Recommended Dietary Allowances (RDA) and Adequate Intakes (AI) for Vitamins

Age (yr)	Thiamin RDA (mg/day)	Riboflavin RDA (mg/day)	Niacin RDA (mg/day)[a]	Biotin AI (µg/day)	Pantothenic acid AI (mg/day)	Vitamin B_6 RDA (mg/day)	Folate RDA (µg/day)[b]	Vitamin B_{12} RDA (µg/day)	Choline AI (mg/day)	Vitamin C RDA (mg/day)	Vitamin A RDA (IU/day)[c]	Vitamin D RDA (IU/day)[d]	Vitamin E RDA (mg/day)[e]	Vitamin K AI (µg/day)
Infants														
0–0.5	0.2	0.3	2	5	1.7	0.1	65	0.4	125	40	400	400 (10 µg)	4	2.0
0.5–1	0.3	0.4	4	6	1.8	0.3	80	0.5	150	50	500	400 (10 µg)	5	2.5
Children														
1–3	0.5	0.5	6	8	2	0.5	150	0.9	200	15	300	600 (15 µg)	6	30
4–8	0.6	0.6	8	12	3	0.6	200	1.2	250	25	400	600 (15 µg)	7	55
Males														
9–13	0.9	0.9	12	20	4	1.0	300	1.8	375	45	600	600 (15 µg)	11	60
14–18	1.2	1.3	16	25	5	1.3	400	2.4	550	75	900	600 (15 µg)	15	75
19–30	1.2	1.3	16	30	5	1.3	400	2.4	550	90	900	600 (15 µg)	15	120
31–50	1.2	1.3	16	30	5	1.3	400	2.4	550	90	900	600 (15 µg)	15	120
51–70	1.2	1.3	16	30	5	1.7	400	2.4	550	90	900	600 (15 µg)	15	120
>70	1.2	1.3	16	30	5	1.7	400	2.4	550	90	900	800 (20 µg)	15	120
Females														
9–13	0.9	0.9	12	20	4	1.0	300	1.8	375	45	600	600 (15 µg)	11	60
14–18	1.0	1.0	14	25	5	1.2	400	2.4	400	65	700	600 (15 µg)	15	75
19–30	1.1	1.1	14	30	5	1.3	400	2.4	425	75	700	600 (15 µg)	15	90
31–50	1.1	1.1	14	30	5	1.3	400	2.4	425	75	700	600 (15 µg)	15	90
51–70	1.1	1.1	14	30	5	1.5	400	2.4	425	75	700	600 (15 µg)	15	90
>70	1.1	1.1	14	30	5	1.5	400	2.4	425	75	700	800 (20 µg)	15	90
Pregnancy														
≤18	1.4	1.4	18	30	6	1.9	600	2.6	450	80	750	600 (15 µg)	15	75
19–30	1.4	1.4	18	30	6	1.9	600	2.6	450	85	770	600 (15 µg)	15	90
31–50	1.4	1.4	18	30	6	1.9	600	2.6	450	85	770	600 (15 µg)	15	90
Lactation														
≤18	1.4	1.6	17	35	7	2.0	500	2.8	550	115	1200	600 (15 µg)	19	75
19–30	1.4	1.6	17	35	7	2.0	500	2.8	550	120	1300	600 (15 µg)	19	90
31–50	1.4	1.6	17	35	7	2.0	500	2.8	550	120	1300	600 (15 µg)	19	90

Note: For all nutrients, values for infants are AI. The glossary on the inside back cover defines units of nutrient measure.
[a] Niacin recommendations are expressed as niacin equivalents (NE), except for recommendations for infants younger than 6 months, which are expressed as preformed niacin.
[b] Folate recommendations are expressed as dietary folate equivalents (DFE).
[c] Vitamin A recommendations are expressed as retinol activity equivalents (RAE).
[d] Vitamin D recommendations are expressed as cholecalciferol and assume an absence of adequate exposure to sunlight.
[e] Vitamin E recommendations are expressed as α-tocopherol.

Recommended Dietary Allowances (RDA) and Adequate Intakes (AI) for Minerals

Age (yr)	Sodium AI (mg/day)	Chloride AI (mg/day)	Potassium AI (mg/day)	Calcium RDA (mg/day)	Phosphorus RDA (mg/day)	Magnesium RDA (mg/day)	Iron RDA (mg/day)	Zinc RDA (mg/day)	Iodine RDA (µg/day)	Selenium RDA (µg/day)	Copper RDA (µg/day)	Manganese AI (mg/day)	Fluoride AI (mg/day)	Chromium AI (µg/day)	Molybdenum RDA (µg/day)
Infants															
0–0.5	120	180	400	200	100	30	0.27	2	110	15	200	0.003	0.01	0.2	2
0.5–1	370	570	700	260	275	75	11	3	130	20	220	0.6	0.5	5.5	3
Children															
1–3	1000	1500	3000	700	460	80	7	3	90	20	340	1.2	0.7	11	17
4–8	1200	1900	3800	1000	500	130	10	5	90	30	440	1.5	1.0	15	22
Males															
9–13	1500	2300	4500	1300	1250	240	8	8	120	40	700	1.9	2	25	34
14–18	1500	2300	4700	1300	1250	410	11	11	150	55	890	2.2	3	35	43
19–30	1500	2300	4700	1000	700	400	8	11	150	55	900	2.3	4	35	45
31–50	1500	2300	4700	1000	700	420	8	11	150	55	900	2.3	4	35	45
51–70	1300	2000	4700	1000	700	420	8	11	150	55	900	2.3	4	30	45
>70	1200	1800	4700	1200	700	420	8	11	150	55	900	2.3	4	30	45
Females															
9–13	1500	2300	4500	1300	1250	240	8	8	120	40	700	1.6	2	21	34
14–18	1500	2300	4700	1300	1250	360	15	9	150	55	890	1.6	3	24	43
19–30	1500	2300	4700	1000	700	310	18	8	150	55	900	1.8	3	25	45
31–50	1500	2300	4700	1000	700	320	18	8	150	55	900	1.8	3	25	45
51–70	1300	2000	4700	1200	700	320	8	8	150	55	900	1.8	3	20	45
>70	1200	1800	4700	1200	700	320	8	8	150	55	900	1.8	3	20	45
Pregnancy															
≤18	1500	2300	4700	1300	1250	400	27	12	220	60	1000	2.0	3	29	50
19–30	1500	2300	4700	1000	700	350	27	11	220	60	1000	2.0	3	30	50
31–50	1500	2300	4700	1000	700	360	27	11	220	60	1000	2.0	3	30	50
Lactation															
≤18	1500	2300	5100	1300	1250	360	10	13	290	70	1300	2.6	3	44	50
19–30	1500	2300	5100	1000	700	310	9	12	290	70	1300	2.6	3	45	50
31–50	1500	2300	5100	1000	700	320	9	12	290	70	1300	2.6	3	45	50

NOTE: For all nutrients, values for infants are AI. The glossary on the inside back cover defines units of nutrient measure.

Tolerable Upper Intake Levels (UL) for Vitamins

Age (yr)	Niacin (mg/day)[a]	Vitamin B$_6$ (mg/day)	Folate (µg/day)[a]	Choline (mg/day)	Vitamin C (mg/day)	Vitamin A (µg/day)[b]	Vitamin D (IU/day)	Vitamin E (mg/day)[c]
Infants								
0–0.5	—	—	—	—	—	600	1000 (25 µg)	—
0.5–1	—	—	—	—	—	600	1500 (38 µg)	—
Children								
1–3	10	30	300	1000	400	600	2500 (63 µg)	200
4–8	15	40	400	1000	650	900	3000 (75 µg)	300
9–13	20	60	600	2000	1200	1700	4000 (100 µg)	600
Adolescents								
14–18	30	80	800	3000	1800	2800	4000 (100 µg)	800
Adults								
19–70	35	100	1000	3500	2000	3000	4000 (100 µg)	1000
>70	35	100	1000	3500	2000	3000	4000 (100 µg)	1000
Pregnancy								
≤18	30	80	800	3000	1800	2800	4000 (100 µg)	800
19–50	35	100	1000	3500	2000	3000	4000 (100 µg)	1000
Lactation								
≤18	30	80	800	3000	1800	2800	4000 (100 µg)	800
19–50	35	100	1000	3500	2000	3000	4000 (100 µg)	1000

[a]The UL for niacin and folate apply to synthetic forms obtained from supplements, fortified foods, or a combination of the two.
[b]The UL for vitamin A applies to the preformed vitamin only.
[c]The UL for vitamin E applies to any form of supplemental α-tocopherol, fortified foods, or a combination of the two.

Tolerable Upper Intake Levels (UL) for Minerals

Age (yr)	Sodium (mg/day)	Chloride (mg/day)	Calcium (mg/day)	Phosphorus (mg/day)	Magnesium (mg/day)[d]	Iron (mg/day)	Zinc (mg/day)	Iodine (µg/day)	Selenium (µg/day)	Copper (µg/day)	Manganese (mg/day)	Fluoride (mg/day)	Molybdenum (µg/day)	Boron (mg/day)	Nickel (mg/day)	Vanadium (mg/day)
Infants																
0–0.5	—	—	1000	—	—	40	4	—	45	—	—	0.7	—	—	—	—
0.5–1	—	—	1500	—	—	40	5	—	60	—	—	0.9	—	—	—	—
Children																
1–3	1500	2300	2500	3000	65	40	7	200	90	1000	2	1.3	300	3	0.2	—
4–8	1900	2900	2500	3000	110	40	12	300	150	3000	3	2.2	600	6	0.3	—
9–13	2200	3400	3000	4000	350	40	23	600	280	5000	6	10	1100	11	0.6	—
Adolescents																
14–18	2300	3600	3000	4000	350	45	34	900	400	8000	9	10	1700	17	1.0	—
Adults																
19–50	2300	3600	2500	4000	350	45	40	1100	400	10,000	11	10	2000	20	1.0	1.8
51–70	2300	3600	2000	4000	350	45	40	1100	400	10,000	11	10	2000	20	1.0	1.8
>70	2300	3600	2000	3000	350	45	40	1100	400	10,000	11	10	2000	20	1.0	1.8
Pregnancy																
≤18	2300	3600	3000	3500	350	45	34	900	400	8000	9	10	1700	17	1.0	—
19–50	2300	3600	2500	3500	350	45	40	1100	400	10,000	11	10	2000	20	1.0	—
Lactation																
≤18	2300	3600	3000	4000	350	45	34	900	400	8000	9	10	1700	17	1.0	—
19–50	2300	3600	2500	4000	350	45	40	1100	400	10,000	11	10	2000	20	1.0	—

[d]The UL for magnesium applies to synthetic forms obtained from supplements or drugs only. NOTE: An Upper Limit was not established for vitamins and minerals not listed and for those age groups listed with a dash (—) because of a lack of data, not because these nutrients are safe to consume at any level of intake. All nutrients can have adverse effects when intakes are excessive.

SOURCE: Adapted with permission from the *Dietary Reference Intakes* series, National Academies Press. Copyright 1997, 1998, 2000, 2001, 2002, 2005, 2011 by the National Academies of Sciences.

Eighth Edition

Nutrition & Diet Therapy

Eighth Edition

Nutrition &
Diet Therapy

Linda Kelly DeBruyne

Kathryn Pinna

Ellie Whitney

WADSWORTH
CENGAGE Learning™

Australia • Brazil • Japan • Korea • Mexico • Singapore • Spain • United Kingdom • United States

Nutrition & Diet Therapy, Eighth Edition
Linda Kelly DeBruyne, Kathryn Pinna,
Ellie Whitney

Publisher/Executive Editor: Yolanda Cossio

Acquisitions Editor: Peggy Williams

Developmental Editor: Elesha Feldman

Assistant Editor: Elesha Feldman

Editorial Assistant: Shana Baldassari

Media Editor: Miriam Myers

Marketing Manager: Laura McGinn

Marketing Communications Manager: Linda Yip

Content Project Manager: Carol Samet

Creative Director: Rob Hugel

Art Director: John Walker

Print Buyer: Becky Cross

Rights Acquisitions Specialist: Roberta Broyer

Production Service: Laura Hakala,
PreMediaGlobal

Text Designer: Tani Hasegawa

Photo Researcher: Carly Bergey,
PreMediaGlobal

Copy Editor: Paula Bonilla

Cover Designer: Riezebos Holzbaur/
Tim Heraldo

Cover Image: Kazuo Ogawa; Sean Justice/
Corbis; Tetra Images; © momentimages/
Tetra Images/Corbis

Compositor: PreMediaGlobal

Library of Congress Control Number: 2011923171

ISBN-13: 978-0-8400-4944-5

ISBN-10: 0-8400-4944-7

Wadsworth
20 Davis Drive
Belmont, CA 94002-3098
USA

Cengage Learning is a leading provider of customized learning solutions with office locations around the globe, including Singapore, the United Kingdom, Australia, Mexico, Brazil, and Japan. Locate your local office at **www.cengage. com/global**

Cengage Learning products are represented in Canada by Nelson Education, Ltd.

To learn more about Wadsworth visit **www.cengage.com/Wadsworth**

Purchase any of our products at your local college store or at our preferred online store **www.cengagebrain.com**

Printed in the United States of America
1 2 3 4 5 6 7 14 13 12 11

Contents in Brief

Contents

How To Features Appear on the Following Pages

Case Study Features Appear on the Following Pages

Preface

Numerous discoveries in nutrition science over the past few years have continued to benefit the field of health care. In revising this eighth edition of *Nutrition and Diet Therapy*, we have been pleased at the number of new findings we have been able to incorporate into these pages. As always, major goals of this textbook are to present both core nutrition information and guidelines about the prevention of, and care during, illness. Another mission is to help nutrition students evaluate information and products available from the media, colleagues, and the marketplace.

As in the previous edition of this book, Chapters 1 through 10 introduce basic concepts in nutrition and explain how sound nutrition supports health. Chapters 11 to 13 apply nutrition principles to individuals during different stages of life, from pregnancy and birth through old age. The second half of the book addresses the concerns of individuals who are at risk of illness or have medical conditions that require medical nutrition therapy. A hallmark of the text is the "Nutrition in Practice" section located at the end of every chapter: these sections provide coverage of current research topics, advanced subjects, or specialty areas.

Changes for This Edition

Each chapter of this book is based on current nutrition knowledge and the latest clinical practice guidelines. Some major changes in this edition include the following:

Chapter 1

- Introduced the 2010 *Dietary Guidelines for Americans*
- Added the new protein foods subgroups to the USDA Food Guide
- Expanded the table of recommended weekly amounts from vegetable subgroups to include protein foods subgroups
- Added a new table of metric/household measures

Chapter 2

- Revised and updated the figure about added sugar consumption
- Added a new figure on sugar alternatives on food labels
- Changed a table about fiber to a figure, adding photos and more explanation
- Added information about the herbal sweetener stevia
- Expanded and updated the table of U.S. approved sweeteners
- Added a new Nutrition in Practice about the glycemic index and its clinical applications

Chapter 3

▌ Reorganized several topics

▌ Increased the coverage of *trans* fat

▌ Rewrote fish oil supplement section

▌ Expanded tables of functions of fats in the body and in food

▌ Added a ranked list of seafood sources of omega-3 fatty acids to Table 3-3: Food Sources of Omega-6 and Omega-3 Fatty Acids

Chapter 4

▌ Regorganized "Roles of Body Proteins" section; added a new section on proteins as structural components

▌ Moved and rewrote section on protein turnover and nitrogen balance

▌ Enhanced discussion of protein excess

▌ Added new illustrated table on vegetarian sources of key nutrients

Chapter 5

▌ Added key terms and revised glossaries of GI terms and digestive glands

▌ Revised Figure 5-1: The Gastrointestinal Tract

Chapter 6

▌ Added more information about the health hazards of central obesity

▌ Included new figure on abdominal obesity

▌ Expanded the section on the health risks of obesity and included health risks of underweight

▌ Defined several new terms—insulin resistance, adipokines, subcutaneous fat, and inflammation

Chapter 7

▌ Added brief discussion of adiponectin

▌ Added brief discussion of cognitive skills and weight management

▌ Added new table of physical activity strategies for weight management from the American College of Sports Medicine 2009 position paper

▌ Added strategies for successful weight maintenance

▌ Shortened and simplified the Nutrition in Practice on fad diets

Chapter 8

▌ Added margin note on minimizing nutrient losses (during food preparation and storage)

▌ Enhanced and clarified discussion of vitamin A's role in vision

▌ Enhanced vitamin–D deficiency symptom figure

▌ Added list of factors that may limit vitamin D synthesis in the skin

▌ Incorporated updated DRI recommendations for vitamin D

Chapter 9

▌ Added new margin list of percentages of water in selected foods

▌ Included updated DRI recommendations for calcium

▌ Added brief discussion of iron deficiency and obesity

▌ Enhanced the discussion of zinc

Chapter 10

▌ Introduced *Physical Activity Guidelines for Americans, 2008*

▌ Updated and enhanced the fitness pyramid to include the *Physical Activity Guidelines for Americans, 2008*

▌ Added a short section on enhanced waters

Chapter 11

▌ Enhanced the discussion of weight gain during pregnancy, addressing current concerns about obesity during pregnancy; included new pregnancy weight gain guidelines for obese women

▌ Added a discussion of weight loss after pregnancy

▌ Redesigned the figure on the neural tube defect spina bifida

▌ Added a discussion about frequency and duration of breastfeeding

Chapter 12

▌ Updated sample meal plan for a one-year-old

▌ Expanded guidelines for mealtimes with one-year-olds

▌ Enhanced the discussion of nutrient needs during childhood

▌ Moved into the chapter and expanded the discussion of childhood obesity from the Nutrition in Practice; the Practice now focuses on later diseases related to obesity, including CHD, hypertension, and diabetes

▌ Added MyPyramid for Preschoolers to the MyPyramid for Kids figure

▌ Added a new figure of a physical activity pyramid

▌ Included new information about food skills for preschoolers

Chapter 13

▌ Added a new table listing signs of dementia

▌ Expanded the Alzheimer's section and added a comparison of signs of normal memory loss versus memory loss with Alzheimer's

▌ Added a new box: How to Turn Convenience Foods into Nutritious Meals

Chapter 14

▌ Expanded and reorganized the section on nutrition screening; revised the table on Subjective Global Assessment and moved it to this section

▌ Revised the discussion and table about historical information used in nutrition assessment

▌ Revised the discussion about determination of healthy body weight

- Revised the discussion, table, and "How to" about determining energy needs in hospital patients

- Expanded the section on documenting nutrition care, including a discussion of the ADIME format; updated the figure of the SOAP note to include nutrition diagnoses

- Added information about epigenetics to the Nutrition in Practice

Chapter 15

- Revised the introductory section about medications in disease treatment

- Simplified the table on terms prohibited in clinical documentation

- Revised the tables listing grapefruit juice–drug interactions and foods high in tyramine

- Moved the section about herbal products to the end of the chapter; simplified the table on herb-drug interactions

Chapter 16

- Revised and reorganized the section about enteral formulas

- Added photos showing transnasal and gastrostomy tube feedings, open feeding systems and closed feeding systems, and an example of an infusion pump

- Revised the section about the administration of tube feedings

- Revised and reorganized the table about the causes and management of tube feeding complications

- Added a figure showing a parenteral nutrition order form

Chapter 17

- Revised the table on dysphagia diets, including new listings of foods to avoid in each type of diet

- Eliminated descriptions of the rarely performed vagotomy and pyloroplasty procedures

- Revised the table on the postgastrectomy diet

- Expanded the section on bariatric surgery, including a revision of the material on nutrition care after bariatric surgery

- Moved the Nutrition in Practice about nutrition and oral (dental) health from Chapter 2 to here, and added discussions about periodontal disease and the relationships between dental diseases and other illnesses

Chapter 18

- Modified the section on diarrhea and revised the table listing foods that affect diarrhea

- Revised the section on irritable bowel syndrome

- Added a photo of a case of diverticulosis as seen in a colonoscopy

Chapter 19

- Modified the section about evaluating malabsorption

- Revised the table listing foods to include and restrict in a fat-controlled diet, and added a new figure showing a sample menu for a fat-controlled diet

- Revised the table describing a gluten-free diet and added a sample menu
- Modified the sections on pancreatitis, cystic fibrosis, and short bowel syndrome

Chapter 20

- Expanded descriptions of the different types of hepatitis viruses
- Revised the sections on the medical treatment and nutrition therapy for cirrhosis

Chapter 21

- Added the glycated hemoglobin measure to the section about diagnosis of diabetes
- Revised the discussion about type 2 diabetes in children and adolescents
- Revised the discussion about complications of diabetes, and included a new figure that outlines the acute effects of insulin insufficiency
- Extended the list of antidiabetic drugs to include recently approved medications for diabetes
- Revised the section about physical activity and diabetes management
- Moved the Nutrition in Practice about metabolic syndrome from Chapter 22 to here

Chapter 22

- Reorganized and revised the sections on the causes of atherosclerosis and the treatment of hypertension
- Modified several sections related to therapeutic lifestyle changes
- Moved the section on stroke to follow the section on coronary heart disease; moved the Nutrition in Practice about feeding disabilities from Chapter 17 to here

Chapter 23

- Revised the section on the nephrotic syndrome and the table on foods to include and restrict in a sodium-controlled diet
- Replaced the terms *acute renal failure* and *chronic renal failure* with the more current terms *acute kidney injury* and *chronic kidney disease*
- Added new tables showing the approximate potassium content of common fruits and vegetables, foods high in phosphorus, and an appropriate menu for a person with chronic kidney disease; eliminated the tables on dietary guidelines following a kidney transplant and foods high in purines
- Added a new section about the medical treatment of kidney stones

Chapter 24

- Modified the sections on the energy and nutrient needs of patients undergoing acute stress and approaches to nutrition care
- Clarified the descriptions of chronic bronchitis and emphysema
- Modified the discussion about nutrition care for respiratory failure

Chapter 25

▪ Revised the discussion about nutrition and cancer risk; modified the tables about the nutrition-related factors that influence cancer risk and recommendations for reducing risk

▪ Modified the section about hematopoietic stem cell transplantation; added a section about biological therapies for cancer

▪ Expanded the "How to" about increasing kcalories and protein in meals

▪ Added a section about the low-microbial diet for individuals with suppressed immunity

▪ Expanded the table listing antiretroviral drugs to include recently approved medications for HIV infection

▪ Reorganized and revised the section on nutrition therapy for HIV infection, including a discussion about weight management and overweight/obesity

Features of this Text

Throughout the book, the readable text and pedagogic features should help to facilitate students' understanding and retention of the material. Features include the following:

▪ Definitions of key terms, clarifications, reminders, and cross-references appear in the margins.

▪ "How to" skill boxes help readers work through calculations or give practical suggestions for applying nutrition information.

▪ "In Summary" statements at the end of each major chapter section help students assimilate the material and assess reading comprehension.

▪ The "Self Check" at the end of each chapter helps readers test their understanding of the chapter material.

▪ "Your Diet" exercises ask students to apply nutrition information from each chapter to their own diets.

▪ Case studies in the later chapters challenge readers to apply chapter information to clinical situations.

▪ "Clinical Applications" provide practice with mathematical calculations and help students understand the impact of nutrition-related issues on health care professionals and their clients.

▪ "Nutrition Assessment Checklists" summarize assessment parameters relevant to different stages of the life cycle or groups of disorders.

▪ "Diet-Drug Interaction" boxes point out interactions relevant to the medications described in each chapter.

▪ The appendixes include a wealth of information on the contents of foods and enteral formulas, U.S. nutrient intake recommendations and the exchange system, Canadian guidelines and food guides, physical activity and energy requirements, additional information about nutrition assessment, and aids to calculations.

We hope that as you discover the many fascinating aspects of nutrition, you will enthusiastically apply the concepts in both your professional and your

personal life. To access additional course materials including Course Mate, please visit **www.cengagebrain.com.** At the CengageBrain.com home page, search for the ISBN of your title (from the back cover of your book) using the search box at the top of the page. This will take you to the product page where these resources can be found.

Ancillaries

▌ New for this edition, the **Course Mate** provides online access to study tools and complementary multimedia for students, and allows instructors to monitor their activities on the website.

▌ The **instructor's manual with test bank** has been thoroughly updated to reflect changes to the text, and provides assignment worksheets; activity suggestions; answer keys for the "Clinical Applications" and case study exercises; chapter outlines with summaries; and a robust assortment of multiple-choice, true/false, matching, and essay questions. The test bank has been expanded with many new items, including new case study–based multiple-choice and essay questions.

▌ The **ExamView** test creation software suite is available with the test bank questions preloaded.

▌ The **Power Lecture** features PowerPoint-ready figures and tables from the text and relevant video clips—many new for this edition—as well as updated ready-to-use PowerPoint lecture notes.

▌ Book-specific content for learning management systems is available in a **WebTutor Toolbox** cartridge.

Acknowledgments

Among the most difficult words to write are those that express the depth of our gratitude to the many dedicated people whose efforts have made this book possible. A special note of appreciation to Sharon Rolfes for her numerous contributions to the chapters and Nutrition in Practice sections as well as to the Dietary Reference Intakes on the inside front cover and the appendices. Many thanks to Fran Webb for sharing her knowledge, ideas, and resources about the latest nutrition developments. Elesha Feldman's critical eye, numerous suggestions, and unceasing support were especially helpful in revising both the normal nutrition and the clinical chapters. Special thanks to Alexandra Rodriguez for help in manuscript preparation. We also wish to acknowledge the efforts of the folks at Axxya for their assistance in creating the food composition appendix. We are indebted to our editorial team, Peggy Williams and Elesha Feldman, and our production team, especially Carol Samet, for seeing this project through from start to finish. We would also like to acknowledge Laura McGinn for her marketing efforts. To the many others involved in designing, indexing, typesetting, dummying, and marketing, we offer our thanks. We are especially grateful to our associates, family, and friends for their continued encouragement and support and to our reviewers who consistently offer excellent suggestions for improving the text.

Reviewers

Katie Clark, M.P.H., R.D., C.D.E.
University of San Diego

Dr. Rachelle D. Duncan
Oklahoma State University

Lisa C. Gibson, M.S., R.D./L.D.
Rose State College

Jorunn Gran-Henriksen
State University of New York College at Plattsburgh

Amy Roy Huggins
Greenville Technical College

Dr. Laina Karthikeyan
New York City College of Technology

Dr. Myrtle McCulloch, R.D.
Georgetown University

Michele Taylor McMullen
Montana State University, Billings

Amy Bolinger Snow
Greenville Technical College

M. Eva Weicker, M.D., M.B.A.
Alvernia University

Janet M. Westhoff
Mott Community College

Mary Width, M.S., R.D.
Wayne State University

About the Authors

Linda Kelly DeBruyne, M.S., R.D., received her B.S. in 1980 and her M.S. in 1982 in nutrition and food science at Florida State University. She is a founding member of Nutrition and Health Associates, an information resource center in Tallahassee, Florida, where her specialty areas are life cycle nutrition and fitness. Her other publications include the textbooks *Nutrition for Health and Health Care,* and *Health: Making Life Choices* and a multimedia CD-ROM called *Nutrition Interactive.* As a consultant for a group of Tallahassee pediatricians, she teaches infant nutrition classes to parents. She is a registered dietitian and maintains a professional membership in the American Dietetic Association.

Kathryn Pinna, Ph.D., R.D., received her M.S. and Ph.D. degrees in nutrition from the University of California at Berkeley. She has taught nutrition, food science, and human biology courses in the San Francisco Bay Area for over 20 years and currently teaches nutrition classes at City College of San Francisco. She has also worked as an outpatient dietitian, Internet consultant, and freelance writer. Her other publications include the textbooks *Understanding Normal and Clinical Nutrition* and *Nutrition for Health and Health Care.* She is a registered dietitian and a member of the American Society for Nutrition and the American Dietetic Association.

Ellie Whitney, Ph.D., grew up in New York City and received her B.A. and Ph.D. degrees in English and biology at Radcliffe/Harvard University and Washington Universities, respectively. She has taught at both Florida State University and Florida A&M University, has written newspaper columns on environmental matters for the *Tallahassee Democrat,* and has authored almost a dozen college textbooks on nutrition, health, and related topics, many of which have been revised multiple times over the years. In addition to teaching and writing, she has spent the past three-plus decades exploring outdoor Florida and studying its ecology. Her latest book is *Priceless Florida: The Natural Ecosystems* (Pineapple Press, 2004).

To my creative and talented son and daughter-in-law, Tyler and Shannon: your joy of cooking is inspiring and the food is delicious. —Linda Kelly DeBruyne

To the professors at UC-Berkeley who taught me how to learn: Nancy Amy, Kenneth Carpenter, Marc Hellerstein, Janet King, and Susan Oace. —Kathryn Pinna

To the memory of my parents, Edith Tyler Noss and Henry H. B. Noss, who supported me with love, discipline, and pride. —Ellie Whitney

1

Overview of Nutrition and Health

Every day, several times a day, you make choices that will either improve your **health** or harm it. Each choice may influence your health only a little, but when these choices are repeated over years and decades, their effects become significant.

The choices people make each day affect not only their physical health but also their **wellness**—all the characteristics that make a person strong, confident, and able to function well with family, friends, and others. People who consistently make poor lifestyle choices, on a daily basis, increase their risks of developing diseases. Figure 1-1 shows how a person's health can fall anywhere along a continuum, from maximum wellness on one end to total failure to function (death) on the other.

As health care professionals, when you take responsibility for your own health by making daily choices and practicing behaviors that enhance your well-being, you prepare yourself physically, mentally, and emotionally to meet the demands of your profession. As health care professionals, however, you have a responsibility to your clients as well as to yourselves. You have unique opportunities to make your clients aware of the benefits of positive health choices and behaviors, to show them how to change their behaviors and make daily choices to enhance their own health, and to serve as role models for those behaviors.

This text focuses on how nutrition choices affect health and disease. The early chapters introduce the basics of nutrition to support good health. The later chapters emphasize nutrition therapy and its role in supporting health and treating diseases and symptoms.

Health care professionals generally use either *client* or *patient* when referring to an individual under their care. The first 13 chapters of this text emphasize the nutrition concerns of people in good health; therefore, the term *client* is used in these chapters.

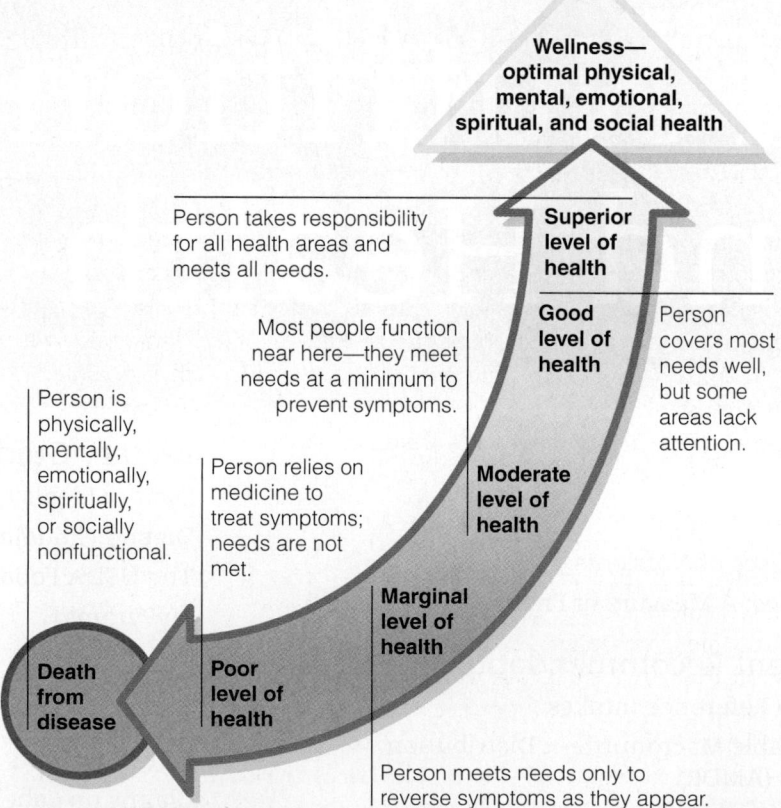

FIGURE 1-1 The Health Line
No matter how well you maintain your health today, you may still be able to improve tomorrow. Likewise, a person who is well today can slip by failing to maintain health-promoting habits.

Food Choices

Sound **nutrition** throughout life does not ensure good health and long life, but it can certainly help to tip the balance in their favor. Nevertheless, most people choose foods for reasons other than their nourishing value. Even people who claim to choose foods primarily for the sake of health or nutrition will admit that other factors also influence their food choices. Because food choices become an integral part of their lifestyles, individuals sometimes find it difficult to change their eating habits. Health care professionals who help clients make diet changes must understand the dynamics of food choices because people will alter their eating habits only if their preferences are honored. Developing **cultural competence** is an important aspect of honoring individual preferences, especially for health care professionals who help clients to achieve a nutritious diet.[1]

Nutrition is only one of the many factors that influence people's food choices.

Nutrition in Practice 8 addresses phytochemicals and their role in disease prevention.

Preference Why do people like certain foods? One reason, of course, is their preference for certain tastes. Some tastes are widely liked, such as the sweetness of sugar and the zest of salt. Research suggests that genetics influence people's taste preferences, a finding that may eventually have implications for clinical nutrition.[2] For example, sensitivity to bitter taste is an inheritable trait. People born with great sensitivity to bitter tastes tend to avoid foods with bitter flavors, such as broccoli, cabbage, brussels sprouts, spinach, and grapefruit juice. These foods, as well as many other fruits and vegetables, contain **bioactive food components—phytochemicals** and nutrients— that may reduce the risk of cancer. Thus, the role that genetics may play in food selection is gaining importance in cancer research.

Habit Sometimes habit dictates people's food choices. People eat a sandwich for lunch or drink orange juice at breakfast simply because they have always done so.

Associations People also like foods with happy associations—foods eaten in the midst of warm family gatherings on traditional holidays or given to them as children by someone who loved them. By the same token, people can attach intense and unalterable dislikes to foods that they ate when they were sick or that were forced on them when they weren't hungry.

Ethnic Heritage and Tradition Every country, and every region of a country, has its own typical foods and ways of combining them into meals. The **foodways** of North America reflect the many different cultural and ethnic backgrounds of its inhabitants. Many foods with foreign origins are familiar items on North American menus: tacos, egg rolls, lasagna, and gyros, to name a few. Still others, such as spaghetti and croissants, are almost staples in the "American diet." North American regional cuisines, like Cajun and TexMex, blend the traditions of several cultures. Table 1-1 (p. 4) presents selected **ethnic diets** and food choices.

Values People's values, environmental ethics, religious beliefs, and political views also influence their food choices. By choosing to eat some foods or to avoid others, people make statements that reflect their values. For example, people may select only foods that come in containers that can be reused or recycled. A concerned consumer may boycott fruit or vegetables picked by migrant workers who have been exploited. People may buy vegetables from local farmers to save the fuel and environmental costs of foods shipped from far away. Labels on some foods carry statements or symbols—known as ecolabels—that imply that the foods have been produced in ways that are considered environmentally favorable. One of the most successful ecolabels is "Certified Organic," meaning the food was grown and processed according to regulations defining the use of synthetic fertilizers, herbicides, pesticides, and other chemical ingredients.

health: a range of states with physical, mental, emotional, spiritual, and social components. At a minimum, health means freedom from physical disease, mental disturbances, emotional distress, spiritual discontent, social maladjustment, and other negative states. At a maximum, health means *wellness*.

wellness: maximum well-being; the top range of health states; the goal of the person who strives toward realizing his or her full potential physically, mentally, emotionally, spiritually, and socially.

nutrition: the science of foods and the nutrients and other substances they contain, and of their ingestion, digestion, absorption, transport, metabolism, interaction, storage, and excretion. A broader definition includes the study of the environment and of human behavior as it relates to these processes.

cultural competence: having an awareness and acceptance of one's own and others' cultures and abilities leading to effective interactions.

bioactive food components: compounds in foods, either nutrients or phytochemicals, that alter physiological processes in the body.

phytochemicals (FIGH-toe-CHEM-ih-cals): compounds in plants that confer color, taste, and other characteristics. Some phytochemicals are bioactive food components in functional foods. Nutrition in Practice 8 provides details.

foodways: the eating habits and culinary practices of a people, region, or historical period.

ethnic diets: foodways and cuisines typical of national origins, races, cultural heritages, or geographic locations.

TABLE 1-1 Selected Ethnic Cuisines and Food Choices

	Grains	Vegetables	Fruits	Meats, Poultry, Fish, Legumes, Eggs, and Nuts	Milk
Asian	Barley, buckwheat, millet, rice, wheat, rice or wheat noodles	Amaranth, baby corn, bamboo shoots, bok choy, cabbages, mung bean sprouts, scallions, seaweed, snow peas, straw mushrooms, water chestnuts, wild yam	Kumquats, loquats, lychee, mandarin oranges, melons, pears, persimmon, plums	Beef, pork, poultry, fish and other seafood, squid, soybeans, tofu, duck eggs, cashews, peanuts	Soymilk
Mediterranean	Bulgur, couscous, focaccia, Italian bread, pastas, pita pocket bread, polenta, rice	Cucumbers, eggplant, grape leaves, onions, peppers, tomatoes	Dates, figs, grapes, lemons, melons, olives, raisins	Beef, gyros, lamb, pork, sausage, chicken, fish and other seafood, fava beans, lentils, almonds, walnuts	Feta, goat, mozzarella, parmesan, provolone, and ricotta cheeses; yogurt
Mexican	Taco shells, tortillas (corn or flour), rice	Cactus, cassava, chayote, chilies, corn, jicama, onions, tomatoes, tomato salsa, yams	Avocado, bananas, guava, lemons, limes, mango, oranges, papaya, plantain	Beef, chorizo, chicken, fish, refried beans, eggs	Cheese, flan (caramel custard)

Religion also influences many people's food choices. Jewish law sets forth an extensive set of dietary rules. Many Christians forgo meat on Fridays during Lent, the period prior to Easter. In Islamic dietary laws, permitted or lawful foods are called *halal*. Other faiths prohibit some dietary practices and promote others. Diet planners can foster sound nutrition practices only if they respect and honor each person's values.

Ethnic meals and family gatherings nourish the spirit as well as the body.

Social Interaction Social interaction is another powerful influence on people's food choices. Meals are social events, and the sharing of food is part of hospitality. It is often considered rude to refuse food or drink being shared by a group or offered by a host. Food brings people together for many different reasons: to celebrate a holiday or special event, to renew an old friendship, to make new friends, to conduct business, and many more. Sometimes food defines status. Lobster and caviar are often associated with wealth; peanut butter and jelly, with children. Sometimes food is used to influence or impress someone. For example, a business executive invites a prospective new client out to dinner in hopes of edging out the competition. In each case, for whatever the purpose, food plays an integral part in the social interaction.

Emotional State People may eat in response to emotional stimuli—for example, to relieve boredom or depression or to calm anxiety. A lonely person may choose to eat rather than to call a friend and risk rejection. A person who has returned home from an exciting evening out may unwind with a late-night snack. Eating in response to emotions can easily lead to overeating and obesity, but it may be appropriate at times. For example, sharing food at times of bereavement serves both the giver's need to provide comfort and the receiver's need to be cared for and to interact with others.

In contrast, some people do not eat at all, or eat very little, in response to emotions. A depressed person may simply have no appetite for food. A person overcome with grief can easily lose all interest in food. As long as a person's appetite or interest in food returns quickly, little harm is done.

Availability, Convenience, and Economy The influence of these factors on people's food selections is clear. You cannot eat foods if they are not available, if you cannot get to the grocery store, if you do not have the time or skill to prepare them, or if you cannot afford them. Consumers today value convenience so highly that they are willing to spend half of their food budget on meals that require little or no preparation. They frequently eat out, bring home ready-to-eat meals, have food delivered, or cook meals ahead at meal assembly kitchens, which are gaining popularity throughout the country.[3] Along with convenience, however, consumers want great taste and quality—foods that are fast, delicious, and nutritious. The demand for fully prepared ready-to-eat foods or foods that can be easily prepared in a microwave demonstrates this influence. Consumer demand for microwavable foods that taste and smell as though they were prepared in a conventional oven has prompted manufacturers to expand the variety of microwavable foods available and to change the formulation and packaging of the foods. Food manufacturers are developing many more microwavable meals and snacks for use at home and the office—rising-crust pizzas, chicken and fish dishes, precooked entrées, and side dishes such as French fries and rice. Many of the new products are packaged in single-serve containers.

Age Age influences people's food choices. Infants, for example, depend on others to choose foods for them. Older children also rely on others but become more active in selecting foods that taste sweet and are familiar to them and rejecting those whose taste or texture they dislike. In contrast, the links between taste preferences and food choices in adults are less direct than they are in children. Adults often choose foods based on health concerns, such as body weight. Indeed, adults may avoid sweet or familiar foods because of such concerns.

Occupation Some people have jobs that keep them away from home for days at a time, require them to conduct business in restaurants or at conventions, or involve hectic schedules that allow little or no time for meals at home. For these people, the kinds of restaurants available to them and the cost of eating out so often may limit food choices.

Eating disorders are the topic of Nutrition in Practice 6.

Body Weight and Image Sometimes people select certain foods and supplements that they believe will improve their physical appearance and avoid those they believe might be detrimental. Such decisions can be beneficial when based on sound nutrition and fitness knowledge, but they may undermine good health when based on fads or carried to extremes.

Medical Conditions Sometimes, medical conditions and their treatments (including medications) limit the foods a person can select. For example, a person with heart disease might need to adopt a diet low in certain types of fats. The chemotherapy needed to treat cancer can interfere with a person's appetite or limit food choices by causing vomiting. Allergy to certain foods can also limit choices. The second half of this text discusses how the diet can be modified to accommodate different medical conditions.

Nutrition in Practice 8 offers more discussion of functional foods.

Health and Nutrition Consumers today cite nutrition as a primary concern in making food choices, yet the foods they choose do not always reflect this concern. Food manufacturers and restaurant chefs have responded to scientific findings linking health with nutrition by offering an abundant selection of health-promoting foods and beverages. Foods that provide health benefits beyond their nutrient contributions are called **functional foods.**[4] Whole foods—as natural and familiar as oatmeal or tomatoes—are the simplest functional foods. In other cases, foods have been modified through fortifications, enrichment, or enhancement. Examples of these functional foods include orange juice fortified with calcium to build strong bones, bread enriched with folate to promote normal fetal development, and margarine enhanced with a plant sterol to lower blood cholesterol.

Consumers typically welcome new foods into their diets, provided that these foods are reasonably priced, clearly labeled, easy to find in the grocery store, and convenient to prepare. These foods must also taste good—as good as the traditional choices. Of course, a person need not eat any "special" foods to enjoy a healthy diet; many "regular" foods provide numerous health benefits as well. In fact, foods such as whole grains; vegetables and legumes; fruits; meats, fish, and poultry; and milk products are among the healthiest choices a person can make.

IN SUMMARY

▌ A person selects foods for many different reasons.

▌ Food choices influence health—both positively and negatively. Individual food selections neither make nor break a diet's healthfulness, but the balance of foods selected over time can make an important difference to health.

▌ In the interest of health, people are wise to think "nutrition" when making their food choices.

functional foods: whole or modified foods that contain bioactive food components believed to provide health benefits, such as reduced disease risks, beyond the benefits that their nutrients contribute. All whole foods are functional in some ways because they provide at least some needed substances, but certain foods stand out as rich sources of bioactive food components.

nutrients: substances obtained from food and used in the body to provide energy and structural materials and to serve as regulating agents to promote growth, maintenance, and repair. Nutrients may also reduce the risks of some diseases.

essential nutrients: nutrients a person must obtain from food because the body cannot make them for itself in sufficient quantities to meet physiological needs.

organic: carbon containing. The four organic nutrients are carbohydrate, fat, protein, and vitamins.

The Nutrients

You are a collection of molecules that move. All these moving parts are arranged in patterns of extraordinary complexity and order—cells, tissues, and organs. Although the arrangement remains constant, the parts are continually changing, using **nutrients** and energy derived from nutrients.

Almost any food you eat is composed of dozens or even hundreds of different kinds of materials. Spinach, for example, is composed mostly of water (95 percent), and most of its solid materials consist of carbohydrates, fats (properly called lipids), and proteins. If you could remove these materials, you would find a tiny residue of minerals, vitamins, and other compounds.

Six Classes of Nutrients

Water, carbohydrates, fats, proteins, vitamins, and minerals are the six classes of nutrients commonly found in spinach and other foods. Some of the other materials in foods, such as pigments and other phytochemicals, are not nutrients but may still be important to health. The body can make some nutrients for itself, at least in limited quantities, but it cannot make them all, and it makes some in insufficient quantities to meet its needs. Therefore, the body must obtain many nutrients from foods. The nutrients that foods must supply are called **essential nutrients.**

Carbohydrates, Fats, and Proteins Four of the six classes of nutrients (carbohydrates, fats, proteins, and vitamins) contain carbon, which is found in all living things. They are therefore **organic** (meaning, literally, "alive"). During metabolism, three of these four (carbohydrates, fats, and proteins) provide energy the body can use. These **energy-yielding nutrients** continually replenish the energy you spend daily. Carbohydrates and fats meet most of the body's energy needs; proteins make a significant contribution only when other fuels are unavailable.

Vitamins, Minerals, and Water Vitamins are organic but do not provide energy to the body. They facilitate the release of energy from the three energy-yielding nutrients. In contrast, minerals and water are **inorganic** nutrients. Minerals yield no energy in the human body, but, like vitamins, they help to regulate the release of energy, among their many other roles. As for water, it is the medium in which all of the body's processes take place.

kCalories: A Measure of Energy

The amount of energy that carbohydrates, fats, and proteins release can be measured in **calories**—units of energy so tiny that a single apple provides tens of thousands of them. To ease calculations, energy is expressed in 1000-calorie metric units known as **kilocalories** (shortened to **kcalories,** but commonly called "calories"). When you read in popular books or magazines that an apple provides "100 calories," understand that it means 100 kcalories. This book uses the term *kcalorie* and its abbreviation *kcal* throughout, as do other scientific books and journals. kCalories are not constituents of foods; they are a measure of the energy foods provide. The energy a food provides depends on how much carbohydrate, fat, and protein the food contains.

Carbohydrate yields 4 kcalories of energy from each gram, and so does protein. Fat yields 9 kcalories per gram. Thus, fat has a greater **energy density** than either carbohydrate or protein. If you know how many grams of each nutrient a food contains, you can derive the number of kcalories potentially available from the food. Simply multiply the carbohydrate grams by 4, the protein grams by 4, and the fat grams by 9, and add the results together (the "How to" (p. 8) describes how to calculate the energy a food provides).

Energy Nutrients in Foods Practically all foods contain mixtures of the energy-yielding nutrients, although foods are sometimes classified by their predominant nutrient. To speak of meat as "a protein" or of bread as "a carbohydrate," however, is inaccurate. Each is rich in a particular nutrient, but a protein-rich food such as beef contains a lot of fat along with the protein, and a carbohydrate-rich food such as cornbread also contains fat (corn oil) and protein. Only a few foods are exceptions to this rule, the common ones being sugar (which is pure carbohydrate) and oil (which is pure fat).

Energy Storage in the Body The body first uses the energy-yielding nutrients to build new compounds and fuel metabolic and physical activities. Excesses are then rearranged into storage compounds, primarily body fat, and put away for later use. Thus,

The six classes of nutrients are water, carbohydrates, fats, proteins, vitamins, and minerals.

Metabolism is the set of processes by which nutrients are rearranged into body structures or broken down to yield energy.

Food energy can also be measured in kilojoules (kJ). The kilojoule is the international unit of energy. One kcalorie equals 4.2 kJ.

Chapter 7 revisits energy density and describes its role in weight management.

energy-yielding nutrients: the nutrients that break down to yield energy the body can use. The three energy-yielding nutrients are carbohydrate, protein, and fat.

inorganic: not containing carbon or pertaining to living things.

calories: units by which energy is measured. Food energy is measured in **kilocalories** (1000 calories equal 1 kilocalorie), abbreviated **kcalories** or kcal. One kcalorie is the amount of heat necessary to raise the temperature of 1 kilogram (kg) of water 1°C. The scientific use of the term *kcalorie* is the same as the popular use of the term *calorie.*

energy density: a measure of the energy a food provides relative to the amount of food (kcalories per gram).

To calculate the energy available from a food, multiply the number of grams of carbohydrate, protein, and fat by 4, 4, and 9, respectively. Then add the results together. For example, one slice of bread with 1 tablespoon of peanut butter on it contains 16 grams of carbohydrate, 7 grams of protein, and 9 grams of fat:

$$16 \text{ g carbohydrate} \times 4 \text{ kcal/g} = 64 \text{ kcal}$$
$$7 \text{ g protein} \times 4 \text{ kcal/g} = 28 \text{ kcal}$$
$$9 \text{ g fat} \times 9 \text{ kcal/g} = 81 \text{ kcal}$$
$$\text{Total} = 173 \text{ kcal}$$

From this information, you can calculate the percentage of kcalories each of the energy nutrients contributes to the total. To determine the percentage of kcalories from fat, for example, divide the 81 fat kcalories by the total 173 kcalories:

$$81 \text{ fat kcal} \div 173 \text{ total kcal} = 0.468 \text{ (rounded to 0.47)}$$

Then multiply by 100 to get the percentage:

$$0.47 \times 100 = 47\%$$

Dietary recommendations that urge people to limit fat intake to 20 to 35 percent of kcalories refer to the day's total energy intake, not to individual foods. Still, if the proportion of fat in each food choice throughout a day exceeds 35 percent of kcalories, then the day's total surely will, too. Knowing that this snack provides 47 percent of its kcalories from fat alerts a person to the need to make lower-fat selections at other times that day.

if you take in more energy than you expend, whether from carbohydrate, fat, or protein, the result is an increase in energy stores and weight gain. Similarly, if you take in less energy than you expend, the result is a decrease in energy stores and weight loss.

Alcohol, Not a Nutrient One other substance contributes energy: alcohol. The body derives energy from alcohol at the rate of 7 kcalories per gram. Alcohol is not a nutrient, however, because it cannot support the body's growth, maintenance, or repair. Nutrition in Practice 20 discusses alcohol's effects on nutrition.

IN SUMMARY

▮ Foods provide nutrients—substances that support the growth, maintenance, and repair of the body's tissues.

▮ The six classes of nutrients are water, carbohydrates, fats, proteins, vitamins, and minerals.

▮ Vitamins, minerals, and water facilitate a variety of activities in the body.

▮ Foods rich in the energy-yielding nutrients (carbohydrates, fats, and proteins) provide the major materials for building the body's tissues and yield energy the body can use or store.

▮ Energy is measured in kcalories.

Nutrient Recommendations

Appendix B presents nutrient recommendations developed by the World Health Organization.

Nutrient recommendations are used as standards to evaluate healthy people's intakes of energy and nutrients. Nutrition experts use the recommendations to assess nutrient intakes and to guide people regarding amounts to consume. Individuals can use them to determine how much of a nutrient they need to consume.

Dietary Reference Intakes

Defining the amounts of energy, nutrients, and other dietary components that best support health is a huge task. Nutrition experts have produced a set of standards that define the amounts of energy, nutrients, other dietary components, and physical

activity that best support health. These recommendations are called **Dietary Reference Intakes (DRI)** and reflect the collaborative efforts of scientists in both the United States and Canada.*[5] The DRI values are presented on the inside front covers.

Setting Nutrient Recommendations: RDA and AI One advantage of the DRI is that they apply to the diets of individuals. The DRI committee offers two sets of values to be used as nutrient intake goals by individuals: a set called the **Recommended Dietary Allowances (RDA)** and a set called **Adequate Intakes (AI).**

Based on solid experimental evidence and other reliable observations, the RDA are the foundation of the DRI. The AI values are based on less extensive scientific findings and rely more heavily on scientific judgment. The committee establishes an AI value whenever scientific evidence is insufficient to generate an RDA.[6] To see which nutrients have an AI and which have an RDA, see the inside front cover of this textbook.

In the last several decades, abundant new research has linked nutrients in the diet with the promotion of health and the prevention of chronic diseases. An advantage of the DRI is that, where appropriate, they take into account disease prevention as well as adequate nutrient intake. For example, the RDA for calcium is based on intakes that are thought to reduce the likelihood of osteoporosis-related fractures later in life.

To ensure that the vitamin and mineral recommendations meet the needs of as many people as possible, the recommendations are set near the top end of the range of the population's estimated average requirements (see Figure 1-2). Small amounts above the daily **requirement** do no harm, whereas amounts below the requirement may lead to health problems. When people's intakes are consistently **deficient,** their nutrient stores decline, and over time this decline leads to deficiency symptoms and poor health.

Facilitating Nutrition Research and Policy: EAR In addition to the RDA and AI, the DRI committee has established another set of values: **Estimated Average Requirements (EAR).** These values establish average requirements for given life-stage and gender groups that researchers and nutrition policymakers use in their work. Nutrition scientists may use the EAR as standards in research. Public health officials may use them to assess nutrient intakes of populations and make recommendations. EAR values form the scientific basis on which RDA are set.

Establishing Safety Guidelines: UL The DRI committee also establishes upper limits of intake for nutrients posing a hazard when consumed in excess. These values, the **Tolerable Upper Intake Levels (UL),** are indispensable to consumers who take supplements. Consumers need to know how much of a nutrient is too much. The UL are also of value to public health officials who set allowances for nutrients that are added to foods and water. The UL values are listed on the inside front cover of this textbook.

Using Nutrient Recommendations Each of the four DRI categories serves a unique purpose. For example, the EAR are most appropriately used to develop and evaluate nutrition programs for *groups,* such as schoolchildren or military personnel. The RDA (or AI, if an RDA is not available) can be used to set goals for *individuals.* The UL help to keep nutrient intakes below the amounts that increase the risk of toxicity. With these understandings, professionals can use the DRI for a variety of purposes.

*The DRI reports are produced by the Food and Nutrition Board, Institute of Medicine of the National Academies, with active involvement of scientists from Canada.

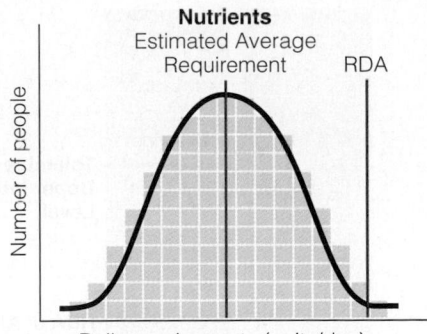

FIGURE 1-2 Nutrient Intake Recommendations
The nutrient intake recommendations are set high enough to cover nearly everyone's requirements (the boxes represent people).

Dietary Reference Intakes (DRI): a set of values for the dietary nutrient intakes of healthy people in the United States and Canada. These values are used for planning and assessing diets.

Recommended Dietary Allowances (RDA): a set of values reflecting the average daily amounts of nutrients considered adequate to meet the known nutrient needs of practically all healthy people in a particular life stage and gender group; a goal for dietary intake by individuals.

Adequate Intakes (AI): a set of values that are used as guides for nutrient intakes when scientific evidence is insufficient to determine an RDA.

requirement: the lowest continuing intake of a nutrient that will maintain a specified criterion of adequacy.

deficient: in regard to nutrient intake, describes the amount below which almost all healthy people can be expected, over time, to experience deficiency symptoms.

Estimated Average Requirements (EAR): the average daily nutrient intake levels estimated to meet the requirements of half of the healthy individuals in a given age and gender group; used in nutrition research and policymaking and as the basis on which RDA values are set.

Tolerable Upper Intake Levels (UL): a set of values reflecting the highest average daily nutrient intake levels that are likely to pose no risk of toxicity to almost all healthy individuals in a particular life stage and gender group. As intake increases above the UL, the potential risk of adverse health effects increases.

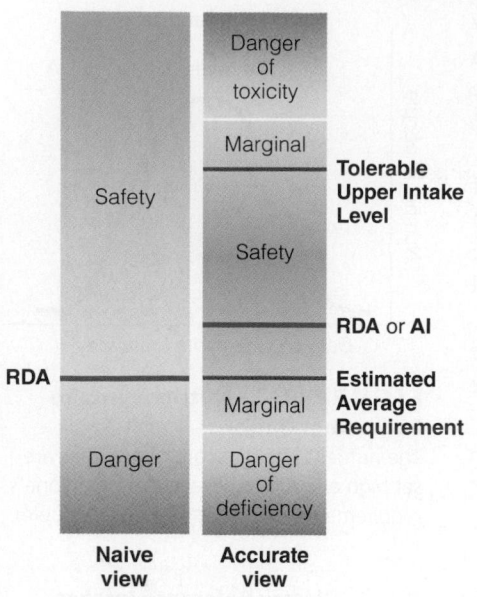

FIGURE 1-3 Naive versus Accurate View of Nutrient Intakes
The RDA or AI for a given nutrient represents a point that lies within a range of appropriate and reasonable intakes between toxicity and deficiency. Both of these recommendations are high enough to provide reserves in times of short-term dietary inadequacies, but not so high as to approach toxicity. Nutrient intakes above or below this range may be equally harmful.

In addition to understanding the unique purposes of the DRI, it is important to keep their uses in perspective. Consider the following:

- The values are recommendations for safe intakes, not minimum requirements; except for energy, they include a generous margin of safety. Figure 1-3 presents an accurate view of how a person's nutrient needs fall within a range, with marginal and danger zones both below and above the range.

- The values reflect daily intakes to be achieved on average, over time. They assume that intakes will vary from day to day, and they are set high enough to ensure that body nutrient stores will meet nutrient needs during periods of inadequate intakes lasting a day or two for some nutrients and up to a month or two for others.

- The values are chosen in reference to specific indicators of nutrient adequacy, such as blood nutrient concentrations, normal growth, and reduction of certain chronic diseases or other disorders when appropriate, rather than prevention of deficiency symptoms alone.

- The recommendations are designed to meet the needs of most healthy people. Medical problems alter nutrient needs, as later chapters describe.

- The recommendations are specific for people of both genders as well as various ages and stages of life: infants, children, adolescents, men, women, pregnant women, and lactating women.

Setting Energy Recommendations In contrast to the vitamin and mineral recommendations, the recommendation for energy, called the **Estimated Energy Requirement (EER),** is not generous because excess energy cannot be excreted and is eventually stored as body fat. Rather, the key to the energy recommendation is balance. For a person whose body weight, body composition, and physical activity level are consistent with good health, energy intake from food should match energy expenditure, so the person achieves energy balance. Enough energy is needed to sustain a healthy, active life, but too much energy leads to obesity. The EER is therefore set at a level of energy intake predicted to maintain energy balance in a healthy adult of a defined age, gender, weight, height, and physical activity level.* Another difference between the requirements for other nutrients and those for energy is that each person has an obvious indicator of whether energy intake is inadequate, adequate, or excessive: body weight. Because *any* amount of energy in excess of need leads to weight gain, the DRI committee did not set a Tolerable Upper Intake Level.

Acceptable Macronutrient Distribution Ranges (AMDR)

As noted earlier, the DRI committee considers prevention of chronic disease as well as nutrient adequacy when establishing recommendations. To that end, the committee established healthy ranges of intakes for the energy-yielding nutrients—carbohydrate, fat, and protein—known as **Acceptable Macronutrient Distribution Ranges (AMDR).** Each of these three energy-yielding nutrients contributes to a person's total energy (kcalorie) intake, and those contributions vary in relation to each other. The DRI committee has determined that a diet that provides the energy-yielding nutrients in the following proportions provides adequate energy and nutrients and reduces the risk of chronic disease:

- 45 to 65 percent of kcal from carbohydrate

- 20 to 35 percent of kcal from fat

- 10 to 35 percent of kcal from protein

*The EER for children, pregnant women, and lactating women includes energy needs associated with the deposition of tissue or the secretion of milk at rates consistent with good health.

- The Dietary Reference Intakes (DRI) are a set of nutrient intake values that can be used to plan and evaluate dietary intakes for healthy people.

- The Estimated Energy Requirement (EER) defines the energy intake level needed to maintain energy balance in a healthy adult of a defined age, gender, weight, height, and physical activity level.

- The Acceptable Macronutrient Distribution Ranges (AMDR) define the proportions contributed by carbohydrate, fat, and protein to a healthy diet.

National Nutrition Surveys

How do nutrition experts know whether people are meeting nutrient recommendations? The Dietary Reference Intakes and other major reports that examine the relationships between diet and health depend on information collected from nutrition surveys. Researchers use nutrition surveys to learn which foods people are eating and which supplements they are taking, to assess people's nutritional health, and to determine people's knowledge, attitudes, and behaviors about nutrition and how these relate to health. The resulting wealth of information can be used for a variety of purposes. For example, Congress uses this information to establish public policy on nutrition education, to assess food assistance programs, and to regulate the food supply. The food industry uses the information to guide decisions in public relations and product development. Scientists use the information to establish research priorities. One of the first nutrition surveys, taken before World War II, suggested that up to a third of the U.S. population might be eating poorly. Programs to correct **malnutrition** have been evolving ever since.

Coordinating Nutrition Survey Data

The National Nutrition Monitoring program coordinates the many nutrition-related activities of various federal agencies. All major reports that examine the contribution of diet and nutrition status to the health of the people of the United States depend on information collected and coordinated by this national program. The integration of two major national surveys provides comprehensive data efficiently. One portion of the survey collects data on the kinds and amounts of foods people eat.* Researchers then calculate the energy and nutrients in the foods and compare the amounts consumed with standards such as the DRI. The other portion of the survey examines the people themselves, using nutrition assessment methods.† The data provide valuable information on several nutrition-related conditions such as growth retardation, heart disease, and nutrient deficiencies. These data also provide the basis for developing and monitoring national health goals.

National Health Goals

The **Healthy People** program identifies the nation's health priorities and guides policies that promote health and prevent disease. At the start of each decade, the program sets goals for improving the nation's health during the following 10 years. Nutrition and Weight Status is one of 38 topic areas of Healthy People 2020, each with numerous objectives (www.healthypeople.gov).

*This survey was formerly called the Continuing Survey of Food Intakes by Individuals (CSFII), conducted by the U.S. Department of Agriculture (USDA).

†This survey is known as the National Health and Nutrition Examination Survey (NHANES).

Estimated Energy Requirement (EER): the dietary energy intake level that is predicted to maintain energy balance in a healthy adult of a defined age, gender, weight, and physical activity level consistent with good health.

Acceptable Macronutrient Distribution Ranges (AMDR): ranges of intakes for the energy-yielding nutrients that provide adequate energy and nutrients and reduce the risk of chronic disease.

malnutrition: any condition caused by deficient or excess energy or nutrient intake or by an imbalance of nutrients.

Healthy People: a national public health initiative under the jurisdiction of the U.S. Department of Health and Human Services (DHHS) that identifies the most significant preventable threats to health and focuses efforts toward eliminating them.

IN SUMMARY

▪ Nutrition surveys measure people's food consumption and evaluate the nutrition status of populations.

▪ Information gathered from nutrition surveys serves as the basis for many major diet and nutrition reports, including Healthy People.

Dietary Guidelines and Food Guides

Today, government authorities are as much concerned about **overnutrition** as they once were about **undernutrition.** Research confirms that dietary excesses, especially of energy, certain fats, and alcohol, contribute to many **chronic diseases,** including heart disease, cancer, stroke, diabetes, and liver disease.[7] Only two common life-style habits have more influence on health than a person's choice of diet: smoking and other tobacco use, and excessive drinking of alcohol. Table 1-2 lists the leading causes of death in the United States; notice that the top three are nutrition related (and related to tobacco use). Note, however, that although diet is a powerful influence on these diseases, they cannot be prevented by a healthy diet alone; genetics, physical activity, and lifestyle also play a role. Within the range set by genetic inheritance, however, disease development is strongly influenced by the foods a person chooses to eat.

So, how can health care professionals help people select foods to create a diet that supplies all the needed nutrients in amounts consistent with good health? The principle is simple enough: encourage clients to eat a variety of foods that supply all the nutrients the body needs. In practice, how do people do this? It helps to keep in mind that a nutritious diet achieves six basic ideals.

Dietary Ideals

A nutritious diet has six characteristics. The first, **adequacy,** was already addressed in the earlier discussion on the DRI. An adequate diet includes enough energy and enough of every nutrient (as well as fiber) to meet the needs of healthy people. Second is **balance:** the food choices do not overemphasize one nutrient or food type at the expense of another.

The essential minerals calcium and iron illustrate the importance of dietary balance. Meats, fish, and poultry are rich in iron but poor in calcium. Conversely, milk and milk products are rich in calcium but poor in iron. Use some meat or meat alternatives for iron; use some milk and milk products for calcium; and save some space for other

A nutritious diet has the following characteristics:
▪ Adequacy
▪ Balance
▪ kCalorie control
▪ Nutrient density
▪ Moderation
▪ Variety

Balance in the diet helps to ensure adequacy.

overnutrition: overconsumption of food energy or nutrients sufficient to cause disease or increased susceptibility to disease; a form of malnutrition.

undernutrition: underconsumption of food energy or nutrients severe enough to cause disease or increased susceptibility to disease; a form of malnutrition.

chronic diseases: diseases characterized by slow progression, long duration, and degeneration of body organs due in part to such personal lifestyle elements as poor food choices, smoking, alcohol use, and lack of physical activity.

TABLE 1-2 Leading Causes of Death in the United States

The diseases in bold italics are nutrition related.

1. *Heart disease*
2. *Cancers*
3. *Strokes*
4. Chronic lung diseases
5. Accidents
6. Alzheimer's disease
7. *Diabetes mellitus*
8. Pneumonia and influenza
9. Kidney disease
10. Infections of the blood

Source: National Center for Health Statistics, 2009.

foods, too, because a diet consisting of milk and meat alone would not be adequate. For other nutrients, people need to consume whole grains, vegetables, and fruit.

The third characteristic is **kcalorie control:** the foods provide the amount of energy needed to maintain a healthy body weight—not more, not less. The key to kcalorie control is to select foods that deliver the most nutrients for the least food energy. This fourth characteristic is known as **nutrient density.** Consider foods containing calcium, for example. You can get about 300 milligrams of calcium from either 1½ ounces of cheddar cheese or 1 cup of fat-free milk, but the cheese delivers about twice as much food energy (kcalories) as the milk. The fat-free milk, then, is twice as calcium dense as the cheddar cheese; it offers the same amount of calcium for half the kcalories. Both foods are excellent choices for adequacy's sake alone, but to achieve adequacy while controlling kcalories, the fat-free milk is the better choice. (Alternatively, a person could select a low-fat cheddar cheese providing kcalories comparable to fat-free milk.)

Just as a financially responsible person pays for rent, food, clothes, and tuition on a limited budget, healthy people obtain iron, calcium, and all the other essential nutrients on a limited energy (kcalorie) allowance. Success depends on getting many nutrients for each kcalorie "dollar." For example, a can of cola and handful of grapes may both provide about the same number of kcalories, but grapes deliver many more nutrients. A person who makes nutrient-dense choices, such as fruit instead of cola, can meet daily nutrient needs on a lower energy budget. Such choices support good health.

Foods that are notably low in nutrient density—such as potato chips, candy, and colas—are sometimes called **empty-kcalorie foods.** The kcalories these foods provide are called "empty" because they deliver energy (from sugar, fat, or both) with little or no protein, vitamins, or minerals.

The fifth characteristic of a nutritious diet is **moderation.** Foods rich in fat and sugar provide enjoyment and energy but relatively few nutrients. In addition, they promote weight gain when eaten in excess. A person who practices moderation eats such foods only on occasion and regularly selects foods low in solid fats and added sugars, a practice that automatically improves nutrient density. Returning to the example of cheddar cheese and fat-free milk, the milk not only offers more calcium for less energy, but it contains far less fat than the cheese.

Finally, the sixth characteristic of a nutritious diet is **variety:** the foods chosen differ from one day to the next. A diet may have all the virtues just described and still lack variety if a person eats the same foods day after day. People should select foods from each of the food groups daily and vary their choices within each food group from day to day for a couple of reasons. First, different foods within the same group contain different arrays of nutrients. Among the fruits, for example, strawberries are especially rich in vitamin C, while apricots are rich in vitamin A. Variety improves nutrient adequacy.[8] Second, no food is guaranteed entirely free of substances that, in excess, could be harmful. The strawberries might contain trace amounts of one contaminant, the apricots another. By alternating fruit choices, a person will ingest very little of either contaminant.

Dietary Guidelines for Americans

The *Dietary Guidelines for Americans 2010* provide evidence-based advice to help people attain and maintain a healthy weight, reduce the risk of chronic diseases, and promote overall health through diet and physical activity.[9] Table 1-3 (p. 14) presents the key recommendations of the *Dietary Guidelines for Americans 2010*, clustered into four major topic areas. The first area focuses on balancing kcalories to manage a healthy body weight by improving eating habits and engaging in regular physical activity. The second area advises people to reduce their intakes of such foods and food components as sodium, solid fats (and the saturated and *trans* fatty acids they contain), cholesterol,

Nutrient density promotes adequacy and kcalorie control.

Moderation contributes to adequacy, balance, and kcalorie control.

adequacy: the characteristic of a diet that provides all the essential nutrients, fiber, and energy necessary to maintain health and body weight.

balance: the dietary characteristic of providing foods of a number of types in proportion to each other, such that foods rich in some nutrients do not crowd out foods that are rich in other nutrients.

kcalorie (energy) control: management of food energy intake.

nutrient density: a measure of the nutrients a food provides relative to the energy it provides. The more nutrients and the fewer kcalories, the higher the nutrient density.

empty-kcalorie foods: a popular term used to denote foods that contribute energy but lack protein, vitamins, and minerals.

moderation: providing enough, but not too much, of a substance.

variety (dietary): eating a wide selection of foods within and among the major food groups (the opposite of monotony).

TABLE 1-3 Key Recommendations of the *Dietary Guidelines for Americans 2010*

Balancing kCalories to Manage Weight
- Prevent and/or reduce overweight and obesity through improved eating and physical activity behaviors.
- Control total kcalorie intake to manage body weight. For people who are overweight or obese, this will mean consuming fewer kcalories from foods and beverages.
- Increase physical activity and reduce time spent in sedentary behaviors.
- Maintain appropriate kcalorie balance during each stage of life—childhood, adolescence, adulthood, pregnancy and breastfeeding, and older age.

Foods and Food Components to Reduce
- Reduce daily sodium intake to less that 2300 milligrams and further reduce intake to 1500 milligrams among persons who are 51 and older and those of any age who are African American or have hypertension, diabetes, or chronic kidney disease. The 1500 milligrams recommendation applies to about half of the U.S. population, including children and the majority of adults.
- Consume less than 10 percent of kcalories from saturated fatty acids by replacing them with monounsaturated and polyunsaturated fatty acids.
- Consume less than 300 milligrams per day of dietary cholesterol.
- Keep *trans* fatty acid consumption as low as possible by limiting foods that contain synthetic sources of *trans* fats, such as partially hydrogenated oils, and by limiting other solid fats.
- Reduce the intake of kcalories from solid fats and added sugars.
- Limit the consumption of foods that contain refined grains, especially refined grain foods that contain solid fats, added sugars, and sodium.
- If alcohol is consumed it should be consumed in moderation—up to one drink per day for women and two drinks per day for men—and only by adults of legal drinking age.

Foods and Nutrients to Increase
- Increase vegetable and fruit intake.
- Eat a variety of vegetables, especially dark-green and red and orange vegetables and beans and peas.
- Consume at least half of all grains as whole grains. Increase whole-grain intake by replacing refined grains with whole grains.
- Increase intake of fat-free or low-fat milk and milk products, such as milk, yogurt, cheese, or fortified soy beverages.
- Choose a variety of protein foods, which include seafood, lean meat and poultry, eggs, beans and peas, soy products, and unsalted nuts and seeds.
- Increase the amount and variety of seafood consumed by choosing seafood in place of some meat and poultry.
- Replace protein foods that are higher in solid fats with choices that are lower in solid fats and kcalories and/or are sources of oils.
- Use oils to replace solid fats where possible.
- Choose foods that provide more potassium, dietary fiber, calcium, and vitamin D, which are nutrients of concern in American diets. These foods include vegetables, fruits, whole grains, and milk and milk products.

Building Healthy Eating Patterns
- Select an eating pattern that meets nutrient needs over time at an appropriate kcalorie level.
- Account for all foods and beverages consumed and assess how they fit within a total healthy eating pattern.
- Follow food safety recommendations when preparing and eating foods to reduce the risk of food-borne illnesses.

Note: These guidelines are intended for adults and children ages 2 and older.
Source: The *Dietary Guidelines for Americans 2010,* available at www.dietaryguidelines.gov

added sugars, refined grain products, and alcoholic beverages (for those who partake). The third area encourages consumers to select a variety of fruits and vegetables, whole grains, and low-fat milk products and protein foods (including seafood). The fourth area helps consumers build healthy eating patterns that meet energy and nutrient needs while reducing the risk of foodborne illnesses. Together, the *Dietary Guidelines for Americans 2010* point the way toward longer, healthier, and more active lives.

Some people might wonder why *dietary* guidelines include recommendations for physical activity. The simple answer is that most people who maintain a healthy body weight do more than eat right. They also exercise—the equivalent of 30 to 60 minutes or more of moderately intense physical activity on most days.[10] As you will see repeatedly throughout this text, food and physical activity choices are integral partners in supporting good health.

▌ A well-planned diet delivers adequate nutrients, a balanced array of nutrients, and an appropriate amount of energy.

▌ A well-planned diet is based on nutrient-dense foods, moderate in substances that can be detrimental to health, and varied in its selections.

▌ The *Dietary Guidelines* apply these principles, offering practical advice on how to eat for good health.

The USDA Food Guide

To help people achieve the goals set forth by the *Dietary Guidelines for Americans* (see Table 1-3), the USDA provides a **food group plan**—the **USDA Food Guide**—that builds a diet from categories of foods that are similar in vitamin and mineral content. Thus, each group provides a set of nutrients that differs somewhat from the nutrients supplied by the other groups. Selecting foods from each of the groups eases the task of creating an adequate and balanced diet.

Figure 1-4 (on pp. 16–17) presents the USDA Food Guide. The USDA Food Guide assigns foods to five major food groups and recommends daily amounts of foods from each group to meet nutrient needs. The judicious use of oils, solid fats, and added sugars is also described in the USDA Food Guide. In addition to presenting the food groups, the figure indicates the most notable nutrients of each group and lists some foods within each group, sorted by nutrient density.

Recommended amounts of food are given in cups for fruits, vegetables, and milk and in ounces for grains and protein foods. Figure 1-4 provides equivalent measures for foods that are not readily measured in cups and ounces. For example, 1 ounce of grains is considered equivalent to 1 slice of bread or ½ cup of cooked rice.

Recommended Daily Food Amounts All food groups offer valuable nutrients, and people should make selections from each group daily. Table 1-4 (p. 18) specifies the amounts of food needed from each group daily to create a healthful diet for several energy (kcalorie) levels. Estimated daily kcalorie needs for sedentary and active

The DASH Eating Plan, presented in Chapter 22, is another dietary pattern that meets the goals of the *Dietary Guidelines for Americans*.

Five food groups:
▌ Fruits
▌ Vegetables
▌ Grains
▌ Protein Foods
▌ Milk and milk products

Chapter 12 provides a food guide for children, and Appendix B presents Canada's food group plan, *Eating Well with Canada's Food Guide*.

Chapter 6 explains how to determine energy needs.

A portion of grains is 1 ounce, yet most bagels today weigh 4 ounces or more—meaning that a single bagel can easily supply four or more portions of grains, not one, as many people assume.

© Matthew Farruggio

food group plan: a diet-planning tool that sorts foods into groups based on nutrient content and then specifies that people should eat certain amounts of food from each group.

USDA Food Guide: the USDA's food group plan for ensuring dietary adequacy that assigns foods to five major food groups.

GRAINS

© Polara Studios, Inc.

Make at least half of the grain selections whole grains.

These foods contribute folate, niacin, riboflavin, thiamin, iron, magnesium, selenium, and fiber.

> 1 oz grains is equivalent to 1 slice bread; ½ c cooked rice, pasta, or cereal;
> 1 oz dry pasta or rice; 1 c ready-to-eat cereal; 3 c popped popcorn.

● Whole grains (amaranth, barley, brown rice, buckwheat, bulgur, millet, oats, quinoa, rye, wheat) and whole-grain, low-fat breads, cereals, crackers, and pastas; popcorn.

● Enriched bagels, breads, cereals, pastas (couscous, macaroni, spaghetti), pretzels, rice, rolls, tortillas.

△ Biscuits, cakes, cookies, cornbread, crackers, croissants, doughnuts, French toast, fried rice, granola, muffins, pancakes, pastries, pies, presweetened cereals, taco shells, waffles.

VEGETABLES

© Polara Studios, Inc.

Choose a variety of vegetables each day, and choose from all five subgroups several times a week.

These foods contribute folate, vitamin A, vitamin C, vitamin K, vitamin E, magnesium, potassium, and fiber.

> 1 c vegetables is equivalent to 1 c cut-up raw or cooked vegetables;
> 1 c cooked legumes; 1 c vegetable juice; 2 c raw, leafy greens.

Vegetable subgroups

● Dark green vegetables: Broccoli and leafy greens such as arugula, beet greens, bok choy, collard greens, kale, mustard greens, romaine lettuce, spinach, and turnip greens.

● Red and orange vegetables: Carrots, carrot juice, pumpkin, red peppers, sweet potatoes, tomatoes, and winter squash (acorn, butternut).

● Legumes: Black beans, black-eyed peas, garbanzo beans (chickpeas), kidney beans, lentils, navy beans, pinto beans, soybeans and soy products such as tofu, and split peas.

● Starchy vegetables: Cassava, corn, green peas, hominy, lima beans, and potatoes.

● Other vegetables: Artichokes, asparagus, bamboo shoots, bean sprouts, beets, Brussels sprouts, cabbages, cactus, cauliflower, celery, cucumbers, eggplant, green beans, iceberg lettuce, mushrooms, okra, onions, peppers, seaweed, snow peas, vegetable juices, zucchini.

△ Baked beans, candied sweet potatoes, coleslaw, French fries, potato salad, refried beans, scalloped potatoes, tempura vegetables.

FRUITS

© Polara Studios, Inc.

Consume a variety of fruits and no more than one-half of the recommended intake as fruit juice.

These foods contribute folate, vitamin A, vitamin C, potassium, and fiber.

> 1 c fruit is equivalent to 1 c fresh, frozen, or canned fruit; ½ c dried fruit;
> 1 c fruit juice.

● Apples, apricots, avocados, bananas, blueberries, cantaloupe, cherries, grapefruit, grapes, guava, kiwi, mango, nectarines, oranges, papaya, peaches, pears, pineapples, plums, raspberries, strawberries, tangerines, watermelon; dried fruit (dates, figs, raisins); unsweetened juices.

△ Canned or frozen fruit in syrup; juices, punches, ades, and fruit drinks with added sugars; fried plantains.

FIGURE 1-4 USDA Food Guide

MILK AND MILK PRODUCTS

© Polara Studios, Inc.

Make fat-free or low-fat choices. Choose lactose-free products or other calcium-rich foods if you don't consume milk.

These foods contribute protein, riboflavin, vitamin B_{12}, calcium, magnesium, potassium, and, when fortified, vitamin A and vitamin D.

> 1 c milk is equivalent to 1 c milk, fortified soy beverage, or yogurt; $1\frac{1}{2}$ oz natural cheese; 2 oz processed cheese.

● Fat-free milk and fat-free milk products such as buttermilk, cheeses, cottage cheese, yogurt; fat-free fortified soy beverage.

△ 1% low-fat milk, 2% reduced-fat milk, and whole milk; low-fat, reduced-fat, and whole-milk products such as cheeses, cottage cheese, and yogurt; milk products with added sugars such as chocolate milk, custard, ice cream, ice milk, milk shakes, pudding, sherbet; fortified soy beverage.

PROTEIN FOODS

© Polara Studios, Inc.

Make lean or low-fat choices. Prepare them with little, or no, added fat.

Meat, poultry, fish, and eggs contribute protein, niacin, thiamin, vitamin B_6, vitamin B_{12}, iron, magnesium, potassium, and zinc; legumes and nuts are notable for their protein, folate, thiamin, vitamin E, iron, magnesium, potassium, zinc, and fiber.

> 1 oz meat is equivalent to 1 oz cooked lean meat, poultry, or seafood; 1 egg; $\frac{1}{4}$ c cooked legumes or tofu; 1 tbs peanut butter; $\frac{1}{2}$ oz nuts or seeds.

● Seafood: Fish, shellfish

● Meat, poultry, eggs: Lean or low-fat meat (fat-trimmed beef, game, ham, lamb, pork), poultry (no skin), eggs

● Nuts, seeds, soy products: Unsalted nuts (almonds, filberts, pistachios, walnuts), seeds (flaxseeds, pumpkin seeds, sunflower seeds), legumes, low-fat tofu, tempeh, peanuts, peanut butter

△ Bacon; baked beans; fried meat, fish, poultry, eggs, or tofu; refried beans; ground beef; hot dogs; luncheon meats; marbled steaks; poultry with skin; sausages; spare ribs.

OILS

Matthew Farruggio

Select the recommended amounts of oils from among these sources.

These foods contribute vitamin E and essential fatty acids (see Chapter 4), along with abundant kcalories.

> 1 tsp oil is equivalent to 1 tbs low-fat mayonnaise; 2 tbs light salad dressing; 1 tsp vegetable oil; 1 tsp soft margarine.

● Liquid vegetable oils such as canola, corn, flaxseed, nut, olive, peanut, safflower, sesame, soybean, and sunflower oils; mayonnaise, oil-based salad dressing, soft *trans*-free margarine.

● Unsaturated oils that occur naturally in foods such as avocados, fatty fish, nuts, olives, seeds (flaxseeds, sesame seeds), and shellfish.

SOLID FATS AND ADDED SUGARS

Matthew Farruggio

Limit intakes of food and beverages with solid fats and added sugars.

Solid fats deliver saturated fat and *trans* fat, and intake should be kept low. Solid fats and added sugars contribute abundant kcalories but few nutrients, and intakes should not exceed the discretionary kcalorie allowance—kcalories to meet energy needs after all nutrient needs have been met with nutrient-dense foods. Alcohol also contributes abundant kcalories but few nutrients, and its kcalories are counted among discretionary kcalories. See Table 1-4 for some discretionary kcalorie allowances.

△ Solid fats that occur in foods naturally such as milk fat and meat fat (see △ in previous lists).

△ Solid fats that are often added to foods such as butter, cream cheese, hard margarine, lard, sour cream, and shortening.

△ Added sugars such as brown sugar, candy, honey, jelly, molasses, soft drinks, sugar, and syrup.

△ Alcoholic beverages such as beer, wine, and liquor.

FIGURE 1-4 (continued)

TABLE 1-4 Recommended Daily Amounts from Each Food Group

FOOD GROUP	1600 kcal	1800 kcal	2000 kcal	2200 kcal	2400 kcal	2600 kcal	2800 kcal	3000 kcal
Fruits	1½ c	1½ c	2 c	2 c	2 c	2 c	2½ c	2½ c
Vegetables	2 c	2½ c	2½ c	3 c	3 c	3½ c	3½ c	4 c
Grains	5 oz	6 oz	6 oz	7 oz	8 oz	9 oz	10 oz	10 oz
Protein foods	5 oz	5 oz	5½ oz	6 oz	6½ oz	6½ oz	7 oz	7 oz
Milk	3 c	3 c	3 c	3 c	3 c	3 c	3 c	3 c
Oils	5 tsp	5 tsp	6 tsp	6 tsp	7 tsp	8 tsp	8 tsp	10 tsp
Discretionary kcalorie allowance	121 kcal	161 kcal	258 kcal	266 kcal	330 kcal	362 kcal	345 kcal	459 kcal

men and women are shown in Table 1-5. For example, a person needing 2000 kcalories a day would select 2 cups of fruit; 2½ cups of vegetables (dispersed among the vegetable subgroups); 6 ounces of grain foods (with at least half coming from whole grains); 5½ ounces of protein foods such as meat, poultry, or fish, or the equivalent of **legumes,** eggs, seeds, or nuts; and 3 cups of milk or yogurt, or the equivalent of cheese or fortified soy products. Additionally, a small amount of unsaturated oil, such as vegetable oil or the oils of nuts, olives, or fatty fish, is required to supply needed nutrients.

All vegetables provide an array of vitamins, fiber, and the mineral potassium, but some vegetables are especially good sources of certain nutrients and beneficial phytochemicals. For this reason, the USDA Food Guide sorts the vegetable group into subgroups. The dark green vegetables deliver the B vitamin folate; the orange vegetables provide abundant vitamin A; legumes supply iron and protein; the starchy vegetables contribute abundant carbohydrate energy; and the other vegetables fill in the gaps and add more of these same nutrients.

In a 2000-kcalorie diet, then, the recommended 2½ cups of daily vegetables should be spread among the subgroups over a week's time. In other words, eating 2½ cups of potatoes or even spinach every day for seven days would *not* meet the recommended vegetable intakes. Potatoes and spinach make excellent choices when consumed in balance with other vegetables from other subgroups. Intakes of vegetables are appropriately averaged over a week's time—it isn't necessary to include every subgroup every day.

For similar reasons, the USDA Food Guide sorts protein foods into three subgroups. Perhaps most notably, each of these subgroups contributes a different assortment of fats. Table 1-6 presents the recommended weekly amounts for each of the subgroups for vegetables and protein foods.

Notable Nutrients As Figure 1-4 (pp. 16–17) notes, each food group contributes key nutrients. This feature provides flexibility in diet planning: a person can select any food from a food group and receive similar nutrients. For example, a person can choose milk, cheese, or yogurt and receive the same key nutrients the milk group offers. Legumes are included in the protein foods group because these foods are notable for their contributions of protein, iron, and zinc. Legumes are unique, however, because they are also grouped with the vegetables by virtue of their fiber and vitamins. Thus legumes can count in either the protein foods group or the vegetable group. In general, people who regularly eat meat, poultry, and fish count legumes as vegetables, and vegetarians and others who seldom eat meat, poultry, or fish, count legumes in the protein foods group.

TABLE 1-5 Estimated Daily kCalorie Needs for Adults

	Sedentary[a]	Active[b]
Women		
19–30 yr	1900	2400
31–50 yr	1800	2200
51+ yr	1600	2100
Men		
19–30 yr	2500	3000
31–50 yr	2300	2900
51+ yr	2100	2600

Note: kCalorie values reflect the midpoint of the range appropriate for age and gender, but within each group older adults may need fewer kcalories and younger adults may need more. In addition to gender, age, and activity level, energy needs vary with height and weight (see Chapter 6).

[a]*Sedentary* describes a lifestyle that includes only the activities typical of day-to-day life.
[b]*Active* describes a lifestyle that includes physical activity equivalent to walking more than 3 miles per day at a rate of 3 to 4 miles per hour, in addition to the activities typical of day-to-day life.

TABLE 1-6 Recommended Weekly Amounts from Vegetable and Protein Foods Subgroups

Table 1-4 specifies the recommended amounts of total vegetables and protein foods per *day*. This table shows those amounts dispersed among five vegetable and three protein foods subgroups per *week*.

Protein Foods Subgroup	1600 kcal	1800 kcal	2000 kcal	2200 kcal	2400 kcal	2600 kcal	2800 kcal	3000 kcal
Seafood	8 oz	8 oz	8 oz	9 oz	10 oz	10 oz	11 oz	11 oz
Meat, poultry, eggs	24 oz	24 oz	26 oz	29 oz	31 oz	31 oz	34 oz	34 oz
Nuts, seeds, soy products	4 oz	4 oz	4 oz	4 oz	5 oz	5 oz	5 oz	5 oz
Vegetable Subgroup								
Dark green	1½ c	1½ c	1½ c	2 c	2 c	2½ c	2½ c	2½ c
Red and orange	4 c	5½ c	5½ c	6 c	6 c	7 c	7 c	7½ c
Legumes	1 c	1½ c	1½ c	2 c	2 c	2½ c	2½ c	3 c
Starchy	4 c	5 c	5 c	6 c	6 c	7 c	7 c	8 c
Other	3½ c	4 c	4 c	5 c	5 c	5½ c	5½ c	7 c

The USDA Food Guide encourages greater consumption from certain food groups to provide the nutrients most often lacking in the diets of Americans. In general, most people need to eat:

- *More* dark green vegetables, red and orange vegetables, legumes, fruits, whole grains, and low-fat milk and milk products.

- *Fewer* refined grains, total fats (especially saturated fat, *trans* fat, and cholesterol), added sugars, and total kcalories.

> The USDA nutrient of concern is vitamin D; the minerals calcium and potassium; and dietary fiber.

Nutrient Density The USDA Food Guide provides a foundation for a healthy diet by emphasizing nutrient-dense options within each food group. By consistently selecting nutrient-dense foods, a person can both obtain all the nutrients needed and keep kcalories under control. In contrast, eating foods that are low in nutrient density makes it difficult to get enough nutrients without exceeding energy needs and gaining weight. For this reason, consumers should select low-fat foods from each group and foods without added fats or sugars—for example, fat-free milk instead of whole milk, baked chicken without the skin instead of hot dogs, green beans instead of french fries, orange juice instead of fruit punch, and whole-wheat bread instead of biscuits. Notice that Figure 1-4 (pp. 16–17) provides a key indicating which foods *within each group* are high or low in nutrient density. Oil is a notable exception: even though oil is pure fat and therefore rich in kcalories, a small amount of oil from sources such as nuts, fish, or vegetable oils is necessary every day to provide nutrients lacking from other foods. Consequently, these high-fat foods are listed among the nutrient-dense foods (see Nutrition in Practice 3 to learn why).

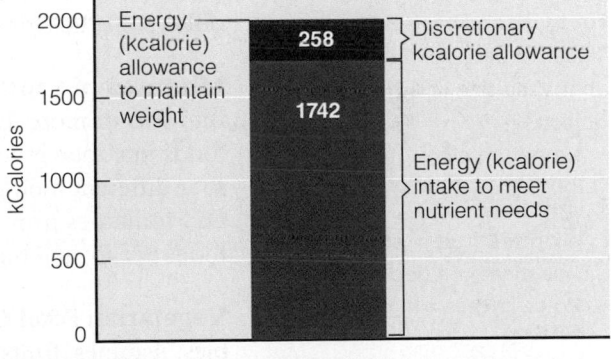

FIGURE 1-5 Discretionary kCalorie Allowance in a 2000-kCalorie Diet

Discretionary kCalorie Allowance At each kcalorie level, people who consistently choose nutrient-dense foods may be able to meet their nutrient needs without consuming their full allotment of kcalories. This difference between the kcalories needed to supply nutrients and those needed for energy—known as the **discretionary kcalorie allowance**—is illustrated in Figure 1-5.

Table 1-4 includes the discretionary kcalorie allowance for several kcalorie levels. A person with discretionary kcalories available might choose to:

- Eat additional nutrient-dense foods, such as an extra portion of skinless chicken or a second ear of corn.

- Select a few foods with fats or added sugars, such as reduced-fat milk or sweetened cereal.

> **legumes** (lay-GYOOMS, LEG-yooms): plants of the bean and pea family with seeds that are rich in protein compared with other plant-derived foods.
>
> **discretionary kcalorie allowance:** the kcalories remaining in a person's energy allowance after consuming enough nutrient-dense foods to meet all nutrient needs for a day.

- Add a little fat or sugar to foods, such as butter or jelly on toast.
- Consume some alcohol. (Nutrition in Practice 20 explains why this may not be a good choice for some individuals.)

Alternatively, a person wanting to lose weight might choose to:

- Omit the kcalories available from the discretionary kcalorie allowance.

Compared with physically active adults, sedentary adults have lower kcalorie needs, so they have a lower discretionary kcalorie allowance. Physically active adults expend more energy and therefore have a higher energy need and a greater discretionary allowance.

Added fats and sugars are always counted as discretionary kcalories. The kcalories from the fat in higher-fat milks and meats are also counted among discretionary kcalories. It helps to think of fat-free milk as "milk," and whole milk or reduced-fat milk as "milk with added fat." Similarly, "meats" should be the leanest; other cuts are "meats with added fat." Puddings and other desserts made from whole milk provide discretionary kcalories from both the sugar added to sweeten them and the naturally occurring fat in the whole milk they contain. Even fruits, vegetables, and grains can carry discretionary kcalories into the diet in the form of peaches canned in syrup, French fries, or high-fat crackers.

Discretionary kcalories must be counted separately from the kcalories of the nutrient-dense foods of which they may be a part. A fried chicken leg, for example, provides discretionary kcalories from two sources: the naturally occurring fat of the chicken skin and the added fat absorbed during frying. The kcalories of the skinless chicken underneath are not discretionary kcalories—they are necessary to provide the nutrients of chicken.

Portion Control To control kcalories, the diet planner must also learn to control food portions (USDA serving equivalents are listed in Figure 1-4 on p. 18). The trend in the United States has been toward consuming larger food portions, especially of foods rich in fat and sugar. At the same time, body weights have been steadily creeping upward, suggesting an increasing need to control portion sizes. In contrast to the random-size (and often large) helpings offered in restaurants, fast-food franchises, and elsewhere, the quantities recommended in the USDA Food Guide are specific, precise, and reliable for delivering certain amounts of key nutrients in food. The margin offers some tips for estimating portion sizes.

Mixtures of Foods Some foods—such as casseroles, soups, and sandwiches—fall into two or more food groups. With a little practice, users can learn to divide these foods into food groups. From the USDA Food Guide point of view, a taco represents four different food groups: the taco shell from the grains group; the onions, lettuce, and tomatoes from the "other vegetable" group; the ground beef from the protein foods group; and the cheese from the milk group.

Vegetarian Food Guide Vegetarian diets rely mainly on plant foods: grains, vegetables, legumes, fruits, seeds, and nuts. Some vegetarian diets include eggs, milk products, or both. People who do not eat meats or milk products can still use the USDA Food Guide to create an adequate diet.[11] The food groups are similar, and the recommended amounts of foods remain the same. Vegetarians select protein foods such as legumes, seeds, nuts, tofu, and, for those who eat them, eggs. Legumes and at least one cup of dark green leafy vegetables help to supply the iron that meats usually provide. Vegetarians who do not drink cow's milk can use soy "milk"—a product made from soybeans that provides similar nutrients if it has been fortified with calcium, vitamin D, and vitamin B_{12}. Nutrition in Practice 4 presents a Food Guide for Vegetarians, defines vegetarian terms, and provides more information on vegetarian diet planning.

Ethnic Food Choices People can use the USDA Food Guide and still enjoy a diverse array of culinary styles by sorting ethnic foods into their appropriate food groups. For example, a person eating Mexican foods would find tortillas in the grains

For quick and easy estimates, visualize each portion as being about the size of a common object:

- 1 medium piece of fruit or potato = a regular (60-watt) light bulb
- ¼ c dried fruit = a golf ball
- 3 oz meat = a deck of cards
- 1½ oz cheese = a nine-volt battery
- ½ c ice cream = a racquetball
- 2 Tbsp peanut butter = a ping-pong ball

group, jicama in the vegetable group, and guava in the fruit group. Table 1-1 (p. 4) features ethnic food choices.

MyPyramid

The USDA created an educational tool called MyPyramid to illustrate the concepts presented in the *Dietary Guidelines* and the USDA Food Guide. Figure 1-6 presents a graphic image of MyPyramid, which was designed to encourage consumers to make healthy food and physical activity choices every day.

An abundance of material supporting MyPyramid is available to consumers who want to find the kinds and amounts of foods to eat each day (www.MyPyramid.gov). In addition to creating a personal plan, consumers can find tips to help them improve their diet and lifestyle by taking small steps each day.

IN SUMMARY

▌ Food group plans, such as the USDA Food Guide, serve as the basis for planning adequate, balanced, and varied diets.

▌ Each food group contributes key nutrients, a feature that provides flexibility in diet planning.

▌ The USDA Food Guide emphasizes nutrient-dense foods within each group.

▌ The discretionary kcalorie allowance is the difference between the kcalories needed to meet nutrient needs and those needed for energy.

▌ MyPyramid is an educational tool used to illustrate the concepts presented in the Dietary Guidelines and the USDA Food Guide.

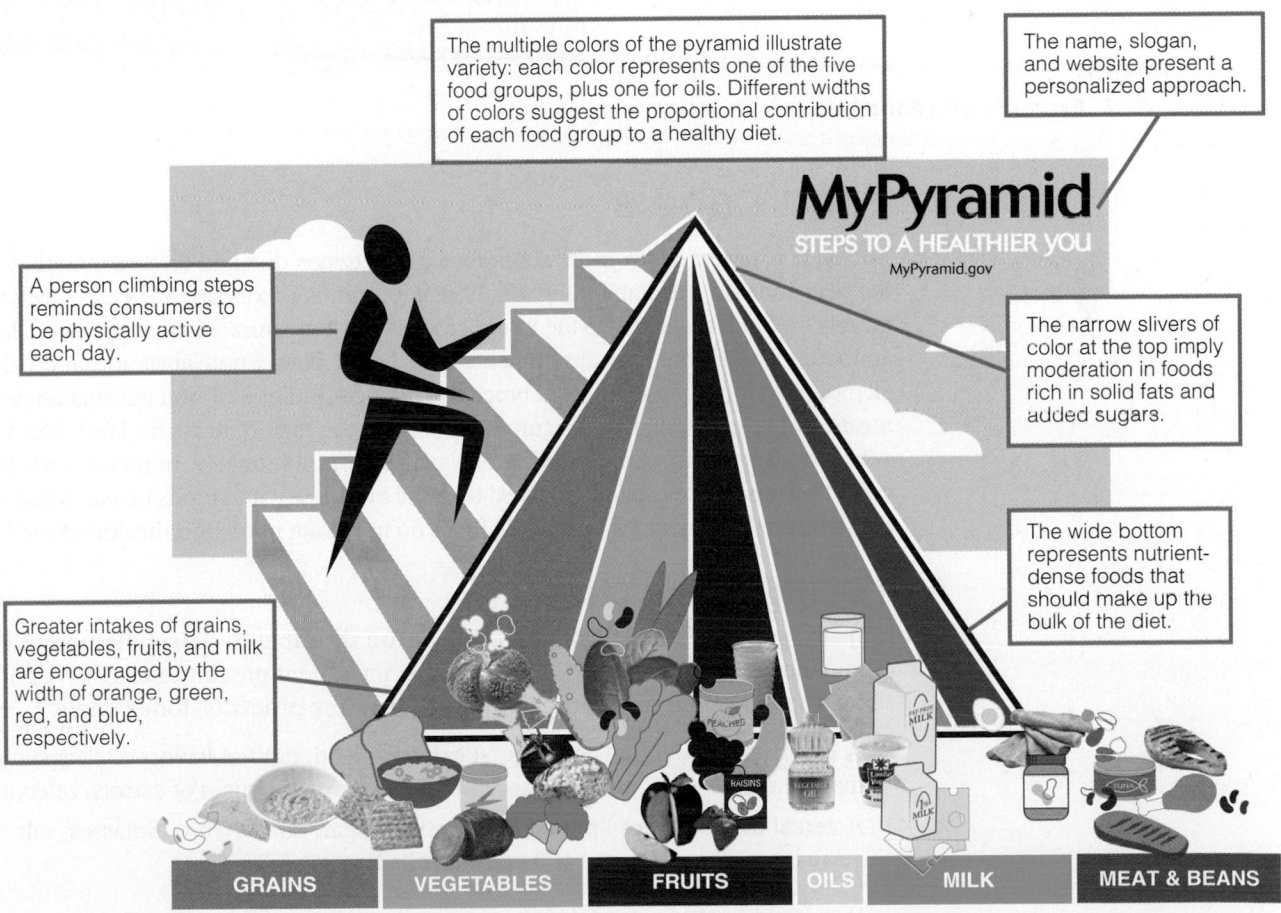

The multiple colors of the pyramid illustrate variety: each color represents one of the five food groups, plus one for oils. Different widths of colors suggest the proportional contribution of each food group to a healthy diet.

The name, slogan, and website present a personalized approach.

A person climbing steps reminds consumers to be physically active each day.

The narrow slivers of color at the top imply moderation in foods rich in solid fats and added sugars.

Greater intakes of grains, vegetables, fruits, and milk are encouraged by the width of orange, green, red, and blue, respectively.

The wide bottom represents nutrient-dense foods that should make up the bulk of the diet.

MyPyramid
STEPS TO A HEALTHIER YOU
MyPyramid.gov

GRAINS   VEGETABLES   FRUITS   OILS   MILK   MEAT & BEANS

FIGURE 1-6 **MyPyramid: Steps to a Healthier You**

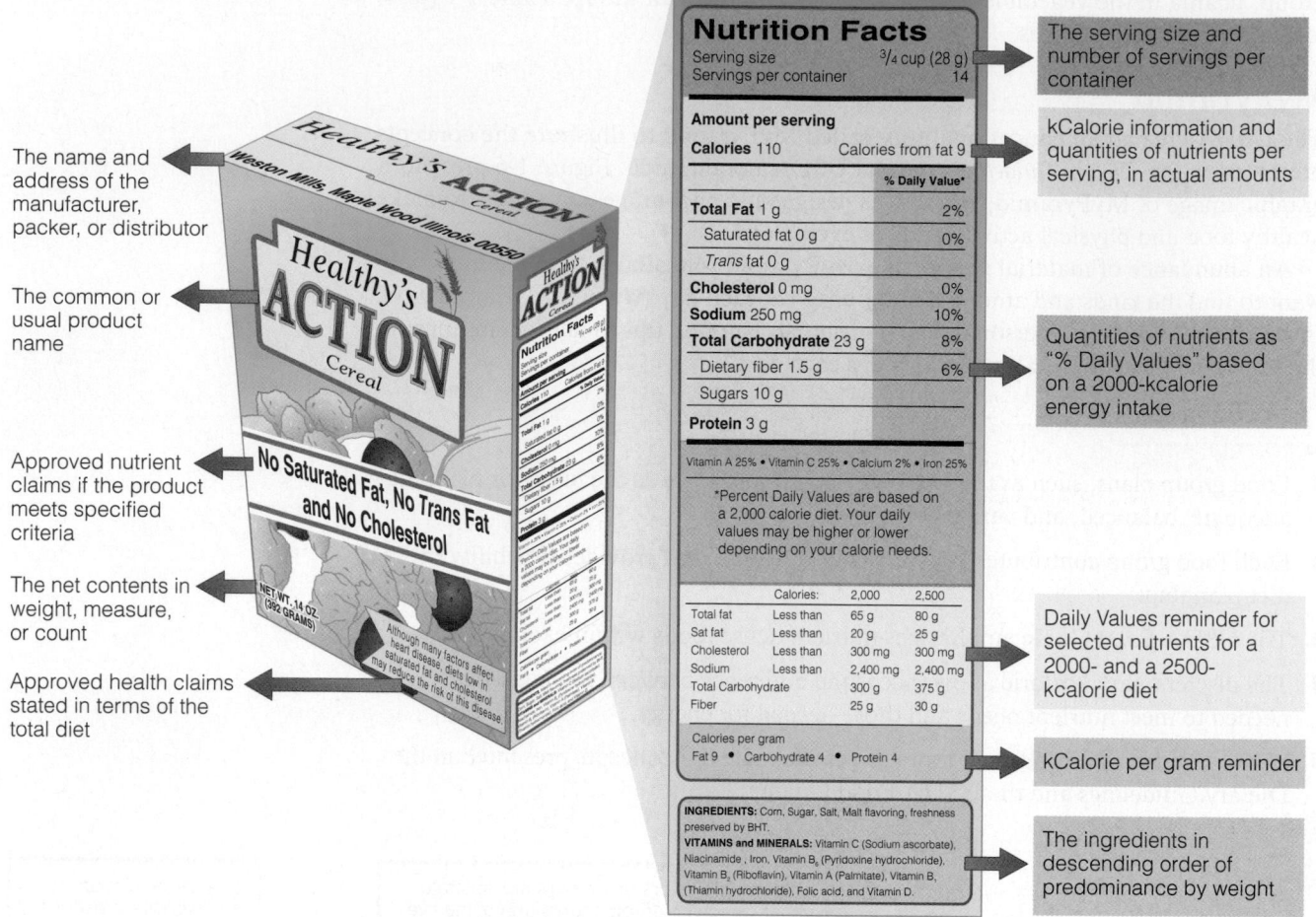

The name and address of the manufacturer, packer, or distributor

The common or usual product name

Approved nutrient claims if the product meets specified criteria

The net contents in weight, measure, or count

Approved health claims stated in terms of the total diet

Healthy's ACTION Cereal
Weston Mills, Maple Wood Illinois 00550

No Saturated Fat, No Trans Fat and No Cholesterol

NET. WT. 14 OZ. (392 GRAMS)

Although many factors affect heart disease, diets low in saturated fat and cholesterol may reduce the risk of this disease.

Nutrition Facts

Serving size ¾ cup (28 g)
Servings per container 14

Amount per serving

Calories 110   Calories from fat 9

	% Daily Value*
Total Fat 1 g	2%
Saturated fat 0 g	0%
Trans fat 0 g	
Cholesterol 0 mg	0%
Sodium 250 mg	10%
Total Carbohydrate 23 g	8%
Dietary fiber 1.5 g	6%
Sugars 10 g	
Protein 3 g	

Vitamin A 25% • Vitamin C 25% • Calcium 2% • Iron 25%

*Percent Daily Values are based on a 2,000 calorie diet. Your daily values may be higher or lower depending on your calorie needs.

		Calories:	2,000	2,500
Total fat	Less than		65 g	80 g
Sat fat	Less than		20 g	25 g
Cholesterol	Less than		300 mg	300 mg
Sodium	Less than		2,400 mg	2,400 mg
Total Carbohydrate			300 g	375 g
Fiber			25 g	30 g

Calories per gram
Fat 9 • Carbohydrate 4 • Protein 4

INGREDIENTS: Corn, Sugar, Salt, Malt flavoring, freshness preserved by BHT.
VITAMINS and MINERALS: Vitamin C (Sodium ascorbate), Niacinamide , Iron, Vitamin B₆ (Pyridoxine hydrochloride), Vitamin B₂ (Riboflavin), Vitamin A (Palmitate), Vitamin B₁ (Thiamin hydrochloride), Folic acid, and Vitamin D.

The serving size and number of servings per container

kCalorie information and quantities of nutrients per serving, in actual amounts

Quantities of nutrients as "% Daily Values" based on a 2000-kcalorie energy intake

Daily Values reminder for selected nutrients for a 2000- and a 2500-kcalorie diet

kCalorie per gram reminder

The ingredients in descending order of predominance by weight

FIGURE 1-7 Example of a Food Label

Food Labels

Today, consumers know more about the links between diet and disease than they did in the past, and they are demanding still more information on disease prevention. Many people rely on food labels to help them select foods with less saturated fat, *trans* fat, cholesterol, and sodium and more vitamins, minerals, and dietary fiber. Food labels appear on virtually all processed foods, and posters or brochures provide similar nutrition information for fresh meats, fruits, and vegetables. Figure 1-7 illustrates the requirements for label information. A few foods need not carry nutrition labels: those contributing few nutrients, such as plain coffee, tea, and spices; foods produced by small businesses; and foods prepared and sold in the same establishment, as long as the foods do not make nutrient or health claims.*

The Ingredient List

All packaged foods must list *all* ingredients on the label in descending order of predominance by weight. Knowing that the first ingredient predominates by weight, consumers can glean much information. Compare these products, for example:

▌ A beverage powder that contains "sugar, citric acid, natural flavors . . ." versus a juice that contains "water, tomato concentrate, concentrated juices of carrots, celery. . . ."

▌ A cereal that contains "puffed milled corn, sugar, corn syrup, molasses, salt . . ." versus one that contains "100 percent rolled oats. . . ."

*For example, restaurants need not provide complete nutrition information unless they are making "heart-healthy" claims for menu items.

In each comparison, consumers can tell that the second product is the more nutrient dense.

Nutrition Facts Panel

The FDA requires food labels to include key nutrition facts. The "Nutrition Facts" panel provides such information as serving sizes, Daily Values, and nutrient quantities.

Serving Sizes Because labels present nutrient information per serving, they must identify the size of a serving. The Food and Drug Administration (FDA) has established specific serving sizes for various foods and requires that all labels for a given product use the same serving size. For example, the serving size for all ice creams is ½ cup and for all beverages, 8 fluid ounces. This facilitates comparison shopping. Consumers can see at a glance which brand has more or fewer kcalories or grams of fat, for example. Standard serving sizes are expressed in both common household measures, such as cups, and metric measures, such as milliliters, to accommodate users of both types of measures (see Table 1-7).

When examining the nutrition information on a food label, consumers need to consider how the serving size compares with the actual quantity eaten. If it is not the same, they will need to adjust the quantities accordingly. For example, if the serving size is four cookies and you only eat two, then you need to cut the nutrient and kcalorie values in half; similarly, if you eat eight cookies, then you need to double the values. Notice, too, that small bags or individually wrapped items, such as chips or candy bars, may contain more than a single serving. The number of servings per container is listed just below the serving size.

The Daily Values To help consumers evaluate the information found on labels, the FDA created a set of nutrient standards, called the **Daily Values,** specifically for use on food labels. The Daily Values do two things: they set adequacy standards for nutrients that are desirable in the diet, such as protein, vitamins, minerals, and fiber; and they also set moderation standards for other nutrients that must be limited, such as fat, saturated fat, cholesterol, and sodium.

The "% Daily Value" column on a label provides a ballpark estimate of how individual foods contribute to the total diet. It compares key nutrients in a serving of food with the daily goals of a person consuming 2000 kcalories. Most labels list, at the bottom, Daily Values for both a 2000-kcalorie and a 2500-kcalorie diet, but the "% Daily Value" column on all labels applies only to a 2000-kcalorie diet. Although the Daily Values are based on a 2000-kcalorie diet, people's actual energy intakes vary widely; some people need fewer kcalories, and some people need more. This makes the Daily Values very useful for comparing one food with another but less useful as nutrient intake targets for individuals. By examining a food's general nutrient profile, however, a person can determine whether the food contributes a little or a lot of a nutrient, whether it contributes more or less than another food, and how well it fits into the consumer's overall diet.

Nutrient Quantities In addition to the serving size and the servings per container, the FDA requires that the "Nutrition Facts" panel on a label present nutrient information in two ways—in quantities (such as grams) and as percentages of the Daily Values. The Nutrition Facts panel must provide the nutrient amount, the percent Daily Value, or both, for the following:

- Total food energy (kcalories)
- Food energy from fat (kcalories)
- Total fat (grams and percent Daily Value)
- Saturated fat (grams and percent Daily Value)

TABLE 1-7 Household and Metric Measures

- 1 teaspoon (tsp) = 5 milliliters (mL)
- 1 tablespoon (tbs) = 15 mL
- 1 cup (c) = 240 mL
- 1 fluid ounce fl (fl oz) = 30 mL
- 1 ounce (oz) = 28 grams (g)

Note: The Aids to Calculation section at the back of the book provides additional weights and measures.

Daily Values: reference values developed by the FDA specifically for use on food labels.

- *Trans* fat (grams)
- Cholesterol (milligrams and percent Daily Value)
- Sodium (milligrams and percent Daily Value)
- Total carbohydrate, including starch, sugar, and fiber (grams and percent Daily Value)
- Dietary fiber (grams and percent Daily Value)
- Sugars (grams), including both those naturally present in and those added to the food
- Protein (grams)

The labels must also present nutrient content information as a percentage of the Daily Values for the following vitamins and minerals:

- Vitamin A
- Vitamin C
- Iron
- Calcium

The FDA developed the Daily Values for use on food labels because comparing nutrient amounts against a standard helps make them meaningful to consumers. A person might wonder, for example, whether 1 milligram of iron or calcium is a little or a lot. As Table 1-8 shows, the Daily Value for iron is 18 milligrams, so 1 milligram of iron is enough to take notice of: it is more than 5 percent. But the Daily Value for calcium on food labels is 1000 milligrams, so 1 milligram of calcium is a negligible amount.

Claims on Labels

In addition to the Nutrition Facts panel, consumers may find various claims on labels. These claims include nutrient claims, health claims, and structure-function claims.

Nutrient Claims The FDA defines the **nutrient claims** that may be used a label to describe the contents of a product (see Table 1-9). Definitions include the conditions under which each term can be used. For example, in addition to having fewer than 2 milligrams of cholesterol, a product labeled "cholesterol free" may not contain more than 2 grams of saturated fat and *trans* fat combined per serving.

Some descriptions imply that a food contains, or does not contain, a nutrient. Implied claims are prohibited unless they meet specified criteria. For example, a claim

nutrient claims: statements that characterize the quantity of a nutrient in a food.

TABLE 1-8 **Daily Values for Food Labels**

Food labels must present the "% Daily Value" for these nutrients.

Food Component	Daily Value	Calculation Factors
Fat	65 g	30% of kcalories
Saturated fat	20 g	10% of kcalories
Cholesterol	300 mg	—
Carbohydrate (total)	300 g	60% of kcalories
Fiber	25 g	11.5 g per 1000 kcalories
Protein	50 g	10% of kcalories
Sodium	2400 mg	—
Potassium	3500 mg	—
Vitamin C	60 mg	—
Vitamin A	1500 µg	—
Calcium	1000 mg	—
Iron	18 mg	—

Note: Daily Values were established for adults and children over 4 years old. The values for energy-yielding nutrients are based on 2000 kcalories a day. For fiber, the Daily Value was rounded up from 23.

TABLE 1-9 Terms Used on Food Labels

GENERAL TERMS

free: "nutritionally trivial" and unlikely to have a physiological consequence; synonyms include *without, no,* and *zero.* A food that does not contain a nutrient naturally may make such a claim but only as it applies to all similar foods (for example, "applesauce, a fat-free food").

good source of: the product provides between 10 and 19 percent of the Daily Value for a given nutrient per serving.

healthy: a food that is low in fat, saturated fat, cholesterol, and sodium and that contains at least 10 percent of the Daily Value for vitamin A, vitamin C, iron, calcium, protein, or fiber.

high: 20 percent or more of the Daily Value for a given nutrient per serving; synonyms include *rich in* or *excellent source.*

less: at least 25 percent less of a given nutrient or kcalories than the comparison food (see individual nutrients); synonyms include *fewer* and *reduced.*

light or **lite:** one-third fewer kcalories than the comparison food; 50 percent or less of the fat or sodium than the comparison food; any use of the term other than as defined must specify what it is referring to (for example, "light in color" or "light in texture").

low: an amount that would allow frequent consumption of a food without exceeding the Daily Value for the nutrient. A food that is naturally low in a nutrient may make such a claim but only as it applies to all similar foods (for example, "fresh cauliflower, a low-sodium food"); synonyms include *little, few,* and *low source of.*

more: at least 10 percent more of the Daily Value for a given nutrient than the comparison food; synonyms include *added* and *extra.*

organic: on food labels, indicates that at least 95 percent of the product's ingredients have been grown and processed according to USDA regulations defining the use of fertilizers, herbicides, insecticides, fungicides, preservatives, and other chemical ingredients.

ENERGY

kcalorie-free: fewer than 5 kcalories per serving.

low kcalorie: 40 kcalories or less per serving.

reduced kcalorie: at least 25 percent fewer kcalories per serving than the comparison food.

FAT AND CHOLESTEROL[a]

percent fat free: may be used only if the product meets the definition of low fat or fat free and must reflect the amount of fat in 100 grams (for example, a food that contains 2.5 grams of fat per 50 grams can claim to be "95 percent fat free").

fat free: less than 0.5 gram of fat per serving (and no added fat or oil); synonyms include *zero-fat, no-fat,* and *nonfat.*

low fat: 3 grams or less fat per serving.

less fat: 25 percent or less fat than the comparison food.

saturated fat free: less than 0.5 gram of saturated fat and 0.5 gram of *trans* fat per serving.

low saturated fat: 1 gram or less saturated fat and less than 0.5 gram of *trans* fat per serving.

less saturated fat: 25 percent or less saturated fat and *trans* fat combined than the comparison food.

trans fat free: less than 0.5 gram of *trans* fat and less than 0.5 gram of saturated fat per serving.

cholesterol-free: less than 2 milligrams cholesterol per serving and 2 grams or less saturated fat and *trans* fat combined per serving.

low cholesterol: 20 milligrams or less cholesterol per serving and 2 grams or less saturated fat and *trans* fat combined per serving.

less cholesterol: 25 percent or less cholesterol than the comparison food (reflecting a reduction of at least 20 milligrams per serving), and 2 grams or less saturated fat and *trans* fat combined per serving.

extra lean: less than 5 grams of fat, 2 grams of saturated fat and *trans* fat combined, and 95 milligrams of cholesterol per serving and per 100 grams of meat, poultry, and seafood.

lean: less than 10 grams of fat, 4.5 grams of saturated fat and *trans* fat combined, and 95 milligrams of cholesterol per serving and per 100 grams of meat, poultry, and seafood. For mixed dishes such as burritos and sandwiches, less than 8 grams of fat, 3.5 grams of saturated fat, and 80 milligrams of cholesterol per reference amount customarily consumed.

CARBOHYDRATES: FIBER AND SUGAR

high fiber: 5 grams or more fiber per serving. A high-fiber claim made on a food that contains more than 3 grams fat per serving and per 100 grams of food must also declare total fat.

sugar-free: less than 0.5 gram of sugar per serving.

SODIUM

sodium-free and **salt-free:** less than 5 milligrams of sodium per serving.

low sodium: 140 milligrams or less per serving.

very low sodium: 35 milligrams or less per serving.

[a]Foods containing more than 13 grams total fat per serving or per 50 grams of food must indicate those contents immediately after a cholesterol claim. As you can see, all cholesterol claims are prohibited when the food contains more than 2 grams saturated fat and *trans* fat combined per serving.

that a product "contains no oil" implies that the food contains no fat. If the product is truly fat free, then it may make the no-oil claim, but if it contains another source of fat, such as butter, it may not.

Health Claims Until recently, the FDA held manufacturers to the highest standards of scientific evidence before allowing them to place **health claims** on food labels.[12] When a label stated, "diets low in sodium may reduce the risk of high blood pressure," for example, consumers could be sure that the FDA had examined much scientific evidence and found substantial support for the claim. Such reliable health claims still appear on food labels, and they have a high degree of scientific validity (see Table 1-10).

Today, however, the FDA also allows other claims backed by weaker evidence to appear on labels. These are *qualified* claims, in the sense that labels bearing them must also state the strength of the scientific evidence backing them up. Unfortunately, most people are not knowledgeable enough to distinguish between scientifically reliable claims and those that are best ignored.[13]

Structure-Function Claims Consumers need to be aware that a different kind of claim, known as a "structure-function" claim, may also appear on food or dietary supplement labels. **Structure-function claims** are statements about a food substance's effect on a structure or function of the body—for example, "antioxidants support heart health." Structure-function claims are not required to have FDA approval, but the claims may not refer to the reduction of disease risk. The claims must be carefully worded to avoid any mention of a specific disease. Typical claims are "slows aging," "improves memory," and "builds strong bones." To make a more specific claim such as "prevents osteoporosis," the manufacturer would have to submit to the rigorous requirements for health claims or meet the even stricter safety and efficacy standards applied to drugs. These rules ensure that consumers can have confidence that when a claim names a specific disease, there is substantial scientific agreement that the food, in the context of a healthy diet, may help protect against that disease.

IN SUMMARY

▌ Food labels provide consumers with information they need to select foods that will help them meet their nutrition and health goals.

▌ Daily Values are a set of nutrient standards created by the FDA for use on food labels.

▌ Reliable health claims are backed by the highest standards of scientific evidence.

TABLE 1-10 Reliable Health Claims on Food Labels

▌ Calcium and reduced risk of osteoporosis

▌ Sodium and reduced risk of hypertension

▌ Dietary saturated fat and cholesterol and reduced risk of coronary heart disease

▌ Dietary fat and reduced risk of cancer

▌ Fiber-containing grain products, fruits, and vegetables and reduced risk of cancer

▌ Fruits, vegetables, and grain products that contain fiber, particularly soluble fiber, and reduced risk of coronary heart disease

▌ Fruits and vegetables and reduced risk of cancer

▌ Folate and reduced risk of neural tube defects

▌ Sugar alcohols and reduced risk of tooth decay

▌ Soluble fiber from whole oats and from psyllium seed husk and reduced risk of heart disease

▌ Soy protein and reduced risk of heart disease

▌ Whole grains and reduced risk of heart disease and certain cancers

▌ Plant sterol and plant stanol esters and reduced risk of heart disease

▌ Potassium and reduced risk of hypertension and stroke

health claims: statements that characterize the relationship between a nutrient or other substance in food and a disease or health-related condition.

structure-function claims: statements that describe how a product may affect a structure or function of the body; for example, "calcium builds strong bones." Structure-function claims do not require FDA authorization.

Your Diet

The secret to making healthy food choices is learning to incorporate the *Dietary Guidelines* (Table 1-3 on p. 14) and the USDA Food Guide (Figure 1-4 on p. 16) into your decision-making process. Before completing this assignment, you may want to review the section called "Mixtures of Foods" on p. 20 to see where foods such as casseroles or soups fall in the food group plan.

▪ Keep a record of the foods you eat for 24 hours. Record both the types and amounts of foods eaten.

▪ Turn to Table 1-5 on p. 18 to determine your estimated kcalorie needs (Chapter 6 offers a more individualized approach to estimating your kcalorie needs). What is your estimated daily kcalorie need? _____ kcalories.

▪ Compare the foods you ate with the USDA Food Guide recommendations for your energy needs (see Table 1-4 on p. 18) by listing them in the accompanying table. (Keep a copy of this for later reference.) For example, if you need 2400 kcalories a day, the recommended amount of fruits is 2 cups, while that of grains is 8 ounces. The USDA Food Guide on p. 16 provides equivalent measures for foods that are not readily measured in cups or ounces. Thus, 1 slice of bread counts as 1 ounce of grains, while 1 egg counts as 1 ounce of meat.

▪ Do your food choices include foods from each of the food groups?

▪ Are some food groups over- or underrepresented? If so, what can you do to ensure a more balanced diet?

▪ Do you eat a variety of foods within each group? If not, suggest ways to enhance the variety in your diet.

▪ Did you choose at least some whole-grain foods from the grain group, and fat-free or low-fat foods from the milk and milk products group and the protein foods group? In your list below (foods consumed), circle the foods you selected that are whole grain (grain group), fat free, and low fat (milk and protein foods groups).

Food Group	Recommended Amount	Foods Consumed	Your Diet: Total Amount Consumed
Grains	_____ oz		_____ oz
Vegetables	_____ cups		_____ cups
Fruits	_____ cups		_____ cups
Milk and milk products	_____ cups		_____ cups
Protein foods	_____ oz		_____ oz
Oils	_____ tsp		_____ tsp

Clinical Applications

1. Make a list of the foods and beverages you've consumed in the past two days. Look at each item on your list and consider why you chose that particular food or beverage. Did you eat cereal for breakfast because that's what you always eat (habit), or because it was the easiest, quickest food to prepare (convenience)? Did you put fat-free milk on the cereal because you want to control your energy intake (nutrition)? In going down your list, you may be surprised to discover exactly why you chose certain foods.

2. As a health care professional, you can uncover clues about a client's food choices by paying close attention. You may be surprised to discover why a client chooses certain foods, but you can then use this knowledge to serve the best interests of the client. For example, an elderly, undernourished widower may eat the same sandwich for lunch every day. In talking with the client, you discover that this is what he and his wife fixed together each day. Consider ways you might be able to help the client learn to eat other foods and vary his choices.

3. Using the list of foods and beverages from question #1, compare your day's intake with the USDA Food Guide. Did you vary your choices within each food group? Did your intake match the daily recommended amounts from each group? If not, list some changes you could have made to meet the recommendations.

Self Check

1. When people eat the foods typical of their families or geographic region, their choices are influenced by:
   a. occupation.
   b. nutrition.
   c. emotional state.
   d. ethnic heritage or tradition.

2. The energy-yielding nutrients are:
   a. fats, minerals, and water.
   b. minerals, proteins, and vitamins.
   c. carbohydrates, fats, and vitamins.
   d. carbohydrates, fats, and proteins.

3. The inorganic nutrients are:
   a. proteins and fats.
   b. vitamins and minerals.
   c. minerals and water.
   d. vitamins and proteins.

4. Alcohol is not a nutrient because:
   a. the body derives no energy from it.
   b. it is organic.
   c. it is converted to body fat.
   d. it does not contribute to the body's growth or repair.

5. DRI stands for:
   a. Daily Recommended Intakes.
   b. Dietary Requirements for Individuals.
   c. Dietary Reference Intakes.
   d. Daily Recommendations for Individuals.

6. Which of the following is consistent with the *Dietary Guidelines for Americans*?
   a. Limit intakes of fruits, vegetables, and whole grains.
   b. Engage in regular physical activity.
   c. Choose a diet with plenty of whole-milk products.
   d. Eat an abundance of foods to ensure nutrient adequacy.

7. In a food group plan such as the USDA Food Guide, foods within a given food group are similar in their contents of:
   a. energy.
   b. proteins and fibers.
   c. vitamins and minerals.
   d. carbohydrates and fats.

8. A slice of apple pie supplies 350 kcalories with 3 grams of fiber; an apple provides 80 kcalories and the same 3 grams of fiber. This is an example of:
   a. kcalorie control.
   b. nutrient density.
   c. variety.
   d. essential nutrients.

9. Which of the following statements does *not* apply to a person's discretionary kcalorie allowance?
   a. Compared with physically active adults, sedentary adults have a lower discretionary kcalorie allowance.
   b. It is the difference between the kcalories needed to supply nutrients and those needed for energy.
   c. Compared to physically active adults, sedentary adults have a higher discretionary kcalorie allowance.
   d. A person with discretionary kcalories available can, if he or she chooses to, select a few foods with added fat or sugar.

10. Food labels list ingredients in:
    a. alphabetical order.
    b. ascending order of predominance by weight.
    c. descending order of predominance by weight.
    d. the manufacturer's order of preference.

Answers to these questions can be found in Appendix H.

Notes

1. K. Stein, Navigating cultural competency: In preparation for an expected standard in 2010, *Journal of the American Dietetic Association* 110 (2010): S13–S20.

2. J. R. Krebs, The gourmet ape: Evolution and human food preferences, *American Journal of Clinical Nutrition* 90 (2009): 707S–711S; Q. Y. Chen and coauthors, Perceptual variation in umami taste and polymorphisms in *TAS1R* taste receptor genes, *American Journal of Clinical Nutrition* 90 (2009): 770S–779S; B. J. Tepper, Nutritional implications of genetic taste variation: The role of PROP sensitivity and other taste phenotypes, *Annual Review of Nutrition* 28 (2008): 367–388; A. A. Bachmanov and G. K. Beauchamp, Taste receptor genes, *Annual Review of Nutrition* 27 (2007): 389–414; A. El-Sohemy and coauthors, Nutrigenomics of taste—Impact on food preferences and food production, *Forum of Nutrition* 60 (2007): 176–182; K. Keskitalo and coauthors, Same genetic components underlie different measures of sweet taste preference, *American Journal of Clinical Nutrition* 86 (2007): 1663–1669; M. R. Yeomans and coauthors, Human

hedonic responses to sweetness: Role of taste genetics and anatomy, *Physiology and Behavior* 91 (2007): 264–273; A. Knaapila and coauthors, Food neophobia shows heritable variation in humans, *Physiology and Behavior* 91 (2007): 573–578.

3. J. E. Tillotson, Americans' food shopping in today's lousy economy (part 2), *Nutrition Today* 44 (2009): 218–221; J. E. Tillotson, Fast food—through the ages: Part 1, *Nutrition Today* 43 (2008): 70–74; J. Mathieu, Hey everyone, dinner's ready . . . two weeks ago, *Journal of the American Dietetic Association* 107 (2007): 26–27.

4. Position of the American Dietetic Association: Functional foods, *Journal of the American Dietetic Association* 109 (2009): 735–746.

5. Standing Committee on the Scientific Evaluation of Dietary Reference Intakes, Food and Nutrition Board, Institute of Medicine, *Dietary Reference Intakes for Water, Potassium, Sodium, Chloride, and Sulfate* (Washington, D.C.: National Academies Press, 2004); Standing Committee on the Scientific Evaluation of Dietary Reference Intakes, Food and Nutrition Board, Institute of Medicine, *Dietary Reference Intakes for Energy, Carbohydrate, Fiber, Fat, Fatty Acids, Cholesterol, Protein, and Amino Acids* (Washington, D.C.: National Academies Press, 2005); Standing Committee on the Scientific Evaluation of Dietary Reference Intakes, Food and Nutrition Board, Institute of Medicine, *Dietary Reference Intakes for Vitamin A, Vitamin K, Arsenic, Boron, Chromium, Copper, Iodine, Iron, Manganese, Molybdenum, Nickel, Silicon, Vanadium, and Zinc* (Washington, D.C.: National Academy Press, 2001); Standing Committee on the Scientific Evaluation of Dietary

Reference Intakes, Food and Nutrition Board, Institute of Medicine, *Dietary Reference Intakes for Vitamin C, Vitamin E, Selenium, and Carotenoids* (Washington, D.C.: National Academy Press, 2000); Standing Committee on the Scientific Evaluation of Dietary Reference Intakes, Food and Nutrition Board, Institute of Medicine, *Dietary Reference Intakes for Thiamin, Riboflavin, Niacin, Vitamin B6, Folate, Vitamin B12, Pantothenic Acid, Biotin, and Choline* (Washington, D.C.: National Academy Press, 1998); Standing Committee on the Scientific Evaluation of Dietary Reference Intakes, Food and Nutrition Board, Institute of Medicine, *Dietary Reference Intakes for Calcium, Phosphorus, Magnesium, Vitamin D, and Fluoride* (Washington, D.C.: National Academy Press, 1997).

6. Institute of Medicine, *Dietary Reference Intakes: The Essential Guide to Nutrient Requirements*, eds., J. J. Otten, J. P. Hellwig, and L. D. Meyers (Washington, D.C.: The National Academies Press, 2006), pp. 5–18.

7. F. Magkos and coauthors, Management of the metabolic syndrome and type 2 diabetes through lifestyle modification, *Annual Review of Nutrition* 29 (2009): 223–256; S. S. Gidding and coauthors, Implementing American Heart Association pediatric and adult nutrition guidelines: A scientific statement from the American Heart Association Nutrition Committee of the Council on Nutrition, Physical Activity and Metabolism, Council on Cardiovascular Disease in the Young, Council on Arteriosclerosis, Thrombosis and Vascular Biology, Council on Cardiovascular Nursing, Council on Epidemiology and Prevention, and Council for High Blood Pressure Research, *Circulation* 119

(2009): 1161–1175; A. Galimanis and coauthors, Lifestyle and stroke risk: A review, *Current Opinion in Neurology* 22 (2009): 60–68; World Cancer Research Fund/American Institute for Cancer Research, *Food, Nutrition, Physical Activity, and the Prevention of Cancer: A Global Perspective* (Washington, D.C.: World Cancer Research Fund/American Institute for Cancer Research, 2007), pp. 66–196.

8. S. P. Murphy and coauthors, Simple measures of dietary variety are associated with improved dietary quality, *Journal of the American Dietetic Association* 106 (2006): 425–429.

9. U.S. Department of Agriculture and U.S. Department of Health and Human Services, *Dietary Guidelines for Americans 2010*, available at www.dietaryguidelines.gov

10. U.S. Department of Health and Human Services, *2008 Physical Activity Guidelines for Americans*, available at www.health.gov/paguidelines; U.S. Department of Health and Human Services, *Dietary Guidelines for Americans, 2010*, available at www.dietaryguidelines.gov.

11. Position of the American Dietetic Association: Vegetarian diets, *Journal of the American Dietetic Association* 109 (2009): 1266–1282.

12. C. L. Taylor and V. L. Wilkening, How the nutrition food label was developed, part 2: The purpose and promise of nutrition claims, *Journal of the American Dietetic Association* 108 (2008): 618–623.

13. Position of the American Dietetic Association, Functional foods, *Journal of the American Dietetic Association* 109 (2009): 735–746.

Nutrition in Practice

Nutrition and health receive so much attention on television, in the popular press, and on the Internet that it is easy to be overwhelmed with conflicting, confusing information. Determining whether nutrition information is accurate can be a challenging task. It is also an important task because nutrition affects a person both professionally and personally.

A person watches a nutrition report on television and then reads a conflicting report in the newspaper. Why do nutrition news reports and claims for nutrition products seem to contradict each other so often?

The problem of conflicting messages arises for several reasons:

▌ Popular media, often faced with tight deadlines and limited time or space to report new information, rush to present the latest breakthrough in a headline or a 60-second spot. They can hardly help omitting important facts about the study or studies that the breakthrough is based on.

▌ Despite tremendous advances in the past few decades, scientists still have much to learn about the human body and nutrition. Scientists themselves often disagree on their first tentative interpretations of new research findings, yet these are the very findings that the public hears most about.

▌ The popular media often broadcast preliminary findings in hopes of grabbing attention and boosting readership or television ratings.

▌ Commercial promoters turn preliminary findings into advertisements for products or supplements long before the findings have been validated—or disproved. The scientific process requires many experiments or trials to confirm a new finding. Seldom do promoters wait as long as they should to make their claims.

▌ Promoters are aware that consumers like to try new products or treatments, even though they probably will not withstand the tests of time and scientific scrutiny.

So how can a person tell what claims to believe?

Valid nutrition information derives from scientific research, which has the following characteristics:

▌ Scientists test their ideas by conducting properly designed scientific experiments. They report their methods and procedures in detail so that other scientists can verify the findings through replication.

▌ Scientists recognize the inadequacy of personal testimonials.

▌ Scientists who use animals in their research do not apply their findings directly to human beings.

▌ Scientists may use specific segments of the population in their research. When they do, they are careful not to generalize the findings to all people.

▌ Scientists report their findings in respected scientific journals. Their work must survive a screening review by their peers before it is accepted for publication.

With each report from scientists, the field of nutrition changes a little—each finding contributes another piece to the whole body of knowledge. Table NP1-1 features 10 red flags of junk science to help consumers distinguish valid from misleading nutrition information.[1]

Because nutrition misinformation harms the health and economic status of consumers, the American Dietetic Association (ADA) works with health care professionals and educators to present sound nutrition information to the public and to actively confront nutrition misinformation.[2] Table NP1-2 offers a list of credible sources of nutrition information.

What about nutrition and health information found on the Internet? How does a person know whether the websites are reliable?

With hundreds of millions of websites on the Internet, searching for nutrition and health information can be daunting. The Internet offers no guarantee of the accuracy of the information found there, and much of it is pure fiction. Websites must be evaluated for their accuracy, just like every other source. Table NP1-3 provides clues to identifying reliable nutrition information sites and lists some credible sites.

TABLE NP1-1 Red Flags of Junk Science

1. Promises of a quick and easy fix.
2. Dire warnings of danger from a single product or regimen.
3. Too good to be true. The claim says what most people want to hear.
4. Enticingly simple conclusions drawn from a complex study.
5. Recommendations based on a single study.
6. Dramatic statements that are refuted by reputable scientific organizations.
7. Lists of "good" and "bad" foods.
8. Claims are made to help sell a product.
9. Claims or recommendations based on studies published without peer review.
10. Recommendations from studies that ignore differences among individuals or groups.

Source: Adapted from Position of the American Dietetic Association: Food and nutrition misinformation, *Journal of the American Dietetic Association* 106 (2006): 601–607. Used by permission of Elsevier, Inc.

TABLE NP1-2 Credible Sources of Nutrition Information

Government agencies, volunteer associations, consumer groups, and professional organizations provide consumers with reliable health and nutrition information. Credible sources of nutrition information include:

- Nutrition and food science departments at a university or community college
- Local agencies such as the health department or County Cooperative Extension Service
- Government health agencies such as:
  - Department of Agriculture (USDA) — www.usda.gov
  - Department of Health and Human Services (DHHS) — www.os.dhhs.gov
  - Food and Drug Administration (FDA) — www.fda.gov
  - Health Canada — www.hc-sc.gc.ca/fn-an
- Volunteer health agencies such as:
  - American Cancer Society — www.cancer.org
  - American Diabetes Society — www.diabetes.org
  - American Heart Association — www.americanheart.org
- Reputable consumer groups such as:
  - American Council on Science and Health — www.acsh.org
  - Federal Citizen Information Center — www.pueblo.gsa.gov
  - International Food Information Council — www.ific.org
- Professional health organizations such as:
  - American Dietetic Association — www.eatright.org
  - American Medical Association — www.ama-assn.org
  - Dietitians of Canada — www.dietitians.ca
- Journals such as:
  - *American Journal of Clinical Nutrition* — www.ajcn.org
  - *Journal of the American Dietetic Association* — www.adajournal.org
  - *New England Journal of Medicine* — www.nejm.org
  - *Nutrition Reviews* — www.blackwellpublishing.com/nure

TABLE NP1-3 Evaluating the Reliability of Websites

To judge whether an Internet site offers reliable nutrition information, answer the following questions.

Who is responsible for the site? Clues can be found in the three-letter "tag" that follows the dot in the site's name. For example, "gov" and "edu" indicate government and educational institution sites, respectively, which are usually reliable sources of information.

Do the names and credentials of information providers appear? Is an editorial board identified? Many legitimate sources provide e-mail addresses or other ways to obtain more information about the site and the information providers behind it.

Are links with other reliable information sites provided? Reputable organizations almost always provide links with other similar sites because they want you to know of other experts in their area of knowledge. Caution is needed when you evaluate a site by its links, however. Anyone, even a quack, can link a web page to a reputable site without the organization's permission. Doing so may give the quack's site the appearance of legitimacy, just the effect the quack is hoping for.

Is the site updated regularly? Nutrition information changes rapidly, and sites should be updated often.

Is the site selling a product or service? Commercial sites may provide accurate information, but they also may not. Their profit motive increases the risk of bias.

Does the site charge a fee to gain access to it? Many academic and government sites offer the best information, usually for free. Some legitimate sites do charge fees, but before paying up, check the free sites. Chances are good you will find what you are looking for without paying.

Some credible websites include:
National Council against Health Fraud
www.ncahf.org
Stephen Barrett's Quackwatch
www.quackwatch.com

One of the most trustworthy sites used by scientists and others is the National Library of Medicine's PubMed (www.ncbi.nlm.nih.gov), which provides free access to more than 10 million abstracts (short descriptions) of research papers published in scientific journals around the world. Many abstracts provide links to websites where full articles are available.

Promoters of fraudulent health products use the Internet as a primary means to sell their wares. Agencies such as the Food and Drug Administration (FDA) take action against fraudulent marketing of supplements and health products on the Internet.[3] The latest actions target unscrupulous companies that use the Internet to promote products to the most vulnerable consumers—those with diseases such as cancer or AIDS. Of greatest concern are those products that not only make false promises but also are potentially dangerous. For example, herbal products touted as safe treatments for serious illnesses such as cancer may interact with and impair the effectiveness of medications. FDA advises consumers to be suspicious of:

- Claims that a product is "natural" or "nontoxic." Natural or nontoxic does not always mean safe.
- Claims that a product is a "scientific breakthrough," "miraculous cure," "secret ingredient," or "ancient remedy."
- Claims that a product cures a wide range of illnesses.
- Claims that use impressive-sounding medical terms.
- Claims of a money-back guarantee.

Consumers with questions or suspicions about fraud can contact the FDA on the Internet at www.FDA.gov or by telephone at (888) INFO-FDA.

Everyone seems to be giving advice on nutrition. How can a person tell whom to listen to?

Registered dietitians (RDs) and nutrition professionals with advanced degrees (MS, PhD) are experts (see the glossary). These professionals are probably in the best position to answer nutrition questions. On the other hand, a **nutritionist** may be an expert or a quack, depending on which state the person practices in.

Some states require people who use this title to meet strict standards. In other states, a nutritionist may be any individual who claims a career connection with the nutrition field. There is no accepted national definition for the term *nutritionist*.

Other purveyors of nutrition information may also lack credentials. A health food store owner may be in the nutrition business simply because it is a lucrative market. The owner may have a background in business or sales and no education in nutrition at all. Such a person is not qualified to provide nutrition information to customers. For accurate nutrition information, seek out a trained professional with a college education in nutrition—an expert in the field of **dietetics.**

What about nurses and other health care professionals?

All members of the health care team share responsibility for helping each client to achieve optimal health, but the registered dietitian is usually the primary nutrition expert. Each of the other team members has a related specialty. Some physicians are specialists in clinical nutrition and are also experts in the field. Other physicians, nurses, and **dietetic technicians** often assist dietitians in providing nutrition information and may help to administer direct nutrition care. Nurses play central roles in client care management and client relationships. Visiting nurses and home health care nurses may become intimately involved in clients' nutrition care at home, teaching them both theory and cooking techniques. Physical therapists can provide individualized exercise programs related to nutrition—for example, to help control obesity. Social workers may provide practical and emotional support.

What roles might these other health care professionals play in nutrition care?

Some of the responsibilities of the health care professional might be:

▎ Helping people understand why nutrition is important to them.

▎ Answering questions about food and diet.

▎ Explaining to clients how modified diets work.

▎ Collecting information about clients that may influence their nutritional health.

▎ Identifying clients at risk for poor nutrition status (see Chapter 14) and recommending or taking appropriate action.

▎ Recognizing when clients need extra help with nutrition problems (in such cases, the problems should be referred to a dietitian or physician).

Health care professionals may routinely perform these nutrition-related tasks:

▎ Obtaining diet histories.

▎ Taking weight and height measurements.

▎ Feeding clients who cannot feed themselves.

GLOSSARY OF TERMS ASSOCIATED WITH NUTRITION EXPERTS

dietetic technicians: persons who have completed a minimum of an associate's degree from an accredited college or university and an approved dietetic technician program. A **dietetic technician, registered (DTR)** has also passed a national examination and maintains registration through continuing professional education.

dietetics: the application of nutrition principles to achieve and maintain optimal human health.

nutritionists: all registered dietitians are nutritionists, but not all nutritionists are registered dietitians. Some state licensing boards set specific qualifications for holding the title. For states that regulate this title, the definition varies from state to state. There are "nutritionist" credentials that require little more than payment to obtain.

registered dietitians (RDs): dietitians who have graduated from a university or college after completing a program of dietetics that has been accredited by the American Dietetic Association (or Dietitians of Canada). The dietitians must serve in an approved internship or coordinated program to practice the necessary skills, pass the association registration examination, and maintain competency through continuing education. Many states require licensing for practicing dietitians. Licensed dietitians (LDs) have met all state requirements to offer nutrition advice.

▎ Recording what clients eat or drink.

▎ Observing clients' responses and reactions to foods.

▎ Helping clients mark menus.

▎ Monitoring weight changes.

▎ Monitoring food and drug interactions.

▎ Encouraging clients to eat.

▎ Assisting clients at home in planning their diets and managing their kitchen chores.

▎ Alerting the physician or dietitian when nutrition problems are identified.

▎ Charting actions taken and communicating on these matters with other professionals as needed.

Thus, although the dietitian assumes the primary role as the nutrition expert on a health care team, other health care professionals play important roles in administering nutrition care.

Notes

1. Position of the American Dietetic Association: Food and nutrition misinformation, *Journal of the American Dietetic Association* 106 (2006): 601–607.

2. Position of the American Dietetic Association, 2006.

3. Food and Drug Administration, *FDA 101: Health Fraud Awareness*, FDA Consumer Health Information, May 2009, available at www.fda.gov.

2

Carbohydrates

Grains, vegetables, legumes, fruits, and milk offer ample carbohydrate.

Chapter 3 describes the roles of fats in health and disease.

Most people would like to feel good all the time. Part of the secret of feeling good is replenishing the body's energy supply with food. That means choosing foods that contain the energy nutrients—carbohydrate and fat, primarily. But which to choose?

Carbohydrate is the preferred energy source for many of the body's functions. As long as carbohydrate is available, the human brain depends exclusively on it as an energy source. Athletes eat a "high-carb" diet to store as much muscle fuel as possible, and dietary recommendations urge people to eat carbohydrate-rich foods for better health. Many people, however, mistakenly think of carbohydrate-rich foods as "fattening" and avoid them. In truth, people who wish to lose fat and to maintain lean tissue and health can best do so by being physically active, paying close attention to portion sizes, and designing a diet based on foods that supply carbohydrate in balance with other energy nutrients.[1] Most unrefined plant foods—grains, vegetables, legumes, and fruits—provide ample carbohydrate and fiber with little or no fat. Milk is the only animal-derived food that contains significant amounts of carbohydrate.

Carbohydrate shares its fuel-providing responsibility with fat. Fat, however, normally is not used as fuel by the brain and central nervous system, and diets high in certain types of fat are associated with chronic diseases. The other energy sources available to the body—protein and alcohol—offer no advantage as fuels. Protein is best left to serve its own diverse functions, as discussed in Chapter 4. Alcohol, of course, has well-known undesirable side effects when used in excess. Alcohol and its relationships with health and disease are the subject of Nutrition in Practice 20.

carbohydrates: energy nutrients composed of monosaccharides.
carbo = carbon
hydrate = water

monosaccharides (mon-oh-SACK-uh-rides): single sugar units.
mono = one
saccharide = sugar

disaccharides (dye-SACK-uh-rides): pairs of sugar units bonded together.
di = two

polysaccharides: long chains of monosaccharide units arranged as starch, glycogen, or fiber.
poly = many

glucose: a monosaccharide, the sugar common to all disaccharides and polysaccharides; also called *blood sugar* or *dextrose.*

fructose: a monosaccharide; sometimes known as *fruit sugar.* It is abundant in fruits, honey, and saps.
fruct = fruit

galactose: a monosaccharide; part of the disaccharide lactose.

homeostasis (HOME-ee-oh-STAY-sis): the maintenance of constant internal conditions (such as chemistry, temperature, and blood pressure) by the body's control systems.
homeo = the same
stasis = staying

The Chemist's View of Carbohydrates

The dietary **carbohydrates** include the sugars, starch, and fiber.[2] Chemists describe the sugars as:

▮ **Monosaccharides** (single sugars)

▮ **Disaccharides** (double sugars)

Starch and fiber are:*

▮ **Polysaccharides**—compounds composed of chains of monosaccharide units

All of these carbohydrates are composed of the single sugar **glucose** and other compounds that are much like glucose in composition and structure. Figure 2-1 shows the chemical structure of glucose.

Monosaccharides

Three monosaccharides are important in nutrition: glucose, **fructose,** and **galactose.** All three monosaccharides have the same number and kinds of atoms, but in different arrangements.

Glucose Most cells depend on glucose for their fuel to some extent, and the cells of the brain and the rest of the nervous system depend almost exclusively on glucose for their energy. The body can obtain this glucose from carbohydrates. To function optimally, the body must maintain blood glucose within limits that allow the cells to nourish themselves. If blood glucose falls below normal, the person may become dizzy and weak; if it rises substantially above normal, the person may become

*Monosaccharides and disaccharides (sugars) are sometimes called *simple carbohydrates*, and the polysaccharides (starch and fiber) are sometimes called *complex carbohydrates*.

fatigued. Left untreated, fluctuations to the extremes—either high or low—can be fatal. Blood glucose **homeostasis** is regulated primarily by two hormones: **insulin,** which moves glucose from the blood into the cells, and **glucagon,** which brings glucose out of storage when blood glucose falls (as occurs between meals).

Fructose Fructose is the sweetest of the sugars. Fructose occurs naturally in fruits, honey, and saps. Other sources include soft drinks, ready-to-eat cereals, and other products sweetened with high-fructose corn syrup. Glucose and fructose are the most common monosaccharides in nature.

Galactose The third single sugar, galactose, occurs mostly as part of lactose, a disaccharide also known as milk sugar. During digestion, galactose is freed as a single sugar.

Disaccharides

In disaccharides, pairs of single sugars are linked together. Three disaccharides are important in nutrition: maltose, sucrose, and lactose. All three have glucose as one of their single sugars. As Table 2-1 shows, the other monosaccharide is either another glucose (in maltose), fructose (in sucrose), or galactose (in lactose). The shapes of the sugars in Table 2-1 reflect their chemical structures as they are drawn on paper.

Sucrose Sucrose (table, or white, sugar) is the most familiar of the three disaccharides and is what people generally mean when they speak of "sugar." This sugar is usually obtained by refining the juice from sugar beets or sugarcane to provide the brown, white, and powdered sugars available in the supermarket, but it occurs naturally in many fruits and vegetables. When a person eats a food containing sucrose, enzymes in the digestive tract split the sucrose into its glucose and fructose components.

Lactose Lactose is the principal carbohydrate of milk. Most human infants are born with the digestive enzymes necessary to split lactose into its two monosaccharide parts, glucose and galactose, so as to absorb it. Breast milk thus provides a simple, easily digested carbohydrate that meets an infant's energy needs; many formulas do, too, because they are made from milk.

Maltose The third disaccharide, **maltose,** is a plant sugar that consists of two glucose units. Maltose is produced whenever starch breaks down—as happens in plants when they break down their stored starch for energy and start to sprout, and in human beings during carbohydrate digestion.

Polysaccharides

Unlike the sugars, which contain the three monosaccharides—glucose, fructose, and galactose—in different combinations, the polysaccharides are composed almost entirely of glucose (and, in some cases, other monosaccharides). Three types of polysaccharides are important in nutrition: glycogen, starch, and fibers.

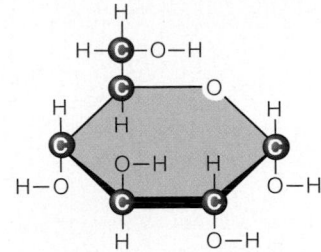

FIGURE 2-1 Chemical Structure of Glucose
On paper, the structure of glucose has to be drawn flat, but in nature, the five carbons and oxygen are roughly in a plane, with the H, OH, and CH_2OH extending out above and below it.

Diabetes, a disorder characterized by elevated blood glucose, is the topic of Chapter 21.

Many people lose the ability to digest lactose after infancy. This condition, known as lactose intolerance, is discussed in Chapter 19.

insulin: a hormone secreted by the pancreas in response to high blood glucose. It promotes cellular glucose uptake for use or storage.

glucagon (GLOO-ka-gon): a hormone that is secreted by special cells in the pancreas in response to low blood glucose concentration and elicits release of glucose from storage.

sucrose: a disaccharide composed of glucose and fructose; commonly known as *table sugar*, *beet sugar*, or *cane sugar*.

  sucro = sugar

lactose: a disaccharide composed of glucose and galactose; commonly known as *milk sugar*.

  lact = milk

maltose: a disaccharide composed of two glucose units; sometimes known as *malt sugar*.

TABLE 2-1 The Major Sugars

Monosaccharides	Disaccharides
Glucose	Sucrose (glucose + fructose)
Fructose	Lactose (glucose + galactose)
Galactose	Maltose (glucose + glucose)
(found primarily as part of lactose)	

Glycogen is a storage form of energy for human beings and animals; starch plays that role in plants; and fibers provide structure in stems, trunks, roots, leaves, and skins of plants. Both glycogen and starch are built of entirely glucose units; fibers are composed of a variety of monosaccharides and other carbohydrate derivatives.

Glycogen **Glycogen** molecules, which are made of chains of glucose, are more highly branched than starch molecules. Glycogen is found in meats only to a limited extent and not at all in plants.* For this reason, glycogen is not a significant food source of carbohydrate, but it does play an important role in the body. The human body stores much of its glucose as glycogen in the liver and muscles.

After a meal, as blood glucose rises, the pancreas is the first organ to respond. It releases the hormone insulin, which signals the body's tissues to take up surplus glucose. Muscle and liver cells use some of this excess glucose to build glycogen. The muscles hoard two-thirds of the body's total glycogen and use it just for themselves during physical activity. The brain stores a tiny fraction of the total, thought to provide an emergency glucose reserve sufficient to fuel the brain for an hour or two in severe glucose deprivation.[3] The liver stores the remainder and is generous with its glycogen, making it available as blood glucose for the brain or other tissues when the supply runs low.

Starch **Starch** is a long, straight or branched chain of hundreds or thousands of glucose units linked together. These giant molecules are packed side by side in grains such as rice or wheat, in root crops and tubers such as yams and potatoes, and in legumes such as peas and beans. When a person eats the plant, the body splits the starch into glucose units and uses the glucose for energy.

All starchy foods come from plants. Grains are the richest food source of starch. In most human societies, people depend on a staple grain for much of their food energy: rice in Asia; wheat in Canada, the United States, and Europe; corn in much of Central and South America; and millet, rye, barley, and oats elsewhere. A second important source of starch is the legume (bean and pea) family. Legumes include peanuts and dry beans such as butter beans, kidney beans, navy beans, black-eyed peas (cowpeas), chickpeas (garbanzo beans), and soybeans. Root vegetables (tubers) such as potatoes and yams are a third major source of starch, and, in many non-Western societies, they are the primary starch sources. Grains, legumes, and tubers not only are rich in starch but may also contain abundant dietary fiber, protein, and other nutrients.

Fibers **Dietary fibers** are the structural parts of plants and thus are found in all plant-derived foods—vegetables, fruits, whole grains, and legumes. Most dietary fibers are polysaccharides—chains of sugars—just as starch is, but in fibers the sugar units are held together by bonds that human digestive enzymes cannot break. Consequently, most dietary fibers pass through the body, providing little or no energy for its use. Figure 2-2 shows the difference between starch and the plant fiber cellulose. In addition to cellulose, fibers include the polysaccharides hemicellulose, pectins, gums, and mucilages, as well as the nonpolysaccharide lignins.

The short chains of glucose units that result from the breakdown of starch are known as *dextrins*. The word sometimes appears on food labels because dextrins can be used as thickening agents in foods.

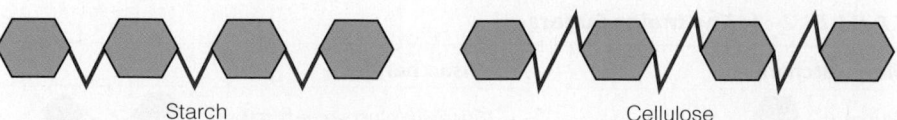

Starch   Cellulose

FIGURE 2-2 **Starch and Cellulose Molecules Compared (Small Segments)**
The bonds that link the glucose units together in cellulose are different from the bonds in starch (and glycogen). Human enzymes cannot digest cellulose.

*Glycogen in animal muscles rapidly breaks down after slaughter.

Cellulose is the main constituent of plant cell walls, so it is found in all vegetables, fruits, and legumes. Hemicellulose is the main constituent of cereal fibers. Pectins are abundant in vegetables and fruits, especially citrus fruits and apples. The food industry uses pectins to thicken jelly and keep salad dressing from separating. Gums and mucilages have similar structures and are used as additives or stabilizers by the food industry. Lignins are the tough, woody parts of plants; few foods people eat contain much lignin.

A few starches are classified as fibers. Known as **resistant starches,** these starches escape digestion and absorption in the small intestine. Starch may resist digestion for several reasons, including an individual's efficiency in digesting starches and the food's physical properties. Resistant starch is common in whole or partially milled grains, legumes, raw potatoes, and unripe bananas.[4] Cooked potatoes, pasta, and rice that have been chilled also contain resistant starch. Similar to some fibers, resistant starch may support a healthy colon.[5]

Although cellulose and other dietary fibers are not broken down by human enzymes, some fibers can be digested by bacteria in the human digestive tract. Bacterial digestion of fibers can generate some absorbable products that can yield energy when metabolized. Food fibers, therefore, can contribute some energy (1.5 to 2.5 kcalories per gram), depending on the extent to which they break down in the body.

Fibers are divided into two general groups by their chemical and physical properties.* In the first group are fibers that dissolve in water (**soluble fibers**). These form gels (are **viscous**) and are easily digested by bacteria in the human large intestine (are easily fermented). Commonly found in barley, legumes, fruits, oats, and vegetables, these fibers are often associated with lower risks of chronic diseases (as discussed in a later section). In foods, soluble fibers add a pleasing consistency; for example, pectin puts the gel in jelly, and gums are added to salad dressings and other foods to thicken them.

Other fibers do not dissolve in water (**insoluble fibers**), do not form gels (are not viscous), and are less readily fermented. Insoluble fibers, such as cellulose and many hemicelluloses, are found in the outer layers of whole grains (bran), the strings of celery, the hulls of seeds, and the skins of corn kernels. These fibers retain their structure and rough texture even after hours of cooking. In the body, they aid the digestive system by easing elimination.[6]

IN SUMMARY

▪ Carbohydrate is the body's preferred energy source. Six sugars are important in nutrition: the three monosaccharides (glucose, fructose, and galactose) and the three disaccharides (sucrose, lactose, and maltose).

▪ The three disaccharides are pairs of monosaccharides; each contains glucose paired with one of the three monosaccharides. The polysaccharides (chains of monosaccharides) are glycogen, starches, and fibers.

▪ Both glycogen and starch are storage forms of glucose—glycogen in the body and starch in plants—and both yield energy for human use.

▪ The dietary fibers also contain glucose (and other monosaccharides), but their bonds cannot be broken by human digestive enzymes, so they yield little, if any, energy.

glycogen (GLY-co-gen): a polysaccharide composed of glucose, made and stored by liver and muscle tissues of human beings and animals as a storage form of glucose. Glycogen is not a significant food source of carbohydrate and is not counted as one of the polysaccharides in foods.

starch: a plant polysaccharide composed of glucose and digestible by human beings.

dietary fibers: a general term denoting in plant foods the polysaccharides cellulose, hemicellulose, pectins, gums and mucilages, as well as the nonpolysaccharide lignins, which are not digested by human digestive enzymes, although some are digested by GI tract bacteria.

resistant starches: starches that escape digestion and absorption in the small intestine of healthy people.

soluble fibers: indigestible food components that readily dissolve in water and often impart gummy or gel-like characteristics to foods. An example is pectin from fruit, which is used to thicken jellies.

viscous: having a gel-like consistency.

insoluble fibers: the tough, fibrous structures of fruits, vegetables, and grains; indigestible food components that do not dissolve in water.

*The DRI committee has proposed these fiber definitions: the term *dietary fibers* refers to naturally occurring fibers in intact foods, and *functional fibers* refers to added fibers that have health benefits; *total fiber* refers to the sum of fibers from both sources.

Health Effects of Sugars and Alternative Sweeteners

The *USDA Food Guide* distinguishes between naturally occurring and added sugars, designating the kcalories from added sugars as discretionary kcalories.

Fiber-rich carbohydrate foods such as vegetables, whole grains, legumes, and fruits should predominate in people's diets; concentrated sweets such as candy, cola beverages and other soft drinks, cookies, pies, cakes, and other foods with **added sugars** supply kcalories, but few, if any, other nutrients. These foods therefore contribute discretionary kcalories to the diet. Estimated intakes of added sugars in the United States far exceed the allowance for discretionary kcalories, regardless of energy intake.[7]

The **naturally occurring sugars** in vegetables, fruits, grains, and milk are acceptable because they are accompanied by many nutrients. People who want to limit their use of sugar may choose from two sets of alternative sweeteners: sugar alcohols and artificial sweeteners.

Sugary soft drinks are the leading source of added sugars in the United States; cakes, cookies, pies, and other baked goods come next; and sweetened fruit drinks and punches follow closely behind.

Sugars

Estimates are that each man, woman, and child in the United States consumes more than 100 pounds of sugar per year, or a little less than 2 pounds per week. Figure 2-3 depicts the increasing availability of added sugars in the U.S. diet and offers the American Heart Association prudent upper intake limits for added sugars for comparison. (Note that the columns in the figure represent sugars in the food supply and do not account for waste, such as the syrup drained from sweet pickles or jam that molds and is thrown away.) The steady upward trend in consumption of added sugars shown in Figure 2-3 is largely the result of a dramatic increase in consumption of commercially prepared foods and beverages that contain sugars added by food manufacturers during processing. In contrast, people are adding less sugar in the kitchen. Soft drinks and other sugar-sweetened beverages are the main source of added sugars in the diets of U.S. consumers.[8] The usual amount of added sugars consumed in a day averages more than 22 teaspoons.[9] The committee that drafted the *Dietary Guidelines for Americans* offers clear advice on added sugars: treat them as discretionary kcalories.[10]

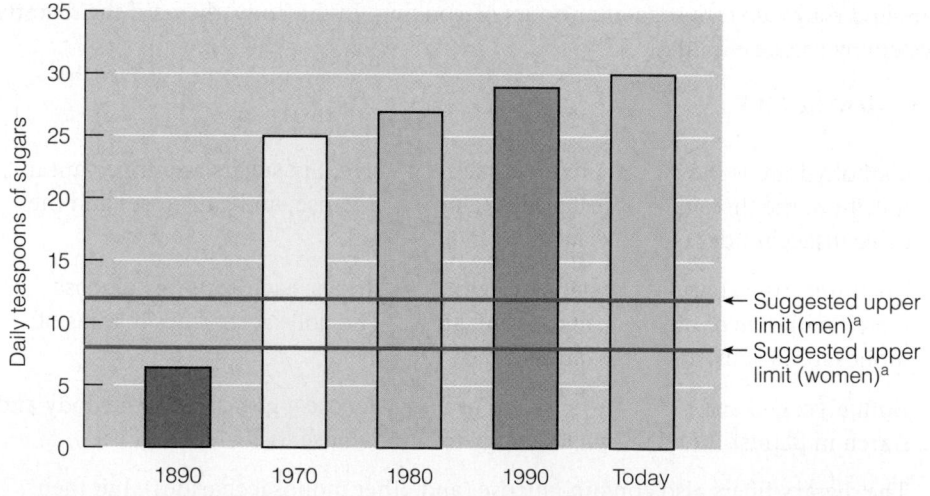

[a]These USDA suggested upper limits for added sugars reflect one-half of the average discretionary kcalorie allowance for sedentary men aged 19–30 consuming 2400 kcal/day and sedentary women aged 19–30 consuming 2000 kcal/day. The American Heart Association sets a prudent upper limit of 6 teaspoons of sugar for most women, and 9 for most men.

FIGURE 2-3 Added Sugars: Average U.S. Supply per Person Compared with Suggested Upper Intake Limits

Sources: USDA Food Guide; R. K. Johnson and coauthors, Dietary sugars intake and cardiovascular health: A Scientific Statement from the American Heart Association, *Circulation* 120 (2009): 1011–1020.

In other words, fulfill the day's nutrient needs with foods contributing little or no added sugar. Most people can afford only a little added sugar in their diets if they are to meet nutrient needs within kcalorie limits.

The U.S. population is not alone in its increasing sugar consumption. The trend is a worldwide phenomenon. In response, the World Health Organization has also taken a stand on sugar intake: consume no more than 10 percent of total kcalories from added sugars.*

The increase in sugar consumption has raised many questions about sugar's effects on health. Many accusations have been made against sugar, but the Food and Drug Administration (FDA) and the National Academy of Sciences have concluded that, in moderate amounts, sugars pose no major health risk. In excess, however, they can be detrimental in two ways. One, sugars can contribute to nutrient deficiencies by supplying energy (kcalories) without providing nutrients. Two, sugars contribute to tooth decay, or **dental caries.**

High intakes of added sugars, along with dramatically increasing rates of obesity, have raised concerns about the relationship between obesity and excessive sugar intake.[11] A later section in this discussion addresses this concern.

Sugar and Nutrient Deficiencies Empty-kcalorie foods that contain lots of added sugars such as cakes, candies, and sodas provide the body with glucose and energy, but few, if any, other nutrients.[12] In comparison, foods such as whole grains, vegetables, legumes, and fruits that contain some natural sugars and lots of starches and fibers also provide protein, vitamins, and minerals.

A person spending 200 kcalories of a day's energy allowance on a 16-ounce soda gets little of value for those kcalories. In contrast, a person using 200 kcalories on three slices of whole-wheat bread gets 9 grams of protein, 6 grams of fiber, plus several of the B vitamins with those kcalories. For the person who wants something sweet, a reasonable compromise might be two slices of bread with a teaspoon of jam on each. The amount of sugar a person can afford to eat depends on how many discretionary kcalories are available beyond those needed to deliver indispensable vitamins and minerals.

By following MyPyramid and making careful food selections, a typical adult can obtain all the needed nutrients within an allowance of about 1500 kcalories. Some people have more generous energy allowances. For example, an active teenage boy may need as many as 3000 kcalories a day. If he eats mostly nutritious foods, then he may have discretionary kcalories available for cola beverages and other extras. In contrast, an inactive older woman who is limited to fewer than 1500 kcalories a day can afford to eat only the most nutrient-dense foods—with few, or no, discretionary kcalories available. As mentioned earlier, intakes of added sugars far exceed discretionary kcalorie allowances, regardless of energy needs.[13]

Sugar and Dental Caries Does sugar contribute to dental caries? The evidence says yes. Any carbohydrate-containing food, including bread, bananas, or milk, as well as sugar, can support bacterial growth in the mouth. These bacteria produce the acid that eats away tooth enamel. Of major importance is the length of time the food stays in the mouth. This, in turn, depends on the composition of the food, how sticky the food is, how often a person eats the food, and especially whether the teeth are brushed afterward.[14] Total sugar intake still plays a major role in caries incidence; populations whose diets provide no more than 10 percent of kcalories from sugar have a low prevalence of dental caries.

Sugar and Obesity Over the past several decades, as obesity rates increased sharply, consumption of added sugars reached an all-time high—much of it because high-fructose corn syrup use, especially in beverages, surged.[15] High-fructose corn syrup is

Nutrition in Practice 17 discusses nutrition and oral health.

added sugars: sugars and syrups added to a food for any purpose, such as to add sweetness or bulk or to aid in browning (baked foods). Also called *carbohydrate sweeteners,* they include glucose, fructose, corn syrup, concentrated fruit juice, and other sweet carbohydrates.

naturally occurring sugars: sugars that are not added to a food but are present as its original constituents, such as the sugars of fruit or milk.

dental caries: the gradual decay and disintegration of a tooth.

---

*The World Health Organization uses the term *free sugars* to mean all monosaccharides and disaccharides added to foods by the manufacturer, cook, or consumer plus the naturally occurring sugars in honey, syrups, and fruit juices.

composed of fructose and glucose in a ratio of about 50:50. Compared with sucrose, high-fructose corn syrup is less expensive, easier to use, and more stable. In addition to being used in beverages, high-fructose corn syrup sweetens candies, baked goods, and hundreds of other foods. The use of high-fructose corn syrup sweetener parallels unprecedented increases in the incidence of obesity, but does this mean that the increasing sugar intakes are responsible for the increase in body fat and its associated health problems?[16] Excess sugar in the diet may be associated with more fat on the body.[17] When eaten in excess of need, energy from added sugars contributes to body fat stores, just as excess energy from other sources does.[18] Added sugars provide excess energy, raising the risk of weight gain.[19] When total energy intake is controlled, however, *moderate* amounts of sugar do not *cause* obesity. A recent meta-analysis of sugar-sweetened beverage consumption and body mass index in children and adolescents showed no significant relationship.[20] The authors point out that sugar-sweetened beverages are a source of energy, and excess energy intake promotes weight gain. Thus, to the extent that sugar contributes to an excessive energy intake, it can play a role in the development of obesity.

Recommended Sugar Intakes Moderate sugar intakes—enough for pleasure but not enough to displace more nutritious foods—are not harmful. Sugar is a delicious, concentrated source of food energy, but it contains no protein, vitamins, or minerals. Eaten in the place of nutrient- and fiber-rich foods, it makes malnutrition likely.

The *Dietary Guidelines for Americans* urge people to reduce consumption of foods containing added sugars. The USDA Food Guide suggests that, within kcalorie limits, small amounts of added sugars can be enjoyed as part of the discretionary kcalorie allowance in a nutrient-dense diet. The USDA Food Guide recommendations represent about 5 to 10 percent of the day's total energy intake. The American Heart Association suggests a prudent upper limit of not more than half the discretionary kcalorie allowance (no more than 100 kcalories of added sugars for most women or 150 kcalories for most men). As noted earlier, the World Health Organization agrees that people should restrict their consumption of added sugars to 10 percent or less of total energy.

The DRI committee did not set a Tolerable Upper Intake Level for added sugars, but, as mentioned, excessive intakes can interfere with sound nutrition and dental health. Few people can eat lots of sugary treats and still meet all of their nutrient needs without exceeding their kcalorie allowance. Instead, the DRI committee suggests, as a rather high maximum, that added sugars should account for no more than 25 percent of the day's total energy intake.[21] For a person consuming 2000 kcalories a day, 25 percent represents 500 kcalories from added sugars—quite a lot of sugar. Perhaps an athlete in training whose energy needs are high can afford the added sugars from sports drinks without compromising nutrient intake, but most people would do better following recommendations to limit added sugar consumption to less than 10 percent of the day's total energy intake.

Recognizing Sugars People often fail to recognize sugar in all its forms and so do not realize how much they consume. To help your clients estimate how much sugar they consume, tell them to treat all of the following concentrated sweets as equivalent to 1 teaspoon of white sugar (4 grams of carbohydrate):

▌ 1 teaspoon brown sugar, candy, jam, jelly, any corn sweetener, syrup, honey, molasses, or maple sugar

▌ 1 tablespoon catsup

▌ 1½ ounces carbonated soft drink

*The amounts of added sugars suggested here are not specific recommendations for amounts of added sugars to consume but rather represent the amounts that can be included in the diet at each kcalorie level.

The USDA Food Guide suggests:*
▐ 3 tsp for 1,600 kcal
▐ 5 tsp for 1,800 kcal
▐ 8 tsp for 2,000 kcal
▐ 9 tsp for 2,200 kcal
▐ 12 tsp for 2,400 kcal

For perspective, each of these sources of concentrated sugars provides about 500 kcal:
▐ 40 oz cola
▐ ½ cup honey
▐ 125 jelly beans
▐ 23 marshmallows
▐ 30 tsp sugar

brown sugar: white sugar with molasses added; 95 percent pure sucrose.

concentrated fruit juice sweetener: a concentrated sugar syrup made from dehydrated, deflavored fruit juice, commonly grape juice; used to sweeten products that can then claim to be "all fruit."

confectioner's sugar: finely powdered sucrose; 99.9 percent pure.

corn sweeteners: corn syrup and sugar solutions derived from corn.

corn syrup: a syrup, mostly glucose, partly maltose, produced by the action of enzymes on cornstarch.

dextrose: an older name for glucose.

evaporated cane juice: raw sugar from which impurities have been removed.

fructose, galactose, glucose: the monosaccharides.

granulated sugar: common table sugar, crystalline sucrose, 99.9 percent pure.

high-fructose corn syrup (HFCS): a mixture of primarily fructose and glucose; the predominant sweetener used in processed foods today.

honey: a concentrated solution primarily composed of glucose and fructose; produced by enzymatic digestion of the sucrose in nectar by bees.

invert sugar: a mixture of glucose and fructose formed by the splitting of sucrose in an industrial process. Sold only in liquid form and sweeter than sucrose, invert sugar forms during certain cooking procedures and works to prevent crystallization of sucrose in soft candies and sweets.

lactose, maltose, sucrose: the disaccharides.

levulose: an older name for fructose.

maple sugar: a concentrated solution of sucrose derived from the sap of the sugar maple tree, mostly sucrose. This sugar was once common but is now usually replaced by sucrose and artificial maple flavoring.

molasses: a syrup left over from the refining of sucrose from sugarcane; a thick, brown syrup. The major nutrient in molasses is iron, a contaminant from the machinery used in processing it.

raw sugar: the first crop of crystals harvested during sugar processing. Raw sugar cannot be sold in the United States because it contains too much filth (dirt, insect fragments, and the like). Sugar sold as "raw sugar" is actually evaporated cane juice.

turbinado (ter-bih-NOD-oh) **sugar:** raw sugar from which the filth has been washed; legal to sell in the United States.

white sugar: pure sucrose, produced by dissolving, concentrating, and recrystallizing raw sugar.

These portions of sugar all provide about the same number of kcalories. Some are closer to 10 kcalories (for example, 14 kcalories for molasses), while some are more than 20 (22 kcalories for honey), so an average figure of 16 kcalories is an acceptable approximation. The accompanying glossary presents the multitude of names that denote sugar on food labels.

People often ask: What is the difference between honey and white sugar? Is honey, by virtue of being natural, more nutritious? Honey, like white sugar, contains glucose and fructose. The difference is that, in white sugar, the glucose and fructose are bonded together in pairs, whereas in honey some of them are paired and some are free single sugars. When you eat either white sugar or honey, though, your body breaks all of the sugars apart into single sugars. It ultimately makes no difference, then, whether you eat single sugars linked together, as in white sugar, or the same sugars unlinked, as in honey; they will end up as single sugars in your body.

Honey does contain a few vitamins and minerals, but not many. Honey is denser than crystalline sugar, too, so it provides more energy per spoonful. Table 2-2 (p. 42) shows that honey and white sugar are similar nutritionally—and both fall short of milk, legumes, fruits, grains, and vegetables. Honey may offer some health benefits, however: It seems to relieve nighttime coughing in children and reduce the severity of mouth ulcers in cancer patients undergoing chemotherapy or radiation.[22]

Some sugar sources, however, are more nutritious than others. Consider a fruit such as an orange. The orange provides the same sugars and about the same energy as a tablespoon of sugar or honey, but the packaging makes a big difference in nutrient density. The sugars of the orange are diluted in a large volume of fluid that contains valuable vitamins and minerals, and the flesh and skin of the orange are supported by fibers that also offer health benefits. A tablespoon of honey offers no such bonuses, and neither, of course, does a cola beverage containing many teaspoons of sugar.

You receive about the same amount and kinds of sugars from an orange as from a tablespoon of honey, but the packaging makes a big nutrition difference.

A 12 oz can of cola contains the equivalent of 8 tsp of sugar.

TABLE 2-2 **Sample Nutrients in Sugars and Other Foods**

The indicated portion of any of these foods provides approximately 100 kcalories. Notice that—for a similar number of kcalories and grams of carbohydrate—milk, legumes, fruits, grains, and vegetables offer more of the other nutrients than do the sugars.

	Size of 100 kCal Portion	Carbohydrate (g)	Protein (g)	Calcium (mg)	Iron (mg)	Vitamin A (µg)	Vitamin C (mg)
Foods							
Milk, 1% low-fat	1 c	12	8	300	0.1	144	2
Kidney beans	½ c	20	7	30	1.6	0	2
Apricots	6	24	2	30	1.1	554	22
Bread, whole wheat	1½ slices	20	4	30	1.9	0	0
Broccoli, cooked	2 c	20	12	188	2.2	696	148
Sugars							
Sugar, white	2 tbs	24	0	trace	trace	0	0
Molasses, blackstrap	2½ tbs	28	0	343	12.6	0	0.1
Cola beverage	1 c	26	0	6	trace	0	0
Honey	1½ tbs	26	trace	2	0.2	0	trace

IN SUMMARY

▌ Sugars pose no major health threat except for an increased risk of dental caries.

▌ Excessive sugar intakes may displace needed nutrients and fiber and may contribute to obesity.

▌ A person deciding to limit daily sugar intake should recognize that not all sugars need to be restricted, just concentrated sweets with added sugars, which are high in kcalories and relatively lacking in other nutrients. Sugars that occur naturally in fruits, vegetables, and milk are acceptable.

Alternative Sweeteners

To control weight gain, blood glucose, and dental caries, many consumers turn to alternative sweeteners to help them limit kcalories and minimize sugar intake. In doing so, they encounter three types of alternative sweeteners: sugar alcohols, artificial sweeteners, and herbal products known as **stevia sweeteners.**

Sugar Alcohols The **sugar alcohols** are carbohydrates, but they yield slightly less energy (2 to 3 kcalories per gram) than sucrose (4 kcalories per gram) because they are not absorbed completely.[23] The sugar alcohols are sometimes called **nutritive sweeteners** because they do yield some energy.

The sugar alcohols occur naturally in fruits and vegetables; they are also used by manufacturers to provide sweetness and bulk to cookies, sugarless gum, hard candies, and jams and jellies. Their sweetness relative to sugar is shown in Table 2-3. Unlike sucrose, sugar alcohols are fermented in the large intestine by intestinal bacteria. Consequently, side effects such as gas, abdominal discomfort, and diarrhea make the sugar alcohols less attractive than the artificial sweeteners.

The advantage of using sugar alcohols is that they do not contribute to dental caries. Bacteria in the mouth metabolize sugar alcohols much more slowly than sucrose, thereby inhibiting the production of acids that promote caries' formation. They are therefore valuable in chewing gums, breath mints, and other products that people

TABLE 2-3 **Sugar Alcohols**

Sugar Alcohols	Relative Sweetness*	Energy (kcal/g)	Approved Uses
Erythritol	0.7	0.2	Bulk sweetener in low-kcalorie foods
Isomalt	0.5	2.0	Candies, chewing gum, ice cream, jams and jellies, frostings, beverages, baked goods
Lactitol	0.4	2.0	Candies, chewing gum, frozen dairy desserts, jams and jellies, frostings, baked goods
Maltitol	0.9	2.1	Particularly good for candy coating
Mannitol	0.7	1.6	Bulking agent, chewing gum
Sorbitol	0.5	2.6	Special dietary foods, candies, gums, frozen desserts, baked goods
Xylitol	1.0	2.4	Chewing gum, candies, pharmaceutical and oral health products

*The relative sweetness depends on the temperature, acidity, and other flavors of the foods in which the substance occurs. The sweetness of pure sucrose is the standard with which the approximate sweetness of sugar substitutes is compared.

keep in their mouths awhile. The FDA allows food labels to carry a health claim (see p. 26 in Chapter 1) about the relationship between sugar alcohols and the non-promotion of dental caries as long as the FDA criteria for sugar-free status and other criteria are met. Figure 2-4 presents labeling information for products using sugar alternatives.

Artificial Sweeteners The **artificial sweeteners** are not carbohydrates. They yield virtually no energy in the amounts typically used and are sometimes called nonnutritive sweeteners. Like the sugar alcohols, artificial sweeteners make foods taste sweet without promoting tooth decay. The FDA has set **Acceptable Daily Intake (ADI)**

stevia sweeteners: zero-kcalorie sweeteners derived from the sweetest part of the native South American stevia plant. Stevia sweeteners are "generally recognized as safe" by the FDA for use as general-purpose sweeteners.

sugar alcohols: sugar-like compounds. Like sugars, they are sweet to taste but yield 2 to 3 kcal per gram, slightly less than sucrose. Examples are maltitol, mannitol, sorbitol, isomalt, lactitol, and xylitol.

nutritive sweeteners: sweeteners that yield energy, including both the sugars and the sugar alcohols.

artificial sweeteners: noncarbohydrate, non-kcaloric synthetic sweetening agents; sometimes called *nonnutritive sweeteners*.

Acceptable Daily Intake (ADI): the amount of an artificial sweetener that individuals can safely consume each day over the course of a lifetime without adverse effect. It includes a 100-fold safety factor.

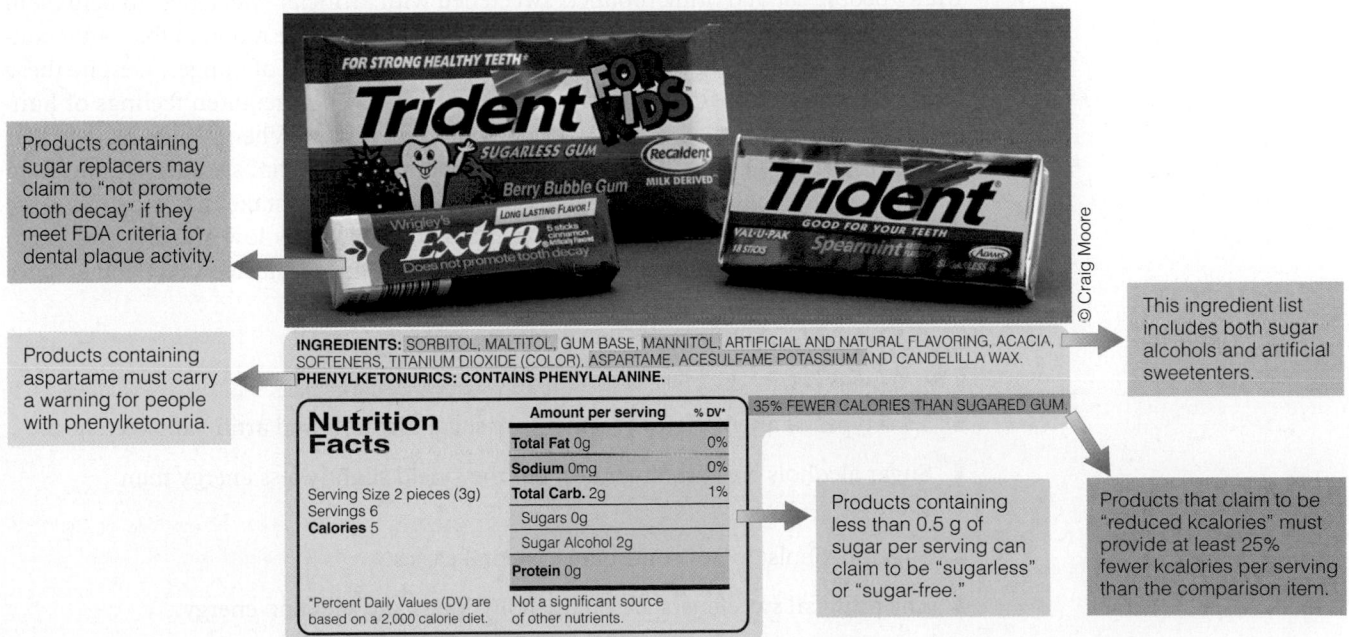

Products containing sugar replacers may claim to "not promote tooth decay" if they meet FDA criteria for dental plaque activity.

Products containing aspartame must carry a warning for people with phenylketonuria.

INGREDIENTS: SORBITOL, MALTITOL, GUM BASE, MANNITOL, ARTIFICIAL AND NATURAL FLAVORING, ACACIA, SOFTENERS, TITANIUM DIOXIDE (COLOR), ASPARTAME, ACESULFAME POTASSIUM AND CANDELILLA WAX. PHENYLKETONURICS: CONTAINS PHENYLALANINE.

This ingredient list includes both sugar alcohols and artificial sweetenters.

35% FEWER CALORIES THAN SUGARED GUM.

Nutrition Facts

Serving Size 2 pieces (3g)
Servings 6
Calories 5

Amount per serving	% DV*
Total Fat 0g	0%
Sodium 0mg	0%
Total Carb. 2g	1%
Sugars 0g	
Sugar Alcohol 2g	
Protein 0g	

*Percent Daily Values (DV) are based on a 2,000 calorie diet. Not a significant source of other nutrients.

Products containing less than 0.5 g of sugar per serving can claim to be "sugarless" or "sugar-free."

Products that claim to be "reduced kcalories" must provide at least 25% fewer kcalories per serving than the comparison item.

© Craig Moore

FIGURE 2-4 **Sugar Alternatives on Food Labels**

acesulfame (AY-sul-fame) **potassium,** also called **acesulfame-K:** a zero-kcalorie sweetener approved by the FDA and Health Canada.

aspartame: a compound of phenylalanine and aspartic acid that tastes like the sugar sucrose but is much sweeter. It is used in both the United States and Canada.

cyclamate: a zero-kcalorie sweetener under consideration for use in the United States and used with restrictions in Canada.

neotame (NEE-oh-tame): an artificial sweetener composed of two amino acids (phenylalanine and aspartic acid) linked in such a way as to make them indigestible by human enzymes.

saccharin: a zero-kcalorie sweetener used freely in the United States but restricted in Canada.

sucralose: a non-kcaloric sweetener derived from a chlorinated form of sugar that travels through the digestive tract unabsorbed. Approved for use in the United States and Canada.

tagatose: an incompletely absorbed monosaccharide sweetener derived from lactose with a kcaloric value of 1.5 kcalories per gram. About 80 percent of the ingested tagatose travels to the large intestine where bacterial colonies ferment it. Tagatose is not readily used by mouth bacteria and so does not promote dental caries.

levels for the artificial sweeteners used in the United States. Table 2-4 and the glossary of artificial sweeteners offer details on artificial sweeteners of interest.

Through the years, questions have emerged about the safety of some artificial sweeteners. For example, early research indicated that large quantities of saccharin caused bladder tumors in laboratory animals, but these issues have since been resolved. Common sense dictates that consuming large amounts of saccharin is probably not safe, but, at moderate intake levels, saccharin is currently assumed to be safe for most people.

Aspartame, a sweetener made from two amino acids (phenylalanine and aspartic acid), is one of the most thoroughly studied food additives ever approved. For most people, long-term consumption of aspartame is safe and is not associated with any adverse health effects.[24] Individuals with the metabolic disorder phenylketonuria (PKU) are the exception, and the labels of products that contain aspartame must include information for individuals with PKU. People with PKU cannot dispose of the phenylalanine contained in aspartame efficiently (see Nutrition in Practice 16).

Artificial Sweeteners and Weight Control

Many people eat and drink products sweetened with artificial sweeteners to help them control weight. Does this work? Ironically, a few studies have reported that, after consuming such products, people experience heightened feelings of hunger. Despite these reports, most studies find that artificial sweeteners do not heighten feelings of hunger, enhance food intake, or cause weight gain in people.[25] When people reduce their energy intakes by replacing sugar in their diets with artificial sweeteners and then compensate for the reduced energy at later meals, energy intake may stay the same or increase. Using artificial sweeteners will not automatically lower energy intake; to successfully control energy intake, a person needs to make informed diet and activity decisions throughout the day.

IN SUMMARY

▌ Two types of alternative sweeteners are sugar alcohols and artificial sweeteners.

▌ Sugar alcohols are carbohydrates, but they yield slightly less energy than sucrose.

▌ Sugar alcohols do not contribute to dental caries.

▌ The artificial sweeteners are not carbohydrates and yield no energy.

▌ Like the sugar alcohols, artificial sweeteners do not promote tooth decay.

TABLE 2-4 U.S.-Approved Sweeteners

Sweetener	Chemical Composition	Body's Response	Sweetness Relative to Sucrose[a]	Energy (kcal/g)	Acceptable Daily Intake (ADI) and (Estimated Equivalent[b])	Approved Uses
Artificial Sweeteners						
Acesulfame potassium or acesulfame-K (Sunette, Sweet One)	Potassium salt	Not digested or absorbed	200	0	15 mg/kg body weight[c] (30 cans diet soda)	Alcoholic beverages, baked goods, candies, chewing gum, desserts, gelatins, puddings, tabletop sweeteners
Aspartame (NutraSweet, Equal, others)	Amino acids (phenylalanine and aspartic acid) and a methyl group	Digested and absorbed	180	4[d]	50 mg/kg body weight[e] (18 cans diet soda)	General-purpose sweetener in all foods and beverages; warning to population with PKU
Neotame	Aspartame with an additional side group attached	Not digested or absorbed	7000	0	18 mg/day	Baked goods, beverages (nonalcoholic), candies, chewing gum, frosting, frozen desserts, gelatins, puddings, jams and jellies, syrups
Saccharin (SugarTwin, Sweet'N Low, others)	Benzoic sulfimide	Rapidly absorbed and excreted	300	0	5 mg/kg body weight (10 packets of sweetener)	Tabletop sweeteners, wide range of foods, beverages, cosmetics, and pharmaceutical products
Sucralose (Splenda)	Sucrose with Cl atoms instead of OH groups	Not digested or absorbed	600	0	5 mg/kg body weight (6 cans diet soda)	Baked goods, carbonated beverages, chewing gum, coffee and tea, dairy products, frozen desserts, fruit spreads, salad dressing, syrups, tabletop sweeteners
Tagatose (Nutralose, Nutrilatose, Tagatesse)	Monosaccharide similar in structure to fructose; naturally occurring or derived from lactose	Not well absorbed	0.9	1.5	7.5 g/day	Bakery products, beverages, cereals, chewing gum, confections, dairy products, dietary supplements, health bars, tabletop sweetener
Herbal Sweeteners						
Stevia (Sweetleaf, Truvia, PurVia)	Glyosides found in the leaves of the Stevia rebaudiana herb	Digested and absorbed	200–300	0	4 mg/kg body weight	Tabletop sweeteners, a variety of foods and beverages

[a]Relative sweetness is determined by comparing the approximate sweetness of a sugar substitute with the sweetness of pure sucrose, which has been defined as 1.0. Chemical structure, temperature, acidity, and other flavors of the foods in which the substance occurs all influence relative sweetness.

[b]Based on a person weighing 70 kg (154 lbs).

[c]Recommendations from the World Health Organization limit acesulfame-K intake to 9 mg per kilogram of body weight per day.

[d]Aspartame provides 4 kcal per gram, as does protein, but because so little is used, its energy contribution is negligible. In powdered form, it is sometimes mixed with lactose, however, so a 1-g packet may provide 4 kcal.

[e]Recommendations from the World Health Organization and in Europe and Canada limit aspartame intake to 40 mg per kilogram of body weight per day.

Health Effects of Starch and Dietary Fibers

Despite dietary recommendations that people should eat generous servings of starch and fiber-rich carbohydrate foods for their health, many people still believe that carbohydrate is the "fattening" component of foods. Gram for gram, carbohydrates contribute fewer kcalories to the body than do dietary fats, so a moderate diet based on starch- and fiber-rich carbohydrate foods is likely to be lower in kcalories than a diet based on high-fat foods.

For health's sake, most people should increase their intakes of carbohydrate-rich foods such as whole grains, vegetables, legumes, and fruits—foods noted for their starch, fiber, and naturally occurring sugars. In addition, most people should also limit their intakes of foods with added sugars and the types of fats associated with heart disease (see Chapter 3). A diet that emphasizes whole grains, vegetables, legumes, and fruits is almost invariably moderate in food energy, low in fats that can harm health, and high in dietary fiber, vitamins, and minerals. All these factors working together can help reduce the risks of obesity, cancer, cardiovascular disease, diabetes, dental caries, gastrointestinal disorders, and malnutrition.

It is difficult to sort out which carbohydrates contribute to which health benefits. Starch and fibers almost always occur together in foods (except refined foods), so it is hard to distinguish their effects. Some health effects appear to be especially closely associated with fibers, however.

Carbohydrates: Disease Prevention and Recommendations

Fiber-rich carbohydrate foods benefit health in many ways. Foods such as whole grains, legumes, vegetables, and fruits supply valuable vitamins, minerals, and phytochemicals, along with abundant dietary fiber and little or no fat. The following paragraphs describe some of the health benefits of diets that emphasize a variety of these foods each day.

Heart Disease Diets rich in whole grains, legumes, and vegetables, especially those rich in whole grains, may protect against heart disease and stroke by lowering blood pressure, improving blood lipids, and reducing inflammation.[26] Such diets are low in animal fat and cholesterol and high in dietary fibers, vegetable proteins, and phytochemicals—all factors associated with a lower risk of heart disease. Foods rich in soluble fibers (such as oat bran, barley, and legumes) lower blood cholesterol by binding cholesterol compounds and carrying them out of the body with the feces.[27] High-fiber foods may also lower blood cholesterol indirectly by displacing fatty, cholesterol-raising foods from the diet. Even when dietary fat intake is low, research shows that high intakes of soluble fiber exert separate and significant blood cholesterol–lowering effects.

The role of animal fat and cholesterol in heart disease is discussed in Chapter 3. The role of vegetable proteins in heart disease is presented in Chapter 4. The benefits of phytochemicals in disease prevention are presented in Nutrition in Practice 8.

Diabetes High-fiber foods—and especially whole grains—play a key role in reducing the risk of **type 2 diabetes** (see Chapter 21).[28] The soluble fibers of foods such as oats and legumes can help regulate the blood glucose following a carbohydrate-rich meal. Soluble fibers trap nutrients and delay their transit through the digestive tract, slowing glucose absorption and preventing the glucose surge and rebound often associated with diabetes onset.

The term **glycemic response** refers to how quickly glucose is absorbed after a person eats, how high the blood glucose rises, and how quickly it returns to normal. Slow absorption, a modest rise in blood glucose, and a smooth return to normal are desirable (a low glycemic response). Fast absorption, a surge in blood glucose, and an overreaction that plunges glucose below normal are less desirable (a high glycemic response). Different foods have different effects on blood glucose. The **glycemic index** is a method of classifying foods according to their potential to raise blood glucose and is the topic of Nutrition in Practice 2.

GI Health Dietary fibers may enhance the health of the large intestine. The healthier the intestinal walls, the better they can block absorption of unwanted constituents. Fibers that both enlarge and soften stools such as cellulose (in cereal brans, fruits, and vegetables) ease elimination for the rectal muscles and thereby alleviate or prevent constipation and hemorrhoids. Other fibers help to solidify watery stools.

Some fibers (again, such as cereal bran) help keep the contents of the intestinal tract moving easily. This action helps prevent compaction of the intestinal contents, which could obstruct the appendix and permit bacteria to invade and infect it. In addition, fibers stimulate the muscles of the gastrointestinal (GI) tract so that they retain their strength and resist bulging out in places, as occurs in diverticulosis.[29] Insoluble fiber seems to be most beneficial in lowering the risk of diverticulosis, which is described in Chapter 18.

Cancer Many studies show that, as people increase their dietary fiber intakes, their risk for colon cancer declines.[30] When the largest study of diet and cancer to date examined the diets of more than a half million people in 10 countries, the researchers found an inverse association between dietary fiber and colon cancer.[31] People who ate the most dietary fiber (35 grams per day) reduced their risk of colon cancer by 40 percent compared with those who ate the least fiber (15 grams per day). Importantly, the study focused on dietary fiber, not fiber supplements or additives. Fiber supplements and additives lack valuable nutrients and phytochemicals that also help protect against cancer.

How fiber may help prevent colon cancer is under investigation. One focus of research is fiber's ability to dilute and speed removal of potential cancer-causing agents from the large intestine (colon). In addition, the colon's bacteria ferment soluble fibers, forming small fatlike molecules that lower the pH. These small fatlike molecules activate cancer-killing enzymes and inhibit inflammation in the colon.[32]

Other processes may also be at work. When researchers examine other lifestyle factors, obesity, red and processed meat intake, cigarette smoking, and other factors emerge as associated with colon and rectal cancers, too.[33] As research progresses, cancer experts recommend that fiber in the diet come from five to nine half-cup servings of vegetables and fruit, along with generous portions of whole grains and legumes.

Weight Management Fiber-rich foods tend to be low in fats and added sugars and, therefore, promote weight loss by delivering less energy per bite.[34] In addition, as fibers absorb water from the digestive juices, they swell, creating feelings of fullness, lowering food intake, and delaying hunger.[35] Recent research suggests that dietary fiber intake may not only protect against excess body fat but may offer specific protection against abdominal obesity, a risk factor for type 2 diabetes. In a study of overweight adolescents, a modest *decrease* (3 grams per 1,000 kcalories per day) in dietary fiber intake over one year was associated with a 21 percent increase in abdominal obesity compared with those who increased dietary fiber intake by the same modest amount.[36] The researchers concluded that public health messages and strategies focusing on increasing dietary fiber intake are especially urgent for young people at risk of obesity and other metabolic disorders.

Many weight-loss products on the market today contain bulk-inducing fibers such as methylcellulose, but buying pure fiber compounds like this is not advised. Besides adding bulk to the diet, high-fiber foods supply other nutrients as well. Most experts agree that the health benefits attributed to fiber may come from other constituents of fiber-containing foods and not from the fiber alone.[37] Therefore, to use fiber in a weight-loss plan, select fresh fruits, vegetables, legumes, and whole-grain foods. As a bonus, fiber-rich foods are often economical as well as nutritious. Figure 2-5 (p. 48) summarizes fibers and their health benefits.

Harmful Effects of Excessive Fiber Intake Despite fiber's benefits to health, when too much fiber is consumed, some minerals may bind to it and be excreted with it without becoming available for the body to use. When mineral intake is adequate, however, a reasonable intake of high-fiber foods does not seem to compromise mineral balance.

People with marginal intakes who eat mostly high-fiber foods may not be able to take in enough food to meet energy or nutrient needs. The malnourished, the elderly,

pH is the unit of measure expressing a substance's acidity or alkalinity. (Chapter 4 provides a more detailed definition.)

type 2 diabetes: the type of diabetes that accounts for 90 to 95 percent of diabetes cases and usually results from insulin resistance coupled with insufficient insulin secretion.

glycemic response: the extent to which a food raises the blood glucose concentration and elicits an insulin response.

glycemic index: a method of classifying foods according to their potential for raising blood glucose.

People who eat these foods...		obtain these types of fibers...	with these actions in the body...	and these probable health benefits...
		Viscous, soluble, more fermentable		
	• Barley, oats, oat bran, rye, fruits (apples, citrus), legumes (especially young green peas and black-eyed peas), seaweeds, seeds and husks, many vegetables, fibers used as food additives	• Gums • Pectins • Psyllium[a] • Some hemicellulose	• Lower blood cholesterol by binding bile • Slow glucose absorption • Slow transit of food through upper GI tract • Hold moisture in stools, softening them • Yield small fat molecules after fermentation that the colon can use for energy • Increase satiety	• Lower risk of heart disease • Lower risk of diabetes • Lower risk of colon and rectal cancer • Increased satiety, and may help with weight management
		Nonviscous, insoluble, less fermentable		
	• Brown rice, fruits, legumes, seeds, vegetables (cabbage, carrots, Brussels sprouts), wheat bran, whole grains, extracted fibers used as food additives	• Cellulose • Lignins • Resistant starch • Hemicellulose	• Increase fecal weight and speed fecal passage through colon • Provide bulk and feelings of fullness	• Alleviate constipation • Lower risk of diverticulosis, hemorrhoids, and appendicitis • Lower risk of colon and rectal cancer

[a]Psyllium, a soluble fiber derived from seeds, is used as a laxative and food additive.

FIGURE 2-5 **Characteristics, Sources, and Health Effects of Fibers**

45% of 2000 kcal:

2000 × 0.45 = 900 kcal.

900 kcal ÷ 4 kcal/g = 225 g

65% of 2000 kcal:

2000 × 0.65 = 1300 kcal.

1300 kcal ÷ 4 kcal/g = 325 g

RDA for carbohydrate:

130 g/day

45 to 65% of energy intake

Daily Value:

300 g carbohydrate (based on 60% of 2000-kcal diet)

Daily Values:

25 g fiber (based on 11.5 g/1000 kcal)

Fiber AI:

14 g/1000 kcal/day

Men:

19–50 yr: 38 g/day

51+ yr: 30 g/day

Women:

19–50 yr: 25 g/day

51+ yr: 21 g/day

---

and young children adhering to all-plant (vegan) diets are especially vulnerable to this problem. Fibers also carry water out of the body and can cause dehydration. Advise clients to add an extra glass or two of water to go along with the fiber added to their diets. Athletes may want to avoid bulky, fiber-rich foods just prior to competition.

Adding purified fibers, such as oat bran or wheat bran, to foods can be taken to extremes. Too much fiber and too little fluid can obstruct the GI tract. Also, purified fiber may not affect the body the same way as the fiber in its original food product.

Recommended Intakes of Starches and Fibers The DRI committee advises that carbohydrates should contribute about half (45 to 65 percent) of the energy requirement. A person consuming 2000 kcalories a day should therefore have 900 to 1300 kcalories of carbohydrate, or between 225 and 325 grams. This amount is more than adequate to meet the RDA for carbohydrate, which is set at 130 grams per day based on the average minimum amount of glucose used by the brain.[38]

When it established the Daily Values that appear on food labels, the FDA used a guideline of 60 percent of kcalories in setting the Daily Value for carbohydrate at 300 grams per day. For most people, this means increasing total carbohydrate intake. To this end, the *Dietary Guidelines for Americans* encourage people to choose fiber-rich whole grains, vegetables, fruits, and legumes daily.

Recommendations for fiber encourage the same foods just mentioned: whole grains, vegetables, fruits, and legumes, which also provide vitamins, minerals, and phytochemicals. The FDA set the Daily Value for fiber at 25 grams, rounding up from the recommended 11.5 grams per 1000 kcalories for a 2000-kcalorie intake. The DRI recommendation is slightly higher, at 14 grams per 1000-kcalorie intake—roughly 25 to 35 grams of dietary fiber daily. These recommendations are about two times higher than the usual intake in the United States.[39]

As health care professionals, you can advise your clients that an effective way to add dietary fiber while lowering fat is to substitute plant sources of proteins (legumes)

for some of the animal sources of protein (meats and cheeses) in the diet. Another way to add fiber is to encourage clients to consume the recommended amounts of fruits and vegetables each day. People choosing high-fiber foods are wise to seek out a variety of fiber sources and to drink extra fluids to help the fiber do its job. Many foods provide fiber in varying amounts, as Figure 2-6 (p. 50) shows.

As mentioned earlier, too much fiber is no better than too little. The World Health Organization recommends an upper limit of 40 grams of dietary fiber a day.

Reminder: An *Adequate Intake (AI)* is used as a guide for nutrient intake when an RDA cannot be established (see Chapter 1).

IN SUMMARY

▌ A diet rich in starches and dietary fibers helps prevent heart disease, diabetes, GI disorders, and possibly some types of cancer. It also supports efforts to manage body weight.

▌ For these reasons, recommendations urge people to eat plenty of whole grains, vegetables, legumes, and fruits—enough to provide 45 to 65 percent of the daily energy from carbohydrate.

Carbohydrates: Food Sources

A day's meals based on the USDA Food Guide not only meet carbohydrate recommendations but provide abundant fiber, too. Grains, vegetables, fruits, and legumes deliver fiber to the diet and are noted for their valuable energy-yielding starches and dilute sugars. Each class of foods makes its own typical carbohydrate contribution. The USDA Food Guide in Chapter 1 can help you and your clients obtain carbohydrate-rich foods.

Grains Most foods in this group—a slice of whole-wheat bread, half an English muffin or bagel, a 6-inch tortilla, or ½ cup of rice, pasta, or cooked cereal—provide about 15 grams of carbohydrate, mostly as starch.* Most grain choices should be low in fat and sugar. When extra kcalories are needed to meet energy needs, some selections higher in fat, preferably unsaturated fat (see Chapter 3), and sugar can supply discretionary kcalories. These choices include biscuits, muffins, and snack crackers.

Vegetables Some vegetables are major contributors of starch in the diet. Just a small white or sweet potato or ½ cup of cooked dry beans, corn, peas, plantain, or winter squash provides 15 grams of carbohydrate, as much as in a slice of bread, though as a mixture of sugars and starch. A ½-cup portion of carrots, okra, onions, tomatoes, cooked greens, and most other nonstarchy vegetables or a cup of salad greens provides about 5 grams as a mixture of starch and sugars. Each of these foods also contributes a little protein, some fiber, and no fat.

Fruits The size of a typical serving of fruit varies depending on the form of the fruit: ½ cup of juice; a small banana, apple, or orange; ½ cup of most canned or fresh fruit; or ¼ cup of dried fruit. A typical fruit serving contains an average of about 15 grams of carbohydrate, mostly as sugars, including the fruit sugar fructose. Fruits vary greatly in their water and fiber contents; therefore, their sugar concentrations vary also. No more than one-half of the day's fruit should come from juice. With the exception of avocado, which is high in fat, the fruits contain insignificant amounts of fat and protein.

---

*Gram values in this section are adapted from the Exchange Lists for Diabetes.

Grains

Whole-grain products provide 1 to 2 g of fiber or more per serving:

- 1 slice whole wheat or rye bread (1 g)
- 1 slice pumpernickel bread (2 g)
- 1/2 c ready-to-eat 100% bran cereal (10 g)
- 1/2 c cooked barley, bulgur, grits, oatmeal (2 to 3 g)

Vegetables

Most vegetables contain 2 to 3 g of fiber per serving:

- 1 c raw bean sprouts
- 1/2 c cooked broccoli, Brussels sprouts, cabbage, carrots, cauliflower, collards, corn, eggplant, green beans, green peas, kale, mushrooms, okra, parsnips, potatoes, pumpkin, spinach, sweet potatoes, swiss chard, winter squash
- 1/2 c chopped raw carrots, peppers

Fruits

Fresh, frozen, and dried fruits have about 2 g of fiber per serving:

- 1 medium apple, banana, kiwi, nectarine, orange, pear
- 1/2 c applesauce, blackberries, blueberries, raspberries, strawberries
- Fruit juices contain very little fiber

Legumes and Nuts

Many legumes provide about 8 g of fiber per serving:

- 1/2 c cooked baked beans, black beans, black-eyed peas, kidney beans, navy beans, pinto beans

Some legumes provide about 5 g of fiber per serving:

- 1/2 c cooked garbanzo beans, great northern beans, lentils, lima beans, split peas

Most nuts and seeds provide 1 to 3 g of fiber per serving:

- 1 oz almonds, cashews, hazelnuts, peanuts, pecans, pumpkin seeds, sunflower seeds

FIGURE 2-6 **Fiber in Selected Foods**

Milk and Milk Products One cup of milk or yogurt or the equivalent (1 cup of buttermilk, ⅓ cup of dry milk powder, or ½ cup of evaporated milk) provides a generous 12 grams of carbohydrate. Among cheeses, cottage cheese provides about 6 grams of carbohydrate per cup, while most other types contain little, if any, carbohydrate. These foods also contribute high-quality protein as well as several important vitamins and minerals. Calcium-fortified soy beverages are options for providing calcium and about the same amount of carbohydrate as milk. All milk products vary in fat content, an important consideration in choosing among them; Chapter 3 provides details.

Cream and butter, although dairy products, are not equivalent to milk because they contain little or no carbohydrate and insignificant amounts of the other nutrients important in milk. They are appropriately placed with the solid fats and added sugars.

Protein Foods With two exceptions, foods of this group provide almost no carbohydrate to the diet. The exceptions are nuts, which provide a little starch and fiber along with their abundant fat, and dry beans, which are excellent sources of both starch and fiber. Just a ½-cup serving of beans provides 15 grams of carbohydrate, an amount equal to the richest carbohydrate sources. Among sources of fiber, beans and other legumes are outstanding, providing as much as 8 grams in ½ cup. The carbohydrate content of a diet can be determined by using a nutrient composition table such as that found in Appendix A, the exchange list system described in Chapter 21, or a computer diet analysis program.

Carbohydrates: Food Labels and Health Claims

Food labels list the amount, in grams, of total carbohydrate—including starch, fibers, and sugars—per serving. Fiber grams are also listed separately, as are the grams of sugars. (With this information, consumers can calculate starch grams by subtracting the grams of fibers and sugars from the total carbohydrate.) Sugars on the Nutrition Facts panel of a food label reflect both added sugars and those that occur naturally in foods. Total carbohydrate and dietary fiber are also expressed in the "% Daily Values" column for a person consuming 2000 kcalories; there is no Daily Value for sugars.

The FDA authorizes four health claims on food labels concerning fiber-rich carbohydrate foods. One is for "fiber-containing grain products, fruits, and vegetables and reduced risk of cancer." Another is for "fruits, vegetables, and grain products that contain fiber, and reduced risk of coronary heart disease." A third is for "soluble fiber from whole oats and from psyllium seed husk and reduced risk of coronary heart disease," and a fourth is for "whole grains and reduced risk of heart disease and certain cancers." Chapter 1 describes the criteria foods must meet to bear these health claims.

IN SUMMARY

▌ Grains, vegetables, fruits, and legumes contribute dietary fiber and energy-yielding starches and dilute sugars to people's diets. One cup of milk or yogurt or the equivalent contributes 12 grams of carbohydrate as well as protein and other important nutrients.

▌ Food labels list grams of total carbohydrate and also provide separate listings of grams of fiber and sugar.

Your Diet

Carbohydrates

Most of the energy people receive from foods comes from carbohydrates. Healthy choices provide carbohydrates rich in fiber, starches, vitamins, minerals, and naturally occurring sugars. A diet that is consistently low in dietary fiber and high in added sugar can lead to health problems. The table below displays related foods in categories that roughly indicate their fiber and sugar contents.

Look back to the food record you completed for the Your Diet activity at the end of Chapter 1. Draw a table like the one below, listing *five* of your carbohydrate choices in the appropriate column(s). If your food is not listed in the first column (High in Fiber/Low in Added Sugar), suggest an alternative selection to place in this column. As an example, if you consumed a Rice Krispies Treat® (low in fiber, high in added sugar), you could list the food in the third column and write "brown rice" in the first column. If the food has no obvious alternatives, like candy or sweetened soda, circle the food and list an alternative choice below the table (for example, you can replace regular soda with diet soda or low-fat milk).

High in Fiber/Low in Added Sugar	Intermediate Fiber and/or Sugar	Low in Fiber/High in Added Sugar
Apple with peel	Applesauce, sweetened	Fruit drink, 10% apple juice
Oatmeal	Granola	Granola breakfast bar
Whole-wheat bread	Bagel, plain	Danish pastry
Corn on the cob	Creamed corn	Corn flakes, added sugar
Baked sweet potato	Candied sweet potato casserole	Sweet potato pie

Clinical Applications

1. Considering the health benefits of carbohydrate-rich foods, especially those that provide starch and fiber, what suggestions would you offer to a client who reports the following:

   ▌ Eats three servings of refined, sugary breads or cereals each day.

   ▌ Eats one serving of vegetables (usually French fries) each day.

   ▌ Drinks fruit juice once a day but never eats fruit.

   ▌ Eats cheese at least twice a day but does not drink milk.

   ▌ Eats large servings of meat at least twice a day.

   ▌ Eats hard candy two or three times a day.

Self Check

1. Carbohydrates are found in virtually all foods **except**:
   a. milks.
   b. meats.
   c. breads.
   d. vegetables.

2. Polysaccharides include:
   a. galactose, starch, and glycogen.
   b. starch, glycogen, and fiber.
   c. lactose, maltose, and glycogen.
   d. sucrose, fructose, and glucose.

3. The chief energy source of the body is:
   a. sucrose.
   b. starch.
   c. glucose.
   d. fructose.

4. The primary form of stored glucose in animals is:
   a. glycogen.
   b. cellulose.
   c. starch.
   d. lactose.

5. The polysaccharide that helps form the cell walls of plants is:
   a. cellulose.
   b. starch.
   c. glycogen.
   d. lactose.

6. Which of the following items may denote sugar on food labels?
   a. Corn syrup
   b. Aspartame
   c. Xylitol
   d. Cellulose

7. The two types of alternative sweeteners are:
   a. saccharin and cyclamate.
   b. sugar alcohols and artificial sweeteners.
   c. sorbitol and xylitol.
   d. sucrose and fructose.

8. A diet high in carbohydrate-rich foods such as whole grains, vegetables, fruits, and legumes is:
   a. most likely low in fat.
   b. most likely low in fiber.
   c. most likely poor in vitamins and minerals.
   d. most likely disease promoting.

9. A fiber-rich diet may help to prevent or control:
   a. diabetes.
   b. heart disease.
   c. constipation.
   d. all of the above.

10. The DRI fiber recommendation is:
    a. 10 grams per 1000 kcalories.
    b. 15 to 25 grams per day.
    c. 14 grams per 1000 kcalories.
    d. 40 to 55 grams per day.

Answers to these questions can be found in Appendix H.

Notes

1. A. T. Merchant and coauthors, Carbohydrate intake and overweight and obesity among healthy adults, *Journal of the American Dietetic Association* 109 (2009): 1165–1172; Standing Committee on the Scientific Evaluation of Dietary Reference Intakes, Food and Nutrition Board, Institute of Medicine, *Dietary Reference Intakes for Energy, Carbohydrate, Fiber, Fat, Fatty Acids, Cholesterol, Protein, and Amino Acids* (Washington, D.C.: National Academies Press, 2005), pp. 265–338.

2. J. H. Cummings and A. M. Stephen, Carbohydrate terminology and classification, *European Journal of Clinical Nutrition* 61 (2007): S5–S18.

3. A. M. Brown and B. R. Ransom, Astrocyte glycogen and brain energy metabolism, *Glia* 55 (2007): 1263–1271; M. K. Dalsgaard and coauthors, High glycogen levels in the hippocampus of patients with epilepsy, *Journal of Cerebral Blood Flow and Metabolism* 27 (2007): 1137–1141.

4. M. M. Murphy, J. S. Douglass, and A. Birkett, Resistant starch intakes in the United States, *Journal of the American Dietetic Association* 108 (2008): 67–78.

5. S. S. Dronamraju and coauthors, Cell kinetics and gene expression changes in colorectal cancer patients given resistant starch: A randomised controlled trial, *Gut* 58 (2009): 413–420; M. Nofrarias and coauthors, Long-term intake of resistant starch improves colonic mucosal integrity and reduces gut apoptosis and blood immune cells, *Nutrition* 23 (2007): 861–870.

6. Position of the American Dietetic Association: Health implications of dietary fiber, *Journal of the American Dietetic Association* 108 (2008): 1716–1731.

7. R. K. Johnson and coauthors, Dietary sugars intake and cardiovascular health: A scientific statement from the American Heart Association, *Circulation* 120 (2009): 1011–1020.

8. Johnson and coauthors, 2009.

9. National Cancer Institute, Usual intake of added sugars, available at http://riskfactor.cancer.gov/diet/usualintakes/pop/t35.html.

10. U.S. Department of Agriculture and U.S. Department of Health and Human Services, *Dietary Guidelines for Americans, 2010* available at www.dietaryguidelines.gov.

11. Johnson and coauthors, 2009.

12. A. Bhargava and A. Amialchuk, Added sugars displaced the use of vital nutrients in the National Food Stamp Program Survey, *Journal of Nutrition* 137 (2007): 453–460.

13. Johnson and coauthors, 2009; U. S. Department of Agriculture, MyPyramid.gov, available at www.mypramid.gov, accessed October 15, 2009.

14. Position of the American Dietetic Association: Oral health and nutrition, *Journal of the American Dietetic Association* 107 (2007): 1418–1428.

15. V. S. Malik, M. B. Schulze, and F. B. Hu, Intake of sugar-sweetened beverages and weight gain: A systematic review, *American Journal of Clinical Nutrition* 84 (2006): 274–288.

16. G. A. Bray, How bad is fructose? *American Journal of Clinical Nutrition* 86 (2007): 895–896; R. Dhingra and coauthors, Soft drink consumption and risk of developing cardiometabolic risk factors and the metabolic syndrome in middle-aged adults in the community, *Circulation* 116 (2007): 480–488; R. J. Johnson and coauthors, Potential role of sugar (fructose) in the epidemic of hypertension, obesity and the metabolic syndrome, diabetes, kidney disease, and cardiovascular disease, *American Journal of Clinical Nutrition* 86 (2007): 899–906; S. C. Larsson, L. Bergkvist, and A. Wolk, Consumption of sugar and sugar-sweetened foods and the risk of pancreatic cancer in a prospective study, *American Journal of Clinical Nutrition* 84 (2006): 1171–1176.

17. J. N. Davis and coauthors, Associations of dietary sugar and glycemic index with adiposity and insulin dynamics in overweight Latino youth, *American Journal of Clinical Nutrition* 86 (2007): 1331–1338.

18. R. A. Forshee and coauthors, A critical examination of the evidence relating high fructose corn syrup and weight gain, *Critical Reviews in Food Science and Nutrition* 47 (2007): 561–582.

19. L. R. Vartanian, M. B. Schwartz, and K. D. Brownell, Effects of soft drink consumption on nutrition and health: A systematic review and meta-analysis, *American Journal of Public Health* 97 (2007): 667–675.

20. R. A. Forshee, P. A. Anderson, and M. L. Storey, Sugar-sweetened beverages and body mass index in children and adolescents: A meta-analysis, *American Journal of Clinical Nutrition* 87 (2008): 1662–1671.

21. Standing Committee on the Scientific Evaluation of Dietary Reference Intakes, 2005, p. 770.

22. I. M. Paul and coauthors, Effect of honey, dextromethorphan, and no treatment on nocturnal cough and sleep quality for coughing children and their parents, *Archives of Pediatric and Adolescent Medicine* 161 (2007): 1140–1146; H. V. Worthington, J. E. Clarkson, and O. B. Eden, Interventions for preventing oral mucositis for patients with cancer receiving treatment, *Cochrane Database of Systematic Reviews* 4 (2007): CD000978.

23. Position of the American Dietetic Association: Use of nutritive and nonnutritive sweeteners, *Journal of the American Dietetic Association* 104 (2004): 255–275.

24. Position of the American Dietetic Association, 2004.

25. R. D. Mattes and B. M. Popkin, Nonnutritive sweetener consumption in humans: Effects on appetite and food intake and their putative mechanisms, *American Journal of Clinical Nutrition* 89 (2009): 1–14.

26. K. C. Maki and coauthors, Whole-grain ready-to-eat oat cereal, as part of a dietary program for weight loss, reduces low-density lipoprotein cholesterol in adults with overweight and obesity more than a dietary program including low-fiber control foods, *Journal of the American Dietetic Association* 110 (2010): 205–214; Position of the American Dietetic Association, 2008; M. T. Streppel and coauthors, Dietary fiber intake in relation to coronary heart disease and all-cause mortality over 40 y: The Zutphen Study, *American Journal of Clinical Nutrition* 88 (2008): 1119–1125; M. F. Chong, B. A. Fielding, and K. N. Frayn, Metabolic interaction of dietary sugars and plasma lipids with a focus on mechanisms and de novo lipogenesis, *Proceedings of the Nutrition Society* 66 (2007): 52–59; P. B. Mellen, T. F. Walsh, and D. M. Herrington, Whole grain intake and cardiovascular disease: A meta-analysis, *Nutrition, Metabolism and Cardiovascular Diseases* 18 (2008): 283–290; R. Solà and coauthors, Effects of soluble fiber (*Plantago ovata* husk) on plasma lipids, lipoproteins, and apolipoproteins in men with ischemic heart disease, *American Journal*

of *Clinical Nutrition* 85 (2007): 1157–1163; P. B. Mellen and coauthors, Whole-grain intake and carotid artery atherosclerosis in a multiethnic cohort: The Insulin Resistance Atherosclerosis Study, *American Journal of Clinical Nutrition* 85 (2007): 1495–1502.

27. J. A. Nettleton and coauthors, Dietary patterns and incident cardiovascular disease in the MultiEthnic Study of Atherosclerosis, *American Journal of Clinical Nutrition* 90 (2009): 647–654; A. Mente and coauthors, A systematic review of the evidence supporting a causal link between dietary factors and coronary heart disease, *Archives of Internal Medicine* 169 (2009): 659–669; Streppel and coauthors, 2008; Position of the American Dietetic Association, 2008.

28. V. Vuksan and coauthors, Fiber facts: Benefits and recommendations for individuals with type 2 diabetes, *Current Diabetes Reports* 9 (2009): 405–411; H. Kim and coauthors, Glucose and insulin responses to whole grain breakfasts varying in soluble fiber, beta-glucan: A dose response study in obese women with increased risk for insulin resistance, *European Journal of Nutrition* 48 (2009): 170–175; P. L. Lutsey and coauthors, Whole grain intake and its cross-sectional association with obesity, insulin resistance, inflammation, diabetes, and subclinical CVD: The MESA Study, *British Journal of Nutrition* 98 (2007): 397–405; M. B. Schulze and coauthors, Fiber and magnesium intake and incidence of type 2 diabetes, *Archives of Internal Medicine* 167 (2007): 956–965; J. S. de Munter and coauthors, Whole grain, bran,

and germ intake and risk of type 2 diabetes: A prospective cohort study and systematic review, *PLoS Medicine* 4 (2007): e261.

29. J. R. Korzenik, Diverticulitis: New frontiers for an old country: Risk factors and pathogenesis, *Journal of Gastroenterology* 42 (2008): 1128–1129; Position of the American Dietetic Association, 2008; D. O. Jacobs, Diverticulitis, *New England Journal of Medicine* 357 (2007): 2057–2066.

30. L. B. Sansbury and coauthors, The effect of strict adherence to a high-fiber, high-fruit and -vegetable, and low-fat eating pattern on adenoma recurrence, *American Journal of Epidemiology* 170 (2009): 576–584; World Cancer Research Fund/American Institute for Cancer Research, *Food, Nutrition, Physical Activity, and the Prevention of Cancer: A Global Perspective* (Washington, D.C.: AICR, 2007), pp. 280–288.

31. S. Bingham, The fibre-folate debate in colo-rectal cancer, *Proceedings of the Nutrition Society* 65 (2006): 19–23; S. A. Bingham and coauthors, Dietary fibre in food and protection against colorectal cancer in the European Prospective Investigation into Cancer and Nutrition (EPIC): An observational study, *Lancet* 361 (2003): 1496–1501.

32. D. J. Rose and coauthors, Influence of dietary fiber on inflammatory bowel disease and colon cancer: Importance of fermentation pattern, *Nutrition Reviews* 65 (2007): 51–62.

33. K. Wallace and coauthors, The association of lifestyle and dietary factors with the risk for

serrated polyps of the colorectum, *Cancer Epidemiology, Biomarkers, and Prevention* 18 (2009): 2310–2317.

34. L. A. Tucker and K. S. Thomas, Increasing total fiber intake reduces of weight and fat gains in women, *Journal of Nutrition* 139 (2009): 567–581.

35. M. Lyly and coauthors, Fiber in beverages can enhance perceived satiety, *European Journal of Nutrition* 48 (2009): 251–258; K. R. Juvonen and coauthors, Viscosity of oat bran-enriched beverages influences gastrointestinal hormonal responses in healthy humans, *Journal of Nutrition* 139 (2009): 461–466; R. A. Samra and G. H. Anderson, Insoluble cereal fiber reduces appetite and short-term food intake and glycemic response to food consumed 75 min later by healthy men, *American Journal of Clinical Nutrition* 86 (2007): 972–979.

36. J. N. Davis and coauthors, Inverse relation between dietary fiber intake and visceral adiposity in overweight Latino youth, *American Journal of Clinical Nutrition* 90 (2009): 1160–1166.

37. Davis and coauthors, 2009; Standing Committee on the Scientific Evaluation of Dietary Reference Intakes, 2005, pp. 391–399.

38. Standing Committee on the Scientific Evolution of Dietary Reference Intakes, 2005, p. 265.

39. Position of the American Dietetic Association, 2008.

Carbohydrate-rich foods vary in the degree to which they elevate both blood glucose and insulin concentrations. Chapter 2 introduces the *glycemic index (GI),* a ranking of carbohydrate foods based on their glycemic effect after ingestion. The glycemic index may be of interest to people with diabetes who must regulate their blood glucose to protect their health. In diabetes treatment, however, the total amount of carbohydrate is more important than the type of carbohydrate consumed.[1] Thus, dietetics experts have debated the usefulness of the glycemic index for diabetes treatment. Despite some controversy, however, the American Diabetes Association encourages low-glycemic foods that are rich in fiber and other nutrients.[2] Furthermore, because recent research shows that a low-glycemic index diet can improve blood glucose control in type 2 diabetes, the use of low-glycemic diets for this purpose may be gaining credibility.[3]

Researchers are also trying to determine whether low-GI diets may be helpful for improving risk factors for a number of other chronic diseases.[4] This Nutrition in Practice will describe the factors that contribute to a food's glycemic effect and the results of research studies that have examined the potential benefits of selecting mainly low-GI foods.

How is the glycemic index measured?

The glycemic index is essentially a measure of how quickly the carbohydrate in a food is digested and absorbed. Although testing methods vary to some degree, the most common protocol is to feed the test food—which contains a measured quantity of digestible carbohydrate—to research subjects and then measure blood glucose levels for two or three hours after the feeding. The increase in blood glucose over the two- or three-hour period is then compared to the blood glucose rise after an identical amount of digestible carbohydrate is ingested from a reference food, such as pure glucose or white bread. Figure NP2-1 illustrates the difference in the blood glucose response to a low-GI food and a high-GI food. The blood glucose curve displays the surge in blood glucose above normal fasting levels after the food is consumed and the subsequent fall over several hours. Table NP2-1 (p. 56) lists the GIs of various carbohydrate-containing foods, arranged from highest to lowest within each food group listed.

The *amount* of carbohydrate consumed also influences the glycemic response. A food's total glycemic effect—expressed as the *glycemic load (GL)*—is the product of its GI and the amount of available carbohydrate from the portion consumed, divided by 100.[5]

What factors influence a food's glycemic effect?

Table NP2-1 shows that starchy foods such as bread and potatoes tend to have high GI values, whereas many fruits and legumes have low GI values. The main factors that influence the GI value of a food include the following:

▌ *Starch structure.* Starch is present in foods as either a straight chain or branched chain of glucose molecules. Whereas digestion of the branched form tends to release glucose quickly, the straight chain is resistant to digestion. Thus, foods that contain mainly the branched form of starch tend to raise blood glucose levels more quickly and have a high GI value. Due to the subtle differences in starch among foods, different species of the same foods can have substantially different GI values; for example, current GI values for rice range from low (38 for parboiled white rice) to high (85 for Japanese sushi-style white rice).[6]

▌ *Fiber content.* Certain types of dietary fibers (primarily soluble fibers) increase the viscosity of chyme (semiliquid, partially digested food), slowing the passage of food in the stomach and upper intestine and making it more difficult for enzymes to digest the food. Therefore, foods such as beans, fruits, and vegetables, which contain soluble fibers, tend to have lower GI values.

▌ *Presence of fat and protein.* The fat in foods tends to slow stomach emptying, thus reducing the rate of digestion and absorption; hence, the presence of fat usually reduces a food's GI value. The protein in foods can also influence the GI because protein promotes insulin secretion, increasing the rate at which glucose is taken up from the blood.[7]

FIGURE NP2-1 **Blood Glucose Response to High-GI and Low-GI Foods**

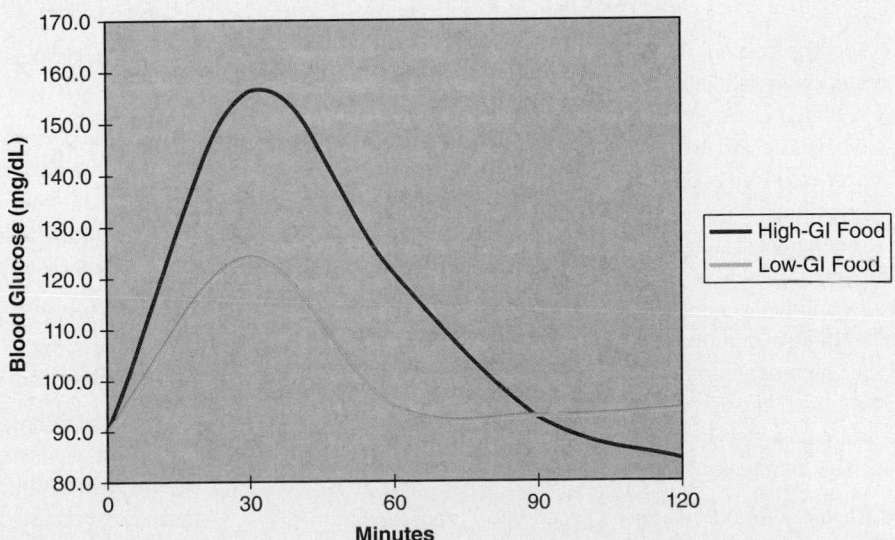

Foods that are digested slowly have a low glycemic index. These foods cause a gradual and more moderate response in blood glucose than do foods with a high glycemic index.

TABLE NP2-1 Glycemic Index (GI) of Selected Foods[a]

Value	GI
High	≥70
Medium	56–69
Low	≤55

Food Item	Glycemic Index
Grains	
Cornflakes	81
Instant oatmeal, cooked	79
White bread, enriched	75
Whole-wheat bread	74
White rice	73
Bagel, white	69
Brown rice	68
Couscous	65
Popcorn	65
Bran flakes	63
Oatmeal, cooked	55
Spaghetti, white, boiled	49
Spaghetti, whole meal, boiled	48
Corn tortilla	46
Oat bran bread; 50% oat bran	44
Barley	28
Milk products	
Ice cream	51
Yogurt, fruit	41
Whole milk	39
Nonfat milk	37
Soy milk	34
Legumes	
Lentils	32
Garbanzo beans	28

Food Item	Glycemic Index
Kidney beans	24
Soy beans	16
Vegetables	
Russet potato, baked	111
Potato, instant mash	87
Potato, boiled	78
Pumpkin, boiled	64
Potato, french fries	63
Sweet potato, boiled	63
Taro	53
Sweet corn	52
Green peas, boiled	51
Carrots, boiled	39
Fruits	
Watermelon	76
Pineapple	59
Bananas	51
Mangoes	51
Orange juice	50
Oranges	43
Apple juice	41
Apples	36
Snack Foods/Beverages	
Fruit punch	67
Soft drink/soda	59
Chocolate	56
Potato chips	40
Sugars	
Sucrose	65
Honey	61
Fructose	15

[a]Reference food: glucose = 100

Source: F. S. Atkinson, K. Foster-Powell, and J. C. Brand-Miller, International tables of glycemic index and glycemic load values: 2008. *Diabetes Care* (2008): 2281–2283.

■ *Food processing.* The manner in which a food is processed and cooked influences the interactions among starch, protein, and fiber and thus affects the final GI value. For example, both pasta and bread are prepared from wheat flour, but pasta (cooked *al dente*) has a lower GI because the starch granules in pasta are surrounded by a sturdy protein barrier that hampers starch digestion. Cooking the pasta for longer periods can break down its structure and raise the GI value. As another example, the GI values for oatmeal vary according to the size and thickness of the oats used to prepare it: oatmeal prepared from steel-cut oats has a lower GI value than oatmeal prepared from quick oats. This is because the steel-cut oats are solid particles of grain, whereas "quick oats" are small, thin flakes.

■ *Mixture of foods in a meal.* Because foods are rarely consumed in isolation, the GI value of an individual food may be less important than the combination of foods consumed at a meal. For example, in a cheese sandwich, the high GI of the bread is lowered by the addition of fat and protein in the cheese.

■ *Individual glucose tolerance.* Cellular responses to insulin vary; thus, individual variability affects the glycemic response to foods. Persons with diabetes or prediabetes exhibit higher blood glucose levels after ingesting carbohydrate foods than do healthy individuals.[8]

What evidence suggests that a low-GI diet may influence chronic disease risk?

Some research shows that low-GI diets may reduce the risks of developing diabetes, heart disease, and obesity and help individuals lose weight.[9] Other studies, however, do not support such findings.[10] Because of mixed findings so far, health practitioners do not routinely recommend that patients consume low-GI diets to prevent or treat disease. An abundance of ongoing research to reveal specific relationships between low-GI diets and chronic

Pasta cooked *al dente* has a lower glycemic index than many other starchy foods.

Scott Vickers/iStockphoto.com

disease risk, however, may change such thinking. Examples of research include the following:

- *Diabetes prevention*. Some researchers have proposed that a high glycemic load can increase the body's demand for insulin and eventually reduce pancreatic function, resulting in inadequate insulin secretion. Indeed, results of several studies suggest that low-GI diets might prevent or delay the onset of type 2 diabetes in those at risk.[11] In fact, a summary of an international workshop on the glycemic index stated that "lower GI-diets are beneficial for those with impaired glucose metabolism, but that it is as yet unclear what such diets mean for healthy people."[12]

- *Heart disease risk*. Although some research suggests that low-GI diets may improve blood lipid levels, the majority of research shows no consistent effects of low-GI diets on heart disease risk.[13] One author concludes that there is a minimal role for the implementation of low GI diets in the prevention or treatment of heart disease.[14]

- *Appetite and weight loss*. Research shows that low-GI foods can slow the after-meal rise in blood glucose, supporting the hypothesis that such foods may promote satiety and suppress hunger.[15] Research results are inconsistent, however, on whether low-GI diets assist in weight loss.[16]

Given the mixed results of research studies on chronic disease prevention, are there any benefits associated with consuming low-GI foods?

Yes, if the low-GI foods are nutrient-dense, high-fiber foods. Not all low-GI foods meet these criteria: note that cakes, cookies, and candy bars may have a low GI due to their high fat content. Thus, a food's GI should be considered along with other nutrient criteria when assessing the health benefits.

The GI can be a helpful tool for choosing the most healthful food from a food group. For example, low-GI breakfast cereals tend to be high in fiber and low in added sugars, whereas high-GI cereals tend to be those that contain refined flours and significant amounts of added sugars. In other words, low GI foods are often wholesome foods that have been minimally processed.

In general, should people avoid consuming high-GI foods?

Some people assume that starchy foods, such as breads and potatoes, should be avoided due to their high GI values. As mentioned earlier, these foods are rarely consumed in isolation, and their GI values are reduced in a mixed meal. For example, breads often have a GI greater than 70, but adding cheese or peanut butter reduces the GI to 55 or 59, respectively. Also worth considering is that GI values often vary considerably. For example, published values for white potatoes range from 24 to 101, and many samples have values in the mid-50s.[17] For these reasons and others, some researchers believe more studies are needed to confirm whether the GI is practical or beneficial for healthy people.[18]

Given the complexity of the GI, what are the current recommendations?

The potential benefits associated with consuming low-GI diets are still under investigation. As discussed earlier, people with type 2 diabetes may benefit from limiting high-GI foods—those that produce too great a rise, or too sudden a fall, in blood glucose.[19] Additional research is needed to justify the use of diets based on the GI for preventing or treating diseases such as heart disease, obesity, or other medical problems.

At present, many nutrition scientists advocate consuming a plant-based diet that contains minimally processed grains, legumes, vegetables, and fruits. Such a diet would include abundant fiber and limited amounts of solid fats and added sugars. Undoubtedly, meals consisting of these foods would tend to have low or medium GI values.

Notes

1. American Diabetes Association, Position statement: Nutrition recommendations and interventions for diabetes, *Diabetes Care* 31 (2008): S61–78.

2. American Diabetes Association, 2008.

3. D. Thomas and E. J. Elliott, Low glycaemic index, or low glycaemic load, diets for diabetes mellitus, *Cochrane Database of Systematic Reviews* January 21, 2009, CD006296; Low glycaemic index diet and disposition index in type 2 diabetes (the Canadian trial of carbohydrates in diabetes): A randomized controlled trial, *Diabetologia* 51 (2009): 1607–1615; D. J. A. Jenkins and coauthors, Effect of a low-glycemic index or a high-cereal fiber diet on type 2 diabetes, *Journal of the American Medical Association* 300 (2008): 2742–2753; G. Riccardi, A. A. Rivellese, and R. Giacco, Role of glycemic index and glycemic load in the healthy state, in prediabetes, and in diabetes, *American Journal of Clinical Nutrition* 87 (2008): 269S–274S.

4. G. Radulian and coauthors, Metabolic effects of low glycaemic index diets, *Nutrition Journal* 8 (2009): 5; A. W. Barclay and coauthors, Glycemic index, glycemic load, and chronic disease

risk—A meta-analysis of observational studies, *American Journal of Clinical Nutrition* 87 (2008): 627–637; J. Brand-Miller and coauthors, Carbohydrates—the good, the bad and the whole grain, *Asia Pacific Journal of Clinical Nutrition* 17 (2008): 16–19.

5. H. Hare-Bruun and coauthors, Should glycemic index and glycemic load be considered in dietary recommendations? *Nutrition Reviews* 66 (2008): 569–590.

6. F. S. Atkinson, K. Foster-Powell, and J. C. Brand-Miller, International tables of glycemic index and glycemic load values: 2008, *Diabetes Care* 31 (2008): 228–2283.

7. P. Small and J. Brand-Miller, From complex carbohydrate to glycemic index, *Nutrition Today* 44 (2009): 236–243.

8. Riccardi, Rivellese, and Giacco, 2008.

9. Radulian and coauthors, 2009; Brand-Miller and coauthors, 2008; Barclay and coauthors, 2008; Jenkins and coauthors, 2008; K. C. Maki and coauthors, Effects of a reduced-glycemic-load diet on body weight, body composition, and cardiovascular disease risk markers in overweight and obese adults, *American Journal of Clinical Nutrition* 85 (2007): 724–734; M. A. Pereira, Weighing in on glycemic index and body weight, *American Journal of Clinical Nutrition* 84 (2006): 677–679.

10. N. R. Sahyoun and coauthors, Dietary glycemic index and glycemic load and the risk of type 2 diabetes in older adults, *American Journal of Clinical Nutrition* 87 (2008): 126–131; A. Mosdol and coauthors, Dietary glycemic index and glycemic load are associated with high-density-lipoprotein cholesterol at baseline but not with increased risk of diabetes in the Whitehall II Study, *American Journal of Clinical Nutrition* 86 (2007): 988–994; E. B. Levitan and coauthors, Dietary glycemic index, dietary glycemic load, and cardiovascular disease in middle-aged and older Swedish men, *American Journal of Clinical Nutrition* 85 (2007): 1521–1526; R. Sichieri and coauthors, An 18-mo randomized trial of a low-glycemic-index diet and weight change in Brazilian women, *American Journal of Clinical Nutrition* 86 (2007): 707–713.

11. Radulian and coauthors, 2009; Brand-Miller and coauthors, 2008.

12. J. Howlett and M. Ashwell, Glycemic response and health: Summary of a workshop, *American Journal of Clinical Nutrition* 87 (2008): 212S–216S.

13. J. M. Shikany and coauthors, Effects of low- and high-glycemic index/glycemic load diets on coronary heart disease risk factors in overweight/obese men, *Metabolism* 58 (2009): 1793-1801; E. B. Levitan and coauthors, Dietary glycemic index, dietary glycemic load, blood lipids, and C-reactive protein, *Metabolism* 57 (2008): 437–443; M. J. Franz, Is there a role for the glycemic index in coronary heart disease prevention or treatment? *Current Atherosclerosis Reports* 10 (2008): 497–502; J. E. Milton and coauthors, Relationship of glycaemic index with cardiovascular risk factors: Analysis of the National Diet and Nutrition Survey for people aged 65 and older, *Public Health Nutrition* 10 (2007): 1321–1335.

14. Franz, 2008.

15. R. C. Reynolds and coauthors, Effect of the glycemic index of carbohydrates on day-long (10 h) profiles of plasma glucose, insulin, cholecystokinin and ghrelin, *European Journal of Clinical Nutrition* 63 (2009): 872–878; S. Pal, S. Lim, and G. Egger, The effect of a low glycaemic index breakfast on blood glucose, insulin, lipid profiles, blood pressure, body weight, body composition and satiety in obese and overweight individuals: A pilot study, *Journal of the American College of Nutrition* 27 (2008): 387–393.

16. M. A. Mendez and coauthors, Glycemic load, glycemic index, and body mass index in Spanish adults, *American Journal of Clinical Nutrition* 89 (2009): 316–322; D. E. Thomas, E. J. Elliott, and L. Baur, Low glycaemic index or low glycaemic load diets for overweight and obesity, *Cochrane Database of Systematic Reviews*, July 18, 2007: CD005105; C. B. Ebbeling and coauthors, Effects of a low-glycemic load vs. low-fat diet in obese young adults, *Journal of the American Medical Association* 297 (2007): 2092–2102; R. Sichieri and coauthors, An 18-mo randomized trial of a low-glycemic-index diet and weight change in Brazilian women, *American Journal of Clinical Nutrition* 86 (2007): 707–713; S. K. Das and coauthors, Long-term effects of 2 energy-restricted diets differing in glycemic load on dietary adherence, body composition, and metabolism in CALERIE: A 10y randomized controlled trial, *American Journal of Clinical Nutrition* 85 (2007): 1023–1030.

17. Atkinson, Foster-Powell, and Brand-Miller, 2008.

18. H. Hare-Bruun and coauthors, Should glycemic index and glycemic load be considered in dietary recommendations? *Nutrition Reviews* 66 (2008): 569–590.

19. Riccardi, Rivellese, and Giacco, 2008.

3

Lipids

Olive Oil

Most people know that too much fat—especially certain kinds of fat—in the diet incurs health risks, but they may be surprised to learn that too little fat poses risks, too. People in the United States, however, are more likely to eat too much fat than too little.

Fat is a member of the class of compounds called **lipids.** The lipids in foods and in the human body include triglycerides (**fats** and **oils**), phospholipids, and sterols.

Roles of Body Fat

Lipids perform many tasks in the body, but most importantly they provide energy. A constant flow of energy is so vital to life that, in a pinch, any other function is sacrificed to maintain it. Chapter 2 described one safeguard against such an emergency—the stores of glycogen in the liver that provide glucose to the blood whenever the supply runs short. The body's stores of glycogen are limited, however. In contrast, the body's capacity to store fat for energy is virtually unlimited due to the fat-storing cells of the **adipose tissue.** The fat cells of the adipose tissue readily take up and store fat, growing in size as they do so. When extra energy storage is needed, new fat cells are readily produced. Fat cells are more than just storage depots, however; fat cells secrete hormones that help to regulate the appetite and influence other body functions.[1] Figure 3-1 shows a fat cell.

The fat stored in fat cells supplies 60 percent of the body's ongoing energy needs during rest. During some types of physical activity or prolonged periods of food deprivation, fat stores may make an even greater energy contribution. The brain and nerves, however, need their energy as glucose, and the body cannot convert fat to glucose. After a long period of glucose deprivation (during fasting or starvation), brain and nerve cells develop the ability to derive about half of their energy from a special form of fat known as **ketones,** but they still require glucose as well. This means that people wanting to lose weight need to eat a certain minimum amount of carbohydrate to meet their energy needs, even when they are limiting their food intakes.

Chapter 6 discusses fat use during fasting.

Cultura/Jupiter Images

Body fat supplies much of the fuel that muscles need to do their work.

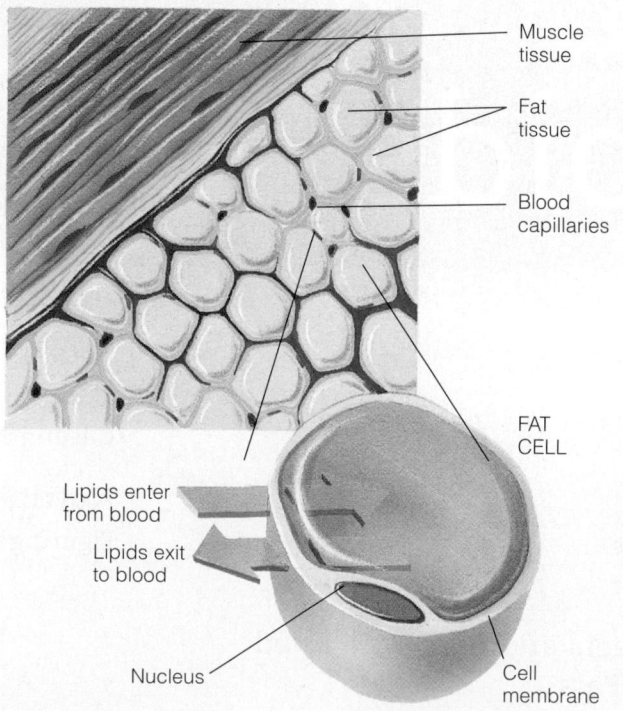

Muscle tissue

Fat tissue

Blood capillaries

FAT CELL

Lipids enter from blood

Lipids exit to blood

Nucleus

Cell membrane

FIGURE 3-1 A Fat Cell
Within the fat, or adipose, cell, lipid is stored in a droplet. This droplet can greatly enlarge, and the fat cell membrane will expand to accommodate its swollen contents.

60

In addition to supplying energy, fat serves other roles in the body. Natural oils in the skin provide a radiant complexion; in the scalp, they help nourish the hair and make it glossy. The layer of fat beneath the skin insulates the body from extremes of temperature. A pad of hard fat beneath each kidney protects it from being jarred and damaged, even during a motorcycle ride on a bumpy road. The soft fat in a woman's breasts protects her mammary glands from heat and cold and cushions them against shock. The fat embedded in muscle tissue shares with muscle glycogen the task of providing energy when the muscles are active. Phospholipids and the sterol cholesterol are cell membrane constituents that help maintain the structure and health of all cells. Table 3-1 summarizes the major functions of fats in the body.

IN SUMMARY

▪ Lipids in the body not only serve as energy reserves but also protect the body from temperature extremes, cushion the vital organs, and provide the major material of cell membranes.

The Chemist's View of Lipids

The diverse and vital functions that lipids play in the body reveal why eating too little fat can be harmful. As mentioned earlier, though, too much fat in the diet seems to be the greater problem for most people. To understand both the beneficial and harmful effects that fats exert on the body, a closer look at the structure and function of members of the lipid family is in order.

Triglycerides

When people talk about fat—for example, "I'm too fat" or "That meat is fatty"—they are usually referring to triglycerides. Among lipids, **triglycerides** predominate—both in the diet and in the body. The name *triglyceride* almost explains itself: three *(tri)* **fatty acids** attached to a **glycerol** "backbone." Figure 3-2 shows how three fatty acids combine with glycerol to make a triglyceride.

Fatty Acids

When energy from any energy-yielding nutrient is to be stored as fat, the nutrient is first broken into small fragments. Then the fragments are linked together into chains known as fatty acids. The fatty acids are then packaged, three at a time, with glycerol to make triglycerides.

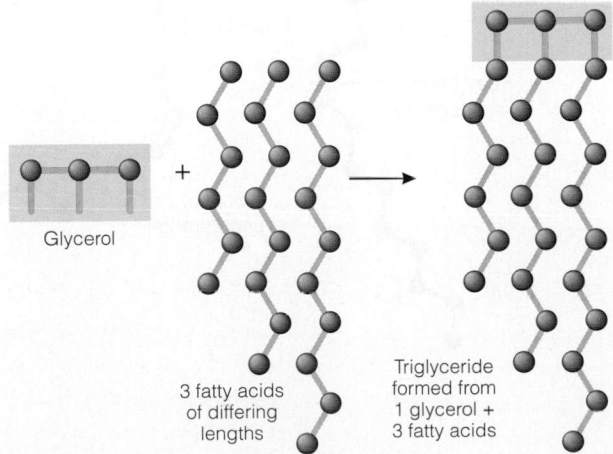

Glycerol

3 fatty acids of differing lengths

Triglyceride formed from 1 glycerol + 3 fatty acids

FIGURE 3-2 Triglyceride Formation
Glycerol, a small, water-soluble compound, plus three fatty acids, equals a triglyceride.

TABLE 3-1

The Functions of Fats in the Body

▪ **Energy stores.** Fats are the body's chief form of stored energy.

▪ **Muscle fuel.** Fats provide most of the energy to fuel muscular work.

▪ **Emergency reserve.** Fats serve as an emergency fuel supply in times of illness and diminished food intake.

▪ **Padding.** Fats protect the internal organs from shock through fat pads inside the body cavity.

▪ **Insulation.** Fats insulate against temperature extremes through a fat layer under the skin.

▪ **Cell membranes.** Fats form the major material of cell membranes.

▪ **Raw materials.** Fats are converted to other compounds, such as hormones, bile, and vitamin D, as needed.

lipids: a family of compounds that includes triglycerides (fats and oils), phospholipids, and sterols. Lipids are characterized by their insolubility in water.

fats: lipids that are solid at room temperature (70°F or 21°C).

oils: lipids that are liquid at room temperature (70°F or 21°C).

adipose tissue: the body's fat, which consists of masses of fat-storing cells called adipose cells.

ketones (KEY-tones): acidic, fat-related compounds formed from the incomplete breakdown of fat when carbohydrate is not available; technically known as *ketone* bodies.

triglycerides (try-GLISS-er-rides): one of the main classes of lipids; the chief form of fat in foods and the major storage form of fat in the body; composed of glycerol with three fatty acids attached.

fatty acids: organic compounds composed of a chain of carbon atoms with hydrogen atoms attached and an acid group at one end.

glycerol (GLISS-er-ol): an organic compound, three carbons long, that can form the backbone of triglycerides and phospholipids.

61

Chain Length and Saturation Fatty acids may differ from one another in two ways—in chain length and in degree of saturation. The chain length refers to the number of carbons in a fatty acid. Saturation also refers to its chemical structure—specifically, to the number of hydrogen atoms the carbons in the fatty acid are holding. If every available carbon is filled to capacity with hydrogen atoms, the chain is called a **saturated fatty acid.** A saturated fatty acid is fully loaded with hydrogen atoms and has only single bonds between the carbons. The first zigzag structure in Figure 3-3 represents a saturated fatty acid.

Unsaturated Fatty Acids In some fatty acids, including most of those in plants and fish, hydrogen atoms are missing in the fatty acid chains. The places where the hydrogen atoms are missing are called points of unsaturation, and a chain containing such points is called an **unsaturated fatty acid.** An unsaturated fatty acid has at least one double bond between its carbons. If there is one point of unsaturation, the chain is a **monounsaturated fatty acid.** The second structure in Figure 3-3 is an example. If there are two or more points of unsaturation, then the fatty acid is a **polyunsaturated fatty acid** (see the third structure in Figure 3-3).

Hard and Soft Fat A triglyceride can contain any combination of fatty acids—long chain or short chain; saturated, monounsaturated, or polyunsaturated. The degree of saturation of the fatty acids in a fat influences health (discussed in a later section) and food characteristics. Fats that contain the shorter-chain or the more unsaturated fatty acids are softer at room temperature and melt more readily. A comparison of three fats—lard (which comes from pork), chicken fat, and safflower oil—illustrates these differences: lard is the most saturated and the hardest; chicken fat is less saturated and somewhat soft; and safflower oil, which is the most unsaturated, is a liquid at room temperature.

Stability Saturation also influences stability. Fats can become **rancid** when exposed to oxygen. Polyunsaturated fatty acids spoil most readily because their double bonds are unstable. The **oxidation** of unsaturated fats produces a variety of compounds that smell

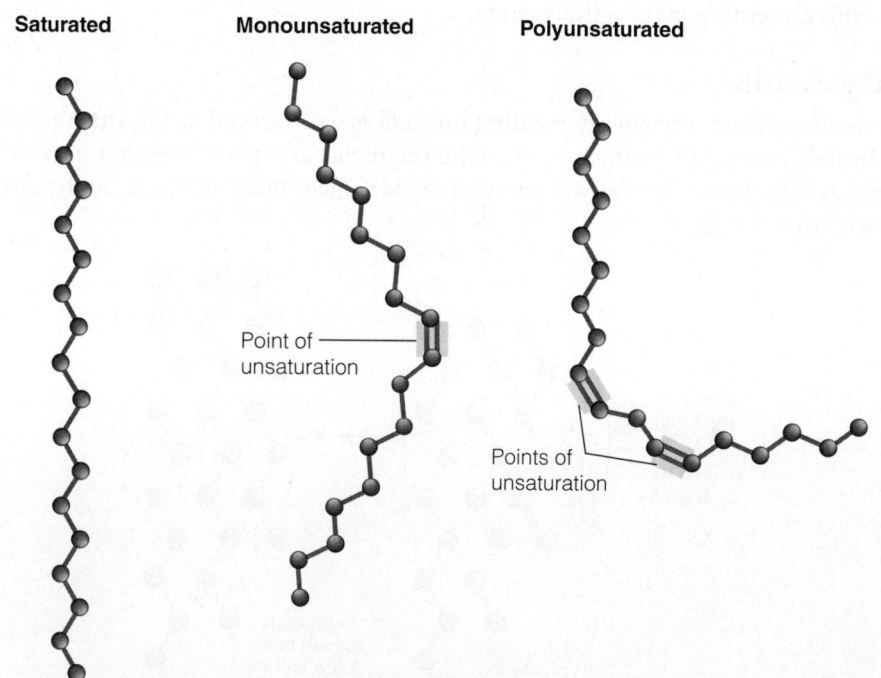

Saturated   Monounsaturated   Polyunsaturated

Point of unsaturation

Points of unsaturation

FIGURE 3-3 Three Types of Fatty Acids
The more carbon atoms in a fatty acid, the longer it is. The more hydrogen atoms attached to those carbons, the more saturated the fatty acid is.

and taste rancid; saturated fats are more resistant to oxidation and are thus less likely to become rancid. Other types of spoilage can occur due to microbial growth, however.

Manufacturers can protect fat-containing products against rancidity in three ways—none of them perfect. First, products may be sealed airtight and refrigerated— an expensive and inconvenient storage system. Second, manufacturers may add **antioxidants** to compete for the oxygen and thus protect the oil (examples are the additives **BHA** and **BHT** and vitamins C and E).* Third, manufacturers may saturate some or all of the points of unsaturation by adding hydrogen atoms—a process known as hydrogenation.

Hydrogenation **Hydrogenation** offers two advantages: it protects against oxidation (thereby prolonging shelf life) and also alters the texture of foods by increasing the solidity of fats. When partially hydrogenated, vegetable oils become spreadable margarine. Hydrogenated fats make pie crusts flaky and puddings creamy. A disadvantage is that hydrogenation makes polyunsaturated fats more saturated. Consequently, any health advantages of using polyunsaturated fats instead of saturated fats are lost in hydrogenation.

Trans-Fatty Acids Another disadvantage of hydrogenation is that some of the molecules that remain unsaturated after processing change shape from *cis* to *trans*. In nature, most unsaturated fatty acids are *cis*-fatty acids—meaning that the hydrogen atoms next to the double bonds are on the same side of the carbon chain. Only a few fatty acids in nature (notably a small percentage of those found in milk and meat products) are *trans*-**fatty acids**—meaning that the hydrogen atoms next to the double bonds are on opposite sides of the carbon chain (see Figure 3-4). These arrangements result in different configurations for the fatty acids, and this difference affects function: in the body, *trans*-fatty acids behave more like saturated fats than like unsaturated fats. The relationship between *trans*-fatty acids and heart disease is the subject of much ongoing research, as a later section describes. In contrast, naturally occurring fatty acids that have a *trans* configuration, such as **conjugated linoleic acid,** may have health benefits.[2]

Essential Fatty Acids The human body can synthesize all the fatty acids it needs from carbohydrate, fat, or protein except for two—**linoleic acid** and **linolenic acid.** Both linoleic acid and linolenic acid are polyunsaturated fatty acids. Because they cannot be made from other substances in the body, they must be obtained from food and are therefore called **essential fatty acids.** Linoleic acid and linolenic acid are found in small amounts in plant oils, and the body readily stores them, making deficiencies unlikely. From both of these essential fatty acids, the body makes important substances that help regulate a wide range of body functions: blood pressure, clot formation, blood lipid concentration, the immune response, the inflammatory response to injury,

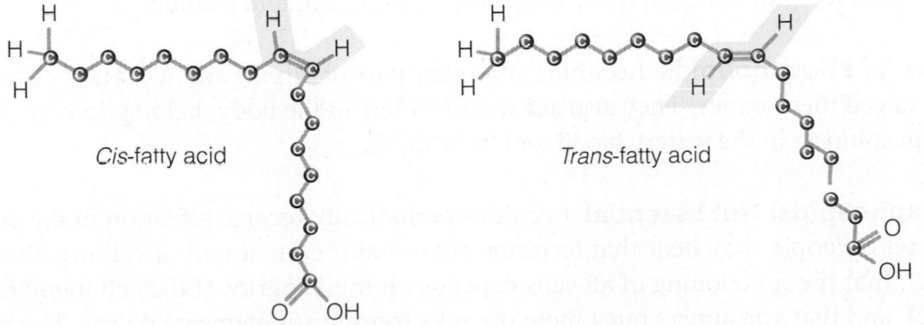

Cis-fatty acid   Trans-fatty acid

FIGURE 3-4 *Cis-* and *Trans-*Fatty Acids Compared

*BHA is butylated hydroxyanisole; BHT is butylated hydroxytoluene.

saturated fatty acid: a fatty acid carrying the maximum possible number of hydrogen atoms (having no points of unsaturation).

unsaturated fatty acid: a fatty acid with one or more points of unsaturation where hydrogen atoms are missing (includes monounsaturated and polyunsaturated fatty acids).

monounsaturated fatty acid (MUFA): a fatty acid that has one point of unsaturation; for example, the oleic acid found in olive oil.

polyunsaturated fatty acids (PUFA): fatty acids with two or more points of unsaturation. For example, linoleic acid has two such points, and linolenic acid has three. Thus, polyunsaturated *fat* is composed of triglycerides containing a high percentage of PUFA.

rancid: the term used to describe fats when they have deteriorated, usually by oxidation. Rancid fats often have an off odor.

oxidation (OKS-ee-day-shun): the process of a substance combining with oxygen.

antioxidants: as a food additive, preservatives that delay or prevent rancidity of foods and other damage to food caused by oxygen.

BHA, BHT: preservatives commonly used to slow the development of off flavors, odors, and color changes caused by oxidation.

hydrogenation (high-drah-gen-AY-shun): a chemical process by which hydrogen atoms are added to monounsaturated or polyunsaturated fats to reduce the number of double bonds, making the fats more saturated (solid) and more resistant to oxidation (protecting against rancidity). Hydrogenation produces *trans*-fatty acids.

***trans*-fatty acids:** fatty acids in which the hydrogen atoms next to the double bond are on opposite sides of the carbon chain.

conjugated linoleic acid: a collective term for several fatty acids that have the same chemical formulas as linoleic acid but with different configurations.

linoleic acid, linolenic acid: polyunsaturated fatty acids that are essential for human beings.

essential fatty acids: fatty acids that the body requires but cannot make in amounts sufficient to meet its physiological needs.

and many others.[3] These two essential nutrients also serve as structural components of cell membranes.

TABLE 3-2
The Lipid Family

Triglycerides (fats and oils)
- Glycerol (1 per triglyceride)
- Fatty acids (3 per triglyceride)
  Saturated
  Monounsaturated
  Polyunsaturated
    Omega-6
    Omega-3

Phospholipids (such as the lecithins)

Sterols (such as cholesterol)

omega-6 fatty acid: a polyunsaturated fatty acid with its endmost double bond located six carbons back from the end of its carbon chain; long recognized as important in nutrition. Linoleic acid is an example.

omega-3 fatty acids: polyunsaturated fatty acids in which the endmost double bond is three carbons back from the end of the carbon chain; relatively newly recognized as important in nutrition. Linolenic acid is an example.

EPA, DHA: omega-3 fatty acids made from linolenic acid. The full name for EPA is *eicosapentaenoic* (EYE-cosa-PENTA-ee-NO-ick) *acid*. The full name for DHA is *docosahexaenoic* (DOE-cosa-HEXA-ee-NO-ick) *acid*.

phospholipids: one of the three main classes of lipids; compounds that are similar to triglycerides but have *choline* (or another compound) and a phosphorus-containing acid in place of one of the fatty acids.

lecithins: one type of phospholipid.

choline: a nonessential nutrient that can be made in the body from an amino acid.

emulsifiers: substances that mix with both fat and water and that disperse the fat in the water, forming an emulsion.

Linoleic Acid: An Omega-6 Fatty Acid Linoleic acid is an **omega-6 fatty acid,** found in the seeds of plants and in the oils produced from the seeds. Any diet that contains vegetable oils, seeds, nuts, and whole-grain foods provides enough linoleic acid to meet the body's needs. Researchers have long known and appreciated the importance of the omega-6 fatty acid family.

Linolenic Acid and Other Omega-3 Fatty Acids Linolenic acid belongs to a family of polyunsaturated fatty acids known as **omega-3 fatty acids,** a family that also includes **EPA** and **DHA.** EPA and DHA are found primarily in fish oils. As mentioned, the human body cannot make linolenic acid, but given dietary linolenic acid, it can make EPA and DHA, although the process is slow.

The importance of omega-3 fatty acids was first recognized during the 1980s when research began to unveil impressive roles for EPA and DHA in metabolism and disease prevention. The brain has a high content of DHA, and both EPA and DHA are needed for normal brain development.[4] DHA is also especially active in the rods and cones of the retina of the eye.[5] Today, researchers know that these omega-3 fatty acids are essential for normal growth and development and that they may play an important role in the prevention and treatment of heart disease, diabetes, hypertension, arthritis, and cancer.[6]

Phospholipids

Up to now, this discussion has focused on one class of lipids, the triglycerides (fats and oils), and their component parts, the fatty acids (see Table 3-2). Two other classes of lipids, the **phospholipids** and sterols, make up only 5 percent of the lipids in the diet, but they are nevertheless worthy of attention. Among the phospholipids, the lecithins are of particular interest.

Structure of Phospholipids Like the triglycerides, **lecithins** and other phospholipids have a backbone of glycerol; they differ from triglycerides in having only two fatty acids attached to the glycerol. In place of the third fatty acid, they have a phosphate group (a phosphorus-containing acid) and a molecule of **choline** or a similar compound. The fatty acids make phospholipids soluble in fat; the phosphate group enables them to dissolve in water. Such versatility benefits the food industry, which uses phospholipids as **emulsifiers** to mix fats with water in such products as mayonnaise and candy bars.

Phospholipids in Foods In addition to the phospholipids used by the food industry as emulsifiers, phospholipids are also found naturally in foods. The richest food sources of lecithin are eggs, liver, soybeans, wheat germ, and peanuts.

Roles of Phospholipids Lecithins and other phospholipids are important constituents of cell membranes. They also act as emulsifiers in the body, helping to keep other fats in solution in the watery blood and body fluids.

Phospholipids: Not Essential Lecithins periodically receive attention in the popular press. People may hear that lecithins are a major constituent of cell membranes (true), that the functioning of all cells depends on the integrity of the cell membranes (true), and that consumers must therefore take lecithin supplements (false). The body digests lecithins before it absorbs them, so the lecithins people eat do not reach the body tissues intact. Instead, the lecithins used for building cell membranes are made from scratch by the liver. In other words, the lecithins are not essential nutrients.

Sterols

Sterols are large, complex molecules consisting of interconnected rings of carbon. Cholesterol is the most familiar sterol, but others, such as vitamin D and the sex hormones (for example, testosterone), are important, too.

Sterols in Foods Foods derived from both plants and animals contain sterols, but only those from animals—meats, eggs, fish, poultry, and dairy products—contain significant amounts of cholesterol. Organ meats, such as liver and kidneys, and eggs are richest in cholesterol; cheeses and other meats have less. Shellfish contain many sterols but much less cholesterol than was previously thought.

Sterols other than cholesterol are naturally found in plants. Being structurally similar to cholesterol, plant sterols interfere with cholesterol absorption. By inhibiting cholesterol absorption, a diet rich in plant sterols lowers blood cholesterol levels.[7] Food manufacturers have fortified foods such as margarine with plant sterols, creating a functional food that helps to reduce blood cholesterol.

Cholesterol Synthesis Like the lecithins, cholesterol can be made by the body, so it is not an essential nutrient. Your liver is manufacturing it now, as you read. The raw materials that the liver uses to make cholesterol can all be taken from glucose or fatty acids. In other words, cholesterol can be made from either carbohydrate or fat. Most of the body's cholesterol ends up in the membranes of cells, where it performs vital structural and metabolic functions.

Cholesterol's Two Routes in the Body After being made, cholesterol leaves the liver by two routes:

1. It may be made into **bile,** stored in the gallbladder, and delivered to the intestine.
2. It may travel, via the bloodstream, to all the body's cells.

The bile that is made from cholesterol in the liver is released into the intestine to aid in the digestion and absorption of fat. After bile does its job, most of it is reabsorbed into the body and recycled; the rest is excreted in the feces.

Cholesterol Excreted While bile is in the intestine, some of it may be trapped by soluble fibers or by some medications, which carry it out of the body in feces. The excretion of bile reduces the total amount of cholesterol remaining in the body.

Cholesterol Transport Some cholesterol, packaged with other lipids and protein, leaves the liver via the arteries and is transported to the body tissues by the blood. These packages of lipids and proteins are called **lipoproteins.** As the lipoproteins travel through the body, tissues can extract lipids from them. Cholesterol's harmful effects in the body occur when it forms deposits in the artery walls. These deposits contribute to **atherosclerosis,** a disease that can cause heart attacks and strokes.

Both the intestine and the liver make lipoproteins. Chapter 5 tells the story of lipid transport.

sterols: one of the main classes of lipids; includes cholesterol, vitamin D, and the sex hormones (such as testosterone).

bile: a compound that prepares fats and oils for digestion; made by the liver, stored in the gallbladder, and released into the small intestine.

lipoproteins: clusters of lipids associated with proteins that serve as transport vehicles for lipids in the lymph and blood.

atherosclerosis (ath-er-oh-scler-OH-sis): a type of artery disease characterized by accumulations of lipid-containing material on the inner walls of the arteries (see Chapter 22).

IN SUMMARY

▌ Table 3-2 summarizes the members of the lipid family.

▌ The predominant lipids, both in foods and in the body, are triglycerides, which have glycerol backbones with three fatty acids attached.

▌ Fatty acids vary in the length of their carbon chains and their degree of saturation. Those that are fully loaded with hydrogen atoms are saturated; those that are missing hydrogen atoms and therefore have double bonds are unsaturated (monounsaturated or polyunsaturated).

▌ Most triglycerides contain more than one type of fatty acid.

- Fatty acid saturation affects the physical characteristics and storage properties of fats.

- Hydrogenation, which makes polyunsaturated fats more saturated, gives rise to *trans*-fatty acids, altered fatty acids that may have health effects similar to those of saturated fatty acids.

- Linoleic acid and linolenic acid are essential nutrients. In addition to serving as structural parts of cell membranes, they make powerful substances that help regulate blood pressure, blood clot formation, and the immune response.

- Phospholipids, including the lecithins, have a unique chemical structure that allows them to be soluble in both water and fat.

- In the body, phospholipids are major constituents of cell membranes; the food industry uses phospholipids as emulsifiers.

- Sterols include cholesterol, bile, vitamin D, and the sex hormones.

- Only animal-derived foods contain significant amounts of cholesterol.

Health Effects and Recommended Intakes of Fats

Of all the dietary factors related to chronic diseases prevalent in developed countries, high intakes of certain fats are by far the most significant. The person who chooses a diet too high in saturated fats or *trans* fats invites the risk of **cardiovascular disease (CVD)**, and heart disease is the number one killer of adults in the United States and Canada. As for cancer, evidence is less compelling than for heart disease, but it does suggest that a diet high in certain kinds of fat is associated with a greater-than-average risk of developing some types of cancer.[8] Conversely, some research suggests that omega-3 fatty acids from fish may protect against some cancers.[9] Obesity carries serious risks to health, and the high energy density of fatty foods makes it easy for people to exceed their energy needs and so gain unneeded weight. Some points about fats and heart health are presented here because they underlie dietary recommendations concerning fats.

Nutrition and heart disease is the topic of Chapter 22.

Nutrition and cancer is a topic of Chapter 25.

The health risks of obesity are described in Chapter 6, and Chapter 7 focuses on weight management.

Fats and Heart Health

As noted earlier and described in Chapter 5, cholesterol travels in the blood within lipoproteins. Two of the lipoproteins, LDL and HDL, play major roles with regard to heart health and are the focus of most recommendations made for reducing the risk of heart disease. A high blood LDL cholesterol concentration is a predictor of the likelihood of suffering a fatal heart attack or stroke, and the higher the LDL, the earlier the episode is expected to occur. Conversely, high HDL cholesterol signifies a *lower* disease risk.

Most people realize that elevated blood cholesterol is an important risk factor for heart disease. Most people may not realize, though, that cholesterol in *food* is not the main influential factor in raising *blood* cholesterol.

Saturated Fats and Blood Cholesterol The main dietary factors associated with elevated blood LDL cholesterol are high saturated fat and high *trans* fat intakes.*[10] LDL cholesterol indicates a risk of heart disease because high LDL concentrations promote cholesterol uptake in blood vessel walls.

*It should be noted that not all saturated fatty acids have the same cholesterol-raising effect. Stearic acid, an 18-carbon fatty acid, does not seem to raise blood cholesterol.

Fats from animal sources are the main contributors of saturated fats in most people's diets. Some vegetable fats (coconut oil, palm kernel oil, and palm oil) and hydrogenated fats such as shortening or stick margarine provide smaller amounts of saturated fats. To minimize intake of saturated fat, most people need to eat less meat. When eating meat, choose the leanest cuts, and trim away the visible fat.[11] Selecting fat-free milk and using nonhydrogenated margarine and unsaturated cooking oil such as olive oil, safflower oil, or canola oil are other simple changes that can dramatically lower saturated fat intake and heart disease risk.

Trans-Fatty Acids and Blood Cholesterol Consuming fats with *trans*-fatty acids poses a risk to the health of the heart and arteries by raising LDL and lowering HDL cholesterol, and by producing inflammation.[12] When news of *trans*-fatty acids' effects on heart health was first emerging, some people hastily switched from using margarine back to butter, believing oversimplified reports that margarine provided no heart health advantage over butter. It is true that hard margarines and virtually all shortenings are made from mostly hydrogenated fats and therefore contain substantial *trans*-fatty acids. Since the requirement that *trans* fats be listed on the Nutrition Facts panel of food labels took effect in 2006, however, margarine makers have reformulated their products to contain much less *trans* fat.[13] Soft or liquid varieties are made from unhydrogenated oils, which are mostly unsaturated and so are less likely to elevate blood cholesterol than the saturated fats of butter. Some margarines contain olive oil, omega-3 fatty acids, or plant sterols (mentioned earlier), making these products preferable to butter and other margarines for the heart.*

With more than 50 types of margarines and spreads on the market in sticks, tubs, sprays, and liquids, consumers must educate themselves, read labels, and select margarines that provide the preferred flavor with the least saturated and *trans* fat. When oils (but not hydrogenated oils) are the first ingredient listed on a margarine label, that margarine is probably low in both *trans*-fatty acids and saturated fat. In general, margarine contributes less *trans* fat to the average diet than do other contributors, such as processed foods.

Simply adding margarine that contains plant sterols, olive oil, or omega-3 fatty acids to a high-fat diet is unlikely to bring benefits. These substances cannot undo the damage from a diet consistently high in saturated fats. Margarines with plant sterols have drawbacks, too. They cost three or four times more than regular margarine, they equal regular margarine in kcalories, and they are not proven safe for use by certain populations, such as growing children.

Many foods other than margarine also contribute *trans*-fatty acids. Fast foods, chips, baked goods, and other commercially prepared foods are high in fats containing up to 50 percent *trans*-fatty acids. The 2006 labeling requirement to list *trans* fat on packaged food has prompted many food manufacturers to eliminate or greatly reduce *trans* fats in foods. Many restaurants have also taken steps to eliminate or reduce *trans* fats from their menu items.[14] The average daily intake of *trans*-fatty acids in the United States is about 6 grams per day—mostly from products that have been hydrogenated.

Dietary Cholesterol and Blood Cholesterol Dietary cholesterol has also been implicated in raising blood cholesterol and increasing the risk of heart disease, although its effect is not as strong as that of saturated fat or *trans* fat. Still, health experts advise limiting cholesterol intake.

Recall that cholesterol is found primarily in foods derived from animals. Consequently, eating less fat from meats, eggs, and milk products helps lower dietary cholesterol intake (as well as total and saturated fat intakes).

*Two brand names of margarines with plant sterols currently on the market are *Benecol* and *Take Control*.

Major sources of saturated fats:

- Whole milk (and even reduced-fat milk), cheese, butter, cream, cream cheese, sour cream, and ice cream
- Fatty cuts of beef and pork, and processed meats such as bacon, sausage, and hot dogs
- Tropical oils (coconut, palm, and palm kernel) and products that contain them such as cakes, cookies, doughnuts, and pastries
- Shortening and lard

Nutrition in Practice 3 examines various types of fats and their roles in supporting or harming heart health.

The words *hydrogenated vegetable oil* or *shortening* in an ingredients list indicate *trans*-fatty acids in the product.

Major food sources of *trans* fats:

- Cakes, cookies, pies, doughnuts, and crackers
- Meat and dairy products
- Hard margarine
- Fried potatoes and other commercial fried foods such as chicken
- Potato chips and corn chips
- Shortening

Major sources of cholesterol:

- Eggs
- Meat, poultry, and shellfish
- Cheese
- Milk

cardiovascular disease (CVD): a general term for all diseases of the heart and blood vessels (see Chapter 22).

Monounsaturated Fatty Acids and Blood Cholesterol Replacing saturated and *trans* fats with monounsaturated fat such as olive oil may be an effective dietary strategy to prevent heart disease. The lower rates of heart disease among people in the Mediterranean region are often attributed to their liberal use of olive oil, a rich source of monounsaturated fatty acids.[15] Olive oil also delivers valuable phytochemicals that help to protect against heart disease.[16] Nutrition in Practice 3 examines the role of olive oil and other fats in supporting or harming heart health.

Polyunsaturated Fatty Acids, Blood Cholesterol, and Heart Disease Risk Polyunsaturated fatty acids (PUFA) of the omega-6 and omega-3 families are potent protectors against heart disease. The primary omega-6 fatty acid, linoleic acid, which is found in vegetable oils such as corn and sunflower oil, exerts most of its beneficial effect by lowering both total blood cholesterol and LDL cholesterol.[17]

The omega-3 fatty acids EPA and DHA, which are found mainly in fatty fish, exert their beneficial effects by influencing the function of both the heart and blood vessels. Specifically, EPA and DHA protect heart health by:[18]

▮ Lowering blood triglycerides

▮ Preventing blood clots

▮ Protecting against irregular heartbeats

▮ Lowering blood pressure

▮ Defending against inflammation

The primary member of the omega-3 family, linolenic acid, may benefit heart health as well, but evidence for this effect is much less certain than for EPA and DHA. Table 3-3 provides sources of omega-6 and omega-3 fatty acids.

TABLE 3-3 Food Sources of Omega-6 and Omega-3 Fatty Acids

Omega-6	
Linoleic acid	Seeds, nuts, vegetable oils (corn, cottonseed, safflower, sesame, soybean, sunflower), poultry fat
Omega-3	
Linolenic acid[a]	Oils (canola, flaxseed, soybean, walnut, wheat germ; liquid or soft margarine made from canola or soybean oil) Nuts and seeds (flaxseeds, walnuts, soybeans) Vegetables (soybeans)
EPA and DHA	Human milk Fish and seafood: **>500 mg per 3.5 oz serving:** European seabass (bronzini), herring (Atlantic and Pacific), mackerel, oyster (Pacific wild), salmon (wild and farmed), sardines, toothfish (includes Chilian seabass), trout (wild and farmed) **150–500 mg per 3.5 oz serving:** black bass, catfish (wild and farmed), clam, cod (Atlantic), crab (Alaskan king), croakers, flounder, haddock, hake, halibut, oyster (eastern and farmed), perch, scallop, shrimp (mixed varieties), sole, swordfish, tilapia (farmed) **<150 mg per 3.5 oz serving:** cod (Pacific), grouper, lobster, mahimahi, monkfish, red snapper, skate, triggerfish, tuna, wahoo

[a]Alpha-linolenic acid. Also found in the seed oil of the herb evening primrose.

Source for fish data: K. L. Weaver and coauthors, The content of favorable and unfavorable polyunsaturated fatty acids found in commonly eaten fish, *Journal of the American Dietetic Association* 108 (2008): 1178–1185; P. M. Kris-Etherton, W. S. Harris, and L. J. Appel, Fish consumption, fish oil, omega-3 fatty acids, and cardiovascular disease, *Circulation* 106 (2002): 2747–2757.

Balance Omega-6 and Omega-3 Intakes The U.S. diet is high in omega-6 fatty acids (due to the increased production and use of vegetable cooking oils such as corn and cottonseed oils) and low in omega-3 fatty acids. Experts recommend a more balanced intake. The best way for people to increase their intakes of omega-3 fatty acids is to follow the advice of the American Heart Association: eat at least two servings of fatty fish (see Table 3-3) each week.[19] Even one fish meal per week has been associated with a reduced risk of heart disease.[20]

Greater heart health benefits can be expected when fish is grilled, baked, or broiled, partly because the varieties prepared this way often contain more EPA and DHA than species used for fried fish in fast-food restaurants and frozen products. Additionally, benefits are attained by avoiding commercial frying fats, which are often laden with *trans* fat and saturated fat. Further benefits arise when fish replaces high-fat meats or other foods rich in saturated fats in several meals each week.

Some species of fish and shellfish, however, may contain significant levels of mercury or other environmental contaminants. Most healthy people can safely consume most species of ocean fish several times a week, but some people face greater risks. Women who may become pregnant, pregnant and lactating women, and children are more sensitive to contaminants than others, but even they can benefit from safer fish varieties within recommended limits (see Chapter 11 for details).

For everyone, consuming a variety of different types of fish is a good idea to minimize exposure to any single toxin that may accumulate in a favored species. The margin lists the species most heavily contaminated with mercury and those that are low in mercury. As for freshwater fish in some areas, consumers should check local advisories about the safety of fish caught by family and friends.

Omega-3 Supplements Fish, not fish oil supplements, is the preferred source of omega-3 fatty acids, except for persons with cardiovascular disease who cannot obtain enough EPA and DHA from fish, and then only under a physician's supervision.[21] Supplements of 2 grams a day of EPA or more than 3 grams of fish oil interfere with blood clotting and cause prolonged bleeding times.[22] Supplements also lack the *other* beneficial nutrients that fish provides. Some fish oil supplements also taste fishy, and people often complain of a lingering or repeating taste for hours after taking them.

In addition, fish oil supplements are made from fish skins and livers, which can accumulate toxic concentrations of dioxins and other industrial contaminants from polluted waters.*[23] As stated earlier, mercury contamination is also a problem in the flesh of both freshwater and saltwater species. For supplements, newer processing methods can rid fish oil of most contaminants, but not all supplements are refined in this way. In addition to contamination, fish liver oil naturally contains high levels of the two most potentially toxic vitamins, A and D. Lastly, supplements of fish oil are expensive—they must be purchased in addition to food, while fatty fish replaces other protein foods that would have been purchased instead.

Recommendations

Some fat in the diet is essential for good health, but too much fat, especially saturated fat and *trans* fat, increases the risks of chronic diseases. Defining the exact amount of fat, saturated fat, or cholesterol that benefits health or begins to harm health, however, is not possible. For this reason, no RDA or UL has been set.[24] Instead, the

Grilling or broiling fish, instead of frying them, preserves their beneficial omega-3 fatty acids while adding little or no saturated fat.

▪ **Fish most heavily contaminated with mercury:** king mackerel, shark, swordfish, and tilefish (also called golden bass or golden snapper).

▪ **Fish or shellfish lower in mercury:** catfish, pollock, salmon, sardines, shrimp, and canned light tuna. Canned albacore ("white") tuna generally contains more mercury than light tuna.

---

*Contaminants include organically bound arsenic, organochlorines, polychlorinated biphenyls (PCBs), and others.

DRI for fat:

▌ 20 to 35 percent of energy intake

Daily Value:

▌ 65 g fat (based on 30 percent of 2000-kcal diet)

Linoleic acid (omega-6) AI:

▌ 5 to 10 percent of energy intake

Men:

▌ 19–50 yr: 17 g/day
▌ 51+ yr: 14 g/day

Women:

▌ 19–50 yr: 12 g/day
▌ 51+ yr: 11 g/day

Linolenic acid (omega-3) AI:

▌ 0.6 to 1.2 percent of energy intake

Men: 1.6 g/day

Women: 1.1 g/day

Dietary Guidelines for Americans:

▌ Consume less than 10 percent of kcalories from saturated fatty acids and less than 300 mg per day of cholesterol, and keep *trans*-fatty acid consumption as low as possible.

Maximum saturated fat intakes set by the *Dietary Guidelines for Americans:*

1600-kcal diet: 18 g
2000-kcal diet: 20 g
2200-kcal diet: 24 g
2500-kcal diet: 28 g
2800-kcal diet: 31 g

Daily Values:

▌ 20 g saturated fat (based on 10 percent of 2000-kcal diet)
▌ 300 mg cholesterol

DRI and *Dietary Guidelines* suggest a diet that is low in saturated fat, *trans* fat, and cholesterol and provides 20 to 35 percent of the daily energy intake from fat. These recommendations recognize that diets with up to 35 percent of kcalories from fat can be compatible with good health if energy intake is reasonable and saturated and *trans* fat intakes are low. When total fat intake exceeds 35 percent of kcalories, saturated fat intakes increase to unhealthy levels.[25] Fat and oil intakes below 20 percent of kcalories increase the risk of inadequate essential fatty acid intakes. The FDA established Daily Values on food labels using 30 percent of energy intake as the guideline for fat.

Part of the allowance for total fat should provide for the essential fatty acids— linoleic acid and linolenic acid. Recommendations suggest that linoleic acid provide 5 to 10 percent of the daily energy intake and linolenic acid, 0.6 to 1.2 percent.

Saturated fats, *trans* fats, and cholesterol increase total blood cholesterol and LDL cholesterol and, therefore, the risk of heart disease. Even low intakes of each may elevate heart disease risk.[26] Thus, recommendations urge people to eat diets that are low in saturated fat, *trans* fat, and cholesterol. Specifically, consume less than 10 percent of kcalories from saturated fat, keep *trans* fat intakes as low as possible, and consume less than 300 milligrams of cholesterol each day.[27] To help consumers meet these goals, the FDA established Daily Values on food labels, using 10 percent of energy intake for saturated fat; the Daily Value for cholesterol is 300 milligrams, regardless of energy intake. There is no Daily Value for *trans* fat.

Nutrition in Practice 3 suggests substituting monounsaturated or polyunsaturated fats for harmful saturated fats. No beneficial change in blood lipids occurs when monounsaturated or polyunsaturated fat is *added* to a diet rich in saturated or *trans* fat, however. The best diet for heart health is also rich in fruits, vegetables, nuts, and whole grains that offer many health advantages by supplying abundant nutrients, fiber, and phytochemicals.

IN SUMMARY

▌ High intakes of saturated or *trans* fats contribute to heart disease, obesity, and other health problems.

▌ High blood cholesterol, specifically, poses a risk of heart disease, and high intakes of saturated fat contribute most to high blood cholesterol. High intakes of *trans*-fatty acids also appear to raise blood cholesterol. Cholesterol in foods presents less of a risk.

▌ When monounsaturated fat, such as olive oil, replaces saturated and *trans* fats in the diet, the risk of heart disease may be lessened.

▌ Polyunsaturated fatty acids of the omega-6 and omega-3 families protect against heart disease.

▌ Though some fat in the diet is necessary, health authorities recommend a diet moderate in total fat and low in saturated fat, *trans* fat, and cholesterol.

Fats in Foods

Fats are important in foods as well as in the body. Many of the compounds that give foods their flavors and aromas are found in fats and oils. The delicious aromas associated with bacon, ham, and other meats, as well as with onions being sautéed, come from fats. Fats also influence the texture of many foods, enhancing smoothness, creaminess, moistness, or crispness. In addition, four vitamins—A, D, E, and K—are soluble in fat. When the fat is removed from a food, many fat-soluble

compounds, including these vitamins, are also removed. Table 3-4 summarizes the roles of fats in foods.

Finding the Fats in Foods

The remainder of this chapter and the Nutrition in Practice show you how to choose fats wisely with the goals of providing optimal health and pleasure in eating. To achieve such goals, you need to know which foods offer unsaturated fats that provide the essential fatty acids and which offer harmful saturated and *trans* fats. Perhaps most important for many people is learning to control portion sizes, particularly portions of fatty foods that can pack hundreds of kcalories into just a few bites.

The *Dietary Guidelines for Americans* urge that, beyond a healthy minimum, people should limit their fat intakes in order to limit intakes of saturated fat, *trans* fat, and kcalories. To do this requires consistently making nutrient-dense choices, such as fat-free milk, fat-free cheese, and the leanest meats, and refraining from adding solid fats, such as butter, hard margarine, or shortening, to foods during preparation or at the table. Higher-fat foods may be included in the diet, but the fat kcalories they provide must fit within a person's discretionary kcalorie allowance. Exceptions are the fats of fatty fish, nuts, and vegetable oils: they provide EPA and DHA, linoleic acid and linolenic acid, and vitamin E needed for a healthful diet. Within kcalorie limits, the kcalories these fats provide are necessary, not discretionary. Many people, especially sedentary people, have few or no discretionary kcalories to spend on high-fat foods.

Added Fats in Foods A dollop of dessert topping, a spread of butter on bread, oil or shortening in a recipe, dressing on a salad—all of these are examples of *added* fats. Indeed, all sorts of fats can be added to foods during commercial or home preparation or at the table. The following amounts of these fats contain about 5 grams of pure fat, providing 45 kcalories and negligible protein and carbohydrate:

- 1 teaspoon of oil or shortening
- 1½ teaspoons of mayonnaise, butter, or margarine
- 1 tablespoon of regular salad dressing, cream cheese, or heavy cream
- 1½ tablespoons of sour cream

The majority of added fats in the diet are invisible. They are the hidden fats of fried foods and baked goods, sauces and mixed dishes, and dips and spreads. Other invisible fats include the fats in the marbling of meat and the fat ground into lunch meats and hamburger.

Milk and Milk Products The fat in whole milk is about 63 percent saturated fat; the cholesterol content is 24 milligrams per cup for whole milk or 5 milligrams for fat-free milk. Thus, choosing fat-free in place of whole milk reduces your intake of cholesterol as well as saturated fat.

Note that cream and butter do not appear in the milk group. Milk and yogurt are rich in calcium and protein, but cream and butter are not. Cream and butter are fats, as are whipped cream, sour cream, and cream cheese, so they are grouped together with the solid fats. That is why in this textbook, the food group that includes milk is carefully labeled the "milk and milk products group," not the "dairy group." Cheeses are major contributors of saturated fat in people's diets.

Protein Foods Meats conceal a good deal of the fat—and much of the saturated fat—that people consume. To help "see" the fat in meats, it is useful to think of them

---

TABLE 3-4
The Functions of Fats in Foods

- **Nutrient.** Fats provide essential fatty acids.
- **Energy.** Fats provide a concentrated energy source in foods.
- **Transport.** Fats carry fat-soluble vitamins A, D, E, and K along with some phytochemicals and assist in their absorption.
- **Sensory appeal.** Fats contribute to the taste and smell of foods.
- **Appetite.** Fats stimulate the appetite.
- **Satiety.** Fats contribute to feelings of fullness.
- **Texture.** Fats help make foods tender.

Table 1-4 of Chapter 1 (p. 18) specifies how much oil is required to meet nutrient needs at several kcalorie levels.

The concept of discretionary kcalories is discussed in Chapter 1.

1 c whole milk:
- 8 g fat
- 5 g saturated fat
- 24 mg cholesterol

1 c reduced-fat milk:
- 5 g fat
- 2 g saturated fat
- 20 mg cholesterol

1 c low-fat milk:
- 2 g fat
- 1.5 g saturated fat
- 10 mg cholesterol

1 c fat-free milk:
- 0 g fat
- 0 g saturated fat
- 5 mg cholesterol

as falling into four categories: very lean, lean, medium-fat, and high-fat meats, as the exchange lists do in Appendix C. Meats in all four categories contain about equal amounts of protein, but their fat contents differ, and their saturated fat and kcalorie amounts vary significantly. Table 1-9 on p. 25 in Chapter 1 provides some definitions concerning the fat contents of meats.

The USDA Food Guide suggests that most adults limit a day's intake of meats or equivalents to about 5 to 7 ounces. For comparison, the smallest fast-food hamburger weighs about 3 ounces. A steak served in a restaurant often runs 8, 12, or 16 ounces, more than a whole day's meat allowance. You may have to weigh a serving or two of meat to see how much you are eating.

People think of meat as protein food, but calculation of its nutrient content reveals a surprising fact. A big (4-ounce) fast-food hamburger sandwich contains 23 grams of protein and 20 grams of fat. Because protein offers 4 kcalories per gram and fat offers 9, the sandwich provides 92 kcalories from protein and about twice that amount from fat. The kcalorie total, counting carbohydrates from the bun and condiments, is more than 400 kcalories, with more than 50 percent of them from fat. Hot dogs, fried chicken sandwiches, and fried fish sandwiches are also high-fat choices. Because so much of the energy in a meat eater's diet is hidden from view, people can easily overeat on high-fat food, making weight control difficult.

When choosing beef or pork, look for lean cuts named *loin* or *round,* from which the fat can be trimmed. Eat small portions, too. As for chicken and turkey, these meats are naturally lean, but commercial processing and frying add fats, especially in patties, nuggets, fingers, or wings. Chicken wings are mostly skin, and a chicken stores most of its fat just under its skin. The tastiest wing snacks have also been fried in cooking fat (often a hydrogenated, saturated type with *trans*-fatty acids); smothered with a buttery, spicy sauce; and then dipped in blue cheese dressing, making wings an extraordinarily high-fat snack. People who snack on wings may want to plan on eating low-fat foods at several other meals to balance them out.

Vegetables, Fruits, and Grains Choosing vegetables, fruits, whole grains, and legumes also helps lower the saturated fat, cholesterol, and total fat content of the diet. Most vegetables and fruits naturally contain little or no fat; avocados and olives are exceptions, but most of their fat is unsaturated, which is not harmful to heart health. Most grains contain only small amounts of fat. Some grain products such as fried taco shells, croissants, and biscuits are high in saturated fat, so consumers need to read food labels. Similarly, many people add butter, margarine, or cheese sauce to grains and vegetables, which raises their saturated and *trans* fat contents. Because fruits are often eaten without added fat, a diet that includes several servings of fruit daily can help a person meet the dietary recommendations for fat.

Cutting Fat Intake and Choosing Unsaturated Fats

Knowing which foods contain the most fat is the first step toward meeting the recommendation to limit dietary fat in general and saturated fat in particular. As a general rule, a person who eats meat and wishes to reduce both saturated fat and cholesterol intake can eat fewer high-fat meats and dairy foods; fewer eggs; and more poultry (without the skin), fish, and fat-free dairy products. A vegetarian who eats dairy products and eggs can shift to fat-free milk and fat-free cheeses and limit butter and egg intake. Vegetarians who omit animal-derived foods generally eat less saturated fat and little or no cholesterol. The accompanying "How to" offers strategies for making heart-healthy choices, food group by food group.

Breads and Cereals

▌ Select breads, cereals, and crackers that are low in saturated and *trans* fat (for example, bagels instead of croissants).

▌ Prepare pasta with a tomato sauce instead of a cheese or cream sauce.

Vegetables and Fruits

▌ Enjoy the natural flavor of steamed vegetables (without butter) for dinner and fruits for dessert.

▌ Eat at least two vegetables (in addition to a salad) with dinner.

▌ Snack on raw vegetables or fruits instead of high-fat items like potato chips.

▌ Buy frozen vegetables without sauce.

Milk and Milk Products

▌ Switch from whole milk to reduced-fat, from reduced-fat to low-fat, and from low-fat to fat-free (nonfat).

▌ Use fat-free and low-fat cheeses (such as part-skim ricotta and low-fat mozzarella) instead of regular cheeses.

▌ Use fat-free or low-fat yogurt or sour cream instead of regular sour cream.

▌ Use evaporated fat-free milk instead of cream.

▌ Enjoy fat-free frozen yogurt, sherbet, or ice milk instead of ice cream.

Protein Foods

▌ Fat adds up quickly, even with lean meat; limit intake to about 6 ounces (cooked weight) daily.

▌ Eat at least two servings of fish per week (particularly fish such as mackerel, lake trout, herring, sardines, and salmon).

▌ Choose fish, poultry, or lean cuts of pork or beef; look for unmarbled cuts named round or loin (eye of round, top round, bottom round, round tip, tenderloin, sirloin, center loin, and top loin).

▌ Choose processed meats such as lunch meats and hot dogs that are low in saturated fat and cholesterol.

▌ Trim the fat from pork and beef; remove the skin from poultry.

▌ Grill, roast, broil, bake, stir-fry, stew, or braise meats; don't fry. When possible, place food on a rack so that fat can drain.

▌ Use lean ground turkey or lean ground beef in recipes; brown ground meats without added fat, then drain off fat.

▌ Select tuna, sardines, and other canned meats packed in water; rinse oil-packed items with hot water to remove much of the fat.

▌ Fill kabob skewers with lots of vegetables and slivers of meat; create main dishes and casseroles by combining a little meat, fish, or poultry with a lot of pasta, rice, or vegetables.

▌ Use legumes often.

▌ Eat a meatless meal or two daily.

▌ Use egg substitutes in recipes instead of whole eggs or use two egg whites in place of each whole egg.

Fats and Oils

▌ Use butter or stick margarine sparingly; select soft margarines instead of hard margarines.

▌ Use fruit butters, reduced-kcalorie margarines, or butter replacers instead of butter.

▌ Use low-fat or fat-free mayonnaise and salad dressing instead of regular.

▌ Limit use of lard and meat fat.

▌ Limit use of products made with coconut oil, palm kernel oil, and palm oil (read labels on bakery goods, processed foods, popcorn oils, and nondairy creamers).

▌ Reduce use of hydrogenated shortenings and stick margarines and products that contain them (read labels on crackers, cookies, and other commercially prepared baked goods); use vegetable oils instead.

Miscellaneous

▌ Use a nonstick pan, or coat the pan lightly with vegetable oil.

▌ Refrigerate soups and stews; when the fat solidifies, remove it before reheating.

▌ Use wine; lemon, orange, or tomato juice; herbs; spices; fruits; or broth instead of butter or margarine when cooking.

▌ Stir-fry in a small amount of oil; add moisture and flavor with broth, tomato juice, or wine.

▌ Use variety to enhance enjoyment of the meal: vary colors, textures, and temperatures—hot cooked versus cool raw foods—and use garnishes to complement food.

▌ Omit high-fat meat gravies and cheese sauces.

▌ Order pizzas with lots of vegetables, a little lean meat, and half the cheese.

Source: Adapted from *Expert Panel on Detection, Evaluation, and Treatment of High Blood Cholesterol in Adults (Adult Treatment Panel III), Third Report of the National Cholesterol Education Program (NCEP)*, NIH publication no. 02-5215 (Bethesda, MD: National Heart, Lung, and Blood Institute, 2002), pp. V-25–V-27.

Fats and kCalories   Removing fat from food also removes energy, as Figure 3-5 (p. 74) shows. A pork chop with the fat trimmed to within a half-inch of the lean provides 290 kcalories; with the fat trimmed off completely, it supplies 174 kcalories. A baked potato with butter and sour cream (1 tablespoon each) has 315 kcalories; a plain baked potato has 188 kcalories. The single most effective step you can take to reduce the energy value of a food is to eat it with less fat.

	kCalories	Fat	Saturated Fat
Pork chop (3 oz) with fat	290 kcal	24 g	9 g
Pork chop (3 oz) with fat trimmed	174 kcal	8 g	3 g

	kCalories	Fat	Saturated Fat
Potato (7 oz) with 1 tbsp each sour cream and butter	315 kcal	13.5 g	8.5 g
Plain potato (7 oz)	188 kcal	0 g	0 g

	kCalories	Fat	Saturated Fat
Whole milk (1 cup)	150 kcal	8 g	5 g
Fat-free milk (1 cup)	90 kcal	<1 g	<1 g

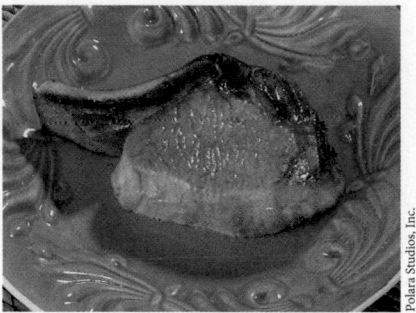  

FIGURE 3-5 Cutting Fat Cuts kCalories—and Saturated Fat

At room temperature, unsaturated fats (such as those found in oil) are usually liquid, whereas saturated fats (such as those found in butter) are solid.

Choosing Unsaturated Fats People who wish to make choices consistent with current recommendations need to learn how to read food labels, limit fat in general, and seek out the polyunsaturated and monounsaturated fats in preference to the saturated ones. Remember, the softer a fat is, the more unsaturated it is. Generally speaking, vegetable and fish oils are rich in polyunsaturates, olive oil and canola oil are rich in monounsaturates, and the harder fats—animal fats—are more saturated (see Figure 3-6). Nonetheless, vegetable fat or vegetable oil doesn't always mean unsaturated fat. Both coconut oil and palm oil, for example, which are often used in nondairy creamers, are saturated fats, and both raise blood cholesterol.

Don't Overdo Fat Restriction Some people actually manage to eat too *little* fat—to their detriment. Among them are young women and men with eating disorders, described in Nutrition in Practice 6. As a practical guideline, it is wise to include the equivalent of at least a teaspoon of fat in every meal.

Fat Replacers Today, consumers can choose from thousands of fat-reduced products. Many bakery goods, lunch meats, cheeses, spreads, frozen desserts, and other products made with **fat replacers** offer less than half a gram of fat, saturated fat, and *trans* fat in a serving. Some of these products contain **artificial fats,** and others use conventional ingredients in unconventional ways to reduce fats and kcalories. Among the latter, manufacturers can:

▌ Add water or whip air into foods.

▌ Add fat-free milk to creamy foods.

74

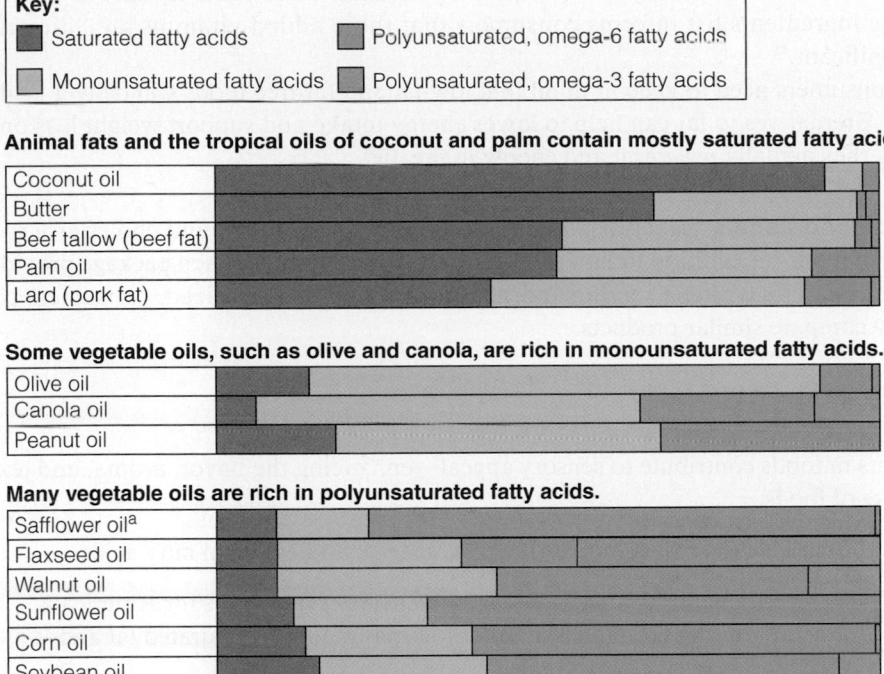

Key:
- ■ Saturated fatty acids
- ■ Monounsaturated fatty acids
- ■ Polyunsaturated, omega-6 fatty acids
- ■ Polyunsaturated, omega-3 fatty acids

Animal fats and the tropical oils of coconut and palm contain mostly saturated fatty acids.

Coconut oil
Butter
Beef tallow (beef fat)
Palm oil
Lard (pork fat)

Some vegetable oils, such as olive and canola, are rich in monounsaturated fatty acids.

Olive oil
Canola oil
Peanut oil

Many vegetable oils are rich in polyunsaturated fatty acids.

Safflower oil[a]
Flaxseed oil
Walnut oil
Sunflower oil
Corn oil
Soybean oil
Cottonseed oil

[a] Salad or cooking type, over 70% linoleic acid.

FIGURE 3-6 Comparison of Dietary Fats
Most fats are a mixture of saturated, monounsaturated, and polyunsaturated fatty acids.

▌ Use lean meats and soy protein to replace high-fat meats.

▌ Bake foods instead of frying them.

Common food ingredients such as fibers, sugars, or proteins can also take the place of fats in some foods. In particular, fat replacers made from oats or barley not only cut down on fats in foods but also introduce beneficial viscous fibers, while imparting desirable tastes and textures associated with real fats.[28] Products made from sugars or proteins still provide kcalories but far fewer kcalories from fats. Manufactured fat replacers consist of chemical derivatives of carbohydrate, protein, or fat, or modified versions of foods rich in those constituents.

A familiar example of an artificial fat that has been approved for use in snack foods such as potato chips, crackers, and tortilla chips is **olestra.** Olestra's chemical structure is similar to that of a regular fat (a triglyceride), but with important differences. A triglyceride is composed of a glycerol molecule with three fatty acids attached, whereas olestra is made of a sucrose molecule with six to eight fatty acids attached. Enzymes in the digestive tract cannot break the bonds of olestra, so, unlike sucrose or fatty acids, olestra passes through the system unabsorbed.

The FDA's evaluation of olestra's safety addressed two questions. First, is olestra toxic? Research on both animals and humans supports olestra's safety as a partial replacement for dietary fats and oils, with no reports of cancer or birth defects. Second, does olestra affect either nutrient absorption or the health of the digestive tract? When olestra passes through the digestive tract unabsorbed, it binds with the fat-soluble vitamins A, D, E, and K and carries them out of the body, robbing the person of these valuable nutrients. To compensate for these losses, the FDA requires the manufacturer to fortify olestra with vitamins A, D, E, and K. Saturating olestra with these vitamins does not make the product a good source of vitamins, but it

fat replacers: ingredients that replace some or all of the functions of fat in foods and may or may not provide energy.

artificial fats: zero-energy fat replacers that are chemically synthesized to mimic the sensory and cooking qualities of naturally occurring fats but are totally or partially resistant to digestion.

olestra: a synthetic fat made from sucrose and fatty acids that provides zero kcalories per gram; also known as *sucrose polyester*.

does block olestra's ability to bind with the vitamins from other foods. An asterisk in the ingredients list informs consumers that these added vitamins are "dietarily insignificant."

Consumers need to keep in mind that low-fat and fat-free foods still deliver kcalories. Alternatives to fat can help to lower energy intake and support weight loss only when they actually *replace* fat and energy in the diet.

Read Food Labels Labels list total fat, saturated fat, *trans* fat, and cholesterol contents of foods, in addition to fat kcalories, per serving. Because each package provides information for a single serving and serving sizes are standardized, consumers can easily compare similar products.

IN SUMMARY

▌ Fats in foods contribute to sensory appeal—enhancing the flavor, aroma, and texture of foods.

▌ Fats in foods deliver fat-soluble vitamins, energy, and essential fatty acids.

▌ While some fat in the diet is necessary, the Dietary Guidelines for Americans recommend limiting fat intakes in order to limit intakes of saturated fat and *trans* fat.

▌ Fats added to foods during preparation or at the table are a major source of fat in the diet.

▌ The choice between whole and fat-free milk products can make a big difference to the fat, saturated fat, and cholesterol content of a diet.

▌ Meats account for a large proportion of the hidden fat and saturated fat in many people's diets.

▌ Most people consume more meat than is recommended.

▌ Most vegetables and fruits naturally contain little or no fat.

▌ Grain products such as croissants and biscuits can be high in saturated fat, so consumers need to read food labels to learn which foods in this group contain fats.

▌ Consumers today can choose from an array of fat-reduced products, and many bakery goods and other foods made with fat replacers offer less than half a gram of fat, saturated fat, and *trans* fat per serving.

▌ Some products use artificial fats such as olestra, whereas others use conventional ingredients, such as water or fat-free milk, to reduce fat and kcalories.

▌ Food labels list total fat, saturated fat, cholesterol, and *trans* fat, as well as fat kcalories per serving.

Chapters 2 and 3 look briefly at the two major energy fuels in the body—carbohydrate and fat. When used for energy, each has desirable characteristics. The glucose derived from carbohydrate is needed by the brain and nerve tissues and is easily used for energy in other cells. Fat is a particularly useful fuel because the body stores it efficiently and in generous amounts. Chapter 4 looks at protein, a nutrient that can be used as fuel, but whose primary role is to provide machinery for getting things done.

Your Diet

Lipids

To maintain good health, eat enough—but not too much—fat and select the right kinds of fats. For example, moderate intakes of foods rich in polyunsaturated and monounsaturated fatty acids such as olive oil, nuts, and fish lower the risk of heart disease. Conversely, diets high in saturated fats and *trans* fats increase the risk of heart disease. Table NP3-1 (p. 81) in the Nutrition in Practice that follows this chapter lists food sources of healthful and harmful fatty acids. As Chapter 3 describes, the major sources of saturated fats in most people's diets are whole milk and whole milk products such as cheese and ice cream, fatty cuts of beef and pork, processed meats such as bacon and sausage, and products that contain tropical oils (coconut oil, palm kernel oil, and palm oil) such as cookies, pastries, and doughnuts. Major sources of *trans* fats include fast foods, chips, baked goods, and other commercially prepared foods.

■ Make a list of all the fast foods and packaged foods such as chips and cookies that you ate today, or that you might typically eat in a day.

■ Use Appendix A to look up the saturated and *trans* fat contents of the fast foods you ate and add this information to your list of foods.

■ Check the labels of the packaged foods you ate and write down their saturated and *trans* fat contents.

■ The recommended saturated fat intake is less than 10 percent of total kcalories per day. The recommendation for *trans* fat is to keep it as low as possible. Thus, for a 2000-kcalorie diet, the recommended saturated fat intake is 22 grams or less. (Recall that 1 gram of fat is equal to 9 kcalories.)

Example:

$$2000 \text{ kcal} \times 0.10 = 200 \text{ kcal}$$
$$200 \text{ kcal} \div 9 \text{ kcal/gram fat} = 22 \text{ grams saturated fat}$$

■ Based on your own estimated kcalorie needs (see Table 1-5 on p. 18), is your saturated fat intake 10 percent or less of total kcalories?

Estimated kcalorie need: _____ kcalories
Saturated fat recommendation: _____ grams
Your saturated fat intake: _____ grams

■ If not, list several foods you consumed that supplied large amounts of saturated fat and suggest alternative foods in the same food group that contain less.

■ If your food choices included packaged or fast foods that contain significant amounts of *trans* fats, suggest alternative foods in the same food group that contain minimal amounts.

Clinical Applications

1. The connection between the overconsumption of fats, especially saturated and *trans* fat, and chronic diseases (obesity, diabetes, cancer, and cardiovascular disease) underscores the importance of being alert to a client's fat intake. What advice would you offer a client who reports the following?

   ■ Eats two or more 6-ounce servings of meat each day.

   ■ Drinks whole milk and eats regular cheddar cheese each day.

   ■ Eats four to five servings of breads and cereals each day, including a bagel with cream cheese for breakfast, a bologna sandwich on white bread for lunch, and biscuits or cornbread with butter to accompany dinner.

   ■ Eats one serving of fruit each day and eats vegetables only on occasion.

2. Make a list of foods, beverages, and seasonings that your client can substitute for foods high in saturated fat.

Self Check

1. Three classes of lipids in the body are:
   a. triglycerides, fatty acids, and cholesterol.
   b. triglycerides, phospholipids, and sterols.
   c. fatty acids, phospholipids, and cholesterol.
   d. glycerol, fatty acids, and triglycerides.

2. A triglyceride consists of:
   a. three glycerols attached to a lipid.
   b. three fatty acids attached to a glucose.
   c. three fatty acids attached to a glycerol.
   d. three phospholipids attached to a cholesterol.

3. A fatty acid that has the maximum possible number of hydrogen atoms is known as a(n):
   a. saturated fatty acid.
   b. monounsaturated fatty acid.
   c. PUFA.
   d. essential fatty acid.

4. The difference between *cis-* and *trans-*fatty acids is:
   a. the number of double bonds.
   b. the length of their carbon chains.
   c. the location of the first double bond.
   d. the configuration around the double bond.

5. Essential fatty acids:
   a. are used to make substances that regulate blood pressure, among other functions.
   b. can be made from carbohydrates.
   c. include lecithin and cholesterol.
   d. cannot be found in commonly eaten foods.

6. Lecithins and other phospholipids in the body function as:
   a. emulsifiers.
   b. enzymes.
   c. temperature regulators.
   d. shock absorbers.

7. To minimize saturated fat intake and lower the risk of heart disease, most people need to:
   a. eat less meat.
   b. select fat-free milk.
   c. use nonhydrogenated margarines and cooking oils such as olive oil or canola oil.
   d. all of the above.

8. To include omega-3 fatty acids in the diet, the American Heart Association recommends eating:
   a. cholesterol-free margarine.
   b. fish oil supplements.
   c. hydrogenated margarine.
   d. at least two fish meals per week.

9. Some examples of foods with hidden fats are:
   a. cheese, lettuce, and fruit juices.
   b. fried foods, sauces, dips, and lunch meats.
   c. fish, rice, and potatoes.
   d. baked potatoes, vegetables, and fruits.

10. Generally speaking, vegetable and fish oils are rich in:
    a. polyunsaturated fat.
    b. saturated fat.
    c. cholesterol.
    d. *trans-*fatty acids.

Answers to these questions can be found in Appendix H.

Notes

1. G. Govindarajan, M. A. Alpert, and L. Tejwani, Endocrine and metabolic effects of fat: Cardiovascular implications, *American Journal of Medicine* 121 (2008): 366–370; T. Yamauchi and T. Kadowaki, Physiological and pathophysiological roles of adiponectin and adiponectin receptors in the integrated regulation of metabolic and cardiovascular diseases, *International Journal of Obesity* 32 (2008): S13–S18; T. Yamada and H. Katagiri, Avenues of communication between the brain and tissues/organs involved in energy homeostasis, *Endocrine Journal* 54 (2007): 497–505.

2. L. A. Smit, A. Baylin, and H. Campos, Conjugated linoleic acid in adipose tissue and risk of myocardial infarction, *American Journal of Clinical Nutrition* 92 (2010): 34–40; J. M. Chardigny and coauthors, Do *trans* fatty acids from industrially produced sources and from natural sources have the same effect on cardiovascular disease risk factors in healthy subjects? Results of the *trans* Fatty Acids Collaboration (TRANSFACT) study, *American Journal of Clinical Nutrition* 87 (2008): 558–566.

3. W. S. Harris and coauthors, Omega-6 fatty acids and risk for cardiovascular disease: A science advisory from the American Heart Association Nutrition Subcommittee of the Council on Nutrition, Physical Activity, and Metabolism; Council on Cardiovascular Nursing; and Council on Epidemiology and Prevention, *Circulation* 119 (2009): 902–907; P. C. Calder, n-3 Polyunsaturated fatty acids, inflammation, and inflammatory diseases, *American Journal of Clinical Nutrition* 83 (2006): 1505S–1519S.

4. S. M. Innis, Dietary (n-3) fatty acids and brain development, *Journal of Nutrition* 137 (2007): 855–859.

5. C. Agostoni, Role of long-chain polyunsaturated fatty acids in the first year of life, *Journal of Pediatric Gastroenterology and Nutrition* 47 (2008): S41–S44; E. E. Birch and coauthors, Visual acuity and cognitive outcomes at 4 years of age in a double-blind, randomized trial of long-chain polyunsaturated fatty acid-supplemented infant formula, *Early Human Development* 83 (2007): 279–284.

6. P. J. Smith and coauthors, Association between n-3 fatty acid consumption and ventricular ectopy after myocardial infarction, *American Journal of Clinical Nutrition* 89 (2009): 1315–1320; J. K. Virtanen and coauthors, Fish consumption and risk of major chronic disease in men, *American Journal of Clinical Nutrition* 88 (2008): 1618–1625; J. M. Norris and coauthors, Omega-3 polyunsaturated fatty acid intake and islet autoimmunity in children at increased risk for type 1 diabetes, *Journal of the American Medical Association* 298 (2007): 1420–1428; J. Shannon and coauthors, Erythrocyte fatty acids and breast cancer risk: A case-control study in Shanghai, China, *American Journal of Clinical Nutrition* 85 (2007): 1090–1097; H. E. Theobald and coauthors, Low-dose docosahexaenoic acid lowers diastolic blood pressure in middle-aged men and women, *Journal of Nutrition* 137 (2007): 973–978; W. C. Willett, The role of dietary n-6 fatty acids in the prevention of cardiovascular disease, *Journal of Cardiovascular Medicine* 8

(2007): S42–S45; C. Chrysohoou and coauthors, Long-term fish consumption is associated with protection against arrhythmia in healthy persons in a Mediterranean region—the ATTICA study, *American Journal of Clinical Nutrition* 85 (2007): 1385–1391; J. L. Breslow, n-3 fatty acids and cardiovascular disease, *American Journal of Clinical Nutrition* 83 (2006): 1477S–1482S.

7. S. B. Racette and coauthors, Dose effects of dietary phytosterols on cholesterol metabolism: A controlled feeding study, *American Journal of Clinical Nutrition* 91 (2010): 32–38; S. Klingberg and coauthors, Inverse relation between dietary intake of naturally occurring plant sterols and serum cholesterol in northern Sweden, *American Journal of Clinical Nutrition* 87 (2008): 993–1001.

8. S. Sieri and coauthors, Dietary fat and breast cancer risk in the European Prospective Investigation into Cancer and Nutrition, *American Journal of Clinical Nutrition* 88 (2008): 1304–1312; A. C. Thiébaut and coauthors, Dietary fat and postmenopausal invasive breast cancer in the National Institutes of Health—AARP Diet and Health Study cohort, *Journal of the National Cancer Institute* 99 (2007): 451–462; R. Prentice and coauthors, Low-fat dietary pattern and cancer incidence in the Women's Health Initiative Dietary Modification Randomized Controlled Trial, *Journal of the National Cancer Institute* 99 (2007): 1534–1543.

9. J. Shannon and coauthors, Erythrocyte fatty acids and breast cancer risk: A case-control study in Shanghai, China, *American Journal of Clinical*

Nutrition 85 (2007): 1090–1097; E. Theodoratou and coauthors, Dietary fatty acids and colorectal cancer: A case-control study, *American Journal of Epidemiology* 166 (2007): 181–195; R. S. Chapkin, D. N. McMurray, and J. R. Lupton, Colon cancer, fatty acids and anti-inflammatory compounds, *Current Opinion in Gastroenterology* 23 (2007): 48–54.

10. L. Van Horn and coauthors, The evidence for dietary prevention and treatment of cardiovascular disease, *Journal of the American Dietetic Association* 108 (2008): 287–331; Position of the American Dietetic Association and Dietitians of Canada: Dietary fatty acids, *Journal of the American Dietetic Association* 107 (2007): 1599–1611.

11. Standing Committee on the Scientific Evaluation of Dietary Reference Intakes, Food and Nutrition Board, Institute of Medicine, *Dietary Reference Intakes for Energy, Carbohydrate, Fiber, Fat, Fatty Acids, Cholesterol, Protein, and Amino Acids* (Washington, D.C.: National Academies Press, 2005), pp. 835–836.

12. S. K. Wallace and D. Mozaffarian, Trans-fatty acids and nonlipid risk factors, *Current Atherosclerosis Reports* 11 (2009): 423–433; D. Mozaffarian, A. Aro, and W. C. Willett, Health effects of trans-fatty acids: Experimental and observational evidence, *European Journal of Clinical Nutrition* 63 (2009): S5–S21; Standing Committee on the Scientific Evaluation of Dietary Reference Intakes, 2005, pp. 494–505.

13. D. B. Haytowitz, P. R. Pehrsson, and J. M. Holden, The National Food and Nutrient Analysis Program: A decade of progress, *Journal of Food Composition and Analysis* 21 (2008): S94–S102; M. J. Albers and coauthors, 2006 marketplace survey of trans-fatty acid content of margarines and butters, cookies and snack cakes, and savory snacks, *Journal of the American Dietetic Association* 108 (2008): 367–370.

14. S. Borra and coauthors, An update of trans-fat reduction in the American diet, *Journal of the American Dietetic Association* 107 (2007): 2048–2050.

15. S. Cicerale and coauthors, Chemistry and health of olive oil phenolics, *Critical Reviews in Food Science and Nutrition* 49 (2009): 218–236.

16. Cicerale and coauthors, 2009; J. Ruano and coauthors, Intake of phenol-rich virgin olive oil improves the postprandial prothrombotic profile in hypercholesterolemic patients, *American Journal of Clinical Nutrition* 86 (2007): 341–346; M. I. Covas and coauthors, The effect of polyphenols in olive oil on heart disease risk factors, *Annals of Internal Medicine* 145 (2006): 333–341.

17. Harris and coauthors, 2009.

18. N. D. Riediger and coauthors, A systemic review of the roles of n-3 fatty acids in health and disease, *Journal of the American Dietetic Association* 109 (2009): 668–679; P. P. Dimitrow and M. Jawien Pleiotropic, Cardioprotective effects of omega-3 polyunsaturated fatty acids, *Mini Reviews in Medicinal Chemistry* 9 (2009): 1030–1039; Van Horn and coauthors, 2008; Breslow, 2006.

19. A. H. Lichtenstein and coauthors, Diet and lifestyle recommendations revision 2006: A scientific statement from the American Heart Association Nutrition Committee, *Circulation* 114 (2006): 82–96.

20. Breslow, 2006.

21. P. M. Kris-Etherton and A. M. Hill, n-3 Fatty acids: Foods or supplements? *American Dietetic Association* 108 (2008): 1125–1130; Lichtenstein and coauthors, 2006.

22. Fish and Omega-3 Fatty Acids 2009, a document available at www.americanheart.org; R. Emsley and coauthors, Safety of the omega-3 fatty acid, eicosapentaenoic acid (EPA in psychiatric patients: Results from a randomized, placebo controlled trial, *Psychiatry Research* 161 (2008): 284–291.

23. Riediger and coauthors, 2009.

24. Standing Committee on the Scientific Evaluation of Dietary Reference Intakes, 2005, pp. 769–770.

25. Standing Committee on the Scientific Evaluation of Dietary Reference Intakes, 2005, pp. 797–802.

26. Standing Committee on the Scientific Evaluation of Dietary Reference Intakes, 2005, pp. 835–836.

27. U.S. Department of Agriculture and U.S. Department of Health and Human Services, *Dietary Guidelines for Americans 2010*, available online at www.dietaryguidelines.gov.

28. Position of the American Dietetic Association: Fat replacers, *Journal of the American Dietetic Association* 105 (2005): 266–275.

To consumers, advice about dietary fat appears to change almost daily. "Eat less fat." "Eat more fatty fish." "Give up butter—use margarine instead." "Give up margarine—replace it with olive oil." "Steer clear of saturated fats." "Seek out omega-3." "Stay away from *trans* fats." "Stick with mono- and polyunsaturated fats." No wonder people feel confused about dietary fat. This Nutrition in Practice begins with a look at the dietary guidelines for fat intake. It continues by identifying which foods provide which fats. It closes with strategies to help consumers choose the right amounts of the right kinds of fats for a healthy diet.

Why do today's fat messages seem to change constantly and become more confusing?

The confusion stems in part from the complexities of fat and in part from the nature of recommendations. As Chapter 3 explained, "dietary fat" refers to several kinds of fats; some fats support health, whereas others damage it; and foods typically provide a mixture of fats in varying proportions. It has taken researchers decades to sort through the relationships among the various kinds of fat and their roles in supporting or harming health. Translating these research findings into dietary recommendations is a challenging process. Too little information can mislead consumers, but too much detail can overwhelm them. As scientific understanding has grown, recommendations have evolved to become more specific. Recommendations may seem to "change constantly and become more confusing," but in fact they are becoming more meaningful.

How exactly have dietary recommendations for fat changed to become more meaningful for consumers?

Dietary recommendations for fat have changed by shifting the emphasis from lowering total fat, in general, to limiting saturated and *trans* fat, specifically. For decades, health experts urged consumers to limit total fat intake to 30 percent or less of energy intake. This advice was straightforward—cut the fat, improve your health. Health experts recognized that saturated fats and *trans* fats were the ones that raise blood cholesterol, but they reasoned that when total fat was limited, saturated and *trans* fat intake would decline as well. People were simply advised to cut back on all fat so that they would cut back on saturated and *trans* fat. Such advice may have oversimplified the message and unnecessarily restricted total fat.

But low-fat diets have been recommended for years to help people manage weight and reduce the risk of heart disease. Are you saying that low-fat diets are no longer recommended?

Low-fat diets remain a key recommendation in treatment plans for people with elevated blood lipids or heart disease and therefore are important in nutrition.[1] As for healthy people, evidence from around the world has led researchers to change population-wide recommendations from a "low-fat" to a "wise-fat" approach. Several problems accompany low-fat diets. For one, many people find low-fat diets difficult to maintain over time. For another, low-fat diets are not necessarily low-kcalorie diets; if energy intake exceeds energy needs, weight gain follows, and obesity brings a host of health problems, including heart disease. For another, diets high in refined carbohydrates, even if low in fat, can cause blood triglycerides to rise and HDL to fall, a deleterious combination for heart health.[2] Finally, taken to the extreme, a low-fat diet may exclude fatty fish, nuts, seeds, and vegetable oils—all valuable sources of many essential fatty acids, phytochemicals, vitamins, and minerals. Importantly, the fats from these sources protect against heart disease, as later sections explain.

How have today's recommendations for fat been revised?

Today, health experts have revised dietary recommendations to acknowledge that not all fats have damaging health consequences.[3] In fact, higher intakes of some kinds of fats (for example, the omega-3 fatty acids) support good health. Instead of urging people to cut back on all fats, current recommendations suggest carefully replacing the "bad" saturated and *trans* fats with the "good" unsaturated fats and enjoying these fats within kcalorie limits.[4] The goal is to create a diet moderate in kcalories that provides enough of the fats that support good health, but not too much of those that harm health. (Turn to pp. 66–69 for a review of the health consequences of each type of fat.)

With these findings and goals in mind, the DRI committee concluded that a diet containing 20 to 35 percent of energy intake from fat, but reduced in saturated fat and *trans* fat and moderate in energy, is compatible with low rates of heart disease, diabetes, obesity, and cancer.[5] The *Dietary Guidelines for Americans* make clear that the human body has no need of dietary saturated fat or *trans* fat, so the less consumed the better, as long as the diet is adequate in nutrients, including the essential fatty acids.[6]

How can people distinguish between the fats in foods that support health and those that might harm it?

Asking consumers to limit their total fat intake was less than perfect advice, but it was straightforward—find the fat and cut back. Asking consumers to keep their intakes of saturated fats, *trans* fats, and cholesterol low and to use monounsaturated and polyunsaturated fats instead may be more on target with heart health, but it also makes diet planning more complicated. To make appropriate selections, consumers must first learn which foods contain which fats. For example, avocados, bacon, walnuts, potato chips, and mackerel are all high-fat foods, yet some of these foods have detrimental effects on heart health when consumed in excess, and others seem neutral or even beneficial.

Is there evidence to clarify why some high-fat foods are compatible with a heart-healthy diet and others are not?

Yes. The traditional diets of Greece and other countries in the Mediterranean region are exemplary in their use of "good" fats, especially olives and olive oil. A classic study of the world's people, the Seven Countries Study, found that death rates from heart disease were strongly associated with diets high in saturated fats, but only weakly linked with total fat.[7] In fact, the two countries with the highest fat intakes, Finland and the Greek island of Crete, had the highest (Finland) and lowest (Crete) rates of heart disease deaths. In both countries, the people consumed 40 percent or more of their kcalories from fat. Clearly, a high-fat diet was not the primary problem, so researchers refocused their attention on the type of fat. They found that Cretans ate diets high in olive oil but low in saturated fat (less than 10 percent of kcalories), a pattern they linked with relatively low disease risks. Many studies that followed yielded similar results—people who eat "Mediterranean-type" diets have low rates of heart disease, some cancers, and other chronic diseases, and their life expectancy is high.[8] Unfortunately, many busy Mediterranean people today, especially the young, are trading labor-intensive traditional diets for convenient and fast Western-style foods. At the same time, their health advantages are rapidly disappearing.[9]

When olive oil replaces saturated fats, such as those of butter, coconut or palm oil, hydrogenated stick margarine, lard, or shortening, it may offer numerous health benefits.[10] Olive oil helps to protect against heart disease by:

▌ Lowering total and LDL cholesterol and not lowering HDL cholesterol or raising triglycerides.[11]

▌ Reducing LDL cholesterol's susceptibility to oxidation.[12]

▌ Lowering blood-clotting factors.[13]

▌ Providing phytochemicals that act as antioxidants (see Nutrition in Practice 8).[14]

▌ Lowering blood pressure.[15]

▌ Interfering with the inflammatory response.[16]

When compared with other fats, olive oil seems to be a wise choice, but controlled clinical trials are too scarce to support population-wide recommendations to switch to a high-fat diet rich in olive oil. Importantly, olive oil is not a magic potion; drizzling it on foods does not make them healthier. Like other fats, olive oil delivers 9 kcalories per gram, which can contribute to weight gain in people who fail to balance their energy intake with their energy output. Its role in a healthy diet is to *replace* the saturated fats. When choosing olive oils, select the darker "extra virgin" kind because it contains the highest levels of potentially beneficial phytochemicals.

Other vegetable oils, such as canola oil or safflower oil, in their liquid unhydrogenated states are also generally low in saturated fats and high in unsaturated fats. Such oils, when they replace solid, saturated fats in the diet, preserve heart health.

Good, olive oil may help protect against heart disease; are there other food fats that may also be protective?

Possibly so. People who eat a 1-ounce serving of nuts on five or more days a week have lower LDL cholesterol and a reduced risk

Olives and their oil may benefit heart health.

of heart disease compared with people who consume no nuts.[17] Even in women with diabetes, whose risks are high, nuts (5 ounces per week) or peanut butter (5 tbs per week) were associated with reduced heart disease risk.[18]

The nuts under study are those commonly eaten in the United States: almonds, Brazil nuts, cashews, hazelnuts, macadamia nuts, pecans, pistachios, walnuts, and, as mentioned, peanut butter.[19] On average, these nuts contain mostly monounsaturated fat (59 percent), some polyunsaturated fat (27 percent), and little saturated fat (14 percent).

Research has shown a benefit from walnuts and almonds in particular. In study after study, walnuts, when substituted for other fats in the diet, produce favorable effects on blood lipids—even in people with elevated total and LDL cholesterol.[20] Results are similar for almonds.[21]

Studies on peanuts, macadamia nuts, pecans, and pistachios follow suit, indicating that including nuts may be a wise strategy against heart disease. Nuts may protect against heart disease because they provide:

▌ Monounsaturated and polyunsaturated fats in abundance but few saturated fats.

▌ Fiber, vegetable protein, and other valuable nutrients, including the antioxidant vitamin E.

▌ Phytochemicals that act as antioxidants (see Nutrition in Practice 8).

▌ Plant sterols.

In addition to their heart benefits, nuts may also benefit other body organs—people who frequently consume nuts and other healthy fats suffer fewer gallbladder problems.[22]

Tree nuts and peanuts once had no place in a low-fat or low-kcalorie diet—and for good reason. Nuts provide up to 80 percent of their kcalories from fat, and a quarter cup (about an ounce) of mixed nuts provides more than 200 kcalories. However, today's research does not support greater weight gain in those who consume nuts, so even dieters may choose to include a few nuts each day for their potential health benefits.[23]

81

Stay mindful of kcalories when snacking on nuts.

If olive oil, nuts, and fatty fish are protective against heart disease, which fats are harmful?

The number one dietary determinant of LDL cholesterol is saturated fat. Figure NP3-1 shows that each 1 percent increase in energy from saturated fatty acids in the diet may produce a 2 percent jump in heart disease risk by elevating blood LDL cholesterol. Conversely, reducing saturated fat intake by 1 percent can be expected to produce a 2 percent drop in heart disease risk by the same mechanism. Even a 2 percent drop in LDL represents a significant improvement for the health of the heart.[26] Like saturated fats, *trans* fats also raise heart disease risk by elevating LDL cholesterol. A heart-healthy diet limits foods that are rich in these two types of fat.

Which foods are highest in saturated and *trans* fats?

The major sources of saturated fats in the U.S. diet are fatty meats, whole-milk products, tropical oils, and products made from any of these foods. More than a third of the fat in most meats is saturated. Similarly, more than half of the fat is saturated in whole milk and other high-fat dairy products, such as cheese, butter, cream, half-and-half, cream cheese, sour cream, and ice cream. Consumers rarely use the tropical oils of palm, palm kernel, and coconut in the kitchen, but these oils are used heavily by food manufacturers and so are commonly found in many commercially prepared foods.

When choosing meats, milk products, and commercially prepared foods, look for those lowest in saturated fat. Labels provide a useful guide for comparing products in this regard, and Appendix A lists the saturated fat in several thousand foods.

Even with careful selections, a nutritionally adequate diet will provide some saturated fat. Zero saturated fat is not possible even when experts design menus with the mission of keeping saturated fat as low as possible.[27] However, diets based on fruits, vegetables, legumes, nuts, soy products, and whole grains can, and often do, deliver less saturated fat than diets that depend heavily on animal-derived foods.

As for *trans* fats, Chapter 3 explained that solid shortening and margarine are made from vegetable oil that has been hardened through hydrogenation. This process both saturates some of the

What about fish? I have a friend whose doctor told her that eating fish is good for the heart. Is this true?

Yes. The preceding chapter made clear that fish oils hold the potential to improve health, particularly heart health. Research studies have provided strong evidence that increasing omega-3 fatty acids in the diet supports heart health and lowers the risk of death from heart disease.[24] For this reason, the American Heart Association and other authorities recommend including two fatty fish servings a week in a heart-healthy diet. People who eat some fish each week can lower their risks of heart attack and stroke.[25] Table 3-3 on page 68 ranks commonly eaten fish by their omega-3 fatty acid content.

Fish is the best source of EPA and DHA in the diet, but it is also a major source of mercury and other environmental contaminants. Most fish contain at least trace amounts of mercury, but tilefish, swordfish, king mackerel, marlin, and shark have especially high levels. Freshwater fish may contain PCBs and other pollutants, so local advisories warn sport fishers of species that can pose problems. The chapter listed safer species of fish. To minimize risks while obtaining fish benefits, vary your choices among fatty fish species often.

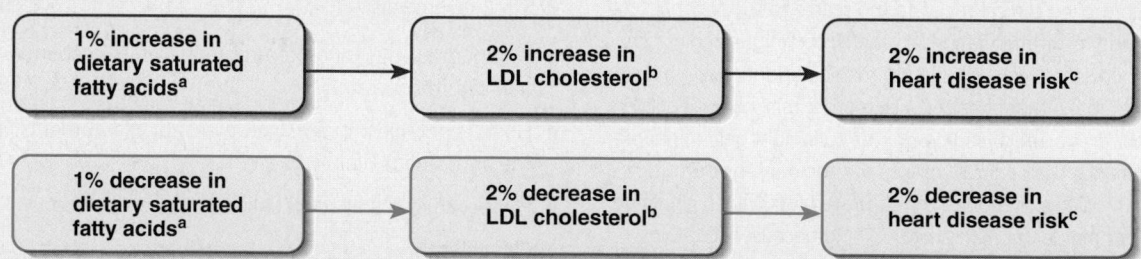

[a] Percentage of change in total dietary energy from saturated fatty acids.
[b] Percentage of change in blood LDL cholesterol.
[c] Percentage of change in an individual's risk of heart disease; the percentage of change in risk may increase when blood lipid changes are sustained over time.

FIGURE NP3-1 Potential Relationships among Dietary Saturated Fatty Acids, LDL Cholesterol, and Heart Disease Risk

Source: Third Report of the National Cholesterol Education Program (NCEP) Expert Panel on Detection, Evaluation, and Treatment of High Blood Cholesterol in Adults (Adult Treatment Panel III), NIH publication no. 02-5215 (Bethesda, Md.: National Heart, Lung, and Blood Institute, 2002), p. V-8 and II-4.

unsaturated fatty acids and introduces *trans*-fatty acids. Many convenience foods contain *trans* fats. Table NP3-1 summarizes which foods provide which fats. Substituting unsaturated fats for saturated fats at each meal and snack can help protect against heart disease. Table NP3-2 provides several examples and shows how such substitutions can lower saturated fat and raise unsaturated fat—even when total fat and kcalories remain unchanged.

So it seems that some fats are "good" and others are "bad" from the body's point of view. Is that right?

The saturated and *trans* fats do indeed seem mostly bad for the heart. Aside from providing energy, which unsaturated fats can do equally well, saturated and *trans* fats bring no indispensable benefits to the body. Furthermore, no harm can come from consuming diets low in them.

In contrast, the unsaturated fats are mostly good for the heart when consumed in moderation. To date, their one proven fault seems to be that they, like all fats, provide abundant energy to the body and so may promote obesity if they drive kcalorie intakes higher than energy needs.[28] Obesity, in turn, often begets many body ills, as Chapter 6 makes clear.

When judging foods by their fatty acids, keep in mind that the fat in foods is a mixture of "good" and "bad," providing both saturated and unsaturated fatty acids. Even predominantly monounsaturated olive oil delivers some saturated fat. Consequently, even when a person chooses foods with mostly unsaturated fats, saturated fat can still add up if total fat is high. For this reason, fat

---

TABLE NP3-1 Food Sources of Fatty Acids

Healthful Fatty Acids

Monounsaturated	Omega-6 polyunsaturated	Omega-3 polyunsaturated
Avocado	Margarine (nonhydrogenated)	Fatty fish (herring, mackerel, salmon, tuna)
Oils (canola, olive, peanut, sesame)	Oils (corn, cottonseed, safflower, soybean)	Flaxseed
Nuts (almonds, cashews, filberts, hazelnuts, macadamia nuts, peanuts, pecans, pistachios)	Nuts (pine nuts, walnuts)	Nuts (walnuts)
Olives	Mayonnaise	
Peanut butter	Salad dressing	
Seeds (sesame)	Seeds (pumpkin, sunflower)	

Harmful Fatty Acids

Saturated	Trans
Bacon	Fried foods (hydrogenated shortening)
Butter	Margarine (hydrogenated or partially hydrogenated)
Chocolate	Nondairy creamers
Coconut	Many fast foods
Cream cheese	Shortening
Cream, half-and-half	Commercial baked goods (including doughnuts, cakes, cookies)
Lard	Many snack foods (including microwave popcorn, chips, crackers)
Meat	
Milk and milk products (whole)	
Oils (coconut, palm, palm kernel)	
Shortening	
Sour cream	

Note: Keep in mind that foods contain a mixture of fatty acids. See Figure 3-6 on page 75.

TABLE NP3-2 Choosing Unsaturated Fat Instead of Saturated Fat

Examples of ways to replace saturated fats with unsaturated fats include sautéing foods in olive oil instead of butter, garnishing salads with sunflower seeds instead of bacon, snacking on mixed nuts instead of potato chips, using avocado instead of cheese on a sandwich, and eating salmon instead of steak. Portion sizes have been adjusted so that each of these foods provides approximately 100 kcalories. Notice that, for a similar number of kcalories and grams of fat, the first choices offer less saturated fat and more unsaturated fat.

Foods (100 kcal portions)	Saturated Fat (g)	Unsaturated Fat (g)	Total Fat (g)
Olive oil 1 tbs vs. butter 1 tbs	2 vs. 7	9 vs. 4	11 vs. 11
Sunflower seeds 2 tbs vs. bacon 2 slices	1 vs. 3	7 vs. 6	8 vs. 9
Mixed nuts 2 tbs vs. potato chips 10 chips	1 vs. 2	8 vs. 5	9 vs. 7
Avocado 6 slices vs. cheese 1 slice	2 vs. 4	8 vs. 4	10 vs. 8
Salmon 2 oz vs. steak 1 oz	1 vs. 2	3 vs. 3	4 vs. 5
Totals	**7 vs. 18**	**35 vs. 22**	**42 vs. 40**

must be kept below 35 percent of total kcalories if the diet is to be moderate in saturated fat.

Basing one's diet on vegetables, fruits, and legumes as part of a balanced daily diet is a good idea, as is *replacing* saturated fats such as butter, shortening, and meat fat with unsaturated fats like olive oil and the oils from nuts and fish. These foods provide vitamins, minerals, and phytochemicals—all valuable in protecting the body's health. To further protect your health, you may want to reduce fats from convenience foods and fast foods; choose small portions of meats, fish, and poultry; and include fresh foods from all the groups each day. Take care to select portion sizes that will best meet your energy needs. Also, be physically active each day.

Notes

1. A. H. Lichtenstein and coauthors, Diet and lifestyle recommendations revision 2006: A scientific statement from the American Heart Association Nutrition Committee, *Circulation* 114 (2006): 82–96.

2. A. H. Lichtenstein, Dietary fat, carbohydrate, and protein: Effects on plasma lipoprotein patterns, *Journal of Lipid Research* 47 (2006): 1661–1667.

3. S. S. Gidding and coauthors, Implementing American Heart Association Pediatric and Adult Nutrition Guidelines: A scientific statement from the American Heart Association Nutrition Committee of the Council on Nutrition, Physical Activity and Metabolism, Council on Cardiovascular Disease in the Young, Council on Arteriosclerosis, Thrombosis and Vascular Biology, Council on Cardiovascular Nursing, Council on Epidemiology and Prevention, and Council for High Blood Pressure Research, *Circulation* 119 (2009): 1161–1175; M. U. Jakobsen and coauthors, Major types of dietary fat and risk of coronary heart disease: A pooled analysis of 11 cohort studies, *American Journal of Clinical Nutrition* 89 (2009): 1425–1432; L. Van Horn and coauthors, The evidence for dietary prevention and treatment of cardiovascular disease, *Journal of the American Dietetic Association* 108 (2008): 287–331.

4. Lichtenstein and coauthors, 2006; U.S. Department of Agriculture and U.S. Department of Health and Human Services, *Dietary Guidelines for Americans 2010*, available online at www.dietaryguidelines.gov; Standing Committee on the Scientific Evaluation of Dietary Reference Intakes, Food and Nutrition Board, Institute of Medicine, *Dietary Reference Intakes for Energy, Carbohydrate, Fiber, Fat, Fatty Acids, Cholesterol, Protein, and Amino Acids* (Washington, D.C.: National Academies Press, 2005).

5. Standing Committee on the Scientific Evaluation of Dietary Reference Intakes, 2005, pp. 769–770.

6. U.S. Department of Agriculture and U.S. Department of Health and Human Services, 2010.

7. A. Keys, *Seven Countries: A Multivariate Analysis of Death and Coronary Heart Disease* (Cambridge, MA: Harvard University Press, 1980).

8. A. Trichopoulou, C. Bamia, and D. Trichopoulos, Anatomy of health effects of Mediterranean diet: Greek EPIC prospective cohort study, *British Medical Journal* 338 (2009), doi:10.1136/bmj.b2337; F. Sofi, The Mediterranean diet revisited: Evidence of its effectiveness grows, *Current Opinion in Cardiology* 24 (2009): 442–446; M. C. Sparling and J. J. B. Anderson, The Mediterranean diet and cardiovascular diseases, *Nutrition Today* 44 (2009): 124–133.

9. F. Bellisle, Infrequently asked questions about the Mediterranean diet, *Public Health Nutrition* 12 (2009): 1644–1647; P. A. Gilbert and S. Khokhar, Changing dietary habits of ethnic groups in Europe and implications for health, *Nutrition Reviews* 66 (2008): 203–215.

10. M. I. Covas, V. Konstantinidou, and M. Fitó, Olive oil and cardiovascular health, *Journal of Cardiovascular Pharmacology* 54 (2009): 477–482; S. Cicerale and coauthors, Chemistry and health of olive oil phenolics, *Critical Reviews in Food Science and Nutrition* 49 (2009): 218–236.

11. E. Waterman and B. Lockwood, Active components and clinical applications of olive oil, *Alternative Medicine Review* 12 (2007): 331–342; M. I. Covas and coauthors, The effect of polyphenols in olive oil on heart disease risk factors, *Annals of Internal Medicine* 145 (2006): 333–341.

12. M. Fitó and coauthors, Effect of a traditional Mediterranean diet on lipoprotein oxidation, *Archives of Internal Medicine* 167 (2007): 1195–1203.

13. J. López-Miranda, Monounsaturated fat and cardiovascular risk, *Nutrition Reviews* 64 (2006): S2–S12.

14. Cicerale and coauthors, 2009; J. Ruano and coauthors, Intake of phenol-rich virgin olive oil improves the postprandial prothrombotic profile in hypercholesterolemic patients, *American Journal of Clinical Nutrition* 86 (2007): 341–346; Covas and coauthors, 2006.

15. Waterman and Lockwood, 2007; B. M. Rasmussen and coauthors, Effects of dietary saturated, monounsaturated, and n-3 fatty acids on blood pressure in healthy subjects, *American Journal of Clinical Nutrition* 83 (2006): 221–226.

16. F. Pérez-Jiménez and coauthors, The influence of olive oil on human health: Not a question of fat alone, *Molecular Nutrition and Food Research* 51 (2007): 1199–1208.

17. A. E. Griel and P. M. Kris-Etherton, Tree nuts and the lipid profile: A review of clinical studies, *British Journal of Nutrition* 96 (2006): S68–S78; J. H. Kelly and J. Sabaté, Nuts and coronary heart disease: An epidemiological perspective, *British Journal of Nutrition* 96 (2006): S61–S67.

18. T. Y. Li and coauthors, Regular consumption of nuts is associated with a lower risk of cardiovascular disease in women with type 2 diabetes, *Journal of Nutrition* 139 (2009): 1333–1338.

19. J. Sabaté and Y. Ang, Nuts and health outcomes: New epidemiologic evidence, *American Journal of Clinical Nutrition* 89 (2009): 1643S–1648S; S. K. Gebauer and coauthors, Effects of pistachios on cardiovascular disease risk factors and potential mechanisms of action: A dose-response study, *American Journal of Clinical Nutrition* 88 (2008): 651–659.

20. S. Rajaram and coauthors, Walnuts and fatty fish influence different serum lipid fractions in normal to mildly hyperlipidemic individuals: A randomized controlled study, *American Journal of Clinical Nutrition* 89 (2009): 1657S–1663S; D. K. Banel and F. B. Hu, Effects of walnut consumption on blood lipids and other cardiovascular risk factors: A meta-analysis and systematic review, *American Journal of Clinical Nutrition* 90 (2009): 56–63.

21. P. M. Kris-Etherton, W. Karmally, and R. Ramakrishnan, Almonds lower LDL cholesterol, *Journal of the American Dietetic Association* 109 (2009): 1521–1522; O. J. Phung and coauthors, Almonds have a neutral effect on serum lipid profile: A meta-analysis of randomized trials, *Journal of the American Dietetic Association* 109 (2009): 865–873; S. Rajaram, K. M. Connell, and J. Sabaté, Effect of almond-enriched high-monounsaturated fat diet on selected markers of inflammation: A randomized, controlled, crossover study, *British Journal of Nutrition,* 103 (2010): 907-912.

22. Sabaté and Ang, 2009.

23. M. Bes-Rastrollo and coauthors, Prospective study of nut consumption, long-term weight change, and obesity risk in women, *American Journal of Clinical Nutrition* 89 (2009): 1913–1919; Sabaté and Ang, 2009; S. Rajaram and J. Sabaté, Nuts, body weight and insulin resistance, *British Journal of Nutrition* 96 (2006): S79–S86.

24. N. D. Riediger and coauthors, A systemic review of the roles of n-3 fatty acids in health and disease, *Journal of the American Dietetic Association* 109 (2009): 668–679; J. K. Virtanen and coauthors, Fish consumption and risk of major chronic disease in men, *American Journal of Clinical Nutrition* 88 (2008): 1618–1625; M. K. Duda and coauthors, Fish oil, but not flaxseed oil, decreases inflammation and prevents pressure overload-induced cardiac dysfunction, *Cardiovascular Research* 81 (2009): 319–327; C. Chrysohoou and coauthors, Long-term fish consumption is associated with protection against arrhythmia in healthy persons in Mediterranean region—the ATTICA study, *American Journal of Clinical Nutrition* 85 (2007): 1385–1391; D. S. Kelley and coauthors, Docosahexaenoic acid supplementation improves fasting and postprandial lipid profiles in hypertriglyceridemic men, *American Journal of Clinical Nutrition* 86 (2007): 324–333; J. L. Breslow, n-3 fatty acids and cardiovascular disease, *American Journal of Clinical Nutrition* 86 (2006): 1477S–1482S.

25. T. L. Psota, S. K. Gebauer, and P. Kris-Etherton, Dietary omega-3 fatty acid intake and cardiovascular risk, *American Journal of Cardiology* 98 (2006): 3i–18i; D. Mozaffarian, R. N. Lemaitre, and L. H. Kuller, Fish consumption and stroke risk in elderly individuals, *Archives of Internal Medicine* 165 (2005): 200–206.

26. *Third Report of the National Cholesterol Education Program (NCEP) Expert Panel on Detection, Evaluation, and Treatment of High Blood Cholesterol in Adults (Adult Treatment Panel III)*, NIH publication no. 025215 (Bethesda, MD.: National Heart, Lung, and Blood Institute, 2002), p. V-8.

27. Standing Committee on the Scientific Evaluation of Dietary Reference Intakes, 2005, p. 835.

28. Standing Committee on the Scientific Evaluation of Dietary Reference Intakes, 2005, pp. 796–797.

4

Protein

People think of proteins as bodybuilding nutrients, the material of strong muscles, and rightly so. No new living tissue can be built without them. Some proteins form structures such as muscle, bone, skin, and other tissues. Other proteins do the cells' work. The energy to fuel that work comes primarily from carbohydrates and fats.

The Chemist's View of Proteins

Proteins are chemical compounds that contain the same atoms as carbohydrates and lipids—carbon (C), hydrogen (H), and oxygen (O)—but proteins are different in that they also contain nitrogen (N) atoms. These nitrogen atoms give the name *amino* (nitrogen containing) to the amino acids that form the links in the chains we call proteins.

The Structure of Proteins

About 20 different **amino acids** may appear in proteins.* All amino acids share a common chemical backbone, and it is these backbones that are linked together to form proteins. Each amino acid also carries a side group, which varies from one amino acid to another and is responsible for the molecule's characteristic size, shape, and electrical charge (see Figure 4-1). The side groups on amino acids are what make proteins so varied in comparison with either carbohydrates or lipids.

Protein Chains The 20 amino acids can be linked end-to-end in a virtually infinite variety of sequences to form proteins. When two amino acids bond together, the resulting structure is known as a **dipeptide.** Three amino acids bonded together form a **tripeptide.** As additional amino acids join the chain, the structure becomes a **polypeptide.** Most proteins are a few dozen to several hundred amino acids long.

Protein Shapes Polypeptide chains twist into complex shapes. Each amino acid has special characteristics that attract it to, or repel it from, the surrounding fluids and other amino acids. Because of these interactions, polypeptide chains fold and intertwine into intricate coils (see Figure 4-2 on p. 88) and other shapes. The amino acid sequence of a protein determines the specific way the chain will fold.

Protein Functions Dramatically different shapes of proteins enable them to perform different tasks in the body. Some, such as hemoglobin in the blood (see Figure 4-3 on p. 88), are globular in shape; some are hollow balls that can carry and store materials within them; and some, such as those that form tendons, are more than 10 times as long as they are wide, forming stiff, sturdy, rodlike structures.

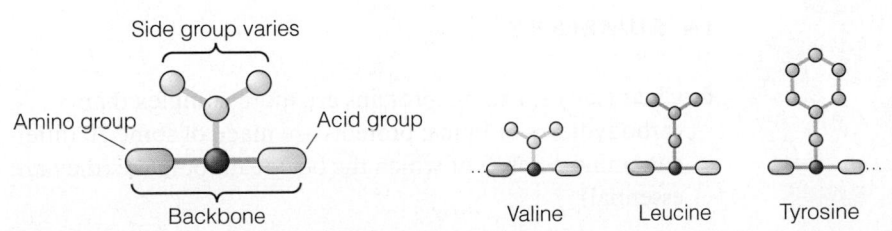

Side group varies

Amino group   Acid group

Backbone

All amino acids have a "backbone" made of an amino acid group (which contains nitrogen) and an acid group. The side group varies from one amino acid to the next.

Valine   Leucine   Tyrosine

Note that the side group is a unique structure that differentiates one amino acid from another.

FIGURE 4-1 Amino Acid Structure and Examples of Amino Acids

*Besides the 20 common amino acids, which can all be components of proteins, others occur individually (for example, ornithine).

proteins: compounds made from strands of amino acids composed of carbon, hydrogen, oxygen, and nitrogen atoms. Some amino acids also contain sulfur atoms.

amino (a-MEEN-oh) **acids:** building blocks of protein. Each contains an amino group, an acid group, a hydrogen atom, and a distinctive side group, all attached to a central carbon atom.
  amino = containing nitrogen

dipeptide: two amino acids bonded together.
  di = two
  peptide = amino acid

tripeptide: three amino acids bonded together.
  tri = three

polypeptide: 10 or more amino acids bonded together. An intermediate strand of between 4 and 10 amino acids is an *oligopeptide.*
  poly = many
  oligo = few

FIGURE 4-2 **The Coiling and Folding of a Protein Molecule**

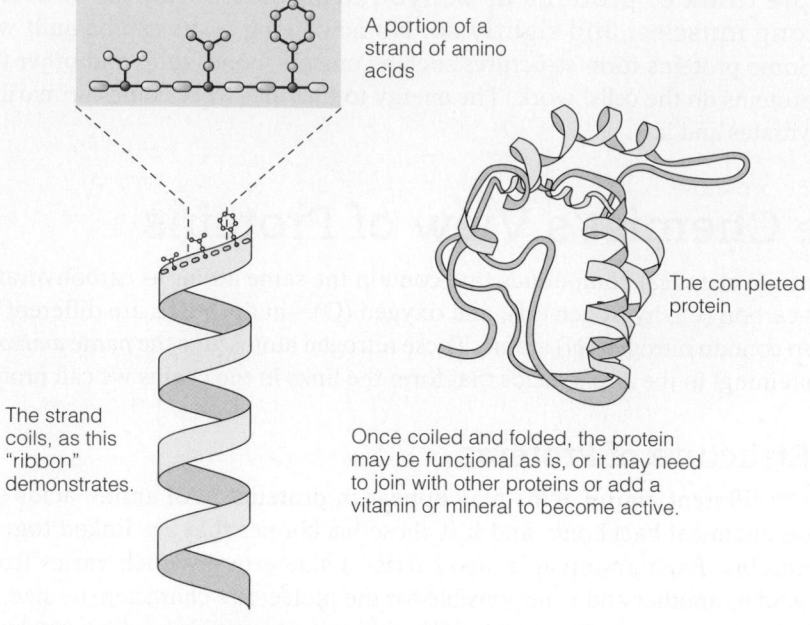

A portion of a strand of amino acids

The strand coils, as this "ribbon" demonstrates.

The completed protein

Once coiled and folded, the protein may be functional as is, or it may need to join with other proteins or add a vitamin or mineral to become active.

The strand of amino acids takes on a spring-like shape as its side groups variously attract and repel each other.

Essential Amino Acids

Proteins in foods do not provide body proteins directly, but rather they supply the amino acids from which the body makes its own proteins. More than half of the amino acids are **nonessential,** meaning that the body can make them for itself. Proteins in foods usually deliver these amino acids, but it is not essential that they do so. There are other amino acids that the body cannot make at all, however, and some that it cannot make fast enough to meet its needs. The proteins in foods must supply these nine amino acids to the body; they are therefore called **essential amino acids.**

Under special circumstances, a nonessential amino acid can become essential. For example, the body normally makes tyrosine (a nonessential amino acid) from the essential amino acid phenylalanine. If the diet fails to supply enough phenylalanine or if the body cannot make the conversion for some reason (as happens in the inherited disease phenylketonuria), then tyrosine becomes a **conditionally essential amino acid.**

Some researchers refer to essential amino acids as *indispensable* and to nonessential amino acids as *dispensable.*

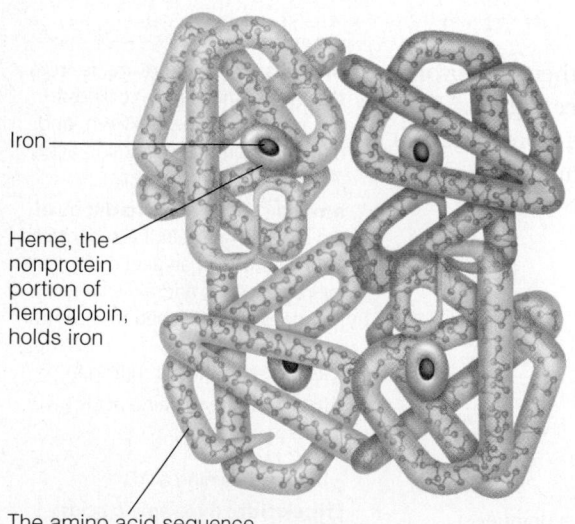

Iron

Heme, the nonprotein portion of hemoglobin, holds iron

The amino acid sequence determines the shape of the polypeptide chain

FIGURE 4-3 The Structure of Hemoglobin
Four highly folded polypeptide chains form the globular hemoglobin protein.

IN SUMMARY

▪ Chemically speaking, proteins are more complex than carbohydrates or lipids; proteins are made of some 20 different amino acids, 9 of which the body cannot make (they are essential).

▪ Each amino acid contains a central carbon atom with an amino group, an acid group, a hydrogen atom, and a unique side group attached to it.

▪ The distinctive sequence of amino acids in each protein determines its shape and function.

Protein Turnover and Nitrogen Balance

Within each cell of the body, proteins are continually being made and broken down, a process known as **protein turnover.** Amino acids must be continuously available to build the proteins of new tissues. The new tissues may be in an embryo, in the muscles of an athlete in training, in a growing child, in the scar tissue that heals wounds, or in new hair and nails.

Less obvious is the protein that helps replace worn-out cells and internal cell structures. For example, the millions of cells that line the intestinal tract live for three to five days; they are constantly being shed and must be replaced. The cells of the skin die and rub off, and new ones grow from underneath.

Protein Turnover

When proteins break down, their component amino acids are liberated within the cells or released into the bloodstream. Some of these amino acids are promptly recycled into other proteins. By reusing amino acids to build proteins, the body conserves and recycles a valuable commodity. Other amino acids are stripped of their nitrogen and used for energy. Each day, about a quarter of the body's available amino acids are irretrievably broken down and used for energy. For this reason, amino acids from food are needed daily to support the new growth and maintenance of cells.

Nitrogen Balance

Researchers use **nitrogen balance** studies to estimate protein requirements. In healthy adults, protein synthesis balances with protein degradation, and nitrogen intake from protein in food balances with nitrogen excretion in the urine, feces, and sweat. When nitrogen intake equals nitrogen output, a person is in nitrogen equilibrium, or zero nitrogen balance.

If the body synthesizes more than it degrades and adds protein, nitrogen status becomes positive. Nitrogen status is positive in growing infants, children, and adolescents; pregnant women; and people recovering from protein deficiency or illness; their nitrogen intake exceeds their nitrogen output. They are retaining protein in new tissues as they add blood, bone, skin, and muscle to their bodies.

If the body degrades more than it synthesizes and loses protein, nitrogen status becomes negative. Nitrogen status is negative in people who are starving or suffering other severe stresses such as burns, injuries, infections, and fever; their nitrogen output exceeds their nitrogen intake. During these times, the body loses nitrogen as it breaks down muscle and other body proteins for energy.

IN SUMMARY

▌ The process by which proteins are continually being made and broken down is known as protein turnover.

▌ The body needs dietary amino acids to grow new cells and to replace worn-out ones.

▌ When nitrogen intake equals nitrogen output, a person is in nitrogen equilibrium, or zero nitrogen balance.

Nitrogen balance:

▌ Nitrogen equilibrium (zero nitrogen balance): N in = N out

▌ Positive nitrogen: N in > N out

▌ Negative nitrogen: N in < N out

nonessential amino acids: amino acids that the body can synthesize.

essential amino acids: amino acids that the body cannot synthesize in amounts sufficient to meet physiological need. Nine amino acids are known to be essential for human adults:

histidine (HISS-tuh-deen)
isoleucine (eye-so-LOO-seen)
leucine (LOO-seen)
lysine (LYE-seen)
methionine (meh-THIGH-oh-neen)
phenylalanine (fen-il-AL-uh-neen)
threonine (THREE-oh-neen)
tryptophan (TRIP-toe-fane, TRIP-toe-fan)
valine (VAY-leen)

conditionally essential amino acid: an amino acid that is normally nonessential but must be supplied by the diet in special circumstances when the need for it becomes greater than the body's ability to produce it.

protein turnover: the continuous breakdown and synthesis of body proteins involving the recycling of amino acids.

nitrogen balance: the amount of nitrogen consumed (N in) as compared with the amount of nitrogen excreted (N out) in a given period of time. The laboratory scientist can estimate the protein in a sample of food, body tissue, or excreta by measuring the nitrogen in it.

Growing children end each day with more bone, blood, muscle, and skin cells than they had at the beginning of the day.

Roles of Body Proteins

What distinguishes you chemically from any other human being are minute differences in your particular body proteins (enzymes, antibodies, and others). These differences are determined by your proteins' amino acid sequences, which are written into the genes you inherited from your parents and ancestors. The genes direct the making of all the body's proteins.

The human body contains an estimated 30,000 or more different kinds of proteins. The roles of more than 3000 of these proteins are now known, although the number is growing rapidly with the recent surge in knowledge gained from sequencing the human genome. Only a few of the many roles proteins play are described here, but these should serve to illustrate proteins' versatility, uniqueness, and importance.

The *human genome* is the full set of chromosomes, including all of the genes and associated DNA. Nutritional genomics is the topic of Nutrition in Practice 14.

As Structural Components A great deal of the body's protein exists in muscle tissue, which allows the body to move. The amino acids of muscle protein can also be released when the need is dire, as in starvation. These amino acids are integral parts of the muscle structure, and their loss exacts a cost of functional protein, as a later section makes clear.

Other structural proteins confer shape and strength on bones, teeth, tendons, cartilage, blood vessels, and other tissues. These proteins exist in a stable form and are more resistant to breakdown than are the proteins of muscles.

As Enzymes Enzymes are catalysts that help assemble and disassemble compounds in the body. For example, enzymes in plant or animal cells put together pairs of sugars to make disaccharides and strands of sugars to make starch, cellulose, and glycogen. Different enzymes dismantle these and other compounds to free their constituent parts and release energy. As Figure 4-4 shows, enzymes themselves are not altered by the reactions they facilitate. All enzymes are proteins, and when amino acids have to be put together to make proteins, it is enzymes that put them together, too. In other words, these proteins can even make other proteins.

The protein story moves in a circle. To follow the circle in nutrition, start with a person eating food proteins. The food proteins are broken down by digestive enzymes,

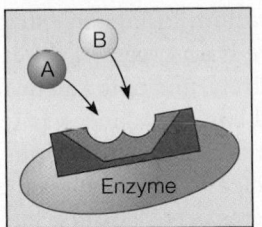 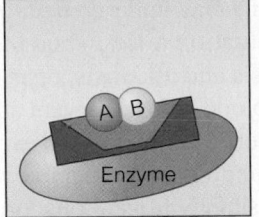 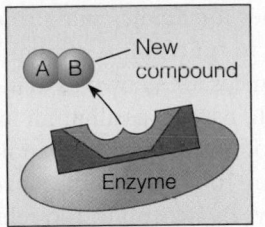

The separate compounds, A and B, are attracted to the enzyme's active site, making a reaction likely.

The enzyme forms a complex with A and B.

The enzyme is unchanged, but A and B have formed a new compound, AB.

FIGURE 4-4 Enzyme Action
Each enzyme facilitates a specific chemical reaction. In this diagram, an enzyme enables two compounds to make a more complex structure, but the enzyme itself remains unchanged.

proteins themselves, into amino acids. The amino acids enter the cells of the body, where other proteins (enzymes) put the amino acids together in long chains whose sequences are specified by the genes. The chains fold and twist back on themselves to form proteins, and some of these proteins become enzymes themselves. Some of these enzymes break apart compounds; others put compounds together. Day by day, in billions of reactions, these processes repeat themselves, and life goes on.

As Transporters A large group of proteins specialize in transporting other substances, such as lipids, vitamins, and minerals, around the body. To do their jobs, those substances must move from place to place within the blood, into and out of cells, or around the cellular interiors. Two familiar examples: the protein hemoglobin carries oxygen from the lungs to the cells, and the lipoproteins transport lipids in the watery blood.

As Regulators of Fluid and Electrolyte Balance Proteins help maintain the body's **fluid and electrolyte balance.** As Figure 4-5 shows, the body's fluids are contained in three major body compartments: (1) the spaces inside the blood vessels; (2) the spaces within the cells; and (3) the spaces between the cells (the interstitial spaces outside the blood vessels). Fluids flow back and forth between these compartments, and proteins in the fluids, together with minerals, help to maintain the needed distribution of these fluids.

Proteins are able to help determine the distribution of fluids in living systems for two reasons: first, proteins cannot pass freely across the membranes that separate the body compartments, and second, they are attracted to water. A cell that "wants" a certain amount of water in its interior space cannot move the water around directly, but it can manufacture proteins, and these proteins will hold water. Thus, the cell can use proteins to help regulate the distribution of water indirectly. Similarly, the body makes proteins for the blood and the interstitial (intercellular) spaces. These proteins help maintain the fluid volume in those spaces. Excess fluid accumulation in the interstitial spaces is called **edema.**

Not only is the quantity of the body fluids vital to life, but so is their composition. Special transport proteins in the membranes of cells continuously transfer substances into and out of cells to maintain balance. For example, sodium is concentrated outside the cells, and potassium is concentrated inside. The balance of these two minerals is critical to nerve transmission and muscle contraction. Any disturbance in this balance triggers a major medical emergency: imbalances can cause irregular heartbeats, kidney failure, muscular weakness, and even death.

As Regulators of Acid-Base Balance Proteins also help maintain the balance between **acids** and **bases** within the body's fluids. Normal body processes continually produce acids and bases, which must be carried by the blood to the kidneys and lungs for excretion. The blood must do this without upsetting its own **acid-base balance.** Blood **pH** is one of the most tightly controlled conditions in the body. If the blood

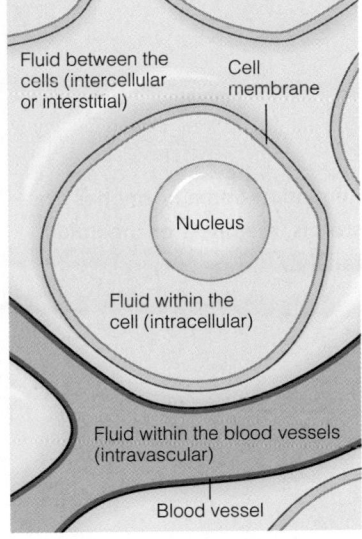

Fluid between the cells (intercellular or interstitial)

Cell membrane

Nucleus

Fluid within the cell (intracellular)

Fluid within the blood vessels (intravascular)

Blood vessel

FIGURE 4-5 One Cell and Its Associated Fluids

Minerals are helper nutrients. The attraction of protein and mineral particles to water is due to osmotic pressure (see Chapter 9).

enzymes: protein catalysts. A catalyst is a compound that facilitates chemical reactions without itself being changed in the process

fluid and electrolyte balance: maintenance of the necessary amounts and types of fluid and minerals in each compartment of the body fluids.

edema (eh-DEEM-uh): the swelling of body tissue caused by leakage of fluid from the blood vessels and accumulation of the fluid in the interstitial spaces.

acids: compounds that release hydrogen ions in a solution.

bases: compounds that accept hydrogen ions in a solution.

acid-base balance: the balance maintained between acid and base concentrations in the blood and body fluids.

pH: the concentration of hydrogen ions. The lower the pH, the stronger the acid. Thus, pH 2 is a strong acid; pH 6 is a weak acid; pH 7 is neutral; and a pH above 7 is alkaline.

becomes too acidic, vital proteins may undergo **denaturation,** losing their shape and ability to function. A similar situation arises when there is an excess of base. These imbalances are known as **acidosis** and **alkalosis,** respectively, and both can be fatal.

Proteins such as albumin in blood help to prevent acid-base imbalances. In a sense, the proteins protect one another by gathering up extra acid (hydrogen) ions when there are too many in the surrounding medium and by releasing them when there are too few. By accepting and releasing hydrogen ions, proteins act as **buffers,** maintaining the acid-base balance of the blood and body fluids.

As Antibodies Other proteins in the blood—the **antibodies**—defend against viruses, bacteria, and other disease agents. The antibodies work so efficiently that if a million bacterial cells are injected into the skin of a healthy person, fewer than 10 of those cells are likely to survive for five hours, which explains why most diseases never have a chance to get started. Without sufficient protein, however, the body cannot maintain its resistance to disease.

As Hormones The blood also carries messenger molecules known as **hormones,** and *some* hormones are proteins. Among the proteins that act as hormones are glucagon and insulin. Hormones have many profound effects, which will become evident in subsequent chapters.

As a Source of Energy and Glucose Without energy, cells die; without glucose, the brain and nervous system falter. Even though amino acids are needed to do the work that only they can perform—build vital proteins—they will be sacrificed to provide energy and glucose during times of starvation or insufficient carbohydrate intake.[1] When glucose or fatty acids are limited, cells are forced to use amino acids for energy and glucose. The body does not make a specialized storage form of protein as it does for carbohydrate and fat. Glucose is stored as glycogen in the liver and muscles, fat as triglycerides in the adipose tissue, but body protein is available only as the working and structural components of the tissues. When the need arises, the body dismantles its tissue proteins and uses them for energy. Thus, over time, energy deprivation (starvation) always incurs wasting of lean body tissue as well as fat loss.

IN SUMMARY

▌ As summarized in Table 4-1, an immense variety of proteins play numerous important roles in the body.

TABLE 4-1 Summary of Functions of Proteins

Structural components. Proteins form integral parts of most body tissues and confer shape and strength on bones, skin, tendons, and other tissues. Structural proteins of muscles allow movement.

Enzymes. Proteins facilitate chemical reactions.

Transporters. Proteins transport substances such as lipids, vitamins, minerals, and oxygen around the body.

Fluid and electrolyte balance. Proteins help to maintain the distribution and composition of various body fluids.

Acid-base balance. Proteins help maintain the acid-base balance of body fluids by acting as buffers.

Antibodies. Proteins act against infectious agents to fight diseases.

Hormones. Proteins regulate body processes. Some, but not all, hormones are made of protein.

Energy and glucose. Proteins provide some fuel, and glucose if needed, for the body's energy needs.

Reminder: Some hormones are sterols, members of the lipid family.

denaturation (dee-nay-cher-AY-shun): the change in a protein's shape brought about by heat, acid, or other agents. Past a certain point, denaturation is irreversible.

acidosis: acid accumulation in the blood and body fluids; depresses the central nervous system and can lead to disorientation and, eventually, coma.

alkalosis: excessive base in the blood and body fluids.

buffers: compounds that can reversibly combine with hydrogen ions to help keep a solution's acidity or alkalinity constant.

antibodies: large proteins of the blood and body fluids, produced in response to invasion of the body by unfamiliar molecules (mostly proteins) called *antigens*. Antibodies inactivate the invaders and so protect the body.

anti = against

hormones: chemical messengers. Hormones are secreted by a variety of glands in the body in response to altered conditions. Each travels to one or more target tissues or organs and elicits specific responses to restore normal conditions.

Protein and Health

During the time that scientists have been studying nutrition, no nutrient has been more intensely scrutinized than protein. As you know by now, it is indispensable to life. And it should come as no surprise that protein deficiency can have devastating effects on people's health. But, as with the other nutrients, protein in excess can be harmful, too, so this section also discusses the consequences of protein excess.

Protein-Energy Malnutrition

When people are deprived of food and suffer an energy deficiency, they degrade their own body protein for energy and indirectly suffer a protein deficiency as well. This combination—**protein-energy malnutrition (PEM)**—is the most widespread form of malnutrition in the world today. PEM is prevalent in Africa, Central America, South America, the Middle East, and East and Southeast Asia, but developed countries, including the United States, are not immune to it. PEM is common among some population groups in the United States: impoverished people living in inner cities and in rural areas; many elderly people; homeless children; and those suffering from the eating disorder anorexia nervosa.

Of all population groups, children are most seriously affected by malnutrition. Worldwide, three-fourths of those who die each year from starvation and related illnesses are children.[2] Children who are thin for their heights may have recently developed PEM, whereas children who are short for their ages may have experienced PEM for extended periods of time. Stunted growth due to PEM is easy to overlook because a small child may look quite normal, but it may be the most common sign of malnutrition in the developing countries. PEM takes two different forms, with some cases exhibiting a combination of the two. In one form, the person is shriveled and emaciated—this disease is called **marasmus.** In the second, a swollen belly and skin rash are present, and the disease is named **kwashiorkor.** In the combination, some features of each type are present. Marasmus reflects a severe deprivation of food over a long time (chronic PEM). Put simply, the person is starving and suffering from an inadequate energy *and* protein intake (and inadequate essential fatty acids, vitamins, and minerals as well). Kwashiorkor may result from even acute malnutrition, with too little protein to support body functions.

Marasmus Marasmus commonly occurs in children from 6 to 18 months of age in all the overpopulated and impoverished areas of the world. Children in impoverished nations subsist on a weak cereal drink that supplies scant energy and protein of low quality: such food can barely sustain life, much less support growth. Consequently, marasmic children look like little old people—just skin and bones.

Without adequate nutrition, muscles, including the heart muscle, waste and weaken. Because the brain normally grows to almost its full adult size within the first two years of life, marasmus impairs brain development and learning ability. Reduced synthesis of key hormones leads to a metabolism so slow that body temperature drops below normal. There is little or no fat under the skin to insulate against cold. Some hospital workers find that the primary need of marasmic children is to be clothed, covered, and kept warm.

The starving child faces this threat to life by engaging in as little activity as possible—not even crying for food. The body gathers all its forces to meet the crisis, so it cuts down on any expenditure of energy not needed for the heart, lungs, and brain to function. Growth ceases; the child is no larger at age four than at age two. Digestive enzymes are in short supply, the digestive tract lining deteriorates, and absorption fails. The child cannot assimilate what little food is eaten.

Blood proteins, including hemoglobin, are no longer synthesized. Antibodies to fight off invading bacteria are degraded to provide amino acids for other uses, rendering the child vulnerable to infection. Then **dysentery,** an infection of the digestive

PEM can be a consequence of many different conditions. PEM has been recognized in people with many chronic diseases such as cancer and AIDS and in those who have severe stresses such as burns or extensive infections (see Chapter 24). The consequences of PEM as a world malnutrition problem are considered here; the problems associated with PEM and illness are described throughout later chapters.

protein-energy malnutrition (PEM): a deficiency of protein and food energy; the world's most widespread malnutrition problem, including both marasmus and kwashiorkor.

mal = bad, poor

marasmus (ma-RAZZ-mus): the most common form of severe PEM before one year of age. Marasmus is characterized by generalized muscle wasting associated with extreme deprivation, or impaired absorption, of energy, protein, vitamins, and minerals.

kwashiorkor (kwash-ee-OR-core or kwash-ee-or-CORE): a severe form of PEM that occurs more frequently after 18 months of age. Kwashiorkor is characterized by failure to grow and develop, changes in the pigmentation of the hair and skin, edema, and fatty liver. Kwashiorkor is associated with inadequate protein intake and infections.

dysentery (DIS-en-terry): an infection of the gastrointestinal tract caused by an amoeba or bacterium that gives rise to severe diarrhea.

dys = bad
entery = intestine

tract, causes diarrhea, further depleting the body of nutrients. In the marasmic child, once infection has set in, kwashiorkor often follows and the immune response weakens further. The infection that occurs with malnutrition is responsible for two-thirds of the deaths of young children in developing countries.

If the condition is caught in time, the life of a starving child may be saved with rehydration and nutrition intervention.[3] In severe cases, children with diarrhea will have incurred dramatic fluid and mineral losses that need to be replaced during the first 24 to 48 hours to help raise the blood pressure and strengthen the heartbeat. After that, protein and energy may be given in *small* quantities several times a day, with intakes *gradually* increased as tolerated.[4] Years after PEM is corrected, however, a child may still experience deficits in thinking and achievement in school compared with well-nourished peers.

Kwashiorkor Kwashiorkor typically reflects a sudden and recent deprivation of food (acute PEM). *Kwashiorkor* is a Ghanaian word meaning a "sickness that infects the first child when the second child is born." If you consider how kwashiorkor often develops, you can easily see how the Ghanaians arrived at this name for the disease. When a mother who has been nursing her first child bears a second child, she weans the first child and puts the second one on the breast. The first child, suddenly switched from nutrient-dense, protein-rich breast milk to a starchy, protein-poor gruel, soon begins to sicken and die. Kwashiorkor typically sets in between 18 months and 2 years. Kwashiorkor usually develops rapidly as a result of protein deficiency or, more commonly, is precipitated by an illness such as measles or other infection.[5]

Some symptoms of kwashiorkor resemble those of marasmus (see Table 4-2). Proteins and hormones that previously maintained fluid balance diminish, and fluid leaks into the interstitial spaces. The child's limbs and belly become swollen with edema, a distinguishing feature of kwashiorkor; a **fatty liver** develops due to a lack of the protein carriers that transport fat out of the liver. The child's hair loses its color; the skin becomes patchy and scaly, sometimes with ulcers and sores that fail to heal.

Clearly, preventing marasmus and kwashiorkor from developing in the first place is more desirable than attempting to treat the conditions after they have set in. Experts

TABLE 4-2 Features of Marasmus and Kwashiorkor in Children

Separating PEM into two classifications oversimplifies the condition, but at the extremes, marasmus and kwashiorkor exhibit marked differences. Marasmus-kwashiorkor mix presents symptoms common to both marasmus and kwashiorkor. In all cases, children are likely to develop diarrhea, infections, and multiple nutrient deficiencies.

Marasmus	Kwashiorkor
Infancy (less than 2 yr)	Older infants and young children (1 to 3 yr)
Severe deprivation, or impaired absorption, of protein, energy, vitamins, and minerals	Inadequate protein intake or, more commonly, infections
Develops slowly; chronic PEM	Rapid onset; acute PEM
Severe weight loss	Some weight loss
Severe muscle wasting, with no body fat	Some muscle wasting, with retention of some body fat
Growth: <60% weight-for-age	Growth: 60 to 80% weight-for-age
No detectable edema	Edema
No fatty liver	Enlarged fatty liver
Anxiety, apathy	Apathy, misery, irritability, sadness
Good appetite possible	Loss of appetite
Hair is sparse, thin, and dry; easily pulled out	Hair is dry and brittle; easily pulled out; changes color; becomes straight
Skin is dry, thin, and easily wrinkles	Skin develops lesions

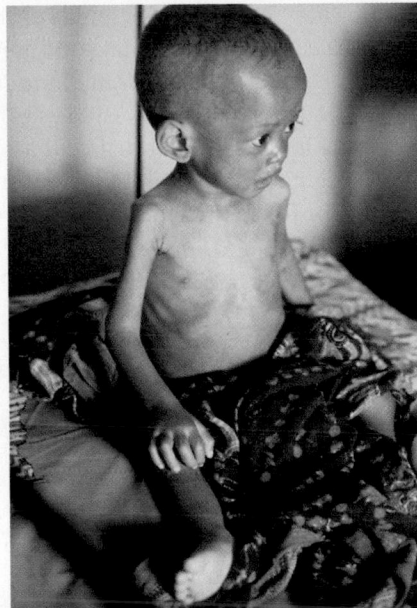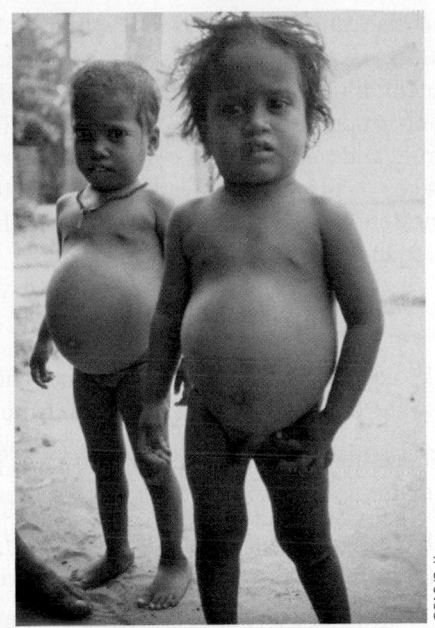

In the photo on the left, the extreme loss of muscle and fat characteristic of marasmus is apparent in the child's matchstick arms and legs. In contrast, the edema and enlarged liver characteristic of kwashiorkor are apparent in the swollen bellies of the children in the photo on the right.

assure us that we possess the knowledge, technology, and resources to end the hunger that leads to PEM. Programs that have involved the local people in the process of identifying the problem and devising its solution have met with some success. But until those who have the food, technology, and resources make fighting hunger a priority, the war on hunger will not be won.

Protein Excess

While many of the world's people struggle to obtain enough food and enough protein to survive, in the developed nations protein is so abundant that problems of protein excess are observed. Overconsumption of protein offers no benefits and may pose health risks for the heart, weakened kidneys, and the bones.[6] Selecting too many protein-rich foods, such as red meats and fat-containing milk products, adds saturated fat and crowds out fruits, vegetables, and whole grains, a problem from the standpoint of chronic disease risks.

Heart Disease Foods rich in animal protein can also be rich in saturated fats. Consequently, it is not surprising that people who take in a great deal of animal protein (red meats and dairy products) have a greater risk of heart disease.[7] In the United States, meat consumption often accompanies higher energy (kcalorie) intakes and increased abdominal body fatness, and excess abdominal fat raises the risk for heart disease.[8] As the Nutrition in Practice points out, people who substitute vegetable protein for animal protein lower their risk of dying from heart disease.[9]

Kidney Disease Excretion of the end products of protein metabolism depends, in part, on an adequate fluid intake and healthy kidneys. A high protein intake increases the work of the kidneys but does not appear to damage healthy kidneys or cause kidney disease.[10] A high-protein diet does worsen existing kidney disease and may accelerate a decline in only mildly impaired kidneys.[11] One of the most effective ways to slow the progression of kidney disease is to restrict dietary protein.

Adult Bone Loss When human subjects are given increasing doses of purified protein, they release increasingly larger amounts of calcium from the body into the urine.

fatty liver: an accumulation of fat in the liver. In PEM, fat accumulates in the liver because no protein is available to form the lipoproteins that normally escort fat molecules in the blood (see Chapter 19).

This fact raises concerns that a high-protein diet may cause or worsen adult bone loss. In a recent study, however, researchers concluded that, unlike intakes of purified protein, high protein intakes from whole food sources, such as meat or milk products, do not appear to increase calcium losses.[12] Other factors, such as low intakes of vitamin D and bone minerals, ultimately harm bone health to a greater degree than high intakes of protein.[13] In establishing protein recommendations, the DRI committee considered protein's effect on bone health but did not find sufficient evidence to set a Tolerable Upper Intake Level.[14]

IN SUMMARY

▌ Protein deficiencies arise from both energy-poor and protein-poor diets and lead to the devastating diseases of marasmus and kwashiorkor.

▌ Together, these diseases are known as protein-energy malnutrition (PEM), a major form of malnutrition causing death in children worldwide.

▌ Excesses of protein offer no advantage; in fact, overconsumption of protein-rich foods may cause health problems as well.

Protein and Amino Acid Supplements

Why do people take protein or amino acid supplements? Athletes often take them when trying to build muscle. Dieters may take them to spare their bodies' protein while losing weight. People take individual amino acid supplements too—to cure herpes, to make themselves sleep better, and to relieve pain and depression. Do protein and amino acid supplements really do these things? Probably not. Are they safe? Not always.

Protein Supplements Well-fed athletes do not need protein supplements. Dietary protein is necessary for building muscle tissue, and consuming protein in conjunction with resistance exercise helps muscles build new proteins.[15] Protein supplements, however, do not improve athletic performance beyond the gains from well-timed meals of ordinary foods.

Weight-loss dieters may benefit from consistently consuming protein-rich foods because protein often satisfies the appetite, while insufficient protein may increase it.[16] However, extra protein from powders, pills, or beverages is unlikely to dampen the appetite further, although it contributes unneeded kcalories. Evidence does not support taking protein supplements for weight loss.

Amino Acid Supplements Enthusiastic popular reports have led to widespread use of individual amino acids. Lysine, for example, is promoted to prevent or relieve the infections that cause herpes sores on the mouth or genital organs, but it does not cure herpes infections. Tryptophan—which plays a role as a precursor for the brain neurotransmitter serotonin, an important regulator of sleep, appetite, mood, and sensory perception—is advertised to relieve pain, depression, and insomnia. The DRI committee concludes that high doses of tryptophan may induce sleepiness, but they may also cause side effects, such as nausea and skin disorders.

The body is designed to handle whole proteins best. It breaks them into manageable pieces (dipeptides and tripeptides), then splits these, a few at a time, simultaneously releasing them into the blood. This slow bit-by-bit assimilation is ideal because groups of chemically similar amino acids compete for the carriers that absorb them into the blood. An excess of one amino acid can produce such a demand for a carrier that it limits the absorption of another amino acid, creating a temporary imbalance.[17]

The DRI committee reviewed the available research on amino acids, but with next to no safety research in existence, the committee was unable to set Tolerable Upper Intake Levels for supplemental doses. Until research becomes available, no level of

amino acid supplementation can be assumed to be safe for all people. Anyone considering taking amino acid supplements should be cautious not to exceed levels normally found in foods.[18]

Protein Recommendations and Intakes

The committee that established the RDA states that a generous daily protein allowance for a healthy adult is 0.8 gram per kilogram (2.2 pounds) of healthy body weight. The RDA covers the needs for replacing worn-out tissue, so it increases for larger people; it also covers the needs for building new tissue during growth, so is slightly higher for infants, children, and pregnant and lactating women.

In setting the RDA, the committee assumes that the protein eaten will be of good quality, that it will be consumed together with adequate energy from carbohydrate and fat, and that other nutrients in the diet will be adequate. The committee also assumes that the RDA will be applied only to healthy individuals with no unusual alteration of protein metabolism.

Research demonstrates that a median protein intake for U.S. adult males is about 16 percent of total kcalories, an amount that falls within the DRI suggested range of between 10 and 35 percent of kcalories.[19] Women, children, and some elderly people may typically take in less protein—13 to 15 percent. A small percentage of adolescent girls and elderly women consume insufficient protein or barely enough to meet their needs.[20]

IN SUMMARY

▮ Normal, healthy people do not need amino acid or protein supplements.

▮ Optimally, an adult's diet will be adequate in energy from carbohydrate and fat and will deliver at least 0.8 grams of protein per kilogram of healthy body weight each day.

Protein in Foods

In the United States and Canada, where nutritious foods are abundant, most people easily obtain enough protein to receive all the amino acids that they need. In countries where food is scarce and the people eat only marginal amounts of protein-rich foods, however, the *quality* of the protein becomes crucial.

Protein Quality

The protein quality of the diet determines, in large part, how well children grow and how well adults maintain their health. Put simply, **high-quality proteins** provide enough of all the essential amino acids needed to support the body's work, and low-quality proteins don't. Two factors influence protein quality—the protein's digestibility and its amino acid composition.

Digestibility As explained earlier, proteins must be digested before they can provide amino acids. **Protein digestibility** depends on such factors as the protein's source and the other foods eaten with it. The digestibility of most animal proteins is high (90 to 99 percent); plant proteins are less digestible (70 to 90 percent for most, but more than 90 percent for soy).

Amino Acid Composition To make proteins, cells must have all the needed amino acids available simultaneously. The liver can produce any nonessential amino acid that may be in short supply so that the cells can continue linking amino acids into protein strands. If an essential amino acid is missing, however, a cell must

---

The following groups of people are especially likely to suffer harm from amino acid supplements:

▮ All women of childbearing age
▮ Pregnant or lactating women
▮ Infants, children, and adolescents
▮ Elderly people
▮ People with certain inborn errors of metabolism
▮ Smokers
▮ People on low-protein diets
▮ People who are ill and take amino acids without medical supervision

RDA for protein (adult) = 0.8 g/kg
▮ 10 to 35 percent of energy intake

To calculate your protein need:
1. Find your body weight in pounds (or kilograms).
2. Convert pounds to kilograms, if necessary (pounds divided by 2.2).
3. Multiply kilograms by 0.8 to find total grams of protein recommended.

For example:
1. Weight = 150 lb
2. 150 lb ÷ 2.2 lb/kg = 68 kg
3. 68 kg × 0.8 g/kg = 54 g

high-quality proteins: dietary proteins containing all the essential amino acids in relatively the same amounts that human beings require. They may also contain nonessential amino acids.

protein digestibility: a measure of the amount of amino acids absorbed from a given protein intake.

Vegetarians obtain their protein from whole grains, legumes, nuts, vegetables, and, in some cases, eggs and milk products.

dismantle its own proteins to obtain it. Therefore, to prevent protein breakdown, dietary protein must supply at least the nine essential amino acids plus enough nitrogen-containing amino groups and energy for the synthesis of the others. If the diet supplies too little of any essential amino acid, protein synthesis will be limited. The body makes whole proteins only; if one amino acid is missing, the others cannot form a "partial" protein. An essential amino acid that is available in the shortest supply relative to the amount needed to support protein synthesis is called a **limiting amino acid.**

High-Quality Proteins A high-quality protein contains all the essential amino acids in amounts adequate for human use; it may or may not contain all the others. Generally, proteins derived from animal foods (meats, fish, poultry, cheese, eggs, yogurt, and milk) are high quality, although gelatin is an exception. Proteins derived from plant foods (legumes, grains, and vegetables) tend to be limiting in one or more essential amino acids. Some plant proteins are notoriously low quality—for example, corn protein. Others are high quality—for example, soy protein. As discussed in Nutrition in Practice 4, the educated vegetarian can design a diet that is adequate in protein by choosing a variety of legumes, whole grains, nuts, and vegetables. Table 4-3 lists the protein contents of foods based on the food groups of the USDA Food Guide in Chapter 1. Fruits are not included in Table 4-3 because they contribute only small amounts of protein.

TABLE 4-3 Protein-Containing Foods

Milk and Milk Products

Each of the following provides about 8 grams of protein:

- 1 c milk, buttermilk, or yogurt (choose low fat or fat free)
- 1 oz regular cheese (for example, cheddar or Swiss; choose low fat)
- ¼ c cottage cheese (choose low fat or fat free)

Protein Foods

Each of the following provides about 7 grams of protein:

- 1 oz meat, poultry, or fish (choose lean meats to limit saturated fat intake)
- ½ c legumes (navy beans, pinto beans, black beans, lentils, soybeans, and other dried beans and peas)
- 1 egg
- ½ c tofu (soybean curd)
- 2 tbs peanut butter
- 1 to 2 oz nuts or seeds

Grains

Each of the following provides about 3 grams of protein:

- 1 slice of bread
- ½ c cooked rice, pasta, cereals, or other grain foods

Vegetables

Each of the following provides about 2 grams of protein:

- ½ c cooked vegetables
- 1 c raw vegetables

Complementary Proteins If the body does not receive all the essential amino acids it needs, the supply of essential amino acids will dwindle until body organs are compromised. Obtaining enough essential amino acids presents no problem to people who regularly eat high-quality proteins, such as those of animal-derived foods and many soybean products, which contain ample amounts of all the essential amino acids. An equally sound choice is to eat two different protein foods from plants so that each supplies the amino acids missing in the other. In this strategy, the two protein-rich foods are combined to yield **complementary proteins** (see Figure 4-6)—proteins containing all the essential amino acids in amounts sufficient to support health. The two proteins need not even be eaten together, as long as the day's meals supply them both, and the diet provides enough energy and total protein from a variety of sources.

	Ile	Lys	Met	Trp
Legumes	✓	✓		
Grains			✓	✓
Together	✓	✓	✓	✓

FIGURE 4-6 Complementary Proteins

In general, legumes provide plenty of isoleucine (Ile) and lysine (Lys) but fall short in methionine (Met) and tryptophan (Trp). Grains have the opposite strengths and weaknesses, making them a perfect match for legumes.

Protein Sparing

Dietary protein—no matter how high the quality—will not be used efficiently and will not support growth when energy from carbohydrate and fat is lacking. The body assigns top priority to meeting its energy need and, if necessary, will break down protein to meet this need. After stripping off and excreting the nitrogen from the amino acids, the body will use the remaining carbon skeletons in much the same way it uses those from glucose or fat. A major reason why people must have ample carbohydrate and fat in the diet is to prevent this wasting of protein.

Reminder: Carbohydrate and fat allow amino acids to be used to build body proteins. This is known as the *protein-sparing effect* of carbohydrate and fat.

IN SUMMARY

▪ A diet inadequate in any of the essential amino acids limits protein synthesis.

▪ The best guarantee of amino acid adequacy is to eat foods containing high-quality proteins or combinations of foods containing complementary proteins so that each can supply the amino acids missing in the other.

▪ Vegetarians who consume no foods of animal origin can meet their protein needs by eating a variety of whole grains, legumes, seeds, nuts, and vegetables.

▪ Ample carbohydrate and fat in the diet allow amino acids to be used to build body proteins.

limiting amino acid: an essential amino acid that is present in dietary protein in the shortest supply relative to the amount needed for protein synthesis in the body.

complementary proteins: two or more proteins whose amino acid assortments complement each other in such a way that the essential amino acids missing from one are supplied by the other.

Your Diet

Proteins and Amino Acids

Many people in the United States and Canada consume more protein than they need. This is not surprising, considering the abundance of food eaten and the central role meats hold in the North American diet. Foods derived from animals—meats, fish, poultry, eggs, and milk products—provide plenty of protein but are often accompanied by fat. Those derived from plants—whole grains, vegetables, and legumes—may provide less protein but also less fat.

▪ Calculate your daily protein needs: Look up the healthy weight range for a person of your height in the body mass index (BMI) table on the inside back cover of this textbook. If your present weight falls within that range, use it for the following calculations. If your present weight falls outside the range, use the midpoint of the healthy weight range as your reference weight.

▪ Convert pounds to kilograms, if necessary (pounds divided by 2.2 equals kilograms).

▪ Multiply kilograms by 0.8 to get your RDA in grams per day. (Older teens 14 to 18 years old, multiply by 0.85.)

▪ Protein RDA: ___ grams

▪ Using your 24-hour food record (from the diet exercise in Chapter 1) and Table 4-3, estimate your protein intake for the day. Make a list of the foods you ate, and multiply the number of portions (as defined in the table) of each food by the protein grams per portion to estimate the protein grams provided by each food. (You can skip fruits and oils because they contribute little, if any, protein.) Add these numbers together to obtain the total intake.

Total protein intake: ___ grams

- Do you eat enough protein to meet your protein RDA?

- Next, determine whether your protein intake came mostly from plant-based foods or animal-based foods. To estimate your intake from plant-based foods, add up the grams of protein from grains, vegetables, legumes, tofu, nuts, and other plant protein sources on your list. Your total protein intake from plant-based foods: ___ grams

- What percentage of your total protein intake derived from plant-based foods? For example, if your total protein intake for the day was 60 grams and your protein intake from plant-based foods was 20 grams:

$$20 \div 60 = 0.33 \times 100 = 33\% \text{ protein intake from}$$
$$\text{plant-based foods}$$

- Does your diet consist of mostly plant-based or animal-based protein foods? ___

- If the majority of protein in your diet derives from animal-based foods, suggest some strategies to include more plant-based protein foods in your diet. Legumes, whole grains, vegetables, nuts, and seeds are rich in fiber and lower in saturated fat than some animal-based protein foods such as red meat, whole milk, and cheese.

Clinical Applications

1. Considering the health effects of too little dietary protein, what suggestions would you have for a teenage girl who reports the following information about her food intake:

   - She never eats any meat or other animal-derived foods because she is a vegan. On a typical day, she consumes toast and juice for breakfast; chips, a soft drink, and a piece of fruit for lunch; and a small serving of plain pasta with tomato sauce or steamed vegetables for dinner, along with a glass of water or tea.

   - She takes amino acid supplements because a friend told her that the only way to get amino acids if she doesn't eat meat is to take them as supplements.

2. Considering the health effects of excess dietary protein, what advice would you have for a college athlete who tells you he wants to bulk up his muscles and reports the following information about his food intake:

   - He eats large servings of meat (usually red meat) at least twice a day. He drinks whole milk two or three times a day and eats eggs and bacon for breakfast almost every day.

   - He avoids breads, cereals, and pasta in order to save room for protein-rich foods such as meat, milk, and eggs.

   - He eats a piece of fruit once in a while but seldom eats vegetables because he thinks they are too time consuming to prepare.

Self Check

1. Proteins are chemically different from carbohydrates and fats because they also contain:
   a. iron.
   b. sodium.
   c. nitrogen.
   d. phosphorus.

2. The basic building blocks for protein are:
   a. side groups.
   b. amino acids.
   c. glucose units.
   d. saturated bonds.

3. Enzymes are proteins that, among other things:
   a. defend the body against disease.
   b. regulate fluid and electrolyte balance.
   c. facilitate chemical reactions by changing themselves.
   d. help assemble disaccharides into starch, cellulose, or glycogen.

4. Functions of proteins in the body include:
   a. supplying omega-3 fatty acids for growth, lowering serum cholesterol, and helping with weight control.
   b. supplying fiber to aid digestion, digesting cellulose, and providing the main fuel source for muscles.
   c. protecting organs against shock, helping the body use carbohydrate efficiently, and providing triglycerides.
   d. serving as structural components, supplying hormones to regulate body processes, and maintaining fluid and electrolyte balance.

5. The swelling of body tissue caused by the leakage of fluid from the blood vessels into the interstitial spaces is called:
   a. edema.
   b. anemia.
   c. acidosis.
   d. sickle-cell anemia.

6. Major proteins in the blood that protect against bacteria and other disease agents are called:
   a. acids.
   b. buffers.
   c. antigens.
   d. antibodies.

7. Marasmus can be distinguished from kwashiorkor because in marasmus:
   a. only adults are victims.
   b. the cause is usually an infection.
   c. severe wasting of body fat and muscle occurs.
   d. the limbs and face swell with edema, and the belly bulges with a fatty liver.

8. The RDA for protein for a healthy adult is ___ gram(s) per kilogram of appropriate body weight for height.
   a. 0.5
   b. 0.8
   c. 1.1
   d. 1.4

9. Generally speaking, from which of the following foods are complete proteins derived?
   a. Milk, gelatin, and soy
   b. Rice, potatoes, and eggs
   c. Meats, fish, and poultry
   d. Vegetables, grains, and fruits

10. An incomplete protein lacks one or more:
    a. hydrogen bonds.
    b. essential fatty acids.
    c. saturated fatty acids.
    d. essential amino acids.

Answers to these questions can be found in Appendix H.

Notes

1. Standing Committee on the Scientific Evaluation of Dietary Reference Intakes, Food and Nutrition Board, Institute of Medicine, *Dietary Reference Intakes for Energy, Carbohydrate, Fiber, Fat, Fatty Acids, Cholesterol, Protein, and Amino Acids* (Washington, D.C.: National Academies Press, 2005), p. 605.

2. Food and Agriculture Organization of the United Nations, *The State of Food Insecurity in the World, 2009*, Food Security Statistics, available from www.fao.org.

3. J. F. Desjeux, Recent issues in energy-protein malnutrition in children, *Nestle Nutrition Workshop Series: Pediatric Program* 58 (2006): 177–184.

4. D. R. Brewster, Critical appraisal of the management of severe malnutrition: 2. Dietary management, *Journal of Paediatrics and Child Health* 42 (2006): 575–582; A. A. Jackson, A. Ashworth, and S. Khanum, Improving child survival: Malnutrition Task Force and the paediatrician's responsibility, *Archives of Diseases in Childhood* 91 (2006): 706–710.

5. N. S. Scrimshaw, Fifty-five-year personal experience with human nutrition worldwide, *Annual Review of Nutrition* 27 (2007): 1–18.

6. H. Frank and coauthors, Effect of short term high-protein compared with normal-protein diets on renal hemodynamics and associated variables in healthy young men, *American Journal of Clinical Nutrition* 90 (2009): 1509–1516; Standing Committee on the Scientific Evaluation of Dietary Reference Intakes, 2005, p. 694.

7. R. Sinha and coauthors, Meat intake and mortality: A prospective study of over half a million people, *Archives of Internal Medicine* 169 (2009): 562–571; L. E. Kelemen and coauthors, Associations of dietary protein with disease and mortality in a prospective study of postmenopausal women, *American Journal of Epidemiology* 161 (2005): 239–249.

8. Y. Wang and M. A. Beydoun, Meat consumption is associated with obesity and central obesity among US adults, *International Journal of Obesity* 33 (2009): 621–628.

9. N. R. Matthan and coauthors, Effect of soy protein from differently processed products on cardiovascular disease risk factors and vascular endothelial function in hypercholesterolemic subjects, *American Journal of Clinical Nutrition* 85 (2007): 960–966; B. L. McVeigh and coauthors, Effect of soy protein varying in isoflavone content on serum lipids in healthy young men, *American Journal of Clinical Nutrition* 83 (2006): 244–251.

10. R. Pecoits-Filho, Dietary protein intake and kidney disease in Western diet, *Contributions to Nephrology* 155 (2007): 102–112.

11. A. M. Bernstein, L. Treyzon, and Z. Li, Are high-protein, vegetable-based diets safe for kidney function? A review of the literature, *Journal of the American Dietetic Association* 107 (2007): 644–650.

12. A. L. Darling and coauthors, Dietary protein and bone health: A systematic review and meta-analysis, *American Journal of Clinical Nutrition* 90 (2009): 1674–1692; K. Rafferty and R. P. Heaney, Nutrient effects on the calcium economy: Emphasizing the potassium controversy, *Journal of Nutrition* 138 (2008): 166S–171S.

13. Rafferty and Heaney, 2008.

14. Standing Committee on the Scientific Evaluation of Dietary Reference Intakes, 2005, p. 694.

15. M. K. C. Hesselink, R. Minnaard, and P. Schrauwen, Eat the meat or feed the meat: Protein turnover in remodeling muscle, *Current Opinion in Clinical Nutrition and Metabolic Care* 9 (2006): 672–676; R. R. Wolfe, Skeletal muscle protein metabolism and resistance exercise, *Journal of Nutrition* 136 (2006): 525S–528S.

16. J. W. Apolzan and coauthors, Inadequate dietary protein increases hunger and desire to eat in younger and older men, *Journal of Nutrition* 137 (2007): 1478–1482.

17. G. Wu, Amino acids: Metabolism, functions, and nutrition, *Amino Acids*, 37 (2009): 1–17.

18. *Dietary Reference Intakes—The Essential Guide to Nutrient Requirements* (Washington, D.C.: National Academies Press, 2006), p. 152.

19. V. L. Fulgoni III, Current protein intake in America: Analysis of the National Health and Nutrition Examination survey, 2003–2004, *American Journal of Clinical Nutrition* 87 (2008): 1554S–1557S.

20. Fulgoni III, 2008.

CHAPTER 4

Nutrition in Practice

Eating patterns all along the continuum of dietary choices—from one end, where people eat no foods of animal origin, to the other end, where they eat generous quantities of meat every day—can support or compromise nutritional health. The quality of the diet does not depend on whether it consists of all plant foods or centers on meat. Rather, it depends on whether the eater's food choices are based on sound nutrition principles: adequacy of nutrient intakes, balance and variety of foods chosen, appropriate energy intake, and moderation in intakes of substances such as saturated fat, *trans* fat, added sugars, sodium, and alcohol that are harmful when consumed in excess. As mentioned in Chapter 4, however, because vegetarian diets exclude at least some animal-derived foods, they are usually lower in saturated fat and cholesterol than many meat-based diets.

People choose to exclude meat and other animal-derived foods from their diets for various reasons—concern for animal welfare, beliefs about health, religious convictions, or convenience. Some believe that vegetarianism is better for the environment; some, that it is healthier; and some, that it is less costly than the meat-eating alternative. Some just prefer it. Whatever the reasons, vegetarians and the health professionals who work with them should be aware of the nutrition and health implications of vegetarian diets.

Because vegetarian diets vary in both the types and the amounts of animal-derived foods they include, these differences must be considered when evaluating the health status of vegetarians. The accompanying glossary defines the various kinds of vegetarian diets.

Are vegetarian diets nutritionally sound?

The American Dietetic Association takes the position that well-planned vegetarian diets offer nutrition and health benefits to adults in general.[1] Research suggests that meat-eating adults who switch to vegetarian diets reduce their risks of heart disease, hypertension, diabetes, some types of cancer, and obesity.[2]

GLOSSARY OF VEGETARIAN TERMS

lacto-ovo vegetarians: people who include milk or milk products and eggs but exclude meat, fish, shellfish, and poultry from their diets.

lacto-vegetarians: people who include milk or milk products but exclude meat, poultry, fish, shellfish, and eggs from their diets.

ovo-vegetarians: people who include eggs, vegetables, grains, legumes, fruits, and nuts but exclude flesh, seafood, and milk products.

partial vegetarians: people who include some, but not all, groups of animal-derived foods in their diets; they usually exclude meat and may occasionally include poultry, fish, and shellfish.

vegans: people who exclude all animal-derived foods (including meat, poultry, fish, shellfish, eggs, cheese, and milk) from their diets; also called strict vegetarians or total vegetarians.

vegetarians: people who include plant-based foods and eliminate some or all animal-derived foods.

What should be my main concerns when planning a nutritionally sound vegetarian diet?

A vegetarian diet planner faces the same task as other diet planners—obtaining a variety of foods that provide all the needed nutrients within an energy allowance that maintains a healthy body weight. The challenge is to do so using at least one fewer food group. Because all vegetarians omit meat and some omit other animal-derived foods, protein, the nutrient that meat is famous for, merits some discussion here.

Isn't protein a problem in vegetarian diets?

No, protein is not the problem it was once thought to be in vegetarian diets. People who include animal-derived foods such as milk and eggs in their diets need not worry at all about protein deficiency. Even for **vegans**—those who eat only plant-based foods—protein intakes are usually satisfactory as long as energy intakes are adequate and protein sources are varied.[3] A mixture of proteins from whole grains, legumes, seeds, nuts, and vegetables can provide adequate amounts of high-quality protein. An advantage of many vegetarian sources of protein is that they are generally lower in saturated fat than meat and are often high in fiber and richer in some vitamins and minerals.

Some vegetarians may use meat replacements made of textured vegetable protein (soy protein). These foods are formulated to look and taste like meat, fish, or poultry. Many of these products are fortified to provide the vitamins and minerals found in animal sources of protein. Some products, however, may fall short of providing the nutrients of meat, and they may be high in salt, sugar, or other additives. Labels list all of the ingredients in such foods. A wise vegetarian learns to use a variety of whole, unrefined foods often and commercially prepared foods less frequently. Vegetarians may also use soybeans in other forms, such as plain tofu (soybean curd), edamame (cooked green soybeans), or soy flour, to bolster protein intake without consuming unwanted salt, sugar, or other additives.

What sorts of food energy intakes do vegetarian diets provide?

Researchers find that vegetarians as a group are closer to a healthy body weight than nonvegetarians.[4] Because obesity impairs health in a number of ways, vegetarians therefore have a health advantage. Vegetarian diets tend to be high in starch- and fiber-rich carbohydrates and low in saturated fat, characteristics that are consistent with current dietary recommendations aimed at reducing the incidence of obesity and other chronic diseases in this country.

Not all vegetarians fit the average pattern, however. Obesity can threaten vegetarians who include milk, eggs, and cheese in their diets. They can easily consume excess saturated fat and food energy and so must be careful to select fat-free and low-fat dairy foods and to avoid relying too heavily on these foods in general.

In contrast, people who exclude all animal-derived foods (vegans) may have trouble obtaining *enough* food energy. This is especially true for children. Vegan diets can fail to provide food energy sufficient to support the growth of a child within a bulk of food small enough for the child to eat. Frequent meals of fortified breads, cereals, or pastas with legumes, nuts, nut butters, and sources of unsaturated fats can help to meet protein and energy needs in a smaller volume at each sitting.[5] The MyPyramid resources, introduced in Chapter 1, include tips for planning vegetarian diets using the USDA Food Guide.

How can vegetarians and health professionals who plan vegetarian meals help ensure adequate nutrient intakes?

Most vegetarians easily obtain large quantities of the nutrients that are abundant in plant foods: carbohydrate, fiber, thiamin, folate, vitamin B_6, vitamin C, vitamin A, and vitamin E. Vegetarian food guides help to ensure adequate intakes of the main nutrients vegetarian diets might otherwise lack: protein, iron, zinc, calcium, vitamin B_{12}, vitamin D, and omega-3 fatty acids. Table NP4-1 on p. 104 presents good vegetarian sources of these key nutrients.

Tell me more about vitamins and minerals. Does a person eating a vegetarian diet need to take vitamin supplements?

That depends on the kind of vegetarian diet. The diet of **lacto-ovo vegetarians** can be complete in all vitamins, but for vegans, several vitamins may be a problem. One such vitamin is B_{12}. Because vitamin B_{12} occurs only in animal-derived foods, regular use of vitamin B_{12}-fortified foods, such as fortified soy and rice beverages, some breakfast cereals, and meat replacements, or supplements, is necessary to prevent deficiency.[6] Women who have adhered to all-plant diets for many years are especially likely to have low vitamin B_{12} stores. Pregnant vegan women, whose needs for vitamin B_{12} are especially high, find it virtually impossible to maintain adequate vitamin B_{12} status without taking supplements or including a reliable food source of the nutrient.

People who stop eating animal-derived foods containing vitamin B_{12} may take several years to develop deficiency symptoms because the body recycles much of its vitamin B_{12}, reabsorbing it over and over again. Even when the body fails to absorb vitamin B_{12}, deficiency may take up to three years to develop because the body conserves its supply. Neurological degeneration, a sign of vitamin B_{12} deficiency, appears rapidly in infants born to mothers with unsupplemented vegan diets or those with impaired absorption of the vitamin.[7] All vegan mothers must be sure to take the appropriate supplements or to use vitamin B_{12}–fortified products.

Because vegans do not drink vitamin D–fortified cow's milk, do they get enough vitamin D?

Overall, vegetarians are similar to nonvegetarians in their vitamin D status; factors such as taking supplements, skin color, and sun exposure have a greater influence on vitamin D than diet.[8] People who do not use vitamin D–fortified foods and do not receive enough exposure to sunlight to synthesize adequate vitamin D may need supplements to fend off bone loss. Of particular concern are infants, children, and older adults in northern climates during winter months.

So, on a vegan diet, vitamin B_{12} and vitamin D can be problems if a person is not careful. What about minerals?

For *all* vegetarians, not just vegans, two minerals may be of concern—iron and zinc. The iron in plant foods such as legumes, dark green leafy vegetables, iron-fortified cereals, and whole-grain breads and cereals is not as absorbable as that in meat. For this reason, the iron recommendation for adult vegetarian men, premenopausal women, and adolescent girls is almost double the recommendation for meat eaters of the same gender and age.[9]

Fortunately, the body seems to adapt to a vegetarian diet by absorbing iron more efficiently. Furthermore, iron absorption is enhanced by vitamin C, and vegetarians typically eat many vitamin C–rich fruits and vegetables. Consequently, vegetarians suffer no more iron deficiency than other people do.

Zinc is similar to iron in that meat is its richest food source, and zinc from plant sources is not as well absorbed. Zinc absorption from plant foods is hindered by phytate, a mineral binder found in grains, legumes, and seeds. Thus, although cereals and legumes contain zinc, the presence of phytate reduces zinc absorption from these foods. The zinc needs of vegetarians and the effects of mineral binders are subjects of intensive study at the present time. While research continues, vegetarians are advised to eat varied diets that include legumes such as navy beans and kidney beans, zinc-enriched cereals, and whole-grain breads well leavened with yeast, which improves the availability of their minerals. For partial vegetarians who include seafood in their diets, oysters, crabmeat, and shrimp are rich in zinc.

What about calcium for the vegan?

Yes, calcium is of concern. The milk-drinking vegetarian is protected from deficiency, but the vegan must find other sources of calcium. Some good calcium sources are regular and ample servings of dark green leafy vegetables such as kale and collard; legumes; calcium-fortified foods such as breakfast cereals, soy milk, and orange juice; some nuts such as almonds; and certain seeds such as sesame seeds. The choices should be varied because binders in some of these foods may hinder calcium absorption. The vegan is urged to use calcium-fortified soy milk in ample quantities regularly. This is especially important for children. Infant formula based on soy is fortified with calcium and can easily be used in food preparation, even for adults.

Do vegetarian diets provide adequate amounts of the essential fatty acids?

Vegetarian diets typically provide enough of the essential fatty acids linoleic acid and linolenic acid, but they lack a dietary source of EPA and DHA.[10] Fatty fish and products fortified with their oils can provide EPA and DHA, but all of these sources ultimately derive from fish, an unacceptable food for vegans. Certain marine algae and their oils provide an alternative DHA source, and more foods are being fortified with such oils, which are listed among the ingredients on a food's label.[11] A vegetarian's daily diet should include small amounts of flaxseed, walnuts, and their oils, as well as soybeans and canola oil to provide essential fatty acids.

TABLE NP4-1 Vegetarian Sources of Key Nutrients

Nutrients	Food Groups					
	Grains	Vegetables	Fruits	Protein-Rich Foods	Milk or Soy milk	Oils
Protein	Whole grains[a]			Legumes, seeds, nuts, nut butters, soy products (tempeh, tofu, veggie burgers)[a] Eggs (for ovo-vegetarians)	Milk, cheese, yogurt (for lacto-vegetarians) Soy milk, soy yogurt, soy cheeses	
Iron	Fortified cereals, enriched and whole grains	Dark green leafy vegetables (spinach, turnip greens)	Dried fruits (apricots, prunes, raisins)	Legumes (black-eyed peas, kidney beans, lentils), soy products		
Zinc	Fortified cereals, whole grains			Legumes (garbanzo beans, kidney beans, navy beans), nuts, seeds (pumpkin seeds)	Milk, cheese, yogurt (for lacto-vegetarians) Soy milk, soy yogurt, soy cheeses	
Calcium	Fortified cereals	Dark green leafy vegetables (bok choy, broccoli, collard greens, kale, mustard greens, turnip greens, watercress)	Fortified juices, figs	Fortified soy products, nuts (almonds), seeds (sesame seeds)	Milk, cheese, yogurt (for lacto-vegetarians) Fortified soy milk, fortified soy yogurt, fortified soy cheese	
Vitamin B_{12}	Fortified cereals			Eggs (for ovo-vegetarians) Fortified soy products	Milk, cheese, yogurt (for lacto-vegetarians) Fortified soy milk, fortified soy yogurt, fortified soy cheese	
Vitamin D	Fortified cereals				Milk, cheese, yogurt (for lacto-vegetarians) Fortified soy milk, fortified soy yogurt, fortified soy cheese	
Omega-3 Fatty Acids		Marine algae and its oils		Flaxseed, walnuts, soybeans Fortified margarine Fortified eggs (for ovo-vegetarians)	Fortified soy milk	Flaxseed oil, walnut oil, soybean oil

[a]As Chapter 4 explains, many plant proteins do not contain all the essential amino acids in the amounts and proportions needed by human beings. To improve protein quality, vegetarians can eat grains and legumes together, for example, although it is not necessary if protein intake is varied and energy intake is sufficient.

Are there any other health advantages to the vegetarian diet?

Yes. Vegetarian protein foods are often higher in fiber, richer in certain vitamins and minerals, and lower in fat—especially saturated fat—than meats. Vegetarians can enjoy a nutritious diet low in saturated fat provided that they limit foods such as butter, cream cheese, and sour cream. If vegetarians follow the guidelines presented here and plan carefully, they can support their health as well as, or perhaps better than, nonvegetarians.

Abundant evidence supports the idea that vegetarians may actually be healthier than meat eaters. Informed vegetarians are not only more likely to be at the desired weights for their heights, but they are also more likely to have lower blood cholesterol levels, lower rates of certain kinds of cancer, better digestive function, and more. Even among people who are health conscious, generally vegetarians experience fewer deaths from cardiovascular disease than meat eaters do. Because many vegetarians also abstain from smoking and the consumption of alcohol, dietary practices alone probably do not account for all the aspects of improved health. Clearly, however, they contribute significantly to it.

Notes

1. Position of the American Dietetic Association, Vegetarian diets, *Journal of the American Dietetic Association* 109 (2009): 1266–1282.

2. G. E. Fraser, Vegetarian diets: What do we know of their effects on common chronic diseases? *American Journal of Clinical Nutrition* 89 (2009): 1607S–1612S; N. D. Barnard and coauthors, A low-fat vegan diet and a conventional diabetes diet in the treatment of type 2 diabetes: A randomized, controlled, 74-wk clinical trial, *American Journal of Clinical Nutrition* 89 (2009): 1588S–1596S; W. J. Craig, Health effects of vegan diets, *American Journal of Clinical Nutrition* 89 (2009): 1627S–1633S; T. J. Key, P. N. Appleby, and M. S. Rosell, Health effects of vegetarian and vegan diets, *Proceedings of the Nutrition Society* 65 (2006): 35–41.

3. Position of the American Dietetic Association, 2009.

4. Fraser, 2009; R. Robinson-O'Brien and coauthors, Adolescent and young adult vegetarianism: Better dietary intake and weight outcomes but increased risk of disordered eating behaviors, *Journal of the American Dietetic Association* 109 (2009): 648–655; S. E. Berkow and N. Barnard, Vegetarian diets and weight status, *Nutrition Reviews* 64 (2006): 175–188; M. Rosell and coauthors, Weight gain over 5 years in 21,966 meat-eating, fish-eating, vegetarian, and vegan men and women in EPIC-Oxford, *International Journal of Obesity* 30 (2006): 1389–1396.

5. Position of the American Dietetic Association, 2009.

6. Position of the American Dietetic Association, 2009.

7. D. K. Dror and L. H. Allen, Effect of vitamin B12 deficiency on neurodevelopment in infants: Current knowledge and possible mechanisms, *Nutrition Reviews* 66 (2008): 250–255.

8. J. Chan, K. Jaceldo-Siegl, and Gary E. Fraser, Serum 25-hydroxyvitamin D status of vegetarians, partial vegetarians, and nonvegetarians: The Adventist Health Study, *American Journal of Clinical Nutrition* 89 (2009): 1686S–1692S.

9. Standing Committee on the Scientific Evaluation of Dietary Reference Intakes, Food and Nutrition Board, Institute of Medicine, *Dietary Reference Intakes for Vitamin A, Vitamin K, Arsenic, Boron, Chromium, Copper, Iodine, Iron, Manganese, Molybdenum, Nickel, Silicon, Vanadium, and Zinc* (Washington, D.C.: National Academy Press, 2001), pp. 9–45.

10. I. Mangat, Do vegetarians have to eat fish for optimal cardiovascular protection? *American Journal of Clinical Nutrition* 89 (2009): 1597–1601.

11. J. Whelan and C. Rust, Innovative dietary sources of n-3 fatty acids, *Annual Review of Nutrition* 26 (2006): 75–103.

Digestion and Absorption

The human body's ability to transform the foods a person eats into the nutrients that fuel the body's work is quite remarkable. Yet most people probably give little, if any, thought to what their bodies do with food once it is eaten. This chapter offers the reader the opportunity to learn how the body digests, absorbs, and transports the nutrients and how it excretes the unwanted substances in foods.

One of the beauties of the digestive tract is that it is selective. Materials that are nutritive for the body are broken down into particles that can be absorbed into the bloodstream. Most of the nonnutritive materials are left undigested and pass out the other end of the digestive tract.

Anatomy of the Digestive Tract

The **gastrointestinal (GI) tract** is a flexible, muscular tube extending from the mouth to the anus. Figure 5-1 on pp. 108–109 traces the path followed by food from one end of the GI tract to the other. The accompanying glossary defines GI anatomical terms. In a sense, the human body surrounds the GI tract. Only when a nutrient or other substance passes through the cells of the digestive tract wall does it actually enter the body.

The Digestive Organs

The process of **digestion** begins in the **mouth.** As you chew, your teeth crush and soften the food, while saliva mixes with the food mass and moistens it for comfortable swallowing. Saliva also helps dissolve the food so that you can taste it; only particles in solution can react with taste buds.

The tongue allows you not only to taste food but also to move food around the mouth, facilitating chewing and swallowing. When you swallow a mouthful of food, it passes through the **pharynx,** a short tube that is shared by both the **digestive system** and the respiratory system.

Mouth to the Esophagus Once a mouthful of food has been chewed and swallowed, it is called a **bolus.** Each bolus first slides across your **epiglottis,** bypassing the entrance to your lungs. During each swallow, the epiglottis closes off your trachea, the air passageway to the lungs, so that you do not choke.

Esophagus to the Stomach The **esophagus** has a **sphincter** muscle at each end. During a swallow, the upper **esophageal sphincter** opens. The bolus then slides down the esophagus, which conducts it through the diaphragm to the **stomach.** The lower esophageal sphincter closes behind the bolus so that it cannot slip back. The stomach retains the bolus for a while, adds juices to it (gastric juices are discussed on p. 113), and transforms it into a semiliquid mass called **chyme.** Then, bit by bit, the stomach releases the chyme through another sphincter, the **pyloric sphincter,** which opens into the **small intestine** and then closes after the chyme passes through.

The Small Intestine At the beginning of the small intestine, the chyme passes by an opening from the common bile duct, which secretes digestive fluids into the small intestine from two organs outside the GI tract—the **gallbladder** and the **pancreas.** The chyme travels on down the small intestine through its three segments—the **duodenum,** the **jejunum,** and the **ileum.** Together, the segments amount to a total of about 10 feet of tubing coiled within the abdomen.* Digestion is completed within the small intestine.

---

*The small intestine is almost two and a half times shorter in living adults than it is at death, when muscles are relaxed and elongated.

gastrointestinal (GI) tract: the digestive tract. The principal organs are the stomach and intestines.
   gastro = stomach

digestion: the process by which complex food particles are broken down to smaller, absorbable particles.

digestive system: all the organs and glands associated with the ingestion and digestion of food.

bolus (BOH-lus): the portion of food swallowed at one time.

chyme (KIME): the semiliquid mass of partly digested food expelled by the stomach into the duodenum (the top portion of the small intestine).

FIBER	CARBOHYDRATE

Mouth and salivary glands
The mechanical action of the mouth and teeth crushes and tears fiber in food and mixes it with saliva to moisten it for swallowing.

The salivary glands secrete saliva into the mouth to moisten the food. The salivary enzyme amylase begins digestion:

$$\text{Starch} \xrightarrow{\text{amylase}} \text{small polysaccharides, maltose.}$$

Stomach
Fiber is unchanged.

Stomach acid inactivates salivary enzymes, halting starch digestion. To a small extent, stomach acid hydrolyzes maltose and sucrose.

Small intestine
Fiber is unchanged.

The pancreas produces enzymes and releases them through the pancreatic duct into the small intestine:

$$\text{Polysaccharides} \xrightarrow[\text{amylase}]{\text{pancreatic}} \text{disaccharides.}$$

Then enzymes on the surfaces of the small intestinal cells break disaccharides into monosaccharides, and the cells absorb them:

$$\text{Maltose} \xrightarrow{\text{maltase}} \text{glucose + glucose.}$$

$$\text{Sucrose} \xrightarrow{\text{sucrase}} \text{fructose + glucose.}$$

$$\text{Lactose} \xrightarrow{\text{lactase}} \text{galactose + glucose.}$$

Large intestine (colon)
Most fiber passes intact through the digestive tract to the large intestine. Here, bacterial enzymes digest some fiber:

$$\text{Some fiber} \xrightarrow[\text{enzymes}]{\text{bacterial}} \text{fatty acids, gas.}$$

Fiber holds water; regulates bowel activity; and binds cholesterol and some minerals, carrying them out of the body as it is excreted with feces.

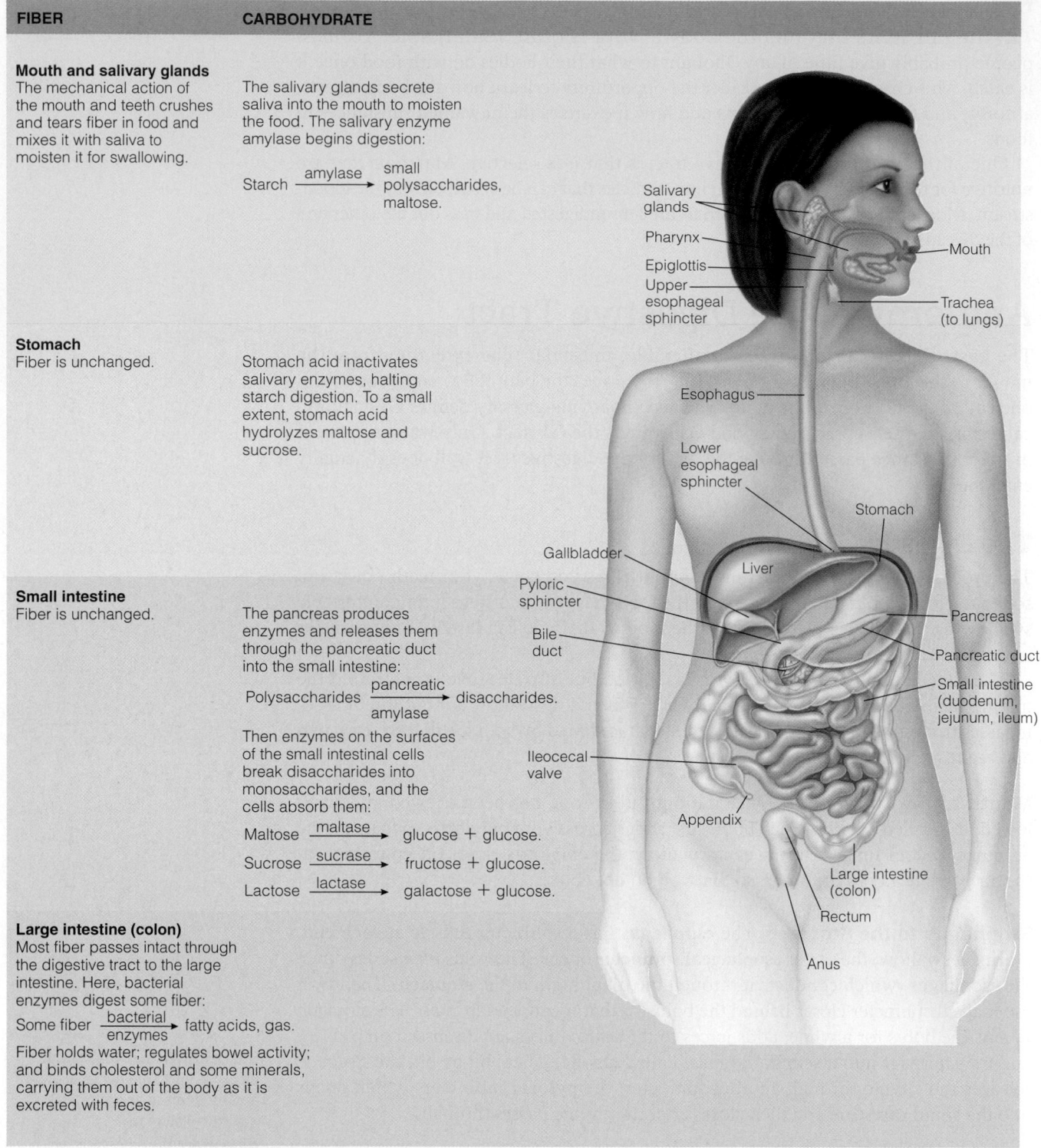

Salivary glands
Pharynx
Epiglottis
Upper esophageal sphincter
Mouth
Trachea (to lungs)
Esophagus
Lower esophageal sphincter
Stomach
Gallbladder
Liver
Pyloric sphincter
Bile duct
Pancreas
Pancreatic duct
Small intestine (duodenum, jejunum, ileum)
Ileocecal valve
Appendix
Large intestine (colon)
Rectum
Anus

FIGURE 5-1 The Gastrointestinal Tract

The Large Intestine (Colon) Having traveled the length of the small intestine, what remains of the intestinal contents passes through another sphincter, the **ileocecal valve,** into the beginning of the **large intestine (colon)** in the lower right-hand side of the abdomen. Upon entering the colon, the contents pass another opening. Should any intestinal contents slip into this opening, they would end up in the **appendix,** a blind sac about the size of your little finger. Normally, the contents bypass this opening,

FAT	PROTEIN	VITAMINS	MINERALS AND WATER
Mouth and salivary glands The sublingual salivary gland in the base of the tongue secretes a fat-digesting enzyme known as lingual lipase. Some hard fats begin to melt as they reach body temperature.	Chewing and crushing moisten protein-rich foods and mix them with saliva to be swallowed.	No action.	The salivary glands add water to disperse and carry food.
Stomach The acid stable lingual lipase splits one bond of triglycerides to produce diglycerides and fatty acids. The degree of hydrolysis is slight for most fats but may be appreciable for milk fats. The stomach's churning action mixes fat with water and acid. A gastric lipase accesses and hydrolyzes a very small amount of fat.	Hydrochloric acid (HCL) uncoils protein strands and activates stomach enzymes: Protein $\xrightarrow[\text{HCl}]{\text{pepsin}}$ smaller polypeptides.	Intrinsic factor (see Chapter 8) attaches to vitamin B_{12}.	Stomach acid (HCl) acts on iron to reduce it, making it more absorbable. The stomach secretes enough watery fluid to turn a moist, chewed mass of solid food into liquid chyme.
Small intestine Bile flows in from the liver and gallbladder (via the common bile duct): Fat $\xrightarrow{\text{bile}}$ emulsified fat. Pancreatic lipase flows in from the pancreas (via the pancreatic duct): Emulsified fat $\xrightarrow{\text{pancreatic lipase}}$ monoglycerides, glycerol, fatty acids (absorbed).	Polypeptides $\xrightarrow[\text{and intestinal proteases}]{\text{pancreatic}}$ tripeptides, dipeptides, amino acids Then enzymes on the surface of the small intestinal cells hydrolyze these peptides and the cells absorb them. Peptides $\xrightarrow[\text{dipeptidases and tripeptidases}]{\text{intestinal}}$ amino acids (absorbed)	Bile emulsifies fat-soluble vitamins and aids in their absorption with other fats. Water-soluble vitamins are absorbed.	The small intestine, pancreas, and liver add enough fluid so that approximately 2 gallons are secreted into the intestine in a day. Many minerals are absorbed. Vitamin D aids in the absorption of calcium.
Large intestine Some fat and cholesterol, trapped in fiber, exit in feces.		Bacteria produce vitamin K, which is absorbed.	More minerals and most of the water are absorbed.

FIGURE 5-1 (continued)

however, and travel up the right-hand side of the abdomen, across the front to the left-hand side, down to the lower left-hand side, and finally below the other folds of the intestines to the back side of the body above the **rectum.**

The Rectum As the intestinal contents pass to the rectum, the colon withdraws water, leaving semisolid waste. The strong muscles of the rectum hold back this waste until it is time to defecate. Then the rectal muscles relax, and the last sphincter in

These terms are listed in order from start to end of the digestive system.

mouth: the oral cavity containing the tongue and teeth.

pharynx (FAIR-inks): the passageway leading from the nose and mouth to the larynx and esophagus, respectively.

epiglottis (epp-ih-GLOTT-iss): cartilage in the throat that guards the entrance to the trachea and prevents fluid or food from entering it when a person swallows.

   epi = upon (over)
   glottis = back of tongue

esophagus (ee-SOFF-ah-gus): the food pipe; the conduit from the mouth to the stomach.

sphincter (SFINK-ter): a circular muscle surrounding, and able to close, a body opening. Sphincters are found at specific points along the GI tract and regulate the flow of food particles.

   sphincter = band (binder)

esophageal (ee-SOF-a-GEE-al) **sphincter:** a sphincter muscle at the upper or lower end of the esophagus. The lower esophageal sphincter is also called the **cardiac sphincter.**

stomach: a muscular, elastic, saclike portion of the digestive tract that grinds and churns swallowed food, mixing it with acid and enzymes to form chyme.

pyloric (pie-LORE-ic) **sphincter:** the circular muscle that separates the stomach from the small intestine and regulates the flow of partially digested food into the small intestine; also called **pylorus** or **pyloric valve.**

   pylorus = gatekeeper

small intestine: a 10-foot length of small-diameter intestine that is the major site of digestion of food and absorption of nutrients. Its segments are the duodenum, jejunum, and ileum.

duodenum (doo-oh-DEEN-um, doo-ODD-num): the top portion of the small intestine (about "12 fingers' breadth long" in ancient terminology).
   duodecim = twelve

jejunum (je-JOON-um): the first two-fifths of the small intestine beyond the duodenum.

ileum (ILL-ee-um): the last segment of the small intestine.

gallbladder: the organ that stores and concentrates bile. When it receives the signal that fat is present in the duodenum, the gallbladder contracts and squirts bile through the bile duct into the duodenum.

pancreas: a gland that secretes digestive enzymes and juices into the duodenum. (The pancreas also secretes hormones into the blood that help to maintain glucose homeostasis.)

ileocecal (ill-ee-oh-SEEK-ul) **valve:** the sphincter separating the small and large intestines.

large intestine or **colon** (COAL-un): the lower portion of intestine that completes the digestive process. Its segments are the ascending colon, the transverse colon, the descending colon, and the sigmoid colon.
   sigmoid = shaped like the letter S (sigma in Greek)

appendix: a narrow blind sac extending from the beginning of the colon that stores lymph cells.

rectum: the muscular terminal part of the intestine, extending from the sigmoid colon to the anus.

anus (AY-nus): the terminal outlet of the GI tract.

The path of food through the digestive tract:
- Mouth
- Esophagus
- Lower esophageal sphincter (or cardiac sphincter)
- Stomach
- Pyloric sphincter
- Duodenum (common bile duct enters here), jejunum, ileum
- Ileocecal valve
- Large intestine (colon)
- Rectum
- Anus

the system, the **anus,** opens to allow the wastes to pass. Thus, food follows the path shown in the margin.

The Involuntary Muscles and the Glands

You are usually unaware of all the activity that goes on between the time you swallow and the time you defecate. As is the case with so much else that happens in the body, the muscles and **glands** of the digestive tract meet internal needs without your having to exert any conscious effort to get the work done.

People consciously chew and swallow, but even in the mouth there are some processes over which you have no control. The salivary glands secrete just enough saliva to moisten each mouthful of food so that it can pass easily down your esophagus.

Gastrointestinal Motility Once you have swallowed, materials are moved through the rest of the GI tract by involuntary muscular contractions. This motion, known as **gastrointestinal motility,** consists of two types of movement, peristalsis and segmentation (see Figure 5-2). Peristalsis propels, or pushes; segmentation mixes, with more gradual pushing.

Peristalsis Peristalsis begins when the bolus enters the esophagus. The entire GI tract is ringed with circular muscles, which are surrounded by longitudinal muscles. When the rings tighten and the long muscles relax, the tube is constricted. When the rings relax and the long muscles tighten, the tube bulges. These actions alternate continually and push the intestinal contents along. If you have ever watched a bolus of food pass along the body of a snake, you have a good picture of how these muscles work. The waves of contraction ripple through the GI tract at varying rates and intensities depending on the part of the GI tract and on whether food is present. Peristalsis,

FIGURE 5-2
**Peristalsis and
Segmentation**

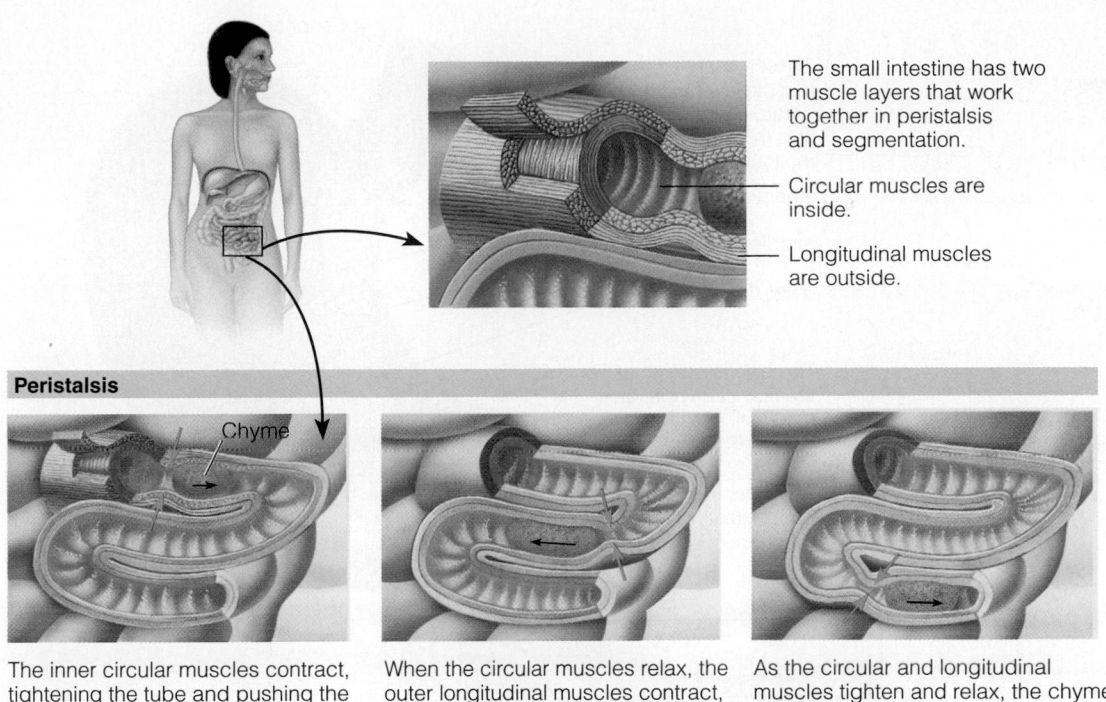

The small intestine has two muscle layers that work together in peristalsis and segmentation.

Circular muscles are inside.

Longitudinal muscles are outside.

Peristalsis

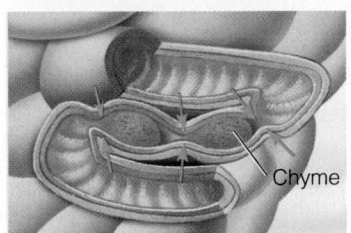

Chyme

The inner circular muscles contract, tightening the tube and pushing the food forward in the intestine.

When the circular muscles relax, the outer longitudinal muscles contract, and the intestinal tube is loose.

As the circular and longitudinal muscles tighten and relax, the chyme moves ahead of the constriction.

Segmentation

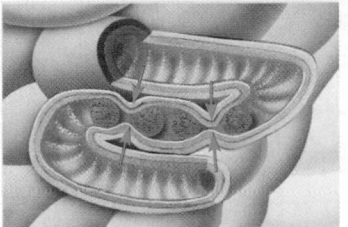

Chyme

Circular muscles contract, creating segments within the intestine.

As each set of circular muscles relaxes and contracts, the chyme is broken up and mixed with digestive juices.

These alternating contractions, occurring 12 to 16 times per minute, continue to mix the chyme and bring the nutrients into contact with the intestinal lining for absorption.

aided by the sphincter muscles located at key places, keeps things moving along. However, factors such as stress, medicines, and medical conditions may interfere with normal GI tract contractions.[1]

Segmentation The intestines not only push but also periodically squeeze their contents as if a string tied around the intestines were being pulled tight. This motion, called **segmentation,** forces the contents back a few inches, mixing them and promoting close contact with the digestive juices and the absorbing cells of the intestinal walls before letting the contents slowly move along again.

Liquefying Process Besides forcing the intestinal contents along, the muscles of the GI tract help to liquefy them to chyme so that the digestive juices will have access to all their nutrients. The mouth initiates this liquefying process by chewing, adding saliva, and stirring with the tongue to reduce the food to a coarse mash suitable for swallowing. The stomach then further mixes and kneads the food.

Stomach Action The stomach has the thickest walls and strongest muscles of all the GI tract organs. In addition to circular and longitudinal muscles, the stomach has a third layer of diagonal muscles that also alternately contract and relax (see Figure 5-3 on p. 112). These three sets of muscles work to force the chyme downward, but the pyloric

gastrointestinal motility: spontaneous motion in the digestive tract accomplished by involuntary muscular contractions.

peristalsis (peri-STALL-sis): successive waves of involuntary muscular contractions passing along the walls of the GI tract that push the contents along.
peri = around
stellein = wrap

segmentation: a periodic squeezing or partitioning of the intestine by its circular muscles that both mixes and slowly pushes the contents along.

FIGURE 5-3
Stomach Muscles
The stomach has three layers
of muscles.

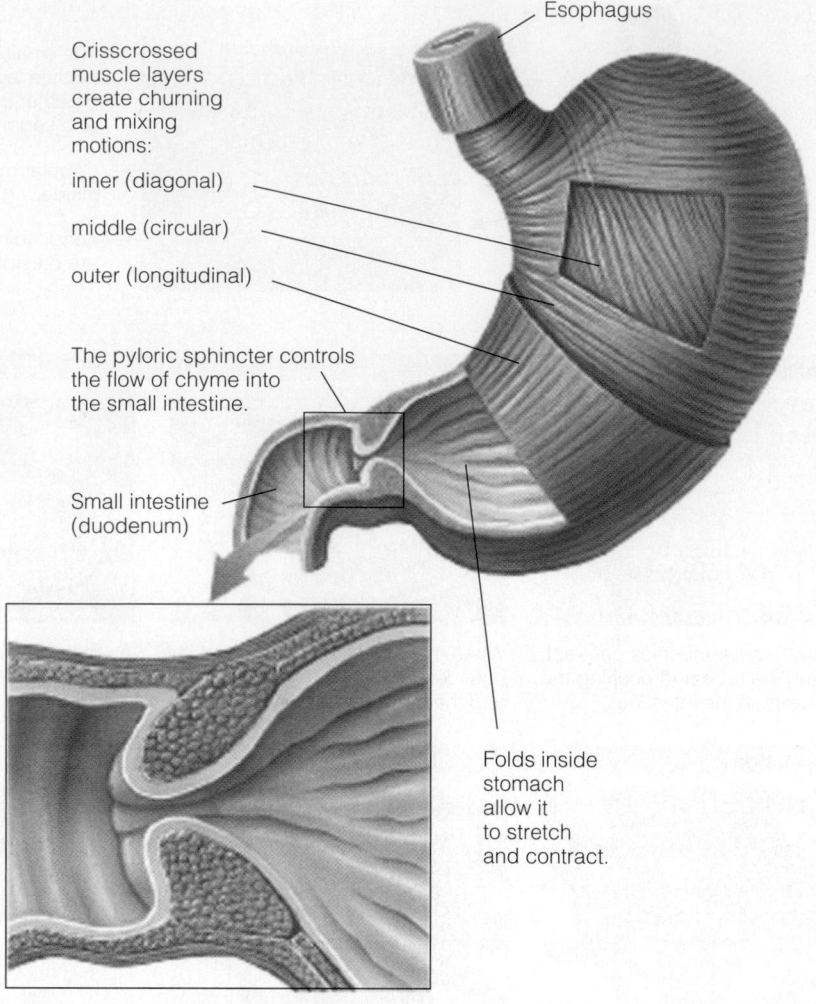

Esophagus

Crisscrossed
muscle layers
create churning
and mixing
motions:

inner (diagonal)

middle (circular)

outer (longitudinal)

The pyloric sphincter controls
the flow of chyme into
the small intestine.

Small intestine
(duodenum)

Folds inside
stomach
allow it
to stretch
and contract.

sphincter usually remains tightly closed, so that the stomach's contents are thoroughly mixed and squeezed before being released. Meanwhile, the gastric glands are adding juices. When the chyme is thoroughly liquefied, the pyloric sphincter opens briefly, about three times a minute, to allow small portions through. At this point, the intestinal contents no longer resemble food in the least.

IN SUMMARY

▌ As Figure 5-1 shows, food enters the mouth and travels down the esophagus and through the lower esophageal sphincter to the stomach, then through the pyloric sphincter to the small intestine, on through the ileocecal valve to the large intestine, past the appendix to the rectum, and ending at the anus.

▌ The wavelike contractions of peristalsis and the periodic squeezing of segmentation keep things moving at a reasonable pace.

The Process of Digestion

One person eats nothing but vegetables, fruits, and nuts; another, nothing but meat, milk, and potatoes. How is it that both people wind up with essentially the same body composition? It all comes down to the body rendering food—whatever it is to start with—into the basic units that make up carbohydrate, fat, and protein. The body absorbs these units and builds its tissues from them.

To digest food, five different body organs secrete digestive juices: the salivary glands, the stomach, the small intestine, the liver (via the gallbladder), and the pancreas. These secretions enter the GI tract at various points along the way, bringing an abundance of water and a variety of **enzymes.** Each of the juices has a turn to mix with the food and promote its breakdown to small units that can be absorbed into the body. The accompanying glossary defines some of the digestive glands and their juices.

Digestion in the Mouth

Digestion of carbohydrate begins in the mouth, where the **salivary glands** secrete **saliva,** which contains water, salts, and enzymes (including salivary **amylase**) that break the bonds in the chains of starch. Saliva also protects the tooth surfaces and linings of the mouth, esophagus, and stomach from attack by molecules that might harm them. The enzymes in the mouth do not, for the most part, affect the fats, proteins, vitamins, minerals, and fiber that are present in the foods people eat (review Figure 5-1 on pp. 108–109).

Digestion in the Stomach

Gastric juice, secreted by the **gastric glands,** is composed of water, enzymes, and **hydrochloric acid.** The acid is so strong that it burns the throat if it happens to reflux into the upper esophagus and mouth. The stomach's strong acidity prevents bacterial growth and kills most bacteria that enter the body along with food. You might expect that the stomach's acid would attack the stomach itself, but the cells of the stomach wall secrete **mucus,** a thick, slimy, white polysaccharide that coats and protects the stomach's lining.

The major digestive event in the stomach is the initial breakdown of proteins. Other than being crushed and mixed with saliva in the mouth, nothing happens to protein until it comes in contact with the gastric juices in the stomach. There, the acid helps to

enzymes: protein catalysts. A catalyst is a compound that facilitates chemical reactions without itself being changed in the process.

GLOSSARY OF DIGESTIVE GLANDS AND THEIR SECRETIONS

These terms are listed in order from the beginning of the digestive tract to the end.

glands: cells or groups of cells that secrete materials for special uses in the body. Glands may be *exocrine* (EKS-oh-crin) *glands*, secreting their materials "out" (into the digestive tract or onto the surface of the skin), or *endocrine* (EN-doe-crin) *glands*, secreting their materials "in" (into the blood).

  exo = outside
  endo = inside
  krine = to separate

salivary glands: exocrine glands that secrete saliva into the mouth.

saliva: the secretion of the salivary glands. The principal enzyme is salivary amylase.

amylase (AM-uh-lace): an enzyme that splits amylose (a form of starch). Amylase is a carbohydrase. The ending *–ase* indicates an enzyme; the root tells what it digests. Other examples: protease, lipase.

gastric glands: exocrine glands in the stomach wall that secrete gastric juice into the stomach.
  gastro = stomach

gastric juice: the digestive secretion of the gastric glands containing a mixture of water, hydrochloric acid, and enzymes. The principal enzymes are pepsin (acts on proteins) and lipase (acts on emulsified fats).

hydrochloric acid (HCl): an acid composed of hydrogen and chloride atoms; normally produced by the gastric glands.

mucus (MYOO-cuss): a mucopolysaccharide (a relative of carbohydrate) secreted by cells of the stomach wall that protects the cells from exposure to digestive juices (and other destructive agents). The cellular lining of the stomach wall with its coat of mucus is known as the mucous membrane. (The noun is *mucus*; the adjective is *mucous*.)

pepsin: a protein-digesting enzyme (gastric protease) in the stomach. It circulates as a precursor, pepsinogen, and is converted to pepsin by the action of stomach acid.

intestinal juice: the secretion of the intestinal glands; contains enzymes for the digestion of carbohydrate and protein and a minor enzyme for fat digestion.

liver: the organ that manufactures bile. (The liver's other functions are described in Chapter 20.)

bile: an emulsifier that prepares fats and oils for digestion; made by the liver, stored in the gallbladder, and released into the small intestine when needed.

pancreatic (pank-ree-AT-ic) **juice:** the exocrine secretion of the pancreas, containing enzymes for the digestion of carbohydrate, fat, and protein. Juice flows from the pancreas into the small intestine through the pancreatic duct. The pancreas also has an endocrine function, the secretion of insulin and other hormones.

bicarbonate: an alkaline secretion of the pancreas; part of the pancreatic juice. (Bicarbonate also occurs widely in all cell fluids.)

uncoil (denature) the protein's tangled strands so that the stomach enzymes can attack the bonds. Both the enzyme **pepsin** and the stomach acid itself act as catalysts in the process. Minor events are the digestion of some fat by a gastric lipase, the digestion of sucrose (to a very small extent) by the stomach acid, and the attachment of a protein carrier to vitamin B_{12}.

The stomach enzymes work most efficiently in the stomach's strong acid, but salivary amylase, which is swallowed with food, does not work in acid this strong. Consequently, the digestion of starch gradually ceases as the acid penetrates the bolus. In fact, salivary amylase becomes just another protein to be digested. The amino acids in amylase end up being absorbed and recycled into other body proteins.

Digestion in the Small and Large Intestines

By the time food leaves the stomach, digestion of all three energy-yielding nutrients has begun, but the process gains momentum in the small intestine. There, the pancreas and the liver contribute additional digestive juices through the duct leading into the duodenum, and the small intestine adds **intestinal juice.** These juices contain digestive enzymes, bicarbonate, and bile.

Digestive Enzymes Pancreatic juice contributes enzymes that digest fats, proteins, and carbohydrates. Glands in the intestinal wall also secrete digestive enzymes. (Review the glossary of digestive glands and their secretions on p. 113 for details.)

Bicarbonate Pancreatic juice also contains sodium **bicarbonate,** which neutralizes the acidic chyme as it enters the small intestine. From this point on, the digestive tract contents are neutral or slightly alkaline. The enzymes of both the intestine and the pancreas work best in this environment.

Bile Bile is secreted continuously by the **liver** and is concentrated and stored in the gallbladder. The gallbladder squirts bile into the duodenum whenever fat arrives there. Bile is not an enzyme but an **emulsifier** that brings fats into suspension in water (see Figure 5-4). After the fats are emulsified, enzymes can work on them, and they can be absorbed. Thanks to all these secretions, all three energy-yielding nutrients are digested in the small intestine.

Mayonnaise, made from vinegar and oil, would separate as other vinegar-and-oil salad dressings do if food chemists did not blend in a third ingredient—an emulsifier. The emulsifier mixes well with the fatty oil and the watery vinegar. In the case of mayonnaise, the emulsifier is lecithin from egg yolks.

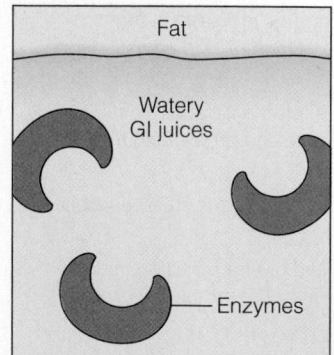 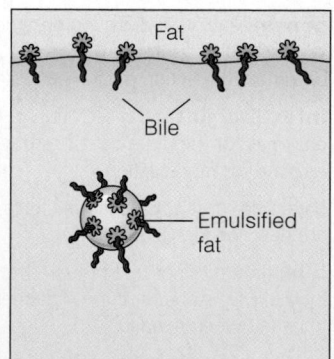 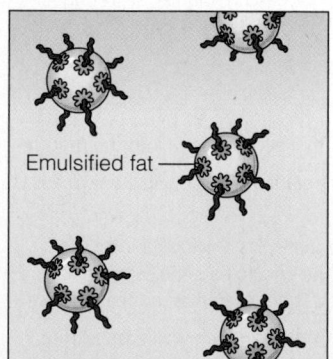 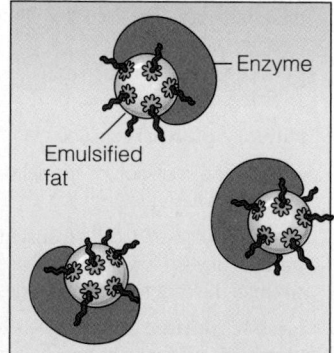

In the stomach, the fat and watery GI juices tend to separate. The enzymes are in the water and cannot get at the fat.	When fat enters the small intestine, the gallbladder secretes bile. Bile has an affinity for both fat and water, so it can bring the fat into the water.	Bile's emulsifying action converts large fat globules into small droplets that repel each other.	After emulsification, the enzymes have easy access to the fat droplets.

FIGURE 5-4 Emulsification of Fat by Bile
Like bile, detergents are emulsifiers and work the same way, which is why they are effective at removing grease spots from clothes. Molecule by molecule, the grease is dissolved out of the spot and suspended in water, where it can be rinsed away.

114

The Rate of Digestion The rate of digestion of the energy nutrients depends on the meal contents. If the meal is high in simple sugars, digestion proceeds fairly rapidly. On the other hand, if it is rich in fat, digestion is slower.

Protective Factors The intestines contain bacteria that produce a variety of vitamins, including biotin and vitamin K (although bacteria alone cannot meet the need for these vitamins). The GI bacteria also protect people from infections. Provided that the normal **intestinal flora** are thriving, infectious bacteria have a hard time getting established and launching an attack on the system. In addition, the small intestine and the entire GI tract manufacture and maintain a strong arsenal of defenses against foreign invaders. Several different types of defending cells are present in the GI tract, and they confer specific immunity against intestinal diseases.

Probiotics and intestinal health is the topic of Nutrition in Practice 18.

The Final Stage The story of how food is broken down into nutrients that can be absorbed is now nearly complete. The three energy-yielding nutrients—carbohydrate, fat, and protein—are disassembled to basic building blocks before they are absorbed. Most of the other nutrients—vitamins, minerals, and water—are absorbed as they are. Undigested residues, such as some fibers, are not absorbed but continue through the digestive tract as a semisolid mass that stimulates the tract's muscles, helping them remain strong and able to perform peristalsis efficiently. Fiber also retains water, keeping the stools soft, and carries some bile acids, sterols, and fat out of the body. Drinking plenty of water in conjunction with eating foods high in fiber supplies fluid for the fiber to take up. This is the basis for the recommendation to drink water and eat fiber-rich foods to relieve constipation.

The process of absorbing the nutrients into the body is discussed in the next section. For the moment, let us assume that the digested nutrients simply disappear from the GI tract as they are ready. Virtually all nutrients are gone by the time the contents of the GI tract reach the end of the small intestine. Little remains but water, a few salts and body secretions, and undigested materials such as fiber. These enter the large intestine (colon).

In the colon, intestinal bacteria degrade some of the fiber to simpler compounds. The colon itself retrieves from its contents the materials that the body is designed to recycle—water and dissolved salts. The waste that is finally excreted has little or nothing of value left in it. The body has extracted all that it can use from the food.

IN SUMMARY

▪ To digest food, the salivary glands, stomach, pancreas, liver (via the gallbladder), and small intestine deliver fluids and digestive enzymes.

The Absorptive System

Within three or four hours after you have eaten a meal, your body must find a way to absorb millions of molecules one by one. The absorptive system is ingeniously designed to accomplish this task.

The Small Intestine

Most absorption takes place in the small intestine. The small intestine is a tube about 10 feet long and about an inch across, yet it provides a surface comparable in area to a tennis court. When nutrient molecules make contact with this surface, they are absorbed and carried off to the liver and other parts of the body.

Villi and Microvilli How does the intestine manage to provide such a large absorptive surface area? Its inner surface looks smooth, but viewed through a

emulsifier: a substance that mixes with both fat and water and that disperses the fat in the water, forming an emulsion.

intestinal flora: the bacterial inhabitants of the GI tract.
flora = plant growth

FIGURE 5-5
The Small Intestinal Villi

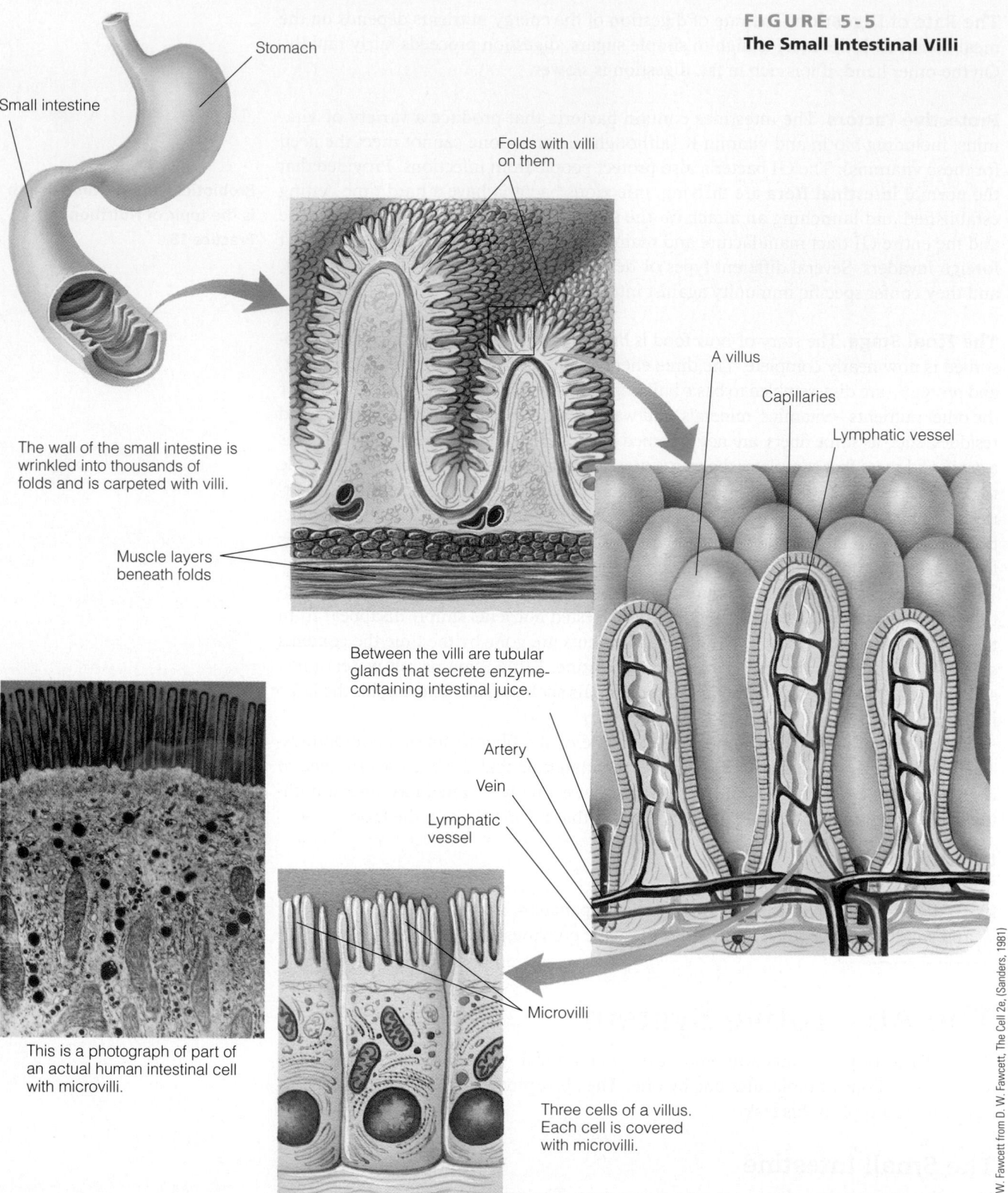

Small intestine

Stomach

The wall of the small intestine is wrinkled into thousands of folds and is carpeted with villi.

Folds with villi on them

A villus

Capillaries

Lymphatic vessel

Muscle layers beneath folds

Between the villi are tubular glands that secrete enzyme-containing intestinal juice.

Artery

Vein

Lymphatic vessel

Microvilli

This is a photograph of part of an actual human intestinal cell with microvilli.

Three cells of a villus. Each cell is covered with microvilli.

D. W. Fawcett from D. W. Fawcett, The Cell 2e, (Sanders, 1981)

microscope, it turns out to be wrinkled into hundreds of folds. Each fold is covered with thousands of fingerlike projections called **villi.** The villi are as numerous as the fibers on velvet fabric. A single villus, magnified still more, turns out to be composed of several hundred cells, each covered with microscopic hairs called **microvilli** (see Figure 5-5).

The villi are in constant motion. A thin sheet of muscle lines each villus so that it can wave, squirm, and wiggle like a sea anemone's tentacles. Any nutrient molecule small enough to be absorbed is trapped among the microvilli and drawn into a cell beneath them. Some partially digested nutrients are caught in the microvilli, digested further by enzymes there, and then absorbed into the cells.

Specialization in the Intestinal Tract As you can see, the intestinal tract is beautifully designed to perform its functions. A further refinement of the system is that the cells of successive portions of the tract are specialized to absorb different nutrients. The nutrients that are ready for absorption early are absorbed near the top of the tract; those that take longer to be digested are absorbed farther down. The rate at which the nutrients travel through the GI tract is finely adjusted to maximize their availability to the appropriate absorptive segment of the tract when they are ready. The lowly "gut" turns out to be one of the most elegantly designed organ systems in the body.

The Myth of "Food Combining" Some popular fad diets advocate the idea that people should not eat certain food combinations (for example, fruit and meat) at the same meal, because the digestive system cannot handle more than one task at a time. This is a myth. The art of "food combining" (which actually emphasizes "food separating") is based on this idea, and it represents faulty logic and a gross underestimation of the body's capabilities. In fact, the opposite is often true: foods eaten together can enhance each other's use by the body. For example, vitamin C in a pineapple or citrus fruit can enhance the absorption of iron from a meal of chicken and rice or other iron-containing foods. Many other instances of mutually beneficial interactions are presented in later chapters.

Absorption of Nutrients

Once a molecule has entered a cell in a villus, the next step is to transmit it to a destination elsewhere in the body by way of the body's two transport systems—the bloodstream and the **lymphatic system.** As Figure 5-5 shows, both systems supply vessels to each villus. Through these vessels, the nutrients leave the cell and enter either the **lymph** or the blood. In either case, the nutrients end up in the blood, at least for a while. The water-soluble nutrients (and the smaller products of fat digestion) are released directly into the bloodstream by way of the capillaries, but the larger fats and the fat-soluble vitamins find direct access into the capillaries impossible because these nutrients are insoluble in water (and blood is mostly water). They require some packaging before they are released.

The intestinal cells assemble the products of fat digestion into larger molecules called **triglycerides.** These triglycerides, fat-soluble vitamins (when present), and other large lipids (cholesterol and the phospholipids) are then packaged for transport. They cluster together with special proteins to form **chylomicrons,** one kind of **lipoproteins** (lipoproteins are described beginning on p. 118). Finally, the cells release the chylomicrons into the lymphatic system. They can then glide through the lymph vessels until they arrive at a point of entry into the bloodstream near the heart.

IN SUMMARY

▪ The many folds and villi of the small intestine dramatically increase its surface area, facilitating nutrient absorption.

▪ Nutrients pass through the cells of the villi and enter either the blood (if they are water soluble or small fat fragments) or the lymph (if they are fat soluble).

villi (VILL-ee or VILL-eye): fingerlike projections from the folds of the small intestine. The singular form is **villus.**
  villus = shaggy hair

microvilli (MY-cro-VILL-ee or MY-cro-VILL-eye): tiny, hairlike projections on each cell of every villus that can trap nutrient particles and transport them into the cells. The singular form is **microvillus.**

lymphatic system: a loosely organized system of vessels and ducts that conveys the products of digestion toward the heart.

lymph (LIMF): the body fluid found in lymphatic vessels. Lymph consists of all the constituents of blood except red blood cells.

triglycerides (try-GLISS-er-rides): one of the main classes of lipids: the chief form of fat in foods and the major storage form of fat in the body; composed of glycerol with three fatty acids attached.
  tri = three
  glyceride = a compound of glycerol

chylomicrons (kye-lo-MY-crons): the lipoproteins that transport lipids from the intestinal cells into the body. The cells of the body remove the lipids they need from the chylomicrons, leaving chylomicron remnants to be picked up by the liver cells.

lipoproteins: clusters of lipids associated with proteins that serve as transport vehicles for lipids in the lymph and blood.

Transport of Nutrients

Once a nutrient has entered the bloodstream or the lymphatic system, it may be transported to any part of the body, from the tips of the toes to the roots of the hair, where it becomes available to any of the cells. The circulatory systems are arranged to deliver nutrients wherever they are needed.

The Vascular System

The vascular or blood circulatory system is a closed system of vessels through which blood flows continuously in a figure eight, with the heart serving as a pump at the crossover point. On each loop of the figure eight, blood travels a simple route: heart to arteries to capillaries to veins to heart.

The routing of blood through the digestive system is different, however. Blood is carried to the digestive system (as it is to all organs) by way of an **artery,** which (as in all organs) branches into **capillaries** to reach every cell. Blood leaving the digestive system goes by way of a **vein.** The **hepatic portal vein,** however, directs blood not back to the heart but to another organ—the liver. This vein *again* branches into *capillaries* so that every cell of the liver has access to the newly absorbed nutrients that the blood is carrying. Blood leaving the liver then *again* collects into a vein, called the **hepatic vein,** which returns the blood to the heart. The route is thus heart to arteries to capillaries (in intestines) to hepatic portal vein to capillaries (in liver) to hepatic vein to heart.

An anatomist studying this system knows there must be a reason for this special arrangement. The liver is located in the circulation system at the point where it will have the first chance at the materials absorbed from the GI tract. In fact, the liver is the body's major metabolic organ (see Figure 5-6) and must prepare the absorbed nutrients for use by the rest of the body. Furthermore, the liver stands as a gatekeeper to waylay intruders that might otherwise harm the heart or brain. Chapter 20 offers more information about this crucial organ.

The Lymphatic System

The lymphatic system is a one-way route for fluids to travel from tissue spaces into the blood. The lymphatic system has no pump; instead, lymph is squeezed from one portion of the body to another like water in a sponge, as muscles contract and create pressure here and there. Ultimately, the lymph flows into a large duct behind the heart. This duct terminates in a vein that conducts the lymph into the heart. In this way, fat-soluble nutrients absorbed into the lymphatic system from the GI tract finally enter the bloodstream.

Transport of Lipids: Lipoproteins

Within the circulatory system, lipids always travel from place to place bundled with protein, that is, as lipoproteins. When physicians measure a person's blood lipid profile, they are interested in both the types of fat present (such as triglycerides and cholesterol) and the types of lipoproteins that carry them.

VLDL, LDL, and HDL As mentioned earlier, chylomicrons transport newly absorbed (*diet-derived*) lipids from the intestinal cells to the rest of the body. As chylomicrons circulate through the body, cells remove their lipid contents, so the chylomicrons gradually become smaller. The liver picks up these chylomicron remnants. When necessary, the liver can assemble different lipoproteins, which are known as **very-low-density lipoproteins (VLDL).** As the body's cells remove triglycerides from the VLDL, the proportions of their lipid and protein contents shift. As this occurs, VLDL become cholesterol-rich **low-density lipoproteins (LDL).** Cholesterol returning to the liver from other parts of the body for metabolism or excretion is packaged in lipoproteins known as **high-density lipoproteins (HDL).** HDL are synthesized primarily in the liver.

The artery that delivers oxygen-rich blood from the heart and lungs to the liver is the *hepatic artery.*

The duct that conveys lymph toward the heart is the *thoracic (thor-ASS-ic) duct.* The *subclavian vein* connects this duct with the right upper chamber of the heart, providing a passageway by which lymph can be returned to the vascular system.

artery: a vessel that carries blood away from the heart.

capillaries: small vessels that branch from an artery. Capillaries connect arteries to veins. Oxygen, nutrients, and waste materials are exchanged across capillary walls.

vein: a vessel that carries blood back to the heart.

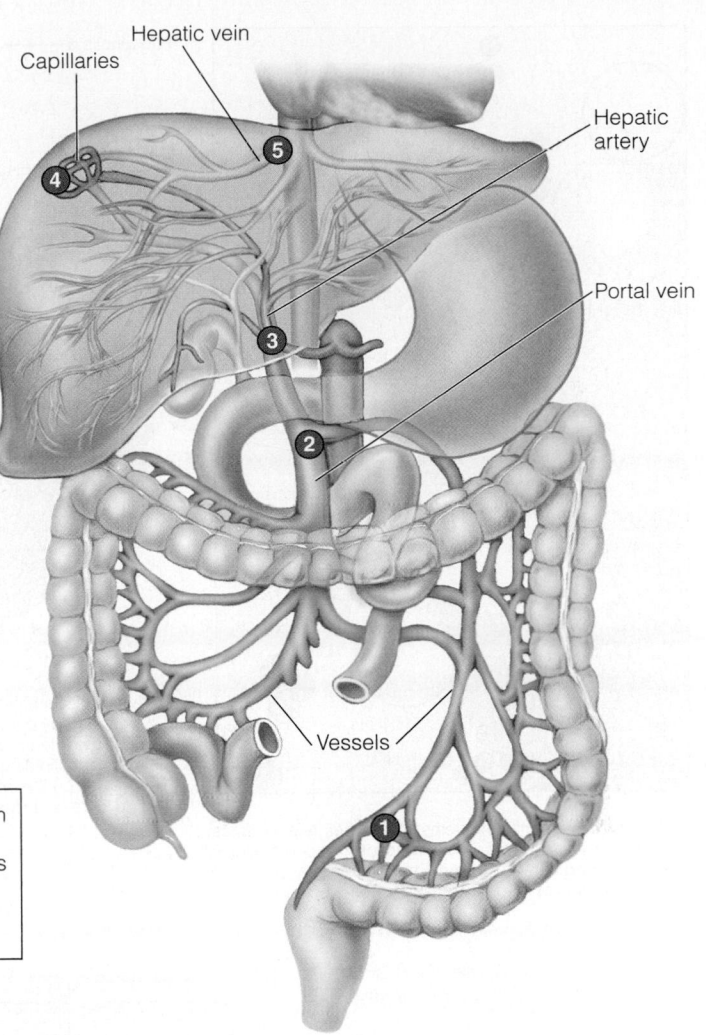

1 Vessels gather up nutrients and reabsorbed water and salts from all over the digestive tract.

Not shown here:
Parallel to these vessels (veins) are other vessels (arteries) that carry oxygen-rich blood from the heart to the intestines.

2 The vessels merge into the portal vein, which conducts all absorbed materials to the liver.

3 The hepatic artery brings a supply of freshly oxygenated blood (not loaded with nutrients) from the lungs to supply oxygen to the liver's own cells.

4 Capillaries branch all over the liver, making nutrients and oxygen available to all its cells and giving the cells access to blood from the digestive system.

5 The hepatic vein gathers up blood in the liver and returns it to the heart.

In contrast, nutrients absorbed into lymph do not go to the liver first. They go to the heart, which pumps them to all the body's cells. The cells remove the nutrients they need, and the liver then has to deal only with the remnants.

Hepatic vein

Capillaries

Hepatic artery

Portal vein

Vessels

FIGURE 5-6 The Liver and Its Circulatory System

The density of lipoproteins varies according to the proportion of lipids and protein they contain. The more lipids in the lipoprotein molecule, the lower the density; the more protein, the higher the density. Both LDL and HDL carry lipids around in the blood, but LDL are larger, lighter, and filled with more lipid; HDL are smaller, denser, and packaged with more protein. LDL deliver cholesterol and triglycerides from the liver to the tissues; HDL scavenge excess cholesterol from the tissues and return it to the liver for metabolism or disposal. Figure 5-7 (p. 120) shows the relative sizes and compositions of the lipoproteins.

Health Implications of LDL and HDL The distinction between LDL and HDL has implications for the health of the heart and blood vessels. Elevated LDL concentrations in the blood are associated with a high risk of heart disease, and elevated HDL concentrations are associated with a low risk.[2] These associations explain why some people refer to LDL as "bad" cholesterol and HDL as "good" cholesterol. Keep in mind, though, that there is only *one* kind of cholesterol molecule; the differences between LDL and HDL reflect *proportions* of lipids and proteins within them—not the type of cholesterol. Factors that improve the LDL-to-HDL ratio include:

▮ Weight management (see Chapter 7).

▮ Polyunsaturated or monounsaturated, instead of saturated, fatty acids in the diet (see Chapter 3).

▮ Soluble fibers (see Chapter 2).

hepatic portal vein: the vein that collects blood from the GI tract and conducts it to capillaries in the liver.
 portal = gateway

hepatic vein: the vein that collects blood from the liver capillaries and returns it to the heart.
 hepatic = liver

very-low-density lipoproteins (VLDL): the type of lipoproteins made primarily by liver cells to transport lipids to various tissues in the body; composed primarily of triglycerides.

low-density lipoproteins (LDL): the type of lipoproteins derived from VLDL as cells remove triglycerides from them. LDL carry cholesterol and triglycerides from the liver to the cells of the body and are composed primarily of cholesterol.

high-density lipoproteins (HDL): the type of lipoproteins that transport cholesterol back to the liver from peripheral cells; composed primarily of protein.

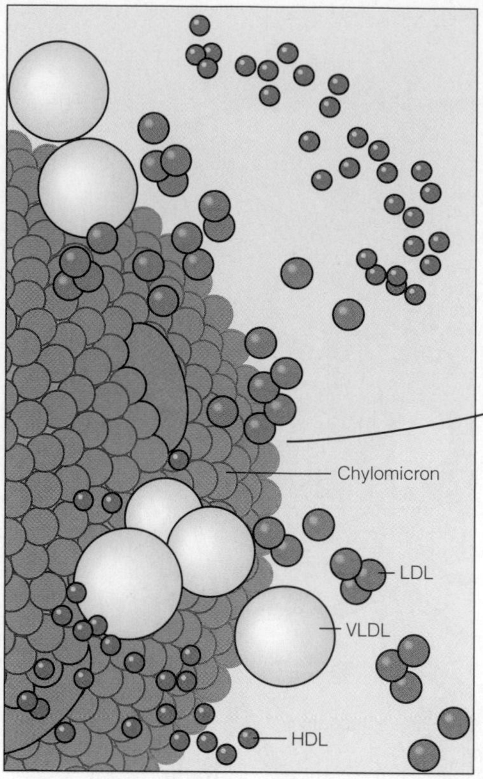

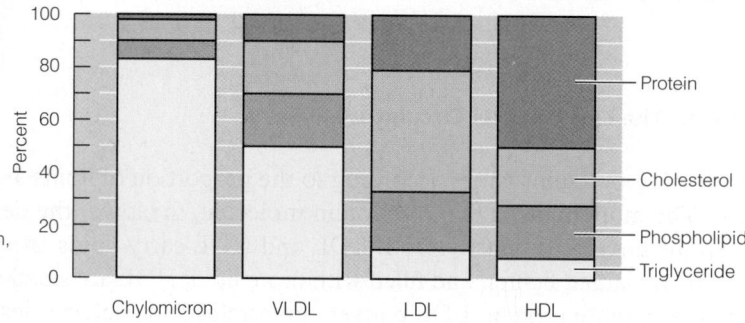

Phospholipid — — Protein

Cholesterol —

Triglyceride —

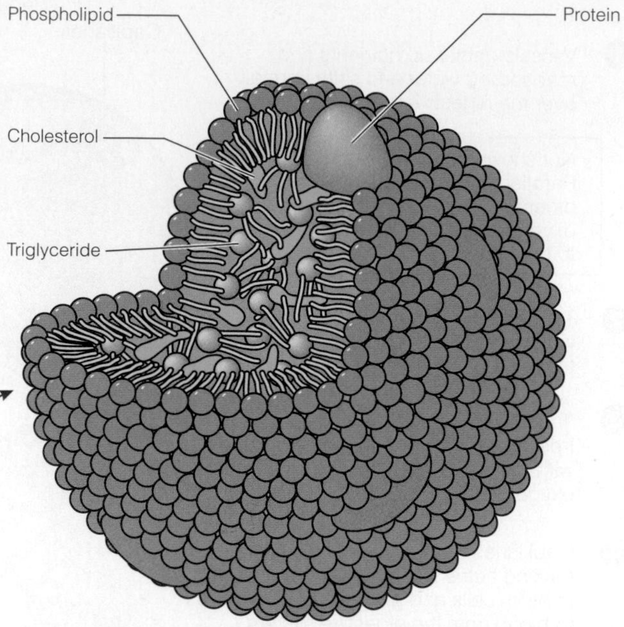

A typical lipoprotein contains an interior of triglycerides and cholesterol surrounded by phospholipids. The phospholipids' fatty-acid "tails" point toward the interior, where the lipids are. Proteins near the outer ends of the phospholipids cover the structure. This arrangement of hydrophobic molecules on the inside and hydrophilic molecules on the outside allows lipids to travel through the watery fluids of the blood.

— Chylomicron

— LDL

— VLDL

— HDL

This solar system of lipoproteins shows their relative sizes. Notice how large the fat-filled chylomicron is compared with the others and how the others get progressively smaller as their proportion of fat declines and protein increases.

Chylomicrons contain so little protein and so much triglyceride that they are the lowest in density.

Very-low-density lipoproteins (VLDL) are half triglycerides, accounting for their low density.

Low-density lipoproteins (LDL) are half cholesterol, accounting for their implication in heart disease.

High-density lipoproteins (HDL) are half protein, accounting for their high density.

Percent

100
80
60
40
20
0

Chylomicron   VLDL   LDL   HDL

— Protein
— Cholesterol
— Phospholipid
— Triglyceride

FIGURE 5-7 The Lipoproteins

Physical activity (see Chapter 10).

Lipoproteins and heart disease are discussed in Chapter 22.

IN SUMMARY

Nutrients leaving the digestive system via the blood are routed directly to the liver before being transported to the body's cells. Those leaving via the lymphatic system eventually enter the vascular system, but they bypass the liver at first.

Within the circulatory system, lipids travel bundled with proteins as lipoproteins. Different types of lipoproteins include chylomicrons, very-low-density lipoproteins (VLDL), low-density lipoproteins (LDL), and high-density lipoproteins (HDL).

■ Elevated blood concentrations of LDL are associated with a high risk of heart disease. Elevated HDL are associated with a low risk of heart disease.

The System at Its Best

The GI tract is the first organ in the body to deal with the nutrients that will ultimately maintain the health and nutrition status of the whole body. Its intricate architecture makes it sensitive and responsive to conditions in its environment. One condition indispensable to the GI tract's performance is its own good health. Such lifestyle factors as sleep, physical activity, state of mind, and nutrition affect GI tract health. Adequate sleep allows for repair and maintenance of tissue. Physical activity promotes healthy muscle tone and may protect against cancer of the colon.[3] Mental state profoundly affects digestion and absorption through the activity of nerves and hormones that help regulate these processes. A relaxed, peaceful attitude during a meal enhances digestion and absorption.

Clinical Applications

1. People who experience malabsorption frequently have the most difficulty digesting fat. Considering the differences in fat, carbohydrate, and protein digestion and absorption, can you offer an explanation?

2. How might you explain the importance of dietary fiber to a client who frequently experiences constipation?

Self Check

1. Once food is swallowed, it travels through the digestive tract in this order:
   a. esophagus, stomach, large intestine, liver.
   b. esophagus, stomach, small intestine, large intestine.
   c. small intestine, stomach, esophagus, large intestine.
   d. small intestine, large intestine, stomach, esophagus.

2. Once chyme travels the length of the small intestine, it passes through the ileocecal valve at the beginning of the:
   a. large intestine.
   b. stomach.
   c. esophagus.
   d. jejunum.

3. The periodic squeezing or partitioning of the intestine by its circular muscles that both mixes and slowly pushes the contents along is known as:
   a. secretion.
   b. absorption.
   c. peristalsis.
   d. segmentation.

4. An enzyme in saliva begins the digestion of:
   a. starch.
   b. vitamins.
   c. protein.
   d. minerals.

5. Bile is:
   a. an enzyme that splits starch.
   b. an alkaline secretion of the pancreas.
   c. an emulsifier made by the liver that prepares fats and oils for digestion.
   d. a stomach secretion containing water, hydrochloric acid, and the enzymes pepsin and lipase.

6. Which nutrient passes through the large intestine mostly unabsorbed?
   a. Fiber
   b. Vitamins
   c. Minerals
   d. Starch

7. The two major nutrient transport systems in the body are:
   a. LDL and HDL.
   b. digestion and absorption.
   c. lipoproteins and chylomicrons.
   d. the vascular and lymphatic systems.

8. Within the circulatory system, lipids always travel from place to place bundled with proteins as:
   a. microvilli.
   b. chylomicrons.
   c. lipoproteins.
   d. phospholipids.

9. Elevated LDL concentrations in the blood are associated with:
   a. a high-protein diet.
   b. a low risk of diabetes.
   c. too much physical activity.
   d. a high risk of heart disease.

10. Three factors that improve the LDL-to-HDL ratio include:
    a. polyunsaturated fat, rest, and dietary HDL.
    b. antioxidants, insoluble fibers, and dietary HDL.
    c. saturated fat, antioxidants, and insoluble fibers.
    d. weight control, soluble fibers, and physical activity.

Answers to these questions can be found in Appendix H.

Notes

1. A. Stengel and Y. Taché, Neuroendocrine control of the gut during stress: Corticotropin-releasing factor signaling pathways in the spotlight, *Annual Review of Physiology* 71 (2009): 219–239; P. Holzer, Opioid receptors in the gastrointestinal tract, *Regulatory Peptides* 155 (2009): 11–17; J. López-Herce, Gastrointestinal complications in critically ill patients: What differs between adults and children? *Current Opinion in Clinical Nutrition and Metabolic Care* 12 (2009): 180–185; National Digestive Diseases Information Clearinghouse, *Irritable Bowel Syndrome,* NIH publication no. 07-693 (Bethesda, MD: National Institute of Diabetes and Digestive and Kidney Diseases, 2007).

2. A. K. Chhatriwalla and coauthors, Low levels of low-density lipoprotein cholesterol and blood pressure and progression of coronary atherosclerosis, *Journal of the American College of Cardiology* 53 (2009): 1110–1115; A. H. Lichtenstein and coauthors, Diet and lifestyle recommendations revision 2006: A scientific statement from the American Heart Association Nutrition Committee, *Circulation* 114 (2006): 82–96; Expert Panel on Detection, Evaluation, and Treatment of High Blood Cholesterol in Adults (Adult Treatment Panel III), *Third Report of the National Cholesterol Education Program (NCEP)*, NIH publication no. 02-5215 (Bethesda, MD: National Heart, Lung, and Blood Institute, 2002) pp. II-1–II-61.

3. World Cancer Research Fund/American Institute for Cancer Research, *Food, Nutrition, Physical Activity, and the Prevention of Cancer: A Global Perspective* (Washington, D.C.: American Institute for Cancer Research, 2007), pp. 244–321.

CHAPTER 5

Nutrition in Practice

The Food and Drug Administration (FDA) lists **foodborne illness** as the leading food safety concern in the United States because episodes of food poisoning far outnumber episodes of any other kind of food contamination. The **CDC** estimates that 48 million people experience foodborne illness each year in the United States.[1] For some 3000 people each year, the symptoms (Table NP5-1) can be so severe as to cause death. Most vulnerable are pregnant women; very young, very old, sick, or malnourished people; and those with a weakened immune system (as in AIDS). By taking the proper precautions, people can minimize their chances of contracting foodborne illnesses. The accompanying glossary defines related terms.

What causes foodborne illness?

Foodborne illness can be caused by either an infection or an intoxication. Table NP5-2 (on p. 124) summarizes the most common or severe foodborne illnesses, their food sources, general symptoms, and prevention methods.

What is the difference between foodborne infections and food intoxications?

Foodborne infections are caused by eating foods contaminated by infectious microbes. Two of the most common foodborne **pathogens** are *Campylobacter jejuni* and *Salmonella*, which enter the GI tract in contaminated foods such as undercooked poultry and unpasteurized milk. Symptoms generally include abdominal cramps, fever, vomiting, and diarrhea.

Norovirus, another leading cause of foodborne illness, illustrates the importance of personal hygiene.[2] Norovirus is passed in the stool and vomit of infected people. Thus, infected people who do not wash their hands adequately can pass the virus directly to other people, or they can pass it indirectly by way of contaminated food or water. Outbreaks in the United States are often linked to food touched by infected food handlers or to person-to-person contact in day care centers, nursing homes, and cruise ships.[3] People can also be infected with norovirus by eating raw shellfish such as oysters and clams that are grown in sewage-contaminated waters.

TABLE NP5-1 Symptoms of Foodborne Illness

Get medical help when these symptoms occur:

- Bloody diarrhea
- Diarrhea lasting more than three days
- Prolonged vomiting that prevents keeping liquids down and can lead to dehydration
- Difficulty breathing
- Difficulty swallowing
- Double vision
- Fever lasting more than 24 hours
- Headache accompanied by muscle stiffness and fever
- Numbness, muscle weakness, and tingling sensations in the skin
- Rapid heart rate, fainting, and dizziness

Food intoxications are caused by eating foods containing natural toxins or, more likely, microbes that produce toxins. The most common food toxin is produced by *Staphylococcus aureus;* it affects more than 1 million people each year. Less common, but more infamous, is *Clostridium botulinum,* an organism that produces a deadly toxin in anaerobic conditions such as improperly canned (especially home-canned) foods and homemade garlic or herb-flavored oils stored at room temperature. Botulism paralyzes muscles; consequently, seeing, speaking, swallowing, and breathing become difficult. Because death can occur within 24 hours of onset, botulism demands immediate medical attention. Even then, survivors may suffer the effects for months or years.

How do people get foodborne illness?

Transmission of foodborne illness has changed as our food supply and lifestyles have changed.[4] In the past, foodborne illness was caused by one person's error in a small setting, such as improperly refrigerated egg salad at a family picnic, and affected only a few victims. Today, we are eating more foods prepared and packaged by others. Consequently, when a food manufacturer or restaurant chef makes an error, foodborne illness can become epidemic. An estimated 80 percent of reported foodborne illnesses are caused by errors in a commercial setting, such as the improper **pasteurization** of milk at a large dairy.

TABLE NP5-2 Foodborne Illnesses

Disease and Organism That Causes It	Most Frequent Food Sources	Onset and General Symptoms	Prevention Methods[a]
Foodborne Infections			
Campylobacteriosis (KAM-pee-loh-BAK-ter-ee-OH-sis) *Campylobacter* bacterium	Raw and undercooked poultry, unpasteurized milk, contaminated water	Onset: 2 to 5 days. Diarrhea, vomiting, abdominal cramps, fever; sometimes bloody stools; lasts 2 to 10 days.	Cook foods thoroughly; use pasteurized milk; use sanitary food-handling methods.
Cryptosporidiosis (KRIP-toe-spo-rid-ee-OH-sis) *Crytosporidium parvum* parasite	Contaminated swimming or drinking water, even from treated sources; highly chlorine-resistant; contaminated raw produce and unpasteurized juices and ciders	Onset: 2 to 10 days. Diarrhea, stomach cramps, upset stomach, slight fever; symptoms may come and go for weeks or months.	Wash all raw vegetables and fruits before peeling; use pasteurized milk and juice; do not swallow drops of water while using pools, hot tubs, ponds, lakes, rivers, or streams for recreation.
Cyclosporiasis (sigh-clo-spore-EYE-uh-sis) *Cyclospora cayetanensis* parasite	Contaminated water, contaminated fresh produce	Onset: 1 to 14 days. Diarrhea, loss of appetite, weight loss, stomach cramps, nausea, vomiting, fatigue; symptoms may come and go for weeks or months.	Use treated, boiled, or bottled water; cook foods thoroughly; peel fruits.
E. coli infection *Escherichia coli*[b] bacterium	Undercooked ground beef, unpasteurized milk and juices, raw fruits and vegetables, contaminated water, and person-to-person contact	Onset: 1 to 8 days. Severe bloody diarrhea, abdominal cramps, vomiting; lasts 5 to 10 days.	Cook ground beef thoroughly; use pasteurized milk; use sanitary food-handling methods; use treated, boiled, or bottled water.
Gastroenteritis[c] Norovirus	Person-to-person contact; raw foods, salads, sandwiches	Onset: 1 to 2 days. Vomiting; lasts 1 to 2 days.	Use sanitary food-handling methods.
Giardiasis (JYE-are-DYE-ah-sis) *Giardia intestinalis* parasite	Contaminated water; uncooked foods	Onset: 7 to 14 days. Diarrhea (but occasionally constipation), abdominal pain, gas.	Use sanitary food-handling methods; avoid raw fruits and vegetables where parasites are endemic; dispose of sewage properly.
Hepatitis (HEP-ah-TIE-tis) Hepatitis A virus	Undercooked or raw shellfish	Onset: 15 to 50 days (28 days average). Diarrhea, dark urine, fever, headache, nausea, abdominal pain, jaundice (yellowed skin and eyes from buildup of wastes); lasts 2 to 12 weeks.	Cook foods thoroughly.
Listeriosis (lis-TER-ee-OH-sis) *Listeria monocytogenes* bacterium	Unpasteurized milk; fresh soft cheeses; luncheon meats, hot dogs	Onset: 1 to 21 days. Fever, muscle aches; nausea, vomiting, blood poisoning, complications in pregnancy, and meningitis (stiff neck, severe headache, and fever).	Use sanitary food-handling methods; cook foods thoroughly; use pasteurized milk.
Perfringens (per-FRINGE-enz) **food poisoning** *Clostridium perfringens* bacterium	Meats and meat products stored at between 120°F and 130°F	Onset: 8 to 16 hours. Abdominal pain, diarrhea, nausea; lasts 1 to 2 days.	Use sanitary food-handling methods; use pasteurized milk; cook foods thoroughly; refrigerate foods promptly and properly.
Salmonellosis (sal-moh-neh-LOH-sis) *Salmonella* bacteria (>2300 types)	Raw or undercooked eggs, meats, poultry, raw milk and other dairy products, shrimp, frog legs, yeast, coconut, pasta, and chocolate	Onset: 1 to 3 days. Fever, vomiting, abdominal cramps, diarrhea; lasts 4 to 7 days; can be fatal.	Use sanitary food-handling methods; use pasteurized milk; cook foods thoroughly; refrigerate foods promptly and properly.

continued

Disease and Organism That Causes It	Most Frequent Food Sources	Onset and General Symptoms	Prevention Methods[a]
Shigellosis (shi-gel-LOH-sis) *Shigella* bacteria (>30 types)	Person-to-person contact, raw foods, salads, sandwiches, and contaminated water	Onset: 1 to 2 days. Bloody diarrhea, cramps, fever; lasts 4 to 7 days.	Use sanitary food-handling methods; cook foods thoroughly; use proper refrigeration.
Vibrio (VIB-ree-oh) **infection** *Vibrio vulnificus*[d] bacterium	Raw or undercooked seafood, contaminated water	Onset: 1 to 7 days. Diarrhea, abdominal cramps, nausea, vomiting; lasts 2 to 5 days; can be fatal.	Use sanitary food-handling methods; cook foods thoroughly.
Yersiniosis (yer-SIN-ee-OH-sis) *Yersinia enterocolitica* bacterium	Raw and undercooked pork, unpasteurized milk	Onset: 1 to 2 days. Diarrhea, vomiting, fever, abdominal pain; lasts 1 to 3 weeks.	Cook foods thoroughly; use pasteurized milk; use treated, boiled, or bottled water.
Food Intoxications			
Botulism (BOT-chew-lizm) Botulinum toxin produced by *Clostridium botulinum* bacterium, which grows without oxygen, in low-acid foods, and at temperatures between 40°F and 120°F; the **botulinum** (BOT-chew-line-um) **toxin** responsible for botulism is called **botulin** (BOT-chew-lin)	Anaerobic environment of low acidity (canned corn, peppers, green beans, soups, beets, asparagus, mushrooms, ripe olives, spinach, tuna, chicken, chicken liver, liver pâté, luncheon meats, ham, sausage, stuffed eggplant, lobster, and smoked and salted fish)	Onset: 4 to 36 hours. Nervous system symptoms, including double vision, inability to swallow, speech difficulty, and progressive paralysis of the respiratory system; often fatal; leaves prolonged symptoms in survivors.	Use proper canning methods for low-acid foods; refrigerate homemade garlic and herb oils; avoid commercially prepared foods with leaky seals or with bent, bulging, or broken cans. Do not give infants honey because it may contain spores of *Clostridium botulinum*, which is a common source of infection for infants.
Staphylococcal (STAF-il-oh-KOK-al) **food poisoning** Staphylococcal toxin (produced by *Staphylococcus aureus* bacterium)	Toxin produced in improperly refrigerated meats; egg, tuna, potato, and macaroni salads; cream-filled pastries	Onset: 1 to 6 hours. Diarrhea, nausea, vomiting, abdominal cramps, fever; lasts 1 to 2 days.	Use sanitary food-handling methods; cook food thoroughly; refrigerate foods promptly and properly; use proper home-canning methods.

Note: Travelers' diarrhea is most commonly caused by *E. coli, Campylobacter jejuni, Shigella,* and *Salmonella.*

[a]Table NP5-3 on p. 127 provides more details on the proper handling, cooking, and refrigeration of foods.

[b]The most serious strain is *E. coli* STEC O157.

[c]Gastroenteritis refers to an inflammation of the stomach and intestines but is the most common name used for illnesses caused by noroviruses.

[d]Most cases of *Vibrio vulnificus* infection occur in persons with underlying illness, particularly those with liver disorders, diabetes, cancer, and AIDS, and those who require long-term steroid use. The fatality rate is 50 percent for this population.

In the early 2000s, a national food company had to recall more than 4 million pounds of poultry products after *Listeria* poisoning killed 7 people and made more than 50 others sick. In the 2006 *E. coli* outbreak from fresh spinach, nearly 200 people became sick, and two elderly women and a two-year-old boy died before consumers got the FDA message not to eat fresh spinach. In 2009, *Salmonella* was found in peanut butter that had been used in more than 2100 products made by more than 200 companies. These incidents and others focus the national spotlight on two important safety issues: disease-causing organisms are commonly found in foods, and safe food-handling practices can minimize harm from most of these foodborne pathogens.

What kinds of programs are in place to help keep foods safe?

To make our food supply safe for consumers, the U.S. Department of Agriculture (USDA), the FDA, and the food-processing industries have developed and implemented programs to control foodborne illness.* The **Hazard Analysis Critical Control Points (HACCP)** system requires food manufacturers to identify points of contamination and implement controls to prevent foodborne disease. For example, after tracing two large outbreaks of salmonellosis to imported cantaloupe, producers began using chlorinated water to wash the melons and to make ice for packing and shipping. Safety procedures such as this prevent hundreds of thousands of foodborne illnesses each year and are responsible for the decline in infections over the past decade.

This example raises another issue regarding the safety of imported foods. Because FDA inspectors cannot keep pace with the increasing numbers of imported foods, the FDA is working

*In addition to HACCP, other programs initiated under the Food Safety Initiative include FoodNet, PulseNet, the Environmental Health Specialists Network (EHS-Net), and Fight BAC!

with other countries to help them adopt the safe food-handling practices used in the United States.

Are foods bought in grocery stores and foods eaten in restaurants safe?

Canned and packaged foods sold in grocery stores are easily controlled, but rarely, accidents do happen. Batch numbering makes it possible to recall contaminated foods through public announcements via newspapers, television, and radio. In the grocery store, consumers can buy items before the "sell by" date and inspect the safety seals and wrappers of packages. A broken seal, bulging can lid, or mangled package fails to protect the consumer against microbes, insects, spoilage, or even vandalism.

State and local health regulations provide guidelines on the cleanliness of facilities and the safe preparation of foods for restaurants, cafeterias, and fast-food establishments. Even so, consumers can also take these actions to help prevent foodborne illnesses when dining out:

▌ Wash hands with hot, soapy water before meals.

▌ Expect clean tabletops, dinnerware, utensils, and food preparation areas.

▌ Expect cooked foods to be served piping hot and salads to be fresh and cold.

▌ Refrigerate take-home items within two hours.

Improper handling of foods can occur anywhere along the line from commercial manufacturers to large supermarkets to small restaurants to private homes. Maintaining a safe food supply requires everyone's efforts.

What can people do to protect themselves from foodborne illness?

Whether microbes multiply and cause illness depends, in part, on a few key food-handling behaviors in the kitchen—whether the kitchen is in your home, a school cafeteria, a gourmet restaurant, or a canned goods manufacturer. Figure NP5-1 summarizes the four simple things that can help most to prevent foodborne illness:

▌ *Keep a clean, safe kitchen.* Wash countertops, cutting boards, hands, sponges, and utensils in hot, soapy water before and after each step of food preparation.

▌ *Avoid cross-contamination.* Keep raw eggs, meat, poultry, and seafood separate from other foods. Wash all utensils and surfaces (such as cutting boards or platters) that have been in contact with these foods with hot, soapy water before using them again. Bacteria inevitably left on the surfaces from the raw meat can recontaminate the cooked meat or other foods—a problem known as **cross-contamination.** Washing raw eggs, meat, and poultry is not recommended as the extra handling increases the risk of cross-contamination.

▌ *Keep hot foods hot.* Cook foods long enough to reach internal temperatures that will kill microbes, and maintain adequate temperatures to prevent bacterial growth until the foods are served.

FIGURE NP5-1 Fight Bac!
Four ways to keep food safe. The Fight Bac! website is at www.fightbac.org.

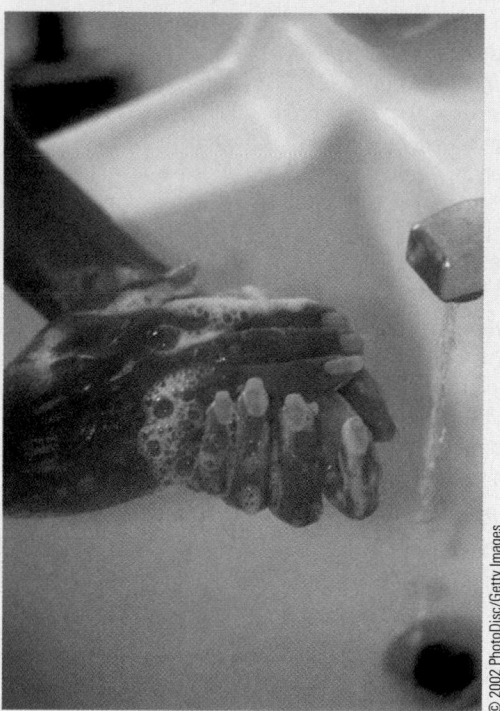

Wash your hands with warm water and soap for at least 20 seconds before preparing or eating food to reduce the chance of microbial contamination.

▌ *Keep cold foods cold.* Go directly home upon leaving the grocery store and immediately unpack foods into the refrigerator or freezer upon arrival. After a meal, refrigerate any leftovers immediately.

Unfortunately, consumers commonly fail to follow these simple food-handling recommendations. See Table NP5-3 for additional food safety tips.

What precautions need to be taken when preparing meat and poultry?

Figure NP5-2 (on p. 128) presents label instructions for the safe handling of meat and poultry and two types of USDA seals. Meats and poultry contain bacteria and provide a moist, nutrient-rich

126

Most foodborne illnesses can be prevented by following four simple rules: keep a clean kitchen, avoid cross-contamination, keep hot foods hot, and keep cold foods cold.

Keep a Clean Kitchen

▪ Wash fruits and vegetables in a clean sink with a scrub brush and warm water; store washed and unwashed produce separately.

▪ Use hot, soapy water to wash hands, utensils, dishes, nonporous cutting boards, and countertops before handling food and between tasks when working with different foods. Use a bleach solution on cutting boards (one capful per gallon of water).

▪ Cover cuts with clean bandages before food preparation; dirty bandages carry harmful microorganisms.

▪ Mix foods with utensils, not hands; keep hands and utensils away from mouth, nose, and hair.

▪ Anyone may be a carrier of bacteria and should avoid coughing or sneezing over food. A person with a skin infection or infectious disease should not prepare food.

▪ Wash or replace sponges and towels regularly.

▪ Clean up food spills and crumb-filled crevices.

Avoid Cross-Contamination

▪ Wash all surfaces that have been in contact with raw meats, poultry, eggs, fish, and shellfish before reusing.

▪ Serve cooked foods on a clean plate with a clean utensil. Separate raw foods from those that have been cooked.

▪ Don't use marinade that was in contact with raw meat for basting or sauces.

Keep Hot Foods Hot

▪ When cooking meats or poultry, use a thermometer to test the internal temperature. Insert the thermometer between the thigh and the body of a turkey or into the thickest part of other meats, making sure the tip of the thermometer is not in contact with bone or the pan. Cook to the temperature indicated for that particular meat (see Figure NP5-3); cook hamburgers to at least medium well done. If you have safety questions, call the USDA Meat and Poultry Hotline: (800) 535-4555.

▪ Cook stuffing separately, or stuff poultry just prior to cooking.

▪ Do not cook large cuts of meat or turkey in a microwave oven; it leaves some parts undercooked while overcooking others.

▪ Cook eggs before eating them (soft-boiled for at least 3½ minutes; scrambled until set, not runny; fried for at least 3 minutes on one side and 1 minute on the other).

▪ Cook seafood thoroughly. If you have safety questions about seafood, call the FDA hotline: (800) FDA-4010.

▪ When serving foods, maintain temperatures at 140°F or higher.

▪ Heat leftovers thoroughly to at least 165°F.

Keep Cold Foods Cold

▪ When running errands, stop at the grocery store last. When you get home, refrigerate the perishable groceries (such as meats and dairy products) immediately. Do not leave perishables in the car any longer than it takes for ice cream to melt.

▪ Put packages of raw meat, fish, or poultry on a plate before refrigerating to prevent juices from dripping on food stored below.

▪ Buy only foods that are solidly frozen in store freezers.

▪ Keep cold foods at 40°F or less; keep frozen foods at 0°F or less (keep a thermometer in the refrigerator).

▪ Marinate meats in the refrigerator, not on the counter.

▪ Look for "Keep Refrigerated" or "Refrigerate after Opening" on food labels.

▪ Refrigerate leftovers promptly; use shallow containers to cool foods faster; use leftovers within three to four days.

▪ Thaw meats or poultry in the refrigerator, not at room temperature. If you must hasten thawing, use cool water (changed every 30 minutes) or a microwave oven.

▪ Freeze meat, fish, or poultry immediately if not planning to use within a few days.

In General

▪ Do not reuse disposable containers; use nondisposable containers or recycle instead.

▪ Do not taste food that is suspect. "If in doubt, throw it out."

▪ Throw out foods with danger-signaling odors. Be aware, though, that most food-poisoning bacteria are odorless, colorless, and tasteless.

▪ Do not buy or use items that have broken seals or mangled packaging. Check safety seals, buttons, and expiration dates.

▪ Follow label instructions for storing and preparing packaged and frozen foods; throw out foods that have been thawed or refrozen.

▪ Discard foods that are discolored, moldy, or decayed or that have been contaminated by insects or rodents.

For Specific Food Items

▪ *Canned goods.* Carefully discard food from cans that leak or bulge so that other people and animals will not accidentally ingest it; before canning, seek professional advice from the USDA Extension Service (check your phone book under U.S. government listings, or go online: www.crees.usda.gov).

▪ *Milk and cheeses.* Use only pasteurized milk and milk products. Aged cheeses, such as cheddar and Swiss, do well for an hour or two without refrigeration, but they should be refrigerated or stored in an ice chest for longer periods.

▪ *Eggs.* Use clean eggs with intact shells. Do not eat eggs, even pasteurized eggs, raw; raw eggs are commonly found in Caesar salad dressing, eggnog, cookie dough, hollandaise sauce, and key lime pie. Cook eggs until whites are firmly set and yolks begin to thicken.

▪ *Honey.* Honey may contain dormant bacterial spores, which can awaken in the human body to produce botulism. In adults, this poses little hazard, but infants younger than 1 year should never be fed honey. Honey can accumulate enough toxin to kill an infant; it has been implicated in several cases of sudden infant death. (Honey can also be contaminated with environmental pollutants picked up by the bees.)

▪ *Mayonnaise.* Commercial mayonnaise may actually help a food to resist spoilage because of the acid content. Still, keep it refrigerated after opening.

▪ *Mixed salads.* Mixed salads of chopped ingredients spoil easily because they have extensive surface area for bacteria to invade, and they have been in contact with cutting boards, hands, and kitchen utensils that easily transmit bacteria to food (regardless of their mayonnaise content). Chill them well before, during, and after serving.

▪ *Picnic foods.* Choose foods that last without refrigeration, such as fresh fruits and vegetables, breads and crackers, and canned spreads and cheeses that can be opened and used immediately. Pack foods cold, layer ice between foods, and keep foods out of water.

▪ *Seafood.* Buy only fresh seafood that has been properly refrigerated or iced. Cooked seafood should be stored separately from raw seafood to avoid cross-contamination.

The voluntary "Graded by USDA" seal indicates that the product has been graded for tenderness, juiciness, and flavor. Beef is graded Prime (abundant marbling of the meat muscle), Choice (less marbling), and Select (lean). Similarly, poultry is graded A, B, and C.

The mandatory "Inspected and Passed by the USDA" seal ensures that meat and poultry products are safe, wholesome, and correctly labeled. Inspection does not guarantee that the meat is free of potentially harmful bacteria.

Safe Handling Instructions

THIS PRODUCT WAS PREPARED FROM INSPECTED AND PASSED MEAT AND/OR POULTRY. SOME FOOD PRODUCTS MAY CONTAIN BACTERIA THAT CAN CAUSE ILLNESS IF THE PRODUCT IS MISHANDLED OR COOKED IMPROPERLY. FOR YOUR PROTECTION, FOLLOW THESE SAFE HANDLING INSTRUCTIONS.

KEEP REFRIGERATED OR FROZEN. THAW IN REFRIGERATOR OR MICROWAVE.

KEEP RAW MEAT AND POULTRY SEPARATE FROM OTHER FOODS. WASH WORKING SURFACES (INCLUDING CUTTING BOARDS), UTENSILS, AND HANDS AFTER TOUCHING RAW MEAT OR POULTRY.

COOK THOROUGHLY.

KEEP HOT FOODS HOT. REFRIGERATE LEFTOVERS IMMEDIATELY OR DISCARD.

The USDA requires that safe handling instructions appear on all packages of meat and poultry.

FIGURE NP5-2 Meat and Poultry Safety, Grading, and Inspection Seals
Inspection is mandatory; grading is voluntary. Neither guarantees that the product will not cause foodborne illnesses, but consumers can help to prevent foodborne illnesses by following the safe handling instructions.

environment that favors microbial growth. Ground meat is especially susceptible because it receives more handling than other kinds of meat and has more surface exposed to bacterial contamination. Consumers cannot detect the harmful bacteria in or on meat. For safety's sake, cook meat thoroughly, using a thermometer to test the internal temperature (see Figure NP5-3).

How can a person enjoy seafood safely?

Most seafood available in the United States and Canada is safe, but eating it undercooked or raw can cause severe illnesses— hepatitis, worms, parasites, viral intestinal disorders, and other diseases.* Rumor has it that freezing fish will make it safe to eat raw, but this is only partly true. Commercial freezing will kill mature parasitic worms, but only cooking can kill all worm eggs and other microorganisms that can cause illness. For safety's sake, all seafood should be cooked until it is opaque. Even **sushi** delicacies can be safe to eat when prepared with cooked seafood.

Eating raw oysters can be dangerous for anyone, but people with liver disease and weakened immune systems are most vulnerable. At least 10 species of bacteria found in raw oysters can cause serious illness and even death.† Raw oysters may also carry the hepatitis A virus, which can cause liver disease. Some hot sauces can kill many of these bacteria but not the virus; alcohol may also protect some people against some oyster-borne illnesses but not enough to guarantee protection (or to recommend drinking alcohol). Pasteurization of raw oysters—holding them at a specified temperature for a specified time—holds promise for killing bacteria without cooking the oyster or altering its texture or flavor.

As population density increases along the shores of seafood-harvesting waters, pollution inevitably invades the sea life there.

Preventing seafood-borne illness is in large part a task of controlling water pollution. To help ensure a safe seafood market, the FDA requires processors to adopt food safety practices based on the HACCP system mentioned earlier.

Chemical pollution and microbial contamination lurk not only in the water but also in the boats and warehouses where seafood is cleaned, prepared, and refrigerated. Seafood is one of the most perishable foods: time and temperature are critical to its freshness and flavor. To keep seafood as fresh as possible, people in the industry "keep it cold, keep it clean, and keep it moving." Wise consumers eat it cooked.

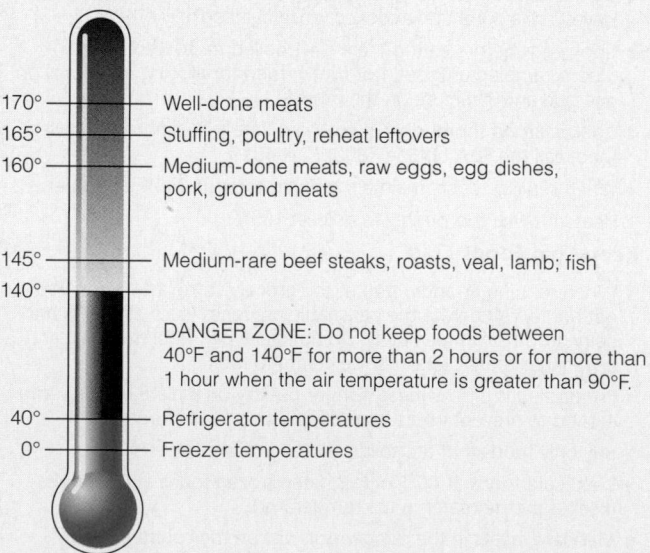

170° — Well-done meats
165° — Stuffing, poultry, reheat leftovers
160° — Medium-done meats, raw eggs, egg dishes, pork, ground meats
145° — Medium-rare beef steaks, roasts, veal, lamb; fish
140°
DANGER ZONE: Do not keep foods between 40°F and 140°F for more than 2 hours or for more than 1 hour when the air temperature is greater than 90°F.
40° — Refrigerator temperatures
0° — Freezer temperatures

FIGURE NP5-3 Recommended Safe Temperatures (Fahrenheit)
Bacteria multiply rapidly at temperatures between 40°F and 140°F. Cook foods to the temperatures shown on this thermometer and hold them at 140°F or higher.

*Diseases caused by toxins from the sea include ciguatera poisoning, scombroid poisoning, and paralytic and neurotoxic shellfish poisoning.

†Raw oysters can carry the bacterium *Vibrio vulnificus;* see Table NP5-2 for details.

Eric O'Connell/Taxi/Getty Images

Cook hamburgers to 160°F; color alone cannot determine doneness. Some burgers will turn brown before reaching 160°F, whereas others may retain some pink color, even when cooked to 175°F.

Davic Chasey/Photodisc Red/Getty Images

Eating raw seafood is a risky proposition.

Do foods that are unsafe to eat smell bad?

Fresh food generally smells fresh. Not all types of food poisoning are detectable by odor, but some bacterial wastes produce off odors. If an abnormal odor exists, the food is spoiled. Throw it out or, if it was recently purchased, return it to the grocery store. Do not taste it. Table NP5-4 lists safe refrigerator storage times for selected foods.

Local health departments and the USDA Extension Service can provide additional information about food safety. Should precautions fail and mild foodborne illness develop, drink clear liquids to replace fluids lost through vomiting and diarrhea. If serious foodborne illness is suspected, first call a physician.

TABLE NP5-4 Safe Refrigerator Storage Times (≤40°F)

One to Two Days

Raw ground meats, breakfast or other raw sausages, raw fish or poultry; gravies

Three to Five Days

Raw steaks, roasts, or chops; cooked meats, poultry, vegetables, and mixed dishes; lunch meats (packages opened); mayonnaise salads (chicken, egg, pasta, tuna)

One Week

Hard-cooked eggs, bacon, or hot dogs (opened packages); smoked sausages or seafood

Two to Four Weeks

Raw eggs (in shells); lunch meats, bacon, or hot dogs (packages unopened); dry sausages (pepperoni, hard salami); most aged and processed cheeses (Swiss, brick)

Two Months

Mayonnaise (opened jar); most dry cheeses (Parmesan, Romano)

Then wrap the remainder of the suspected food and label the container so that the food cannot be mistakenly eaten, place it in the refrigerator, and hold it for possible inspection by health authorities.

How can a person defend against foodborne illness when traveling to foreign countries?

People who travel to other countries have an estimated 40 percent chance of contracting a foodborne illness, commonly described as **travelers' diarrhea**.[5] Like many other foodborne illnesses, travelers' diarrhea is a sometimes serious, always annoying bacterial infection of the digestive tract. The risk is high because some countries' cleanliness standards for food and water may be lower than those in the United States and Canada. Also, every region's microbes are different, and although people are immune to those in their own neighborhoods, they have had no chance to develop immunity to the pathogens in places they are visiting for the first time. In addition to the food safety tips outlined on p. 127, precautions while traveling include:

▌ Wash hands frequently with soap and hot water, especially before handling food or eating. Use antiseptic gel or hand wipes regularly.

▌ Eat only well-cooked and hot or canned foods. Eat raw fruits or vegetables only if washed in purified water and peeled with clean hands.

▌ Use purified, bottled water for drinking, making ice cubes, and brushing teeth. Alternatively, use disinfecting tablets or boil water.

▌ Refuse dairy products that have not been pasteurized and refrigerated properly.

▌ Travel with antidiarrheal medication in case efforts to avoid illness fail.

To sum up these recommendations, "Boil it, cook it, peel it, or forget it."

Notes

1. Centers for Disease Control and Prevention, Press release: New estimates more precise, available at www.cdc.gov/eid, December 15, 2010. Position of the American Dietetic Association: Food and water safety, *Journal of the American Dietetic Association* 109 (2009): 1449–1460.

2. Centers for Disease Control and Prevention, Surveillance for foodborne disease outbreaks—United States, 2006, *Morbidity and Mortality Weekly Report* 58 (2009): 609–614.

3. L. Verhoef and coauthors, Emergence of new norovirus variants on spring cruise ships and prediction of winter epidemics, *Emerging Infectious Diseases* 14 (2008): 238–243.

4. Position of the American Dietetic Association, 2009.

5. H. L. Koo and H. L. DuPont, Current and future developments in travelers' diarrhea therapy, *Expert Review of Anti-Infective Therapy* 4 (2006): 417–427.

6

Metabolism, Energy Balance, and Body Composition

Every organ, every tissue, and every cell of the body engages in **metabolism,** the chemical reactions involved in releasing energy, breaking down compounds, and making new compounds. Much like a factory, the body works efficiently to manufacture needed products and dispose of wastes. All these processes are regulated by hormonal signals that coordinate supply and demand.

In disease, metabolic processes can become disturbed, and some diseases are caused by metabolic disturbances. This chapter introduces the principal organs and their metabolic roles, the metabolism of the energy nutrients, energy balance, body composition, and the health risks associated with too much or too little body fat. The next chapter offers strategies toward solving the problems of too much or too little body fat.

The Organs and Their Metabolic Roles

The metabolic reactions of every organ contribute to the body's ability to function normally and maintain health. Metabolic reactions also use or release energy and therefore affect body weight, with consequences for health.

The Principal Organs

Of particular concern to metabolism are the digestive organs, liver, pancreas, circulatory system, and kidneys. Together, they perform much of the work of breaking down compounds, making new ones, transporting nutrients and oxygen throughout the body, and removing the wastes generated by metabolic processes.

The Digestive Organs As Chapter 5 describes, the digestive system transports foods through the gastrointestinal (GI) tract, produces digestive juices and enzymes, absorbs nutrients, provides transport proteins to carry lipids and vitamins to other sites in the body, and reabsorbs salts and fluids. The digestive system also possesses the body's most rapidly multiplying cells: when healthy, they replace themselves every few days. Disorders affecting the GI tract interfere with the ingestion, digestion, absorption, and metabolism of nutrients, as described in Chapters 17 and 18.

The Liver Nutrients absorbed into the bloodstream are taken first to the liver, as described in Chapter 5. The liver is one of the body's most active metabolic factories. It receives nutrients and metabolizes, packages, stores, or ships them out for use by other organs. It manufactures bile, which the body uses to emulsify fat for digestion and absorption. It metabolizes and detoxifies drugs, prepares waste products for excretion, and participates in iron recycling and blood cell manufacture. It also makes many proteins necessary for health, including immune factors, transport proteins, and clotting factors. When liver disorders disrupt metabolism, they profoundly affect both nutrition and health status, as described in Chapter 20.

The Pancreas The pancreas contributes digestive juices to the GI tract, but also has another metabolic function: it produces the hormones insulin and glucagon that regulate the body's use of glucose. After a meal, as blood glucose rises, the pancreas secretes insulin. Insulin prompts cells to take up glucose and use it as fuel; insulin also prompts liver cells to store glucose as glycogen. When blood glucose falls (as occurs between meals), the pancreas responds by secreting glucagon into the blood. Glucagon raises blood glucose by signaling the liver to dismantle its glycogen stores and release glucose into the blood for use by all the other body cells. Glucose is an indispensable fuel for brain cells, nerve cells, and red blood cells. Its availability is therefore crucial to normal nervous system activity and blood chemistry. Abnormalities associated with the digestive functions of the pancreas

are described in Chapter 19, and those associated with its hormonal functions are described in Chapter 21.

The Heart and Blood Vessels The heart and blood vessels conduct blood, with its cargo of nutrients and oxygen, to all the other body cells and carry wastes away from them. Diseases of the heart and arteries therefore affect the health of the whole body. Metabolic reactions that affect the heart and blood vessels include, most importantly, the making and transport of lipoproteins, which are the carriers of cholesterol and other lipids from the liver to the tissues and back again. High blood levels of low-density lipoproteins (LDL) and very-low-density lipoproteins (VLDL) promote atherosclerosis, which increases the risk of disability or death from heart attacks and strokes. Chapter 22 is devoted to these conditions.

The Kidneys The kidneys are also active metabolic organs. Unceasingly, for 24 hours of every day, they filter waste products from the blood to be excreted in the urine and reabsorb needed nutrients, thereby maintaining the blood's delicate chemical balances. The kidneys' cells also produce compounds that help to regulate blood pressure and convert a precursor compound to active vitamin D, thereby helping to maintain the bones. Thus, disorders of the kidneys nearly always involve the heart and the skeleton. Kidney disorders are the subject of Chapter 23.

The Body's Metabolic Work

The metabolic work that the body's cells do, like all work, requires energy, and foods supply that energy. Foods, in turn, get their energy from the sun, either directly (in the case of photosynthesizing plants that make carbohydrate) or indirectly (in the case of animals that eat plants). When chemical reactions in cells release stored energy from energy-yielding nutrients, that energy becomes available to do the cells' work.

Heat Energy and Body Temperature The cells of each organ conduct metabolic activities specific to that organ. In addition, all cells must maintain themselves, and many must reproduce. To do this, they must have all the essential nutrients available to them: energy nutrients, vitamins, and minerals, as well as water. As cells do their metabolic work, the chemical reactions that are involved release heat, and this heat keeps the body warm. By regulating the rates at which these metabolic reactions release heat energy, the body maintains its constant normal temperature of about 98.6°F.

Accelerated Metabolism During severe stress to the body, metabolism speeds up. Fever sometimes develops. An accelerated metabolism signifies that fuels are being used at a rate more rapid than normal; this may lead to wasting of body organs and loss of weight, including loss of vital lean tissue. Chapters 24 and 25 describe the metabolic consequences of severe stress and the wasting syndrome, respectively.

Building Up and Breaking Down Compounds When not needed by the cells for energy, the basic units of the energy-yielding nutrients are used to build body compounds. The building up of body compounds is known as **anabolism;** this book represents anabolic reactions, wherever possible, by "up" arrows in chemical diagrams (such as those shown in Figure 6-1 on p. 134). Glucose units can be strung together to make glycogen chains. Glycerol and fatty acids can be assembled into triglycerides. Amino acids can be linked together to make proteins. These anabolic reactions, in which simple compounds are put together to form larger, more complex structures, involve doing work and so require the energy provided by ATP.

The breaking down of body compounds is known as **catabolism.** Catabolic reactions usually release energy and are represented, wherever possible, by "down" arrows in chemical diagrams (review Figure 6-1). Glycogen can be broken down to glucose, triglycerides to fatty acids and glycerol, and protein to amino acids.

metabolism: the sum total of all the chemical reactions that go on in living cells.

anabolism (an-AB-o-lism): reactions in which small molecules are put together to build larger ones. Anabolic reactions require energy.
  ana = up

catabolism (ca-TAB-o-lism): reactions in which large molecules are broken down to smaller ones. Catabolic reactions release energy.
  kata = down

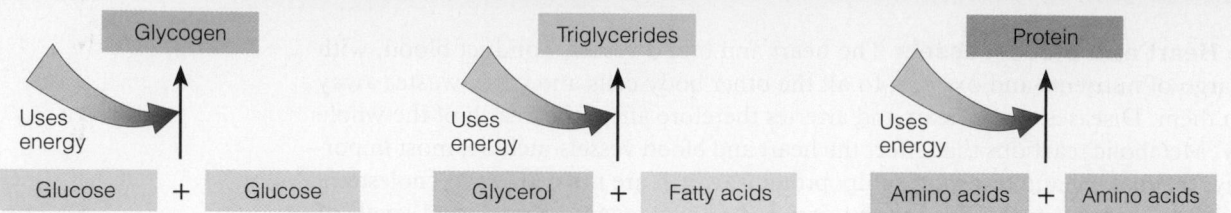

Anabolic reactions include the making of glycogen, triglycerides, and protein; these reactions require differing amounts of energy.

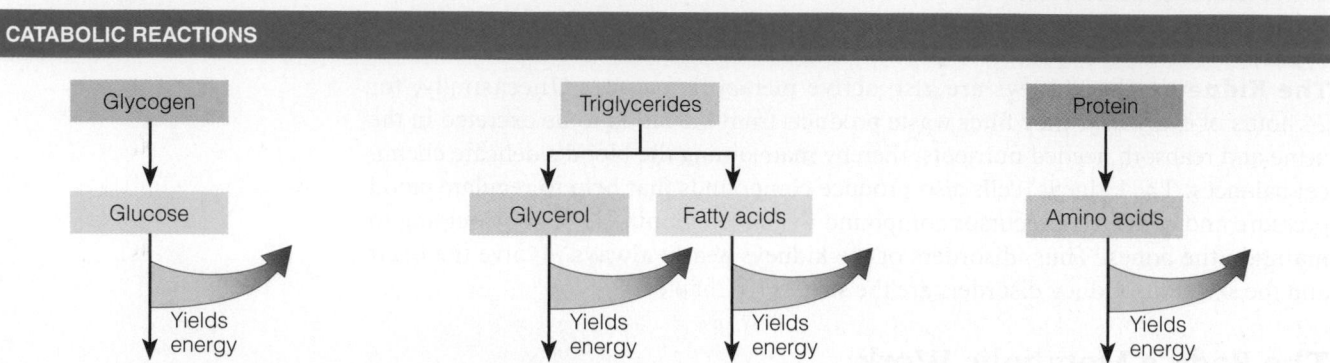

Catabolic reactions include the breakdown of glycogen, triglycerides, and protein; the further catabolism of glucose, glycerol, fatty acids, and amino acids releases differing amounts of energy. Much of the energy released is captured in the bonds of adenosine triphosphate (ATP).

FIGURE 6-1 Anabolic and Catabolic Reactions Compared

IN SUMMARY

▌ The digestive organs, the liver, the pancreas, the circulatory system, and the kidneys perform much of the metabolic work of the body.

▌ Metabolism occurs throughout the body, all the time, and supports normal health. Foods supply the energy required for the metabolic work of the body.

▌ When not needed for energy, the basic units of the energy-yielding nutrients are used by the body's cells to build body compounds (anabolism).

▌ When the body needs energy, the cells break the basic units of the energy-yielding nutrients down further to release energy (catabolism).

The Body's Use of Fuels

How does your body get the energy needed to maintain all of its cellular activities from the foods you eat? The answer to this question lies in an understanding of metabolism, the chemical reactions that occur in all living cells.

Energy Metabolism

Although every aspect of our lives depends on energy, the concept of energy can be difficult to grasp because it cannot be seen or touched, and it manifests in various forms, including heat, mechanical, electrical, and chemical energy. In the body, heat energy maintains a constant body temperature, and electrical energy sends nerve impulses, for example. Energy is stored in foods and in the body as chemical energy. **Energy metabolism** is the sum total of all the chemical reactions that the body uses to obtain or expend energy from foods. Earlier chapters described how the energy-yielding

nutrients—carbohydrate, fat, and protein—are broken down into basic units that are absorbed into the blood:

▌ From carbohydrates: glucose

▌ From fat (triglycerides): glycerol and fatty acids

▌ From proteins: amino acids

This section picks up from there, describing what becomes of these nutrients.

The high-energy compound **ATP (adenosine triphosphate)** is able to transfer small amounts of usable energy to move our muscles and supply our enzymes with the energy they need to catalyze chemical reactions. Figure 6-2 illustrates how ATP does this. When ATP breaks down and releases one of its phosphate groups, a small amount of energy is released and used in the body to build compounds. With the loss of the phosphate group, the ATP becomes ADP (adenosine diphosphate). During energy metabolism, ATP is re-created by attaching a phosphate group to ADP. ATP is produced continuously throughout the day, using the energy from the breakdown of the energy-yielding nutrients.

Breaking Down Nutrients for Energy To produce ATP, the body breaks down any or all of the four basic units—glucose, fatty acids, glycerol, and amino acids—into even smaller units. Each of these nutrients travels down a different pathway, but all can eventually become acetyl CoA, enter the TCA cycle, and provide hydrogens for the electron transport chain. Details follow.

The breakdown of glucose (a 6-carbon compound) into two molecules of **pyruvate** (a 3-carbon compound) is called **glycolysis,** and it produces just two usable ATP. As the carbons in glucose are broken apart to produce pyruvate, the hydrogen atoms attached to the carbons are transferred by **coenzymes** to the electron transport chain. Thus, the reactions of glycolysis produce a small amount of ATP, pyruvate, and hydrogen-rich coenzymes that are used later in energy metabolism. After glycolysis, pyruvate is converted to **acetyl CoA,** which consists of a 2-carbon fragment and a coenzyme called **CoA.**

Acetyl CoA can be produced not only from pyruvate, but also from the other energy-yielding nutrients. Fatty acids can be broken down into 2-carbon fragments that combine with CoA to form acetyl CoA. As the carbons in fatty acids are broken apart to produce acetyl CoA, hydrogen atoms are released to coenzymes that transfer them to the electron transport chain. Glycerol can easily be converted to pyruvate and therefore can also produce acetyl CoA. The amino acids have various pathways; some can be converted to pyruvate, others can be converted to acetyl CoA, and a few can enter the TCA cycle directly.

The coenzymes used in energy metabolism contain B vitamins; hence, B vitamins play critical roles in ATP production.

energy metabolism: all the reactions by which the body obtains and expends the energy from food.

ATP or adenosine (ah-DEN-oh-seen) **triphosphate** (tri-FOS-fate): a common high-energy compound that contains three phosphate groups. The bonds between the phosphate groups are often described as "high-energy" because of their readiness to release energy.

pyruvate (PIE-roo-vate): a 3-carbon compound that plays a key role in energy metabolism.

glycolysis (gligh-COLL-ih-sis): the metabolic breakdown of glucose to pyruvate.
  glyco = glucose
  lysis = breakdown

coenzymes (co-EN-zime): small molecules that work with enzymes to facilitate the enzymes' activity. Most coenzymes have B vitamins as part of their structures.

acetyl CoA (ASS-eh-teel, or ah-SEET-il, coh-AY): a 2-carbon compound (acetate, or acetic acid) with a molecule of CoA attached to it.

CoA (coh-AY): coenzyme A; the coenzyme derived from the B vitamin pantothenic acid and central to energy metabolism.

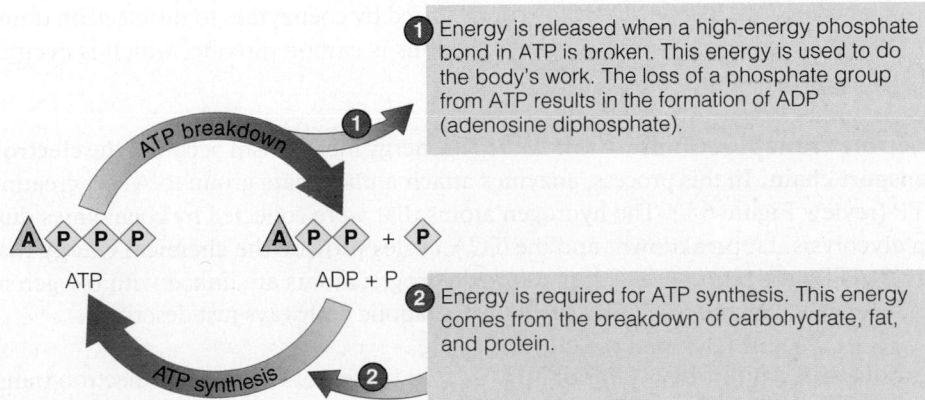

① Energy is released when a high-energy phosphate bond in ATP is broken. This energy is used to do the body's work. The loss of a phosphate group from ATP results in the formation of ADP (adenosine diphosphate).

② Energy is required for ATP synthesis. This energy comes from the breakdown of carbohydrate, fat, and protein.

FIGURE 6-2 ATP

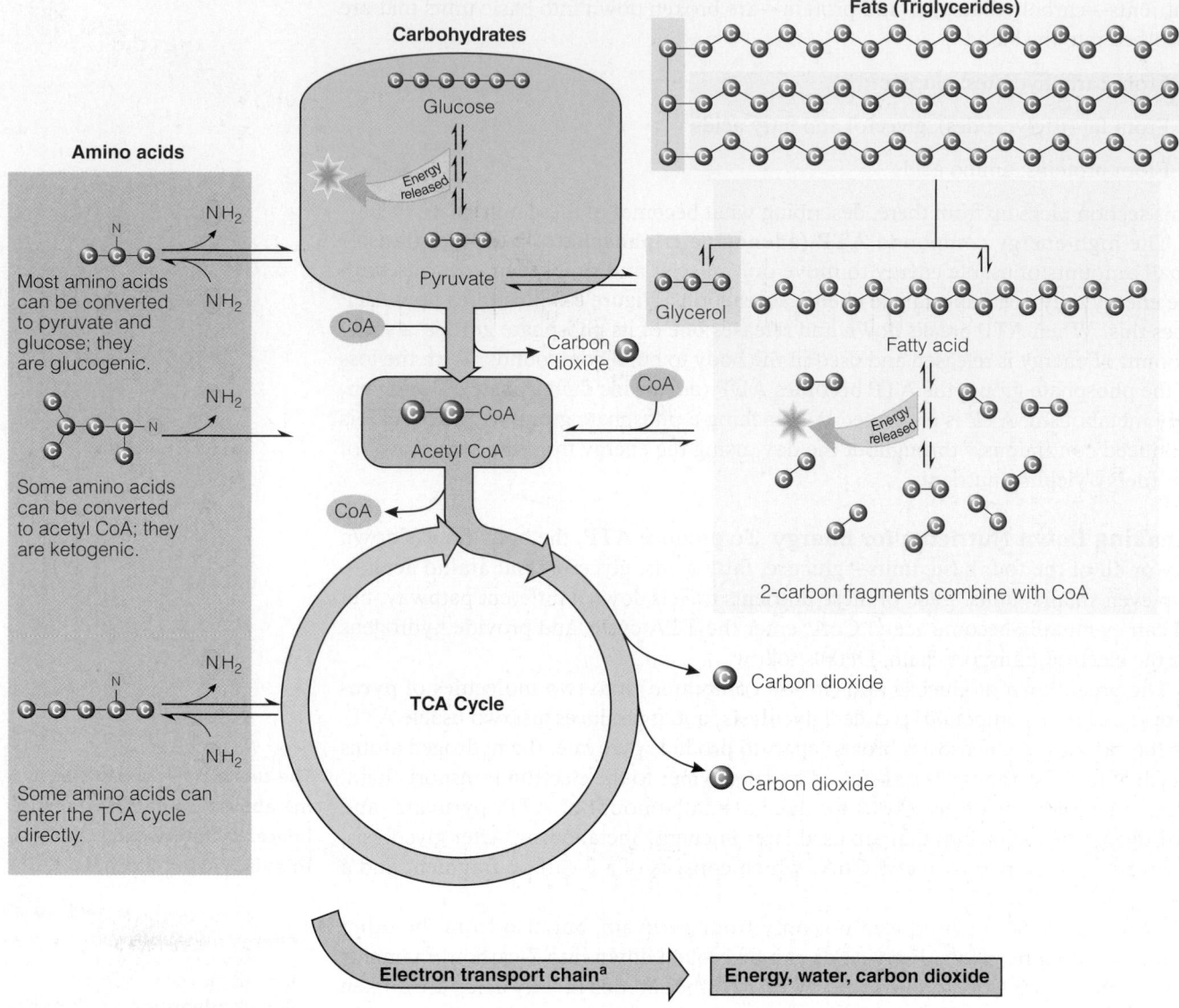

Carbohydrates

Glucose

Energy released

Pyruvate

Amino acids

NH_2

Most amino acids can be converted to pyruvate and glucose; they are glucogenic.

NH_2

NH_2

N

Some amino acids can be converted to acetyl CoA; they are ketogenic.

NH_2

N

NH_2

Some amino acids can enter the TCA cycle directly.

CoA

Carbon dioxide

CoA

CoA—Acetyl CoA

CoA

TCA Cycle

Fats (Triglycerides)

Glycerol

Fatty acid

Energy released

2-carbon fragments combine with CoA

Carbon dioxide

Carbon dioxide

Electron transport chain[a] → **Energy, water, carbon dioxide**

[a] Coenzymes carry hydrogens from the breakdown of the energy nutrients to the electron transport chain.

FIGURE 6-3 The Central Pathways of Energy Metabolism

TCA is the abbreviation for *tricarboxylic acid,* named after the chemical structures of the compounds in this pathway.

TCA Cycle The breakdown of energy nutrients continues in the **TCA cycle (tricarboxylic acid cycle),** as enzymes break down acetyl CoA molecules. With each turn of the TCA cycle, hydrogen atoms are carried by coenzymes to the electron transport chain. The waste product of these reactions is carbon dioxide, which is eventually exhaled.

Electron Transport Chain The final step in energy metabolism occurs at the **electron transport chain.** In this process, enzymes attach a phosphate group to ADP, creating ATP (review Figure 6-2). The hydrogen atoms that were collected by coenzymes during glycolysis, fat breakdown, and the TCA cycles provide the chemical energy that drives ATP production. Finally, the same hydrogen atoms are linked with oxygen to produce water. Figure 6-3 summarizes the metabolic pathways just described.

Aerobic and Anaerobic Metabolism The production of ATP via the electron transport chain requires oxygen in the final step and is called *aerobic* metabolism. Glycolysis produces ATP without the need for oxygen and is therefore called *anaerobic* metabolism.

When exercise intensity requires more ATP than can be provided by the electron transport chain (due to limited oxygen or other factors), ATP production from glycolysis is stepped up, making glucose a critical fuel for the exercising muscles. Chapter 10 provides information about anaerobic and aerobic activities and their effects on fuel use.

Glucose Production

When glucose levels drop, glucose can be produced from several other compounds in a process called **gluconeogenesis.** In Figure 6-3, you can see two-way arrows between glucose and pyruvate, but only a one-way arrow between pyruvate and acetyl CoA. The arrows show that pyruvate can be reconverted to glucose, but that acetyl CoA cannot. Any compound that can be converted to pyruvate can be used to make glucose. Any compound that has been converted to acetyl CoA cannot be used to make glucose.

Glucose $\longleftrightarrow$ pyruvate $\longrightarrow$ acetyl CoA $\longrightarrow$ energy

Triglycerides and Glucose Production Recall that triglycerides (the primary form of fat in the body) consist of three fatty acids and a glycerol. Because fatty acids break down to acetyl CoA, they cannot be used to make glucose. The glycerol portion of a triglyceride, however, can be converted to pyruvate and thus can yield glucose. Because glycerol represents only about 5 percent of the weight of a triglyceride molecule, fat is an inefficient source of glucose.* About 95 percent of a triglyceride cannot be converted to glucose at all. The task of producing glucose is left to amino acids, which are obtained by breaking down the body's proteins.

Amino Acids and Glucose Production As Chapter 4 explains, the primary role of amino acids is to maintain supplies of needed body proteins. If amino acids are needed for energy or if they are consumed in excess, they first undergo **deamination,** a reaction in which they are stripped of their nitrogen. The nitrogen can be used to make other compounds, including the nonessential amino acids, or it can be excreted. With nitrogen removed, most of the amino acids can be converted to pyruvate and can therefore provide glucose. Several of the amino acids can only be converted to acetyl CoA and therefore cannot supply glucose. Thus protein, unlike fat, is a fairly efficient source of glucose when carbohydrate is not available.

The principal nitrogen-excretion product of protein metabolism is **urea.**

IN SUMMARY

■ ATP provides chemical energy for our bodies' cells and is produced by breaking down the energy-yielding nutrients.

■ The chemical pathways that produce ATP are glycolysis, the TCA cycle, and the electron transport chain.

■ Glycolysis breaks down glucose to pyruvate, which can be converted to acetyl CoA. Fatty acids can be broken down to acetyl CoA, as can many amino acids.

■ The TCA cycle breaks down acetyl CoA molecules, producing carbon dioxide as waste and releasing hydrogen atoms to coenzymes that deliver them to the electron transport chain.

■ When the hydrogen atoms that originate from energy-yielding nutrients enter the electron transport chain, ATP is made.

■ Glucose can be produced from compounds that can be converted to pyruvate, including glycerol and most amino acids. Fatty acids can be used only for energy and cannot make glucose.

TCA cycle or **tricarboxylic** (try-car-box-ILL-ick) **acid cycle:** a series of metabolic reactions that break down molecules of acetyl CoA to carbon dioxide and hydrogen atoms; also called the **Kreb's cycle** after the biochemist who elucidated its reactions.

electron transport chain: the final pathway in energy metabolism that transports electrons from hydrogen to oxygen and captures the energy released in the bonds of a high-energy compound, ATP.

gluconeogenesis (gloo-co-nee-oh-GEN-ih-sis): the making of glucose from a noncarbohydrate source.

deamination: removal of the amino group (NH_2) from a compound such as an amino acid.

urea (yoo-REE-uh): the principal nitrogen-excretion product of protein metabolism.

*Figure 3-2 in Chapter 3 showed that glycerol (3 carbons) plus 3 fatty acids (most often 16 to 18 carbons) equals a triglyceride. Thus the small glycerol molecule represents only 3 of the 50 or so carbons in the triglyceride.

Energy Imbalance

When a person takes in and expends roughly the same number of kcalories, body weight remains stable. In other words, the body's energy is in balance. Many people, however, overeat and gain weight; others eat too little and lose weight. This section examines the two extremes of energy imbalance—feasting and fasting.

Feasting

When people consume more energy than they expend, much of the excess is stored as body fat. Fat can be made from an excess of any energy-yielding nutrient. In addition, excess energy from alcohol is also stored as fat.[1] Alcohol has also been shown to slow down the body's use of fat for fuel, causing more fat to be stored. The fat cells of the adipose tissue enlarge as they fill with fat, as Figure 6-4 shows.

Excess Carbohydrate Surplus carbohydrate (glucose) is first stored as glycogen in the liver and muscles, but the glycogen-storing cells have limited capacity. Once glycogen stores are filled, most of the additional carbohydrate is used for energy, displacing the body's use of fat for energy and allowing body fat to accumulate. Thus excess carbohydrate can contribute to obesity.

Excess Fat Surplus dietary fat contributes easily to the body's fat stores. After a meal, fat is sent to the body's adipose tissue, where it is stored until needed for energy. Thus, excess fat from food easily adds to body fat.

Excess Protein Surplus protein may also contribute to body fat. If not needed to build body protein or to meet energy needs, amino acids will lose their nitrogens and be converted, through a series of reactions, to triglycerides. These, too, swell the fat cells and add to body weight. Figure 6-5 shows the metabolic events of feasting.

IN SUMMARY

▍ Excess energy from carbohydrate, fat, protein, and alcohol is stored as body fat in adipose tissue.

Fasting

The body expends energy all the time. Even when a person is asleep and totally relaxed, the cells of many organs are hard at work.

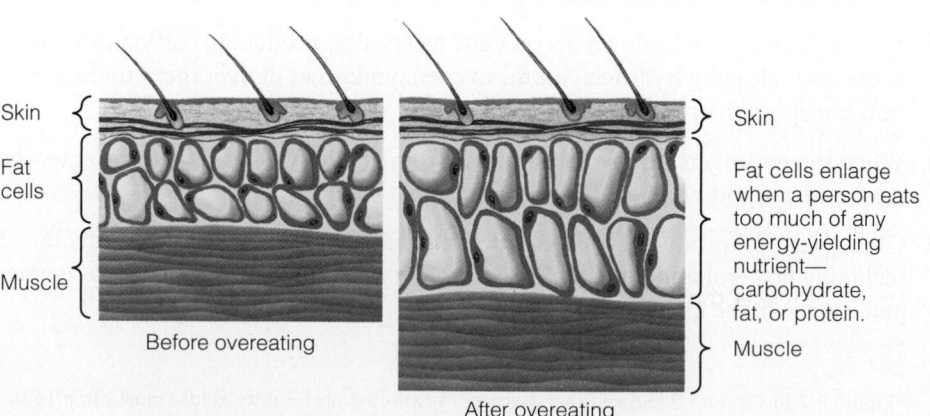

Skin

Fat cells

Muscle

Before overeating

Skin

Fat cells enlarge when a person eats too much of any energy-yielding nutrient—carbohydrate, fat, or protein.

Muscle

After overeating

FIGURE 6-4 Fat Cell Enlargement

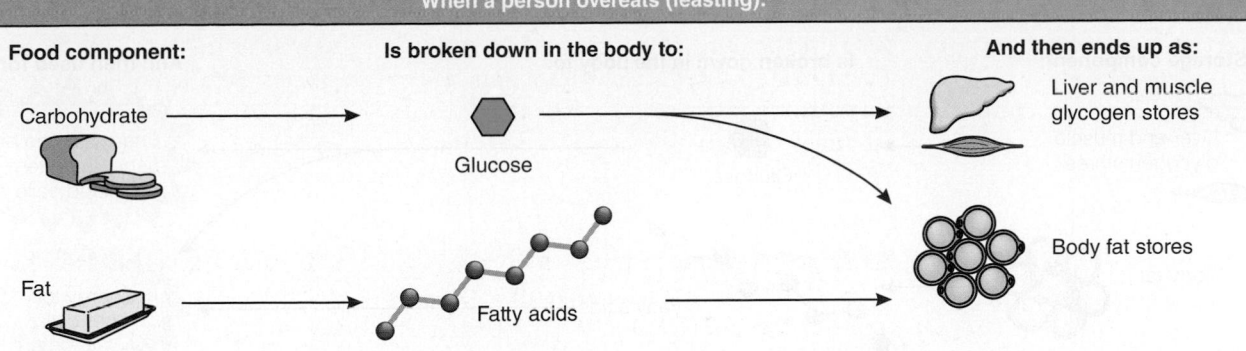

Food component: | **Is broken down in the body to:** | **And then ends up as:**

Carbohydrate → Glucose → Liver and muscle glycogen stores

Fat → Fatty acids → Body fat stores

Protein → Amino acids (first used to replace body proteins) → Nitrogen lost in urine

FIGURE 6-5 Feasting
When people overeat, they store energy.

Energy Deficit The body's top priority is to meet the energy needs for its ongoing cellular activity. Its normal way of meeting its energy needs is by periodic refueling, that is, by eating several times per day. When food is not available, the body uses fuel reserves from its own tissues. If people voluntarily choose not to eat, we say they are fasting; if they have no choice (as in a famine), we say they are starving. The body, however, makes no distinction between the two—metabolically, fasting and starvation are identical. In either case, the body switches to a wasting metabolism, drawing on its stores of carbohydrate and fat and, within a day or so, on its vital protein tissues as well.

Glycogen Used First As fasting begins, glucose from the liver's glycogen stores and fatty acids from the body's adipose tissue flow into the cells to fuel their work. Several hours later, liver glycogen is exhausted, and most of the glucose is used up. Low blood glucose concentrations serve as a signal to promote further fat breakdown.

Glucose Needed for the Brain At this point, a few hours into a fast, most of the cells depend on fatty acids to continue providing fuel. But as mentioned earlier, the nervous system (brain and nerves) and red blood cells cannot use fatty acids; they need glucose. Glucose has to be present to permit the brain's energy-metabolizing machinery to work. Normally, the nervous system consumes a little more than half of the total glucose used each day—about 400 to 600 kcalories' worth.

Protein Breakdown and Ketosis Because fat stores cannot provide the glucose needed by the brain, nerves, and red blood cells, body protein tissues (such as liver and muscle) always break down to some extent during fasting. In the first few days of a fast, body protein provides about 90 percent of the needed glucose, and glycerol provides about 10 percent. If body protein losses were to continue at this rate, death would ensue within about three weeks. As the fast continues, however, the body finds a way to use its fat to fuel the brain. It adapts by condensing together fragments derived from fatty acids to produce ketone bodies, which can serve as fuel for some brain cells. Ketone body production rises until, after several weeks of fasting, it is meeting much of the nervous system's energy needs. Still, many areas of the brain rely exclusively on glucose, and body protein continues to be sacrificed to produce it. Figure 6-6 (p. 140) shows the metabolic events that occur during fasting.

Slowed Metabolism As fasting continues and the nervous system shifts to partial dependence on ketone bodies for energy, the body simultaneously reduces its energy

Reminder: Glycerol represents only about 5 percent of the weight of a triglyceride molecule. About 95 percent of a triglyceride (the fatty acids attached to the glycerol) cannot be converted to glucose.

Reminder: Ketone bodies are acidic, fat-related compounds formed from the incomplete breakdown of fat when carbohydrate is not available. Small amounts of ketone bodies are normally produced during energy metabolism, but when their blood concentration rises, they spill into the urine. The combination of a high blood concentration of ketone bodies (ketonemia) and ketone bodies in the urine (ketonuria) is called ketosis.

Fasting = living on the body's fat and protein

In fasting, muscle and lean tissues give up protein to supply amino acids for conversion to glucose. This glucose, with ketone bodies produced from fat, fuels the brain's activities.

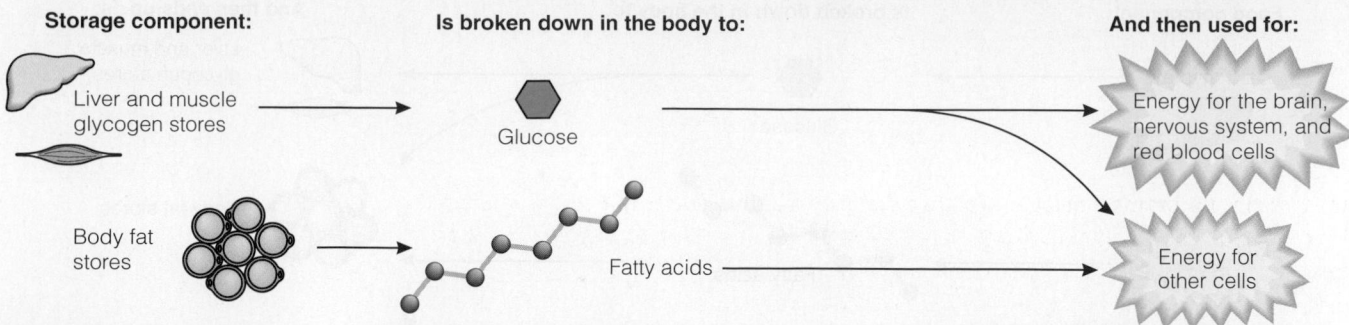

When a person draws on stores (fasting):

Storage component: **Is broken down in the body to:**   **And then used for:**

Liver and muscle glycogen stores → Glucose → Energy for the brain, nervous system, and red blood cells

Body fat stores → Fatty acids → Energy for other cells

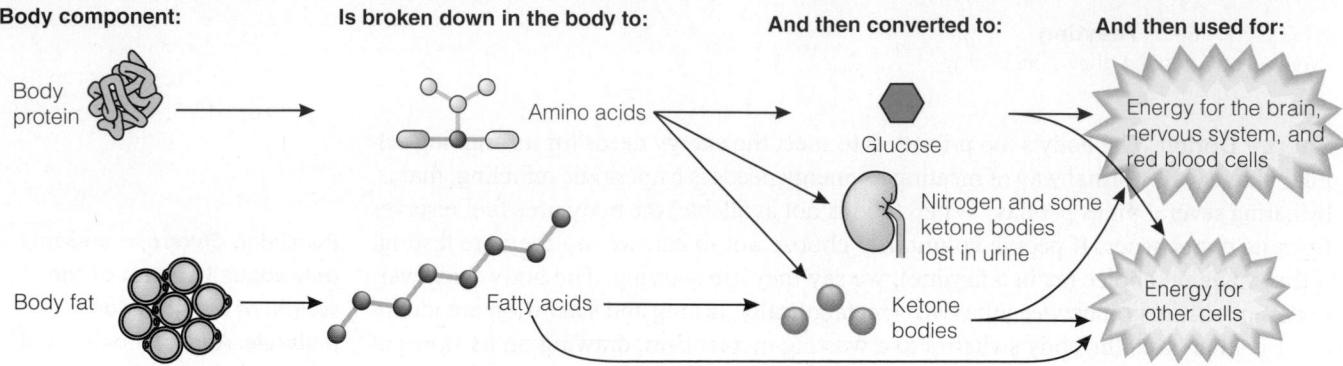

If the fast continues beyond glycogen depletion:

Body component: **Is broken down in the body to:**   **And then converted to:**   **And then used for:**

Body protein → Amino acids → Glucose → Energy for the brain, nervous system, and red blood cells

Nitrogen and some ketone bodies lost in urine

Body fat → Fatty acids → Ketone bodies → Energy for other cells

FIGURE 6-6 Fasting

When people are fasting, they draw on stored energy.

output (metabolic rate) and conserves both fat and lean tissue. Because of the slowed metabolism, energy use falls to a bare minimum.

Hazards of Fasting The body's adaptations to fasting are sufficient to maintain life for a long period. Mental alertness need not be diminished. Even physical energy can remain sufficient for a surprisingly long time. Still, fasting is not without its hazards. Among the many changes that take place in the body are:

▌ Wasting of lean tissues

▌ Impairment of disease resistance

▌ Lowering of body temperature

▌ Disturbances of the body's fluid and electrolyte balances

For the person who wants to lose weight, fasting is not the best way to go. The body's lean tissue continues to be degraded, sometimes amounting to as much as 50 percent of the weight lost over the first week. Over the long term, a diet only moderately restricted in energy promotes primarily *fat* loss and the retention of more lean tissue than a severely restricted fast.

IN SUMMARY

▌ When fasting, the body makes a number of adaptations: increasing the breakdown of fat to provide energy for most of the cells, using glycerol and amino acids to make glucose for the red blood cells and central nervous system, producing ketones to fuel the brain, and slowing metabolism.

- All of these measures conserve energy and minimize losses.

- Over the long term, a diet moderately restricted in energy promotes primarily fat loss and the retention of lean tissue.

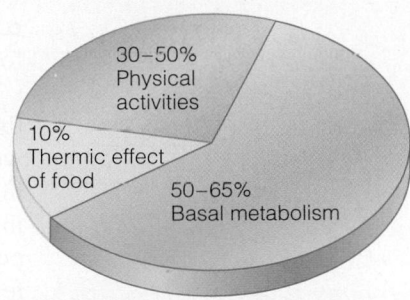

Energy Balance

If a person maintains a healthy weight over time, that person is in energy balance. Food energy intake equals energy expenditure: deposits of fat made at one time are compensated for by withdrawals made at another. In other words, the body uses fat as a savings account for energy. But, unlike money, having more fat is not better; there is an optimum.

A day's energy balance can be stated like this: Change in energy stores equals the food energy taken in (kcalories) minus the energy expended on metabolism and physical activities (kcalories). More simply:

Change in energy stores = energy in (kcalories) – energy out (kcalories)

Energy In

The energy provided by foods and beverages is the only contributor to the "energy in" side of the energy balance equation. One way to become familiar with the amounts of energy provided by foods and beverages is to look up the kcalories in the Table of Food Composition (Appendix A). Alternatively, diet analysis computer programs can readily provide this information for those with computer access.

Food composition data would reveal that an apple provides about 70 kcalories from carbohydrate and a candy bar supplies about 250 kcalories, mostly from fat and carbohydrate. You may already know that for each 3500 kcalories you eat in excess of expenditures, you store approximately 1 pound of body fat. Similarly, a pound of fat is lost for each 3500 kcalories expended beyond those consumed. The fat stores of even a healthy-weight adult represent an ample reserve of energy—50,000 to 200,000 kcalories.

Energy Out

The body expends energy in two major ways: to fuel its **basal metabolism** and to fuel its **voluntary activities.** People can change their voluntary activities to expend more or less energy in a day, and over time they can also change their basal metabolism by building up the body's metabolically active lean tissue, as explained in Chapter 7.

Energy for Basal Metabolism Basal metabolism supports the body's work that goes on all the time without the person's conscious awareness. The beating of the heart, inhaling and exhaling of air, maintenance of body temperature, and transmission of nerve and hormonal messages to direct these activities are the basal processes that maintain life. As Figure 6-7 shows, basal metabolism represents about two-thirds of the total energy a sedentary person spends in a day. In practical terms, a person whose total energy needs are 2000 kcalories per day may expend 1000 to 1300 of them to support basal metabolism.

The **basal metabolic rate (BMR)** is the rate at which the body expends energy for these activities. This rate varies from person to person and may vary for an individual with a change in circumstance or physical condition. The rate is slowest when a person is sleeping undisturbed, but it is usually measured when the person is awake and lying still, in a room with a comfortable temperature after a restful sleep and an overnight (12- to 14-hour) fast. A similar measure of energy output—called the

FIGURE 6-7 Components of Energy Expenditure
The amount of energy expended in a day differs for each individual, but, in general, basal metabolism is the largest component of energy expenditure, and the thermic effect of food is the smallest. The amount expended in voluntary physical activities has the greatest variability, depending on a person's activity patterns. For a sedentary person, physical activities may account for less than half as much energy as basal metabolism, whereas an extremely active person may expend as much on activity as for basal metabolism.

1 lb body fat = 3500 kcal
Body fat, or adipose tissue, is composed of a mixture of mostly fat, some protein, and water. A pound of body fat (454 g) is approximately 87% fat, or (454 × 0.87) 395 g, and 395 g × 9 kcal/g = 3555 kcal.

basal metabolism: the energy needed to maintain life when a person is at complete digestive, physical, and emotional rest. Basal metabolism is normally the largest part of a person's daily energy expenditure.

voluntary activities: the component of a person's daily energy expenditure that involves conscious and deliberate muscular work—walking, lifting, climbing, and other physical activities. Voluntary activities normally require less energy per day than basal metabolism does.

basal metabolic rate (BMR): the rate of energy use for metabolism under specified conditions: after a 12-hour fast and restful sleep, without any physical activity or emotional excitement, and in a comfortable setting. It is usually expressed as kcalories per kilogram of body weight per hour.

resting metabolic rate (RMR)—is slightly higher than the BMR because its criteria for recent food intake and physical activity are not as strict.

Table 6-1 summarizes the factors that raise and lower the BMR. For the most part, the BMR is highest in people who are growing (children, adolescents, and pregnant women) and in those with considerable lean body mass (physically fit people and males). One way to increase the BMR is to maximize lean body tissue by participating regularly in endurance and strength-building activities. The BMR is also fast in people who are tall and so have a large surface area for their weight, in people with fever or under stress, in people taking certain medications, and in people with highly active thyroid glands. The BMR is slowed by loss of lean tissue and by depression of thyroid hormone activity due to disease, inactivity, fasting, or malnutrition.

Energy for Physical Activities The number of kcalories expended on voluntary physical activities depends on three factors: muscle mass, body weight, and activity. The larger the muscle mass required for the activity and the heavier the body part being moved, the more kcalories are expended. The activity's duration, frequency, and intensity also influence energy costs: the longer, the more frequent, and the more intense the activity, the more kcalories expended. Table 6-2 gives average energy expenditures for people of different body weights engaged in various activities and shows that a heavy person uses more energy per minute than a light person does in the same activity.

TABLE 6-1 Factors That Affect the BMR

Factor	Effect on BMR
Age	Lean body mass diminishes with age, slowing the BMR.[a]
Height	In tall, thin people, the BMR is higher.[b]
Growth	In children and pregnant women, the BMR is higher.
Body composition (gender)	The more lean tissue, the higher the BMR (which is why males usually have a higher BMR than females). The more fat tissue, the lower the BMR.
Fever	Fever raises the BMR.[c]
Stresses	Stresses (including many diseases and certain drugs) raise the BMR.
Environmental temperature	Both heat and cold raise the BMR.
Fasting/starvation	Fasting/starvation lowers the BMR.[d]
Malnutrition	Malnutrition lowers the BMR.
Hormones (gender)	The thyroid hormone thyroxin, for example, can speed up or slow down the BMR.[e] Premenstrual hormones slightly raise the BMR.
Smoking	Nicotine increases energy expenditure.
Caffeine	Caffeine increases energy expenditure.
Sleep	BMR is lowest when sleeping.

[a]The BMR begins to decrease in early adulthood (after growth and development cease) at a rate of about 2 percent a decade. A reduction in voluntary activity as well brings the total decline in energy expenditure to 5 percent a decade.
[b]If two people weigh the same, the taller, thinner person will have the faster metabolic rate, reflecting the greater skin surface, through which heat is lost by radiation, in proportion to the body's volume.
[c]Fever raises the BMR by 7 percent for each degree Fahrenheit.
[d]Prolonged starvation reduces the total amount of metabolically active lean tissue in the body, although the decline occurs sooner and to a greater extent than body losses alone can explain. More likely, the neural and hormonal changes that accompany fasting are responsible for changes in the BMR.
[e]The thyroid gland releases hormones that travel to the cells and influence cellular metabolism. Thyroid hormone activity can speed up or slow down the rate of metabolism by as much as 50 percent.

resting metabolic rate (RMR): a measure of the energy use of a person at rest in a comfortable setting; similar to the BMR but with less stringent criteria for recent food intake and physical activity. Consequently, the RMR is slightly higher than the BMR.

TABLE 6-2 **Energy Spent on Various Activities**
The values listed in this table reflect both the energy spent in physical activity and the amount used for BMR.

Activity	kcal/lb/min[a]	Activity	kcal/lb/min[a]	Activity	kcal/lb/min[a]
Aerobic dance (vigorous)	.062	Horseback riding (trot)	.052	Table tennis (skilled)	.045
Basketball (vigorous, full court)	.097	Rowing (vigorous)	.097	Tennis (beginner)	.032
Bicycling		Running		Walking (brisk pace)	
13 mph	.045	5 mph	.061	3.5 mph	.035
15 mph	.049	6 mph	.074	4.5 mph	.048
17 mph	.057	7.5 mph	.094	Weight lifting	
19 mph	.076	9 mph	.103	light-to-moderate effort	.024
21 mph	.090	10 mph	.114	vigorous effort	.048
23 mph	.109	11 mph	.131	Wheelchair basketball	.084
25 mph	.139	Soccer (vigorous)	.097	Wheeling self in wheelchair	.030
Canoeing, flat water, moderate pace	.045	Studying	.011		
Cross-country skiing		Swimming			
8 mph	.104	20 yd/min	.032		
Golf (carrying clubs)	.045	45 yd/min	.058		
Handball	.078	50 yd/min	.070		

[a]To calculate kcalories spent per minute of activity for your own body weight, multiply kcal/lb/min by your exact weight and then multiply that number by the number of minutes spent in the activity. For example, if you weigh 142 pounds, and you want to know how many kcalories you spent doing 30 minutes of vigorous aerobic dance: 0.062 × 142 = 8.8 kcalories per minute; 8.8 × 30 (minutes) = 264 total kcalories spent.

Energy for Thermic Effect of Food When food is taken into the body, many cells that have been dormant become active. The muscles that move the food through the intestinal tract speed up their rhythmic contractions, and the cells that manufacture and secrete digestive juices begin their tasks. All these and other cells need extra energy as they participate in the digestion, absorption, and metabolism of food. This cellular activity produces heat and is known as the **thermic effect of food.** The thermic effect of food is generally thought to represent about 10 percent of the total food energy taken in. For purposes of rough estimates, the thermic effect of food is not always included.

Estimating Energy Requirements

In estimating energy requirements, the DRI committee developed equations that consider how the following factors influence energy expenditure:

▌ *Gender.* In general, women have a lower BMR than men, in large part because men typically have more lean body mass. In addition, menstrual hormones influence the BMR in women, raising it just prior to menstruation. Two sets of energy equations—one for men and one for women—were developed to accommodate the influence of gender on energy expenditure.

▌ *Growth.* The BMR is high in people who are growing. For this reason, pregnant women, children, and adolescents have their own sets of energy equations.

▌ *Age.* The BMR declines during adulthood as lean body mass diminishes. Physical activities tend to decline as well, bringing the average reduction in energy expenditure to about 5 percent per decade. The decline in the BMR that occurs when

Physical activity expends energy and benefits health in many ways.

Note that Table 6-1 listed these factors among those that influence BMR and consequently energy expenditure.

thermic effect of food: an estimation of the energy required to process food (digest, absorb, transport, metabolize, and store ingested nutrients).

a person becomes less active reflects the loss of lean body mass and may be prevented with ongoing physical activity. Because age influences energy expenditure, it is also factored into the energy equations.

▌ *Physical activity.* Using individual values for various physical activities (as in Table 6-2) is time-consuming and impractical for estimating the energy needs of a population. Instead, various activities are clustered according to the typical intensity of a day's efforts. Energy equations include a physical activity factor for various levels of intensity for each gender (see the accompanying "How to").

▌ *Body composition and body size.* The BMR is high in people who are tall and so have a large surface area. Similarly, the more a person weighs, the more energy is expended on basal metabolism. For these reasons, the energy equations include a factor for both height and weight.

Appendix D presents tables that provide a shortcut to estimating total energy expenditure and instructions to help you determine the appropriate physical activity factor to use in the equations.

As just explained, energy needs vary between individuals depending on such factors as gender, growth, age, physical activity, and body composition and body size. Even when two people are similarly matched, however, their energy needs will still differ because of genetic differences. The "How to" provides instructions on calculating estimated energy requirements using the DRI equations and physical activity factors.

How To ESTIMATE ENERGY REQUIREMENTS

To determine your estimated energy requirements (EER), use the appropriate equation (of the two options below), inserting your age in years, weight (wt) in kilograms, height (ht) in meters, and physical activity (PA) factor from the accompanying table. (To convert pounds to kilograms, divide by 2.2; to convert inches to meters, divide by 39.37.)

▌ For men 19 years and older:

$$EER = [662 - (9.53 \times age)] + PA \times [(15.91 \times wt) + (539.6 \times ht)]$$

▌ For women 19 years and older:

$$EER = [354 - (6.91 \times age)] + PA \times [(9.36 \times wt) + (726 \times ht)]$$

For example, consider an active 30-year-old male who is 5 feet 11 inches tall and weighs 178 pounds. First, he converts his weight from pounds to kilograms and his height from inches to meters, if necessary:

$$178 \text{ lb} \div 2.2 = 80.9 \text{ kg}$$
$$71 \text{ in} \div 39.37 = 1.8 \text{ m}$$

Next, he considers his level of daily physical activity and selects the appropriate PA factor from the accompanying table (in this example, 1.25 for an active male). Then, he inserts his age, PA factor, weight, and height into the appropriate equation:

$$EER = [662 - (9.53 \times 30)] + 1.25 \times [(15.91 \times 80.9) + (539.6 \times 1.8)]$$

(A reminder: do calculations within the parentheses first, and multiply before adding or subtracting.) He calculates:

$$EER = [662 - (9.53 \times 30)] + 1.25 \times (1287 + 971)$$
$$EER = [662 - (9.53 \times 30)] + (1.25 \times 2258)$$
$$EER = 662 - 286 + 2823$$
$$EER = 3199$$

The estimated energy requirement for an active 30-year-old male who is 5 feet 11 inches tall and weighs 178 pounds is about 3200 kcalories/day. His actual requirement probably falls within a range of 200 kcalories above and below this estimate.

Physical Activity (PA) Factors for EER Equations

	Men	Women	Physical Activity
Sedentary	1.0	1.0	Typical daily living activities
Low active	1.11	1.12	Plus 30 to 60 minutes moderate activity
Active	1.25	1.27	Plus ≥60 minutes moderate activity
Very active	1.48	1.45	Plus ≥60 minutes moderate activity and 60 minutes vigorous activity or 120 minutes moderate activity

Note: Moderate activity is equivalent to walking at 3½ to 4½ miles per hour.

■ A person takes in energy from food and, on average, expends most of it on basal metabolic activities, some of it on physical activities, and about 10 percent on the thermic effect of food.

■ Because energy requirements vary from person to person, such factors as gender, growth, age, and body composition and body size must be considered when calculating energy expended on basal metabolism, and the intensity and duration of the activity must be taken into account when calculating expenditures on physical activities.

Body Weight, Body Composition, and Health

The body's weight reflects its composition—the proportions of its bone, muscle, fat, fluid, and other tissue. All of these body components can vary in quantity and quality: the bones can be dense or porous, the muscles can be well developed or underdeveloped, fat can be abundant or scarce, and so on. By far the most variable tissue is body fat. For health's sake, weight management efforts should focus on eating and activity habits to improve body composition.

Defining Healthy Body Weight

How much should a person weigh? How can a person know if her weight is appropriate for her height and age? How can a person know if his weight is jeopardizing his health? Questions such as these seem so simple, yet the answers can be complex—and different depending on whom you ask.

The Criterion of Fashion In asking what an ideal body weight is, people often mistakenly turn to fashion for the answer. Without a doubt, our society sets unrealistic ideals for body weight, especially for women.[2] Magazines, movies, and television all convey the message that to be thin is to be beautiful. As a result, the media have a great influence on the weight concerns and dieting patterns of people of all ages, but most tragically on young, impressionable children and adolescents.[3]

Importantly, perceived body image has little to do with actual body weight or size. People of all shapes, sizes, and ages—including extremely thin fashion models with anorexia nervosa and fitness instructors with ideal body composition—have learned to be unhappy with their bodies. Such dissatisfaction can lead to damaging behaviors such as starvation diets and diet pill abuse, and to the development of eating disorders.[4] The first step toward making healthy changes may be self-acceptance. Keep in mind that fashion is fickle; the body shapes valued by our society change with time and differ from those valued by other societies. The standards defining "ideal" are subjective and frequently have little in common with health. Table 6-3 (p. 146) offers some tips for adopting health as an ideal, rather than society's misconceived image of beauty.

The Criterion of Health Even if our society were to accept fat as beautiful, obesity would still be a major risk factor for several life-threatening diseases as discussed later in this chapter. For this reason, the most important criterion for determining how much a person should weigh and how much body fat a person needs is not appearance but good health and longevity. A range of healthy body weights has been identified using a common measure of weight and height—the body mass index.

Body Mass Index The **body mass index (BMI)** describes relative weight for height and often correlates with degree of body fatness and disease risks. The equation that is used

body mass index (BMI): an index of a person's weight in relation to height; determined by dividing the weight (in kilograms) by the square of the height (in meters).

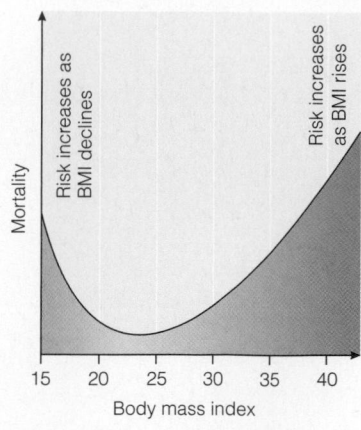

FIGURE 6-8 Body Mass Index and Mortality

This J-shaped curve describes the relationship between body mass index (BMI) and mortality and shows that both underweight and overweight present risks of a premature death.

$$BMI = \frac{weight\ (kg)}{height\ (m)^2}$$

- To convert pounds to kilograms, divide by 2.2.
- To convert inches to meters, divide by 39.37.

visceral fat: fat stored within the abdominal cavity in association with the internal abdominal organs, as opposed to fat stored directly under the skin (subcutaneous fat); also called *intra-abdominal fat*.

central obesity: excess fat around the trunk of the body; also called *abdominal fat* or *upper- body fat*.

adipokines (AD-ih-poh-kynz): protein hormones made and released by adipose tissue (fat) cells.

inflammation: an immunological response to cellular injury characterized by an increase in white blood cells.

subcutaneous fat: fat stored directly under the skin.
  sub = beneath
  cutaneous = skin

waist circumference: a measurement used to assess a person's abdominal fat.

skinfold measure: a clinical estimate of total body fatness in which the thickness of a fold of skin on the back of the arm (over the triceps muscle), below the shoulder blade (subscapular), or in other places is measured with a caliper.

The inflammatory response is described in detail in Chapter 24. Insulin resistance is a central feature of the metabolic syndrome, which is discussed in Nutrition in Practice 21.

TABLE 6-3 Tips for Accepting a Healthy Body Weight

- Value yourself and others for human attributes other than body weight. Realize that prejudging people by weight is as harmful as prejudging them by race, religion, or gender.
- Use positive, nonjudgmental descriptions of your body.
- Accept positive comments from others.
- Focus on your whole self, including your intelligence, social grace, and professional and scholastic achievements.
- Accept that no magic diet exists.
- Stop dieting to lose weight. Adopt a lifestyle of healthy eating and physical activity permanently.
- Follow the USDA Food Guide. Never restrict food intake below the minimum levels that meet nutrient needs.
- Become physically active, not because it will help you get thin but because it will make you feel good and enhance your health.
- Seek support from loved ones. Tell them of your plan for a healthy life in the body you have been given.
- Seek professional counseling, not from a weight-loss counselor but from someone who can help you make gains in self-esteem without weight as a factor.

to derive BMI values is provided in the margin. A person who takes measurements in pounds and inches can convert them to metric units or can use this modified equation:

$$BMI = \frac{weight\ (lb)}{height\ (in)^2} \times 703$$

The table on the inside back cover of this textbook shows weights for various heights using the BMI to define underweight, healthy weight, overweight, and obesity. As the table shows, healthy weight falls between a BMI of 18.5 and 24.9. Most people with a BMI within this range have few of the health risks typically associated with too-low or too-high body weight. Risks increase as BMI falls below 18.5 or rises above 24.9 (see Figure 6-8), reflecting the reality that both underweight and overweight impair health status.

The BMI values are most accurate in assessing degrees of obesity and are less useful for evaluating nonobese people's body fatness. BMI values fail to reveal two pieces of information valuable in assessing disease risk: how much of the weight is fat, and where the fat is located. For this knowledge, measures of body composition are needed.

Body Composition

For many people, being overweight compared with the standard means that they are over*fat*. This is not the case for athletes with dense bones and well-developed muscles; they may be over*weight* but carry little body fat. Conversely, inactive people may seem to have acceptable weights but still carry too much body fat. In addition, the distribution of fat on the body may be even more critical than overfatness alone.

Central Obesity Even more than total body fat, fat that collects deep within the central abdominal area of the body, called **visceral fat,** may be especially likely to lead to diabetes, stroke, hypertension, and coronary artery disease (see Figure 6-9).[5] The risk of death from all causes may be higher for those with **central obesity** than for those whose fat accumulates elsewhere in the body.[6]

One possible explanation for why fat in the abdomen increases the risk of disease involves **adipokines,** hormones released by visceral adipose tissue. Adipokines help to regulate **inflammation** and energy metabolism in the tissues.[7] In central obesity, a shift occurs in the balance of adipokines, favoring those that increase both inflammation and insulin resistance of tissues.[8] The resulting chronic inflammation and insulin resistance contribute to diabetes, atherosclerosis (a cause of heart disease), and other chronic diseases.[9] With weight loss, the adipokine balance is restored, and inflammation and insulin resistance are relieved.[10] Chronic disease risks drop in response.

Abdominal fat creates the "apple" profile of central obesity. **Subcutaneous fat** around the hips and thighs creates more of a "pear" profile (see Figure 6-10 on p. 148). Abdominal fat is common in women past menopause and even more common in men. Even when total body fat is similar, men have more abdominal fat than either premenopausal or postmenopausal women. For those women with abdominal fat, the risks of cardiovascular disease and mortality are increased, just as they are for men. Smokers, too, may carry more of their body fat centrally. Interestingly, smokers tend to have more visceral fat than nonsmokers, even though they typically have a lower BMI.[11] Two other factors that may affect body fat distribution are intakes of alcohol and physical activity. Moderate-to-high alcohol consumption may favor central obesity.[12] In contrast, regular physical activity seems to prevent abdominal fat accumulation.[13]

Waist Circumference A person's **waist circumference** is a good indicator of fat distribution and central obesity (see Figure 6-11 on p. 148).[14] In general, women with a waist circumference greater than 35 inches and men with a waist circumference greater than 40 inches have a high risk of central obesity-related health problems.

Skinfold Measures Skinfold measurements provide an accurate estimate of total body fat and a fair assessment of the fat's location. About half of the fat in the body lies directly beneath the skin, so the thickness of this subcutaneous fat is assumed to reflect total body fat. Measures taken from central-body sites (around the abdomen) better reflect changes in fatness than those taken from upper sites (arm and back). A skilled assessor can obtain an accurate **skinfold measure** and then compare the measurement with standards (see Appendix E).

How Much Body Fat Is Too Much? People often ask exactly how much fat is too fat for health. Ideally, a person has enough fat to meet basic needs but not so much as to incur health risks. The ideal amount of body fat depends partly on the person. A man with a BMI within the recommended range may have between 13 and 21 percent body fat; a woman, because of her greater quantity of essential fat, 23 to 31 percent.

Many athletes have a lower percentage of body fat—just enough fat to provide fuel, insulate and protect the body, assist in nerve impulse transmissions, and support normal hormone activity, but not so much as to burden the muscles. For athletes, body fat might be 5 to 10 percent for men and 15 to 20 percent for women.

For an Alaskan fisherman, a higher-than-average percentage of body fat is probably beneficial because fat helps prevent heat loss in cold weather. A woman starting a pregnancy needs sufficient body fat to support conception and fetal growth. Below a certain threshold for body fat, individuals may become infertile, develop depression, experience abnormal hunger regulation, or be unable to keep warm. These thresholds differ for each function and for each individual; much remains to be learned about them.

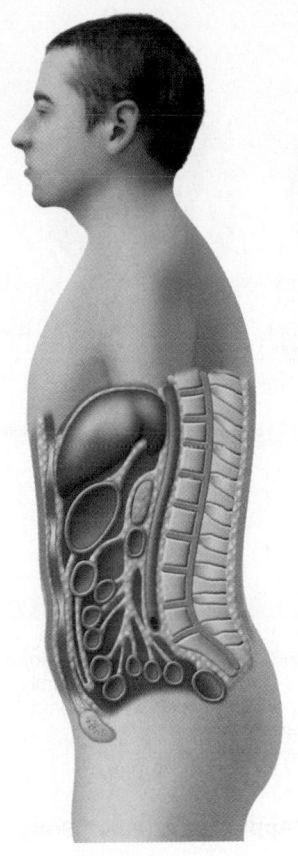
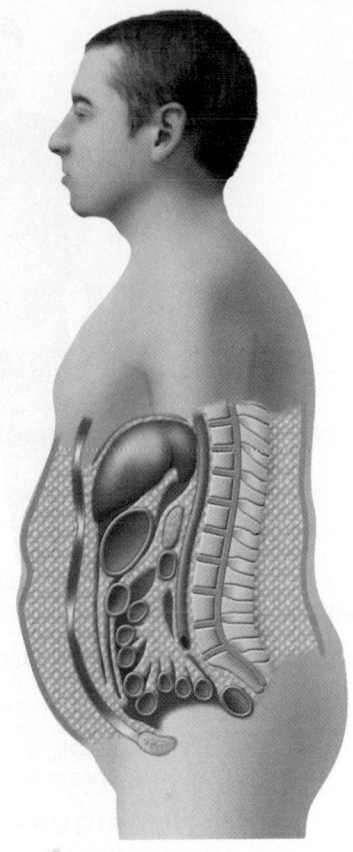

In healthy-weight people, some fat is stored around the organs of the abdomen.

In overweight people, excess abdominal fat increases the risks of diseases.

FIGURE 6-9 Abdominal Fat

The higher percentage of body fat in women compared to men is normal and necessary for reproduction.

At 6 feet 4 inches tall and 250 pounds, this runner would be considered overweight by most standards. Yet he is clearly not overfat.

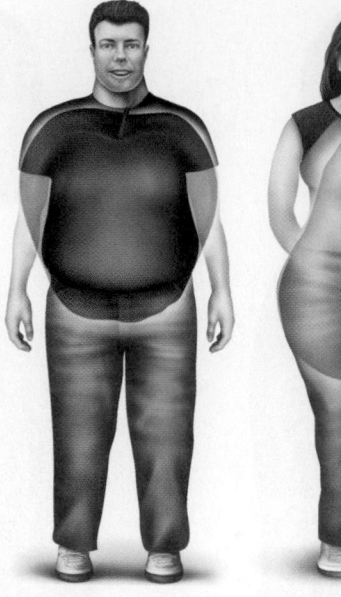

Upper-body fat is more common in men than in women and is closely associated with heart disease, stroke, diabetes, hypertension, and some types of cancer.

Lower-body fat is more common in women than in men and is not usually associated with chronic diseases.

FIGURE 6-10 "Apple" and "Pear" Body Shapes Compared
Popular articles sometimes call bodies with upper-body fat "apples" and those with lower-body fat "pears."

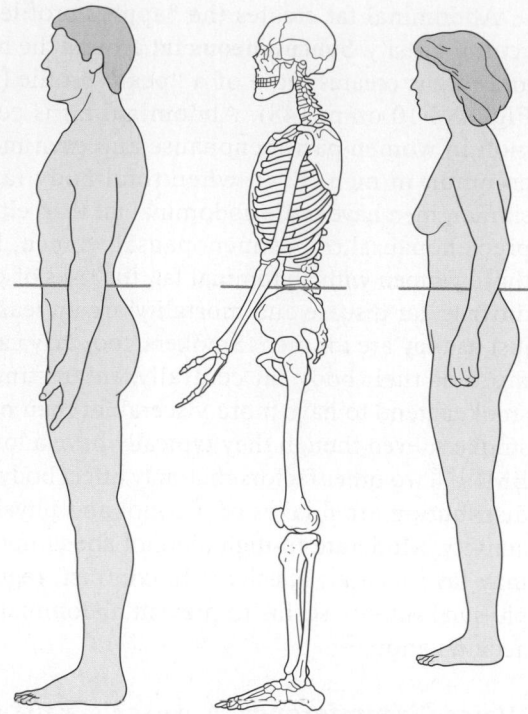

FIGURE 6-11 Measuring Waist Circumference
Using a non-stretching tape measure, measure around the body near the belly button. (The skeleton shows the tape position relative to the hip bone.) Take the measure of the end of a normal expiration. A healthy waist circumference for men is no larger than 102 centimeters (40 inches); for women, no larger than 88 centimeters (35 inches).

IN SUMMARY

▮ Clearly, the most important criterion of appropriate fatness is health.

▮ Current standards for body weight are based on the BMI (body mass index), which describes a person's weight in relation to height.

▮ Health risks increase with a BMI below 18.5 or above 24.9.

▮ Central obesity, in which excess fat is distributed around the trunk of the body, presents greater health risks than excess fat distributed on the lower body.

▮ Researchers use a number of techniques to assess body composition including waist circumference and skinfold measures.

Health Risks Associated with Body Weight and Body Fat

As mentioned earlier and shown in Figure 6-8 (p. 146), health risks increase as BMI falls below 18.5 or rises above 24.9. People who are extremely underweight or extremely obese carry higher risks of early death than those whose weights fall within the healthy, or even the slightly overweight, range.[15] These mortality risks decline with age.[16] Independently of BMI, factors such as smoking habits raise health risks, and physical fitness lowers them.

148

Health Risks of Underweight Some underweight people enjoy an active, healthy life, but others are underweight because of malnutrition, smoking habits, substance abuse, or illnesses. Weight and fat measures alone would not reveal these underlying causes, but a complete assessment that includes a diet and medical history, physical examination, and biochemical analysis would.

Underweight people, especially older adults, may be unable to preserve lean tissue when fighting a wasting disease such as cancer. Overly thin people are also at a disadvantage in the hospital, where nutrient status can easily deteriorate if they have to go without food for days at a time while undergoing tests or surgery.[17] Underweight and significant weight loss are also associated with osteoporosis and bone fractures. For all these reasons, underweight people may benefit from enough of a weight gain to provide an energy reserve and protective amounts of all the nutrients that can be stored.

An extreme underweight condition known as anorexia nervosa is sometimes seen in young people who exercise unreasonable self-denial in order to control their weight. Anorexia nervosa is a major eating disorder seen in our society today. Eating disorders are the subject of the Nutrition in Practice that follows this chapter.

Health Risks of Overweight and Obesity Despite our nation's preoccupation with body image and weight loss, the prevalence of overweight and obesity continues to rise dramatically (see Figure 6-12).[18] During the previous three decades, obesity increased in every state, in both genders, and across all ages, races, and education levels. Today, 68 percent of U.S. adults are **overweight** (BMI of 25 or greater) or dangerously obese (BMI of 30 or greater).[19] The problem reaches around the globe, in urban and rural areas alike.[20] In short, obesity is a major public health problem that is becoming more prevalent.

The growing prevalence of overweight and obesity is a matter of concern because both conditions present risks to health. Indeed, the health risks of **obesity** are so many that it has been declared a disease. Excess weight contributes to up to half of all cases of hypertension, thereby increasing the risk of heart attack and stroke.[21] Obesity raises blood pressure in part by altering kidney function and promoting fluid retention.[22] Often weight loss alone can normalize the blood pressure of an overfat person.

Excess body weight also increases the risk of type 2 diabetes. Most adults with type 2 diabetes are overweight or obese, and obesity itself can directly cause some degree of **insulin resistance**.[23] Diabetes (type 2) is three times more likely to develop in an obese person than in a nonobese person. Furthermore, the person with type 2 diabetes often has central obesity. Central-body fat cells appear to be larger and more insulin resistant than lower-body fat cells.[24]

In addition to diabetes and hypertension, other risks threaten obese adults. Among them are high blood lipids, cardiovascular disease, sleep apnea (abnormal ceasing of breathing during sleep), osteoarthritis, abdominal hernias, some cancers, varicose veins, gout, gallbladder disease, kidney stones, respiratory problems (including Pickwickian syndrome, a breathing blockage linked with sudden death), nonalcoholic

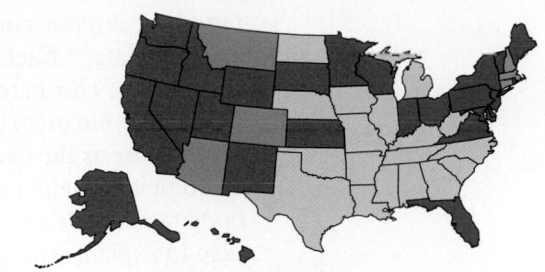

1999: Most states had prevalence rates less than 20 percent.

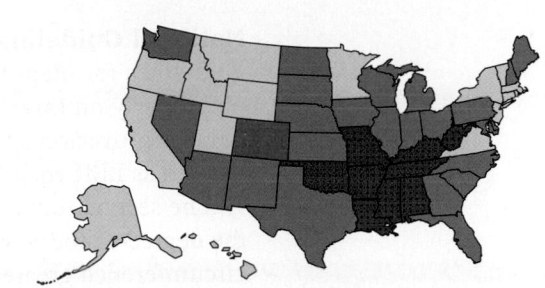

2009: Most states had prevalence rates greater than 25 percent, with nine states reporting prevalence rates greater than or equal to 30 percent.

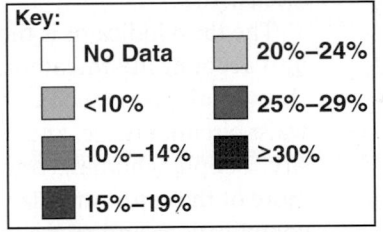

Key:
☐ No Data	☐ 20%–24%
☐ <10%	☐ 25%–29%
☐ 10%–14%	☐ ≥30%
☐ 15%–19%	

FIGURE 6-12 The Increasing Prevalence of Obesity among U.S. Adults by State
Source: Behavioral Risk Factor Surveillance System, Centers for Disease Control and Prevention; http://www.cdc.gov/obesity/data/trends.html.

overweight: overfatness of a moderate degree; defined as a body mass index (BMI) of 25.0 through 29.9.

obesity: overfatness with adverse health effects, as determined by reliable measures and interpreted with good medical judgment. Obesity is officially defined as a body mass index (BMI) of 30 or higher.

insulin resistance: the condition in which a normal amount of insulin produces a subnormal effect in muscle, adipose, and liver cells, resulting in an elevated fasting glucose; a metabolic consequence of obesity that precedes type 2 diabetes.

fatty liver disease, complications in pregnancy and surgery, flat feet, and even a high accident rate.[25] Each year these obesity-related illnesses cost our nation billions of dollars. The cost in terms of lives is also great. People with lifelong obesity are twice as likely to die prematurely as others. In the United States, obesity is second only to tobacco use as the most significant cause of preventable death.

Some overweight people, however, remain healthy and live long lives despite their body fatness. Genetic inheritance, smoking habits, and level of physical activity may help to explain why some overweight individuals stay well while others become ill.[26] To help in identifying those most at risk, obesity experts have developed guidelines, described next.

National Guidelines for Identifying Those at Risk from Obesity The U.S. guidelines for identifying and evaluating the risks to health from overweight and obesity rely on three indicators. The first indicator is a person's BMI. As a general guideline, overweight for adults is defined as BMI of 25.0 through 29.9, and obesity is defined as BMI equal to or greater than 30.

The second indicator is waist circumference, which, as discussed earlier, reflects the degree of abdominal fatness in proportion to body fatness. Women with a waist circumference greater than 35 inches and men with a waist circumference greater than 40 inches are at greater risk of type 2 diabetes, hypertension, and cardiovascular disease than women or men with waist circumferences equal to, or below, these measures. In other words, waist circumference is an independent predictor of disease risk.

The third indicator is the person's disease risk profile. The disease risk profile takes into account life-threatening diseases, family history, and common risk factors for chronic diseases (such as blood lipid profile). The higher the BMI, the greater the waist circumference, and the more risk factors—the greater the urgency to treat obesity. People who have one or more of the diseases listed in the margin, or three or more of the cardiovascular disease risk factors listed, have a very high risk for disease complications and mortality that requires aggressive treatment to manage the disease or modify the risk factors.

Other Risks of Obesity Although some obese people seem to escape health problems, few in our society can avoid the social and economic handicaps. Our society places enormous value on thinness. Obese people are less sought after for romance, less often hired, and less often admitted to college.[27] They pay higher insurance premiums and more for clothing. This is especially true for women. In contrast, people with other chronic conditions such as asthma, diabetes, and epilepsy do not differ socially or economically from non-overweight people.

Prejudice defines people by their appearance rather than by their abilities and characters. Obese people suffer emotional pain when others treat them with insensitivity, hostility, and contempt, and they may internalize a sense of guilt and self-deprecation. Health care professionals, even dietitians, can be among the offenders without realizing it.[28] To free our society of its obsession with body fatness and prejudice against obese people, activists are promoting respect for individuals of all body weights.

IN SUMMARY

▌ The health risks of underweight and obesity are many and serious.

▌ Guidelines for identifying the health risks of overweight and obesity are based on a person's BMI, waist circumference, and disease risk profile.

▌ Obesity also incurs social, economic, and psychological risks.

The National Heart, Lung, and Blood Institute states that aggressive treatment is urgently needed for a clinically obese person (BMI ≥ 30) who also has any one of the following:

▌ Heart disease
▌ Diabetes (type 2)
▌ Sleep apnea

The same urgency for treatment exists for an obese person with any *three* of the following CVD risk factors:

▌ Hypertension
▌ Cigarette smoking
▌ High LDL cholesterol
▌ Low HDL cholesterol
▌ Impaired glucose tolerance
▌ Age older than 45 years (men) or 55 years (women)
▌ Heart disease of an immediate family member before age 55 (male) or 65 (female)

Source: National Heart, Lung, and Blood Institute, National Institutes of Health, *The Practical Guide: Identification, Evaluation, and Treatment of Overweight and Obesity in Adults,* NIH publication no. 00-4084 (Washington, D.C.: Government Printing Office, 2000).

This chapter described energy metabolism, energy balance, and the health risks of underweight and obesity. The next chapter discusses the causes of obesity and strategies to lose, maintain, or gain weight.

Clinical Applications

1. Compare the energy a person might spend on various physical activities. Refer to Table 6-2 on p. 143, and compute how much energy a person who weighs 142 pounds would spend doing each of the following activities. An example using aerobic dance has been provided for you. You may want to compare various activities based on your own weight.

   30 minutes of vigorous aerobic dance:

   0.062 kcal/lb/min × 142 lb = 8.8 kcal/min
   8.8 kcal/min × 30 min = 264 kcal

   a. 2 hours of golf, carrying clubs
   b. 20 minutes running at 9 mph
   c. 45 minutes of swimming at 20 yd/min
   d. 1 hour of walking at 3.5 mph

2. Using the "How to" on p. 144 as a guide, determine your estimated energy requirements (EER).

   Answers

   1. (a) 0.045 kcal/lb/min × 142 lb = 6.4 kcal/min, 6.4 kcal/min × 120 min = 768 kcal; (b) 0.103 kcal/lb/min × 142 lb = 14.6 kcal/min, 14.6 kcal/min × 20 min = 292 kcal; (c) 0.032 kcal/lb/min × 142 lb = 4.5 kcal/min, 4.5 kcal/min × 45 = 202.5 kcal; (d) 0.035 kcal/lb/min × 142 lb = 5 kcal/min, 5 kcal/min × 60 min = 300 kcal.

Self Check

1. Before entering the TCA cycle, each of the energy-yielding nutrients is broken down to:
   a. ammonia.
   b. pyruvate.
   c. electrons.
   d. acetyl CoA.

2. As carbohydrate and fat stores are depleted during fasting or starvation, the body then uses ___ as its fuel source.
   a. alcohol
   b. protein
   c. glucose
   d. triglycerides

3. When carbohydrate is not available to provide energy for the brain, as in starvation, the body produces ketone bodies from:
   a. glucose.
   b. glycerol.
   c. fatty acid fragments.
   d. amino acids.

4. Three factors that affect the body's basal metabolic rate are:
   a. height, weight, and energy intake.
   b. weight, fever, and environmental temperature.
   c. fever, body composition, and altitude.
   d. age, body composition, and height.

5. The largest component of energy expenditure is:
   a. thermic effect of food.
   b. physical activity.
   c. indirect calorimetry.
   d. basal metabolism.

6. Which of the following reflects height and weight?
   a. body mass index
   b. central obesity
   c. waist circumference
   d. body composition

7. The BMI range that correlates with the fewest health risks is:
   a. 16.5 to 20.9.
   b. 18.5 to 24.9.
   c. 25.5 to 30.9.
   d. 30.5 to 34.9.

8. A person's waist circumference is a good indicator of:
   a. central obesity.
   b. subcutaneous fat.
   c. total body fat.
   d. energy metabolism.

9. The health risks of underweight include:
   a. sleep apnea.
   b. osteoporosis.
   c. hypertension.
   d. type 2 diabetes.

10. Which of the following health risks is *not* associated with being overweight?
    a. hypertension
    b. heart disease
    c. type 1 diabetes
    d. gallbladder disease

Answers to these questions can be found in Appendix H.

Notes

1. S. G. Wannamethee, A. G. Shaper, and P. H. Whincup, Alcohol and adiposity: Effects of quantity and type of drink and time relation with meals, *International Journal of Obesity and Related Metabolic Disorders*, online print, August 2, 2005; R. A. Breslow and B. A. Smothers, Drinking patterns and body mass index in never smokers, *American Journal of Epidemiology* 161 (2005): 368–376.

2. J. B. Martin, The development of ideal body image perceptions in the United States, *Nutrition Today* 45 (2010): 98–110.

3. M. J. Hogan and V. C. Strasburger, Body image, eating disorders, and the media, *Adolescent Medicine: State of the Art Reviews* 19 (2009): 521–546; A. J. Hill, Motivation for eating behavior in adolescent girls: The body beautiful, *Proceedings of the Nutrition Society* 65 (2006): 376–384.

4. Martin, 2010.

5. S. S. Dhaliwal and T. A. Welborn, Central obesity and multivariable cardiovascular risk as assessed by the Framingham Prediction Scores, *American Journal of Cardiology* 103 (2009): 1403–1407; G. A. Bray and coauthors, Relation of central adiposity and body mass index to the development of diabetes in the Diabetes Prevention Program, *American Journal of Clinical Nutrition* 87 (2008): 1212–1218; C. S. Fox and coauthors, Abdominal visceral and subcutaneous adipose tissue compartments: Association with metabolic risk factors in the Framingham Heart Study, *Circulation* 116 (2007): 39–48; A. Tchernof, Visceral adipocytes and the Metabolic Syndrome, *Nutrition Reviews* 65 (2007): S24–S29; C. D. Lee and coauthors, Abdominal obesity and coronary artery calcification in young adults: The Coronary Artery Risk Development in Young Adults (CARDIA) Study, *American Journal of Clinical Nutrition* 86 (2007): 48–54; S. B. Votruba and M. D. Jensen, Regional fat deposition as a factor in FFA metabolism, *Annual Review of Nutrition* 27 (2007): 149–163.

6. J. P. Reis and coauthors, Overall obesity and abdominal adiposity as predictors of mortality in U.S. white and black adults, *Annals of Epidemiology* 19 (2009): 134–142.

7. G. Govindarajan, M. A. Alpert, and L. Tejwani, Endocrine and metabolic effects of fat: Cardiovascular implications, *American Journal of Medicine* 121 (2008): 366–370; G. Fantuzzi and T. Mazzone, Adipose tissue and atherosclerosis: Exploring the connection, *Arteriosclerosis, Thrombosis, and Vascular Biology* 27 (2007): 996–1003; P. Trayhurn, C. Bing, and I. S. Wood, Adipose tissue and adipokines—Energy regulation from the human perspective, *Journal of Nutrition* 136 (2006): 1935S–1939S; C. Bulcão and coauthors, The new adipose tissue and adipocytokines, *Current Diabetes Reviews* 2 (2006): 19–28; T. J. Guzik, D. Mangalat, and R. Korbut, Adipocytokines—Novel link between inflammation and vascular function? *Journal of Physiology and Pharmacology* 57 (2006): 505–528.

8. W. V. Brown and coauthors, Obesity: Why be concerned? *American Journal of Medicine* 122 (2009): S4–S11; J. Korner, S. C. Woods, and K. A. Woodworth, Regulation of energy homeostasis and health consequences in obesity, *American Journal of Medicine* 122 (2009): S12–S18; W. L. Holland and coauthors, Lipid mediators of insulin resistance, *Nutrition Reviews* 65 (2007): S39–S46; R. Weiss, Fat distribution and storage: How much, where, and how? *European Journal of Endocrinology* 157 (2007): S39–S45.

9. Brown and coauthors, 2009; V. Z. Rocha and P. Libby, The multiple facets of the fat tissue, *Thyroid* 18 (2008): 175–183; P. Calabro and E. T. Yeh, Intra-abdominal adiposity, inflammation, and cardiovascular risk: New insight into global cardiometabolic risk, *Current Hypertension Reports* 10 (2008): 32–38.

10. J. P. Bastard and coauthors, Recent advances in the relationship between obesity, inflammation, and insulin resistance, *European Cytokine Network* 17 (2006): 4–12.

11. E. W. Demerath, Causes and consequences of human variation in visceral adiposity, *American Journal of Clinical Nutrition* 91 (2010): 1–2; E. A. Molenaar and coauthors, Association of lifestyle factors with abdominal subcutaneous and visceral adiposity: The Framingham Heart Study, *Diabetes Care* 32 (2009): 505–510; D. Canoy and coauthors, Cigarette smoking and fat distribution in 21,828 British men and women: A population-based study, *Obesity Research* 13 (2005): 1466–1475.

12. Molenaar and coauthors, 2009; Wannamethee, Shaper, and Whincup, 2005.

13. B. J. Arsenault and coauthors, Physical inactivity, abdominal obesity and risk of coronary heart disease in apparently healthy men and women, *International Journal of Obesity* 34 (2010): 340–347; Y. Kim and S. Lee, Physical activity and abdominal obesity in youth, *Applied Physiology, Nutrition and Metabolism* 34 (2009): 571–581.

14. J. P. Després and coauthors, Abdominal obesity and the metabolic syndrome: Contribution to global cardiometabolic risk, *Arteriosclerosis, Thrombosis, and Vascular Biology* 28 (2008): 1039–1049; S. Klein and coauthors, Waist circumference and cardiometabolic risk: A consensus statement from Shaping America's Health: Association for Weight Management and Obesity Prevention; NAASO, The Obesity Society; The American Society for Nutrition; and the American Diabetes Association, *American Journal of Clinical Nutrition* 85 (2007): 1197–1202.

15. K. M. Flegal and coauthors, Cause-specific excess deaths associated with underweight, overweight, and obesity, *Journal of the American Medical Association* 298 (2007): 2028–2037; C. L. Ogden and coauthors, The epidemiology of obesity, *Gastroenterology* 132 (2007): 2087–2102; G. M. Price and coauthors, Weight, shape, and mortality risk in older persons: Elevated waist-hip ratio, not high body mass index, is associated with a greater risk of death, *American Journal of Clinical Nutrition* 84 (2006): 449–460.

16. Price and coauthors, 2006.

17. O. Bouillanne and coauthors, Fat mass protects hospitalized elderly persons against morbidity and mortality, *American Journal of Clinical Nutrition* 90 (2009): 505–510.

18. C. M. Apovian, The causes, prevalence, and treatment of obesity revisited in 2009: What have we learned so far? *American Journal of Clinical Nutrition* 91 (2010): 277S–279S; K. M. Flegal and coauthors, Prevalence and trends in obesity among US adults, 1999–2008, *Journal of the American Medical Association* 303 (2010): 235–241.

19. Flegal and coauthors, 2010.

20. B. M. Popkin, Recent dynamics suggest selected countries catching up to U.S. obesity, *American Journal of Clinical Nutrition* 91 (2010): 284S–288S.

21. G. Hu and coauthors, Body mass index, waist circumference, and waist-hip ratio on the risk of total and type-specific stroke, *Archives of Internal Medicine* 167 (2007): 1420–1427.

22. J. Redon and coauthors, Mechanisms of hypertension in the cardiometabolic syndrome, *Journal of Hypertension* 27 (2009): 441–451; F. W. Visser and coauthors, Rise in extracellular fluid volume during high sodium depends on BMI in healthy men, *Obesity (Silver Spring)* 17 (2009): 1684–1688.

23. American Diabetes Association, Diagnosis and classification of diabetes mellitus, *Diabetes Care* 32 (2009): S62–S67.

24. E. H. Livingston, Lower body subcutaneous fat accumulation and diabetes mellitus risk, *Surgery for Obesity and Related Diseases* 2 (2006): 362–368.

25. L. W. York, S. Puthalapattu, and G. Y. Wu, Nonalcoholic fatty liver disease and low-carbohydrate diets, *Annual Review of Nutrition* 29 (2009): 365–379; G. Whitlock and coauthors, Body-mass index and cause-specific mortality in 900,000 adults: Collaborative analyses of 57 prospective studies, *Lancet* 373 (2009): 1083–1096; C. J. Lavie, R. V. Milani, and H. O. Ventura, Obesity and cardiovascular disease: Risk factor, paradox, and impact of weight loss, *Journal of the American College of Cardiology* 53 (2009): 1925–1932; A. G. Renehan and coauthors, Body-mass index and incidence of cancer: A systematic review and meta-analysis of prospective observational studies, *Lancet* 371 (2008): 569–578; K. F. Adams and coauthors, Body mass and colorectal cancer risk in the NIH-AARP cohort, *American Journal of Epidemiology* 166 (2007): 36–45.

26. N. Stefan and coauthors, Identification and characterization of metabolically benign obesity in humans, *Archives of Internal Medicine* 168 (2008): 1609–1616; R. P. Wildman and coauthors, The obese without cardiometabolic risk factor clustering and the normal weight with cardiometabolic risk factor clustering, *Archives of Internal Medicine* 168 (2008): 1617–1624; P. T. Williams, Maintaining vigorous activity attenuates 7-yr weight gain in 8340 runners, *Medicine & Science in Sports & Exercise* 39 (2007): 801–809.

27. M. S. Arguete, J. L. Edman, and A. Yates, Romantic interest in obese college students, *Eating Behaviors* 10 (2009): 143–145.

28. D. E. Berryman and coauthors, Dietetics students possess negative attitudes toward obesity similar to nondietetics students, *Journal of the American Dietetic Association* 106 (2006): 1678–1682.

Nutrition in Practice   EATING DISORDERS

The exact number of people in the United States afflicted with some form of an **eating disorder** (see the accompanying glossary for the relevant terms) is unknown because many cases are never reported.[1] An estimated 6 percent of females and 3 percent of males have **anorexia nervosa, bulimia nervosa,** or **binge eating disorder.**[2] Many more suffer from other related conditions that do not meet the strict criteria for anorexia nervosa, bulimia nervosa, or binge eating disorder but still imperil a person's well-being. Characteristics of disordered eating, such as restrained eating, binge eating, purging, fear of fatness, and distortion of body image, are common, especially among young middle-class girls. In most other societies, these behaviors and attitudes are much less prevalent.

Why do so many young people in our society suffer from eating disorders?

Most experts agree that the causes are multifactorial: sociocultural, psychological, and probably also neurochemical.[3] Excessive pressure to be thin is at least partly to blame. When low body weight becomes an important goal, people begin to view normal, healthy body weight as too fat. Healthy people then take unhealthy actions to lose weight. Severe restriction of food intake may create intense hunger that leads to binges. Research confirms this theory, showing that unhealthy or dangerous diets often precede binge eating in adolescent girls.[4] Energy restriction followed by bingeing can set in motion a pattern of **weight cycling,** which may make weight loss and maintenance more difficult over time.

People who attempt extreme weight loss are dissatisfied with their bodies to begin with; they may also be depressed or suffer social anxiety.[5] As weight loss becomes increasingly difficult, psychological problems worsen, and the likelihood of developing full-blown eating disorders increases.

People with anorexia nervosa suffer from an extreme preoccupation with weight loss that seriously endangers their health and even their lives. People with bulimia engage in episodes of binge eating alternating with periods of severe dieting or self-starvation. Some bulimics also follow binge eating with self-induced vomiting, laxative abuse, or diuretic abuse in an attempt to undo the perceived damage caused by the binge.

Are there other groups, besides girls and young women, who are vulnerable to anorexia nervosa and bulimia nervosa?

Yes. Athletes who participate in sports that emphasize leanness are at special risk for developing eating disorders.[6] Athletes must often meet stringent weight requirements to compete in their sport. Many athletes report that they engage in behaviors that are typical of people with eating disorders. Female athletes are most vulnerable, but males—especially dancers, wrestlers, skaters,

GLOSSARY

amenorrhea (ay-MEN-oh-REE-ah): the absence of or cessation of menstruation. Primary amenorrhea is menarche delayed beyond 16 years of age. Secondary amenorrhea is the absence of three to six consecutive menstrual cycles.

anorexia nervosa: an eating disorder characterized by a refusal to maintain a minimally normal body weight, self-starvation to the extreme, and a disturbed perception of body weight and shape; seen (usually) in adolescent girls and young women.
> *anorexia* = without appetite
> *nervosa* = of nervous origin

binge eating disorder: an eating disorder whose criteria are similar to those of bulimia nervosa, excluding purging or other compensatory behaviors.

bulimia (byoo-LEEM-ee-uh) **nervosa:** recurring episodes of binge eating combined with a morbid fear of becoming fat, usually followed by self-induced vomiting or purging.

cathartic: a strong laxative.

cognitive therapy: psychological therapy aimed at changing undesirable behaviors by changing underlying thought processes contributing to these behaviors. In anorexia nervosa, a goal is to replace false beliefs about body weight, eating, and self-worth with health-promoting beliefs.

eating disorder: a disturbance in eating behavior that jeopardizes a person's physical and psychological health.

emetic (em-ETT-ic): an agent that causes vomiting.

female athlete triad: a potentially fatal triad of medical problems: disordered eating, amenorrhea, and osteoporosis.

stress fractures: bone damage or breaks caused by stress on bone surfaces during exercise.

weight cycling: repeated rounds of weight loss and subsequent regain, with reduced ability to lose weight with each attempt; also called yo-yo dieting.

jockeys, and gymnasts—suffer from eating disorders, too, and their numbers may be increasing.[7] These athletes may engage in extreme weight-loss practices such as overtraining, prolonged fasting, vomiting, taking diet pills, and using steam baths and saunas to induce sweating. In doing so, they actually compromise their athletic abilities. The diminished anaerobic strength, reduced endurance, decreased oxygen capacity, and general weakness caused by food deprivation and dehydration can impair performance, an effect lasting days after food and water are replenished.

Even among athletes, however, women are most susceptible to developing eating disorders. Many female athletes appear healthy but in fact may develop the three interrelated components of the **female athlete triad:** disordered eating, **amenorrhea** (the absence of three or more consecutive menstrual cycles), and osteoporosis (see Figure NP6-1 on p. 154).

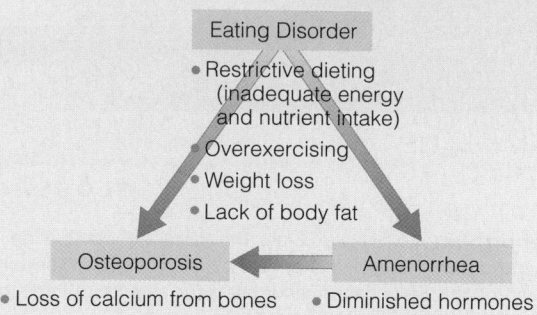

How does the female athlete triad develop?

Many athletic women engage in self-destructive eating behaviors (disordered eating) because they and their coaches have adopted unsuitable weight standards. An athlete's body must be heavier for height than a nonathlete's body because the athlete's bones and muscles are denser. Weight standards that may be appropriate for others are inappropriate for athletes. Techniques such as skinfold measures yield more useful information about body composition.

Many young female athletes severely restrict energy intakes to improve performance, enhance the aesthetic appeal of their performance, or meet the weight guidelines of their specific sports. They fail to realize that the loss of lean tissue that accompanies energy restriction actually impairs their physical performance. Risk factors for the female athlete triad include the following:

▌ Young age (adolescence)

▌ Pressure to excel at a chosen physical activity

▌ Focus on achieving or maintaining an "ideal" body weight or body fat percentage

▌ Participation in sports or competitions that judge performance on aesthetic appeal such as gymnastics, figure skating, or dance

▌ Weight-loss dieting at an early age

▌ Unsupervised dieting

As for amenorrhea, its prevalence among premenopausal women in the United States is about 2 to 5 percent overall, but among female athletes it may be as high as 66 percent. Amenorrhea is *not* a normal adaptation to strenuous physical training: it is a symptom of something going wrong. Amenorrhea is characterized by low blood estrogen, infertility, and often bone mineral losses.

In general, weight-bearing physical activity, dietary calcium, and the hormone estrogen protect against the bone loss of osteoporosis, but in women with disordered eating and amenorrhea, strenuous activity may impair bone health.[8] Vigorous training combined with low food energy intakes disrupts metabolic and hormonal balances. These disturbances compromise bone health, greatly increasing the risks of **stress fractures.**[9] Stress fractures, a serious form of bone injury, commonly occur among dancers and other competitive athletes with amenorrhea, low calcium intakes, and disordered eating. Many underweight young athletes have bones like those of postmenopausal women, and they may never recover their lost bone even after diagnosis and treatment—which makes prevention critical. Young athletes should be encouraged to consume at least 1300 milligrams of calcium each day, to eat nutrient-dense foods, and to obtain enough food energy to support weight gain and the energy expended in physical activity. Nutrition is critical to bone recovery.[10]

What can be done to prevent eating disorders in athletes and dancers?

To prevent eating disorders in athletes and dancers, both the performers and their coaches must be educated about links between inappropriate body weight ideals, improper weight-loss techniques, eating disorder development, adequate nutrition, and safe weight-control methods. Coaches and dance instructors should never encourage unhealthy weight loss to qualify for competition or to conform with distorted artistic ideals. Frequent weighings can push young people who are striving to lose weight into a cycle of starving to confront the scale and then bingeing uncontrollably afterward. The erosion of self-esteem that accompanies these events can interfere with the normal psychological development of the teen years and set the stage for serious problems later on.

Table NP6-1 provides some suggestions to help athletes and dancers protect themselves against developing eating disorders. The next sections describe eating disorders that anyone, athlete or nonathlete, may experience.

TABLE NP6-1 **Strategies to Combat Eating Disorders**

The following guidelines may be useful in combating eating disorders:

▌ Never restrict food intakes to below the amounts suggested for adequacy by the USDA Food Guide.

▌ Eat frequently. People often do not eat frequent meals because of time constraints, but eating can be incorporated into other activities, such as snacking while studying or commuting. The person who eats frequently never gets so hungry as to allow hunger to dictate food choices.

▌ If not at a healthy weight, establish a reasonable weight goal based on a healthy body composition.

▌ Allow a reasonable time to achieve the goal. A reasonable loss of excess fat can be achieved at the rate of about 1 percent of body weight per week.

▌ Learn to recognize media image biases, and reject ultra-thin standards for beauty. Shift focus to health, competencies, and human interactions; bring behaviors in line with those beliefs.

Specific guidelines for athletes and dancers:

▌ Replace weight-based goals with performance-based goals.

▌ Remember that eating disorders impair physical performance. Seek confidential help in obtaining treatment if needed.

▌ Restrict weight-loss activities to the off-season.

▌ Focus on proper nutrition as an important facet of your training— as important as proper technique.

What are the characteristics of anorexia nervosa?

Most anorexia nervosa victims are females who come from middle- or upper-class families. The person with anorexia nervosa is often a perfectionist who works hard to please her parents. She may identify so strongly with her parents' ideals and goals for her that she sometimes feels she has no identity of her own. She is respectful of authority but sometimes feels like a robot, and she may act that way, too: polite but controlled, rigid, and unspontaneous. She earnestly desires to control her own destiny, but she feels controlled by others. When she does not eat, she gains control.

How does a person know when dieting is going too far?

When a person loses weight to well below the average for her height, becoming too slim, and still doesn't stop, she has gone too far. Regardless of how thin she is, she looks in the mirror and sees herself as fat. Central to the diagnosis of anorexia nervosa is a distorted body image that overestimates body fatness. Table NP6-2 shows the criteria that professionals use to diagnose anorexia nervosa. Anorexia nervosa resembles an addiction. The characteristic behavior is obsessive and compulsive. Before drawing conclusions

TABLE NP6-2 Criteria for Diagnosis of Anorexia Nervosa

A person with anorexia nervosa demonstrates the following:

A. Refusal to maintain body weight at or above a minimal normal weight for age and height, for example, weight loss leading to maintenance of body weight less than 85 percent of that expected; or failure to make expected weight gain during period of growth, leading to body weight less than 85 percent of that expected.

B. Intense fear of gaining weight or becoming fat, even though underweight.

C. Disturbance in the way in which one's body weight or shape is experienced; undue influence of body weight or shape on self-evaluation, or denial of the seriousness of the current low body weight.

D. In females past puberty, amenorrhea, that is, the absence of at least three consecutive menstrual cycles. (A woman is considered to have amenorrhea if her periods occur only following hormone administration, e.g., estrogen.)

Two types:

■ **Restricting type:** During the episode of anorexia nervosa, the person does not regularly engage in binge eating or purging behavior (i.e., self-induced vomiting or the misuse of laxatives, diuretics, or enemas).

■ **Binge eating/purging type:** During the episode of anorexia nervosa, the person regularly engages in binge eating or purging behavior (i.e., self-induced vomiting or the misuse of laxatives, diuretics, or enemas).

Source: Reprinted with permission from the *Diagnostic and Statistical Manual of Mental Disorders*, 4th ed., Text Revision, American Psychiatric Association, 2000.

Women with anorexia nervosa see themselves as fat, even when they are dangerously underweight.

about someone who is extremely thin or who eats very little, remember that diagnosis of anorexia nervosa requires professional assessment.

What is the harm in being very thin?

Anorexia nervosa damages the body much as starvation does. In young people, growth ceases, and normal development falters. They lose so much lean tissue that basal metabolic rate slows. Additionally, the heart pumps inefficiently and irregularly, the heart muscle becomes weak and thin, the heart chambers diminish in size, and the blood pressure falls. Minerals that help to regulate the heartbeat become unbalanced. Many deaths in people with anorexia nervosa are due to heart failure. The kidneys often fail as well.[11]

Starvation brings other physical consequences: loss of brain tissue, impaired immune response, anemia, and a loss of digestive function that worsens malnutrition.[12] Digestive functioning becomes sluggish, the stomach empties slowly, and the lining of the intestinal tract shrinks. The ailing digestive tract fails to sufficiently digest any food the victim may eat. The pancreas slows its production of digestive enzymes. The person may suffer from diarrhea, further worsening malnutrition.

What kind of treatment helps people with anorexia nervosa?

Treatment of anorexia nervosa requires a multidisciplinary approach that addresses two sets of issues and behaviors: those relating to food and weight, and those involving relationships with oneself and others. Teams of physicians, nurses, psychiatrists, family therapists, and dietitians work together to treat people with anorexia nervosa. The expertise of a registered dietitian is essential because an appropriate, individually crafted diet is crucial for normalizing body weight, and nutrition counseling is indispensable.[13] Seldom are clients willing to eat for themselves, but if they are, chances are they can recover without other interventions.

Professionals classify clients based on the risks posed by the degree of malnutrition present.* Clients with low risks may benefit from family counseling, **cognitive therapy**, behavior modification, and nutrition guidance; those with greater risks may also need other forms of psychotherapy and supplemental formulas to provide extra nutrients and energy.

Sometimes, intensive behavior management treatment in a live-in facility can help to normalize food intake and exercise.[14] When starvation leads to severe underweight (less than 75 percent of ideal body weight), high medical risks ensue and patients require hospitalization. They must be stabilized and carefully fed to forestall death.[15] However, involuntary feeding through a tube can cause psychological trauma and may not be necessary in all cases.[16] Antidepressants and other drugs are commonly prescribed but are often ineffective in anorexia nervosa.[17]

Stopping weight loss is a first goal; establishing regular eating patterns is next. Because body weight is low and fear of weight gain is high, initial food intake may be small—1200 kcalories per day can be an achievement.[18] A variety of high-energy foods and beverages can help deliver needed kcalories and nutrients.[19] Even after recovery, however, energy intakes and eating behaviors may not fully return to normal.

Almost half of women who are treated can successfully maintain their body weight at just below a healthy weight; at that weight, many of them begin menstruating again. The other half have poor or fair treatment outcomes, relapse into abnormal eating behaviors, or die. Anorexia nervosa has one of the highest mortality rates among psychiatric disorders.[20] An estimated 1000 women die each year of anorexia nervosa—most commonly from cardiac complications due to malnutrition or from suicide.

How does bulimia nervosa differ from anorexia nervosa?

Bulimia nervosa is distinct from anorexia nervosa and is more prevalent. More men suffer from bulimia nervosa than from anorexia, but bulimia is still more common in women. The secretive nature of bulimic behaviors makes recognition of the problem difficult, but once it is recognized, diagnosis is based on the criteria listed in Table NP6-3.

The typical person with bulimia is well educated, in her early twenties, and close to ideal body weight. She is a high achiever

*Indicators of malnutrition include a low percentage of body fat, low blood proteins, and impaired immune response.

TABLE NP6-3 Criteria for Diagnosis of Bulimia Nervosa

A person with bulimia nervosa demonstrates the following:

A. Recurrent episodes of binge eating. An episode of binge eating is characterized by both of the following:
1. Eating, in a discrete period of time (e.g., within any two-hour period), an amount of food that is definitely larger than most people would eat during a similar period of time and under similar circumstances
2. A sense of lack of control over eating during the episode (e.g., a feeling that one cannot stop eating or control what or how much one is eating)

B. Recurrent inappropriate compensatory behavior in order to prevent weight gain, such as self-induced vomiting; misuse of laxatives, diuretics, enemas, or other medications; fasting; or excessive exercise.

C. Binge eating and inappropriate compensatory behaviors that both occur, on average, at least twice a week for three months.

D. Self-evaluation unduly influenced by body shape and weight.

E. The disturbance does not occur exclusively during episodes of anorexia nervosa.

Two types:

- **Purging type:** The person regularly engages in self-induced vomiting or the misuse of laxatives, diuretics, or enemas.
- **Non-purging type:** The person uses other inappropriate compensatory behaviors, such as fasting or excessive exercise, but does not regularly engage in self-induced vomiting or the misuse of laxatives, diuretics, or enemas.

Source: Reprinted with permission from the *Diagnostic and Statistical Manual of Mental Disorders*, 4th ed., Text Revision, American Psychiatric Association, 2000.

but emotionally insecure. She experiences considerable social anxiety and has difficulty establishing personal relationships. She is sometimes depressed and often exhibits impulsive behavior.[21]

Like the person with anorexia nervosa, the person with bulimia spends much time thinking about her body weight and food. Her preoccupation with food manifests itself in secretive binge eating episodes followed by self-induced vomiting, fasting, or the use of laxatives or diuretics. Such behaviors typically begin in late adolescence after a long series of various unsuccessful weight-reduction diets. People with bulimia commonly follow a pattern of restrictive dieting interspersed with bulimic behaviors and experience weight fluctuations of more than 10 pounds over short periods of time.

Unlike the person with anorexia nervosa, the person with bulimia is aware of the consequences of her behavior, feels that it is abnormal, and is deeply ashamed of it. She feels inadequate and unable to control her eating, so she tends to be passive and to look to others for confirmation of her sense of self-worth. When she is rejected, either in reality or in her imagination, her bulimia becomes worse. If her depression deepens, she may seek solace in drug or alcohol abuse or other addictive behaviors. Clinical depression is common in people with bulimia nervosa, and the rates of substance abuse are high.

What exactly is binge eating?

Binge eating is unlike normal eating, and the food is not consumed for its nutritional value. The binge eater has a compulsion to eat. A typical binge occurs periodically, is done in secret, usually at night, and lasts an hour or more. A binge frequently follows a period of rigid dieting, so the binge eating is accelerated by hunger. During a binge, the person with bulimia may consume a thousand kcalories or more. The food typically contains little fiber or water, has a smooth texture, and is high in sugar and fat, so it is easy to consume vast amounts rapidly with little chewing.

What are the consequences of binge eating?

After a binge, the person may use a **cathartic**—a strong laxative that can injure the lower intestinal tract. Or the person may induce vomiting using an **emetic**—a drug intended as first aid for poisoning.

On first glance, purging seems to offer a quick and easy solution to the problems of unwanted kcalories and body weight. Many people perceive such behavior as neutral or even positive, when, in fact, bingeing and purging have serious physical consequences. Fluid and electrolyte imbalances caused by vomiting or diarrhea can lead to abnormal heart rhythms and injury to the kidneys. Vomiting causes irritation and infection of the pharynx, esophagus, and salivary glands; erosion of the teeth; and dental caries. The esophagus may rupture or tear, as may the stomach. Overuse of emetics depletes potassium concentrations and can lead to death by heart failure.

What is the treatment for bulimia nervosa?

As for people with anorexia nervosa, a team approach provides the most effective treatment for people with bulimia nervosa. Bulimia nervosa is easier to treat than anorexia nervosa in many respects because it seems to be more of a chosen behavior. People with bulimia know that their behavior is abnormal, and many are willing to try to cooperate.

The goal of the dietary plan to treat bulimia is to help the client gain control, establish regular eating patterns, and restore nutritional health. Energy intake should not be severely restricted—hunger can be a trigger for a binge.[22] The person needs to learn to eat a quantity of nutritious food sufficient to nourish her body and to satisfy hunger (at least 1600 kcalories a day). Table NP6-4 offers some ways to begin correcting bulimia nervosa. Most people diagnosed with bulimia nervosa recover within 5 to 10 years, with or without treatment, but treatment probably speeds the recovery process.

Anorexia nervosa and bulimia nervosa are distinct eating disorders, yet they sometimes overlap. Anorexia victims may purge, and victims of both conditions share an overconcern with body weight and the tendency to drastically undereat. The two disorders can also appear in the same person, or one can lead to the other. Other people have eating disorders that fall short of anorexia nervosa or bulimia nervosa but share some of their features, such as fear of body fatness. One such condition is binge eating disorder (defined earlier in the glossary).

TABLE NP6-4 Strategies to Combat Bulimia Nervosa

Planning Principles

- Plan meals and snacks; record plans in a food diary prior to eating.
- Plan meals and snacks that require eating at the table and using utensils.
- Refrain from finger foods.
- Refrain from fasting or skipping meals.

Nutrition Principles

- Eat a well-balanced diet and regularly timed meals consisting of a variety of foods.
- Include raw vegetables, salad, or raw fruit at meals to prolong eating times.
- Choose whole-grain, high-fiber breads, pasta, rice, and cereals to increase bulk.
- Consume adequate fluid, particularly water.

Other Tips

- Choose foods that provide protein and fat for satiety and bulky, fiber-rich carbohydrates for immediate feelings of fullness.
- Try including soups and other water-rich foods for satiety.
- Choose portions that meet the definition of a serving according to the USDA Food Guide (pp. 16–17).
- For convenience (and to reduce temptation) select foods that naturally divide into portions. Select one potato, rather than rice or pasta that can be overloaded onto the plate; purchase yogurt and cottage cheese in individual containers; look for small packages of precut steak or chicken; choose frozen dinners with measured portions.
- Include 30 minutes of physical activity every day—exercise may be an important tool in defeating bulimia.

How does binge eating disorder differ from bulimia nervosa?

Up to half of all people who restrict eating to lose weight periodically binge without purging, including about one-third of obese people. Obesity itself, however, does not constitute an eating disorder. Table NP6-5 (p. 158) lists the official diagnostic criteria for binge eating disorder.

Clinicians note differences between people with bulimia nervosa and those with binge eating disorder. People with binge eating disorder consume less during a binge, rarely purge, and exert less restraint during times of dieting. Similarities also exist, including feeling out of control, disgusted, depressed, embarrassed, or guilty because of their self-perceived gluttony. Binge eating behavior responds more readily to treatment than other eating disorders, and resolving such behaviors can be a first step to authentic weight control. Successful treatment also improves physical health, mental health, and the chances of breaking the cycle of rapid weight losses and gains.

Chapter 6 describes how our society sets unrealistic ideals for body weight, especially in women, and devalues those who do not conform to them. Anorexia nervosa and bulimia nervosa are not a form of rebellion against these unreasonable expectations but rather the exaggerated acceptance of them. Body dissatisfaction is a primary factor in the development of eating disorders.[23] Perhaps a person's best defense against these disorders is to learn to appreciate his or her own uniqueness.

TABLE NP6-5 Criteria for Diagnosis of Binge Eating Disorder

A person with a binge eating disorder demonstrates the following:

A. Recurrent episodes of binge eating. An episode of binge eating is characterized by both of the following:

1. Eating, in a discrete period of time (e.g., within any two-hour period), an amount of food that is definitely larger than most people would eat in a similar period of time under similar circumstances.

2. A sense of lack of control over eating during the episode (e.g., a feeling that one cannot stop eating or control what or how much one is eating).

B. Binge eating episodes are associated with at least three of the following:

1. Eating much more rapidly than normal.
2. Eating until feeling uncomfortably full.
3. Eating large amounts of food when not feeling physically hungry.
4. Eating alone because of being embarrassed by how much one is eating.
5. Feeling disgusted with oneself, depressed, or very guilty after overeating.

C. The binge eating causes marked distress.

D. The binge eating occurs, on average, at least twice a week for six months.

E. The binge eating is not associated with the regular use of inappropriate compensatory behaviors (e.g., purging, fasting, excessive exercise) and does not occur exclusively during the course of anorexia nervosa or bulimia nervosa.

Source: Reprinted with permission from the *Diagnostic and Statistical Manual of Mental Disorders*, 4th ed., Text Revision, American Psychiatric Association, 2000.

Notes

1. Position of the American Dietetic Association: Nutrition intervention in the treatment of anorexia nervosa, bulimia. nervosa, and other eating disorders, *Journal of the American Dietetic Association* 106 (2006): 2073–2082.

2. J. L. Hudson and coauthors, The prevalence and correlates of eating disorders in the National Comorbidity Survey Replication, *Biological Psychiatry* 61 (2007): 348–358.

3. S. E. Mazzeo and C. M. Bulik, Environmental and genetic risk factors for eating disorders: What the clinician needs to know, *Child and Adolescent Psychiatric Clinics of North America* 18 (2009): 67–82; S. S. O'Sullivan, A. H. Evans, and A. J. Lees, Dopamine dysregulation syndrome: An overview of its epidemiology, mechanisms, and management, *CNS Drugs* 23 (2009): 157–170; T. D. Müller and coauthors, Leptin-mediated neuroendocrine alterations in anorexia nervosa: Somatic and behavioral implications, *Child and Adolescent Psychiatric Clinics of North America* 18 (2009): 117–129; V. Costarelli, M. Demerzi, and D. Stamou, disordered eating attitudes in relation body image and emotional intelligence in young women, *Journal of Human Nutrition and Dietetics* 22 (2009): 239–245.

4. D. Neumark-Sztainer and coauthors, Obesity, disordered eating, and eating disorders in a longitudinal study of adolescents: How do dieters fare 5 years later? *Journal of the American Dietetic Association* 106 (2006): 559–568; B. A. Spear, does dieting increase the risk for obesity and eating disorders? *Journal of the American Dietetic Association* 106 (2006): 523–525.

5. B. Blaine, Does depression cause obesity? A meta-analysis of longitudinal studies of depression and weight control, *Journal of Health Psychology* 13 (2008): 1190–1197; S. E. Anderson and coauthors, Association of depression and anxiety disorders with weight change in a prospective community-based study of children followed up into adulthood, *Archives of Pediatric and Adolescent Medicine* 160 (2006): 285–291.

6. M. Vertalino and coauthors, Participation in weight-related sports is associated with higher use of unhealthful weight-control behaviors and steroid use, *Journal of the American Dietetic Association* 107 (2007): 434–440; American College of Sports Medicine, Position stand: The female athlete triad, *Medicine and Science in Sports and Exercise* 39 (2007): 1867–1882.

7. J. B. Martin, The development of ideal body image perceptions in the United States, *Nutrition Today* 45 (2010): 98–110; J. L. Glazer, Eating disorders among male athletes, *Current Sports Medicine Reports* 7 (2008): 332–337.

8. M. T. Barrack and coauthors, Dietary restraint and low bone mass in female adolescent endurance runners, *American Journal of Clinical Nutrition* 87 (2008): 36–43; D. L. Nichols, C. F. Sanborn, and E. V. Essery, Bone density and young athletic women, *Sports Medicine* 37 (2007): 1001–1014.

9. J. L. Kelsey and coauthors, Risk factors for stress fracture among young female cross-country runners, *Medicine and Science in Sports and Exercise* 39 (2007): 1457–1463.

10. J. Dominguez and coauthors, Treatment of anorexia nervosa is associated with increases in bone mineral density, and recovery is a biphasic process involving both nutrition and return of menses, *American Journal of Clinical Nutrition* 86 (2007): 92–99.

11. J. M. Torpy, Anorexia nervosa, JAMA patient page, *Journal of the American Medical Association* 295 (2006): 2684.

12. J. Yager and A. E. Andersen, Anorexia nervosa, *New England Journal of Medicine* 353 (2005): 1481–1488.

13. Position of the American Dietetic Association, 2006.

14. E. Attia and B. T. Walsh, Behavioral management for anorexia nervosa, *New England Journal of Medicine* 360 (2009): 500–506.

15. Position of the American Dietetic Association, 2006.

16. C. Theils, Forced treatment of patients with anorexia, *Current Opinion in Psychiatry* 21 (2008): 495–498.

17. B. T. Walsh and coauthors, Fluoxetine after weight restoration in anorexia nervosa: A randomized controlled trial, *Journal of the American Medical Association* 295 (2006): 2605–2612.

18. Yager and Andersen, 2005.

19. J. E. Schebendach and coauthors, Dietary energy density and diet variety as predictors of outcome in anorexia nervosa, *American Journal of Clinical Nutrition* 87 (2008): 810–816.

20. J. M. Holm-Denoma and coauthors, Deaths by suicide among individuals with anorexia as arbiters between competing explanations of the anorexia-suicide link, *Journal of Affective Disorders* 107 (2008): 231–236; G. Murialdo and coauthors, Alterations in the autonomic control of heart rate variability in patients with anorexia or bulimia nervosa: Correlations between sympathovagal activity, clinical features, and leptin levels, *Journal of Endocrinological Investigation* 30 (2007): 356–362; D. Casiero and W. H. Frishman, Cardiovascular complications of eating disorders, *Cardiology in Review* 14 (2006): 227–231; M. Pompili and coauthors, Suicide and attempted suicide in eating disorders, obesity and weight-image concern, *Eating Behaviors* 7 (2006): 384–394.

21. W. Kaye, Neurobiology of anorexia and bulimia nervosa, *Physiology and Behavior* 94 (2008): 121–135.

22. Position of the American Dietetic Association, 2006.

23. Martin, 2010.

7

Weight Management: Overweight and Underweight

160

A re you pleased with your body weight? If you answered yes, you are a rare individual. Nearly all people in our society think they should weigh more or less (mostly less) than they do. Usually, their primary reason is appearance, but they often perceive, correctly, that their weight is also related to physical health. Chapter 6 addressed the health risks of being overweight or underweight.

Overweight and underweight both result from energy imbalance. The simple picture is as follows: Overweight people have consumed more food energy than they have expended and have banked the surplus in their body fat. To reduce body fat, overweight people need to expend more energy than they take in from food. In contrast, underweight people have consumed too little food energy to support their activities and so have depleted their bodies' fat stores and possibly some of their lean tissues as well. To gain weight, they need to take in more food energy than they expend.

This chapter's missions are to present strategies toward solving the problems of excessive and deficient body fatness and to point out how appropriate body composition, once achieved, can be maintained. The chapter emphasizes overweight because it has been more intensively studied and is a more widespread health problem in the developed countries.

Causes of Obesity

Henceforth, this chapter will use the term *obesity* to refer to excess body fat. Excess body fat accumulates when people take in more food energy than they expend. Why do they do this? Is it genetic? Metabolic? Psychological? Behavioral? All of these? Most likely, obesity has many interrelated causes.

Genetics and Weight

A person's genetic makeup influences the body's tendency to consume or store too much energy or to expend too little.[1] Evidence that genes influence eating behavior and body composition comes from family, twin, and adoption studies.[2] Adopted children tend to be more similar in weight to their biological parents than to their adoptive parents.[3] Studies of twins yield similar findings: compared with fraternal twins, identical twins are twice as likely to weigh the same.[4] Genes clearly influence a person's tendency to gain weight or stay lean. The risk of obesity is two to three times higher for a person with a family history of obesity than for a person without such history.[5] Despite such findings, however, only 1 to 5 percent of obesity cases can be explained by a single gene mutation.[6] Furthermore, although genomics researchers have identified hundreds of genes with possible roles in obesity development, they have not yet identified genetic causes of common obesity.[7]

Exceptionally complex relationships exist among the many genes related to energy metabolism and obesity, and they each interact with environmental factors, too. Although an individual's genetic inheritance may make obesity likely, it will not necessarily develop unless given a push by environmental factors that encourage energy consumption and discourage energy expenditure.[8] The following sections describe research involving proteins that might help explain appetite control, energy regulation, and obesity development.

Lipoprotein Lipase Some of the research investigating genetic influence on obesity focuses on the enzyme **lipoprotein lipase (LPL),** which promotes fat storage in fat cells and muscle cells.[9] People with high LPL activity are especially efficient at storing fat. Obese people generally have much more LPL activity in their fat cells than lean people do (their muscle cell LPL activity is similar, though). This high LPL activity makes fat storage especially efficient. Consequently, even modest excesses in energy intake have a more dramatic impact on obese people than on lean people.

lipoprotein lipase (LPL): an enzyme mounted on the surface of fat cells (and other cells). It hydrolyzes triglycerides in the blood into fatty acids and glycerol for absorption into the cells. There they are metabolized or reassembled for storage.

Genes instruct cells to make proteins, and each protein performs a unique function.

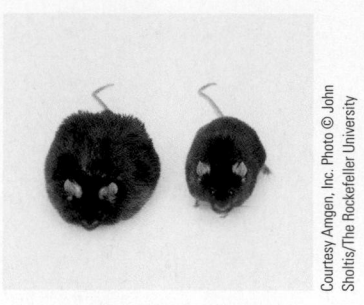

The mouse on the left is genetically obese—it lacks the gene for producing leptin. The mouse on the right is *also* genetically obese, but because it receives leptin, it eats less, expends more energy, and is less obese than it would be had it not received the leptin.

Leptin Researchers have discovered a gene in humans called the obesity *(ob)* gene. The obesity gene codes for the protein **leptin.** Leptin is a hormone primarily produced and secreted by the fat cells in proportion to the amount of fat stored.[10] A gain in body fatness stimulates the production of leptin, which, by way of the hypothalamus, suppresses the appetite, increases energy expenditure, and produces fat loss.[11] Fat loss produces the opposite effect—suppression of leptin production, increased appetite, and decreased energy expenditure. As the accompanying photo shows, mice with a defective obesity gene do not produce leptin and can weigh up to three times as much as normal mice. When injected with leptin, the mice lose weight. (Because leptin is a protein, it would be destroyed during digestion if given orally; consequently, it must be given by injection.)

Although it is extremely rare, researchers have identified a genetic deficiency of leptin in human beings as well.[12] An error in the gene that codes for leptin was discovered in two extremely obese children whose blood levels of leptin were barely detectable. Without leptin, the children had little appetite control; they were constantly hungry and ate considerably more than their siblings or peers. Given daily injections of leptin, these children lost a substantial amount of weight, confirming leptin's role in regulating appetite and body weight.

Most obese people do not have leptin deficiency, however. In fact, most obese people produce plenty of leptin, but they fail to respond to it, a condition called *leptin resistance.*[13] Researchers speculate that blood leptin rises in an effort to suppress appetite and inhibit fat storage when fat cells are ample. Obese people with elevated leptin concentrations may be resistant to its satiating effect. The absence of or resistance to leptin in obesity parallels the scenario of insulin in diabetes: some people have an insulin deficiency (type 1), whereas many others have elevated insulin but are resistant to its glucose-storing effect (type 2).

Adiponectin In addition to leptin, adipose tissue secretes another protein known as **adiponectin.** Unlike leptin, however, adiponectin correlates inversely with body fat: lean people have higher amounts than obese people—which helps to explain some of the relationships between obesity and diseases.[14] Adiponectin seems to have the beneficial effects of inhibiting inflammation and protecting against insulin resistance, type 2 diabetes, and cardiovascular disease.[15] Researchers are hopeful that, if they can find ways to raise adiponectin levels or enhance its activity, they will be able to reduce the disease risks associated with obesity.[16]

Ghrelin Another protein, known as **ghrelin,** is synthesized and secreted primarily by the stomach cells and promotes a positive energy balance by stimulating appetite and promoting efficient energy storage.[17] The role ghrelin plays in regulating food intake and body weight is the subject of much intense research.[18] Pharmaceutical companies are eager to develop products that mimic ghrelin to treat wasting conditions, as well as products that oppose ghrelin's actions to treat obesity.

Ghrelin powerfully triggers the desire to eat. Blood levels of ghrelin typically rise before and fall after a meal in proportion to the kcalories ingested—reflecting the hunger and satiety that precede and follow eating. In general, fasting blood levels correlate inversely with body weight: lean people have high ghrelin levels and obese people have low levels.[19]

Ghrelin fights to maintain a stable body weight. On average, ghrelin levels are high whenever the body is in negative energy balance, as occurs during low-kcalorie diets, for example. This response may help explain why weight loss is so difficult to maintain. Ghrelin levels decline again whenever the body is in positive energy balance, as occurs with weight gains.

Fat Cell Development Another cause of obesity may be the development of excess fat cells during childhood. The amount of fat on a person's body reflects both fat cell *number* and fat cell *size.* The number of fat cells increases most rapidly during the

CHAPTER SEVEN

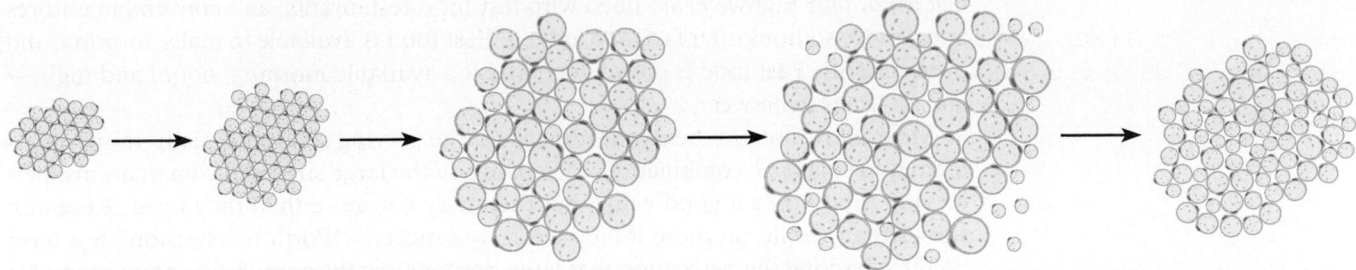

During growth, fat cells increase in number.

When energy intake exceeds expenditure, fat cells increase in size.

When fat cells have enlarged and energy intake continues to exceed energy expenditure, fat cells increase in number again.

With fat loss, the size of the fat cells shrinks, but not the number.

FIGURE 7-1 Fat Cell Development
Fat cells are capable of increasing their size by 20-fold and their number by several thousandfold.

growing years of late childhood and early puberty.[20] Fat cell numbers increase more rapidly in obese children than in lean children, and obese children entering their teen years may already have as many fat cells as do adults of normal weight.

Fat cells can also expand in size. After they reach their maximum size, more cells can develop to store more fat. Thus, obesity develops when a person's fat cells increase in number, in size, or quite often both. Figure 7-1 illustrates fat cell development.

With fat loss, the size of the fat cells shrinks, but the number of fat calls does not. For this reason, people with extra fat cells may tend to regain lost weight rapidly. Prevention of obesity, then, is most critical during the growing years when fat cell number is increasing.

Set-Point Theory One popular theory for why a person may store too much fat is the **set-point theory.** The set-point theory proposes that body weight, like body temperature, is physiologically regulated. Researchers have noted that many people who lose weight quickly regain all their lost weight. This suggests that somehow the body chooses a preferred weight and defends that weight by regulating eating behaviors and hormonal actions. Research confirms that the body adjusts its metabolism whenever it gains or loses weight—in the direction that returns to the initial body weight: energy expenditure increases with weight gain and decreases with weight loss. These changes in energy expenditure are greater than those predicted based on body composition and help to explain why it is so difficult for an obese person to maintain weight losses.[21] An individual's set point for body weight may be adjustable, shifting over the life span in response to physiological changes and to genetic, dietary, and other factors.

Environmental Stimuli

As discussed earlier, genetic factors play a partial role in determining a person's susceptibility to obesity, but they do not fully explain obesity. Obesity rates have risen dramatically during recent decades, but the human gene pool has remained unchanged. The environment must therefore play a role as well. Obesity reflects the interaction between genes and the environment.[22] The *environment* includes all of the circumstances that people encounter daily that push them toward fatness or thinness. Over the past four decades, the demand for physical activity has decreased as the abundance of food has increased.[23]

Overeating People may overeat in response to stimuli in their surroundings—primarily, the availability of many delectable foods. Most people in the United States find high-kcalorie foods readily available, relatively inexpensive, heavily advertised, and reasonably tasty. Food is available everywhere, all the time—thanks largely to

leptin: a hormone produced by fat cells under the direction of the *ob* gene. It decreases appetite and increases energy expenditure.
  leptos = thin

adiponectin: a protein produced by the fat cells that inhibits inflammation and protects against insulin resistance, type 2 diabetes, and cardiovascular disease.

ghrelin (GRELL-in): a hormone produced primarily by the stomach cells. It signals the hypothalamus of the brain to stimulate appetite and food intake.

set-point theory: the theory that the body tends to maintain a certain weight by means of its own internal controls.

fast food. Our highways are lined with fast-food restaurants, and convenience stores and service stations offer fast food as well. Fast food is available in malls, airports, and even schools. Fast food is convenient, and it's available morning, noon, and night—and all times in between.

Most alarming are the extraordinarily large serving sizes and ready-to-go meals offered in supersize combinations.[24] People buy the large sizes and combinations, perceiving them to be a good value, but then they eat more than they need. Research shows that people eat more if they are served more.[25] "Portion distortion" is a term used to describe the perception that large portions are the appropriate amounts to eat at a single sitting.[26] And portion sizes of virtually all foods and beverages (and even plates, glasses, and utensils) have increased markedly in the past several decades, most notably at fast-food restaurants.[27] The increase in portion sizes parallels the growing prevalence of overweight and obesity in the United States, beginning in the 1970s, rising sharply in the 1980s and 1990s, and continuing today.

Research suggests that fast-food consumption contributes significantly to the development of obesity.[28] Fast food is often energy-dense food, which increases energy intake, BMI, and body fatness.[29] The combination of large portions and energy-dense foods strikes a double blow.[30] Reducing portion sizes is helpful, but the real kcalorie savings come from lowering the energy density. Satisfying portions of foods with low energy density such as fruits and vegetables can help with weight loss.[31] Unfortunately, low-energy-density foods tend to be more expensive and less convenient than energy-dense foods.[32] The financial interests of the food industry do not always align with consumer health goals.[33] Consumers' health would benefit from restaurants providing appropriate portion sizes and offering more fruits, vegetables, legumes, and whole grains.

Learned Behavior Psychological stimuli also trigger inappropriate eating behaviors in some people. Appropriate eating behavior is a response to **hunger.** Hunger is a drive programmed into people by their heredity. **Appetite,** in contrast, is learned and can lead people to ignore hunger or to over-respond to it. Hunger is physiological, whereas appetite is psychological, and the two do not always coincide.

Food behavior is also intimately connected to deep emotional needs such as the primitive fear of starvation. Yearnings, cravings, and addictions with profound psychological significance can express themselves in people's eating behavior. An emotionally insecure person might eat rather than call a friend and risk rejection. Another person might eat to relieve boredom or to ward off depression.

Physical Inactivity The possible causes of obesity mentioned so far all relate to the input side of the energy equation. What about output? People may be obese not because they eat too much, but because they move too little. Obese people observed closely are often seen to eat less than lean people, but they are sometimes so extraordinarily inactive that they still manage to accumulate an energy surplus. Reducing their food intake further would jeopardize health and incur nutrient deficiencies. Physical activity, then, is a necessary component of nutritional health. People must be physically active if they are to eat enough food to deliver all the nutrients needed without unhealthy weight gain.

Our environment, however, fosters inactivity. In turn, inactivity contributes to obesity and poor health.[34] Television and sedentary video and computer entertainment have all but replaced outdoor activity for many people.[35] In addition, most people work at sedentary jobs. One hundred years ago, 30 percent of the energy used in farm and factory work came from muscle power; today, only 1 percent does. Modern technology has replaced physical activity at home, at work, at school, and in transportation. The more time spent sitting still, the higher the risk of dying from heart disease and other causes.[36]

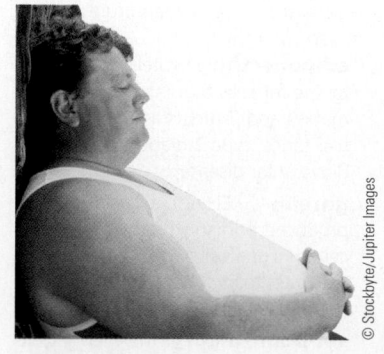

Lack of physical activity fosters obesity.

Health experts urge people to "take the stairs instead of the elevator" or to "walk or bike to work." These are excellent suggestions: climbing stairs provides an impromptu workout, and people who walk or ride a bicycle for transportation most often meet their needs for physical activity.[37]

Some aspects of the **built environment,** including buildings, sidewalks, and transportation opportunities, can discourage physical activity. For example, most stairwells of modern buildings are inconvenient, isolated, and unsafe. Roadways often lack sidewalks, crosswalks, or lanes marked for bicycles. The air on roadways can be dangerously high in carbon monoxide gas and other pollutants from gasoline engine emissions. Hot and cold weather also poses hazards for outdoor commuters. In contrast, those with access to health-promoting foods and built environments more easily make healthy choices.[38] Safe, affordable biking and walking areas and public exercise facilities help maintain health and body leanness. Physical activity strategies for weight loss and weight maintenance are offered in a later section of this chapter.

IN SUMMARY

▍ Genetics, fat cell development, set point, and overeating all offer possible, but still incomplete, explanations of obesity.

▍ Most likely, obesity has not one cause but different causes and combinations of causes in different people.

▍ Some causes may be within a person's control, and some may be beyond it.

▍ Like all the other causes of obesity, inactivity alone fails to explain it fully.

Obesity Treatment: Who Should Lose?

Millions of U.S. adults are trying to lose weight at any given time. Some of these people do not even need to lose weight. Others may benefit from weight loss but are not successful. Relatively few people succeed in losing weight, and even fewer succeed permanently.

Many people assume that every overweight person can achieve slenderness and should pursue that goal. Consider, however, that most overweight people cannot become slender. People vary in their weight tendencies just as they vary in their potentials for height and degrees of health. The question of whether a person should lose weight depends on many factors: the extent of overweight, age, health, and genetics, to name a few. Weight loss advice, then, does not apply equally to all overweight people. Some people may risk more in the process of losing weight than in remaining overweight. Others may reap significant health benefits with just modest weight loss.

IN SUMMARY

▍ Weight-loss advice does not apply equally to all overweight people.

▍ Some people may risk more through misguided efforts to lose weight than by remaining overweight, whereas others may benefit from just a modest weight loss.

Inappropriate Obesity Treatments

The risks people incur in attempting to lose weight often depend on how they go about it. Weight-loss plans and obesity treatments abound—some are adequate, but many are ineffective and possibly dangerous. Fad diets are the topic of Nutrition in Practice 7. This section addresses other inappropriate obesity interventions. Aggressive

hunger: the physiological need to eat, experienced as a drive to obtain food; an unpleasant sensation that demands relief.

appetite: the psychological desire to eat; a learned motivation that is experienced as a pleasant sensation that accompanies the sight, smell, or thought of appealing foods.

built environment: the buildings, roads, utilities, homes, fixtures, parks, and all other manufactured entities that form the physical characteristics of a community.

approaches to treatment for those obese people who face high risks of medical problems and must lose weight rapidly are discussed in the next section. Reasonable approaches for overweight persons seeking safe, gradual weight losses are saved for the last part of the obesity-treatment discussion.

Over-the-Counter Weight-Loss Supplements

Millions of people in the United States use nonprescription weight-loss products, believing them to be safe. Most of the people who use such products are women, especially young overweight women, but almost 10 percent are of normal weight.[39] Promoters and marketers of weight-loss products make all kinds of claims for their products with only one intention—profit. Such claims as "eat all you want and lose weight," "take three pills before bedtime and watch the fat disappear," "blocks carbs," "blocks fat," and many more lure people into believing that maybe this time a product will really work.

In an investigation of over-the-counter weight-loss pills, powders, and other "dietary supplements," the FDA found an alarming number of products to illegally contain prescription medications. Strong diuretics, unproven experimental drugs, psychotropic drugs used to treat mental illnesses, and even drugs deemed unsafe and so banned from U.S. markets were among those discovered, and all pose serious health risks.[40]

In their search for weight-loss magic, some consumers turn to "natural" herbal products and dietary supplements, even though few have proved to be effective.[41] People falsely believe that herbs are not harmful to the body, but many herbs contain toxins. Belladonna and hemlock are infamous examples, but many lesser-known herbs, such as sassafras, contain toxins as well. Furthermore, because herbs are marketed as dietary supplements, manufacturers need not present scientific evidence of their safety or effectiveness to the FDA before marketing them. Evidence about their safety is gathered only through reports of consumers who sicken or die after using the remedies.

A now-familiar example is ephedra (also called *ma huang*), an herb that showed promise as a weight-loss drug in preliminary studies. Immediately, ephedra-containing products for dieters and athletes flooded the market. Many consumers of these products reported ill effects, including cardiac arrest, abnormal heartbeats, hypertension, strokes, and seizures; the supplements have been linked to some deaths as well. For this reason, the FDA has banned the sale of dietary supplements containing ephedra and its active constituent, ephedrine.* Table 7-1 presents the claims and the dangers behind ephedrine and several other weight-loss supplements.

Herbal laxatives containing senna, aloe, rhubarb root, cascara, castor oil, and buckthorn (or various combinations) are commonly sold as "dieter's tea." Such concoctions cause nausea, vomiting, diarrhea, cramping, and fainting and may have contributed to the deaths of four women who had drastically reduced their food intakes. Consumers mistakenly believe that laxatives will diminish nutrient absorption and reduce kcalorie intake, but remember that absorption occurs primarily in the upper small intestine and these laxatives act on the lower large intestine. Nutrition in Practice 15 explores the possible benefits and potential dangers of herbal products and other alternative therapies. Anyone using dietary supplements for weight loss should first consult a physician.

> Ephedrine is an amphetamine-like substance extracted from the Chinese herb *ma huang.*

Other Gimmicks

Other gimmicks don't help with weight loss either. Hot baths do not speed up metabolism so that pounds can be lost in hours. Steam and sauna baths do not melt the fat off the body, although they may dehydrate people so that they lose water weight. Brushes, sponges, wraps, creams, and massages intended to move, burn, or break up **"cellulite"** do nothing of the kind, because there is no such thing as cellulite.

Ma huang (ephedrine) is illegal in Canada.

TABLE 7-1 Selected Herbal and Other Dietary Supplements Marketed for Weight Loss

Product	Claims	Research Findings	Risks
Bitter orange[a] (*Citrus aurantium,* a natural flavoring that contains synephrine, a compound structurally similar to epinephrine)	Stimulates weight loss; provides an alternative to ephedra	Little evidence available	May increase blood pressure; may interact with drugs
Chitosan[b] (pronounced KITE-oh-san; derived from chitin, the substance that forms the hard shells of lobsters, crabs, and other crustaceans)	Binds to dietary fat, preventing digestion and absorption	Ineffective	Impaired absorption of fat-soluble vitamins
Chromium (trace mineral)	Eliminates body fat	Ineffective; weight gain reported when not accompanied by exercise	Headaches, sleep disturbances, and mood swings; hexavalent form is toxic and carcinogenic
Conjugated linoleic acid (CLA; a group of fatty acids related to linoleic acid but with different *cis-* and *trans-*configurations)	Reduces body fat and suppresses appetite	Some evidence in animal studies, modest fat loss in human studies	None known
Ephedrine[c] (amphetamine-like substance derived from the Chinese ephedra herb *ma huang*)	Speeds body's metabolism	Short-term weight loss and dangerous side effects	Insomnia, tremors, heart attacks, strokes, and death; FDA has banned the sale of these products
Fucoxanthin[d] (derived from seaweed)	Speeds metabolism; burns fat	No evidence available	None known
Hoodia (derived from cactus)	Suppresses appetite	Little evidence available	None known
Hydroxycitric acid[e] (active ingredient derived from the rind of the tropical fruit *Garcinia cambogia*)	Inhibits the enzyme that converts citric acid to fat; suppresses appetite	Ineffective	Toxicity symptoms reported in animal studies; headaches, respiratory and gastrointestinal distress in humans
Pyruvate[f] (3-carbon compound produced during glycolysis)	Speeds body's metabolism	Modest weight loss with high doses	GI distress
Yohimbine (derived from the bark of a West African tree)	Promotes weight loss	Ineffective	Nervousness, insomnia, anxiety, dizziness, tremors, headaches, nausea, vomiting, hypertension

Note: The FDA has not approved the use of any of these products; most products are used in conjunction with a 1000- to 1800-kcalorie diet.

[a] Marketed under the trade names Xenadrine EFX, Metabolife Ultra, and NOW Diet Support.
[b] Marketed under the trade names Chitorich, Exofat, Fat Breaker, Fat Blocker, Fat Magnet, Fat Trapper, and Fatsorb.
[c] Marketed under the trade names Diet Fuel, Metabolife, and Nature's Nutrition Formula One.
[d] Marketed under the trade name FucoThin.
[e] Marketed under the trade names Ultra Burn, Citralean, CitriMax, Citrin, Slim Life, Brindleslim, Medislim, and Beer Belly Busters.
[f] Marketed under the trade names Exercise in a Bottle, Pyruvate Punch, Pyruvate-c, and Provate.

IN SUMMARY

▌ When pursued via unwise weight-loss techniques, such as over-the-counter supplements and herbal products, weight-loss efforts can be physically and psychologically damaging.

Aggressive Treatments of Obesity

For some obese people, the medical problems caused by their obesity demand treatment approaches that may, themselves, incur some risks. The health benefits to be gained by weight loss, however, may make these risks worth taking.

cellulite (SELL-you-light or SELL-you-leet): supposedly, a lumpy form of fat; actually, a fraud. Fatty areas of the body may appear lumpy when the strands of connective tissue that attach the skin to underlying muscles pull tight where the fat is thick. The fat itself is the same as fat anywhere else in the body. If the fat in these areas is lost, the lumpy appearance disappears.

Obesity Drugs

Several prescription medications for weight loss have been marketed over the years. When used as part of a long-term, comprehensive weight-loss program, medications can help with modest weight loss. Because weight regain commonly occurs with the discontinuation of drug therapy, treatment is long term—and the long-term use of medications poses risks. Medical experts do not yet know whether a person would benefit more from maintaining 50 pounds of excess weight or from taking a medication for a decade to keep the 50 pounds off.

The challenge, then, is to develop an effective drug—or more likely, a combination of drugs—that can be used over time without adverse side effects or the potential for abuse.[42] The FDA has approved two drugs for long-term treatment of obesity: sibutramine and orlistat. One other drug, phentermine, is approved for short-term treatment of obesity.

Sibutramine Sibutramine is an appetite suppressant that works on the brain's neurotransmitters.* The drug is most effective when used in combination with a reduced-kcalorie diet and increased physical activity. Side effects include dry mouth, rapid heart rate, insomnia, headache, and high blood pressure. The FDA warns those with heart failure, hypertension, irregular heart beats, and other heart problems not to use sibutramine and advises others to monitor their blood pressure. As more information becomes known about the molecular chemistry of appetite control, safer appetite suppressants may be developed.

Orlistat Orlistat takes a different approach to weight loss.† Not an appetite suppressant, orlistat inhibits the action of fat-digesting enzymes in the GI tract and so reduces fat digestion and absorption by about 30 percent.[43] As a result of absorbing less fat, people often lose weight. The drug is taken with meals and is most effective when accompanied by a reduced-kcalorie, low-fat diet. Side effects include gas, frequent bowel movements, and reduced absorption of fat-soluble vitamins. The FDA recently approved an over-the-counter, low-dose version of orlistat, making it the first nonprescription weight-loss drug for use in the United States.

Phentermine Phentermine reduces appetite and is a widely prescribed weight-loss drug. However, phentermine use is approved by the FDA for no longer than three months.[44] Side effects include dry mouth, insomnia, and constipation.

Surgery

Surgery as an approach to weight loss is justified in some specific cases of **clinically severe obesity.** Two procedures, **gastric bypass** and **gastric banding,** have gained wide acceptance. Both procedures limit food intake by effectively reducing the capacity of the stomach and suppressing hunger by reducing the production of ghrelin.[45] The results are significant: depending on the type of surgery, a loss of 20 to 32 percent of body weight initially and 14 to 25 percent after 10 years.[46] Research shows that weight-loss surgery also helps to improve blood lipids, diabetes, sleep apnea, and hypertension.[47]

Laparoscopic weight-loss surgery techniques are now used to perform gastric bypass and other weight-loss surgeries. Laparoscopic weight-loss surgery produces significant, long-term weight loss, shortens recovery, and is less invasive than open surgery.[48]

The long-term safety and effectiveness of gastric surgery depend, in large part, on compliance with dietary instructions. Common immediate postsurgical complications include infections, nausea, vomiting, and dehydration; in the long term, vitamin

---

Drugs may be an option for people with all of the following conditions:

- Unable to achieve adequate weight loss with diet and exercise
- BMI ≥30, or BMI ≥27 with weight-related health problems
- No medical contraindications

Postsurgical diets are discussed in Chapter 17.

---

*Sibutramine's trade name is Meridia.

†Orlistat's trade name is Xenical. The low-dose, over-the-counter version of orlistat is marketed under the trade name Alli (AL-eye).

and mineral deficiencies and psychological problems are common. Lifelong medical supervision is necessary for those who choose the surgical route, but in suitable candidates, the health benefits of weight loss may prove worth the risks.[49]

IN SUMMARY

▌ Obese people with high risks of medical problems may need aggressive treatment, including drugs or surgery.

Reasonable Strategies for Weight Loss

Successful weight-loss strategies embrace small changes, moderate losses, and reasonable goals.[50] People who lose 10 to 20 pounds in a year by consistently choosing nutrient-dense foods and engaging in regular physical activity are much more likely to maintain the loss and reap health benefits than if they were to lose more weight in less time by adopting the latest fad diet. In keeping with this philosophy, the *Dietary Guidelines for Americans* advise that those who need to lose weight should over time, "consume fewer kcalories from foods and beverages, increase physical activity, and reduce time spent in sedentary behaviors." Modest weight loss, even when a person is still overweight, can improve control of diabetes and reduce the risks of heart disease by lowering blood pressure and blood cholesterol, especially for those with abdominal fat.[51]

Of course, the same eating and activity habits that improve health often lead to a healthier body weight and composition as well. Successful weight loss, then, is defined not by pounds lost but by health gained. People less concerned with disease risks may prefer to set goals for personal fitness, such as being able to play with children or climb stairs without becoming short of breath.

Whether the goal is health or fitness, weight-loss expectations need to be reasonable. Unreachable targets ensure frustration and failure. Setting reasonable goals helps to achieve the desired result in managing weight. For example, obese people who must reduce their weight to lower their disease risks might set three broad goals:

1. Reduce body weight by about 10 percent over half a year's time.[52]
2. Maintain a lower body weight over the long term.
3. At a minimum, prevent further weight gain.

Such goals may be achieved or even exceeded, providing a sense of accomplishment instead of disappointment.

A Healthful Eating Plan

Contrary to the claims of many fad diets, no particular eating plan is magical, and no specific food must be either included or avoided for weight management. In developing a plan, people need only consider foods that they like or can learn to like, that are available, and that are within their means.

A Realistic Energy Intake The main characteristic of a weight-loss diet is that it provides less energy than the person needs to maintain present body weight. If food energy is restricted too severely, dieters may not receive sufficient nutrients and may lose lean tissue. Rapid weight loss usually means excessive loss of lean tissue, a lower

Weight management strategies for children and adolescents are discussed in Chapter 12.

Courtesy of author Linda DeBruyne

A healthy body contains enough lean tissue to support health and the right amount of fat to meet body needs.

clinically severe obesity: a BMI of 40 or greater, or a BMI of 35 or greater with one or more serious conditions such as hypertension. Another term used to describe the same condition is *morbid obesity*.

gastric bypass: surgery that restricts stomach size and reroutes food from the stomach to the lower part of the small intestine; creates a chronic, lifelong state of malabsorption by preventing normal digestion and absorption of nutrients.

gastric banding: a surgical means of producing weight loss by restricting stomach size with a constricting band; used in people whose severe obesity brings extreme health risks.

laparoscopic weight-loss surgery: a procedure in which surgeons gain access to the abdomen via several small incisions. A tiny video camera is inserted through one of the incisions and surgical instruments through the others. The surgeons watch their work on a large-screen monitor.

BMR, and a rapid weight gain to follow. Restrictive eating may also set in motion the unhealthy behaviors of eating disorders (described in Nutrition in Practice 6).

Table 7-2 outlines the recommendations of a weight-loss diet. Energy intake should provide nutritional adequacy without excess—that is, somewhere between deprivation and complete freedom to eat whatever, whenever. A reasonable suggestion is that an adult needs to increase activity and reduce food intake to create a deficit of 500 to 1000 kcalories per day. Such a deficit produces a weight loss of 1 to 2 pounds per week—a rate that supports the loss of fat efficiently while retaining lean tissue.[53] In general, weight-loss diets provide about 1200 kcalories per day for women and 1600 kcalories per day for men.[54] Table 7-3 suggests daily food amounts from which to build balanced 1200- to 1600-kcalorie diets.

Some people skip meals, typically breakfast, in an effort to reduce energy intake, but research suggests such a strategy may be counterproductive. Breakfast frequency is inversely associated with obesity—that is, people who frequently eat breakfast have lower BMIs than those who tend to skip breakfast.[55] Furthermore, when people eat breakfast, overall diet quality is better and daily energy density is lower—two factors that support healthy body weight.[56]

Nutritional Adequacy Nutritional adequacy is difficult to achieve on fewer than 1200 kcalories a day, and most healthy adults need never consume any less than that. A plan that provides an adequate intake supports a healthier and more successful weight loss than a restrictive plan that creates feelings of starvation and deprivation, which can lead to an irresistible urge to binge.

1 lb body fat = 3500 kcal. To lose a pound a week, cut 500 kcal/day.

TABLE 7-2 **Recommendations for a Weight-Loss Diet**

Nutrient	Recommended Intake
kCalories	
For people with BMI ≥ 35	Approximately 500 to 1000 kcalories per day reduction from usual intake
For people with BMI between 27 and 35	Approximately 300 to 500 kcalories per day reduction from usual intake
Total fat	30% or less of total kcalories
Saturated fatty acids[a]	8 to 10% of total kcalories
Monounsaturated fatty acids	Up to 15% of total kcalories
Polyunsaturated fatty acids	Up to 10% of total kcalories
Cholesterol[a]	< 300 mg per day
Protein[b]	Approximately 15% of total kcalories
Carbohydrate[c]	55% or more of total kcalories
Sodium chloride[d]	No more than 2300 mg of sodium or approximately 6 g of sodium chloride (salt) per day
Calcium	1000 to 1500 mg per day
Fiber[c]	20 to 30 g per day

[a]People with high blood cholesterol should aim for less than 7 percent kcalories from saturated fat and 200 milligrams cholesterol per day.

[b]Protein should be derived from plant sources and lean sources of animal protein.

[c]Carbohydrates and fiber should be derived from vegetables, fruits, and whole grains.

[d]Value from the *Dietary Guidelines for Americans 2010*; the Tolerable Upper Intake Level for sodium is 2300 milligrams.

Source: National Heart, Lung, and Blood Institute, National Institutes of Health, *The Practical Guide: Identification, Evaluation, and Treatment of Overweight and Obesity in Adults,* NIH publication no. 00-4084 (Washington, DC: Government Printing Office, 2000), p. 27.

TABLE 7-3 Daily Amounts from Each Food Group for 1200- to 1600-kCalorie Diets

Food Group	1200 kCalories	1400 kCalories	1600 kCalories
Fruit	1 c	1½ c	1½ c
Vegetables	1½ c	1½ c	2 c
Grains	4 oz	5 oz	5 oz
Protein foods	3 oz	4 oz	5 oz
Milk	3 c	3 c	3 c
Oils	3 tsp	3 tsp	4 tsp

Note: The USDA Food Guide patterns for 1200 and 1400 kcalories were designed for children and provided 2½ cups milk. They were modified here to include an additional ½ cup of milk, as 3 cups per day is recommended for all adults. The discretionary kcalorie allowance for these patterns is about 100 kcalories.

Take a look at the 1200-kcalorie diet in Table 7-3. Such an intake would allow most people to lose weight and still meet their nutrient needs with careful, nutrient-dense food selections. (Women might need calcium or iron supplements.) Keep in mind that well-balanced diets that emphasize fruits, vegetables, whole grains, lean protein foods, and low-fat or fat-free milk products offer many health rewards even when they don't result in weight loss.[57] A dietary supplement providing vitamins and minerals at or below 100 percent of the Daily Values can help people following low-kcalorie diets to achieve nutrient adequacy.

Small Portions As mentioned earlier, large portion sizes increase energy intakes, and the huge helpings served by restaurants and sold in packages are the enemy of people striving to manage their weight.[58] Similarly, big bowls, plates, utensils, and drinking glasses encourage people to take and consume larger portions. Small plates and bowls, tall and thin drinking glasses, and luncheon-sized plates have the opposite effect.[59]

For health's sake, overweight people may need to learn to eat less food at each meal—one piece of chicken for dinner instead of two, a teaspoon of butter on the vegetables instead of a tablespoon, and one cookie for dessert instead of six. Chew foods slowly and thoroughly. The goal is to eat enough food for energy, nutrients, and pleasure, but not more. This amount should leave a person feeling satisfied—not necessarily full. Keep in mind that even fat-free and low-fat foods can deliver a lot of kcalories when a person eats large quantities.

Balancing Carbohydrates, Fats, and Protein Healthy diets based on abundant fresh fruits and vegetables; low-fat milk products; legumes; lean meats, fish, or poultry; and whole grains are high in carbohydrates, adequate in fiber and protein, and low in the kinds of fats associated with diseases. They are also best for managing weight. As earlier chapters described, each of the energy-yielding nutrients is important to health. Therefore, diets for weight management should provide all three within the ranges recommended by the DRI committee (see the margin).

Wholesome, high-fiber, and unprocessed or lightly processed foods offer bulk and satiety for fewer kcalories than smooth, quickly consumed refined foods. Thus, choosing whole grains and fiber-rich vegetables in place of most refined grains and added fats and sugars benefits both weight and nutrition. Choose fats sensibly by avoiding sources of saturated and *trans* fats and including enough of the health-supporting fats (details in Nutrition in Practice 3) to provide satiety but not so much as to oversupply kcalories. Nuts provide fat and protein, and including a moderate amount in the diet may assist in controlling body weight.[60] Lean meats or other low-fat protein sources also provide satiety. Limit these foods, but don't eliminate them.

The DRI committee recommends:
- 45 to 65% kcalories from carbohydrate
- 20 to 35% kcalories from fat
- 10 to 35% kcalories from protein

Selecting grapes with their high water content instead of raisins increases the volume and cuts the energy intake in half.

Even at the same weight and similar serving sizes, the fiber-rich broccoli delivers twice the fiber of the potatoes for about one-fourth the energy.

By selecting the water-packed tuna (on the right) instead of the oil-packed tuna, a person can enjoy the same amount for fewer kcalories.

FIGURE 7-2 Energy Density
Decreasing the energy density (kcal/g) of foods allows a person to eat satisfying portions while still reducing energy intake. To lower energy density, select foods high in water or fiber and low in fat.

Lower Energy Density As discussed earlier, to lower energy intake, people can choose smaller portion sizes, or they can reduce the energy density of the foods they eat. Research shows that eating satisfying portions of foods that are low in energy density (such as fruits, vegetables, and broth-based soups) maintains satiety while reducing energy intake.[61] Foods containing substantial water or fiber and those low in fat help to lower a meal's energy density (see Figure 7-2). The Clinical Applications feature at the end of this chapter (p. 179) describes how to calculate the energy density of foods.

Sugar and Alcohol A person trying to achieve or maintain a healthy weight needs to pay attention to sugar and alcohol, as well as fat. Using them for pleasure on occasion is compatible with health as long as most daily choices are of nutrient-dense foods.

Meal Spacing Three meals a day is standard in our society, but no law says you can't have four or five—just be sure they are smaller, of course. People who eat small, frequent meals can be as successful at weight loss and maintenance as those who eat three.[62] Make sure that mild hunger, not appetite, is prompting you to eat. Eat regularly, and eat before you become extremely hungry.

Adequate Water Learn to satisfy thirst with water. Water helps with weight management in several ways. For one, foods with high water content (such as broth-based soups) increase fullness, reduce hunger, and consequently reduce energy intake.[63] For another, drinking a large glass of water before a meal may ease hunger, fill the stomach, and reduce energy intake.[64] Importantly, water adds no kcalories. The average U.S. diet delivers an estimated 75 to 150 kcalories a day from sweetened beverages.[65] Consuming large portions of kcalorie-containing beverages increases energy intake.[66] Simply replacing nutrient-poor, energy-dense beverages with water could save a person up to 15 pounds a year. Water also helps the GI tract adapt to a high-fiber diet.

Physical Activity

The best approach to weight management includes physical activity.[67] Either energy restriction or physical activity alone can produce some weight loss. Clearly, however, the combination is most effective.[68] People who combine diet and physical activity are more likely to lose more fat, retain more muscle, and regain less weight than those who only diet.[69] Figure 10-1 in Chapter 10 presents the *Physical Activity Guidelines for Americans, 2008*, which specify the minimum amount of physical activity people need

TABLE 7-4 Physical Activity Strategies for Weight Management

A negative energy balance incurred through physical activity will result in weight loss, and the larger the negative energy balance, the greater the weight loss.

- Weight gain prevention and augmented weight loss occur with at least 2 hours and 30 minutes (150 minutes) per week of at least moderate-intensity physical activity.
- Greater weight loss and improved weight maintenance after weight loss occur with more than 4 hours and 10 minutes (>250 minutes) per week of at least moderate-intensity physical activity.
- Both aerobic (endurance) and muscle-strengthening (resistance) physical activities are beneficial, but kcalorie restriction must accompany resistance training to achieve weight loss.

Source: American College of Sports Medicine, Position stand: Appropriate physical activity intervention strategies for weight loss and prevention of weight regain for adults, *Medicine and Science in Sports and Exercise* 41 (2009): 459–471.

to gain general *health* benefits. Table 7-4 presents physical activity strategies from the American College of Sports Medicine developed specifically to prevent weight gain and promote at least modest weight loss.[70]

Even without weight loss, physical activity may also help counteract some of the negative effects of excess body weight on health.[71] For example, physical activity reduces abdominal obesity, and this change improves blood pressure, insulin resistance, and fitness of the heart and lungs, even without weight loss.[72]

Energy Expenditure Physical activity directly increases energy output by the muscles and cardiovascular system. Table 6-2 in Chapter 6 (p. 143) shows how much energy each of several activities uses. The number of kcalories spent in an activity depends on body weight, intensity, and duration. For example, a 150-pound person who walks 3.5 miles in 60 minutes expends about 315 kcalories. That same person running 3 miles in 30 minutes uses a similar amount. By comparison, a 200-pound person running 3 miles in 30 minutes expends an additional 125 kcalories or so (about 444 kcalories total). The goal is to expend as much energy as time allows. The greater the energy deficit created by physical activity, the greater the fat loss. A word of caution, however: people who reward themselves for "good behavior" with high-kcalorie foods can easily negate any kcalorie deficits incurred with physical activity.[73]

BMR Activity also contributes to energy output in an indirect way—by speeding up basal metabolism. It does this both immediately and over the long term. On any given day, basal metabolism remains slightly elevated for several hours after intense and prolonged exercise.[74] Over the long term, a person who engages in daily vigorous activity gradually develops more lean tissue, which is more active metabolically than fat tissue. Metabolic rate rises accordingly, and this makes a contribution toward continued weight loss or maintenance.

Appetite Control Physical activity also helps to control appetite. People think that exercising will make them want to eat, but this is not entirely true. Active people do have healthy appetites, but appetite is suppressed immediately following a workout and satiation during meals is heightened.[75] The reasons are unclear, but physical activity may help to control appetite by altering the levels of appetite-regulating hormones.[76]

Psychological Benefits Physical activity helps especially to curb the inappropriate appetite that prompts a person to eat when bored, anxious, or depressed. Weight-management programs encourage people to go out and be active when they are tempted to eat but are not really hungry.

Being active—even if overweight—is healthier than being sedentary. Weight loss itself, however, best reduces disease risks.

Physical activity also helps to reduce stress. Because stress itself is a cue to inappropriate eating behavior for many people, activity can help here, too.

Activity offers still more psychological advantages. The fit person looks and feels healthy, and high self-esteem accompanies these benefits. High self-esteem tends to support a person's resolve to persist in a weight-control effort, rounding out a beneficial cycle.

Benefits of physical activity in a weight-management program:

- Improved body composition
- Favorable effects on disease risks
- Short-term increase in energy expenditure (from exercise and from a slight rise in BMR)
- Long-term increase (slight) in BMR
- Appetite control
- Stress reduction and control of stress eating
- Physical, and therefore psychological, well-being
- Improved self-esteem

Choosing Activities What kind of physical activity is best? For health, a combination of moderate-to-vigorous aerobic physical activity along with **resistance training** at a safe level provides benefits. However, any physical activity is better than being sedentary.[77]

People seeking to lose weight should choose activities that they enjoy and are willing to do regularly. Health care professionals frequently advise people who want to manage their body weight and lose fat to engage in activities of low-to-moderate intensity for a long duration, such as an hour-long fast-paced walk. The reasoning behind such advice is that people exercising at low-to-moderate intensity are likely to stick with their activity for longer times and are less likely to injure themselves. People who regularly engage in more *vigorous* physical activities (fast bicycling or endurance running, for example), however, have less body fat than those who engage in moderately intense activities. The conditioned body that is adapted to strenuous and prolonged aerobic activity uses more fat all day long, not just during activity. The bottom line on physical activity and weight and/or fat loss seems to be that total energy expenditure is the main factor, regardless of how a person does it.

In addition to activities such as walking or cycling, there are hundreds of ways to incorporate energy-expending activities into daily routines: take the stairs instead of the elevator, walk to the neighbor's apartment instead of making a phone call, and rake the leaves instead of using a blower. These activities burn only a few kcalories each, but over a year's time they become significant.

Spot Reducing People sometimes ask about "spot reducing." Unfortunately, no one part of the body gives up fat in preference to another. Fat cells all over the body release fat in response to demand, and the fat is then used by whatever muscles are active. No exercise can remove the fat from any one particular area—and, incidentally, neither can a massage machine that claims to break up fat on trouble spots.

Physical activity can help with trouble spots in another way, though. Strengthening muscles in a trouble area can help to improve their tone; stretching to gain flexibility can help with posture problems. Thus, cardiorespiratory endurance, strength, and flexibility workouts all have a place in fitness programs.

Behavior and Attitude

Behavior modification provides ways to overcome barriers to making dietary changes and increasing physical activity. Behavior modification does more than help people decide which behaviors to change: it also teaches them how to change.[78] Behavior and attitude are important supporting factors in achieving and maintaining appropriate body weight and composition. Changing the behaviors of overeating and underexercising that lead to, and perpetuate, obesity requires time and effort. A person must commit to take action.

Becoming Aware of Behaviors A person who is aware of all the behaviors that create a problem has a head start on developing a solution. First, the person needs to establish a baseline (a record of present eating and physical activity behaviors) against which to measure future progress. It is best to keep a diary (see Figure 7-3) that includes the time and place of meals and snacks, the type and amount of foods eaten, the persons present when food is eaten, and a description of the individual's feelings

Time	Place	Activity or food eaten	People present	Mood
10:30–10:40	School vending machine	6 peanut butter crackers and 12 oz. cola	by myself	Starved
12:15–12:30	Restaurant	Sub sandwich and 12 oz. cola	friends	relaxed & friendly
3:00–3:45	Gym	Weight training	workout partner	tired
4:00–4:10	Snack bar	Small frozen yogurt	by myself	OK

FIGURE 7-3 Food and Activity Diary
The entries in a food and activity diary should include the times and places of meals and snacks, the types and amounts of foods eaten, and a description of the individual's feelings when eating. The diary should also record physical activities: the kind, the intensity level, the duration, and the person's feelings about them.

when eating. The diary should also record physical activities: the kind, the intensity level, the duration, and the person's feelings about them. These entries will help the individual identify possible behaviors to change.

Making Small Changes Behavior modification strategies focus on learning desired eating and exercise behaviors and eliminating unwanted behaviors. With so many possible behavior changes, a person can feel overwhelmed. Start with small time-specific goals for each behavior—for example, "I'm going to take a 30-minute walk after dinner every evening" instead of "I'm going to run a marathon someday." Practice desired behaviors until they become routine. Using a reward system seems to effectively support weight-loss efforts.[79] The "How to" (p. 176) describes behavioral strategies to support weight management. A particularly attractive feature of these strategies is that they do not involve blaming oneself or putting oneself down—an important element in fostering self-esteem.

Cognitive Skills Behavior therapists often teach **cognitive skills,** or new ways of thinking, to help overweight people solve problems and correct false thinking that can undermine healthy eating behaviors.[80] Thinking habits are as important as eating and activity habits to achieving a healthy body weight, and thinking habits can be changed.* A paradox of making a change is that it takes belief in oneself and honoring of oneself to lay the foundation for changing that self. That is, self-acceptance predicts success, while self-loathing predicts failure. Positive self-talk is a concept worth cultivating—many people succeed because their mental dialogue supports, rather than degrades, their efforts. Negative thoughts ("I'm not getting thin anyway, so what is the use of continuing?") should be viewed in light of empirical evidence ("my starting weight: 174 pounds; today's weight: 163 pounds").

Take credit for new behaviors; be aware of any physical improvements, too, such as lower blood pressure or less painful knees, even without a noticeable change in pant size. Finally, remember to enjoy your emerging fit and healthy self.

Personal Attitude For many people, overeating and being overweight may have become an integral part of their identity. Changing diet and activity behaviors without attention to a person's self-concept invites failure.

Many people overeat to cope with the stresses of life. To break out of that pattern, they must first identify the particular stressors that trigger their urges to overeat. Then, when faced with these situations, they must learn to practice problem-solving skills. When the

*Psychologists have a term for changing thinking habits: *cognitive restructuring.*

resistance training: the use of free weights or weight machines to provide resistance for developing muscle strength, power, and endurance; also called *weight training.* A person's own body weight may also be used to provide resistance as when a person does push-ups, pull-ups, or abdominal crunches.

behavior modification: the changing of behavior by the manipulation of *antecedents* (cues or environmental factors that trigger behavior), the behavior itself, and *consequences* (the penalties or rewards attached to behavior).

cognitive skills: as taught in behavior therapy, changes to conscious thoughts with the goal of improving adherence to lifestyle modifications; examples are problem-solving skills or the correction of false negative thoughts, termed *cognitive restructuring.*

How To APPLY BEHAVIOR MODIFICATION TO MANAGE BODY FATNESS

1. Eliminate inappropriate cues:
   - Do not buy problem foods.
   - Eat only in one room at the designated time.
   - Shop when not hungry.
   - Avoid vending machines, fast-food restaurants, and convenience stores.
   - Turn off television, video games, and computers.

2. Suppress the cues you cannot eliminate:
   - Serve individual plates; do not serve "family style."
   - Make small portions look large by spreading them over the plate.
   - Create obstacles to consuming problem foods—wrap them and freeze them, making them less quickly accessible.
   - Control deprivation; plan and eat regular meals.

   - Plan the time spent in sedentary activities, such as watching television or using a computer—do not use these activities just to fill time.

3. Strengthen cues to appropriate behaviors:
   - Share appropriate foods with others.
   - Store appropriate foods in convenient spots in the refrigerator.
   - Learn appropriate portion sizes.
   - Plan appropriate snacks.
   - Keep sports and play equipment by the door.

4. Repeat desired behaviors:
   - Slow down eating—put down utensils between bites.
   - Always use utensils.
   - Leave some food on your plate.
   - Move more—shake a leg, pace, stretch often.
   - Join groups of active people and participate.

5. Arrange negative consequences for negative behavior:
   - Ask that others respond neutrally to your deviations (make no comments—even negative attention is a reward).
   - If you slip, do not punish yourself.

6. Reward yourself personally and immediately for positive behaviors:
   - Buy tickets to sports events, movies, concerts, or other nonfood amusement.
   - Indulge in a new small purchase.
   - Get a massage; buy some flowers.
   - Take a hot bath; read a good book.
   - Treat yourself to a lesson in a new active pursuit such as horseback riding, handball, or tennis.
   - Praise yourself; visit friends.
   - Nap; relax.

problems that trigger the urge to overeat are dealt with in alternative ways, people may find that they eat less. The message is that sound emotional health supports the ability to take care of health in all ways—including nutrition, weight management, and fitness.

Weight Maintenance Finally, be aware that it can be hard to maintain weight loss. Millions of people have experienced the frustration of achieving a desired change in weight only to see their hard work visibly slip away in a seemingly never-ending cycle of weight loss and weight regain. Disappointment, frustration, and self-condemnation are common in people who have slipped back to their original weight or even higher. Some hints for successful long-term weight loss can be gleaned from data of the National Weight Control Registry, a self-selected population of 6000 people who lost at least 30 pounds and kept it off for at least a year.[81]

A key to weight maintenance is accepting it as a lifelong endeavor and not a goal to be achieved and then forgotten. People who maintain their weight loss continue to employ the behaviors that reduced their weight in the first place. They cultivate habits of people who maintain a healthy weight, such as eating low-kcalorie meals (averaging 1800 kcalories per day) and being physically active. They must also control susceptibility to overeating regardless of the initial weight-loss method.[82] Those who maintain weight losses over time also generally:

- Believe they have the ability to control their weight, an attribute known as **self-efficacy.**

- Eat breakfast every day.

- Average about one hour of physical activity per day.

- Monitor body weight about once a week.

- Maintain consistent lower-kcalorie eating patterns.

- Quickly address small **lapses** to prevent small gains from turning into major ones.

- Watch fewer than 10 hours of television per week.

- Eat high-fiber food, particularly whole grains, vegetables, and fruit, and consume sufficient water each day.[83]

- Cultivate and honor realistic expectations regarding body size and shape.

IN SUMMARY

- A person who adopts a lifelong "eating plan for good health" rather than a "diet for weight loss" will be more likely to keep the lost weight off. Table 7-5 offers several tips for successful weight management.

- Physical activity should be an integral part of a weight-management program.

- Physical activity can increase energy expenditure, improve body composition, help control appetite, reduce stress and stress eating, and enhance physical and psychological well-being.

- Behavior modification provides ways to overcome barriers to successful weight management.

self-efficacy: a person's belief in his or her ability to succeed in an undertaking.

lapses: periods of returning to old habits.

TABLE 7-5 Weight Management Strategies

In General

- Focus on healthy eating and activity habits, not on weight losses or gains.
- Adopt reasonable expectations about health and fitness goals and about how long it will take to achieve them.
- Make nutritional adequacy a high priority.
- Learn, practice, and follow a healthful eating plan for the rest of your life.
- Participate in some form of physical activity regularly.
- Adopt permanent lifestyle changes to achieve and maintain a healthy weight.

For Weight Loss

- Energy out should exceed energy in by about 500 kcalories/day. Increase your physical activity enough to spend more energy than you consume from foods.
- Emphasize foods with a low energy density and a high nutrient density.
- Eat small portions. Share a restaurant meal with a friend or take home half for lunch tomorrow.
- Eat slowly.
- Limit high-fat foods. Make legumes, whole grains, vegetables, and fruits central to your diet plan.
- Limit low-fat treats to the serving size on the label.
- Limit concentrated sweets and alcoholic beverages.
- Drink a glass of water before you begin to eat and another while you eat. Drink plenty of water throughout the day (eight glasses or more per day).
- Keep a record of diet and exercise habits; it reveals problem areas, the first step toward improving behaviors.
- Learn alternative ways to deal with emotions and stresses.
- Attend support groups regularly or develop supportive relationships with others.

For Weight Gain

- Energy in should exceed energy out by at least 500 kcalories/day. Increase your food intake enough to store more energy than you spend in exercise. Exercise and eat to build muscles.
- Expect weight gain to take time (1 pound per month would be reasonable).
- Emphasize energy-dense foods.
- Eat at least three meals per day.
- Eat large portions of foods and expect to feel full.
- Eat snacks between meals.
- Drink plenty of juice and milk.

Strategies for Weight Gain

Some people are unalterably thin by reasons of genetics or early physical influences. Those who wish to gain weight for appearance's sake or to improve athletic performance should be aware that a healthful weight can be achieved only through physical activity, particularly strength training, combined with a high energy intake. Eating many high-kcalorie foods can bring about weight gain, but it will be mostly fat, and this can be as detrimental to health as being slightly underweight. In an athlete, such a weight gain can impair performance. Therefore, in weight gain, as in weight loss, physical activity and energy intake are essential components of a sound plan.

Physical Activity to Build Muscles The person who wants to gain weight should primarily use resistance training. As activity is increased, energy intake must be increased to support that activity. Eating extra food will then support a gain of both muscle and fat.

Energy-Dense Foods Energy-dense foods (the very ones eliminated from a successful weight-loss diet) hold the key to weight gain. Pick the highest-kcalorie items from each food group—that is, milk shakes instead of fat-free milk, peanut butter instead of lean meat, avocados instead of cucumbers, and whole-wheat muffins instead of whole-wheat bread. Because fat contains more than twice as many kcalories per teaspoon as sugar does, fat adds kcalories without adding much bulk.

Be aware that health experts recommend a moderate-fat diet for the general U.S. population because the general population is overweight and at risk for heart disease. Consumption of excessive fat is not healthy for most people, of course, but it may be essential for an underweight individual who needs to gain weight. An underweight person who is physically active and eating a nutritionally adequate diet can afford a few extra kcalories from fat. For health's sake, it is wise to select foods with monounsaturated and polyunsaturated fats instead of those with saturated or *trans* fats: for example, sautéing vegetables in olive oil instead of butter or hydrogenated margarine.

Three Meals Daily People wanting to gain weight should eat at least three hearty meals a day. Many people who are underweight have simply been too busy (sometimes for months) to eat enough to gain or maintain weight. Therefore, they need to make meals a priority and plan them in advance. Taking time to prepare and eat each meal can help, as can learning to eat more food within the first 20 minutes of a meal, before you begin to feel full. Another suggestion is to eat meaty appetizers or the main course first and leave the soup or salad until later.

Large Portions It is also important for the underweight person to learn to eat more food at each meal: have two sandwiches for lunch instead of one, drink milk from a larger glass, and eat cereal from a larger bowl. Expect to feel full. Most underweight individuals are accustomed to small quantities of food. When they begin eating significantly more, they feel uncomfortable. This is normal and passes over time.

Extra Snacks Because a substantially higher energy intake is needed each day, in addition to eating more food at each meal, it is necessary to eat more frequently. Between-meal snacking offers a solution. For example, a student might make three sandwiches in the morning and eat them between classes in addition to the day's three regular meals.

Juice and Milk Beverages provide an easy way to increase energy intake. Consider that 6 cups of cranberry juice add almost 1000 kcalories to the day's intake. kCalories can be added to milk by mixing in powdered milk or packets of instant breakfast.

For people who are underweight due to illness, concentrated liquid formulas are often recommended because a weak person can swallow them easily. A registered dietitian can recommend high-protein, high-kcalorie formulas to help the underweight person maintain or gain weight. Used in addition to regular meals, these formulas can help considerably.

IN SUMMARY

▌ To gain weight, a person must train physically and increase energy intake by selecting energy-dense foods, eating regular meals, taking larger portions, and consuming extra snacks and beverages. Review the tips for weight gain in Table 7-5 (p. 177).

Your Diet

To enjoy good health and maintain a reasonable body weight, you need to combine sensible eating habits and regular physical activity. This exercise allows you to evaluate your current body weight and consider some lifestyle factors important for weight management.

First, check to see whether your current body weight falls within the "healthy weight" range in the BMI table at the back of this text. Find your height in the left-hand column of the table and look across the row to find your weight. Is your BMI in the healthy range, which is highlighted in green?

Alternatively, you can calculate your BMI using the equation shown on p. 146. A healthy weight usually falls within a BMI range of 18.5 to 24.9.

Your BMI: ___

If your BMI falls within the underweight, overweight, or obese range, you may want to gain or lose weight to improve your health. Even if your BMI is within the healthy range, you may wish to improve your eating habits or fitness level. Review Figure 7-3 (p. 175) and complete your own food and activity record for at least a 24-hour period to reveal information about daily habits that may support or undermine achieving a healthy weight.

Time	Place	Activity or Food Eaten	People Present	Mood

▌ Make a list of the habits that support maintaining or achieving a healthy weight. For example, do you drink water rather than soda throughout the day and limit the time you spend watching television or playing computer games?

▌ Make a list of the habits that do not support maintaining or achieving a healthy weight. For example, do you skip breakfast, and then eat fast food or vending machine food?

▌ What changes would you like to make in your daily habits to improve your health and nutrition?

Clinical Applications

1. Chapter 1 discusses the nutrient density of foods—their nutrient contribution per kcalorie. Another way to evaluate foods is to consider their energy density—their energy contribution per gram:

▌ A carrot weighing 72 grams delivers 31 kcalories.

▌ To calculate the energy density, divide kcalories by grams: 31 kcalories divided by 72 grams = 0.43 kcalories per gram.

2. Do the same for French fries weighing 50 grams and contributing 167 kcalories: 167 kcalories divided by 50 grams = _____ kcalories per gram.

   ▮ The more kcalories per gram, the greater the energy density.

   ▮ Which food is more energy dense? The conclusion is no surprise, but understanding the mathematics may offer valuable insight into the concept of energy density.

   ▮ French fries are more energy dense, providing 3.34 kcalories per gram. They provide more energy per gram—and per bite.

3. Considering a food's energy density is especially useful in planning diets for weight management. Foods with a high energy density can help with weight gain, whereas those with a low energy density can help with weight loss. Give some examples of foods that you might suggest for a client who wants to gain weight and some that might be appropriate for a client who is trying to lose weight.

© Matthew Farruggio

Self Check

1. Two causes of obesity in humans are:
   a. set-point theory and BMI.
   b. genetics and physical inactivity.
   c. genetics and low-carbohydrate diets.
   c. mineral imbalances and fat cell imbalance.

2. The protein produced by the fat cells under the direction of the *ob* gene is called:
   a. leptin.
   b. orlistat.
   c. sibutramine.
   d. lipoprotein lipase.

3. All of the following describe the behavior of fat cells *except:*
   a. the number decreases when fat is lost from the body.
   b. the storage capacity for fat depends on both cell number and cell size.
   c. the size is larger in obese people than in normal-weight people.
   d. the number increases several-fold during the growth years and tapers off when adult status is reached.

4. The obesity theory that suggests the body chooses to be at a specific weight is the:
   a. fat cell theory.
   b. enzyme theory.
   c. set-point theory.
   d. external cue theory.

5. The biggest problem associated with the use of prescription drugs in the treatment of obesity is:
   a. cost.
   b. the necessity for long-term dosage.
   c. ineffectiveness.
   d. adverse side effects.

6. A nutritionally sound weight-loss diet might restrict daily energy intake to create a:
   a. 1000-kcalorie-per-month deficit.
   b. 500-kcalorie-per-month deficit.
   c. 500-kcalorie-per-day deficit.
   d. 3500-kcalorie-per-day deficit.

7. What is the best approach to weight loss?
   a. Avoid foods containing carbohydrates.
   b. Eliminate all fats from the diet and decrease water intake.
   c. Greatly increase protein intake to prevent body protein loss.
   d. Reduce daily energy intake and increase energy expenditure.

8. Physical activity does *not* help a person to:
   a. lose weight.
   b. lose fat in trouble spots.
   c. retain muscle.
   d. maintain weight loss.

9. Suggestions to change behaviors for successful weight control include:
   a. shop only when hungry.
   b. eat in front of the television for distraction.
   c. learn appropriate portion sizes.
   d. eat quickly.

10. Which strategy would *not* help an underweight person to gain weight?
   a. Exercise.
   b. Drink plenty of water.
   c. Eat snacks between meals.
   d. Eat large portions of foods.

Answers to these questions can be found in Appendix H.

Notes

1. M. de Krom and coauthors, Genetic variation and effects on human eating behavior, *Annual Review of Nutrition* 29 (2009): 283–304; T. Rankinen and C. Bouchard, Genetics of food intake and eating behavior phenotypes in humans, *Annual Review of Nutrition* 26 (2006): 413–434.

2. J. Hebebrand and A. Hinney, Environmental and genetic risk factors in obesity, *Child and Adolescent Psychiatric Clinics of North America* 18 (2009): 83–94.

3. K. Silventhoinen and coauthors, The genetic and environment influences on childhood obesity: A systematic review of twin and adoption studies, *International Journal of Obesity* 34 (2010): 29–40.

4. J. Wardle and coauthors, Evidence for a strong genetic influence on childhood adiposity despite the force of the obesogenic environment, *American Journal of Clinical Nutrition* 87 (2008): 398–404.

5. A. Newell and coauthors, Addressing the obesity epidemic: A genomics perspective, *Preventing Chronic Disease* [CDC serial online] April 2007, available at www.cdc.gov/pcd/issues/2007/apr/06_0068.htm.

6. C. Bouchard, Defining the genetic architecture of the predisposition to obesity: A challenging but not insurmountable task, *American Journal of Clinical Nutrition* 91 (2010): 5–6; Newell and coauthors, 2007.

7. Li Shengxu and coauthors, Cumulative effects and predictive value of common obesity-susceptibility variants identified by genome-wide association studies, *American Journal of Clinical Nutrition* 91 (2010): 184–190.

8. I. Romao and J. Roth, Genetic and environmental interactions in obesity and type 2 diabetes, *Journal of the American Dietetic Association* 108 (2008): S24–S28.

9. H. Wang and R. H. Eckel, Lipoprotein lipase: From gene to obesity, *American Journal of Physiology. Endocrinology and Metabolism* 297 (2009): E271–E288.

10. S. C. Woods, R. J. Seeley, and D. Cota, Regulation of food intake through hypothalamic signaling networks involving mTOR, *Annual Review of Nutrition* 28 (2008): 295–311.

11. J. M. Friedman, Leptin at 14 y of age: An ongoing story, *American Journal of Clinical Nutrition* 89 (2009): 973S–979S.

12. I. S. Farooqi and S. O'Rahilly, Leptin: A pivotal regulator of human homeostasis, *American Journal of Clinical Nutrition* 89 (2009): 980S–984S.

13. Friedman, 2009.

14. J. Beltowski, A. Jamroz-Wisniewska, and S. Widomska, Adiponectin and its role in cardiovascular disease, *Cardiovascular & Hematological Disorders Drug Targets* 8 (2008): 7–46.

15. S. Li and coauthors, Adiponectin levels and risk of type 2 diabetes: A systematic review and meta-analysis, *Journal of the American Medical Association* 302 (2009): 179–188; G. Wolf, New insights into thiol-mediated regulation of adiponectin secretion, *Nutrition Reviews* 66 (2008): 642–645;

M. Garaulet and coauthors, Adiponectin, the controversial hormone, *Public Health Nutrition* 10 (2007): 1145–1150; Y. Takemura, K. Walsh, N. Ouchi, Adiponectin and cardiovascular inflammatory responses, *Current Atherosclerosis Report* 9 (2007): 238–243.

16. M. Guerre-Millo, Adiponectin: An update, *Diabetes & Metabolism* 34 (2008): 12–18.

17. D. E. Cummings, K. E. Foster-Schubert, and J. Overduin, Ghrelin and energy balance: Focus on current controversies, *Current Drug Targets* 6 (2005): 153–169.

18. T. R. Castaneda and coauthors, Ghrelin in the regulation of body weight and metabolism, *Frontiers in Neuroendocrinology* November 5, 2009 31 (2010): 44–60; Cummings, Foster-Schubert, and Overduin, 2005.

19. V. Monti and coauthors, Relationship of ghrelin and leptin hormones with body mass index and waist circumference in a random sample of adults, *Journal of the American Dietetic Association* 106 (2006): 822–828.

20. K. L. Spalding and coauthors, Dynamics of fat cell turnover in humans, *Nature* 453 (2008): 783–787.

21. G. C. Major and coauthors, Clinical significance of adaptive thermogenesis, *International Journal of Obesity* 31 (2007): 204–212.

22. M. M. Hetherington and J. E. Cecil, Gene-environment interactions in obesity, *Forum of Nutrition* 63 (2010): 195–203; D. Heber, An integrative view of obesity, *American Journal of Clinical Nutrition* 91 (2010): 280S–283S; L. Qi and Y. A. Cho, Gene-environment interaction and obesity, *Nutrition Reviews* 66 (2008): 684–694; Newell and coauthors, 2007.

23. W. P. James, The fundamental drivers of the obesity epidemic, *Obesity Reviews* 9 (2008): S6–S13.

24. B. Wansink and K. Van Ittersum, Portion size me: Downsizing our consumption, *Journal of the American Dietetic Association* 107 (2007): 1103–1106.

25. B. J. Rolls, Plenary Lecture 1: Dietary strategies for the prevention and treatment of obesity, *Proceedings of the Nutrition Society* 69 (2010): 70–79; I. H. Steenhuis and W. M. Vermeer, Portion size: Review and framework for interventions, *International Journal of Behavioral Nutrition and Physical Activity* 6 (2009): 58.

26. Position of the American Dietetic Association. Weight management, *Journal of the American Dietetic Association* 109 (2009): 330–346.

27. Wansink and Van Ittersum, 2007; L. R. Young and M. Nestle, Portion sizes and obesity: Responses of fast-food companies, *Journal of Public Health Policy* 28 (2007): 238–248.

28. K. J. Duffey and coauthors, Regular consumption from fast food establishments relative to other restaurants is differentially associated with metabolic outcomes in young adults, *Journal of Nutrition* 139 (2009): 2113–2118; R. Rosenheck, Fast food consumption and increased caloric intake: A systematic review of a trajectory towards weight gain and obesity risk, *Obesity Reviews* 9 (2008): 535–547; M. A. Pereira and coauthors, Fast-food

habits, weight gain, and insulin resistance (the CARDIA Study): 15-year prospective analysis, *Lancet* 365 (2005): 36–42.

29. L. Johnson and coauthors, Energy-dense, low-fiber, high-fat dietary pattern is associated with increased fatness in childhood, *American Journal of Clinical Nutrition* 87 (2008): 846–854; N. C. Howarth and coauthors, Dietary energy density is associated with overweight status among 5 ethnic groups in the Multiethnic Cohort Study, *Journal of Nutrition* 136 (2006): 2243–2248.

30. B. J. Rolls, L. S. Roe, and J. S. Meengs, Reductions in portion size and energy density of foods are additive and lead to sustained decreases in energy intake, *American Journal of Clinical Nutrition* 83 (2006): 11–17.

31. Rolls, 2010; J. A. Ello-Martin, J. H. Ledikwe, and B. J. Rolls, The Influence of food portion size and energy density on energy intake: Implications for weight management, *American Journal of Clinical Nutrition* 82 (2005): 236S–241S.

32. A. Drewnowski and P. Eichelsdoefer, Can low-income Americans afford a healthy diet? *Nutrition Today* 44 (2009): 246–249; P. Monsivais and A. Drewnowski, The rising cost of low-energy-density foods, *Journal of the American Dietetic Association* 107 (2007): 2071–2076; L. H. Epstein and coauthors, Price and maternal obesity influence purchasing of low- and high-energy-dense foods, *American Journal of Clinical Nutrition* 86 (2007): 914–922.

33. D. S. Ludwig and M. Nestle, Can the food industry play a constructive role in the obesity epidemic? *Journal of the American Medical Association* 300 (2008): 1808–1811.

34. P. T. Katzmarzyk and coauthors, Sitting time and mortality from all causes, cardiovascular disease, and cancer, *Medicine and Science in Sports and Exercise* 41 (2009): 998–1005; S. Mandic and coauthors, Characterizing differences in mortality at the low end of the fitness spectrum, *Medicine and Science in Sports and Exercise* 41 (2009): 1573–1579; American College of Sports Medicine, Position stand: Appropriate physical activity intervention strategies for weight loss and prevention of weight regain for adults, *Medicine and Science in Sports and Exercise* 41 (2009): 459–471; P. Gordon-Larsen and coauthors, Fifteen-year longitudinal trends in walking patterns and their impact on weight change, *American Journal of Clinical Nutrition* 89 (2009): 19–26.

35. B. Swinburn and A. Shelly, Effects of TV time and other sedentary pursuits, *International Journal of Obesity* 32 (2008): S132–S136; D. A. Raynor and coauthors, Television viewing and long-term maintenance: Results from the National Weight Control Registry, *Obesity* 14 (2006): 1816–1824.

36. Katzmarzyk and coauthors, 2009.

37. D. Berrigan and coauthors, Active transportation increases adherence to activity recommendations, *American Journal of Preventive Medicine* 31 (2006): 210–216.

38. S. K. Kumanyika and coauthors, Population-based prevention of obesity. The need for comprehensive promotion of healthful eating,

physical activity, and energy balance. A scientific statement from the American Heart Association Council on Epidemiology and Prevention, Interdisciplinary Committee for Prevention, *Circulation* 118 (2008): 428–464.

39. J. L. Pilletteri and coauthors, Use of dietary supplements for weight loss in the United States: Results of a national survey, *Obesity (Silver Springs)* 16 (2008): 790–796.

40. U.S. Food and Drug Administration, Questions and answers about FDA's initiative against contaminated weight loss products, April 30, 2009, available at http://www.fda.gov/Drugs/ResourcesForYou/Consumers/QuestionsAnswers/ucm136187.htm.

41. Ethics opinion: Weight loss products and medications, *Journal of the American Dietetic Association* 108 (2008): 2109–2113; H. M. Blanck and coauthors, Use of nonprescription dietary supplements for weight loss is common among Americans, *Journal of the American Dietetic Association* 107 (2007): 441–447.

42. R. F. Kushner, Anti-obesity drugs, *Expert Opinion on Pharmacotherapy* 9 (2008): 1339–1350.

43. Position of the American Dietetic Association, 2009.

44. Position of the American Dietetic Association, 2009.

45. Cummings, Foster-Schubert, and Overduin, 2005.

46. L. Sjostrom and coauthors, Effects of bariatric surgery on mortality in Swedish obese subjects, *New England Journal of Medicine* 357 (2007): 741–752; G. L. Blackburn, Solutions in weight control: Lessons from gastric surgery, *American Journal of Clinical Nutrition* 82 (2005): 248S–252S.

47. J. B. Dixon and coauthors, Adjustable gastric banding and conventional therapy for type 2 diabetes: A randomized controlled trial, *Journal of the American Medical Association* 299 (2008): 316–323; E. N. Hansen, A. Torquati, and N. N. Abumrad, Results of bariatric surgery, *Annual Review of Nutrition* 26 (2006): 481–511.

48. Blackburn, 2005.

49. M. K. Robinson, Surgical Treatment of obesity—weighing the facts, *New England Journal of Medicine* 361 (2009): 520–521; B. R. Smith, P. Schauer, and N. T. Nguyen, surgical approaches to the treatment of obesity: Bariatric surgery, *Endocrinology Metabolism Clinics of North America* 37 (2008): 943–964; J. A. Vogel and coauthors, Reduction in predicted coronary heart disease risk after substantial weight reduction after bariatric surgery, *American Journal of Cardiology* 99 (2007): 222–226.

50. Position of the American Dietetic Association, 2009; J. O. Hill, Can a small-changes approach help address the obesity epidemic? A report of the Joint Task Force of the American Society for Nutrition, Institute of Food Technologists, and International Food Information Council, *American Journal of Clinical Nutrition* 89 (2009): 477–484.

51. D. R. Jacobs and coauthors, Association of 1-y change in diet pattern with cardiovascular disease risk factors and adipokines: Results from the 1-y randomized Oslo Diet and Exercise Study, *American Journal of Clinical Nutrition* 89 (2009): 509–517; M. L. Fernandez, The metabolic syndrome, *Nutrition Reviews* 65 (2007): S30–S34; C. Galani and H. Schneider, Prevention and treatment of

obesity with lifestyle interventions: Review and meta-analysis, *International Journal of Public Health* 52 (2007): 348–359.

52. K. M. Kolasa, D. N. Collier, and K. J. Cable, Weight loss strategies that really work, *Journal of Family Practice* 59 (2010): 378–385.

53. Position of the American Dietetic Association, 2009; National Heart, Lung, and Blood Institute, National Institutes of Health, *The Practical Guide: Identification, Evaluation, and Treatment of Overweight and Obesity in Adults,* NIH publication no. 00-4084 (Washington, D.C.: Government Printing Office, 2000), p. 23.

54. National Heart, Lung, and Blood Institute, National Institutes of Health, 2000, pp. 26–27.

55. M. T. Timlin and coauthors, Breakfast eating and weight change in a 5-year prospective analysis of adolescents: Project EAT (Eating Among Teens), *Pediatrics* 121 (2008): e638; M. T. Timlin and M. A. Pereira, Breakfast frequency and quality in the etiology of adult obesity and chronic diseases, *Nutrition Reviews* 65 (2007): 268–281.

56. A. K. Kant and coauthors, Association of breakfast energy density with diet quality and body mass index in American adults: National Health and Nutrition Examination Surveys, 1999–2004, *American Journal of Clinical Nutrition* 88 (2008): 1396–1404.

57. M. Bulló and coauthors, Inflammation, obesity and comorbidities: The role of diet, *Public Health Nutrition* 10 (2007): 1164–1172.

58. Rolls, Roe, and Meengs, 2006.

59. K. Van Ittersum and B. Wansink, Do children really prefer large portions? Visual illusions bias their estimates and intake, *Journal of the American Dietetic Association* 107 (2007): 1107–1110; B. Wansink, K. Van Ittersum, and J. E. Painter, Ice cream illusions: Bowls, spoons, and self-served portion sizes, *American Journal of Preventive Medicine* 31 (2006): 240–243.

60. M. Bes-Rastrollo and coauthors, Prospective study of nut consumption, long-term weight change, and obesity risk in women, *American Journal of Clinical Nutrition* 89 (2009): 1913–1919.

61. M. C. de Oliveira, R. Sichieri, and R. V. Mozzer, A low-energy-dense diet adding fruit reduces weight and energy intake in women, *Appetite* 51 (2008): 291–295; J. A. Ello-Martin, and coauthors, dietary energy density in the treatment of obesity: A year-long trial comparing 2 weight-loss diets, *American Journal of Clinical Nutrition* 85 (2007): 1465–1477.

62. M. A. Palmer, S. Capra, and S. K. Baines, Association between eating frequency, weight, and health, *Nutrition Reviews* 67 (2009): 379–390.

63. J. E. Flood and B. J. Rolls, Soup preloads in a variety of forms reduce meal energy intake, *Appetite* 49 (2007): 626–634.

64. B. M. Davy and coauthors, Water consumption reduces energy intake at a breakfast meal in obese older adults, *Journal of the American Dietetic Association* 108 (2008): 1236–1239; E. L. Van Walleghen and coauthors, Pre-meal water consumption reduces meal energy intake in older but not younger subjects, *Obesity* 15 (2007): 93–99.

65. B. M. Popkin and coauthors, A new proposed guidance system for beverage consumption in the

United States, *American Journal of Clinical Nutrition* 83 (2006): 529–542.

66. J. E. Flood and coauthors, The effect of increased beverage portion size on energy intake at a meal, *Journal of the American Dietetic Association* 106 (2006): 1984–1990.

67. D. E. Larson-Meyer and coauthors, Caloric restriction with or without exercise: The fitness versus fatness debate, *Medicine and Science in Sports and Exercise* 42 (2010): 152–159; American College of Sports Medicine, 2009; Position of the American Dietetic Association, 2009.

68. A. E. Field and coauthors, Weight control behaviors and subsequent weight change among adolescents and young adult females, *American Journal of Clinical Nutrition* 91 (2010): 147–153; American College of Sports Medicine, 2009.

69. B. J. Nicklas and coauthors, Effect of exercise intensity on abdominal fat loss during calorie restriction in overweight and obese postmenopausal women: A randomized, controlled trial, *American Journal of Clinical Nutrition* 89 (2009): 1043–1052; American College of Sports Medicine, 2009; S. J. Elder and S. B. Roberts, The effects of exercise on food intake and body fatness: A summary of published studies, *Nutrition Reviews* 65 (2007): 1–19.

70. American College of Sports Medicine, 2009.

71. Larson-Meyer and coauthors, 2010; F. Magkos and coauthors, Management of the Metabolic Syndrome and type 2 diabetes through lifestyle modification, *Annual Review of Nutrition* 29 (2009): 223–256.

72. M. Hamer and G. O'Donovan, Cardiorespiratory fitness and metabolic risk factors in obesity, *Current Opinion in Lipidology* 21 (2010): 1–7; B. A. Irving and coauthors, Effect of exercise training intensity on abdominal visceral fat and body composition, *Medicine and Science in Sports and Exercise* 40 (2008): 1863–1872; M. Fogelholm, How physical activity can work? *International Journal of Pediatric Obesity* 3 (2008): 10–14.

73. N. A. King and coauthors, Individual variability following 12 weeks of supervised exercise: Identification and characterization of compensation for exercise-induced weight loss, *International Journal of Obesity (London)* (2008): 177–184.

74. K. Ohkawara and coauthors, Twenty-four-hour analysis of elevated energy expenditure after physical activity in a metabolic chamber: Models of daily total energy expenditure, *American Journal of Clinical Nutrition* 87 (2008): 1268–1276.

75. N. A. King and coauthors, Dual process action of exercise on appetite control: Increase in orexigenic drive but improvement in meal-induced satiety, *American Journal of Clinical Nutrition* 90 (2009): 921–927; C. Martins and coauthors, effects of exercise on gut peptides, energy intake and appetite, *Journal of Endocrinology* 193 (2007): 251–258.

76. C. Martins, L. Morgan, and H. Truby, A review of the effects of exercise on appetite regulation: An obesity perspective, *International Journal of Obesity* 32 (2008): 1337–1347; D. R. Broom and coauthors, Exercise-induced suppression of acylated ghrelin in humans, *Journal of Applied Physiology* 102 (2007): 2165–2171; R. R. Kraemer and V. D. Castracane, Exercise and humoral mediators of peripheral energy balance: Ghrelin and

adiponectin, *Experimental Biology and Medicine (Maywood):* 232 (2007): 184–194.

77. U.S. Department of Agriculture and U.S. Department of Health and Human Services, *2008 Physical Activity Guidelines for Americans,* available at www.health.gov/paguidelines/default.aspx.

78. B. Van Dorsten and E. M. Lindley, Cognitive and behavioral approaches in the treatment of obesity, *Endocrinology and Metabolism Clinics of North America* 37 (2008): 905–922; M. D. Tsiros and coauthors, cognitive behavioral therapy improves diet and body composition in overweight and obese adolescents, *American Journal of Clinical Nutrition* 87 (2008): 1134–1140.

79. K. G. Volpp and coauthors, Financial incentive-based approaches for weight loss: A randomized trial, *Journal of the American Medical Association* 300 (2008): 2631–2637.

80. A. N. Fabricatore, Behavior therapy and cognitive-behavioral therapy for obesity: Is there a difference? *Journal of the American Dietetic Association* 107 (2007): 92–99.

81. R. Wing, Secrets of successful losers, *Nutrition Action Health Letter,* January/February, 2008, p. 8.

82. D. S. Bond, Weight-loss maintenance in successful weight losers: Surgical vs non-surgical methods, *International Journal of Obesity (London)* 33 (2009): 173–180.

83. J. N. Davis and coauthors, Normal-weight adults consume more fiber and fruit than their age- and height-matched overweight/obese counterparts, *Journal of the American Dietetic Association* 106 (2006): 833–840.

Nutrition in Practice

To paraphrase William Shakespeare, "a fad diet by any other name would still be a fad diet." And the names are legion: the Atkins Diet, the Cheater's Diet, the South Beach Diet, the Zone Diet.* Year after year, "new and improved" diets appear on bookstore shelves and circulate among friends. Sometimes fad diets seem to work for a while, but more often than not, their success is short-lived. Then another fad diet takes the spotlight. Here's how Dr. K. Brownell, an obesity researcher at Yale University, describes this phenomenon: "When I get calls about the latest diet fad, I imagine a trick birthday cake candle that keeps lighting up and we have to keep blowing it out."

Why don't health professionals speak out against the relentless promotion of fad diets?

Realizing that many fad diets do not offer a safe and effective plan for weight loss, health professionals speak out, but they never get the candle blown out permanently. New fad diets can keep making outrageous claims because no one requires their advocates to prove what they say. Fad diet gurus do not have to conduct credible research on the benefits or dangers of their diets. They can simply make recommendations and then later, if questioned, search for bits and pieces of research that support the conclusions they have already reached. That's backwards. Diet and health recommendations should *follow* years of sound research that has been reviewed by panels of scientists *before* being offered to the public.

How can the promoters of fad diets get away with exaggerated claims?

Because anyone can publish anything—in books or on the Internet—peddlers of fad diets can make unsubstantiated statements that fall far short of the truth but sound impressive to the uninformed. They often offer distorted bits of legitimate research. They may start with one or more actual facts but then leap from one erroneous conclusion to the next. Anyone who wants to believe these claims has to wonder how the thousands of scientists working on obesity research over the past century could possibly have missed such obvious connections. Table NP7-1 presents some of the claims and truths of fad diets.

Each popular diet features a unique way to make losing weight fast and easy. Which one works best?

Fad diets come in almost as many shapes and sizes as the people who search them out. Some restrict fats or carbohydrates, some limit portion sizes, some focus on food combinations, and some

*The following sources offer comparisons and evaluations of various fad diets for your review: New diet winners, *Consumer Reports*, June 2007, pp. 12–17; Battle of the diet books II, *Nutrition Action Healthletter*, July/August 2006, pp. 10–11; B. Liebman, Weighing the diet books, *Nutrition Action Healthletter*, January/February 2004, pp.1–8.

TABLE NP7-1 The Claims and Truths of Fad Diets

The Claim:	You can lose weight "easily."
The Truth:	Most fad diet plans have complicated rules that require you to calculate protein requirements, count carbohydrate grams, combine certain foods, time meal intervals, purchase special products, plan daily menus, and measure serving sizes.
The Claim:	You can lose weight by eating a specific ratio of carbohydrate, protein, and fat.
The Truth:	Weight loss depends on expending more energy than you take in, not on the proportion of energy nutrients.
The Claim:	This "revolutionary diet" can "reset your genetic code."
The Truth:	You inherited your genes and cannot alter your genetic code.
The Claim:	High-protein diets are popular, selling more than 20 million books, because they work.
The Truth:	Weight-loss books are popular because people grasp for quick fixes and simple solutions to their weight problems. If book sales were an indication of weight-loss success, we would be a lean nation—but they are not, and neither are we.
The Claim:	People gain weight on low-fat diets.
The Truth:	People can gain weight on low-fat diets if they overindulge in carbohydrates and proteins while cutting fat; low-fat diets are not necessarily low-kcalorie diets. But people can also lose weight on low-fat diets if they cut kcalories as well as fat.
The Claim:	High-protein diets energize the brain.
The Truth:	The brain depends on glucose for its energy; the primary dietary source of glucose is carbohydrate, not protein.
The Claim:	Thousands of people have been successful with this plan.
The Truth:	Authors of fad diets have not published their research findings in scientific journals. Success stories are anecdotal, and failures are not reported.
The Claim:	Carbohydrates raise blood glucose levels, triggering insulin production and fat storage.
The Truth:	Insulin promotes fat storage when energy intake exceeds energy needs. Furthermore, insulin is only one of many hormones involved in the complex processes of maintaining the body's energy balance and health.
The Claim:	Eat protein and lose weight.
The Truth:	For every complicated problem, there is a simple—and wrong—solution.

claim that a person's genetic type or blood type determines the foods best suited to manage weight and prevent disease. Despite claims that each new diet is "unique" in its approach to weight loss, most fad diets are designed to ensure a low energy intake. The "magic feature" that best supports weight loss is to limit energy intake to less than energy expenditure. Most of the sample menu plans, especially in the early stages, deliver an average of 1200 kcalories per day. Total kcalories tend to be low simply because food intake is so limited. Table NP7-2 compares some of today's more popular diets.

What is the appeal of fad diets?

Probably the greatest appeal of fad diets is that they tend to ignore current diet recommendations. Foods such as meats and milk products that need to be selected carefully to limit saturated fat

---

TABLE NP7-2 Popular Diets Compared

Diet	Major Premise Promoted	Strong Point(s)	Weak Point(s)
Atkins Diet	▪ People are overweight or obese because they have metabolic imbalances caused by eating too many carbohydrates; by restricting carbohydrates, these imbalances can be corrected. ▪ You can lose weight without lowering kcalorie intake.	▪ Quick, short-term weight loss is achieved.	▪ Restricts carbohydrates to a level that induces ketosis. ▪ Ketosis can cause nausea, lightheadedness, and fatigue. ▪ Ketosis can worsen existing medical problems such as kidney disease. ▪ A diet high in fat such as Atkins can increase the risk of heart disease and some cancers.
Cheater's Diet	▪ Successful weight loss depends on eliminating boredom and allowing indulgences. ▪ Cheating on weekends "stokes your metabolism."	▪ Meals are proportioned one-half fruit or vegetables, one-fourth lean protein, and one-fourth whole grains. ▪ Encourages as much exercise as possible.	▪ No scientific data on cheating to boost metabolism or support weight loss.
Eat Right 4 Your Type	▪ Your blood type determines which foods you should eat or not eat.	▪ None	▪ Food groups or individual foods are excluded, depending on blood type. ▪ No scientific data on the relationship between blood type and food choices.
Glucose Revolution	▪ Low-glycemic index foods satisfy hunger, control blood glucose, and promote weight loss.	▪ Emphasizes fiber-rich vegetables, legumes, fruits, and whole grains. ▪ Minimizes saturated fat intake.	▪ Difficult to know the glycemic index of some foods.
Ornish Diet	▪ By strictly limiting fat (both animal and vegetable), you eat fewer kcalories without eating less food.	▪ High-fiber, low-fat foods in this plan can lower blood cholesterol and blood pressure.	▪ So little fat that essential fatty acids may be lacking. ▪ Limits fish, nuts, and olive oil, which may protect against heart disease.
Pritikin Program	▪ By eating low-fat, mainly plant-based foods, you can eat more food and still feel satisfied.	▪ No food group is completely eliminated in this high-fiber, low-fat diet program. ▪ Some use of foods rich in omega-3 fatty acids is encouraged.	▪ For some people, very low-fat diets may be unsatisfying and therefore difficult to adhere to.
Sonoma Diet	▪ Enjoying portion-controlled Mediterranean-style foods supports weight loss and promotes good health.	▪ Emphasizes nutrient-dense foods.	▪ Initial phase restricts fruits and limits milk products.
South Beach Diet	▪ Eating "good carbohydrates" such as vegetables, whole-wheat pastas, and brown rice will maintain satiety and resist cravings for "bad carbohydrates" such as white rice and potatoes.	▪ Encourages consumption of vegetables, lean meats, and fish, and the use of unsaturated oils when cooking. ▪ Restricts fatty meats and cheeses as well as sweets.	▪ Starchy carbohydrates and all fruits are completely excluded during the first two weeks.

continued

Diet	Major Premise Promoted	Strong Point(s)	Weak Point(s)
Ultimate Weight Solution Diet	▪ Foods that require great effort to prepare and eat are nutrient dense; eating these kinds of foods (raw vegetables, vegetable soups, whole grains, beans, meats, poultry, and fish) will lead to weight loss. ▪ Foods that take little effort to prepare and eat provide excess kcalories relative to nutrients; eating these kinds of foods (fast foods, puddings, high-kcalorie convenience foods, processed foods) leads to uncontrolled eating and weight gain.	▪ Encourages consumption of lean meats and fish; whole grains; vegetables; fruit; and low-fat milk, yogurt, and cheese. ▪ Restricts fatty meats and cheeses as well as sweets. ▪ Encourages exercise.	▪ Confusing as to exactly what to eat or how much.
Zone Diet	▪ Eating the correct proportions of carbohydrates, fat, and protein leads to hormonal balance, weight loss, disease prevention, and increased vitality.	▪ Promotes weight loss because it is a low-kcalorie diet.	▪ The diet is rigid, restrictive, and complicated, making it difficult for most people to follow accurately. ▪ The overblown health claims of the diet's proponents are based on misinterpreted science and remain unsubstantiated.

can now be eaten with abandon. Whole grains, legumes, vegetables, and fruits that should be eaten in abundance can now be bypassed. For some people, this is a dream come true: steaks without the potatoes, ribs without the coleslaw, and meatballs without the pasta. Who can resist the promise of weight loss while eating freely from a list of favorite foods?

Dieters are also lured into fad diets by sophisticated—yet often erroneous—explanations of the metabolic consequences of eating certain foods. Terms such as *eicosanoids* or *adipokines* are scattered about, often intimidating readers into believing that the authors must be right given their brilliance in understanding the body.

Are fad diets adequate?

When food choices are limited, nutrient intakes may be inadequate. To help shore up some of these inadequacies, fad diets often recommend a dietary supplement. Conveniently, many of the companies selling fad diets also sell these supplements, often at an inflated price. As Nutrition in Practice 9 explains, however, foods offer many more health benefits than any supplement can provide.[1] Quite simply, if the diet is inadequate, it needs to be improved, not supplemented.

When researchers evaluated popular weight-loss plans against dietary standards for health, they found most fell short in some ways.[2] Of the eight popular plans examined, Ornish ranked highest and Atkins, which famously eliminates most carbohydrate-containing foods from the diet, ranked lowest.

Are the diets effective?

If fad diets were entirely ineffective, consumers would eventually stop pursuing them. Obviously, this is not the case. Similarly, if the diets were especially effective, then consumers who tried them would lose weight, and the obesity problem would be solved. Clearly, this is not happening either. As mentioned earlier, most fad diets succeed in contriving ways to limit kcalorie intakes, and

so produce weight loss (at least in the short term). Simple kcalorie deficit, as it turns out, is the key to weight loss; an emphasis on protein, carbohydrate, or fat, or any other gimmick does not improve weight loss.[3]

With over half of our nation's adults overweight and many more concerned about their weight, weight-loss books and products are a $33 billion-a-year business. Even a plan that offers only minimal weight-loss success easily attracts a following.

Many people seem confused about what kinds of dietary changes they need to make to lose weight. Isn't it helpful to follow a plan?

Yes. Most people need specific instructions and examples to make dietary changes. Popular diets offer dieters a plan. The user doesn't have to decide what foods to eat, how to prepare them, or how much to eat. Unfortunately, these instructions serve short-term weight-loss needs only. Over the long run, people tire of the monotony of diet plans and most revert to their previous patterns.

By now you can see that the major drawback of most fad weight-loss schemes is that they fail to create lifestyle changes needed to support long-term weight maintenance and improve health.[4] For that, people must do the hard work of learning the facts about nutrition and weight loss, setting their own realistic goals, and then devising a lifestyle plan that can achieve them. A balanced, kcalorie-restricted diet with sufficient physical activity to support it may not be the shortcut that most people wish for, but it assures nutrient adequacy and provides the best chance of long-term success.[5] Table NP7-3 offers guidelines for identifying fad diets and other weight-loss scams; it includes the hallmarks of healthy weight-loss programs as well. Chapter 7 presents reasonable approaches to weight management and concludes that the ideal diet is one you can live with for the rest of your life. Keep that criterion in mind when you evaluate the next "latest and greatest weight-loss diet" that comes along.

TABLE NP7-3 Guidelines for Identifying Fad Diets and Other Weight-Loss Scams

Fad Diets and Weight-Loss Scams	Healthy Diet Guidelines
1. They promise dramatic, rapid weight loss.	1. Weight loss should be gradual and not exceed 2 pounds per week.
2. They promote diets that are nutritionally unbalanced or extremely low in kcalories.	2. Diets should provide: ▪ A reasonable number of kcalories (not fewer than 1200 kcalories per day) ▪ Enough, but not too much, protein (between the RDA and twice the RDA) ▪ Enough, but not too much, fat (between 20 and 35 percent of daily energy intake from fat) ▪ Enough carbohydrates to spare protein and prevent ketosis (at least 100 grams per day) and 20 to 30 grams of fiber from food sources ▪ A balanced assortment of vitamins and minerals from a variety of foods from each of the food groups ▪ At least 1 liter (about 1 quart) of water daily or 1 milliliter per kcalorie daily—whichever is more
3. They use liquid formulas rather than foods.	3. Foods should accommodate a person's ethnic background, taste preferences, and financial means.
4. They attempt to make clients dependent upon special foods or devices.	4. Programs should teach clients how to make good choices from the conventional food supply.
5. They fail to encourage permanent, realistic lifestyle changes.	5. Programs should teach physical activity plans that involve expending at least 300 kcalories a day and behavior-modification strategies that help to correct poor eating habits.
6. They misrepresent salespeople as "counselors" supposedly qualified to give guidance in nutrition and/or general health.	6. Even if adequately trained, such "counselors" would still be objectionable because of the obvious conflict of interest that exists when providers profit directly from products they recommend and sell.
7. They collect large sums of money at the start or require that clients sign contracts for expensive, long-term programs.	7. Programs should be reasonably priced and run on a pay-as-you-go basis.
8. They fail to inform clients of the risks associated with weight loss in general or the specific program being promoted.	8. They should provide information about dropout rates, the long-term success of their clients, and possible diet side effects.
9. They promote unproven or spurious weight-loss aids such as human chorionic gonadotropin hormone (HCG), starch blockers, diuretics, sauna belts, body wraps, passive exercise, ear stapling, acupuncture, electric muscle-stimulating (EMS) devices, spirulina, amino acid supplements (e.g., arginine, ornithine), glucomannan, methylcellulose (a "bulking agent"), "unique" ingredients, and so forth.	9. They should focus on nutrient-rich foods and regular exercise.
10. They fail to provide for weight maintenance after the program ends.	10. They should provide a plan for weight maintenance after successful weight loss.

Sources: Adapted from National Heart, Lung, and Blood Institute, National Institutes of Health, *The Practical Guide: Identification, Evaluation, and Treatment of Overweight and Obesity in Adults,* NIH publication no. 00-4084 (Washington, D.C.: Government Printing Office, 2000), pp. 25–33; American College of Sports Medicine, *ACSM's Guidelines for Exercise Testing and Prescription* (Baltimore: Lippincott Williams & Wilkins, 2006), pp. 133–173.

Notes

1. D. R. Jacobs, M. D. Gross, and L. C. Tapsell, Food Synergy: An operational concept for understanding nutrition, *American Journal of Clinical Nutrition* 89 (2009): 1543S–1548S.

2. Y. Ma and coauthors, A dietary quality comparison of popular weight-loss plans, *Journal of the American Dietetic Association* 107 (2007): 1786–1791.

3. F. M. Sacks and coauthors, Comparison of weight-loss diets with different compositions of fat, protein, and carbohydrates, *New England Journal of Medicine* 360 (2009): 859–873.

4. I. Shai and M. J. Stampfer, Weight-loss diets—can you keep it off? *American Journal of Clinical Nutrition* 88 (2008): 1185–1186.

5. I. Shai and M. J. Stampfer, 2008; B. Strasser, A. Spreitzer, and P. Haber, Fat loss depends on energy deficit only, independently of the method for weight loss, *Annals of Nutrition and Metabolism* 51 (2007): 428–432.

8

The Vitamins

arlier chapters focused primarily on the energy-yielding nutrients—carbohydrate, fat, and protein. This chapter and the next one discuss the nutrients everyone thinks of when nutrition is mentioned—vitamins and minerals.

Vitamins—An Overview

Vitamins occur in foods in much smaller quantities than do energy-yielding nutrients, and they themselves contribute no energy to the body. Instead, they serve mostly as facilitators of body processes. They are a powerful group of substances, as their absence attests: vitamin A deficiency can cause blindness, a lack of niacin can cause dementia, and a lack of vitamin D can retard bone growth. The consequences of deficiencies are so dire and the effects of restoring the needed nutrients so dramatic that people spend billions of dollars each year on **dietary supplements** to cure many different ailments. Vitamins certainly contribute to sound nutritional health, but supplements do not cure all ills. Actually, a vitamin can cure only the disease caused by a deficiency of that vitamin. The vitamins' roles in supporting optimal health extend far beyond preventing deficiency diseases, however. Emerging evidence points to relationships between low intakes of vitamins and chronic diseases such as cancer and heart disease.

A child once defined vitamins as "what, if you don't eat, you get sick." The description is both insightful and accurate. A more prosaic definition is that vitamins are potent, essential, nonkcaloric, organic nutrients needed from foods in trace amounts to perform specific functions that promote growth, reproduction, and the maintenance of health and life. Two characteristics distinguish vitamins from energy nutrients:

- Vitamins do not yield energy when broken down, but rather they assist the enzymes that release energy from carbohydrate, fat, and protein.

- The amounts of vitamins people ingest daily from foods and the amounts they require are measured in *micrograms* (µg) or *milligrams* (mg), rather than grams (g).

The vitamins are similar to the energy-yielding nutrients, however, in that they are vital to life, organic, and available from foods.

As the individual vitamins were discovered, they were named or given letters, numbers, or both. This led to the confusion that still exists today. This chapter uses the names shown in Table 8-1; alternative names are given in Tables 8-2 and 8-3, which appear later in the chapter.

Bioavailability The amount of vitamins available from foods depends on two factors: the quantity provided by a food and the amount absorbed and used by the body (the vitamin's **bioavailability**). Researchers analyze foods to determine their vitamin contents and publish the results in tables of food composition such as Appendix A. Determining the bioavailability of a vitamin is more difficult because it depends on many factors, including:

- The efficiency of digestion and time of transit through the GI tract.

- A person's previous nutrient intake and nutrition status.

- Other foods eaten at the same time.

- The method of food preparation (raw or cooked, for example).

- The source of the nutrient (naturally occurring, synthetic, or fortified).

This chapter and the next describe factors that inhibit or enhance the absorption of individual vitamins and minerals. Experts consider these factors when estimating recommended intakes.

TABLE 8-1
Vitamin Names

Fat-Soluble Vitamins
Vitamin A
Vitamin D
Vitamin E
Vitamin K

Water-Soluble Vitamins
B vitamins
Thiamin
Riboflavin
Niacin
Pantothenic acid
Biotin
Vitamin B_6
Folate
Vitamin B_{12}
Vitamin C

vitamins: essential, nonkcaloric, organic nutrients needed in tiny amounts in the diet.

dietary supplements: products that are added to the diet and contain any of the following ingredients: a vitamin; a mineral; an herb or other botanical; an amino acid; a metabolite; a constituent; or an extract.

bioavailability: the rate and extent to which a nutrient is absorbed and used.

Precursors Some of the vitamins are available from foods in inactive forms known as **precursors,** or provitamins. Once inside the body, the precursor is converted to the active form of the vitamin. Thus, in measuring a person's vitamin intake, it is important to count both the amount of the actual vitamin and the potential amount available from its precursors.

Organic nutrients contain carbon.

Organic Nature Being organic, vitamins can be destroyed during storage and in cooking, and therefore must be handled with care. Prolonged heating may destroy much of the thiamin in food. Because riboflavin can be destroyed by the ultraviolet rays of the sun or by fluorescent light, foods stored in transparent glass containers are most likely to lose riboflavin. Oxygen destroys vitamin C, so losses occur when foods are cut, processed, and stored. The margin lists ways to minimize nutrient losses in the kitchen.

Minimizing nutrient losses:

▌ To slow the degradation of vitamins, refrigerate (most) fruits and vegetables.

▌ To minimize the oxidation of vitamins, store fruits and vegetables that have been cut in airtight wrappers, and store open juices in closed containers (and refrigerate them).

▌ To prevent losses during washing, rinse fruits and vegetables before cutting or peeling.

▌ To prevent heat destruction, cook fruits and vegetables in a microwave oven, or quickly stir fry, or steam them over a small amount of water. Recapture dissolved vitamins by using the cooking water for soups, stews, or gravies. Avoid high temperatures and long cooking times.

Solubility Vitamins fall naturally into two classes—fat soluble and water soluble. The solubility of a vitamin confers on it many characteristics and determines how it is absorbed, transported, stored, and excreted. This discussion of vitamins begins with the fat-soluble vitamins.

IN SUMMARY

▌ Vitamins are essential, nonkcaloric nutrients that are needed in trace amounts in the diet to help facilitate body processes.

The Fat-Soluble Vitamins

The fat-soluble vitamins—A, D, E, and K—usually occur together in the fats and oils of foods, and the body absorbs them in the same way it absorbs lipids. Therefore, any condition that interferes with fat absorption can precipitate a deficiency of the fat-soluble vitamins. Once absorbed, fat-soluble vitamins are stored in the liver and fatty tissues until the body needs them. They are not readily excreted, and, unlike most of the water-soluble vitamins, they can build up to toxic concentrations. Excesses of vitamins A and D from supplements can reach toxic levels especially easily.

The capacity to store fat-soluble vitamins affords a person some flexibility in dietary intake. When blood concentrations begin to decline, the body can retrieve the vitamins from storage. Thus, a person need not eat a day's allowance of each fat-soluble vitamin every day but need only make sure that, over time, average daily intakes approximate recommended intakes. In contrast, most water-soluble vitamins must be consumed more regularly because the body does not store them to any great extent.

Vitamin A and Beta-Carotene

Vitamin A has the distinction of being the first fat-soluble vitamin to be recognized. A century later, vitamin A and its plant-derived precursor, **beta-carotene,** continue to intrigue researchers with their diverse roles and profound effects on health.

Vitamin A is a versatile vitamin, with roles in gene expression, vision, cell differentiation (thereby maintaining the health of body linings and skin), immunity, and reproduction and growth.[1] Three different forms of vitamin A are active in the body: retinol, retinal, and retinoic acid. Each form of vitamin A performs specific tasks. Retinol supports reproduction and is the major transport and storage form of the vitamin; the cells convert retinol to retinal or retinoic acid as needed. Retinal is active in vision, and retinoic acid acts as a hormone, regulating cell differentiation, growth, and embryonic development. A special transport protein, **retinol-binding protein (RBP),** picks up retinol from the liver, where it is stored, and carries it in the blood.

Measurement of the blood concentration of RBP is a sensitive test of vitamin A status.

Vitamin A's Role in Gene Expression Vitamin A exerts considerable influence on body functions through its regulation of the activities of the genes.[2] Genes direct the synthesis of proteins, including enzymes, and enzymes perform the metabolic work of the tissues (see Chapter 5). Hence, factors that influence gene expression also affect the metabolic activities of the tissues and the health of the body. Hundreds of genes have been suggested as regulatory targets of the retinoic acid form of vitamin A.[3]

Researchers have long known that simply possessing the genetic equipment needed to make a particular protein does not guarantee that the protein will be produced, any more than owning a car guarantees you a trip across town. To get the car rolling, you must also use the right key to trigger the events that start up its engine or turn it off at the appropriate time. Some dietary components, including the retinoic acid form of vitamin A, are now known to be such keys—they help to activate or deactivate certain genes and thus affect the production of specific proteins.[4]

Vitamin A's Role in Vision Vitamin A plays two indispensable roles in the eye. It helps maintain a healthy, crystal-clear outer window, the **cornea;** and it participates in the events of light detection at the **retina.** Some of the photosensitive cells of the retina contain **pigment** molecules called **rhodopsin;** each rhodopsin molecule is composed of a protein called **opsin** bonded to a molecule of the retinal form of vitamin A. When light passes through the cornea of the eye and strikes the retina, rhodopsin responds by changing shape and becoming bleached. In turn, this initiates the signal that conveys the sensation of sight to the optic center in the brain. Figure 8-1 shows vitamin A's site of action inside the eye.

When vitamin A is lacking, the eye has difficulty adapting to changing light levels. At night, after the eye has adapted to darkness, a lag occurs before the eye can see again after a flash of bright light. This lag in the recovery of night vision is known as **night blindness.** Because night blindness is easy to test, it aids in the diagnosis of vitamin A deficiency. Night blindness is only a symptom, however, and may indicate a condition other than vitamin A deficiency.

Vitamin A's Role in Protein Synthesis and Cell Differentiation The role that vitamin A plays in vision is undeniably important, but only one-thousandth of the body's vitamin A is in the retina. Much more is in the skin and the linings of organs, where it works behind the scenes at the genetic level to promote protein synthesis and cell **differentiation.** The process of cell differentiation allows each type of cell to mature so that it is capable of performing a specific function.

All body surfaces, both inside and out, are covered by layers of cells known as **epithelial cells.** The **epithelial tissue** on the outside of the body is, of course, the skin. The epithelial tissues inside the body include the linings of the mouth, stomach, and

Photosensitive cells of the retina:
- Rods contain the rhodopsin pigment and respond to faint light.
- Cones contain the iodopsin pigment and function in color vision.

precursors: compounds that can be converted into other compounds; with regard to vitamins, compounds that can be converted into active vitamins; also known as **provitamins.**

vitamin A: a fat-soluble vitamin. Its three chemical forms are *retinol* (the alcohol form), *retinal* (the aldehyde form), and *retinoic acid* (the acid form).

beta-carotene: a vitamin A precursor made by plants and stored in human fat tissue; an orange pigment.

retinol-binding protein (RBP): the specific protein responsible for transporting retinol.

cornea (KOR-nee-uh): the hard, transparent membrane covering the outside of the eye.

retina (RET-in-uh): the layer of light-sensitive nerve cells lining the back of the inside of the eye; consists of rods and cones.

pigment: a molecule capable of absorbing certain wavelengths of light so that it reflects only those that we perceive as a certain color.

rhodopsin (ro-DOP-sin): a light-sensitive pigment of the retina; contains the retinal form of vitamin A and the protein opsin.

hod = red (pigment)
opsin = visual protein

opsin (OP-sin): the protein portion of the visual pigment molecule.

night blindness: the slow recovery of vision after exposure to flashes of bright light at night; an early symptom of vitamin A deficiency.

differentiation: the development of specific functions different from those of the original.

epithelial (ep-i-THEE-lee-ul) **cells:** cells on the surface of the skin and mucous membranes.

epithelial tissue: tissue composing the layers of the body that serve as selective barriers between the body's interior and the environment (examples are the cornea, the skin, the respiratory lining, and the lining of the digestive tract).

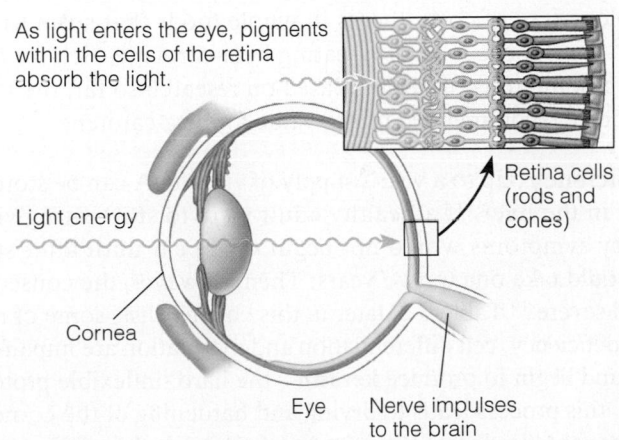

As light enters the eye, pigments within the cells of the retina absorb the light.

Retina cells (rods and cones)

Light energy

Cornea

Eye   Nerve impulses to the brain

FIGURE 8-1 Vitamin A's Role in Vision

intestines; the linings of the lungs and the passages leading to them; the lining of the bladder; the linings of the uterus and vagina; and the linings of the eyelids and sinus passageways. The epithelial tissues on the inside of the body must be kept smooth. To ensure that they are, the epithelial cells on their surfaces secrete a smooth, slippery substance (mucus) that coats the tissues and protects them from invasive microorganisms and other harmful particles. The **mucous membrane** that lines the stomach also shields its cells from digestion by gastric juices. Vitamin A, by way of its role in cell differentiation, helps to maintain the integrity of the epithelial cells.

Vitamin A's Role in Immunity Vitamin A has gained a reputation as an "anti-infective" vitamin because so many of the body's defenses against infection depend on an adequate supply.[5] Much research supports the need for vitamin A in the regulation of the genes involved in immunity. Without sufficient vitamin A, these genetic interactions produce an altered response to infection that weakens the body's defenses.

Vitamin A's Role in Reproduction, Growth, and Development Vitamin A is crucial to normal reproduction and growth. In men, vitamin A participates in sperm development, and, in women, vitamin A promotes normal fetal growth and development.[6] During pregnancy, vitamin A is transferred to the fetus and is essential to the development of the nervous system, lungs, heart, kidneys, skeleton, eyes, and ears.

Nutrition during pregnancy is discussed in Chapter 11.

Beta-Carotene's Role as an Antioxidant Beta-carotene is an extremely effective **antioxidant** in the body. Antioxidants are compounds that protect other compounds (such as lipids in cell membranes) from attack by oxygen. Oxygen triggers the formation of compounds known as **free radicals** that can start chain reactions in cell membranes. If left uncontrolled, these chain reactions can damage cell structures and impair cell functions. Oxidative and free-radical damage to cells is suspected of instigating some early stages of cancer and heart disease.[7] Research has identified links between oxidative damage and the development of many other diseases, including age-related blindness, Alzheimer's disease, arthritis, cataracts, diabetes, and kidney disease.[8]

Studies of populations suggest that people whose diets are low in foods rich in beta-carotene have higher incidences of certain types of cancer than those whose diets contain generous amounts of such foods. Based on these findings, researchers designed a study to determine the effects of beta-carotene supplements on the incidence of lung cancer among smokers. The researchers expected to see a beneficial effect, but instead they found that smokers taking the beta-carotene supplements suffered a *greater* incidence of lung cancer than those taking placebos. Beta-carotene is one of many **dietary antioxidants** present in foods—others include vitamin E, vitamin C, the mineral selenium, and many phytochemicals. Dietary antioxidants are just one class of a complex array of constituents in whole foods that seem to benefit health synergistically.[9] Until more is known, eating beta-carotene–rich foods, not supplements, is in the best interests of health. Based on research so far, the DRI committee has not established a recommended intake value for beta-carotene.

A placebo is an inactive substance used in research studies.

Reminder: *Phytochemicals* are compounds in plants that confer color, taste, and other characteristics. Some phytochemicals are bioactive food components in functional foods. Phytochemicals and functional foods are the topic of Nutrition in Practice 8.

Vitamin A Deficiency Up to a year's supply of vitamin A can be stored in the body, 90 percent of it in the liver. If a healthy adult were to stop eating vitamin A–rich foods, deficiency symptoms would not begin to appear until after stores were depleted, which would take one to two years. Then, however, the consequences would be profound and severe.[10] Table 8-2, later in this chapter, lists some of them.

In vitamin A deficiency, cell differentiation and maturation are impaired. The epithelial cells flatten and begin to produce **keratin**—the hard, inflexible protein of hair and nails. In the eye, this process leads to drying and hardening of the cornea, which may progress to permanent blindness. Vitamin A deficiency is the major cause of preventable blindness in children worldwide, causing as many as half a million children to lose

The progressive blindness caused by vitamin A deficiency is called *xerophthalmia* (zer-off-THAL-mee-uh).

  xero = dry
  ophthalm = eye

their sight every year.[11] As many as 250 million children worldwide endure less severe forms of vitamin A deficiency, making them vulnerable to infectious diseases. Routine vitamin A supplementation and food fortification can be life-saving interventions.[12]

All body surfaces, both inside and out, maintain their integrity with the help of vitamin A. When vitamin A is lacking, cells of the skin harden and flatten, making it dry, rough, scaly, and hard. An accumulation of keratin makes a lump around each hair **follicle** (keratinization).

In the mouth, a vitamin A deficiency results in drying and hardening of the salivary glands, making them susceptible to infection. Secretions of mucus in the stomach and intestines are reduced, hindering normal digestion and absorption of nutrients. Infections of other mucous membranes also become likely.

Vitamin A's role in maintaining the body's defensive barriers may partially explain the relationship between vitamin A deficiency and susceptibility to infection. In developing countries around the world, measles is a devastating infectious disease, killing 450 children each day.[13] Deaths are usually due to related infections such as pneumonia and severe diarrhea. Providing large doses of vitamin A reduces the risk of dying from these infections.

The evidence that vitamin A reduces the severity of measles and measles-related infections and diarrhea has prompted the World Health Organization (WHO) and the United Nations International Children's Emergency Fund (UNICEF) to make control of vitamin A deficiency a major goal in their quest to improve child survival throughout the developing world. They recommend routine vitamin A supplementation for all children with measles in areas where vitamin A deficiency is a problem or where the measles death rate is high. The American Academy of Pediatrics recommends vitamin A supplementation for certain groups of measles-infected infants and children in the United States.

Vitamin A Toxicity Vitamin A toxicity is a real possibility when people consume concentrated amounts of **preformed vitamin A** in foods derived from animals, fortified foods, or supplements.[14] Plant foods contain the vitamin only as beta-carotene, its inactive precursor form. The precursor does not convert to active vitamin A rapidly enough to cause toxicity.

Overdoses of vitamin A damage the same body systems that exhibit symptoms in vitamin A deficiency (see Table 8-2 later in the chapter). Children are most vulnerable to vitamin A toxicity because they need less vitamin A and are more sensitive to overdoses. The availability of breakfast cereals, instant meals, fortified milk, and chewable candy-like supplements, each containing 100 percent or more of the recommended daily intake of vitamin A, makes it possible for a well-meaning parent to provide several times the daily allowance of the vitamin to a child within a few hours.[15] Serious toxicity is seen in infants and young children when they are given more than 10 times the recommended amount every day for weeks at a time.

Excessive vitamin A also poses a **teratogenic** risk.[16] Excessive vitamin A during pregnancy can injure the spinal cord of the developing fetus, increasing the risk of birth defects.[17] The Tolerable Upper Intake Level of 3000 micrograms for women of childbearing age is based on the teratogenic effect of vitamin A.[18]

Excessive amounts of vitamin A over the years may weaken the bones and contribute to fractures and osteoporosis (see the later discussion of vitamin D for more on osteoporosis).[19] More research is needed to clarify the relationship between vitamin A intake and bone health, but such findings suggest that most people should not take vitamin A supplements. Even multivitamin supplements provide more vitamin A than most people need.

Certain vitamin A relatives are available by prescription as acne treatments. When applied directly to the skin surface, these preparations help relieve the symptoms of acne. Taking massive doses of vitamin A internally will *not* cure acne, however, and may cause the symptoms itemized in Table 8-2 (p. 201). In most cases, foods are a

An early sign of xerophthalmia is *xerosis* (drying of the cornea); the last and most severe stage is *keratomalacia* (kerr-uh-to-mal-AY-shuh), or total blindness.

malacia = softening, weakening
The accumulation of the hard material keratin around each hair follicle is *follicular hyperkeratosis*.

mucous membrane: membrane composed of mucus-secreting cells that lines the surfaces of body tissues. (Reminder: *Mucus* is the smooth, slippery substance secreted by these cells.)

antioxidant (anti-OX-ih-dant): a compound that protects other compounds from oxygen by itself reacting with oxygen. *Oxidation* is a potentially damaging effect of normal cell chemistry involving oxygen.
anti = against
oxy = oxygen

free radicals: highly reactive chemical forms that can cause destructive changes in nearby compounds, sometimes setting up a chain reaction.

dietary antioxidants: compounds typically found in plant foods that significantly decrease the adverse effects of oxidation on living tissues. The major antioxidant vitamins are vitamin E, vitamin C, and beta-carotene.

keratin (KERR-uh-lin): a water-insoluble protein; the normal protein of hair and nails. Keratin-producing cells may replace mucus-producing cells in vitamin A deficiency.

follicle (FOLL-i-cul): a group of cells in the skin from which a hair grows.

preformed vitamin A: vitamin A in its active form.

teratogenic (ter-AT-oh-jen-ik): causing abnormal fetal development and birth defects.
terato = monster
genic = to produce

Multivitamin supplements typically provide:
▪ 750 µg (2500 IU).
▪ 1500 µg (5000 IU).
For perspective, the RDA for vitamin A is 700 µg for women and 900 µg for men. The IU (*international unit*) is used to express the amount of some vitamins on supplement labels.

FIGURE 8-2 Symptom of Beta-Carotene Excess—Discoloration of the Skin
The hand on the right shows the skin discoloration that occurs from excess beta-carotene. (Another person's hand is shown on the left for comparison.)

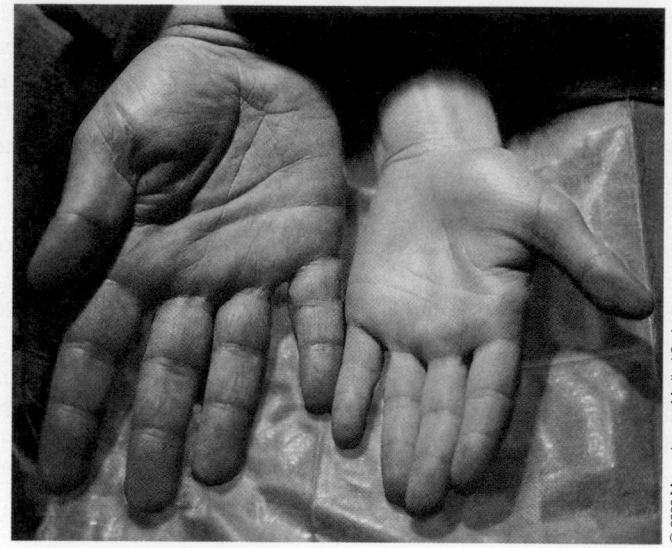

1 µg RAE = 1 µg retinol
= 12 µg beta-carotene from food

Yellowing of the skin caused by excess carotene in the blood is known as *carotenemia* (KAR-oh-teh-NEE-me-ah). Carotenemia can be distinguished from jaundice because the mucous membranes lining the eyelids do not turn yellow as they do in jaundice.

better choice than supplements for needed nutrients. The best way to ensure a safe vitamin A intake is to eat generous servings of vitamin A–rich foods.

Beta-Carotene Conversion and Toxicity Nutrition scientists do not use micrograms to specify the quantity of beta-carotene in foods. Instead, they use a value known as **retinol activity equivalents (RAE),** which express the amount of retinol the body actually derives from a plant food after conversion. The body can make one unit of retinol from about 12 units of beta-carotene.

As mentioned earlier, beta-carotene from plant foods is not converted to the active form of vitamin A rapidly enough to be hazardous. It has, however, been known to turn people bright yellow if they eat too much. Beta-carotene builds up in the fat just beneath the skin and imparts a yellowish cast (see Figure 8-2). As discussed earlier, however, over-consumption of beta-carotene from *supplements* may be harmful, especially to smokers.[20]

Vitamin A in Foods Preformed vitamin A is found only in foods of animal origin. The richest sources of vitamin A are liver and fish oil, but milk, cheese, and fortified cereals are also good sources. Healthy people can eat vitamin A–rich foods in large amounts without risking toxicity, with the possible exception of liver. Eating liver once every week or so is enough. Butter and eggs also provide some vitamin A to the diet.

Because vitamin A is fat soluble, it is lost when milk is skimmed. To compensate, reduced-fat, low-fat, and fat-free milks are often fortified with vitamin A. Margarine is also usually fortified so as to provide the same amount of vitamin A as butter. Snapshot 8-1 shows a sampling of the richest food sources of both preformed vitamin A and beta-carotene.

Beta-Carotene in Foods Many foods from plants contain beta-carotene, the orange pigment responsible for the bold colors of many fruits and vegetables. Carrots, sweet potatoes, pumpkins, cantaloupe, and apricots are all rich sources, and their bright orange color enhances the eye appeal of the plate. Another colorful group, *dark* green vegetables, such as spinach, other greens, and broccoli, owe their color to both chlorophyll and beta-carotene. The orange and green pigments together impart a deep, murky green color to the vegetables. Other colorful vegetables, such as iceberg lettuce, beets, and sweet corn, can fool you into thinking they contain beta-carotene, but these foods derive their color from other pigments and are poor sources of beta-carotene. As for white plant foods such as rice and potatoes, they have little or none. Similarly, fast foods often lack vitamin A. Recommendations to eat *dark* green or *deep* orange vegetables and fruits at least every other day help people to meet their vitamin A needs.

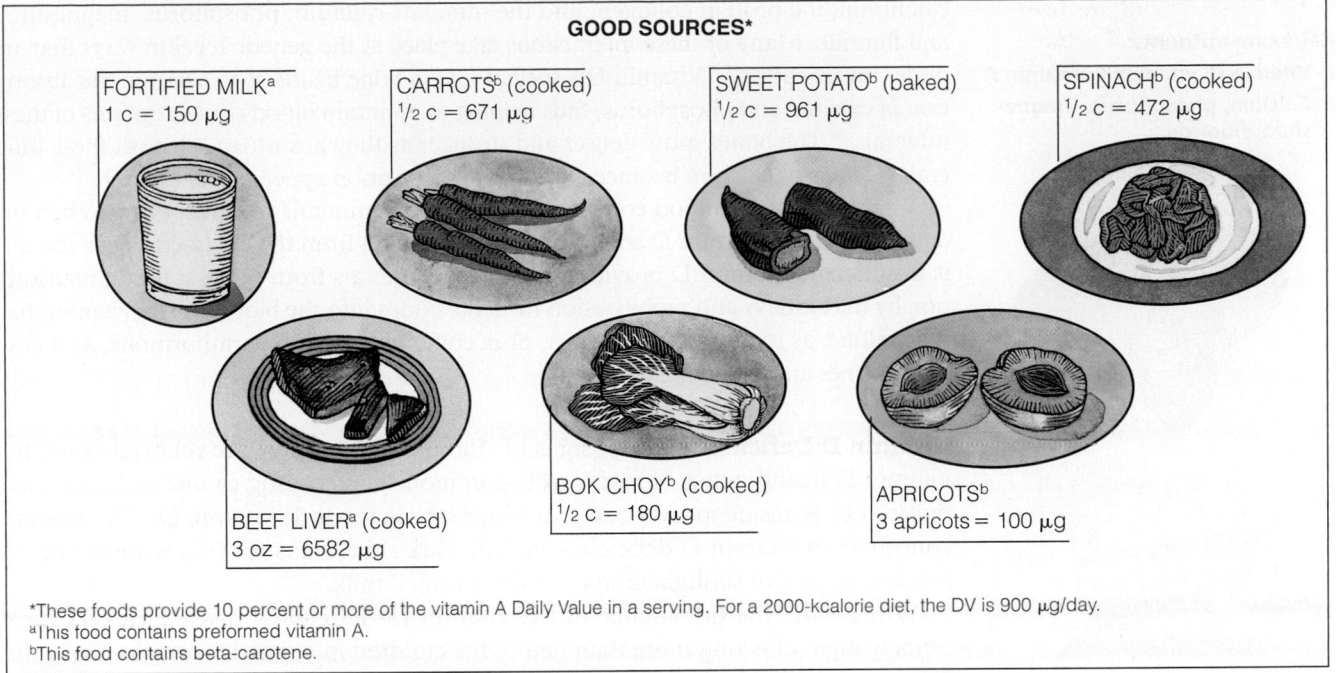

GOOD SOURCES*

FORTIFIED MILK[a]
1 c = 150 µg

CARROTS[b] (cooked)
1/2 c = 671 µg

SWEET POTATO[b] (baked)
1/2 c = 961 µg

SPINACH[b] (cooked)
1/2 c = 472 µg

BEEF LIVER[a] (cooked)
3 oz = 6582 µg

BOK CHOY[b] (cooked)
1/2 c = 180 µg

APRICOTS[b]
3 apricots = 100 µg

*These foods provide 10 percent or more of the vitamin A Daily Value in a serving. For a 2000-kcalorie diet, the DV is 900 µg/day.
[a]This food contains preformed vitamin A.
[b]This food contains beta-carotene.

SNAPSHOT 8-1 Vitamin A and Beta-Carotene

Vitamin D

Vitamin D is different from all the other nutrients in that the body can synthesize it in significant quantities with the help of sunlight. Therefore, in a sense, vitamin D is not an essential nutrient. Given enough sun, people need no vitamin D from foods.

Vitamin D's Metabolic Conversions The liver manufactures a vitamin D precursor, which migrates to the skin, where it is converted to a second precursor with the help of the sun's ultraviolet rays. Next, the liver and then the kidneys alter the second precursor to produce the active vitamin. Whether made from sunlight or obtained from food, vitamin D requires the same two conversions by the liver and kidneys to become active.[21] The biological activity of the active vitamin is 500- to 1000-fold greater than that of its precursor. Diseases that affect either the liver or the kidneys may impair the transformations of precursor vitamin D to active vitamin D and therefore produce symptoms of vitamin D deficiency.

Vitamin D's Actions Although known as a vitamin, vitamin D is actually a hormone—a compound manufactured by one organ of the body that has effects on another. The best-known vitamin D target organs are the small intestine, the kidneys, and the bones, but scientists have discovered more than 30 other vitamin D target tissues, including the brain, the pancreas, the skin, the reproductive organs, and cells of the immune system.[22] In many cases, vitamin D enhances or suppresses the activity of genes that regulate cell growth.[23] As such, it may be valuable in treating a number of diseases.[24]

Some research suggests that a deficit of vitamin D is associated with an increased risk of high blood pressure; cardiovascular diseases; some common cancers; infections such as tuberculosis; inflammatory conditions; autoimmune diseases such as type 1 diabetes, rheumatoid arthritis, and multiple sclerosis; and even premature death.[25] The well-established vitamin D roles, however, concern calcium balance and bone health during growth and throughout life.[26]

Vitamin D's Roles in Bone Growth Vitamin D is a member of a large, cooperative bone-making and maintenance team composed of nutrients and other compounds, including vitamins A, C, and K; the hormones parathormone (parathyroid hormone) and

Vitamin D comes in many forms, but the two most important in the diet are a plant version called vitamin D_2, or *ergocalciferol* (ER-go-kal-SIF-er-ol), and an animal version called vitamin D_3, or *cholecalciferol* (KO-lee-kal-SIF-er-ol).

The precursor of vitamin D made in the liver is 7-dehydrocholesterol, which is made from cholesterol. This is one of the body's many "good" uses for cholesterol.

The final, active vitamin is calcitriol (1-25 dihydroxyvitamin D).

retinol activity equivalents (RAE): a measure of vitamin A activity; the amount of retinol that the body will derive from a food containing preformed retinol or its precursor beta-carotene.

Key bone nutrients:

- Vitamin D, vitamin K, vitamin A
- Calcium, phosphorus, magnesium, fluoride

Sunlight promotes vitamin D synthesis in the skin. Exposure to the sun should be moderate, however; excessive exposure may cause skin cancer.

calcitonin; the protein collagen; and the minerals calcium, phosphorus, magnesium, and fluoride. Many of these interactions take place at the genetic level in ways that are under investigation.[27] Vitamin D's special role in bone health is to assist in the absorption of calcium and phosphorus, thus helping to maintain blood concentrations of these minerals.[28] The bones grow denser and stronger as they absorb and deposit these minerals. Details of calcium balance and mineral deposition appear in Chapter 9.

Vitamin D raises blood concentrations of bone minerals in three ways. When the diet is sufficient, vitamin D enhances their absorption from the GI tract. When the diet is insufficient, vitamin D provides the needed minerals from other sources: reabsorption by the kidneys and mobilization from the bones into the blood.[29] The vitamin may work alone, as it does in the GI tract, or in combination with parathormone, as it does in the bones and kidneys.[30]

Vitamin D Deficiency Overt signs of vitamin D deficiency are relatively rare, but vitamin D insufficiency is remarkably common.[31] According to one estimate, two-thirds of U. S. residents over one year of age take in too little vitamin D.[32] Factors that contribute to vitamin D deficiency include dark skin, breastfeeding without supplementation, lack of sunlight, and not using fortified milk.

Worldwide, the prevalence of the vitamin D–deficiency disease **rickets** is extremely high, affecting more than half of the children in countries such as Mongolia, Tibet, and the Netherlands.[33] In the United States, more than 58 million children are reported to be vitamin D deficient or insufficient, but rickets itself is uncommon.[34] When it occurs, black children and adolescents—especially females and overweight teens—are the ones most likely to be affected.[35] In rickets, the bones fail to calcify normally, causing growth retardation and skeletal abnormalities. The bones become so weak that they bend when they have to support the body's weight (see Figure 8-3). A child with rickets who is old enough to walk characteristically develops bowed legs, often the most obvious sign of the disease. Another sign is the beaded ribs that result from the poorly formed attachments of the bones to the cartilage.

FIGURE 8-3
Vitamin D–Deficiency Symptoms—Bowed Legs and Beaded Ribs of Rickets

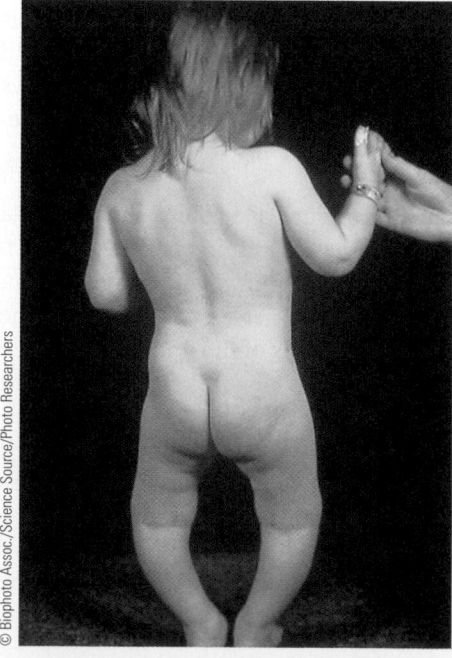

Bowed legs. In rickets, the poorly formed long bones of the legs bend outward as weight-bearing activities such as walking begin.

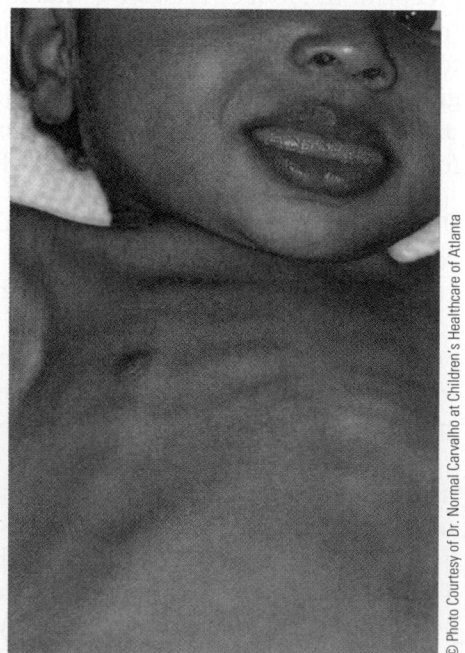

Beaded ribs. In rickets, a series of "beads" develop where the cartilages and bones attach.

Adolescents, who often abandon vitamin D–fortified milk in favor of soft drinks, may also prefer indoor pastimes such as computer games to outdoor activities during daylight hours. Such teens often lack vitamin D and so fail to develop the bone density needed to prevent bone loss in later life.[36]

In adults, the poor mineralization of bone results in the painful bone disease **osteomalacia.** Any failure to synthesize adequate vitamin D or obtain enough from foods sets the stage for a loss of calcium from the bones, which can result in fractures secondary to **osteoporosis** (reduced bone density). The simple act of taking a combined vitamin D and calcium supplement could easily save the life of an elderly person who might otherwise suffer dangerous bone fractures and falls.[37]

Vitamin D Toxicity Whereas vitamin D deficiency depresses calcium absorption, blood calcium, and bone mineralization, an excess of vitamin D does the opposite. It enhances calcium absorption, produces high blood calcium, and promotes return of bone calcium into the blood. The excess calcium then tends to precipitate in the soft tissues, forming stones, including kidney stones.[38] Calcification may also harden the blood vessels and is especially dangerous in the major arteries of the brain, heart, and lungs, where it can cause death.

Vitamin D in excess is the most toxic of all the vitamins. The amounts of vitamin D in foods available in the United States and Canada are well within safe limits, but supplements containing the vitamin in concentrated form are not. Adults should use caution when taking vitamin D supplements and keep them out of the reach of children. The DRI committee has set a Tolerable Upper Intake Level for vitamin D at 100 micrograms per day (4000 IU on supplement labels) for those 9 years of age and older.[39]

Tolerable Upper Intake Levels for infants and young children are on the inside front cover of this text.

Vitamin D from the Sun Most of the world's population relies on natural exposure to sunlight to maintain adequate vitamin D nutrition. The sun imposes no risk of vitamin D toxicity. Prolonged exposure to sunlight degrades the vitamin D precursor in the skin, preventing its conversion to the active vitamin. Even lifeguards on southern beaches are safe from vitamin D toxicity from the sun.

Prolonged exposure to sunlight has other undesirable consequences, however, such as premature wrinkling of the skin and skin cancers. UV radiation from the sun, including UVB that promotes vitamin D synthesis, is carcinogenic, contributing to about a million skin cancers per year in the United States.[40] Sunscreens help reduce the risks of premature wrinkling and skin cancers, but sunscreens with sun protection factors (SPF) of 8 and above also retard vitamin D synthesis. Still, even with an SPF 15–30 sunscreen, sufficient vitamin D synthesis can be obtained in 10 to 20 minutes of sun exposure.[41] A strategy to avoid this dilemma is to apply sunscreen after enough time has elapsed to provide sufficient vitamin D. For most people, exposing hands, face, and arms on a clear summer day for 5 to 10 minutes, a few times a week, should be sufficient to maintain vitamin D nutrition. Dark-skinned people require longer exposure than light-skinned people, but by three hours, vitamin D synthesis in heavily pigmented skin arrives at the same plateau as in fair skin after 30 minutes.

Latitude, season, and time of day also have dramatic effects on vitamin D synthesis and status.[42] Heavy clouds, smoke, or smog block the ultraviolet (UV) rays of the sun that promote vitamin D synthesis. Differences in skin pigmentation, latitude, and smog may account for the finding that African American people, especially those in northern, smoggy cities, are most likely to be vitamin D deficient and develop rickets. Vitamin D deficiency is especially prevalent in the winter.[43] To ensure an adequate vitamin D status, supplements may be needed. The body's vitamin D supplies from summer synthesis alone are insufficient to meet winter needs.[44]

The ultraviolet rays from tanning lamps and tanning booths may also stimulate vitamin D synthesis, but the hazards outweigh any possible benefits. The Food and Drug Administration (FDA) warns that, if the lamps are not properly filtered, people using tanning booths risk burns, damage to the eyes and blood vessels, and skin cancer.

Factors that may limit sun exposure and, therefore, vitamin D synthesis:

- **Geographic location**
- **Season of the year**
- **Time of day**
- **Air pollution**
- **Clothing**
- **Tall buildings**
- **Indoor living**
- **Sunscreens**

rickets: the vitamin D–deficiency disease in children.

osteomalacia (os-tee-oh-mal-AY-shuh): a bone disease characterized by softening of the bones. Symptoms include bending of the spine and bowing of the legs. The disease occurs most often in adults with renal failure or malabsorption disorders.
osteo = bone
mal = bad (soft)

osteoporosis: (os-tee-oh-pore-OH-sis): literally, porous bones; reduced density of the bones, also known as *adult bone loss.*

Vitamin D in Foods Only a few animal foods, notably eggs, liver, butter, some fatty fish, and fortified milk, supply significant amounts of vitamin D. For those who use margarine in place of butter, fortified margarine is a significant source. Infant formulas are fortified with vitamin D in amounts adequate for daily intake as long as infants consume at least one liter (1000 milliliters) or one quart (32 ounces) of formula. Breast milk is low in vitamin D, so vitamin D supplements (400 IU daily) are recommended for infants who are breastfed exclusively and for those who do not receive at least 1000 milliliters of vitamin D–fortified formula per day.[45] These sources, plus any exposure to the sun, provide infants with more than enough of this vitamin.

The fortification of milk with vitamin D is the best guarantee that children will meet their vitamin D needs and underscores the importance of milk in children's diets. Unlike milk, cheese and yogurt are not fortified with vitamin D. Vegans, and especially their children, may have low vitamin D intakes because few fortified plant sources exist. Exceptions include margarine and some soy milks. In the United States, breakfast cereals may be fortified with vitamin D, as their labels indicate.

Vitamin D RDA:
- Adults (19–70 yr): 15 µg/day
- Adults (>70 yr): 20 µg/day

Vitamin D Recommendations In setting the recently revised dietary recommendations, the DRI Committee assumed that no vitamin D was available from skin synthesis.[46] Advancing age increases the risk of vitamin D deficiency, so intake recommendations increase for those over the age of 70 (see the margin).

Vitamin E

More than 80 years ago, researchers discovered a compound in vegetable oils necessary for reproduction in rats. The compound was named **tocopherol,** which means "offspring." Eventually, the compound was named vitamin E. When chemists isolated four tocopherol compounds, they designated them by the first four letters of the Greek alphabet: alpha, beta, gamma, and delta. Of these, alpha-tocopherol is the gold standard for vitamin E activity; recommended intakes are based on it. Table 8-2 later in the chapter summarizes important information about vitamin E.

Reminder: *Oxidation* is a type of chemical reaction, so named because oxygen is one of the agents that often brings it about.

Vitamin E as an Antioxidant Like beta-carotene, vitamin E is a fat-soluble antioxidant. It protects other substances from oxidation by being oxidized itself. If there is plenty of vitamin E in the membranes of cells exposed to an oxidant, chances are this vitamin will take the brunt of any oxidative attack, protecting the lipids and other vulnerable components of the membranes. Vitamin E is especially effective in preventing the oxidation of the polyunsaturated fatty acids (PUFA), but it protects all other lipids (for example, vitamin A) as well.

Vitamin E exerts an especially important antioxidant effect in the lungs, where the cells are exposed to high concentrations of oxygen. Vitamin E also protects the lungs from air pollutants that are strong oxidants.

Some evidence suggests that vitamin E may also offer protection against heart disease by protecting low-density lipoproteins (LDL) from oxidation and reducing inflammation.[47] The oxidation of LDL encourages the development of atherosclerosis. Despite this theoretical understanding of how vitamin E may exert protection against heart disease, the results of controlled clinical studies in which human beings were given vitamin E supplements have been disappointing.[48] Several leading vitamin E researchers point out that clinical trials of vitamin E supplementation differ in many important ways, such as the selection of subjects, the source of the vitamin, the dose of the vitamin, and the outcomes studied. Such differences may partly explain the inconsistent findings.

Further research that addresses such issues may clarify the relationship between vitamin E and heart disease. In the meantime, the American Heart Association supports the consumption of antioxidant-rich fruits and vegetables, as well as whole grains and nuts, to reduce the risk of heart disease.[49]

tocopherol (tuh-KOFF-er-ol): a general term for several chemically related compounds, one of which has vitamin E activity.

Vitamin E Deficiency When blood concentrations of vitamin E fall below a certain critical level, the red blood cells tend to break open and spill their contents, probably because the PUFA in their membranes oxidize. This classic vitamin E–deficiency symptom, known as **erythrocyte hemolysis,** is seen in premature infants born before the transfer of vitamin E from the mother to the fetus that takes place in the last weeks of pregnancy. Vitamin E treatment corrects **hemolytic anemia.**

The few symptoms of vitamin E deficiency that have been observed in adults include loss of muscle coordination and reflexes with impaired movement, vision, and speech. Vitamin E treatment corrects all of these symptoms.

In adults, vitamin E deficiency is usually associated with diseases, notably those that cause malabsorption of fat. These include diseases of the liver, gallbladder, and pancreas, as well as various hereditary diseases involving digestion and use of nutrients.

On rare occasions, vitamin E deficiencies develop in people without diseases. Most likely, such deficiencies occur after years of eating diets extremely low in fat; using fat substitutes, such as diet margarines and salad dressings, as the only sources of fat; or consuming diets composed of highly processed or "convenience" foods. Extensive heating in the processing of foods destroys vitamin E.

Vitamin E Toxicity Vitamin E in foods is safe to consume. Reports of vitamin E toxicity symptoms are rare across a broad range of intakes.[50] Vitamin E supplement use has increased in recent years, however, as its antioxidant action against disease has been recognized. As a result, signs of toxicity are now known or suspected, although vitamin E toxicity is not nearly as common, and its effects are not as serious, as vitamin A or vitamin D toxicity. Extremely high doses of vitamin E interfere with the blood-clotting action of vitamin K and enhance the action of anticoagulant medications, leading to hemorrhage.

Recently, the pooled results from 67 experiments involving almost a quarter-million people suggested that taking vitamin E supplements may slightly increase mortality in both healthy and sick people.[51] Other studies find no effect or a slight decrease in mortality among certain groups.[52] To err on the safe side, until more is known about the safety of vitamin E supplements, intakes should probably be kept low, and they certainly should not exceed the UL of 1000 milligrams of alpha-tocopherol per day. The UL for vitamin E is more than 65 times greater than the recommended intake for adults (15 milligrams).

Vitamin E in Foods Vitamin E is widespread in foods. Much of the vitamin E in the diet comes from vegetable oils and the products made from them, such as margarine, salad dressings, and shortenings. Wheat germ oil is especially rich in vitamin E. Other sources of the vitamin include whole grains, fruits, vegetables, and nuts. Because vitamin E is readily destroyed by heat processing and oxidation, fresh or lightly processed foods are the best sources of this vitamin.

Prior to 2000, values of the vitamin E in food reflected all of the different tocopherols and were expressed in "milligrams of tocopherol equivalents." These measures overestimated the amount of alpha-tocopherol. To estimate the alpha-tocopherol content of foods stated in tocopherol equivalents, multiply by 0.8.

Vitamin K

Vitamin K has long been known for its role in blood clotting, where its presence can make the difference between life and death. The vitamin also participates in the synthesis of several bone proteins. Without vitamin K, the bones produce an abnormal protein that cannot bind to the minerals that normally form bones, resulting in low bone density.[53] An adequate intake of vitamin K helps to decrease bone turnover and protect against fractures.[54] People who consume abundant vitamin K, often in the form of green leafy vegetables, suffer fewer hip fractures than those with lower intakes. Vitamin K supplements seem ineffective against bone loss, however.[55] Vitamin K is also under investigation for roles in heart disease prevention.[56]

Nutrition and upper-GI disorders are discussed in Chapter 17. Nutrition and lower-GI disorders are discussed in Chapter 18. Nutrition and liver disorders are discussed in Chapter 20.

Vegetable oils, some nuts and seeds such as almonds and sunflower seeds, and wheat germ are rich in vitamin E.

K stands for the Danish word *koagulation* (coagulation, or "clotting").

erythrocyte (er-REETH-ro-cite) **hemolysis** (he-MOLL-uh-sis): rupture of the red blood cells, caused by vitamin E deficiency.
erythro = red
cyte = cell
hemo = blood
lysis = breaking

hemolytic (HE-moh-LIT-ick) **anemia:** the condition of having too few red blood cells as a result of erythrocyte hemolysis.

FIGURE 8-4 **Blood-Clotting Process**

When blood is exposed to air, to foreign substances, or to secretions from injured tissues, platelets (small, cell-like structures in the blood) release a phospholipid known as thromboplastin. Thromboplastin catalyzes the conversion of the inactive protein prothrombin to the active enzyme thrombin. Thrombin then catalyzes the conversion of the precursor protein fibrinogen to the active protein fibrin that forms the clot.

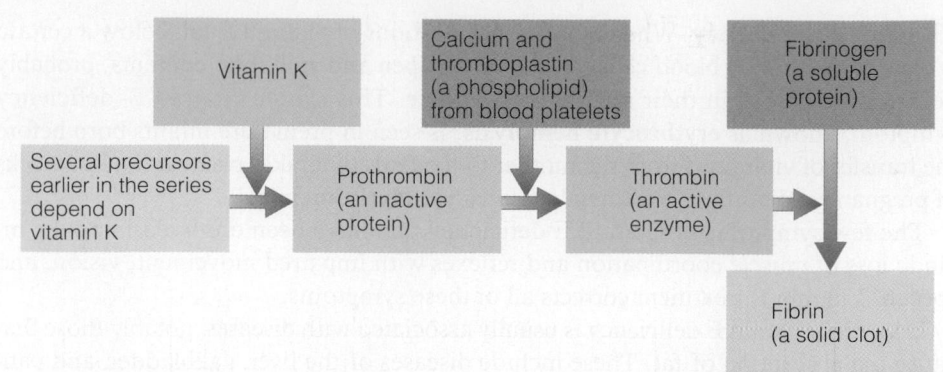

Several precursors earlier in the series depend on vitamin K → Prothrombin (an inactive protein)

Vitamin K

Calcium and thromboplastin (a phospholipid) from blood platelets → Thrombin (an active enzyme)

Fibrinogen (a soluble protein) → Fibrin (a solid clot)

Reminder: The bacterial inhabitants of the digestive tract are known as the *intestinal flora.*

flora = plant inhabitants

Vitamin K AI:
- Men: 120 μg/day
- Women: 90 μg/day

Notable food sources of vitamin K include green vegetables such as collards, spinach, bib lettuce, Brussels sprouts, and cabbage and vegetable oils such as soybean oil and canola oil.

© Matthew Farruggio

Blood Clotting At least 13 different proteins and the mineral calcium are involved in the blood clotting process. Vitamin K is essential for the activation of several of these proteins, among them prothrombin, the precursor of the enzyme thrombin (see Figure 8-4). When any of the blood-clotting factors is lacking, **hemorrhagic disease** results. If an artery or vein is cut or broken, bleeding goes unchecked. Of course, this is not to say that hemorrhaging is always caused by a vitamin K deficiency.

Intestinal Synthesis Like vitamin D, vitamin K can be obtained from a nonfood source. Bacteria in the intestinal tract synthesize vitamin K that the body can absorb, but people cannot depend on this source alone for their vitamin K.

Vitamin K Deficiency Vitamin K deficiency is rare, but it may occur in two circumstances. First, it may arise in conditions of fat malabsorption. Second, some medications interfere with vitamin K's synthesis and action in the body: antibiotics kill the vitamin K–producing bacteria in the intestine, and anticoagulant medications interfere with vitamin K metabolism and activity. When vitamin K deficiency does occur, it can be fatal.

Vitamin K for Newborns Newborn infants present a unique case of vitamin K nutrition. An infant is born with a **sterile** digestive tract, and some weeks pass before the vitamin K–producing bacteria become fully established in the infant's intestines. At the same time, plasma prothrombin concentrations are low (this helps prevent blood clotting during the stress of birth, which might otherwise be fatal). A single dose of vitamin K, usually in a water-soluble form, is recommended at birth to prevent hemorrhagic disease in the newborn.

Vitamin K Toxicity Vitamin K toxicity is rare, and no adverse effects have been reported with high intakes. Therefore, a Tolerable Upper Intake Level has not been established. High doses of vitamin K can reduce the effectiveness of anticoagulant medications used to prevent blood clotting. People taking these medications should eat vitamin K–rich foods in moderation and keep their intakes consistent from day to day.

Vitamin K in Foods Many foods contain ample amounts of vitamin K, notably green leafy vegetables, members of the cabbage family, and some vegetable oils. Other vegetables such as iceberg lettuce and green beans provide smaller amounts.

IN SUMMARY

- The fat-soluble vitamins are vitamins A, D, E, and K.

- Vitamin A is essential to gene expression, vision, cell differentiation and integrity of epithelial tissues, immunity, and reproduction and growth.

- Vitamin A deficiency can cause blindness, sickness, and death and is a major problem worldwide.

- Overdoses of vitamin A are possible and dangerous.

- Vitamin D raises calcium and phosphorus levels in the blood. A deficiency can cause rickets in children or osteomalacia in adults.

- Vitamin D is the most toxic of all the vitamins.

- People exposed to the sun make vitamin D in their skin; fortified milk is an important food source.

- Vitamin E acts as an antioxidant in cell membranes and is especially important in the lungs where cells are exposed to high concentrations of oxygen.

- Vitamin E may protect against heart disease, but the evidence is not conclusive yet.

- Vitamin E deficiency is rare in healthy human beings. The vitamin is widely distributed in plant foods.

- Vitamin K is necessary for blood to clot and for bone health.

- The bacterial inhabitants of the digestive tract produce vitamin K, but people need vitamin K from foods as well.

- Dark green, leafy vegetables are good sources of vitamin K.

hemorrhagic (hem-oh-RAJ-ik) **disease:** the vitamin K–deficiency disease in which blood fails to clot.

sterile: free of microorganisms such as bacteria.

Table 8-2 offers a complete summary of the fat-soluble vitamins.

TABLE 8-2 The Fat-Soluble Vitamins—A Summary

Vitamin Name	Chief Functions	Deficiency Symptoms	Toxicity Symptoms	Significant Sources
Vitamin A (Retinol, retinal, retinoic acid; main precursor is beta-carotene)	Vision, maintenance of cornea, epithelial cells, mucous membranes, skin; bone and tooth growth; reproduction; regulation of gene expression; immunity	Infectious diseases, night blindness, blindness (xerophthalmia), keratinization	Reduced bone mineral density, liver abnormalities, birth defects	Retinol: milk and milk products; eggs; liver Beta-carotene: spinach and other dark, leafy greens; broccoli; deep orange fruits (apricots, cantaloupe) and vegetables (carrots, winter squashes, sweet potatoes, pumpkin)
Vitamin D (Calciferol, cholecalciferol, dihydroxy vitamin D; precursor is cholesterol)	Mineralization of bones (raises blood calcium and phosphorus by increasing absorption from digestive tract, withdrawing calcium from bones, stimulating retention by kidneys)	Rickets, osteomalacia	Calcium imbalance (calcification of soft tissues and formation of stones)	Synthesized in the body with the help of sunshine; fortified milk, margarine, butter, and cereals; eggs; liver; fatty fish (salmon, sardines)
Vitamin E (Alpha-tocopherol, tocopherol)	Antioxidant (stabilization of cell membranes, regulation of oxidation reactions, protection of polyunsaturated fatty acids [PUFA] and vitamin A)	Erythrocyte hemolysis, nerve damage	Hemorrhagic effects	Polyunsaturated plant oils (margarine, salad dressings, shortenings), green and leafy vegetables, wheat germ, whole-grain products, nuts, seeds
Vitamin K (Phylloquinone, menaquinone, naphthoquinone)	Synthesis of blood-clotting proteins and bone proteins	Hemorrhage	None known	Synthesized in the body by GI bacteria; green, leafy vegetables; cabbage-type vegetables; vegetable oils

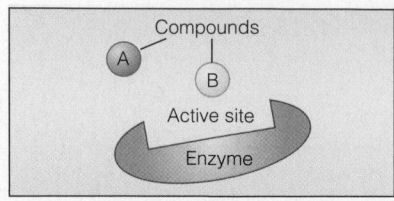

Without the coenzyme, compounds A and B do not respond to the enzyme.

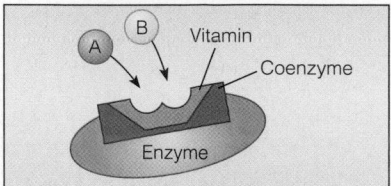

With the coenzyme in place, compounds A and B are attracted to the active site on the enzyme, and they react.

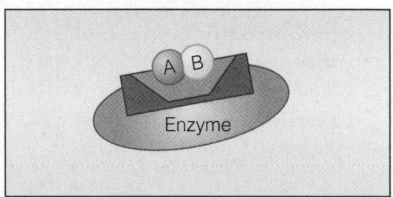

The reaction is completed with the formation of a new product. In this case, the product is AB.

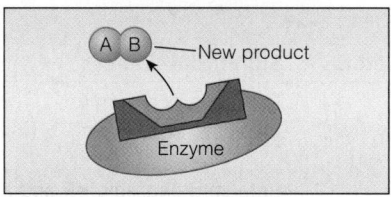

The product AB is released.

FIGURE 8-5
Coenzyme Action

coenzyme (co-EN-zime): a small molecule that works with an enzyme to promote the enzyme's activity. Many coenzymes have B vitamins as part of their structure.
  co = with

beriberi: the thiamin-deficiency disease; characterized by loss of sensation in the hands and feet, muscular weakness, advancing paralysis, and abnormal heart action.

pellagra (pell-AY-gra): the niacin-deficiency disease. Symptoms include the "4 Ds": diarrhea, dermatitis, dementia, and, ultimately, death.
  pellis = skin
  agra = seizure

The Water-Soluble Vitamins

The B vitamins and vitamin C are the water-soluble vitamins. These vitamins, found in the watery compartments of foods, are distributed into water-filled compartments of the body. They are easily absorbed into the bloodstream and are just as easily excreted if their blood concentrations rise too high. Thus, the water-soluble vitamins are less likely to reach toxic concentrations in the body than are the fat-soluble vitamins. Foods never deliver excessive amounts of the water-soluble vitamins, but the large doses concentrated in vitamin supplements can reach toxic levels.

The B Vitamins

Despite advertisements that claim otherwise, the B vitamins do not give people energy. Carbohydrate, fat, and protein—the *energy-yielding* nutrients—supply the fuel for energy. The B vitamins *help* the body use that fuel but do not serve as fuel themselves.

Coenzymes   The eight B vitamins were listed in Table 8-1 (p. 189). Each is part of an enzyme helper known as a **coenzyme.** Some B vitamins have other important functions in the body as well, but the roles these vitamins play as parts of coenzymes are the best understood. A coenzyme is a small molecule that combines with an enzyme to make it active. With the coenzyme in place, a substance is attracted to the enzyme, and the reaction proceeds instantaneously. Figure 8-5 illustrates coenzyme action.

Active forms of five of the B vitamins—thiamin, riboflavin, niacin, pantothenic acid, and biotin—participate in the release of energy from carbohydrate, fat, and protein. A coenzyme containing vitamin B_6 assists enzymes that metabolize amino acids. Folate and vitamin B_{12} help cells to multiply. Among these cells are the red blood cells and the cells lining the GI tract—cells that deliver energy to all the others.

The eight B vitamins play many specific roles in helping the enzymes to perform thousands of different molecular conversions in the body. They must be present in every cell continuously for the cells to function as they should. As for vitamin C, its primary role, discussed later, is as an antioxidant.

B Vitamin Deficiencies   In academic and clinical discussions of the vitamins, different sets of deficiency symptoms are ascribed to each individual vitamin. Such clear-cut symptoms, however, are found only in laboratory animals that have been fed contrived diets that lack just one nutrient. In reality, a deficiency of any single B vitamin seldom shows up in isolation because people do not eat nutrients one by one; they eat foods containing mixtures of many nutrients. If a major class of foods is missing from the diet, all of the nutrients delivered by those foods will be lacking to various extents.

In only two cases have dietary deficiencies associated with single B vitamins been observed on a large scale in human populations. Diseases have been named for these deficiency states. One of them, **beriberi,** was first observed in Southeast Asia when the custom of polishing rice became widespread. Rice contributed 80 percent of the energy intake of the people in these areas, and rice bran was their principal source of thiamin. When the bran was removed to make the rice whiter, beriberi spread like wildfire.

The niacin-deficiency disease, **pellagra,** became widespread in the southern United States in the early part of the 20th century among people who subsisted on a low-protein diet with a staple grain of corn. This diet was unusual in that it supplied neither enough niacin nor enough tryptophan, its amino acid precursor, to make the niacin intake adequate.

Even in the cases of beriberi and pellagra, the deficiencies were probably not pure. When foods were provided containing the one vitamin known to be needed, other vitamins that may have been in short supply came as part of the package.

Major deficiency diseases such as pellagra and beriberi no longer occur in the United States and Canada, but more subtle deficiencies of nutrients, including the B vitamins, are sometimes observed. When they do occur, it is usually in people whose food choices are poor because of poverty, ignorance, illness, or poor health habits such as alcohol abuse.

Interdependent Systems Table 8-3, at the end of this chapter, sums up a few of the better-established facts about B vitamin deficiencies. A look at the table will make another generalization possible. Different body systems depend on these vitamins to different extents. Processes in nerves and in their responding tissues, the muscles, depend heavily on glucose metabolism and hence on thiamin, so paralysis sets in when this vitamin is lacking. But thiamin is important in all cells, not just in nerves and muscles. Similarly, because the red blood cells and GI tract cells divide the most rapidly, two of the first symptoms of a deficiency of folate are a type of anemia and GI tract deterioration—but again, all systems depend on folate, not just these. The list of symptoms in Table 8-3 is far from complete.

B Vitamin Enrichment of Foods If the staple food of a region is made from **refined grain,** vitamin B deficiencies are especially likely. One way to protect people from deficiencies is to add nutrients to their staple food, a process known as **fortification** or **enrichment.** The enrichment of refined breads and cereals has drastically reduced the incidence of iron and B vitamin deficiencies.

The preceding discussion has shown both the great importance of the B vitamins in promoting normal, healthy functioning of all body systems and the severe consequences of deficiency. Now you may want to know how to be sure you and your clients are getting enough of these vital nutrients. The next sections present information on each B vitamin. While reading further, keep in mind that foods can provide all the needed nutrients and that supplements are a poor second choice. Some supplements are absurdly costly, but even if they are inexpensive, most people don't need them. Nutrition in Practice 9 discusses uses and choices of supplements in more detail.

Thiamin

All cells use thiamin, which plays a critical role in their energy metabolism. Thiamin also occupies a special site on nerve cell membranes. Consequently, as mentioned earlier, thiamin is critical to the normal functioning of the nerves and muscles.

Thiamin Need People who fail to eat enough food to meet energy needs risk nutrient deficiencies, including thiamin deficiency. Inadequate thiamin intakes have been reported among the nation's malnourished and homeless people. Similarly, people risk thiamin deficiency when they derive most of their energy from empty-kcalorie foods and beverages. Alcohol is a good example. It contributes energy but provides few, if any, nutrients and often displaces food. In addition, alcohol impairs thiamin absorption and enhances thiamin excretion in the urine, doubling the risk of deficiency. Many alcoholics are thiamin deficient.

Thiamin in Foods Thiamin occurs in small quantities in virtually all nutritious foods, but it is concentrated in only a few foods, of which pork is the most commonly eaten. A useful guideline for meeting thiamin needs is to keep empty-kcalorie foods to a minimum and to include 10 or more different servings of nutritious foods each day, assuming that each serving will contribute, on the average, about 10 percent of needs. Foods chosen from the bread and cereal group should be either whole grain or enriched. Thiamin is not stored in the body to any great extent, so daily intake is important.

refined grain: a product from which the bran, germ, and husk have been removed, leaving only the endosperm.

fortification: the addition to a food of nutrients that were either not originally present or present in insignificant amounts. Fortification can be used to correct or prevent a widespread nutrient deficiency, to balance the total nutrient profile of a food, or to restore nutrients lost in processing.

enrichment: the addition to a food of nutrients to meet a specified standard. In the case of refined bread or cereal, five nutrients have been added: thiamin, riboflavin, niacin, and folate in amounts approximately equivalent to, or higher than, those originally present and iron in amounts to alleviate the prevalence of iron-deficiency anemia.

Note: The terms *fortified* and *enriched* may be used interchangeably.

Nutritious foods such as pork, legumes, sunflower seeds, and enriched and whole-grain breads are valuable sources of thiamin.

Severe thiamin deficiency in alcohol abusers is called *Wernicke-Korsakoff syndrome.* Symptoms include disorientation, loss of short-term memory, jerky eye movements, and staggering gait.

Thiamin RDA:
▪ Men: 1.2 mg/day
▪ Women: 1.1 mg/day

Milk and milk products supply much (about 50 percent) of the riboflavin in people's diets, but meats, eggs, green vegetables, and enriched and whole-grain breads and cereals are good sources, too.

Riboflavin RDA:
▮ Men: 1.3 mg/day
▮ Women: 1.1 mg/day

Niacin-rich foods include meat, fish, poultry, and peanut butter, as well as enriched breads and cereals and certain vegetables.

A food containing 1 mg of niacin and 60 mg of tryptophan contains the niacin equivalent of 2 mg, or 2 mg NE.

Niacin RDA:
▮ Men: 16 mg NE/day
▮ Women: 14 mg NE/day

When a normal dose of a nutrient clears up a deficiency condition, the effect is a *physiological* one. When a large dose of a nutrient overwhelms a body system and acts like a drug, the effect is a *pharmacological* one.

Riboflavin

Like thiamin, riboflavin facilitates energy metabolism in the body. The needs of infants, children, and pregnant women rise rapidly during periods of active growth.

Riboflavin Deficiency and Toxicity When thiamin is deficient, riboflavin may be lacking too, but its deficiency symptoms, such as cracks at the corners of the mouth and sore throat, may go undetected because those of thiamin are more severe. A diet that remedies riboflavin deficiency invariably contains some thiamin and so clears up both deficiencies. Excesses of riboflavin appear to cause no harm, and no UL has been established.

Riboflavin in Foods Unlike thiamin, riboflavin is not evenly distributed among the food groups. The major contributors of riboflavin to people's diets are milk and milk products, followed by enriched breads, cereals, and other grain products. Green vegetables (broccoli, turnip greens, asparagus, and spinach) and meats are also contributors. The riboflavin richness of milk and milk products is a good reason to include these foods in every day's meals. No other commonly eaten food can make such a substantial contribution. People who omit milk and milk products from their diets can substitute generous servings of dark green, leafy vegetables. Among the meats, liver and heart are the richest sources, but all lean meats, as well as eggs, offer some riboflavin.

Effects of Light Riboflavin is light sensitive; the ultraviolet rays of the sun or of fluorescent lamps can destroy it, as can irradiation. For this reason, milk is often sold in cardboard or opaque plastic containers to protect the riboflavin in the milk from light. In contrast, riboflavin is heat stable, so ordinary cooking does not destroy it.

Niacin

Like thiamin and riboflavin, niacin participates in the energy metabolism of every body cell. Niacin is unique among the B vitamins in that the body can make it from protein. The amino acid tryptophan can be converted to niacin in the body: 60 milligrams of tryptophan yield 1 milligram of niacin. Recommended intakes are therefore stated in **niacin equivalents (NE),** reflecting the body's ability to convert tryptophan to niacin.

Naturally occurring niacin from foods causes no harm. Certain forms of niacin supplements taken in doses three to four times the dietary recommendation or larger cause "niacin flush," a dilation of the capillaries of the skin with perceptible tingling that, if intense, can be painful. The Tolerable Upper Intake Level (35 milligrams NE) is based on flushing as the critical adverse effect.

Niacin Used as a Medication Physicians sometimes use diet and large doses of a form of niacin (nicotinic acid) to lower blood cholesterol in the treatment of atherosclerosis. When used this way, niacin leaves the realm of nutrition to become a pharmacological agent, a drug. As with any medication, self-dosing with niacin is ill advised; large doses may injure the liver and produce some symptoms of diabetes.[57]

Niacin in Foods Meat, poultry, fish, legumes, and enriched and whole grains contribute about half the niacin equivalents most people consume. Among the vegetables, mushrooms, asparagus, and potatoes are the richest niacin sources. Niacin is less vulnerable to losses during food preparation and storage than other water-soluble vitamins. Being fairly heat resistant, niacin can withstand reasonable cooking times, but like other water-soluble vitamins, it will leach into cooking water.

Pantothenic Acid and Biotin

Two other B vitamins—pantothenic acid and biotin—are also important in energy metabolism. Pantothenic acid was first recognized as a substance that stimulates growth. It is a component of a key enzyme that makes possible the release of energy from the energy nutrients. Pantothenic acid is involved in more than 100 different steps in the synthesis of lipids, neurotransmitters, steroid hormones, and hemoglobin. Biotin plays an important role in metabolism as a coenzyme that carries carbon dioxide. Emerging evidence indicates that biotin participates in other processes such as gene expression and cell signaling and in the structure of DNA-binding proteins in the cell nucleus.[58]

Pantothenic Acid and Biotin in Foods Both pantothenic acid and biotin are more widespread in foods than the other vitamins discussed so far. There seems to be no danger that people who consume a variety of foods will suffer deficiencies. Claims that pantothenic acid and biotin are needed in pill form to prevent or cure disease conditions are at best unfounded and at worst intentionally misleading.

Biotin Deficiency Biotin deficiencies are rare, but they have been reported in adults fed artificially by vein without biotin supplementation. Researchers can induce biotin deficiency in animals or human beings by feeding them raw egg whites, which contain a protein that binds biotin and prevents its absorption.

Pantothenic acid AI:
- Adults: 5 mg/day

Biotin AI:
- Adults: 30 µg/day

The protein *avidin* in egg whites binds biotin.

Vitamin B$_6$

A surge of recent research interest in vitamin B$_6$ has not only revealed new knowledge but has also raised new questions. For example, unlike the other water-soluble vitamins, vitamin B$_6$ is stored extensively in muscle tissue. Most recently, research interest has centered on a possible role for vitamin B$_6$ in the treatment of disease.[59]

Metabolic Roles of Vitamin B$_6$ Vitamin B$_6$ has long been known to play roles in protein and amino acid metabolism. In the cells, vitamin B$_6$ helps convert one kind of amino acid, which the cells have in abundance, to other nonessential amino acids that the cells lack. It also aids in the conversion of the amino acid tryptophan to niacin and plays important roles in the synthesis of hemoglobin and neurotransmitters, the communication molecules of the brain. Vitamin B$_6$ also assists in releasing stored glucose from glycogen and thus contributes to the regulation of blood glucose. Research suggests roles for vitamin B$_6$ in immune function, cognitive performance, and hormone response.[60] The association between vitamin B$_6$ and immune function is related to the critical role the vitamin plays in protein metabolism. Vitamin B$_6$ deficiency can significantly impair the immune response, perhaps by way of impaired antibody production.

Vitamin B$_6$ status may be related to cardiovascular disease risk. Elevated blood levels of the amino acid homocysteine correlate with a high incidence of heart disease.[61] Evidence suggests that low blood concentrations of vitamin B$_6$, vitamin B$_{12}$, and folate are associated with elevated homocysteine concentrations.[62] The missing link, however, is how, or even whether, a deficiency of B vitamins or elevated homocysteine directly affects processes leading to heart disease.[63]

Vitamin B$_6$ Deficiency Besides a weakening immune response, vitamin B$_6$ deficiency is expressed in general symptoms, such as weakness, depression, confusion, and irritability. Other symptoms include a greasy, flaky dermatitis; anemia; and, in advanced cases, convulsions.

Vitamin B$_6$ Toxicity For years it was believed that vitamin B$_6$, like other water-soluble vitamins, could not reach toxic concentrations in the body. The toxic effects of vitamin B$_6$ became known when a physician reported them in women who had been

Most protein-rich foods such as meat, fish, and poultry provide ample vitamin B$_6$; certain vegetables and fruits are good sources too.

niacin equivalents (NE): the amount of niacin present in food, including the niacin that can theoretically be made from tryptophan, its precursor, present in the food.

taking more than 2 *grams* of vitamin B_6 daily (20 times the current UL of 100 *milligrams*) for two months or more, attempting to cure premenstrual syndrome. The first symptom of toxicity was numb feet; then the women lost sensation in their hands; then they became unable to walk. The women recovered after they discontinued the supplements.

Vitamin B_6 Recommendations Because vitamin B_6 coenzymes play many roles in amino acid metabolism, previous RDA were expressed in terms of protein intakes; the current RDA for vitamin B_6, however, is not. Research does not support claims that large doses of vitamin B_6 enhance muscle strength or physical endurance.

Vitamin B_6 in Foods The richest food sources of vitamin B_6 are protein-rich meat, fish, and poultry. Potatoes, a few other vegetables, and some fruits are good sources, too. Foods lose vitamin B_6 when heated.

Folate

The B vitamin folate is active in cell division. During periods of rapid growth and cell division, such as pregnancy and adolescence, folate needs increase and deficiency is especially likely. When a deficiency occurs, the replacement of the rapidly dividing cells of the blood and the GI tract falters. Not surprisingly, then, two of the first symptoms of a folate deficiency are a type of anemia and GI tract deterioration (see Table 8-3 on p. 212).

Folate, Alcohol, and Drugs Of all the vitamins, folate appears to be the most vulnerable to interactions with alcohol and other drugs.[64] As Nutrition in Practice 20 describes, alcohol-addicted people risk folate deficiency because alcohol impairs folate's absorption and increases its excretion. Furthermore, as people's alcohol intakes rise, their folate intakes decline. Many medications, including aspirin, oral contraceptives, and anticonvulsants, also impair folate status. Smoking exerts a negative effect on folate status as well.

Folate and Neural Tube Defects Research studies confirm the importance of folate in preventing **neural tube defects (NTD).** The brain and spinal cord develop from the neural tube, and defects in its orderly formation during the early weeks of pregnancy may result in various central nervous system disorders and death. Folate supplements taken before conception and continued throughout the first trimester of pregnancy can prevent NTD. For this reason, all women of childbearing age who are capable of becoming pregnant should consume 400 micrograms (0.4 milligrams) of folate daily from supplements, fortified foods, or both, *in addition* to eating folate-rich foods.

Folate status improves more with supplementation or fortification than with a dietary intake that meets recommendations. Neural tube defects arise early in pregnancy before most women realize they are pregnant, and most women eat too few fruits and vegetables to supply even half the folate needed to prevent NTD.[65] For these reasons, the FDA mandated that enriched grain products (flour, cornmeal, pasta, and rice) be fortified with an especially absorbable synthetic form of folate, folic acid.

Fortification has improved folate status in women of childbearing age and lowered the number of neural tube defects that occur each year.[66] Folate fortification also raises safety concerns, however. High doses of folate can complicate the diagnosis of vitamin B_{12} deficiency, as discussed later. Other suspected but unconfirmed potential harms from high blood folic acid levels include suppression of normal immune functioning and increased cancer risks.[67] The DRI committee set a Tolerable Upper Intake Level of 1000 micrograms per day from fortified foods or supplements. Except for individuals who take supplements with more than 400 micrograms of folic acid, few people (less than 3 percent of the U.S. population) exceed the Tolerable Upper Intake Level for folic acid.[68] Thus, researchers conclude that the current level of fortification of the food supply appears to be safe.

Vitamin B_6 RDA:
- Adults (19–50): 1.3 mg/day

The two main types of neural tube defects are *spina bifida* (literally, "split spine") and *anencephaly* ("no brain"). Chapter 11 includes a figure of a neural tube defect and a discussion.

Folate RDA:
- Adults: 400 µg/day

Bread products, flour, corn grits, and pasta must be fortified with 140 µg per 100 g of food (about ½ c cooked food or 1 slice of bread).

Folate in Foods As Snapshot 8-2 shows, the best food sources of folate are liver, legumes, beets, and leafy green vegetables (the vitamin's name suggests the word *foliage*). Among the fruits, oranges, orange juice, and cantaloupe are the best sources. With fortification, grain products are good sources of folate, too. Heat and oxidation during cooking and storage can destroy up to half of the folate in foods.

The difference in absorption between naturally occurring food folate and synthetic folate that enriches foods and is added to supplements necessitated a new unit of measurement for folate: the **dietary folate equivalents,** or **DFE.** The DFE convert all forms of folate into units that are equivalent to the folate in foods. Most food labels and tables of food composition express folate values in micrograms, however, so the accompanying "How to" describes how to estimate dietary folate equivalents.

neural tube defects (NTD): malformations of the brain, spinal cord, or both that occur during embryonic development.

dietary folate equivalents (DFE): the amount of folate available to the body from naturally occurring sources, fortified foods, and supplements, accounting for differences in bioavailability from each source.

How To ESTIMATE DIETARY FOLATE EQUIVALENTS

Folate is expressed in terms of dietary folate equivalents (DFE) because synthetic folate from supplements and fortified foods is absorbed at almost twice (1.7 times) the rate of naturally occurring folate from other foods. Use the following equation to calculate:

$$DFE = \mu g \text{ food folate} + (1.7 \times \mu g \text{ synthetic folate})$$

Consider, for example, a pregnant woman who takes a supplement and eats a bowl of fortified corn flakes, two slices of fortified bread, and one cup of fortified pasta. From the supplement and fortified foods, she obtains synthetic folate:

Supplement	100 μg folate
Fortified corn flakes	100 μg folate
Fortified bread	40 μg folate
Fortified pasta	60 μg folate
	300 μg folate

To calculate the DFE, multiply the amount of synthetic folate by 1.7:

$$300 \ \mu g \times 1.7 = 510 \ \mu g \text{ DFE}$$

Now add the naturally occurring folate from the other foods in her diet—in this example, another 90 micrograms of folate.

$$510 \ \mu g \text{ DFE} + 90 \ \mu g = 600 \ \mu g \text{ DFE}$$

Notice that if we had not converted synthetic folate from supplements and fortified foods to DFE, this woman's intake would appear to fall short of the 600 micrograms recommended for pregnancy (300 μg + 190 μg = 390 μg). But as this example shows, her intake does meet the recommendation. At this time, supplement and fortified food labels list folate in micrograms only, not micrograms DFE, making such calculations necessary.

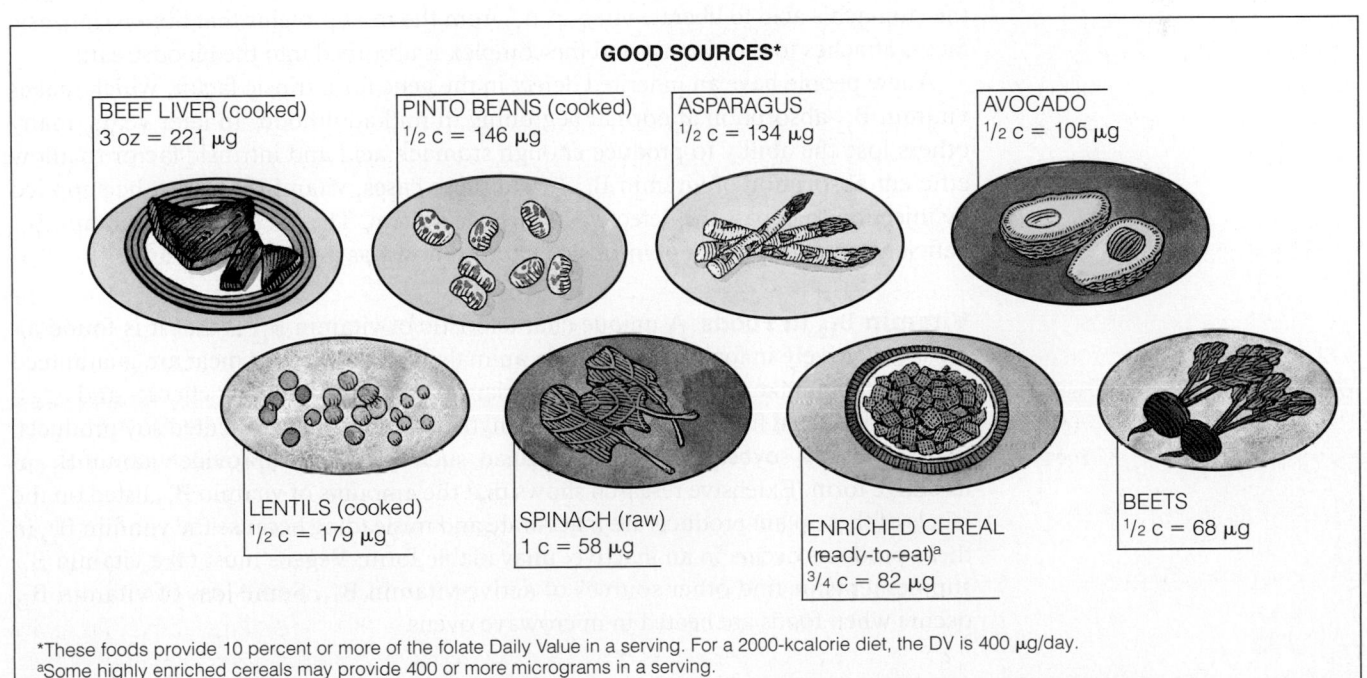

GOOD SOURCES*

BEEF LIVER (cooked)
3 oz = 221 μg

PINTO BEANS (cooked)
1/2 c = 146 μg

ASPARAGUS
1/2 c = 134 μg

AVOCADO
1/2 c = 105 μg

LENTILS (cooked)
1/2 c = 179 μg

SPINACH (raw)
1 c = 58 μg

ENRICHED CEREAL
(ready-to-eat)ª
3/4 c = 82 μg

BEETS
1/2 c = 68 μg

*These foods provide 10 percent or more of the folate Daily Value in a serving. For a 2000-kcalorie diet, the DV is 400 μg/day.
ªSome highly enriched cereals may provide 400 or more micrograms in a serving.

SNAPSHOT 8-2 Folate

Vitamin B$_{12}$

Vitamin B$_{12}$ and folate share a special relationship: vitamin B$_{12}$ assists folate in cell division. Their roles intertwine, but each performs a specific task that the other cannot accomplish.

Vitamin B$_{12}$, Folate, and Cell Division Vitamin B$_{12}$ (in coenzyme form) stands by to accept carbon groups from folate as folate removes them from other compounds. The passing of these carbon groups from folate to vitamin B$_{12}$ regenerates the active form of folate so that it can continue its dismantling tasks. In the absence of vitamin B$_{12}$, folate is trapped in its inactive, metabolically useless form, unable to do its job. When folate is either trapped due to a vitamin B$_{12}$ deficiency or unavailable due to a deficiency of folate itself, cells that are growing most rapidly—notably, the blood cells—are the first to be affected. Thus, a deficiency of either nutrient—vitamin B$_{12}$ or folate—impairs maturation of the blood cells and produces anemia. The anemia is identifiable by microscopic examination of the blood, which reveals many large, immature red blood cells. Either vitamin B$_{12}$ or folate will clear up the anemia.

Vitamin B$_{12}$ and the Nervous System Although either vitamin will clear up the anemia caused by vitamin B$_{12}$ deficiency, if folate is given when vitamin B$_{12}$ is needed, the result is disastrous, not to the blood but to the nervous system. The reason: vitamin B$_{12}$ also helps maintain nerve fibers. A vitamin B$_{12}$ deficiency can ultimately result in devastating neurological symptoms, undetectable by a blood test. A deceptive folate "cure" of the anemia in vitamin B$_{12}$ deficiency allows the nerve deterioration to progress, leading to paralysis and permanent nerve damage. This interaction between folate and vitamin B$_{12}$ raises safety concerns about the use of folate supplements and fortification of foods.[69] In an effort to prevent excessive folate intakes that could mask symptoms of a vitamin B$_{12}$ deficiency, the FDA specifies the exact amounts of folic acid that can be added to enriched foods.

Vitamin B$_{12}$ Absorption Absorption of vitamin B$_{12}$ requires an **intrinsic factor,** a compound made by the stomach with instructions from the genes. With the help of the stomach's acid to liberate vitamin B$_{12}$ from the food proteins that bind it, intrinsic factor attaches to the vitamin and the complex is absorbed into the bloodstream.

A few people have an inherited defect in the gene for intrinsic factor, which makes vitamin B$_{12}$ absorption abnormal beginning in mid-adulthood. In later years, many others lose the ability to produce enough stomach acid and intrinsic factor to allow efficient absorption of vitamin B$_{12}$.[70] * In these cases, vitamin B$_{12}$ must be supplied by injection to bypass the defective absorptive system. The anemia of the vitamin B$_{12}$ deficiency caused by lack of intrinsic factor is known as **pernicious anemia.**

Vitamin B$_{12}$ in Foods A unique characteristic of vitamin B$_{12}$ is that it is found almost exclusively in foods derived from animals. People who eat meat are guaranteed an adequate intake, and lacto-ovovegetarians (who consume milk, cheese, and eggs) are also protected from deficiency. It is a myth, however, that fermented soy products, such as miso (a soybean paste), or sea algae, such as spirulina, provide vitamin B$_{12}$ in its active form. Extensive research shows that the amounts of vitamin B$_{12}$ listed on the labels of these plant products are inaccurate and misleading because the vitamin B$_{12}$ in these products occurs in an inactive, unavailable form. Vegans must take vitamin B$_{12}$ supplements or find other sources of active vitamin B$_{12}$. Some loss of vitamin B$_{12}$ occurs when foods are heated in microwave ovens.

*The condition is atrophic gastritis (a-TROH-fik gas-TRY-tis), a chronic inflammation of the stomach accompanied by a diminished size and functioning of the stomach's mucous membrane and glands.

Large-cell anemia is known as *macrocytic* or *megaloblastic* anemia.
macro = large
cyte = cell
mega = large
blast = immature cell

Vitamin B$_{12}$ RDA:
▌ Adults: 2.4 μg/day

Vitamin B$_{12}$ Deficiency in Vegans Vegans who omit supplements are at special risk for undetected vitamin B$_{12}$ deficiency for two reasons: first, they receive none in their diets, and second, they consume large amounts of folate in the vegetables they eat. Because the body can store many times the amount of vitamin B$_{12}$ used each day, a deficiency may take years to develop in a new vegetarian. When a deficiency does develop, though, it may progress to a dangerous extreme because the deficiency of vitamin B$_{12}$ may be masked by the high folate intake.

Worldwide, vitamin B$_{12}$ deficiency among vegetarians is a growing problem.[71] A pregnant or lactating vegetarian woman who eats no foods of animal origin should be aware that her infant can develop a vitamin B$_{12}$ deficiency, even if the mother appears healthy. Breastfed infants born to vegan mothers with low concentrations of vitamin B$_{12}$ in their breast milk can develop severe neurological symptoms such as seizures and cognitive problems.[72]

Non–B Vitamins

Other compounds are sometimes inappropriately called B vitamins because, like the true B vitamins, they serve as coenzymes in metabolism. Even if they were essential, however, supplements would be unnecessary because these compounds are abundant in foods.

Among the non–B vitamins are a trio of substances known as inositol, choline, and carnitine. Researchers are exploring the possibility that these substances may be essential. Thus far, only choline has been assigned an Adequate Intake value.

Choline AI:
- Men: 550 mg/day
- Women: 425 mg/day

Vitamin C

Three hundred and sixty years ago, any man who joined the crew of a seagoing ship knew he had only half a chance of returning alive—not because he might be slain by pirates or die in a storm, but because he might contract the dread disease **scurvy.** Then, a physician with the British navy found that citrus fruits could cure the disease, and thereafter, all ships were required to carry lime juice for every sailor. (This is why British sailors are still called "limeys" today.) In the 1930s, the antiscurvy factor in citrus fruits was isolated from lemon juice and named **ascorbic acid.** Today, hundreds of millions of vitamin C pills are produced in pharmaceutical laboratories. Vitamin C's action defies a simple, tidy description. It plays many important roles in the body, and its modes of action differ in different situations.

Vitamin C's Role in Collagen Formation The best-understood action of vitamin C is its role in helping to form **collagen,** the single most important protein of connective tissue. Collagen serves as the matrix on which bone is formed, the material of scars, and an important part of the "glue" that attaches one cell to another. This latter function is especially important in the artery walls, which must expand and contract with each beat of the heart, and in the walls of the capillaries, which are thin and fragile. Vitamin C also plays a role in the production of carnitine, important for transporting fatty acids within cells.

Vitamin C as an Antioxidant Vitamin C is also an important antioxidant. Recall that the antioxidants beta-carotene and vitamin E protect fat-soluble substances from oxidizing agents; vitamin C protects water-soluble substances the same way. By being oxidized itself, vitamin C regenerates already-oxidized substances such as iron and copper to their original, active form. In the intestines, it protects iron from oxidation and so enhances iron absorption. In the cells and body fluids, it helps to protect other molecules, including the fat-soluble compounds vitamin A, vitamin E, and the polyunsaturated fatty acids.

Vitamin C in Amino Acid Metabolism Vitamin C is also involved in the metabolism of several amino acids. Some of these amino acids end up being used to make substances of great importance in body functioning, among them the neurotransmitter norepinephrine and the hormone thyroxine.

intrinsic factor: a factor inside the system. Anemia that reflects a vitamin B$_{12}$ deficiency caused by lack of intrinsic factor is known as pernicious anemia.

pernicious (per-NISH-us) **anemia:** a blood disorder that reflects a vitamin B$_{12}$ deficiency caused by lack of intrinsic factor and characterized by large, immature red blood cells and damage to the nervous system (*pernicious* means "highly injurious or destructive").

scurvy: the vitamin C–deficiency disease.

ascorbic acid: one of the two active forms of vitamin C. Many people refer to vitamin C by this name.
   a = without
   scorbic = having scurvy

collagen: the characteristic protein of connective tissue.
   kolla = glue
   gennan = produce

Role of Stress During stress, the adrenal glands release large quantities of vitamin C together with the stress hormones epinephrine and norepinephrine. The vitamin's exact role in the stress reaction remains unclear, but physical stresses (described in a later section) raise vitamin C needs.

Vitamin C as a Possible Antihistamine Newspaper headlines touting vitamin C as a cure for colds and cancer have appeared frequently over the years. Some research suggests that vitamin C (2 grams per day for two weeks) may reduce the severity and duration of cold and allergy symptoms by reducing blood histamine concentrations. In other words, vitamin C acts as an antihistamine. If further research confirms vitamin C's antihistamine effect, its use may permit people to rely less heavily on antihistamine drugs when suffering from cold and allergy symptoms.

Vitamin C's Role in Cancer Prevention and Treatment The role of vitamin C in the prevention and treatment of cancer is still being studied. Evidence to date indicates that foods containing vitamin C probably protect against cancer of the esophagus.[73] The correlation may reflect not just an association with vitamin C but the broader benefits of a diet rich in fruits and vegetables and low in fat. It does not support the taking of vitamin C supplements to prevent or treat cancer.

Vitamin C Deficiency When vitamin C intake is inadequate, the body's vitamin C pool dwindles, and the blood vessels show the first deficiency signs. The gums around the teeth begin to bleed easily, and capillaries under the skin break spontaneously, producing pinpoint hemorrhages. As vitamin C concentrations continue to fall, the symptoms of scurvy appear. Muscles, including the heart muscle, may degenerate. The skin becomes rough, brown, scaly, and dry. Wounds fail to heal because scar tissue will not form without collagen. Bone rebuilding falters; the ends of the long bones become softened, malformed, and painful; and fractures occur. The teeth may become loose in the jawbone and fall out. Anemia and infections are common. Sudden death is likely, perhaps because of massive bleeding into the joints and body cavities.

It takes only 10 or so milligrams of vitamin C a day to prevent scurvy, and not much more than that to cure it. Once diagnosed, scurvy is readily reversible with moderate doses, in the neighborhood of 100 milligrams per day. Such an intake is easily achieved by including vitamin C–rich foods in the diet.

Vitamin C Toxicity The easy availability of vitamin C in pill form and the publication of books recommending vitamin C to prevent everything from the common cold to life-threatening cancer have led thousands of people to take megadoses of vitamin C. Not surprisingly, instances of vitamin C causing harm have surfaced.

Some of the suspected toxic effects of vitamin C megadoses have not been confirmed, but others have been seen often enough to warrant concern. Nausea, abdominal cramps, and diarrhea are often reported. Several instances of interference with medical regimens are known. Large amounts of vitamin C excreted in the urine obscure the results of tests used to detect diabetes. People taking anticoagulants may unwittingly counteract the effect of these medications if they also take massive doses of vitamin C. Vitamin C megadoses can also enhance iron absorption too much, resulting in iron overload (see Chapter 9).

People with sickle-cell anemia may be especially vulnerable to megadoses of vitamin C. Those who have a tendency toward **gout,** as well as those who have a genetic abnormality that alters the way they metabolize vitamin C, are more prone to forming kidney stones if they take megadoses of vitamin C.

Recommended Intakes of Vitamin C The vitamin C RDA is 90 milligrams for men and 75 milligrams for women. These amounts are far higher than the 10 milligrams per day needed to prevent the symptoms of scurvy. In fact, they are close to the amount at which the body's pool of vitamin C is full to overflowing: about 100 milligrams per day.

Doses of 10 to 30 or more times the recommended intake of a nutrient are termed *megadoses.* In the case of vitamin C, current recommendations are 75 mg/day for women and 90 mg/day for men. The Tolerable Upper Intake Level for vitamin C is 2000 mg/day.

The anticoagulants with which vitamin C interferes are warfarin and dicumarol.

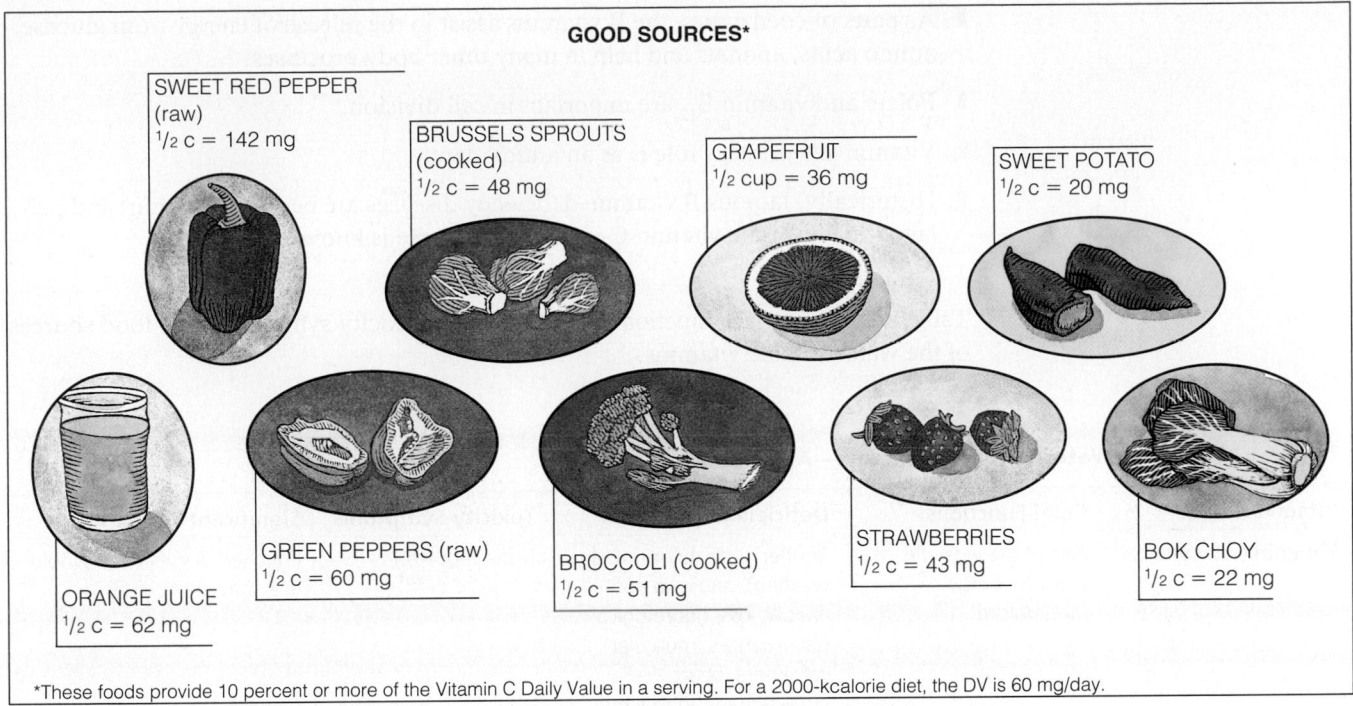

GOOD SOURCES*

SWEET RED PEPPER (raw)
¹/₂ c = 142 mg

BRUSSELS SPROUTS (cooked)
¹/₂ c = 48 mg

GRAPEFRUIT
¹/₂ cup = 36 mg

SWEET POTATO
¹/₂ c = 20 mg

ORANGE JUICE
¹/₂ c = 62 mg

GREEN PEPPERS (raw)
¹/₂ c = 60 mg

BROCCOLI (cooked)
¹/₂ c = 51 mg

STRAWBERRIES
¹/₂ c = 43 mg

BOK CHOY
¹/₂ c = 22 mg

*These foods provide 10 percent or more of the Vitamin C Daily Value in a serving. For a 2000-kcalorie diet, the DV is 60 mg/day.

SNAPSHOT 8-3 **Vitamin C**

Special Needs for Vitamin C As is true of all nutrients, unusual circumstances may raise vitamin C needs. Among the stresses known to do so are infections; burns; surgery; extremely high or low temperatures; toxic doses of heavy metals, such as lead, mercury, and cadmium; and the chronic use of certain medications, including aspirin, barbiturates, and oral contraceptives. Smoking, too, has adverse effects on vitamin C status. Cigarette smoke contains oxidants, which deplete this potent antioxidant. Accordingly, the vitamin C recommendation for smokers is set high, at 125 milligrams for men and 105 milligrams for women.

Vitamin C in Foods The inclusion of intelligently selected fruits and vegetables in the daily diet guarantees a generous intake of vitamin C. Even those who wish to ingest amounts well above the RDA can easily meet their goals by eating certain foods (see Snapshot 8-3). Citrus fruits are rightly famous for their vitamin C contents. Certain other fruits and vegetables are also rich sources: cantaloupe, strawberries, broccoli, and brussels sprouts. No animal foods other than organ meats, such as chicken liver and kidneys, contain vitamin C. The humble potato is an important source of vitamin C in Western countries, where potatoes are eaten so frequently that they make substantial vitamin C contributions overall. They provide about 20 percent of all the vitamin C in the average diet. Vitamin C in foods is easily oxidized, so store cut produce and juices in airtight containers.

Vitamin C and Iron Absorption Eating foods containing vitamin C at the same meal with foods containing iron can double or triple the absorption of iron from those foods. This strategy is highly recommended for women and children, whose energy intakes are not large enough to guarantee that they will get enough iron from the foods they eat.

Iron is discussed in Chapter 9.

IN SUMMARY

▌ The B vitamins and vitamin C are the water-soluble vitamins.

▌ Each B vitamin is part of an enzyme helper known as a coenzyme.

gout (GOWT): a metabolic disease in which crystals of uric acid precipitate in the joints.

- As parts of coenzymes, the B vitamins assist in the release of energy from glucose, amino acids, and fats and help in many other body processes.

- Folate and vitamin B_{12} are important in cell division.

- Vitamin C's primary role is as an antioxidant.

- Historically, famous B vitamin–deficiency diseases are beriberi (thiamin) and pellagra (niacin). The vitamin C–deficiency disease is known as scurvy.

Table 8-3 summarizes functions, deficiency and toxicity symptoms, and food sources of the water-soluble vitamins.

TABLE 8-3 The Water-Soluble Vitamins—A Summary

Vitamin Name	Chief Functions	Deficiency Symptoms	Toxicity Symptoms	Significant Sources
Thiamin (Vitamin B_1)	Part of a coenzyme used in energy metabolism	Beriberi (edema or muscle wasting), anorexia and weight loss, neurological disturbances, muscular weakness, heart enlargement and failure	None reported	Enriched, fortified, or whole-grain products; pork
Riboflavin (Vitamin B_2)	Part of coenzymes used in energy metabolism	Inflammation of the mouth, skin, and eyelids; sensitivity to light; sore throat	None reported	Milk products; enriched, fortified, or whole-grain products; liver
Niacin (Nicotinic acid, nicotinamide, niacinamide, vitamin B_3; precursor is dietary tryptophan, an amino acid)	Part of coenzymes used in energy metabolism	Pellagra (diarrhea, dermatitis, and dementia)	Niacin flush, liver damage, impaired glucose tolerance	Milk, eggs, meat, poultry, fish, whole-grain and enriched breads and cereals, nuts, and all protein-containing foods
Biotin	Part of a coenzyme used in energy metabolism	Skin rash, hair loss, neurological disturbances	None reported	Widespread in foods; GI bacteria synthesis
Pantothenic acid	Part of a coenzyme used in energy metabolism	Digestive and neurological disturbances	None reported	Widespread in foods
Vitamin B_6 (Pyridoxine, pyridoxal, pyridoxamine)	Part of coenzymes used in amino acid and fatty acid metabolism	Scaly dermatitis, depression, confusion, convulsions, anemia	Nerve degeneration, skin lesions	Meats, fish, poultry, potatoes, legumes, noncitrus fruits, fortified cereals, liver, soy products
Folate (Folic acid, folacin, pteroylglutamic acid)	Activates vitamin B_{12}; helps synthesize DNA for new cell growth	Anemia; smooth, red tongue; mental confusion; elevated homocysteine	Masks vitamin B_{12} deficiency	Fortified grains, leafy green vegetables, legumes, seeds, liver
Vitamin B_{12} (Cobalamin)	Activates folate; helps synthesize DNA for new cell growth; protects nerve cells	Anemia; nerve damage and paralysis	None reported	Foods derived from animals (meat, fish, poultry, shellfish, milk, cheese, eggs), fortified cereals
Vitamin C (Ascorbic acid)	Synthesis of collagen, carnitine, hormones, neurotransmitters; antioxidant	Scurvy (bleeding gums, pinpoint hemorrhages, abnormal bone growth, and joint pain)	Diarrhea, GI distress	Citrus fruits, cabbage-type vegetables, dark green vegetables (such as bell peppers and broccoli), cantaloupe, strawberries, lettuce, tomatoes, potatoes, papayas, mangoes

Your Diet

Vitamins

A diet that supplies nutrient-dense foods from each food group can provide ample vitamins. To learn whether your food choices include good sources of vitamins, insert your recommended amounts of each food group/subgroup listed in the table below using information from Tables 1-4 and 1-6 (pp. 18 and 19). Next, estimate your intake for each food group (you can use the food lists generated for previous diet projects to help you assess your intakes). In the third column, calculate the percentages of recommended intakes you consumed (your intake ÷ suggested intake × 100). Circle the percentages that are less than 100 percent.

■ Do you eat an adequate amount of fruits? Citrus fruits offer abundant vitamin C and some folate. Some deep orange fruits such as cantaloupe (not oranges) supply vitamin A.

■ Do you choose whole or enriched grains often? These choices supply substantial amounts of thiamin, riboflavin, niacin, and folate.

■ Do you consume vitamin A- and D-fortified milk products regularly? Milk products supply riboflavin and vitamins A, D, and B_{12}.

■ Do you make frequent choices from the vegetable subgroups? Most dark green vegetables provide significant folate and

Food Choices	Suggested Daily Intake	Your Daily Intake	% of Recommended Intake
Fruits	_____ c	_____ c	_____ %
Grains	_____ oz	_____ oz	_____ %
Milk and milk products	_____ c	_____ c	_____ %

Vegetable Subgroups	Suggested Weekly Intake	Your Weekly Intake	
Dark green vegetables	_____ c	_____ c	_____ %
Orange and deep yellow	_____ c	_____ c	_____ %
Legumes	_____ c	_____ c	_____ %

vitamins A and E. Many deep orange vegetables supply vitamin A. Legumes supply substantial folate.

■ If you take supplements, list the vitamin content of each supplement. Then compare your intake with the UL for vitamins listed on the first page of this book.

Clinical Applications

1. How might a vitamin deficiency weaken a client's resistance to disease?

2. Pull together information from Chapter 1 about the different food groups and the significant sources of vitamins shown in the photos and snapshots throughout this chapter. Consider which vitamins might be lacking in the diet of a client who reports the following:

■ Dislikes leafy, green vegetables

■ Never uses milk, milk products, or cheese

■ Follows a very low-fat diet

■ Eats a fruit or vegetable once a day

What additional information would help you pinpoint problems with vitamin intake?

Self Check

1. Which of the following vitamins are fat soluble?
   a. Vitamins B, C, and E
   b. Vitamins B, C, D, and E
   c. Vitamins A, C, E, and K
   d. Vitamins A, D, E, and K

2. Which of the following describes fat-soluble vitamins?
   a. They include thiamin, vitamin A, and vitamin K.
   b. They cannot be stored to any great extent and so must be consumed daily.
   c. Toxic levels can be reached by consuming citrus fruits and vegetables.
   d. They can be stored in the liver and fatty tissues and can build up toxic concentrations.

3. Night blindness and susceptibility to infection are the result of a deficiency of which vitamin?
   a. Niacin
   b. Vitamin C
   c. Vitamin A
   d. Vitamin K

4. Good sources of vitamin D include:
   a. eggs, fortified milk, and sunlight.
   b. citrus fruits, sweet potatoes, and spinach.
   c. leafy, green vegetables; cabbage; and liver.
   d. breast milk, polyunsaturated plant oils, and citrus fruits.

5. Which of the following describes water-soluble vitamins?
   a. They include vitamins D and E.
   b. They are frequently toxic.
   c. They are stored extensively in tissues.
   d. They are easily absorbed and excreted.

6. A coenzyme is:
   a. a fat-soluble vitamin.
   b. an energy-yielding nutrient.
   c. a source of vitamin K.
   d. a molecule that combines with an enzyme to make it active.

7. Good food sources of folate include:
   a. citrus fruits, dairy products, and eggs.
   b. liver; legumes; and leafy, green vegetables.
   c. dark green vegetables, corn, and cabbage.
   d. potatoes, broccoli, and whole-wheat bread.

8. Which vitamin is present only in foods of animal origin?
   a. Riboflavin
   b. Pantothenic acid
   c. Vitamin B_{12}
   d. The inactive form of vitamin A

9. Which of the following nutrients is an antioxidant that protects water-soluble substances from oxidizing agents?
   a. Beta-carotene
   b. Thiamin
   c. Vitamin C
   d. Vitamin D

10. Eating foods containing vitamin C at the same meal can increase the absorption of which mineral?
    a. Iron
    b. Calcium
    c. Magnesium
    d. Folate

Answers to these questions can be found in Appendix H.

Notes

1. S. M. Ahmad and coauthors, Markers of innate immune function are associated with vitamin A stores in men, *Journal of Nutrition* 139 (2009): 377–385; M. A. Marzinke and coauthors, Calmin expression in embryos and the adult brain, and its regulation by all-trans retinoic acid, *Developmental Dynamics* 239 (2009): 610–619; M. A. Muhit and coauthors, Causes of severe visual impairment and blindness in Bangladesh: A study of 1935 children, *British Journal of Ophthalmology* 91 (2007): 1000–1004; A. C. Ross, Vitamin A and carotenoids, in M. E. Shils and coeditors, *Modern Nutrition in Health and Disease*, 10th ed. (Philadelphia: Lippincott, Williams, & Wilkins, 2006), pp. 351–375.

2. R. Blomhoff and H. K. Blomhoff, Overview of retinoid metabolism and function, *Journal of Neurobiology* 66 (2006): 606–630.

3. Blomhoff and Blomhoff, 2006.

4. G. Wolf, Retinoic acid as cause of cell proliferation or cell growth inhibition depending on activation of one of two different nuclear receptors, *Nutrition Reviews* 66 (2008): 55–59.

5. Ahmad and coauthors, 2009; C. Trottier and coauthors, Retinoids inhibit measles virus through a type I IFN-dependent bystander effect, *FASEB Journal* 23 (2009): 3203–3212; K. Z. Long and coauthors, Impact of vitamin A on selected gastrointestinal pathogen infections and associated diarrheal episodes among children in Mexico City, Mexico, *Journal of Infectious Diseases* 194 (2006): 1217–1225.

6. N. Noy, Between death and survival: Retinoic acid in regulation of apoptosis, *Annual Review of Nutrition* 30 (2010): 201–217; W. Checkley and coauthors, Maternal vitamin A supplementation and lung function in offspring, *New England Journal of Medicine* 362 (2010): 1784–1794.

7. E. Herrera and coauthors, Aspects of antioxidant foods and supplements in health and disease, *Nutrition Reviews* 67 (2009): S140–S144; L. C. Yong and coauthors, High dietary antioxidant intakes are associated with decreased chromosome translocation frequency in airline pilots, *American Journal of Clinical Nutrition* 90 (2009): 1402–1410.

8. Herrera and coauthors, 2009; J. K. Shen and coauthors, Oxidative damage in age-related macular degeneration, *Histology and Histopathology* 12 (2007): 1301–1308.

9. D. R. Jacobs, M. D. Gross, and L. C. Tapsell, Food synergy: An operational concept for understanding nutrition, *American Journal of Clinical Nutrition* 89 (2009): 1543S–1548S.

10. A. Sommer, Vitamin A deficiency and clinical disease: An historical overview, *Journal of Nutrition* 138 (2008): 1835–1839.

11. World Health Organization, Micronutrient deficiencies: Vitamin A deficiency, available at www.who.int/topics/vat/en, accessed February 3, 2010; Standing Committee on the Scientific Evaluation of Dietary Reference Intakes, *Dietary Reference Intakes for Vitamin A, Vitamin D, Arsenic, Boron, Chromium, Copper, Iodine, Iron, Manganese, Molybdenum, Nickel, Silicon, Vanadium, and Zinc* (Washington, D.C.: National Academy Press, 2001), pp. 95–97.

12. R. D. Semba and coauthors, The role of expanded coverage of the national vitamin A program in preventing morbidity and mortality among preschool children in India, *Journal of Nutrition* 140 (2010): 208S–212S; S. A. Abrams and D. C. Hilmers, Postnatal vitamin A supplementation in developing countries: An intervention whose time has come? *Pediatrics* 122 (2008): 180–181; K. Kraemer and coauthors, Are low tolerable upper intake levels for vitamin A undermining effective food fortification efforts? *Nutrition Reviews* 66 (2008): 517–525.

13. World Health Organization, Measles, available at www.who.int/mediacentre/factsheets, accessed February 3, 2010.

14. A. Sheth, R. Khurana, and V. Khurana, Potential liver damage associated with over-the-counter vitamin supplements, *Journal of the American Dietetic Association* 108 (2008): 1536–1537; K. L. Penniston and S. A. Tanumihardjo, The acute and chronic toxic effects of vitamin A, *American Journal of Clinical Nutrition* 83 (2006): 191–201.

15. H. S. Lam and coauthors, Risk of vitamin A toxicity from candy-like chewable vitamin supplements for children, *Pediatrics* 118 (2006): 820–824.

16. Standing Committee on the Scientific Evaluation of Dietary Reference Intakes, 2001, pp. 128–129.

17. J. Zhao and coauthors, Retinoic acid downregulates microRNAs to induce abnormal development of spinal cord in spina bifida rat model, *Child's Nervous System* 24 (2008): 485–492.

18. Standing Committee on the Scientific Evaluation of Dietary Reference Intakes, 2001, pp. 126–133.

19. J. D. Ribaya-Mercado and J. B. Blumberg, Vitamin A: Is it a risk factor for osteoporosis and bone fracture? *Nutrition Reviews* 65 (2007): 425–438; Penniston and Tanumihardjo, 2006.

20. T. Tanvetyanon and G. Bepler, Beta-carotene in multivitamins and the possible risk of lung cancer among smokers versus former smokers: A meta-analysis and evaluation of national brands, *Cancer* 113 (2008): 150–157.

21. A. W. Norman, From vitamin D to hormone D: Fundamentals of the vitamin D endocrine system essential for good health, *American Journal of Clinical Nutrition* 88 (2008): 491S–499S.

22. Norman, 2008.

23. S. Samual and M. D. Sitrin, Vitamin D's role in cell proliferation and differentiation, *Nutrition Reviews* 66 (2008): S116–S124.

24. E. van Etten and coauthors, Regulation of vitamin D homeostasis: Implications for the immune system, *Nutrition Reviews* 66 (2008): S125–S134.

25. S. Cheng and coauthors, Adiposity, cardiometabolic risk, and vitamin status: The Framingham heart study, *Diabetes* 59 (2010): 242–248; A. Kilkkinen and coauthors, Vitamin D status and the risk of cardiovascular death, *American Journal of Epidemiology* 170 (2009): 1032–1039; C. F. Garland and coauthors, Vitamin D for cancer prevention: Global perspective, *Annals of Epidemiology* 19 (2009): 468–483; A. Zitterman, J. Gummert, and J. Börgermann, Vitamin D deficiency and mortality, *Current Opinion in Clinical Nutrition and Metabolic Care* 12 (2009): 634–639; R. P. Heaney, Vitamin D and calcium interactions: Functional outcomes, *American Journal of Clinical Nutrition* 88 (2008): 541S–544S; M. T. Cantorna, Vitamin D and multiple sclerosis: An update, *Nutrition Reviews* 66 (2008): S135–S138; M. F. Holick, Vitamin D: A d-lightful health perspective, *Nutrition Reviews* 66 (2008): S182–S194; S. E. Judd and coauthors, Optimal vitamin D status attenuates the age-associated increase in systolic blood pressure in white Americans: Results from the third National Health and Nutrition Examination Survey, *American Journal of Clinical Nutrition* 87 (2008): 136–141; J. M. Lappe and coauthors, Vitamin D and calcium supplementation reduces cancer risk: Results of a randomized trial, *American Journal of Clinical Nutrition* 85 (2007): 1586–1591; J. Lin and coauthors, Intakes of calcium and vitamin D and breast cancer risk in women, *Archives of Internal Medicine* 167 (2007): 1050–1059; G. E. Mullin and A. Dobs, Vitamin D and its role in cancer and immunity: A prescription for sunlight, *Nutrition in Clinical Practice* 22 (2007): 305–322; Y. Cui and T. E. Rohan, Vitamin D, calcium, and breast cancer risk: A review, *Cancer Epidemiological, Biomarkers and Prevention* 15 (2006): 1427–1437; A. F. Gombart, Q. T. Luong, and H. P. Koeffler, Vitamin D compounds: Activity against microbes and cancer, *Anticancer Research* 26 (2006): 2531–2542; P. T. Liu and coauthors, Toll-like receptor triggering of a vitamin D-mediated human antimicrobial response, *Science* 311 (2006): 1770–1773; L. A. Martini and R. J. Wood, Vitamin D status and the metabolic syndrome, *Nutrition Reviews* 64 (2006): 479–486; K. L. Munger and coauthors, Serum 25-hydroxyvitamin D levels and risk of multiple sclerosis, *Journal of the American Medical Association* 296 (2006): 2832–2838; S. S. Schleithoff and coauthors, Vitamin D supplementation improves cytokine profiles in patients with congestive heart failure: A double-blind, randomized, placebo-controlled trial, *American Journal of Clinical Nutrition* 83 (2006): 754–759.

26. Committee on Dietary Reference Intakes, *Dietary Reference Intakes for Calcium and Vitamin D* (Washington, D.C.: National Academies Press, 2011), p. 3-3.

27. R. P Heaney, Vitamin D and calcium interactions: Functional outcomes, *American Journal of Clinical Nutrition* 88 (2008): 541S–544S.

28. Heaney, 2008.

29. H. F. Deluca, Evolution of our understanding of vitamin D, *Nutrition Reviews* 66 (2008): S73–S87; M. R. Haussler and coauthors, Vitamin D receptor: Molecular signaling and actions of nutritional ligands in disease prevention, *Nutrition Reviews* 66 (2008): S98–S112.

30. R. C. Khanal and I. Nemere, Regulation of intestinal calcium transport, *Annual Review of Nutrition* 28 (2008): 179–196.

31. A. C. Looker and coauthors, Serum 25-hydroxyvitamin D status of the US population: 1988–1994 compared with 2000–2004, *American Journal of Clinical Nutrition* 88 (2008): 1519–1527; S. A. Bowden and coauthors, Prevalence of vitamin D deficiency and insufficiency in children with osteopenia or osteoporosis referred to a pediatric metabolic bone clinic, *Pediatrics* 121 (2008): e1585–e1590; M. L. Neuhouser and coauthors, Vitamin D insufficiency in a multiethnic cohort of breast cancer survivors, *American Journal of Clinical Nutrition* 88 (2008): 133–139.

32. A. Moshfegh and coauthors, *What We Eat in America, NHANES 2005–2006: Usual Nutrient Intakes from Food and Water Compared to 1997 Dietary Reference Intakes for Vitamin D, Calcium, Phosphorus, and Magnesium,* (Beltsville, MD: USDA 2009), available at www.ars.usda.gov/ba/bhnrc/fsrg.

33. A. Prentice, Vitamin D deficiency: A global perspective, *Nutrition Reviews* 66 (2008): S153–S164.

34. J. Kumar and coauthors, Prevalence and associations of 25-hydroxyvitamin D deficiency in US children: NHANES 2001–2004, *Pediatrics* 124 (2009): e362–e370.

35. S. Saintonge, H. Bang, and L. M. Gerber, Implications of a new definition of vitamin D deficiency in a multiracial US adolescent population: The National Health and Nutrition Examination Survey III, *Pediatrics* 123 (2009): 797–803; F. R. Greer, 25-Hydroxyvitamin D: Functional outcome in infants and young children, *American Journal of Clinical Nutrition* (2008): 529S–533S; S. Y. Huh and C. M. Gordon, Vitamin D deficiency in children and adolescents: Epidemiology, impact and treatment, *Reviews in Endocrine and Metabolic Disorders* 9 (2008): 161–170; M. Misra and coauthors, Vitamin D deficiency in children and its management: Review of current knowledge and recommendations, *Pediatrics* 122 (2008): 398–417.

36. K. D. Cashman and coauthors, Low vitamin D status adversely affects bone health parameters in adolescents, *American Journal of Clinical Nutrition* 87 (2008): 1039–1044.

37. H. A. Bischoff-Ferrari and coauthors, Prevention of nonvertebral fractures with oral vitamin D and dose dependency: A meta-analysis of randomized controlled trials, *Archives of Internal Medicine* 169 (2009): 551–561; G. Buhr and C. W. Bales, Nutritional supplements for older adults: Review and recommendations—part I, *Journal of Nutrition for the Elderly* 28 (2009): 5–29; A. Avenell and coauthors, Vitamin D and vitamin D analogues for preventing fractures associated with involutional and post-menopausal osteoporosis, *Cochrane Database of Systematic Reviews* 2 (2009), CD000227; B. Dawson-Hughes, Serum 25-hydroxyvitamin D and functional outcomes in the elderly, *American Journal of Clinical Nutrition* 88 (2008): 537S–540S; S. A. Talwar and coauthors, Dose response to vitamin D supplementation among postmenopausal African American women, *American Journal of Clinical Nutrition* 86 (2007): 1657–1662.

38. R. P. Heaney, Vitamin D: Criteria for safety and efficacy, *Nutrition Reviews* 66 (2008): S178–S181.

39. Committee on Dietary Reference Intakes, 2011.

40. F. Ghissassi and coauthors, A review of human carcinogens—Part D: Radiation, *The Lancet* 10 (2009): 751–752; B. A. Gilchrest, Sun exposure and vitamin D sufficiency, *American Journal of Clinical Nutrition* 88 (2008): 570S–577S.

41. Gilchrest, 2008.

42. A. Prentice, F. R. Goldberg, and I. Schoenmakers, Vitamin D across the lifecycle: Physiology and biomarkers, *American Journal of Clinical Nutrition* 88 (2008): 500S–506S; M. J. Bolland and coauthors, The effects of seasonal variation of 25-hydroxyvitamin D and fat mass on a diagnosis of vitamin D sufficiency, *American Journal of Clinical Nutrition* 86 (2007): 959–964.

43. E. Hyppönen and C. Power, Hypovitaminosis D in British adults at age 45 y: Nationwide cohort study of dietary and lifestyle predictors, *American Journal of Clinical Nutrition* 85 (2007): 860–868; F. L. Weng and coauthors, Risk factors for low serum 25-hydroxyvitamin D concentrations in otherwise healthy children and adolescents, *American Journal of Clinical Nutrition* 86 (2007): 150–158.

44. K. D. Cashman and coauthors, Estimation of the dietary requirement for vitamin D in healthy adults, *American Journal of Clinical Nutrition* 88 (2008): 1535–1542.

45. C. L. Wagner, F. R. Greer, and the Section on Breastfeeding and Committee on Nutrition, Prevention of rickets and vitamin D deficiency in infants, children, and adolescents, *Pediatrics* 122 (2008): 1142-1152.

46. Committee on Dietary Reference Intakes, 2011.

47. D. L. Rainwater and coauthors, Vitamin E dietary supplementation significantly affects multiple risk factors for cardiovascular disease in baboons, *American Journal of Clinical Nutrition* 86 (2007): 597–603; U. Singh, S. Devaraj, and I. Jialal, Vitamin E, oxidative stress, and inflammation, *Annual Review of Nutrition* 25 (2005): 151–174.

48. B. J. Wilcox, J. D. Curb, and B. L. Rodriguez, Antioxidants in cardiovascular health and disease: Key lessons from epidemiologic studies, *American Journal of Cardiology* 101 (2009): 75D–86D.

49. A. H. Lichtenstein and coauthors, Diet and lifestyle recommendations revision 2006: A scientific statement from the American Heart Association Nutrition Committee, *Circulation* 114 (2006): 82–96.

50. J. N. Hathcock and coauthors, Vitamins E and C are safe across a broad range of intakes, *American Journal of Clinical Nutrition* 81 (2005): 736–745.

51. G. Bjelakovic and coauthors, Antioxidant supplements for prevention of mortality in healthy participants and patients with various diseases, *Cochrane Database of Systematic Reviews* April 16, 2008: CD007176.

52. G. Pocobelli and coauthors, Use of supplements of multivitamins, vitamin C and vitamin E in relation to mortality, *American Journal of Epidemiology* 170 (2009): 472–483.

53. H. M. Macdonald and coauthors, Vitamin K$_1$ intake is associated with higher bone mineral density and reduced bone resorption in early postmenopausal Scottish women: No evidence of gene-nutrient interaction with apolipoprotein E polymorphisms, *American Journal of Clinical Nutrition* 87 (2008): 1513–1520; S. Cockayne and coauthors, Vitamin K and the prevention of fractures: Systematic review and meta-analysis of randomized controlled trials, *Archives of Internal Medicine* 166 (2006): 1256–1261.

54. Cockayne and coauthors, 2006; J. Iwamoto, T. Takeda, and Y. Sato, Menatetrenone (vitamin K2) and bone quality in the treatment of postmenopausal osteoporosis, *Nutrition Reviews* 64 (2006): 509–517; K. D. Cashman, Vitamin K status may be an important determinant of childhood bone health, *Nutrition Reviews* 63 (2005): 284–293.

55. K. D. Cashman and E. O'Connor, Does high vitamin K$_1$ intake protect against bone loss in later life? *Nutrition Reviews* 66 (2008): 532–538; M. K. Shea and S. L. Booth, Update on the role of vitamin K in skeletal health, *Nutrition Reviews* 66 (2008): 549–557.

56. M. K. Shea and coauthors, Vitamin K supplementation and progression of coronary artery calcium in older men and women, *American Journal of Clinical Nutrition* 89 (2009): 1799–1807.

57. C. Bourgeois, D. Cercantes-Laurean, and J. Moss, Niacin, in M. E. Shils and coeditors, *Modern Nutrition in Health and Disease*, 10th ed. (Philadelphia: Lippincott, Williams, & Wilkins, 2006), pp. 426–433.

58. Y. L. Hassan and J. Zempleni, A novel, enigmatic histone modification: Biotinylation of histones by holocarboxylase synthetase, *Nutrition Reviews* 66 (2008): 721–725; J. Zempleni, Uptake, localization, and noncarboxylase roles of biotin, *Annual Review of Nutrition* 25 (2005): 175–196.

59. J. Shen and coauthors, Association of vitamin B-6 status with inflammation, oxidative stress, and chronic inflammatory conditions: The Boston Puerto Rican Health Study, *American Journal of Clinical Nutrition* 91 (2010): 337–342.

60. E. S. Wintergerst, S. Maggini, and D. H. Hornig, Contribution of selected vitamins and trace elements to immune function, *Annals of Nutrition and Metabolism* 51 (2007): 301–323; A. D. Mackey, S. R. Davis, and J. F. Gregory, Vitamin B$_6$, in M. E. Shils and coeditors, *Modern Nutrition in Health and Disease*, 10th ed. (Philadelphia: Lippincott, Williams, & Wilkins, 2006), pp. 452–461.

61. L. L. Humphrey and coauthors, Homocysteine level and coronary heart disease incidence: A systematic review and meta-analysis, *Mayo Clinic Proceedings* 83 (2008): 3–16.

62. K. Woolf and M. M. Manore, Elevated plasma homocysteine and low vitamin B-6 status in nonsupplementing older women with rheumatoid arthritis, *Journal of the American Dietetic Association* 108 (2008): 443–453; J. Selhub, Public health significance of elevated homocysteine, *Food and Nutrition Bulletin* 29 (2008): S116–S125.

63. J. H. Page and coauthors, Plasma B(6) and risk of myocardial infarction in women, *Circulation* 120 (2009): 649–655; A. J. Marti-Carvajal and coauthors, Homocysteine lowering interventions for preventing cardiovascular events, *Cochrane Database of Systematic Reviews* October 7, 2009: CD006612; M. Ebbing and coauthors, Mortality and cardiovascular events in patients treated with homocysteine-lowering B vitamins after coronary angiography: A randomized controlled trial, *Journal of the American Medical* Association 300 (2008): 795–804.

64. R. Carmel, Folic acid, in M. E. Shils and coeditors, *Modern Nutrition in Health and Disease*, 10th ed. (Philadelphia: Lippincott, Williams, & Wilkins, 2006), pp. 470–481.

65. R. L. Bailey and coauthors, Total folate and folic acid intake from foods and dietary supplements in the United States: 2003–2006, *American Journal of Clinical Nutrition* 91 (2010): 231–237.

66. P. D. Wals and coauthors, Reduction in neural-tube defects after folic acid fortification in Canada, *New England Journal of Medicine* 357 (2007): 135–142.

67. M. Ebbing and coauthors, Cancer incidence and mortality after treatment with folic acid and vitamin B$_{12}$, *Journal of the American Medical Association* 302 (2009): 2119–2126; U. C. Ericson and coauthors, Increased breast cancer risk at high plasma folate concentrations among women with the MTHFR 677T allele, *American Journal of Clinical Nutrition* 90 (2009): 1380–1389; A. M. Troen and coauthors, Unmetabolized folic acid in plasma is associated with reduced natural killer cell cytotoxicity among postmenopausal women, *Journal of Nutrition* 136 (2006): 189–194.

68. Bailey and coauthors, 2010; Q. Yang and coauthors, Folic acid source, usual intake, and folate and vitamin B-12 status in US adults: National Health and Nutrition Examination Survey (NHANES) 2003–2006, *American Journal of Clinical Nutrition* 91 (2010): 64–72.

69. J. W. Miller and coauthors, Metabolic evidence of vitamin B-12 deficiency, including high homocysteine and methylmalonic acid and low holotranscobalamin, is more pronounced in older adults with elevated plasma folate, *American Journal of Clinical Nutrition* 90 (2009): 1586–1592.

70. S. Park and M. A. Johnson, What is an adequate dose of oral vitamin B$_{12}$ in older people with poor vitamin B$_{12}$ status? *Nutrition Reviews* 64 (2006): 383–388.

71. I. Elmadfa and I. Singer, Vitamin B-12 and homocysteine status among vegetarians, *American Journal of Clinical Nutrition* 89 (2009): 1693S–1698S.

72. Carmel, 2006.

73. World Cancer Research Fund and American Institute for Cancer Research, *Food, Nutrition, Physical Activity, and the Prevention of Cancer: A Global Perspective* (Washington, D.C.: American Institute for Cancer Research, 2007), pp. 253–258.

Nutrition in Practice PHYTOCHEMICALS AND FUNCTIONAL FOODS

The wisdom of the familiar advice, "Eat your vegetables; they're good for you," stands on firmer scientific ground today than ever before as population studies around the world suggest that diets rich in vegetables and fruits protect against heart disease, cancer, and other chronic diseases.[1] We now know that the "goodness" of vegetables, fruits, and other whole foods such as legumes and grains comes not only from the nutrients they contain but also from the **phytochemicals** that they offer.[2] Phytochemicals often act as **bioactive food components**, food constituents with the ability to alter body processes. (Terms are defined in Table NP8-1.)

Vegetables, fruits, and other whole foods are the simplest examples of foods now known as **functional foods.** Functional foods provide health benefits beyond basic nutrition by altering one or more physiological processes. Modified foods, such as those that have been fortified, enriched, or enhanced with nutrients, phytochemicals, herbs, or other food components, also are functional foods.[3] Functional foods that fit this description include orange juice fortified with calcium, folate-enriched cereal, beverages with herbal additives, and margarine enhanced with **phytosterols.** This Nutrition in Practice begins with a look at the evidence concerning the effectiveness and safety of a few selected phytochemicals in the simplest of functional foods—vegetables, fruits, and other whole foods. Then, the discussion turns to examine the most controversial of functional foods—novel foods to which phytochemicals have been added to promote health. How these foods fit into a healthy diet is still unclear.[4]

What are phytochemicals, and what do they do?

Phytochemicals are bioactive compounds found in plants. In foods, phytochemicals impart tastes, aromas, colors, and other characteristics. They give hot peppers their burning sensation, garlic and onions their pungent flavor, chocolate its bitter tang, and tomatoes

TABLE NP8-1 Phytochemical and Functional Food Terms

antioxidants (anti-OX-ih-dants): compounds that protect other compounds from damaging reactions involving oxygen by themselves reacting with oxygen (*anti* means "against"; *oxy* means "oxygen"); *oxidation* is a potentially damaging effect of normal cell chemistry involving oxygen.

bioactive food components: compounds in foods, either nutrients or phytochemicals, that alter physiological processes. Also defined in Chapter 1.

carotenoids (kah-ROT-eh-noyds): pigments commonly found in plants and animals, some of which have vitamin A activity. The carotenoid with the greatest vitamin A activity is beta-carotene.

flavonoids (FLAY-von-oyds): yellow pigments in foods; phytochemicals that may exert physiological effects on the body. *Flavus* means "yellow."

flaxseed: small brown seed of the flax plant; valued in nutrition as a source of fiber, lignans, and the omega-3 fatty acid linolenic acid.

functional foods: whole or modified foods that contain bioactive food components believed to provide health benefits, such as reduced disease risks, beyond the benefits that their nutrients contribute. All whole foods are functional in some ways because they provide at least some needed substances, but certain foods stand out as rich sources of bioactive food components. Also defined in Chapter 1.

genistein (GEN-ih-steen): a phytoestrogen found primarily in soybeans that both mimics and blocks the action of estrogen in the body.

lignans: phytochemicals present in flaxseed, but not flaxseed oil, that are converted to phytoestrogens by intestinal bacteria and are under study as possible anticancer agents.

lutein (LOO-teen): a plant pigment of yellow hue; a phytochemical believed to play roles in eye functioning and health.

lycopene (LYE-koh-peen): a pigment responsible for the red color of tomatoes and other red-hued vegetables; a phytochemical that may act as an antioxidant in the body.

organosulfur compounds: a large group of phytochemicals containing the mineral sulfur. Organosulfur phytochemicals are responsible for the pungent flavors and aromas of foods belonging to the onion, leek, chive, shallot, and garlic family and are thought to stimulate cancer defenses in the body.

phytochemicals (FIGH-toe-CHEM-ih-cals): compounds in plants that confer color, taste, and other characteristics. Some phytochemicals are bioactive food components in functional foods. Also defined in Chapter 1.

phytoestrogens (FIGHT-toe-ESS-troh-gens): phytochemicals structurally similar to human estrogen. Phytoestrogens weakly mimic or modulate estrogen in the body.

phytosterols: phytochemicals that resemble cholesterol in structure but that lower blood cholesterol by interfering with cholesterol absorption in the intestine. Phytosterols include sterol esters and stanol esters.

resveratrol (rez-VER-ah-trol): a flavonoid of grapes under study for potential health benefits.

tofu: white curd made of soybeans, popular in Asian cuisines, and considered to be a functional food.

their dark red color. In the body, phytochemicals can have profound physiological effects—acting as antioxidants, mimicking hormones, stimulating or inhibiting enzymes, interfering with DNA replication, destroying bacteria, and binding physically to cell walls. Any of these actions may suppress the development of diseases.[5]

Notably, cancer and heart disease are linked to processes involving oxygen compounds in the body, and **antioxidants** are thought to oppose these actions. Table NP8-2 introduces the names, possible physiological effects, and food sources of some of the better-known phytochemicals.

TABLE NP8-2 **Phytochemicals—Possible Health Effects and Food Sources**

Chemical Name	Possible Effects	Food Sources
Alkylresorcinols (phenolic lipids)	May contribute to the protective effect of grains in reducing the risks of diabetes, heart disease, and some cancers	Whole-grain wheat and rye
Allicin (organosulfur compound)	Antimicrobial that may reduce ulcers; may lower blood cholesterol	Chives, garlic, leeks, onions
Capsaicin	Modulates blood clotting, possibly reducing the risk of fatal clots in heart and artery disease	Hot peppers
Carotenoids (include beta-carotene, lycopene, lutein, and hundreds of related compounds)	Act as antioxidants, possibly reducing risks of cancer and other diseases	Deeply pigmented fruits and vegetables (apricots, broccoli, cantaloupe, carrots, pumpkin, spinach, sweet potatoes, tomatoes)
Curcumin	Acts as an antioxidant and anti-inflammatory agent; may reduce blood clot formation; may inhibit enzymes that activate carcinogens	Tumeric, a yellow-colored spice
Flavonoids (include flavones, flavonols, isoflavones, catechins, and others)	Act as antioxidants; scavenge carcinogens; bind to nitrates in the stomach, preventing conversion to nitrosamines; inhibit cell proliferation	Berries, black tea, celery, citrus fruits, green tea, olives, onions, oregano, purple grapes, purple grape juice, soybeans and soy products, vegetables, whole wheat, wine
Genistein and daidzein (isoflavones)	Phytoestrogens that inhibit cell replication in GI tract; may reduce or elevate risk of breast, colon, ovarian, prostate, and other estrogen-sensitive cancers; may reduce cancer cell survival; may reduce risk of osteoporosis	Soybeans, soy flour, soy milk, tofu, textured vegetable protein, other legume products
Indoles (organosulfur compound)	May trigger production of enzymes that block DNA damage from carcinogens; may inhibit estrogen action	Cruciferous vegetables such as broccoli, brussels sprouts, cabbage, cauliflower; horseradish, mustard greens, kale
Isothiocyanates (organosulfur compounds that include sulforaphane)	Act as antioxidants; inhibit enzymes that activate carcinogens; activate enzymes that detoxify carcinogens; may reduce risk of breast cancer, prostate cancer	Cruciferous vegetables such as broccoli, brussels sprouts, cabbage, cauliflower; horseradish, mustard greens, kale
Lignans	Phytoestrogens that block estrogen activity in cells, possibly reducing the risk of cancer of the breast, colon, ovaries, and prostate	Flaxseed, whole grains
Monoterpenes (including limonene)	May trigger enzyme production to detoxify carcinogens; inhibit cancer promotion and cell proliferation	Citrus fruit peels and oils
Phenolic acids	May trigger enzyme production to make carcinogens water soluble, facilitating excretion	Coffee beans, fruits (apples, blueberries, cherries, grapes, oranges, pears, prunes), oats, potatoes, soybeans
Phytic acid	Binds to minerals, preventing free-radical formation, possibly reducing cancer risk	Whole grains
Resveratrol	Acts as antioxidant; may inhibit cancer growth; reduces inflammation, LDL oxidation, and blood clot formation	Red wine, peanuts, grapes, raspberries
Saponins (glucosides)	May interfere with DNA replication, preventing cancer cells from multiplying; stimulate immune response	Alfalfa sprouts, other sprouts, green vegetables, potatoes, tomatoes
Tannins	Act as antioxidants; may inhibit carcinogen activation and cancer promotion	Black-eyed peas, grapes, lentils, red and white wine, tea

Why are phytochemicals receiving so much attention these days, and what are some examples of those in the spotlight?

Diets rich in whole grains, legumes, vegetables, and fruits seem to be protective against heart disease and cancer, but identifying *the* specific foods or components of foods that are responsible is difficult. Scientists are conducting extensive research studies to discover phytochemical connections to disease prevention, but, so far, solid evidence is generally lacking. Some of the likeliest candidates include **flavonoids** and **carotenoids** (including **lycopene**).

What are flavonoids, and in which foods are they found?

Flavonoids, a large group of phytochemicals known for their health-promoting qualities, are found in whole grains, soy, vegetables, fruits, herbs, spices, teas, chocolate, nuts, olive oil, and red wine. Flavonoids are powerful antioxidants that may help to protect LDL against oxidation, minimize inflammation, and reduce blood platelet stickiness, thereby slowing the progression of atherosclerosis and making blood clots less likely.[6] Whereas an abundance of flavonoid-containing *foods* in the diet may lower the risks of chronic diseases, no claims can be made for flavonoids themselves as the protective factor, particularly when they are extracted from foods and sold as supplements.

Flavonoids impart a bitter taste to foods, so manufacturers often refine away the natural flavonoids to please consumers, who usually prefer milder flavors. For example, to produce white grape juice or white wine, manufacturers remove the red, flavonoid-rich grape skins to lighten the flavor and color of the product, while greatly reducing its beneficial flavonoid content. Specifically, one flavonoid in purple grape juice and red wine, **resveratrol,** seems to hold promise as a disease fighter, but the amount present in wine or a serving of grape juice may be too small to benefit human health.[7]

What about carotenoids?

In addition to flavonoids, fruits and vegetables are rich in carotenoids—the red and yellow pigments of plants. Some carotenoids, such as beta-carotene, are vitamin A precursors. Some research suggests that a diet rich in carotenoids is associated with a lower risk of heart disease.[8] Among the carotenoids that may defend against heart disease is lycopene. Lycopene may also protect against certain types of cancer.

What is lycopene, and what foods contain it?

Lycopene is a red pigment with powerful antioxidant activity found in guava, papaya, pink grapefruit, tomatoes (especially cooked tomatoes and tomato products), and watermelon. More than 80 percent of the lycopene consumed in the United States comes from tomato products such as tomato sauce, tomato juice, and catsup. Around the world, people who eat five or more tomato-containing meals per week are less likely to suffer from cancers of the esophagus, prostate, or stomach than those who avoid tomatoes. Lycopene is a leading candidate for this protective effect.

Theoretically, the potent antioxidant capability of lycopene plays a role in its action against cancer, but, so far, research does not support the idea.[9] Tomatoes contain many other phytochemicals and nutrients that may contribute to the beneficial health effects of eating tomatoes and tomato products. An evidence-based review by the FDA concluded that little or no credible evidence exists to support an association between lycopene or tomato consumption and reduced cancer rates.[10]

Do foods contain other phytochemicals that may help to protect people from cancer or other diseases?

Foods contain thousands of different phytochemicals, and so far only a few have been researched at all. There are still many questions about the phytochemicals that have been studied and only tentative answers about their roles in human health. For example, compared with people in the West, Asians living in Asia suffer less frequently from cancers, especially of the breast, colon, and prostate; heart disease; and osteoporosis (adult bone loss). Among many differences between the diets of the two regions, Asians consume far more soybeans and soy products such as **tofu** than do Westerners.

Soybeans are rich sources of phytochemicals known as **phytoestrogens.** Researchers suspect that the phytoestrogens of soy foods, their protein content, or a combination of these factors may be responsible for health effects in soy-eating peoples. Nevertheless, research, although ongoing, is limited. So far, we know with certainty that phytoestrogens are plant-derived chemical relatives of the human hormone estrogen and that they weakly mimic or modulate the hormone's effects on some body tissues. With regard to cancer, concerns about breast cancer, colon cancer, and prostate cancer involve estrogen-sensitive varieties—cancers that grow when exposed to estrogen. The age of the person eating soy affects the results: a high soy intake during childhood and adolescence seems to reduce breast cancer risk in women before menopause; soy intake by adults may or may not reduce this risk, but more research is needed to clarify these relationships.[11] With regard to the biological activity of soy, scientific understanding is incomplete.[12]

Low doses of one soy phytoestrogen, **genistein,** appear to speed up division of breast cancer cells in laboratory cultures and in mice, whereas high doses seem to do the opposite.[13] Still under study is whether dietary soy phytoestrogens have similar effects on cancer cells in living people, but it seems unlikely that moderate intakes of soy foods would do so.[14] If they did, soy-eating cultures would have higher, not lower, incidences of these cancers.

The opposing actions of phytoestrogens should raise a red flag against taking supplements, especially by people who have had cancer or whose close relatives have developed cancer. The American Cancer Society recommends that breast cancer survivors and those under treatment for breast cancer should consume only moderate amounts of soy foods as part of a healthy plant-based diet and should not intentionally ingest very high levels of soy products.

Other foods under study for potential health benefits include **flaxseed** and its oil. Flaxseed is found as a whole seed or ground meal, or as flaxseed oil. Flaxseed is of interest for its possible benefits to heart health because it is a good source of soluble fiber, and it is the richest known source of both the omega-3 fatty acid linolenic

acid and **lignans**, compounds converted into biologically active phytoestrogens by bacteria that normally reside in the human intestine.[15] Flaxseed oil, though rich in linolenic acid, does not contain fiber or lignans. Large quantities of flaxseed can cause digestive distress, and severe allergic reactions to flaxseed have been reported.

What about other phytochemical supplements?

Even when people don't eat in the best interest of health, taking supplements of purified phytochemicals is not the way to go. Phytochemicals can alter body functions, sometimes powerfully. Researchers are just beginning to understand how a handful of phytochemicals work, and what is current today may change tomorrow. Foods deliver thousands of bioactive food components, all within a food matrix that maximizes their availability and effectiveness.[16] The best way to reap the benefits of phytochemicals is by eating foods, not taking supplements (see Figure NP8-1).

How do whole foods compare with processed foods that have been enriched with phytochemicals?

Good question. The U.S. food supply is being transformed by a proliferation of functional foods—foods claimed to provide health benefits beyond those of the traditional nutrients. Virtually all whole foods have some special value in supporting health and are therefore functional foods. Cranberries may protect against urinary tract infections because cranberries contain a phytochemical that dislodges bacteria from the tract.[17] Cooked tomatoes, as mentioned, provide lycopene, along with **lutein** (an antioxidant associated with healthy eye function), vitamin C (an antioxidant vitamin), and many other healthful attributes. This has not stopped food manufacturers from trying to create functional foods as well. As consumer demand for healthful foods continues to grow, so will the development of functional foods.[18]

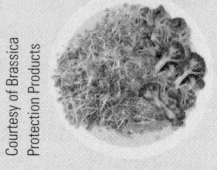

Broccoli and broccoli sprouts contain an abundance of the cancer-fighting phytochemical sulforaphane.

An apple a day—rich in flavonoids—may protect against lung cancer.

The phytoestrogens of soybeans seem to starve cancer cells and inhibit tumor growth; the phytosterols may lower blood cholesterol and protect cardiac arteries.

Garlic, with its abundant organosulfur compounds, may lower blood cholesterol and protect against stomach cancer.

The phytochemical resveratrol found in grapes (and nuts) protects against cancer by inhibiting cell growth and against heart disease by limiting clot formation and inflammation.

The ellagic acid of strawberries may inhibit certain types of cancer.

Tomatoes, with their abundant lycopene, may defend against cancer by protecting DNA from oxidative damage.

The monoterpenes of citrus fruits (and cherries) may inhibit cancer growth.

The flavonoids in black tea may protect against heart disease, whereas those in green tea may defend against cancer.

The flavonoids in cocoa and chocolate defend against oxidation and reduce the tendency of blood to clot.

Spinach and other colorful vegetables contain the carotenoids lutein and zeaxanthin, which help protect the eyes against macular degeneration.

Flaxseed, the richest source of lignans, may prevent the spread of cancer.

Blueberries, a rich source of flavonoids, improve memory in animals.

FIGURE NP8-1 An Array of Phytochemicals in a Variety of Fruits and Vegetables

What are some examples of manufactured functional foods?

Many processed foods become functional foods when they are fortified with nutrients (calcium-fortified orange juice to support bone health, for example). In other cases, processed foods are enhanced with bioactive food components (margarine blended with a phytosterol to lower cholesterol, for example). The creation of some novel manufactured functional foods raises the question—is it a food or a drug?[19]

Isn't the distinction between a food and a drug pretty clear?

Not too long ago, most of us could agree on what was a food and what was a drug. Today, functional foods blur the distinctions. They have characteristics similar to both foods and drugs but do not fit neatly into either category. Consider the margarine example above.

Eating nonhydrogenated margarine sparingly instead of butter generously may lower blood cholesterol slightly over several months and clearly falls into the food category. Taking a statin drug, on the other hand, lowers blood cholesterol significantly within weeks and clearly falls into the drug category. But margarine enhanced with a phytosterol that lowers blood cholesterol is in a gray area between the two.* The margarine looks and tastes like a food, but it acts like a drug.

What about using sterol-enhanced margarines to lower cholesterol or other manufactured functional foods to help prevent cancer?

To achieve a desired health effect, which is the better choice: to eat a food designed to affect some body function or simply to adjust the diet? Does it make more sense to use a margarine enhanced with a phytosterol that lowers blood cholesterol or simply to limit the amount of butter eaten? Is it smarter to eat eggs enriched with omega-3 fatty acids or to restrict egg consumption? Might functional foods offer a sensible solution for improving our nation's health—if done correctly? Perhaps so—but there is a problem with functional foods: the food industry is moving too fast for either scientists or the Food and Drug Administration (FDA) to keep up. Consumers were able to buy soup with St. John's wort that claimed to enhance mood and fruit juice with echinacea that was supposed to fight colds while scientists were still conducting their studies on these ingredients. Research to determine the safety and effectiveness of these substances is still in progress. Until this work is complete, consumers are on their own in finding the answers to the following questions:

▌ *Does it work?* Research is generally lacking, and findings are often inconclusive.

▌ *How much does it contain?* Food labels are not required to list the quantities of added phytochemicals. Even if they were, consumers have no standard for comparison and cannot deduce whether the amounts listed are a little or a lot. Most importantly, until research is complete, food manufacturers do not know what amounts (if any) are most effective—or most toxic.

*Margarine products that lower blood cholesterol contain either sterol esters—from vegetable oils, soybeans, and corn—or stanol esters from wood pulp.

▌ *Is it safe?* Functional foods can act like drugs. They contain ingredients that can alter body functions and cause allergies, drug interactions, drowsiness, and other side effects. Yet, unlike drug labels, food labels do not provide instructions for the dosage, frequency, or duration of treatment.

▌ *Has the FDA issued warnings about any of the ingredients?* Check the FDA's website (www.fda.gov) to find out.

▌ *Is it healthy?* Adding phytochemicals to a food does not magically make it a healthy choice. A candy bar may be fortified with phytochemicals, but it is still made mostly of sugar and fat.

Critics suggest that the designation "functional foods" may be nothing more than a marketing tool. After all, even the most experienced researchers cannot yet identify the perfect combination of nutrients and phytochemicals to support optimal health. Yet manufacturers are freely experimenting with various concoctions as if they possessed that knowledge.

What is the final word regarding phytochemicals and functional foods?

Nature has elegantly designed foods to provide us with a complex array of dozens of nutrients and thousands of additional compounds that may benefit health—most of which we have yet to identify or understand. Over the years, we have taken those foods and first deconstructed them and then reconstructed them in an effort to "improve" them. With new scientific understandings of how nutrients—and the myriad of other compounds in foods—interact with genes, we may someday be able to design foods to meet the *exact* health needs of *each* individual. Indeed, our knowledge of the human genome and of human nutrition may well merge to allow specific recommendations for individuals based on their predisposition to diet-related diseases.

If the present trend continues, physicians may someday be able to prescribe the perfect foods to enhance a person's health, and farmers will be able to grow them. In the meantime, however, it seems clear that a moderate approach to phytochemicals and functional foods is warranted. People who eat the recommended amounts of a variety of fruits and vegetables may cut their risk of many diseases by as much as half. Replacing some meat with soy

Functional foods currently on the market promise to "enhance mood," "promote relaxation and good karma," "increase alertness," and "improve memory," among other claims.

221

foods and other legumes may also lower heart disease and cancer risks. Beneficial constituents are widespread among foods. Take a no-nonsense approach where your health is concerned: choose a wide variety of whole grains, legumes, fruits, and vegetables in the context of an adequate, balanced, and varied diet, and receive all of the health benefits that these foods offer.

Notes

1. S. S. Gidding and coauthors, Implementing American Heart Association pediatric and adult nutrition guidelines: A scientific statement from the American Heart Association Nutrition Committee of the Council on Nutrition, Physical Activity and Metabolism, Council on Cardiovascular Disease in the Young, Council on Arteriosclerosis, Thrombosis, and Vascular Biology, Council on Cardiovascular Nursing, Council on Epidemiology and Prevention, and Council for High Blood Pressure Research, *Circulation* 119 (2009): 1161–11745; F. Sofi, The Mediterranean diet revisited: Evidence of its effectiveness grows, *Current Opinion in Cardiology* 24 (2009): 442–446; J. A. Nettleton and coauthors, Dietary patterns and incident cardiovascular disease in the Multi-Ethnic Study of Atherosclerosis, *American Journal of Clinical Nutrition* 90 (2009): 647–654; F. J. B. van Duijnhoven and coauthors, Fruit, vegetables, and colorectal cancer risk: The European Prospective Investigation into Cancer and Nutrition, *American Journal of Clinical Nutrition* 89 (2009): 1441–1452; C. La Vecchia, Association between Mediterranean dietary patterns and cancer risk, *Nutrition Reviews* 67 (2009): S126–S129.

2. N. Khan, F. Afaq, and H. Mukhtar, Cancer chemoprevention through dietary antioxidants: Progress and promise, *Antioxidants and Redox Signaling* 10 (2008): 475–510; M. Rhone and A. Basu, Phytochemicals and age-related eye diseases, *Nutrition Reviews* 66 (2008): 465–472; C. C. Udenigwe and coauthors, Potential of resveratrol in anticancer and anti-inflammatory therapy, *Nutrition Reviews* 66 (2008): 445–454.

3. Position of the American Dietetic Association: Functional foods, *Journal of the American Dietetic Association* 109 (2009): 735–746.

4. G. Williamson and coauthors, Functional foods for health promotion: State-of-the-science on dietary flavonoids. Extended abstracts from the 12th Annual Conference on Functional Foods for Health Promotion, April, 2009, *Nutrition Reviews* 67 (2009): 736–743; L. R. Ferguson, Nutrigenomics approaches to functional foods, *Journal of the American Dietetic Association* 109 (2009): 452–458.

5. S. Jew, S. S. AbulMweis, and P. J. Jones, Evolution of the human diet: Linking our ancestral diet to modern functional foods as a means of chronic disease prevention, *Journal of Medicinal Foods* 12 (2009): 925–934; M. Kaur, C. Agarwal, and R. Agarwal, Anticancer and cancer chemopreventive potential of grape seed extract and other grape-based products, *Journal of Nutrition* 139 (2009): 1806S–1812S; G. Riccioni and coauthors, Protective effect of lycopene in cardiovascular disease, *European Review for Medical and Pharmacological Sciences* 12 (2008): 183–190.

6. Williamson and coauthors, 2009; M. Monagas and coauthors, Effect of cocoa powder on the modulation of inflammatory biomarkers in patients at high risk of cardiovascular disease, *American Journal of Clinical Nutrition* 90 (2009): 1144–1150; R. di Giuseppe and coauthors, Regular consumption of dark chocolate is associated with low serum concentrations of C-reactive protein in a healthy Italian population, *Journal of Nutrition* 138 (2008): 1939–1945; I. Erlund and coauthors, Favorable effects of berry consumption on platelet function, blood pressure, and HDL cholesterol, *American Journal of Clinical Nutrition* 87 (2008): 323–331; L. Hooper and coauthors, Flavonoids, flavonoid-rich foods, and cardiovascular risk: A meta-analysis of randomized controlled trials, *American Journal of Clinical Nutrition* 88 (2008): 38–50.

7. S. C. Forester and A. L. Waterhouse, Metabolites are key to understanding health effects of wine polyphenolics, *Journal of Nutrition* 138 (2009): 1824S–1831S; M. D. Knutson and C. Leeuwenburgh, Resveratrol and novel potent activators of SIRT1: Effects on aging and age-related diseases, *Nutrition Reviews* 66 (2008): 591–596.

8. A. Hozawa and coauthors, Circulating carotenoid concentrations and incident hypertension: The Coronary Artery Risk Development in Young Adults (CARDIA) Study, *Journal of Hypertension* 27 (2009): 237–242; G. Riccioni, Carotenoids and cardiovascular disease, *Current Atherosclerosis Reports* 11 (2009): 434–439; G. Riccioni and coauthors, Protective effect of lycopene in cardiovascular disease, *European Review for Medical and Pharmacological Sciences* 12 (2008): 183–190.

9. J. R. Mein, F. Lian, and X-D Wang, Biological activity of lycopene metabolites: Implications for cancer prevention, *Nutrition Reviews* 66 (2008): 667–683; U. Peters and coauthors, Serum lycopene, other carotenoids, and prostate cancer risk: A nested case-control study in the Prostate, Lung, Colorectal, and Ovarian Cancer Screening Trial, *Cancer Epidemiology Biomarkers and Prevention* 16 (2007): 962–968.

10. C. J. Kavanaugh, P. R. Trumbo, and K. E. Ellwood, The U.S. Food and Drug Administration's evidence-based review for qualified health claims: Tomatoes, lycopene, and cancer, *Journal of the National Cancer Institute* 99 (2007): 1047–1085.

11. S-A. Lee and coauthors, Adolescent and adult soy food intake and breast cancer risk: Results from the Shanghai Women's Health Study, *American Journal of Clinical Nutrition* 89 (2009): 1920–1926; M. Messina and A. H. Wu, Perspectives on the soy–breast cancer relation, *American Journal of Clinical Nutrition* 89 (2009): 1673S–1679S; A. Warri and coauthors, The role of early life genistein exposures in modifying breast cancer risk, *British Journal of Cancer* 98 (2008): 1485–1493.

12. J. W. Lampe, Is equal the key to the efficacy of soy foods? *American Journal of Clinical Nutrition* 89 (2009): 1664S–1667S.

13. Y. Zhang, and coauthors, Soy isoflavones and their bone protective effects, *Inflammopharmacology* 16 (2008): 213–215.

14. C. K. Taylor and coauthors, The effect of genistein aglycone on cancer and cancer risk: A review of in vitro, preclinical, and clinical studies, *Nutrition Reviews* 67 (2009): 398–415.

15. A. Pan and coauthors, Meta-analysis of the effects of flaxseed interventions on blood lipids, *American Journal of Clinical Nutrition* 90 (2009): 288–297.

16. D. R. Jacobs, M. D. Gross, and L. C. Tapsell, Food synergy: An operational concept for understanding nutrition, *American Journal of Clinical Nutrition* 89 (2009): 1543S–1548S.

17. D. R. Guay, Cranberry and urinary tract infections, *Drugs* 69 (2009): 775–807; J. Jass and G. Reid, Effect of cranberry drink on bacterial adhesion in vitro and vaginal microbiota in healthy females, *Canadian Journal of Urology* 16 (2009): 4901–4907.

18. Position of the American Dietetic Association, 2009.

19. P. J. Jones and K. A. Varady, Are functional foods redefining nutritional requirements? *Applied Physiology, Nutrition, and Metabolism* 33 (2008): 118–123.

9

Water and the Minerals

224

The body's water cannot be considered separately from the minerals dissolved in it. A person can drink pure water, but in the body, water mingles with minerals to become fluids in which all life processes take place. This chapter begins by discussing the body's fluids and their chief minerals. The focus then shifts to other functions of the minerals.

Water and Body Fluids

Water is the most indispensable nutrient of all.

Water constitutes about 60 percent of an adult's body weight and a higher percentage of a child's. Every cell in the body is bathed in a fluid of the exact composition that is best for that cell. The body fluids bring to each cell the ingredients it requires and carry away the end products of the life-sustaining reactions that take place within the cell's boundaries. The water in the body fluids:

▌ Carries nutrients and waste products throughout the body.

▌ Maintains the structure of large molecules such as proteins and glycogen.

▌ Participates in metabolic reactions.

▌ Serves as the solvent for minerals, vitamins, amino acids, glucose, and many other small molecules.

▌ Maintains blood volume.

▌ Aids in the regulation of normal body temperature, as the evaporation of sweat from the skin removes excess heat from the body.

▌ Acts as a lubricant and cushion around joints and inside the eyes, spinal cord, and amniotic sac surrounding a fetus in the womb.

To support these and other vital functions, the body actively regulates its **water balance.**

Water Balance

The cells themselves regulate the composition and amounts of fluids within and surrounding them. The entire system of cells and fluids remains in a delicate but firmly maintained state of dynamic equilibrium. Imbalances such as **dehydration** (see Table 9-1 on p. 226) and **water intoxication** can occur, but the body quickly restores the balance to normal if it can. The body controls both water intake and water excretion to maintain water equilibrium.

Water Intake Regulation The body can survive for only a few days without water. In healthy people, thirst and satiety govern water intake.[1] Thirst is finely adjusted to ensure a water intake that meets the body's needs. When the blood becomes too concentrated (having lost water but not salt and other dissolved substances), the mouth becomes dry, and the brain center known as the **hypothalamus** initiates drinking behavior.

Thirst lags behind the lack of water. A water deficiency that develops slowly can switch on drinking behavior in time to prevent serious dehydration, but a deficiency that develops quickly may not. Also, thirst itself does not remedy a water deficiency; a person must pay attention to the thirst signal and take the time to get a drink. With aging, thirst sensations may diminish. Dehydration can threaten elderly people who do not develop the habit of drinking water regularly.

Water Excretion Regulation Water excretion is regulated by the brain and the kidneys. The cells of the brain's hypothalamus, which monitor blood salts, stimulate the **pituitary gland** to release **antidiuretic hormone (ADH)** whenever the salts are too

water balance: the balance between water intake and water excretion that keeps the body's water content constant.

dehydration: the loss of water from the body that occurs when water output exceeds water input. The symptoms progress rapidly from thirst, to weakness, to exhaustion and delirium, and end in death if not corrected.

water intoxication: the rare condition in which body water contents are too high. The symptoms may include confusion, convulsion, coma, and even death in extreme cases.

hypothalamus (high-poh-THALL-uh-mus): a part of the brain that helps regulate many body balances, including fluid balance.

pituitary (pit-TOO-ih-tary) **gland:** in the brain, the "king gland" that regulates the operation of many other glands.

antidiuretic hormone (ADH): a hormone released by the pituitary gland in response to high salt concentrations in the blood. The kidneys respond by reabsorbing water.

TABLE 9-1 Signs of Mild and Severe Dehydration

Mild Dehydration (Loss of <5% Body Weight)	Severe Dehydration (Loss of >5% Body Weight)
Thirst	Pale skin
Sudden weight loss	Bluish lips and fingertips
Rough, dry skin	Confusion; disorientation
Dry mouth, throat, body linings	Rapid, shallow breathing
Rapid pulse	Weak, rapid, irregular pulse
Low blood pressure	Thickening of blood
Lack of energy; weakness	Shock; seizures
Impaired kidney function	Coma; death
Reduced quantity of urine; concentrated urine	
Decreased mental functioning	
Decreased muscular work and athletic performance	
Fever or increased internal temperature	
Fainting	

Source: Standing Committee on the Scientific Evaluation of Dietary Reference Intakes, Food and Nutrition Board, Institute of Medicine, *Dietary Reference Intakes: Water, Potassium, Sodium, Chloride, and Sulfate* (Washington, D.C.: National Academies Press, 2005), pp. 90–122.

concentrated or the blood volume or blood pressure is too low. ADH stimulates the kidneys to reabsorb water rather than excrete it. Thus, the more water you need, the less you excrete.

If too much water is lost from the body, blood volume and blood pressure fall. Cells in the kidneys respond to the low blood pressure by releasing an enzyme. Through a complex series of events involving the hormone **aldosterone,** this enzyme also causes the kidneys to retain more water. Again, the effect is that, when more water is needed, less is excreted.

Minimum Water Needed These mechanisms can maintain water balance only if a person drinks enough water. The body must excrete a minimum of about 500 milliliters each day as urine—enough to carry away the waste products generated by a day's metabolic activities. Above this amount, excretion adjusts to balance intake, so the more a person drinks, the more diluted the urine becomes. In addition to urine, some water is lost from the lungs as vapor, some is excreted in feces, and some evaporates from the skin. A person's water losses from all of these routes total about 2½ liters (about 2½ quarts) a day on the average. Table 9-2 shows how fluid intake and output naturally balance out.

The enzyme renin (REN-in), released by the kidneys in response to low blood pressure, aids the kidneys in retaining water through the renin–angiotensin mechanism.

500 mL = about ½ qt

aldosterone (al-DOS-ter-own): a hormone secreted by the adrenal glands that stimulates the reabsorption of sodium by the kidneys; also regulates chloride and potassium concentrations.

salts: compounds composed of charged particles (ions). An example of a salt is potassium chloride (K+Cl−).

electrolyte: a salt that dissolves in water and dissociates into charged particles called ions.

electrolyte solutions: solutions that can conduct electricity.

TABLE 9-2 Water Balance

Water Sources	Amount (mL)	Water Losses	Amount (mL)
Liquids	550 to 1500	Kidneys (urine)	500 to 1400
Foods	700 to 1000	Skin (sweat)	450 to 900
Metabolic water	200 to 300	Lungs (breath)	350
		GI tract (feces)	150
Total	1450 to 2800	Total	1450 to 2800

Water Recommendations and Sources Water needs vary greatly depending on the foods a person eats, the environmental temperature and humidity, the person's activity level, and other factors. Accordingly, a general water requirement is difficult to establish. In the past, recommendations for adults were expressed in proportion to the amount of energy expended under normal environmental conditions. For the person who expends about 2000 kcalories a day, this works out to 2 to 3 liters, or about 8 to 12 cups. This recommendation is in line with the Adequate Intake (AI) for *total* water set by the DRI committee. Total water includes not only drinking water but also water in other beverages and in foods as well.[2]

Because a wide range of water intakes will prevent dehydration and its harmful consequences, the AI is based on average intakes. Strenuous physical activity and heat stress can increase water needs considerably, however.[3] In general, you can tell from the color of the urine whether a person needs more water. Pale yellow urine reflects appropriate dilution.[4]

The obvious dietary sources of water are water itself and other beverages, but nearly all foods also contain water. Water constitutes up to 95 percent of the volume of most fruits and vegetables and at least 50 percent of many meats and cheeses. The energy nutrients in foods also yield water during metabolism.

Which beverages are best? Any beverage can readily meet the body's fluid needs, but those with few or no kcalories do so without contributing to weight gain. Given that obesity is a major health problem and that beverages contribute more than 20 percent of the total energy intake in the United States, water is the best choice for most people.[5]

People often ask whether caffeine-containing beverages such as coffee, tea, or soda can help to meet water needs. When people who normally abstain from caffeine drink a caffeine-containing beverage, their urine output increases somewhat more than it would for a similar amount of water. This occurs because caffeine acts as a mild diuretic. Research is mixed on whether any but the highest caffeine intakes (four or five cups of coffee) cause a net water deficit in the body; most people make up for small water losses by drinking additional fluid later. The DRI committee considered such findings in making its recommendations for water intake and concluded that "caffeinated beverages contribute to the daily total water intake similar to that contributed by non-caffeinated beverages."[6] In other words, it doesn't seem to matter whether people rely on caffeine-containing beverages or other beverages to meet their fluid needs.

In contrast, alcohol should not be used to meet fluid needs. As Nutrition in Practice 20 explains, alcohol acts as a diuretic, and it has many adverse effects on health and nutrition status.

Fluid and Electrolyte Balance

When mineral **salts** dissolve in water, they separate (dissociate) into charged particles known as ions, which can conduct electricity. For this reason, a salt that dissociates in water is known as an **electrolyte.** The body fluids, which contain water and partly dissociated salts, are **electrolyte solutions.**

The body's electrolytes are vital to cell survival and therefore must be closely regulated to help maintain the appropriate distribution of body fluids. The major minerals form salts that dissolve in the body fluids; the cells direct where these salts go; and the movement of the salts determines where the fluids flow because water follows salt. Cells use this force to move fluids back and forth across their membranes. Thanks to the electrolytes, water can be held in compartments where it is needed.

Proteins in the cell membranes move ions into or out of the cells. These protein pumps tend to concentrate sodium and chloride outside cells and potassium and other ions inside. By maintaining specific amounts of sodium outside and potassium

AI for *total* water:
- Men: 3.7 L/day
- Women: 2.7 L/day

Percentage of water in selected foods:
- 90–99%: Fat-free milk, cantaloupe, grapefruit, strawberries, watermelon, lettuce, tomato, cabbage, celery, spinach, squash, cucumber
- 80–89%: Fruit juice, yogurt, whole milk, apples, grapes, oranges, carrots, broccoli (cooked), pears, pineapple
- 70–79%: Avocados, bananas, corn, potatoes, cottage cheese, ricotta cheese, shrimp
- 60–69%: Pasta, legumes, salmon, lean steak, chicken breast, ice cream
- 50–59%: Ground beef, hot dogs, pork chop, feta cheese
- 40–49%: Pizza, cheeseburger
- 30–39%: Cheddar cheese, bagels, bread
- 20–29%: Pepperoni sausage, cake, biscuits
- 10–19%: Butter, margarine, raisins
- 1–9%: Crackers, ready-to-eat cereals, pretzels, taco shells, peanut butter, nuts
- 0%: Oils, white sugar

Exceptions: A compound in which the positive ions are hydrogen ions (H^+) is an acid (example: hydrochloric acid, or H^+Cl^-); a compound in which the negative ions are hydroxyl ions (OH^-) is a base (example: potassium hydroxide, or K^+OH^-).

The simple statement that water follows salt describes the force that chemists call *osmosis.*

Water follows salt. Notice the beads of "sweat" formed on the right-hand slices of eggplant, which were sprinkled with salt. Cellular water moves across each cell's membrane (water-permeable divider) toward the higher concentration of salt (dissolved particles) on the surface.

inside, cells can regulate the exact amounts of water inside and outside their boundaries.

Healthy kidneys regulate the body's sodium, as well as its water, with remarkable precision. The intestinal tract absorbs sodium readily, and it travels freely in the blood, but the kidneys excrete unneeded amounts. The kidneys actually filter all of the sodium out of the blood; then, they return to the bloodstream the exact amount the body needs to retain. Thus, the body's total electrolytes remain constant, while the urinary electrolytes fluctuate according to what is eaten.

In some cases, the body's mechanisms for maintaining fluid and electrolyte balances cannot compensate for a sudden loss of large amounts of fluid and electrolytes. Vomiting, diarrhea, heavy sweating, fever, burns, wounds, and the like may incur great fluid and electrolyte losses, precipitating an emergency that demands medical intervention.

The body's responses to severe stress and trauma are discussed in Chapter 24.

Acid–Base Balance

The body uses ions not only to help maintain water balance but also to regulate the acidity (pH) of its fluids. Like proteins, electrolyte mixtures in the body fluids protect the body against changes in acidity by acting as **buffers**—substances that can accommodate excess acids or bases.

The body's buffer systems serve as a first line of defense against changes in the fluids' acid–base balance. The lungs, skin, gastrointestinal (GI) tract, and kidneys provide other defenses. Of these organ systems, the kidneys play the primary role in maintaining acid–base balance. Thus, disorders of the kidneys impair the body's ability to regulate its acid–base balance, as well as its fluid and electrolyte balances.

IN SUMMARY

▌ Water makes up about 60 percent of the body's weight.

▌ Water helps transport nutrients and waste products throughout the body, participates in metabolic reactions, acts as a solvent, assists in maintaining blood volume and body temperature, acts as a lubricant and cushion around joints, and serves as a shock absorber.

▌ To maintain water balance, intake from liquids, foods, and metabolism must equal losses from kidneys, skin, lungs, and feces.

▌ Electrolytes help maintain the appropriate distribution of body fluids and help to maintain acid–base balance as well.

TABLE 9-3 The Major and Trace Minerals

Major Minerals	Trace Minerals
Calcium	Arsenic
Chloride	Boron
Magnesium	Chromium
Phosphorus	Cobalt
Potassium	Copper
Sodium	Fluoride
Sulfur	Iodine
	Iron
	Manganese
	Molybdenum
	Nickel
	Selenium
	Silicon
	Vanadium
	Zinc

The Major Minerals

Table 9-3 lists the major and trace minerals in the body, and Figure 9-1 shows the amounts found in the body. As you can see, the most prevalent minerals are calcium and phosphorus, the chief minerals of bone. The distinction between the **major minerals** and the **trace minerals** does not mean that one group is more important than the other. A deficiency of the few micrograms of iodine needed daily is just as serious as a deficiency of the several hundred milligrams of calcium. The major minerals are so named because they are present, and needed, in larger amounts in the body than the trace minerals.

Although all the major minerals influence the body's fluid balance, sodium, chloride, and potassium are most noted for that role. For this reason, those three minerals are discussed first. Each major mineral also plays other specific roles in the body. Sodium, potassium, calcium, and magnesium are critical to nerve transmission and muscle contractions. Phosphorus and magnesium are involved in energy metabolism. Calcium, phosphorus, and magnesium contribute to the structure of the bones. Sulfur helps determine the shape of proteins. Table 9-6, shown later in the chapter (p. 238), provides a summary of information about the major minerals.

Sodium

Sodium is the principal electrolyte in the **extracellular fluid** (the fluid outside the cells) and the primary regulator of the extracellular fluid volume. When the blood concentration of sodium rises, as when a person eats salted foods, thirst prompts the person to drink water until the appropriate sodium-to-water ratio is restored. Sodium also helps maintain acid–base balance and is essential to muscle contraction and nerve transmission. Too much sodium, however, can contribute to high blood pressure.

Sodium Recommendations and Food Sources Diets rarely lack sodium, and even when intakes are low, the body adapts by reducing sodium losses in urine and sweat, thus making deficiencies unlikely. Sodium recommendations are set low enough to protect against high blood pressure but high enough to allow an adequate intake of other nutrients. Because high sodium intakes correlate with high blood pressure, the Tolerable Upper Intake Level (UL) for adults is set at 2300 milligrams per day, slightly lower than the Daily Value used on food labels (2400 milligrams). The UL corresponds to about 1 teaspoon of salt (sodium chloride).

Today, the average U.S. sodium intake is more than 3400 mg per day, an amount that far exceeds the UL.[7] Three groups of people encompassing about 70 percent of U.S. adults—persons with hypertension, all middle-aged and older adults, and all African Americans—are urged to limit their sodium intakes (to no more than 1500 milligrams of sodium per day) for the sake of their blood pressure and heart health.[8]

Sodium intakes also vary widely. People who eat mostly processed and fast foods have the highest sodium intakes, whereas those who eat mostly whole, unprocessed

Sodium AI:
- 1500 mg/day (19–50 yr)
- 1300 mg/day (51–70 yr)
- 1200 mg/day (>70 yr)

Salt (sodium chloride) is about 40 percent sodium.
1 g salt contributes 400 mg sodium.
6 g salt = 1 tsp.
1 tsp salt contributes about 2300 mg sodium.

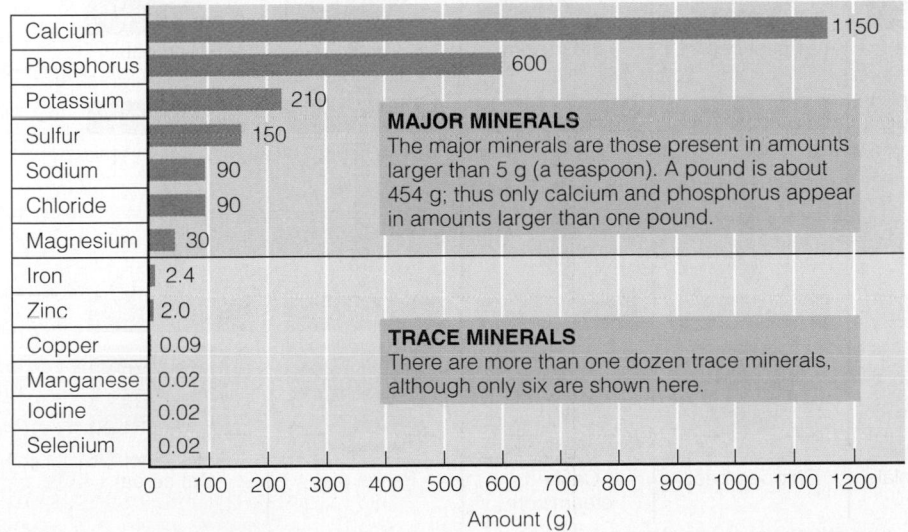

Mineral	Amount (g)
Calcium	1150
Phosphorus	600
Potassium	210
Sulfur	150
Sodium	90
Chloride	90
Magnesium	30
Iron	2.4
Zinc	2.0
Copper	0.09
Manganese	0.02
Iodine	0.02
Selenium	0.02

MAJOR MINERALS
The major minerals are those present in amounts larger than 5 g (a teaspoon). A pound is about 454 g; thus only calcium and phosphorus appear in amounts larger than one pound.

TRACE MINERALS
There are more than one dozen trace minerals, although only six are shown here.

FIGURE 9-1 Amounts of Minerals in a 60-kilogram (132-pound) Human Body
Not only are the major minerals present in the body in larger amounts than the trace minerals but they are also needed by the body in large amounts. Recommended intakes for the major minerals are stated in *hundreds of milligrams* or *grams*, whereas those for the trace minerals are listed in *tens of milligrams* or even *micrograms*.

buffers: compounds that can reversibly combine with hydrogen ions to help keep a solution's acidity or alkalinity constant. Also defined in Chapter 4.

major minerals: essential mineral nutrients required in the adult diet in amounts greater than 100 milligrams per day.

trace minerals: essential mineral nutrients required in the adult diet in amounts less than 100 milligrams per day.

extracellular fluid: fluid residing outside the cells; includes the fluid between the cells (*interstitial fluid*), plasma, and the water of structures such as the skin and bones. Extracellular fluid accounts for about one-third of the body's water.

foods, such as fresh fruits and vegetables, have the lowest intakes. In fact, about three-fourths of the sodium in people's diets comes from salt added to foods by manufacturers. Figure 9-2 shows that processed foods contain not only more sodium but also less potassium than their less processed counterparts.

Chapter 22 presents more details about blood pressure and lists standard ranges for blood pressure readings.

Sodium and Blood Pressure High intakes of salt among the world's people correlate with high rates of **hypertension,** heart disease, and cerebral hemorrhage, a hypertension-related stroke.[9] An estimated 29 percent of U.S. adults have hypertension (that is, high blood pressure), and another 28 percent have **prehypertension.**[10]

In many people, the relationship between salt intake and blood pressure is immediate and direct—as sodium intakes increase, blood pressure rises with them in a stepwise fashion within just days or weeks of exposure.[11] And with high blood pressure, the heart becomes damaged and risks of death from stroke and heart disease climb steeply. In a large 15-year intervention study, when people with prehypertension reduced their sodium intakes, they reduced their risk of developing heart disease or dying from it by 25 to 30 percent.[12]

In fact, a salt-restricted diet lowers blood pressure in people without hypertension as well. Because reducing salt intake causes no harm and diminishes the risk of hypertension and heart disease, the *Dietary Guidelines for Americans* advise limiting daily *sodium* intakes to less than 2300 milligrams (approximately 1 teaspoon of *salt*). Higher intakes seem to be well tolerated in most healthy people, however. The accompanying "How to" offers strategics for cutting salt (and, therefore, sodium) intake.

Chapter 22 offers a complete discussion of hypertension and the dietary recommendations for its prevention and treatment.

One diet plan known as the DASH (Dietary Approaches to Stop Hypertension) diet also lowers blood pressure.[13] The DASH approach emphasizes fruits, vegetables, and low-fat dairy products; includes whole grains, nuts, poultry, and fish; and calls for reduced intakes of red meat, butter, and other high-fat foods. The DASH diet in combination with a reduced sodium intake is even more effective at lowering blood pressure than either strategy alone.

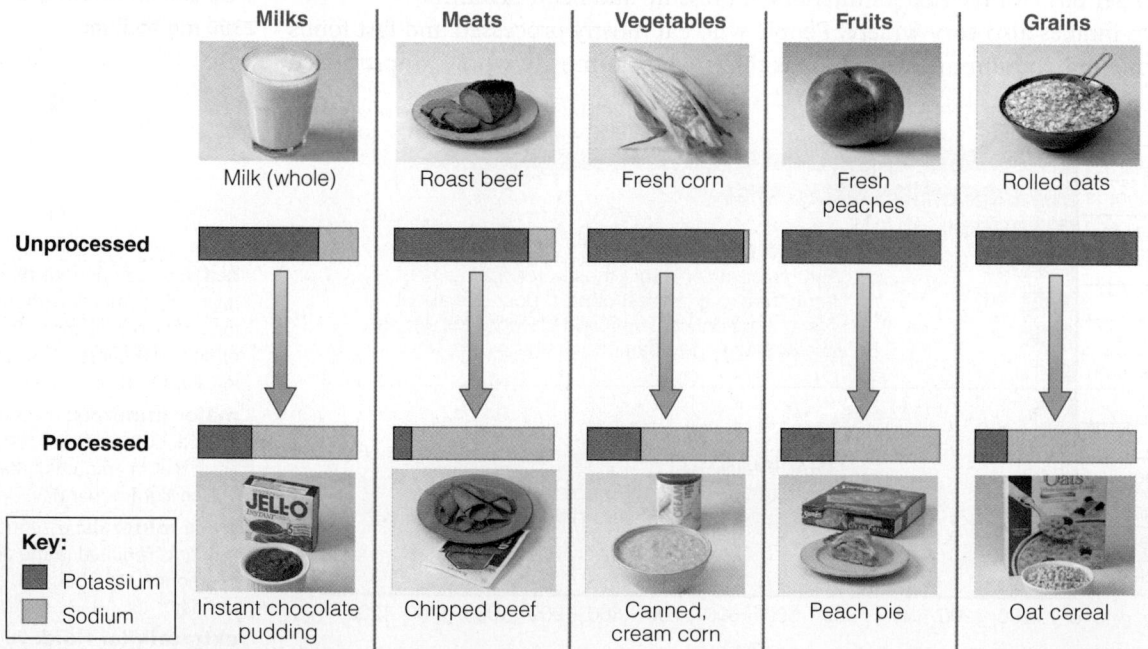

FIGURE 9-2 What Processing Does to the Sodium and Potassium Contents of Foods
People who eat foods high in salt often happen to be eating fewer potassium-containing foods at the same time. Notice how potassium is lost and sodium is gained as foods become more processed, causing the potassium-to-sodium ratio to fall dramatically. Even when potassium is not lost, the addition of sodium still lowers the potassium-to-sodium ratio.

CHAPTER NINE

How To CUT SALT INTAKE

Most people eat more salt (and, therefore, sodium) than they need. Some people can lower their blood pressure by avoiding highly salted foods and removing the saltshaker from the table. Foods eaten without salt may seem less tasty at first, but with repetition, people can learn to enjoy the natural flavors of many unsalted foods. Strategies to cut salt intake include:

- Select fresh, unprocessed foods.
- Cook with little or no added salt.
- Prepare foods with sodium-free spices such as basil, bay leaves, curry, garlic, ginger, mint, oregano, pepper, rosemary, and thyme; lemon juice; vinegar; or wine.
- Add little or no salt at the table; taste foods before adding salt.
- Read labels with an eye open for sodium. (See Table 1-9 on p. 25 for terms used to describe the sodium contents of foods on labels.)
- Select low-salt or salt-free products when available.

Use these foods sparingly:

- Foods prepared in brine, such as pickles, olives, and sauerkraut.
- Salty or smoked meats, such as bologna, corned or chipped beef, bacon, frankfurters, ham, lunch meats, salt pork, sausage, and smoked tongue.
- Salty or smoked fish, such as anchovies, caviar, salted and dried cod, herring, sardines, and smoked salmon.
- Snack items such as potato chips, pretzels, salted popcorn, salted nuts, and crackers.
- Condiments such as bouillon cubes; seasoned salts; MSG; soy, teriyaki, Worcestershire, and barbeque sauces; prepared horseradish, catsup, and mustard.
- Cheeses, especially processed types.
- Canned and instant soups.

Chloride

The chloride ion is the major negative ion of the extracellular fluids, where it occurs primarily in association with sodium. Like sodium, chloride is critical to maintaining fluid, electrolyte, and acid–base balances in the body. In the stomach, the chloride ion is part of hydrochloric acid, which maintains the strong acidity of the gastric fluids.

Salt is a major food source of chloride, and, as with sodium, processed foods are a major contributor of this mineral to people's diets. Because salt contains a higher proportion of chloride (by weight) than sodium, chloride recommendations are slightly higher than, but still equivalent to, those of sodium. In other words, ¾ teaspoon of salt will deliver some sodium, more chloride, and still meet the AI for both.

Chloride AI:

- 2300 mg/day (19–50 yr)
- 2000 mg/day (51–70 yr)
- 1800 mg/day (>70 yr)

Potassium

Potassium is the principal positively charged ion inside the body cells. It plays a major role in maintaining fluid and electrolyte balance and cell integrity. Potassium is also critical to keeping the heartbeat steady. The sudden deaths that occur in severe diarrhea and in children with kwashiorkor or people with eating disorders are likely due to heart failure caused by potassium loss.

Potassium Deficiency and Toxicity Potassium deficiency is characterized by an increase in blood pressure, salt sensitivity, kidney stones, and bone turnover. As deficiency progresses, symptoms include irregular heartbeats, muscle weakness, and glucose intolerance. Potassium deficiency results more often from excessive losses than from deficient intakes. Deficiency arises in abnormal conditions such as diabetic acidosis, dehydration, or prolonged vomiting or diarrhea; potassium deficiency can also result from the regular use of certain medications, including **diuretics, steroids,** and **cathartics.** Inadequate potassium intakes are possible with diets low in fresh fruits and vegetables, but out-and-out deficiencies of potassium are unlikely in healthy people.

Potassium toxicity does not result from overeating foods high in potassium; therefore, a Tolerable Upper Intake Level was not set. Toxicity can result from

hypertension: high blood pressure.

prehypertension: blood pressure values that predict hypertension.

diuretics (dye-yoo-RET-ics): medications that promote the excretion of water through the kidneys. Not all diuretics increase the urinary loss of potassium. Some, called potassium-sparing diuretics, are less likely to result in a potassium deficiency (see Chapter 23).

steroids (STARE-oids): medications used to reduce tissue inflammation, to suppress the immune response, or to replace certain steroid hormones in people who cannot synthesize them.

cathartics (ca-THART-ics): strong laxatives.

Fresh fruits and vegetables provide potassium in abundance.

Potassium AI:
▮ Adults: 4700 mg

overconsumption of potassium salts or supplements and from certain diseases or medications. Given more potassium than the body needs, the kidneys accelerate their excretion. If the GI tract is bypassed, however, and potassium is injected directly into a vein, it can stop the heart.

Potassium Recommendations and Food Sources In healthy people, almost any reasonable diet provides enough potassium to prevent the dangerously low blood potassium that indicates a severe deficiency. Potassium is abundant inside all living cells, both plant and animal, and because cells remain intact until foods are processed, the richest sources of potassium are *fresh* foods of all kinds—especially fruits and vegetables. A typical U.S. diet, however, with its low intakes of fruits and vegetables, provides only about half the recommended intake.[14] Although blood potassium may remain normal on such a diet, chronic diseases are more likely to occur.

Calcium

Calcium occupies more space in this chapter than any other major mineral. Calcium deserves such emphasis, however, because an adequate intake of calcium early in life helps grow a healthy skeleton and prevent bone disease in later life.

Calcium in Bone Calcium owns the distinction of being the most abundant mineral in the body. Ninety-nine percent of the body's calcium is stored in the bones (and teeth), where it plays two important roles.[15] First, it is an integral part of bone structure. Second, it serves as a calcium bank available to the body fluids should a drop in blood calcium occur.

As bones begin to form, calcium salts form crystals on a matrix of the protein collagen. As the crystals become denser, they give strength and rigidity to the maturing bones. As a result, the long leg bones of children can support their weight by the time they have learned to walk. Figure 9-3 shows the lacy network of calcium-containing crystals in the bone.

Many people have the idea that bones are inert, like rocks. Not so. Bones continuously gain and lose minerals in an ongoing process of remodeling. Growing children gain more bone than they lose, and healthy adults maintain a reasonable balance. When withdrawals substantially exceed deposits, however, problems such as osteoporosis develop.

From birth to approximately age 20, the bones are actively growing by modifying their length, width, and shape (see Figure 9-4). This rapid growth phase overlaps with the next period of peak bone mass development, which occurs between the ages of 12 and 30. During this period, skeletal mass increases. Bones grow thicker and denser by remodeling, a maintenance and repair process involving the loss of existing bone and the deposition of new bone. In the final phase, which begins between 30 and 40 years of age and continues throughout the remainder of life, bone loss exceeds new bone formation.

Calcium in Body Fluids The 1 percent of the body's calcium that circulates in the fluids as ionized calcium is vital to life. It plays these major roles:

▮ Regulates the transport of ions across cell membranes and is particularly important in nerve transmission

▮ Helps maintain normal blood pressure

▮ Plays an essential role in the clotting of blood

▮ Is essential for muscle contraction and therefore for the heartbeat

▮ Allows secretion of hormones, digestive enzymes, and neurotransmitters

▮ Activates cellular enzymes that regulate many processes

Because of its importance, blood calcium concentration is tightly controlled.[16]

FIGURE 9-3 Cross Section of Bone
The lacy structural elements are *trabeculae* (tra-BECK-you-lee), which can be drawn on to replenish blood calcium.

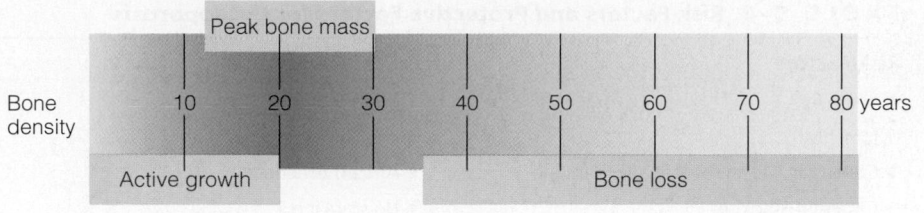

FIGURE 9-4 **Phases of Bone Development throughout Life**

The active growth phase occurs from birth to approximately age 20. The next phase of peak bone mass development occurs between the ages of 12 and 30. The final phase, when bone resorption exceeds formation, begins between the ages of 30 and 40 and continues through the remainder of life.

Calcium Balance Whenever blood calcium rises too high, a system of hormones and vitamin D promotes its deposit into bone. Whenever blood calcium falls too low, the regulatory system acts in three locations to raise it:

1. The small intestine absorbs more calcium.
2. The bones release more calcium into the blood.
3. The kidneys excrete less calcium.

Thus, blood calcium rises to normal.

The calcium stored in bone provides a nearly inexhaustible source of calcium for the blood. Even in a calcium deficiency, blood calcium remains normal. Blood calcium changes only in response to abnormal regulatory control, not to diet.

Although a chronic *dietary* deficiency of calcium or a chronic deficiency due to poor absorption does not change blood calcium, it does deplete the calcium in the bones. Because this is an important concept, we repeat: it is the bones, not the blood, that are robbed by calcium deficiency.

The regulators are hormones from the thyroid and parathyroid glands, as well as vitamin D. One hormone, parathormone, raises blood calcium. Another hormone, calcitonin, lowers blood calcium by inhibiting release of calcium from bone. The hormonelike vitamin D raises blood calcium by acting at the three sites listed.

Calcium and Osteoporosis As mentioned earlier, bone mass peaks at the time of skeletal maturity (about age 30), and a high peak bone mass is the best protection against later age-related bone loss and fracture. Adequate calcium nutrition during the growing years is essential to achieving optimal peak bone mass.[17] Following menopause, women lose about 15 percent of their bone mass, as do middle-aged and older men. When bone loss has reached such an extreme that bones fracture under even common, everyday stresses, the condition is known as osteoporosis. Osteoporosis afflicts more than 40 million people, mostly women 50 years of age or older.[18] Men, however, are not immune to osteoporosis. Each year, more than 2 million people—30 percent of them men—suffer broken hips, pelvises, legs, arms, hands, and ankles attributable to osteoporosis.

Both genetic and environmental factors contribute to osteoporosis. Table 9-4 (p. 234) summarizes risk and protective factors for osteoporosis. Osteoporosis is more prevalent in women than men for several reasons. First, women consume less dietary calcium than men do. Second, at all ages, women's bone mass is lower than men's because women generally have smaller bodies. Finally, bone loss accelerates after menopause and exceeds men's losses for up to 10 years.

In addition to calcium, many other minerals and vitamins, including phosphorus, magnesium, fluoride, and vitamin D, help to form and stabilize the structure of bones. Any or all of these elements are needed to prevent bone loss. The first, most obvious lines of defense, however, are to maintain a lifelong adequate intake of calcium and to "exercise it into place." Active bones are denser than sedentary bones.[19] Weight-bearing physical activity, such as walking, running, dancing, and weight training, prompts the bones to deposit minerals. It has long been known that, when people are confined to bed, both their muscles and their bones lose strength. Muscle strength and bone strength go together: when muscles work, they pull on the bones, and both are stimulated to grow stronger.

Reminder: *Osteoporosis* is a condition characterized by reduced density of the bones. The bones become porous and fragile and fracture easily.

Calcium and Disease Prevention Calcium may protect against hypertension, although research results are inconsistent and inconclusive.[20] Considering the success of the DASH diet in lowering blood pressure, restricting sodium to treat hypertension

TABLE 9-4 Risk Factors and Protective Factors for Osteoporosis

Risk Factors	Protective Factors
▪ Older age	▪ Younger age
▪ Low BMI	▪ High BMI
▪ Caucasian, Asian, or Hispanic heritage	▪ African American heritage
▪ Cigarette smoking	▪ No smoking
▪ Alcohol consumption in excess	▪ Alcohol consumption in moderation
▪ Sedentary lifestyle	▪ Regular weight-bearing exercise
▪ Use of glucocorticoids or anticonvulsants	▪ Use of diuretics
▪ Female gender	▪ Male gender
▪ Maternal history of osteoporosis fracture or personal history of fracture	▪ Bone density assessment and treatment (if necessary)
▪ Estrogen deficiency in women (amenorrhea or menopause, especially early or surgically induced); testosterone deficiency in men	▪ Use of estrogen therapy
▪ Lifetime diet inadequate in calcium and vitamin D	▪ Lifetime diet rich in calcium and vitamin D

may be narrow advice. The DASH diet is not particularly low in sodium, but it is rich in calcium, as well as in potassium and magnesium. As mentioned earlier, the DASH diet, together with a reduced sodium intake, is more effective at lowering blood pressure than either strategy alone. Some research also suggests protective relationships between calcium and blood cholesterol, diabetes, and colon and rectal cancers.[21] Calcium from low-fat milk and milk products (but not from supplements) has been linked with having a healthy body weight in some, but not all, studies.[22] Large, well-designed clinical studies are needed to clarify any effects of dietary calcium on body weight.[23]

Calcium Recommendations As mentioned earlier, blood calcium concentration does not reflect calcium status. Calcium recommendations are therefore based on balance studies, which measure daily intake and excretion. An optimal calcium intake reflects the amount needed to retain the most calcium. The more calcium retained, the greater the bone density (within genetic limits) and, potentially, the lower the risk of osteoporosis. Calcium recommendations during adolescence are set high (1300 milligrams) to help ensure that the skeleton will be strong and dense. Between the ages of 19 and 50, recommendations are lowered slightly, and for women over 50 and all adults over age 70, recommendations are raised again to minimize bone loss. Many people in the United States have calcium intakes below current recommendations.[24] High intakes of calcium from supplements may have adverse effects, such as kidney stone formation. For this reason, a UL has been established (see inside front cover).

Calcium in Foods Calcium is found most abundantly in a single food group—milk and milk products. For this reason, dietary recommendations advise daily consumption of low-fat or fat-free milk products. A cup of milk offers about 300 milligrams of calcium, so an adult who drinks 3 cups of milk a day (or eats the equivalent in yogurt) is well on the way to meeting daily calcium needs (see Table 9-5). The other dairy food that contains comparable amounts of calcium is cheese. One slice of cheese (1 ounce) contains about two-thirds as much calcium as a cup of milk. Cottage cheese, however, contains much less. Snapshot 9-1 shows foods that are rich in calcium, and the accompanying "How to" suggests ways of adding calcium to meals.

Calcium RDA:

- ▪ **Adults (19–50 yr): 1000 mg/day**
- ▪ **Men (51–70): 1000 mg/day**
- ▪ **Women (51–70): 1200 mg/day**
- ▪ **Adults (>70): 1200 mg/day**

TABLE 9-5 Suggested Minimum Daily Milk and Milk Products Intakes

Young children (4 to 8 years of age)	2½ cups
Older children and adolescents	3 cups
Adults	3 cups
Pregnant or lactating women	3 cups
Women past menopause	3 cups

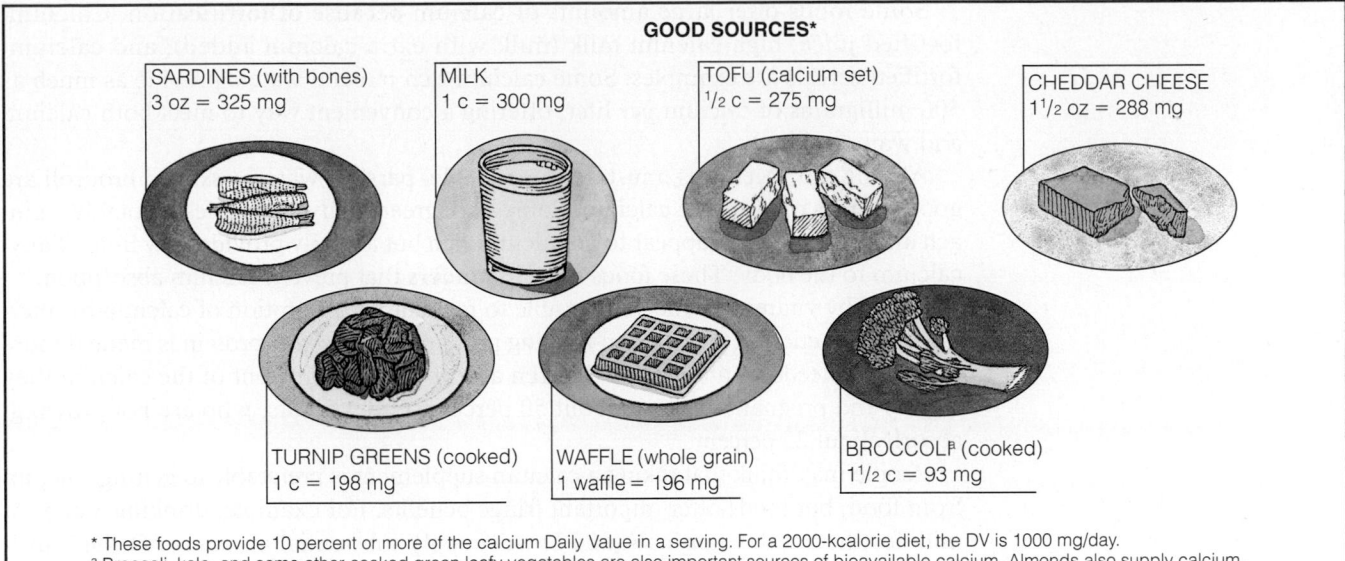

GOOD SOURCES*

SARDINES (with bones)
3 oz = 325 mg

MILK
1 c = 300 mg

TOFU (calcium set)
1/2 c = 275 mg

CHEDDAR CHEESE
1 1/2 oz = 288 mg

TURNIP GREENS (cooked)
1 c = 198 mg

WAFFLE (whole grain)
1 waffle = 196 mg

BROCCOLI[a] (cooked)
1 1/2 c = 93 mg

* These foods provide 10 percent or more of the calcium Daily Value in a serving. For a 2000-kcalorie diet, the DV is 1000 mg/day.
a Broccoli, kale, and some other cooked green leafy vegetables are also important sources of bioavailable calcium. Almonds also supply calcium.
Other greens, such as spinach and chard, contain calcium in an unabsorbable form. Some calcium-rich mineral waters may also be good sources.

SNAPSHOT 9-1 Calcium

How To ADD CALCIUM TO DAILY MEALS

For those who tolerate milk, many cooks slip extra calcium into meals by sprinkling a tablespoon or two of fat-free dry milk into almost everything. The added kcalorie value is small, and changes to the taste and texture of the dish are practically nil. Yet each 2 tablespoons adds about 100 extra milligrams of calcium and moves people closer to meeting the recommendation to obtain 3 cups of milk each day. Here are some more tips for including calcium-rich foods in your meals.

At Breakfast

- Choose calcium-fortified orange or vegetable juice.
- Serve tea or coffee, hot or iced, with milk.
- Choose cereals, hot or cold, with milk.
- Cook hot cereals with milk instead of water; then mix in 2 tablespoons of fat-free dry milk.
- Make muffins or quick breads with milk and extra fat-free powdered milk.
- Add milk to scrambled eggs.
- Moisten cereals with flavored yogurt.

At Lunch

- Add low-fat cheeses to sandwiches, burgers, or salads.
- Use a variety of green vegetables, such as watercress or kale, in salads and on sandwiches.
- Drink fat-free milk or calcium-fortified soy milk as a beverage or in a smoothie.
- Drink calcium-rich mineral water as a beverage (studies suggest significant calcium absorption).

- Marinate cabbage shreds or broccoli spears in low-fat Italian dressing for an interesting salad that provides calcium.
- Choose coleslaw over potato and macaroni salads.
- Mix the mashed bones of canned salmon into salmon salad or patties.
- Eat sardines with their bones.
- Stuff potatoes with broccoli and low-fat cheese.
- Try pasta such as ravioli stuffed with low-fat ricotta cheese instead of meat.
- Sprinkle Parmesan cheese on pasta salads.

At Supper

- Toss a handful of thinly sliced green vegetables, such as kale or young turnip greens, with hot pasta; the greens wilt pleasingly in the steam of the freshly cooked pasta.
- Serve a green vegetable every night and try new ones— how about kohlrabi? It tastes delicious when cooked like broccoli.
- Learn to stir-fry Chinese cabbage and other Asian foods.
- Try tofu (the calcium-set kind); this versatile food has inspired whole cookbooks devoted to its creative uses.
- Add fat-free powdered milk to almost anything—meatloaf, sauces, gravies, soups, stuffings, casseroles, blended beverages, puddings, quick breads, cookies, brownies. Be creative.
- Choose frozen yogurt, ice milk, or custards for dessert.

Some foods offer large amounts of calcium because of fortification. Calcium-fortified juice, high-calcium milk (milk with extra calcium added), and calcium-fortified cereals are examples. Some calcium-rich mineral waters provide as much as 500 milligrams of calcium per liter, offering a convenient way to meet both calcium and water needs.[25]

Among the vegetables, mustard greens, kale, parsley, watercress, and broccoli are good sources of available calcium. Some dark green, leafy vegetables—notably, spinach and Swiss chard—appear to be calcium rich but actually provide very little, if any, calcium to the body. These foods contain **binders** that prevent calcium absorption.

Aided by vitamin D, the body is able to regulate its absorption of calcium by altering its production of the calcium-binding protein. More of this protein is made if more calcium is needed. Infants and children absorb up to 60 percent of the calcium they ingest, and pregnant women, about 50 percent. Other adults, who are not growing, absorb about 25 percent.

People may think that taking a calcium supplement is preferable to getting calcium from food, but foods offer important fringe benefits. For example, drinking 3 cups of milk fortified with vitamins A and D will supply substantial amounts of other nutrients. Furthermore, the vitamin D and possibly other nutrients in the milk enhance calcium absorption. Some people absorb calcium better from milk and milk products than from even the most absorbable supplements. The National Institutes of Health concludes that foods are the best sources of calcium and recommends supplements only when intake from food is insufficient.

Phosphorus

Phosphorus is the second most abundant mineral in the body. About 85 percent of it is found combined with calcium in the crystals of the bones and teeth. As part of one of the body's buffer systems (phosphoric acid), phosphorus is also found in all body tissues. Phosphorus is a part of DNA and RNA, the genetic material present in every cell. Thus, phosphorus is necessary for all growth. Phosphorus also plays many key roles in the transfer of energy that occurs during cellular metabolism. Phosphorus-containing lipids (phospholipids) help transport other lipids in the blood. Phospholipids are also principal components of cell membranes.

Animal protein is the best source of phosphorus because the mineral is so abundant in the cells of animals. Diets that provide adequate energy and protein also supply adequate phosphorus. Dietary deficiencies are rare. A summary of facts about phosphorus appears in Table 9-6 later in the chapter (p. 238).

Phosphorus RDA:

▌ Adults: 700 mg/day

Magnesium

Magnesium barely qualifies as a major mineral. Only about 1 ounce of magnesium is present in the body of a 130-pound person, more than half of it in the bones. Most of the rest is in the muscles, heart, liver, and other soft tissues, with only 1 percent in the body fluids. Bone magnesium seems to be a reservoir to ensure that some will be on hand for vital reactions regardless of recent dietary intake.

Magnesium is critical to the operation of hundreds of enzymes. It acts in all the cells of the soft tissues, where it forms part of the protein-making machinery and is necessary for the release of energy. Magnesium also helps muscles to relax after contraction.

Magnesium Deficiency Magnesium deficiency can result from vomiting, diarrhea, alcohol abuse, or protein malnutrition; in people who have been fed incomplete fluids intravenously for too long after surgery; or in people using diuretics. A severe magnesium deficiency causes tetany, an extreme and prolonged contraction of the muscles.

binders: chemical compounds in foods that combine with nutrients (especially minerals) to form complexes the body cannot absorb. Examples include *phytates* and *oxalates*.

Magnesium deficiency is thought to cause the hallucinations commonly experienced during withdrawal from alcohol intoxication.

Magnesium Toxicity Magnesium toxicity is rare, but it can be fatal. Toxicity occurs only with high intakes from nonfood sources such as supplements or magnesium salts. Accidental poisonings may occur in children with access to medicine cabinets and in older adults who abuse magnesium-containing laxatives, antacids, and other medications. The consequences include diarrhea, abdominal cramps, and, in severe cases, acid–base imbalance and potassium depletion.[26]

Magnesium Intakes and Food Sources Although almost half the U.S. population has magnesium intakes below those recommended, deficiency symptoms are rare in healthy people.[27] In various parts of the country, the water contains both calcium and magnesium. Known as "hard" water, this water can contribute significantly to magnesium intakes.

Magnesium-rich food sources include dark green, leafy vegetables; nuts; legumes; whole-grain breads and cereals; seafood; chocolate; and cocoa (see Snapshot 9-2). Magnesium is easily lost from foods during processing, so unprocessed foods are the best choices.

Magnesium RDA:
- Men (19–30 yr): 400 mg/day
  (31 and older): 420 mg/day
- Women (19–30 yr): 310 mg/day
  (31 and older): 320 mg/day

Sulfate

Sulfate is the oxidized form of sulfur as it exists in food and water. The body requires sulfate for the synthesis of many important sulfur-containing compounds. Sulfur-containing amino acids play an important role in helping to shape strands of protein. The particular shape of a protein enables it to do its specific job, such as enzyme work. Skin, hair, and nails contain some of the body's more rigid proteins, which have high sulfur contents.

There is no recommended intake for sulfur, and no deficiencies are known. Only a person who lacks protein to the point of severe deficiency will lack the sulfur-containing amino acids.

The sulfur-containing amino acids are methionine and cysteine. Cysteine in one part of a protein chain can bind to cysteine in another part of the chain by way of a sulfur-sulfur bridge, thus helping to stabilize the protein structure.

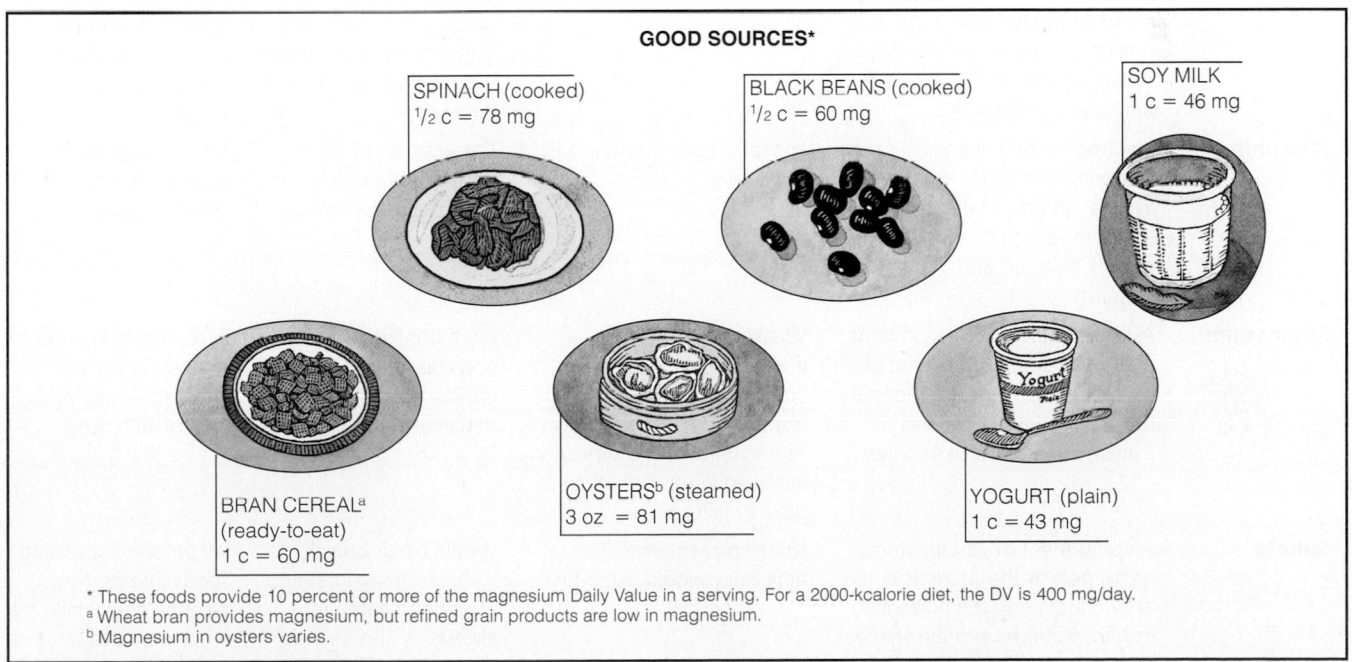

GOOD SOURCES*

SPINACH (cooked)
½ c = 78 mg

BLACK BEANS (cooked)
½ c = 60 mg

SOY MILK
1 c = 46 mg

BRAN CEREAL[a]
(ready-to-eat)
1 c = 60 mg

OYSTERS[b] (steamed)
3 oz = 81 mg

YOGURT (plain)
1 c = 43 mg

* These foods provide 10 percent or more of the magnesium Daily Value in a serving. For a 2000-kcalorie diet, the DV is 400 mg/day.
a Wheat bran provides magnesium, but refined grain products are low in magnesium.
b Magnesium in oysters varies.

SNAPSHOT 9-2 Magnesium

IN SUMMARY

■ All of the major minerals influence the body's fluid balance, but sodium, chloride, and potassium are most noted for this role.

■ Excess sodium in the diet contributes to high blood pressure.

■ Most of the body's calcium is in the bones, where it provides a rigid structure and a reservoir of calcium for the blood.

■ Table 9-6 offers a summary of the major minerals and their functions.

TABLE 9-6 The Major Minerals—A Summary

Mineral Name	Chief Functions in the Body	Deficiency Symptoms	Toxicity Symptoms	Significant Sources
Sodium	With chloride and potassium (electrolytes), maintains cells' normal fluid balance and acid–base balance in the body. Also critical to nerve impulse transmission and muscle contraction.	Muscle cramps, mental apathy, loss of appetite	Hypertension	Salt, soy sauce, processed foods
Chloride	Part of the hydrochloric acid found in the stomach and necessary for proper digestion.	Does not occur under normal circumstances	Normally harmless (the gas chlorine is a poison but evaporates from water); can cause vomiting	Salt, soy sauce; moderate quantities in whole, unprocessed foods; large amounts in processed foods
Potassium	Facilitates reactions, including the making of protein; the maintenance of fluid and electrolyte balance; the support of cell integrity; the transmission of nerve impulses; and the contraction of muscles, including the heart.	Moderate deficiency: elevated blood pressure, increased salt sensitivity, increased risk of kidney stones, increased bone turnover; severe deficiency: cardiac arrhythmias, muscle weakness, glucose intolerance	Causes muscular weakness; triggers vomiting; if given into a vein, can stop the heart	All whole foods: meats, milk, fruits, vegetables, grains, legumes
Calcium	The principal mineral of bones and teeth. Also acts in normal muscle contraction and relaxation, nerve functioning, blood clotting, blood pressure, and immune defenses.	Stunted growth in children; adult bone loss (osteoporosis)	Constipation; increased risk of urinary stone formation and kidney dysfunction; interference with absorption of other minerals	Milk and milk products, oysters, small fish (with bones), tofu (bean curd), greens, legumes
Phosphorus	Involved in the mineralization of bones and teeth. Important in cells' genetic material, in cell membranes as phospholipids, in energy transfer, and in buffering systems.	Phosphorus deficiency unknown	Calcification of nonskeletal tissues, particularly the kidneys	Foods from animal sources (meat, fish, poultry, eggs, milk)
Magnesium	Another factor involved in bone mineralization, the building of protein, enzyme action, normal muscular contraction, and transmission of nerve impulses.	Weakness; confusion; if extreme, convulsions, bizarre movements (especially of eyes and face), hallucinations, and difficulty in swallowing; in children, growth failure[a]	From nonfood sources only; diarrhea; acid–base imbalance; potassium depletion	Nuts, legumes, whole grains, dark green vegetables, seafoods, chocolate, cocoa
Sulfate	A component of certain amino acids; part of the vitamins biotin and thiamin and the hormone insulin; stabilizes protein shape by forming sulfur-sulfur bridges.	None known; protein deficiency would occur first.	Would occur only if sulfur-containing amino acids were eaten in excess; this (in animals) depresses growth	All protein-containing foods (meats, fish, poultry, eggs, milk, legumes, nuts)

[a]A still more severe deficiency causes tetany, an extreme, prolonged contraction of the muscles similar to that caused by low blood calcium.

The Trace Minerals

Each of the trace minerals performs some vital role for which no substitute will do. A deficiency of any of them can be fatal, and an excess of many can be equally deadly.

Recommendations have been established for nine of the trace minerals. Others are recognized as essential nutrients for some animals but have not been proved to be required for human beings (see Table 9-7). Still others are under study to determine whether they, too, perform indispensable roles in the body. Table 9-8, at the end of the chapter (p. 248), provides a summary of the trace minerals.

Iron

Every living cell—both plant and animal—contains iron. Most of the iron in the body is a component of the proteins **hemoglobin** in red blood cells and **myoglobin** in muscle cells. The iron in both hemoglobin and myoglobin helps them carry and hold oxygen and then release it. Hemoglobin in the blood carries oxygen from the lungs to tissues throughout the body. Myoglobin holds oxygen for the muscles to use when they contract. As part of many enzymes, iron is vital to the processes by which cells generate energy. Iron is also needed to make new cells, amino acids, hormones, and neurotransmitters.

The special provisions the body makes for iron's handling show that it is a precious mineral to be tightly hoarded. For example, when a red blood cell dies, the liver saves the iron and returns it to the bone marrow, which uses it to build new red blood cells. The body does lose iron from the digestive tract, in nail and hair trimmings, and in shed skin cells, but only in tiny amounts.[28] Bleeding, however, can cause significant iron loss from the body.

Normally, only about 10 to 15 percent of dietary iron is absorbed, but if the body's supply is diminished or if the need increases for any reason (such as pregnancy), absorption increases. The body makes several provisions for absorbing iron. A special protein in the intestinal cells captures iron and holds it in reserve for release into the body as needed; another protein transfers the iron to a special iron carrier in the blood. The blood protein **(transferrin)** carries the iron to tissues throughout the body. When more iron is needed, more of these special proteins are produced, so that more than the usual amount of iron can be absorbed and carried. If there is a surplus of iron, special storage proteins in the liver, bone marrow, and other organs store it.

Researchers are discovering genes that participate in the regulation of the body's iron at an extraordinary pace.[29] Along the way, new information about how the body controls the uptake, storage, and distribution of iron is emerging. For example, it has long been known that the liver is the main site of iron storage in the body, but only recently has research revealed that the hormone hepcidin, produced by the liver, is central to the regulation of iron balance.[30] Hepcidin helps to maintain blood iron within the normal range by limiting absorption from the small intestine and controlling release from the liver, spleen, and bone marrow.

Iron Deficiency Worldwide, **iron deficiency** is the most common nutrient deficiency, with **iron-deficiency anemia** affecting more than 1.6 billion people—almost half preschool children and pregnant women.[31] In the United States, iron deficiency is less prevalent, but it still affects about 14 percent of toddlers and 10 percent of adolescent girls and women of childbearing age.[32] Iron deficiency is also more prevalent among overweight children and adolescents compared with those who are normal weight.[33] The association between iron deficiency and obesity has yet to be explained, but researchers are currently examining the relationships between the inflammation that develops with excess body fat and reduced iron absorption.[34]

Women are especially prone to iron deficiency during their reproductive years because of blood losses during menstruation. Pregnancy places further iron demands

TABLE 9-7 Trace Minerals

RDA

Copper
Iodine
Iron
Molybdenum
Selenium
Zinc

Adequate Intake (AI)

Chromium
Fluoride
Manganese

Known Essential for Animals; Human Requirements under Study

Arsenic
Boron
Nickel
Silicon
Vanadium

Known Essential for Some Animals; No Evidence That Intake by Humans Is Ever Limiting; No Recommendation Necessary

Cobalt

Note: The evidence for requirements and essentiality is weak for the trace minerals cadmium, lead, lithium, and tin.

The storage proteins are ferritin (FERR-i-tin) and hemosiderin (heem-oh-SID-er-in).

hemoglobin: the oxygen-carrying protein of the red blood cells.
hemo = blood
globin = globular protein

myoglobin: the oxygen-carrying protein of the muscle cells.
myo = muscle

transferrin (trans-FERR-in): the body's iron-carrying protein.

iron deficiency: the condition of having depleted iron stores.

iron-deficiency anemia: a blood iron deficiency characterized by small, pale red blood cells; also called **microcytic hypochromic anemia**.
micro = small
cytic = cells
hypo = too little
chrom = color

on women: iron is needed to support the added blood volume, the growth of the fetus, and blood loss during childbirth. Infants (six months or older) and young children receive little iron from their high-milk diets, yet they need extra iron to support their rapid growth and brain development.[35] The rapid growth of adolescence, especially for males, and the blood losses of menstruation for females also demand extra iron that a typical teen diet may not provide.

Causes of Iron Deficiency The cause of iron deficiency is usually inadequate intake from an ignorance of which foods to choose, from sheer lack of food altogether, or from high consumption of iron-poor foods. In the Western world, high sugar and fat intakes are often associated with low iron intakes. Blood loss is the primary nonnutritional cause, especially in poor regions of the world where parasitic infections of the GI tract may lead to blood loss.

Stages of iron deficiency:
- Iron stores diminish.
- Transport iron decreases.
- Hemoglobin production declines.

Assessment of Iron Deficiency Iron deficiency develops in stages. This section provides a brief overview of how to detect these stages, and Appendix E provides more details. In the first stage of iron deficiency, iron stores diminish, as do levels of ferritin, an iron-storing protein. Measures of serum ferritin (in the blood) reflect iron stores and are most valuable in assessing iron status at this earliest stage.[36]

The second stage of iron deficiency is characterized by a decrease in transport iron: levels of serum iron fall, and the levels of iron-carrying protein transferrin *increase* (an adaptation that enhances iron absorption). Together, these two measures can determine the severity of the deficiency—the more transferrin and the less iron in the blood, the more advanced the deficiency is. Transferrin saturation—the percentage of transferrin that is saturated with iron—decreases as iron stores decline.

The third stage of iron deficiency occurs when the lack of iron limits hemoglobin production. Now the hemoglobin precursor, **erythrocyte protoporphyrin,** begins to accumulate as hemoglobin and **hematocrit** values decline.

Hemoglobin and hematocrit tests are easy, quick, and inexpensive, so they are the tests most commonly used in evaluating iron status. Their usefulness is limited, however, because they are late indicators of iron deficiency. Furthermore, other nutrient deficiencies and medical conditions can influence their values.

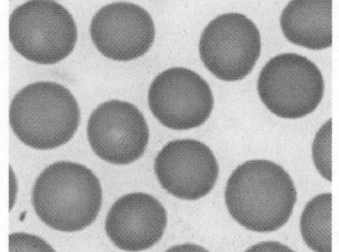

Normal red blood cells. Both size and color are normal.

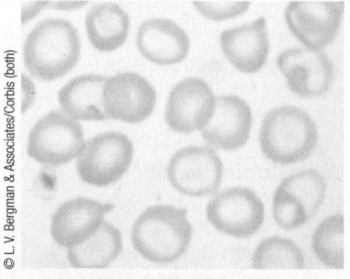

Blood cells in iron-deficiency anemia. These cells are small and pale because they contain less hemoglobin.

FIGURE 9-5 Normal and Anemic Blood Cells

The effects of iron deficiency on children's behavior are revisited in Chapter 12.

Iron Deficiency and Anemia Iron deficiency and iron-deficiency anemia are not the same: people may be iron deficient without being anemic. The term *iron deficiency* refers to depleted body iron stores without regard to the degree of depletion or to the presence of anemia. The term *iron-deficiency anemia* refers to the severe depletion of iron stores that results in a low hemoglobin concentration. In iron-deficiency anemia, red blood cells are pale and small (see Figure 9-5). They can't carry enough oxygen from the lungs to the tissues. Without adequate iron, energy metabolism in the cells falters. The result is fatigue, weakness, headaches, apathy, pallor, and poor resistance to cold temperatures. Because hemoglobin is the bright red pigment of the blood, the skin of a fair person who is anemic may become noticeably pale. In a dark-skinned person, the tongue and eye lining, normally pink, will be very pale.

The fatigue that accompanies iron-deficiency anemia differs from the tiredness a person experiences from a simple lack of sleep. People with anemia feel fatigue only when they exert themselves. Iron supplementation can, over time, relieve the fatigue and improve the body's response to physical activity.

Less severe iron deficiency produces symptoms, too. Long before the red blood cells are affected and anemia is diagnosed, a developing iron deficiency affects behavior.[37] Even at slightly lowered iron levels, energy metabolism is impaired and neurotransmitter synthesis is altered, reducing physical work capacity and mental productivity. Children deprived of iron become irritable, restless, and unable to pay attention. These symptoms are among the first to appear when the body's iron begins to fall and among the first to disappear when iron status is restored.

240

Iron Deficiency and Pica A curious behavior seen in some iron-deficient people, especially in women and children of low-income groups, is **pica**—the craving for and consumption of clay, ice, chalk, starch, and other nonfood substances. These substances contain no iron and cannot remedy a deficiency; in fact, clay actually inhibits iron absorption, which may explain the iron deficiency that accompanies such behavior. The cause of pica is unknown, but researchers hypothesize that it may be motivated by hunger, nutrient deficiencies, or an attempt to protect against toxins or microbes.[38]

Caution on Self-Diagnosis Low hemoglobin may reflect an inadequate iron intake, and, if it does, the physician may prescribe iron supplements. Any nutrient deficiency, disease, or agent that interferes with hemoglobin synthesis, disrupts hemoglobin function, or causes a loss of red blood cells can precipitate anemia, however.

Feeling fatigued, weak, and apathetic is a sign that something is wrong, but it does not indicate that a person should take iron supplements; it means that the person should consult a physician. In fact, taking iron supplements may be the worst possible thing a person can do, because such supplements can mask a serious medical condition, such as hidden bleeding from cancer or an ulcer. Remember, don't self-diagnose.

Iron Overload Normally, the body protects itself against absorbing too much iron by setting up a block in the intestinal cells. The system can be overwhelmed, however, resulting in **iron overload.** Once considered rare, iron overload has increased in frequency over the past few decades.

Iron overload, known as **hemochromatosis,** is usually caused by a genetic failure to prevent unneeded iron in the diet from being absorbed.[39] Other causes of iron overload include repeated blood transfusions, massive doses of supplementary iron, and other rare metabolic disorders.

Some of the signs and symptoms of iron overload are similar to those of iron deficiency: apathy, lethargy, and fatigue. Therefore, taking iron supplements before assessing iron status is clearly unwise; hemoglobin tests alone would fail to make the distinction because excess iron accumulates in storage. Iron overload assessment tests measure transferrin saturation and serum ferritin.

Iron overload is characterized by tissue damage, especially in iron-storing organs such as the liver. Infections are likely because bacteria thrive on iron-rich blood.[40] Symptoms are most severe in alcohol abusers because alcohol damages the intestine, further impairing its defenses against absorbing excess iron. Untreated hemochromatosis aggravates the risk of diabetes, liver cancer, heart disease, and arthritis.

Iron overload is much more common in men than in women and is twice as prevalent among men as iron deficiency.[41] The widespread fortification of foods with iron makes it difficult for people with hemochromatosis to follow an iron-restricted diet.

Iron Poisoning The rapid ingestion of massive amounts of iron can cause sudden death. Iron-containing supplements can easily cause accidental poisonings in young children.[42] Keep iron-containing supplements out of children's reach. If you suspect iron poisoning, call the nearest poison center or a physician immediately.

Iron Recommendations The average diet in the United States provides only about 6 to 7 milligrams of iron for every 1000 kcalories. Men need 8 milligrams of iron each day; most men easily take in more than 2000 kcalories, so a man can meet his iron needs without special effort. The recommendation for women during childbearing years, however, is 18 milligrams. Because women have higher iron needs and typically consume fewer than 2000 kcalories per day, they have trouble achieving appropriate iron intakes. On the average, women receive only 12 to 13 milligrams of iron per day, not enough until after menopause. To meet their iron needs from foods, premenopausal women need to select iron-rich foods at every meal. Vegetarians, because vegetable sources of iron are poorly absorbed, should aim for 1.8 times the normal requirement. Vegetarian diets are the topic of Nutrition in Practice 4.

Binding proteins in the intestinal cells (*mucosal ferritin* and *mucosal transferrin*) capture and hold unneeded iron to be shed with the cells, thereby forming a mucosal block to iron absorption.

Iron RDA:
- Men (19 and older): 8 mg/day
- Women (19–50 yr): 18 mg/day
- Women (>50 yr): 8 mg/day

To calculate the RDA for vegetarians, multiply by 1.8:
- 8 mg × 1.8 = 14 mg/day (vegetarian men)
- 18 mg × 1.8 = 32 mg/day (vegetarian women, 19 to 50 yr)

erythrocyte protoporphyrin (PRO-toh-PORE-fe-rin): a precursor to hemoglobin.

hematocrit (hee-MAT-oh-crit): measurement of the volume of the red blood cells packed by centrifuge in a given volume of blood.

pica (PIE-ka): a craving for nonfood substances; also known as *geophagia* (jee-oh-FAY-jee-uh) when referring to clay-eating behavior.
  picus = woodpecker or magpie
  geo = earth
  phagein = to eat

iron overload: toxicity from excess iron.

hemochromatosis (heem-oh-crome-a-TOH-sis): iron overload characterized by deposits of iron-containing pigment in many tissues, with tissue damage. Hemochromatosis is usually caused by a hereditary defect in iron absorption.

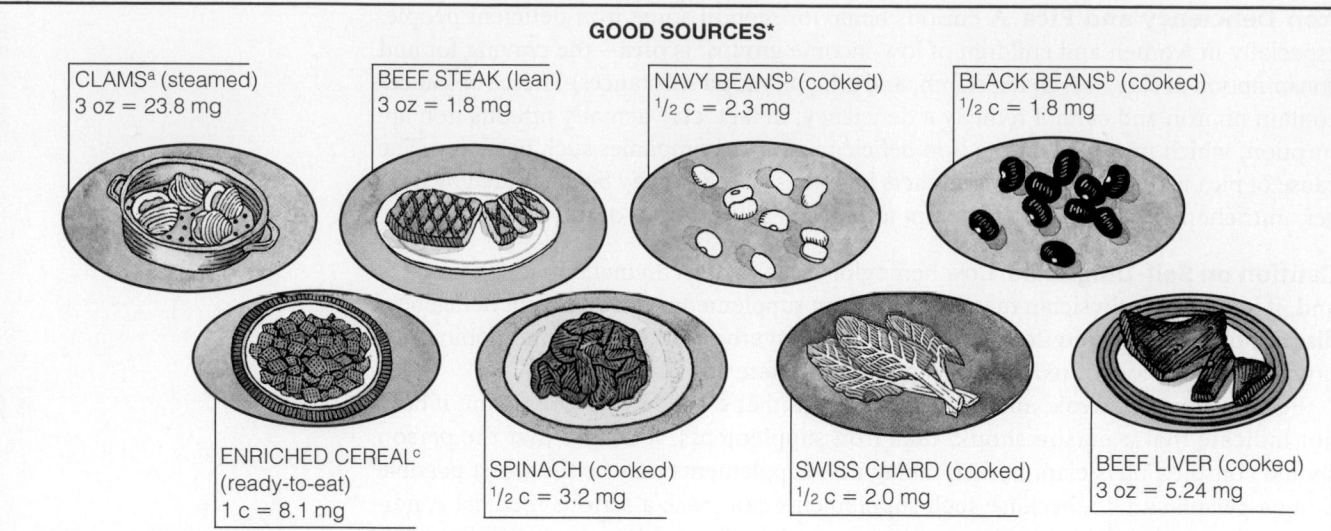

GOOD SOURCES*

CLAMS[a] (steamed)
3 oz = 23.8 mg

BEEF STEAK (lean)
3 oz = 1.8 mg

NAVY BEANS[b] (cooked)
½ c = 2.3 mg

BLACK BEANS[b] (cooked)
½ c = 1.8 mg

ENRICHED CEREAL[c]
(ready-to-eat)
1 c = 8.1 mg

SPINACH (cooked)
½ c = 3.2 mg

SWISS CHARD (cooked)
½ c = 2.0 mg

BEEF LIVER (cooked)
3 oz = 5.24 mg

* These foods provide 10 percent or more of the iron Daily Value in a serving. For a 2000-kcalorie diet, the DV is 18 mg/day.
Note: Dried figs contain 0.6 mg per ¼ cup; raisins contain 0.8 mg per ¼ cup.
[a] Some clams may contain less, but most types are iron-rich foods.
[b] Legumes contain phytates that reduce iron absorption. Soaking in water before cooking reduces phytates, and consuming legumes with vitamin C or meats increases iron absorption.
[c] Enriched cereals vary widely in iron content.

SNAPSHOT 9-3 Iron

About 40 percent of the iron in meat, fish, and poultry is bound into molecules of heme (HEEM), the iron-holding part of the hemoglobin and myoglobin proteins. This heme iron is much more absorbable than nonheme iron.

This chili dinner provides iron and MFP factor from meat, iron from legumes, and vitamin C from tomatoes. The combination of heme iron, nonheme iron, MFP factor, and vitamin C helps to achieve maximum iron absorption.

Iron in Foods Iron occurs in two forms in foods, one of which is up to 10 times more absorbable than the other. The most absorbable form is heme iron, which is bound into the iron-carrying proteins hemoglobin and myoglobin in meat, poultry, and fish. Heme iron contributes a small portion of the iron consumed by most people, but it is absorbed at a fairly constant rate of about 23 percent. The less absorbable form is nonheme iron, found in meats and also in plant foods. People absorb nonheme iron at a lower rate (2 to 20 percent); its absorption depends on several dietary factors and iron stores. Most of the iron people consume is nonheme iron from vegetables, grains, eggs, meat, fish, and poultry. Snapshot 9-3 shows the iron amounts found in usual serving sizes of different foods.

Iron absorption from foods can be maximized by two substances that enhance iron absorption: MFP factor and vitamin C. Meat, fish, and poultry contain a factor (MFP factor) other than heme that promotes the absorption of iron. MFP factor even enhances the absorption of nonheme iron from other foods eaten at the same time. Vitamin C eaten in the same meal also doubles or triples nonheme iron absorption. Additionally, cooking with iron skillets can contribute iron to the diet. Some substances impair iron absorption; they include the **tannins** of tea and coffee, the calcium in milk, and the **phytates** that accompany fiber in legumes and whole-grain cereals.[43] The accompanying "How to" offers suggestions on obtaining adequate iron.

Zinc

Zinc is a versatile trace mineral required as a cofactor by more than 100 enzymes. These zinc-requiring enzymes perform tasks in the eyes, liver, kidneys, muscles, skin, bones, and male reproductive organs. Zinc works with the enzymes that make genetic material; manufacture heme; digest food; metabolize carbohydrate, protein, and fat; liberate vitamin A from storage in the liver; and dispose of damaging free radicals.[44] Zinc also interacts with platelets in blood clotting, affects thyroid hormone function, assists in immune function, and affects behavior and learning performance.[45] Zinc is needed to produce the active form of vitamin A in visual pigments and is essential to wound healing, taste perception, the making of sperm, and fetal development. When zinc deficiency occurs, it impairs all these and other functions.

How To ADD IRON TO DAILY MEALS

The following set of guidelines can be used for planning an iron-rich diet:

▪ *Breads and cereals*. Use only whole-grain, enriched, and fortified products (iron is one of the enrichment nutrients).

▪ *Vegetables*. The dark green, leafy vegetables are good sources of vitamin C and iron. Eat vitamin C–rich vegetables often to enhance absorption of the iron from foods eaten with them.

▪ *Fruits*. Dried fruits, such as raisins, apricots, peaches, and prunes, are high in iron. Eat vitamin C–rich fruits often with iron-containing foods.

▪ *Milk and cheese*. Do not overdo foods from the milk group; they are poor sources of iron. But do not omit them, either, because they are rich in calcium. Drink fat-free milk to free kcalories to be invested in iron-rich foods.

▪ *Meats*. Meat, fish, and poultry are excellent iron sources.

▪ *Legumes*. Include legumes frequently. One cup of peas or beans can supply up to 7 milligrams of iron.

The body's handling of zinc differs from that of iron but with some interesting similarities. For example, like iron, extra zinc that enters the body is held within the intestinal cells, and only the amount needed is released into the bloodstream. As with iron, zinc status influences the percentage of zinc absorbed from the diet; if more is needed, more is absorbed.[46]

Zinc's main transport vehicle in the blood is the protein albumin. This may account for observations that serum zinc concentrations decline in conditions that lower plasma albumin concentrations—for example, pregnancy and malnutrition.

Zinc Deficiency Zinc deficiency in human beings was first reported in the 1960s from studies of growing children and male adolescents in Egypt, Iran, and Turkey. Their diets were typically low in zinc and high in fiber and phytates (which impair zinc absorption). The zinc deficiency was marked by dwarfism, or severe growth retardation, and arrested sexual maturation—symptoms that were responsive to zinc supplementation.

Since that time, zinc deficiency has been recognized elsewhere and is known to affect more than growth. It drastically impairs immune function, causes loss of appetite, and, during pregnancy, may lead to growth and developmental disorders.[47] Zinc deficiency is a substantial contributor to illness throughout the developing world and is responsible for almost half a million deaths each year.[48] A detailed list of symptoms of zinc deficiency is presented later in Table 9-8 (p. 248). Conditions other than poor diet that contribute to the development of zinc deficiency include loss of blood due to parasitic infections, climates that increase sweat losses, and the practice of clay eating.

Clay eating is a form of pica (see p. 241).

Pronounced zinc deficiency is not widespread in developed countries, but deficiencies do occur in the most vulnerable groups of the U.S. population—pregnant women, young children, the elderly, and the poor.[49] Even mild zinc deficiency can result in metabolic changes such as impaired immune response, abnormal taste, and abnormal dark adaptation (zinc is required to produce the active form of vitamin A, retinal, in visual pigments).

Some people are at greater risk of zinc deficiency than others. Pregnant teenagers need zinc for their own growth as well as for the developing fetus. Vegetarians whose diets emphasize whole grains, legumes, and other plant foods may also be at risk. These foods, though rich in zinc, also contain phytate, a potent inhibitor of zinc absorption.[50] Protein enhances zinc absorption, but most plant-protein foods also contain phytate. The DRI committee suggests that the dietary zinc requirement for vegetarians who exclude all animal-derived foods may be as much as 50 percent greater than the RDA, but so far evidence is insufficient to establish zinc recommendations based on the presence of other food components or nutrients.[51] Vegetarians who

tannins: compounds in tea (especially black tea) and coffee that bind iron.

phytates: nonnutrient components of grains, legumes, and seeds. Phytates can bind minerals such as iron, zinc, calcium, and magnesium in insoluble complexes in the intestine, and the body excretes them unused.

include cheese, eggs, or other animal protein in their diet absorb more zinc than those who exclude these foods.

Zinc Toxicity Zinc can be toxic if consumed in large enough quantities. A high zinc intake is known to produce copper-deficiency anemia by inducing the intestinal cells to synthesize large amounts of a protein (metallothionein) that captures copper in a nonabsorbable form. Accidental consumption of high levels of zinc can cause vomiting, diarrhea, headaches, and other symptoms (see Table 9-8, p. 248). The Tolerable Upper Intake Level for zinc for adults is 40 milligrams per day.

Zinc Recommendations and Food Sources Zinc is most abundant in foods high in protein, such as shellfish (especially oysters), meats, and poultry. In general, two ordinary servings a day of animal protein provide most of the zinc a healthy person needs. Milk, eggs, and whole-grain products are good sources of zinc if eaten in large quantities. For infants, breast milk is a good source of zinc. Commercial infant formulas are fortified with zinc. Snapshot 9-4 shows zinc-rich foods.

Most people in the United States have zinc intakes that approximate recommendations.[52] Zinc supplements are not recommended except for an accurately diagnosed zinc deficiency.

The use of zinc lozenges and nasal sprays or gels to treat the common cold has been controversial and inconclusive, with some studies finding them effective and others not.[53] Zinc from lozenges, sprays, and gels may or may not relieve a cold, but these products contribute zinc to the body, and the FDA has reported permanent loss of the sense of smell in people who used nasal zinc products.[54]

Selenium

Selenium is an essential trace mineral that functions as an antioxidant nutrient, working primarily as a part of proteins—most notably, the enzyme glutathione peroxidase.[55] Glutathione peroxidase and vitamin E work in tandem. Glutathione peroxidase prevents free-radical formation, thus blocking the damaging chain reaction before it begins; if free radicals do form, and a chain reaction starts, vitamin E halts it.

Zinc RDA:
- **Men: 11 mg/day**
- **Women: 8 mg/day**

GOOD SOURCES*

OYSTERS[a] (steamed)
3 oz = 72 mg

BEEF STEAK (lean)
3 oz = 4.9 mg

YOGURT (plain)
1 c = 2.2 mg

SHRIMP (cooked)
3 oz = 1.5 mg

ENRICHED CEREAL[b]
(ready-to-eat)
1 c = 3.8 mg

PORK CHOP
3 oz = 2 mg

* These foods provide 10 percent or more of the zinc Daily Value in a serving. For a 2000-kcalorie diet, the DV is 15 mg/day.
[a] Some oysters contain more or less than this amount, but all types are zinc-rich foods.
[b] Enriched cereals vary widely in zinc content.

SNAPSHOT 9-4 Zinc

Selenium-containing enzymes are necessary for the proper functioning of the iodine-containing thyroid hormones that regulate metabolism.[56]

Selenium and Cancer The question of whether selenium protects against the development of some cancers, particularly prostate cancer, is under investigation.[57] Given the potential for harm and the lack of additional evidence, recommendations to take selenium supplements to prevent cancer would be premature.

Selenium Deficiency Selenium deficiency is associated with a heart disease in children and young women living in regions of China where the soil and foods lack selenium. Although the primary cause of this heart disease is probably a virus, selenium deficiency appears to predispose people to it, and adequate selenium seems to prevent it.

Selenium Toxicity Because high doses of selenium are toxic, a UL has been set (see inside front cover). Selenium toxicity causes vomiting, diarrhea, loss of hair and nails, and lesions of the skin and nervous system.

Selenium Recommendations and Intakes Anyone who eats a normal diet composed mostly of unprocessed foods need not worry about meeting selenium recommendations. Selenium is widely distributed in foods such as meats and shellfish and in vegetables and grains grown on selenium-rich soil. Some regions in the United States and Canada produce crops on selenium-poor soil, but people are protected from deficiency because they eat selenium-rich meat and supermarket foods transported from other regions. Eating as few as two Brazil nuts a day effectively improves selenium status.[58]

Iodine

Traces of the iodine ion (called iodide) are indispensible to life. In the GI tract, iodine from foods becomes iodide. This chapter uses the term *iodine* when referring to the nutrient in foods, and *iodide* when referring to it in the body. Iodide occurs in the body in minuscule amounts, but its principal role in human nutrition is well known, and the amount needed is well established. Iodide is an integral part of the thyroid hormones, which regulate body temperature, metabolic rate, reproduction, growth, the making of blood cells, nerve and muscle function, and more.

Iodine Deficiency When the iodide concentration in the blood is low, the cells of the thyroid gland enlarge in an attempt to trap as many particles of iodide as possible. If the gland enlarges until it is visible, the swelling is called a **goiter.** People with iodine deficiency this severe become sluggish and may gain weight. Goiter afflicts about 200 million people the world over, many of them in South America, Asia, and Africa. In most cases of goiter, the cause is iodine deficiency, but some people have goiter because they overconsume foods of the cabbage family and others that contain an antithyroid substance **(goitrogen)** whose effect is not counteracted by dietary iodine.

Goiter may be the earliest and most obvious sign of iodine deficiency, but the most tragic and prevalent damage occurs in the brain. Iodine deficiency is the most common cause of *preventable* mental retardation and brain damage in the world. Children with even a mild iodine deficiency typically have goiters and perform poorly in school. With sustained treatment, however, mental performance in the classroom as well as thyroid function improves.

A severe iodine deficiency during pregnancy causes the extreme and irreversible mental and physical retardation known as **cretinism.** A child with cretinism may have an IQ as low as 20 (100 is normal) and a face and body with many abnormalities. Cretinism can be averted if the pregnant woman's deficiency is detected and treated in time.[59]

The heart disease is named *Keshan disease* for one of the provinces of China where it was studied.

Selenium RDA:
- Adults: 55 µg/day

In iodine deficiency, the thyroid gland enlarges—a condition known as simple goiter.

goiter (GOY-ter): an enlargement of the thyroid gland due to an iodine deficiency, malfunction of the gland, or overconsumption of a thyroid antagonist. Goiter caused by iodine deficiency is sometimes called *simple goiter*.

goitrogen (GOY-troh-jen): a substance that enlarges the thyroid gland and causes *toxic goiter*. Goitrogens occur naturally in such foods as cabbage, kale, brussels sprouts, cauliflower, broccoli, and kohlrabi.

cretinism (CREE-tin-ism): an iodine-deficiency disease characterized by mental and physical retardation.

Iodine Toxicity Excessive intakes of iodine can enlarge the thyroid gland, just as deficiencies can. Intakes in the United States are slightly above the recommended intake of 150 micrograms but still below the Tolerable Upper Intake Level of 1100 micrograms per day for an adult.[60]

Iodine Sources The ocean is the world's major source of iodine. In coastal areas, seafood, water, and even iodine-containing sea mist are important iodine sources. Further inland, the amount of iodine in the diet is variable and generally reflects the amount present in the soil in which plants grow or on which animals graze. In the United States and Canada, the use of iodized salt has largely wiped out the iodine deficiency that once was widespread. In the United States, labels indicate whether salt is iodized; in Canada, all table salt is iodized.

Copper

The body contains about 100 milligrams of copper. About one-fourth is in the muscles; one-fourth is in the liver, brain, and blood; and the rest is in the bones, kidneys, and other tissues. The primary function of copper in the body is to serve as a constituent of enzymes. The copper-containing enzymes have diverse metabolic roles: they catalyze the formation of hemoglobin, help manufacture the protein collagen, assist in the healing of wounds, and help maintain the sheaths around nerve fibers. One of copper's most vital roles is to help cells use iron. Like iron, copper is needed in many reactions related to respiration and energy metabolism.

Copper Deficiency Copper deficiency is rare but not unknown. It has been seen in premature or malnourished infants. High intakes of zinc interfere with copper absorption and can lead to deficiency.

Copper Toxicity Some genetic disorders create a copper toxicity. Copper toxicity from foods, however, is unlikely. The Tolerable Upper Intake Level for copper is set at 10,000 micrograms per day.

Copper Recommendations and Food Sources The RDA for copper is 900 micrograms per day, which is slightly below the average intake for adults in the United States.[61] The best food sources of copper are legumes, whole grains, seafood, nuts, and seeds.

Manganese

The human body contains a tiny 20 milligrams of manganese, mostly in the bones and metabolically active organs such as the liver, kidneys, and pancreas. Manganese is a cofactor for many enzymes, helping to facilitate dozens of different metabolic processes. Deficiencies of manganese have not been noted in people, but toxicity may be severe. Miners who inhale large quantities of manganese dust over prolonged periods show many symptoms of a brain disease, along with abnormalities in appearance and behavior. The Tolerable Upper Intake Level for manganese is 11 milligrams per day.

Manganese requirements are low, and plant foods such as nuts, whole grains, and leafy green vegetables contain significant amounts of this trace mineral. Deficiencies are therefore unlikely.

Manganese AI:
▮ Men: 2.3 mg/day
▮ Women: 1.6 mg/day

fluorapatite (floor-APP-uh-tite): the stabilized form of bone and tooth crystal, in which fluoride has replaced the hydroxy portion of hydroxyapatite.

Fluoride

Only a trace of fluoride occurs in the human body, but research demonstrates that where diets are high in fluoride during the growing years, crystalline deposits in bones and teeth are larger and harder. When bones and teeth become mineralized,

first a crystal called hydroxyapatite forms from calcium and phosphorus. Then fluoride replaces the hydroxy portion of hydroxyapatite, forming **fluorapatite,** which makes the bones stronger and the teeth more resistant to decay. Once the teeth have erupted, the topical application of fluoride by way of toothpaste or mouth rinse continues to exert a caries-reducing effect.

Fluoride Deficiency Where fluoride is lacking in the water supply, the incidence of dental decay is high. Fluoridation of water to raise its fluoride concentration to 1 part per million is recommended as an important public health measure. Those fortunate enough to have had sufficient fluoride during the tooth-forming years of infancy and childhood are protected throughout life from dental decay. Dental problems are of great concern because they can lead to a multitude of other health problems affecting the whole body. Based on the accumulated evidence of its beneficial effects, the CDC named water fluoridation as one of the 10 most important public health measures of the 20th century.[62] About 70 percent of the U.S. population served by public water systems receives optimal levels of fluoride (see Figure 9-6).[63] Most bottled waters lack fluoride.

Fluoride Sources All normal diets include some fluoride. However, drinking water; processed soft drinks and fruit juice made with fluoridated water; and fluoride toothpastes, gels, and oral rinses are the most common fluoride sources in the United States. Fish and tea may supply substantial amounts as well.

In some areas, the natural fluoride concentration in water is high, and too much fluoride can damage teeth, causing **fluorosis.** For this reason, a Tolerable Upper Intake Level has been established. In mild cases, the teeth develop small white specks; in severe cases, the enamel becomes pitted and permanently stained (see Figure 9-7 on p. 248). Fluorosis occurs only during tooth development and cannot be reversed, making its prevention a high priority.

Chromium

Chromium is an essential mineral that participates in carbohydrate and lipid metabolism. Chromium enhances the activity of the hormone insulin. When chromium is lacking, a diabetes-like condition may develop with elevated blood glucose and impaired glucose tolerance, insulin response, and glucagon response. Some research findings suggest that chromium supplements improve glucose or insulin responses in diabetes, but these relationships are uncertain.[64]

Chromium deficiency is unlikely, given the small amount of chromium required and its presence in a variety of foods. The more refined foods people eat, however, the less chromium they obtain from their diets. Unrefined foods such as liver, brewer's yeast, and whole grains are the best sources.

Other Trace Minerals

An RDA has been established for one other trace mineral, molybdenum. **Molybdenum** functions as a working part of several metal-containing enzymes, some of which are giant proteins. Deficiencies or toxicities of molybdenum are unknown.

Several other trace minerals are known or suspected to contribute to the health of the body. Nickel is recognized as important for the health of many body tissues. Silicon is involved in the formation of bones and collagen. Cobalt is found in the large vitamin B_{12} molecule. Boron influences the activity of many enzymes and may play a key role in bone health, brain activities, and immune response.[65]

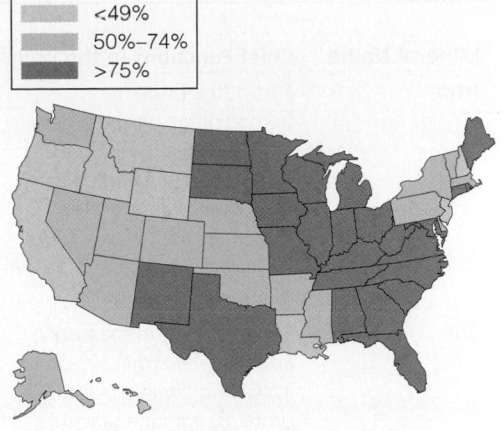

Key:
- <49%
- 50%–74%
- >75%

FIGURE 9-6 Percentage of State Populations with Access to Optimally Fluoridated Water through Public Water Systems

Fluoride AI:
- Men: 4 mg/day
- Women: 3 mg/day

Fluoride Tolerable Upper Intake Level:
- Adults: 10 mg/day

To prevent fluorosis:
- Monitor the fluoride content of the local water supply.
- Supervise children younger than six when they brush their teeth (to ensure that they don't swallow the toothpaste), and use only a pea-size amount of toothpaste.
- Use fluoride supplements only as prescribed by a physician.

Small organic compounds that enhance insulin's activity are called glucose tolerance factors (GTF). Some glucose tolerance factors contain chromium.

Chromium AI:
- Men (19–50 yr): 35 μg/day
  (51 and older): 30 μg/day
- Women (19–50 yr): 25 μg/day
  (51 and older): 20 μg/day

Molybdenum RDA: 45 μg/day

fluorosis (floor-OH-sis): mottling of the tooth enamel from ingestion of too much fluoride during tooth development.
molybdenum (mo-LIB-duh-num): a trace element.

TABLE 9-8 The Trace Minerals—A Summary

Mineral Name	Chief Functions in the Body	Deficiency Symptoms	Toxicity Symptoms	Significant Sources
Iron	Part of the protein hemoglobin, which carries oxygen in the blood; part of the protein myoglobin in muscles, which makes oxygen available for muscle contraction; necessary for the utilization of energy	Anemia: weakness, pallor, headaches, reduced work productivity, inability to concentrate, impaired cognitive function (children), lowered cold tolerance	Iron overload: infections, liver injury, possible increased risk of heart attack, acidosis, bloody stools, shock	Red meats, fish, poultry, shellfish, eggs, legumes, dried fruits
Zinc	Part of the hormone insulin and many enzymes; involved in making genetic material and proteins, immune reactions, transport of vitamin A, taste perception, wound healing, the making of sperm, and normal fetal development	Growth retardation, delayed sexual maturation, impaired immune function, loss of taste, poor wound healing, eye and skin lesions	Loss of appetite, impaired immunity, low HDL, nausea, vomiting, diarrhea, headaches, copper and iron deficiencies	Protein-containing foods: meats, fish, shellfish, poultry, grains, vegetables
Selenium	Assists a group of enzymes that break down reactive chemicals that harm cells	Predisposition to heart disease characterized by cardiac tissue becoming fibrous (uncommon)	Nausea, abdominal pain, nail and hair changes, nerve damage	Seafoods, organ meats, other meats, whole grains and vegetables (depending on soil content)
Iodine	A component of two thyroid hormones, which help to regulate growth, development, and metabolic rate	Goiter, cretinism	Depressed thyroid activity; goiterlike thyroid enlargement	Iodized salt; seafood; bread; plants grown in most parts of the country and animals fed those plants
Copper	Necessary for the absorption and use of iron in the formation of hemoglobin; part of several enzymes	Anemia, bone abnormalities	Vomiting, diarrhea, liver damage	Organ meats, seafood, nuts, seeds, whole grains, drinking water
Manganese	Facilitator, with enzymes, of many cell processes; bone formation	Rare	Nervous system disorders	Nuts, whole grains, leafy vegetables, tea
Fluoride	An element involved in the formation of bones and teeth; helps to make teeth resistant to decay	Susceptibility to tooth decay	Fluorosis (pitting and discoloration of teeth)	Drinking water (if fluoride containing or fluoridated), tea, seafood
Chromium	Enhances insulin action and may improve glucose tolerance	Diabetes-like condition marked by an inability to use glucose normally	None reported	Meats, whole grains

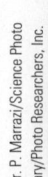
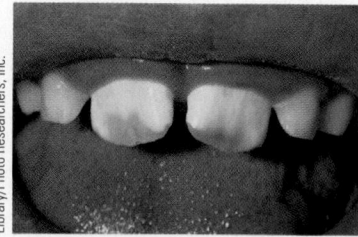

© Dr. P. Marrazzi/Science Photo Library/Photo Researchers, Inc.

FIGURE 9-7 Fluoride Toxicity Symptom—The Mottled Teeth of Fluorosis

IN SUMMARY

▪ The body requires trace minerals in tiny amounts, and they function in similar ways—assisting enzymes all over the body.

▪ Eating a diet that consists of a variety of foods is the best way to ensure an adequate intake of these important nutrients.

▪ Many dietary factors, including the trace minerals themselves, affect the absorption and availability of these nutrients.

▪ Table 9-8 offers a summary of facts about trace minerals in the body.

Your Diet

Minerals

The two minerals most likely to fall short in the diet are calcium and iron. Interestingly, both are found in protein-rich foods, but not in the same foods. Milk and milk products are rich in calcium but poor in iron. Conversely, meats, poultry, fish, and legumes are rich in iron but (with the exception of small fish such as canned sardines and other canned fishes prepared with their bones) poor in calcium. Including food sources of calcium and iron in the diet can help defend against osteoporosis and iron deficiency, respectively. Review the table you completed at the end of Chapter 1, which compares the foods you ate in a day with the USDA Food Guide.

▌ For calcium, did you drink three cups of milk or eat the equivalent in yogurt, low-fat cheese, calcium-set tofu, calcium-fortified foods, or dark green vegetables such as kale, mustard greens, or broccoli? For iron, did you eat a variety of foods, including some meat, poultry, fish, legumes, eggs, iron-fortified cereals, and dark green vegetables? (Review Snapshots 9-1 and 9-3, which show good sources of calcium and iron, or consult Appendix A for the calcium and iron amounts in many foods.)

▌ Using Appendix A, list the foods that supplied substantial amounts of calcium and iron (suggested in the previous question) along with the calcium and iron content of the foods listed. What are your estimated calcium and iron intakes for the day?

▌ Consult the inside front cover of this text to find the recommended calcium and iron intakes for your age and gender group. For example, if you are between the ages of 19 and 30, your calcium RDA is 1000 milligrams. What is your calcium RDA? What is your iron RDA?

▌ Calculate your calcium and iron intakes as a percentage of the recommended intake. For example, if your calcium intake for the day was 600 milligrams and your calcium RDA is 1000 milligrams: 600 mg ÷ 1000 mg × 100 = 60 percent.

▌ Comment on your calcium and iron intakes. Are these intakes typical on a day-to-day basis?

▌ If your calcium and iron intakes were 80 percent or less of the recommendation, which foods were your best sources? Can you eat more of these foods to bring your intake up to the recommended level? If not, list some foods that you might eat to increase your intake.

Clinical Applications

1. Pull together information from Chapter 1 about the different food groups and the significant sources of minerals shown or discussed in this chapter. Consider which minerals might be lacking (or excessive) in the diet of a client who reports the following:

▌ Relies on highly processed foods, snack foods, and fast foods as mainstays of the diet.

▌ Never uses milk, milk products, or cheese.

▌ Dislikes leafy green vegetables.

▌ Never eats meat, fish, poultry, or even meat alternates such as legumes.

What additional information would help you pinpoint problems with mineral intake?

Self Check

1. Which of the following body structures helps to regulate thirst?
   a. Brain stem
   b. Cerebellum
   c. Optic nerve
   d. Hypothalamus

2. Which of the following is *not* a function of water in the body?
   a. Lubricant
   b. Source of energy
   c. Maintains protein structure
   d. Participant in chemical reactions

3. Two situations in which a person may experience fluid and electrolyte imbalances are:
   a. vomiting and burns.
   b. diarrhea and cuts.
   c. broken bones and fever.
   d. heavy sweating and excessive carbohydrate intake.

4. Three-fourths of the sodium in people's diets comes from:
   a. fresh meats.
   b. home-cooked foods.
   c. frozen vegetables and meats.
   d. salt added to food by manufacturers.

5. Which mineral is critical to keeping the heartbeat steady and plays a major role in maintaining fluid and electrolyte balance?
   a. Sodium
   b. Calcium
   c. Potassium
   d. Magnesium

6. The two best ways to prevent age-related bone loss and fracture are to:
   a. take calcium supplements and estrogen.
   b. participate in aerobic activity and drink eight glasses of milk daily.
   c. eat a diet low in fat and salt and refrain from smoking.
   d. maintain a lifelong adequate calcium intake and engage in weight-bearing physical activity.

7. Three good food sources of calcium are:
   a. milk, sardines, and broccoli.
   b. spinach, yogurt, and sardines.
   c. cottage cheese, spinach, and tofu.
   d. Swiss chard, mustard greens, and broccoli.

8. Foods high in iron that help prevent or treat anemia include:
   a. green peas and cheese.
   b. dairy foods and fresh fruits.
   c. homemade breads and most fresh vegetables.
   d. meat and dark green, leafy vegetables.

9. Two groups of people who are especially at risk for zinc deficiency are:
   a. Asians and children.
   b. infants and teenagers.
   c. smokers and athletes.
   d. pregnant adolescents and vegetarians.

10. A deficiency of ___ is the world's most common preventable cause of mental retardation.
   a. zinc
   b. iodine
   c. selenium
   d. magnesium

Answers to these questions can be found in Appendix H.

Notes

1. B. M. Popkin, K. E. D'Anci, and I. H. Rosenberg, Water, hydration, and health, *Nutrition Reviews* 68 (2010): 439–458.

2. Standing Committee on the Scientific Evaluation of Dietary Reference Intakes, Food Nutrition Board, Institute of Medicine, *Dietary Reference Intakes for Water, Potassium, Sodium, Chloride, and Sulfate* (Washington, D.C.: National Academies Press, 2005), p. 73.

3. K. M. Kolasa, C. J. Lackey, and A. C. Grandjean, Hydration and health promotion, *Nutrition Today* 44 (2009): 190–201; American College of Sports Medicine, Position stand: Exercise and fluid replacement, *Medicine and Science in Sports and Exercise* 39 (2007): 377–390.

4. L. E. Armstrong, Assessing hydration status: The elusive gold standard, *Journal of the American College of Nutrition* 26 (2007): 575S–584S.

5. M. C. Daniels and B. M. Popkin, Impact of water intake on energy intake and weight status: A systematic review, *Nutrition Reviews* 68 (2010): 505–521.

6. Standing Committee on the Scientific Evaluation of Dietary Reference Intakes, 2005, pp. 133–134.

7. Centers for Disease Control and Prevention, Application of Lower Sodium Intake Recommendations to Adults—United States, 1999–2006, *Morbidity and Mortality Weekly Report* 58 (2009): 281–283.

8. Centers for Disease Control and Prevention, 2009.

9. R. Takachi and coauthors, Consumption of sodium and salted foods in relation to cancer and cardiovascular disease: The Japan Public Health Center-based prospective study, *American Journal of Clinical Nutrition* 91 (2010): 456–464; F. J. He and G. A. MacGregor, A comprehensive review

on salt and health and current experience of worldwide salt reduction programmes, *Journal of Human Hypertension* 23 (2009): 363–384.

10. Centers for Disease Control and Prevention, 2009.

11. B. N. Van Vliet and J.-P. Montani, The time course of salt-induced hypertension, and why it matters, *International Journal of Obesity* 32 (2008): S35–S47.

12. N. R. Cook and coauthors, Long-term effects of dietary sodium reduction on cardiovascular disease outcomes: Observational follow-up of the trials of hypertension prevention (TOHP), *British Medical Journal* 334 (2007): 885–888.

13. J. A. Blumenthal and coauthors, Effects of the DASH diet alone and in combination with exercise and weight loss on blood pressure and cardiovascular biomarkers in men and women with high blood pressure: The ENCORE study, *Archives of Internal Medicine* 170 (2010): 126–135.

14. Standing Committee on the Scientific Evaluation of Dietary Reference Intakes, 2005, pp. 186–268.

15. Committee on Dietary Reference Intakes, *Dietary Reference Intakes for Calcium and Vitamin D* (Washington, D.C.: National Academies Press, 2011), p. 2–7.

16. R. C. Khanal and I. Nemere, Regulation of intestinal calcium transport, *Annual Review of Nutrition* 28 (2008): 179–196.

17. L. L. Moore and coauthors, Effects of average childhood dairy intake on adolescent bone health, *Journal of Pediatrics* 153 (2008): 667–673.

18. National Osteoporosis Foundation, *Osteoporosis: Fast Facts* (2008), available at www.nof.org/osteoporosis/diseasefacts.htm.

19. A. Guadalupe-Grau and coauthors, Exercise and bone mass in adults, *Sports Medicine* 39 (2009):

439–468; D. C. Wilks, S. F. Gilliver, and J. Rittweger, Forearm and tibial bone measures of distance- and sprint-trained master cyclists, *Medicine and Science in Sports and Exercise* 41 (2009): 566–573.

20. V. Centeno and coauthors, Molecular mechanisms triggered by low-calcium diets, *Nutrition Research Reviews* 22 (2009): 163–174; P. R. Trumbo and K. C. Ellwood, Supplemental calcium and risk reduction of hypertension, pregnancy-induced hypertension, and preeclampsia: An evidence-based review by the US Food and Drug Administration, *Nutrition Reviews* 65 (2007): 78–87.

21. Y. Park and coauthors, Dairy food, calcium, and risk of cancer in the NIH-AARP Diet and Health Study, *Archives of Internal Medicine* 169 (2009): 391–401; J. Ishihara and coauthors, Dietary calcium, vitamin D, and the risk of colorectal cancer, *American Journal of Clinical Nutrition* 88 (2008): 1576–1583.

22. R. P. Heaney and K. Rafferty, Preponderance of the evidence: An example from the issue of calcium intake and body composition, *Nutrition Reviews* 67 (2009): 32–39; G. C. Major and co-authors, Recent developments in calcium-related obesity research, *Obesity Reviews* 9 (2008): 428–445; M. B. Snijder and coauthors, Is higher dairy consumption associated with lower body weight and fewer metabolic disturbances? The Hoorn Study, *American Journal of Clinical Nutrition* 85 (2007): 989–995.

23. M. Bortolotti and coauthors, Dairy calcium supplementation in overweight or obese persons: Its effect on markers of fat metabolism, *American Journal of Clinical Nutrition* 88 (2008): 877–885.

24. A. Moshfegh and coauthors, *What We Eat in America, NHANES 2005–2006: Usual Nutrient Intakes from Food and Water Compared to 1997*

Dietary Reference Intakes for Vitamin D, Calcium, Phosphorus, and Magnesium (Beltsville, MD: USDA, 2009).

25. R. P. Heaney, Absorbability and utility of calcium in mineral water, *American Journal of Clinical Nutrition* 84 (2006): 371–374.

26. *Dietary Reference Intakes: The Essential Guide to Nutrient Requirements*, eds., J. J. Otten, J. P. Hellwig, and L. D. Meyers (Washington, D.C.: National Academies Press, 2006), pp. 341–349.

27. Moshfegh and coauthors, 2009.

28. J. R. Hunt, C. A. Zito, and L. K. Johnson, Body iron excretion by healthy men and women, *American Journal of Clinical Nutrition* 89 (2009): 1792–1798.

29. M. U. Muckenthaler, B. Galy, and M. W. Hentze, Systemic iron homeostasis and the iron-responsive element/iron-regulatory protein (IRE/IRP) regulatory network, *Annual Review of Nutrition* 28 (2008): 197–213.

30. M. D. Knutson, Iron-sensing proteins that regulate hepcidin and enteric iron absorption, *Annual Review of Nutrition* 30 (2010): 149–171.

31. *Worldwide prevalence of anaemia 1993–2005: WHO Global Database on Anaemia*, published 2008, available at www.who.org.

32. M. E. Cogswell and coauthors, Assessment of iron deficiency in US preschool children and nonpregnant females of childbearing age: National Health and Nutrition Examination survey 2003–2006, *American Journal of Clinical Nutrition* 89 (2009): 1334–1342.

33. L. M. Tussing-Humphreys and coauthors, Excess adiposity, inflammations, and iron-deficiency in female adolescents, *Journal of the American Dietetic Association* 109 (2009): 297–302.

34. M. Wessling-Resnick, Iron homeostasis and inflammatory response, *Annual Review of Nutrition* 30 (2010): 105–122; J. P. McClung and J. P. Karl, Iron deficiency and obesity: The contribution of inflammation and diminished iron absorption, *Nutrition Reviews* 67 (2009): 100–104.

35. J. L. Beard, Why iron deficiency is important in infant development, *Journal of Nutrition* 138 (2008): 2534–2536.

36. Z. Yang and coauthors, Comparison of plasma ferritin concentration with the ratio of plasma transferrin receptor to ferritin in estimating body iron stores: Results of 4 intervention trials, *American Journal of Clinical Nutrition* 87 (2008): 1892–1898.

37. J. C. McCann and B. N. Ames, An overview of evidence for a causal relation between iron deficiency during development and deficits in cognitive or behavioral function, *American Journal of Clinical Nutrition* 85 (2007): 931–945.

38. S. L. Young and coauthors, Toward a comprehensive approach to the collection and analysis of pica substances, with emphasis on geophagic materials, *PloS ONE* 3 (2008): e3147.

39. A. Pietrangelo, Hereditary hemochromatosis, *Annual Review of Nutrition* 26 (2006): 251–270.

40. H. Drakesmith and A. Prentice, Viral infection and iron metabolism, *Nature Reviews: Microbiology* 6 (2008): 541–552.

41. K. J. Allen and coauthors, Iron-overload—Related disease in *HFE* hereditary hemochromatosis, *New England Journal of Medicine* 358 (2008): 221–230.

42. M. Carlsson and coauthors, Severe iron intoxication treated with exchange transfusion, *Archives of Disease in Childhood* 93 (2008): 321–322.

43. Standing Committee on the Scientific Evaluation of Dietary Reference Intakes, Food and Nutrition Board, National Institute of Health, *Dietary Reference Intakes for Vitamin A, Vitamin K, Arsenic, Boron, Chromium, Copper, Iodine, Iron, Manganese, Molybdenum, Nickel, Silicon, Vanadium, and Zinc* (Washington, D.C.: National Academy Press, 2001), pp. 311–316.

44. S. G. Bell and B. L. Vallee, The metallothionein/thionein system: An oxidoreductive metabolic zinc link, *ChemBioChem* 10 (2009): 55–62; Y. Song and coauthors, Zinc deficiency affects DNA damage, oxidative stress, antioxidant defenses, and DNA repair in rats, *Journal of Nutrition* 139 (2009): 1626–1631.

45. H. Haase and L. Rink, Functional significance of zinc-related signaling pathways in immune cells, *Annual Review of Nutrition* 29 (2009): 133–152; G. A. Kandhro and coauthors, Effect of zinc supplementation on the zinc level in serum and urine and their relation to thyroid hormone profile in male and female goitrous patients, *Clinical Nutrition* 28 (2009): 162–168.

46. K. M. Hambidge and coauthors, Zinc bioavailability and homeostasis, *American Journal of Clinical Nutrition* 91 (2010): 1478S–1483S.

47. S. Y. Hess and J. C. King, Effects of maternal zinc supplementation on pregnancy and lactation outcomes, *Food and Nutrition Bulletin* 30 (2009): S60–S78.

48. C. L. Fischer Walker, M. Ezzati, and R. E. Black, Global and regional child mortality and burden of disease attributable to zinc deficiency, *European Journal of Clinical Nutrition* 63 (2009): 591–597.

49. S. N. Meydani and coauthors, Serum zinc and pneumonia in nursing home elderly, *American Journal of Clinical Nutrition* 86 (2007): 1167–1173; J. M. Schneider and coauthors, The prevalence of low serum zinc and copper levels and dietary habits associated with serum zinc and copper in 12- to 36-month-old children from low-income families at risk for iron deficiency, *Journal of the American Dietetic Association* 107 (2007): 1924–1929.

50. Hambidge and coauthors, 2010.

51. Standing Committee on the Scientific Evaluation of Dietary Reference Intakes, 2001, pp. 479–480.

52. Standing Committee on the Scientific Evaluation of Dietary Reference Intakes, 2001, p. 442.

53. T. J. Caruso, C. G. Prober, and J. M. Gwaltney, Treatment of naturally acquired common colds with zinc: A structured review, *Clinical Infectious Diseases* 45 (2007): 569–574.

54. FDA, Warnings on three Zicam intranasal zinc products, *For Consumers*, June 2009, available at www.fda.gov/forconsumers/ConsumerUpdates/ucm166931.htm.

55. F. P. Bellinger and coauthors, Regulation and function of selenoproteins in human disease, *Biochemical Journal* 422 (2009): 11–22; X. G. Lei, W. Cheng, and J. P. McClung, Metabolic regulation and function of glutathione peroxidase-1, *Annual Review of Nutrition* 27 (2007): 41–61.

56. D. L. St. Germain, V. A. Galton, and A. Hernandez, Minireview: Defining the roles of the iodothyronine deiodinases: Current concepts and challenges, *Endocrinology* 150 (2009): 1097–1107.

57. S. M. Lippman and coauthors, Effect of selenium and vitamin E on risk of prostate cancer and other cancers: The Selenium and Vitamin E Cancer Prevention Trial (SELECT), *Journal of the American Medical Association* 301 (2009): 39–51; N. Facompre and K. El-Bayoumy, Potential stages for prostate cancer prevention with selenium: Implications for cancer survivors, *Cancer Research* 69 (2009): 2699–2703.

58. C. D. Thomson and coauthors, Brazil nuts: An effective way to improve selenium status, *American Journal of Clinical Nutrition* 87 (2008): 379–384.

59. M. B. Zimmermann, Iodine deficiency in pregnancy and the effects of maternal iodine supplementation on the offspring: A review, *American Journal of Clinical Nutrition* 89 (2009): 668S–672S.

60. Standing Committee on the Scientific Evaluation of Dietary Reference Intakes, 2001, pp. 278–282.

61. Standing Committee on the Scientific Evaluation of Dietary Reference Intakes, 2001, pp. 224–257.

62. Position of the American Dietetic Association: The impact of fluoride on health, *Journal of the American Dietetic Association* 105 (2005): 1620–1628.

63. Populations receiving optimally fluoridated public drinking water: United States, 1992–2006, *Morbidity and Mortality Weekly Report* 57 (2008): 737–741.

64. E. M. Balk and coauthors, Effect of chromium supplementation on glucose metabolism and lipids: A systematic review of randomized controlled trials, *Diabetes Care* 30 (2007): 2154–2163.

65. F. H. Nielsen, Is boron nutritionally relevant? *Nutrition Reviews* 66 (2008): 183–191.

Nutrition in Practice

At least half of U.S. adults collectively spend tens of *billions* of dollars a year on dietary supplements.[1] This trend is accelerating as scientists discover more links between nutrition and disease prevention. Most people take a daily multivitamin and mineral pill to make up for dietary shortfalls; others take single-nutrient supplements in an attempt to ward off diseases.[2] In many cases, taking supplements is a costly but harmless practice; sometimes, it is both costly and harmful to health. The main message of this Nutrition in Practice is that most healthy people can get the nutrients they need from foods. Supplements cannot substitute for a healthy diet. For some people, however, certain nutrient supplements may be desirable. In some cases, they can correct deficiencies; in others, they can reduce disease risks.

© Tom Carter/PhotoEdit

Do foods really contain enough vitamins and minerals to supply all that most people need?

Emphatically, yes, for both healthy adults and children who choose a variety of foods. The USDA Food Guide described in Chapter 1 is the guide to follow to achieve adequate intakes. People who meet their nutrient needs from foods, rather than supplements, have little risk of deficiency or toxicity.

Do some people need supplements?

Yes, some people may suffer marginal nutrient deficiencies due to illness, alcohol or drug addiction, or other conditions that limit food intake.[3] Certain people may benefit from nutrient supplements in amounts consistent with the RDA. For example:

▌ People with specific nutrient deficiencies need specific nutrient supplements.

▌ People whose energy intakes are particularly low (fewer than 1600 kcalories per day) need multivitamin and mineral supplements.

▌ Vegetarians who eat all-plant diets (vegans) and older adults with atrophic gastritis need vitamin B_{12}.

▌ People who have lactose intolerance or milk allergies or who otherwise do not consume enough milk products to forestall extensive bone loss need calcium.

▌ People in certain stages of the life cycle who have increased nutrient requirements need specific nutrient supplements. For example, infants need iron and fluoride supplements, women of childbearing age and pregnant women need folate and iron supplements, and the elderly need supplements of vitamins B_{12} and D.

▌ People who have inadequate milk intakes, limited sun exposure, or heavily pigmented skin need vitamin D.

▌ People who have diseases, infections, or injuries, or who have undergone surgery that interferes with the intake, absorption, metabolism, or excretion of nutrients may need specific nutrient supplements.

▌ People taking medications that interfere with the body's use of specific nutrients may need specific nutrient supplements.

Except for people in these circumstances, most adults can normally get all the nutrients they need by eating a varied diet of nutrient-dense foods. Whenever a health care professional finds a person's diet inadequate, the right corrective step is to improve the person's food choices and eating patterns, not to begin supplementation.

Why do so many people take supplements?

People frequently take supplements for mistaken reasons, such as "They give me energy" or "They make me strong." Other invalid reasons why people may take supplements include:

▌ Their feeling of insecurity about the nutrient content of the food supply.

▌ Their belief that extra vitamins and minerals will help them cope with stress.

▌ Their belief that supplements can enhance athletic performance or build lean body tissue without physical work.

▌ Their desire to prevent, treat, or cure symptoms or diseases ranging from the common cold to cancer.[4]

Ironically, many supplement users eat more nutrient-dense diets than nonusers and therefore need supplements less.[5] In addition, little relationship exists between the nutrients people need and the ones they take in supplements. In fact, an argument against

supplements is that they may lull people into a false sense of security. A person might eat irresponsibly, thinking, "My supplement will cover my needs."

Are there other arguments against taking supplements?

Yes, there are several arguments against taking supplements. First, foods rarely cause nutrient imbalances or toxicities, but supplements can. The higher the dose, the greater the risk of harm. People's tolerances for high doses of nutrients vary, just as their risks of deficiencies do. The Tolerable Upper Intake Levels of the DRI (see inside front cover) answer the question "How much is too much?" by defining the highest amount that appears safe for most healthy people.

Second, supplement users are more likely to have excessive intakes of certain nutrients—notably iron, zinc, vitamin A, and niacin.[6] The true extent of supplement toxicity in this country is unknown, but many adverse effects are reported each year from vitamins, minerals, essential oils, herbs, and other supplements.[7] Some authorities believe supplements should bear warning labels, advising consumers that large doses may be toxic.

Third, the U.S. Food and Drug Administration recently identified more than 140 dietary supplements sold on the U.S. market that were contaminated with pharmaceutical drugs, such as steroid hormones and stimulants. Toxic plant material, toxic heavy metals, bacteria, and other substances have also shown up in a wide variety of dietary supplements. Whereas hazardous products are quickly removed from the market upon discovery, many others remain on store shelves because current regulations make the supplement market difficult to monitor and control.[8] Plain multivitamin and mineral supplements from reputable sources, without herbs or other unnecessary ingredients often test free from contamination.

A fourth argument against taking supplements arises when people who are ill come to believe that high doses of vitamins or minerals can be therapeutic. Not only can high doses be toxic, but the person may take them instead of seeking medical help.

A final argument against taking supplements is that the body absorbs nutrients best from foods that dilute them and disperse them among other substances to facilitate their absorption and use by the body.[9] Taken in pure, concentrated form, nutrients are likely to interfere with one another's absorption or with the absorption of other nutrients from foods eaten at the same time.

Do antioxidant supplements prevent cancer?

Again, it is better advice to eat a very nutritious diet. Convincing evidence links high intakes of antioxidant-rich fruits and vegetables with good health and disease prevention.[10] More than 200 population studies have examined the effects of fruits and vegetables on cancer risk, and many show that people who eat more of these foods are less likely to develop certain cancers. Findings from other types of studies, such as intervention studies and clinical trials, however, show weaker associations between fruit and vegetable intake and reduced risk of cancer. Some researchers speculate that fruit and vegetable intake may play a smaller role in total cancer protection than previously thought, but fruits and vegetables do contain protective factors for specific cancers. For example, research suggests dietary fiber protects against colorectal cancer.[11] When research combines many different fruits and vegetables, specific protective factors may not

stand out. Many experts agree that the antioxidant vitamins in these foods are probably important protective factors, but they also note that other constituents of fruits and vegetables (see Table NP8-2 on p. 218) certainly have not been ruled out as contributing factors.

The way to apply this information is to eat nutritious foods. Before supplementation is recommended as a strategy to prevent cancer or other diseases, researchers must determine the optimal doses to reduce risk and the potential adverse effects of long-term supplementation.

When a person needs a vitamin-mineral supplement, what kind should be used?

Take your health care professional's advice if it is offered. If you are selecting a supplement yourself, a single, balanced vitamin-mineral supplement with no added extras such as herbs should suffice. Choose the kind that provides all the nutrients in amounts less than, equal to, or very close to the RDA (remember, you get some nutrients from foods). Avoid any preparations that, in a daily dose, provide more than the recommended intake of vitamin A, vitamin D, or any mineral or more than the Tolerable Upper Intake Level for any nutrient. In addition, avoid the following:

▌ High doses of iron (more than 10 milligrams per day), except for menstruating women. People who menstruate need more iron; others do not.

▌ "Organic" or "natural" preparations with added substances. They are no better than standard types, but they cost more.

▌ "High-potency" or "therapeutic dose" supplements. More is not better.

▌ Items not needed in human nutrition such as inositol and carnitine. These particular items won't harm you, but they reveal a marketing strategy that makes the whole mix suspect.

Finally, be aware that if you see a USP symbol on the label, it means that the manufacturer has voluntarily paid an independent laboratory to test the product and affirm that it contains the ingredients listed and that it will dissolve or disintegrate in the digestive tract to make the ingredients available for absorption. The symbol does not imply that the supplement has been tested for safety or effectiveness with regard to health, however.

Used with permission of The United States Pharmacopeial Convention

This symbol means that a supplement contains the nutrients stated and that it will dissolve in the digestive system—the symbol does not guarantee safety or health advantages.

253

Can supplement labels help consumers make informed choices?

Yes, to some extent. To enable consumers to make more informed choices about nutrient supplements, the Food and Drug Administration (FDA) published labeling regulations for supplements. The Dietary Supplement Health and Education Act subjects supplements to the same general labeling requirements that apply to foods. Specifically:

▪ Nutrition labeling for dietary supplements is required. The nutrition panel on supplements is called "Supplement Facts" (see Figure NP9-1).

▪ Labels may make nutrient claims (as "high" or "low") according to specific criteria (for example, "an excellent source of vitamin C").

▪ Labels may make health claims that are supported by significant scientific agreement and are not brand specific (for example, "folate protects against neural tube defects").

▪ Labels may claim that the lack of a nutrient can cause a deficiency disease, but if they do, they must also include the prevalence of that deficiency disease in the United States.

▪ Labels may claim to diagnose, treat, cure, or relieve common complaints such as menstrual cramps or memory loss, but they may *not* make claims about specific diseases (except as noted previously).

▪ Labels may make structure-function claims (see Chapter 1) about the role a nutrient plays in the body, explain how the nutrient performs its function, and indicate that consuming the nutrient is associated with general well-being. These claims must be accompanied by an FDA disclaimer statement: "This statement has not been evaluated by the Food and Drug Administration. This product is not intended to diagnose, treat, cure, or prevent any disease."

Note, however, that despite these requirements, the Dietary Supplement Health and Education Act in effect resulted in the deregulation of the supplement industry. Under current law, the FDA is responsible only for taking action against unsafe dietary supplements already on the market. No advance registration or approval by the FDA is needed before a manufacturer can put a supplement on store shelves.[12] To act against unsafe supplements, the FDA must receive manufacturers' reports concerning serious adverse health effects reported to them by consumers.[13] Manufacturers list contact information on supplement labels for this purpose, but many symptoms of adverse reactions are easily mistaken for something else—stomach flu, headache, or fatigue. Consumers can also report adverse reactions from supplements directly to the FDA via its hotline or website, but most people are unaware of these options (see the note on this page).*

People in developed nations are far more likely to suffer from *overnutrition* and poor lifestyle choices than from nutrient deficiencies. People wish that swallowing vitamin pills would boost their health. The truth—that they need to improve their eating and exercise habits—is harder to swallow.

*Consumers should report suspected harm from dietary supplements to their healthcare providers or to the FDA's MedWatch program at (800) FDA-1088 or on the Internet at www.fda.gov/medwatch/.

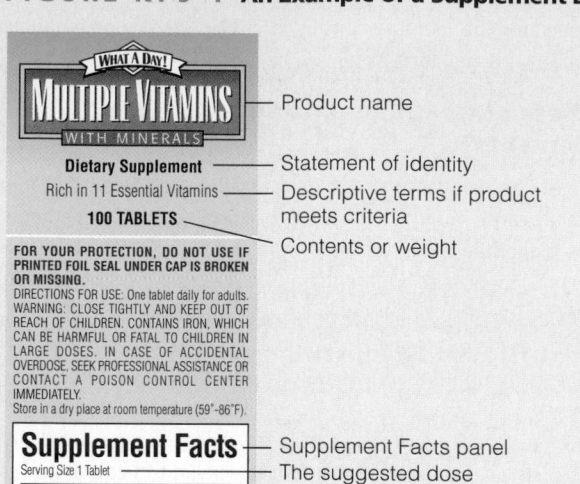

FIGURE NP9-1 An Example of a Supplement Label

— Product name

— Statement of identity

— Descriptive terms if product meets criteria

— Contents or weight

— Supplement Facts panel

— The suggested dose

— The name, quantity per tablet, and "% Daily Value" for all nutrients listed; nutrients without a Daily Value may be listed below.

— All ingredients must be listed on the label, but not necessarily in the ingredient list nor in descending order of predominance; ingredients named in the Supplement Facts panel need not be repeated here.

— Name and address of manufacturer

Notes

1. Position of the American Dietetic Association: Nutrient Supplementation, *Journal of the American Dietetic Association* 109 (2009): 2073–2085; National Institutes of Health State-of-the-Science Conference Statement: Multivitamin/mineral supplements and chronic disease prevention, *American Journal of Clinical Nutrition* 85 (2007): 257S–264S.

2. National Institutes of Health State-of-the-Science Conference Statement, 2007; E. Sloan, Why people use vitamin and mineral supplements, *Nutrition Today* 42 (2007): 55–61.

3. Position of the American Dietetic Association, 2009.

4. Sloan, 2007.

5. National Institutes of Health State-of-the-Science Conference Statement, 2007.

6. S. P. Murphy and coauthors, Multivitamin-multimineral supplements' effect on total nutrient intake, *American Journal of Clinical Nutrition* 85 (2007): 280S–284S.

7. A. C. Bronstein and coauthors, 2007 Annual report of the American Association of Poison Control Centers' national poisoning and exposure database, *Clinical Toxicology* 46 (2008): 927–1057.

8. P. A. Cohen, American roulette—contaminated dietary supplements, *New England Journal of Medicine* 361 (2009): 1523–1525.

9. D. R. Jacobs, M. D. Gross, and L. C. Tapsell, Food synergy: An operational concept for understanding nutrition, *American Journal of Clinical Nutrition* 89 (2009): 1543S–1548S.

10. L. Djousse, J. A. Driver, and J. M. Gazianao, Relation between modifiable lifestyle factors and lifetime risk of heart failure, *Journal of the American Medical Association* 302 (2009): 394–400; F. J. B. van Duijnhoven and coauthors, Fruit, vegetables, and colorectal cancer risk: The European Prospective Investigation into Cancer and Nutrition, *American Journal of Clinical Nutrition* 89 (2009): 1441–1452; World Cancer Research Fund/American Institute for Cancer Research, *Food, Nutrition, Physical Activity and the Prevention of Cancer: A Global Perspective* (Washington, D.C.: AICR, 2007), pp. 75–115.

11. L. B. Sansbury and coauthors, The effect of strict adherence to a high-fiber, high-fruit–and–vegetable, and low-fat eating pattern on adenoma recurrence, *American Journal of Epidemiology* 170 (2009): 576–584; World Cancer Research Fund/American Institute for Cancer Research, 2007, pp. 280–288.

12. U.S. Food and Drug Administration, Center for Food Safety and Applied Nutrition, *Dietary Supplements,* available at www.cfsan.fda.gov.

13. DSHEA slightly strengthened, *Consumer Health Digest,* January 30, 2007, available at www.quackwatch.org.

10

Fitness and Nutrition

In the body, nutrition and physical activity are interactive—each influences the other. The working body demands all three energy-yielding nutrients—carbohydrate, lipids, and protein—to fuel activity. The body also needs protein and a host of supporting nutrients to build lean tissue. Physical activity benefits the body's nutrition by helping to regulate the use of fuels, by pushing the body composition toward the lean, and by increasing the daily kcalorie allowance. With more kcalories come more nutrients and other beneficial constituents of foods.

For those just beginning to increase **fitness,** be assured that improvement is not only possible, but also an inevitable result of becoming more active. As you improve your physical fitness, you not only *feel* better and stronger, but you also *look* better. Physically fit people walk with confidence and purpose because posture and self-image improve along with physical fitness.

If you are already physically fit, the following description applies. You move with ease and balance. You have endurance, and your energy lasts for hours. You are strong and meet daily physical challenges without strain. What is more, you are prepared to meet mental and emotional challenges because physical fitness also supports mental and emotional energy and resilience.

This chapter is written for athletes and for active people who train like athletes. Casual athletes (those who compete only with their own goals) and competitive athletes (those who compete with others) need the right foods and fluids at the right time to support their efforts. The chapter refers to "you" to make the connection between general principles and personal choices. To understand the interactions between physical activity and nutrition, you must first know a few things about fitness, its benefits, and **training** to develop fitness.

Fitness

Fitness depends on a certain minimum amount of **physical activity** or **exercise.** Physical activity and exercise both involve bodily movement, muscle contraction, and enhanced energy expenditure, but by definition, a distinction is made between the two terms. Exercise is often considered to be a vigorous, structured, and planned type of physical activity. Because this chapter focuses on the active body's use of energy nutrients—whether that body is pedaling a bike across campus or pedaling a stationary bike in a gym—for our purposes, the terms *physical activity* and *exercise* will be used interchangeably.

Benefits of Fitness

People who regularly engage in just moderate physical activity live longer on average than those who are physically inactive.[1] Yet, despite an increasing awareness of the health benefits that physical activity confers, about half of adults in the United States are not regularly active and about 20 percent are completely inactive.[2] A **sedentary** lifestyle ranks with smoking and obesity as a powerful risk factor for developing the major killer diseases of our time—cardiovascular disease, some forms of cancer, stroke, diabetes, and hypertension.[3]

As a person becomes physically active, the health of the entire body improves. Compared with unfit people, physically fit people enjoy:

▮ *More restful sleep.* Rest and sleep occur naturally after periods of physical activity. During rest, the body repairs injuries, disposes of wastes generated during activity, and builds new physical structures.

▮ *Improved nutritional health.* Physical activity expends energy and thus allows people to eat more food. If they choose wisely, active people will consume more nutrients and be less likely to develop nutrient deficiencies.

© Zoran Milich/Masterfile

Physical activity, or its lack, exerts a significant and pervasive influence on everyone's nutrition and overall health.

Similar differences in risks for chronic disease and death exist between:

▮ **Vigorous exercise versus minimal exercise**

▮ **Healthy weight versus 20 percent overweight**

▮ **Nonsmoking versus smoking (one pack per day)**

fitness: the characteristics that enable the body to perform physical activity; more broadly, the ability to meet routine physical demands with enough reserve energy to rise to a physical challenge; the body's ability to withstand stress of all kinds.

training: regular practice of an activity, which leads to physical adaptations of the body with improvement in flexibility, strength, or endurance.

physical activity: bodily movement produced by muscle contractions that substantially increase energy expenditure.

exercise: planned, structured, and repetitive bodily movement that promotes or maintains physical fitness.

sedentary: physically inactive (literally "sitting down a lot").

- *Improved body composition.* A balanced program of physical activity limits body fat and increases or maintains lean tissue. Thus, physically active people have relatively less body fat than sedentary people at the same body weight.[4]

- *Improved bone density.* Weight-bearing physical activity builds bone strength and protects against osteoporosis.[5]

- *Enhanced resistance to colds and other infectious diseases.* Fitness enhances immunity.*[6]

- *Lower risks of some types of cancers.* Lifelong physical activity may help to protect against colon cancer, breast cancer, and some other cancers.[7]

- *Stronger circulation and lung function.* Physical activity that challenges the heart and lungs strengthens both the circulatory and the respiratory systems.

- *Lower risks of cardiovascular disease.* Physical activity lowers blood pressure, slows resting pulse rate, lowers total blood cholesterol, and raises HDL cholesterol, thus reducing the risks of heart attacks and strokes.[8] Some research suggests that physical activity may reduce the risk of cardiovascular disease in another way as well—by reducing intra-abdominal fat stores.[9]

- *Lower risks of type 2 diabetes.* Physical activity normalizes glucose tolerance.[10] Regular physical activity reduces the risk of developing type 2 diabetes and benefits those who already have the condition.

- *Reduced risk of gallbladder disease (women).* Regular physical activity reduces women's risk of gallbladder disease—perhaps by facilitating weight control and lowering blood lipid levels.[11]

- *Lower incidence and severity of anxiety and depression.* Physical activity may improve mood and enhance the quality of life by reducing depression and anxiety.[12]

- *Stronger self-image.* The sense of achievement that comes from meeting physical challenges promotes self-confidence.

- *Long life and high quality of life in the later years.* Active people live longer, healthier lives than sedentary people do.[13] Even a two-mile walk daily can add years to a person's life. In addition to extending longevity, physical activity supports independence and mobility in later life by reducing the risk of falls and minimizing the risk of injury should a fall occur.[14]

What does a person have to do to reap the health rewards of physical activity? The *Physical Activity Guidelines for Americans, 2008* (see Figure 10-1) specify the minimum amount of **aerobic physical activity** people need to gain substantial *health* benefits.[15] The guidelines recommend longer or shorter times for activity depending on whether the activity is **moderate-intensity physical activity** or **vigorous-intensity physical activity.** For clarity and effectiveness, a minimum length of 10 minutes for short bouts of aerobic physical activity is recommended.[16] The *Physical Activity Guidelines 2008*, the American College of Sports Medicine (ACSM), and the DRI Committee, however, all advise that exceeding the minimum for intensity and duration of aerobic physical activity brings more extensive health benefits such as maintaining a healthy body weight (BMI of 18.5 to 24.9) and further reducing the risk of chronic diseases.

For many people, the health benefits of regular, moderate physical activity are reward enough. Others seek the kinds and amount of physical activity that will not only benefit health, but will also improve their physical fitness or sports performance. To develop and maintain *fitness,* the American College of Sports Medicine (ACSM) recommends the types and amounts of physical activity presented in Table 10-1.[17]

aerobic physical activity: activity in which the body's large muscles move in a rhythmic manner for a sustained period of time. Aerobic activity, also called *endurance activity,* improves cardiorespiratory fitness. Brisk walking, running, swimming, and bicycling are examples.

moderate-intensity physical activity: physical activity that requires some increase in breathing and/or heart rate and expends 3.5 to 7 kcalories per minute. Walking at a speed of 3 to 4.5 miles per hour (about 15 to 20 minutes to walk one mile) is an example.

vigorous-intensity physical activity: physical activity that requires a large increase in breathing and/or heart rate and expends more than 7 kcalories per minute. Walking at a very brisk pace (>4.5 miles per hour) or running at a pace of at least 5 miles per hour are examples.

*Moderate physical activity can stimulate immune function. Intense, vigorous, prolonged activity such as marathon running, however, may compromise immune function.

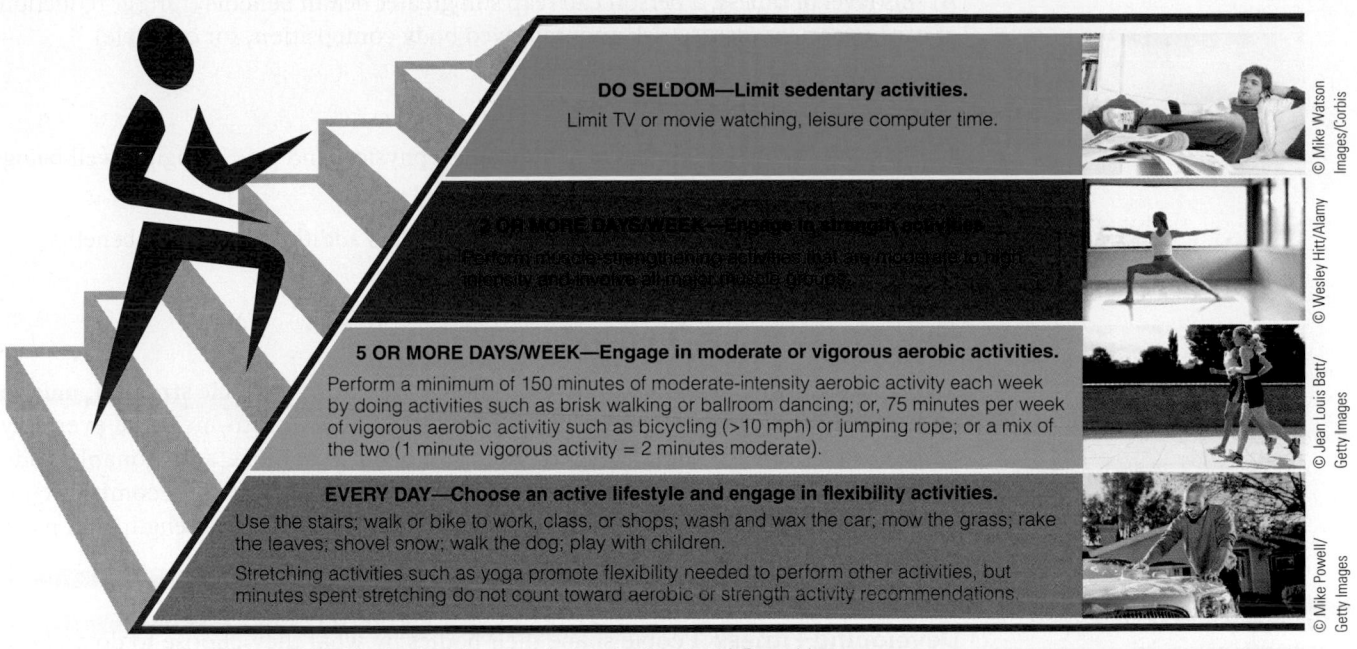

Note: Tips for increasing physical activity every day can be found at MyPyramid.gov.

FIGURE 10-1 *Physical Activity Guidelines for Americans, 2008*

TABLE 10-1 **American College of Sports Medicine's Guidelines for Physical Activity**

	Cardiorespiratory	Strength	Flexibility
Type of Activity	Aerobic activity that uses large-muscle groups and can be maintained continuously	Resistance activity that is performed at a controlled speed and through a full range of motion	Stretching activity that uses the major muscle groups
Frequency	5 to 7 days per week	2 or more nonconsecutive days per week	2 to 7 days per week
Intensity	Moderate (equivalent to walking at a pace of 3 to 4 miles per hour)[a]	Enough to enhance muscle strength and improve body composition	Enough to develop and maintain a full range of motion
Duration	At Least 30 minutes	8 to 12 repetitions of 8 to 10 different exercises (minimum)	2 to 4 repetitions of 15 to 30 seconds per muscle group
Examples	Running, cycling, swimming, inline skating, rowing, power walking, crosscountry skiing, kickboxing, jumping rope; sports activities such as basketball, soccer, racquetball, tennis, volleyball	Pull-ups, push-ups, weight lifting, Pilates	Yoga

[a]For those who prefer vigorous-intensity aerobic activity such as walking, at a very brisk pace (>4.5 mph) or running (≥5 mph) a minimum of 20 minutes per day, 3 days per week is recommended.

Source: Adapted from W. L. Haskell and coauthors, Physical, activity and public health: Updated recommendation for adults from the American College of Sports Medicine and the American Heart Association, *Medicine & Science in Sports & Exercise* 39 (2007): 1423–1434.

At this level of fitness, a person can reap still greater health benefits (further reduction of cardiovascular disease risk and improved body composition, for example).[18]

IN SUMMARY

▌ Physical activity and fitness benefit people's physical and psychological well-being and improve their resistance to disease.

▌ Physical activity to improve physical fitness offers additional personal benefits.

The Essentials of Fitness

To be physically fit, you need to develop enough **flexibility, muscle strength, muscle endurance,** and **cardiorespiratory endurance** to allow you to meet the everyday demands of life with some to spare, and you need to achieve a reasonable body composition. A person who practices a physical activity *adapts* by becoming better able to perform it after each session—with more flexibility, more strength, and more endurance.

Developing Fitness People shape their bodies by what they choose to do and not do. Muscle cells and tissues respond to an **overload** of physical activity by gaining strength and size, a response called **hypertrophy.** The opposite is also true: if not called on to perform, muscles diminish in size and weaken, a response called **atrophy.** Thus, cyclists often have well-developed legs but less arm or chest strength; a tennis player may have one superbly strong arm, while the other is just average. A variety of physical activities produces the best overall fitness, and to this end, people need to work different muscle groups from day to day. In a balanced fitness program (see Table 10-2), aerobic activity improves cardiorespiratory fitness, stretching enhances flexibility, and resistance training develops muscle strength, **muscle power,** and muscle endurance. It makes sense to give muscles a rest, too, because it takes a day or two to replenish muscle fuel supplies and to repair wear and tear incurred through physical activity.

Periodic rest also gives muscles time to adapt to an activity. During rest, muscles build more of the cellular structures required to perform the activity that preceded the rest. For example, the muscle cells of a superbly trained weight lifter store extra granules of glycogen, build up strong connective tissues, and add bulk to the special

TABLE 10-2 A Sample Balanced Fitness Program

Monday, Tuesday, Wednesday, Thursday, Friday:

▌ 5 minutes of warm-up activity
▌ 30–60 minutes of aerobic activity
▌ 10 minutes of cool-down activity and stretching

Tuesday, Thursday, Saturday

▌ 5 minutes of warm-up activity
▌ 30 minutes of resistance training
▌ 10 minutes of cool-down activity and stretching

Saturday and/or Sunday

▌ Sports, walking, hiking, biking, or swimming

flexibility: the capacity of the joints to move through a full range of motion; the ability to bend and recover without injury.

muscle strength: the ability of muscles to work against resistance.

muscle endurance: the ability of a muscle to contract repeatedly within a given time without becoming exhausted.

cardiorespiratory endurance: the ability to perform large-muscle dynamic exercise of moderate-to-high intensity for prolonged periods.

overload: an extra physical demand placed on the body; an increase in the frequency, duration, or intensity of an activity. A principle of training is that for a body system to improve, it must be worked at frequencies, durations, or intensities that increase by increments.

© Ryan McVay/PhotoDisc/Getty Images

© David Madison/Stone/Getty Images

People's bodies are shaped by the activities they perform.

proteins that contract the muscles, thereby increasing the muscles' ability to perform.* In the same way, the muscle cells of a distance runner adapt to burn fat and to sustain prolonged exertion. Therefore, if you wish to become a better jogger, swimmer, or biker, you should train mostly by jogging, swimming, or biking. Your performance will improve as your muscles adapt to do the activity.

IN SUMMARY

▌ The components of fitness are flexibility, muscle strength, muscle endurance, and cardiorespiratory endurance.

▌ To build fitness, a person must engage in physical activity. Muscles adapt to activities they are called upon to perform repeatedly.

Resistance Training **Resistance training** has long been recognized as a method to build lean body mass and develop and maintain muscle strength, muscle power, and muscle endurance. Additional benefits of resistance training, however, have also emerged. Progressive resistance training helps prevent and manage several chronic diseases, including cardiovascular disease, and enhances psychological well-being.[19]

By promoting strong muscles in the back and abdomen, weight training can improve posture and reduce the risk of back injury. Resistance training can also help prevent the decline in physical mobility that often accompanies aging.[20] Older adults, even those in their eighties, who participate in resistance training programs not only gain muscle strength, but also improve their muscle endurance, which enables them to walk significantly longer before exhaustion. Leg strength and walking endurance are powerful indicators of an older adult's physical abilities.

Resistance training also helps to maximize and maintain bone mass, especially in combination with adequate dietary calcium and vitamin D intakes.[21] Even in women past menopause (when most women are losing bone), a one-year program of resistance training improves bone density.

Resistance training builds muscle strength, muscle power, and muscle endurance. To emphasize muscle strength, combine high resistance (heavy weight) with few (3 to 6) repetitions.[22] To emphasize muscle power, combine moderate resistance (light to medium weight) with high velocity (as fast as safely possible). To emphasize muscle endurance, combine less resistance (lighter weight) with more (15 to 20) repetitions. Resistance training enhances performance in other sports, too. Swimmers can develop a more efficient stroke and tennis players, a more powerful serve, when they train with weights.

IN SUMMARY

▌ Resistance training offers health and fitness benefits to adults.

▌ Resistance training reduces the risk of cardiovascular disease, improves older adults' physical mobility, and helps maximize and maintain bone mass.

---

*All muscles contain a variety of muscle fibers, but there are two main types—slow-twitch (also called *red fibers*) and fast-twitch (also called *white fibers*). Slow-twitch fibers contain extra metabolic equipment to perform fat-burning aerobic work; the fast-twitch type store extra glycogen for anaerobic work. Muscle fibers of one type take on some of the characteristics of the other as an adaptation to exercise.

Extremes in physical activity, together with severely restricted energy intakes, may be detrimental to bone health in some young women. Such women risk developing the "female athlete triad," discussed in Nutrition in Practice 6.

hypertrophy (high-PURR-tro-fee): an increase in size (for example, of a muscle) in response to use.

atrophy (AT-tro-fee): a decrease in size (for example, of a muscle) because of disuse.

muscle power: the efficiency of a muscle contraction, measured by force and time.

resistance training: physical activity that develops muscle strength, power, endurance, and mass. Resistance can be provided by free weights, weight machines, elastic bands, or other objects. A person's own body weight may also be used to provide resistance, as when a person does push-ups, pull-ups, or sit-ups; also called *weight training*.

Cardiorespiratory Endurance Resistance training provides some cardiovascular benefits, but the kind of physical activity most beneficial to the health of the heart is cardiorespiratory endurance training. Cardiorespiratory endurance determines how long you can remain active with an elevated heart rate—it is the ability of the heart and lungs to sustain a given physical demand. Cardiorespiratory training enhances the capacity of the heart, lungs, and blood to deliver oxygen to, and remove wastes from, the body's cells.[23] Cardiorespiratory endurance training, therefore, is aerobic.

The body's adaptation to the demands of regular aerobic activity involves a complex sequence of heart-healthy events. As cardiorespiratory endurance improves, the body delivers oxygen more efficiently. In fact, the accepted measure of a person's cardiorespiratory fitness is maximal oxygen uptake **(VO_2max).**[24] With cardiorespiratory endurance, the total blood volume and the number of red blood cells increase, so the blood can carry more oxygen. The heart muscle becomes stronger, and its **cardiac output** increases. Each beat empties the heart's chambers more completely, so the heart pumps more blood per beat—its **stroke volume** increases. This makes fewer beats necessary, so the pulse rate falls. The muscles that inflate and deflate the lungs gain strength and endurance, so breathing becomes more efficient. Circulation through the blood vessels improves. Blood moves easily, and blood pressure falls.[25] Figure 10-2 shows the major relationships among the heart, lungs, and muscles. The improvements that come with cardiorespiratory endurance also raise blood HDL, the lipoprotein associated with lower heart disease risk.

Cardiorespiratory endurance is characterized by:

- Increased cardiac output and oxygen delivery
- Increased heart strength and stroke volume
- Slowed resting pulse
- Increased breathing efficiency
- Improved circulation
- Reduced blood pressure

FIGURE 10-2

Delivery of Oxygen by the Heart and Lungs to the Muscles

The cardiorespiratory system responds to the muscles' demand for oxygen by building up its capacity to deliver oxygen. Researchers can measure cardiorespiratory fitness by measuring the maximum amount of oxygen a person consumes per minute while working out, a measure called VO_2max.

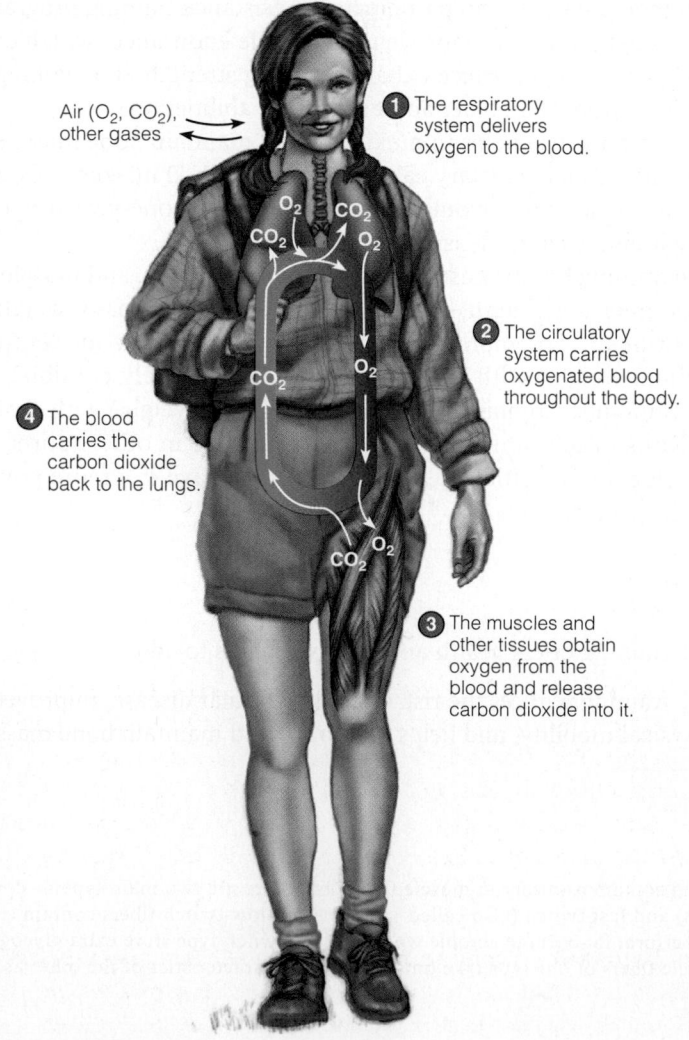

Air (O_2, CO_2), other gases

1 The respiratory system delivers oxygen to the blood.

2 The circulatory system carries oxygenated blood throughout the body.

4 The blood carries the carbon dioxide back to the lungs.

3 The muscles and other tissues obtain oxygen from the blood and release carbon dioxide into it.

Which activities produce these beneficial changes? Effective activities elevate the heart rate, are sustained for longer than 20 minutes, and use most of the large-muscle groups of the body (legs, buttocks, and abdomen). Examples are swimming, cross-country skiing, rowing, fast walking, jogging, fast bicycling, soccer, hockey, basketball, in-line skating, lacrosse, and rugby. The ACSM guidelines for developing and maintaining cardiorespiratory fitness were given in Table 10-1 on page 259.

An informal pulse check can give you some indication of how conditioned your heart is. The average resting pulse rate for adults is around 70 beats per minute. Active people can have resting pulse rates of 50 or even lower. To take your pulse, follow the directions in the margin.

© Corbis

The key to regular physical activity is finding an activity that you enjoy.

IN SUMMARY

▮ Cardiorespiratory endurance training enhances the ability of the heart and lungs to deliver oxygen to the muscles.

▮ With cardiorespiratory endurance training, the heart becomes stronger, breathing becomes more efficient, and the health of the entire body improves.

The rest of this chapter describes the interactions between nutrients and physical activity. Nutrition alone cannot endow you with fitness or athletic ability, but along with the right mental attitude, it complements your effort to obtain them. Conversely, unwise food selections can stand in your way.

The Active Body's Use of Fuels

The fuels that support physical activity are glucose (from carbohydrate), fatty acids (from fat), and, to a small extent, amino acids (from protein). The body uses different mixtures of fuels depending on the intensity and duration of its activities and depending on its own prior training.

During rest, the body derives a little more than half of its energy from fatty acids, most of the rest from glucose, and a little from amino acids. During physical activity, the body adjusts its fuel mix to use the stored glucose of muscle glycogen. In the early minutes of an activity, muscle glycogen provides the majority of energy the muscles use to go into action. As activity continues, messenger molecules, including the hormone **epinephrine**, flow into the bloodstream to signal the liver and fat cells to liberate their stored energy nutrients, primarily glucose and fatty acids.

Glucose Use during Physical Activity

Both the liver and skeletal muscles store glucose as glycogen. Glucose comes from dietary sources, but the liver can also make glucose from fragments of other nutrients. Muscles retain their glycogen stores for their own use—they do not release their glucose into the bloodstream to share with other body tissues, as the liver does. This is fortunate, since a muscle that conserves its glycogen is prepared to act quickly in emergencies because muscle glucose fuels quick action. As activity continues, glucose from the liver's stored glycogen and dietary glucose absorbed from the digestive tract also become important sources of fuel for muscle activity.

Diet Affects Glycogen Storage and Use The body constantly uses and replenishes its glycogen. The more carbohydrate a person eats, the more glycogen muscles store (up to a limit) and the longer the stores will last to support physical activity.

To take your resting pulse:
Sit down and relax for five minutes before you begin. Using a watch or clock with a second hand, place your hand over your heart or your finger firmly over an artery at the underside of the wrist or side of the throat under the jawbone. Start counting your pulse at a convenient second, and continue counting for 10 seconds. If a heartbeat occurs exactly on the 10th second, count it as one-half beat. Multiply by 6 to obtain the beats per minute.

To ensure a true count:
▮ Use only fingers, not your thumb, on the pulse point. (The thumb has a pulse of its own.).
▮ Press just firmly enough to feel the pulse. Too much pressure can interfere with the pulse rhythm.

VO$_2$ max: the maximum rate of oxygen consumption by an individual (measured at sea level).

cardiac output: the volume of blood discharged by the heart each minute.

stroke volume: the amount of oxygenated blood ejected from the heart toward body tissues at each beat.

epinephrine (EP-ih-NEFF-rin): the major hormone that elicits the stress response.

FIGURE 10-3

The Effect of Diet on Physical Endurance

A high-carbohydrate diet can increase an athlete's endurance. In this study, the high-fat diet provided 94 percent of kcalories from fat and 6 percent from protein; the normal mixed diet provided 55 percent of kcalories from carbohydrate; and the high-carbohydrate diet provided 83 percent of kcalories from carbohydrate.

Maximum endurance time:

Fat and protein diet — 57 min

Normal mixed diet — 114 min

High-carbohydrate diet — 167 min

Jupiter Images

A classic report compared fuel use during physical activity by three groups of runners, each on a different diet. For several days before testing, one of the groups ate a normal mixed diet (55 percent of kcalories from carbohydrate), a second group ate a high-carbohydrate diet (83 percent of kcalories from carbohydrate), and the third group ate a high-fat diet (94 percent of kcalories from fat). As Figure 10-3 shows, the high-carbohydrate diet enabled the athletes to work longer before exhaustion. This study and many others established that a high-carbohydrate diet enhances an athlete's endurance by ensuring ample glycogen stores.[26]

Intensity of Activity Affects Glycogen Use The body stores far less glycogen than fat. How long a person's glycogen will last during physical activity depends not only on diet but also on the intensity of the activity.

Moderate-intensity activities such as easy jogging use glycogen slowly. The lungs and circulatory system have no trouble keeping up with the muscles' need for oxygen. The individual breathes easily, and the heart beats steadily—the activity is aerobic. The muscles derive their energy from both glucose and fatty acids. By depending partly on fatty acids, moderate aerobic activity conserves glycogen stores. Joggers still use glycogen, however, and eventually they can run out of it.

Intense activities—the kind that make it difficult to catch your breath, such as a quarter-mile race—use glycogen quickly. Muscles must begin to rely more heavily on glucose, which can be partially broken down by **anaerobic** metabolism. Thus, the muscles begin drawing more heavily on their limited glycogen supply. Anaerobic breakdown of glycogen yields energy to muscle tissue when energy demands outstrip the body's ability to provide energy aerobically, but it does so by spending the muscles' glycogen reserves.

Lactate During intense activity, anaerobic breakdown of glucose produces **lactate.** Muscles release lactate formed during exercise into the blood, and it travels to the liver. There, liver enzymes convert the lactate back into glucose. Glucose can then return to the muscles to fuel additional activity. At low intensities, lactate is readily cleared from the blood by the liver, but at higher intensities, lactate accumulates. When the rate of lactate production exceeds the rate of clearance, intense activity can be maintained for only one to three minutes (as in a 400- or 800-meter race or a boxing match). Accumulation of lactate was long blamed for a type of muscle fatigue, but recent thought disputes this idea. It is true that muscles produce lactate during a type of fatigue, but the lactate does not cause the fatigue.[27]

Activity Duration Affects Glycogen Use The *duration* of a physical activity as well as its *intensity* affect glucose use. In the first 10 minutes or so of an activity, the active muscles rely almost entirely on their own stores of glycogen for glucose. Within the first 20 minutes of moderate activity, a person uses up about one-fifth of the available glycogen. As the muscles devour their own glycogen, they become ravenous for more glucose, and the liver responds by emptying out its glycogen stores.

A person who exercises moderately for longer than 20 minutes begins to use less glucose and more fat for fuel. Still, glucose use continues, and if the activity goes on long enough and at a high enough intensity, blood glucose declines and muscle and liver glycogen stores are depleted. Physical activity can continue for a short time thereafter only because the liver produces, from lactate and certain amino acids, the minimum amount of glucose needed to briefly forestall total depletion. Glycogen depletion generally occurs after about two hours of vigorous exercise.* When glycogen depletion occurs, it brings nervous system function to a near halt, making continued exertion at the same intensity almost impossible. Marathon runners call this "hitting the wall."

Maintaining Blood Glucose for Activity To delay the onset of fatigue and enhance endurance performance, athletes try to maintain their blood glucose concentrations for as long as they can. The following guidelines can help endurance athletes maximize glucose supply:

Those who compete in endurance activities require fluid and carbohydrate fuel.

- Eat a high-carbohydrate diet (approximately 8 grams of carbohydrate per kilogram of body weight or about 70 percent of energy intake) regularly.†

- Take glucose (usually in sports drinks) periodically during activities that last for one hour or more.

- Eat carbohydrate-rich foods (approximately 60 grams of carbohydrate) immediately following activity.

Another strategy, **carbohydrate loading,** involves training the muscles to store as much glycogen as they can, while supplying the dietary glucose to enable them to do so. This strategy benefits endurance athletes who exercise at high intensity for longer than 90 minutes and who cannot meet their carbohydrate needs during competition. Although carbohydrate-loading can increase an individual's glycogen stores, performance may or may not improve.[28] Those who exercise for shorter times or at a slower pace are better served by a regular high-carbohydrate diet. The last section of this chapter, "Diets for Physically Active People," discusses how to design a high-carbohydrate diet for performance.

For perspective, snack ideas providing 60 grams of carbohydrate include:

- **16 oz sports drink and a small bagel**
- **16 oz milk and four oatmeal-raisin cookies**
- **8 oz pineapple juice and a granola bar**

Glucose during Activity Muscles can obtain the glucose they need not only from glycogen stores but also from foods and beverages consumed during activity.[29] Consuming carbohydrate is especially useful during exhausting endurance activities (lasting more than one hour) and during games such as soccer or hockey, which last for hours and demand repeated bursts of intense activity.[30]

Endurance athletes often run short of glucose by the end of competitive events. To ensure optimal carbohydrate intake, sports nutrition experts recommend 30 to 60 grams of carbohydrate per hour during prolonged events.[31] Carbohydrate-based sports drinks

anaerobic (AN-air-ROH-bic): not requiring oxygen. Anaerobic activity may require strength but does not work the heart and lungs very hard for a sustained period.

lactate: a compound produced during the breakdown of glucose in anaerobic metabolism.

carbohydrate loading: a regimen of moderate exercise, followed by eating a high-carbohydrate diet, which enables muscles to temporarily store glycogen beyond their normal capacity; also called *glycogen loading* or *glycogen supercompensation*.

*Here "vigorous exercise" means exercise at 75 percent of VO_2 max.

†Percentage of energy intake is meaningful only when total energy intake is known. Consider that at high energy intakes (5000 kcalories/day), even a moderate carbohydrate diet (50 percent of energy intake) supplies 625 grams of carbohydrate—enough for a 165-pound (75 kilogram) athlete in heavy training. By comparison, at a moderate energy intake (2000 kcalories/day), a high-carbohydrate intake (70 percent of energy intake) supplies 350 grams—plenty of carbohydrate for most people, but not enough for athletes in heavy training.

offer a convenient way to meet this recommendation and also help replace water and electrolyte losses. Thus, to ensure optimal hydration and carbohydrate intake, endurance athletes are advised to drink one-half to one liter of a 4 to 8 percent carbohydrate-based sports drink per hour, in small, frequent doses during activity.[32] During the last stages of an endurance competition, when glycogen is running low, glucose consumed during the event can slowly make its way from the digestive tract to the muscles and augment the body's supply of glucose enough to forestall exhaustion.

Glucose after Activity Eating high-carbohydrate foods *after* physical activity also enlarges glycogen stores. Train normally; then, within two hours after physical activity, consume a high-carbohydrate meal, such as a glass of orange juice and some graham crackers, toast, or cereal. This method accelerates the rate of glycogen storage.[33] This is especially important to athletes who train hard more than once per day. For athletes who do not feel like eating right after exercise, **high-carbohydrate energy drinks** are available. These fruit-flavored drinks are higher in kcalories and carbohydrate than the regular sports drinks discussed later in this chapter (p. 273).

Nutrition in Practice 2 discusses the possible health benefits of eating a diet ranking low on the glycemic index. Such a diet may also benefit endurance performance. Some research indicates that foods with a low glycemic index enhance fatty acid availability and use during subsequent activity, thereby reducing reliance on the muscles' own lipid and glycogen stores.[34] More research is needed to confirm these findings.

Degree of Training Affects Glycogen Use Training affects glycogen use during activity in at least two ways. First, muscles that deplete their glycogen stores through work adapt to store greater amounts of glycogen to support that work. Second, trained muscles burn more fat and at higher intensities than untrained muscles, so they require less glucose to perform the same amount of work. A person attempting an activity for the first time uses up much more glucose per minute than an athlete trained to perform it.

IN SUMMARY

<table>
<tr><td>■</td><td>Glucose is supplied by dietary carbohydrate or made by the liver.</td></tr>
<tr><td>■</td><td>Glucose is stored in both liver and muscle tissue as glycogen.</td></tr>
<tr><td>■</td><td>Total glycogen stores affect an athlete's endurance.</td></tr>
<tr><td>■</td><td>The more intense an activity, the more glucose it demands.</td></tr>
<tr><td>■</td><td>During anaerobic metabolism, the body spends glucose rapidly and accumulates lactate.</td></tr>
<tr><td>■</td><td>Physical activity of long duration places demands on the body's glycogen stores.</td></tr>
<tr><td>■</td><td>Carbohydrate ingested before and during long-duration activity may help to forestall fatigue.</td></tr>
<tr><td>■</td><td>Carbohydrate loading is a regimen of physical activity and diet that enables an athlete's muscles to store larger-than-normal amounts of glycogen to extend endurance.</td></tr>
<tr><td>■</td><td>Highly trained muscles use less glucose and more fat than do untrained muscles to perform the same work, so their glycogen lasts longer.</td></tr>
</table>

Fat Use during Activity

As Figure 10-3 (p. 264) showed, researchers have long recognized the importance of a high-carbohydrate diet for endurance performance. When endurance athletes "fat load" by consuming high-fat, low-carbohydrate diets for one to three days, their

Factors that affect glucose use during physical activity:

I Carbohydrate intake

I Intensity and duration of the activity

I Degree of training

performance is impaired because their small glycogen stores are depleted quickly.[35] Endurance athletes who adhere to a high-fat, low-carbohydrate diet for more than one week, however, adapt by relying more on fat to fuel activity. Nonetheless, even with fat adaptation, performance benefits are not consistently evident.[36]

Diets high in saturated fat carry risks of heart disease, too. Physical activity offers some protection against cardiovascular disease, but athletes, like everyone else, can suffer heart attacks and strokes. Most nutrition experts agree that the potential for adverse health effects from prolonged high-fat diets makes them an unwise choice for athletes.

A diet that overly restricts fat is not recommended either. Athletes who restrict fat below 20 percent of total energy intake may fail to consume adequate energy and nutrients. Sports nutrition experts recommend that endurance athletes consume 20 to 35 percent of their energy from fat.[37]

In contrast to *dietary* fat, *body* fat stores are extremely important during physical activity, as long as the activity is not too intense. Unlike the body's glycogen stores, which are limited to less than 2000 kcalories of energy, fat stores can total 70,000 kcalories or more and can fuel hours of activity without running out. Early in activity, muscles begin to draw on fatty acids from two sources—fats stored within the working muscles and fats from fat deposits such as the fat under the skin. Areas with the most fat to spare donate the greatest amounts of fatty acids to the blood (although they may not be the areas from which one might choose to lose fat). This is why "spot reducing" does not work: muscles do not own the fat that surrounds them. Fat cells release fatty acids into the blood for all the muscles to share. Proof is found in a tennis player's arms: the fat folds measure the same in both arms, even though one arm has better-developed muscles than the other. A balanced fitness program that includes strength training, however, will tighten muscles underneath the fat, improving the overall appearance. Keep in mind that some body fat is essential to good health.

Intensity and Duration Affect Fat Use   The *intensity* of physical activity affects fat use. As the intensity of activity increases, fat makes less and less of a contribution to the fuel mixture. Remember that fat can be broken down for energy only by aerobic metabolism. For fat to fuel activity, oxygen must be abundantly available. If a person is breathing easily during activity, the muscles are getting all the oxygen they need and are able to use more fat in the fuel mixture.

The *duration* of activity also influences fat use. Early in an activity, as the muscles draw on fatty acids, blood levels fall. If the activity continues for more than a few minutes, the hormone epinephrine signals the fat cells to begin breaking down their stored triglycerides and liberating fatty acids into the blood. After about 20 minutes of activity, the blood fatty acid concentration surpasses the normal resting concentration. Thereafter, sustained, moderate activity uses body fat stores as its major fuel.

Degree of Training Affects Fat Use   Training—repeated aerobic activity—produces the adaptations that permit the body to draw more heavily on fat for fuel. Aerobically trained muscles burn fat more readily than untrained muscles. With aerobic training, the heart and lungs also become stronger and better able to deliver oxygen to the muscles during high-intensity activities. In turn, this improved oxygen supply enables the muscles to burn more fat. These adaptations reward not only trained athletes, but also all active people; a person who trains by way of aerobic activities such as distance running or cycling becomes well suited to the activity.

IN SUMMARY

▌ Athletes who eat high-fat diets may burn more fat during endurance activity, but the risks to health outweigh any possible performance benefits.

▌ The intensity and duration of activity, as well as the degree of training, affect fat use.

Low- to moderate-intensity aerobic exercises that can be sustained for a long time (more than 20 minutes) use some glucose but more fat for fuel.

Factors that affect fat use during physical activity:
▌ Fat intake
▌ Intensity and duration of the activity
▌ Degree of training

high-carbohydrate energy drinks: fruit-flavored commercial beverages used to restore muscle glycogen after exercise or as a pre-game beverage.

Protein Use during Activity

Athletes use protein to build and maintain muscle and other lean tissue structures and, to a small extent, to fuel activity. The body handles protein differently during activity than during rest.

Protein for Building Muscle Tissue In the hours of rest that follow physical activity, muscles speed up their rate of protein synthesis—they build more of the proteins they need to perform the activity. Research shows that eating high-quality protein, either by itself or together with carbohydrate, immediately following physical activity stimulates muscle protein synthesis.[38] Whenever the body rebuilds a part of itself, it also tears down old structures to make way for new ones. Repeated activity, with just a slight overload, triggers the protein-dismantling and the protein-synthesizing equipment of each muscle cell to make needed changes—that is, to adapt.

Dietary protein provides the needed amino acids for synthesis of new muscle proteins. As Chapter 4 pointed out, the true director of synthesis of muscle protein is physical activity itself.

The intensity and pattern of muscle contractions initiate signals that direct the muscles' genetic material to make particular proteins. For example, a weight lifter's workout sends the information that muscle fibers need added bulk for strength and more enzymes for making and using glycogen. A jogger's workout stimulates production of proteins needed for aerobic oxidation of fat and glucose. Muscle cells are exquisitely responsive to the need for proteins, and they build them conservatively, only as needed.

Finally, after muscle cells have made all the decisions about which proteins to build and when, protein nutrition comes into play. During active muscle-building phases of training, a weight lifter might add between ¼ ounce and 1 ounce (between 7 and 28 grams) of protein to existing muscle mass each day. This extra protein comes from ordinary food.

Protein for Fuel Not only do athletes retain more protein, but they also use a little more protein as fuel. Muscles speed up their use of amino acids for energy during physical activity, just as they speed up their use of glucose and fatty acids. Still, protein contributes at most about 10 percent of the total fuel used, both during activity and during rest.

For perfect functioning, every nutrient is needed.

Diet Affects Protein Use during Activity The factors that regulate how much protein is used during activity seem to be the same ones that regulate the use of glucose and fat—one factor is diet. People who consume diets adequate in energy and rich in *carbohydrate* use less protein than those who eat protein- and fat-rich diets. Recall that carbohydrates spare proteins from being broken down to make glucose when needed. Because physical activity requires glucose, a diet lacking in carbohydrate necessitates the conversion of amino acids to glucose.

Intensity and Duration Affect Protein Use The intensity and duration of the activity also affect protein use. Endurance athletes who train for more than one hour per day, engaging in aerobic activity of moderate intensity and long duration, may deplete their glycogen stores by the end of their training and become more dependent on body protein for energy. In contrast, anaerobic strength training does not use more protein for energy but does demand more protein to build muscle. Thus, the protein needs of both endurance and strength athletes are higher than those of sedentary people, but not as high as the protein intakes many athletes consume.

Factors that affect protein use during physical activity:

■ Carbohydrate intake
■ Intensity and duration of the activity
■ Degree of training

Degree of Training Affects Protein Use Finally, the extent of training also affects the use of protein. Particularly in strength athletes such as bodybuilders, the higher the degree of training, the less protein a person uses during activity at a given intensity.

TABLE 10-3 **Recommended Protein Intakes for Athletes**

	Recommendations (g/kg/day)	Protein Intakes (g/day)	
		Males	Females
RDA for adults	0.8	56	44
Recommended intake for power (strength or speed) athletes	1.2–1.7	84–119	66–94
Recommended intake for endurance athletes	1.2–1.4	84–98	66–77
U.S. average intake		102	70

Note: Daily protein intakes are based on a 70-kilogram (154-pound) man and 55-kilogram (121-pound) woman.

Sources: Position of the American Dietetic Association, Dietitians of Canada, and the American College of Sports Medicine: Nutrition and athletic performance, *Journal of the American Dietetic Association* 109 (2009): 509–527; U.S. Department of Agriculture, Agricultural Research Service, 2008, *Nutrient Intakes from Food: Mean Amounts Consumed per Individual One Day, 2005–2006*. Available at www.ars.usda.gov/ba/bhnrc/fsrg; Committee on Dietary Reference Intakes, *Dietary Reference Intakes for Energy, Carbohydrate, Fiber, Fat, Fatty Acids, Cholestrol, Protein and Amino Acids* (Washington, D.C.: National Academies Press, 2005), pp. 660–661.

Protein Recommendations for Active People Although most athletes need somewhat more protein than do sedentary people, average protein intakes in the United States are high enough to cover those needs. Therefore, athletes in training should attend to protein needs but should back up the protein with ample carbohydrate. Otherwise, they will burn off as fuel the very protein they wish to retain in muscle.

The DRI Committee does not recommend greater-than-normal protein intakes for athletes, but other authorities do.[39] These recommendations specify different protein intakes for athletes pursuing different activities (see Table 10-3).[40] A later section translates protein recommendations into a diet plan and shows that no one needs protein supplements or even large servings of meat to obtain the highest recommended protein intakes.

IN SUMMARY

▌ Physical activity stimulates muscle cells to break down and synthesize protein, resulting in muscle adaptation to activity.

▌ Athletes use protein both for building muscle tissue and for energy. Diet, intensity and duration of activity, and training affect protein use during activity.

▌ Although athletes need more protein than sedentary people, a balanced, high-carbohydrate diet provides sufficient protein to cover an athlete's needs.

Vitamins and Minerals to Support Activity

Many vitamins and minerals assist in releasing energy from fuels and transporting oxygen. This knowledge has led many people to believe, mistakenly, that vitamin and mineral *supplements* offer physically active people both health benefits and athletic advantages.

Supplements

Although many athletes take supplements in the hope of improving their performance, nutrient supplements do not enhance the performance of well-nourished athletes or active people. However, deficiencies of vitamins and minerals do impede performance. Regular, strenuous physical activity increases the demand for energy, but athletes and

Nutrition in Practice 9 offers a discussion of vitamin and mineral supplements and Nutrition in Practice 10 explores supplements and other products people use in the hope of enhancing athletic performance.

active people who eat enough nutrient-dense foods to meet energy needs also meet their vitamin and mineral needs. Active people eat more food; it stands to reason that with the right choices, they will get more nutrients.

Athletes who lose weight to meet low body-weight requirements, however, may consume so little food that they fail to obtain all the nutrients they need.[41] The practice of "making weight" is opposed by many health and fitness organizations, but for athletes who choose this course of action, a single daily multivitamin–mineral tablet that provides no more than the DRI recommendations for nutrients can be beneficial. In addition, some athletes do not eat enough food to maintain body weight during times of intense training or competition. For these athletes, too, a daily multivitamin–mineral supplement can be helpful.

Some athletes believe that taking vitamin or mineral supplements just before competition will enhance performance. These beliefs are contrary to scientific reality. Most vitamins and minerals function as small parts of larger working units. After entering the blood, they have to wait for the cells to combine them with their appropriate other parts so that they can do their work. This takes time—hours or days. Vitamins or minerals taken right before an event do not improve performance, even if the person is actually suffering deficiencies of those nutrients.

Nutrients of Special Concern

In general, active people who eat well-balanced meals do not need vitamin or mineral supplements. Two nutrients, vitamin E and iron, do merit special attention, however, for different reasons. Vitamin E is addressed because so many athletes take supplements of it. Iron is discussed because some female athletes may be unaware that they need supplements.

Vitamin E During prolonged, high-intensity physical activity, the muscles' consumption of oxygen increases tenfold or more, enhancing the production of damaging free radicals in the body.[42] Vitamin E is a potent fat-soluble antioxidant that vigorously defends cell membranes against oxidative damage.

Does vitamin E supplementation protect against exercise-induced oxidative stress? Some studies find that it does; others show no effect, and still others report enhanced oxidative stress after vitamin E supplementation.[43] Recent research may offer some insight into these inconsistencies. Although free radicals are usually damaging, during repeated episodes of endurance activities, they may actually be beneficial. Free radicals activate powerful antioxidant enzymes, which may enhance the athlete's tolerance to such oxidative stresses.[44] Researchers speculate that antioxidant supplements such as vitamin E interfere with this protective adaptive response. This may explain why, in some studies, athletes taking vitamin E show signs of increased oxidative stress. Clearly, more research is needed, but in the meantime, physically active people can benefit by using vegetable oils and eating generous servings of antioxidant-rich fruits and vegetables regularly.

Iron and Performance Physically active young women, especially those who engage in endurance activities such as distance running, are prone to iron deficiency. Habitually low intakes of iron-rich foods, high iron losses through menstruation, and the high demands of muscles for the iron-containing molecules of aerobic metabolism and the muscle protein myoglobin can contribute to iron deficiency in young female athletes.

Adolescent female athletes who eat vegetarian diets may be particularly vulnerable to iron deficiency. The bioavailability of iron is often poor in plant-based diets because such diets are high in fiber and phytic acid and because the nonheme iron in plant foods is not absorbed as well as the heme iron in animal-derived foods. Vegetarian diets are usually rich in vitamin C, however, which enhances iron absorption.

Stringent weight requirements pose a risk of developing eating disorders. See Nutrition in Practice 6.

The Tolerable Upper Intake Level (UL) for vitamin E is 1000 mg per day.

To protect against iron deficiency, vegetarian athletes need to pay close attention to their intake of good dietary sources of iron (fortified cereals, legumes, nuts, and seeds) and include vitamin C–rich foods with each meal.[45] As long as vegetarian athletes, like all athletes, consume enough nutrient-dense foods, they can perform as well as anyone.

Iron deficiency impairs performance because iron helps deliver the muscles' oxygen. Insufficient oxygen delivery reduces aerobic work capacity, so the person tires easily. Whether marginal deficiency without clinical signs of anemia hinders physical performance is less clear.[46]

Early in training, athletes may develop low blood hemoglobin. This condition, sometimes called "sports anemia," is not a true iron-deficiency condition. Strenuous training promotes destruction of the more fragile, older red blood cells, and the resulting cleanup work reduces the blood's iron content temporarily. Strenuous activity also promotes increases in the fluid portion of the blood; with more fluid, the red blood cell count in a unit of blood drops. Most researchers view sports anemia as an *adaptive,* temporary response to endurance training. True iron-deficiency anemia requires treatment with prescribed iron supplements, but sports anemia goes away by itself, even with continued training.

The best strategy concerning iron is to determine individual needs. Many menstruating women border on iron deficiency even without the additional iron demand and losses incurred by physical activity. Active teens of both genders have high iron needs because they are growing. For women and teens, prescribed supplements may be needed to correct a deficiency of iron that is confirmed by tests. (Medical testing is needed to eliminate nondietary causes of anemia, such as internal bleeding or cancer.)

Female athletes may be at special risk of iron deficiency.

Rocky Widner/WNBAE via Getty Images

IN SUMMARY

- With the possible exception of iron, well-nourished active people and athletes do not need nutrient supplements.

- Female athletes need to pay special attention to their iron needs.

- Iron-deficiency anemia impairs physical performance because iron is the blood's oxygen handler.

- Sports anemia is a harmless temporary adaptation to physical activity.

Fluids and Electrolytes in Physical Activity

The body's need for water far surpasses its need for any other nutrient. If the body loses too much water, its life-supporting chemistry is compromised.

The exercising body loses water primarily via sweat; second to that, breathing uses water, exhaled as vapor. During physical activity, both routes can be significant, and dehydration is a real threat. The first symptom of dehydration is fatigue. A water loss of greater than 2 percent of body weight can reduce a person's capacity to do muscular work.[47] A person with a water loss of about 7 percent is likely to collapse. The athlete who arrives at an event even slightly dehydrated starts out at a competitive disadvantage.

Temperature Regulation

As Chapter 9 pointed out, sweat cools the body. The conversion of water to vapor uses up a great deal of heat, so as sweat evaporates, it cools the skin's surface and the blood flowing beneath it.

Symptoms of heat stroke:[48]

■ Clumsiness
■ Confusion, other mental changes, loss of consciousness
■ Dizziness
■ Headache
■ Internal (rectal) temperature above 104° Fahrenheit
■ Nausea
■ Stumbling
■ Hot, flushed, dry skin, or pale, sweat-soaked skin

Hyperthermia In hot, humid weather, sweat may fail to evaporate because the surrounding air is already laden with water. In **hyperthermia,** body heat builds up and triggers maximum sweating, but without sweat evaporation, little cooling takes place. In such conditions, active people must take precautions to avoid **heat stroke.** Heat stroke is an especially dangerous accumulation of body heat with accompanying loss of body fluid. To reduce the risk of heat stroke, drink enough fluid before and during the activity, rest in the shade when tired, and wear lightweight clothing that allows sweat to evaporate. The rubber or heavy suits sold with promises of weight loss during physical activity are dangerous because they promote profuse sweating, prevent sweat evaporation, and invite heat stroke. If you experience any of the symptoms of heat stroke listed in the margin, stop your activity, sip cold fluids, seek shade, and ask for help. The condition demands medical attention—it can kill.

Hypothermia In cold weather, **hypothermia,** or loss of body heat, can pose as serious a threat as heat stroke does in hot weather. Inexperienced runners participating in long races on cold or wet, chilly days are especially vulnerable to hypothermia. Slow runners can produce too little heat to keep warm, especially if their clothing is inadequate. Early symptoms of hypothermia include feeling cold, shivering, apathy, and social withdrawal.[49] As body temperature continues to fall, shivering stops, and disorientation, slurred speech, and change in behavior or appearance set in. People with these symptoms soon become helpless to protect themselves from further body heat losses. Even in cold weather, the body still sweats and needs fluids, but the fluids should be warm or at room temperature to help prevent hypothermia.

Fluid Needs during Physical Activity

Endurance athletes can lose 1.5 quarts or more of fluid during *each hour* of activity. To prepare for fluid losses, the athlete must hydrate before activity. To replace fluid losses, the person must rehydrate during and after activity. (Table 10-4 presents one schedule of hydration for physical activity.) Even then, in hot weather, the digestive tract may not be able to absorb enough water fast enough to keep up with an athlete's sweat losses, and some degree of dehydration may be inevitable. Athletes who know their body's **hourly sweat rate** can strive to replace the total amount of fluid lost during activity to prevent dehydration.

Athletes who are preparing for competition are often advised to drink extra fluids in the last few days of training before the event. The extra fluid is not stored in the body, but drinking extra ensures maximum tissue hydration at the start of the event. Full hydration is imperative for every athlete both in training and in competition. Any

TABLE 10-4 Hydration Schedule for Physical Activity

When to Drink	Amount of Fluid
2 to 3 hr before activity	2 to 3 c
15 min before activity	1 to 2 c
Every 15 min during activity	½ to 1 c (Drink enough to minimize loss of body weight, but don't overdrink.)
After activity	2 c for each pound of body weight lost[a]

[a]Drinking 2 cups of fluid every 20 to 30 minutes after exercise until the total amount required is consumed is more effective for rehydration than drinking the needed amount all at once. Rapid fluid replacement after exercise stimulates urine production and results in less body water retention.

Sources: Adapted from American College of Sports Medicine, Position stand, Exercise and fluid replacement, *Medicine & Science in Sports & Exercise* 39 (2007): 377–390; C. K. Seto, D. Way, and N. O'Connor, Environmental illness in athletes, *Clinics in Sports Medicine* 24 (2005): 695–718; R. Murray, Fluid, electrolytes, and exercise, in *Sports Nutrition: A Practice Manual for Professionals.* 4th ed., ed. M. Dunford (Chicago: American Dietetic Association, 2006), pp. 94–115; D. J. Casa, P. M. Clarkson, and W. O. Roberts, American College of Sports Medicine Roundtable on Hydration and Physical Activity: Consensus statements, *Current Sports Medicine Reports* 4 (2005): 115–127.

coach or athlete who withholds fluids during practice for any reason takes a great risk and is subject to sanctions by the American College of Sports Medicine.

Athletes who rely on thirst to govern fluid intake can easily become dehydrated. During activity thirst becomes detectable only *after* fluid stores are depleted. Do not wait to feel thirsty before drinking.

Active people need extra fluid, even in cold weather.

Water What is the best fluid to support physical activity? For noncompetitive, everyday active people, plain cool water is recommended for two reasons: (1) water rapidly leaves the digestive tract to enter the tissues, and (2) it cools the body from the inside out. Endurance athletes are an exception: they need more from their fluids than water alone. The first priority for endurance athletes should always be replacement of fluids to prevent life-threatening heat stroke. But endurance athletes also need carbohydrate to supplement their limited glycogen stores, so glucose is important, too. A later section compares water and sports drinks as fluid sources for endurance athletes.

Electrolyte Losses and Replacement During physical activity, the body loses electrolytes—the minerals sodium, potassium, and chloride—in sweat. Beginners lose these electrolytes to a much greater extent than do trained athletes. The body's adaptation to physical activity includes better conservation of these electrolytes. To replenish lost electrolytes, a person ordinarily needs only to eat a regular diet that meets energy and nutrient needs. In events lasting more than one hour, sports drinks may be needed to replace fluids and electrolytes. Salt tablets can worsen dehydration and impair performance; they increase potassium losses, irritate the stomach, and cause vomiting. Athletes should avoid them.

Sodium Depletion When athletes compete in endurance sports lasting longer than three hours, replenishing electrolytes becomes crucial. If athletes sweat profusely over a long period of time and do not replace lost sodium, a dangerous condition of sodium depletion, known as **hyponatremia,** may result. The symptoms of hyponatremia are similar to, but not the same as, those of dehydration (see the margin). Recent research shows that some athletes who sweat profusely may also lose more sodium in their sweat than others—and are prone to debilitating heat cramps.[50] These athletes lose twice as much sodium in sweat as athletes who do not cramp, making them more susceptible to hyponatremia. Depending on individual variation, exercise intensity, and changes in ambient temperature and humidity, sweat rates for these athletes can exceed 2 quarts per hour.[51]

Hyponatremia may also occur when endurance athletes drink such large amounts of water over the course of a long event that they overhydrate, diluting the body fluids to such an extent that the sodium concentration becomes extremely low. During long competitions, when athletes lose sodium through heavy sweating *and* consume excessive amounts of liquids, especially water, hyponatremia becomes likely.

Some athletes may be vulnerable to hyponatremia even when they drink sports drinks during an event. Sports drinks do contain sodium, but as a later section points out, the sodium content of sports drinks is low, and in some cases, too low to replace sweat losses. Still, sports drinks do offer more sodium than plain water.

To prevent hyponatremia, endurance athletes need to replace sodium during prolonged events. They should favor sports drinks over water and eat pretzels in the last half of a long race. Some may need beverages with higher sodium concentrations than commercial sports drinks. In the days before the event, especially an event in the heat, athletes should not restrict salt in their diets.

Sports Drinks

Hydration is critical to optimal performance. As stated earlier, water best meets the fluid needs of most people, yet manufacturers market many good-tasting sports drinks that deliver both fluid and carbohydrate for active people. The term *sports drink*

Symptoms of hyponatremia:
▮ **Severe headache**
▮ **Vomiting**
▮ **Bloating, puffiness from water retention (shoes tight, rings tight)**
▮ **Confusion**
▮ **Seizure**

hyperthermia: an above-normal body temperature.

heat stroke: an acute and life-threatening reaction to heat buildup in the body.

hypothermia: a below-normal body temperature.

hourly sweat rate: the amount of weight lost plus fluid consumed during exercise per hour.

hyponatremia (HIGH-poh-na-TREE-mee-ah): a decreased concentration of sodium in the blood (*hypo* means "below"; *natrium* means "sodium"; *emia* means "blood").

generally refers to beverages that contain carbohydrates and electrolytes in specific concentrations, and they are the focus of this discussion.

Many sports drinks compete for their share of the more than $1 billion market. What do sports drinks have to offer?

Fluid Sports drinks offer fluids to help offset the loss of fluids during physical activity, but plain water can do this, too. Alternatively, diluted fruit juices or flavored water can be used if preferred to plain water.

Glucose Sports drinks offer simple sugars or **glucose polymers** that help maintain hydration and blood glucose and enhance performance as effectively as, or in some circumstances even better than, water. Such measures are especially beneficial for strenuous endurance activities lasting longer than one hour, during intense activities, or during prolonged competitive games that demand repeated intermittent activity.[52] Sports drinks are also suitable for events lasting less than one hour, although plain water is appropriate as well. Fluid transport to the tissues from beverages containing up to 8 percent glucose is rapid. Most sports drinks contain about 7 percent carbohydrate (about half the sugar of ordinary soft drinks, or about 5 teaspoons in each 12 ounces). Less than 6 percent may not enhance performance, and more than 8 percent may cause abdominal cramps, nausea, and diarrhea. Although glucose does enhance endurance and performance in strenuous competitive events, for the moderate exerciser, it can be counterproductive if weight loss is the goal. Glucose is sugar, and like candy, it provides only empty kcalories—no vitamins or minerals. Most sports drinks provide between 50 and 100 kcalories per cup.

Sodium and Other Electrolytes Sports drinks offer sodium and other electrolytes to help replace those lost during physical activity. Sodium in sports drinks also helps to increase the rate of fluid absorption from the GI tract and maintain plasma volume during activity and recovery. Most physically active people do not need to replace the minerals lost in sweat immediately; a meal eaten within hours of competition replaces these minerals soon enough. Most sports drinks are relatively low in sodium, however, so those who choose to use these beverages run little risk of excessive intake.

Good Taste Manufacturers reason that if a drink tastes good, people will drink more, thereby ensuring adequate hydration. For athletes who prefer the flavors of sports drinks over water, it may be worth paying for good taste to replace lost fluids.

For athletes who exercise for one hour or more, sports drinks provide an advantage over water. For most physically active people, though, water is the best fluid to replenish lost fluids. The most important thing to do is drink—even if you don't feel thirsty.

Enhanced Water

Another beverage often marketed to athletes and active people is **enhanced water.** Enhanced waters are lightly flavored waters with lower carbohydrate and electrolyte contents than traditional sports drinks. Marketers promote these beverages for the added vitamins, minerals, and in some cases, protein, they contain. In fact, most enhanced waters contain small amounts of only a few minerals, some of the B vitamins, and sometimes vitamin C or vitamin E. In the context of daily needs, the vitamins and minerals in these drinks do not add up to much. For example, it takes a quart of most of these beverages to provide only 10 percent of the RDA for iron or calcium. Quite simply, enhanced waters are not a substitute for eating nutrient-rich fruits and vegetables. Enhanced waters may not be harmful (except maybe to the wallet), but most people do not need them. Plain water can meet fluid needs. If the flavor of enhanced waters encourages greater fluid intake, then they may offer some advantage. Serious endurance athletes need the carbohydrate-electrolyte sports drinks discussed earlier.

Other Beverages

Some beverages, such as soft drinks or the increasingly popular energy drinks, deliver caffeine along with fluid. Caffeine is a stimulant, and athletes sometime use it to enhance performance, as this chapter's Nutrition in Practice explains. Carbonated soft drinks, with or without caffeine, may not be a wise choice for athletes: bubbles make a person feel full quickly and so may limit fluid intake. Some energy drinks contain amounts of caffeine equivalent to a cup or two of coffee. When used in excess or in combination with stimulants or other unregulated substances, energy drinks can hinder performance and are potentially dangerous.[53] Another reason energy drinks should not be used for fluid replacement during athletic events is that the carbohydrate concentrations are too high for optimal fluid absorption.

Athletes sometimes drink beverages that contain alcohol, but these beverages are inappropriate as fluid replacements. Alcohol is a diuretic. It promotes the excretion of water; of vitamins such as thiamin, riboflavin, and folate; and of minerals such as calcium, magnesium, and potassium—exactly the wrong effects for fluid balance and nutrition. Alcohol has many detrimental effects on physical activity. It impairs temperature regulation, making hypothermia or heat stroke much more likely. It alters perceptions and slows reaction time. It depletes strength and endurance and deprives people of their judgment, thereby compromising their safety in sports. Many sports-related fatalities and injuries each year involve alcohol.

IN SUMMARY

▌ Hyperthermia and heat stroke can be a threat to physically active people in hot, humid weather.

▌ Hypothermia threatens those who exercise in the cold.

▌ Physically active people lose fluids and must replace them to avoid dehydration.

▌ Water is the best drink for most physically active people, but endurance athletes need beverages that supply glucose as well as fluids.

▌ During events lasting longer than three hours, athletes need to pay special attention to replacing sodium losses to prevent hyponatremia.

▌ Caffeine-containing drinks within limits may not impair performance, but water and fruit juice are preferred.

▌ Alcohol use can impair performance in many ways and is not recommended.

Diets for Physically Active People

No one diet best supports physical performance. Active people who choose foods within the framework of the diet-planning principles presented in Chapter 1 can design many excellent diets.

Nutrient Density A physically active person needs a diet composed mostly of nutrient-dense foods, the kind that supply a maximum of vitamins and minerals for the energy they provide. When athletes eat mostly refined, processed foods that have suffered nutrient losses and contain too much added sugar and solid fat, their nutrition status suffers. Even if foods are fortified or enriched, manufacturers cannot replace the whole range of nutrients and phytochemicals lost in refining. For example, manufacturers mill out much of a food's original magnesium and chromium but do not replace them. This does not mean that athletes can never choose a white bread, bologna, and mayonnaise sandwich, but only that later they should eat a large salad or big

The caffeine contents of selected beverages (including energy drinks) and foods are listed in the table at the beginning of Appendix A.

Beer facts:

▪ Beer is not carbohydrate-rich. Beer is kcalorie-rich, but only one-third of its kcalories are from carbohydrates. The other two-thirds are from alcohol.

▪ Beer is mineral-poor. Beer contains a few minerals, but to replace those lost in sweat, athletes need good sources such as fruit juices.

▪ Beer is vitamin-poor. Beer contains tiny traces of some B vitamins, but it cannot compete with rich food sources.

▪ Beer causes fluid losses. Beer is a fluid, but alcohol is a diuretic and causes the body to lose more fluid in urine than is provided by the beer.

glucose polymers: compounds that supply glucose, not as single molecules, but linked in chains somewhat like starch. The objective is to attract less water from the body into the digestive tract.

enhanced water: water that is fortified with ingredients such as vitamins, minerals, protein, oxygen, or herbs. Enhanced water is marketed as *vitamin water, sports water, oxygenated water,* and *protein water.*

portions of vegetables and whole grains and drink a glass of milk to compensate. The nutrient-dense foods will provide the magnesium and chromium; the bologna sandwich provides extra energy, mostly from fats.

Carbohydrate Athletes must eat for energy, and their energy needs can be immense. Athletes need full glycogen stores, and they need to strive to prevent heart disease and cancer by limiting fat, especially saturated fat. To serve these special needs, a diet that is high in carbohydrate (60 to 70 percent of total kcalories), moderate in fat (20 to 35 percent), and adequate in protein (10 to 20 percent) works best. Even if you do not compete in glycogen-depleting events, such a diet provides adequate fiber while supplying abundant nutrients and energy.

With these principles in mind, compare the two 500-calorie sandwich meals in the margin. The trick to getting enough carbohydrate energy is easy, at least in theory: just reduce the amount of fat and meat in a meal, and let carbohydrate-rich foods fill in for them.

Adding carbohydrate-rich foods is a sound and reasonable option for increasing energy intake, up to a point. It becomes unreasonable when the person cannot eat enough food to meet energy needs. At that point, the person can add more food energy into the diet by adding concentrated carbohydrate foods such as dried fruits, sweet potatoes, and nectars, and even high-fat foods such as avocados and nuts. Still, these energy-rich additions must be superimposed on nutrient-rich choices; energy alone is not enough.

Some athletes use commercial high-carbohydrate liquid supplements to obtain the carbohydrate and energy needed for heavy training and top performance. Most of these products contain glucose polymers and about 18 to 24 percent carbohydrate. These supplements do not *replace* regular food; they are meant to be used in *addition* to it. Unlike the sports beverages discussed earlier, these high-carbohydrate supplements are too concentrated in carbohydrate to be used for fluid replacement.

Protein In addition to carbohydrate and some fat (and the energy they provide), athletes need protein. Meats and milk products head the list of protein-rich foods, but suggesting that active people eat more than the recommended servings of meat is narrow advice. As mentioned repeatedly, active people need diets rich in carbohydrate, and meats have none to offer. Legumes, whole grains, and vegetables provide some protein with abundant carbohydrate. Table 10-3 (p. 269) shows recommended protein intakes.

A Performance Diet Example A person weighing 70 kilograms who engages in vigorous daily physical activity could easily require more than 3000 kcalories per day. To meet this need, the person can choose a variety of nutrient-dense foods. Figure 10-4 shows examples of meals that provide about 3000 kcalories. These meals supply about 125 grams of protein, equivalent to the highest recommended intake for an athlete weighing 160 pounds. Obviously, the higher a person's energy intake, the more protein that person will receive, assuming the foods chosen are nutrient dense. This relationship breaks down only when people meet their energy needs with high-fat, high-sugar confections.

The meals in Figure 10-4 provide 63 percent of their kcalories from carbohydrate. Athletes who train exhaustively for endurance events may want to aim for somewhat higher carbohydrate intakes.

Meals before Competition Science indicates that the **pregame meal** include plenty of fluids and be easy to digest. It should provide between 300 and 800 kcalories, primarily from carbohydrate-rich foods that are familiar and well-tolerated by the athlete. The meal should end three to four hours before competition to allow time for the stomach to empty before exertion.

Compare and decide which best meets your needs:

- 1 sandwich of 2 slices bologna, 2 slices white bread, 2 tbsp mayonnaise (525 kcal, 9% protein, 23% carbohydrate, 68% fat)

or

- 2 sandwiches of 2 slices lean ham, 4 slices whole-wheat bread, 2 tsp mayonnaise (503 kcal, 20% protein, 51% carbohydrate, 29% fat)

Breakfast
1 c shredded wheat with
  low-fat milk and banana
2 slices whole-wheat toast
  with jelly
1½ c orange juice

Lunch
2 turkey sandwiches
1½ c low-fat milk
Large bunch of grapes

Snack
3 c plain popcorn
A smoothie made from:
  1½ c apple juice
  1½ frozen bananas

Dinner
Salad: 1 c spinach, carrots,
  and mushrooms with
  ½ c garbanzo beans,
  1 tbs sunflower seeds, and
  1 tbs ranch salad dressing
1 c spaghetti with meat sauce
1 c green beans
1 corn on the cob
2 slices Italian bread
4 tsp butter
1 piece angel food cake with
  fresh strawberries and
  whipping cream
1 c low-fat milk

Total kcal: about 3000

63% kcal from carbohydrate
22% kcal from fat
15% kcal from protein

All vitamin and mineral intakes exceed
the RDA for both men and women.

FIGURE 10-4 High-Carbohydrate Meals for Athletes
This sample menu provides about 3000 kcalories, with almost 520 grams of carbohydrate (63 percent of total kcalories) and about 125 grams of protein (15 percent of total kcalories). In addition to meeting the carbohydrate and protein needs of an athlete who has a 3000-kcalorie intake, these meals also meet or exceed recommendations for all vitamins and minerals.

Breads, potatoes, pasta, and fruit juices—carbohydrate-rich foods low in fat, protein, and fiber—form the basis of the pregame meal (see Figure 10-5 for some examples). Bulky, fiber-rich foods such as raw vegetables and high-fiber cereals, although usually desirable, are best avoided just before competition. Such foods can cause stomach discomfort during performance.

What about drinks or sport bars claiming to provide "complete" nutrition? These mixtures of carbohydrate, protein (usually amino acids), fat, some fiber, and certain vitamins and minerals may taste good and provide additional food energy for a game or for weight gain. They fall short of providing complete nutrition, however, because

pregame meal: a meal eaten three to four hours before athletic competition.

300-kcalorie meal
1 large apple
4 saltine crackers
1½ tbs reduced-fat
  peanut butter

500-kcalorie meal
1 large whole-wheat bagel
2 tbs jelly
1½ c low-fat milk

750-kcalorie meal
1 large baked potato
2 tsp soft margarine
1 c steamed broccoli
1 c mixed carrots and green
  peas
5 vanilla wafers
1½ c apple or pineapple juice

FIGURE 10-5 Examples of High-Carbohydrate Pregame Meals
Pregame meals should be eaten three to four hours before the event and provide 300 to 800 kcalories, primarily from carbohydrate-rich foods. Each of these sample meals provides at least 65 percent of total kcalories from carbohydrate.

they lack many of the beneficial nutrients and phytochemicals that real food offers. These products may provide one advantage for active people—they are easy to eat in the hours before competition.

Liquid meals are easy to digest, and many such meals are commercially available. Alternatively, athletes can mix fat-free milk or juice, frozen fruits, and flavorings in a blender. Do not drop a raw egg in the blender because raw eggs often carry bacteria that cause food poisoning.

High-carbohydrate, liquid pregame meal ideas:
- Apple juice, frozen banana, and cinnamon
- Papaya juice, frozen strawberries, and mint
- Fat-free milk, frozen banana, and vanilla

IN SUMMARY

- The person who wants to excel physically will apply accurate nutrition knowledge along with dedication to rigorous training.

- A diet that provides ample fluid and consists of a variety of nutrient-dense foods in quantities to meet energy needs will enhance not only athletic performance, but overall health as well.

- Carbohydrate-rich foods that are light and easy to digest are recommended for the pregame meal.

Training and genetics being equal, who would win a competition—the person who habitually consumes less than the amounts of nutrients needed or the one who arrives at the event with a long history of full nutrient stores and well-met metabolic needs?

Your Diet

Fitness and Nutrition

Fitness depends on a certain minimum amount of physical activity. Ideally, the quantity and quality of the physical activity you select will improve your cardiorespiratory endurance, body composition, strength, and flexibility. Examine your activity choices by keeping an activity diary for one week. For each physical activity, be sure to record the type of activity, the level of intensity, and the duration. In addition, record the times and places of beverage consumption and the types and amounts of beverages consumed. Now compare the choices you made in your one-week activity diary to the American College of Sports Medicine's guidelines for physical fitness (Table 10-1, p. 259).

- How often were you engaged in aerobic activity to improve cardiorespiratory endurance? Was the intensity of the activity moderate or vigorous? Did each session last at least

30 minutes if of moderate intensity and at least 20 minutes if of vigorous intensity?

- How often did you participate in resistance activities to develop strength? Was the intensity enough to enhance muscle strength and improve body composition? Did you perform 8 to 10 different exercises, repeating each one 8 to 12 times?

- How often did you stretch to improve your flexibility? Was the intensity enough to develop and maintain a full range of motion? Did you hold each stretch 15 to 30 seconds and repeat each stretch two to four times?

- Do you drink plenty of fluids daily, especially water, before, during, and after physical activity?

- What changes could you make to improve your fitness?

Clinical Applications

1. During her freshman year in college, Kim spent much of her free time bonding with new friends over pizzas, burgers, and fried chicken. Now in her sophomore year, she has decided that it is time to shed the "freshman 15" she gained and get fit quickly. She is replacing her pizzas and burgers with protein shakes and diet sodas; she avoids carbohydrates; and she takes vitamin supplements because she has heard they will give her energy. Kim's demanding sophomore academic load leaves little time during the week to do anything but go to class, study, and work on assignments. She has therefore decided to spend much of her weekend time running, working out with weights, swimming, and playing pick-up soccer on Sunday

afternoons. After a few weeks of her new eating and fitness plan, however, Kim is feeling tired and run down much of the time, her muscles ache, and she falls asleep when she is trying to study.

▌ What dietary advice would you suggest to help Kim feel healthier and more energetic?

▌ What fitness strategies would you offer Kim to prevent her fatigue and sore muscles?

2. Zak, a junior in college, has been weight training four days a week for the past year to gain strength and bulk up his muscles. Along with his training, he has changed his diet from a normal mixed diet to one that emphasizes large servings of meat, eggs, and milk. Zak also takes amino acid supplements in hopes of building more muscle, faster. What advice would you offer Zak about his dietary habits? What would you tell him about the supplements he is taking?

Self Check

1. Regular physical activity helps protect against:
   a. backaches, cancer, and emphysema.
   b. cancer, diabetes, and heart disease.
   c. obesity, kidney disease, and anemia.
   d. high blood pressure, cancer, and osteopenia.

2. Fitness benefits health by:
   a. increasing lean body tissue and enhancing resistance to colds and other infectious diseases.
   b. lowering the risk of heart disease, decreasing muscle mass, and improving nutritional health.
   c. building bone strength, lowering the risk of some cancers, and increasing anxiety.
   d. reducing diabetes risk, compromising lung function, and promoting a strong self-image.

3. Which of the following characteristics is not a component of fitness?
   a. muscle endurance
   b. conditioning
   c. flexibility
   d. muscle strength

4. Which of the following provides most of the energy the muscles use in the early minutes of activity?
   a. fat
   b. protein
   c. glycogen
   d. vitamins

5. "Hitting the wall" is a term runners sometimes use to describe:
   a. dehydration.
   b. competition.
   c. indigestion.
   d. glucose depletion.

6. Conditioned muscles rely less on ___ and more on ___ for energy.
   a. protein; fat
   b. fat; protein
   c. glycogen; fat
   d. fat; glycogen

7. Vitamin or mineral supplements taken right before an event are useless for improving performance because the:
   a. athlete sweats the nutrients out during the event.
   b. stomach can't digest supplements during physical activity.
   c. nutrients are diluted by all the fluids the athlete drinks.
   d. body needs hours or days for the nutrients to do their work.

8. Physically active young women, especially those who are endurance athletes, are prone to:
   a. energy excess.
   b. iron deficiency.
   c. protein overload.
   d. iodine deficiency.

9. Plain, cool water is the best fluid for everyday active people because it:
   a. rapidly leaves the digestive tract to enter the tissues and cool the body.
   b. tastes good.
   c. provides carbohydrate.
   d. leaves the digestive tract slowly.

10. A recommended pregame meal includes plenty of fluids and provides between:
    a. 300 and 800 kcalories, mostly from fat-rich foods.
    b. 50 and 100 kcalories, mostly from fiber-rich foods.
    c. 1000 and 2000 kcalories, mostly from protein-rich foods.
    d. 300 and 800 kcalories, mostly from carbohydrate-rich foods.

Answers to these questions can be found in Appendix H.

Notes

1. Q. Sun and coauthors, Physical activity at midlife in relation to successful survival in women at age 70 years or older, *Archives of Internal Medicine* 170 (2010): 194–201; W. J. Nusselder and coauthors, Living healthier for longer: Comparative effects of three heart-healthy behaviors on life expectancy with and without cardiovascular disease, *BMC Public Health* 24 (2009): 487;T. M. Manini and coauthors, Daily activity energy expenditure and mortality among older adults, *Journal of the American Medical Association* 296 (2006): 171–179.

2. Centers for Disease Control and Prevention, Surveillance of certain health behaviors and conditions among states and selected local areas: Behavioral Risk Factor Surveillance System (BRFSS), United States, 2006, *Morbidity and Mortality Weekly Report* 57 (2008): 1–188; Centers for

Disease Control and Prevention, Prevalence of regular physical activity among adults: United States, 2001 and 2005, *Morbidity and Mortality Weekly Report* 56 (2007): 1209–1212.

3. T. Y. Warren and coauthors, Sedentary behaviors increase risk of cardiovascular disease mortality in men, *Medicine and Science in Sports and Exercise* 42 (2010): 879–885; P. T. Katzmarzyk and coauthors, Sitting time and mortality from all causes, cardiovascular disease, and cancer, *Medicine and Science in Sports and Exercise* 41 (2009): 998–1005; W. Kemmler and coauthors, Exercise decreases the risk of metabolic syndrome in elderly females, *Medicine and Science in Sports and Exercise* 41 (2009): 297–305; K. Y. Wolin and coauthors, Physical activity and colon cancer prevention: A meta-analysis, *British Journal of Cancer* 100 (2009): 611–616; J. B. Peel and coauthors, A prospective study of cardiorespiratory fitness and breast cancer mortality, *Medicine and Science in Sports and Exercise* 41 (2009): 742–748; D. E. Warburton, C. W. Nicol, and S. S. Bredin, Health benefits of physical activity: The evidence, *Canadian Medical Association Journal* 174 (2006): 801–809.

4. American College of Sports Medicine, Position paper, Appropriate physical activity intervention strategies for weight loss and prevention of weight regain for adults, *Medicine and Science in Sports and Exercise* 41 (2009): 459–471; K. S. Vimaleswaran and coauthors, Physical activity attenuates the body mass index-increasing influence of genetic variation in the *FTO* gene, *American Journal of Clinical Nutrition* 90 (2009): 425–428; S. J. Elder and S. B. Roberts, The effects of exercise on food intake and body fatness: A summary of published studies, *Nutrition Reviews* 65 (2007): 1–19; P. T. Williams, Maintaining vigorous activity attenuates 7-yr weight gain in 8340 runners, *Medicine and Science in Sports and Exercise* 39 (2007): 801–809.

5. R. S. Rector and coauthors, Lean body mass and weight-bearing activity in the prediction of bone mineral density in physically active men, *Journal of Strength and Conditioning Research* 23 (2009): 427–435; A. Guadalupe-Grau and coauthors, Exercise and bone mass in adults, *Sports Medicine* 39 (2009): 439–468.

6. J. Romeo and coauthors, Physical activity, immunity and infection, *Proceedings of the Nutrition Society* 69 (2010): 390–399.

7. X. Sui and coauthors, Influence of cardiorespiratory fitness on lung cancer mortality, *Medicine and Science in Sports and Exercise* 42 (2010): 872–878; J. B. Peel and coauthors, A prospective study of cardiorespiratory fitness and breast cancer mortality, *Medicine and Science in Sports and Exercise* 41 (2009): 742–748; S. Y. Pan and M. DesMeules, energy intake, physical activity energy balance, and cancer: Epidemiologic evidence, *Methods in Molecular Biology* 472 (2009): 191–215.

8. M. Hamer and E. Stamatakis, Physical activity and risk of cardiovascular disease events: Inflammatory and metabolic mechanisms, *Medicine and Science in Sports and Exercise* 41 (2009) 1206–1211; N. L. Chase and coauthors, The association of cardiorespiratory fitness and physical activity with incidence of hypertension in men, *American Journal of Hypertension* 22 (2009): 417–424.

9. M. Fogelhom, How physical activity can work? *International Journal of Pediatric Obesity* 3 (2008): 10–14; B. A. Irving and coauthors, Effect of exercise training intensity on abdominal visceral fat and body composition, *Medicine and Science in Sports and Exercise* 40 (2008): 1863–1872.

10. J. Ralph and coauthors, Low-intensity exercise reduces the prevalence of hyperglycemia in type 2 diabetes, *Medicine and Science in Sports and Exercise* 42 (2010): 219–225; C. Y Jeon and coauthors, Physical activity of moderate intensity and risk of type 2 diabetes: A systematic review, *Diabetes Care* 30 (2007): 744–752.

11. P. J. Banim and coauthors, Physical activity reduces the risk of symptomatic gallstones: A prospective cohort study, *European Journal of Gastroenterology and Hepatology* 22 (2010): 983–988; A. M. Kriska and coauthors, Physical activity and gallbladder disease determined by ultrasonography, *Medicine and Science in Sports and Exercise* 39 (2007): 1927–1932.

12. D. B. Nelson and coauthors, Effect of physical activity on menopausal symptoms among urban women, *Medicine and Science in Sports and Exercise* 40 (2008): 50–58; J. A. Blumenthal and coauthors, Exercise and pharmacotherapy in the treatment of major depressive disorder, *Psychosomatic Medicine* 69 (2007): 587–596; D. I. Galper and coauthors, Inverse association between physical inactivity and mental health in men and women, *Medicine and Science in Sports and Exercise* 38 (2006): 173–178.

13. P. Kokkinos and coauthors, Exercise capacity and mortality in black and white men, *Circulation* 117 (2008): 614–622; L. B. Yates and coauthors, Exceptional longevity in men, *Archives of Internal Medicine* 168 (2008): 284–290; X. Sui and coauthors, Cardiorespiratory fitness and adiposity as mortality predictors in older adults, *Journal of the American Medical Association* 298 (2007): 2507–2516; T. M. Manini and coauthors, 2006.

14. P. Aagaard and coauthors, Mechanical muscle function, morphology, and fiber type in lifelong trained elderly, *Medicine and Science in Sports and Exercise* 39 (2007): 1989–1996; M. E. Nelson and coauthors, Physical activity and public health in older adults: Recommendation from the American College of Sports Medicine and the American Heart Association, *Medicine and Science in Sports and Exercise* 39 (2007): 1435–1445; N. Takeshima and coauthors, Functional fitness gain varies in older adults depending on exercise mode, *Medicine and Science in Sports and Exercise* 39 (2007): 2036–2043.

15. U.S. Department of Health and Human Services, 2008 Physical Activity Guidelines for Americans, available at www.health.gov/paguidelines/default.aspx, site updated October 16, 2008, and accessed October 07, 2010.

16. W. L. Haskell and coauthors, Physical activity and public health: Updated recommendation for adults from the American College of Sports Medicine and the American Heart Association, *Medicine and Science in Sports and Exercise* 39 (2007): 1423–1434.

17. Haskell and coauthors, 2007.

18. U.S. Department of Health and Human Services, 2008 Physical Activity Guidelines for Americans.

19. American College of Sports Medicine, Position stand: Progression models in resistance training for healthy adults, *Medicine and Science in Sports and Exercise* 41 (2009): 687–708.

20. S. Karikanta and coauthors, A multi-component exercise regimen to prevent functional decline and bone fragility in home-dwelling elderly women: Randomized, controlled trial, *Osteoporosis International* 18 (2007): 453–462; Nelson and coauthors, 2007; J. A. Katula and coauthors, Strength training in older adults: An empowering intervention, *Medicine and Science in Sports and Exercise* 38 (2006): 106–111.

21. A. Guadalupe-Grau and coauthors, Exercise and bone mass in adults, *Sports Medicine* 39 (2009): 439–468; American College of Sports Medicine, Position Stand: Physical activity and bone health, *Medicine and Science in Sports and Exercise* 36 (2004): 1985–1996.

22. American College of Sports Medicine, 2009.

23. S. Marwood and coauthors, Faster pulmonary oxygen uptake kinetics in trained versus untrained male adolescents, *Medicine and Science in Sports and Exercise* 42 (2010): 127–134.

24. P. G. Snell and coauthors, Maximal oxygen uptake as a parametric measure of cardiorespiratory capacity, *Medicine and Science in Sports and Exercise* 39 (2007): 103–107.

25. T. Rankinen and coauthors, Cardiorespiratory fitness, BMI, and risk of hypertension: The HYPGENE Study, *Medicine and Science in Sports and Exercise* 39 (2007): 1687–1692.

26. Position of the American Dietetic Association, Dietitians of Canada, and the American College of Sports Medicine: Nutrition and athletic performance, *Journal of the American Dietetic Association* 109 (2009): 509–527.

27. A. M. Bellinger and coauthors, Remodeling of ryanodine receptor complex causes "leaky" channels: a molecular mechanism for decreased exercise capacity, *Proceedings of the National Academy of Sciences* 105 (2008): 2198–2202; S. P. Cairns, Lactic acid and exercise performance: Culprit or friend? *Sports Medicine* 36 (2006): 279–291.

28. D. A. Sedlock, The latest on carbohydrate loading: A practical approach, *Current Sports Medicine Reports* 7 (2008): 209–213.

29. Position of the American Dietetic Association, Dietitians of Canada, and the American College of Sports Medicine, 2009; G. A. Wallis and coauthors, Dose-response effects of ingested carbohydrate on exercise metabolism in women, *Medicine and Science in Sports and Exercise* 39 (2007): 131–138.

30. L. M. Burke, Fueling strategies to optimize performance: Training high or training low? *Scandinavian Journal of Medicine and Science in Sports* 20 (2010): 48–58; A. Ali and C. Williams, Carbohydrate ingestion and soccer skill performance during prolonged intermittent exercise, *Journal of Sport Science* 27 (2009): 1499–1508; A. Foskett and coauthors, Carbohydrate availability and muscle energy metabolism during intermittent running, *Medicine and Science in Sports and Exercise* 40 (2008): 96–103.

31. American College of Sports Medicine, Position stand: Exercise and fluid replacement, *Medicine and Science in Sports and Exercise* 39 (2007): 377–390.

32. American College of Sports Medicine, 2007, pp. 377–390.

33. Position of the American Dietetic Association, Dietitians of Canada, and the American College of Sports Medicine, 2009; G. A. Wallis and coauthors, Postexercise muscle glycogen synthesis with combined glucose and fructose ingestion, *Medicine and Science in Sports and Exercise* 40 (2008): 1789–1794.

34. C. M. Perry, T. L. Perry, and M. C. Rose, Glycemic index and endurance performance, *International Journal of Sport Nutrition and Exercise Metabolism* 20 (2010): 154–165; E. J. Stevenson and coauthors, Dietary glycemic index influences lipid oxidation but not muscle or liver glycogen oxidation during exercise, *American Journal of Physiology, Endocrinology and Metabolism* 296 (2009): E-1140–1147; M. I. Trenell and coauthors, Effect of high and low glycaemic index recovery diets on intramuscular lipid oxidation during aerobic exercise, *British Journal of Nutrition* 99 (2008): 326–332.

35. C. M. Cook and M. D. Haud, Low-carbohydrate diets and performance, *Current Sports Medicine Reports* 6 (2007): 225–229.

36. L. Havemann and coauthors, Fat adaptation followed by carbohydrate loading compromises high-intensity sprint performance, *Journal of Applied Physiology* 100 (2006): 194–202.

37. Position of the American Dietetic Association, Dietitians of Canada, and the American College of Sports Medicine, 2009.

38. T. B. Symons and coauthors, A moderate serving of high-quality protein maximally stimulates skeletal muscle protein synthesis in young and elderly subjects, *Journal of the American Dietetic Association* 109 (2009): 1582–1586; K. D. Tipton and A. A. Ferrando, Improving muscle mass: response of muscle metabolism to exercise, nutrition, and anabolic agents, *Essays in Biochemistry* 44 (2008): 85–98; T. G. Anthony and coauthors, Feeding meals containing soy or whey protein after exercise stimulates protein synthesis and translation initiation in the skeletal muscle

of male rats, *Journal of Nutrition* 137 (2007): 357–362; N. R. Rodriguez, L. M. Vislocky, and P. Courtney Gaine, Dietary protein, endurance exercise, and human skeletal-muscle protein turnover, *Current Opinion in Clinical and Metabolic Care* 10 (2007): 40–45.

39. Standing Committee on the Scientific Evaluation of Dietary Reference Intakes, 2005, pp. 660–661.

40. Position of the American Dietetic Association, Dietitians of Canada, and the American College of Sports Medicine, 2009.

41. Position of the American Dietetic Association, Dietitians of Canada, and the American College of Sports Medicine, 2009.

42. M. J. Jackson, Free radicals generated by contracting muscle; By-products of metabolism or key regulators of muscle function? *Free Radical Biology & Medicine* 44 (2008): 132–141; S. Sachdev and K. J. Davies, Production, detection, and adaptive responses to free radicals in exercise, *Free Radical Biology & Medicine* 44 (2008): 215–223.

43. V. H. Teixeira and coauthors, Antioxidants do not prevent postexercise peroxidation and may delay muscle recovery, *Medicine and Science in Sports and Exercise* 41 (2009): 1752–1760; W. L. Knex, D. G. Jenkins, and J. S. Coombes, Oxidative stress in half and full ironman triathletes, *Medicine and Science in Sports and Exercise* 39 (2007): 283–288; S. L. Williams and coauthors, Antioxidant requirements of endurance athletes: Implications for health, *Nutrition Reviews* 64 (2006): 93–108.

44. Teixeira, 2009; M. C. Gomez-Cabrera, E. Domenech, and J. Vina, Moderate exercise is an antioxidant: Upregulation of antioxidant genes by training, *Free Radical Biology and Medicine* 44 (2008): 126–131; Jackson, 2008; Sachdev and Davies, 2008; J. Padilla and T. D. Mickleborough, Does antioxidant supplementation prevent favorable adaptations to exercise training? *Medicine and Science in Sports and Exercise* 39 (2007): 1887.

45. Position of the American Dietetic Association: Vegetarian diets, *Journal of the American Dietetic Association* 109 (2009): 1266–1282.

46. J. P. McClung and coauthors, Randomized, double-blind, placebo-controlled trial of iron supplementation in female soldiers during military training: Effects on iron status, physical performance, and mood, *American Journal of Clinical Nutrition* 90 (2009): 124–131; P. S. Hinton and L. M. Sinclair, Iron supplementation maintains ventilatory threshold and improves energetic efficiency in iron-deficient nonanemic athletes, *European Journal of Clinical Nutrition* 61 (2007): 30–39.

47. American College of Sports Medicine, Position stand: Exercise and fluid replacement, *Medicine & Science in Sports & Exercise* 39 (2007): 377–390.

48. American College of Sports Medicine, Position stand: Exertional heat illness during training and competition, *Medicine & Science in Sports & Exercise* 39 (2007): 556–572.

49. American College of Sports Medicine, Position stand: Prevention of cold injuries during exercise, *Medicine and Science in Sports and Exercise* 38 (2006): 2012–2029.

50. American College of Sports Medicine, Position stand: Exercise and fluid replacement, *Medicine & Science in Sports & Exercise* 39 (2007): 377–390.

51. Standing Committee on the Scientific Evaluation of Dietary Reference Intakes, Food and Nutrition Board, Institute of Medicine, *Dietary Reference Intakes for Water, Potassium, Sodium, Chloride, and Sulfate* (Washington, D.C.: National Academies Press, 2005), pp. 127–132.

52. American College of Sports Medicine, Position stand: Exercise and fluid replacement, *Medicine & Science in Sports & Exercise* 39 (2007): 377–390.

53. Position of the American Dietetic Association, Dietitians of Canada, and the American College of Sports Medicine, 2009.

Nutrition in Practice

Athletes gravitate to promises that they can easily improve their performance by taking pills, powders, or potions. Unfortunately, they often hear such promises from their coaches and peers, who advise them to use nutrient supplements, take drugs, or follow procedures that claim to deliver results with little effort.[1] The wish to win is strong, but no amount of wishing can change the fact that a large majority of supplements sold for athletes are frauds. When products sold as performance-enhancing aids have no effect and are harmless, they are only a waste of money. When they are harmful or actually impair performance, they are a waste of athletic potential as well. This Nutrition in Practice looks at scientific evidence for and against some of the most common dietary supplements and hormonal preparations available to athletes, and the accompanying Glossary defines them. Chapter 4 included a discussion on protein powders and amino acid supplements (see p. 96).

GLOSSARY

The Glossary on p. 284 includes additional supplements commonly used to enhance performance.

anabolic steroids: drugs related to the male sex hormone, testosterone, that stimulate the development of lean body mass.

> *anabolic* = promoting growth
> *sterols* = compounds chemically related to cholesterol

caffeine: a natural stimulant found in many common foods and beverages, including coffee, tea, and chocolate. High doses cause headaches, trembling, rapid heart rate, and other undesirable side effects.

carnitine: a nonessential nonprotein amino acid made in the body from lysine that helps transport fatty acids across the mitochondrial membrane. Carnitine supposedly "burns" fat and spares glycogen during endurance events, but in reality it does neither.

chromium picolinate (CROW-mee-um pick-oh-LYN-ate): a trace mineral supplement; falsely promoted as building muscle, enhancing energy, and burning fat. Picolinate is a derivative of the amino acid tryptophan that seems to enhance chromium absorption.

conjugated linoleic acid: a naturally occurring *trans*-fatty acid with 18 carbons and 2 double bonds; sometimes taken as a supplement to improve body composition.

creatine (KREE-ah-tin): a nitrogen-containing compound that combines with phosphate to form the high-energy compound creatine phosphate (or phosphocreatine) in muscles.

DHEA (dehydroepiandrosterone) and **androstenedione:** hormones made in the adrenal glands that serve as precursors to the male hormone testosterone; falsely promoted as burning fat, building muscle, and slowing aging. Side effects include acne, aggressiveness, and liver enlargement.

ergogenic (ER-go-JEN-ick) **aids:** substances or techniques used in an attempt to enhance physical performance.

> *ergo* = work
> *genic* = gives rise to

What does *ergogenic* mean?

Ergogenic means work enhancing or work producing. In connection with athletic performance, ergogenic aids are substances or treatments that purportedly improve athletic performance above and beyond what is possible through training alone. Research findings do not, for the most part, support the claims made for **ergogenic aids.**[2] When you hear a claim that a product is ergogenic, remember to consider the source of the claim and ask who may gain from the sale. Nutrition in Practice 1 describes ways to recognize misinformation and quackery.

I know that some athletes, especially endurance athletes, are taking carnitine supplements. What is carnitine?

Carnitine is a nonessential nutrient. Endurance athletes believe carnitine will help them burn more fat, thereby sparing glycogen during endurance events. Carnitine is also promoted to bodybuilders as a "fat burner."

In the body, carnitine facilitates the transfer of fatty acids across the mitochrondrial membrane. Supplement manufacturers suggest that with more carnitine available, fat oxidation will be enhanced, but this does not seem to be the case. Carnitine supplementation neither raises muscle carnitine concentrations nor enhances athletic performance.[3] Milk and meat products are good sources of carnitine, and supplements are not needed.

Advertisements in magazines and health food stores make all kinds of impressive claims about a supplement called chromium picolinate. Are any of the claims true?

Chapter 9 introduces chromium as an essential trace mineral involved in carbohydrate and lipid metabolism. Advertisements in bodybuilding magazines claim that **chromium picolinate,** which is more easily absorbed than chromium alone, builds muscle, enhances energy, and burns fat. Such claims derive from one or two initial studies on this mineral. Most subsequent studies, however, show no effects of chromium picolinate supplementation on strength, lean body mass, or body fat.[4]

A lot of my friends take creatine. Why is it so popular?

Interest in and use of **creatine** supplements to enhance energy production during intense activity has grown dramatically in the last few years. Power athletes such as weight lifters use creatine supplements to increase stores of the high-energy compound creatine phosphate (or phosphocreatine) in muscles. Theoretically, the more creatine phosphate in muscles, the higher the intensity at which an athlete can train. High-intensity training stimulates the muscles to adapt, which, in turn, improves performance.

The results of some studies suggest that creatine supplementation enhances performance of short-term, high-intensity strength activity such as weight lifting or repeated sprinting.[5] In contrast, creatine supplementation has not been shown to benefit endurance activity.[6] The question of whether short-term use of creatine supplements is safe continues to be studied, but so far, the supplements are viewed to be safe for healthy adults.[7]

Some medical and fitness experts voice concern that, like many performance enhancement supplements before it, creatine is being taken in huge doses (up to 30 grams per day) before evidence of its value or safety has been ascertained. Even people who eat red meat, which is a creatine-rich food, do not consume near the amount some athletes are taking. Despite the uncertainties, creatine supplements are not illegal in international competition. The American Academy of Pediatrics strongly discourages the use of creatine supplements, as well as the use of any performance-enhancing substance in adolescents less than 18 years of age.[8]

What is conjugated linoleic acid, and why do some athletes take it?

Conjugated linoleic acid (CLA) derives from the essential fatty acid, linoleic acid. CLA is part of a group of naturally occurring polyunsaturated fatty acids found in beef, lamb, and dairy products. In animal studies, CLA has been shown to reduce body fat and increase lean body mass—findings that have sparked interest in CLA as a performance-enhancing aid.[9] Only a few studies of the effects of supplemental CLA on body composition in human beings have been conducted, and the results seem less promising.[10] When researchers studied the combined effects of supplemental CLA and resistance training on body composition in men and women, they found small increases in lean body mass and reductions in body fat, but no improvements in strength.[11] The researchers noted that although the effects were statistically significant, they were nevertheless small and should be weighed against the relatively high cost of supplemental CLA.

What about caffeine? I have heard that it can improve endurance performance.

Some research supports the use of **caffeine** to enhance endurance and, to some extent, to enhance short-term, high-intensity exercise performance.[12] Caffeine is a stimulant that elicits a number of physiological and psychological effects in the body. Caffeine enhances alertness and reduces fatigue.[13] The possible benefits of caffeine use must be weighed against its adverse effects—stomach upset, nervousness, irritability, headaches, and diarrhea. Caffeine-containing beverages should be used in moderation, if at all, and *in addition* to other fluids, not as a substitute for them. In 2004, the World Anti-Doping Agency (WADA), established by the International Olympic Committee, removed caffeine from its list of prohibited substances. Caffeine is a *restricted* substance by the National Collegiate Athletic Association, which allow urine concentrations of 15 milligrams or less. This is equivalent to about 5 cups of coffee consumed within a few hours before testing. Urine tests that detect more caffeine than this disqualify athletes from competition. (The table at the start of Appendix A provides a list of common caffeine-containing items and the doses they deliver.)

I have heard that it is dangerous to take anabolic steroids. Is this true?

Anabolic steroids have dangerous side effects and are illegal. Technically called androgenic-anabolic steroid drugs, they are derivatives of the male sex hormone testosterone. Testosterone promotes the development of male characteristics (androgenic) and lean body mass (anabolic). Athletes take steroids to stimulate muscle bulking. In professional circles such as Major League Baseball and the National Football League, where monetary rewards for excellence are sky-high, steroid use is common despite its illegality and side effects.

The American College of Sports Medicine and the American Academy of Pediatrics condemn athletes' use of anabolic steroids, and the International Olympic Committee bans their use. These authorities cite the known toxic side effects and maintain that steroid use is a form of cheating. Competitors who use the drugs put other athletes in the difficult position of either conceding an unfair advantage to abusing competitors or taking steroids and accepting the risk of harmful side effects (see Table NP10-1 on p. 284). Young athletes should never be forced to make such a choice.

The dangers of steroid use cannot be overemphasized. Health care professionals are obligated to warn athletes of these dangers. Speak simply and emphatically: the price for the potential competitive edge that steroids confer is high—sometimes life itself. Steroids are not simple pills that build bigger muscles. They are complex chemicals to which the body reacts in many ways, particularly when bodybuilders and other athletes take large amounts.[14] Among the side effects and adverse reactions that steroids produce are cancerous liver tumors; testicular shrinkage in men and masculinization of women; cardiovascular problems; and sterility. The safest effective way to build muscle has always been through hard training, and always will be.

What are DHEA and androstenedione, and why do some athletes use them?

Some athletes use **DHEA** (dehydroepiandrosterone) and **androstenedione** as alternatives to anabolic steroids. Androstenedione, or "andro," made headlines in the late 1990s when the media reported its use by baseball great Mark McGwire.

DHEA and androstenedione are hormones made in the adrenal glands that serve as precursors to the male hormone testosterone. Advertisements claim the hormones "burn fat," "build muscle," and "slow aging," but evidence to support such claims is lacking.

Short-term side effects of DHEA and androstenedione include oily skin, acne, body hair growth, liver enlargement, testicular shrinkage, and aggressive behavior. Long-term effects, such as serious liver damage, may take years to become evident. The potential for harm from these supplements is great, and athletes, as well as others, should avoid them.

Effective in 2005, androstenedione was reclassified from a dietary supplement to an anabolic steroid in the United States.[15] The Food and Drug Administration (FDA) sent letters to producers of dietary supplements warning that products containing androstenedione are considered to be adulterated and therefore are illegal to sell and that criminal penalties could result from continued sales. The National Collegiate Athletic Association, the National Football League, and the International Olympic Committee have banned the use of

TABLE NP10-1 Anabolic Steroids: Side Effects and Adverse Reactions

Mind

- Extreme aggression with hostility ("steroid rage"); mood swings; anxiety; dizziness; drowsiness; unpredictability; insomnia; psychotic depression; personality changes, suicidal thoughts

Face and Hair

- Swollen appearance; greasy skin; severe, scarring acne; mouth and tongue soreness; yellowing of whites of eyes (jaundice)
- In females, male-pattern hair loss and increased growth of face and body hair

Voice

- In females, irreversible deepening of voice

Chest

- In males, breathing difficulty, breast development
- In females, breast atrophy

Heart

- Heart disease; elevated or reduced heart rate; heart attack; stroke; hypertension; increased LDL; reduced HDL

Abdominal Organs

- Nausea; vomiting; bloody diarrhea; pain; edema; liver tumors (possibly cancerous); liver damage, disease, or rupture leading to fatal liver, liver failure; kidney stones and damage; gallstones; frequent urination; possible rupture of aneurysm or hemorrhage

Blood

- Blood clots: high risk of blood poisoning; those who share needles risk contracting HIV (the AIDS virus) or other disease-causing organisms; septic shock (from injections)

Reproductive System

- In males, permanent shrinkage of testes; prostate enlargement with increased risk of cancer; sexual dysfunction; loss of fertility, excessive and painful erections
- In females, loss of menstruation and fertility; permanent enlargement of external genitalia; fetal damage, if pregnant

Muscles, Bones, and Connective Tissues

- Increased susceptibility to injury with delayed recovery times; cramps; tremors; seizure-like movements; injury at injection site
- In adolescents, failure to grow to normal height

Other

- Fatigue; increased risk of cancer

GLOSSARY OF SUBSTANCES PROMOTED AS ERGOGENIC AIDS

The glossary on p. 282 includes supplements mentioned in the text. Chapter 4 includes a discussion on protein and amino acid supplements.

arginine: a nonessential amino acid falsely promoted as enhancing the secretion of human growth hormone, the breakdown of fat, and the development of muscle.

boron: a nonessential mineral that is falsely promoted as a "natural" steroid replacement.

coenzyme Q10: a lipid found in cells (mitochondria) shown to improve exercise performance in heart disease patients, but not effective in improving the performance of healthy athletes.

gamma-oryzanol: a plant sterol that supposedly provides the same physical responses as anabolic steroids without the adverse side effects; also known as ferulic acid, lerulate, or FRAC.

ginseng: a plant whose extract supposedly boosts energy. Side effects of chronic use include headache, nervousness, and sleeplessness.

glycerol: a 3-carbon molecule that is part of triglycerides and phospholipids. Studies show glycerol is ineffective for enhancing performance, but findings are inconsistent regarding its efficacy in improving hydration and temperature regulation during exercise.

HMB (beta-hydroxy-beta-methylbutyrate): a metabolite of the branched-chain amine acid leucine. Some research shows that HMB may increase muscle mass and strength in untrained individuals. More research is needed, however, to determine whether HMB enhances training adaptations in athletes.

pyruvate: a 3-carbon compound that plays a key role in energy metabolism. Supplements claim to burn fat and enhance endurance performance.

ribose: a 5-carbon sugar falsely promoted as improving the regeneration of ATP and thereby the speed of recovery after high-intensity exercise.

royal jelly: the substance produced by worker bees and fed to the queen bee; falsely promoted as enhancing stamina and reducing fatigue.

sodium bicarbonate: baking soda; an alkaline salt believed to neutralize blood lactic acid and thereby to reduce pain and enhance possible workload. "Soda loading" may cause intestinal bloating and diarrhea.

androstenedione and DHEA in competition. The American Academy of Pediatrics and many other medical professional groups have spoken out against the use of these and other "hormone replacement" substances.

Carnitine and chromium picolinate do not enhance athletic performance, and experts are concerned about the long-term effects of large doses of creatine. Caffeine may be effective, but can have adverse side effects. Steroids pose serious health risks and are illegal. Do any of the substances athletes use to boost performance work?

For the most part, no. Many of these substances have been studied and found to be worthless. The Glossary lists and describes more substances promoted as ergogenic aids.

Health professionals can positively influence athletes and others interested in boosting athletic performance by stressing the measures that do help to enhance performance. They are regular training and sound nutrition.

Notes

1. K. A. Erdman, T. S. Fung, and R. A. Reimer, Influence of performance level on dietary supplementation in elite Canadian athletes, *Medicine and Science in Sports and Exercise* 38 (2006): 349–356.

2. L. Di Luigi, Supplements and the endocrine system in athletes, *Clinics in Sports Medicine* 27 (2008): 131–151; R. Calfee and P. Fadale, Popular ergogenic drugs and supplements in young athletes, *Pediatrics* 117 (2006): e577–589.

3. L. L. Spriet, C. G. Perry, and J. L. Talanian, Legal pre-event nutritional supplements to assist energy metabolism, *Essays in Biochemistry* 44 (2008): 27–43.

4. Di Luigi, 2008; M. L. Diaz and coauthors, Chromium picolinate and conjugated linoleic acid do not synergistically influence diet- and exercise-induced changes in body composition and health indexes in overweight women, *Journal of Nutritional Biochemistry* 19 (2008): 61–68; H. C. Lukaski, W. A. Siders, and J. G. Penland, Chromium picolinate supplementation in women: Effects on body weight, composition, and iron status, *Nutrition* 23 (2007): 187–195.

5. D. H. Fukuda and coauthors, The effects of creatine loading and gender on anaerobic running capacity, *Journal of Strength and Conditioning Research* 24 (2010): 1826–1833; Spriet, Perry, and Talanian, 2008; K. D. Tipton and A. A. Ferrando, Improving muscle mass: response of muscle metabolism to exercise, nutrition, and anabolic agents, *Essays in Biochemistry* 44 (2008): 85–98; J. T. Brosnan and M. E. Brosnan, Creatine: endogenous metabolite, dietary, and therapeutic supplement, *Annual Review of Nutrition* 27 (2007): 241–261.

6. Position of the American Dietetic Association, Dietitians of Canada, and American College of Sports Medicine: Nutrition and Athletic Performance, *Journal of the American Dietetic Association* 109 (2009): 509–526; Brosnan and Brosnan, 2007.

7. Position of the American Dietetic Association, Dietitians of Canada, and American College of Sports Medicine, 2009; A. Shao and J. N. Hathcock, Risk assessment for creatine monohydrate, *Regulatory Toxicology and Pharmacology* 45 (2006): 242–251.

8. American Academy of Pediatrics, Policy Statement, Committee on Sports Medicine and Fitness, Use of performance-enhancing substances, *Pediatrics* 115 (2005): 1103–1106.

9. B. A. Cori and coauthors, Conjugated linoleic acid reduces body fat accretion and lipogenic gene expression in neonatal pigs fed low-or high-fat formulas, *Journal of Nutrition* 138 (2008): 449–454.

10. B. Campbell and R. Kreider, Conjugated linoleic acids, *Current Sports Medicine Reports* 7 (2008): 237–241; M. Plourde and coauthors, Conjugated linoleic acids: Why the discrepancy between animal and human studies? *Nutrition Reviews* 66 (2008): 415–421; S. E. Steck and coauthors, Conjugated linoleic acid supplementation for twelve weeks increases lean body mass in obese humans, *Journal of Nutrition* 137 (2007): 1188–1193; J. M. Gaullier and coauthors, Six months supplementation with conjugated linoleic acid (CLA) induces regional-specific fat mass decreases in overweight and obese, *British Journal of Nutrition* 97 (2007): 50–60.

11. C. Pinkoski and coauthors, The effects of conjugated linoleic acid supplementation during resistance training, *Medicine and Science in Sports and Exercise* 38 (2006): 339–348.

12. E. Hogervorst and coauthors, Caffeine improves physical and cognitive performance during exhaustive exercise, *Medicine & Science in Sports & Exercise* 40 (2008): 1841–1851; M. Glaister and coauthors, Caffeine supplementation and multiple sprint running performance, *Medicine & Science in Sports & Exercise* 40 (2008):1835–1840; G. Jones, Caffeine and other sympathomimetic stimulants: Modes of action and effects on sports performance, *Essays in Biochemistry* 44 (2008): 109–123.

13. Jones, 2008.

14. Di Luigi, 2008; A. B. Parkinson and N. A. Evans, Anabolic androgenic steroids: A survey of 500 users, *Medicine and Science in Sports and Exercise* 38 (2006): 644–651.

15. A. S. Fragakis, and C. Thomson, *The Health Professional's Guide to Popular Dietary Supplements*, 3rd ed. (Chicago: American Dietetic Association, 2007), pp. 23–26.

11

Nutrition through the Life Span: Pregnancy and Lactation

286

All people need the same nutrients, but the amounts they need vary depending on their stage of life. This chapter focuses on nutrition in preparation for, and support of, pregnancy and lactation. The next two chapters address the needs of infants, children, adolescents, and older adults.

Pregnancy: The Impact of Nutrition on the Future

The woman who enters pregnancy with full nutrient stores, sound eating habits, and a healthy body weight has done much to ensure an optimal pregnancy. Then, if she eats a variety of nutrient-dense foods during pregnancy, her own and her infant's health will benefit further.

Nutrition Prior to Pregnancy

A discussion of nutrition prior to pregnancy must, by its nature, focus mainly on women. A man's nutrition may affect his **fertility** and possibly the genetic contributions he makes to his children, but nutrition exerts its primary influence through the woman. Her body provides the environment for the growth and development of a new human being. Full nutrient stores *before* pregnancy are essential both to conception and to healthy infant development during pregnancy. In the early weeks of pregnancy, before many women are even aware that they are pregnant, significant developmental changes occur that depend on a woman's nutrient stores. In preparation for a healthy pregnancy, a woman can establish the following habits:[1]

▌ *Achieve and maintain a healthy body weight.* Both underweight and overweight women, and their newborns, face increased risks of complications.

▌ *Choose an adequate and balanced diet.* Malnutrition reduces fertility and impairs the early development of an infant should a woman become pregnant.

▌ *Be physically active.* A woman who wants to be physically active *when* she is pregnant needs to become physically active *beforehand*.

▌ *Receive regular medical care.* Regular health care visits can help ensure a healthy start to pregnancy.

▌ *Avoid harmful influences.* Both maternal and paternal ingestion of harmful substances (such as cigarettes, alcohol, drugs, or environmental contaminants) can cause miscarriage or abnormalities, alter genes or their expression, and interfere with fertility.

Young adults who nourish and protect their bodies do so not only for their own sakes but also for future generations.

Prepregnancy Weight

Appropriate weight prior to pregnancy benefits pregnancy outcome. Being either underweight or overweight (see p. 297) presents medical risks during pregnancy and childbirth. Underweight women are therefore advised to gain weight before becoming pregnant, and overweight women are advised to lose excess weight.

Underweight Infant birthweight correlates with prepregnancy weight and weight gain during pregnancy and is the most potent single predictor of the infant's future health and survival. An underweight woman has a high risk of having a **low-birthweight** infant, especially if she is unable to gain sufficient weight during pregnancy.[2] Compared with normal-weight infants, low-birthweight infants are more likely to contract

Both parents can prepare in advance for a healthy pregnancy.

fertility: the capacity of a woman to produce a normal ovum periodically and of a man to produce normal sperm; the ability to reproduce.

low birthweight (LBW): a birthweight less than 5½ lb (2500 g); indicates probable poor health in the newborn and poor nutrition status of the mother during pregnancy. Optimal birthweight for a full-term infant is 6.8 to 7.9 lb (about 3100 to 3600 g). Low-birthweight infants are of two different types. Some are **premature:** they are born early and are of a weight **appropriate for gestational age (AGA).** Others have suffered growth failure in the uterus; they may or may not be born early, but they are **small for gestational age (SGA).**

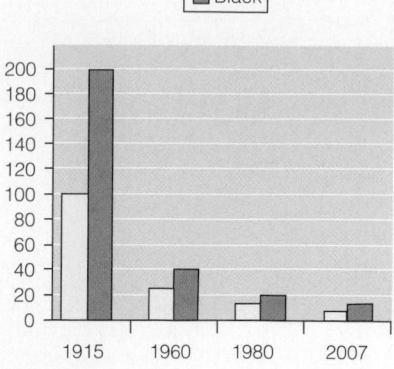

200
180
160
140
120
100
80
60
40
20
0

1915   1960   1980   2007

FIGURE 11-1 Infant
Mortality Decline over Time
The graph shows infant deaths per 1000
live births.

Source: M. Heron and coauthors, Annual
summary of vital statistics: 2007, *Pediatrics* 125
(2010): 4–15.

Neural tube defects and gestational diabetes are discussed in
later sections.

uterus (YOO-ter-us): the womb,
the muscular organ within which the
infant develops before birth.

placenta (pla-SEN-tuh): an organ
that develops inside the uterus early
in pregnancy, in which maternal and
fetal blood circulate in close proximity and exchange materials. The
fetus receives nutrients and oxygen
across the placenta; the mother's
blood picks up carbon dioxide and
other waste materials to be excreted
via her lungs and kidneys.

gestation: the period of about
40 weeks (three trimesters) from
conception to birth; the term of a
pregnancy.

diseases and nearly 40 times more likely to die in the first month of life. Impaired
growth and development during pregnancy may have long-term health effects as
well. Research suggests that, when nutrient supplies fail to meet demands, permanent adaptations take place that may make obesity or chronic diseases such as heart
disease and hypertension more likely in later life (see Nutrition in Practice 12).[3]
Other potential problems of low birthweight may include lower adult IQ and other
brain impairments, short stature, and educational disadvantages.[4] Underweight
women are therefore advised to gain weight before becoming pregnant and to
strive to gain adequately during pregnancy.

Nutritional deficiency, coupled with low birthweight, is the underlying cause
of more than half of all the deaths worldwide of children under five years of age.
In the United States, the infant mortality rate in 2007 was 6.8 deaths per 1000 live
births.[5] This rate, though higher than that of some other developed countries, has
seen a significant steady decline over the past four decades and stands as a tribute
to public health efforts aimed at reducing infant deaths (see Figure 11-1).

Not all cases of low birthweight reflect poor nutrition. Heredity, disease conditions, smoking, and drug use (including alcohol) during pregnancy all contribute.[6] Even with optimal nutrition and health during pregnancy, some women
give birth to small infants for unknown reasons. But poor nutrition is the major
factor in low birthweight—and an avoidable one, as later sections make clear.[7]

Overweight and Obesity Obese women are also urged to strive for healthy weights
before pregnancy. The infant of an obese mother may be larger than normal and may
be large even if born prematurely. The large early infant may not be recognized as premature and thus may not receive the special medical care required. Maternal obesity
may double the risk for neural tube defects.[8] Folate's role has been examined, but a
more likely explanation seems to be poor glycemic control.[9] Obese women are more
likely to require drugs to induce labor or require surgical intervention for the birth,
and they suffer gestational diabetes, hypertension, and infections after the birth more
often than do women of healthy weight.[10] In addition, both overweight and obese
women have a greater risk of giving birth to infants with heart defects and other abnormalities.[11] An appropriate goal for the obese woman who wishes to become pregnant is to strive to attain a healthy prepregnancy body weight so as to minimize her
medical risks and those of her future child.

Healthy Support Tissues

A major reason that the mother's prepregnancy nutrition is so crucial is that it determines
whether her **uterus** will be able to support the growth of a healthy **placenta** during the
first month of **gestation.** The placenta is both a supply depot and a waste-removal system
for the fetus. If the placenta works perfectly, the fetus wants for nothing; if it doesn't, no
alternative source of sustenance is available, and the fetus will fail to thrive. Figure 11-2
shows the placenta, a mass of tissue in which maternal and fetal blood vessels intertwine
and exchange materials. The two bloods never mix, but the barrier between them is notably thin. To grasp how thin, picture your hands as fetal blood vessels, skintight surgical
gloves as the tissue-thin placenta, and finally, your gloved hands immersed in water as
the pool of maternal blood. Across this thin barrier, nutrients and oxygen move from
the mother's blood into the fetus's blood, and wastes move out of the fetal blood to be
excreted by the mother. The **umbilical cord** is the pipeline from the placenta to the fetus.
The **amniotic sac** surrounds and cradles the fetus, cushioning it with fluids.

The placenta is a versatile, metabolically active organ. It actively gathers up hormones, nutrients, and protein molecules such as antibodies and transfers them into the
fetal bloodstream.[12] The placenta also produces a broad range of hormones that act in
many ways to maintain pregnancy and prepare the mother's breasts for **lactation.** A
healthy placenta is essential for the developing fetus to attain its full potential.[13]

288

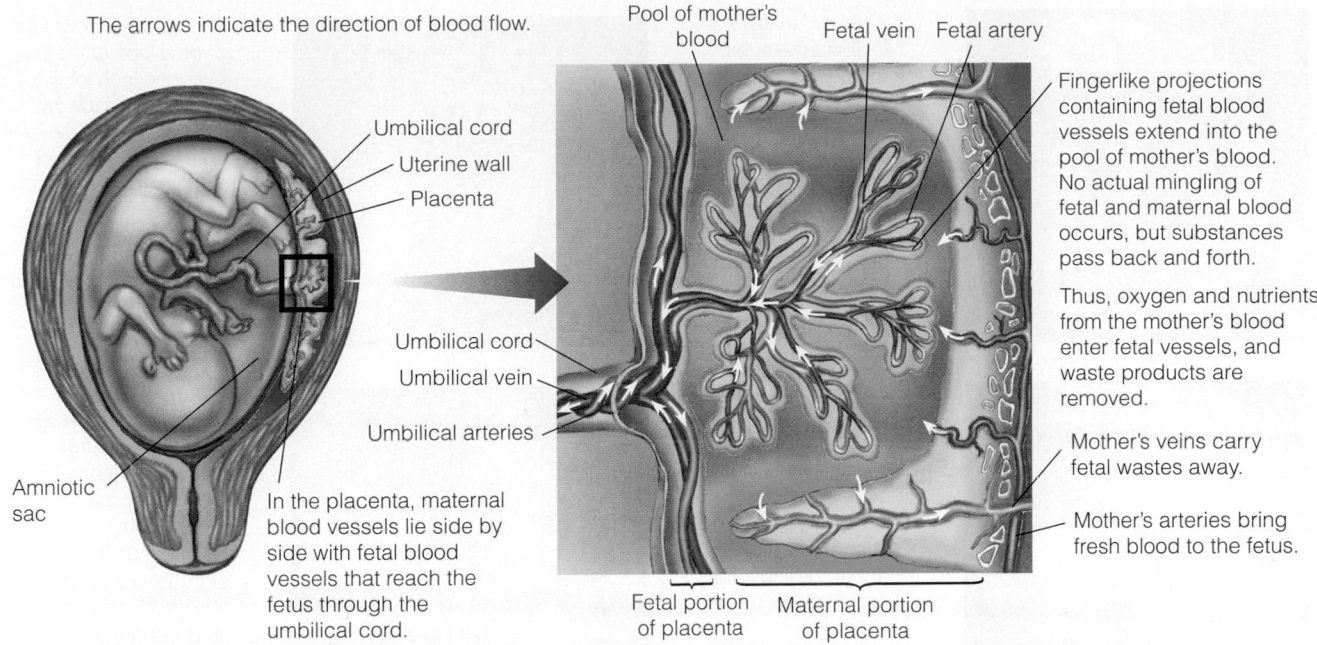

The arrows indicate the direction of blood flow.

Umbilical cord
Uterine wall
Placenta

Amniotic sac

In the placenta, maternal blood vessels lie side by side with fetal blood vessels that reach the fetus through the umbilical cord.

Pool of mother's blood

Fetal vein   Fetal artery

Fingerlike projections containing fetal blood vessels extend into the pool of mother's blood. No actual mingling of fetal and maternal blood occurs, but substances pass back and forth.

Thus, oxygen and nutrients from the mother's blood enter fetal vessels, and waste products are removed.

Mother's veins carry fetal wastes away.

Mother's arteries bring fresh blood to the fetus.

Umbilical cord
Umbilical vein
Umbilical arteries

Fetal portion of placenta   Maternal portion of placenta

FIGURE 11-2 **The Placenta**

IN SUMMARY

▌ Adequate nutrition before pregnancy establishes physical readiness and nutrient stores to support fetal growth.

▌ Both underweight and overweight women should strive for appropriate body-weights before pregnancy.

▌ Newborns who weigh less than 5½ pounds face greater health risks than normal-weight infants.

▌ The healthy development of the placenta depends on adequate nutrition before pregnancy.

The Events of Pregnancy

The newly fertilized **ovum** is called a **zygote.** It begins as a single cell and rapidly divides to become a **blastocyst.** During the first week, the blastocyst floats down into the uterus where it will embed itself in the inner uterine wall—a process known as **implantation.** Minimal growth in size takes place at this point, but it is a crucial period in development. Adverse influences such as smoking, drug abuse, and malnutrition at this time lead to failure to implant or to abnormalities such as neural tube defects that can cause the loss of the developing embryo, possibly before the woman knows she is pregnant.

The Embryo and Fetus During the next six weeks of development, the **embryo** registers astonishing physical changes (see Figure 11-3 on p. 290). At eight weeks, the **fetus** has a complete central nervous system, a beating heart, a fully formed digestive system, well-defined fingers and toes, and the beginnings of facial features.

In the last seven months of pregnancy, the fetal period, the fetus grows 50 times heavier and 20 times longer. Critical periods of cell division and development occur in organ after organ. Most successful pregnancies last 38 to 42 weeks and produce a healthy infant weighing between 6.8 and 7.9 pounds.[14] The 40 or so weeks of pregnancy are divided into thirds, each of which is called a **trimester.**

umbilical (um-BIL-ih-cul) **cord:** the ropelike structure through which the fetus's veins and arteries reach the placenta; the route of nourishment and oxygen into the fetus and the route of waste disposal from the fetus.

amniotic (am-nee-OTT-ic) **sac:** the "bag of waters" in the uterus in which the fetus floats.

lactation: production and secretion of breast milk for the purpose of nourishing an infant.

ovum (OH-vum): the female reproductive cell, capable of developing into a new organism upon fertilization; commonly referred to as an egg.

zygote (ZY-goat): the product of the union of ovum and sperm; a fertilized ovum.

blastocyst (BLASS-toe-sist): the developmental stage of the zygote when it is about five days old and ready for implantation.

implantation: the stage of development in which the blastocyst embeds itself in the wall of the uterus and begins to develop; occurs during the first two weeks after conception.

embryo (EM-bree-oh): the developing infant from two to eight weeks after conception.

fetus (FEET-us): the developing infant from eight weeks after conception until its birth.

trimester: a period representing one-third of the term of gestation. A trimester is about 13 to 14 weeks.

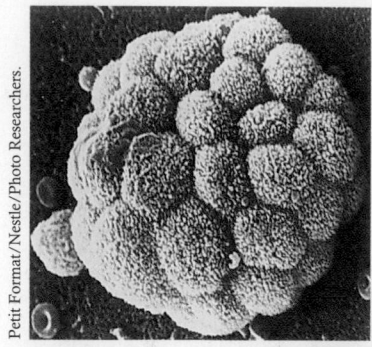

(1) A newly fertilized ovum is about the size of the period at the end of this sentence. Less than one week after fertilization and multiple cell divisions, implantation occurs.

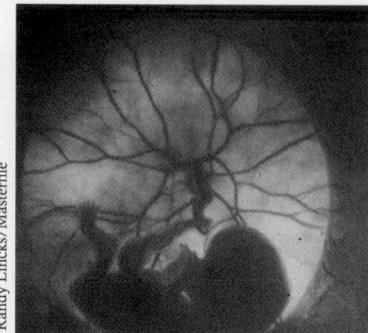

(3) A fetus after 11 weeks of development is just over one inch long. Notice the umbilical cord and blood vessels connecting the fetus with the placenta.

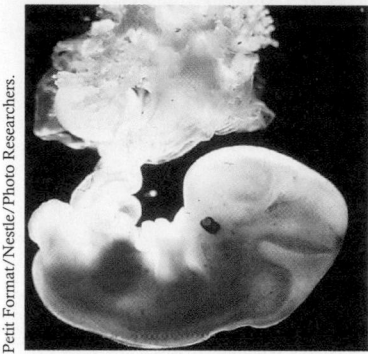

(2) After implantation, the placenta develops and begins to provide nourishment to the developing embryo. An embryo five weeks after fertilization is about 1/2 inch long.

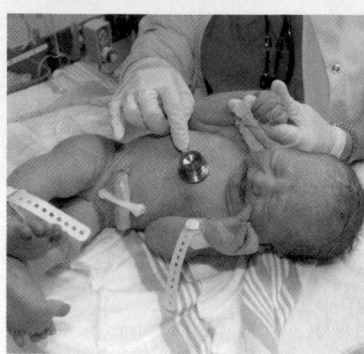

(4) A newborn infant after nine months of development measures close to 20 inches in length. The average birth weight is about 7 1/2 pounds. From eight weeks to term, this infant grew 20 times longer and 50 times heavier.

FIGURE 11-3 Stages of Embryonic and Fetal Development

A Note about Critical Periods Each organ and tissue type grows with its own characteristic pattern and timing. The development of each takes place only at a certain time—the **critical period.** Whatever nutrients and other environmental conditions are necessary during this period must be supplied on time if the organ is to reach its full potential. If the development of an organ is limited during a critical period, recovery is impossible. For example, the fetus's heart and brain are well developed at 14 weeks; the lungs, 10 weeks later. Therefore, early malnutrition impairs the heart and brain; later malnutrition impairs the lungs.

The effects of malnutrition during critical periods of pregnancy are seen in defects of the nervous system of the embryo (explained later), in the child's poor dental health, and in the adolescent's and adult's vulnerability to infections and possibly higher risks of diabetes, hypertension, stroke, or heart disease.[15] The effects of malnutrition during critical periods are irreversible: abundant and nourishing food, consumed after the critical time, cannot remedy harm already done.

IN SUMMARY

▌ Placental development, implantation, and early critical periods depend on maternal nutrition before and during pregnancy.

Nutrient Needs during Pregnancy

Nutrient and energy intake recommendations for pregnant women are listed on the inside front cover.

Nutrient needs during pregnancy increase more for certain nutrients than for others. Figure 11-4 shows the percentage increase in nutrient intakes recommended for pregnant women compared with nonpregnant women. To meet the high nutrient demands of pregnancy, a woman must make careful food choices, but her body will also do its part by maximizing nutrient absorption and minimizing losses.

FIGURE 11-4

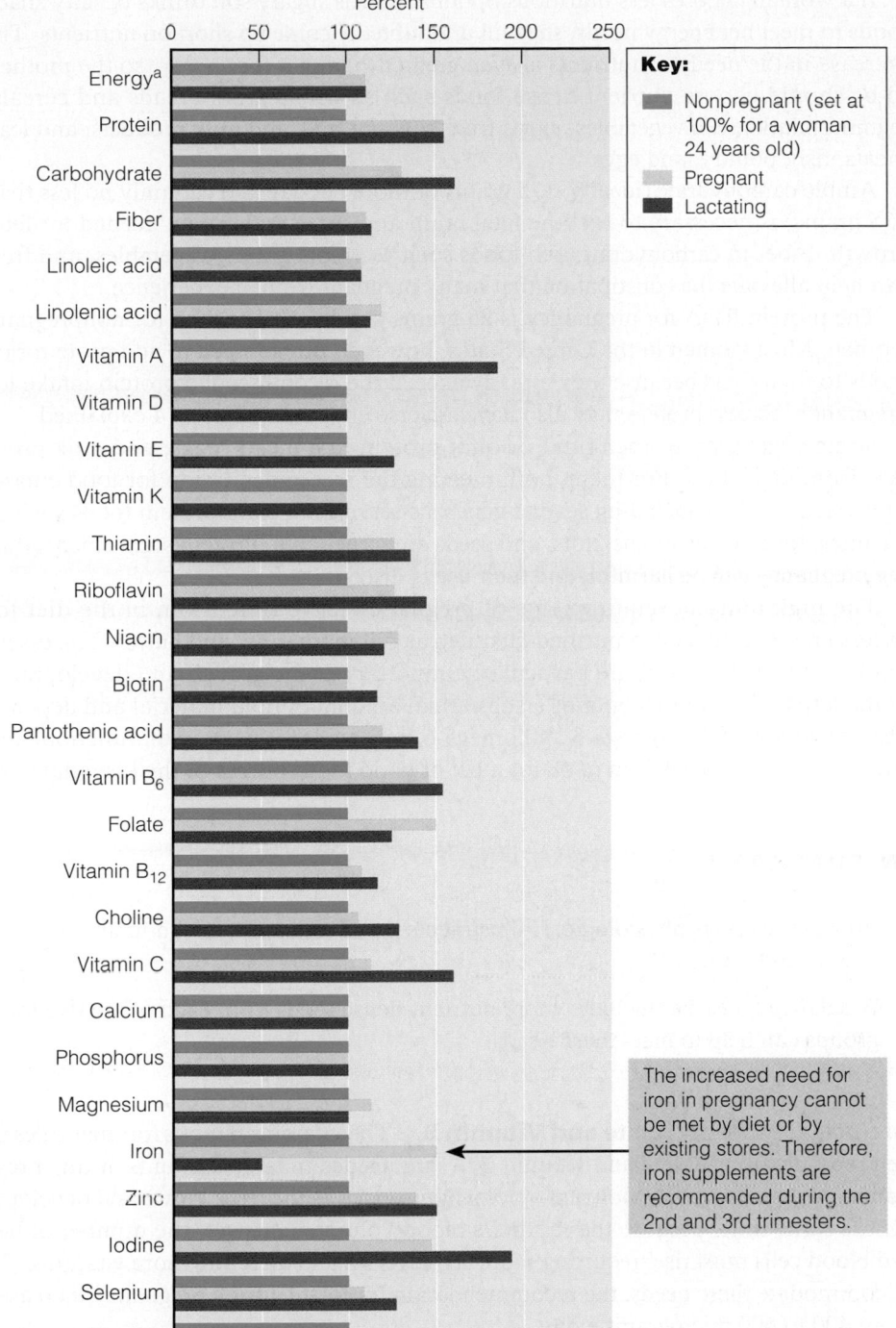

The increased need for iron in pregnancy cannot be met by diet or by existing stores. Therefore, iron supplements are recommended during the 2nd and 3rd trimesters.

Energy, Carbohydrate, Protein, and Fat Energy needs vary with the progression of pregnancy. In the first trimester, the pregnant woman needs no additional energy, but as pregnancy progresses, her energy needs rise. She requires an additional 340 kcalories daily during the second trimester and an extra 450 kcalories each day during the third trimester.[16] Well-nourished pregnant women meet these demands for more energy in several ways: some eat more food, some reduce their activity, and some store less of their food energy as fat. A woman can easily meet the need for extra kcalories by selecting more nutrient-dense foods from the five food groups. Table 1-4 (on p. 18) provides suggested eating patterns for several kcalorie levels, and Table 11-1 (p. 292) offers a sample menu for pregnant and lactating women.

critical period: a finite period during development in which certain events occur that will have irreversible effects on later developmental stages; usually a period of rapid cell division.

TABLE 11-1 Daily Food Choices for Pregnancy (Second and Third Trimesters) and Lactation

Breakfast

1 whole-wheat English muffin

2 tbs peanut butter

1 c low-fat vanilla yogurt

½ c fresh strawberries

1 c orange juice

Midmorning snack

½ c cranberry juice

1 oz pretzels

Lunch

Sandwich (tuna salad on whole-wheat bread)

½ carrot (sticks)

1 c low-fat milk

Dinner

Chicken cacciatore

   3 oz chicken

   ½ c stewed tomatoes

1 c rice

½ c summer squash

1½ c salad (spinach, mushrooms, carrots)

1 tbs salad dressing

1 slice Italian bread

2 tsp soft margarine

1 c low-fat milk

Note: This sample meal plan provides about 2500 kcalories (55 percent from carbohydrate, 20 percent from protein, and 25 percent from fat) and meets most of the vitamin and mineral needs of pregnant and lactating women.

Folate RDA during pregnancy:

▌ **600 µg/day**

Reminder: Neural tube defects (NTD) are malformations of the brain, spinal cord, or both during embryonic development.

A pregnancy affected by a neural tube defect can occur in any woman, but these factors make it more likely:[19]

▌ **A previous pregnancy affected by a neural tube defect**

▌ **Maternal diabetes**

▌ **Maternal use of certain anti-seizure medications**

▌ **Mutations in folate-related enzymes**

▌ **Maternal obesity**

If a woman chooses less nutritious options such as sugary soft drinks or fatty snack foods to meet her energy needs, she will undoubtedly come up short on nutrients. The increase in the need for nutrients is even greater than that for energy, so the mother-to-be should choose nutrient-dense foods such as whole-grain breads and cereals, legumes, dark green vegetables, citrus fruits, low-fat milk and milk products, and lean meats, fish, poultry, and eggs.

Ample carbohydrate (ideally, 175 grams or more per day and certainly no less than 135 grams) is necessary to fuel the fetal brain and spare the protein needed for fetal growth. Fiber in carbohydrate-rich foods such as whole grains, vegetables, and fruit can help alleviate the constipation that many pregnant women experience.

The protein RDA for pregnancy is 25 grams per day higher than for nonpregnant women. Most women in the United States, however, do not need to add protein-rich foods to their diets because they already exceed the recommended protein intake for pregnancy. Excess protein may also have adverse effects, as Chapter 4 explained.

Some vegetarian women limit or omit protein-rich meats, eggs, and milk products from their diets. For them, both meeting the recommendation for food energy each day as well as including several generous servings of plant-protein foods such as legumes, tofu, whole grains, nuts, and seeds are imperative. Protein supplements during pregnancy can be harmful, and their use is discouraged.

The high nutrient requirements of pregnancy leave little room in the diet for excess energy from added purified fats such as oil, margarine, and butter. The essential fatty acids, however, are particularly important to the growth and development of the fetus.[17] The brain contains a substantial amount of lipid material and depends heavily on long-chain omega-3 and omega-6 fatty acids for its growth, function, and structure. (See Table 3-3 on p. 68 for a list of good food sources of the essential fatty acids.)

IN SUMMARY

▌ Pregnancy brings physiological adjustments that demand increased intakes of energy and nutrients.

▌ A balanced diet that includes more nutrient-dense foods from each of the five food groups can help to meet these needs.

Of Special Interest: Folate and Vitamin B$_{12}$ The vitamins famous for their roles in cell reproduction—folate and vitamin B$_{12}$—are needed in large amounts during pregnancy. New cells are laid down at a tremendous pace as the fetus grows and develops. At the same time, because the mother's blood volume increases, the number of her red blood cells must rise, requiring more cell division and therefore more vitamins. To accommodate these needs, the recommendation for folate during pregnancy increases from 400 to 600 micrograms a day.

As described in Chapter 8, folate plays an important role in preventing neural tube defects. To review, the early weeks of pregnancy are a critical period for the formation and closure of the **neural tube** that will later develop to form the brain and spinal cord. By the time a woman suspects she is pregnant, usually around the sixth week of pregnancy, the embryo's neural tube normally has closed. A **neural tube defect (NTD)** occurs when the tube fails to close properly. In the United States, an estimated 2500 infants with NTD are born each year.[18] When the neural tube fails to close properly and brain development fails, a rare but lethal defect known as **anencephaly** occurs. All infants with anencephaly die shortly after birth.

In a more common NTD, the spinal cord and backbone do not develop normally—and the result is **spina bifida** (see Figure 11-5). The membranes covering the spinal cord often protrude from the spine as a sac, and sometimes a portion of the spinal

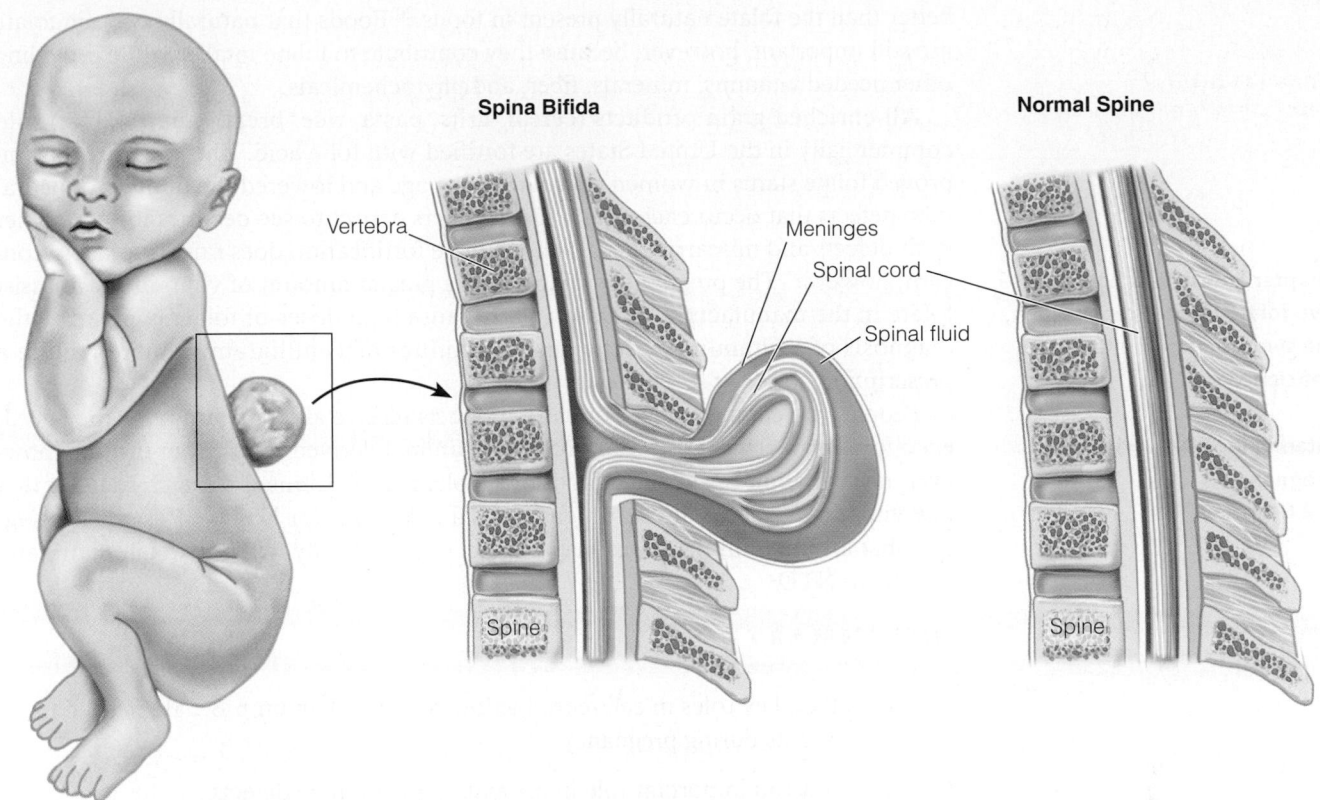

Spina Bifida

Vertebra

Normal Spine

Meninges

Spinal cord

Spinal fluid

Spine

Spine

FIGURE 11-5 Spina Bifida—A Neural Tube Defect
Spina bifida, a common neural tube defect, occurs when the vertebrae of the spine fail to close around the spinal cord, leaving it unprotected. The B vitamin folate helps prevent spina bifida and other neural tube defects.

Source: From the *Journal of the American Medical Association*, June 20, 2001, Vol. 285, No. 23 p. 3050. Reprinted with permission of the American Medical Association.

cord is contained in the sac. Spina bifida is often accompanied by varying degrees of paralysis, depending on the extent of spinal cord damage. Mild cases may not be noticed. Moderate cases may involve curvature of the spine, muscle weakness, mental handicaps, and other ills, while severe cases can lead to death.

To reduce the risk of neural tube defects, *the Dietary Guidelines for Americans, 2010* advise women who are capable of becoming pregnant to obtain 400 micrograms of folic acid daily from supplements, fortified foods, or both, *in addition* to eating folate-rich foods (see Table 11-2). The DRI committee recommends intake of synthetic folate, called folic acid, in supplements and fortified food because it is absorbed

neural tube: the embryonic tissue that later forms the brain and spinal cord.

neural tube defect (NTD): a serious central nervous system birth defect that often results in lifelong disability or death.

anencephaly (AN-en-SEF-a-lee): an uncommon and always fatal type of neural tube defect; characterized by the absence of a brain.
   an = not (without)
   encephalus = brain

spina (SPY-nah) **bifida** (BIFF-ih-dah): one of the most common types of neural tube defects; characterized by the incomplete closure of the spinal cord and its bony encasement.
   spina = spine
   bifida = split

TABLE 11-2 Rich Folate Sources[a]

Natural Folate Sources	Fortified Folate Sources
Liver (3 oz): 221 µg	Multi-Grain Cheerios Plus cereal (1 c): 400 µg[b]
Lentils (½ c): 179 µg	Product 19 cereal (1 c): 400 µg[b]
Chickpeas or pinto beans (½ c): 145 µg	Total cereal (1 c): 400 µg[b]
Asparagus (½ c): 134 µg	Pasta, cooked (1 c): 110 µg
Spinach (1 c raw): 58 µg	Rice, cooked (1 c): 134 µg
Avocado (½ c): 45 µg	Bagel (1 small whole): 75 µg
Orange juice (1 c): 74 µg	Waffles, frozen (2): 36 µg
Beets (½ c): 68 µg	Bread, white (1 slice): 28 µg

[a]Folate amounts for these and 2000 other foods are listed in the Table of Food Composition in Appendix A.
[b]Folate in cereals varies; read the Nutrition Facts panel of the label.

better than the folate naturally present in foods.[20] Foods that naturally contain folate are still important, however, because they contribute to folate intakes while providing other needed vitamins, minerals, fiber, and phytochemicals.

All enriched grain products (cereal, grits, pasta, rice, bread, and the like) sold commercially in the United States are fortified with folic acid. This measure has improved folate status in women of childbearing age and lowered the number of neural tube defects that occur each year.[21] Researchers expect to see declines in some other birth defects and miscarriages as well.[22] Folate fortification does raise one safety concern, however. The pregnant woman needs a greater amount of vitamin B_{12} to assist folate in the manufacture of new cells. Because high doses of folate complicate the diagnosis of a vitamin B_{12} deficiency, quantities of 1 milligram or more require a prescription.

People who eat meat, eggs, or dairy products receive all the vitamin B_{12} they need, even for pregnancy. Those who exclude all animal-derived foods from the diet, however, need vitamin B_{12}-fortified foods or supplements.[23] Limited research suggests that low vitamin B_{12} status during pregnancy is a risk factor for NTD.[24] Researchers suggest that low vitamin B_{12} status may interact synergistically with low folate status and lead to an NTD.

IN SUMMARY

▌ Due to their key roles in cell reproduction, folate and vitamin B_{12} are needed in large amounts during pregnancy.

▌ Folate plays an important role in preventing neural tube defects.

Vitamin D and Calcium for Bones Vitamin D and the minerals involved in building the skeleton—calcium, phosphorus, magnesium, and fluoride—are in great demand during pregnancy. Insufficient intakes may produce abnormal fetal bones and teeth.

Vitamin D plays a vital role in calcium absorption and utilization. Consequently, severe maternal vitamin D deficiency interferes with normal calcium metabolism, which, in rare cases, may cause rickets in the infant.[25] Regular exposure to sunlight and consumption of vitamin D–fortified milk are usually sufficient to support normal calcium metabolism during pregnancy, although some researchers question whether current recommendations, even with prenatal supplements, are adequate.[26]

Intestinal absorption of calcium doubles early in pregnancy, when the mother's bones store the mineral. Later, as the fetal bones begin to calcify, there is a dramatic shift of calcium across the placenta. Whether the calcium added to the mother's bones early in pregnancy is withdrawn to build the fetus's bones later is unclear. In the final weeks of pregnancy, more than 300 milligrams a day are transferred to the fetus. Recommendations to ensure an adequate calcium intake during pregnancy are aimed at conserving the mother's bone mass while supplying fetal needs.

For women whose prepregnancy calcium intakes are below recommendations, as most are, increased calcium intakes may be especially important. Milk products offer many advantages over supplements, as emphasized in earlier chapters. Because bones are still actively depositing minerals until about age 25, adequate calcium is especially important for young women. Pregnant women under age 25 who consume less than 600 milligrams of calcium a day need to increase their intakes of milk, cheese, yogurt, and other calcium-rich foods. Alternatively, and less preferably, they may need a daily supplement of 600 milligrams of calcium.

Women who exclude milk products need calcium-fortified foods such as soymilk, but not all soymilk is fortified with calcium. Products fortified with calcium and

Chapter 8 describes how excessive folate intakes can mask the symptoms of a vitamin B_{12} deficiency.

Vitamin B_{12} RDA during pregnancy:
▌ 2.6 µg/day

Calcium RDA during pregnancy:
▌ 1300 mg/day (14 to 18 yr)
▌ 1000 mg/day (19 to 50 yr)
Phosphorus RDA during pregnancy:
▌ 1250 mg/day (14 to 18 yr)
▌ 700 mg/day (19 to 50 yr)
Magnesium RDA during pregnancy:
▌ 400 mg/day (14 to 18 yr)
▌ 350 mg/day (19 to 30 yr)
▌ 360 mg/day (31 to 50 yr)
Fluoride AI during pregnancy:
▌ 3 mg/day

Three cups of milk a day will supply 900 mg of calcium. For other food sources of calcium, see Chapter 9.

vitamin D are recommended. Calcium-fortified orange juice offers folate and vitamin C as well as calcium.

IN SUMMARY

▮ Maternal vitamin D deficiency interferes with calcium metabolism in the infant.

▮ All pregnant women, but especially those who are younger than 25 years of age, need to pay special attention to calcium to ensure adequate intakes.

Iron The body conserves iron especially well during pregnancy: menstruation ceases, and absorption of iron increases up to threefold due to a rise in the blood's iron-absorbing and iron-carrying protein transferrin. Still, iron needs are so high that stores dwindle during pregnancy.

The developing fetus draws heavily on the mother's iron stores to create stores of its own to last through the first four to six months of life. Even women with inadequate iron stores transfer significant amounts of iron to the fetus, suggesting that the iron needs of the fetus have priority over those of the mother. Iron losses also occur with the bleeding that is inevitable at birth.

Few women enter pregnancy with adequate iron stores. Women who enter pregnancy with iron-deficiency anemia have a greater-than-normal risk of delivering low-birthweight or preterm infants.[27] For all women not taking supplements containing iron, a daily iron supplement containing 30 milligrams is recommended during the second and third trimesters of pregnancy. When a low hemoglobin or hematocrit is confirmed by a repeat test, more than 30 milligrams of iron may be prescribed. To enhance iron absorption, the supplement should be taken between meals and with liquids other than milk, coffee, or tea, which inhibit iron absorption.

Zinc Zinc is required for DNA and RNA synthesis and thus for protein synthesis. Typical zinc intakes for pregnant women are lower than recommendations, but, fortunately, zinc absorption increases when zinc intakes are low. Routine supplementation during pregnancy is not advised.[28] Women taking iron supplements (more than 30 milligrams per day), however, may need zinc supplements because large doses of iron can interfere with the body's absorption and use of zinc. Most supplements for pregnancy provide about 30 to 60 milligrams of iron a day. Zinc is most abundant in protein-rich foods such as shellfish, meat, and nuts.

IN SUMMARY

▮ A daily iron supplement is recommended for all pregnant women during the second and third trimesters.

▮ Iron interferes with zinc absorption, so women taking iron supplements (more than 30 milligrams per day) may need zinc supplements as well.

Nutrient Supplements Physicians often recommend daily multivitamin-mineral supplements for pregnant women. These prenatal supplements typically provide more folate, iron, and calcium than regular supplements. Prenatal supplements are especially beneficial for women who do not eat adequately and for those in high-risk groups: women carrying twins or triplets, cigarette smokers, and alcohol and drug abusers.[29] For these women, prenatal supplements may be of some help in reducing the risks of preterm delivery, low birthweights, and birth defects (see Figure 11-6 on p. 296). Be aware, however, that supplements cannot prevent the vast majority of fetal harm from tobacco, alcohol, and drugs, which continues unopposed, as later sections explain.

Iron RDA during pregnancy:
▮ 27 mg/day

In pregnancy, hemoglobin values of 12 g/dL are not unusual, and 11 g/dL is where the line defining "too low" is often drawn. Appendix E discusses more sensitive measures of iron status.

Food sources of iron:
▮ Liver, oysters
▮ Red meat, fish, other meat
▮ Dried fruits (raisins, prunes)
▮ Legumes (dried beans, peas, lima beans)
▮ Dark green vegetables

Dietary Guidelines for Americans, 2010:
Women capable of becoming pregnant should choose foods that supply heme iron, which is more readily absorbed by the body, additional iron sources, and enhancers of iron absorption, such as vitamin C.

Zinc RDA during pregnancy:
▮ 12 mg/day (≤18 yr)
▮ 11 mg/day (19 to 50 yr)

Supplement Facts
Serving Size 1 Tablet

Amount Per Tablet	% Daily Value for Pregnant/ Lactating Women
Vitamin A 4000 IU	50%
Vitamin C 100 mg	167%
Vitamin D 400 IU	100%
Vitamin E 11 IU	37%
Thiamin 1.84 mg	108%
Riboflavin 1.7 mg	85%
Niacin 18 mg	90%
Vitamin B_6 2.6 mg	104%
Folate 800 mcg	100%
Vitamin B_{12} 4 mcg	50%
Calcium 200 mg	15%
Iron 27 mg	150%
Zinc 25 mg	167%

INGREDIENTS: calcium carbonate, microcrystalline cellulose, dicalcium phosphate, ascorbic acid, ferrous fumarate, zinc oxide, acacia, sucrose ester, niacinamide, modified cellulose gum, di-alpha tocopheryl acetate, hydroxypropyl methylcellulose, hydroxypropyl cellulose, artificial colors (FD&C blue no. 1 lake, FD&C red no. 40 lake, FD&C yellow no. 6 lake, titanium dioxide), polyethylene glycol, starch, pyridoxine hydrochloride, vitamin A acetate, riboflavin, thiamin mononitrate, folic acid, beta carotene, cholecalciferol, maltodextrin, gluten, cyanocobalamin, sodium bisulfite.

FIGURE 11-6 Example of a Prenatal Supplement Label
Notice that vitamin A is reduced to guard against birth defects, while extra amounts of folate, iron, and other nutrients are provided to meet the specific needs of pregnant women.

high-risk pregnancy: a pregnancy characterized by risk factors that make it likely the birth will be surrounded by problems such as premature delivery, difficult birth, retarded growth, birth defects, and early infant death.

low-risk pregnancy: a pregnancy characterized by factors that make it likely the birth will be normal and the infant healthy.

Special Supplemental Nutrition Program for Women, Infants, and Children (WIC): a high-quality, cost-effective health care and nutrition services program administered by the U.S. Department of Agriculture for low-income women, infants, and children who are nutritionally at risk. WIC provides supplemental foods, nutrition education, and referrals to health care and other social services.

TABLE 11-3 High-Risk Pregnancy Factors

Prepregnancy BMI either <18.5 or ≥25
Insufficient or excessive pregnancy weight gain
Nutrient deficiencies or toxicities; eating disorders
Poverty, lack of family support, low level of education, limited food available
Smoking, alcohol, or other drug use
Teens, especially 15 years or younger; women 35 years or older
Many previous pregnancies (three or more to mothers under age 20; four or more to mothers age 20 or older)
Short or long intervals between pregnancies (<18 months or >59 months)
Previous history of problems
Twins or triplets
Low- or high-birthweight infants
Development of gestational hypertension
Development of gestational diabetes
Diabetes; hypertension; heart, respiratory, and kidney disease; certain genetic disorders; special diets and medications

Table 11-3 identifies characteristics of a **high-risk pregnancy.** A woman with none of these factors is said to have a **low-risk pregnancy.** The more factors that apply, the higher the risk. All pregnant women, especially those in high-risk categories, need prenatal medical care, including dietary advice.

IN SUMMARY

▌ Women most likely to benefit from multivitamin-mineral supplements during pregnancy include those who do not eat adequately, those carrying twins or triplets, and those who smoke cigarettes or are alcohol or drug abusers.

Food Assistance Programs

Women of limited financial means may eat diets too low in calcium, iron, vitamins A and C, and protein. Often, they and their children need help in obtaining food and benefit from nutrition counseling. At the federal level, the **Special Supplemental Nutrition Program for Women, Infants, and Children (WIC)** provides vouchers redeemable for nutritious foods, nutrition education, and referrals to health and social services to low-income pregnant and lactating women and their children.[30] The foods offered include milk and cheese, iron-fortified cereals, fruit or vegetable juices, carrots, eggs, dried beans, tuna fish, and peanut butter. Soy-based beverages, tofu, fruits and vegetables, baby foods, whole-wheat bread, and other whole-grain options were recently added. These foods provide nutrients often lacking in diets of low-income women and children. For infants given infant formula, WIC also provides iron-fortified formula. WIC encourages mothers to breastfeed their infants, however, and offers incentives to those who do.[31]

More than 8 million people—most of them infants and young children—receive WIC benefits each month. Participation in the WIC program benefits both the nutrient status and the growth and development of infants and children. WIC participation during pregnancy can effectively reduce infant mortality, low birthweight, and maternal and newborn medical costs.

The Supplemental Nutrition Assistance Program (SNAP) provides a debit card that can also help to stretch the low-income pregnant woman's grocery dollars. Many

communities provide educational services and materials, including nutrition, food budgeting, and shopping information through the local agricultural extension service. Organizations such as the American Dietetic Association and local hospitals also provide nutrition information.

IN SUMMARY

■ Food assistance programs, such as WIC, can provide nutritious food for pregnant women of limited financial means.

Weight Gain

Women must gain weight during pregnancy—fetal and maternal well-being depend on it. Ideally, a woman will have begun her pregnancy at a healthy weight, but even more importantly, she will gain within the recommended weight range based on her prepregnancy body mass index (BMI). Table 11-4 presents recommended weight gains for pregnancy. Pregnancy weight gains within the recommended ranges are associated with fewer surgical births, a greater number of healthy birthweights, and other positive outcomes for both mothers and infants, but many women do not gain within these ranges.[32] Among women in the United States, excessive weight gain during pregnancy is more prevalent than inadequate weight gain. To improve pregnancy outcomes, researchers and health care providers are placing greater emphasis on preventing excessive weight gains during pregnancy than in the past.[33]

Weight loss during pregnancy is not recommended.[34] Even obese women are advised to gain between 11 and 20 pounds for the best chance of delivering a healthy infant.[35] Ideally, overweight women will achieve a healthy body weight before becoming pregnant, avoid excessive weight gain during pregnancy, and postpone weight loss until after childbirth.[36]

According to current recommendations, pregnant adolescents who are still growing themselves should strive for gains at the upper end of the target range. Concerns about obesity raise questions as to whether such recommendations are always appropriate. Compared with older mothers, for example, the risk of lifetime weight retention after pregnancy may be far greater for young adolescents.[37]

The ideal weight gain pattern for a woman who begins pregnancy at a healthy weight is 3½ pounds during the first trimester and 1 pound per week thereafter. If a woman gains more than is recommended early in pregnancy, she should not restrict her energy intake later on in order to lose weight. A sudden, large weight gain is a danger signal, however, because it may indicate the onset of preeclampsia (discussed later in the chapter).

The weight the pregnant woman gains is nearly all lean tissue: placenta, uterus, blood, milk-producing glands, and the fetus itself (see Figure 11-7 on p. 298). The fat she gains is needed later for lactation. Physical activity can help a pregnant woman cope with the extra weight, as a later section explains.

A prenatal weight-gain grid (Appendix E) plots the rate of weight gain during pregnancy.

TABLE 11-4 Recommended Weight Gains Based on Prepregnancy Weight

Prepregnancy Weight	Recommended Weight Gain	
	For Single Birth	For Twin Birth
Underweight (BMI <18.5)	28 to 40 lb (12.5 to 18.0 kg)	Insufficient data to make recommendation
Healthy weight (BMI 18.5 to 24.9)	25 to 35 lb (11.5 to 16.0 kg)	37 to 54 lb (17 to 25 kg)
Overweight (BMI 25.0 to 29.9)	15 to 25 lb (7.0 to 11.5 kg)	31 to 50 lb (14 to 23 kg)
Obese (BMI ≥30)	11 to 20 lb (5 to 9 kg)	25 to 42 lb (11 to 19 kg)

Source: Institute of Medicine, *Weight Gain During Pregnancy: Reexamining the Guidelines.* Reprinted with permission from The National Academies Press, Copyright © 2009, National Academy of Sciences.

FIGURE 11-7
Components of Weight Gain during Pregnancy

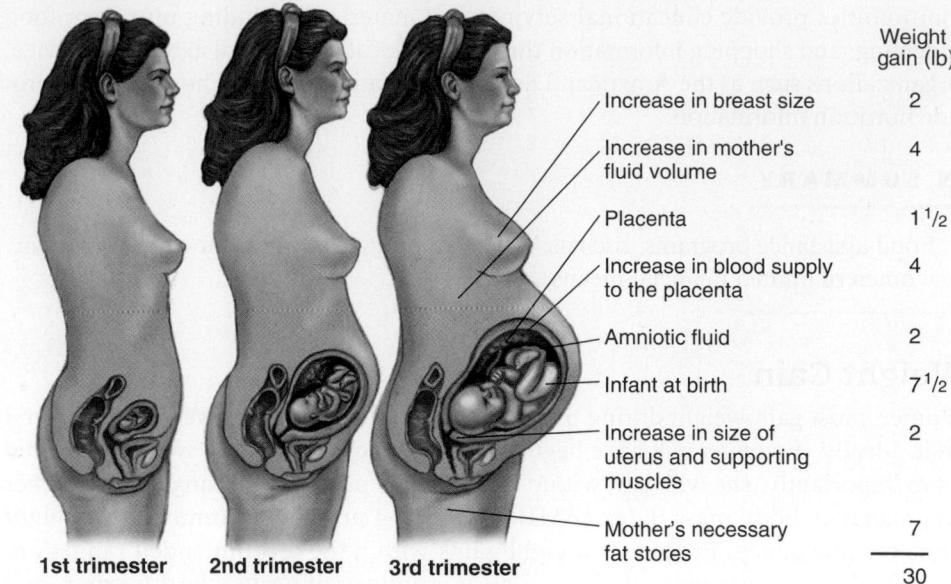

	Weight gain (lb)
Increase in breast size	2
Increase in mother's fluid volume	4
Placenta	1 1/2
Increase in blood supply to the placenta	4
Amniotic fluid	2
Infant at birth	7 1/2
Increase in size of uterus and supporting muscles	2
Mother's necessary fat stores	7
	30

1st trimester   2nd trimester   3rd trimester

Weight Loss after Pregnancy

The pregnant woman loses some weight at delivery. In the following weeks, she loses more as her blood volume returns to normal and she gets rid of accumulated fluids. The typical woman does not, however, return to her prepregnancy weight. In general, the more weight a woman gains beyond the needs of pregnancy, the more she retains—mostly as body fat.[38] Even with an average weight gain during pregnancy, most women tend to retain a couple of pounds with each pregnancy. When those couple of pounds becomes 7 or more and BMI increases by a unit or more, complications such as diabetes and hypertension in future pregnancies as well as chronic diseases later in life can increase—even for women who are not overweight.[39] Those who are successful in losing their pregnancy weight are more likely to limit weight gains through middle adulthood.[40]

IN SUMMARY

▌ Weight gain is essential for a healthy pregnancy. Excessive weight gain during pregnancy, however, can impair pregnancy outcome.

▌ A woman's prepregnancy BMI, her own nutrient needs, and the number of fetuses she is carrying help to determine appropriate weight gain.

Physical Activity

Physical activity is important to the pregnant woman, not only to help her carry the extra weight of pregnancy without strain, but also to help ease her upcoming childbirth. Staying active during the course of a normal, healthy pregnancy improves the fitness of the mother-to-be, facilitates labor, helps to prevent or manage gestational diabetes, and reduces psychological stress.[41] Women who remain active during pregnancy report fewer discomforts throughout their pregnancies and retain habits that help in losing excess weight and regaining fitness after the birth.

Pregnant women must take care in choosing their physical activities, however. They should participate in low-impact activities and avoid sports in which they might fall or be hit by other people or objects. Swimming and water aerobics are particularly beneficial because they allow the body to remain cool and move freely with the water's support, thus reducing back pain.[42]

DO		DON'T
Do exercise regularly (most, if not all, days of the week).		Don't exercise vigorously after long periods of inactivity.
Do warm up with 5 to 10 minutes of light activity.		Don't exercise in hot, humid weather.
Do 30 minutes or more of moderate physical activity.		Don't exercise when sick with fever.
Do cool down with 5 to 10 minutes of slow activity and gentle stretching.		Don't exercise while lying on your back after the first trimester of pregnancy or stand motionless for prolonged periods.
Do drink water before, after, and during exercise.		Don't exercise if you experience any pain or discomfort.
Do eat enough to support the additional needs of pregnancy plus exercise.		Don't participate in activities that may harm the abdomen or involve jerky, bouncy movements.
	Pregnant women can enjoy the benefits of physical activity.	Don't scuba dive.

FIGURE 11-8 **Guidelines for Physical Activity during Pregnancy**

As is true for everyone, the frequency, duration, and intensity of the activity affect the likelihood of the benefits or risks. A pregnant woman should consult her health care provider before taking up additional activity. A few guidelines are offered in Figure 11-8. Several of the guidelines are aimed at preventing excessively high internal body temperatures and dehydration, both of which can harm fetal development. To this end, a pregnant woman should also stay out of saunas, steam rooms, and hot whirlpools.

IN SUMMARY

▌ By remaining active throughout pregnancy, a woman can develop the strength she needs to carry the extra weight and maintain habits that will help her lose it after the birth.

Common Nutrition-Related Concerns of Pregnancy

Food sensitivities, nausea, heartburn, and constipation are common during pregnancy. A few simple strategies can help alleviate maternal discomforts (see Table 11-5 on p. 300).

Food Cravings and Aversions Some women develop cravings for, or aversions to, certain foods and beverages during pregnancy. Individual **food cravings** during pregnancy do not seem to reflect real physiological needs. In other words, a woman who craves pickles does not necessarily need salt. Similarly, cravings for ice cream are common during pregnancy but do not signify a calcium deficiency. **Food aversions** and cravings that arise during pregnancy are probably due to hormone-induced changes in taste and sensitivities to smells.

Nonfood Cravings Some pregnant women develop cravings for nonfood items such as laundry starch, clay, soil, or ice—a practice known as pica.[43] Pica may be practiced for cultural reasons that reflect a society's folklore; it is especially common among African American women. Pica is often associated with iron deficiency, but whether iron deficiency leads to pica or pica leads to iron deficiency is unclear.[44] Eating clay or soil may interfere with iron absorption and displace iron-rich foods from the diet. Furthermore, if the soil or clay contains environmental contaminants such as lead or parasites, health and nutrition suffer.

Physical Activity Guidelines for Americans, 2008

▌ Healthy women who are not already highly active or doing vigorous-intensity activity should get at least 150 minutes (2 hours and 30 minutes) of moderate-intensity aerobic activity per week during pregnancy and the postpartum period. Preferably, this activity should be spread throughout the week.

▌ Pregnant women who habitually engage in vigorous-intensity aerobic activity or are highly active can continue physical activity during pregnancy and the postpartum period, provided that they remain healthy and discuss with their health care provider how and when activity should be adjusted over time.

food cravings: deep longings for particular foods.

food aversions: strong desires to avoid particular foods.

TABLE 11-5 Strategies to Alleviate Maternal Discomforts

To alleviate the nausea of pregnancy:
On waking, get up slowly.
Eat dry toast or crackers.
Chew gum or suck hard candies.
Eat small, frequent meals whenever hunger strikes.
Avoid foods with offensive odors.
When nauseated, do not drink citrus juice, water, milk, coffee, or tea.

To prevent or alleviate constipation:
Eat foods high in fiber.
Exercise daily.
Drink at least 8 glasses of liquids per day.
Respond promptly to the urge to defecate.
Use laxatives only as prescribed by a physician; avoid mineral oil—it carries needed fat-soluble vitamins out of the body.

To prevent or relieve heartburn:
Relax and eat slowly.
Eat small, frequent meals.
Drink liquids between meals.
Avoid spicy or greasy foods.
Sit up while eating.
Wait 1 hour after eating before lying down.
Wait 2 hours after eating before exercising.

Morning Sickness The nausea of "morning" (actually, anytime) sickness is usually benign, although it is distressing to some women. It arises from the hormonal changes taking place early in pregnancy, ranges from mild queasiness to debilitating nausea, and afflicts more than half of all pregnant women. Many women complain that smells, especially cooking smells, make them sick. Thus, minimizing odors is a key to alleviating morning sickness. Traditional strategies for quelling nausea are listed in Table 11-5, but many women benefit most from simply eating the foods they want when they feel like eating.

Heartburn Heartburn, a burning sensation in the lower esophagus near the heart, is common during pregnancy and is also benign. As the growing fetus puts increasing pressure on the woman's stomach, acid may back up and create a burning sensation in her throat. Tips to relieve heartburn are also listed in Table 11-5.

Constipation As the hormones of pregnancy alter muscle tone and the thriving infant crowds intestinal organs, an expectant mother may complain of constipation, another harmless but annoying condition. A high-fiber diet, physical activity, and plentiful fluids will help relieve this condition. Also, responding promptly to the urge to defecate can help. Laxatives should be used only as prescribed by the physician. Mineral oil should not be used because it interferes with the absorption of fat-soluble vitamins.

IN SUMMARY

▋ Food cravings typically do not reflect physiological needs.

▋ The nausea, heartburn, and constipation that sometimes accompany pregnancy can usually be alleviated with a few simple strategies.

Problems in Pregnancy

Just as adequate nutrition and normal weight gain support the health of the mother and growth of the fetus, maternal diseases can have an adverse effect. If discovered early, many diseases can be controlled—another reason why early prenatal care is recommended. Some nutrition measures can help alleviate the most common problems encountered during pregnancy.

Preexisting Diabetes Pregnancy presents special challenges for the management of diabetes.[45] Insulin-induced hypoglycemia has a more rapid onset during pregnancy and is a danger to the mother, especially in those with type 1 diabetes. Women with type 2 diabetes often start pregnancy with insulin resistance and obesity, making optimal glycemic control difficult. The risks of diabetes during pregnancy depend on how well it is managed before, during, and after. Excellent glycemic control in the first trimester and throughout the pregnancy is associated with the lowest frequency of maternal, fetal, and newborn complications.[46] Without proper management, women face high infertility rates, and those who do conceive may experience episodes of severe hypoglycemia or hyperglycemia, preterm labor, and pregnancy-related hypertension. Infants may be large, suffer physical and mental abnormalities, and experience other complications such as severe hypoglycemia or respiratory distress, both of which can be fatal. Signs of fetal health problems are apparent even when maternal glucose is above normal but still below the diagnosis of diabetes.[47] Ideally, a woman will receive the prenatal care needed to achieve glucose control before conception and continued glucose control throughout pregnancy. For optimal long-term outcomes, continuation of intensified diabetes management after pregnancy is in the best interest of the mother's health.[48]

Gestational Diabetes Some women are prone to developing a pregnancy-related form of diabetes, **gestational diabetes.** Gestational diabetes usually resolves after the infant is born, but some women go on to develop diabetes (type 2) later in life, especially if they are overweight.[49] Gestational diabetes can lead to fetal or infant sickness or death. When it is identified early and managed properly, however, the most serious risks fall dramatically.[50] More commonly, gestational diabetes leads to surgical birth and high infant birthweight.[51] To ensure that the problems of gestational diabetes are dealt with promptly, physicians screen for the risk factors listed in the margin of this page, and they test high-risk women for glucose intolerance immediately and average-risk women between 24 and 28 weeks gestation.[52] Chapter 21 provides information about medical nutrition therapy for gestational diabetes.

Hypertension Hypertension (blood pressure $\geq 140/90$ millimeters mercury) complicates pregnancy and affects its outcome in different ways, depending on how severe it becomes. Hypertension during pregnancy is classified as **chronic hypertension** or **gestational hypertension.**[53] Chronic hypertension can be a preexisting condition that develops before a woman becomes pregnant. In women whose prepregnancy blood pressure is unknown, diagnosis of chronic hypertension is based on the presence of sustained hypertension before 20 weeks of gestation.[54] In contrast, gestational hypertension develops after the 20th week of gestation. In women with gestational hypertension, blood pressure usually returns to normal during the first few weeks after childbirth.

Both types of hypertension pose risks to the mother and fetus. In addition to the health risks normally imposed by hypertension (heart attack and stroke), high blood pressure increases the risks of having a low-birthweight infant or of having the placenta separate from the wall of the uterus before the birth, resulting in stillbirth. Both chronic hypertension and gestational hypertension also increase the risk of preeclampsia.

Risk factors for gestational diabetes:
- Age 25 or older
- BMI ≥ 25 or excessive weight gain
- Complications in previous pregnancies, including gestational diabetes or high birthweight
- Prediabetes or symptoms of diabetes
- Family history of diabetes
- Hispanic American, African American, Native American, Asian American, Pacific Islander

gestational diabetes: glucose intolerance with first onset or first recognition during pregnancy.

chronic hypertension: in pregnant women, hypertension that is present and documented before pregnancy; in women whose prepregnancy blood pressure is unknown, the presence of sustained hypertension before 20 weeks of gestation.

gestational hypertension: high blood pressure that develops in the second half of pregnancy and usually resolves after childbirth.

▮ Hypertension
▮ Protein in the urine
▮ Upper abdominal pain
▮ Severe and constant headaches
▮ Swelling, especially of the face
▮ Dizziness
▮ Blurred vision
▮ Sudden weight gain (1 lb/day)

Preeclampsia **Preeclampsia** is a condition characterized not only by high blood pressure but also by protein in the urine.[55] Preeclampsia usually occurs with first pregnancies and almost always appears after 20 weeks' gestation. Symptoms typically regress within 48 hours of delivery.

Preeclampsia affects almost all of the woman's organs—the circulatory system, liver, kidneys, and brain. If it progresses, she may experience seizures; when this occurs, the condition is called **eclampsia.** Maternal mortality during pregnancy is rare in developed countries, but eclampsia is the most common cause. Preeclampsia demands prompt medical attention. Treatment focuses on regulating blood pressure and preventing seizures.

IN SUMMARY

▮ Conditions such as gestational diabetes, hypertension, and preeclampsia can threaten the health and life of both mother and infant.

▮ Such conditions require medical and nutrition treatment.

Practices to Avoid

A general guideline for the pregnant woman is to eat a normal, healthy diet and practice moderation. A woman's daily choices during pregnancy take on enormous importance. Forewarned, pregnant women can choose to abstain from or avoid potentially harmful practices.

Cigarette Smoking One practice to be avoided during pregnancy is cigarette smoking. A surgeon general's warning states that parental smoking can kill an otherwise healthy fetus or newborn. Unfortunately, an estimated 10 percent of pregnant women in the United States smoke, and rates are even higher for unmarried women and those who have not graduated from high school.[56] Constituents of cigarette smoke, such as nicotine, carbon monoxide, arsenic, and cyanide, are toxic to a fetus.[57] Research shows that smoking during pregnancy can cause damage to fetal chromosomes, which could lead to developmental defects or genetic disorders, including cancer.[58] Smoking also restricts the blood supply to the growing fetus and so limits the delivery of oxygen and nutrients and the removal of wastes. It slows growth, thus retarding physical development of the fetus. Smoking during pregnancy may reduce brain size and impair the intellectual and behavioral development of the child later in life.[59]

A mother who smokes is more likely to have a complicated birth and a low-birthweight infant.[60] The more a mother smokes, the smaller her infant will be. Of all preventable causes of low birthweight in the United States, smoking has the greatest impact. Sudden infant death syndrome (SIDS), the unexplained death that sometimes occurs in an otherwise healthy infant, has been linked to the mother's cigarette smoking during pregnancy.[61] Research suggests that even in women who do not smoke, exposure to **environmental tobacco smoke (ETS,** or secondhand smoke) during pregnancy increases the risk of low birthweight and the likelihood of SIDS.[62]

▮ Fetal growth retardation
▮ Low birthweight
▮ Complications at birth (prolonged final stage of labor)
▮ Mislocation of the placenta
▮ Premature separation of the placenta
▮ Vaginal bleeding
▮ Spontaneous abortion
▮ Fetal death
▮ Sudden infant death syndrome (SIDS)
▮ Middle ear diseases
▮ Cardiac and respiratory diseases

In addition to the harms already described, cigarette (and cigar) smoking adversely affects the pregnant woman's nutrition status and thus impairs fetal nutrition and development. The margin lists complications of smoking during pregnancy.

Medicinal Drugs and Herbal Supplements Medicinal drugs taken during pregnancy can cause serious birth defects. Pregnant women should not take over-the-counter drugs or any medications not prescribed by a physician. Drug labels warn pregnant and lactating women of possible harmful consequences. Such warnings should be taken seriously.

Some pregnant women mistakenly consider herbal supplements to be safe alternatives to medicinal drugs and take them to relieve nausea, promote water loss, alleviate

depression, help them sleep, or for other reasons. Some herbal products may be safe, but very few have been tested for safety or effectiveness during pregnancy. Pregnant women should stay away from herbal supplements, teas, or other products unless their safety during pregnancy has been ascertained.[63] Chapter 15 offers more information about herbal supplements and other alternative therapies.

Drugs of Abuse Drugs of abuse such as cocaine easily cross the placenta and impair fetal growth and development. Furthermore, such drugs are responsible for preterm births, low-birthweight infants, and sudden infant deaths. If these newborns survive, central nervous system damage is evident: their cries, sleep, and behaviors early in life are abnormal, and their cognitive development later in life is impaired.[64] They may be hypersensitive or underaroused; those who test positive for drugs suffer the greatest effects of toxicity and withdrawal. Their growth throughout childhood continues at a slow rate.[65]

Environmental Contaminants Infants and young children of pregnant women exposed to environmental contaminants such as lead and mercury show signs of impaired mental and psychomotor development. During pregnancy, lead readily moves across the placenta, inflicting severe damage on the developing fetal nervous system. In addition, infants exposed to even low levels of lead during gestation weigh less at birth and consequently struggle to survive. For these reasons, it is particularly important that pregnant women consume foods and beverages free of contamination. A diet high in calcium will help to defend against lead contamination.

Mercury is among the contaminants of concern. As Chapter 3 mentioned, fatty fish are a good source of omega-3 fatty acids, but some fish contain large amounts of the pollutant mercury, which can impair fetal growth and harm the developing brain and nervous system.[66] Because the benefits of seafood consumption seem to outweigh the risks, *the Dietary Guidelines for Americans, 2010* advise pregnant (and lactating) women to do the following:[67]

▌ Avoid shark, swordfish, king mackerel, and tilefish (also called golden snapper or golden bass).

▌ Limit average weekly consumption to 12 ounces (cooked or canned) of seafood *or* to 6 ounces (cooked or canned) of white (albacore) tuna.

Fish-oil supplements are not recommended because they may contain concentrated toxins and because their effects on pregnancy remain unknown.

Foodborne Illness The vomiting and diarrhea caused by many foodborne illnesses can leave a pregnant woman exhausted and dangerously dehydrated. Particularly threatening, however, is **listeriosis,** which can cause miscarriage, stillbirth, or severe brain or other infections to fetuses and newborns. Pregnant women are about 20 times more likely than other healthy adults to get listeriosis. A woman with listeriosis may develop symptoms such as fever, vomiting, and diarrhea in about 12 hours after eating a contaminated food, and serious symptoms may develop a week to six weeks later. A blood test can reliably detect listeriosis, and antibiotics given promptly to the pregnant sufferer can often prevent infection of the fetus or newborn. The margin lists preventive measures pregnant women can take to avoid contracting listeriosis.

Vitamin-Mineral Megadoses Many vitamins and minerals are toxic when taken in excess. Excessive vitamin A is particularly infamous for its role in fetal malformations of the cranial nervous system. Intakes before the seventh week of pregnancy appear to be the most damaging. For this reason, vitamin A supplements are not given during pregnancy, unless there is specific evidence of deficiency, which is rare.

Dieting Weight-loss dieting, even for short periods, can be hazardous during pregnancy. Low-carbohydrate diets or fasts that cause ketosis deprive the growing fetal brain of needed glucose and may impair its development. Such diets are also likely

preeclampsia (PRE-ee-KLAMP-see-ah): a condition characterized by hypertension and protein in the urine.

eclampsia (eh-KLAMP-see-ah): a severe complication during pregnancy in which seizures occur.

environmental tobacco smoke (ETS): the combination of exhaled smoke (mainstream smoke) and smoke from lighted cigarettes, pipes, or cigars (sidestream smoke) that enters the air and may be inhaled by other people.

listeriosis: a serious foodborne infection that can cause severe brain infection or death in a fetus or newborn; caused by the bacterium *Listeria monocytogenes*, which is found in soil and water.

To protect their fetuses and newborns from listeriosis, pregnant women should:

▌ Avoid the following Mexican soft cheeses: *queso blanco, queso fresco, queso de hoja, queso de crema,* and *asadero.* Also avoid feta cheese, brie, Camembert, and blue-veined cheeses like Roquefort.

▌ Use only pasteurized dairy products and juices.

▌ Eat only thoroughly cooked meat, poultry, and seafood.

▌ Thoroughly reheat until steaming hot all hot dogs, luncheon meats, and deli meats, including cured meats like salami, before eating them.

▌ Wash all fruits and vegetables.

▌ Do not eat refrigerated smoked seafood such as salmon or trout, or any fish labeled "nova-style," "lox," or "kippered," unless it is an ingredient in a cooked dish.

▌ Do not eat refrigerated pâté or meat spreads. Canned or shelf-stable pâté and meat spreads are safer.

to be deficient in other nutrients vital to fetal growth. Regardless of prepregnancy weight, pregnant women need an adequate diet and sufficient weight gain to support healthy fetal development.

Sugar Substitutes Artificial sweeteners have been studied extensively and found to be acceptable during pregnancy if used within the FDA's guidelines (see Chapter 2).[68] Still, it would be prudent for pregnant women to use sweeteners in moderation and within an otherwise nutritious and well-balanced diet. Women with phenylketonuria should not use aspartame, as Chapter 2 explains.

Caffeine Caffeine crosses the placenta, and the fetus has only a limited ability to metabolize it. Research studies have not proved that caffeine (even in high doses) causes birth defects in human infants (as it does in animals), but limited evidence suggests that heavy use increases the risk of miscarriage and fetal death.[69] (In these studies, heavy caffeine use is defined as the equivalent of three or more cups of coffee a day.) Depending on the quantities consumed and the mother's metabolism, caffeine may also interfere with fetal growth.[70] In light of this evidence, the most sensible course is to limit caffeine consumption to the equivalent of one cup of coffee or two 12-ounce cola beverages a day. Caffeine amounts in food and beverages are listed in Appendix A.

Alcohol Drinking alcohol during pregnancy threatens the fetus with irreversible brain damage, growth retardation, mental retardation, facial abnormalities, vision abnormalities, and many more health problems—a spectrum of symptoms known as **fetal alcohol spectrum disorders,** or **FASD.** Children at the most severe end of the spectrum (those with all of the symptoms) are defined as having **fetal alcohol syndrome (FAS).**[71] The fetal brain is extremely vulnerable to a glucose or oxygen deficit, and alcohol causes both by disrupting placental functioning. The lifelong mental retardation and other tragedies of FAS can be prevented by abstaining from drinking alcohol during pregnancy. Once the damage is done, however, the child remains impaired.

The accompanying photo shows the facial abnormalities of FAS, which are easy to depict. A visual picture of the internal harm is impossible, but that damage seals the fate of the child. An estimated five to twenty of every 10,000 children are victims of FAS, making it one of the leading known preventable causes of mental retardation in the world.[72]

One out of eight pregnant women drinks alcohol sometime during her pregnancy; one out of fifty pregnant women reports binge drinking (five or more drinks on one occasion).[73] Almost half of all pregnancies are unintended, and many are conceived during a binge-drinking episode.

For women who know they are pregnant and choose to drink alcohol, the question is how much alcohol is too much. Even one drink a day threatens neurological development and behaviors. Low birthweight is reported among infants born to women who drink 1 ounce (two drinks) of alcohol per day during pregnancy, and FAS is also known to occur with as few as two drinks a day. Birth defects have been reliably observed among the children of women who drink 2 ounces (four drinks) of alcohol daily during pregnancy. The most severe impact is likely to occur in the first two months, before the woman may even be aware that she is pregnant.

For every child diagnosed with full-blown FAS, many more with FASD go undiagnosed until problems develop in the preschool years. Upon reaching adulthood, such children are ill equipped for employment, relationships, and the other facets of life most adults take for granted. Prenatal exposure to alcohol may permanently alter a person's response to it, and also to certain drugs, making addictions likely.

The American Academy of Pediatrics takes the position that women should stop drinking as soon as they *plan* to become pregnant. Researchers have looked for a safe alcohol intake limit during pregnancy and have found none. Their conclusion: abstinence from alcohol is the best policy for pregnant women.

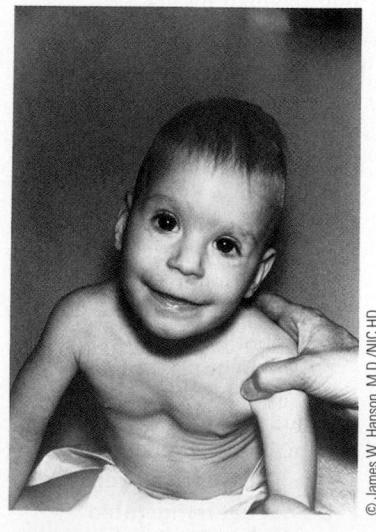

© James W. Hanson, M.D./NIC HD

These facial traits are typical of fetal alcohol syndrome, caused by drinking alcohol during pregnancy—low nasal bridge, short eyelid opening, underdeveloped groove in the center of the upper lip, small midface, short nose, and small head circumference.

For pregnant women who have already drunk alcohol, the advice is "stop now." A woman who has drunk heavily during the first two-thirds of her pregnancy can still prevent some organ damage by stopping heavy drinking during the third trimester.

IN SUMMARY

▌ Abstaining from smoking and other drugs, including alcohol, limiting intake of foods known to contain unsafe levels of contaminants such as mercury, taking precautions against foodborne illness, avoiding large doses of nutrients, refraining from dieting, using artificial sweeteners in moderation, and limiting caffeine use are recommended during pregnancy.

Adolescent Pregnancy

Many adolescents become sexually active before age 19, and more than 400,000 infants are born to teenage mothers each year in the United States.[74] A pregnant adolescent presents a special case of intense nutrient needs. Young teenage girls have a hard enough time meeting nutrient needs for their own rapid growth and development, let alone those of pregnancy. Many teens enter pregnancy deficient in vitamins B_{12} and D, folate, and iron, which increases the risk of impaired fetal growth.[75] Pregnant adolescents are less likely to receive early prenatal care and are more likely to smoke during pregnancy—two factors that predict low birthweight and infant death.[76]

The rates of stillbirths, preterm births, and low-birthweight infants are high for teenagers—both for teen moms and for teen dads.[77] Their greatest risk, however, is death of the infant: mothers under age 16 bear more infants who die within the first year than do women in any other age group. These factors combine to make adolescent pregnancy a major public health problem.

Adequate nutrition is an indispensable component of prenatal care for adolescents and can substantially improve the outlook for both mother and infant. To support the needs of both mother and fetus, a pregnant teenager with a BMI in the normal range is encouraged to gain about 35 pounds to reduce the likelihood of a low-birthweight infant. As mentioned earlier, however, compared with older mothers, the lifetime risk of postpartum weight retention in young adolescents may be far greater. Research shows that women who give birth during adolescence have higher body weights, BMIs, and percent body fat later on than adolescents who do not experience pregnancy.[78] Researchers agree that optimal weight gain recommendations for pregnant adolescents need focused attention. Meanwhile, pregnant and lactating adolescents would do well to follow the eating pattern presented in Table 1-4 (on p. 18), making sure to choose a kcalorie level high enough to support adequate, but not excessive, weight gain.

IN SUMMARY

▌ Proper nutrition and adequate weight gain are especially important in reducing the risk of poor pregnancy outcome in adolescents.

Breastfeeding

The American Academy of Pediatrics (AAP) recommends that infants receive breast milk for at least the first 12 months of life and beyond for as long as mutually desired by mother and child.[79] The American Dietetic Association (ADA) advocates breastfeeding for the nutritional health it confers on the infant as well as for the physiological, social, economic, and other benefits it offers the mother.[80] The AAP and the ADA recognize **exclusive breastfeeding** for 6 months and breastfeeding

fetal alcohol spectrum disorders (FASD): a spectrum of physical, behavioral, and cognitive disabilities caused by prenatal alcohol exposure.

fetal alcohol syndrome (FAS): the cluster of symptoms seen in an infant or child whose mother consumed excessive alcohol during her pregnancy. FAS includes, but is not limited to, brain damage, growth retardation, mental retardation, and facial abnormalities.

exclusive breastfeeding: an infant's consumption of human milk with no supplementation of any type (no water, no juice, no nonhuman milk, and no foods) except for vitamins, minerals, and medications.

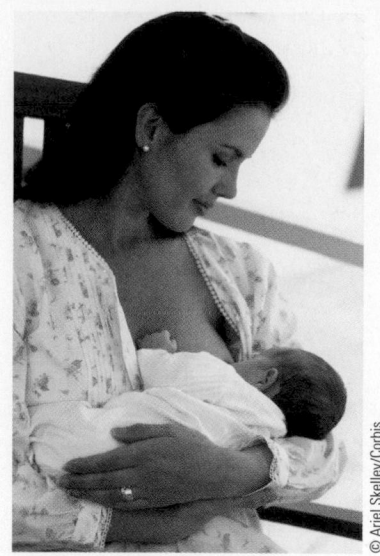

A woman who decides to breastfeed offers her infant a full array of nutrients and protective factors to support optimal health and development.

Energy requirement during lactation:

▌ **First 6 months:** +330 kcal/day

▌ **Second 6 months:** +400 kcal/day

with complementary foods for at least 12 months as an optimal feeding pattern for infants.[81] Breast milk's unique nutrient composition and protective factors promote optimal infant health and development. The only acceptable alternative to breast milk is iron-fortified formula. Adequate nutrition of the mother supports successful lactation, and without it, lactation is likely to falter or fail.

Nutrition during Lactation

By continuing to eat nutrient-dense foods, not restricting weight gain unduly, and enjoying ample food and fluid at frequent intervals throughout lactation, the mother who chooses to breastfeed her infant will be nutritionally prepared to do so. An inadequate diet does not support the stamina, patience, and self-confidence that nursing an infant demands. Figure 11-4 (on p. 291) shows how a lactating woman's nutrient needs differ from those of a nonpregnant woman, and Table 11-1 (on p. 292) presents a sample menu that meets those needs.

Energy A nursing woman produces about 25 ounces of milk a day, with considerable variation from woman to woman and in the same woman from time to time, depending primarily on the infant's demand for milk. Producing this milk costs a woman almost 500 kcalories per day above her regular need during the first six months of lactation. To meet this energy need, the woman is advised to eat an extra 330 kcalories of food each day. The other 170 kcalories can be drawn from the fat stores she accumulated during pregnancy. The food energy consumed by the nursing mother should carry with it abundant nutrients. Severe energy restriction hinders milk production and can compromise the mother's health.

Weight Loss After the birth of the infant, many women actively try to lose the extra weight and body fat they accumulated during pregnancy. How much weight a woman retains after pregnancy depends on her gestational weight gain and the duration and intensity of breastfeeding. Many women who follow recommendations for gestational weight gain and breastfeeding can readily return to prepregnancy weight by six months.[82] Neither the quality nor the quantity of breast milk is adversely affected by moderate weight loss, and infants grow normally.

Women often choose to be physically active to lose weight and improve fitness, and this is compatible with breastfeeding and infant growth.[83] A gradual weight loss (1 pound per week) is safe and does not reduce milk output. Too large an energy deficit, however, especially soon after birth, will inhibit lactation.

Vitamins and Minerals A question often raised is whether a mother's milk may lack a nutrient if she fails to get enough in her diet. The answer differs from one nutrient to the next, but, in general, nutritional deprivation of the mother reduces the *quantity,* not the *quality,* of her milk. Women can produce milk with adequate protein, carbohydrate, fat, folate, and most minerals, even when their own supplies are limited. For these nutrients, milk quality is maintained at the expense of maternal stores. This is most evident in the case of calcium: dietary calcium has no effect on the calcium concentration of breast milk, but maternal bones lose some of their density during lactation if calcium intakes are inadequate.[84] Such losses are generally made up quickly when lactation ends, and breastfeeding has no long-term harmful effects on women's bones. The nutrients in breast milk that are most likely to decline in response to prolonged inadequate intakes are the vitamins—especially vitamins B_6, B_{12}, A, and D. Vitamin supplementation of undernourished women appears to help normalize the vitamin concentrations in their milk and may be beneficial.

Water The volume of breast milk produced depends on how much milk the infant demands, not on how much fluid the mother drinks. The nursing mother is nevertheless advised to drink plenty of liquids each day (about 13 cups) to protect herself from

dehydration. To help themselves remember to drink enough liquid, many women make a habit of drinking a glass of milk, juice, or water each time the infant nurses as well as at mealtimes.

The DRI recommendation for *total* water intake during lactation is 3.8 L/day. This includes 3.1 L or about 13 c as total beverages, including drinking water.

Particular Foods Foods with strong or spicy flavors (such as onions or garlic) may alter the flavor of breast milk. A sudden change in the taste of the milk may annoy some infants. Familiar flavors may enhance enjoyment. A mother who is breastfeeding her infant is advised to eat whatever nutritious foods she chooses. Then, if a particular food seems to cause the infant discomfort, she can try eliminating that food from her diet for a few days and see if the problem resolves.

Current evidence does not support a major role for maternal dietary restrictions during lactation to prevent or delay the onset of food allergy in infants.[85] Infants who develop symptoms of food allergy, however, may be more comfortable if the mother's diet excludes the most common offenders—cow's milk, eggs, fish, peanuts, and tree nuts. Generally, infants with a strong family history of food allergies benefit from breastfeeding.[86]

The accompanying Case Study presents a woman who is four months pregnant. Answering the questions offers practice in thinking through some of the issues related to pregnancy and breastfeeding.

Contraindications to Breastfeeding

Some substances impair maternal milk production or enter the breast milk and interfere with infant development. Some medical conditions prohibit breastfeeding.

Alcohol Alcohol easily enters breast milk and can adversely affect the production, volume, composition, and ejection of breast milk as well as overwhelm an infant's immature alcohol-degrading system.[87] Alcohol concentration in breast milk peaks within one hour after ingestion of even moderate amounts (equivalent to a can of beer). It may alter the taste of the milk in such a way that the nursing infant may drink less than normal.

Caffeine Caffeine can make an infant jittery and wakeful. As during pregnancy, caffeine consumption should be moderate.

Cigarette Smoke Health care professionals should actively discourage smoking by lactating women. Research shows that lactating women who smoke produce less milk, and milk with a lower fat content, than mothers who do not smoke. Consequently, their infants gain less weight than infants of nonsmokers. Furthermore, infants of breastfeeding mothers who smoke sleep less than infants of those who do not smoke.[88]

Case Study WOMAN IN HER FIRST PREGNANCY

Ellen Cassidy is a 24-year-old woman who is four months pregnant. This is her first pregnancy, and she is eager to learn how to feed herself during pregnancy as well as her infant after birth. She is 5 feet 3 inches tall and currently weighs 150 pounds. Her prepregnancy weight was 148 pounds. Ellen is very concerned about her 2-pound weight gain.

1. Consult the BMI table (inside back cover), and using the "Healthy Weight" section, find a healthy weight in the middle of the range appropriate for a woman of Ellen's height.

2. Do you think that Ellen's weight at the start of her pregnancy was appropriate for her height? Why or why not? Should Ellen be concerned about her 2-pound weight gain? Why or why not?

3. What advice should you give Ellen about her weight gain during pregnancy? What other dietary advice would you give her?

4. Discuss methods of infant feeding with Ellen and describe some of the advantages breastfeeding would offer her. What advice will you give Ellen if she decides to breastfeed?

A lactating woman who smokes not only exposes her infant to nicotine and other chemicals via her breast milk but may also expose the infant to sidestream smoke. Infants who are "smoked over" experience a wide array of health problems—poor growth, hearing impairment, vomiting, breathing difficulties, and even unexplained death.

Medications and Illicit Drugs If a nursing mother must take medication that is secreted in breast milk and is known to affect the infant, then breastfeeding must be put off for the duration of treatment. Meanwhile, the flow of milk can be sustained by pumping the breasts and discarding the milk. Many prescription medications do not reach nursing infants in sufficient quantities to affect them adversely and so have no impact on breastfeeding. Other drugs are not at all compatible with breastfeeding, either because they are secreted into the milk and can harm the infant or because they suppress lactation. A nursing mother should consult with the prescribing physician before taking medicines. Breastfeeding is also contraindicated if the mother uses illicit drugs. Breast milk can deliver such high doses of illicit drugs as to cause irritability, tremors, hallucinations, and even death in infants.

Many women wonder about using oral contraceptives during lactation. One type that combines the hormones estrogen and progestin seems to suppress milk output, lower the nitrogen content of the milk, and shorten the duration of breastfeeding. In contrast, progestin-only pills have no effect on breast milk or breastfeeding and are considered appropriate for lactating women.

Maternal Illness If a woman has an ordinary cold, she can go on nursing without worry. If susceptible, the infant will catch it from her anyway, and, thanks to immunological protection, a breastfed baby may be less susceptible than a formula-fed infant would be. If a woman has active untreated tuberculosis or is receiving therapeutic radioactive isotopes, breastfeeding is contraindicated.[89]

The human immunodeficiency virus (HIV), responsible for causing AIDS, can be passed from an infected mother to her infant during pregnancy, at birth, or through breast milk, especially during the early months of breastfeeding. In developed countries such as the United States, where safe alternatives are available, HIV-positive women should not breastfeed their infants.[90]

Throughout the world, breastfeeding prevents millions of infant deaths each year. In developing countries, where the feeding of inappropriate or contaminated formulas causes 1.5 million infant deaths each year, breastfeeding can be critical to infant survival. Thus, the question of whether HIV-infected women in developing countries should breastfeed comes down to a delicate balance between risks and benefits. For HIV-positive women in developing countries, the most appropriate infant-feeding option depends on individual circumstances, including the health status of the mother and the local situation, as well as the health services, counseling, and support available. The World Health Organization (WHO) recommends exclusive breastfeeding for infants of HIV-infected women for the first six months of life unless replacement feeding is acceptable, feasible, affordable, sustainable, and safe for mothers and their infants before that time.[91]

IN SUMMARY

■ The lactating woman needs enough energy and nutrients to produce about 25 ounces of milk a day. She also needs extra fluid.

■ Alcohol, caffeine, smoking, and drugs may reduce milk production or enter breast milk and impair infant development.

■ Some maternal illnesses are incompatible with breastfeeding.

Nutrition Assessment Checklist FOR PREGNANT AND LACTATING WOMEN

Medical History

Check the medical record for:

- Alcohol or illicit drug abuse
- Chronic diseases
- Gestational diabetes
- History of previous pregnancies (number, intervals, outcomes, multiple births, gestational age, and birth weights)
- Hypertension
- Neural tube defect in an infant born previously
- Preeclampsia

Note risk factors for complications during pregnancy, including:

- Cigarette smoking
- Food faddism
- Lactose intolerance
- Low socioeconomic status
- Significant or prolonged vomiting
- Very young or old age
- Weight-loss dieting

Note any complaints of:

- Constipation
- Heartburn
- Morning sickness

Medications

For pregnant women who are using drug therapy for medical conditions, note:

- Potential for contraindication to breastfeeding
- GI tract side effects that might reduce food intake or change nutrient needs

Dietary Intake

For all pregnant and lactating women, especially those considered at risk nutritionally, assess the diet for:

- Total energy
- Protein
- Calcium, phosphorus, magnesium, iron, and zinc
- Folate and vitamin B_{12}
- Vitamin D

Anthropometric Data

Measure baseline height and weight:

- Prepregnancy weight

Reassess weight at each medical checkup and determine whether gains are appropriate. Note:

- Weight gain during pregnancy
- Gestational age

Laboratory Tests

Monitor the following laboratory tests for pregnant women:

- Hemoglobin, hematocrit, or other tests of iron status
- Blood glucose

Physical Signs

Blood pressure measurement is routine in physical exams but is especially important for pregnant women. Look for physical signs of:

- Iron deficiency
- Edema
- Protein-energy malnutrition
- Folate deficiency

This chapter has focused on nutrition during pregnancy and lactation. The Nutrition Assessment Checklist helps to identify nutrition-related factors that may help prevent or correct potential problems in pregnant or lactating women. Chapter 12 explores the dietary needs of infants, children, and adolescents.

Clinical Applications

1. Consider the different factors in a pregnant woman's history that can affect her nutrition status and the outcome of her pregnancy. Describe what steps you would take to remedy potential problems for the following clients:

   a. A 15-year-old adolescent of low socioeconomic status is in her first trimester of pregnancy. She began the pregnancy at a normal, healthy weight, but her weight gain during pregnancy so far has been less than expected. Her favorite beverages are soft drinks; her favorite foods are French fries and boxed macaroni and cheese.

   b. A lactose-intolerant, 22-year-old pregnant woman has been eating a vegan diet for the past year or so. She began the pregnancy slightly underweight (BMI = 18.4), but her weight gain has been adequate and consistent during the four months of her pregnancy. She complains of feeling tired all the time.

2. What information would you give to a pregnant woman who is considering breastfeeding her infant, but isn't quite sure why she should?

Self Check

1. The most important single predictor of an infant's future health and survival is:
   a. the infant's birth weight.
   b. the infant's iron status at birth.
   c. the mother's weight at delivery.
   d. the mother's prepregnancy weight.

2. A mother's prepregnancy nutrition is important to a healthy pregnancy because it determines the development of:
   a. the largest baby possible.
   b. adequate maternal iron stores.
   c. an adequate fat supply for the mother.
   d. healthy support tissues—the placenta, amniotic sac, umbilical cord, and uterus.

3. A pregnant woman needs an extra 450 calories above the allowance for nonpregnant women during which trimester(s)?
   a. First
   b. Second
   c. Third
   d. First, second, and third

4. Two nutrients needed in large amounts during pregnancy for rapid cell proliferation are:
   a. vitamin B_{12} and vitamin C.
   b. calcium and vitamin B_6.
   c. folate and vitamin B_{12}.
   d. copper and zinc.

5. For a woman who is at the appropriate weight for height and is carrying a single fetus, the recommended weight gain during pregnancy is:
   a. 40 to 60 pounds.
   b. 25 to 35 pounds.
   c. 10 to 20 pounds.
   d. 20 to 40 pounds.

6. Rewards of physical activity during pregnancy may include:
   a. weight loss.
   b. decreased incidence of pica.
   c. relief from morning sickness.
   d. reduced stress and easier labor.

7. During pregnancy, the combination of high blood pressure, protein in the urine, and edema signals:
   a. jaundice.
   b. preeclampsia.
   c. gestational diabetes.
   d. gestational hypertension.

8. Which of the following preventative measures should pregnant women take to avoid contracting listeriosis?
   a. Avoid feta cheese.
   b. Avoid pasteurized milk.
   c. Thoroughly heat hot dogs.
   d. (a) and (c)

9. To facilitate lactation, a mother needs:
   a. about 5000 kcalories a day.
   b. adequate nutrition and fluid intake.
   c. vitamin and mineral supplements.
   d. a glass of wine or beer before each feeding.

10. A woman who breastfeeds her infant should drink plenty of water to:
    a. produce more milk.
    b. suppress lactation.
    c. prevent dehydration.
    d. dilute nutrient concentrations.

Answers to these questions can be found in Appendix H.

Notes

1. Recommendations to improve preconception health and health care—United States, *Morbidity and Mortality Weekly Report* 55 (2006): 1–23.

2. M. Viswanathan and coauthors, Outcomes of maternal weight gain, *Evidence Report/Technology Assessment* 168 (2008): 1–223; R. L. Goldenberg and J. F. Culhane, Low birthweight in the United States, *American Journal of Clinical Nutrition* 85 (2007): 584S–590S.

3. C. Bouchard, Childhood obesity: Are genetic differences involved? *American Journal of Clinical Nutrition* 89 (2009): 1494S–1501S; M. E. Symonds, T. Stephenson, and H. Budge, Early determinants of cardiovascular disease: The role of early diet in later blood pressure control, *American Journal of Clinical Nutrition* 89 (2009): 1518S–1522S; P. D. Gluckman and coauthors,

Effect of in utero and early-life conditions on adult health and disease, *New England Journal of Medicine* 359 (2008): 61–73.

4. D. S. Alam, Prevention of low birthweight, *Nestle Nutrition Workshop Series: Pediatric Program* 63 (2009): 209–225; E. M. Lundgren and T. Tuvemo, Effects of being born small for gestational age on long-term intellectual performance, *Best Practice and Research: Clinical Endocrinology and Metabolism* 22 (2008): 477–488.

5. M. Heron and coauthors, Annual summary of vital statistics: 2007, *Pediatrics* 125 (2010): 4–15.

6. L. M. McCowan and coauthors, Spontaneous preterm birth and small for gestational age infants in women who stop smoking early in

pregnancy: Prospective cohort study, *British Medical Journal* 338 (2009): b1081.

7. Position of the American Dietetic Association: Nutrition and lifestyle for a healthy pregnancy outcome, *Journal of the American Dietetic Association* 108 (2008): 553–561; T. O. Scholl, Maternal nutrition before and during pregnancy, Nestle Nutrition Workshop Series; *Pediatrics Program* 61 (2008): 79–89.

8. D. K. Waller and coauthors, Prepregnancy obesity as a risk factor for structural birth defects, *Archives of Pediatrics and Adolescent Medicine* 161 (2007): 745–750.

9. Position of the American Dietetic Association and American Society for Nutrition: Obesity, reproduction, and pregnancy outcomes, *Journal*

of the American Dietetic Association 109 (2009): 918–927.

10. Position of the American Dietetic Association, 2008; E. A. Nohr and coauthors, Combined associations of prepregnancy body mass index and gestational weight gain with the outcome of pregnancy, American Journal of Clinical Nutrition 87 (2008): 1750–1759; S. Y. Chu and coauthors, Association between obesity during pregnancy and increased use of health care, New England Journal of Medicine 358 (2008): 1444–1453.

11. K. J. Stothard and coauthors, Maternal overweight and obesity and the risk of congenital anomalies: A systematic review and meta-analysis, Journal of the American Medical Association 301 (2009): 636–650; Waller and coauthors, 2007.

12. V. E. Murphy and coauthors, Endocrine regulation of human fetal growth: The role of the mother, placenta, and fetus, Endocrine Reviews 27 (2006): 141–169.

13. M. Desforges and C. P. Sibley, Placental nutrient supply and fetal growth, International Journal of Developmental Biology 54 (2010): 377–390.

14. Position of the American Dietetic Association, 2008.

15. S. H. Zeisel, Epigenetic mechanisms for nutrition determinants of later health outcomes, American Journal of Clinical Nutrition 89 (2009): 1488S–1493S; P. D. Gluckman and coauthors, Effect of in utero and early-life conditions on adult health and disease, New England Journal of Medicine 359 (2008): 61–73.

16. Standing Committee on the Scientific Evaluation of Dietary Reference Intakes, Food and Nutrition Board, Institute of Medicine, Dietary Reference Intakes for Energy, Carbohydrate, Fiber, Fat, Fatty Acids, Cholesterol, Protein, and Amino Acids (Washington, D.C.: National Academies Press, 2005), pp. 185–194.

17. P. Haggarty, Fatty acid supply to the human fetus, Annual Review of Nutrition 30 (2010): 237–255; S. M. Innis and R. W. Freisen, Essential n-3 fatty acids in pregnant women and early visual acuity maturation in term infants, American Journal of Clinical Nutrition 87 (2008): 548–557; M. von Eijsden and coauthors, Maternal n-3, n-6 and trans fatty acid profile early in pregnancy and term birth weight: A prospective cohort study, American Journal of Clinical Nutrition 87 (2008): 887–895.

18. R. M. Pitkin, Folate and neural tube defects, American Journal of Clinical Nutrition 85 (2007): 285S–288S.

19. U.S. Preventive Services Task Force, Folic acid for the prevention of neural tube defects: U.S. Preventive Services Task Force recommendation statement, Annals of Internal Medicine 150 (2009): 626–631.

20. V. F. Ohrvik and coauthors, Folate bioavailability from breads and a meal assessed with a human stable-isotope area under the curve and ileostomy model, American Journal of Clinical Nutrition 92 (2010): 532–538.

21. P. De Wals and coauthors, Reduction in neural-tube defects after folic acid fortification in Canada, New England Journal of Medicine 357 (2007): 135–142; J. I. Rader and B. O. Schneeman, Prevalence of neural tube defects, folate status, and folate fortification of enriched cereal-grain products in the United States, Pediatrics 117 (2006): 1394–1399.

22. A. J. Wilcox and coauthors, Folic acid supplements and risk of facial clefts: National population based case-control study, British Medical Journal 334 (2007): 464.

23. I. Elmadfa and I. Singer, Vitamin B-12 and homocysteine status among vegetarians: A global perspective, American Journal of Clinical Nutrition 89 (2009): 1693S–1698S.

24. A. M. Molloy and coauthors, Maternal vitamin B_{12} status and risk of neural tube defects in a population with high neural tube defect prevalence and no folic acid fortification, Pediatrics 123 (2009): 917–923.

25. C. L. Wagner, F. R. Greer, and the Section on Breastfeeding and Committee on Nutrition, Prevention of rickets and vitamin D deficiency in infants, children, and adolescents, Pediatrics 122 (2008): 1142–1152; C. S. Kovacs, Vitamin D in pregnancy and lactation: Maternal, fetal, and neonatal outcomes from human and animal studies, American Journal of Clinical Nutrition 88 (2008): 520S–528S.

26. Committee on Dietary Reference Intakes, Dietary Reference Intakes for Calcium and Vitamin D (Washington, D.C.: National Academies Press, 2011).

27. S. Mahajan and coauthors, Nutritional anaemia dysregulates endocrine control of fetal growth, British Journal of Nutrition 100 (2008): 408–417; F. M. Rioux and C. P. LeBlanc, Iron supplementation during pregnancy: What are the risks and benefits of current practices? Applied Physiology, Nutrition and Metabolism 32 (2007): 282–288.

28. D. Shah and H. P. S. Sachdev, Zinc deficiency in pregnancy and fetal outcome, Nutrition Reviews 64 (2006):15–30.

29. Position of the American Dietetic Association, 2008.

30. WIC, The Special Supplemental Nutrition Program for Women, Infants, and Children, available at www.fns.usda.gov/fns. Updated April 2009.

31. A. Jacknowitz, D. Novillo, and L. Tiehen, Special supplemental Nutrition Program for Women, Infants, and Children and infant feeding practices, Pediatrics 119 (2007): 281–288.

32. Position of the American Dietetic Association and American Society for Nutrition, 2009; S. Y. Chu and coauthors, Gestational weight gain by body mass index among US women delivering live births, 2004–2005: Fueling future obesity, American Journal of Obstetrics and Gynecology 200 (2009): 271, e1–e7; C. M. Olson, Achieving a healthy weight gain during pregnancy, Annual Review of Nutrition 28 (2008): 411–423.

33. Position of the American Dietetic Association and American Society for Nutrition, 2009.

34. Position of the American Dietetic Association and American Society for Nutrition, 2009.

35. Institute of Medicine, Weight Gain during Pregnancy: Reexamining the Guidelines (Washington, D.C.: The National Academies Press, 2009).

36. J. H. Cohen and H. Kim, Sociodemographic and health characteristics associated with attempting weight loss during pregnancy, Preventing Chronic Disease 6 (2009): A07.

37. National Research Council, Institute of Medicine, Influence of Pregnancy Weight on Maternal and Child Health (Washington, D.C.: National Academies Press, 2007).

38. Position of the American Dietetic Association and American Society for Nutrition, 2009; Viswanathan and coauthors, 2008; J. L. Baker and coauthors, Breastfeeding reduces postpartum weight retention, American Journal of Clinical Nutrition 88 (2008): 1543–1551.

39. D. A. Krummel, Postpartum weight control: A vicious cycle, Journal of the American Dietetic Association 107 (2007): 37–40; E. Villamor and S. Cnattingius, Interpregnancy weight change and risk of adverse pregnancy outcomes: A population-based study, Lancet 368 (2006): 1164–1170.

40. A. R. Amorium and coauthors, Does excess pregnancy weight gain constitute a major risk for increasing long-term BMI? Obesity 15 (2007): 1278–1286.

41. Position of the American Dietetic Association, 2008.

42. A. B. Granath, M. S. Hellgren, and R. K. Gunnarsson, Water aerobics reduces sick leave due to low back pain during pregnancy, Journal of Obstetrics, Gynecology, and Neonatal Nursing 35 (2006): 465–471.

43. S. L. Young, Pica in Pregnancy: New ideas about an old condition, Annual Review of Nutrition 30 (2010): 403–422.

44. S. L. Young, 2010.

45. J. L. Kitzmiller and coauthors, Managing preexisting diabetes for pregnancy: Summary of evidence and consensus recommendations for care, Diabetes Care 31 (2008): 1060–1079.

46. J. L. Kitzmiller and coauthors, 2008.

47. The HAPO Study Cooperative Research Group, Hyperglycemia and adverse pregnancy outcomes, New England Journal of Medicine 358 (2008): 1991–2002; V. Seshiah and coauthors, "Abnormal" fasting plasma glucose during pregnancy, Diabetes Care 31 (2008): e92.

48. J. L. Kitzmiller and coauthors, 2008.

49. Position statement, Standards of medical care in diabetes—2008, Diabetes Care 31 (2008): S12–S54.

50. C. Mulholland and coauthors, Comparison of guidelines available in the United States for diagnosis and management of diabetes before, during, and after pregnancy, Journal of Women's Health 16 (2007): 790–801.

51. The HAPO study Cooperative Research Group, 2008.

52. American Diabetes Association, Diagnosis and classification of diabetes mellitus, Diabetes Care 31 (2008): S55–S60.

53. P. E. Marik, Hypertensive disorders of pregnancy, Postgraduate Medicine 121 (2009): 69–76.

54. B. M. Sibai, Caring for women with hypertension, Journal of the American Medical Association 298 (2007): 1566–1568.

55. Position of the American Dietetic Association, 2008.

56. V. T. Tong and coauthors, Trends in smoking before, during, and after pregnancy—Pregnancy Risk Assessment Monitoring System (PRAMS), United States, 31 sites, 2000–2005, Morbidity and Mortality Weekly Report 58 (2009): SS-4, pp. 1–31.

57. J. M. Rogers, Tobacco and pregnancy: Overview of exposures and effects, *Birth Defects Research (Part C)* 84 (2008): 1–15.

58. E. Jauniaux and G. J. Burton, Morphological and biological effects of maternal exposure to tobacco smoke on the feto-placental unit, *Early Human Development* 83 (2007): 699–706.

59. M. J. Rivkin and coauthors, Volumetric MRI study of brain in children with intrauterine exposure to cocaine, alcohol, tobacco, and marijuana, *Pediatrics* 121 (2008): 741–750.

60. J. M. Rogers, 2008.

61. J. M. Rogers, 2008; American Academy of Pediatrics, Task Force on Sudden Infant Death Syndrome, The changing concept of sudden infant death syndrome: Diagnostic coding shifts, controversies regarding the sleeping environment, and new variables to consider in reducing risk, *Pediatrics* 116 (2005): 1245–1255.

62. J. M. Rogers, 2008.

63. Position of the American Dietetic Association, 2008.

64. M. J. Rivkin and coauthors, 2008; H. S. Bada and coauthors, Impact of prenatal cocaine exposure on child behavior problems through school age, *Pediatrics* 119 (2007): e348; B. A. Lewis and coauthors, Prenatal cocaine and tobacco effects on children' language trajectories, *Pediatrics* 120 (2007): e78.

65. G. A. Richardson, L. Goldschmidt, and C. Larkby, Effects of prenatal cocaine exposure on growth: A longitudinal analysis, *Pediatrics* 120 (2007): e1017.

66. T. I. Halldorsson and coauthors, Is high consumption of fatty fish during pregnancy a risk factor for fetal growth retardation? A study of 44,824 Danish pregnant women, *American Journal of Epidemiology* 166 (2007): 687–696.

67. E. Oken and coauthors, Maternal fish intake during pregnancy, blood mercury levels, and child cognition at age 3 years in a US cohort, *American Journal of Epidemiology* 167 (2008): 1171–1181; D. Mozaffarian and E. B. Rimm, Fish intake, contaminants, and human health: Evaluating the risks and the benefits, *Journal of the American Medical Association* 296 (2006): 1885–1899; Institute of Medicine report brief, *Seafood Choices: Balancing Benefits and Risks,* October 2006.

68. Position of the American Dietetic Association: Use of nutritive and nonnutritive sweeteners, *Journal of the American Dietetic Association* 104 (2004): 255–275.

69. X. Weng, R. Odouli, and D. Li, Maternal caffeine consumption during pregnancy and the risk of miscarriage: A prospective cohort study, *American Journal of Obstetrics and Gynecology* 198 (2008): 279.e1–279.e8; M. L. Browne, Maternal exposure to caffeine and risk of congenital anomalies: A systematic review, *Epidemiology* 17 (2006): 324–331; A. Matijasevich and coauthors, Maternal caffeine consumption and fetal death: A case-control study in Uruguay, *Paediatric and Perinatal Epidemiology* 20 (2006): 100–109.

70. CARE Study Group, Maternal caffeine intake during pregnancy and risk of fetal growth restriction: A large prospective observational study, *British Medical Journal* 337 (2008): a2332.

71. H. E. Hoyme and coauthors, A practical approach to diagnosis of fetal alcohol spectrum disorders: Clarification of the 1996 Institute of Medicine criteria, *Pediatrics* 115 (2005): 39–47.

72. C. H. Denny and coauthors, Alcohol use among pregnant and nonpregnant women of childbearing age—United States, 1991–2005, *Morbidity and Mortality Weekly Report* 58 (2009): 529–532; Centers for Disease Control, Fetal alcohol syndrome, www.cdc.gov/ncbddd/fas/fasask.htm, accessed May 29, 2009.

73. C. H. Denny and coauthors, 2009.

74. J. A. Martin and coauthors, Births: Final data for 2006, *National Vital Statistics Reports* 57 (2009): 1–102.

75. P. N. Baker and coauthors, A prospective study of micronutrient status in adolescent pregnancy, *American Journal of Clinical Nutrition* 89 (2009): 1114–1124; S. Saintonge, H. Bang, and L. M. Gerber, Implications of a new definition of vitamin D deficiency in a multiracial US adolescent population: The National Health and Nutrition examination Survey III, *Pediatrics* 123 (2009): 797–803.

76. Quickstats—Birthrates among females aged 15-19 years, by state—United States, 2004, *Morbidity and Mortality Weekly Report* 55 (2007): 1383.

77. X. Chen and coauthors, Paternal age and adverse birth outcomes: Teenager or 40+, who is at risk? *Human Reproduction* 23 (2008): 1290–1296.

78. E. P. Gunderson and coauthors, Longitudinal study of growth and adiposity in parous compared with nulligravid adolescents, *Archives of Pediatrics and Adolescent Medicine* 163 (2009): 349–356.

79. Breastfeeding, in *Pediatric Nutrition Handbook,* 6th ed., ed. R. E. Kleinman (Elk Grove Village, IL: American Academy of Pediatrics, 2009), pp. 29–59; American Academy of Pediatrics, Policy statement: Breastfeeding and the use of human milk, *Pediatrics* 115 (2005): 496–506.

80. Position of the American Dietetic Association: Promoting and supporting breastfeeding, *Journal of the American Dietetic Association* 109 (2009): 1926–1942.

81. Breastfeeding, in *Pediatric Nutrition Handbook,* 2009; Position of the American Dietetic Association, 2009; American Academy of Pediatrics, 2005.

82. J. Baker and coauthors, 2008.

83. D. Su and coauthors, Breast-feeding mothers can exercise: Results of a cohort study, *Public Health Nutrition* 10 (2007): 1089–1093.

84. K. O. O'Brien and coauthors, Bone calcium turnover during pregnancy and lactation in women with low calcium diets is associated with calcium intake and circulating insulin-like growth factor 1 concentrations, *American Journal of Clinical Nutrition* 83 (2006): 317–323.

85. F. R. Greer, S. H. Sicherer, A. Wesley Burks, and the Committee on Nutrition and Section on Allergy and Immunology, Effects of early nutritional interventions on the development of atopic disease in infants and children: The role of maternal dietary restriction, breastfeeding, timing of introduction of complementary foods, and hydrolyzed formulas, *Pediatrics* 121 (2008): 183–191.

86. F. R. Greer, S. H. Sicherer, A. Wesley Burks, and the Committee on Nutrition and Section on Allergy and Immunology, 2008.

87. American Academy of Pediatrics, 2005.

88. J. A. Mennella, L. M. Yourshaw, and L. K. Morgan, Breastfeeding and smoking: Short-term effects on infant feeding and sleep, *Pediatrics* 120 (2007): 497–502.

89. American Academy of Pediatrics, 2005.

90. P. L. Havens, L. M. Mofenson, and the Committee on Pediatric AIDS, *Pediatrics* 123 (2009): 175–187.

91. World Health Organization, *HIV and Infant Feeding,* available at HTTP://www.who.int/child_adolescent_health/topics/prevention_care/child/nutrition/hivif/en/, accessed June 1, 2009; M. W. Kline, Early exclusive breastfeeding: Still the cornerstone of child survival, *American Journal of Clinical Nutrition* 89 (2009): 1281–1282.

Nutrition in Practice

As discussed in Chapters 11 and 12, breastfeeding offers benefits to both mother and infant. The American Academy of Pediatrics (AAP), the American Dietetic Association, and the Canadian Paediatric Society all advocate breastfeeding as the preferred means of infant feeding.[1] Promotion of breastfeeding is an integral part of the WIC program's nutrition education component. National efforts to promote breastfeeding seem to be working, at least to some extent: the percentage of infants who were ever breastfed rose from 60 percent among those born in 1994 to 77 percent among infants born in 2006.[2] Despite this encouraging trend, the rate of breastfeeding at six months of age did not change for infants born between 1993 and 2004.[3] Only about one in three infants is still being breastfed at six months of age. The AAP and many other health organizations recommend exclusive breastfeeding for the first six months of life.[4] Exclusive breastfeeding is defined as an infant's consumption of human milk with no supplementation of any kind (no water, no other type of milk, no juice, and no other foods) except for vitamins, minerals, and medications. The AAP recommends that breastfeeding continue for at least a year and thereafter for as long as mutually desired.[5] In the United States, only about one in five infants is still breastfeeding at one year of age. Increasing the rates of breastfeeding initiation and duration is one of the goals of *Healthy People 2020*:

> Increase the proportion of mothers who breastfeed immediately after birth, for the first six months, and preferably, through the infant's first year of life. Increase the proportion of mothers who breastfeed exclusively.[6]

Despite the trend toward increasing breastfeeding, the percentage of mothers choosing to breastfeed their infants and continuing to do so still falls short of goals.

Why don't more women choose to breastfeed their infants?

Many experts cite two major deterrents: infant formula manufacturers' public advertising and promotion of their products and the medical community's failure to encourage breastfeeding. Infant formula manufacturers spend millions of dollars each year marketing their products, often claiming that formula "is like breast milk." Such advertising efforts seem to be working. According to a recent survey, one out of four people of various ages, races, and socioeconomic backgrounds agree with the statement "infant formula is as good as breast milk."[7] Infant formula is an appropriate substitute for breast milk when breastfeeding is specifically contraindicated, but for most infants, the benefits of breast milk outweigh those of formula. Despite medical evidence of the benefits of breast milk, medical practice is not always supportive.

As an example of the medical lack of encouragement, some hospitals routinely separate mother and infant soon after birth. The child's first feeding then comes from the bottle rather than the breast. Furthermore, many hospitals send new mothers home with free samples of infant formula.[8] The World Health Organization opposes this practice because it sends a misleading message that medical authorities favor infant formula over breast milk for infants. Even in hospitals where women are encouraged to breastfeed and are supported in doing so, little, if any, assistance is available after hospital discharge when many breastfeeding women still need assistance. Up to half of mothers who initially breastfeed their infants stop within three months—seemingly due to lack of knowledge and support.

Women who receive early and repeated breastfeeding information and support breastfeed their infants longer than other women do.[9] Information and instruction are especially important during the *prenatal* period, when most women decide whether to breastfeed or to feed formula. Nurses and other health care professionals can play a crucial role in encouraging successful breastfeeding by offering women adequate, accurate information about breastfeeding that permits them to make informed choices. Table NP11-1 lists ten steps that maternity facilities and health care professionals can take to promote successful breastfeeding among new mothers.

If breastfeeding is a natural process, what do mothers need to learn?

Although lactation is an automatic physiological process, breastfeeding requires some learning. This learning is most successful in a supportive environment. It begins with preparatory steps taken before the infant is born.

TABLE NP11-1 Ten Steps to Successful Breastfeeding

To promote breastfeeding, every maternity facility should:

- Develop a written breastfeeding policy that is routinely communicated to all health care staff.
- Train all health care staff in the skills necessary to implement the breastfeeding policy.
- Inform all pregnant women about the benefits and management of breastfeeding.
- Help mothers initiate breastfeeding within half an hour of birth.
- Show mothers how to breastfeed and how to maintain lactation, even if they need to be separated from their infants.
- Give newborn infants no food or drink other than breast milk, unless medically indicated.
- Practice rooming-in, allowing mothers and infants to remain together 24 hours per day.
- Encourage breastfeeding on demand.
- Give no artificial nipples or pacifiers to breastfeeding infants.
- Foster the establishment of breastfeeding support groups and refer mothers to them at discharge from the facility.

Sources: U.S. Department of Health and Human Services, Office on Women's Health, *Breastfeeding: HHS Blueprint for Action on Breastfeeding* (Washington, DC: U.S. Department of Health and Human Services, 2000).

What are these preparatory steps?

Toward the end of pregnancy and throughout lactation, a woman who intends to breastfeed should stop using soap and lotions on her breasts. The natural secretions of the breasts themselves lubricate the nipple area best. A woman who plans to breastfeed should also acquire at least two nursing bras before her infant is born. The bras should provide good support and have drop-flaps so that either breast can be freed for nursing.

How soon after birth should breastfeeding start?

It should start as soon as possible. Immediately after the delivery, for a short period, the infant is intensely alert and intent on suckling. This is the ideal time to initiate breastfeeding, and doing so facilitates successful lactation.

What does the new mother need to know to continue breastfeeding her infant successfully?

She needs to learn how to relax and position herself so that she and the infant will be comfortable and the infant can breathe freely while nursing. She also needs to understand that infants have a **rooting reflex** that makes them turn toward any touch on the face (see the glossary). Consequently, she should touch her nipple to the infant's cheek, so that the infant will turn the correct way in order to nurse. The mother can then place four fingers under the breast and her thumb on top to support the breast and present the nipple to the infant. The mother's fingers and thumb should be behind the areola, the colored ring around the nipple, so as not to interfere with the infant latching onto the breast (see Figure NP11-1). With the breast supported, the mother tickles the infant's lips with the breast until the infant's mouth opens wide. The mother can then gently bring the infant forward onto the breast. The nipple must rest well back on the infant's tongue so that the infant's gums will squeeze on the glands that release the milk and swallowing will be effortless. To break the suction, if necessary, the mother can slip a finger between the infant's mouth and her breast.

Does it hurt to have the infant sucking so hard on the breast?

Breastfeeding should not be painful if the infant is positioned correctly. The mother has a **letdown reflex** that forces milk to the front of her breast when the infant begins to nurse, virtually propelling the milk into the infant's mouth. Letdown is necessary for the infant to obtain milk easily, and the mother needs to relax for

---

FIGURE NP11-1 Infant's Grasp on Mother's Breast

The mother supports the breast with her fingers and thumb behind the areola to present the nipple to the infant. Once the infant latches onto the breast, the infant's lips and gums pump the areola, releasing milk from the mammary glands into the milk ducts that lie beneath the areola.

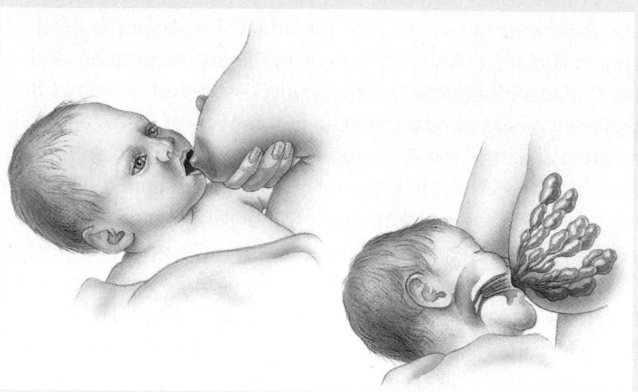

letdown to occur. The mother who assumes a comfortable position in an environment without interruptions will find it easiest to relax.

How long should the infant be allowed to nurse at each feeding?

Although the infant receives half the milk from the breast within the first 2 minutes, and 80 to 90 percent of it within 4 minutes, suckling on each breast for 10 to 15 minutes is encouraged. The suckling itself, as well as the complete removal of milk from the breast, stimulates the mammary glands to produce milk for the next nursing session. Successive sessions should start on alternate breasts to ensure that each breast is emptied regularly. This pattern maintains the same supply and demand for each breast and thus prevents either breast from overfilling.

Infants should be fed on demand and not be held to a rigid schedule. The breastfed infant may average 8 to 12 feedings per 24-hour period during the first month or so. Once the mother's milk supply is well established and the infant's capacity has increased, the intervals between feedings will become longer.

What if a mother wants to skip one or two feedings daily—for example, because she works outside the home?

The mother can express breast milk into a bottle ahead of time, freeze the breast milk, and, when needed, substitute the expressed breast milk for a nursing session. Breast milk can be kept refrigerated for 48 hours or frozen. Frozen milk can be kept for one month in a freezer attached to a refrigerator or for three to six months in a zero-degree deep freezer.[10]

The mother can hand express her breast milk or use one of several different breast pumps available. The bicycle-horn type of manual breast pump is difficult to keep clean and is not recommended. Cylinder-type manual pumps or electric breast pumps are safer and are also more efficient. Alternatively, a mother can substitute formula for those missed feedings and continue to breastfeed at other times.

What about problems associated with breast-feeding such as sore nipples or infection of the breast?

Most problems associated with breastfeeding can be resolved. Many mothers experience sore nipples during the initial days of breastfeeding. Sore nipples need to be treated kindly, but nursing can continue. Improper feeding position is a frequent cause of sore nipples: the mother should make sure the infant is taking the entire nipple and part of the areola onto the tongue. She should nurse on the less sore breast first to get letdown going while the infant is sucking hardest; then she can switch to the sore breast. Between times, she should expose her nipples to light and air to heal them.

Before lactation is well established, when the schedule changes, or when a feeding is missed, the breasts may become full and hard—an uncomfortable condition known as **engorgement.** The infant cannot grasp an engorged nipple and so cannot provide relief by nursing. A gentle massage or warming the breasts with a cloth soaked in warm water or in a shower helps to initiate letdown and to release some of the accumulated milk; then the mother can pump out some of her milk and allow the infant to nurse.

Infection of the breast, known as **mastitis,** is best managed by continuing to breastfeed. By drawing off the milk, the infant helps to relieve pressure in the infected area. The infant is safe because the infection is between the milk-producing glands, not inside them.

Even if everything is going smoothly, the nursing mother should ideally have enough help and support so that she can rest in bed a few hours each day for the first week or so. Successful breastfeeding requires the support of all those who care. This, plus adequate nutrition, ample fluids, fresh air, and physical activity, will do much to enhance the well-being of mother and infant.

Notes

1. Position of the American Dietetic Association: Promoting and Supporting Breastfeeding, *Journal of the American Dietetic Association* 109 (2009): 1926–1942; Breastfeeding, in *Pediatric Nutrition Handbook,* 6th ed., R. E. Kleinman (Elk Grove Village, IL: American Academy of Pediatrics, 2009), pp. 29–59; American Academy of Pediatrics, Policy statement: Breastfeeding and the use of human milk, *Pediatrics* 115 (2005): 496–506; M. Boland, Exclusive breastfeeding should continue to six months, *Paediatrics and Child Health* 10 (2005): 148–149; reaffirmed February, 2009.

2. Centers for Disease Control and Prevention, National Center for Health Statistics, M. M. McDowell, C. Wang, and J. Kennedy-Stephenson, Breastfeeding in the United States: Findings from the National Health and Nutrition Examination Surveys, 1999–2006, *NCHS Data Brief,* No. 5, April 2008, available at www.cdc.gov/nchs/fastats/Default.htm.

3. Centers for Disease Control and Prevention, National Center for Health Statistics, M. M. McDowell, C. Wang, and J. Kennedy-Stephenson, 2008.

4. Position of the American Dietetic Association, 2009; Breastfeeding, in *Pediatric Nutrition Handbook,* 2009.

5. Breastfeeding, in *Pediatric Nutrition Handbook,* 2009; American Academy of Pediatrics, 2005.

6. *Healthy People 2020,* Objectives retained as is from *Healthy People 2010,* available at www.healthypeople.gov/hp2020/OObjectives/TopicArea.aspx?id=32&TopicArea=M; accessed October 28, 2010.

7. L. Rowell, V. J. Rock, and L. Grummer-Strawn, Changes in public attitudes toward breastfeeding in the United States, *Journal of the American Dietetic Association* 107 (2007): 122–127.

8. Centers for Disease Control and Prevention, Breastfeeding-related maternity practices at hospitals and birth centers—United States, 2007, *Morbidity and Mortality Weekly Report* 57 (2008): 621–625.

9. M. Chung and coauthors, Interventions in primary care to promote breastfeeding: An evidence review for the U.S. Preventive Services Task Force, *Annals of Internal Medicine* 149 (2008): 565–582; I. Mannan and coauthors, Can early postpartum home visits by trained community health workers improve breastfeeding of newborns? *Journal of Perinatology* 28 (2008): 632–640; J. Clifford and E. McIntyre, Who supports breastfeeding? *Breastfeeding Review* 16 (2008): 9–19.

10. Breastfeeding beyond infancy, in *American Academy of Pediatrics, New Mother's Guide to Breastfeeding,* ed. J. Y. Meek and S. Tippins (New York: Bantam Books, 2002), pp. 158–188.

12

Nutrition through the Life Span: Infancy, Childhood, and Adolescence

The first year of life is a time of phenomenal growth and development. After the first year, a child continues to grow and change, but more slowly. Sound nutrition throughout infancy and childhood promotes normal growth and development; facilitates academic and physical performance; and helps prevent obesity, diabetes, heart disease, cancer, and other degenerative diseases in adulthood. As children enter their teens, a foundation built by years of eating nutritious foods best prepares them to meet the upcoming demands of rapid growth. This chapter examines the special nutrient needs of infants, children, and adolescents.

Nutrition of the Infant

Early nutrition affects later development, and early feeding sets the stage for eating habits that will influence nutrition status for a lifetime. Trends change, and experts argue about the fine points, but properly nourishing an infant is relatively simple, overall. The infant initially drinks only breast milk or formula but later begins to eat some foods, as appropriate. Common sense in the selection of infant foods and a nurturing, relaxed environment go far to promote an infant's health and well-being.

Nutrient Needs during Infancy

An infant grows faster during the first year than ever again, as Figure 12-1 shows. The growth of infants and children directly reflects their nutritional well-being and is an important parameter in assessing their nutrition status. Health care professionals use growth charts to evaluate the growth and development of children from birth to age 20 (see Appendix E).

Nutrients to Support Growth An infant's birthweight doubles by about five months and triples by the age of one year, typically reaching 20 to 25 pounds. (Consider that if an adult, starting at 120 pounds, were to do this, the person's weight would increase to 360 pounds in a single year.) The infant's length changes more slowly than weight, increasing about 10 inches from birth to one year. By the end of the first year, the growth rate slows considerably. An infant typically gains fewer than than 10 pounds during the second year and grows about 5 inches in height.

Not only do infants grow rapidly, but, in proportion to body weight, their basal metabolic rate is remarkably high—about twice that of an adult. The rapid growth and metabolism of the infant demand an ample supply of all the nutrients. Of special importance during infancy are the energy nutrients and the vitamins and minerals critical to the growth process, such as vitamin A, vitamin D, and calcium.

Because they are small, infants need smaller *total* amounts of these nutrients than adults do, but as a percentage of body weight, infants need more than twice as much of most nutrients. Infants require about 100 kcalories per kilogram of body weight per day; most adults require fewer than 40 (see Table 12-1). Figure 12-2 (p. 318) compares a five-month-old infant's needs (per unit of body weight) with those of an adult man. You can see that differences in vitamin D

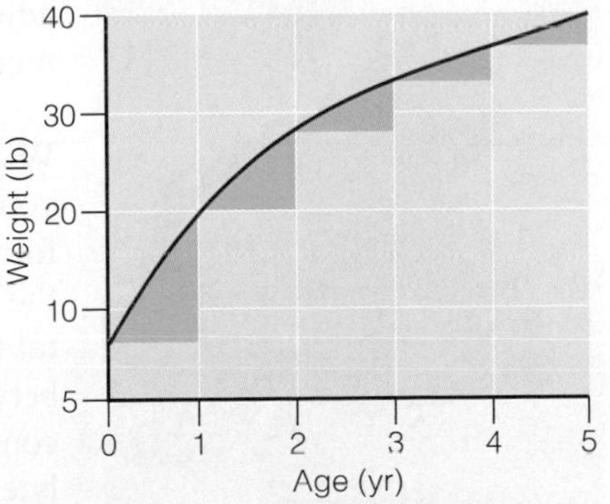

FIGURE 12-1 **Weight Gain of Human Infants in Their First Five Years of Life**
In the first year, an infant's birthweight may triple, but over the following several years the rate of weight gain gradually diminishes.

TABLE 12-1 **Infant and Adult Heart Rate, Respiration Rate, and Energy Needs Compared**

	Infants	Adults
Heart rate (beats/minute)	120 to 140	70 to 80
Respiration rate (breaths/minute)	20 to 40	15 to 20
Energy needs (kcal/body weight)	45/lb (100/kg)	<18/lb (<40/kg)

FIGURE 12-2 Nutrient Recommendations for a Five-Month-Old Infant and an Adult Male Compared on the Basis of Body Weight

Infants may be relatively small and inactive, but they use large amounts of energy and nutrients in proportion to their body size to keep all their metabolic processes going.

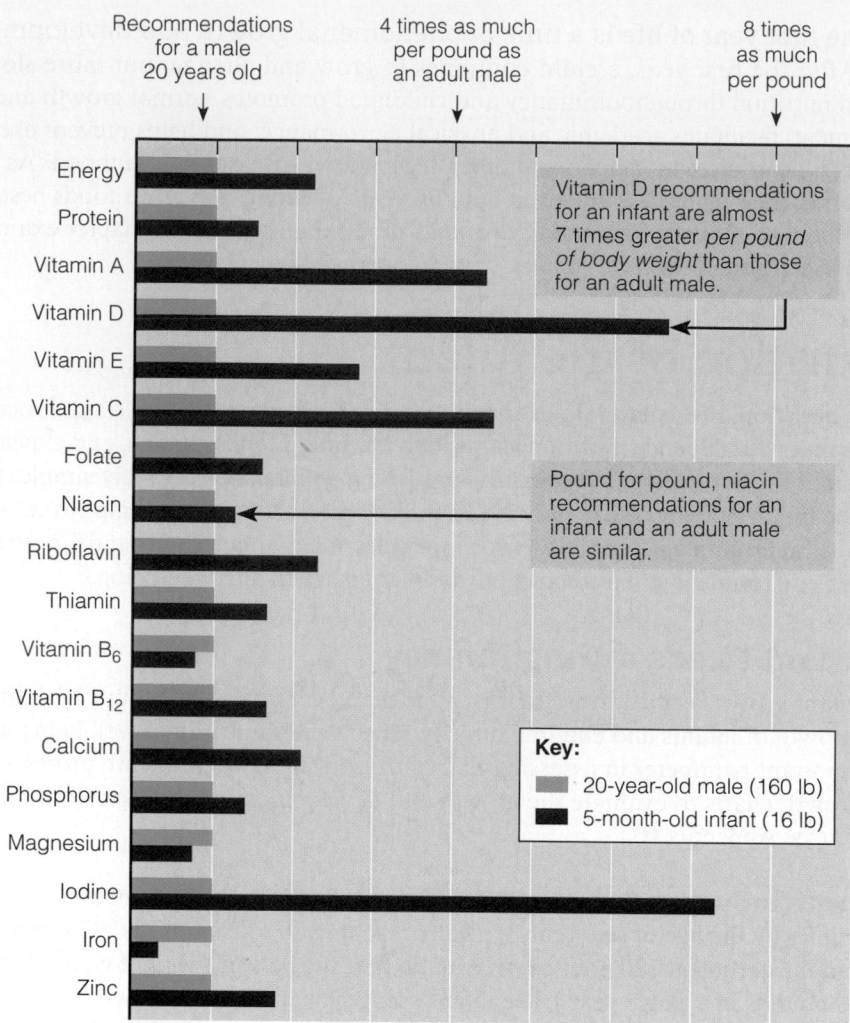

Recommendations for a male 20 years old

4 times as much per pound as an adult male

8 times as much per pound

Energy
Protein
Vitamin A
Vitamin D
Vitamin E
Vitamin C
Folate
Niacin
Riboflavin
Thiamin
Vitamin B$_6$
Vitamin B$_{12}$
Calcium
Phosphorus
Magnesium
Iodine
Iron
Zinc

Vitamin D recommendations for an infant are almost 7 times greater *per pound of body weight* than those for an adult male.

Pound for pound, niacin recommendations for an infant and an adult male are similar.

Key:
20-year-old male (160 lb)
5-month-old infant (16 lb)

After six months of age, the energy saved by slower growth is spent on increased activity.

and iodine, for instance, are extraordinary. Around six months of age, energy needs begin to increase less rapidly as the growth rate begins to slow, but some of the energy saved by slower growth is spent on increased activity. When their growth slows, infants spontaneously reduce their energy intakes. Parents should expect their infants to adjust their food intakes downward when appropriate and should not force or coax them to eat more.

Water One of the most important nutrients for infants, as for everyone, is water. The younger a child is, the more of the child's body weight is water. Breast milk or infant formula normally provides enough water to replace fluid losses in a healthy infant. If the environmental temperature is extremely high, however, infants need supplemental water.[1] Because proportionately more of an infant's body water than an adult's is between the cells and in the vascular space, this water is easy to lose. Consequently, conditions that cause rapid fluid loss, such as vomiting or diarrhea, require an electrolyte solution designed for infants.

IN SUMMARY

Infants' rapid growth and development depend on adequate nutrient supplies, including water from breast milk and formula.

Breast Milk

Breast milk excels as a source of nutrients for the young infant. With the possible exception of vitamin D (discussed later), breast milk provides all the nutrients a healthy infant needs for the first six months of life.[2] It provides many other health benefits as well.

Frequency of Breastfeeding Breast milk is more easily and completely digested than formula, so breastfed infants usually need to eat more frequently than formula-fed infants do. During the first few weeks, approximately 8 to 12 feedings a day, on demand, as soon as the infant shows early signs of hunger such as increased alertness, activity, or suckling motions, promote optimal milk production and infant growth.[3] Crying is a late indicator of hunger. An infant who nurses every two to three hours and sleeps contentedly between feedings is adequately nourished. As the infant gets older, stomach capacity enlarges, and the mother's milk production increases, allowing for longer intervals between feedings.

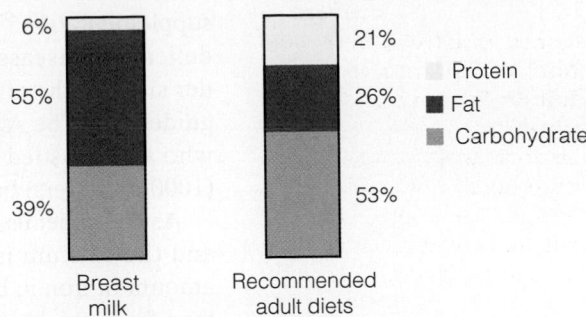

FIGURE 12-3 Percentages of Energy-Yielding Nutrients in Breast Milk and in Recommended Adult Diets

The proportions of energy-yielding nutrients in human breast milk differ from those recommended for adults.[a]

[a]The values listed for adults represent approximate midpoints of the acceptable ranges for protein (10 to 35 percent), fat (20 to 35 percent), and carbohydrate (45 to 65 percent).

Energy Nutrients The balance of energy nutrients in breast milk differs dramatically from the balance recommended for adults (see Figure 12-3). Yet, for infants, breast milk is nature's most nearly perfect food, providing the clear lesson that people at different stages of life have different nutrient needs.

The carbohydrate in breast milk (and standard infant formula) is lactose. In addition to being easily digested, lactose enhances calcium absorption. The carbohydrate component of breast milk also contains abundant oligosaccharides, which are present only in trace amounts in cow's milk and infant formula made from cow's milk.[4] Human milk oligosaccharides help protect the infant from infection by preventing the binding of pathogens to the infant's intestinal cells.[5]

Breast milk contains less protein than cow's milk, but this is actually beneficial because it places less stress on the infant's immature kidneys to excrete the major end product of protein metabolism, urea. The protein in breast milk is largely **alpha-lactalbumin,** which is efficiently digested and absorbed.

The lipids in breast milk—and infant formula—provide the main source of energy in the infant's diet. Breast milk contains a generous proportion of the essential fatty acids linoleic acid and linolenic acid, as well as their longer-chain derivatives arachidonic acid and docosahexaenoic acid (DHA). Most formulas today also contain added arachidonic acid and DHA (read the label). Infants can produce some arachidonic acid and DHA from linoleic and linolenic acid, respectively, but some infants may need more than they can make.

As Chapter 4 mentioned, DHA is the most abundant fatty acid in the brain and is also present in the retina of the eye. DHA accumulation in the brain is greatest during fetal development and early infancy.[6] Research has focused on the mental and visual development of breastfed infants and infants fed standard formula with and without DHA added.[7] Most studies show no beneficial effect of DHA supplementation of formula for term infants.[8]

Vitamins and Minerals With the exception of vitamin D, the vitamin content of the breast milk of a well-nourished mother is ample. Even vitamin C, for which cow's milk is a poor source, is supplied generously. The concentration of vitamin D in breast milk is low, however, and vitamin D deficiency impairs bone mineralization. Vitamin D deficiency is most likely in infants who are not exposed to sunlight daily, have darkly pigmented skin, and receive breast milk without vitamin D

alpha-lactalbumin (lackt-AL-byoo-min): the chief protein in human breast milk, as **casein** (CAY-seen) is the chief protein in cow's milk.

supplementation.[9] Reports of infants in the United States developing the vitamin D–deficiency disease rickets and recommendations by the AAP to keep infants under six months of age out of direct sunlight have prompted revisions in vitamin D guidelines. The AAP currently recommends a vitamin D supplement for all infants who are breastfed exclusively and for any infants who do not receive at least 1 liter (1000 milliliters) or 1 quart (32 ounces) of vitamin D–fortified formula daily.[10]

As for minerals, the calcium content of breast milk is ideal for infant bone growth, and the calcium is well absorbed. Breast milk is also low in sodium. The limited amount of iron in breast milk is highly absorbable, and its zinc, too, is absorbed better than from cow's milk, thanks to the presence of a zinc-binding protein.

Supplements for Infants Pediatricians may prescribe supplements containing vitamin D, iron, and fluoride (after six months of age), as outlined in Table 12-2. Vitamin K nutrition for newborns presents a unique case: the AAP recommends giving a single dose of vitamin K to infants at birth to prevent uncontrolled bleeding.

See Chapter 8 for a description of vitamin K's role in blood clotting.

Immunological Protection In addition to its nutritional benefits, breast milk offers immunological protection. Not only is breast milk sterile, but it actively fights disease and protects infants from illnesses.[11]

During the first two or three days after delivery, the breasts produce **colostrum,** a premilk substance containing mostly serum with antibodies and white blood cells. Colostrum (like breast milk) helps protect the newborn from infections against which the mother has developed immunity. The maternal antibodies in the breast milk inactivate disease-causing bacteria within the infant's digestive tract before they can start infections.[12] This explains, in part, why breastfed infants have fewer intestinal infections than formula-fed infants.

In addition to antibodies, colostrum and breast milk provide other powerful agents that help to fight against bacterial infection. Among them are the oligosaccharides, described earlier, that prevent pathogens from binding to intestinal cells. Also present are **bifidus factors,** which favor the growth of the "friendly" bacterium *Lactobacillus bifidus* in the infant's digestive tract, so that other, harmful bacteria cannot become established. An iron-binding protein in breast milk, **lactoferrin,** keeps bacteria from getting the iron they need to grow, helps absorb iron into the infant's bloodstream, and kills some bacteria directly.[13] The protein **lactadherin** in breast milk binds to, and inhibits replication of, the virus that causes most infant diarrhea.[14] Breastfeeding also

Protective factors in breast milk:

- Antibodies
- Oligosaccharides
- Bifidus factors
- Lactoferrin
- Lactadherin
- Growth factor
- Lipase enzyme

colostrum (co-LAHS-trum): a milk-like secretion from the breasts that is rich in protective factors. Colostrum is present during the first day or so after delivery, before milk appears.

bifidus (BIFF-id-us, by-FEED-us) **factors:** factors in colostrum and breast milk that favor the growth of the "friendly" bacterium *Lactobacillus* (lack-toe-ba-SILL-us) *bifidus* in the infant's intestinal tract. These bacteria prevent other, less desirable intestinal inhabitants from flourishing.

TABLE 12-2 Supplements for Full-Term Infants

	Vitamin D[a]	Iron[b]	Fluoride[c]
Breastfed infants:			
Birth to six months	✓	–	–
Six months to one year	✓	✓	✓
Formula-fed infants:			
Birth to six months	–	–	–
Six months to one year	–	✓	✓

[a]Vitamin D supplements are recommended for all infants who are exclusively breastfed and for any infants who do not receive at least 1 liter (1000 milliliters) or 1 quart (32 ounces) of vitamin D–fortified formula per day.
[b]All infants six months of age need additional iron, preferably in the form of iron-fortified infant cereal and/or infant meats. Formula-fed infants need iron-fortified infant formula.
[c]At six months of age, breastfed infants and formula-fed infants who receive ready-to-use formulas (these are prepared with water low in fluoride) or formula mixed with water that contains little or no fluoride (less than 0.3 ppm) need supplements.

Source: Adapted from Committee on Nutrition, American Academy of Pediatrics, *Pediatric Nutrition Handbook,* 6th ed., ed. R. E. Kleinman (Elk Grove Village, IL: American Academy of Pediatrics, 2009).

protects against other common illnesses of infancy such as middle ear infection and respiratory illness.[15] In addition, a growth factor that is present in breast milk stimulates the development and maintenance of the infant's digestive tract and its protective factors. Several breast milk enzymes, such as lipase, also help protect the infant against infection. Nutrition in Practice 11 offers suggestions for successful breastfeeding.

Allergy and Disease Protection In addition to protection against infection, breast milk may offer protection against the development of allergies.[16] Compared with formula-fed infants, breastfed infants have a lower incidence of allergic reactions, such as recurrent wheezing and skin rashes.[17] This protection is especially noticeable among infants with a family history of allergies.[18] Similarly, breast milk may offer protection against the development of cardiovascular disease. Compared with formula-fed infants, breastfed infants have lower blood pressure and lower blood cholesterol as adults.[19]

Other Potential Benefits Breastfeeding may offer some protection against excessive weight gain in later life, although findings are inconsistent.[20] One review reports that various studies have shown a protective effect, a protective effect only in certain groups, or no effect.[21] Some research suggests that the longer the duration of breastfeeding, the lower the risk of overweight in childhood, while other research conflicts with such findings.[22] Researchers note that many other factors—socioeconomic status, other infant and child feeding practices, and especially the mother's weight—strongly predict a child's body weight.

Some research suggests a beneficial effect of breastfeeding on intelligence, but when subjected to strict standards of methodology (for example, large sample size and appropriate intelligence testing), the evidence is less convincing.[23] Most likely, a combination of factors is involved. More large, well-controlled studies are needed to confirm the effects, if any, of breastfeeding on later intelligence.

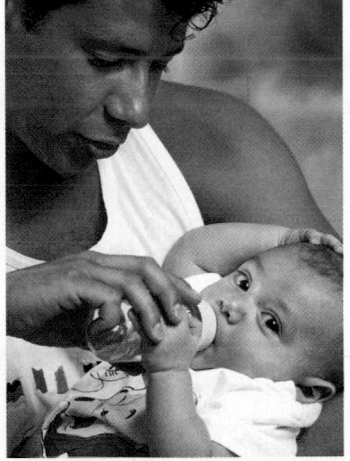

The infant thrives on infant formula offered with affection.

Formula preparation:

- Liquid concentrate (moderately expensive, relatively easy)—mix with equal part water.
- Powdered formula (least expensive, lightest for travel)—read label directions.
- Ready-to-feed (easiest, most expensive)—pour directly into clean bottles.

Infant Formula

Breastfeeding offers many benefits to both mother and infant, and it should be encouraged whenever possible. The mother who has decided to use formula, however, should be supported in her choice, just as the breastfeeding mother should be. She can offer the same closeness, warmth, and stimulation during feedings as the breastfeeding mother can.

Many mothers choose to breastfeed at first but **wean** their children within the first 1 to 12 months. If infants are less than a year of age, mothers must wean them onto *infant formula*, not onto plain cow's milk of any kind—whole, reduced fat, low fat, or fat free.

Infant Formula Composition Manufacturers can prepare formulas from cow's milk in such a way that they do not differ significantly from human milk in nutrient content. Figure 12-4 (p. 322) illustrates the energy nutrient balance of breast milk, infant formula, and cow's milk. Notice the higher protein concentration of cow's milk, which stresses the infant's kidneys. The AAP recommends that all formula-fed infants receive iron-fortified infant formulas.[24]

Infant Formula Standards National and international standards have been set for the nutrient contents of infant formulas. United States standards are based on AAP recommendations, and the FDA mandates quality control procedures to ensure that these standards are met. All standard formulas are therefore nutritionally similar. Small differences in nutrient content are sometimes confusing but usually unimportant.

Special Formulas Standard cow's milk–based formulas are inappropriate for some infants. Special formulas have been designed to meet the dietary needs of infants with specific conditions such as prematurity or inherited diseases. Most infants who are

lactoferrin (lack-toe-FERR-in): protein in breast milk that binds iron and keeps it from supporting the growth of the infant's intestinal bacteria.

lactadherin (lack-tad-HAIR-in): a protein in breast milk that attacks diarrhea-causing viruses.

wean: to gradually replace breast milk with infant formula or other foods appropriate to an infant's diet.

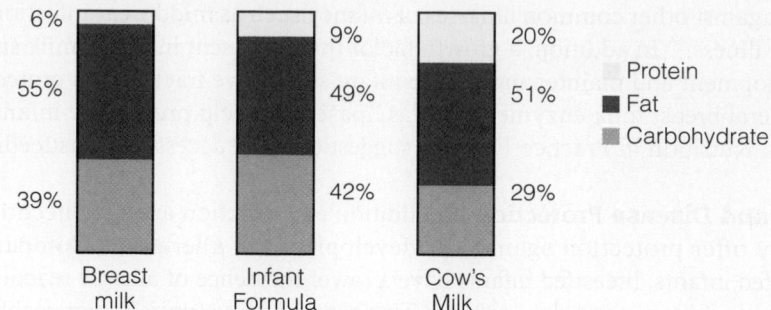

allergic to milk protein can drink formulas based on soy protein.[25] Soy formulas also use cornstarch and sucrose instead of lactose and so are recommended for infants with lactose intolerance as well. They are also useful as an alternative to milk-based formulas for vegan families.

Some infants who are allergic to cow's milk protein may also be allergic to soy protein.[26] For these infants, special formulas based on hydrolyzed protein are available.

Risks of Formula Feeding Infant formulas contain no protective antibodies for infants, but, in general, vaccinations, purified water, and clean environments in developed countries help protect infants from infections. Formulas can be prepared safely by following the rules of proper food handling and by using water that is free of contamination.

In developing countries and in poor areas of the United States, formula may be unavailable, overdiluted in an attempt to save money, or prepared with contaminated water. Overdilution of formula can cause malnutrition and growth failure. Contaminated formula often causes infections leading to diarrhea, dehydration, and malabsorption. Wherever sanitation is poor, breastfeeding should take priority over feeding formula. Breast milk is sterile, and its antibodies strengthen an infant's resistance to disease.

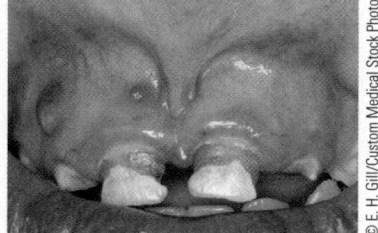

Nursing bottle tooth decay, an extreme example. The upper teeth have decayed all the way to the gum line.

Nursing Bottle Tooth Decay Dentists advise against putting an infant to bed with a bottle. Salivary flow, which normally cleanses the mouth, diminishes as the infant falls asleep. Prolonged sucking on a bottle of formula, milk, or juice bathes the upper teeth in a carbohydrate-rich fluid that nourishes decay-producing bacteria. (The tongue covers and protects most of the lower teeth, but they, too, may be affected.) The result is extensive and rapid tooth decay. To prevent **nursing bottle tooth decay,** no child should be put to bed with a bottle as a pacifier.

The Transition to Cow's Milk

The age at which whole cow's milk should be introduced to the infant's diet has long been a source of controversy. The AAP advises that whole cow's milk is not appropriate during the first year.[27] Children one to two years of age should not be given reduced-fat, low-fat, or fat-free milk routinely; they need the fat contained in whole milk. Between the ages of two and five, a gradual transition from whole milk to the lower-fat milks can take place, but care should be taken to avoid excessive restriction of dietary fat.

In some infants, particularly those younger than six months, whole cow's milk causes intestinal bleeding, which can lead to iron deficiency. Cow's milk is also a poor source of iron. Consequently, it both causes iron loss and fails to replace iron. Furthermore, the bioavailability of iron from infant cereal and other foods is reduced when cow's milk replaces breast milk or iron-fortified formula during the first year. Compared with breast milk or iron-fortified formula, cow's milk is higher in calcium and lower in vitamin C, characteristics that reduce iron absorption. In short, cow's milk is a poor choice during the first year of life; infants need breast milk or iron-fortified infant formula.

TABLE 12-3 Infant Development and Recommended Foods

Age (mo)	Feeding Skill	Foods Introduced into the Diet
0–4	Turns head toward any object that brushes cheek.	Feed breast milk or infant formula.
	Initially swallows using back of tongue; gradually begins to swallow using front of tongue as well.	
	Strong reflex (extrusion) to push food out during first 2 to 3 months.	
4–6	Extrusion reflex diminishes, and the ability to swallow nonliquid foods develops.	Begin iron-fortified cereal mixed with breast milk, formula, or water. Begin pureed meats, legumes, vegetables, and fruits.
	Indicates desire for food by opening mouth and leaning forward.	
	Indicates satiety or disinterest by turning away and leaning back.	
	Sits erect with support at 6 months.	
	Begins chewing action.	
	Brings hand to mouth.	
	Grasps objects with palm of hand.	
6–8	Able to self-feed finger foods.	Begin textured vegetables and fruits.
	Develops pincher (finger to thumb) grasp.	Begin unsweetened, diluted fruit juices from cup.
	Begins to drink from cup.	
8–10	Begins to hold own bottle.	Begin breads and cereals from table.
	Reaches for and grabs food and spoon.	Begin yogurt.
	Sits unsupported.	Begin pieces of soft, cooked vegetables and fruit from table.
		Gradually begin finely cut meats, fish, casseroles, cheese, eggs, and mashed legumes.
10–12	Begins to master spoon but still spills some.	Add variety.
		Gradually increase portion sizes.[a]

Note: Because each stage of development builds on the previous stage, the foods from an earlier stage continue to be included in all later stages.

[a]Portion sizes for infants and young children are smaller than those for an adult. For example, a grain serving might be ½ slice of bread instead of 1 slice, or ¼ cup rice instead of ½ cup.

Source: Adapted in part from Committee on Nutrition, American Academy of Pediatrics, *Pediatric Nutrition Handbook*, 6th ed., ed. R. E. Kleinman (Elk Grove Village, IL: American Academy of Pediatrics, 2009), pp. 113–142.

Introducing First Foods

Changes in the body organs during the first year affect the infant's readiness to accept solid foods. Until the child is several months old, the immature stomach and intestines can digest milk sugar (lactose) but not starch. This is one of the many reasons why breast milk and formula are such good foods for an infant; they provide simple, easily digested carbohydrate that supplies energy for the infant's growth and activity.

When to Introduce Solid Food The AAP supports exclusive breastfeeding for approximately six months but recognizes that infants are often developmentally ready to accept complementary foods between four and six months of age.[28] The main purpose of introducing solid foods is to provide needed nutrients that are no longer supplied adequately by breast milk or formula alone. The foods chosen must be those that the infant is developmentally capable of handling both physically and metabolically. The exact timing depends on the individual infant's needs and developmental readiness (see Table 12-3), which vary from infant to infant because of differences in growth rates, activities, and environmental conditions. The addition of foods to an infant's diet should be governed by three considerations: the infant's nutrient needs, the infant's physical readiness to handle different forms of foods, and the need to detect and control allergic reactions. With respect to nutrient needs, iron is needed earliest; next is vitamin C.

nursing bottle tooth decay: extensive tooth decay due to prolonged tooth contact with formula, milk, fruit juice, or other carbohydrate-rich liquid offered to an infant in a bottle.

Foods such as iron-fortified cereals and formulas, mashed legumes, and strained meats provide iron.

© Polara Studios, Inc.

Foods to Provide Iron and Vitamin C Rapid growth demands iron. At about four to six months, the infant begins to need more iron than body stores plus breast milk or iron-fortified formula can provide. In addition to breast milk or iron-fortified formula, infants can receive iron from iron-fortified cereals and, once they readily accept solid foods, from protein foods such as meat and legumes. Iron-fortified cereals contribute a significant amount of iron to an infant's diet, but the iron's bioavailability is poor.[29] Caregivers can enhance iron absorption from iron-fortified cereals by serving vitamin C–rich foods with meals.

The best sources of vitamin C are fruits and vegetables (see p. 211 in Chapter 8). Fruit juice is a source of vitamin C, but excessive juice intake can lead to diarrhea in infants and young children.[30] Furthermore, too much fruit juice contributes excessive kcalories and displaces other nutrient-rich foods. AAP recommendations limit juice consumption for infants and young children (one to six years of age) to between 4 and 6 ounces per day.[31] Fruit juices should be diluted and served in a cup, not a bottle, once the infant is six months of age or older.

Physical Readiness for Solid Foods The ability to swallow solid food develops at around four to six months, and food offered by spoon helps to develop swallowing ability. At eight months to a year, an infant can sit up, can handle finger foods, and begins to teethe. At that time, hard crackers and other hard finger foods may be introduced to promote the development of manual dexterity and control of the jaw muscles. These feedings must occur under the watchful eye of an adult because the infant can also choke on such foods.

Some parents want to feed solids at an earlier age, mistakenly believing that "stuffing the baby" at bedtime promotes sleeping through the night. On average, infants start to sleep through the night at about the same age (three to four months) regardless of when solid foods are introduced.

Allergy-Causing Foods To prevent allergy and to facilitate its prompt identification should it occur, experts recommend introducing single-ingredient foods, one at a time, in small portions, and waiting three to five days before introducing the next new food.[32] For example, rice cereal is usually the first cereal introduced because it is the least allergenic. When it is clear that rice cereal is not causing an allergy, another grain, perhaps barley or oats, is introduced. Wheat cereal is offered last because it is the most common offender. If a cereal causes an allergic reaction such as a skin rash, digestive upset, or respiratory discomfort, it should be discontinued before introducing the next food. A later section offers more information about food allergies.

Choice of Infant Foods Infant foods should be selected to provide variety, balance, and moderation. Commercial baby foods offer a wide variety of palatable, nutritious foods in a safe and convenient form. Homemade infant foods can be as nutritious as commercially prepared ones, as long as the cook minimizes nutrient losses during preparation. Ingredients for homemade foods should be fresh whole foods without added salt, sugar, or seasonings. Pureed food can be frozen in ice cube trays, providing convenient-sized blocks of food that can be thawed, warmed, and fed to the infant. To guard against foodborne illnesses, hands and equipment must be kept clean.

Dietary fat should not be restricted in children under two. To prevent parents from restricting fat in infants' diets, manufacturers are required to omit information about fat from labels on foods for children under two (such as infant meats and cereals). Infants and young children, because of their rapid growth, need more fat than older children and adults.

Foods to Omit Sweets of any kind (including baby-food desserts) have no place in an infant's diet. The added food energy conveys few, if any, nutrients to support growth and contributes to obesity. Canned vegetables are also inappropriate for infants; they

324

often contain too much sodium. Honey and corn syrup should never be fed to infants because of the risk of botulism. Infants and young children are vulnerable to foodborne illnesses. An infant's caregiver must be on guard against food poisoning and take precautions against it, as described in Nutrition in Practice 5.

Infants, or even young children, cannot safely chew and swallow any of the foods listed in the margin; they can easily choke on these foods, a risk not worth taking. Nonfood items of small size should always be kept out of the infant's reach to prevent choking.

Foods at One Year Whole milk is the best food to supply most of the nutrients the infant needs at one year of age; 2 to 3 cups a day meet those needs sufficiently. More milk than this displaces iron-rich foods and can lead to the iron-deficiency anemia known as milk anemia. Other foods—meat and meat alternates, iron-fortified cereals, whole-grain or enriched bread, fruits, and vegetables—should be supplied in variety and in amounts sufficient to round out total energy needs. Ideally, a one-year-old will sit at the table, eat many of the same foods everyone else eats, and drink liquids from a cup—not a bottle. Table 12-4 shows a sample menu that meets a one-year-old's requirements.

Looking Ahead

Probably the most important single measure to undertake during the first year is to encourage eating habits that will support continued normal weight as the child grows. This means introducing a variety of nutritious foods in an inviting way, not forcing the infant to finish drinking the bottle or eating the jar of baby food, avoiding concentrated sweets and empty-kcalorie foods, and encouraging physical activity. Parents should avoid teaching infants to seek food as a reward, to expect food as comfort for unhappiness, or to associate food deprivation with punishment. Normal dental development is also promoted by supplying nutritious foods, avoiding sweets, and discouraging the association of food with reward or comfort. Oral health is the subject of Nutrition in Practice 17.

Mealtimes

Nurturing a young child involves more than nutrition. Those who care for young children are responsible for providing not only nutritious foods, milk, and water but also a safe, loving, secure environment in which the children may grow and develop. The person feeding a one-year-old should be aware that exploring and experimenting are

---

To prevent choking, do not give infants or young children the following:
- Gum
- Popcorn
- Whole grapes
- Cherries
- Raw celery
- Carrots
- Whole beans
- Hot dog slices
- Hard or gel-type candies
- Marshmallows
- Nuts
- Peanut butter

Keep these nonfood items out of their reach:
- Coins
- Balloons
- Small balls
- Pen tops
- Other items of similar size

Reminder: *Milk anemia* develops when an excessive milk intake displaces iron-rich foods from the diet.

TABLE 12-4 **Sample Menu for a One-Year-Old**

Breakfast	1 scrambled egg 1 slice whole-wheat toast ½ c whole milk
Morning Snack	½ c yogurt ¼ c fruit[a]
Lunch	½ grilled cheese sandwich: 1 slice whole-wheat bread with 1 slice cheese ½ c vegetables[b] (steamed carrots) ¼ c 100% fruit juice
Afternoon snack	½ c fruit[a] ½ c toasted oat cereal
Dinner	1 oz chopped meat or ¼ c well-cooked mashed legumes ½ c rice or pasta ½ c vegetables[b] (chopped broccoli) ½ c whole milk

Note: This sample menu provides about 1000 kcalories.

[a]Include citrus fruits, melons, and berries.

[b]Include dark green, leafy, and orange vegetables.

Ideally, a one-year-old eats many of the same healthy foods as the rest of the family.

normal and desirable behaviors at this time in a child's life. The child is developing a sense of autonomy that, if allowed to develop, will lay the foundation for later confidence and effectiveness as an individual. In light of the developmental and nutrient needs of one-year-olds, and in the face of their often contrary and willful behavior, a few feeding guidelines may be helpful:

▌ *Discourage unacceptable behavior (such as standing at the table or throwing food) by removing the child from the table to wait until later to eat.* Be consistent and firm, not punitive. For example, instead of saying "You make me mad when you don't sit down," say "The fruit salad tastes good—please sit down and eat some with me." The child will soon learn to sit and eat.

▌ *Let young children explore and enjoy food.* This may mean eating with fingers for a while. Learning to use a spoon will come in time. Children who are allowed to touch, mash, and smell their food while exploring it are more likely to accept it.

▌ *Don t force food on children.* Rejecting new foods is normal, and acceptance is more likely as children become familiar with new foods through repeated opportunities to taste them. Instead of saying "You cannot go outside to play until you taste your carrots," say "You can try the carrots again another time."

▌ *Provide nutritious foods, and let children choose which ones, and how much, they will eat.* Gradually, they will acquire a taste for different foods.

▌ *Limit sweets.* Infants and young children have little room for empty-kcalorie foods in their daily energy allowance. Do not use sweets as a reward for eating meals.

These recommendations reflect a spirit of tolerance that serves the best interests of the child.

IN SUMMARY

▌ The primary food for infants during the first 12 months is either breast milk or iron-fortified formula.

▌ In addition to nutrients, breast milk also offers immunological protection.

▌ At about four to six months, infants should gradually begin eating solid foods. By one year, they are drinking from a cup and eating many of the same foods as the rest of the family.

Nutrition during Childhood

After age one, growth rate slows, but the body continues to change dramatically (see Figure 12-5). At one, infants have just learned to stand and toddle; by two, they walk confidently and are learning to run, jump, and climb. Nutrition and physical activity have helped them prepare for these new accomplishments by adding to the mass and density of their bone and muscle tissue. Thereafter, their bones continue to grow longer and their muscles to gain size and strength, though unevenly and more slowly, until adolescence.

Energy and Nutrient Needs

Children's appetites begin to diminish around the first birthday, consistent with the slowed growth rate. Thereafter, the appetite fluctuates. At times, children seem to be insatiable, and at other times they seem to live on air and water. Parents and other caregivers need not worry about this—a child will need and demand more food during periods of rapid growth than during slow periods. The perfect regulation of appetite in children of healthy weight guarantees that their food energy intakes will be right for each stage of growth.

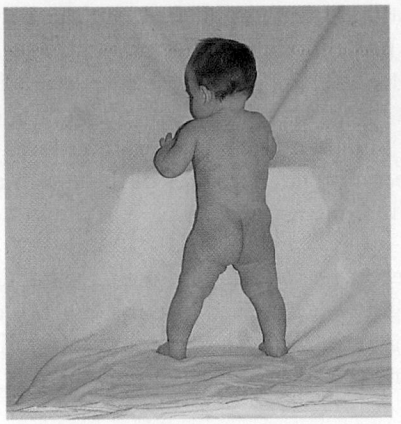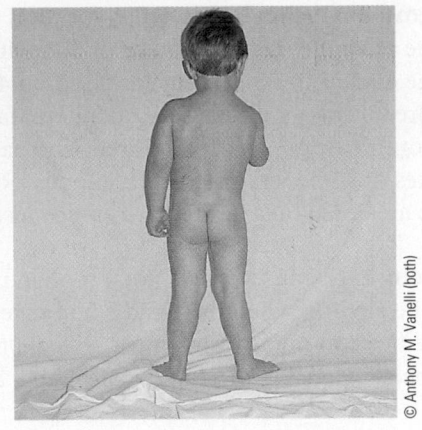

© Anthony M. Vanelli (both)

FIGURE 12-5 Body Shape of a One-Year-Old and a Two-Year-Old Compared
The body shape of a one-year-old (left) changes dramatically by age two (right). The two-year-old has lost much of the baby fat; the muscles (especially in the back, buttocks, and legs) have firmed and strengthened; and the leg bones have lengthened.

Ideally, children accumulate stores of nutrients before adolescence. Then, when they take off on the adolescent growth spurt and their nutrient intakes cannot keep pace with the demands of rapid growth, they can draw on the nutrient stores accumulated earlier. This is especially true of calcium; the denser the bones are in childhood, the better prepared they will be to support teen growth and still withstand the inevitable bone losses of later life.[33] Consequently, the way children eat influences their nutritional health during childhood, during their teen years, and for the rest of their lives.

Children's Appetites Many people mistakenly believe that they must make their children eat the right amounts of food, and children's erratic appetites often reinforce this belief. Although children's food energy intakes vary widely from meal to meal, total daily energy intake remains remarkably constant.[34] If children eat less at one meal, they typically eat more at the next, and vice versa.

Parents do, however, need to help children choose the right foods. With overweight children, parents may need to help more, as described later. Overweight children may not adjust their energy intakes appropriately and, disregarding appetite-regulation signals, may eat in response to external cues, such as television commercials.[35]

Energy Individual children's energy needs vary widely, depending on their growth and physical activity. A one-year-old child needs approximately 800 kcalories a day; an active six-year-old needs twice as many kcalories a day. By age ten, an active child needs about 2000 kcalories a day. Total energy needs increase gradually with age, but energy needs per kilogram of body weight actually decline. Physically active children of any age need more energy because they expend more, and inactive children can become obese even when they eat less food than the average.

Children who adhere to a vegan diet may have difficulty meeting their energy needs. Grains, vegetables, and fruits provide plenty of fiber, adding bulk, but these foods may provide too few kcalories to support growth. Soy products, other legumes, and nut or seed butters offer more concentrated sources of energy to support optimal growth and development.[36]

Carbohydrate and Fiber Carbohydrate recommendations are based on the brain's glucose use. After one year of age, brain glucose use remains fairly constant and is within the adult range. Carbohydrate recommendations for children older than one year are therefore the same as for adults (see inside front cover of this textbook).[37]

Fiber recommendations derive from adult intakes shown to reduce the risk of coronary heart disease and are based on energy intakes. Consequently, fiber recommendations for younger children with low energy intakes are less than those for older ones with high energy intakes.[38]

Fiber recommendations for children:

Age (yr)	DRI (g/day)
1–3	19
4–8	25
9–13	
Boys	31
Girls	26
14–18	
Boys	38
Girls	26

TABLE 12-5 Iron-Rich Foods Children Like[a]

Breads, cereals, and grains

Canned macaroni (½ c)

Canned spaghetti (½ c)

Cream of wheat (¼ c)

Fortified dry cereals (1 oz)[b]

Noodles, rice, or barley (½ c)

Tortillas (1 flour or whole wheat, 2 corn)

Whole-wheat, enriched, or fortified bread (1 slice)

Vegetables

Baked flavored potato skins (½ skin)

Cooked mung bean sprouts or snow peas (½ c)[c]

Cooked mushrooms (½ c)

Green peas (½ c)

Mixed vegetable juice (1 c)

Fruits

Canned plums (3 plums)

Cooked dried apricots (¼ c)

Dried peaches (4 halves)

Raisins (1 tbsp)

Protein Foods

Bean dip (¼ c)

Canned pork and beans (⅓ c)

Lean chopped roast beef or cooked ground beef (1 oz)

Liverwurst on crackers (½ oz)

Meat casseroles (½ c)

Mild chili or other bean/meat dishes (¼ c)

Peanut butter and jelly sandwich (½ sandwich)

Sloppy joes (½ sandwich)

[a]Each serving provides at least 1 milligram of iron. Vitamin C–rich foods included with these snacks increase iron absorption.

[b]Some fortified breakfast cereals contain more than 10 milligrams of iron per half-cup serving (read the labels).

[c]Raw sprouts may pose a bacterial hazard to young children.

Vitamin D RDA for children:

▌ 15 µg/day

▌ www.MyPyramid.gov/preschoolers

▌ www.MyPyramid.gov/kids

Fat and Fatty Acids No RDA for total fat has been established, but the DRI committee recommends a fat intake of 30 to 40 percent of energy for children 1 to 3 years of age and 25 to 35 percent for children 4 to 18 years of age.[39] However, as long as children's energy intakes are adequate, fat intakes below 30 percent of total energy do not impair growth.[40] Children who eat low-fat diets, however, tend to have low intakes of some vitamins and minerals. Recommended intakes of the essential fatty acids are based on average intakes (see inside front cover).

Protein Like energy needs, total protein needs increase slightly with age but actually decline slightly when the child's body weight is considered (see inside front cover). Protein recommendations must address the requirements for maintaining nitrogen balance, the quality of protein consumed, and the added needs of growth.

Vitamins and Minerals The vitamin and mineral needs of children increase with age (see inside front cover). A balanced diet of nutritious foods can meet children's needs for these nutrients, with the notable exception of iron, and possibly vitamin D.[c] Iron-deficiency anemia is a major problem worldwide, and is prevalent among U.S. and Canadian children, especially toddlers one to three years of age.[41] During the second year of life, toddlers progress from a diet of iron-rich infant foods such as breast milk, iron-fortified formula, and iron-fortified infant cereal to a diet of adult foods and iron-poor cow's milk. In addition, their appetites often fluctuate—some become finicky about the foods they eat, and others prefer milk and juice to solid foods. These situations can interfere with children eating iron-rich foods at a critical time for brain growth and development.

To prevent iron deficiency, children's foods must deliver 7 to 10 milligrams of iron per day. To achieve this goal, snacks and meals should include iron-rich foods, and milk intake should be reasonable so that it will not displace lean meats, fish, poultry, eggs, legumes, and whole-grain or enriched products. That means 2 cups of milk per day for ages two and three, 2½ cups for ages four through eight, and 3 cups per day from age nine on. After age two, if reduced-fat or low-fat milk is used instead of whole milk, the saved kcalories can be invested in iron-rich foods such as lean meats, fish, poultry, eggs, and legumes. Whole-grain or enriched breads and cereals also contribute iron. Table 12-5 lists iron-rich foods children like.

The DRI committee recently revised their recommendations for vitamin D intakes for healthy Americans.[42] Children obtain vitamin D from fortified milk (2.5 micrograms per 1-cup serving) and dry cereals (1 microgram per ½-cup serving), but children who do not meet their RDA should receive a vitamin D supplement.[43] Remember that sunlight is also a source of vitamin D, especially in warm climates and warm seasons.

Supplements With the exception of specific recommendations for fluoride, iron, and vitamin D during infancy and childhood, the AAP and other professional groups agree that well-nourished children do not need vitamin and mineral supplements. Despite this, many children and adolescents take supplements.[44] Ironically, children with poor nutrient intakes typically do not receive supplements, while those who do take them usually receive extra nutrients they do not need.[45] Furthermore, researchers are still studying the safety of supplement use by children. Dietary supplements on the market today include many herbal products that have not been tested for safety and effectiveness in children.

Food Patterns for Children To provide all the needed nutrients, a child's meals and snacks should include a variety of foods from each food group—in amounts suited to the child's appetite and needs. Figure 12-6 presents MyPyramid for Preschoolers, designed for children 2 to 5 years of age, and MyPyramid for Kids, designed for children 6 to 11 years of age. The figure includes the recommended

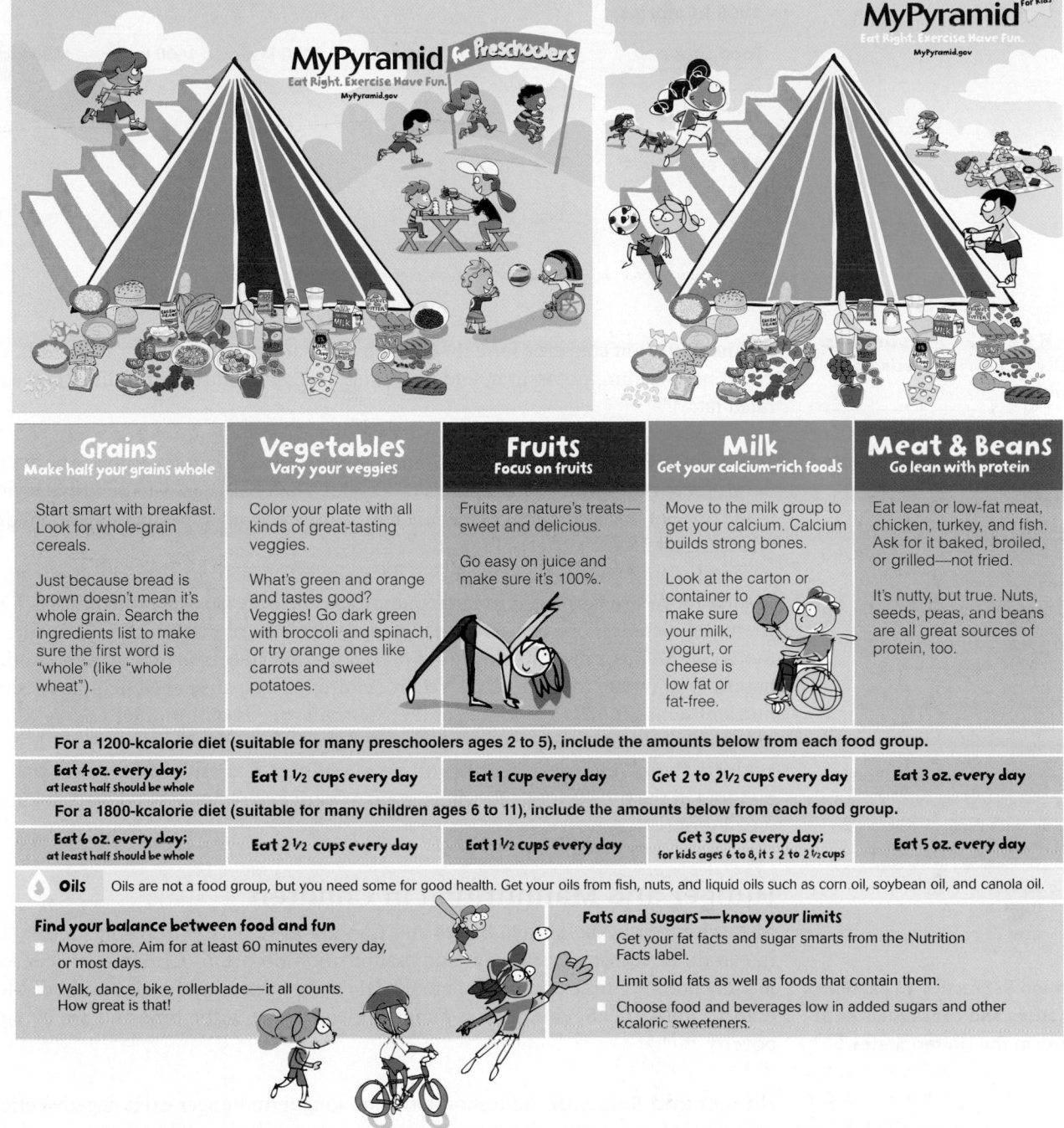

Grains
Make half your grains whole

Vegetables
Vary your veggies

Fruits
Focus on fruits

Milk
Get your calcium-rich foods

Meat & Beans
Go lean with protein

Grains	Vegetables	Fruits	Milk	Meat & Beans
Start smart with breakfast. Look for whole-grain cereals. Just because bread is brown doesn't mean it's whole grain. Search the ingredients list to make sure the first word is "whole" (like "whole wheat").	Color your plate with all kinds of great-tasting veggies. What's green and orange and tastes good? Veggies! Go dark green with broccoli and spinach, or try orange ones like carrots and sweet potatoes.	Fruits are nature's treats—sweet and delicious. Go easy on juice and make sure it's 100%.	Move to the milk group to get your calcium. Calcium builds strong bones. Look at the carton or container to make sure your milk, yogurt, or cheese is low fat or fat-free.	Eat lean or low-fat meat, chicken, turkey, and fish. Ask for it baked, broiled, or grilled—not fried. It's nutty, but true. Nuts, seeds, peas, and beans are all great sources of protein, too.

For a 1200-kcalorie diet (suitable for many preschoolers ages 2 to 5), include the amounts below from each food group.

Eat 4 oz. every day; at least half should be whole	Eat 1½ cups every day	Eat 1 cup every day	Get 2 to 2½ cups every day	Eat 3 oz. every day

For a 1800-kcalorie diet (suitable for many children ages 6 to 11), include the amounts below from each food group.

Eat 6 oz. every day; at least half should be whole	Eat 2½ cups every day	Eat 1½ cups every day	Get 3 cups every day; for kids ages 6 to 8, it's 2 to 2½ cups	Eat 5 oz. every day

Oils Oils are not a food group, but you need some for good health. Get your oils from fish, nuts, and liquid oils such as corn oil, soybean oil, and canola oil.

Find your balance between food and fun
- Move more. Aim for at least 60 minutes every day, or most days.
- Walk, dance, bike, rollerblade—it all counts. How great is that!

Fats and sugars—know your limits
- Get your fat facts and sugar smarts from the Nutrition Facts label.
- Limit solid fats as well as foods that contain them.
- Choose food and beverages low in added sugars and other kcaloric sweeteners.

FIGURE 12-6 MyPyramid for Preschoolers and for Kids
Source: MyPyramid.gov.

amounts of food for a 1200-kcalorie intake (appropriate for many preschoolers) and for an 1800-kcalorie intake (appropriate for many older children). Table 12-6 (p. 330) lists recommended amounts from each food group for several kcalorie levels. Estimated daily kcalorie needs for active and sedentary children of various ages are shown in Table 12-7 (p. 330).

▸ For kcalorie levels over 1800, see Table 1-4 on p. 18.

Children whose diets follow the pattern presented in Figure 12-6 meet their nutrient needs fully, but few children eat according to these recommendations. One analysis of the quality of children's diets found that most children (up to 88 percent) between two and nine years of age have diets that need substantial improvement.[46]

TABLE 12-6 Recommended Daily Amounts for Each Food Group (1000 to 1800 kCalories)

Food Group	1000 kcal	1200 kcal	1400 kcal	1600 kcal	1800 kcal
Fruit	1 c	1 c	1½ c	1½ c	1½ c
Vegetables	1 c	1½ c	1½ c	2 c	2½ c
Grains	3 oz	4 oz	5 oz	5 oz	6 oz
Meat and legumes	2 oz	3 oz	4 oz	5 oz	5 oz
Milk	2 c	2½ c	2½ c	3 c	3 c
Oils	3 tsp	3 tsp	3 tsp	4 tsp	5 tsp

Note: The discretionary kcalorie allowance for these patterns is about 100 kcalories.

TABLE 12-7 Estimated Daily kCalorie Needs for Children

Children	Sedentary[a]	Active[b]
2 to 3 yr	1000	1400
Females		
4 to 8 yr	1200	1800
9 to 13 yr	1600	2200
Males		
4 to 8 yr	1400	2000
9 to 13 yr	1800	2600

[a]Sedentary describes a lifestyle that includes only the activities typical of day-to-day life.

[b]Active describes a lifestyle that includes at least 60 minutes per day of moderate physical activity (equivalent to walking more than 3 miles per day at 3 to 4 miles per hour) in addition to the activities of day-to-day life.

Nutrition in Practice 13 examines the causes and consequences of hunger in the United States.

Among nutrition concerns for U.S. children are inadequate intakes of vitamin E, calcium, magnesium, potassium, and fiber, and excessive intakes of sodium and saturated fat.[47]

Children's Food Choices Parents and other caregivers can do much to foster the development of healthy eating habits in a child. The challenge is to deliver nutrients in the form of meals and snacks that are both nutritious and appealing so that children will learn to enjoy a variety of nutritious foods.

Candy, cola, and other concentrated sweets must be limited in children's diets. If such foods are permitted in large quantities, the only possible outcomes are nutrient deficiencies, obesity, or both. Children can't be trusted to choose nutritious foods on the basis of taste alone; the preference for sweets is innate, and children naturally gravitate to them. Overweight children, especially, need help in selecting nutrient-dense foods that will meet their nutrient needs within their energy allowances. Underweight children or active, healthy-weight children can enjoy higher-kcalorie foods, but these should still be nutritious. Examples are ice cream and pudding in the milk group and whole-grain or enriched pancakes and crackers in the bread group.

Hunger and Malnutrition in Children

Most children in the United States and Canada have access to regular meals, but in certain circumstances hunger and malnutrition do occur. Children in very low-income families, for example, are more likely to be hungry and malnourished. More than 12 million U.S. children are hungry at least some of the time and are living in poverty.[48]

Hunger and Behavior Both short-term and long-term hunger exert negative effects on behavior and health. Short-term hunger, such as when a child misses a meal, impairs the child's ability to pay attention and to be productive. Hungry children are irritable, apathetic, and uninterested in their environment. Long-term hunger impairs growth and immune defenses. Food assistance programs such as the WIC program (discussed in Chapter 11) and the School Breakfast and National School Lunch Programs (discussed later in this chapter) are designed to protect against hunger and improve children's health.[49]

Children who eat nutritious breakfasts improve their school performance and are tardy or absent significantly less often than their peers who do not.[50] A nutritious breakfast is a central feature of a diet that meets children's needs and supports their healthy growth and development. Children who skip breakfast typically do not make up the deficits at later meals—they simply have lower intakes of energy, vitamins, and minerals than those who eat breakfast. Without breakfast, children perform poorly

in tasks requiring concentration, their attention spans are shorter, and they even score lower on intelligence tests than their well-fed peers. Malnourished children are particularly vulnerable. Common sense dictates that it is unreasonable to expect anyone to learn and perform without fuel.

Iron Deficiency and Behavior Iron deficiency has well-known and widespread effects on children's behavior and intellectual performance.[51] In addition to carrying oxygen in the blood, iron transports oxygen within cells, which use it for energy metabolism. Iron is also used to make neurotransmitters—most notably, those that regulate the ability to pay attention, which is crucial to learning. Consequently, iron deficiency not only causes an energy crisis but also directly impairs attention span and learning ability.

Iron deficiency is often diagnosed by a quick, easy, inexpensive hemoglobin or hematocrit test that detects a deficit of iron in the *blood*. A child's *brain*, however, is sensitive to low iron concentrations long before the blood effects appear. Iron deficiency lowers the "motivation to persist in intellectually challenging tasks" and impairs overall intellectual performance. Anemic children perform poorly on tests and are disruptive in the classroom; iron supplementation improves learning and memory. When combined with other nutrient deficiencies, iron-deficiency anemia has synergistic effects that are especially detrimental to learning. Furthermore, children who had iron-deficiency anemia *as infants* continue to perform poorly as they grow older, even if their iron status improves.[52] The long-term damaging effects on mental development make prevention and treatment of iron deficiency during infancy and early childhood a high priority.

Healthy, well-nourished children are alert in the classroom and energetic at play.

Other Nutrient Deficiencies A child with any of several nutrient deficiencies may be irritable, aggressive, and disagreeable, or sad and withdrawn. Such a child may be labeled "hyperactive," "depressed," or "unlikable," when in fact these traits may be due to simple, even marginal, malnutrition. Though parents and medical practitioners often overlook the possibility that malnutrition may account for these abnormalities, any departure from normal, healthy appearance and behavior is a sign of possible poor nutrition. In any such case, inspection of the child's diet by a registered dietitian or other qualified health care professional is in order. Any suspicion of dietary inadequacies, *no matter what other causes may be implicated*, should prompt steps to correct those inadequacies immediately.

Lead Poisoning in Children

Children who are malnourished are vulnerable to lead poisoning. They absorb more lead if their stomachs are empty; if they have low intakes of calcium, zinc, vitamin C, or vitamin D; and, of greatest concern because it is so common, if they have an iron deficiency. Iron deficiency weakens the body's defenses against lead absorption, and lead poisoning can cause iron deficiency.[53] Common to both iron deficiency and lead poisoning are a low socioeconomic background and a lack of immunizations against infectious diseases. Another common factor is pica—a craving for nonfood items. Many children with lead poisoning eat dirt or chips of old paint, two common sources of lead.

The anemia brought on by lead poisoning may be mistaken for a simple iron deficiency and therefore may be incorrectly treated. Like iron deficiency, mild lead toxicity has nonspecific symptoms, including diarrhea, irritability, and fatigue. Adding

Old lead-based paint can threaten the health of an exploring child.

- Ask a pediatrician whether your child should be tested for lead poisoning.
- Clean floors, window frames, window sills, and other surfaces regularly. Use a mop or sponge with warm water and a general all-purpose cleaner.
- Prevent children from putting dirty or old painted objects in their mouths, and make sure children wash their hands before eating.
- Feed children balanced, timely meals with ample iron and calcium.
- Wash children's bottles, pacifiers, and toys often.
- Do not use lead-contaminated water to make infant formula.
- Have your water tested for lead. If the cold water hasn't been used for more than a few hours, let it run for 15 to 30 seconds before drinking it.
- Be aware that other countries do not have the same regulations protecting consumers against lead. Children have been poisoned by eating crayons made in China, putting toys from China in their mouth, and drinking fruit juice canned in Mexico.

Reminder: *Antigens* are substances foreign to the body that elicit the formation of antibodies or an inflammation reaction from immune system cells. Food antigens are usually glycoproteins (large proteins with glucose molecules attached).

Reminder: *Antibodies* are large proteins that are produced in response to antigens and then inactivate the antigens.

iron to the diet does not reverse the symptoms; exposure to lead must stop and treatment for lead poisoning must begin. With further exposure, the symptoms become more pronounced, and children develop learning disabilities and behavior problems. Still more severe lead toxicity can cause irreversible nerve damage, paralysis, mental retardation, and death.

More than 300,000 children in the United States—most of them under age six—have blood lead concentrations high enough to cause mental, behavioral, and other health problems.[54] Lead toxicity in young children comes from their own behaviors and activities—putting their hands in their mouths, playing in dirt and dust, and chewing on nonfood items.[55] Unfortunately, the body readily absorbs lead during times of rapid growth and hoards it possessively thereafter. Lead is not easily excreted and accumulates mainly in the bones but also in the brain, teeth, and kidneys. Tragically, a child's neuromuscular system is also maturing during these first few years of life. No wonder children with elevated lead levels experience impairment of balance, motor development, and the relaying of nerve messages to and from the brain. Deficits in intellectual development are only partially reversed when lead levels decline.

Federal laws mandating reductions in leaded gasoline, lead-based solder, and other products over the past four decades have helped to reduce the amounts of lead in food and in the environment in the United States. As a consequence, the prevalence of lead toxicity in children has declined dramatically for most of the nation. Nevertheless, lead exposure is still a threat in certain communities. Strategies for defending children against lead toxicity are listed in the margin. The Environmental Protection Agency (EPA) offers many more tips to protect against lead poisoning (www.epa.gov/lead).

Food Allergy

Food allergy is frequently blamed for physical and behavioral abnormalities in children, but just 6 to 8 percent of children under four years of age are diagnosed with true food allergies.[56] Food allergies diminish with age, until, in adulthood, they affect less than 4 percent of the population. The prevalence of food allergy, especially peanut allergy, is on the rise, however.[57] Reasons for an increase in peanut allergy are not yet clear, but possible contributing factors include genetics, food preparation methods (roasting peanuts at very high temperatures makes them more allergenic), and exposure to medicinal skin creams containing peanut oil.[58]

A true food allergy occurs when fractions of a food protein or other large molecule are absorbed into the blood and elicit an immunologic response. (Recall that proteins are normally dismantled in the digestive tract to amino acids that are absorbed without such a reaction.) The body's immune system reacts to these large food molecules as it does to other antigens—by producing antibodies or other defensive agents.

Asymptomatic and Symptomatic Allergies Allergies may have one or two components. They always involve antibodies; they may or may not involve symptoms. A person may produce antibodies without having any symptoms (known as *asymptomatic allergy*) or may produce antibodies and have symptoms (known as *symptomatic allergy*). A person who experiences symptoms without producing antibodies, however, does not have an allergy. This means that allergies have to be diagnosed by testing for antibodies. Once a food allergy has been diagnosed, therapy requires strict elimination of the offending food. Children with allergies, like all children, need all of their nutrients, so it is important to include other foods that offer the same nutrients as the omitted foods.[59]

Immediate and Delayed Reactions Allergic reactions to food may be immediate or delayed. In either case, the antigen interacts immediately with the immune system,

but symptoms may appear within minutes or after several (up to 24) hours. Identifying the food that causes an immediate allergic reaction is easy because symptoms correlate closely with the time of eating the food. Identifying the food that caused a delayed reaction is more difficult because the symptoms may not appear until a day after the offending food was eaten; by this time, many other foods will have been eaten, too, complicating the picture. Allergic reactions to single foods are common. Reactions to multiple foods are the exception, not the rule.

Anaphylactic Shock The life-threatening food allergy reaction of **anaphylactic shock** is most often caused by peanuts, tree nuts, milk, eggs, wheat, soybeans, fish, or shellfish. Among these foods, eggs, milk, soy, and peanuts most often cause problems in children.[60] Children are more likely to outgrow allergies to eggs, milk, and soy than allergies to peanuts. Peanuts cause more life-threatening reactions than do all other food allergies combined. Research is currently under way to help those with peanut allergies tolerate small doses, thus saving lives and minimizing reactions.[61]

Families of children with a life-threatening food allergy and school personnel who supervise them must guard them against any exposure to the allergen. Parents of allergic children can pack safe foods for lunches and snacks and ask school officials to strictly enforce a "no swapping" policy in the lunchroom. The child must learn to identify which foods pose a problem and then learn and use refusal skills for all foods that may contain the allergen. The child must be able to recognize the symptoms of impending anaphylactic shock. Any person with food allergies severe enough to cause anaphylactic shock should wear a medical alert bracelet or necklace. Finally, the responsible child and the school staff should be prepared to administer injections of **epinephrine,** which prevents anaphylaxis after exposure to the allergen. Many preventable deaths occur each year when people with food allergies accidentally ingest the allergen but have no epinephrine.

Technology may soon offer new solutions. New drugs are being developed that may interfere with the immune response that causes allergic reactions.[62] Also, through genetic engineering, scientists may one day banish allergens from peanuts, soybeans, and other foods to make them safer.

Food Labeling Food labels must list the presence of common allergens in plain language, using the names of the eight most common allergy-causing foods.[63] For example, a food containing "textured vegetable protein" must say "soy" on its label. Similarly, "casein" must be identified as "milk," and so forth. Food producers must also prevent cross-contamination during production and clearly label the foods in which it is likely to occur. For example, equipment used for making peanut butter must be scrupulously clean before being used to pulverize cashew nuts for cashew butter to protect unsuspecting cashew butter consumers from peanut allergens.

Other Adverse Reactions to Foods Not all **adverse reactions** to foods are food allergies, although even physicians may describe them as such. Signs of adverse reactions to foods include stomachaches, headaches, rapid pulse rate, nausea, wheezing, hives, bronchial irritation, coughs, and other such discomforts. Among the causes may be reactions to chemicals in foods, such as the flavor enhancer monosodium glutamate (MSG), the natural laxative in prunes, or the mineral sulfur; digestive diseases, obstructions, or injuries; enzyme deficiencies, such as lactose intolerance; and even psychological aversions. These reactions involve symptoms but no antibody production. Therefore, they are **food intolerances,** not allergies.

Pesticides on produce may also cause adverse reactions. Pesticides that were applied in the fields may linger on the foods. Health risks from pesticide exposure may be low for healthy adults, but children are vulnerable. Therefore, government agencies have set a **tolerance level** for each pesticide by first identifying foods that children

These eight normally wholesome foods—milk, shellfish, fish, peanuts, tree nuts, eggs, wheat, and soybeans (and soy products)—may cause life-threatening symptoms in people with allergies.

Symptoms of impending anaphylactic shock:

- Tingling sensation in mouth
- Swelling of the tongue and throat
- Irritated, reddened eyes
- Difficulty breathing, asthma
- Hives, swelling, rashes
- Vomiting, abdominal cramps, diarrhea
- Drop in blood pressure
- Loss of consciousness

food allergy: an adverse reaction to food that involves an immune response; also called *food-hypersensitivity reactions.*

anaphylactic (AN-ah-feh-LAC-tic) **shock:** a life-threatening whole-body allergic reaction to an offending substance.

epinephrine: one of the stress hormones secreted whenever emergency action is needed, prescribed therapeutically to relax the bronchioles during allergy or asthma attacks.

adverse reactions: unusual responses to food (including intolerances and allergies).

food intolerances: adverse reactions to foods or food additives that do not involve the immune system.

tolerance level: the maximum amount of residue permitted in a food when a pesticide is used according to the label directions.

commonly eat in large amounts and then considering the effects of pesticide exposure during each developmental stage.

Food Dislikes Parents are advised to watch for signs of food dislikes and take them seriously. Children's food aversions may be the result of nature's efforts to protect them from allergic or other adverse reactions. Test for allergies, and then apply nutrition knowledge conscientiously in deciding how to alter the diet.

Hyperactivity

Hyperactivity affects behavior and learning in about 5 to 10 percent of young school-aged children.[64] Left untreated, it can interfere with a child's social development and ability to learn. Treatment focuses on relieving the symptoms and controlling the associated problems. Physicians often manage hyperactivity through behavior modification, special educational techniques, and psychological counseling. In many cases, they prescribe medication.[65]

Parents of hyperactive children often blame sugar as the cause and mistakenly believe that simply eliminating candy and other sweet treats will solve the problem. Despite such speculation, studies have consistently found no convincing evidence that sugar causes hyperactivity or worsens behavior.

Food additives have also been blamed for hyperactivity and other behavior problems in children, but scientific evidence to substantiate the connection has been elusive—until recently. A well-controlled study of close to 300 children suggests that food additives such as artificial colors or sodium benzoate preservative (or both) exacerbate hyperactive symptoms such as inattention and impulsivity.[66] Additional studies are needed to confirm the findings and to determine which additives studied might be responsible for negative behaviors.

Children can become excitable, rambunctious, and unruly as a result of a desire for attention, lack of sleep, overstimulation, watching too much television or playing too many computer games, too much caffeine from colas or chocolate, or a lack of physical activity. Such behaviors may suggest that more consistent care is needed. It helps to insist on regular hours of sleep, regular mealtimes, and regular outdoor activity. The accompanying Case Study offers an opportunity to think about these issues in relation to a specific child.

IN SUMMARY

▌ Children's appetites and nutrient needs reflect their stage of growth.

▌ Long-term hunger and malnutrition impair growth and health.

▌ Short-term hunger exerts more subtle effects on children's health and behavior—such as poor academic performance.

▌ Iron deficiency is widespread and has many physical and behavioral consequences.

▌ Lead toxicity is prevalent among young children and can have irreversible effects on health and behavior.

▌ True food allergies are somewhat rare in children, and children can outgrow some food allergies.

▌ Some allergies, however, can cause dangerous, life-threatening reactions in both children and adults.

▌ "Hyper" behavior is not caused by poor nutrition; misbehavior may reflect inconsistent care.

Freddie Willis is a six-year-old boy who seldom sits still, often misbehaves, and is frequently sick. Freddie's eating habits are erratic and poor, as is his appetite. He often misses breakfast because he is too tired to get up in time to eat before school. By midmorning, Freddie is irritable and disruptive in the classroom. At lunchtime, he trades the peanut butter and banana sandwich his mother packed in his lunchbox for a piece of cake. After school, he hurries home to watch television while he eats his favorite snack—cola and potato chips. At dinnertime, Freddie picks at his food because he isn't very hungry. Later on, when it's time for bed, Freddie complains that he's hungry. His parents let him stay up to have a bowl of cereal (the kind with marshmallows) before he finally falls asleep.

1. What factors in Freddie's daily routine might be contributing to his restless behavior?

2. Discuss some changes in diet that might improve Freddie's health and disposition.

Childhood Obesity

The number of overweight children has increased dramatically over the past three decades.[67] Like their parents, children in the United States are becoming fatter. An estimated 32 percent of U.S. children and adolescents 2 to 19 years of age are overweight, and 17 percent are obese.[68] Based on data from the BMI-for-age growth charts, children and most adolescents are categorized as *overweight* above the 85th percentile and as *obese* at the 95th percentile and above.[69] Because BMI at the 95th percentile for older adolescents is higher than 30, the adult obesity cutoff point, obesity in young people is defined as BMI at the 95th percentile or BMI of 30 or greater, whichever is lower. For children younger than 2 years of age, BMI values are not available. For this age group, weight-for-height values above the 95th percentile are classified as overweight. Figures E-5A and E-5B in Appendix E present the BMI-for-age percentiles for children and adolescents.

Genetic and Environmental Factors Parental obesity predicts an early increase in a young child's BMI, and it more than doubles the chances that a young child will become an obese adult. Children with neither parent obese have a less than 10 percent chance of becoming obese in adulthood, whereas overweight teens with at least one obese parent have a greater than 80 percent chance of being obese adults. The chances of an obese child becoming an obese adult grow greater as the child grows older.[70] The link between parental and child obesity reflects both genetic and environmental factors (as described in Chapter 7).

Diet and physical inactivity must also play a role in explaining why children are heavier today than they were 30 or so years ago. The prevalence of childhood obesity throughout the United States has more than doubled for young children and more than tripled for children 6 to 11 years of age and for adolescents; meanwhile, the society our children live in has changed considerably.[71] In many families today, both parents work outside the home and work longer hours; more emphasis is placed on convenience foods and foods eaten away from home; meal choices at school are more diverse and often less nutritious; sedentary activities such as watching television and playing computer games occupy much of children's free time; and opportunities for physical activity and outdoor play both during and after school have declined.[72] All of these factors—and many others—influence children's eating and activity patterns.

Children learn food behaviors from their families, and research confirms the significant roles parents play in teaching their children about healthy food choices, providing nutrient-dense foods, and serving as role models.[73] When parents eat fruits and vegetables frequently, their children do, too.[74] The more fruits and vegetables children eat, the more vitamins, minerals, and fiber, and the less saturated fat, in their diets.[75]

hyperactivity: inattentive and impulsive behavior that is more frequent and severe than is typical of others of a similar age; professionally called **attention deficit/ hyperactivity disorder (ADHD).**

Chapter 2 points out that as the prevalence of obesity among both children and adults has surged over the past few decades, so has the consumption of added sugars and, especially, high-fructose corn syrup—the easily consumed, energy-dense liquid sugar added to soft drinks.[76] Research shows that soft drink consumption is associated with increased energy intake and body weight.[77]

No doubt, the tremendous increase in soft drink consumption plays a role, but much of the obesity epidemic can be explained by lack of physical activity. Children have become more sedentary, and sedentary children are more often overweight.[78] Television watching may contribute most to physical inactivity. Excessive television time is linked with overweight in children.[79] A child who spends more than an hour or two each day in front of a television, computer monitor, or other media can become overweight even while eating fewer kcalories than a more active child.

Children who have television sets in their bedrooms spend more time watching TV, spend less time being physically active, and are more likely to be overweight than children who do not have televisions in their rooms.[80] Children who watch a great deal of television are most likely to be overweight and least likely to eat family meals or fruits and vegetables.[81] They often snack on the nutrient-poor, energy-dense foods that are advertised.[82] The average child sees an estimated 40,000 TV commercials a year—many peddling foods high in sugar, saturated fat, and salt such as sugar-coated breakfast cereals, candy bars, chips, fast foods, and carbonated beverages.[83] More than half of all food advertisements are aimed specifically at children and market their products as fun and exciting.[84]

Simply reducing the amount of time spent watching television (and playing computer games) can improve a child's BMI. The American Academy of Pediatrics (AAP) now recommends no television viewing before two years of age and thereafter limiting television and video time to two hours per day as a strategy to help prevent childhood obesity.[85]

Physical Health Like overweight adults, overweight children display a blood lipid profile indicating that atherosclerosis is beginning to develop: high levels of total cholesterol, triglycerides, and LDL cholesterol. Overweight children also tend to have high blood pressure; in fact, obesity is a leading cause of pediatric hypertension.[86] Their risks for developing type 2 diabetes and respiratory diseases (such as asthma) are also exceptionally high.[87] These relationships between childhood obesity and chronic diseases are discussed fully in Nutrition in Practice 12.

Psychological Development In addition to the physical consequences, childhood obesity brings a host of emotional and social problems.[88] Because people frequently judge others on appearance more than on character, overweight children are often victims of prejudice. Many suffer discrimination by adults and rejection by their peers. They may have poor self-images, a sense of failure, and a passive approach to life.

Prevention and Treatment of Obesity Medical science has worked wonders in preventing or curing many of even the most serious childhood diseases, but obesity remains a challenge.[89] Once excess fat has been stored, it is challenging to lose. In light of all this, parents are encouraged to make major efforts to prevent childhood obesity, starting at birth, or to begin treatment early—before adolescence.[90] The Expert Committee of the American Medical Association recommends specific eating and physical activity behaviors to prevent obesity for all children (see Table 12-8).

Weight-loss programs that involve parents and other caregivers in treatment report greater success than those without parental involvement. Because obesity in parents and their children tends to be positively correlated, both benefit when parents participate in a weight-loss program. The main goal of obesity treatment is to improve long-term physical health through permanent healthy lifestyle habits.[91] The most

TABLE 12-8 Recommended Eating and Physical Activity Behaviors to Prevent Obesity

The Expert Committee of the American Medical Association recommends the following healthy habits for children 2 to 18 years of age to help prevent childhood obesity:

- Limit consumption of sugar-sweetened beverages, such as soft drinks and fruit-flavored punches.
- Eat the recommended amounts of fruits and vegetables every day (2 to 4.5 cups per day based on age).
- Learn to eat age-appropriate portions of foods.
- Eat foods low in energy density such as those high in fiber and/or water and modest in fat.
- Eat a nutritious breakfast every day.
- Eat a diet rich in calcium.
- Eat a diet balanced in recommended proportions for carbohydrate, fat, and protein.
- Eat a diet high in fiber.
- Eat together as a family as often as possible.
- Limit the frequency of restaurant meals.
- Limit television watching or other screen time to no more than two hours per day, and do not have televisions or computers in sleeping areas.
- Engage in at least 60 minutes of moderate to vigorous physical activity every day.

Source: S. E. Barlow, Expert Committee recommendations regarding the prevention, assessment, and treatment of child and adolescent overweight and obesity: Summary report, *Pediatrics* 120 (2007): S164–S192. Used by permission.

successful approach integrates diet, physical activity, psychological support, and behavioral changes.[92] In contrast to traditional weight-loss programs that focus on *what* to eat, behavioral programs focus on *how* to eat.

As a first step, the Expert Committee recommends that overweight and obese children and their families adopt the same healthy eating and activity behaviors presented in Table 12-8 for obesity prevention. The goal for overweight and obese children is to improve BMI. If the child's BMI does not improve after several months, the Expert Committee recommends increasing the intensity of the treatment. The level of intensity depends on treatment response, age, degree of obesity, health risks, and the family's readiness to change. Advanced treatment involves close follow-up monitoring by a health care provider and greater support and structure for the child.[93]

Diet The initial goal for overweight children is to reduce the rate of weight gain, that is, to maintain weight as the child grows taller. Continued growth will then accomplish the desired change in BMI. Weight loss is usually not recommended because diet restriction can interfere with growth and development. Intervention for some overweight children with accompanying medical conditions may warrant weight loss, but this treatment requires an individualized approach based on the degree of overweight and severity of the medical conditions.[94] Dietary strategies begin with those listed in Table 12-8 and progress to more structured family meal plans when necessary.

Physical Activity The many benefits of physical activity are well known but often are not enough to motivate overweight people, especially children. Yet regular vigorous activity can improve a child's weight, body composition, and physical fitness.[95] The *Physical Activity Guidelines for Americans 2008* make specific recommendations for children that can be used as a guide for how much activity children need. Opportunities to be physically active can include team, individual, and recreational activities (see Figure 12-7 on p. 338). Physical activity is a natural and lifelong behavior of healthy living. The AAP supports the efforts of schools to include more physical activity in the curriculum and encourages parents to support their children's participation.[96]

Physical Activity Guidelines for Americans 2008
- **Children and adolescents should engage in 60 minutes (1 hour) or more of physical activity daily.**
- **Aerobic: Most of the 60 or more minutes a day should be either moderate- or vigorous-intensity aerobic activity (bicycle riding, running) and should include vigorous-intensity physical activity at least three days a week.**
- **Muscle strengthening: As part of their 60 or more minutes of daily physical activity, children and adolescents should include muscle-strengthening physical activity (rope climbing, sit-ups) on at least three days of the week.**
- **Bone strengthening: As part of their 60 or more minutes of daily physical activity, children and adolescents should include bone-strengthening (jumping rope, gymnastics) physical activity on at least three days of the week.**

Preschoolers (2 to 5 years)	Older children (6 to 12 years)

Games in the yard or park
Family walks after dinner
Playing with the dog
Dancing freestyle
Tumbling and gymnastics
T-ball
Playing catch
Family bike rides
Building a snowman
Family swimming at the pool
    or beach
Playing hide and seek
	Throwing a Frisbee
	Jumping rope
	Bicycling
	Playing games and sports such as soccer,
	    softball, baseball, and basketball
	Rollerblading
	Running
	Weight training with light weights
	Dancing
	Competitive swimming
	Snowboarding or skiing
	Family kayaking, canoeing, or surfing

FIGURE 12-7 Physical Activity Pyramid for Kids

Drugs The use of weight-loss drugs to treat obesity in children merits special concern because the long-term effects of these drugs on growth and development have not been studied.[97] The drugs may be used in addition to structured lifestyle changes for carefully selected children or adolescents who are at high risk for severe obesity in adulthood. Only two obesity drugs, orlistat and sibutramine (see Chapter 7), have been approved for limited use in children and adolescents.

Surgery The use of surgery to treat severe obesity in adults (see Chapter 7) has created interest in surgery for obese adolescents. Limited research shows that, after surgery, extremely obese adolescents lose significant weight and experience improvements in type 2 diabetes and cardiovascular risk factors.[98] The selection criteria for surgery to treat obesity in adolescents are based on recommendations of a panel of pediatricians and surgeons.[99]

Surgery may be an option for adolescents who meet the following criteria:

▮ **Have reached physical maturity**

▮ **BMI ≥50, or BMI ≥40 with significant weight-related health problems**

▮ **Have experienced failure in a formal, six-month weight-loss program**

▮ **Are capable of adhering to the long-term lifestyle changes required after surgery**

IN SUMMARY

▮ Childhood obesity has become a major health problem.

▮ Genetics, diet, and physical inactivity play roles in childhood obesity.

▮ Childhood obesity can impair physical and psychological health.

▮ The main goal of obesity treatment in children is to improve long-term physical health through permanent healthy lifestyle habits.

Mealtimes at Home

The childhood years are the parents' last chance to influence their children's food choices. Parents who want to promote nutritious choices and healthful habits provide access to nutrient-dense, delicious foods and opportunities for active play at home. Food choices and regular physical activity can not only promote healthy growth but, as mentioned earlier, can also help prevent the degenerative diseases of later life. Because the interactions between parents and children can set the stage for lifelong attitudes and habits, a child's preferences should be treated with respect, even when nutrient needs must take precedence.

Honoring Children's Preferences Researchers attempting to explain children's food preferences encounter many contradictions. Children say they like colorful foods, yet most often they reject green and yellow vegetables while favoring brown

338

peanut butter and white potatoes, apple wedges, and bread. They do like raw vegetables better than cooked ones, however, so it is wise to offer vegetables that are raw or slightly undercooked, crunchy, and bright in color. They should be warm, not hot, because a child's mouth is much more sensitive than an adult's. The flavor should be mild (a child has more taste buds), and smooth foods such as mashed potatoes or pea soup should have no lumps (a child wonders, with some disgust, what the lumps might be).

Young children like to eat at little tables and to be served little portions of food. They also love to eat with other children and have been observed to stay at the table longer and eat more food when in the company of their peers. Parents who serve food in a relaxed and casual manner, without anxiety, provide an environment in which a child's negative emotions will be minimized.

Eating is more fun when friends are there.

Avoiding Power Struggles Problems over food often arise during the second or third year, when children begin asserting their independence. Many of these problems stem from the conflict between children's developmental stages and capabilities and parents who, in attempting to do what they think is best for their children, try to control every aspect of eating. Such conflicts can disrupt children's abilities to regulate their own food intakes or to determine their own likes and dislikes. For example, many people share the misconception that children must be persuaded or coerced to try new foods. In fact, the opposite is true. When children are forced to try new foods, especially when offered rewards for eating a particular food (if you eat your vegetables, you can watch TV), they are less likely to try those foods again than are children who are left to decide for themselves. Similarly, when children are restricted from eating their favorite foods, they are more likely to want those foods.[100] As dietitian and family therapist Ellyn Satter notes, the parent is responsible for *what* the child is offered to eat, but the child is responsible for *how much* and even *whether* to eat.

When introducing new foods at the table, parents are advised to offer them one at a time and only in small amounts at first. The more often a food is presented to a young child, the more likely the child will accept that food.[101] Between 5 and 10 exposures to a new food are necessary before a toddler shows an enhanced preference for the food. Offer the new food at the beginning of the meal, when the child is hungry, and allow the child to make the decision to accept or reject it.

Choking Prevention Parents must always be alert to the dangers of choking. A choking child is silent, so an adult should be present whenever a child is eating. Make sure the child sits when eating; choking is more likely when a child is running or falling (see p. 325 for a list of foods and nonfood items most likely to cause choking).

Play First Ideally, each meal is preceded, not followed, by the activity the child looks forward to the most. A number of schools have discovered that children eat a much better lunch if it is served after, rather than before, recess. Otherwise, children eat hurriedly so that they can go play.

Child Participation Allowing children to help plan and prepare the family's meals provides enjoyable learning experiences and encourages children to eat the foods they have prepared. Fresh vegetables can provide opportunities for children to learn about color, growing things and their seeds, and shapes and textures—all of which are fascinating to young children. Measuring, stirring, decorating, and arranging foods are skills that even a very young child can practice with enjoyment and pride (see Table 12-9).

TABLE 12-9 Food Skills of Preschool Children[a]

Age one to two years, when large muscles develop:

- Uses a spoon
- Helps feed self
- Lifts and drinks from a cup
- Helps scrub fruits and vegetables, tear lettuce or greens, snap green beans, or dip foods
- Wipes tables
- Places empty paper and plastic containers in recycle bins

Age three years, when medium hand muscles develop:

- Spears food with a fork
- Feeds self independently
- Adds ingredients to pancake batters, cookie recipes, salads, or other mixed dishes
- Helps wrap, pour, mix, shake, stir, or spread foods
- Helps crack nuts with supervision

Age four years, when small finger muscles develop:

- Uses all utensils and napkin
- Helps roll, juice, or mash foods
- Helps measure dry ingredients
- Cracks egg shells
- Helps make sandwiches and toss salads
- Peels foods such as hard-boiled eggs and bananas

Age five years, when fine coordination of fingers and hands develops:

- Measures liquids
- Helps grind, grate, and cut (soft foods with dull knife)
- Uses hand mixer with supervision

[a]These ages are approximate. Healthy, normal children develop at their own pace.

Source: Adapted from MyPyramid for Preschoolers, Kitchen activities, available at www.mypyramid.gov

Snacks Parents may find that their children often snack so much that they aren't hungry at mealtimes. Instead of teaching children *not* to snack, teach them *how* to snack. Provide snacks that are as nutritious as the foods served at mealtime. Snacks can even be mealtime foods that are served individually over time, instead of all at once on one plate. When providing snacks to children, think of the food groups and offer such snacks as pieces of cheese, sliced strawberries, cooked baby carrots, and egg salad on whole-wheat crackers (see Table 12-10).

Preventing Dental Caries Children frequently snack on sticky, sugary foods that stay on the teeth and provide an ideal environment for the growth of bacteria that cause dental caries. Teach children to brush and floss after meals, to brush or rinse after eating snacks, to avoid sticky foods, and to select crisp or fibrous foods frequently.

Serving as Role Models In an effort to practice these many tips, parents may overlook perhaps the single most important influence on their children's food habits—themselves.[102] Parents who do not eat oranges should not be surprised when their children refuse to eat oranges. Likewise, parents who dislike the smell of brussels sprouts may not be able to persuade children to try them. Children learn much through imitation. Parents, older siblings, and other caregivers set an irresistible example by sitting with younger children, eating the same foods, and having pleasant conversations during mealtime.

Nutrition at School

While parents are doing what they can to establish good eating habits in their children at home, others are preparing and serving foods to their children at day-care centers and schools. In addition, children begin learning about food and nutrition in the classroom. Meeting the nutrition and education needs of children is critical to supporting their healthy growth and development.[103]

The U.S. government funds programs to provide nutritious, high-quality meals for children at school. Both the School Breakfast Program and the National School Lunch Program provide meals for free or at reduced cost to children from low-income families, and at a reasonable cost to other children.

School Breakfast The School Breakfast Program is available in more than 80 percent of the nation's schools that offer school lunch, and 10 million children participate in it.[104] Nevertheless, many children who need the School Breakfast Program either do not have access to the program or do not participate in it.[105] The majority of children who eat school breakfasts are from low-income families. As research results continue to emphasize the positive impact breakfast has on school performance and health, vigorous campaigns to expand school breakfast programs are under way.[106]

School Lunch More than 30 million children receive lunches through the National School Lunch Program—more than half of them free or at a reduced price.[107] School lunches offer a variety of food choices and are designed to provide at least a third of the recommendation for energy, protein, vitamin A, vitamin C, iron, and calcium. In an effort to help reduce cardiovascular disease risk, all government-funded meals served at schools must follow the *Dietary Guidelines for Americans*.

Parents often rely on school lunches to meet a significant part of their children's nutrient needs on school days. Indeed, students who regularly eat school lunches have higher intakes of many nutrients and fiber than students who do not.[108] Children don't always like what they are served, however, and school lunch programs must strike a balance between what children want to eat and what will nourish them and guard their health.

The school breakfast must contain, at a minimum:

- One serving of fluid milk.
- One serving of fruit or vegetable or full-strength juice.
- Two servings of bread or bread alternatives, or two servings of meat or meat alternatives, or one of each.

The American Dietetic Association has set nutrition standards for child-care programs. Among them, meal plans should:

- Be nutritionally adequate and consistent with the *Dietary Guidelines for Americans*.
- Involve parents in planning.
- Follow recommended meal patterns that balance energy and nutrients with children's ages, appetites, activity levels, and special needs while respecting cultural and ethnic differences.
- Minimize added fat, sugar, and sodium.
- Emphasize fresh fruit, fresh and frozen vegetables, and whole grains.
- Provide furniture and eating utensils that are age appropriate and developmentally suitable to encourage children to accept and enjoy mealtime.

Source: Position of the American Dietetic Association: Benchmarks for nutrition programs in child care settings, *Journal of the American Dietetic Association* 105 (2005): 979–986.

Competing Influences at School Serving nutritious lunches is only half the battle; students need to eat them, too. Short lunch periods and long lines prevent some students from eating a school lunch and leave others with too little time to finish eating their meals.[109] Nutrition efforts at schools are also undermined when students can buy what the USDA labels "competitive foods"—meals from fast-food restaurants or à la carte foods such as pizza or snack foods and carbonated beverages from snack bars, school stores, and vending machines.[110] These foods and beverages compete with nutritious school lunches. Children receive a mixed message when they are left on their own to choose between the health-supporting school lunch and the high-fat, high-salt foods with low nutrient density that their taste buds may prefer.

Increasingly, school-based nutrition issues are being addressed by legislation. Some states restrict the sale of competitive foods and have higher rates of participation in school meal programs than the national average. Federal legislation mandates that all school districts that participate in the USDA's National School Lunch Program develop and put in place a local wellness policy.[111] By law, wellness policies must:[112]

▌ Set goals for nutrition education, physical activity, and other school-based activities.

▌ Establish nutrition guidelines for all foods available on school campuses during the school day.

▌ Develop a plan to measure policy implementation.

School districts across the nation have made progress toward meeting these goals, but implementation is inconsistent, and, because wellness policies are established locally, a great deal of variety exists among them. Some are well defined and detailed, whereas others are vague.[113] To enhance local wellness policies, standards for competitive foods and beverages served in schools spell out appropriate fat, saturated fat, kcalorie, sugar, and sodium contents.[114] Establishing and implementing these nutrition standards for competitive foods and beverages helps to ensure that all foods served in schools are consistent and comply with the *Dietary Guidelines for Americans*.

IN SUMMARY

▌ Adults at home and at school need to provide children with nutrient-dense foods and teach them how to make healthful choices.

▌ Adults also need to provide ample opportunity for children to be physically active.

Nutrition during Adolescence

As children pass through **adolescence** on their way to becoming adults, they change in many ways. Their physical changes make their nutrient needs high, and their emotional, intellectual, and social changes make meeting those needs a challenge.

Teenagers make many more choices for themselves than they did as children. They are not fed; they eat. Food choices made during the teen years profoundly affect health, both now and in the future. At the same time, social pressures thrust choices at them: whether to drink alcoholic beverages and whether to develop their bodies to meet extreme ideals of slimness or athletic prowess. Their interest in nutrition—both valid information and misinformation—derives from personal, immediate experiences. They are concerned with how diet can improve their lives now—they try the latest fad diet to fit into a new bathing suit, avoid greasy foods in an effort to clear acne, or eat a plate of pasta to prepare for a big sporting event.

TABLE 12-10 Healthful Snack Ideas—Think Food Groups, Alone and in Combination

Grain Products
▪ Cereal with fruit and milk
▪ Crackers and cheese
▪ Whole-grain toast with peanut butter
▪ Popcorn with grated cheese
▪ Oatmeal raisin cookies with milk

Vegetables
▪ Red bell pepper strips with hummus dip
▪ Sliced cherry tomatoes and mozzarella cheese bites
▪ Celery with peanut butter
▪ Broccoli, cauliflower, and carrot sticks with a flavored cottage cheese dip

Fruits
▪ Apples and cheese
▪ Bananas and peanut butter
▪ Peaches with yogurt
▪ Raisins mixed with sunflower seeds or nuts

Protein Foods
▪ Refried beans with nachos and cheese
▪ Tuna on crackers
▪ Luncheon meat on whole-grain bread

Milk and Milk Products
▪ Milk (as a beverage with any snack)
▪ Yogurt and cheese, alone or with other foods

adolescence: the period of growth from the beginning of puberty until full maturity. Timing of adolescence varies from person to person.

Growth and Development during Adolescence

With the onset of adolescence, the steady growth of childhood speeds up abruptly and dramatically, and the growth patterns of females and males become distinct. Hormones direct the intensity and duration of the adolescent growth spurt, profoundly affecting every organ of the body, including the brain. After two to three years of intense growth and a few more at a slower pace, physically mature adults emerge.

In general, the adolescent growth spurt begins at age 10 or 11 for females and at age 12 or 13 for males. It lasts about two and a half years. Before **puberty,** male and female body compositions differ only slightly, but during the adolescent spurt, differences between the genders become apparent in the skeletal system, lean body mass, and fat stores. In females, fat assumes a larger percentage of total body weight, and in males, the lean body mass—principally muscle and bone—increases much more than in females. On average, males grow 8 inches taller, and females, 6 inches taller. Males gain approximately 45 pounds, and females, about 35 pounds.

Energy and Nutrient Needs

The energy needs of adolescents vary greatly, depending on the current rate of growth, gender, body composition, and physical activity.[115] Boys' energy needs may be especially high; they typically grow faster than girls and, as mentioned, develop a greater proportion of lean body mass. An exceptionally active boy of 15 may need 3500 kcalories or more a day just to maintain his weight. Girls start growing earlier than boys and attain shorter heights and lower weights, so their energy needs peak sooner and decline earlier than those of their male peers. An inactive girl of 15 whose growth is nearly at a standstill may need fewer than 1800 kcalories a day if she is to avoid excessive weight gain. Thus, teenage girls need to pay special attention to being physically active and selecting foods of high nutrient density so that they will meet their nutrient needs without exceeding their energy needs.

Obesity The insidious problem of obesity becomes ever more apparent in adolescence and often continues into adulthood. The problem is most evident in African American females and in Hispanic children of both genders. The consequences of obesity are so dramatic, and our society's attitude toward obese people is so negative, that even healthy-weight or underweight teens may perceive a need to lose weight. When taken to extremes, restrictive diets bring dramatic physical consequences of their own, as Nutrition in Practice 7 explains.

Vitamin D Recommendations for most vitamins increase during the teen years (see the tables on the inside front cover). Several of the vitamin recommendations for adolescents are similar to those for adults, including the recently revised recommendations for vitamin D.[116] Vitamin D is essential for bone growth and development. Recent studies of vitamin D status in adolescents show that as many as half of adolescents may be vitamin D deficient.[117] Adolescents who do not receive enough vitamin D from fortified foods such as milk and dry cereals, or from sun exposure each day, should take a vitamin D supplement.[118]

Iron The need for iron increases during adolescence for both females and males—but for different reasons. Iron requirements increase for females as they start to menstruate and for males as their lean body mass develops. Hence, the RDA increases at age 14 for both males and females. Because menstruation continues throughout a woman's childbearing years, the RDA for iron remains high for women into late adulthood. For males, the RDA returns to preadolescent values in early adulthood.

Vitamin D RDA for adolescents:
■ 15 μg/day

In addition, iron needs increase when the adolescent growth spurt begins, whether that occurs before or after age 14. Therefore, boys in a growth spurt need an additional 2.9 milligrams of iron per day above the RDA for their age; girls need an additional 1.1 milligrams per day.

Furthermore, iron recommendations for girls before age 14 do not reflect the iron losses of menstruation, even though the average age of menarche (first menstruation) in the United States is 12.5 years. Therefore, for girls under the age of 14 who have started to menstruate, an additional 2.5 milligrams of iron per day is recommended. Thus, the RDA for iron depends not only on age and gender but also on whether the individual is in a growth spurt or has begun to menstruate, as listed in the margin.

Iron intakes often fail to keep pace with increasing needs, especially for females, who typically consume fewer iron-rich foods such as meat and fewer total kcalories than males. Not surprisingly, iron deficiency is most prevalent among adolescent girls. Iron-deficient children and teens score lower on standardized tests than those who are not iron deficient.

Calcium Adolescence is a crucial time for bone development, and the requirement for calcium reaches its peak during these years.[119] Unfortunately, low calcium intakes among adolescents have reached crisis proportions: 90 percent of females and 70 percent of males ages 12 to 19 years have calcium intakes below recommendations.[120] Low calcium intakes during times of active growth, especially if paired with physical inactivity, can compromise the development of peak bone mass, which is considered the best protection against adolescent fractures and adult osteoporosis. Increasing milk products in the diet to meet calcium recommendations greatly increases bone density.[121] Once again, however, teenage girls are most vulnerable, for their milk intakes—and therefore their calcium intakes—begin to decline at the time when their calcium needs are greatest. Furthermore, women have much greater bone losses than men in later life. In addition to dietary calcium, physical activity makes bones grow stronger. However, because some high schools do not require students to attend physical activity classes, many adolescents must make a point to be physically active during leisure time.

Food Choices and Health Habits

Teenagers like the freedom to come and go as they choose. They eat what they want if it is convenient and if they have the time. With a multitude of after-school, social, and job activities, they almost inevitably fall into irregular eating habits. At any given time, on any given day, a teenager may be skipping a meal, eating a snack, preparing a meal, or consuming food prepared by a parent or restaurant. Adolescents who frequently eat meals with their families, however, eat more fruits, vegetables, grains, and calcium-rich foods, and drink fewer soft drinks, than those who seldom eat with their families.[122]

Many adolescents also begin to skip breakfast on a regular basis, missing out on important nutrients that are not made up at later meals during the day. Teenagers who eat breakfast are therefore more likely to meet their nutrient recommendations.

Breakfast skipping may also lead to weight gain in adolescents. Research shows a dose-response, inverse relationship between breakfast eating and BMI.[123] As adolescents make the transition to adulthood, not only do they skip breakfast more often, they also eat fast food more often. Both skipping breakfast and eating fast foods lead to weight gain.[124]

Ideally, in light of adolescents' busy schedules and desire for freedom, the adult becomes a **gatekeeper,** controlling the type and availability of food in the teenager's environment. Teenage sons and daughters and their friends should find plenty of nutritious, easy-to-grab food in the refrigerator (meats for sandwiches; low-fat cheeses;

Iron RDA for males:
- 9–13 yr.: 8 mg/day
- 9–13 yr. in growth spurt: 10.9 mg/day
- 14–18 yr.: 11 mg/day
- 14–18 yr. in growth spurt: 13.9 mg/day

Iron RDA for females:
- 9–13 yr.: 8 mg/day
- 9–13 yr. in menarche: 10.5 mg/day
- 9–13 yr. in menarche and growth spurt: 11.6 mg/day
- 14–18 yr.: 15 mg/day
- 14–18 yr. in growth spurt: 16.1 mg/day

puberty: the period in life in which a person becomes physically capable of reproduction.

gatekeeper: with respect to nutrition, a key person who controls other people's access to foods and thereby exerts a profound impact on their nutrition. Examples are the spouse who buys and cooks the food, the parent who feeds the children, and the caregiver in a day-care center.

Nutritious snacks contribute valuable nutrients to an active teen's diet.

Appendix A provides a table of the caffeine contents of beverages, foods, and medications.

fresh, raw vegetables and fruits; fruit juices; and milk) and more in the cabinets (whole-grain breads, peanut butter, nuts, popcorn, and cereal). In many households today, all the adults work outside the home, and teenagers perform some of the gatekeeper's roles, such as shopping for groceries or choosing fast or prepared foods.

Snacks On average, about a fourth of an adolescent's total daily energy intake comes from snacks, which, if chosen carefully, can contribute some of the needed nutrients (see Table 12-10 on p. 341). A survey of more than 4000 adolescents found that those who ate snacks more often had higher intakes of fruit compared with those who ate snacks less often.[125] However, adolescents frequently choose foods that are too high in added sugars, saturated fat, and sodium and too low in fiber to support the future health of their arteries.[126]

Beverages Most frequently, adolescents drink soft drinks instead of fruit juice or milk with lunch, supper, and snacks. About the only time they select fruit juices is at breakfast. When teens drink milk, they are more likely to consume it with a meal (especially breakfast) than as a snack.

Soft drinks, when chosen as the primary beverage, may affect bone density, partly because they displace milk from the diet.[127] Over the past three decades, teens (especially girls) have been drinking more soft drinks and less milk. Adolescents who drink soft drinks regularly have a higher energy intake and a lower calcium intake than those who do not; they are also more likely to be overweight.[128]

Soft drinks containing caffeine present a different problem if caffeine intake becomes excessive. Caffeine seems to be relatively harmless, however, when used in moderate doses (the equivalent of fewer than, say, three 12-ounce cola beverages a day). In greater amounts, it can cause the symptoms associated with anxiety—sweating, tenseness, and inability to concentrate.

Eating Away from Home Adolescents eat about one-third of their meals away from home, and their nutritional welfare is enhanced or hindered by the choices they make. A lunch consisting of a hamburger, a chocolate shake, and French fries supplies substantial quantities of many nutrients at a kcalorie cost of about 800, an energy intake some adolescents can afford. When they eat this sort of lunch, teens can adjust their breakfast and dinner choices to include fruits and vegetables for vitamin A, vitamin C, folate, and fiber, and lean meats and legumes for iron and zinc. Fortunately, many fast-food restaurants are offering more nutritious choices than the standard hamburger meal.

Peer Influence Physical maturity and growing independence present adolescents with new choices. The consequences of those choices will influence their health and nutrition status both today and throughout life. Many of the food and health choices adolescents make reflect the opinions and actions of their peers. When others perceive milk as babyish, a teen may choose soft drinks instead; when others skip lunch and hang out in the parking lot, a teen may join in for the camaraderie, regardless of hunger. Some teenagers begin using drugs, alcohol, and tobacco; others wisely refrain. Adults can set up the environment so that nutritious foods are available and can stand by with reliable information and advice about health and nutrition, but the rest is up to the adolescents. Ultimately, they make the choices.

IN SUMMARY

▌ Nutrient needs rise dramatically as children enter the rapid growth phase of the teen years.

▌ The busy lifestyles of teenagers add to the challenge of meeting their nutrient needs, especially for iron and calcium.

344

Nutrition Assessment Checklist FOR INFANTS, CHILDREN, AND ADOLESCENTS

Medical History

Check the medical record for:

- Alcohol, tobacco, or illicit drug abuse
- Attention deficit/hyperactivity disorder (ADHD)
- Diabetes or other chronic disorders
- Eating disorders
- Food allergies
- Lactose intolerance
- Obesity
- Pregnancy

Medications

For children or adolescents being treated with drug therapy for medical conditions, note:

- Side effects that might reduce food intake or change nutrient needs
- Proper administration of medication with respect to food intake

Dietary Intake

For infants, note:

- Method of feeding (breastfeeding, formula, or both)
- Frequency and duration of breastfeeding
- Amount of infant formula
- Practice of putting infant to bed with bottle
- Solid foods the infant is fed, if any
- Amount of food the infant is fed

For all children and adolescents, especially those considered at risk nutritionally, assess the diet for:

- Total energy
- Protein
- Calcium and iron
- Vitamin A, vitamin C, and folate
- Fiber

Note the following:

- Number of days each week a nutritious breakfast is eaten
- Number of hours the child or teen sleeps each day

- Number of soft drinks the child or teen drinks each day
- Number of fast-food meals eaten each day
- Number and type of snacks eaten each day
- Type and amount of physical activity
- Amount of caffeine consumed

Anthropometric Data

Measure baseline height and weight.

- Note infant birth weight.
- Reassess height, weight, and growth patterns at each medical checkup.
- Note weight, length, and head circumference of infants.
- Note significant obesity or underweight and any intervention strategies employed.

Laboratory Tests

Monitor the following laboratory tests for infants:

- Blood glucose of infants born to mothers with gestational diabetes
- Results of tests for inborn errors of metabolism (see Nutrition in Practice 16)

Monitor the following laboratory tests for children and adolescents:

- Hemoglobin, hematocrit, or other tests of iron status
- Blood glucose for children or adolescents with diabetes
- Blood lead concentrations

Physical Signs

Look for physical signs of:

- Protein-energy malnutrition
- Iron deficiency
- Vitamin A deficiency
- Vitamin C deficiency
- Folate deficiency

Clinical Applications

1. At 2½ years old, Travis is healthy, though slightly underweight, and headstrong. Travis's mother hovers over him at every meal and insists that he take several bites of every food on his plate, even if he dislikes the food or is not familiar with it. Even though Travis is hungry when he sits down to a meal, his mother's constant urging to get him to eat quickly quells any interest he had in eating. Travis simply folds his arms across his chest, closes his mouth tightly, and refuses to eat any more food.

After more begging, pleading, and nagging, Travis's mother becomes angry and sends him away from the table. Travis is not allowed to snack between meals because his mother is concerned that snacks will ruin his appetite.

- What factors might be contributing to Travis's refusal to eat?

- Travis's mother is concerned about her son's underweight. What strategies would you suggest to help Travis gain weight?

- What advice would you offer Travis's mother to help her improve mealtimes with her son?

2. Loni is a physically inactive, overweight 15-year-old girl who enjoys watching television and playing computer games in her spare time. She typically buys a cola and bag of chips from the vending machine for lunch and then stops by a fast-food restaurant after school for chicken nuggets or a hamburger with French fries. She has noticed that she has gained weight and lacks energy.

- What nutrients might be excessive or lacking in her current diet plan?
- What dietary advice would you offer to help Loni look and feel healthier?
- What would you tell Loni to motivate her to become physically active?

Self Check

1. Breast milk is recommended for the first 12 months of life because it offers complete nutrition and ___ to the infant.
   a. fluoride
   b. fructose
   c. immunological protection
   d. pica

2. Which of the following is a characteristic of iron deficiency in children?
   a. It rarely develops in those with high intakes of milk.
   b. It affects brain function before anemia sets in.
   c. It is a primary factor in hyperactivity.
   d. Mild deficiency enhances mental performance by lowering physical activity level, thereby leading to increased attention span.

3. Three symptoms of lead toxicity are:
   a. diarrhea, irritability, and fatigue.
   b. low blood sugar, hair loss, and skin rash.
   c. increased heart rate, hyperactivity, and dry skin.
   d. bleeding gums, brittle fingernails, and swollen glands.

4. Allergic reactions to foods are most often caused by:
   a. corn, rice, or meats.
   b. eggs, peanuts, or milk.
   c. red meats, milk, or MSG.
   d. seafood, dark greens, or lactose.

5. When introducing new foods to children:
   a. reward children as they try new foods.
   b. offer many choices to encourage variety.
   c. offer one new food at the end of the meal.
   d. offer one new food at the beginning of the meal.

6. Which of the following is *not* true? Children who watch a lot of television are likely to:
   a. become obese.
   b. spend less time being physically active.
   c. learn healthy eating tips from programs.
   d. eat the foods most often advertised on television.

7. Which of the following strategies is *not* effective?
   a. Play first, eat later.
   b. Provide small portions.
   c. Encourage children to help prepare meals.
   d. Use dessert as a reward for eating vegetables.

8. During the growth spurt of adolescence:
   a. females gain more weight than males.
   b. males gain more fat, proportionately, than females.
   c. differences in body composition between males and females become apparent.
   d. similarities in body composition between males and females become apparent.

9. Two nutrients that are usually lacking in adolescents' diets are:
   a. zinc and fat.
   b. iron and calcium.
   c. protein and thiamin.
   d. vitamin A and riboflavin.

10. To help teenagers consume a balanced diet, parents can:
    a. monitor the teens' food intake.
    b. give up—parents can't influence teenagers.
    c. keep the cabinets and refrigerator well stocked.
    d. forbid snacking and insist on regular, well-balanced meals.

Answers to these questions can be found in Appendix H.

Notes

1. Formula feeding of term infants, in *Pediatric Nutrition Handbook,* 6th ed., ed. R. E. Kleinman (Elk Grove Village, IL: American Academy of Pediatrics, 2009), pp. 61–78.

2. Position of the American Dietetic Association: Promoting and supporting breastfeeding, *Journal of the American Dietetic Association* 109 (2009):

1926–1942; American Academy of Pediatrics, Policy statement: Breastfeeding and the use of human milk, *Pediatrics* 115 (2005): 496–506.

3. American Academy of Pediatrics, 2005.

4. L. Bode, Recent advances on structure, metabolism, and function of human milk

oligosaccharides, *Journal of Nutrition* 136 (2006): 2127–2130.

5. S. M. Donovan, Human milk oligosaccharides— The plot thickens, *British Journal of Nutrition* 101 (2009): 1267–1269.

6. S. M. Innis, Dietary (n-3) fatty acids and brain development, *Journal of Nutrition* 137 (2007): 855–859.

7. S. E. Carlson, Early determinants of development: A lipid perspective, *American Journal of Clinical Nutrition* 89 (2009): 1523S–1529S; K. Simmer, S. K. Patole, and S. C. Rao, Longchain polyunsaturated fatty acid supplementation in infants born at term, *Cochrane Database of Systematic Reviews* January 23, 2008, CD000376; S. M. Innis, 2007; M. S. Fewtrell, Long-chain polyunsaturated fatty acids in early life: Effects on multiple health outcomes, in *Primary Prevention by Nutrition Intervention in Infancy and Childhood,* eds., A. Lucas and H. A. Sampson, *Nestle Nutrition Workshop Series Pediatric Program* 57 (2006): 203–221.

8. K. Simmer, S. K. Patole, and S. C. Rao, 2008.

9. Fat-soluble vitamins, in *Pediatric Nutrition Handbook,* 6th ed., ed. R. E. Kleinman (Elk Grove Village, IL: American Academy of Pediatrics, 2009), pp. 461–474.

10. C. L. Wagner, F. R. Greer, and the Section on Breastfeeding and Committee on Nutrition, Prevention of rickets and vitamin D deficiency in infants, children, and adolescents, *Pediatrics* 122 (2008): 1142–1152.

11. Breastfeeding, in *Pediatric Nutrition Handbook,* 6th ed., R. E. Kleinman, ed. (Elk Grove Village, IL: American Academy of Pediatrics, 2009), pp. 29–59; Position of the American Dietetic Association, 2009.

12. K. Sadeharju and coauthors, Maternal antibodies in breast milk protect the child from enterovirus infections, *Pediatrics* 119 (2007): 941–946.

13. L. A. Hanson, Session 1: Feeding and infant development breast-feeding and immune function, *Proceedings of the Nutrition Society* 66 (2007): 384–396.

14. D. S. Newburg, G. M. Ruiz-Palacios, and A. L. Morrow, Human milk glycans protect infants against enteric pathogens, *Annual Review of Nutrition* 25 (2005): 37–58.

15. Breastfeeding, in *Pediatric Nutrition Handbook,* 2009; Position of the American Dietetic Association, 2009; L. A. Hanson, Session 1: Feeding and infant development breast-feeding and immune function, *Proceedings of the Nutrition Society* 66 (2007): 384–396.

16. F. R. Greer, S. H. Sicherer, A. W. Burks, and the Committee on Nutrition and Section on Allergy and Immunology, Effect of early nutritional interventions on the development of atopic disease in infants and children: The role of maternal dietary restriction, breastfeeding, timing of introduction of complementary foods, and hydrolyzed formulas, *Pediatrics* 121 (2008): 183–191.

17. F. R. Greer, S. H. Sicherer, A. W. Burks, and the Committee on Nutrition and Section on Allergy and Immunology, 2008.

18. F. R. Greer, S. H. Sicherer, A. W. Burks, and the Committee on Nutrition and Section on Allergy and Immunology, 2008; A. C. Krakowski and coauthors, Management of atopic dermatitis in the pediatric population, *Pediatrics* 122 (2008): 812–824.

19. D. A. Leon and G. Ronalds, Breast-feeding influences on later life—cardiovascular disease, *Advances in Experimental Medicine and Biology*

639 (2009): 153–166; C. G. Owen and coauthors, Does initial breastfeeding lead to lower blood cholesterol in adult life? A quantitative review of the evidence, *American Journal of Clinical Nutrition* 88 (2008): 305–314.

20. L. Shields and coauthors, Breastfeeding and obesity at 21 years: A cohort study, *Journal of Clinical Nursing* 19 (2010): 1612–1617; L. Schack-Nielsen and coauthors, Late introduction of complementary feeding, rather than duration breastfeeding, may protect against adult overweight, *American Journal of Clinical Nutrition* 91 (2010): 619–627; L. Twells and L. A. Newhook, Can exclusive breastfeeding reduce the likelihood of childhood obesity in some regions of Canada? *Canadian Journal of Public Health* 101 (2010): 36–39; P. Chivers and coauthors, Body mass index, adiposity rebound and early feeding in a longitudinal cohort (Raine Study), *International Journal of Obesity* 34 (2010): 1169–1176; A. S. Ryan, Breastfeeding and the risk of childhood obesity, *Collegium Antropologicum* 31 (2007): 19–28; S. Scholtens and coauthors, Breastfeeding, weight gain in infancy, and overweight at seven years of age: The prevention and incidence of asthma and mite allergy birth cohort study, *American Journal of Epidemiology* 165 (2007): 919–926; A. M. Toschke and coauthors, Infant feeding method and obesity: Body mass index and dual-energy X-ray absorptiometry measurements at 9-10 y of age from the Avon Longitudinal Study of Parents and Children (ALSPAC), *American Journal of Clinical Nutrition* 85 (2007): 1578–1585.

21. A. S. Ryan, 2007.

22. P. Chivers and coauthors, 2010; L. Shields and coauthors, 2010. T. Harder and coauthors, Duration of breastfeeding and risk of overweight: A meta-analysis, *Journal of Epidemiology* 162 (2005): 397–403.

23. L. Schack-Nielsen and K. F. Michaelsen, Advances in our understanding of the biology of human milk and its effects on the offspring, *Journal of Nutrition* 137 (2007): 503S–510S; G. Der, G. D. Batty, and I. J. Deary, Effect of breast feeding on intelligence in children; prospective study, sibling pairs analysis, and meta-analysis, *British Medical Journal* 333 (2006): 929–930.

24. Formula feeding of term infants, in *Pediatric Nutrition Handbook,* 2009.

25. Formula feeding of term infants, in *Pediatric Nutrition Handbook,* 2009.

26. Formula feeding of term infants, in *Pediatric Nutrition Handbook,* 2009.

27. Formula feeding of term infants, in *Pediatric Nutrition Handbook,* 2009.

28. Complementary feeding, in *Pediatric Nutrition Handbook,* 6th ed., ed. R. E. Kleinman (Elk Grove Village, IL: American Academy of Pediatrics, 2009), pp. 113–142.

29. Iron, in *Pediatric Nutrition Handbook,* 6th ed., ed. R. E. Kleinman (Elk Grove Village, IL: American Academy of Pediatrics, 2009), pp. 403–422.

30. Feeding the child, in *Pediatric Nutrition Handbook,* 6th ed., ed. R. E. Kleinman (Elk Grove Village, IL: American Academy of Pediatrics, 2009), pp. 145–174.

31. Feeding the child, in *Pediatric Nutrition Handbook,* 2009.

32. Complementary feeding, in *Pediatric Nutrition Handbook,* 2009; A. Fiocchi, A. Assa'ad, and S. Bahna, Food allergy and the introduction of solid foods to infants: A consensus document, *Annals of Allergy, Asthma and Immunology* 97 (2006): 10–21.

33. L. L. Moore and coauthors, Effects of average childhood dairy intake on adolescent bone health, *Journal of Pediatrics* 153 (2008): 667–673; K. S. Ondrak and D. W. Morgan, Physical activity, calcium intake and bone health in children and adolescents, *Sports Medicine* 37 (2007): 587–600.

34. M. K. Fox and coauthors, Relationship between portion size and energy intake among infants and toddlers: Evidence of self-regulation, *Journal of the American Dietetic Association* 106 (2006): S77–S83.

35. Position of the American Dietetic Association: Nutrition guidance for healthy children ages 2 to 11 years, *Journal of the American Dietetic Association* 108 (2008): 1038–1047; Position of the American Dietetic Association: Individual-, family-, school-, and community-based interventions for pediatric overweight, *Journal of the American Dietetic Association* 106 (2006): 925–945.

36. Nutritional aspects of vegetarian diets, in *Pediatric Nutrition Handbook,* 6th ed., ed. R. E. Kleinman (Elk Grove Village, IL: American Academy of Pediatrics, 2009), pp. 201–224.

37. Committee on the Scientific Evaluation of Dietary Reference Intakes, Food and Nutrition Board, Institute of Medicine, *Dietary Reference Intakes for Energy, Carbohydrate, Fiber, Fat, Fatty Acids, Cholesterol, Protein, and Amino Acids* (Washington, D.C.: National Academies Press, 2005), pp. 265–338.

38. Committee on the Scientific Evaluation of Dietary Reference Intakes, 2005, pp. 339–421.

39. Committee on the Scientific Evaluation of Dietary Reference Intakes, 2005, pp. 769–879.

40. Committee on the Scientific Evaluation of Dietary Reference Intakes, 2005, pp. 422–541.

41. J. M. Brotanek and coauthors, Iron deficiency in early childhood in the United States: Risk factors and racial/ethnic disparities, *Pediatrics* 120 (2007): 568–575.

42. Committee on Dietary Reference Intakes, *Dietary Reference Intakes for Calcium and Vitamin D* (Washington, D.C.: National Academies Press, 2011).

43. C. L. Wagner, F. R. Greer, and the Section on Breastfeeding and Committee on Nutrition, Prevention of rickets and vitamin D deficiency in infants, children, and adolescents, *Pediatrics* 122 (2008): 1142–1152.

44. Feeding the child, in *Pediatric Nutrition Handbook,* 6th ed., ed. R. E. Kleinman (Elk Grove Village, IL: American Academy of Pediatrics, 2009), pp. 145–174.

45. U. Shaikh, R. S. Byrd, and P. Auinger, Vitamin and mineral supplement use by children and adolescents in the 1999–2004 National Health and Nutrition Examination Survey: Relationship with nutrition, food security, physical activity, and health care access, *Archives of Pediatrics and Adolescent Medicine* 163 (2009): 150–157; R. Briefel and coauthors, Feeding Infants and Toddlers Study: Do vitamin and mineral

supplements contribute to nutrient adequacy or excess among US infants and toddlers? *Journal of the American Dietetic Association* 106 (2006): S52–S65.

46. Position of the American Dietetic Association, 2008.

47. M. Vadiveloo, L. Zhu, and P. A. Quatromoni, Diet and physical activity patterns of school-aged children, *Journal of the American Dietetic Association* 109 (2009): 145–151; Position of the American Dietetic Association, 2008.

48. Bread for the World, Hunger facts: Domestic, available at www.bread.org/learn/hunger-basics/hunger-facts-domestic.html, updated February, 2008; accessed January 23, 2009.

49. Position of the American Dietetic Association: Child and adolescent nutrition assistance programs, *Journal of the American Dietetic Association* 110 (2010): 791–799.

50. K. Widenhorn-Muller and coauthors, Influence of having breakfast on cognitive performance and mood in 13- to 20-year-old high school students: Results of a crossover trial, *Pediatrics* 122 (2008): 279–284.

51. J. C. McCann and B. N. Ames, An overview of evidence for a causal relation between iron deficiency during development and deficits in cognitive or behavioral function, *American Journal of Clinical Nutrition* 85 (2007): 931–945.

52. B. Lozhoff and coauthors, Long-lasting neural and behavioral effects of iron deficiency in infancy, *Nutrition Reviews* 64 (2006): S34–S43.

53. K. Kordas, Iron, lead, and children's behavior and cognition, *Annual Review of Nutrition* 30 (2010): 123–148.

54. Centers for Disease Control and Prevention, General lead information: Questions and answers, available at www.cdc.gov/nceh/lead/faq/about.htm, accessed January 23, 2009.

55. Centers for Disease Control and Prevention, Interpreting and managing blood lead levels <10μg/dL in children and reducing childhood exposures to lead: Recommendations of CDC's Advisory Committee on Childhood Lead Poisoning Prevention, *Morbidity and Mortality Weekly Report* 56/RR-8 (2007): 1–16.

56. National Institutes of Health, National Institute of Allergy and Infectious Diseases, *Food Allergy: Report of the NIH Expert Panel on Food Allergy Research*, March 13-14, 2006, available at www.naid.nih.gov.

57. U.S. Food and Drug Administration, Food allergies: Reducing the risks, January 23, 2009, available at www.fda.gov/consumer/updates/foodallergies012209.html.

58. A. Boulay and coauthors, A EuroPrevall review of factors affecting incidence of peanut allergy: Priorities for research and policy, *Allergy* 63 (2008): 797–809; L. A. Lee and A. W. Burks, Food allergies: Prevalence, molecular characterization, and treatment/prevention strategies, *Annual Review of Nutrition* 26 (2006): 539–565.

59. M. Boguniewica, N. Moore, and K. Paranto, Allergic diseases, quality of life, and the role of the dietitian, *Nutrition Today* 43 (2008): 6–10.

60. S. Ramesh, Food allergy overview in children, *Clinical Reviews in Allergy and Immunology* 34 (2008): 217–230.

61. L. A. Lee and A. W. Burks, 2006.

62. National Institutes of Health, National Institute of Allergy and Infectious Diseases, 2006.

63. Lee and Burks, 2006.

64. P. N. Pastor and C. A. Reuben, Diagnosed attention deficit hyperactivity disorder and learning disability: United States, 2004-2006, *Vital and Health Statistics, Series 10, Data from the National Health Survey* 237 (2008): 1–14.

65. M. L. Wolraich and coauthors, Attention-deficit/hyperactivity disorder among adolescents: A review of the diagnosis, treatment, and clinical implications, *Pediatrics* 115 (2005): 1734–1746.

66. D. McCann and coauthors, Food additives and hyperactive behaviour in 3-year-old and 8/9-year-old children in the community: A randomized, double-blinded, placebo-controlled trial, *Lancet* 370 (2007): 1560–1567.

67. C. L. Ogden and coauthors, Prevalence of high body mass index in US children and adolescents, 2007-2008, *Journal of the American Medical Association* 303 (2010): 242–249.

68. C. L. Ogden, and coauthors, 2010.

69. S. E. Barlow and the Expert Committee, Expert Committee recommendations regarding the prevention, assessment, and treatment of child and adolescent overweight and obesity: Summary report, *Pediatrics* 120 (2007): S164–S192.

70. N. F. Krebs and coauthors, Assessment of child and adolescent overweight and obesity, *Pediatrics* 120 (2007): S193–S228.

71. Barlow and the Expert Committee, 2007.

72. B. A. Spear and coauthors, Recommendations for treatment of child and adolescent overweight and obesity, *Pediatrics* 120 (2007): S254–S287; S. L. Martin, S. M. Lee, and R. Lowry, National prevalence and correlates of walking and bicycling to school, *American Journal of Preventive Medicine* 33 (2007): 98–105; American Academy of Pediatrics, Council on Sports Medicine and Fitness and Council on School Health, Active healthy living: Prevention of childhood obesity through increased physical activity, *Pediatrics* 117 (2006): 1834–1842; J. P. Koplan, C. T. Liverman, and V. I. Kraak, eds., *Preventing Childhood Obesity: Health in the Balance* (Washington, D.C.: National Academies Press, 2005), pp. 79–123.

73. K. J. Campbell and coauthors, Associations between the home food environment and obesity-promoting eating behaviors in adolescence, *Obesity* 15 (2007): 719–730; Barlow and the Expert Committee, 2007.

74. K. S. Geller and D. A. Dzewaltowski, Longitudinal and cross-sectional influences on youth fruit and vegetable consumption, *Nutrition Reviews* 67 (2009): 65–76; J. Brug and coauthors, Taste preferences, liking and other factors related to fruit and vegetable intakes among schoolchildren: Results from observational studies, *British Journal of Nutrition* 99 (2008): S7–S14; C. A. Forestell and J. A. Mennella, Early determinants of fruit and vegetable acceptance, *Pediatrics* 120 (2007): 1247–1254; C. Arcan and coauthors, Parental eating behaviours, home food environment and adolescent intakes of fruits, vegetables and dairy foods: Longitudinal finding form Project EAT, *Public Health Nutrition* 11 (2007): 1257–1265; M. Wind and coauthors,

Correlates of fruit and vegetable consumption among 11-year-old Belgina-Flemish and Dutch schoolchildren, *Journal of Nutrition Education and Behavior* 38 (2006): 211–221.

75. K. E. Leahy, L. L. Birch, and B. Rolls, Reducing the energy density of multiple meals decreases the energy intake of preschool-age children, *American Journal of Clinical Nutrition* 88 (2008): 1459–1468.

76. S. N. Bleich and coauthors, Increasing consumption of sugar-sweetened beverages among US adults: 1988–1994 to 1999–2004, *American Journal of Clinical Nutrition* 89 (2009): 372–381; Spear and coauthors, 2007; L. DuBois and coauthors, Regular sugar-sweetened beverage consumption between meals increases risk of overweight among preschool-aged children, *Journal of the American Dietetic Association* 107 (2007): 924–934.

77. L. R. Vartanian, M. B. Schwartz, and K. D. Brownell, Effects of soft drink consumption on nutrition and health: A systematic review and meta-analysis, *American Journal of Public Health* 97 (2007): 667–675.

78. American Academy of Pediatrics, Council on Sports Medicine and Fitness and Council on School Health, 2006.

79. S. Gable, Y. Chang, and J. L. Drull, Television watching and frequency of family meals are predictive of overweight onset and persistence in a national sample of school-aged children, *Journal of the American Dietetic Association* 107 (2007): 53–61; Spear and coauthors, 2007.

80. D. J. Barr-Anderson and coauthors, Characteristics associated with older adolescents who have a television in their bedrooms, *Pediatrics* 121 (2008): 718–724; A. M. Adachi-Mejia and coauthors, Children with a TV in their bedroom at higher risk for being overweight, *International Journal of Obesity* 31 (2007): 644–651.

81. L. Dubois and coauthors, Social factors and television use during meals and snacks is associated with higher BMI among pre-school children, *Public Health Nutrition* 11 (2008): 1267–1279; S. Gable, Y. Chang, and J. L. Krull, Television watching and frequency of family meals are predictive of overweight onset and persistence in a national sample of school-aged children, *Journal of the American Dietetic Association* 107 (2007): 53–61.

82. J. L. Wiecha and coauthors, When children eat what they watch: Impact of television viewing on dietary intake in youth, *Archives of Pediatrics & Adolescent Medicine* 160 (2006): 436–442; S. C. Folta and coauthors, Food advertising targeted at school-age children: A content analysis, *Journal of Nutrition Education and Behavior* 38 (2006): 244–248.

83. A. Batada and coauthors, Nine out of 10 food advertisements shown during Saturday morning children's television programming are for foods high in fat, sodium, or added sugars, or low in nutrients, *Journal of the American Dietetic Association* 108 (2008): 673–678; L. M. Powell and coauthors, Nutritional content of television food advertisements seen by children and adolescents in the United States, *Pediatrics* 120 (2007): 576–583.

84. S. M. Connor, Food-related advertising on preschool television: Building brand recognition in

young viewers, *Pediatrics* 118 (2006): 1478–1485; Folta and coauthors, 2006.

85. Barlow and the Expert Committee, 2007.

86. M. Salvadori and coauthors, Elevated blood pressure in relation to overweight and obesity among children in a rural Canadian community, *Pediatrics* 122 (2008): e821–e827; R. Jago and coauthors, Prevalence of abnormal lipid and blood pressure values among an ethnically diverse population of eighth-grade adolescents and screening implications, *Pediatrics* 117 (2006): 2065–2073.

87. K. L. Jones, Role of obesity in complicating and confusing the diagnosis and treatment of diabetes in children, *Pediatrics* 121 (2008): 361–368; S. Cook and coauthors, Metabolic Syndrome rates in United States adolescents, from the National Health and Nutrition Examination Survey, 1999–2002, *Journal of Pediatrics* 152 (2008): 165–170; C. L. Carroll and coauthors, Childhood overweight increases hospital admission rates for asthma, *Pediatrics* 120 (2007): 734–740.

88. D. S. Ludwig, Childhood obesity—The shape of things to come, *New England Journal of Medicine* 357 (2007): 2325–2326; J. Franklin and coauthors, Obesity and risk of low self-esteem: A statewide survey of Australian children, *Pediatrics* 118 (2006): 2481–2487; A. J. Daley and coauthors, Exercise therapy as a treatment for psychopathologic conditions in obese and morbidly obese adolescents: A randomized, controlled trial, *Pediatrics* 118 (2006): 2126–2134.

89. Barlow and the Expert Committee, 2007.

90. M. M. Davis and coauthors, Recommendations for prevention of childhood obesity, *Pediatrics* 120 (2007): S229–S253; Position of the American Dietetic Association: Individual-, family-, school-, and community-based interventions for pediatric overweight, 2006.

91. Barlow and the Expert Committee, 2007.

92. Spear and coauthors, 2007.

93. Barlow and the Expert Committee, 2007.

94. Barlow and the Expert Committee, 2007.

95. American Academy of Pediatrics, Council on Sports Medicine and Fitness and Council on School Health, 2006.

96. American Academy of Pediatrics, Council on Sports Medicine and Fitness and Council on School Health, 2006.

97. Spear and coauthors, 2007.

98. T. H. Inge and coauthors, Reversal of type 2 diabetes mellitus and improvements in cardiovascular risk factors after surgical weight loss in adolescents, *Pediatrics* 123 (2009): 214–222.

99. Spear and coauthors, 2007.

100. E. Jansen and coauthors, From the Garden of Eden to the land of plenty: Restriction of fruit and sweets intake leads to increased fruit and sweets consumption in children, *Appetite* 51 (2008): 570–575; E. Jansen, S. Mulkens, and A. Jansen, Do not eat the red food: Prohibition of snacks leads to their relatively higher consumption in children, *Appetite* 49 (2007): 572–577.

101. Position of the American Dietetic Association, Nutrition guidance for healthy children ages 2 to 11 years, 2008.

102. Position of the American Dietetic Association, Nutrition guidance for healthy children ages 2 to 11 years, 2008; J. Wardle, S. Carnell, and L. Cooke, Parental control over feeding and children's fruit and vegetable intake: How are they related? *Journal of the American Dietetic Association* 105 (2005): 227–232.

103. Position of the American Dietetic Association, School Nutrition Association, and Society for Nutrition Education: Comprehensive school nutrition services, *Journal of the American Dietetic Association* 110 (2010): 1738–1749.

104. Position of the American Dietetic Association: Local support for nutrition integrity in schools, *Journal of the American Dietetic Association* 110 (2010): 1244–1254.

105. Position of the American Dietetic Association: Local support for nutrition integrity in schools, 2010.

106. M. K. Crepinsek and coauthors, Meals offered and served in US public schools: Do they meet nutrient standards? *Journal of the American Dietetic Association* 109 (2009): S31–S43; K. Widenhorn-Muller and coauthors, Influence of having breakfast on cognitive performance and mood in 13- to 20-year-old high school students: Results of a crossover trial, *Pediatrics* 122 (2008): 279–284.

107. Position of the American Dietetic Association: Local support for nutrition integrity in schools, 2010.

108. M. Story, The Third School Nutrition Dietary Assessment Study: Findings and policy implications for improving the health of US children, *Journal of the American Dietetic Association* 109 (2009): S7–S13; Position of the American Dietetic Association, Nutrition guidance for healthy children ages 2 to 11 years, 2008.

109. Position of the American Dietetic Association: Local support for nutrition integrity in schools, 2010.

110. Position of the American Dietetic Association: Local support for nutrition integrity in schools, 2010; Centers for Disease Control and Prevention, Competitive foods and beverages available for purchase in secondary schools—Selected sites, United States, 2006, *Morbidity and Mortality Weekly Report* 57 (2008): 935–938.

111. Position of the American Dietetic Association: Local support for nutrition integrity in schools, 2010.

112. Institute of Medicine, Food and Nutrition Board, Committee on Nutrition Standards for Foods in Schools, eds. V. A. Stallings and A. L. Yaktine, *Nutrition Standards for Foods in Schools: Leading the Way toward Healthier Youth* (Washington, D.C.: National Academies Press, 2007).

113. Institute of Medicine, Food and Nutrition Board, Committee on Nutrition Standards for Foods in Schools, 2007.

114. Institute of Medicine, Food and Nutrition Board, Committee on Nutrition Standards for Foods in Schools, 2007.

115. Committee on the Scientific Evaluation of Dietary Reference Intakes, 2005, pp. 177–182.

116. Committee on Dietary Reference Intakes, 2011.

117. Wagner, Greer, and the Section on Breastfeeding and Committee on Nutrition, 2008.

118. C. L. Wagner, F. R. Greer, and the Section on Breastfeeding and Committee on Nutrition, 2008.

119. F. R. Greer, N. F. Krebs, and the Committee on Nutrition, American Academy of Pediatrics, Optimizing bone health and calcium intakes of infants, children, and adolescents, *Pediatrics* 117 (2006): 578–585.

120. Greer, Krebs, and the Committee on Nutrition, 2006.

121. L. Esterie and coauthors, Milk, rather than other foods, is associated with vertebral bone mass and circulating IGF-1 in female adolescents, *Osteoporosis International* 20 (2009): 567–575; M. M. Murphy and coauthors, Drinking flavored or plain milk is positively associated with nutrient intake and is not associated with adverse effects on weight status in US children and adolescents, *Journal of the American Dietetic Association* 108 (2008): 631–639.

122. N. I. Larson and coauthors, Family meals during adolescence are associated with higher diet quality and healthful meal patterns during young adulthood, *Journal of the American Dietetic Association* 107 (2007): 1502–1510.

123. M. T. Timlin and coauthors, Breakfast eating and weight change in a 5-year prospective analysis of adolescents: Project EAT (Eating Among Teens), *Pediatrics* 121 (2008): e638–e645.

124. C. M. McDonald and coauthors, Overweight is more prevalent than stunting and is associated with socioeconomic status, maternal obesity, and a snacking dietary pattern in school children from Bogota, Colombia, *Journal of Nutrition* 139 (2009): 370–376; H. M. Niemeier and coauthors, Fast food consumption and breakfast skipping: Predictors of weight gain from adolescence to adulthood in a nationally representative sample, *Journal of Adolescent Health* 39 (2006): 842–849.

125. R. S. Sebastian, L. E. Cleveland, and J. D. Goldman, Effect of snacking frequency on adolescents' dietary intakes and meeting national recommendations, *Journal of Adolescent Health* 42 (2008): 503–511.

126. American Heart Association, S. S. Gidding, and coauthors, Dietary recommendations for children and adolescents: A guide for practitioners, *Pediatrics* 117 (2006): 544–559.

127. L. Libuda and coauthors, Association between long-term consumption of soft drinks and variables of bone modeling and remodeling in a sample of healthy German children and adolescents, *American Journal of Clinical Nutrition* 88 (2008): 1670–1677; Greer, Krebs, and the Committee on Nutrition, 2006.

128. L. R. Vartanian, M. B. Schwartz, and K. D. Brownell, Effects of soft drink consumption on nutrition and health: A systematic review and meta-analysis, *American Journal of Public Health* 97 (2007): 667–675.

Nutrition in Practice

When people think about the health problems of children and adolescents, they typically think of ear infections, colds, and acne—not heart disease, diabetes, or hypertension. Today, however, unprecedented numbers of U.S. children are being diagnosed with obesity and the serious "adult diseases," such as type 2 diabetes, that accompany overweight.[1] When type 2 diabetes develops before the age of 20, the incidence of diabetic kidney disease and death in middle age increases dramatically, largely because of the long duration of the disease.[2] For children born in the United States in 2000, the risk of developing type 2 diabetes sometime in their lives is estimated to be 30 percent for boys and 40 percent for girls. U.S. children are not alone—rapidly rising rates of obesity threaten the health of an alarming number of children around the globe.[3] Without immediate intervention, millions of children are destined to develop type 2 diabetes and hypertension in childhood followed by **cardiovascular disease (CVD)** in early adulthood.[4]

Over the past three decades, researchers have been observing how changes in body weight, blood lipids, blood pressure, and individual behaviors correlate with the development of CVD over time—from infancy to childhood through adolescence and into young adulthood. Some major findings have emerged from this research:

▌ Changes inside the arteries—changes predictive of CVD—are evident in childhood.

▌ Obesity in children affects these changes.

▌ Behaviors that influence the development of obesity and of CVD are learned and begin early in life. These behaviors include overeating, physical inactivity, and cigarette smoking.

This Nutrition in Practice focuses on efforts to prevent childhood obesity, type 2 diabetes, and CVD (see the accompanying glossary for definitions of the relevant terms), but the benefits extend to other obesity-related diseases as well. The childhood years (ages 2 to 18) are emphasized here, for the earlier in life health-promoting habits become established, the better they will stick.

What about genetics? Do some people inherit the tendency to become obese or develop diabetes or CVD regardless of the lifestyle habits they adopt?

For obesity, as well as for CVD, hypertension, and type 2 diabetes, genetics does not appear to play a *determining* role; that is, a person is not simply destined at birth to develop them. Instead, genetics appears to play a *permissive* role—the potential is inherited and will then be realized, if given a push by factors in the environment such as poor diet, sedentary lifestyle, and cigarette smoking.[5] Researchers note that the relationship between genes and the environment is a synergistic one—their combined effects are greater than the sum of their individual effects.

Many experts agree that preventing or treating obesity in childhood will reduce the rate of chronic diseases in adulthood. Without intervention, most overweight children become overweight adolescents, who become overweight adults, and being overweight exacerbates every chronic disease that adults face.[6] Fatty liver, a condition that correlates directly with BMI, was not even recognized in pediatric research until recently. Today, fatty liver disease affects about one in three obese children.[7]

What about events that take place during fetal development—malnutrition for example? Can they affect a person's tendency to develop diseases later in life?

A theory called *fetal programming* or *fetal origins of disease* states that maternal malnutrition or other harmful conditions at a critical period of fetal development may have lifelong effects on an individual's pattern of genetic expression and therefore on the tendency to develop obesity and certain diseases.[8] Poor maternal diet or health during pregnancy may alter the infant's bodily functions such as blood pressure, cholesterol metabolism, and immune functions that influence disease development.[9] For example, maternal diet may alter blood-vessel growth and program lipid metabolism and lean body mass development in such a way that the infant will develop risk factors for cardiovascular disease as an adult.[10]

Why has type 2 diabetes become so prevalent?

Type 2 diabetes, a chronic disease closely linked with obesity, has been on the rise among children and adolescents as the prevalence of obesity in U.S. youth has increased in recent years. Obesity is the most important risk factor for type 2 diabetes—most of

the children diagnosed with type 2 diabetes are obese.[11] Most are diagnosed during puberty, but as children become more obese and less active, the disease is appearing in younger and younger children. Type 2 diabetes is most likely to occur in those who are obese and sedentary and have a family history of diabetes.

How does type 2 diabetes develop?

In type 2 diabetes, the body's cells become insulin resistant—that is, the cells become less sensitive to insulin, reducing the amount of glucose entering the cells from the blood. The combination of obesity and insulin resistance produces a cluster of symptoms, including high blood pressure and high blood lipids, which in turn promotes the development of atherosclerosis and the early development of CVD.[12] Other common problems evident by early adulthood include kidney disease, blindness, and miscarriages. The complications of diabetes, especially when encountered at a young age, can shorten life expectancy. Chapter 21 offers a detailed discussion of diabetes.

Prevention and treatment of type 2 diabetes depend on weight management, which can be particularly difficult in a young person's world of food advertising, computer games, and pocket money for candy bars. The activity and dietary suggestions to help defend against heart disease later in this discussion apply to type 2 diabetes as well.

How does CVD develop, and when does its development begin?

Most CVD involves **atherosclerosis**—the accumulation of cholesterol and other blood lipids along the walls of the arteries. Frequently, atherosclerosis and its complications interfere with the flow of blood to the heart and can lead to coronary heart disease (CHD), which, in turn, raises the likelihood of a heart attack. When atherosclerosis interferes with blood flow to the brain, a stroke can result. Infants are born with healthy, smooth, clear arteries, but within the first decade of life, **fatty streaks** may begin to appear. During adolescence, these fatty streaks may begin to turn into **plaques** (Figure 22-2 in Chapter 22 shows the formation of plaques in atherosclerosis). By early adulthood, the fibrous plaques may begin to calcify and become raised lesions, especially in boys and young men. As the lesions grow more numerous and thicken, the heart disease rate begins to rise, and the rise becomes dramatic at about age 45 in men and 55 in women. From this point on, arterial damage and blockage progress rapidly, and heart attacks and strokes threaten life. In short, the consequences of atherosclerosis, which become apparent only in adulthood, have their beginnings in the first decades of life.[13]

Children with the highest risks of developing heart disease are sedentary and obese, with diabetes, high blood pressure, and high blood cholesterol. In contrast, children with the lowest risks of heart disease are physically active and of normal weight, with low blood pressure and favorable lipid profiles.

Parents do not need to worry about their children's blood cholesterol, do they?

Atherosclerotic lesion development reflects blood cholesterol: as blood cholesterol increases, lesion coverage increases. Cholesterol values at birth are similar in all populations; differences emerge in early childhood. Standard values for cholesterol screening in

TABLE NP12-1 Cholesterol Values for Children and Adolescents

Disease Risk	Total Cholesterol (mg/dL)	LDL Cholesterol (mg/dL)
Acceptable	<170	<110
Borderline	170–199	110–129
High	≥200	≥130

Note: Adult values appear in Table 22-2 on p. 572.

children and adolescents are listed in Table NP12-1. Cholesterol concentrations change with age in children and adolescents, however, and are especially variable during puberty.[14] Thus, use of a single cut point for all pediatric age groups has limitations.

In general, blood cholesterol tends to rise as dietary saturated fat intakes increase. Blood cholesterol also correlates with childhood obesity, especially abdominal obesity.[15] LDL cholesterol rises with obesity, and HDL declines. These relationships are apparent throughout childhood, and their magnitude increases with age.

Children who are overweight and have high blood cholesterol are likely to have parents who develop heart disease early. For this reason, selective screening is recommended for children and adolescents who are overweight or obese; those whose parents (or grandparents) have premature (≤55 years of age for men and ≤65 years of age for women) heart disease; those whose parents have elevated blood cholesterol; those who have other risk factors for heart disease such as hypertension, cigarette smoking, or diabetes; and those whose family history is unavailable.[16] Because blood cholesterol in children is a good predictor of adult values, some experts recommend universal screening for all children, and particularly for those who are overweight, smoke, are sedentary, or consume diets high in saturated fat.

Early—but not advanced—atherosclerotic lesions are reversible, making screening and education a high priority. Both those with family histories of heart disease and those with multiple risk factors need intervention.

Is hypertension a concern for children and adolescents?

Pediatricians routinely monitor blood pressure in children and adolescents. High blood pressure may signal an underlying disease or the early onset of hypertension. Hypertension accelerates the development of atherosclerosis.[17]

Like atherosclerosis and high blood cholesterol, hypertension may develop in the first decades of life, especially among obese children, and worsen with time. Children can control their hypertension by participating in regular aerobic activity and by losing weight or maintaining their weight as they grow taller. Restricting dietary sodium also causes an immediate drop in most children's and adolescents' blood pressure.[18]

Regular physical activity lowers risks for heart disease and hypertension in adults; does it do so in children as well?

Yes. Research has also confirmed an association between blood lipids and physical activity in children, similar to that seen in adults.

Physically active children have a better lipid profile and lower blood pressure than physically inactive children, and these positive findings often persist into adulthood. The *Physical Activity Guidelines for Americans, 2008* recommendations for children and adolescents are listed in Chapter 12 on p. 337.

Just as blood cholesterol and obesity track over the years, so does a child's level of physical activity. Those who are inactive now are likely to still be inactive years later. Similarly, those who are physically active now tend to remain so. Compared with inactive teens, those who are physically active weigh less, smoke less, eat a diet lower in saturated fats, and have better blood lipid profiles. Both obesity and blood cholesterol correlate with the inactive pastime of watching television. The message is clear: physical activity offers numerous health benefits, and children who are active today are most likely to be active for years to come.

Are adult dietary recommendations appropriate for children?

Regardless of family history, all children over age two should eat a variety of foods and maintain desirable weight (see Table NP 12-2). Children (4 to 18 years of age) should receive at least 25 percent and no more than 35 percent of total energy from fat, less than 10 percent from saturated fat, and less than 300 milligrams of cholesterol per day.[19]

Recommendations limiting fat and cholesterol are not intended for infants or children under two years old. Infants and toddlers need a higher percentage of fat to support their rapid growth. For children between 1 year of age and 2 years of age who are overweight or obese, or have a family history of heart disease, obesity, or abnormal blood lipids, however, the use of reduced-fat milk is appropriate.[20]

Healthy children over age two can begin the transition to eating according to recommendations by selecting fewer foods high in saturated fat and more fruits and vegetables. Healthy meals can occasionally include moderate amounts of a child's favorite food, such as ice cream, even if it is high in saturated fat. A steady diet from the children's menus in some restaurants—which feature chicken nuggets, hot dogs, and French fries—easily exceeds a prudent intake of saturated fat, *trans* fat, and kcalories, however, and invites both nutrient shortages and weight gains.[21] Fortunately, most restaurant chains are changing children's menus to include steamed vegetables, fruit cups, and broiled or grilled chicken—additions welcomed by busy parents who often dine out or purchase take-out foods.

Other fatty foods, such as nuts, vegetable oils, and some varieties of fish such as tuna or salmon, contribute essential fatty acids. Low-fat milk and milk products also deserve special attention in a child's diet for the needed calcium and other nutrients they supply.[22]

Parents and caregivers play a key role in helping children establish healthy eating habits. Balanced meals need to provide lean meat, poultry, fish, and legumes; fruits and vegetables; whole grains; and low-fat milk products. Such meals can provide enough energy and nutrients to support growth and maintain blood cholesterol within a healthy range.

Pediatricians warn parents to avoid extremes. Although intentions may be good, excessive food restriction may create nutrient deficiencies and impair growth. Furthermore, parental control over eating may instigate battles and foster attitudes about foods that can lead to inappropriate eating behaviors.

Do pediatricians ever prescribe drugs to lower blood cholesterol in children?

Experts agree that children with high blood cholesterol should first be treated with diet. If high blood cholesterol persists despite dietary intervention in children eight years of age and older, then

---

TABLE NP12-2 **American Heart Association Dietary Guidelines and Strategies for Children**[a]

▪ Balance dietary kcalories with physical activity to maintain normal growth.

▪ Every day, engage in 60 minutes of moderate to vigorous play or physical activity.

▪ Eat vegetables and fruits daily. Use fresh, frozen, and canned vegetables and fruits and serve at every meal; limit those with added fats, salt, and sugar.

▪ Limit juice intake (4 to 6 ounces per day for children 1 to 6 years of age, 8 to 12 ounces for children 7 to 18 years of age).

▪ Use vegetable oils (canola, soybean, olive, safflower, or other unsaturated oils) and soft margarines low in saturated fat and *trans* fatty acids instead of butter or most other animal fats in the diet.

▪ Choose whole-grain breads and cereals rather than refined products; read labels and make sure that "whole grain" is the first ingredient.

▪ Reduce the intake of sugar-sweetened beverages and foods.

▪ Consume low-fat and nonfat milk and milk products daily.

▪ Include two servings of fish per week, especially fatty fish such as broiled or baked salmon.

▪ Choose legumes and tofu in place of meat for some meals.

▪ Choose only lean cuts of meat and reduced-fat meat products; remove the skin from poultry.

▪ Use less salt, including salt from processed foods. Breads, breakfast cereals, and soups may be high in salt and/or sugar so read food labels and choose high-fiber, low-salt, low-sugar alternatives.

▪ Limit the intake of high-kcalorie add-ons such as gravy, Alfredo sauce, cream sauce, cheese sauce, and hollandaise sauce.

▪ Serve age-appropriate portion sizes on appropriately sized plates and bowls.

[a]These guidelines are for children three years of age and older.

Source: Adapted from American Heart Association, S. S. Gidding, and coauthors, Dietary recommendations for children and adolescents: A guide for practitioners, *Pediatrics* 117 (2006): 544–559.

drug treatment may be necessary. Drugs can effectively lower blood cholesterol without interfering with adolescent growth or development.[23]

Can parents or caregivers do anything else to help children reduce their risks of CVD?

Even though the focus of this text is nutrition, another risk factor for heart disease that starts in childhood and carries over into adulthood must also be addressed: cigarette smoking. Each day, 3000 children light up for the first time—typically in elementary school. Among high school students, almost half have tried smoking, and one in five smokes regularly.[24] Approximately 80 percent of all adult smokers began smoking before age 18.

Of those teenagers who continue smoking, half will eventually die of smoking-related causes. Efforts to teach children about the dangers of smoking need to be aggressive. Children are not likely to consider the long-term health consequences of tobacco use. They are more likely to be struck by the immediate health consequences, such as shortness of breath when playing sports, or social consequences, such as having bad breath. Whatever the context, the message to all children and teens should be clear: don't start smoking. If you have already started, quit.

In conclusion, *adult* heart disease is a major *pediatric* problem. Without intervention, some 60 million children are destined to suffer its consequences within the next 30 years. Optimal prevention efforts focus on children, especially on those who are overweight.[25] Just as young children receive vaccinations against infectious diseases, they need screening for, and education about, chronic diseases. Many health education programs have been implemented in schools around the country. These programs are most effective when they include education in the classroom, heart-healthy meals in the lunchroom, fitness activities on the playground, and parental involvement at home.

Notes

1. Childhood obesity, *Lancet* 375 (2010): 1737–1748; F. M. Biro and M. Wien, Childhood obesity and adult morbidities, *American Journal of Clinical Nutrition* 91 (2010): 1499S–1505S; C. L. Ogden and coauthors, Prevalences of high body mass index in US children and adolescents, 2007–2008, *Journal of the American Medical Association* 303 (2010): 242–249; D. S. Freedman and coauthors, Risk factors and adult body mass index among overweight children: The Bogalusa Heart Study, *Pediatrics* 123 (2009): 750–757; K. L. Jones, Role of obesity in complicating and confusing the diagnosis and treatment of diabetes in children, *Pediatrics* 121 (2008): 361–368; M. Gardner, D. W. Gardner, and J. R. Sowers, The cardiometabolic syndrome in the adolescent, *Pediatric Endocrinology Reviews* Suppl. 4 (2008): 964–968; S. Cook and coauthors, Metabolic Syndrome rates in United States adolescents, from the National Health and Nutrition Examination survey, 1999–2002, *Journal of Pediatrics* 152 (2008): 165–170.

2. M. E. Pavkov and coauthors, Effect of youth-onset type 2 diabetes mellitus on incidence of end-stage renal disease and mortality in young and middle-aged Pima Indians, *Journal of the American Medical Association* 296 (2006): 421–426.

3. C. Bouchard, Childhood obesity: Are genetic differences involved? *American Journal of Clinical Nutrition* 89 (2009): 1494S–1501S; W. Maziak, K. D. Ward, and M. B. Stockton, Childhood obesity: Are we missing the big picture? *Obesity Reviews* 9 (2008): 35–42; Y. Wang and I. Lobstein, Worldwide trends in childhood overweight and obesity, *International Journal of Pediatric Obesity* 1 (2006): 11–25.

4. Freedman and coauthors, 2009; H. Zhu and coauthors, Relationships of cardiovascular phenotypes with healthy weight, at risk of overweight, and overweight in US youths, *Pediatrics* 121 (2008): 115–122.

5. C. Bouchard, 2009; J. M. Ordovas, Genetic influences on blood lipids and cardiovascular disease risk: Tools for primary prevention, *American Journal of Clinical Nutrition* 89 (2009): 1509S–1517S.

6. S. E. Barlow and the Expert Committee, Expert Committee recommendations regarding the prevention, assessment, and treatment of child and adolescent overweight and obesity: Summary report, *Pediatrics* 120 (2007): S164–S192; D. S. Ludwig, Childhood obesity—The shape of things to come, *New England Journal of Medicine* 357 (2007): 2325–2327.

7. D. S. Ludwig, 2007.

8. M. E. Symonds, T. Stephenson, and H. Budge, Early determinants of cardiovascular disease: The role of early diet in later blood pressure control, *American Journal of Clinical Nutrition* 89 (2009): 1518S–1522S; S. H. Zeisel, Epigenetic mechanisms for nutrition determinants of later health outcomes, *American Journal of Clinical Nutrition* 89 (2009): 1488S–1493S; P. D. Gluckman and coauthors, Effect of in utero and early-life conditions on adult health and disease, *New England Journal of Medicine* 359 (2008): 61.

9. W. Palinski and coauthors, Developmental programming: Maternal hypercholesterolemia and immunity influence susceptibility to atherosclerosis, *Nutrition Reviews* 65 (2007): S182–S187.

10. R. C. Painter and coauthors, Early onset of coronary artery disease after prenatal exposure to the Dutch famine, *American Journal of Clinical Nutrition* 84 (2006): 322–327.

11. Jones, 2008; T. S. Hannon, F. Rao, and S. A. Arslanian, Childhood obesity and type 2 diabetes mellitus, *Pediatrics* 116 (2005): 473–480.

12. F. M. Biro and M. Wien, 2010; G. S. Boyd and coauthors, Effect of obesity and high blood pressure on plasma lipid levels in children and adolescents, *Pediatrics* 116 (2005): 442–446.

13. G. Raghuveer, Lifetime cardiovascular risk of childhood obesity, *American Journal of Clinical Nutrition* 91 (2010): 1514S–1519S; H. Zhu and coauthors, 2008; D. S. Freedman and coauthors, Cardiovascular risk factors and excess adiposity among overweight children and adolescents: The Bogalusa Heart Study, *Journal of Pediatrics* 150 (2007): 12–17; D. R. Thompson and coauthors, Childhood overweight and

cardiovascular disease risk factors: The National Heart, Lung, and Blood Institute Growth and Health Study, *Journal of Pediatrics* 150 (2007): 18–25.

14. S. R. Daniels, F. R. Greer, and the Committee on Nutrition, Lipid screening and cardiovascular health in childhood, *Pediatrics* 122 (2008): 198–208.

15. D. S. Freedman and coauthors, Risk factors and adult body mass index among overweight children: The Bogalusa Heart Study, *Pediatrics* 123 (2009): 750–757; S. Cook and coauthors, Metabolic Syndrome rates in United States Adolescents, from the National Health and Nutrition Examination Survey, 1999–2002, *Journal of Pediatrics* 152 (2008): 165–170; J. Botton and coauthors, Cardiovascular risk factor levels and their relationships with overweight and fat distribution in children: The Fleurbaix Laventie Ville Sante II Study, *Metabolism* 56 (2007): 614–622; D. R. Thompson and coauthors, 2007.

16. S. R. Daniels, F. R. Greer, and the Committee on Nutrition, 2008.

17. National High Blood Pressure Education Program Working Group on High Blood Pressure in Children and Adolescents, The fourth report on the diagnosis, evaluation, and treatment of high blood pressure in children and adolescents, *Pediatrics* 114 (2004): 555S–576S.

18. F. J. He and G. A. MacGregor, Importance of salt in determining blood pressure in children: Meta-analysis of controlled trials, *Hypertension* 48 (2006): 861–869.

19. Standing Committee on the Scientific Evaluation of Dietary Reference Intakes, Food and Nutrition Board, Institute of Medicine, *Dietary Reference Intakes for Energy, Carbohydrate, Fiber, Fat, Fatty Acids, Cholesterol, Protein, and Amino Acids* (Washington, D.C.: National Academies Press, 2005), pp. 769–879.

20. S. R. Daniels, F. R. Greer, and the Committee on Nutrition, 2008.

21. L. Johnson and coauthors, Energy-dense, low-fiber, high-fat dietary pattern is associated with increased fatness in childhood, *American Journal of Clinical Nutrition* 87 (2008): 846–854.

22. F. R. Greer, N. F. Krebs, and the Committee on Nutrition, American Academy of Pediatrics, Optimizing bone health and calcium intakes of infants, children, and adolescents, *Pediatrics* 117 (2006): 578–585.

23. S. R. Daniels, F. R. Greer, and the Committee on Nutrition, 2008.

24. Centers for Disease Control and Prevention, Cigarette use among high school students—United States, 1991–2009, *Morbidity and Mortality Weekly Report* 59 (2010): 797–801.

25. S. R. Daniels, F. R. Greer, and the Committee on Nutrition, 2008; S. E. Barlow and the Expert Committee, 2007.

13

Nutrition through the Life Span: Later Adulthood

The last two chapters were devoted to stages of the life cycle that require special nutrition attention: pregnancy, lactation, infancy, childhood, and adolescence. Much of the text before that focused on nutrition to support wellness during adulthood. This chapter describes the unique nutrition needs of the later adult years.

The most urgent nutrition need of older people, however, is to have made good food choices in the past! All of life's nutrition choices incur health consequences, for the better or for the worse. A single day's intakes of nutrients may exert only a minute effect on body organs and their functions, but, over years and decades, the repeated effects accumulate to have major impacts. This being the case, it is of great importance for everyone, of every age, to pay close attention today to nutrition.

The "graying" of America is a continuing trend.[1] The majority of citizens are now middle aged, and the ratio of old people to young is increasing, as Figure 13-1 shows. Our society uses the arbitrary age of 65 to define the transition point between middle age and old age, but growing old happens day by day, with changes occurring gradually over time. Since 1950, the population aged 65 and over has almost tripled. Remarkably, the fastest-growing age group has been people over 85 years; since 1950, their numbers have increased sevenfold. The number of people in the United States age 100 or older doubled in the past decade. Similar trends are occurring in populations worldwide. The U.S. Bureau of the Census projects that by the year 2040 more than a million Americans will be 100 years old or older.

Life expectancy in the United States is 78 years, up from about 47 years in 1900.[2] Advances in medical science—antibiotics and other treatments—are largely responsible for almost doubling the life expectancy in the 20th century. Improved nutrition and an abundant food supply have also contributed to lengthening life expectancy. Ironically, this abundant food supply has also jeopardized the chances of further lengthening life expectancy as obesity rates increase.[3]

The human **life span,** currently estimated at 130 years, is the upper limit of human **longevity,** even given optimal nutrition. With work progressing in medical and genetic technologies, however, the human life span may be extended significantly.

Research in the field of aging is active—and difficult. Researchers are challenged by the diversity of older adults. When older adults experience health problems, it is hard to know whether to attribute these problems to genetics, aging, or other environmental factors such as nutrition. The idea that nutrition can influence the way the human body ages is particularly appealing because diet is a factor that people can control and change.

Nutrition and Longevity

What has been learned so far about the effects of nutrition and environment on longevity provides incentive for researchers to keep asking questions about how and why human beings age. Among their questions are:

▮ To what extent is aging inevitable, and can it be slowed through changes in lifestyle and environment?

▮ What role does nutrition play in aging, and what role can it play in slowing aging?

With respect to the first question, it seems that aging is an inevitable, natural process, programmed into the genes at conception. People can, however, slow the process within the natural limits set by heredity by adopting healthy lifestyle habits such as eating nutritious food and engaging in physical activity. In fact, an estimated 70 to 80 percent of the average person's life expectancy may depend on individual health-related behaviors; genes determine the remaining 20 to 30 percent.

With respect to the second question, good nutrition helps to maintain a healthy body and can therefore ease the aging process in many significant ways.[4] Clearly, nutrition can improve the **quality of life** in the later years.

Commonly used age groups:

▮ **Young old (65–74 years)**

▮ **Old (75–84 years)**

▮ **Oldest old (≥85 years)**

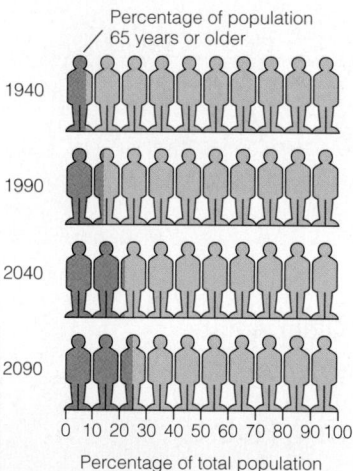

Percentage of population 65 years or older

1940

1990

2040

2090

0 10 20 30 40 50 60 70 80 90 100

Percentage of total population

FIGURE 13-1 The Aging of the U.S. Population
In 1940, 6.8 percent of the population was 65 or older. In 1990, 12.7 percent of us had reached age 65; by 2040, 21.7 percent will have reached age 65; and by 2090, nearly one out of four Americans will be 65 or older.

356

Slowing the Aging Process

One approach researchers use to search out the secret of long life is to study older people. Some people are young for their ages, whereas others are old for their ages. What makes the difference?

Healthy Habits Six lifestyle habits seem to have a profound influence on people's health and therefore on their **physiological age:**[5]

▌ Sleeping regularly and adequately

▌ Eating well-balanced meals rich in fruits and vegetables regularly

▌ Maintaining a healthy body weight

▌ Engaging in regular physical activity

▌ Not smoking

▌ Not using alcohol, or using it in moderation

Over the years, the effects of these lifestyle choices accumulate—that is, those who follow all of these practices live longer and have fewer disabilities as they age. They are in better health, even if older in **chronological age,** than people who do not adopt these behaviors.[6] Even though people cannot alter their birth dates, they may be able to add years to, and enhance the quality of, their lives.[7] Physical activity seems to be most influential in preventing or slowing the many changes that many people seem to accept as an inevitable consequence of old age. In other words, physical activity and long life seem to go together.[8]

Physical Activity The many and remarkable benefits of regular physical activity are not limited to the young. Compared with those who are inactive, older adults who are active weigh less; have greater flexibility, more endurance, better balance, and better health; and live longer.[9] They reap additional benefits from various types of activities as well: aerobic activities improve cardiorespiratory endurance, blood pressure, and blood lipid concentrations; moderate endurance activities improve the quality of sleep; and strength training significantly improves posture and mobility. In fact, regular physical activity is a powerful predictor of a person's mobility in the later years. Physical activity also increases blood flow to the brain, thereby preserving mental ability, alleviating depression, and supporting independence.[10]

Muscle mass and muscle strength tend to decline with aging, making older people vulnerable to falls and immobility.[11] Falls are a major cause of fear, injury, disability, dependence, and even death among older adults. Regular physical activity tones, firms, and strengthens muscles, helping to improve confidence, reduce the risk of falling, and minimize the risk of injury should a fall occur. Strength training, even in frail, elderly people over 85 years of age, has been shown not only to improve balance, muscle strength, and mobility but also to increase energy expenditure and energy intake. This finding highlights another reason to be physically active: a person spending energy on physical activity can afford to eat more food and, with it, more nutrients. People who are committed to an ongoing fitness program have higher energy and nutrient intakes than more sedentary people.

Activities of all kinds are recommended to maintain and promote health.[12] Strength training improves muscle strength, which enhances a person's ability to perform many of life's daily tasks such as climbing stairs and carrying packages.[13] Muscle strength during midlife (between the ages of 45 and 65) predicts health and disability 25 years later. Greater muscle strength during midlife may act as a strength *reserve* later on, protecting older adults from disability even when chronic conditions develop. In short, improving overall strength during early and middle adulthood could potentially lower the risk of later physical disability.

© R.W. Jones/Corbis

Regular physical activity promotes a healthy, independent lifestyle.

life expectancy: the average number of years lived by people in a given society.

life span: the maximum number of years of life attainable by a member of a species.

longevity: long duration of life.

quality of life: a person's perceived physical and mental well-being.

physiological age: a person's age as estimated from her or his body's health and probable life expectancy.

chronological age: a person's age in years from his or her date of birth.

Ideally, physical activity should be part of each day's schedule and should be intense enough to prevent muscle atrophy and to speed up the heartbeat and respiration rate. Although aging affects both speed and endurance to some degree, older adults can still train and achieve exceptional performances. Healthy older adults who have not been active can ease into a suitable routine. They can start by walking short distances until they are walking at least 10 minutes continuously and then gradually increase their distance to a 30-minute walk at least five days a week.[14] With persistence, people can achieve great improvements at any age. Table 13-1 provides exercise guidelines for seniors; people with medical conditions should check with a physician before beginning an exercise routine, as should sedentary men over 40 and women over 50 who want to participate in a vigorous program.

Restriction of kCalories In their efforts to understand longevity, researchers have not only observed people but have also manipulated influencing factors, such as diet, in animals. This research has produced some interesting and suggestive findings. For example, animals live longer and have fewer age-related diseases when their energy intakes are restricted. These life-prolonging benefits become evident when the diet provides enough food to prevent malnutrition and energy intake of about 70 percent of normal; benefits decline as the age of starting the energy restriction is delayed.[15]

Exactly how energy restriction prolongs life remains unexplained, although gene activity appears to play a key role. Energy restriction in animals prevents alterations

TABLE 13-1 Exercise Guidelines for Older Adults

	Aerobic	Strength	Balance	Flexibility
Examples	© Geoff Manasse/Photodisc/PictureQuest	© IT Stock Free/PictureQuest	© IT Stock Free/PictureQuest	© Ron Chapple/Thinkstock/PictureQuest
Start easy and progress gradually	Be active 5 minutes on most or all days	Using 0- to 2-pound weights, do one set of 8–12 repetitions twice a week	Hold onto table or chair with one hand, then with one finger	Hold stretch for 10 seconds; do each stretch three times
Frequency	At least five days per week of moderate activity or at least three days per week of vigorous activity	At least two (nonconsecutive) days per week	Three days each week	At least two days per week, preferably on all days that aerobic or strength activities are performed
Intensity[a]	Moderate, vigorous, or combination	Moderate to high; 10 to 15 repetitions per exercise		At least 10 minutes per day
Duration	At least 30 minutes of moderate activity in bouts of at least 10 minutes each or at least 20 minutes of continuous vigorous activity	8 to 10 exercises involving the major muscle groups		Stretch major muscle groups for 10–30 seconds, repeating each stretch three to four times
Cautions and comments	Stop if you are breathing so hard you can't talk or if you feel dizziness or chest pain	Breathe out as you contract and in as you relax (do not hold breath); use smooth, steady movements	Incorporate balance techniques with strength exercises as you progress	Stretch after strength and endurance exercises for 20 minutes, three times a week; use slow, steady movements; bend joints slightly

Note: Activity recommendations are in addition to routine activities of daily living (such as getting dressed, cooking, and grocery shopping) and moderate activities lasting less than 10 minutes.

[a]On a 10-point scale, where sitting = 0 and maximum effort = 10, moderate intensity = 5 to 6 and vigorous intensity = 7 to 8.

Source: M. E. Nelson and coauthors, Physical activity and public health in older adults: Recommendation from the American College of Sports Medicine and the American Heart Association, *Medicine and Science in Sports and Exercise* 39 (2007): 1435–1445.

in gene expression that are associated with aging.[16] Food restriction may also extend the life span by preventing damaging lipid oxidation, thereby delaying the onset of age-related diseases, such as atherosclerosis.[17] Other research suggests that energy restriction beneficially alters several aspects of fat cell metabolism, which may play a role in reducing chronic disease risk and extending life.[18] Experiments with food restriction and longevity in animals have *not* suggested any direct applications to human nutrition.

Moderate energy restriction (80 percent of usual intake) in human beings may be valuable. When people restrict energy intake moderately, body weight, body fat, and blood pressure drop, and HDL cholesterol rises—favorable changes for preventing chronic diseases.[19] The reduction in oxidative damage that occurs with energy restriction in animals also occurs in people whose diets include antioxidant nutrients and phytochemicals. Diets such as the Mediterranean diet that include an abundance of fruits, vegetables, olive oil, whole grains, and legumes—with their array of antioxidants and phytochemicals—support good health and long life.

Nutrition and Disease Prevention

Nutrition alone, even if ideal, cannot ensure a long and robust life. Nevertheless, nutrition clearly affects aging and longevity in human beings by way of its role in disease prevention. Among the better-known relationships between nutrition and disease are the following:

▌ Appropriate energy intake helps prevent *obesity, diabetes,* and related *cardiovascular diseases* such as atherosclerosis and hypertension (Chapters 7, 21, and 22) and may influence the development of some forms of *cancer* (Chapter 25).

▌ Adequate intakes of essential nutrients prevent *deficiency diseases* such as scurvy, goiter, anemia, and the like (Chapters 8 and 9).

▌ Variety in food intake, as well as ample intakes of certain fruits and vegetables, may be protective against certain types of *cancer* (Chapter 25).

▌ Moderation in sugar intake helps prevent *dental caries* (Nutrition in Practice 17).

▌ Appropriate fiber intakes help prevent disorders of the digestive tract such as *constipation, diverticulosis,* and possibly *colon cancer* (Chapters 2, 18, and 25).

▌ Moderate sodium intake and adequate intakes of potassium, calcium, and other minerals help prevent *hypertension* (Chapters 9 and 22).

▌ An adequate calcium intake throughout life helps protect against *osteoporosis* (Chapter 9).

Other, less well-established links between nutrition and disease are being discovered each day. Research that focuses on how life factors affect aging and disease processes is vital to ensuring that increasing numbers of people can look forward to long, healthy lives.

IN SUMMARY

▌ Life expectancy in the United States increased dramatically in the 20th century.

▌ Factors that enhance longevity include limited or no alcohol use, regular balanced meals rich in fruits and vegetables, weight control, adequate sleep, abstinence from smoking, and regular physical activity.

▌ Nutrition alone, even if ideal, cannot guarantee a long and robust life. At the very least, however, nutrition—especially when combined with regular physical activity—can influence aging and longevity in human beings by supporting good health and preventing disease.

Nutrition-Related Concerns during Late Adulthood

Nutrition through the prime years may play a greater role than has been realized in preventing many changes once thought to be inevitable consequences of growing older. The following discussions of cataracts and macular degeneration, arthritis, and the aging brain show that nutrition may provide at least some protection against some of the conditions commonly associated with aging.

Cataracts and Macular Degeneration

Cataracts are age-related thickenings in the lenses of the eye that impair vision. If not surgically removed, they ultimately lead to blindness. Oxidative stress appears to play a significant role in the development of cataracts, and the antioxidant nutrients and phytochemicals in fruits and vegetables may help minimize the damage. Some studies suggest that a diet providing ample carotenoids, vitamin C, and vitamin E may be especially important for preventing early onset of cataracts.[20] Also, people who follow the *Dietary Guidelines for Americans* are reported to have fewer cataracts, though cataracts can occur even in well-nourished individuals due to exposure to ultraviolet light, oxidative damage, viral infections, toxic substances, genetic disorders, injury, or other trauma. Most cataracts are vaguely called "senile" cataracts, meaning "caused by aging." In the United States, more than half of all adults age 65 and older have a cataract.

One other diet-related factor may play a role in cataract development: obesity. How obesity may influence the development of cataracts is not known, but researchers speculate that conditions that typically accompany obesity such as glucose intolerance and insulin resistance may provide clues.

The leading cause of vision loss among older people is **macular degeneration,** a deterioration of the macular region of the eye.[21] As with cataracts, risk factors for age-related macular degeneration include oxidative stress from sunlight. Preventive factors for macular degeneration may include supplements of the omega-3 fatty acid DHA and the carotenoids lutein and zeaxanthin.[22]

Arthritis

An estimated 50 million people in the United States have some form of **arthritis.**[23] As the population ages, it is expected that the prevalence will increase to 60 million by 2020. The most common type of arthritis that disables older people is **osteoarthritis,** a painful swelling of the joints. During movement, the ends of bones are normally protected from wear by cartilage and by small sacs of fluid that lubricate the joint. With age, the cartilage sometimes disintegrates, and the joints become malformed and painful to move.

One known connection between osteoarthritis and nutrition is overweight. Weight loss can help overweight people with osteoarthritis, partly because the joints affected are often weight-bearing joints that are stressed and irritated by having to carry excess poundage. Interestingly, however, weight loss often relieves the worst pain of osteoarthritis in the hands as well, even though they are not weight-bearing joints. Jogging and other weight-bearing activities do not worsen osteoarthritis. In fact, both aerobic activity and weight training offer modest improvements in physical performance and pain relief, especially when accompanied by even modest weight loss.[24]

Nutrition quackery to treat arthritis is abundant, but no single universally effective diet for arthritis relief is known. Table 13-2 presents some of the many ineffective dietary treatments for osteoarthritis. Traditional medical intervention for arthritis includes medication and surgery. Two popular supplements for treating osteoarthritis—glucosamine and chondroitin—may alleviate pain and improve mobility, but mixed

TABLE 13-2 Ineffective Dietary Strategies for Arthritis

- Alfalfa tea
- Aloe vera liquid
- Amino acid supplements
- Blackstrap molasses
- Burdock root
- Calcium
- Celery juice
- Cod liver oil
- Copper supplements
- Dimethyl sulfoxide (DMSO)
- Fasting
- Fresh fruit
- Garlic
- Honey
- Inositol
- Kelp
- Lecithin
- Para-amino benzoic acid (PABA)
- Raw liver
- Superoxide dismutase (SOD)
- Vitamin D
- Vitamin megadoses
- Watercress
- Yeast

reports from studies emphasize the need for additional research.[25] Drugs and supplements used to relieve arthritis can impose nutrition risks; some affect appetite and alter the body's use of nutrients, as Chapter 15 explains.

Another type of arthritis, known as **rheumatoid arthritis,** has a possible link to diet through the immune system. In rheumatoid arthritis, the immune system mistakenly attacks the bone coverings as if they were made of foreign tissue. In some individuals, certain foods, notably a Mediterranean-type diet of fish, vegetables, and olive oil, may moderate the inflammatory responses and provide some relief.[26]

The omega-3 fatty acids commonly found in fish oil reduce joint tenderness and improve mobility in some people with rheumatoid arthritis. The same diet recommended for heart health—one low in saturated fat from meats and milk products and high in omega-3 fats from fish—helps prevent or reduce the inflammation in the joints that makes arthritis so painful. Fish, not fish oil supplements, is the preferred source of omega-3 fatty acids; supplements are appropriate only for persons with cardiovascular disease who cannot obtain enough EPA and DHA from fish, and then only under a physician's supervision.[27]

The Aging Brain

The brain, like all of the body's organs, responds to both inherited and environmental factors that can enhance or diminish its amazing capacities. One of the challenges researchers face when studying the aging of the human brain is to distinguish among changes caused by normal, age-related, physiological processes; changes caused by diseases; and changes caused by cumulative, extrinsic factors such as diet.

The brain normally changes in some characteristic ways as it ages. For one thing, its blood supply decreases. For another, the number of **neurons,** the brain cells that specialize in transmitting information, diminishes as people age. When the number of nerve cells in one part of the **cerebral cortex** diminishes, hearing and speech are affected. Losses of neurons in other parts of the cortex can impair memory and cognitive function. When the number of neurons in the hindbrain diminishes, balance and posture are affected. Losses of neurons in other parts of the brain affect still other functions.

Nutrient Deficiencies and Brain Function Clinicians now recognize that much of the cognitive loss and forgetfulness generally attributed to aging is due in part to extrinsic, and therefore controllable, factors such as nutrient deficiencies. The ability of neurons to synthesize specific neurotransmitters depends in part on the availability of precursor nutrients that are obtained from the diet. The neurotransmitter serotonin, for example, derives from the amino acid tryptophan. In addition, the enzymes involved in neurotransmitter synthesis require vitamins and minerals to function properly. Thus, nutrient deficiencies may contribute to the loss of memory and cognition that some older adults experience.[28] Such losses may be preventable or at least diminished or delayed through diet. Table 13-3 summarizes some of the better-known connections between brain function and nutrients.

Both foods and mental challenges nourish the brain.

Factors that protect brain function:

- Physical activities
- Intellectual challenges
- Social interactions
- Balanced diet rich in antioxidants

cataracts: thickenings of the eye lenses that impair vision and can lead to blindness.

macular (MACK-you-lar) **degeneration:** deterioration of the macular area of the eye that can lead to loss of central vision and eventual blindness. The **macula** is a small, oval, yellowish region in the center of the retina that provides the sharp, straight-ahead vision so critical to reading and driving.

arthritis: inflammation of a joint, usually accompanied by pain, swelling, and structural changes.

osteoarthritis: a painful, chronic disease of the joints that occurs when the cushioning cartilage in a joint breaks down; joint structure is usually altered, with loss of function; also called *degenerative arthritis.*

rheumatoid arthritis: a disease of the immune system involving painful inflammation of the joints and related structures.

neurons: nerve cells; the structural and functional units of the nervous system. Neurons initiate and conduct nerve transmissions.

cerebral cortex: the outer surface of the cerebrum, which is the largest part of the brain.

TABLE 13-3 Summary of Nutrient–Brain Relationships

Brain Function	Depends on an Adequate Intake of:
Short-term memory	Vitamin B_{12}, vitamin C, vitamin E
Performance in problem-solving tests	Riboflavin, folate, vitamin B_{12}, vitamin C
Mental health	Thiamin, niacin, zinc, folate
Cognition	Folate, vitamin B_6, vitamin B_{12}, iron, vitamin E
Vision	Essential fatty acids, vitamin A
Neurotransmitter synthesis	Tyrosine, tryptophan, choline

A person with typical benign memory loss of aging may:

▪ Forget names or places temporarily.

▪ Need more time to remember a task the person set out to do.

▪ Misplace items in typical places, such as leaving reading glasses in the car.

▪ Joke about apparent forgetfulness.

▪ Experience little interference in living a full and productive life.

A person in the early stages of Alzheimer's disease may:

▪ Ask the same questions repeatedly.

▪ Have difficulty remembering common words—saying "that thing for my mouth" instead of "toothbrush," for example.

▪ Mix words up—saying "bed" instead of "table," for example.

▪ Be unable to complete familiar tasks, such as cooking a meal or making a phone call.

▪ Misplace items in uncommon places, such as putting the phone in the refrigerator.

▪ Get lost while driving on familiar streets.

▪ Have sudden, unexplained mood or behavior changes.

▪ Become less able to follow directions.

▪ Be unaware of memory loss or forgetfulness.

▪ Experience worsening memory loss that progressively diminishes the quality of life.

TABLE 13-4 Common Signs of Dementia

▪ Agitated behavior
▪ Becoming lost in familiar surroundings or circumstances
▪ Confusion
▪ Delusions
▪ Loss of interest in daily activities
▪ Loss of memory
▪ Loss of problem-solving skills
▪ Unclear thinking

In some instances, the degree of cognitive loss is extensive. Such **senile dementia** may be attributable to a specific disorder such as a brain tumor or Alzheimer's disease. Table 13-4 lists common signs of dementia.

Alzheimer's Disease In **Alzheimer's disease,** the most prevalent form of senile dementia, brain cell death occurs in the areas of the brain that coordinate memory and cognition (see the margin). Alzheimer's disease afflicts an estimated 5 million people in the United States, and that number is expected to almost triple by the year 2050.[29] Diagnosis of Alzheimer's depends on its characteristic symptoms: the victim gradually loses memory and reasoning ability, the ability to communicate, physical capabilities, and eventually life itself.

Researchers are closing in on the cause of Alzheimer's disease.* Clearly, genetic factors are involved.[30] Free radicals and oxidative stress also seem to be involved.[31] Nerve cells in the brains of people with Alzheimer's disease show evidence of free-radical attack—damage to DNA, cell membranes, and proteins—and of the minerals that trigger these attacks—iron, copper, zinc, and aluminum.[32] Some research suggests that the antioxidant nutrients can limit free-radical damage and delay or prevent Alzheimer's disease.[33]

Increasing evidence also suggests that overweight and obesity in middle age are associated with dementia, in general, and with Alzheimer's disease, in particular.[34] The possible relationship between obesity and Alzheimer's disease is disturbing given the current obesity epidemic. Efforts to prevent and treat obesity, however, may also help prevent Alzheimer's disease.

In Alzheimer's disease, the brain develops **senile plaques** and **neurofibrillary tangles.** Senile plaques are clumps of a protein fragment called beta-amyloid, whereas neurofibrillary tangles are snarls of the fibers that extend from the nerve cells. Both seem to occur in response to oxidative stress.[35] Researchers question whether these characteristics are the cause or the result of Alzheimer's disease.[36] In fact, scientists are unsure whether these plaques and tangles are causing the damage, serving as markers, or even protecting the brain by sequestering the proteins that begin the dementia process.[37] In any case, treatment research focuses on lowering beta-amyloid levels.[38]

Research suggests that cardiovascular disease risk factors such as high blood pressure, diabetes, and elevated levels of homocysteine may be related to the development of Alzheimer's disease.[39] Diets designed to support a healthy heart, which include the omega-3 fatty acids of oily fish, may benefit brain health as well.[40] Similarly, physical activity supports heart health and slows the cognitive decline of Alzheimer's disease.[41]

Treatment for Alzheimer's disease involves providing care to clients and support to their families. Drugs are used to improve or at least to slow the loss of short-term memory and cognition, but they do not cure the disease. Other drugs may be used to control depression, anxiety, and behavior problems.

Maintaining appropriate body weight may be the most important nutrition concern for the person with Alzheimer's disease. Depression and forgetfulness can lead to changes in eating behaviors and poor food intake. Furthermore, changes in the body's weight-regulation system may contribute to weight loss. Perhaps the best that a caregiver can do nutritionally for a person with Alzheimer's disease is to supervise food planning and mealtimes. Providing well-liked and well-balanced meals and snacks in a cheerful atmosphere encourages food consumption. To minimize confusion, offer a few ready-to-eat foods, in bite-size pieces, with seasonings and sauces. To avoid mealtime disruptions, control distractions such as music, television, children, and the telephone.

*Information on the genetic and other aspects of Alzheimer's is available from the Alzheimer's Disease Education and Referral Center, www.nia.nih.gov/Alzheimers.

■ Cataracts, age-related macular degeneration, and arthritis afflict millions of older adults, while others face senile dementia and other losses of brain function.

■ Some of these problems may be inevitable, but others are preventable, and good nutrition may play a key preventive role.

Energy and Nutrient Needs during Late Adulthood

Knowledge about the nutrient needs and nutrition status of older adults has grown considerably in recent years. The Dietary Reference Intakes (DRI) cluster people over the age of 50 into two age categories—51 to 70 years old, and 71 years and older. Research is showing that the nutrition needs of people 50 to 70 years old may be very different from those of people over 70—a pattern that makes sense considering the wide age span involved.

Setting standards for older people is difficult, though, because individual differences become more pronounced as people grow older. People start out with different genetic predispositions and ways of handling nutrients, and the effects of these differences become magnified with years of unique dietary habits. For example, one person may tend to omit fruits and vegetables from his diet, and by the time he is old, he may have a set of nutrition problems associated with a lack of fiber and antioxidants. Another person who omitted milk and milk products all her life may have nutrition problems related to a lack of calcium. Also, as people age, they may suffer different chronic diseases and take different medications—both have impacts on nutrient needs. Table 13-5 lists some changes of aging that can affect nutrition. For all these reasons, researchers have difficulty even defining "healthy aging," a prerequisite to developing recommendations that are designed to meet the "needs of practically all healthy persons." Still, some generalizations are valid. The next sections give special attention to a few nutrients of concern.

Energy and Energy Nutrients

Energy needs decline with advancing age. As a general rule, an adult's energy needs decline an estimated 5 percent per decade. One reason is that people usually reduce their physical activity as they age, although they need not do so. Another reason is that

TABLE 13-5 Examples of Physical Changes of Aging That Affect Nutrition

Mouth	Tooth loss, gum disease, and reduced salivary output impede chewing and swallowing. Swallowing disorders and choking may become likely. Discomfort and pain associated with eating may reduce food intake.
Digestive tract	Intestines lose muscle strength, resulting in sluggish motility that leads to constipation (see Chapter 18). Stomach inflammation, abnormal bacterial growth, and greatly reduced acid output impair digestion and absorption. Pain may cause food avoidance or reduced intake.
Hormones	For example, the pancreas secretes less insulin and cells become less responsive, causing abnormal glucose metabolism.
Sensory organs	Diminished senses of smell and taste can reduce appetite; diminished sight can make food shopping and preparation difficult.
Body composition	Weight loss and decline in lean body mass lead to lowered energy requirements. May be preventable or reversible through physical activity.
Urinary tract	Increased frequency of urination may limit fluid intake.

senile dementia: the loss of brain function beyond the normal loss of physical adeptness and memory that occurs with aging.

Alzheimer's disease: a progressive, degenerative disease that attacks the brain and impairs thinking, behavior, and memory.

senile plaques: clumps of the protein fragment beta-amyloid on the nerve cells, commonly found in the brains of people with Alzheimer's dementia.

neurofibrillary tangles: snarls of the threadlike strands that extend from the nerve cells, commonly found in the brains of people with Alzheimer's dementia.

FIGURE 13-2 **Sarcopenia**

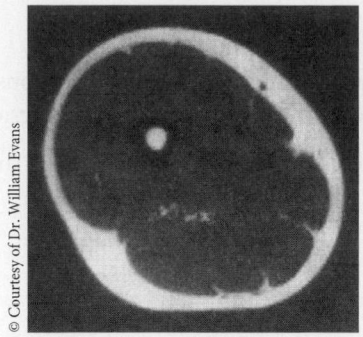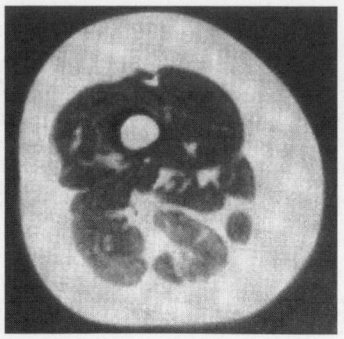

These cross sections of two women's thighs may appear to be about the same size from the outside, but the 20-year-old woman's thigh (left) is dense with muscle tissue. The 64-year-old woman's thigh (right) has lost muscle and gained fat, changes that may be largely preventable with strength-building physical activities.

basal metabolic rate declines 1 to 2 percent per decade, in part because lean body mass and thyroid hormones diminish. Loss of muscle mass, known as **sarcopenia,** can be significant in the later years (its prevalence is more than 50 percent among those older than 80), and its consequences can be dramatic (see Figure 13-2). As skeletal muscle mass diminishes, people lose their ability to move and to maintain balance, making falls likely. The limitations that accompany the loss of muscle mass and strength play a key role in the diminishing health that often accompanies aging.[42] To some extent, however, declines in lean body mass and energy needs may not be entirely inevitable. Optimal nutrition with sufficient protein and regular physical activity can help maintain muscle mass and strength and minimize the changes in body composition associated with aging.[43] Physical activity not only increases energy expenditure but, along with sound nutrition, enhances bone density and supports many body functions as well.[44]

The lower energy expenditures of many older adults require that they consume less food energy to maintain their weights. Accordingly, the estimated energy requirements for adults decrease steadily after age 19. Energy intakes typically decline in parallel with needs. Still, many older adults are overweight, indicating that their food intakes do not decline enough to compensate for their reduced energy expenditures. Overweight and obesity in older adults increase the risk for many diseases and for disabilities as well.[45]

On limited energy allowances, people must select mostly nutrient-dense foods. There is little leeway for added sugars, solid fats, or alcohol. Older adults can follow the USDA Food Guide (see pp. 16–17), making sure to choose the recommended daily amounts of food from each food group that are appropriate to their energy needs (see Table 1-4 on p. 18).

Protein The protein needs of older adults appear to be about the same as those of younger people. Protein is especially important for older adults in order to support a healthy immune system and to prevent muscle wasting. Because energy needs decrease, however, the protein has to be obtained from low-kcalorie sources of high-quality protein, such as lean meats, poultry, fish, and eggs; fat-free and low-fat milk products; and legumes.

Underweight or malnourished older adults need protein- and energy-dense snacks such as hard-boiled eggs, tuna fish and crackers, peanut butter on wheat toast, and hearty soups. Drinking liquid nutritional formulas between meals can also boost energy and nutrient intakes.

Carbohydrate and Fiber As always, abundant carbohydrate is needed to protect protein from being used as an energy source. The recommendation to obtain ample amounts of mostly whole-grain breads, cereals, rice, and pasta holds true for older

sarcopenia *(SAR-koh-PEE-nee-ah):* loss of skeletal muscle mass, strength, and quality.

people. As for younger people, a steady supply of carbohydrate is essential for optimal brain functioning. With age, fiber takes on extra importance for its role in alleviating constipation, a common complaint among older adults and among nursing home residents in particular. Fruits and vegetables supply viscous (soluble) fibers thought to help ward off diseases of aging, but factors such as transportation problems, limited cooking facilities, and chewing problems limit some elderly people's intakes of fresh fruits and vegetables.[46] Most older adults do not obtain the recommended daily 25 or more grams of fiber (14 grams per 1000 kcalories).[47] When low fiber intakes are combined with low fluid intakes, inadequate physical activity, and constipating medications, constipation becomes inevitable.

Fat As is true for people of all ages, fat intake needs to be moderate in the diets of most older adults—enough to enhance flavors and provide valuable essential fatty acids and other nutrients, but not so much as to raise risks of cancer, atherosclerosis, and other degenerative diseases. The recommendation should not be taken too far; limiting fat too severely may lead to nutrient deficiencies and weight loss—two problems that carry even greater health risks in the elderly than overweight.

Water

Dehydration is a risk for older adults, who may not notice or pay attention to their thirst or who find it difficult and bothersome to get a drink or to get to a bathroom.[48] Older adults who have lost bladder control may be afraid to drink too much water. Despite real fluid needs, older people do not seem to feel as thirsty or notice mouth dryness as readily as younger people. Many nursing home employees say it is hard to persuade their elderly clients to drink enough water and fruit juices.

Total body water decreases as people age, so even mild stresses such as fever or hot weather can precipitate rapid dehydration in older adults. Dehydrated older adults seem to be more susceptible to urinary tract infections, pneumonia, pressure ulcers, confusion, and disorientation.[49] An intake of 9 cups a day of total beverages, including water, is recommended for women; for men, the recommendation is 13 cups a day of total beverages, including water.[50]

Vitamins and Minerals

As research reveals more about how specific vitamins and minerals influence disease prevention and how age-related physiological changes affect nutrient metabolism, optimal intakes of vitamins and minerals for different groups of older adults are being defined. This section highlights the vitamins and minerals of greatest concern to older adults.

Vitamin D Older adults face a greater risk of vitamin D deficiency than younger people do. Vitamin D–fortified milk is the most reliable source of vitamin D, but many older adults drink little or no milk. Consequently, the vitamin D intakes of many older adults are less than half of recommendations. Further compromising the vitamin D status of many older people, especially those in nursing homes, is their limited exposure to sunlight. Finally, aging reduces the skin's capacity to make vitamin D and the kidneys' ability to convert it to its active form. Not only are older adults not getting enough vitamin D, but they may actually need more to improve both muscle and bone strength.[51] The margin lists vitamin D recommendations to prevent bone loss and to maintain vitamin D status in older people, especially those who engage in minimal outdoor activity.

Vitamin D RDA during late adulthood:
- 15 µg/day (51–70 yr)
- 20 µg/day (>70 yr)

Vitamin B$_{12}$ The DRI committee recommends that adults aged 51 years and older obtain 2.4 micrograms of vitamin B$_{12}$ daily *and* that vitamin B$_{12}$–fortified foods (such as fortified cereals) or supplements be used to meet much of the DRI recommended intake. The committee's recommendation reflects the finding that an estimated 10 to 30 percent of adults older than 50 years lose the ability to produce enough stomach

Vitamin B$_{12}$ RDA during late adulthood:
- 2.4 µg/day

acid to make the protein-bound form of vitamin B_{12} available for absorption. Synthetic vitamin B_{12} is reliably absorbed, however. Given the poor cognition, anemia, and devastating neurological effects associated with a vitamin B_{12} deficiency, an adequate intake is imperative.[52]

One cause of the malabsorption of protein-bound vitamin B_{12} is a condition known as **atrophic gastritis.** The prevalence of atrophic gastritis among those 60 years of age and older is high.

Folate As is true of vitamin B_{12}, folate intakes of older adults typically fall short of recommendations. The elderly are also more likely to have medical conditions or to take medications that can compromise folate status.

Iron Iron-deficiency anemia is less common in older adults than in younger people, but it still occurs in some older adults, especially in those with low food-energy intakes. Aside from diet, other factors in many older people's lives make iron deficiency likely: chronic blood loss from disease conditions and medicines and poor iron absorption due to reduced secretion of stomach acid and antacid use. Anyone concerned with older people's nutrition should keep these possibilities in mind.

Zinc Zinc intake is commonly low in older people. Zinc deficiency can depress the appetite and blunt the sense of taste, thereby leading to low food intakes and worsening of zinc status. Zinc deficiency may also increase the likelihood of infection.[53] Many medications that older adults commonly use can impair zinc absorption or enhance its excretion and thus lead to deficiency.

Calcium The importance of abundant dietary calcium throughout life to protect against osteoporosis has been emphasized throughout this book. The calcium intakes of many people, especially women, in the United States are well below the recommended amount. If milk causes stomach discomfort, as many older adults report, then lactose-modified milk or other calcium-rich foods should take its place.

Nutrient Supplements for Older Adults

People judge for themselves how to manage their nutrition, and more than half of older adults turn to dietary supplements. When recommended by a physician or registered dietitian, vitamin D and calcium supplements for osteoporosis or vitamin B_{12} for pernicious anemia may be beneficial. Advertisers target older people with appeals to take supplements and eat "health" foods, claiming that these products prevent disease and promote longevity. Quite often, those who take supplements are not deficient in the nutrients being supplemented.

Elderly people often benefit from a balanced low-dose vitamin and mineral supplement, however.[54] Such supplements supply many of the needed minerals along with the vitamins often lacking in older people's diets without providing too much of any one nutrient.

Food is still the best source of nutrients for everybody, however. Supplements are just that—supplements to foods, not substitutes for them. For anyone who is motivated to obtain the best possible health, it is never too late to learn to eat well, become physically active, and adopt other lifestyle changes such as quitting smoking, moderating alcohol use, and the like. Table 13-6 summarizes the nutrient concerns of aging. Table 13-7 offers strategies for growing old healthfully.

The Effects of Drugs on Nutrients

As people grow older, the use of medicines—from over-the-counter types such as aspirin and laxatives to prescription medications of all kinds—becomes commonplace. Most drugs interact with one or more nutrients in several ways, usually resulting in greater-than-normal needs for these nutrients. Chapter 15 offers a discussion

TABLE 13-6 Summary of Nutrient Concerns of Aging

Nutrient	Effect of Aging	Comments
Water	Lack of thirst and decreased total body water make dehydration likely.	Mild dehydration is a common cause of confusion. Difficulty obtaining water or getting to the bathroom may compound the problem.
Energy	Need decreases as muscle mass decreases (sarcopenia).	Physical activity moderates the decline.
Fiber	Likelihood of constipation increases with low intakes and changes in the GI tract.	Inadequate water intakes and lack of physical activity, along with some medications, compound the problem.
Protein	Needs may stay the same or increase slightly.	Low-fat, high-fiber legumes and grains meet both protein and other nutrient needs.
Vitamin B_{12}	Atrophic gastritis is common.	Deficiency causes neurological damage; supplements may be needed.
Vitamin D	Increased likelihood of inadequate intake; skin synthesis declines.	Daily sunlight exposure in moderation or supplements may be beneficial.
Calcium	Intakes may be low; osteoporosis is common.	Stomach discomfort commonly limits milk intake; calcium substitutes or supplements may be needed.
Iron	In women, status improves after menopause; deficiencies are linked to chronic blood losses and low stomach acid output.	Adequate stomach acid is required for absorption; antacid or other medicine use may aggravate iron deficiency; vitamin C and meat increase absorption.

TABLE 13-7 Strategies for Growing Old Healthfully

- Choose nutrient-dense foods.
- Be physically active. Walk, run, dance, swim, bike, or row for aerobic activity. Lift weights, do calisthenics, or pursue some other activity to tone, firm, and strengthen muscles. Practice balancing on one foot or doing simple movements with your eyes closed. Modify activities to suit changing abilities and tastes.
- Maintain appropriate body weight.
- Reduce stress (cultivate self-esteem, maintain a positive attitude, manage time wisely, know your limits, practice assertiveness, release tension, and take action).
- For women, discuss with a physician the risks and benefits of estrogen replacement therapy.
- For people who smoke, discuss with a physician strategies and programs to help you quit.
- Expect to enjoy sex and learn new ways of enhancing it.
- Use alcohol only moderately, if at all; use drugs only as prescribed.
- Take care to prevent accidents.
- Expect good vision and hearing throughout life; obtain glasses and hearing aids if necessary.
- Take care of your teeth; obtain dentures if necessary.
- Be alert to confusion as a disease symptom and seek diagnosis.
- Take medications as prescribed; see a physician before self-prescribing medicines or herbal remedies and a registered dietitian before self-prescribing supplements.
- Control depression through activities and friendships; seek professional help if necessary.
- Drink plenty of water every day.
- Practice mental skills. Keep on solving math problems and crossword puzzles, playing cards or other games, reading, writing, imagining, and creating.
- Make financial plans early to ensure security.
- Accept change. Work at recovering from losses; make new friends.
- Cultivate spiritual health. Cherish personal values. Make life meaningful.
- Go outside for sunshine and fresh air as often as possible.
- Be socially active—play bridge, join an exercise or dance group, take a class, teach a class, eat with friends, volunteer time to help others.
- Stay interested in life—pursue a hobby, spend time with grandchildren, take a trip, read, grow a garden, or go to the movies.
- Enjoy life.

of diet-drug interactions and describes the many reasons why elderly people are vulnerable to such interactions.

The most common drug that can affect nutrition in older people is alcohol. The effects of alcohol on people of all ages are explained in Nutrition in Practice 20.

IN SUMMARY

- Table 13-6 summarizes the nutrient concerns of aging.

- The ever-growing number of older people creates an urgent need to know more about how their nutrient requirements differ from those of others and how such knowledge can enhance their health.

atrophic gastritis (a-TRO-fik gas-TRI-tis): a condition characterized by chronic inflammation of the stomach accompanied by a diminished size and functioning of the mucosa and glands.

Food Choices and Eating Habits of Older Adults

To provide any benefit, strategies and interventions to improve a person's nutrition status must be based on knowledge of food preferences and eating patterns. Menus and feeding programs for older adults must take into consideration not only the food likes and dislikes but also the living conditions, economic status, and medical conditions of this diverse group of people. If nutrition intervention is to be successful, it is essential to know what foods people will eat, in what settings they like to eat these foods, and whether they can buy and prepare meals.

Older people are, for the most part, independent, socially sophisticated, mentally lucid, fully participating members of society who report themselves to be happy and healthy. In fact, chronic disabilities among the elderly have declined dramatically in recent years. Older people spend more money per person on foods to eat at home than other age groups and less money on foods away from home. Older adults appreciate good-tasting, nutritious foods in easy-to-open, single-serving packages with labels that are easy to read. Such products enable older adults to maintain their independence; most of them want to take care of themselves and need to feel a sense of control and involvement in their own lives. As discussed earlier, another way older adults can take care of themselves is by remaining or becoming physically active. Physical activity helps preserve one's ability to perform daily tasks and so promotes independence.[55]

Individual Preferences

Familiarity, taste, and health beliefs are most influential on older people's food choices. Eating foods that are familiar, especially ethnic foods that recall family meals and pleasant times, can be comforting. Older adults are choosing poultry and fish, low-fat milk and milk products, and high-fiber breads and grains, indicating that they recognize the importance of diet in supporting good health. Few older adults, however, consume the recommended amounts of milk products.

Meal Setting

The food choices and eating habits of older adults are also affected by the changes in lifestyle that often accompany aging in this society. Whether people live alone, with others, or in institutions affects the way they eat. For example, men living alone are most likely to be poorly nourished. Older adults who live alone do not make poorer food choices than those who live with companions; rather, they consume too little food: loneliness is directly related to inadequacies, especially of energy intakes.

Shared meals can brighten the day and enhance the appetite.

Depression

Another factor affecting food intake and appetite in older people is depression. Loss of appetite and motivation to cook or even to eat frequently accompanies depression. An overwhelming feeling of grief and sadness at the death of a spouse, friend, or family member may leave many people, particularly elderly people, with a feeling of powerlessness to overcome the depression. The support and companionship of family and friends, especially at mealtimes, can help overcome depression and enhance appetite. The accompanying Case Study presents a man who has several of these problems. Use the suggestions here, and in the last section of this chapter, to help develop solutions. The Nutrition Assessment Checklist for Older Adults at the end of this chapter helps

Case Study ELDERLY MAN WITH A POOR DIET

Mr. Brezenoff is 75 years old and lives alone. He has slowly been losing weight since his wife died a year ago. At 5 · feet 8 inches tall, he currently weighs 124 pounds. His previous weight was 150 pounds. In talking with Mr. Brezenoff, you realize that he doesn't even like to talk about food, let alone eat it. "My wife always did the cooking before, and I ate well. Now I just don't feel like eating." You manage to find out that he skips breakfast, has soup and bread for lunch, and sometimes eats a cold-cut sandwich or a frozen dinner for supper. He seldom sees friends or relatives. Mr. Brezenoff has also lost several teeth and doesn't eat any raw fruits or vegetables because he finds them hard to chew. He lives on a meager but adequate income.

1. Consult the BMI table (inside back cover), and judge whether Mr. Brezenoff is at a healthy weight. What other assessments might you use to back up your judgment? Is his weight loss significant?

2. What factors are contributing to Mr. Brezenoff's poor food intake? What nutrients are probably deficient in his diet?

3. Look at Mr. Brezenoff as an individual and suggest ways he can improve his diet and his lifestyle.

4. What other aspects of Mr. Brezenoff's physical and mental health should you consider in helping him improve his food intake?

to pinpoint nutrition-related factors to look for when working with older adults. To *determine* the risk of malnutrition in older clients, health care providers can keep in mind the characteristics listed in the margin.

Food Assistance Programs

Federally funded programs can provide food and nutrition services for older adults.[56] The Older Americans Act (OAA) provides many different services and support to help older adults remain independent. An integral component of the OAA is the OAA Nutrition Program, which offers services that promote health to older adults. Table 13-8 summarizes food assistance programs available to the elderly.

Risk factors for malnutrition in older adults:

- **Disease**
- **Eating poorly**
- **Tooth loss or oral pain**
- **Economic hardship**
- **Reduced social contact**
- **Multiple medications**
- **Involuntary weight loss or gain**
- **Needs assistance with self-care**
- **Elderly person older than 80 years**

TABLE 13-8 Food Assistance Programs for Older Adults

OAA Nutrition Program

Services: Provides congregate and home-delivered meals to improve older people's nutrition status. Includes transportation to congregate meal sites; shopping assistance; information and referral; and, to some extent, nutrition counseling and education.

Impact: Improves the nutrient content of high-risk older adults' diets and offers socialization and recreation. Many of the nutrition programs around the country go above and beyond federal requirements of congregate and home meals by offering lunch clubs, ethnic meals, and meals for older homeless people.

Supplemental Nutrition Assistance Program (SNAP)

Services: Supplements income for low-income households by means of a card similar to a debit card that can be used to purchase food.

Impact: Serves more as an income supplement for some elderly participants than as a device to improve nutrition status. For other elderly participants, nutrient intakes are higher than those of nonparticipants with similar incomes.

Meals on Wheels

Services: Delivers meals directly to the homebound elderly; integrated into the meal delivery services provided by the OAA Program.

Impact: Focuses on filling the need for weekend and holiday meals for the homebound elderly people, a service that is limited in the OAA Nutrition Program.

Senior Farmers Market Nutrition Program

Services: Provides low-income older adults with coupons that can be exchanged for fresh fruits, vegetables, and herbs at community-supported farmers' markets and roadside stands; administered by the USDA. State agencies may limit sales to specific foods that are locally grown to encourage recipients to support farmers in their own states.

Impact: Increases fresh fruit and vegetable consumption, provides nutrition information, and even reaches the homebound elderly, a group of people who normally do not have access to farmers' markets.

Meals for Singles

Many older adults live alone, and singles of all ages face challenges in purchasing, storing, and preparing food. Large packages of meat and vegetables are often intended for families of four or more, and even a head of lettuce can spoil before one person can use it all. Many singles live in small dwellings and have little storage space for foods. A limited income presents additional obstacles. This section offers suggestions that can help to solve some of the problems singles face, beginning with a note about the dangers of foodborne illness.

Foodborne Illness The risk of older adults getting a foodborne illness is greater than for other adults. The consequences of an upset stomach, diarrhea, fever, vomiting, abdominal cramps, and dehydration are oftentimes more severe, sometimes leading to paralysis, meningitis, or even death. For these reasons, older adults need to carefully follow the food safety suggestions presented in Nutrition in Practice 5.

Spend Wisely People who have the means to shop and cook for themselves can cut their food bills just by being wise shoppers. Large supermarkets are usually less expensive than convenience stores. A grocery list helps reduce impulse buying, and specials and coupons can save money when the items featured are those that the shopper needs and uses.

Buying the right amount so as not to waste any food is a challenge for people eating alone. They can buy fresh milk in the size best suited for personal needs. Pint-size and even cup-size boxes of milk are available and can be stored unopened on a shelf for up to three months without refrigeration.

Many foods that offer a variety of nutrients for practically pennies have a long shelf life; staples such as rice, pastas, dry powdered milk, and dried legumes can be purchased in bulk and stored for months at room temperature. Other foods that are usually a good buy include whole pieces of cheese rather than sliced or shredded cheese, fresh produce in season, variety meats such as chicken livers, and cereals that require cooking instead of ready-to-serve cereals.

A person who has ample freezing space can buy large packages of meat, such as pork chops, ground beef, or chicken, when they are on sale. Then the meat can be immediately wrapped in individual servings for the freezer. All the individual servings can be put in a bag marked appropriately with the contents and the date.

Frozen vegetables are more economical in large bags than in small boxes. The amount needed can be taken out, and the bag closed tightly with a twist tie or rubber band. If the package is returned quickly to the freezer each time, the vegetables will stay fresh for a long time.

Finally, breads and cereals usually must be purchased in larger quantities. Again the amount needed for a few days can be taken out and the rest stored in the freezer.

People who do not have freezer space can ask the grocer to break open a package of wrapped meat and rewrap the portion needed. Similarly, eggs can be purchased by the half-dozen. Eggs can keep for long periods, however, if stored properly in the refrigerator.

Fresh fruits and vegetables can be purchased individually. A person can buy fresh fruit at various stages of ripeness: a ripe one to eat right away, a semiripe one to eat soon after, and a green one to ripen on the windowsill. If vegetables are packaged in large quantities, the grocer can break open the package so that a smaller amount can be purchased. Small cans of fruits and vegetables, even though they are more expensive per unit, are a reasonable alternative, considering that it is expensive to buy a regular-size can and let the unused portion spoil.

Be Creative Creative chefs think of various ways to use foods when only large amounts are available. For example, a head of cauliflower can be divided into thirds. Then one-third is cooked and eaten hot, another third is put into a vinegar and oil marinade for use in a salad, and the last third can be used in a casserole or stew.

Boxes of milk that can be stored at room temperature have been exposed to temperatures above those of pasteurization just long enough to sterilize the milk—a process called *ultrahigh temperature (UHT)*.

A variety of vegetables and meats can be enjoyed stir-fried; inexpensive vegetables such as cabbage, celery, and onion are delicious when crisp cooked in a little oil with herbs or lemon added. Interesting frozen vegetable mixtures are available in larger grocery stores. Cooked, leftover vegetables can be dropped in at the last minute. A bonus of a stir-fried meal is that there is only one pan to wash. Similarly, a microwave oven allows a home chef to use fewer pots and pans. Meals and leftovers can also be frozen or refrigerated in microwavable containers to reheat as needed.

Many frozen dinners or grocery store take-out foods offer nutritious options. Adding a fresh salad, a whole-wheat roll, and a glass of milk can make a nutritionally balanced meal. The "How to" offers time-saving tips to turn convenience foods into nutritious meals.

Also, single people shouldn't hesitate to invite someone to share meals with them whenever there is enough food. It's likely that the person will return the invitation, and both parties will get to enjoy companionship and a meal prepared by others.

IN SUMMARY

▮ Food choices of older adults are affected by health status and changed life circumstances.

▮ Older people can benefit from both the nutrients provided and the social interaction available at congregate meals. Other government programs deliver meals to those who are homebound.

▮ With creativity and careful shopping, those living alone can prepare nutritious, inexpensive meals.

How To   TURN CONVENIENCE FOODS INTO NUTRITIOUS MEALS

These time-saving tips can turn convenience foods into nutritious meals:

▮ Add extra nutrients and a fresh flavor to canned stews and soups by tossing in some frozen ready-to-use mixed vegetables. Choose vegetables frozen without salty, fatty sauces—prepared foods generally contain enough salt to season the whole dish, including added vegetables.

▮ Buy frozen vegetables in a bag, toss in a variety of herbs, and use as needed.

▮ When grilling burgers or chicken, wrap a mixture of frozen broccoli, onion, and carrots in a foil packet with a tablespoon of Italian dressing and grill alongside the meat for seasoned grilled vegetables.

▮ Use canned fruits in their own juices as desserts. Toss in some frozen berries or peach slices and top with flavored yogurt for an instant fruit salad.

▮ Prepared rice or noodle dishes are convenient, but those claiming to contain broccoli, spinach, or other vegetables seldom contain enough to qualify as a serving of vegetables.

Pump up the nutrient value by adding a half-cup of frozen vegetables per serving of pasta or rice just before cooking.

▮ Purchase frozen onion, mushroom, and pepper mixtures to embellish jarred spaghetti sauce or small frozen pizzas. Top with parmesan cheese.

▮ Use frozen shredded potatoes, sold for hash browns, in soups or stews or mix with a handful of shredded reduced-fat cheese or a can of fat-free "cream of anything" soup and bake for a quick and hearty casserole.

▮ Purchase a bag of triple-washed, ready-to-eat salad and add any or all of the following to make a hearty, healthy salad: a small can of garbanzo beans or a package of frozen edamame (immature soy beans); a handful of shredded reduced-fat cheese; a hard-boiled egg; a handful of toasted almonds or other nuts.

▮ Open a can of pinto beans or black beans, heat a tablespoon of olive oil, and sauté a little bit of frozen onion and pepper mixture. Add the beans and mash them in the oil and vegetables. Serve the bean mixture in a soft taco shell with shredded reduced-fat cheese, lettuce, and salsa for a tasty, nutritious bean taco.

Nutrition Assessment Checklist FOR OLDER ADULTS

Medical History

Check the medical record for:

- Alcohol abuse
- Alzheimer's disease or other dementia or confusion
- Arthritis
- Cataracts
- Chronic diseases (cancer, heart disease, hypertension, diabetes)
- Cigarette, cigar, or pipe smoking; use of other tobacco products
- Constipation
- Dehydration
- Dental disease or tooth loss
- Depression
- Inflammation of the stomach (gastritis)
- Swallowing disorders

Medications

For older adults being treated with drug therapy for medical conditions, note:

- Use of multiple medications—prescription and/or over-the-counter medications, such as laxatives and pain relievers
- Side effects that might reduce food intake or change nutrient needs
- Proper administration of medication with respect to food intake
- Malnutrition—is the person's nutrition status questionable even before considering side effects of medications that worsen nutrition status?
- Diminished mental capacity that might interfere with taking correct medications and doses
- Dehydration (can alter effects of medications)

Dietary Intake

For all older adults, especially those at risk nutritionally, assess the diet for:

- Total energy
- Protein
- Calcium, iron, and zinc
- Vitamin B_6, vitamin B_{12}, folate, and vitamin D

Note the following:

- Number of meals eaten each day
- Number and ages of people in household
- Amount of milk consumed each day
- Type and frequency of outdoor activity
- Type and frequency of physical activity
- Financial resources
- Transportation resources
- Physical disabilities
- Mental alertness

Anthropometric Data

Measure baseline height and weight.

- Reassess height and weight at each medical checkup.
- Note significant overweight or underweight, which warrants intervention.
- Use skinfold measures to reveal altered body composition that may indicate malnutrition and loss of lean tissue.

Laboratory Tests

- Hemoglobin, hematocrit, or other tests of iron status
- Serum albumin or other measures of protein status
- Serum folate
- Serum B_{12}

Physical Signs

Look for physical signs of:

- Protein-energy malnutrition
- Iron and zinc deficiency
- Folate deficiency

Clinical Applications

1. Ms. Hamilton is an 80-year-old woman in excellent health who lives alone, eats a well-balanced diet, enjoys an active social life, and walks every day. Consider the way Ms. Hamilton's health and nutrition status might be affected by the following situations:

   - Many of Ms. Hamilton's friends pass away or move into extended care facilities.

   - Ms. Hamilton falls and breaks her hip.

   - Ms. Hamilton fails her vision test and is not permitted to renew her driver's license.

   Describe interventions the health care professional can take to help Ms. Hamilton deal with each situation to prevent her from falling into a downward spiral.

Self Check

1. The fastest-growing age group in the United States is:
   a. under 21 years of age.
   b. 30 to 45 years of age.
   c. 50 to 70 years of age.
   d. over 85 years of age.

2. Which of the following lifestyle habits can enhance the length and quality of people's lives?
   a. Moderate smoking
   b. Six hours of sleep daily
   c. Regular physical activity
   d. Skipping breakfast

3. Which of the following is among the better-known relationships between nutrition and disease prevention?
   a. Appropriate fiber intake helps prevent goiter.
   b. Moderate sodium intake helps prevent obesity.
   c. Moderate sugar intake helps prevent hypertension.
   d. Appropriate energy intake helps prevent diabetes and cardiovascular disease.

4. A disease of the immune system that involves painful inflammation of the joints is:
   a. sarcopenia.
   b. osteoarthritis.
   c. senile dementia.
   d. rheumatoid arthritis.

5. Examples of low-kcalorie, high-quality protein foods include:
   a. cottage cheese, sour cream, and eggs.
   b. green and yellow vegetables and citrus fruits.
   c. potatoes, rice, pasta, and whole-grain breads.
   d. lean meats, poultry, fish, legumes, fat-free milk, and eggs.

6. For malnourished and underweight people, protein- and energy-dense snacks include:
   a. fresh fruits and vegetables.
   b. yogurt and cottage cheese.
   c. whole grains and high-fiber legumes.
   d. hard-boiled eggs and peanut butter and graham crackers.

7. Which of the following does not explain why dehydration is a risk for older adults?
   a. They do not seem to feel thirsty.
   b. Total body water increases with age.
   c. They may find it difficult to get a drink.
   d. They may have difficulty swallowing liquids.

8. Inadequate milk intake and limited exposure to sunlight contribute to older adults' risk of:
   a. vitamin A deficiency.
   b. vitamin D deficiency.
   c. riboflavin deficiency.
   d. vitamin B_{12} deficiency.

9. Two risk factors for malnutrition in older adults are:
   a. loneliness and multiple medication use.
   b. increased energy needs and lack of fiber.
   c. decreased mineral absorption and antioxidant intake.
   d. high carbohydrate intake and lack of physical activity.

10. Two strategies to improve nutrition status when growing old include:
    a. increase vitamin A intake and exercise 30 minutes daily.
    b. choose nutrient-dense foods and maintain appropriate weight.
    c. avoid high-fiber foods and take a daily vitamin-mineral supplement.
    d. eat at least one big meal per day and drink at least 10 glasses of water daily.

Answers to these questions can be found in Appendix H.

Notes

1. National Center for Health Statistics, Life expectancy at age 65 years, by sex and race—United States, 2000–2006, *Morbidity and Mortality Weekly Report* 58 (2009): 473.

2. J. Xu and coauthors, Deaths: Final data for 2007, *National Vital Statistics Reports* 58 (2010): 1–136.

3. G. Whitlock and coauthors, Body-mass index and cause-specific mortality in 900,000 adults: Collaborative analyses of 57 prospective studies, *Lancet* 373 (2009): 1083–1096; C. J. Lavie and coauthors, Obesity and cardiovascular disease, *Journal of the American College of Cardiology* 53 (2009): 1925–1932.

4. Position of the American Dietetic Association: Nutrition across the spectrum of aging, *Journal of the American Dietetic Association* 105 (2005): 616–633.

5. L. B. Yates and coauthors, Exceptional longevity in men: Modifiable factors associated with survival and function to age 90 years, *Archives of Internal Medicine* 168 (2008): 284–290; R. S. Rivlin, Keeping the young-elderly healthy: Is it too late to improve our health through nutrition? *American Journal of Clinical Nutrition* 86 (2007): 1572S–1576S.

6. Yates and coauthors, 2008; K. Khaw and coauthors, Combined impact of health behaviours and mortality in men and women: The EPIC-Norfolk Prospective Population Study, *PLoS Medicine* 5 (2008): e12.

7. Yates and coauthors, 2008.

8. Q. Sun and coauthors, Physical activity at midlife in relation to successful survival in women at age 70 years or older, *Archives of Internal Medicine* 170 (2010): 194–201; Yates and coauthors, 2008; X. Sui and coauthors, Cardiorespiratory fitness and adiposity as mortality predictors in older adults, *Journal of the American Medical Association* 298 (2007): 2507–2516.

9. American College of Sports Medicine, Position stand: Exercise and physical activity for older adults, *Medicine and Science in Sports and Exercise* 41 (2009): 1510–1530; P. Kokkinos and coauthors, Exercise capacity and mortality in black and white men, *Circulation* 117 (2008): 614–622.

10. R. C. Cassilhas and coauthors, The impact of resistance exercise on the cognitive function of

the elderly, *Medicine and Science in Sports and Exercise* 39 (2007): 1401–1407; W. R. Bixby and coauthors, The unique relation of physical activity to executive function in older men and women, *Medicine and Science in Sports and Exercise* 39 (2007): 1408–1416.

11. Rivlin, 2007.

12. M. E. Nelson and coauthors, Physical activity and public health in older adults: Recommendation for the American College of Sports Medicine and the American Heart Association, *Medicine and Science in Sports and Exercise* 39 (2007): 1435–1445.

13. R. Koopman and L. J. van Loon, Aging, exercise, and muscle protein metabolism, *Journal of Applied Physiology,* 106 (2009): 2040–2048.

14. Nelson and coauthors, 2007.

15. J. R. Speakman and C. Hambly, Starving for life: What animal studies can and cannot tell us about the use of caloric restriction to prolong human lifespan, *Journal of Nutrition* 137 (2007): 1078–1086.

16. L. Fontana and S. Klein, Aging, adiposity, and calorie restriction, *Journal of the American Medical Association* 297 (2007): 986–994; Position of the American Dietetic Association, 2005.

17. Fontana and Klein, 2007.

18. K. A. Varady and M. K. Hellerstein, Do calorie restriction or alternate-day fasting regimens modulate adipose tissue physiology in a way that reduces chronic disease risk? *Nutrition Reviews* 66 (2008): 333–342.

19. Fontana and Klein, 2007.

20. W. G. Christen and coauthors, Dietary carotenoids, vitamins C and E, and risk of cataract in women. A prospective study, *Archives of Ophthalmology* 126 (2008): 102–109; A. G. Tan and coauthors, Antioxidant nutrient intake and the long-term incidence of age-related cataract: The Blue Mountains Eye Study, *American Journal of Clinical Nutrition* 87 (2008): 1899–1905.

21. R. D. Jager, W. F. Mieler, and J. W. Miller, Age-related macular degeneration, *New England Journal of Medicine* 358 (2008): 2606–2617.

22. J. P. SanGiovanni and coauthors, ω-3 long-chain polyunsaturated fatty acid intake and 12-y incidence of neovascular age-related macular degeneration and central geographic atrophy: AREDS report 30, a prospective cohort study for the Age-Related Eye Disease Study, *American Journal of Clinical Nutrition* 90 (2009): 1601–1607; C. Augood and coauthors, Oily fish consumption, dietary docosahexaenoic acid and eicosapentaenoic acid intakes, and associations with neovascular age-related macular degeneration, *American Journal of Clinical Nutrition* 88 (2008): 398–406; E. J. Johnson and coauthors, The influence of supplemental lutein and docosahexaenoic acid on serum, lipoproteins, and macular pigmentation, *American Journal of Clinical Nutrition* 87 (2008): 1521–1529; E. D. O'Connell and coauthors, Diet and risk factors for age-related maculopathy, *American Journal of Clinical Nutrition* 87 (2008): 712–722.

23. Prevalence of doctor-diagnosed arthritis and arthritis-attributable activity limitation—United States, 2007-2009, *Morbidity and Mortality Weekly* 59 (2010): 1261–1265.

24. L. E. Hart and coauthors, The relationship between exercise and osteoarthritis in the elderly, *Clinical Journal of Sport Medicine* 18 (2008): 508–521; L. Devos-Comby, T. Cronan, and S. C. Roesch, Do exercise and self-management interventions benefit patients with osteoarthritis of knee? A metaanalytic review, *Journal of Rheumatology* 33 (2006): 744–756.

25. C. T. Vangsness, W. Spiker, and J. Erickson, A review of evidence-based medicine for glucosamine and chondroitin sulfate use in knee osteoarthritis, *Arthroscopy* 25 (2009): 86–94; National Center for Complementary and Alternative Medicine, The NIH Glucosamine/ Chondroitin Arthritis Intervention Trial (GAIT), *Journal of Pain and Palliative Care Pharmacotherapy* 22 (2008): 39–43; S. Dahmer and R. M. Schiller, Glucosamine, *American Family Physician* 78 (2008): 471–476; O. Bruyere and J. Y. Reginster, Glucosamine and chondroitin sulfate as therapeutic agents for knee and hip osteoarthritis, *Drugs and Aging* 24 (2007): 573–580.

26. G. McKellar and coauthors, A pilot study of a Mediterranean-type diet intervention in female patients with rheumatoid arthritis living in areas of social deprivation in Glasgow, *Annals of the Rheumatic Diseases* 66 (2007): 1239–1243.

27. P. M. Kris-Etherton and A. M. Hill, n-3 fatty acids: Foods or supplements? *Journal of the American Dietetic Association* 108 (2008): 1125–1130.

28. A. D. Smith and H. Refsum, Vitamin B-12 and cognition in the elderly, *American Journal of Clinical Nutrition* 89 (2009): 707S–711S; D. M. Lee and coauthors, Association between 25-hydroxyvitamin D levels and cognitive performance in middle-aged and older European men, *Journal of Neurology, Neurosurgery, and Psychiatry* 80 (2009): 722–729; W. J. Lukiw and N. G. Bazan, Docosahexaenoic acid and the aging brain, *Journal of Nutrition* 138 (2008): 2510–2514; B. M. van Gelder and coauthors, Fish consumption, n-3 fatty acids, and subsequent 5-y cognitive decline in elderly men: The Zutphen Elderly Study, *American Journal of Clinical Nutrition* 85 (2007): 1142–1147.

29. Centers for Disease Control and Prevention, Alzheimer's Disease, available at www.cdc.gov/ Features/Alzheimerís, accessed August 17, 2009.

30. L. M. Bekris and coauthors, Genetics of Alzheimer disease, *Journal of Geriatric Psychiatry Neurology* 23 (2010): 213–227; A. Burns and S. Liffe, Clinical review—Alzheimer's disease, *British Medical Journal* 338 (2009): b158.

31. D. Praticó, Evidence of oxidative stress in Alzheimer's disease brain and antioxidant therapy: Lights and shadows, *Annals of the New York Academy of Sciences* 1147 (2008): 70–78; I. G. Onyango and S. M. Khan, Oxidative stress, mitochondrial dysfunction, and stress signaling in Alzheimer's disease, *Current Alzheimer Research* 3 (2006): 339–349.

32. Praticó, 2008; P. I. Moreira and coauthors, Alzheimer disease and the role of free radicals in the pathogenesis of the disease, *CNS and Neurological Disorders Drug Targets* 7 (2008): 3–10; A. Nunomura and coauthors, Involvement of oxidative stress in Alzheimer disease, *Journal of Neuropathology and Experimental Neurology* 65 (2006): 631–641.

33. W. Wang and coauthors, Nutritional biomarkers in Alzheimer's disease: The association between carotenoids, n-3 fatty acids, and dementia severity, *Journal of Alzheimer's Disease* 13 (2008): 31–38; H. J. Wengreen and coauthors, Antioxidant intake and cognitive function of elderly men and women: The Cache County Study, *Journal of Nutrition, Health, and Aging,* 11 (2007): 230–237; F. C. Lau, B. Shukitt-Hale, and J. A. Joseph, Nutritional intervention in brain aging: Reducing the effects of inflammation and oxidative stress, *Sub-cellular Biochemistry* 42 (2007): 299–318; L. Letenneur and coauthors, Flavonoid intake and cognitive decline over a 10-year period, *American Journal of Epidemiology* 165 (2007): 1364–1371.

34. J. A. Luchsinger and D. R. Gustafson, Adiposity and Alzheimer's disease, *Current Opinion in Clinical Nutrition and Metabolic Care* 12 (2009): 15–21; D. B. Miller and J. P. O'Callaghan, Do early-life insults contribute to the late-life development of Parkinson and Alzheimer diseases? *Metabolism* 57 (2008): S44–S49; R. A. Whitmer, The epidemiology of adiposity and dementia, *Current Alzheimer Research* 4 (2007): 117–122.

35. Moreira and coauthors, 2008; R. J. Castellani and coauthors, Antioxidant protection and neurodegenerative disease: The role of amyloid-beta and tau, *American Journal of Alzheimer's Disease and Other Dementias* 21 (2006): 126–130; P. Zafrilla and coauthors, Oxidative stress in Alzheimer patients in different stages of the disease, *Current Medicinal Chemistry* 13 (2006): 1075–1083.

36. R. A. Armstrong, Plaques and tangles and the pathogenesis of Alzheimer's disease, *Folia Neuropathologica* 44 (2006): 1–11; G. L. Wenk, Neuropathologic changes in Alzheimer's disease: Potential targets for treatment, *Journal of Clinical Psychiatry* 67 (2006): 3–7.

37. R. J. Castellani and coauthors, Reexamining Alzheimer's disease; Evidence for a protective role for amyloid-beta protein precursor and amyloid-beta, *Journal of Alzheimer's Disease* 18 (2009): 447-452; A. Nunomura and coauthors, Neuropathology in Alzheimer's disease: Awaking from a hundred-year-old dream, *Science of Aging Knowledge Environment* (2006): pe10

38. D. J. Selkoe, Developing preventive therapies for chronic diseases: Lessons learned from Alzheimer's disease, *Nutrition Reviews* 65 (2007): S239–S243.

39. J. A. Luchsinger and D. R. Gustafson, Adiposity, type 2 diabetes, and Alzheimer's disease, *Journal of Alzheimer's Disease* 16 (2009): 693–704; M. N. Haan and coauthors, Homocysteine, B vitamins, and the incidence of dementia and cognitive impairment: Results from the Sacramento Area Latino Study on Aging, *American Journal of Clinical Nutrition* 85 (2007): 511–517.

40. L. J. Whalley and coauthors, n-3 Fatty acid erythrocyte membrane content, APOE ε4, and cognitive variation: An observational follow-up study in late adulthood, *American Journal of Clinical Nutrition* 87 (2008): 449–454; M. A. Beydoun and coauthors, Plasma n-3 fatty acids and the risk of cognitive decline in older adults: The Atherosclerosis Risk in Communities study, *American Journal of Clinical Nutrition* 85 (2007): 1103–1111; W. E. Connor and S. L. Connor, The importance of fish and docosahexaenoic acid in Alzheimer

disease, *American Journal of Clinical Nutrition* 85 (2007): 929–930; C. Dullemeijer and coauthors, n-3 Fatty acid proportions in plasma and cognitive performance in older adults, *American Journal of Clinical Nutrition* 86 (2007): 1479–1485; B. M. V. Gelder and coauthors, Fish consumption, n-3 fatty acids, and subsequent 5-y cognitive decline in elderly men: The Zutphen Elderly study, *American Journal of Clinical Nutrition* 85 (2007): 1142–1147; E. Nurk and coauthors, Cognitive performance among the elderly and dietary fish intake: The Hordaland Health study, *American Journal of Clinical Nutrition* 86 (2007): 1470–1478.

41. N. Scarmeas and coauthors, Physical activity, diet, and risk of Alzheimer disease, *Journal of the American Medical Association* 302 (2009): 627–637; T. Liu-Ambrose and M. G. Donaldson, Exercise and cognition in older adults: Is there a role for resistance training programmes? *British Journal of Sports Medicine* 43 (2009): 25–27; E. B. Larson, Physical activity for older adults at risk for Alzheimer disease, *Journal of the American Medical Association* 300 (2008): 1077–1079; N. T. Lautenschlager and coauthors, Effect of physical activity on cognitive function in older adults at risk for Alzheimer disease, *Journal of the American Medical Association* 300 (2008): 1027–1037.

42. Koopman and van Loon, 2009; M. Cesari and coauthors, Frailty syndrome and skeletal muscle: Results from the Invecchiare in Chianti Study, *American Journal of Clinical Nutrition* 83 (2006): 1142–1148.

43. Koopman and van Loon, 2009; D. K. Houston and coauthors, Dietary protein intake is associated with lean mass change in older, community-dwelling adults: The Health, Aging, and Body Composition (Health ABC) Study, *American Journal of Clinical Nutrition* 87 (2008): 150–155; D. Paddon-Jones and coauthors, Role of dietary protein in the sarcopenia of aging,

American Journal of Clinical Nutrition 87 (2008): 1562S–1566S; H. B. Iglay and coauthors, Resistance training and dietary protein: Effects on glucose tolerance and contents of skeletal muscle insulin signaling proteins in older persons, *American Journal of Clinical Nutrition* 85 (2007): 1005–1013; W. W. Campbell, Synergistic use of higher-protein diets or nutritional supplements with resistance training to counter sarcopenia, *Nutrition Reviews* 65 (2007): 416–422.

44. S. Karinkanta and coauthors, A multi-component exercise regimen to prevent functional decline and bone fragility in home-dwelling elderly women: Randomized, controlled trial, *Osteoporosis International* 18 (2007): 453–462; Rivlin, 2007.

45. Rivlin, 2007; Position of the American Dietetic Association, 2005.

46. R. B. Ervin and B. A. Dye, The effect of functional dentition on Healthy Eating Index scores and nutrient intakes in a nationally representative sample of older adults, *Journal of Public Health Dentistry* 69 (2009): 207–216; R. B. Ervin, Healthy Eating Index scores among adults 60 years of age and over by sociodemographic and health characteristics: United States, 1999–2002, *Advance Data* 20 (2008): 1–16; Position of the American Dietetic Association, 2005.

47. Standing Committee on the Scientific Evaluation of Dietary Reference Intakes, Food and Nutrition Board, Institute of Medicine, *Dietary Reference Intakes for Energy, Carbohydrate, Fiber, Fat, Fatty Acids, Cholesterol, Protein, and Amino Acids* (Washington D.C.: National Academies Press, 2005), pp. 387–389.

48. Effect of aging on regional cerebral blood flow responses associated with osmotic thirst and its satiation by water drinking: A PET study, *Proceedings of the National Academy of Sciences of the United States of America* 105 (2008): 382–387.

49. Standing Committee on the Scientific Evaluation of Dietary Reference Intakes, Food and Nutrition Board, Institute of Medicine, *Dietary Reference Intakes for Water, Potassium, Sodium, Chloride, and Sulfate* (Washington, D.C.: National Academies Press, 2005), pp. 118–127.

50. Standing Committee on the Scientific Evaluation of Dietary Reference Intakes, 2005, pp. 149–150.

51. B. Dawson-Hughes, Serum 25-hydroxyvitamin D and functional outcomes in the elderly, *American Journal of Clinical Nutrition* 88 (2008): 537S–540S.

52. A. D. Smith and H. Refsum, Vitamin B-12 and cognition in the elderly, *American Journal of Clinical Nutrition* 89 (2009): 707S–711S; R. Clarke and coauthors, Low vitamin B-12 status and risk of cognitive decline in older adults, *American Journal of Clinical Nutrition* 86 (2007): 1384–1391.

53. A. S. Prasad and coauthors, Zinc supplementation decreases incidence of infections in the elderly: Effect of zinc on generation of cytokines and oxidative stress, *American Journal of Clinical Nutrition* 85 (2007): 837–844.

54. R. S. Sebastian and coauthors, Older adults who use vitamin/mineral supplements differ from nonusers in nutrient intake adequacy and dietary attitudes, *Journal of the American Dietetic Association* 107 (2007): 1322–1332; Position of the American Dietetic Association, 2005.

55. Karinkanta and coauthors, 2007.

56. Position of the American Dietetic Association, American Society for Nutrition, and Society for Nutrition Education: Food and nutrition programs for community-residing older adults, *Journal of the American Dietetic Association* 110 (2010): 463–472.

Nutrition in Practice
HUNGER AND COMMUNITY NUTRITION

Worldwide, nearly one billion people experience persistent hunger—not the healthy appetite triggered by anticipation of a hearty meal, but the painful sensation caused by a lack of food.[1] Hunger deprives a person of the physical and mental energy needed to enjoy a full life. It often leads to severe malnutrition and death. Tens of thousands die of starvation each day—one child every five seconds.

In the United States, where most people enjoy a life of relative abundance, about one in every nine households has one or more members who experience pain from hunger caused by lack of food. In these households, almost 13 million children do not know where their next meal is coming from or when it will come.[2] Given the agricultural bounty and enormous wealth in this country, do these numbers surprise you? The limited or uncertain availability of nutritionally adequate and safe foods is known as **food insecurity** and is a major social problem in our nation today.[3] Table NP13-1 presents the questions used in national surveys to identify food insecurity in the United States, and Figure NP13-1 presents the most recent findings. Surveys like these provide crude but necessary data to estimate the degree of hunger in this country. The glossary defines related terms.

Why is hunger a problem in developed countries such as the United States where food is abundant?

Hunger has many causes, but in developed countries, the primary cause is **food poverty.** People are hungry not because there is no food nearby to purchase but because they lack sufficient money

---

TABLE NP13-1 Questions to Identify Food Insecurity in a U.S. Household

To determine the extent of food insecurity in a household, surveys ask questions about behaviors and conditions known to characterize households having difficulty meeting basic food needs during the past 12 months. Most often, adults tend to protect their children from hunger. In the most severe cases, children also suffer from hunger and eat less.

1. Did you worry whether your food would run out before you got money to buy more?
2. Did you find that the food you bought just didn't last, and you didn't have money to buy more?
3. Were you unable to afford to eat balanced meals?
4. Did you or other adults in your household ever cut the size of your meals or skip meals because there wasn't enough money for food?
5. Did this happen in three or more months during the previous year?
6. Did you ever eat less than you felt you should because there wasn't enough money for food?
7. Were you ever hungry but didn't eat because you couldn't afford enough food?
8. Did you lose weight because you didn't have enough money for food?
9. Did you or other adults in your household ever not eat for a whole day because there wasn't enough money for food?
10. Did this happen in three or more months during the previous year?
11. Did you rely on only a few kinds of low-cost food to feed your children because you were running out of money to buy food?
12. Were you unable to feed your children a balanced meal because you couldn't afford it?
13. Were your children not eating enough because you just couldn't afford enough food?
14. Did you ever cut the size of your children's meals because there wasn't enough money for food?
15. Were your children ever hungry but you just couldn't afford more food?
16. Did your children ever skip a meal because there wasn't enough money for food?
17. Did this happen in three or more months during the previous year?
18. Did your children ever not eat for a whole day because there wasn't enough money for food?

The more positive responses, the greater the food insecurity. Households with children answer all of the questions and are categorized as follows:

   ≤2 positive responses = food secure
   3–7 positive responses = low food security
   ≥8 positive responses = very low food security

Households without children answer the first 10 questions and are categorized as follows:

   ≤2 positive responses = food secure
   3–5 positive responses = low food security
   ≥6 positive responses = very low food security

Source: United States Department of Agriculture, *Household Food Security in the United States, 2008,* available at www.ers.usda.gov

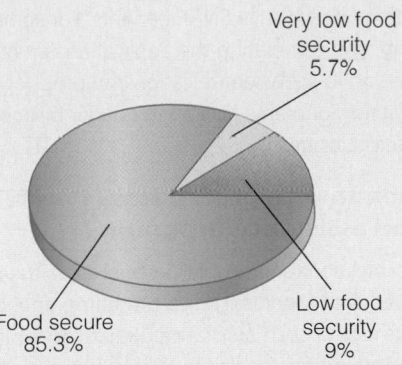

Very low food
security
5.7%

Low food
security
9%

Food secure
85.3%

FIGURE NP13-1 **Prevalence of Food Security, Food Insecurity, and Hunger in U.S. Households, 2009**

Source: USDA Economic Research Service; available at www.ers.usda.gov/publication/, update December 2009.

with which to buy nutritious food and pay for other necessities, such as housing, clothing, medicines, and utilities. More than 12 percent of the population of the United States lives in poverty. Even those above the poverty line may not enjoy **food security.**

Physical and mental illnesses and disabilities, sudden job losses, and high living expenses threaten their financial stability. Further contributing to food poverty are other problems such as abuse of alcohol and other drugs; lack of awareness of available food assistance programs; and the reluctance of people, particularly the elderly, to accept what they perceive as welfare or charity. Lack of resources remains the major cause of food poverty in developed countries, and solving this problem would go a long way toward relieving hunger.

In the United States, food poverty and hunger reach into many segments of society, affecting not only the chronic poor (migrant workers, the unskilled and unemployed, the homeless, and some elderly) but also the so-called working poor. Some are displaced farm families. Some are former blue-collar and white-collar workers forced out of their trades and professions into minimum-wage jobs. These people outnumber the chronic poor, and they are not on welfare—they have jobs, but the pay is too low to meet their needs. Families with incomes below a certain level are simply unable to buy sufficient amounts of nourishing foods, even if they are skilled in food shopping. Consequently, their diets tend to be inadequate.[4] For many children in these families, school lunch (and breakfast, where available) may be the only nourishment for the day. Otherwise, they go hungry, waiting for an adult to find money for food. Not surprisingly, these children are more likely to have health problems than those who eat regularly.[5] They also tend to perform poorly in school and in social situations.[6]

What U.S. food programs are directed at relieving hunger in the United States?

The American Dietetic Association (ADA) calls for aggressive action to bring an end to domestic hunger and to achieve food and nutrition security for all residents of the United States.[7] Many federal and local programs aim to prevent or relieve malnutrition and hunger in the United States.

An extensive network of federal assistance programs provides life-giving food daily to millions of U.S. citizens. One out of every five Americans receives food assistance of some kind, at a total

© Charles Gupton/Stock, Boston/PictureQuest

These people and many others like them in the United States face food insecurity daily.

377

cost of almost $60 billion per year. Even so, the programs are not fully successful in preventing hunger, but they nevertheless seem to improve the nutrient intakes of those who participate. Programs described in the life cycle chapters include the WIC program for low-income pregnant women, breastfeeding mothers, and their young children (Chapter 11); the school lunch and breakfast programs for children (Chapter 12); and the food assistance programs for older adults such as congregate meals and Meals on Wheels (Chapter 13).

The centerpiece of food programs for low-income people in the United States is the Supplemental Nutrition Assistance Program (SNAP), administered by the U.S. Department of Agriculture (USDA). The USDA issues debit cards through state agencies to households—people who buy and prepare food together. The amount a household receives depends on its size, resources, and income. Recipients may use the cards to purchase food and food-bearing plants and seeds but not to buy tobacco, cleaning items, alcohol, or other nonfood items. The accompanying "How to" offers shopping tips for those on a limited budget.

SNAP is the largest of the federal food assistance programs, both in amount of money spent and in number of people participating. It provides assistance to more than 28 million people at a cost of more than $34 billion per year; about half of the recipients are children.[8]

Although food assistance programs improve nutrient intakes significantly, hunger continues to plague the United States. Of the estimated 2 million homeless people in the United States who are eligible for food assistance, only 15 percent of single adults and 50 percent of families receive SNAP benefits. For some, reading, understanding, and completing the application can be difficult. For others, having to show identification and proof of homelessness can be frustrating. For many, accepting hunger is simply easier than meeting these challenges.

Why do health care professionals need to know about food assistance programs?

Health care professionals who work in public health are generally well acquainted with food assistance programs, and, often, many of their clients receive such assistance. Regardless of the setting in which health care professionals see clients, however, it is important to encourage those who may be having financial problems to talk with a social worker who can assess their eligibility for food assistance programs. The subject of food assistance must be approached in a nonjudgmental and tactful manner—the client may feel uncomfortable about seeking assistance.

Are there other programs aimed at reducing hunger in the United States?

Efforts to resolve the problem of hunger in the United States do not depend solely on federal assistance programs. National **food recovery** programs have made a dramatic difference; the largest program, Feeding America, coordinates the efforts of more than 40,000 **food pantries, emergency shelters,** and **soup kitchens** that feed an estimated 24 to 27 million people a year. Table NP13-2 provides mission statements and websites for Feeding America and other hunger relief organizations.

How To PLAN HEALTHY, THRIFTY MEALS

Chapter 1 introduces the USDA MyPyramid Food Guide and principles for planning a healthy diet. Meeting that goal on a limited budget adds to the challenge. To save money and spend wisely, plan and shop for healthy meals with the following tips in mind:

Planning

▪ Make a grocery list before going to the store to avoid expensive impulse items.

▪ Do not shop when hungry.

▪ Use leftovers.

▪ Center meals on rice, noodles, and other grains.

▪ Use small quantities of meat, poultry, fish, or eggs.

▪ Use legumes instead of meat, poultry, fish, or eggs several times a week.

▪ Use cooked cereals such as oatmeal instead of ready-to-eat breakfast cereals.

▪ Cook large quantities when time and money allow.

▪ Check for sales and clip coupons for products you need; plan meals to take advantage of sale items.

Shopping

▪ Buy day-old bread and other products from a bakery outlet.

▪ Select whole foods instead of convenience foods (potatoes instead of instant mashed potatoes, for example).

▪ Try store brands.

▪ Buy fresh produce that is in season; buy canned or frozen items at other times.

▪ Buy only the amount of fresh foods that you will eat before it spoils. Buy large bags of frozen items or dry goods; when cooking, take out the amount needed and store the remainder.

▪ Buy fat-free dry milk; mix and refrigerate quantities needed for a day or two. Buy fresh milk by the gallon or half-gallon.

▪ Buy less expensive cuts of meat. Chuck and bottom round roast are usually inexpensive; cover during cooking and cook long enough to make meat tender. Buy whole chickens instead of pieces.

▪ Compare the unit price (cost per ounce, for example) of similar foods so that you can select the least expensive brand or size.

▪ Buy nonfood items such as toilet paper and laundry detergent at discount stores instead of grocery stores.

For daily menus and recipes for healthy, thrifty meals, visit the USDA Center for Nutrition Policy and Promotion: **www.cnpp.usda.gov/.**

TABLE NP13-2 Hunger Relief Organizations

Organization	Mission Statement
Action without Borders www.idealist.org	International organization seeking to connect people, organizations, and resources to help build a world where all people can live free and dignified lives.
Bread for the World www.bread.org	Nonpartisan, Christian citizens' movement seeking to influence reform in policies, programs, and conditions that allow hunger and poverty to persist globally.
Catholic Relief Services www.crs.org	Humanitarian service agency assisting the impoverished and disadvantaged through community-based, sustainable development initiatives.
Community Food Security Coalition www.foodsecurity.org	North American coalition of diverse people and organizations working from the local to the international levels to build community food security.
Congressional Hunger Center www.hungercenter.org	Bipartisan organization training and inspiring leaders with the intent to end hunger and advocating public policies to create a food-secure world.
Feeding America www.feedingamerica.org	Domestic charity organization providing food assistance through a nationwide network of member food banks and facilitating education to end hunger nationally.
Food and Agriculture Organization (FAO) of the United Nations www.fao.org	International organization leading efforts to defeat hunger by helping to develop and modernize countries' agriculture, forestry, and fishery practices.
Oxfam America www.oxfamamerica.org	International relief and development organization aiming to create lasting solutions to poverty, hunger, and injustice.
Pan American Health Organization www.paho.org	International public health agency aiming to strengthen national and local health systems with the purpose of improving the quality of, and lengthening, the lives of the peoples in the Americas.
Society of St. Andrew www.endhunger.org	Ecumenical Christian ministry salvaging and redirecting large amounts of fresh produce to hunger agencies for distribution to the poor.
The Hunger Project www.thp.org	International relief organization emphasizing sustainable solutions such as rural development and self-reliance to facilitate food security.
United Nations Children's Fund (UNICEF) www.unicef.org	International organization advocating for the protection of children's rights, to help meet their basic needs and to expand their opportunities to reach their full potentials.
World Food Programme www.wfp.org	Food aid branch of the United Nations aiming to prepare for, protect during, and provide assistance after emergencies, as well as reduce hunger and undernutrition.
World Health Organization (WHO) www.who.int	United Nations agency acting as the authority on international public health by influencing policy, setting research agendas, establishing standards, and providing technical support to monitor and assess health trends.
World Hunger Year (WHY) www.whyhunger.org	Domestic organization supporting and funding community-based organizations intent on empowering individuals and building self-reliance to provide long-term solutions to hunger and poverty.

Feeding the hungry in the United States.

© Skjold/The Image Works

Each year, an estimated one-fifth of our food supply is wasted in fields, commercial kitchens, grocery stores, and restaurants—enough food to feed 49 million people. Food recovery programs collect and distribute good food that would otherwise go to waste. Volunteers might pick corn left in an already harvested field, a grocer might deliver ripe bananas to a local **food bank,** and a caterer might take leftover chicken salad to a community shelter, for example. All of these efforts help to feed the hungry in the United States.

What about local efforts and community nutrition programs?

Food recovery programs depend on volunteers. Concerned citizens work through local agencies and churches to feed the hungry. Community-based food pantries provide groceries, and soup kitchens serve prepared meals. Meals often deliver

Community-based efforts to feed citizens include food pantries that provide groceries.

adequate nourishment, but most homeless people receive fewer than one and a half meals a day, so many are still inadequately nourished. Health care professionals can serve as valuable members of community groups seeking to provide food assistance.

Notes

1. *State of Food Insecurity in the World*, 2010, Food Security Statistics available from Food and Agriculture Organization, www.fao.org.

2. United States Department of Agriculture, *Household Food Security in the United States, 2007*, ERS Report Summary, November 2008, available at www.ers.usda.gov/Publications/ERR66.

3. Position of the American Dietetic Association: Food insecurity in the United States, *Journal of the American Dietetic Association* 110 (2010): 1368–1377.

4. C. M. Champagne and coauthors, Poverty and food intake in rural America: Diet quality is lower in food insecure adults in the Mississippi Delta, *Journal of the American Dietetic Association* 107 (2007): 1886–1894; M. L. Kropf and coauthors, Food security status and produce intake and behaviors of Special Supplemental Nutrition Program for Women, Infants, and Children and Farmer's Market Nutrition Program participants, *Journal of the American Dietetic Association* 107 (2007): 1903–1908.

5. R. Rose-Jacobs and coauthors, Household food insecurity: Associations with at-risk infant and toddler development, *Pediatrics* 121 (2008): 65–72.

6. Position of the American Dietetic Association, 2010.

7. Position of the American Dietetic Association, 2010.

8. *Leading the Fight against Hunger—Federal Nutrition Assistance*, June 2008, www.fns.usda.gov/fns.hunger.pdf; P. S. Landers, The Food Stamp Program: History, nutrition education, and impact, *Journal of the American Dietetic Association* 107 (2007): 1945–1951.

14

Illness and Nutrition Care

Previous chapters of this book introduced the nutrients and described how the appropriate dietary choices can support good health. Turning to clinical nutrition, this chapter describes the nutrition care process and the implementation of nutrition care in clinical practice. Ensuring that nutrient needs are met is a key part of this process, so the chapter also describes typical methods for estimating energy requirements and provides examples of common nutrition modifications.

Nutrition in Health Care

Many medical problems can alter nutrient needs and cause malnutrition. Malnutrition has been reported in 38 to 62 percent of patients hospitalized with acute illness,[1] and patients with no nutrition problems often exhibit a decline in nutrition status within a few weeks of admission. Poor nutrition status can weaken immune function and compromise a person's healing ability, which influences both the course of disease and the body's response to treatment. Thus, preventing and correcting nutrition problems may improve the outcome of medical treatments and help to prevent complications. In addition, patients are often concerned about the impact their diet has on their medical condition.

Effects of Illness on Nutrition Status

An illness, its symptoms, and its treatments may lead to malnutrition by reducing food intake, interfering with digestion and absorption, or altering nutrient metabolism and excretion (see Figure 14-1). For example, the nausea caused by some illnesses and disease treatments can diminish appetite and reduce food intake. Inflammation in the mouth or esophagus can cause discomfort when an individual consumes food. Some medications may cause **anorexia** (loss of appetite) or gastrointestinal (GI) discomfort or interfere with nutrient function and metabolism. Prolonged bed rest may result in **pressure sores,** which increase metabolic stress and raise protein and energy needs.

The dietary changes required during an acute illness are usually temporary and can be tailored to accommodate an individual's preferences and lifestyle. However, chronic illnesses may require long-term modifications. For example, diabetes treatment requires lifelong changes in diet and lifestyle that some people may find difficult to maintain.

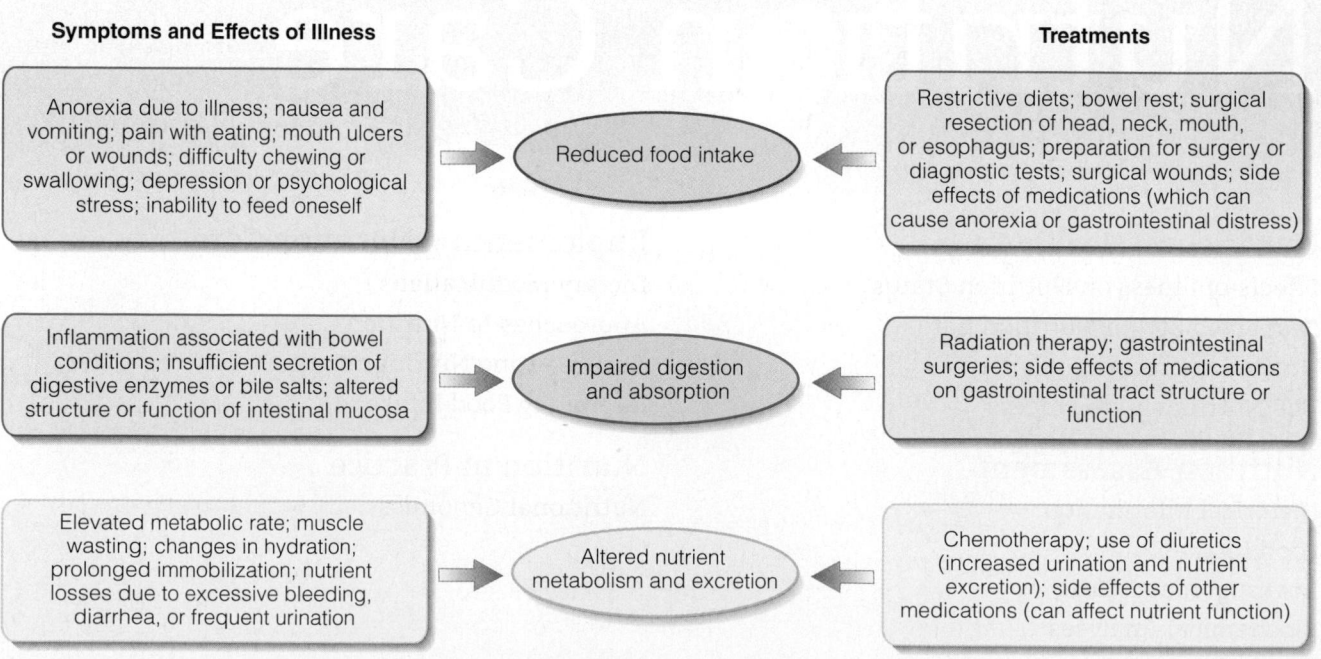

Symptoms and Effects of Illness

Treatments

Anorexia due to illness; nausea and vomiting; pain with eating; mouth ulcers or wounds; difficulty chewing or swallowing; depression or psychological stress; inability to feed oneself → Reduced food intake ← Restrictive diets; bowel rest; surgical resection of head, neck, mouth, or esophagus; preparation for surgery or diagnostic tests; surgical wounds; side effects of medications (which can cause anorexia or gastrointestinal distress)

Inflammation associated with bowel conditions; insufficient secretion of digestive enzymes or bile salts; altered structure or function of intestinal mucosa → Impaired digestion and absorption ← Radiation therapy; gastrointestinal surgeries; side effects of medications on gastrointestinal tract structure or function

Elevated metabolic rate; muscle wasting; changes in hydration; prolonged immobilization; nutrient losses due to excessive bleeding, diarrhea, or frequent urination → Altered nutrient metabolism and excretion ← Chemotherapy; use of diuretics (increased urination and nutrient excretion); side effects of other medications (can affect nutrient function)

FIGURE 14-1 Ways in Which Illness Can Affect Nutrition Status

The challenge for health professionals is to help their patients appreciate the potential benefits of treatment and accept dietary changes that can improve their health.

Responsibility for Nutrition Care

The members of a health care team work together to ensure that the nutritional needs of patients are met during illness. The roles of health professionals may vary somewhat in different institutions, and their responsibilities can sometimes overlap. In some cases, nutrition care is incorporated into the medical care plan developed by the entire health care team. Such plans, called **critical pathways,** outline coordinated plans of care for specific medical diagnoses, treatments, or procedures.

Physicians Physicians are responsible for meeting all of a patient's medical needs, including nutrition. They prescribe **nutrition prescriptions** (also called *diet orders*) and other instructions pertaining to nutrition care, including recommendations for nutrition assessment and dietary counseling. Physicians rely on nurses, registered dietitians, and other health professionals to alert them to nutrition problems, suggest strategies for handling nutrition care, and provide nutrition services.

Registered Dietitians Registered dietitians are food and nutrition experts who are qualified to provide **medical nutrition therapy.** They conduct nutrition and dietary assessments; diagnose nutritional problems; develop, implement, and evaluate **nutrition care plans** (described in a later section); plan and approve menus; and provide dietary counseling and nutrition education services. Registered dietitians may also manage food and cafeteria services in health care institutions.

Registered Dietetic Technicians Registered dietetic technicians often work in partnership with registered dietitians and assist in the implementation and monitoring of nutrition services. Depending on their background and experience, they may screen patients for nutrition problems, provide patient education and counseling, develop menus and recipes, ensure appropriate meal delivery, and monitor patients' food choices and intakes. Dietetic technicians sometimes supervise foodservice operations and may have roles in purchasing, inventory, quality control, sanitation, or safety.

Nurses Nurses interact closely with patients and thus are in an ideal position to identify people who would benefit from nutrition services. Nurses often screen patients for nutrition problems and may participate in nutrition assessments. Nurses also provide direct nutrition care, such as encouraging patients to eat, finding practical solutions to food-related problems, recording a patient's food intake, and answering questions about special diets. As members of **nutrition support teams,** nurses are responsible for administering tube and intravenous feedings. In facilities that do not employ registered dietitians, nurses often assume responsibility for much of the nutrition care.

Other Health Care Professionals Other heath care professionals may also assist with nutrition care. Pharmacists, physical therapists, occupational therapists, speech therapists, social workers, nursing assistants, and home health care aides can be instrumental in alerting dietitians or nurses to nutrition problems or may share relevant information about a patient's health status or personal needs.

Nutrition Screening

A **nutrition screening,** an assessment procedure that helps to identify patients who are malnourished or at risk for malnutrition, may lead to a referral for nutrition care. The screening should be sensitive enough to identify patients who require nutrition care but simple enough to be completed within 10 to 15 minutes. The information used in screening includes the admitting diagnosis, information from the medical record, physical measurements and laboratory test results collected during the admission process, and responses given by the patient or caregiver in an interview or

Reminder: A *registered dietitian* has met the academic and professional requirements for the RD credential conferred by the American Dietetic Association (or Dietitians of Canada), including a bachelor's degree in nutrition or dietetics, a supervised internship, and the successful completion of a national registration examination.

anorexia: loss of appetite.

pressure sores: regions of damaged skin and tissue due to prolonged pressure on the affected area by an external object, such as a bed, wheelchair, or cast; vulnerable areas of the body include buttocks, hips, and heels. Also called *decubitus* (deh-KYU-bih-tus) *ulcers.*

critical pathways: coordinated programs of treatment that merge the care plans of different health practitioners; also called *clinical pathways.*

nutrition prescriptions: specific instructions concerning nutrition care; also referred to as *diet orders* or *diet prescriptions.*

medical nutrition therapy: nutrition care provided by a registered dietitian; includes assessing nutrition status, diagnosing nutrition problems, developing nutrition care plans, and providing dietary counseling and nutrition education.

nutrition care plans: strategies for meeting an individual's nutritional needs.

nutrition support teams: health care professionals responsible for the provision of nutrients by tube feeding or intravenous infusion.

nutrition screening: a brief assessment of health-related variables to identify patients who are malnourished or at risk for nutrition.

TABLE 14-1 **Criteria for Identifying Malnutrition Risk**

- Age, medical diagnosis, severity of illness
- Height and weight, body mass index (BMI), unintentional weight changes
- Tissue wasting, loss of subcutaneous fat
- Changes in appetite or food intake
- Problems that interfere with food intake (such as chewing or swallowing difficulty, or nausea and vomiting)
- Food allergies or intolerances, extensive dietary restrictions
- Laboratory test results that indicate poor health status
- History of diabetes, renal disease, or other chronic illness
- Presence of anemia or pressure sores
- Use of medications that can impair nutrition status
- Depression, social isolation, dementia

questionnaire. Table 14-1 lists examples of information collected during screening. The **Joint Commission,** a nonprofit organization that confers accreditation to health care institutions, recommends that a nutrition screening be conducted within 24 hours of a patient's admission to a hospital or other extended-care facility. Nutrition screening is also frequently included in outpatient services and community health programs.

Nutrition screening methods often combine different types of health-related information to improve accuracy; one such example is the Subjective Global Assessment, which includes weight and dietary changes, gastrointestinal symptoms, work capacity, level of metabolic stress, degree of muscle wasting and fat loss, and presence of edema or ascites (see Table 14-2). More rapid screening techniques may use just two or three variables; for example, several tools screen for malnutrition risk by evaluating unintentional weight changes and reduced appetite or food intake.[2]

Ascites, the abnormal accumulation of fluid in the abdominal cavity, is discussed in Chapter 20.

TABLE 14-2 **Subjective Global Assessment**

The Subjective Global Assessment evaluates a person's risk of malnutrition by ranking key variables of the medical history and physical examination. These variables are each given an A, B, or C rating: A for well nourished, B for potential or mild malnutrition, and C for severe malnutrition. Patients are classified according to the final numbers of A, B, and C ratings.

Medical History

- Body weight changes: percentage change in past six months; weight change in past two weeks
- Dietary changes: suboptimal, low kcalorie, liquid diet, or starvation
- GI symptoms: nausea, diarrhea, vomiting, or anorexia for more than two weeks
- Functional ability: full capacity versus suboptimal, walking versus bedridden
- Degree of disease-related metabolic stress: low, medium, or high

Physical Examination

- Subcutaneous fat loss (triceps or chest)
- Muscle loss (quadriceps or deltoids)
- Ankle edema
- Sacral (lower spine) edema
- Ascites (abdominal edema)

Classification:

A: Well nourished: if no significant loss of weight, fat, or muscle tissue and no dietary difficulties, functional impairments, or GI symptoms; also applies to patients with recent weight gain and improved appetite, functioning, or medical prognosis

B: Moderate malnutrition: if 5 to 10 percent weight loss, mild loss of muscle or fat tissue, decreased food intake, and digestive or functional difficulties that impair food intake; the B classification usually applies to patients with an even mix of A, B, and C ratings

C: Severe malnutrition: if more than 10 percent weight loss, severe loss of muscle or fat tissue, edema, multiple GI symptoms, and functional impairments

Sources: R. S. Gibson, *Principles of Nutritional Assessment* (New York: Oxford University Press, 2005), pp. 809–826; A. S. Detsky and coauthors, What is subjective global assessment of nutritional status? *Journal of Parenteral and Enteral Nutrition* 11 (1987): 8–13.

The Nutrition Care Process

Registered dietitians use a systematic approach to medical nutrition therapy called the **nutrition care process.**[3] Figure 14-2 presents the four distinct, yet interrelated, steps of the nutrition care process:

1. Nutrition assessment
2. Nutrition diagnosis
3. Nutrition intervention
4. Nutrition monitoring and evaluation

Although the nutrition care process is easiest to visualize as a series of steps, earlier steps are frequently revisited to reassess and revise diagnoses and intervention strategies. Each step must be documented in the medical record, providing a record for future reference and facilitating communication among members of the health care team.

Nutrition Assessment Nutrition assessment involves the collection and analysis of health-related information in order to identify specific nutrition problems and their underlying causes. The information may be obtained from the medical record, physical examination, laboratory analyses, medical procedures, interview with the patient or caregiver, and consultation with other health professionals. The assessment data are used to develop a plan of action to prevent or correct energy or nutrient imbalances, or to determine whether a care plan is working. The next section of this chapter describes the components of nutrition assessment in more detail.

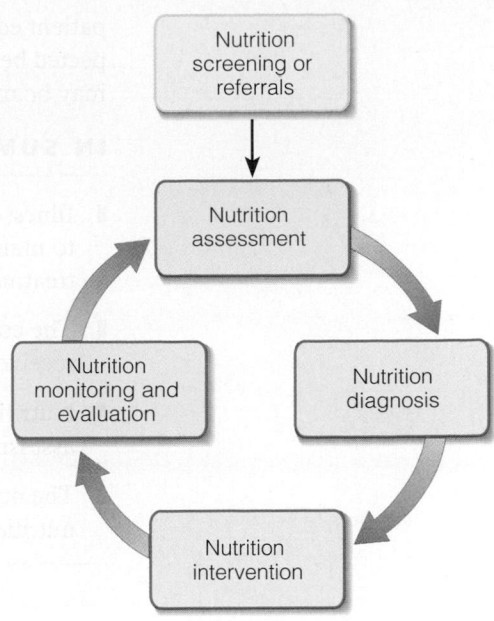

FIGURE 14-2 The Nutrition Care Process

Nutrition Diagnosis Each nutrition problem receives a separate diagnosis, which is formatted to include the specific nutrition problem, etiology or cause, and signs and symptoms that provide evidence of the problem.[4] For example, a nutrition diagnosis might state, "Unintentional weight loss *(the problem)* related to insufficient kcaloric intake *(the etiology or cause)* as evidenced by a 10-pound weight loss (8 percent of body weight) in the past few months *(the sign or symptom)*." A nutrition diagnosis can change during the course of an illness.

This format is called a *PES statement* because it includes the Problem, the Etiology, and the Signs and symptoms.

Nutrition Intervention After nutrition problems are identified, the appropriate nutrition therapy can be determined. An intervention may include counseling or education about appropriate dietary and lifestyle practices, changes in a medication or other treatment, or adjustments in the meals provided to a hospital patient. Ideally, the intervention considers an individual's food habits, lifestyle, and other personal factors. Goals are specified in terms of measurable outcomes; for example, goals for an overweight person with diabetes might include improved blood glucose levels and body weight. Other goals may be changes in the patient's dietary behaviors and lifestyle; for example, a diabetes patient may learn how to control carbohydrate intake and begin a regular exercise program.

Nutrition Monitoring and Evaluation The patient's progress should be monitored closely. Updated assessment data or diagnoses may require adjustments in goals or outcome measures. Sometimes a new situation alters nutritional needs; for example, a change in the medical treatment or a new medication may alter a person's tolerance to certain foods. The nutrition care plan must be flexible enough to adapt to the new situation.

If a patient is unable or unwilling to make the suggested changes, the care plan should be redesigned and take into account the reasons why the earlier plan was not successful. This new plan may need to include motivational techniques or additional

Joint Commission: a nonprofit organization that sets standards for health care performance and safety and confers accreditation to health care organizations and programs that meet those standards.

nutrition care process: a problem-solving method that dietetics professionals use to evaluate and treat nutrition-related problems.

patient education. If the patient remains unwilling to modify behaviors despite the expected benefits, the health care provider can try again at a later time when the patient may be more receptive.

IN SUMMARY

▪ Illnesses and their treatments can affect food intake and nutrient needs, leading to malnutrition. In turn, poor nutrition can reduce the effectiveness of medical treatment.

▪ The combined efforts of each member of the health care team ensure that patients receive optimal nutrition care.

▪ Nutrition screening identifies individuals who can benefit from nutrition assessment and follow-up nutrition care.

▪ The nutrition care process includes four interrelated steps: nutrition assessment, nutrition diagnosis, nutrition intervention, and nutrition monitoring and evaluation.

Nutrition Assessment

As described earlier, a nutrition assessment provides necessary information for identifying nutrition problems and designing a nutrition care plan; follow-up assessments help to determine if the care plan has been effective. Ideally, the assessment should be sensitive enough to detect subtle nutrition problems and specific enough to identify problem nutrients. This section describes the types of information and measures that are most commonly obtained in a nutrition assessment.

Historical Information

Historical information provides clues about the patient's nutrition status and nutrient needs. Historical information also uncovers personal preferences that should be considered when developing a nutrition care plan. Table 14-3 summarizes the various types of historical data that contribute to a nutrition assessment. This information can be obtained from the medical record or by interviewing the patient or caregiver.

Medical History The medical history helps the practitioner identify health problems that may affect food intake or require dietary changes. Table 14-4 lists medical problems that are frequently associated with nutrition problems. Many other conditions can lead to malnutrition, especially if nutrition problems are not recognized and corrected before they become serious.

TABLE 14-3 Historical Information Used in Nutrition Assessment[a]

Medical History	Medication and Supplement History	Personal and Social History	Food and Nutrition History
Age	Prescription drugs	Employment status	Food intake
Current complaint(s)	Over-the-counter drugs	Educational level	Alcohol consumption
Past medical conditions	Dietary and herbal supplements	Socioeconomic status	Dietary restrictions
Surgical history		Cultural/ethnic identity	Food allergies or intolerances
Family medical history		Religious beliefs	Nutrition and health knowledge
Chronic disease risk		Home/family situation	Food availability
Allergies		Cognitive abilities	Physical activity and exercise habits
Mental/emotional health status		Use of tobacco or illegal drugs	

[a]Historical information is classified in different ways among medical institutions.

Medication and Supplement History Certain medications can have detrimental effects on nutrition status, and some dietary components can alter the absorption and metabolism of drugs. Chapter 15 provides examples of diet-drug interactions that need consideration when planning nutrition care.

Personal and Social History Personal and social factors can influence both food choices and the ability to deal with health and nutrition problems. For example, an individual's cultural background can affect food preferences, whereas financial concerns may restrict access to health care and nutritious foods. Some individuals depend on others to prepare or procure food. A person who lives alone or is depressed may eat poorly or be unable to follow complex dietary instructions.

Food and Nutrition History A food and nutrition history (or *diet history*) is a detailed account of a person's dietary practices. It includes information about food intake, lifestyle habits, and other factors that affect food choices. Although various methods exist for obtaining data, the procedure often includes an interview about recent food intake (for example, a *24-hour recall*) and a survey of usual food choices (such as a *food frequency questionnaire*). The following section describes the most common methods of gathering food intake information.

Food Intake Data

Obtaining accurate food intake data is challenging, as results may vary depending on both the individual's memory and honesty and the assessor's skill and training. In addition, each method has its own strengths and weaknesses, so best results are obtained from using a combination of methods. Table 14-5 summarizes the methods commonly used and each method's advantages and disadvantages.

The 24-Hour Recall The **24-hour recall** is a guided interview in which an individual recounts all of the foods and beverages consumed in the past 24 hours or during the previous day. The interviewer includes questions about what times meals or snacks were eaten, amounts consumed, and ways in which foods were prepared. Accuracy can be improved by prompting the respondent to recall items that are often forgotten, such as condiments, snacks, and beverages.[5]

In a typical interview, the assessor may begin by asking: "What is the first thing you ate or drank yesterday morning?" After the respondent describes the first items, follow-up questions might be: "What time was that?" and "How much did you eat?" Questioning continues until the intake record for the day is complete. Food models or measuring cups and spoons can be used to help the individual visualize and describe the amounts consumed. After the day's intake is recounted, the interviewer asks whether the intake that day is fairly typical and, if not, how it varies from the person's usual intake. A recall interview may be conducted on several nonconsecutive days to get a better representation of a person's usual diet.

A recall interview can yield useful data for developing an acceptable nutrition care plan and identifying food items that may need to be restricted due to illness. However, it is a poor technique for determining the adequacy of a diet because it does not take into account fluctuations in food intake or seasonal variations. Moreover, food intakes are often underestimated because the process relies on an individual's memory and reporting accuracy. People often forget to mention beverages, condiments, and snack foods unless specifically prompted to do so,

Food models and measuring utensils can help an individual visualize portion sizes.

© Nathan Benn/Corbis

TABLE 14-4 Medical Problems Often Associated with Malnutrition

- Acquired immune deficiency syndrome (AIDS)
- Alcoholism
- Anorexia nervosa or bulimia
- Burns (extensive or severe)
- Cancer and cancer treatments
- Cardiovascular diseases
- Celiac disease
- Chewing or swallowing difficulties
- Chronic kidney disease
- Dementia or other mental illness
- Diabetes mellitus
- Feeding disabilities
- Infections
- Inflammatory bowel diseases
- Liver disease
- Pressure sores
- Surgery (major)
- Vomiting (prolonged or severe)

24-hour recall: a record of foods consumed in the previous 24 hours; sometimes modified to include foods consumed in a typical day.

TABLE 14-5 Methods for Obtaining Food Intake Data

Method	Description	Advantages	Disadvantages
24-hour recall	Guided interview in which the foods and beverages consumed in a 24-hour period are described in detail.	▪ Results are not dependent on literacy or educational level of respondent. ▪ Interview occurs after food is consumed, so method does not influence dietary choices. ▪ Results are obtained quickly; method is relatively easy to conduct.	▪ Process is reliant on memory. ▪ Underestimation and overestimation of food intakes are common. ▪ Food items that cause embarrassment (alcohol, desserts) may be omitted. ▪ Data from a single day cannot accurately represent the respondent's usual intake. ▪ Seasonal variations may not be addressed. ▪ Skill of interviewer affects outcome.
Food frequency questionnaire	Written survey of food consumption during a specific period of time, often a one-year period.	▪ Process examines long-term food intake, so day-to-day and seasonal variability should not affect results. ▪ Questionnaire is completed after food is consumed, so method does not influence food choices. ▪ Method is inexpensive to administer.	▪ Process is reliant on memory. ▪ Food lists often include common foods only. ▪ Serving sizes are often difficult for respondents to evaluate without assistance. ▪ Calculated nutrient intakes may not be accurate. ▪ Food lists for the general population are of limited value in special populations. ▪ Method is not effective for monitoring short-term changes in food intake.
Food record	Written account of food consumed during a specified period, usually several consecutive days. Accuracy is improved by including weights or measures of foods.	▪ Process does not rely on memory. ▪ Recording foods as they are consumed may improve accuracy of food intake data. ▪ Process is useful for controlling intake because keeping records increases awareness of food choices.	▪ Recording process itself influences food intake. ▪ Underreporting and portion size errors are common. ▪ Process is time-consuming and burdensome for respondent; requires high degree of motivation. ▪ Method requires literacy and the physical ability to write. ▪ Seasonal changes in diet are not taken into account.
Direct observation	Observation of meal trays or shelf inventories before and after eating; possible only in residential facilities.	▪ Process does not rely on memory. ▪ Method does not influence food intake. ▪ Method can be used to evaluate the acceptability of a prescribed diet.	▪ Process is possible only in residential situations. ▪ Method is labor intensive.

and some find it embarrassing to report consumption of foods such as chocolate, butter, and red meat.[6]

Food Frequency Questionnaire A **food frequency questionnaire** surveys the foods and beverages regularly consumed during a specific time period. Some questionnaires are qualitative only: food lists contain common foods, organized by food group, with boxes to check to indicate frequency of consumption. Other types of questionnaires provide semiquantitative information by including portion sizes as well. Figure 14-3 shows a sample section of a semiquantitative questionnaire that surveys fruit intake over the previous year. Because the respondent is often asked to estimate food intakes over a one-year period, the results should not be affected by seasonal changes in diet. Conversely, a disadvantage of this method is its inability to determine recent changes in food intake.

FRUIT	HOW OFTEN								HOW MUCH			
	Never or less than once per month	1 per mon.	2–3 per mon.	1 per week	2 per week	3–4 per week	5–6 per week	Every day	MEDIUM SERVING	YOUR SERVING SIZE		
										S	M	L
EXAMPLE: Bananas	○	○	○	●	○	○	○	○	1 medium	○ 1/2	● 1	○ 2
Bananas	○	○	○	○	○	○	○	○	1 medium	○ 1/2	○ 1	○ 2
Apples, applesauce	○	○	○	○	○	○	○	○	1 medium or 1/2 cup	○ 1/2	○ 1	○ 2
Oranges (not including juice)	○	○	○	○	○	○	○	○	1 medium	○ 1/2	○ 1	○ 2
Grapefruit (not including juice)	○	○	○	○	○	○	○	○	1/2 medium	○ 1/4	○ 1/2	○ 1
Cantaloupe	○	○	○	○	○	○	○	○	1/4 medium	○ 1/8	○ 1/4	○ 1/2
Peaches, apricots (fresh, in season)	○	○	○	○	○	○	○	○	1 medium	○ 1/2	○ 1	○ 2
Peaches, apricots (canned or dried)	○	○	○	○	○	○	○	○	1 medium or 1/2 cup	○ 1/2	○ 1	○ 2
Prunes, or prune juice	○	○	○	○	○	○	○	○	1/2 cup	○ 1/4	○ 1/2	○ 1
Watermelon (in season)	○	○	○	○	○	○	○	○	1 slice	○ 1/2	○ 1	○ 2
Strawberries, other berries (in season)	○	○	○	○	○	○	○	○	1/2 cup	○ 1/4	○ 1/2	○ 1
Any other fruit, including kiwi, fruit cocktail, grapes, raisins, mangoes	○	○	○	○	○	○	○	○	1/2 cup	○ 1/4	○ 1/2	○ 1

FIGURE 14-3 Sample Section of a Food Frequency Questionnaire

Simple versions of food frequency questionnaires focus on food categories relevant to a person's medical condition. For example, a questionnaire designed to evaluate calcium intake may include only milk products, fortified foods, certain fruits and vegetables, and dietary supplements that contain calcium. A computer analysis can then quickly estimate the individual's calcium intake and compare it to recommendations.

Food Record A **food record** is a written account of foods and beverages consumed during a specified time period, usually several consecutive days. Foods are recorded as they are consumed in order to obtain the most complete and accurate record possible; thus, the process does not rely on memory. A detailed food record includes the types and amounts of foods and beverages consumed, times of consumption, and methods of preparation. It provides valuable information about food intake as well as a person's response to and compliance with medical nutrition therapy. Unfortunately, food records require a great deal of time to complete, and people need to be highly motivated to keep accurate records. Another drawback is that the recording process itself may influence food intake. Furthermore, it is often difficult to obtain accurate estimates of nutrient intake in just a few days or even a week due to day-to-day and seasonal variations in food intake.

Direct Observation In facilities that serve meals, food intakes can be directly observed and analyzed. This method can also reveal a person's food preferences, changes in appetite, and any problems with a prescribed diet. Health practitioners use direct observation to conduct patients' **kcalorie counts,** which are estimates of the food energy (and often protein) consumed by patients during a single day or several consecutive days. To perform a kcalorie count, the clinician records the dietary items that a patient is given at meals and subtracts the amounts remaining after meals are completed; this procedure allows an estimate of the kcaloric content of foods and beverages actually consumed. Although a useful means of discerning patients' intakes, direct observation requires regular and careful documentation and can be labor-intensive and costly.

food frequency questionnaire: a survey of foods routinely consumed. Some questionnaires ask about the types of food eaten and yield only qualitative information, whereas others include questions about portions consumed and yield semiquantitative data as well.

food record: a detailed log of food eaten during a specified time period, usually several days; also called a *food diary*. A food record may also include information regarding disease symptoms, physical activity, and medication use.

kcalorie counts: the determination of food energy (and often protein) consumed by patients for one or more days.

Anthropometric Data

Reminder: *Protein-energy malnutrition* is a deficiency of protein and food energy characterized by loss of weight and muscle mass.

Appendix E provides information about methods for determining the percentage of body fat.

Length is measured while a person is recumbent (lying down), whereas *height* is measured while a person is standing upright.

Measures of body size, known as **anthropometric** measurements, can reveal problems related to both overnutrition and protein-energy malnutrition (PEM). Height (or length) and weight are the most common anthropometric measurements and are used to evaluate growth in children and nutrition status in adults. Other helpful data include body fat percentages and circumferences of the head, waist, and limbs.

Height (or Length) Poor growth in children can signify malnutrition. In adults, height measurements alone do not reflect current nutrition status but can be used for estimating a person's energy needs or appropriate body weight. Length is measured in infants and children younger than 24 months, and height is usually measured in older children and adults. Length can also be measured in adults and children who cannot stand unassisted due to physical or medical reasons. The "How to" describes some standard techniques for measuring length and height.

In adults who are bedridden or unable to stand, height can be estimated from equations that include either the knee height or the full arm span, both of which correlate well with height.[7] Knee height, which extends from the heel to the top of the knee when the leg is bent at a 90-degree angle, can be measured in either a sitting or supine position with a knee-height caliper; specific formulas are available for various age, gender, and ethnic groups. The full arm span is the distance from the tip of one middle finger to the other while the arms are extended horizontally. In children with disabilities that affect stature, alternative measures of linear growth include the full arm span, lower-leg lengths (knee to heel, similar to the knee height measure), and upper-arm lengths (shoulder to elbow), all of which can be compared with reference percentiles.

Body Weight During clinical care, health care providers monitor body weights closely: weight changes may reflect changes in hydration status, and an involuntary weight loss can be a sign of PEM. Body weights can be compared with healthy ranges

How To MEASURE LENGTH AND HEIGHT

To improve the accuracy of length and height measurements, keep the following in mind:

▌ Always measure—never ask! Self-reported heights are less accurate than measured heights. If height is not measured, document that the height is self-reported.

▌ Measure the length of infants and young children by using a measuring board with a fixed headboard and a movable footboard. It generally takes two people to measure length: one person gently holds the infant's head against the headboard; the other straightens the infant's legs and moves the footboard to the bottom of the infant's feet.

▌ Measure height next to a wall on which a nonstretchable measuring tape or board has been fixed. Ask the person to stand erect without shoes and with

heels together. The person's eyes and head should be facing forward, with heels, buttocks, and shoulder blades touching the wall. Place a ruler or other flat, stiff object on the top of the head at a right angle to the wall and carefully note the height measurement. Immediately record length and height measurements to the nearest 1/8 inch or 0.1 centimeter.

▌ For evaluating growth rate in young children, use the appropriate growth chart (Appendix E) when plotting results. If length is measured, use the growth chart for children between 0 and 36 months; if height is measured, use the chart for individuals between 2 and 20 years.

▌ Higher values are obtained from supine measurements than from vertical height measurements due to gravity.

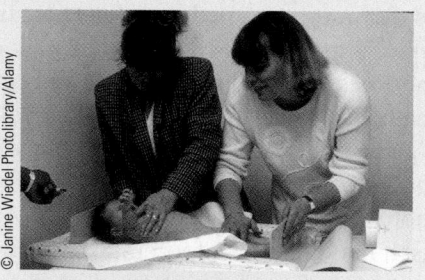

It generally takes two people to measure the length of an infant.

Standing erect allows for an accurate height measurement.

How To MEASURE WEIGHT

Tips for measuring weight include:

■ Always measure—never ask! Self-reported weights are often inaccurate. If weight is not measured, document that the weight is self-reported.

■ Valid weight measurements require scales that have been carefully maintained, calibrated, and checked for accuracy at regular intervals. Beam balance and electronic scales are the most accurate. Bathroom scales are inaccurate and inappropriate for clinical use.

■ Measure an infant's weight with a scale that allows the infant to sit or lie down. The tray should be large enough to support an infant or young child up to 40 pounds, and weight graduations should be in ½-ounce or 10-gram increments. For accurate results, weigh infants without clothes or diapers. Excessive movement by the infant can reduce accuracy.

■ Children who can stand are weighed in the same way as adults, using beam balance or electronic scales with platforms large enough for standing comfortably. If repeated weight measurements are needed, each weighing should take place at the same time of day (preferably before breakfast), in the same amount of clothing, after the person has voided, and using the same scale. Record weights to the nearest ¼ pound or 0.1 kilogram.

■ Special scales and hospital beds with built-in scales are available for weighing people who are bedridden.

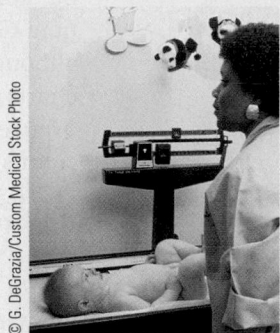

Infants are weighed on scales that allow them to sit or lie down.

Beam balance scales allow accurate weight measurements for older children and adults.

on height-weight tables and growth charts or used to calculate the body mass index (BMI). A healthy body weight typically falls within a BMI range of 18.5 to 25; thus, an appropriate body weight can usually be estimated by using a BMI table or graph (see the inside back cover of this book).[8] The "How to" includes suggestions for improving the accuracy of weight measurements.

Reminder: BMI = $\dfrac{\text{weight (kg)}}{\text{height (m)}^2}$

Head Circumference A measurement of head circumference helps to assess brain growth and malnutrition in children up to three years of age, although this measure is not necessarily reduced in a malnourished child. Head circumference values can also track brain development in premature and small-for-gestational-age infants. To measure head circumference, the assessor encircles the largest circumference measure of a child's head with a nonstretchable measuring tape: the tape is placed just above the eyebrows and ears and around the occipital prominence at the back of the head (see the photo). The measurement is read to the nearest $\frac{1}{8}$ inch or 0.1 centimeter.

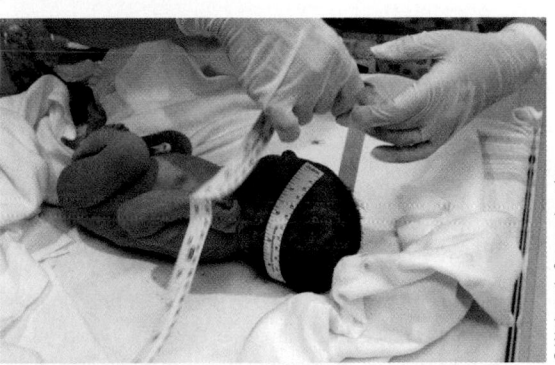

Head circumference measurements help to assess brain growth.

Circumferences of Waist and Limbs Circumferences of the waist and limbs are useful for evaluating body fat and muscle mass, respectively. Waist circumference correlates with intra-abdominal fat and can help in assessing overnutrition. Circumferences of the mid-upper arm, mid-thigh, and mid-calf regions can help in evaluating the effects of illness, aging, and PEM on skeletal muscle content. For improved accuracy, circumference measurements are often used together with skinfold measurements to correct for the subcutaneous fat in limbs.

Anthropometric Assessment in Infants and Children To evaluate growth patterns, periodic measurements of height (or length), weight, and head circumference can be plotted on growth charts, such as those provided in Appendix E. The most

anthropometric (AN-throw-poe-MEH-trik): related to physical measurements of the human body, such as height, weight, body circumferences, and percentage of body fat.

commonly used growth charts compare height (or length) to age, weight to age, head circumference to age, weight to length, and BMI to age. Although individual growth patterns vary, a child's growth will generally stay at about the same percentile throughout childhood; a sharp drop in a previously steady growth pattern suggests malnutrition. Growth patterns that fall below the 5th percentile may also be cause for concern, although genetic influences must be considered when interpreting low values. Growth charts with BMI-for-age percentiles can be used to assess risk of underweight and overweight in children over two years of age: the 5th and 85th percentiles are used as cutoffs to identify children who may be malnourished or overweight, respectively.[9]

Anthropometric Assessment in Adults To evaluate the nutritional risks associated with illness, clinicians frequently monitor both the total reduction in weight and rate of weight loss over time.[10] Weight changes must be evaluated carefully: although unintentional weight *loss* can indicate malnutrition, weight *gain* may result from fluid retention rather than overnutrition. Fluid retention often accompanies worsening disease in patients with heart failure, liver cirrhosis, and kidney failure and can mask the weight loss associated with PEM. Some medications can also lead to weight loss or gain. As mentioned, the rate of weight loss should be considered as well as the amount; for example, an adult is at risk of PEM if there is an involuntary weight loss of more than 5 percent within one month or more than 10 percent within six months.[11]

Body weight data can be expressed as a percentage of usual body weight (%UBW) or ideal body weight (%IBW). The %UBW is more effective than %IBW for interpreting weight changes that occur in underweight, overweight, or obese individuals. In overweight persons, the %IBW may fail to identify significant weight loss. Conversely, in underweight individuals, the %IBW can overstate the degree of weight loss due to illness. The "How to" describes how to estimate %UBW and %IBW, and Table 14-6 shows how to interpret these values.

Some illnesses discussed in later chapters are associated with lean tissue losses that resist nutrition intervention. In aging individuals, losses in both lean tissue and height are common even though body weights may remain stable. Including measures such as skinfold measurements and limb circumferences can help the clinician identify changes in body composition that need to be addressed in the treatment plan.

$$\text{Percent weight loss} = \frac{\text{usual weight} - \text{current weight}}{\text{usual weight}} \times 100$$

TABLE 14-6 Body Weight and Nutritional Risk

%UBW	%IBW	Nutritional Risk
85–95	80–90	Risk of mild malnutrition
75–84	70–79	Risk of moderate malnutrition
<75	<70	Risk of severe malnutrition

How To ESTIMATE AND EVALUATE %UBW AND %IBW

%UBW: To estimate %UBW, compare an individual's current weight with the weight that the person generally maintains:

$$\%UBW = \frac{\text{current weight}}{\text{usual weight}} \times 100$$

For example, if a man loses 32 pounds during illness and his usual weight is 180 pounds, his current weight would be 148 pounds. These values can be incorporated into the previous equation:

$$\%UBW = \frac{148}{180} \times 100 = 82.2\%$$

The man in this example weighs 82.2 percent of his usual weight. A look at Table 14-6 shows that a person who is at 82 percent of UBW may be moderately malnourished.

%IBW: To estimate %IBW, compare an individual's current weight with a reasonable (ideal) weight from a BMI table or other appropriate reference:

$$\%IBW = \frac{\text{current weight}}{\text{ideal weight}} \times 100$$

For example, suppose you wish to calculate the %IBW for a woman who is 5 feet 8 inches tall and weighs 116 pounds. The midpoint of the healthy BMI range is approximately 22, so using a BMI table (as shown on the inside back cover of this book), you estimate that a reasonable weight for this woman would be about 144 pounds:

$$\%IBW = \frac{116}{144} \times 100 = 80.6\%$$

The woman in this example weighs about 80.6 percent of her ideal body weight. A look at Table 14-6 suggests that, at 80.6 percent of IBW, she may be mildly malnourished. Keep in mind that the calculation of "ideal body weight" is somewhat arbitrary, because the BMI table and various other references provide a range of weights for individuals of a given height.

Biochemical Analyses

Biochemical data provide information about protein-energy nutrition, vitamin and mineral status, fluid and electrolyte balance, and organ function. Most tests are based on analyses of blood and urine samples, which contain proteins, nutrients, and metabolites that reflect nutrition and health status. Table 14-7 lists and describes common blood tests with nutritional implications. Laboratory tests relevant to specific diseases will be discussed in the chapters that follow.

Either *plasma* or *serum* levels of blood components may be reported. *Plasma* is the yellow fluid that remains after cells are removed and still contains clotting factors. *Serum* is the fluid remaining after both cells and clotting factors are removed.

TABLE 14-7 **Routine Laboratory Tests with Nutritional Implications**

This table presents a partial listing of some uses of commonly performed lab tests that have implications for nutritional problems.

Laboratory Test	Acceptable Range	Description
Hematology		
Red blood cell (RBC) count	Male: 4.3–5.7 million/μL Female: 3.8–5.1 million/μL	Number of RBC; aids anemia diagnosis.
Hemoglobin (Hb)	Male: 13.5–17.5 g/dL Female: 12.0–16.0 g/dL	Hemoglobin content of RBC; aids anemia diagnosis.
Hematocrit (Hct)	Male: 39–49% Female: 35–45%	Percentage RBC in total blood volume; aids anemia diagnosis.
Mean corpuscular volume (MCV)	80–100 fL	RBC size; helps to distinguish between microcytic and macrocytic anemia.
Mean corpuscular hemoglobin concentration (MCHC)	31–37% Hb/cell	Hb concentration within RBC; helps to distinguish iron-deficiency anemia.
White blood cell (WBC) count	4500–11,000 cells/μL	Number of WBC; general assessment of immunity.
Serum Proteins		
Total protein	6.4–8.3 g/dL	Protein levels are not specific to disease or highly sensitive; they can reflect body protein, illness or infections, changes in hydration or metabolism, pregnancy, or use of certain medications.
Albumin	3.4–4.8 g/dL	May reflect illness or PEM; slow to respond to improvement or worsening of disease.
Transferrin	200–400 mg/dL >60 yr: 180–380 mg/dL	May reflect illness, PEM, or iron deficiency; slightly more sensitive to health status changes than albumin.
Prealbumin (transthyretin)	10–40 mg/dL	May reflect illness or PEM; more responsive to health status changes than albumin or transferrin.
C-reactive protein	68–8200 ng/mL	Indicator of inflammation or disease.
Serum Enzymes		
Creatine kinase (CK)	Male: 38–174 U/L Female: 26–140 U/L	Different forms of CK are found in muscle, brain, and heart. High levels in blood may indicate heart attack, brain tissue damage, or skeletal muscle injury.
Lactate dehydrogenase (LDH)	208–378 U/L	LDH is found in many tissues. Specific types may be elevated after heart attack, lung damage, or liver disease.
Alkaline phosphatase	25–100 U/L	Found in many tissues; often measured to evaluate liver function.
Aspartate aminotransferase (AST, formerly SGOT)	10–30 U/L	Usually monitored to assess liver damage; elevated in most liver diseases. Levels are somewhat increased after muscle injury.
Alanine aminotransferase (ALT, formerly SGPT)	Male: 10–40 U/L Female: 7–35 U/L	Usually monitored to assess liver damage; elevated in most liver diseases. Levels are somewhat increased after muscle injury.

(continued)

TABLE 14-7 (continued)

Laboratory Test	Acceptable Range	Description
Serum Electrolytes		
Sodium	136–146 mEq/L	Helps to evaluate hydration status or neuromuscular, kidney, and adrenal functions.
Potassium	3.5–5.1 mEq/L	Helps to evaluate acid-base balance and kidney function; can detect potassium imbalances.
Chloride	98–106 mEq/L	Helps to evaluate hydration status and detect acid-base and electrolyte imbalances.
Other		
Glucose (fasting)[a]	74–106 mg/dL >60 yr: 80–115 mg/dL	Detects risk of glucose intolerance, diabetes mellitus, and hypoglycemia; helps to monitor diabetes treatment.
Glycated hemoglobin (HbA$_{1c}$)	5.0–7.5% of total Hb	Used to monitor long-term blood glucose control (approximately 1 to 3 months prior).
Blood urea nitrogen (BUN)	6–20 mg/dL	Primarily used to monitor kidney function; value is altered by liver failure, dehydration, or shock.
Uric acid	Male: 3.5–7.2 mg/dL Female: 2.6–6.0 mg/dL	Used for detecting gout or changes in kidney function; levels affected by age and diet; varies among different ethnic groups.
Creatinine (serum or plasma)	Male: 0.7–1.3 mg/dL Female: 0.6–1.1 mg/dL	Used to monitor renal function.

Note: μL = microliter; dL = deciliter; fL = femtoliter; ng = nanogram; U/L = units per liter; mEq = milliequivalents.

[a]Fasting glucose levels that repeatedly exceed 100 mg/dL suggest prediabetes.

Source: L. Goldman and D. Ausiello, coeditors, *Cecil Medicine* (Philadelphia: Saunders, 2008).

Fluid retention can cause lab results to be deceptively low. Dehydration may cause lab results to be deceptively high.

Interpreting laboratory values can be challenging because a number of factors may influence test results. For example, serum protein values can be affected by fluid imbalances, infections, pregnancy, and some medications. Similarly, serum levels of vitamins and minerals are often poor indicators of nutrient deficiency because the values are affected by multiple variables; therefore, a variety of tests are generally needed to diagnose a nutrition problem. Taken together with other assessment data, however, laboratory test results can present a clearer picture than is possible otherwise.

Plasma Proteins Plasma protein levels can help in the assessment of protein-energy status, but the levels may fluctuate for other reasons as well.[12] For example, plasma proteins are synthesized in the liver, so plasma levels of these proteins can reflect liver function. Metabolic stress alters plasma proteins because the liver responds by increasing its synthesis of some proteins and reducing synthesis of others. Values are also influenced by pregnancy, kidney function, zinc status, and some medications. Because plasma proteins are affected by so many factors, their values must be considered along with other data in order to evaluate nutrition status.

Albumin Albumin is the most abundant plasma protein, and its levels are routinely measured during illness. Although many medical conditions influence albumin, it is slow to reflect changes in nutrition status because of its large body pool and slow rate of degradation. In people with chronic PEM, albumin levels remain normal for long periods of time despite depletion of body proteins; levels fall only after prolonged malnutrition. Likewise, albumin levels increase slowly when malnutrition is treated, so albumin is not a sensitive indicator of effective treatment.

In blood tests, the term *half-life* describes the length of time that a substance remains in plasma. The albumin in plasma has a half-life of 14 to 20 days, meaning that half of the amount circulating in plasma is degraded in this time period.

Transferrin Transferrin is an iron-transport protein, and its concentrations respond to both iron status and PEM. Transferrin levels rise as iron status worsens and fall as

iron status improves, so using transferrin values to evaluate protein-energy status is difficult if iron deficiency is also present. Transferrin degrades more rapidly than albumin but its levels change relatively slowly in response to nutrition therapy.

Transferrin's half-life in plasma is approximately 8 to 10 days.

Prealbumin and Retinol-Binding Protein Levels of prealbumin (also called transthyretin) and retinol-binding protein decrease rapidly during PEM and respond quickly to improved protein intakes. Thus, these proteins are more sensitive than albumin to changes in protein status. Like other plasma proteins, their usefulness in nutrition assessment is limited because they are affected by a number of different factors, including metabolic stress, zinc deficiency, and various medical conditions. Prealbumin and retinol-binding protein are more expensive to measure than albumin, so they are not routinely included in a nutrition assessment.

Half-lives of prealbumin and retinol-binding protein are 2 to 3 days and 12 hours, respectively.

Physical Examinations

As with other assessment methods, interpreting physical signs of malnutrition requires skill and clinical judgment. Most physical signs are nonspecific; they can reflect any of several nutrient deficiencies as well as conditions unrelated to nutrition. For example, cracked lips may be caused by several B vitamin deficiencies but they can also result from sunburn, windburn, or dehydration. Dietary and laboratory data are usually needed as additional evidence to confirm suspected nutrient deficiencies.

Clinical Signs of Malnutrition Signs of malnutrition tend to appear most often in parts of the body where cells are replaced at a rapid rate, such as the hair, skin, and digestive tract (including the mouth and tongue). Table 14-8 lists some clinical signs of nutrient deficiencies. Many of the symptoms listed occur only in advanced stages of deficiency. Chapters 8 and 9 provide additional examples of the clinical signs of nutrient imbalances.

TABLE 14-8 Clinical Signs of Nutrient Deficiencies

Body System	Acceptable Appearance	Signs of Malnutrition	Other Possible Causes
Hair	Shiny, firm in scalp	Dull, brittle, dry, loose; falls out (PEM); corkscrew hair (vitamin C)	Excessive hair bleaching, hair loss from aging, chemotherapy, or radiation therapy
Eyes	Bright; clear; shiny; pink, moist membranes; adjust easily to light	Pale membranes (iron); spots, dryness, night blindness (vitamin A); redness at corners of eyes (B vitamins)	Anemia that is unrelated to nutrition; eye disorders; allergies
Lips	Smooth	Dry, cracked, or with sores in the corners of the lips (B vitamins)	Sunburn, windburn, excessive salivation from ill-fitting dentures or various disorders
Mouth and gums	Oral tissues without lesions, swelling, or bleeding; red tongue; normal sense of taste; teeth without caries; ability to chew and swallow	Bleeding gums (vitamin C); smooth or magenta tongue (B vitamins), poor taste sensation (zinc)	Medications, periodontal disease (poor oral hygiene)
Skin	Smooth, firm, good color	Poor wound healing (PEM, vitamin C, zinc); dry, rough, lack of fat under skin (essential fatty acids, PEM, vitamin A, B vitamins); bruising or bleeding under skin (vitamins C and K); pale (iron)	Poor skin care, diabetes mellitus, aging, medications
Nails	Smooth, firm, pink	Ridged (PEM); spoon shaped, pale (iron)	—
Other	—	Dementia, peripheral neuropathy (B vitamins); swollen glands at front of neck (PEM, iodine); bowed legs (vitamin D)	Disorders of aging (dementia), diabetes mellitus (peripheral neuropathy)

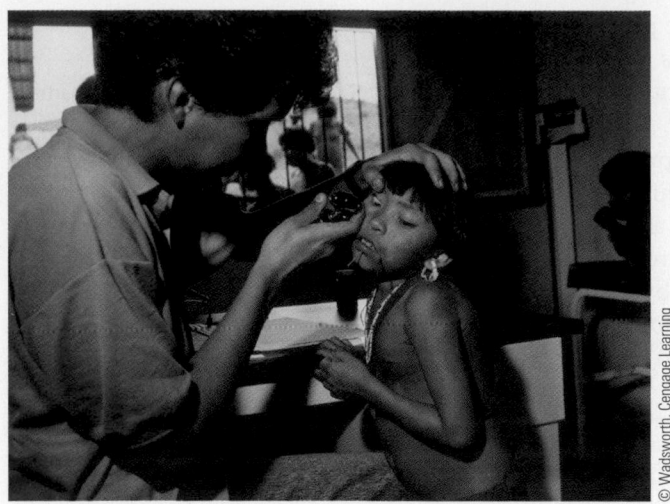

Physical signs of malnutrition are often evident in parts of the body where cells are replaced at a rapid rate.

Fluid Imbalances As mentioned previously, fluid imbalances may accompany some illnesses and can also result from the use of certain medications. Recognizing the physical signs of fluid retention or dehydration is necessary for the correct interpretation of blood test results and the body weight measurement.

Fluid retention (also called *edema*) can accompany malnutrition, infection, or injury or the use of certain medications. It can be caused by impaired blood circulation and frequently accompanies disorders of the heart and blood vessels, kidneys, liver, and lungs. Physical signs of fluid retention include weight gain, facial puffiness, swelling of limbs, abdominal distention, and tight-fitting shoes.

Dehydration can result from vomiting, diarrhea, fever, sweating, excessive urination, and skin injury or burns (due to fluid loss through skin lesions). Dehydration risk is greatest in elderly individuals, who have a reduced thirst response and various other impairments in fluid regulation.[13] Symptoms of dehydration include thirst, dry skin or mouth, reduced skin tension, dark yellow or amber urine, and low urine volume.

Functional Assessment Nutrient deficiencies sometimes impair normal physiological functions, so functional tests or procedures can be used to evaluate some aspects of malnutrition. For example, both PEM and zinc deficiency can depress immunity, which can be evaluated by testing the skin's response to antigens that cause redness and swelling when immune function is adequate. Muscle weakness due to **wasting,** or loss of muscle tissue, may be assessed by testing hand-grip strength. Exercise tolerance, which is reduced in heart and lung disorders, is sometimes evaluated using a treadmill or cycle ergometer.

Determining Energy Requirements

To determine the energy requirements of hospital patients, clinicians typically measure or estimate the resting metabolic rate (RMR), and then adjust the RMR value with "stress factors" that account for medical problems, or, in some cases, medical treatments. In ambulatory patients, a factor for activity level may also be applied. Table 14-9 presents examples of RMR equations in common use, and the "How to" presents an example of this method. Note that in individuals who are overweight or obese, the Mifflin–St. Jeor equation has been found to yield the most accurate results.[14]

In critical-care patients, energy needs may be considerably higher than normal due to fever, mechanical ventilation, restlessness, or the presence of open wounds. Patients who are critically ill are usually bedridden and inactive, however, so the energy needed for physical activity is minimal. Energy requirements for critical-care patients are described in Chapter 24.

TABLE 14-9 **Selected Equations for Estimating Resting Metabolic Rate (RMR)**

Harris–Benedict[a]

Women: RMR = 655.1 + [9.563 × weight (kg)] + [1.85 × height (cm)] − [4.676 × age (years)].
Men: RMR = 66.5 + [13.75 × weight (kg)] + [5.003 × height (cm)] − [6.755 × age (years)].

Mifflin–St. Jeor

Women: RMR = [9.99 × weight (kg)] + [6.25 × height (cm)] − [4.92 × age (years)] − 161
Men: RMR = [9.99 × weight (kg)] + [6.25 × height (cm)] − [4.92 × age (years)] + 5

[a]Although these equations are sometimes used for estimating basal metabolic rate (BMR), they were derived from data measured during resting conditions in most cases.

How To ESTIMATE THE ENERGY NEEDS OF A HOSPITAL PATIENT

To determine the energy needs of a hospital patient, the health practitioner calculates the patient's resting metabolic rate (RMR) and then applies a "stress factor" to accommodate the additional energy needs imposed by illness. The stress factor 1.25 has been shown to be reasonably accurate for many hospitalized patients; other examples are listed in Table 24-2 on p. 624.

The following example uses the Mifflin–St. Jeor equation (shown in Table 14-9) and the stress factor 1.25 to determine the energy needs of a 57-year-old female patient who is 5 feet 3 inches tall, weighs 115 pounds, and is confined to bed.

Step 1: The patient's weight and height are converted to the units used in the equation:

> Weight in kilograms = 115 lb ÷ 2.2 lb/kg = 52.3 kg
> Height in centimeters = 63 in × 2.54 cm/in = 160 cm

Step 2: Using the Mifflin–St. Jeor equation for estimating RMR in women:

$$RMR = [9.99 \times weight\ (kg)] + [6.25 \times height\ (cm)]$$
$$- [4.92 \times age\ (years)] - 161$$
$$= (9.99 \times 52.3) + (6.25 \times 160) - (4.92 \times 57) - 161$$
$$= 522 + 1000 - 280 - 161 = 1081\ kcal$$

Step 3: The RMR value is multiplied by the appropriate stress factor:

> RMR × stress factor = 1081 × 1.25 = 1351 kcal

Thus, an appropriate energy intake for this patient would be approximately 1351 kcal. Her weight should be monitored to determine if her actual needs are higher or lower.

For a patient who is not confined to bed, an additional activity factor can be applied to accommodate the extra energy needs. For example, if the patient in the example begins limited activity while in the hospital, an activity factor of 1.2 can be multiplied by the results obtained in Step 3:

> 1351 × activity factor = 1351 × 1.2 = 1621 kcal

The activity factor for a hospitalized patient often falls between 1.1 and 1.4, and it is likely to change as the patient's condition improves.

IN SUMMARY

▌ Nutrition assessments include historical information, anthropometric data, biochemical analyses, and physical examinations. Health care providers assess food intake using 24-hour recall interviews, food frequency questionnaires, food records, and direct observation.

▌ Anthropometric measurements help clinicians evaluate growth patterns, overnutrition and undernutrition, and body composition.

▌ Biochemical analyses help in the assessment of nutrient imbalances but are also influenced by various other medical problems.

▌ Physical examinations allow the detection of nutrient deficiencies, fluid imbalances, and functional impairments related to nutritional problems.

▌ The energy needs of hospital patients can be estimated by multiplying the resting metabolic rate (RMR) by factors that account for the medical condition, medical treatments, and activity level. The accompanying Case Study can help you review the different components of a nutrition assessment.

Implementing Nutrition Care

Once the health care professional has collected assessment information, the next steps of the nutrition care process can be carried out. A nutrition care plan often includes both dietary adjustments and nutrition education. Note that some aspects of nutrition care fall within the scope of dietetics practice, whereas others require the assistance of other health professionals.

wasting: the gradual atrophy (loss) of body tissues; associated with protein-energy malnutrition or chronic illness.

Elise Sawrey is an 80-year-old retired businesswoman who has been a widow for 10 years. She uses a walker and has poorly fitting dentures. She was recently admitted to the hospital with pneumonia and also has congestive heart failure and diabetes. She routinely takes several medications to control her blood glucose levels, hypertension, and heart function. In addition to these medications, the physician has recently ordered antibiotics to treat the pneumonia. During an initial nutrition screening, Mrs. Sawrey stated that she had been eating very poorly over the past two weeks. She said that she usually weighs about 125 pounds—a fact that was documented in her medical chart from a previous visit. Although she felt she was losing weight, she didn't know how much weight she may have lost or when she started losing weight. Upon admission to the hospital, Mrs. Sawrey weighed 110 pounds and was 5 feet 3 inches tall. Her serum albumin level was 3.0 grams per deciliter. A physical exam revealed edema, and several other laboratory tests confirmed that she was retaining fluid. As a result of the nutrition screening, Mrs. Sawrey was referred to a registered dietitian for a nutrition assessment.

1. From the brief description provided, which items in Mrs. Sawrey's medical history, personal and social history, and food and nutrition history might alert the dietitian that this patient is at risk of malnutrition?

2. Identify a healthy body weight for Mrs. Sawrey, and calculate her %UBW and %IBW. What do the results reveal? What effect does fluid retention have on Mrs. Sawrey's weight?

3. How might fluid retention alter Mrs. Sawrey's serum protein levels? What physical symptoms may have suggested that she was retaining excess fluid?

4. What tools can be used to estimate Mrs. Sawrey's usual food intake? What medical, physical, and personal factors are likely to influence her diet?

5. Describe other types of assessment information the dietitian may need before developing a nutrition care plan.

Dietary Modifications

During many illnesses, a person can meet energy and nutrient needs by following a **regular diet.** In other cases, adjustments in energy or nutrient intakes or a **modified diet** may be necessary. Modified diets contain foods with altered texture, consistency, or nutrient content, or they may include or omit specific foods.

Table 14-10 lists examples of modified diets that are often prescribed during illness.[15] Diets that contain foods with altered texture and consistency may be recommended for individuals with chewing or swallowing impairments. Other dietary modifications may relieve the symptoms of disease; for example, restricting dietary fat or fiber may be helpful for people with certain kinds of intestinal disorders, and controlling sodium intake can help prevent fluid accumulation. A high-kcalorie, high-protein diet may help to prevent or reverse malnutrition. Note that a person may have several medical problems and a number of modifications may be needed.

Modified diets should be adjusted to satisfy individual preferences and tolerances and may also need to be altered as a patient's condition changes. Later chapters include more specific information about modified diets and additional dietary strategies for treating nutritional problems.

Diet Progression A change in diet as a patient's food tolerance improves is called **diet progression.** For example, the diet order may read, "progress diet from clear liquids to a regular diet as tolerated." In practice, this means that the patient would be given clear beverages initially, and then gradually be provided with other beverages or solid foods that are unlikely to cause discomfort. As another example, the diet may progress from small, frequent feedings to larger meals as tolerance improves.

Diet Manual When designing menus for modified diets, the dietary and foodservice personnel refer to a **diet manual,** which details the exact foods or preparation methods to include or exclude in a modified diet. The diet manual may also outline the rationale and indications for use of the diets and include sample menus. The manual

TABLE 14-10 Examples of Modified Diets

Type of Diet	Description of Diet	Appropriate Uses
Modified Texture and Consistency		
Mechanically altered diets	Contain foods that are modified in texture. Pureed diets include only pureed foods; mechanical soft diets may include solid foods that are mashed, minced, ground, or soft.	Pureed diets are used for people with swallowing difficulty, poor lip and tongue control, or oral hypersensitivity. Mechanical soft diets are appropriate for people with limited chewing ability or certain swallowing impairments.
Blenderized liquid diet	Contains fluids and foods that are blenderized to liquid form.	For people who cannot chew, swallow easily, or tolerate solid foods.
Clear liquid diet	Contains clear fluids or foods that are liquid at room temperature and leave minimal residue in the colon.	For preparation for bowel surgery or colonoscopy, for acute GI disturbances (such as after GI surgeries), or as a transition diet after intravenous feeding. For short-term use only.
Modified Nutrient or Food Content		
Fat-controlled diet	Limits dietary fat to low (<50 g/day) or very low (<25 g/day) intakes.	For people who have certain malabsorptive disorders or symptoms of diarrhea, flatulence, or steatorrhea (fecal fat) resulting from dietary fat intolerance.
Fiber-restricted diet	Limits dietary fiber; degree of restriction depends on the patient's condition and reason for restriction.	For acute phases of intestinal disorders or to reduce fecal output before surgery. Not recommended for long-term use.
Sodium-controlled diet	Limits dietary sodium; degree of restriction depends on symptoms and disease severity.	To help lower blood pressure or prevent fluid retention; used in hypertension, congestive heart failure, renal disease, and liver disease.
High-kcalorie, high-protein diet	Contains foods that are kcalorie and protein dense.	Used for patients with high kcalorie and protein requirements (due to cancer, AIDS, burns, trauma, and other conditions); also used to reverse malnutrition, improve nutritional status, or promote weight gain.

Source: American Dietetic Association, *Nutrition Care Manual* (Chicago: American Dietetic Association, 2010).

may be compiled by the dietetics staff or adopted from another health care facility or a dietetics organization.

Alternative Feeding Routes In most cases, patients meet their nutrient needs by consuming regular foods. If their nutrient needs are high or their appetites poor, liquid formulas can be added to their diets to supplement their intakes. Sometimes, however, a person's medical condition makes it difficult to meet nutrient needs orally. Two options remain: **tube feedings** and **parenteral nutrition,** described more fully in Chapter 16.

▌ *Tube feedings.* Nutritionally complete formulas can be delivered through a tube placed directly into the stomach or intestine. Tube feedings are preferred to parenteral nutrition if the GI tract is functioning.

▌ *Parenteral nutrition.* A person's medical condition sometimes prohibits the use of the GI tract to deliver nutrients. If the person is malnourished and the GI tract cannot be used for a significant period of time, parenteral nutrition, in which nutrients are supplied intravenously, can meet nutritional needs.

Nothing by Mouth (NPO) An order to not give a patient anything at all—food, beverages, or medications—is indicated by NPO, an abbreviation for *non per os,* meaning "nothing by mouth." For example, an order may read "NPO for 24 hours" or "NPO

regular diet: a diet that includes all foods and meets the nutrient needs of healthy people; also called a *standard diet* or *house diet.*

modified diet: a diet that contains foods altered in texture, consistency, or nutrient content or that includes or omits specific foods; also called a *therapeutic diet.*

diet progression: a change in diet as a patient's tolerances permit.

diet manual: a resource that specifies the foods or preparation methods to include or exclude in modified diets and provides sample menus.

tube feedings: liquid formulas delivered through a tube placed in the stomach or intestine.

parenteral nutrition: the provision of nutrients through a vein, bypassing the intestine.

until after X-ray." The NPO order is commonly used during certain acute illnesses or diagnostic tests involving the GI tract.

Approaches to Nutrition Care

A nutrition care plan often involves significant dietary modifications. To ensure better compliance, the plan must be compatible with the desires and abilities of the person it is designed to help. The challenge is greater if dietary changes are required for extended periods.

Long-Term Dietary Intervention When long-term changes are necessary, a care plan must take into account the person's current food practices, lifestyle, and degree of motivation. Behavior change is a process that occurs in stages; therefore, more than one consultation is usually necessary. The following approaches may be helpful in implementing long-term dietary changes:[16]

■ *Determine the individual's readiness for change.* Some people have little desire to change their dietary behaviors, and even those who are willing may not be fully prepared to take the necessary steps. The health practitioner needs to consider a patient's readiness to adopt new dietary behaviors before attempting to implement an ambitious care plan.

■ *Emphasize what to eat, rather than what not to eat.* Emphasizing foods to include in the diet, rather than those to restrict, can make dietary changes more appealing. For example, encouraging additional fruits and vegetables is a more attractive message than telling the patient to restrict butter, cream sauces, and ice cream.

■ *Suggest only one or two changes at a time.* People are more likely to adopt a dietary plan that does not deviate too much from their usual diet. If they succeed in adopting one or two changes, they are more likely to stick to the plan and be open to additional suggestions. Stricter plans may yield quicker results but are useful only for highly motivated people.

Nutrition Education Nutrition education allows patients to learn about the dietary factors that affect their particular medical condition. Ideally, this knowledge can motivate them to change their diet and lifestyle to improve their health status.

Dietary counseling requires sensitivity to cultural orientation, educational background, and motivation for change.

© Wadsworth, Cengage Learning

A nutrition education program should be tailored to a person's age, level of literacy, and cultural background. Learning style should also be considered: some people learn best by discussion supplemented with written materials, whereas others prefer visual examples, such as food models and measuring devices.[17] Information can be provided in one-on-one sessions or group discussions. The meeting should include an assessment of the person's understanding of the material and commitment to making changes. Follow-up sessions can reveal whether the person has successfully adopted a dietary plan. For example, a dietitian who counsels a woman who is lactose intolerant and hesitant to use milk products might proceed as follows:

■ The dietitian provides sample menus of a nutritionally adequate diet that limits milk and milk products. Together, the dietitian and the patient design menus that consider her food preferences.

■ The dietitian describes the types and amounts of milk products that are unlikely to cause symptoms and explains how to gradually incorporate these foods into the diet.

- Using diet analysis software, the dietitian demonstrates how altering intakes of calcium-containing foods changes a meal's calcium content.

- The dietitian explains how to use the Daily Values information on food labels to estimate the calcium content of packaged foods.

- The dietitian provides information about the advantages and disadvantages of different calcium supplements.

- The dietitian assesses the patient's understanding by having her identify nonmilk products that are high in calcium.

Ideally, the dietitian would be able to monitor the patient's progress in a subsequent session.

Documenting Nutrition Care

Each step of the nutrition care process must be documented in the patient's medical record. The entries should be as succinct as possible so that they can be easily read and understood by other members of the health care team. In addition, electronic (computerized) data systems, which have been widely adopted in the past decade, have standardized templates that require concise language. Before making entries in medical records, health care professionals need to learn the particular charting methods preferred by their medical facility. Although a variety of charting styles are in use, the content is more relevant than the particular format used.

ADIME Format The ADIME format closely reflects the steps of the nutrition care process. The letters represent the different steps: *Assessment*, *Diagnosis*, *Intervention*, and *Monitoring* and *Evaluation*. Using this format, the nutrition care plan would be recorded as follows:

- *Assessment*. The assessment section summarizes relevant assessment results, such as the medical problem, historical information, anthropometric and laboratory test results, and relevant symptoms.

- *Diagnosis*. The diagnosis section lists and prioritizes the nutrition diagnoses.

- *Intervention*. The intervention section describes treatment goals and expected outcomes, specific interventions, and the patient's responses to treatment.

- *Monitoring and Evaluation*. The monitoring and evaluation section records the patient's progress, changes in the patient's condition, and adjustments in the care plan.

SOAP Format The SOAP format is the oldest method used for documenting nutrition care and is still in popular use. The letters represent the types of information included in each section: *Subjective*, *Objective*, *Assessment*, and the *Plan* for care.

- *Subjective* information is obtained in an interview with the patient or caregiver and includes the chief medical problem and relevant symptoms.

- *Objective* information includes nutrition screening or assessment data, such as the results of anthropometric and laboratory tests and the physical examination.

- The *Assessment* section contains a brief evaluation of the subjective and objective data and provides concise diagnoses of the nutrition problems.

- The *Plan* includes recommendations that can help solve the problem, including the nutrition prescription, plan for nutrition education and counseling, and referrals to other professionals.

FIGURE 14-4 **Example of a SOAP Note**

SOAP NOTE

Patient Name: *James Steiner*   Date: *Sept. 15, 2011*

Age: *58*   Gender: *Male*   Medical diagnosis: *Hypercholesterolemia*

Subjective:

Mr. Steiner recently learned of his hypercholesterolemia; wants to try dietary/ lifestyle changes to reduce need for the medication. Reports frequent snacking and little time for exercise. Willing to attempt weight loss.

Objective:

Total cholesterol: 288 mg/dl   *Height: 6'1"; Weight: 268 lb.*
LDL-C: 214 mg/dl; HDL-C: 48 mg/dl   *BMI: 35.4*
Triglycerides: 132 mg/dl   *Waist circumference: 45"*

Assessment:

Abdominal obesity; analysis reveals intake of approximately 4200 kcal per day, about 1500 kcal above estimated needs; snack food choices are high in kcal and saturated fat.
Nutrition Diagnoses:
1. Obesity related to excess energy intake of 1500 kcal/day and physical inactivity as evidenced by BMI of 35.4
2. Undesirable food choices related to inadequate access to appropriate foods at work as evidenced by elevated body weight and LDL cholesterol

Plan:

Goal: 15 lb. weight loss over next 6 months.
Mr. Steiner to start 45-minute walking program, evenings.
Nutrition prescription: reduction of food intake to about 2400 kcal per day with about 30% kcal from fat, and 7% of kcal from saturated fat.
Initial education: appropriate food portions, low-kcal foods and snacks, food sources of saturated fat, pre-planning lunches at work.
Referral: Heart-healthy workshop on Sept. 22 (one week); Mr. Steiner to attend with wife.
Follow-up visit: Oct 15 (one month); Mr. Steiner to keep 3-day food record before visit; will identify appropriate food portions and between-meal snacks.

Form completed by: *Genevieve Johnson, MPH, RD* Position: *Dietitian, Nutrition Services*

Figure 14-4 shows an example of a SOAP note, although there are many possible variations.

Improving Food Intake

People in hospitals and other medical facilities often lose their appetites as a result of their medical condition, treatment, or emotional distress. In addition, some medications and other treatments can dramatically alter taste perceptions. Patients usually receive meals at specified times whether they are hungry or not and often must eat in bed without companionship; under these conditions, eating can be more of a chore than a pleasurable experience. Meals may also be unwelcome if the person is in pain or is under sedation.

Nurses and dietetic technicians often have central roles in helping patients to eat. If appetite or the sense of taste is affected by illness, the health practitioner should work with the patient to identify foods that are the most enjoyable. When meals are served, the nurse can help the patient wash up before a meal and check to see that foods and utensils are arranged attractively. The "How to" lists additional suggestions that may help to improve food intake at mealtimes. The accompanying Case Study provides an opportunity for you to review the implementation of nutrition care.

1. Empathize with the patient. Show that you understand how difficult eating may be when a person feels too sick to move or too tired to sit up. Help to motivate the patient by explaining how important good nutrition is to recovery.

2. Help patients select the foods they like and mark menus appropriately. When appropriate and permissible, let friends or family members bring favorite foods from outside the hospital.

3. For patients who are weak, suggest foods that require little effort to eat. Eating a roast beef sandwich, for example, requires less effort than cutting and eating a steak. Drinking soup from a cup may be easier than eating it with a spoon.

4. During mealtimes, make sure the patient's room is quiet and has sufficient lighting for viewing the food. See that the room is free of odors that may interfere with the appetite.

5. Help patients prepare for meals. Help them get comfortable, either in bed or in a chair. Adjust the extension table to a comfortable distance and height and make sure it is clean. Take these steps before the tray arrives so that the meal can be served promptly and at the right temperature.

6. When the food cart arrives, check the patient's tray. Confirm that the patient is receiving the right diet, the foods on the tray are those selected from the menu, and the foods look appealing. Order a new tray if the foods are not appropriate.

7. Help with eating, if necessary. Help patients open containers or cut foods, and assist with feeding if patients cannot feed themselves. Encourage patients with little appetite to eat the most nutritious foods first and to drink liquids between meals.

8. Take a positive attitude toward the hospital's food. Never say something like "I couldn't eat this either." Instead, say, "The foodservice department really tries to make foods appetizing. I'm sure we can find a solution."

Case Study IMPLEMENTING NUTRITION CARE

Max is an 11-year-old boy who was admitted to the hospital after he passed out while playing with friends. Tests confirm a diagnosis of type 1 diabetes mellitus. Max remains in the hospital for several days until his blood glucose and ketone levels are under control. During this time, he and his family learn about diabetes, the diet Max needs to follow, the use of insulin, the monitoring of blood glucose levels, and the required coordination of diet, insulin, and physical activity. The details of diabetes mellitus are reserved for Chapter 21, but for now you can consider the steps that are necessary for implementing Max's nutrition care.

1. Given the chronic nature of Max's illness and his age, what approaches should be used when discussing the required dietary and medical treatments with Max and his family?

2. What factors need consideration when designing a nutrition education program for Max and his parents?

3. Max will need additional care to learn more about diabetes and to make the adjustments that will allow him to cope with his condition. Why is it important to address follow-up care before Max leaves the hospital?

IN SUMMARY

▌ Dietary modifications prescribed during illness include changes in food texture or consistency, modified energy or nutrient content, or the inclusion or exclusion of certain foods. Some medical conditions may require the use of tube feedings or parenteral nutrition.

▌ A nutrition intervention should take into account a person's food practices, lifestyle, and degree of motivation. Nutrition education should be individualized to accommodate a patient's needs and learning style.

▌ Each step of nutrition care is documented in the medical record; the ADIME and SOAP formats are popular styles of documentation.

▌ Hospital patients may need assistance at mealtime and encouragement to consume adequate amounts of food.

Clinical Applications

1. Describe the potential nutritional implications of these findings from a patient's medical, personal, and social histories: age 78, lives alone, recently lost spouse, uses a walker, has no natural teeth or dentures, has a history of hypertension and diabetes, uses medications that cause frequent urination.

2. Calculate the %UBW and %IBW for a man who is 5 feet 11 inches tall with a current weight of 150 pounds and a usual body weight of 180 pounds. What additional information do you need to interpret the implications of his weight loss?

3. Jonathan is a 49-year-old male who is 6 feet 2 inches tall and has a usual body weight of 180 pounds. He was admitted to the hospital following an automobile accident and was treated for minor injuries. Using the method described in the "How to" on p. 397, estimate his energy requirement using the Mifflin–St. Jeor equation and the stress factor 1.25.

4. A healthy 70-year-old woman, admitted to the hospital for a hip replacement surgery, develops an infection after the surgery and recovery is slower than expected. You notice that she seems uninterested in meals and has eaten only small amounts of food for several days. What steps can be taken to uncover and address problems that the woman might be having with food?

Self Check

1. Mr. Hom experiences loss of appetite, difficulty swallowing, and mouth pain as a consequence of illness. Mr. Hom is at risk of malnutrition due to:
   a. altered metabolism.
   b. reduced food intake.
   c. altered excretion of nutrients.
   d. altered digestion and absorption.

2. The member of the health care team best qualified to provide medical nutrition therapy is the:
   a. physician.
   b. nurse.
   c. registered dietitian.
   d. pharmacist.

3. The nutrition care process is a systematic approach for:
   a. identifying the nutrient content of foods.
   b. ordering special diets.
   c. conducting nutrition screening.
   d. identifying and meeting the nutrition needs of patients.

4. Which dietary assessment method does a nurse use to conduct a kcalorie count?
   a. Direct observation
   b. 24-hour recall interview
   c. Food frequency questionnaire
   d. Food record

5. The %UBW of a person who weighs 135 pounds and has a usual body weight of 150 pounds is:
   a. 111 percent.
   b. 90 percent.
   c. 86 percent.
   d. 74 percent.

6. A malnourished, acutely ill patient has just begun to eat after days without significant amounts of food. Which of the following blood test results would change most quickly as the patient's nutrition and health status improves?
   a. Albumin
   b. Transferrin
   c. Serum electrolytes
   d. Retinol-binding protein

7. Fluid retention can cause all of the following effects *except*:
   a. lab results that are deceptively high.
   b. facial puffiness.
   c. lab results that are deceptively low.
   d. tight-fitting shoes.

8. A nurse notices a food on a patient's tray that she thinks should be excluded from the patient's prescribed diet. What resource can she check to learn whether the food is appropriate?
   a. Nutrition care plan
   b. Diet order
   c. Diet manual
   d. Medical record

9. A successful nutrition intervention would include a long list of:
   a. dietary changes that the patient should consider making.
   b. foods that the patient should avoid.
   c. appetizing meals and foods that the patient can include in his or her diet.
   d. reasons why the patient should make dietary changes.

10. The most important factor(s) that affect(s) how nutrition education is presented is (are):
    a. the person's nutrient needs and nutrition status.
    b. the person's abilities and motivation.
    c. the person's medical history.
    d. the entries in the medical record.

Answers to these questions can be found in Appendix H.

Notes

1. D. C. Heimburger, Adulthood, in M. E. Shils and coeditors, *Modern Nutrition in Health and Disease* (Baltimore: Lippincott Williams & Wilkins, 2006), pp. 830–842.

2. P. Charney and M. Marian, Nutrition screening and nutrition assessment, in P. Charney and A. M. Malone, eds., *ADA Pocket Guide to Nutrition Assessment* (Chicago: American Dietetic Association, 2009), pp. 1–19.

3. Writing Group of the Nutrition Care Process/ Standardized Language Committee, Nutrition care process and model part I: The 2008 update, *Journal of the American Dietetic Association* 108 (2008): 1113–1117; K. Lacey and E. Pritchett, Nutrition care process and model: ADA adopts road map to quality care and outcomes management, *Journal of the American Dietetic Association* 103 (2003): 1061–1072.

4. Writing Group of the Nutrition Care Process/ Standardized Language Committee, 2008.

5. F. E. Thompson and A. F. Subar, Dietary assessment methodology, in A. M. Coulston and C. J. Boushey, eds., *Nutrition in the Prevention and Treatment of Disease* (Burlington, MA: Elsevier Academic Press, 2008), pp. 3–39.

6. L. C. Tapsell, V. Brenninger, and J. Barnard, Applying conversation analysis to foster accurate reporting in the diet history interview, *Journal of the American Dietetic Association* 100 (2000): 818–824.

7. E. Saltzman and M. A. McCrory, Physical assessment of nutritional status, in A. M. Coulston and C. J. Boushey, eds., *Nutrition in the Prevention and Treatment of Disease* (Burlington, MA: Elsevier Academic Press, 2008), pp. 57–73.

8. B. Shah, K. Sucher, and C. B. Hollenbeck, Comparison of ideal body weight equations and published height-weight tables with body mass index tables for healthy adults in the United States, *Nutrition in Clinical Practice* 21 (2006): 312–319.

9. S. E. Barlow and the Expert Committee, Expert Committee recommendations regarding the prevention, assessment, and treatment of child and adolescent overweight and obesity: Summary Report, *Pediatrics* 120 (2007): S164–S192.

10. S. B. Heymsfield and R. N. Baumgartner, Body composition and anthropometry, in M. E. Shils and coeditors, *Modern Nutrition in Health and Disease* (Baltimore: Lippincott Williams & Wilkins, 2006), pp. 751–770.

11. American Dietetic Association, *Nutrition Care Manual* (Chicago: American Dietetic Association, 2010).

12. C. W. Thompson, Laboratory assessment, in P. Charney and A. M. Malone, eds., *ADA Pocket Guide to Nutrition Assessment* (Chicago: American Dietetic Association, 2009), pp. 62–153.

13. Standing Committee on the Scientific Evaluation of Dietary Reference Intakes, Food and Nutrition Board, Institute of Medicine, *Dietary Reference Intakes for Water, Potassium, Sodium, Chloride, and Sulfate* (Washington, D.C.: National Academies Press, 2005), pp. 147–150.

14. J. Boullata and coauthors, Accurate determination of energy needs in hospitalized patients, *Journal of the American Dietetic Association* 107 (2007): 393–401.

15. American Dietetic Association, *Nutrition Care Manual* (Chicago: American Dietetic Association, 2010).

16. K. Glanz, Current theoretical bases for nutrition intervention and their uses, in A. M. Coulston and C. J. Boushey, eds., *Nutrition in the Prevention and Treatment of Disease* (Burlington, MA: Elsevier Academic Press, 2008), pp. 127–138.

17. L. M. Delahanty and J. M. Heins, Tools and techniques to facilitate nutrition intervention, in A. M. Coulston and C. J. Boushey, eds., *Nutrition in the Prevention and Treatment of Disease* (Burlington, MA: Elsevier Academic Press, 2008), pp. 149–167.

CHAPTER 14

Nutrition in Practice NUTRITIONAL GENOMICS

Imagine this scenario: A physician scrapes a sample of cells from inside your cheek and submits it to a **genomics** lab. In a short time, you receive a report that reveals your disease susceptibilities and recommends dietary and lifestyle changes that can maintain your health. You may even be given a prescription for a dietary supplement that will best meet your personal nutrient requirements. Unlikely? Perhaps, but these possibilities are being explored by scientists working in the field of **nutritional genomics,** the study of dietary effects on **gene expression.** Recent research suggests that some dietary factors may be more helpful (or more harmful) in people who have particular genetic variations. The promise of nutritional genomics is a custom-designed dietary prescription that fits each person's specific needs. The accompanying glossary defines genomics and related terms.

What is a genome?

Genetic information is encoded in DNA molecules within the nuclei of almost all of the cells in our bodies. Figure NP14-1 shows how the genetic material is organized within the **genome,** the complete set of genetic information within our cells. The DNA molecules are tightly packed along with associated proteins within the 46 **chromosomes.** Segments of a DNA strand that can eventually be translated into proteins are called **genes.** The sequence of **nucleotides** within each gene encodes the amino acid sequence of a particular protein. Scientists estimate that there are between 20,000 and 24,000 genes in the human genome.[1] However, only a small percentage of the genome codes for proteins: most DNA consists of **noncoding sequences,** which may help to regulate gene expression or have other functions.

When proteins are made, the information in the DNA sequence is first transcribed (copied) to messenger RNA molecules, which carry the genetic information out of the nucleus. Gene expression can be measured by determining the amounts of messenger RNA in a tissue sample. The expression of thousands of genes can be measured simultaneously using **microarray technology** (see photo).

FIGURE NP14-1 The Human Genome

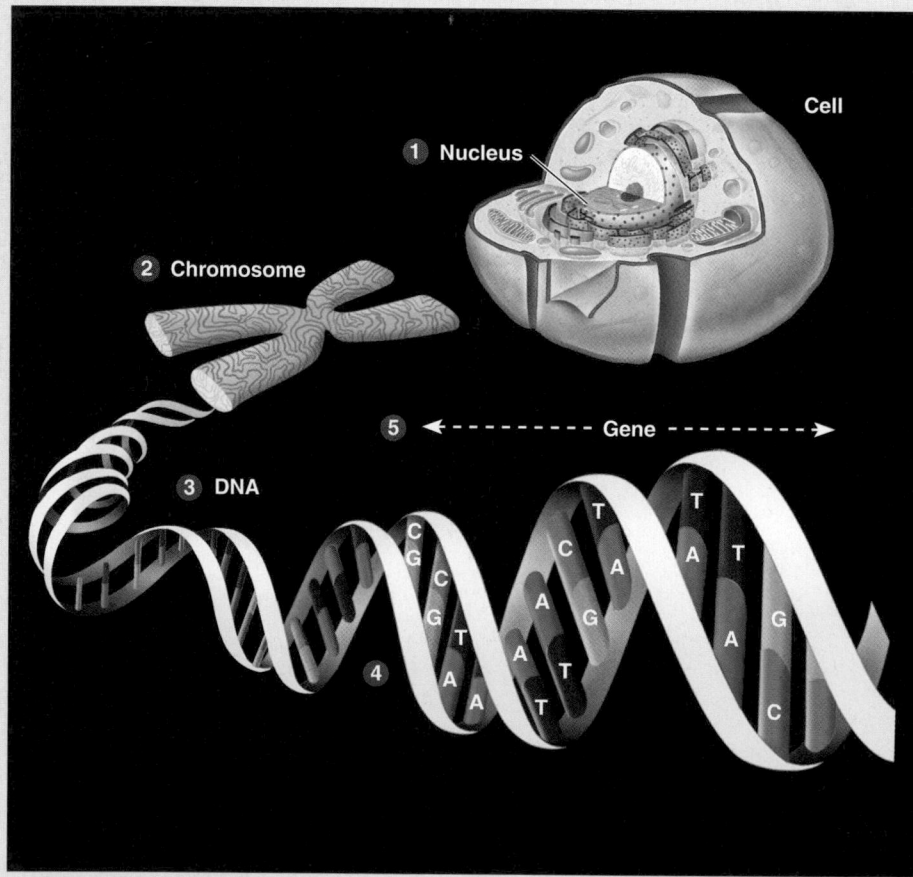

1. The human genome is a complete set of genetic material organized into 46 chromosomes, located within the nucleus of a cell.

2. A chromosome is made of DNA and associated proteins.

3. The double helical structure of a DNA molecule is made up of two long chains of nucleotides. Each nucleotide is composed of a phosphate group, a 5-carbon sugar, and a base.

4. The sequence of nucleotide bases (C, G, A, T) determines the amino acid sequence of proteins. These bases are connected by hydrogen bonding to form base pairs: adenine (A) with thymine (T) and guanine (G) with cytosine (C).

5. A gene is a segment of DNA that includes the information needed to synthesize one or more proteins.

Source: Adapted from "A Primer: From DNA to Life," Human Genome Project, U.S. Department of Energy Office of Science, www.ornl.gov/sci/techresources/Human_Genome/primer_pic.shtml.

chromosomes: structures within the nucleus of a cell that contain the cell's DNA and associated proteins.

epigenetics: the study of heritable changes in DNA structure that do not change the DNA nucleotide sequence.

gene expression: the process by which a cell converts the genetic code into RNA and protein.

genes: segments of DNA that contain the information needed to make proteins.

genome (JEE-nome): the full complement of genetic material in the chromosomes of a cell.

genomics (jee-NO-miks): the study of genomes.

inherited disorders: medical conditions resulting from genetic defects.

methylation: the addition of methyl ($-CH_3$) groups.

microarray technology: research technology that monitors the expression of thousands of genes simultaneously.

multigene or **polygenic:** involving a number of genes, rather than a single gene.

noncoding sequences: regions of DNA that do not code for proteins. Some noncoding sequences may have regulatory or structural properties, but most have no known function.

nucleotides: the subunits of DNA and RNA molecules. These compounds—cytosine (C), thymine (T), uracil (U), guanine (G), and adenine (A)—are each composed of a phosphate group, a 5-carbon sugar (ribose), and a nitrogen-containing base. A DNA molecule is made up of two long chains of nucleotides held together by hydrogen bonding between nucleotide bases on opposing strands; each hydrogen-bonded nucleotide couple is called a *base pair*.

nutritional genomics: the study of dietary effects on genetic expression; also known as **nutrigenomics.**

polymorphisms: differences in the DNA sequences among individuals. **A single-nucleotide polymorphism** involves a single nucleotide at a particular area in the DNA strand.
- *poly* = many
- *morph* = form
- *ism* = condition

promoter: a region of DNA involved with gene activation.

transcription factors: proteins that bind DNA at specific sequences to regulate gene expression.

A DNA microarray allows researchers to monitor the expression of thousands of genes simultaneously.

How do nutrients alter gene expression?

Some nutrients can switch gene expression on or off.[2] The **promoter** region of a gene (a DNA region involved with gene activation) acts as the master switch. A large variety of proteins known as **transcription factors** can bind to areas on the promoter and either enhance or inhibit gene expression. A combination of dietary factors and hormones influences the types of transcription factors that reach the nucleus and their tendency to bind to DNA. Specific examples of how nutrients can influence transcription factors include:

- The transcription factor that enhances the gene expression of enzymes required for cholesterol synthesis enters the nucleus only when the cellular cholesterol content is low.

- The transcription factor that inhibits the expression of ferritin, an iron-storage protein, changes its affinity for DNA based on the iron content of the cell.

Gene expression is also influenced by modifications in the structure of DNA and its packaging in chromosomes. For example, **methylation** of DNA molecules (the addition of methyl groups) alters the expression of numerous genes and depends on both inherited factors and the availability of various nutrients. In individuals who are susceptible for a disease that is influenced by DNA methylation, nutrient supplementation may reduce the risk of disease.[3] The study of heritable changes in DNA structure that are separate from the underlying DNA nucleotide sequence is called **epigenetics.**

How much genetic variation is there among people?

Except for identical twins, no two individuals are genetically identical. However, the variation in the genomes of any two persons is only about 0.1 percent, a difference of only one base in every 1000.[4] The most common genetic differences, known as **polymorphisms,** are changes in single nucleotides. A **single-nucleotide polymorphism** is due to a nucleotide insertion, deletion, or substitution within the DNA molecule. Such variations are significant only if they affect a protein's amino acid sequence in a way that alters the protein's function, or if the change in the DNA molecule changes how a particular gene is regulated.

Genetic variation gives rise to the diversity among human beings—it explains most of the differences in our physical appearances and metabolic characteristics. Along with environmental factors, it also determines our susceptibilities to disease. Diseases affected by a single gene tend to be relatively rare and usually exert their effects early in life. In contrast, common diseases such as heart disease and cancer are influenced by many genes and typically develop over several decades or even longer. In these more complex **multigene,** or **polygenic,** disorders, many genes can contribute to disease risk, but no single gene may be sufficient to cause the disease on its own.

What are some examples of single-gene disorders?

Examples of single-gene disorders include phenylketonuria (discussed in Nutrition in Practice 16), sickle-cell anemia, and the iron-overload disease hemochromatosis. Single-gene disorders may seriously disrupt metabolism and often require significant dietary or medical intervention. However, not all single-gene disorders have life-threatening ramifications. For example, lactose intolerance can result from an alteration in the promoter of the lactase gene; it may cause gastrointestinal discomfort, but it is readily managed by simple dietary changes.

How are multigene disorders different from single-gene disorders?

Multigene disorders are usually sensitive to a number of environmental influences, including diet and lifestyle. These environmental factors can directly influence the expression of the genes involved. Multigene disorders tend to develop over many years, so determining genetic susceptibility may allow a person to modify diet and lifestyle appropriately and reduce the risk of developing a disease.

Heart disease is an example of a disease with multiple gene influences. Its many risk factors represent the involvement of an assortment of genes, which affect disparate aspects of physiology and metabolism. Consider that the major risk factors for heart disease include elevated blood cholesterol levels, obesity, diabetes, and hypertension. The underlying cause of any of these risk factors is rarely known; currently, clinicians screen for the presence of risk factors but not for the reasons why they occur. Should genomic research prove successful, a future assessment approach might be to identify specific genetic alterations and changes in metabolism that lead to the development of individual risk factors. For example, tests may determine whether blood cholesterol levels are high due to excessive cholesterol absorption, excessive cholesterol production in the liver, or reduced cholesterol degradation.[5] This information could then guide health care providers to the most appropriate intervention, allowing a better match between treatment recommendations and a person's genetic profile.

Can genomic research be used to explore the differences in nutrient needs among people?

Even though most people apparently can meet their nutrient needs by consuming nutrients at recommended levels, it would be useful to learn more about genetic variations within healthy populations. The techniques that have emerged from genomic research may provide a means for fine-tuning nutrient recommendations for different individuals.[6] Moreover, ideal indicators of nutrient status are still lacking for several of the minerals, such as zinc, magnesium, and chromium. Scientists hope to eventually produce genomic maps that will indicate how various nutrient deficiencies and combinations of deficiencies affect gene expression. These maps may eventually provide data that can help diagnose nutrient deficiencies.

Will knowledge about the human genome substantially change the manner in which health care is provided?

The enthusiasm surrounding genomic research must be put into perspective in terms of both the status of clinical medicine at present and people's willingness to make difficult lifestyle choices. Critics have questioned whether genetic markers for disease would be more useful than simple and inexpensive clinical measurements, which reflect both genetic *and* environmental influences. In other words, knowing that a person is genetically predisposed toward high cholesterol levels is not necessarily more useful than knowing the person's actual blood cholesterol level.[7] Furthermore, a person's family history is already a simple genomic tool that indicates a higher genetic risk for certain illnesses. Understanding the family's medical history prompts the clinician to use risk-reduction strategies that are appropriate to the culture and literacy of the patient, rather than focusing on genetic risk alone.[8]

Obtaining additional knowledge about disease risk is not necessarily useful unless people are motivated to make serious lifestyle changes. For example, despite the present abundance of disease prevention recommendations, many people seem unwilling to make the changes known to improve health. Researchers have estimated that heart disease and type 2 diabetes are 80 percent and 90 percent preventable, respectively, by changing one's lifestyle to include an appropriate diet, a healthy body weight, and regular exercise, among other factors.[9] Given the difficulty that people have with current recommendations, it is unlikely that they will enthusiastically adopt an even more detailed list of dietary and lifestyle modifications.

What ethical concerns are raised by having extensive knowledge about an individual's genome?

The ability to obtain detailed genetic information raises a number of concerns. A primary consideration is confidentiality: should information about a person's susceptibility to disease be released to others (including other family members at risk) without that person's consent? Because environmental factors play such an important role in disease risk, genetic predisposition alone usually cannot predict whether a person will develop a particular disease. Concern about privacy issues has led to federal and state legislation that prevents discrimination by group health plans, health insurers, and employers on the basis of an individual's genetic predisposition to disease.[10]

Another important concern is whether genetic testing is always in the best interest of children. Although early knowledge of a child's predisposition to illnesses may be useful for parents who want to provide optimal care, the release of this information could threaten the child's privacy and increase the potential for genetic discrimination in the future.

Although genomic research has the potential to improve our ability to diagnose and treat disease, it is still unclear how knowledge of the genome will be translated into useful medical treatments. Still, health care professionals will need to keep informed of the ethical, legal, and social implications of nutritional and medical genomics as this remarkable research continues.

Notes

1. S. Barnes, Nutritional genomics, polyphenols, diets, and their impact on dietetics, *Journal of the American Dietetic Association* 108 (2008): 1888–1895; International Human Genome Sequencing Consortium, Finishing the euchromatic sequence of the human genome, *Nature* 431 (2004): 931–945.

2. R. J. Cousins, Nutritional regulation of gene expression and nutritional genomics, in M. E. Shils and coeditors, *Modern Nutrition in Health and Disease* (Baltimore: Lippincott Williams & Wilkins, 2006), pp. 615–626.

3. P. J. Stover and M. A. Caudill, Genetic and epigenetic contributions to human nutrition and health: Managing genome–diet interactions, *Journal of the American Dietetic Association* 108 (2008): 1480–1487.

4. P. J. Stover and C. Garza, Polymorphisms: Effect on nutrient utilization and metabolism, in M. E. Shils and coeditors, *Modern Nutrition in Health and Disease* (Baltimore: Lippincott Williams & Wilkins, 2006), pp. 627–635.

5. J. B. German, M.-A. Roberts, and S. M. Watkins, Personal metabolomics as a next generation nutritional assessment, *Journal of Nutrition* 133 (2003): 4260–4266.

6. Stover and Caudill, 2008.

7. W. C. Willett, Balancing life-style and genomics research for disease prevention, *Science* 296 (2002): 695–698.

8. M. J. Khoury and coauthors, Do we need genomic research for the prevention of common diseases with environmental causes? *American Journal of Epidemiology* 161 (2005): 799–805.

9. Khoury and coauthors, 2005; Willett, 2002.

10. P. R. Reilly and R. M. DeBusk, Ethical and legal issues in nutritional genomics, *Journal of the American Dietetic Association* 108 (2008): 36–40.

15

Medications, Diet-Drug Interactions, and Herbal Supplements

Although people rely on medications to prevent and treat their health problems, these helpful substances may also alter metabolism and potentially disrupt body processes. In addition, serious side effects may occur if the drugs interact with nutrients or other dietary components. This chapter reviews the use of medications in clinical care, describes potential diet-drug interactions, and introduces herbal supplements, which many individuals use in hope of treating their medical problems. Nutrition in Practice 15 describes the various categories and uses of complementary and alternative therapies.

Medications in Disease Treatment

Drugs must be proved to be safe and effective before they can be marketed in the United States. The Food and Drug Administration (FDA) is responsible for approving sales of new drugs and inspecting facilities where drugs are manufactured. By law, drugs are divided into two categories:[1]

▌ *Prescription drugs* are usually given to treat serious conditions and may cause severe side effects. For these reasons, they are sold by prescription only, which ensures that a physician has evaluated the patient's medical condition and determined that the benefits of using the medication outweigh the risks of incurring side effects.

▌ *Over-the-counter (OTC) drugs* are those that can generally be used safely and effectively without medical supervision. People use them to treat less serious illnesses that are easily self-diagnosed. Examples include aspirin to treat headaches or pain and antacids to combat indigestion. The FDA regulates labels on OTC drugs to make sure they provide accurate information about the drugs' appropriate uses and dosages and potential adverse effects. Prescription drugs considered safe enough for self-medication are frequently switched to OTC status, sometimes in smaller doses than are available by prescription.

Brand-name drugs are usually given patent protection for 20 years after the patent is submitted. After the patent expires, less-expensive generic versions of the drugs can be sold. To gain FDA approval, a generic drug must have similar biological effects as compared to the original drug: it must contain the same active ingredients; be identical in strength, dosage form, and route of administration; and meet the same requirements for purity and quality. In some cases, the bioavailability (amount absorbed) of a brand-name drug and generic drug may differ due to differences in the drugs' solubility or the types of inactive ingredients present; thus, greater benefit may be obtained by using the brand-name drug.[2] Most often, however, consumers can be confident that generic drugs are as safe and effective as the brand-name products they replace.

Generic vs. brand-name drugs:
Generic name: diazepam;
brand name: Valium (a sedative)

Generic name: furosemide;
brand name: Lasix (a diuretic)

Risks from Medications

The risk of an adverse reaction always accompanies the use of a medicine. Thus, a medication should be used only when the benefits of using it outweigh the potential risks. The risks become greater when a drug is incorrectly prescribed or administered. This section discusses the types of risks associated with medications and suggests some steps for managing risk.

Side Effects By the time a drug reaches the marketplace, large-scale clinical trials have revealed the majority of side effects associated with its use. However, rare side effects are sometimes detected only after a drug has been more widely used. In some instances, these effects occur because drugs are used for longer periods or in different circumstances than originally anticipated.

Although over-the-counter drugs are considered safe enough for self-medication, they can cause adverse effects when used inappropriately.

The FDA's MedWatch program encourages health professionals and consumers to report medication problems by mail, fax, telephone, or the Internet (www.fda.gov/medwatch).

The FDA monitors adverse events after drugs are marketed. Manufacturers are required to submit periodic reports, and individuals using the drugs are encouraged to report unexpected effects directly to the FDA. The FDA may decide to change labeling information or even withdraw drugs from the marketplace if they are believed to cause unacceptable risks to health.

Because OTC drugs are available without a prescription, patients may not realize that adverse effects can occur if the drugs are used inappropriately. Under certain circumstances, the active ingredients in these drugs may worsen a medical condition, produce complications, or interact with other medications. Furthermore, people using products with several active ingredients may inadvertently take toxic amounts of a substance when using several drugs simultaneously. For example, a person with a cold may take one medication to treat a cough and another medication for a headache without realizing that both contain an analgesic (pain medication).

Drug-Drug Interactions When a person uses multiple drugs, one drug may alter the effects of another drug, and the risk of side effects increases. These problems are common in older adults, who often use several medications daily over long periods. Although primary care physicians may supervise medication use, some individuals use drugs prescribed by a number of different physicians. Others may use OTC medications and dietary supplements in addition to prescription drugs without being aware of the risks associated with certain combinations.

Diet-Drug Interactions Substances in the diet may alter the effectiveness of drugs, and drugs may affect food intake or the digestion, absorption, metabolism, or excretion of nutrients. Later sections of this chapter describe these interactions in detail.

Medication Errors A medication error is any preventable action that causes inappropriate drug use or patient harm due to mistakes made by a health professional or patient. Many errors leading to patient harm involve the use of incorrect drugs or improper dosages.[3] The wrong drug is sometimes administered when two different drugs have names that look or sound alike or have similar packaging. In other cases, the physician's prescription is misread or misinterpreted; for example, one patient died after receiving 10 milliliters of morphine solution instead of 10 milligrams—a 20-fold overdose.

Drugs that have similar names:

- Bupropion (for depression) and buspirone (for anxiety)
- Hydralazine (for hypertension) and hydroxyzine (for anxiety)

Several policy changes and programs are helping to reduce medication errors. The bar codes currently used on medications and patient identification bracelets allow health practitioners to verify that the correct medication and dosage are administered: error messages alert personnel if the drug, dose, or timing of administration is inappropriate. In addition, a national education campaign is attempting to eliminate one of the most common but preventable sources of medication errors—the use of ambiguous medical abbreviations (see examples in Table 15-1). Because terms such as these are easily misread or misinterpreted, they can no longer be used in clinical documentation related to patient care.

Patients at High Risk of Adverse Effects

Health care professionals should be aware that some patients are more vulnerable than others to adverse effects from drugs. This category includes the populations that rarely participate in clinical trials that determine product safety: pregnant and lactating women, children, and people with medical conditions that are not the main focus of the study. In these groups, side effects may be discovered only after a drug has been marketed. Children may react in different ways to drugs than adults do, and the appropriate dosage for their age and size is often unknown. Also, limited data are available on drug safety in older adults. Elderly people with chronic diseases that require multiple medications are especially susceptible to adverse effects. They are also more

TABLE 15-1 Terms Prohibited on Clinical Documentation

Prohibited Terms	Intended Meaning	Potential Problem	Correct Term for Documentation
U	Unit	Can be misread as the number 0 or 4; may cause 10-fold overdose or higher.	Write out "unit."
IU	International unit	Can be misread as IV (intravenous) or 10.	Write out "international unit."
Trailing zero (1.0 mg) or lack of leading zero (.1 mg)	1 mg; 0.1 mg	Decimal point can be missed, leading to 10-fold error in dosages.	Never use zero by itself after a decimal. Always use zero before a decimal point.
μg	Microgram	Can be misread as mg (milligram).	Write out "microgram."
Q.D. (q.d.), Q.O.D. (q.o.d.)	*Q.D.* means "every day"; *Q.O.D.* means "every other day"	Can be mistaken for one another or misread as "q.i.d." (four times daily).	Write out "daily" or "every other day."

likely to have impaired function of the liver or kidneys—the two organs critical to metabolizing and eliminating drugs from the body.

To reduce the likelihood of adverse effects, health professionals should discuss with patients the potential benefits and risks of using medications before prescribing them. These suggestions may help:

▪ Advise the patient that drugs should not be taken unless absolutely necessary. Discuss dietary or lifestyle practices that have benefits similar to those of drugs. For example, laxatives may not be necessary if an individual increases consumption of foods high in fiber and begins exercising regularly.

▪ Request a complete list of prescription medications, OTC drugs, and dietary supplements that the patient is taking. Ensure that at least one physician is coordinating the patient's drug use. Encourage the patient to purchase all medications at the same pharmacy so that the pharmacist can alert physicians and patients to potential problems.

▪ Verify that the patient understands how to take medications properly. Alert the patient to potential drug-drug and diet-drug interactions.

▪ Encourage the patient to keep track of side effects. Inform the patient that new or unusual symptoms may be due to a new medication rather than the medical condition. In some cases, another medication that treats the condition may have fewer side effects.

IN SUMMARY

▪ Both prescription and OTC drugs must be shown to be safe and effective before they are sold. A medication's benefits should be greater than the risks associated with its use. Potential risks include side effects, drug-drug and diet-drug interactions, and medication errors.

▪ Medication errors often involve incorrect dosing or using the wrong drug. Bar coding is required on medications sold to health institutions, and confusing terms are being eliminated from documents related to patient care.

▪ Patients at highest risk of experiencing adverse drug effects include pregnant and nursing women, children, and the elderly.

Elderly people using multiple medications are especially susceptible to adverse effects from drugs.

© 2002 Macduff Everton/The Image Bank/Getty Images

- Health professionals should discuss the risks and benefits of medications with patients and alert them to potential dangers and possible solutions.

Diet-Drug Interactions

When working with patients, health practitioners should remain alert to possible interactions between drugs and dietary substances. These interactions can raise health care costs and result in serious, and sometimes fatal, complications. Accordingly, health professionals must learn to take steps to prevent or lessen their adverse consequences. Diet-drug interactions generally fall into the following categories:

- Drugs may alter food intake by reducing the appetite or by causing complications that make food consumption difficult or unpleasant. Other drugs may increase the appetite and cause weight gain.

- Drugs may alter the absorption, metabolism, or excretion of nutrients. Conversely, nutrients and other food components may alter the absorption, metabolism, and excretion of drugs.

- Some interactions between dietary components and drugs can be toxic.

Examples of these types of diet-drug interactions are shown in Table 15-2.[4]

Drug Effects on Food Intake

Some drugs can make food intake difficult or unpleasant: they may suppress the appetite, induce nausea or vomiting, cause mouth dryness, alter the sense of taste, or lead to inflammation or lesions in the mouth or GI tract. Certain side effects of drugs, including abdominal discomfort, constipation, and diarrhea, may be aggravated by food consumption. Medications that cause drowsiness, such as sedatives and some painkillers, can make a person too tired to eat.

Drug complications that reduce food intake are significant only when they continue for a long period. Although many drugs can cause nausea in some individuals, the nausea often subsides after the first few doses of the medication and therefore has little effect on nutrition status. If side effects persist, other medications are sometimes used to treat them; for example, antinauseants and antiemetics may help to reduce nausea and vomiting and thereby improve food intake.

Some medications stimulate the appetite and encourage weight gain. Unintentional weight gain may result from the use of some antidepressants, antipsychotics, antidiabetic drugs, and corticosteroids (such as prednisone).[5] For some conditions, however, weight gain is desirable. Patients with diseases that cause wasting, such as cancer or AIDS, are sometimes prescribed appetite enhancers such as megestrol acetate (Megace), a progesterone analog, or dronabinol (Marinol), which is derived from the active ingredient in marijuana.

Drug Effects on Nutrient Absorption

The medications that most often cause nutrient malabsorption are those that upset gastrointestinal function or damage the intestinal mucosa. Antineoplastic and antiretroviral drugs are especially detrimental, whereas nonsteroidal anti-inflammatory drugs (NSAIDs) and some antibiotics can have similar, though milder, effects. This section describes additional ways in which medications may alter nutrient absorption.

Antineoplastic drugs combat tumor growth. *Antiretroviral drugs* treat HIV infection.

Drug-Nutrient Binding Some medications bind to nutrients in the GI tract, preventing their absorption. For example, bile acid binders, which are used to reduce cholesterol levels, may bind to fat-soluble vitamins. Some antibiotics, notably tetracycline and ciprofloxacin (Cipro), bind to the calcium in foods and supplements, reducing the

414

CHAPTER FIFTEEN

TABLE 15-2 Examples of Diet-Drug Interactions

Drugs may alter food intake by:

Altering the appetite (amphetamines suppress appetite; corticosteroids increase appetite).

Interfering with taste or smell (amphetamines change taste perceptions).

Inducing nausea or vomiting (digitalis may do both).

Interfering with oral function (some antidepressants may cause dry mouth).

Causing sores or inflammation in the mouth (methotrexate may cause painful mouth ulcers).

Drugs may alter nutrient absorption by:

Changing the acidity of the digestive tract (antacids may interfere with iron and folate absorption).

Damaging mucosal cells (cancer chemotherapy may damage mucosal cells).

Binding to nutrients (bile acid binders bind to fat-soluble vitamins).

Foods and nutrients may alter drug absorption by:

Stimulating the secretion of gastric acid (the antifungal agent ketoconazole is absorbed better with meals due to increased acid secretion).

Altering the rate of gastric emptying (intestinal absorption of drugs may be delayed when they are taken with food).

Binding to drugs (calcium binds to tetracycline, reducing the absorption of both substances).

Competing for absorption sites in the intestine (dietary amino acids interfere with levodopa absorption).

Drugs and nutrients may interact and alter metabolism by:

Acting as structural analogs (as do warfarin and vitamin K).

Using similar enzyme systems (phenobarbital induces liver enzymes that increase the metabolism of folate, vitamin D, and vitamin K).

Competing for transport on plasma proteins (fatty acids and drugs may compete for the same sites on the plasma protein albumin).

Drugs may alter nutrient excretion by:

Altering nutrient reabsorption in the kidneys (some diuretics increase the excretion of sodium and potassium).

Causing diarrhea or vomiting (diarrhea and vomiting may cause electrolyte losses).

Food substances may alter drug excretion by:

Inducing the activities of liver enzymes that metabolize drugs, increasing drug excretion (components of charcoal-broiled meats increase the metabolism of warfarin, theophylline, and acetaminophen).

Food substances and drugs may interact and cause toxicity by:

Increasing side effects of the drug (the caffeine in beverages can increase the adverse effects of stimulants).

Increasing drug action to excessive levels (grapefruit components inhibit the enzymes that degrade certain drugs, increasing drug concentrations in the body).

To help prevent diet-drug interactions, find out about all of the drugs and supplements the patient takes, including prescription and over-the-counter medications, herbs, and other dietary supplements.

absorption of both the calcium and the antibiotic. Other antibiotics can bind to minerals such as iron, magnesium, and zinc. Consumers are generally advised to use dairy products and all mineral supplements at least two hours apart from these medications.

Altered Stomach Acidity Medications that reduce stomach acid can impair the absorption of vitamin B_{12}, folate, and iron. Examples include antacids, which neutralize stomach acid by acting as weak bases, and antiulcer drugs (proton pump inhibitors and H2 blockers), which interfere with acid secretion.

Direct Inhibition Several drugs impede nutrient absorption by interfering with their intestinal metabolism or transport into mucosal cells. For example, the antibiotics trimethoprim (Proloprim) and pyrimethamine (Daraprim) compete with folate for absorption into intestinal cells.

Dietary Effects on Drug Absorption

Major influences on drug absorption include the stomach-emptying rate, the level of acidity in the stomach, and direct interactions with dietary components. The drug's formulation may also influence its absorption. The instructions included with medications typically advise whether food should be included or avoided with use.

Stomach-Emptying Rate Drugs reach the small intestine more quickly when the stomach is empty. Therefore, taking a medication with meals may delay its absorption, even though the total amount absorbed may not be lower. As an example, aspirin works faster when taken on an empty stomach, although taking it with food is often encouraged to reduce stomach irritation.

Slow stomach emptying can sometimes enhance drug absorption because the drug's absorption sites in the small intestine are less likely to become saturated. However, a slow drug absorption rate (due to slow stomach emptying) can be a problem if high drug concentrations are needed for effectiveness, as when a hypnotic is taken to induce sleep.

Stomach Acidity Some drugs are better absorbed in an acidic medium, whereas others are better absorbed under alkaline conditions. For example, reduced stomach acidity (due to secretory disorders or antacid medications) can reduce the absorption of ketoconazole (an antifungal medication) and atazanavir (an antiretroviral medication), but increase the absorption of digoxin (Lanoxin, which treats heart failure) and alendronate (Fosamax, which treats osteoporosis).[6] Some drugs can be damaged by acid and are available in coated forms that resist the stomach's acidity.

Binding between Food Components and Drugs Some dietary substances can bind to drugs and inhibit their absorption. For example, the phytates in foods can bind to digoxin. High-fiber diets can decrease the absorption of some tricyclic antidepressants due to binding between the fiber and the drugs. As mentioned earlier, minerals can bind to some antibiotics, reducing absorption of both the minerals and the drugs.

Drug Effects on Nutrient Metabolism

Drugs and nutrients share similar enzyme systems in the small intestine and liver. Consequently, some drugs may enhance or inhibit the activities of enzymes needed for nutrient metabolism. For example, the anticonvulsants phenobarbital and phenytoin increase levels of the liver enzymes that metabolize folate, vitamin D, and vitamin K; therefore, persons using these drugs may require supplements of these vitamins.

The drug methotrexate, which treats cancer (and some inflammatory conditions), acts by interfering with folate metabolism and thus depriving rapidly dividing cancer cells of the folate they need to multiply. Methotrexate resembles folate in structure (see Figure 15-1) and competes with folate for the enzyme that converts folate to its

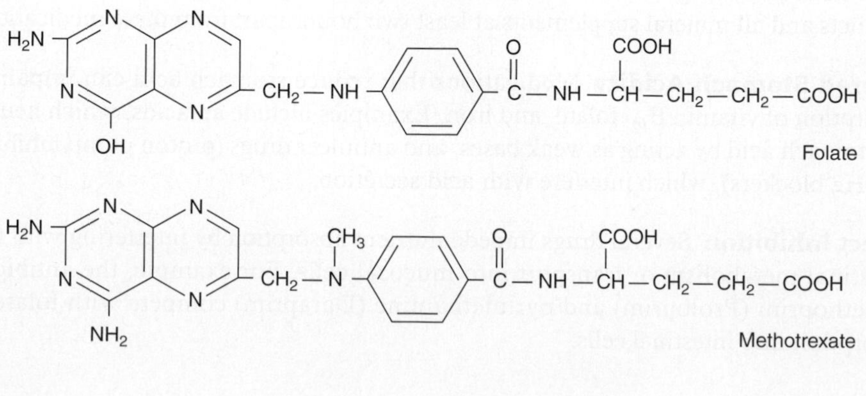

Reminder: *Phytates* are compounds found in many plant foods, including whole grains and legumes. Phytates can bind to minerals and reduce their absorption.

FIGURE 15-1
Folate and Methotrexate
By competing for the enzyme that activates folate, methotrexate prevents cancer cells from obtaining the folate they need to multiply. In the process, normal cells are also deprived of the folate they need.

active form. The adverse effects of using methotrexate therefore include symptoms of folate deficiency. These adverse effects can be reduced by using a preactivated form of folate (called leucovorin), which is often prescribed along with methotrexate to ensure that the body's rapidly dividing cells (cells of the digestive tract, skin cells, and red blood cells) receive adequate folate.

Corticosteroids, used as anti-inflammatory agents and immunosuppressants, have actions that mimic those of the hormone cortisol. Long-term corticosteroid use can have broad effects on nutritional health and may cause weight gain, muscle wasting, bone loss, and hyperglycemia, with eventual development of osteoporosis and diabetes.

> Cortisol is a steroid hormone secreted by the adrenal glands as part of the body's stress response.

Dietary Effects on Drug Metabolism

Some food components alter the activities of enzymes that metabolize drugs or may counteract drug effects in other ways. Compounds in grapefruit juice (or whole grapefruit) have been found to inhibit or inactivate enzymes that metabolize a number of different drugs. As a result of the reduced enzyme action, blood levels of the drugs increase, leading to stronger physiological effects. The effect of the grapefruit juice lasts for a substantial period after the juice is consumed; for example, in experiments with a drug prescribed for heart disease, the juice's effect had an estimated half-life of 12 hours.[7] Table 15-3 provides examples of drugs that interact with grapefruit juice, as well as some common drugs that are unaffected.

A number of dietary substances can alter the activity of the anticoagulant drug warfarin (Coumadin). One important interaction is with vitamin K, which is structurally similar to warfarin. Warfarin acts by blocking the enzyme that activates vitamin K, thereby preventing the synthesis of blood-clotting factors. The amount of warfarin prescribed is dependent, in part, on how much vitamin K is in the diet. If vitamin K consumption from foods or supplements changes substantially, it can alter the effect of the drug. Individuals using warfarin are advised to consume similar amounts of vitamin K daily to keep warfarin activity stable. The dietary sources highest in vitamin K are green, leafy vegetables.

Several popular herbs contain natural compounds that enhance the activity of warfarin and therefore should be avoided during warfarin treatment. These herbs include St. John's wort, garlic, ginseng, dong quai, danshen, and others.[8]

> Reminder: the term *half-life* defines the time period of a chemical effect. If the grapefruit effect has a 12-hour half-life, this means that after 12 hours, its biological effect is half of the maximum effect measured.

> Reminder: Vitamin K is required for the synthesis of prothrombin and various other blood-clotting proteins.

TABLE 15-3 Examples of Grapefruit Juice–Drug Interactions

Drug Category	Drugs Affected by Grapefruit Juice	Drugs Unaffected by Grapefruit Juice
Cardiovascular drugs	Amiodarone Felodipine Nicardipine	Amlodipine Digoxin Diltiazem
Cholesterol-lowering drugs	Atorvastatin Lovastatin Simvastatin	Fluvastatin Pravastatin Rosuvastatin
Central nervous system drugs	Buspirone Carbamazepine Diazepam	Alprazolam Haloperidol Lorazepam
Anti-infective drugs	Erythromycin Saquinavir	Clarithromycin Quinine
Estrogens	Ethinylestradiol	17-β-estradiol
Anticoagulants	—	Acenocoumarol Warfarin
Immunosuppressants	Cyclosporine Tacrolimus	Prednisone

Drug Effects on Nutrient Excretion

Drugs that increase urine production may reduce nutrient reabsorption in the kidneys and consequently increase urinary losses of these nutrients. For example, some diuretics increase losses of calcium, potassium, magnesium, and thiamin; thus, dietary supplements may be necessary to avoid deficiency. Risk of nutrient depletion is higher if multiple drugs with the same effect are used, if kidney function is impaired, or if the medications are used for a long time. Note that some diuretics can cause certain minerals to be retained, rather than excreted.[9]

A number of drugs can increase the excretion of vitamin B_6. One example is isoniazid (INH), an antituberculosis drug similar in structure to vitamin B_6. This drug induces excretion of vitamin B_6 and therefore may lead to a vitamin B_6 deficiency.[10] Because the drug must be taken for at least six months to treat infection, vitamin B_6 supplements are often given simultaneously to prevent deficiency.

Dietary Effects on Drug Excretion

Inadequate excretion of medications can cause toxicity, whereas excessive losses may reduce the amount available for therapeutic effect. Some food components influence drug excretion by altering the amount reabsorbed in the kidneys. For example, the amount of lithium (a mood stabilizer) reabsorbed in the kidneys is similar to the amount of sodium that is reabsorbed. Consequently, both dehydration and sodium depletion, which promote sodium reabsorption, can result in lithium retention. Similarly, a person with a high sodium intake will excrete more sodium in the urine and, therefore, more lithium. Individuals using lithium are advised to maintain a consistent sodium intake from day to day to maintain stable blood concentrations of lithium.

Urine acidity can affect drug excretion due to the effects of pH on a compound's ionic (chemical) form. The medication quinidine, used to treat arrhythmias, is excreted more readily in acidic urine. Foods or drugs that cause urine to become more alkaline may reduce quinidine excretion and raise blood levels of the medication.

Diet-Drug Interactions and Toxicity

Interactions between food components and drugs can cause toxicity or exacerbate a drug's side effects. The combination of tyramine, a food component, and monoamine oxidase (MAO) inhibitors, which treat depression and Parkinson's disease, can be fatal. MAO inhibitors block an enzyme that normally inactivates tyramine, as well as the hormones epinephrine and norepinephrine. When people who take MAO inhibitors consume excessive tyramine, the increased tyramine in the blood can induce a sudden release of stored norepinephrine. This surge in norepinephrine results in severe headaches, rapid heartbeat, and a dangerous rise in blood pressure. For this reason, people taking MAO inhibitors are advised to restrict their intakes of foods rich in tyramine. Foods that often contain substantial amounts of tyramine are listed in Table 15-4.

Considering the many ways in which drugs and dietary substances can interact, health professionals should attempt to understand the mechanisms underlying diet-drug

When the kidneys reabsorb a substance, they retain it in the blood. Substances that are not reabsorbed are excreted in urine.

TABLE 15-4 Examples of Foods with a High Tyramine Content[a]

▪ Aged cheeses (cheddar, Gruyère)	▪ Mushrooms
▪ Aged or cured meats (sausage, salami)	▪ Prepared soy foods (miso, tempeh, tofu)
▪ Beer	▪ Soy sauce
▪ Fermented vegetables (sauerkraut, kim chee)	▪ Wine (red)
▪ Fish, smoked or pickled	▪ Yeast extract (Marmite, Vegemite)

[a]Although tyramine occurs naturally in foods, it is also formed when bacteria degrade food proteins; thus, the tyramine content increases when a food ages or spoils. Individuals at risk of tyramine toxicity are advised to buy mainly fresh foods and consume them promptly.

The Joint Commission, an accreditation agency for health care organizations, has recommended that all patients be educated about potential diet-drug interactions. Health professionals can help by informing patients of precautions related to medications and watching for signs of problems that may arise.

To prevent diet-drug interactions, first list the types and amounts of over-the-counter drugs, prescription drugs, and dietary supplements that the patient uses on a regular basis. Look up each drug in a drug reference and make a note of:

▪ The appropriate method of administration (twice daily or at bedtime, for example).

▪ How the drug should be administered with respect to foods, beverages, and specific nutrients (for example, take on an empty stomach, take with food, do not take with milk, or do not drink alcoholic beverages while using the medication).

▪ How the drug should be used with respect to other medications.

▪ The side effects that may influence food intake (nausea and vomiting, diarrhea, constipation, or sedation, for example) or nutrient needs (interference with nutrient absorption or metabolism, for example).

A similar process can be used to review the dietary supplements that a person is taking. A reliable reference may list their appropriate uses, possible side effects, and potential interactions with food and medications.

Patients who take multiple medications may need to time their intakes carefully to avoid drug-drug or diet-drug interactions. The health professional can use information from a patient's food and nutrition history (see Chapter 14) to help the patient coordinate meals and drugs so as to avoid interactions.

Some medications have well-known effects on nutritional status. The health professional should remain alert for signs of problems, especially when:

▪ Nutritional problems are a frequent result of using the medication.

▪ A patient requires multiple medications.

▪ The patient is in a high-risk group; for example, a child, a pregnant or lactating woman, an older adult, or a person who is malnourished, abuses alcohol, or has impaired liver or kidney function.

▪ The patient needs to use the medications for an extended period.

Check with the pharmacist for additional information about drugs and their potential adverse effects.

interactions, identify them when they occur, and prevent them whenever possible. The accompanying "How to" offers some practical advice about preventing diet-drug interactions.

IN SUMMARY

▪ Medications can alter food intake and affect the absorption, metabolism, and excretion of nutrients. Components of foods can similarly affect drug activity.

▪ Drugs can alter food intake by increasing or decreasing the appetite, altering the sense of taste, causing GI discomfort, or damaging the lining of the GI tract.

▪ Drugs can affect nutrient absorption by binding to nutrients, changing stomach acidity, or interfering with nutrient transport into intestinal cells. Dietary substances can influence drug absorption by altering the stomach emptying rate, changing stomach acidity, or directly binding to drugs.

▪ Drugs and nutrients may interfere with each other's metabolism because they use similar enzymes in the small intestine and liver.

▪ Diet-drug interactions may increase nutrient losses in urine and alter the urinary excretion of drugs. Some interactions between food components and drugs may result in toxicity.

Herbal Supplements

The use of herbal supplements has grown rapidly in the past decade. Currently, about 17 percent of adults in the United States report using herbal supplements regularly.[11] Consumers use these products in the hope of improving their general health and

TABLE 15-5 Popular Herbs, Their Common Uses, and Adverse Effects

Herb	Scientific Name	Common Uses	Adverse Effects
Black cohosh	*Cimicifuga racemosa*	Relief of menopausal symptoms	Rare; occasional stomach upset, headache, weight gain
Chaparral	*Larrea tridentata*	General tonic; treatment of infection, cancer, and arthritis	Hepatitis, liver failure
Comfrey	*Symphytum officinale*	Wound healing (topical use), treatment of lung and GI disorders	Liver damage
Echinacea	*Echinacea augustifolia, E. pallida, E. purpurea*	Prevention and treatment of upper respiratory infections	Rare; stomach upset, headache, occasional allergic reactions
Feverfew	*Tanacetum parthenium*	Prevention of migraine headache	Mouth and tongue sores, swelling of lips, GI upset
Garlic	*Allium sativum*	Reduction of blood clotting, atherosclerosis, blood pressure, and blood cholesterol	Bad breath, body odor, occasional stomach upset or flatulence, excessive bleeding
Ginger	*Zingiber officinale*	Prevention and treatment of nausea and motion sickness	Rare; occasional heartburn
Ginkgo	*Ginkgo biloba*	Treatment of dementia, memory defects, and circulatory impairment	Rare; occasional stomach upset, headache, skin hypersensitivity, excessive bleeding
Ginseng	*Panax ginseng, P. quinquefolius*	General tonic, reduction of blood glucose levels	Rare
Kava kava	*Piper methysticum*	Treatment of anxiety, stress, and insomnia	Dyspepsia, restlessness, drowsiness, tremor, headache, dermatitis (with heavy use), occasional hepatitis and liver failure
St. John's wort	*Hypericum perforatum*	Treatment of mild to moderate depression	Rare; occasional stomach upset, fatigue, dizziness, headache, dry mouth, dermatitis, skin photosensitivity
Saw palmetto	*Serenoa repens*	Reduction of symptoms associated with enlarged prostate	Rare
Valerian	*Valeriana officinalis*	Sedation, treatment of insomnia	Rare
Yohimbe	*Pausinystalia yohimbe*	Treatment of erectile dysfunction	Anxiety, headache, dizziness, nausea, rapid heartbeat, hypertension, increased urinary frequency; isolated reports of renal failure, blood disorders, and airway constriction

An *herb* is a non-woody, seed-producing plant; *herbal supplements* include other types of plant products, such as garlic and ginkgo.

preventing or treating specific diseases. Top-selling herbal supplements include echinacea, garlic, ginkgo biloba, ginseng, and St. John's wort.[12] Table 15-5 lists these and other popular herbal products along with their common uses and the potential risks associated with their use.

Effectiveness and Safety of Herbal Products

Despite the popularity of herbal products, the benefits of their use are uncertain. There is no question that many medicinal herbs contain naturally occurring compounds that can exert physiological effects. Few herbal products, however, have been rigorously tested, many make unfounded claims, and some contain contaminants or produce toxic effects.[13]

Efficacy Herbs have been used for centuries to treat medical conditions, and many have acquired reputations for being beneficial for people with specific diseases. Unfortunately, only a limited number of clinical studies support the traditional uses, and the

results of studies that suggest little or no benefit are rarely publicized by the supplement industry. The National Center for Complementary and Alternative Medicine (a division of the National Institutes of Health) is currently funding large, controlled trials of several popular herbal treatments in an effort to obtain reliable efficacy and safety data.

Although labels on herbal products cannot make claims about preventing or treating specific diseases, suggestive statements are common. For example, a label may claim that an herb "promotes restful sleep" but cannot state that it cures insomnia. Stores often shelve herbal products by health condition; for example, posted signs may indicate the supplements suggested for "liver health" or "men's health." The reading materials positioned close to those shelves may also suggest that the products can improve one's health.

Consistency of Herbal Ingredients Herbs contain numerous compounds, and it is often unclear which of these ingredients, if any, might produce the implied beneficial effects. Because the compounds in herbs vary among species and are affected by a plant's growing conditions, different samples of an herb can have different chemical compositions. The preparation method may also cause variations in the composition of an herbal product. Some manufacturers attempt to standardize the herbal extracts they sell so that the compound believed to be beneficial is more likely to be obtained from each dose.

Even when the active ingredients in an herbal supplement have been shown to be effective, the dosage suggested on the label might not provide the quantity of active ingredients required for benefit. For example, a consumer group (ConsumerLab.com) tested seven ginkgo biloba products and found that four of the products, when consumed at the recommended dosage, lacked the amounts of the compounds indicated on their labels.[14] In a university study of echinacea preparations, 10 percent of the 59 products tested contained no measurable echinacea, and only 52 percent of the samples contained the variety of echinacea listed on the label.[15]

Safety Issues Consumers often assume that because plants are "natural," herbal products must be harmless. Many herbal remedies have toxic effects, however. The most common adverse effects of herbs include diarrhea, nausea, and vomiting. The popular herbs chaparral, comfrey, and kava have caused liver damage. The use of yohimbe (promoted for bodybuilding) has been linked to alterations in blood pressure and heart arrhythmias.[16] Note that the adverse effects of herbs are seldom listed on supplement labels.

Contamination of herbal products is another safety concern.[17] Some products have been found to contain lead and other toxic metals in excessive amounts. Other contaminants frequently found in herbal supplements include molds, bacteria, and pesticides that have been banned for use on food crops.[18] Adulteration of imported products is a serious concern: several studies found that some herbal products imported from China and India contained synthetic drugs that were not declared on the label.[19] There have also been reports of illnesses and fatalities occurring from the intentional or accidental substitutions of one plant species for another.

Unlike drugs, herbal products do not need FDA approval before they are marketed. According to the Dietary Supplement Health and Education Act (DSHEA) of 1994, the companies that produce or distribute dietary (including herbal) supplements are responsible for determining their safety, yet these companies are not required to provide any evidence or conduct safety studies. If a company receives reports of illness or injury related to the use of its products, it is not required to submit this information to the FDA. In addition, the FDA must show that a supplement is unsafe before it can take action to remove the product from the marketplace.

Herb-Drug Interactions Like drugs, herbs may either intensify or interfere with the effects of other herbs and drugs or they may raise the risk of toxicity.[20] For example, garlic, ginkgo, and ginseng may increase the risk of bleeding when used

Despite the popularity of echinacea, its benefits for treating the common cold have not been supported by some well-designed clinical studies.

TABLE 15-6 Examples of Herb-Drug Interactions

Herb	Drugs	Interaction
Echinacea	Immunosuppressant drugs	Suppresses drug effects
Feverfew	Anticoagulants, aspirin	Enhances drug effects
Garlic, ginkgo, ginseng	Anticoagulants, antiplatelet drugs	Increases risk of hemorrhaging
Kava kava	Acetaminophen, antifungal drugs, methotrexate, steroids	Increases risk of liver damage
St. John's wort	Various	Suppresses drug effects
Valerian	Antidepressants, antiepileptics, anesthetics, analgesics, barbiturates	May enhance drug effects, increase sedation

with anticoagulant drugs. St. John's wort has been found to inhibit the actions of oral contraceptives, anticoagulants, and other drugs. Ginseng contains compounds that raise blood pressure and may increase the toxicity of drugs that have a similar side effect.[21] Unfortunately, information about herb-drug interactions is limited, and much of what is known has been obtained from case studies rather than controlled clinical trials. Table 15-6 provides some examples of herb-drug interactions.

Use of Herbal Products in Illness

When people self-medicate or ask the advice of store clerks instead of seeking effective medical treatment, the consequences are sometimes serious and irreversible. Purchasing an herbal remedy may be less stressful than a visit to the doctor, but it may delay getting an appropriate treatment and allow an illness to progress. Although retailers are not legally permitted to provide medical advice, a 2010 investigation found that sellers of herbal products routinely made improper claims that their herbal products were able to treat, prevent, or cure specific illnesses.[22]

Patients are often unaware that herbal products may be unsafe or can interact with medications. Elderly individuals (65 years old and older) are at highest risk of herb-drug interactions because most individuals in this age group take three or more prescription drugs over the course of a year.[23] Some pharmacology textbooks and handbooks now contain information about herbal supplements and potential herb-drug interactions, and various consumer websites and periodicals provide information about the safety of brand-name herbal products. Health professionals should turn to these resources to help patients who plan to use herbal supplements.

IN SUMMARY

▌ Herbal products are not reliable treatments for medical conditions; there is little evidence demonstrating their effectiveness, and the concentrations of active ingredients in products can vary greatly. Safety concerns include adverse effects, contamination, and herb-drug interactions.

▌ Manufacturers and distributors of herbal supplements are responsible for determining product safety but are not required to conduct safety studies. The FDA must prove that a supplement is unsafe before removing it from the market.

▌ Consumers using herbs may delay getting an appropriate treatment for their condition and may receive questionable advice from supplement retailers.

Clinical Applications

1. An elderly woman in a residential home has been losing weight since her arrival there. She has been taking several medications to treat both a heart problem and a mild case of bronchitis. You notice that she eats only a few bites at mealtimes and seems uninterested in food. Describe several steps you can take to learn whether the medications are interfering with her food intake in some way.

2. A patient mentions that he regularly takes six to seven dietary and herbal supplements and that he has not told the physician that he uses them. His prescription medications include an antihypertensive agent (to reduce blood pressure) and warfarin. What approach might you take to learn the details about his supplement use and his reasons for taking them? If you discover that some of the supplements may pose a risk for diet-drug or herb-drug interactions with his prescription medications, what steps should you take?

Self Check

1. Over-the-counter drugs are:
   a. unlikely to cause adverse effects.
   b. unlikely to interact with dietary components.
   c. generally used for longer periods of time than prescription medications.
   d. used to treat illnesses that are normally self-diagnosed and self-treated.

2. Recommendations for reducing the incidence of medication errors include:
   a. avoiding use of confusing terms on clinical documents.
   b. advising patients to take only one medication at a time.
   c. requiring that prescriptions be handwritten instead of typed.
   d. physician supervision whenever drugs are administered.

3. Adverse drug effects are most likely when:
   a. multiple medications are used.
   b. generic drugs are substituted for brand-name drugs.
   c. patients begin using a new medication.
   d. medications are taken for just one or two days.

4. Examples of medication-related symptoms that can significantly limit food intake include:
   a. ringing in the ears.
   b. persistent nausea and vomiting.
   c. insomnia.
   d. skin rash.

5. Factors that may influence drug absorption include:
   a. binding between drugs and food components.
   b. use of antacid therapies.
   c. rapid stomach emptying rate.
   d. all of the above.

6. Compounds in grapefruit juice:
   a. bind to antibiotics, reducing absorption.
   b. cause excessive drug excretion.
   c. strengthen the effects of certain drugs.
   d. alter acidity in the stomach, impairing drug absorption.

7. Vitamin K consumption should be consistent in patients using:
   a. tetracycline.
   b. warfarin.
   c. lithium.
   d. isoniazid.

8. People who use MAO inhibiters must limit consumption of:
   a. whole milk and yogurt.
   b. aged cheeses.
   c. dark green leafy vegetables.
   d. grapefruit juice.

9. An initial step that health practitioners can take to limit risk of medication-related side effects is to:
   a. recommend use of over-the-counter drugs instead of prescription medications.
   b. encourage use of herbal supplements rather than prescription medications.
   c. advise patients to take medications separately from meals.
   d. ask patients to fully describe the types and amounts of medications and dietary supplements they are using.

10. An important difference between medications and herbal products that reach the marketplace is that:
    a. medications that cause adverse effects cannot be sold.
    b. medications are subject to contamination with toxic metals, molds, and bacteria.
    c. herbal products are not required to prove safety and effectiveness.
    d. herbal products must provide standard amounts of active ingredients.

Answers to these questions can be found in Appendix H.

Notes

1. R. L. Corelli, Therapeutic and toxic potential of over-the-counter agents, in B. G. Katzung, ed., *Basic and Clinical Pharmacology* (New York: Lange Medical Books/McGraw-Hill, 2007), pp. 1041–1049.

2. P. W. Lofholm and B. G. Katzung, Rational prescribing and prescription writing, in B. G. Katzung, ed., *Basic and Clinical Pharmacology* (New York: Lange Medical Books/McGraw-Hill, 2007), pp. 1063–1072.

3. Institute of Medicine, Committee on Identifying and Preventing Medication Errors, *Preventing Medication Errors* (Washington, D.C.: National Academy Press, 2007).

4. L.-N. Chan, Drug–nutrient interactions, in M. E. Shils and coeditors, *Modern Nutrition in Health and Disease* (Baltimore: Lippincott Williams & Wilkins, 2006), pp. 1539–1553.

5. W. S. Leslie, C. R. Hankey, and M. E. J. Lean, Weight gain as an adverse effect of some commonly prescribed drugs: A systematic review, *QJM: An International Journal of Medicine* 100 (2007): 395–404.

6. E. Lahner and coauthors, Systematic review: Impaired drug absorption related to the co-administration of antisecretory therapy, *Alimentary Pharmacology and Therapeutics* 29 (2009): 1219–1229.

7. D. G. Bailey, Grapefruit juice–drug interaction issues, in J. I. Boullata and V. T. Armenti, eds., *Handbook of Drug-Nutrient Interactions* (Totowa, NJ: Humana Press, 2004), pp. 175–194.

8. M. L. Chavez, M. A. Jordan, and P. I. Chavez, Evidence-based drug–herbal interactions, *Life Sciences* 78 (2006): 2146–2157.

9. S. A. Shapses, Y. R. Schlussel, and M. Cifuentes, Drug-nutrient interactions that impact mineral status, in J. I. Boullata and V. T. Armenti, eds., *Handbook of Drug-Nutrient Interactions* (Totowa, NJ: Humana Press, 2004), pp. 301–328; B. J. McCabe, E. H. Frankel, and J. J. Wolfe, Monitoring nutritional status in drug regimens, in B. J. McCabe, E. H. Frankel, and J. J. Wolfe, eds., *Handbook of Food-Drug Interactions* (Boca Raton, FL: CRC Press, 2003), pp. 73–108.

10. H. F. Chambers, Antimycobacterial drugs, in B. G. Katzung, ed., *Basic and Clinical Pharmacology* (New York: Lange Medical Books/McGraw-Hill, 2007), pp. 771–780.

11. M. E. Gershwin and coauthors, Public safety and dietary supplementation, *Annals of the New York Academy of Sciences* 1190 (2010): 104–117.

12. U.S. Government Accountability Office (GAO), Herbal dietary supplements: Examples of deceptive or questionable marketing practices and potentially dangerous advice, GAO-10-662T (Washington, D.C.: May 26, 2010); J. Kennedy, Herb and supplement use in the U.S. adult population, *Clinical Therapeutics* 27 (2005): 1847–1858.

13. Gershwin and coauthors, 2010; U.S. Government Accountability Office (GAO), 2010.

14. ConsumerLab.com, Product review: Supplements for memory and cognition enhancement (ginkgo, huperzine-A, and acetyl-L-carnitine), available at www.consumerlab.com, posted December 30, 2009; accessed June 8, 2010.

15. C. M. Gilroy and coauthors, Echinacea and truth in labeling, *Archives of Internal Medicine* 163 (2003): 699–704.

16. J. A. Rindfleisch and B. Barrett, Herbs and other dietary supplements, in R. E. Rakel, ed., *Textbook of Family Medicine* (Philadelphia: Saunders, 2007), pp. 243–266.

17. Gershwin and coauthors, 2010; ConsumerLab. com, 2010.

18. V. H. Tournas, E. Katsoudas, and E. J. Miracco, Moulds, yeasts, and aerobic plate counts in ginseng supplements, *International Journal of Food Microbiology* 108 (2006): 178–181; K. S. Leung and coauthors, Systematic evaluation of organochlorine pesticide residues in Chinese materia medica, *Phytotherapy Research* 19 (2005): 514–518.

19. R. J. Ko, A U.S. perspective on the adverse reactions from traditional Chinese medicines, *Journal of the Chinese Medical Association* 67 (2004): 109–116; E. Ernst, Toxic heavy metals and undeclared drugs in Asian herbal medicines, *Trends in Pharmacological Sciences* 23 (2002): 136–139.

20. C. E. Dennehy and C. Tsourounis, Botanicals ("herbal medications") and nutritional supplements, in B. G. Katzung, ed., *Basic and Clinical Pharmacology* (New York: McGraw-Hill, 2007), pp. 1050–1062; Rindfleisch and Barrett, 2007; J. J. Mucksavage and L.-N. Chan, Dietary supplement interactions with medication, in J. I. Boullata and V. T. Armenti, eds., *Handbook of Drug-Nutrient Interactions* (Totowa, NJ: Humana Press, 2004), pp. 217–233.

21. J. Barnes, L. A. Anderson, and J. D. Phillipson, *Herbal Medicines: A Guide for Healthcare Professionals* (Chicago: Pharmaceutical Press, 2002).

22. U.S. Government Accountability Office (GAO), 2010.

23. U.S. Government Accountability Office (GAO), 2010.

Nutrition in Practice

The medical treatments described in the clinical chapters are based upon current scientific understanding of human physiology and biochemistry and are generally supported by well-conducted clinical research. This Nutrition in Practice examines therapies that have *not* been scientifically validated and are therefore not currently promoted by conventional health professionals; these therapies fall into a category called *complementary and alternative medicine (CAM)*. When the therapies are used together with conventional medicine, they are called *complementary*; when used in place of conventional medicine, they are called *alternative*.[1] The term "alternative" may be misleading, because it inappropriately implies that unproven methods of treatment are valid alternatives to conventional treatments.

How popular is CAM in the United States?

An estimated 38 percent of adults in the United States use some form of CAM (excluding the use of prayer).[2] CAM is most prevalent among people with chronic, debilitating diseases; for example, 84 percent of AIDS patients reportedly use CAM.[3] Many patients use CAM as an adjunct to conventional medicine—often for symptoms or illnesses that are not sufficiently helped by conventional treatments. CAM therapies remain popular despite the dearth of evidence demonstrating their effectiveness. Reasons for their popularity include consumers' growing interest in self-help measures, the noninvasive nature of many CAM therapies, and the positive interactions consumers have with CAM practitioners.[4]

In response to the enormous popularity of CAM in the United States, in 1998 Congress established the National Center for Complementary and Alternative Medicine (NCCAM), which is now one of the 27 institutes that make up the National Institutes of Health (NIH). NCCAM's missions are to investigate complementary and alternative therapies by funding well-designed scientific studies and to provide authoritative information for consumers and health professionals. If enough evidence is found to support the use of a complementary or alternative therapy, it will likely become a treatment offered by conventional health practitioners.[5]

Why should mainstream health professionals learn more about CAM?

Due to the enormous consumer interest in trying novel treatments, health professionals should be familiar with CAM therapies so that they can better communicate with patients regarding their medical care and advise them when an alternative approach conflicts with standard therapy or presents a danger to health. To provide medical students with objective information about CAM, many medical schools in the United States now offer elective courses about alternative forms of treatment. Physicians who practice *integrative medicine* may refer patients for complementary therapies while continuing to provide standard treatments.

What kinds of practices are considered CAM therapies?

CAM encompasses all therapies that are not normally part of conventional medicine. Consequently, the list of CAM approaches includes hundreds of advertised therapies purchased and used by consumers. Unfortunately, CAM has become a marketing buzzword and is used by unscrupulous sellers of worthless treatments. NCCAM categorizes CAM therapies as shown in Table NP15-1 and defined in the Glossary of Alternative Therapies on p. 426. Several popular examples are described in this Nutrition in Practice; other examples are discussed on the NCCAM website (http://nccam.nih.gov/health).

How do alternative medical systems differ from conventional medicine?

Alternative medical systems are based on beliefs that lack the scientific basis of the theories underlying conventional medicine. The alternative systems were developed well over 100 years ago, before our bodies' biochemical and physiological processes

TABLE NP15-1 Examples of Complementary and Alternative Medicine

Alternative Medical Systems
- Naturopathic medicine
- Homeopathic medicine
- Traditional Chinese medicine
- Ayurveda

Mind-Body Interventions
- Biofeedback
- Meditation
- Faith healing (prayer)
- Mental healing (including hypnotherapy)
- Music, art, and dance therapy

Biologically Based Therapies
- Dietary supplements
- Foods and special diets
- Herbal products
- Hormones
- Aromatherapy

Manipulative and Body-Based Methods
- Chiropractic
- Massage therapy
- Osteopathic manipulation
- Reflexology

Energy Therapies
- Biofield therapies (including therapeutic touch, acupuncture, and qi gong
- Bioelectrical therapies (including electrical and magnetic fields)

acupuncture (AK-you-PUNK-chur): a therapy that involves inserting thin needles into the skin at specific anatomical points, allegedly to correct disruptions in the flow of energy within the body.

bioelectrical or **bioelectromagnetic therapies:** therapies that involve the unconventional use of electric or magnetic fields to cure illness.

biofeedback: a technique in which individuals are trained to gain voluntary control of certain physiological processes, such as skin temperature or brain wave activity, to help reduce stress and anxiety.

biofield therapies: healing methods based on the belief that illnesses can be cured by manipulating energy fields that purportedly surround and penetrate the body. Examples include *acupuncture, qi gong,* and *therapeutic touch.*

chiropractic (KYE-roh-PRAK-tic): a method of treatment based on the unproven theory that spinal manipulation can restore health.
▌ A *subluxation* is a misaligned vertebra or other spinal alteration that may cause illness.
▌ *Adjustment* is the manipulative therapy practiced by chiropractors.

faith healing: the use of prayer or belief in divine intervention to promote healing.

homeopathic (HO-mee-oh-PATH-ic) **medicine:** a practice based on the theory that "like cures like"; that is, substances believed to cause certain symptoms are prescribed for curing the same symptoms, but are given in extremely diluted amounts.
▌ **homeo** = like
▌ **pathos** = suffering

imagery: the use of mental images of things or events to aid relaxation or promote self-healing.

massage therapy: manual manipulation of muscles to reduce tension, increase blood circulation, improve joint mobility, and promote healing of injuries.

meditation: a self-directed technique of calming the mind and relaxing the body.

naturopathic (NAY-chur-oh-PATH-ic) **medicine:** an approach to health care using practices alleged to enhance the body's natural healing abilities. Treatments may include a variety of alternative therapies including dietary supplements, herbal remedies, exercise, and homeopathy.

qi gong (chee-GUNG): a traditional Chinese system that combines movement, meditation, and breathing techniques and allegedly cures illness by enhancing the flow of qi (energy) within the body.

therapeutic touch: a technique of passing hands over a patient to purportedly identify energy imbalances and transfer healing power from therapist to patient; also called *laying on of hands.*

traditional Chinese medicine (TCM): an approach to health care based on the concept that illness can be cured by enhancing the flow of qi (energy) within a person's body. Treatments may include herbal therapies, physical exercises, meditation, acupuncture, and remedial massage.

---

were well understood. The alternative treatments may appeal to consumers because the interventions are nontechnical and seem nonthreatening. In general, however, the alternative theories and practices remain rooted in the past and have not been updated to include current knowledge. Examples of alternative medical systems include the following:

▌ **Naturopathic medicine** proposes that a person's natural "life force" can foster self-healing. This life force is allegedly stimulated by certain health-promoting factors and suppressed by excesses and deficiencies. Naturopathic therapies aim to enhance the natural healing powers of the body and may include special diets or fasting, herbal remedies and other dietary supplements, **acupuncture,** homeopathy, massage, and various other interventions.

▌ **Homeopathic medicine** is based on the dubious theory that "like cures like." Homeopaths believe that a substance that causes a particular set of symptoms can be used to cure a disease that has similar symptoms. Homeopathic medicines are usually natural substances that are substantially diluted in the belief that dilution increases potency, and most remedies are so extremely diluted that the original substance is no longer present. Homeopaths theorize that even though their remedies no longer contain the diluted substance, they still have powerful healing effects because the water structure is somehow altered during the dilution process used to prepare homeopathic medicines. This theory, however, conflicts with scientific understanding of water structure and properties.

▌ **Traditional Chinese medicine (TCM)** includes a large number of folk practices that originated in China. TCM is based on the theory that the body has pathways (called *meridians*) that conduct energy (called *qi;* pronounced "chee"). The interrupted flow of qi is believed to cause illness. TCM practices allegedly improve the flow of qi and include acupuncture, **qi gong,** herbal remedies, dietary practices, and massage. Ironically, TCM is used by relatively few in the Chinese population, as Chinese physicians have largely adopted the Western approach to managing illness.[6]

What is the theory underlying mind-body interventions?

Mind-body interventions attempt to improve a person's sense of well-being despite the presence of illness. The treatments are also used in the hope of reducing stress, dealing with pain, or lowering blood pressure. Some of these therapies have been incorporated into mainstream medicine for stress reduction or relaxation. For example, **biofeedback** training, in which individuals learn to monitor skin temperature, muscle tension, or brain wave activity while practicing relaxation techniques, is frequently taught by behavioral

Biofeedback training is a stress reduction and relaxation technique.

medicine specialists to help patients reduce stress or anxiety. Similar benefits may be achieved using **meditation,** art and music therapy, and prayer.

The clinical applications of other mind-body therapies are far more questionable. One example is guided **imagery,** in which a person tries to reverse the disease process (for example, shrink a tumor) by using mental pictures. Another example is the use of **faith healing** in place of proven conventional treatments to cure disease.

Which alternative practices involve physical manipulation, and how do they work?

Manipulative interventions include physical touch, forceful movement of different parts of the body, and the application of pressure. Some practitioners maintain that special energy fields are manipulated during the physical treatment and that proper energy flow induces healing. Popular practices include the following:

▎ **Chiropractic** theory proposes that keeping the nervous system free from obstruction allows the body to heal itself, because the healing process stems from the brain and is conducted via the spinal cord and nerves to all parts of the body. Chiropractors claim to diagnose illnesses by detecting subluxations in the spine, which are variously described as misaligned vertebrae or pinched nerves that allegedly cause subtle interferences within the nervous system. The main treatment is the adjustment, a manual manipulation that is said to correct a subluxation and restore the body's natural healing ability. Although spinal manipulation has mainly been found to be helpful for improving back pain, many chiropractors still assert that chiropractic can cure disease rather than simply relieve symptoms.[7] For example, many chiropractors promote spinal manipulation to treat infectious diseases and prevent cancer, even though the nervous system and spinal alignment do not play roles in the pathology of these conditions.

▎ **Massage therapy** is the manipulation of muscle and connective tissue to improve muscle function, reduce pain, or promote relaxation. Massage therapists may also apply heat or cold and give advice about exercises that may improve muscle tone and range of motion. Massage is often integrated into conventional physical therapy, although some massage therapists may incorrectly suggest that massage is a valid treatment for a wide range of medical conditions.

What are the alleged effects of "energy" therapies?

Two categories of therapies involve the alleged curative power of "energy":

▎ **Biofield therapies** are said to influence the energy that surrounds or pervades the human body, and proponents claim that an energy therapy can strengthen or restore a person's "energy flow" and induce healing. Acupuncture, qi gong, and **therapeutic touch** are among the therapies that subscribe to these theories. Note that CAM adherents often use the term "energy" unscientifically and that there is no objective evidence of this sort of energy flow.

▎ **Bioelectrical** or **bioelectromagnetic therapies** use electric or magnetic fields to allegedly promote healing; for example, magnets have been marketed with claims that they can improve circulation and reduce inflammation.

Do conventional health practitioners consider CAM treatments to be safe and effective?

CAM treatments are generally excluded from mainstream medical practice because there is no evidence proving that they are effective for treating the diseases and medical conditions for which they are used. Many consumers think otherwise and seem satisfied that these treatments work. How is this dichotomy to be explained?

Surveys suggest that consumers perceive their visits to CAM therapists as far more pleasant than visits to conventional health practitioners. As explained earlier, CAM therapists spend more time with patients, are more attentive, and use less invasive interventions.[8] Self-help measures are encouraged, so the consumer has more control over the treatment. The therapies appear to be more "natural" and to have fewer side effects. Possible explanations for "cures" include the following:

▎ A person may seem cured because of misdiagnosis; that is, the condition diagnosed by the CAM practitioner may not have actually existed.

▎ The condition may have been self-limiting, or it may have gone into temporary remission after the treatment.

▎ Undue credit may be inappropriately assigned to the CAM therapy when the improvement was actually due to a previous or concurrent conventional treatment.

▎ The placebo effect may have had an influence on the course of disease.

The question remains: Do the CAM therapies merely make people *feel* better, or do they really *get* better? This question can be answered only by well-controlled research studies.

Are any potential dangers associated with the use of CAM?

One of the attractions of alternative therapies is the assumption that they are safe. Recall, however, the concerns associated with the use of herbal products discussed in Chapter 15, which include the potential toxicity of herbal ingredients, product contamination or adulteration, and interactions with conventional medications. Another concern is that use of CAM therapies may delay the use of reliable treatments that have demonstrable benefits.[9] Various reports have described people with treatable medical conditions who suffered permanent disability or death when they were misdiagnosed or improperly treated by CAM practitioners. For example, a rare but well-known risk of spinal cord injury or stroke is associated with a type of cervical manipulation performed by chiropractors.[10] Unfortunately, because most CAM therapies are not regulated or monitored, there are no accurate estimates of their adverse effects.

What should health practitioners do if they think their patients are using CAM?

Health practitioners should routinely inquire about the use of CAM therapies and educate patients about the hazards of postponing or stopping conventional treatment. Patients should also be told about potential interactions between conventional treatments and CAM therapies. Some patients may want to learn about differences between evidence-based medical practices and untested CAM theories and may be interested in the integrative medicine options available.

All alternative therapies have one characteristic in common: their effectiveness is, for the most part, unproven.[11] Because patients often choose CAM therapies because of positive interactions with alternative practitioners, health practitioners should realize that empathizing with patients may go a long way toward winning their trust and improving compliance with therapy. Furthermore, health practitioners should use reliable, objective resources to update their knowledge about unconventional practices so that they can discuss these options with patients.

Notes

1. S. E. Straus, Complementary and alternative medicine, in L. Goldman and D. Ausiello, eds., *Cecil Medicine* (Philadelphia: Saunders, 2008), pp. 206–209.

2. P. M. Barnes, B. Bloom, and R. Nahin, Complementary and alternative medicine use among adults and children: United States, 2007, *CDC National Health Statistics Report 12* (December 10, 2008): 1–24.

3. J. D. Berman and S. E. Straus, Implementing a research agenda for complementary and alternative medicine, *Annual Review of Medicine* 55 (2004): 239–254.

4. A. M. McCaffrey, G. F. Pugh, and B. B. O'Connor, Understanding patient preference for integrative medical care: Results from patient focus groups, *Journal of General Internal Medicine* 22 (2007): 1500–1505; E. Ernst, The role of complementary and alternative medicine, *British Medical Journal* 321 (2000): 1133–1135.

5. Straus, 2008.

6. D. Normile, The new face of Chinese medicine, *Science* 299 (2003): 188–190.

7. J. C. Keating and coauthors, Subluxation: Dogma or science? *Chiropractic and Osteopathy* epub 10 August 2005 (DOI 10.1186/1746-1340-13-17).

8. McCaffrey, Pugh, and O'Connor, 2007; B. Barrett and coauthors, What complementary and alternative medicine practitioners say about health and health care, *Annals of Family Medicine* 2 (2004): 253–259.

9. Straus, 2008.

10. W.-L. Chen and coauthors, Vertebral artery dissection and cerebellar infarction following chiropractic manipulation, *Emergency Medical Journal* 23 (2006): e1.doi:10.1136/emj.2004.015636; W. S. Smith and coauthors, Spinal manipulative therapy is an independent risk factor for vertebral artery dissection, *Neurology* 13 (2003): 1424–1428; R. Dziewas and coauthors, Cervical artery dissection—Clinical features, risk factors, therapy and outcome in 126 patients, *Journal of Neurology* 250 (2003): 1179–1184.

11. Barnes, Bloom, and Nahin, 2008.

Specialized Nutrition Support: Enteral and Parenteral Nutrition

Patients are often too sick to consume a diet that provides the energy and nutrients they need. Moreover, some illnesses may interfere with eating, digestion, or absorption to such a degree that conventional foods cannot supply the necessary nutrients. In such cases, **nutrition support**—the delivery of nutrients using a feeding tube or intravenous infusions—can meet a patient's nutritional needs. **Enteral nutrition** provides nutrients using the gastrointestinal (GI) tract. Enteral nutrition includes oral diets or supplements but most often refers to the use of tube feedings, which supply nutrients directly to the stomach or intestine via a thin, flexible tube. **Parenteral nutrition** provides nutrients intravenously to patients who do not have adequate gastrointestinal function to handle enteral feedings. If the GI tract remains functional, enteral nutrition support is preferred, partly to avoid the expense and complications associated with intravenous infusions and partly to preserve healthy GI function. Figure 16-1 summarizes the decision-making process for selecting the most appropriate feeding method.

Enteral Nutrition Support

If gastrointestinal function is normal and the primary nutrition problem is a poor appetite, patients can drink enteral formulas to supplement their usual diet. If patients cannot consume enough food or drink enough formula to meet nutrient needs, tube feedings can deliver the required nutrients.

Enteral Formulas

Most enteral formulas can supply all of an individual's nutrient requirements when consumed in sufficient volume, a necessity for the patient using a tube feeding or oral liquid diet for more than a few days. Formulas can be used alone or provided along with other foods.

FIGURE 16-1
Selecting a Feeding Route

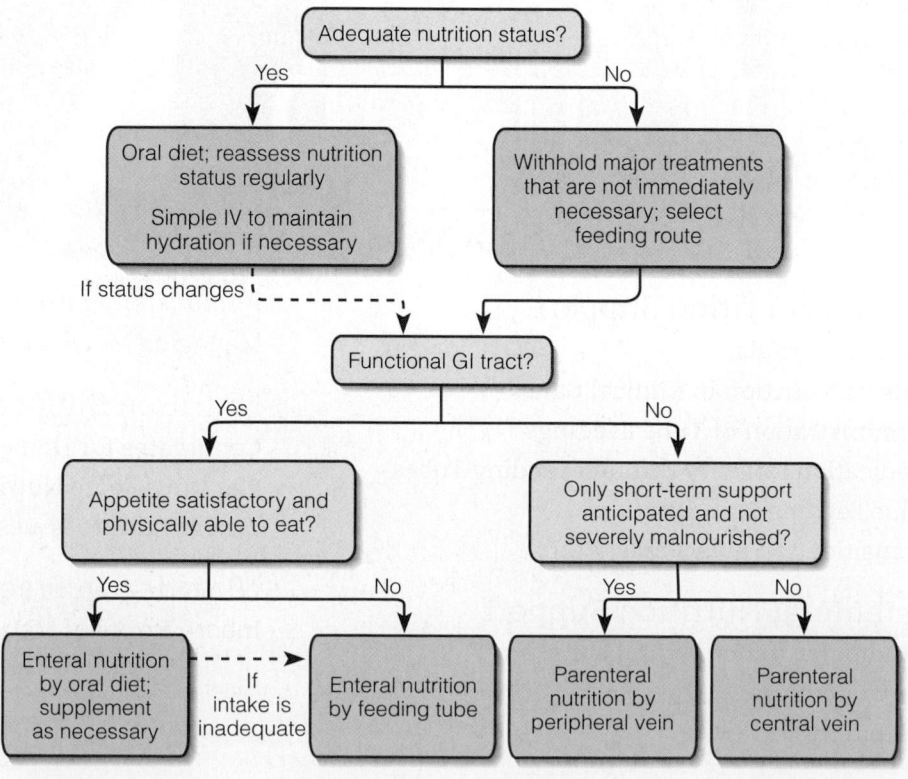

Types of Enteral Formulas More than 100 enteral formulas are currently marketed;[1] Appendix G lists examples. The main types of formulas include the following:

- **Standard formulas,** also called *polymeric formulas*, are provided to individuals who can digest and absorb nutrients without difficulty. They contain intact proteins extracted from milk or soybeans (called **protein isolates**) or a combination of such proteins. The carbohydrate sources include modified starches, glucose polymers (such as maltodextrin), and sugars. A few formulas, called **blenderized formulas,** are made from whole foods and derive their protein primarily from pureed meat or poultry.

- **Elemental formulas,** also called *hydrolyzed*, *chemically defined*, or *monomeric formulas*, are prescribed for patients who have compromised digestive or absorptive functions. Elemental formulas contain proteins and carbohydrates that have been broken down into fragments that require little (if any) digestion. The formulas are often low in fat or may provide fat from **medium-chain triglycerides (MCT)** to ease digestion and absorption.

- **Specialized formulas,** also called *disease-specific formulas*, can meet the specific nutrient needs of patients with particular illnesses. Products are available for individuals with liver, kidney, and lung diseases; glucose intolerance; and metabolic stress (later chapters provide details). Specialized formulas are generally expensive, and their effectiveness is controversial.[2]

- **Modular formulas,** created from individual macronutrient preparations called *modules*, are prepared for patients who require specific nutrient combinations. Vitamin and mineral mixtures are also included in these formulas so that they can meet all of a person's nutrient needs. In some cases, one or more modules may be added to other enteral formulas to adjust their nutrient composition.

Macronutrient Composition The amounts of protein, carbohydrate, and fat in enteral formulas vary substantially (see Appendix G for details). Protein content ranges from 12 to 20 percent of total kcalories;[3] note that protein needs are high in patients with severe metabolic stress, whereas protein restrictions are necessary for patients with chronic kidney disease. Carbohydrate and fat provide most of the energy in enteral formulas; standard formulas generally provide 40 to 60 percent of kcalories from carbohydrate and 30 to 40 percent of kcalories from fat.

Energy Density The energy density of enteral formulas ranges from 0.5 to 2.0 kcalories per milliliter of fluid. Standard formulas provide 1.0 to 1.2 kcalories per milliliter; formulas with higher energy densities can meet energy and nutrient needs in a smaller volume of fluid and therefore benefit patients with high nutrient needs or fluid restrictions.

Fiber Content Fiber-containing formulas can be helpful for normalizing intestinal function, treating diarrhea or constipation, and maintaining blood glucose control. Conversely, fiber-containing formulas may be avoided in patients with acute intestinal conditions or pancreatitis, and before or after some intestinal examinations and surgeries.

Osmolality Osmolality refers to the moles of osmotically active solutes (or *osmoles*) per kilogram of solvent. An enteral formula with an osmolality similar to that of blood serum (about 300 milliosmoles per kilogram) is an **isotonic formula,** whereas a **hypertonic formula** has an osmolality greater than that of blood serum.

Most enteral formulas have osmolalities between 300 and 700 milliosmoles per kilogram; generally, hydrolyzed formulas and nutrient-dense formulas have higher

---

Reminder: The *macronutrients* are proteins, carbohydrates, and fats.

Osmotically active solutes affect *osmosis*, the movement of water across semipermeable membranes.

nutrition support: the delivery of nutrients using a feeding tube or intravenous infusions.

enteral (EN-ter-al) **nutrition:** the provision of nutrients using the GI tract, including the use of tube feedings and oral diets.

parenteral (par-EN-ter-al) **nutrition:** the intravenous provision of nutrients that bypasses the GI tract.
par = beside
entero = intestine

standard formulas: enteral formulas that contain mostly intact proteins and polysaccharides; also called *polymeric formulas.*

protein isolates: proteins that have been isolated from foods.

blenderized formulas: enteral formulas that are prepared by using a food blender to mix and puree whole foods.

elemental formulas: enteral formulas that contain carbohydrates and proteins that are partially or fully hydrolyzed; also called *hydrolyzed, chemically defined,* or *monomeric formulas.*

medium-chain triglycerides (MCT): triglycerides that contain fatty acids that are 8 to 10 carbons in length. MCTs do not require digestion and can be absorbed in the absence of lipase or bile.

specialized formulas: enteral formulas designed to meet the nutrient needs of patients with specific illnesses; also called *disease-specific formulas.*

modular formulas: enteral formulas prepared in the hospital from *modules* that contain single macronutrients; used for people with unique nutrient needs.

osmolality (OZ-moe-LAL-ih-tee): the concentration of osmotically active solutes in a solution, expressed as milliosmoles (mOsm) per kilogram of solvent.

isotonic formula: a formula with an osmolality similar to that of blood serum (about 300 milliosmoles per kilogram).
iso = equal
tono = pressure

hypertonic formula: a formula with an osmolality greater than that of blood serum.

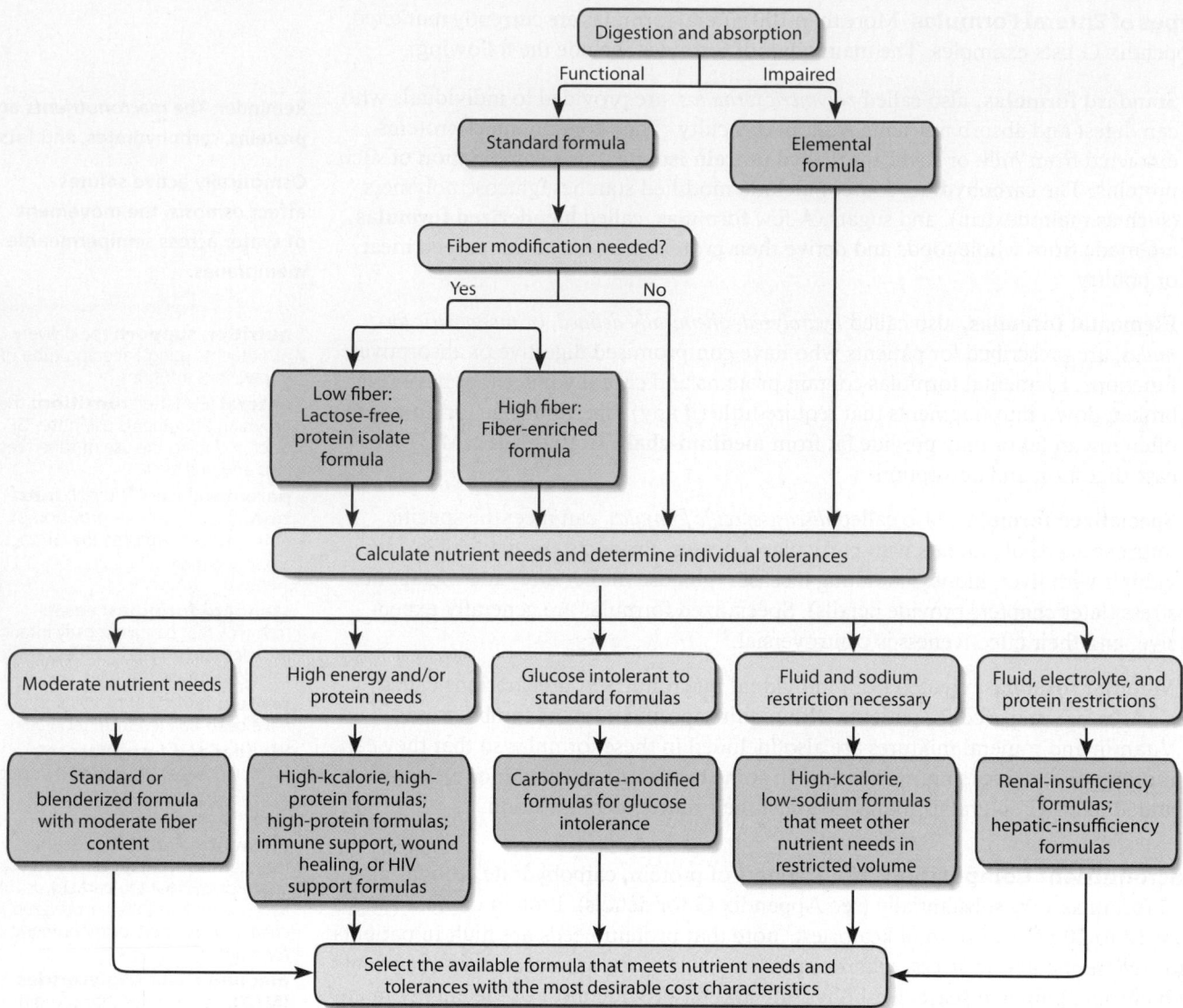

FIGURE 16-2 **Selecting a Formula**

osmolalities than standard formulas. Most people are able to tolerate both isotonic and hypertonic feedings without difficulty. When medications are infused along with enteral feedings, however, the osmotic load increases substantially and may contribute to the diarrhea experienced by many tube-fed patients.

Formula Selection Some of the factors considered when choosing a formula are shown in Figure 16-2. Generally, the best formula is one that meets the patient's medical and nutrient needs with the lowest risk of complications and the lowest cost. The vast majority of patients can use standard formulas. A person with a functional, but impaired, GI tract may require an elemental formula. Factors that influence formula selection include:

▌ *Nutrient and energy needs.* As with patients consuming regular diets, adjustments in nutrient and energy intakes are sometimes necessary for tube-fed patients. If fluids are restricted, the formula should have adequate nutrient and energy densities to deliver the required nutrients in the volume prescribed.

▌ *Fiber modifications.* The choice of formulas is narrower if fiber intake needs to be low or high.

Individual tolerances (food allergies and sensitivities). Most formulas are lactose-free because many patients using enteral formulas have some degree of lactose intolerance. Many formulas are also gluten-free and can accommodate the needs of individuals with gluten sensitivity.

Health care facilities stock a limited number of formulas, so formula selection is influenced by availability. The medical staff may initially choose a formula based on the criteria previously mentioned and then reevaluate the decision according to the patient's response to the formula.

Enteral Nutrition in Clinical Care

A patient with a functional GI tract who cannot meet nutrient needs with a regular diet may be a candidate for enteral nutrition support. Enteral feedings are preferred over parenteral nutrition because they help to stimulate or maintain gut function, cause fewer complications, and are less costly.[4] Similarly, oral feedings are preferred to tube feedings when the person is able to drink enteral formulas, to avoid the stress, complications, and expense associated with tube feedings.

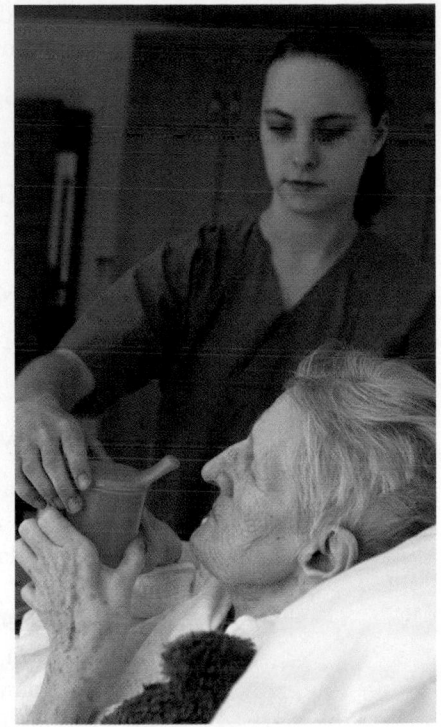

Patients can drink enteral formulas when they are unable to consume enough food from a conventional diet.

Oral Use of Enteral Formulas Some patients drink enteral formulas to supplement their diets when they are unable to consume enough energy or nutrients to meet their needs. In addition, individuals who are weak or debilitated may find formulas easier to manage than meals. Some enteral products are sold in pharmacies and grocery stores for home use; examples include Ensure, Boost, and Carnation Instant Breakfast.

Taste is an important consideration when a patient drinks a formula, so allowing patients to select the products they prefer helps to promote acceptance. The "How to" offers additional suggestions for helping patients accept and enjoy oral formulas.

Candidates for Tube Feedings Tube feedings are typically recommended for patients at risk of protein-energy malnutrition who are unable to consume adequate

How To   HELP PATIENTS ACCEPT ORAL FORMULAS

People using enteral formulas are often quite ill and have poor appetites. Even when a person enjoys a formula, the taste can become monotonous in time. Elemental formulas are usually less palatable than standard formulas, and patients may find them difficult to drink. Health practitioners can motivate patients to drink formulas by trying these suggestions:

∎ Let the patient sample different formulas that are appropriate for his or her needs, and use only those that the patient enjoys.

∎ Serve formulas attractively, and remind patients to drink them. Formulas offered in a glass on an attractive plate may be more appealing than those served from a can with an unfamiliar name.

∎ If a patient finds the smell of a formula unappealing, it may help to cover the top of the glass with plastic wrap or a lid, leaving just enough room for a straw.

∎ Provide easy access. Keep the formula close to the patient's bed where it can be reached with little effort and within sight so that the patient is reminded to drink it. Patients who are very ill may lack the motivation to reach for the formula, let alone drink it.

∎ Try keeping the formula in an ice bath so that it will be cool and refreshing when the patient drinks it. Check with the patient to make sure the colder temperature is suitable.

∎ For patients with little appetite, offer the formula in small amounts that are easy to tolerate, and serve it more frequently during the day.

∎ If the patient stops enjoying the formula, recommend different flavors or try other formulas. Check with the pharmacy to see if alternative enteral products are available, such as milkshakes, puddings, or snack bars.

food or formula for at least seven days.[5] The following medical conditions or treatments may indicate the need for tube feedings:

- Severe swallowing disorders

- Impaired motility in the upper GI tract

- Gastrointestinal obstructions and **fistulas** that can be bypassed with a feeding tube

- Certain types of intestinal surgeries

- Mechanical ventilation

- Extremely high nutrient requirements

- Little or no appetite for extended periods, especially if the patient is malnourished

- Mental incapacitation due to confusion, neurological disorders, or coma

Contraindications for tube feedings include severe GI bleeding, high-output fistulas, **intractable** vomiting or diarrhea, complete intestinal obstruction, and severe malabsorption.[6]

Feeding Routes The feeding route chosen depends on the patient's medical condition, expected duration of tube feeding, and potential complications of a particular route. Figure 16-3 illustrates the main feeding routes, and the Glossary of Tube Feeding Routes describes each route.

For short-term tube feedings (less than four weeks), the feeding tube is routed into the GI tract via the nose (**nasogastric** or **nasoenteric** routes). The patient is frequently awake during **transnasal** (through-the-nose) placement of a feeding tube. While the

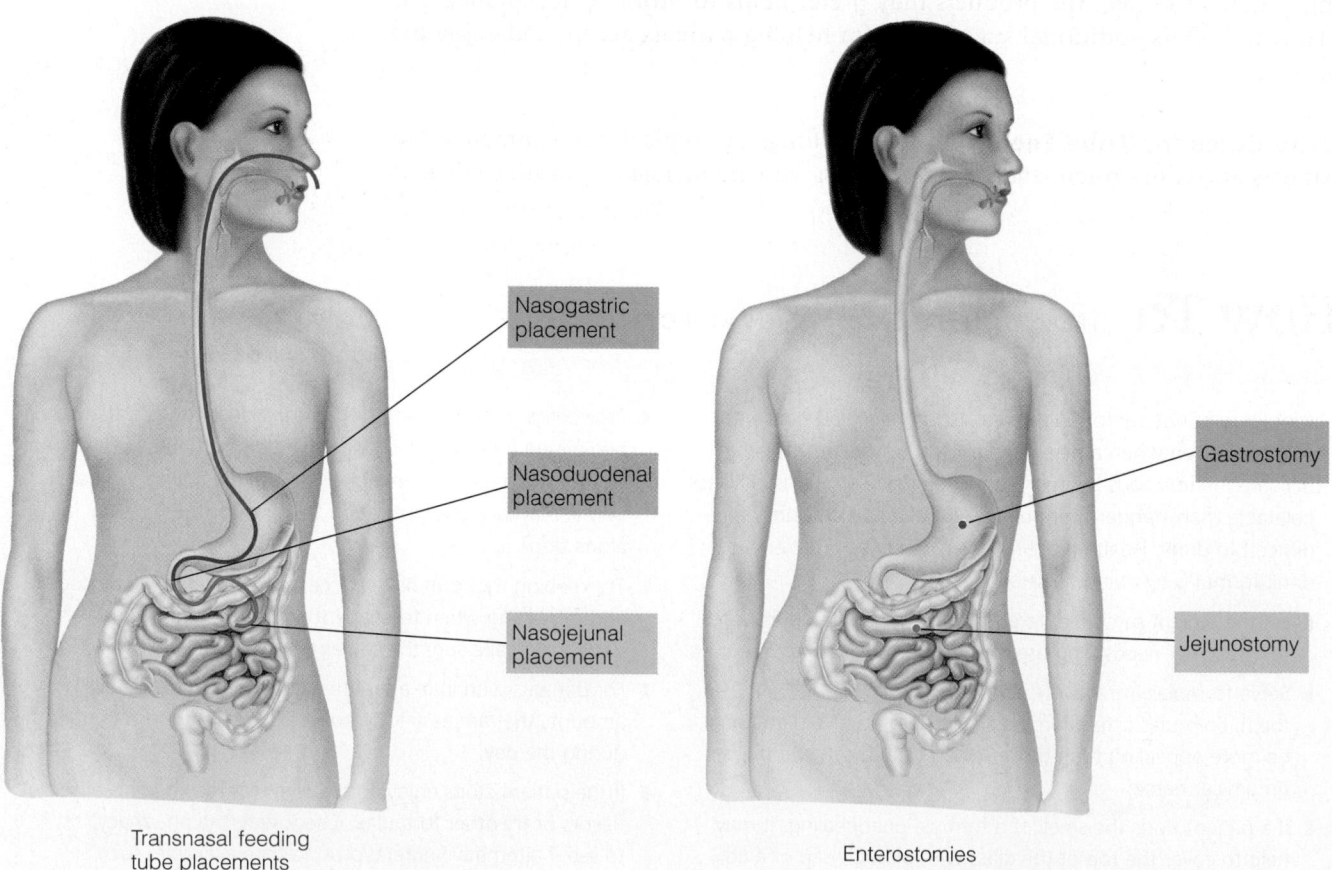

Nasogastric placement

Nasoduodenal placement

Nasojejunal placement

Gastrostomy

Jejunostomy

Transnasal feeding tube placements

Enterostomies

FIGURE 16-3 Tube Feeding Routes

For each type of tube placement, the terms are listed in order from the upper to lower organs of the digestive system.

transnasal: a *transnasal feeding tube* is one that is inserted through the nose.

- ▌ **nasogastric (NG):** tube is placed into the stomach via the nose.
- ▌ **nasoenteric:** tube is placed into the GI tract via the nose. (*Nasoenteric feedings* usually refer to *nasoduodenal* and *nasojejunal* feedings.)
- ▌ **nasoduodenal (ND):** tube is placed into the duodenum via the nose.
- ▌ **nasojejunal (NJ):** tube is placed into the jejunum via the nose.

orogastric: tube is inserted into the stomach through the mouth. This method is often used to feed infants because a nasogastric tube may hinder the infant's breathing.

enterostomy (EN-ter-AH-stoe-mee): an opening into the GI tract through the abdominal wall.

- ▌ **gastrostomy** (gah-STRAH-stoe-mee): an opening into the stomach through which a feeding tube can be passed. A nonsurgical technique for creating a gastrostomy under local anesthesia is called *percutaneous endoscopic gastrostomy (PEG)*.
- ▌ **jejunostomy** (JEH-ju-NAH-stoe-mee): an opening into the jejunum through which a feeding tube can be passed. A nonsurgical technique for creating a jejunostomy is called *percutaneous endoscopic jejunostomy (PEJ)*. The tube can either be guided into the jejunum via a gastrostomy or passed directly into the jejunum (*direct PEJ*).

patient is in a slightly upright position with head tilted, the tube is inserted into a nostril and passed into the stomach (nasogastric route), duodenum (**nasoduodenal** route), or jejunum (**nasojejunal** route). The final position of the feeding tube tip is verified by abdominal X-ray or other means. In infants, **orogastric** placement, in which the feeding tube is passed into the stomach via the mouth, is sometimes preferred over transnasal routes; this placement allows the infant to breathe more normally during feedings.

When a patient will be tube fed for longer than four weeks or if the nasoenteric route is inaccessible due to an obstruction or other medical reasons, a direct route to the stomach or intestine may be created by passing the tube through an **enterostomy,** an opening in the abdominal wall that leads to the stomach (**gastrostomy**) or jejunum (**jejunostomy**). An enterostomy can be made by either surgical incision or needle puncture.

Gastric feedings (nasogastric and gastrostomy routes) are preferred whenever possible. These feedings are more easily tolerated and less complicated to deliver than intestinal feedings because the stomach controls the rate at which nutrients enter the intestine. Gastric feedings are not possible, however, if patients have gastric obstructions or motility disorders that interfere with the stomach's ability to empty. Gastric

fistulas (FIST-you-luz): abnormal passages between organs or tissues (or between an internal organ and the body's surface) that permit the passage of fluids or secretions.

intractable: not easily managed or controlled.

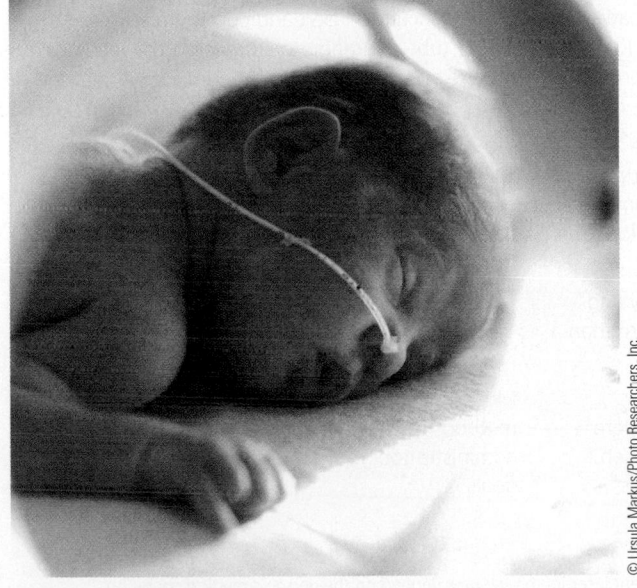

A transnasal feeding tube accesses the GI tract via the nose.

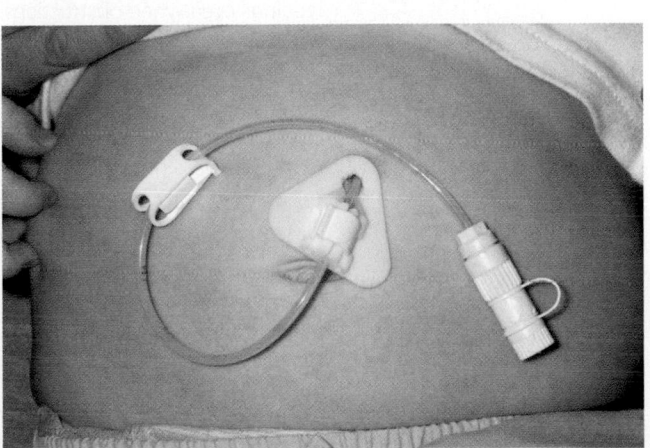

In a gastrostomy, the feeding tube accesses the GI tract through the abdominal wall.

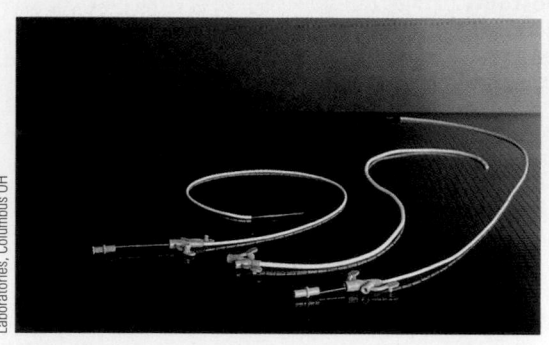

© Flexiflo® Over-The-Counter Nasojejunal Feeding Tube, Courtesy of Ross Products Div., Abbott Laboratories, Columbus OH

The thin wires protruding from the ends of these feeding tubes are stylets, which stiffen the tubes to ease insertion and are discarded thereafter. The orange Y-connectors provide ports for administering water or medications without disrupting the feeding.

feedings are also avoided in patients at high risk of **aspiration,** a complication in which formula or GI secretions enter the lungs, possibly from the backflow of stomach contents (note that studies have not consistently shown that gastric feedings increase aspiration risk[7]). Table 16-1 summarizes the advantages and disadvantages of the various tube feeding routes.

Feeding Tubes Feeding tubes are made from soft, flexible materials (such as silicone or polyurethane) and come in a variety of lengths and diameters. The tube selected is often the smallest-diameter tube through which the formula will flow without clogging.

The outer diameter of a feeding tube is measured in **French units,** in which each unit equals 1/3 millimeter; thus, a "12 French" feeding tube has a 4-millimeter diameter. The inner diameter depends on the thickness of the tubing material. Double-lumen tubes are also available; these allow a single tube to be used for both intestinal feedings and **gastric decompression,** a procedure in which the stomach contents of patients with motility disorders are removed by suction.

Administration of Tube Feedings

After the feeding route and formula have been selected, the formula must be safely delivered. The methods of tube feeding administration vary somewhat from one health care facility to the next. The procedures presented in the following sections are suggested guidelines.

TABLE 16-1 Comparison of Tube Feeding Routes[a]

Insertion Method or Feeding Site	Advantages	Disadvantages
Transnasal	Does not require surgery or incisions for placement; tubes can be placed by a nurse or trained dietitian.	Easy to remove by disoriented patients; long-term use may irritate the nasal passages, throat, and esophagus.
Nasogastric	Easiest to insert and confirm placement; least expensive method; feedings can often be given intermittently and without an infusion pump.	Highest risk of aspiration in compromised patients;[b] risk of tube migration to small intestine.
Nasoduodenal and nasojejunal	Lower risk of aspiration in compromised patients;[b] allows for earlier tube feedings than gastric feedings during severe stress; may allow enteral feedings even when obstructions, fistulas, or other medical conditions prevent gastric feedings.	More difficult to insert and confirm placement; risk of tube migration to stomach; tube feedings require an infusion pump for administration; may take longer to reach nutrition goals.
Tube enterostomies	Allow the lower esophageal sphincter to remain closed, reducing the risk of aspiration;[b] more comfortable than transnasal insertion for long-term use; site is not visible under clothing.	Tubes must be placed by physician or surgeon; general anesthesia may be required for surgically placed tubes; risk of complications from the insertion procedure; risk of infection at insertion site.
Gastrostomy	Feedings can often be given intermittently and without a pump; easier insertion procedure than a jejunostomy.	Moderate risk of aspiration in high-risk patients.[b]
Jejunostomy	Lowest risk of aspiration;[b] allows for earlier tube feedings than gastrostomy during severe stress; may allow enteral feedings even when obstructions, fistulas, or medical conditions prevent gastric feedings.	Most difficult insertion procedure; most costly method; feedings require an infusion pump for administration; may take longer to reach nutrition goals.

[a]Relative to other tube feeding routes. The actual advantages and disadvantages of different insertion procedures depend on the person's medical condition.
[b]The risk of aspiration associated with the different feeding routes is controversial and still under investigation.

Safe Handling Individuals who are ill or malnourished often have suppressed immune systems, making them vulnerable to infection from foodborne illness. Thus, the personnel involved with preparing or delivering formula—usually individuals in the foodservice department or pharmacy—must follow specific protocols at their facility that prevent formula contamination.

Formulas are available in open feeding systems and closed feeding systems. With an **open feeding system,** the formula needs to be transferred from its original packaging to a feeding container. Examples include formulas that are packaged in cans or bottles, concentrates that need to be diluted, and powders that require reconstitution. In a **closed feeding system,** the formula is prepackaged in a container that can be connected directly to a feeding tube. Closed systems are less likely to become contaminated, require less nursing time, and can hang for longer periods of time than open systems. Although closed systems cost more initially, they may be less expensive in the long run because they prevent bacterial contamination and thus avoid the costs of treating infections.

Safety Guidelines After the formula reaches the nursing station, the nursing staff assumes responsibility for its safe handling. Hands should be carefully washed before handling formulas and feeding containers. The following steps can reduce the risk of formula contamination when using open feeding systems:

- Before opening a can of formula, clean the lid with a disposable alcohol wipe and wash the can opener with detergent and hot water. (Check HACCP protocols for details.) If you do not use the entire can at one feeding, label the can with the date and time it was opened.

- Store opened cans or mixed formulas in clean, closed containers. Refrigerate unused portions promptly.

- Discard unlabeled or improperly labeled containers and all opened containers of formula that are not used within 24 hours.

- Hang no more than an 8-hour supply of formula when using an open feeding system. Discard any formula that remains, rinse out the feeding bag and tubing, and add fresh formula to the feeding bag. Use a new feeding container and tubing (except for the feeding tube itself) every 24 hours.[8]

Reminder: Health care facilities have protocols for handling food products and formulas based on the potential hazards and critical control points in food preparation, referred to as *Hazard Analysis and Critical Control Points (HACCP)*.

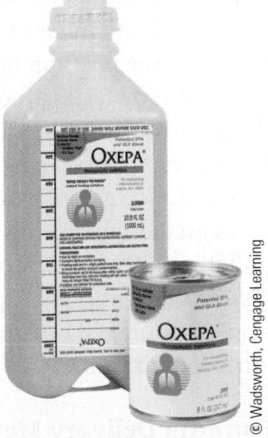

In an open feeding system, the formula is transferred from its original packaging to a feeding container.

In a closed feeding system, the formula is prepackaged in a container that can be attached directly to a feeding tube, such as the bottle shown on the left. The formula in the can at right can be used in an open feeding system.

aspiration: drawing in by suction or breathing; a common complication of enteral feedings in which foreign material enters the lungs, often from GI secretions or the reflux of stomach contents.

French units: units of measure used to indicate a feeding tube's outer diameter; 1 French unit equals one-third of a millimeter.

gastric decompression: the removal of the stomach contents (including swallowed saliva, stomach secretions, and gas) of patients with motility disorders or obstructions that prevent stomach emptying.

open feeding system: a delivery system that requires the formula to be transferred from its original packaging to a feeding container before being administered through the feeding tube.

closed feeding system: a delivery system in which the formula comes prepackaged in a container that can be attached directly to the feeding tube for administration.

How To HELP PATIENTS COPE WITH TUBE FEEDINGS

Patients may be less apprehensive about tube feedings once they understand the insertion procedure, the expected duration of the tube feeding, and the strategic role that nutrition plays in recovery from disease. The pointers that follow can help health practitioners prepare patients for transnasal tube feedings:

■ Allow the patient to see and touch the feeding tube. Understanding that the tube is soft and narrow (only about half the diameter of a pencil) often alleviates anxiety.

■ Show the patient how the feeding equipment is attached to the feeding tube, and explain how the feeding will work. For young children, use dolls or stuffed toys to demonstrate tube insertion and feeding procedures.

■ Explain that the patient remains fully alert during the procedure and helps pass the tube by swallowing. A numbing solution sprayed on the back of the throat minimizes discomfort and prevents gagging during the procedure.

■ Inform the patient that after the tube has been inserted, most people become accustomed to its presence within a few hours. In most cases, the patient can continue to swallow foods and beverages with the tube in place.

Tube feedings may cause some patients to feel that they have lost control over an important aspect of their lives. They may also feel self-conscious about how the feeding tube looks or feel awkward when moving around with the equipment. A few measures can help:

■ Involve patients in the decision-making and care process whenever possible. Patients can help to arrange their daily feeding schedules and can perform some of the feeding procedures themselves.

■ Show patients how to manipulate the feeding equipment so that they can get out of bed and move around.

For closed feeding systems, the hang time should be no longer than 24 to 48 hours. Contamination is more likely with the longer time periods.

Preparing for Tube Feedings Before starting a tube feeding, health practitioners can ease fears by fully discussing the procedure with the patient and family members. The discussion should address the reasons why tube feeding is appropriate as well as the benefits and risks of the procedure. The "How to" offers suggestions that may help ease the concerns of patients who may benefit from tube feeding.

Serious complications can develop if a transnasal tube is accidentally inserted into the respiratory tract or if formula or GI secretions are aspirated into the lungs. To check for incorrect tube placement, clinicians use X-rays to verify the position of the feeding tube before a feeding is initiated. After the tube's placement has been confirmed, the nurse secures the tube to the patient's nose and cheek with tape and monitors the position of the tubing throughout the day. Tube placement can also be monitored by testing the pH of a fluid sample drawn into the feeding tube; recall that the pH of stomach fluid is much lower than the pH of fluid obtained from the intestine or respiratory tract.

> **A fasting gastric sample usually has a pH of 5 or lower. A sample from the intestine or respiratory tract has a pH of about 7 or higher.**

To reduce the risk of aspiration, the patient's upper body is elevated to a 30- to 45-degree angle during the feeding and for 30 minutes after the feeding whenever possible. The addition of blue food coloring to formula was formerly suggested as a means of identifying aspirated formula in lung secretions; however, this practice is now discouraged due to its lack of reliability and because several deaths have been attributed to its use.[9]

Formula Delivery Methods Formulas may be delivered continuously during the day **(continuous feedings)** or in relatively large amounts several times per day **(intermittent feedings).** A patient may also start with continuous feedings and gradually transition to intermittent feedings. Each method has specific uses, advantages, and disadvantages.

Continuous feedings are delivered slowly and at a constant rate over a period of 8 to 24 hours. Continuous feedings are used to deliver intestinal feedings, and are generally recommended for critically ill patients because the slower delivery rate is

How To PLAN A TUBE FEEDING SCHEDULE

After selecting a suitable formula, the clinician must determine the volume of formula that meets the patient's nutritional needs. Consider a patient who needs 2000 kcalories daily and is using a standard formula that provides 1.0 kcalorie per milliliter. The total volume of formula required would be 2000 milliliters per day:

$$x \text{ mL} \times 1.0 \text{ kcal/mL} = 2000 \text{ kcal}$$

$$x \text{ mL} = \frac{2000 \text{ kcal}}{1.0 \text{ kcal/mL}} = 2000 \text{ mL}$$

If the patient is to receive intermittent feedings six times a day, he will need about 333 milliliters of formula at each feeding:

$$2000 \text{ mL} \div 6 \text{ feedings} = 333 \text{ mL/feeding}$$

Alternatively, if he is to receive intermittent feedings eight times a day, he will need 250 milliliters (or about one can of ready-to-feed formula) at each feeding:

$$2000 \text{ mL} \div 8 \text{ feedings} = 250 \text{ mL/feeding}$$

If the patient is to receive the formula continuously over 24 hours, he will need about 83 milliliters of formula each hour:

$$2000 \text{ mL} \div 24 \text{ hours} = 83 \text{ mL/hr}$$

easier to tolerate. An infusion pump is required to ensure accurate and steady flow rates; consequently, the feedings can limit the patient's freedom of movement and are also more costly.

Intermittent feedings are best tolerated when they are delivered into the stomach (not the intestine). Generally, a total of about 250 to 400 milliliters of formula is delivered over 30 to 45 minutes using a gravity drip method or an infusion pump. The exact amount is determined by dividing the required volume of formula into several daily feedings, as shown in the accompanying "How to." Due to the relatively high volume of formula delivered at one time, intermittent feedings may be difficult for some patients to tolerate, and the risk of aspiration may be higher than with continuous feedings. An advantage of intermittent feedings is that they are similar to the usual pattern of eating and allow the patient freedom of movement between meals.

Rapid delivery of a large volume of formula into the stomach (250 to 500 milliliters in less than 20 minutes) is called a **bolus feeding.** This type of feeding may be given every 3 to 4 hours using a syringe. Bolus feedings are convenient for patients and staff because they are rapidly administered, do not require an infusion pump, and allow greater independence for patients. However, bolus feedings can cause abdominal discomfort, nausea, and cramping in some patients, and the risk of aspiration is greater

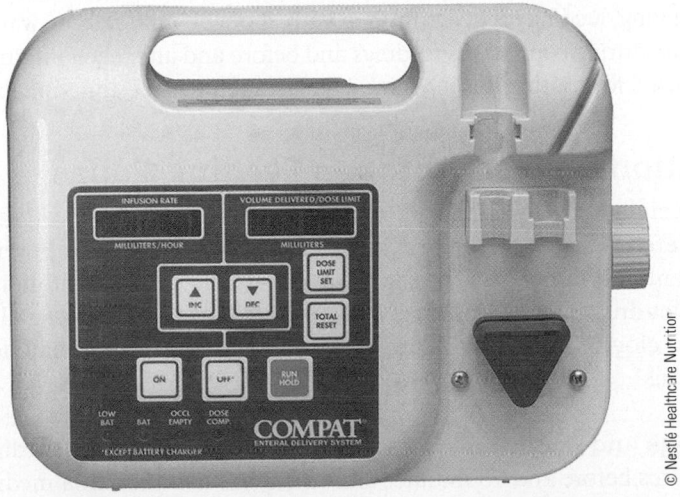

© Nestlé Healthcare Nutrition

The delivery of intermittent and continuous feedings can be controlled with an infusion pump.

continuous feedings: slow delivery of formula at a constant rate over an 8- to 24-hour period.

intermittent feedings: delivery of about 250 to 400 milliliters of formula over 20 to 40 minutes.

bolus (BOH-lus) **feeding:** delivery of about 250 to 500 milliliters of formula in less than 20 minutes.

than with other methods of feeding. For these reasons, bolus feedings are used only in patients who are not critically ill.

Initiating and Advancing Tube Feedings Formula administration techniques vary widely among institutions, so protocols should be reviewed carefully before working with patients. In addition, patient tolerance must be considered when adjusting formula delivery rates. Keep in mind that few studies have evaluated the various methods for initiating and advancing enteral feedings.

Formulas are typically provided full-strength, although they may occasionally be diluted if the patient's fluid requirements are high and water needs cannot be met by other means.[10] Continuous feedings may start at about 10 to 40 milliliters per hour and be raised by 10 to 20 milliliters per hour every 8 to 12 hours as tolerated. Intermittent feedings may start with 60 to 120 milliliters at the initial feeding and be increased by 60 to 120 milliliters at each feeding until the goal volume is reached.[11] Concentrated formulas are often started at the slower rates. If the patient cannot tolerate an increased rate of delivery, the feeding rate is slowed until the person adapts. In some patients, formula delivery can be started at the goal rate immediately, which significantly improves the patient's overall caloric intake.[12]

Checking the Gastric Residual Volume In patients who are given gastric feedings, the nurse regularly measures the **gastric residual volume** (the volume of formula remaining in the stomach after feeding) to ensure that the stomach is emptying properly. The gastric residual volume is measured by gently withdrawing the gastric contents through the feeding tube using a syringe, usually before each intermittent feeding and every 4 to 6 hours during continuous feedings.[13] Although opinions vary, some experts recommend that an evaluation be conducted if the gastric residual volume exceeds 200 milliliters and that feedings be withheld if it exceeds 500 milliliters.[14] If the tendency to accumulate fluids persists, the physician may recommend intestinal feedings or begin drug therapy to improve gastric emptying.

Meeting Water Needs Although water needs vary, many adults require about 2000 milliliters (about 2 quarts) of water daily. Fluids may be restricted in persons with kidney, liver, or heart disease. Additional water is required in patients with fever, high urine output, diarrhea, excessive sweating, severe vomiting, fistula drainage, high-output ostomies, blood loss, or open wounds.

The water in formulas provides most of a person's water needs; standard formulas contain about 85 percent water, or about 850 milliliters of water per liter of formula. Water can also be provided by flushing water separately through the feeding tube. To prevent clogging, feeding tubes are flushed with about 30 milliliters of warm water about every 4 hours during continuous feedings and before and after each intermittent feeding. The water used for routine flushes should be included when estimating fluid intakes.

Medication Delivery through Feeding Tubes

Patients receiving tube feedings sometimes require one or more medications that need to be delivered through feeding tubes. Because medications can interact with substances in enteral formulas in the same ways that they interact with food substances, potential diet-drug interactions must be considered. Medications can also cause feeding tubes to clog. The "How to" provides some guidelines that may help to prevent complications.

Medications and Continuous Feeding A continuous feeding is ordinarily stopped for 15 minutes before and 15 minutes after the administration of a medication so that the components of enteral formulas do not interfere with the medication's absorption.

To estimate fluid requirements in adults and children:

▌ Adults: allow 30 to 40 mL/kg; 25 to 30 mL/kg in adults ≥65 years old

▌ Children: allow 50 to 60 mL/kg

▌ Infants: allow 100 to 150 mL/kg

The pharmacist is your best resource for learning how and when medications can be administered via feeding tubes, especially when you are dealing with an unfamiliar drug. Check with the pharmacist to learn the following:

- Whether a particular medication is known to be incompatible with formulas.

- The proper timing of medication administration to avoid drug-nutrient interactions.

- For patients using intestinal feedings, whether a medication can be absorbed without exposure to stomach acid.

- Whether a liquid form of a medication is available and, if so, the appropriate dosage of the liquid form.

- If only tablets are available, whether the tablets can be crushed and mixed with water. Enteric-coated and sustained-release medications should not be crushed due to the potential for adverse effects.

In general, it is best to give medications by mouth instead of by tube whenever possible. In some cases, the injectable form of a medication may be the best option. For medications that must be given by feeding tube:

- Do not mix medications with enteral formulas. Do not mix medications together.

- Before administering medications, ensure that the feeding tube is placed correctly, that it is not clogged, and that the gastric residual volume is not excessive.

- Position the patient in a semi-upright position (30 degrees or higher) to prevent aspiration.

- Flush the feeding tube with 30 milliliters of warm water before and after administering a medication. When more than one medication is administered, flush the feeding tube with water between medications.

- Use liquid forms of medications whenever possible. Dilute viscous or hypertonic liquid medications with at least 30 milliliters of water before administering them through the feeding tube.

- If tablets are used, crush tablets to a fine powder and mix with about 30 milliliters of warm water before administering.

Some medications may require a longer formula-free interval; for example, feedings need to be stopped for at least 1 hour before and after administering phenytoin, a medication that controls seizures.[15] In such cases, the formula's delivery rate needs to be increased so that the correct amount of formula can be delivered.

Diarrhea Medications are a major cause of the diarrhea that frequently accompanies tube feedings. Diarrhea is especially associated with the administration of sorbitol-containing medications, laxatives, and some types of antibiotics.[16] The high osmolality of many liquid medications can also cause diarrhea, so dilution of hypertonic medications may be helpful.

Tube Feeding Complications

Complications of tube feedings include gastrointestinal problems, such as constipation and diarrhea; mechanical problems related to the tube feeding process; and metabolic problems, such as biochemical alterations and nutrient deficiencies. Examples of the most common complications, along with some preventive and corrective measures, are summarized in Table 16-2 on p. 442.

Transition to Table Foods

After the patient's condition improves, the volume of formula can be tapered off as the patient gradually shifts to an oral diet. Individuals receiving continuous feedings may be switched to intermittent feedings initially. Patients using elemental formulas may begin the transition by using a standard formula, either orally or via tube feeding. Oral intake should supply about two-thirds of estimated nutrient needs before the tube feeding is discontinued completely.[17] The Case Study allows you to consider the many factors involved in tube feedings.

gastric residual volume: the volume of formula remaining in the stomach from a previous feeding.

TABLE 16-2 Causes and Management of Tube Feeding Complications

Complications	Possible Causes	Preventive/Corrective Measures
Aspiration of formula	Inappropriate tube placement	Ensure correct placement of feeding tube.
	Delayed gastric emptying	Elevate head of bed during and after feeding; decrease formula delivery rate if gastric residual volume is excessive; consider using intestinal feedings in high-risk patients.
	Excessive sedation	Minimize use of medications that cause sedation.
Clogged feeding tube	Excessive formula viscosity	Ensure that tube size is appropriate; flush tubing with water before and after giving formula. Remedies to unclog feeding tubes include flushes with warm water or solutions that contain pancreatic enzymes and sodium bicarbonate; consult pharmacist for more options.
	Improper administration of medications	Use oral, liquid, or injectable medications whenever possible; flush tubing with water before and after a medication is given; avoid mixing medications with formula; dilute thick or sticky liquid medications before administering; crush tablets to a fine powder and mix with water (except enteric-coated or sustained-release medications).
Constipation	Inadequate dietary fiber	Use a formula with appropriate fiber content.
	Dehydration	Provide additional fluids.
	Lack of exercise	Encourage walking and other activities, if appropriate.
	Medication side effect	Consult physician about minimizing or replacing medications that cause constipation.
Diarrhea	Medication intolerance	Dilute hypertonic medications before administering; avoid using poorly tolerated medications.
	Infection in GI tract	Consult physician about specific diagnosis and appropriate treatment.
	Formula contamination	Review safety guidelines for formula preparation and delivery.
	Excessively rapid formula administration	Decrease formula delivery rate or use continuous feedings.
	Lactose or gluten intolerance	Use lactose-free or gluten-free formula in patients with intolerances.
Fluid and electrolyte imbalances	Diarrhea	See items under *Diarrhea*.
	Inappropriate fluid intake or excessive losses	Monitor daily weights, intake and output records, serum electrolyte levels, and clinical signs that indicate dehydration or overhydration; ensure that water intake and formula delivery rates are appropriate.
	Inappropriate insulin, diuretic, or other therapy	Ensure that medication doses are appropriate.
	Inappropriate nutrient intake	Use a formula with appropriate nutrient content; ensure that malnourished patients do not receive excessive nutrients.[a]
Nausea and vomiting, cramps	Delayed stomach emptying	Decrease formula delivery rate or use continuous feedings; halt feeding if gastric residual volume is excessive (>500 mL); evaluate for obstruction; consider use of medications to improve emptying rate.
	Formula intolerance	Ensure that formula is at room temperature, delivery rate is appropriate, and formula odor is not objectionable; consider using formula that is low in fat, low in fiber, or elemental.
	Medication intolerance	Consult physician about replacing medications that are poorly tolerated.
	Response to disease or disease treatment	Consider use of medications that control nausea and vomiting.

[a]An excessive nutrient intake in malnourished patients may cause *refeeding syndrome,* a disorder that can lead to fluid and electrolyte imbalances; see p. 449 for details.

Sharyn Eschler is a 24-year-old student who suffered multiple fractures when she fell from a cliff while hiking. She has been in the hospital for two weeks and has no appetite. Due to her injuries, she is in traction and is immobile, although the head of her bed can be elevated 45 degrees. Sharyn has lost 8 pounds over the course of her hospitalization, and the health care team agrees that nasoduodenal tube feeding should be instituted before her nutrition status deteriorates further. The standard formula selected for the feeding is lactose free, and Sharyn's nutrient requirements can be met with 2200 milliliters of the formula per day.

1. What steps can be taken to prepare Sharyn for tube feeding? What are some general reasons why nasoduodenal placement of the feeding tube might be preferred over nasogastric placement?

2. The physician's orders specify that the feeding should be given continuously over 18 hours. Develop an appropriate tube feeding schedule.

3. Describe precautions that should be taken if Sharyn is to receive medications through the feeding tube.

4. After three days of feeding, Sharyn develops diarrhea. Check Table 16-2 to determine the possible causes. What measures can be taken to correct the diarrhea?

IN SUMMARY

▪ Enteral formulas can supplement conventional diets or be used for tube feedings. They vary in macronutrient composition, energy density, fiber content, and osmolality.

▪ A nasoenteric feeding route is preferred for short-term tube feedings, whereas enterostomies are used for longer-term feedings. Gastric feedings are preferred over intestinal feedings.

▪ The formula can be delivered continuously, intermittently, or in bolus feedings. Medications should be given separately and accompanied by water flushes to prevent tube clogging.

▪ Complications of tube feedings can be gastrointestinal, mechanical, or metabolic in nature.

Parenteral Nutrition Support

The first half of this chapter described how enteral formulas can supplement or replace a conventional diet. Because enteral formulas cannot be used if intestinal function is inadequate, the ability to meet nutrient needs intravenously is a lifesaving option for critically ill persons. The procedure is costly, however, and is associated with a number of potentially dangerous complications. As previous sections suggested, enteral nutrition support is preferred if the GI tract is functional.

Candidates for Parenteral Nutrition

Parenteral nutrition is usually recommended for patients who are unable to use the GI tract and who are either malnourished or likely to become so (review Figure 16-1 on p. 430). The following conditions may require use of parenteral nutrition:

▪ Intestinal obstructions or fistulas

▪ Paralytic ileus (intestinal paralysis)

▪ Short bowel syndrome (part of the small intestine has been removed)

▪ Intractable vomiting or diarrhea

▪ Bone marrow transplants

▪ Severe malnutrition and intolerance to enteral nutrition

Venous Access

The access sites for parenteral nutrition fall into two main categories: the **peripheral veins** located in the arms and legs, and the large-diameter **central veins** located near the heart.

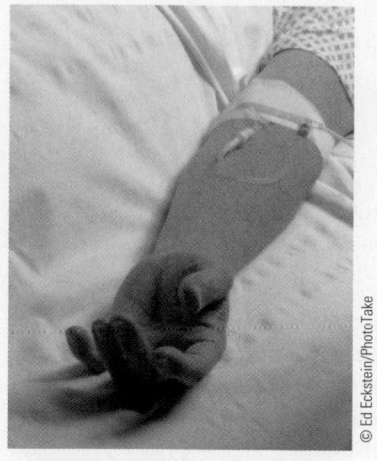

The peripheral veins can provide access to the blood for the delivery of parenteral solutions.

peripheral veins: the small-diameter veins that carry blood from the arms and legs.

central veins: the large-diameter veins located close to the heart.

peripheral parenteral nutrition (PPN): the infusion of nutrient solutions into peripheral veins, usually a vein in the arm or back of the hand.

phlebitis (fleh-BYE-tiss): inflammation of a vein.

osmolarity: the concentration of osmotically active solutes in a solution, expressed as milliosmoles per liter of solution (mOsm/L). *Osmolality* (mOsm/kg) is an alternative term used to describe a solution's osmotic properties.

Peripheral Parenteral Nutrition In **peripheral parenteral nutrition (PPN),** nutrients are delivered using only the peripheral veins. Peripheral veins can be damaged by overly concentrated solutions, however—**phlebitis** may develop, characterized by redness, swelling, and tenderness at the infusion site. To prevent phlebitis, the **osmolarity** of parenteral solutions used for PPN is generally kept below 900 milliosmoles per liter,[18] a concentration that limits the amounts of energy and protein the solution can provide. PPN is used most often in patients who require short-term nutrition support (about 7 to 10 days) and who do not have high nutrient needs or fluid restrictions. The use of PPN is not possible if the peripheral veins are too weak to tolerate the procedure. In many cases, clinicians must rotate venous access sites to avoid damaging veins.

Total Parenteral Nutrition Most patients meet their nutrient needs using the larger central veins, where blood volume is greater and nutrient concentrations do not need to be limited. Because this method can reliably meet a person's nutrient requirements, it is called **total parenteral nutrition (TPN).** Central veins lie close to the heart, where the large volume of blood rapidly dilutes parenteral solutions. Therefore, patients with very high nutrient needs or fluid restrictions are able to receive the nutrient-dense solutions they require. TPN is also preferred for patients who require long-term parenteral nutrition.

To access central veins, the tip of a central venous **catheter** can either be placed directly into a large-diameter central vein or threaded into a central vein through a peripheral vein (see Figure 16-4). Peripheral insertion of central catheters is less invasive and lower in cost than direct insertion into central veins.

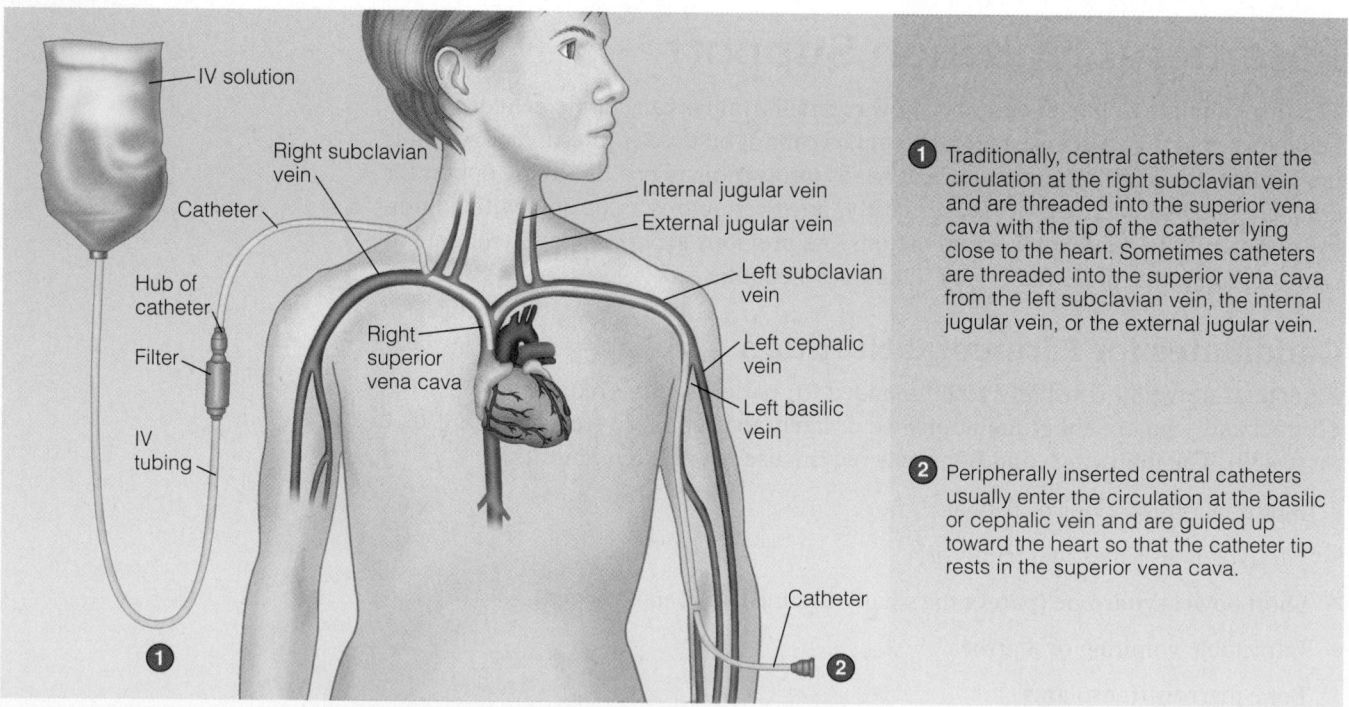

1 Traditionally, central catheters enter the circulation at the right subclavian vein and are threaded into the superior vena cava with the tip of the catheter lying close to the heart. Sometimes catheters are threaded into the superior vena cava from the left subclavian vein, the internal jugular vein, or the external jugular vein.

2 Peripherally inserted central catheters usually enter the circulation at the basilic or cephalic vein and are guided up toward the heart so that the catheter tip rests in the superior vena cava.

FIGURE 16-4 Accessing Central Veins for Total Parenteral Nutrition

Parenteral Solutions

The pharmacies located within health care institutions are often responsible for preparing parenteral solutions. This arrangement is convenient because the pharmacist can customize the solutions to meet patients' needs and because the solutions have a limited shelf life. The physician typically submits an order form such as the one shown in Figure 16-5 to the pharmacy. Prescriptions for parenteral solutions are highly individualized and may need to be recalculated daily until the patient's condition is stable.

> **total parenteral nutrition (TPN):** the infusion of nutrient solutions into a central vein.
>
> **catheter:** a thin tube placed within a narrow lumen (such as a blood vessel) or body cavity; can be used to infuse or withdraw fluids or keep a passage open.

Physician Orders
PARENTERAL NUTRITION (PN) – ADULT

Primary Diagnosis: _____ Ht: _____ cm Dosing Wt: _____ kg

PN Indication: _____ Allergies _____

Instructions: This form must be completed for a new order or continuation of PN and faxed to the Pharmacy by [Insert Time] to receive same day preparation. PN administration begins at [Insert Time]. Contact the Nutrition Support Service at (XXX) XXX-XXXX for additional information.

Administration Route: CVC or PICC   *Note: Proper tip placement of the CVC or PICC must be confirmed prior to PN infusion*

Peripheral IV (PIV) (*Final PN Osmolarity* ≤ _____ *mOsm/L*)

Monitoring: Daily weights, Strict input & output, Bedside glucose monitoring every _____ hours

Na, K, Cl, CO_2, Glucose, BUN, Scr, Mg, PO_4 every _____

T, Bili, Alk Phos, AST, ALT, Albumin, Triglycerides, Calcium every _____

Base Solution: *Select one*	*Parenteral nutrition MUST be administered through a dedicated infusion port and filtered with a 1.2-micron in-line filter at all times. Discard any unused volume after 24 hours.*	
PERIPHERAL 2-in-1 Dextrose _____ g Amino Acids (*Brand* _____) _____ g *For patients with PIV and established glucose tolerance; Provides _____ kcal; Maximum Rate not to exceed _____ mL/hour*	**CENTRAL 2-in-1** Dextrose _____ g Amino Acids (*Brand* _____) _____ g *For patients with CVC or PICC and established glucose tolerance; Provides _____ kcal; Maximum Rate not to exceed _____ mL/hour*	**CENTRAL 3-in-1** Dextrose _____ g Amino Acids (*Brand* _____) _____ g Fat Emulsion (*Brand* _____) _____ g *For patients with CVC or PICC and established glucose/fat emulsion tolerance; Provides _____ kcal; Maximum Rate not to exceed _____ mL/hour* *Use of additional fat emulsion not required with 3-in-1 base solution*
RATE & VOLUME: _____ mL/hour for _____ hours = _____ mL/day ***Must specify***		
or **CYCLIC INFUSION:** _____ mL/hour for _____ hours, then _____ mL/hour for _____ hours = _____ mL/day		

Fat Emulsion (Brand _____) – via PIV or CVC with 2-in-1 base solutions	(*Select caloric density & volume*)	
10% 250 mL 20% 500 mL	Infuse at _____ mL/hour over _____ hours (*Note: infusions < 4 or > 12 hours not recommended*)	Frequency _____ *Discard any unused volume after 12 hours.*

Additives: (*per day*)			**Normal Dosages**	**Additives:** (*per day*)
Sodium Chloride	_____	mEq	1-2 mEq Sodium/kg/day	**Regular Insulin** _____ units
as Acetate	_____	mEq	pH or CO_2 dependent	*Recommend if hyperglycemic, start*
as Phosphate	_____	mmol of PO_4	Consider if hyperkalemic	*with 1 unit for every 10 g of dextrose*
Potassium Chloride	_____	mEq	1-2 mEq Potassium/kg/day	
as Acetate	_____	mEq	pH or CO_2 dependent	
as Phosphate	_____	mmol of PO_4	20-40 mmol/day (1 mmol Phos = 1.5 mEq K)	***Pharmacy Use Only:*** Ca/PO_4
Calcium **Gluconate**	_____	mEq	5-15 mEq/day	**Limit Checked** _____
				(*Note: Some brands of amino acids contain phosphate*)
Magnesium **Sulfate**	_____	mEq	8-24 mEq/day	
Adult **Multivitamins**	_____	mL/day	Contains Vitamin K 150 mcg	
Adult **Trace Elements**	_____	mL/day	Zn ___ mg, Cu ___ mg, Mn ___ mg, Cr ___ mcg, Se ___ mcg (with normal hepatic function)	
H_2 **Antagonist** _____	_____	mg	____ mg/day with normal renal function	
Other:				

Physician's Signature: _____ Pager Number: _____ Date/time: _____

Orders transcribed by: _____ Date/time: _____ Orders verified by: _____ Date/time: _____

SEND COMPLETED ORDERS TO PHARMACY

FIGURE 16-5 Sample Parenteral Nutrition Order Form

Because the nutrients are provided intravenously, they must be given in forms that are safe to inject directly into the bloodstream.

Amino Acids Parenteral solutions contain all of the essential amino acids and various combinations of the nonessential amino acids. Amino acid concentrations range from 3.5 to 15 percent; the more concentrated solutions are used only for TPN. Just as in regular foods, the amino acids provide 4 kcalories per gram. Disease-specific amino acid solutions are available for patients with liver disease, kidney disease, and metabolic stress.

Carbohydrate Glucose is the main source of energy in parenteral solutions. It is provided in the form of dextrose monohydrate, in which each glucose molecule is associated with a single water molecule. Dextrose monohydrate provides 3.4 kcalories per gram, slightly less than pure glucose, which provides 4 kcalories per gram. Commercial dextrose solutions are available in concentrations between 2.5 and 70 percent. Concentrations greater than 12.5 percent are used only in TPN solutions.[19]

A 10 percent dextrose solution provides 10 g of dextrose monohydrate per 100 mL of solution.

Lipids Lipid emulsions supply essential fatty acids and are a significant source of energy. The emulsions usually contain triglycerides from soybean oil and safflower oil, phospholipids to serve as emulsifying agents, and glycerol to make the solutions isotonic. Lipid emulsions are available in 10, 20, and 30 percent solutions, containing 1.1, 2.0, and 3.0 kcalories per milliliter, respectively. Therefore, a 500-milliliter container of 10 percent lipid emulsion would provide 550 kcalories; the same volume of a 20 percent lipid emulsion would provide 1000 kcalories. In the United States, the 30 percent lipid emulsion can be used for preparing mixed parenteral solutions but cannot be directly infused into patients.[20]

500 mL of a 10% lipid emulsion:
500 mL × 1.1 kcal/mL = 550 kcal
500 mL of a 20% lipid emulsion:
500 mL × 2 kcal/mL = 1000 kcal

Lipid emulsions are often provided daily and may supply 20 to 30 percent of total kcalories. Including lipids as an energy source reduces the need for energy from dextrose and lowers the risk of hyperglycemia in glucose-intolerant patients. Lipid infusions must be restricted in patients with hypertriglyceridemia, however. There is also some concern that lipid emulsions that contain excessive linoleic acid can suppress some aspects of the immune response.

Milliequivalents are determined by dividing an ion's molecular weight (MW) by its number of charges. For example:

- For calcium, MW = 40, and the ion has 2 positive charges: 40 ÷ 2 = 20. Thus, 1 mEq of Ca^{++} is equivalent to 20 mg of calcium.
- For sodium, MW = 23, and the ion has 1 positive charge: 23 ÷ 1 = 23. Thus, 1 mEq of Na^+ is equivalent to 23 mg of sodium.
- 1 mEq of Ca^{++} has the same number of charges as 1 mEq of Na^+.

Fluids and Electrolytes Daily fluid needs average between 1500 and 2500 milliliters for most people (see p. 440). The amounts are adjusted according to daily fluid losses and the results of hydration assessment.

The electrolytes added to parenteral solutions include sodium, potassium, chloride, calcium, magnesium, and phosphate. The amounts infused differ from DRI values because the nutrients are delivered into the blood and therefore are not influenced by absorption, as they are when consumed orally. Because electrolyte imbalances can be lethal, electrolyte management by experienced professionals is necessary whenever intravenous therapies are used. Blood tests are administered daily to monitor electrolyte levels until patients have stabilized.

The electrolyte content of parenteral solutions is expressed in *milliequivalents (mEq)*, which are units indicating the number of ionic charges provided by electrolytes. The body's fluids and parenteral solutions are neutral solutions that contain equal numbers of positive and negative charges.

Vitamins and Trace Minerals All vitamins are usually included in parenteral solutions, although a preparation without vitamin K is available for patients using warfarin therapy.[21] The trace minerals typically added to parenteral solutions include zinc, copper, chromium, selenium, and manganese. Iron is excluded because it alters the stability of other ingredients in parenteral mixtures; therefore, special forms of iron need to be injected separately.

Reminder: The anticoagulant warfarin works by interfering with vitamin K's blood-clotting function (see Chapter 15).

Osmolarity Recall that the osmolarity of PPN solutions is limited to 900 milliosmoles per liter, whereas TPN solutions may be as nutrient dense as necessary. Amino acids, dextrose, and electrolytes contribute the most to a solution's osmolarity, whereas the lipid content has little effect. Thus, lipid emulsions are often used to increase the energy provided in PPN solutions.

Medications To avoid the need for a separate infusion site, medications are occasionally added directly to parenteral solutions or infused through a separate port in the catheter.

Parenteral Formulations When a parenteral solution contains dextrose, amino acids, and lipids, it is called a **total nutrient admixture (TNA)**, a **3-in-1 solution,** or an **all-in-one solution.** A **2-in-1 solution** excludes lipids, and the lipid emulsion is administered separately, often using a second port in the catheter. Although the administration of TNA solutions is simpler because only one infusion pump is required, the addition of lipid emulsion to solutions reduces their stability, a major concern when TNA solutions are compounded. Thus, lipids are usually provided separately when they are not a major energy source and are used only to provide essential fatty acids. The "How to" describes a method for calculating the macronutrient and energy content of a parenteral solution.

A lipid emulsion gives a parenteral solution a milky-white color.

Administering Parenteral Nutrition

Parenteral nutrition is a complex treatment that requires skills from a variety of disciplines. Many hospitals organize nutrition support teams, consisting of physicians, nurses, dietitians, and pharmacists, that specialize in the provision of both intravenous and tube feedings. The nurse, who performs direct patient care, plays a central role in administering and monitoring parenteral infusions.

Care of Intravenous Catheters Catheter-related problems frequently cause complications (see Table 16-3 on p. 448). Catheters may be improperly positioned or may dislodge after placement. Air can leak into catheters and escape into the bloodstream, obstructing blood flow. Catheters in peripheral veins may cause phlebitis, necessitating

total nutrient admixture (TNA): a parenteral solution that contains dextrose, amino acids, and lipids; also called a **3-in-1 solution** or an **all-in-one solution.**

2-in-1 solution: a parenteral solution that contains dextrose and amino acids, but excludes lipids.

How To   CALCULATE THE MACRONUTRIENT AND ENERGY CONTENT OF A PARENTERAL SOLUTION

Suppose a patient is receiving 1.25 liters (1250 milliliters) of a parenteral solution that contains 5 percent amino acids and 30 percent dextrose, supplemented with 250 milliliters of a 20 percent lipid emulsion daily. How many grams of protein and carbohydrate is the person receiving, and what is the total energy intake for the day?

Amino acids:

$$5\% \text{ amino acids} = \frac{5 \text{ g amino acids}}{100 \text{ mL}}$$

$$\frac{5 \text{ g amino acids}}{100 \text{ mL}} \times 1250 \text{ mL} = 62.5 \text{ g of amino acids}$$

62.5 g amino acids × 4.0 kcal/g = 250 kcal

Carbohydrate:

$$30\% \text{ dextrose} = \frac{30 \text{ g dextrose}}{100 \text{ mL}}$$

$$\frac{30 \text{ g dextrose}}{100 \text{ mL}} \times 1250 \text{ mL} = 375 \text{ g of dextrose}$$

375 g dextrose × 3.4 kcal/g = 1275 kcal

Lipids:

Recall that a 20 percent lipid emulsion provides 2.0 kcalories per milliliter. If the patient is given 250 milliliters of the emulsion:

250 mL × 2.0 kcal/mL = 500 kcal

Total energy intake:

250 kcal + 1275 kcal + 500 kcal = 2025 kcal

TABLE 16-3 Potential Complications of Parenteral Nutrition

Catheter-Related	Metabolic
Air embolism	Abnormal liver function
Blood clotting at catheter tip	Electrolyte imbalances
Clogging of catheter	Gallbladder disease
Dislodgment of catheter	Hyperglycemia, hypoglycemia
Improper placement	Hypertriglyceridemia
Infection, sepsis	Metabolic bone disease
Phlebitis	Nutrient deficiencies
Tissue injury	Refeeding syndrome

reinsertion at an alternate site. A catheter may become clogged from blood clotting or from a buildup of scar tissue around the catheter tip. Catheters are also a leading cause of infection: contamination may be introduced during insertion or may develop at the placement site.

To reduce the risk of complications, nurses use aseptic techniques when inserting catheters, changing tubing, or changing a dressing that covers the catheter site. Unusual bleeding or a wet dressing suggests a problem with catheter placement. A change in infusion rate may indicate a clogged catheter. Infection may be indicated by redness or swelling around the catheter site or by an unexplained fever. Routine inspections of equipment and frequent monitoring of patients' symptoms help to minimize the problems associated with catheter use.

Administration of Parenteral Solutions Infusion protocols vary among institutions. One approach is to start the infusion at a slow rate (with a solution that is either full strength or nutrient dilute) and increase the rate gradually over a 2- to 3-day period. For example, 40 milliliters per hour can be infused during the first 24 hours of administration (supplying 960 milliliters), and the volume can be increased to the goal rate on the second day. Another method is to give the full volume of a nutrient-dilute solution on the first day and advance nutrient concentrations as tolerated. Solutions can often be started at full volume and full strength, unless there is a risk of hyperglycemia or other complications.[22]

Parenteral solutions may be infused continuously over 24 hours **(continuous parenteral nutrition)** or during 8- to 16-hour periods only **(cyclic parenteral nutrition).** Continuous infusions are given to critically ill and malnourished patients who cannot receive adequate nutrients in the shorter time periods. Cyclic infusions are sometimes provided at night so that patients can participate in routine activities during the day. This method is especially suited to patients who require long-term parenteral support or who will be infusing parenteral solutions at home. Patients may begin with continuous parenteral nutrition and transition to cyclic parenteral nutrition as their condition improves.

Regular monitoring helps to prevent complications. The parenteral solution and tubing are checked frequently for signs of contamination. Routine testing of glucose, lipid, and electrolyte levels helps to determine tolerance to solutions. Frequent reassessment of nutrition status may be necessary until a patient has stabilized. Rapid changes in infusion rate are discouraged in some patients due to a risk of developing hyperglycemia or hypoglycemia.[23]

Discontinuing Parenteral Nutrition The patient must have adequate GI function before parenteral nutrition can be tapered off and enteral feedings begun. During the transition to oral feedings, a combination of methods is often necessary. Parenteral infusions are usually tapered off at the same time that tube feedings or oral feedings are

begun, such that the two methods can together supply the needed nutrients. Clear liquids are generally the first foods offered and include pulp-free fruit juices, soft drinks, and clear broths; small amounts are given initially to determine tolerance. Later feedings include beverages and solid foods that are unlikely to cause discomfort; a low-fat, lactose-free diet may be recommended. If gastrointestinal symptoms (such as nausea, vomiting, bloating, or diarrhea) develop, oral feedings are limited in size or frequency until the intestines adapt. Once about two-thirds to three-fourths of nutrient needs can be provided enterally, the parenteral infusions may be discontinued. Transitioning to an oral diet is sometimes difficult because a person's appetite remains suppressed for several weeks after parenteral nutrition is terminated.

Chapter 14 provides more information about the clear liquid diet.

Managing Metabolic Complications

As discussed previously, the catheters used for intravenous infusions may cause a number of serious complications. This section discusses some metabolic complications that may result from parenteral nutrition (review Table 16-3).[24]

Hyperglycemia Hyperglycemia most often occurs in patients who are glucose intolerant or undergoing severe metabolic stress. It can be prevented by providing insulin along with parenteral solutions or by restricting the amount of dextrose in a solution. Dextrose infusions are generally limited to less than 5 milligrams per kilogram of body weight per minute in critically ill adult patients so that the carbohydrate intake does not exceed the maximum glucose oxidation rate.

For most patients receiving parenteral nutrition, blood glucose levels should not exceed 200 mg/dL.

Hypoglycemia Hypoglycemia occasionally occurs when parenteral nutrition is interrupted or discontinued or if excessive insulin is given. In patients at risk, infusions may be tapered off over several hours before discontinuation. Another option is to infuse a dextrose solution at the same time that parenteral nutrition is stopped.[25]

Hypertriglyceridemia Hypertriglyceridemia may develop in certain patients who cannot tolerate the amount of lipid supplied. If blood triglyceride levels exceed 500 milligrams per deciliter, lipid infusions should be reduced or stopped.[26]

Refeeding Syndrome Severely malnourished patients who are fed aggressively (parenterally or otherwise) may develop **refeeding syndrome,** characterized by electrolyte and fluid imbalances and hyperglycemia. These effects occur because dextrose infusions raise circulating insulin levels, which promote anabolic processes that quickly remove potassium, phosphate, and magnesium from the blood. The altered electrolyte levels can lead to fluid retention and life-threatening changes in organ systems. To prevent refeeding syndrome, health practitioners may start parenteral infusions slowly and monitor electrolyte and glucose levels when malnourished patients begin receiving nutrition support.

Abnormal Liver Function Fatty liver often results from parenteral nutrition, but it is usually corrected after the parenteral infusions are discontinued. Long-term parenteral nutrition, however, can result in progressive liver disease. To minimize the risk, clinicians may avoid giving the patient excess energy, dextrose, or lipids. Cyclic infusions may be less problematic than continuous infusions.[27]

Gallbladder Disease When parenteral nutrition continues for more than four weeks, sludge (thickened bile) may build up in the gallbladder and eventually lead to gallstone formation. Patients requiring long-term parenteral nutrition may be given medications to stimulate gallbladder contraction or improve bile flow or may have their gallbladders removed surgically.

continuous parenteral nutrition: continuous administration of parenteral solutions over a 24-hour period.

cyclic parenteral nutrition: administration of parenteral solutions over an 8- to 16-hour period each day.

refeeding syndrome: a condition that sometimes develops when a severely malnourished person is aggressively fed; characterized by electrolyte and fluid imbalances and hyperglycemia.

Metabolic Bone Disease Long-term parenteral nutrition is associated with lower bone density, which may be related to altered calcium, phosphorus, magnesium, and sodium metabolism. Inappropriate intakes of vitamin D, vitamin K, and phosphorus have also been implicated. Interventions may include dietary adjustments, medications, and physical activity.[28]

IN SUMMARY

■ Peripheral parenteral nutrition is provided to patients who need short-term parenteral support and do not have high nutrient needs or fluid restrictions. Total parenteral nutrition supplies nutrient-dense solutions for long-term parenteral support.

■ Parenteral solutions include amino acids, dextrose, electrolytes, vitamins, and minerals. Lipid emulsions may be included in the mixture or may be administered separately.

■ Catheters are frequently the cause of complications, which include improper placement or dislodgment, infection, clotting, embolism, and phlebitis.

■ Critically ill patients may require continuous parenteral nutrition, whereas stable patients and long-term users may prefer cyclic parenteral nutrition.

■ Metabolic complications include hyperglycemia, hypoglycemia, hypertriglyceridemia, refeeding syndrome, and diseases affecting the liver, gallbladder, and bone. The Case Study checks your understanding of the concepts introduced in this section.

Nutrition Support at Home

Some individuals may require nutrition support—either tube feedings or parenteral nutrition—after a medical condition has stabilized and they no longer require hospital services. For such a person, home nutrition support may be a suitable option. Current technology allows for the safe administration of nutrition support in the home, and insurance coverage often pays a substantial portion of the costs. Users of home nutrition services (or their caregivers) must be capable of learning the required procedures and managing any complications that arise.

Case Study PATIENT WITH INTESTINAL DISEASE REQUIRING PARENTERAL NUTRITION

Jerry Huang, a 27-year-old man with an inflammatory intestinal disease, underwent a surgical procedure in which a substantial portion of his small intestine was removed. He had received TPN prior to surgery and continued to receive it afterward. After 10 days, tube feeding was begun and initially delivered very small feedings.

1. List some reasons why the nutrition support team initially chose TPN as a means of nutrition support for this patient. How would you explain the need for parenteral nutrition to Jerry?

2. Describe the components of a typical TPN solution. Calculate the energy content of 1 liter of a solution that provides 140 grams of dextrose monohydrate, 45 grams of amino acids, and 90 milliliters of 20 percent lipid emulsion. If Jerry's energy requirement is 2100 kcalories per day, how many liters of solution will he need each day?

3. Why is it important that Jerry begin enteral feedings as soon as possible? Assuming that Jerry eventually tolerates a tube feeding, in what ways can the health care team help Jerry make the

transition from parenteral nutrition to tube feedings? Consider some of the physiological problems that Jerry might face when he begins an oral diet.

4. If Jerry is unable to meet his nutrient needs orally, he may need to continue tube feeding or TPN at home. As you read through the section on nutrition support at home, consider the factors that would make Jerry a good candidate for a home nutrition support program. Consider both the benefits of a proposed program and the problems he could encounter.

Candidates for Home Nutrition Support

Individuals referred for home nutrition support usually need long-term nutrition care for chronic medical conditions. Candidates for home nutrition support include the following:[29]

▮ For home enteral nutrition, individuals who have disorders that prevent food from reaching the intestines or interfere with nutrient absorption. Examples include people with head and neck cancers, severe dysphagia, gastric outlet obstructions, and pancreatic or intestinal conditions that cause malabsorption.

▮ For home parenteral nutrition, individuals who have disorders that severely impede nutrient absorption or interfere with intestinal motility. Examples include people with short bowel syndrome, inflammatory bowel diseases, and intestinal obstructions.

Portable pumps and convenient carrying cases allow people who require home nutrition support to move about freely.

Planning Home Nutrition Care

As with the nutrition support provided in health care facilities, planning for home nutrition care involves decisions about access sites, nutrient delivery methods, and formulas. Users of home services should be involved in the decision making to ensure long-term compliance and satisfaction.

Home Enteral Nutrition Access to the GI tract is possible using either nasal tubes or enterostomies. Although people can learn to place nasogastric tubes themselves, active children and adults often prefer low-profile gastrostomy or jejunostomy tubes, which allow them to lead a more normal lifestyle. For gastric feedings, bolus infusions are simplest and can be quickly delivered. For intermittent feedings that require slow or reliable delivery rates, infusion pumps may be necessary. Portable pumps can free individuals from the need to infuse formula at home.

Insurance reimbursements do not always include the cost of enteral formulas, which are considered food products. For this reason, some people choose to prepare simple formulas at home. Blenderizing home-cooked foods is possible, but the foods need to be strained to remove particles or clumps that may obstruct the tube. Closed feeding systems are useful for avoiding contamination risk but are not appropriate for intermittent feedings that require smaller amounts of formula.

Home Parenteral Nutrition Although both peripheral parenteral nutrition and total parenteral nutrition (TPN) can be provided at home, long-term therapy requires access to the larger central veins that are appropriate for TPN. The catheter's exit site is generally placed in a region accessible to the patient. Most people prefer cyclic infusions over continuous infusions and transition to cyclic infusions before discharge from the hospital. As infusion pumps are required for home TPN, sufficient battery backup is necessary in case electrical service is interrupted. Portable pumps are useful for individuals who lead active lifestyles.

Parenteral solutions need to be aseptically prepared, and individuals who mix their own solutions must be carefully trained. Ready-made parenteral solutions require refrigeration and are stable for limited periods; for example, 3-in-1 solutions may be stable for only one week when refrigerated.

Quality-of-Life Issues

Individuals who depend on home nutrition support face unique challenges that can affect quality of life. Activities and work schedules must be planned around feedings. People receiving nocturnal feedings often cite disturbed sleep as a major problem.

Among social issues, the inability to share meals with family and friends is often a great concern.[30] The Oley Foundation (www.oley.org) is an excellent source of current information and emotional support for people who require home nutrition services.

IN SUMMARY

▌ Candidates for home enteral nutrition services have disorders that interfere with swallowing ability, GI motility, or nutrient absorption. Candidates for home parenteral nutrition have disorders that severely impair nutrient absorption or cause intestinal motility problems.

▌ Patients and caregivers should participate in decisions about access sites, nutrient delivery methods, and formulas. Enteral formulas and parenteral solutions can be purchased or prepared in the home.

▌ The use of portable pumps may help individuals lead a normal lifestyle. Nevertheless, lifestyle adjustments to nutrition support may be difficult and stressful.

Nutrition Assessment Checklist FOR PEOPLE RECEIVING ENTERAL NUTRITION SUPPORT

Medical History

Check the medical record for medical conditions that:

☐ Alter nutrient needs and influence the formula selection
☐ Influence the selection of tube placement sites and feeding routes
☐ Suggest the length of time that the tube feeding will be needed

Monitor the medical record for complications or risks that may influence the formula selection or delivery technique, including:

☐ Aspiration
☐ Constipation
☐ Fluid and electrolyte imbalances
☐ Diarrhea
☐ Hyperglycemia
☐ Nausea and vomiting
☐ Skin irritation

Medications

Check medications for those that can cause side effects similar to the complications of tube feeding, such as:

☐ Nausea and vomiting
☐ Diarrhea
☐ Constipation
☐ GI discomfort

For medications delivered through the feeding tube, check:

☐ Form of medication and possible alternatives
☐ Viscosity of liquid medications
☐ Potential for diet-drug interactions

Dietary Intake

To assess nutritional adequacy, check to see whether:

☐ The formula is appropriate for patient's needs
☐ Supplemental water is provided to meet needs
☐ The formula is administered as prescribed

Anthropometric Data

Measure baseline height and weight, and monitor body weight regularly. If weight is not appropriate:

☐ Determine whether energy needs have been correctly assessed
☐ Check to see if the formula is being delivered as prescribed
☐ Check for signs of dehydration or overhydration

Laboratory Tests

Check serum and urine tests for signs of:

☐ Fluid and electrolyte imbalances
☐ Glucose intolerance
☐ Inadequate protein status (serum protein levels)
☐ Improvement or deterioration of the medical condition

Physical Signs

Look for physical signs of:

☐ Dehydration or overhydration
☐ Delayed gastric emptying
☐ Malnutrition

Nutrition Assessment Checklist FOR PEOPLE RECEIVING PARENTERAL NUTRITION SUPPORT

Medical History

Check the medical record for medical conditions that:

- Prevent the use of enteral nutrition
- Suggest the time period that parenteral nutrition will be required

Monitor the medical record for complications or risks that may influence the parenteral solution formulation or delivery technique, including:

- Acid-base imbalances
- Fluid and electrolyte imbalances
- Hyperglycemia or hypoglycemia
- Hypertriglyceridemia
- Preexisting liver disease
- Refeeding syndrome

Medications

For medications added to the parenteral solution, determine:

- Medications' compatibility with the parenteral solution
- Length of time that the medication can remain stable in solution

For medications infused separately, determine:

- Length of time that the infusion may need to be stopped
- Necessary adjustments in parenteral infusions to compensate for medication delivery

Dietary Intake

To assess nutritional adequacy, check to see whether:

- Patient's nutrient needs were correctly determined
- Solution is administered as prescribed
- Infusion pump is operating correctly

Anthropometric Data

Measure baseline height and weight, and monitor daily weights. If weight is not appropriate:

- Determine whether energy needs have been correctly assessed
- Check to see if the parenteral solution is being delivered as prescribed
- Check for signs of dehydration or overhydration

Laboratory Tests

Check serum and urine tests for signs of:

- Fluid, electrolyte, and acid-base imbalances
- Hyperglycemia or hypoglycemia
- Hypertriglyceridemia
- Abnormal liver function
- Inadequate protein status (serum protein levels)
- Improvement or deterioration of medical condition

Physical Signs

Routinely monitor the following:

- Catheter insertion site for signs of infection or inflammation
- Blood pressure, temperature, pulse, and respiration for signs of fluid, electrolyte, and acid-base imbalances

Look for physical signs of:

- Dehydration or overhydration
- Protein-energy malnutrition
- Malnutrition

Clinical Applications

1. Appendix G provides examples of enteral formulas on the market and lists their energy and macronutrient contents. Select one standard formula and one elemental formula from Tables G-1 and G-2, respectively. For the two formulas you selected, calculate the volume of formula that would meet the energy needs of a patient who requires about 1750 kcalories daily. Use these results in answering the following questions:

   a. What is the amount of protein, carbohydrate, and fat that the patient would obtain in a typical day? Determine the percentages of kcalories that come from carbohydrate and fat. Do these percentages fall within the Acceptable Macronutrient Distribution Ranges described in Chapter 1 (p. 10)?

   b. Tables G-1 and G-2 show the formula volumes that would meet the Reference Daily Intakes (RDI). Would the volumes you obtained meet typical vitamin and mineral needs?

2. A liter of a TPN solution contains 500 milliliters of 50 percent dextrose solution and 500 milliliters of 5 percent amino acid solution. Determine the daily energy and protein intakes of a person who receives 2 liters per day of such a solution. Calculate the average daily energy intake if the person also receives 500 milliliters of a 20 percent fat emulsion three times a week.

3. Consider the social, psychological, clinical, and financial ramifications of using home parenteral nutrition, with no foods allowed by mouth, in answering the following questions:

   a. What would be the advantages of living at home instead of in a hospital or other residential facility? Can you think of some disadvantages?

   b. Think about how you, as the patient, might manage daily infusions; consider the time, cost, and commitment required to maintain the therapy.

   c. If you were not allowed to consume foods, what possible difficulties might you encounter? How would you handle holidays and special occasions that center around food?

Self Check

1. An important measure that may prevent bacterial contamination in tube feeding formulas is:
   a. nonstop feeding of formula.
   b. using the same feeding bag and tubing each day.
   c. discarding opened containers of formula not used within 24 hours.
   d. adding formula to the feeding container before it empties completely.

2. Compared with intermittent tube feedings, continuous feedings:
   a. require an infusion pump.
   b. allow greater freedom of movement.
   c. are more similar to normal patterns of eating.
   d. are associated with more GI side effects.

3. A patient needs 1800 milliliters of formula a day. If the patient is to receive formula intermittently every 4 hours, how many milliliters of formula will she need at each feeding?
   a. 225
   b. 300
   c. 400
   d. 425

4. The term that describes the amount of enteral formula remaining in the stomach from a previous feeding is:
   a. residue.
   b. osmolar load.
   c. gastric residual volume.
   d. intermittent feeding.

5. The nurse using a feeding tube to deliver medications recognizes that:
   a. medications given by feeding tube generally do not cause GI complaints.
   b. medications can usually be added directly to the feeding container.
   c. enteral formulas do not interact with medications in the same way that foods do.
   d. thick or sticky liquid medications and crushed tablets can clog feeding tubes.

6. TPN is preferred over PPN for a patient who:
   a. does not have high nutrient requirements.
   b. needs long-term parenteral nutrition support.
   c. has strong peripheral veins and moderate nutrient needs.
   d. needs parenteral infusions as a supplement to tube feedings.

7. For a patient receiving central TPN who also receives intravenous lipid emulsions two or three times a week, the lipid emulsions serve primarily as a source of:
   a. essential fatty acids.
   b. cholesterol.
   c. fat-soluble vitamins.
   d. concentrated energy.

8. Which nutrient is often omitted from parenteral solutions because it may destabilize other ingredients in the solution?
   a. Calcium
   b. Vitamin K
   c. Iron
   d. Chromium

9. Refeeding syndrome causes dangerous fluctuations in:
   a. serum electrolytes.
   b. serum liver enzyme levels.
   c. blood triglyceride levels.
   d. ketone bodies.

10. Patients using home parenteral nutrition:
   a. are unable to use TNA solutions.
   b. are usually given continuous rather than cyclic infusions.
   c. are generally unable to work out of the home or travel.
   d. require infusion pumps for use at home.

Answers to these questions can be found in Appendix H.

Notes

1. A. M. Malone, Enteral formula selection, in P. Charney and A. Malone, eds., *ADA Pocket Guide to Enteral Nutrition* (Chicago: American Dietetic Association, 2006), pp. 63–122.

2. J. L. Rombeau, Enteral nutrition, in L. Goldman and D. Ausiello, eds., *Cecil Medicine* (Philadelphia: Saunders, 2008), pp. 1617–1621; M. Shike, Enteral feeding, in M. E. Shils and coeditors, *Modern*

Nutrition in Health and Disease (Baltimore: Lippincott Williams & Wilkins, 2006), pp. 1554–1566.

3. Shike, 2006.

4. N. Gupta and R. G. Martindale, Parenteral vs. enteral nutrition, in G. Cresci, ed., *Nutrition Support for the Critically Ill Patient: A Guide to Practice* (Boca Raton, FL: Taylor & Francis Group, 2005), pp. 193–208.

5. Rombeau, 2008.

6. M. Marian and P. Charney, Patient selection and indications for enteral feedings, in P. Charney and A. Malone, eds., *ADA Pocket Guide to Enteral Nutrition* (Chicago: American Dietetic Association, 2006), pp. 1–25; Shike, 2006.

7. K. K. Kattelmann and coauthors, Preliminary evidence for a medical nutrition therapy protocol: Enteral feedings for critically ill patients, *Journal of the American Dietetic Association* 106 (2006): 1226–1241; B. Taylor and J. E. Mazuski, Enteral feeding access in the critically ill, in G. Cresci, ed., *Nutrition Support for the Critically Ill Patient: A Guide to Practice* (Boca Raton, FL: Taylor & Francis Group, 2005), pp. 235–252.

8. P. D. Wohlt and coauthors, Recommendations for the use of medications with continuous enteral nutrition, *American Journal of Health-Systems Pharmacy* 66 (2009): 1458–1467.

9. Kattelmann, 2006.

10. C. Thompson, Initiation, advancement, and transition of enteral feedings, in P. Charney and A. Malone, eds., *ADA Pocket Guide to Enteral Nutrition* (Chicago: American Dietetic Association, 2006), pp. 123–154.

11. Wohlt and coauthors, 2009.

12. A. Desachy and coauthors, Initial efficacy and tolerability of early enteral nutrition with immediate or gradual introduction in intubated patients, *Intensive Care Medicine* 34 (2008): 1054–1059.

13. M. K. Russell, Monitoring complications of enteral feedings, in P. Charney and A. Malone, eds., *ADA Pocket Guide to Enteral Nutrition* (Chicago: American Dietetic Association, 2006), pp. 155–192.

14. S. A. McClave and coauthors, Guidelines for the provision and assessment of nutrition support therapy in the adult critically ill patient: Society of Critical Care Medicine (SCCM) and American Society for Parenteral and Enteral Nutrition (A.S.P.E.N.), *Journal of Parenteral and Enteral Nutrition* 33 (2009): 277–316; Russell, 2006.

15. Wohlt and coauthors, 2009.

16. Russell, 2006.

17. Thompson, 2006.

18. C. Hamilton, Vascular access, in P. Charney and A. Malone, eds., *ADA Pocket Guide to Parenteral Nutrition* (Chicago: American Dietetic Association, 2007), pp. 33–51.

19. M. L. Christensen, Parenteral formulations, in G. Cresci, ed., *Nutrition Support for the Critically Ill Patient: A Guide to Practice* (Boca Raton, FL: Taylor & Francis Group, 2005), pp. 279–302.

20. R. O. Brown and G. Minard, Parenteral nutrition, in M. E. Shils and coeditors, *Modern Nutrition in Health and Disease* (Baltimore: Lippincott Williams & Wilkins, 2006), pp. 1567–1597.

21. A. M. Malone, Parenteral nutrients and formulations, in P. Charney and A. Malone, eds., *ADA Pocket Guide to Parenteral Nutrition* (Chicago: American Dietetic Association, 2007), pp. 52–63.

22. S. Roberts, Initiation, advancement, and acute complications, in P. Charney and A. Malone, eds., *ADA Pocket Guide to Parenteral Nutrition* (Chicago: American Dietetic Association, 2007), pp. 76–102.

23. E. E. Szeszycki and S. Benjamin, Complications of parenteral feeding, in G. Cresci, ed., *Nutrition Support for the Critically Ill Patient: A Guide to Practice* (Boca Raton, FL: Taylor & Francis Group, 2005), pp. 303–319.

24. M. P. Fuhrman, Complications of long-term parenteral nutrition, in P. Charney and A. Malone, eds., *ADA Pocket Guide to Parenteral Nutrition* (Chicago: American Dietetic Association, 2007), pp. 103–117; Roberts, 2007; Szeszycki and Benjamin, 2005.

25. A. Wilmer and G. Van den Berghe, Parenteral nutrition, in L. Goldman and D. Ausiello, eds., *Cecil Medicine* (Philadelphia: Saunders, 2008), pp. 1621–1626; Szeszycki and Benjamin, 2005.

26. Roberts, 2007.

27. Roberts, 2007.

28. Fuhrman, 2007; Brown and Minard, 2006.

29. C. Hamilton and T. Austin, Home parenteral nutrition, in P. Charney and A. Malone, eds., *ADA Pocket Guide to Parenteral Nutrition* (Chicago: American Dietetic Association, 2007), pp. 118–146; A. Pattinson and J. Buchholtz, Home enteral nutrition, in P. Charney and A. Malone, eds., *ADA Pocket Guide to Enteral Nutrition* (Chicago: American Dietetic Association, 2006), pp. 193–227.

30. M. F. Winkler, Living with enteral and parenteral nutrition: How food and eating contribute to quality of life, *Journal of the American Dietetic Association* 110 (2010): 169–177.

Nutrition in Practice

An **inborn error of metabolism** is an inherited trait, caused by a genetic **mutation,** that results in the absence, deficiency, or malfunction of a protein that has a critical metabolic role.[1] The severity of the inborn error's effects are ultimately related to the degree of impairment caused by the altered or missing protein. This Nutrition in Practice describes some inborn errors of metabolism and discusses the role of diet in two of these disorders: phenylketonuria and galactosemia. The accompanying glossary defines the relevant terms.

What problems can result from inborn errors of metabolism?

The protein affected by an inborn error may function as an enzyme, receptor, transport protein, or structural protein. When the body fails to make a protein, the functions that depend on that protein are impaired. For example, when an enzyme is missing or malfunctioning in a metabolic pathway that typically converts compound A to compound B, compound A will accumulate and compound B will not be made. The excess of compound A and the lack of compound B may have harmful effects. Furthermore, the imbalances in one pathway may affect other pathways and ultimately cause a number of metabolic and physiologic disturbances. Table NP16-1 lists some examples of inborn errors related to defects in nutrient metabolism.

GLOSSARY

cystic fibrosis: an inherited disorder that affects the transport of chloride across epithelial cell membranes; primarily affects the gastrointestinal and respiratory systems.

galactosemia (ga-LAK-toe-SEE-me-ah): an inherited disorder that affects galactose metabolism. Accumulated galactose causes damage to the liver, kidneys, and brain in untreated patients.

gene therapy: treatment for inherited disorders, in which DNA sequences are introduced into the chromosomes of affected cells, prompting the cells to express the protein needed to correct the disease.

genetic counseling: support for families at risk of genetic disorders; involves diagnosis of disease, identification of inheritance patterns within the family, and review of reproductive options.

hemophilia (HE-moh-FEEL-ee-ah): an inherited bleeding disorder characterized by deficiency or malfunction of a plasma protein needed for clotting blood.

inborn error of metabolism: an inherited trait (one that is present at birth) that causes the absence, deficiency, or malfunction of a protein that has a critical metabolic role.

metabolites: products of metabolism; compounds produced by a biochemical pathway.

mutation: an inheritable alteration in the DNA sequence of a gene.

phenylketonuria (FEN-il-KEY-toe-NU-ree-ah) or **PKU:** an inherited disorder that affects the conversion of the essential amino acid phenylalanine to the amino acid tyrosine.

What role can diet play in treating inborn errors of metabolism?

Medical nutrition therapy is the primary treatment for many inborn errors that involve nutrient metabolism. Once the biochemical pathway affected by a mutation is identified, a health practitioner may be able to manipulate elements of the diet to compensate for deficiencies and excesses. Dietary intervention generally involves restricting substances that cannot be properly metabolized and supplying substances that cannot be produced. Thus, dietary changes may be able to improve outcomes of some inborn errors by:

▐ Preventing the accumulation of toxic metabolites

▐ Replacing nutrients that are deficient as a result of a defective metabolic pathway

▐ Providing a diet that supports normal growth and development and maintains health

Successful treatment for an inborn error of metabolism depends on the ability to screen newborns and diagnose metabolic diseases before irreversible damage can occur. After a genetic defect is identified, family members undergo **genetic counseling** to evaluate the likelihood that they may pass on the disorder to future offspring. During counseling, couples may learn about reproductive options such as artificial insemination, *in vitro* fertilization, or prenatal monitoring after conception.

Are there treatments for inborn errors that don't involve dietary changes?

Nondietary therapies can treat some inborn errors of metabolism, although the options are somewhat limited. In some cases, the missing protein is infused; this is the primary means of treating **hemophilia,** caused by deficiency of one of the plasma proteins needed for clotting blood. Drug therapy is the main treatment for some inborn errors, including **cystic fibrosis** (discussed in Chapter 19), which is characterized by a defect that prevents normal chloride transport across cell membranes. Future approaches may include **gene therapy,** a treatment that introduces DNA sequences into the chromosomes of affected cells, prompting the cells to express the protein needed to correct the abnormality.

What is an example of an inborn error that benefits from dietary treatment?

A classic example is **phenylketonuria (PKU),** a metabolic disorder that affects amino acid metabolism. PKU occurs in approximately 1 out of every 10,000 births in the United States each year.[2] In PKU, the missing or defective protein is a liver enzyme that converts the essential amino acid phenylalanine to the amino acid tyrosine (see Figure NP16-1 on p. 458). Without this enzyme, phenylalanine and its **metabolites** (metabolic products) accumulate and damage the

TABLE NP16-1 Nutrition-Related Inborn Errors of Metabolism

Disorder	Affected Nutrient(s) or Substance	Metabolic Defect	Nutritional Treatment
Amino acid metabolism			
Maple syrup urine disease	Branched-chain amino acids (isoleucine, leucine, and valine)	Impaired metabolism of branched-chain amino acids	Restriction of branched-chain amino acids; thiamin supplementation
Phenylketonuria	Phenylalanine	Impaired conversion of phenylalanine to tyrosine	Phenylalanine-restricted diet; tyrosine supplementation
Carbohydrate metabolism			
Galactosemia	Galactose	Impaired conversion of galactose to glucose	Galactose-restricted diet
Glycogen storage disease	Glycogen	Impaired metabolism or transport of glycogen, resulting in glycogen accumulation in tissues	Varies; may require frequent feedings, cornstarch supplementation, high-protein diet
Lipid metabolism			
Carnitine transporter deficiency	Fatty acids	Impaired transport of fatty acids into mitochondria for oxidation	Carnitine supplementation; avoidance of fasting and strenuous exercise
X-adrenoleukodystrophy[a]	Very long-chain fatty acids	Impaired breakdown of very long-chain fatty acids in peroxisomes	Under investigation; limited benefit from restriction of very long-chain fatty acids and supplementation with various fatty acid mixtures[a]
Mineral metabolism			
Hemochromatosis	Iron	Excessive iron absorption (causes iron accumulation)	Avoidance of iron and vitamin C supplements and alcoholic beverages (routine blood draws remove excess iron from the body)
Wilson's disease	Copper	Impaired copper excretion (causes copper accumulation)	Avoidance of copper-rich foods; zinc therapy (reduces copper absorption)

[a]The disease X-adrenoleukodystrophy was featured in the 1992 film *Lorenzo's Oil*.

developing nervous system. The impairment in the metabolic pathway also prevents liver synthesis of tyrosine and tyrosine-derived compounds (such as the neurotransmitter epinephrine). Under these conditions, tyrosine becomes essential: the body cannot produce tyrosine, and therefore the diet must supply it.

Although PKU's most debilitating effect is on brain development, other symptoms may manifest if the condition is untreated. Infants with PKU may have poor appetites and grow slowly. They may be irritable or have tremors or seizures. Their bodies and urine may have a musty odor. Their skin may be unusually pale, and they may develop skin rashes. In older children and adults who discontinue treatment, neurological and psychological problems are common.

How is PKU diagnosed?

PKU must be diagnosed soon after birth so that early treatment can prevent its devastating effects. For this reason, newborns are screened for PKU in all 50 states.[3] A standard blood test for phenylalanine is typically conducted by heel puncture after the infant has consumed several meals containing protein. Abnormal results require further testing.

The screening of newborns for PKU is one of the most common genetic tests in the United States and many other countries. Before widespread newborn screening, infants with PKU demonstrated developmental delays (for example, inability to crawl) by 6 to 9 months of age. By the time parents recognized the problem, the damage was irreversible. Most of the damaging consequences of this disorder are now prevented due to the early detection and treatment of PKU.

What is the treatment for PKU?

The only current treatment for PKU is a diet that restricts phenylalanine and supplies tyrosine so that the blood levels of these amino acids are maintained within safe ranges. Because phenylalanine is an essential amino acid, the diet cannot exclude it completely. Children with PKU need phenylalanine to grow, but they cannot handle excesses without detrimental effects. Therefore, their diets must provide enough phenylalanine to support growth and health, but not so much as to cause harm. The diets must also provide tyrosine, which is an essential nutrient for individuals with PKU. To ensure that blood concentrations of phenylalanine and tyrosine are close to normal, blood tests are performed periodically, and diets

FIGURE NP16-1 Biochemical Alterations in PKU

Normal:

Normally, the amino acid phenylalanine follows two pathways, one in the liver, the other in the kidneys. In the liver, the enzyme phenylalanine hydroxylase adds a hydroxyl group (OH) to produce the amino acid tyrosine. Tyrosine, in turn, produces melanin, the pigmented compound found in skin and brain cells; the neurotransmitters epinephrine and norepinephrine; and the hormone thyroxine. In the kidneys, enzymes convert phenylalanine to by-products that are excreted.

In the liver:

Phenylalanine →(Phenylalanine hydroxylase)→ Tyrosine →
Melanin
Epinephrine
Norepinephrine
Thyroxine

In the kidneys:

Phenylalanine → Phenylpyruvic acid (a ketone body) → Other phenyl acids (excreted)

In PKU:

Individuals with PKU lack the liver enzyme phenylalanine hydroxylase, impairing conversion of phenylalanine to tyrosine. Phenylalanine accumulates in the liver and blood, reaching the kidneys in abnormally high concentrations. In the kidneys, an aminotransferase enzyme converts phenylalanine to the ketone body phenylpyruvic acid, which spills into the urine—thus the name phenylketonuria.

In the liver:

Phenylalanine (accumulates) →(Phenylalanine hydroxylase (deficient))→ Tyrosine (deficient)

In the kidneys:

Phenylalanine (accumulates) → Phenylpyruvic acid (accumulates) → Other phenyl acids (accumulate)

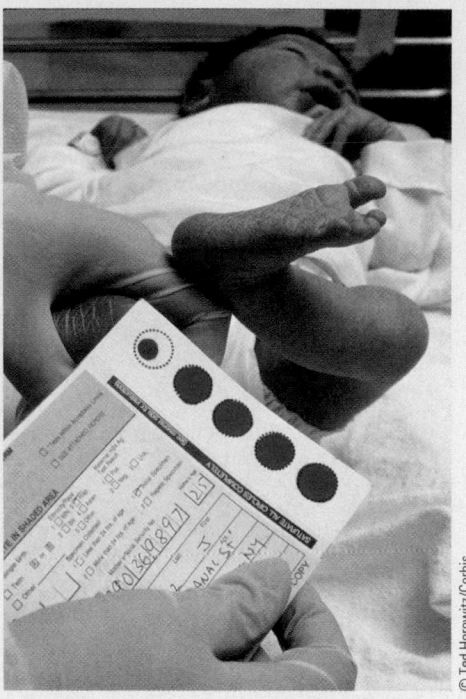

A simple blood test screens newborns for PKU—a common inborn error of metabolism.

are adjusted when necessary. If the dietary treatment is conscientiously followed, it can prevent the symptoms described earlier. Adults with PKU must continue to follow the PKU diet, as well, in order to prevent deterioration in brain function.

What are the main features of the PKU diet?

Central to the PKU diet is the use of an enteral formula that is phenylalanine-free yet supplies energy, amino acids, vitamins, and minerals. For infants, the phenylalanine-free formula can be supplemented with measured amounts of breast milk or regular infant formula to provide the phenylalanine needed for growth. Low-phenylalanine formulas are available for infants who must meet all of their nutrient needs with formula. Formula requirements need to be recalculated periodically to accommodate the growing infant's shifting needs for protein, phenylalanine, tyrosine, and energy.

Once food consumption begins, a phenylalanine-free formula supplies the needed amino acids, and foods that contain phenylalanine are carefully monitored. All proteins contain some phenylalanine; therefore, high-protein foods such as meat, fish, poultry, milk, cheese, legumes, and nuts (including peanut butter) are omitted. Fruits, vegetables, and cereals also contain phenylalanine, so only limited amounts are allowed. Low-protein flours and mixes are available for making low-phenylalanine breads, pasta, cakes, and cookies. Foods that do not contain phenylalanine, such as jams, jellies, and most sweeteners, can be used freely. Growth

rates and nutrition status are monitored to ensure that the diet is adequate. Note that older children, teens, and adults with PKU should continue to use the phenylalanine-free formulas to help meet their protein and energy needs.

Individuals with PKU should be encouraged to develop creative ways to make their diets enjoyable. The formula can be flavored or combined with fruits or juices to make smoothies or frozen juice bars. Sandwiches can include low-phenylalanine breads and fillings such as mashed bananas or avocados, shredded carrots and olives, or tomato slices with mayonnaise. Children often enjoy creating special recipes with permitted foods to make their choices more varied and to share meals with friends.

How long should a person with PKU continue the dietary treatment?

Lifelong adherence to a phenylalanine-restricted diet is currently recommended for all individuals with PKU. Elevated phenylalanine levels can adversely affect cognitive function at any age. Case studies have suggested that individuals with PKU who discontinue dietary management may have problems with attention span, concentration, and memory. It is especially important that women with PKU maintain safe phenylalanine concentrations during pregnancy. Elevated phenylalanine levels, especially during the first trimester, have been associated with mental retardation and organ malformations in the offspring of PKU mothers who have discontinued dietary treatment.[4]

What is another example of an inborn error that requires dietary changes?

Galactosemia is an example of an inborn error of carbohydrate metabolism. Individuals with galactosemia are deficient in one of the enzymes needed to metabolize galactose, a sugar that is primarily found in milk products (recall that each lactose molecule contains galactose). An accumulation of galactose can cause damage in multiple tissues. Infants with galactosemia who are given milk react with severe vomiting and liver jaundice within days of the initial feeding. Serious liver damage can develop and progress to symptomatic cirrhosis. Other complications may include kidney failure, cataracts, and brain damage. Treatment in the first weeks of life can prevent the most detrimental effects of galactose accumulation, but if treatment is delayed, the damage to the brain is irreversible.[5]

What is the dietary treatment for galactosemia?

The diet for galactosemia is much simpler than the diet for PKU. For one thing, galactose is not an essential nutrient. The galactosemia diet essentially eliminates galactose from the diet and does not need to provide a carefully determined amount of any nutrient, as the PKU diet does. In addition, dietary galactose is primarily obtained from lactose (the milk sugar), so the main focus of dietary treatment is the exclusion of milk and milk products. A number of other foods that contain galactose in substantial amounts, such as organ meats and some legumes, fruits, and vegetables, must also be avoided or restricted. Patients receive food lists that identify the galactose content of common foods.

Infants diagnosed with galactosemia are given lactose-free formulas to meet their nutrient needs. Once a child can consume adequate amounts of regular foods, special formulas are unnecessary. However, care must be taken to ensure that the diet supplies adequate calcium.

How effective is the dietary treatment for galactosemia?

Although the early introduction of a galactose-restricted diet can eliminate the acute toxic effects of galactosemia, complications of the disease may develop despite an individual's compliance with diet therapy. For example, most patients experience delays in speech and language development. Ovarian failure occurs in up to 85 percent of women who have galactosemia.[6] In addition, some evidence suggests that IQ declines as a person with galactosemia ages. The reasons for these long-term complications are not fully understood.

As our scientific understanding of human genetics and biochemistry increases, more inborn errors of metabolism are being recognized. Mainstays of management for these diseases include effective diagnosis, early treatment, and control of environmental factors that cause toxicity. In some cases, dietary changes are central to treatment and can prevent serious complications. Not all inborn errors are easily treated, however. Future developments in biotechnology may someday allow medical practitioners to correct genetic errors using gene therapy.

Notes

1. L. J. Elsas II, Approach to inborn errors of metabolism, in L. Goldman and D. Ausiello, eds., *Cecil Medicine* (Philadelphia: Saunders, 2008), pp. 1539–1546.

2. S. D. Cederbaum, Disorders of phenylalanine and tyrosine metabolism, in L. Goldman and D. Ausiello, eds., *Cecil Medicine* (Philadelphia: Saunders, 2008), pp. 1573–1576.

3. L. J. Elsas II and P. B. Acosta, Inherited metabolic disease: Amino acids, organic acids, and galactose, in M. E. Shils and coeditors, *Modern Nutrition in Health and Disease* (Baltimore: Lippincott Williams & Wilkins, 2006), pp. 909–959.

4. Cederbaum, 2008.

5. L. J. Elsas II, Galactosemia, in L. Goldman and D. Ausiello, eds., *Cecil Medicine* (Philadelphia: Saunders, 2008), pp. 1555–1558.

6. Elsas, Galactosemia, 2008.

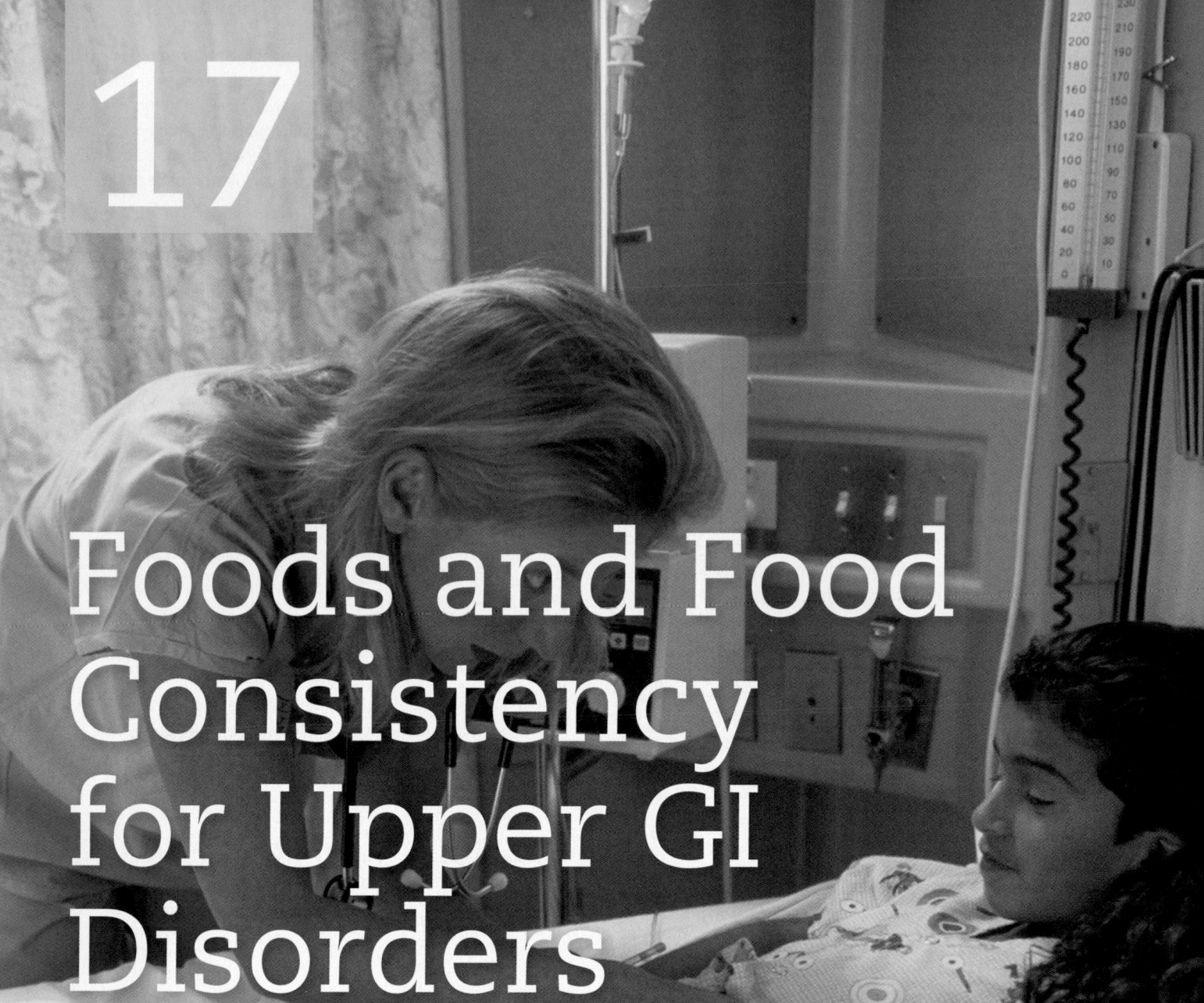

17

Foods and Food Consistency for Upper GI Disorders

Gastrointestinal (GI) illnesses account for a significant fraction of hospital admissions and visits to health practitioners each year. Diagnosis is not always straightforward, however, because many patients with GI symptoms exhibit no physical abnormalities. Evaluation requires a detailed review of a patient's symptoms and responses to dietary adjustments. Because GI complications frequently accompany other illnesses, the medical history can sometimes uncover the underlying source of distress.

Modifications in Food Texture and Consistency

Mechanically altered diets, which contain foods that are modified in texture or consistency, are frequently recommended for individuals with chewing or swallowing impairments. The foods are typically liquid, pureed, ground, chopped, or soft in texture. A **clear liquid diet** may be prescribed before or after certain types of medical procedures.

Mechanically Altered Diets

Table 17-1 and Figure 17-1 provide examples of foods that are usually included in mechanically altered diets. Although the names for these diets vary, a more restrictive diet may contain mostly pureed foods *(pureed diet)*, whereas a less restrictive diet may include ground or minced foods or moist, soft-textured foods that easily form a bolus *(mechanical soft diet,* or simply *soft diet)*. Because chewing and swallowing difficulties vary in severity and fluctuate over time, individual tolerances ultimately determine whether the modifications are acceptable. A later section provides additional information about the various diets used for treating swallowing difficulties (see Dysphagia, pp. 464–466).

A *blenderized diet* is most often recommended following oral or facial surgeries (for example, jaw wiring). Foods that can be blenderized (often with added liquid) are available from all food groups; they include cereals and breads; cooked vegetables; fresh or cooked fruits without skins or seeds; cooked, tender meats and fish; and potatoes, rice, and pasta. Foods that do not blend well should be excluded; examples include hard or rubbery foods such as nuts and seeds, coconut, dried fruits, hard cheeses, sausages and frankfurters, and some raw vegetables.

Clear Liquid Diet

Clear liquids, which require minimal digestion and are easily tolerated by the GI tract, are often the foods recommended before some GI procedures (including GI

mechanically altered diets: diets that contain foods that have been mechanically altered in some way; foods may be liquid, pureed, ground, chopped, minced, or tenderized.

clear liquid diet: a diet that consists of foods that are liquid at room temperature and leave almost no residue (undigested material) in the intestines after digestion and absorption.

TABLE 17-1 Foods Included in Mechanically Altered Diets

Depending on the feeding problem, a mechanically altered diet may include foods that are pureed, mashed, ground, minced, or soft textured. Foods vary according to tolerance.

Pureed Diets	Mechanical Soft Diets
Milk products: Milk, smooth yogurt, pudding, custard	**Milk products:** Milk, yogurt with soft fruit, pudding, cottage cheese
Fruits: Pureed fruits and juices without pulp, seeds, skins, or chunks; well-mashed fresh bananas; applesauce	**Fruits:** Canned or cooked fruits without seeds or skin, fruit juices with small amounts of pulp, ripe bananas
Vegetables: Pureed cooked vegetables without seeds, skins, or chunks; mashed potatoes, pureed potatoes with gravy	**Vegetables:** Soft, well-cooked vegetables that are not rubbery or fibrous; well-cooked, moist potatoes
Meats and meat substitutes: Pureed meats; smooth, homogenous souffles; hummus or other pureed legume spreads	**Meats and meat substitutes:** Ground, minced, or tender meat, poultry, or fish with gravy or sauce; tofu; well-cooked, moist legumes; scrambled or soft-cooked eggs
Breads and cereals: Smooth cooked cereals such as Cream of Wheat, slurried breads and pancakes,[a] pureed rice and pasta	**Breads and cereals:** Cooked cereals or moistened dry cereals with minimal texture, soft pancakes or breads, well-cooked noodles or dumplings in sauce or gravy

[a]Slurried foods are mixed with liquid until the consistency is appropriate; they may be gelled and shaped to improve their appearance.

FIGURE 17-1 **Diets for People with Chewing or Swallowing Problems**

A meal of pureed foods can include meats, beans, breads, cereals, fruits, and vegetables that are pureed to a pudding-like consistency. Foods should be homogeneous and cohesive. Sticky foods, such as melted cheese or peanut butter, are generally not included.

Most foods can be ground or minced so that chewing is not required. The diet can include ground meats; soft, moist casseroles; scrambled eggs and egg salad; bite-sized pasta; smooth cooked cereals; and mashed or minced fruits and vegetables.

A diet with soft-textured, easy-to-swallow foods may include moist, shaved, tender meats; baked fish; cooked or cold cereals with milk; well-cooked vegetables; thinly sliced cucumber without skin or seeds; soft breads and cookies without nuts or seeds; and fruits without seeds or tough skins.

© Polara Studios, Inc. (all)

examinations, X-rays, or surgeries), after GI surgery, or after fasting or intravenous feeding. The *clear liquid diet* consists of clear fluids and foods that are liquid at room temperature and leave little undigested material (called **residue**) in the colon. Permitted foods include clear or pulp-free fruit juices, carbonated beverages, clear meat and vegetable broths (such as consommé and bouillon), fruit-flavored gelatin, fruit ices made from clear juices, frozen juice bars, and plain hard candy. Although the clear liquid diet provides fluid and electrolytes, its nutrient and energy contents are extremely limited. If used for longer than a day or two, the diet should be supplemented with commercially prepared low-residue formulas that provide required nutrients. Figure 17-2 gives an example of a one-day clear liquid menu.

A liquid diet that includes milk and other opaque liquids (such as fruit nectars, yogurt, and liquid supplements) is called a **full liquid diet.** Although rarely necessary, it is sometimes used as a transitional diet between the clear liquid diet and diets that contain solid foods.

Reminder: A change in diet as a patient's food tolerance improves is called *diet progression.*

- Mechanically altered diets are often prescribed for people with swallowing and chewing difficulties.

- Clear liquid diets may be used briefly before some GI procedures, after GI surgery, or after fasting or intravenous feeding.

Conditions Affecting the Mouth and Esophagus

This section describes how people with dry mouth can relieve discomfort and examines the causes and treatments of the two most common problems affecting the esophagus: dysphagia (difficulty swallowing) and gastroesophageal reflux disease.

Dry Mouth

Dry mouth **(xerostomia),** caused by reduced salivary flow, is a side effect of many medications and is associated with a number of diseases and disease treatments. Antihistamines, antihypertensive agents, antidepressants, decongestants, and other medications can cause dry mouth. Poorly controlled diabetes is often associated with dry mouth, as are conditions that directly affect salivary gland function, such as **Sjögren's syndrome.** Radiation therapy that treats head and neck cancers often damages salivary glands, sometimes permanently. Mouth breathing is also a common cause of dry mouth.[1]

A reduction in salivary flow can impair health in a variety of ways. Dry mouth can interfere with speech and cause bad breath. Mouth infections and dental diseases are more common. Chewing and swallowing are more difficult, and taste sensation is diminished. Dentures may be uncomfortable to wear, and ulcerations may develop where they contact the mouth. Dry mouth may cause a person to reduce food intake and may thereby increase malnutrition risk. Table 17-2 offers suggestions that may help to manage dry mouth.

SAMPLE MENU

Breakfast	Strained orange juice Flavored gelatin Ginger ale Coffee or tea, sugar
Lunch	Bouillon or consommé Flavored gelatin Frozen juice bars Apple or grape juice Coffee or tea, sugar
Dinner	Bouillon or consommé Flavored gelatin Fruit ice Cranberry juice Coffee or tea, sugar
Snacks	Soft drinks Fruit ices Hard candy

FIGURE 17-2 Menu—Clear Liquid Diet

TABLE 17-2 Suggestions for Managing Dry Mouth

- Take frequent sips of water or another sugarless beverage.
- Chew sugarless gum to help stimulate salivary flow.
- Suck on ice cubes or frozen fruit juice bars (unless their coldness causes discomfort).
- Avoid citrus juices and spicy or salty foods if they cause mouth irritation.
- Avoid dry foods like toast, chips, and crackers.
- Avoid caffeine, alcohol, and smoking, which may dry the mouth.
- Consume foods that have a high fluid content, such as soups, stews, sauces and gravies, yogurt, and pureed fruits.
- Try over-the-counter saliva substitutes (available as gels, sprays, and tablets), especially just before meals and at bedtime.
- Try rinsing the mouth with small amounts of vegetable oil or softened margarine.
- Use a humidifier during the night.
- Pay strict attention to oral hygiene, brushing and flossing at least twice daily. Try to brush immediately after each meal.
- Avoid alcohol- and detergent-containing mouthwashes that may dry and irritate the mouth.
- If dry mouth is caused by a medication, ask your physician about possible alternatives.
- Ask your physician if using a medication to stimulate saliva secretion may be of benefit; examples include nicotinic acid tablets and pilocarpine.

residue: material left in the intestine after digestion; includes mostly dietary fiber and undigested starches and proteins.

full liquid diet: a liquid diet that includes milk and other opaque liquids.

xerostomia (ZEE-roh-STOE-me-ah): dry mouth caused by reduced salivary flow.
  xero = dry
  stoma = mouth

Sjögren's (SHOW-grenz) **syndrome:** an autoimmune disease characterized by the destruction of secretory glands, resulting in dry mouth and dry eyes.

Can you tell that the foods in this photo are pureed foods shaped with food molds?

dysphagia (dis-FAY-ja): difficulty swallowing.

oropharyngeal dysphagia (OR-oh-fah-ren-JEE-al diss-FAY-jee-ah): an inability to transfer food from the mouth and pharynx to the esophagus; usually caused by a neurological or muscular disorder.

esophageal dysphagia: an inability to move food through the esophagus; usually caused by an obstruction or a motility disorder.

stricture: abnormal narrowing of a passageway; often due to inflammation, scarring, or a congenital abnormality.

achalasia (ack-ah-LAY-zhah): an esophageal disorder characterized by weakened peristalsis and impaired relaxation of the lower esophageal sphincter.

  a = without
  chalasia = relaxation

Dysphagia

Dysphagia may result from neurological diseases, surgical procedures involving the head and neck, or abnormalities that restrict the movement of food within the throat or esophagus. Dysphagia can be categorized according to the phase of swallowing that is impaired (see Table 17-3).

Oropharyngeal dysphagia, which inhibits the transfer of food from the mouth and pharynx to the esophagus, is often due to a neuromuscular condition that upsets the swallowing reflex or impairs the mobility of the tongue and other oral tissues. Symptoms include an inability to initiate swallowing, coughing during or after swallowing (due to aspiration), and nasal regurgitation. Other signs include bad breath, a gurgling noise after swallowing, a hoarse or "wet" voice, or a speech disorder. Oropharyngeal dysphagia is common in elderly persons and frequently follows a stroke.[2]

Esophageal dysphagia interferes with the passage of materials through the esophageal lumen and into the stomach and is usually caused by an obstruction in the esophagus or a motility disorder. The main symptom is the sensation of food "sticking" in the esophagus after it is swallowed. An obstruction can be caused by a **stricture** (abnormal narrowing), tumor, or compression of the esophagus by surrounding tissues. Whereas an obstruction can prevent the passage of solid foods but may not affect liquids, a motility disorder hinders the passage of both solids and liquids. **Achalasia,** the most common motility disorder, is a degenerative nerve condition affecting the esophagus; it is characterized by impaired peristalsis and incomplete relaxation of the lower esophageal sphincter when swallowing.[3]

Complications of Dysphagia Clinicians should be alert to the various complications that frequently accompany dysphagia. If the condition restricts food consumption, malnutrition and weight loss may occur. Individuals who cannot swallow liquids are at increased risk of dehydration. If aspiration occurs, it may cause choking, airway obstruction, or respiratory infections, including pneumonia.

Nutrition Intervention for Dysphagia The National Dysphagia Diet, developed in 2002 by a panel of dietitians, speech and language therapists, and a food scientist, has helped to standardize the nutrition care of dysphagia patients.[4] Table 17-4 presents brief descriptions of the different levels of the diet and some sample meals. After the appropriate dietary level is selected, it must be adjusted to suit the person's swallowing abilities and tolerances. In many cases, the most appropriate foods may be determined only by trial and error. Furthermore, a person's swallowing ability can fluctuate over time, so the dietary plan needs frequent reassessment. A consultation with a swallowing expert, such as a speech and language therapist, is often necessary.

Food Properties and Preparation Foods included in dysphagia diets should have easy-to-manage textures and consistencies. Soft, cohesive foods are easier to swallow than hard or crumbly foods. Moist foods are better tolerated than dry foods. Some foods within a category may be acceptable and others may not; for example, some cookies are soft and tender, whereas others are hard and brittle. Sticky or gummy foods, such as peanut butter and cream cheese, may be difficult to clear from the mouth and throat.

The textures of foods are typically altered to make them easier to swallow. Foods are often pureed, mashed, ground, or minced (review Table 17-1 on p. 461). Foods that have more than one texture, such as vegetable soup or cereal with milk, are harder to handle, so ingredients may be blended to a single consistency with items such as nuts and seeds omitted.

TABLE 17-4 National Dysphagia Diet

Level 1: Dysphagia Pureed

Foods should be pureed or well mashed, homogeneous, and cohesive. This diet is for patients with moderate to severe dysphagia and poor oral or chewing ability.

Sample menus:

▪ *Breakfast:* Cream of wheat, slurried muffins or pancakes,[a] pureed scrambled eggs, plain or vanilla yogurt, well-mashed bananas, fruit juice without pulp (thickened as needed), coffee or tea (if thin liquids are acceptable).

▪ *Lunch or dinner:* Pureed tomato soup, slurried crackers, pureed meat or poultry, zucchini soufflé, mashed potatoes with gravy, pureed carrots or green beans, smooth applesauce, pureed peaches, chocolate pudding.

Foods to avoid: Dry breads and cereals, oatmeal, rice, fruit yogurt, cheese (including cottage cheese), peanut butter, nuts and seeds, raw fruits and vegetables, chunky applesauce, fruit preserves with chunks or seeds, tomato sauce with seeds, beverages with pulp, coarsely ground pepper, herbs.

Level 2: Dysphagia Mechanically Altered

Foods should be moist, cohesive, and soft textured and should easily form a bolus. This diet is for patients with mild to moderate dysphagia; some chewing ability is required.

Sample menus:

▪ *Breakfast:* Moist oatmeal, cornflakes or puffed rice cereal with milk (thickened as needed), moist pancakes or muffins (with butter, margarine, or jam; without nuts or seeds), soft scrambled eggs, cottage cheese, ripe bananas or cooked fruit without skin or seeds, fruit juice (thickened as needed), coffee or tea (if thin liquids are allowed).

▪ *Lunch or dinner:* Soup with easy-to-chew meat and vegetables; slurried bread or crackers; minced, tender-cooked meat; well-cooked pasta with moist meatballs and meat sauce; baked potato with gravy; soft, tender-cooked vegetables (not fibrous or rubbery); canned-peach slices; soft fruit pie (with bottom crust only); soft, smooth chocolate bar.

Foods to avoid: Dry or coarse foods; breads and cereals with nuts, seeds, or dried fruit; frankfurters and sausages; hard-cooked eggs; corn and clam chowders; sandwiches; pizza; sliced cheese; rice; potato skins; French fries; raw vegetables; fibrous, rubbery, or non-tender cooked vegetables such as asparagus, broccoli, brussels sprouts, cabbage, celery, corn, and peas; peanut butter; coconut; nuts and seeds; raw fruit (except banana); cooked fruits with skin or seeds; pineapple; mango; uncooked dried fruit; popcorn; chewy candies (such as caramel or licorice).

Level 3: Dysphagia Advanced

Foods should be moist and in bite-sized pieces when swallowed; foods with mixed textures are included. This diet is for patients with mild dysphagia and adequate chewing ability.

Sample menus:

▪ *Breakfast:* Cereal with milk, moist pancakes or muffins (with butter, margarine, or jam; without nuts or seeds), poached or scrambled eggs, fruit yogurt, soft fresh fruit (peeled) or berries, coffee or tea (if thin liquids are tolerated).

▪ *Lunch or dinner:* Chicken noodle soup; moistened crackers or moist bread; thin-sliced tender meat; cheese; moist, soft-cooked potatoes or rice; tender-cooked vegetables; shredded lettuce with dressing; fresh, peeled peach or melon; canned fruit salad; moist chocolate chip cookie (without nuts).

Foods to avoid: Dry or coarse foods; breads and cereals with nuts, seeds, or dried fruit; corn and clam chowders; potato skins; raw vegetables (except shredded lettuce); corn; chunky peanut butter; coconut; nuts and seeds; hard fruit (such as apples or pears); fruit with skin, seeds, or stringy textures (such as mango or pineapple); uncooked dried fruit; fruit leathers; popcorn; chewy candies (such as caramel or licorice).

Liquid Consistencies (only those tolerated are allowed in the diet)

▪ *Thin:* Watery fluids; may include milk, coffee, tea, juices, carbonated beverages.
▪ *Nectarlike:* Fluids thicker than water that can be sipped through a straw; may include buttermilk, eggnog, tomato juice.
▪ *Honeylike:* Fluids that can be eaten with a spoon but do not hold their shape; may include honey, tomato sauce, yogurt.
▪ *Spoon-thick:* Thick fluids that must be eaten with a spoon and can hold their shape; may include milk pudding, thickened applesauce.

[a]Slurried foods are mixed with liquid until the consistency is appropriate; they may be gelled and shaped to improve appearance.

Consuming foods that have a similar consistency can quickly become monotonous. By using commercial thickeners and food molds, pureed foods can be formed into attractive shapes. Including a variety of flavors and colors can also make a meal more appealing. The "How to" offers additional suggestions for improving the acceptance of pureed and other mechanically altered foods.

Take a moment to think about a meal of pureed or ground foods. A typical dinner of baked chicken, potatoes, carrots, and green beans can look like mounds of differently colored mush. The foods may taste great, but a person may have little appetite before trying a first bite. To improve appetite, be creative when preparing and serving meals:

▌ Help to stimulate the appetite by preparing favorite foods and foods with pleasant smells. Enliven food flavors with seasonings and spices.

▌ Use attractive plates and silverware to improve the visual appeal of a meal. Consider colors and shapes when arranging foods on a plate; colorful garnishes can add color and eye appeal. Substitute brightly colored vegetables for white vegetables; for example, replace mashed potatoes with mashed sweet potatoes.

▌ Try layering ingredients so that the food looks like a fancy casserole or popular hors d'oeuvre. For example, food items can resemble lasagna, moussaka, tamales, or sushi.

▌ Shape pureed and ground foods to resemble traditional dishes; for example, flatten a spoonful of pureed meat to make a patty, or use small scoops so that meat resembles meatballs. Use food molds to restore slurried breads and pureed meats to their traditional shapes.

Efforts to improve the visual appearance of foods can go a long way toward helping people eat nourishing meals and maintain a healthy weight.

Properties of Liquids Thickened liquids are easier to swallow than thin liquids such as water or juice. Table 17-4 describes the four levels of liquid consistencies prescribed for dysphagia patients, referred to as thin, nectarlike, honeylike, and spoon-thick. To increase viscosity, commercial starch thickeners can be stirred into beverages and soup broths. Some beverages can lose their appeal when thickened, however; for example, individuals may find thickened coffee and tea unacceptable.

Alternative Feeding Strategies for Dysphagia Some patients may be able to learn alternative feeding techniques to help them compensate for their swallowing problem. For example, changing the position of the head and neck while eating and drinking can minimize some swallowing difficulties. (As an example, cups designed for dysphagia patients allow drinking without tilting the head back.) Individuals with oropharyngeal dysphagia can be taught exercises that strengthen the jaws, tongue, or larynx, or they can learn new methods of swallowing that allow them to consume a normal diet. Speech and language therapists are often responsible for teaching patients these techniques.

Gastroesophageal Reflux Disease

Gastroesophageal reflux disease (GERD) is a condition in which the stomach's acidic contents back up into the esophagus, causing discomfort and, sometimes, tissue damage. People who suffer from GERD often refer to these symptoms as *heartburn* or *acid indigestion*. Reflux does not necessarily cause symptoms or injury—it occurs occasionally in healthy people and is a problem only if it creates complications and requires lifestyle changes or medical treatment.

Causes of GERD The lower esophageal sphincter is the main barrier to gastric reflux, so GERD can result if the sphincter muscle is weak or relaxes inappropriately. Medical conditions that either interfere with the sphincter's function or prevent rapid clearance of acid from the esophagus can predispose a person to GERD.

Conditions associated with high rates of GERD include pregnancy, obesity, asthma, and **hiatal hernia,** a condition in which a portion of the stomach protrudes above the diaphragm (see Figure 17-3). Pregnancy is the most common predisposing condition—as many as two-thirds of pregnant women report heartburn, which usually worsens during the third trimester.[5] Some medications can increase the risk of reflux, as does the use of nasogastric tubes in tube feedings. Various other conditions can exacerbate GERD by increasing stomach distention or weakening the lower esophageal sphincter; Table 17-5 lists examples.

Normal	Acid reflux	Hiatal hernia

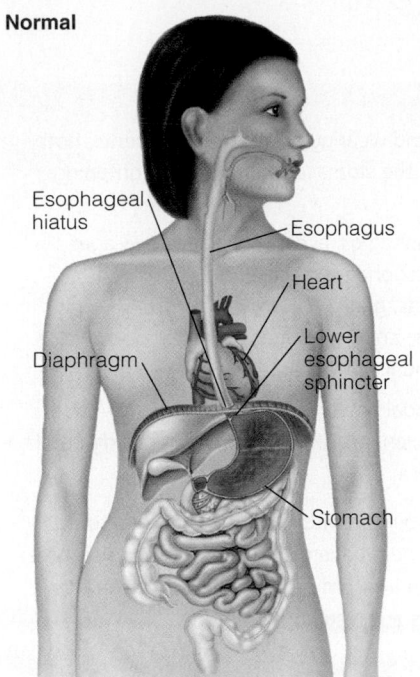
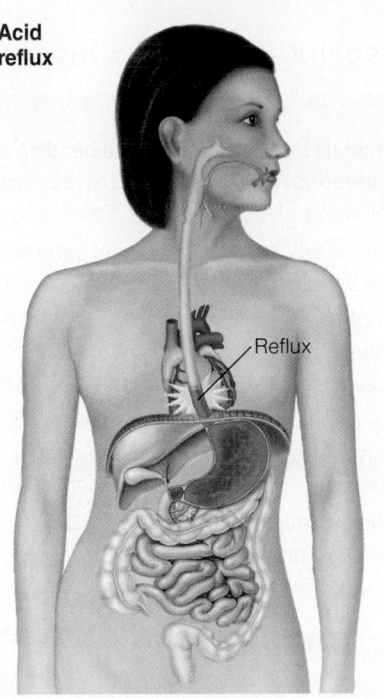
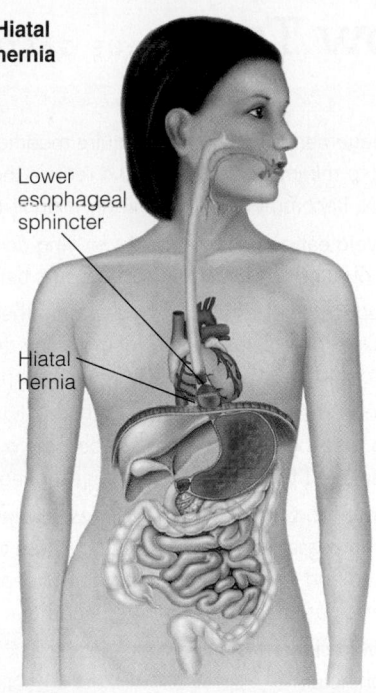

Normal — labels: Esophageal hiatus, Esophagus, Heart, Lower esophageal sphincter, Diaphragm, Stomach

Acid reflux — label: Reflux

Hiatal hernia — labels: Lower esophageal sphincter, Hiatal hernia

The stomach normally lies below the diaphragm, and the esophagus passes through the esophageal hiatus. The lower esophageal sphincter prevents reflux of stomach contents.

Whenever the pressure in the stomach exceeds the pressure in the esophagus, as can occur with overeating and overdrinking, the chance of reflux increases. The resulting "heartburn" is so-named because it is felt in the area of the heart.

Risk of acid reflux may increase as a consequence of a hiatal hernia. A "sliding" hiatal hernia occurs when part of the stomach, along with the lower esophageal sphincter, rises above the diaphragm.

FIGURE 17-3 The Upper GI Tract, Acid Reflux, and Hiatal Hernia

Consequences of GERD If gastric acid remains in the esophagus long enough to damage the esophageal lining, the resulting inflammation is called **reflux esophagitis.** Severe and chronic inflammation may lead to esophageal ulcers, with consequent bleeding. Healing and scarring of ulcerated tissue may narrow the inner diameter of the esophagus, causing esophageal stricture. A slowly progressive dysphagia for solid foods sometimes results, and swallowing occasionally becomes painful. Pulmonary disease can develop if gastric contents are aspirated into the lungs. Chronic reflux is also associated with **Barrett's esophagus,** a condition in which damaged esophageal

TABLE 17-5 Conditions and Substances Associated with Esophageal Reflux

Conditions That Increase Pressure within the Stomach	Substances That Weaken the Lower Esophageal Sphincter
Ascites (abdominal fluid accumulation)	Alcohol
Delayed gastric emptying	Anticholinergic drugs
Eating large meals	Antihistamines
Lying down after eating	Caffeinated beverages
Obesity	Calcium channel blockers
Pregnancy	Chocolate
Wearing tight clothing around the waist or abdomen	Cigarette smoking
	Diazepam
	Fatty foods
	Peppermint and spearmint
	Progesterone
	Theophylline
	Tricyclic antidepressants

gastroesophageal reflux disease (GERD): condition characterized by the backward flow (reflux) of the stomach's acidic contents into the esophagus.

hiatal hernia: a condition in which the upper portion of the stomach protrudes above the diaphragm; most cases are asymptomatic.

reflux esophagitis: inflammation in the esophagus related to the reflux of acidic stomach contents.

Barrett's esophagus: a condition in which esophageal cells damaged by chronic exposure to stomach acid are replaced by cells that resemble those in the stomach or small intestine, sometimes becoming cancerous.

Management of GERD may require modifications in diet and lifestyle to minimize discomfort and reduce the recurrence of acid reflux. Recommendations typically include the following:

▌ Avoid eating bedtime snacks or lying down after meals. Meals should be consumed at least 3 hours before bedtime.

▌ Reduce nighttime reflux by elevating the head of the bed on 6-inch blocks, inserting a foam wedge under the mattress, or propping pillows under the head and upper torso.

▌ Consume only small meals, and drink liquids between meals so that the stomach does not become overly distended, which can exert pressure on the lower esophageal sphincter.

▌ Limit foods that weaken lower esophageal sphincter pressure or increase gastric acid secretion; these include chocolate, fried and fatty foods, spearmint and peppermint, coffee (both caffeinated and decaffeinated), and tea.

▌ Avoid cigarettes and alcohol; both relax the lower esophageal sphincter.

▌ Avoid bending over and wearing tight-fitting garments; both can cause pressure in the stomach to increase, heightening the risk of reflux.

▌ During periods of esophagitis, avoid foods and beverages that may irritate the esophagus, such as citrus fruits and juices, tomato products, garlic, onions, pepper, spicy foods, carbonated beverages, and very hot or very cold foods (depending on individual tolerances).

▌ Avoid using nonsteroidal anti-inflammatory drugs (NSAIDs) such as aspirin, naproxen, and ibuprofen, which can damage the esophageal mucosa.

Food tolerances among people with GERD can vary markedly. Health practitioners can help patients pinpoint food intolerances by advising them to keep a record of the foods and beverages they consume, as well as any resulting symptoms.

cells are gradually replaced by cells that resemble those in gastric or intestinal tissue; such cellular changes increase the risk of developing esophageal cancer. GERD can also damage tissues in the mouth, pharynx, and larynx, resulting in eroded tooth enamel, sore throat, and laryngitis.[6]

Treatment of GERD Treatment objectives are to alleviate symptoms and facilitate the healing of damaged tissue. Severe ulcerative disease may require immediate acid-suppressing medication, whereas a mild case may be managed with dietary and lifestyle modifications. The "How to" lists lifestyle modifications that may help to prevent the recurrence of gastrointestinal reflux.

Medications that suppress gastric acid secretion help the healing process by reducing the damaging effects of acid on esophageal tissue. **Proton-pump inhibitors** are the most effective of the antisecretory agents and are used both for rapid healing of esophagitis and as a maintenance treatment. Other antisecretory drugs include **histamine-2 receptor blockers** (often referred to as *H2 blockers*) and antacids, which neutralize gastric acid. Antacids are frequently used to relieve occasional heartburn, but they have only short-term effects and may cause some nutrient deficiencies when used over the long term. The Case Study includes questions that review the treatments available for a patient with GERD.

IN SUMMARY

▌ Dry mouth can increase the risk of developing oral infections, diminish taste sensation, and lead to reduced food intake. It can be managed with oral hygiene, dietary changes, and saliva substitutes.

▌ Dysphagia may interfere with food intake and increase the risk of aspiration. Treatment may include dietary adjustments, strengthening exercises, and using different swallowing techniques.

▌ Gastroesophageal reflux disease (GERD) may lead to esophageal ulcers, inflammation, bleeding, and stricture. Treatment includes the use of acid-suppressing drugs and lifestyle changes.

Elyssa Rinaldi is a 39-year-old accountant who is 5 feet 4 inches tall and weighs 165 pounds. During a recent physical examination, she mentioned to her physician that she had been feeling fairly well until she began experiencing heartburn, which has progressively become more frequent and painful. The heartburn often occurs after she eats a large meal and is particularly bad after she goes to bed at night. By directly examining the esophageal lumen using an endoscope (a thin, flexible tube equipped with an optical device), the physician found evidence of reflux esophagitis and a slight narrowing throughout the length of the esophagus.

Ms. Rinaldi's medical history does not indicate any significant health problems. During her last physical exam, her physician advised her to stop smoking cigarettes and to lose 20 pounds, but she has not attempted to do either. The nutrition assessment reveals that Ms. Rinaldi is feeling stressed because it is the middle of the tax season. She usually has little time for breakfast, eats a lunch of fast foods while continuing to work at her desk, and eats a large dinner at around 8 P.M. She generally has wine with dinner and another alcoholic beverage later in the evening.

1. Explain to Ms. Rinaldi the meaning of the medical diagnoses *reflux esophagitis* and *esophageal stricture.*

2. From the brief history provided, list the factors and behaviors that increase Ms. Rinaldi's risks of experiencing reflux. What recommendations can you make to help her change these behaviors?

3. What medications might the physician prescribe, and why?

Conditions Affecting the Stomach

Stomach disorders range from occasional bouts of discomfort to severe conditions that require surgery. This section begins with a discussion of *dyspepsia* (often called "indigestion"), the feeling of pain or discomfort in the upper abdomen that occurs after food consumption. More serious stomach conditions that may benefit from dietary adjustments include *gastritis* and *peptic ulcers,* which often result from bacterial infection or the use of medications that damage the stomach lining.

Dyspepsia

Dyspepsia refers to general symptoms of indigestion in the upper abdominal region, which may include stomach pain, gnawing sensations, early satiety, nausea, vomiting, and bloating. These symptoms sometimes indicate the presence of more serious illnesses, including GERD or peptic ulcer disease. Although about 25 percent of the population experiences dyspepsia, only one person in four seeks medical attention.[7]

Causes of Dyspepsia Abdominal symptoms don't always lead to a clear diagnosis. Various medical problems can cause abdominal discomfort, including peptic ulcers, GERD, gastric motility disorders, gallbladder and pancreatic diseases, and tumors in the upper GI tract. Chronic diseases such as diabetes mellitus, heart disease, and hypothyroidism can sometimes be accompanied by gastric symptoms. Some medications and dietary supplements can cause gastrointestinal distress. Intestinal conditions such as irritable bowel syndrome or lactose intolerance can mimic dyspepsia. Although pinpointing the cause of gastric symptoms can be difficult, a complete examination is in order if the individual experiences unintentional weight loss, persistent vomiting, dysphagia, anemia, or GI bleeding, which suggest the presence of serious illness.[8]

Bloating and Stomach Gas The feeling of bloating may be caused by excessive gas in the stomach, which accumulates when air is swallowed. Air swallowing often accompanies gum chewing, smoking, rapid eating, drinking carbonated beverages, and using a straw. Omitting these practices generally helps to correct the problem.

Potential Food Intolerances Some people report that their symptoms worsen after eating certain foods or spices. Coffee can induce symptoms in about 50 percent of

proton-pump inhibitors: a class of drugs that inhibits the enzyme that pumps hydrogen ions (protons) into the stomach. Examples include omeprazole (Prilosec) and lansoprazole (Prevacid).

histamine-2 receptor blockers: a class of drugs that suppresses acid secretion by inhibiting receptors on acid-producing cells; commonly called *H2 blockers.* Examples include cimetidine (Tagamet), ranitidine (Zantac), and famotidine (Pepcid).

dyspepsia: a feeling of pain, bloating, or discomfort in the upper abdominal area, often called *indigestion;* a symptom of illness rather than a disease itself.

   dys = bad; impaired
   pepsis = digestion

dyspepsia patients, and it also increases gastric acid production and acid reflux.[9] Spicy foods may cause some injury to the mucosal lining and exacerbate the pain from a preexisting ulcer. High-fat meals can slow gastric emptying and thereby exacerbate dyspepsia. To minimize symptoms, people with dyspepsia are typically advised to avoid consuming excessively large meals, fatty or highly spiced foods, and specific foods believed to trigger symptoms.[10]

Nausea and Vomiting

Nausea and vomiting accompany many illnesses and are common side effects of medications. Although occasional vomiting is not dangerous, prolonged vomiting can cause fluid and electrolyte imbalances and may require medical care. A chronic vomiting problem can reduce food intake and lead to malnutrition and nutrient deficiencies.

The timing of vomiting gives clues to its cause. Vomiting that occurs within an hour after a meal suggests a peptic ulcer or a psychological cause. If it occurs more than one hour after a meal, possible causes include food poisoning, an obstruction that prevents stomach emptying, or a stomach motility disorder.

Treatment of Nausea and Vomiting Most cases are short lived and require no treatment. When treatment is necessary, the main goal is to find and correct the underlying disorder. Restoring hydration and electrolyte balance may also be necessary in some individuals. If a medication is the cause, taking it with food may help. If the cause is unknown or the underlying disorder cannot be corrected, medications that suppress nausea and vomiting may be prescribed. People with **intractable vomiting**—vomiting that is not easily controlled—may require intravenous nutrition support.

Dietary Interventions Sometimes nausea can be prevented or improved with dietary measures.[11] To minimize stomach distention, patients should consume small meals and drink beverages between meals rather than during a meal. Dry, starchy foods such as toast, crackers, and pretzels may help to reduce nausea, whereas fatty or spicy foods and foods with strong odors may worsen symptoms. Foods that are cold or at room temperature may be better tolerated than hot foods. Individuals often have strong food aversions when nauseated, and tolerances vary greatly.

Gastritis

The suffix *-itis* refers to the presence of inflammation in an organ or tissue.

Gastritis is a general term that refers to inflammation of the stomach mucosa. Acute cases of gastritis typically result from irritating substances or treatments that damage the gastric mucosa, resulting in tissue erosions, ulcers, or hemorrhaging (severe bleeding). Chronic cases may be caused by long-term infections or autoimmune disease and can progress to widespread gastric inflammation and tissue atrophy. Most often, gastritis results from *Helicobacter pylori* infection or the use of aspirin or NSAIDs, which are primary causes of peptic ulcer disease as well. Table 17-6 lists some potential causes of gastritis.

Complications of Gastritis The extensive tissue damage that sometimes develops in chronic gastritis can disrupt gastric secretory functions. If hydrochloric acid secretions become abnormally low **(hypochlorhydria)** or absent **(achlorhydria),** absorption of nonheme iron and vitamin B_{12} can be impaired, increasing the risk of deficiency. Pernicious anemia, a condition characterized by the autoimmune destruction of stomach cells that produce intrinsic factor, is a late complication of atrophic gastritis that can result in the macrocytic anemia of vitamin B_{12} deficiency (see p. 208).

Dietary Interventions for Gastritis Dietary recommendations depend on an individual's symptoms. In asymptomatic cases, no dietary adjustments are needed. If pain

TABLE 17-6 Potential Causes of Gastritis

Infection	Internal (Bodily) Causes
▪ Bacterial: *Helicobacter pylori*, *Actinomyces israelii*	▪ Autoimmune disease
▪ Fungal: *Candida albicans*	▪ Bile reflux
▪ Parasitic: *C*ryptosporidiosis, nematode infection	▪ Severe stress
▪ Viral: Cytomegalovirus, herpes simplex virus	▪ Systemic illness or sepsis
Chemical Substances	**Miscellaneous**
▪ Alcohol	▪ Food allergy
▪ Cancer chemotherapy	▪ Foreign bodies
▪ Drugs (especially aspirin and other NSAIDs)	▪ High salt intake
▪ Ingestion of toxins or corrosive materials	▪ Radiation therapy

or discomfort is present, the patient should avoid irritating foods and beverages; these often include alcohol, coffee (including decaffeinated), tea, cola beverages, spicy foods, and fried or fatty foods. If food consumption causes intense pain or nausea and vomiting, food intake should be avoided for 24 to 48 hours to rest the stomach. If hypochlorhydria or achlorhydria is present, supplementation of iron and vitamin B$_{12}$ may be warranted.

Peptic Ulcer Disease

A **peptic ulcer** is an open sore that develops in the GI mucosa when gastric acid and pepsin overwhelm mucosal defenses and destroy mucosal tissue. It is most often caused by *H. pylori* infection, which is present in approximately 30 to 60 percent of patients with gastric ulcers and 70 to 90 percent of those with duodenal ulcers.[12] Another major factor is the use of NSAIDs, which have both topical and systemic effects that can damage mucosal tissue. In rare cases, ulcers may develop from disorders that cause excessive acid secretion. Ulcer risk can be increased by cigarette smoking and psychological stress.

Effects of Emotional Stress Although most ulcers are associated with *H. pylori* infection or NSAID use, about a quarter of ulcers develop for other reasons.[13] Psychological stress is not believed to cause ulcers per se, but it has effects on physiological processes and behaviors that may increase a person's vulnerability. For example, stress may increase secretions of hydrochloric acid and pepsin, promote more rapid stomach emptying (increasing the acid load in the duodenum), and cause hormonal changes that impair immune responses and wound healing. Stress may also lead to behavioral changes, including the increased use of cigarettes, alcohol, and NSAIDs— all potential risk factors for ulcers. Thus, stress may contribute to ulcer development although its precise effects are not fully understood.

Symptoms of Peptic Ulcers Peptic ulcer symptoms vary. Some people are asymptomatic or experience only mild discomfort. In others, ulcer pain may be experienced as a hunger pain, a sensation of gnawing, or a burning pain in the stomach region. The pain or discomfort of duodenal ulcers may be relieved by food and recur several hours after a meal. Gastric ulcers, however, are often aggravated by food and can cause loss of appetite and eventual weight loss. Ulcer symptoms tend to go into remission regularly and recur every few weeks or months.

Complications of Peptic Ulcers Peptic ulcers are a major cause of GI bleeding, which occurs in up to 15 percent of ulcer cases.[14] If bleeding is severe, surgical intervention may be necessary; signs of severe bleeding include black, tarry stool samples

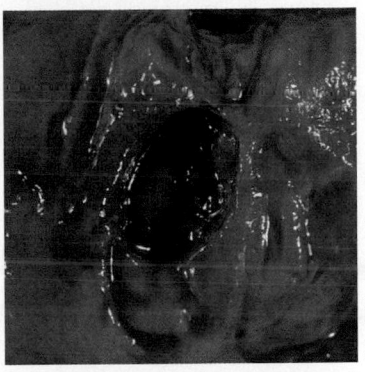

© Dr. E. Walker/Photo Researchers

A peptic ulcer, such as the gastric ulcer shown here, damages mucosal tissue and may cause pain and bleeding.

The specific reasons why ulcers develop are unknown; only 10 to 15 percent of people with chronic *H. pylori* infection develop a peptic ulcer.

In the United States, up to 80 percent of ulcers occur in the duodenum.

intractable vomiting: vomiting that is not easily managed or controlled.

gastritis: inflammation of stomach tissue.

Helicobacter pylori (H. pylori): a species of bacterium that colonizes gastric mucosa; a primary cause of gastritis and peptic ulcer disease.

hypochlorhydria (HIGH-poe-clor-HIGH-dree-ah): abnormally low gastric acid secretions.

achlorhydria (AY-clor-HIGH-dree-ah): absence of gastric acid secretions.

peptic ulcer: an open sore in the gastrointestinal mucosa; may develop in the esophagus, stomach, or duodenum.

peptic = related to digestion

Diet-Drug Interactions

Check this table for notable nutrition-related effects of the medications discussed in this chapter.

Antacids (aluminum hydroxide, magnesium hydroxide, calcium carbonate)	**Gastrointestinal effects:** Constipation (aluminum- or calcium-containing antacids), diarrhea (magnesium-containing antacids) **Dietary interactions:** May decrease iron, folate, or vitamin B_{12} absorption **Metabolic effects:** Electrolyte imbalances
Antibiotics (for *H. pylori* infection; include amoxicillin, metronidazole, tetracycline)	**Gastrointestinal effects:** Diarrhea (amoxicillin, tetracycline), nausea and vomiting (tetracycline), altered taste sensation (metronidazole) **Dietary interactions:** Avoid alcohol with metronidazole; tetracycline decreases iron absorption and binds calcium in the GI tract, reducing absorption of both the tetracycline and the calcium
Antisecretory drugs (proton-pump inhibitors, H2 blockers)	**Gastrointestinal effects:** Constipation, diarrhea, nausea and vomiting, abdominal pain (proton-pump inhibitors) **Dietary interactions:** May decrease iron, folate, and vitamin B_{12} absorption

or, occasionally, vomit that resembles coffee grounds. Other serious complications of ulcers include perforations of the stomach or duodenum and **gastric outlet obstruction** due to scarring or inflammation.

Drug Therapy for Peptic Ulcers The goals of ulcer treatment are to relieve pain, promote healing, and prevent recurrence. In most cases, treatment requires using a combination of antibiotics to eradicate *H. pylori* infection and/or discontinuing the use of aspirin and other NSAIDs, which can irritate the gastric mucosa and delay healing. Antisecretory drugs are prescribed to relieve pain and allow healing; these include proton-pump inhibitors, H2 blockers, and antacids (as used in GERD; see the earlier discussion on p. 468). Bismuth preparations (such as Pepto-Bismol) and the medication sucralfate can also help to promote healing by coating the mucosal lining and preventing further tissue erosion. See the Diet-Drug Interactions feature for significant nutrition-related effects of these medications.

Nutrition Care for Peptic Ulcers The goals of nutrition care are to correct nutrient deficiencies, if necessary, and encourage dietary and lifestyle practices that minimize symptoms. Patients should avoid substances that increase acid secretion or irritate the gastrointestinal lining; examples include alcohol, coffee and other caffeine-containing beverages, chocolate, and pepper, although individual tolerances vary. Small meals may be better tolerated than large ones. Patients should avoid food consumption for at least two hours before bedtime. There is no evidence that dietary adjustments alter the rate of healing.[15]

IN SUMMARY

▌ *Dyspepsia* refers to general symptoms of indigestion such as abdominal pain, nausea, and vomiting, which can be caused by a variety of medical conditions.

▌ Gastritis and peptic ulcer disease are most often associated with *H. pylori* infection, which can be eradicated by antibiotic therapy. NSAID use may promote gastritis and peptic ulcer disease by damaging the mucosal lining.

▌ Nutrition care for gastritis and peptic ulcer disease includes correcting any nutritional deficiencies that develop and eliminating dietary substances that cause pain or discomfort.

Gastric Surgery

Gastric surgery is sometimes necessary for treating stomach cancer, ulcer complications, and ulcers that are resistant to drug therapy. In recent years, gastric surgeries have also become popular treatments for severe obesity. This section describes **gastrectomy,** the surgery that removes diseased areas of the stomach, and **bariatric surgery,** the type of surgery that treats severe obesity.

Gastrectomy

Figure 17-4 illustrates some gastrectomy procedures. In a partial gastrectomy, only part of the stomach is removed, and the remaining portion is connected to the duodenum or jejunum. In a total gastrectomy, the surgeon removes the entire stomach and connects the esophagus directly to the small intestine.

Nutrition Care after Gastrectomy Following a gastrectomy, oral intake of fluids and foods is suspended until some healing has occurred, and fluids are supplied intravenously. Small sips of water, ice chips (melted in the mouth), and broth are usually the first fluids given orally. Once fluids are tolerated, patients are offered liquid meals (with no sugars) at first, and they usually progress to a soft food diet by the fourth or fifth day after surgery. Tube feedings may be necessary if complications prevent a normal progression to solid foods.[16]

Dietary measures after gastrectomy are influenced by the size of the remaining stomach, which influences meal size, and the stomach emptying rate, which affects food tolerances. Initial meals and snacks may include only one or two food items. Depending on the amount of food tolerated, the patient may require as many as six to eight small meals and snacks per day. The patient must avoid sweets and sugars because they increase osmolarity in the small intestine and potentiate the **dumping syndrome** (discussed below). Soluble fibers may be added to meals to help delay stomach emptying and reduce diarrhea. Although tolerances vary, patients may have difficulty with fatty foods, highly spiced foods, carbonated drinks, caffeine-containing beverages, alcohol, extremely hot or cold foods, peppermint, chocolate, and milk products (if lactose intolerant). Liquids are restricted during meals due to limited stomach capacity and because liquids can speed stomach emptying. Table 17-7 lists foods that are often permitted or limited in postgastrectomy diets.

Dumping Syndrome The dumping syndrome, a common complication of gastrectomy, is characterized by a group of symptoms resulting from rapid gastric emptying. Ordinarily, the pyloric sphincter controls the rate of flow from the stomach into the duodenum. After some types of stomach surgery, the hypertonic gastric contents are no longer regulated and rush into the small intestine more quickly after meals, causing a number of unpleasant effects. Early symptoms can occur within 30 minutes and may include nausea, vomiting, abdominal cramping, diarrhea, lightheadedness, rapid

gastric outlet obstruction: an obstruction that prevents the normal emptying of stomach contents into the duodenum.

gastrectomy (gah-STREK-ta-mee): the surgical removal of part of the stomach (partial gastrectomy) or the entire stomach (total gastrectomy).

bariatric (BAH-ree-AH-trik) **surgery:** surgery that treats severe obesity.

baros = weight

dumping syndrome: a cluster of symptoms that result from the rapid emptying of an osmotic load from the stomach into the small intestine.

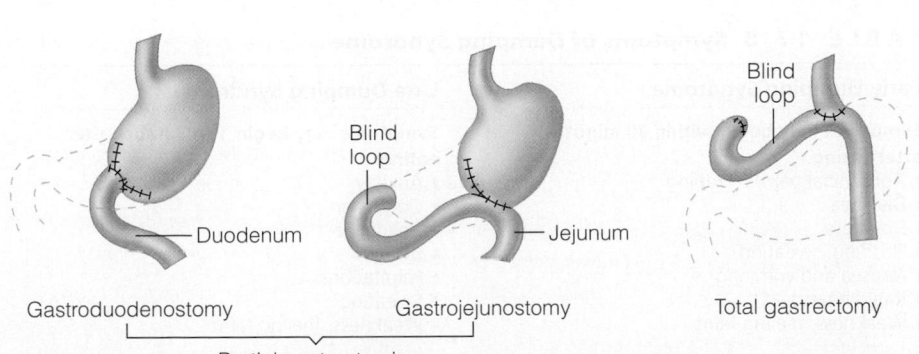

Gastroduodenostomy   Gastrojejunostomy   Total gastrectomy

Partial gastrectomies

FIGURE 17-4
Gastrectomy Procedures
In a gastrectomy, part or all of the stomach is surgically removed. The dashed lines show the removed section.

TABLE 17-7 Postgastrectomy Diet

Food Category	Foods Recommended (as tolerated)	Foods to Limit (unless tolerated)
Meat and meat alternates	Lean tender meats, fish, poultry, shellfish, eggs, smooth nut butters	Fried, tough, or chewy meats; frankfurters and sausages; bacon; luncheon meats; dried peas and beans
Milk and milk products	Milk, plain yogurt, mild cheeses	Milk shakes, chocolate milk, fruit yogurt
Breads and cereals	Breads, crackers, bagels, pasta, and breakfast cereals made from enriched white flour (cereals should contain no added sugars)	Breads and cereals with more than 2 grams of fiber per serving; baked goods with dried fruits, nuts, or seeds; granola; frosted cereals; pastries, doughnuts
Vegetables	Tender-cooked vegetables without peels, skins, or seeds; raw lettuce	Raw vegetables (except lettuce), beets, broccoli, brussels sprouts, cabbage, cauliflower, collard and mustard greens, corn, potato skins
Fruit	Canned fruit without added sugars, bananas, melon	Canned fruits in syrup, raw fruits (except bananas and melons), dried fruits, fruit juices
Beverages	Decaffeinated coffee and tea, beverages sweetened with artificial sweeteners	Caffeinated beverages; alcoholic beverages; beverages sweetened with sugars, corn syrup, or honey

heartbeat, and others (see Table 17-8). These symptoms may be due to a shift of fluid from blood vessels to the intestine that lowers blood volume and increases intestinal distention, and/or the accelerated release of GI hormones that induce strong intestinal contractions. Several hours later, symptoms of hypoglycemia may occur (see Table 17-8) because the unusually large spike in blood glucose following the meal (due to rapid nutrient influx and absorption) can result in an excessive insulin response.

Dietary adjustments can greatly minimize or prevent dumping syndrome. The goals are to limit the amount of food material that reaches the intestine, slow the rate of gastric emptying, and reduce foods that increase hypertonicity. Therefore, meal size is limited, fluids are restricted during meals, and sugars (including milk sugar) are restricted. In some cases, drugs that inhibit gastrointestinal motility may help. The "How to" lists practical suggestions for reducing the occurrence of dumping syndrome. The Case Study provides the opportunity to design a menu for a gastrectomy patient who is at risk for dumping syndrome.

Postsurgical Complications and Nutrition Status After a gastrectomy, it may take time for the patient to learn the amount of food that can be consumed without discomfort. The symptoms associated with meals may lead to food avoidance, substantial weight loss, and eventually malnutrition. Other nutrition problems that may occur after gastrectomy include the following:

▪ *Fat malabsorption.* Fat digestion may be impaired due to the accelerated transit of food material, which prevents normal mixing of fat with lipase and bile. If the

TABLE 17-8 Symptoms of Dumping Syndrome

Early Dumping Syndrome	Late Dumping Syndrome
Symptoms may begin within 30 minutes after eating.	**Symptoms may begin 1 to 3 hours after eating.**
▪ Abdominal pain, cramping	▪ Anxiety
▪ Diarrhea	▪ Confusion
▪ Dizziness	▪ Headache
▪ Flushing, sweating	▪ Hunger
▪ Nausea and vomiting	▪ Palpitations
▪ Rapid heartbeat	▪ Sweating
▪ Weakness, feeling faint	▪ Weakness, feeling faint

How To ALTER THE DIET TO REDUCE SYMPTOMS OF DUMPING SYNDROME

Dietary adjustments can greatly minimize or prevent symptoms of dumping syndrome. The following suggestions may help:

- Eat smaller meals that suit the reduced capacity of the stomach. Increase the number of meals consumed daily so that energy intake is adequate.

- Eat in a relaxed setting. Eat slowly, and chew food thoroughly.

- Limit the amount of fluid included in meals. Avoid drinking beverages within 30 minutes before and after meals, but be sure to consume adequate fluid to avoid dehydration.

- Avoid juices and sweetened beverages and foods that contain high amounts of sugar. Use artificial sweeteners to sweeten beverages and desserts.

- Avoid carbonated beverages if they cause bloating.

- Avoid foods and beverages that are very hot or very cold, unless tolerated.

- Include fiber-rich foods in each meal. Adding soluble fibers like pectin or guar gum to meals may help to control symptoms.

- Avoid milk and most milk products, which are high in lactose. Avoid enzyme-treated milk as well, because the breakdown products of lactose (glucose and galactose) can also cause dumping symptoms. Cheese may be better tolerated than milk because its lactose content is low. Make an effort to consume nonmilk calcium sources such as green leafy vegetables and tofu.

- If symptoms of hypoglycemia continue, try including a protein-rich food in each meal.

- Lie down for 20 to 30 minutes (or longer) after eating to help slow the transit of food to the small intestine. While eating a meal, sit upright.

Case Study NUTRITION CARE FOR PATIENT AFTER GASTRIC SURGERY

Dirk Hanson, a 58-year-old biology teacher, was admitted to the hospital for gastric surgery after numerous medical treatments failed to manage severe complications related to his peptic ulcer disease. A gastrojejunostomy was performed, and after about 24 hours, Mr. Hanson was able to take small sips of warm water. The health care team anticipates multiple nutrition-related problems and is taking measures to prevent them.

1. Review Figure 17-4 to better understand Mr. Hanson's surgical procedure. Consider the possibilities that he might experience the following symptoms: early satiety, nausea and vomiting, weight loss, dumping syndrome, fat malabsorption, anemia, and bone disease. Explain why each of these conditions may occur.

2. What type of diet will the physician prescribe for Mr. Hanson after he begins eating solid foods? Create a day's worth of menus, using foods from Table 17-7.

3. What advice can you give Mr. Hanson that will help to prevent dumping syndrome? List several foods from each major food group that may cause symptoms of dumping syndrome.

duodenum has been removed or bypassed, less lipase is available for fat digestion. Bacterial overgrowth, a common consequence of gastric surgeries, can lead to changes in bile acids that upset bile function. The fat malabsorption that results from these changes can eventually cause deficiencies of fat-soluble vitamins and some minerals. Supplemental pancreatic enzymes are sometimes provided to improve fat digestion.

Bone disease. Fat malabsorption can lead to malabsorption of both calcium and vitamin D; furthermore, patients at risk of dumping syndrome may need to avoid milk products, which are among the best sources of these nutrients. Bone density should be monitored during the years following surgery, and supplementation of calcium and vitamin D is often recommended.

Anemia. After gastrectomy, the reduction in gastric secretions impairs the absorption of both iron and vitamin B_{12}, often leading to anemia. If the duodenum has been removed or is bypassed, the risk of iron deficiency increases because the duodenum is a major site of iron absorption. Supplementation of both iron and vitamin B_{12} is often warranted after surgery.

Bacterial overgrowth may be a consequence of reduced gastric acid secretions, altered motility of intestinal contents, or changes in intestinal anatomy due to surgical reconstruction. Chapter 19 describes bacterial overgrowth in detail.

Fat malabsorption reduces calcium absorption because the negatively charged fatty acids combine with calcium (which is positively charged) and prevent its absorption.

Bariatric Surgery

Bariatric surgery was introduced in Chapter 7 (see p. 168).

Reminder: a healthy BMI usually falls between 18.5 and 25.

The gastric bypass surgery is also known as a *Roux-en-Y gastric bypass:* the Swiss surgeon *Roux* developed the procedure, which reconstructs the small intestine such that it resembles the letter *Y*.

Bariatric surgery is currently considered the most effective and durable treatment for morbid obesity.[17] Candidates for bariatric surgery are obese individuals who have a body mass index (BMI) greater than 40, or a BMI between 35 and 40 accompanied by severe weight-related problems such as diabetes, hypertension, or debilitating osteoarthritis. The individual who seeks bariatric surgery should have attempted at least six months of medically supervised weight-loss measures prior to seeking surgery. Bariatric surgery patients should have realistic expectations about the amount of weight they are likely to lose, the diet they will need to follow, and the complications that may ensue. Some types of bariatric surgery can dramatically affect health and nutrition status, and many patients require lifelong management.

Bariatric Surgical Procedures Figure 17-5 illustrates the most popular surgical options for weight reduction. The gastric bypass operation, which accounts for about 88 percent of bariatric surgeries,[18] creates a small gastric pouch that reduces stomach capacity and thereby restricts meal size. In addition, the gastric pouch is connected directly to the jejunum, resulting in significant nutrient malabsorption because the flow of food bypasses a significant portion of the small intestine. In the gastric banding procedure, a gastric pouch is created using a fluid-filled inflatable band; adjusting the band's fluid level can tighten or loosen the band and alter the size of the opening to the rest of the stomach. A smaller opening slows the pouch emptying rate and prolongs the sense of fullness after a meal. Whereas the gastric bypass operation is usually permanent, the gastric banding procedure is fully reversible.

Clinical studies indicate that the gastric bypass surgery generally results in greater weight loss than the gastric banding procedure. Although study results vary, gastric bypass patients typically lose 60 to 70 percent of their excess body weight, whereas those who undergo gastric banding lose about 50 percent of their excess weight.[19]

Nutrition Care after Bariatric Surgery The gastric pouch created by surgery may eventually expand to hold up to one-half cup of food, but its initial capacity is only a few tablespoons. Only sugar-free clear liquids are given during the first two days following bariatric surgery. Afterward, patients consume a full liquid diet (low in sugars) at first, followed by pureed foods and then a soft food diet; the diet is advanced as

FIGURE 17-5 Surgical Procedures for Severe Obesity

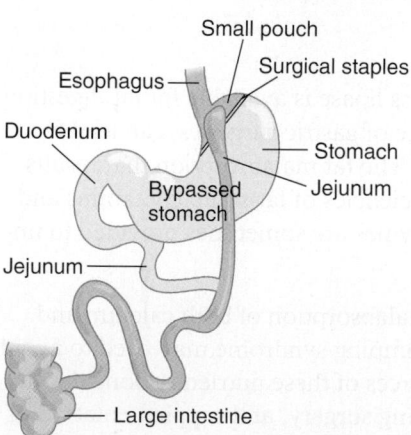

In a gastric bypass operation, the surgeon constructs a small gastric pouch and connects it directly to the jejunum so that the flow of food bypasses a substantial portion of the small intestine. The dark pink area highlights the altered flow of food through the GI tract. The pale pink area indicates the bypassed sections.

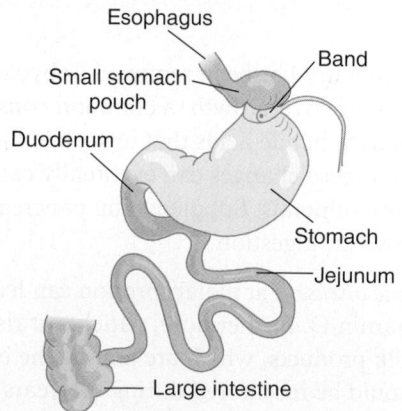

In a gastric banding procedure, the surgeon creates a small gastric pouch using an inflatable band placed near the top of the stomach. The band is tightened or loosened by adding or removing fluid via an access port placed under the skin.

476

tolerated.[20] Once the diet progresses to solid foods, patients may consume between three and six small meals per day. Only small portions of food can be consumed at each meal because overeating can stretch the gastric pouch or result in vomiting or regurgitation. Similarly, fluids must be consumed separately from meals to avoid excessive distention. Other nutrition-related concerns include the following:

- *Protein.* Recommendations range from 1.0 to 1.5 grams of protein per kilogram of ideal body weight per day[21]; high-protein foods should be eaten before consuming other foods in a meal. Protein supplements are used to improve protein intakes.

- *Vitamins and minerals.* Supplemental vitamin B_{12}, vitamin D, iron, and calcium are recommended after surgery. A daily multivitamin/mineral supplement ensures that patients meet their needs for other nutrients.

- *Foods to avoid.* Some foods may obstruct the gastric outlet; these include doughy or sticky breads, pasta products, and rice; fibrous vegetables such as asparagus and celery; foods with seeds, peels, or skins; nuts; popcorn; and tough, dry meats.

- *Dumping syndrome.* To avoid symptoms of dumping syndrome, gastric bypass patients must carefully control food portions, avoid foods high in sugars, and consume liquids between meals.

After bariatric surgery, patient education and counseling are critical for weight loss and weight management, and patients also need to learn the elements of a healthy diet. The "How to" includes additional dietary suggestions for patients who have undergone bariatric surgery.

Nutrition Problems following Bariatric Surgery   Common complaints after bariatric surgery include nausea, vomiting, and constipation.[22] Although the cause of these problems varies among patients, dietary noncompliance and inadequate fluid intake are often contributing factors, and improved dietary intakes can help in resolving these conditions. The long-term complications that may develop after gastric bypass surgery are similar to those that arise after gastrectomy and include fat malabsorption, bone disease, and anemia. Rapid weight loss increases a person's risk of developing

How To   ALTER DIETARY HABITS TO ACHIEVE AND MAINTAIN WEIGHT LOSS AFTER BARIATRIC SURGERY

Patients need to learn new dietary habits after bariatric surgery. The following recommendations may help:

- Consume only small portions of food and chew food thoroughly. Use a small spoon, and take small bites. Relax and enjoy the meal, taking at least 20 minutes to eat.

- Understand that, at first, the appropriate amount of each food served at mealtime may be only a few spoonfuls. Learn to recognize the sensations that occur when the gastric pouch is full. Signs of fullness may include pressure in the stomach region, a slight feeling of nausea, or pain in the upper chest or shoulder.

- To control vomiting, try eating smaller volumes of food, eating more slowly, and avoiding foods known to cause difficulty. Continued vomiting may be a sign that some food choices or amounts are inappropriate.

- Consume food only during designated mealtimes (usually three to six small meals per day), and avoid consuming foods

at other times of the day. Snacking throughout the day can become a bad habit that causes weight to be regained.

- Avoid consuming liquids within 30 minutes of mealtime. Avoid high-kcalorie drinks like sweetened soda, milk shakes, and alcoholic beverages. Avoid drinking carbonated beverages or using a straw, as these practices can increase stomach gas and cause bloating.

- Sip water and other beverages throughout the day to obtain sufficient fluids. Aim to consume between 6 and 8 cups of water and other noncaloric beverages daily. Remember that most people meet a significant fraction of their fluid needs by eating food, but a person who has undergone bariatric surgery must limit food intake.

- Engage in regular physical activity. Activity is a valuable aid to weight maintenance and can help to maintain lean tissue while weight is being lost.

gallbladder disease; patients at especially high risk sometimes have their gallbladders removed while undergoing bariatric surgery.

IN SUMMARY

▌ Gastric surgeries, used to treat cancer, peptic ulcer complications, and obesity, require dietary adjustments after surgery and are associated with complications that may affect nutrition status.

▌ Dumping syndrome, a common complication of gastric surgery, results from the rapid influx of nutrients from the stomach into the small intestine.

▌ After bariatric surgery, patients must learn to consume appropriate food portions; use dietary supplements to prevent nutrient deficiencies; and choose foods that are unlikely to cause discomfort, vomiting, or dumping syndrome.

Nutrition Assessment Checklist FOR PEOPLE WITH UPPER GI TRACT DISORDERS

Medical History

Check the medical history to uncover conditions or treatments that may:

☐ Interfere with chewing or swallowing
☐ Lead to dry mouth
☐ Lead to dyspepsia, nausea, or vomiting

Check for a medical diagnosis of:

☐ Gastritis or peptic ulcer
☐ GERD
☐ Hiatal hernia
☐ Pernicious anemia

For a patient who has undergone gastric surgery, check for the following complications:

☐ Anemia
☐ Bone disease
☐ Dumping syndrome
☐ Fat malabsorption

Medications

Record all medications and note:

☐ Aspirin or NSAID use in patients with gastritis or peptic ulcer disease
☐ Medications that may cause dry mouth
☐ Medications that may cause nausea and vomiting

To help alleviate nausea, suggest that medications be taken with food, when possible.

Dietary Intake

To devise an acceptable meal plan, obtain:

☐ An accurate and thorough record of food intake
☐ A record of foods that provoke symptoms of GERD, dyspepsia, gastritis, peptic ulcers, or dumping syndrome

For patients on long-term dysphagia diets, monitor:

☐ Appetite
☐ Food tolerances
☐ Variety of foods regularly consumed

Anthropometric Data

Measure baseline height and weight. Address weight loss early to prevent malnutrition in patients with:

☐ Dumping syndrome
☐ Dyspepsia or long-term nausea
☐ Dysphagia or difficulty chewing
☐ Malabsorption

Laboratory Tests

Check laboratory tests for signs of dehydration for patients with:

☐ Constipation
☐ Dumping syndrome
☐ Persistent vomiting

Check laboratory tests for nutrition-related anemia in patients with:

☐ Gastritis
☐ Long-term use of antisecretory drugs
☐ Previous gastric surgeries

Physical Signs

Look for physical signs of:

☐ Dehydration—in patients with constipation, dumping syndrome, or persistent vomiting
☐ Iron and vitamin B_{12} deficiencies—in patients with hypochlorhydria or achlorhydria

Clinical Applications

1. Although some individuals require a mechanically altered diet for just a few weeks, others have medical problems that require long-term use of such diets. Consider the difference between working with a person who has had a swallowing problem for years and a person who recently had mouth surgery and is just beginning to eat again.

   ▮ Explain how the needs of these individuals may differ. What nutrition-related problems may develop if a person has been following a restrictive dysphagia diet for several years?

   ▮ Using Tables 17-1 and 17-4 and the "How to" on p. 466, create a day's worth of menus for a person who requires long-term use of a pureed dysphagia diet and tolerates only liquids that have a honeylike consistency.

2. Jillian, a 35-year-old woman who is 5 feet 4 inches tall and weighs 227 pounds, has had severe hip and knee osteoarthritis for several years and was recently diagnosed with type 2 diabetes. After trying numerous diet programs without success, she finally visits a bariatric surgeon to learn about the surgical options for treating her obesity.

   ▮ Calculate Jillian's BMI, and explain why or why not Jillian would be a good candidate for bariatric surgery.

   ▮ Should Jillian decide to undergo gastric bypass surgery, she will need to permanently change her dietary habits. Outline the diet progression typically recommended after bariatric surgery, and describe the dietary measures necessary for preventing vomiting, distention of the gastric pouch, gastric outlet obstruction, and dumping syndrome.

   ▮ Explain why dehydration is a frequent complication following bariatric surgery. What tips can you give Jillian to help her avoid this problem?

Self Check

1. Foods permitted on the clear liquid diet include all of the following *except:*
   a. milk.
   b. fruit ices.
   c. flavored gelatin.
   d. consommé.

2. If a patient with dysphagia has difficulty swallowing solids but can easily swallow liquids:
   a. the problem is probably a motility disorder.
   b. the patient most likely has achalasia.
   c. the problem is probably an esophageal obstruction.
   d. the problem is most likely oropharyngeal dysphagia.

3. The clinican working with a patient with dysphagia should recognize that:
   a. only pureed foods should be given to minimize the risk of aspiration.
   b. the patient can have any food that can be comfortably and safely chewed and swallowed.
   c. highly seasoned foods are often restricted.
   d. conventional diets are unable to meet total nutrient needs, so supplements are required.

4. Conditions associated with an increased risk of developing GERD include:
   a. hiatal hernia.
   b. asthma.
   c. pregnancy.
   d. all of the above.

5. For the patient with persistent vomiting, the major nutrition-related concern(s) is/are:
   a. dehydration and malnutrition.
   b. reflux esophagitis.
   c. dyspepsia.
   d. peptic ulcers.

6. Chronic gastritis frequently leads to:
   a. dumping syndrome.
   b. bone disease.
   c. iron and vitamin B_{12} deficiencies.
   d. excessive hydrochloric acid secretion.

7. The primary cause of most peptic ulcers is:
   a. consumption of spicy foods.
   b. hypochlorhydria.
   c. smoking cigarettes.
   d. *Helicobacter pylori* infection.

8. Foods discouraged for patients with gastritis or active ulcers include those that:
   a. are high in fiber.
   b. irritate the gastric mucosa.
   c. are easy to swallow.
   d. contain simple sugars.

9. People at risk of dumping syndrome should generally avoid:
   a. high-fiber foods.
   b. sweets and sugars.
   c. beverages.
   d. bread and potatoes.

10. The clinician assessing a patient who underwent a gastrectomy several years ago should be alert to signs of:
   a. dysphagia.
   b. GERD.
   c. anemia.
   d. gastritis.

Answers to these questions can be found in Appendix H.

Notes

1. T. E. Daniels, Diseases of the mouth and salivary glands, in L. Goldman and D. Ausiello, eds., *Cecil Medicine* (Philadelphia: Saunders, 2008), pp. 2867–2874; D. P. DePaola and coauthors, Nutrition and dental medicine, in M. E. Shils and coeditors, *Modern Nutrition in Health and Disease* (Baltimore: Lippincott Williams & Wilkins, 2006), pp. 1152–1178.

2. R. K. Goyal, A. Chaudhury, and H. Mashimo, Oropharyngeal and esophageal motility disorders, in N. J. Greenberger, ed., *Current Diagnosis and Treatment: Gastroenterology, Hepatology, and Endoscopy* (New York: McGraw-Hill Companies, 2009), pp. 155–171.

3. R. C. Orlando, Diseases of the esophagus, in L. Goldman and D. Ausiello, eds., *Cecil Medicine* (Philadelphia: Saunders, 2008), pp. 998–1009.

4. American Dietetic Association, *Nutrition Care Manual* (Chicago: American Dietetic Association, 2010).

5. J. R. Agrawal and S. Friedman, Gastrointestinal and biliary complications of pregnancy, in N. J. Greenberger, ed., *Current Diagnosis and Treatment: Gastroenterology, Hepatology, and Endoscopy* (New York: McGraw-Hill Companies, 2009), pp. 75–97.

6. Agrawal and Friedman, 2009; Orlando, 2008.

7. N. J. Talley, Functional gastrointestinal disorders: Irritable bowel syndrome, dyspepsia, and noncardiac chest pain, in L. Goldman and D. Ausiello, eds., *Cecil Medicine* (Philadelphia: Saunders, 2008), pp. 990–998.

8. V. S. Wang and R. Burakoff, Functional (nonulcer) dyspepsia, in N. J. Greenberger, ed., *Current Diagnosis and Treatment: Gastroenterology, Hepatology, and Endoscopy* (New York: McGraw-Hill Companies, 2009), pp. 189–199.

9. Talley, 2008.

10. Wang and Burakoff, 2009; S. Escott-Stump, *Nutrition and Diagnosis-Related Care* (Baltimore: Lippincott Williams & Wilkins, 2008).

11. American Dietetic Association, 2010; Escott-Stump, 2008.

12. E. Lew, Peptic ulcer disease, in N. J. Greenberger, ed., *Current Diagnosis and Treatment: Gastroenterology, Hepatology, and Endoscopy* (New York: McGraw-Hill Companies, 2009), pp. 175–183.

13. K. Ramakrishnan and R. C. Salinas, Peptic ulcer disease, *American Family Physician* 76 (2007): 1005–1012.

14. Lew, 2009.

15. F. F. Ferri, *Ferri's Clinical Advisor 2009: Instant Diagnosis and Treatment* (Philadelphia: Mosby, 2009), pp. 693–694; W. F. Stenson, The esophagus and stomach, in M. E. Shils and coeditors, *Modern Nutrition in Health and Disease* (Baltimore: Lippincott Williams & Wilkins, 2006), pp. 1179–1188.

16. American Dietetic Association, 2010.

17. M. K. Robinson and N. J. Greenberger, Treatment of obesity: The impact of bariatric surgery, in N. J. Greenberger, ed., *Current Diagnosis and Treatment: Gastroenterology, Hepatology, and Endoscopy* (New York: McGraw-Hill Companies, 2009), pp. 210–221.

18. G. Woodard and J. Morton, Bariatric surgery, in M. Feldman and coeditors, *Sleisenger and Fordtran's Gastrointestinal and Liver Disease* (Philadelphia: Saunders, 2010), pp. 115–119.

19. Weight Management Dietetic Practice Group, *ADA Pocket Guide to Bariatric Surgery* (Chicago: American Dietetic Association, 2009), pp. 1–16.

20. J. I. Mechanick and coauthors, American Association of Clinical Endocrinologists, the Obesity Society, and American Society for Metabolic and Bariatric Surgery medical guidelines for clinical practice for the perioperative nutrition, metabolic, and nonsurgical support of the bariatric surgery patients, *Obesity* 17 (2009): S1–S41.

21. L. Aills and coauthors, ASMBS allied health nutritional guidelines for the surgical weight loss patient, *Surgery for Obesity and Related Diseases* 4 (2008): S73–S108.

22. Robinson and Greenberger, 2009.

CHAPTER 17

Nutrition in Practice
NUTRITION AND ORAL HEALTH

Various aspects of nutrition and oral health were discussed earlier in this book. Chapters 2 and 9 described the effects of carbohydrate and fluoride, respectively, on the development of dental caries. Chapter 12 explained how the prolonged use of nursing bottles increases the risk of dental caries in babies. This Nutrition in Practice provides additional information about dental caries and introduces other problems related to oral health. Related terms are defined in the accompanying glossary.

What is dental caries, and what factors influence its development?

Dental caries is an oral infectious disease that affects the structures and integrity of the teeth. Caries develops when the bacteria that reside in **dental plaque** metabolize dietary carbohydrates and produce acids that attack tooth enamel. If allowed to progress, the decay can penetrate the dentin and destroy other structures that support and maintain the tooth (see Figure NP17-1). The development of dental caries is influenced by the type of carbohydrate consumed, the frequency of carbohydrate intake, the stickiness of the foods that contain carbohydrate, and the availability of saliva to rinse the teeth and neutralize acid. Other factors include oral hygiene, fluoride intake, and the composition of tooth enamel. Poor nutrition during pregnancy, infancy, or early childhood can impair the development of healthy teeth and increase vulnerability to dental caries.[1] Table NP17-1 lists examples of nutrient deficiencies that influence the development of dental caries.

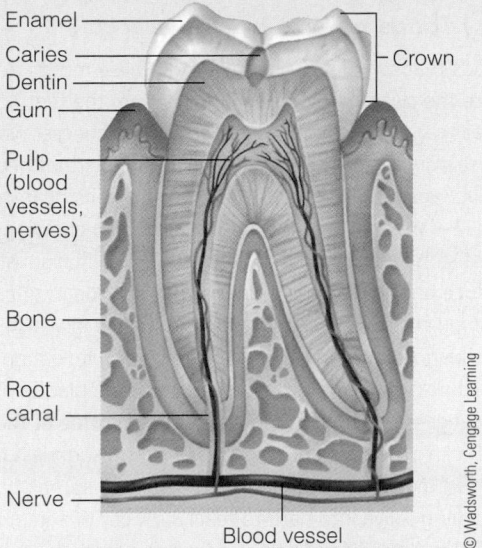

FIGURE NP17-1 Development of Dental Caries
Dental caries begins when acid dissolves the enamel that covers the tooth. If not repaired, the decay may penetrate the dentin and spread into the pulp of the tooth, causing inflammation, abscess, and possible loss of the tooth.

Which foods promote caries development?

The most **cariogenic** foods are carbohydrate-containing foods that remain in contact with the teeth for prolonged periods, are difficult to clear from the mouth, or are consumed frequently or over an extended time period. Examples include hard candies or lozenges that dissolve slowly in the mouth; sticky or chewy foods such as dried fruits, jelly beans, or chewy breads; starchy snack foods such as pretzels or chips, and sweetened beverages that are repeatedly sipped. These foods can be eaten without provoking tooth decay if they are consumed quickly and removed from tooth surfaces promptly.

GLOSSARY

cariogenic (KAH-ree-oh-JEN-ik): conducive to development of dental caries.

dental calculus: mineralized dental plaque, often associated with inflammation and bleeding.

dental caries (KAH-reez): infectious disease of the teeth that causes the gradual decay and disintegration of tooth structures.

dental plaque (PLACK): a film of bacteria and bacterial by-products that accumulates on the tooth surface.

gingiva (jin-JYE-va, JIN-jeh-va): the gums.

gingivitis (jin-jeh-VYE-tus): inflammation of the gums, characterized by redness, swelling, and bleeding.

periodontal disease: disease that involves the connective tissue that supports the teeth.

periodontitis: inflammation or degeneration of the tissues that support the teeth.

periodontium: the tissues that support the teeth, including the gums, cementum (bonelike material covering the dentin layer of the tooth), periodontal ligament, and underlying bone.
▍ *peri* = around, surrounding
▍ *odont* = tooth

TABLE NP17-1 Nutrient Deficiencies Affecting Development of Dental Caries

Nutrient Deficiency	Effect on Dental Tissues
Protein	Increased solubility of tooth enamel
Vitamin A	Impaired formation of tooth enamel
Vitamin D	Impaired formation of tooth enamel, reduced enamel mineralization
Calcium	Reduced enamel mineralization
Phosphorus	Reduced enamel mineralization
Fluoride	Increased susceptibility to enamel demineralization, reduced protection against decay-causing bacteria

Acidic foods and beverages, such as cola drinks, citrus fruits and juices, pickles, and some herbal teas, also contribute to the erosion of tooth enamel.[2] Acidic food items are more cariogenic when consumed after or between meals because they are less likely to be rinsed away by saliva or neutralized by alkaline foods that may be consumed during the meal.

Do any foods help to prevent caries?

Yes, foods that stimulate saliva flow, neutralize mouth acidity, or induce the clearance of food particles from the teeth can help to prevent caries formation. Examples include cheese, which increases salivary secretions and contains nutrients that neutralize acid; milk, which reduces mouth acidity due to its nearly neutral pH; and raw vegetables, which require vigorous chewing and therefore stimulate saliva flow. Some foods contain substances that reduce enamel solubility, such as the fluoride in tea and an unidentified compound in cocoa. The sequence in which foods are eaten influences caries development; for example, eating cheese after consuming an acidic fruit dessert can raise plaque pH and reduce caries risk, whereas drinking sugared coffee at the end of the meal can prolong plaque acidity. Chewing sugarless gum after or between meals can help to prevent caries by stimulating salivary flow, pushing saliva into hard-to-reach crevices in and around teeth, and removing food particles from the teeth.[3]

How does saliva protect against caries formation?

In addition to rinsing away the sugars and food particles that remain on the teeth, saliva contains substances (such as proteins, bicarbonate, and phosphates) that dilute and neutralize mouth acidity. Furthermore, saliva contains antimicrobial proteins (immunoglobulins and lysozyme) that defend against bacteria and fungi. Finally, the calcium, phosphate, and fluoride ions in saliva help to prevent dissolution of enamel and promote remineralization. Thus, if salivary secretions are low or absent, the risk of developing dental caries and other dental diseases greatly increases.

Does dental plaque contribute to other dental diseases?

Yes, the deposits of plaque can thicken and lead to other dental problems. As plaque accumulates, it fills with calcium and phosphate, eventually forming **dental calculus.** Calculus develops either at the gum surface or in the crevice between the gum and a tooth; its presence may lead to additional plaque retention. The buildup of plaque and calculus increases the likelihood of infection and subsequent inflammation.

Periodontal disease is the name given to inflammatory conditions that involve the **periodontium,** the structures that support the tooth in its bony socket. The periodontium includes the gums (called **gingiva**), other connective tissues surrounding the tooth, and the bone underneath. Inflammation of the gums, called **gingivitis,** is characterized by redness, bleeding, and swelling of gum tissue. **Periodontitis** is an inflammation of the other tissues surrounding the tooth. As plaque invades the space below the gum line, the combination of toxic bacterial by-products and the body's immune response can damage the tissues holding a tooth in place. Left untreated, the tissues and bone of the

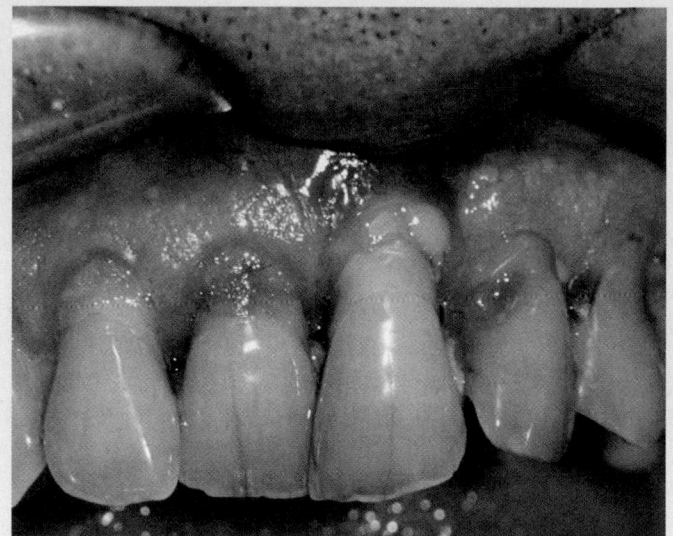

Periodontal disease destroys the tissues and bones that hold teeth in place.

© CNRI/Photo Researchers, Inc.

peridontium may ultimately be destroyed, leading to permanent tooth loss.

What are the risk factors for periodontal disease, and how is it treated?

Dental plaque is the major risk factor associated with periodontal disease, and the severity of disease is related to the amount of plaque present. Tobacco smoking is another factor, possibly due to its destructive effects on cellular immune responses. The likelihood of developing periodontal disease is increased if a person has a chronic illness that impairs immune status, such as diabetes mellitus or HIV infection. Other risk factors include stress, pregnancy, use of certain medications (including oral contraceptives, antiepileptic drugs, and anticancer drugs), and dental conditions that increase plaque accumulation, such as poorly aligned teeth or ill-fitting bridges.[4] Strategies for reducing risk focus on improving oral hygiene (proper brushing and flossing) and encouraging smoking cessation.

Treatment of periodontal disease depends on the extent of damage. In mild cases, deep cleaning and proper oral hygiene may reverse the condition. Antimicrobial mouth rinses and topical antibiotics may be prescribed to control infection. Surgical approaches may be used to remove plaque or calculus deposits underneath gum tissue or to replace tissues that have been destroyed.

How do chronic illnesses influence a person's risk for dental and oral diseases?

Some chronic illnesses can alter the structure and function of dental tissues, impair immune responses, or cause reduced salivary flow. Furthermore, many medications can reduce salivary secretions, along with the immune protection that saliva provides. Examples of conditions that contribute to dental and oral diseases include the following:

▌ *Diabetes mellitus.* Periodontal disease is more prevalent among people with diabetes mellitus, especially those whose diabetes is poorly controlled. People with diabetes often have

impaired immune responses and a greater susceptibility to infections. Diabetes also favors the growth of bacteria that tend to infect periodontal tissues. People with diabetes tend to have higher plaque accumulations and dry mouth. In addition, the damaging effects of hyperglycemia weaken the collagen structure of dental tissues, making them more vulnerable to destruction.[5]

▌ *Human immunodeficiency virus (HIV)/AIDS.* HIV infection is characterized by compromised immunity, and the risk of developing periodontal disease is closely linked to the extent of HIV infection. Those at greatest risk include smokers and patients in the advanced stages of disease. HIV-infected individuals often develop dry mouth as a result of medications or salivary gland dysfunction.[6] In untreated persons, fungal and viral infections are common and may cause burning in the mouth and painful ulcerations.

▌ *Oral cancers.* Radiation treatment of oral cancers often causes serious oral and dental complications. Inflammation and tissue damage can be so severe that the radiation treatment may need to be halted or the intensity significantly reduced. Other complications include dry mouth, fungal and viral infections, changes in taste sensation, and tissue and muscle scarring (which often reduces chewing ability). To minimize complications, dental care is often initiated before radiation therapy begins.

Can dental diseases have adverse effects on health beyond their effects on the teeth?

Yes, the bacteria that reside on dental tissues can enter the bloodstream and travel to other tissues; therefore, they may be able to trigger immune responses or cause infections elsewhere in the body. Evidence supports a link between dental bacteria and other medical conditions, including the following:[7]

▌ *Atherosclerosis and heart disease.* The inflammatory process induced by periodontal pathogens may increase levels of cytokines and other mediators that accelerate the progression of atherosclerosis. Periodontal bacteria may also initiate plaque formation and induce blood clotting.

▌ *Diabetes mellitus.* The chronic inflammation caused by periodontal disease can exacerbate insulin resistance and provoke events leading to type 2 diabetes. Severe periodontal disease has also been linked to poor glycemic control in persons with diabetes.

▌ *Respiratory illnesses.* The teeth of hospitalized individuals can become colonized with bacteria that cause respiratory illnesses. In addition, mortality rates associated with pneumonia have been found to be four times higher in individuals with periodontal disease.[8]

Research studies are in progress to confirm cause-and-effect relationships between oral bacteria and the medical conditions described above, as well as the specific mechanisms involved.

Developing sound eating habits and maintaining good dental hygiene are practices that can promote dental health and possibly reduce the risk of developing other medical problems. Additional studies will help to clarify the complex interactions between dental disease and chronic illnesses.

Notes

1. D. P. DePaola and coauthors, Nutrition and dental medicine, in M. E. Shils and coeditors, *Modern Nutrition in Health and Disease* (Baltimore: Lippincott Williams & Wilkins, 2006), pp. 1152–1178.

2. S. Wongkhantee and coauthors, Effect of acidic food and drinks on surface hardness of enamel, dentine, and tooth-coloured filling materials, *Journal of Dentistry* 34 (2006): 214–220.

3. S. K. Stookey, The effect of saliva on dental caries, *Journal of the American Dental Association* 139 (2008): 11S–17S; DePaola and coauthors, 2006.

4. J. Kim and S. Amar, Periodontal disease and systemic conditions: A bidirectional relationship, *Odontology* 94 (2006): 10–21.

5. American Dietetic Association, Position of the American Dietetic Association: Oral health and nutrition, *Journal of the American Dietetic Association* (2007): 1418–1428; DePaola and coauthors, 2006.

6. J. C. C. Filho and E. M. Giovani, Xerostomy, dental caries and periodontal disease in HIV+ patients, *Brazilian Journal of Infectious Diseases* 13(2009): 13–17.

7. P. Weidlich and coauthors, Association between periodontal diseases and systemic diseases, *Brazilian Oral Research* 22 (2008): 32–43.

8. Weidlich and coauthors, 2008.

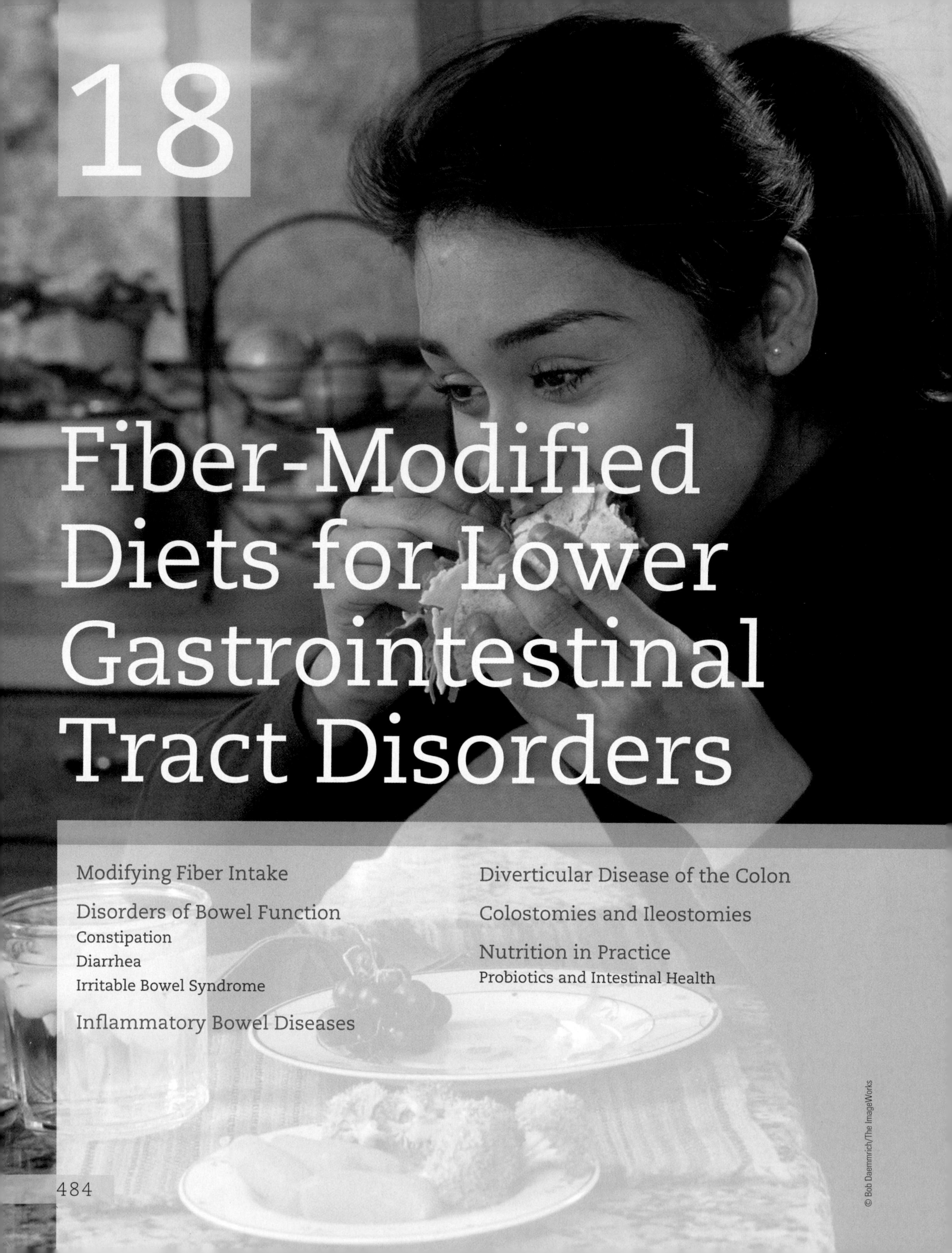

Fiber-Modified Diets for Lower Gastrointestinal Tract Disorders

Modifying Fiber Intake

Disorders of Bowel Function
Constipation
Diarrhea
Irritable Bowel Syndrome

Inflammatory Bowel Diseases

Diverticular Disease of the Colon

Colostomies and Ileostomies

Nutrition in Practice
Probiotics and Intestinal Health

The lower gastrointestinal (GI) tract, which includes the small intestine and large intestine, continues the digestion of food that arrives from the stomach and facilitates nutrient absorption. This chapter discusses the potential benefits of adjusting fiber intake when treating some lower GI disorders.

Modifying Fiber Intake

Diets can be modified by either increasing or decreasing dietary fiber. As discussed in Chapter 2, *insoluble fibers* increase fecal weight and speed the passage of wastes through the large intestine. *Soluble, viscous fibers* slow the passage of food through the GI tract, which increases satiety and delays glucose absorption in the small intestine. Table 18-1 lists some medical problems that may be treated with modifications in fiber intake.

Chapter 2 and Appendix A provide information about the fiber content of foods. People can increase their intake of fiber-rich foods by emphasizing the consumption of whole grains, legumes, nuts and seeds, and fruits and vegetables. Figure 18-1 shows a one-day menu that includes about 38 grams of fiber, which is the DRI for an adult male. People who must restrict dietary fiber can limit consumption of the high-fiber foods listed in these resources.

Increased intestinal gas **(flatulence)** may be an unpleasant side effect of consuming a high-fiber diet. The undigested fibers pass into the colon, where they are fermented by bacteria, which produce gas as a by-product. Therefore, health practitioners advise that fiber-containing foods be added gradually at first and portions increased as tolerance improves. Intestinal gas also develops when other carbohydrates are incompletely digested or absorbed; these include fructose, sugar alcohols (sorbitol, xylitol, mannitol), the indigestible carbohydrates in beans (raffinose and stachyose), and some forms of resistant starch, found in grain products and potatoes.[1] Table 18-2 lists foods commonly associated with excessive gas production, although individual responses vary. Malabsorption disorders (discussed in Chapter 19) also cause flatulence because the undigested nutrients can be metabolized by colonic bacteria. Swallowed air that is not expelled by belching may travel to the intestines and be a source of intestinal gas as well. Note that people often attribute abdominal bloating and pain to excessive gas, but these symptoms do not correlate well with increased gas production.[2]

IN SUMMARY

- Insoluble dietary fibers affect stool volume and transit time in the large intestine. Soluble fibers slow the passage of food in the GI tract and help to reduce blood cholesterol levels.

- Because fiber-containing foods can increase intestinal gas, they should be added gradually and portions increased as tolerance improves.

Disorders of Bowel Function

Many different factors can alter the fluid content of stools or transit time and lead to constipation or diarrhea. Consequently, these symptoms accompany a wide variety of health problems. Because dietary fiber affects GI transit time as well as stool volume and composition, modifying fiber intake can often serve in the treatment of these problems.

TABLE 18-1 Indications for Modifying Fiber Intake

Increasing Fiber Intake
- Constipation
- Diabetes mellitus
- Diverticulosis (prevention)
- Heart disease
- Irritable bowel syndrome
- Prediabetes
- Weight management

Reducing Fiber Intake
- Inflammatory bowel diseases (active episodes)
- Post-surgical diet (depending on tolerance)

The fiber DRI for women and men aged 19 to 50 years are 25 grams and 38 grams, respectively.

TABLE 18-2 Foods That May Increase Intestinal Gas

Apples
Artichokes
Asparagus
Beer
Broccoli
Brussels sprouts
Cabbage
Carbonated beverages
Carrots
Corn
Dried beans and peas
Fructose-sweetened products
Fruit juices
Green beans
Leeks
Milk products (if lactose intolerant)
Onions
Peanuts
Pears
Turnips
Wheat

flatulence: the condition of having excessive intestinal gas, which causes abdominal discomfort.

FIGURE 18-1 Sample One-Day Menu Containing 38 g Fiber

Constipation

Diagnosis of constipation is based, in part, on a defecation frequency of fewer than three bowel movements per week. Other symptoms may include the passage of hard stool and excessive straining during defecation. In some cases, a person's perception of constipation may be due to a mistaken notion of what constitutes "normal" bowel habits, so the person's expectations about bowel function may need to be addressed. Constipation affects up to 28 percent of the population in Western countries and is particularly prevalent among women and older adults.[3]

Causes of Constipation The risk of constipation is increased in individuals with a low-fiber diet, low food intake, inadequate fluid intake, or low level of physical activity. All of these factors can extend transit time, leading to increased water reabsorption within the colon and dry, hard stools that are difficult to pass. Medical conditions often associated with constipation include diabetes mellitus and hypothyroidism. Neurological conditions such as Parkinson's disease, spinal cord lesions, and multiple sclerosis may cause motor problems that lead to constipation. During pregnancy, women often experience constipation because the enlarged uterus presses against the rectum and colon. Constipation is also a common side effect of several classes of medications and some dietary supplements, including opiate-containing analgesics, tricyclic antidepressants, anticonvulsants, calcium channel blockers, diuretics, aluminum-containing antacids, and iron and calcium supplements.

Treatment of Constipation The primary treatment for constipation is a gradual increase in fiber intake to about 25 grams per day.[4] High-fiber diets increase stool weight and fecal water content and promote a more rapid transit of materials through the colon. Foods that increase stool weight the most are wheat bran, fruits, and vegetables.[5] Bran intake can be increased by adding bran cereals and whole-wheat bread to the diet or by mixing bran powder with beverages or foods. The transition to a high-fiber diet may be difficult for some people because it can increase intestinal gas, so high-fiber foods should be added gradually, as tolerated. Fiber supplements like methylcellulose (Citrucel), psyllium (Metamucil, Fiberall), and polycarbophil (Fiber-Lax) are also effective (see Table 18-3). Unlike other fibers, methylcellulose and polycarbophil do not increase intestinal gas.

Several other measures may also help constipation. Consuming adequate fluid (usually 1.5 to 2 liters daily) helps to increase stool frequency in people who are already consuming a high-fiber diet.[6] An appropriate fluid intake prevents excessive reabsorption of water from the colon, resulting in wetter stools. Adding prunes or prune juice to the diet is often recommended because prunes contain compounds that have a mild laxative effect.

High-fiber foods promote regular bowel movements.

Laxatives Laxatives work by increasing stool weight, increasing the water content of the stool, or stimulating peristaltic contractions. Table 18-3 includes examples of common laxatives and describes their modes of action. Enemas and suppositories (chemicals introduced into the rectum) can also promote defecation; they work by distending and stimulating the rectum or by lubricating the stool.

Medical Interventions For patients with severe constipation who do not respond to dietary or laxative treatments, medications that stimulate colonic contractions may be prescribed. Physical therapy and biofeedback techniques are sometimes successful in training patients to relax their pelvic muscles more effectively. Surgical interventions are a last resort and include colonic resections and colostomy operations, which are discussed later in this chapter.

TABLE 18-3 Laxatives and Bulk-Forming Agents

Laxative Type	Active Ingredients	Product Examples	Method of Action	Cautions
Fiber (bulk-forming agents)	Malt soup extract, methylcellulose, polycarbophil, psyllium	Citrucel, Fiberall, Fiber-Lax, Metamucil	Fiber supplements increase stool weight and aid in formation of soft, bulky stools. Similar effects are achieved by adding bran to the diet. For mild constipation. Safe for long-term use.	Some fiber supplements may increase flatulence. Psyllium may cause an allergic reaction.
Osmotic laxatives: nonabsorbable salts	Magnesium citrate, magnesium hydroxide, sodium phosphate	Epsom salts, milk of magnesia	Unabsorbed salts attract and retain water in the large intestine and stimulate contractions.	May cause bloating and watery stools or diarrhea. Should be used with caution. Avoid using in renal patients and children.
Osmotic laxatives: nonabsorbable sugars	Lactulose, polyethylene glycol, sorbitol	Cephulac, Chronulac	Unabsorbed sugars attract water to the large intestine and promote softer stools. Must be used for several days to take effect. Safe for long-term use.	May cause flatulence and cramps. Can lose effectiveness over time.
Stimulant or irritant laxatives	Aloe, bisacodyl, cascara, castor oil, senna	Correctol, Dulcolax, Ex-Lax	Act as local irritants to colonic tissue; stimulate peristalsis and mucosal secretions. For moderate-to-severe constipation. Long-term use is discouraged.	Usually given only after milder treatments fail. May alter fluid and electrolyte balances. May lead to laxative dependency.
Stool surfactant agents (stool softeners)	Docusate sodium	Colace, Kaopectate, Surfak	Detergent action promotes the mixing of water with stools. Prevents formation of dry, hard stools.	Do not increase stool weight. Limited effectiveness.

Diarrhea

Diarrhea is characterized by the passage of frequent, watery stools. In most cases, it lasts for only a day or two and subsides without complication. Severe or persistent diarrhea, however, can cause dehydration and electrolyte imbalances. Diarrhea may be accompanied by other symptoms, such as fever, abdominal cramps, dyspepsia, or bleeding, which help in diagnosing the cause.

Causes of Diarrhea Diarrhea is a complication of various GI disorders and may also be caused by infections, medications, or dietary substances. It is due to inadequate fluid reabsorption in the intestines, sometimes in conjunction with an increase in intestinal secretions.[7] *Osmotic diarrhea* results from nutrient malabsorption, as when poorly absorbed sugars like sorbitol or fructose attract water to the colon and increase fecal water content. In *secretory diarrhea,* the fluid secreted by the intestines exceeds the amount that can be reabsorbed by intestinal cells. Secretory diarrhea may be caused by bacterial food poisoning, chemical substances, or the inflammation associated with some intestinal diseases. *Motility disorders* can result in diarrhea because they accelerate the transit of colonic residue, reducing the contact time available for fluid reabsorption.

Treatment of Diarrhea Correcting the underlying medical disorder is the first step in treating diarrhea. Antibiotics are prescribed to treat intestinal infections. If a medication is the cause of diarrhea, a different drug may be suggested. If certain foods are responsible, they can be omitted from the diet. Bulk-forming agents such as psyllium (Metamucil) or methylcellulose (Citrucel) can help to reduce the liquidity of the stool. If chronic diarrhea does not respond to treatment, antidiarrheal drugs may be given to slow GI motility or reduce intestinal secretions.

Probiotics—live bacteria provided in foods and dietary supplements for the purpose of preventing or treating disease—may be beneficial for certain types of diarrhea (especially diarrhea caused by infections), but standard treatment protocols have not been developed. Nutrition in Practice 18 describes the potential health benefits of probiotics.

An oral rehydration solution can be mixed from the following ingredients:

▪ ½ tsp sodium chloride (table salt)

▪ ⅓ tsp potassium chloride (salt substitute)

▪ ¾ tsp sodium bicarbonate (baking soda)

▪ 1⅓ tbs sugar

▪ 1 qt water

Oral Rehydration Therapy Severe diarrhea requires the replacement of lost fluid and electrolytes. Oral rehydration solutions can be purchased or mixed using water, salts, and glucose or sucrose (see the recipe in the margin). The addition of carbohydrate to the solution facilitates sodium and water absorption. Commercial sports drinks are not ideal fluids for rehydration because their sodium content is too low, but they can be used if accompanied by salty snack foods.[8] When diarrhea results in extreme dehydration, intravenous solutions can quickly replenish fluid and electrolytes.

Nutrition Therapy for Diarrhea Nutrition care depends on the cause of diarrhea and its severity and duration. The dietary treatment often recommended is a low-fiber, low-fat, lactose-free diet,[9] which limits foods that contribute to colonic residue or worsen symptoms, such as those with significant amounts of fiber, resistant starch, fructose, sugar alcohols, and lactose (in lactose-intolerant individuals). Avoidance of fatty foods is advised because they may aggravate diarrhea. Gas-producing foods (those with poorly digested or absorbed carbohydrates) can increase intestinal distention and cause additional discomfort. Patients should avoid caffeinated coffee and tea because caffeine stimulates GI motility and can thereby reduce water reabsorption. In the treatment of formula-fed infants, apple pectin or banana flakes are sometimes added to formulas to help thicken stool consistency. Table 18-4 lists foods that may worsen diarrhea, although individual tolerances vary.

Irritable Bowel Syndrome

People with **irritable bowel syndrome** experience chronic and recurring intestinal problems that cannot be explained by specific physical abnormalities. The symptoms usually include disturbed defecation (diarrhea and/or constipation), flatulence, and abdominal discomfort or pain; the pain is often aggravated by eating and relieved by defecation. In some patients, symptoms are mild; in others, the disturbances in colonic function can interfere with work and social activities enough to dramatically alter the person's lifestyle and sense of well-being.[10]

Although the causes of irritable bowel syndrome remain elusive, people with the disorder tend to have excessive colonic responses to meals, GI hormones, and stress. Many individuals exhibit hypersensitivity to a normal degree of intestinal distention

TABLE 18-4 **Foods That May Worsen Diarrhea[a]**

Foods to Avoid	Rationale	Selected Examples
High-fiber foods	They increase colonic residue.	Breads and cereals with more than 2 g fiber per serving, fruits and vegetables with peels or skins
Foods with indigestible carbohydrates	They contribute to osmotic diarrhea.	Artichokes, asparagus, brussels sprouts, cabbage, dried beans and peas, fruit, garlic, green beans, leeks, onions, wheat, zucchini
Foods that contain fructose or sugar alcohols	They contribute to osmotic diarrhea.	Dried fruits, fresh fruits (except bananas), fruit juices, fructose-sweetened soft drinks, sugar-free gums and candies
Milk products, if person is lactose intolerant	They contribute to osmotic diarrhea.	Milk and milk products
Caffeine-containing beverages	They increase intestinal motility.	Coffee, tea, colas, energy drinks

[a]Individual tolerances vary; the foods to avoid are best determined by trial and error.

and feel discomfort when experiencing normal meal transit or typical amounts of intestinal gas. Intestinal motility after meals may be excessive, leading to diarrhea, or be reduced, causing constipation. Some patients show signs of low-grade intestinal inflammation; others may have had a bacterial infection that initiated their GI problems. Symptoms often worsen during periods of psychological stress. Diagnosing irritable bowel syndrome is difficult because its symptoms are typical of other GI disorders and laboratory tests for the condition are nonexistent.[11]

Treatment of Irritable Bowel Syndrome Medical treatment of irritable bowel syndrome often includes dietary adjustments, stress management, and behavioral therapies. Medications may be prescribed to manage symptoms, but they are not always helpful. The drugs prescribed may include antidiarrheal drugs, anticholinergics (which affect GI motility), antidepressants, and laxatives.

Nutrition Therapy for Irritable Bowel Syndrome Although dietary adjustments may reduce symptoms, responses among patients can vary considerably. The most common recommendation is to gradually increase fiber intake (from food or supplements) to relieve constipation and improve stool bulk. However, clinical studies suggest that additional fiber has only marginal effectiveness in improving symptoms and may worsen flatulence.[12] Psyllium supplementation may be helpful for individuals with constipation. Some individuals have fewer symptoms when they consume small, frequent meals instead of larger ones. Foods that aggravate symptoms may include fried or fatty foods, gas-producing foods, milk products (not necessarily due to lactose intolerance), wheat products, coffee (with or without caffeine), and alcoholic beverages; however, individual tolerances are best determined by trial and error. A careful evaluation of the dietary patterns that exacerbate symptoms may uncover the foods and habits most closely associated with intestinal discomfort.

Treatments under investigation for irritable bowel syndrome include peppermint oil, which relaxes smooth muscle within the GI tract, and various types of probiotics (see Nutrition in Practice 18). The Case Study allows you to apply your knowledge about irritable bowel syndrome to a clinical situation.

irritable bowel syndrome: an intestinal disorder of unknown cause that disturbs the functioning of the lower bowel; symptoms include abdominal pain, flatulence, diarrhea, and constipation.

Case Study YOUNG ADULT WITH IRRITABLE BOWEL SYNDROME

Shannon Calloway is a 22-year-old recent college graduate who began her first professional job in a bank one month ago. As a college student, she occasionally experienced abdominal pain and cramping after eating. She also had frequent bouts of diarrhea and felt somewhat better after bowel movements. Once Shannon began her new job, her symptoms occurred more frequently. At first, she attributed her symptoms to job stress, but when the symptoms continued for several months, she decided to see her physician. After taking a careful history and conducting tests to rule out other bowel disorders, the physician diagnosed irritable bowel

syndrome. The physician prescribed bulk-forming fibers and advised Shannon to keep a record of her food intake and symptoms for one week. Shannon was then referred to a dietitian for a review of her dietary record. The dietitian noticed that Shannon routinely drank several cups of coffee in the morning and had large meals for lunch and dinner. Shannon often ate out in Mexican restaurants and favored spicy foods, refried beans, and fatty desserts. Between meals, she snacked on low-carbohydrate foods sweetened with sugar alcohols and drank several cans of soda daily. Her dietary fiber intake, however, totaled only about 13 grams daily.

1. Describe the characteristics of irritable bowel syndrome to Shannon, and indicate the role that stress might play in her illness.

2. Explain how the record of food intake and symptoms might be helpful in devising an appropriate dietary plan for Shannon. Could any of the foods that are currently in Shannon's diet be aggravating her symptoms?

3. What dietary measures might benefit individuals with irritable bowel syndrome? What problems might some of the dietary changes cause?

- Constipation may be caused by many different health problems; a high-fiber diet or fiber supplements are often beneficial.

- Diarrhea may result from malabsorption, intestinal infections, motility disorders, and dietary substances; a low-fiber, low-fat, lactose-free diet may be recommended.

- Irritable bowel syndrome is characterized by recurring intestinal symptoms such as diarrhea, constipation, abdominal pain, and flatulence; dietary measures may include increasing fiber intake, consuming small meals, and omitting certain foods from the diet.

Inflammatory Bowel Diseases

Inflammatory bowel diseases are chronic inflammatory illnesses characterized by abnormal immune responses in the GI tract.[13] Both genetic and environmental factors are believed to contribute to the development of these diseases, although the exact triggers are unknown. Table 18-5 compares the two major forms of inflammatory bowel disease, **Crohn's disease** and **ulcerative colitis.** Crohn's disease usually involves the small intestine and may lead to nutrient malabsorption, whereas ulcerative colitis affects the large intestine, where little nutrient absorption occurs. Both diseases may cause nutrient losses due to tissue damage, bleeding, and diarrhea.

Complications of Crohn's Disease Crohn's disease may occur in any region of the GI tract, but most cases involve the ileum and/or large intestine. Lesions may develop in different areas in the intestine, with normal tissue separating affected regions (called "skip" lesions). During exacerbations, the inflammation may extend deeply into intestinal tissue and be accompanied by ulcerations, fissures, and **fistulas** (abnormal passages between tissues). The resultant scar tissue can eventually thicken, narrowing the lumen and possibly causing strictures or obstructions. About three-fourths of patients require surgery within 20 years of diagnosis.[14] Patients with Crohn's disease are also at increased risk of developing intestinal cancers.

TABLE 18-5 Comparison of Crohn's Disease and Ulcerative Colitis

	Crohn's Disease	Ulcerative Colitis
Location of inflammation	Approximately 40% of cases involve the ileum and cecum, 30% are in the small intestine only, and 20% are in the colon	Inflammation is confined to the rectum and colon; it begins at the rectum and spreads into the colon
Pattern of inflammation	Discrete areas separated by normal tissue ("skip" lesions)	Continuous inflammation that begins at the rectum and ends abruptly within the colon
Depth of damage	Damage throughout all layers of tissue; causes deep fissures that give intestinal tissue a "cobblestone" appearance	Damage primarily in the mucosa and submucosa (layers of intestinal tissue closest to the lumen)
Fistulas	Common	Usually do not occur
Cancer risk	Increased	Greatly increased

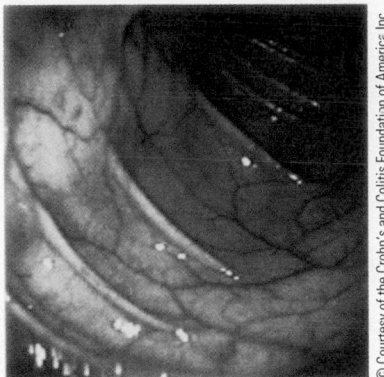

The healthy colon has a smooth surface with a visible pattern of fine blood vessels.

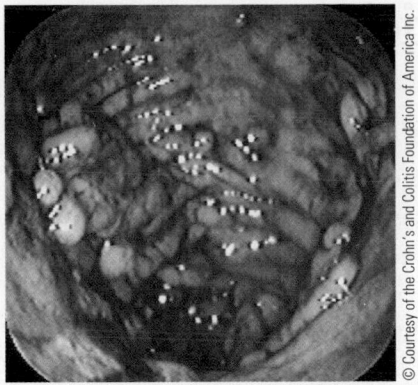

In Crohn's disease, the mucosa has a "cobblestone" appearance due to deep fissuring in the inflamed mucosal tissue.

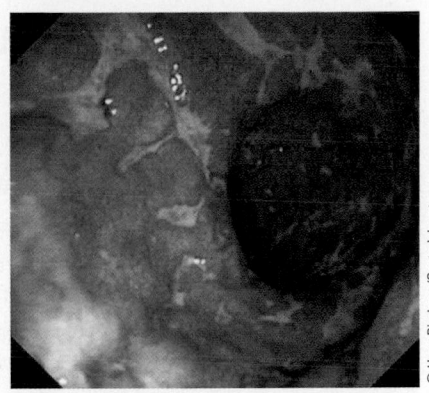

In ulcerative colitis, the colon appears inflamed and reddened, and ulcers are visible.

Malnutrition may result from malabsorption, nutrient losses associated with the tissue damage, reduced food intake, and surgical resections that shorten the small intestine. If the ileum is affected, bile acids may become depleted, causing malabsorption of fat, fat-soluble vitamins, calcium, magnesium, and zinc (the minerals bind to unabsorbed fatty acids). Because the ileum is the site of vitamin B_{12} absorption, patients often require vitamin B_{12} injections. Anemia may result from bleeding, nutrient deficiencies, or the metabolic effects of chronic illness (see Nutrition in Practice 19).

Most of the bile used during digestion is eventually reabsorbed in the ileum and returned to the liver.

Complications of Ulcerative Colitis Ulcerative colitis always involves the rectum and usually extends into the colon. Inflammation is continuous along the length of intestine affected, ending abruptly at the area where healthy tissue begins. Tissue erosion or ulceration develops primarily in the mucosa and submucosa (the layers of tissue closest to the lumen). During active episodes, patients have frequent, urgent bowel movements that are small in volume. Stools are often streaked with blood.

Although mild disease may cause few complications, weight loss, fever, and weakness are common when much of the colon is involved. Severe disease is often associated with anemia (due to blood loss), dehydration, and electrolyte imbalances. Protein losses from the inflamed tissue can be substantial. A **colectomy** (removal of the colon) is performed in 20 to 25 percent of patients and prevents future recurrence.[15] Colon cancer risk is substantially increased in ulcerative colitis patients.

Drug Treatment of Inflammatory Bowel Diseases Medications help to control symptoms, reduce inflammation, and minimize complications. The drugs prescribed include antidiarrheal agents, immunosuppressants, anti-inflammatory drugs (usually corticosteroids and salicylates), and antibiotics. The Diet-Drug Interactions feature (p. 492) lists some nutrition-related effects of these medications.

Nutrition Therapy for Inflammatory Bowel Diseases Crohn's disease often requires aggressive dietary management because it can lead to protein-energy malnutrition (PEM), nutrient deficiencies, and growth failure in children. Specific dietary measures depend on the symptoms and complications that develop.

During disease exacerbations, a low-fiber, low-fat diet provided in small, frequent feedings can minimize stool output and reduce symptoms of malabsorption. High-kcalorie, high-protein diets may be prescribed to prevent or treat malnutrition.

Crohn's disease: an inflammatory bowel disease that usually occurs in the lower portion of the small intestine and the colon. Inflammation may pervade the entire intestinal wall.

ulcerative colitis (ko-LYE-tis): an inflammatory bowel disease that involves the colon. Inflammation affects the mucosa and submucosa of the intestinal wall.

fistulas (FIST-you-luz): abnormal passages between organs or tissues that permit the passage of fluids or secretions.

colectomy: removal of a portion or all of the colon.

Diet-Drug Interactions

Check this table for notable nutrition-related effects of the medications discussed in this chapter.

Antidiarrheal drugs	**Gastrointestinal effect:** Constipation
Anti-inflammatory drugs (sulfasalazine, corticosteroids)	**Gastrointestinal effects:** Nausea, heartburn (sulfasalazine)
	Dietary interactions: Sulfasalazine may decrease folate absorption; supplementation is recommended
	Metabolic effects: Anemia (sulfasalazine); fluid retention, hyperglycemia, hypocalcemia, hypokalemia, hypophosphatemia, increased appetite, protein catabolism (corticosteroids)
Laxatives	**Gastrointestinal effects:** Diarrhea, flatulence, abdominal discomfort
	Metabolic effects: Dehydration, electrolyte imbalances, laxative dependency

Liquid supplements may help to increase energy intake and improve weight gain. Vitamin and mineral supplements are usually necessary, especially if nutrient malabsorption is present. In some instances, tube feedings are used to supplement the diet or may be the sole means of providing nutrients; elemental formulas are most easily tolerated. Table 18-6 includes some other adjustments that may be beneficial.

In most cases, the diet for ulcerative colitis requires few adjustments. As in Crohn's disease, the symptoms and complications that arise are managed with the appropriate dietary measures. During disease exacerbations, emphasis is given to restoring fluid and electrolyte balances and correcting deficiencies that result from

TABLE 18-6 Management of Symptoms and Complications in Crohn's Disease

Symptom or Complication	Possible Dietary Measures
Growth failure, weight loss, or muscle wasting	High-kcalorie, high-protein diet Liquid supplements Elemental tube feedings
Anorexia or pain with eating	Small, frequent meals Liquid supplements If long-term (>5 to 7 days): elemental tube feedings
Malabsorption	High-kcalorie diet Nutrient supplementation
Steatorrhea (fat malabsorption)	Fat restriction Medium-chain triglycerides Nutrient supplementation
Diarrhea	Fluid and electrolyte replacement Nutrient supplementation
Lactose intolerance	Avoidance of lactose-containing foods
Nutrient deficiencies	Nutrient-dense diet Nutrient supplementation
Strictures/fistulas	Low-fiber diet
Severe bowel obstruction, high-output fistulas, or severe exacerbations of disease	Total parenteral nutrition

protein and blood losses. A low-fiber diet may reduce irritation by minimizing fecal volume. If colon function becomes severely impaired, food and fluids may be withheld and fluids and electrolytes supplied intravenously until colon function returns.[16]

IN SUMMARY

▌ Crohn's disease can damage all layers of intestinal tissue, and it impairs the intestine's absorptive function. Ulcerative colitis affects the mucosa and submucosa of the rectum and colon.

▌ The treatment for inflammatory bowel diseases includes drug therapies, surgical removal of diseased areas, and correction of malnutrition.

▌ Dietary measures for inflammatory bowel diseases aim to reduce symptoms, manage complications, and correct deficiencies. A low-fiber diet can reduce irritation during active episodes of illness.

Diverticular Disease of the Colon

Diverticulosis refers to the presence of pebble-sized herniations (outpockets) in the intestinal wall, known as diverticula (see Figure 18-2). Most people with diverticulosis are symptom free and remain unaware of the condition until a complication develops. The prevalence of diverticulosis increases with age, occurring in 50 to 65 percent of 80-year-old individuals.[17]

The development of diverticula is strongly influenced by the amount of dietary fiber a person consumes. By increasing stool weight and bulk, high-fiber diets reduce the workload of the circular muscles that move wastes through the colon. Low-fiber diets require more vigorous muscle contractions, increasing pressure within the segments immediately adjacent to the circular muscles. This increase in pressure may induce small areas of intestinal tissue to weaken and balloon outward over time.

Diverticulitis Inflammation or infection sometimes develops in the area around a diverticulum. This condition, called **diverticulitis,** is the most common complication of diverticulosis, affecting 10 to 25 percent of individuals with the condition.[18] It is thought to result from erosion of the diverticular wall due to high pressure within the lumen, causing inflammation and eventually a microperforation that leads to subsequent infection.[19] If the infection spreads to adjacent organs, fistulas may develop. Symptoms of diverticulitis may include persistent abdominal pain, tenderness in the affected area, fever, constipation, diarrhea, nausea, and vomiting.

Treatment for Diverticular Disease Medical treatment for diverticulosis is necessary only if symptoms develop. Because a high-fiber diet may prevent disease progression and intestinal symptoms, patients are advised to increase fiber intake, with an emphasis on insoluble fiber sources. Fiber should be increased gradually to ensure tolerance. Bulk-forming agents, such as psyllium, can help to increase fiber intake if food sources are insufficient. Avoiding nuts, seeds, and popcorn is

diverticulosis (dye-ver-tic-you-LOH-sis): an intestinal condition characterized by the presence of small herniations (called diverticula) in the intestinal wall.

diverticulitis (dye-ver-tic-you-LYE-tis): an inflammation or infection involving diverticula.

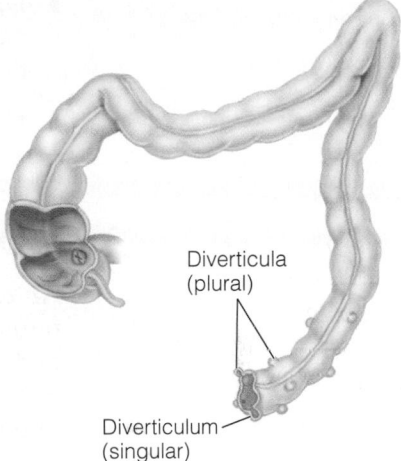

Diverticula (plural)

Diverticulum (singular)

FIGURE 18-2
Diverticula in the Colon
Diverticula are small pouches that develop in weakened areas of the intestinal wall. The condition of having diverticula is known as *diverticulosis.*

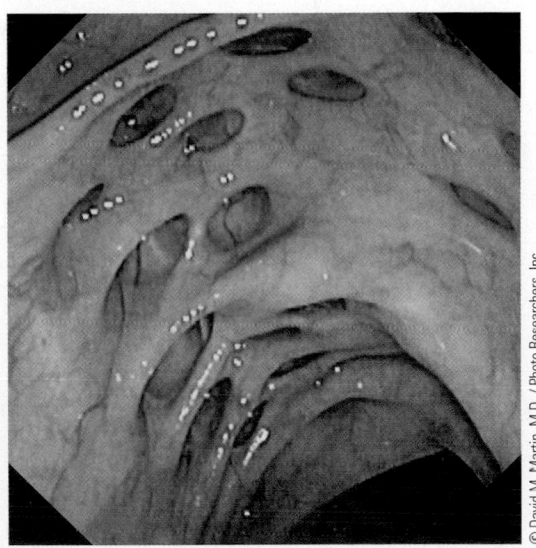

© David M. Martin, M.D. / Photo Researchers, Inc.

Diverticula are frequently seen during a colonoscopy, a procedure that uses a flexible, lighted tube to examine the inside of the colon.

sometimes suggested to prevent complications, but no evidence is available to justify the recommendation.[20]

Patients with diverticulitis may need antibiotics to treat infections and, possibly, pain-control medications. In mild cases, a clear liquid diet may be advised initially, with progression to solid foods as tolerated. In more severe cases, bowel rest is necessary (oral fluids and food are withheld), and fluids are given intravenously. Afterwards, an oral diet is gradually reintroduced as the condition improves, beginning with clear liquids and progressing to a restricted-fiber diet until inflammation and bleeding subside. After recovery, a high-fiber diet is recommended to prevent disease progression and symptom recurrence.

IN SUMMARY

▮ Diverticulosis is characterized by the presence of small herniations in the intestinal wall and is usually asymptomatic. A high-fiber diet may help to prevent the development and progression of diverticulosis.

▮ Diverticulitis is an inflammation or infection involving the diverticula.

Colostomies and Ileostomies

An *ostomy* is a surgically created opening (called a **stoma**) in the abdominal wall through which dietary waste can be eliminated. Whereas a permanent ostomy is necessary after a partial or total colectomy, a temporary ostomy is sometimes constructed to bypass the colon after injury or extensive surgery. To create the stoma, the cut end of the remaining segment of functioning intestine is routed through an opening in the abdominal wall so that it empties to the exterior. The stoma can be formed from a section of the colon **(colostomy)** or ileum **(ileostomy),** as shown in Figure 18-3. Conditions that may require these procedures include inflammatory bowel diseases, diverticulitis, and colorectal cancers.

To collect wastes, a disposable bag is affixed to the skin around the stoma and emptied during the day as needed. Alternatively, an interior pouch can be constructed

FIGURE 18-3
Colostomy and Ileostomy

Colostomy

Ileostomy

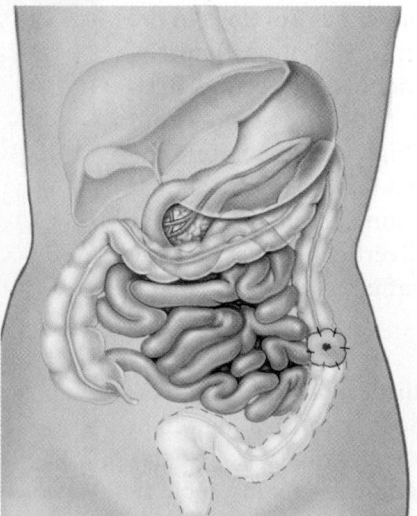

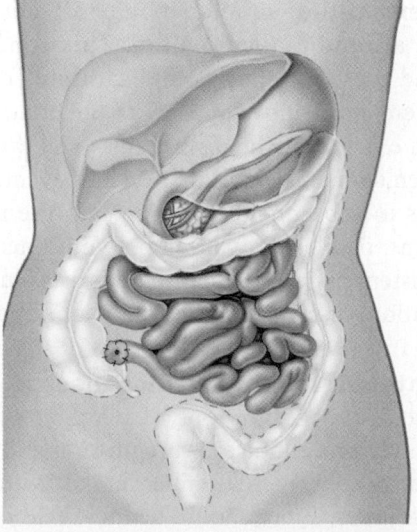

In a colostomy, a portion of the colon is removed or bypassed, and the stoma is formed from the remaining section of functional colon.

In an ileostomy, the entire colon is removed or bypassed, and the stoma is formed from the ileum.

behind the stoma using intestinal tissue; the pouch can be emptied with a catheter when convenient. Stool consistency varies according to the length of colon that is functional. If a small portion of the colon is absent or bypassed, the stools may continue to be semisolid. If the entire colon has been removed or is bypassed, the output is liquid.

Nutrition Care for Patients with Ostomies Following surgery, the diet progresses from clear liquids (low in sugars) to regular foods as tolerated. To reduce stool output, a low-fiber diet may be recommended. Small, frequent meals may be more acceptable than larger ones. To determine food tolerances, patients should try small amounts of questionable foods and assess their effects; a food that causes problems can be tried again later.

People with ileostomies need to chew thoroughly to ensure that foods are adequately digested and to prevent obstructions, a common complication due to the small diameter of the ileal lumen. Foods high in insoluble fibers are sometimes avoided because they reduce transit time, may cause obstructions, and increase stool output. Because less fluid is reabsorbed following an ileostomy, the diet should provide adequate liquid to prevent dehydration. To replace electrolyte losses, patients are encouraged to use salt liberally and to ingest beverages with added electrolytes (such as sports drinks and rehydration beverages), if necessary. Ileostomy patients are also at risk of fat malabsorption (due to bile acid depletion) and vitamin B_{12} deficiency (vitamin B_{12} is absorbed in the ileum).

Dietary concerns after colostomies depend on the length of colon remaining. Most patients have no dietary restrictions and can return to a regular diet. Patient concerns typically include stool odors, excessive gas production, and diarrhea. If a large portion of colon was removed, recommendations may be similar to those given to ileostomy patients.

Obstructions As mentioned, foods that are incompletely digested may cause obstructions in ileostomy patients. Although almost any food can be consumed if carefully chewed, the following foods may cause difficulty: celery, coconut, corn, dried fruit, grapes, nuts, raw cabbage (for example, in coleslaw), and unpeeled apples.[21]

Reducing Gas and Odors Persons with ostomies are often concerned about foods that may increase gas production or cause strong odors. Foods and practices that contribute to gas formation were discussed earlier (p. 485). Foods that may produce unpleasant odors include asparagus, beer, broccoli, brussels sprouts, cabbage, dried beans and peas, eggs, fish, garlic, and onions. Foods that may help to reduce odors include buttermilk, cranberry juice, parsley, and yogurt.[22]

Diarrhea Examples of foods that may aggravate diarrhea were listed in Table 18-4 (p. 488). Foods that may thicken stool include applesauce, banana, cheese, smooth peanut butter, pasta, potatoes, tapioca, and white rice.[23] What works may differ for each individual, however, and is best determined by trial and error.

IN SUMMARY

▌ Colostomies and ileostomies are surgically created openings in the abdominal wall using the colon or ileum. Fluid and electrolyte requirements are greater after an ostomy because colon function is reduced or absent.

▌ Poorly digested foods may cause obstructions in people with ostomies, although thorough chewing can reduce risk. Other concerns include excessive gas production, food odors, and diarrhea.

stoma (STOE-ma): a surgically created opening in a body tissue or organ.

colostomy (co-LAH-stoe-me): a surgical passage through the abdominal wall into the colon.

ileostomy (ill-ee-AH-stoe-me): a surgical passage through the abdominal wall into the ileum.

Nutrition Assessment Checklist FOR PEOPLE WITH LOWER GI TRACT DISORDERS

Medical History

Check the medical record for diseases that:

- Cause chronic GI symptoms, such as irritable bowel syndrome or ulcerative colitis
- Interfere with nutrient absorption, such as Crohn's disease

Check for surgical procedures involving the lower GI tract, such as:

- Intestinal resections
- Ileostomy
- Colostomy

Check for the following symptoms or complications:

- Anemia
- Constipation
- Diarrhea, dehydration
- Fistulas
- Nutrient deficiencies
- Strictures or obstructions
- Poor growth (in children)

Medications

Check for medications or dietary supplements that may:

- Cause constipation or diarrhea
- Alter appetite or nutrient needs

Dietary Intake

Note the following problems, and contact the dietitian if you suspect difficulties such as:

- Poor appetite or food intake
- Food intolerances
- Inadequate fiber intake (in patients with constipation)
- Lactose intolerance (in patients with diarrhea)
- Inadequate fluid intake

Anthropometric Data

Measure baseline height and weight. Address weight loss early to prevent malnutrition in patients with:

- Severe or persistent diarrhea
- Nutrient malabsorption

Laboratory Tests

Check laboratory tests for signs of dehydration, electrolyte imbalances, nutrient deficiencies, and anemia in patients with:

- Severe or persistent diarrhea
- Nutrient malabsorption
- Intestinal resections

Physical Signs

Look for physical signs of:

- Dehydration
- Fat-soluble vitamin deficiencies
- Vitamin B_{12} deficiencies
- Mineral deficiencies
- Protein-energy malnutrition

Clinical Applications

1. A health practitioner working with a patient with a constipation problem provides him with detailed information about a high-fiber diet. At a follow-up appointment, the patient reports no change in symptoms. His food diary for that day shows that he consumed an omelet and toast for breakfast and a sandwich with juice for lunch.

   ▪ Considering these two meals only, what additional information would help the health practitioner evaluate the man's compliance with the diet he was given?

   ▪ Review the discussion about fiber in Chapter 2 and create a one-day menu that provides the DRI for fiber for an adult male, using the fiber values listed in Appendix A.

2. A number of symptoms and disorders of the GI tract, as described in this chapter and the previous one, are associated with aging. Review both chapters and list these symptoms and disorders. Referring back to Chapter 13, describe some effects of aging on the GI tract and explain how these changes might relate to the GI symptoms and disorders you listed.

3. The diets described in this chapter are highly individualized: a particular food may cause discomfort for one person and have no effect on another. Describe some practical ways to keep track of food intolerances.

Self Check

1. Foods that do *NOT* supply fiber include:
   a. cabbage and broccoli.
   b. beef and pork.
   c. oats and corn.
   d. beans and peas.

2. The health practitioner advising an elderly patient with constipation encourages the patient to:
   a. consume a low-fat diet low in sodium.
   b. consume a high-protein diet rich in calcium.
   c. gradually add high-fiber foods to the diet.
   d. eliminate gas-forming foods from the diet.

3. Laxatives help to promote bowel movements by:
   a. stimulating peristalsis.
   b. increasing stool weight.
   c. increasing the water content of stools.
   d. all of the above.

4. Osmotic diarrhea often results from:
   a. excessive colonic contractions.
   b. excessive fluid secretion by the intestines.
   c. nutrient malabsorption.
   d. viral, bacterial, or protozoal infections.

5. An oral rehydration solution can be mixed at home using:
   a. water, vinegar, honey, and table salt.
   b. water, table salt, and salt substitute.
   c. water, milk, sugar, and table salt.
   d. water, table salt, baking soda, potassium chloride, and sugar.

6. Symptoms of irritable bowel syndrome most often include:
   a. constipation, diarrhea, and flatulence.
   b. weight loss and malnutrition.
   c. strong odors and obstructions.
   d. nauea and vomiting.

7. A patient with Crohn's disease may develop all of the following nutrition problems, *except:*
   a. fat malabsorption.
   b. dumping syndrome.
   c. vitamin B_{12} deficiency.
   d. anemia.

8. Ulcerative colitis can afflict which region of the digestive tract?
   a. ileum, rectum, and colon
   b. rectum and colon
   c. stomach and duodenum
   d. most regions of the GI tract can be affected

9. Diverticulosis is most often associated with:
   a. a low-fiber diet.
   b. inadequate exercise.
   c. intestinal surgery.
   d. a high-fiber diet.

10. After an ileostomy, the most serious concern is that:
    a. the diet is too restrictive to meet nutrient needs.
    b. waste disposal causes frequent daily interruptions.
    c. incompletely digested foods may cause obstructions.
    d. fluid restrictions prevent patients from drinking beverages freely.

Answers to these questions can be found in Appendix H.

Notes

1. J. S. Barrett and P. R. Gibson, Clinical ramifications of malabsorption of fructose and other short-chain carbohydrates, *Practical Gastroenterology* 31 (August, 2007): 51–65.

2. F. Azpiroz and M. D. Levitt, Intestinal gas, in M. Feldman, L. S. Friedman, and L. J. Brandt, eds., *Sleisenger and Fordtran's Gastrointestinal and Liver Disease* (Philadelphia: Saunders, 2010), pp. 233–240.

3. A. J. Lembo and S. P. Ullman, Constipation, in M. Feldman, L. S. Friedman, and L. J. Brandt, eds., *Sleisenger and Fordtran's Gastrointestinal and Liver Disease* (Philadelphia: Saunders, 2010), pp. 259–284.

4. C. A. Ternent and coauthors, Practice parameters for the evaluation and management of constipation, *Diseases of the Colon and Rectum* 50 (2007): 2013–2022.

5. Standing Committee on the Scientific Evaluation of Dietary Reference Intakes, Food and Nutrition Board, Institute of Medicine, *Dietary Reference Intakes for Energy, Carbohydrate, Fiber, Fat, Fatty Acids, Cholesterol, Protein, and Amino Acids* (Washington, DC: National Academies Press, 2002).

6. Ternent and coauthors, 2007.

7. L. R. Schiller and J. H. Sellin, Diarrhea, in M. Feldman, L. S. Friedman, and L. J. Brandt, eds., *Sleisenger and Fordtran's Gastrointestinal and Liver Disease* (Philadelphia: Saunders, 2010), pp. 211–232.

8. Schiller and Sellin, 2010.

9. American Dietetic Association, *Nutrition Care Manual* (Chicago: American Dietetic Association, 2010).

10. S. Friedman, Irritable bowel syndrome, in N. J. Greenberger, ed., *Current Diagnosis and Treatment: Gastroenterology, Hepatology, and Endoscopy* (New York: McGraw-Hill Companies, 2009), pp. 279–290.

11. N. J. Talley, Irritable bowel syndrome, in M. Feldman, L. S. Friedman, and L. J. Brandt, eds., *Sleisenger and Fordtran's Gastrointestinal and Liver Disease* (Philadelphia: Saunders, 2010), pp. 2091–2104.

12. W. D. Heizer, S. Southern, and S. McGovern, The role of diet in symptoms of irritable bowel syndrome in adults: A narrative review, *Journal of the American Dietetic Association* 109 (2009): 1204–1214; A. Sanjeevi and D. F. Kirby, The role of food and dietary intervention in the irritable bowel syndrome, *Practical Gastroenterology* 32 (July 2008): 33–42.

13. B. E. Sands and C. A. Siegel, Crohn's disease, in M. Feldman, L. S. Friedman, and L. J. Brandt, eds., *Sleisenger and Fordtran's Gastrointestinal and Liver Disease* (Philadelphia: Saunders, 2010), pp. 1941–1973.

14. Sands and Siegel, 2010.

15. W. F. Stenson, Inflammatory bowel disease, in L. Goldman and D. Ausiello, eds., *Cecil Medicine* (Philadelphia: Saunders, 2008), pp. 1042–1050.

16. Stenson, 2008.

17. A. C. Travis and R. S. Blumberg, Diverticular disease of the colon, in N. J. Greenberger, ed., *Current Diagnosis and Treatment: Gastroenterology, Hepatology, and Endoscopy* (New York: McGraw-Hill Companies, 2009), pp. 243–255.

18. C. Prather, Inflammatory and anatomic diseases of the intestine, peritoneum, mesentery, and omentum, in L. Goldman and D. Ausiello, eds., *Cecil Medicine* (Philadelphia: Saunders, 2008), pp. 1050–1061.

19. Travis and Blumberg, 2009.

20. Travis and Blumberg, 2009; L. L. Strate and co-authors, Nut, corn, and popcorn consumption and the incidence of diverticular disease, *Journal of the American Medical Association* 300 (2008): 907–914.

21. American Dietetic Association, 2010.

22. American Dietetic Association, 2010.

23. American Dietetic Association, 2010.

CHAPTER 18

Nutrition in Practice

Soon after birth, the warm, nutrient-rich environment within the gastrointestinal tract is colonized by a wide variety of bacterial species. The approximately 10 trillion bacterial cells inhabiting our bodies **(flora)** make up more than 90 percent of all our cells. Most bacterial cells reside in our colon, which harbors 400 to 500 different species.[1] Although the exact composition of intestinal bacteria varies among individuals, the pattern within an individual tends to remain constant over time, fluctuating somewhat due to illness, antibiotic treatment, and, to some extent, dietary factors. Table NP18-1 lists the predominant types of bacteria that colonize the human intestines, and Table NP18-2 shows how the bacterial populations vary within different regions of the GI tract.

Over the past several decades, nutritional scientists and microbiologists have tried to determine whether **probiotics**—live, **nonpathogenic** bacteria supplied in sufficient numbers to possibly benefit our health—can be useful for preventing or treating various medical conditions. Although the diseases of interest include gastrointestinal disorders, researchers have also been studying the effects of bacterial cells on cancer, immune system disorders, and other illnesses. This Nutrition in Practice discusses some of the research and explains some of the issues involved in selecting and consuming probiotic bacteria. The accompanying glossary defines the relevant terms.

How do our intestinal bacteria influence health?

Intestinal bacteria can benefit our health in a number of different ways. First, the bacteria degrade much of our undigested or unabsorbed dietary carbohydrate, including dietary fibers, starch that is

TABLE NP18-1 Intestinal Flora

Predominant Types	Subdominant Types
Bacteroides	Enterobacteria
Bifidobacteria	Enterococci
Clostridia	Escherichia
Eubacteria	Klebsiella
Peptococci	Lactobacilli
Peptostreptococci	Micrococci
Ruminococci	Staphylococci

TABLE NP18-2 Bacterial Populations in the Gastrointestinal Tract

Organ	Total Bacteria (per mL of contents)
Stomach, duodenum	10 to 1000
Jejunum, ileum	10^4 to 10^8
Colon	10^{10} to 10^{12}

GLOSSARY

bacterial translocation: movement of bacteria across the intestinal mucosa, allowing access to body tissues.

flora: the bacteria that normally reside in a person's body.

nonpathogenic: not capable of causing disease.

pathogenic: capable of causing disease.

prebiotics: indigestible substances in foods that stimulate the growth of nonpathogenic bacteria within the large intestine.

probiotics: live bacteria provided in foods and dietary supplements for the purpose of preventing or treating disease.

resistant to digestion, and poorly absorbed sugars and sugar alcohols. In turn, the bacteria produce some vitamins, as well as short-chain fatty acids that our colonic epithelial cells and other body cells can use as an energy source. Intestinal bacteria also assist in the development and maintenance of mucosal tissue, protect intestinal tissue from **pathogenic** bacteria, and stimulate immune defenses in mucosal cells and other body tissues.[2]

Certain indigestible substances in food, called **prebiotics,** can stimulate the growth or activity of resident bacteria within the large intestine. Prebiotics include some of the carbohydrates found in asparagus, chicory root, garlic, Jerusalem artichokes, onions, and other foods.[3] Because the intestinal bacteria that degrade these substances produce gas as a by-product, people who consume high amounts of these foods may experience more flatulence than usual.

Why are certain types of bacteria considered "probiotic"?

For bacteria to be "probiotic"—that is, beneficial to health—they must be nonpathogenic when consumed. They must survive their transit through the digestive tract; therefore, they must be resistant to destruction by stomach acid, bile, and other digestive substances. They should be able to alter the intestinal environment in some way that is beneficial to the human host, either by producing antimicrobial substances, altering immune defenses, metabolizing undigested foodstuffs, or protecting the intestinal walls.[4]

Probiotic bacteria must be consumed in large amounts—between 100 million and 100 billion live bacteria per day—to survive in sufficient numbers to influence the bacterial populations in the large intestine; a serving of yogurt usually provides these amounts. Carefully controlled studies have not found that probiotic bacteria actually *colonize* the intestine, however, as they are no longer detected in fecal or intestinal samples once ingestion of the probiotic product stops.[5] Note that only a few different types of bacteria are used in foods, and the relatively small amounts consumed cannot compete with the huge populations that normally populate our digestive tract.

What types of medical problems are helped by probiotic bacteria?

Although the results of research studies vary, probiotic bacteria may help to prevent and treat some gastric and intestinal disorders, alter susceptibility to food allergens and alleviate some allergy symptoms, and improve the availability and digestibility of various nutrients.[6] Other potential benefits include improved immune responses, reduced symptoms of lactose intolerance, and reduced cancer risk.[7]

Much of the research investigating probiotics and intestinal illness has focused on the prevention and treatment of infectious diarrhea. For example, controlled trials have suggested that certain strains of probiotic bacteria may shorten the duration of diarrhea caused by rotavirus infection in infants and children, decrease the incidence of traveler's diarrhea in tourists visiting high-risk areas, and prevent the recurrence of infectious diarrhea in hospitalized patients.[8] In studies of children and adults using antibiotics, some strains of probiotic bacteria have been shown to reduce the incidence and duration of antibiotic-associated diarrhea. As another example, some studies have suggested that probiotic treatment may help to reduce the recurrence of *pouchitis,* an inflammation of the surgical pouch created in patients who have had an ileostomy or colostomy.[9]

Despite promising research results thus far, there are no clear conclusions about the appropriate probiotic doses or durations of treatment for many of these conditions. Moreover, the beneficial effects of one bacterial strain cannot be extrapolated to other strains of the same species.[10] Thus, individuals who decide to consume probiotic-containing foods and supplements to benefit their health cannot be certain that the substances they use will help their condition. At best, probiotics should be considered an adjunct therapy rather than a primary treatment for an illness.

What are the main dietary sources of probiotic bacteria?

Probiotics are provided mainly by fermented foods. In the United States, yogurt and acidophilus milk are produced using various species of lactobacilli and bifidobacteria, although the species are chosen for their ability to produce desirable food products rather than their potential health benefits. In Europe and Asia, food products containing probiotic bacteria include yogurt, milk, ice cream, oatmeal gruel, and soft drinks. Although lactobacilli are used to produce various other fermented food products, such as sauerkraut, pickles, brined olives, and sausages, these foods retain few, if any, live bacteria after they undergo typical food processing methods.[11]

A number of companies market probiotic supplements, which are available in capsules, tablets, and powders. Because probiotic bacteria are living organisms, storage conditions may affect their viability—heat, moisture, and oxygen can reduce survival times—and therefore consumers should check the expiration date before purchasing a product. When a consumer group (ConsumerLab.com) tested 13 probiotic supplements, they found that 5 of the products contained substantially fewer live bacteria than was claimed on the label.[12] Thus, there is no guarantee that a dietary supplement will contain the amount of bacteria expected.

Various species of *Lactobacillus* are used in the production of fermented food products, such as the foods shown in this photo.

Are there any potential problems associated with the use of probiotic bacteria?

Yes. One major concern is the possibility that probiotic bacteria may cause infection in immune-compromised individuals. Various species of probiotic bacteria, including *Lactobacillus* species, have been isolated from the infection sites of severely ill individuals who were consuming the probiotic.[13] Risk is increased by the use of antibiotic therapy (which reduces intestinal flora populations), illnesses or medications that suppress immunity, and illnesses that increase risk of **bacterial translocation** (including inflammatory bowel illnesses and intestinal infections). Care should be taken to inquire about probiotic use in these patients.

Other safety concerns are related to the lack of industry standards for probiotics in foods and supplements: the concentrations and strains of probiotic bacteria in foods may vary substantially.[14] Thus, a consumer who wishes to try probiotics would find it difficult to determine how much of a product to consume in order to achieve the desired effect.

In recent years, the contributions of our intestinal flora to health have been increasingly recognized. Preliminary research suggests that altering our bacterial populations by consuming probiotics or prebiotics may help to improve our defenses against certain illnesses. Additional studies are needed to verify the beneficial effects of probiotics and prebiotics and to develop standard protocols that can be used for treating illness.

Notes

1. S. O'Mahony and F. Shanahan, Enteric microbiota and small intestinal bacterial overgrowth, in M. Feldman, L. S. Friedman, and L. J. Brandt, eds., *Sleisenger and Fordtran's Gastrointestinal and Liver Disease* (Philadelphia: Saunders, 2010), pp. 1769–1778.

2. A. M. O'Hara and F. Shanahan, The gut flora as a forgotten organ, *EMBO Reports* 7 (2006): 688–693.

3. S. Kolida and G. R. Gibson, Prebiotic capacity of inulin-type fructans, *Journal of Nutrition* 137 (2007): 2503S–2506S.

4. P. Winkler and coauthors, Molecular and cellular basis of microflora host interactions, *Journal of Nutrition* 137 (2007): 756S–772S.

5. B. Corthésy, H. R. Gaskins, and A. Mercenier, Cross-talk between probiotic bacteria and the host immune system, *Journal of Nutrition* 137 (2007): 781S–790S.

6. M. de Vrese and P. R. Marteau, Probiotics and prebiotics: Effects on diarrhea, *Journal of Nutrition* 137 (2007): 803S–811S; A. C. Ouwehand, Antiallergic effects of probiotics, *Journal of Nutrition* 137 (2007): 794S–797S.

7. S. Parvez and coauthors, Probiotics and their fermented food products are beneficial for health, *Journal of Applied Microbiology* 100 (2006): 1171–1185.

8. de Vrese and Marteau, 2007.

9. J. J. Jones and A. E. Foxx-Orenstein, Probiotics in inflammatory bowel disease, *Practical Gastroenterology* (March 2006): 44–50.

10. L. C. Douglas and M. E. Sanders, Probiotics and prebiotics in dietetics practice, *Journal of the American Dietetic Association* (2008): 510–521.

11. Douglas and Sanders, 2008.

12. ConsumerLab.com, Product review: Probiotic supplements (including *Lactobacillus acidophilus*, *Bifidobacterium*, and others), available at www.consumerlab.com, accessed August 19, 2010.

13. K. Whelan and C. E. Myers, Safety of probiotics in patients receiving nutritional support: A systematic review of case reports, randomized controlled trials, and nonrandomized trials, *American Journal of Clinical Nutrition* 91 (2010): 687–703.

14. E. R. Farnworth, The evidence to support health claims for probiotics, *Journal of Nutrition* 138 (2008): 1250S–1254S.

Carbohydrate- and Fat-Modified Diets for Malabsorption Disorders

The gastrointestinal (GI) illnesses that are the most detrimental to nutrition status are those that cause nutrient malabsorption. Malabsorption can lead to nutrient deficiencies and weight loss and cause serious complications. This chapter describes disorders that require carbohydrate- and fat-modified diets to treat malabsorption.

Malabsorption Syndromes

To digest and absorb nutrients, we depend on normal digestive secretions and healthy intestinal mucosa. Malabsorption can therefore be caused by pancreatic disorders that result in enzyme or bicarbonate deficiencies, conditions that lead to bile deficiency, and inflammatory conditions or medical treatments that damage intestinal tissue. In some cases, the treatment of an intestinal disorder requires surgical removal of a section (resection) of the small intestine, leaving minimal absorptive capacity in the portion that remains. In addition, various medications can damage the mucosa and impair the digestive and absorptive functions of the small intestine. Table 19-1 lists examples of diseases and treatments that are frequently associated with malabsorption.

Malabsorption rarely involves a single nutrient. When malabsorption is caused by pancreatic enzyme deficiencies, all macronutrients—protein, carbohydrate, and fat—may be affected. When fat is malabsorbed, fat-soluble nutrients and some minerals are usually malabsorbed as well. Malabsorption disorders and their treatments can tax nutritional status further by causing complications that alter food intake, raise nutrient needs, and incur additional nutrient losses.

Evaluating Malabsorption

A number of clinical procedures and laboratory tests are used to determine whether an individual has a malabsorption problem. Examples include the following:[1]

- *Endoscopy or biopsy.* Direct examination of the duodenal mucosa with an **endoscope** may reveal physical changes characteristic of intestinal diseases that cause malabsorption. A **biopsy** can be taken during the procedure for later analysis.

- *Hydrogen breath test.* When carbohydrate is malabsorbed, colonic bacteria digest the carbohydrate and produce hydrogen gas, which is absorbed and later can be measured in the breath. The hydrogen breath test is often used to diagnose lactose intolerance, but it can also diagnose malabsorption of other types of carbohydrate as well.

- *Fecal fat analysis.* Fat malabsorption can be determined by placing the patient on a high-fat diet (100 grams per day), performing a 48- to 72-hour stool collection, and measuring the stool's fat content. In healthy individuals, fecal fat excretion is usually less than 7 grams per day under these conditions. Excessive fat in the stools is known as **steatorrhea.**

- *Schilling test.* The Schilling test helps to diagnose malabsorption of vitamin B_{12}. After the patient takes an oral dose of radioactive vitamin B_{12}, a urine test can reveal whether the vitamin B_{12} was absorbed. More extensive testing can determine whether vitamin B_{12} malabsorption is caused by intrinsic factor deficiency or pancreatic enzyme insufficiency.

TABLE 19-1 Potential Causes of Malabsorption

Genetic disorders
- Enzyme deficiencies

Intestinal disorders
- AIDS-related enteropathy
- Bacterial overgrowth
- Celiac disease
- Crohn's disease
- Radiation enteritis

Intestinal infections
- Giardiasis
- Nematode (roundworm) infections

Liver disease (bile insufficiency)

Pancreatic disorders
- Chronic pancreatitis
- Cystic fibrosis

Surgeries
- Gastric or intestinal bypass surgery
- Intestinal resection (short bowel syndrome)

Reminder: Intrinsic factor is a stomach protein that binds to vitamin B_{12} and facilitates its absorption in the ileum.

resection: the surgical removal of part of an organ or body structure.

endoscope: a flexible fiber optic tube used for viewing internal cavities of the body.

biopsy: removal of a tissue sample to determine the cause of an illness.

steatorrhea (stee-AH-tor-REE-ah): excessive fat in the stools due to fat malabsorption; characterized by stools that are loose, frothy, and foul smelling due to a high fat content.
- *steat* = fat
- *rheo* = flow

Fat Malabsorption

Fat is the nutrient most frequently malabsorbed because both digestive enzymes and bile must be present for its digestion. Thus, fat malabsorption often develops when an illness interferes with pancreatic secretions or bile production. For example, both pancreatitis and cystic fibrosis can reduce the secretion of pancreatic lipase, whereas severe liver disease can cause bile insufficiency. Motility disorders that accelerate gastric emptying or intestinal transit can cause fat malabsorption because they prevent the normal mixing of dietary fat with lipase and bile. Fat malabsorption may also be caused by conditions or treatments that damage the intestinal mucosa, such as inflammatory bowel diseases, AIDS, and radiation treatments for cancer.

Consequences of Fat Malabsorption Fat malabsorption is associated with losses of food energy, essential fatty acids, fat-soluble vitamins, and some minerals (see Figure 19-1). Weight loss may result if the individual does not consume alternative sources of energy. Deficiencies of fat-soluble vitamins and essential fatty acids are common in chronic conditions. Malabsorption of some minerals, including calcium, magnesium, and zinc, often occurs because the minerals form **soaps** with the unabsorbed fatty acids. Calcium deficiency may lead to bone loss, which is further aggravated by the vitamin D deficiency that may be present due to fat malabsorption.

Another consequence of fat malabsorption is an increased risk of kidney stones, which are most often composed of calcium oxalate. The oxalates in foods ordinarily bind to calcium in the small intestine and are excreted in the stool. If calcium instead binds to fatty acids or bile acids, the oxalates are free to be absorbed into the blood and are ultimately excreted in the urine. The risk of developing oxalate stones increases when urinary oxalate levels are high. Kidney stones are discussed further in Chapter 23.

Oxalates are plant compounds found in green leafy vegetables and some other foods. The oxalates can bind to minerals in the GI tract and form complexes that the body cannot absorb.

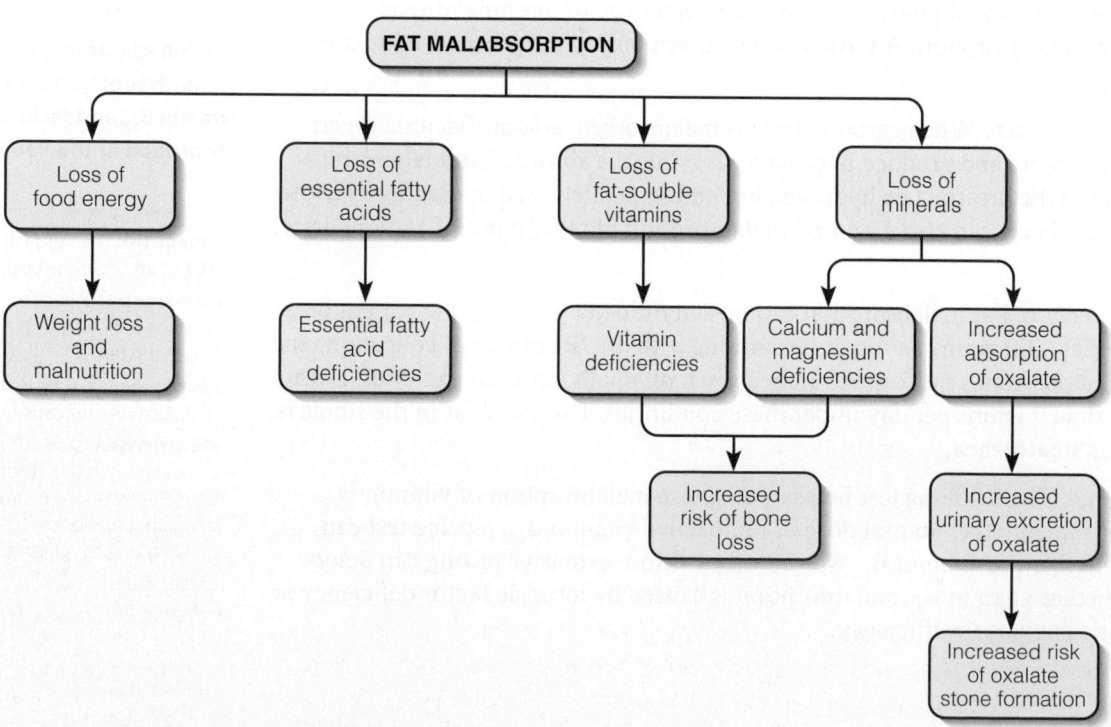

FIGURE 19-1 The Consequences of Fat Malabsorption

TABLE 19-2 Fat-Controlled Diet

A fat-controlled diet includes mostly low-fat and fat-free foods. For a fat intake of 50 grams per day, limit meats and meat substitutes to 6 ounces per day, and limit fats and oils to 8 teaspoons per day.[a] Foods from other food groups should provide less than 1 gram of fat per serving.

Food Category	Foods Recommended	Foods to Avoid
Meat and meat alternates	Lean meats, fish, and skinless poultry prepared by broiling, roasting, grilling, or boiling; low-fat luncheon meats such as sliced turkey breast; meat alternates such as dried beans or peas; low-fat egg substitutes. Limit whole eggs to two per week.	Meat with visible fat, ground beef (unless extra lean), sausage, bacon, frankfurters, spareribs, duck, tuna packed in oil
Milk and milk products	Fat-free milk, fat-free yogurt, fat-free sour cream substitutes, fat-free half-and-half and cream substitutes, fat-free cheeses. Low-fat milk products can be used in moderation.	Milk products that are not fat free or low fat
Breads, cereals, rice, and pasta	Whole-grain and enriched breads, cooked cereals and most cold breakfast cereals, plain tortillas, bagels, English muffins, fat-free muffins, saltine crackers, graham crackers, plain rice, plain noodles and pasta.	Biscuits, pancakes, waffles, granola, snack crackers made with fat, cornbread, doughnuts, corn chips, fried rice
Vegetables	All vegetables prepared without added fat.	Buttered, creamed, breaded, or fried vegetables; vegetables prepared au gratin style; french-fried potatoes; olives
Fruits	All fruits except avocado.	Avocado; fruits prepared with fats, nuts, or coconut
Desserts	Sherbet; fruit ices; flavored gelatin; angel food cake; meringue; fat-free puddings; fat-free bakery products; fat-free ice cream or frozen yogurt; fat-free candies such as marshmallows, jelly beans, and hard candy.	Cakes, cookies, pies, and pastries made with fat; puddings made with whole milk or eggs; ice cream; candies made with fat such as caramel and chocolates
Fats	For a total fat intake of 50 grams per day, limit intake of the following foods to 8 teaspoons daily: vegetable oil, butter, margarine, mayonnaise, lard (each has 3½ to 4½ grams of fat). Each of these foods can replace 1 teaspoon of fat in the amounts specified: 1 tbs salad dressing, 2 tbs low-fat salad dressing, ½ tbs peanut butter, 1 tbs chopped nuts, 2 tbs mashed avocado.	Dietary fat that exceeds the amount specified in the nutrition prescription
Beverages	Fruit juices, soft drinks, fat-free milk, coffee, tea, coffee substitutes.	Beverages made with milk (unless fat free) or added cream, chocolate milk, eggnog, milk shakes

Sample menu (about 50 grams of fat):

Breakfast: 6 oz orange juice, 1 c oatmeal with nonfat milk and raisins, 1 slice whole-wheat toast with 1 tsp margarine, coffee with fat-free half-and-half

Lunch: Turkey breast sandwich (2 slices whole-wheat bread, 2 oz lean turkey breast, 2 tomato slices, lettuce leaf, and 2 tsp mayonnaise), 2 c salad greens with 1 tbs salad dressing, fruit cup (1 c peaches and ½ c berries) with ½ c orange sherbet

Snack: 6 oz nonfat fruit yogurt, 6 saltine crackers with 1 tbs peanut butter and ½ tbs honey

Dinner: 4 oz cod with sliced lemon, 1 slice French bread with 1 tsp butter, 1 c steamed rice with herbs and walnut oil (includes ½ tsp oil), 1 c steamed broccoli and carrots with ½ tsp margarine, 1 piece angel food cake with fat-free whipped cream

[a]For a fat intake of 25 grams per day, limit meats and meat substitutes to 4 ounces per day, and limit fats and oils to 2 teaspoons per day.

Nutrition Therapy for Fat Malabsorption If fat malabsorption does not improve, a fat-controlled diet may be recommended (see Table 19-2). The diet may relieve intestinal symptoms that are aggravated by fat intake (such as diarrhea and flatulence) and reduce vitamin and mineral losses. Because fat is a primary energy source, it should not be restricted more than necessary. **Medium-chain triglycerides (MCT),** which do not require lipase or bile for digestion and absorption, can be used as an alternative source of dietary fat, although MCT oil does not provide essential fatty acids. The "How to" offers suggestions for following a fat-controlled diet and for using MCT oil.

soaps: chemical compounds formed from fatty acids and positively charged minerals.

medium-chain triglycerides (MCT): triglycerides with fatty acids that are 8 to 10 carbons in length. MCT do not require digestion and can be absorbed in the absence of lipase or bile.

For some individuals, fat-controlled diets may be difficult to follow. Fats add flavors, aromas, and textures to foods—characteristics that make foods more enjoyable. Unlike some diets that can be introduced gradually, a fat-controlled diet is often implemented immediately, allowing little time for adaptation. These suggestions may help:

▪ Fat is better tolerated if provided in small portions. Divide the day's allotment into several servings that can be consumed throughout the day.

▪ Use variety to enhance enjoyment of meals: vary flavors, textures, colors, and seasonings.

▪ Look for fat-free items when grocery shopping. Incorporate fat-free ingredients when preparing favorite recipes.

▪ Try fat-free and low-fat condiments to improve the diet's palatability. Experiment with herbs and spices. Instead of butter, use fruit butters on toast. Use butter-flavored granules on vegetables. Replace mayonnaise in sandwiches with spicy mustard. Replace salad dressings with flavored vinegars.

▪ Avoid products that contain the fat substitute olestra, which may aggravate GI symptoms.

If patients are interested in using medium-chain triglyceride (MCT) oil:

▪ Explain that MCT products are expensive but that the cost is sometimes covered by medical insurance.

▪ Advise patients to add MCT oil to the diet gradually. Diarrhea and abdominal cramps may result if too much is used at once. Tolerance to MCT oil may improve in time.

▪ Advise patients that MCT oil may have an unpleasant taste when used alone. Suggest using MCT oil in recipes as a substitute for regular oil. MCT oil can replace oil in salad dressings, be incorporated into sauces, and be used in cooking or baking. It can also be added to fat-free milk products to make milk shakes.

▪ Point out that MCT oil should not be used to fry foods because it decomposes at lower temperatures than most cooking oils.

Bacterial Overgrowth

Ordinarily, the GI tract is protected from **bacterial overgrowth** by gastric acid, which destroys bacteria; peristalsis, which flushes bacteria through the small intestine before they multiply; and immunoglobulins secreted into the GI lumen.[2] When bacterial overgrowth does occur, it can lead to fat malabsorption because the bacteria dismantle the bile acids needed for fat emulsification. Deficiencies of the fat-soluble vitamins A, D, and E may eventually develop. The bacteria also produce enzymes and toxins that disturb the intestinal mucosa, destroying some mucosal enzymes (especially lactase). Some types of bacteria metabolize vitamin B_{12}, reducing its absorption and increasing the risk of deficiency. Although symptoms of bacterial overgrowth are often minor and nonspecific, severe cases may lead to chronic diarrhea, steatorrhea, flatulence, bloating, and weight loss.

Causes of Bacterial Overgrowth Conditions that impair intestinal motility and allow material to stagnate can greatly increase susceptibility to bacterial overgrowth. For example, in some types of gastric surgery, a portion of the small intestine is bypassed, preventing the flow of material in the bypassed region and allowing bacteria to flourish (see the "blind loop" shown in Figure 17-4 on p. 473). Intestinal motility can also be reduced by strictures, obstructions, and diverticula in the small intestine, as well as by some chronic illnesses, including diabetes mellitus, scleroderma, and chronic kidney disease.

Reduced secretions of gastric acid can also lead to bacterial overgrowth. Possible causes include atrophic gastritis, use of acid-suppressing medications, and some gastrectomy procedures.

Treatment for Bacterial Overgrowth Treatment may include antibiotics to suppress bacterial growth and surgical correction of the anatomical defects that contribute to a motility disorder. Medications may be given to stimulate peristalsis. A lactose-restricted diet may reduce flatulence and diarrhea in some individuals.[3] Dietary supplements can correct nutrient deficiencies, especially deficiencies of fat-soluble vitamins, calcium (which binds to unabsorbed fatty acids), and vitamin B_{12}.

IN SUMMARY

▌ Nutrient malabsorption can result from an undersupply of digestive secretions, motility disorders, and damaged intestinal mucosa. Clinical procedures or laboratory tests may reveal the underlying conditions causing a malabsorption problem.

▌ Fat malabsorption can cause losses of food energy and deficiencies of essential fatty acids, fat-soluble vitamins, and some minerals. In severe cases, nutrition therapy may include a fat-controlled diet and use of MCT oil.

▌ Bacterial overgrowth may result from conditions that reduce gastric acidity or intestinal motility; it typically causes malabsorption of fat and some essential nutrients.

Most people with lactose intolerance can drink milk, especially if they drink it along with other foods and limit the amount they consume at any one time.

Lactose Intolerance

Approximately 75 percent of people worldwide have some degree of **lactose intolerance,** which is caused by the loss or reduction of lactase, the intestinal enzyme that digests the lactose in milk products.[4] Lactose intolerance is especially prevalent among individuals of certain ethnic groups, including Asians, African Americans, Native Americans, Ashkenazi Jews, and Latinos. It can also result from GI disorders that damage the small intestinal mucosa. The primary symptoms of lactose intolerance are diarrhea and increased intestinal gas.

Lactose intolerance is rarely serious and is easily managed by simple dietary adjustments. Although people with the condition are sometimes reluctant to consume milk products, clinical studies have found that individuals with lactose intolerance can tolerate up to 2 cups of milk daily without significant symptoms.[5] People who avoid milk for fear of intestinal discomfort can be urged to gradually increase consumption of lactose-containing foods. They may more readily tolerate milk when intake is divided throughout the day and the milk is taken with food. Some people tolerate chocolate milk better than plain milk. Most aged cheeses are well tolerated because they contain little lactose. Yogurts that contain live bacterial cultures are usually acceptable because the bacteria contain lactase, which may aid in lactose digestion. Other options are to add a lactase preparation to milk or to take enzyme tablets before consuming lactose-containing foods. Lactose-free milk is also commercially available.

People who develop lactose intolerance as a result of intestinal illness may be advised to temporarily restrict milk and milk products. Foods that contain lactose can be reintroduced in small amounts once the condition improves. Individuals who restrict milk products should be encouraged to consume alternative food sources of calcium and vitamin D (see Chapters 8 and 9).

IN SUMMARY

▌ Lactose intolerance, caused by lactase deficiency, can be managed by adjusting the amount and timing of lactose consumption.

▌ Most people with lactose intolerance can consume up to 2 cups of milk daily without significant symptoms; tolerance may be improved by consuming small portions of milk at one time and consuming the milk with foods.

bacterial overgrowth: excessive bacterial colonization of the stomach and small intestine; may be due to low gastric acidity, altered GI motility, mucosal damage, or contamination.

lactose intolerance: a condition characterized by the inability to digest the lactose in milk products; symptoms include flatulence, abdominal discomfort, and diarrhea.

Disorders of the Pancreas

As mentioned previously, pancreatic disorders can lead to maldigestion and malabsorption due to the impaired secretion of digestive enzymes. This section describes several pancreatic illnesses that are characterized by widespread malabsorption.

Pancreatitis

Pancreatitis is an inflammatory disease of the pancreas. Although mild cases may subside in a few days, other cases can persist for weeks or months. Chronic pancreatitis can lead to irreversible damage to pancreatic tissue and permanent loss of function.

Acute Pancreatitis In acute pancreatitis, the digestive enzymes within pancreatic cells become prematurely activated, causing destruction of pancreatic tissue and subsequent inflammation. About 70 to 80 percent of acute cases are caused by gallstones or alcohol abuse; less frequent causes include high blood triglyceride levels or exposure to toxins.[6] Common symptoms include severe abdominal pain, nausea and vomiting, and abdominal distention. In most patients, the condition resolves within a week with no complications. More serious cases may lead to chronic pancreatitis, infection, the systemic inflammatory response syndrome (see p. 622), or multiple organ failure.

(see p. 622)

Nutrition Therapy for Acute Pancreatitis The initial treatment for acute pancreatitis is supportive and includes pain control, intravenous hydration, and supplementary oxygen, if necessary. Oral fluids and food are withheld until the patient is pain free and experiences no nausea or vomiting.[7] Afterwards, patients may consume a liquid diet or small low-fat meals, as tolerated (fat stimulates the pancreas more than other nutrients). In severe pancreatitis, tube feedings may be necessary; either standard formulas or elemental formulas may be used, depending on patient tolerance. Protein and energy needs are high in severe cases due to the catabolic and hypermetabolic effects of inflammation. Patients with acute pancreatitis require nutrient supplementation until food intake can meet nutritional needs.[8]

Reminder: An *elemental formula* contains hydrolyzed nutrients that require minimal digestion and are easily absorbed.

Chronic Pancreatitis Chronic pancreatitis is characterized by progressive, permanent damage to pancreatic tissue, resulting in the impaired secretion of digestive enzymes and bicarbonate. About 70 percent of cases are caused by excessive alcohol consumption.[9] In chronic pancreatitis, abdominal pain is often severe and unrelenting and may worsen with eating. Fat maldigestion develops sooner than maldigestion of protein or carbohydrate, and steatorrhea is common in advanced cases. Long-term illness is associated with reductions in both insulin and glucagon secretions, and diabetes eventually develops in 40 to 80 percent of patients.

Nutrition Therapy for Chronic Pancreatitis The objectives of nutrition therapy are to correct malnutrition, reduce malabsorption, and prevent symptom recurrence. Dietary supplements are used to correct nutrient deficiencies, which may be due to malabsorption or to the alcohol abuse that caused the disease. To improve food tolerance, patients should consume small, low-fat meals. They should also avoid alcohol completely and quit smoking cigarettes, as these substances can exacerbate illness and interfere with healing.[10]

Steatorrhea is usually treated with pancreatic enzyme replacement. Pancreatic enzymes are often **enteric coated** to resist the acidity of the stomach and do not dissolve until the pH is above 5.5. If nonenteric-coated preparations are used, acid-suppressing drugs are also required (see the Diet-Drug Interactions feature for nutrition-related side effects of these medications). Fecal fat concentrations must be monitored to determine whether the enzyme treatment has been effective. In some cases, a fat-controlled diet may help to reduce symptoms, and MCT oil can be used as an alternative source of fat kcalories.

Diet-Drug Interactions

Check this table for notable nutrition-related effects of the medications discussed in this chapter.

Antisecretory drugs (proton-pump inhibitors, H2 blockers)	**Gastrointestinal effects:** Nausea and vomiting, diarrhea, constipation, abdominal pain (proton-pump inhibitors)
	Dietary interactions: May decrease iron, folate, and vitamin B$_{12}$ absorption
Pancreatic enzyme replacements	**Gastrointestinal effects:** Nausea and vomiting, stomach cramping, diarrhea, constipation, irritation of GI mucosa
	Metabolic effects: Elevated serum or urinary uric acid levels (with high doses), allergic reactions (rare)

Cystic Fibrosis

Cystic fibrosis is the most common life-threatening genetic disorder among Caucasians, with an incidence of about 1 in 2500 to 1 in 3200 white births.[11] It is caused by a mutation that disturbs the transport of chloride ions in **exocrine** glands. This defect results in thickened glandular secretions and a broad range of serious complications. Until a few decades ago, few infants born with cystic fibrosis survived to adulthood. Now, with early detection and advances in medical treatment, the median life span has extended beyond 36 years of age, with many patients surviving into their 50s.

Consequences of Cystic Fibrosis Cystic fibrosis is characterized by abnormal chloride and sodium levels in exocrine secretions. The secretions are unusually viscous and can block the ducts that normally allow their passage. The major complications of cystic fibrosis involve the lungs, pancreas, and sweat glands:

- *Lungs.* Abnormally thick mucus secretions cause obstructions in many of the small airways of the lungs. The obstructions lead to chronic coughing and persistent respiratory infections, both of which contribute to progressive inflammation in bronchial tissues. The eventual lung damage causes breathing difficulties and lower exercise tolerance.

- *Pancreas.* Thickened pancreatic secretions obstruct the pancreatic ducts, leading to progressive damage and scarring within pancreatic tissue. Fewer pancreatic enzymes reach the small intestine, causing malabsorption of protein, fat, and fat-soluble vitamins.

- *Sweat glands.* Salt losses in sweat are usually excessive, increasing the risk of dehydration.

Children with cystic fibrosis are chronically undernourished, grow poorly, and have difficulty maintaining normal body weight. Complications that may develop over time include pancreatitis, glucose intolerance or diabetes (due to destruction of insulin-producing cells), and gallbladder and liver diseases.

Nutrition Therapy for Cystic Fibrosis Patients with cystic fibrosis have high energy and protein requirements, which range from 110 to 200 percent of DRI values.[12] To compensate for fat malabsorption, about 35 to 40 percent of kcalories should come from fat. Patients are typically encouraged to eat high-kcalorie and high-fat foods, eat frequent meals and snacks, and supplement meals with milk shakes or liquid dietary supplements. Supplemental tube feedings can help to improve nutrition status if energy intakes are inadequate.

Nutrition problems associated with chronic obstructive lung diseases are described in Chapter 24.

enteric coated: refers to medications or enzyme preparations that are coated to withstand gastric acidity and dissolve only at the higher pH of the small intestine.

cystic fibrosis: an inherited disease characterized by the production of abnormally viscous exocrine secretions; often leads to respiratory illness and pancreatic insufficiency.

exocrine: pertains to external secretions, such as those of the mucous membranes or the skin. Opposite of *endocrine*, which pertains to hormonal secretions into the blood.

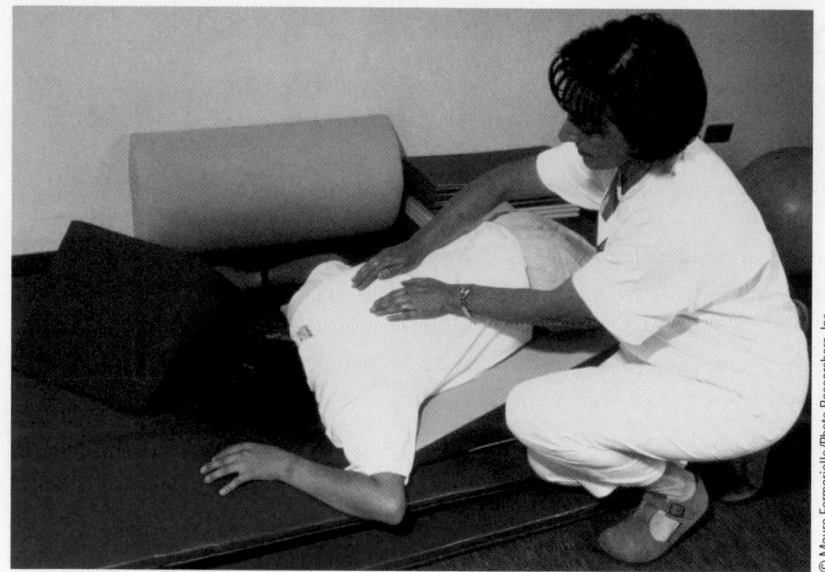
Postural drainage, a type of physical therapy used in cystic fibrosis, helps to clear the thick, sticky secretions that block airways and increase infection risk.

Pancreatic enzyme replacement therapy is a central feature of cystic fibrosis treatment. Supplemental enzymes must be included with every meal or snack. For young children, the contents of capsules are mixed in small amounts of liquid or a soft food (such as applesauce) and fed with a spoon. Enzyme dosages may need to be increased if malabsorption continues, as evidenced by poor growth or GI symptoms such as steatorrhea, intestinal gas, or abdominal pain.

The risk of deficiency depends on the degree of malabsorption; nutrients of greatest concern include the fat-soluble vitamins, essential fatty acids, and calcium. Multivitamin and fat-soluble vitamin supplements are routinely recommended. The liberal use of table salt and salty foods is encouraged to make up for losses of sodium in sweat. The accompanying Case Study describes a patient with cystic fibrosis.

IN SUMMARY

■ Whereas acute pancreatitis is short lived and does not cause permanent damage, chronic pancreatitis leads to digestive enzyme deficiencies and requires pancreatic enzyme replacement therapy.

■ Cystic fibrosis can cause obstructive lung disease and pancreatic damage; the nutrition treatment includes a high-kcalorie, high-protein diet and pancreatic enzyme replacement therapy.

Disorders of the Small Intestine

Reminder: *Crohn's disease* is a chronic, progressive inflammatory illness characterized by substantial damage to all three layers of intestinal tissue.

Malabsorption is a frequent consequence when the intestinal mucosa is damaged due to inflammation, infection, or other causes. One intestinal illness characterized by severe inflammation—Crohn's disease—was previously described in Chapter 18. This section discusses *celiac disease,* an intolerance to wheat, barley, and rye that can

Case Study CHILD WITH CYSTIC FIBROSIS

Julie is a 7-year-old girl diagnosed with cystic fibrosis. Symptoms of steatorrhea and poor growth during infancy prompted the tests that led to the diagnosis. She is currently 45 inches tall and weighs 42 pounds. Her height-for-age and weight-for-age fall near the 10th percentile (see Appendix E). Julie eats regular foods during the day and receives additional nutrients by tube feedings delivered overnight.

1. What do the height and weight percentiles tell you about Julie's nutritional status? Why is growth failure common in children with cystic fibrosis?

2. Explain why Julie's energy needs are so much higher than normal. Describe the elements of the diet that Julie should follow in order to improve growth.

3. Explain to Julie's parents how to use enzyme replacement therapy effectively.

4. Julie's parents are hoping to discontinue the nightly tube feedings. Do you think the tube feedings are necessary? Why or why not?

damage the intestinal mucosa, and *short bowel syndrome,* the malabsorption disorder that results when a significant portion of the small intestine is surgically removed.

Celiac Disease

Celiac disease is an immune disorder characterized by an abnormal response to a protein fraction in **wheat gluten** and to related proteins in barley and rye. The reaction to gluten causes severe damage to the intestinal mucosa and subsequent malabsorption. Celiac disease may affect approximately 1 percent of individuals in the United States.[13]

Consequences of Celiac Disease The immune reaction to gluten can cause striking changes in intestinal tissue. In affected areas, the villi may be shortened or absent, resulting in a significant reduction in mucosal surface area (and, therefore, in intestinal digestive enzymes). The damage may be restricted to the duodenum or may involve the full length of the small intestine. Individuals with severe disease may malabsorb all nutrients to some degree, especially the macronutrients, calcium, iron, folate, the fat-soluble vitamins, and vitamin B_{12}.[14]

Symptoms of celiac disease include GI disturbances such as diarrhea, steatorrhea, and flatulence. Because lactase deficiency can result from the mucosal damage, milk products may exacerbate GI symptoms. Children with celiac disease often exhibit poor growth, low body weights, muscle wasting, and anemia. Adults may develop anemia, bone disorders, neurological symptoms, and fertility problems.

Some gluten-sensitive individuals may have few GI symptoms but react to gluten by developing a severe rash. This condition is called **dermatitis herpetiformis** and requires dietary adjustments similar to those for celiac disease.

Nutrition Therapy for Celiac Disease The treatment for celiac disease is lifelong adherence to a gluten-free diet. Improvement in symptoms often occurs within several weeks, although mucosal healing can sometimes take years. If lactase deficiency is suspected, patients should avoid lactose-containing foods until the intestine has recovered.

celiac (SEE-lee-ack) **disease:** an immune disorder characterized by an abnormal immune response to the dietary protein gluten; also called *gluten-sensitive enteropathy* or *celiac sprue.*

wheat gluten (GLU-ten): a family of water-insoluble proteins in wheat; includes the gliadin (GLY-ah-din) fractions that are toxic to persons with celiac disease.

dermatitis herpetiformis (DERM-ah-TYE-tis HER-peh-tih-FOR-mis): a gluten-sensitive disorder characterized by a severe skin rash.

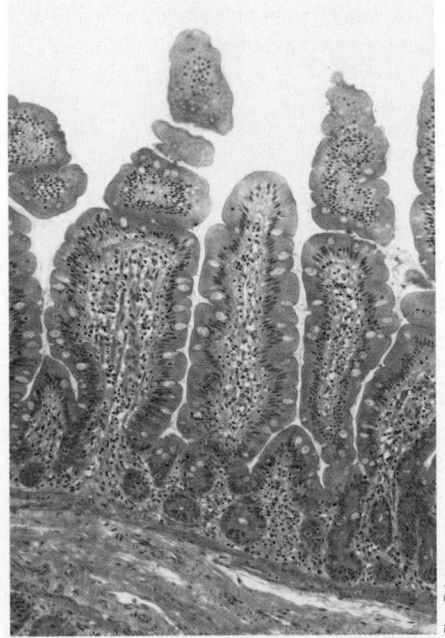

In the healthy intestine, the villi greatly increase the absorptive surface area.

In celiac disease, the villi may be shortened or absent, resulting in substantial reductions in nutrient absorption.

TABLE 19-3 Gluten-Free Diet

Food Category	Gluten-Free Choices	Potential Gluten Sources
Meat and meat alternates	Fresh meat, fish, or poultry; shellfish; dried peas and beans; tofu; nuts and seeds; eggs	Luncheon meats, sandwich spreads, meatloaf, meatballs, frankfurters, sausages, poultry injected with broth, imitation meat products, imitation seafood, meat extenders, miso, egg substitutes, dried egg products, dry roasted nuts, peanut butter. *Avoid:* products made with hydrolyzed vegetable protein (HVP), marinades, and soy sauce; breaded foods; foods prepared with cream sauces or gravies.
Milk and milk products	Milk, buttermilk, half-and-half, cream, plain yogurt, cheese, cottage cheese, cream cheese	Chocolate milk, milk shakes, frozen yogurt, flavored yogurt, cheese spreads, cheese sauces. *Avoid:* malted milk, malted milk powders.
Breads, cereals, rice, and pasta	Breads, bakery products, and cereals made with amaranth, arrowroot, buckwheat, corn, flax, hominy grits, millet, potato flour or potato starch, quinoa, rice, sorghum, soybean flour, tapioca, and teff; pasta and noodles made with the grains or starches listed above; corn tacos and corn tortillas	Oatmeal and oat bran (due to contamination), rice crackers, rice cakes, corn cakes. *Avoid:* breads, bakery products, cereals, tortillas, pastas, and pancake or baking mixes made with wheat, rye, barley, and triticale. *Wheat products* include bulghur, couscous, durum flour, einkorn, emmer, farina, graham flour, kamut, semolina, spelt, wheat bran, wheat germ. *Barley products* include malt, malt flavoring, and malt extract.
Fruits and vegetables	Any fresh, frozen, or canned fruits or vegetables	French fries from fast-food restaurants, commercial salad dressings, fruit pie fillings, dried fruits (may be dusted with flour). *Avoid:* scalloped potatoes (usually made with wheat flour), creamed vegetables, vegetables dipped in batters.
Desserts	Bakery products made with gluten-free flours, most ice creams, sherbet, sorbet, Italian ices, popsicles, gelatin desserts, egg custards, most chocolate bars, chocolate chips, hard candies, whipped toppings	Some ice creams (especially if made with cookie dough, brownies, nuts, and other added ingredients), icing or frosting, candies and candy bars, marshmallows. *Avoid:* bakery products or doughnuts made with wheat, rye, or barley; puddings made with wheat flour; ice cream or sherbets that contain gluten stabilizers; ice cream cones; licorice.
Beverages	Coffee; tea; cocoa made with pure cocoa powder; soft drinks; wine; distilled alcoholic beverages such as rum, gin, whiskey, and vodka	Instant tea or coffee, coffee substitutes, chocolate drinks, hot cocoa mixes. *Avoid:* beer, ale, lager, malted beverages, cereal beverages, beverages that contain nondairy cream substitutes.

Gluten-free products help people with celiac disease enjoy a wider variety of foods.

The gluten-free diet eliminates foods that contain wheat, barley, and rye (see Table 19-3). Because many foods contain ingredients derived from these grains, foods that are problematic are not always obvious. Even small amounts of gluten may cause symptoms in some people, so patients should check ingredient lists carefully. Gluten-containing products that may be overlooked include beer, caramel coloring, coffee substitutes, communion wafers, imitation meats, malt syrup, medications, salad dressings, and soy sauce. Special gluten-free products can be purchased to replace common food items such as bread, pasta, and cereals. Figure 19-2 shows an example of a menu for a gluten-free diet.

Although most people with celiac disease can safely consume moderate amounts of oats, most oats grown in the United States are contaminated with wheat, barley, or rye.

Oats are usually grown in rotation with other grains and may become contaminated during harvesting or processing. However, several oat manufacturers in the United States produce oats in dedicated facilities and test the products to ensure that they are gluten free. Individuals who wish to include oats in their diet should be advised to purchase only uncontaminated oats and to limit intakes to the amounts found to be safe (about ½ cup of dry rolled oats per day).

Short Bowel Syndrome

The treatment of Crohn's disease, cancers of the small intestine, and other intestinal disorders may include the surgical resection of a major portion of the small intestine. **Short bowel syndrome** is the malabsorption syndrome that results when the absorptive capacity of the remaining intestine is insufficient for meeting nutritional needs. Without appropriate dietary adjustments, short bowel syndrome can result in fluid and electrolyte imbalances and multiple nutrient deficiencies. Symptoms include severe diarrhea, steatorrhea, dehydration, weight loss, and growth impairment in children.

Figure 19-3 reviews nutrient absorption in the GI tract and describes how absorption is affected by surgical resections. Generally, up to 50 percent of the small intestine can be resected without serious nutritional consequences.[15] More extensive resections lead to generalized malabsorption, and patients may need lifelong parenteral nutrition to supplement oral intakes.

Intestinal Adaptation After an intestinal resection, the remaining intestine undergoes **intestinal adaptation,** which dramatically improves the intestine's absorptive capacity. Adaptation begins soon after surgery and continues for several years. During this period, the remaining section of intestine develops taller villi and deeper crypts and also grows in length and diameter; these changes dramatically increase the absorptive surface area. The ileum has a greater capacity for adaptation than the jejunum; thus, removal of the ileum has more severe consequences than removal of the jejunum. Loss of the ileum permanently disrupts vitamin B_{12} and bile acid absorption, and depletion of bile acids worsens both fat malabsorption and diarrhea. Adaptation is achieved more easily if the colon is present, because the colon's bacteria metabolize unabsorbed carbohydrates and produce some usable nutrients. An intact colon also helps to reduce losses of fluids and electrolytes.

Nutrition Therapy for Short Bowel Syndrome Total parenteral nutrition meets nutritional needs after surgery. To promote intestinal adaptation, oral feedings may be started within a week after surgery, after diarrhea subsides somewhat and some bowel function is restored. Initial oral intake may consist of sips of clear, sugar-free liquids, progressing to larger amounts of liquid formulas and then to solid foods, as tolerated. Very small, frequent feedings can utilize the remaining intestine most efficiently. Parenteral nutrition can be discontinued once oral intakes supply adequate nourishment. Some patients require tube feedings in addition to oral feedings to meet nutrient needs.

The exact diet prescribed depends on the portion of intestine removed, the length of remaining intestine, and whether the colon is still intact.[16] A high-kcalorie diet helps to compensate for malabsorption; thus, a high-fat, low-carbohydrate diet may be recommended if fat is sufficiently tolerated. Conversely, a high-complex-carbohydrate, low-fat diet may be suggested for patients who have an intact colon, because the colon bacteria can produce short-chain fatty acids from unabsorbed carbohydrates (which are readily absorbed in the colon) and the low fat intake can improve steatorrhea. Patients should avoid concentrated sweets (which attract fluids) if they cause additional diarrhea. Lactose intolerance, if present, can be managed by consuming limited amounts of milk at one time. Vitamin and mineral supplementation can help to

SAMPLE MENU

Breakfast
- Orange juice
- Gluten-free pancake with maple syrup
- Plain yogurt with banana and strawberries
- Coffee with half-and-half

Lunch
- Grilled chicken breast with cranberry chutney
- Baked potato topped with grated cheddar cheese
- Sliced tomato with chopped basil
- Raspberry sherbet

Snack
- Tortilla chips and guacamole
- Hot cocoa (made with cocoa powder)

Dinner
- Sauteed catfish with sliced lemon and dill
- Wild rice pilaf
- Collard greens and garlic sauteed in olive oil
- Green salad with oil and vinegar dressing
- Vanilla egg custard

FIGURE 19-2 Sample Menu—Gluten-Free Diet

The presence of nutrients in the lumen stimulates the growth of intestinal tissue, thereby promoting adaptation.

short bowel syndrome: the malabsorption syndrome that follows resection of the small intestine, resulting in inadequate absorptive capacity in the remaining intestine.

intestinal adaptation: the process of intestinal recovery following resection that leads to improved absorptive capacity.

WHAT IS ABSORBED

Duodenum/jejunum
- Simple carbohydrates
- Fats
- Amino acids
- Vitamins[a]
- Minerals[a]
- Water

Ileum
- Bile salts
- Vitamin B_{12}
- Water
(Assumes absorptive function of duodenum and jejunum with adaptation)

Colon
- Water
- Electrolytes
- Short-chain fatty acids

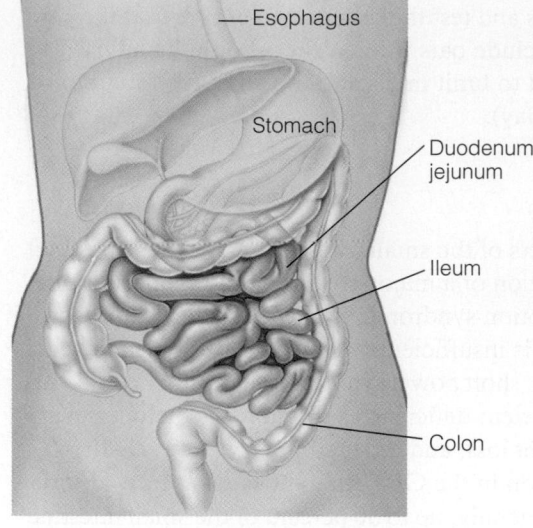

— Esophagus

Stomach

— Duodenum/jejunum

— Ileum

— Colon

POSSIBLE CONSEQUENCES OF RESECTION

Duodenum/jejunum
- Minimal consequences if the ileum remains intact
- Calcium and iron malabsorption if duodenum resected

Ileum
- Fat malabsorption
- Protein malabsorption
- Malabsorption of fat-soluble vitamins and vitamin B_{12}
- Reduced calcium, magnesium, and zinc absorption
- Fluid losses
- Diarrhea/steatorrhea

Colon
- Fluid and electrolyte losses
- Diarrhea
(Losses are compounded if ileum is also resected)

© Wadsworth, Cengage Learning

[a] The absorption of vitamins and minerals begins in the duodenum and continues throughout the length of the small intestine.

FIGURE 19-3 Nutrient Absorption and Consequences of Intestinal Surgeries

prevent deficiencies. The Case Study can help you review the material on short bowel syndrome.

IN SUMMARY

▌ Celiac disease is characterized by an abnormal immune response to gluten, resulting in malabsorption; a gluten-free diet is the primary treatment.

▌ Short bowel syndrome may result from major intestinal resections. Although intestinal adaptation can improve absorptive capacity, individuals with extensive resections may require lifelong parenteral nutrition support.

Case Study PATIENT WITH SHORT BOWEL SYNDROME

Judi Morel is a 28-year-old economist with an eight-year history of Crohn's disease. Judi is 5 feet 7 inches tall. Three years ago, she underwent a small bowel resection and remained free of active disease for two years. During that time, her symptoms subsided; she was able to tolerate most foods without any problem and gained weight. Ten months ago, Judi experienced a severe flare-up of her Crohn's disease. Since that time, she has lost 15 pounds and currently weighs 118 pounds. She has experienced severe abdominal pain and fatigue that have persisted despite aggressive medical management that included

intravenous nutrition. Five days ago, Judi underwent another resection, which left her with 40 percent of healthy small intestine. Her colon is intact. She is experiencing extensive diarrhea.

1. Describe the manifestations of Crohn's disease, and explain why surgery is sometimes performed as part of the treatment (review Chapter 18). Describe the complications of disease that may affect nutrient needs.

2. Using the BMI table in the back of the book, check the ideal weight range for a person of Judi's height. What

nutrition-related concerns are suggested by Judi's recent weight loss? What other nutrition problems did Judi probably experience as a consequence of Crohn's disease?

3. Discuss the complications that may follow an extensive intestinal resection. What factors may affect a person's ability to meet nutrient needs with an oral diet?

4. Describe the dietary progression recommended after an extensive intestinal resection. After Judi is able to eat solid foods, what factors may affect the type of diet that is recommended for her?

Nutrition Assessment Checklist FOR PEOPLE WITH MALABSORPTION DISORDERS

Medical History

Check the medical record for medical diagnoses of:

- Celiac disease
- Crohn's disease
- Chronic pancreatitis
- Cystic fibrosis
- Lactose intolerance
- Short bowel syndrome

Check for surgical procedures involving the GI tract, such as:

- Gastrectomy
- Gastric bypass surgery
- Intestinal bypass surgery
- Intestinal resections

Check for the following symptoms or complications:

- Anemia
- Bacterial overgrowth
- Bone disease
- Diarrhea, dehydration
- Fistulas
- Kidney stones
- Lactose intolerance
- Obstructions
- Poor growth (in children)
- Steatorrhea

Medications

Check for medications or dietary supplements that may:

- Interfere with appetite or food intake
- Irritate the mucosal lining
- Reduce gastric acid secretion

Dietary Intake

Note the following problems and contact the dietitian if you suspect difficulties such as:

- Food intolerances
- Inadequate fluid intake
- Poor appetite or food intake

Anthropometric Data

Measure baseline height and weight. Address weight loss early to prevent malnutrition in patients with:

- Gastric or intestinal bypass surgeries
- Gastric or intestinal resections
- Malabsorption
- Severe or persistent diarrhea
- Steatorrhea

Laboratory Tests

Check laboratory tests for signs of dehydration, electrolyte imbalances, and nutrient deficiencies in patients with:

- Intestinal resections
- Malabsorption
- Severe or persistent diarrhea
- Steatorrhea

Physical Signs

Look for physical signs of:

- Dehydration
- Essential fatty acid and fat-soluble vitamin deficiencies
- Folate and vitamin B_{12} deficiencies
- Mineral deficiencies
- Protein-energy malnutrition

Clinical Applications

1. Using Table 19-2 on p. 505 as a guide, plan a day's menus for a diet containing approximately 50 grams of fat. Take care to make the meals both palatable and nutritious. How can these menus be improved using the suggestions in the "How to" on p. 506?

2. As stated in this chapter, treatment of celiac disease is deceptively simple—eliminate wheat, barley, and rye, and possibly oats. Remaining on a gluten-free diet is more challenging than it appears, however.

   - Randomly select 10 of your favorite snack and convenience foods. Take a trip to the grocery store, and check the labels of the products you selected to see if they would be allowed on a gluten-free diet. Keep in mind that the labels may not list all offending ingredients.

   - Find acceptable substitutes for the products that are not allowed, either by substituting other foods or by checking for gluten-free products in the grocery store. If you have access to the Internet, you may want to investigate websites that advertise gluten-free products to get an idea of what's available.

Self Check

1. Possible causes of malabsorption include all of the following *except:*
   a. inflammatory bowel disease.
   b. pancreatic dysfunction.
   c. liver disease.
   d. flatulence.

2. The Schilling test helps to diagnose:
   a. lactose intolerance.
   b. fat malabsorption.
   c. malabsorption of vitamin B_{12}.
   d. pancreatic disorders.

3. Nutrition problems that may result from fat malabsorption include all of the following *except:*
   a. weight loss.
   b. essential amino acid deficiencies.
   c. bone loss.
   d. oxalate kidney stones.

4. Common nutrition problems associated with bacterial overgrowth in the stomach and small intestine include:
   a. sensitivity to gluten.
   b. fat malabsorption and vitamin B_{12} deficiency.
   c. constipation.
   d. permanent loss of digestive enzymes.

5. Lactose intolerance is a direct consequence of:
   a. insufficient lactase.
   b. milk allergy.
   c. fluid imbalance.
   d. pancreatic dysfunction.

6. The majority of chronic pancreatitis cases can be attributed to:
   a. bacterial and viral infections.
   b. gallstones.
   c. excessive alcohol use.
   d. elevated triglyceride levels.

7. The most appropriate diet for a person with cystic fibrosis is a:
   a. high-kcalorie diet.
   b. high-fiber diet.
   c. gluten-free diet.
   d. fat-controlled diet.

8. Chronic pancreatitis and cystic fibrosis are both treated with:
   a. intestinal resection.
   b. postural drainage.
   c. enzyme replacement therapy.
   d. stool softeners.

9. A person on a gluten-free diet must avoid products containing:
   a. wheat, barley, and rye.
   b. barley, soybeans, and corn.
   c. wheat, corn, and rice.
   d. buckwheat, rice, and oats.

10. If 60 percent of the small intestine remains after a jejunal resection:
    a. lifelong parenteral nutrition is the only option available.
    b. bile salts must be taken orally.
    c. pancreatic enzyme replacement is required.
    d. oral diets may eventually be able to meet nutrient needs.

Answers to these questions can be found in Appendix H.

Notes

1. C. Högenauer and H. F. Hammer, Maldigestion and malabsorption, in M. Feldman, L. S. Friedman, and L. J. Brandt, eds., *Sleisenger and Fordtran's Gastrointestinal and Liver Disease* (Philadelphia: Saunders, 2010), pp. 1735–1767.

2. J. S. Trier, Intestinal malabsorption, in N. J. *Greenberger, ed., Current Diagnosis and Treatment: Gastroenterology, Hepatology, and Endoscopy* (New York: McGraw-Hill Companies, 2009), pp. 223–242.

3. J. K. DiBaise, Nutritional consequences of small intestinal bacterial overgrowth, *Practical Gastroenterology* 32 (December, 2008): 15–28.

4. S. Hertzler and coauthors, Nutrient considerations in lactose intolerance, in A. M. Coulston and C. J. Boushey, eds., *Nutrition in the Prevention and Treatment of Disease* (Burlington, MA: Elsevier, 2008), pp. 755–770.

5. Hertzler and coauthors, 2008.

6. V. Singh, D. L. Conwell, and P. A. Banks, Acute pancreatitis, in N. J. Greenberger, ed., *Current Diagnosis and Treatment: Gastroenterology,*

Hepatology, and Endoscopy (New York: McGraw-Hill Companies, 2009), pp. 291–298.

7. S. Tenner and W. M. Steinberg, Acute pancreatitis, in M. Feldman, L. S. Friedman, and L. J. Brandt, eds., *Sleisenger and Fordtran's Gastrointestinal and Liver Disease* (Philadelphia: Saunders, 2010), pp. 959–983; Singh, Conwell, and Banks, 2009.

8. American Dietetic Association, *Nutrition Care Manual* (Chicago: American Dietetic Association, 2010).

9. C. E. Forsmark, Chronic pancreatitis, in M. Feldman, L. S. Friedman, and L. J. Brandt, eds., *Sleisenger and Fordtran's Gastrointestinal and Liver Disease* (Philadelphia: Saunders, 2010), pp. 985–1015.

10. Forsmark, 2010; B. Wu, D. Conwell, and P. Banks, Chronic pancreatitis, in N. J. Greenberger, ed., *Current Diagnosis and Treatment: Gastroenterology, Hepatology, and Endoscopy* (New York: McGraw-Hill Companies, 2009), pp. 299–304.

11. D. C. Whitcomb and M. E. Lowe, Hereditary, familial, and genetic disorders of the pancreas and pancreatic disorders in childhood, in M. Feldman, L. S. Friedman, and L. J. Brandt, eds., *Sleisenger and Fordtran's Gastrointestinal and Liver Disease* (Philadelphia: Saunders, 2010), pp. 931–957.

12. Whitcomb and Lowe, 2010; American Dietetic Association, 2010.

13. M. M. Niewinski, Advances in celiac disease and gluten-free diet, *Journal of the American Dietetic Association* (2008): 661–672.

14. Niewinski, 2008.

15. Trier, 2009.

16. A. L. Buchman, Short bowel syndrome, in M. Feldman, L. S. Friedman, and L. J. Brandt, eds., *Sleisenger and Fordtran's Gastrointestinal and Liver Disease* (Philadelphia: Saunders, 2010), pp. 1779–1795; C. R. Parrish, The clinician's guide to short bowel syndrome, *Practical Gastroenterology* (September 2005): 67–106.

Nutrition in Practice | ANEMIA IN ILLNESS

Anemia—a reduction of red blood cells that lowers the oxygen-carrying capacity of the blood—is frequently the first sign of illness and may be the disorder that initially drives an individual to seek medical attention. Anemia is associated with a great number of diseases and is common among hospital patients: some 20 to 40 percent exhibit some degree of anemia.[1] Earlier chapters in this textbook described some of the relationships between nutrient deficiencies and anemia. This Nutrition in Practice explains how and why anemia develops during the course of illness. The accompanying glossary defines relevant terms.

How does anemia develop?

Anemia develops when red blood cells (also called *erythrocytes*) are not produced in sufficient numbers, are too quickly destroyed, or are lost due to bleeding. Because red blood cells contain the hemoglobin that supplies oxygen to tissues, their absence can result in fatigue and reduced stamina. The deficiency of oxygen in tissues is the main stimulus for the production of additional red blood cells. Table NP19-1 provides an overview of some different categories of anemia and their underlying causes.

How are red blood cells produced?

The production of red blood cells **(erythropoiesis)** takes place in the bone marrow, a soft tissue found in certain types of bone. The process begins when kidney cells sense the low oxygen

TABLE NP19-1 Types of Anemia

Type of Anemia	General Mechanism
Anemia of chronic disease	Reduced iron availability due to inflammatory processes; results in reduced red blood cell (RBC) production
Aplastic anemia	Failure of stem cells to develop into RBCs; may be due to immune disease, viruses, drugs and toxins, or genetic defects
Hemolytic anemia	Premature destruction of red blood cells; results in shortened RBC life span and fewer RBCs
Hemorrhagic anemia	Blood loss; causes reduction in circulating RBCs
Iron-deficiency anemia	Reduced iron availability due to dietary deficiency; interferes with hemoglobin production and results in small, hypochromic RBCs
Megaloblastic anemia	Reduced availability of nutrients required for DNA synthesis and cell division; results in large, immature RBCs
Sickle-cell anemia	Genetic mutation that results in altered hemoglobin molecule; causes production of abnormal, sickle-shaped RBCs
Thalassemia	Genetic mutation that reduces hemoglobin synthesis; results in reduced RBC production

GLOSSARY

anemia of chronic disease: anemia that develops in persons with chronic illness; may resemble iron-deficiency anemia even though iron stores are often adequate. Also called *anemia of chronic inflammation.*

aplastic anemia: anemia characterized by the inability of bone marrow to produce adequate numbers of blood cells. Causes include drug toxicity, viruses, and genetic defects.

erythropoiesis (eh-RIH-throh-poy-EE-sis): production of red blood cells within the bone marrow.

erythropoietin (eh-RIH-throh-POY-eh-tin): a hormone produced by kidney cells that stimulates red blood cell production.

hemolytic (hee-moe-LIH-tic) **anemia:** anemia characterized by the breakdown of red blood cells.

megaloblastic anemia: anemia characterized by large, immature red blood cells, as occurs in folate and vitamin B_{12} deficiency; also called *macrocytic anemia.*

microcytic anemia: anemia characterized by small, hypochromic (pale) red blood cells, as occurs in iron deficiency.

peripheral blood smear: a blood sample spread on a glass slide and stained for analysis under a microscope. *Peripheral* refers to the use of circulating blood rather than tissue blood.

reticulocytes: immature red blood cells released into blood by bone marrow.

content of blood and release the hormone **erythropoietin** (see Figure NP19-1). Erythropoietin travels to bone marrow, where it stimulates precursor cells (stem cells) to divide and differentiate into red blood cells. The cells that are released from the bone marrow are immature red blood cells called **reticulocytes.** Reticulocytes develop into mature red blood cells over a 24- to 48-hour period while they circulate in the bloodstream.

Which nutrient deficiencies can interfere with red blood cell production?

The nutrient deficiencies that most frequently upset red blood cell production are those of iron, folate, and vitamin B_{12}. Iron is required for hemoglobin production, and deficiency results in **microcytic anemia,** characterized by small, hypochromic cells (see pp. 239–240). Vitamin B_{12} and folate participate in DNA synthesis, and deficiency of either nutrient leads to **megaloblastic anemia** (also called *macrocytic anemia*), characterized by large, immature cells (see p. 208). Other nutrient deficiencies may cause anemia, although not as frequently. Vitamin E helps to maintain cell membrane integrity, and its deficiency is associated with **hemolytic anemia** (red blood cell breakdown). Vitamin B_6 plays a

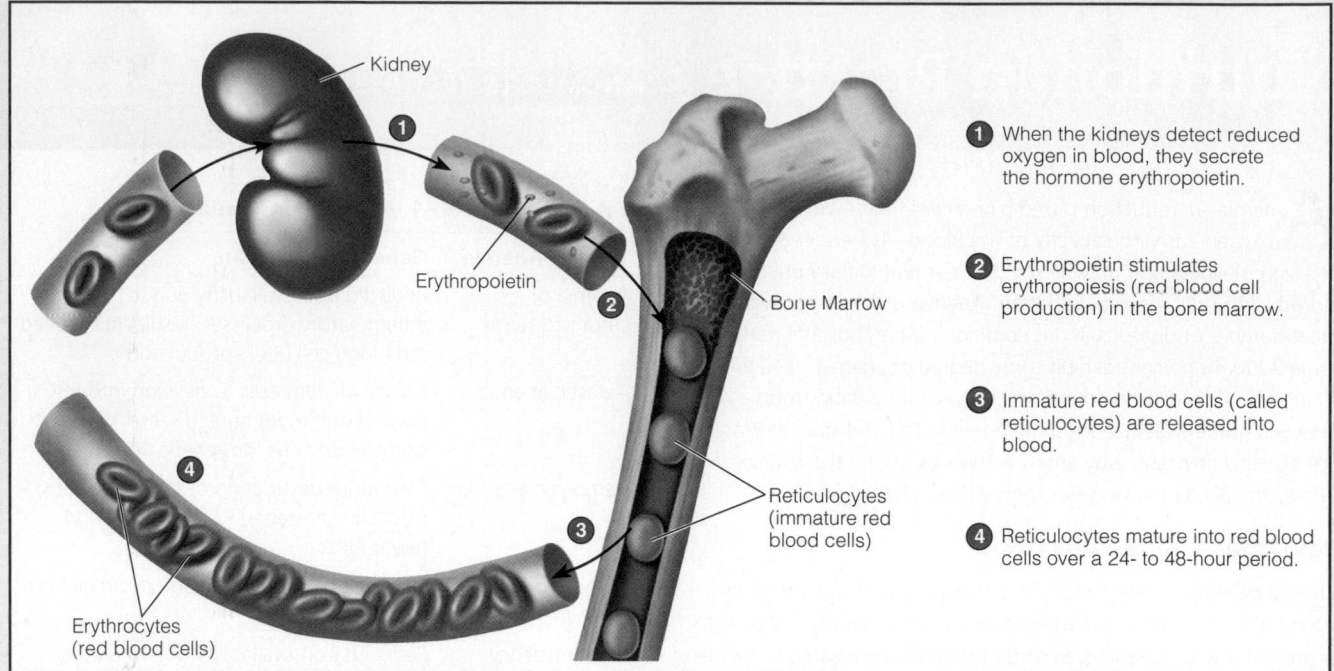

FIGURE NP19-1 **Erythropoiesis**
Source: Reprinted with permission from L. Sherwood, *Human Physiology,* 5th ed., (Brooks/Cole, 2004), Figure 11-4, p. 395.

The following labels appear in the figure:

Kidney

1 When the kidneys detect reduced oxygen in blood, they secrete the hormone erythropoietin.

Erythropoietin

2 Erythropoietin stimulates erythropoiesis (red blood cell production) in the bone marrow.

Bone Marrow

3 Immature red blood cells (called reticulocytes) are released into blood.

Reticulocytes (immature red blood cells)

4 Reticulocytes mature into red blood cells over a 24- to 48-hour period.

Erythrocytes (red blood cells)

role in hemoglobin production, and a deficiency may occasionally cause microcytic anemia. Vitamin C supports blood vessel integrity; fragile and bleeding capillaries may result from its deficiency. Protein-energy malnutrition leads to anemia because red blood cell development depends on protein synthesis. Although nutrient deficiencies may result from dietary inadequacy, they can also arise during the course of illness due to the effects of disease on intestinal absorption, nutrient metabolism, and nutrient losses.

How can illnesses cause the nutrient deficiencies that lead to anemia?

There are numerous ways in which illnesses can lead to deficiencies of iron, folate, or vitamin B_{12}, the main causes of the nutritional anemias. Blood loss, common to many illnesses, is a primary cause of iron deficiency. Some illnesses may result in a reduction in food intake, as discussed throughout the clinical chapters. The liver's stores of iron and vitamin B_{12} are often adequate to prevent deficiencies during transient illnesses, but reserves of folate are limited; thus, a folate deficiency can develop within a few months if dietary intakes are low. If several deficiencies occur simultaneously, it may be difficult to identify the cause of anemia using standard blood tests (see Appendix E) because both megaloblastic anemia and microcytic anemia may be present.

Which illnesses are associated with blood loss?

Conditions that involve the gastrointestinal tract often cause bleeding; examples include peptic ulcers, inflammatory bowel conditions, and gastrointestinal varices (enlarged veins) that develop in advanced liver disease. Excessive bleeding can also accompany coagulation disorders, which are often due to liver disease, genetic defects, or vitamin K deficiency. Frequent blood draws or surgical procedures can contribute to blood loss and result in iron

deficiency. Unfortunately, slow, chronic bleeding is sometimes difficult to identify before anemia develops.

How do malabsorption disorders contribute to the development of anemia?

Chapters 18 and 19 explained how disorders that damage the small intestine can lead to nutrient malabsorption. Diseases like Crohn's disease and celiac disease can destroy intestinal mucosa and reduce the absorption of all nutrients. Iron is primarily absorbed in the duodenum and upper jejunum, and its absorption is impaired by conditions that reduce hydrochloric acid secretion or result in surgical resection (removal) of the upper intestine. Resection of the stomach or ileum can hasten the onset of vitamin B_{12} deficiency because both organs have roles in vitamin B_{12} absorption: you may recall from Chapter 8 that the stomach produces a protein called *intrinsic factor* that is needed for vitamin B_{12} absorption, and that the ileum is the site of vitamin B_{12} absorption.

Why is anemia present during many chronic illnesses, even when the illnesses are not accompanied by nutrient deficiencies or bleeding disorders?

Chronic disease itself can cause anemia, and anemia is sometimes the initial sign that chronic disease is present. In fact, the **anemia of chronic disease** is the most common type of anemia affecting hospitalized patients and patients with chronic illnesses.[2] This type of anemia usually occurs in individuals who have inflammatory conditions, chronic infections, autoimmune disorders, or cancer. Although often a mild form of anemia, it can progress and become severe enough to require blood transfusions.

The anemia of chronic disease is characterized by alterations both in the distribution of iron among tissues and in the rates of

red blood cell production and destruction.[3] During chronic illness, inflammatory mediators induce the production of the protein *hepcidin,* which blocks the release of iron from storage and thereby renders iron unavailable for red blood cell production. Furthermore, hepcidin inhibits iron's release from intestinal cells into the blood and therefore interferes with iron absorption. Finally, inflammatory processes cause red blood cells to be degraded more quickly than usual, and the reduced production of red blood cells cannot keep pace. Eventually, outright iron deficiency may be a consequence of the impaired iron absorption.

How is the anemia of chronic disease diagnosed?

Blood tests help to distinguish between the anemia of chronic disease and iron-deficiency anemia (see Table NP19-2). The combination of low serum iron and low total iron-binding capacity suggests the anemia of chronic disease rather than iron deficiency. In addition, serum ferritin levels are normal or elevated, whereas they are typically low in iron deficiency. Diagnosis is more complicated if both types of anemia are present.

Can medications cause anemia?

Yes, anemia is among the adverse effects that can result from medication use. Various medications may disrupt nutrient metabolism, impair blood coagulation and erythropoiesis, and increase red blood cell destruction. Because the life span of red blood cells is about 120 days, the long-term use of such medications is more likely to result in anemia than short-term use.

How do medications alter nutrient metabolism?

There are numerous ways in which medications can alter nutrient metabolism; the most common are listed in Table 15-2 on p. 415. As an example, a number of medications are known to influence the absorption or metabolism of folate and lead to megaloblastic anemia. Sulfasalazine (used for ulcerative colitis) and some anticonvulsant drugs inhibit folate absorption, and methotrexate (an immunosuppressive) and pyrimethamine (an antimalarial) interfere with folate metabolism.[4] If a medication is known to result in deficiency, nutrient supplementation is usually recommended as an adjunct therapy.

How do medications impair blood coagulation?

Anticoagulants, which are prescribed specifically to reduce blood clotting, may sometimes lead to excessive bleeding. These medications work by interfering with one of the steps involved in blood clotting, such as platelet function, vitamin K function, or the synthesis of clotting proteins. Other than anticoagulants, a large number of other drugs can impair coagulation (including commonly used drugs such as aspirin and other nonsteroidal anti-inflammatory drugs). The anticoagulant effects may be augmented if several of these drugs are used simultaneously. The slow, chronic bleeding that sometimes develops may go unnoticed until excessive blood loss has occurred.

Which categories of drugs inhibit erythropoiesis?

The categories of drugs that can inhibit erythropoiesis include anticonvulsants, antibiotics, antidiabetic drugs, diuretics, antithyroid drugs, and anticancer agents.[5] The anemia that occurs when the bone marrow fails to produce adequate numbers of blood cells is called **aplastic anemia.** Aplastic anemia can also be caused by viral infections, exposure to toxins, and genetic defects.

How can medications cause blood cell destruction?

Some patients may develop hemolytic anemia as a result of drug interactions with red blood cells. For example, a drug may alter the red blood cell membrane in such a way that a component of the membrane becomes an antigen and induces an antibody response that destroys the cell.[6] Several types of antibiotics, including penicillin and cephalosporin, may cause this type of response. Withdrawal of the drug can eventually reverse the anemia, and sometimes medications are given to suppress the immune response.

With all the possible ways in which anemia can develop, how can its cause be determined?

Identifying the cause of anemia is sometimes quite challenging. In some cases, anemia may be a well-known consequence of disease, as when renal failure impairs the synthesis of the hormone erythropoietin. When anemia develops rapidly, blood loss is often the cause, whereas a more gradual onset suggests malnutrition, chronic illness, or slow, chronic bleeding. As mentioned earlier, the results of laboratory tests provide valuable clues, although

TABLE NP19-2 **Laboratory Tests for Evaluating Iron Deficiency and Anemia of Chronic Disease**

Laboratory Test	Effect of Iron Deficiency	Effect of Chronic Disease
Red blood cell (RBC) size and number	Microcytic; reduced RBC count	Normocytic or microcytic; reduced RBC count
Serum iron	Low	Low
Serum ferritin	Low	Normal or elevated
Serum transferrin	Elevated	Low
Total iron-binding capacity	High	Normal or low
Bone marrow iron	Low	Normal or elevated

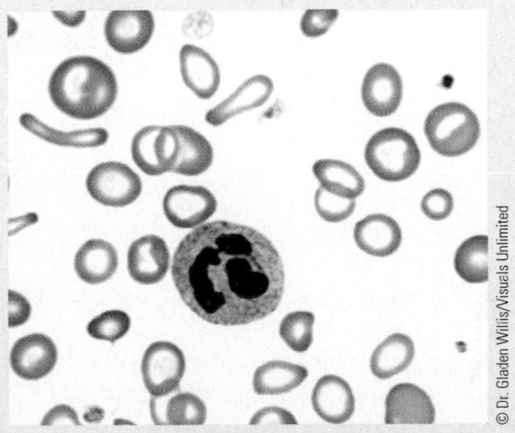

© Dr. Gladen Willis/Visuals Unlimited

A peripheral blood smear provides information about the number and shape of blood cells.

conditions such as dehydration and inflammation can influence the values. Laboratory results are especially difficult to analyze if several disturbances are present simultaneously. A **peripheral blood smear** (see photo) is often used to study abnormalities in red blood cell shape and may reveal an underlying cause.

Anemia is a disorder associated with many different diseases, and it may also be caused by disease treatment. When it occurs during illness, its causes must be investigated before it leads to complications that worsen prognosis. The medical history, blood tests, and peripheral blood smears may all help to determine the reasons why anemia has developed.

Notes

1. K. S. Zuckerman, Approach to the anemias, in L. Goldman and D. Ausiello, eds., *Cecil Textbook of Medicine* (Philadelphia: Saunders, 2008), pp. 1179–1187.

2. G. D. Ginder, Microcytic and hypochromic anemias, in L. Goldman and D. Ausiello, eds., *Cecil Textbook of Medicine* (Philadelphia: Saunders, 2008), pp. 1187–1194.

3. Ginder, 2008.

4. A. C. Antony, Megaloblastic anemias, in L. Goldman and D. Ausiello, eds., *Cecil Textbook of Medicine* (Philadelphia: Saunders, 2008), pp. 1231–1241.

5. H. Castro-Malaspina and R. J. O'Reilly, Aplastic anemia and related disorders, in L. Goldman and D. Ausiello, eds., *Cecil Textbook of Medicine* (Philadelphia: Saunders, 2008), pp. 1241–1248.

6. R. S. Schwartz, Autoimmune and intravascular hemolytic anemias, in L. Goldman and D. Ausiello, eds., *Cecil Textbook of Medicine* (Philadelphia: Saunders, 2008), pp. 1194–1203.

20

Nutrition Therapy for Liver and Gallbladder Diseases

The liver is the most metabolically active organ in the body. It plays a central role in processing, storing, and redistributing the nutrients provided by the foods we eat. The liver also produces bile, which emulsifies fat during digestion; between meals, this bile is stored and concentrated in the gallbladder. The liver manufactures most of the proteins that circulate in plasma, including albumin, clotting proteins, and transport proteins. In addition, the liver detoxifies drugs and alcohol and processes excess nitrogen so that it can be safely excreted as urea. If the liver's numerous roles are upset by liver damage or disease, the effects on health and nutrition status can be profound.

Liver disease progresses slowly. Its symptoms are sometimes so mild that complications may develop before liver disease is diagnosed. Once liver disease is recognized, preserving the remaining liver function becomes a primary concern because the liver can regenerate some healthy tissue, improving the prognosis. Preventing additional damage is the principal means of avoiding liver failure or transplantation.

Fatty Liver and Hepatitis

Fatty liver and hepatitis are the two most common disorders affecting the liver. Although both conditions may be mild and are usually reversible, each may progress to more serious illness and eventually cause liver damage.

Fatty Liver

Fatty liver is an accumulation of fat in liver tissue. Ordinarily, the liver's excess triglycerides are packaged into very-low-density lipoproteins (VLDL) and exported to the bloodstream. Fatty liver represents an imbalance between the amount of fat synthesized or picked up from the blood and the amount exported to the blood via VLDL. Fatty liver is estimated to affect 20 percent or more of the adult population in the United States.[1]

Fatty liver is a clinical finding that is common to many conditions. It is present in the majority of patients who have alcoholic liver disease (discussed in Nutrition in Practice 20) and can also result from exposure to drugs and toxic metals. It often accompanies diabetes mellitus, metabolic syndrome, obesity, and diseases of malnutrition, including kwashiorkor and marasmus. Fatty liver may also follow gastrointestinal bypass surgery and long-term total parenteral nutrition.[2]

Consequences of Fatty Liver In many individuals, fatty liver is asymptomatic and causes no harm. In other cases, it may be associated with liver enlargement **(hepatomegaly),** inflammation **(steatohepatitis),** and fatigue. If liver damage and scarring develop, fatty liver may progress to cirrhosis (discussed in a later section), liver failure, or liver cancer.[3]

Fatty liver is a frequent cause of abnormal levels of liver enzymes in the blood. Laboratory findings may include elevated blood concentrations of the liver enzymes ALT and AST, as well as increased levels of triglycerides, cholesterol, and glucose. Table 20-3 on p. 525 provides normal ranges for these liver enzymes.

The liver enzymes ALT (alanine aminotransferase) and AST (aspartate aminotransferase) are involved in amino acid catabolism.

Treatment of Fatty Liver The usual treatment for fatty liver is to eliminate the factors that cause it. For example, if fatty liver is due to alcohol abuse or drug treatment, it may improve after the patient discontinues use of the substance. In patients with elevated blood lipids, fatty liver may improve after blood lipid levels are lowered. An appropriate treatment for obese or diabetic patients might be weight reduction, increased physical activity, or medications that improve insulin sensitivity. Rapid weight loss should be discouraged, however, because it may accelerate the progression of liver disease.[4] Note that lifestyle modifications are not always successful in reversing fatty liver, especially in patients who lack the usual risk factors.

TABLE 20-1 Features of Hepatitis Viruses

Hepatitis Virus	Major Mode of Transmission	New Cases (United States, 2007)	Chronic Disease Rate (% of cases)	Chronic Cases (United States)	Vaccination Available
A	Fecal-oral	25,000	None	0	Yes
B	Bloodborne, sexual transmission	43,000	2–7	1.25 million	Yes
C	Bloodborne	17,000	50–85	3.2 million	No

Note: There are fewer new cases of hepatitis C virus (HCV) infection than of hepatitis B virus (HBV) infection each year, but more HCV cases become chronic. Therefore, there are more HCV carriers than HBV carriers.

Hepatitis

Hepatitis, a condition of liver inflammation, can result from any factor that causes damage to liver tissue. Most often, the damage is caused by infection with specific viruses, designated by the letters A, B, C, D, E, and G. Other causes include alcohol abuse, fatty liver disease, autoimmune disease, and exposure to some drugs, herbal substances, and toxic chemicals. Long-term hepatitis can lead to cirrhosis (discussed in a later section) and liver cancer.

Viral Hepatitis Acute hepatitis is most often caused by infection with hepatitis virus A, B, or C (see Table 20-1). Specific features of these viruses include the following:

▌ *Hepatitis A virus* (HAV) is primarily spread via fecal-oral transmission, which usually involves the ingestion of foods or beverages that have been contaminated with fecal matter. Outbreaks are often associated with floods and other natural disasters, when untreated sewage contaminates water supplies. In the United States, vaccinations against HAV are routinely provided to children and high-risk individuals. HAV infection usually resolves within a few months and does not cause chronic illness or permanent liver damage.

▌ *Hepatitis B virus* (HBV) is transmitted by infected blood or needles or by sexual contact. A major global health concern, HBV has infected one-third of the world population, although chronic illness develops in less than 10 percent of cases.[5] Vaccinations are currently recommended for newborn infants and children, health-care workers, recipients of blood products, dialysis patients, sexually active adults, and users of injected drugs.

▌ *Hepatitis C virus* (HCV) is spread via infected blood or needles but is not readily spread by sexual contact. Most cases progress to chronic illness, and currently HCV is the most common cause of chronic liver disease in the United States.[6] Preventive measures include blood donor screening, viral inactivation of blood products, and infection control practices in health care settings. No vaccine is available to protect against HCV infection.

Symptoms of Hepatitis The effects of hepatitis depend on the cause and severity of the disease. Individuals with mild or chronic cases are often asymptomatic. The onset of acute hepatitis may be accompanied by fatigue, nausea, vomiting, anorexia, and pain in the liver area. The liver is often slightly enlarged and tender. **Jaundice** (yellow discoloration of tissues) may develop, causing yellowing of the skin, urine, and the whites of the eyes. Other symptoms may include fever, muscle weakness, joint pain, and skin rashes. Serum levels of the liver enzymes ALT and AST are typically elevated. Chronic hepatitis can be associated with complications that are typical of liver cirrhosis.

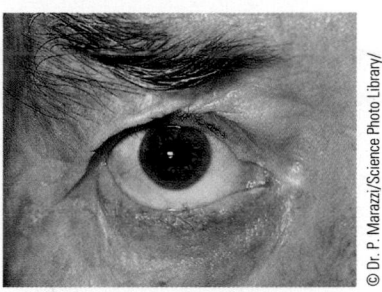

© Dr. P. Marazzi/Science Photo Library/ Photo Researchers, Inc.

Jaundice is a yellow discoloration of the tissues that is most easily seen in the whites of the eyes.

Jaundice results when liver dysfunction impairs the metabolism of bilirubin, a breakdown product of hemoglobin that is normally eliminated in bile. Accumulation of bilirubin in the bloodstream leads to yellow discoloration of tissues.

fatty liver: an accumulation of fat in liver tissue; also called *hepatic steatosis* (STEE-ah-TOE-sis).
 hepatic = pertaining to the liver
 steato = fat

hepatomegaly (HEP-ah-toe-MEG-ah-lee): enlargement of the liver.

steatohepatitis (STEE-ah-to-HEP-ah-TYE-tis): liver inflammation that is associated with fatty liver.

hepatitis (hep-ah-TYE-tis): inflammation of the liver.

jaundice (JAWN-dis): yellow discoloration of the skin and eyes due to an accumulation of bilirubin, a breakdown product of hemoglobin that normally exits the body via bile secretions.

Treatment of Hepatitis Hepatitis is treated with supportive care, such as bed rest (if necessary) and an appropriate diet. Hepatitis patients should avoid substances that irritate the liver, such as alcohol, drugs, and dietary supplements that cause liver damage. Hepatitis A infection usually resolves without the use of medications, whereas antiviral drugs may be used to treat HBV and HCV infections. Nonviral forms of hepatitis may be treated with anti-inflammatory and immunosuppressant drugs.

Nutrition care varies according to a patient's symptoms and nutrition status.[7] Most individuals require no dietary changes. Those with anorexia or abdominal discomfort may find small, frequent meals easier to tolerate. Malnourished individuals need to consume adequate protein and energy to replenish nutrient stores. A low-fat diet may be recommended for those with steatorrhea. Patients with persistent vomiting may require fluid and electrolyte replacement. Liquid supplements can be helpful for improving nutrient intakes.

IN SUMMARY

▌ Fatty liver can result from excessive alcohol intake, drug toxicity, and chronic disorders such as diabetes and obesity. Hepatitis is frequently caused by viral infection but can also result from alcohol abuse, drug toxicity, and other causes.

▌ Although fatty liver is often benign, hepatitis can become chronic and lead to cirrhosis and liver cancer. Hepatitis treatment involves supportive care, such as bed rest, elimination of liver toxins, and dietary measures that maintain or improve nutrition status.

Cirrhosis

Cirrhosis is the final phase of chronic liver disease. Long-term liver disease gradually destroys liver tissue, leading to scarring (fibrosis) in some regions and small areas of regenerated, healthy tissue in others. As the disease progresses, the scarring becomes more extensive, leaving fewer areas of healthy tissue. A cirrhotic liver is often shrunken in size and has an irregular, nodular appearance. Cirrhosis impairs liver function and can eventually lead to liver failure. It is currently the 12th leading cause of death in the United States.[8]

The chief causes of cirrhosis in the United States are alcoholic liver disease and chronic hepatitis C infection, followed by fatty liver disease and chronic hepatitis B infection.[9] Additional causes include other types of chronic hepatitis, drug-induced liver injury, inherited disorders, and bile duct blockages, which cause bile acids to accumulate to toxic levels in the liver (see Table 20-2).

TABLE 20-2 Causes of Cirrhosis

Alcoholic liver disease
Autoimmune hepatitis
Bile duct obstructions
▌ Complications of gallbladder surgery
▌ Cystic fibrosis
▌ Diseases that cause bile duct injury
Drug-induced liver injury
Inherited disorders
▌ Galactosemia
▌ Glycogen storage disease
▌ Hemochromatosis (causes excessive liver iron)
▌ Wilson's disease (causes excessive liver copper)
Nonalcoholic fatty liver disease
Viral hepatitis
▌ Hepatitis B
▌ Hepatitis C

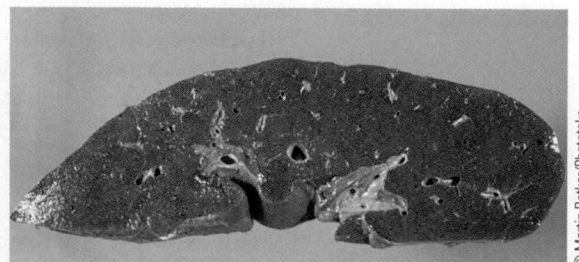

Normal liver tissue is smooth and has a regular texture.

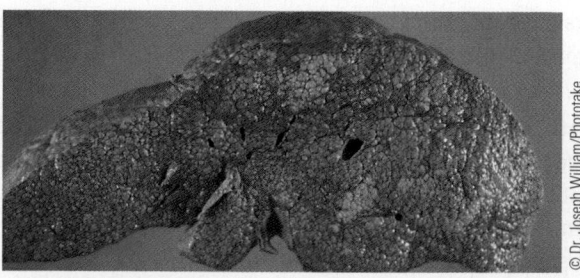

A cirrhotic liver has an irregular, nodular appearance. The nodules represent clusters of regenerating cells within the damaged liver tissue.

Consequences of Cirrhosis

About 40 percent of cirrhosis patients are asymptomatic.[10] Because liver disease progresses slowly, the effects of cirrhosis may be subtle at first. Initial symptoms are usually nonspecific and may include fatigue, weakness, anorexia, and weight loss. Later, patients may develop anemia, bruise easily, and be more susceptible to infections. If bile obstruction occurs, jaundice and fat malabsorption are likely. The physical changes in liver tissue may interfere with blood flow, causing fluid to accumulate in blood vessels and body tissues. Advanced cirrhosis can disrupt kidney and lung function. Figure 20-1 illustrates some of the clinical effects of liver cirrhosis, and later sections describe some of these complications in more detail.

Table 20-3 lists laboratory tests that are used to monitor the extent of liver damage. Serum liver enzyme levels are elevated in liver disease because the injured liver tissue releases the enzymes into the bloodstream. Serum levels of bilirubin may be increased if the liver is too damaged to process it or if bile ducts are blocked and prevent its excretion. The impaired synthesis of plasma proteins by the liver lowers albumin levels and extends blood-clotting time. Liver damage also reduces the conversion of ammonia to urea, causing ammonia levels in the blood to rise.

Portal Hypertension A large volume of blood normally flows through the liver. The portal vein and hepatic artery together supply approximately 1500 milliliters (about 1.5 quarts) of blood each minute to the extensive network of vessels in the liver. The scarred tissue of a cirrhotic liver impedes the flow of blood, three-fourths of which is supplied by the portal vein.[11] The resistance to blood flow within the liver causes a rise in blood pressure within the portal vein, called **portal hypertension.**

Collaterals and Gastroesophageal Varices When blood flow through the portal vein is impeded, the blood is diverted to the smaller blood vessels surrounding the liver. These **collaterals** develop throughout the gastrointestinal (GI) tract and in regions near the abdominal wall. As pressure builds, the collateral vessels become enlarged and engorged, forming **varices** (see the photo). The varices that develop in the esophagus *(esophageal varices)* and stomach *(gastric varices)* are vulnerable to rupture because they have thin walls and often bulge into the lumen. If ruptured, they can cause massive bleeding

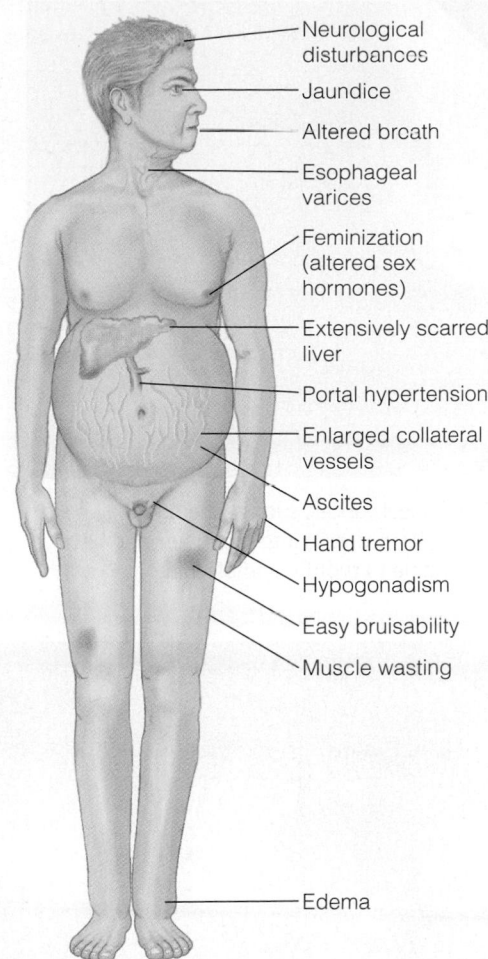

- Neurological disturbances
- Jaundice
- Altered breath
- Esophageal varices
- Feminization (altered sex hormones)
- Extensively scarred liver
- Portal hypertension
- Enlarged collateral vessels
- Ascites
- Hand tremor
- Hypogonadism
- Easy bruisability
- Muscle wasting
- Edema

FIGURE 20-1 Clinical Effects of Liver Cirrhosis

Reminder: The *portal vein* is the large blood vessel that carries nutrient-rich blood from the digestive tract to the liver.

TABLE 20-3 Laboratory Tests for Evaluation of Liver Disease

Laboratory Test	Normal Ranges (serum)	Values in Liver Disease
Alanine aminotransferase (ALT)	Male: 10–40 U/L Female: 7–35 U/L	Elevated
Albumin	3.4–4.8 g/dL	Decreased
Alkaline phosphatase	25–100 U/L	Normal or elevated
Ammonia	15–45 µg N/dL	Elevated
Aspartate aminotransferase (AST)	10–30 U/L	Elevated
Bilirubin (total)	0.3–1.2 mg/dL	Elevated
Blood urea nitrogen (BUN)	6–20 mg/dL	Normal or decreased
Prothrombin time[a]	10–13 seconds	Prolonged

[a]The test for prothrombin time evaluates the clotting ability of blood.
Note: U/L = units per liter; dL = deciliter; µg = micrograms; N = nitrogen.

cirrhosis (sih-ROE-sis): an advanced stage of liver disease in which extensive scarring replaces healthy liver tissue, causing impaired liver function and liver failure.

portal hypertension: elevated blood pressure in the portal vein due to obstructed blood flow through the liver.

collaterals: blood vessels that enlarge to allow an alternative pathway for diverted blood.

varices (VAH-rih-seez): abnormally dilated blood vessels (singular: *varix*).

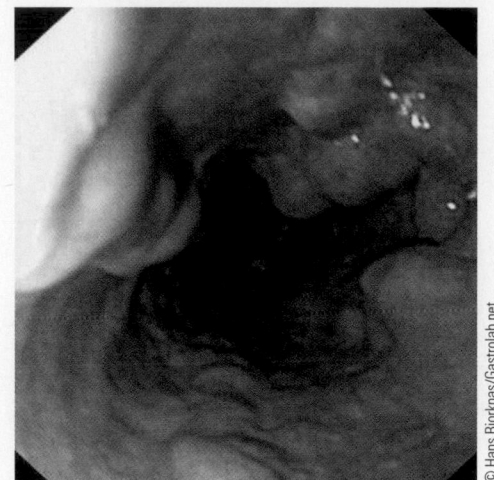

Esophageal varices, such as the one shown here, may protrude into the lumen and be vulnerable to rupture and bleeding.

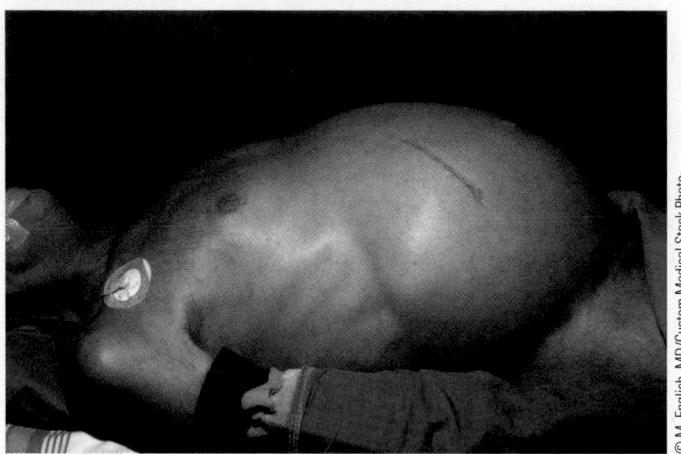

Ascites is caused by various illnesses, but cirrhosis is the underlying cause in most patients with the condition.

The aromatic amino acids— phenylalanine, tyrosine, and tryptophan—have carbon rings in their side groups. The branched-chain amino acids are leucine, isoleucine, and valine; their side groups have a branched structure.

that is sometimes fatal. The blood loss is exacerbated by the liver's reduced production of blood-clotting factors.

Ascites Within 10 years of disease onset, about 50 percent of cirrhosis patients develop **ascites,** a large accumulation of fluid in the abdominal cavity. The development of ascites indicates that liver damage has reached a critical stage, as half of patients with ascites die within 2 years.[12] Ascites is primarily a consequence of portal hypertension, sodium and water retention in the kidneys, and reduced albumin synthesis in the diseased liver. The increased pressure within the portal vein triggers the release of chemical factors (such as nitric oxide) that dilate local blood vessels and lower the blood volume elsewhere; these changes activate sodium and water retention in the kidneys and cause an accumulation of body fluid. Elevated pressure within the liver's small blood vessels **(sinusoids)** causes leaking of fluid into lymphatic vessels and, ultimately, the abdominal cavity. The water pooling is exacerbated by low levels of serum albumin, a protein that helps to retain fluid in blood vessels. Ascites can cause abdominal discomfort and early satiety, which contribute to malnutrition. Because ascites can raise body weight considerably, weight changes may be difficult to interpret.

Hepatic Encephalopathy Advanced liver disease sometimes leads to **hepatic encephalopathy,** a disorder characterized by abnormal neurological functioning. Symptoms of hepatic encephalopathy include changes in personality, behavior, mental abilities, and motor functions (see Table 20-4). At worst, amnesia, seizures, and **hepatic coma** may develop. Although hepatic encephalopathy is fully reversible with medical treatment, the prognosis is poor when it progresses to the advanced stages.

The exact causes of hepatic encephalopathy remain elusive, although elevated blood ammonia levels are thought to play a key role in its development due to ammonia's neurotoxicity. Other substances that may accumulate in brain tissue and disturb brain function include sulfur compounds, short-chain fatty acids, and manganese.[13] Another theory is that the brain's neurotransmitters are altered by an increased ratio of aromatic amino acids to branched-chain amino acids in brain tissue, a result of disordered amino acid metabolism in the liver. Most likely, a combination of abnormalities contributes to the disruption in neurological functioning.

Elevated Blood Ammonia Levels Much of the body's free ammonia is produced by bacterial action on unabsorbed dietary protein in the colon. Normally, the liver

TABLE 20-4 Symptoms of Hepatic Encephalopathy

Early Stages	Middle Stages	Later Stages
Short attention span	Disorientation, impaired memory	Confusion, amnesia
Depression, irritability	Anxiety, impaired judgment	Anger, paranoia
Lack of coordination, tremor	Slurred speech, abnormal reflexes	Muscular rigidity, abnormal reflexes
Sleep disorders	Lethargy	Semi-stupor, coma

TABLE 20-5 Possible Causes of Malnutrition in Liver Disease

Mechanism	Examples
Reduced nutrient intake	Abdominal discomfort, altered mental status, anorexia, early satiety (due to ascites), effects of medications (including gastrointestinal disturbances and taste changes), fasting for medical procedures, fatigue, nausea and vomiting, restrictive diets
Malabsorption or nutrient losses	Diarrhea, effects of medications (including malabsorption and nutrient losses from diuretic use), fat malabsorption (due to reduced bile flow), gastrointestinal bleeding, vomiting
Altered metabolism or increased nutrient needs	Hypermetabolism, impaired protein synthesis, infections or inflammation, muscle catabolism, reduced nutrient storage and metabolism in the liver

extracts this ammonia from portal blood and converts it to urea, which is then excreted by the kidneys. In advanced liver disease, the liver is unable to process the ammonia sufficiently. In addition, the ammonia-laden portal blood bypasses the liver by way of collateral vessels and reaches the general blood circulation, causing a substantial increase in the ammonia that reaches brain tissue. Although ammonia levels do not correlate well with the degree of neurological impairment in hepatic encephalopathy, ammonia-reducing medications can successfully reverse the neurological symptoms.[14]

Malnutrition and Wasting Most patients with cirrhosis develop protein-energy malnutrition (PEM) and experience some degree of wasting. Malnutrition is usually caused by a combination of factors (see Table 20-5). Patients may consume less food due to reduced appetite, GI symptoms, early satiety associated with ascites, or fatigue. If the diet is sodium restricted (to treat ascites), foods may seem unpalatable. Fat malabsorption is common due to reduced bile flow, potentially leading to deficiencies of fat-soluble vitamins and some minerals. Additional nutrient losses may result from diarrhea, vomiting, and GI bleeding. If cirrhosis is a consequence of alcohol abuse, multiple nutrient deficiencies may be present.

Treatment of Cirrhosis

Medical treatment for cirrhosis aims to correct both the underlying cause of illness and any complications that develop. Antiviral medications may be prescribed to treat viral infections. Avoidance of alcohol and liver toxins is critical for preserving the remaining liver function and extending survival. Patients should be screened and treated for life-threatening complications, such as gastroesophageal varices and liver cancer. Liver transplantation may be necessary in advanced cirrhosis.

Medications can effectively treat many of the complications that accompany cirrhosis.[15] Individuals with portal hypertension and varices may be given propranolol (Inderal) or octreotide (Sandostatin), which reduce both portal blood pressure and bleeding risk. Diuretics can help to control portal hypertension and ascites; common examples include spironolactone (Aldactone) and furosemide (Lasix). Lactulose, a nonabsorbable disaccharide, treats hepatic encephalopathy by reducing ammonia production and absorption in the colon. The antibiotic rifaximin is an alternative treatment for elevated ammonia that works by altering bacterial populations. To stimulate the appetite and promote weight gain, megestrol acetate (Megace) or dronabinol (Marinol) may be prescribed. The Diet-Drug Interactions feature lists potential nutritional problems associated with these medications.

ascites (ah-SIGH-teez): an abnormal accumulation of fluid in the abdominal cavity.

sinusoids: the small capillary-like passages that carry blood through liver tissue.

hepatic encephalopathy (en-sef-ah-LOP-ah-thie): a condition that develops in advanced liver disease that is characterized by altered neurological functioning, including changes in personality and behavior, reduced mental abilities, and disturbances in motor function.
  encephalo = brain
  pathy = disease

hepatic coma: loss of consciousness resulting from severe liver disease.

Diet-Drug Interactions

Check this table for notable nutrition-related effects of the medications discussed in this chapter.

Appetite stimulants (megestrol acetate, dronabinol)	**Gastrointestinal effects:** Nausea, vomiting, diarrhea **Metabolic effects:** Hyperglycemia (megestrol acetate)
Diuretics (furosemide, spironolactone[a])	**Gastrointestinal effects:** Dry mouth, anorexia, decreased taste perception **Dietary interactions:** Furosemide's bioavailability is reduced when taken with food **Metabolic effects:** Fluid and electrolyte imbalances,[a] hyperglycemia (spironolactone), hyperlipidemia (spironolactone), thiamin and zinc deficiencies
Immunosuppressants (cyclosporine, tacrolimus)	**Gastrointestinal effects:** Nausea, vomiting, diarrhea, anorexia (tacrolimus) **Dietary interactions:** Cyclosporine potentiates the effects of alcohol. The bioavailability of tacrolimus is reduced when the drug is taken with food. Grapefruit juice can raise serum concentrations of these drugs to toxic levels. **Metabolic effects:** Electrolyte imbalances, hypertension, hyperglycemia, hyperlipidemia
Lactulose	**Gastrointestinal effects:** Diarrhea **Metabolic effects:** Fluid and electrolyte imbalances

[a] *Furosemide* is a "potassium-wasting" diuretic; patients should increase intakes of potassium-rich foods. *Spironolactone* is a "potassium-sparing" diuretic; patients should avoid supplemental potassium and potassium-containing salt substitutes.

Nutrition Therapy for Cirrhosis

Nutrition care for cirrhosis is customized to each patient's needs, which vary considerably and depend on the accompanying complications. Common problems include PEM and muscle wasting; thus, protein and energy intakes must be sufficient for maintaining nitrogen balance. Dietary substances that may cause liver injury should be avoided; examples include alcohol, some herbal supplements, and vitamin or mineral megadoses. Table 20-6 lists the general dietary guidelines for cirrhosis.

Energy Energy requirements may range from 20 to 40 percent above resting metabolic rate (RMR); if possible, indirect calorimetry should be used to determine RMR. Hypermetabolism, infection, nutrient malabsorption, and recent unintentional weight

TABLE 20-6 Nutrition Therapy for Liver Cirrhosis

Energy	▪ Energy needs may range from 20 to 40 percent above resting metabolic rate (RMR); if possible, use indirect calorimetry to determine RMR. ▪ Use estimated dry body weight for RMR calculations in patients with ascites. ▪ Energy requirements may be higher in patients with hypermetabolism, infection, malabsorption, or malnutrition. Energy requirements may be lower in patients who would benefit from weight loss.
Meal frequency	▪ To improve food intake, patients should consume small meals four to six times daily.
Protein	▪ Provide 0.8 to 1.2 g protein per kilogram of dry body weight per day to maintain nitrogen balance and prevent wasting.
Carbohydrate	▪ No carbohydrate restrictions unless the patient has insulin resistance or diabetes. ▪ For persons with insulin resistance or diabetes, monitor carbohydrate intakes and provide a diet that maintains blood glucose control.
Fat	▪ No fat restrictions unless fat malabsorption is present. ▪ If fat is malabsorbed, restrict fat to 30 percent of total kcalories or as necessary to control steatorrhea; use medium-chain triglycerides (MCT) to increase kcalories.
Sodium and fluid	▪ Restrict sodium as necessary to control ascites; 2000 mg sodium per day is adequate restriction in most cases. ▪ If ascites is accompanied by low serum sodium levels (less than 128 mEq/L), restrict fluids to 1200 to 1500 mL per day. In severe cases (serum sodium less than 125 mEq/L), restrict fluids to 1000 to 1200 mL per day.
Vitamins and minerals	▪ Ensure adequate intake from diet or supplements based on individual needs.

loss can increase energy needs. For patients with ascites, RMR calculations should use either the patient's desirable weight or an estimated dry weight (weight without ascites). A value for dry weight can be obtained after diuretic therapy or after a medical procedure that directly removes excess abdominal fluid.

Many patients with cirrhosis have difficulty consuming enough food, and some individuals may find four to six small meals easier to tolerate than three large meals each day. Nutritional supplements, including liquid formulas and energy bars, can be used to improve energy intakes. The "How to" offers additional suggestions that can help a patient meet energy needs.

Protein The protein recommendation is 0.8 to 1.2 grams of protein per kilogram of body weight per day, based on dry weight or an appropriate weight for height.[16] Patients with hepatic encephalopathy should avoid excessive protein consumption, and their protein intake should be spread throughout the day so that they consume only modest amounts at each meal. Protein restriction is not helpful, however, because an inadequate protein intake can worsen malnutrition and wasting.

Reminder: The protein RDA for healthy adults is 0.8 g/kg.

In an attempt to normalize amino acid ratios in blood plasma and brain tissue and possibly improve the mental status of patients with hepatic encephalopathy, some health care providers may prescribe enteral formulas enriched with branched-chain amino acids. Clinical studies testing the use of these formulas have yielded mixed

How To   HELP THE CIRRHOSIS PATIENT EAT ENOUGH FOOD

Individuals with cirrhosis often have difficulty consuming enough food to prevent malnutrition and its consequences. Ascites and gastrointestinal symptoms such as nausea and vomiting may interfere with food intake. Fatigue may cause a lack of interest in food preparation. Sodium restrictions may make foods unpalatable. To improve food intake:

- If nutrient restrictions are necessary, make sure the patient fully understands how to modify the diet so that food intake is not restricted unnecessarily. Provide lists of acceptable foods and menus. Explain how recipes can be altered so that favorite foods can still be incorporated into the diet.

- Suggest between-meal snacks during the day and a snack at bedtime. A liquid supplement like Ensure can substitute for a snack and requires no preparation. Snacks should not be consumed within two hours of meals, or they may reduce appetite at mealtime.

- If the patient has little appetite or is quickly satiated, suggest foods that are higher in food energy, such as whole milk instead of reduced-fat milk or canned fruit that is packed in heavy syrup instead of fruit juice. Suggest that beverages be consumed separately from meals.

- Recommend energy boosters. Cream sauces and gravies can add kcalories to entrées. Fruit juices and fruit nectars can substitute for drinking water. The following additions can boost the energy content of meals:

  - Sour cream and butter—on vegetables and potatoes

  - Mayonnaise—in sandwiches and salads

- Half-and-half and light cream—in soups and on cereals

- Hard-cooked eggs—in casseroles and meat loaf

- Cheese—in salads and casseroles and melted on steamed vegetables

- Peanut butter, nut butters, and cream cheese—on crackers or celery and in milk shakes

- Chopped nuts—in salads, cooked cereals, and bakery products

Sodium-controlled diets are recommended for treating ascites and other medical conditions, including kidney and heart disorders. The "How to" on p. 585 offers suggestions to help patients implement sodium restrictions. To improve the palatability of low-sodium meals:

- Suggest that patients replace the salt they use for cooking and seasoning with strong-flavored herbs and spices such as chili powder, coriander, cumin, curry powder, garlic, ginger, lemon, mint, and parsley.

- Advise patients to check food labels to learn the sodium content of the foods they eat. Similar products may be available that are lower in sodium. (Persons using potassium-sparing diuretics should be cautioned to avoid salt substitutes that replace sodium with potassium.)

Offer support and encouragement to the patient with cirrhosis. Significant weight loss is less likely to occur if dietary advice is provided before problems progress.

results, however, and their use is recommended only in patients who do not respond to conventional treatment.[17]

Carbohydrate Carbohydrate provides a substantial proportion of energy needs. Many patients with cirrhosis are insulin resistant, however, and require medications or insulin to manage their hyperglycemia. These individuals should follow the dietary guidelines for diabetes: monitor carbohydrate intakes and consume a diet that maintains blood glucose levels within a normal range (see Chapter 21). Carbohydrate intakes should be fairly consistent from day to day for improved blood glucose control.

Fat Fat provides both energy and essential fatty acids. In patients with fat malabsorption, fat intake may be restricted to less than 30 percent of total kcalories or as necessary to control steatorrhea. Medium-chain triglycerides (MCT) may be used to provide additional energy. Severe steatorrhea warrants supplementation of fat-soluble vitamins, calcium, magnesium, and zinc (see Chapter 19).

Sodium and Fluid Patients with ascites are typically advised to restrict sodium. Because ascites is partly caused by sodium and water retention in the kidneys, treatment usually includes both a moderate sodium restriction (to no more than 2000 milligrams of sodium per day) and diuretic therapy to promote fluid loss. Potassium intake should be monitored if a potassium-wasting diuretic (such as furosemide) is used.

Many patients find low-sodium diets unpalatable, so some health practitioners may allow a more liberal sodium intake and depend on diuretics to mobilize excess fluids. If patients do not respond to sodium restriction and diuretic therapy, fluid may be removed directly by surgical puncture **(paracentesis)** or may be diverted to the bloodstream using a catheter **(peritoneovenous shunt).**

Fluid restriction may be necessary when ascites is accompanied by a low serum sodium concentration. If the sodium level falls below 128 milliequivalents per liter, the fluid intake should be limited to 1200 to 1500 milliliters daily; with a sodium level below 125 milliequivalents per liter, fluids should be restricted to 1000 to 1200 milliliters per day.[18]

Vitamins and Minerals Vitamin and mineral deficiencies are common in patients with cirrhosis due to the effects of illness, disease complications, or the alcohol abuse that may have induced the liver disease. Therefore, multivitamin supplementation is often necessary. If steatorrhea is present, fat-soluble nutrients can be provided in water-soluble forms. Patients with esophageal varices may find it easier to ingest supplements in liquid form.

Enteral and Parenteral Nutrition Support In patients who are unable to consume enough food, tube feedings may be infused overnight as a supplement to oral intakes or may replace oral feedings entirely. Although standard formulas are often appropriate, an energy-dense, moderate-protein, low-electrolyte formula may be necessary for patients with ascites or fluid restrictions. In patients with esophageal varices, the feeding tube should be as narrow and flexible as possible to prevent rupture and bleeding. Parenteral nutrition support should be considered for patients who are unable to tolerate enteral feedings due to intestinal obstruction, gastrointestinal bleeding, or uncontrollable vomiting. To avoid excessive fluid delivery, patients with ascites typically require concentrated parenteral solutions, which are infused into central veins. The Case Study allows you to apply your knowledge of cirrhosis to a clinical situation.

Table 23-1 (p. 598) and the "How to" on p. 585 provide information about following a sodium-controlled diet.

Case Study MAN WITH CIRRHOSIS

Lenny Levitt, a 49-year-old carpenter, has just been diagnosed with cirrhosis, which is a consequence of his alcohol abuse over the past 25 years. Although he understands that he has an alcohol problem and recently entered an alcohol rehabilitation program, he is still drinking. At 5 feet 8 inches tall, Mr. Levitt, who formerly weighed 160 pounds, now weighs 130 pounds. According to family members, he is showing signs of mental deterioration, such as forgetfulness and an inability to concentrate. He is jaundiced and appears thin, although his abdomen is distended with ascites. Laboratory findings indicate elevated serum concentrations of

AST, ALT, and ammonia; reduced albumin levels; and hyperglycemia.

1. Do Mr. Levitt's laboratory values suggest liver disease? Compare the results of his laboratory tests with the values shown in Table 20-3 on p. 525.

2. From the limited information available, evaluate Mr. Levitt's nutrition status. What medical problem makes it difficult to interpret his present weight? Describe the development of that type of problem in liver disease, and explain how the diet is usually adjusted for such a patient.

3. Estimate Mr. Levitt's energy and protein needs. Describe the general diet you might recommend for him. What suggestions do you have for increasing his energy intake?

4. Explain the significance of Mr. Levitt's elevated blood ammonia levels. What are some signs that would indicate that he is undergoing mental decline?

5. Describe each of the following complications of liver disease: portal hypertension, jaundice, and gastroesophageal varices. What complication may result if the esophageal varices are not treated?

IN SUMMARY

▌ Liver cirrhosis is characterized by extensive fibrosis and progressive liver dysfunction. The primary causes of cirrhosis in the United States are alcoholic liver disease and chronic hepatitis C infection.

▌ Symptoms of cirrhosis include fatigue, anorexia, and weight loss. Eventually, patients may bruise easily, be more susceptible to infections, and develop complications such as portal hypertension, gastroesophageal varices, ascites, and hepatic encephalopathy.

▌ Treatment of cirrhosis is highly individualized and depends on the accompanying symptoms and complications. Both drug therapies and dietary adjustments are usually necessary. If warranted, the diet may need to be restricted in fat, sodium, or fluids.

Liver Transplantation

Acute or chronic liver disease can lead to liver failure, in which case liver transplantation is the only remaining treatment option. The most common illnesses that precede liver transplantation are chronic hepatitis C infection and alcoholic liver disease, which account for about 50 percent of liver transplant cases.[19] The 5-year survival rate among transplant recipients ranges from 58 to 81 percent, depending on the cause of illness.[20] Complications such as ascites and hepatic encephalopathy worsen the prognosis.

Nutrition Status of Transplant Patients As mentioned earlier, advanced liver disease is usually associated with malnutrition, which can increase the risk of complications following a liver transplant. Evaluating nutrition status in transplant candidates can be difficult, however, because liver dysfunction and malnutrition often have similar metabolic effects. If fluid retention is present, it can mask weight loss and alter

paracentesis (pah-rah-sen-TEE-sis): a surgical puncture of a body cavity with an aspirator to draw out excess fluid.

peritoneovenous (PEH-rih-toe-NEE-oh-VEE-nus) **shunt:** a surgical passage created between the peritoneum and the jugular vein to divert fluid and relieve ascites. The peritoneum is the membrane that surrounds the abdominal cavity.

anthropometric and laboratory values. Correcting malnutrition prior to transplant surgery can help speed recovery after the surgery.

Posttransplantation Concerns The immediate concerns following a transplant are organ rejection and infection. Immunosuppressive drugs, including prednisone, tacrolimus, and cyclosporine, help to reduce the immune responses that cause rejection, but they also raise the risk of infection. Infections are a potential cause of death following a liver transplant; therefore, antibiotics and antiviral medications are prescribed to reduce infection risk.

Immunosuppressive drugs can affect nutrition status in numerous ways. Gastrointestinal side effects include nausea, vomiting, diarrhea, abdominal pain, and mouth sores. Some medications may alter appetite and taste perception. Some of the drugs may cause hyperglycemia or outright diabetes, which may need to be controlled with insulin. Electrolyte and fluid imbalances are common. Other possible effects include hypertension, hyperlipidemias, protein catabolism, and increased osteoporosis risk.

Protein and energy requirements are increased after transplantation due to the stress of surgery. High-kcalorie, high-protein snacks and enteral supplements can help the transplant patient meet postsurgical needs. Vitamin and mineral supplementation is also an integral part of nutrition care. To help transplant patients avoid developing foodborne illnesses, health practitioners can provide information about food safety measures, such as cooking meats adequately, washing fresh produce, and avoiding foods that may be contaminated. Nutrition in Practice 5 (pp. 123–130) provides additional information about food safety.

IN SUMMARY

▌ Liver transplantation has improved the long-term outlook for patients with advanced liver disease. Transplant patients are usually malnourished and may have medical problems that affect transplant success.

▌ Due to the potential for organ rejection, immunosuppressive drugs are prescribed following surgery. Use of these drugs increases the risk of infection, and the drugs have side effects that can impair nutrition status and general health.

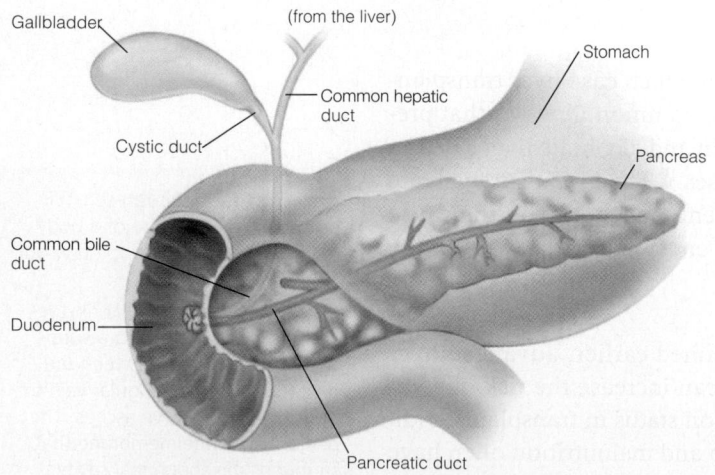

Gallbladder

(from the liver)

Stomach

Common hepatic duct

Cystic duct

Pancreas

Common bile duct

Duodenum

Pancreatic duct

FIGURE 20-2 The Gallbladder and Bile Ducts

Gallbladder Disease

The gallbladder concentrates and stores the bile produced by the liver until the bile is needed for fat digestion (see Figure 20-2). Disorders that obstruct the liver's release of bile can damage the liver. More commonly, disorders of the **biliary system**—the gallbladder and bile ducts—involve the formation of **gallstones.** Gallstones affect an estimated 20 million people in the United States, or about 12 percent of the adult population.[21]

Types of Gallstones

The formation of gallstones, or **cholelithiasis,** results from the excessive concentration and crystallization of compounds in bile. Bile is a solution of bile salts, cholesterol, phospholipids (primarily lecithin), proteins, and bile pigment (bilirubin). While stored in

the gallbladder, bile concentration increases approximately tenfold as its water content is extracted. Factors that raise bile's cholesterol concentration, promote crystal formation and development, or reduce gallbladder motility favor gallstone formation.[22]

Most gallstones are made primarily of cholesterol; they can be as small as a pea or as large as a ping-pong ball.

Cholesterol Gallstones In about 80 percent of cases, the gallstones are composed primarily of cholesterol, although they also contain calcium salts and bilirubin.[23] The cholesterol in bile can precipitate out of solution and form small crystals, which eventually coalesce to form stones. The stones can be as small as a pea or as large as a ping-pong ball. Some people tend to form many small stones, while others may form only one or two large ones. Cholesterol gallstones often develop after the bile concentrate thickens and forms a type of **sludge** that cannot be easily expelled by gallbladder contraction. Biliary sludge may develop after rapid weight loss or fasting, gastric bypass surgery, or long-term total parenteral nutrition, and it can also occur during pregnancy.

Pigment Gallstones Although pigment stones account for a minority of cases in the United States, they are the predominant type of gallstone in some Asian countries. Pigment stones are primarily made up of the calcium salt of bilirubin (calcium bilirubinate). They often develop as a result of bacterial infection, which alters the structure of bilirubin and causes it to precipitate out of bile and form stones. Other cases result from excessive red blood cell breakdown, leading to an abnormal accumulation of bilirubin. Conditions associated with pigment stone formation include biliary tract infections, pancreatitis, and red blood cell disorders, such as sickle-cell anemia. Pigment stones may form in either the gallbladder or the bile duct. Unlike the crystalline cholesterol stones, pigment stones are soft and easily crushed.

Consequences of Gallstones

About 80 percent of gallstones are asymptomatic and are discovered accidentally while testing for other conditions.[24] In other cases, patients may experience an aggressive course of illness with recurring symptoms.

Gallstone Symptoms Gallstone pain usually arises when a gallstone temporarily blocks the cystic duct, which leads from the gallbladder to the common bile duct (review Figure 20-2). The pain is steady and severe and may last for several minutes or several hours. Although the pain is usually located in the upper abdomen, it may radiate to the chest or to the back. Nausea, vomiting, and bloating may also be present. Symptoms usually develop after meals, especially after eating fatty foods. Pain may also occur during the night and awaken a person from sleep.

Complications of Gallstones If a gallstone remains lodged in the cystic duct, it can obstruct bile flow to the duodenum and cause **cholecystitis**—distention and inflammation of the gallbladder. Cholecystitis can lead to infection or to more severe complications, including perforation of the gallbladder, **peritonitis,** and fistulas. If gallstones obstruct the common bile duct, they can block bile flow from the liver and lead to jaundice or damage to liver tissue. An impacted stone within the bile ducts may lead to infection and the condition known as **bacterial cholangitis,** which causes severe pain, sepsis, and fever and is often a medical emergency. Gallstones can block the pancreatic duct as well—a primary cause of acute pancreatitis. Due to the potential danger of these complications, individuals should seek medical attention if gallstone pain does not resolve over time or if fever, jaundice, or persistent nausea and vomiting develop.

biliary system: the gallbladder and ducts that deliver bile from the liver and gallbladder to the small intestine.

gallstones: stones that form in the gallbladder from crystalline deposits of cholesterol or bilirubin.

cholelithiasis (KOH-leh-lih-THIGH-ah-sis): formation of gallstones.
chole = bile
lithiasis = formation of stones

sludge: literally, a semisolid mass. Biliary sludge is made up of mucus, cholesterol crystals, and bilirubin granules.

cholecystitis (KOH-leh-sih-STY-tis): inflammation of the gallbladder, usually caused by obstruction of the cystic duct by gallstones.

peritonitis: inflammation of the peritoneal membrane, which lines the abdominal cavity.

bacterial cholangitis (KOH lan-JYE-tis): bacterial infection involving the bile ducts.

Risk Factors for Cholesterol Gallstones

The risk of developing cholesterol gallstones is influenced by a number of genetic and lifestyle factors. As described in the sections that follow, the risk factors may either cause an increase in bile's cholesterol concentration or a reduction in gallbladder motility, thereby promoting gallstone crystallization or subsequent stone growth.

Ethnicity Although the genetic factors related to gallstone formation are not well understood, ethnicity strongly influences gallstone formation. The Pima Indians are an exceptionally high-risk population; gallstones develop in about 70 percent of adult women. Other high-risk populations include Scandinavians, Chileans, Hispanic Americans, and various Native American groups in North and South America. In the United States, African Americans have a lower prevalence of gallstones than white populations.[25]

Aging Because gallstones do not dissolve spontaneously, gallstone prevalence increases with age. Moreover, bile composition tends to change with aging: the cholesterol concentration increases while bile salts decrease, leading to a greater likelihood of cholesterol crystallization.

Gender The prevalence of cholesterol gallstones is about twice as high in women as in men.[26] The reason for the gender difference is that estrogen alters cholesterol metabolism, causing an increased secretion of cholesterol into bile. The use of estrogen replacement therapy after menopause increases gallstone risk by promoting additional cholesterol secretion into bile.

Pregnancy Some women experience their first gallstone symptoms during pregnancy. Gallstone risk is increased in pregnancy due to hormonal changes: the higher serum estrogen levels raise bile's cholesterol concentration, and the increase in progesterone levels reduces gallbladder motility. The risk of gallstones worsens as the pregnancy progresses and is especially high during the third trimester.

Obesity and Weight Loss Obesity is associated with increased cholesterol synthesis in the liver, leading to higher cholesterol concentrations in bile. In a large study of disease risk in women of different weights, researchers found that the incidence of gallstones was seven times higher in women with body mass indexes (BMIs) greater than 45 as compared with nonobese women.[27]

Gallstones frequently develop as a result of rapid weight loss, occurring in about one-quarter of obese persons who undergo severe kcalorie restriction and up to half of those who undergo gastric bypass surgery.[28] Rapid weight loss increases the secretion of cholesterol into bile and may also decrease gallbladder motility. Another effect of rapid weight loss is the increased production of the gallbladder's *mucin* proteins, which are a major component of biliary sludge and also serve as a matrix for cholesterol crystals during stone growth. The oral ingestion of bile salts has been shown to reduce the risk of gallstone formation during rapid weight loss.

Other Risk Factors Long-term total parenteral nutrition usually reduces gallbladder motility, which promotes the development of biliary sludge. Some medications (such as octreotide) may have similar effects. The medication clofibrate, used for heart disease, increases the cholesterol concentration of bile, promoting gallstone crystallization. High triglyceride levels in blood are also associated with increased gallstone risk, as are spinal cord injuries and diseases affecting the ileum.

Reminder: A healthy weight usually falls between a BMI of 18.5 and 25. Overweight and obesity are usually defined by BMIs above 25 and 30, respectively.

Treatment of Gallstones

Asymptomatic gallstones generally do not require treatment. Gallstones that cause symptoms or complications are usually treated with gallbladder surgery or nonsurgical procedures that dissolve or fragment the stones. To minimize symptoms before the gallbladder or gallstones are removed, a low-fat diet (with less than 30 percent of total kcalories from fat) may be prescribed; some individuals may tolerate small, frequent meals better than large meals.[29]

Surgery Gallbladder removal, or **cholecystectomy,** is the primary treatment for patients with recurring gallstones.[30] The preferred surgical approach is a **laparoscopic** method, which relies on narrow surgical telescopes (laparoscopes) to view and perform the necessary procedures via small incisions in the abdomen. The procedure takes only one or two hours, and many patients are discharged on the same day as the surgery. In patients with complications that make organ removal difficult, open cholecystectomy—which requires opening the abdominal cavity—may be necessary.

Once the gallbladder has been removed, the common bile duct collects bile between meals and releases it into the duodenum at mealtimes; thus, patients can usually tolerate a regular diet. Some individuals may experience diarrhea due to an increased amount of bile in the large intestine, which has a laxative effect. Abdominal pain is sometimes caused by the presence of residual stones within the common bile duct that were overlooked during surgery or that formed within the duct itself. Bile duct injuries occasionally result from the surgical procedure.

Nonsurgical Procedures Nonsurgical methods are used primarily in patients who have small cholesterol stones and transient conditions associated with gallstone formation. The gallstones can be treated by oral intake of ursodeoxycholic acid (ursodiol), a bile acid that reduces cholesterol secretion by the liver and eventually causes the cholesterol crystals in gallstones to dissolve. Ursodeoxycholic acid must be used for 6 to 12 months and is best suited for stones that are 5 millimeters (about ¼ inch) in diameter or smaller. Recurrence rates after dissolution are as high as 50 percent.[31]

Cholesterol gallstones can be fragmented using **shock-wave lithotripsy,** a procedure that is also used to fragment kidney stones. This technique uses high-amplitude sound waves (called shock waves) to break gallstones into pieces that are small enough to either pass into the intestine without causing symptoms or be dissolved with ursodeoxycholic acid. Shock-wave lithotripsy can be performed only in patients with few gallstones. Success is highest in patients with solitary stones that are less than 20 millimeters (¾ inch) in diameter. Recurrence of gallstones has been reported in up to 54 percent of patients using this procedure.[32]

IN SUMMARY

▌ Gallstones are the most common disorder affecting the gallbladder. They are formed by the concentration of compounds in bile, especially cholesterol and the bile pigment bilirubin.

▌ Although most gallstones are asymptomatic, some gallstones can cause recurring pain and GI problems that often appear after meals and persist for several hours.

▌ The risk of gallstone disease is influenced by ethnicity, gender, pregnancy, obesity, rapid weight loss, and other factors.

▌ Treatments for gallstones include gallbladder removal and gallstone dissolution or fragmentation.

cholecystectomy (KOH-leh-sis-TEK-toe-mee): surgical removal of the gallbladder.

laparoscopic: pertaining to procedures that use a laparoscope for internal examination or surgery. A laparoscope is a narrow surgical telescope that is inserted into the abdominal cavity through a small incision. A video camera is usually attached so that the procedure can be viewed on a television monitor.

shock-wave lithotripsy: a nonsurgical procedure that uses high-amplitude sound waves to fragment gallstones or kidney stones.

Nutrition Assessment Checklist FOR PEOPLE WITH DISORDERS OF THE LIVER AND GALLBLADDER

Medical History

Check the medical record to determine:

- ☐ Type of liver disorder
- ☐ Cause of the liver disorder
- ☐ If the patient has received a liver transplant
- ☐ If the patient has a history of gallstones

Review the medical record for complications that may alter nutritional needs, including:

- ☐ Abdominal pain
- ☐ Anemia
- ☐ Ascites
- ☐ Esophageal varices
- ☐ Hepatic encephalopathy
- ☐ Impaired kidney or lung function
- ☐ Infections
- ☐ Insulin resistance or diabetes mellitus
- ☐ Malabsorption
- ☐ Malnutrition
- ☐ Pancreatitis

Medications

In patients with liver dysfunction, the risk of diet-drug interactions is high because most drugs are metabolized in the liver. Risk of interactions is intensified for patients with:

- ☐ Ascites (medications may take a long time to reach the liver)
- ☐ Renal failure (medications are often metabolized further in the kidneys and excreted in the urine)
- ☐ Malnutrition
- ☐ Multiple prescriptions
- ☐ Long-term medication use

Dietary Intake

For patients with fatty liver, pay special attention to:

- ☐ Energy intake, if the patient is overweight or malnourished, has diabetes, or is receiving total parenteral nutrition
- ☐ Carbohydrate intake, if the patient has diabetes or is receiving total parenteral nutrition
- ☐ Alcohol abuse

For patients with hepatitis, cirrhosis, or ascites:

- ☐ Check appetite.
- ☐ Ensure that energy and nutrient intakes are adequate.
- ☐ Determine whether alcohol is being consumed.
- ☐ Determine whether sodium or fluid restriction is warranted.
- ☐ Base energy intakes on desirable weight or an estimated dry weight to avoid overfeeding.

Anthropometric Data

Take baseline height and weight measurements, and monitor weight regularly. For patients with ascites and edema:

- ☐ Monitor weight changes to evaluate the degree of fluid retention.
- ☐ Remember that the patient may be malnourished, and weight may be deceptively high.

Laboratory Tests

Note that albumin and serum proteins are often reduced in people with liver disease and are not appropriate indicators of nutrition status. Review the following laboratory test results to assess liver function:

- ☐ Albumin
- ☐ Alkaline phosphatase
- ☐ ALT and AST
- ☐ Ammonia
- ☐ Bilirubin
- ☐ Prothrombin time

Check laboratory test results for complications associated with liver failure, including:

- ☐ Anemia
- ☐ Decreased renal function
- ☐ Fluid retention
- ☐ Hyperglycemia

Physical Signs

Look for physical signs of:

- ☐ Fluid retention (ascites and edema)
- ☐ PEM (muscle wasting and unintentional weight loss)
- ☐ Nutrient deficiencies

Clinical Applications

1. Vijaya Reddy is a college student who visited relatives near her parents' birthplace in Anantapur, India, during summer vacation. Although her relatives provided boiled or purified water at their home, they occasionally took Vijaya to local restaurants, where she drank tap water. Several weeks after Vijaya returned home, she developed flu-like symptoms and started feeling extremely tired. She also experienced upper abdominal pain and felt nauseous after meals. After her roommate told her that her eyes and skin appeared yellow, she knew something was definitely wrong. A physician at the student health center diagnosed hepatitis.

   ▮ Which type of hepatitis did Vijaya most likely have?

   ▮ What additional symptoms may develop? Is Vijaya's condition likely to become chronic?

■ What medical treatment is suggested for Vijaya's condition? Describe the dietary modifications that may be necessary in some cases.

2. As discussed in the section on cirrhosis, many patients develop protein-energy malnutrition and wasting during the course of illness. Review Table 20-5 on p. 527 to find examples of problems that may lead to malnutrition. Select three nutrition or medical problems (from the "Examples" column), and discuss the complications of liver disease that may cause the problems you selected. What dietary or medical treatments can help in managing these problems?

Self Check

1. Which of the following dietary strategies would be the most appropriate for reversing fatty liver associated with diabetes mellitus?
   a. Following a low-protein diet
   b. Following a fat-controlled diet
   c. Following a fluid- and sodium-controlled diet
   d. Modifying energy to achieve a desirable weight and modifying carbohydrates to attain blood glucose control

2. Which of the following statements about hepatitis is true?
   a. Chronic hepatitis can progress to cirrhosis.
   b. Whatever the cause of hepatitis, symptoms are typically severe.
   c. Vaccines are available to protect against hepatitis A, B, and C viruses.
   d. HCV infection can be spread through contaminated foods and water.

3. Esophageal varices are a dangerous complication of liver disease primarily because they:
   a. interfere with food intake.
   b. can lead to massive bleeding.
   c. divert blood flow from the GI tract.
   d. contribute to hepatic encephalopathy.

4. A complication of liver disease that contributes to the development of ascites is:
   a. portal hypertension.
   b. rising blood ammonia levels.
   c. elevated serum albumin levels.
   d. insulin resistance.

5. A patient with cirrhosis may develop personality changes and motor dysfunction, which are signs of:
   a. jaundice.
   b. hepatic encephalopathy.
   c. hyperammonemia.
   d. hepatic coma.

6. With respect to protein intake, patients with cirrhosis should:
   a. consume no more than the protein RDA.
   b. restrict protein intake to 0.6 grams per kilogram of body weight.
   c. maintain nitrogen balance by consuming 0.8 to 1.2 grams of protein per kilogram of body weight per day.
   d. use formulas enriched with aromatic amino acids to meet their protein needs.

7. People with ascites must often restrict dietary intake of:
   a. fat.
   b. protein.
   c. sugars.
   d. sodium.

8. After a liver transplant, a major health concern is:
   a. jaundice.
   b. ascites.
   c. infection.
   d. hepatic coma.

9. Regarding risk factors for gallstone disease:
   a. prevalence is much higher in men than in women.
   b. gallstone risk is increased during pregnancy.
   c. rapid weight loss can temporarily shrink gallstones.
   d. risk is generally similar among ethnic groups.

10. Nonsurgical approaches to gallstone treatment include:
    a. cholecystectomy.
    b. weight loss.
    c. dissolution and fragmentation.
    d. immunosuppressant drug therapy.

Answers to these questions can be found in Appendix H.

Notes

1. A. M. Diehl, Alcoholic and nonalcoholic steatohepatitis, in L. Goldman and D. Ausiello, eds., *Cecil Medicine* (Philadelphia: Saunders, 2008), pp. 1135–1139.

2. D. E. Cohen and F. A. Anania, Nonalcoholic fatty liver disease, in N. J. Greenberger, ed., *Current Diagnosis and Treatment: Gastroenterology, Hepatology, and Endoscopy* (New York: McGraw-Hill Companies, 2009), pp. 467–472; Diehl, 2008.

3. A. E. Reid, Nonalcoholic fatty liver disease, in M. Feldman, L. S. Friedman, and L. J. Brandt, eds., *Sleisenger and Fordtran's Gastrointestinal and Liver Disease* (Philadelphia: Saunders, 2010), pp. 1401–1411.

4. Reid, 2010.

5. J. M. Crawford and C. Liu, Liver and biliary tract, in V. Kumar and coeditors, *Robbins and Cotran Pathologic Basis of Disease* (Philadelphia: Saunders, 2010), pp. 833–890.

6. Crawford and Liu, 2010.

7. American Dietetic Association, *Nutrition Care Manual* (Chicago: American Dietetic Association, 2010).

8. J. Q. Xu and coauthors, Deaths: Final data for 2007, *National Vital Statistics Reports* 58 (May 2010).

9. G. Garcia-Tsao, Cirrhosis and its sequelae, in L. Goldman and D. Ausiello, eds., *Cecil Medicine* (Philadelphia: Saunders, 2008), pp. 1140–1147.

10. J. J. Heidelbaugh and coauthors, Gastroenterology, in R. E. Rakel, ed., *Textbook of Family Medicine* (Philadelphia: Saunders, 2007), pp. 1115–1171.

11. V. H. Shah and P. S. Kamath, Portal hypertension and gastrointestinal bleeding, in M. Feldman, L. S. Friedman, and L. J. Brandt, eds., *Sleisenger and Fordtran's Gastrointestinal and Liver Disease* (Philadelphia: Saunders, 2010), pp. 1489–1516.

12. B. A. Runyon, Ascites and spontaneous bacterial peritonitis, in M. Feldman, L. S. Friedman, and L. J. Brandt, eds., *Sleisenger and Fordtran's Gastrointestinal and Liver Disease* (Philadelphia: Saunders, 2010), pp. 1517–1541.

13. N. J. Greenberger, Portal systemic encephalopathy and hepatic encephalopathy, in N. J. Greenberger, ed., *Current Diagnosis and Treatment: Gastroenterology, Hepatology, and Endoscopy* (New York: McGraw-Hill Companies, 2009), pp. 473–477.

14. Greenberger, 2009.

15. Garcia-Tsao, 2008.

16. American Dietetic Association, 2010.

17. S. A. McClave and coauthors, Guidelines for the provision and assessment of nutrition support therapy in the adult critically ill patient: Society of Critical Care Medicine (SCCM) and American Society for Parenteral and Enteral Nutrition (A.S.P.E.N.), *Journal of Parenteral and Enteral Nutrition* 33 (2009): 277–316.

18. American Dietetic Association, 2010.

19. E. B. Keeffe, Hepatic failure and liver transplantation, in L. Goldman and D. Ausiello, eds., *Cecil Medicine* (Philadelphia: Saunders, 2008), pp. 1147–1152.

20. Keeffe, 2008.

21. D. Q.-H. Wang and N. H. Afdhal, Gallstone disease, in M. Feldman, L. S. Friedman, and L. J. Brandt, eds., *Sleisenger and Fordtran's Gastrointestinal and Liver Disease* (Philadelphia: Saunders, 2010), pp. 1089–1120.

22. Wang and Afdhal, 2010.

23. G. Paumgartner and N. J. Greenberger, Gallstone disease, in N. J. Greenberger, ed., *Current Diagnosis and Treatment: Gastroenterology, Hepatology, and Endoscopy* (New York: McGraw-Hill Companies, 2009), pp. 537–546.

24. Paumgartner and Greenberger, 2009.

25. Wang and Afdhal, 2010.

26. Paumgartner and Greenberger, 2009.

27. Wang and Afdhal, 2010.

28. Wang and Afdhal, 2010.

29. American Dietetic Association, 2010.

30. R. E. Glasgow and S. J. Mulvihill, Treatment of gallstone disease, in M. Feldman, L. S. Friedman, and L. J. Brandt, eds., *Sleisenger and Fordtran's Gastrointestinal and Liver Disease* (Philadelphia: Saunders, 2010), pp. 1121–1138.

31. Glasgow and Mulvihill, 2010.

32. Glasgow and Mulvihill, 2010.

Nutrition in Practice   ALCOHOL IN HEALTH AND DISEASE

As Chapter 20 described, excessive alcohol consumption is a primary cause of liver disease. Alcohol can be toxic to other organs as well, including the brain, gastrointestinal (GI) tract, and pancreas. In addition, **alcohol abuse** can lead to a number of nutrient deficiencies. Moderate use of alcohol, however, has been associated with a number of health benefits, including the reduction of heart disease risk in middle-aged and older adults. This Nutrition in Practice discusses current recommendations concerning alcohol and the health problems and benefits associated with its use. The glossary defines the relevant terms.

What are the current dietary guidelines for alcohol consumption?

For those who choose to drink alcoholic beverages, the *Dietary Guidelines for Americans* recommend that women and men limit their average daily intakes of alcohol to one drink and two drinks per day, respectively. In addition, women should avoid consuming more than three drinks in a single day, whereas men should consume no more than four drinks in one day.[1] One **drink** is defined as 12 ounces of beer, 5 ounces of wine, 10 ounces of wine cooler, or 1½ ounces of 80 proof distilled spirits such as gin, rum, vodka, and whiskey. Some individuals should not consume alcohol at all; these include pregnant and lactating women, women of childbearing age who may become pregnant, children and adolescents, people using medications that can interact with alcohol, and people who are unable to voluntarily restrict their alcohol intake. Alcohol should also be avoided by anyone who is involved in an activity that requires attention or coordination, such as driving or operating machinery.

About 64 percent of adults in the United States drink alcoholic beverages.[2] The prevalence of alcohol abuse is estimated to be between 7.4 and 9.7 percent in the general population.

12 oz beer | 10 oz wine cooler | 1½ oz hard liquor (80 proof whiskey, gin, brandy, rum, vodka) | 5 oz wine

© Polara Studios, Inc.

Each of the amounts shown is equivalent to one drink.

What happens to alcohol in the body?

Recall that alcohol is a source of food energy, providing 7 kcalories per gram. Although a small amount of alcohol is metabolized in the stomach, most of the alcohol that is consumed is quickly absorbed in the stomach or small intestine and passes readily into the body's cells, where alcohol concentrations reach levels similar to those in the blood. The liver is the site of most alcohol metabolism, although its ability to metabolize alcohol is somewhat limited: the average adult metabolizes only 7 to 10 grams of alcohol per hour, about the amount in one drink.[3] The main product of alcohol metabolism is **acetate,** which can be used as a source of energy by most tissues.

Because there is no storage pool for alcohol and it can be toxic to cells, the metabolism of alcohol in the liver takes priority over that of other substances. Thus, alcohol suppresses the breakdown of fat for energy, leading to fat accumulation in the liver and the increased release of triglyceride-carrying lipoproteins (VLDL). Excessive alcohol intakes inhibit both the storage of glycogen and the liver's production of glucose between meals. Heavy drinkers are at risk of developing hypoglycemia, an effect that is accentuated in people with diabetes who use insulin or medication to reduce glucose levels. Alcohol ingestion can also impair the synthesis and secretion of a number of liver proteins, including albumin and fibrinogen (a blood-clotting protein).

How is alcohol toxic to cells?

Alcohol can damage cells either directly, or indirectly via the metabolite **acetaldehyde** that is created during alcohol metabolism. High alcohol concentrations change the structure of cell membranes, increasing membrane permeability. Cell membrane proteins may become damaged, interfering with the transport of substances across the membrane. The presence of alcohol also activates inflammatory responses in cells, which can eventually lead to cell death and tissue damage. Acetaldehyde can cause

GLOSSARY

acetaldehyde (ah-set-AL-deh-hide): an intermediate in alcohol metabolism that, in excess, can damage the body's tissues and interfere with cellular functions.

acetate (AH-seh-tate): the product of alcohol metabolism that is used as a source of energy by many tissues in the body.

alcohol abuse: the continued use of alcohol despite the development of social, legal, or health problems. Individuals who abuse alcohol are called *alcoholics.*

alcoholic liver disease: liver disease that is related to excessive alcohol consumption. Disorders that may develop include fatty liver, hepatitis, and cirrhosis, which may lead to liver failure.

drink: an alcoholic beverage that contains about half an ounce of pure alcohol. One drink is equivalent to 12 ounces of beer, 5 ounces of wine, 10 ounces of wine cooler, or 1½ ounces of 80 proof distilled spirits.

numerous adverse effects: it binds to proteins and interferes with their functions, alters immune responses, causes oxidative damage, inhibits DNA repair, and prevents the formation of microtubules within cells.

How does alcohol affect brain function?

Alcohol acts as a central nervous system depressant. It can cause sedation, slow reaction times, and relieve anxiety. In excess, it impairs judgment, reduces inhibitions, and impairs speech and motor functions. Extremely high blood alcohol levels can lead to coma, respiratory depression, and death.

Chronic heavy drinking can cause certain types of neurological damage. The most common abnormality is injury to the peripheral nerves, which is evidenced by tingling in the hands and feet, lack of muscular coordination, and changes in a person's manner of walking. Visual impairments, such as blurred vision and optic nerve degeneration, can also occur.

Chronic drinkers who are forced to discontinue alcohol use may experience withdrawal symptoms, which include anxiety, irritability, and palpitations. In severe cases, symptoms may include nausea and vomiting, delirium, and seizures.[4]

What are some other long-term consequences of drinking too much alcohol?

Alcoholic liver disease is the most common complication of alcohol abuse, occurring in about 15 to 30 percent of chronic heavy drinkers.[5] The disease develops in a manner similar to the disease progression described in Chapter 20: fatty liver, hepatitis, and, eventually, cirrhosis and liver failure. In addition to liver damage, alcohol can cause damage to the GI tract, pancreas, and heart (see Figure NP20-1). In the GI tract, alcohol's eventual effects may include gastritis, ulcers, nutrient malabsorption, diarrhea, and GI cancers. Chronic alcohol ingestion frequently injures pancreatic tissue and may lead to chronic pancreatitis. Heavy drinking is also

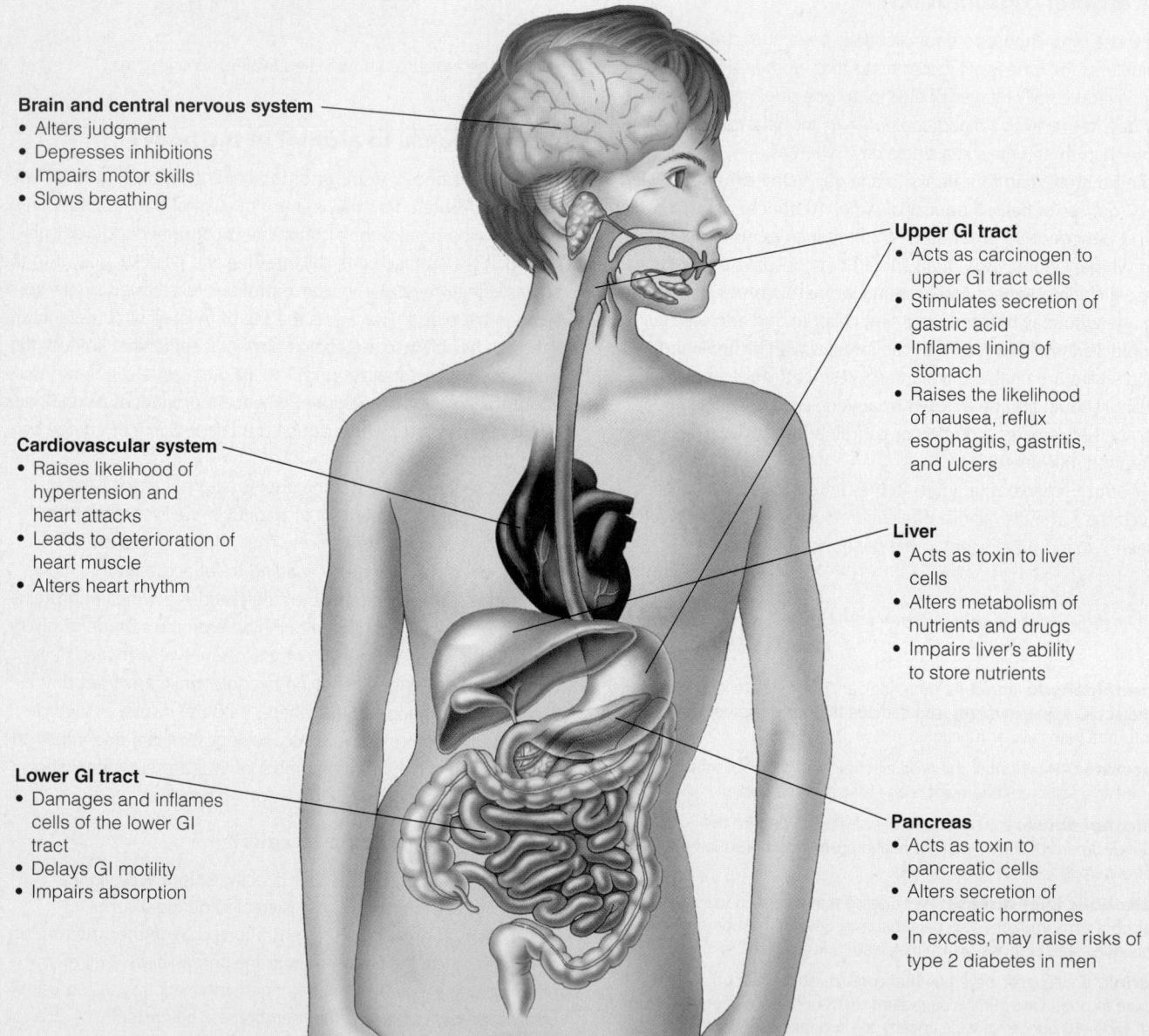

Brain and central nervous system
- Alters judgment
- Depresses inhibitions
- Impairs motor skills
- Slows breathing

Cardiovascular system
- Raises likelihood of hypertension and heart attacks
- Leads to deterioration of heart muscle
- Alters heart rhythm

Lower GI tract
- Damages and inflames cells of the lower GI tract
- Delays GI motility
- Impairs absorption

Upper GI tract
- Acts as carcinogen to upper GI tract cells
- Stimulates secretion of gastric acid
- Inflames lining of stomach
- Raises the likelihood of nausea, reflux esophagitis, gastritis, and ulcers

Liver
- Acts as toxin to liver cells
- Alters metabolism of nutrients and drugs
- Impairs liver's ability to store nutrients

Pancreas
- Acts as toxin to pancreatic cells
- Alters secretion of pancreatic hormones
- In excess, may raise risks of type 2 diabetes in men

FIGURE NP20-1 Effects of Excessive Alcohol Intake on Organ Systems

associated with heart arrhythmias, impaired heart muscle contractility, and hypertension.

What are the effects of excessive alcohol consumption on nutrition status?

An excessive intake of alcohol often causes multiple nutrient deficiencies. Because alcohol supplies 7 kcalories per gram, it displaces other energy sources along with the essential nutrients such foods would provide. As mentioned earlier, alcohol can cause nutrient malabsorption as a result of direct damage to the GI mucosa. Alcohol also interferes with the way the body processes nutrients. Examples of common deficiencies in persons who abuse alcohol include:[6]

- *Vitamin A.* Alcohol and vitamin A are metabolized by similar enzymes, so an increase in the enzymes that break down alcohol—induced by heavy drinking—can increase degradation of vitamin A.

- *Thiamin.* Alcohol abuse is the most common cause of thiamin deficiency in the United States. Heavy drinking is associated with low thiamin intakes, and alcohol ingestion dramatically reduces thiamin absorption.

- *Folate.* Alcohol reduces the absorption of folate in the small intestine, alters folate metabolism, and increases folate losses in urine. Because folate has a central role in cell division (and the cells of the GI tract are rapidly dividing cells), folate deficiency contributes to the malabsorption of other nutrients as well.

Can alcohol have disruptive effects on the metabolism of medications?

Yes. Some of the liver enzymes that degrade alcohol have the additional role of metabolizing drugs, which prepares the drugs for excretion. When alcohol is consumed, the enzymes involved in alcohol degradation are unavailable for drug metabolism; thus, the effects of the drugs may be prolonged. It is especially problematic to combine alcohol and drugs if they have similar effects in the body; examples of such drugs include sedatives and tricyclic antidepressants. Some drugs can even become liver toxins if their metabolism is disrupted; for example, consuming three or more drinks daily increases the risk of toxicity from therapeutic doses of acetaminophen (Tylenol).[7]

Does alcohol have any beneficial effects on health?

Yes. Epidemiological studies in Western countries have consistently shown that a moderate alcohol intake reduces total mortality among middle-aged and older men and women. In addition, researchers have learned that moderate alcohol drinking can help reduce risks of developing the following chronic illnesses.[8]

- *Heart disease.* Alcohol increases levels of HDL cholesterol (which protects against the development of atherosclerosis) and also reduces the tendency for blood clotting. These effects occur when any type of alcoholic beverage is consumed, and are seen most frequently when alcohol is consumed several times per week. Alcohol's protective effects are seen mainly in older persons who have one or more classic risk factors for heart disease.[9]

- *Stroke.* Moderate alcohol consumption may reduce the risk of stroke, especially ischemic stroke (caused by blood clotting in arteries that supply blood to the brain). Conversely, heavy drinking increases the risk of all forms of stroke.

- *Diabetes.* Risk of type 2 diabetes may be lower in moderate drinkers than in individuals who abstain from alcohol. In addition, some studies have found that a moderate alcohol intake may improve insulin sensitivity and lower fasting glucose concentrations.

Note that unlike other recommendations for disease prevention, medical personnel rarely recommend that a nondrinker begin drinking to reduce disease risk, due to the potential for addiction and the adverse effects caused by excessive intakes.

The benefits or harm associated with alcohol consumption depend on a person's health status, age, and the amount consumed. Most people in the United States who consume alcoholic beverages are not heavy drinkers; however, health professionals should be alert to those who may be abusing alcohol and potentially endangering their health.

Notes

1. *Report of the Dietary Guidelines Advisory Committee on the Dietary Guidelines for Americans, 2010,* available at http://www.cnpp.usda.gov/DGAs2010-DGACReport.htm; accessed September 21, 2010.

2. P. G. O'Connor, Alcohol abuse and dependence, in L. Goldman and D. Ausiello, eds., *Cecil Medicine* (Philadelphia: Saunders, 2008), pp. 167–174.

3. S. B. Masters, The alcohols, in B. G. Katzung, ed., *Basic and Clinical Pharmacology* (New York: McGraw-Hill Companies, 2007), pp. 363–373.

4. O'Connor, 2008; Masters, 2007.

5. Masters, 2007.

6. C. S. Lieber, Nutrition in liver disorders and the role of alcohol, in M. E. Shils and coeditors, *Modern Nutrition in Health and Disease* (Baltimore: Lippincott Williams & Wilkins, 2006), pp. 1235–1266.

7. Masters, 2007.

8. *Report of the Dietary Guidelines Advisory Committee on the Dietary Guidelines for Americans, 2010,* 2010.

9. A. Lichtenstein and coauthors, Diet and lifestyle recommendations revision 2006: A scientific statement from the American Heart Association Nutrition Committee, *Circulation* 114 (2006): 82–96.

Carbohydrate-Controlled Diets for Diabetes Mellitus

The incidence of **diabetes mellitus** is steadily increasing in the United States and many other countries. It now affects an estimated 10.2 percent of adults aged 20 and older in the United States, or about 24 million people.[1] About 25 percent of persons with diabetes are unaware that they have it,[2] a danger because its damaging effects often occur before symptoms develop. Diabetes ranks seventh among the leading causes of death in the United States. It also contributes to the development of other life-threatening diseases, including heart disease and kidney failure.

Overview of Diabetes Mellitus

The term *diabetes mellitus* refers to metabolic disorders characterized by elevated blood glucose concentrations and disordered insulin metabolism. People with diabetes may have defects in insulin secretion or insulin effectiveness, or both. These impairments result in inadequate glucose utilization in muscle and adipose cells and unrestrained glucose production in the liver. The result is **hyperglycemia,** a marked elevation in blood glucose levels that can ultimately cause damage to blood vessels, nerves, and tissues. The glossary on p. 544 defines diabetes-related symptoms and complications.

Reminder: *Insulin* is a pancreatic hormone that regulates blood glucose concentrations. Its actions are countered mainly by the hormone *glucagon.*

Symptoms of Diabetes

Symptoms of diabetes are usually related to the degree of hyperglycemia present (see Table 21-1). When the plasma glucose level rises above about 200 milligrams per deciliter (mg/dL), it exceeds the **renal threshold,** the concentration at which the kidneys begin to pass glucose into the urine **(glycosuria).** The presence of glucose in the urine draws additional water from the blood, increasing the amount of urine produced. Thus, the symptoms that arise in diabetes typically include frequent urination **(polyuria),** dehydration, and increased thirst **(polydipsia).** Other possible symptoms include weight loss, an increased appetite **(polyphagia),** blurred vision (caused by the exposure of eye tissues to hyperosmolar fluids), frequent infections (due to weakened immune responses and impaired circulation), and fatigue (due to altered energy metabolism, dehydration, or other effects of illness).

Reminder: *Osmolarity* refers to the concentration of osmotically active particles in solution. Hyperglycemia increases the osmolarity of the body's fluids.

Normal fasting plasma glucose levels are approximately 75 to 100 mg/dL (published values vary).

Diagnosis of Diabetes

The diagnosis of diabetes is based primarily on plasma glucose levels, which may be measured under fasting conditions or at random times during the day. In some cases, an **oral glucose tolerance test** is given: the individual ingests a 75-gram glucose load, and plasma glucose is measured at one or more time intervals following glucose ingestion. **Glycated hemoglobin (HbA$_{1c}$)** levels, which reflect hemoglobin's exposure to glucose over prolonged periods, are an indirect assessment of blood glucose levels. The following criteria are currently used to diagnose diabetes:[3]

diabetes (DYE-ah-BEE-teez) **mellitus:** a group of metabolic disorders characterized by hyperglycemia and disordered insulin metabolism.

renal threshold: the blood concentration of a substance that exceeds the kidneys' capacity for reabsorption, causing the substance to be passed into the urine.

oral glucose tolerance test: a test that evaluates a person's ability to tolerate an oral glucose load.

glycated hemoglobin (HbA$_{1c}$): hemoglobin that has non-enzymatically attached to glucose; the level of HbA$_{1c}$ in blood helps to diagnose diabetes and evaluate long-term glycemic control. Also called *glycosylated hemoglobin.*

- The plasma glucose concentration of a blood sample obtained at a random time during the day (without regard to food intake) is 200 mg/dL or higher, and classic symptoms of diabetes are present.

- The plasma glucose concentration is 126 mg/dL or higher after a fast of at least eight hours.

- The plasma glucose concentration measured two hours after a 75-gram glucose load is 200 mg/dL or higher.

- The HbA$_{1c}$ level is 6.5 percent or higher.

Overt symptoms of hyperglycemia help to confirm the diagnosis. Otherwise, diagnosis of diabetes is confirmed only if a subsequent test yields similar results.

GLOSSARY OF DIABETES-RELATED SYMPTOMS AND COMPLICATIONS

acetone breath: a distinctive fruity odor on the breath of a person with ketosis.

claudication (CLAW-dih-KAY-shun): pain in the legs while walking; usually due to an inadequate supply of blood to muscles.

diabetic coma: a coma that occurs in uncontrolled diabetes; may be due to diabetic ketoacidosis, the hyperosmolar hyperglycemic syndrome, or severe hypoglycemia.

diabetic nephropathy (neh-FRAH-pah-thee): damage to the kidneys that results from long-term diabetes.

diabetic neuropathy (nur-RAH-pah-thee): complications of diabetes that cause damage to nerves.

diabetic retinopathy (REH-tih-NAH-pah-thee): retinal damage that results from long-term diabetes.

gangrene: death of tissue due to a deficient blood supply and/or infection.

gastroparesis (GAS-troe-pah-REE-sis): delayed stomach emptying caused by nerve damage in stomach tissue.

glycosuria (GLY-co-SOOR-ee-ah): the presence of glucose in the urine.

hyperglycemia: elevated blood glucose concentrations. Normal fasting plasma glucose levels are less than 100 mg/dL. Fasting plasma glucose levels between 100 and 125 mg/dL suggest prediabetes; values of 126 mg/dL and above suggest diabetes.

hyperosmolar hyperglycemic syndrome: extreme hyperglycemia associated with dehydration, hyperosmolar blood, and altered mental status; sometimes called the *hyperosmolar hyperglycemic nonketotic state*.

hypoglycemia: abnormally low concentrations of blood glucose. In diabetes, hypoglycemia is treated when plasma glucose levels fall below 70 mg/dL.

ketoacidosis (KEY-toe-ass-ih-DOE-sis): an acidosis (lowering of blood pH) that results from the excessive production of ketone bodies.

ketonuria (KEY-toe-NOOR-ee-ah): the presence of ketone bodies in the urine.

macrovascular complications: disorders that affect the large blood vessels, including the coronary arteries and arteries of the limbs.

microalbuminuria: the presence of albumin (a blood protein) in the urine, a sign of diabetic nephropathy.

microvascular complications: disorders that affect the small blood vessels and capillaries, including those in the retina and kidneys.

peripheral vascular disease: disorders characterized by impaired blood circulation in the limbs.

polydipsia (POL-ee-DIP-see-ah): excessive thirst.

polyphagia (POL-ee-FAY-jee-ah): excessive appetite or food intake.

polyuria (POL-ee-YOOR-ree-ah): excessive urine production.

The term **prediabetes** pertains to individuals who have blood glucose levels between normal and diabetic. A person with prediabetes is at risk of developing diabetes and cardiovascular diseases.

Types of Diabetes

Table 21-2 lists the features of the two main types of diabetes, type 1 and type 2 diabetes. Pregnancy can lead to abnormal glucose tolerance and the condition known as *gestational diabetes*, which often resolves after pregnancy but is a risk factor for type 2 diabetes. Diabetes can also be caused by medical conditions that either damage the pancreas or interfere with insulin function.

TABLE 21-2 Features of Type 1 and Type 2 Diabetes Mellitus

Feature	Type 1	Type 2
Prevalence in diabetic population	5 to 10 percent of cases	90 to 95 percent of cases
Age of onset	<30 years	>40 years[a]
Associated conditions	Autoimmune diseases, viral infection, inherited factors	Obesity, aging, inactivity, inherited factors
Major defect	Destruction of pancreatic beta cells; insulin deficiency	Insulin resistance; insulin deficiency relative to needs
Insulin secretion	Little or none	Varies; may be normal, increased, or decreased
Requirement for insulin therapy	Always	Sometimes
Former names	Juvenile-onset diabetes Insulin-dependent diabetes	Adult-onset diabetes Noninsulin-dependent diabetes

[a]Incidence of type 2 diabetes is increasing in children and adolescents; in more than 90 percent of these cases, it is associated with overweight or obesity and a family history of type 2 diabetes.

Type 1 Diabetes **Type 1 diabetes** accounts for 5 to 10 percent of diabetes cases. It is caused by **autoimmune** destruction of the pancreatic beta cells, which produce and secrete insulin. By the time symptoms develop, damage to the beta cells has progressed so far that insulin must be provided, most often by injection. Although the reason for the autoimmune attack is usually unknown, environmental toxins or infections are likely triggers. People with type 1 diabetes often have a genetic susceptibility for the disorder and an increased risk of developing other autoimmune diseases.

Type 1 diabetes usually develops during childhood or adolescence, and symptoms may appear abruptly in previously healthy children.[4] Classic symptoms are polyuria, polydipsia, weight loss, and weakness or fatigue. **Ketoacidosis**—acidosis due to excessive production of ketone bodies—is sometimes the first sign of disease. Disease onset tends to be more gradual in individuals who develop type 1 diabetes in later years. Blood tests that detect antibodies to insulin, pancreatic islet cells, and pancreatic enzymes can confirm the diagnosis and help to predict development of the disease in close relatives.

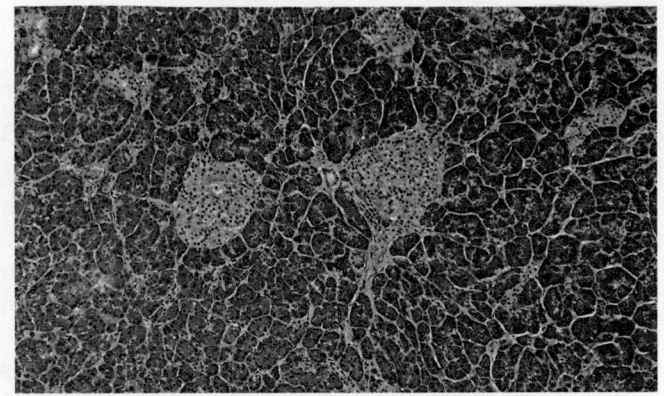

Cross-sections of the pancreas reveal small clusters of cells known as the islets of Langerhans; these regions contain the beta cells that produce insulin.

Reminder: *Ketone bodies* are products of fat metabolism that are produced in the liver; they accumulate in tissues when fatty acids are released in abnormally high amounts from adipose tissue.

Type 2 Diabetes **Type 2 diabetes** is the most prevalent form of diabetes, accounting for 90 to 95 percent of cases, and it is often asymptomatic. The primary defect in type 2 diabetes is **insulin resistance,** the reduced sensitivity to insulin in muscle, adipose, and liver cells. To compensate, the pancreas increases its secretion of insulin, and plasma insulin concentrations may rise to abnormally high levels **(hyperinsulinemia).** Over time, the pancreas becomes less able to compensate for the cells' reduced sensitivity to insulin, and hyperglycemia worsens. The high demand for insulin can eventually exhaust the beta cells of the pancreas and lead to impaired insulin production and reduced plasma insulin concentrations. Type 2 diabetes is therefore associated with both insulin resistance and relative insulin deficiency; that is, the amount of insulin secreted is insufficient to compensate for its diminished effect in cells.

Although the precise causes of type 2 diabetes are unknown, risk is substantially increased by obesity (especially abdominal obesity), aging, and physical inactivity. An estimated 80 to 90 percent of individuals with type 2 diabetes are obese, and obesity itself can directly cause some degree of insulin resistance.[5] Prevalence increases with age and approaches 23 percent in persons over 60 years of age; however, many of these cases remain undiagnosed.[6] Inherited factors strongly influence risk, and type 2 diabetes is more common in certain ethnic groups, including African Americans, Asian Americans, Hispanic Americans, Mexican Americans, Native Americans, and Pacific Islanders.

Type 2 Diabetes in Children and Adolescents Although most cases of type 2 diabetes are diagnosed in individuals over 40 years old, children and teenagers who are overweight or obese or have a family history of diabetes are at increased risk. Because type 2 diabetes is frequently asymptomatic, it is generally identified in youths only when high-risk groups are screened for the disease.

Increased rates of both type 1 and type 2 diabetes have been documented in children in past decades and correlate with the rise in childhood obesity. Type 1 and type 2 diabetes are sometimes difficult to distinguish in children, however, and a few studies suggest that some children diagnosed with type 1 diabetes may actually have had type 2 diabetes.[7] One multicenter study found that about half of the cases of newly diagnosed diabetes in African American and Hispanic individuals between 10 and 19 years old were classified as type 2 diabetes.[8] Although type 2 diabetes is still relatively rare in children, its increasing prevalence indicates that

prediabetes: the condition in which blood glucose levels are higher than normal but not high enough to be diagnosed as diabetes.

type 1 diabetes: the type of diabetes that accounts for 5 to 10 percent of diabetes cases and usually results from autoimmune destruction of pancreatic beta cells.

autoimmune: an immune response directed against the body's own tissues.

auto = self

type 2 diabetes: the type of diabetes that accounts for 90 to 95 percent of diabetes cases and usually results from insulin resistance coupled with insufficient insulin secretion.

insulin resistance: reduced sensitivity to insulin in muscle, adipose, and liver cells.

hyperinsulinemia: abnormally high levels of insulin in the blood.

routine screening and diabetes prevention programs may be important safeguards for children at risk.

Acute Complications of Diabetes

Untreated diabetes may result in life-threatening complications. Insulin deficiency can significantly disturb energy metabolism, and severe hyperglycemia can lead to dehydration and electrolyte imbalances. In treated diabetes, **hypoglycemia** is a possible complication of inappropriate disease management. Figure 21-1 shows an overview of some of the acute effects of insulin insufficiency on energy metabolism.

Diabetic Ketoacidosis in Type 1 Diabetes A severe lack of insulin causes diabetic ketoacidosis. Without insulin, glucagon's effects become more pronounced, leading to the unrestrained breakdown of the triglycerides in adipose tissue and the protein in muscle. As a result, an increased supply of fatty acids and amino acids arrives in the liver, fueling the production of ketone bodies and glucose. Ketone bodies, which are acidic, can reach dangerously high levels in the blood (ketoacidosis) and spill into the urine **(ketonuria).** Blood pH typically falls below 7.30. Blood glucose levels exceed 250 mg/dL and rise above 1000 mg/dL in severe cases. The main features of diabetic ketoacidosis therefore include ketosis, acidosis, and hyperglycemia.[9]

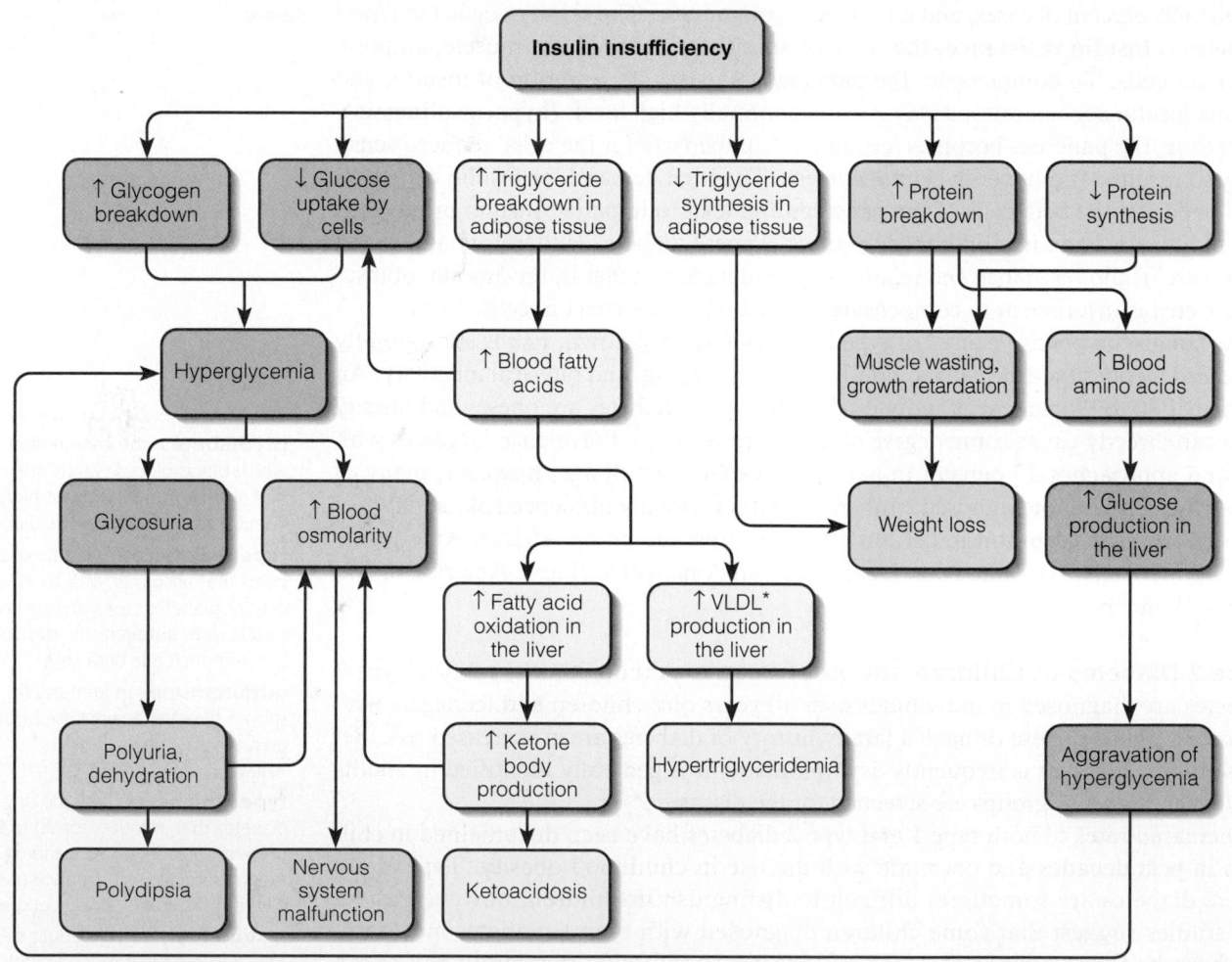

*Very low density lipoproteins; these lipoproteins transport triglycerides from the liver to other tissues.

FIGURE 21-1 Acute Effects of Insulin Insufficiency
The effects of insulin insufficiency can be grouped according to its effects on carbohydrate, protein, and fat metabolism.

Patients with ketoacidosis exhibit symptoms of both acidosis and dehydration. Acidosis is partially corrected by exhalation of carbon dioxide, so rapid or deep breathing is characteristic. Ketone accumulation is sometimes evident by a fruity odor on the breath **(acetone breath).** Hyperglycemia leads to polyuria, which lowers blood volume and blood pressure and depletes electrolytes. In response, patients may demonstrate marked fatigue, lethargy, nausea, and vomiting. Mental state may vary from alertness to comatose **(diabetic coma).** Treatment of ketoacidosis includes insulin therapy to correct hyperglycemia, fluid and electrolyte replacement, and, possibly, bicarbonate therapy to treat acidosis.

Diabetic ketoacidosis may result from inadequate diabetes treatment (such as missed insulin injections), illness or infection, or other physiological stressors.[10] The condition usually develops quickly, within one or two days. Although diabetic ketoacidosis can occur in type 2 diabetes—usually due to severe stressors such as infection, trauma, or surgery—it rarely develops because even relatively low insulin concentrations can suppress ketone body production.

Hyperosmolar Hyperglycemic Syndrome in Type 2 Diabetes The **hyperosmolar hyperglycemic syndrome** is a condition of severe hyperglycemia and dehydration that develops in the absence of ketosis. It is usually precipitated by infection or illness, and is sometimes the first sign of diabetes in older persons. The condition evolves slowly, over one week or longer, when patients are unable to recognize thirst or adequately replace fluids due to age, illness, or incapacity. The profound dehydration that eventually develops exacerbates the rise in blood glucose levels, which often exceed 600 mg/dL and may climb above 1000 mg/dL. Blood plasma may become so hyperosmolar as to cause neurological abnormalities, such as abnormal reflexes, motor impairments, and seizures; about 10 percent of patients lapse into coma.[11] Treatment includes insulin therapy and fluid and electrolyte replacement.

Hypoglycemia Hypoglycemia, or low blood glucose, is due to the inappropriate management of diabetes rather than the disease itself. It is usually caused by excessive dosages of insulin or antidiabetic drugs, prolonged exercise, skipped or delayed meals, inadequate food intake, or the consumption of alcohol without food. Hypoglycemia is the most frequent cause of coma in insulin-treated patients and is believed to account for 3 to 4 percent of deaths in this population.[12]

Symptoms of hypoglycemia include hunger, shakiness, dizziness, irritability, sweating, and heart palpitations. Mental confusion may prevent a person from recognizing the problem and taking such corrective action as ingesting glucose tablets, juice, or candy. If hypoglycemia occurs during the night, patients may be completely unaware of its presence.

Chronic Complications of Diabetes

Prolonged hyperglycemia can damage cells and tissues. Glucose combines with proteins to form reactive compounds known as **advanced glycation end products (AGEs);** in diabetes, these AGEs accumulate to such high levels that they alter the structures of proteins and blood vessels. Chronic complications of diabetes typically affect the large blood vessels **(macrovascular complications),** smaller vessels such as arterioles and capillaries **(microvascular complications),** and the nervous system **(diabetic neuropathy).** Complications may appear 15 to 20 years after diabetes onset.[13]

Macrovascular Complications The damage caused by diabetes accelerates the development of atherosclerosis in the coronary arteries. Cardiovascular disease is the leading cause of death in people with diabetes, accounting for up to 70 percent of deaths.[14] Moreover, type 2 diabetes is often accompanied by multiple risk factors for coronary

Bicarbonate is a buffer in the blood that corrects acidosis. Acid (H^+) combines with bicarbonate (HCO_3^-) to form carbonic acid (H_2CO_3), which breaks down to water (H_2O) and carbon dioxide (CO_2). The carbon dioxide is then exhaled.

Diabetic coma was a frequent cause of death before insulin was routinely used to manage diabetes.

advanced glycation end products (AGEs): reactive compounds formed after glucose combines with protein; AGEs can damage tissues and lead to diabetic complications.

The *metabolic syndrome,* a cluster of symptoms often associated with insulin resistance (including hyperglycemia, hypertension, and abnormal blood lipids), substantially increases heart disease risk (see Nutrition in Practice 21).

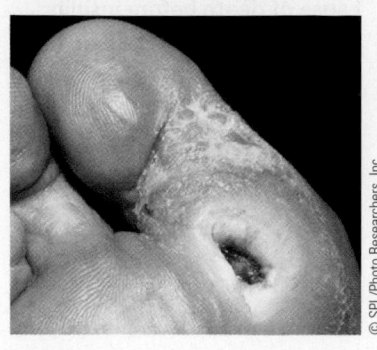

Foot ulcers are a common complication of diabetes because blood circulation is impaired, which slows healing, and nerve damage dampens foot pain, delaying recognition and treatment of cuts and bruises.

heart disease, including hypertension, abnormal blood lipids, and obesity. People with diabetes also have increased tendencies for thrombosis (blood clot formation) and abnormal ventricle function, both of which can worsen the clinical course of heart disease.

Peripheral vascular disease (impaired blood circulation in the limbs) increases the risk of **claudication** (pain while walking) and contributes to the development of foot ulcers (see the photo). Left untreated, foot ulcers can lead to **gangrene** (tissue death), and some patients require foot amputation, a major cause of disability in diabetes. About 15 percent of diabetic patients experience clinically significant foot ulcers during the course of illness.[15]

Microvascular Complications Long-term diabetes causes progressive damage to capillaries in the retina **(diabetic retinopathy),** leading to visual impairments and, possibly, blindness. Damage to the specialized capillaries in the kidneys **(diabetic nephropathy)** impairs filtration of the blood, and kidney failure often develops, requiring the use of dialysis (artificial filtration) for survival. Retinopathy and nephropathy progress most rapidly when diabetes is poorly controlled, and intensive management substantially reduces the risks of developing these conditions.

Diabetic Neuropathy Symptoms of diabetic neuropathy vary and may be experienced as deep pain or burning in the legs and feet, weakness of the arms and legs, or numbness and tingling in the hands and feet. Pain and cramping, especially in the legs, are often severe during the night and may interrupt sleep. Neuropathy also contributes to the development of foot ulcers because cuts and bruises may go unnoticed until wounds are severe. Other manifestations of neuropathy include sweating abnormalities, disturbances in bladder and bowel function, sexual dysfunction, constipation, and delayed stomach emptying **(gastroparesis).**

IN SUMMARY

▌ Diabetes mellitus is a chronic condition characterized by inadequate insulin secretion and/or impaired insulin action; diagnosis is based on indicators of hyperglycemia.

▌ In type 1 diabetes, the pancreas secretes little or no insulin, and insulin therapy is necessary for survival. Type 2 diabetes is characterized by insulin resistance coupled with relative insulin deficiency.

▌ Acute complications of diabetes include diabetic ketoacidosis, in which hyperglycemia is accompanied by ketosis and acidosis; the hyperosmolar hyperglycemic syndrome, characterized by severe hyperglycemia, dehydration, and mental impairments; and hypoglycemia, which results from inappropriate disease management.

▌ Chronic complications of diabetes include macrovascular disorders such as cardiovascular disease and peripheral vascular disease, microvascular conditions such as retinopathy and nephropathy, and neuropathy.

Treatment of Diabetes Mellitus

Diabetes is a chronic and progressive illness that requires lifelong treatment. Managing blood glucose levels is a delicate balancing act that involves meal planning, proper timing of medications, and physical exercise. Frequent adjustments in treatment are often necessary to establish good **glycemic** control. Individuals with type 1 diabetes require insulin therapy for survival. Type 2 diabetes may be treated initially with nutrition therapy and exercise, but most patients eventually need antidiabetic medications

548

	Conventional Therapy	Intensive Therapy
Blood glucose monitoring	Monitored daily	Monitored at least three times daily
Insulin therapy	One or two daily injections; no daily adjustments	Three or more daily injections or use of an external insulin pump; dosage adjusted according to the results of blood glucose monitoring and expected carbohydrate intake
Advantages	Fewer incidences of severe hypoglycemia; less weight gain	Delayed progression of retinopathy, nephropathy, and neuropathy
Disadvantages	More rapid progression of retinopathy, nephropathy, and neuropathy	Twofold to threefold increase in severe hypoglycemia; weight gain; increased risk of becoming overweight

or insulin. Although the health care team must determine the appropriate therapy, the individual with diabetes ultimately assumes much of the responsibility for treatment and therefore requires education in self-management of the disease.

Treatment Goals

The main goal of diabetes treatment is to maintain blood glucose levels within a desirable range to prevent or reduce the risk of complications. Several multicenter clinical trials have shown that *intensive* diabetes treatment, which keeps blood glucose levels tightly controlled, can reduce the incidence and severity of diabetic nephropathy, retinopathy, and neuropathy.*[16] Therefore, maintenance of near-normal glucose levels has become the fundamental objective of diabetes care plans. Other goals of treatment include maintaining healthy blood lipid concentrations, controlling blood pressure, and managing weight—measures that can help to prevent or delay diabetes complications as well. Table 21-3 provides examples of the major differences between conventional and intensive therapies for type 1 diabetes. Although intensive therapy is also associated with some risks, including an increased risk of hypoglycemia, its benefits outweigh these disadvantages.

Diabetes education provides an individual with the knowledge and skills necessary to implement treatment. The primary instructor is often a **Certified Diabetes Educator (CDE),** a health care professional (often a nurse or dietitian) who has specialized knowledge about diabetes treatment and the health education process. To manage diabetes, patients must learn about appropriate meal planning, medication administration, blood glucose monitoring, weight management, appropriate physical activity, and prevention and management of diabetic complications.

Evaluating Diabetes Treatment

Diabetes treatment is largely evaluated by monitoring glycemic status. Good glycemic control requires frequent home monitoring of blood glucose levels using a glucose meter, referred to as **self-monitoring of blood glucose.** Glucose testing provides valuable feedback when the patient adjusts food intake, medications, and physical activity and is helpful for preventing hypoglycemia. Ideally, patients with type 1 diabetes should monitor blood glucose three or more times daily—and more frequently when therapy is adjusted.[17] Some patients may achieve better glycemic control by also using a **continuous glucose monitoring** system, which measures tissue glucose levels every few minutes using a tiny sensor placed under the skin. Self-monitoring of blood glucose is also useful in type 2 diabetes, but the recommended frequency varies according to the specific needs of individual patients.

*Intensive treatment may be inappropriate for individuals with limited life expectancies, a history of hypoglycemia, or previous heart disease.

Studies that evaluated the benefits of intensive diabetes treatment include the *Diabetes Control and Complications Trial* and the *United Kingdom Prospective Diabetes Study*.

Goals for glycemic control:
- Fasting glucose: 70–130 mg/dL
- One to two hours after mealtime: <180 mg/dL
- Bedtime glucose: 90–150 mg/dL

glycemic (gly-SEE-mic): pertaining to blood glucose.

Certified Diabetes Educator (CDE): a health care professional who specializes in diabetes management education. Certification is obtained from the National Certification Board for Diabetes Educators.

self-monitoring of blood glucose: home monitoring of blood glucose levels using a glucose meter.

continuous glucose monitoring: continuous monitoring of tissue glucose levels using a small sensor placed under the skin.

Long-Term Glycemic Control The amount of HbA$_{1c}$ (glycated hemoglobin) in blood reflects glycemic control over the preceding two or three months (the average age of circulating blood cells). The goal of diabetes treatment is an HbA$_{1c}$ value under 7 percent,[18] but it is often markedly higher in people with diabetes, even those who are maintaining near-normal blood glucose levels. Less stringent HbA$_{1c}$ goals may be suitable for certain patients, including those with limited life expectancy, advanced diabetic complications, or a history of severe hypoglycemia.

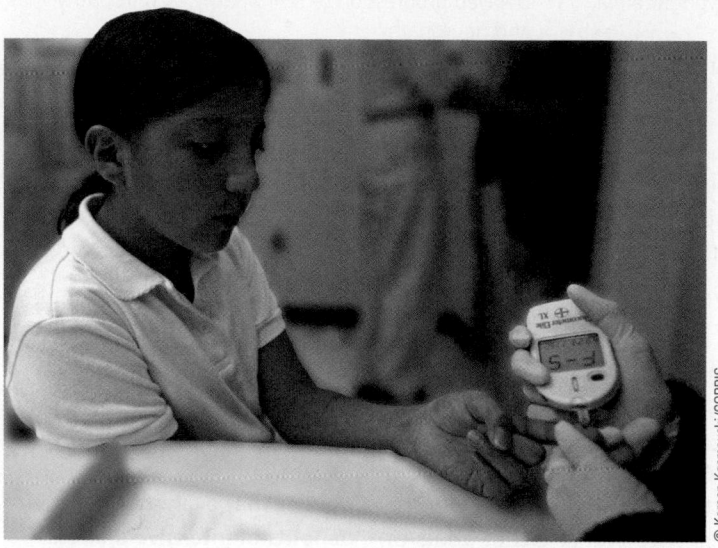

Self-monitoring of blood glucose can help individuals with diabetes learn how to maintain blood glucose levels within a desirable range.

Monitoring for Long-Term Complications Individuals with diabetes are routinely monitored for signs of long-term complications. Blood pressure is measured at each checkup. Annual lipid screening is suggested for adult patients. Routine checks for urinary protein (**microalbuminuria**) can determine if nephropathy has developed. Physical examinations generally screen for signs of retinopathy, neuropathy, and foot problems.

Ketone Testing Ketone testing, which checks for the development of ketoacidosis, should be performed if symptoms are present or if risk has increased due to acute illness, stress, or pregnancy. Both blood and urine tests are available for home use, although the blood tests are generally more reliable. Ketone testing is most useful for patients with type 1 diabetes or gestational diabetes. Individuals with type 2 diabetes may produce excessive ketone bodies when severely stressed by infection or trauma.

Body Weight Concerns

Whereas individuals with newly diagnosed type 1 diabetes are likely to be thin, most people with type 2 diabetes are overweight or obese. In children, body weight and growth patterns are monitored to evaluate whether energy intakes are appropriate.

Body Weight in Type 1 Diabetes In general, people with type 1 diabetes are less likely to be overweight than those in the general population. However, excessive weight gain is sometimes an unintentional side effect of improved glycemic control, especially in those undergoing intensive insulin therapy. Although the cause of the weight gain is unclear, it is possibly related to the insulin treatment, which may stimulate fat synthesis or induce energy intake in some way.[19] In addition, the insulin treatment eliminates urinary glucose losses and may indirectly contribute to energy excess.[20] Although patients should try to prevent excessive weight gain, concerns about weight should not discourage the use of intensive therapy, which is associated with longer life expectancy and fewer complications than occur with conventional therapy (review Table 21-3). It is also important to ensure that growing children receive sufficient energy for normal growth and development.

Body Weight in Type 2 Diabetes Because excessive body fat can worsen insulin resistance, weight loss is recommended for overweight or obese individuals who have diabetes.[21] Even moderate weight loss (10 to 20 pounds) can help to improve glycemic control, blood lipid levels, and blood pressure. Weight loss is most beneficial early in the course of diabetes, before insulin secretion has diminished.

Not all persons with type 2 diabetes are overweight or obese. Older adults and those in long-term care facilities are often underweight and may need to gain weight. Low body weight increases risks of morbidity and mortality in these individuals.

Nutrition Therapy: Nutrient Recommendations

Nutrition therapy can both improve blood glucose levels and slow the progression of diabetic complications. As always, the nutrition care plan must consider personal preferences and lifestyle habits. In addition, dietary intakes must be modified to accommodate growth, lifestyle changes, aging, and any complications that develop. Although all members of the diabetes care team should understand the principles of dietary treatment, a registered dietitian is best suited to design and implement the nutrition therapy provided to diabetes patients. This section presents the nutrient recommendations for diabetes; a later section describes meal-planning strategies.

Total Carbohydrate Intake The amount of carbohydrate consumed has the greatest influence on blood glucose levels after meals—the more grams of carbohydrate ingested, the greater the glycemic response. The carbohydrate recommendation is based on the person's metabolic needs (which are related to the type of diabetes, degree of glucose tolerance, and blood lipid levels), the type of insulin or other medications used to manage the diabetes, and individual preferences. Low-carbohydrate diets, which restrict carbohydrate intake to less than 130 grams per day, are not recommended.[22]

Carbohydrate Sources Different carbohydrate-containing foods have different effects on blood glucose levels; for example, consuming a portion of white rice causes blood glucose to rise more than would a similar portion of barley. A food's glycemic effect is influenced by the type of carbohydrate in a food, the food's fiber content, the preparation method, the other foods included in a meal, and individual tolerances. For individuals with diabetes, using the glycemic index may provide some additional benefit for achieving glycemic control, compared with that obtained by considering only the amount of carbohydrate consumed.[23] In addition, high-fiber, minimally processed foods—which typically have more moderate effects on blood glucose than do highly processed, starchy foods—are among the foods frequently recommended for persons with diabetes.

Reminder: Nutrition in Practice 2 discusses the glycemic index. The website **www.glycemicindex.com** provides glycemic index values for a wide variety of common foods.

Fiber Fiber recommendations for diabetes are similar to those for the general population; thus, people with diabetes are encouraged to include fiber-rich foods such as legumes, whole-grain cereals, fruits, and vegetables in their diet. Although some studies have suggested that very high intakes of fiber (50 grams or more per day) may improve glycemic control, many individuals have difficulty enjoying or tolerating such large amounts of fiber.[24]

The fiber DRI for adult women and men ranges from 21 to 38 g; the DRI table on the inside front cover provides specific values.

Sugars A common misperception is that people with diabetes need to avoid sugar and sugar-containing foods. In reality, table sugar (sucrose), made up of glucose and fructose, has a lower glycemic effect than starch. Because moderate consumption of sugar has not been shown to adversely affect glycemic control,[25] sugar recommendations for people with diabetes are similar to those for the general population, which suggest minimizing foods and beverages that contain added sugars. However, sugars and sugary foods must be counted as part of the daily carbohydrate allowance.

Although fructose has a minimal glycemic effect, its use as an added sweetener is not advised because excessive dietary fructose may adversely affect blood lipid levels. (Note that it is not necessary to avoid the naturally occurring fructose in fruits and vegetables.) Sugar alcohols (such as sorbitol and maltitol) provide about half the kcalories and carbohydrate grams of sucrose and other sugars, so the amounts consumed need to be considered when determining the total carbohydrate intake. Artificial sweeteners (such as aspartame, saccharin, and sucralose) contain no digestible carbohydrate and can be safely used in place of sugar.

Dietary Fat Because people with diabetes are at high risk of developing cardiovascular diseases, guidelines for dietary fat are similar to those suggested for other persons at risk: saturated fat should be less than 7 percent of total kcalories, *trans* fat should be minimized, and cholesterol intake should be less than 200 milligrams daily. In addition, two or more servings of fish are recommended weekly because fish supplies omega-3 fatty acids that reduce heart disease risk (see Chapter 22).

Protein Protein recommendations for people with diabetes are similar to those for the general population. In the United States, the average protein intake is about 15 percent of the energy intake. Although small, short-term studies have suggested that protein intakes above 20 percent of kcalories may improve glycemic control, increase satiety, and help with weight loss, the long-term effects of such diets on diabetes management are unknown.[26] In addition, high protein intakes are discouraged because they may be detrimental to kidney function in some individuals.

Alcohol Use in Diabetes Guidelines for alcohol intake are similar to those for the general population, which recommend that women and men limit their average daily intakes of alcohol to one drink and two drinks per day, respectively. In addition, individuals using insulin or medications that promote insulin secretion should consume food when they ingest alcoholic beverages to avoid hypoglycemia. Alcohol can cause hypoglycemia by interfering with glucose production in the liver. Conversely, an excessive alcohol intake (three or more drinks per day) can worsen hyperglycemia and raise triglyceride levels in susceptible persons. People who should avoid alcohol include pregnant women and individuals with advanced neuropathy, abnormally high triglyceride levels, or a history of alcohol abuse.

Micronutrients Micronutrient recommendations for people with diabetes are the same as for the general population. Vitamin and mineral supplementation is not recommended unless nutrient deficiencies develop; those at risk include the elderly, pregnant or lactating women, strict vegetarians, and individuals on kcalorie-restricted diets.

Nutrition Therapy: Meal-Planning Strategies

Dietitians provide a number of meal-planning strategies to help people with diabetes maintain glycemic control. These strategies emphasize control of carbohydrate intake and portion sizes. Initial dietary instructions may include guidelines for maintaining a healthy diet, improving blood lipids, and reducing cardiovascular risk factors. Sample menus that include commonly eaten foods can help to illustrate general principles.

Carbohydrate Counting Carbohydrate counting techniques are simpler and more flexible than other menu-planning approaches and are widely used for planning diabetes diets. Carbohydrate counting works as follows: After a dietitian determines a person's nutrient and energy needs, the individual is given a daily carbohydrate allowance, often divided into a pattern of meals and snacks according to individual preferences. The carbohydrate allowance can be expressed in grams or as the number of carbohydrate portions allowed per meal (see Table 21-4). The plan user need only be concerned about meeting carbohydrate goals and can select from any of the carbohydrate-containing food groups when planning meals (see Table 21-5 and Figure 21-2). Although encouraged to make healthy food choices, the individual has the freedom to choose the foods desired at each meal without risking loss of glycemic control. Some people may also need guidance about noncarbohydrate foods to help them choose a healthy diet that improves blood lipids or energy intakes. The "How to" on pp. 553–554 shows how to implement carbohydrate counting in clinical practice.

How To USE CARBOHYDRATE COUNTING IN CLINICAL PRACTICE

1. The first step in basic carbohydrate counting is to determine an appropriate carbohydrate intake and suitable distribution pattern; an example is shown in Table 21-4. A nutrition assessment can help to estimate a person's usual energy and carbohydrate intakes. The carbohydrate level should be acceptable to the person using the plan. Frequent monitoring of blood glucose levels can help determine whether additional carbohydrate restriction would be helpful.

   The example given in Table 21-4 illustrates a meal pattern for a person consuming 2000 kcalories daily with a carbohydrate allowance of 50 percent of kcalories. This is calculated as follows:

   $50\% \times 2000$ kcal = 1000 kcal of carbohydrate

   $$\frac{1000 \text{ kcal carbohydrate}}{4 \text{ kcal/g carbohydrate}} = 250 \text{ g carbohydrate/day}$$

   $$\frac{250 \text{ g carbohydrate}}{15 \text{ g/1 carbohydrate portion}} = 16.7 \text{ carbohydrate portions/day}$$

2. The distribution of carbohydrates among meals and snacks is based on both individual preferences and metabolic needs. In type 1 diabetes, the insulin regimen must coordinate with the individual's dietary and lifestyle choices. People using conventional insulin therapy must maintain a consistent carbohydrate intake from day to day to match their particular insulin prescription, whereas those using intensive therapy can alter insulin dosages when carbohydrate intakes change. People with type 2 diabetes are encouraged to develop dietary patterns that suit their lifestyle and medication schedules. For all types of diabetes, the carbohydrate recommendation may need to be altered periodically to improve blood glucose control.

3. Carbohydrate counting can be done in one of two ways:

   ■ Count the grams of carbohydrate provided by foods.

   ■ Count carbohydrate portions, expressed in terms of servings that contain approximately 15 grams each.

TABLE 21-4 Sample Carbohydrate Distribution for a 2000-kCalorie Diet

Meals	Carbohydrate Allowance	
	Grams	Portions[a]
Breakfast	60	4
Lunch	60	4
Afternoon snack	30	2
Dinner	75	5
Evening snack	30	2
Totals	**255 g**	**17**

Note: The carbohydrate allowance in this example is approximately 50 percent of total kcalories.

[a]1 portion = 15 g carbohydrate = 1 portion of starchy food, milk, or fruit.

Success with carbohydrate counting requires knowledge about the food sources of carbohydrates and an understanding of portion control. As shown in Table 21-5, food selections that contain about 15 grams of carbohydrate are interchangeable. The portions of foods that contain 15 grams may vary substantially, however, even among foods in a single food group. Accurate carbohydrate counting often requires instruction and practice in portion control using measuring

TABLE 21-5 Carbohydrate-Containing Food Groups and Sample Portion Sizes

Bread, cereal, rice, and pasta: 1 portion = 15 g carbohydrate
- 1 slice of bread or 1 tortilla
- $1/2$ English muffin
- $3/4$ c unsweetened, ready-to-eat cereal
- $1/2$ c cooked oatmeal
- $1/3$ c cooked rice or pasta

Starchy vegetables: 1 portion = 15 g carbohydrate
- 1 small (3 oz) potato
- $1/2$ c canned or frozen corn
- $1/2$ c cooked beans
- 1 c winter squash, cubed

Fruit: 1 portion = 15 g carbohydrate
- 1 medium apple, orange, or peach
- 1 small banana
- $3/4$ c blueberries or chopped pineapple
- $1/2$ c apple or orange juice

Milk products: 1 portion = 12 g carbohydrate; may be rounded up to 15 g for ease in counting carbohydrate portions
- 1 c milk (whole, low fat, or fat free)
- 1 c buttermilk
- 6 oz plain yogurt

Sweets and desserts: Considerable variation in carbohydrate content; portions listed contain approximately 15 g
- $1/2$ c ice cream
- 2 sandwich cookies (with cream filling)
- $1/2$ frosted cupcake
- 1 granola bar (1 oz)
- 1 tbs honey

Nonstarchy vegetables: 1 portion = 3 to 6 g carbohydrate; 3 servings are equivalent to 1 carbohydrate portion; can be disregarded if less than 3 servings are consumed
- $1/2$ c cooked cauliflower
- $1/2$ c cooked cabbage, collards, or kale
- $1/2$ c cooked okra
- $1/2$ c diced or raw tomatoes

Note: Unprocessed meats, fish, and poultry contain negligible amounts of carbohydrate.

(continued)

cups, spoons, and a food scale. Food lists that indicate the carbohydrate contents of common foods are available from the American Diabetes Association and the American Dietetic Association; these are helpful resources for learning carbohydrate-counting methods.

When using packaged foods, individuals should check the Nutrition Facts panel of food labels to find the carbohydrate content of a serving. If the fiber content is greater than 5 grams per serving, it should be subtracted from the *Total Carbohydrate* value, as fiber does not contribute to blood glucose. If the sugar alcohol content is greater than 5 grams per serving, half of the grams of sugar alcohol can be subtracted from the *Total Carbohydrate* value.

4. Once they have learned the basic carbohydrate counting method, individuals can select whatever foods they wish as long as they do not exceed their carbohydrate goals. Figure 21-2 shows a day's menu that provides the carbohydrate allowance shown in Table 21-4. Although carbohydrate counting focuses on a single macronutrient, people using this technique should be encouraged to follow a healthy eating plan that meets other dietary objectives as well.

❊ SAMPLE MENU ❊

	Carbohydrate Portions			Carbohydrate Portions
Breakfast:			**Afternoon snack:**	
Carbohydrate goal = 4 portions or 60 g.			**Carbohydrate goal = 2 portions or 30 g.**	
$3/4$ c unsweetened, ready-to-eat cereal	1		2 sandwich cookies	1
$1/2$ c low-fat milk	$1/2$		1 c low-fat milk	1
1 scrambled egg	—			
1 slice whole wheat toast (with margarine or butter)	1		**Dinner:**	
6 oz orange juice	$1^1/2$		**Carbohydrate goal = 5 portions or 75 g.**	
Coffee (without milk or sugar)	—		4 oz grilled steak	—
Lunch:			1 small baked potato (with margarine or butter)	1
			Corn on cob, 1 large ear	2
Carbohydrate goal = 4 portions or 60 g.			$1/2$ c steamed collard greens[a] ⎤	
1 tuna salad sandwich (includes 2 slices whole-grain bread, mayonnaise)	2		1 c sliced, raw tomatoes[a] ⎦	1
6 oz yogurt (plain) with 3/4 c blueberries and artificial sweetener	2		$1/2$ c ice cream	1
Diet cola	—		**Evening snack:**	
			Carbohydrate goal = 2 portions or 30 g.	
			1 medium apple	1
			1 oz granola bar	1

[a] Three servings of nonstarchy vegetables are equivalent to 1 carbohydrate portion.

FIGURE 21-2 Translating Carbohydrate Portions into a Day's Meals

---

Carbohydrate counting is taught at different levels of complexity depending on a person's needs and abilities. The basic carbohydrate counting method just described can be helpful for most people, although it requires a consistent carbohydrate intake from day to day to match the medication or insulin regimen. Advanced carbohydrate counting allows more flexibility, but it is best suited for patients using intensive insulin therapy. With this method, a person can determine the specific dosage of insulin needed to cover the amount of carbohydrate consumed in a meal. The person is then free to choose the types and portions of food desired without sacrificing glycemic control. Advanced carbohydrate counting requires some training and should be attempted only after an individual has mastered more basic methods.

Exchange Lists for Meal Planning The exchange list system is an alternative meal-planning method, although it may be more difficult for patients to learn than carbohydrate counting. The exchange system sorts foods according to their proportions of carbohydrate, fat, and protein so that each item in a food group

(or "exchange list") has a similar macronutrient and energy content (see pp. C-1 to C-2). Thus, any food on a list can be exchanged, or traded, for any other food on the same list without affecting the macronutrient balance in a day's meals. Although the exchange list system can be helpful for individuals who want to maintain a diet with specific percentages of protein, carbohydrate, and fat, it offers no advantages for maintaining glycemic control and is less flexible than carbohydrate counting. This system of meal planning is described further in Appendix C (Appendix B for Canadians).

The exchange lists can be helpful resources for individuals using carbohydrate-counting methods because the portions in the exchange lists are interchangeable with the portions used in carbohydrate counting. Foods listed in the starch, fruit, and milk lists, for example, are equivalent to carbohydrate portions, as each item contains approximately 15 grams of carbohydrate (see pp. C-3 to C-6; note that the carbohydrate in the milk exchanges can be rounded up to 15 grams). In the list labeled "Sweets, Desserts, and Other Carbohydrates" (pp. C-6 to C-8), the carbohydrate portions are indicated in the far-right column.

Insulin Therapy

Insulin therapy is necessary for individuals who cannot produce enough insulin to meet their metabolic needs. It is therefore required by all persons with type 1 diabetes and those with type 2 diabetes who cannot maintain glycemic control with medications, diet, and exercise. The pancreas normally secretes insulin in relatively low amounts between meals and during the night (called *basal insulin*) and in much higher amounts when meals are ingested. Ideally, the insulin treatment should reproduce the natural pattern of insulin secretion as closely as possible.

Insulin Preparations The forms of insulin that are commercially available differ by their onset of action, timing of peak action, and duration of effects. Table 21-6 and Figure 21-3 show how insulin preparations are classified: they may be rapid acting (lispro, aspart, and glulisine), short acting (regular), intermediate acting (NPH), or long acting (glargine and detemir), thereby allowing substantial flexibility in establishing a suitable insulin regimen.[27] The rapid- and short-acting insulins are used at mealtimes, whereas the intermediate- and long-acting insulins provide basal insulin for the periods between meals and during the night. Thus, mixtures of several types of insulin can produce greater glycemic control than any one type alone. Several premixed formulations are also available; some examples are listed in Table 21-6.

subcutaneous (sub-cue-TAY-nee-us): beneath the skin.

syringes: devices used for injecting medications. A syringe consists of a hypodermic needle attached to a hollow tube with a plunger inside.

Insulin Delivery Insulin is most often administered by **subcutaneous** injection, either self-administered or provided by caregivers. Disposable **syringes,** which are filled from vials that contain multiple doses of insulin, are the most common devices

Because insulin is a protein, it would be destroyed by digestive processes if taken orally.

TABLE 21-6 Insulin Preparations

Form of Insulin	Common Preparations	Onset of Action	Peak Action	Duration of Action
Rapid acting	Lispro (Humalog) Aspart (Novolog) Glulisine (Apidra)	5 to 15 minutes	60 to 90 minutes	3 to 5 hours
Short acting	Regular	30 minutes	2 to 3 hours	5 to 8 hours
Intermediate acting	NPH	2 to 4 hours	6 to 10 hours	10 to 16 hours
Long acting	Glargine (Lantus) Detemir (Levemir)	1 to 2 hours	Steady effects	24 hours
Insulin mixtures (with sample ratios)	NPH/regular (70:30) NPH/regular (50:50)	Variable; depends on formulation	Variable; depends on formulation	Variable; depends on formulation

FIGURE 21-3 **Effects of Insulin Preparations**

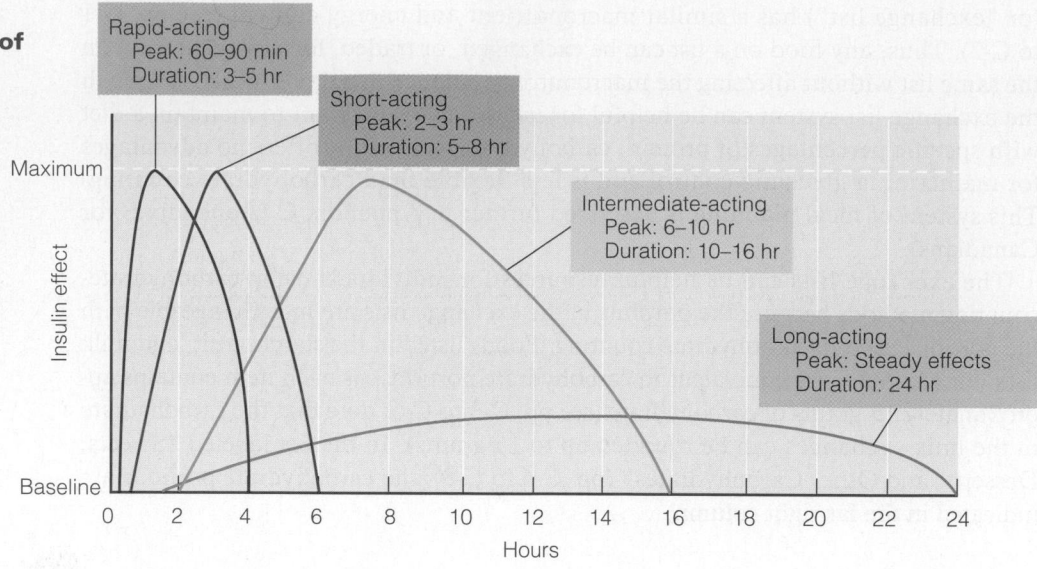

Rapid-acting
Peak: 60–90 min
Duration: 3–5 hr

Short-acting
Peak: 2–3 hr
Duration: 5–8 hr

Intermediate-acting
Peak: 6–10 hr
Duration: 10–16 hr

Long-acting
Peak: Steady effects
Duration: 24 hr

Maximum

Insulin effect

Baseline

0   2   4   6   8   10   12   14   16   18   20   22   24

Hours

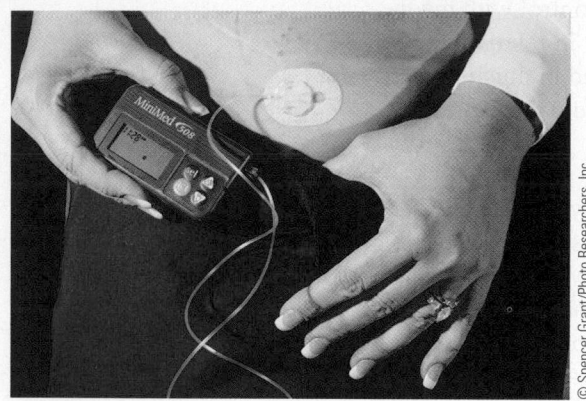

External insulin pumps deliver insulin continuously through thin, flexible tubing inserted into the skin.

© Spencer Grant/Photo Researchers, Inc.

Rapid-acting insulin begins working within 15 minutes, so it can be injected right before a meal. Short-acting insulin requires a half-hour wait before the meal can begin.

used. Another option is to use insulin pens, injection devices that resemble permanent marking pens. Disposable insulin pens are prefilled with insulin and used one time only, whereas reusable pens can be fitted with prefilled insulin cartridges and replaceable needles. Some individuals use insulin pumps, computerized devices that can be programmed to deliver basal insulin continuously and bolus doses at mealtimes. These pumps infuse insulin through thin, flexible tubing that remains in the skin. An insulin pump can be worn under clothes, attached to a belt, or kept in a pocket.

Insulin Regimen for Type 1 Diabetes Type 1 diabetes is best managed with intensive insulin therapy, which involves multiple daily injections of several types of insulin or the use of an insulin pump. Usually, intermediate- or long-acting insulin meets basal insulin needs, and rapid- or short-acting insulin is injected before meals. Three or more daily injections are required for good glycemic control. Simpler regimens involve twice-daily injections of a mixture of intermediate- and short-acting insulin. Regimens that include three or more injections allow for greater flexibility in carbohydrate intake and meal timing. With fewer injections, the timing of both meals and injections must be similar from day to day to avoid periods of insulin deficiency or excess.

A person using intensive therapy must learn to accurately determine the amount of insulin to inject before each meal. The amount required depends on the pre-meal blood glucose level, the carbohydrate content of the meal, and the person's body weight and sensitivity to insulin. To determine insulin sensitivity, the individual keeps careful records of food intake, insulin dosages, and blood glucose levels. Eventually, these records are analyzed by medical personnel to determine the appropriate **carbohydrate-to-insulin ratio** for that individual, which assists in calculating insulin dosages at mealtime. Intensive therapy allows for substantial variation in food intake and lifestyle, but it requires frequent testing of blood glucose levels and a good understanding of carbohydrate counting.

After insulin therapy is initiated, persons with type 1 diabetes may experience a temporary remission of disease symptoms and a reduced need for insulin, known as

the "honeymoon period." The remission is due to a temporary improvement in pancreatic function and may last for several weeks or months.[28] It is important to anticipate this period of remission to avoid insulin excess. After the honeymoon period ends, the patient will need to reinstate full insulin treatment.

Insulin Regimen for Type 2 Diabetes Approximately 30 percent of people diagnosed with type 2 diabetes can benefit from insulin therapy.[29] Although initial treatment of type 2 diabetes may involve nutrition therapy, physical activity, and antidiabetic medications, long-term results with these treatments are often disappointing. As the disease progresses, pancreatic function worsens, and many individuals require insulin therapy to maintain glycemic control.

Many possible regimens can be used to control type 2 diabetes. Some persons may be treated with insulin alone, whereas others may use insulin in combination with other antidiabetic drugs. Many patients need only one or two daily injections. Some regimens involve a mixture of rapid- and intermediate-acting insulin in the morning and an injection of intermediate- or long-acting insulin at dinner or before bedtime. In other cases, only a single injection of intermediate- or long-acting insulin may be needed at bedtime.[30] Dosages and timing are adjusted according to the results of blood glucose self-monitoring.

Insulin Therapy and Hypoglycemia Hypoglycemia is the most common complication of insulin treatment, although it may also result from the use of some antidiabetic drugs. It most often results from intensive insulin therapy, because the attempt to attain near-normal blood glucose levels increases the risk of overtreatment. Hypoglycemia can be corrected with the immediate intake of glucose or a carbohydrate-containing food. Usually, 15 to 20 grams of carbohydrate can relieve hypoglycemia in about 15 minutes, although patients should retest their blood glucose levels after 15 minutes in case additional treatment is necessary. Foods that provide pure glucose yield a better response than foods that contain other sugars, such as sucrose or fructose. Individuals using insulin are usually advised to carry glucose tablets or a source of carbohydrate that can be easily ingested. Those at risk of severe hypoglycemia are often given prescriptions for the hormone glucagon, which can be injected by caregivers in case of unconsciousness.

Fasting Hyperglycemia Insulin therapy must sometimes be adjusted to prevent **fasting hyperglycemia,** which has three possible causes. The usual cause is a waning of insulin action during the night due to insufficient insulin. A second possibility, known as the **dawn phenomenon,** occurs when blood glucose levels increase in the morning due to the early morning secretion of growth hormone, which counteracts insulin's actions. Less frequently, fasting hyperglycemia develops as a result of nighttime hypoglycemia, which causes hormonal responses that stimulate glucose production; the resulting condition is known as **rebound hyperglycemia** (also called the **Somogyi effect**). Whatever the cause, fasting hyperglycemia can be treated by adjusting the dosage or formulation of insulin administered in the evening.

Antidiabetic Drugs

Treatment of type 2 diabetes often requires the use of oral medications and injectable drugs other than insulin. As shown in Table 21-7, these drugs can improve hyperglycemia by various modes of action. Treatment may involve the use of a single medication (monotherapy) or a combination of several medications (combination therapy). By relying on several mechanisms at once, combination therapy achieves more rapid and sustained glycemic control than is possible with monotherapy. The Diet-Drug Interactions feature lists some nutrition-related effects of antidiabetic drugs.

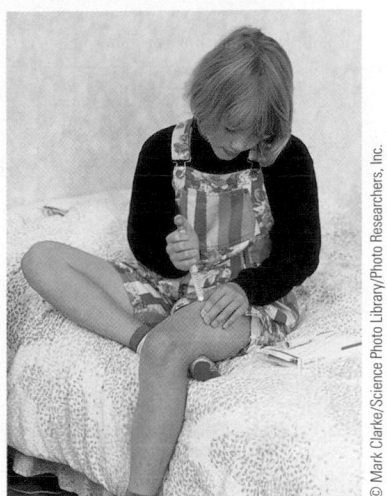

Children often become adept at administering the insulin they require.

Each of the following sources provides about 15 g of carbohydrate:
- Glucose tablets: 4 tablets
- Table sugar: 4 tsp
- Honey: 1 tbs
- Jelly beans: 15 small
- Orange juice: ½ c

carbohydrate-to-insulin ratio: the amount of carbohydrate that can be handled per unit of insulin. On average, every 15 grams of carbohydrate requires about 1 unit of rapid- or short-acting insulin.

fasting hyperglycemia: hyperglycemia that develops after an overnight fast of at least eight hours.

dawn phenomenon: morning hyperglycemia that is caused by the early-morning release of growth hormone, which counteracts insulin's glucose-lowering effects.

rebound hyperglycemia: hyperglycemia that results from the release of counterregulatory hormones following nighttime hypoglycemia; also called the **Somogyi effect.**

TABLE 21-7 Antidiabetic Drugs

Drug Category	Common Examples	Mode of Action
Alpha-glucosidase inhibitors	Acarbose (Precose) Miglitol (Glyset)	Delay carbohydrate digestion and absorption
Amylin analogs (injected)	Pramlintide (Smylin)	Suppress glucagon secretion, delay stomach emptying, suppress appetite
Biguanides	Metformin (Glucophage)	Inhibit liver glucose production, improve glucose utilization
Dipeptidyl peptidase 4 (DPP-4) inhibitors	Saxagliptin (Onglyza) Sitagliptin (Januvia)	Improve insulin secretion, suppress glucagon secretion, delay stomach emptying
Incretin mimetics (injected)	Exenatide (Byetta) Liraglutide (Victoza)	Improve insulin secretion, suppress glucagon secretion, delay stomach emptying
Meglitinides	Nateglinide (Starlix) Repaglinide (Prandin)	Stimulate insulin secretion by the pancreas
Sulfonylureas	Glipizide (Glucotrol) Glyburide (Glynase PresTab)	Stimulate insulin secretion by the pancreas
Thiazolidinediones	Pioglitazone (Actos) Rosiglitazone (Avandia)	Decrease insulin resistance

Physical Activity and Diabetes Management

Regular physical activity can improve glycemic control considerably and is therefore a central feature of disease management. A regular exercise program improves insulin sensitivity, which reduces insulin requirements. Physical activity also benefits other aspects of health, including blood lipid levels, blood pressure, body weight, and cardiovascular functioning. People with diabetes are advised to perform at least 150 minutes of moderate-intensity activity or 75 minutes of vigorous aerobic activity each week (or an equivalent combination). In addition, they should participate in a resistance exercise program at least two or three times weekly, unless contraindicated by a medical condition that increases risk of injury.[31]

Medical Evaluation before Exercise Before a person with diabetes begins a new exercise program, a medical evaluation should screen for problems that may be aggravated by certain activities. Complications involving the heart and blood vessels, eyes, kidneys, feet, and nervous system may limit the types of activity recommended. For individuals with a low level of fitness who have been relatively inactive, only mild or moderate exercise may be prescribed at first; for example, a short walk at a comfortable pace may be the first activity suggested. People with severe retinopathy should avoid vigorous aerobic or resistance exercise, which may lead to retinal detachment and damage to eye tissue. Peripheral neuropathy may preclude repetitive weight-bearing exercises such as jogging and step exercises because these activities may lead to foot ulcerations. To prevent dehydration, which can adversely affect blood glucose levels and heart function, proper hydration should be encouraged before and during exercise.

Maintaining Glycemic Control Individuals using insulin or medications that increase insulin secretion must carefully adjust food intake and medication dosages to prevent hypoglycemia during physical activity. Medication dosages that precede exercise often need to be reduced substantially. Blood glucose levels should be checked both before and after an activity. If blood glucose is below 100 mg/dL, carbohydrate should be consumed before exercise begins. Additional carbohydrate may be needed during

Diet-Drug Interactions

Check this table for notable nutrition-related effects of the medications discussed in this chapter.

Sulfonylureas	**Gastrointestinal effects:** Nausea, vomiting, cramps, diarrhea **Dietary interactions:** Avoid using with alcohol due to a toxic reaction that causes flushing, throbbing head and neck pain, shortness of breath, palpitations, and sweating. Avoid using with dietary supplements that contain ginseng, garlic, fenugreek, coriander, and celery, as they may increase risk of hypoglycemia. **Metabolic effects:** Hypoglycemia, weight gain, allergic skin reactions
Biguanides (metformin)	**Gastrointestinal effects:** Abdominal pain and cramps, nausea, vomiting, gas, diarrhea, metallic taste, anorexia **Dietary interactions:** Avoid using with alcohol. **Metabolic effects:** Decreased folate and vitamin B_{12} absorption
Thiazolidinediones	**Metabolic effects:** Hypoglycemia, weight gain, fluid retention, edema, anemia
Alpha-glucosidase inhibitors	**Gastrointestinal effects:** Abdominal pain and cramps, nausea, gas, diarrhea **Metabolic effects:** Elevated liver enzymes, hyperbilirubinemia

or after prolonged activity or even several hours after the activity is completed. The individual should avoid strenuous exercise if blood glucose levels exceed 250 mg/dL and should avoid all types of physical activity when blood glucose levels exceed 300 mg/dL or ketosis is present.[32]

Sick-Day Management

Illness, infection, or injury can cause hormonal changes that raise blood glucose levels and increase the risk of complications. During illness, individuals with diabetes should measure blood glucose and ketone levels several times daily. They should continue to use antidiabetic drugs, including insulin, as prescribed; adjustments in dosages may be necessary if hyperglycemia persists. If appetite is poor, patients should select easy-to-manage foods and beverages that provide the prescribed amount of carbohydrate at each meal. To prevent dehydration, especially if vomiting or diarrhea is present, patients should make sure they consume adequate amounts of liquids throughout the day.

IN SUMMARY

▌ Diabetes treatment includes nutrition therapy, the use of insulin or other antidiabetic medications, and appropriate physical activity. Glycemic control is evaluated by monitoring blood glucose levels and glycated hemoglobin.

▌ The quantity of carbohydrate consumed has the greatest influence on blood glucose levels after meals. The total amount of carbohydrate ingested is more important than the type of carbohydrate consumed.

▌ Carbohydrate counting is widely used in menu planning and can be taught at different levels of complexity depending on individual needs and abilities.

▌ Insulin therapy is required for patients who are unable to produce sufficient insulin. Antidiabetic drugs improve hyperglycemia by various modes of action.

▌ Physical activity can improve glycemic control and enhance various aspects of general health. Illness can worsen glycemic control and may require medication adjustments. The Case Study provides an opportunity to review the treatment for diabetes.

Case Study CHILD WITH TYPE 1 DIABETES

Nora is a 12-year-old girl who was diagnosed with type 1 diabetes two years ago. She practices intensive therapy and has had the support of her parents and an excellent diabetes management team. With their help, Nora has been able to assume the bulk of the responsibility for her diabetes care and has managed to control her blood glucose remarkably well. In the past few months, however, Nora has been complaining bitterly about the impositions diabetes has placed on her life and her interactions with friends. Sometimes she refuses to monitor her blood glucose levels, and she has skipped insulin injections a

few times. Recently, Nora was admitted to the emergency room complaining of fever, nausea, vomiting, and intense thirst. The physician noted that Nora was confused and lethargic. A urine test was positive for ketones, and her blood glucose levels were 400 mg/dL. The diagnosis was diabetic ketoacidosis.

1. Describe the metabolic events that lead to ketoacidosis. Were Nora's symptoms and laboratory tests consistent with the diagnosis?

2. Review Table 21-3 on p. 549, and consider the advantages and disadvantages

that intensive therapy might have for Nora.

3. Discuss how Nora's age might influence her ability to cope with and manage her diabetes. Why might she feel that diabetes is disrupting her life? What suggestions may help?

4. Review the complications associated with long-term diabetes. How might you explain the importance of glycemic control to a 12-year-old girl?

Diabetes Management in Pregnancy

Women with diabetes face new challenges during pregnancy. Due to hormonal changes, pregnancy increases insulin resistance and the need for insulin, so maintaining glycemic control may be more difficult. In addition, about 7 percent of nondiabetic women develop gestational diabetes and require treatment during pregnancy.[33] Women with gestational diabetes are at greater risk of developing type 2 diabetes later in life, and their children are at increased risk of developing obesity and type 2 diabetes as they enter adulthood.

A pregnancy complicated by diabetes increases health risks for both mother and fetus. Uncontrolled diabetes is linked with increased incidences of miscarriage, birth defects, and fetal deaths. Newborns are more likely to suffer from respiratory distress and to develop metabolic problems such as hypoglycemia, jaundice, and hypocalcemia. Women with diabetes often deliver babies with **macrosomia** (abnormally large bodies), which makes delivery more difficult and can result in birth trauma or the need for a cesarean section. Macrosomia results because maternal hyperglycemia induces excessive insulin production by the fetal pancreas, which stimulates growth and fat deposition.[34]

Glycemic control during pregnancy offers the best chance of a safe delivery and a healthy infant.

© EHproductions Ltd/Jupiter Images

Pregnancy in Type 1 or Type 2 Diabetes

Women with diabetes who achieve glycemic control at conception and during the first trimester of their pregnancy substantially reduce

the risks of birth defects and spontaneous abortion. For this reason, it is recommended that women contemplating pregnancy receive preconception care to avoid the complications associated with poorly controlled diabetes. Maintaining glycemic control during the second and third trimesters can minimize the risks of macrosomia and morbidity in newborn infants.

Nutrient requirements during pregnancy are generally similar for women with and without diabetes. The dietary adjustments suggested for improving glycemic control should be based on a woman's dietary habits and the results of blood glucose monitoring. Regular meals and snacks help to avoid hypoglycemia, which is more likely to occur during pregnancy because glucose is continuously supplied to the fetus. An evening snack is usually required to prevent overnight hypoglycemia and ketosis. Insulin and medication changes are often needed during pregnancy, and the woman may have to adjust her dietary habits further as a result.

Gestational Diabetes

Risk of gestational diabetes is highest in women who have a family history of diabetes, are obese, are in a high-risk ethnic group (Hispanic American, Native American, Asian American, African American, or Pacific Islander), or have previously given birth to an infant weighing over 9 pounds. To ensure that problems are dealt with promptly, physicians routinely test all women for gestational diabetes between 24 and 28 weeks of gestation. In high-risk women, screening should begin prior to pregnancy or soon after conception. Even mild hyperglycemia can have adverse effects on a developing fetus and may lead to complications during pregnancy.[35]

Women with gestational diabetes who are overweight or obese may need to adjust their energy intakes during pregnancy. Although adequate energy is needed for fetal development, a modest kcaloric reduction (about 30 percent less than total energy needs) may improve glycemic control without increasing the risk of ketosis.[36] Restricting carbohydrate to 40 to 45 percent of total energy intake may improve blood glucose levels after meals. Carbohydrate is usually poorly tolerated in the morning; therefore, restricting carbohydrate (to about 30 grams) at breakfast is often necessary.[37] The remaining carbohydrate intake should be spaced throughout the day in several meals and snacks, including an evening snack to prevent ketosis during the night. Regular aerobic activity is often recommended because it can help to improve glycemic control. Women who fail to achieve glycemic goals by diet and exercise alone may need to use insulin or an antidiabetic drug that is safe during pregnancy.[38] The Case Study reviews the connections between gestational diabetes and type 2 diabetes.

IN SUMMARY

▮ Careful management of blood glucose levels before and during pregnancy may reduce complications in mother and infant. Most nutrient requirements during pregnancy are similar for women with and without diabetes.

▮ Carbohydrate intake should be distributed into several meals and snacks, including an evening snack to prevent overnight ketosis. Carbohydrate restriction may be recommended, especially in women with gestational diabetes.

▮ Moderate energy restriction may help to improve glycemic control in overweight and obese women with gestational diabetes.

macrosomia (MAK-roh-SOH-mee-ah): the condition of having an abnormally large body; in infants, refers to birth weights of 4000 grams (8 pounds 13 ounces) and above.

Case Study WOMAN WITH TYPE 2 DIABETES

Alicia Cordova is a 41-year-old Mexican American woman recently diagnosed with type 2 diabetes. Mrs. Cordova developed gestational diabetes while she was pregnant with her second child. Her blood glucose levels returned to normal following pregnancy, and she was advised to get regular checkups, maintain a desirable weight, and engage in regular physical activity. Although she reports that she does not overeat and that she exercises regularly, she has been unable to maintain a healthy weight. At 5 feet 3 inches tall, Mrs. Cordova currently weighs 155 pounds. She has decided to lose weight and join a gym because she is concerned about the long-term effects of diabetes and the possibility that she may need insulin injections. She is also concerned about her husband and children because they are overweight and not very active. The physician refers Mrs. Cordova to a dietitian to help her plan a diet.

1. What factors in Mrs. Cordova's medical history increase her risk for diabetes? Are her husband and children also at risk?

2. Describe the general characteristics of a diet and exercise program that would be appropriate for Mrs. Cordova. How might weight loss and physical activity benefit her diabetes?

3. If Mrs. Cordova is unable to control her blood glucose with diet and physical activity, what treatment might be suggested? Can you explain to Mrs. Cordova why she would probably not require insulin at this time?

4. What dietary and lifestyle changes may help to prevent diabetes in Mrs. Cordova's husband and children?

Nutrition Assessment Checklist FOR PEOPLE WITH DIABETES

Medical History

Check the medical record to determine:

- Type of diabetes
- Duration of diabetes
- Acute and chronic complications
- Conditions, including pregnancy, that may alter treatment

Medications

For people with preexisting diabetes who use antidiabetic drugs (including insulin), note:

- Type of medication
- Administration schedule

Check for use of other medications, including:

- Medications that affect blood glucose levels
- Cholesterol- and triglyceride-lowering medications
- Antihypertensive medications

Dietary Intake

To devise an acceptable meal plan and coordinate medications, obtain:

- An accurate and thorough record of food intake and meal patterns
- An account of usual physical activities

At medical checkups, reassess the person's ability to:

- Maintain an appropriate carbohydrate intake
- Maintain an appropriate energy intake

- Monitor blood glucose levels at home
- Adjust insulin and diet to accommodate sick days
- Use appropriate foods to treat hypoglycemia

Anthropometric Data

Take accurate baseline height and weight measurements as a basis for:

- Appropriate energy intake
- Initial insulin therapy

Periodically reassess height and weight for children and weight for adults and pregnant women to ensure that the meal plan provides an appropriate energy intake.

Laboratory Tests

Monitor the success of diabetes treatment using these tests:

- Blood lipid concentrations
- Blood or urinary ketones
- Glycated hemoglobin
- Urinary protein (microalbuminuria)

Physical Signs

Look for physical signs of:

- Dehydration, especially in older adults
- Foot ulcers
- Nerve damage
- Vision problems

Clinical Applications

1. Using the carbohydrate-counting method described in the "How to" on pp. 553–554, determine an appropriate carbohydrate intake (in both grams and portions) for a man with type 2 diabetes who requires approximately 2600 kcalories daily. Assume he would benefit from a carbohydrate allowance that is 50 percent of his energy intake. Using information from Tables 21-4 and 21-5, develop a one-day sample menu that is likely to meet his carbohydrate goals. Use the exchange lists in Appendix C to find additional examples of foods to include in your menu.

2. Take a trip to a pharmacy or use information from an online drugstore to price these items: blood glucose meter, test strips for the glucose meter selected, lancets, insulin, and syringes. Determine the approximate cost of insulin and syringes for a person who uses 12 units of short-acting insulin (regular) and 18 units of intermediate-acting insulin (NPH) taken in three injections daily (thus, the insulin requirement is 30 units per day). Also estimate the cost of testing blood glucose three times daily. Approximately how much would these supplies cost per month?

Self Check

1. Which of the following is characteristic of type 1 diabetes?
   a. Abdominal obesity increases risk.
   b. The pancreas makes little or no insulin.
   c. It is the predominant form of diabetes.
   d. It often arises during pregnancy.

2. Which of the following describes type 2 diabetes?
   a. It is usually an autoimmune disease.
   b. The pancreas makes little or no insulin.
   c. Diabetic ketoacidosis is a common complication.
   d. Chronic complications may develop before it is diagnosed.

3. Most chronic complications associated with diabetes result from:
   a. altered kidney function.
   b. infections that deplete nutrient reserves.
   c. weight gain and hypertension.
   d. damage to blood vessels and nerves.

4. Long-term glycemic control is usually evaluated by:
   a. self-monitoring of blood glucose.
   b. testing urinary ketone levels.
   c. measuring glycated hemoglobin.
   d. testing urinary protein levels (microalbuminuria).

5. Regarding dietary carbohydrate, a patient with diabetes should be most concerned about:
   a. consuming the correct quantity of carbohydrate at each meal or snack.
   b. consuming the correct proportion of sugars, starches, and fiber in meals.
   c. avoiding added sugars and kcaloric sweeteners.
   d. choosing meals with ideal proportions of protein, carbohydrate, and fat.

6. Which of the following is true regarding the general use of alcohol in diabetes?
   a. A serving of alcohol is considered part of the carbohydrate allowance.
   b. Alcohol contributes to hyperglycemia and should be avoided completely.
   c. Alcohol can cause hypoglycemia and should therefore be consumed with food if patients use insulin or medications that stimulate insulin secretion.
   d. Patients can use alcohol in unlimited quantities unless they are pregnant.

7. The meal-planning strategy best suited to all people with diabetes is:
   a. carbohydrate counting.
   b. the exchange list system.
   c. following menus and recipes provided by a registered dietitian.
   d. any approach that best helps the patient control blood glucose levels.

8. A patient using intensive insulin therapy is likely to follow a regimen that involves:
   a. twice-daily injections that combine short-, intermediate-, and long-acting insulin in each injection.
   b. a mixture of intermediate- and long-acting insulin injected between meals.
   c. multiple daily injections that supply basal insulin and precise insulin doses at each meal.
   d. the use of both insulin and oral antidiabetic agents.

9. In a person who has previously maintained good glycemic control, hyperglycemia can be precipitated by:
   a. infections or illnesses.
   b. chronic alcohol ingestion.
   c. undertreatment of hypoglycemia.
   d. prolonged exercise.

10. Women with pregnancies complicated by diabetes:
    a. generally benefit from larger meals and a snack at bedtime.
    b. often need less carbohydrate at breakfast.
    c. need more carbohydrate than women with diabetes who are not pregnant.
    d. need more kcalories to support the pregnancy than women without diabetes.

Answers to these questions can be found in Appendix H.

Notes

1. National Center for Health Statistics, *Health, United States, 2009—With Special Feature on Medical Technology* (Hyattsville, MD: author, 2010), p. 258.

2. National Center for Health Statistics, 2010.

3. American Diabetes Association, Diagnosis and classification of diabetes mellitus, *Diabetes Care* 34 (2011): S62–S69.

4. S. E. Inzucchi and R. S. Sherwin, Type 1 diabetes mellitus, in L. Goldman and D. Ausiello, eds., *Cecil Medicine* (Philadelphia: Saunders, 2008), pp. 1727–1747.

5. A. Maitra, The endocrine system, in V. Kumar and coeditors, *Robbins and Cotran Pathologic Basis of Disease* (Philadelphia: Saunders, 2010), pp. 1097–1164.

6. National Center for Health Statistics, 2010.

7. R. B. Lipton, Incidence of diabetes in children and youth—tracking a moving target, *Journal of the American Medical Association* 297 (2007): 2760–2762.

8. The Writing Group for the SEARCH for Diabetes in Youth Study Group, Incidence of diabetes in youth in the United States, *Journal of the American Medical Association* 297 (2007): 2716–2724.

9. U. Masharani and M. S. German, Pancreatic hormones and diabetes mellitus, in D. G. Gardner and D. Shoback, eds., *Greenspan's Basic and Clinical Endocrinology* (New York: McGraw-Hill/ Lange, 2007), pp. 661–747.

10. Inzucchi and Sherwin, Type 1 diabetes mellitus, 2008.

11. S. E. Inzucchi and R. S. Sherwin, Type 2 diabetes mellitus, in L. Goldman and D. Ausiello, eds., *Cecil Medicine* (Philadelphia: Saunders, 2008), pp. 1748–1760.

12. Inzucchi and Sherwin, Type 1 diabetes mellitus, 2008.

13. Maitra, The endocrine system, 2010.

14. Inzucchi and Sherwin, Type 2 diabetes mellitus, 2008.

15. Inzucchi and Sherwin, Type 1 diabetes mellitus, 2008.

16. American Diabetes Association, Implications of the United Kingdom Prospective Diabetes Study, *Diabetes Care* 21 (1998): 2180–2184; Diabetes Control and Complications Trial Research Group, The effect of intensive treatment of diabetes on the development and progression of long-term complications in insulin-dependent diabetes mellitus, *New England Journal of Medicine* 329 (1993): 977–986.

17. American Diabetes Association, Standards of medical care in diabetes—2011, *Diabetes Care* 34 (2011): S11–S61.

18. American Diabetes Association, Standards of medical care in diabetes—2011, 2011.

19. M. Ryan and coauthors, Is a failure to recognize an increase in food intake a key to understanding insulin-induced weight gain? *Diabetes Care* epub 17 December 2007 (DOI 10.2337/dc07-1171); A. N. Jacob and coauthors, Potential causes of weight gain in type 1 diabetes mellitus, *Diabetes, Obesity and Metabolism* 8 (2006): 404–411.

20. A. Daly, Use of insulin and weight gain: Optimizing diabetes nutrition therapy, *Journal of the American Dietetic Association* 107 (2007): 1386–1393.

21. American Diabetes Association, Standards of medical care in diabetes—2011, 2011; M. J. Franz and coauthors, The evidence for medical nutrition therapy for type 1 and type 2 diabetes in adults, *Journal of the American Dietetic Association* 110 (2010): 1852–1889.

22. American Diabetes Association, Standards of medical care in diabetes—2011, 2011; American Diabetes Association, Nutrition recommendations and interventions for diabetes, *Diabetes Care* 31 (2008): S61–S78.

23. American Diabetes Association, Standards of medical care in diabetes—2011, 2011.

24. American Diabetes Association, Nutrition recommendations and interventions for diabetes, 2008.

25. Franz and coauthors, 2010; American Diabetes Association, Nutrition recommendations and interventions for diabetes, 2008.

26. Franz and coauthors, 2010; American Diabetes Association, Nutrition recommendations and interventions for diabetes, 2008.

27. M. S. Nolte and J. H. Karam, Pancreatic hormones and antidiabetic drugs, in B. G. Katzung, ed., *Basic and Clinical Pharmacology* (New York: McGraw-Hill/ Lange, 2007), pp. 683–705; Masharani and German, 2007.

28. Inzucchi and Sherwin, Type 1 diabetes mellitus, 2008.

29. Nolte and Karam, 2007.

30. Inzucchi and Sherwin, Type 2 diabetes mellitus, 2008.

31. American Diabetes Association, Standards of medical care in diabetes—2011, 2011.

32. Inzucchi and Sherwin, Type 1 diabetes mellitus, 2008; *Diabetes Mellitus: A Guide to Patient Care* (Philadelphia: Lippincott Williams & Wilkins, 2007), p. 74.

33. American Diabetes Association, Diagnosis and classification of diabetes mellitus, 2011.

34. A. Maitra, Diseases of infancy and childhood, in V. Kumar and coeditors, *Robbins and Cotran Pathologic Basis of Disease* (Philadelphia: Saunders, 2010), pp. 447–483.

35. Z. Hussain and L. Jovanovic, Nutritional strategies in pregestational, gestational, and postpartum diabetic patients, in J. I. Mechanick and E. M. Brett, eds., *Nutritional Strategies for the Diabetic and Prediabetic Patient* (Boca Raton, FL: CRC Press, 2006), pp. 133–148.

36. American Diabetes Association, Nutrition recommendations and interventions for diabetes, 2008.

37. American Dietetic Association, *Nutrition Care Manual* (Chicago: American Dietetic Association, 2011).

38. O. Langer, Oral antidiabetic drugs in pregnancy: The other alternative, *Diabetes Spectrum* 20 (2007): 101–105.

Nutrition in Practice

THE METABOLIC SYNDROME

Chapter 21 described how insulin resistance—a reduced sensitivity to insulin in muscle, adipose, and liver cells—can contribute to hyperglycemia and hyperinsulinemia and, eventually, to type 2 diabetes. Insulin resistance is also a central feature of several other conditions, including the **metabolic syndrome,** a cluster of metabolic abnormalities that raise the risk of developing cardiovascular diseases (CVD) and type 2 diabetes. This Nutrition in Practice describes how the metabolic syndrome is diagnosed, how and why it might develop, its consequences, and current treatment approaches. The glossary defines the relevant terms.

How is the metabolic syndrome diagnosed, and how many people in the United States does it affect?

Table NP21-1 lists the laboratory values used to identify the metabolic syndrome, which is diagnosed when at least three of the following disorders are present: hyperglycemia, obesity, **hypertriglyceridemia** (high blood triglyceride levels), low HDL cholesterol levels, and hypertension (high blood pressure). An estimated 34 percent of adults in the United States may meet the criteria for the metabolic syndrome.[1] As Figure NP21-1 shows, prevalence increases with age. Risk also varies among ethnic groups: Mexican Americans have the highest incidence of the metabolic syndrome in the United States, with an overall prevalence of about 37 percent.[2]

What causes the metabolic syndrome?

Both genetic and environmental factors probably contribute to the development of the metabolic syndrome.[3] However, the close relationship between abdominal obesity and insulin resistance suggests that the current obesity crisis may be partly responsible for its high prevalence. Obesity is thought to induce a number of

TABLE NP21-1 Features of the Metabolic Syndrome

Metabolic syndrome is diagnosed when three or more of the following abnormalities are present.

Measure	Diagnostic Cut Point
Hyperglycemia	Fasting plasma glucose ≥100 mg/dL
Abdominal obesity	Waist circumference >40" in men, >35" in women
Hypertriglyceridemia	≥150 mg/dL
Reduced HDL cholesterol	<40 mg/dL in men, <50 mg/dL in women
Hypertension	≥130/85 mm Hg

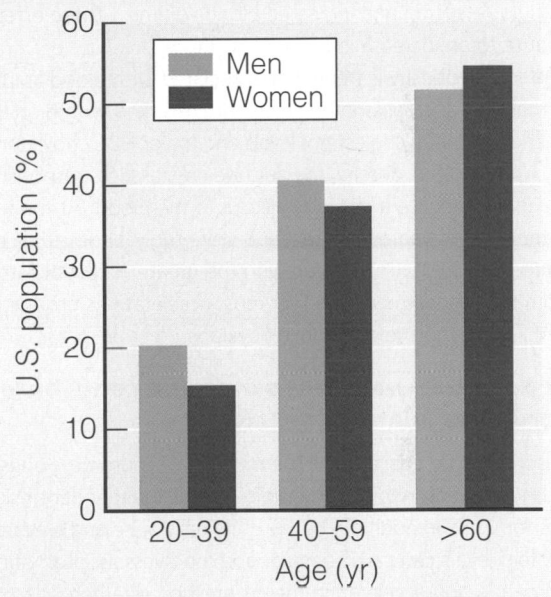

FIGURE NP21-1 Prevalence of the Metabolic Syndrome in the U.S. Population

Source: R. B. Ervin, Prevalence of metabolic syndrome among adults 20 years of age and over, by sex, age, race and ethnicity, and body mass index: United States, 2003–2006, *National Health Statistics Reports;* no. 13 (Hyattsville, MD: National Center for Health Statistics, 2009).

metabolic changes that lead to insulin resistance, which then leads to hyperglycemia and other abnormalities. Note that overweight individuals (BMI from 25 to 29.9) have about a sixfold increased risk of developing metabolic syndrome compared with underweight and normal-weight persons (BMI less than 25).[4]

How can obesity lead to insulin resistance?

Various theories have been proposed to explain the relationship between obesity and insulin resistance. Much of the research suggests that obesity leads to an increase in fatty acid concentrations

GLOSSARY

adiponectin (AH-dih-poe-NECK-tin): a hormone produced by adipose cells that improves insulin sensitivity.

fibrinogen (fye-BRIN-oh-jen): a liver protein that promotes blood clot formation.

hypertriglyceridemia (HYE-per-try-gliss-er-rye-DEE-me-ah): high blood triglyceride levels.

metabolic syndrome: a cluster of interrelated clinical disorders, including obesity, insulin resistance, high blood pressure, and abnormal blood lipids, which together increase risk of diabetes and cardiovascular disease; also known as *insulin resistance syndrome* or *syndrome X.*

nitric oxide: a compound produced by blood vessel cells that helps to regulate blood vessel function, including dilation and constriction.

plasminogen activator inhibitor-1: a protein that promotes blood clotting by inhibiting blood clot degradation within blood vessels.

resistin (re-ZIST-in): a hormone produced by adipose cells that induces insulin resistance.

in the blood, resulting in the abnormal deposition of triglycerides in the muscle, liver, and abdominal region.[5] The abnormally high fat content of these tissues may alter cellular responses to insulin and result in insulin resistance. In addition, adipose cells produce hormones and proteins that influence metabolic processes and fuel use in the body.[6] Obesity is associated with the reduced secretion of **adiponectin,** a hormone involved in sensitivity to insulin. Conversely, the hormone **resistin,** which promotes insulin resistance, is released in greater amounts. The enlarged adipose cells also secrete a variety of cytokines (signaling proteins) that induce inflammation and blood coagulation. The inflammatory process activates multiple metabolic changes that antagonize insulin responsiveness.

Can obesity lead to other problems related to metabolic syndrome?

Obesity increases the risk of developing high blood pressure, a common component of the metabolic syndrome. Both the insulin resistance and hyperinsulinemia that develop in obesity may be implicated in raising blood pressure. Insulin resistance interferes with the normal relaxation and dilation of blood vessels. Hyperinsulinemia promotes reabsorption of sodium in the kidneys, resulting in fluid retention and increased blood volume.[7] These effects contribute to an increase in blood pressure.

Abdominal obesity is frequently associated with blood lipid abnormalities.[8] Increases in body weight are linked with higher triglyceride and LDL cholesterol levels and lower HDL cholesterol levels. As a result of obesity, the adipose cells are less responsive to insulin and release more fatty acids into the blood. At the same time, they are less able to extract and store triglycerides from chylomicrons and VLDL. To keep up with the greater influx of fatty acids from the bloodstream, the liver must accelerate its production of VLDL, and hypertriglyceridemia develops.

How does the metabolic syndrome contribute to cardiovascular disease risk?

The disorders that characterize the metabolic syndrome—obesity, lipid abnormalities, and hypertension—are all independent risk factors for CVD. In addition, both insulin resistance and elevated lipoprotein levels can cause damage to blood vessels, promoting inflammation and accelerating the progression of atherosclerosis.[9] Blood vessel inflammation induces liver secretion of **fibrinogen,** a protein that promotes blood clot formation (increasing risk of a heart attack or stroke). C-reactive protein, which is elevated by both inflammation and obesity, inhibits **nitric oxide** production by blood vessel cells, an effect that impairs blood vessel activity and also promotes blood clotting.[10] Another procoagulant factor— **plasminogen activator inhibitor-1**— is overproduced as a consequence of both obesity and hyperinsulinemia. The combined effect of these multiple abnormalities can worsen atherosclerosis and increase the risks of developing heart attack and stroke. Individuals with metabolic syndrome are also at increased risk of developing diabetes, which is another major risk factor for CVD.

What is the treatment for the metabolic syndrome?

Metabolic syndrome is primarily treated with dietary and lifestyle changes, with the goal of correcting the abnormalities that increase CVD risk.[11] In most individuals, a combination of weight loss and physical activity can improve insulin resistance, blood pressure, and blood lipid levels. Even a moderate weight loss (about 7 percent of body weight) can improve symptoms, although many people find this difficult to achieve. Additional dietary strategies depend on a patient's specific symptoms. If dietary and lifestyle changes are not successful, medications may be prescribed. Because effective treatment requires lifelong commitment, health care providers should work with patients to develop a treatment plan that they are willing to adopt.

What dietary strategies, other than weight loss, are suggested for people with metabolic syndrome?

In individuals with hypertriglyceridemia, the general recommendation is to reduce intake of added sugars and refined grain products (soda, juices, white bread, sweetened cereal, and desserts) and increase servings of whole grains and foods high in fiber (whole-wheat bread, oatmeal, legumes, fruits, and vegetables). In some people, carbohydrate restriction may help to reduce blood triglyceride levels and improve hyperglycemia.[12] Including fish in the diet each week may also improve triglyceride levels. Individuals with hypertension are encouraged to reduce sodium intake and increase consumption of fruits and vegetables and low-fat milk products. A diet low in saturated fat, *trans* fats, and cholesterol can help to reduce LDL cholesterol levels. Chapter 22 describes additional dietary modifications that may reduce CVD risk.

© Rolf Bruderer/Corbis

Regular exercise can reduce the risks of developing the metabolic syndrome, cardiovascular diseases, and type 2 diabetes.

Why is physical activity recommended for people with the metabolic syndrome?

Regular physical activity helps with weight management and may also improve blood lipid concentrations, hypertension, and insulin resistance—all changes that can reduce the risk of developing CVD. A regular exercise program can also prevent or delay the onset of diabetes in persons at risk.[13] A program that includes both aerobic exercise and strength training is best. A minimum of 30 minutes of moderate aerobic activity (brisk walking, jogging, or cycling) daily is suggested, although longer periods (1 hour daily) are recommended for weight control. A sedentary lifestyle can worsen the progression of metabolic syndrome and should be discouraged.

What types of medications are used to treat the metabolic syndrome?

If dietary and lifestyle changes are unsuccessful, medications may be prescribed to correct hypertriglyceridemia and hypertension (Chapter 22 provides details). At present, antidiabetic drugs are not routinely used to treat insulin resistance in patients with the metabolic syndrome due to insufficient evidence that the drugs can improve long-term outcomes better than lifestyle changes.

As explained in this Nutrition in Practice, the metabolic syndrome consists of a cluster of interrelated disorders that increase the risk for developing CVD and type 2 diabetes. Whereas individually the features of the metabolic syndrome are risk factors for CVD, in combination they may raise risk twofold to threefold. Treatment of the metabolic syndrome emphasizes dietary and lifestyle changes.

Notes

1. R. B. Ervin, Prevalence of metabolic syndrome among adults 20 years of age and over, by sex, age, race and ethnicity, and body mass index: United States, 2003–2006, *National Health Statistics Reports,* no. 13 (Hyattsville, MD: National Center for Health Statistics, 2009).

2. Ervin, 2009.

3. Z. T. Bloomgarden, The 6th Annual World Congress on the Insulin Resistance Syndrome, *Diabetes Care* 32 (2009): e104–e111; R. L. Pollex and R. A. Hegele, Genetic determinants of the metabolic syndrome, *Nature Clinical Practice Cardiovascular Medicine* 3 (2006): 482–489.

4. Ervin, 2009.

5. J.-P. Després and I. Lemieux, Abdominal obesity and metabolic syndrome, *Nature* 444 (2006): 881–887.

6. A. Maitra, The endocrine system, in V. Kumar and coeditors, *Robbins and Cotran Pathologic Basis of Disease* (Philadelphia: Saunders, 2010), pp. 1097–1164; M. D. Jensen, Obesity, in L. Goldman and D. Ausiello, eds., *Cecil Medicine* (Philadelphia: Saunders, 2008), pp. 1643–1652.

7. V. Kumar and coauthors, Environmental and nutritional diseases, in V. Kumar and coeditors, *Robbins and Cotran Pathologic Basis of Disease* (Philadelphia: Saunders, 2010), pp. 399–445.

8. Jensen, 2008.

9. R. F. Kushner and J. L. Roth, Nutritional strategies for patients with obesity and the metabolic syndrome, in J. I. Mechanick and E. M. Brett, eds., *Nutritional Strategies for the Diabetic and Prediabetic Patient* (Boca Raton, FL: CRC Press, 2006), pp. 55–80.

10. S. M. Grundy and coauthors, Clinical management of metabolic syndrome: Report of the American Heart Association/National Heart, Lung, and Blood Institute/American Diabetes Association Conference on Scientific Issues Related to Management, *Circulation* 109 (2004): 551–556.

11. F. Magkos and coauthors, Management of the metabolic syndrome and type 2 diabetes through lifestyle modification, *Annual Review of Nutrition* 29 (2009): 223–256; Kushner and Roth, 2006.

12. Kushner and Roth, 2006.

13. Magkos and coauthors, 2009; *Diabetes Mellitus: A Guide to Patient Care* (Philadelphia: Lippincott Williams & Wilkins, 2007), pp. 37–49.

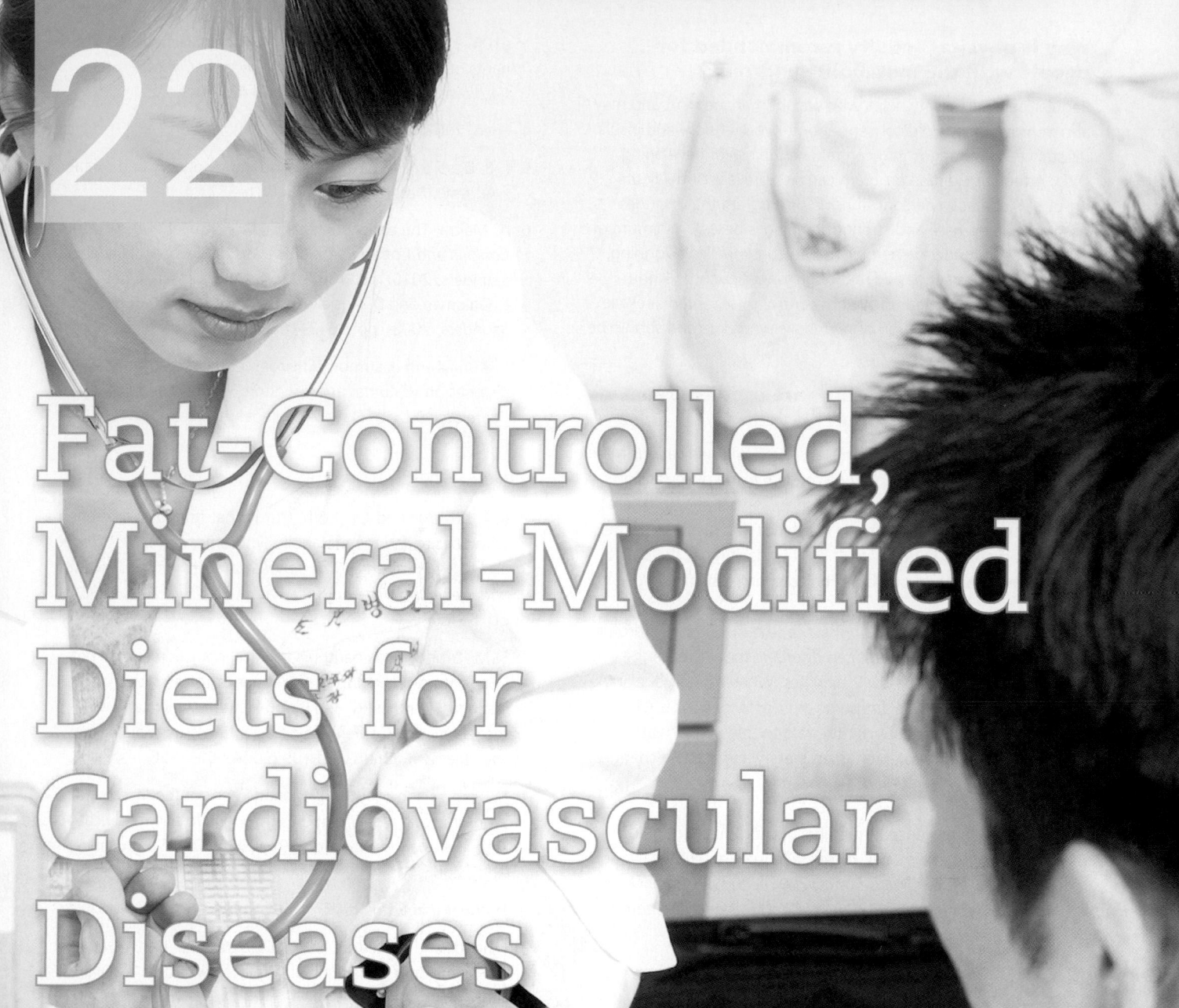

22

Fat-Controlled, Mineral-Modified Diets for Cardiovascular Diseases

Cardiovascular disease (CVD) is a general term describing diseases of the heart and blood vessels. **Coronary heart disease (CHD),** the most common form of CVD, is usually caused by **atherosclerosis** in the coronary arteries that supply blood to the heart muscle. If atherosclerosis restricts blood flow in these arteries, the resulting deprivation of oxygen and nutrients can destroy heart tissue and cause a **myocardial infarction (MI)**—a **heart attack.** When the blood supply to brain tissue is blocked, a **stroke** occurs.

Both heart attack and stroke may result in disablement or death. This chapter describes these and other cardiovascular disorders. Figure 22-1 shows the percentages of deaths resulting from all types of CVD. The accompanying glossary defines common terms related to CVD.

Cardiovascular disease is responsible for nearly 34 percent of deaths in the United States.[1] Although many people assume that heart conditions are men's diseases, more women than men die each year from the various types of CVD. Furthermore, CVD is a global health issue; it is the leading cause of death in Europe and contributes to 30 percent of deaths worldwide.[2]

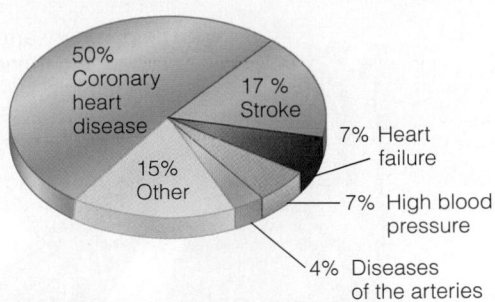

FIGURE 22-1 Percentage Breakdown of Deaths from Cardiovascular Diseases in the United States

Source: V. L. Roger and coauthors, Heart disease and stroke statistics 2011 update: A report from the American Heart Association, *Circulation* 123 (2011): DOI: 10.1161/ CIR.0b013e3182009701.

Atherosclerosis

In atherosclerosis, the artery walls become progressively thickened due to an accumulation of fatty deposits, smooth muscle cells, and fibrous connective tissue, collectively known as **plaque.** Atherosclerosis initially arises in response to minimal but chronic injuries that damage the inner arterial wall. The first lesions tend to develop in regions where the arteries branch or bend because the blood flow is disturbed in those areas (see Figure 22-2). Damage to the artery elicits an inflammatory response, attracting immune cells and increasing the permeability of artery walls. Low-density lipoproteins (LDL) slip under the thin layer of blood vessel cells, become oxidized by local enzymes, and accumulate. Arterial macrophages engulf this altered LDL and become **foam cells;**

C-reactive protein, an acute-phase protein secreted during the inflammatory response, is associated with increased heart disease risk (see Chapter 24).

Macrophages are immune cells that engulf pathogens and cellular debris; they are derived from white blood cells called monocytes.

GLOSSARY OF TERMS RELATED TO CARDIOVASCULAR DISEASES

aneurysm (AN-you-rih-zum): an abnormal enlargement or bulging of a blood vessel (usually an artery) caused by weakness in the blood vessel wall.

angina (an-JYE-nah or AN-ji-nah) **pectoris:** a condition caused by ischemia in the heart muscle that results in discomfort or dull pain in the chest region. The pain often radiates to the left shoulder and arm or to the back, neck, and lower jaw.

atherosclerosis (ATH-er-oh-scler-OH-sis): a type of artery disease characterized by accumulations of fatty material on the inner walls of arteries.

cardiovascular disease (CVD): a general term describing diseases of the heart and blood vessels.

▌ *cardio* = heart
▌ *vascular* = blood vessels

coronary heart disease (CHD): a chronic, progressive disease characterized by obstructed blood flow in the coronary arteries; also called *coronary artery disease.*

embolism (EM-boh-lizm): the obstruction of a blood vessel by an embolus, causing sudden tissue death.

▌ *embol* = to insert, plug

embolus (EM-boh-lus): an abnormal particle, such as a blood clot or air bubble, that travels in the blood.

fatty streaks: initial lesions of atherosclerosis that form on the artery wall, characterized by accumulations of foam cells, lipid material, and connective tissue.

foam cells: swollen cells in the artery wall that accumulate lipids.

intermittent claudication (claw-dih-KAY-shun): severe pain and weakness in the legs (especially the calves) caused by inadequate blood supply to the muscles; it usually occurs with walking and subsides during rest.

ischemia (iss-KEE-mee-a): inadequate blood supply within tissues due to obstructed blood flow in the arteries.

myocardial (MY-oh-CAR-dee-al) **infarction** (in-FARK-shun), or **MI:** death of heart muscle caused by a sudden reduction in coronary blood flow; also called a **heart attack** or *cardiac arrest.*

▌ *myo* = muscle
▌ *cardial* = heart
▌ *infarct* = tissue death

plaque (PLACK): an accumulation of fatty deposits, smooth muscle cells, and fibrous connective tissue in blood vessels.

stroke: a sudden injury to brain tissue resulting from impaired blood flow through an artery that supplies blood to the brain; also called a *cerebrovascular accident.*

▌ *cerebro* = brain

thrombosis (throm-BOH-sis): the formation or presence of a blood clot in blood vessels. A *coronary thrombosis* occurs in a coronary artery, and a *cerebral thrombosis* occurs in an artery that supplies blood to the brain.

▌ *thrombo* = clot

thrombus: a blood clot formed within a blood vessel that remains attached to its place of origin.

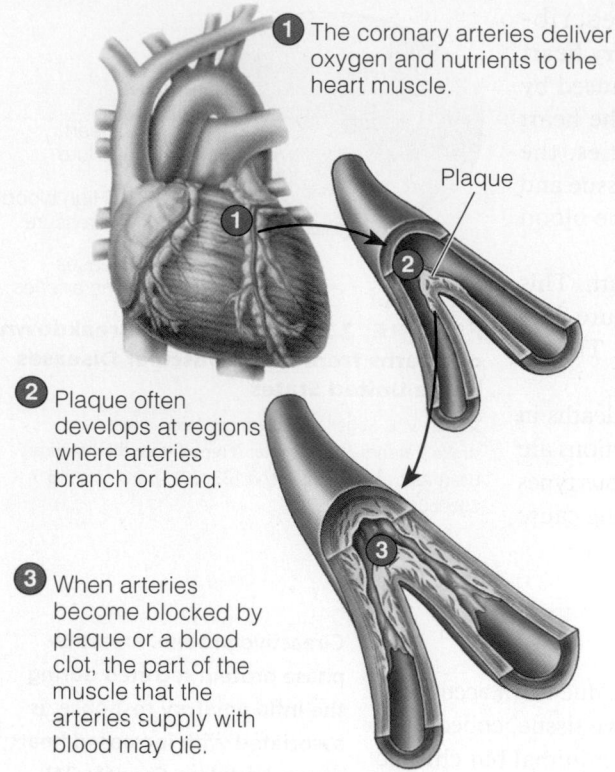

1 The coronary arteries deliver oxygen and nutrients to the heart muscle.

Plaque

2 Plaque often develops at regions where arteries branch or bend.

3 When arteries become blocked by plaque or a blood clot, the part of the muscle that the arteries supply with blood may die.

FIGURE 22-2 **Plaque Formation in Atherosclerosis**

these fat-laden cells are visible as fatty deposits along artery walls, known as **fatty streaks** (see Figure 22-3). Eventually, the plaque thickens and hardens as additional lipids, connective tissue, and calcium accumulate in the region. Atherosclerosis begins to develop as early as childhood or adolescence and typically progresses over several decades before symptoms develop.

Consequences of Atherosclerosis

As atherosclerosis worsens, it can eventually narrow the lumen of an artery and interfere with blood flow. Some types of plaque are highly susceptible to rupture, which promotes blood clotting within the artery **(thrombosis).** A blood clot **(thrombus)** may enlarge in time and ultimately obstruct blood flow. A portion of a clot can also break free **(embolus)** and travel through the circulatory system until it lodges in a narrowed artery and shuts off blood flow to the surrounding tissue **(embolism).** Most complications of atherosclerosis result from the deficiency of blood and oxygen within the tissue served by an artery **(ischemia).**

Atherosclerosis can affect almost any organ or tissue in the body and, accordingly, is a major cause of disablement or death. Obstructed blood flow in the coronary arteries can cause pain or discomfort in the chest and surrounding regions **(angina pectoris)** or lead to a heart attack. As mentioned earlier, obstructed blood flow to the brain can cause injury to brain tissue, or a stroke. Impaired blood circulation in the legs can cause fatigue and pain while walking, known as **intermittent claudication.** Blockage of the arteries that supply the kidneys can result in kidney disease or even acute kidney failure.

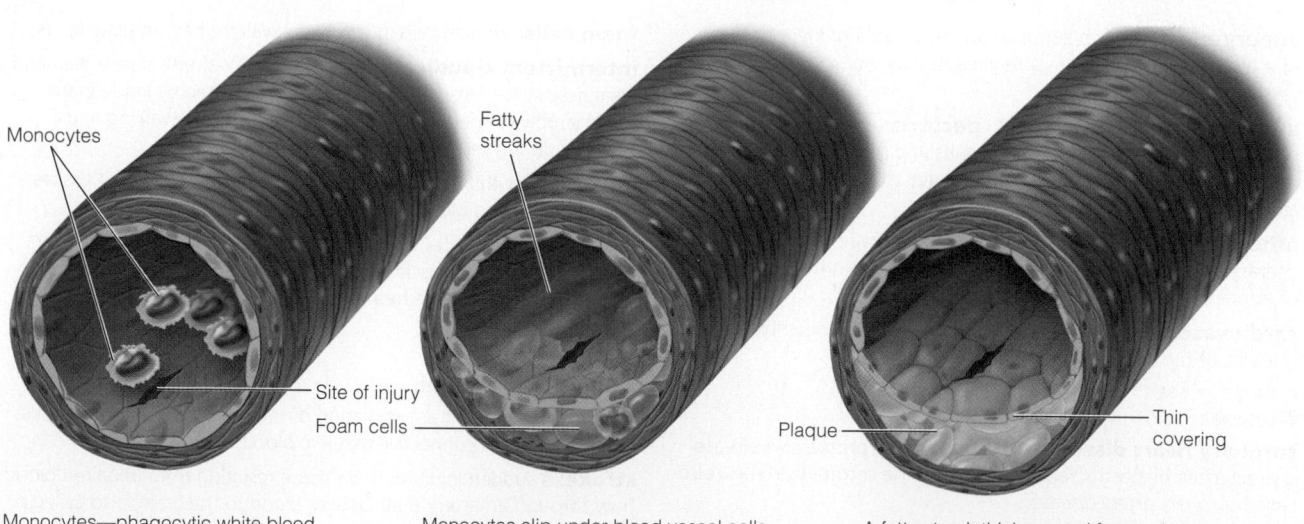

Monocytes

Fatty streaks

Site of injury

Foam cells

Plaque

Thin covering

Monocytes—phagocytic white blood cells—circulate in the bloodstream and respond to injury on the artery wall.

Monocytes slip under blood vessel cells and engulf LDL cholesterol, becoming foam cells. The thin layers of foam cells that develop on artery walls are known as *fatty streaks.*

A fatty streak thickens and forms plaque as it accumulates additional lipids, smooth muscle cells, connective tissue, and cellular debris.

The artery may expand to accommodate plaque. When this occurs, the plaque that develops often contains a large lipid core with a thin fibrous covering and is vulnerable to rupture and thrombosis.

FIGURE 22-3 **Stages of Plaque Progression**

Atherosclerosis is the most common cause of an **aneurysm**—the abnormal dilation of a blood vessel. This occurs because plaque can weaken the blood vessel wall, allowing the damaged region to stretch and balloon outward. Aneurysms can rupture and lead to massive bleeding and death, particularly when a large vessel such as the aorta is affected. In the arteries of the brain, an aneurysm may lead to bleeding within the brain, coma, or a stroke.

Causes of Atherosclerosis

The factors that initiate atherosclerosis either cause direct damage to the artery wall or allow lipid materials to penetrate its surface. Factors that generally worsen atherosclerosis are those that lead to plaque rupture or blood coagulation. The development of advanced atherosclerosis is a long-term process that involves recurrent plaque rupture, thrombosis, and healing at sites in the artery wall.[3]

Shear Stress/Hypertension The stress of blood flow along the artery walls—called *shear stress*—can cause physical damage to arteries.[4] Hypertension (high blood pressure) intensifies the stress of blood flow on arterial tissue.

Elevated LDL and VLDL High levels of LDL and very-low-density lipoproteins (VLDL) can both initiate and worsen atherosclerosis. When levels are high, LDL and VLDL are actively taken up and retained in susceptible regions in the artery wall.[5] Because high-density lipoproteins (HDL) help to prevent oxidation and also remove cholesterol from circulation, low levels of HDL contribute to the development of atherosclerosis as well.

Elevated concentrations of a variant form of LDL called *lipoprotein(a)* have been found to speed the progression of atherosclerosis and to raise the risk of various types of CVD.[6] Lipoprotein(a) levels are primarily genetically determined and are influenced to only a minor degree by age and environmental factors.

Cigarette Smoking Chemicals in cigarette smoke are toxic to blood vessel cells, and the resulting damage can initiate or contribute to arterial injury. Other effects of smoking include chronic inflammation, enhanced blood coagulation, increased LDL cholesterol, and decreased HDL cholesterol—all effects that can promote the progression of atherosclerosis.[7]

Diabetes Mellitus Diabetes can both initiate and accelerate the development of atherosclerosis. Chronic hyperglycemia leads to the accumulation of advanced glycation end products (AGEs), which can directly damage blood vessel cells and disturb blood vessel function.[8] AGEs also promote inflammation and oxidative stress, which contribute to plaque's progression.

Age and Gender As a person ages, arterial cells tend to degenerate and risk factors for CVD accumulate. Risk of atherosclerosis increases substantially in men aged 45 or older and women aged 55 or older. After menopause, women's risk increases because the reduction in estrogen levels increases the tendency for plaque formation. Levels of the amino acid homocysteine, which may damage artery walls and promote blood clotting, rise with age and are generally higher in men; however, researchers have not determined whether homocysteine is a cause or an effect of the disease process.[9]

IN SUMMARY

- Atherosclerosis, characterized by the plaque buildup in artery walls, can lead to complications such as angina pectoris, heart attack, stroke, intermittent claudication, kidney disease, and aneurysms.

- Leading causes of plaque formation and progression include shear stress, hypertension, elevated LDL and VLDL, cigarette smoking, diabetes, and aging.

Shear stress is a stress that occurs sideways against a surface rather than perpendicular to a surface.

Reminder: LDL transport cholesterol in the blood, whereas VLDL transport triglycerides. In clinical practice, VLDL levels are commonly referred to as *blood triglycerides*.

Reminder: *Advanced glycation end products* are reactive compounds formed after glucose nonenzymatically attaches to proteins (see Chapter 21).

Estrogen replacement therapy after menopause has mixed effects on heart disease risk; it can improve blood vessel function, lower LDL, and raise HDL, but it also promotes blood clotting.

Reminder: Blood homocysteine levels are influenced by intakes of folate, vitamin B$_{12}$, and vitamin B$_6$ (see Chapter 8).

TABLE 22-1 Risk Factors for CHD

Major Risk Factors for CHD (not modifiable)

- Increasing age
- Male gender
- Family history of heart disease

Major Risk Factors for CHD (modifiable)

- High LDL cholesterol
- High blood triglyceride (VLDL) levels
- Low HDL cholesterol
- Hypertension (high blood pressure)
- Diabetes
- Obesity (especially abdominal obesity)
- Physical inactivity
- Cigarette smoking
- Alcohol overconsumption (≥3 drinks per day)
- An atherogenic diet (includes high saturated fat, cholesterol, and trans fat intakes; low fruit and vegetable intakes)

Note: Risk factors highlighted in yellow have relationships with diet.

Sources: V. L. Roger and coauthors, Heart disease and stroke statistics 2011 update: A report from the American Heart Association, *Circulation* 123 (2011): DOI: 10.1161/CIR.0b013e3182009701; M. H. Criqui, Epidemiology of cardiovascular disease, in L. Goldman and D. Ausiello, eds., *Cecil Medicine* (Philadelphia: Saunders, 2008), pp. 301–305.

Abdominal obesity is suggested by a waist circumference of >40 inches for men and >35 inches for women.

Coronary Heart Disease (CHD)

Coronary heart disease (CHD), also called *coronary artery disease*, is the most common type of cardiovascular disease and the leading cause of death in the United States.[10] As described earlier, CHD is usually caused by atherosclerosis, which reduces blood flow in the coronary arteries and deprives the heart muscle of oxygen and nutrients. The most common symptom of CHD is pain or discomfort in the chest region; the pain may radiate to the shoulders, left arm, back, neck, jaw, or teeth. Other possible symptoms include shortness of breath, weakness, lightheadedness, sweating, nausea, and vomiting.

Evaluating Risk for Coronary Heart Disease

Because CHD develops over many years, prevention should begin well before symptoms appear. Population studies have suggested that about 90 percent of people with CHD have at least one of the four classic risk factors: cigarette smoking, high LDL cholesterol, hypertension, or diabetes.[11] These and other major risk factors for CHD are listed in Table 22-1; most of the risk factors listed can be modified by changes in diet and lifestyle.

CHD Risk Assessment Risk assessment requires several key laboratory measures (see Table 22-2) and a thorough medical history. A complete lipoprotein profile, which includes measures of total cholesterol, LDL and HDL cholesterol, and blood triglycerides (VLDLs), should be obtained every 5 years starting at 20 years of age. Sometimes the ratio of total cholesterol to HDL cholesterol is used to predict CHD risk: a high total cholesterol value suggests elevated LDL cholesterol levels, and a low HDL value is often linked with other lipid abnormalities. Overweight and obesity predispose to CHD, particularly when abdominal obesity is present. Hypertension is a major risk factor; for people over 50 years of age, a high systolic blood pressure is more predictive of CHD risk than is high diastolic blood pressure. Finally, cigarette smoking and the presence of diabetes strongly contribute to CHD risk. The "How to" on p. 573 presents a screening method for assessing a person's 10-year risk of developing CHD that includes some of these risk factors.

Blood Cholesterol Levels and CHD Risk Once a person's level of risk has been identified, much of the treatment focuses on lowering LDL cholesterol. Elevated LDL levels are directly related to the development of atherosclerosis, and clinical studies have confirmed that LDL-lowering treatments can successfully reduce CHD mortality rates. CHD is seldom seen in populations that maintain desirable LDL levels.

TABLE 22-2 Laboratory Measures for CHD Risk Assessment

Clinical Measures	Desirable	Borderline Risk	High Risk
Total blood cholesterol (mg/dL)	<200	200–239	≥240
LDL cholesterol (mg/dL)	<100[a]	130–159	160–189[b]
HDL cholesterol (mg/dL)	≥60	40–59	<40
Triglycerides, fasting (mg/dL)	<150	150–199	200–499[c]
Body mass index (BMI)[d]	18.5–24.9	25–29.9	≥30
Blood pressure (systolic and diastolic pressure)	<120/<80	120–139/80–89[e]	≥140/≥90[f]

[a]LDL levels of 100–129 mg/dL indicate a near or above optimal level; <70 mg/dL is a desirable goal for very high-risk persons.
[b]LDL levels ≥190 mg/dL indicate a very high risk.
[c]Triglyceride levels ≥500 mg/dL indicate a very high risk.
[d]Body mass index (BMI) was defined in Chapter 6; BMI standards are found on the inside back cover.
[e]These values indicate prehypertension.
[f]These values indicate stage one hypertension; ≥160/≥100 indicates stage two hypertension. Physicians use these classifications to determine medical treatment.

This assessment estimates a person's 10-year risk for experiencing a major coronary event associated with CHD, such as a heart attack.[a] A high score does not mean that the person *will* have a heart attack, but it warns of the possibility and suggests the need to consult a physician. To use this tool, you need to know a person's age, total and HDL cholesterol levels, and blood pressure.

Age (years):

	Men	Women
20–34	−9	−7
35–39	−4	−3
40–44	0	0
45–49	3	3
50–54	6	6
55–59	8	8
60–64	10	10
65–69	11	12
70–74	12	14
75–79	13	16

HDL (mg/dL):

	Men	Women
≥60	−1	−1
50–59	0	0
40–49	1	1
<40	2	2

Systolic Blood Pressure (mm Hg):

	Untreated		Treated	
	Men	Women	Men	Women
<120	0	0	0	0
120–129	0	1	1	3
130–139	1	2	2	4
140–159	1	3	2	5
≥160	2	4	3	6

Total Cholesterol (mg/dL):

	Age 20–39		Age 40–49		Age 50–59		Age 60–69		Age 70–79	
	Men	Women	Men	Women	Men	Women	Men	Women	Men	Women
<160	0	0	0	0	0	0	0	0	0	0
160–199	4	4	3	3	2	2	1	1	0	1
200–239	7	8	5	6	3	4	1	2	0	1
240–279	9	11	6	8	4	5	2	3	1	2
≥280	11	13	8	10	5	7	3	4	1	2

Smoking (any cigarette smoking in the past month):

	Men	Women	Men	Women	Men	Women	Men	Women	Men	Women
Smoker	8	9	5	7	3	4	1	2	1	1
Nonsmoker	0	0	0	0	0	0	0	0	0	0

Scoring Heart Disease Risk

Add up the total points: _____. Using the table at the right, find the total in the first column for the appropriate gender, and then check the second column to learn the percentage risk of developing severe CHD within the next 10 years. A person's risk for an acute coronary event (such as a heart attack) may be identified as low (<10%), moderate (10–20%), or high (>20%). Treatment strategies vary according to a person's risk category.

Men		Women	
Total	Risk (%)	Total	Risk (%)
<0	<1	<9	<1
0–4	1	9–12	1
5–6	2	13–14	2
7	3	15	3
8	4	16	4
9	5	17	5
10	6	18	6
11	8	19	8
12	10	20	11
13	12	21	14
14	16	22	17
15	20	23	22
16	25	24	27
≥17	≥30	≥25	≥30

[a]A link to the electronic version of this assessment is available on the ATP III page of the National Heart, Lung, and Blood Institute's website (www.nhlbi.nih.gov/guidelines/cholesterol).

Source: Adapted from Expert Panel on Detection, Evaluation, and Treatment of High Blood Cholesterol in Adults (Adult Treatment Panel III), *Third Report of the National Cholesterol Education Program (NCEP),* NIH publication no. 02-5216 (Bethesda, MD: National Heart, Lung, and Blood Institute, 2002), section III.

As mentioned earlier, HDL help to protect against atherosclerosis, and low HDL levels often coexist with other lipid abnormalities (such as high VLDL); thus, a low HDL level is highly predictive of CHD risk. In addition, some risk factors for CHD—obesity, smoking, inactivity, and male gender—are associated with reduced HDL levels. It is not known whether raising HDL will help to reduce CHD risk, but weight loss, smoking cessation, and regular physical activity can all independently help to lower risk.

Therapeutic Lifestyle Changes for Lowering CHD Risk

People who have CHD or multiple risk factors for CHD are often advised to make dietary and lifestyle changes before considering drug treatment. An approach to risk reduction promoted by the National Cholesterol Education Program, known as Therapeutic Lifestyle Changes (TLC), is summarized in Table 22-3.[12] The main features of the TLC plan include a cholesterol-lowering diet, regular physical activity, and weight reduction. If the recommendations are followed carefully, substantial progress may be seen after six weeks. People with a high risk of CHD should try to lower LDL cholesterol with at least a three-month trial of TLC before starting drug therapy.

Saturated Fat Of the dietary lipids, saturated fat has the strongest effect on blood cholesterol levels, and replacing saturated fat with monounsaturated and polyunsaturated fats can generally lower LDL levels. The TLC recommendation is to consume less than 7 percent of total kcalories as saturated fat. The average saturated fat intake in the United States is about 11 percent of total kcalories consumed.[13]

For most people, cutting down on saturated fat involves more than just switching from butter to vegetable oil, as the main sources of saturated fat in the United States are whole-milk products, high-fat meats, and baked goods. Choosing lean meats or fish, using fat-free or low-fat milk products, and avoiding certain types of bakery products are usually more effective ways of reducing saturated fat.

Replacing saturated fat with carbohydrate can also lower LDL cholesterol, but such a change may lower HDL cholesterol as well and also raise blood triglyceride (VLDL)

The National Cholesterol Education Program was developed by a division of the National Institutes of Health.

For updated guidelines related to CHD risk reduction, access this website: www.nhlbi.nih.gov/guidelines

TABLE 22-3 Reducing Risk of CHD with Therapeutic Lifestyle Changes

Dietary Strategies

- Limit saturated fat to less than 7 percent of total kcalories and cholesterol to less than 200 milligrams per day. Maintaining a fat intake that is 25 to 35 percent of total kcalories may help with this goal.
- Replace saturated fats with unsaturated fats from fish, vegetable oils, and nuts or with carbohydrates from whole grains, legumes, fruits, and vegetables.
- Avoid food products that contain *trans*-fatty acids. The *trans* fat content in packaged foods is shown on the Nutrition Facts panel.
- Choose foods high in soluble fibers, including oats, barley, legumes, and fruit. Food supplements that contain psyllium seed husks can be used to help lower LDL cholesterol levels.
- Regularly consume food products that contain added plant sterols or stanols.
- To reduce blood pressure, limit sodium intake to 2400 milligrams per day,[a] and choose a diet that is high in fruits, vegetables, and whole grains and includes low-fat milk products and nuts.
- Fish can be consumed regularly as part of a CHD risk-reduction diet.
- If alcohol is consumed, it should be limited to one drink daily for women and two drinks daily for men.

Lifestyle Choices

- Physical activity: At least 30 minutes of moderate-intensity endurance activity should be undertaken on most days of the week. The eventual goal should be an expenditure of at least 2000 kcalories weekly.
- Smoking cessation: Exposure to any form of tobacco smoke should be minimized.

Weight Reduction

- Weight reduction may improve other CHD risk factors. The general goals of a weight-management program should be to prevent weight gain, reduce body weight, and maintain a lower body weight over the long term. The initial goal of a weight-loss program should be to lose no more than 10 percent of original body weight.

[a]According to DRI recommendations, sodium intake should be limited to 2300 milligrams daily.

levels. The effect on triglyceride levels can be minimized by limiting added sugars and including fiber-rich foods; ideally, the diet should include generous amounts of whole grains, legumes, fruits, and vegetables. The TLC plan recommends a carbohydrate intake in the range of 50 to 60 percent of total kcalories.

Total Fat For people whose total fat intake includes substantial saturated fat, limiting total fat may indirectly reduce saturated fat. Therefore, the TLC recommendation for total fat is 25 to 35 percent of kcalories. Individuals with elevated blood triglycerides may benefit from a fat intake at the upper end of this range (30 to 35 percent) so that their carbohydrate intakes are not excessive. Fat intakes higher than 35 percent of kcalories are discouraged because they may promote weight gain in some people.

Trans Fats *Trans*-fatty acids can raise LDL cholesterol levels, and when they replace saturated fats in the diet (as when stick margarine replaces butter), they can also reduce HDL cholesterol. The TLC recommendation is to keep *trans* fats as low as possible.

Most sources of *trans* fats are products made with partially hydrogenated oils; examples include baked goods such as cookies, crackers, and doughnuts; snack foods such as potato chips and corn chips; and fried foods such as French fries and fried chicken. In the past few years, food manufacturers have reformulated many food products so that they contain little or no *trans* fat. In some cases, unfortunately, the *trans* fats have been replaced with saturated fat sources, so consumers should read labels carefully to avoid both types of cholesterol-raising fats.

Dietary Cholesterol A high cholesterol intake can raise LDL levels, and reducing dietary cholesterol lowers LDL cholesterol in most people. The TLC recommendation is a cholesterol intake of less than 200 milligrams per day. Currently, the cholesterol intakes of women and men in the United States average about 230 and 362 milligrams per day, respectively.[14] Eggs contribute about one-quarter of the cholesterol in the U.S. diet, followed by chicken, beef, and cheese.[15]

Soluble Fibers Soluble, viscous fibers can reduce LDL cholesterol levels by inhibiting cholesterol and bile absorption in the small intestine and reducing cholesterol synthesis in the liver. Dietary sources of soluble fibers include oats, barley, legumes, and fruits. The soluble fiber in psyllium seed husks, frequently used to treat constipation, is effective for lowering cholesterol levels when used as a dietary supplement.

Plant Sterols and Stanols Foods or supplements that contain significant amounts of plant sterols or plant stanols can help to lower LDL cholesterol levels. Plant sterols or stanols are added to various food products, such as margarines and orange juice, or supplied in dietary supplements. These plant compounds work by interfering with cholesterol and bile absorption. About 2 grams of plant sterols daily (provided by 2 to 2½ tablespoons of sterol-enriched margarines) can lower LDL cholesterol by up to 15 percent without lowering HDL cholesterol.[16]

Sodium and Potassium Intakes Excessive dietary sodium may raise blood pressure, whereas potassium can help to lower blood pressure. A low-sodium diet that contains generous amounts of fruits and vegetables, low-fat milk products, nuts, and whole grains has been found to substantially reduce blood pressure, largely due to the diet's content of potassium and several other minerals that have blood pressure–lowering effects. This diet (the *DASH Eating Plan*) and other factors that influence blood pressure are discussed in a later section (see pp. 582–585).

Fish and Omega-3 Fatty Acids The omega-3 fatty acids in fatty fish, known as EPA and DHA, may benefit people who have had a heart attack by lowering blood triglyceride levels, reducing blood clotting, and stabilizing heart rhythm. In addition,

Regular aerobic exercise can strengthen the cardiovascular system, promote weight loss, reduce blood pressure, and improve blood glucose and lipid levels.

Cigar and pipe smoking can also increase the risk of CHD, but the risk may not be quite as great because the smoke is less likely to be inhaled.

fish typically replaces entrées that contain animal fat (a source of saturated fat). The American Heart Association recommends consuming two servings of fish per week, with an emphasis on fatty fish.[17] Chapter 3 provides additional information about omega-3 fatty acids, as well as a discussion about the use of fish oil supplements.

Alcohol Moderate consumption of alcohol—from beer, wine, or liquor—has favorable effects on atherosclerosis, blood-clotting activity, HDL cholesterol levels, inflammation, and insulin resistance.[18] These benefits are most apparent in men and women who are at least 45 and 55 years old, respectively. Of note, only low or moderate amounts of alcohol—no more than one drink daily for women and two for men—have been found to lower CHD risk, and higher intakes are associated with higher mortality rates. For some people, alcohol's negative effects can offset its health advantages (see Nutrition in Practice 20).

Physical Activity Regular aerobic activity reverses a number of risk factors for CHD: it can lower blood triglycerides, raise HDL levels, lower blood pressure, promote weight loss, improve insulin sensitivity, strengthen heart muscle, and increase coronary artery size and tone. The American Heart Association recommends that all adults participate in at least 30 minutes of moderate-intensity physical activity on most days of the week.[19] Activities that use large muscle groups have the greatest benefits; such activities include brisk walking, running, swimming, cycling, stair-stepping, and cross-country skiing. If preferred, physical activity can be divided into several sessions during the day.

Smoking Cessation In addition to promoting atherosclerosis, cigarette smoking decreases the oxygen-carrying capacity of the blood, raises the heart rate, inhibits vasodilation, reduces exercise tolerance, and promotes blood clotting, among other effects.[20] Quitting smoking improves CHD risk quickly; the incidence of CHD drops to levels near those of nonsmokers in just two years.[21]

Weight Reduction In persons who are obese, weight reduction can improve such CHD risk factors as insulin resistance, hypertension, and blood lipid abnormalities. However, individuals should focus on weight reduction only *after* they have adopted other dietary measures to lower LDL.[22] This approach ensures that LDL reduction is given priority and that the individual does not receive a multitude of dietary suggestions at one time. For some, avoiding additional weight gain may be a desirable starting point.

Successful Adherence to Lifestyle Changes Adopting multiple lifestyle changes at once can be challenging. Health practitioners can help to motivate patients by explaining the reasons for each change, setting obtainable goals, and providing practical suggestions. In some individuals, high LDL cholesterol levels may persist despite adherence to a TLC program; drug therapy may be the only effective treatment. Review Table 22-3 on p. 574 for a summary of the suggestions discussed in this section. The "How to" offers suggestions for implementing a heart-healthy diet.

Lifestyle Changes for Hypertriglyceridemia

Hypertriglyceridemia (elevated blood triglycerides) affects nearly one-third of adults in the United States.[23] It is common in people with diabetes and the metabolic syndrome and may also result from other disorders. Elevated blood triglyceride levels may coexist with elevated LDL cholesterol or occur separately. Whereas mild or moderate hypertriglyceridemia is often associated with increased risk of CHD, severe hypertriglyceridemia can cause additional complications, including fatty deposits in the skin and soft tissues and acute pancreatitis.[24]

Blood triglycerides:
- Borderline high: 150–199 mg/dL
- High: 200–499 mg/dL
- Very high: ≥500 mg/dL

For some people, following a heart-healthy diet may require significant changes in food choices. It is often easier to adopt a new diet if only a few changes are made at a time. Discussing positive choices (what to eat) first, rather than negative ones (what not to eat), may improve compliance. These suggestions can help patients implement their diet:

Breads, Cereals, and Pasta

▪ Choose whole-grain breads and cereals. Make sure the first ingredient on bread and cereal labels is "whole wheat" rather than "enriched wheat flour." Consume oats and barley regularly, as they are good sources of soluble fibers.

▪ Bakery products and snack foods often contain *trans*-fatty acids. Choose products whose labels list 0 grams of *trans* fat on the Nutrition Facts panel; the ingredient lists should not include any "partially hydrogenated vegetable oil," the main source of *trans* fatty acids.

▪ Avoid products that contain tropical oils (coconut, palm, or palm kernel oil), which are high in saturated fat.

Fruits and Vegetables

▪ Incorporate at least one or two servings of fruits and vegetables into each meal. Keep the refrigerator stocked with a variety of colorful fruits and vegetables (baby carrots, blueberries, grapes) that can be eaten when the urge to nibble arises.

▪ Check food labels on canned products carefully. Canned vegetables (especially tomato-based products) are often high in sodium. Fruits that are canned in juice are higher in nutrient density than those canned in syrup.

▪ Avoid French fries from fast-food restaurants, which are often prepared with *trans* fats. Restrict high-sodium foods such as pickles, olives, sauerkraut, and kimchee.

Lunch and Dinner Entrées

▪ Limit meat, fish, and poultry intake to 5 ounces per day. Plan to eat fish twice a week, preferably fatty fish such as salmon, tuna, and mackerel.

▪ Select lean cuts of beef, such as sirloin tip and round steak; lean cuts of pork, such as loin chops and tenderloin; and skinless poultry pieces. Trim visible fat before cooking.

▪ Select extra-lean ground meat and drain well after cooking. Use lean ground turkey, without skin added, in place of ground beef.

▪ Prepare pasta and vegetable stir-fry dishes several times weekly to help reduce meat intake and increase vegetable intake. Use soybean products and other legumes as sources of protein.

▪ Limit cholesterol-rich organ meats (liver, brain, sweetbreads) and shrimp. Limit intake of whole eggs to two per week, as the yolks are high in cholesterol (about 210 milligrams per yolk). Replace whole eggs in recipes with egg whites or commercial egg substitutes or similar reduced-cholesterol products.

▪ Restrict these high-sodium foods: cured or smoked meats such as beef jerky, bologna, corned beef, frankfurters, ham, luncheon meats, salt pork, and sausage; salty or smoked fish such as anchovies, caviar, salted or dried cod, herring, and smoked salmon; and canned, frozen, or packaged soups, sauces, and entrées.

Milk Products

▪ Select fat-free or low-fat milk products only. Use yogurt or fat-free sour cream to make dips or salad dressings. Substitute evaporated fat-free milk for heavy cream.

▪ Restrict foods high in saturated fat or sodium, such as cheese, processed cheeses, and ice cream or other milk-based desserts.

Fats and Oils

▪ Prepare salad dressings and other foods with vegetable oils rich in omega-3 fatty acids, such as canola, soybean, flaxseed, and walnut oils. Select other unsaturated vegetable oils, such as corn, olive, peanut, safflower, sesame, and sunflower oils, instead of saturated fat sources such as butter and lard.

▪ Select margarines that indicate 0 grams of *trans* fat on the Nutrition Facts panel; these products should contain little or no "partially hydrogenated vegetable oil." Tub margarines are less likely to contain *trans* fat than stick margarines. To help lower LDL cholesterol levels, use margarines with added plant sterols or stanols.

▪ Add unsalted nuts or avocados to meals to make them more appetizing; these foods are good sources of unsaturated fats.

▪ Avoid tropical oils (coconut, palm, or palm kernel oil), which are high in saturated fat.

Spices and Seasonings

▪ Use salt only at the end of cooking, and you will need to add much less. Use salt substitutes at the table.

▪ Spices and herbs can improve food flavor without adding sodium. Try using more garlic, ginger, basil, curry or chili powder, cumin, pepper, lemon, mint, oregano, rosemary, and thyme.

▪ Check the sodium content on food labels. Flavorings and sauces that are usually high in sodium include bouillon cubes, soy sauce, steak and barbecue sauces, relishes, mustard, and catsup.

Snacks and Desserts

▪ Select low-sodium and low–saturated fat snacks such as unsalted pretzels and nuts, plain popcorn, and unsalted chips and crackers. Check labels on snack foods and desserts to ensure that they do not include *trans* fats.

▪ Enjoy angel food cake, which is made without egg yolks and added fat. Select low-fat frozen desserts such as sherbet, sorbet, fruit bars, and some low-fat ice creams.

▪ Choose canned or dried fruits and crunchy raw vegetables to boost fruit and vegetable intake.

Nutrition Therapy for Hypertriglyceridemia Dietary and lifestyle changes can improve most cases of hypertriglyceridemia.[25] Overweight and obesity, a sedentary lifestyle, and cigarette smoking all can raise triglyceride levels. Dietary factors that increase triglyceride levels include high intakes of alcohol and refined carbohydrates; sucrose and fructose are the carbohydrates with the strongest effect. Thus, controlling body weight, being physically active, quitting smoking, restricting alcohol, and limiting intakes of refined carbohydrates (especially foods made with white flour or added sugars) are basic treatments for hypertriglyceridemia. As mentioned earlier, high triglyceride levels are often associated with low HDL, and the lifestyle changes listed here are likely to improve HDL levels as well.

Severe Hypertriglyceridemia In addition to dietary and lifestyle changes, medications are usually necessary for lowering blood triglyceride levels that exceed 500 milligrams per deciliter. If blood triglycerides exceed 1000 milligrams per deciliter, a very low-fat diet, providing less than 15 percent of kcalories from fat, may be required.[26] Patients must also eliminate consumption of alcoholic beverages.

Fish Oil Supplements and Hypertriglyceridemia Fish oil supplements are sometimes recommended for treating hypertriglyceridemia. Clinical trials suggest that a daily intake of 2 to 4 grams of EPA and DHA (combined) can lower triglyceride levels by 30 to 50 percent.[27] Fish oil therapy should be monitored by a physician due to the potential for adverse effects.

Vitamin Supplementation and CHD Risk

People often ask about the possible benefits of using certain types of vitamin supplements for reducing CHD risk, particularly B vitamin and antioxidant supplements. Most clinical trials have not been able to confirm any benefits from using these supplements, as described below:

- *B vitamin supplements.* As mentioned earlier, elevated blood homocysteine is a risk factor for CHD, but whether homocysteine itself is directly damaging or is simply an indicator of other abnormalities remains unknown. Although increased intakes of folate, vitamin B_6, and vitamin B_{12} can lower homocysteine levels, clinical trials have not demonstrated that supplementation with these vitamins can reduce the incidence of heart attacks in those at risk.[28] Hence, B vitamin supplements are not currently recommended for patients at risk for CHD.

- *Antioxidant vitamin supplements.* Because oxidized LDL promote atherosclerosis, researchers have hypothesized that antioxidant supplements may help to prevent atherosclerosis and reduce CHD risk. However, most studies that tested supplementation with either single antioxidants (such as vitamins C or E) or combinations of antioxidants have produced weak or inconsistent results, and several studies suggested possible harm.[29] Until more data are available, the use of antioxidant supplements is not recommended for heart disease prevention.[30]

Drug Therapies for CHD Prevention

Individuals who cannot reach LDL goals with dietary and lifestyle changes alone may be prescribed one or more medications. The most common drugs prescribed are *statins* (such as Lipitor or Zocor), which reduce cholesterol synthesis in the liver. *Bile acid sequestrants* (such as Colestid or Questran) reduce LDL cholesterol levels by interfering with bile acid reabsorption in the small intestine. For lowering triglyceride levels and increasing HDL, both *fibrates* (such as Lopid) and *nicotinic acid* (a form of niacin) are effective; nicotinic acid can also reduce LDL and lipoprotein(a) levels. Individuals using these medications should continue using dietary and lifestyle modifications so that they can use the lowest possible doses of the drugs.[31]

In addition to lipid-lowering medications, some people may require drugs that suppress blood clotting (such as anticoagulants and aspirin) or reduce blood pressure. Nitroglycerin (a vasodilator) may be given to alleviate angina as needed. Some medications may affect nutrition status or food intake (see the Diet-Drug Interactions feature); the interactions can be even more complicated when multiple medications are used.

Diet-Drug Interactions

Check this table for notable nutrition-related effects of the medications discussed in this chapter.

Anticoagulants (warfarin)	**Dietary interactions:** Require consistent vitamin K intake to maintain effectiveness. Drug effects are enhanced with supplementation of vitamin E, dong quai, danshen, fish oils, garlic, and ginkgo. Drug effects are reduced with coenzyme Q, ginseng, and green tea. Avoid alcohol.
Antihypertensives	
Beta-blockers	**Metabolic effects:** Elevated serum potassium levels, hypoglycemia
Calcium channel blockers	**Gastrointestinal effects:** Nausea, constipation
	Dietary interactions: Avoid herbal supplements that contain natural licorice. Avoid grapefruit juice, which may enhance drug effects.
ACE inhibitors[a]	**Gastrointestinal effects:** Reduced taste sensation
	Dietary interactions: Avoid herbal supplements that contain natural licorice. Avoid potassium supplements and salt substitutes that contain potassium.
	Metabolic effects: Elevated serum potassium levels
Antilipemics	
Statins	**Gastrointestinal effects:** Constipation, flatulence, GI discomfort
	Dietary interactions: Avoid grapefruit juice, which may enhance drug effects.
	Metabolic effects: Elevated serum liver enzymes
Bile acid sequestrants	**Gastrointestinal effects:** Constipation, flatulence, diarrhea
	Dietary interactions: Cause reduced absorption of fat-soluble vitamins
	Metabolic effects: Electrolyte imbalances, iron deficiency
Nicotinic acid	**Gastrointestinal effects:** GI discomfort
	Dietary interactions: Avoid alcoholic beverages, coffee, and tea, which may increase side effects.
	Metabolic effects: Elevated serum liver enzymes, elevated uric acid levels, hyperglycemia, low blood pressure
Digoxin	**Gastrointestinal effects:** Anorexia, nausea, stomach cramps, diarrhea
	Dietary interactions: High-fiber foods and magnesium supplements can reduce drug absorption. St. John's wort may reduce drug efficacy.
	Metabolic effects: Elevated serum potassium and reduced serum magnesium levels. Drug toxicity can develop if body potassium levels are low.
Diuretics (furosemide, spironolactone[b])	**Gastrointestinal effects:** Dry mouth, anorexia, decreased taste perception
	Dietary interactions: Furosemide bioavailability is reduced when the drug is taken with food.
	Metabolic effects: Fluid and electrolyte imbalances,[b] hyperglycemia (spironolactone), hyperlipidemia (spironolactone), thiamin and zinc deficiencies
Nitroglycerin	**Gastrointestinal effects:** Decreased taste perception
	Dietary interactions: Increases effects of alcohol.

[a]ACE is an abbreviation for *angiotensin-converting enzyme*. An ACE inhibitor interferes with the conversion of angiotensin I to angiotensin II, a peptide that helps to regulate blood pressure.

[b]*Furosemide* is a potassium-wasting diuretic; patients taking furosemide should increase intakes of potassium-rich foods. *Spironolactone* is a potassium-sparing diuretic; patients taking spironolactone should avoid supplemental potassium and salt substitutes that contain potassium.

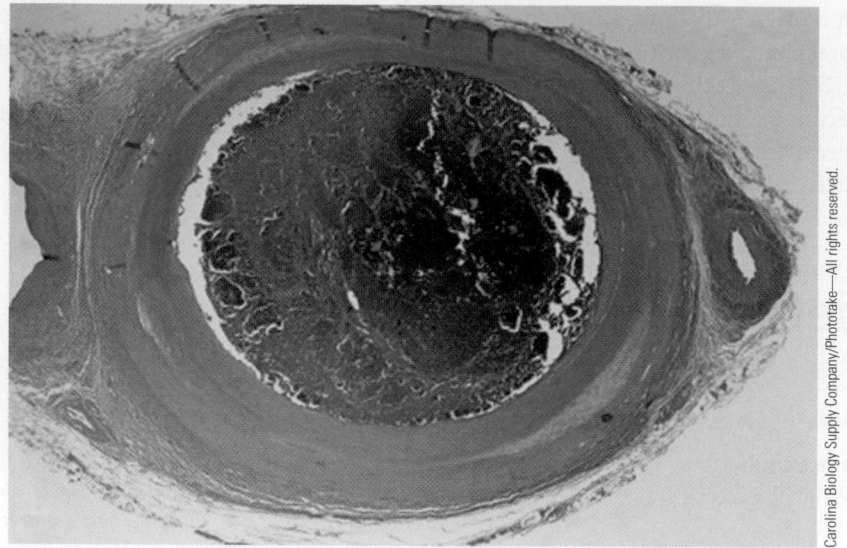

A heart attack occurs when a blood clot blocks the flow of blood in a coronary artery narrowed by atherosclerosis.

Treatment for Heart Attack

As explained earlier, a heart attack occurs when the blood supply to heart muscle is blocked, causing damage or death to heart tissue. Drug therapies given immediately after a heart attack may include thrombolytic drugs (clot-busting drugs), anticoagulants, aspirin, painkillers, and medications that regulate heart rhythm and reduce blood pressure.[32] Patients are not given food or beverages, except for sips of water or clear liquids, until their condition stabilizes. Once able to eat, they are initially offered small portions of foods that are low in sodium, saturated fat, and cholesterol. The sodium restriction helps to limit fluid retention but may be lifted after several days if the patient shows no signs of heart failure.

A heart attack patient needs to regain strength and learn strategies that can reduce the risk of a future heart attack; such strategies are similar to the lifestyle changes described earlier. Thus, the cardiac rehabilitation programs in hospitals and outpatient clinics include exercise therapy, instruction about heart-healthy food choices, help with smoking cessation, and medication counseling.

IN SUMMARY

▌ CHD management involves risk reduction, with an emphasis on lowering LDL cholesterol.

▌ The Therapeutic Lifestyle Changes (TLC) approach can help patients reduce LDL levels and eliminate other risk factors. Dietary recommendations are to reduce saturated fat, *trans* fats, and cholesterol; increase soluble fiber; and incorporate plant sterols, plant stanols, and fish into the diet.

▌ Treatment for mild hypertriglyceridemia emphasizes weight control, regular physical activity, smoking cessation, alcohol restriction, and avoiding a high carbohydrate intake. Severe hypertriglyceridemia requires drug therapies and dietary fat restriction.

▌ Medications given after a heart attack suppress blood clotting, regulate heart rhythm, and reduce blood pressure. To reduce the risk of a future heart attack, patients must learn strategies similar to the TLC approach.

Stroke

Stroke is the fourth most common cause of death in the United States[33] and a leading cause of serious long-term disability in adults. About 87 percent of strokes are **ischemic strokes**,[34] caused by the obstruction of blood flow to brain tissue. **Hemorrhagic strokes** occur in 13 percent of cases and result from bleeding within the brain, which damages brain tissue. Most strokes are a consequence of atherosclerosis, hypertension, or both.

Strokes that occur suddenly and are short-lived (lasting several minutes to an hour) are called **transient ischemic attacks (TIAs).** These brief strokes are a warning sign that a heart attack or a more severe stroke may follow.

Stroke Prevention

Stroke is largely preventable by recognizing its risk factors and making lifestyle choices that reduce risk. Many of the risk factors are similar to those for heart disease, and include hypertension, elevated LDL cholesterol, diabetes mellitus, cigarette smoking, and a history of cardiovascular disease. Medications that suppress blood clotting reduce the risk of ischemic stroke, especially in people who have suffered a first stroke or a transient ischemic attack. The drugs typically prescribed include antiplatelet drugs (including aspirin) or anticoagulants such as warfarin (Coumadin). Anticoagulant therapy requires regular follow-up and occasional adjustments in dosage to prevent excessive bleeding.

Reminder: Warfarin acts by interfering with vitamin K's blood-clotting function (see Chapter 15).

Stroke Management

The effects of a stroke vary according to the area of the brain that has been injured. Body movements, senses, and speech are often impaired, and one side of the body may be weakened or paralyzed. Early diagnosis and treatment are necessary to preserve brain tissue and minimize long-term disability. Ideally, thrombolytic (clot-busting) drugs are used within the first few hours following an ischemic stroke to restore blood flow and prevent further brain damage.[35]

The focus of nutrition care is to help patients maintain nutrition status and overall health despite the disabilities caused by the stroke. Dysphagia (difficulty swallowing) is a frequent complication of stroke and is associated with a poorer prognosis. Difficulty with speech prevents patients from communicating food preferences or describing the problems they may be having with eating. Coordination problems can make it hard for patients to grasp utensils or bring food from table to mouth. In some cases, tube feedings may be necessary until the patient has regained these skills. Nutrition in Practice 22 describes various options for feeding people with disabilities, such as those that follow stroke.

IN SUMMARY

▌ The two major types of strokes, ischemic and hemorrhagic stroke, may be a consequence of atherosclerosis, hypertension, or both. Transient ischemic attacks, which are short-lived ischemic strokes, are a warning sign that a heart attack or more severe stroke may follow.

▌ Strokes are largely preventable by reversing modifiable risk factors. Treatment includes the use of anticlotting drugs and anticoagulants. A patient who has had a major stroke may have problems eating due to difficulty swallowing or lack of coordination.

Hypertension

Hypertension (high blood pressure) affects about one-third of adults in the United States.[36] Prevalence is especially high in African Americans, who develop hypertension earlier in life and sustain higher average blood pressures throughout their lives

ischemic strokes: strokes caused by the obstruction of blood flow to brain tissue.

hemorrhagic strokes: strokes caused by bleeding within the brain, which destroys or compresses brain tissue.

transient ischemic attacks (TIAs): brief ischemic strokes that cause short-term neurological symptoms.

Blood pressure is measured both when heart muscle contracts (*systolic* blood pressure) and when it relaxes (*diastolic* blood pressure). Measurements are expressed as millimeters of mercury (mm Hg).

	Systolic	Diastolic
Desirable blood pressure	<120	<80
Prehypertension	120–139	80–89
Hypertension	≥140	≥90

The equation describing this relationship is: blood pressure (BP) = cardiac output (CO) × peripheral resistance (PR).

FIGURE 22-4
Determinants of Blood Pressure

than other ethnic groups. An estimated 20 percent of people with hypertension are unaware that they have it.[37]

Although people cannot feel the physical effects of hypertension, it is a primary risk factor for atherosclerosis and cardiovascular diseases. Elevated blood pressure forces the heart to work harder to eject blood into the arteries; this effort weakens heart muscle and increases the risk of developing heart arrhythmias, heart failure, and even sudden death. Hypertension is also a primary cause of stroke and kidney failure; reducing blood pressure can dramatically reduce the incidence of these diseases.

Factors That Influence Blood Pressure

As shown in Figure 22-4, blood pressure depends on the volume of blood pumped by the heart (cardiac output) and the resistance the blood encounters in the arterioles (peripheral resistance). When either cardiac output or peripheral resistance increases, blood pressure rises. Cardiac output is raised when heart rate or blood volume increases; peripheral resistance is affected mostly by the diameters of the arterioles and blood viscosity. Blood pressure is therefore influenced by the nervous system, which regulates heart muscle contractions and arteriole diameters, and hormonal signals, which may cause fluid retention or blood vessel constriction. The kidneys also play a role in regulating blood pressure by controlling the secretion of the hormones involved in vasoconstriction and retention of sodium and water.

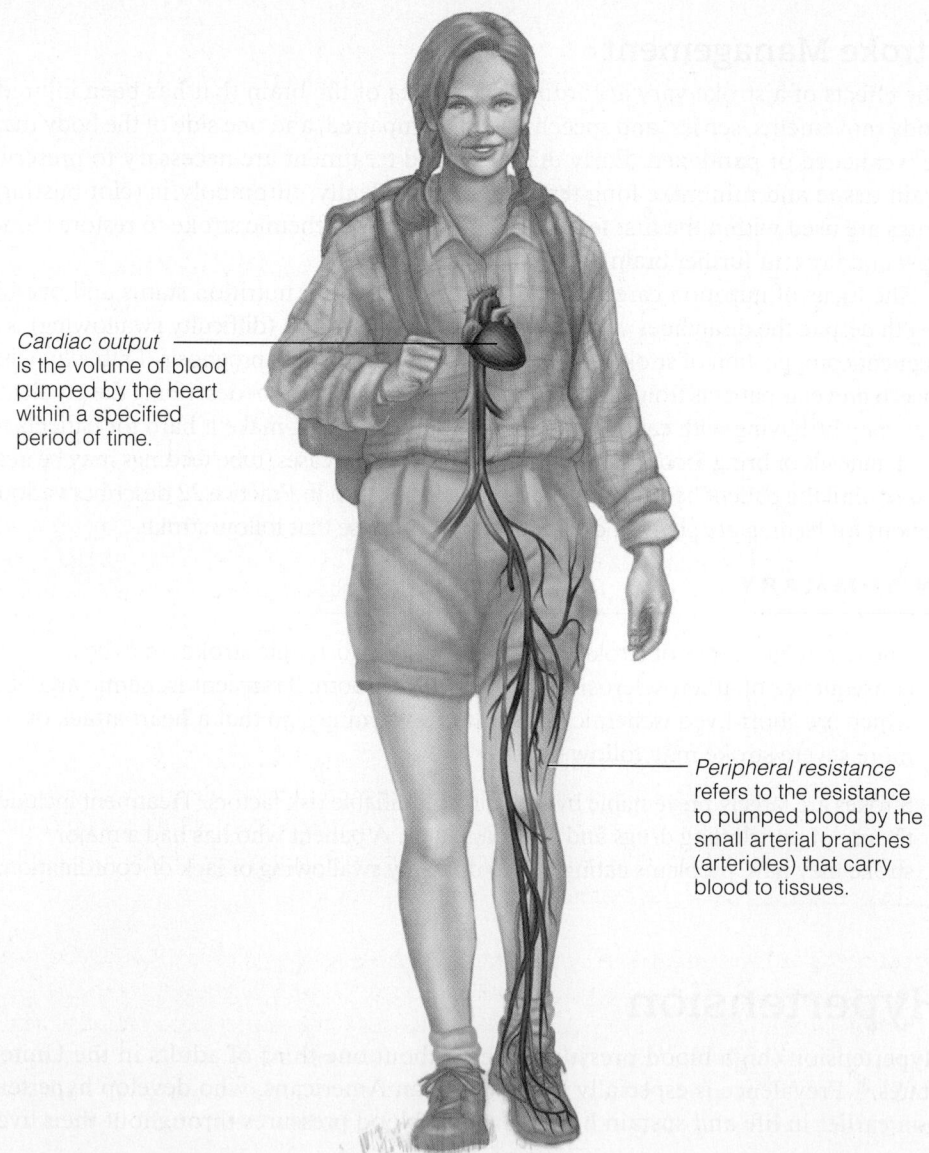

Cardiac output is the volume of blood pumped by the heart within a specified period of time.

Peripheral resistance refers to the resistance to pumped blood by the small arterial branches (arterioles) that carry blood to tissues.

Factors That Contribute to Hypertension

In 90 to 95 percent of hypertension cases, the cause is unknown.[38] In other cases, hypertension is caused by a known physical or metabolic disorder, such as an abnormality in an organ or hormone involved in blood pressure regulation. For example, conditions characterized by the narrowing of renal arteries often result in the increased production of proteins and hormones that stimulate water retention and vasoconstriction, thereby raising blood pressure.

A number of risk factors for hypertension have been identified. These include the following:

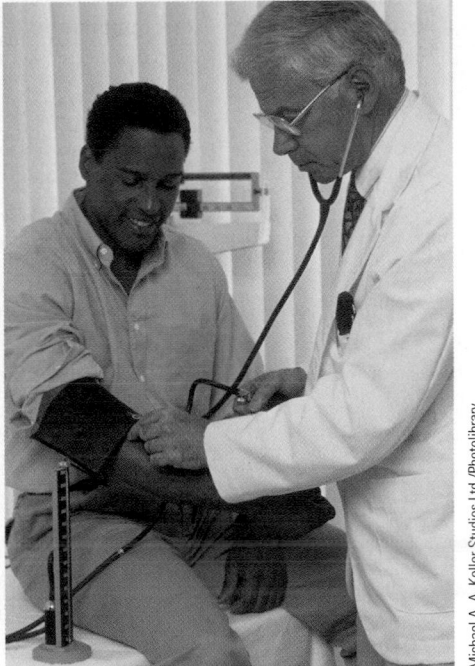

Michael A. A. Keller Studios Ltd./Photolibrary

- *Aging.* Hypertension risk increases with age. More than two-thirds of persons older than 65 years have hypertension.[39] Moreover, individuals who have normal blood pressure at age 55 still have a 90 percent risk of developing high blood pressure during their lifetimes.[40]

- *Genetics.* Risk of hypertension is similar among family members. It is also more prevalent and severe in certain ethnic groups; for example, the prevalence in African American adults is about 44 percent, compared with a prevalence of about 33 percent in whites and 28 percent in Mexican Americans.[41]

- *Obesity.* Most people with hypertension—an estimated 60 percent—are obese.[42] Obesity raises blood pressure, in part, by increasing blood volume, promoting vasoconstriction, and increasing activity of the sympathetic nervous system.[43]

- *Salt sensitivity.* Approximately 50 percent of those with hypertension have blood pressure that is sensitive to salt.[44] These people can improve their blood pressure by reducing salt in their diets.

Screening people for hypertension is a first step toward early detection and prevention of complications.

- *Alcohol.* Heavy drinking (three or more drinks daily) increases the incidence and severity of hypertension by stimulating the sympathetic nervous system.[45] Reducing alcohol consumption reverses this effect.

- *Dietary factors.* A person's diet may affect hypertension risk. As explained later, dietary modifications that increase intakes of potassium, calcium, and magnesium have been shown to reduce blood pressure.

Treatment of Hypertension

Both lifestyle modifications and medications are used to treat hypertension. Table 22-4 lists lifestyle modifications that can reduce blood pressure and the expected reduction in systolic blood pressure for each change. The recommendations include weight reduction if overweight or obese; a diet low in sodium and rich in potassium, calcium, and magnesium; regular physical activity; and a moderate alcohol intake, if one chooses to drink. Combining two or more of these modifications can enhance results.

Weight Reduction For obese individuals, weight reduction may reduce blood pressure significantly. Clinical studies suggest that systolic blood pressure can be reduced by about 1 mm Hg for each kilogram of weight loss and that the blood pressure reduction may be sustained for several years.[46] In the long term, however (more than three years), blood pressure tends to revert to initial levels, even when weight loss is partially maintained. Weight reduction is most beneficial for blood pressure control during periods of weight loss and weight maintenance.

Dietary Approaches for Reducing Blood Pressure Several research studies have shown that a significant reduction in blood pressure can be achieved by following a diet that emphasizes fruits, vegetables, and low-fat dairy products and includes whole grains, poultry, fish, and nuts.[47] The diet tested in these studies, now known as the *DASH Eating Plan,* provides more fiber, potassium, magnesium, and calcium than the typical American diet. The diet also limits red meat, sweets, sugar-containing

The goal of hypertension treatment is to reduce blood pressure to <140/<90 mm Hg. For people with CHD, diabetes, or kidney disease, the blood pressure goal is <130/<80 mm Hg.

For updated guidelines related to hypertension treatment, access this website: www.nhlbi.nih.gov/guidelines

The DASH Eating Plan is based on the test diet used in a study called Dietary Approaches to Stop Hypertension.

TABLE 22-4 Lifestyle Modifications for Blood Pressure Reduction

Modification	Recommendation	Expected Reduction in Systolic Blood Pressure
Weight reduction	Maintain healthy body weight (BMI below 25).	5–20 mm Hg/10 kg weight loss
DASH Eating Plan[a]	Adopt a diet rich in fruits, vegetables, and low-fat milk products with reduced saturated fat intake.	8–14 mm Hg
Sodium restriction	Reduce dietary sodium intake to less than 2400 milligrams sodium (less than 6 grams salt) per day.[b]	2–8 mm Hg
Physical activity	Perform aerobic physical activity for at least 30 minutes per day, most days of the week.	4–9 mm Hg
Moderate alcohol consumption	Men: Limit to two drinks per day. Women and lighter-weight men: Limit to one drink per day.	2–4 mm Hg

[a]The DASH Eating Plan was tested in a study called Dietary Approaches to Stop Hypertension.
[b]According to DRI recommendations, sodium intake should be limited to 2300 milligrams daily.

Source: Adapted from *Reference Card from the Seventh Report of the Joint National Committee on Prevention, Detection, Evaluation, and Treatment of High Blood Pressure (JNC 7)*, NIH publication no. 03-5231 (Bethesda, MD: National Institutes of Health, National Heart, Lung, and Blood Institute, and National High Blood Pressure Education Program, May 2003).

beverages, saturated fat (to 7 percent of kcalories), and cholesterol (to 150 milligrams per day), so it is beneficial for reducing CHD risk as well. The DASH Eating Plan, shown in Table 22-5, is a dietary pattern that meets the goals specified in the *Dietary Guidelines for Americans* (see Chapter 1).

The DASH Eating Plan is even more effective when accompanied by a low sodium intake. In a research study that tested the blood pressure–lowering effects of the

TABLE 22-5 The DASH Eating Plan

Food Group	Recommended Servings for Different Energy Intakes *(servings per day except as noted)*			
	1600 kcal	2000 kcal	2600 kcal	3100 kcal
Grains and grain products[a] (1 serving = 1 slice bread, 1 oz dry cereal,[b] or ½ c cooked rice, pasta, or cereal)	6	6–8	10–11	12–13
Vegetables (1 serving = ½ c cooked vegetables, 1 c raw leafy vegetables, or ½ c vegetable juice)	3–4	4–5	5–6	6
Fruits (1 serving = 1 medium fruit; ½ c fresh, frozen, or canned fruit; ¼ c dried fruit; or ½ c fruit juice)	4	4–5	5–6	6
Milk products (low fat or fat free) (1 serving = 1 c milk or yogurt, or 1½ oz cheese)	2–3	2–3	3	3–4
Meat, poultry, and fish (1 serving = 3 oz cooked lean meat, poultry, or fish)	1–2	1–2	2	2–3
Nuts, seeds, and legumes (1 serving = ⅓ c nuts, 2 tbsp peanut butter, 2 tbsp seeds, or ½ c cooked dry beans or peas)	3 per week	4–5 per week	1	1
Fats and oils (1 serving = 1 tsp vegetable oil or soft margarine, ½ tbsp mayonnaise, or 1 tbsp salad dressing)	2	2–3	3	4
Sweets (1 serving = 1 tbsp sugar, jelly, or jam; ½ c sorbet; or 1 c lemonade)	0	up to 5 per week	≤2	≤2

[a]Whole grains are recommended for most servings consumed.
[b]One ounce of dry cereal may be equivalent to ½ to 1¼ cups, depending on the cereal. Check the food label for the portion size.

Source: National Heart, Lung, and Blood Institute, National Institutes of Health, *Your Guide to Lowering Your Blood Pressure with DASH*, NIH publication no. 06-4082 (Bethesda, MD: National Heart, Lung, and Blood Institute, 2006).

DASH dietary pattern in combination with sodium restriction, the best results were achieved when sodium was reduced to 1500 milligrams daily—a lower level than the maximum intake of 2400 milligrams* recommended for people with hypertension.[48] These results suggest that the optimal sodium intake for people with hypertension may be far lower than is typically recommended. The "How to" lists practical suggestions for restricting sodium intake; additional detail is provided in Table 23-1 on p. 598.

Drug Therapies for Reducing Blood Pressure People with hypertension usually require two or more medications to meet their blood pressure goals. Using a combination of drugs with different modes of action can reduce the doses of each drug needed and minimize side effects. Most treatments include diuretics, which lower blood pressure by reducing blood volume. Other drugs commonly prescribed include angiotensin-converting enzyme (ACE) inhibitors, beta-blockers, and calcium channel blockers; these drugs are also used to treat other heart conditions. Drug dosages may need regular adjustment until the blood pressure goal is reached.

IN SUMMARY

- About one in three persons in the United States has hypertension, which increases the risk of developing CHD, stroke, heart failure, and kidney failure.

- Blood pressure is elevated by factors that increase blood volume, heart rate, or resistance to blood flow. Although the cause of most hypertension cases is unknown, risk factors include aging, family history, ethnicity, obesity, and various dietary factors.

- Treatment of hypertension usually includes a combination of lifestyle modifications and drug therapies.

How To REDUCE SODIUM INTAKE

- Select fresh, unprocessed foods. Packaged foods, canned goods, and frozen meals are often high in sodium.
- Do not use salt at the table or while cooking. Salt substitutes may be useful for some people. Salt substitutes often contain potassium, however, and are not appropriate for people using diuretics that promote potassium retention in the blood.
- Avoid eating in fast-food restaurants; most menu choices are very high in sodium.
- Check food labels. The labeling term *low sodium* is a better guide than the terms *reduced sodium* (contains 25 percent less sodium than the regular product) or *light in sodium* (contains 50 percent less sodium). To be labeled *low sodium,* a food product must contain less than 140 milligrams of sodium per serving. Keep your sodium goal in mind when you read labels.
- Recognize the high-sodium foods in each food category, and purchase only unsalted or low-sodium varieties of these products if they are available. High-sodium foods include the following:
  - Snack foods made with added salt, such as tortilla chips, popcorn, and nuts.

- Processed meats, such as ham, corned beef, bologna, salami, sausage, bacon, frankfurters, and pastrami.
- Processed fish, such as salted fish and canned fish.
- Tomato-based products, such as tomato sauce, tomato juice, pizza, canned tomatoes, and catsup.
- Canned soups and broths; note that even reduced-sodium varieties may contain excessive sodium.
- Cheese, such as cottage cheese, American cheese, and Parmesan and most other hard cheeses.
- Bakery products made with baking powder or baking soda (sodium bicarbonate), such as cakes, cookies, doughnuts, and muffins.
- Condiments and relishes, such as bouillon cubes, olives, and pickled vegetables.
- Flavoring sauces, such as soy sauce, barbecue sauce, and steak sauce.
- Check for the word *sodium* on medication labels. Sodium is often an ingredient in some types of antacids and laxatives.

*The maximum sodium intake recommended for people with hypertension (2400 mg) is similar to the Tolerable Upper Intake Level (UL) for sodium, which was set at 2300 mg to prevent the adverse effects of high sodium intakes on blood pressure in the general population.

Case Study PATIENT WITH CARDIOVASCULAR DISEASE

Robert Reid, a 48-year-old African American computer programmer, is 5 feet 9 inches tall and weighs 240 pounds. He sits for long hours at work and is too tired to exercise when he gets home at night. His meals usually include fatty meats, eggs, and cheese, and he likes dairy desserts such as pudding and ice cream. He has a family history of CHD and hypertension. His recent laboratory tests show that his blood pressure is 160/100 mm Hg, and his LDL and HDL levels are 160 mg/dL and 35 mg/dL, respectively. He smokes a pack of cigarettes each day and usually has two glasses of wine at both lunch and dinner.

1. Identify Mr. Reid's major risk factors for CHD and hypertension. Which can be modified? What complications might occur if he delays treatment for his blood lipids and blood pressure?

2. What dietary changes would you recommend that could help to improve Mr. Reid's blood pressure and his LDL and HDL cholesterol? Explain the rationale for each dietary change. Prepare a day's menu for Mr. Reid using the DASH Eating Plan as an outline for your choices.

3. What other laboratory tests or measurements would you need to better assess Mr. Reid's condition? Why?

4. Describe several benefits that Mr. Reid might obtain from a program that includes weight reduction and regular physical activity. Explain why the use of alcohol can be both a protective and a damaging lifestyle habit.

5. Assuming that Mr. Reid does not make any changes in his diet and lifestyle and suffers a heart attack, identify the elements of a cardiac rehabilitation program that would be critical for his long-term survival.

The Case Study above provides an opportunity to review the risk factors and treatments for CHD and hypertension.

Heart Failure

Heart failure, also called *congestive heart failure,* is characterized by the heart's inability to pump adequate blood, resulting in inadequate blood delivery and a buildup of fluids in the veins and tissues. Heart failure has various causes, but it is often a consequence of chronic disorders that create extra work for the heart muscle, such as hypertension or CHD. To accommodate the extra workload, the heart enlarges or pumps faster or harder, but it eventually may weaken enough to fail completely. Heart failure develops mostly in older adults and the elderly: approximately 75 percent of persons with heart failure in the United States are age 65 or older.[49]

Consequences of Heart Failure

In heart failure, fluid may accumulate in the liver, abdomen, and lower extremities, causing chest pain, difficulty with digestion and absorption, and swelling in the legs,

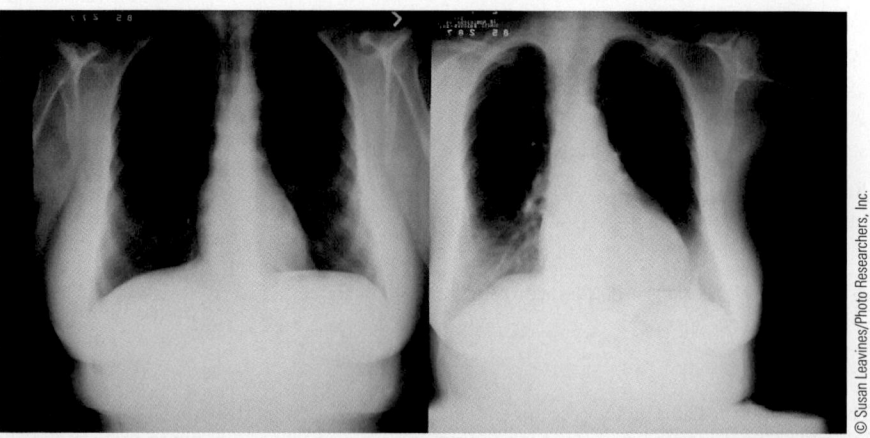

An overburdened heart enlarges in an effort to supply blood to the body's tissues.

© Susan Leavines/Photo Researchers, Inc.

ankles, and feet. Fluid can sometimes build up in the lungs (called *pulmonary edema*), resulting in extreme shortness of breath and limited oxygen for activity; in severe cases, it can lead to respiratory failure. With inadequate blood flow, the functions of various organs, such as the liver and kidneys, may become impaired. Note that the effects of heart failure depend on the severity of illness: mild cases may be asymptomatic, but severe cases may cause considerable damage to health.

Heart failure often affects a person's food intake and level of physical activity. In persons with abdominal bloating and liver enlargement, pain and discomfort may worsen with meals. Limb weakness and fatigue can limit physical activity. End-stage heart failure is often accompanied by **cardiac cachexia,** a condition of severe malnutrition characterized by severe weight loss and tissue wasting. Cardiac cachexia may develop due to increased levels of pro-inflammatory cytokines (which promote catabolism), elevated metabolic rate, loss of appetite, and reduced food intake.[50] The resultant weakness further lowers the person's strength, functional capacity, and activity levels.

Medical Management of Heart Failure

Heart failure is a chronic, progressive illness that may require frequent hospitalizations. Many patients face a combination of debilitating symptoms, complex treatments, and an uncertain outcome. Important goals of medical therapy are to slow disease progression and enhance the patient's quality of life.

The specific treatment for heart failure depends on the nature and severity of the illness. Medications help to manage fluid retention and improve heart function. Dietary sodium and fluid restrictions can help to prevent fluid accumulation. Vaccinations for influenza and pneumonia reduce the risk of developing respiratory infections. Treatment of CHD risk factors, such as hypertension and lipid disorders, can help to slow disease progression. Heart failure patients are also encouraged to participate in exercise programs to avoid becoming physically disabled and to improve endurance.

Nutrition Therapy for Heart Failure The main dietary recommendation for heart failure is a sodium restriction of 2000 milligrams or less daily to reduce the likelihood of fluid retention.[51] In patients with persistent or recurrent fluid retention, fluid intakes may be restricted to 2 liters per day or less. Individuals who have difficulty eating due to abdominal or chest pain may tolerate small, frequent meals better than large meals.

Patients with heart failure may be prone to constipation due to diuretic use and reduced physical activity. Maintaining an adequate fiber intake can help to minimize constipation problems. Because of alcohol's deleterious effects on blood pressure and heart function, most patients need to restrict or avoid alcoholic beverages.

No known therapies can reverse cardiac cachexia, and prognosis is poor. For some patients, liquid supplements, tube feedings, or parenteral nutrition support can be supportive additions to treatment.

The recommendation to limit sodium intake to 2000 mg is not very restrictive. The current DRI recommendation is to limit sodium intake to 2300 mg daily.

IN SUMMARY

- Heart failure is usually a chronic, progressive condition that results from other cardiovascular illnesses.

- In heart failure, the heart is unable to pump adequate blood to tissues. Consequences may include fluid accumulation in the veins, lungs, and other organs and impaired organ function.

- Treatment of heart failure usually includes drug therapies that reduce fluid accumulation and improve heart function. Nutrition therapy may include sodium and fluid restrictions.

heart failure: a condition with various causes that is characterized by the heart's inability to pump adequate blood to the body's cells, resulting in fluid accumulation in the tissues; also called *congestive heart failure*.

cardiac cachexia: a condition of severe malnutrition that develops in heart failure patients; characterized by weight loss and tissue wasting.

Nutrition Assessment Checklist FOR PEOPLE WITH CARDIOVASCULAR DISEASES

Medical History

Check the medical record for a diagnosis of:

- Coronary heart disease
- Stroke
- Hypertension
- Heart failure

Review the medical record for complications related to cardiovascular diseases:

- Heart attack
- Transient ischemic attack
- Cardiac cachexia

Note risk factors for CHD or stroke that are related to diet, including:

- Elevated LDL or triglyceride levels
- Obesity or overweight
- Diabetes
- Hypertension

Medications

For patients using drug treatments for cardiovascular diseases, note:

- Side effects that may alter food intake
- Medications that may interact with grapefruit juice
- Use of warfarin, which requires a consistent vitamin K intake
- Use of diuretics or other drugs associated with potassium imbalances
- Potential diet-drug or herb-drug interactions

Dietary Intake

For patients with CHD, a previous stroke, or hypertension, assess the diet for:

- Energy intake
- Saturated fat, *trans* fat, cholesterol, and sodium content
- Soluble fiber and plant sterol or plant stanol content
- Intake of whole grains, fruits, vegetables, legumes, and nuts
- Alcohol content

For patients with complications resulting from cardiovascular diseases:

- Check physical disabilities that may interfere with food preparation or consumption following a stroke.
- Check adequacy of food intake in patients with heart failure.

Anthropometric Data

Measure baseline height and weight, and reassess weight at each medical checkup. Note whether patients are meeting weight goals, including:

- Weight loss or maintenance in patients who are overweight
- Weight maintenance in patients with advanced heart failure

Remember that weight may be deceptively high in people who are retaining fluids, especially individuals with heart failure.

Laboratory Tests

Monitor the following laboratory tests in people with cardiovascular diseases:

- LDL cholesterol, blood triglycerides, and HDL cholesterol
- Blood glucose in patients with diabetes
- Serum potassium in patients using diuretics, antihypertensive medications, or digoxin
- Blood-clotting times in patients using anticoagulants
- Indicators of fluid retention in patients with heart failure

Physical Signs

Blood pressure measurement is routine in physical exams but is especially important for people who:

- Have cardiovascular diseases
- Have experienced a heart attack or stroke
- Have risk factors for CHD or hypertension

Look for signs of:

- Potassium imbalances (muscle weakness, numbness and tingling, irregular heartbeat) in those using diuretics, anti-hypertensive medications, or digoxin
- Fluid overload in patients with heart failure

Clinical Applications

1. List risk factors for coronary heart disease, and identify possible interrelationships among the factors. For example, a woman over age 55 is also at risk for diabetes; a person with diabetes is more likely to have hypertension.

2. Review the DASH Eating Plan shown in Table 22-5 on p. 584. As the chapter describes, the DASH dietary pattern is helpful for lowering blood pressure and for reducing CHD risk as well.

- List elements of the DASH Eating Plan that are consistent with the TLC recommendations.

- Suggest ways in which a person following the DASH Eating Plan might accomplish the following additional dietary modifications: consume a higher percentage of fat from monounsaturated sources, reduce intake of *trans*-fatty acids, and include EPA/DHA and plant sterols in the diet.

Self Check

1. Ischemia in the coronary arteries is a frequent cause of:
   a. angina pectoris.
   b. hemorrhagic stroke.
   c. aneurysm.
   d. hypertension.

2. Risk factors for atherosclerosis include all of the following *except:*
   a. smoking.
   b. hypertension.
   c. diabetes mellitus.
   d. elevated HDL cholesterol.

3. The dietary lipids with the strongest LDL cholesterol–raising effects are:
   a. monounsaturated fats.
   b. polyunsaturated fats.
   c. saturated fats.
   d. plant sterols.

4. The omega-3 fatty acids EPA and DHA, which may improve some risk factors for heart disease, are obtained by consuming:
   a. fatty fish.
   b. soy products.
   c. egg yolks and organ meats.
   d. nuts and seeds.

5. Moderate alcohol consumption can improve heart disease risk, in part, because it:
   a. lowers blood pressure.
   b. improves nutrition status.
   c. offsets the damage from smoking.
   d. increases HDL cholesterol levels.

6. Patients with mild hypertriglyceridemia may improve their triglyceride levels by:
   a. reducing sodium intake.
   b. consuming moderate amounts of alcohol.
   c. avoiding a high carbohydrate intake.
   d. reducing cholesterol intake.

7. Hemorrhagic stroke:
   a. is the most common type of stroke.
   b. results from obstructed blood flow within brain tissue.
   c. comes on suddenly and usually lasts for up to 30 minutes.
   d. results from bleeding within the brain, which damages brain tissue.

8. In most cases of hypertension, the cause is:
   a. excessive alcohol use.
   b. atherosclerosis.
   c. hormonal imbalances.
   d. unknown.

9. Hypertensive patients can benefit from all of the following dietary and lifestyle modifications *except:*
   a. including fat-free or low-fat milk products in the diet.
   b. reducing total fat intake.
   c. consuming generous amounts of fruits, vegetables, legumes, and nuts.
   d. reducing sodium intake.

10. Nutrition therapy for a patient with heart failure usually includes:
    a. weight loss.
    b. reducing total fat intake.
    c. sodium restriction.
    d. cholesterol restriction.

Answers to these questions can be found in Appendix H.

Notes

1. V. L. Roger and coauthors, Heart disease and stroke statistics 2011 update: A report from the American Heart Association, *Circulation* 123 (2011): DOI: 10.1161/CIR.0b013e3182009701.

2. American Heart Association, International Cardiovascular Disease Statistics (2009), available at http://www.americanheart.org/presenter.jhtml?identifier=3001008; accessed November 26, 2010.

3. R. N. Mitchell and F. J. Schoen, Blood vessels, in V. Kumar and coeditors, *Robbins and Cotran Pathologic Basis of Disease* (Philadelphia: Saunders, 2010), pp. 487–528; W. Insull, The pathology of atherosclerosis: Plaque development and plaque responses to medical treatment, *American Journal of Medicine* 122 (2009): S3–S14.

4. C. Cheng and coauthors, Atherosclerotic lesion size and vulnerability are determined by patterns of fluid shear stress, *Circulation* 113 (2006): 2744–2753.

5. Insull, 2009; I. Tabas, K. J. Williams, and J. Boren, Subendothelial lipoprotein retention as the initiating process in atherosclerosis: Update and therapeutic implications, *Circulation* 116 (2007): 1832–1844.

6. G. T. Jones and coauthors, Plasma lipoprotein(a) indicates risk for 4 distinct forms of vascular disease, *Clinical Chemistry* 53 (2007): 679–685.

7. N. L. Benowitz, Tobacco, in L. Goldman and D. Ausiello, eds., *Cecil Medicine* (Philadelphia: Saunders, 2008), pp. 162–166.

8. S. Vasdev, V. Gill, and P. Singal, Role of advanced glycation end products in hypertension and atherosclerosis: Therapeutic implications, *Cell Biochemistry and Biophysics* 49 (2007): 48–63.

9. K. S. McCully, Homocysteine, vitamins, and vascular disease prevention, *American Journal of Clinical Nutrition* 86 (2007): 1563S–1568S.

10. Roger and coauthors, 2011.

11. Roger and coauthors, 2011.

12. National Cholesterol Education Program, *Third Report of the National Cholesterol Education Program (NCEP) Expert Panel on Detection, Evaluation, and Treatment of High Blood Cholesterol in Adults (Adult Treatment Panel III): Final Report,* NIH publication no. 02-5215 (Bethesda, MD: National Heart, Lung, and Blood Institute, 2002).

13. U.S. Department of Agriculture, Agricultural Research Service, Nutrient intakes from food: Mean amounts consumed per individual, by gender and age, *What We Eat in America, NHANES 2007–2008* (2010); available at www.ars.usda.gov/ba/bhnrc/fsrg; accessed December 3, 2010.

14. U.S. Department of Agriculture, Agricultural Research Service, 2010.

15. National Cancer Institute, Applied Research Program, *Sources of Cholesterol Among the U.S. Population, 2005–06* (updated July 21, 2010); available at http://riskfactor.cancer.gov/diet/

foodsources/cholesterol; accessed December 4, 2010.

16. A. H. Lichtenstein and coauthors, Diet and lifestyle recommendations revision 2006: A scientific statement from the American Heart Association Nutrition Committee, *Circulation* 114 (2006): 82–96.

17. Lichtenstein and coauthors, 2006.

18. Lichtenstein and coauthors, 2006; I. Gigleux and coauthors, Moderate alcohol consumption is more cardioprotective in men with the metabolic syndrome, *Journal of Nutrition* 136 (2006): 3027–3032.

19. Lichtenstein and coauthors, 2006.

20. Benowitz, 2008.

21. M. H. Criqui, Epidemiology of cardiovascular disease, in L. Goldman and D. Ausiello, eds., *Cecil Medicine* (Philadelphia: Saunders, 2008), pp. 301–305.

22. National Cholesterol Education Program, 2002.

23. E. S. Ford and coauthors, Hypertriglyceridemia and its pharmacologic treatment among US Adults, *Archives of Internal Medicine* 169 (2009): 572–578.

24. G. Yuan, K. Z. Al-Shali, and R. A. Hegele, Hypertriglyceridemia: Its etiology, effects, and treatment, *Canadian Medical Association Journal* 176 (2007): 1113–1120; M. J. Malloy and J. P. Kane, Disorders of lipoprotein metabolism, in D. G. Gardner and D. Shoback, eds., *Greenspan's Basic and Clinical Endocrinology* (New York: McGraw-Hill/Lange, 2007), pp. 770–795.

25. American Dietetic Association, *Nutrition Care Manual* (Chicago: American Dietetic Association, 2011); Yuan, Al-Shali, and Hegele, 2007.

26. R. C. Oh and J. B. Lanier, Management of hypertriglyceridemia, in *American Family Physician* 75 (2007): 1365–1371.

27. Oh and Lanier, 2007.

28. J. M. Armitage and coauthors, Study of the Effectiveness of Additional Reductions in Cholesterol and Homocysteine (SEARCH) Collaborative Group, Effects of homocysteine-lowering with folic acid plus vitamin B_{12} vs. placebo on mortality and major morbidity in myocardial infarction survivors: A randomized trial, Journal of the American Medical Association 303 (2010): 2486–2494; A. J. Martí-Carvajal and coauthors, Homocysteine lowering interventions for preventing cardiovascular events, *Cochrane Database of Systematic Reviews* Oct 7 (2009): CD006612; Van Horn and coauthors, 2008.

29. C. Hatzigeorgiou and coauthors, Antioxidant vitamin intake and subclinical coronary atherosclerosis, *Preventive Cardiology* 9 (2006): 75–81; D. H. Lee and coauthors, Does supplemental vitamin C increase cardiovascular disease risk in women with diabetes? *American Journal of Clinical Nutrition* 80 (2004): 1194–1200.

30. Van Horn and coauthors, 2008; L. Mosca and coauthors, Evidence-based guidelines for cardiovascular disease prevention in women: 2007 update, *Circulation* 115 (2007): 1481–1501.

31. Toth and coauthors, 2007.

32. J. L. Anderson, ST segment elevation acute myocardial infarction and complications of myocardial infarction, in L. Goldman and D. Ausiello, eds., *Cecil Medicine* (Philadelphia: Saunders, 2008), pp. 500–518.

33. A. M. Miniño, J. Xu, and K. D. Kochanek, Deaths: Preliminary data for 2008, *National Vital Statistics Reports* vol 59 no 2 (Hyattsville, MD: National Center for Health Statistics, 2010).

34. Roger and coauthors, 2011.

35. J. A. Zivin, Ischemic cerebrovascular disease, in L. Goldman and D. Ausiello, eds., *Cecil Medicine* (Philadelphia: Saunders, 2008), pp. 2708–2719.

36. Roger and coauthors, 2011.

37. Roger and coauthors, 2011.

38. R. G. Victor, Arterial hypertension, in L. Goldman and D. Ausiello, eds., *Cecil Medicine* (Philadelphia: Saunders, 2008), pp. 430–450.

39. Roger and coauthors, 2011.

40. M. H. Beers and coeditors, *The Merck Manual of Diagnosis and Therapy* (Whitehouse Station, NJ: Merck Research Laboratories, 2006), pp. 570–772.

41. Roger and coauthors, 2011.

42. T. A. Kotchen and J. M. Kotchen, Nutrition, diet, and hypertension, in M. E. Shils and coeditors, *Modern Nutrition in Health and Disease* (Philadelphia: Lippincott Williams & Wilkins, 2006), pp. 1095–1107.

43. M. D. Jensen, Obesity, in L. Goldman and D. Ausiello, eds., *Cecil Medicine* (Philadelphia: Saunders, 2008), pp.1643–1652.

44. B. Rodriguez-Iturbe and N. D. Vaziri, Salt-sensitive hypertension—Update on novel findings, *Nephrology Dialysis Transplantation* 22 (2007): 992–995.

45. Victor, 2008.

46. L. Aucott and coauthors, Long-term weight loss from lifestyle intervention benefits blood pressure? A systematic review, *Hypertension* 54 (2009): 756–762.

47. F. M. Sacks and coauthors, Effects on blood pressure of reduced dietary sodium and the Dietary Approaches to Stop Hypertension (DASH) Diet, *New England Journal of Medicine* 344 (2001): 3–10; L. J. Appel and coauthors, A clinical trial on the effects of dietary patterns on blood pressure, *New England Journal of Medicine* 336 (1997): 1117–1124.

48. Sacks and coauthors, 2001.

49. B. M. Massie, Heart failure: Pathophysiology and diagnosis, in L. Goldman and D. Ausiello, eds., *Cecil Medicine* (Philadelphia: Saunders, 2008), pp. 345–354.

50. Massie, 2008.

51. American Dietetic Association, 2011.

CHAPTER TWENTY TWO

Nutrition in Practice

Chapter 22 referred to difficulties following a stroke that can interfere with the ability to eat independently. This Nutrition in Practice discusses the problems faced by individuals who must cope with various disabilities that interfere with the process of eating, including those that interfere with chewing, swallowing, or bringing food to the mouth. These obstacles can arise at any time during a person's life and from any number of causes. An infant may be born with a physical impairment such as cleft palate; an adolescent may lose motor control following injuries sustained in an automobile accident; an older adult may struggle with the pain of arthritis or the mental deterioration of dementia. Table NP22-1 lists some of the conditions that may lead to feeding problems.

In what ways can disabilities impair a person's ability to eat?

Eating and drinking require a considerable number of individual coordinated motions. Consider an infant learning the skills required for feeding: each step—sitting, grasping cups and utensils, bringing food to the mouth, biting, chewing, and swallowing—requires coordinated movements. An injury or disability that interferes with any of these movements can lead to feeding problems and inadequate food intake. Total food intake is often significantly reduced when individuals with inefficient motor function take a long time to eat.[1] Difficulties that affect procurement of food, such as the inability to drive, walk, or carry groceries, can also lower food intake and lead to malnutrition and weight loss.

Can disabilities alter a person's energy needs?

Yes, certain disabilities can either increase or decrease energy requirements.[2] Disabilities that affect muscle tension or mobility can reduce physical activity and, consequently, energy requirements. Other disabilities, such as certain forms of cerebral palsy, cause involuntary muscle activity that raises energy requirements. Loss of a limb due to amputation reduces energy needs in proportion to the weight and metabolism represented by the missing limb, but

energy needs may be greater if an individual increases activity to compensate for the loss, such as by propelling a wheelchair. Because the effects of disabilities are often unpredictable, the health care practitioner may find it difficult to assess energy requirements until weight gain or loss has occurred.

Overweight and obesity often accompany conditions that limit mobility or result in short stature; examples include Down syndrome and spina bifida. Obesity may also develop because the family or caregiver provides an inappropriate amount of food, sometimes out of sympathy for the disabled individual. In these cases, the health practitioner may need to counsel the family or caregiver about appropriate food choices and portion sizes.

How may disease symptoms or medications alter food intake or nutrition status?

Examples of disease symptoms that can alter food intake include nausea, frequent coughing or choking, difficulty breathing, and gastroesophageal reflux. Individuals with speech or hearing problems may have difficulty communicating with caregivers about thirst and hunger. Mobility problems can influence nutrition status by leading to bone demineralization and pressure sores.

Conditions that require the use of multiple medications can also have a significant impact on nutrition status (see Chapter 15). Medications may increase or decrease appetite, interfere with nutrient metabolism, or have gastrointestinal effects that cause pain or discomfort with eating.

Which health professionals typically work with people who have feeding problems?

Evaluating and treating feeding problems often involve the joint efforts of health care professionals from a variety of disciplines, including nurses, dietitians, occupational and physical therapists, speech-language pathologists, and dentists. Together, these professionals evaluate each patient's dietary needs and assess abilities to chew, sip, swallow, grasp utensils, use utensils to pick up foods, and bring foods from the plate to the mouth. A speech-language pathologist most often evaluates chewing and swallowing abilities and trains patients to use lips, tongue, and throat for eating and speaking. An occupational therapist can demonstrate alternative feeding strategies, including changes in body position that improve feeding, techniques for handling utensils and food, and use of special feeding devices.

Direct observation of a patient during mealtimes allows health care professionals to assess current eating behaviors, demonstrate feeding techniques, monitor the patient's and caregiver's understanding of the techniques, and evaluate how well the care plan is working. To illustrate, consider a child with a feeding problem caused by hypersensitivity to oral stimulation. The health care professional may start by teaching the caregiver to gently stroke the child's face with a hand, washcloth, or soft toy. Once the child tolerates touch on less sensitive areas of the face, the health care

TABLE NP22-1 Conditions That May Lead to Feeding Problems

The following conditions may lead to feeding problems by interfering with a person's ability to suck, bite, chew, swallow, or coordinate hand-to-mouth movements.

▪ Accidents	▪ Language, visual, or hearing impairments
▪ Amputations	▪ Multiple sclerosis
▪ Arthritis	▪ Muscle weakness
▪ Birth defects	▪ Muscular dystrophy
▪ Cerebral palsy	▪ Neuromotor dysfunction
▪ Cleft palate	▪ Parkinson's disease
▪ Down syndrome	▪ Polio
▪ Head injuries	▪ Spinal cord injuries
▪ Huntington's chorea	▪ Stroke

Adaptive feeding equipment can help patients with feeding disabilities gain independence.

professional may encourage the caregiver to slowly begin to rub the child's lips, gums, palate, and tongue. With time, the child may be better able to tolerate the presence of food in the mouth. Examples of other strategies that can help feeding problems are listed in Table NP22-2.

Can special equipment be used to help people with certain feeding difficulties?

Yes. Adaptive feeding devices can make a remarkable difference in a person's ability to eat independently. Figure NP22-1 shows a few of the many special feeding devices that are available and describes their uses. Other examples of adaptive equipment include specialized chairs to improve posture, bolsters inserted under arms to improve elbow stability, and raised trays or eating surfaces to simplify hand-to-mouth movements.

Sometimes, despite the best efforts of all involved, a patient is unable to consume enough food by mouth. In these cases, tube feedings can help to improve nutrition status. Tube feedings are typically recommended for patients with severe dysphagia (difficulty swallowing), aspiration pneumonia, recurrent malnutrition, or failure to thrive.[3]

In what ways can feeding difficulties affect family life?

Mealtimes are a critical time for social interaction, and therefore individuals with feeding problems may encounter emotional and social problems if they are unable to participate. Children may fail to develop social skills, whereas adults may miss the social stimulation that mealtimes provide. Individuals should be encouraged to sit with family and friends during meals so that they are not deprived of the social and cultural aspects of eating.

The responsibility of caring for a person with a feeding problem can frequently overwhelm a caregiver.[4] Caring for a person with disabilities requires time and patience—and many new therapies to be learned and administered. The caregiver may

TABLE NP22-2 Interventions for Feeding-Related Problems

Inability to Suck
- Use squeeze bottles, which do not require sucking, to express liquids into the mouth.
- Place a spoon on the center of the tongue, and apply downward pressure to stimulate sucking.
- Apply rhythmic, slow strokes on the tongue to alter tongue position and improve the sucking response.

Inability to Chew
- Place foods between gums and teeth to promote chewing.
- Improve chewing skills with foods of different textures; for example, fruit leathers stimulate jaw movements but dissolve quickly enough to minimize choking.
- Provide soft foods that require minimal chewing or are easily chewed.

Inability to Swallow
- Provide thickened liquids, pureed foods, and moist foods that form boluses easily.
- Provide cold formulas, frozen fruit juice bars, and ice; cold substances promote swallowing movements by the tongue and soft palate.
- Make sure the patient's jaws and lips are closed to facilitate swallowing action.
- Correct posture and head position if they interfere with swallowing ability.

Inability to Grasp or Coordinate Movements
- Provide utensils that have modified handles, or are smaller or larger as necessary.
- Encourage the use of hands for feeding if utensils are difficult to maneuver.
- Provide plates with food guards to prevent spilling.
- Supply clothing protection.

Impaired Vision
- Place foods (meats, vegetables) in similar locations on the plate at each meal.
- Provide plates with food guards to prevent spilling.

spend many hours preparing special foods, monitoring the use of adaptive feeding equipment, and helping with feedings. Moreover, a person with disabilities may need help with other tasks as well, and all may require a considerable amount of time. In many cases, a caregiver receives little or no assistance. These conditions may lead to strained interactions between caregiver and patient and cause stress and frustration. Psychologists can offer counseling to patients or caregivers to help them adjust, and all members of the health care team can offer emotional support and practical suggestions to ease caregivers' responsibilities and frustrations.

Successful therapy for people with feeding disabilities requires the involvement of many health care professionals and depends on accurate identification of impaired feeding skills and determination of appropriate interventions. Ideally, with training, people with disabilities attain total independence—they are able to prepare, serve, and eat nutritionally adequate food daily without help. In some cases, these goals can be met with the help of caregivers. The combined efforts of the health care team can support both patients and caregivers in enhancing quality of life and in achieving independence to the greatest degree possible.

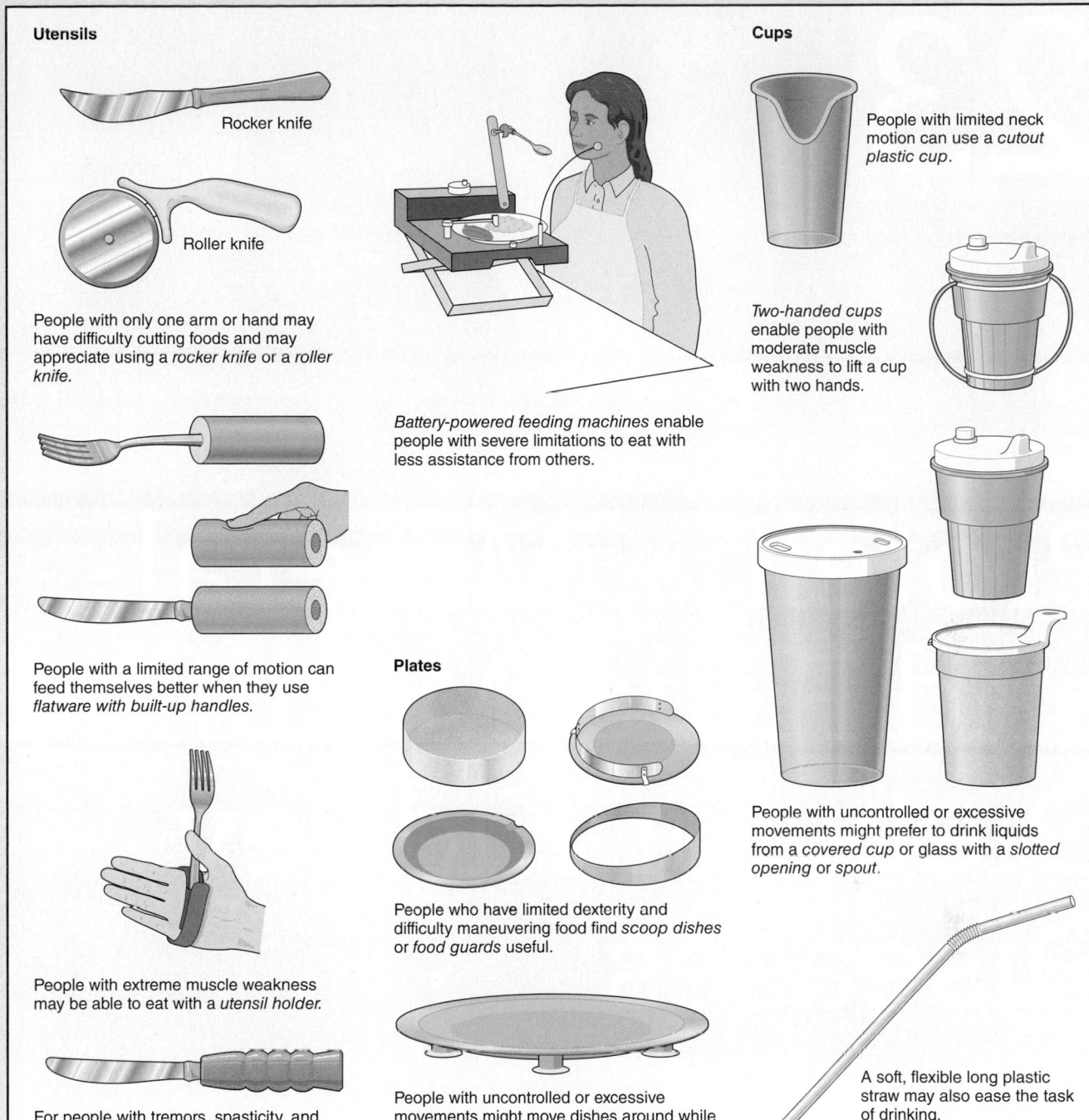

Utensils

Rocker knife

Roller knife

People with only one arm or hand may have difficulty cutting foods and may appreciate using a *rocker knife* or a *roller knife.*

People with a limited range of motion can feed themselves better when they use *flatware with built-up handles.*

People with extreme muscle weakness may be able to eat with a *utensil holder.*

For people with tremors, spasticity, and uneven jerky movements, *weighted utensils* can aid the feeding process.

Battery-powered feeding machines enable people with severe limitations to eat with less assistance from others.

Plates

People who have limited dexterity and difficulty maneuvering food find *scoop dishes* or *food guards* useful.

People with uncontrolled or excessive movements might move dishes around while eating and may benefit from using *unbreakable dishes with suction cups.*

Cups

People with limited neck motion can use a *cutout plastic cup.*

Two-handed cups enable people with moderate muscle weakness to lift a cup with two hands.

People with uncontrolled or excessive movements might prefer to drink liquids from a *covered cup* or glass with a *slotted opening* or *spout.*

A soft, flexible long plastic straw may also ease the task of drinking.

FIGURE NP22-1 Examples of Adaptive Feeding Devices

Notes

1. Position of the Canadian Paediatric Society: Nutrition in neurologically impaired children, *Paediatrics and Child Health* 14 (2009): 395–401.

2. Position of the American Dietetic Association: Providing nutrition services for people with developmental disabilities and special health care needs, *Journal of the American Dietetic Association* 110 (2010): 296–307.

3. Position of the American Dietetic Association, 2010.

4. P. Raina and coauthors, The health and well-being of caregivers of children with cerebral palsy, *Pediatrics* 115 (2005): e626–e636.

23

Protein-, Mineral-, and Fluid-Modified Diets for Kidney Diseases

The two kidneys sit just above the waist on each side of the spinal column. As part of the urinary system, they are responsible for filtering the blood and removing excess fluid and wastes for elimination in urine. Figure 23-1 shows the kidneys' placement and structure and one of their functional units, the **nephron.** Within each nephron, the **glomerulus,** a ball-shaped tuft of capillaries, serves as a gateway through which the blood components must pass to form **filtrate.** The glomerulus and surrounding capsule function like a sieve, retaining blood cells and most plasma proteins in blood while allowing fluid and small solutes to enter the nephron's system of **tubules.** As the filtrate passes through the tubules, its composition changes as some components are reabsorbed and returned to the blood, while the remaining substances contribute to the final urine product. In this manner, the kidneys regulate the extracellular fluid volume and osmolarity, electrolyte concentrations, and acid-base balance. They also excrete metabolic waste products such as urea and creatinine, as well as various drugs and toxicants. Other roles of the kidneys include the following:

▮ Secretion of the enzyme *renin*, which helps to regulate blood pressure

▮ Production of the hormone **erythropoietin,** which stimulates red blood cell production in the bone marrow

▮ Conversion of vitamin D to its active form, thereby helping to regulate calcium balance and bone formation

As this chapter explains, **renal** diseases can interfere with the kidneys' various functions and severely disrupt health.

nephron (NEF-ron): the functional unit of the kidneys, consisting of a glomerulus and tubules.

▮ *nephros* = kidney

glomerulus (gloh-MEHR-yoo-lus): a tuft of capillaries within the nephron that filters water and solutes from the blood as urine production begins (plural: *glomeruli*).

filtrate: the substances that pass through the glomerulus and travel through the nephron's tubules, eventually forming urine.

tubules: tubelike structures of the nephron that process filtrate during urine production. The tubules are surrounded by capillaries that reabsorb the substances retained by tubule cells.

erythropoietin (eh-RITH-ro-POY-eh-tin): a hormone made by the kidneys that stimulates red blood cell production.

renal (REE-nil): pertaining to the kidneys.

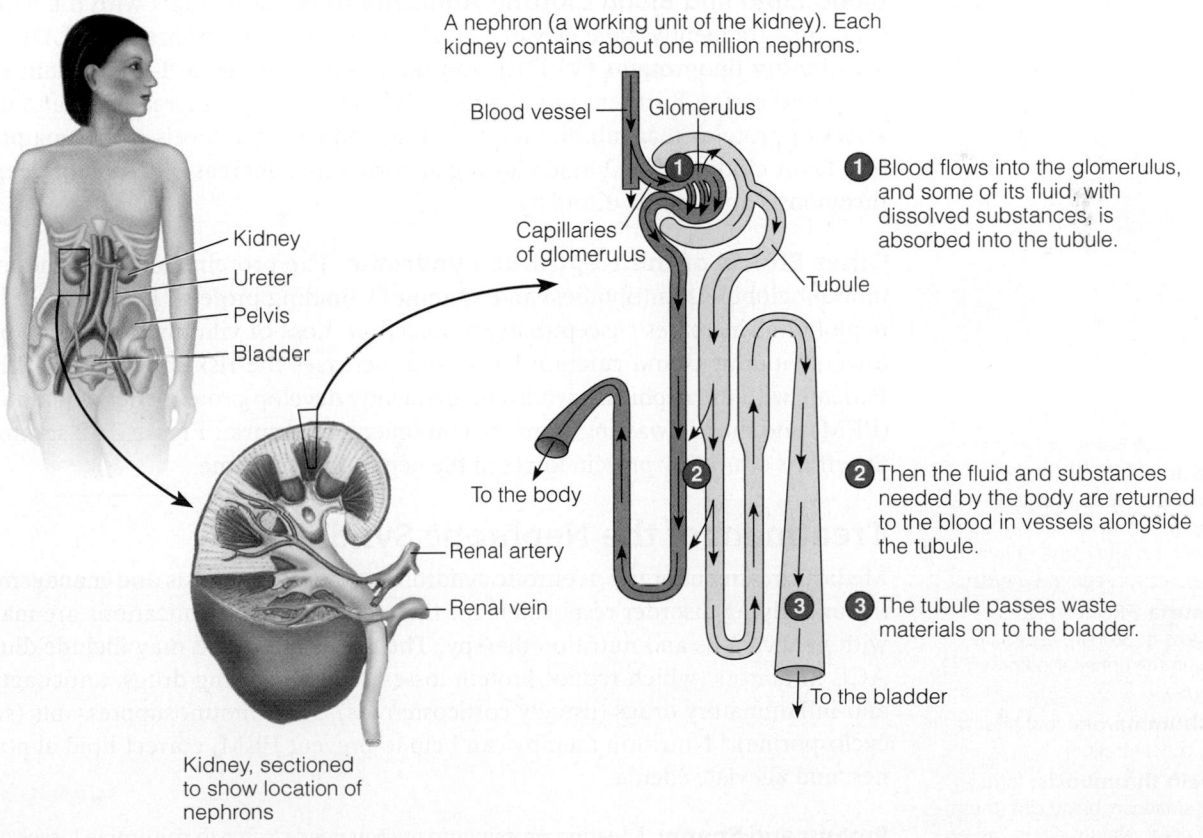

A nephron (a working unit of the kidney). Each kidney contains about one million nephrons.

Blood vessel — Glomerulus

Kidney
Ureter
Pelvis
Bladder

Capillaries of glomerulus

Tubule

❶ Blood flows into the glomerulus, and some of its fluid, with dissolved substances, is absorbed into the tubule.

To the body

Renal artery
Renal vein

❷ Then the fluid and substances needed by the body are returned to the blood in vessels alongside the tubule.

❸ The tubule passes waste materials on to the bladder.

To the bladder

Kidney, sectioned to show location of nephrons

FIGURE 23-1 The Kidneys and Nephron Function

Nephrotic Syndrome

The **nephrotic syndrome** is not a specific disease; rather, the term refers to kidney disorders that result in significant urinary protein losses **(proteinuria)** due to severe glomerular damage. The loss of plasma proteins (typically more than 3½ grams per day) causes serious consequences, including edema, blood lipid abnormalities, blood coagulation disorders, and infections. In some cases, the nephrotic syndrome can progress to renal failure.

Causes of the nephrotic syndrome include glomerular disorders, diabetic nephropathy, immunological and hereditary diseases, infections (involving the kidneys or elsewhere in the body), chemical damage (from medications or illicit drugs), and some cancers.[1] Depending on the underlying condition, some patients may experience one or more relapses and require additional treatment to prevent proteinuria from recurring.

Consequences of the Nephrotic Syndrome

In the nephrotic syndrome, urinary protein losses generally average about 8 grams daily.[2] The liver attempts to compensate by increasing its synthesis of various plasma proteins, but some of the proteins are produced in excessive amounts. The imbalance in plasma protein concentrations contributes to a number of complications.

Edema Albumin is the most abundant plasma protein, and it is the protein subject to the most significant urinary losses as well. The **hypoalbuminemia** characteristic of the nephrotic syndrome contributes to a fluid shift from blood plasma to the interstitial spaces and, thus, edema. Impaired sodium excretion also contributes to edema: the nephrotic kidney tends to reabsorb sodium in greater amounts than usual, causing sodium and water retention within the body.[3]

Blood Lipid and Blood Clotting Abnormalities Individuals with the nephrotic syndrome frequently have elevated levels of low-density lipoproteins (LDL), very-low-density lipoproteins (VLDL), and the more atherogenic LDL variant known as lipoprotein(a). Furthermore, the risk of blood clotting is increased due to urinary losses of proteins that inhibit blood clotting and elevated levels of plasma proteins that favor clotting. The blood clotting abnormalities increase the risk of **deep vein thrombosis** and similar disorders.

Other Effects of the Nephrotic Syndrome The proteins lost in urine include immunoglobulins (antibodies) and vitamin D–binding protein. Depletion of immunoglobulins increases susceptibility to infection. Loss of vitamin D–binding protein lowers vitamin D and calcium levels and increases the risk of rickets in children. Patients with the nephrotic syndrome frequently develop protein-energy malnutrition (PEM) and muscle wasting from the continued proteinuria. Figure 23-2 summarizes the effects of urinary protein losses in the nephrotic syndrome.

Treatment of the Nephrotic Syndrome

Medical treatment of the nephrotic syndrome requires diagnosis and management of the underlying disorder responsible for the proteinuria. Complications are managed with medications and nutrition therapy. The drugs prescribed may include diuretics, ACE inhibitors (which reduce protein losses), lipid-lowering drugs, anticoagulants, anti-inflammatory drugs (usually corticosteroids), and immunosuppressants (such as cyclosporine).[4] Nutrition therapy can help to prevent PEM, correct lipid abnormalities, and alleviate edema.

Protein and Energy Meeting protein and energy needs helps to minimize losses of muscle tissue. High-protein diets are not advised, however, because they can exacerbate urinary protein losses and result in further damage to the kidneys. Instead, the protein intake

Reminder: Plasma proteins, such as albumin, help to maintain fluid balance within the blood.

nephrotic (neh-FROT-ik) **syndrome:** a syndrome associated with kidney disorders that damage the glomerulus and cause urinary protein losses exceeding 3½ g/day.

proteinuria (PRO-teen-NOO-ree-ah): loss of protein, especially albumin, in the urine; also known as *albuminuria*.

hypoalbuminemia: low plasma albumin concentrations.

deep vein thrombosis: formation of a stationary blood clot (thrombus) in a deep vein, usually in the leg, which causes inflammation, pain, and swelling, and is potentially fatal.

```
                    Nephrotic syndrome
                            │
                            ▼
              Glomerular permeability increases
                            │
                            ▼
                    Plasma proteins spill
                         into urine
                            │
                            ▼
                     Plasma protein
                       levels fall
                            │
    ┌──────────┬───────────┼───────────┬──────────┐
    ▼          ▼           ▼           ▼          ▼
 Lean body   Albumin   Immuno-      Vitamin     Blood
 tissues     declines  globulins    D–binding   clearance
 break down            decline      protein     of lipids
                                    lost        declines
    │          │           │           │          │
    ▼          ▼           ▼           ▼          ▼
 Wasting     Edema      Infection    Rickets    Altered
                                                blood lipids
```

should fall between 0.8 and 1.0 gram per kilogram of body weight per day, and at least half of the protein consumed should be from high-quality protein sources.[5] An adequate energy intake (about 35 kcalories per kilogram of body weight daily) sustains weight and spares protein. Weight loss or infections suggest the need for additional energy.

Lipids As Chapter 22 explains, a diet low in saturated fat, *trans* fats, cholesterol, and refined sugars helps to control elevated LDL and VLDL. Dietary measures are usually inadequate for controlling blood lipids, however, so physicians usually prescribe lipid-lowering medications as well. In some cases, treating the underlying cause of nephrotic syndrome is sufficient for correcting the lipid disorders.[6]

Sodium and Potassium Controlling sodium intake helps to control edema; therefore, sodium intake is often limited to 1 to 2 grams daily.[7] Table 23-1 provides guidelines for following a diet restricted to 2 grams of sodium. If diuretics prescribed for the edema cause potassium losses, patients are encouraged to select foods rich in potassium (see Chapter 9).

Vitamins and Minerals Multivitamin-mineral supplementation is often advised to avoid nutrient deficiencies; nutrients at risk include vitamin B_6, vitamin B_{12}, folate, iron, copper, and zinc.[8] Patients with the nephrotic syndrome also require calcium (about 1000 to 1500 milligrams per day) and vitamin D supplementation to help prevent bone loss and rickets.

Nutrient deficiencies may develop if the carrier proteins for nutrients are lost in the urine.

IN SUMMARY

■ The nephrotic syndrome is characterized by proteinuria due to glomerular damage. Complications include edema, lipid and blood coagulation abnormalities, infections, and PEM.

■ Medications treat the underlying cause of proteinuria and manage complications. The diet should provide sufficient protein and energy to maintain health. Other dietary adjustments can help to correct lipid disorders, edema, and nutrient deficiencies.

TABLE 23-1 **Sodium-Controlled Diet**

General Guidelines

About 75 percent of the sodium in a typical diet comes from processed foods, about 10 percent from unprocessed natural foods, and about 15 percent from table salt. With this in mind:

▪ Whenever possible, select fresh foods, which are usually low in sodium.

▪ Select frozen and canned food products that have been prepared without added salt.

▪ Avoid adding salt to foods while cooking or at the table.

▪ When dining in restaurants, ask that meals be prepared without salt.

Sodium in Foods

All foods contain sodium, but some contain more than others. Use the information in the table below to plan meals that are low in sodium.

Food Group	Serving Size	Sodium per Serving (mg)
Milk products	1 cup milk or yogurt; 1 oz hard cheese (cheddar, Swiss, jack)	150–200
	Avoid: buttermilk, cottage cheese, cheese spreads, prepared cheeses (such as American cheese)	
Meat, fish, poultry, and eggs	3 oz fresh meat, fish, or poultry; 1 large egg	60
	Avoid: luncheon meats, corned beef, salt pork, sausage, frankfurters, bacon, canned meats or fish, fresh meats prepared with injected broth	
Fruits and vegetables	½ cup fresh vegetables, ½ cup fresh or frozen fruit, 6 oz fruit juice; 6 oz tomato or vegetable juice without added salt	10–20
	Avoid: pickled vegetables, olives, tomato or vegetable juices with added salt; dried fruits with added sodium sulfite	
Breads and cereals	½–⅔ cup dry or cooked cereal without added salt, ½ cup cooked rice or pasta	0–10
	1 slice bread, 1 roll or tortilla, ½–⅔ cup dry or cooked cereal prepared with salt	150
	Avoid: pancakes, waffles, muffins, biscuits, and quick breads made with baking powder or baking soda; instant and ready-to-eat cereals with >175 mg sodium; salted snack foods	
Condiments	Unsalted butter; low-sodium salad dressings, mayonnaise, sauces, and gravies; low-sodium catsup, mustard, and hot sauces; garlic or onion powders without added salt; lemon or lime juice, vinegar	Varies; check labels
	Avoid: commercial salad dressings, gravy or soup mixes, barbeque sauces, soy sauce, steak sauces, spices or herb products made with salt, bouillon, meat tenderizers, monosodium glutamate	

A Sample Diet Restricted to 2 Grams (2000 mg) of Sodium

Using the guidelines presented here, an individual can develop a variety of sample menus. A possible plan for a day might look like this:

Food Group	Sodium (mg)
Meat, 6 oz (2 servings × 60 mg)	120
Milk, 3 c (3 servings × 150 mg)	450
Fruit, 2 servings	negligible
Vegetables, 3 servings	45
Whole-grain bread, 4 slices (4 × 150 mg)	600
Salt, ¼ tsp (used lightly at meals)	600
Total	**1815**

Individuals can use the remainder of the sodium allowance for whatever foods they desire. The sodium content of most foods can be determined by reading food labels or using food composition tables (such as Appendix A). See additional information about reducing sodium intake in Chapter 22 (p. 585).

Acute Kidney Injury

In **acute kidney injury,** kidney function deteriorates rapidly, over hours or days. The loss of kidney function reduces urine output and allows nitrogenous wastes to build up in the blood. The degree of renal dysfunction varies from mild to severe. With prompt treatment, acute kidney injury is often reversible, although mortality rates are high, ranging from 36 to 86 percent.[9] Up to 7 percent of hospital patients develop acute kidney injury.[10]

Causes of Acute Kidney Injury

Many different disorders can lead to acute kidney injury, and it often develops as a consequence of severe illness, sepsis, or injury. To aid in diagnosis and treatment, its causes are commonly classified as prerenal, intrarenal, or postrenal. *Prerenal* factors are those that cause a sudden reduction in blood flow to the kidneys; they often involve a severe stressor such as heart failure, shock, or blood loss. Factors that damage kidney tissue, such as infections, toxicants, drugs, or direct trauma, are classified as *intrarenal* causes of acute kidney injury. *Postrenal* factors are those that prevent excretion of urine due to urinary tract obstructions. Table 23-2 provides examples of specific disorders that may cause acute kidney injury.

Consequences of Acute Kidney Injury

A decline in renal function alters the composition of blood and urine. The kidneys become unable to regulate the levels of electrolytes, acid, and nitrogenous wastes in the blood. Urine may be diminished in quantity **(oliguria)** or absent **(anuria),** leading to fluid retention. Diagnosis is often a complex task because the clinical effects can be subtle and vary according to the underlying cause of disease.

Fluid and Electrolyte Imbalances About half of patients with acute kidney injury experience oliguria, producing less than 400 milliliters of urine per day.[11] The reduced excretion of fluids and electrolytes results in sodium retention and elevated levels of potassium, phosphate, and magnesium in the blood. Elevated potassium **(hyperkalemia)** is of particular concern because potassium imbalances can alter heart rhythm and lead to heart failure. Elevated serum phosphate levels **(hyperphosphatemia)** promote excessive secretion of parathyroid hormone, leading to losses of bone calcium. Due to the sodium retention and reduced urine production, edema is a common symptom of acute kidney injury and may be apparent as puffiness in the face and hands and swelling of the feet and ankles.

Uremia As a result of impaired kidney function, nitrogen-containing compounds and various other waste products may accumulate in the blood—a condition often

A sudden rise in blood urea nitrogen (BUN) and/or serum creatinine suggests the presence of acute kidney injury. Refer to Table 14-7 on p. 393–394 for normal laboratory values.

Normal urine volume typically exceeds 800 mL per day, or about 3½ cups.

To measure serum phosphate, the phosphorus content of the blood is analyzed; thus the terms *serum phosphate* and *serum phosphorus* are often used interchangeably.

acute kidney injury: a condition associated with the abrupt decline of kidney function over a period of hours or days; potentially a cause of acute renal failure.

oliguria (OL-lih-GOO-ree-ah): an abnormally low amount of urine, often less than 400 mL/day.

anuria (ah-NOO-ree-ah): the absence of urine; clinically identified as urine output less than 50 mL/day.

hyperkalemia (HIGH-per-ka-LEE-me-ah): elevated serum potassium levels.

hyperphosphatemia (HIGH-per-fos-fa-TEE-me-ah): elevated serum phosphate levels.

TABLE 23-2 Causes of Acute Kidney Injury

Prerenal Factors (60 to 70% of cases)	Intrarenal Factors (25 to 40% of cases)	Postrenal Factors (5 to 10% of cases)
▪ **Low blood volume or pressure:** hemorrhage, burns, sepsis or shock, anaphylactic reactions, nephrotic syndrome, gastrointestinal losses, diuretics, antihypertensive medications ▪ **Renal artery disorders:** blood clots or emboli, stenosis, aneurysm, trauma ▪ **Heart disorders:** heart failure, heart attack, arrhythmias	▪ **Vascular disorders:** sickle-cell disease, diabetes mellitus, transfusion reactions ▪ **Obstructions (within kidney):** inflammation, tumors, stones, scar tissue ▪ **Renal injury:** infections, environmental contaminants, drugs, medications, *E. coli* food poisoning	▪ **Obstructions (ureter or bladder):** strictures, tumors, stones, trauma ▪ **Prostate disorders:** cancer or hyperplasia ▪ **Renal vein thrombosis** ▪ **Bladder disorders:** neurological conditions, bladder rupture ▪ **Pregnancy**

The related term *azotemia* refers specifically to the accumulation of nitrogenous wastes in the blood.

Nutrition in Practice 23 describes common dialysis procedures, including *continuous renal replacement therapy,* the approach usually used for treating acute kidney injury.

referred to as **uremia.** The clinical outcome, called the *uremic syndrome*, includes a cluster of symptoms caused by impairments in multiple body systems. Although the clinical signs vary considerably among patients, the symptoms may reflect hormonal imbalances, electrolyte and acid-base imbalances, disturbed heart and gastrointestinal functioning, neuromuscular disturbances, and depressed immunity, among other conditions. The uremic syndrome is described in more detail later in this chapter.

Treatment of Acute Kidney Injury

Treatment of acute kidney injury involves a combination of drug therapy, **dialysis,** and nutrition therapy to restore fluid and electrolyte balances and minimize blood concentrations of toxic waste products. Both medical care and dietary measures must be highly individualized to suit each patient's needs. Correcting the underlying illness is necessary to prevent further damage to the kidneys.

In oliguric patients (those with reduced urine output), recovery from kidney injury sometimes begins with a period of **diuresis,** in which large amounts of fluid (up to 3 liters daily) are excreted.[12] Because tubular function is minimal at this stage, electrolytes may not be sufficiently reabsorbed; consequently, both fluid depletion and electrolyte losses become a concern. Patients with this pattern of recovery (generally those with tubular injury) require close monitoring in case they need fluid and electrolyte replacement.

Drug Treatment in Acute Kidney Injury Because kidney function is required for drug excretion, patients may need to use lower doses of their usual medications to compensate for limited urine output. Conversely, dialysis treatment may increase losses of some drugs, and doses may need to be increased. Drugs that are **nephrotoxic** (including some antibiotics and nonsteroidal anti-inflammatory drugs) must be avoided until kidney function improves.

The medications prescribed for acute kidney injury depend on the cause of illness and complications that develop. Edema is treated with diuretics to mobilize fluids; furosemide (Lasix) is the usual choice. Patients with hyperkalemia are given potassium-exchange resins that bind potassium ions in the gastrointestinal (GI) tract, ensuring the elimination of potassium in the stool. Rapid correction of hyperkalemia requires the use of insulin, which causes a temporary shift of extracellular potassium into the cells. (Glucose must be supplied along with insulin to prevent hypoglycemia.) If acidosis is present, bicarbonate may be administered orally or intravenously.

Energy and Protein Although its effects are highly variable, acute kidney injury is often a catabolic condition associated with hypermetabolism and muscle wasting. Therefore, sufficient energy and protein must be ingested to preserve muscle mass. Initially, the patient may be provided with 25 to 35 kcalories per kilogram of body weight per day, while body weight is monitored to ensure that energy intake is adequate.[13] If available, indirect calorimetry provides the best estimate of energy needs.

Protein contributes nitrogen, increasing the kidneys' workload, but intake should be sufficient to maintain nitrogen balance (to the extent possible) and prevent additional wasting. Protein recommendations are influenced by kidney function, the degree of catabolism, and the use of dialysis (dialysis removes nitrogenous wastes). For patients who are not treated with dialysis, protein intakes should be limited to 0.8 to 1.2 grams per kilogram body weight per day. Higher intakes (1.2 to 1.5 grams per kilogram daily) may be recommended if kidney function improves or the treatment includes dialysis. Patients who are catabolic or septic typically have high protein needs, but they require dialysis to accommodate the additional nitrogen load.

Fluids Health practitioners can assess fluid status by monitoring weight fluctuations, blood pressure, pulse rates, and appearance of the skin and mucous membranes. Another assessment method is the measurement of serum sodium concentrations: a low level of sodium often indicates excessive fluid intake, and a high level suggests inadequate intake.

Fluid balance must be restored in patients who are either overhydrated or dehydrated. Thereafter, fluid needs can be estimated by measuring urine output and adding about 500 milliliters to account for the water lost from skin, lungs, and perspiration. An individual with fever, vomiting, or diarrhea requires additional fluid. Patients undergoing dialysis can ingest fluids more freely.

Electrolytes Serum electrolyte levels are monitored closely to determine appropriate electrolyte intakes. Depending on the results of laboratory tests and the clinical assessment, restrictions may be necessary for potassium (2000 to 3000 milligrams per day), phosphorus (8 to 15 milligrams per kilogram body weight per day), and sodium (2000 to 3000 milligrams per day).[14] Patients undergoing dialysis may be allowed more liberal intakes. As mentioned previously, oliguric patients who experience diuresis at the beginning of the recovery period may need electrolyte replacement to compensate for urinary losses.

Enteral and Parenteral Nutrition Some patients need nutrition support to obtain adequate energy. Enteral support (tube feeding) is preferred over parenteral nutrition because it is less likely to cause infection and sepsis. Enteral formulas for patients with acute kidney injury are more kcalorically dense and may have lower protein and electrolyte concentrations than standard formulas. Total parenteral nutrition is necessary only if patients are severely malnourished or cannot consume food or tolerate tube feedings for more than 14 days.[15]

IN SUMMARY

▐ Acute kidney injury is characterized by a rapid loss in kidney function, causing a buildup of waste products in the blood. Causes include prerenal, intrarenal, or postrenal factors.

▐ Consequences of acute kidney injury may include fluid and electrolyte imbalances and uremia. If hyperkalemia develops, it can alter heart rhythm and lead to heart failure.

▐ Acute kidney injury is treated with medications, dialysis, and dietary modifications.

The accompanying Case Study checks your understanding of acute kidney injury.

uremia (you-REE-me-ah): the accumulation of nitrogenous and various other waste products in the blood; often associated with symptoms that reflect impairments in multiple body systems (literally, "urine in the blood"). The term may also be used to indicate the toxic state that results when wastes are retained in the blood.

dialysis (dye-AH-lih-sis): a treatment that removes wastes and excess fluid from the blood after the kidneys have stopped functioning.

diuresis (DYE-uh-REE-sis): increased urine production.

nephrotoxic: toxic to the kidneys.

Case Study WOMAN WITH ACUTE KIDNEY INJURY

Catherine Garber is a 42-year-old office manager admitted to the hospital's intensive care unit. She was first seen in the emergency room with severe edema, headache, nausea and vomiting, and a rapid heart rate. She reported an inability to pass more than minimal amounts of urine in the past two days. Her son, who drove her to the emergency room, reported that she had missed work for several days and seemed confused and unusually tired. Laboratory tests revealed elevated serum creatinine, BUN, and potassium levels. After learning from her medical history that Mrs. Garber had begun taking penicillin earlier in the week, the physician diagnosed acute kidney

injury, probably caused by a reaction to the medication. Mrs. Garber is 5 feet 3 inches tall and weighs 125 pounds.

1. Describe the probable reason for Mrs. Garber's inability to produce urine. Is her reaction to penicillin considered a prerenal, intrarenal, or postrenal cause of kidney injury? Give examples of other medical problems that can cause acute kidney injury.

2. What medications can the physician prescribe to treat Mrs. Garber's edema and hyperkalemia? What recommendation is likely regarding her continued use of penicillin?

3. What concerns should be kept in mind when determining Mrs. Garber's energy, protein, fluid, and electrolyte needs during acute kidney injury? How would dialysis treatment alter recommendations?

4. After treatment begins, Mrs. Garber suddenly begins producing copious amounts of urine. How should this development alter dietary treatment?

As you read through the discussion of chronic kidney disease, consider how Mrs. Garber's diet should change if her kidney problems become chronic.

TABLE 23-3 Clinical Effects of Chronic Kidney Disease

Early Stages

- Anorexia
- Exercise Intolerance
- Fatigue
- Headache
- Hypercoagulation
- Hypertension
- Proteinuria, hematuria (blood in urine)

Advanced Stages

- Anemia, bleeding tendency
- Cardiovascular disease
- Confusion, mental impairments
- Electrolyte imbalances
- Fluid retention, edema
- Hormonal abnormalities
- Itching
- Metabolic acidosis
- Nausea and vomiting
- Peripheral neuropathy
- Protein-energy malnutrition
- Reduced immunity
- Renal osteodystrophy

The kidneys' ability to function despite loss of nephrons is referred to as *renal reserve.*

Reminder: *Creatinine* is a waste product of creatine, a nitrogen-containing compound in muscle cells.

Reminder: *Aldosterone* promotes sodium (and therefore water) retention and potassium excretion.

Reminder: *Parathyroid hormone* helps to regulate serum concentrations of calcium and phosphorus.

Chronic Kidney Disease

Unlike acute kidney injury, in which kidney function declines suddenly and rapidly, **chronic kidney disease** is characterized by gradual, irreversible deterioration. Because the kidneys have a large functional reserve, the disease typically progresses over many years without causing symptoms. Patients are typically diagnosed late in the course of illness after most kidney function has been lost.

The most common causes of chronic kidney disease are diabetes mellitus and hypertension, which are estimated to cause 45 and 27 percent of cases, respectively.[16] Other conditions that lead to chronic kidney disease include inflammatory, immunological, or hereditary diseases that directly involve the kidneys. Chronic kidney disease affects approximately 13 percent of the U.S. population.[17]

Consequences of Chronic Kidney Disease

In the early stages of chronic kidney disease, the nephrons compensate by enlarging so that they can handle the extra workload. As the nephrons deteriorate, however, there is additional work for the remaining nephrons. The overburdened nephrons continue to degenerate until finally the kidneys are unable to function adequately, resulting in kidney failure. Once the extent of kidney damage necessitates active treatment—either dialysis or a kidney transplant—the condition is classified as **end-stage renal disease (ESRD).** Table 23-3 lists common clinical effects of the early and advanced stages of chronic kidney disease.

Renal disease is evaluated using the **glomerular filtration rate (GFR),** the rate at which the kidneys form filtrate. The GFR can be estimated using predictive equations that are based on serum creatinine levels, age, gender, race, and body size. Table 23-4 shows how chronic kidney disease is classified according to estimated GFR. Other laboratory measures used to assess kidney function include urinary protein levels, BUN, and the ratio of albumin to creatinine in a urine sample.[18]

Altered Electrolytes and Hormones As the GFR falls, the increased activity by the remaining nephrons is often sufficient to maintain electrolyte excretion. Thus, fluid and electrolyte disturbances may not develop until the third or fourth stage of chronic kidney disease. A number of hormonal adaptations also help to regulate electrolyte levels, but these changes may cause complications of their own. The increased secretion of aldosterone helps to prevent increases in serum potassium but contributes to fluid overload and the development of hypertension (in patients who were not previously hypertensive). Increased secretion of parathyroid hormone helps to prevent elevations in serum phosphorus but contributes to bone loss and development of **renal osteodystrophy,** a bone disorder common in renal patients. Electrolyte imbalances

TABLE 23-4 Evaluation of Chronic Kidney Disease

Stage of Disease	Description	GFR[a] (mL/min per 1.73 m^2)
1	Kidney damage with normal or increased GFR	≥90
2	Kidney damage with mildly decreased GFR	60–89
3	Moderately decreased GFR	30–59
4	Severely decreased GFR	15–29
5	Kidney failure	<15 (or undergoing dialysis)

[a]Glomerular filtration rate, or GFR, is estimated from the Modification of Diet in Renal Disease study equation and is based on age, gender, race, and calibration for serum creatinine. Normal GFR is approximately 125 milliliters per minute.

Source: A. S. Levey and coauthors, National Kidney Foundation practice guidelines for chronic kidney disease: Evaluation, classification, and stratification, *Annals of Internal Medicine* 139 (2003): 137–147. Reprinted by permission.

are likely when the GFR becomes extremely low (less than 5 milliliters per minute), when hormonal adaptations are inadequate, or when intakes of water and electrolytes are either very restricted or excessive.

Because the kidneys are responsible for maintaining acid-base balance, acidosis often develops in chronic kidney disease. Although usually mild, the acidosis exacerbates renal bone disease because compounds in bone (for example, protein and phosphates) are released to buffer the acid in blood.

Uremic Syndrome Uremia usually develops during the final stages of chronic kidney disease, when the GFR falls below about 15 milliliters per minute.[19] As mentioned previously, the numerous symptoms and complications that result from uremia are collectively known as the **uremic syndrome.** Clinical effects may include the following:[20]

■ *Hormonal imbalances.* Diseased kidneys are unable to produce erythropoietin, causing anemia. Reduced production of active vitamin D contributes to bone disease.

■ *Altered heart function/increased heart disease risk.* Fluid and electrolyte imbalances result in hypertension, arrhythmias, and heart muscle enlargement. Excessive parathyroid hormone secretion leads to calcification of arteries and heart tissue.

■ *Neuromuscular disturbances.* Symptoms may include altered thought processes, sleep disorders, muscle cramping, sensory deficits, tremor, and seizures.

■ *Other effects.* Defects in platelet function and clotting factors prolong bleeding time and contribute to bruising and gastrointestinal bleeding. Skin changes include increased pigmentation and severe pruritus (itchiness). Patients with uremia typically have suppressed immune responses.

Protein-Energy Malnutrition Patients with chronic kidney disease often develop PEM and wasting. Anorexia is thought to contribute to the poor food intake of kidney patients and may result from hormonal disturbances, nausea and vomiting, restrictive diets, uremia, and medications. Nutrient losses also contribute to malnutrition and may be a consequence of vomiting, diarrhea, gastrointestinal bleeding, and dialysis. In addition, many of the illnesses that lead to chronic kidney disease induce a catabolic state that contributes to protein losses.[21]

Treatment of Chronic Kidney Disease

The goals of treatment for chronic kidney disease are to slow disease progression and prevent or alleviate symptoms. Nutrition therapy helps to prevent PEM and weight loss. Once kidney disease reaches the final stages, dialysis or a kidney transplant is necessary to sustain life.

Drug Therapy Medications help to control some of the complications associated with chronic kidney disease. Treatment of hypertension is critical for slowing disease progression and reducing cardiovascular disease risk; thus, antihypertensive drugs are usually prescribed (see Chapter 22). Some antihypertensive drugs (such as ACE inhibitors) can reduce proteinuria, helping to prevent additional kidney damage. Anemia is treated by injection or intravenous administration of erythropoietin (epoetin). Other drug treatments include phosphate binders (taken with food) to reduce serum phosphorus levels, sodium bicarbonate to reverse acidosis, and cholesterol-lowering medications. Supplementation with active vitamin D (called *calcitriol*) helps to raise serum calcium and reduce parathyroid hormone levels.

Dialysis Dialysis replaces kidney function by removing excess fluid and wastes from the blood. In **hemodialysis,** the blood is circulated through a **dialyzer** (artificial kidney), where it is bathed by a **dialysate,** a solution that selectively removes fluid and

Reminder: A screening method sometimes used for assessing PEM risk is the *Subjective Global Assessment,* described in Ch. 14 (Table 14-2 on p. 384).

chronic kidney disease: kidney disease characterized by gradual, irreversible deterioration of the kidneys; also known as *chronic renal failure.*

end-stage renal disease (ESRD): an advanced stage of chronic kidney disease in which dialysis or a kidney transplant is necessary to sustain life.

glomerular filtration rate (GFR): the rate at which filtrate is formed within the kidneys, normally about 125 mL/min.

renal osteodystrophy: a bone disorder that develops in patients with chronic kidney disease as a consequence of the increased secretion of parathyroid hormone, reduced serum calcium, acidosis, and impaired vitamin D activation by the kidneys.

uremic syndrome: the cluster of symptoms associated with inadequate kidney function; the symptoms reflect fluid, electrolyte, and hormonal imbalances; altered heart function; neuromuscular disturbances; and other metabolic derangements.

hemodialysis (HE-moe-dye-AL-ih-sis): a treatment that removes fluids and wastes from the blood by passing the blood through a dialyzer.

dialyzer (DYE-ah-LYE-zer): a machine used in hemodialysis to filter the blood; also called an *artificial kidney.*

dialysate (dye-AL-ih-sate): the solution used in dialysis to draw wastes and fluids from the blood.

TABLE 23-5 Dietary Recommendations for Chronic Kidney Disease

Nutrient	Predialysis	Hemodialysis	Peritoneal Dialysis
Energy[a] (kcal/kg body weight)	35 for <60 years old 30–35 for ≥60 years old	35 for <60 years old 30–35 for ≥60 years old	35 for <60 years old 30–35 for ≥60 years old (total energy intake includes kcalories absorbed from the dialysate)
Protein (g/kg body weight)	0.60–0.75 (≥50% high-quality proteins)	≥1.2 (≥50% high-quality proteins)	≥1.2–1.3 (≥50% high-quality proteins)
Fat	As necessary to maintain a healthy lipid profile	As necessary to maintain a healthy lipid profile	As necessary to maintain a healthy lipid profile
Fluid (mL/day)	Unrestricted if urine output is normal	500–1000 plus daily urine output	As necessary to maintain fluid balance
Sodium (mg/day)	1000–3000	1000–3000	2000–4000
Potassium (mg/day)	Unrestricted unless hyperkalemia is present	2000–3000; adjust according to serum potassium levels	3000–4000; adjust according to serum potassium levels
Calcium (mg/day)	1000–1500	≤2000 from diet and medications	≤2000 from diet and medications
Phosphorus (mg/day)	800–1000 if serum phosphorus or parathyroid hormone is elevated	800–1000 if serum phosphorus or parathyroid hormone is elevated	800–1000 if serum phosphorus or parathyroid hormone is elevated

[a]Values listed apply to adults; recommendations for children should not fall below DRI levels.

Sources: American Dietetic Association, *Nutrition Care Manual* (Chicago: American Dietetic Association, 2011); D. J. Goldstein-Fuchs and C. M. Goeddeke-Merickel, Nutrition and kidney disease, in A. Greenberg, ed., *Primer on Kidney Diseases* (Philadelphia: Saunders, 2009), pp. 479–486.

wastes. In **peritoneal dialysis,** the dialysate is infused into a person's peritoneal cavity, and blood is filtered by the peritoneum (the membrane that surrounds the abdominal cavity). After several hours, the dialysate is drained, removing unneeded fluid and wastes. Nutrition in Practice 23 provides additional information about dialysis.

Nutrition Therapy for Chronic Kidney Disease The patient's diet strongly influences disease progression and the development of complications. Because the dietary measures for kidney disease are complex and nutrient needs change frequently during the course of illness, a dietitian who specializes in renal disease is best suited to provide nutrition therapy. Table 23-5 summarizes the general dietary guidelines for patients in the different stages of disease. Because patients' needs vary considerably, actual recommendations should be based on the results of a careful and complete nutrition assessment.

Energy The energy intake should be high enough to allow patients to maintain a healthy weight and to prevent wasting. Foods and beverages with a high energy density are typically recommended. Malnourished patients may require oral supplements or tube feedings to maintain weight. The "How to" on p. 645 in Chapter 25 includes suggestions for increasing the energy content of meals.

The dialysate used in peritoneal dialysis contains glucose in order to draw fluid from the blood to the peritoneal cavity by osmosis; about 60 percent of this glucose is absorbed. The kcalories from glucose (as many as 800 kcalories daily) must be included in estimates of energy intake. Weight gain is sometimes a problem when peritoneal dialysis continues for a long period.[22]

Protein A low-protein diet is usually prescribed to reduce the amount of nitrogenous waste produced. In addition, low-protein diets supply less phosphorus than high-protein diets, reducing the risks associated with hyperphosphatemia. Because

Reminder: Foods with a high energy density contain a high number of kcalories per unit weight; these foods are generally high in fat and low in water content.

The protein RDA for adults is 0.8 g/kg.

renal patients often develop PEM, however, their diet must provide enough protein to meet needs and prevent wasting. During the predialysis period, the recommended protein intake is 0.60 to 0.75 grams per kilogram of body weight per day.[23] At least 50 percent of the protein consumed should come from high-quality protein sources (such as eggs, milk products, meat, poultry, fish, and soybeans) to ensure that the patient consumes adequate amounts of the essential amino acids. Low-protein breads, pastas, and other grain-based products are commercially available to help renal patients improve energy intakes without increasing protein consumption.

Because of the high risk of wasting and compliance difficulties associated with low-protein diets, some dietitians suggest that patients consume higher amounts of protein to preserve health.[24] Once dialysis has begun, protein restrictions can be relaxed because dialysis removes nitrogenous wastes and results in some amino acid losses as well.

In a renal diet, at least half of the protein consumed should be from high-quality protein sources such as eggs, milk products, meat, poultry, and fish.

Lipids To control elevated blood lipids and reduce heart disease risk, patients with chronic kidney disease are advised to restrict their intakes of saturated fat, *trans* fat, and cholesterol. Although patients are often encouraged to consume high-fat foods to improve their energy intakes, the foods they select should provide mostly unsaturated fats. Good choices include nuts and seeds, oil-based salad dressings, mayonnaise, avocados, and soybean products.

Sodium and Fluids As kidney disease progresses, patients excrete less urine and cannot handle normal amounts of sodium and fluids. Recommendations depend on the total urine output, changes in body weight and blood pressure, and serum sodium levels. A rise in body weight and blood pressure suggests that the person is retaining sodium and fluid; conversely, declines in these measurements indicate fluid loss. Most persons with kidney disease tend to retain sodium and may benefit from mild restriction; less frequently, a patient may have a salt-wasting condition that requires additional dietary sodium.

People on a renal diet can consume most fruits and vegetables in limited amounts.

Fluids are not restricted until urine output decreases. For a person who is neither dehydrated nor overhydrated, the daily fluid intake should match the daily urine output. (Obligatory water losses—from skin and lungs—are replaced by the water contained in the solid foods that are consumed.) Once a person is on dialysis, sodium and fluid intakes should be controlled so that only about 2 pounds of water weight are gained daily—this excess fluid is then removed during the next dialysis treatment. Patients on fluid-restricted diets should be advised that foods such as flavored gelatin, soups, frozen fruit juice bars, and ice milk contribute to the fluid allowance.

Potassium Before dialysis treatments begin, most patients can handle typical intakes of potassium; restrictions are usually necessary only in those with elevated potassium levels. Individuals with diabetic nephropathy are at high risk of hyperkalemia and may also need to limit dietary potassium during the early stages of disease.

Dialysis patients must control potassium intakes to prevent hyperkalemia or, more rarely, **hypokalemia.** Restriction is necessary for persons treated with hemodialysis, whereas those undergoing peritoneal dialysis can consume potassium more freely.

peritoneal (PEH-rih-toe-NEE-al) **dialysis:** a treatment that removes fluids and wastes from the blood by using the body's peritoneal membrane as a filter.

hypokalemia (HIGH-po-ka-LEE-me-ah): low serum potassium levels.

Recommended intakes are based on serum potassium levels, renal function, medications, and the dialysis procedure used.

All fresh foods provide potassium, but some fruits and vegetables contain such high amounts that some patients must restrict intakes. Table 23-6 shows the potassium content of some common fruits and vegetables. Foods in other food groups may be high in potassium as well; examples include dried beans, fish, milk and milk products, molasses, nuts and nut butters, and wheat bran.

Calcium, Phosphorus, and Vitamin D To prevent bone disease, calcium and phosphorus intakes may need adjustment, even during the early stages of kidney disease. Laboratory values usually help to guide dietary recommendations for these nutrients. Serum calcium levels must be monitored to guard against **hypercalcemia,** which can develop in response to simultaneous calcium and vitamin D supplementation. Elevated serum phosphorus levels indicate the need for dietary phosphorus restriction. Vitamin D supplementation is a standard treatment for many renal patients, but the amount prescribed depends on the serum levels of calcium, phosphorus, and parathyroid hormone.

High-protein foods are also high in phosphorus, so protein-restricted diets curb phosphorus intakes as well. After dialysis treatments begin and protein intakes are liberalized, phosphate binders (taken with meals) become essential for phosphorus control. Because foods that are rich in calcium (such as milk and milk products) are usually high in phosphorus and are therefore restricted, patients must rely on calcium supplements to meet their calcium needs. Table 23-7 lists examples of foods that are high in phosphorus.

Vitamins and Minerals The restrictive renal diet interferes with vitamin and mineral intakes, increasing the risk of deficiencies. In addition, patients treated with dialysis lose water-soluble vitamins and some trace minerals into the dialysate. Dietary supplements for dialysis patients typically supply generous amounts of folic acid and vitamin B_6—0.8 to 1.0 milligram and 5 to 10 milligrams per day, respectively—along with recommended amounts of the other water-soluble vitamins.[25] Supplemental vitamin C should be limited to 100 milligrams per day, because excessive intakes can

TABLE 23-6 Potassium Guide—Fruits and Vegetables

This table lists common fruits and vegetables according to their potassium content. One serving is ½ cup raw fruit or cooked vegetable unless otherwise noted. Keep in mind that the portion size may determine how a food is categorized. Check Appendix A for additional information about the potassium content of foods.

High Potassium (>250 mg per serving)	Medium Potassium (150–250 mg per serving)	Low Potassium (<150 mg per serving)
Avocado	Apple (1 medium)	Blueberries
Banana	Apricots (2 whole)	Cabbage
Beets	Asparagus	Carrots (1 medium)
Chard	Broccoli	Cauliflower
Dates (3 whole)	Cantaloupe	Cucumbers
Nectarine (1 small)	Celery	Eggplant
Orange (1 medium)	Corn	Grapes
Parsnips	Grapefruit (½ fruit)	Green beans
Potatoes	Honeydew melon	Green pepper
Pumpkin	Kale	Lettuce (4 leaves, raw)
Raisins	Peach (1 small)	Onions (1 small)
Spinach	Pear (1 medium)	Plum (1 small)
Sweet potatoes	Peas	Strawberries
Tomato	Zucchini	Watermelon

contribute to kidney stone formation in those at risk (see p. 610). Vitamin A supplements are not recommended because vitamin A levels tend to rise as kidney function worsens.

Iron deficiency is common in hemodialysis patients and may be due to inadequate erythropoietin, gastrointestinal bleeding, reduced iron absorption, or blood losses associated with the dialysis treatment.[26] Intravenous administration of iron, in conjunction with erythropoietin therapy, is more effective than oral iron supplementation for improving iron status.

Enteral and Parenteral Nutrition Nutrition support is sometimes necessary for renal patients who cannot consume adequate amounts of food. The enteral formulas suitable for chronic kidney disease are more kcalorically dense and have lower protein and electrolyte concentrations than standard formulas. **Intradialytic parenteral nutrition** is an option for supplying supplemental nutrients to dialysis patients; this technique combines parenteral infusions with hemodialysis treatments. An advantage of this approach is that the volume of parenteral solution infused can be simultaneously removed (recall that fluid intake is controlled in dialysis patients). However, clinical studies have not shown intradialytic parenteral nutrition to be more successful than oral supplementation in malnourished dialysis patients.[27]

Dietary Compliance Adhering to a renal diet is probably the most challenging aspect of treatment for renal disease patients. These patients often require extensive counseling once multiple dietary restrictions become necessary. Depending on the stage of illness and the patient's laboratory values, the renal diet may limit protein, fluids, sodium, potassium, and phosphorus, thereby affecting food selections from all major food groups. In addition, adjustments in nutrient intake are required as the disease progresses. Because these diets have so many restrictions, patient compliance is often a problem. The "How to" provides suggestions to help patients comply with renal diets, and Table 23-8 shows an example of a one-day menu that includes some of the restrictions discussed in the previous paragraphs. The accompanying Case Study allows you to apply your knowledge about chronic kidney disease and hemodialysis.

TABLE 23-7 Foods High in Phosphorus[a]

- Barley
- Bran (oat, wheat)
- Buckwheat groats
- Bulgur
- Canned iced teas
- Canned lemonade
- Coconut
- Cola beverages
- Cornmeal
- Couscous
- Dried peas and beans
- Fish
- Milk products
- Nuts and seeds
- Organ meats
- Peanut butter
- Processed meats
- Soybeans, tofu

[a]For a complete list, visit the USDA's Nutrient Database at www.nal.usda. gov/fnic/foodcomp/search. Click on "Nutrient lists," and then find the list of foods sorted in descending order by phosphorus content (click on the letter "W" to the right of the word "Phosphorus").

TABLE 23-8 **Chronic Kidney Disease—One-Day Menu**

The menu below provides 2028 kcalories, 46 g protein, 784 mg phosphorus, 2190 mg potassium, and 1510 mg sodium.[a] The energy and protein content would be appropriate for a 135-pound predialysis patient.

Breakfast
- Corn flakes with milk (1 cup cereal, ½ cup whole milk)
- Apricot nectar (1 cup)
- Caffé latte (brewed coffee, 2 tsp sugar, ½ cup cream substitute)

Lunch
- Turkey sandwich (2 slices white bread, 1½ oz dark meat, 5 slices cucumber, 1 tbs mayonnaise)
- Grape juice (1 cup)
- Orange sherbet (½ cup)

Dinner
- Spaghetti with tomato sauce (1 cup cooked spaghetti, ½ cup bottled tomato sauce, ½ tbs grated cheese)
- Green beans with olive oil (1 cup cooked green beans, 1 tbs olive oil)
- Biscuit with margarine (2½-inch biscuit, ½ tbs margarine)
- Baked apple with nondairy sour cream (1 large apple, ¼ cup nondairy sour cream)

[a]Energy and nutrient values were obtained from the USDA National Nutrient Database for Standard Reference: www.nal.usda.gov/fnic/foodcomp/search

hypercalcemia (HIGH-per-kal-SEE-me-ah): elevated serum calcium levels.

intradialytic parenteral nutrition: the infusion of nutrients during hemodialysis, often providing amino acids, dextrose, lipids, and some trace minerals.

Patients with renal disease and their caregivers face considerable challenges as they learn to manage a renal diet. The following suggestions may help:

1. *To keep track of fluid intake:*

   ▌ Fill a container with an amount of water equal to your total fluid allowance. Each time you consume a liquid food or beverage, discard an equivalent amount of water from the container. The amount remaining in the container will show you how much fluid you have left for the day.

   ▌ Be sure to save enough fluid to take medications.

2. *To help control thirst:*

   ▌ Chew gum or suck hard candy.

   ▌ Suck on frozen grapes.

   ▌ Freeze beverages to a semisolid state so that they take longer to consume.

   ▌ Add lemon juice or crumpled mint leaves to water to make it more refreshing.

   ▌ Gargle with refrigerated mouthwash.

3. *To increase the energy content of meals:*

   ▌ Add extra margarine or a flavored oil to rice, noodles, breads, crackers, and cooked vegetables. Add extra salad dressing or mayonnaise to salads.

   ▌ Add nondairy whipped toppings to desserts.

   ▌ Include fried foods in your diet.

4. *To include more of your favorite vegetables in meals:*

   ▌ Consult your nurse or dietitian to learn whether you can safely use the process of leaching to remove some of the potassium from vegetables.

   ▌ To leach potassium from vegetables: Cut the vegetables into $1/8$-inch slices and rinse. Soak the vegetables in a large amount of warm water for two hours—about 10 parts of water to 1 part of vegetables. Rinse vegetables well. Boil vegetables using 5 parts of water to 1 part of vegetables.

5. *To prevent the diet from becoming monotonous:*

   ▌ Experiment with new combinations of allowed foods.

   ▌ Substitute nondairy products for milk products. Nondairy products, which are lower in protein, phosphorus, and potassium, can substitute for milk and add energy to the diet.

   ▌ Add flavor to foods by seasoning with garlic, onion, chili powder, curry powder, oregano, mint, basil, parsley, pepper, or lemon juice.

   ▌ Consult a nurse or dietitian when you want to eat restricted foods. Many restricted foods can be used occasionally and in small amounts if the menu is carefully adjusted.

Case Study MAN WITH CHRONIC KIDNEY DISEASE

Thomas Stone is a 55-year-old banker who developed chronic kidney disease as a result of hypertension. His condition was discovered several years ago, when routine laboratory tests revealed elevated serum creatinine and BUN levels. Since then, he has been taking antihypertensive medications and restricting dietary sodium; he reported difficulty following the low-protein diet that was also prescribed. Mr. Stone recently visited his doctor with complaints of low urine output and reduced sensation in his hands and feet. He also reported feeling drowsy at work and mentioned that he was bruising more than usual. The examination revealed a 9-pound weight gain since his last visit and swelling in his ankles and feet. Tests revealed that his GFR had fallen to 10 milliliters per minute. Mr. Stone is 5 feet 8 inches tall and normally weighs 160 pounds.

1. Explain how chronic kidney disease progresses. What happens to GFR, serum creatinine levels, and BUN as renal function declines?

2. Describe the clinical effects you would expect during the final stage of disease, when kidney failure develops. Explain the significance of each of Mr. Stone's physical complaints.

3. Explain why a low-sodium, low-protein diet was prescribed for Mr. Stone at a former visit. What energy and protein intakes were probably recommended at that time?

4. The physician determines that Mr. Stone's kidney disease has reached the final stage and prescribes hemodialysis. How will dialysis alter Mr. Stone's diet? Calculate his new protein recommendation, and compare it to the amount of protein recommended before dialysis. What other changes in nutrient intake may be necessary?

Kidney Transplants

A preferred alternative to dialysis in patients with end-stage renal disease is kidney transplantation.[28] A successful kidney transplant restores kidney function, allows a more liberal diet, and frees the patient from routine dialysis. Given the choice, many patients would prefer transplants, but the demand for suitable kidneys far exceeds the supply. Other barriers to transplantation include advanced age, poor health, and financial difficulties. Approximately 30 percent of patients who develop end-stage renal disease receive a kidney transplant.[29]

Immunosuppressive Drug Therapy To prevent tissue rejection following transplant surgery, patients require high doses of immunosuppressive drugs. These drugs have multiple effects that can alter nutrition status, including nausea, vomiting, diarrhea, glucose intolerance, altered blood lipids, fluid retention, hypertension, and increased risk of infection. Because immunosuppressive drugs increase the risk of foodborne infection, food safety guidelines should be provided to patients and caregivers. The Diet-Drug Interactions feature summarizes the nutrition-related effects of the drugs mentioned in this chapter.

Examples of immunosuppressive drugs used after a kidney transplant:
- Azathioprine (Imuran)
- Corticosteroids (prednisone)
- Cyclosporine (Sandimmune)
- Tacrolimus (Prograf)

Nutrition Therapy after a Kidney Transplant Energy and protein requirements increase after surgery due to stress and the catabolic effects of drug therapy. Once recovery is under way, recommendations for energy and most nutrients are similar to those suggested for the general population. Patients should attempt to maintain a healthy body weight and consume a diet that reduces their risk for cardiovascular diseases.

For most transplant patients, the side effects of drugs are the primary reason that dietary adjustments may be required. Although sodium, potassium, phosphorus, and fluid intakes are usually liberalized following a transplant, serum electrolyte levels must be monitored because some drug therapies can cause electrolyte imbalances or fluid retention. If corticosteroids are used as immunosuppressants, calcium supplementation is recommended because the medication increases urinary calcium losses. If drug treatment leads to hyperglycemia, patients should limit intakes of refined carbohydrates and concentrated sweets; for some individuals, oral medications or insulin therapy may be necessary. As noted earlier, patients must carefully follow food safety guidelines to avoid foodborne illness.

Diet-Drug Interactions

Check this table for notable nutrition-related effects of the medications discussed in this chapter.

Immunosuppressants

Cyclosporine, tacrolimus	**Gastrointestinal effects:** Nausea, vomiting, diarrhea, anorexia (tacrolimus)
	Dietary interactions: Cyclosporine potentiates the effects of alcohol. The bioavailability of tacrolimus is reduced when the drug is taken with food. Grapefruit juice can raise serum concentrations of these drugs to toxic levels.
	Metabolic effects: Electrolyte imbalances, hypertension, hyperglycemia, hyperlipidemia
Corticosteroids	**Metabolic effects:** Glucose intolerance, sodium retention, negative nitrogen balance, appetite stimulation, weight gain, growth suppression in children
Phosphate binders (calcium-containing)	**Gastrointestinal effect:** Constipation **Metabolic effects:** Electrolyte imbalances
Potassium-exchange resins (sodium polystyrene sulfonate)	**Metabolic effects:** Fluid retention, hypokalemia, hypocalcemia
Potassium citrate	**Gastrointestinal effects:** Nausea, vomiting, stomach pain, diarrhea **Metabolic effect:** Hyperkalemia

- Chronic kidney disease causes gradual loss of kidney function and often results from long-standing diabetes mellitus or hypertension.

- Depending on the stage of illness, complications of chronic kidney disease may include fluid and electrolyte imbalances, hypertension, renal osteodystrophy, mental impairments, bleeding abnormalities, anemia, increased risk for cardiovascular disease, and reduced immunity.

- Treatment of chronic kidney disease can slow disease progression and correct complications and includes drug therapies, dialysis, and nutrition therapy. Dietary measures usually feature a low-protein diet, controlled fluid and sodium intakes, phosphorus restrictions, and calcium and vitamin D supplementation; potassium restrictions are usually necessary after dialysis treatment begins.

- In patients with chronic kidney disease, kidney transplantation can restore renal function and liberalize dietary restrictions.

Kidney Stones

Approximately 12 percent of men and 6 percent of women in the United States develop one or more **kidney stones** during their lifetimes.[30] A kidney stone is a crystalline mass that forms within the urinary tract. Although stones are often asymptomatic, their passage can cause severe pain or block the urinary tract. Stones tend to recur, but they can be prevented with dietary measures and medical treatment.

The most common type of kidney stone is composed of calcium oxalate crystals, as shown here. Kidney stones may be as small as a bread crumb or as large as a golf ball.

Reminder: Fat malabsorption promotes oxalate absorption, thereby increasing the risk of forming calcium oxalate stones (see Chapter 19).

Formation of Kidney Stones

Kidney stones develop when stone constituents become concentrated in urine, allowing crystals to form and grow. About 70 percent of kidney stones are composed primarily of calcium oxalate. Less commonly, stones are composed of calcium phosphate, uric acid, the amino acid cystine, or magnesium ammonium phosphate (the latter are known as *struvite* stones). Stone formation is promoted by factors that reduce urine volume, block urine flow, or increase concentrations of stone-forming substances.

Calcium Oxalate Stones The most common abnormality in people with calcium oxalate stones is **hypercalciuria** (elevated urinary calcium levels). Hypercalciuria can result from excessive calcium absorption, impaired calcium reabsorption in kidney tubules, or elevated serum levels of parathyroid hormone or vitamin D. However, some people with calcium oxalate stones excrete normal amounts of calcium in the urine, and the reason they form stones is unknown.

Elevated urinary oxalate levels, or **hyperoxaluria**, also promote the formation of calcium oxalate crystals. Oxalate is a normal product of metabolism that readily binds to calcium. Hyperoxaluria may reflect an increase in the body's synthesis of oxalate or increased absorption from dietary sources. People who form calcium oxalate stones are advised to reduce their dietary intake of oxalate (see Table 23-9) and to avoid supplementation with vitamin C, which degrades to oxalate in the body.[31]

Uric Acid Stones Uric acid stones develop when the urine is abnormally acidic, contains excessive uric acid, or both. These stones are frequently associated with **gout**, a metabolic disorder characterized by elevated uric acid levels in the blood and urine. A diet rich in **purines** also contributes to high uric acid levels; purines are abundant

TABLE 23-9 Foods High in Oxalate

Vegetables	Fruits	Other
Beets*	Apricots, dried	Barley
Chard	Blackberries	Buckwheat
Collard greens	Blueberries	Chocolate*
Dried beans	Currants, red	Cocoa
Eggplant	Figs	Cornmeal, grits
Escarole	Grapes, Concord	Miso
Green beans	Kiwi	Nuts, nut butters*
Kale	Lemon peel	Peanut butter*
Leeks	Oranges, orange peel	Sesame seeds, tahini
Mustard greens	Raspberries	Soybean products
Parsley	Rhubarb*	Tea*
Spinach*	Strawberries*	Wheat bran*
Sweet potatoes		Whole-wheat flour

Note: The oxalate content of many foods has not been analyzed, and few studies have been conducted to determine which foods raise urinary oxalate levels.

*The foods marked with an asterisk have been documented to raise urinary oxalate levels and should be avoided by people who form calcium oxalate stones.

in animal proteins (meat, poultry, seafood) and degrade to uric acid in the body. In addition, a high intake of animal protein increases urine acidity, which promotes the crystallization of uric acid.

Cystine and Struvite Stones Cystine stones can form in people with the inherited disorder **cystinuria,** in which the renal tubules are unable to reabsorb the amino acid cystine. This abnormality results in high concentrations of cystine in the urine, leading to subsequent crystallization and stone formation. **Struvite** stones, composed primarily of magnesium ammonium phosphate, form in alkaline urine; the urinary pH is elevated due to the bacterial degradation of urea to ammonia. Struvite stones can accompany chronic urinary infections or disorders that interfere with urinary flow.

Consequences of Kidney Stones

In most cases, kidney stones do not pose serious medical problems. Small stones can readily pass through the ureters and out of the body with minimal treatment.

Renal Colic A stone passing through the ureter can produce severe, stabbing pain, called **renal colic.** The pain can be severe enough to cause nausea and vomiting and sometimes requires medication. Blood may appear in the urine (**hematuria**) as a result of damage to the kidney or ureter lining.

Urinary Tract Complications Depending on the location of the stone, symptoms may include urination urgency, frequent urination, or inability to urinate. Stones that are unable to pass through the ureter can cause a urinary tract obstruction and possibly lead to infection or acute kidney injury.

Prevention and Treatment of Kidney Stones

Solutes are less likely to crystallize and form stones in dilute urine. Therefore, people who form kidney stones are advised to drink 12 to 16 cups of fluid throughout the day in order to maintain urine volumes of least 2 to 2½ liters per day.[32] Additional fluid may be needed in hot weather or if an individual is extremely active. For some patients, dietary modifications, medications, or surgical stone removal may be necessary.

Calcium Oxalate Stones Most dietary measures and drug treatments for calcium oxalate stones aim to reduce urinary calcium and oxalate levels. Dietary measures may

kidney stones: crystalline masses that form in the urinary tract; also called *renal calculi* or *nephrolithiasis.*

hypercalciuria (HIGH-per-kal-see-YOO-ree-ah): elevated urinary calcium levels.

hyperoxaluria (HIGH-per-ox-ah-LOO-ree-ah): elevated urinary oxalate levels.

gout (GOWT): a metabolic disorder characterized by elevated uric acid levels in the blood and urine and the deposition of uric acid in and around the joints, causing acute joint inflammation.

purines (PYOO-reens): products of nucleotide metabolism that degrade to uric acid.

cystinuria (SIS-tin-NOO-ree-ah): an inherited disorder characterized by the elevated urinary excretion of several amino acids, including cystine.

struvite (STROO-vite): crystals of magnesium ammonium phosphate.

renal colic: the intense pain that occurs when a kidney stone passes through the ureter.

hematuria (HE-mah-TOO-ree-ah): blood in the urine.

include adjustments in calcium, oxalate, protein, and sodium intakes.[33] Patients should consume adequate calcium from food sources (about 800 and 1200 milligrams per day for men and women, respectively) because dietary calcium combines with oxalate in the intestines, reducing oxalate absorption and helping to control hyperoxaluria. Conversely, low-calcium diets promote oxalate absorption and higher urinary oxalate levels. Some individuals with hyperoxaluria may benefit from dietary oxalate restriction.

High protein and sodium intakes increase urinary calcium excretion, so moderate protein consumption (0.8 to 1.0 gram per kilogram of body weight per day) and a controlled sodium intake (no more than 3450 milligrams daily) are also advised. Vitamin C intakes should not exceed the RDA (90 and 75 milligrams for men and women, respectively) because vitamin C degrades to oxalate. Medications used to prevent calcium oxalate stones include thiazide diuretics, which reduce urinary calcium levels; cholestyramine (Questran), which reduces oxalate absorption; potassium citrate (a base), which inhibits crystal formation; and allopurinol (Zyloprim), which reduces uric acid production in the body and may have other effects.[34]

Uric Acid Stones Diets restricted in purines may help to control urinary uric acid levels. Because all animal proteins contain purines, strict dietary control over a long period may be difficult to achieve. In addition, the benefits of purine restriction are unclear. Drug treatments for uric acid stones include allopurinol to reduce uric acid levels and potassium citrate to reduce urine acidity.

Cystine and Struvite Stones High fluid intakes may prevent the formation of cystine stones in some patients, whereas other individuals require drug therapy to reduce cystine production in the body. Medications frequently prescribed include penicillamine (Cuprimine) and tiopronin (Thiola), which increase the solubility of cystine, and potassium citrate, which reduces urine acidity. For preventing struvite stones, preventing or promptly treating urinary tract infections is a central strategy.

Medical Treatment for Kidney Stones Medical treatment may be necessary for a kidney stone that is too large to pass, blocks urine flow, or causes severe pain or bleeding. Medications that may be given to facilitate stone passage include alpha-blockers and calcium channel blockers, which relax the ureter and increase urine flow. Sometimes a *stent* (a thin, flexible tube) is placed in the ureter to promote stone passage, although the stent may be uncomfortable and cause excessive bleeding. Some kidney stones can be fragmented into pieces that are small enough to pass in the urine; the most common method is *extracorporeal shock wave lithotripsy*, a procedure that uses high-amplitude sound waves to break down the kidney stone. Various other surgical methods—which involve physical removal of the stone—have a higher success rate but are also more invasive.

IN SUMMARY

▌ Kidney stones form when stone constituents—calcium oxalate, calcium phosphate, uric acid, cystine, or magnesium ammonium phosphate—crystallize in urine. Complications include renal colic, difficulty with urination, and obstruction.

▌ Kidney stones may be prevented by maintaining urine volumes of 2 to 2½ liters daily. Dietary measures include the consumption of appropriate amounts of calcium, oxalates, protein, sodium, and purines.

▌ Symptomatic kidney stones are sometimes treated with medications that facilitate stone passage or surgeries that fragment or remove stones.

© James Darell/Getty Images

Drinking plenty of water throughout the day is the most important measure for preventing kidney stones.

Nutrition Assessment Checklist FOR PEOPLE WITH KIDNEY DISEASES

Medical History

Check the medical record to determine:

- Degree of kidney function
- Cause of the nephrotic syndrome or kidney disease
- Type of dialysis, if appropriate
- Whether the patient has received a kidney transplant
- Type of kidney stone

Review the medical record for complications that may alter nutritional needs:

- Anemia
- Diabetes mellitus
- Edema or oliguria
- Hyperlipidemia
- Hypertension
- Metabolic stress or infection
- Protein-energy malnutrition

Medications

Assess risks for medication-related malnutrition related to:

- Long-term use of medications
- Multiple medication use, especially if medications affect nutrition status

For all patients with kidney diseases, note:

- Whether medications or supplements contain electrolytes that must be controlled
- Use of drugs or herbs that may be toxic to the kidneys

Dietary Intake

For patients with the nephrotic syndrome, kidney disease, or kidney transplant, assess intakes of:

- Protein and energy
- Fluid
- Vitamins, especially vitamin D
- Minerals, especially calcium, phosphorus, iron, and electrolytes

For patients with kidney stones or a history of kidney stones:

- Stress the need to drink plenty of fluids throughout the day.
- Assess intake of calcium, oxalate, sodium, protein, purines, or vitamin C, as appropriate for the type of stone.

Anthropometric Data

Take accurate baseline height and weight measurements. Keep in mind that:

- Fluid retention due to the nephrotic syndrome or kidney failure can mask malnutrition.
- For dialysis patients, the weight measured immediately after the dialysis treatment (called the *dry weight*) most accurately reflects the person's true weight. Rapid weight gain between dialysis treatments reflects fluid retention. If fluid retention is excessive, review fluid intake to determine if the patient understands and is complying with diet recommendations.

Laboratory Tests

Note that serum protein levels are often low in patients with nephrotic syndrome or advanced kidney disease. Review the following laboratory test results to assess the degree of kidney function and response to treatments:

- Blood urea nitrogen (BUN)
- Creatinine
- Glomerular filtration rate (GFR)
- Serum electrolytes
- Urinary protein

Check laboratory test results for complications associated with kidney disease, including:

- Anemia
- Hyperglycemia
- Hyperlipidemia
- Hyperparathyroidism (related to bone disease)

Physical Signs

For patients with nephrotic syndrome or kidney disease, look for physical signs of:

- Bone disease
- Dehydration or fluid retention
- Hyperkalemia
- Iron deficiency
- Uremia

Clinical Applications

1. A person with chronic kidney disease may need multiple medications to control disease progression and treat symptoms and complications. For people with diabetes and hyperlipidemias who develop chronic kidney disease, medications might include insulin, oral hypoglycemic drugs, antihypertensives, diuretics, lipid-lowering medications, and phosphate binders. Review the nutrition-related side effects of these medications. Describe the ways in which these medications may make it harder for people to maintain nutrition status.

2. Because the diet for chronic kidney disease is so restrictive, patients often find the diet difficult to manage and maintain over the long term. Review the suggestions in the "How to" on p. 608. Can you think of additional suggestions that may help? List some ideas that may help patients adjust to each of the different aspects of their renal diets.

Self Check

1. Which of the following is *not* a function of the kidneys?
   a. Activation of vitamin K
   b. Maintenance of acid-base balance
   c. Elimination of metabolic waste products
   d. Maintenance of fluid and electrolyte balances

2. The nephrotic syndrome frequently results in:
   a. the uremic syndrome.
   b. oliguria.
   c. edema.
   d. renal colic.

3. Dietary recommendations for patients with the nephrotic syndrome include:
   a. a high protein intake.
   b. sodium restriction.
   c. potassium and phosphorus restrictions.
   d. fluid restriction.

4. Hyperkalemia is often treated by:
   a. eliminating potassium from the diet.
   b. using diuretics to increase potassium losses.
   c. increasing fluid consumption.
   d. using potassium-exchange resins, which bind potassium in the GI tract.

5. Fluid requirements for oliguric patients are estimated by adding about ____ milliliters to the volume of urine output.
   a. 100
   b. 300
   c. 500
   d. 750

6. The most common cause of chronic kidney disease is:
   a. diabetes mellitus.
   b. hypertension.
   c. autoimmune disease.
   d. exposure to toxins.

7. A person with chronic kidney disease who has been following a renal diet for several years begins hemodialysis treatment. An appropriate dietary adjustment would be to:
   a. reduce protein intake.
   b. consume protein more liberally.
   c. increase intakes of sodium and water.
   d. consume potassium and phosphorus more liberally.

8. Which of the following nutrients may be unintentionally restricted when a patient restricts phosphorus intake?
   a Fluid
   b. Calcium
   c. Potassium
   d. Sodium

9. Most kidney stones are made primarily from:
   a. struvite.
   b. uric acid.
   c. calcium oxalate.
   d. cystine.

10. Treatment for all kidney stones includes:
    a. dietary oxalate restriction.
    b. dietary protein restriction.
    c. vitamin C supplementation.
    d. a fluid intake that maintains a urine volume of at least 2 to 2½ liters per day.

Answers to these questions can be found in Appendix H.

Notes

1. G. B. Appel, Glomerular disorders and nephrotic syndromes, in L. Goldman and D. Ausiello, eds., *Cecil Medicine* (Philadelphia: Saunders, 2008), pp. 866–876.

2. B. R. Don and G. A. Kaysen, Proteinuria and nephrotic syndrome, in R. W. Schrier, ed., *Renal and Electrolyte Disorders* (Philadelphia: Lippincott Williams & Wilkins, 2010), pp. 519–558.

3. Don and Kaysen, 2010.

4. Appel, 2008; J. A. Charlesworth, D. M. Gracey, and B. A. Pussell, Adult nephrotic syndrome: Non-specific strategies for treatment, *Nephrology* 13 (2008): 45–50.

5. American Dietetic Association, *Nutrition Care Manual* (Chicago: American Dietetic Association, 2011); Don and Kaysen, 2010.

6. Don and Kaysen, 2010.

7. American Dietetic Association, 2011.

8. American Dietetic Association, 2011; G. M. Podda and coauthors, Abnormalities of homocysteine and B vitamins in the nephrotic syndrome, *Thrombosis Research* 120 (2007): 647–652.

9. B. A. Molitoris, Acute kidney injury, in L. Goldman and D. Ausiello, eds., *Cecil Medicine* (Philadelphia: Saunders, 2008), pp. 862–866.

10. R. W. Schrier and C. L. Edelstein, Acute kidney injury: Pathogenesis, diagnosis, and management, in R. W. Schrier, ed., *Renal and Electrolyte Disorders* (Philadelphia: Lippincott Williams & Wilkins, 2010), pp. 325–388.

11. Schrier and Edelstein, 2010.

12. C. E. Alpers, The kidney, in V. Kumar and coeditors, *Robbins and Cotran Pathologic Basis of Disease* (Philadelphia: Saunders, 2010), pp 905–969.

13. American Dietetic Association, 2011.

14. American Dietetic Association, 2011.

15. Schrier and Edelstein, 2010.

16. W. E. Mitch, Chronic kidney disease, in L. Goldman and D. Ausiello, eds., *Cecil Medicine* (Philadelphia: Saunders, 2008), pp. 921–930.

17. J. Coresh and coauthors, Prevalence of chronic kidney disease in the United States, *Journal of the American Medical Association* 298 (2007): 2038–2047.

18. Mitch, 2008.

19. G. T. Obrador, Chronic renal failure and the uremic syndrome, in E. V. Lerma and coeditors, *Current Diagnosis and Treatment: Nephrology and Hypertension* (New York: McGraw Hill/Lange, 2009), pp. 149–154.

20. M. Chonchol and L. Chan, Chronic kidney disease: Manifestations and pathogenesis, in R. W. Schrier, ed., *Renal and Electrolyte Disorders* (Philadelphia: Lippincott Williams & Wilkins, 2010), pp. 389–425; A. Schieppati, R. Pisoni, and G. Remuzzi, Pathophysiology of chronic kidney disease, in A. Greenberg, ed., *Primer on Kidney Diseases* (Philadelphia: Saunders, 2009), pp. 422–445.

21. J. D. Kopple, Nutrition, diet, and the kidney, in M. E. Shils and coeditors, *Modern Nutrition in Health and Disease* (Baltimore, MD: Lippincott Williams & Wilkins, 2006), pp. 1475–1511.

22. A. J. Hutchison and A. Vardhan, Peritoneal dialysis, in A. Greenberg, ed., *Primer on Kidney Diseases* (Philadelphia: Saunders, 2009), pp. 459–471.

23. American Dietetic Association, 2011.

24. J. A. Beto and V. K. Bansal, Medical nutrition therapy in chronic kidney failure: Integrating clinical practice guidelines, *Journal of the American Dietetic Association* 104 (2004): 404–409.

25. D. J. Goldstein-Fuchs and C. M. Goeddeke-Merickel, Nutrition and kidney disease, in A. Greenberg, ed., *Primer on Kidney Diseases* (Philadelphia: Saunders, 2009), pp. 478–486.

26. N. Tolkoff-Rubin, Treatment of irreversible renal failure, in L. Goldman and D. Ausiello, eds., *Cecil Medicine* (Philadelphia: Saunders, 2008), pp. 936–947.

27. N. J. Cano and coauthors, Intradialytic parenteral nutrition does not improve survival in malnourished hemodialysis patients: A 2-year multicenter, prospective, randomized study, *Journal of the American Society of Nephrology* 18 (2007): 2583–2591.

28. Tolkoff-Rubin, 2008.

29. S. Beddhu, Outcome of end-stage renal disease therapies, in A. Greenberg, ed., *Primer on Kidney Diseases* (Philadelphia: Saunders, 2009), pp. 472–477.

30. G. C. Curhan, Nephrolithiasis, in A. Greenberg, ed., *Primer on Kidney Diseases* (Philadelphia: Saunders, 2009), pp. 382–388.

31. L. K. Massey, M. Liebman, and S. A. Kynast-Gales, Ascorbate increases human oxaluria and kidney stone risk, *Journal of Nutrition* 135 (2005): 1673–1677.

32. American Dietetic Association, 2011; Curhan, 2009.

33. American Dietetic Association, 2011.

34. Curhan, 2009.

CHAPTER 23

Nutrition in Practice

Although there is no perfect substitute for one's own kidneys, dialysis offers a life-sustaining treatment option for people with chronic kidney disease who develop renal failure. Dialysis can serve as a permanent treatment or as a temporary measure to sustain life until a suitable kidney donor can be found. Dialysis can also restore fluid and electrolyte balances in patients with acute kidney injury. Clinicians who routinely work with renal patients should understand how dialysis procedures work. This Nutrition in Practice describes the process of dialysis and outlines the various types of procedures that are available. The accompanying glossary defines the relevant terms.

How does dialysis work?

Dialysis removes excess fluids and wastes from the blood by employing the processes of **diffusion, osmosis,** and **ultrafiltration** (see Figure NP23-1). The dialysate, a solution similar in composition to normal blood plasma, is delivered to a compartment beside a **semipermeable membrane;** the person's blood flows along the other side of the membrane. The semipermeable membrane acts like a filter: small molecules such as urea and glucose can pass through microscopic pores in the membrane, whereas large molecules are unable to cross.

In *hemodialysis,* the tiny tubes that carry blood through the dialyzer are made of materials that serve as semipermeable membranes. In *peritoneal dialysis,* the body's peritoneal membrane, rich with blood vessels, is used to filter the blood.

How are solutes separated from the blood in dialysis?

The chemical composition of the dialysate affects the movement of solutes across the semipermeable membrane. When the concentration of a substance is lower in the dialysate than in the blood, the substance—provided it can cross the membrane—will diffuse out of the blood. For example, the goal is to remove as much as

continuous ambulatory peritoneal dialysis (CAPD): the most common method of peritoneal dialysis; involves frequent exchanges of dialysate, which remains in the peritoneal cavity throughout the day.

continuous renal replacement therapy (CRRT): a slow, continuous method of removing solutes and/or fluids from blood by gently pumping blood across a filtration membrane over a prolonged time period.

diffusion: movement of solutes from an area of high concentration to one of low concentration.

hemofiltration: removal of fluid and solutes by pumping blood across a membrane; no osmotic gradients are created during the process.

oncotic pressure: the pressure exerted by fluid on one side of a membrane as a result of osmosis.

osmosis: movement of water across a membrane toward the side where solutes are more concentrated.

peritonitis: inflammation of the peritoneal membrane.

pressure gradient: the change in pressure over a given distance. In dialysis, a pressure gradient is created between the blood and the dialysate.

semipermeable membrane: a membrane that allows some particles to pass through but not others.

ultrafiltration: removal of fluids and solutes from blood by using pressure to transfer the blood across a semipermeable membrane.

urea kinetic modeling: a method of determining the adequacy of dialysis treatment by calculating the urea clearance from blood.

possible of the waste product urea from the blood, so the dialysate contains no urea. For many other solutes, the dialysate is adjusted so that only excesses will be removed. Potassium can be removed from the blood, for example, by providing a dialysate that has a lower concentration of potassium than is found in the

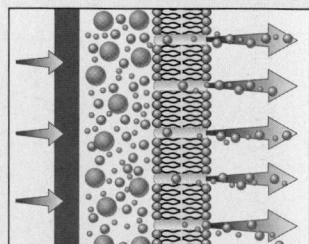

Diffusion

Small molecules (electrolytes and waste products) move from an area of high concentration to an area of low concentration by diffusion.

Osmosis

Water moves from an area of high water concentration to an area of low water concentration. In other words, water moves toward the side where solutes are more concentrated.

Ultrafiltration

Pressure squeezes water and small molecules through the pores of a semipermeable membrane during ultrafiltration.

person's blood. The dialysate must contain some potassium, how-ever; otherwise, the blood potassium would fall too low.

The dialysate can also be used to add needed components back into the blood. For a person with acidosis, for example, bases such as bicarbonate are added to the dialysate; the bases then move by diffusion into the blood to alleviate the acidosis.

How is fluid removed from the blood?

Because albumin and other plasma proteins are so adept at retaining fluids in blood, osmosis alone is not an efficient process for removing fluid. In hemodialysis, a **pressure gradient** is created between the blood and the dialysate. Most modern dialyzers produce *positive* pressure in the blood compartment and *negative* pressure in the dialysate compartment, establishing a pressure gradient that "pushes" water (and accompanying solutes) through the pores of the membrane.[1] This process, called ultrafiltration, relies on pumps to establish an appropriate flow rate between the blood and the dialysate.

How does the health practitioner know if the dialysis treatment has been effective?

A number of methods have been devised for gauging the adequacy of dialysis treatment. The most common method is **urea kinetic modeling,** a technique that evaluates the amount of urea cleared from the blood. The formula used most often is Kt/V, where K is the amount of urea cleared, t is the time spent on dialysis, and V is the blood volume. The value obtained indicates whether the patient has undergone sufficient dialysis; the goal is a Kt/V result of approximately 1.2. Because technical data (such as dialyzer clearance data, blood flow rate, and dialysate flow rate) need to be incorporated into the calculation, the computation is usually done by computer analysis. Current treatment guidelines recommend that hemodialysis adequacy be evaluated at least monthly, or more often if problems develop or patients are noncompliant.[2]

How long does a hemodialysis treatment last?

As described previously, hemodialysis employs a dialyzer to cleanse the patient's blood. Although dialyzers vary in efficiency, the treatment usually lasts 3 to 4 hours and is required at least 3 times weekly. Other options include short daily dialysis,

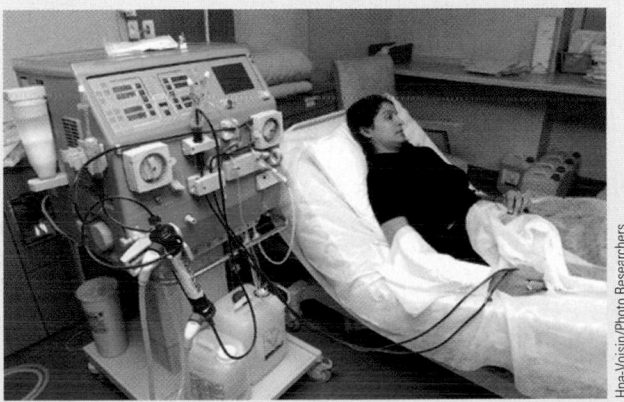
During hemodialysis, blood passes through a dialyzer where wastes are extracted, and the cleansed blood is returned to the body.

performed for about 2 hours each day, and daily nocturnal he-modialysis, in which dialysis is done at home overnight while the patient is sleeping. Although some studies have reported improved outcomes in patients who undergo daily dialysis, these approaches have not been widely adopted.[3] Note that most patients must visit dialysis centers to obtain treatment; few patients have access to a dialysis machine at home.

Are any complications associated with hemodialysis?

Yes. Although lifesaving, hemodialysis is associated with a substantial number of complications.[4] Problems at the vascular access site include infections and blood clotting. Hypotension can develop while blood is circulated through the dialyzer. Muscle cramping often occurs during the procedure, especially in the hands, legs, and feet. Blood losses can worsen anemia, which is already severe in two-thirds of patients beginning hemodialysis treatment.[5] Patients may also experience headaches, weakness, nausea, vomiting, restlessness, and agitation. Many patients experience extreme fatigue after a hemodialysis treatment, and some may require rest or sleep.

How does peritoneal dialysis work?

In peritoneal dialysis, the peritoneal membrane surrounding the abdominal organs serves as a semipermeable membrane. The dialysate is infused into a catheter that empties into the peritoneal space—the space within the abdomen near the intestines (see Figure NP23-2). In the most common procedure, **continuous ambulatory peritoneal dialysis (CAPD),** the dialysate remains in the peritoneal cavity for 4 to 6 hours, after which it is drained and replaced with fresh dialysate (about 2 to 3 liters in adults). Generally, the dialysate solution is exchanged four times daily and requires only about 30 minutes to drain and replace.

Because a pressure gradient cannot be created in the peritoneal cavity as it can in a dialyzer, the glucose concentration in the dialysate must be high enough to create enough **oncotic pressure** to draw fluid from the blood. As indicated in Chapter 23, a substantial amount of glucose can be absorbed into the patient's blood and may contribute to weight gain over time. The high glucose load may also cause hyperglycemia and hypertriglyceridemia in some patients.

What are the advantages and disadvantages of peritoneal dialysis?

Peritoneal dialysis offers a number of advantages over hemodialysis: vascular access is not required, dietary restrictions are fewer, and the procedure can be scheduled when convenient. The most common complication is infection, which can occur at the catheter site or within the peritoneal cavity **(peritonitis).** Other problems that may arise include blood clotting in the catheter, catheter migration, and abdominal hernia due to the dialysate volume.

What are the features of continuous renal replacement therapy?

In people with acute kidney injury, **continuous renal replacement therapy (CRRT)** removes fluids and wastes. CRRT uses the process of **hemofiltration,** in which blood is gently pumped

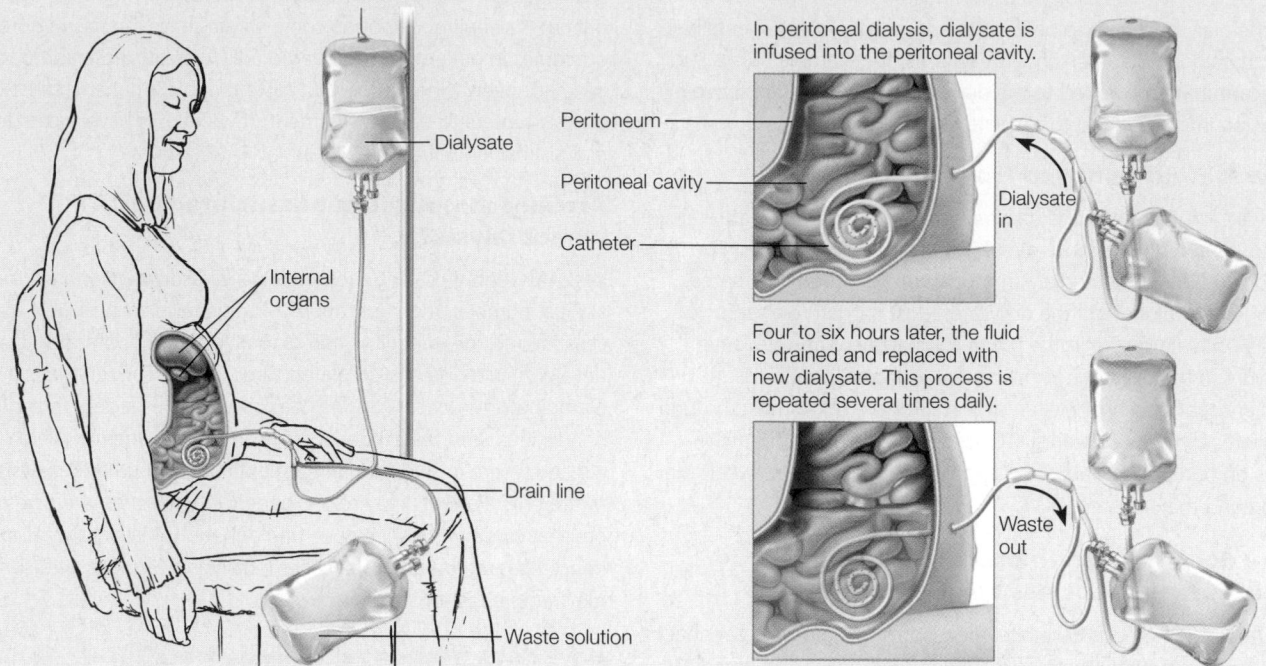

In peritoneal dialysis, dialysate is infused into the peritoneal cavity.

Peritoneum

Peritoneal cavity

Catheter

Dialysate in

Four to six hours later, the fluid is drained and replaced with new dialysate. This process is repeated several times daily.

Waste out

Dialysate

Internal organs

Drain line

Waste solution

across a filtration membrane over a prolonged time period. (This process differs from dialysis treatments that rely on the diffusion of wastes across a membrane into the dialysate.) Either a pump or the patient's own blood pressure moves the blood across the membrane. The procedure can be used to remove fluids, solutes, or both. Some patients require fluid replacement during the procedure to maintain adequate blood volume, so hydration status must be closely monitored.

The use of CRRT is advantageous in acute care situations because it corrects imbalances without causing sudden shifts in blood volume, which are poorly tolerated in acute care patients. In addition, replacement fluids can include parenteral feedings without upsetting fluid balance. Complications include clotting problems, damage to arteries, and inadequate blood flow rates in hypotensive patients.

Dialysis and CRRT help to remove the wastes and fluids that are normally removed by healthy kidneys. Although these procedures cannot restore the kidneys' hormonal functions, they provide a lifesaving means of alleviating symptoms of uremia, hypertension, and edema.

Notes

1. J. Z. Kallenbach and coauthors, *Review of Hemodialysis for Nurses and Dialysis Personnel* (St. Louis: Elsevier/Mosby, 2005), pp. 61–70.

2. National Kidney Foundation, K/DOQI clinical practice guidelines for hemodialysis adequacy: Update 2006, www.kidney. org/PROFESSIONALS/kdoqi/guideline_upHD_PD_VA/hd_ guide2.htm; accessed December 23, 2010.

3. K. L. Johansen and coauthors, Survival and hospitalization among patients using nocturnal and short daily compared to conventional hemodialysis: A USRDS study, *Kidney International* 76 (2009): 984–990.

4. N. Tolkoff-Rubin, Treatment of irreversible renal failure, in L. Goldman and D. Ausiello, eds., *Cecil Medicine* (Philadelphia: Saunders, 2008), pp. 936–947.

5. Tolkoff-Rubin, 2008.

24

Energy- and Protein-Modified Diets for Metabolic and Respiratory Stress

The body's response to severe stress can alter metabolism enough to threaten survival. Stress also raises nutritional needs considerably—increasing the risk of malnutrition even in previously healthy individuals. **Metabolic stress,** a disruption in the body's internal chemical environment, can result from uncontrolled infections or extensive tissue damage. As the first part of this chapter explains, the body's stress response is an attempt to restore balance, but it can have both helpful and harmful effects. Later sections of this chapter describe **respiratory stress,** characterized by inadequate oxygen and excessive carbon dioxide in the blood and tissues. Both metabolic and respiratory stress can lead to **hypermetabolism** (above-normal metabolic rate), **wasting** (loss of muscle tissue), and, in severe circumstances, life-threatening complications.

The Body's Responses to Stress and Injury

The **stress response** is the body's *nonspecific* response to a variety of stressors, such as infection, fractures, surgery, and burns. During stress, metabolic processes that support immediate survival are given priority, while those of lesser consequence are delayed. Energy is of primary importance, and therefore the energy nutrients are mobilized from storage and made available in the blood. Heart rate and respiration (breathing rate) increase to deliver oxygen and nutrients to cells more quickly, and blood pressure rises. Meanwhile, energy is diverted from processes that are not life sustaining, such as growth, reproduction, and long-term immunity. If stress continues for a long period, interference with these processes may begin to cause damage or illness.

Hormonal Responses to Stress

The stress response is mediated by several hormones, which are released into the blood soon after the onset of injury (see Table 24-1).[1] The catecholamines (epinephrine and norepinephrine), often called the *fight-or-flight hormones*, stimulate heart muscle, raise blood pressure, and increase metabolic rate. Epinephrine also promotes glucagon secretion from the pancreas, prompting the release of nutrients from storage. The steroid hormone cortisol enhances protein degradation, raising amino acid levels in the blood and making amino acids available for conversion to glucose. All of these hormones have similar effects on glucose and fat metabolism, causing the breakdown of glycogen, the production of glucose from amino acids, and the breakdown of triglycerides in adipose tissue. Thus, the combined effects of these

The catecholamines, glucagon, and cortisol have actions that oppose those of insulin and are therefore referred to as *counter-regulatory hormones.*

metabolic stress: a disruption in the body's chemical environment due to the effects of disease or injury. Metabolic stress is characterized by changes in metabolic rate, heart rate, blood pressure, hormonal status, and nutrient metabolism.

respiratory stress: abnormal gas exchange between the air and blood, resulting in lower-than-normal oxygen levels and higher-than-normal carbon dioxide levels.

hypermetabolism: a higher-than-normal metabolic rate.

wasting: the breakdown of muscle tissue that results from disease or malnutrition.

stress response: the chemical and physical changes that occur within the body during stress.

inflammatory response: a group of nonspecific immune responses to infection or injury.

phagocytes (FAG-oh-sites): white blood cells (neutrophils and macrophages) that have the ability to engulf and destroy antigens.

▮ *phagein* = to eat

TABLE 24-1 Metabolic Effects of Hormones Released during the Stress Response

Hormone	Metabolic Effects
Catecholamines	▮ Increase in metabolic rate ▮ Glycogen breakdown in liver and muscle ▮ Glucose production from amino acids ▮ Release of fatty acids from adipose tissue ▮ Glucagon secretion from pancreas
Glucagon	▮ Glycogen breakdown in liver ▮ Glucose production from amino acids ▮ Release of fatty acids from adipose tissue
Cortisol	▮ Protein degradation ▮ Enhancement of glucagon's action on liver glycogen ▮ Glucose production from amino acids ▮ Release of fatty acids from adipose tissue
Aldosterone	▮ Sodium reabsorption in kidneys
Antidiuretic hormone	▮ Water reabsorption in kidneys

hormones contribute to hyperglycemia, which often accompanies critical illness. Two other hormones induced by stress, aldosterone and antidiuretic hormone, help to maintain blood volume by stimulating the kidneys to reabsorb more sodium and water, respectively.

Cortisol's effects can be detrimental when stress is prolonged. In excess, cortisol causes the depletion of protein in muscle, bone, connective tissue, and skin. It impairs wound healing, so high cortisol levels may be especially dangerous for a patient with severe injuries. Because cortisol inhibits protein synthesis, consuming more protein cannot easily reverse tissue losses. Excess cortisol also leads to insulin resistance, contributing to hyperglycemia. In addition, cortisol suppresses immune responses, increasing susceptibility to infection. Note that pharmaceutical forms of cortisol are common anti-inflammatory medications (such as *cortisone* and *prednisone*); their long-term use can cause undesirable side effects such as muscle wasting, thinning of the skin, diabetes, and early osteoporosis.

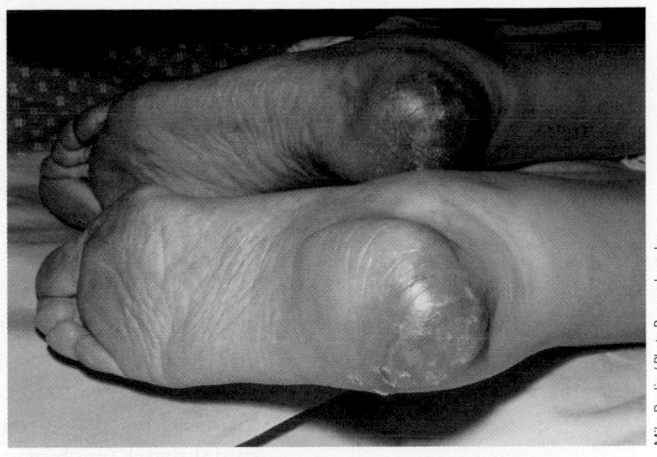

Mike Devlin / Photo Researchers, Inc

Pressure sores, wounds that develop when prolonged pressure cuts off blood circulation to the skin and underlying tissues, are a frequent source of metabolic stress in bedridden and wheelchair-bound patients.

The Inflammatory Response

Cells of the immune system mount a quick, nonspecific response to infection or tissue injury. This so-called **inflammatory response** serves to contain and destroy infectious agents (and their products) and prevent further tissue damage. As in the stress response, however, there is a delicate balance between a response that protects tissues from further injury and an excessive response that can cause additional damage to tissue.

The Inflammatory Process The inflammatory response begins with the dilation of arterioles and capillaries at the site of injury, which increases the blood flow to the affected area. The capillaries within the damaged tissue become more permeable, allowing some blood plasma to escape into the tissue and cause local edema (see Figure 24-1). The various changes in blood vessels also encourage the entry of immune cells that can destroy foreign agents. Among the first cells to arrive are the **phagocytes,** which slip

Classic signs of inflammation that accompany altered blood flow are:

- *Redness*—due to dilation of blood vessels in the injured area
- *Heat*—due to the influx of warm arterial blood
- *Swelling*—due to the accumulation of fluid at the site of injury
- *Pain*—due to the pressure of edema within damaged tissue and the actions of certain chemical mediators on pain receptors

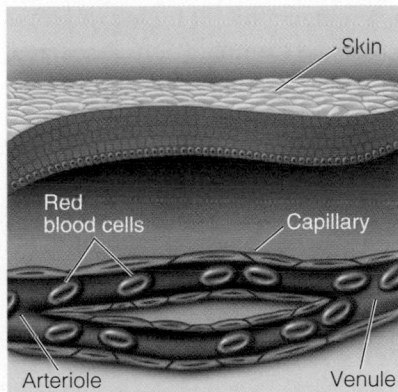

Cells lining the blood vessels lie close together, and normally do not allow the contents to cross into tissue.

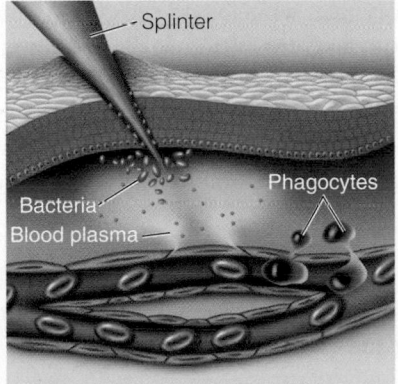

When tissues are damaged, immune cells release histamine, which dilates some blood vessels, increasing blood flow to the damaged area. Fluid leaks out of capillaries (causing swelling), and phagocytes escape between the small gaps in the blood vessel walls.

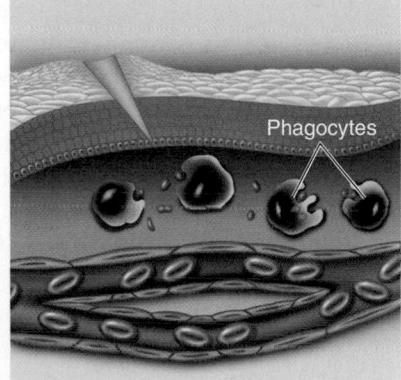

Phagocytes engulf bacteria and disable them with hydrolytic enzymes and reactive forms of oxygen.

FIGURE 24-1 **The Inflammatory Process**

through gaps between the endothelial cells that form the vessel walls. The phagocytes engulf microorganisms and destroy them with hydrolytic enzymes and reactive forms of oxygen. When inflammation becomes chronic, these normally useful products of phagocytes can damage healthy tissue.

Mediators of Inflammation Numerous chemical substances control the inflammatory process. These *mediators* are released from damaged tissue, blood vessel cells, and activated immune cells. Many of them help to regulate more than one step in the process. Histamine, a small molecule similar to an amino acid in structure, is released from granules within **mast cells,** causing vasodilation and capillary permeability. Other compounds that participate in the inflammatory process include **cytokines,** produced by white blood cells (and some other cells), and **eicosanoids,** which are derived from dietary fatty acids. Note that most anti-inflammatory medications, including steroidal drugs (such as cortisone and prednisone) and nonsteroidal anti-inflammatory drugs (such as aspirin and ibuprofen), act by blocking eicosanoid synthesis.

Changing dietary fat sources can have subtle effects on the inflammatory process.[2] The major precursor for the eicosanoids is arachidonic acid, which derives from the omega-6 fatty acids in vegetable oils. Some omega-3 fatty acids compete with arachidonic acid and inhibit the production of the most powerful inflammatory mediators. Partially replacing vegetable oils rich in omega-6 fatty acids with food sources high in omega-3 fatty acids (such as fish oil) helps to suppress inflammation, but it is not a reliable treatment.

Systemic Effects of Inflammation Cytokines released during the inflammatory process induce **systemic** effects as well as the localized effects described earlier. Within hours after inflammation, infection, or severe injury, the liver steps up its production of certain proteins in an effort known as the **acute-phase response.**[3] These acute-phase proteins include **C-reactive protein, complement,** blood-clotting proteins such as fibrinogen and prothrombin, and others. At the same time, plasma concentrations of albumin, iron, and zinc fall. The acute-phase response is accompanied by muscle catabolism to make amino acids available for glucose production, tissue repair, and immune protein synthesis; consequently, negative nitrogen balance (and wasting) frequently results. Other clinical features include hypermetabolism, increased blood neutrophil levels, lethargy, anorexia, and, often, fever.

If inflammation does not resolve, the continued production of pro-inflammatory cytokines may lead to the **systemic inflammatory response syndrome (SIRS),** which is diagnosed when the patient's symptoms include substantial increases in heart rate, respiratory rate, white blood cell counts, and/or body temperature. If these symptoms result from a severe infection, the condition is called **sepsis.** Complications associated with severe cases of SIRS or sepsis include fluid retention and tissue edema, low blood pressure, and impaired blood flow. If the reduction in blood flow is severe enough to deprive the body's tissues of oxygen and nutrients (a condition known as **shock**), multiple organs may fail simultaneously, as discussed in Nutrition in Practice 24.

IN SUMMARY

- The stress and inflammatory responses are nonspecific responses to stressors that cause infection and injury.

- The stress response is mediated by the catecholamine hormones, cortisol, and glucagon, which together raise nutrient levels in blood, stimulate heart rate, raise blood pressure, and increase metabolic rate. Aldosterone and antidiuretic hormone help to maintain adequate blood volume.

- The inflammatory process can lead to systemic effects that alter nutrient metabolism, heart rate, blood pressure, body temperature, and immune cell functions.

Antihistamines are medications taken to reduce the effects of histamine.

C-reactive protein, the best clinical indicator of the acute-phase response, is elevated during many chronic illnesses.

mast cells: cells within connective tissue that produce and release histamine.

cytokines (SIGH-toe-kines): signaling proteins produced by the body's cells; those produced by white blood cells regulate various aspects of immune function.

eicosanoids (eye-KO-sa-noids): 20-carbon molecules derived from dietary fatty acids that help to regulate blood pressure, blood clotting, and other body functions.
- *eicosa* = twenty

systemic (sih-STEM-ic): relating to the entire body.

acute-phase response: changes in body chemistry resulting from infection, inflammation, or injury; characterized by alterations in plasma proteins.

C-reactive protein: an acute-phase protein released from the liver during acute inflammation or stress.

complement: a group of plasma proteins that assist the activities of antibodies.

Nutrition Treatment of Acute Stress

As described earlier, an excessive response to metabolic stress can worsen illness and even threaten survival. Therefore, medical personnel must manage both the medical condition that initiated stress and complications that result from the stress and inflammatory responses. Immediate concerns are to restore lost fluids and electrolytes and remove underlying stressors. Thus, initial treatments include correcting fluid and electrolyte imbalances, treating infections, repairing wounds, draining **abscesses** (accumulated pus), and removing dead tissue (**debridement**). After stabilization, nutrient needs can be assessed and nutrition therapy provided.

Determining Nutritional Requirements

Notable metabolic changes in patients undergoing metabolic stress include hypermetabolism, negative nitrogen balance, insulin resistance, and hyperglycemia. Hypermetabolism and negative nitrogen balance can lead to wasting, which may impair organ function and delay recovery. Hyperglycemia increases the risk of infection. Therefore, the principal goals of nutrition care are to preserve lean (muscle) tissue, maintain immune defenses, and promote healing.

Feeding an acutely stressed patient is often challenging. Overfeeding increases the risks of refeeding syndrome and its associated hyperglycemia. Underfeeding can worsen negative nitrogen balance and increase lean tissue losses. Assessing nutritional needs can be complicated, however, because fluid imbalances prevent accurate weight measurements, and laboratory data may reflect metabolic alterations of illness rather than the person's nutrition status.

The amounts of protein and energy to provide during acute illness are sometimes difficult to determine. Because various conditions can lead to metabolic stress, each patient's situation may be somewhat different.[4] Moreover, nutritional needs can vary substantially over the course of illness. To help guide their decisions about treatment, clinicians need to closely observe patients' responses to feedings and readjust nutrient intakes as necessary.

Estimating Energy Needs in Acute Stress A common method for determining the energy needs of acutely stressed individuals is to estimate or measure the resting metabolic rate (RMR) and then multiply the result by a stress factor to account for the increased energy requirements of stress and healing. This method was introduced in Chapter 14 (see pp. 396–397); Table 24-2 reviews the use of stress factors and provides examples. Generally, energy needs are increased by fever, mechanical ventilation, infections, burns, and the presence of open wounds. Patients undergoing acute stress are usually bedridden, so the energy needed for physical activity is minimal. Note that several predictive equations developed for use in critical care populations include "built-in" stress factors to account for various injuries or stressors.*

Daily energy requirements for nonobese critical care patients often fall within the range of 25 to 30 kcalories per kilogram of body weight;[5] a patient weighing 160 pounds (72.7 kilograms) may therefore require between 1818 and 2181 kcalories per day. The energy intake is sometimes started within this range and then adjusted as the patient's body weight and other determinants of nutrition status change. For critically ill obese patients (BMI >30), energy needs may range from 22 to 25 kcalories per kilogram of ideal body weight per day.[6]

---

*One example is the Ireton–Jones equation, developed for ventilator-dependent patients: Energy needs (kcal/day) = 1925 + [5 × wt (kg)] − [10 × age (yr)] + [281 × sex] + [292 × trauma] + [851 × burn]; where sex is male (× 1) or female (× 0), trauma is the presence of physical injury (× 1) or not (× 0), and burn is the presence of a burn injury (× 1) or not (× 0).

---

systemic inflammatory response syndrome (SIRS): a whole-body inflammatory response caused by severe illness or trauma; characterized by raised heart and respiratory rates, abnormal white blood cell counts, and elevated body temperature.

sepsis: a whole-body inflammatory response caused by infection; characterized by symptoms similar to those of SIRS.

shock: a severe reduction in blood flow that deprives the body's tissues of oxygen and nutrients; characterized by reduced blood pressure, raised heart and respiratory rates, and muscle weakness.

abscesses (AB-sess-es): accumulations of pus.

debridement: the surgical removal of dead, damaged, or contaminated tissue resulting from burns or wounds; helps to prevent infection and hasten healing.

Reminder: *Refeeding syndrome* can develop when a severely malnourished person is aggressively fed; it is associated with fluid and electrolyte imbalances and hyperglycemia.

72.7 kg × 25 kcal/kg = 1818 kcal
72.7 kg × 30 kcal/kg = 2181 kcal

Reminder: Ideal body weights typically fall within the BMI range of 18.5 to 25.

TABLE 24-2 Disease-Specific Stress Factors for Estimating Energy Needs in Hospital Patients

Method:
Step 1. Estimate the energy needed to support resting metabolic rate (RMR) using indirect calorimetry or a predictive equation (see Table 14-9, p. 396).
Step 2. Multiply the estimated RMR by an appropriate stress factor for acute illness (see the "How to" box on p. 397).

Examples of stress factors:

- Advanced liver disease: 1.0 to 1.16
- Pancreatitis: 1.13 to 1.21
- Surgery: 1.2 to 1.4
- Mechanical ventilation: 1.32 to 1.34
- Burns (20 to 30 percent of body surface): 1.6 to 1.7
- Burns (30 to 40 percent of body surface): 1.8 to 1.9

Adjustment for obesity: In some individuals with BMI >30, using an adjusted body weight in RMR equations may improve results.[a] One method is to use a weight that falls between the person's actual body weight and an ideal body weight (IBW):

$$\text{Adjusted weight} = \text{IBW} + [0.5 \times (\text{actual body weight} - \text{IBW})]$$

[a]The use of adjusted body weights for determining energy requirements in obese individuals is controversial and under investigation.

Sources: American Dietetic Association, *Nutrition Care Manual* (Chicago: American Dietetic Association, 2010); N. Barak, E. Wall-Alonso, and M. D. Sitrin, Evaluation of stress factors and body weight adjustments currently used to estimate energy expenditure in hospitalized patients, *Journal of Parenteral and Enteral Nutrition* 26 (2002): 231–238.

Protein Requirements in Acute Stress To maintain lean tissue, the protein intakes recommended during acute stress are higher than DRI values. In nonobese critically ill patients, protein needs may range from 1.2 to 2.0 grams per kilogram body weight per day;[7] burn patients may require between 2 and 3 grams per kilogram body weight each day due to the substantial losses of protein associated with burn wounds.[8] For most obese patients, protein requirements may range from 2.0 to 2.5 grams per kilogram of ideal body weight per day.[9] Even with adequate protein, however, negative nitrogen balance cannot be prevented during acute stress because hormonal changes encourage protein catabolism. The bed rest required during illness also contributes substantially to muscle breakdown.

> **Reminder:** The protein RDA for adults is 0.8 gram per kilogram body weight.

The amino acids glutamine and arginine are sometimes added to the diets of acutely stressed and immune-compromised patients. Studies have suggested that glutamine supplementation may improve immune function, preserve muscle mass, and reduce mortality rates in critically ill patients.[10] Arginine supplementation may benefit the immune responses and nitrogen balance of critically ill and postoperative patients.[11] Although glutamine and arginine are often added to enteral formulas promoted for wound healing and enhanced immunity, their use remains controversial.[12]

Carbohydrate and Fat Intakes in Acute Stress The bulk of energy needs are supplied from carbohydrate and fat. Carbohydrate is usually the main source of energy, providing 50 to 60 percent of total energy requirements.[13] In patients with severe hyperglycemia, fat may provide up to 50 percent of kcalories, although high fat intakes may suppress immune function and increase the risks of developing infections and hypertriglyceridemia. Patients with blood triglyceride levels above 300 to 400 milligrams per deciliter may require fat restriction.[14]

> **Reminder:** When parenteral infusions are necessary, dextrose is usually limited to 5 milligrams per kilogram body weight per minute.

Micronutrient Needs in Acute Stress Acutely stressed patients are believed to have increased micronutrient needs, but specific requirements remain unknown.[15]

In hypermetabolic patients, the need for B vitamins may be higher to support the increase in energy metabolism. Several micronutrients, such as vitamin A, vitamin C, and zinc, have critical roles in immunity and wound healing, and their supplementation may speed recovery under certain circumstances. Patients with burns and tissue injuries may have increased requirements for trace minerals due to tissue losses.

The acute-phase response causes a redistribution in the tissue content of some micronutrients that either raises or lowers their blood levels; therefore, micronutrient status is sometimes difficult to interpret. Blood concentrations of trace minerals are monitored in patients receiving parenteral nutrition support to ensure that excessive amounts are not given intravenously.

During the acute-phase response, plasma levels of iron and zinc fall, whereas plasma copper levels rise.

Approaches to Nutrition Care in Acute Stress

As mentioned earlier, the initial care following acute stress focuses on maintaining fluid and electrolyte balances. Simple intravenous solutions often contain dextrose, providing minimal kcalories. Once patients are stable, nutrition support may be necessary if poor appetite, the medical condition, or a medical procedure (such as mechanical ventilation) interferes with food intake. For patients with a functional GI tract, early enteral feedings—started in the first 24 to 48 hours after hospitalization—are associated with fewer complications and shorter hospital stays compared with delayed feedings. If enteral nutrition is not possible, malnourished patients may receive parenteral nutrition support soon after admission to the hospital. In previously healthy patients, parenteral nutrition support may be withheld during the first seven days of hospitalization to avoid the risk of infectious complications.[16]

Once patients can tolerate oral feedings, a high-kcalorie, high-protein diet is often prescribed, although care must be taken not to overfeed patients at risk of developing refeeding syndrome or hyperglycemia. Because meeting protein and energy needs may be difficult, enteral formulas are often provided to supplement the diet. Many such formulas have high nutrient density, and some contain extra amounts of nutrients believed to promote healing or benefit immune function, such as the amino acids arginine and glutamine, omega-3 fatty acids, and the antioxidant nutrients. Nutrient needs should be reassessed frequently until the patient's condition improves. The accompanying Case Study reviews the nutrition care of a patient undergoing acute metabolic stress.

Case Study PATIENT WITH A SEVERE BURN

David Bray, a 42-year-old man, has been admitted to intensive care. He suffered a severe burn covering 35 percent of his body when he was trapped inside a burning building. His wife told the nurse that Mr. Bray's height is 6 feet and that he usually weighs about 175 pounds. The physician ordered lab tests, including serum protein concentrations, but the results are not yet available.

1. Identify Mr. Bray's immediate needs after the injury. Describe the initial concerns of the health care team and the measures they might take soon after Mr. Bray's arrival at the hospital.

2. Considering Mr. Bray's condition, what problems might the health care team encounter when they attempt to obtain information that can help them assess his nutrition status?

3. Estimate Mr. Bray's energy and protein needs (use a protein factor of 2.5 grams per kilogram). What problems may interfere with Mr. Bray's ability to meet his nutrient needs?

4. After Mr. Bray transitions to oral feedings, he is able to obtain only 65 percent of his energy requirements. What other feeding options may be considered?

■ The objective of nutrition care during metabolic stress is to provide a diet that preserves muscle tissue, maintains immune defenses, and promotes healing.

■ Energy needs during acute stress can be estimated by modifying RMR values with stress factors that account for the increased demands of the medical condition or treatment. Protein recommendations are higher than DRI levels to help prevent tissue losses and allow healing of damaged tissue.

■ Nutrition care during acute stress focuses on maintaining fluid and electrolyte balances, meeting the patient's nutritional needs, and providing enteral or parenteral nutrition support when necessary.

Nutrition and Respiratory Stress

Some medical problems upset the process of gas exchange between the air and blood and result in respiratory stress, which is characterized by a reduction in the blood's oxygen supply and an increase in carbon dioxide levels. Excessive carbon dioxide in the blood may disturb the breathing pattern enough to interfere with food intake. Moreover, the labored breathing caused by many respiratory disorders entails a higher energy cost than normal breathing does, raising energy needs and increasing carbon dioxide production further. Lung diseases make physical activity difficult and can lead to muscle wasting. Weight loss and malnutrition therefore become dangerous outcomes of some types of respiratory illnesses.

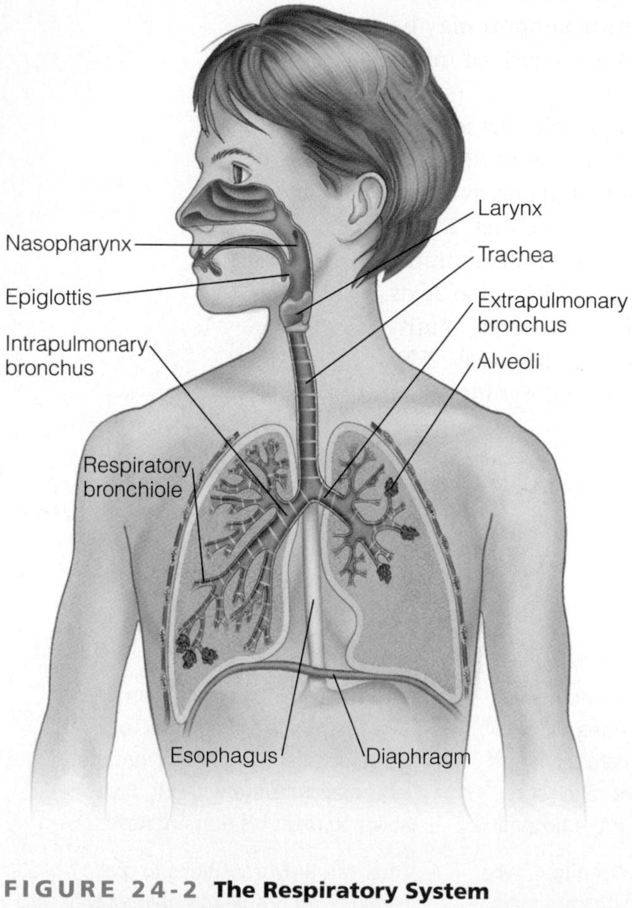

Chronic Obstructive Pulmonary Disease

Chronic obstructive pulmonary disease (COPD) refers to a group of conditions characterized by the persistent obstruction of airflow through the lungs. Figure 24-2 illustrates the main airways **(bronchi** and **bronchioles)** and air sacs **(alveoli)** of the normal respiratory system, and Figure 24-3 shows how they are altered in COPD. The two main types of COPD are **chronic bronchitis** and **emphysema,** and many patients display features of both conditions:[17]

■ *Chronic bronchitis* is characterized by persistent inflammation and excessive secretions of mucus in the main airways of the lungs, which may ultimately thicken and become too narrow for adequate mucus clearance. Chronic bronchitis is diagnosed when a chronic, productive cough persists for at least three months of the year for two consecutive years.

■ *Emphysema* is characterized by the breakdown of the lungs' elastic structure and destruction of the walls of the bronchioles and alveoli, changes that significantly reduce the surface area available for respiration. Emphysema is diagnosed on the basis of clinical signs and the results of lung function tests.

Both chronic bronchitis and emphysema are associated with abnormal levels of oxygen and carbon dioxide in the blood and shortness of breath **(dyspnea).** COPD may eventually lead to respiratory or heart failure, and together with

FIGURE 24-2 The Respiratory System
Inhaled air travels via the trachea to the bronchi and bronchioles, the major airways of the lungs. Oxygen and carbon dioxide are exchanged across the thin-walled alveoli, which are surrounded by capillaries.

Source: Based on a drawing in Carol Mattson Porth, *Pathophysiology,* 5th ed. (Philadelphia: Lippincott Williams and Wilkins, 1998).

Labels: Nasopharynx, Epiglottis, Intrapulmonary bronchus, Respiratory bronchiole, Esophagus, Larynx, Trachea, Extrapulmonary bronchus, Alveoli, Diaphragm

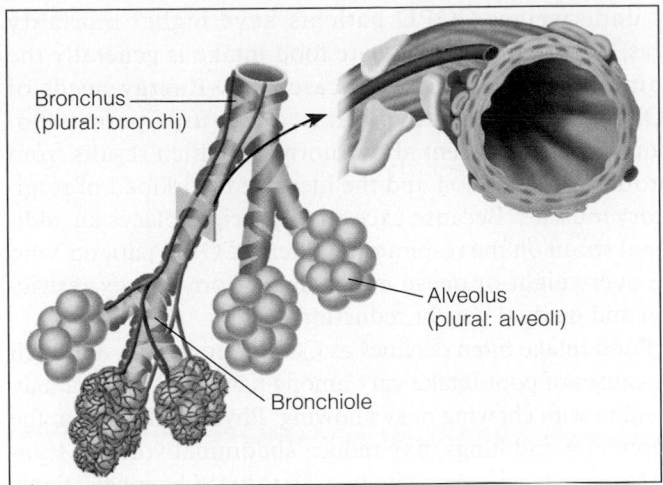

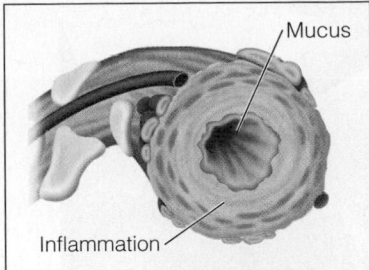

Chronic bronchitis is characterized by inflammation, excessive secretion of mucus, and narrowing of the bronchi—factors that reduce normal airflow.

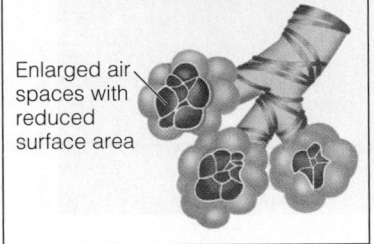

Emphysema is characterized by gradual destruction of the walls separating the alveoli and reduced lung elasticity.

Healthy bronchi provide an open passageway for air. Healthy alveoli permit gas exchange between the air and blood.

FIGURE 24-3 Chronic Obstructive Pulmonary Disease

other chronic respiratory illnesses, ranks as the third leading cause of death in the United States.[18]

COPD is a debilitating condition. Generally, dyspnea worsens as the disease progresses, resulting in dramatic reductions in physical activity and quality of life. Activities of daily living such as bathing or dressing may cause exhaustion or breathlessness. Weight loss and wasting are common in the advanced stages of disease and may result from hypermetabolism, poor food intake, and the actions of various inflammatory proteins.

Causes of COPD Cigarette smoking is the primary risk factor in 90 percent of COPD cases[19] and is especially damaging when combined with respiratory infections or an occupational exposure to dusts or chemicals. Only a minority of smokers (about 15 percent) develop COPD, however; thus, genetic susceptibility also contributes to its development. Genetic factors are especially likely in patients with early-onset COPD. Alpha-1-antitrypsin deficiency, an inherited disorder, accounts for 1 to 2 percent of COPD cases.[20] Individuals with this defect have inadequate blood levels of a plasma protein (alpha-1 antitrypsin) that normally inhibits enzymatic breakdown of the lungs' connective tissue.

Treatment of COPD The primary objectives of COPD treatment are to prevent the disease from progressing and relieve major symptoms (dyspnea and coughing). Individuals with COPD are encouraged to quit smoking to prevent disease progression and to get vaccinated against influenza and pneumonia to avoid complications. The most frequently prescribed medications are bronchodilators, which improve airflow, and corticosteroids (anti-inflammatory medications), which help to prevent symptom recurrence; note that corticosteroids promote catabolic processes and can exacerbate muscle loss. For people with severe COPD, supplemental oxygen therapy (12 hours daily) can maintain normal oxygen levels in the blood and reduce mortality risk. The Diet-Drug Interactions feature lists nutrition-related effects of the medications used to treat COPD.

Nutrition Therapy for COPD The main goals of nutrition therapy for COPD are to correct malnutrition (which affects up to 60 percent of COPD patients[21]), promote the maintenance of a healthy body weight, and prevent muscle wasting.

chronic obstructive pulmonary disease (COPD): a group of lung diseases characterized by persistent obstructed airflow through the lungs and airways; includes chronic bronchitis and emphysema.

bronchi (BRON-key), **bronchioles** (BRON-key-oles): the main airways of the lungs. The singular form of bronchi is *bronchus.*

alveoli (al-VEE-oh-lie): air sacs in the lungs. One air sac is an *alveolus.*

chronic bronchitis (bron-KYE tis): a lung disorder characterized by persistent inflammation and excessive secretions of mucus in the main airways of the lungs.

emphysema (EM-fih-ZEE-mah): a progressive lung condition characterized by the breakdown of the lungs' elastic structure and destruction of the walls of the bronchioles and alveoli, reducing the surface area involved in respiration.

dyspnea (DISP-nee-ah): shortness of breath.

Patients who need supplemental oxygen can use lightweight, portable equipment that allows them to move about freely.

As underweight COPD patients have higher mortality rates,[22] encouraging adequate food intake is generally the main focus of the nutrition care plan. Energy needs of COPD patients are usually raised due to hypermetabolism (about 20 percent above normal), which results from chronic inflammation and the increased workload of respiratory muscles. Because excess body weight places an additional strain on the respiratory system, COPD patients who are overweight or obese may benefit from energy restriction and gradual weight reduction.[23]

Food intake often declines as COPD progresses, although the causes of poor intake vary among patients. Dyspnea may interfere with chewing or swallowing. Physical changes in the diaphragm and lungs may reduce abdominal volume, leading to early satiety. Appetite may be reduced by medications, depression or anxiety, or altered taste perception. Some patients may become too disabled to shop or prepare food or may lack adequate support at home. The clinician must assess the unique needs of a COPD patient before proposing a nutrition care plan.

The altered sense of taste in COPD patients may be due to chronic mouth breathing, which dries the mouth, or the use of certain medications, including some bronchodilators.

Some patients may benefit from eating frequent, small meals spaced throughout the day rather than two or three large ones. The lower energy content of small meals reduces the carbon dioxide load, and the smaller meals may produce less abdominal discomfort and dyspnea. Some individuals may eat better if they receive supplemental oxygen at mealtimes. Consuming adequate fluids should be encouraged to help prevent the secretion of overly thick mucus; however, some patients should consume liquids between meals so as not to interfere with food intake. For undernourished persons, a high-kcalorie, high-protein diet may be helpful, but high energy intakes increase the amount of carbon dioxide produced and can increase respiratory stress. Liquid supplements may be recommended as between-meal snacks to improve weight gain or endurance, but patients should be cautioned not to consume amounts that reduce energy intake at mealtime.

Pulmonary Formulas Enteral formulas designed for use in COPD provide more kcalories from fat and fewer from carbohydrate than standard formulas. The ratio of carbon dioxide production to oxygen consumption is lower when fat is consumed, so theoretically these formulas should lower respiratory requirements. However, research studies have not confirmed that the reduced-carbohydrate formulas improve clinical outcomes more than moderate energy intakes.[24]

Diet-Drug Interactions

Check this table for notable nutrition-related effects of the medications discussed in this chapter.

Bronchodilators (theophylline, dyphylline)	**Gastrointestinal effects:** Increased gastric acid secretion, acid reflux
	Dietary interactions: Caffeine enhances drug effects (daily caffeine intake should be consistent); excessive alcohol intake may reduce drug clearance
Corticosteroids (prednisone)	**Metabolic effects:** Glucose intolerance, sodium retention, negative nitrogen balance, decreased bone mineral density, appetite stimulation, weight gain, growth suppression in children

Case Study

John Norback is an 82-year-old man who has emphysema that severely affects both lungs. He is 5 feet 9 inches tall and currently weighs 150 pounds, about 20 pounds less than his weight in earlier years. He lives with a daughter and son-in-law and eats meals with their family. He becomes breathless when eating and when walking around the house, and he feels tired much of the time. A medical clinic recently ordered oxygen therapy for home use, but supplies have not yet arrived. Mr. Norback's daughter is concerned about her father's recent weight loss and breathlessness.

1. Assess Mr. Norback's risk of malnutrition, using information from Table 14-6 in Chapter 14 (p. 392). What factors may have contributed to his weight loss?

2. What are possible reasons for Mr. Norback's difficulty with eating?

3. Based on the history given, what factors may account for Mr. Norback's tiredness? What suggestions would you give Mr. Norback and his daughter regarding physical activity?

List some dietary suggestions that may help to improve his appetite and food intake. How might the use of oxygen therapy help?

Incorporating an Exercise Program Loss of muscle can be more readily prevented or reversed if the treatment plan includes an effective exercise program. With exercise, patients are likely to see improvements in their endurance and become less fearful of their physical limitations. For some patients, the combination of an exercise plan and oral supplements may be better for maintaining weight and improving muscle mass than either component of treatment alone.[25] The accompanying Case Study allows you to review the nutrition care for a patient with COPD.

Respiratory Failure

In respiratory failure, the gas exchange between the air and circulating blood is severely impaired, causing abnormal levels of tissue gases that can be life threatening. Any condition that causes lung injury or impairs lung function can be the underlying cause of failure; examples include infection (such as pneumonia or sepsis), physical trauma, neuromuscular disorders, smoke inhalation, and airway obstruction.[26] Severe lung damage can lead to **acute respiratory distress syndrome (ARDS)**, a life-threatening condition that requires the use of mechanical ventilation to restore normal oxygen and carbon dioxide levels. In ARDS, the lungs exhibit extensive inflammation and fluid buildup (called *pulmonary edema*) that interfere with lung ventilation and gas exchange in the alveoli. Later stages are associated with a proliferation of lung cells, which causes fibrosis and disrupts lung structure. A dangerous complication of ARDS is the progression to multiple organ dysfunction syndrome, which is described in the Nutrition in Practice following this chapter.

Consequences of Respiratory Failure Respiratory failure is characterized by severe **hypoxemia** (low oxygen levels in the blood) and **hypercapnia** (excessive carbon dioxide in the blood). The low oxygen content of tissues **(hypoxia)** impedes cellular function and may lead to cell death. Severe hypercapnia can cause **acidosis,** which interferes with normal functioning of the central nervous system. To compensate for respiratory failure, a person breathes more rapidly, and the heart rate increases. The skin may become sweaty and develop a bluish cast **(cyanosis).** Headache, confusion, and drowsiness may occur. Severe cases of respiratory failure can cause heart arrhythmias and, ultimately, coma.

Treatment of Respiratory Failure The treatment of respiratory failure focuses on supporting lung function and correcting the underlying disorder. Because respiratory failure can be caused by a number of different conditions, treatment plans vary

acute respiratory distress syndrome (ARDS): respiratory failure triggered by severe lung injury; a medical emergency that causes dyspnea and pulmonary edema and usually requires mechanical ventilation.

hypoxemia (high-pock-SEE-me-ah): a low level of oxygen in the blood.

hypercapnia (high-per-CAP-nee-ah): excessive carbon dioxide in the blood.

hypoxia (high-POCK-see-ah): a low amount of oxygen in body tissues.

acidosis: acid accumulation in body tissues; depresses the central nervous system and may lead to disorientation and, eventually, coma.

cyanosis (sigh-ah-NOH-sis): a bluish cast in the skin due to the color of deoxygenated hemoglobin. Cyanosis is most evident in individuals with lighter, thinner skin; it is mostly seen on lips, cheeks, and ears, and under the nails.

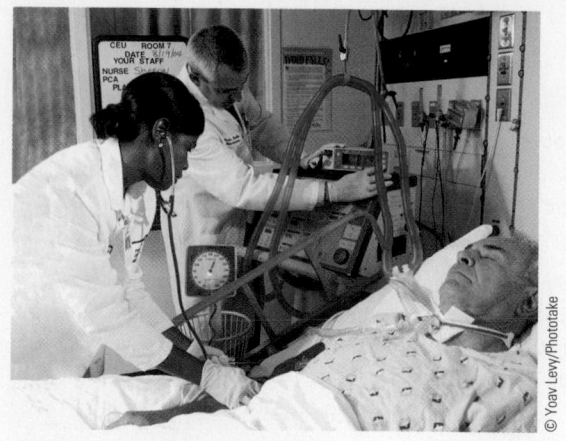

Mechanical ventilation controls the rate and amount of oxygen supplied to a person's airways.

considerably. Individuals with chronic lung disorders may be provided with oxygen therapy via a face mask or nasal tubing to relieve symptoms, whereas patients with ARDS receive mechanical ventilation until they are able to breathe independently. Diuretics may be prescribed to mobilize the fluid that has accumulated in lung tissue. Medications are given to treat infections, keep airways open, or relieve inflammation. Complications are common in ARDS and must be forestalled to prevent multiple organ dysfunction.

Nutrition Therapy for Respiratory Failure Patients with lung injuries or ARDS are frequently hypermetabolic and/or catabolic and at high risk of muscle wasting. The primary concerns are therefore to provide enough energy and protein to sustain muscle mass and lung function without overtaxing the respiratory system. Specific recommendations include the following:

▌ *Energy.* Energy needs typically range from 25 to 35 kilocalories per kilgram.[27] The body weight used for determining energy needs may need to be corrected for pulmonary edema. Energy needs can also be estimated using predictive equations such as those described on p. 623. Overfeeding should be avoided because it can cause excessive carbon dioxide production and worsen respiratory function.

▌ *Protein.* Protein needs are increased in patients with lung inflammation or ARDS. For mild or moderate lung injury, protein recommendations range from 1.0 to 1.5 grams of protein per kilogram body weight per day. Patients with ARDS may require 1.5 to 2.0 grams of protein per kilogram body weight daily.

▌ *Fluids.* Although most patients have normal fluid requirements, fluid status should be monitored daily to prevent fluid imbalance. Some patients may require fluid restriction to prevent edema in lung tissue, whereas others may become dehydrated due to diuretic therapy.

Nutrition Support in Respiratory Failure Patients with severe cases of respiratory failure may be unable to eat meals and may require nutrition support. Tube feedings are used if the intestine is functional, and intestinal feedings may be preferred over gastric feedings because they reduce the risk of aspiration. Patients with acute lung injuries or ARDS may benefit from enteral formulas designed to reduce inflammation and promote healing; these formulas are typically fortified with omega-3 fatty acids and antioxidant nutrients.[28] Nutrient-dense formulas (1.5 to 2.0 kcalories per milliliter) are prescribed for patients with fluid restrictions. If the risk of aspiration is too high to continue enteral feedings, parenteral nutrition support may be considered.

IN SUMMARY

▌ Chronic obstructive pulmonary diseases are debilitating, progressive illnesses that can lead to malnutrition, muscle wasting, and activity intolerance. The goals of nutrition therapy are to improve food intake, maintain proper weight, preserve muscle tissue, and improve exercise endurance.

▌ Respiratory failure can result from conditions that cause lung injury or impair lung function. Acute respiratory distress syndrome is a severe form of respiratory failure that requires mechanical ventilation. Goals of nutrition therapy are to provide enough energy and protein to support lung function without burdening the respiratory system.

Nutrition Assessment Checklist
FOR PEOPLE UNDERGOING METABOLIC OR RESPIRATORY STRESS

Medical History

Check the medical record to determine:

- [ ] Cause of stress
- [ ] Severity of stress
- [ ] Whether any organ system is compromised
- [ ] Whether nutrition support is required

For patients with COPD, check to determine:

- [ ] Degree of breathing difficulty
- [ ] Use of oxygen therapy
- [ ] Activity tolerance

Review the medical record for complications related to underfeeding or overfeeding, such as:

- [ ] Dehydration or fluid overload
- [ ] Electrolyte imbalances
- [ ] Fatty liver
- [ ] Hyperglycemia
- [ ] Hypertriglyceridemia

Medications

Record all medications and note:

- [ ] Side effects that may alter food intake or nutrition status
- [ ] Use of theophylline in patients who may need to monitor caffeine intake

Dietary Intake

If the patient is not meeting nutrition goals:

- [ ] Monitor intakes to ensure that the patient is receiving the diet prescribed.
- [ ] Investigate appetite problems or difficulties with eating.
- [ ] Consider interventions to improve food intake.
- [ ] Consider the need for supplementation.
- [ ] In patients with COPD, consider problems that may hamper the patient's ability to prepare or consume foods.

Anthropometric Data

Measure baseline height and weight, and monitor daily weights. Remember that body weight can fluctuate in acutely ill patients who undergo fluid resuscitation. After the patient's weight has stabilized:

- [ ] Reevaluate protein and energy needs.
- [ ] Consider the need to alter the energy prescription to meet weight goals.

Laboratory Tests

Laboratory tests that may be affected by stress and therefore require careful interpretation include:

- [ ] Albumin
- [ ] C-reactive protein
- [ ] Prealbumin
- [ ] Serum iron and zinc
- [ ] Transferrin
- [ ] White blood cell count

Monitor laboratory tests for signs of:

- [ ] Dehydration or fluid overload
- [ ] Electrolyte and acid-base imbalances
- [ ] Hyperglycemia
- [ ] Hypertriglyceridemia
- [ ] Nutrient deficiencies
- [ ] Negative nitrogen balance
- [ ] Organ dysfunction or organ function that has normalized

Physical Signs

Regularly assess vital signs, including:

- [ ] Blood pressure
- [ ] Body temperature
- [ ] Pulse
- [ ] Respiratory rate

Look for physical signs of:

- [ ] Protein-energy malnutrition
- [ ] Dehydration or fluid overload
- [ ] Nutrient deficiencies and excesses

Clinical Applications

1. Adam is a 29-year-old male who is 6 feet 2 inches tall and has a usual body weight of 180 pounds. He underwent emergency surgery following an accident and is now being cared for in the intensive care unit. Using the method described in Table 24-2 on p. 624, estimate Adam's energy requirement. Estimate his protein requirement, using the factor 1.5 grams per kilogram of body weight.

2. Ayla is a 23-year-old law student who was admitted to the hospital following an automobile accident in which she broke several bones and ruptured part of her small intestine. She has been in the hospital for several weeks and has just begun eating table foods. Her brother, who was driving the vehicle, was also seriously injured and nearly lost his life. Aside from the increased nutritional needs imposed by the stress of the accident, discuss how the following factors might interfere with Ayla's ability to improve her nutrition status:

   ▌ Ayla's injuries are painful.

   ▌ Ayla's medications cause drowsiness.

   ▌ Ayla is depressed.

   ▌ Ayla is often out of her room for X-rays and other diagnostic tests when the menus and food trays arrive.

   ▌ Ayla's food intake is sometimes restricted due to the procedures she is undergoing.

   How might these problems be resolved to improve Ayla's food intake?

Self Check

1. Which of the following metabolic changes accompanies acute stress?
   a. Reduced plasma concentrations of glucose and fatty acids
   b. Reduced blood volume and blood pressure
   c. Increased insulin effectiveness
   d. Catabolism of protein in muscle and connective tissue

2. Tissue injury is followed by:
   a. fluid accumulation in damaged tissue.
   b. reduced blood flow to injured tissue.
   c. reduced capillary permeability.
   d. decreased body temperature.

3. What is a possible effect of replacing vegetable oils rich in omega-6 fatty acids with oils rich in omega-3 fatty acids?
   a. Improvement in blood circulation
   b. Suppression of inflammation
   c. Protection against sepsis
   d. Hypertriglyceridemia

4. The acute-phase response results in an increased plasma concentration of:
   a. albumin.
   b. iron.
   c. C-reactive protein.
   d. zinc.

5. Which of the following statements concerning protein and energy recommendations during acute metabolic stress is true?
   a. Protein and energy recommendations are similar to those for healthy people.
   b. Protein and energy recommendations are reduced because a stressed individual cannot metabolize nutrients normally.
   c. Acutely stressed individuals can benefit from as much protein and energy as can be provided.
   d. Protein and energy recommendations are increased to minimize muscle tissue losses.

6. The amount of protein recommended for a nonobese acutely stressed patient who weighs 150 pounds ranges from about _____ grams of protein per day.
   a. 55 to 68
   b. 82 to 136
   c. 102 to 170
   d. 136 to 205

7. The primary risk factor for COPD is:
   a. alpha-1 antitrypsin deficiency.
   b. occupational exposure to dusts or chemicals.
   c. cigarette smoking.
   d. respiratory infections.

8. A primary feature of emphysema is:
   a. obstruction within the bronchi.
   b. obstruction within the bronchioles.
   c. destruction of the walls separating the alveoli.
   d. excessive lung elasticity.

9. The weight loss and wasting that often occur in COPD can be caused by:
   a. reduced food intake.
   b. increased metabolic rate.
   c. reduced exercise tolerance.
   d. all of the above.

10. Nutrition therapy for a person with respiratory failure includes:
    a. careful attention to providing enough, but not too much, energy.
    b. a generous fluid intake to facilitate mucus clearance.
    c. a high fat intake to prevent weight loss.
    d. a high carbohydrate intake to increase carbon dioxide production.

Answers to these questions can be found in Appendix H.

Notes

1. S. F. Lowry and J. M. Perez, The hypercatabolic state, in M. E. Shils and coeditors, *Modern Nutrition in Health and Disease* (Baltimore: Lippincott Williams & Wilkins, 2006), pp. 1381–1400.

2. N. D. Riediger and coauthors, A systemic review of the roles of n-3 fatty acids in health and disease, *Journal of the American Dietetic Association* 109 (2009): 668–679; P. C. Calder, Polyunsaturated fatty acids, inflammatory processes and inflammatory bowel diseases, *Molecular Nutrition and Food Research* 8 (2008): 885–897.

3. V. Kumar and coauthors, Acute and chronic inflammation, in V. Kumar and coeditors, *Robbins and Cotran Pathologic Basis of Disease* (Philadelphia: Saunders, 2010), pp. 43–77; D. S. Pisetsky, Laboratory testing in the rheumatic diseases, in L. Goldman and D. Ausiello, eds., *Cecil Medicine* (Philadelphia: Saunders, 2008), pp. 1966–1970.

4. Y. Debaveye and G. Van den Berghe, Risks and benefits of nutritional support during critical illness, *Annual Review of Nutrition* 26 (2006): 513–538.

5. K. A. Kudsk and G. S. Sacks, Nutrition in the care of the patient with surgery, trauma, and sepsis, in M. E. Shils and coeditors, *Modern Nutrition in Health and Disease* (Baltimore: Lippincott Williams & Wilkins, 2006), pp. 1414–1435.

6. S. A. McClave and coauthors, Guidelines for the provision and assessment of nutrition support therapy in the adult critically ill patient: Society of Critical Care Medicine (SCCM) and American Society for Parenteral and Enteral Nutrition (A.S.P.E.N.), *Journal of Parenteral and Enteral Nutrition* 33 (2009): 277–316.

7. McClave and coauthors, 2009.

8. American Dietetic Association, *Nutrition Care Manual* (Chicago: American Dietetic Association, 2010).

9. McClave and coauthors, 2009.

10. McClave and coauthors, 2009; Lowry and Perez, 2006.

11. M. Zhou and R. G. Martindale, Arginine in the critical care setting, *Journal of Nutrition* 137 (2007): 1687S–1692S; P. Furst, Protein and amino acid metabolism: Comparison of stressed and nonstressed states, in G. Cresci, ed., *Nutrition Support for the Critically Ill Patient: A Guide to Practice* (Boca Raton, FL: Taylor & Francis Group, 2005), pp. 27–47.

12. Debaveye and Van den Berghe, 2006.

13. Kudsk and Sacks, 2006.

14. J. Lefton and P. P. Lopez, Macronutrient requirements: Carbohydrate, protein, and lipid, in G. Cresci, ed., *Nutrition Support for the Critically Ill Patient: A Guide to Practice* (Boca Raton, FL: Taylor & Francis Group, 2005), pp. 99–108.

15. K. Sriram and J. I. Cué, Micronutrient and antioxidant therapy in critically ill patients, in G. Cresci, ed., *Nutrition Support for the Critically Ill Patient: A Guide to Practice* (Boca Raton, FL: Taylor & Francis Group, 2005), pp. 109–123.

16. McClave and coauthors, 2009.

17. A. N. Husain, The lung, in V. Kumar and coeditors, *Robbins and Cotran Pathologic Basis of Disease* (Philadelphia: Saunders, 2010), pp. 677–737; N. Anthonisen, Chronic obstructive pulmonary disease, in L. Goldman and D. Ausiello, eds., *Cecil Medicine* (Philadelphia: Saunders, 2008), pp. 619–627.

18. A. M. Miniño, J. Xu, and K. D. Kochanek, Deaths: Preliminary data for 2008, *National Vital Statistics Reports* 59, no. 2 (Hyattsville, MD: National Center for Health Statistics, 2010).

19. T. J. Prendergast and S. J. Ruoss, Pulmonary disease, in S. J. McPhee and W. F. Ganong, eds., *Pathophysiology of Disease: An Introduction to Clinical Medicine* (New York: McGraw-Hill/Lange, 2006), pp. 218–258.

20. M. H. Beers and coeditors, *The Merck Manual of Diagnosis and Therapy* (Whitehouse Station, NJ: Merck Research Laboratories, 2006), pp. 400–422.

21. B. Suckling, M. M. Johnson, and R. Chin, Jr., Nutrition, respiratory function, and disease, in M. E. Shils and coeditors, *Modern Nutrition in Health and Disease* (Baltimore: Lippincott Williams & Wilkins, 2006), pp. 1462–1474.

22. D. A. King, F. Cordova, and S. M. Scharf, Nutritional aspects of chronic obstructive pulmonary disease, *Proceedings of the American Thoracic Society* 5 (2008): 519–523.

23. M. Poulain and coauthors, The effect of obesity on chronic respiratory diseases: Pathophysiology and therapeutic strategies, *Canadian Medical Association Journal* 174 (2006): 1293–1299.

24. McClave and coauthors, 2009, A. Malone, Enteral formula selection, in P. Charney and A. Malone, eds., *ADA Pocket Guide to Enteral Nutrition* (Chicago: American Dietetic Association, 2006), pp. 63–122.

25. M. C. Steiner and coauthors, Nutritional enhancement of exercise performance in chronic obstructive pulmonary disease: A randomised controlled trial, *Thorax* 58 (2003): 745–751.

26. L. D. Hudson and A. S. Slutsky, Acute respiratory failure, in L. Goldman and D. Ausiello, eds., *Cecil Medicine* (Philadelphia: Saunders, 2008), pp. 723–734.

27. American Dietetic Association, 2010.

28. McClave and coauthors, 2009.

Nutrition in Practice

Multiple organ dysfunction syndrome (MODS), also called *multiple organ failure,* is a frequent cause of death in intensive care patients. Described as the progressive dysfunction of two or more of the body's organ systems, MODS most often involves the lungs, liver, kidneys, and gastrointestinal (GI) tract. MODS is not a disease per se, but rather a late stage of severe illness or injury that results from a severe inflammatory response (discussed in Chapter 24).[1] MODS can be initiated by a number of very different critical illnesses and conditions, including respiratory failure, sepsis, burn injuries, extensive surgery, and pancreatitis. This Nutrition in Practice discusses how MODS develops, the manner in which it is treated, and the importance of its prevention.

How long has MODS been a major clinical problem?

MODS was recognized as a clinical entity only after World War II. Prior to the mid-20th century, patients with severe illnesses or multiple injuries frequently died of shock or circulatory failure. After fluid replacement and blood transfusions became standard treatments, the kidneys became the organs at highest risk, and kidney failure became the most common cause of death. Eventually, physicians learned to better support kidney function by providing appropriate electrolyte solutions and improving urine output. With improved kidney care, the lungs became the most vulnerable organ after severe injury. Improved treatment of respiratory failure eventually led to the current situation: advances in critical care allow patients to survive severe illnesses and injuries, but the body's defenses often overburden organs that were not originally injured.

Why does critical illness sometimes lead to MODS?

As discussed earlier in Chapter 24, injury and infection cause the release of chemical mediators that have systemic (whole-body) effects. A severe, persistent inflammatory response can lead to systemic inflammatory response syndrome (SIRS), which is associated with a constellation of symptoms including fever, increased heart and respiratory rates, and abnormal white blood cell counts. SIRS is a normal adaptive response to a severe insult, but if not reversed quickly enough it can progress to shock, which is characterized by extremely low blood pressure and an inadequate blood supply for the tissues and organs of the body.[2]

As might be expected from a systemic reduction in blood availability, shock can impair numerous organ systems. The abnormal delivery of oxygen and nutrients to tissues and insufficient removal of wastes result in irreversible injury to cells and tissues. Although each organ system is affected differently, ultimately one or more organs may begin to fail. The failure of one organ may place excessive demands on another, causing the second to fail as

well. The progression of SIRS to MODS reflects the inability of the body's defenses and medical treatments to counter the detrimental effects of a sustained and potent inflammatory response.

The specific pathophysiology of MODS is poorly understood. Although early reports attempted to link the development of MODS directly to sepsis, sepsis is not present in all cases. Infection often results from impaired immune function and therefore is a frequent consequence of MODS, but it is not necessarily the underlying trigger of organ dysfunction. Recall from this chapter that sepsis gives rise to symptoms identical to those seen in SIRS. Figure NP24-1 illustrates the relationships among SIRS, infection, sepsis, and MODS.

Do organs fail in a specific pattern?

Although the clinical course differs substantially among patients, the sequence of organ dysfunction often follows a similar pattern: first the lungs fail, then the liver, and finally the kidneys, GI tract, or heart. Other organs or systems may also become involved, and each additional failure reduces the likelihood of survival. Table NP24-1 lists the organs and systems most often involved in MODS and the potential consequences of their failure.

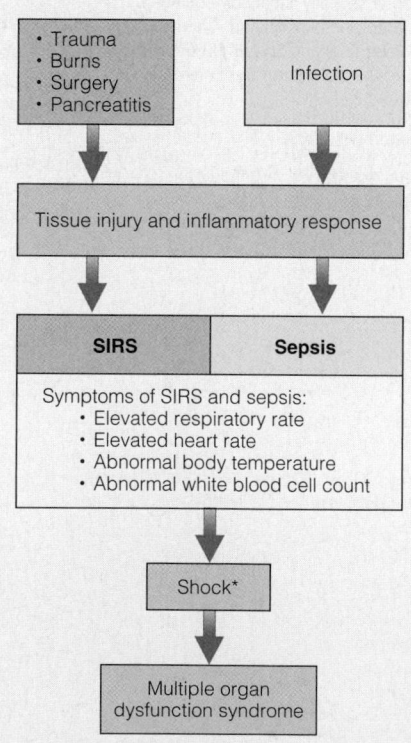

*After critical injury, shock may sometimes precede and be the cause of SIRS.

FIGURE NP24-1 Relationships among SIRS, Sepsis, and Multiple Organ Dysfunction Syndrome

TABLE NP24-1 Physiological Effects of Organ or System Failure

Organ or System	Effects of Failure
Lungs	Inability to maintain gas exchange
Liver	Altered metabolic processes
Kidneys	Inability to regulate blood volume, maintain electrolytes, remove wastes
Heart	Low cardiac output, low blood pressure, inadequate circulation, shock
GI tract	Impaired digestion and absorption, abnormal bleeding, bacterial translocation
Immune system	Infection, sepsis
Coagulation system	Excessive bleeding or blood clotting
Central nervous system	Decreased perceptions, brain injury, coma

TABLE NP24-2 Factors That Influence Risk of Multiple Organ Dysfunction Syndrome

Age over 55 years
Prior chronic disease
Persistent SIRS
Major infection
Blood transfusions
Severity of tissue injury
Length of time between injury and arrival at hospital
Malnutrition

Are there any risk factors for MODS?

Epidemiological studies have identified a number of factors that increase risk. For example, people who develop MODS are often older, have multiple or severe injuries, and develop severe infections. Table NP24-2 lists the major risk factors associated with MODS, some of which are discussed below:

▌ *Age.* Patients over 55 years old are several times more likely to develop MODS than are younger patients. In elderly patients, the increased risk may be due to the presence of chronic illnesses that directly affect organ function, such as heart disease, lung disease, diabetes, or liver damage. Aging also decreases the functional reserve of organs, thereby reducing an older patient's ability to deal with the additional stress that arises during critical illness.

▌ *Severity of SIRS.* The length of time that SIRS persists is related to the development of MODS. Patients who have SIRS that persists for more than three days are more likely to develop MODS than patients who have SIRS for less than two days.

▌ *Infection.* Prolonged SIRS can suppress immune function and increase the risk of developing an infection. During hospital stays, critically ill patients often contract pneumonia—the principal infection associated with MODS. The risks of infection and sepsis greatly increase with the use of invasive catheters, which are frequently needed during intensive care to provide oxygen support, intravenous fluid resuscitation, nutrition support, and urine clearance.

▌ *Blood transfusions.* Blood transfusions are immunosuppressive and may increase a patient's risks of developing infection or sepsis. Blood transfusions frequently have adverse effects that can add further stress; they may cause acute lung injury, allergic reactions, red blood cell hemolysis (breakdown), and other complications.

What is the treatment for MODS?

Once MODS has developed, extensive medical support is needed until the inflammatory response has abated. Unfortunately, aggressive treatments can have damaging effects of their own and may cause further injury to organs that are already weakened by illness. Health practitioners should be aware of the adverse effects of aggressive therapies and remain alert to a patient's responses to treatments. Therapies that are often used to manage MODS include:[3]

▌ *Lung support.* Mechanical ventilation is used to assist injured lungs and sustain gas exchange.

▌ *Fluid resuscitation.* Fluids and electrolytes are supplied to restore blood volume and maintain electrolyte balance.

▌ *Support of heart and blood vessel function.* Medications help to sustain or increase cardiac output and maintain adequate blood pressure.

▌ *Kidney support.* Hemofiltration or dialysis helps to prevent the buildup of toxic metabolites in blood.

▌ *Protection against infection.* Antibiotic therapy may reverse or prevent infections.

▌ *Nutrition support.* Enteral and parenteral nutrition support provide nutrients, help to prevent excessive wasting, and promote recovery.

What can be done to reduce the incidence of MODS?

Because mortality rates for MODS are so high, prevention must be considered at the earliest stages of injury and treatment, before an excessive inflammatory response can cause further damage. Health practitioners have learned to identify the conditions that can increase organ stress whether they are due to a disease process, an inflammatory response, or an aggressive treatment that is intended to provide organ support. Although improvements in care over the past few decades have reduced some of the complications that arise during intensive care, rates of mortality from MODS have not changed. Thus, a focus on prevention is critical until a better understanding of the pathophysiology of MODS is achieved, which may lead to additional therapeutic options.

Notes

1. J. Parrillo, Approach to the patient with shock, in L. Goldman and D. Ausiello, eds., *Cecil Medicine* (Philadelphia: Saunders, 2008), pp. 742–750.

2. J. A. Russell, Shock syndromes related to sepsis, in L. Goldman and D. Ausiello, eds., *Cecil Medicine* (Philadelphia: Saunders, 2008), pp. 755–763.

3. M. H. Oltermann, Systemic inflammatory response and sepsis: A multidisciplinary approach to the nutritional considerations, in G. Cresci, ed., *Nutrition Support for the Critically Ill Patient: A Guide to Practice* (Boca Raton, FL: Taylor & Francis Group, 2005), pp. 565–577.

25

Energy- and Protein-Modified Diets for Cancer and HIV Infection

Although **cancers** and **human immunodeficiency virus (HIV)** infections are distinct disorders, from a nutritional standpoint they share some similarities. Both disorders have debilitating effects that influence nutritional needs, and both can lead to severe wasting in advanced cases. These illnesses require nutrition therapy that is highly individualized based on the symptoms manifested and the organ systems involved.

Cancers are classified by the tissues or cells from which they develop:

- *Adenocarcinomas* (ADD-eh-no-CAR-sih-NO-muz) arise from glandular tissues.
- *Carcinomas* (CAR-sih-NO-muz) arise from epithelial tissues.
- *Leukemias* (loo-KEY-mee-uz) arise from white blood cell precursors.
- *Lymphomas* (lim-FOE-muz) arise from lymphoid tissue.
- *Melanomas* (MEL-ah-NO-muz) arise from pigmented skin cells.
- *Myelomas* (MY-ah-LOE-muz) arise from plasma cells in the bone marrow.
- *Sarcomas* (sar-KO-muz) arise from connective tissues, such as muscle or bone.

An abnormal mass of cells that is noncancerous is called a *benign* tumor.

Cancer

Cancer, the growth of **malignant** tissue, is the second most common cause of death in the United States, ranking just below cardiovascular disease. Cancer is not a single disorder, however; there are many different kinds of malignant growths. The different types of cancer have different characteristics, occur in different locations in the body, take different courses, and require different treatments. Whereas an isolated, non-spreading type of skin cancer may be removed in a physician's office with no effect on nutrition status, advanced cancers—especially those of the gastrointestinal (GI) tract and pancreas—can seriously impair nutrition status. In the United States, the most common cancers are breast cancer (in women), prostate cancer (in men), lung cancer, and colorectal cancers.[1]

How Cancer Develops

The development of cancer, called **carcinogenesis,** often proceeds slowly and continues for several decades. A cancer arises from mutations in the genes that control cell division in a single cell.[2] These mutations may promote cellular growth, interfere with growth restraint, inhibit genetic repair, or prevent cellular death. The affected cell thereby loses its built-in capacity for halting cell division and produces daughter cells with the same genetic defects. As the abnormal mass of cells, called a **tumor** (or *neoplasm*), grows, a network of blood vessels forms to supply the tumor with the nutrients it needs to support its growth. The tumor can disrupt the functioning of the normal tissue around it, and some tumor cells may **metastasize** to other regions in the body. Figure 25-1 illustrates the steps in cancer development. In leukemia (cancer of the white blood cells), the abnormal cells do not form a tumor; they accumulate in the blood and other tissues.

The reasons why cancers develop are numerous and varied. Vulnerability to cancer is sometimes inherited, as when a person is born with a genetic defect that alters DNA structure, function, or repair. Certain metabolic processes may initiate carcinogenesis, as when phagocytes (immune cells) produce oxidants that cause DNA damage, or when chronic inflammation increases the rate of cell division and the risk of a damaging mutation. More often, cancers are caused by interactions between a person's

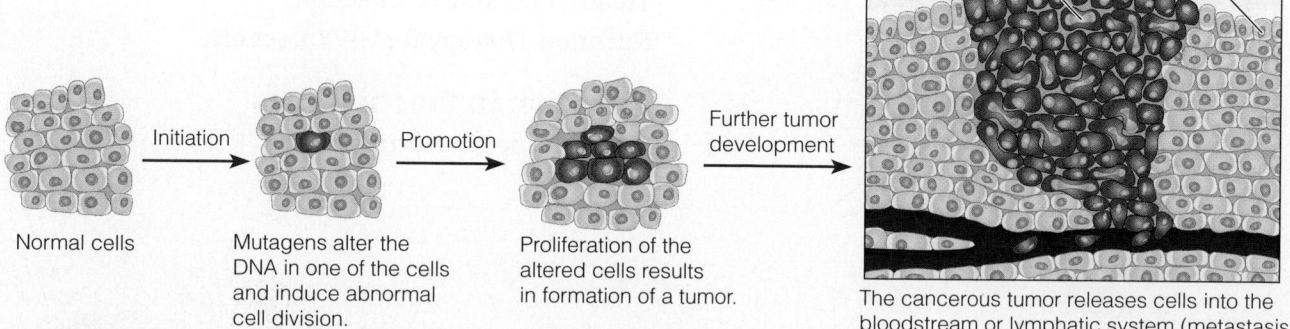

Normal cells → Initiation → Mutagens alter the DNA in one of the cells and induce abnormal cell division. → Promotion → Proliferation of the altered cells results in formation of a tumor. → Further tumor development →

Malignant cells   Normal cells

The cancerous tumor releases cells into the bloodstream or lymphatic system (metastasis)

FIGURE 25-1 **Cancer Development**

TABLE 25-1 Environmental Factors That Increase Cancer Risk

Environmental Factors	Cancer Sites
Aflatoxins (toxins in moldy peanuts or grains)	Liver
Asbestos[a]	Lung, pleura, peritoneum
Chromium (hexavalent) compounds	Nasal cavity, lung
Estrogen-progesterone replacement therapy	Breast
Immunosuppressive medications	Lymphoid tissues, liver
Infection with *Helicobacter pylori*	Stomach
Infection with hepatitis B and hepatitis C viruses	Liver
Infection with human papillomavirus (HPV)	Cervix
Ionizing radiation (X-rays, radioactive isotopes, and other sources)	White blood cells (leukemia), esophagus, stomach, colon, thyroid, lung, bladder, breast
Tobacco[b]	Nasal cavity, lung, mouth, pharynx, larynx, esophagus, stomach, colon, rectum, liver, pancreas, kidney, renal pelvis, bladder
Ultraviolet radiation (sun exposure)	Skin

[a]Risk is greatly increased in cigarette smokers.
[b]A combined exposure to tobacco and alcohol multiplies the risks of developing cancers of the oral cavity, pharynx, larynx, and esophagus.

Sources: M. J. Thun, Epidemiology of cancer, in L. Goldman and D. Ausiello, eds., *Cecil Medicine* (Philadelphia: Saunders, 2008), pp. 1335–1340; World Cancer Research Fund/American Institute for Cancer Research, *Food, Nutrition, Physical Activity, and the Prevention of Cancer: A Global Perspective* (Washington, D.C.: American Institute for Cancer Research, 2007), pp. 157–171.

genes and the environment. Exposure to cancer-causing substances, or **carcinogens,** may induce genetic mutations that lead to cancer; other substances may stimulate division of the altered cells. Table 25-1 provides examples of environmental factors that increase cancer risk.

Nutrition and Cancer Risk

Like other environmental factors, diet and lifestyle strongly influence cancer risk. Certain food components may directly damage DNA, alter the metabolism of carcinogens by liver enzymes, or inhibit the formation of carcinogens in the body.[3] In addition, energy balance and growth rates affect the rate of cell division and consequently influence the rates at which mutations form and are replicated. Table 25-2 includes examples of nutrition-related factors that may increase or decrease the risk of developing cancer.

Nutrition and Increased Cancer Risk As shown in Table 25-2, obesity is a risk factor for a number of different cancers, including some relatively common cancers such as colon cancer and postmenopausal breast cancer. Obesity increases cancer risk, in part, by altering levels of hormones that influence cell growth, such as the sex hormones, insulin, and certain growth factors.[4] For example, in the case of breast cancer in postmenopausal women, the hormone estrogen is likely involved: obese women have higher estrogen levels than lean women do because adipose tissue is the primary source of estrogen after menopause.

Alcohol consumption correlates strongly with cancers of the head and neck, colon, rectum, and breast. For head and neck cancers, the risk is multiplied when alcohol drinkers also smoke tobacco.[5] In addition, alcohol abuse may damage the liver and instigate the development of liver cancer. These findings illustrate why the potential benefits of moderate alcohol consumption on cardiovascular disease risk must be weighed against the potential dangers.

cancers: diseases characterized by the uncontrolled growth of a group of cells, which can destroy adjacent tissues and spread to other areas of the body via lymph or blood.

human immunodeficiency virus (HIV): the virus that causes acquired immune deficiency syndrome (AIDS). HIV destroys immune cells and progressively impedes the body's ability to fight infections and certain cancers.

malignant (ma-LIG-nent): describes a cancerous cell or tumor, which can injure healthy tissue and spread cancer to other regions of the body.

carcinogenesis (CAR-sin-oh-JEN-eh-sis): the process of cancer development.

tumor: an abnormal tissue mass that has no physiological function; also called a *neoplasm* (NEE-oh-plazm).

metastasize (meh-TAS-tah-size): to spread from one part of the body to another; refers to cancer cells.

carcinogens (CAR-sin-oh-jenz or car-SIN-oh-jenz): substances that can cause cancer (the adjective is *carcinogenic*).

TABLE 25-2 Nutrition-Related Factors That Influence Cancer Risk

Nutrition-Related Factors[a]	Cancer Sites
Factors that may increase cancer risk:	
Obesity	Esophagus, colon, rectum, pancreas, gallbladder, kidney, breast (postmenopausal), endometrium
Red meat, processed meats	Colon, rectum
Salted and salt-preserved foods	Stomach
Beta-carotene supplements	Lung[b]
High-calcium diets (over 1500 mg daily)	Prostate
Alcohol[c]	Mouth, pharynx, larynx, esophagus, colon, rectum, liver, breast
Low level of physical activity[d]	Colon, breast (postmenopausal), endometrium
Factors that may decrease cancer risk:	
Fruits and nonstarchy vegetables	Lung, mouth, pharynx, larynx, esophagus, stomach
Carotenoid-containing foods	Lung, mouth, pharynx, larynx, esophagus
Tomato products	Prostate
Allium vegetables (onion, garlic)	Stomach, colon, rectum
Vitamin C–containing foods	Esophagus
Folate-containing foods	Pancreas
Fiber-containing foods	Colon, rectum
Milk and calcium supplements	Colon, rectum
High level of physical activity[d]	Colon, breast (postmenopausal), endometrium

[a]Altered cancer risk is associated with high intakes of the dietary substances listed.

[b]Cancer risk is increased in tobacco smokers and may not apply to other groups.

[c]A combined exposure to alcohol and tobacco multiplies the risks of developing cancers of the oral cavity, pharynx, larynx, and esophagus.

[d]Physical activity may influence cancer risk by altering body fatness, intestinal transit time, insulin sensitivity, hormone levels, enzyme activities, and immune responses.

Source: World Cancer Research Fund/American Institute for Cancer Research, *Food, Nutrition, Physical Activity, and the Prevention of Cancer: A Global Perspective* (Washington, D.C.: American Institute for Cancer Research, 2007).

Food preparation methods are responsible for producing certain types of carcinogens. Cooking meat, poultry, and fish at high temperatures (frying, broiling) causes the amino acids and creatine in these foods to react together and form carcinogens.[6]* Carcinogens also accompany the smoke that adheres to foods during grilling and are present in the charred surfaces of grilled meat and fish.† However, the cancer risk from eating such foods is unclear because the biological actions of these carcinogens are modulated by other dietary components, including compounds in vegetables and other plant foods.

Nutrition and Decreased Cancer Risk The consumption of fruits and vegetables may provide some benefits in protecting against the development of cancer (see Table 25-2). Fruits and vegetables contain both nutrients and phytochemicals with antioxidant activity, and these substances may prevent or reduce the oxidative reactions in cells that cause DNA damage. Phytochemicals may also help to inhibit carcinogen production in the body, enhance immune functions that protect against cancer development, or promote enzyme reactions that inactivate carcinogens.[7] The B vitamin folate, which is provided by certain fruits and vegetables, plays roles in DNA synthesis and repair; thus, inadequate folate intakes may allow DNA damage to accumulate. Fruits and vegetables also contribute dietary fiber, which may help to protect against colon and

To minimize carcinogen formation during cooking:

- Marinate meats before cooking.
- Use lower-heat options such as roasting, stewing, or microwaving.
- Choose lean meats for grilling, and take care not to blacken surfaces.
- To reduce smoke formation, prevent fat from dripping onto the heat source.

anorexia: lack of appetite.

cancer cachexia (ka-KEK-see-ah): a wasting syndrome associated with cancer that is characterized by anorexia, muscle wasting, weight loss, and fatigue.

*These carcinogens are *heterocyclic amines*.

†These carcinogens are *polycyclic aromatic hydrocarbons*.

TABLE 25-3 **Recommendations for Reducing Cancer Risk**

Maintain a healthy body weight throughout life.

▮ Be as lean as possible within the normal range of body weight for your height.

▮ Avoid weight gain and increases in waist circumference throughout adulthood.

Be physically active as part of everyday life.

▮ For adults: engage in moderate to vigorous physical activity for at least 30 minutes on at least 5 days per week; 45 to 60 minutes is preferable.

▮ For children and adolescents: engage in moderate to vigorous activity for 60 minutes on at least 5 days of the week.

▮ Limit sedentary habits such as watching television.

Choose a healthy diet that emphasizes plant sources.

▮ Limit consumption of energy-dense foods (>225 kcal per 100 g food) and sugary drinks that contribute to weight gain.

▮ Consume relatively unprocessed grains and/or legumes with every meal. Choose whole-grain products instead of processed (refined) grains.

▮ Consume five or more servings of non-starchy vegetables and fruits every day.

▮ Limit consumption of red meats (such as beef, pork, and lamb) and processed meats (such as those preserved by smoking, curing, or salting).

▮ Avoid salt-preserved, salted, and salty foods.

▮ Avoid moldy grains and legumes.

Limit consumption of alcoholic beverages.

▮ For women: drink no more than one drink daily.

▮ For men: drink no more than two drinks daily.

Aim to meet nutritional needs through the diet.

▮ Obtain necessary nutrients from the diet. Dietary supplements are not recommended for cancer prevention, and they may have unexpected adverse effects.

Avoid using tobacco in any form.

Sources: World Cancer Research Fund/American Institute for Cancer Research, *Food, Nutrition, Physical Activity, and the Prevention of Cancer: A Global Perspective* (Washington, D.C.: American Institute for Cancer Research, 2007);
L. H. Kushi and coauthors, American Cancer Society guidelines on nutrition and physical activity for cancer prevention: Reducing the risk of cancer with healthy food choices and physical activity, *CA: A Cancer Journal for Clinicians* 56 (2006): 254–281.

rectal cancers by diluting potential carcinogens in fecal matter and accelerating their removal from the GI tract. Table 25-3 summarizes the dietary and lifestyle practices that may help to reduce the risk of developing cancer.

Consequences of Cancer

Once cancer develops, its consequences depend on the location of the tumor, its severity, and the treatment. The complications that develop are often due to the tumor's impingement on surrounding tissues. Nonspecific effects of cancer include **anorexia,** lethargy, weight loss, night sweats, and fever.[8] During the early stages, many cancers produce no symptoms, and the person may be unaware of the threat to health.

Wasting Associated with Cancer Anorexia, muscle wasting, weight loss, anemia, and fatigue typify **cancer cachexia,** a condition of severe malnutrition that develops in up to 50 percent of cancer patients.[9] Without adequate energy and nutrients, the body is poorly equipped to maintain organ function, support immune defenses, and mend damaged tissues. An involuntary weight loss of more than 10 percent, which indicates significant malnutrition, is cause for concern.[10] Care must be taken not to overlook unintentional weight loss in patients who are overweight or obese.

Many factors play a role in the wasting associated with cancer. Cytokines, released by both tumor cells and immune cells, induce a catabolic state. The combined effects of a poor appetite, accelerated and abnormal metabolism, and the diversion of nutrients to support tumor growth result in a lower supply of energy and nutrients at a time when demands are high. Appetite and food intake are further disturbed by the effects of treatments and medications prescribed for cancer patients. Unlike in starvation, nutrition intervention alone is unable to reverse cachexia.[11]

Cruciferous vegetables, such as cauliflower, broccoli, and brussels sprouts, contain nutrients and phytochemicals that inhibit cancer development.

Reminder: The immune responses to tissue injury include the release of *cytokines* into the bloodstream (described in Chapter 24). The cytokines that induce cachexia include tumor necrosis factor-α, interleukin-1, interleukin-6, and γ-interferon.

Metabolic Changes The metabolic changes that arise in cancer exacerbate the wasting described in the previous section.[12] Cancer patients exhibit an increased rate of protein turnover but reduced muscle protein synthesis. Muscle contributes amino acids for glucose production, further depleting the body's supply of protein. Triglyceride breakdown increases, elevating serum lipids. Many patients develop insulin resistance. These metabolic abnormalities help to explain why people with cancer fail to regain lean tissue or maintain healthy body weights even when they are consuming adequate energy and nutrients.

Anorexia and Reduced Food Intake Anorexia is a major contributor to the wasting associated with cancer. Some factors that contribute to anorexia or otherwise reduce food intake include:

▌ *Chronic nausea and early satiety.* People with cancer frequently experience nausea and a premature feeling of fullness after eating small amounts of food.

▌ *Fatigue.* People with cancer often tire easily and lack the energy to prepare and eat meals. Once cachexia develops, these tasks become even more difficult.

▌ *Pain.* People in pain may have little interest in eating, particularly if eating makes the pain worse.

▌ *Mental stress.* A cancer diagnosis can cause distress, anxiety, and depression, all of which may reduce appetite. Facing and undergoing cancer treatments induces additional psychological stress.

▌ *Gastrointestinal obstructions.* A tumor may partially or completely obstruct a portion of the GI tract, causing complications such as nausea and vomiting, early satiety, delayed gastric emptying, and bacterial overgrowth. Some patients with obstructions are unable to tolerate oral diets.

▌ *Effects of cancer therapies.* Chemotherapy and radiation treatments for cancer frequently have side effects that make food consumption difficult, such as nausea, vomiting, altered taste perceptions, mouth sores, inflammation of mucosal tissue, abdominal pain or discomfort, and diarrhea.

Treatments for Cancer

The primary medical treatments for cancer—surgery, chemotherapy, radiation therapy, or any combination of the three—aim to remove cancer cells, prevent further tumor growth, and alleviate symptoms.[13] The likelihood of effective treatment is highest with early detection and intervention. Because treatment decisions are difficult and cancer therapies have considerable side effects, patients rely on health care providers to help them make informed decisions.

Surgery Surgery is performed to remove tumors, determine the extent of cancer, and protect nearby tissues. Often, surgery must be followed by other cancer treatments to prevent growth of new tumors. The acute metabolic stress caused by surgery raises protein and energy needs and can exacerbate wasting. Surgery also contributes to pain, fatigue, and anorexia, all of which can reduce food intake at a time when nutritional needs are substantial. Blood loss contributes to nutrient losses and further exacerbates malnutrition. Some surgeries may have long-term effects on nutritional status (see Table 25-4).

Chemotherapy Chemotherapy relies on the use of drugs to treat cancer, and is used to inhibit tumor growth, shrink tumors before surgery, and prevent or eradicate metastasis. Some cancer drugs interfere with the process of cell division; others sterilize cells that are in a resting phase and are not actively dividing. Unfortunately, most of these drugs are toxic to normal cells as well as cancer cells, and they are especially harmful to rapidly dividing cells, such as those of the GI tract, skin, and bone

---

TABLE 25-4 Nutrition-Related Side Effects of Cancer Surgeries

Head and Neck Surgeries

Aspiration

Dry or sore mouth

Reduced chewing or swallowing ability

Reduced sense of taste or smell

Esophageal Resection

Acid reflux

Altered gastric motility

Reduced swallowing ability

Gastric Resection

Dumping syndrome

Early satiety

Inadequate gastric acid secretion

Malabsorption of iron, folate, and vitamin B_{12}

Intestinal Resection

Bile insufficiency

Diarrhea

Fluid and electrolyte imbalances

General malabsorption

Pancreatic Resection

Diabetes mellitus

General malabsorption

One drug that inhibits cell division is *methotrexate*, which closely resembles the B vitamin folate (see Figure 15-1 on p. 416). Folate is required for cell division because it is needed for DNA synthesis. Methotrexate works by blocking activity of the enzyme that converts folate to its active form.

TABLE 25-5 Nutrition-Related Side Effects of Chemotherapy and Radiation Therapy

	Reduced Nutrient Intake	Accelerated Nutrient Losses	Altered Metabolism
Chemotherapy	Abdominal pain Anorexia Mouth ulcers Nausea and vomiting Reduced taste sensation	Diarrhea Gastrointestinal inflammation Malabsorption Vomiting	Anemia, neutropenia Fluid and electrolyte imbalances as a consequence of vomiting, diarrhea, or malabsorption Hyperglycemia Interference with vitamins or body compounds Negative nitrogen and micronutrient balances Secondary effects of malnutrition, infection, or inflammation
Radiation therapy	Anorexia Damage to teeth, jaws, or salivary glands Dysphagia Esophagitis Mouth sores Nausea and vomiting Reduced salivary secretions Reduced taste sensation	Blood loss from intestine and bladder Diarrhea Fistulas Intestinal obstructions Malabsorption Radiation enteritis Vomiting	Fluid and electrolyte imbalances as a consequence of vomiting, diarrhea, or malabsorption Secondary effects of malnutrition, infection, or inflammation

marrow. The bone marrow damage can suppress the production of red blood cells (causing anemia) and white blood cells (causing **neutropenia**). Some of the newer drugs are able to target properties specific to cancer cells and are better tolerated by the body's tissues. Table 25-5 includes a summary of the nutrition-related side effects that may result from chemotherapy.

Radiation Therapy **Radiation therapy** treats cancer by bombarding cancer cells with X-rays, gamma rays, or various atomic particles. These treatments generate reactive forms of oxygen, such as superoxide and hydroxyl radicals, which can damage cellular DNA and cause cell death. Newer techniques can focus radiation directly at tumors and minimize damage to nearby tissues. An advantage of radiation therapy over surgery is that it can shrink tumors while preserving organ structure and function. Compared with chemotherapy, radiation therapy is better able to target specific regions of the body, rather than involving all body cells. Nonetheless, radiation therapy can damage healthy tissues and sometimes has long-term detrimental effects on nutrition status (see Table 25-5). Radiation to the head and neck area can damage the salivary glands and taste buds, causing inflammation, dry mouth, and a reduced sense of taste; in severe cases, the damage may be permanent. Radiation treatment in the lower abdominal area can cause **radiation enteritis,** an inflammatory condition of the small intestine that causes nausea, vomiting, and diarrhea; the condition may persist for months or years and lead to chronic malabsorption in some individuals.

Hematopoietic Stem Cell Transplantation **Hematopoietic stem cell transplantation** replaces the blood-forming stem cells that have been destroyed by high-dose chemotherapy or radiation therapy. These procedures may be used to treat leukemia, lymphomas, and multiple myeloma.[14] If possible, stem cells are collected from the patient before chemotherapy or radiation treatment begins so that it is not necessary to find a separate donor. If another person's cells are used, the patient must take immunosuppressant drugs to prevent **tissue rejection.** The high-dose chemotherapy or radiation therapy preceding the transplant and the immunosuppressant drugs required afterward can impair immune function substantially and increase the risk of foodborne illness. Patients are often unable to consume adequate food during or after these procedures and usually require nutrition support.

chemotherapy: the use of drugs to arrest or destroy cancer cells; these drugs are called *antineoplastic agents.*

neutropenia: a low white blood cell (neutrophil) count, which increases susceptibility to infection.

radiation therapy: the use of X-rays, gamma rays, or atomic particles to destroy cancer cells.

radiation enteritis: inflammation of intestinal tissue caused by radiation therapy.

hematopoietic stem cell transplantation: transplantation of the stem cells that produce red blood cells and white blood cells; the stem cells are obtained from the bone marrow (*bone marrow transplantation*) or circulating blood.

tissue rejection: destruction of donor tissue by the recipient's immune system, which recognizes the donor cells as foreign.

Biological Therapies Newer therapies for cancer include the use of biological molecules that stimulate immune responses against cancer cells (also called *immunotherapy*). These substances include antibodies, cytokines, and other proteins that strengthen the body's immune defenses, enable the destruction of cancer cells, or interfere with cancer development in some way. Although side effects vary, many of these treatments can cause anorexia, GI symptoms, and general discomfort, reducing a person's ability or desire to consume adequate amounts of food.

Medications to Combat Anorexia and Wasting To help patients combat anorexia, medications may be prescribed to stimulate the appetite and promote weight gain. Examples include megestrol acetate (Megace), a synthetic compound similar in structure to the hormone progesterone, and dronabinol (Marinol), which resembles the psychoactive ingredient in marijuana and stimulates the appetite at doses that have minimal mental effects. Under investigation are medications that promote muscle protein synthesis, induce the secretion of growth hormone or growth factors, or inhibit some pro-inflammatory cytokines (which have catabolic effects).[15]

Alternative Therapies Many patients turn to *complementary and alternative medicine (CAM)* to assist them in their fight against cancer. Although few abandon conventional medicine, up to 80 percent of cancer patients combine one or more CAM approaches with standard treatment.[16] Many patients do not discuss their use of CAM with physicians.

Multivitamin and herbal supplements are among the most frequently used CAM therapies. Although many supplements can be used without risk, some may have adverse effects or interfere with conventional treatments. Use of the herb St. John's wort, for example, can reduce the effectiveness of some anticancer drugs.[17] As another example, some studies suggest that antioxidant supplements interfere with chemotherapy and radiation treatments.[18] Clinical trials of several popular supplements are in progress and are expected to reveal more about their potential effects and interactions with cancer treatments.

> Reminder: The term *complementary and alternative medicine (CAM)* refers to health care practices that have not been proved to be effective and consequently are not included as part of conventional treatment.

Nutrition Therapy for Cancer

The objectives of nutrition therapy for cancer patients are to minimize loss of weight and muscle tissue, correct nutrient deficiencies, and provide a diet that patients can tolerate and enjoy despite the complications of illness. Appropriate nutrition care helps patients preserve their strength and improves recovery after stressful cancer treatments. Moreover, malnourished cancer patients develop more complications and have shorter survival times than patients who maintain good nutrition status.[19]

Because there are many forms of cancer and a variety of potential treatments, nutritional needs among cancer patients vary considerably. Furthermore, a person's needs may change at different stages of illness. Patients should be screened for malnutrition when cancer is diagnosed and reassessed during the treatment and recovery periods.

Protein and Energy For patients at risk of weight loss and wasting, the focus of nutrition care is to ensure appropriate intakes of protein and energy. Health practitioners regularly monitor weight changes and adjust intake recommendations as necessary. Patients who cannot eat adequate food may be able to meet their needs by supplementing the diet with nutrient-dense formulas. The "How to" provides suggestions that can help to increase the energy and protein content of meals.

Although weight loss is a problem for many cancer patients, breast cancer patients often gain weight.[20] The weight gain occurs during the first two years after breast cancer diagnosis and is associated with an increase in total body fat. By discussing weight maintenance soon after diagnosis and encouraging physical activity, health practitioners can help patients avoid unnecessary weight gain.

- For maintenance of body weight and lean tissue, suggest 1.0 to 1.5 g protein/kg body weight and 25 to 35 kcal/kg body weight daily.

- For weight regain and repletion of muscle tissue, suggest 1.5 to 2.0 g protein/kg body weight and 35 to 45 kcal/kg body weight daily.

To increase the energy content of a meal, try these suggestions:

- *Meats.* Choose high-fat meats instead of lean meats. Sauté or pan-fry meat instead of baking or roasting it, and use sauces or gravies liberally. Sprinkle bacon bits or sausage pieces on vegetable dishes.

- *Cheese.* Include cheese slices or cream cheese in sandwiches made with luncheon meats. Spread cream cheese on raw vegetables, toast, and crackers or mix into dishes that contain chopped fruit.

- *Half-and-half and cream.* Replace milk or water with half-and-half or cream in soups, sauces, hot chocolate, desserts, mashed potatoes, and cold cereals. Use sour cream or cream sauces on potato dishes, vegetable dishes, and soups. Add whipped cream to fruit salads and desserts.

- *Breads and cereals.* Choose high-fat grain products such as granola, pancakes, waffles, French toast, and biscuits. Prepare hot cereals with whole milk or cream or added fat.

- *Fruits.* Mash avocados to make guacamole, or use mashed avocado as a sandwich spread. Add chopped dried fruits to salads and baked goods. Snack on dried fruits between meals.

- *Nuts.* Add chopped nuts to pasta dishes, stir-fried vegetables, fruit salads, and green salads. Use nut meats in baked products. Spread nut butters on bread and crackers.

- *Butter or margarine.* Melt on pasta, potatoes, rice, and cooked vegetables. Add to hot cereals, casseroles, and soups. Spread liberally on bread, crackers, and rolls.

- *Mayonnaise or salad dressings.* Add to pasta, tuna, and potato salads. Use as dressings for raw or cooked vegetables.

- *Beverages.* Replace water and non-kcaloric beverages with sweetened drinks, fruit juices, and milk shakes. Drink whole milk instead of low-fat or nonfat milk. Add strawberry or chocolate syrup to plain milk to boost kcalories.

These suggestions can help to add protein to a meal:

- *Meats.* Add small chunks of meat to soups, egg dishes, casseroles, bean dishes, and pasta sauces. Add minced meats to vegetable dishes. Add chunks of cooked chicken or turkey to salads.

- *Eggs.* Add raw eggs when preparing casseroles, meatballs, and hamburgers. Add chopped hard-cooked eggs to salads, vegetable dishes, sandwich fillings, and pasta and potato salads.

- *Cheese.* Melt on burgers, meat loaf, cooked vegetables, scrambled eggs, casseroles, and potatoes. Add cottage cheese to casseroles, egg dishes, pasta recipes, and salad dressings. Grate hard cheeses and sprinkle on soups, salads, and cooked vegetable dishes. Avoid using reduced-fat cheeses.

- *Milk.* Use in place of water when preparing cereals and soups. Use cream sauces (which are made with milk) to flavor vegetable and pasta dishes.

- *Powdered milk (use full-fat milk powder if available).* Add to recipes that include milk. Dissolve extra milk powder into milk-containing beverages. Stir into hot cereals, potato dishes, casseroles, and sauces. Add to scrambled eggs, hamburgers, and meat loaf.

- *Protein supplements.* Snack on protein bars between meals. Add protein powders to beverages and shakes. Drink meal replacement formulas, such as Ensure or Boost, instead of juices or soda.

Managing Symptoms and Complications A thorough nutrition assessment often uncovers specific problems that interfere with food consumption. Table 25-6 lists dietary considerations related to cancers affecting different sites in the body. The "How to" outlines a variety of dietary measures that may alleviate symptoms and improve food intake. Patients' responses to these strategies vary considerably, and in some cases a number of adjustments may be necessary.

Low-Microbial Diet Patients with suppressed immunity or neutropenia may be prescribed a **low-microbial diet** (also called a *neutropenic diet*), which includes only foods that are unlikely to be contaminated with bacteria or other microbes.[21] Generally, patients should consume only well-cooked meats and eggs, pasteurized milk products, well-washed fruits and vegetables, and shelf-stable packaged foods. Foods that must be avoided include unwashed raw fruits and vegetables, unpasteurized juices and milk products, leftover luncheon meats, and foods from salad bars or street vendors. In addition, patients should be instructed to follow safe food-handling practices to minimize the risk of foodborne illness (see Nutrition in Practice 5).

Enteral and Parenteral Nutrition Support Nutrition support is used in limited situations during cancer treatment. Generally, tube feedings and parenteral nutrition are provided to patients who have long-term or permanent gastrointestinal impairment or

low-microbial diet: a diet that contains foods that are unlikely to be contaminated with bacteria and other microbes.

TABLE 25-6 Dietary Considerations for Specific Cancers

Cancer Sites	Common Complications[a]	Possible Dietary Measures
Brain and nervous system	Chewing or swallowing difficulty, headache, altered taste or smell sensation, difficulty feeding oneself	Mechanically altered diet, use of adaptive feeding devices (see Nutrition in Practice 22)
Head and neck[b]	Chewing or swallowing difficulty, aspiration, inflamed mucosa, dry mouth, altered taste or smell sensation	Tube feeding, mechanically altered diet
Esophagus	Swallowing difficulty, aspiration, obstruction, acid reflux, inflamed mucosa	Tube feeding, mechanically altered diet
Stomach	Anorexia, early satiety, reduced secretion of gastric acid and intrinsic factor, delayed stomach emptying, dumping syndrome, malabsorption, nutrient deficiencies	Tube feeding (for obstruction or unmanageable dumping syndrome); postgastrectomy diet; small, frequent meals; limited sugars and insoluble fibers (see Chapter 17); nutrient supplementation
Intestine	Inflamed mucosa, bacterial overgrowth, obstruction, lactose intolerance, general malabsorption, bile insufficiency, nutrient deficiencies, short bowel syndrome (if resected), altered bowel function, fluid and electrolyte imbalances	Tube feeding or total parenteral nutrition for obstruction, enteritis, or short bowel syndrome; fat- and lactose-restricted diet (see Chapter 19); nutrient supplementation
Pancreas	Reduced secretion of digestive enzymes, bile insufficiency, general malabsorption, nutrient deficiencies, hyperglycemia	Fat-restricted diet; enzyme replacement (see Chapter 19); small, frequent meals; carbohydrate-controlled diet (Chapter 21); nutrient supplementation

[a]Actual complications depend on the exact location of the cancer and the specific methods used for treating the cancer.
[b]Includes cancers of the salivary glands, oral and nasal cavities, pharynx, and larynx.

How To HELP PATIENTS HANDLE FOOD-RELATED PROBLEMS

In people with cancer or HIV infection, various complications can interfere with food intake. Health care providers can try to identify a patient's specific problems and offer appropriate solutions. Not every suggestion will work for each person; encourage patients to experiment and find strategies that work best.

I just don't have an appetite.

▌ Eat small meals and snacks at regular times each day.

▌ Eat the largest meal at the time of day when you feel the best.

▌ Include nutrient-dense foods in meals, and consume them before other foods.

▌ Indulge in favorite foods throughout the day. Serve foods attractively.

▌ Avoid drinking large amounts of liquids before or with meals.

▌ Eat in a pleasant and relaxed environment. Eat with family and friends when possible.

▌ Listen to your favorite music or enjoy a TV or radio program while you eat.

▌ Ask your doctor about appetite-enhancing medications.

I am too tired to fix meals and eat.

▌ Let family members and friends prepare food for you.

▌ Obtain foods that are easy to prepare and easy to eat, such as sandwiches, frozen dinners, take-out meals from restaurants, instant breakfast drinks, liquid formulas, and energy bars.

▌ Find time to rest before you attempt to prepare a large meal.

▌ Prepare soups, stews, and casserole dishes in sufficient quantity to provide enough for several meals, so that you will have enough to eat at times when you are too tired to cook.

Foods just don't taste right.

▌ Brush your teeth or use mouthwash before you eat.

▌ Consume foods chilled or at room temperature. Use plastic, rather than metal, eating utensils.

▌ Choose eggs, fish, poultry, and milk products instead of meats.

▌ Experiment with sauces, seasonings, herbs, spices, and sweeteners to improve food taste and flavor.

▌ Save your favorite foods for times when you are not feeling nauseated.

I am nauseated a lot of the time, and sometimes I need to vomit.

▌ Consume liquids throughout the day to replace fluids.

▌ If you become nauseated from chemotherapy treatments, avoid eating for at least two hours before treatments.

▌ Consume your largest meal at a time when you are least likely to feel nauseous.

▌ Try consuming smaller meals, and eat slowly. Experiment with foods to see if some foods cause nausea more than others.

▌ Avoid foods and meals that have strong odors or are fatty, greasy, or gas forming.

I have problems chewing and swallowing food.

▌ Experiment with food consistencies to find the ones you can manage best. Thin liquids, dry foods, and sticky foods (such as peanut butter) are often difficult to swallow.

▌ Add sauces and gravies to dry foods.

▌ Drink fluids during meals to ease chewing and swallowing.

- Try using a straw to drink liquids. Experiment with beverage thickeners if you cannot tolerate thin beverages.
- Tilt your head forward and backward to see if you can swallow more easily when your head is positioned differently.

I have sores in my mouth, and they hurt when I eat.

- Try eating chilled or frozen foods; they are often soothing.
- Try soft foods such as ice cream, milk shakes, bananas, applesauce, mashed potatoes, cottage cheese, and macaroni and cheese. Mix dry foods with sauces or gravies.
- Cut foods into smaller pieces, so they are less likely to irritate the mouth.
- Avoid foods that irritate mouth sores, such as citrus fruits and juices, tomatoes and tomato-based products, spicy foods, foods that are very salty, foods with seeds (such as poppy seeds and sesame seeds) that can scrape the sores, and coarse foods such as raw vegetables, crackers, corn chips, and toast.
- Ask your doctor about using a local anesthetic solution such as lidocaine before eating to reduce pain.
- Use a straw for drinking liquids, in order to bypass the sores.

My mouth is really dry.

- Rinse your mouth with warm salt water or mouthwash frequently. Avoid using mouthwash that contains alcohol.
- Drink small amounts of liquid frequently between meals.
- Ask your doctor or pharmacist about medications or saliva substitutes that can help a dry mouth condition.
- Use sour candy or chewing gum to stimulate the flow of saliva.
- Sip fluids frequently while eating. Add broth, sauces, gravies, mayonnaise, butter, or margarine to dry foods.
- Make sure you brush your teeth and floss regularly to prevent tooth decay and oral infections.

I am having trouble with constipation.

- Drink plenty of fluids. Try warm fluids, especially in the morning.
- Eat whole-grain breads and cereals, nuts, fresh fruits and vegetables, prunes, and prune juice. Avoid refined carbohydrate foods such as white bread, white rice, and pasta.
- Engage in physical activity regularly.
- Try an over-the-counter bulk-forming agent, such as methylcellulose (Citrucel),

psyllium (Metamucil or Fiberall), or polycarbophil (Fiber-Lax).

I am having trouble with diarrhea.

- To avoid dehydration, drink plenty of fluids throughout the day. Salty broths and soups, diluted fruit juices, and sports drinks are good choices. Avoid caffeine- and alcohol-containing beverages. For severe diarrhea, try oral rehydration formulas that are commercially prepared.
- Avoid foods and beverages that increase gas, such as legumes, onions, vegetables of the cabbage family, foods that contain sorbitol or mannitol, and carbonated beverages.
- Try using lactase enzyme replacements when you use milk products in case you are experiencing lactose intolerance. Yogurt and aged cheeses may be easier to tolerate than milk and fresh cheeses.
- Avoid fatty foods if you are fat intolerant. Try reducing your intake of whole-grain breads and cereals if they worsen the diarrhea.
- Eat small, frequent meals instead of large ones. Try consuming cool or lukewarm foods instead of very cold or hot foods.
- Ask your doctor about using a bulk-forming agent or antidiarrhea medication.

are experiencing complications that interfere with food intake.[22] For example, many patients undergoing radiation treatment for head and neck cancers require long-term tube feeding and may need to continue tube feedings at home. Parenteral nutrition is reserved for patients who have inadequate GI function, such as individuals with chronic radiation enteritis. Whenever possible, enteral nutrition is strongly preferred over parenteral nutrition, to preserve GI function and avoid infection.

Radiation to the head and neck regions often causes dysphagia and mouth sores.

IN SUMMARY

- Cancer arises from mutations in the genes that control cell division. Some dietary substances promote carcinogenesis, while others may help to prevent cancer.

- Cancer's effects on nutrition status depend on the type of cancer a person has, its severity, and the methods used to treat the cancer. Cancer cachexia is a frequent complication of cancer and may be related to anorexia, altered metabolism, and responses to cancer treatment.

- Medical treatments for cancer include surgery, chemotherapy, radiation therapy, and biological therapies. Nutrition therapy aims to minimize weight loss and wasting, correct deficiencies, and manage complications that impair food intake.

The Case Study allows you to apply information about nutrition and cancer to a clinical situation.

Case Study WOMAN WITH CANCER

Sheri Nixon is a 58-year-old public relations consultant who was recently diagnosed with colon cancer after a routine colonoscopy, a procedure in which the colon is examined using a flexible tube attached to an optical device. Mrs. Nixon is scheduled to have surgery to remove the segment of colon that contains the tumor and to determine if the cancer has spread to the surrounding lymph nodes and, possibly, other organs. The nurse completing the nutrition assessment finds that Mrs. Nixon is 5 feet 5 inches tall and weighs 178 pounds. Mrs. Nixon usually spends most of the day sitting and has little time to engage in recreational exercise. Her diet typically includes red meat at both lunch and dinner, and she consumes one or two glasses of wine with both meals. She eats two or three servings of fruits and vegetables each day, although she does not like green leafy vegetables very much. She rarely drinks milk or consumes milk products.

1. Review Table 25-2 on p. 640, and describe the factors in Mrs. Nixon's diet and lifestyle that may have contributed to the development of colon cancer.

2. What symptoms and complications may arise after colon surgery and impair nutrition status? If the cancer team decides that Mrs. Nixon needs follow-up chemotherapy, how might the chemotherapy affect her nutrition status?

3. If Mrs. Nixon is unresponsive to treatment and her cancer progresses, she may develop cancer cachexia. Describe this syndrome, its causes, and its consequences.

4. Provide suggestions that may help Mrs. Nixon handle the following problems should they develop: poor appetite, fatigue, taste alterations, nausea and vomiting, chewing and swallowing difficulties, mouth sores, dry mouth, diarrhea, constipation, and weight loss.

HIV Infection

Possibly the most infamous infectious disease today is **acquired immune deficiency syndrome (AIDS).** AIDS develops from infection with human immunodeficiency virus (HIV), which attacks the immune system and disables a person's defenses against other diseases, including infections and certain cancers. Then these diseases—which would cause few, if any, symptoms in people with healthy immune systems—destroy health and life.

The HIV/AIDS epidemic continues to sweep across countries, especially in sub-Saharan Africa (see Table 25-7). For many years, the destructive effects of HIV infection seemed unstoppable, but in the mid to late 1990s the death rate from AIDS began to decline in the United States, and the progression from HIV infection to AIDS slowed dramatically. Although AIDS is still incurable, remarkable progress has been made in understanding and treating HIV infection.

Without a cure for AIDS, the best course is prevention. HIV is most often sexually transmitted and can be spread by direct contact with contaminated body fluids, such as blood, semen, vaginal secretions, and breast milk. Because many people remain symptom-free during the early stages of infection, they may not realize that they can pass the infection to others. To reduce the spread of HIV infection, individuals at risk (see Table 25-8) are encouraged to undergo testing. A blood test can usually detect HIV antibodies within several months after exposure and, often, after two or three weeks. An estimated 21 percent of persons in the United States who have HIV infection are unaware that they are infected.[23]

acquired immune deficiency syndrome (AIDS): the late stage of illness caused by infection with the human immunodeficiency virus (HIV); characterized by severe damage to immune function.

helper T cells: lymphocytes that have a specific protein called CD4 on their surfaces and therefore are also known as *CD4+ T cells;* these are the cells most affected in HIV infection.

TABLE 25-7 The HIV and AIDS Epidemic at a Glance, 2008

Stage of Epidemic	World	Sub-Saharan Africa	North America
Individuals living with HIV infection or AIDS	33,400,000	22,400,000	1,400,000
Individuals newly infected with HIV	2,700,000	1,900,000	55,000
AIDS-related deaths	2,000,000	1,400,000	25,000

Source: Joint United Nations Program on HIV/AIDS and World Health Organization, *AIDS epidemic update: December 2009,* available at http://data.unaids.org/pub/Report/2009/2009_epidemic_update_en.pdf; site visited November 21, 2010.

Consequences of HIV Infection

HIV infection destroys immune cells that have a protein called CD4 on their surfaces. The cells most affected are the **helper T cells,** also called *CD4+ T cells* because the presence of CD4 is a primary characteristic. Other cells that have the CD4 protein (and are infected by HIV) include tissue macrophages, blood monocytes, and certain cells of the central nervous system.[24] Early symptoms of HIV infection are nonspecific and may include fever, sore throat, malaise, swollen lymph nodes, skin rashes, muscle and joint pain, and diarrhea. After these symptoms subside, many people remain symptom-free for 5 to 10 years or even longer. If the HIV infection is not treated, however, the depletion of T cells eventually increases the person's susceptibility to **opportunistic infections**—that is, infections caused by microorganisms that normally do not cause disease in healthy individuals.

The term *AIDS* applies to the advanced stages of HIV infection, in which the inability to fight illness allows a number of serious diseases and complications to develop; such **AIDS-defining illnesses** include severe infections, certain cancers, and wasting of muscle tissue. Without treatment, AIDS develops in 26 to 36 percent of HIV-infected persons within 7 years.[25] Health practitioners evaluate disease progression by measuring the concentrations of helper T cells and circulating virus (called the *viral load*) and by monitoring clinical symptoms. Although current drug therapies can dramatically slow the progression of HIV infection, the drugs' side effects may make it difficult for patients to adhere to treatments, as discussed in some of the following sections.

Lipodystrophy About one-third or more of patients using drug therapies to suppress HIV infection develop abnormalities in glucose and fat metabolism.[26] These complications, collectively known as the **HIV-lipodystrophy syndrome,** include body fat redistribution, abnormal blood lipid levels, and insulin resistance. Patients may accumulate abdominal fat, lose fat from the face and extremities, or both.[27] Also observed are breast enlargement (in both men and women), fat accumulation at the base of the neck (called a **buffalo hump**), and benign growths composed of fat tissue (called **lipomas**). The changes in body composition are often disfiguring and may cause physical discomfort; moreover, patients often develop hypertriglyceridemia, elevated low-density lipoprotein (LDL) cholesterol levels, low high-density lipoprotein (HDL) cholesterol levels, glucose intolerance, and hyperinsulinemia. The reasons for the development of lipodystrophy are unknown.

Weight Loss and Wasting Even with effective treatment of HIV infection, weight loss and wasting are ongoing problems for many HIV-infected patients. *HIV-associated wasting* is diagnosed in patients who unintentionally lose 7.5 percent of body weight within 6 months, or 10 percent of body weight within 12 months.[28] The wasting has been linked with accelerated disease progression, reduced strength, and fatigue. In the later stages of AIDS, the wasting is severe and increases the risk of death. Much as in cancer, wasting associated with HIV infection has many causes: anorexia and inadequate food intake, altered metabolism, malabsorption, chronic diarrhea, and diet-drug interactions.

Anorexia and Reduced Food Intake Inadequate food intake is a key factor in the development of wasting. Poor food intake may result from various factors, including the following:

- *Emotional distress, pain, and fatigue.* The physical and social problems that accompany chronic illness may cause fear, anxiety, and depression, which contribute to anorexia. Pain and fatigue, which may be associated with some disease complications, can cause anorexia and difficulty with eating.

- *Oral infections.* The oral infections associated with HIV infection can cause discomfort and interfere with food consumption. Common infections include **candidiasis**

TABLE 25-8 Risk Factors for HIV Infection

- History of receiving blood transfusions or blood components before 1985
- Infant born to mother with HIV infection
- Intravenous drug use in which syringes are shared among users
- Sexual contact with intravenous drug users, prostitutes, or individuals with a history of HIV or other sexually transmitted diseases
- Sexual contact with multiple partners
- Unsafe sexual practices

T cells are lymphocytes that develop in the thymus gland. The other lymphocytes are the *B cells* (which develop in bone marrow) and *natural killer cells.*

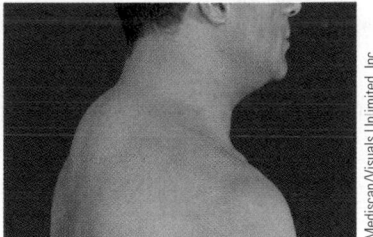

Mediscan/Visuals Unlimited, Inc.

HIV lipodystrophy is sometimes evident by the accumulation of fatty tissue at the base of the neck, referred to as *buffalo hump.*

opportunistic infections: infections from microorganisms that normally do not cause disease in healthy people but are damaging to persons with compromised immune function.

AIDS-defining illnesses: diseases and complications associated with the later stages of an HIV infection, including wasting, recurrent bacterial pneumonia, opportunistic infections, and certain cancers.

HIV-lipodystrophy (LIP-oh-DIS-tro-fee) **syndrome:** a collection of abnormalities in fat and glucose metabolism that may result from drug treatments for HIV infection; changes include body fat redistribution, abnormal blood lipid levels, and insulin resistance. The accumulation of abdominal fat is sometimes called *protease paunch.*

buffalo hump: the accumulation of fatty tissue at the base of the neck.

lipomas (lih-POE-muz): benign tumors composed of fatty tissue.

candidiasis: a fungal infection on the mucous membranes of the oral cavity and elsewhere; usually caused by *Candida albicans.*

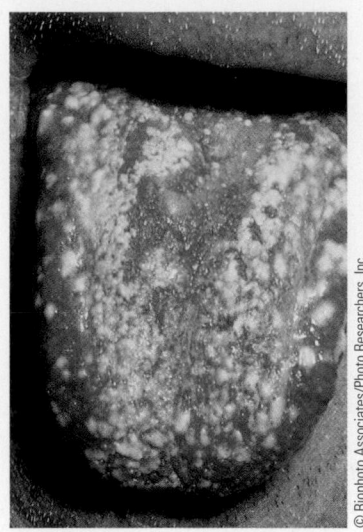

The oral infection *thrush* is easily identified by the characteristic milky white patches that appear on the tongue.

The AIDS-related abnormalities that develop in the intestinal mucosa are sometimes referred to as *AIDS enteropathy* (EN-ter-OP-ah-thy).

and **herpes simplex virus** infection. Candidiasis (commonly called *thrush*) can cause mouth pain, dysphagia (difficulty swallowing), and altered taste sensation; infection with herpes simplex virus may cause painful lesions around the lips and in the mouth.

▌ *Respiratory disorders.* Respiratory infections, including pneumonia and tuberculosis, are common in people with HIV infection. Symptoms often include chest pain, shortness of breath, and cough, which interfere with eating and contribute to anorexia.

▌ *Cancer.* As described earlier in this chapter, cancer leads to anorexia for numerous reasons. In addition, **Kaposi's sarcoma,** a type of cancer frequently associated with HIV infection, can cause lesions in the mouth and throat that make eating painful.

▌ *Medications.* The medications given to treat HIV infection, other infections, and cancer often cause anorexia, nausea and vomiting, altered taste sensation, food aversions, and diarrhea.

GI Tract Complications Complications of HIV infection involving the GI tract may result from opportunistic infections, the HIV infection itself, or medications.[29] In addition to the oral infections described previously, infections may develop in the esophagus, stomach, and intestines. Furthermore, advanced AIDS is often accompanied by characteristic changes in the small intestinal lining: the villi appear shortened and flattened, and the absorptive area is substantially reduced; these changes can cause malabsorption, steatorrhea, and diarrhea.

As described earlier, many patients are unable to tolerate the medications used to suppress HIV and develop nausea, vomiting, and diarrhea. Furthermore, the medications that treat infections in the GI tract contribute to bacterial overgrowth. Thus, HIV-infected patients using standard treatments face an extremely high risk of malnutrition due to the combination of intestinal discomfort, bacterial overgrowth, malabsorption, and nutrient losses from vomiting, steatorrhea, and diarrhea.

Other Complications Patients with HIV infection can develop anemia due to nutrient malabsorption, blood loss, disturbed bone marrow function, or medication side effects. HIV infection may also lead to neurological complications (dementia, muscle weakness, gait disturbances), skin disorders (rashes, infections, cancers), kidney diseases (nephrotic syndrome, chronic kidney disease), eye disorders (retinal infection or detachment), and coronary heart disease.

Treatments for HIV Infection

Although there is no cure for HIV infection, treatments can help to slow its progression, reduce complications, and alleviate pain. The standard treatment for suppressing HIV infection, called *highly active antiretroviral therapy (HAART)*, combines three or more antiretroviral drugs.[30] Table 25-9 lists the major drug categories included in antiretroviral therapy and describes the drugs' modes of action. These antiretroviral agents have multiple adverse effects that make their long-term use difficult to tolerate. In addition to the GI effects discussed previously, side effects include skin rashes, headache, anemia, tingling and numbness, hepatitis, pancreatitis, and kidney stones. Thus, although HAART has improved life span and quality of life for many patients, the drug regimens are difficult to adhere to and cause complications that require continual management. The Diet-Drug Interactions feature summarizes the nutrition-related effects of some of the antiretroviral agents and other drugs mentioned in this chapter.

Control of Anorexia and Wasting Anabolic hormones, appetite stimulants, and regular physical activity have been successful in reversing weight loss and increasing muscle mass in HIV-infected patients. Testosterone and human growth hormone

TABLE 25-9 **Antiretroviral Drugs for Treatment of HIV Infection**

Category	Examples	Mode of Action
CCR5 antagonist	Maraviroc	CCR5 antagonists prevent HIV from entering cells by blocking a membrane receptor on the host cell.
Fusion inhibitor	Enfuvirtide	Fusion inhibitors prevent HIV from entering cells by binding a viral protein needed for its entry.
Integrase inhibitor	Raltegravir	Integrase inhibitors impair the function of HIV's integrase enzyme, which incorporates viral DNA into the host cell's genome.
Non-nucleoside reverse transcriptase inhibitor (NNRTI)	Delavirdine, efavirenz, nevirapine	NNRTI bind active sites on HIV's reverse transcriptase enzyme, blocking the ability of HIV to produce DNA copies of its genetic material.
Nucleoside reverse transcriptase inhibitor (NRTI)	Didanosine, lamivudine, zidovudine	As analogs of the nucleosides needed for DNA synthesis, NRTI impair the ability of HIV's reverse transcriptase enzyme to produce usable copies of DNA.
Protease inhibitor (PI)	Saquinavir, ritonavir, indinavir	PI inhibit HIV's protease enzyme, which cleaves HIV's gene products into usable structural proteins.

Sources: U.S. Department of Health and Human Services, Panel on Antiretroviral Guidelines for Adults and Adolescents, *Guidelines for the Use of Antiretroviral Agents in HIV-1-Infected Adults and Adolescents,* December 1, 2009, pp. 1–168, available at http://aidsinfo.nih.gov/contentfiles/AdultandAdolescentGL.pdf; accessed November 21, 2010; S. Safrin, Antiviral agents, in B. G. Katzung, ed., *Basic and Clinical Pharmacology* (New York: Lange/McGraw-Hill, 2007), pp. 790–818.

Diet-Drug Interactions

Check this table for notable nutrition-related effects of the medications discussed in this chapter.

Appetite stimulants (megestrol acetate, dronabinol)	**Gastrointestinal effects:** Nausea, vomiting, diarrhea **Metabolic effects:** Hyperglycemia (megestrol acetate)
Didanosine	**Gastrointestinal effects:** Nausea, vomiting, dry mouth, altered taste perception, anorexia, diarrhea, constipation **Dietary interactions:** Take medication either one-half hour before or two hours after a meal; avoid alcohol and aluminum- and magnesium-containing antacids. **Metabolic effects:** Pancreatitis; anemia; reduced serum copper, zinc, and vitamin B_{12} levels
Methotrexate	**Gastrointestinal effects:** Nausea, vomiting, diarrhea, reduced absorption of vitamin B_{12} and calcium **Dietary interactions:** Milk may reduce methotrexate absorption if the milk and methotrexate are ingested together. **Metabolic effects:** Increased serum uric acid levels, anemia, liver toxicity
Ritonavir	**Gastrointestinal effects:** Nausea, vomiting, altered taste perception, anorexia, diarrhea **Metabolic effects:** Pancreatitis, hyperglycemia, reduced serum copper and zinc levels; increased levels of blood triglycerides, LDL cholesterol, uric acid, liver enzymes, and creatine kinase
Zidovudine	**Gastrointestinal effects:** Nausea, vomiting, altered taste perception, anorexia, mouth sores, constipation **Dietary interactions:** Do not take medication with a high-fat meal. **Metabolic effects:** Anemia; reduced serum copper, zinc, and vitamin B_{12} levels

Note: Most antiretroviral drugs that treat HIV infection have gastrointestinal and metabolic side effects; only a few are listed here as examples.

have demonstrated positive effects on nitrogen balance and lean tissue content, especially in combination with resistance training. A regular program of resistance exercise improves muscle mass and strength and corrects some of the metabolic abnormalities (altered blood lipids and insulin resistance) that are common in HIV-infected patients. The medications megestrol acetate and dronabinol are sometimes prescribed to stimulate appetite and improve weight gain, although much of the weight increase is attributable to a gain of fat rather than lean tissue.[31]

herpes simplex virus: a common virus that can cause blisterlike lesions on the lips and in the mouth.

Kaposi's (kah-POH-seez) **sarcoma:** a common cancer in HIV-infected persons that is characterized by lesions in the skin, lungs, and GI tract.

Resistance training can help a person with HIV infection maintain muscle mass and strength.

Control of Lipodystrophy Treatment strategies for lipodystrophy are under investigation. Both aerobic activity and resistance training may help to reduce abdominal fat, although some patients opt for cosmetic surgery. Patients may be given alternative antiretroviral drugs to alleviate symptoms. Medications may be prescribed to treat abnormal blood lipids and insulin resistance.

Alternative Therapies Like cancer patients, people with HIV infection and AIDS are frequently tempted to try unconventional methods of treatment. Although many alternative therapies are harmless, some may have side effects that worsen complications or interfere with treatment. The herb St. John's wort, for example, can reduce the effectiveness of some antiretroviral drugs.[32] Monitoring patients' use of dietary supplements is essential to reduce the possibility of nutrient-drug and herb-drug interactions.

Nutrition Therapy for HIV Infection

HIV-infected individuals must learn how to maintain body weight and muscle mass, prevent malnutrition, and cope with nutrition-related side effects of medications. Nutrition assessment and counseling should begin soon after diagnosis. The initial assessment should include an evaluation of body weight and body composition.

Weight Management Since the development of successful drug therapies for HIV infection, obesity and overweight have become more prevalent than wasting among HIV-infected individuals in the United States.[33] Because excessive body weight can increase risks for cardiovascular disease and diabetes, moderate weight loss is recommended for patients with HIV infection who are overweight or obese.

Individuals who experience weight loss and wasting may benefit from a high-kcalorie, high-protein diet. Daily energy needs may be 30 to 40 kcalories per kilogram body weight, and protein requirements may be as high as 1.2 to 2.0 grams per kilogram body weight.[34] If food consumption is difficult, small, frequent feedings may be better tolerated than several large meals. The addition of nutrient-dense snacks, protein or energy bars, and oral supplements can improve intakes (review the "How to" on p. 645). Liquid formulas may be useful for the person who is too tired to eat or prepare meals.

Metabolic Complications To improve insulin resistance and blood lipid abnormalities, patients are advised to achieve or maintain a desirable weight, replace saturated fats with monounsaturated and polyunsaturated fats, increase fiber intake, and limit intakes of *trans*-fatty acids, cholesterol, added sugars, and alcohol. Regular physical activity may improve both insulin resistance and blood lipid levels.

Additional suggestions for managing insulin resistance and blood lipid abnormalities are available in Chapters 21 and 22, respectively.

Vitamins and Minerals Because nutrient deficiencies are likely to result from reduced food intake, malabsorption, diet-drug interactions, and nutrient losses, multivitamin-mineral supplements are typically recommended. Patients should be cautioned to maintain intakes that are close to DRI recommendations, however, due to the risk of adverse interactions between excessive amounts of vitamins and minerals, and antiretroviral drugs.[35]

Symptom Management The discomfort associated with antiretroviral therapy, opportunistic GI infections, and malabsorption symptoms can make food consumption difficult, and problems such as vomiting and diarrhea contribute to fluid and electrolyte losses. The "How to" on pp. 646–647 describes measures that can improve food and fluid intakes and alleviate discomfort in individuals with these problems.

Food Safety The depressed immunity of people with HIV infections places them at extremely high risk of developing foodborne illnesses. Health practitioners should provide detailed instructions about the safe handling and preparation of foods; for some individuals, a low-microbial diet may be suggested (see p. 645). Water can also be a source of foodborne illness and is a common cause of **cryptosporidiosis** in HIV-infected individuals. In places where water quality is questionable, patients should consult their local health departments to determine whether the tap water is safe to drink. If not, or to take additional safety measures, water used for drinking and making ice cubes should be boiled for one minute.

Enteral and Parenteral Nutrition Support In later stages of illness, people with HIV infections may be unable to consume enough food and may need aggressive nutrition support. Tube feedings are preferred whenever the GI tract is functional; they can be provided at night to supplement oral diets consumed during the day. Parenteral nutrition is reserved for patients who are unable to tolerate enteral nutrition, such as those with GI obstructions that prevent food intake.

IN SUMMARY

▌ By attacking immune cells, HIV causes progressive damage to immune function and may eventually lead to AIDS.

▌ Improved drug therapies have slowed the progression of HIV infection; however, these drugs may promote the HIV-lipodystrophy syndrome, characterized by body fat redistribution, abnormal lipid levels, and insulin resistance.

▌ HIV infection may lead to weight loss and wasting, anorexia, and various complications that affect food intake. Dietary adjustments, resistance training, and medications can help patients maintain their weight and prevent wasting.

▌ People with HIV infections must pay strict attention to food safety guidelines to prevent foodborne illnesses.

cryptosporidiosis (KRIP-toe-spor-ih-dee-OH-sis): a foodborne illness caused by the parasite *Cryptosporidium parvum*.

The Case Study provides an opportunity to review the nutrition concerns of a person with HIV infection.

Case Study MAN WITH HIV INFECTION

Three years ago, George Judd, a 37-year-old financial planner, sought medical help when he began feeling run-down and developed a painful white fungal infection over his mouth and tongue. The presence of thrush, recent weight loss, and anemia alerted Mr. Judd's physician to the possibility of an HIV infection. When Mr. Judd tested positive for HIV, he and his family and friends were devastated by the news, but those close to him have remained supportive. During the three years

since Mr. Judd began antiretroviral drug therapy, he has maintained his weight but has also developed lipodystrophy and hypertriglyceridemia. Mr. Judd is 6 feet tall and currently weighs 185 pounds. He occasionally develops diarrhea and sometimes anorexia.

1. Describe lipodystrophy, and discuss its typical pattern in people who have an HIV infection. What adjustments in treatment and lifestyle may be helpful for Mr. Judd?

2. Describe an appropriate diet for Mr. Judd. What strategies may improve his problems with diarrhea and anorexia? Suggest reasons why diarrhea and anorexia may develop in people with HIV infections.

3. Explain why an HIV infection can lead to wasting as the disease progresses to the later stages. What recommendations may be helpful for maintaining weight and health if wasting becomes a problem?

Nutrition Assessment Checklist FOR PEOPLE WITH CANCER OR HIV INFECTIONS

Medical History

Check the medical record to determine:

- Type and stage of cancer
- Stage of HIV infection

Review the medical record for complications that may alter nutrition therapy, including:

- Altered organ function
- Altered taste perception
- Anorexia
- Dry mouth and oral infections
- GI symptoms and infections
- Hyperlipidemias
- Insulin resistance
- Malnutrition and wasting

Medications

For patients with cancer or HIV infections:

- Check medications to identify potential diet-drug interactions.
- Recommend the use of antinauseants at mealtime, if needed.
- Ask about the use of dietary supplements, including herbal products.

For cancer patients who require chemotherapy:

- Recommend strategies to prevent food aversions.
- Offer suggestions for managing drug-related complications.

For HIV-infected patients using antiretroviral drug therapy:

- Remind patients that some drugs are better absorbed with foods and that others must be taken on an empty stomach.
- Help patients work out a medication schedule that suits their lifestyle and is timed appropriately in regard to food intake.
- Offer suggestions for managing drug-related complications.

Dietary Intake

For patients with poor food intakes and weight loss:

- Determine the reasons for reduced food intake.
- Offer appropriate suggestions to improve food intake.
- Provide interventions before weight loss progresses too far.

For patients with HIV infections who experience weight gain, elevated triglyceride or LDL cholesterol levels, or hyperglycemia:

- Assess the diet for energy, total fat, types of fat, carbohydrates, fiber, and sugars.
- For patients with hyperlipidemias, recommend a diet low in saturated fat, *trans*-fatty acids, cholesterol, and sugars.
- For patients with hyperglycemia, recommend a consistent carbohydrate intake at meals and snacks that emphasizes complex carbohydrates and limits concentrated sweets.
- Recommend regular physical activity for weight control and for improving blood lipid levels and insulin resistance.

Anthropometric Data

Take baseline height and weight measurements, monitor weight regularly, and suggest dietary adjustments for weight maintenance, if necessary. Remember that body composition may change without affecting body weight. Perform baseline and periodic body composition measurements in HIV-infected patients who are using antiretroviral drug therapy.

Laboratory Tests

Note that albumin and other serum proteins may be reduced in patients with cancer or HIV infections, especially in those experiencing wasting. Check laboratory tests for indications of:

- Anemia
- Dehydration
- Elevated LDL cholesterol levels
- Elevated triglyceride levels
- Hyperglycemia

For patients with HIV infections, evaluate disease progression by checking:

- Helper T cell counts
- Viral load

Physical Signs

Look for physical signs of:

- Dehydration (especially for patients with fever, vomiting, or diarrhea)
- Kaposi's sarcoma
- Oral infections
- Protein-energy malnutrition and wasting

Clinical Applications

1. Consider the nutrition problems that may develop in a 36-year-old woman with a malignant brain tumor that affects her ability to move the right side of her body (including the tongue) and to speak coherently. She is taking a pain medication that makes her nauseated and sleepy. Her expected survival time is only about six months.

   ▪ If she is right-handed, how might her impairment interfere with eating? What suggestions do you have for overcoming this problem?

   ▪ How might her nutrition status be affected by her inability to communicate effectively? What suggestions may help?

   ▪ In what ways might the pain medication she is taking affect her nutrition status?

2. Various types of chronic conditions can lead to weight loss and wasting. For some of these conditions, such as Crohn's disease and celiac disease (Chapters 18 and 19), diet is a cornerstone of treatment. For others, such as cancer and HIV infection, nutrition plays a supportive role. What determines whether nutrition plays a primary role or a supportive role in the treatment of disease?

Self Check

1. Which of these dietary substances may help to protect against cancer?
   a. Alcohol
   b. Well-cooked meats, poultry, and fish
   c. Animal fats
   d. Phytochemicals from fruits and vegetables

2. The metabolic changes that often accompany cancer include all of the following except:
   a. increased triglyceride breakdown.
   b. increased protein turnover.
   c. increased muscle protein synthesis.
   d. insulin resistance.

3. An advantage of radiation therapy over chemotherapy is that:
   a. radiation is not damaging to rapidly dividing cells.
   b. side effects of radiation therapy do not include malnutrition.
   c. radiation can be directed toward the regions affected by cancer.
   d. the radiation used is too weak to damage GI tissues.

4. Although many cancer patients lose weight, which type of cancer is often associated with weight gain?
   a. Kidney cancer
   b. Breast cancer
   c. Colon cancer
   d. Lung cancer

5. Oral diets after hematopoietic stem cell transplants may restrict:
   a. fiber.
   b. carbohydrates.
   c. high-protein foods.
   d. raw fruits and vegetables.

6. HIV can enter and destroy:
   a. epithelial cells.
   b. helper T cells.
   c. liver cells.
   d. intestinal cells.

7. The HIV-lipodystrophy syndrome may result in all of these changes except:
   a. increased abdominal fat.
   b. increased fat in the arms and legs.
   c. fat accumulation at the base of the neck.
   d. hypertriglyceridemia.

8. Mouth sores in people with HIV infections are most frequently due to:
   a. oral infections.
   b. dehydration.
   c. nutrient deficiency.
   d. foodborne illnesses.

9. Megestrol acetate and dronabinol are:
   a. medications used to promote weight gain.
   b. protease inhibitors that fight HIV infection.
   c. medications that treat common opportunistic infections.
   d. anabolic hormones that promote gain of muscle tissue.

10. To prevent cryptosporidiosis, a person with HIV infection may need to:
    a. wash hands carefully before meals.
    b. avoid consuming undercooked meat, poultry, and eggs.
    c. consume a high-kcalorie, high-protein diet.
    d. boil drinking water for one minute.

Answers to these questions can be found in Appendix H.

Notes

1. National Center for Health Statistics, *Health, United States, 2009: With Special Feature on Medical Technology* (Hyattsville, MD: U.S. Department of Health and Human Services, 2009), pp. 254–257.

2. R. S. K. Chaganti, Genetics of cancer, in L. Goldman and D. Ausiello, eds., *Cecil Medicine* (Philadelphia: Saunders, 2008), pp. 1340–1344.

3. W. C. Willett and E. Giovannucci, Epidemiology of diet and cancer risk, in M. E. Shils and co-editors, *Modern Nutrition in Health and Disease* (Philadelphia: Lippincott Williams & Wilkins, 2006), pp.1267–1279.

4. World Cancer Research Fund/American Institute for Cancer Research, *Food, Nutrition, Physical Activity, and the Prevention of Cancer: A Global Perspective* (Washington, D.C.: American Institute for Cancer Research, 2007).

5. World Cancer Research Fund/American Institute for Cancer Research, 2007.

6. R. J. Turesky, Formation and biochemistry of carcinogenic heterocyclic aromatic amines in cooked meats, *Toxicology Letters* 168 (2007): 219–227; T. Sugimura and coauthors, Heterocyclic amines: Mutagens/carcinogens produced during cooking of meat and fish, *Cancer Science* 95 (2004): 290–299.

7. World Cancer Research Fund/American Institute for Cancer Research, 2007.

8. H. S. Rugo, Paraneoplastic syndromes and other non-neoplastic effects of cancer, in L. Goldman and D. Ausiello, eds., *Cecil Medicine* (Philadelphia: Saunders, 2008), pp. 1353–1362.

9. M. J. Tisdale, Mechanisms of cancer cachexia, *Physiological Reviews* 89 (2009): 381–410.

10. M. Schattner and M. Shike, Nutrition support of the patient with cancer, in M. E. Shils and coeditors, *Modern Nutrition in Health and Disease* (Philadelphia: Lippincott Williams & Wilkins, 2006), pp. 1290–1313.

11. S. Dodson and coauthors, Muscle wasting in cancer cachexia: Clinical implications, diagnosis, and emerging treatment strategies, *Annual Review of Medicine* 62 (2011): 8.1–8.15.

12. T. Agustsson and coauthors, Mechanism of increased lipolysis in cancer cachexia, *Cancer Research* 67 (2007): 5531–5537; L. G. Melstrom and coauthors, Mechanisms of skeletal muscle degradation and its therapy in cancer cachexia, *Histology and Histopathology* 22 (2007): 805–814.

13. M. C. Perry, Principles of cancer therapy, in L. Goldman and D. Ausiello, eds., *Cecil Medicine* (Philadelphia: Saunders, 2008), pp. 1370–1387.

14. J. M. Vose and S. Z. Pavletic, Hematopoietic stem cell transplantation, in L. Goldman and D. Ausiello, eds., *Cecil Medicine* (Philadelphia: Saunders, 2008), pp. 1328–1332.

15. Dodson and coauthors, 2011.

16. C. S. Roberts, Patient-physician communication regarding use of complementary therapies during cancer treatment, *Journal of Psychosocial Oncology* 23 (2005): 35–60.

17. C. E. Dennehy and C. Tsourounis, Botanicals ("herbal medications") and nutritional supplements, in B. G. Katzung, ed., *Basic and Clinical Pharmacology* (New York: Lange/McGraw-Hill, 2007), pp. 1050–1062.

18. M. L. Heaney and coauthors, Vitamin C antagonizes the cytotoxic effects of antineoplastic drugs, *Cancer Research* 68 (2008): 8031–8038; B. Bruemmer and coauthors, The association between vitamin C and vitamin E supplement use before hematopoietic stem cell transplant and outcomes to two years, *Journal of the American Dietetic Association* 103 (2003): 982–990.

19. Schattner and Shike, 2006.

20. N. Saquib and coauthors, Weight gain and recovery of pre-cancer weight after breast cancer treatments: Evidence from the women's healthy eating and living (WHEL) study, *Breast Cancer Research and Treatment* 105 (2007): 177–186; G. Makari-Judson, C. H. Judson, and W. C. Mertens, Longitudinal patterns of weight gain after breast cancer diagnosis: Observations beyond the first year, *Breast Journal* 13 (2007): 258–265.

21. American Dietetic Association, *Nutrition Care Manual* (Chicago: American Dietetic Association, 2011).

22. Schattner and Shike, 2006.

23. UNAIDS/WHO, *AIDS Epidemic Update: December 2009,* available at http://data.unaids.org/pub/Report/2009/2009_epidemic_update_en.pdf; accessed November 21, 2010.

24. G. M. Shaw, Biology of human immunodeficiency viruses, in L. Goldman and D. Ausiello, eds., *Cecil Medicine* (Philadelphia: Saunders, 2008), pp. 2557–2561.

25. Shaw, 2008.

26. H. Masur, L. Healey, and C. Hadigan, Treatment of human immunodeficiency virus infection and acquired immunodeficiency syndrome, in L. Goldman and D. Ausiello, eds., *Cecil Medicine* (Philadelphia: Saunders, 2008), pp. 2571–2582.

27. K. R. Dong, HIV-associated fat deposition, in K. M. Hendricks, K. R. Dong, and J. L. Gerrior, eds., *Nutrition Management of HIV and AIDS* (Chicago: American Dietetic Association, 2009), pp. 57–65; Masur, Healey, and Hadigan, 2008.

28. J. L. Gerrior, Unintentional weight loss and wasting in HIV infection, in K. M. Hendricks, K. R. Dong, and J. L. Gerrior, eds., *Nutrition Management of HIV and AIDS* (Chicago: American Dietetic Association, 2009), pp. 41–55.

29. J. G. Bartlett, Gastrointestinal manifestations of human immunodeficiency virus and acquired immunodeficiency syndrome, in L. Goldman and D. Ausiello, eds., *Cecil Medicine* (Philadelphia: Saunders, 2008), pp. 2582–2585.

30. U.S. Department of Health and Human Services, Panel on Antiretroviral Guidelines for Adults and Adolescents, *Guidelines for the Use of Antiretroviral Agents in HIV-1-Infected Adults and Adolescents,* December 1, 2009, pp. 1–168, available at http://aidsinfo.nih.gov/contentfiles/AdultandAdolescentGL.pdf; accessed November 21, 2010; S. Safrin, Antiviral agents, in B. G. Katzung, ed., *Basic and Clinical Pharmacology* (New York: Lange/McGraw-Hill, 2007), pp. 790–818.

31. K. Mulligan and coauthors, Testosterone supplementation of megestrol therapy does not enhance lean tissue accrual in men with human immunodeficiency virus-associated weight loss: A randomized, double-blind, placebo-controlled, multicenter trial, *Journal of Clinical Endocrinology and Metabolism* 92 (2007): 563–570.

32. Dennehy and Tsourounis, 2007.

33. L. Vining, General nutrition issues for healthy living with HIV infection, in K. M. Hendricks, K. R. Dong, and J. L. Gerrior, eds., *Nutrition Management of HIV and AIDS* (Chicago: American Dietetic Association, 2009), pp. 23–40.

34. Gerrior, 2009.

35. A. Howard, K. M. Hendricks, and J. Dwyer, Dietary supplement use in HIV infection, in K. M. Hendricks, K. R. Dong, and J. L. Gerrior, eds., *Nutrition Management of HIV and AIDS* (Chicago: American Dietetic Association, 2009), pp. 109–127; P. K. Drain and coauthors, Micronutrients in HIV-positive persons receiving highly active antiretroviral therapy, *American Journal of Clinical Nutrition* 85 (2007).

Nutrition in Practice

s with other medical technologies, the availability of specialized nutrition support forces health care professionals and members of our society to face difficult **ethical** issues. When medical treatments prolong life by merely delaying death, the lifetime that remains may be of extremely low quality. This Nutrition in Practice examines the ethical dilemmas that clinicians must face when dealing with patients in critical care. The accompanying glossary defines the relevant terms.

If providing nutrition care can do little to promote recovery, is it appropriate to withhold or withdraw nutrition support?

In attempting to answer questions such as these, health professionals must consider the following ethical principles:[1]

▮ A patient has the right to make decisions concerning his or her own well-being **(patient autonomy),** even if refusing treatment could result in death. It is generally accepted that a patient's preferences should take precedence over the desires of others.[2]

▮ A patient should be fully informed of a treatment's benefits and risks in a fair and honest manner **(disclosure).** A patient's acceptance of a treatment that has been adequately disclosed is considered **informed consent.**

▮ A patient must have the mental capacity to make appropriate health care decisions **(decision-making capacity).** If a patient is mentally incapable of doing so, a person designated by the patient should serve as a **surrogate** decision maker.

▮ The potential benefits **(beneficence)** of any treatment should outweigh its potential harm **(maleficence).**

▮ Health care providers must determine whether the provision of health care to one patient would unfairly limit the care of other patients **(distributive justice).**

Although these principles may seem simple and logical, it is often difficult to determine the appropriate action to take during intensive care.[3] When clinicians and families disagree, the courts may be asked to decide.

GLOSSARY

advance medical directive: written or oral instruction regarding one's preferences for medical treatment to be used in the event of becoming incapacitated; also called an *advance health care directive.*

beneficence (be-NEF-eh-sense): the act of performing beneficial services rather than harmful ones.

cardiopulmonary resuscitation (CPR): life-sustaining treatment that supplies oxygen and restores a person's ability to breathe and pump blood.

decision-making capacity: the ability to understand pertinent information and make appropriate decisions; known as *decision-making competency* within the legal system.

defibrillation: life-sustaining treatment in which an electronic device is used to shock the heart and reestablish a pattern of normal contractions. Defibrillation is used when the heart has arrhythmias or has experienced arrest.

dialysis: life-sustaining treatment in which a patient's blood is filtered using selective diffusion through a semipermeable membrane; substitutes for kidney function.

disclosure: the act of revealing pertinent information. For example, clinicians should accurately describe proposed tests and procedures, their benefits and risks, and alternative approaches.

distributive justice: the equitable distribution of resources.

do-not-resuscitate (DNR) order: a request by a patient or surrogate to withhold cardiopulmonary resuscitation.

durable power of attorney: a legal document (sometimes called a *health care proxy)* that gives legal authority to another *(a health care agent)* to make medical decisions in the event of incapacitation.

ethical: in accordance with accepted principles of right and wrong.

futile: describes medical care that will not improve the medical circumstances of a patient.

health care agent: a person given legal authority to make medical decisions for another in the event of incapacitation.

informed consent: a patient's or caregiver's agreement to undergo a treatment that has been adequately disclosed. Persons must be mentally competent in order to make the decision.

living will: a written statement that specifies the medical procedures desired or not desired in the event that a person is unable to communicate or is incapacitated; also called an *advance medical directive.*

maleficence (mah-LEF-eh-sense): the act of doing evil or harm.

mechanical ventilation: life-sustaining treatment in which a mechanical ventilator is used to substitute for a patient's failing lungs.

patient autonomy: a principle of self-determination, such that patients (or surrogate decision makers) are free to choose the medical interventions that are acceptable to them, even if they choose to refuse interventions that may extend their lives.

persistent vegetative state: a vegetative mental state resulting from brain injury that persists for at least one month. Individuals lose awareness and the ability to think but retain noncognitive brain functions, such as motor reflexes and normal sleep patterns.

surrogate: a substitute; a person who takes the place of another.

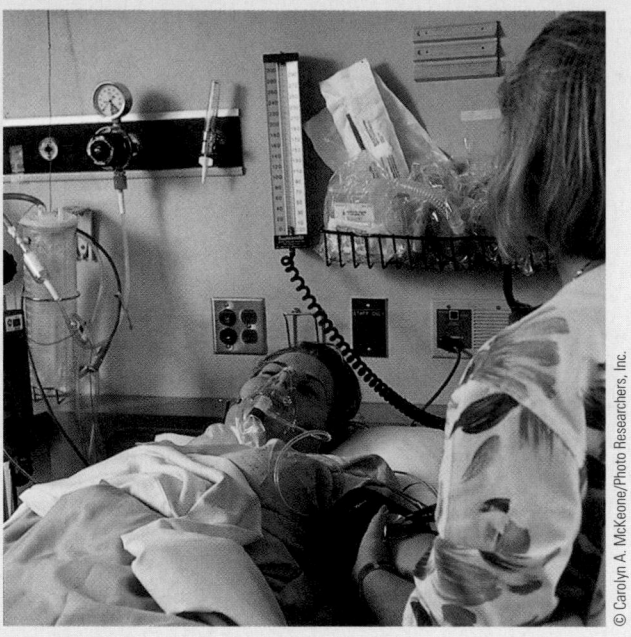

Is it ever morally or legally appropriate to withhold or withdraw nutrition support?

What kinds of treatments can help to sustain a patient's life?

Nutrition support and hydration are both considered life-sustaining treatments because withholding or withdrawing either can result in death. Other life-sustaining treatments include **cardiopulmonary resuscitation (CPR),** which supplies oxygen and restores a person's ability to breathe and pump blood; **defibrillation,** in which an electronic device shocks the heart and reestablishes normal contractions; **mechanical ventilation,** which substitutes for lung function; and **dialysis,** which substitutes for kidney function.

Do patients have a right to life-sustaining treatments?

Although life-sustaining treatments are readily provided to patients who have a reasonable chance of recovering from illness, it is sometimes difficult to determine the best course of action for patients who are dying or who are unlikely to regain consciousness. Under such circumstances, such treatments may be considered **futile** because they are unable to improve the outcome of disease or increase the patient's comfort and well-being. If patients or caregivers demand a treatment that health practitioners have determined to be useless, a legal resolution may be required. Conversely, medical personnel may find it objectionable to withdraw life support when they know that the inevitable consequence will be the patient's death.

How have the courts resolved conflicts involving nutrition support?

One of the landmark cases involving nutrition support concerned Nancy Cruzan, who suffered permanent and irreversible brain damage after a car crash in 1983 when she was 26 years of age.[4]

After she had been in a **persistent vegetative state** for 5 years, her parents requested permission to discontinue tube feeding, but hospital staff refused to honor the request, and the matter was taken to court. The Missouri Supreme Court determined that Nancy had never definitively stated her "right to die" wishes and that her parents were unable to make such a request for her. The court also stated that preserving life, no matter what its quality, should take precedence over all other considerations. Nancy's parents appealed the ruling, but in 1990, the U.S. Supreme Court upheld the Missouri Supreme Court in a five-to-four decision. Three witnesses were eventually found who could testify that Nancy would not desire life-sustaining treatment under the circumstances, and the court finally granted permission to remove the feeding tube. This case illustrates the importance of having an **advance medical directive** (discussed in a later section) that clearly indicates one's preferences for medical treatment in the event of incapacitation.

In a more recent case that received widespread media attention, the spouse and parents of a patient in a persistent vegetative state fought a 10-year legal battle over her medical care. In 1990, at the age of 25, Terri Schiavo suffered a full cardiac arrest.[5] She initially fell into a coma, but her condition evolved into a persistent vegetative state that was considered irreversible. Despite the neurologists' diagnosis and a series of computed tomography (CT) and magnetic resonance imaging (MRI) scans showing extensive brain atrophy, her parents maintained that she was minimally conscious and could improve somewhat with rigorous treatment. Her husband, who was legally responsible for her care, insisted that she would never have wanted to be kept alive in a vegetative state. Like Nancy Cruzan, Terri had never expressed her wishes in an advance directive.

In 1998, Terri's husband filed a petition to have her feeding tube removed, and a Florida court approved the motion in February 2000. Although Terri's parents appealed, an appeals court affirmed the decision, and the Florida Supreme Court declined to review the case. In April 2001, Terri's physicians removed her feeding tube, but within days, a federal circuit court judge ordered it to be reinserted and reopened the case. Eventually, the various motions filed by the parents were dismissed and Terri's feeding tube was removed for the second time in October 2003. Within days, the Florida legislature passed a bill known as "Terri's Law" that gave the governor the authority to intervene, and Governor Jeb Bush ordered the feeding tube reinserted. A year later, Florida's Supreme Court declared Terri's Law to be unconstitutional. Although the governor appealed the decision, his appeal was rejected in January 2005. Terri's feeding tube was removed for the third time in March 2005. Despite emergency petitions by her parents and an attempt by the U.S. Congress to have her case reconsidered, the courts refused to grant a restraining order, and Terri died 13 days after her feeding tube was removed.

How can people ensure that their wishes will be considered in the event that they become incapacitated?

A person can declare preferences about medical treatments in a **living will,** sometimes called an *advance medical directive.* Living wills include instructions about life-sustaining procedures that a

person does or does not want. These directives are incorporated into the medical record and updated when appropriate; they take effect only if a physician determines that a patient lacks the ability to understand and make decisions about available treatments. If a person's preferences are unknown, decisions are based on a patient's best interests as determined by a caregiver or family member.[6]

Another important directive is a **durable power of attorney** (sometimes called a *health care proxy*), in which another person (a **health care agent**) is appointed to act as decision maker in the event of incapacitation. The agent should understand one's medical preferences and be absolutely trustworthy. Only one person can be designated, although one or two alternates may also be listed. If an agent is given comprehensive power to supervise care, he or she may make decisions about medical staff, health care facilities, and medical procedures.

Laws regarding advance directives vary from state to state. In some states, nutrition and hydration are not considered life-sustaining treatments, and a person's instructions about them may need to be indicated separately. Some states restrict the use of advance directives to terminal illness or disallow them if a woman is pregnant. Generally, advance directives created in one state are honored in another.

How does a "do-not-resuscitate" order differ from other advance directives?

A **do-not-resuscitate (DNR) order** is frequently used to withhold CPR in the event of cardiopulmonary arrest, which occurs too suddenly for deliberate decision making.[7] A DNR order is written in the medical record as other directives are, but it does not exclude the use of other life-prolonging measures. A DNR order is most often used in patients with serious illnesses or advanced age. Some institutions allow a physician to write a DNR order for a patient who has a poor prognosis, but the physician must inform the patient or surrogate if this is done.

Have advance directives changed the way that medical care is provided?

Not really. Despite the availability of advance directives, only about 20 percent of people in the United States have completed one.[8] Furthermore, advance directives are often unavailable when intensive care decisions are made: one study found that only 57.5 percent of patient charts indicating the existence of an advance directive actually contained a copy.[9] In addition, advance directives are sometimes too general or vague to guide specific treatment decisions.

Physicians must often provide patient care before they have a chance to discuss treatments with patients or caregivers.[10] In many cases, life-sustaining treatments are begun without the prior knowledge of patients or their decision makers, or the treatments continue even if patients want them stopped. Patients who are fully aware of treatment options and clearly state their preferences are more likely to be successful at obtaining the care they desire.

Notes

1. M. A. Grippi, Ethics in critical care, in A. P. Fishman and coeditors, *Fishman's Manual of Pulmonary Diseases and Disorders* (New York: McGraw-Hill, 2002), pp. 1111–1114.

2. E. J. Emanuel, Bioethics in the practice of medicine, in L. Goldman and D. Ausiello, eds., *Cecil Medicine* (Philadelphia: Saunders, 2008), pp. 6–11.

3. Emanuel, 2008.

4. J. O. Maillet, Position of the American Dietetic Association: Ethical and legal issues in nutrition, hydration, and feeding, *Journal of the American Dietetic Association* 108 (2008): 873–882.

5. R. Cranford, Facts, lies, and videotapes: The permanent vegetative state and the sad case of Terri Schiavo, *Journal of Law, Medicine, and Ethics* 33 (2005): 363–372.

6. L. Snyder and C. Leffler, Position paper: Ethics manual, *Annals of Internal Medicine* (2005): 560–582.

7. Snyder and Leffler, 2005.

8. Emanuel, 2008.

9. Institute of Medicine, *Approaching Death: Improving Care at the End of Life* (Washington, D.C.: National Academy Press, 1997), pp. 202–203.

10. Emanuel, 2008.

Appendices

Appendix A

Table of Food Composition

This edition of the table of food composition includes a wide variety of foods. It is updated with each edition to reflect current nutrient data for foods, to remove outdated foods, and to add foods that are new to the marketplace.* The nutrient database for this appendix is compiled from a variety of sources, including the USDA Nutrient Database and manufacturers' data. The USDA database provides data for a wider variety of foods and nutrients than other sources. Because laboratory analysis for each nutrient can be quite costly, manufacturers tend to provide data only for those nutrients mandated on food labels. Consequently, data for their foods are often incomplete; any missing information on this table is designated as a dash. Keep in mind that a dash means only that the information is unknown and should not be interpreted as a zero. A zero means that the nutrient is not present in the food.

Whenever using nutrient data, remember that many factors influence the nutrient contents of foods. These factors include the mineral content of the soil, the diet fed to the animal or the fertilizer used on the plant, the season of harvest, the method of processing, the length and method of storage, the method of cooking, the method of analysis, and the moisture content of the sample analyzed. With so many influencing factors, users should view nutrient data as a close approximation of the actual amount.

For updates, corrections, and a list of more than 8000 foods and codes found in the diet analysis software that accompanies this text, visit www.cengage.com/nutrition and click on Diet Analysis Plus.

- *Fats.* Total fats, as well as the breakdown of total fats to saturated, monounsaturated, polyunsaturated, and *trans* fats, are listed in the table. The fatty acids seldom add up to the total, in part due to rounding, but also because values may include some non-fatty acids, such as glycerol, phosphate, or sterols.

- *Trans Fats.* *Trans* fat data have been listed in the table. Because food manufacturers have only been required to report *trans* fats on food labels since January 2006, much of the data is incomplete. Missing *trans* fat data are designated with a dash. As additional *trans* fat data become available, the table will be updated.

- *Vitamin A and Vitamin E.* In keeping with the 2001 RDA for vitamin A, this appendix presents data for vitamin A in micrograms (µg) RAE. Similarly, because the 2000 RDA for vitamin E is based only on the alpha-tocopherol form of vitamin E, this appendix reports vitamin E data in milligrams (mg) alpha-tocopherol, listed on the table as Vit E (mg α).

- *Bioavailability.* Keep in mind that the availability of nutrients from foods depends not only on the quantity provided by a food, but also on the amount absorbed and used by the body—the bioavailability. The bioavailability of folate from fortified foods, for example, is greater than from naturally occurring sources. Similarly, the body can make niacin from the amino acid tryptophan, but niacin values in this table (and most databases) report preformed niacin only. Chapter 8 provides conversion factors and additional details.

*This food composition table has been prepared by Cengage Learning. The nutritional data are supplied by Axxya Systems.

Appendix A

A GRAS substance is one that is "generally recognized as safe."

■ *Using the Table.* The foods and beverages in this table are organized into several categories, which are listed at the head of each right-hand page. Page numbers are provided, and each group is color-coded to make it easier to find individual foods.

■ *Caffeine Sources.* Caffeine occurs in several plants, including the familiar coffee bean, the tea leaf, and the cocoa bean from which chocolate is made. Most human societies use caffeine regularly, most often in beverages, for its stimulant effect and flavor. Caffeine contents of beverages vary depending on the plants they are made from, the climates and soils where the plants are grown, the grind or cut size, the method and duration of brewing, and the amounts served. The accompanying table shows that, in general, a cup of coffee contains the most caffeine; a cup of tea, less than half as much; and cocoa or chocolate, less still. As for cola beverages, they are made from kola nuts, which contain caffeine, but most of their caffeine is added, using the purified compound obtained from decaffeinated coffee beans. The FDA lists caffeine as a multipurpose GRAS substance that may be added to foods and beverages. Drug manufacturers use caffeine in many products.

TABLE Caffeine Content of Selected Beverages, Foods, and Medications

Beverages and Foods	Serving Size	Average (mg)	Beverages and Foods	Serving Size	Average (mg)
Coffee			**Soft Drinks**		
Brewed	8 oz	95	A&W Creme Soda	12 oz	29
Decaffeinated	8 oz	2	Barq's Root Beer	12 oz	18
Instant	8 oz	64	Coca-Cola	12 oz	30
			Dr. Pepper, Mr. Pibb, Sunkist Orange	12 oz	36
Tea			A&W Root Beer, club soda, Fresca, ginger ale, 7-Up, Sierra Mist, Sprite, Squirt, tonic water, caffeine-free soft drinks	12 oz	0
Brewed, green	8 oz	30			
Brewed, herbal	8 oz	0			
Brewed, leaf or bag	8 oz	47			
Instant	8 oz	26	Mello Yello	12 oz	51
Lipton Brisk iced tea	12 oz	7	Mountain Dew	12 oz	45
Nestea Cool iced tea	12 oz	12	Pepsi	12 oz	32
Snapple iced tea (all flavors)	16 oz	42			

TABLE Caffeine Content of Selected Beverages, Foods, and Medications *(continued)*

Beverages and Foods	Serving Size	Average (mg)
Energy Drinks		
Amp	8.4 oz	70
Aqua Blast	.5 L	90
Aqua Java	.5 L	55
E Maxx	8.4 oz	74
Java Water	.5 L	125
KMX	8.4 oz	33
Krank	.5 L	100
Red Bull	8.3 oz	67
Red Devil	8.4 oz	42
Sobe Adrenaline Rush	8.3 oz	77
Sobe No Fear	16 oz	141
Water Joe	.5 L	65
Other Beverages		
Chocolate milk or hot cocoa	8 oz	5
Starbucks Frappuccino Mocha	9.5 oz	72
Starbucks Frappuccino Vanilla	9.5 oz	64
Yoohoo chocolate drink	9 oz	3
Candies		
Baker's chocolate	1 oz	26
Dark chocolate covered coffee beans	1 oz	235
Dark chocolate, semisweet	1 oz	18
Milk chocolate	1 oz	6
Milk chocolate covered coffee beans	1 oz	224
White chocolate	1 oz	0

Beverages and Foods	Serving Size	Average (mg)
Foods		
Frozen yogurt, Ben & Jerry's coffee fudge	1 cup	85
Frozen yogurt, Häagen-Dazs coffee	1 cup	40
Ice cream, Starbucks coffee	1 cup	50
Ice cream, Starbucks Frappuccino bar	1 bar	15
Yogurt, Dannon coffee-flavored	1 cup	45

Drugs[a]	Serving Size	Average (mg)
Cold Remedies		
Coryban-D, Dristan	1 tablet	30
Diuretics		
Aqua-Ban	1 tablet	100
Pre-Mens Forte	1 tablet	100
Pain Relievers		
Anacin, BC Fast Pain Reliever	1 tablet	32
Excedrin, Midol, Midol Max Strength	1 tablet	65
Stimulants		
Awake, NoDoz	1 tablet	100
Awake Maximum Strength, Caffedrine, NoDoz Maximum Strength, Stay Awake, Vivarin	1 tablet	200
Weight-Control Aids		
Dexatrim	1 tablet	200

[a]A pharmacologically active dose of caffeine is defined as 200 milligrams.

NOTE: The FDA suggests a maximum of 65 milligrams per 12-ounce cola beverage but does not regulate the caffeine contents of other beverages. Because products change, contact the manufacturer for an update on products you use regularly.

SOURCE: Adapted from USDA database Release 18 (www.nal.usda.gov/fnic/foodcomp/Data/), Caffeine content of foods and drugs, Center for Science and the Public Interest (www.cspinet.org/new/cafchart.htm), and R. R. McCusker, B. A. Goldberger, and E. J. Cone, Caffeine content of energy drinks, carbonated sodas, and other beverages, *Journal of Analytical Toxicology* 30 (2006): 112–114.

Appendix A

TABLE A-1 Table of Food Composition (Computer code is for Cengage Diet Analysis program) (For purposes of calculations, use "0" for t, <1, <.1, <.01, etc.)

DA+ Code	Food Description	Quantity	Measure	Wt (g)	H₂0 (g)	Ener (kcal)	Prot (g)	Carb (g)	Fiber (g)	Fat (g)	Sat	Mono	Poly	Trans
	**Breads, Baked Goods, Cakes, Cookies, Crackers, Chips, Pies**													

Bagels

DA+ Code	Food Description	Quantity	Measure	Wt (g)	H₂0 (g)	Ener (kcal)	Prot (g)	Carb (g)	Fiber (g)	Fat (g)	Sat	Mono	Poly	Trans
8534	Cinnamon and raisin	1	item(s)	71	22.7	194	7.0	39.2	1.6	1.2	0.2	0.1	0.5	—
14395	Multigrain	1	item(s)	61	—	170	6.0	35.0	1.0	1.5	0.5	0.1	0.4	—
8538	Oat bran	1	item(s)	71	23.4	181	7.6	37.8	2.6	0.9	0.1	0.2	0.3	—
4910	Plain, enriched	1	item(s)	71	25.8	182	7.1	35.9	1.6	1.2	0.3	0.4	0.5	0
4911	Plain, enriched, toasted	1	item(s)	66	18.7	190	7.4	37.7	1.7	1.1	0.2	0.3	0.6	0

Biscuits

25008	Biscuits	1	item(s)	41	15.8	121	2.6	16.4	0.5	4.9	1.4	1.4	1.8	—
16729	Scone	1	item(s)	42	11.5	148	3.8	19.1	0.6	6.2	2.0	2.5	1.3	—
25166	Wheat biscuits	1	item(s)	55	21.0	162	3.6	21.9	1.4	6.7	1.9	1.9	2.5	—

Bread

325	Boston brown, canned	1	slice(s)	45	21.2	88	2.3	19.5	2.1	0.7	0.1	0.1	0.3	—
8716	Bread sticks, plain	4	item(s)	24	1.5	99	2.9	16.4	0.7	2.3	0.3	0.9	0.9	—
25176	Cornbread	1	piece(s)	55	25.9	141	4.7	18.3	0.9	5.4	2.1	1.4	1.5	0
327	Cracked wheat	1	slice(s)	25	9.0	65	2.2	12.4	1.4	1.0	0.2	0.5	0.2	—
9079	Croutons, plain	¼	cup(s)	8	0.4	31	0.9	5.5	0.4	0.5	0.1	0.2	0.1	—
8582	Egg	1	slice(s)	40	13.9	113	3.8	19.1	0.9	2.4	0.6	0.9	0.4	—
8585	Egg, toasted	1	slice(s)	37	10.5	117	3.9	19.5	0.9	2.4	0.6	1.1	0.4	—
329	French	1	slice(s)	32	8.9	92	3.8	18.1	0.8	0.6	0.2	0.1	0.3	—
8591	French, toasted	1	slice(s)	23	4.7	73	3.0	14.2	0.7	0.5	0.1	0.1	0.2	—
42096	Indian fry, made with lard (Navajo)	3	ounce(s)	85	26.9	281	5.7	41.0	—	10.4	3.9	3.8	0.9	—
332	Italian	1	slice(s)	30	10.7	81	2.6	15.0	0.8	1.1	0.3	0.2	0.4	—
1393	Mixed grain	1	slice(s)	26	9.6	69	3.5	11.3	1.9	1.1	0.2	0.2	0.5	0
8604	Mixed grain, toasted	1	slice(s)	24	7.6	69	3.5	11.3	1.9	1.1	0.2	0.2	0.5	0
8605	Oat bran	1	slice(s)	30	13.2	71	3.1	11.9	1.4	1.3	0.2	0.5	0.5	—
8608	Oat bran, toasted	1	slice(s)	27	10.4	70	3.1	11.8	1.3	1.3	0.2	0.5	0.5	—
8609	Oatmeal	1	slice(s)	27	9.9	73	2.3	13.1	1.1	1.2	0.2	0.4	0.5	—
8613	Oatmeal, toasted	1	slice(s)	25	7.8	73	2.3	13.2	1.1	1.2	0.2	0.4	0.5	—
1409	Pita	1	item(s)	60	19.3	165	5.5	33.4	1.3	0.7	0.1	0.1	0.3	—
7905	Pita, whole wheat	1	item(s)	64	19.6	170	6.3	35.2	4.7	1.7	0.3	0.2	0.7	—
338	Pumpernickel	1	slice(s)	32	12.1	80	2.8	15.2	2.1	1.0	0.1	0.3	0.4	—
334	Raisin, enriched	1	slice(s)	26	8.7	71	2.1	13.6	1.1	1.1	0.3	0.6	0.2	—
8625	Raisin, toasted	1	slice(s)	24	6.7	71	2.1	13.7	1.1	1.2	0.3	0.6	0.2	—
10168	Rice, white, gluten free, wheat free	1	slice(s)	38	—	130	1.0	18.0	0.5	6.0	0	—	—	0
8653	Rye	1	slice(s)	32	11.9	83	2.7	15.5	1.9	1.1	0.2	0.4	0.3	—
8654	Rye, toasted	1	slice(s)	29	9.0	82	2.7	15.4	1.9	1.0	0.2	0.4	0.3	—
336	Rye, light	1	slice(s)	25	9.3	65	2.0	12.0	1.6	1.0	0.2	0.3	0.3	—
8588	Sourdough	1	slice(s)	25	7.0	72	2.9	14.1	0.6	0.5	0.1	0.1	0.2	—
8592	Sourdough, toasted	1	slice(s)	23	4.7	73	3.0	14.2	0.7	0.5	0.1	0.1	0.2	—
491	Submarine or hoagie roll	1	item(s)	135	40.6	400	11.0	72.0	3.8	8.0	1.8	3.0	2.2	—
8596	Vienna, toasted	1	slice(s)	23	4.7	73	3.0	14.2	0.7	0.5	0.1	0.1	0.2	—
8670	Wheat	1	slice(s)	25	8.9	67	2.7	11.9	0.9	0.9	0.2	0.2	0.4	—
8671	Wheat, toasted	1	slice(s)	23	5.6	72	3.0	12.8	1.1	1.0	0.2	0.2	0.4	—
340	White	1	slice(s)	25	9.1	67	1.9	12.7	0.6	0.8	0.2	0.2	0.3	—
1395	Whole wheat	1	slice(s)	46	15.0	128	3.9	23.6	2.8	2.5	0.4	0.5	1.4	—

Cakes

386	Angel food, prepared from mix	1	piece(s)	50	16.5	129	3.1	29.4	0.1	0.2	0	0	0.1	—
8772	Butter pound, ready to eat, commercially prepared	1	slice(s)	75	18.5	291	4.1	36.6	0.4	14.9	8.7	4.4	0.8	—
28517	Carrot	1	slice(s)	131	56.6	339	4.8	56.5	1.9	11.1	1.0	5.7	3.8	—
4931	Chocolate with chocolate icing, commercially prepared	1	slice(s)	64	14.7	235	2.6	34.9	1.8	10.5	3.1	5.6	1.2	—
8756	Chocolate, prepared from mix	1	slice(s)	95	23.2	352	5.0	50.7	1.5	14.3	5.2	5.7	2.6	—
393	Devil's food cupcake with chocolate frosting	1	item(s)	35	8.4	120	2.0	20.0	0.7	4.0	1.8	1.6	0.6	—
8757	Fruitcake, ready to eat, commercially prepared	1	piece(s)	43	10.9	139	1.2	26.5	1.6	3.9	0.5	1.8	1.4	—
1397	Pineapple upside down, prepared from mix	1	slice(s)	115	37.1	367	4.0	58.1	0.9	13.9	3.4	6.0	3.8	—
411	Sponge, prepared from mix	1	slice(s)	63	18.5	187	4.6	36.4	0.3	2.7	0.8	1.0	0.4	—
8817	White with coconut frosting, prepared from mix	1	slice(s)	112	23.2	399	4.9	70.8	1.1	11.5	4.4	4.1	2.4	—
8819	Yellow with chocolate frosting, ready to eat, commercially prepared	1	slice(s)	64	14.0	243	2.4	35.5	1.2	11.1	3.0	6.1	1.4	—

PAGE KEY: A-6 = Breads/Baked Goods A-12 = Cereal/Rice/Pasta A-16 = Fruit A-20 = Vegetables/Legumes A-30 = Nuts/Seeds A-32 = Vegetarian A-34 = Dairy A-42 = Eggs A-42 = Seafood A-44 = Meats A-48 = Poultry
A-50 = Processed Meats A-50 = Beverages A-54 = Fats/Oils A-56 = Sweets A-58 = Spices/Condiments/Sauces A-62 = Mixed Foods/Soups/Sandwiches A-66 = Fast Food A-86 = Convenience A-88 = Baby Foods

Chol (mg)	Calc (mg)	Iron (mg)	Magn (mg)	Pota (mg)	Sodi (mg)	Zinc (mg)	Vit A (µg)	Thia (mg)	Vit E (mg α)	Ribo (mg)	Niac (mg)	Vit B_6 (mg)	Fola (µg)	Vit C (mg)	Vit B_{12} (µg)	Sele (µg)
0	13	2.69	19.9	105.1	228.6	0.80	14.9	0.27	0.22	0.19	2.18	0.04	78.8	0.5	0	22.0
0	60	1.08	—	—	310.0	—	0	—	—	—	—	—	—	0	—	—
0	9	2.18	22.0	81.7	360.0	0.63	0.7	0.23	0.23	0.24	2.10	0.03	69.6	0.1	0	24.3
0	63	4.29	15.6	53.3	318.1	1.34	0	0.42	0.07	0.18	2.82	0.04	103.0	0.7	0	16.2
0	65	2.97	15.8	56.1	316.8	0.85	0	0.39	0.07	0.17	2.88	0.04	86.5	0	0	16.6
0	38	0.94	6.0	47.4	206.0	0.20	—	0.16	0.01	0.12	1.20	0.01	31.7	0.1	0.1	7.1
49	79	1.35	7.1	48.7	277.2	0.29	64.7	0.14	0.42	0.15	1.19	0.02	32.3	0	0.1	10.9
0	57	1.21	16.1	81.0	321.1	0.42	—	0.19	0.01	0.14	1.65	0.03	35.3	0.1	0.1	0
0	32	0.94	28.4	143.1	284.0	0.22	11.3	0.01	0.14	0.05	0.50	0.03	5.0	0	0	9.9
0	5	1.02	7.7	29.8	157.7	0.21	0	0.14	0.24	0.13	1.26	0.01	38.9	0	0	9.0
21	94	0.91	10.5	71.5	209.8	0.48	—	0.14	0.32	0.15	1.03	0.04	34.6	1.7	0.2	6.2
0	11	0.70	13.0	44.3	134.5	0.31	0	0.09	—	0.06	0.92	0.08	15.3	0	0	6.3
0	6	0.30	2.3	9.3	52.4	0.06	0	0.04	—	0.02	0.40	0.00	9.9	0	0	2.8
20	37	1.21	7.6	46.0	196.8	0.31	25.2	0.17	0.10	0.17	1.93	0.02	42.0	0	0	12.0
21	38	1.23	7.8	46.6	199.8	0.31	25.5	0.14	0.10	0.16	1.77	0.02	36.3	0	0	12.2
0	14	1.16	9.0	41.0	208.0	0.29	0	0.13	0.05	0.09	1.52	0.03	47.4	0.1	0	8.7
0	11	0.89	7.1	32.2	165.6	0.24	0	0.10	0.04	0.09	1.24	0.02	32.2	0	0	6.8
6	48	3.43	15.3	65.5	279.8	0.29	0	0.36	0.00	0.18	3.91	0.03	103.8	—	0	15.8
0	23	0.88	8.1	33.0	175.2	0.25	0	0.14	0.08	0.08	1.31	0.01	57.3	0	0	8.2
0	27	0.65	20.3	59.8	109.2	0.44	0	0.07	0.09	0.03	1.05	0.06	19.5	0	0	8.6
0	27	0.65	20.4	60.0	109.7	0.44	0	0.06	0.10	0.03	1.05	0.07	16.8	0	0	8.6
0	20	0.93	10.5	44.1	122.1	0.26	0.6	0.15	0.13	0.10	1.44	0.02	24.3	0	0	9.0
0	19	0.92	9.2	33.2	121.0	0.28	0.5	0.12	0.13	0.09	1.29	0.01	18.6	0	0	8.9
0	18	0.72	10.0	38.3	161.7	0.27	1.4	0.10	0.13	0.06	0.84	0.01	16.7	0	0	6.6
0	18	0.74	10.3	38.5	162.8	0.28	1.3	0.09	0.13	0.06	0.77	0.02	13.3	0.1	0	6.7
0	52	1.57	15.6	72.0	321.6	0.50	0	0.35	0.18	0.19	2.77	0.02	64.2	0	0	16.3
0	10	1.95	44.2	108.8	340.5	0.97	0	0.21	0.39	0.05	1.81	0.17	22.4	0	0	28.2
0	22	0.91	17.3	66.6	214.7	0.47	0	0.10	0.13	0.09	0.98	0.04	29.8	0	0	7.8
0	17	0.75	6.8	59.0	101.4	0.18	0	0.08	0.07	0.10	0.90	0.01	27.6	0	0	5.2
0	17	0.76	6.7	59.0	101.8	0.19	0	0.07	0.07	0.09	0.81	0.02	23.5	0.1	0	5.2
0	100	1.08	—	—	140	—	—	0.15	—	0.10	1.20	—	32.0	0	—	—
0	23	0.90	12.8	53.1	211.2	0.36	0	0.13	0.10	0.10	1.21	0.02	35.2	0	0	9.9
0	23	0.89	12.5	53.1	210.3	0.36	0	0.11	0.10	0.09	1.09	0.02	29.9	0.1	0	9.9
0	20	0.70	3.9	51.0	175.0	0.18	0	0.10	—	0.08	0.80	0.01	5.3	0	0	8.0
0	11	0.91	7.0	32.0	162.5	0.23	0	0.11	0.05	0.07	1.19	0.03	37.0	0.1	0	6.8
0	11	0.89	7.1	32.2	165.6	0.24	0	0.10	0.04	0.09	1.24	0.02	32.2	0	0	6.8
0	100	3.80	—	128.0	683.0	—	0	0.54	—	0.33	4.50	0.04		0	—	42.0
0	11	0.89	7.1	32.2	165.6	0.24	0	0.10	0.04	0.09	1.24	0.02	32.2	0	0	6.8
0	36	0.87	12.0	46.0	130.3	0.30	0	0.09	0.05	0.08	1.30	0.03	21.3	0.1	0	7.2
0	38	0.94	13.6	51.3	140.5	0.34	0	0.10	0.06	0.09	1.44	0.04	19.8	0	0	7.7
0	38	0.94	5.8	25.0	170.3	0.19	0	0.11	0.06	0.08	1.10	0.02	27.8	0	0	4.3
0	15	1.42	37.3	144.4	159.2	0.69	0	0.13	0.35	0.10	1.83	0.09	29.9	0	0	17.8
0	42	0.11	4.0	67.5	254.5	0.06	0	0.04	0.01	0.10	0.08	0.00	9.5	0	0	7.7
166	26	1.03	8.3	89.3	298.5	0.34	111.8	0.10	—	0.17	0.98	0.03	30.8	0	0.2	6.6
0	65	2.18	23.0	279.6	367.7	0.44	—	0.25	0.01	0.19	1.73	0.10	43.4	4.6	0	14.7
27	28	1.40	21.8	128.0	213.8	0.44	16.6	0.01	0.62	0.08	0.36	0.02	10.9	0.1	0.1	2.1
55	57	1.53	30.4	133.0	299.3	0.65	38.0	0.13	—	0.20	1.08	0.03	25.7	0.2	0.2	11.3
19	21	0.70	—	46.0	92.0	—	—	0.04	—	0.05	0.30	—	2.1	0	—	2.0
2	14	0.89	6.9	65.8	116.1	0.11	3.0	0.02	0.38	0.04	0.34	0.02	8.6	0.2	0	0.9
25	138	1.70	15.0	128.8	366.9	0.35	71.3	0.17	—	0.17	1.36	0.03	29.9	1.4	0.1	10.8
107	26	0.99	5.7	88.8	143.6	0.37	48.5	0.10	—	0.19	0.75	0.03	24.6	0	0.2	11.7
1	101	1.29	13.4	110.9	318.1	0.37	13.4	0.14	0.13	0.21	1.19	0.03	34.7	0.1	0.1	12.0
35	24	1.33	19.2	113.9	215.7	0.39	21.1	0.07	—	0.10	0.79	0.02	14.1	0	0.1	2.2

TABLE A-1 Table of Food Composition *(continued)* (Computer code is for Cengage Diet Analysis program) (For purposes of calculations, use "0" for t, <1, <.1, <.01, etc.)

DA+ Code	Food Description	Quantity	Measure	Wt (g)	H₂O (g)	Ener (kcal)	Prot (g)	Carb (g)	Fiber (g)	Fat (g)	Fat Breakdown (g)			
											Sat	Mono	Poly	*Trans*
Breads, Baked Goods, Cakes, Cookies, Crackers, Chips, Pies—continued														
8822	Yellow with vanilla frosting, ready to eat, commercially prepared	1	slice(s)	64	14.1	239	2.2	37.6	0.2	9.3	1.5	3.9	3.3	—
	Snack cakes													
8791	Chocolate snack cake, crème filled, with frosting	1	item(s)	50	9.3	200	1.8	30.2	1.6	8.0	2.4	4.3	0.9	—
25010	Cinnamon coffee cake	1	piece(s)	72	22.6	231	3.6	35.8	0.7	8.3	2.2	2.6	3.0	—
16777	Funnel cake	1	item(s)	90	37.6	276	7.3	29.1	0.9	14.4	2.7	4.7	6.1	—
8794	Sponge snack cake, crème filled	1	item(s)	43	8.6	155	1.3	27.2	0.2	4.8	1.1	1.7	1.4	—
	Snacks, chips, pretzels													
29428	Bagel chips, plain	3	item(s)	29	—	130	3.0	19.0	1.0	4.5	0.5	—	—	—
29429	Bagel chips, toasted onion	3	item(s)	29	—	130	4.0	20.0	1.0	4.5	0.5	—	—	—
38192	Chex traditional snack mix	1	cup(s)	45	—	197	3.0	33.3	1.5	6.1	0.8	—	—	—
654	Potato chips, salted	1	ounce(s)	28	0.6	155	1.9	14.1	1.2	10.6	3.1	2.8	3.5	—
8816	Potato chips, unsalted	1	ounce(s)	28	0.5	152	2.0	15.0	1.4	9.8	3.1	2.8	3.5	—
5096	Pretzels, plain, hard, twists	5	item(s)	30	1.0	114	2.7	23.8	1.0	1.1	0.2	0.4	0.4	—
4632	Pretzels, whole wheat	1	ounce(s)	28	1.1	103	3.1	23.0	2.2	0.7	0.2	0.3	0.2	—
4641	Tortilla chips, plain	6	item(s)	11	0.2	53	0.8	7.1	0.6	2.5	0.3	0.8	0.5	0.3
	Cookies													
8859	Animal crackers	12	item(s)	30	1.2	134	2.1	22.2	0.3	4.1	1.0	2.3	0.6	—
8876	Brownie, prepared from mix	1	item(s)	24	3.0	112	1.5	12.0	0.5	7.0	1.8	2.6	2.3	—
25207	Chocolate chip cookies	1	item(s)	30	3.7	140	2.0	16.2	0.6	7.9	2.1	3.3	2.1	—
8915	Chocolate sandwich cookie with extra crème filling	1	item(s)	13	0.2	65	0.6	8.9	0.4	3.2	0.7	2.1	0.3	1.1
14145	Fig Newtons cookies	1	item(s)	16	—	55	0.5	11.0	0.5	1.3	0	—	—	0
8920	Fortune cookie	1	item(s)	8	0.6	30	0.3	6.7	0.1	0.2	0.1	0.1	0	—
25208	Oatmeal cookies	1	item(s)	69	12.3	234	5.7	45.1	3.1	4.2	0.7	1.3	1.8	—
25213	Peanut butter cookies	1	item(s)	35	4.1	163	4.2	16.9	0.9	9.2	1.7	4.7	2.3	—
33095	Sugar cookies	1	item(s)	16	4.1	61	1.1	7.4	0.1	3.0	0.6	1.3	0.9	—
9002	Vanilla sandwich cookie with crème filling	1	item(s)	10	0.2	48	0.5	7.2	0.2	2.0	0.3	0.8	0.8	—
	Crackers													
9012	Cheese cracker sandwich with peanut butter	4	item(s)	28	0.9	139	3.5	15.9	1.0	7.0	1.2	3.6	1.4	—
9008	Cheese crackers (mini)	30	item(s)	30	0.9	151	3.0	17.5	0.7	7.6	2.8	3.6	0.7	—
33362	Cheese crackers, low sodium	1	serving(s)	30	0.9	151	3.0	17.5	0.7	7.6	2.9	3.6	0.7	—
8928	Honey graham crackers	4	item(s)	28	1.2	118	1.9	21.5	0.8	2.8	0.4	1.1	1.1	—
9016	Matzo crackers, plain	1	item(s)	28	1.2	112	2.8	23.8	0.9	0.4	0.1	0	0.2	—
9024	Melba toast	3	item(s)	15	0.8	59	1.8	11.5	0.9	0.5	0.1	0.1	0.2	—
9028	Melba toast, rye	3	item(s)	15	0.7	58	1.7	11.6	1.2	0.5	0.1	0.1	0.2	—
14189	Ritz crackers	5	item(s)	16	0.5	80	1.0	10.0	0	4.0	1.0	—	—	0
9014	Rye crispbread crackers	1	item(s)	10	0.6	37	0.8	8.2	1.7	0.1	0	0	0.1	—
9040	Rye wafer	1	item(s)	11	0.6	37	1.1	8.8	2.5	0.1	0	0	0	—
432	Saltine crackers	5	item(s)	15	0.8	64	1.4	10.6	0.5	1.7	0.2	1.1	0.2	0.5
9046	Saltine crackers, low salt	5	item(s)	15	0.6	65	1.4	10.7	0.5	1.8	0.4	1.0	0.3	—
9052	Snack cracker sandwich with cheese filling	4	item(s)	28	1.1	134	2.6	17.3	0.5	5.9	1.7	3.2	0.7	—
9054	Snack cracker sandwich with peanut butter filling	4	item(s)	28	0.8	138	3.2	16.3	0.6	6.9	1.4	3.9	1.3	—
9048	Snack crackers, round	10	item(s)	30	1.1	151	2.2	18.3	0.5	7.6	1.1	3.2	2.9	—
9050	Snack crackers, round, low salt	10	item(s)	30	1.1	151	2.2	18.3	0.5	7.6	1.1	3.2	2.9	—
9044	Soda crackers	5	item(s)	15	0.8	64	1.4	10.6	0.5	1.7	0.2	1.1	0.2	0.5
9059	Wheat cracker sandwich with cheese filling	4	item(s)	28	0.9	139	2.7	16.3	0.9	7.0	1.2	2.9	2.6	—
9061	Wheat cracker sandwich with peanut butter filling	4	item(s)	28	1.0	139	3.8	15.1	1.2	7.5	1.3	3.3	2.5	—
9055	Wheat crackers	10	item(s)	30	0.9	142	2.6	19.5	1.4	6.2	1.6	3.4	0.8	—
9057	Wheat crackers, low salt	10	item(s)	30	0.9	142	2.6	19.5	1.4	6.2	1.6	3.4	0.8	—
9022	Whole wheat crackers	7	item(s)	28	0.8	124	2.5	19.2	2.9	4.8	1.0	1.6	1.8	—
	Pastry													
16754	Apple fritter	1	item(s)	17	6.4	61	1.0	5.5	0.2	3.9	0.9	1.7	1.1	—
41565	Cinnamon rolls with icing, refrigerated dough	1	serving(s)	44	12.3	145	2.0	23.0	0.5	5.0	1.5	—	—	2.0
4945	Croissant, butter	1	item(s)	57	13.2	231	4.7	26.1	1.5	12.0	6.6	3.1	0.6	—
9096	Danish, nut	1	item(s)	65	13.3	280	4.6	29.7	1.3	16.4	3.8	8.9	2.8	—
9115	Doughnut with crème filling	1	item(s)	85	32.5	307	5.4	25.5	0.7	20.8	4.6	10.3	2.6	—

Chol (mg)	Calc (mg)	Iron (mg)	Magn (mg)	Pota (mg)	Sodi (mg)	Zinc (mg)	Vit A (μg)	Thia (mg)	Vit E (mg α)	Ribo (mg)	Niac (mg)	Vit B_6 (mg)	Fola (μg)	Vit C (mg)	Vit B_{12} (μg)	Sele (μg)
35	40	0.68	3.8	33.9	220.2	0.16	12.2	0.06	—	0.04	0.32	0.01	17.3	0	0.1	3.5
0	58	1.80	18.0	88.0	194.5	0.52	0.5	0.01	0.54	0.03	0.46	0.07	13.0	1.0	0	1.7
26	55	1.36	9.9	91.9	277.6	0.30	—	0.17	0.23	0.16	1.29	0.02	36.1	0.3	0.1	9.6
62	126	1.90	16.2	152.1	269.1	0.65	49.5	0.23	1.54	0.32	1.86	0.04	50.4	0	0.3	17.7
7	19	0.54	3.4	37.0	155.1	0.12	2.1	0.06	0.50	0.05	0.52	0.01	17.0	0	0	1.3
0	0	0.72	—	45.0	70.0	—	0	—	—	—	—	—	—	0	0	—
0	0	0.72	—	50.0	300.0	—	0	—	—	—	—	—	—	0	0	—
0	0	0.55	—	75.8	621.2	—	0	0.09	—	0.05	1.21	—	12.1	0	—	—
0	7	0.45	19.8	465.5	148.8	0.67	0	0.01	1.91	0.06	1.18	0.20	21.3	5.3	0	2.3
0	7	0.46	19.0	361.5	2.3	0.30	0	0.04	2.58	0.05	1.08	0.18	12.8	8.8	0	2.3
0	11	1.29	10.5	43.8	514.5	0.25	0	0.13	0.10	0.18	1.57	0.03	51.3	0	0	1.7
0	8	0.76	8.5	121.9	57.6	0.17	0	0.12	—	0.08	1.85	0.07	15.3	0.3	0	—
0	19	0.25	15.8	23.2	45.5	0.26	0	0.00	0.46	0.01	0.13	0.02	2.2	0	0	0.7
0	13	0.82	5.4	30.0	117.9	0.19	0	0.10	0.03	0.09	1.04	0.01	30.9	0	0	2.1
18	14	0.44	12.7	42.2	82.3	0.23	42.2	0.03	—	0.05	0.24	0.02	7.0	0.1	0	2.8
13	11	0.69	12.4	62.1	108.8	0.24	—	0.08	0.54	0.06	0.87	0.01	17.8	0	0	4.1
0	2	1.01	4.7	17.8	45.6	0.10	0	0.02	0.25	0.02	0.25	0.00	6.0	0	0	1.1
0	10	0.36	—	—	57.5	—	0	—	—	—	—	—	—	0	—	—
0	1	0.12	0.6	3.3	21.9	0.01	0.1	0.01	0.00	0.01	0.15	0.00	5.3	0	0	0.2
0	26	1.93	48.8	176.7	311.1	1.42	—	0.26	0.23	0.13	1.35	0.09	34.9	0.3	0	17.4
13	27	0.65	21.1	112.8	154.1	0.46	—	0.08	0.73	0.09	1.85	0.05	23.5	0.1	0.1	4.8
18	5	0.30	1.7	12.2	49.4	0.08	—	0.04	0.28	0.05	0.31	0.01	9.5	0	0	3.1
0	3	0.22	1.4	9.1	34.9	0.04	0	0.02	0.16	0.02	0.27	0.00	5.0	0	0	0.3
0	14	0.76	15.7	61.0	198.8	0.29	0.3	0.15	0.66	0.08	1.63	0.04	26.3	0	0.1	2.3
4	45	1.43	10.8	43.5	298.5	0.33	8.7	0.17	0.01	0.12	1.40	0.16	45.6	0	0.1	2.6
4	45	1.43	10.8	31.8	137.4	0.33	5.1	0.17	0.09	0.12	1.40	0.16	26.7	0	0.1	2.6
0	7	1.04	8.4	37.8	169.4	0.22	0	0.06	0.09	0.08	1.15	0.01	12.9	0	0	2.9
0	4	0.89	7.1	31.8	0.6	0.19	0	0.11	0.01	0.08	1.10	0.03	4.8	0	0	10.5
0	14	0.55	8.9	30.3	124.4	0.30	0	0.06	0.06	0.04	0.61	0.01	18.6	0	0	5.2
0	12	0.55	5.9	29.0	134.9	0.20	0	0.07	—	0.04	0.70	0.01	12.8	0	0	5.8
0	20	0.72	—	10.0	135.0	—	—	—	—	—	—	—	—	0	—	—
0	3	0.24	7.8	31.9	26.4	0.23	0	0.02	0.08	0.01	0.10	0.02	4.7	0	0	3.7
0	4	0.65	13.3	54.5	87.3	0.30	0	0.04	0.08	0.03	0.17	0.03	5.0	0	0	2.6
0	10	0.84	3.3	23.1	160.8	0.12	0	0.01	0.14	0.06	0.78	0.01	20.9	0	0	1.5
0	18	0.81	4.1	108.6	95.4	0.11	0	0.08	0.01	0.06	0.78	0.01	18.6	0	0	2.9
1	72	0.66	10.1	120.1	392.3	0.17	4.8	0.12	0.06	0.19	1.05	0.01	28.0	0	0	6.0
0	23	0.77	15.4	60.2	201.0	0.31	0.3	0.13	0.57	0.07	1.71	0.04	24.1	0	0	3.0
0	36	1.08	8.1	39.9	254.1	0.20	0	0.12	0.60	0.10	1.21	0.01	27.0	0	0	2.0
0	36	1.08	8.1	106.5	111.9	0.20	0	0.12	0.60	0.10	1.21	0.01	27.0	0	0	2.0
0	10	0.84	3.3	23.1	160.8	0.12	0	0.01	0.14	0.06	0.78	0.01	20.9	0	0	1.5
2	57	0.73	15.1	85.7	255.6	0.24	4.8	0.10	—	0.12	0.89	0.07	17.9	0.4	0	6.8
0	48	0.74	10.6	83.2	226.0	0.23	0	0.10	—	0.08	1.64	0.03	19.6	0	0	6.1
0	15	1.32	18.6	54.9	238.5	0.48	0	0.15	0.15	0.09	1.48	0.04	35.1	0	0	1.9
0	15	1.32	18.6	60.9	84.9	0.48	0	0.15	0.15	0.09	1.48	0.04	15.0	0	0	10.1
0	14	0.86	27.7	83.2	184.5	0.60	0	0.05	0.24	0.02	1.26	0.05	7.8	0	0	4.1
14	9	0.26	2.2	22.4	6.8	0.09	7.1	0.03	0.07	0.04	0.23	0.01	6.3	0.2	0.1	2.6
0	—	0.72	—	—	340.1	—	0	—	—	—	—	—	—	—	—	—
38	21	1.15	9.1	67.3	424.1	0.42	117.4	0.22	0.47	0.13	1.24	0.03	50.2	0.1	0.1	12.9
30	61	1.17	20.8	61.8	236.0	0.56	5.9	0.14	0.53	0.15	1.49	0.06	54.0	1.1	0.1	9.2
20	21	1.55	17.0	68.0	262.7	0.68	9.4	0.28	0.24	0.12	1.90	0.05	59.5	0	0.1	9.2

TABLE A-1 Table of Food Composition (continued) (Computer code is for Cengage Diet Analysis program) (For purposes of calculations, use "0" for t, <1, <.1, <.01, etc.)

DA+ Code	Food Description	Quantity	Measure	Wt (g)	H₂0 (g)	Ener (kcal)	Prot (g)	Carb (g)	Fiber (g)	Fat (g)	Fat Breakdown (g) Sat	Mono	Poly	Trans
Breads, Baked Goods, Cakes, Cookies, Crackers, Chips, Pies—continued														
9117	Doughnut with jelly filling	1	item(s)	85	30.3	289	5.0	33.2	0.8	15.9	4.1	8.7	2.0	—
4947	Doughnut, cake	1	item(s)	47	9.8	198	2.4	23.4	0.7	10.8	1.7	4.4	3.7	—
9105	Doughnut, cake, chocolate glazed	1	item(s)	42	6.8	175	1.9	24.1	0.9	8.4	2.2	4.7	1.0	—
437	Doughnut, glazed	1	item(s)	60	15.2	242	3.8	26.6	0.7	13.7	3.5	7.7	1.7	—
10617	Toaster pastry, brown sugar cinnamon	1	item(s)	50	5.3	210	3.0	35.0	1.0	6.0	1.0	4.0	1.0	—
30928	Toaster pastry, cream cheese	1	item(s)	54	—	200	3.0	23.0	0	11.0	4.5	—	—	1.5
	Muffins														
25015	Blueberry	1	item(s)	63	29.7	160	3.4	23.0	0.8	6.0	0.9	1.5	3.3	—
9189	Corn, ready to eat	1	item(s)	57	18.6	174	3.4	29.0	1.9	4.8	0.8	1.2	1.8	—
9121	English muffin, plain, enriched	1	item(s)	57	24.0	134	4.4	26.2	1.5	1.0	0.1	0.2	0.5	—
29582	English muffin, toasted	1	item(s)	50	18.6	128	4.2	25.0	1.5	1.0	0.1	0.2	0.5	—
9145	English muffin, wheat	1	item(s)	57	24.1	127	5.0	25.5	2.6	1.1	0.2	0.2	0.5	—
8894	Oat bran	1	item(s)	57	20.0	154	4.0	27.5	2.6	4.2	0.6	1.0	2.4	—
	Granola bars														
38161	Kudos milk chocolate granola bars w/fruit and nuts	1	item(s)	28	—	90	2.0	15.0	1.0	3.0	1.0	—	—	—
38196	Nature Valley banana nut crunchy granola bars	2	item(s)	42	—	190	4.0	28.0	2.0	7.0	1.0	—	—	—
38187	Nature Valley fruit 'n' nut trail mix bar	1	item(s)	35	—	140	3.0	25.0	2.0	4.0	0.5	—	—	—
1383	Plain, hard	1	item(s)	25	1.0	115	2.5	15.8	1.3	4.9	0.6	1.1	3.0	—
4606	Plain, soft	1	item(s)	28	1.8	126	2.1	19.1	1.3	4.9	2.1	1.1	1.5	—
	Pies														
454	Apple pie, prepared from home recipe	1	slice(s)	155	73.3	411	3.7	57.5	2.3	19.4	4.7	8.4	5.2	—
470	Pecan pie, prepared from home recipe	1	slice(s)	122	23.8	503	6.0	63.7	—	27.1	4.9	13.6	7.0	—
33356	Pie crust mix, prepared, baked	1	slice(s)	20	2.1	100	1.3	10.1	0.4	6.1	1.5	3.5	0.8	—
9007	Pie crust, ready to bake, frozen, enriched, baked	1	slice(s)	16	1.8	82	0.7	7.9	0.2	5.2	1.7	2.5	0.6	—
472	Pumpkin pie, prepared from home recipe	1	slice(s)	155	90.7	316	7.0	40.9	—	14.4	4.9	5.7	2.8	—
	Rolls														
8555	Crescent dinner roll	1	item(s)	28	9.7	78	2.7	13.8	0.6	1.2	0.3	0.3	0.6	—
489	Hamburger roll or bun, plain	1	item(s)	43	14.9	120	4.1	21.3	0.9	1.9	0.5	0.5	0.8	—
490	Hard roll	1	item(s)	57	17.7	167	5.6	30.0	1.3	2.5	0.3	0.6	1.0	—
5127	Kaiser roll	1	item(s)	57	17.7	167	5.6	30.0	1.3	2.5	0.3	0.6	1.0	—
5130	Whole wheat roll or bun	1	item(s)	28	9.4	75	2.5	14.5	2.1	1.3	0.2	0.3	0.6	—
	Sports bars														
37026	Balance original chocolate bar	1	item(s)	50	—	200	14.0	22.0	0.5	6.0	3.5	—	—	—
37024	Balance original peanut butter bar	1	item(s)	50	—	200	14.0	22.0	1.0	6.0	2.5	—	—	—
36580	Clif Bar chocolate brownie energy bar	1	item(s)	68	—	240	10.0	45.0	5.0	4.5	1.5	—	—	0
36583	Clif Bar crunchy peanut butter energy bar	1	item(s)	68	—	250	12.0	40.0	5.0	6.0	1.5	—	—	0
36589	Clif Luna Nutz over Chocolate energy bar	1	item(s)	48	—	180	10.0	25.0	3.0	4.5	2.5	—	—	0
12005	PowerBar apple cinnamon	1	item(s)	65	—	230	9.0	45.0	3.0	2.5	0.5	1.5	0.5	0
16078	PowerBar banana	1	item(s)	65	—	230	9.0	45.0	3.0	2.5	0.5	1.0	0.5	0
16080	PowerBar chocolate	1	item(s)	65	6.4	230	10.0	45.0	3.0	2.0	0.5	0.5	1.0	0
29092	PowerBar peanut butter	1	item(s)	65	—	240	10.0	45.0	3.0	3.5	0.5	—	—	0
	Tortillas														
1391	Corn tortillas, soft	1	item(s)	26	11.9	57	1.5	11.6	1.6	0.7	0.1	0.2	0.4	—
1669	Flour tortilla	1	item(s)	32	9.7	100	2.7	16.4	1.0	2.5	0.6	1.2	0.5	—
	Pancakes, waffles														
8926	Pancakes, blueberry, prepared from recipe	3	item(s)	114	60.6	253	7.0	33.1	0.8	10.5	2.3	2.6	4.7	—
5037	Pancakes, prepared from mix with egg and milk	3	item(s)	114	60.3	249	8.9	32.9	2.1	8.8	2.3	2.4	3.3	—
1390	Taco shells, hard	1	item(s)	13	1.0	62	0.9	8.3	0.6	2.8	0.6	1.6	0.5	0.6
30311	Waffle, 100% whole grain	1	item(s)	75	32.3	200	6.9	25.0	1.9	8.4	2.3	3.3	2.1	—
9219	Waffle, plain, frozen, toasted	2	item(s)	66	20.2	206	4.7	32.5	1.6	6.3	1.1	3.2	1.5	—
500	Waffle, plain, prepared from recipe	1	item(s)	75	31.5	218	5.9	24.7	1.7	10.6	2.1	2.6	5.1	—

PAGE KEY: A-6 = Breads/Baked Goods A-12 = Cereal/Rice/Pasta A-16 = Fruit A-20 = Vegetables/Legumes A-30 = Nuts/Seeds A-32 = Vegetarian A-34 = Dairy A-42 = Eggs A-42 = Seafood A-44 = Meats A-48 = Poultry
A-50 = Processed Meats A-50 = Beverages A-54 = Fats/Oils A-56 = Sweets A-58 = Spices/Condiments/Sauces A-62 = Mixed Foods/Soups/Sandwiches A-66 = Fast Food A-86 = Convenience A-88 = Baby Foods

Chol (mg)	Calc (mg)	Iron (mg)	Magn (mg)	Pota (mg)	Sodi (mg)	Zinc (mg)	Vit A (μg)	Thia (mg)	Vit E (mg α)	Ribo (mg)	Niac (mg)	Vit B_6 (mg)	Fola (μg)	Vit C (mg)	Vit B_{12} (μg)	Sele (μg)
22	21	1.49	17.0	67.2	249.1	0.63	14.5	0.26	0.36	0.12	1.81	0.08	57.8	0	0.2	10.6
17	21	0.91	9.4	59.7	256.6	0.25	17.9	0.10	0.90	0.11	0.87	0.02	24.4	0.1	0.1	4.4
24	89	0.95	14.3	44.5	142.8	0.23	5.0	0.01	0.08	0.02	0.19	0.01	18.9	0	0	1.7
4	26	0.36	13.2	64.8	205.2	0.46	2.4	0.53	—	0.04	0.39	0.03	13.2	0.1	0.1	5.0
0	0	1.80	—	70.0	190.0	—	—	0.15	—	0.17	2.00	0.20	40.0	0	0	—
10	100	1.80	—	—	220.0	—	—	0.15	—	0.17	2.00	—	40.0	0	0.6	—
20	56	1.02	7.8	70.2	289.4	0.28	—	0.17	0.75	0.15	1.25	0.02	34.0	0.4	0.1	8.8
15	42	1.60	18.2	39.3	297.0	0.30	29.6	0.15	0.45	0.18	1.16	0.04	45.6	0	0.1	8.7
0	30	1.42	12.0	74.7	264.5	0.39	0	0.25	—	0.16	2.21	0.02	42.2	0	0	—
0	95	1.36	11.0	71.5	252.0	0.38	0	0.19	0.16	0.14	1.90	0.02	43.5	0.1	0	13.5
0	101	1.63	21.1	106.0	217.7	0.61	0	0.24	0.25	0.16	1.91	0.05	36.5	0	0	16.6
0	36	2.39	89.5	289.0	224.0	1.04	0	0.14	0.37	0.05	0.23	0.09	50.7	0	0	6.3
0	200	0.36	—	—	60.0	—	0	—	—	—	—	—	—	0	0	—
0	20	1.08	—	120.0	160.0	—	0	—	—	—	—	—	—	0	—	—
0	0	0.00	—	—	95.0	—	0	—	—	—	—	—	—	0	—	—
0	15	0.72	23.8	82.3	72.0	0.50	0	0.06	—	0.03	0.39	0.02	5.6	0.2	0	4.0
0	30	0.72	21.0	92.3	79.0	0.42	0	0.08	—	0.04	0.14	0.02	6.8	0	0.1	4.6
0	11	1.73	10.9	122.5	327.1	0.29	17.1	0.22	—	0.16	1.90	0.05	37.2	2.6	0	12.1
106	39	1.80	31.7	162.3	319.6	1.24	100.0	0.22	—	0.22	1.03	0.07	31.7	0.2	0.2	14.6
0	12	0.43	3.0	12.4	145.8	0.07	0	0.06	—	0.03	0.47	0.01	14.0	0	0	4.4
0	3	0.36	2.9	17.6	103.5	0.05	0	0.04	0.42	0.06	0.39	0.01	8.8	0	0	0.5
65	146	1.96	29.5	288.3	348.8	0.71	660.3	0.14	—	0.31	1.21	0.07	32.6	2.6	0.1	11.0
0	39	0.93	5.9	26.3	134.1	0.18	0	0.11	0.02	0.09	1.16	0.02	31.1	0	0.1	5.5
0	59	1.42	9.0	40.4	206.0	0.28	0	0.17	0.03	0.13	1.78	0.03	47.7	0	0.1	8.4
0	54	1.87	15.4	61.6	310.1	0.53	0	0.27	0.23	0.19	2.41	0.02	54.2	0	0	22.3
0	54	1.86	15.4	61.6	310.1	0.53	0	0.27	0.23	0.19	2.41	0.01	54.2	0	0	22.3
0	30	0.69	24.1	77.1	135.5	0.57	0	0.07	0.26	0.04	1.04	0.06	8.5	0	0	14.0
3	100	4.50	40.0	160.0	180.0	3.75	—	0.37	—	0.42	5.00	0.50	100.0	60.0	1.5	17.5
3	100	4.50	40.0	130.0	230.0	3.75	—	0.37	—	0.42	5.00	0.50	100.0	60.0	1.5	17.5
0	250	4.50	100.0	370.0	150.0	3.00	—	0.37	—	0.25	3.00	0.40	80.0	60.0	0.9	14.0
0	250	4.50	100.0	230.0	250.0	3.00	—	0.37	—	0.25	3.00	0.40	80.0	60.0	0.9	14.0
0	350	5.40	80.0	190.0	190.0	5.25	—	1.20	—	1.36	16.00	2.00	400.0	60.0	6.0	24.5
0	300	6.30	140.0	125.0	100.0	5.25	—	1.50	—	1.70	20.00	2.00	400.0	60.0	6.0	—
0	300	6.30	140.0	190.0	100.0	5.25	0	1.50	—	1.70	20.00	2.00	400.0	60.0	6.0	—
0	300	6.30	140.0	200.0	95.0	5.25	0	1.50	—	1.70	20.00	2.00	400.0	60.0	6.0	5.1
0	300	6.30	140.0	130.0	120.0	5.25	0	1.50	—	1.70	20.00	2.00	400.0	60.0	6.0	—
0	21	0.32	18.7	48.4	11.7	0.34	0	0.02	0.07	0.02	0.39	0.06	1.3	0	0	1.6
0	41	1.06	7.0	49.6	203.5	0.17	0	0.17	0.06	0.08	1.14	0.01	33.3	0	0	7.1
64	235	1.96	18.2	157.3	469.7	0.61	57.0	0.22	—	0.31	1.73	0.05	41.0	2.5	0.2	16.0
81	245	1.48	25.1	226.9	575.7	0.85	82.1	0.22	—	0.35	1.40	0.12	104.6	0.7	0.4	—
0	13	0.25	11.3	29.7	51.7	0.21	0.1	0.03	0.09	0.01	0.25	0.03	9.2	0	0	0.6
71	194	1.60	28.5	171.0	371.3	0.87	48.8	0.15	0.32	0.25	1.47	0.08	28.5	0	0.4	20.0
10	203	4.56	15.8	95.0	481.8	0.35	262.7	0.34	0.64	0.46	5.86	0.68	49.5	0	1.9	8.3
52	191	1.73	14.3	119.3	383.3	0.51	48.8	0.19	—	0.26	1.55	0.04	34.5	0.3	0.2	34.7

TABLE A-1 **Table of Food Composition** *(continued)*   (Computer code is for Cengage Diet Analysis program)   (For purposes of calculations, use "0" for t, <1, <.1, <.01, etc.)

DA+ Code	Food Description	Quantity	Measure	Wt (g)	H₂O (g)	Ener (kcal)	Prot (g)	Carb (g)	Fiber (g)	Fat (g)	Fat Breakdown (g)			
											Sat	Mono	Poly	*Trans*
Cereal, Flour, Grain, Pasta, Noodles, Popcorn														
	Grain													
2861	Amaranth, dry	½	cup(s)	98	9.6	365	14.1	64.5	9.1	6.3	1.6	1.4	2.8	—
1953	Barley, pearled, cooked	½	cup(s)	79	54.0	97	1.8	22.2	3.0	0.3	0.1	0	0.2	—
1956	Buckwheat groats, cooked, roasted	½	cup(s)	84	63.5	77	2.8	16.8	2.3	0.5	0.1	0.2	0.2	—
1957	Bulgur, cooked	½	cup(s)	91	70.8	76	2.8	16.9	4.1	0.2	0	0	0.1	—
1963	Couscous, cooked	½	cup(s)	79	57.0	88	3.0	18.2	1.1	0.1	0	0	0.1	—
1967	Millet, cooked	½	cup(s)	120	85.7	143	4.2	28.4	1.6	1.2	0.2	0.2	0.6	—
1969	Oat bran, dry	½	cup(s)	47	3.1	116	8.1	31.1	7.2	3.3	0.6	1.1	1.3	—
1972	Quinoa, dry	½	cup(s)	85	11.3	313	12.0	54.5	5.9	5.2	0.6	1.4	2.8	—
	Rice													
129	Brown, long grain, cooked	½	cup(s)	98	71.3	108	2.5	22.4	1.8	0.9	0.2	0.3	0.3	—
2863	Brown, medium grain, cooked	½	cup(s)	98	71.1	109	2.3	22.9	1.8	0.8	0.2	0.3	0.3	—
37488	Jasmine, saffroned, cooked	½	cup(s)	280	—	340	8.0	78.0	0	0	0	0	0	0
30280	Pilaf, cooked	½	cup(s)	103	74.0	129	2.1	22.2	0.6	3.3	0.6	1.5	1.0	—
28066	Spanish, cooked	½	cup(s)	244	184.2	241	5.7	50.2	3.3	1.9	0.4	0.6	0.7	0
2867	White glutinous, cooked	½	cup(s)	87	66.7	84	1.8	18.3	0.9	0.2	0	0.1	0.1	—
484	White, long grain, boiled	½	cup(s)	79	54.1	103	2.1	22.3	0.3	0.2	0.1	0.1	0.1	—
482	White, long grain, enriched, instant, boiled	½	cup(s)	83	59.4	97	1.8	20.7	0.5	0.4	0	0.1	0	—
486	White, long grain, enriched, par-boiled, cooked	½	cup(s)	79	55.6	97	2.3	20.6	0.7	0.3	0.1	0.1	0.1	—
1194	Wild brown, cooked	½	cup(s)	82	60.6	83	3.3	17.5	1.5	0.3	0	0	0.2	—
	Flour and grain fractions													
505	All purpose flour, self-rising, enriched	½	cup(s)	63	6.6	221	6.2	46.4	1.7	0.6	0.1	0	0.2	—
503	All purpose flour, white, bleached, enriched	½	cup(s)	63	7.4	228	6.4	47.7	1.7	0.6	0.1	0	0.2	—
1643	Barley flour	½	cup(s)	56	5.5	198	4.2	44.7	2.1	0.8	0.2	0.1	0.4	—
383	Buckwheat flour, whole groat	½	cup(s)	60	6.7	201	7.6	42.3	6.0	1.9	0.4	0.6	0.6	—
504	Cake wheat flour, enriched	½	cup(s)	69	8.6	248	5.6	53.5	1.2	0.6	0.1	0.1	0.3	—
426	Cornmeal, degermed, enriched	½	cup(s)	69	7.8	255	5.0	54.6	2.8	1.2	0.1	0.2	0.5	0
424	Cornmeal, yellow whole grain	½	cup(s)	61	6.2	221	4.9	46.9	4.4	2.2	0.3	0.6	1.0	—
1978	Dark rye flour	½	cup(s)	64	7.1	207	9.0	44.0	14.5	1.7	0.2	0.2	0.8	—
1644	Masa corn flour, enriched	½	cup(s)	57	5.1	208	5.3	43.5	5.5	2.1	0.3	0.6	1.0	—
1976	Rice flour, brown	½	cup(s)	79	9.4	287	5.7	60.4	3.6	2.2	0.4	0.8	0.8	—
1645	Rice flour, white	½	cup(s)	79	9.4	289	4.7	63.3	1.9	1.1	0.3	0.3	0.3	—
1980	Semolina, enriched	½	cup(s)	84	10.6	301	10.6	60.8	3.2	0.9	0.1	0.1	0.4	—
2827	Soy flour, raw	½	cup(s)	42	2.2	185	14.7	14.9	4.1	8.8	1.3	1.9	4.9	—
1990	Wheat germ, crude	2	tablespoon(s)	14	1.6	52	3.3	7.4	1.9	1.4	0.2	0.2	0.9	—
506	Whole wheat flour	½	cup(s)	60	6.2	203	8.2	43.5	7.3	1.1	0.2	0.1	0.5	—
	Breakfast bars													
39230	Atkins Morning Start apple crisp breakfast bar	1	item(s)	37	—	170	11.0	12.0	6.0	9.0	4.0	—	—	—
10571	Nutri-Grain apple cinnamon cereal bar	1	item(s)	37	—	140	2.0	27.0	1.0	3.0	0.5	2.0	0.5	—
10647	Nutri-Grain blueberry cereal bar	1	item(s)	37	5.4	140	2.0	27.0	1.0	3.0	0.5	2.0	0.5	—
10648	Nutri-Grain raspberry cereal bar	1	item(s)	37	5.4	140	2.0	27.0	1.0	3.0	0.5	2.0	0.5	—
10649	Nutri-Grain strawberry cereal bar	1	item(s)	37	5.4	140	2.0	27.0	1.0	3.0	0.5	2.0	0.5	—
	Breakfast cereals, hot													
1260	Cream of Wheat, instant, prepared	½	cup(s)	121	—	388	12.9	73.3	4.3	0	0	0	0	0
365	Farina, enriched, cooked w/water and salt	½	cup(s)	117	102.4	56	1.7	12.2	0.3	0.1	0	0	0	—
363	Grits, white corn, regular and quick, enriched, cooked w/water and salt	½	cup(s)	121	103.3	71	1.7	15.6	0.4	0.2	0	0.1	0.1	—
8636	Grits, yellow corn, regular and quick, enriched, cooked w/salt	½	cup(s)	121	103.3	71	1.7	15.6	0.4	0.2	0	0.1	0.1	—
8657	Oatmeal, cooked w/water	½	cup(s)	117	97.8	83	3.0	14.0	2.0	1.8	0.4	0.5	0.7	0
5500	Oatmeal, maple and brown sugar, instant, prepared	1	item(s)	198	150.2	200	4.8	40.4	2.4	2.2	0.4	0.7	0.8	—
5510	Oatmeal, ready to serve, packet, prepared	1	item(s)	186	158.7	112	4.1	19.8	2.7	2.0	0.4	0.7	0.8	—
	Breakfast cereals, ready to eat													
1197	All-Bran	1	cup(s)	62	1.3	160	8.1	46.0	18.2	2.0	0.4	0.4	1.3	0
1200	All-Bran Buds	1	cup(s)	91	2.7	212	6.4	72.7	39.1	1.9	0.4	0.5	1.2	0

PAGE KEY: A-6 = Breads/Baked Goods A-12 = Cereal/Rice/Pasta A-16 = Fruit A-20 = Vegetables/Legumes A-30 = Nuts/Seeds A-32 = Vegetarian A-34 = Dairy A-42 = Eggs A-42 = Seafood A-44 = Meats A-48 = Poultry
A-50 = Processed Meats A-50 = Beverages A-54 = Fats/Oils A-56 = Sweets A-58 = Spices/Condiments/Sauces A-62 = Mixed Foods/Soups/Sandwiches A-66 = Fast Food A-86 = Convenience A-88 = Baby Foods

Chol (mg)	Calc (mg)	Iron (mg)	Magn (mg)	Pota (mg)	Sodi (mg)	Zinc (mg)	Vit A (µg)	Thia (mg)	Vit E (mg α)	Ribo (mg)	Niac (mg)	Vit B$_6$ (mg)	Fola (µg)	Vit C (mg)	Vit B$_{12}$ (µg)	Sele (µg)
0	149	7.40	259.3	356.8	20.5	3.10	0	0.06	—	0.20	1.24	0.20	47.8	4.1	0	—
0	9	1.04	17.3	73.0	2.4	0.64	0	0.06	0.01	0.04	1.61	0.09	12.6	0	0	6.8
0	6	0.67	42.8	73.9	3.4	0.51	0	0.03	0.07	0.03	0.79	0.06	11.8	0	0	1.8
0	9	0.87	29.1	61.9	4.6	0.51	0	0.05	0.01	0.02	0.91	0.07	16.4	0	0	0.5
0	6	0.30	6.3	45.5	3.9	0.20	0	0.05	0.10	0.02	0.77	0.04	11.8	0	0	21.6
0	4	0.75	52.8	74.4	2.4	1.09	0	0.12	0.02	0.09	1.59	0.13	22.8	0	0	1.1
0	27	2.54	110.5	266.0	1.9	1.46	0	0.55	0.47	0.10	0.43	0.07	24.4	0	0	21.2
0	40	3.88	167.4	478.5	4.2	2.62	0.8	0.30	2.06	0.26	1.28	0.40	156.4	0	0	7.2
0	10	0.41	41.9	41.9	4.9	0.61	0	0.09	0.02	0.02	1.49	0.14	3.9	0	0	9.6
0	10	0.51	42.9	77.0	1.0	0.60	0	0.09	—	0.01	1.29	0.14	3.9	0	0	38.0
0	—	2.16	—	—	780.0	—	—	—	—	—	—	—	—	—	—	—
0	11	1.16	9.3	54.6	390.4	0.37	33.0	0.13	0.28	0.02	1.23	0.06	44.3	0.4	0	4.3
0	37	1.52	95.4	330.5	97.1	1.40	—	0.27	0.12	0.05	3.24	0.38	19.0	22.6	0	14.3
0	2	0.12	4.4	8.7	4.4	0.35	0	0.01	0.03	0.01	0.25	0.02	0.9	0	0	4.9
0	8	0.94	9.5	27.7	0.8	0.38	0	0.12	0.03	0.01	1.16	0.07	45.8	0	0	5.9
0	7	1.46	4.1	7.4	3.3	0.40	0	0.06	0.01	0.01	1.43	0.04	57.8	0	0	4.0
0	15	1.43	7.1	44.2	1.6	0.29	0	0.16	0.01	0.01	1.82	0.12	64.0	0	0	7.3
0	2	0.49	26.2	82.8	2.5	1.09	0	0.04	0.19	0.07	1.05	0.11	21.3	0	0	0.7
0	211	2.90	11.9	77.5	793.7	0.38	0	0.42	0.02	0.24	3.64	0.02	122.5	0	0	21.5
0	9	2.90	13.7	66.9	1.2	0.42	0	0.48	0.02	0.30	3.68	0.02	114.4	0	0	21.2
0	16	0.70	45.4	185.9	4.5	1.04	0	0.06	—	0.02	2.56	0.14	12.9	0	0	2.0
0	25	2.42	150.6	346.2	6.6	1.86	0	0.24	0.18	0.10	3.68	0.34	32.4	0	0	3.4
0	10	5.01	11.0	71.9	1.4	0.42	0	0.61	0.01	0.29	4.65	0.02	127.4	0	0	3.4
0	2	2.98	24.1	104.9	4.8	0.48	7.6	0.42	0.10	0.28	3.66	0.12	148.3	0	0	8.0
0	4	2.10	77.5	175.1	21.3	1.10	6.7	0.22	0.24	0.12	2.20	0.18	15.2	0	0	9.4
0	36	4.12	158.7	467.2	0.6	3.58	0.6	0.20	0.90	0.16	2.72	0.28	38.4	0	0	22.8
0	80	4.10	62.7	169.9	2.8	1.00	0	0.80	0.08	0.42	5.60	0.20	132.8	0	0	8.5
0	9	1.56	88.5	228.3	6.3	1.92	0	0.34	0.94	0.06	5.00	0.58	12.6	0	0	—
0	8	0.26	27.6	60.0	0	0.62	0	0.10	0.08	0.02	2.04	0.34	3.2	0	0	11.9
0	14	3.64	39.2	155.3	0.8	0.86	0	0.66	0.20	0.46	5.00	0.08	152.8	0	0	74.6
0	87	2.70	182.0	1067.0	5.5	1.65	2.5	0.24	0.82	0.48	1.83	0.18	146.4	0	0	3.2
0	6	0.90	34.4	128.2	1.7	1.76	0	0.27	—	0.07	0.97	0.18	40.4	0	0	11.4
0	20	2.32	82.8	243.0	3.0	1.74	0	0.26	0.48	0.12	3.82	0.20	26.4	0	0	42.4
0	200	—	—	90.0	70.0	—	—	0.22	—	0.25	3.00	—	—	9.0	—	—
0	200	1.80	8.0	75.0	110.0	1.50	—	0.37	—	0.42	5.00	0.50	40.0	0	—	—
0	200	1.80	8.0	75.0	110.0	1.50	—	0.37	—	0.42	5.00	0.50	40.0	0	—	—
0	200	1.80	8.0	70.0	110.0	1.50	—	0.37	—	0.42	5.00	0.50	40.0	0	—	—
0	200	1.80	8.0	55.0	110.0	1.50	—	0.37	—	0.42	5.00	0.50	40.0	0	—	—
0	862	34.91	21.4	150.8	732.6	0.86	—	1.59	—	1.47	21.55	2.15	431.0	0	0	—
0	5	0.58	2.3	15.1	383.3	0.09	0	0.07	0.01	0.05	0.57	0.01	39.6	0	0	10.6
0	4	0.73	6.1	25.4	269.8	0.08	0	0.10	0.02	0.07	0.87	0.03	39.9	0	0	3.8
0	4	0.73	6.1	25.4	269.8	0.08	2.4	0.10	0.02	0.07	0.87	0.03	39.9	0	0	3.3
0	11	1.05	31.6	81.9	4.7	1.17	0	0.09	0.09	0.02	0.26	0.01	7.0	0	0	6.3
0	26	6.83	49.9	126.4	403.5	1.03	0	1.02	—	0.05	1.56	0.30	42.2	0	0	11.1
0	21	3.96	44.7	112.4	240.9	0.92	0	0.60	—	0.04	0.77	0.18	18.7	0	0	3.8
0	241	10.90	224.4	632.4	150.0	3.00	300.1	1.40	—	1.68	9.16	7.44	800.0	12.4	12.0	5.8
0	57	13.64	186.4	909.1	614.5	4.55	464.5	1.09	1.42	1.27	15.45	6.09	1222.7	18.2	18.2	26.3

TABLE A-1 **Table of Food Composition (*continued*)** (Computer code is for Cengage Diet Analysis program) (For purposes of calculations, use "0" for t, <1, <.1, <.01, etc.)

DA+ Code	Food Description	Quantity	Measure	Wt (g)	H₂O (g)	Ener (kcal)	Prot (g)	Carb (g)	Fiber (g)	Fat (g)	Fat Breakdown (g) Sat	Mono	Poly	*Trans*
Cereal, Flour, Grain, Pasta, Noodles, Popcorn—continued														
1199	Apple Jacks	1	cup(s)	33	0.9	130	1.0	30.0	0.5	0.5	0	—	—	0
1204	Cap'n Crunch	1	cup(s)	36	0.9	147	1.3	30.7	1.3	2.0	0.5	0.4	0.3	—
1205	Cap'n Crunch Crunchberries	1	cup(s)	35	0.9	133	1.3	29.3	1.3	2.0	0.5	0.4	0.3	—
1206	Cheerios	1	cup(s)	30	1.0	110	3.0	22.0	3.0	2.0	0	0.5	0.5	—
3415	Cocoa Puffs	1	cup(s)	30	0.6	120	1.0	26.0	0.2	1.0	—	—	—	—
1207	Cocoa Rice Krispies	1	cup(s)	41	1.0	160	1.3	36.0	1.3	1.3	0.7	0	0	—
5522	Complete wheat bran flakes	1	cup(s)	39	1.4	120	4.0	30.7	6.7	0.7	—	—	—	0
1211	Corn Flakes	1	cup(s)	28	0.9	100	2.0	24.0	1.0	0	0	0	0	0
1247	Corn Pops	1	cup(s)	31	0.9	120	1.0	28.0	0.3	0	0	0	0	0
1937	Cracklin' Oat Bran	1	cup(s)	65	2.3	267	5.3	46.7	8.0	9.3	4.0	4.7	1.3	0
1220	Froot Loops	1	cup(s)	32	0.8	120	1.0	28.0	1.0	1.0	0.5	0	0	—
38214	Frosted Cheerios	1	cup(s)	37	—	149	2.5	31.1	1.2	1.2	—	—	—	—
372	Frosted Flakes	1	cup(s)	41	1.1	160	1.3	37.3	1.3	0	0	0	0	0
38215	Frosted Mini Chex	1	cup(s)	40	—	147	1.3	36.0	0	0	0	0	0	0
10268	Frosted Mini-Wheats	1	cup(s)	59	3.1	208	5.8	47.4	5.8	1.2	0	0	0.6	0
38216	Frosted Wheaties	1	cup(s)	40	—	147	1.3	36.0	1.3	0	0	0	0	0
1223	Granola, prepared	½	cup(s)	61	3.3	298	9.1	32.5	5.5	14.7	2.5	5.8	5.6	0
2415	Honey Bunches of Oats honey roasted	1	cup(s)	40	0.9	160	2.7	33.3	1.3	2.0	0.7	1.2	0.1	—
1227	Honey Nut Cheerios	1	cup(s)	37	0.9	149	3.7	29.9	2.5	1.9	0	0.6	0.6	—
2424	Honeycomb	1	cup(s)	22	0.3	83	1.5	19.5	0.8	0.4	0	—	—	—
10286	Kashi whole grain puffs	1	cup(s)	19	—	70	2.0	15.0	1.0	0.5	0	—	—	0
41142	Kellogg's Mueslix	1	cup(s)	83	7.2	298	7.6	60.8	6.1	4.6	0.7	2.4	1.5	0
1231	Kix	1	cup(s)	24	0.5	96	1.6	20.8	0.8	0.4	—	—	—	—
30569	Life	1	cup(s)	43	1.7	160	4.0	33.3	2.7	2.0	0.3	0.6	0.6	—
1233	Lucky Charms	1	cup(s)	24	0.6	96	1.6	20.0	0.8	0.8	—	—	—	—
38220	Multi-Grain Cheerios	1	cup(s)	30	—	110	3.0	24.0	3.0	1.0	—	—	—	—
1201	Multi-Bran Chex	1	cup(s)	63	1.3	216	4.3	52.9	8.6	1.6	0	0	0.5	0
13633	Post Bran Flakes	1	cup(s)	40	1.5	133	4.0	32.0	6.7	0.7	0	—	—	—
1241	Product 19	1	cup(s)	30	1.0	100	2.0	25.0	1.0	0	0	0	0	0
32432	Puffed rice, fortified	1	cup(s)	14	0.4	56	0.9	12.6	0.2	0.1	0	—	—	—
32433	Puffed wheat, fortified	1	cup(s)	12	0.4	44	1.8	9.6	0.5	0.1	0	—	—	—
13334	Quaker 100% natural granola oats and honey	½	cup(s)	48	—	220	5.0	31.0	3.0	9.0	3.8	4.1	1.2	—
13335	Quaker 100% natural granola oats, honey, and raisins	½	cup(s)	51	—	230	5.0	34.0	3.0	9.0	3.6	3.8	1.1	—
2420	Raisin Bran	1	cup(s)	59	5.0	190	4.0	46.0	8.0	1.0	0	0.1	0.4	—
1244	Rice Chex	1	cup(s)	31	0.8	120	2.0	27.0	0.3	0	0	0	0	0
1245	Rice Krispies	1	cup(s)	26	0.8	96	1.6	23.2	0	0	0	0	0	0
5593	Shredded Wheat	1	cup(s)	49	0.4	177	5.8	40.9	6.9	1.1	0.1	0	0.2	0
1248	Smacks	1	cup(s)	36	1.1	133	2.7	32.0	1.3	0.7	—	—	—	—
1246	Special K	1	cup(s)	31	0.9	110	7.0	22.0	0.5	0	0	0	0	0
3428	Total corn flakes	1	cup(s)	23	0.6	83	1.5	18.0	0.6	0	0	0	0	0
1253	Total whole grain	1	cup(s)	40	1.1	147	2.7	30.7	4.0	1.3	—	—	—	—
1254	Trix	1	cup(s)	30	0.6	120	1.0	27.0	1.0	1.0	—	—	—	—
382	Wheat germ, toasted	2	tablespoon(s)	14	0.8	54	4.1	7.0	2.1	1.5	0.3	0.2	0.9	—
1257	Wheaties	1	cup(s)	36	1.2	132	3.6	28.8	3.6	1.2	—	—	—	—
Pasta, noodles														
449	Chinese chow mein noodles, cooked	½	cup(s)	23	0.2	119	1.9	12.9	0.9	6.9	1.0	1.7	3.9	—
1995	Corn pasta, cooked	½	cup(s)	70	47.8	88	1.8	19.5	3.4	0.5	0.1	0.1	0.2	—
448	Egg noodles, enriched, cooked	½	cup(s)	80	54.2	110	3.6	20.1	1.0	1.7	0.3	0.5	0.4	0
1563	Egg noodles, spinach, enriched, cooked	½	cup(s)	80	54.8	106	4.0	19.4	1.8	1.3	0.3	0.4	0.3	—
440	Macaroni, enriched, cooked	½	cup(s)	70	43.5	111	4.1	21.6	1.3	0.7	0.1	0.1	0.2	0
2000	Macaroni, tricolor vegetable, enriched, cooked	½	cup(s)	67	45.8	86	3.0	17.8	2.9	0.1	0	0	0	—
1996	Plain pasta, fresh-refrigerated, cooked	½	cup(s)	64	43.9	84	3.3	16.0	—	0.7	0.1	0.1	0.3	—
1725	Ramen noodles, cooked	½	cup(s)	114	94.5	104	3.0	15.4	1.0	4.3	0.2	0.2	0.2	—
2878	Soba noodles, cooked	½	cup(s)	95	69.4	94	4.8	20.4	—	0.1	0	0	0	—
2879	Somen noodles, cooked	½	cup(s)	88	59.8	115	3.5	24.2	—	0.2	0	0	0.1	—
493	Spaghetti, al dente, cooked	½	cup(s)	65	41.6	95	3.5	19.5	1.0	0.5	0.1	0.1	0.2	—
2884	Spaghetti, whole wheat, cooked	½	cup(s)	70	47.0	87	3.7	18.6	3.2	0.4	0.1	0.1	0.1	—
Popcorn														
476	Air popped	1	cup(s)	8	0.3	31	1.0	6.2	1.2	0.4	0	0.1	0.2	—
4619	Caramel	1	cup(s)	35	1.0	152	1.3	27.8	1.8	4.5	1.3	1.0	1.6	—

PAGE KEY: A-6 = Breads/Baked Goods A-12 = Cereal/Rice/Pasta A-16 = Fruit A-20 = Vegetables/Legumes A-30 = Nuts/Seeds A-32 = Vegetarian A-34 = Dairy A-42 = Eggs A-42 = Seafood A-44 = Meats A-48 = Poultry A-50 = Processed Meats A-50 = Beverages A-54 = Fats/Oils A-56 = Sweets A-58 = Spices/Condiments/Sauces A-62 = Mixed Foods/Soups/Sandwiches A-66 = Fast Food A-86 = Convenience A-88 = Baby Foods

Chol (mg)	Calc (mg)	Iron (mg)	Magn (mg)	Pota (mg)	Sodi (mg)	Zinc (mg)	Vit A (µg)	Thia (mg)	Vit E (mg α)	Ribo (mg)	Niac (mg)	Vit B$_6$ (mg)	Fola (µg)	Vit C (mg)	Vit B$_{12}$ (µg)	Sele (µg)
0	0	4.50	8.0	30.0	130.0	1.50	150.2	0.37	—	0.42	5.00	0.50	100.0	15.0	1.5	2.4
0	5	6.80	20.0	73.3	266.7	5.00	2.5	0.51	—	0.57	6.68	0.67	133.5	0	0	6.7
0	7	6.53	18.7	73.3	240.0	5.13	2.4	0.51	—	0.57	6.68	0.67	133.7	0	0	6.7
0	100	8.10	40.0	95.0	280.0	3.75	150.3	0.37	—	0.42	5.00	0.50	200.0	6.0	1.5	11.3
0	100	4.50	8.0	50.0	170.0	3.75	0	0.37	—	0.42	5.00	0.50	100.0	6.0	1.5	2.0
0	53	6.00	10.7	66.7	253.3	2.00	200.1	0.49	—	0.56	6.67	0.67	133.3	20.0	2.0	5.8
0	0	24.00	53.3	226.7	280.0	20.00	300.1	2.00	—	2.27	26.67	2.67	533.3	80.0	8.0	4.1
0	0	8.10	3.4	25.0	200.0	0.16	149.8	0.37	—	0.42	5.00	0.50	100.0	6.0	1.5	1.4
0	0	1.80	2.5	25.0	120.0	1.50	150.0	0.37	—	0.42	5.00	0.50	100.0	6.0	1.5	2.0
0	27	2.40	80.0	293.3	200.0	2.00	299.9	0.49	—	0.56	6.67	0.67	133.3	20.0	2.0	14.4
0	0	4.50	8.0	35.0	150.0	1.50	150.1	0.37	—	0.42	5.00	0.50	100.0	15.0	1.5	2.3
0	124	5.60	19.9	68.4	261.3	4.67	—	0.46	—	0.52	6.22	0.62	124.4	7.5	1.9	—
0	0	6.00	3.7	26.7	200.0	0.20	200.1	0.49	—	0.56	6.67	0.67	133.3	8.0	2.0	1.8
0	133	12.00	—	33.3	266.7	4.00	—	0.49	—	0.56	6.67	0.67	266.7	8.0	2.0	—
0	0	16.66	69.4	196.7	5.8	1.74	0	0.43	—	0.49	5.78	0.58	115.7	0	1.7	2.4
0	133	10.80	0	46.7	266.7	10.00	—	1.00	—	1.13	13.33	1.33	533.3	8.0	4.0	—
0	48	2.58	106.8	329.4	15.3	2.45	0.6	0.44	6.77	0.17	1.30	0.17	50.0	0.7	0	17.0
0	0	10.80	21.3	0	253.3	0.40	—	0.49	—	0.56	6.67	0.67	133.0	0	2.0	—
0	124	5.60	39.8	112.0	336.0	4.67	—	0.46	—	0.52	6.22	0.62	248.9	7.5	1.9	8.8
0	0	2.03	6.0	26.3	165.4	1.13	—	0.28	—	0.32	3.74	0.37	75.0	0	1.1	—
0	0	0.36	—	60.0	0	—	—	0.03	—	0.03	0.80	0.00	—	0	—	—
0	48	6.83	74.2	363.3	257.5	5.67	136.7	0.67	6.00	0.67	8.33	3.08	615.0	0.3	9.2	14.4
0	120	6.48	6.4	28.0	216.0	3.00	120.2	0.30	—	0.34	4.00	0.40	160.0	4.8	1.2	4.8
0	149	11.87	41.3	120.0	213.3	5.33	0.9	0.53	—	0.60	7.12	0.71	142.4	0	0	10.7
0	80	3.60	12.8	48.0	168.0	3.00	—	0.30	—	0.34	4.00	0.40	160.0	4.8	1.2	4.8
0	100	18.00	24.0	85.0	200.0	15.00	—	1.50	—	1.70	20.00	2.00	400.0	15.0	6.0	—
0	108	17.50	64.8	237.7	410.6	4.05	171.1	0.40	—	0.45	5.40	0.54	432.2	6.5	1.6	4.9
0	0	10.80	80.0	266.7	280.0	2.00	—	0.49	—	0.56	6.67	0.67	133.3	0	2.0	—
0	0	18.00	16.0	50.0	210.0	15.00	225.3	1.50	—	1.70	20.00	2.00	400.0	60.0	6.0	3.6
0	1	4.43	3.5	15.8	0.4	0.14	0	0.36	—	0.25	4.94	0.01	2.7	0	0	1.5
0	3	3.80	17.4	41.8	0.5	0.28	0	0.31	—	0.21	4.23	0.02	3.8	0	0	14.8
0	61	1.20	51.0	220.0	20.0	1.05	0.5	0.13	—	0.12	0.82	0.07	15.0	0.2	0.1	8.3
0	59	1.20	49.0	250.0	20.0	0.99	0.5	0.13	—	0.12	0.80	0.08	14.1	0.4	0.1	8.8
0	20	10.80	80.0	360.0	360.0	2.25	—	0.37	—	0.42	5.00	0.50	100.0	0	2.1	—
0	100	9.00	9.3	35.0	290.0	3.75	—	0.37	—	0.42	5.00	0.50	200.0	6.0	1.5	1.2
0	0	1.44	12.8	32.0	256.0	0.48	120.1	0.30	—	0.34	4.80	0.40	80.0	4.8	1.2	4.1
0	18	2.90	60.3	179.3	1.1	1.37	0	0.14	—	0.12	3.47	0.18	21.1	0	0	2.0
0	0	0.48	10.7	53.3	66.7	0.40	200.2	0.49	—	0.56	6.67	0.67	133.3	8.0	2.0	17.5
0	0	8.10	16.0	60.0	220.0	0.90	225.1	0.52	—	0.59	7.00	2.00	400.0	21.0	6.0	7.0
0	752	13.53	0	22.6	157.9	11.28	112.8	1.13	22.56	1.28	15.04	1.50	300.8	45.1	4.5	1.2
0	1333	24.00	32.0	120.0	253.3	20.00	200.4	2.00	31.32	2.27	26.67	2.67	533.3	80.0	8.0	1.9
0	100	4.50	0	15.0	190.0	3.75	150.3	0.37	—	0.42	5.00	0.50	100.0	6.0	1.5	6.0
0	6	1.28	45.2	133.8	0.6	2.35	0.7	0.23	2.25	0.11	0.79	0.13	49.7	0.8	0	9.2
0	24	9.72	38.4	126.0	264.0	9.00	180.4	0.90	—	1.02	12.00	1.20	240.0	7.2	3.6	1.7
0	5	1.06	11.7	27.0	98.8	0.31	0	0.13	0.78	0.09	1.33	0.02	20.3	0	0	9.7
0	1	0.18	25.2	21.7	0	0.44	2.1	0.04	—	0.02	0.39	0.04	4.2	0	0	2.0
23	10	1.17	16.8	30.4	4.0	0.52	4.8	0.23	0.13	0.11	1.66	0.03	67.2	0	0.1	19.1
26	15	0.87	19.2	29.6	9.6	0.50	8.0	0.19	0.46	0.09	1.17	0.09	51.2	0	0.1	17.4
0	5	0.90	12.6	30.8	0.7	0.36	0	0.19	0.04	0.10	1.18	0.03	51.1	0	0	18.5
0	7	0.33	12.7	20.8	4.0	0.30	3.4	0.08	0.14	0.04	0.72	0.02	43.6	0	0	13.3
21	4	0.73	11.5	15.4	3.8	0.36	3.8	0.13	—	0.10	0.64	0.02	41.0	0	0.1	—
18	9	0.89	8.5	34.5	414.5	0.30	—	0.08	—	0.04	0.71	0.03	4.0	0.1	0	—
0	4	0.45	8.5	33.2	57.0	0.11	0	0.09	—	0.02	0.48	0.03	6.6	0	0	—
0	7	0.45	1.8	25.5	141.7	0.19	0	0.01	—	0.03	0.08	0.01	1.8	0	0	—
0	7	1.00	12.4	51.5	0.5	0.35	0	0.12	0.04	0.07	0.90	0.04	7.8	0	0	40.0
0	11	0.74	21.0	30.8	2.1	0.57	0	0.08	0.21	0.03	0.50	0.06	3.5	0	0	18.1
0	1	0.25	11.5	26.3	0.6	0.25	0.8	0.01	0.02	0.01	0.18	0.01	2.5	0	0	0
2	15	0.61	12.3	38.4	72.5	0.20	0.7	0.02	0.42	0.02	0.77	0.01	1.8	0	0	1.3

TABLE A-1 Table of Food Composition (*continued*) (Computer code is for Cengage Diet Analysis program) (For purposes of calculations, use "0" for t, <1, <.1, <.01, etc.)

DA+ Code	Food Description	Quantity	Measure	Wt (g)	H₂0 (g)	Ener (kcal)	Prot (g)	Carb (g)	Fiber (g)	Fat (g)	Fat Breakdown (g) Sat	Mono	Poly	Trans
Cereal, Flour, Grain, Pasta, Noodles, Popcorn—continued														
4620	Cheese flavored	1	cup(s)	36	0.9	188	3.3	18.4	3.5	11.8	2.3	3.5	5.5	—
477	Popped in oil	1	cup(s)	11	0.1	64	0.8	5.0	0.9	4.8	0.8	1.1	2.6	—
Fruit and Fruit Juices														
	Apples													
952	Juice, prepared from frozen concentrate	½	cup(s)	120	105.0	56	0.2	13.8	0.1	0.1	0	0	0	—
225	Juice, unsweetened, canned	½	cup(s)	124	109.0	58	0.1	14.5	0.1	0.1	0	0	0	—
224	Slices	½	cup(s)	55	47.1	29	0.1	7.6	1.3	0.1	0	0	0	—
946	Slices without skin, boiled	½	cup(s)	86	73.1	45	0.2	11.7	2.1	0.3	0	0	0.1	—
223	Raw medium, with peel	1	item(s)	138	118.1	72	0.4	19.1	3.3	0.2	0	0	0.1	—
948	Dried, sulfured	¼	cup(s)	22	6.8	52	0.2	14.2	1.9	0.1	0	0	0	—
226	Applesauce, sweetened, canned	½	cup(s)	128	101.5	97	0.2	25.4	1.5	0.2	0	0	0.1	—
227	Applesauce, unsweetened, canned	½	cup(s)	122	107.8	52	0.2	13.8	1.5	0.1	0	0	0	—
38492	Crabapples	1	item(s)	35	27.6	27	0.1	7.0	0.9	0.1	0	0	0	—
	Apricot													
228	Fresh without pits	4	item(s)	140	120.9	67	2.0	15.6	2.8	0.5	0	0.2	0.1	—
229	Halves with skin, canned in heavy syrup	½	cup(s)	129	100.1	107	0.7	27.7	2.1	0.1	0	0	0	—
230	Halves, dried, sulfured	¼	cup(s)	33	10.1	79	1.1	20.6	2.4	0.2	0	0	0	—
	Avocado													
233	California, whole, without skin or pit	½	cup(s)	115	83.2	192	2.2	9.9	7.8	17.7	2.4	11.3	2.1	—
234	Florida, whole, without skin or pit	½	cup(s)	115	90.6	138	2.5	9.0	6.4	11.5	2.2	6.3	1.9	—
2998	Pureed	⅛	cup(s)	28	20.2	44	0.5	2.4	1.8	4.0	0.6	2.7	0.5	—
	Banana													
4580	Dried chips	¼	cup(s)	55	2.4	285	1.3	32.1	4.2	18.5	15.9	1.1	0.3	—
235	Fresh whole, without peel	1	item(s)	118	88.4	105	1.3	27.0	3.1	0.4	0.1	0	0.1	—
	Blackberries													
237	Raw	½	cup(s)	72	63.5	31	1.0	6.9	3.8	0.4	0	0	0.2	—
958	Unsweetened, frozen	½	cup(s)	76	62.1	48	0.9	11.8	3.8	0.3	0	0	0.2	—
	Blueberries													
959	Canned in heavy syrup	½	cup(s)	128	98.3	113	0.8	28.2	2.0	0.4	0	0.1	0.2	—
238	Raw	½	cup(s)	73	61.1	41	0.5	10.5	1.7	0.2	0	0	0.1	—
960	Unsweetened, frozen	½	cup(s)	78	67.1	40	0.3	9.4	2.1	0.5	0	0.1	0.2	—
	Boysenberries													
961	Canned in heavy syrup	½	cup(s)	128	97.6	113	1.3	28.6	3.3	0.2	0	0	0.1	—
962	Unsweetened, frozen	½	cup(s)	66	56.7	33	0.7	8.0	3.5	0.2	0	0	0.1	—
35576	**Breadfruit**	1	item(s)	384	271.3	396	4.1	104.1	18.8	0.9	0.2	0.1	0.3	—
	Cherries													
967	Sour red, canned in water	½	cup(s)	122	109.7	44	0.9	10.9	1.3	0.1	0	0	0	—
3000	Sour red, raw	½	cup(s)	78	66.8	39	0.8	9.4	1.2	0.2	0.1	0.1	0.1	—
3004	Sweet, canned in heavy syrup	½	cup(s)	127	98.2	105	0.8	26.9	1.9	0.2	0	0.1	0.1	—
969	Sweet, canned in water	½	cup(s)	124	107.9	57	1.0	14.6	1.9	0.2	0	0	0	—
240	Sweet, raw	½	cup(s)	73	59.6	46	0.8	11.6	1.5	0.1	0	0	0	—
	Cranberries													
3007	Chopped, raw	½	cup(s)	55	47.9	25	0.2	6.7	2.5	0.1	0	0	0	—
1717	Cranberry apple juice drink	½	cup(s)	123	102.6	77	0	19.4	0	0.1	0	0	0.1	—
1638	Cranberry juice cocktail	½	cup(s)	127	109.0	68	0	17.1	0	0.1	0	0	0.1	—
241	Cranberry juice cocktail, low calorie, with saccharin	½	cup(s)	119	112.8	23	0	5.5	0	0	0	0	0	—
242	Cranberry sauce, sweetened, canned	¼	cup(s)	69	42.0	105	0.1	26.9	0.7	0.1	0	0	0	—
	Dates													
244	Domestic, chopped	¼	cup(s)	45	9.1	125	1.1	33.4	3.6	0.2	0	0	0	—
243	Domestic, whole	¼	cup(s)	45	9.1	125	1.1	33.4	3.6	0.2	0	0	0	—
	Figs													
975	Canned in heavy syrup	½	cup(s)	130	98.8	114	0.5	29.7	2.8	0.1	0	0	0.1	—
974	Canned in water	½	cup(s)	124	105.7	66	0.5	17.3	2.7	0.1	0	0	0.1	—
973	Raw, medium	2	item(s)	100	79.1	74	0.7	19.2	2.9	0.3	0.1	0.1	0.1	—
	Fruit cocktail and salad													
245	Fruit cocktail, canned in heavy syrup	½	cup(s)	124	99.7	91	0.5	23.4	1.2	0.1	0	0	0	—
978	Fruit cocktail, canned in juice	½	cup(s)	119	103.6	55	0.5	14.1	1.2	0	0	0	0	—
977	Fruit cocktail, canned in water	½	cup(s)	119	107.6	38	0.5	10.1	1.2	0.1	0	0	0	—
979	Fruit salad, canned in water	½	cup(s)	123	112.1	37	0.4	9.6	1.2	0.1	0	0	0	—

PAGE KEY: A-6 = Breads/Baked Goods A-12 = Cereal/Rice/Pasta A-16 = Fruit A-20 = Vegetables/Legumes A-30 = Nuts/Seeds A-32 = Vegetarian A-34 = Dairy A-42 = Eggs A-42 = Seafood A-44 = Meats A-48 = Poultry
A-50 = Processed Meats A-50 = Beverages A-54 = Fats/Oils A-56 = Sweets A-58 = Spices/Condiments/Sauces A-62 = Mixed Foods/Soups/Sandwiches A-66 = Fast Food A-86 = Convenience A-88 = Baby Foods

Chol (mg)	Calc (mg)	Iron (mg)	Magn (mg)	Pota (mg)	Sodi (mg)	Zinc (mg)	Vit A (µg)	Thia (mg)	Vit E (mg α)	Ribo (mg)	Niac (mg)	Vit B$_6$ (mg)	Fola (µg)	Vit C (mg)	Vit B$_{12}$ (µg)	Sele (µg)
4	40	0.79	32.5	93.2	317.4	0.71	13.6	0.04	—	0.08	0.52	0.08	3.9	0.2	0.2	4.3
0	0	0.22	8.7	20.0	116.4	0.34	0.9	0.01	0.27	0.00	0.13	0.01	2.8	0	0	0.2
0	7	0.31	6.0	150.6	8.4	0.05	0	0.00	0.01	0.02	0.05	0.04	0	0.7	0	0.1
0	9	0.46	3.7	147.6	3.7	0.04	0	0.03	0.01	0.02	0.12	0.04	0	1.1	0	0.1
0	3	0.06	2.7	58.8	0.5	0.02	1.6	0.01	0.10	0.01	0.05	0.02	1.6	2.5	0	0
0	4	0.16	2.6	75.2	0.9	0.03	1.7	0.01	0.04	0.01	0.08	0.04	0.9	0.2	0	0.3
0	8	0.16	6.9	147.7	1.4	0.05	4.1	0.02	0.24	0.03	0.12	0.05	4.1	6.3	0	0
0	3	0.30	3.4	96.8	18.7	0.04	0	0.00	0.11	0.03	0.20	0.03	0	0.8	0	0.3
0	5	0.44	3.8	77.8	3.8	0.05	1.3	0.01	0.26	0.03	0.24	0.03	1.3	2.2	0	0.4
0	4	0.14	3.7	91.5	2.4	0.03	1.2	0.01	0.25	0.03	0.22	0.03	1.2	1.5	0	0.4
0	6	0.12	2.5	67.9	0.4	—	0.7	0.01	0.20	0.01	0.03	—	2.0	2.8	0	—
0	18	0.54	14.0	362.6	1.4	0.28	134.4	0.04	1.24	0.05	0.84	0.07	12.6	14.0	0	0.1
0	12	0.38	9.0	180.6	5.2	0.14	80.0	0.02	0.77	0.02	0.48	0.07	2.6	4.0	0	0.1
0	18	0.87	10.5	381.5	3.3	0.12	59.1	0.00	1.42	0.02	0.85	0.05	3.3	0.3	0	0.7
0	15	0.66	33.3	583.0	9.2	0.78	8.0	0.08	2.23	0.16	2.19	0.31	102.3	10.1	0	0.4
0	12	0.19	27.6	403.6	2.3	0.45	8.0	0.02	3.03	0.04	0.76	0.08	40.3	20.0	0	—
0	3	0.14	8.0	134.1	1.9	0.17	1.9	0.01	0.57	0.03	0.48	0.07	22.4	2.8	0	0.1
0	10	0.69	41.8	294.8	3.3	0.40	2.2	0.04	0.13	0.01	0.39	0.14	7.7	3.5	0	0.8
0	6	0.30	31.9	422.4	1.2	0.17	3.5	0.03	0.11	0.08	0.78	0.43	23.6	10.3	0	1.2
0	21	0.45	14.4	116.6	0.7	0.38	7.9	0.01	0.84	0.02	0.47	0.02	18.0	15.1	0	0.3
0	22	0.60	16.6	105.7	0.8	0.19	4.5	0.02	0.88	0.03	0.91	0.05	25.7	2.3	0	0.3
0	6	0.42	5.1	51.2	3.8	0.09	2.6	0.04	0.49	0.07	0.14	0.05	2.6	1.4	0	0.1
0	4	0.20	4.4	55.8	0.7	0.12	2.2	0.03	0.41	0.03	0.30	0.04	4.4	7.0	0	0.1
0	6	0.14	3.9	41.9	0.8	0.05	1.6	0.03	0.37	0.03	0.40	0.05	5.4	1.9	0	0.1
0	23	0.55	14.1	115.2	3.8	0.24	2.6	0.03	—	0.04	0.29	0.05	43.5	7.9	0	0.5
0	18	0.56	10.6	91.7	0.7	0.15	2.0	0.04	0.57	0.02	0.51	0.04	41.6	2.0	0	0.1
0	65	2.07	96.0	1881.6	7.7	0.46	0	0.42	0.38	0.11	3.45	0.38	53.8	111.4	0	2.3
0	13	1.67	7.3	119.6	8.5	0.09	46.4	0.02	0.28	0.05	0.22	0.05	9.8	2.6	0	0
0	12	0.25	7.0	134.1	2.3	0.08	49.6	0.02	0.05	0.03	0.31	0.03	6.2	7.8	0	0
0	11	0.44	11.4	183.4	3.8	0.12	10.1	0.02	0.29	0.05	0.50	0.03	5.1	4.6	0	0
0	14	0.45	11.2	162.4	1.2	0.10	9.9	0.03	0.29	0.05	0.51	0.04	5.0	2.7	0	0
0	9	0.26	8.0	161.0	0	0.05	2.2	0.02	0.05	0.02	0.11	0.04	2.9	5.1	0	0
0	4	0.13	3.3	46.8	1.1	0.05	1.7	0.01	0.66	0.01	0.05	0.03	0.6	7.3	0	0.1
0	4	0.09	1.2	20.8	2.5	0.02	0	0.00	0.15	0.00	0.00	0.00	0	48.4	0	0
0	4	0.13	1.3	17.7	2.5	0.04	0	0.00	0.28	0.00	0.05	0.00	0	53.5	0	0.3
0	11	0.05	2.4	29.6	3.6	0.02	0	0.00	0.06	0.00	0.00	0.00	0	38.2	0	0
0	3	0.15	2.1	18.0	20.1	0.03	1.4	0.01	0.57	0.01	0.06	0.01	0.7	1.4	0	0.2
0	17	0.45	19.1	291.9	0.9	0.12	0	0.02	0.02	0.02	0.56	0.07	8.5	0.2	0	1.3
0	17	0.45	19.1	291.9	0.9	0.12	0	0.02	0.02	0.02	0.56	0.07	8.5	0.2	0	1.3
0	35	0.36	13.0	128.2	1.3	0.14	2.6	0.03	0.16	0.05	0.55	0.09	2.6	1.3	0	0.3
0	35	0.36	12.4	127.7	1.2	0.15	2.5	0.03	0.10	0.05	0.55	0.09	2.5	1.2	0	0.1
0	35	0.36	17.0	232.0	1.0	0.14	7.0	0.06	0.10	0.04	0.40	0.10	6.0	2.0	0	0.2
0	7	0.36	6.2	109.1	7.4	0.09	12.4	0.02	0.49	0.02	0.46	0.06	3.7	2.4	0	0.6
0	9	0.25	8.3	112.6	4.7	0.11	17.8	0.01	0.47	0.02	0.48	0.06	3.6	3.2	0	0.6
0	6	0.30	8.3	111.4	4.7	0.11	15.4	0.02	0.47	0.01	0.43	0.06	3.6	2.5	0	0.6
0	9	0.37	6.1	95.6	3.7	0.10	27.0	0.02	—	0.03	0.46	0.04	3.7	2.3	0	1.0

TABLE A-1 **Table of Food Composition** *(continued)* (Computer code is for Cengage Diet Analysis program) (For purposes of calculations, use "0" for t, <1, <.1, <.01, etc.)

DA+ Code	Food Description	Quantity	Measure	Wt (g)	H₂O (g)	Ener (kcal)	Prot (g)	Carb (g)	Fiber (g)	Fat (g)	Fat Breakdown (g) Sat	Mono	Poly	Trans
Fruit and Fruit Juices—continued														
	Gooseberries													
982	Canned in light syrup	½	cup(s)	126	100.9	92	0.8	23.6	3.0	0.3	0	0	0.1	—
981	Raw	½	cup(s)	75	65.9	33	0.7	7.6	3.2	0.4	0	0	0.2	—
	Grapefruit													
251	Juice, pink, sweetened, canned	½	cup(s)	125	109.1	57	0.7	13.9	0.1	0.1	0	0	0	—
249	Juice, white	½	cup(s)	124	111.2	48	0.6	11.4	0.1	0.1	0	0	0	—
3022	Pink or red, raw	½	cup(s)	114	100.8	48	0.9	12.2	1.8	0.2	0	0	0	—
248	Sections, canned in light syrup	½	cup(s)	127	106.2	76	0.7	19.6	0.5	0.1	0	0	0	—
983	Sections, canned in water	½	cup(s)	122	109.6	44	0.7	11.2	0.5	0.1	0	0	0	—
247	White, raw	½	cup(s)	115	104.0	38	0.8	9.7	1.3	0.1	0	0	0	—
	Grapes													
255	American, slip skin	½	cup(s)	46	37.4	31	0.3	7.9	0.4	0.2	0.1	0	0	—
256	European, red or green, adherent skin	½	cup(s)	76	60.8	52	0.5	13.7	0.7	0.1	0	0	0	—
3159	Grape juice drink, canned	½	cup(s)	125	106.6	71	0	18.2	0.1	0	0	0	0	0
259	Grape juice, sweetened, with added vitamin C, prepared from frozen concentrate	½	cup(s)	125	108.6	64	0.2	15.9	0.1	0.1	0	0	0	—
3060	Raisins, seeded, packed	¼	cup(s)	41	6.8	122	1.0	32.4	2.8	0.2	0.1	0	0.1	—
987	**Guava, raw**	1	item(s)	55	44.4	37	1.4	7.9	3.0	0.5	0.2	0	0.2	—
35593	**Guavas, strawberry**	1	item(s)	6	4.8	4	0	1.0	0.3	0	0	0	0	—
3027	**Jackfruit**	½	cup(s)	83	60.4	78	1.2	19.8	1.3	0.2	0	0	0.1	—
990	**Kiwi fruit or Chinese gooseberries**	1	item(s)	76	63.1	46	0.9	11.1	2.3	0.4	0	0	0.2	—
	Lemon													
262	Juice	1	tablespoon(s)	15	13.8	4	0.1	1.3	0.1	0	0	0	0	0
993	Peel	1	teaspoon(s)	2	1.6	1	0	0.3	0.2	0	0	0	0	—
992	Raw	1	item(s)	108	94.4	22	1.3	11.6	5.1	0.3	0	0	0.1	—
	Lime													
269	Juice	1	tablespoon(s)	15	14.0	4	0.1	1.3	0.1	0	0	0	0	—
994	Raw	1	item(s)	67	59.1	20	0.5	7.1	1.9	0.1	0	0	0	—
995	**Loganberries, frozen**	½	cup(s)	74	62.2	40	1.1	9.6	3.9	0.2	0	0	0.1	—
	Mandarin orange													
1038	Canned in juice	½	cup(s)	125	111.4	46	0.8	11.9	0.9	0	0	0	0	—
1039	Canned in light syrup	½	cup(s)	126	104.7	77	0.6	20.4	0.9	0.1	0	0	0	—
999	**Mango**	½	cup(s)	83	67.4	54	0.4	14.0	1.5	0.2	0.1	0.1	0	—
1005	**Nectarine, raw, sliced**	½	cup(s)	69	60.4	30	0.7	7.3	1.2	0.2	0	0.1	0.1	—
	Melons													
271	Cantaloupe	½	cup(s)	80	72.1	27	0.7	6.5	0.7	0.1	0	0	0.1	—
1000	Casaba melon	½	cup(s)	85	78.1	24	0.9	5.6	0.8	0.1	0	0	0	—
272	Honeydew	½	cup(s)	89	79.5	32	0.5	8.0	0.7	0.1	0	0	0	—
318	Watermelon	½	cup(s)	76	69.5	23	0.5	5.7	0.3	0.1	0	0	0	—
	Orange													
14412	Juice with calcium and vitamin D	½	cup(s)	120	—	55	1.0	13.0	0	0	0	0	0	0
29630	Juice, fresh squeezed	½	cup(s)	124	109.5	56	0.9	12.9	0.2	0.2	0	0	0	—
14411	Juice, not from concentrate	½	cup(s)	120	—	55	1.0	13.0	0	0	0	0	0	0
278	Juice, unsweetened, prepared from frozen concentrate	½	cup(s)	125	109.7	56	0.8	13.4	0.2	0.1	0	0	0	—
3040	Peel	1	teaspoon(s)	2	1.5	2	0	0.5	0.2	0	0	0	0	—
273	Raw	1	item(s)	131	113.6	62	1.2	15.4	3.1	0.2	0	0	0	—
274	Sections	½	cup(s)	90	78.1	42	0.8	10.6	2.2	0.1	0	0	0	—
	Papaya, raw													
16830	Dried, strips	2	item(s)	46	12.0	119	1.9	29.9	5.5	0.4	0.1	0.1	0.1	—
282	Papaya	½	cup(s)	70	62.2	27	0.4	6.9	1.3	0.1	0	0	0	—
35640	**Passion fruit, purple**	1	item(s)	18	13.1	17	0.4	4.2	1.9	0.1	0	0	0.1	—
	Peach													
285	Halves, canned in heavy syrup	½	cup(s)	131	103.9	97	0.6	26.1	1.7	0.1	0	0	0.1	—
286	Halves, canned in water	½	cup(s)	122	113.6	29	0.5	7.5	1.6	0.1	0	0	0	—
290	Slices, sweetened, frozen	½	cup(s)	125	93.4	118	0.8	30.0	2.3	0.2	0	0.1	0.1	—
283	Raw, medium	1	item(s)	150	133.3	59	1.4	14.3	2.3	0.4	0	0.1	0.1	—
	Pear													
8672	Asian	1	item(s)	122	107.7	51	0.6	13.0	4.4	0.3	0	0.1	0.1	—
293	D'Anjou	1	item(s)	200	168.0	120	1.0	30.0	5.2	1.0	0	0.2	0.2	—
294	Halves, canned in heavy syrup	½	cup(s)	133	106.9	98	0.3	25.5	2.1	0.2	0	0	0	—
1012	Halves, canned in juice	½	cup(s)	124	107.2	62	0.4	16.0	2.0	0.1	0	0	0	—
291	Raw	1	item(s)	166	139.0	96	0.6	25.7	5.1	0.2	0	0	0	—

Chol (mg)	Calc (mg)	Iron (mg)	Magn (mg)	Pota (mg)	Sodi (mg)	Zinc (mg)	Vit A (µg)	Thia (mg)	Vit E (mg α)	Ribo (mg)	Niac (mg)	Vit B_6 (mg)	Fola (µg)	Vit C (mg)	Vit B_{12} (µg)	Sele (µg)
0	20	0.42	7.6	97.0	2.5	0.14	8.8	0.03	—	0.07	0.19	0.02	3.8	12.6	0	0.5
0	19	0.23	7.5	148.5	0.8	0.09	11.3	0.03	0.28	0.02	0.23	0.06	4.5	20.8	0	0.5
0	10	0.45	12.5	202.2	2.5	0.08	0	0.05	0.05	0.03	0.40	0.03	12.5	33.6	0	0.1
0	11	0.25	14.8	200.1	1.2	0.06	1.2	0.05	0.27	0.02	0.25	0.05	12.4	46.9	0	0.1
0	25	0.09	10.3	154.5	0	0.07	66.4	0.04	0.14	0.03	0.23	0.06	14.9	35.7	0	0.1
0	18	0.50	12.7	163.8	2.5	0.10	0	0.04	0.11	0.02	0.30	0.02	11.4	27.1	0	1.1
0	18	0.50	12.2	161.0	2.4	0.11	0	0.05	0.11	0.03	0.30	0.02	11.0	26.6	0	1.1
0	14	0.07	10.4	170.2	0	0.08	2.3	0.04	0.15	0.02	0.30	0.05	11.5	38.3	0	1.6
0	6	0.13	2.3	87.9	0.9	0.02	2.3	0.04	0.09	0.02	0.14	0.05	1.8	1.8	0	0
0	8	0.27	5.3	144.2	1.5	0.05	2.3	0.05	0.14	0.05	0.14	0.07	1.5	8.2	0	0.1
0	9	0.16	7.5	41.3	11.3	0.04	0	0.28	0.00	0.44	0.18	0.04	1.3	33.1	0	0.1
0	5	0.13	5.0	26.3	2.5	0.05	0	0.02	0.00	0.03	0.16	0.05	1.3	29.9	0	0.1
0	12	1.06	12.4	340.3	11.6	0.07	0	0.04	—	0.07	0.46	0.07	1.2	2.2	0	0.2
0	10	0.14	12.1	229.4	1.1	0.12	17.1	0.03	0.40	0.02	0.59	0.06	27.0	125.6	0	0.3
0	1	0.01	1.0	17.5	2.2	—	0.3	0.00	—	0.00	0.03	0.00	—	2.2	0	
0	28	0.49	30.5	250.0	2.5	0.35	12.4	0.02	—	0.09	0.33	0.09	11.5	5.5	0	0.5
0	26	0.23	12.9	237.1	2.3	0.10	3.0	0.02	1.11	0.01	0.25	0.04	19.0	70.5	0	0.2
0	1	0.00	0.9	18.9	0.2	0.01	0.2	0.00	0.02	0.00	0.02	0.01	2.0	7.0	0	0
0	3	0.01	0.3	3.2	0.1	0.01	0.1	0.00	0.01	0.00	0.01	0.00	0.3	2.6	0	0
0	66	0.75	13.0	156.6	3.2	0.10	2.2	0.05	—	0.04	0.21	0.11	—	83.2	0	1.0
0	2	0.02	1.2	18.0	0.3	0.01	0.3	0.00	0.03	0.00	0.02	0.01	1.5	4.6	0	0
0	22	0.40	4.0	68.3	1.3	0.07	1.3	0.02	0.14	0.01	0.13	0.02	5.4	19.5	0	0.3
0	19	0.47	15.4	106.6	0.7	0.25	1.5	0.04	0.64	0.03	0.62	0.05	19.1	11.2	0	0.1
0	14	0.34	13.7	165.6	6.2	0.64	53.5	0.10	0.12	0.04	0.55	0.05	6.2	42.6	0	0.5
0	9	0.47	10.1	98.3	7.6	0.30	52.9	0.07	0.13	0.06	0.56	0.05	6.3	24.9	0	0.5
0	8	0.10	7.4	128.7	1.7	0.03	31.4	0.05	0.92	0.04	0.48	0.11	11.6	22.8	0	0.5
0	4	0.19	6.2	138.7	0	0.12	11.7	0.02	0.53	0.02	0.78	0.02	3.5	3.7	0	0
0	7	0.17	9.6	213.6	12.8	0.14	135.2	0.03	0.04	0.01	0.59	0.05	16.8	29.4	0	0.3
0	9	0.29	9.4	154.7	7.7	0.06	0	0.01	0.04	0.03	0.20	0.14	6.8	18.5	0	0.3
0	5	0.15	8.8	201.8	15.9	0.07	2.7	0.03	0.01	0.01	0.37	0.07	16.8	15.9	0	0.6
0	5	0.18	7.6	85.1	0.8	0.07	21.3	0.02	0.04	0.01	0.13	0.03	2.3	6.2	0	0.3
0	175	0.00	12.0	225.0	0	—	0	0.08	—	0.03	0.40	0.06	30.0	36.0	0	—
0	14	0.25	13.6	248.0	1.2	0.06	12.4	0.11	0.05	0.04	0.50	0.05	37.2	62.0	0	0.1
0	10	0.00	12.5	225.0	0	0.06	0	0.08	—	0.03	0.40	0.06	30.0	36.0	0	0.1
0	11	0.12	12.5	236.6	1.2	0.06	6.2	0.10	0.25	0.02	0.25	0.06	54.8	48.4	0	0.1
0	3	0.01	0.4	4.2	0.1	0.01	0.4	0.00	0.01	0.00	0.01	0.00	0.6	2.7	0	0
0	52	0.13	13.1	237.1	0	0.09	14.4	0.11	0.23	0.05	0.36	0.07	39.3	69.7	0	0.7
0	36	0.09	9.0	162.9	0	0.06	9.9	0.07	0.16	0.03	0.25	0.05	27.0	47.9	0	0.4
0	73	0.30	30.4	782.9	9.2	0.21	83.7	0.06	2.22	0.08	0.93	0.05	58.0	37.7	0	1.8
0	17	0.07	7.0	179.9	2.1	0.05	38.5	0.02	0.51	0.02	0.24	0.01	26.6	43.3	0	0.4
0	2	0.28	5.2	62.6	5.0	0.01	11.5	0.00	0.00	0.00	0.27	0.01	2.5	5.4	0	0.1
0	4	0.35	6.6	120.5	7.9	0.11	22.3	0.01	0.64	0.03	0.80	0.02	3.9	3.7	0	0.4
0	2	0.39	6.1	120.8	3.7	0.11	32.9	0.01	0.59	0.02	0.63	0.02	3.7	3.5	0	0.4
0	4	0.46	6.3	162.5	7.5	0.06	17.5	0.01	0.77	0.04	0.81	0.02	3.8	117.8	0	0.5
0	9	0.37	13.5	285.0	0	0.25	24.0	0.03	1.09	0.04	1.20	0.03	6.0	9.9	0	0.2
0	5	0.00	9.8	147.6	0	0.02	0	0.01	0.14	0.01	0.26	0.02	9.8	4.6	0	0.1
0	22	0.50	12.0	250.0	0	0.24	—	0.04	1.00	0.08	0.20	0.03	14.6	8.0	0	1.0
0	7	0.29	5.3	86.5	6.7	0.10	0	0.01	0.10	0.02	0.32	0.01	1.3	1.5	0	0
0	11	0.36	8.7	119.0	5.0	0.11	0	0.01	0.10	0.01	0.25	0.02	1.2	2.0	0	0
0	15	0.28	11.6	197.5	1.7	0.16	1.7	0.02	0.19	0.04	0.26	0.04	11.0	7.0	0	0.2

TABLE A-1 Table of Food Composition (*continued*) (Computer code is for Cengage Diet Analysis program) (For purposes of calculations, use "0" for t, <1, <.1, <.01, etc.)

DA+ Code	Food Description	Quantity	Measure	Wt (g)	H₂0 (g)	Ener (kcal)	Prot (g)	Carb (g)	Fiber (g)	Fat (g)	Fat Breakdown (g) Sat	Mono	Poly	*Trans*
Fruit and Fruit Juices—continued														
1017	**Persimmon**	1	item(s)	25	16.1	32	0.2	8.4	—	0.1	0	0	0	—
	Pineapple													
3053	Canned in extra heavy syrup	½	cup(s)	130	101.0	108	0.4	28.0	1.0	0.1	0	0	0	—
1019	Canned in juice	½	cup(s)	125	104.0	75	0.5	19.5	1.0	0.1	0	0	0	—
296	Canned in light syrup	½	cup(s)	126	108.0	66	0.5	16.9	1.0	0.2	0	0	0.1	—
1018	Canned in water	½	cup(s)	123	111.7	39	0.5	10.2	1.0	0.1	0	0	0	—
299	Juice, unsweetened, canned	½	cup(s)	125	108.0	66	0.5	16.1	0.3	0.2	0	0	0.1	—
295	Raw, diced	½	cup(s)	78	66.7	39	0.4	10.2	1.1	0.1	0	0	0	—
1024	**Plantain, cooked**	½	cup(s)	77	51.8	89	0.6	24.0	1.8	0.1	0.1	0	0	—
300	**Plum, raw, large**	1	item(s)	66	57.6	30	0.5	7.5	0.9	0.2	0	0.1	0	—
1027	**Pomegranate**	1	item(s)	154	124.7	105	1.5	26.4	0.9	0.5	0.1	0.1	0.1	—
	Prunes													
5644	Dried	2	item(s)	17	5.2	40	0.4	10.7	1.2	0.1	0	0	0	—
305	Dried, stewed	½	cup(s)	124	86.5	133	1.2	34.8	3.8	0.2	0	0.1	0	—
306	Juice, canned	1	cup(s)	256	208.0	182	1.6	44.7	2.6	0.1	0	0.1	0	—
	Raspberries													
309	Raw	½	cup(s)	62	52.7	32	0.7	7.3	4.0	0.4	0	0	0.2	—
310	Red, sweetened, frozen	½	cup(s)	125	90.9	129	0.9	32.7	5.5	0.2	0	0	0.1	—
311	**Rhubarb, cooked with sugar**	½	cup(s)	120	81.5	140	0.5	37.5	2.7	0.1	0	0	0.1	—
	Strawberries													
313	Raw	½	cup(s)	72	65.5	23	0.5	5.5	1.4	0.2	0	0	0.1	—
315	Sweetened, frozen, thawed	½	cup(s)	128	99.5	99	0.7	26.8	2.4	0.2	0	0	0.1	—
16828	**Tangelo**	1	item(s)	95	82.4	45	0.9	11.2	2.3	0.1	0	0	0	—
	Tangerine													
1040	Juice	½	cup(s)	124	109.8	53	0.6	12.5	0.2	0.2	0	0	0	—
316	Raw	1	item(s)	88	74.9	47	0.7	11.7	1.6	0.3	0	0.1	0.1	—
Vegetables, Legumes														
	Amaranth													
1043	Leaves, boiled, drained	½	cup(s)	66	60.4	14	1.4	2.7	—	0.1	0	0	0.1	—
1042	Leaves, raw	1	cup(s)	28	25.7	6	0.7	1.1	—	0.1	0	0	0	—
8683	**Arugula leaves, raw**	1	cup(s)	20	18.3	5	0.5	0.7	0.3	0.1	0	0	0.1	—
	Artichoke													
1044	Boiled, drained	1	item(s)	120	100.9	64	3.5	14.3	10.3	0.4	0.1	0	0.2	—
2885	Hearts, boiled, drained	½	cup(s)	84	70.6	45	2.4	10.0	7.2	0.3	0.1	0	0.1	—
	Asparagus													
566	Boiled, drained	½	cup(s)	90	83.4	20	2.2	3.7	1.8	0.2	0	0	0.1	—
568	Canned, drained	½	cup(s)	121	113.7	23	2.6	3.0	1.9	0.8	0.2	0	0.3	—
565	Tips, frozen, boiled, drained	½	cup(s)	90	84.7	16	2.7	1.7	1.4	0.4	0.1	0	0.2	—
	Bamboo shoots													
1048	Boiled, drained	½	cup(s)	60	57.6	7	0.9	1.2	0.6	0.1	0	0	0.1	—
1049	Canned, drained	½	cup(s)	66	61.8	12	1.1	2.1	0.9	0.3	0.1	0	0.1	—
	Beans													
1801	Adzuki beans, boiled	½	cup(s)	115	76.2	147	8.6	28.5	8.4	0.1	0	—	—	—
511	Baked beans with franks, canned	½	cup(s)	130	89.8	184	8.7	19.9	8.9	8.5	3.0	3.7	1.1	—
513	Baked beans with pork in sweet sauce, canned	½	cup(s)	127	89.3	142	6.7	26.7	5.3	1.8	0.6	0.6	0.5	0
512	Baked beans with pork in tomato sauce, canned	½	cup(s)	127	93.0	119	6.5	23.6	5.1	1.2	0.5	0.7	0.3	—
1805	Black beans, boiled	½	cup(s)	86	56.5	114	7.6	20.4	7.5	0.5	0.1	0	0.2	—
14597	Chickpeas, garbanzo beans or bengal gram, boiled	½	cup(s)	82	49.4	134	7.3	22.5	6.2	2.1	0.2	0.5	0.9	—
569	Fordhook lima beans, frozen, boiled, drained	½	cup(s)	85	62.0	88	5.2	16.4	4.9	0.3	0.1	0	0.1	—
1806	French beans, boiled	½	cup(s)	89	58.9	114	6.2	21.3	8.3	0.7	0.1	0	0.4	—
2773	Great northern beans, boiled	½	cup(s)	89	61.1	104	7.4	18.7	6.2	0.4	0.1	0	0.2	—
2736	Hyacinth beans, boiled, drained	½	cup(s)	44	37.8	22	1.3	4.0	—	0.1	0.1	0.1	0	—
570	Lima beans, baby, frozen, boiled, drained	½	cup(s)	90	65.1	95	6.0	17.5	5.4	0.3	0.1	0	0.1	—
515	Lima beans, boiled, drained	½	cup(s)	85	57.1	105	5.8	20.1	4.5	0.3	0.1	0	0.1	—
579	Mung beans, sprouted, boiled, drained	½	cup(s)	62	57.9	13	1.3	2.6	0.5	0.1	0	0	0	—
510	Navy beans, boiled	½	cup(s)	91	58.1	127	7.5	23.7	9.6	0.6	0.1	0.1	0.4	0
32816	Pinto beans, boiled, drained, no salt added	½	cup(s)	85	58.8	122	7.7	22.4	7.7	0.6	0.1	0.1	0.2	—

PAGE KEY: A-6 = Breads/Baked Goods A-12 = Cereal/Rice/Pasta A-16 = Fruit A-20 = Vegetables/Legumes A-30 = Nuts/Seeds A-32 = Vegetarian A-34 = Dairy A-42 = Eggs A-42 = Seafood A-44 = Meats A-48 = Poultry
A-50 = Processed Meats A-50 = Beverages A-54 = Fats/Oils A-56 = Sweets A-58 = Spices/Condiments/Sauces A-62 = Mixed Foods/Soups/Sandwiches A-66 = Fast Food A-86 = Convenience A-88 = Baby Foods

Chol (mg)	Calc (mg)	Iron (mg)	Magn (mg)	Pota (mg)	Sodi (mg)	Zinc (mg)	Vit A (µg)	Thia (mg)	Vit E (mg α)	Ribo (mg)	Niac (mg)	Vit B$_6$ (mg)	Fola (µg)	Vit C (mg)	Vit B$_{12}$ (µg)	Sele (µg)
0	7	0.62	—	77.5	0.3	—	0	—	—	—	—	—	—	16.5	0	—
0	18	0.49	19.5	132.6	1.3	0.14	1.3	0.11	—	0.03	0.36	0.09	6.5	9.5	0	—
0	17	0.35	17.4	151.9	1.2	0.12	2.5	0.12	0.01	0.02	0.35	0.09	6.2	11.8	0	0.5
0	18	0.49	20.2	132.3	1.3	0.15	2.5	0.11	0.01	0.03	0.36	0.09	6.3	9.5	0	0.5
0	18	0.49	22.1	156.2	1.2	0.15	2.5	0.11	0.01	0.03	0.37	0.09	6.2	9.5	0	0.5
0	16	0.39	15.0	162.5	2.5	0.14	0	0.07	0.03	0.03	0.25	0.13	22.5	12.5	0	0.1
0	10	0.22	9.3	84.5	0.8	0.09	2.3	0.06	0.02	0.03	0.39	0.09	14.0	37.0	0	0.1
0	2	0.45	24.6	358.1	3.9	0.10	34.7	0.04	0.10	0.04	0.58	0.19	20.0	8.4	0	1.1
0	4	0.11	4.6	103.6	0	0.06	11.2	0.02	0.17	0.02	0.27	0.02	3.3	6.3	0	0
0	5	0.46	4.6	398.9	4.6	0.18	7.7	0.04	0.92	0.04	0.46	0.16	9.2	9.4	0	0.9
0	7	0.16	6.9	123.0	0.3	0.07	6.6	0.01	0.07	0.03	0.32	0.03	0.7	0.1	0	0
0	24	0.51	22.3	398.0	1.2	0.24	21.1	0.03	0.24	0.12	0.90	0.27	0	3.6	0	0.1
0	31	3.02	35.8	706.6	10.2	0.53	0	0.04	0.30	0.17	2.01	0.55	0	10.5	0	1.5
0	15	0.42	13.5	92.9	0.6	0.26	1.2	0.02	0.54	0.02	0.37	0.03	12.9	16.1	0	0.1
0	19	0.81	16.3	142.5	1.3	0.22	3.8	0.02	0.90	0.05	0.28	0.04	32.5	20.6	0	0.4
0	174	0.25	16.2	115.0	1.0	—	—	0.02	—	0.03	0.25	—	—	4.0	0	—
0	12	0.30	9.4	110.2	0.7	0.10	0.7	0.02	0.21	0.02	0.28	0.03	17.3	42.3	0	0.3
0	14	0.59	7.7	125.0	1.3	0.06	1.3	0.01	0.30	0.09	0.37	0.03	5.1	50.4	0	0.9
0	38	0.09	9.5	172.0	0	0.06	10.5	0.08	0.17	0.03	0.26	0.05	28.5	50.5	0	0.5
0	22	0.25	9.9	219.8	1.2	0.04	16.1	0.07	0.16	0.02	0.12	0.05	6.2	38.3	0	0.1
0	33	0.13	10.6	146.1	1.8	0.06	29.9	0.05	0.18	0.03	0.33	0.07	14.1	23.5	0	0.1
0	138	1.49	36.3	423.1	13.9	0.58	91.7	0.01	—	0.09	0.37	0.12	37.6	27.1	0	0.6
0	60	0.65	15.4	171.1	5.6	0.25	40.9	0.01	—	0.04	0.18	0.05	23.8	12.1	0	0.3
0	32	0.29	9.4	73.8	5.4	0.09	23.8	0.01	0.09	0.02	0.06	0.01	19.4	3.0	0	0.1
0	25	0.73	50.4	343.2	72.0	0.48	1.2	0.06	0.22	0.10	1.33	0.09	106.8	8.9	0	0.2
0	18	0.51	35.3	240.2	50.4	0.33	0.8	0.04	0.16	0.07	0.93	0.06	74.8	6.2	0	0.2
0	21	0.81	12.6	201.6	12.6	0.54	45.0	0.14	1.35	0.12	0.97	0.07	134.1	6.9	0	5.5
0	19	2.21	12.1	208.1	347.3	0.48	49.6	0.07	1.47	0.12	1.15	0.13	116.2	22.3	0	2.1
0	16	0.50	9.0	154.8	2.7	0.36	36.0	0.05	1.08	0.09	0.93	0.01	121.5	22.0	0	3.5
0	7	0.14	1.8	319.8	2.4	0.28	0	0.01	—	0.03	0.18	0.06	1.2	0	0	0.2
0	5	0.21	2.6	52.4	4.6	0.43	0.7	0.02	0.41	0.02	0.09	0.09	2.0	0.7	0	0.3
0	32	2.30	59.8	611.8	9.2	2.03	0	0.13	—	0.07	0.82	0.11	139.2	0	0	1.4
8	62	2.24	36.3	304.3	556.9	2.42	5.2	0.08	0.21	0.07	1.17	0.06	38.9	3.0	0.4	8.4
9	75	2.08	41.7	326.4	422.5	1.73	0	0.05	0.03	0.07	0.44	0.07	10.1	3.5	0	6.3
9	71	4.09	43.0	373.2	552.8	6.93	5.1	0.06	0.12	0.05	0.62	0.08	19.0	3.8	0	5.9
0	23	1.80	60.2	305.3	0.9	0.96	0	0.21	—	0.05	0.43	0.05	128.1	0	0	1.0
0	40	2.36	39.4	238.6	5.7	1.25	0.8	0.09	0.28	0.05	0.43	0.11	141.0	1.1	0	3.0
0	26	1.54	35.7	258.4	58.7	0.62	8.5	0.06	0.24	0.05	0.90	0.10	17.9	10.9	0	0.5
0	56	0.95	49.6	327.5	5.3	0.56	0	0.11	—	0.05	0.48	0.09	66.4	1.1	0	1.1
0	60	1.88	44.3	346.0	1.8	0.77	0	0.14	—	0.05	0.60	0.10	90.3	1.2	0	3.6
0	18	0.33	18.3	114.0	0.9	0.16	3.0	0.02	—	0.03	0.20	0.01	20.4	2.2	0	0.7
0	25	1.76	50.4	369.9	26.1	0.49	7.2	0.06	0.57	0.04	0.69	0.10	14.4	5.2	0	1.5
0	27	2.08	62.9	484.5	14.5	0.67	12.8	0.11	0.11	0.08	0.88	0.16	22.1	8.6	0	1.7
0	7	0.40	8.7	62.6	6.2	0.29	0.6	0.03	0.04	0.06	0.51	0.03	18.0	7.1	0	0.4
—	63	2.14	48.2	354.0	0	0.93	0	0.21	0.01	0.06	0.59	0.12	127.4	0.8	0	2.6
0	39	1.79	43.0	373.0	1.0	0.84	0	0.17	0.80	0.05	0.27	0.19	147.0	0.7	0	5.3

Appendix A

TABLE A-1 **Table of Food Composition (continued)** (Computer code is for Cengage Diet Analysis program)   (For purposes of calculations, use "0" for t, <1, <.1, <.01, etc.)

DA+ Code	Food Description	Quantity	Measure	Wt (g)	H₂0 (g)	Ener (kcal)	Prot (g)	Carb (g)	Fiber (g)	Fat (g)	Sat	Mono	Poly	Trans
	Vegetables, Legumes—continued													
1052	Pinto beans, frozen, boiled, drained	½	cup(s)	47	27.3	76	4.4	14.5	4.0	0.2	0	0	0.1	—
514	Red kidney beans, canned	½	cup(s)	128	99.0	108	6.7	19.9	6.9	0.5	0.1	0.2	0.2	—
1810	Refried beans, canned	½	cup(s)	127	96.1	119	6.9	19.6	6.7	1.6	0.6	0.7	0.2	—
1053	Shell beans, canned	½	cup(s)	123	111.1	37	2.2	7.6	4.2	0.2	0	0	0.1	—
1670	Soybeans, boiled	½	cup(s)	86	53.8	149	14.3	8.5	5.2	7.7	1.1	1.7	4.4	—
1108	Soybeans, green, boiled, drained	½	cup(s)	90	61.7	127	11.1	9.9	3.8	5.8	0.7	1.1	2.7	—
1807	White beans, small, boiled	½	cup(s)	90	56.6	127	8.0	23.1	9.3	0.6	0.1	0.1	0.2	—
575	Yellow snap, string or wax beans, boiled, drained	½	cup(s)	63	55.8	22	1.2	4.9	2.1	0.2	0	0	0.1	—
576	Yellow snap, string or wax beans, frozen, boiled, drained	½	cup(s)	68	61.7	19	1.0	4.4	2.0	0.1	0	0	0.1	—
	Beets													
584	Beet greens, boiled, drained	½	cup(s)	72	64.2	19	1.9	3.9	2.1	0.1	0	0	0.1	—
2730	Pickled, canned with liquid	½	cup(s)	114	92.9	74	0.9	18.5	3.0	0.1	0	0	0	—
581	Sliced, boiled, drained	½	cup(s)	85	74.0	37	1.4	8.5	1.7	0.2	0	0	0.1	—
583	Sliced, canned, drained	½	cup(s)	85	77.3	26	0.8	6.1	1.5	0.1	0	0	0	—
580	Whole, boiled, drained	2	item(s)	100	87.1	44	1.7	10.0	2.0	0.2	0	0	0.1	—
585	**Cowpeas or black-eyed peas, boiled, drained**	½	cup(s)	83	62.3	80	2.6	16.8	4.1	0.3	0.1	0	0.1	—
	Broccoli													
588	Chopped, boiled, drained	½	cup(s)	78	69.6	27	1.9	5.6	2.6	0.3	0.1	0	0.1	—
590	Frozen, chopped, boiled, drained	½	cup(s)	92	83.5	26	2.9	4.9	2.8	0.1	0	0	0.1	—
587	Raw, chopped	½	cup(s)	46	40.6	15	1.3	3.0	1.2	0.2	0	0	0	—
16848	**Broccoflower, raw, chopped**	½	cup(s)	32	28.7	10	0.9	1.9	1.0	0.1	0	0	0	—
	Brussels sprouts													
591	Boiled, drained	½	cup(s)	78	69.3	28	2.0	5.5	2.0	0.4	0.1	0	0.2	—
592	Frozen, boiled, drained	½	cup(s)	78	67.2	33	2.8	6.4	3.2	0.3	0.1	0	0.2	—
	Cabbage													
595	Boiled, drained, no salt added	1	cup(s)	150	138.8	35	1.9	8.3	2.8	0.1	0	0	0	—
35611	Chinese (pak choi or bok choy), boiled with salt, drained	1	cup(s)	170	162.4	20	2.6	3.0	1.7	0.3	0	0	0.1	—
16869	Kim chee	1	cup(s)	150	137.5	32	2.5	6.1	1.8	0.3	0	0	0.2	—
594	Raw, shredded	1	cup(s)	70	64.5	17	0.9	4.1	1.7	0.1	0	0	0	—
596	Red, shredded, raw	1	cup(s)	70	63.3	22	1.0	5.2	1.5	0.1	0	0	0.1	—
597	Savoy, shredded, raw	1	cup(s)	70	63.7	19	1.4	4.3	2.2	0.1	0	0	0	—
35417	**Capers**	1	teaspoon(s)	4	—	2	0	0	0	0	0	0	0	0
	Carrots													
8691	Baby, raw	8	item(s)	80	72.3	28	0.5	6.6	2.3	0.1	0	0	0.1	—
601	Grated	½	cup(s)	55	48.6	23	0.5	5.3	1.5	0.1	0	0	0.1	0
1055	Juice, canned	½	cup(s)	118	104.9	47	1.1	11.0	0.9	0.2	0	0	0.1	—
600	Raw	½	cup(s)	61	53.9	25	0.6	5.8	1.7	0.1	0	0	0.1	0
602	Sliced, boiled, drained	½	cup(s)	78	70.3	27	0.6	6.4	2.3	0.1	0	0	0.1	—
32725	**Cassava or manioc**	½	cup(s)	103	61.5	165	1.4	39.2	1.9	0.3	0.1	0.1	0	—
	Cauliflower													
606	Boiled, drained	½	cup(s)	62	57.7	14	1.1	2.5	1.4	0.3	0	0	0.1	—
607	Frozen, boiled, drained	½	cup(s)	90	84.6	17	1.4	3.4	2.4	0.2	0	0	0.1	—
605	Raw, chopped	½	cup(s)	50	46.0	13	1.0	2.6	1.2	0	0	0	0	—
	Celery													
609	Diced	½	cup(s)	51	48.2	8	0.3	1.5	0.8	0.1	0	0	0	—
608	Stalk	2	item(s)	80	76.3	13	0.6	2.4	1.3	0.1	0	0	0.1	—
	Chard													
1057	Swiss chard, boiled, drained	½	cup(s)	88	81.1	18	1.6	3.6	1.8	0.1	0	0	0	—
1056	Swiss chard, raw	1	cup(s)	36	33.4	7	0.6	1.3	0.6	0.1	0	0	0	—
	Collard greens													
610	Boiled, drained	½	cup(s)	95	87.3	25	2.0	4.7	2.7	0.3	0	0	0.2	—
611	Frozen, chopped, boiled, drained	½	cup(s)	85	75.2	31	2.5	6.0	2.4	0.3	0.1	0	0.2	—
	Corn													
29614	Yellow corn, fresh, cooked	1	item(s)	100	69.2	107	3.3	25.0	2.8	1.3	0.2	0.4	0.6	—
615	Yellow creamed sweet corn, canned	½	cup(s)	128	100.8	92	2.2	23.2	1.5	0.5	0.1	0.2	0.3	—
612	Yellow sweet corn, boiled, drained	½	cup(s)	82	57.0	89	2.7	20.6	2.3	1.1	0.2	0.3	0.5	—

A-22

APPENDIX A

PAGE KEY: A-6 = Breads/Baked Goods A-12 = Cereal/Rice/Pasta A-16 = Fruit A-20 = Vegetables/Legumes A-30 = Nuts/Seeds A-32 = Vegetarian A-34 = Dairy A-42 = Eggs A-42 = Seafood A-44 = Meats A-48 = Poultry
A-50 = Processed Meats A-50 = Beverages A-54 = Fats/Oils A-56 = Sweets A-58 = Spices/Condiments/Sauces A-62 = Mixed Foods/Soups/Sandwiches A-66 = Fast Food A-86 = Convenience A-88 = Baby Foods

Chol (mg)	Calc (mg)	Iron (mg)	Magn (mg)	Pota (mg)	Sodi (mg)	Zinc (mg)	Vit A (µg)	Thia (mg)	Vit E (mg α)	Ribo (mg)	Niac (mg)	Vit B$_6$ (mg)	Fola (µg)	Vit C (mg)	Vit B$_{12}$ (µg)	Sele (µg)
0	24	1.27	25.4	303.6	39.0	0.32	0	0.12	—	0.05	0.29	0.09	16.0	0.3	0	0.7
0	32	1.62	35.8	327.7	330.2	2.09	0	0.13	0.02	0.11	0.57	0.10	25.6	1.4	0	0.6
10	44	2.10	41.7	337.8	378.2	1.48	0	0.03	0.00	0.02	0.39	0.18	13.9	7.6	0	1.6
0	36	1.21	18.4	133.5	409.2	0.33	13.5	0.04	0.04	0.07	0.25	0.06	22.1	3.8	0	2.6
0	88	4.42	74.0	442.9	0.9	0.98	0	0.13	0.30	0.24	0.34	0.20	46.4	1.5	0	6.3
0	131	2.25	54.0	485.1	12.6	0.82	7.2	0.23	—	0.14	1.13	0.05	99.9	15.3	0	1.3
0	65	2.54	60.9	414.4	1.8	0.97	0	0.21	—	0.05	0.24	0.11	122.6	0	0	1.2
0	29	0.80	15.6	186.9	1.9	0.23	2.5	0.05	0.28	0.06	0.38	0.04	20.6	6.1	0	0.3
0	33	0.59	16.2	85.1	6.1	0.32	4.1	0.02	0.03	0.06	0.26	0.04	15.5	2.8	0	0.3
0	82	1.36	49.0	654.5	173.5	0.36	275.8	0.08	1.30	0.20	0.35	0.09	10.1	17.9	0	0.6
0	12	0.46	17.0	168.0	299.6	0.29	1.1	0.01	—	0.05	0.28	0.05	30.6	2.6	0	1.1
0	14	0.67	19.6	259.3	65.5	0.30	1.7	0.02	0.03	0.03	0.28	0.05	68.0	3.1	0	0.6
0	13	1.54	14.5	125.8	164.9	0.17	0.9	0.01	0.02	0.03	0.13	0.04	25.5	3.5	0	0.4
0	16	0.79	23.0	305.0	77.0	0.35	2.0	0.02	0.04	0.04	0.33	0.06	80.0	3.6	0	0.7
0	106	0.92	42.9	344.9	3.3	0.85	33.0	0.08	0.18	0.12	1.15	0.05	104.8	1.8	0	2.1
0	31	0.52	16.4	228.5	32.0	0.35	60.1	0.04	1.13	0.09	0.43	0.15	84.2	50.6	0	1.2
0	30	0.56	12.0	130.6	10.1	0.25	46.9	0.05	1.21	0.07	0.42	0.12	51.5	36.9	0	0.6
0	21	0.33	9.6	143.8	15.0	0.19	14.1	0.03	0.36	0.05	0.29	0.08	28.7	40.6	0	1.1
0	11	0.23	6.4	96.0	7.4	0.20	2.6	0.02	0.01	0.03	0.23	0.07	18.2	28.2	0	0.2
0	28	0.93	15.6	247.3	16.4	0.25	30.4	0.08	0.33	0.06	0.47	0.13	46.8	48.4	0	1.2
0	20	0.37	14.0	224.8	11.6	0.18	35.7	0.08	0.39	0.08	0.41	0.22	78.3	35.4	0	0.5
0	72	0.24	22.5	294.0	12.0	0.30	6.0	0.08	0.20	0.04	0.36	0.16	45.0	56.2	0	0.9
0	158	1.76	18.7	630.7	459.0	0.28	360.4	0.04	0.14	0.10	0.72	0.28	69.7	44.2	0	0.7
0	144	1.26	27.0	379.5	996.0	0.36	288.0	0.06	0.36	0.10	0.80	0.32	88.5	79.6	0	1.5
0	28	0.33	8.4	119.0	12.6	0.12	3.5	0.04	0.10	0.02	0.16	0.08	30.1	25.6	0	0.2
0	31	0.56	11.2	170.1	18.9	0.15	39.2	0.04	0.07	0.05	0.29	0.14	12.6	39.9	0	0.4
0	24	0.28	19.6	161.0	19.6	0.18	35.0	0.05	0.11	0.02	0.21	0.13	56.0	21.7	0	0.6
0	0	0.00	—	—	140	—	0	—	—	—	—	—	—	0	—	—
0	26	0.71	8.0	189.6	62.4	0.13	552.0	0.02	—	0.02	0.44	0.08	21.6	2.1	0	0.7
0	18	0.16	6.6	176.0	37.9	0.13	459.2	0.03	0.36	0.03	0.54	0.07	10.4	3.2	0	0.1
0	28	0.54	16.5	344.6	34.2	0.21	1128.1	0.11	1.37	0.07	0.46	0.26	4.7	10.0	0	0.7
0	20	0.18	7.3	195.2	42.1	0.15	509.4	0.04	0.40	0.04	0.60	0.08	11.6	3.6	0	0.1
0	23	0.26	7.8	183.3	45.2	0.15	664.6	0.05	0.80	0.03	0.50	0.11	10.9	2.8	0	0.5
0	16	0.27	21.6	279.1	14.4	0.35	1.0	0.08	0.19	0.04	0.87	0.09	27.8	21.2	0	0.7
0	10	0.19	5.6	88.0	9.3	0.10	0.6	0.02	0.04	0.03	0.25	0.10	27.3	27.5	0	0.4
0	15	0.36	8.1	125.1	16.2	0.11	0	0.03	0.05	0.04	0.27	0.07	36.9	28.2	0	0.5
0	11	0.22	7.5	151.5	15.0	0.14	0.5	0.03	0.04	0.03	0.26	0.11	28.5	23.2	0	0.3
0	20	0.10	5.6	131.3	40.4	0.07	11.1	0.01	0.14	0.03	0.16	0.04	18.2	1.6	0	0.2
0	32	0.16	8.8	208.0	64.0	0.10	17.6	0.01	0.21	0.04	0.25	0.05	28.8	2.5	0	0.3
0	51	1.98	75.3	480.4	156.6	0.29	267.8	0.03	1.65	0.08	0.32	0.07	7.9	15.8	0	0.8
0	18	0.64	29.2	136.4	76.7	0.13	110.2	0.01	0.68	0.03	0.14	0.03	5.0	10.8	0	0.3
0	133	1.10	19.0	110.2	15.2	0.21	385.7	0.03	0.83	0.10	0.54	0.12	88.4	17.3	0	0.5
0	179	0.95	25.5	213.4	42.5	0.22	488.8	0.04	1.06	0.09	0.54	0.09	64.6	22.4	0	1.3
0	2	0.61	32.0	248.0	242.0	0.48	13.0	0.20	0.09	0.07	1.60	0.06	46.0	6.2	0	0.2
0	4	0.48	21.8	171.5	364.8	0.67	5.1	0.03	0.09	0.06	1.22	0.08	57.6	5.9	0	0.5
0	2	0.36	21.3	173.8	0	0.50	10.7	0.17	0.07	0.05	1.32	0.04	37.7	5.1	0	0.2

TABLE A-1 **Table of Food Composition** *(continued)*   (Computer code is for Cengage Diet Analysis program)   (For purposes of calculations, use "0" for t, <1, <.1, <.01, etc.)

DA+ Code	Food Description	Quantity	Measure	Wt (g)	H₂0 (g)	Ener (kcal)	Prot (g)	Carb (g)	Fiber (g)	Fat (g)	Fat Breakdown (g) Sat	Mono	Poly	Trans
Vegetables, Legumes—continued														
614	Yellow sweet corn, frozen, boiled, drained	½	cup(s)	82	63.2	66	2.1	15.8	2.0	0.5	0.1	0.2	0.3	—
618	**Cucumber**	¼	item(s)	75	71.7	11	0.5	2.7	0.4	0.1	0	0	0	—
16870	**Cucumber, kim chee**	½	cup(s)	75	68.1	16	0.8	3.6	1.1	0.1	0	0	0	—
	Dandelion greens													
620	Chopped, boiled, drained	½	cup(s)	53	47.1	17	1.1	3.4	1.5	0.3	0.1	0	0.1	—
2734	Raw	1	cup(s)	55	47.1	25	1.5	5.1	1.9	0.4	0.1	0	0.2	—
1066	**Eggplant, boiled, drained**	½	cup(s)	50	44.4	17	0.4	4.3	1.2	0.1	0	0	0	—
621	**Endive or escarole, chopped, raw**	1	cup(s)	50	46.9	8	0.6	1.7	1.5	0.1	0	0	0	—
8784	**Jicama or yambean**	½	cup(s)	65	116.5	49	0.9	11.4	6.3	0.1	0	0	0.1	—
	Kale													
623	Frozen, chopped, boiled, drained	½	cup(s)	65	58.8	20	1.8	3.4	1.3	0.3	0	0	0.2	—
29313	Raw	1	cup(s)	67	56.6	33	2.2	6.7	1.3	0.5	0.1	0	0.2	—
	Kohlrabi													
1072	Boiled, drained	½	cup(s)	83	74.5	24	1.5	5.5	0.9	0.1	0	0	0	—
1071	Raw	1	cup(s)	135	122.9	36	2.3	8.4	4.9	0.1	0	0	0.1	—
	Leeks													
1074	Boiled, drained	½	cup(s)	52	47.2	16	0.4	4.0	0.5	0.1	0	0	0	—
1073	Raw	1	cup(s)	89	73.9	54	1.3	12.6	1.6	0.3	0	0	0.1	—
	Lentils													
522	Boiled	¼	cup(s)	50	34.5	57	4.5	10.0	3.9	0.2	0	0	0.1	—
1075	Sprouted	1	cup(s)	77	51.9	82	6.9	17.0	—	0.4	0	0.1	0.2	—
	Lettuce													
625	Butterhead leaves	11	piece(s)	83	78.9	11	1.1	1.8	0.9	0.2	0	0	0.1	—
624	Butterhead, Boston or Bibb	1	cup(s)	55	52.6	7	0.7	1.2	0.6	0.1	0	0	0.1	—
626	Iceberg	1	cup(s)	55	52.6	8	0.5	1.6	0.7	0.1	0	0	0	—
628	Iceberg, chopped	1	cup(s)	55	52.6	8	0.5	1.6	0.7	0.1	0	0	0	—
629	Looseleaf	1	cup(s)	36	34.2	5	0.5	1.0	0.5	0.1	0	0	0	—
1665	Romaine, shredded	1	cup(s)	56	53.0	10	0.7	1.8	1.2	0.2	0	0	0.1	—
	Mushrooms													
15585	Crimini (about 6)	3	ounce(s)	85	—	28	3.7	2.8	1.9	0	0	0	0	0
8700	Enoki	30	item(s)	90	79.7	40	2.3	6.9	2.4	0.3	0	0	0.1	—
1079	Mushrooms, boiled, drained	½	cup(s)	78	71.0	22	1.7	4.1	1.7	0.4	0	0	0.1	—
1080	Mushrooms, canned, drained	½	cup(s)	78	71.0	20	1.5	4.0	1.9	0.2	0	0	0.1	—
630	Mushrooms, raw	½	cup(s)	48	44.4	11	1.5	1.6	0.5	0.2	0	0	0.1	—
15587	Portabella, raw	1	item(s)	84	—	30	3.0	3.9	3.0	0	0	0	0	0
2743	Shiitake, cooked	½	cup(s)	73	60.5	41	1.1	10.4	1.5	0.2	0	0.1	0	—
	Mustard greens													
2744	Frozen, boiled, drained	½	cup(s)	75	70.4	14	1.7	2.3	2.1	0.2	0	0.1	0	—
29319	Raw	1	cup(s)	56	50.8	15	1.5	2.7	1.8	0.1	0	0	0	—
	Okra													
16866	Batter-coated, fried	11	piece(s)	83	55.6	156	2.1	12.7	2.0	11.2	1.5	3.7	5.5	—
32742	Frozen, boiled, drained, no salt added	½	cup(s)	92	83.8	26	1.9	5.3	2.6	0.3	0.1	0	0.1	—
632	Sliced, boiled, drained	½	cup(s)	80	74.1	18	1.5	3.6	2.0	0.2	0	0	0	—
	Onions													
635	Chopped, boiled, drained	½	cup(s)	105	92.2	46	1.4	10.7	1.5	0.2	0	0	0.1	—
2748	Frozen, boiled, drained	½	cup(s)	106	97.8	30	0.8	7.0	1.9	0.1	0	0	0	—
1081	Onion rings, breaded and pan fried, frozen, heated	10	piece(s)	71	20.2	289	3.8	27.1	0.9	19.0	6.1	7.7	3.6	—
633	Raw, chopped	½	cup(s)	80	71.3	32	0.9	7.5	1.4	0.1	0	0	0	—
16850	Red onions, sliced, raw	½	cup(s)	57	50.7	24	0.5	5.8	0.8	0	0	0	0	—
636	Scallions, green or spring onions	2	item(s)	30	26.9	10	0.5	2.2	0.8	0.1	0	0	0	—
16860	**Palm hearts, cooked**	½	cup(s)	73	50.7	84	2.0	18.7	1.1	0.1	0	0	0.1	—
637	**Parsley, chopped**	1	tablespoon(s)	4	3.3	1	0.1	0.2	0.1	0	0	0	0	—
638	**Parsnips, sliced, boiled, drained**	½	cup(s)	78	62.6	55	1.0	13.3	2.8	0.2	0	0.1	0	—
	Peas													
639	Green peas, canned, drained	½	cup(s)	85	69.4	59	3.8	10.7	3.5	0.3	0.1	0	0.1	—
641	Green peas, frozen, boiled, drained	½	cup(s)	80	63.6	62	4.1	11.4	4.4	0.2	0	0	0.1	—
35694	Pea pods, boiled with salt, drained	½	cup(s)	80	71.1	32	2.6	5.2	2.2	0.2	0	0	0.1	—

PAGE KEY: A-6 = Breads/Baked Goods A-12 = Cereal/Rice/Pasta A-16 = Fruit A-20 = Vegetables/Legumes A-30 = Nuts/Seeds A-32 = Vegetarian A-34 = Dairy A-42 = Eggs A-42 = Seafood A-44 = Meats A-48 = Poultry
A-50 = Processed Meats A-50 = Beverages A-54 = Fats/Oils A-56 = Sweets A-58 = Spices/Condiments/Sauces A-62 = Mixed Foods/Soups/Sandwiches A-66 = Fast Food A-86 = Convenience A-88 = Baby Foods

Chol (mg)	Calc (mg)	Iron (mg)	Magn (mg)	Pota (mg)	Sodi (mg)	Zinc (mg)	Vit A (µg)	Thia (mg)	Vit E (mg α)	Ribo (mg)	Niac (mg)	Vit B$_6$ (mg)	Fola (µg)	Vit C (mg)	Vit B$_{12}$ (µg)	Sele (µg)
0	2	0.38	23.0	191.1	0.8	0.51	8.2	0.02	0.05	0.05	1.07	0.08	28.7	2.9	0	0.6
0	12	0.20	9.8	110.6	1.5	0.14	3.8	0.01	0.01	0.01	0.07	0.03	5.3	2.1	0	0.2
0	7	3.61	6.0	87.8	765.8	0.38	—	0.02	—	0.02	0.34	0.08	17.3	2.6	0	—
0	74	0.95	12.6	121.8	23.1	0.15	179.6	0.07	1.28	0.09	0.27	0.08	6.8	9.5	0	0.2
0	103	1.70	19.8	218.3	41.8	0.22	279.4	0.10	1.89	0.14	0.44	0.13	14.8	19.2	0	0.3
0	3	0.12	5.4	60.9	0.5	0.06	1.0	0.04	0.20	0.01	0.30	0.04	6.9	0.6	0	0
0	26	0.41	7.5	157.0	11.0	0.39	54.0	0.04	0.22	0.03	0.20	0.01	71.0	3.2	0	0.1
0	16	0.78	15.5	194.0	5.2	0.20	1.3	0.02	0.59	0.04	0.25	0.05	15.5	26.1	0	0.9
0	90	0.61	11.7	208.7	9.8	0.11	477.8	0.02	0.59	0.07	0.43	0.05	9.1	16.4	0	0.6
0	90	1.14	22.8	299.5	28.8	0.29	515.2	0.07	—	0.08	0.66	0.18	19.4	80.4	0	0.6
0	21	0.33	15.7	280.5	17.3	0.26	1.7	0.03	0.43	0.02	0.32	0.13	9.9	44.6	0	0.7
0	32	0.54	25.7	472.5	27.0	0.04	2.7	0.06	0.64	0.02	0.54	0.20	21.6	83.7	0	0.9
0	16	0.56	7.3	45.2	5.2	0.02	1.0	0.01	—	0.01	0.10	0.04	12.5	2.2	0	0.3
0	53	1.86	24.9	160.2	17.8	0.10	73.9	0.05	0.81	0.02	0.35	0.20	57.0	10.7	0	0.9
0	9	1.65	17.8	182.7	1.0	0.63	0	0.08	0.05	0.04	0.52	0.09	89.6	0.7	0	1.4
0	19	2.47	28.5	247.9	8.5	1.16	1.5	0.17	—	0.09	0.86	0.14	77.0	12.7	0	0.5
0	29	1.02	10.7	196.4	4.1	0.16	137.0	0.04	0.14	0.05	0.29	0.06	60.2	3.1	0	0.5
0	19	0.68	7.1	130.9	2.7	0.11	91.3	0.03	0.09	0.03	0.19	0.04	40.1	2.0	0	0.3
0	10	0.22	3.8	77.5	5.5	0.08	13.7	0.02	0.09	0.01	0.07	0.02	15.9	1.5	0	0.1
0	10	0.22	3.8	77.5	5.5	0.08	13.7	0.02	0.09	0.01	0.07	0.02	15.9	1.5	0	0.1
0	13	0.31	4.7	69.8	10.1	0.06	133.2	0.02	0.10	0.02	0.13	0.03	13.7	6.5	0	0.2
0	18	0.54	7.8	138.3	4.5	0.13	162.4	0.04	0.07	0.03	0.17	0.04	76.2	13.4	0	0.2
0	0	0.67	—	—	32.6	—	0	—	—	—	—	—	—	0	0	—
0	1	0.98	14.4	331.2	2.7	0.54	0	0.16	0.01	0.14	5.31	0.07	46.8	0	0	2.0
0	5	1.35	9.4	277.7	1.6	0.67	0	0.05	0.01	0.23	3.47	0.07	14.0	3.1	0	9.3
0	9	0.61	11.7	100.6	331.5	0.56	0	0.06	0.01	0.01	1.24	0.04	9.4	0	0	3.2
0	1	0.24	4.3	152.6	2.4	0.25	0	0.04	0.01	0.19	1.73	0.05	7.7	1.0	0	4.5
0	39	0.35	—	—	9.9	—	0	—	—	—	—	—	—	0	0	—
0	2	0.31	10.2	84.8	2.9	0.96	0	0.02	0.02	0.12	1.08	0.11	15.2	0.2	0	18.0
0	76	0.84	9.8	104.3	18.8	0.15	265.5	0.03	1.01	0.04	0.19	0.08	52.5	10.4	0	0.5
0	58	0.81	17.9	198.2	14.0	0.11	294.0	0.04	1.12	0.06	0.45	0.10	104.7	39.2	0	0.5
2	54	1.13	32.2	170.8	109.7	0.44	14.0	0.16	1.50	0.12	1.29	0.11	39.6	9.2	0	3.6
0	88	0.61	46.9	215.3	2.8	0.57	15.6	0.09	0.29	0.11	0.72	0.04	134.3	11.2	0	0.6
0	62	0.22	28.8	108.0	4.8	0.34	11.2	0.10	0.21	0.04	0.69	0.15	36.8	13.0	0	0.3
0	23	0.24	11.5	174.3	3.1	0.21	0	0.03	0.02	0.02	0.17	0.12	15.7	5.5	0	0.6
0	17	0.32	6.4	114.5	12.7	0.06	0	0.02	0.01	0.02	0.14	0.06	13.8	2.8	0	0.4
0	22	1.20	13.5	91.6	266.3	0.29	7.8	0.19	—	0.09	2.56	0.05	46.9	1.0	0	2.5
0	18	0.16	8.0	116.8	3.2	0.13	0	0.03	0.01	0.02	0.09	0.09	15.2	5.9	0	0.4
0	13	0.10	5.7	82.4	1.7	0.09	0	0.02	0.01	0.01	0.04	0.08	10.9	3.7	0	0.3
0	22	0.44	6.0	82.8	4.8	0.11	15.0	0.01	0.16	0.02	0.15	0.01	19.2	5.6	0	0.2
0	13	1.23	7.3	1318.4	10.2	2.72	2.2	0.03	0.36	0.12	0.62	0.53	14.6	5.0	0	0.5
0	5	0.23	1.9	21.1	2.1	0.04	16.0	0.00	0.02	0.00	0.05	0.00	5.8	5.1	0	0
0	29	0.45	22.6	286.3	7.8	0.20	0	0.06	0.78	0.04	0.56	0.07	45.2	10.1	0	1.3
0	17	0.80	14.5	147.1	214.2	0.60	23.0	0.10	0.02	0.06	0.62	0.05	37.4	8.2	0	1.4
0	19	1.21	17.6	88.0	57.6	0.53	84.0	0.22	0.02	0.08	1.18	0.09	47.2	7.9	0	0.8
0	34	1.57	20.8	192.0	192.0	0.29	41.6	0.10	0.31	0.06	0.43	0.11	23.2	38.3	0	0.6

TABLE A-1 Table of Food Composition *(continued)*   (Computer code is for Cengage Diet Analysis program)   (For purposes of calculations, use "0" for t, <1, <.1, <.01, etc.)

DA+ Code	Food Description	Quantity	Measure	Wt (g)	H₂O (g)	Ener (kcal)	Prot (g)	Carb (g)	Fiber (g)	Fat (g)	Fat Breakdown (g) Sat	Mono	Poly	Trans
Vegetables, Legumes—continued														
1082	Peas and carrots, canned with liquid	½	cup(s)	128	112.4	48	2.8	10.8	2.6	0.3	0.1	0	0.2	—
1083	Peas and carrots, frozen, boiled, drained	½	cup(s)	80	68.6	38	2.5	8.1	2.5	0.3	0.1	0	0.2	—
2750	Snow or sugar peas, frozen, boiled, drained	½	cup(s)	80	69.3	42	2.8	7.2	2.5	0.3	0.1	0	0.1	—
640	Snow or sugar peas, raw	½	cup(s)	32	28.0	13	0.9	2.4	0.8	0.1	0	0	0	—
29324	Split peas, sprouted	½	cup(s)	60	37.4	77	5.3	16.9	—	0.4	0.1	0	0.2	—
	Peppers													
644	Green bell or sweet, boiled, drained	½	cup(s)	68	62.5	19	0.6	4.6	0.8	0.1	0	0	0.1	—
643	Green bell or sweet, raw	½	cup(s)	75	69.9	15	0.6	3.5	1.3	0.1	0	0	0	—
1664	Green hot chili	1	item(s)	45	39.5	18	0.9	4.3	0.7	0.1	0	0	0	—
1663	Green hot chili, canned with liquid	½	cup(s)	68	62.9	14	0.6	3.5	0.9	0.1	0	0	0	—
1086	Jalapeño, canned with liquid	½	cup(s)	68	60.4	18	0.6	3.2	1.8	0.6	0.1	0	0.3	—
8703	Yellow bell or sweet	1	item(s)	186	171.2	50	1.9	11.8	1.7	0.4	0.1	0	0.2	—
1087	**Poi**	½	cup(s)	120	86.0	134	0.5	32.7	0.5	0.2	0	0	0.1	—
	Potatoes													
1090	Au gratin mix, prepared with water, whole milk and butter	½	cup(s)	124	97.7	115	2.8	15.9	1.1	5.1	3.2	1.5	0.2	
1089	Au gratin, prepared with butter	½	cup(s)	123	90.7	162	6.2	13.8	2.2	9.3	5.8	2.6	0.3	—
5791	Baked, flesh and skin	1	item(s)	202	151.3	188	5.1	42.7	4.4	0.3	0.1	0	0.1	—
645	Baked, flesh only	½	cup(s)	61	46.0	57	1.2	13.1	0.9	0.1	0	0	0	—
1088	Baked, skin only	1	item(s)	58	27.4	115	2.5	26.7	4.6	0.1	0	0	0	—
5795	Boiled in skin, flesh only, drained	1	item(s)	136	104.7	118	2.5	27.4	2.1	0.1	0	0	0.1	—
5794	Boiled, drained, skin and flesh	1	item(s)	150	115.9	129	2.9	29.8	2.5	0.2	0	0	0.1	—
647	Boiled, flesh only	½	cup(s)	78	60.4	67	1.3	15.6	1.4	0.1	0	0	0	—
648	French fried, deep fried, prepared from raw	14	item(s)	70	32.8	187	2.7	23.5	2.9	9.5	1.9	4.2	3.0	—
649	French fried, frozen, heated	14	item(s)	70	43.7	94	1.9	19.4	2.0	3.7	0.7	2.3	0.2	—
1091	Hashed brown	½	cup(s)	78	36.9	207	2.3	27.4	2.5	9.8	1.5	4.1	3.7	—
652	Mashed with margarine and whole milk	½	cup(s)	105	79.0	119	2.1	17.7	1.6	4.4	1.0	2.0	1.2	0.7
653	Mashed, prepared from dehydrated granules with milk, water, and margarine	½	cup(s)	105	79.8	122	2.3	16.9	1.4	5.0	1.3	2.1	1.4	—
2759	Microwaved	1	item(s)	202	145.5	212	4.9	49.0	4.6	0.2	0.1	0	0.1	—
2760	Microwaved in skin, flesh only	½	cup(s)	78	57.1	78	1.6	18.1	1.2	0.1	0	0	0	—
5804	Microwaved, skin only	1	item(s)	58	36.8	77	2.5	17.2	4.2	0.1	0	0	0	—
1097	Potato puffs, frozen, heated	½	cup(s)	64	38.2	122	1.3	17.8	1.6	5.5	1.2	3.9	0.3	—
1094	Scalloped mix, prepared with water, whole milk and butter	½	cup(s)	124	98.4	116	2.6	15.9	1.4	5.3	3.3	1.5	0.2	—
1093	Scalloped, prepared with butter	½	cup(s)	123	99.2	108	3.5	13.2	2.3	4.5	2.8	1.3	0.2	—
	Pumpkin													
1773	Boiled, drained	½	cup(s)	123	114.8	25	0.9	6.0	1.3	0.1	0	0	0	—
656	Canned	½	cup(s)	123	110.2	42	1.3	9.9	3.6	0.3	0.2	0	0	—
	Radicchio													
8731	Leaves, raw	1	cup(s)	40	37.3	9	0.6	1.8	0.4	0.1	0	0	0	—
2498	Raw	1	cup(s)	40	37.3	9	0.6	1.8	0.4	0.1	0	0	0	—
657	**Radishes**	6	item(s)	27	25.7	4	0.2	0.9	0.4	0	0	0	0	—
1099	**Rutabaga, boiled, drained**	½	cup(s)	85	75.5	33	1.1	7.4	1.5	0.2	0	0	0.1	—
658	**Sauerkraut, canned**	½	cup(s)	118	109.2	22	1.1	5.1	3.4	0.2	0	0	0.1	—
	Seaweed													
1102	Kelp	½	cup(s)	40	32.6	17	0.6	3.8	0.5	0.2	0.1	0	0	—
1104	Spirulina, dried	½	cup(s)	8	0.4	22	4.3	1.8	0.3	0.6	0.2	0.1	0.2	—
1106	**Shallots**	3	tablespoon(s)	30	23.9	22	0.8	5.0	—	0	0	0	0	—
	Soybeans													
1670	Boiled	½	cup(s)	86	53.8	149	14.3	8.5	5.2	7.7	1.1	1.7	4.4	—
2825	Dry roasted	½	cup(s)	86	0.7	388	34.0	28.1	7.0	18.6	2.7	4.1	10.5	—
2824	Roasted, salted	½	cup(s)	86	1.7	405	30.3	28.9	15.2	21.8	3.2	4.8	12.3	—
8739	Sprouted, stir fried	½	cup(s)	63	42.3	79	8.2	5.9	0.5	4.5	0.6	1.0	2.5	0
	Soy products													
1813	Soy milk	1	cup(s)	240	211.3	130	7.8	15.1	1.4	4.2	0.5	1.0	2.3	0
2838	Tofu, dried, frozen (koyadofu)	3	ounce(s)	85	4.9	408	40.8	12.4	6.1	25.8	3.7	5.7	14.6	

PAGE KEY: A-6 = Breads/Baked Goods A-12 = Cereal/Rice/Pasta A-16 = Fruit A-20 = Vegetables/Legumes A-30 = Nuts/Seeds A-32 = Vegetarian A-34 = Dairy A-42 = Eggs A-42 = Seafood A-44 = Meats A-48 = Poultry
A-50 = Processed Meats A-50 = Beverages A-54 = Fats/Oils A-56 = Sweets A-58 = Spices/Condiments/Sauces A-62 = Mixed Foods/Soups/Sandwiches A-66 = Fast Food A-86 = Convenience A-88 = Baby Foods

Chol (mg)	Calc (mg)	Iron (mg)	Magn (mg)	Pota (mg)	Sodi (mg)	Zinc (mg)	Vit A (µg)	Thia (mg)	Vit E (mg α)	Ribo (mg)	Niac (mg)	Vit B$_6$ (mg)	Fola (µg)	Vit C (mg)	Vit B$_{12}$ (µg)	Sele (µg)
0	29	0.96	17.9	127.5	331.5	0.74	368.5	0.09	—	0.07	0.74	0.11	23.0	8.4	0	1.1
0	18	0.75	12.8	126.4	54.4	0.36	380.8	0.18	0.41	0.05	0.92	0.07	20.8	6.5	0	0.9
0	47	1.92	22.4	173.6	4.0	0.39	52.8	0.05	0.37	0.09	0.45	0.13	28.0	17.6	0	0.6
0	14	0.65	7.6	63.0	1.3	0.08	17.0	0.04	0.12	0.02	0.19	0.05	13.2	18.9	0	0.2
0	22	1.34	33.6	228.6	12.0	0.62	4.8	0.12	—	0.08	1.84	0.14	86.4	6.2	0	0.4
0	6	0.31	6.8	112.9	1.4	0.08	15.6	0.04	0.34	0.02	0.32	0.15	10.9	50.6	0	0.2
0	7	0.25	7.5	130.4	2.2	0.09	13.4	0.04	0.27	0.02	0.35	0.16	7.5	59.9	0	0
0	8	0.54	11.3	153.0	3.2	0.13	26.6	0.04	0.31	0.04	0.42	0.12	10.4	109.1	0	0.2
0	5	0.34	9.5	127.2	797.6	0.10	24.5	0.01	0.46	0.02	0.54	0.10	6.8	46.2	0	0.2
0	16	1.28	10.2	131.2	1136.3	0.23	57.8	0.03	0.47	0.03	0.27	0.13	9.5	6.8	0	0.3
0	20	0.85	22.3	394.3	3.7	0.31	18.6	0.05	—	0.04	1.65	0.31	48.4	341.3	0	0.6
0	19	1.06	28.8	219.6	14.4	0.26	3.6	0.16	2.76	0.05	1.32	0.33	25.2	4.8	0	0.8
19	103	0.39	18.6	271.0	543.3	0.29	64.4	0.02	—	0.10	1.16	0.05	8.7	3.8	0	3.3
28	146	0.78	24.5	485.1	530.4	0.85	78.4	0.08	—	0.14	1.22	0.21	13.5	12.1	0	3.3
0	30	2.18	56.6	1080.7	20.2	0.72	2.0	0.12	0.08	0.09	2.84	0.62	56.6	19.4	0	0.8
0	3	0.21	15.3	238.5	3.1	0.18	0	0.06	0.02	0.01	0.85	0.18	5.5	7.8	0	0.2
0	20	4.08	24.9	332.3	12.2	0.28	0.6	0.07	0.02	0.06	1.77	0.35	12.8	7.8	0	0.4
0	7	0.42	29.9	515.4	5.4	0.40	0	0.14	0.01	0.02	1.95	0.40	13.6	17.7	0	0.4
0	13	1.27	34.1	572.0	7.4	0.46	0	0.14	0.01	0.03	2.13	0.44	15.0	18.4	0	—
0	6	0.24	15.6	255.8	3.9	0.21	0	0.07	0.01	0.01	1.02	0.21	7.0	5.8	0	0.2
0	16	1.05	30.8	567.0	8.4	0.39	0	0.08	0.09	0.03	1.34	0.37	16.1	21.2	0	0.4
0	8	0.51	18.2	315.7	271.6	0.26	0	0.09	0.07	0.02	1.55	0.12	19.6	9.3	0	0.1
0	11	0.43	27.3	449.3	266.8	0.37	0	0.13	0.01	0.03	1.80	0.37	12.5	10.1	0	0.4
1	23	0.27	19.9	344.4	349.6	0.31	43.0	0.09	0.44	0.04	1.23	0.25	9.4	11.0	0.1	0.8
2	36	0.21	21.0	164.8	179.5	0.26	49.3	0.09	0.53	0.09	0.90	0.16	8.4	6.8	0.1	5.9
0	22	2.50	54.5	902.9	16.2	0.72	0	0.24	—	0.06	3.46	0.69	24.2	30.5	0	0.8
0	4	0.31	19.4	319.0	5.4	0.25	0	0.10	—	0.01	1.26	0.25	9.3	11.7	0	0.3
0	27	3.44	21.5	377.0	9.3	0.29	0	0.04	0.01	0.04	1.28	0.28	9.9	8.9	0	0.3
0	9	0.41	10.9	199.7	307.2	0.21	0	0.08	0.15	0.02	0.97	0.08	9.0	4.0	0	0.4
14	45	0.47	17.4	252.2	423.7	0.31	43.5	0.02	—	0.06	1.28	0.05	12.4	4.1	0	2.0
15	70	0.70	23.3	463.1	410.4	0.49	0	0.08	—	0.11	1.29	0.22	13.5	13.0	0	2.0
0	18	0.69	11.0	281.8	1.2	0.28	306.3	0.03	0.98	0.09	0.50	0.05	11.0	5.8	0	0.2
0	32	1.70	28.2	252.4	6.1	0.20	953.1	0.02	1.29	0.06	0.45	0.06	14.7	5.1	0	0.5
0	8	0.23	5.2	120.8	8.8	0.25	0.4	0.01	0.90	0.01	0.10	0.02	24.0	3.2	0	0.4
0	8	0.23	5.2	120.8	8.8	0.25	0.4	0.01	0.90	0.01	0.10	0.02	24.0	3.2	0	0.4
0	7	0.09	2.7	62.9	10.5	0.07	0	0.00	0.00	0.01	0.06	0.01	6.8	4.0	0	0.2
0	41	0.45	19.6	277.1	17.0	0.30	0	0.07	0.27	0.04	0.61	0.09	12.8	16.0	0	0.6
0	35	1.73	15.3	200.6	780.0	0.22	1.2	0.03	0.17	0.03	0.17	0.15	28.3	17.3	0	0.7
0	67	1.12	48.4	35.6	93.2	0.48	2.4	0.02	0.32	0.04	0.16	0.00	72.0	1.2	0	0.3
0	9	2.14	14.6	102.2	78.6	0.15	2.2	0.18	0.38	0.28	0.96	0.03	7.1	0.8	0	0.5
0	11	0.36	6.3	100.2	3.6	0.12	18.0	0.02	—	0.01	0.06	0.09	10.2	2.4	—	0.4
0	88	4.42	74.0	442.9	0.9	0.98	0	0.13	0.30	0.24	0.34	0.20	46.4	1.5	0	6.3
0	120	3.39	196.1	1173.0	1.7	4.10	0	0.36	—	0.64	0.90	0.19	176.3	4.0	0	16.6
0	119	3.35	124.7	1264.2	140.2	2.70	8.6	0.08	0.78	0.12	1.21	0.17	181.5	1.9	0	16.4
0	52	0.25	60.4	356.6	8.8	1.32	0.6	0.26	—	0.12	0.69	0.10	79.9	7.5	0	0.4
0	60	1.53	60.0	283.2	122.4	0.28	0	0.14	0.26	0.16	1.23	0.18	43.2	0	0	11.5
0	310	8.27	50.2	17.0	5.1	4.16	22.1	0.42	—	0.27	1.01	0.24	78.2	0.6	0	46.2

TABLE A-1 **Table of Food Composition** *(continued)*   (Computer code is for Cengage Diet Analysis program)   (For purposes of calculations, use "0" for t, <1, <.1, <.01, etc.)

DA+ Code	Food Description	Quantity	Measure	Wt (g)	H₂0 (g)	Ener (kcal)	Prot (g)	Carb (g)	Fiber (g)	Fat (g)	Fat Breakdown (g) Sat	Mono	Poly	Trans
Vegetables, Legumes—continued														
13844	Tofu, extra firm	3	ounce(s)	85	—	86	8.6	2.2	1.1	4.3	0.5	0.9	2.8	—
13843	Tofu, firm	3	ounce(s)	85	—	75	7.5	2.2	0.5	3.2	0	0.9	2.3	—
1816	Tofu, firm, with calcium sulfate and magnesium chloride (nigari)	3	ounce(s)	85	72.2	60	7.0	1.4	0.8	3.5	0.7	1.0	1.5	—
1817	Tofu, fried	3	ounce(s)	85	43.0	230	14.6	8.9	3.3	17.2	2.5	3.8	9.7	—
13841	Tofu, silken	3	ounce(s)	85	—	42	3.7	1.9	0	2.3	0.5	—	—	—
13842	Tofu, soft	3	ounce(s)	85	—	65	6.5	1.1	0.5	3.2	0.5	1.1	2.2	—
1671	Tofu, soft, with calcium sulfate and magnesium chloride (nigari)	3	ounce(s)	85	74.2	52	5.6	1.5	0.2	3.1	0.5	0.7	1.8	—
	Spinach													
663	Canned, drained	½	cup(s)	107	98.2	25	3.0	3.6	2.6	0.5	0.1	0	0.2	—
660	Chopped, boiled, drained	½	cup(s)	90	82.1	21	2.7	3.4	2.2	0.2	0	0	0.1	—
661	Chopped, frozen, boiled, drained	½	cup(s)	95	84.5	32	3.8	4.6	3.5	0.8	0.1	0	0.4	—
662	Leaf, frozen, boiled, drained	½	cup(s)	95	84.5	32	3.8	4.6	3.5	0.8	0.1	0	0.4	—
659	Raw, chopped	1	cup(s)	30	27.4	7	0.9	1.1	0.7	0.1	0	0	0	—
8470	Trimmed leaves	1	cup(s)	32	27.5	3	0.9	0	2.8	0.1	—	—	—	—
	Squash													
1662	Acorn winter, baked	½	cup(s)	103	85.0	57	1.1	14.9	4.5	0.1	0	0	0.1	—
29702	Acorn winter, boiled, mashed	½	cup(s)	123	109.9	42	0.8	10.8	3.2	0.1	0	0	0	—
29451	Butternut, frozen, boiled	½	cup(s)	122	106.9	47	1.5	12.2	1.8	0.1	0	0	0	—
1661	Butternut winter, baked	½	cup(s)	102	89.5	41	0.9	10.7	3.4	0.1	0	0	0	—
32773	Butternut winter, frozen, boiled, mashed, no salt added	½	cup(s)	121	106.4	47	1.5	12.2	—	0.1	0	0	0	—
29700	Crookneck and straightneck summer, boiled, drained	½	cup(s)	65	60.9	12	0.6	2.6	1.2	0.1	0	0	0.1	—
29703	Hubbard winter, baked	½	cup(s)	102	86.8	51	2.5	11.0	—	0.6	0.1	0	0.3	—
1660	Hubbard winter, boiled, mashed	½	cup(s)	118	107.5	35	1.7	7.6	3.4	0.4	0.1	0	0.2	—
29704	Spaghetti winter, boiled, drained, or baked	½	cup(s)	78	71.5	21	0.5	5.0	1.1	0.2	0	0	0.1	—
664	Summer, all varieties, sliced, boiled, drained	½	cup(s)	90	84.3	18	0.8	3.9	1.3	0.3	0.1	0	0.1	—
665	Winter, all varieties, baked, mashed	½	cup(s)	103	91.4	38	0.9	9.1	2.9	0.4	0.1	0	0.2	—
1112	Zucchini summer, boiled, drained	½	cup(s)	90	85.3	14	0.6	3.5	1.3	0	0	0	0	—
1113	Zucchini summer, frozen, boiled, drained	½	cup(s)	112	105.6	19	1.3	4.0	1.4	0.1	0	0	0.1	—
	Sweet potatoes													
666	Baked, peeled	½	cup(s)	100	75.8	90	2.0	20.7	3.3	0.2	0	0	0.1	—
667	Boiled, mashed	½	cup(s)	164	131.4	125	2.2	29.1	4.1	0.2	0.1	0	0.1	—
668	Candied, home recipe	½	cup(s)	91	61.1	132	0.8	25.4	2.2	3.0	1.2	0.6	0.1	—
670	Canned, vacuum pack	½	cup(s)	100	76.0	91	1.7	21.1	1.8	0.2	0	0	0.1	—
2765	Frozen, baked	½	cup(s)	88	64.5	88	1.5	20.5	1.6	0.1	0	0	0	—
1136	Yams, baked or boiled, drained	½	cup(s)	68	47.7	79	1.0	18.7	2.7	0.1	0	0	0	—
32785	**Taro shoots, cooked, no salt added**	½	cup(s)	70	66.7	10	0.5	2.2	—	0.1	0	0	0	—
	Tomatillo													
8774	Raw	2	item(s)	68	62.3	22	0.7	4.0	1.3	0.7	0.1	0.1	0.3	—
8777	Raw, chopped	½	cup(s)	66	60.5	21	0.6	3.9	1.3	0.7	0.1	0.1	0.3	—
	Tomato													
16846	Cherry, fresh	5	item(s)	85	80.3	15	0.7	3.3	1.0	0.2	0	0	0.1	—
671	Fresh, ripe, red	1	item(s)	123	116.2	22	1.1	4.8	1.5	0.2	0	0	0.1	—
675	Juice, canned	½	cup(s)	122	114.1	21	0.9	5.2	0.5	0.1	0	0	0	—
75	Juice, no salt added	½	cup(s)	122	114.1	21	0.9	5.2	0.5	0.1	0	0	0	—
1699	Paste, canned	2	tablespoon(s)	33	24.1	27	1.4	6.2	1.3	0.2	0	0	0.1	—
1700	Puree, canned	¼	cup(s)	63	54.9	24	1.0	5.6	1.2	0.1	0	0	0.1	—
1118	Red, boiled	½	cup(s)	120	113.2	22	1.1	4.8	0.8	0.1	0	0	0.1	—
3952	Red, diced	½	cup(s)	90	85.1	16	0.8	3.5	1.1	0.2	0	0	0.1	—
1120	Red, stewed, canned	½	cup(s)	128	116.7	33	1.2	7.9	1.3	0.2	0	0	0.1	—
1125	Sauce, canned	¼	cup(s)	61	55.6	15	0.8	3.3	0.9	0.1	0	0	0	—
8778	Sun dried	½	cup(s)	27	3.9	70	3.8	15.1	3.3	0.8	0.1	0.1	0.3	—
8783	Sun dried in oil, drained	¼	cup(s)	28	14.8	59	1.4	6.4	1.6	3.9	0.5	2.4	0.6	—
	Turnips													
678	Turnip greens, chopped, boiled, drained	½	cup(s)	72	67.1	14	0.8	3.1	2.5	0.2	0	0	0.1	—

Chol (mg)	Calc (mg)	Iron (mg)	Magn (mg)	Pota (mg)	Sodi (mg)	Zinc (mg)	Vit A (µg)	Thia (mg)	Vit E (mg α)	Ribo (mg)	Niac (mg)	Vit B_6 (mg)	Fola (µg)	Vit C (mg)	Vit B_{12} (µg)	Sele (µg)
0	65	1.16	84.1	—	0	—	0	—	—	—	—	—	—	0	0	—
0	108	1.16	56.1	—	0	—	0	—	—	—	—	—	—	0	0	—
0	171	1.36	31.5	125.9	10.2	0.70	0	0.05	0.01	0.05	0.08	0.06	16.2	0.2	0	8.4
0	316	4.14	51.0	124.2	13.6	1.69	0.9	0.14	0.03	0.04	0.08	0.08	23.0	0	0	24.2
0	56	0.34	33.1	—	4.7	—	0	—	—	—	—	—	—	0	1.7	—
0	108	1.16	35.5	—	0	—	0	—	—	—	—	—	—	0	1.9	—
0	94	0.94	23.0	102.1	6.8	0.54	0	0.04	0.01	0.03	0.45	0.04	37.4	0.2	0	7.6
0	136	2.45	81.3	370.2	28.9	0.48	524.3	0.02	2.08	0.14	0.41	0.11	104.8	15.3	0	1.5
0	122	3.21	78.3	419.4	63.0	0.68	471.6	0.08	1.87	0.21	0.44	0.21	131.4	8.8	0	1.4
0	145	1.86	77.9	286.9	92.2	0.46	572.9	0.07	3.36	0.16	0.41	0.12	115.0	2.1	0	5.2
0	145	1.86	77.9	286.9	92.2	0.46	572.9	0.07	3.36	0.16	0.41	0.12	115.0	2.1	0	5.2
0	30	0.81	23.7	167.4	23.7	0.16	140.7	0.02	0.61	0.06	0.22	0.06	58.2	8.4	0	0.3
0	25	2.13	25.5	134.1	38.0	0.18	—	0.03	—	0.05	0.18	0.07	0	7.5	0	—
0	45	0.95	44.1	447.9	4.1	0.17	21.5	0.17	—	0.01	0.90	0.19	19.5	11.1	0	0.7
0	32	0.68	31.9	322.2	3.7	0.13	50.2	0.12	—	0.01	0.65	0.14	13.5	8.0	0	0.5
0	23	0.70	10.9	161.9	2.4	0.14	203.3	0.06	0.14	0.05	0.56	0.08	19.5	4.3	0	0.6
0	42	0.61	29.6	289.6	4.1	0.13	569.1	0.07	1.31	0.01	0.99	0.12	19.4	15.4	0	0.5
0	23	0.70	10.9	161.2	2.4	0.14	202.4	0.06	—	0.05	0.56	0.08	19.4	4.2	0	0.6
0	14	0.31	13.6	137.1	1.3	0.19	5.2	0.03	—	0.02	0.29	0.07	14.9	5.4	0	0.1
0	17	0.48	22.4	365.1	8.2	0.15	308.0	0.07	—	0.04	0.57	0.17	16.3	9.7	0	0.6
0	12	0.33	15.3	252.5	5.9	0.11	236.0	0.05	0.14	0.03	0.39	0.12	11.8	7.7	0	0.4
0	16	0.26	8.5	90.7	14.0	0.15	4.7	0.02	0.09	0.01	0.62	0.07	6.2	2.7	0	0.2
0	24	0.32	21.6	172.8	0.9	0.35	9.9	0.04	0.12	0.03	0.46	0.05	18.0	5.0	0	0.2
0	23	0.45	13.3	247.0	1.0	0.23	267.5	0.02	0.12	0.07	0.51	0.17	20.5	9.8	0	0.4
0	12	0.32	19.8	227.7	2.7	0.16	50.4	0.04	0.11	0.04	0.39	0.07	15.3	4.1	0	0.2
0	19	0.54	14.5	216.3	2.2	0.22	10	0.05	0.13	0.04	0.43	0.05	8.9	4.1	0	0.2
0	38	0.69	27.0	475.0	36.0	0.32	961.0	0.10	0.71	0.10	1.48	0.28	6.0	19.6	0	0.2
0	44	1.18	29.5	377.2	44.3	0.33	1290.7	0.09	1.54	0.08	0.88	0.27	9.8	21.0	0	0.3
7	24	1.03	10.0	172.6	63.9	0.13	0	0.01	—	0.03	0.36	0.03	10.0	6.1	0	0.7
0	22	0.89	22.0	312.0	53.0	0.18	399.0	0.04	1.00	0.06	0.74	0.19	17.0	26.4	0	0.7
0	31	0.47	18.4	330.1	7.0	0.26	913.3	0.05	0.67	0.04	0.49	0.16	19.3	8.0	0	0.5
0	10	0.35	12.2	455.6	5.4	0.13	4.1	0.06	0.23	0.01	0.37	0.15	10.9	8.2	0	0.5
0	10	0.28	5.6	240.8	1.4	0.37	2.1	0.02	—	0.03	0.56	0.07	2.1	13.2	0	0.7
0	5	0.42	13.6	182.2	0.7	0.15	4.1	0.03	0.25	0.02	1.25	0.03	4.8	8.0	0	0.3
0	5	0.41	13.2	176.9	0.7	0.15	4.0	0.03	0.25	0.02	1.22	0.04	4.6	7.7	0	0.3
0	9	0.22	9.4	201.5	4.3	0.14	35.7	0.03	0.45	0.01	0.50	0.06	12.8	10.8	0	0
0	12	0.33	13.5	291.5	6.2	0.20	51.7	0.04	0.66	0.02	0.73	0.09	18.5	15.6	0	0
0	12	0.52	13.4	278.2	326.8	0.18	27.9	0.06	0.39	0.04	0.82	0.14	24.3	22.2	0	0.4
0	12	0.52	13.4	278.2	12.2	0.18	27.9	0.06	0.39	0.04	0.82	0.14	24.3	22.2	0	0.4
0	12	0.97	13.8	332.6	259.1	0.20	24.9	0.02	1.41	0.05	1.00	0.07	3.9	7.2	0	1.7
0	11	1.11	14.4	274.4	249.4	0.22	16.3	0.01	1.23	0.05	0.91	0.07	6.9	6.6	0	0.4
0	13	0.82	10.8	261.6	13.2	0.17	28.8	0.04	0.67	0.03	0.64	0.10	15.6	27.4	0	0.6
0	9	0.24	9.9	213.3	4.5	0.15	37.8	0.03	0.48	0.01	0.53	0.07	13.5	11.4	0	0
0	43	1.70	15.3	263.9	281.8	0.22	11.5	0.06	1.06	0.04	0.91	0.02	6.4	10.1	0	0.8
0	8	0.62	9.8	201.9	319.6	0.12	10.4	0.01	0.87	0.04	0.59	0.06	6.7	4.3	0	0.1
0	30	2.45	52.4	925.3	565.7	0.53	11.9	0.14	0.00	0.13	2.44	0.09	18.4	10.6	0	1.5
0	13	0.73	22.3	430.4	73.2	0.21	17.6	0.05	—	0.10	0.99	0.08	6.3	28.0	0	0.8
0	99	0.58	15.8	146.2	20.9	0.10	274.3	0.03	1.35	0.05	0.30	0.13	85.0	19.7	0	0.6

TABLE A-1 Table of Food Composition (continued) (Computer code is for Cengage Diet Analysis program) (For purposes of calculations, use "0" for t, <1, <.1, <.01, etc.)

DA+ Code	Food Description	Quantity	Measure	Wt (g)	H₂0 (g)	Ener (kcal)	Prot (g)	Carb (g)	Fiber (g)	Fat (g)	Fat Breakdown (g) Sat	Mono	Poly	Trans
Vegetables, Legumes—continued														
679	Turnip greens, frozen, chopped, boiled, drained	½	cup(s)	82	74.1	24	2.7	4.1	2.8	0.3	0.1	0	0.1	—
677	Turnips, cubed, boiled, drained	½	cup(s)	78	73.0	17	0.6	3.9	1.6	0.1	0	0	0	—
	Vegetables, mixed													
1132	Canned, drained	½	cup(s)	82	70.9	40	2.1	7.5	2.4	0.2	0	0	0.1	—
680	Frozen, boiled, drained	½	cup(s)	91	75.7	59	2.6	11.9	4.0	0.1	0	0	0.1	—
7489	V8 100% vegetable juice	½	cup(s)	120	—	25	1.0	5.0	1.0	0	0	0	0	0
7490	V8 low sodium vegetable juice	½	cup(s)	120	—	25	0	6.5	1.0	0	0	0	0	0
7491	V8 spicy hot vegetable juice	½	cup(s)	120	—	25	1.0	5.0	0.5	0	0	0	0	0
	Water chestnuts													
31073	Sliced, drained	½	cup(s)	75	70.0	20	0	5.0	1.0	0	0	0	0	0
31087	Whole	½	cup(s)	75	70.0	20	0	5.0	1.0	0	0	0	0	0
1135	**Watercress**	1	cup(s)	34	32.3	4	0.8	0.4	0.2	0	0	0	0	—
Nuts, Seeds, and Products														
	Almonds													
32940	Almond butter with salt added	1	tablespoon(s)	16	0.2	101	2.4	3.4	0.6	9.5	0.9	6.1	2.0	—
1137	Almond butter, no salt added	1	tablespoon(s)	16	0.2	101	2.4	3.4	0.6	9.5	0.9	6.1	2.0	—
32886	Blanched	¼	cup(s)	36	1.6	211	8.0	7.2	3.8	18.3	1.4	11.7	4.4	—
32887	Dry roasted, no salt added	¼	cup(s)	35	0.9	206	7.6	6.7	4.1	18.2	1.4	11.6	4.4	—
29724	Dry roasted, salted	¼	cup(s)	35	0.9	206	7.6	6.7	4.1	18.2	1.4	11.6	4.4	—
29725	Oil roasted, salted	¼	cup(s)	39	1.1	238	8.3	6.9	4.1	21.7	1.7	13.7	5.3	—
508	Slivered	¼	cup(s)	27	1.3	155	5.7	5.9	3.3	13.3	1.0	8.3	3.3	0
1138	**Beechnuts, dried**	¼	cup(s)	57	3.8	328	3.5	19.1	5.3	28.5	3.3	12.5	11.4	—
517	**Brazil nuts, dried, unblanched**	¼	cup(s)	35	1.2	230	5.0	4.3	2.6	23.3	5.3	8.6	7.2	—
1166	**Breadfruit seeds, roasted**	¼	cup(s)	57	28.3	118	3.5	22.8	3.4	1.5	0.4	0.2	0.8	—
1139	**Butternuts, dried**	¼	cup(s)	30	1.0	184	7.5	3.6	1.4	17.1	0.4	3.1	12.8	—
	Cashews													
32931	Cashew butter with salt added	1	tablespoon(s)	16	0.5	94	2.8	4.4	0.3	7.9	1.6	4.7	1.3	—
32889	Cashew butter, no salt added	1	tablespoon(s)	16	0.5	94	2.8	4.4	0.3	7.9	1.6	4.7	1.3	—
1140	Dry roasted	¼	cup(s)	34	0.6	197	5.2	11.2	1.0	15.9	3.1	9.4	2.7	—
518	Oil roasted	¼	cup(s)	32	1.1	187	5.4	9.6	1.1	15.4	2.7	8.4	2.8	—
	Coconut, shredded													
32896	Dried, not sweetened	¼	cup(s)	23	0.7	152	1.6	5.4	3.8	14.9	13.2	0.6	0.2	—
1153	Dried, shredded, sweetened	¼	cup(s)	23	2.9	116	0.7	11.1	1.0	8.3	7.3	0.4	0.1	—
520	Shredded	¼	cup(s)	20	9.4	71	0.7	3.0	1.8	6.7	5.9	0.3	0.1	—
	Chestnuts													
1152	Chinese, roasted	¼	cup(s)	36	14.6	87	1.6	19.0	—	0.4	0.1	0.2	0.1	—
32895	European, boiled and steamed	¼	cup(s)	46	31.3	60	0.9	12.8	—	0.6	0.1	0.2	0.2	—
32911	European, roasted	¼	cup(s)	36	14.5	88	1.1	18.9	1.8	0.8	0.1	0.3	0.3	—
32922	Japanese, boiled and steamed	¼	cup(s)	36	31.0	20	0.3	4.5	—	0.1	0	0	0	—
32923	Japanese, roasted	¼	cup(s)	36	18.1	73	1.1	16.4	—	0.3	0	0.1	0.1	—
4958	**Flax seeds or linseeds**	¼	cup(s)	43	3.3	225	8.4	12.3	11.9	17.7	1.7	3.2	12.6	0
32904	**Ginkgo nuts, dried**	¼	cup(s)	39	4.8	136	4.0	28.3	—	0.8	0.1	0.3	0.3	—
	Hazelnuts or filberts													
32901	Blanched	¼	cup(s)	30	1.7	189	4.1	5.1	3.3	18.3	1.4	14.5	1.7	—
32902	Dry roasted, no salt added	¼	cup(s)	30	0.8	194	4.5	5.3	2.8	18.7	1.3	14.0	2.5	—
1156	**Hickory nuts, dried**	¼	cup(s)	30	0.8	197	3.8	5.5	1.9	19.3	2.1	9.8	6.6	—
	Macadamias													
32905	Dry roasted, no salt added	¼	cup(s)	34	0.5	241	2.6	4.5	2.7	25.5	4.0	19.9	0.5	—
32932	Dry roasted, with salt added	¼	cup(s)	34	0.5	240	2.6	4.3	2.7	25.5	4.0	19.9	0.5	—
1157	Raw	¼	cup(s)	34	0.5	241	2.6	4.6	2.9	25.4	4.0	19.7	0.5	—
	Mixed nuts													
1159	With peanuts, dry roasted	¼	cup(s)	34	0.6	203	5.9	8.7	3.1	17.6	2.4	10.8	3.7	—
32933	With peanuts, dry roasted, with salt added	¼	cup(s)	34	0.6	203	5.9	8.7	3.1	17.6	2.4	10.8	3.7	—
32906	Without peanuts, oil roasted, no salt added	¼	cup(s)	36	1.1	221	5.6	8.0	2.0	20.2	3.3	11.9	4.1	—
	Peanuts													
2807	Dry roasted	¼	cup(s)	37	0.6	214	8.6	7.9	2.9	18.1	2.5	9.0	5.7	—
2806	Dry roasted, salted	¼	cup(s)	37	0.6	214	8.6	7.9	2.9	18.1	2.5	9.0	5.7	—
1763	Oil roasted, salted	¼	cup(s)	36	0.5	216	10.1	5.5	3.4	18.9	3.1	9.4	5.5	—
1884	Peanut butter, chunky	1	tablespoon(s)	16	0.2	94	3.8	3.5	1.3	8.0	1.3	3.9	2.4	—
30303	Peanut butter, low sodium	1	tablespoon(s)	16	0.2	95	4.0	3.1	0.9	8.2	1.8	3.9	2.2	—
30305	Peanut butter, reduced fat	1	tablespoon(s)	18	0.2	94	4.7	6.4	0.9	6.1	1.3	2.9	1.8	—

Chol (mg)	Calc (mg)	Iron (mg)	Magn (mg)	Pota (mg)	Sodi (mg)	Zinc (mg)	Vit A (µg)	Thia (mg)	Vit E (mg α)	Ribo (mg)	Niac (mg)	Vit B_6 (mg)	Fola (µg)	Vit C (mg)	Vit B_{12} (µg)	Sele (µg)
0	125	1.59	21.3	183.7	12.3	0.34	441.2	0.04	2.18	0.06	0.38	0.06	32.0	17.9	0	1.0
0	26	0.14	7.0	138.1	12.5	0.09	0	0.02	0.02	0.02	0.23	0.05	7.0	9.0	0	0.2
0	22	0.86	13.0	237.2	121.4	0.33	475.1	0.04	0.24	0.04	0.47	0.06	19.6	4.1	0	0.2
0	23	0.74	20.0	153.8	31.9	0.44	194.7	0.06	0.34	0.10	0.77	0.06	17.3	2.9	0	0.3
0	20	0.36	12.9	260.0	310.0	0.24	100.0	0.05	—	0.03	0.87	0.17	—	30.0	0	—
0	20	0.36	—	450.0	70.0	—	100.0	0.02	—	0.02	0.75	—	—	30.0	0	—
0	20	0.36	12.9	240.0	360.0	0.24	50.0	0.05	—	0.03	0.88	0.17	—	15.0	0	—
0	7	0.00	—	—	5.0	—	0	—	—	—	—	—	—	2.0	—	—
0	7	0.00	—	—	5.0	—	0	—	—	—	—	—	—	2.0	—	—
0	41	0.06	7.1	112.2	13.9	0.03	54.4	0.03	0.34	0.04	0.06	0.04	3.1	14.6	0	0.3
0	43	0.59	48.5	121.3	72.0	0.49	0	0.02	4.16	0.10	0.46	0.01	10.4	0.1	0	0.8
0	43	0.59	48.5	121.3	1.8	0.48	0	0.02	—	0.09	0.46	0.01	10.4	0.1	0	—
0	78	1.34	99.7	249.0	10.2	1.13	0	0.07	8.95	0.20	1.32	0.04	10.9	0	0	1.0
0	92	1.55	98.7	257.4	0.3	1.22	0	0.02	8.97	0.29	1.32	0.04	11.4	0	0	1.0
0	92	1.55	98.7	257.4	117.0	1.22	0	0.02	8.97	0.29	1.32	0.04	11.4	0	0	1.0
0	114	1.44	107.5	274.4	133.1	1.20	0	0.03	10.19	0.30	1.43	0.04	10.6	0	0	1.1
0	71	1.00	72.4	190.4	0.3	0.83	0	0.05	7.07	0.27	0.91	0.03	13.5	0	0	0.7
0	1	1.39	0	579.7	21.7	0.20	0	0.16	—	0.20	0.48	0.38	64.4	8.8	0	4.0
0	56	0.85	131.6	230.7	1.1	1.42	0	0.21	2.00	0.01	0.10	0.03	7.7	0.2	0	671.0
0	49	0.50	35.3	616.7	15.9	0.58	8.5	0.22	—	0.12	4.20	0.22	33.6	4.3	0	8.0
0	16	1.21	71.1	126.3	0.3	0.94	1.8	0.12	—	0.04	0.31	0.17	19.8	1.0	0	5.2
0	7	0.81	41.3	87.4	98.2	0.83	0	0.05	0.15	0.03	0.26	0.04	10.9	0	0	1.8
0	7	0.81	41.3	87.4	2.4	0.83	0	0.05	—	0.03	0.26	0.04	10.9	0	0	1.8
0	15	2.06	89.1	193.5	5.5	1.92	0	0.07	0.32	0.07	0.48	0.09	23.6	0	0	4.0
0	14	1.95	88.0	203.8	4.2	1.73	0	0.12	0.30	0.07	0.56	0.10	8.1	0.1	0	6.5
0	6	0.76	20.7	125.2	8.5	0.46	0	0.01	0.10	0.02	0.13	0.07	2.1	0.3	0	4.3
0	3	0.45	11.6	78.4	60.9	0.42	0	0.01	0.09	0.00	0.11	0.06	1.9	0.2	0	3.9
0	3	0.48	6.4	71.2	4.0	0.21	0	0.01	0.04	0.00	0.11	0.01	5.2	0.7	0	2.0
0	7	0.54	32.6	173.0	1.4	0.33	0	0.05	—	0.03	0.54	0.15	26.1	13.9	0	2.6
0	21	0.80	24.8	328.9	12.4	0.11	0.5	0.06	—	0.03	0.32	0.10	17.5	12.3	0	—
0	10	0.32	11.8	211.6	0.7	0.20	0.4	0.08	0.18	0.05	0.48	0.18	25.0	9.3	0	0.4
0	4	0.19	6.5	42.8	1.8	0.14	0.4	0.04	—	0.01	0.19	0.03	6.1	3.4	0	—
0	13	0.75	23.2	154.8	6.9	0.51	1.4	0.15	—	—	0.24	0.14	21.4	10.1	0	—
142	142	2.13	156.1	354.0	11.9	1.83	0	0.06	0.14	0.06	0.59	0.39	118.4	0.5	0	2.3
0	8	0.62	20.7	390.2	5.1	0.26	21.5	0.17	—	0.07	4.58	0.25	41.4	11.4	0	—
0	45	0.98	48.0	197.4	0	0.66	0.6	0.14	5.25	0.03	0.46	0.17	23.4	0.6	0	1.2
0	37	1.31	51.9	226.5	0	0.74	0.9	0.10	4.58	0.03	0.61	0.18	26.4	1.1	0	1.2
0	18	0.64	51.9	130.8	0.3	1.29	2.1	0.26	—	0.04	0.27	0.06	12.0	0.6	0	2.4
0	23	0.88	39.5	121.6	1.3	0.43	0	0.23	0.19	0.02	0.76	0.12	3.4	0.2	0	3.9
0	23	0.88	39.5	121.6	88.8	0.43	0	0.23	0.19	0.02	0.76	0.12	3.4	0.2	0	3.9
0	28	1.24	43.6	123.3	1.7	0.44	0	0.40	0.18	0.05	0.83	0.09	3.7	0.4	0	1.2
0	24	1.27	77.1	204.5	4.1	1.30	0.3	0.07	—	0.07	1.61	0.10	17.1	0.1	0	1.0
0	24	1.26	77.1	204.5	229.1	1.30	0	0.06	3.74	0.06	1.61	0.10	17.1	0.1	0	2.6
0	38	0.92	90.4	195.8	4.0	1.67	0.4	0.18	—	0.17	0.70	0.06	20.2	0.2	0	—
0	20	0.82	64.2	240.2	2.2	1.20	0	0.16	2.52	0.03	4.93	0.09	52.9	0	0	2.7
0	20	0.82	64.2	240.2	296.7	1.20	0	0.16	2.84	0.03	4.93	0.09	52.9	0	0	2.7
0	22	0.54	63.4	261.4	115.2	1.18	0	0.03	2.49	0.03	4.97	0.16	43.2	0.3	0	1.2
0	7	0.30	25.6	119.2	77.8	0.45	0	0.02	1.01	0.02	2.19	0.07	14.7	0	0	1.3
0	6	0.29	25.4	107.0	2.7	0.47	0	0.01	1.23	0.02	2.14	0.07	11.8	0	0	1.2
0	6	0.34	30.6	120.4	97.2	0.50	0	0.05	1.20	0.01	2.63	0.06	10.8	0	0	1.4

TABLE A-1 **Table of Food Composition *(continued)*** (Computer code is for Cengage Diet Analysis program) (For purposes of calculations, use "0" for t, <1, <.1, <.01, etc.)

DA+ Code	Food Description	Quantity	Measure	Wt (g)	H₂0 (g)	Ener (kcal)	Prot (g)	Carb (g)	Fiber (g)	Fat (g)	Sat	Mono	Poly	*Trans*
											Fat Breakdown (g)			

Nuts, Seeds, and Products—continued

DA+ Code	Food Description	Quantity	Measure	Wt (g)	H₂0 (g)	Ener (kcal)	Prot (g)	Carb (g)	Fiber (g)	Fat (g)	Sat	Mono	Poly	*Trans*
524	Peanut butter, smooth	1	tablespoon(s)	16	0.3	94	4.0	3.1	1.0	8.1	1.7	3.9	2.3	—
2804	Raw	¼	cup(s)	37	2.4	207	9.4	5.9	3.1	18.0	2.5	8.9	5.7	—
	Pecans													
32907	Dry roasted, no salt added	¼	cup(s)	28	0.3	198	2.6	3.8	2.6	20.7	1.8	12.3	5.7	—
32936	Dry roasted, with salt added	¼	cup(s)	27	0.3	192	2.6	3.7	2.5	20.0	1.7	11.9	5.6	—
1162	Oil roasted	¼	cup(s)	28	0.3	197	2.5	3.6	2.6	20.7	2.0	11.3	6.5	—
526	Raw	¼	cup(s)	27	1.0	188	2.5	3.8	2.6	19.6	1.7	11.1	5.9	—
12973	**Pine nuts or pignolia, dried**	1	tablespoon(s)	9	0.2	58	1.2	1.1	0.3	5.9	0.4	1.6	2.9	—
	Pistachios													
1164	Dry roasted	¼	cup(s)	31	0.6	176	6.6	8.5	3.2	14.1	1.7	7.4	4.3	—
32938	Dry roasted, with salt added	¼	cup(s)	32	0.6	182	6.8	8.6	3.3	14.7	1.8	7.7	4.4	—
1167	**Pumpkin or squash seeds, roasted**	¼	cup(s)	57	4.0	296	18.7	7.6	2.2	23.9	4.5	7.4	10.9	
	Sesame													
32912	Sesame butter paste	1	tablespoon(s)	16	0.3	94	2.9	3.8	0.9	8.1	1.1	3.1	3.6	—
32941	Tahini or sesame butter	1	tablespoon(s)	15	0.5	89	2.6	3.2	0.7	8.0	1.1	3.0	3.5	—
1169	Whole, roasted, toasted	3	tablespoon(s)	10	0.3	54	1.6	2.4	1.3	4.6	0.6	1.7	2.0	—
	Soy nuts													
34173	Deep sea salted	¼	cup(s)	28	—	119	11.9	8.9	4.9	4.0	1.0	—	—	—
34174	Unsalted	¼	cup(s)	28	—	119	11.9	8.9	4.9	4.0	0	—	—	—
	Sunflower seeds													
528	Kernels, dried	1	tablespoon(s)	9	0.4	53	1.9	1.8	0.8	4.6	0.4	1.7	2.1	—
29721	Kernels, dry roasted, salted	1	tablespoon(s)	8	0.1	47	1.5	1.9	0.7	4.0	0.4	0.8	2.6	—
29723	Kernels, toasted, salted	1	tablespoon(s)	8	0.1	52	1.4	1.7	1.0	4.8	0.5	0.9	3.1	—
32928	Sunflower seed butter with salt added	1	tablespoon(s)	16	0.2	93	3.1	4.4	—	7.6	0.8	1.5	5.0	—
	Trail mix													
4646	Trail mix	¼	cup(s)	38	3.5	173	5.2	16.8	2.0	11.0	2.1	4.7	3.6	—
4647	Trail mix with chocolate chips	¼	cup(s)	38	2.5	182	5.3	16.8	—	12.0	2.3	5.1	4.2	—
4648	Tropical trail mix	¼	cup(s)	35	3.2	142	2.2	23.0	—	6.0	3.0	0.9	1.8	—
	Walnuts													
529	Dried black, chopped	¼	cup(s)	31	1.4	193	7.5	3.1	2.1	18.4	1.1	4.7	11.0	—
531	English or Persian	¼	cup(s)	29	1.2	191	4.5	4.0	2.0	19.1	1.8	2.6	13.8	—

Vegetarian Foods

DA+ Code	Food Description	Quantity	Measure	Wt (g)	H₂0 (g)	Ener (kcal)	Prot (g)	Carb (g)	Fiber (g)	Fat (g)	Sat	Mono	Poly	*Trans*
	Prepared													
34222	Brown rice and tofu stir-fry (vegan)	8	ounce(s)	227	244.4	302	16.5	18.0	3.2	21.0	1.7	4.7	13.4	0
34368	Cheese enchilada casserole (lacto)	8	ounce(s)	227	80.3	385	16.6	38.4	4.1	17.8	9.5	6.1	1.1	—
34247	Five bean casserole (vegan)	8	ounce(s)	227	175.8	178	5.9	26.6	6.0	5.8	1.1	2.5	1.9	0
34261	Lentil stew (vegan)	8	ounce(s)	227	227.9	188	11.5	35.9	11.0	0.7	0.1	0.1	0.3	0
34397	Macaroni and cheese (lacto)	8	ounce(s)	227	352.1	391	18.1	37.1	1.0	18.7	9.8	6.0	1.8	0
34238	Steamed rice and vegetables (vegan)	8	ounce(s)	227	222.9	587	11.2	87.9	5.8	23.1	4.1	8.7	9.1	0
34308	Tofu rice burgers (ovo-lacto)	1	piece(s)	218	77.6	435	22.4	68.6	5.6	8.4	1.7	2.4	3.5	—
34276	Vegan spinach enchiladas (vegan)	1	piece(s)	82	59.2	93	4.9	14.5	1.8	2.4	0.3	0.6	1.3	—
34243	Vegetable chow mein (vegan)	8	ounce(s)	227	163.3	166	6.5	22.1	2.0	6.4	0.7	2.7	2.5	0
34454	Vegetable lasagna (lacto)	8	ounce(s)	227	178.9	208	13.7	29.9	2.6	4.1	2.3	1.1	0.3	—
34339	Vegetable marinara (vegan)	8	ounce(s)	252	200.7	104	3.0	16.7	1.4	3.1	0.4	1.4	1.0	0
34356	Vegetable rice casserole (lacto)	8	ounce(s)	227	178.9	238	9.7	24.4	4.0	12.5	4.9	3.5	3.1	—
34311	Vegetable strudel (ovo-lacto)	8	ounce(s)	227	63.1	478	12.0	32.4	2.5	33.8	11.5	16.7	3.9	0
34371	Vegetable taco (lacto)	1	item(s)	85	46.5	117	4.2	13.6	2.9	5.6	2.1	1.9	1.3	—
34282	Vegetarian chili (vegan)	8	ounce(s)	227	191.4	115	5.6	21.4	7.1	1.5	0.2	0.3	0.7	0
34367	Vegetarian vegetable soup (vegan)	8	ounce(s)	227	257.9	111	3.2	16.0	3.2	5.0	1.0	2.1	1.6	0
	Boca burger													
32067	All American flamed grilled patty	1	item(s)	71	—	90	14.0	4.0	3.0	3.0	1.0	—	—	0
32074	Boca chik'n nuggets	4	item(s)	87	—	180	14.0	17.0	3.0	7.0	1.0	—	—	0
32075	Boca meatless ground burger	½	cup(s)	57	—	60	13.0	6.0	3.0	0.5	0	—	—	0
32072	Breakfast links	2	item(s)	45	—	70	8.0	5.0	2.0	3.0	0.5	—	—	0
32071	Breakfast patties	1	item(s)	38	—	60	7.0	5.0	2.0	2.5	0	—	—	0
35780	Cheeseburger meatless burger patty	1	item(s)	71	—	100	12.0	5.0	3.0	5.0	1.5	—	—	0
33958	Original meatless chik'n patties	1	item(s)	71	—	160	11.0	15.0	2.0	6.0	1.0	—	—	0

PAGE KEY: A-6 = Breads/Baked Goods A-12 = Cereal/Rice/Pasta A-16 = Fruit A-20 = Vegetables/Legumes A-30 = Nuts/Seeds A-32 = Vegetarian A-34 = Dairy A-42 = Eggs A-42 = Seafood A-44 = Meats A-48 = Poultry A-50 = Processed Meats A-50 = Beverages A-54 = Fats/Oils A-56 = Sweets A-58 = Spices/Condiments/Sauces A-62 = Mixed Foods/Soups/Sandwiches A-66 = Fast Food A-86 = Convenience A-88 = Baby Foods

Chol (mg)	Calc (mg)	Iron (mg)	Magn (mg)	Pota (mg)	Sodi (mg)	Zinc (mg)	Vit A (µg)	Thia (mg)	Vit E (mg α)	Ribo (mg)	Niac (mg)	Vit B$_6$ (mg)	Fola (µg)	Vit C (mg)	Vit B$_{12}$ (µg)	Sele (µg)
0	7	0.30	24.6	103.8	73.4	0.47	0	0.01	1.44	0.02	2.14	0.09	11.8	0	0	0.9
0	34	1.67	61.3	257.3	6.6	1.19	0	0.23	3.04	0.04	4.40	0.12	87.6	0	0	2.6
0	20	0.78	36.8	118.3	0.3	1.41	1.9	0.12	0.35	0.03	0.32	0.05	4.5	0.2	0	1.1
0	19	0.75	35.6	114.5	103.4	1.36	1.9	0.11	0.34	0.03	0.31	0.05	4.3	0.2	0	1.1
0	18	0.68	33.3	107.8	0.3	1.23	1.4	0.13	0.70	0.03	0.33	0.05	4.1	0.2	0	1.7
0	19	0.69	33.0	111.7	0	1.23	0.8	0.18	0.38	0.04	0.32	0.06	6.0	0.3	0	1.0
0	1	0.47	21.6	51.3	0.2	0.55	0.1	0.03	0.80	0.02	0.37	0.01	2.9	0.1	0	0.1
0	34	1.29	36.9	320.4	3.1	0.71	4.0	0.26	0.59	0.05	0.44	0.39	15.4	0.7	0	2.9
0	35	1.34	38.4	333.4	129.6	0.73	4.2	0.26	0.61	0.05	0.45	0.40	16.0	0.7	0	3.0
0	24	8.48	303.0	457.4	10.2	4.22	10.8	0.12	0.00	0.18	0.99	0.05	32.3	1.0	0	3.2
0	154	3.07	57.9	93.1	1.9	1.17	0.5	0.04	—	0.03	1.07	0.13	16.0	0	0	0.9
0	21	0.66	14.3	68.9	5.3	0.69	0.5	0.24	—	0.02	0.85	0.02	14.7	0.6	0	0.3
0	94	1.40	33.8	45.1	1.0	0.68	0	0.07	—	0.02	0.43	0.07	9.3	0	0	0.5
0	59	1.07	—	—	148.1	—	0	—	—	—	—	—	—	0	—	—
0	59	1.07	—	—	9.9	—	0	—	—	—	—	—	—	0	—	—
0	7	0.47	29.3	58.1	0.8	0.45	0.3	0.13	2.99	0.03	0.75	0.12	20.4	0.1	0	4.8
0	6	0.30	10.3	68.0	32.8	0.42	0	0.01	2.09	0.02	0.56	0.06	19.0	0.1	0	6.3
0	5	0.57	10.8	41.1	51.3	0.44	0	0.03	—	0.02	0.35	0.07	19.9	0.1	0	5.2
0	20	0.76	59.0	11.5	83.2	0.85	0.5	0.05	—	0.05	0.85	0.13	37.9	0.4		
0	29	1.14	59.3	256.9	85.9	1.20	0.4	0.17	—	0.07	1.76	0.11	26.6	0.5	0	—
2	41	1.27	60.4	243.0	45.4	1.17	0.8	0.15	—	0.08	1.65	0.09	24.4	0.5	0	—
0	20	0.92	33.6	248.2	3.5	0.41	0.7	0.15	—	0.04	0.51	0.11	14.7	2.7	0	—
0	19	0.97	62.8	163.4	0.6	1.05	0.6	0.01	0.56	0.04	0.14	0.18	9.7	0.5	0	5.3
0	29	0.85	46.2	129.0	0.6	0.90	0.3	0.10	0.20	0.04	0.32	0.15	28.7	0.4	0	1.4
0	353	6.34	118.3	501.4	142.2	2.03	—	0.23	0.07	0.14	1.49	0.36	51.8	24.8	0	14.8
39	441	2.44	34.6	191.2	1139.7	1.84	—	0.31	0.05	0.35	2.23	0.11	87.1	20.4	0.4	20.0
0	48	1.78	40.8	364.1	613.6	0.61	—	0.10	0.52	0.07	0.93	0.11	39.4	8.3	0	3.3
0	34	3.23	50.0	548.8	436.5	1.42	—	0.24	0.14	0.16	2.31	0.29	91.4	26.4	0	12.1
43	415	1.71	45.4	267.8	1641.0	2.32	—	0.32	0.27	0.48	2.18	0.13	74.2	0.9	0.8	33.3
0	91	3.31	153.1	810.1	3117.8	2.04	—	0.37	3.03	0.21	6.16	0.64	61.7	35.2	0	18.8
52	467	9.01	89.7	455.6	2449.5	2.06	—	0.27	0.12	0.26	3.43	0.29	106.6	2.0	0.1	43.0
0	117	1.13	40.4	170.5	134.2	0.68	—	0.07	—	0.07	0.53	0.10	52.1	1.8	0	5.1
0	189	3.70	28.0	310.3	372.7	0.76	—	0.13	0.05	0.11	1.43	0.14	45.6	8.0	0	6.5
10	176	1.86	41.9	470.0	759.4	1.14	—	0.26	0.05	0.25	2.49	0.22	64.7	19.0	0.4	21.8
0	17	0.94	19.1	189.9	439.6	0.42	—	0.15	0.55	0.08	1.36	0.12	40.6	23.5	0	10.8
17	190	1.28	29.3	414.2	626.0	1.24	—	0.16	0.35	0.29	2.00	0.19	72.6	56.0	0.2	5.8
29	200	2.15	24.5	181.0	512.1	1.24	—	0.28	0.20	0.31	2.88	0.11	88.3	17.4	0.2	19.7
7	77	0.88	26.3	174.1	280.7	0.59	—	0.08	0.04	0.06	0.49	0.08	38.7	4.6	0	3.0
0	65	1.98	41.0	543.1	390.7	0.74	—	0.14	0.15	0.10	1.31	0.18	59.3	20.3	0	4.4
0	46	1.87	34.9	550.3	729.5	0.56	—	0.13	0.55	0.09	1.99	0.27	47.8	29.9	0	1.4
5	150	1.80	—	—	280.0	—	0	—	—	—	—	—	—	0	—	—
0	40	1.44	—	—	500.0	—	0	—	—	—	—	—	—	0	—	—
0	60	1.80	—	—	270.0	—	0	—	—	—	—	—	—	0	—	—
0	20	1.44	—	—	330.0	—	0	—	—	—	—	—	—	0	—	—
0	20	1.08	—	—	280.0	—	0	—	—	—	—	—	—	0	—	—
5	80	1.80	—	—	360.0	—	—	—	—	—	—	—	—	0	—	—
0	40	1.80	—	—	430.0	—	—	—	—	—	—	—	—	0	—	—

TABLE A-1 **Table of Food Composition (*continued*)** (Computer code is for Cengage Diet Analysis program) (For purposes of calculations, use "0" for t, <1, <.1, <.01, etc.)

DA+ Code	Food Description	Quantity	Measure	Wt (g)	H₂0 (g)	Ener (kcal)	Prot (g)	Carb (g)	Fiber (g)	Fat (g)	Fat Breakdown (g)			
											Sat	Mono	Poly	*Trans*
Vegetarian Foods—continued														
32066	Original patty	1	item(s)	71	—	70	13.0	6.0	4.0	0.5	0	—	—	0
32068	Roasted garlic patty	1	item(s)	71	—	70	12.0	6.0	4.0	1.5	0	—	—	0
37814	Roasted onion meatless burger patty	1	item(s)	71	—	70	11.0	7.0	4.0	1.0	0	—	—	0
	Gardenburger													
37810	BBQ chik'n with sauce	1	item(s)	142	—	250	14.0	30.0	5.0	8.0	1.0	—	—	0
39661	Black bean burger	1	item(s)	71	—	80	8.0	11.0	4.0	2.0	0	—	—	0
39666	Buffalo chik'n wing	3	item(s)	95	—	180	9.0	8.0	5.0	12.0	1.5	—	—	0
39665	Country fried chicken with creamy pepper gravy	1	item(s)	142	—	190	9.0	16.0	2.0	9.0	1.0	—	—	0
37808	Flamed grilled chik'n	1	item(s)	71	—	100	13.0	5.0	3.0	2.5	0	—	—	0
37803	Garden vegan	1	item(s)	71	—	100	10.0	12.0	2.0	1.0	—	—	—	0
39663	Homestyle classic burger	1	item(s)	71	—	110	12.0	6.0	4.0	5.0	0.5	—	—	0
37807	Meatless breakfast sausage	1	item(s)	43	—	50	5.0	2.0	2.0	3.5	0	—	—	0
37809	Meatless meatballs	6	item(s)	85	—	110	12.0	8.0	4.0	4.5	1.0	—	—	0
37806	Meatless riblets with sauce	1	item(s)	142	—	160	17.0	11.0	4.0	5.0	0	—	—	0
29913	Original	1	item(s)	71	—	90	10.0	8.0	3.0	2.0	0.5	—	—	0
39662	Sun-dried tomato basil burger	1	item(s)	71	—	80	10.0	11.0	3.0	1.5	0.5	—	—	0
29915	Veggie medley	1	item(s)	71	—	90	9.0	11.0	4.0	2.0	0	—	—	0
	Loma Linda													
9311	Big franks, canned	1	item(s)	51	—	110	11.0	3.0	2.0	6.0	1.0	1.5	3.5	0
9323	Fried chik'n with gravy	2	piece(s)	80	45.9	150	12.0	5.0	2.0	10	1.5	2.5	5.0	0
9326	Linketts, canned	1	item(s)	35	21.0	70	7.0	1.0	1.0	4.0	0.5	1.0	2.5	0
9336	Redi-Burger patties, canned	1	slice(s)	85	50.5	120	18.0	7.0	4.0	2.5	0.5	0.5	1.5	0
9350	Swiss Stake pattie with gravy, frozen	1	piece(s)	92	65.7	130	9.0	9.0	3.0	6.0	1.0	1.5	3.5	0
9354	Tender Rounds meatball substitute, canned in gravy	6	piece(s)	80	53.9	120	13.0	6.0	1.0	4.5	0.5	1.5	2.5	0
	Morningstar Farms													
33707	America's Original Veggie Dog links	1	item(s)	57	—	80	11.0	6.0	1.0	0.5	0	—	—	0
9362	Better'n Eggs egg substitute	¼	cup(s)	57	50.3	20	5.0	0	0	0	0	0	0	0
9371	Breakfast bacon strips	2	item(s)	16	6.8	60	2.0	2.0	0.5	4.5	0.5	1.0	3.0	0
9368	Breakfast sausage links	2	item(s)	45	26.8	80	9.0	3.0	2.0	3.0	0.5	1.5	1.0	0
33705	Chik'n nuggets	4	piece(s)	86	—	190	12.0	18.0	2.0	7.0	1.0	2.0	4.0	0
11587	Chik patties	1	item(s)	71	36.3	150	9.0	16.0	2.0	6.0	1.0	1.5	2.5	0
2531	Garden veggie patties	1	item(s)	67	40.1	100	10.0	9.0	4.0	2.5	0.5	0.5	1.5	0
33702	Spicy black bean veggie burger	1	item(s)	78	—	140	12.0	15.0	3.0	4.0	0.5	1.0	2.5	0
9412	Vegetarian chili, canned	1	cup(s)	230	172.6	180	16.0	25.0	10.0	1.5	0.5	0.5	0.5	0
	Worthington													
9424	Chili, canned	1	cup(s)	230	167.0	280	24.0	25.0	8.0	10.0	1.5	1.5	7.0	0
9436	Diced chik, canned	¼	cup(s)	55	42.7	50	9.0	2.0	1.0	0	0	0	0	0
9440	Dinner roast, frozen	1	slice(s)	85	53.2	180	14.0	6.0	3.0	11.0	1.5	4.5	5.0	0
9420	Meatless chicken slices, frozen	3	slice(s)	57	38.9	90	9.0	2.0	0.5	4.5	1.0	1.0	2.5	0
36702	Meatless chicken style roll, frozen	1	slice(s)	55	—	90	9.0	2.0	1.0	4.5	1.0	1.0	2.5	0
9428	Meatless corned beef, sliced, frozen	3	slice(s)	57	31.2	140	10.0	5.0	0	9.0	1.0	2.0	5.0	0
9470	Meatless salami, sliced, frozen	3	slice(s)	57	32.4	120	12.0	3.0	2.0	7.0	1.0	1.0	5.0	0
9480	Meatless smoked turkey, sliced	3	slice(s)	57	—	140	10.0	4.0	0	9.0	1.5	2.0	5.0	0
9462	Prosage links	2	item(s)	45	26.8	80	9.0	3.0	2.0	3.0	0.5	0.5	2.0	0
9484	Stakelets patty beef steak substitute, frozen	1	piece(s)	71	41.5	150	14.0	7.0	2.0	7.0	1.0	2.5	3.5	0
9486	Stripples bacon substitute	2	item(s)	16	6.8	60	2.0	2.0	0.5	4.5	0.5	1.0	3.0	0
9496	Vegetable Skallops meat substitute, canned	½	cup(s)	85	—	90	17.0	4.0	3.0	1.0	0	0	0.5	0
Dairy														
	Cheese													
1433	Blue, crumbled	1	ounce(s)	28	12.0	100	6.1	0.7	0	8.1	5.3	2.2	0.2	—
884	Brick	1	ounce(s)	28	11.7	105	6.6	0.8	0	8.4	5.3	2.4	0.2	—
885	Brie	1	ounce(s)	28	13.7	95	5.9	0.1	0	7.8	4.9	2.3	0.2	—
34821	Camembert	1	ounce(s)	28	14.7	85	5.6	0.1	0	6.9	4.3	2.0	0.2	—
5	Cheddar, shredded	¼	cup(s)	28	10.4	114	7.0	0.4	0	9.4	6.0	2.7	0.3	—
888	Cheddar or Colby	1	ounce(s)	28	10.8	112	6.7	0.7	0	9.1	5.7	2.6	0.3	—
32096	Cheddar or Colby, low fat	1	ounce(s)	28	17.9	49	6.9	0.5	0	2.0	1.2	0.6	0.1	—
889	Edam	1	ounce(s)	28	11.8	101	7.1	0.4	0	7.9	5.0	2.3	0.2	—

PAGE KEY: A-6 = Breads/Baked Goods A-12 = Cereal/Rice/Pasta A-16 = Fruit A-20 = Vegetables/Legumes A-30 = Nuts/Seeds A-32 = Vegetarian A-34 = Dairy A-42 = Eggs A-42 = Seafood A-44 = Meats A-48 = Poultry
A-50 = Processed Meats A-50 = Beverages A-54 = Fats/Oils A-56 = Sweets A-58 = Spices/Condiments/Sauces A-62 = Mixed Foods/Soups/Sandwiches A-66 = Fast Food A-86 = Convenience A-88 = Baby Foods

Chol (mg)	Calc (mg)	Iron (mg)	Magn (mg)	Pota (mg)	Sodi (mg)	Zinc (mg)	Vit A (µg)	Thia (mg)	Vit E (mg α)	Ribo (mg)	Niac (mg)	Vit B$_6$ (mg)	Fola (µg)	Vit C (mg)	Vit B$_{12}$ (µg)	Sele (µg)
0	60	1.80	—	—	280.0	—	0	—	—	—	—	—	—	0	—	—
0	60	1.80	—	—	370.0	—	0	—	—	—	—	—	—	0	—	—
0	100	2.70	—	—	300.0	—	—	—	—	—	—	—	—	0	—	—
0	150	1.08	—	—	890.0	—	—	—	—	—	—	—	—	0	—	—
0	40	1.44	—	—	330.0	—	—	—	—	—	—	—	—	0	—	—
0	40	0.72	—	—	1000.0	—	—	—	—	—	—	—	—	0	—	—
5	40	1.44	—	—	550.0	—	—	—	—	—	—	—	—	0	—	—
0	60	3.60	—	—	360.0	—	—	—	—	—	—	—	—	0	—	—
0	40	4.50	—	—	230.0	—	—	—	—	—	—	—	—	0	—	—
0	80	1.44	—	—	380.0	—	—	—	—	—	—	—	—	0	—	—
0	20	0.72	—	—	120.0	—	—	—	—	—	—	—	—	0	—	—
0	60	1.80	—	—	400.0	—	—	—	—	—	—	—	—	0	—	—
0	60	1.80	—	—	720.0	—	—	—	—	—	—	—	—	3.6	—	—
0	80	1.08	30.4	193.4	490.0	0.89	—	0.10	—	0.15	1.08	0.08	10.1	1.2	0.1	7.0
5	60	1.44	—	—	260.0	—	—	—	—	—	—	—	—	3.6	—	—
0	40	1.44	27.0	182.0	290.0	0.46	—	0.07	—	0.08	0.90	0.09	10.6	9.0	0	4.0
0	0	0.77	—	50.0	220.0	—	0	0.22	—	0.10	2.00	0.70	—	0	2.4	—
0	20	1.80	—	70.0	430.0	0.33	0	1.05	—	0.34	4.00	0.30	—	0	2.4	—
0	0	0.36	—	20.0	160.0	0.46	0	0.12	—	0.20	0.80	0.16	—	0	0.9	—
0	0	1.06	—	140.0	450.0	—	0	0.15	—	0.25	4.00	0.40	—	0	1.2	—
0	0	0.72	—	200.0	430.0	—	0	0.45	—	0.25	10.00	1.00	—	0	5.4	—
0	20	1.08	—	80.0	340.0	0.66	0	0.75	—	0.17	2.00	0.16	—	0	1.2	—
0	0	0.72	—	60.0	580.0	—	0	—	—	—	—	—	—	0	—	—
0	20	0.72	—	75.0	90.0	0.60	37.5	0.03	—	0.34	0.00	0.08	24.0	—	0.6	—
0	0	0.36	—	15.0	220.0	0.05	0	0.75	—	0.04	0.40	0.07	—	0	0.2	—
0	0	1.80	—	50.0	300.0	—	0	0.37	—	0.17	7.00	0.50	—	0	3.0	—
0	20	2.70	—	320.0	490.0	—	0	0.52	—	0.25	5.00	0.30	—	0	1.5	—
0	0	1.80	—	210.0	540.0	—	0	1.80	—	0.17	2.00	0.20	—	0	1.2	—
0	40	0.72	—	180.0	350.0	—	—	—	—	—	—	—	—	0	—	—
0	40	1.80	—	320.0	470.0	—	0	—	—	—	0.00	—	—	0	—	—
0	40	3.60	—	660.0	900.0	—	—	—	—	—	—	—	—	0	—	—
0	40	3.60	—	330.0	1130.0	—	0	0.30	—	0.13	2.00	0.70	—	0	1.5	—
0	0	1.08	—	100.0	220.0	0.24	0	0.06	—	0.10	4.00	0.08	—	0	0.2	—
0	20	1.80	—	120.0	580.0	0.64	0	1.80	—	0.25	6.00	0.60	—	0	1.5	—
0	250	1.80	—	250.0	250.0	0.26	0	0.37	—	0.13	4.00	0.30	—	0	1.8	—
0	100	1.08	—	240.0	240.0	—	0	0.37	—	0.13	4.00	0.30	—	0	1.8	—
0	0	1.80	—	130.0	460.0	0.26	0	0.45	—	0.17	5.00	0.30	—	0	1.8	—
0	0	1.08	—	95.0	800.0	0.30	0	0.75	—	0.17	4.00	0.20	—	0	0.6	—
0	60	2.70	—	60.0	450.0	0.23	0	1.80	—	0.17	6.00	0.40	—	0	3.0	—
0	0	1.44	—	50.0	320.0	0.36	0	1.80	—	0.17	2.00	0.30	—	0	3.0	—
0	40	1.08	—	130.0	480.0	0.50	0	1.20	—	0.13	3.00	0.30	—	0	1.5	—
0	0	0.36	—	15.0	220.0	0.05	0	0.75	—	0.03	0.40	0.08	—	0	0.2	—
0	0	0.36	—	10.0	390.0	0.67	0	0.03	—	0.03	0.00	0.01	—	0	0	—
21	150	0.08	6.5	72.6	395.5	0.75	56.1	0.01	0.07	0.10	0.28	0.04	10.2	0	0.3	4.1
27	191	0.12	6.8	38.6	158.8	0.73	82.8	0.00	0.07	0.10	0.03	0.01	5.7	0	0.4	4.1
28	52	0.14	5.7	43.1	178.3	0.67	49.3	0.02	0.06	0.14	0.10	0.06	18.4	0	0.5	4.1
20	110	0.09	5.7	53.0	238.7	0.67	68.3	0.01	0.06	0.14	0.18	0.06	17.6	0	0.4	4.1
30	204	0.19	7.9	27.7	175.4	0.87	74.9	0.01	0.08	0.10	0.02	0.02	5.1	0	0.2	3.9
27	194	0.21	7.4	36.0	171.2	0.87	74.8	0.00	0.07	0.10	0.02	0.02	5.1	0	0.2	4.1
6	118	0.11	4.5	18.7	173.5	0.51	17.0	0.00	0.01	0.06	0.01	0.01	3.1	0	0.1	4.1
25	207	0.12	8.5	53.3	273.6	1.06	68.9	0.01	0.06	0.11	0.02	0.02	4.5	0	0.4	4.1

TABLE A-1 **Table of Food Composition** *(continued)*   (Computer code is for Cengage Diet Analysis program)   (For purposes of calculations, use "0" for t, <1, <.1, <.01, etc.)

DA+ Code	Food Description	Quantity	Measure	Wt (g)	H₂0 (g)	Ener (kcal)	Prot (g)	Carb (g)	Fiber (g)	Fat (g)	Sat	Mono	Poly	Trans
Dairy—continued														
890	Feta	1	ounce(s)	28	15.7	75	4.0	1.2	0	6.0	4.2	1.3	0.2	—
891	Fontina	1	ounce(s)	28	10.8	110	7.3	0.4	0	8.8	5.4	2.5	0.5	—
8527	Goat cheese, soft	1	ounce(s)	28	17.2	76	5.3	0.3	0	6.0	4.1	1.4	0.1	—
893	Gouda	1	ounce(s)	28	11.8	101	7.1	0.6	0	7.8	5.0	2.2	0.2	—
894	Gruyère	1	ounce(s)	28	9.4	117	8.5	0.1	0	9.2	5.4	2.8	0.5	—
895	Limburger	1	ounce(s)	28	13.7	93	5.7	0.1	0	7.7	4.7	2.4	0.1	—
896	Monterey jack	1	ounce(s)	28	11.6	106	6.9	0.2	0	8.6	5.4	2.5	0.3	—
13	Mozzarella, part skim milk	1	ounce(s)	28	15.2	72	6.9	0.8	0	4.5	2.9	1.3	0.1	—
12	Mozzarella, whole milk	1	ounce(s)	28	14.2	85	6.3	0.6	0	6.3	3.7	1.9	0.2	—
897	Muenster	1	ounce(s)	28	11.8	104	6.6	0.3	0	8.5	5.4	2.5	0.2	—
898	Neufchatel	1	ounce(s)	28	17.6	74	2.8	0.8	0	6.6	4.2	1.9	0.2	—
14	Parmesan, grated	1	tablespoon(s)	5	1.0	22	1.9	0.2	0	1.4	0.9	0.4	0.1	—
17	Provolone	1	ounce(s)	28	11.6	100	7.3	0.6	0	7.5	4.8	2.1	0.2	—
19	Ricotta, part skim milk	¼	cup(s)	62	45.8	85	7.0	3.2	0	4.9	3.0	1.4	0.2	—
18	Ricotta, whole milk	¼	cup(s)	62	44.1	107	6.9	1.9	0	8.0	5.1	2.2	0.2	—
20	Romano	1	tablespoon(s)	5	1.5	19	1.6	0.2	0	1.3	0.9	0.4	0	—
900	Roquefort	1	ounce(s)	28	11.2	105	6.1	0.6	0	8.7	5.5	2.4	0.4	—
21	Swiss	1	ounce(s)	28	10.5	108	7.6	1.5	0	7.9	5.0	2.1	0.3	—
	Imitation cheese													
42245	Imitation American cheddar cheese	1	ounce(s)	28	15.1	68	4.7	3.3	0	4.0	2.5	1.2	0.1	—
53914	Imitation cheddar	1	ounce(s)	28	15.1	68	4.7	3.3	0	4.0	2.5	1.2	0.1	—
	Cottage cheese													
9	Low fat, 1% fat	½	cup(s)	113	93.2	81	14.0	3.1	0	1.2	0.7	0.3	0	—
8	Low fat, 2% fat	½	cup(s)	113	89.6	102	15.5	4.1	0	2.2	1.4	0.6	0.1	—
	Cream cheese													
11	Cream cheese	2	tablespoon(s)	29	15.6	101	2.2	0.8	0	10.1	6.4	2.9	0.4	—
17366	Fat-free cream cheese	2	tablespoon(s)	30	22.7	29	4.3	1.7	0	0.4	0.3	0.1	0	—
10438	Tofutti Better than Cream Cheese	2	tablespoon(s)	30	—	80	1.0	1.0	0	8.0	2.0	—	6.0	—
	Processed cheese													
24	American cheese food, processed	1	ounce(s)	28	12.3	94	5.2	2.2	0	7.1	4.2	2.0	0.3	—
25	American cheese spread, processed	1	ounce(s)	28	13.5	82	4.7	2.5	0	6.0	3.8	1.8	0.2	—
22	American cheese, processed	1	ounce(s)	28	11.1	106	6.3	0.5	0	8.9	5.6	2.5	0.3	—
9110	Kraft deluxe singles pasteurized process American cheese	1	ounce(s)	28	—	108	5.4	0	0	9.5	5.4	—	—	—
23	Swiss cheese, processed	1	ounce(s)	28	12.0	95	7.0	0.6	0	7.1	4.5	2.0	0.2	—
	Soy cheese													
10437	Galaxy Foods vegan grated parmesan cheese alternative	1	tablespoon(s)	8	—	23	3.0	1.5	0	0	0	0	0	0
10430	Nu Tofu cheddar flavored cheese alternative	1	ounce(s)	28	—	70	6.0	1.0	0	4.0	0.5	2.5	1.0	—
	Cream													
26	Half and half cream	1	tablespoon(s)	15	12.1	20	0.4	0.6	0	1.7	1.1	0.5	0.1	—
32	Heavy whipping cream, liquid	1	tablespoon(s)	15	8.7	52	0.3	0.4	0	5.6	3.5	1.6	0.2	—
28	Light coffee or table cream, liquid	1	tablespoon(s)	15	11.1	29	0.4	0.5	0	2.9	1.8	0.8	0.1	—
30	Light whipping cream, liquid	1	tablespoon(s)	15	9.5	44	0.3	0.4	0	4.6	2.9	1.4	0.1	—
34	Whipped cream topping, pressurized	1	tablespoon(s)	3	1.8	8	0.1	0.4	0	0.7	0.4	0.2	0	—
	Sour cream													
30556	Fat-free sour cream	2	tablespoon(s)	32	25.8	24	1.0	5.0	0	0	0	0	0	0
36	Sour cream	2	tablespoon(s)	24	17.0	51	0.8	1.0	0	5.0	3.1	1.5	0.2	—
	Imitation cream													
3659	Coffeemate nondairy creamer, liquid	1	tablespoon(s)	15	—	20	0	2.0	0	1.0	0	0.5	0	—
40	Cream substitute, powder	1	teaspoon(s)	2	0	11	0.1	1.1	0	0.7	0.7	0	0	—
904	Imitation sour cream	2	tablespoon(s)	29	20.5	60	0.7	1.9	0	5.6	5.1	0.2	0	—
35972	Nondairy coffee whitener, liquid, frozen	1	tablespoon(s)	15	11.7	21	0.2	1.7	0	1.5	0.3	1.1	0	—
35976	Nondairy dessert topping, frozen	1	tablespoon(s)	5	2.4	15	0.1	1.1	0	1.2	1.0	0.1	0	—
35975	Nondairy dessert topping, pressurized	1	tablespoon(s)	4	2.7	12	0	0.7	0	1.0	0.8	0.1	0	—

PAGE KEY: A-6 = Breads/Baked Goods A-12 = Cereal/Rice/Pasta A-16 = Fruit A-20 = Vegetables/Legumes A-30 = Nuts/Seeds A-32 = Vegetarian A-34 = Dairy A-42 = Eggs A-42 = Seafood A-44 = Meats A-48 = Poultry
A-50 = Processed Meats A-50 = Beverages A-54 = Fats/Oils A-56 = Sweets A-58 = Spices/Condiments/Sauces A-62 = Mixed Foods/Soups/Sandwiches A-66 = Fast Food A-86 = Convenience A-88 = Baby Foods

Chol (mg)	Calc (mg)	Iron (mg)	Magn (mg)	Pota (mg)	Sodi (mg)	Zinc (mg)	Vit A (µg)	Thia (mg)	Vit E (mg α)	Ribo (mg)	Niac (mg)	Vit B_6 (mg)	Fola (µg)	Vit C (mg)	Vit B_{12} (µg)	Sele (µg)
25	140	0.18	5.4	17.6	316.4	0.81	35.4	0.04	0.05	0.23	0.28	0.12	9.1	0	0.5	4.3
33	156	0.06	4.0	18.1	226.8	0.99	74.0	0.01	0.07	0.05	0.04	0.02	1.7	0	0.5	4.1
13	40	0.53	4.5	7.4	104.3	0.26	81.6	0.02	0.05	0.10	0.12	0.07	3.4	0	0.1	0.8
32	198	0.06	8.2	34.3	232.2	1.10	46.8	0.01	0.06	0.09	0.01	0.02	6.0	0	0.4	4.1
31	287	0.04	10.2	23.0	95.3	1.10	76.8	0.01	0.07	0.07	0.03	0.02	2.8	0	0.5	4.1
26	141	0.03	6.0	36.3	226.8	0.59	96.4	0.02	0.06	0.14	0.04	0.02	16.4	0	0.3	4.1
25	211	0.20	7.7	23.0	152.0	0.85	56.1	0.00	0.07	0.11	0.02	0.02	5.1	0	0.2	4.1
18	222	0.06	6.5	23.8	175.5	0.78	36.0	0.01	0.04	0.08	0.03	0.02	2.6	0	0.2	4.1
22	143	0.12	5.7	21.5	177.8	0.82	50.7	0.01	0.05	0.08	0.02	0.01	2.0	0	0.6	4.8
27	203	0.11	7.7	38.0	178.0	0.79	84.5	0.00	0.07	0.09	0.02	0.01	3.4	0	0.4	4.1
22	21	0.07	2.3	32.3	113.1	0.14	84.5	0.00	—	0.05	0.03	0.01	3.1	0	0.1	0.9
4	55	0.04	1.9	6.3	76.5	0.19	6.0	0.00	0.01	0.02	0.01	0.00	0.5	0	0.1	0.9
20	214	0.14	7.9	39.1	248.3	0.91	66.9	0.01	0.06	0.09	0.04	0.02	2.8	0	0.4	4.1
19	167	0.27	9.2	76.9	76.9	0.82	65.8	0.01	0.04	0.11	0.04	0.01	8.0	0	0.2	10.3
31	127	0.23	6.8	64.6	51.7	0.71	73.8	0.01	0.06	0.12	0.06	0.02	7.4	0	0.2	8.9
5	53	0.03	2.1	4.3	60	0.12	4.8	0.00	0.01	0.01	0.00	0.00	0.4	0	0.1	0.7
26	188	0.15	8.5	25.8	512.9	0.59	83.3	0.01	—	0.16	0.20	0.03	13.9	0	0.2	4.1
26	224	0.05	10.8	21.8	54.4	1.23	62.4	0.01	0.10	0.08	0.02	0.02	1.7	0	0.9	5.2
10	159	0.08	8.2	68.6	381.3	0.73	32.3	0.01	0.07	0.12	0.03	0.03	2.0	0	0.1	4.3
10	159	0.09	8.2	68.6	381.3	0.73	32.3	0.01	0.07	0.12	0.04	0.03	2.0	0	0.1	4.3
5	69	0.15	5.7	97.2	458.8	0.42	12.4	0.02	0.01	0.18	0.14	0.07	13.6	0	0.7	10.2
9	78	0.18	6.8	108.5	458.8	0.47	23.7	0.02	0.02	0.20	0.16	0.08	14.7	0	0.8	11.5
32	23	0.34	1.7	34.5	85.8	0.15	106.1	0.01	0.08	0.05	0.02	0.01	3.8	0	0.1	0.7
2	56	0.05	4.2	48.9	163.5	0.26	83.7	0.01	0.00	0.05	0.04	0.01	11.1	0	0.2	1.5
0	0	0.00	—	—	135.0	—	0	—	—	—	—	—	—	0	—	—
23	162	0.16	8.8	87.5	358.6	0.90	57.0	0.01	0.06	0.14	0.04	0.02	2.0	0	0.4	4.6
16	159	0.09	8.2	68.6	381.3	0.73	49.0	0.01	0.05	0.12	0.03	0.03	2.0	0	0.1	3.2
27	156	0.05	7.7	47.9	422.1	0.80	72.0	0.01	0.07	0.10	0.02	0.02	2.3	0	0.2	4.1
27	338	0.00	0	33.8	459.0	1.22	114.0	—	—	0.14	—	—	—	0	0.2	—
24	219	0.17	8.2	61.2	388.4	1.02	56.1	0.00	0.09	0.07	0.01	0.01	1.7	0	0.3	4.5
0	60	0.00	—	75.0	97.5	—	—	—	—	—	—	—	—	—	—	—
0	200	0.36	—	—	190.0	—	—	—	—	—	—	—	—	0	—	—
6	16	0.01	1.5	19.5	6.2	0.08	14.6	0.01	0.05	0.02	0.01	0.01	0.5	0.1	0	0.3
21	10	0.00	1.1	11.3	5.7	0.03	61.7	0.00	0.15	0.01	0.01	0.00	0.6	0.1	0	0.1
10	14	0.01	1.4	18.3	6.0	0.04	27.2	0.01	0.08	0.02	0.01	0.01	0.3	0.1	0	0.1
17	10	0.00	1.1	14.6	5.1	0.03	41.9	0.00	0.13	0.01	0.01	0.00	0.6	0.1	0	0.1
2	3	0.00	0.3	4.4	3.9	0.01	5.6	0.00	0.01	0.00	0.00	0.00	0.1	0	0	0
3	40	0.00	3.2	41.3	45.1	0.16	23.4	0.01	0.00	0.04	0.02	0.01	3.5	0	0.1	1.7
11	28	0.01	2.6	34.6	12.7	0.06	42.5	0.01	0.14	0.03	0.01	0.00	2.6	0.2	0.1	0.5
0	0	0.00	—	30.0	0	—	0	0.01	—	0.01	0.20	—	—	0	—	—
0	0	0.02	0.1	16.2	3.6	0.01	0	0.00	0.01	0.00	0.00	0.00	0	0	0	0
0	1	0.11	1.7	46.3	29.3	0.34	0	0.00	0.21	0.00	0.00	0.00	0	0	0	0.7
0	1	0.00	0	28.9	12.0	0.00	0.2	0.00	0.12	0.00	0.00	0.00	0	0	0	0.2
0	0	0.00	0.1	0.9	1.2	0.00	0.3	0.00	0.05	0.00	0.00	0.00	0	0	0	0.1
0	0	0.00	0	0.8	2.8	0.00	0.2	0.00	0.04	0.00	0.00	0.00	0	0	0	0.1

TABLE A-1 **Table of Food Composition (continued)** (Computer code is for Cengage Diet Analysis program) (For purposes of calculations, use "0" for t, <1, <.1, <.01, etc.)

DA+ Code	Food Description	Quantity	Measure	Wt (g)	H₂O (g)	Ener (kcal)	Prot (g)	Carb (g)	Fiber (g)	Fat (g)	Fat Breakdown (g) Sat	Mono	Poly	Trans
Dairy—continued														
	Fluid milk													
60	Buttermilk, low fat	1	cup(s)	245	220.8	98	8.1	11.7	0	2.2	1.3	0.6	0.1	—
54	Low fat, 1%	1	cup(s)	244	219.4	102	8.2	12.2	0	2.4	1.5	0.7	0.1	—
55	Low fat, 1%, with nonfat milk solids	1	cup(s)	245	220.0	105	8.5	12.2	0	2.4	1.5	0.7	0.1	—
57	Nonfat, skim or fat free	1	cup(s)	245	222.6	83	8.3	12.2	0	0.2	0.1	0.1	0	—
58	Nonfat, skim or fat free with nonfat milk solids	1	cup(s)	245	221.4	91	8.7	12.3	0	0.6	0.4	0.2	0	—
51	Reduced fat, 2%	1	cup(s)	244	218.0	122	8.1	11.4	0	4.8	3.1	1.4	0.2	—
52	Reduced fat, 2%, with nonfat milk solids	1	cup(s)	245	217.7	125	8.5	12.2	0	4.7	2.9	1.4	0.2	—
50	Whole, 3.3%	1	cup(s)	244	215.5	146	7.9	11.0	0	7.9	4.6	2.0	0.5	—
	Canned milk													
62	Nonfat or skim evaporated	2	tablespoon(s)	32	25.3	25	2.4	3.6	0	0.1	0	0	0	—
63	Sweetened condensed	2	tablespoon(s)	38	10.4	123	3.0	20.8	0	3.3	2.1	0.9	0.1	—
61	Whole evaporated	2	tablespoon(s)	32	23.3	42	2.1	3.2	0	2.4	1.4	0.7	0.1	—
	Dried milk													
64	Buttermilk	¼	cup(s)	30	0.9	117	10.4	14.9	0	1.8	1.1	0.5	0.1	—
65	Instant nonfat with added vitamin A	¼	cup(s)	17	0.7	61	6.0	8.9	0	0.1	0.1	0	0	—
5234	Skim milk powder	¼	cup(s)	17	0.7	62	6.1	9.1	0	0.1	0.1	0	0	—
907	Whole dry milk	¼	cup(s)	32	0.8	159	8.4	12.3	0	8.5	5.4	2.5	0.2	—
909	**Goat milk**	1	cup(s)	244	212.4	168	8.7	10.9	0	10.1	6.5	2.7	0.4	—
	Chocolate milk													
33155	Chocolate syrup, prepared with milk	1	cup(s)	282	227.0	254	8.7	36.0	0.8	8.3	4.7	2.1	0.5	—
33184	Cocoa mix with aspartame, added sodium and vitamin A, no added calcium or phosphorus, prepared with water	1	cup(s)	192	177.4	56	2.3	10.8	1.2	0.4	0.3	0.1	0	—
908	Hot cocoa, prepared with milk	1	cup(s)	250	206.4	193	8.8	26.6	2.5	5.8	3.6	1.7	0.1	0.2
69	Low fat	1	cup(s)	250	211.3	158	8.1	26.1	1.3	2.5	1.5	0.8	0.1	—
68	Reduced fat	1	cup(s)	250	205.4	190	7.5	30.3	1.8	4.8	2.9	1.1	0.2	—
67	Whole	1	cup(s)	250	205.8	208	7.9	25.9	2.0	8.5	5.3	2.5	0.3	—
70	**Eggnog**	1	cup(s)	254	188.9	343	9.7	34.4	0	19.0	11.3	5.7	0.9	—
	Breakfast drinks													
10093	Carnation Instant Breakfast classic chocolate malt, prepared with skim milk, no sugar added	1	cup(s)	243	—	142	11.1	21.3	0.7	1.3	0.7	—	—	—
10092	Carnation Instant Breakfast classic French vanilla, prepared with skim milk, no sugar added	1	cup(s)	273	—	150	12.9	24.0	0	0.4	0.4	—	—	—
10094	Carnation Instant Breakfast stawberry sensation, prepared with skim milk, no sugar added	1	cup(s)	243	—	142	11.1	21.3	0	0.4	0.4	—	—	—
10091	Carnation Instant Breakfast strawberry sensation, prepared with skim milk	1	cup(s)	273	—	220	12.5	38.8	0	0.4	0.4	—	—	—
1417	Ovaltine rich chocolate flavor, prepared with skim milk	1	cup(s)	258	—	170	8.5	31.0	0	0	0	0	0	0
8539	**Malted milk, chocolate mix, fortified, prepared with milk**	1	cup(s)	265	215.8	223	8.9	28.9	1.1	8.6	5.0	2.2	0.5	—
	Milkshakes													
73	Chocolate	1	cup(s)	227	164.0	270	6.9	48.1	0.7	6.1	3.8	1.8	0.2	—
3163	Strawberry	1	cup(s)	226	167.8	256	7.7	42.8	0.9	6.3	3.9	—	—	—
74	Vanilla	1	cup(s)	227	169.2	254	8.8	40.3	0	6.9	4.3	2.0	0.3	—
	Ice cream													
4776	Chocolate	½	cup(s)	66	36.8	143	2.5	18.6	0.8	7.3	4.5	2.1	0.3	—
12137	Chocolate fudge, no sugar added	½	cup(s)	71	—	100	3.0	16.0	2.0	3.0	1.5	—	—	0
16514	Chocolate, soft serve	½	cup(s)	87	49.9	177	3.2	24.1	0.7	8.4	5.2	2.4	0.3	—
16523	Sherbet, all flavors	½	cup(s)	97	63.8	139	1.1	29.3	3.2	1.9	1.1	0.5	0.1	—
4778	Strawberry	½	cup(s)	66	39.6	127	2.1	18.2	0.6	5.5	3.4	—	—	—
76	Vanilla	½	cup(s)	72	43.9	145	2.5	17.0	0.5	7.9	4.9	2.1	0.3	—
12146	Vanilla chocolate swirl, fat-free, no sugar added	½	cup(s)	71	—	100	3.0	14.0	2.0	3.0	2.0	—	—	0
82	Vanilla, light	½	cup(s)	76	48.3	125	3.6	19.6	0.2	3.7	2.2	1.0	0.2	—
78	Vanilla, light, soft serve	½	cup(s)	88	61.2	111	4.3	19.2	0	2.3	1.4	0.7	0.1	—

PAGE KEY: A-6 = Breads/Baked Goods A-12 = Cereal/Rice/Pasta A-16 = Fruit A-20 = Vegetables/Legumes A-30 = Nuts/Seeds A-32 = Vegetarian A-34 = Dairy A-42 = Eggs A-42 = Seafood A-44 = Meats A-48 = Poultry
A-50 = Processed Meats A-50 = Beverages A-54 = Fats/Oils A-56 = Sweets A-58 = Spices/Condiments/Sauces A-62 = Mixed Foods/Soups/Sandwiches A-66 = Fast Food A-86 = Convenience A-88 = Baby Foods

Chol (mg)	Calc (mg)	Iron (mg)	Magn (mg)	Pota (mg)	Sodi (mg)	Zinc (mg)	Vit A (µg)	Thia (mg)	Vit E (mg α)	Ribo (mg)	Niac (mg)	Vit B_6 (mg)	Fola (µg)	Vit C (mg)	Vit B_{12} (µg)	Sele (µg)
10	284	0.12	27.0	370.0	257.3	1.02	17.2	0.08	0.12	0.37	0.14	0.08	12.3	2.5	0.5	4.9
12	290	0.07	26.8	366.0	107.4	1.02	141.5	0.04	0.02	0.45	0.22	0.09	12.2	0	1.1	8.1
10	314	0.12	34.3	396.9	127.4	0.98	144.6	0.09	—	0.42	0.22	0.11	12.3	2.5	0.9	5.6
5	306	0.07	27.0	382.2	102.9	1.02	149.5	0.11	0.02	0.44	0.23	0.09	12.3	0	1.3	7.6
5	316	0.12	36.8	419.0	129.9	1.00	149.5	0.10	0.00	0.42	0.22	0.11	12.3	2.5	1.0	5.4
20	285	0.07	26.8	366.0	100.0	1.04	134.2	0.09	0.07	0.45	0.22	0.09	12.2	0.5	1.1	6.1
20	314	0.12	34.3	396.9	127.4	0.98	137.2	0.09	—	0.42	0.22	0.11	12.3	2.5	0.9	5.6
24	276	0.07	24.4	348.9	97.6	0.97	68.3	0.10	0.14	0.44	0.26	0.08	12.2	0	1.1	9.0
1	93	0.09	8.6	105.9	36.7	0.28	37.6	0.01	0.00	0.09	0.05	0.01	2.9	0.4	0.1	0.8
13	109	0.07	9.9	141.9	48.6	0.36	28.3	0.03	0.06	0.16	0.08	0.02	4.2	1.0	0.2	5.7
9	82	0.06	7.6	95.4	33.4	0.24	20.5	0.01	0.04	0.10	0.06	0.01	2.5	0.6	0.1	0.7
21	359	0.09	33.3	482.5	156.7	1.21	14.9	0.11	0.03	0.48	0.27	0.10	14.2	1.7	1.2	6.2
3	209	0.05	19.9	289.9	93.3	0.75	120.5	0.07	0.00	0.30	0.15	0.06	8.5	1.0	0.7	4.6
3	214	0.05	20.3	296.0	95.3	0.76	123.1	0.07	0.00	0.30	0.15	0.06	8.7	1.0	0.7	4.7
31	292	0.15	27.2	425.6	118.7	1.06	82.2	0.09	0.15	0.38	0.20	0.09	11.8	2.8	1.0	5.2
27	327	0.12	34.2	497.8	122.0	0.73	139.1	0.11	0.17	0.33	0.67	0.11	2.4	3.2	0.2	3.4
25	251	0.90	50.8	408.9	132.5	1.21	70.5	0.11	0.14	0.46	0.38	0.09	14.1	0	1.1	9.6
0	92	0.74	32.6	405.1	138.2	0.51	0	0.04	0.00	0.20	0.16	0.04	1.9	0	0.2	2.5
20	263	1.20	57.5	492.5	110.0	1.57	127.5	0.09	0.07	0.45	0.33	0.10	12.5	0.5	1.1	6.8
8	288	0.60	32.5	425.0	152.5	1.02	145.0	0.09	0.05	0.41	0.31	0.10	12.5	2.3	0.9	4.8
20	273	0.60	35.0	422.5	165.0	0.97	160.0	0.11	0.10	0.45	0.41	0.06	5.0	0	0.8	8.5
30	280	0.60	32.5	417.5	150.0	1.02	65.0	0.09	0.15	0.40	0.31	0.10	12.5	2.3	0.8	4.8
150	330	0.50	48.3	419.1	137.2	1.16	116.8	0.08	0.50	0.48	0.26	0.12	2.5	3.8	1.1	10.7
9	444	4.00	88.9	631.1	195.6	3.38	—	0.33	—	0.45	4.44	0.44	4.0	26.7	1.3	8.0
9	500	4.50	100.0	665.0	192.0	3.75	—	0.37	—	0.51	5.00	0.49	100.0	30.0	1.5	9.0
9	444	4.00	88.9	568.9	186.7	3.38	—	0.33	—	0.45	4.44	0.44	88.9	26.7	1.3	8.0
9	500	4.47	100.0	665.0	288.0	3.75	—	0.37	—	0.51	5.07	0.50	100.0	30.0	1.5	8.8
5	350	3.60	100.0	—	270.0	3.75	—	0.37	—	—	4.00	0.40	—	12.0	1.2	—
27	339	3.76	45.1	577.7	230.6	1.16	903.7	0.75	0.15	1.31	11.08	1.01	18.6	31.8	1.1	12.5
25	300	0.70	36.4	508.9	252.2	1.09	40.9	0.10	0.11	0.50	0.28	0.05	11.4	0	0.7	4.3
25	256	0.24	29.4	412.0	187.9	0.81	58.9	0.10	—	0.44	0.39	0.10	6.8	1.8	0.7	4.8
27	332	0.22	27.3	415.8	215.8	0.88	56.8	0.06	0.11	0.44	0.33	0.09	15.9	0	1.2	5.2
22	72	0.61	19.1	164.3	50.2	0.38	77.9	0.02	0.19	0.12	0.14	0.03	10.6	0.5	0.2	1.7
10	100	0.36	—	—	65.0	—	—	—	—	—	—	—	—	0	—	—
22	103	0.32	19.0	192.0	43.3	0.45	66.6	0.03	0.22	0.13	0.11	0.03	4.3	0.5	0.3	2.5
0	52	0.13	7.7	92.6	44.4	0.46	9.7	0.02	0.02	0.08	0.07	0.02	6.8	5.6	0.1	1.3
19	79	0.13	9.2	124.1	39.6	0.22	63.4	0.03	—	0.16	0.11	0.03	7.9	5.1	0.2	1.3
32	92	0.06	10.1	143.3	57.6	0.49	85.0	0.03	0.21	0.17	0.08	0.03	3.6	0.4	0.3	1.3
10	100	0.00	—	—	65.0	—	—	—	—	—	—	—	—	0	—	—
21	122	0.14	10.6	158.1	56.2	0.55	97.3	0.04	0.09	0.19	0.10	0.03	4.6	0.9	0.4	1.5
11	138	0.05	12.3	194.5	61.6	0.46	25.5	0.04	0.05	0.17	0.10	0.04	4.4	0.8	0.4	3.2

TABLE A-1 Table of Food Composition (continued) (Computer code is for Cengage Diet Analysis program) (For purposes of calculations, use "0" for t, <1, <.1, <.01, etc.)

DA+ Code	Food Description	Quantity	Measure	Wt (g)	H₂0 (g)	Ener (kcal)	Prot (g)	Carb (g)	Fiber (g)	Fat (g)	Fat Breakdown (g) Sat	Mono	Poly	Trans
Dairy—continued														
	Soy desserts													
10694	Tofutti low fat vanilla fudge nondairy frozen dessert	½	cup(s)	70	—	140	2.0	24.0	0	4.0	1.0	—	—	—
15721	Tofutti premium chocolate supreme nondairy frozen dessert	½	cup(s)	70	—	180	3.0	18.0	0	11.0	2.0	—	—	—
15720	Tofutti premium vanilla nondairy frozen dessert	½	cup(s)	70	—	190	2.0	20.0	0	11.0	2.0	—	—	—
	Ice milk													
16517	Chocolate	½	cup(s)	66	42.9	94	2.8	16.9	0.3	2.1	1.3	0.6	0.1	—
16516	Flavored, not chocolate	½	cup(s)	66	41.4	108	3.5	17.5	0.2	2.6	1.7	0.6	0.1	—
	Pudding													
25032	Chocolate	½	cup(s)	144	109.7	155	5.1	22.7	0.7	5.4	3.1	1.7	0.2	0
1923	Chocolate, sugar free, prepared with 2% milk	½	cup(s)	133	—	100	5.0	14.0	0.3	3.0	1.5	—	—	—
1722	Rice	½	cup(s)	113	75.6	151	4.1	29.9	0.5	1.9	1.1	0.5	0.1	—
4747	Tapioca, ready to eat	1	item(s)	142	102.0	185	2.8	30.8	0	5.5	1.4	3.6	0.1	—
25031	Vanilla	½	cup(s)	136	109.7	116	4.7	17.6	0	2.8	1.6	0.9	0.2	0
1924	Vanilla, sugar free, prepared with 2% milk	½	cup(s)	133	—	90	4.0	12.0	0.2	2.0	1.5	—	—	—
	Frozen yogurt													
4785	Chocolate, soft serve	½	cup(s)	72	45.9	115	2.9	17.9	1.6	4.3	2.6	1.3	0.2	—
1747	Fruit varieties	½	cup(s)	113	80.5	144	3.4	24.4	0	4.1	2.6	1.1	0.1	—
4786	Vanilla, soft serve	½	cup(s)	72	47.0	117	2.9	17.4	0	4.0	2.5	1.1	0.2	—
	Milk substitutes													
	Lactose free													
16081	Fat-free, calcium fortified [milk]	1	cup(s)	240	—	80	8.0	13.0	0	0	0	0	0	0
36486	Low fat milk	1	cup(s)	240	—	110	8.0	13.0	0	2.5	1.5	—	—	—
36487	Reduced fat milk	1	cup(s)	240	—	130	8.0	12.0	0	5.0	3.0	—	—	—
36488	Whole milk	1	cup(s)	240	—	150	8.0	12.0	0	8.0	5.0	—	—	—
	Rice													
10083	Rice Dream carob rice beverage	1	cup(s)	240	—	150	1.0	32.0	0	2.5	0	—	—	—
17089	Rice Dream original rice beverage, enriched	1	cup(s)	240	—	120	1.0	25.0	0	2.0	0	—	—	—
10087	Rice Dream vanilla enriched rice beverage	1	cup(s)	240	—	130	1.0	28.0	0	2.0	0	—	—	—
	Soy													
34750	Soy Dream chocolate enriched soy beverage	1	cup(s)	240	—	210	7.0	37.0	1.0	3.5	0.5	—	—	—
34749	Soy Dream vanilla enriched soy beverage	1	cup(s)	240	—	150	7.0	22.0	0	4.0	0.5	—	—	—
13840	Vitasoy light chocolate soymilk	1	cup(s)	240	—	100	4.0	17.0	0	2.0	0.5	0.5	1.0	—
13839	Vitasoy light vanilla soymilk	1	cup(s)	240	—	70	4.0	10.0	0	2.0	0.5	0.5	1.0	—
13836	Vitasoy rich chocolate soymilk	1	cup(s)	240	—	160	7.0	24.0	1.0	4.0	0.5	1.0	2.5	—
13835	Vitasoy vanilla delite soymilk	1	cup(s)	240	—	120	7.0	13.0	1.0	4.0	0.5	1.0	2.5	—
	Yogurt													
3615	Custard style, fruit flavors	6	ounce(s)	170	127.1	190	7.0	32.0	0	3.5	2.0	—	—	—
3617	Custard style, vanilla	6	ounce(s)	170	134.1	190	7.0	32.0	0	3.5	2.0	0.9	0.1	—
32101	Fruit, low fat	1	cup(s)	245	184.5	243	9.8	45.7	0	2.8	1.8	0.8	0.1	—
29638	Fruit, nonfat, sweetened with low-calorie sweetener	1	cup(s)	241	208.3	123	10.6	19.4	1.2	0.4	0.2	0.1	0	—
93	Plain, low fat	1	cup(s)	245	208.4	154	12.9	17.2	0	3.8	2.5	1.0	0.1	—
94	Plain, nonfat	1	cup(s)	245	208.8	137	14.0	18.8	0	0.4	0.3	0.1	0	—
32100	Vanilla, low fat	1	cup(s)	245	193.6	208	12.1	33.8	0	3.1	2.0	0.8	0.1	—
5242	Yogurt beverage	1	cup(s)	245	199.8	172	6.2	32.8	0	2.2	1.4	0.6	0.1	—
38202	Yogurt smoothie, nonfat, all flavors	1	item(s)	325	—	290	10.0	60.0	6.0	0	0	0	0	0
	Soy yogurt													
34617	Stonyfield Farm O'Soy strawberry-peach pack organic cultured soy yogurt	1	item(s)	113	—	100	5.0	16.0	3.0	2.0	0	—	—	0
34616	Stonyfield Farm O'Soy vanilla organic cultured soy yogurt	1	item(s)	170	—	150	7.0	26.0	4.0	2.0	0	—	—	0
10453	White Wave plain silk cultured soy yogurt	8	ounce(s)	227	—	140	5.0	22.0	1.0	3.0	0.5	—	—	0

Chol (mg)	Calc (mg)	Iron (mg)	Magn (mg)	Pota (mg)	Sodi (mg)	Zinc (mg)	Vit A (µg)	Thia (mg)	Vit E (mg α)	Ribo (mg)	Niac (mg)	Vit B$_6$ (mg)	Fola (µg)	Vit C (mg)	Vit B$_{12}$ (µg)	Sele (µg)
0	0	0.00	—	8.0	90.0	—	0	—	—	—	—	—	—	0	—	—
0	0	0.00	—	7.0	180.0	—	0	—	—	—	—	—	—	0	—	—
0	0	0.00	—	2.0	210.0	—	0	—	—	—	—	—	—	0	—	—
6	94	0.15	13.1	155.2	40.6	0.36	15.7	0.03	0.05	0.11	0.08	0.02	3.9	0.5	0.3	2.2
16	76	0.05	9.2	136.2	48.5	0.47	90.4	0.02	0.05	0.11	0.06	0.01	3.3	0.1	0.2	1.3
35	149	0.46	31.3	226.7	137.0	0.71	—	0.05	0.00	0.22	0.15	0.06	8.3	1.2	0.5	4.9
10	150	0.72	—	330.0	310.0	—	—	0.06	—	0.26	—	—	—	0	—	—
7	113	0.28	15.8	201.4	66.4	0.52	41.6	0.03	0.05	0.17	0.34	0.06	4.5	0.2	0.2	4.8
1	101	0.15	8.5	130.6	205.9	0.31	0	0.03	0.21	0.13	0.09	0.03	4.3	0.4	0.3	0
35	146	0.17	17.2	188.9	136.4	0.52	—	0.04	0.00	0.22	0.10	0.05	8.0	1.2	0.5	4.6
10	150	0.00	—	190.0	380.0	—	—	0.03	—	0.17	—	—	—	0	—	—
4	106	0.90	19.4	187.9	70.6	0.35	31.7	0.02	—	0.15	0.22	0.05	7.9	0.2	0.2	1.7
15	113	0.52	11.3	176.3	71.2	0.31	55.4	0.04	0.10	0.20	0.07	0.04	4.5	0.8	0.1	2.1
1	103	0.21	10.1	151.9	62.6	0.30	42.5	0.02	0.07	0.16	0.20	0.05	4.3	0.6	0.2	2.4
3	500	0.00	—	—	125.0	—	100.0	—	—	—	—	—	—	0	0	—
10	300	0.00	—	—	125.0	—	100.0	—	—	—	—	—	—	0	—	—
20	300	0.00	—	—	125.0	—	98.2	—	—	—	—	—	—	0	—	—
35	300	0.00	—	—	125.0	—	58.1	—	—	—	—	—	—	0	—	—
0	20	0.72	—	82.5	100.0	—	—	—	—	—	—	—	—	1.2	—	—
0	300	0.00	13.3	60.0	90.0	0.24	—	0.06	—	0.00	0.84	0.07	—	0	1.5	—
0	300	0.00	—	53.0	90.0	—	—	—	—	—	—	—	—	0	1.5	—
0	300	1.80	60.0	350.0	160.0	0.60	33.3	0.15	—	0.06	0.80	0.12	60.0	0	3.0	—
0	300	1.80	40.0	260.0	140.0	0.60	33.3	0.15	—	0.06	0.80	0.12	60.0	0	3.0	—
0	300	0.72	24.0	200.0	140.0	0.90	—	0.09	—	0.34	—	—	24.0	0	0.9	—
0	300	0.72	24.0	200.0	120.0	0.90	—	0.09	—	0.34	—	—	24.0	0	0.9	—
0	300	1.08	40.0	320.0	150.0	0.90	—	0.15	—	0.34	—	—	60.0	0	0.9	—
0	40	0.72	—	320.0	115.0	—	0	—	—	—	—	—	—	0	—	—
15	300	0.00	16.0	310.0	100.0	—	—	—	—	0.25	—	—	—	0	—	—
15	300	0.00	16.0	310.0	100.0	—	—	—	—	0.25	—	—	—	0	—	—
12	338	0.14	31.9	433.7	129.9	1.64	27.0	0.08	0.04	0.39	0.21	0.09	22.1	1.5	1.1	6.9
5	369	0.62	41.0	549.5	139.8	1.83	4.8	0.10	0.16	0.44	0.49	0.10	31.3	26.5	1.1	7.0
15	448	0.19	41.7	573.3	171.5	2.18	34.3	0.10	0.07	0.52	0.27	0.12	27.0	2.0	1.4	8.1
5	488	0.22	46.6	624.8	188.7	2.37	4.9	0.11	0.00	0.57	0.30	0.13	29.4	2.2	1.5	8.8
12	419	0.17	39.2	536.6	161.7	2.03	29.4	0.10	0.04	0.49	0.26	0.11	27.0	2.0	1.3	12.0
13	260	0.22	39.2	399.4	98.0	1.10	14.7	0.11	0.00	0.51	0.30	0.14	29.4	2.1	1.5	—
5	300	2.70	100.0	580.0	290.0	2.25	—	0.37	—	0.42	5.00	0.50	100.0	15.0	1.5	—
0	100	1.08	24.0	5.0	20.0	—	0	0.22	—	0.10	—	0.04	—	0	0	—
0	150	1.44	40.0	15.0	40.0	—	—	0.30	—	0.13	—	0.08	—	0	0	—
0	400	1.44	—	0	30.0	—	0	—	—	—	—	—	—	0	—	—

TABLE A-1 Table of Food Composition (continued) (Computer code is for Cengage Diet Analysis program) (For purposes of calculations, use "0" for t, <1, <.1, <.01, etc.)

DA+ Code	Food Description	Quantity	Measure	Wt (g)	H₂O (g)	Ener (kcal)	Prot (g)	Carb (g)	Fiber (g)	Fat (g)	Fat Breakdown (g) Sat	Mono	Poly	Trans
Eggs														
	Eggs													
99	Fried	1	item(s)	46	31.8	90	6.3	0.4	0	7.0	2.0	2.9	1.2	—
100	Hard boiled	1	item(s)	50	37.3	78	6.3	0.6	0	5.3	1.6	2.0	0.7	—
101	Poached	1	item(s)	50	37.8	71	6.3	0.4	0	5.0	1.5	1.9	0.7	—
97	Raw, white	1	item(s)	33	28.9	16	3.6	0.2	0	0.1	0	0	0	—
96	Raw, whole	1	item(s)	50	37.9	72	6.3	0.4	0	5.0	1.5	1.9	0.7	—
98	Raw, yolk	1	item(s)	17	8.9	54	2.7	0.6	0	4.5	1.6	2.0	0.7	—
102	Scrambled, prepared with milk and butter	2	item(s)	122	89.2	204	13.5	2.7	0	14.9	4.5	5.8	2.6	0.7
	Egg substitute													
4028	Egg Beaters	¼	cup(s)	61	—	30	6.0	1.0	0	0	0	0	0	0
920	Frozen	¼	cup(s)	60	43.9	96	6.8	1.9	0	6.7	1.2	1.5	3.7	—
918	Liquid	¼	cup(s)	63	51.9	53	7.5	0.4	0	2.1	0.4	0.6	1.0	—
Seafood														
	Cod													
6040	Atlantic cod or scrod, baked or broiled	3	ounce(s)	85	64.6	89	19.4	0	0	0.7	0.1	0.1	0.2	—
1573	Atlantic cod, cooked, dry heat	3	ounce(s)	85	64.6	89	19.4	0	0	0.7	0.1	0.1	0.2	—
2905	**Eel, raw**	3	ounce(s)	85	58.0	156	15.7	0	0	9.9	2.0	6.1	0.8	—
	Fish fillets													
25079	Baked	3	ounce(s)	84	79.9	99	21.7	0	0	0.7	0.1	0.1	0.3	—
8615	Batter coated or breaded, fried	3	ounce(s)	85	45.6	197	12.5	14.4	0.4	10.5	2.4	2.2	5.3	—
25082	Broiled fish steaks	3	ounce(s)	85	68.1	128	24.2	0	0	2.6	0.4	0.9	0.8	—
25083	Poached fish steaks	3	ounce(s)	85	67.1	111	21.1	0	0	2.3	0.3	0.8	0.7	—
25084	Steamed	3	ounce(s)	85	72.2	79	17.2	0	0	0.6	0.1	0.1	0.2	—
25089	**Flounder, baked**	3	ounce(s)	85	64.4	113	14.8	0.4	0.1	5.5	1.1	2.2	1.4	—
1825	**Grouper, cooked, dry heat**	3	ounce(s)	85	62.4	100	21.1	0	0	1.1	0.3	0.2	0.3	—
	Haddock													
6049	Baked or broiled	3	ounce(s)	85	63.2	95	20.6	0	0	0.8	0.1	0.1	0.3	—
1578	Cooked, dry heat	3	ounce(s)	85	63.1	95	20.6	0	0	0.8	0.1	0.1	0.3	—
1886	**Halibut, Atlantic and Pacific, cooked, dry heat**	3	ounce(s)	85	61.0	119	22.7	0	0	2.5	0.4	0.8	0.8	—
1582	**Herring, Atlantic, pickled**	4	piece(s)	60	33.1	157	8.5	5.8	0	10.8	1.4	7.2	1.0	—
1587	**Jack mackerel, solids, canned, drained**	2	ounce(s)	57	39.2	88	13.1	0	0	3.6	1.1	1.3	0.9	—
8580	**Octopus, common, cooked, moist heat**	3	ounce(s)	85	51.5	139	25.4	3.7	0	1.8	0.4	0.3	0.4	—
1831	**Perch, mixed species, cooked, dry heat**	3	ounce(s)	85	62.3	100	21.1	0	0	1.0	0.2	0.2	0.4	—
1592	**Pacific rockfish, cooked, dry heat**	3	ounce(s)	85	62.4	103	20.4	0	0	1.7	0.4	0.4	0.5	—
	Salmon													
2938	Coho, farmed, raw	3	ounce(s)	85	59.9	136	18.1	0	0	6.5	1.5	2.8	1.6	—
1594	Broiled or baked with butter	3	ounce(s)	85	53.9	155	23.0	0	0	6.3	1.2	2.3	2.3	—
29727	Smoked chinook (lox)	2	ounce(s)	57	40.8	66	10.4	0	0	2.4	0.5	1.1	0.6	—
154	**Sardine, Atlantic with bones, canned in oil**	3	ounce(s)	85	50.7	177	20.9	0	0	9.7	1.3	3.3	4.4	—
	Scallops													
155	Mixed species, breaded, fried	3	item(s)	47	27.2	100	8.4	4.7	—	5.1	1.2	2.1	1.3	—
1599	Steamed	3	ounce(s)	85	64.8	90	13.8	2.0	0	2.6	0.4	1.0	0.8	—
1839	**Snapper, mixed species, cooked, dry heat**	3	ounce(s)	85	59.8	109	22.4	0	0	1.5	0.3	0.3	0.5	—
	Squid													
1868	Mixed species, fried	3	ounce(s)	85	54.9	149	15.3	6.6	0	6.4	1.6	2.3	1.8	—
16617	Steamed or boiled	3	ounce(s)	85	63.3	89	15.2	3.0	0	1.3	0.4	0.1	0.5	—
1570	**Striped bass, cooked, dry heat**	3	ounce(s)	85	62.4	105	19.3	0	0	2.5	0.6	0.7	0.9	—
1601	**Sturgeon, steamed**	3	ounce(s)	85	59.4	111	17.0	0	0	4.3	1.0	2.0	0.7	—
1840	**Surimi, formed**	3	ounce(s)	85	64.9	84	12.9	5.8	0	0.8	0.2	0.1	0.4	—
1842	**Swordfish, cooked, dry heat**	3	ounce(s)	85	58.5	132	21.6	0	0	4.4	1.2	1.7	1.0	—
1846	**Tuna, yellowfin or ahi, raw**	3	ounce(s)	85	60.4	92	19.9	0	0	0.8	0.2	0.1	0.2	—
	Tuna, canned													
159	Light, canned in oil, drained	2	ounce(s)	57	33.9	112	16.5	0	0	4.6	0.9	1.7	1.6	—
355	Light, canned in water, drained	2	ounce(s)	57	42.2	66	14.5	0	0	0.5	0.1	0.1	0.2	—
33211	Light, no salt, canned in oil, drained	2	ounce(s)	57	33.9	112	16.5	0	0	4.7	0.9	1.7	1.6	—
33212	Light, no salt, canned in water, drained	2	ounce(s)	57	42.6	66	14.5	0	0	0.5	0.1	0.1	0.2	—

PAGE KEY: A-6 = Breads/Baked Goods A-12 = Cereal/Rice/Pasta A-16 = Fruit A-20 = Vegetables/Legumes A-30 = Nuts/Seeds A-32 = Vegetarian A-34 = Dairy A-42 = Eggs A-42 = Seafood A-44 = Meats A-48 = Poultry
A-50 = Processed Meats A-50 = Beverages A-54 = Fats/Oils A-56 = Sweets A-58 = Spices/Condiments/Sauces A-62 = Mixed Foods/Soups/Sandwiches A-66 = Fast Food A-86 = Convenience A-88 = Baby Foods

Chol (mg)	Calc (mg)	Iron (mg)	Magn (mg)	Pota (mg)	Sodi (mg)	Zinc (mg)	Vit A (µg)	Thia (mg)	Vit E (mg α)	Ribo (mg)	Niac (mg)	Vit B$_6$ (mg)	Fola (µg)	Vit C (mg)	Vit B$_{12}$ (µg)	Sele (µg)
210	27	0.91	6.0	67.6	93.8	0.55	91.1	0.03	0.56	0.23	0.03	0.07	23.5	0	0.6	15.7
212	25	0.59	5.0	63.0	62.0	0.52	84.5	0.03	0.51	0.25	0.03	0.06	22.0	0	0.6	15.4
211	27	0.91	6.0	66.5	147.0	0.55	69.5	0.02	0.48	0.20	0.03	0.06	17.5	0	0.6	15.8
0	2	0.02	3.6	53.8	54.8	0.01	0	0.00	0.00	0.14	0.03	0.00	1.3	0	0	6.6
212	27	0.91	6.0	67.0	70.0	0.55	70.0	0.03	0.48	0.23	0.03	0.07	23.5	0	0.6	15.9
210	22	0.46	0.9	18.5	8.2	0.39	64.8	0.03	0.43	0.09	0.00	0.06	24.8	0	0.3	9.5
429	87	1.46	14.6	168.4	341.6	1.22	174.5	0.06	1.33	0.53	0.09	0.14	36.6	0.2	0.9	27.5
0	20	1.08	4.0	85.0	115.0	0.60	112.5	0.15	—	0.85	0.20	0.08	60.0	0	1.2	—
1	44	1.18	9.0	127.8	119.4	0.58	6.6	0.07	0.95	0.23	0.08	0.08	9.6	0.3	0.2	24.8
1	33	1.32	5.6	207.1	111.1	0.82	11.3	0.07	0.17	0.19	0.07	0.00	9.4	0	0.2	15.6
47	12	0.41	35.7	207.5	66.3	0.49	11.9	0.07	0.68	0.06	2.13	0.24	6.8	0.8	0.9	32.0
47	12	0.41	35.7	207.5	66.3	0.49	11.9	0.07	0.68	0.06	2.13	0.24	6.8	0.9	0.9	32.0
107	17	0.42	17.0	231.3	43.4	1.37	887.0	0.13	3.40	0.03	2.97	0.05	12.8	1.5	2.6	5.5
44	8	0.31	29.1	489.0	86.1	0.48	—	0.02	—	0.05	2.47	0.46	8.1	3.0	1.0	44.3
29	15	1.79	20.4	272.2	452.5	0.37	9.4	0.09	—	0.09	1.78	0.08	14.5	0	0.9	7.7
37	55	0.97	96.7	524.3	62.9	0.49	—	0.05	—	0.08	6.47	0.36	12.6	0	1.2	42.5
32	48	0.85	84.0	455.6	54.7	0.42	—	0.05	—	0.07	5.92	0.33	11.5	0	1.1	37.0
41	12	0.29	24.7	319.3	41.7	0.34	—	0.06	—	0.06	1.89	0.21	6.1	0.8	0.8	32.0
44	19	0.34	47.3	224.7	280.2	0.20	—	0.06	0.40	0.07	2.02	0.18	7.4	2.8	1.6	33.5
40	18	0.96	31.5	404.0	45.1	0.43	42.5	0.06	—	0.01	0.32	0.29	8.5	0	0.6	39.8
63	36	1.15	42.5	339.4	74.0	0.40	16.2	0.03	0.42	0.03	3.94	0.29	6.8	0	1.2	34.4
63	36	1.14	42.5	339.3	74.0	0.40	16.2	0.03	—	0.03	3.93	0.29	11.1	0	1.2	34.4
35	51	0.91	91.0	489.9	58.7	0.45	45.9	0.05	—	0.07	6.05	0.33	11.9	0	1.2	39.8
8	46	0.73	4.8	41.4	522.0	0.31	154.8	0.02	1.02	0.08	1.98	0.10	1.2	0	2.6	35.1
45	137	1.15	21.0	110.0	214.9	0.57	73.7	0.02	0.58	0.12	3.50	0.11	2.8	0.5	3.9	21.4
82	90	8.11	51.0	535.8	391.2	2.85	76.5	0.04	1.02	0.06	3.21	0.55	20.4	6.8	30.6	76.2
98	87	0.98	32.3	292.6	67.2	1.21	8.5	0.06	—	0.10	1.61	0.11	5.1	1.4	1.9	13.7
37	10	0.45	28.9	442.3	65.5	0.45	60.4	0.03	1.32	0.07	3.33	0.22	8.5	0	1.0	39.8
43	10	0.29	26.4	382.7	40.0	0.36	47.6	0.08	—	0.09	5.79	0.56	11.1	0.9	2.3	10.7
40	15	1.02	26.9	376.6	98.6	0.56	—	0.13	1.14	0.05	8.33	0.18	4.2	1.8	2.3	41.0
13	6	0.48	10.2	99.2	1134.0	0.17	14.7	0.01	—	0.05	2.67	0.15	1.1	0	1.8	21.6
121	325	2.48	33.2	337.6	429.5	1.10	27.2	0.04	1.70	0.18	4.43	0.14	10.2	0	7.6	44.8
28	20	0.38	27.4	154.8	215.8	0.49	10.7	0.02	—	0.05	0.70	0.06	17.2	1.1	0.6	12.5
27	20	0.22	45.9	238.0	358.7	0.78	32.3	0.01	0.16	0.05	0.84	0.11	10.2	2.0	1.1	18.2
40	34	0.20	31.5	444.0	48.5	0.37	29.8	0.04	—	0.00	0.29	0.39	5.1	1.4	3.0	41.7
221	33	0.85	32.3	237.3	260.3	1.48	9.4	0.04	—	0.39	2.21	0.04	11.9	3.6	1.0	44.1
227	31	0.62	28.9	192.1	356.2	1.49	8.5	0.01	1.17	0.32	1.69	0.04	3.4	3.2	1.0	43.7
88	16	0.91	43.4	279.0	74.8	0.43	26.4	0.09	—	0.03	2.17	0.29	8.5	0	3.8	39.8
63	11	0.59	29.8	239.7	388.5	0.35	198.9	0.06	0.52	0.07	8.30	0.19	14.5	0	2.2	13.3
26	8	0.22	36.6	95.3	121.6	0.28	17.0	0.01	0.53	0.01	0.18	0.02	1.7	0	1.4	23.9
43	5	0.88	28.9	313.8	97.8	1.25	34.9	0.03	—	0.09	10.02	0.32	1.7	0.9	1.7	52.5
38	14	0.62	42.5	377.6	31.5	0.44	15.3	0.37	0.42	0.04	8.33	0.77	1.7	0.8	0.4	31.0
10	7	0.79	17.6	117.3	200.6	0.51	13.0	0.02	0.49	0.07	7.03	0.06	2.8	0	1.2	43.1
17	6	0.87	15.3	134.3	191.5	0.43	9.6	0.01	0.19	0.04	7.52	0.19	2.3	0	1.7	45.6
10	7	0.78	17.6	117.4	28.3	0.51	0	0.02	—	0.06	7.03	0.06	2.8	0	1.2	43.1
17	6	0.86	15.3	134.4	28.3	0.43	0	0.01	—	0.04	7.52	0.19	2.3	0	1.7	45.6

TABLE A-1 **Table of Food Composition (*continued*)** (Computer code is for Cengage Diet Analysis program) (For purposes of calculations, use "0" for t, <1, <.1, <.01, etc.)

DA+ Code	Food Description	Quantity	Measure	Wt (g)	H₂0 (g)	Ener (kcal)	Prot (g)	Carb (g)	Fiber (g)	Fat (g)	Sat	Mono	Poly	Trans
Seafood—continued														
2961	White, canned in oil, drained	2	ounce(s)	57	36.3	105	15.0	0	0	4.6	0.7	1.8	1.7	—
351	White, canned in water, drained	2	ounce(s)	57	41.5	73	13.4	0	0	1.7	0.4	0.4	0.6	—
33213	White, no salt, canned in oil, drained	2	ounce(s)	57	36.3	105	15.0	0	0	4.6	0.9	1.4	1.9	—
33214	White, no salt, canned in water, drained	2	ounce(s)	57	42.0	73	13.4	0	0	1.7	0.4	0.4	0.6	—
	Yellowtail													
8548	Mixed species, cooked, dry heat	3	ounce(s)	85	57.3	159	25.2	0	0	5.7	1.4	2.2	1.5	—
2970	Mixed species, raw	2	ounce(s)	57	42.2	83	13.1	0	0	3.0	0.7	1.1	0.8	—
	Shellfish, meat only													
1857	Abalone, mixed species, fried	3	ounce(s)	85	51.1	161	16.7	9.4	0	5.8	1.4	2.3	1.4	—
16618	Abalone, steamed or poached	3	ounce(s)	85	40.7	177	28.8	10.1	0	1.3	0.3	0.2	0.2	—
	Crab													
1851	Blue crab, canned	2	ounce(s)	57	43.2	56	11.6	0	0	0.7	0.1	0.1	0.2	—
1852	Blue crab, cooked, moist heat	3	ounce(s)	85	65.9	87	17.2	0	0	1.5	0.2	0.2	0.6	—
8562	Dungeness crab, cooked, moist heat	3	ounce(s)	85	62.3	94	19.0	0.8	0	1.1	0.1	0.2	0.3	—
1860	**Clams, cooked, moist heat**	3	ounce(s)	85	54.1	126	21.7	4.4	0	1.7	0.2	0.1	0.5	—
1853	**Crayfish, farmed, cooked, moist heat**	3	ounce(s)	85	68.7	74	14.9	0	0	1.1	0.2	0.2	0.4	—
	Oysters													
8720	Baked or broiled	3	ounce(s)	85	68.6	89	5.6	3.2	0	5.8	1.3	2.1	1.9	—
152	Eastern, farmed, raw	3	ounce(s)	85	73.3	50	4.4	4.7	0	1.3	0.4	0.1	0.5	—
8715	Eastern, wild, cooked, moist heat	3	ounce(s)	85	59.8	117	12.0	6.7	0	4.2	1.3	0.5	1.6	—
8584	Pacific, cooked, moist heat	3	ounce(s)	85	54.5	139	16.1	8.4	0	3.9	0.9	0.7	1.5	—
1865	Pacific, raw	3	ounce(s)	85	69.8	69	8.0	4.2	0	2.0	0.4	0.3	0.8	—
1854	**Lobster, northern, cooked, moist heat**	3	ounce(s)	85	64.7	83	17.4	1.1	0	0.5	0.1	0.1	0.1	—
1862	**Mussel, blue, cooked, moist heat**	3	ounce(s)	85	52.0	146	20.2	6.3	0	3.8	0.7	0.9	1.0	—
	Shrimp													
158	Mixed species, breaded, fried	3	ounce(s)	85	44.9	206	18.2	9.8	0.3	10.4	1.8	3.2	4.3	—
1855	Mixed species, cooked, moist heat	3	ounce(s)	85	65.7	84	17.8	0	0	0.9	0.2	0.2	0.4	—
Beef, Lamb, Pork														
	Beef													
4450	Breakfast strips, cooked	2	slice(s)	23	5.9	101	7.1	0.3	0	7.8	3.2	3.8	0.4	—
174	Corned beef, canned	3	ounce(s)	85	49.1	213	23.0	0	0	12.7	5.3	5.1	0.5	—
33147	Cured, thin siced	2	ounce(s)	57	32.9	100	15.9	3.2	0	2.2	0.9	1.0	0.1	—
4581	Jerky	1	ounce(s)	28	6.6	116	9.4	3.1	0.5	7.3	3.1	3.2	0.3	—
	Ground beef													
5898	Lean, broiled, medium	3	ounce(s)	85	50.4	202	21.6	0	0	12.2	4.8	5.3	0.4	—
5899	Lean, broiled, well done	3	ounce(s)	85	48.4	214	23.8	0	0	12.5	5.0	5.7	0.3	0.4
5914	Regular, broiled, medium	3	ounce(s)	85	46.1	246	20.5	0	0	17.6	6.9	7.7	0.6	—
5915	Regular, broiled, well done	3	ounce(s)	85	43.8	259	21.6	0	0	18.4	7.5	8.5	0.5	0.6
	Beef rib													
4241	Rib, small end, separable lean, 0" fat, broiled	3	ounce(s)	85	53.2	164	25.0	0	0	6.4	2.4	2.6	0.2	—
4183	Rib, whole, lean and fat, ¼" fat, roasted	3	ounce(s)	85	39.0	320	18.9	0	0	26.6	10.7	11.4	0.9	—
	Beef roast													
16981	Bottom round, choice, separable lean and fat, ⅛" fat, braised	3	ounce(s)	85	46.2	216	27.9	0	0	10.7	4.1	4.6	0.4	—
16979	Bottom round, separable lean and fat, ⅛" fat, roasted	3	ounce(s)	85	52.4	185	22.5	0	0	9.9	3.8	4.2	0.4	—
16924	Chuck, arm pot roast, separable lean and fat, ⅛" fat, braised	3	ounce(s)	85	42.9	257	25.6	0	0	16.3	6.5	7.0	0.6	—
16930	Chuck, blade roast, separable lean and fat, ⅛" fat, braised	3	ounce(s)	85	40.5	290	22.8	0	0	21.4	8.5	9.2	0.8	—
5853	Chuck, blade roast, separable lean, 0" trim, pot roasted	3	ounce(s)	85	47.4	202	26.4	0	0	9.9	3.9	4.3	0.3	—
4296	Eye of round, choice, separable lean, 0" fat, roasted	3	ounce(s)	85	56.5	138	24.4	0	0	3.7	1.3	1.5	0.1	—
16989	Eye of round, separable lean and fat, ⅛" fat, roasted	3	ounce(s)	85	52.2	180	24.2	0	0	8.5	3.2	3.6	0.3	—

Chol (mg)	Calc (mg)	Iron (mg)	Magn (mg)	Pota (mg)	Sodi (mg)	Zinc (mg)	Vit A (µg)	Thia (mg)	Vit E (mg α)	Ribo (mg)	Niac (mg)	Vit B_6 (mg)	Fola (µg)	Vit C (mg)	Vit B_{12} (µg)	Sele (µg)
18	2	0.36	19.3	188.8	224.5	0.26	2.8	0.01	1.30	0.04	6.63	0.24	2.8	0	1.2	34.1
24	8	0.55	18.7	134.3	213.6	0.27	3.4	0.00	0.48	0.02	3.28	0.12	1.1	0	0.7	37.2
18	2	0.36	19.3	188.8	28.3	0.26	0	0.01	—	0.04	6.63	0.24	2.8	0	1.2	34.1
24	8	0.54	18.7	134.4	28.3	0.27	3.4	0.00	—	0.02	3.28	0.12	1.1	0	0.7	37.3
60	25	0.53	32.3	457.6	42.5	0.56	26.4	0.14	—	0.04	7.41	0.15	3.4	2.5	1.1	39.8
31	13	0.28	17.0	238.1	22.1	0.29	16.4	0.08	—	0.02	3.86	0.09	2.3	1.6	0.7	20.7
80	31	3.23	47.6	241.5	502.6	0.80	1.7	0.18	—	0.11	1.61	0.12	11.9	1.5	0.6	44.1
144	50	4.84	68.9	295.0	980.1	1.38	3.4	0.28	6.74	0.12	1.89	0.21	6.0	2.6	0.7	75.6
50	57	0.47	22.1	212.1	188.8	2.27	1.1	0.04	1.04	0.04	0.77	0.08	24.4	1.5	0.3	18.0
85	88	0.77	28.1	275.6	237.3	3.58	1.7	0.08	1.56	0.04	2.80	0.15	43.4	2.8	6.2	34.2
65	50	0.36	49.3	347.0	321.5	4.65	26.4	0.04	—	0.17	3.08	0.14	35.7	3.1	8.8	40.5
57	78	23.78	15.3	534.1	95.3	2.32	145.4	0.12	—	0.36	2.85	0.09	24.7	18.8	84.1	54.4
117	43	0.94	28.1	202.4	82.5	1.25	12.8	0.03	—	0.06	1.41	0.11	9.4	0.4	2.6	29.1
43	36	5.30	37.4	125.0	403.8	72.22	60.4	0.07	0.98	0.06	1.04	0.04	7.7	2.8	14.7	50.7
21	37	4.91	28.1	105.4	151.3	32.23	6.8	0.08	—	0.05	1.07	0.05	15.3	4.0	13.8	54.1
89	77	10.19	80.8	239.0	358.9	154.45	45.9	0.16	—	0.15	2.11	0.10	11.9	5.1	29.8	60.9
85	14	7.82	37.4	256.8	180.3	28.27	124.2	0.10	0.72	0.37	3.07	0.07	12.8	10.9	24.5	131.0
43	7	4.34	18.7	142.9	90.1	14.13	68.9	0.05	—	0.20	1.70	0.04	8.5	6.8	13.6	65.5
61	52	0.33	29.8	299.4	323.2	2.48	22.1	0.01	0.85	0.05	0.91	0.06	9.4	0	2.6	36.3
48	28	5.71	31.5	227.9	313.8	2.27	77.4	0.25	—	0.35	2.55	0.08	64.6	11.6	20.4	76.2
150	57	1.07	34.0	191.3	292.4	1.17	0	0.11	—	0.11	2.60	0.08	15.3	1.3	1.6	35.4
166	33	2.62	28.9	154.8	190.5	1.32	57.8	0.02	1.17	0.02	2.20	0.10	3.4	1.9	1.3	33.7
27	2	0.71	6.1	93.1	509.2	1.44	0	0.02	0.06	0.05	1.46	0.07	1.8	0	0.8	6.1
73	10	1.76	11.9	115.7	855.6	3.03	0	0.01	0.12	0.12	2.06	0.11	7.7	0	1.4	36.5
23	6	1.53	10.8	243.2	815.9	2.25	0	0.04	0.00	0.10	2.98	0.19	6.2	0	1.5	16.0
14	6	1.53	14.5	169.2	627.4	2.29	0	0.04	0.13	0.04	0.49	0.05	38.0	0	0.3	3.0
58	6	2.00	17.9	266.2	59.5	4.63	0	0.05	—	0.23	4.21	0.23	7.6	0	1.8	16.0
69	12	2.21	18.4	250.0	62.4	5.86	0	0.08	—	0.23	5.10	0.16	9.4	0	1.7	19.0
62	9	2.07	17.0	248.3	70.6	4.40	0	0.02	—	0.16	4.90	0.23	7.6	0	2.5	16.2
71	12	2.30	18.5	242.4	72.4	5.18	0	0.08	—	0.23	4.93	0.17	8.5	0	1.6	18.0
65	16	1.59	21.3	319.8	51.9	4.64	0	0.06	0.34	0.12	7.15	0.53	8.5	0	1.4	29.2
72	9	1.96	16.2	251.7	53.6	4.45	0	0.06	—	0.14	2.85	0.19	6.0	0	2.1	18.7
68	6	2.29	17.9	223.7	35.7	4.59	0	0.05	0.41	0.15	5.05	0.36	8.5	0	1.7	29.3
64	5	1.83	14.5	182.0	29.8	3.76	0	0.05	0.34	0.12	3.92	0.29	6.8	0	1.3	23.0
67	14	2.15	17.0	205.8	42.5	5.93	0	0.05	0.45	0.15	3.63	0.25	7.7	0	1.9	24.1
88	11	2.66	16.2	198.2	55.3	7.15	0	0.06	0.17	0.20	2.06	0.22	4.3	0	1.9	20.9
73	11	3.12	19.6	223.7	60.4	8.73	0	0.06	—	0.23	2.27	0.24	5.1	0	2.1	22.7
49	5	2.16	16.2	200.7	32.3	4.28	0	0.05	0.30	0.15	4.69	0.34	8.5	0	1.4	28.0
54	5	1.98	15.3	193.1	31.5	3.95	0	0.05	0.34	0.13	4.37	0.31	7.7	0	1.5	25.2

TABLE A-1 **Table of Food Composition** *(continued)* (Computer code is for Cengage Diet Analysis program) (For purposes of calculations, use "0" for t, <1, <.1, <.01, etc.)

DA+ Code	Food Description	Quantity	Measure	Wt (g)	H₂O (g)	Ener (kcal)	Prot (g)	Carb (g)	Fiber (g)	Fat (g)	Fat Breakdown (g) Sat	Mono	Poly	Trans
Beef, Lamb, Pork—continued														
	Beef steak													
4348	Short loin, T-bone steak, lean and fat, ¼" fat, broiled	3	ounce(s)	85	43.2	274	19.4	0	0	21.2	8.3	9.6	0.8	—
4349	Short loin, T-bone steak, lean, ¼" fat, broiled	3	ounce(s)	85	52.3	174	22.8	0	0	8.5	3.1	4.2	0.3	—
4360	Top loin, prime, lean and fat, ¼" fat, broiled	3	ounce(s)	85	42.7	275	21.6	0	0	20.3	8.2	8.6	0.7	—
	Beef variety													
188	Liver, pan fried	3	ounce(s)	85	52.7	149	22.6	4.4	0	4.0	1.3	0.5	0.5	0.2
4447	Tongue, simmered	3	ounce(s)	85	49.2	242	16.4	0	0	19.0	6.9	8.6	0.6	0.7
	Lamb chop													
3275	Loin, domestic, lean and fat, ¼" fat, broiled	3	ounce(s)	85	43.9	269	21.4	0	0	19.6	8.4	8.3	1.4	—
	Lamb leg													
3264	Domestic, lean and fat, ¼" fat, cooked	3	ounce(s)	85	45.7	250	20.9	0	0	17.8	7.5	7.5	1.3	—
	Lamb rib													
182	Domestic, lean and fat, ¼" fat, broiled	3	ounce(s)	85	40.0	307	18.8	0	0	25.2	10.8	10.3	2.0	—
183	Domestic, lean, ¼" fat, broiled	3	ounce(s)	85	50.0	200	23.6	0	0	11.0	4.0	4.4	1.0	—
	Lamb shoulder													
186	Shoulder, arm and blade, domestic, choice, lean and fat, ¼" fat, roasted	3	ounce(s)	85	47.8	235	19.1	0	0	17.0	7.2	6.9	1.4	—
187	Shoulder, arm and blade, domestic, choice, lean, ¼" fat, roasted	3	ounce(s)	85	53.8	173	21.2	0	0	9.2	3.5	3.7	0.8	—
3287	Shoulder, arm, domestic, lean and fat, ¼" fat, braised	3	ounce(s)	85	37.6	294	25.8	0	0	20.4	8.4	8.7	1.5	—
3290	Shoulder, arm, domestic, lean, ¼" fat, braised	3	ounce(s)	85	41.9	237	30.2	0	0	12.0	4.3	5.2	0.8	—
	Lamb variety													
3375	Brain, pan fried	3	ounce(s)	85	51.6	232	14.4	0	0	18.9	4.8	3.4	1.9	—
3406	Tongue, braised	3	ounce(s)	85	49.2	234	18.3	0	0	17.2	6.7	8.5	1.1	—
	Pork, cured													
29229	Bacon, Canadian style, cured	2	ounce(s)	57	37.9	89	11.7	1.0	0	4.0	1.3	1.8	0.4	—
161	Bacon, cured, broiled, pan fried or roasted	2	slice(s)	16	2.0	87	5.9	0.2	0	6.7	2.2	3.0	0.7	0
35422	Breakfast strips, cured, cooked	3	slice(s)	34	9.2	156	9.8	0.4	0	12.5	4.3	5.6	1.9	—
189	Ham, cured, boneless, 11% fat, roasted	3	ounce(s)	85	54.9	151	19.2	0	0	7.7	2.7	3.8	1.2	—
29215	Ham, cured, extra lean, 4% fat, canned	2	2 ounce(s)	57	41.7	68	10.5	0	0	2.6	0.9	1.3	0.2	—
1316	Ham, cured, extra lean, 5% fat, roasted	3	ounce(s)	85	57.6	123	17.8	1.3	0	4.7	1.5	2.2	0.5	—
16561	Ham, smoked or cured, lean, cooked	1	slice(s)	42	27.6	66	10.5	0	0	2.3	0.8	1.1	0.3	—
	Pork chop													
32671	Loin, blade, chops, lean and fat, pan fried	3	ounce(s)	85	42.5	291	18.3	0	0	23.6	8.6	10	2.6	—
32672	Loin, center cut, chops, lean and fat, pan fried	3	ounce(s)	85	45.1	236	25.4	0	0	14.1	5.1	6.0	1.6	—
32682	Loin, center rib, chops, boneless, lean and fat, braised	3	ounce(s)	85	49.5	217	22.4	0	0	13.4	5.2	6.1	1.1	—
32603	Loin, center rib, chops, lean, broiled	3	ounce(s)	85	55.4	158	21.9	0	0	7.1	2.4	3.0	0.8	0.1
32478	Loin, whole, lean and fat, braised	3	ounce(s)	85	49.6	203	23.2	0	0	11.6	4.3	5.2	1.0	—
32481	Loin, whole, lean, braised	3	ounce(s)	85	52.2	174	24.3	0	0	7.8	2.9	3.5	0.6	—
	Pork leg or ham													
32471	Pork leg or ham, rump portion, lean and fat, roasted	3	ounce(s)	85	48.3	214	24.6	0	0	12.1	4.5	5.4	1.2	—
32468	Pork leg or ham, whole, lean and fat, roasted	3	ounce(s)	85	46.8	232	22.8	0	0	15.0	5.5	6.7	1.4	—
	Pork ribs													
32693	Loin, country style, lean and fat, roasted	3	ounce(s)	85	43.3	279	19.9	0	0	21.6	7.8	9.4	1.7	—
32696	Loin, country style, lean, roasted	3	ounce(s)	85	49.5	210	22.6	0	0	12.6	4.5	5.5	0.9	—

Chol (mg)	Calc (mg)	Iron (mg)	Magn (mg)	Pota (mg)	Sodi (mg)	Zinc (mg)	Vit A (µg)	Thia (mg)	Vit E (mg α)	Ribo (mg)	Niac (mg)	Vit B_6 (mg)	Fola (µg)	Vit C (mg)	Vit B_{12} (µg)	Sele (µg)
58	7	2.56	17.9	233.9	57.8	3.56	0	0.07	0.18	0.17	3.29	0.27	6.0	0	1.8	10.0
50	5	3.11	22.1	278.1	65.5	4.34	0	0.09	0.11	0.21	3.93	0.33	6.8	0	1.9	8.5
67	8	1.88	19.6	294.3	53.6	3.85	0	0.06	—	0.15	3.96	0.31	6.0	0	1.6	19.5
324	5	5.24	18.7	298.5	65.5	4.44	6586.3	0.15	0.39	2.91	14.86	0.87	221.1	0.6	70.7	27.9
112	4	2.22	12.8	156.5	55.3	3.47	0	0.01	0.25	0.25	2.96	0.13	6.0	1.1	2.7	11.2
85	17	1.53	20.4	278.1	65.5	2.96	0	0.08	0.11	0.21	6.03	0.11	15.3	0	2.1	23.3
82	14	1.59	19.6	263.7	61.2	3.79	0	0.08	0.11	0.21	5.66	0.11	15.3	0	2.2	22.5
84	16	1.59	19.6	229.5	64.6	3.40	0	0.07	0.10	0.18	5.95	0.09	11.9	0	2.2	20.3
77	14	1.87	24.7	266.1	72.3	4.47	0	0.08	0.15	0.21	5.56	0.12	17.9	0	2.2	26.4
78	17	1.67	19.6	213.4	56.1	4.44	0	0.07	0.11	0.20	5.22	0.11	17.9	0	2.2	22.3
74	16	1.81	21.3	225.3	57.8	5.13	0	0.07	0.15	0.22	4.89	0.12	21.3	0	2.3	24.2
102	21	2.03	22.1	260.3	61.2	5.17	0	0.06	0.12	0.21	5.66	0.09	15.3	0	2.2	31.6
103	22	2.29	24.7	287.5	64.6	6.20	0	0.06	0.15	0.23	5.38	0.11	18.7	0	2.3	32.1
2130	18	1.73	18.7	304.5	133.5	1.70	0	0.14	—	0.31	3.87	0.19	6.0	19.6	20.5	10.2
161	9	2.23	13.6	134.4	57.0	2.54	0	0.06	—	0.35	3.13	0.14	2.6	6.0	5.4	23.8
28	5	0.38	9.6	195.0	798.9	0.78	0	0.42	0.11	0.09	3.53	0.22	2.3	0	0.4	14.2
18	2	0.22	5.3	90.4	369.6	0.56	1.8	0.06	0.04	0.04	1.76	0.04	0.3	0	0.2	9.9
36	5	0.67	8.8	158.4	713.7	1.25	0	0.25	0.08	0.12	2.58	0.11	1.4	0	0.6	8.4
50	7	1.13	18.7	347.7	1275.0	2.09	0	0.62	0.26	0.28	5.22	0.26	2.6	0	0.6	16.8
22	3	0.53	9.6	206.4	711.6	1.09	0	0.47	0.09	0.13	3.00	0.25	3.4	0	0.5	8.2
45	7	1.25	11.9	244.1	1023.1	2.44	0	0.64	0.21	0.17	3.42	0.34	2.6	0	0.6	16.6
23	3	0.39	9.2	132.7	557.3	1.07	0	0.28	0.10	0.10	2.10	0.19	1.7	0	0.3	10.7
72	26	0.74	17.9	282.4	57.0	2.71	1.7	0.52	0.17	0.25	3.35	0.28	3.4	0.5	0.7	29.7
78	23	0.77	24.7	361.5	68.0	1.96	1.7	0.96	0.21	0.25	4.76	0.39	5.1	0.9	0.6	33.2
62	4	0.78	14.5	329.1	34.0	1.76	1.7	0.44	—	0.20	3.66	0.26	3.4	0.3	0.4	28.4
56	22	0.57	21.3	291.7	48.5	1.91	0	0.48	0.08	0.18	6.68	0.57	0	0	0.4	38.6
68	18	0.91	16.2	318.1	40.8	2.02	1.7	0.53	0.20	0.21	3.75	0.31	2.6	0.5	0.5	38.5
67	15	0.96	17.0	329.1	42.5	2.10	1.7	0.56	0.17	0.22	3.90	0.32	3.4	0.5	0.5	41.0
82	10	0.89	23.0	318.1	52.7	2.39	2.6	0.63	0.18	0.28	3.95	0.26	2.6	0.2	0.6	39.8
80	12	0.85	18.7	299.4	51.0	2.51	2.6	0.54	0.18	0.26	3.89	0.34	8.5	0.3	0.6	38.5
78	21	0.90	19.6	292.6	44.2	2.00	2.6	0.75	—	0.29	3.67	0.37	4.3	0.3	0.7	31.6
79	25	1.09	20.4	296.8	24.7	3.24	1.7	0.48	—	0.29	3.96	0.37	4.3	0.3	0.7	36.0

TABLE A-1 **Table of Food Composition** *(continued)* (Computer code is for Cengage Diet Analysis program) (For purposes of calculations, use "0" for t, <1, <.1, <.01, etc.)

DA+ Code	Food Description	Quantity	Measure	Wt (g)	H₂O (g)	Ener (kcal)	Prot (g)	Carb (g)	Fiber (g)	Fat (g)	Sat	Mono	Poly	Trans
Beef, Lamb, Pork—continued														
Pork shoulder														
32626	Shoulder, arm picnic, lean and fat, roasted	3	ounce(s)	85	44.3	270	20.0	0	0	20.4	7.5	9.1	2.0	—
32629	Shoulder, arm picnic, lean, roasted	3	ounce(s)	85	51.3	194	22.7	0	0	10.7	3.7	5.1	1.0	—
Rabbit														
3366	Domesticated, roasted	3	ounce(s)	85	51.5	168	24.7	0	0	6.8	2.0	1.8	1.3	—
3367	Domesticated, stewed	3	ounce(s)	85	50.0	175	25.8	0	0	7.2	2.1	1.9	1.4	—
Veal														
3391	Liver, braised	3	ounce(s)	85	50.9	163	24.2	3.2	0	5.3	1.7	1.0	0.9	0.3
3319	Rib, lean only, roasted	3	ounce(s)	85	55.0	151	21.9	0	0	6.3	1.8	2.3	0.6	—
1732	Deer or venison, roasted	3	ounce(s)	85	55.5	134	25.7	0	0	2.7	1.1	0.7	0.5	—
Poultry														
Chicken														
29562	Flaked, canned	2	ounce(s)	57	39.3	97	10.3	0.1	0	5.8	1.6	2.3	1.3	—
Chicken, fried														
29632	Breast, meat only, breaded, baked or fried	3	ounce(s)	85	44.3	193	25.3	6.9	0.2	6.6	1.6	2.7	1.7	—
35327	Broiler breast, meat only, fried	3	ounce(s)	85	51.2	159	28.4	0.4	0	4.0	1.1	1.5	0.9	—
36413	Broiler breast, meat and skin, flour coated, fried	3	ounce(s)	85	48.1	189	27.1	1.4	0.1	7.5	2.1	3.0	1.7	—
36414	Broiler drumstick, meat and skin, flour coated, fried	3	ounce(s)	85	48.2	208	22.9	1.4	0.1	11.7	3.1	4.6	2.7	—
35389	Broiler drumstick, meat only, fried	3	ounce(s)	85	52.9	166	24.3	0	0	6.9	1.8	2.5	1.7	—
35406	Broiler leg, meat only, fried	3	ounce(s)	85	51.5	177	24.1	0.6	0	7.9	2.1	2.9	1.9	—
35484	Broiler wing, meat only, fried	3	ounce(s)	85	50.9	179	25.6	0	0	7.8	2.1	2.6	1.8	—
29580	Patty, fillet or tenders, breaded, cooked	3	ounce(s)	85	40.2	256	14.5	12.2	0	16.5	3.7	8.4	3.7	—
Chicken, roasted, meat only														
35409	Broiler leg, meat only, roasted	3	ounce(s)	85	55.0	162	23.0	0	0	7.2	1.9	2.6	1.7	—
35486	Broiler wing, meat only, roasted	3	ounce(s)	85	53.4	173	25.9	0	0	6.9	1.9	2.2	1.5	—
35138	Roasting chicken, dark meat, meat only, roasted	3	ounce(s)	85	57.0	151	19.8	0	0	7.4	2.1	2.8	1.7	—
35136	Roasting chicken, light meat, meat only, roasted	3	ounce(s)	85	57.7	130	23.1	0	0	3.5	0.9	1.3	0.8	—
35132	Roasting chicken, meat only, roasted	3	ounce(s)	85	57.3	142	21.3	0	0	5.6	1.5	2.1	1.3	—
Chicken, stewed														
1268	Gizzard, simmered	3	ounce(s)	85	57.8	124	25.8	0	0	2.3	0.6	0.4	0.3	0.1
1270	Liver, simmered	3	ounce(s)	85	56.8	142	20.8	0.7	0	5.5	1.8	1.2	1.7	0.1
3174	Meat only, stewed	3	ounce(s)	85	56.8	151	23.2	0	0	5.7	1.6	2.0	1.3	—
Duck														
1286	Domesticated, meat and skin, roasted	3	ounce(s)	85	44.1	287	16.2	0	0	24.1	8.2	11.0	3.1	—
1287	Domesticated, meat only, roasted	3	ounce(s)	85	54.6	171	20.0	0	0	9.5	3.5	3.1	1.2	—
Goose														
35507	Domesticated, meat and skin, roasted	3	ounce(s)	85	44.2	259	21.4	0	0	18.6	5.8	8.7	2.1	—
35524	Domesticated, meat only, roasted	3	ounce(s)	85	48.7	202	24.6	0	0	10.8	3.9	3.7	1.3	—
1297	Liver pate, smoked, canned	4	tablespoon(s)	52	19.3	240	5.9	2.4	0	22.8	7.5	13.3	0.4	—
Turkey														
3256	Ground turkey, cooked	3	ounce(s)	85	50.5	200	23.3	0	0	11.2	2.9	4.2	2.7	—
3263	Patty, batter coated, breaded, fried	1	item(s)	94	46.7	266	13.2	14.8	0.5	16.9	4.4	7.0	4.4	—
219	Roasted, dark meat, meat only	3	ounce(s)	85	53.7	159	24.3	0	0	6.1	2.1	1.4	1.8	—
222	Roasted, fryer roaster breast, meat only	3	ounce(s)	85	58.2	115	25.6	0	0	0.6	0.2	0.1	0.2	—
220	Roasted, light meat, meat only	3	ounce(s)	85	56.4	134	25.4	0	0	2.7	0.9	0.5	0.7	—
1303	Turkey roll, light and dark meat	2	slice(s)	57	39.8	84	10.3	1.2	0	4.0	1.2	1.3	1.0	—
1302	Turkey roll, light meat	2	slice(s)	57	42.5	56	8.4	2.9	0	0.9	0.2	0.2	0.1	0

Chol (mg)	Calc (mg)	Iron (mg)	Magn (mg)	Pota (mg)	Sodi (mg)	Zinc (mg)	Vit A (µg)	Thia (mg)	Vit E (mg α)	Ribo (mg)	Niac (mg)	Vit B$_6$ (mg)	Fola (µg)	Vit C (mg)	Vit B$_{12}$ (µg)	Sele (µg)
80	16	1.00	14.5	276.4	59.5	2.93	1.7	0.44	—	0.25	3.33	0.29	3.4	0.2	0.6	28.6
81	8	1.20	17.0	298.5	68.0	3.46	1.7	0.49	—	0.30	3.66	0.34	4.3	0.3	0.7	32.7
70	16	1.93	17.9	325.7	40.0	1.93	0	0.07	—	0.17	7.17	0.40	9.4	0	7.1	32.7
73	17	2.01	17.0	255.1	31.5	2.01	0	0.05	0.37	0.14	6.09	0.28	7.7	0	5.5	32.7
435	5	4.34	17.0	279.8	66.3	9.55	8026	0.15	0.57	2.43	11.18	0.78	281.5	0.9	72.0	16.4
98	10	0.81	20.4	264.5	82.5	3.81	0	0.05	0.30	0.24	6.37	0.23	11.9	0	1.3	9.4
95	6	3.80	20.4	284.9	45.9	2.33	0	0.15	—	0.51	5.70	—	—	0	—	11.0
35	8	0.89	6.8	147.4	408.2	0.79	19.3	0.01	—	0.07	3.58	0.19	2.3	0	0.2	—
67	19	1.05	24.7	222.6	450.2	0.84	—	0.08	—	0.09	10.97	0.46	4.3	0	0.3	—
77	14	0.96	26.4	234.7	67.2	0.91	6.0	0.06	0.35	0.10	12.57	0.54	3.4	0	0.3	22.3
76	14	1.01	25.5	220.3	64.6	0.93	12.8	0.06	0.39	0.11	11.68	0.49	5.1	0	0.3	20.3
77	10	1.13	19.6	194.8	75.7	2.45	21.3	0.06	0.65	0.19	5.13	0.29	8.5	0	0.3	15.6
80	10	1.12	20.4	211.8	81.6	2.73	15.3	0.06	—	0.20	5.22	0.33	7.7	0	0.3	16.7
84	11	1.19	21.3	216.0	81.6	2.53	17.0	0.07	0.38	0.21	5.68	0.33	7.7	0	0.3	16.0
71	13	0.96	17.9	176.9	77.4	1.80	15.3	0.03	0.40	0.10	6.15	0.50	3.4	0	0.3	21.6
49	11	0.75	19.6	244.8	411.4	0.79	4.3	0.09	1.04	0.12	5.99	0.24	24.7	0	0.2	13.9
80	10	1.11	20.4	205.8	77.4	2.43	16.2	0.06	0.22	0.19	5.37	0.31	6.8	0	0.3	18.8
72	14	0.98	17.9	178.6	78.2	1.82	15.3	0.03	0.22	0.10	6.21	0.50	3.4	0	0.3	21.0
64	9	1.13	17.0	190.5	80.8	1.81	13.6	0.05	—	0.16	4.87	0.26	6.0	0	0.2	16.7
64	11	0.91	19.6	200.7	43.4	0.66	6.8	0.05	0.22	0.07	8.90	0.45	2.6	0	0.3	21.9
64	10	1.02	17.9	194.8	63.8	1.29	10.2	0.05	—	0.12	6.70	0.34	4.3	0	0.2	20.9
315	14	2.71	2.6	152.2	47.6	3.75	0	0.02	0.17	0.17	2.65	0.06	4.3	0	0.9	35.0
479	9	9.89	21.3	223.7	64.6	3.38	3385.8	0.24	0.69	1.69	9.39	0.64	491.6	23.7	14.3	70.1
71	12	0.99	17.9	153.1	59.5	1.69	12.8	0.04	0.22	0.13	5.20	0.22	5.1	0	0.2	17.8
71	9	2.29	13.6	173.5	50.2	1.58	53.6	0.14	0.59	0.22	4.10	0.15	5.1	0	0.3	17.0
76	10	2.29	17.0	214.3	55.3	2.21	19.6	0.22	0.59	0.39	4.33	0.21	8.5	0	0.3	19.1
77	11	2.40	18.7	279.8	59.5	2.22	17.9	0.06	1.47	0.27	3.54	0.31	1.7	0	0.3	18.5
82	12	2.44	21.3	330.0	64.6	2.69	10.2	0.07	—	0.33	3.47	0.39	10.2	0	0.4	21.7
78	36	2.86	6.8	71.8	362.4	0.47	520.5	0.04	—	0.15	1.30	0.03	31.2	0	4.9	22.9
87	21	1.64	20.4	229.6	91.0	2.43	0	0.04	0.28	0.14	4.09	0.33	6.0	0	0.3	31.6
71	13	2.06	14.1	258.5	752.0	1.35	9.4	0.09	0.87	0.17	2.16	0.18	38.5	0	0.2	20.8
72	27	1.98	20.4	246.6	67.2	3.79	0	0.05	0.54	0.21	3.10	0.30	7.7	0	0.3	34.8
71	10	1.30	24.7	248.3	44.2	1.48	0	0.03	0.07	0.11	6.37	0.47	5.1	0	0.3	27.3
59	16	1.14	23.8	259.4	54.4	1.73	0	0.05	0.07	0.11	5.81	0.45	5.1	0	0.3	27.3
31	18	0.76	10.2	153.1	332.3	1.13	0	0.05	0.19	0.16	2.72	0.15	2.8	0	0.1	16.6
19	4	0.21	10.8	242.1	590.8	0.50	0	0.01	0.07	0.08	4.05	0.23	2.3	0	0.2	7.4

TABLE A-1 **Table of Food Composition** *(continued)* (Computer code is for Cengage Diet Analysis program) (For purposes of calculations, use "0" for t, <1, <.1, <.01, etc.)

DA+ Code	Food Description	Quantity	Measure	Wt (g)	H₂0 (g)	Ener (kcal)	Prot (g)	Carb (g)	Fiber (g)	Fat (g)	Fat Breakdown (g) Sat	Mono	Poly	*Trans*
Processed Meats														
	Beef													
1331	Corned beef loaf, jellied, sliced	2	slice(s)	57	39.2	87	13.0	0	0	3.5	1.5	1.5	0.2	—
	Bologna													
13459	Beef	1	slice(s)	28	15.1	90	3.0	1.0	0	8.0	3.5	4.3	0.3	—
13461	Light, made with pork and chicken	1	slice(s)	28	18.2	60	3.0	2.0	0	4.0	1.0	2.0	0.4	—
13458	Made with chicken and pork	1	slice(s)	28	15.0	90	3.0	1.0	0	8.0	3.0	4.1	1.1	—
13565	Turkey bologna	1	slice(s)	28	19.0	50	3.0	1.0	0	4.0	1.0	1.1	1.0	0
	Chicken													
7125	Breast, smoked	1	slice(s)	10	—	10	1.8	0.3	0	0.2	0	—	—	—
	Ham													
7127	Deli-sliced, honey	1	slice(s)	10	—	10	1.7	0.3	0	0.3	0.1	—	—	—
7126	Deli-sliced, smoked	1	slice(s)	10	—	10	1.7	0.2	0	0.3	0.1	—	—	—
8614	**Beef and pork mortadella, sliced**	2	slice(s)	46	24.1	143	7.5	1.4	0	11.7	4.4	5.2	1.4	—
1323	**Pork olive loaf**	2	slice(s)	57	33.1	133	6.7	5.2	0	9.4	3.3	4.5	1.1	—
1324	**Pork pickle and pimento loaf**	2	slice(s)	57	34.2	128	6.4	4.8	0.9	9.1	3.0	4.0	1.6	—
	Sausages and frankfurters													
37296	Beerwurst beef, beer salami (bierwurst)	1	slice(s)	29	16.6	74	4.1	1.2	0	5.7	2.5	2.7	0.2	—
37257	Beerwurst pork, beer salami	1	slice(s)	21	12.9	50	3.0	0.4	0	4.0	1.3	1.9	0.5	—
35338	Berliner, pork and beef	1	ounce(s)	28	17.3	65	4.3	0.7	0	4.9	1.7	2.3	0.4	—
37298	Bratwurst pork, cooked	1	piece(s)	74	42.3	181	10.4	1.9	0	14.3	5.1	6.7	1.5	—
37299	Braunschweiger pork liver sausage	1	slice(s)	15	8.2	51	2.0	0.3	0	4.5	1.5	2.1	0.5	—
1329	Cheesefurter or cheese smokie, beef and pork	1	item(s)	43	22.6	141	6.1	0.6	0	12.5	4.5	5.9	1.3	—
1330	Chorizo, beef and pork	2	ounce(s)	57	18.1	258	13.7	1.1	0	21.7	8.2	10.4	2.0	—
8600	Frankfurter, beef	1	item(s)	45	23.4	149	5.1	1.8	0	13.3	5.3	6.4	0.5	—
202	Frankfurter, beef and pork	1	item(s)	45	25.2	137	5.2	0.8	0	12.4	4.8	6.2	1.2	—
1293	Frankfurter, chicken	1	item(s)	45	28.1	100	7.0	1.2	0.2	7.3	1.7	2.7	1.7	0.1
3261	Frankfurter, turkey	1	item(s)	45	28.3	100	5.5	1.7	0	7.8	1.8	2.6	1.8	0.4
37275	Italian sausage, pork, cooked	1	item(s)	68	32.0	234	13.0	2.9	0.1	18.6	6.5	8.1	2.2	—
37307	Kielbasa or kolbassa, pork and beef	1	slice(s)	30	18.5	67	5.0	1.0	0	4.7	1.7	2.2	0.5	—
1333	Knockwurst or knackwurst, beef and pork	2	ounce(s)	57	31.4	174	6.3	1.8	0	15.7	5.8	7.3	1.7	—
37285	Pepperoni, beef and pork	1	slice(s)	11	3.4	51	2.2	0.4	0.2	4.4	1.8	2.1	0.3	—
37313	Polish sausage, pork	1	slice(s)	21	11.4	60	2.8	0.7	0	5.0	1.8	2.3	0.5	—
206	Salami, beef, cooked, sliced	2	slice(s)	52	31.2	136	6.5	1.0	0	11.5	5.1	5.5	0.5	—
37272	Salami, pork, dry or hard	1	slice(s)	13	4.6	52	2.9	0.2	0	4.3	1.5	2.0	0.5	—
40987	Sausage, turkey, cooked	2	ounce(s)	57	36.9	111	13.5	0	0	5.9	1.3	1.7	1.5	0.2
8620	Smoked sausage, beef and pork	2	ounce(s)	57	30.6	181	6.8	1.4	0	16.3	5.5	6.9	2.2	0
8619	Smoked sausage, pork	2	ounce(s)	57	32.0	178	6.8	1.2	0	16.0	5.3	6.4	2.1	0.1
37273	Smoked sausage, pork link	1	piece(s)	76	29.8	295	16.8	1.6	0	24.0	8.6	11.1	2.8	—
1336	Summer sausage, thuringer, or cervelat, beef and pork	2	ounce(s)	57	25.6	205	9.9	1.9	0	17.3	6.5	7.4	0.7	—
37294	Vienna sausage, cocktail, beef and pork, canned	1	piece(s)	16	10.4	37	1.7	0.4	0	3.1	1.1	1.5	0.2	—
	Spreads													
1318	Ham salad spread	¼	cup(s)	60	37.6	130	5.2	6.4	0	9.3	3.0	4.3	1.6	—
32419	Pork and beef sandwich spread	4	tablespoon(s)	60	36.2	141	4.6	7.2	0.1	10.4	3.6	4.6	1.5	—
	Turkey													
13604	Breast, fat free, oven roasted	1	slice(s)	28	—	25	4.0	1.0	0	0	0	0	0	0
13606	Breast, hickory smoked fat free	1	slice(s)	28	—	25	4.0	1.0	0	0	0	0	0	0
16049	Breast, hickory smoked slices	1	slice(s)	56	—	50	11.0	1.0	0	0	0	0	0	0
16047	Breast, honey roasted slices	1	slice(s)	56	—	60	11.0	3.0	0	0	0	0	0	0
16048	Breast, oven roasted slices	1	slice(s)	56	—	50	11.0	1.0	0	0	0	0	0	0
7124	Breast, oven roasted	1	slice(s)	10	—	10	1.8	0.3	0	0.1	0	0	0	0
13567	Turkey ham, 10% water added	2	slice(s)	56	40.9	70	10.0	2.0	0	3.0	0	0.4	0.6	—
37270	Turkey pastrami	1	slice(s)	28	20.3	35	4.6	1.0	0	1.2	0.3	0.4	0.3	—
3262	Turkey salami	2	slice(s)	57	39.1	98	10.9	0.9	0.1	5.2	1.6	1.8	1.4	0
37318	Turkey salami, cooked	1	slice(s)	28	20.4	43	4.3	0.1	0	2.7	0.8	0.9	0.7	—
Beverages														
	Beer													
866	Ale, mild	12	fluid ounce(s)	360	332.3	148	1.1	13.3	0.4	0	0	0	0	0
686	Beer	12	fluid ounce(s)	356	327.7	153	1.6	12.7	0	0	0	0	0	0

PAGE KEY: A-6 = Breads/Baked Goods A-12 = Cereal/Rice/Pasta A-16 = Fruit A-20 = Vegetables/Legumes A-30 = Nuts/Seeds A-32 = Vegetarian A-34 = Dairy A-42 = Eggs A-42 = Seafood A-44 = Meats A-48 = Poultry
A-50 = Processed Meats A-50 = Beverages A-54 = Fats/Oils A-56 = Sweets A-58 = Spices/Condiments/Sauces A-62 = Mixed Foods/Soups/Sandwiches A-66 = Fast Food A-86 = Convenience A-88 = Baby Foods

Chol (mg)	Calc (mg)	Iron (mg)	Magn (mg)	Pota (mg)	Sodi (mg)	Zinc (mg)	Vit A (µg)	Thia (mg)	Vit E (mg α)	Ribo (mg)	Niac (mg)	Vit B_6 (mg)	Fola (µg)	Vit C (mg)	Vit B_{12} (µg)	Sele (µg)
27	6	1.15	6.2	57.3	540.4	2.31	0	0.00	—	0.06	0.99	0.06	4.5	0	0.7	9.8
20	0	0.36	3.9	47.0	310.0	0.56	0	0.01	—	0.03	0.67	0.04	3.6	0	0.4	—
20	40	0.36	5.6	45.6	300.0	0.45	0	—	—	—	—	—	—	0	—	—
30	20	0.36	5.9	43.1	300.0	0.39	0	—	—	—	—	—	—	0	—	—
20	40	0.36	6.2	42.6	270.0	0.51	0	—	—	—	—	—	—	0	—	—
4	0	0.00	—	—	100.0	—	0	—	—	—	—	—	—	0	—	—
4	0	0.12	—	—	100.0	—	0	—	—	—	—	—	—	0.6	—	—
4	0	0.12	—	—	103.3	—	0	—	—	—	—	—	—	0.6	—	—
26	8	0.64	5.1	75.0	573.2	0.96	0	0.05	0.10	0.07	1.23	0.06	1.4	0	0.7	10.4
22	62	0.30	10.8	168.7	842.9	0.78	34.1	0.16	0.14	0.14	1.04	0.13	1.1	0	0.7	9.3
33	62	0.75	19.3	210.7	740.7	0.95	44.3	0.22	0.22	0.06	1.41	0.23	21.0	4.4	0.3	4.5
18	3	0.44	3.5	66.5	264.9	0.71	0	0.02	0.05	0.03	0.98	0.04	0.9	0	0.6	4.7
12	2	0.15	2.7	53.3	261.0	0.36	0	0.11	0.03	0.04	0.68	0.07	0.6	0	0.2	4.4
13	3	0.32	4.3	80.2	367.7	0.70	0	0.10	—	0.06	0.88	0.05	1.4	0	0.8	4.0
44	33	0.95	11.1	156.9	412.2	1.70	0	0.37	0.01	0.13	2.36	0.15	1.5	0.7	0.7	15.7
24	1	1.42	1.7	27.5	131.5	0.42	641.0	0.03	0.05	0.23	1.27	0.05	6.7	0	3.1	8.8
29	25	0.46	5.6	88.6	465.3	0.96	20.2	0.10	0.10	0.06	1.24	0.05	1.3	0	0.7	6.8
50	5	0.90	10.2	225.7	700.2	1.93	0	0.35	0.12	0.17	2.90	0.30	1.1	0	1.1	12.0
24	6	0.67	6.3	70.2	513.0	1.10	0	0.01	0.09	0.06	1.06	0.04	2.3	0	0.8	3.7
23	5	0.51	4.5	75.2	504.0	0.82	8.1	0.09	0.11	0.05	1.18	0.05	1.8	0	0.6	6.2
43	33	0.52	9.0	90.9	379.8	0.50	0	0.02	0.09	0.11	2.10	0.14	3.2	0	0.2	10.4
35	67	0.66	6.3	176.4	485.1	0.82	0	0.01	0.27	0.08	1.65	0.06	4.1	0	0.4	6.8
39	14	0.97	12.2	206.7	820.8	1.62	6.8	0.42	0.17	0.15	2.83	0.22	3.4	0.1	0.9	15.0
20	13	0.44	4.9	84.4	283.0	0.61	0	0.06	0.06	0.06	0.87	0.05	1.5	0	0.5	5.4
34	6	0.37	6.2	112.8	527.3	0.94	0	0.19	0.32	0.07	1.55	0.09	1.1	0	0.7	7.7
13	2	0.15	2.0	34.7	196.7	0.30	0	0.05	0.00	0.02	0.59	0.04	0.7	0.1	0.2	2.4
15	2	0.29	2.9	37.3	199.3	0.40	0	0.10	0.04	0.03	0.71	0.03	0.4	0.2	0.2	3.7
37	3	1.14	6.8	97.8	592.8	0.92	0	0.04	0.08	0.08	1.68	0.08	1.0	0	1.6	7.6
10	2	0.16	2.8	48.4	289.3	0.53	0	0.11	0.02	0.04	0.71	0.07	0.3	0	0.4	3.3
52	12	0.84	11.9	169.0	377.1	2.19	7.4	0.04	0.10	0.14	3.24	0.18	3.4	0.4	0.7	0
33	7	0.42	7.4	101.5	516.5	0.71	7.4	0.10	0.07	0.06	1.66	0.09	1.1	0	0.3	0
35	6	0.33	6.2	273.9	468.9	0.74	0	0.12	0.14	0.10	1.59	0.10	0.6	0	0.4	10.4
52	23	0.87	14.4	254.6	1136.6	2.13	0	0.53	0.18	0.19	3.43	0.26	3.8	1.5	1.2	16.4
42	5	1.15	7.9	147.4	737.1	1.45	0	0.08	0.12	0.18	2.44	0.14	9.4	3.1	11.5
14	2	0.14	1.1	16.2	155.0	0.25	0	0.01	0.03	0.01	0.25	0.01	0.6	0	0.2	2.7
22	5	0.35	6.0	90.0	547.2	0.66	0	0.26	1.04	0.07	1.25	0.09	0.6	0	0.5	10.7
23	7	0.47	4.8	66.0	607.8	0.61	15.6	0.10	1.04	0.08	1.03	0.07	1.2	0	0.7	5.8
10	0	0.00	—	—	340.0	—	0	—	—	—	—	—	—	0	—	—
10	0	0.00	—	—	300.0	—	0	—	—	—	—	—	—	0	—	—
25	0	0.72	—	—	720.0	—	0	—	—	—	—	—	—	0	—	—
20	0	0.72	—	—	660.0	—	0	—	—	—	—	—	—	0	—	—
20	0	0.72	—	—	660.0	—	0	—	—	—	—	—	—	0	—	—
4	0	0.06	—	—	103.3	—	0	—	—	—	—	—	—	0	—	—
40	0	0.72	12.3	162.4	700.0	1.44	0	—	—	—	—	—	—	0	—	—
19	3	1.19	4.0	97.8	278.1	0.61	1.1	0.01	0.06	0.07	1.00	0.07	1.4	4.6	0.1	4.6
43	23	0.70	12.5	122.5	569.3	1.31	1.1	0.24	0.13	0.17	2.25	0.24	5.7	0	0.6	15.0
22	11	0.35	6.2	61.2	284.6	0.65	0.6	0.12	0.06	0.08	1.12	0.12	2.8	0	0.3	7.5
0	18	0.07	21.6	90.0	14.4	0.03	0	0.03	0.00	0.10	1.62	0.18	21.6	0	0.1	2.5
0	14	0.07	21.4	96.2	14.3	0.03	0	0.01	0.00	0.08	1.82	0.16	21.4	0	0.1	2.1

Appendix A

TABLE A-1 **Table of Food Composition** *(continued)* (Computer code is for Cengage Diet Analysis program) (For purposes of calculations, use "0" for t, <1, <.1, <.01, etc.)

DA+ Code	Food Description	Quantity	Measure	Wt (g)	H₂0 (g)	Ener (kcal)	Prot (g)	Carb (g)	Fiber (g)	Fat (g)	Fat Breakdown (g) Sat	Mono	Poly	*Trans*
Beverages—continued														
16886	Beer, nonalcoholic	12	fluid ounce(s)	360	328.1	133	0.8	29.0	0	0.4	0.1	0	0.2	0
31609	Bud Light beer	12	fluid ounce(s)	355	335.5	110	0.9	6.6	0	0	0	0	0	0
31608	Budweiser beer	12	fluid ounce(s)	355	327.7	145	1.3	10.6	0	0	0	0	0	0
869	Light beer	12	fluid ounce(s)	354	335.9	103	0.9	5.8	0	0	0	0	0	0
31613	Michelob beer	12	fluid ounce(s)	355	323.4	155	1.3	13.3	0	0	0	0	0	0
31614	Michelob Light beer	12	fluid ounce(s)	355	329.8	134	1.1	11.7	0	0	0	0	0	0
	Gin, rum, vodka, whiskey													
857	Distilled alcohol, 100 proof	1	fluid ounce(s)	28	16.0	82	0	0	0	0	0	0	0	0
687	Distilled alcohol, 80 proof	1	fluid ounce(s)	28	18.5	64	0	0	0	0	0	0	0	0
688	Distilled alcohol, 86 proof	1	fluid ounce(s)	28	17.8	70	0	0	0	0	0	0	0	0
689	Distilled alcohol, 90 proof	1	fluid ounce(s)	28	17.3	73	0	0	0	0	0	0	0	0
856	Distilled alcohol, 94 proof	1	fluid ounce(s)	28	16.8	76	0	0	0	0	0	0	0	0
	Liqueurs													
33187	Coffee liqueur, 53 proof	1	fluid ounce(s)	35	10.8	113	0	16.3	0	0.1	0	0	0	—
3142	Coffee liqueur, 63 proof	1	fluid ounce(s)	35	14.4	107	0	11.2	0	0.1	0	0	0	—
736	Cordials, 54 proof	1	fluid ounce(s)	30	8.9	106	0	13.3	0	0.1	0	0	0	—
	Wine													
861	California red wine	5	fluid ounce(s)	150	133.4	125	0.3	3.7	0	0	0	0	0	0
858	Domestic champagne	5	fluid ounce(s)	150	—	105	0.3	3.8	0	0	0	0	0	0
690	Sweet dessert wine	5	fluid ounce(s)	147	103.7	235	0.3	20.1	0	0	0	0	0	0
1481	White wine	5	fluid ounce(s)	148	128.1	121	0.1	3.8	0	0	0	0	0	0
1811	Wine cooler	10	fluid ounce(s)	300	267.4	159	0.3	20.2	0	0.1	0	0	0	—
	Carbonated													
31898	7 Up	12	fluid ounce(s)	360	321.0	140	0	39.0	0	0	0	0	0	0
692	Club soda	12	fluid ounce(s)	355	354.8	0	0	0	0	0	0	0	0	0
12010	Coca-Cola Classic cola soda	12	fluid ounce(s)	360	319.4	146	0	40.5	0	0	0	0	0	0
693	Cola	12	fluid ounce(s)	368	332.7	136	0.3	35.2	0	0.1	0	0	0	—
2391	Cola or pepper-type soda, low calorie with saccharin	12	fluid ounce(s)	355	354.5	0	0	0.3	0	0	0	0	0	0
9522	Cola soda, decaffeinated	12	fluid ounce(s)	372	333.4	153	0	39.3	0	0	0	0	0	0
9524	Cola, decaffeinated, low calorie with aspartame	12	fluid ounce(s)	355	354.3	4	0.4	0.5	0	0	0	0	0	0
1415	Cola, low calorie with aspartame	12	fluid ounce(s)	355	353.6	7	0.4	1.0	0	0.1	0	0	0	—
1412	Cream soda	12	fluid ounce(s)	371	321.5	189	0	49.3	0	0	0	0	0	0
31899	Diet 7 Up	12	fluid ounce(s)	360	—	0	0	0	0	0	0	0	0	0
12031	Diet Coke cola soda	12	fluid ounce(s)	360	—	2	0	0.2	0	0	0	0	0	0
29392	Diet Mountain Dew soda	12	fluid ounce(s)	360	—	0	0	0	0	0	0	0	0	0
29389	Diet Pepsi cola soda	12	fluid ounce(s)	360	—	0	0	0	0	0	0	0	0	0
12034	Diet Sprite soda	12	fluid ounce(s)	360	—	4	0	0	0	0	0	0	0	0
695	Ginger ale	12	fluid ounce(s)	366	333.9	124	0	32.1	0	0	0	0	0	0
694	Grape soda	12	fluid ounce(s)	372	330.3	160	0	41.7	0	0	0	0	0	0
1876	Lemon lime soda	12	fluid ounce(s)	368	330.8	147	0.2	37.4	0	0.1	0	0	0	—
29391	Mountain Dew soda	12	fluid ounce(s)	360	314.0	170	0	46.0	0	0	0	0	0	0
3145	Orange soda	12	fluid ounce(s)	372	325.9	179	0	45.8	0	0	0	0	0	0
1414	Pepper-type soda	12	fluid ounce(s)	368	329.3	151	0	38.3	0	0.4	0.3	0	0	—
29388	Pepsi regular cola soda	12	fluid ounce(s)	360	318.9	150	0	41.0	0	0	0	0	0	0
696	Root beer	12	fluid ounce(s)	370	330.0	152	0	39.2	0	0	0	0	0	0
12044	Sprite soda	12	fluid ounce(s)	360	321.0	144	0	39.0	0	0	0	0	0	0
	Coffee													
731	Brewed	8	fluid ounce(s)	237	235.6	2	0.3	0	0	0	0	0	0	0
9520	Brewed, decaffeinated	8	fluid ounce(s)	237	234.3	5	0.3	1.0	0	0	0	0	0	0
16882	Cappuccino	8	fluid ounce(s)	240	224.8	79	4.1	5.8	0.2	4.9	2.3	1.0	0.2	—
16883	Cappuccino, decaffeinated	8	fluid ounce(s)	240	224.8	79	4.1	5.8	0.2	4.9	2.3	1.0	0.2	—
16880	Espresso	8	fluid ounce(s)	237	231.8	21	0	3.6	0	0.4	0.2	0	0.2	0
16881	Espresso, decaffeinated	8	fluid ounce(s)	237	231.8	21	0	3.6	0	0.4	0.2	0	0.2	0
732	Instant, prepared	8	fluid ounce(s)	239	236.5	5	0.2	0.8	0	0	0	0	0	0
	Fruit drinks													
29357	Crystal Light sugar-free lemonade drink	8	fluid ounce(s)	240	—	5	0	0	0	0	0	0	0	0
6012	Fruit punch drink with added vitamin C, canned	8	fluid ounce(s)	248	218.2	117	0	29.7	0.5	0	0	0	0	0
31143	Gatorade Thirst Quencher, all flavors	8	fluid ounce(s)	240	—	50	0	14.0	0	0	0	0	0	0
260	Grape drink, canned	8	fluid ounce(s)	250	210.5	153	0	39.4	0	0	0	0	0	0
17372	Kool-Aid (lemonade/punch/fruit drink)	8	fluid ounce(s)	248	220.0	108	0.1	27.8	0.2	0	0	0	0	—

PAGE KEY: A-6 = Breads/Baked Goods A-12 = Cereal/Rice/Pasta A-16 = Fruit A-20 = Vegetables/Legumes A-30 = Nuts/Seeds A-32 = Vegetarian A-34 = Dairy A-42 = Eggs A-42 = Seafood A-44 = Meats A-48 = Poultry
A-50 = Processed Meats A-50 = Beverages A-54 = Fats/Oils A-56 = Sweets A-58 = Spices/Condiments/Sauces A-62 = Mixed Foods/Soups/Sandwiches A-66 = Fast Food A-86 = Convenience A-88 = Baby Foods

Chol (mg)	Calc (mg)	Iron (mg)	Magn (mg)	Pota (mg)	Sodi (mg)	Zinc (mg)	Vit A (μg)	Thia (mg)	Vit E (mg α)	Ribo (mg)	Niac (mg)	Vit B$_6$ (mg)	Fola (μg)	Vit C (mg)	Vit B$_{12}$ (μg)	Sele (μg)
0	25	0.21	25.2	28.8	46.8	0.07	—	0.07	0.00	0.18	3.99	0.10	50.4	1.8	0.1	4.3
0	18	0.14	17.8	63.9	9.0	0.10	0	0.03	—	0.10	1.39	0.12	14.6	0	0	4.0
0	18	0.10	21.3	88.8	9.0	0.07	0	0.02	—	0.09	1.60	0.17	21.3	0	0.1	4.0
0	14	0.10	17.7	74.3	14.2	0.03	0	0.01	0.00	0.05	1.38	0.12	21.2	0	0.1	1.4
0	18	0.10	21.3	88.8	9.0	0.07	0	0.02	—	0.09	1.60	0.17	21.3	0	0.1	4.0
0	18	0.14	17.8	63.9	9.0	0.10	0	0.03	—	0.10	1.39	0.12	14.6	0	0	4.0
0	0	0.01	0	0.6	0.3	0.01	0	0.00	—	0.00	0.00	0.00	0	0	0	0
0	0	0.01	0	0.6	0.3	0.01	0	0.00	0.00	0.00	0.00	0.00	0	0	0	0
0	0	0.01	0	0.6	0.3	0.01	0	0.00	0.00	0.00	0.00	0.00	0	0	0	0
0	0	0.01	0	0.6	0.3	0.01	0	0.00	0.00	0.00	0.00	0.00	0	0	0	0
0	0	0.01	0	0.6	0.3	0.01	0	0.00	—	0.00	0.00	0.00	0	0	0	0
0	0	0.02	1.0	10.4	2.8	0.01	0	0.00	0.00	0.00	0.05	0.00	0	0	0	0.1
0	0	0.02	1.0	10.4	2.8	0.01	0	0.00	—	0.00	0.05	0.00	0	0	0	0.1
0	0	0.02	0.6	4.5	2.1	0.01	0	0.00	0.00	0.00	0.02	0.00	0	0	0	0.1
0	12	1.43	16.2	170.6	15.0	0.14	0	0.01	0.00	0.04	0.11	0.05	1.5	0	0	—
0	—	—	—	—	—	—	—	—	—	—	—	—	—	—	0	—
0	12	0.34	13.2	135.4	13.2	0.10	0	0.01	0.00	0.01	0.30	0.00	0	0	0	0.7
0	13	0.39	14.8	104.7	7.4	0.18	0	0.01	0.00	0.01	0.15	0.06	1.5	0	0	0.1
0	18	0.75	15.0	129.0	24.0	0.18	—	0.01	0.03	0.03	0.13	0.03	3.0	5.4	0	0.6
0	—	—	—	0.6	75.0	—	—	—	—	—	—	—	—	—	—	—
0	18	0.03	3.5	7.1	74.6	0.35	0	0.00	0.00	0.00	0.00	0.00	0	0	0	0
0	—	—	—	0	49.5	—	0	—	—	—	—	—	—	0	—	—
0	7	0.41	0	7.4	14.7	0.06	0	0.00	0.00	0.00	0.00	0.00	0	0	0	0.4
0	14	0.06	3.5	14.2	56.8	0.11	0	0.00	0.00	0.00	0.00	0.00	0	0	0	0.3
0	7	0.08	0	11.2	14.9	0.03	0	0.00	0.00	0.00	0.00	0.00	0	0	0	0.4
0	11	0.06	0	24.9	14.2	0.03	0	0.02	0.00	0.08	0.00	0.00	0	0	0	0.3
0	11	0.39	3.5	28.4	28.4	0.03	0	0.02	0.00	0.08	0.00	0.00	0	0	0	0
0	19	0.18	3.7	3.7	44.5	0.26	0	0.00	0.00	0.00	0.00	0.00	0	0	0	0
0	—	—	—	77.0	45.0	—	—	—	—	—	—	—	—	—	—	—
0	—	—	—	18.0	42.0	—	0	—	—	—	—	—	—	0	—	—
0	—	—	—	70.0	35.0	—	—	—	—	—	—	—	—	—	—	—
0	—	—	—	30.0	35.0	—	—	—	—	—	—	—	—	—	—	—
0	—	—	—	109.5	36.0	—	0	—	—	—	—	—	—	0	—	—
0	11	0.65	3.7	3.7	25.6	0.18	0	0.00	0.00	0.00	0.00	0.00	0	0	0	0.4
0	11	0.29	3.7	3.7	55.8	0.26	0	0.00	—	0.00	0.00	0.00	0	0	0	0
0	7	0.41	3.7	3.7	33.2	0.14	0	0.00	0.00	0.00	0.05	0.00	0	0	0	0
0	—	—	—	0	70.0	—	—	—	—	—	—	—	—	—	—	—
0	19	0.21	3.7	7.4	44.6	0.36	0	0.00	—	0.00	0.00	0.00	0	0	0	0
0	11	0.14	0	3.7	36.8	0.14	0	0.00	—	0.00	0.00	0.00	0	0	0	0.4
0	—	—	—	0	35.0	—	—	—	—	—	—	—	—	—	—	—
0	18	0.18	3.7	3.7	48.0	0.26	0	0.00	0.00	0.00	0.00	0.00	0	0	0	0.4
0	—	—	—	0	70.5	—	0	—	—	—	—	—	—	0	—	—
0	5	0.02	7.1	116.1	4.7	0.04	0	0.03	0.02	0.18	0.45	0.00	4.7	0	0	0
0	7	0.14	11.8	108.9	4.7	0.00	0	0.00	0.00	0.03	0.66	0.00	0	0	0	0.5
12	144	0.19	14.4	232.8	50.4	0.50	33.6	0.04	0.09	0.27	0.13	0.04	7.2	0	0.4	4.6
12	144	0.19	14.4	232.8	50.4	0.50	33.6	0.04	0.09	0.27	0.13	0.04	7.2	0	0.4	4.6
0	5	0.30	189.6	272.6	33.2	0.11	0	0.00	0.04	0.42	12.34	0.00	2.4	0.5	0	0
0	5	0.30	189.6	272.6	33.2	0.11	0	0.00	0.04	0.42	12.34	0.00	2.4	0.5	0	0
0	10	0.09	9.5	71.6	9.5	0.01	0	0.00	0.00	0.00	0.56	0.00	0	0	0	0.2
0	0	0.00	—	160.0	40.0	—	0	—	—	—	—	—	—	0	—	—
0	20	0.22	7.4	62.0	94.2	0.02	5.0	0.05	0.04	0.05	0.05	0.02	9.9	89.3	0	0.5
0	0	0.00	—	30.0	110.0	—	0	—	—	—	—	—	—	0	—	—
0	130	0.17	2.5	30.0	40.0	0.30	0	0.00	0.00	0.01	0.02	0.01	0	78.5	0	0.3
0	14	0.45	5.0	49.6	31.0	0.19	—	0.03	—	0.05	0.04	0.01	4.3	41.6	0	1.0

TABLE A-1 Table of Food Composition (continued) (Computer code is for Cengage Diet Analysis program) (For purposes of calculations, use "0" for t, <1, <.1, <.01, etc.)

DA+ Code	Food Description	Quantity	Measure	Wt (g)	H₂0 (g)	Ener (kcal)	Prot (g)	Carb (g)	Fiber (g)	Fat (g)	Sat	Mono	Poly	Trans
Beverages—continued														
17225	Kool-Aid sugar free, low calorie tropical punch drink mix, prepared	8	fluid ounce(s)	240	—	5	0	0	0	0	0	0	0	0
266	Lemonade, prepared from frozen concentrate	8	fluid ounce(s)	248	221.6	99	0.2	25.8	0	0.1	0	0	0	—
268	Limeade, prepared from frozen concentrate	8	fluid ounce(s)	247	212.6	128	0	34.1	0	0	0	0	0	—
14266	Odwalla strawberry C monster smoothie blend	8	fluid ounce(s)	240	—	160	2.0	38.0	0	0	0	0	0	0
10080	Odwalla strawberry lemonade quencher	8	fluid ounce(s)	240	—	110	0	28.0	0	0	0	0	0	0
10099	Snapple fruit punch fruit drink	8	fluid ounce(s)	240	—	110	0	29.0	0	0	0	0	0	0
10096	Snapple kiwi strawberry fruit drink	8	fluid ounce(s)	240	211.2	110	0	28.0	0	0	0	0	0	0
Slim Fast ready-to-drink shake														
16054	French vanilla ready to drink shake	11	fluid ounce(s)	325	—	220	10.0	40.0	5.0	2.5	0.5	1.5	0.5	—
40447	Optima rich chocolate royal ready-to-drink shake	11	fluid ounce(s)	330	—	180	10.0	24.0	5.0	5.0	1.0	3.5	0.5	0
16055	Strawberries n cream ready to drink shake	11	fluid ounce(s)	325	—	220	10.0	40.0	5.0	2.5	0.5	1.5	0.5	—
Tea														
33179	Decaffeinated, prepared	8	fluid ounce(s)	237	236.3	2	0	0.7	0	0	0	0	0	0
1877	Herbal, prepared	8	fluid ounce(s)	237	236.1	2	0	0.5	0	0	0	0	0	0
735	Instant tea mix, lemon flavored with sugar, prepared	8	fluid ounce(s)	259	236.2	91	0	22.3	0.3	0.2	0	0	0	—
734	Instant tea mix, unsweetened, prepared	8	fluid ounce(s)	237	236.1	2	0.1	0.4	0	0	0	0	0	0
733	Tea, prepared	8	fluid ounce(s)	237	236.3	2	0	0.7	0	0	0	0	0	0
Water														
1413	Mineral water, carbonated	8	fluid ounce(s)	237	236.8	0	0	0	0	0	0	0	0	0
33183	Poland spring water, bottled	8	fluid ounce(s)	237	237.0	0	0	0	0	0	0	0	0	0
1821	Tap water	8	fluid ounce(s)	237	236.8	0	0	0	0	0	0	0	0	0
1879	Tonic water	8	fluid ounce(s)	244	222.3	83	0	21.5	0	0	0	0	0	0
Fats and Oils														
Butter														
104	Butter	1	tablespoon(s)	14	2.3	102	0.1	0	0	11.5	7.3	3.0	0.4	—
2522	Butter Buds, dry butter substitute	1	teaspoon(s)	2	—	5	0	2.0	0	0	0	0	0	0
921	Unsalted	1	tablespoon(s)	14	2.5	102	0.1	0	0	11.5	7.3	3.0	0.4	—
107	Whipped	1	tablespoon(s)	9	1.5	67	0.1	0	0	7.6	4.7	2.2	0.3	—
944	Whipped, unsalted	1	tablespoon(s)	11	2.0	82	0.1	0	0	9.2	5.9	2.4	0.3	—
Fats, cooking														
2671	Beef tallow, semisolid	1	tablespoon(s)	13	0	115	0	0	0	12.8	6.4	5.4	0.5	—
922	Chicken fat	1	tablespoon(s)	13	0	115	0	0	0	12.8	3.8	5.7	2.7	—
5454	Household shortening with vegetable oil	1	tablespoon(s)	13	0	115	0	0	0	13.0	3.4	5.5	2.7	2.2
111	Lard	1	tablespoon(s)	13	0	115	0	0	0	12.8	5.0	5.8	1.4	—
Margarine														
114	Margarine	1	tablespoon(s)	14	2.3	101	0	0.1	0	11.4	2.1	5.5	3.4	2.1
5439	Soft	1	tablespoon(s)	14	2.3	103	0.1	0.1	0	11.6	1.7	4.4	2.1	3.0
32329	Soft, unsalted, with hydrogenated soybean and cottonseed oils	1	tablespoon(s)	14	2.5	101	0.1	0.1	0	11.3	2.0	5.4	3.5	—
928	Unsalted	1	tablespoon(s)	14	2.6	101	0.1	0.1	0	11.3	2.1	5.2	3.5	—
119	Whipped	1	tablespoon(s)	9	1.5	64	0.1	0.1	0	7.2	1.2	3.2	2.5	—
Spreads														
54657	I Can't Believe It's Not Butter! tub, soya oil (non-hydrogenated)	1	tablespoon(s)	14	2.3	103	0.1	0.1	0	11.6	2.8	2.0	5.1	0.1
2708	Mayonnaise with soybean and safflower oils	1	tablespoon(s)	14	2.1	99	0.2	0.4	0	11.0	1.2	1.8	7.6	—
16157	Promise vegetable oil spread, stick	1	tablespoon(s)	14	4.2	90	0	0	0	10.0	2.5	2.0	4.0	—
Oils														
2681	Canola	1	tablespoon(s)	14	0	120	0	0	0	13.6	1.0	8.6	3.8	0.1
120	Corn	1	tablespoon(s)	14	0	120	0	0	0	13.6	1.8	3.8	7.4	0
122	Olive	1	tablespoon(s)	14	0	119	0	0	0	13.5	1.9	9.9	1.4	—

Chol (mg)	Calc (mg)	Iron (mg)	Magn (mg)	Pota (mg)	Sodi (mg)	Zinc (mg)	Vit A (µg)	Thia (mg)	Vit E (mg α)	Ribo (mg)	Niac (mg)	Vit B$_6$ (mg)	Fola (µg)	Vit C (mg)	Vit B$_{12}$ (µg)	Sele (µg)
0	0	0.00	—	10.1	10.1	—	0	—	—	—	—	—	—	6.0	—	—
0	10	0.39	5.0	37.2	9.9	0.05	0	0.01	0.02	0.05	0.04	0.01	2.5	9.7	0	0.2
0	5	0.00	4.9	24.7	7.4	0.02	0	0.01	0.00	0.01	0.02	0.01	2.5	7.7	0	0.2
0	20	0.72	—	0	20.0	—	0	—	—	—	—	—	—	600.0	0	—
0	0	0.00	—	70.0	10.0	—	0	—	—	—	—	—	—	54.0	0	—
0	0	0.00	—	20.0	10.0	—	0	—	—	—	—	—	—	0	0	—
0	0	0.00	—	40.0	10.0	—	0	—	—	—	—	—	—	0	0	—
5	400	2.70	140.0	600.0	220.0	2.25	—	0.52	—	0.59	7.00	0.70	120.0	60.0	2.1	17.5
5	1000	2.70	140.0	600.0	220.0	2.25	—	0.52	—	0.59	7.00	0.70	120.0	30.0	2.1	17.5
5	400	2.70	140.0	600.0	220.0	2.25	—	0.52	—	0.59	7.00	0.70	120.0	60.0	2.1	17.5
0	0	0.04	7.1	87.7	7.1	0.04	0	0.00	0.00	0.03	0.00	0.00	11.9	0	0	0
0	5	0.18	2.4	21.3	2.4	0.09	0	0.02	0.00	0.01	0.00	0.00	2.4	0	0	0
0	5	0.05	2.6	38.9	5.2	0.02	0	0.00	0.00	0.00	0.02	0.00	0	0	0	0.3
0	7	0.02	4.7	42.7	9.5	0.02	0	0.00	0.00	0.01	0.07	0.00	0	0	0	0
0	0	0.04	7.1	87.7	7.1	0.04	0	0.00	0.00	0.03	0.00	0.00	11.9	0	0	0
0	33	0.00	0	0	2.4	0.00	0	0.00	—	0.00	0.00	0.00	0	0	0	0
0	2	0.02	2.4	0	2.4	0.00	0	0.00	—	0.00	0.00	0.00	0	0	0	0
0	7	0.00	2.4	2.4	7.1	0.00	0	0.00	0.00	0.00	0.00	0.00	0	0	0	0
0	2	0.02	0	0	29.3	0.24	0	0.00	0.00	0.00	0.00	0.00	0	0	0	0
31	3	0.00	0.3	3.4	81.8	0.01	97.1	0.00	0.32	0.01	0.01	0.00	0.4	0	0	0.1
0	0	0.00	0	1.6	120.0	0.00	0	0.00	0.00	0.00	0.00	0.00	0	0	0	—
31	3	0.00	0.3	3.4	1.6	0.01	97.1	0.00	0.32	0.01	0.01	0.00	0.4	0	0	0.1
21	2	0.01	0.2	2.4	77.7	0.01	64.3	0.00	0.21	0.00	0.00	0.00	0.3	0	0	0.1
25	3	0.00	0.2	2.7	1.3	0.01	78.0	0.00	0.26	0.00	0.00	0.00	0.3	0	0	0.1
14	0	0.00	0	0	0	0.00	0	0.00	0.34	0.00	0.00	0.00	0	0	0	0
11	0	0.00	0	0	0	0.00	0	0.00	0.34	0.00	0.00	0.00	0	0	0	—
0	0	0.00	0	0	0	0.00	0	0.00	—	0.00	0.00	0.00	0	0	0	—
12	0	0.00	0	0	0	0.01	0	0.00	0.07	0.00	0.00	0.00	0	0	0	0
0	4	0.01	0.4	5.9	133.0	0.00	115.5	0.00	1.26	0.01	0.00	0.00	0.1	0	0	0
0	4	0.00	0.3	5.5	155.4	0.00	142.7	0.00	1.00	0.00	0.00	0.00	0.1	0	0	0
0	4	0.00	0.3	5.4	3.9	0.00	103.1	0.00	0.98	0.00	0.00	0.00	0.1	0	0	0
0	2	0.00	0.3	3.5	0.3	0.00	115.5	0.00	1.80	0.00	0.00	0.00	0.1	0	0	0
0	2	0.00	0.2	3.4	97.1	0.00	73.7	0.00	0.45	0.00	0.00	0.00	0.1	0	0	0
0	4	0.00	0.3	5.5	155.3	0.00	142.6	0.00	0.72	0.00	0.00	0.00	0.1	0	0	0
8	2	0.06	0.1	4.7	78.4	0.01	11.6	0.00	3.03	0.00	0.00	0.08	1.1	0	0	0.2
0	10	0.18	—	8.7	90.0	—	—	0.00	—	0.00	0.00	—	—	0.6	—	
0	0	0.00	0	0	0	0.00	0	0.00	2.37	0.00	0.00	0.00	0	0	0	0
0	0	0.00	0	0	0	0.00	0	0.00	1.94	0.00	0.00	0.00	0	0	0	0
0	0	0.07	0	0.1	0.3	0.00	0	0.00	1.93	0.00	0.00	0.00	0	0	0	0

TABLE A-1 Table of Food Composition *(continued)* (Computer code is for Cengage Diet Analysis program) (For purposes of calculations, use "0" for t, <1, <.1, <.01, etc.)

											Fat Breakdown (g)			
DA+ Code	Food Description	Quantity	Measure	Wt (g)	H_2O (g)	Ener (kcal)	Prot (g)	Carb (g)	Fiber (g)	Fat (g)	Sat	Mono	Poly	*Trans*
Fats and Oils—continued														
124	Peanut	1	tablespoon(s)	14	0	119	0	0	0	13.5	2.3	6.2	4.3	—
2693	Safflower	1	tablespoon(s)	14	0	120	0	0	0	13.6	0.8	10.2	2.0	—
923	Sesame	1	tablespoon(s)	14	0	120	0	0	0	13.6	1.9	5.4	5.7	—
128	Soybean, hydrogenated	1	tablespoon(s)	14	0	120	0	0	0	13.6	2.0	5.8	5.1	—
130	Soybean, with soybean and cottonseed oil	1	tablespoon(s)	14	0	120	0	0	0	13.6	2.4	4.0	6.5	—
2700	Sunflower	1	tablespoon(s)	14	0	120	0	0	0	13.6	1.8	6.3	5.0	—
357	**Pam original no stick cooking spray**	1	serving(s)	0	0.2	0	0	0	0	0	0	0	0	—
	Salad dressing													
132	Blue cheese	2	tablespoon(s)	30	9.7	151	1.4	2.2	0	15.7	3.0	3.7	8.3	—
133	Blue cheese, low calorie	2	tablespoon(s)	32	25.4	32	1.6	0.9	0	2.3	0.8	0.6	0.8	—
1764	Caesar	2	tablespoon(s)	30	10.3	158	0.4	0.9	0	17.3	2.6	4.1	9.9	—
29654	Creamy, reduced calorie, fat-free, cholesterol-free, sour cream and/or buttermilk and oil	2	tablespoon(s)	32	23.9	34	0.4	6.4	0	0.9	0.2	0.2	0.5	—
29617	Creamy, reduced calorie, sour cream and/or buttermilk and oil	2	tablespoon(s)	30	22.2	48	0.5	2.1	0	4.2	0.6	1.0	2.4	—
134	French	2	tablespoon(s)	32	11.7	146	0.2	5.0	0	14.3	1.8	2.7	6.7	—
135	French, low fat	2	tablespoon(s)	32	17.4	74	0.2	9.4	0.4	4.3	0.4	1.9	1.6	—
136	Italian	2	tablespoon(s)	29	16.6	86	0.1	3.1	0	8.3	1.3	1.9	3.8	—
137	Italian, diet	2	tablespoon(s)	30	25.4	23	0.1	1.4	0	1.9	0.1	0.7	0.5	—
139	Mayonnaise-type	2	tablespoon(s)	29	11.7	115	0.3	7.0	0	9.8	1.4	2.6	5.3	—
942	Oil and vinegar	2	tablespoon(s)	32	15.2	144	0	0.8	0	16.0	2.9	4.7	7.7	—
1765	Ranch	2	tablespoon(s)	30	11.6	146	0.1	1.6	0	15.8	2.3	5.2	7.6	—
3666	Ranch, reduced calorie	2	tablespoon(s)	30	20.5	62	0.1	2.2	0	6.1	1.1	1.8	2.9	—
940	Russian	2	tablespoon(s)	30	11.6	107	0.5	9.3	0.7	7.8	1.2	1.8	4.4	—
939	Russian, low calorie	2	tablespoon(s)	32	20.8	45	0.2	8.8	0.1	1.3	0.2	0.3	0.7	—
941	Sesame seed	2	tablespoon(s)	30	11.8	133	0.9	2.6	0.3	13.6	1.9	3.6	7.5	—
142	Thousand Island	2	tablespoon(s)	32	14.9	118	0.3	4.7	0.3	11.2	1.6	2.5	5.8	—
143	Thousand Island, low calorie	2	tablespoon(s)	30	18.2	61	0.3	6.7	0.4	3.9	0.2	1.9	0.8	—
	Sandwich spreads													
138	Mayonnaise with soybean oil	1	tablespoon(s)	14	2.1	99	0.1	0.4	0	11.0	1.6	2.7	5.8	0
140	Mayonnaise, low calorie	1	tablespoon(s)	16	10.0	37	0	2.6	0	3.1	0.5	0.7	1.7	—
141	Tartar sauce	2	tablespoon(s)	28	8.7	144	0.3	4.1	0.1	14.4	2.2	3.8	7.7	—
Sweets														
4799	**Butterscotch or caramel topping**	2	tablespoon(s)	41	13.1	103	0.6	27.0	0.4	0	0	0	0	—
	Candy													
1786	Almond Joy candy bar	1	item(s)	45	4.3	220	2.0	27.0	2.0	12.0	8.0	3.3	0.7	0
1785	Bit-O-Honey candy	6	item(s)	40	—	190	1.0	39.0	0	3.5	2.5	—	—	—
33375	Butterscotch candy	2	piece(s)	12	0.6	47	0	10.8	0	0.4	0.2	0.1	0	—
1701	Chewing gum, stick	1	item(s)	3	0.1	7	0	2.0	0.1	0	0	0	0	—
33378	Chocolate fudge with nuts, prepared	2	piece(s)	38	2.9	175	1.7	25.8	1.0	7.2	2.5	1.5	2.9	0.1
1787	Jelly beans	15	item(s)	43	2.7	159	0	39.8	0.1	0	0	0	0	—
1784	Kit Kat wafer bar	1	item(s)	42	0.8	210	3.0	27.0	0.5	11.0	7.0	3.5	0.3	0
4674	Krackel candy bar	1	item(s)	41	0.6	210	2.0	28.0	0.5	10.0	6.0	3.9	0.4	0
4934	Licorice	4	piece(s)	44	7.3	154	1.1	35.1	0	1.0	0	0.1	0	—
1780	Life Savers candy	1	item(s)	2	—	8	0	2.0	0	0	0	0	0	0
1790	Lollipop	1	item(s)	28	—	108	0	28.0	0	0	0	0	0	0
4679	M & Ms peanut chocolate candy, small bag	1	item(s)	49	0.9	250	5.0	30.0	2.0	13.0	5.0	5.4	2.1	—
1781	M & Ms plain chocolate candy, small bag	1	item(s)	48	0.8	240	2.0	34.0	1.0	10.0	6.0	3.3	0.3	—
4673	Milk chocolate bar, Symphony	1	item(s)	91	0.9	483	7.7	52.8	1.5	27.8	16.7	7.2	0.6	—
1783	Milky Way bar	1	item(s)	58	3.7	270	2.0	41.0	1.0	10.0	5.0	3.5	0.3	—
1788	Peanut brittle	1½	ounce(s)	43	0.3	207	3.2	30.3	1.1	8.1	1.8	3.4	1.9	—
1789	Reese's peanut butter cups	2	piece(s)	51	0.8	280	6.0	19.0	2.0	15.5	6.0	7.2	2.7	0
4689	Reese's pieces candy, small bag	1	item(s)	43	1.1	220	5.0	26.0	1.0	11.0	7.0	0.9	0.4	0
33399	Semisweet chocolate candy, made with butter	½	ounce(s)	14	0.1	68	0.6	9.0	0.8	4.2	2.5	1.4	0.1	—
1782	Snickers bar	1	item(s)	59	3.2	280	4.0	35.0	1.0	14.0	5.0	6.1	2.9	—
4694	Special Dark chocolate bar	1	item(s)	41	0.4	220	2.0	25.0	3.0	12.0	8.0	4.6	0.4	0
4695	Starburst fruit chews, original fruits	1	package(s)	59	3.9	240	0	48.0	0	5.0	1.0	2.1	1.8	—

Chol (mg)	Calc (mg)	Iron (mg)	Magn (mg)	Pota (mg)	Sodi (mg)	Zinc (mg)	Vit A (µg)	Thia (mg)	Vit E (mg α)	Ribo (mg)	Niac (mg)	Vit B_6 (mg)	Fola (µg)	Vit C (mg)	Vit B_{12} (µg)	Sele (µg)
0	0	0.00	0	0	0	0.00	0	0.00	2.11	0.00	0.00	0.00	0	0	0	0
0	0	0.00	0	0	0	0.00	0	0.00	4.63	0.00	0.00	0.00	0	0	0	0
0	0	0.00	0	0	0	0.00	0	0.00	0.19	0.00	0.00	0.00	0	0	0	0
0	0	0.00	0	0	0	0.00	0	0.00	1.10	0.00	0.00	0.00	0	0	0	0
0	0	0.00	0	0	0	0.00	0	0.00	1.64	0.00	0.00	0.00	0	0	0	0
0	0	0.00	0	0	0	0.00	0	0.00	5.58	0.00	0.00	0.00	0	0	0	0
0	0	0.00	0	0.3	1.5	0.01	0.1	0.00	0.00	0.00	0.00	0.00	0	0	0	0
5	24	0.06	0	11.1	328.2	0.08	20.1	0.00	1.80	0.03	0.03	0.01	7.8	0.6	0.1	0.3
0	28	0.16	2.2	1.6	384.0	0.08	—	0.01	0.08	0.03	0.01	0.01	1.0	0.1	0.1	0.5
1	7	0.05	0.6	8.7	323.4	0.03	0.6	0.00	1.56	0.00	0.01	0.00	0.9	0	0	0.5
0	12	0.08	1.6	42.6	320.0	0.05	0.3	0.00	0.21	0.01	0.01	0.01	1.9	0	0	0.5
0	2	0.03	0.6	10.8	306.9	0.01	—	0.00	0.71	0.00	0.01	0.01	0	0.1	0	0.5
0	8	0.25	1.6	21.4	267.5	0.09	7.4	0.01	1.60	0.01	0.06	0.00	0	0	0	0
0	4	0.27	2.6	34.2	257.3	0.06	8.6	0.01	0.09	0.01	0.14	0.01	0.6	0	0	0.5
0	2	0.18	0.9	14.1	486.3	0.03	0.6	0.00	1.47	0.01	0.00	0.01	0	0	0	0.6
2	3	0.19	1.2	25.5	409.8	0.05	0.3	0.00	0.06	0.00	0.00	0.02	0	0	0	2.4
8	4	0.05	0.6	2.6	209.0	0.05	6.2	0.00	0.60	0.01	0.00	0.01	1.8	0	0.1	0.5
0	0	0.00	0	2.6	0.3	0.00	0	0.00	1.46	0.00	0.00	0.00	0	0	0	0.5
1	4	0.03	1.2	8.4	354.0	0.01	5.4	0.00	1.84	0.01	0.00	0.00	0.3	0.1	0	0.1
0	5	0.01	1.5	8.4	413.7	0.01	0.9	0.00	0.72	0.01	0.00	0.00	0.3	0.1	0	0.1
0	6	0.20	3.0	51.9	282.3	0.06	13.2	0.01	0.98	0.01	0.16	0.02	1.5	1.4	0	0.5
2	6	0.18	0	50.2	277.8	0.02	0.6	0.00	0.12	0.00	0.00	0.00	1.0	1.9	0	0.5
0	6	0.18	0	47.1	300.0	0.02	0.6	0.00	1.50	0.00	0.00	0.00	0	0	0	0.5
8	5	0.37	2.6	34.2	276.2	0.08	4.5	0.46	1.28	0.01	0.13	0.00	0	0	0	0.5
0	5	0.27	2.1	60.6	249.3	0.05	4.8	0.01	0.30	0.01	0.13	0.00	0	0	0	0
5	1	0.03	0.1	1.7	78.4	0.02	11.2	0.01	0.72	0.01	0.00	0.08	0.7	0	0	0.2
4	0	0.00	0	1.6	79.5	0.01	0	0.00	0.32	0.00	0.00	0.00	0	0	0	0.3
8	6	0.20	0.8	10.1	191.5	0.05	20.2	0.00	0.97	0.00	0.01	0.07	2.0	0.1	0.1	0.5
0	22	0.08	2.9	34.4	143.1	0.07	11.1	0.01	—	0.03	0.01	0.01	0.8	0.1	0	0
0	18	0.33	30.3	126.5	65.0	0.36	0	0.01	—	0.06	0.21	—	—	0	—	—
0	20	0.00	—	—	150.0	—	0	—	—	—	—	—	—	0	—	—
1	0	0.00	0	0.4	46.9	0.01	3.4	0.00	0.01	0.00	0.00	0.00	0	0	0	0.1
0	0	0.00	0	0.1	0	0.00	0	0.00	0.00	0.00	0.00	0.00	0	0	0	0
5	22	0.74	20.9	69.5	14.8	0.54	14.4	0.02	0.09	0.03	0.12	0.03	6.1	0.1	0	1.1
0	1	0.05	0.9	15.7	21.3	0.02	0	0.00	0.00	0.01	0.00	0.00	0	0	0	0.5
3	60	0.36	16.4	126.0	30.0	0.51	0	0.07	—	0.22	1.07	0.05	59.6	0	0.1	2.0
3	40	0.36	—	168.8	50.0	—	0	—	—	—	—	—	—	0	—	—
0	0	0.22	2.6	28.2	126.3	0.07	0	0.01	0.07	0.01	0.04	0.00	0	0	0	—
0	0	0.00	—	0	0	—	0	0.00	—	0.00	0.00	—	0	—	—	0
0	0	0.00	—	—	10.8	—	0	0.00	—	0.00	0.00	0.00	0	—	0	1.0
5	40	0.36	36.5	170.6	25.0	1.13	14.8	0.03	—	0.06	1.60	0.04	17.3	0.6	0.1	1.9
5	40	0.36	19.6	127.4	30.0	0.46	14.8	0.02	—	0.06	0.10	0.01	2.9	0.6	0.1	1.4
22	228	0.82	61.0	398.6	91.9	1.00	0	0.06	—	0.25	0.14	0.10	10.9	2.0	0.4	—
5	60	0.18	19.8	140.1	95.0	0.41	15.1	0.02	—	0.06	0.20	0.02	5.8	0.6	0.2	3.3
5	11	0.51	17.9	71.4	189.2	0.37	16.6	0.05	1.08	0.01	1.12	0.03	19.6	0	0	1.1
3	40	0.72	45.4	217.4	180.0	0.93	0	0.12	—	0.08	2.35	0.07	28.1	0	0.1	2.3
0	20	0.00	18.9	169.9	80.0	0.32	0	0.04	—	0.06	1.22	0.03	12.0	0	0.1	0.8
3	5	0.44	16.3	51.7	1.6	0.23	0.4	0.01	—	0.01	0.06	0.01	0.4	0	0	0.5
5	40	0.36	42.3	—	140.0	1.37	15.3	0.03	—	0.06	1.60	0.05	23.5	0.6	0.1	2.7
0	0	1.80	45.5	136.0	50.0	0.59	0	0.01	—	0.02	0.16	0.01	0.8	0	0	1.2
0	10	0.18	0.6	1.2	0	0.00	—	0.00	—	0.00	0.00	0.00	0	30	0	0.5

TABLE A-1 Table of Food Composition *(continued)* (Computer code is for Cengage Diet Analysis program) (For purposes of calculations, use "0" for t, <1, <.1, <.01, etc.)

DA+ Code	Food Description	Quantity	Measure	Wt (g)	H_2O (g)	Ener (kcal)	Prot (g)	Carb (g)	Fiber (g)	Fat (g)	Fat Breakdown (g)			
											Sat	Mono	Poly	*Trans*
Sweets—continued														
4698	Taffy	3	piece(s)	45	2.2	179	0	41.2	0	1.5	0.9	0.4	0.1	0.1
4699	Three Musketeers bar	1	item(s)	60	3.5	260	2.0	46.0	1.0	8.0	4.5	2.6	0.3	—
4702	Twix caramel cookie bars	2	item(s)	58	2.4	280	3.0	37.0	1.0	14.0	5.0	7.7	0.5	—
4705	York peppermint pattie	1	item(s)	39	3.9	160	0.5	32.0	0.5	3.0	1.5	1.2	0.1	0
	Frosting, icing													
4760	Chocolate frosting, ready to eat	2	tablespoon(s)	31	5.2	122	0.3	19.4	0.3	5.4	1.7	2.8	0.6	—
4771	Creamy vanilla frosting, ready to eat	2	tablespoon(s)	28	4.2	117	0	19.0	0	4.5	0.8	1.4	2.2	0
17291	Dec-A-Cake variety pack candy decoration	1	teaspoon(s)	4	—	15	0	3.0	0	0.5	0	—	—	—
536	White icing	2	tablespoon(s)	40	3.6	162	0.1	31.8	0	4.2	0.8	2.0	1.2	—
	Gelatin													
13697	Gelatin snack, all flavors	1	item(s)	99	96.8	70	1.0	17.0	0	0	0	0	0	0
2616	Sugar free, low calorie mixed fruit gelatin mix, prepared	½	cup(s)	121	—	10	1.0	0	0	0	0	0	0	0
548	**Honey**	1	tablespoon(s)	21	3.6	64	0.1	17.3	0	0	0	0	0	0
	Jams, jellies													
550	Jam or preserves	1	tablespoon(s)	20	6.1	56	0.1	13.8	0.2	0	0	0	0	—
42199	Jams, preserves, dietetic, all flavors, w/sodium saccharin	1	tablespoon(s)	14	6.4	18	0	7.5	0.4	0	0	0	0	—
552	Jelly	1	tablespoon(s)	21	6.3	56	0	14.7	0.2	0	0	0	0	—
545	**Marshmallows**	4	item(s)	29	4.7	92	0.5	23.4	0	0.1	0	0	0	—
4800	**Marshmallow cream topping**	2	tablespoon(s)	40	7.9	129	0.3	31.6	0	0.1	0	0	0	—
555	**Molasses**	1	tablespoon(s)	20	4.4	58	0	14.9	0	0	0	0	0	—
4780	**Popsicle or ice pop**	1	item(s)	59	47.5	47	0	11.3	0	0.1	0	0	0	—
	Sugar													
559	Brown sugar, packed	1	teaspoon(s)	5	0.1	17	0	4.5	0	0	0	0	0	0
563	Powdered sugar, sifted	⅓	cup(s)	33	0.1	130	0	33.2	0	0	0	0	0	—
561	White granulated sugar	1	teaspoon(s)	4	0	16	0	4.2	0	0	0	0	0	0
	Sugar substitute													
1760	Equal sweetener, packet size	1	item(s)	1	—	0	0	0.9	0	0	0	0	0	0
13029	Splenda granular no calorie sweetener	1	teaspoon(s)	1	—	0	0	0.5	0	0	0	0	0	0
1759	Sweet N Low sugar substitute, packet	1	item(s)	1	0.1	4	0	0.5	0	0	0	0	0	0
	Syrup													
3148	Chocolate syrup	2	tablespoon(s)	38	11.6	105	0.8	24.4	1.0	0.4	0.2	0.1	0	—
29676	Maple syrup	¼	cup(s)	80	25.7	209	0	53.7	0	0.2	0	0.1	0.1	—
4795	Pancake syrup	¼	cup(s)	80	30.4	187	0	49.2	0	0	0	0	0	0
Spices, Condiments, Sauces														
	Spices													
807	Allspice, ground	1	teaspoon(s)	2	0.2	5	0.1	1.4	0.4	0.2	0	0	0	—
1171	Anise seeds	1	teaspoon(s)	2	0.2	7	0.4	1.1	0.3	0.3	0	0.2	0.1	—
729	Bakers' yeast, active	1	teaspoon(s)	4	0.3	12	1.5	1.5	0.8	0.2	0	0.1	0	—
683	Baking powder, double acting with phosphate	1	teaspoon(s)	5	0.2	2	0	1.1	0	0	0	0	0	0
1611	Baking soda	1	teaspoon(s)	5	0	0	0	0	0	0	0	0	0	0
8552	Basil	1	teaspoon(s)	1	0.8	0	0	0	0	0	0	0	0	0
34959	Basil, fresh	1	piece(s)	1	0.5	0	0	0	0	0	0	0	0	—
808	Basil, ground	1	teaspoon(s)	1	0.1	4	0.2	0.9	0.6	0.1	0	0	0	—
809	Bay leaf	1	teaspoon(s)	1	0	2	0	0.5	0.2	0.1	0	0	0	—
11720	Betel leaves	1	ounce(s)	28	—	17	1.8	2.4	0	0	—	—	—	—
730	Brewers' yeast	1	teaspoon(s)	3	0.1	8	1.0	1.0	0.8	0	0	0	0	0
11710	Capers	1	teaspoon(s)	5	—	0	0	0	0	0	0	0	0	0
1172	Caraway seeds	1	teaspoon(s)	2	0.2	7	0.4	1.0	0.8	0.3	0	0.2	0.1	—
1173	Celery seeds	1	teaspoon(s)	2	0.1	8	0.4	0.8	0.2	0.5	0	0.3	0.1	—
1174	Chervil, dried	1	teaspoon(s)	1	0	1	0.1	0.3	0.1	0	0	0	0	—
810	Chili powder	1	teaspoon(s)	3	0.2	8	0.3	1.4	0.9	0.4	0.1	0.1	0.2	—
8553	Chives, chopped	1	teaspoon(s)	1	0.9	0	0	0	0	0	0	0	0	—
51420	Cilantro (coriander)	1	teaspoon(s)	0	0.3	0	0	0	0	0	0	0	0	—
811	Cinnamon, ground	1	teaspoon(s)	2	0.2	6	0.1	1.9	1.2	0	0	0	0	0
812	Cloves, ground	1	teaspoon(s)	2	0.1	7	0.1	1.3	0.7	0.4	0.1	0	0.1	—
1175	Coriander leaf, dried	1	teaspoon(s)	1	0	2	0.1	0.3	0.1	0	0	0	0	—
1176	Coriander seeds	1	teaspoon(s)	2	0.2	5	0.2	1.0	0.8	0.3	0	0.2	0	—
1706	Cornstarch	1	tablespoon(s)	8	0.7	30	0	7.3	0.1	0	0	0	0	—

PAGE KEY: A-6 = Breads/Baked Goods A-12 = Cereal/Rice/Pasta A-16 = Fruit A-20 = Vegetables/Legumes A-30 = Nuts/Seeds A-32 = Vegetarian A-34 = Dairy A-42 = Eggs A-42 = Seafood A-44 = Meats A-48 = Poultry
A-50 = Processed Meats A-50 = Beverages A-54 = Fats/Oils A-56 = Sweets A-58 = Spices/Condiments/Sauces A-62 = Mixed Foods/Soups/Sandwiches A-66 = Fast Food A-86 = Convenience A-88 = Baby Foods

Chol (mg)	Calc (mg)	Iron (mg)	Magn (mg)	Pota (mg)	Sodi (mg)	Zinc (mg)	Vit A (µg)	Thia (mg)	Vit E (mg α)	Ribo (mg)	Niac (mg)	Vit B_6 (mg)	Fola (µg)	Vit C (mg)	Vit B_{12} (µg)	Sele (µg)
4	4	0.00	0	1.4	23.4	0.09	12.2	0.01	0.04	0.01	0.00	0.00	0	0	0	0.3
5	20	0.36	17.5	80.3	110.0	0.33	14.5	0.01	—	0.03	0.20	0.01	0	0.6	0.1	1.5
5	40	0.36	18.5	116.8	115.0	0.45	15.0	0.09	—	0.13	0.69	0.01	13.9	0.6	0.1	1.2
0	0	0.33	23.4	66.1	10.0	0.28	0	0.01	—	0.03	0.31	0.01	1.5	0	0	—
0	2	0.44	6.4	60.0	56.1	0.09	0	0.00	0.48	0.00	0.03	0.00	0.3	0	0	0.2
0	1	0.04	0.3	9.5	51.5	0.01	0	0.00	0.43	0.08	0.06	0.00	2.2	0	0	0
0	0	0.00	—	—	15.0	—	0	—	—	—	—	—	—	0	—	—
0	4	0.01	0.4	5.6	76.4	0.01	44.4	0.00	0.32	0.01	0.00	0.00	0	0	0	0.3
0	0	0.00	—	0	40.0	—	0	—	—	—	—	—	—	0	—	—
0	0	0.00	0	0	50.0	0	0	0.00	0.00	0.00	0.00	0.00	0	0	0	0
0	1	0.08	0.4	10.9	0.8	0.04	0	0.00	0.00	0.01	0.02	0.01	0.4	0.1	0	0.2
0	4	0.10	0.8	15.4	6.4	0.01	0	0.00	0.02	0.02	0.01	0.00	2.2	1.8	0	0.4
0	1	0.56	0.7	9.7	0	0.01	0	0.00	0.01	0.00	0.00	0.00	1.3	0	0	0.2
0	1	0.04	1.3	11.3	6.3	0.01	0	0.00	0.00	0.01	0.01	0.00	0.4	0.2	0	0.1
0	1	0.06	0.6	1.4	23.0	0.01	0	0.00	0.00	0.00	0.02	0.00	0.3	0	0	0.5
0	1	0.08	0.8	2.0	32.0	0.01	0	0.00	0.00	0.00	0.03	0.00	0.4	0	0	0.7
0	41	0.94	48.4	292.8	7.4	0.05	0	0.01	0.00	0.00	0.18	0.13	0	0	0	3.6
0	0	0.31	0.6	8.9	4.1	0.08	0	0.00	0.00	0.00	0.00	0.00	0	0.4	0	0.1
0	4	0.03	0.4	6.1	1.3	0.00	0	0.00	0.00	0.00	0.01	0.00	0	0	0	0.1
0	0	0.01	0	0.7	0.3	0.00	0	0.00	0.00	0.00	0.00	0.00	0	0	0	0.2
0	0	0.00	0	0.1	0	0.00	0	0.00	0.00	0.00	0.00	0.00	0	0	0	0
0	0	0.00	0	0	0	0.00	0	0.00	0.00	0.00	0.00	0.00	0	0	0	0
0	0	0.00	—	—	0	—	—	0.00	—	0.00	0.00	—	—	0	0	—
0	0	0.00	—	—	0	—	0	—	0.00	—	—	—	—	0	—	—
0	5	0.79	24.4	84.0	27.0	0.27	0	0.00	0.01	0.01	0.12	0.00	0.8	0.1	0	0.5
0	54	0.96	11.2	163.2	7.2	3.32	0	0.01	0.00	0.01	0.02	0.00	0	0	0	0.5
0	2	0.02	1.6	12.0	65.6	0.06	0	0.01	0.00	0.01	0.00	0.00	0	0	0	0
0	13	0.13	2.6	19.8	1.5	0.01	0.5	0.00	—	0.00	0.05	0.00	0.7	0.7	0	0.1
0	14	0.77	3.6	30.3	0.3	0.11	0.3	0.01	—	0.01	0.06	0.01	0.2	0.4	0	0.1
0	3	0.66	3.9	80.0	2.0	0.25	0	0.09	0.00	0.21	1.59	0.06	93.6	0	0	1.0
0	339	0.51	1.8	0.2	363.1	0.00	0	0.00	0.00	0.00	0.00	0.00	0	0	0	0
0	0	0.00	0	0	1258.6	0.00	0	0.00	0.00	0.00	0.00	0.00	0	0	0	0
0	2	0.02	0.6	2.6	0	0.01	2.3	0.00	0.01	0.00	0.01	0.00	0.6	0.2	0	0
0	1	0.01	0.4	2.3	0	0.00	1.3	0.00	—	0.00	0.00	0.00	0.3	0.1	0	0
0	30	0.58	5.9	48.1	0.5	0.08	6.6	0.00	0.10	0.00	0.09	0.03	3.8	0.9	0	0
0	5	0.25	0.7	3.2	0.1	0.02	1.9	0.00	—	0.00	0.01	0.01	1.1	0.3	0	0
0	110	2.29	—	155.9	2.0	—	—	0.04	—	0.07	0.19	—	—	0.9	0	—
0	6	0.46	6.1	50.7	3.3	0.21	0	0.41	—	0.11	1.00	0.06	104.3	0	0	0
0	—	—	—	—	105.0	—	—	—	—	—	—	—	—	—	0	—
0	14	0.34	5.4	28.4	0.4	0.11	0.4	0.01	0.05	0.01	0.07	0.01	0.2	0.4	0	0.3
0	35	0.89	8.8	28.0	3.2	0.13	0.1	0.01	0.02	0.01	0.06	0.01	0.2	0.3	0	0.2
0	8	0.19	0.8	28.4	0.5	0.05	1.8	0.00	—	0.00	0.03	0.01	1.6	0.3	0	0.2
0	7	0.37	4.4	49.8	26.3	0.07	38.6	0.01	0.75	0.02	0.20	0.09	2.6	1.7	0	0.2
0	1	0.01	0.4	3.0	0	0.01	2.2	0.00	0.00	0.00	0.01	0.00	1.1	0.6	0	0
0	0	0.01	0.1	1.7	0.2	0.00	1.1	0.00	0.01	0.00	0.00	0.00	0.2	0.1	0	0
0	23	0.19	1.4	9.9	0.2	0.04	0.3	0.00	0.05	0.00	0.03	0.00	0.1	0.1	0	0.1
0	14	0.18	5.5	23.1	5.1	0.02	0.6	0.00	0.17	0.01	0.03	0.01	2.0	1.7	0	0.1
0	7	0.25	4.2	26.8	1.3	0.02	1.8	0.01	0.01	0.01	0.06	0.00	1.6	3.4	0	0.2
0	13	0.29	5.9	22.8	0.6	0.08	0	0.00	—	0.01	0.03	—	0	0.4	0	0.5
0	0	0.03	0.2	0.2	0.7	0.01	0	0.00	0.00	0.00	0.00	0.00	0	0	0	0.2

TABLE A-1 **Table of Food Composition** *(continued)* (Computer code is for Cengage Diet Analysis program)   (For purposes of calculations, use "0" for t, <1, <.1, <.01, etc.)

DA+ Code	Food Description	Quantity	Measure	Wt (g)	H₂0 (g)	Ener (kcal)	Prot (g)	Carb (g)	Fiber (g)	Fat (g)	Fat Breakdown (g)			
											Sat	Mono	Poly	*Trans*
Spices, Condiments, Sauces—continued														
1177	Cumin seeds	1	teaspoon(s)	2	0.2	8	0.4	0.9	0.2	0.5	0	0.3	0.1	—
11729	Cumin, ground	1	teaspoon(s)	5	—	11	0.4	0.8	0.8	0.4	—	—	—	—
1178	Curry powder	1	teaspoon(s)	2	0.2	7	0.3	1.2	0.7	0.3	0	0.1	0.1	—
1179	Dill seeds	1	teaspoon(s)	2	0.2	6	0.3	1.2	0.4	0.3	0	0.2	0	—
1180	Dill weed, dried	1	teaspoon(s)	1	0.1	3	0.2	0.6	0.1	0	0	0	0	—
34949	Dill weed, fresh	5	piece(s)	1	0.9	0	0	0.1	0	0	0	0	0	—
4949	Fennel leaves, fresh	1	teaspoon(s)	1	0.9	0	0	0.1	0	0	—	—	—	—
1181	Fennel seeds	1	teaspoon(s)	2	0.2	7	0.3	1.0	0.8	0.3	0	0.2	0	—
1182	Fenugreek seeds	1	teaspoon(s)	4	0.3	12	0.9	2.2	0.9	0.2	0.1	—	—	—
11733	Garam masala, powder	1	ounce(s)	28	—	107	4.4	12.8	0	4.3	—	—	—	—
1067	Garlic clove	1	item(s)	3	1.8	4	0.2	1.0	0.1	0	0	0	0	—
813	Garlic powder	1	teaspoon(s)	3	0.2	9	0.5	2.0	0.3	0	0	0	0	—
1068	Ginger root	2	teaspoon(s)	4	3.1	3	0.1	0.7	0.1	0	0	0	0	—
1183	Ginger, ground	1	teaspoon(s)	2	0.2	6	0.2	1.3	0.2	0.1	0	0	0	—
35497	Leeks, bulb and lower-leaf, freeze-dried	¼	cup(s)	1	0	3	0.1	0.6	0.1	0	0	0	0	—
1184	Mace, ground	1	teaspoon(s)	2	0.1	8	0.1	0.9	0.3	0.6	0.2	0.2	0.1	—
1185	Marjoram, dried	1	teaspoon(s)	1	0	2	0.1	0.4	0.2	0	0	0	0	—
1186	Mustard seeds, yellow	1	teaspoon(s)	3	0.2	15	0.8	1.2	0.5	0.9	0	0.7	0.2	—
814	Nutmeg, ground	1	teaspoon(s)	2	0.1	12	0.1	1.1	0.5	0.8	0.6	0.1	0	—
2747	Onion flakes, dehydrated	1	teaspoon(s)	2	0.1	6	0.1	1.4	0.2	0	0	0	0	—
1187	Onion powder	1	teaspoon(s)	2	0.1	7	0.2	1.7	0.1	0	0	0	0	—
815	Oregano, ground	1	teaspoon(s)	2	0.1	5	0.2	1.0	0.6	0.2	0	0	0.1	—
816	Paprika	1	teaspoon(s)	2	0.2	6	0.3	1.2	0.8	0.3	0	0	0.2	—
817	Parsley, dried	1	teaspoon(s)	0	0	1	0.1	0.2	0.1	0	0	0	0	—
818	Pepper, black	1	teaspoon(s)	2	0.2	5	0.2	1.4	0.6	0.1	0	0	0	—
819	Pepper, cayenne	1	teaspoon(s)	2	0.1	6	0.2	1.0	0.5	0.3	0.1	0	0.2	—
1188	Pepper, white	1	teaspoon(s)	2	0.3	7	0.3	1.6	0.6	0.1	0	0	0	—
1189	Poppy seeds	1	teaspoon(s)	3	0.2	15	0.5	0.7	0.3	1.3	0.1	0.2	0.9	—
1190	Poultry seasoning	1	teaspoon(s)	2	0.1	5	0.1	1.0	0.2	0.1	0	0	0	—
1191	Pumpkin pie spice, powder	1	teaspoon(s)	2	0.1	6	0.1	1.2	0.3	0.2	0.1	0	0	—
1192	Rosemary, dried	1	teaspoon(s)	1	0.1	4	0.1	0.8	0.5	0.2	0.1	0	0	—
11723	Rosemary, fresh	1	teaspoon(s)	1	0.5	1	0	0.1	0.1	0	0	0	0	—
2722	Saffron powder	1	teaspoon(s)	1	0.1	2	0.1	0.5	0	0	0	0	0	—
11724	Sage	1	teaspoon(s)	1	—	1	0	0.1	0	0	—	—	—	—
1193	Sage, ground	1	teaspoon(s)	1	0.1	2	0.1	0.4	0.3	0.1	0	0	0	—
30189	Salt substitute	¼	teaspoon(s)	1	—	0	0	0	0	0	0	0	0	0
30190	Salt substitute, seasoned	¼	teaspoon(s)	1	—	1	0	0.1	0	0	0	—	—	—
822	Salt, table	¼	teaspoon(s)	2	0	0	0	0	0	0	0	0	0	0
1194	Savory, ground	1	teaspoon(s)	1	0.1	4	0.1	1.0	0.6	0.1	0	—	—	—
820	Sesame seed kernels, toasted	1	teaspoon(s)	3	0.1	15	0.5	0.7	0.5	1.3	0.2	0.5	0.6	—
11725	Sorrel	1	teaspoon(s)	3	—	1	0.1	0.1	0	0	0	—	—	—
11721	Spearmint	1	teaspoon(s)	2	1.6	1	0.1	0.2	0.1	0	0	0	0	—
35498	Sweet green peppers, freeze-dried	¼	cup(s)	2	0	5	0.3	1.1	0.3	0	0	0	0	—
11726	Tamarind leaves	1	ounce(s)	28	—	33	1.6	5.2	0	0.6	—	—	—	—
11727	Tarragon	1	ounce(s)	28	—	14	1.0	1.8	0	0.3	—	—	—	—
1195	Tarragon, ground	1	teaspoon(s)	2	0.1	5	0.4	0.8	0.1	0.1	0	0	0.1	—
11728	Thyme, fresh	1	teaspoon(s)	1	0.5	1	0	0.2	0.1	0	0	0	0	—
821	Thyme, ground	1	teaspoon(s)	1	0.1	4	0.1	0.9	0.5	0.1	0	0	0	—
1196	Turmeric, ground	1	teaspoon(s)	2	0.3	8	0.2	1.4	0.5	0.2	0.1	0	0	—
11995	Wasabi	1	tablespoon(s)	14	10.7	10	0.7	2.3	0.2	0	—	—	—	—
	Condiments													
674	Catsup or ketchup	1	tablespoon(s)	15	10.4	15	0.3	3.8	0	0	0	0	0	—
703	Dill pickle	1	ounce(s)	28	26.7	3	0.2	0.7	0.3	0	0	0	0	—
138	Mayonnaise with soybean oil	1	tablespoon(s)	14	2.1	99	0.1	0.4	0	11.0	1.6	2.7	5.8	0
140	Mayonnaise, low calorie	1	tablespoon(s)	16	10.0	37	0	2.6	0	3.1	0.5	0.7	1.7	—
1682	Mustard, brown	1	teaspoon(s)	5	4.1	5	0.3	0.3	0	0.3	—	—	—	—
700	Mustard, yellow	1	teaspoon(s)	5	4.1	3	0.2	0.3	0.2	0.2	0	0.1	0	—
706	Sweet pickle relish	1	tablespoon(s)	15	9.3	20	0.1	5.3	0.2	0.1	0	0	0	—
141	Tartar sauce	2	tablespoon(s)	28	8.7	144	0.3	4.1	0.1	14.4	2.2	3.8	7.7	—
	Sauces													
685	Barbecue sauce	2	tablespoon(s)	31	18.9	47	0	11.3	0.2	0.1	0	0	0.1	0
834	Cheese sauce	¼	cup(s)	63	44.4	110	4.2	4.3	0.3	8.4	3.8	2.4	1.6	—
32123	Chili enchilada sauce, green	2	tablespoon(s)	57	53.0	15	0.6	3.1	0.7	0.3	0	0	0.1	0
32122	Chili enchilada sauce, red	2	tablespoon(s)	32	24.5	27	1.1	5.0	2.1	0.8	0.1	0	0.4	0

PAGE KEY: A-6 = Breads/Baked Goods A-12 = Cereal/Rice/Pasta A-16 = Fruit A-20 = Vegetables/Legumes A-30 = Nuts/Seeds A-32 = Vegetarian A-34 = Dairy A-42 = Eggs A-42 = Seafood A-44 = Meats A-48 = Poultry A-50 = Processed Meats A-50 = Beverages A-54 = Fats/Oils A-56 = Sweets A-58 = Spices/Condiments/Sauces A-62 = Mixed Foods/Soups/Sandwiches A-66 = Fast Food A-86 = Convenience A-88 = Baby Foods

Chol (mg)	Calc (mg)	Iron (mg)	Magn (mg)	Pota (mg)	Sodi (mg)	Zinc (mg)	Vit A (µg)	Thia (mg)	Vit E (mg α)	Ribo (mg)	Niac (mg)	Vit B$_6$ (mg)	Fola (µg)	Vit C (mg)	Vit B$_{12}$ (µg)	Sele (µg)
0	20	1.39	7.7	37.5	3.5	0.10	1.3	0.01	0.07	0.01	0.09	0.01	0.2	0.2	0	0.1
0	20	—	—	43.6	4.8	—	—	—	—	—	—	—	—	—	—	—
0	10	0.59	5.1	30.9	1.0	0.08	1.0	0.01	0.44	0.01	0.06	0.02	3.1	0.2	0	0.3
0	32	0.34	5.4	24.9	0.4	0.10	0.1	0.01	—	0.01	0.05	0.01	0.2	0.4	0	0.3
0	18	0.48	4.5	33.1	2.1	0.03	2.9	0.00	—	0.00	0.02	0.01	1.5	0.5	0	—
0	2	0.06	0.6	7.4	0.6	0.01	3.9	0.00	0.01	0.00	0.01	0.00	1.5	0.9	0	—
0	1	0.02	—	4.0	0.1	—	—	0.00	—	0.00	0.01	0.00	—	0.3	0	—
0	24	0.37	7.7	33.9	1.8	0.07	0.1	0.01	—	0.01	0.12	0.01	—	0.4	0	—
0	7	1.24	7.1	28.5	2.5	0.09	0.1	0.01	—	0.01	0.06	0.02	2.1	0.1	0	0.2
0	215	9.24	93.6	411.1	27.5	1.07	—	0.09	—	0.09	0.70	—	0	0	0	—
0	5	0.05	0.8	12.0	0.5	0.03	0	0.01	0.00	0.00	0.02	0.03	0.1	0.9	0	0.4
0	2	0.07	1.6	30.8	0.7	0.07	0	0.01	0.01	0.00	0.01	0.08	0.1	0.5	0	1.1
0	1	0.02	1.7	16.6	0.5	0.01	0	0.00	0.01	0.00	0.02	0.01	0.4	0.2	0	0
0	2	0.20	3.3	24.2	0.6	0.08	0.1	0.00	0.32	0.00	0.09	0.01	0.7	0.1	0	0.7
0	3	0.06	1.3	19.2	0.3	0.01	0.1	0.01	—	0.00	0.02	0.01	2.9	0.9	0	0
0	4	0.23	2.8	7.9	1.4	0.03	0.7	0.01	—	0.01	0.02	0.00	1.3	0.4	0	0
0	12	0.49	2.1	9.1	0.5	0.02	2.4	0.00	0.01	0.00	0.02	0.01	1.6	0.3	0	0
0	17	0.32	9.8	22.5	0.2	0.18	0.1	0.01	0.09	0.01	0.26	0.01	2.5	0.1	0	4.4
0	4	0.06	4.0	7.7	0.4	0.04	0.1	0.01	0.00	0.00	0.02	0.00	1.7	0.1	0	0
0	4	0.02	1.5	27.1	0.4	0.03	0	0.01	0.00	0.00	0.01	0.02	2.8	1.3	0	0.1
0	8	0.05	2.6	19.8	1.1	0.04	0	0.01	0.01	0.00	0.01	0.02	3.5	0.3	0	0
0	24	0.66	4.1	25.0	0.2	0.06	5.2	0.01	0.28	0.01	0.09	0.01	4.1	0.8	0	0.1
0	4	0.49	3.9	49.2	0.7	0.08	55.4	0.01	0.62	0.03	0.32	0.08	2.2	1.5	0	0.1
0	4	0.29	0.7	11.4	1.4	0.01	1.5	0.00	0.02	0.00	0.02	0.00	0.5	0.4	0	0.1
0	9	0.60	4.1	26.4	0.9	0.03	0.3	0.00	0.01	0.01	0.02	0.01	0.2	0.4	0	0.1
0	3	0.14	2.7	36.3	0.5	0.04	37.5	0.01	0.53	0.01	0.15	0.04	1.9	1.4	0	0.2
0	6	0.34	2.2	1.8	0.1	0.02	0	0.00	—	0.00	0.01	0.00	0.2	0.5	0	0.1
0	41	0.26	9.3	19.6	0.6	0.28	0	0.02	0.03	0.01	0.02	0.01	1.6	0.1	0	0
0	15	0.53	3.4	10.3	0.4	0.04	2.0	0.00	0.02	0.00	0.04	0.02	2.1	0.2	0	0.1
0	12	0.33	2.3	11.3	0.9	0.04	0.2	0.00	0.01	0.00	0.03	0.01	0.9	0.4	0	0.2
0	15	0.35	2.6	11.5	0.6	0.03	1.9	0.01	—	0.01	0.01	0.02	3.7	0.7	0	0.1
0	2	0.04	0.6	4.7	0.2	0.01	1.0	0.00	—	0.00	0.01	0.00	0.8	0.2	0	—
0	1	0.07	1.8	12.1	1.0	0.01	0.2	0.00	—	0.00	0.01	0.01	0.7	0.6	0	0
0	4	—	1.1	2.7	0	0.01	—	0.00	—	—	—	—	—	—	0	—
0	12	0.19	3.0	7.5	0.1	0.03	2.1	0.01	0.05	0.00	0.04	0.01	1.9	0.2	0	0
0	7	0.00	0	603.6	0.1	—	0	—	—	—	—	—	—	0	—	—
0	0	0.00	—	476.3	0.1	—	0	—	—	—	—	—	—	0	—	—
0	0	0.01	0	0.1	581.4	0.00	0	0.00	0.00	0.00	0.00	0.00	0	0	0	0
0	30	0.53	5.3	14.7	0.3	0.06	3.6	0.01	—	—	0.05	0.02	—	0.7	0	0.1
0	3	0.21	9.2	10.8	1.0	0.27	0.1	0.03	0.01	0.01	0.15	0.00	2.6	0	0	0
0	—	—	—	—	0.1	—	—	—	—	—	—	—	—	—	—	—
0	4	0.22	1.2	8.7	0.6	0.02	3.9	0.00	—	0.00	0.01	0.00	2.0	0.3	0	—
0	2	0.16	3.0	50.7	3.1	0.03	4.5	0.01	0.06	0.01	0.11	0.03	3.7	30.4	0	0.1
0	85	1.48	20.2	—	—	—	—	0.06	—	0.02	1.16	—	—	0.9	0	—
0	48	—	14.5	128.1	2.6	0.17	—	0.04	—	—	—	—	—	0.6	0	—
0	18	0.51	5.6	48.3	1.0	0.06	3.4	0.00	—	0.02	0.14	0.03	4.4	0.8	0	0.1
0	3	0.14	1.3	4.9	0.1	0.01	1.9	0.00	—	0.00	0.01	0.00	0.4	1.3	0	—
0	26	1.73	3.1	11.4	0.8	0.08	2.7	0.01	0.10	0.01	0.06	0.01	3.8	0.7	0	0.1
0	4	0.91	4.2	55.6	0.8	0.09	0	0.00	0.06	0.01	0.11	0.04	0.9	0.6	0	0.1
0	13	0.11	—	—	—	—	—	0.02	—	0.01	0.07	—	—	11.2	0	—
0	3	0.07	2.9	57.3	167.1	0.03	7.1	0.00	0.21	0.02	0.21	0.02	1.5	2.3	0	0
0	12	0.10	2.0	26.1	248.1	0.03	2.6	0.01	0.02	0.01	0.03	0.01	0.3	0.2	0	0
5	1	0.03	0.1	1.7	78.4	0.02	11.2	0.01	0.72	0.01	0.00	0.08	0.7	0	0	0.2
4	0	0.00	0	1.6	79.5	0.01	0	0.00	0.32	0.00	0.00	0.00	0	0	0	0.3
0	6	0.09	1.0	6.8	68.1	0.01	0	0.00	0.09	0.00	0.01	0.00	0.2	0.1	0	—
0	3	0.07	2.5	6.9	56.8	0.03	0.2	0.01	0.01	0.00	0.02	0.00	0.4	0.1	0	1.6
0	0	0.13	0.8	3.8	121.7	0.02	9.2	0.00	0.08	0.01	0.03	0.00	0.2	0.2	0	0
8	6	0.20	0.8	10.1	191.5	0.05	20.2	0.00	0.97	0.00	0.01	0.07	2.0	0.1	0.1	0.5
0	4	0.06	3.8	65.0	349.7	0.04	3.8	0.00	0.20	0.01	0.15	0.01	0.6	0.2	0	0.4
18	116	0.13	5.7	18.9	521.6	0.61	50.4	0.00	—	0.07	0.01	0.01	2.5	0.3	0.1	2.0
0	5	0.36	9.5	125.7	61.9	0.11	—	0.02	0.00	0.02	0.63	0.06	5.7	43.9	0	0
0	7	1.05	11.1	231.3	113.8	0.14	—	0.01	0.00	0.21	0.61	0.34	6.6	0.3	0	0.3

Appendix A

TABLE A-1 Table of Food Composition (continued) (Computer code is for Cengage Diet Analysis program) (For purposes of calculations, use "0" for t, <1, <.1, <.01, etc.)

DA+ Code	Food Description	Quantity	Measure	Wt (g)	H₂0 (g)	Ener (kcal)	Prot (g)	Carb (g)	Fiber (g)	Fat (g)	Sat	Mono	Poly	Trans
Spices, Condiments, Sauces—continued														
29688	Hoisin sauce	1	tablespoon(s)	16	7.1	35	0.5	7.1	0.4	0.5	0.1	0.2	0.3	—
1641	Horseradish sauce, prepared	1	teaspoon(s)	5	3.3	10	0.1	0.2	0	1.0	0.6	0.3	0	—
16670	Mole poblano sauce	½	cup(s)	133	102.7	156	5.3	11.4	2.7	11.3	2.6	5.1	3.0	—
29689	Oyster sauce	1	tablespoon(s)	16	12.8	8	0.2	1.7	0	0	0	0	0	—
1655	Pepper sauce or Tabasco	1	teaspoon(s)	5	4.8	1	0.1	0	0	0	0	0	0	—
347	Salsa	2	tablespoon(s)	32	28.8	9	0.5	2.0	0.5	0.1	0	0	0	—
52206	Soy sauce, tamari	1	tablespoon(s)	18	12.0	11	1.9	1.0	0.1	0	0	0	0	—
839	Sweet and sour sauce	2	tablespoon(s)	39	29.8	37	0.1	9.1	0.1	0	0	0	0	—
1613	Teriyaki sauce	1	tablespoon(s)	18	12.2	16	1.1	2.8	0	0	0	0	0	0
25294	Tomato sauce	½	cup(s)	150	132.8	63	2.2	11.9	2.6	1.8	0.2	0.4	0.9	0
728	White sauce, medium	¼	cup(s)	63	46.8	92	2.4	5.7	0.1	6.7	1.8	2.8	1.8	—
1654	Worcestershire sauce	1	teaspoon(s)	6	4.5	4	0	1.1	0	0	0	0	0	0
	Vinegar													
30853	Balsamic	1	tablespoon(s)	15	—	10	0	2.0	0	0	0	0	0	0
727	Cider	1	tablespoon(s)	15	14.0	3	0	0.1	0	0	0	0	0	0
1673	Distilled	1	tablespoon(s)	15	14.3	2	0	0.8	0	0	0	0	0	0
12948	Tarragon	1	tablespoon(s)	15	13.8	2	0	0.1	0	0	0	0	0	0
Mixed Foods, Soups, Sandwiches														
	Mixed dishes													
16652	Almond chicken	1	cup(s)	242	186.8	281	21.8	15.8	3.4	14.7	1.8	6.3	5.6	—
25224	Barbecued chicken	1	serving(s)	177	99.3	327	27.1	15.7	0.5	17.1	4.8	6.8	3.8	0
25227	Bean burrito	1	item(s)	149	81.8	326	16.1	33.0	5.6	14.8	8.3	4.7	0.9	—
9516	Beef and vegetable fajita	1	item(s)	223	143.9	397	22.4	35.3	3.1	18.0	5.9	8.0	2.5	—
16796	Beef or pork egg roll	2	item(s)	128	85.2	225	9.9	18.4	1.4	12.4	2.9	6.0	2.6	—
177	Beef stew with vegetables, prepared	1	cup(s)	245	201.0	220	16.0	15.0	3.2	11.0	4.4	4.5	0.5	—
30233	Beef stroganoff with noodles	1	cup(s)	256	190.1	343	19.7	22.8	1.5	19.1	7.4	5.7	4.4	—
16651	Cashew chicken	1	cup(s)	242	186.8	281	21.8	15.8	3.4	14.7	1.8	6.3	5.6	—
30274	Cheese pizza with vegetables, thin crust	2	slice(s)	140	76.6	298	12.7	35.4	2.5	12.0	4.9	4.7	1.6	—
30330	Cheese quesadilla	1	item(s)	54	18.3	190	7.7	15.3	1.0	10.8	5.2	3.6	1.3	—
215	Chicken and noodles, prepared	1	cup(s)	240	170.0	365	22.0	26.0	1.3	18.0	5.1	7.1	3.9	—
30239	Chicken and vegetables with broccoli, onion, bamboo shoots in soy based sauce	1	cup(s)	162	125.5	180	15.8	9.3	1.8	8.6	1.7	3.0	3.1	—
25093	Chicken cacciatore	1	cup(s)	244	175.7	284	29.9	5.7	1.3	15.3	4.3	6.2	3.3	0
28020	Chicken fried turkey steak	3	ounce(s)	492	276.2	706	77.1	68.7	3.6	12.0	3.4	2.9	3.9	—
218	Chicken pot pie	1	cup(s)	252	154.6	542	22.6	41.4	3.5	31.3	9.8	12.5	7.1	—
30240	Chicken teriyaki	1	cup(s)	244	158.3	364	51.0	15.2	0.7	7.0	1.8	2.0	1.7	—
25119	Chicken waldorf salad	½	cup(s)	100	67.2	179	14.0	6.8	1.0	10.8	1.8	3.1	5.2	—
25099	Chili con carne	¾	cup(s)	215	174.4	198	13.7	21.4	7.5	6.9	2.5	2.8	0.5	0
1062	Coleslaw	¾	cup(s)	90	73.4	70	1.2	11.2	1.4	2.3	0.3	0.6	1.2	—
1574	Crab cakes, from blue crab	1	item(s)	60	42.6	93	12.1	0.3	0	4.5	0.9	1.7	1.4	—
32144	Enchiladas with green chili sauce (enchiladas verdes)	1	item(s)	144	103.8	207	9.3	17.6	2.6	11.7	6.4	3.6	1.0	0
2793	Falafel patty	3	item(s)	51	17.7	170	6.8	16.2	—	9.1	1.2	5.2	2.1	—
28546	Fettuccine alfredo	1	cup(s)	244	88.7	279	13.1	46.1	1.4	4.2	2.2	1.0	0.4	0
32146	Flautas	3	item(s)	162	78.0	438	24.9	36.3	4.1	21.6	8.2	8.8	2.3	—
29629	Fried rice with meat or poultry	1	cup(s)	198	128.5	333	12.3	41.8	1.4	12.3	2.2	3.5	5.7	—
16649	General Tso chicken	1	cup(s)	146	91.0	296	18.7	16.4	0.9	17.0	4.0	6.3	5.3	—
1826	Green salad	¾	cup(s)	104	98.9	17	1.3	3.3	2.2	0.1	0	0	0	—
1814	Hummus	½	cup(s)	123	79.8	218	6.0	24.7	4.9	10.6	1.4	6.0	2.6	—
16650	Kung pao chicken	1	cup(s)	162	87.2	434	28.8	11.7	2.3	30.6	5.2	13.9	9.7	—
16622	Lamb curry	1	cup(s)	236	187.9	257	28.2	3.7	0.9	13.8	3.9	4.9	3.3	—
25253	Lasagna with ground beef	1	cup(s)	237	158.4	284	16.9	22.3	2.4	14.5	7.5	4.9	0.8	—
442	Macaroni and cheese, prepared	1	cup(s)	200	122.3	390	14.9	40.6	1.6	18.6	7.9	6.4	2.9	—
29637	Meat filled ravioli with tomato or meat sauce, canned	1	cup(s)	251	198.7	208	7.8	36.5	1.3	3.7	1.5	1.4	0.3	—
25105	Meat loaf	1	slice(s)	115	84.5	245	17.0	6.6	0.4	16.0	6.1	6.9	0.9	0
16646	Moo shu pork	1	cup(s)	151	76.8	512	18.9	5.3	0.6	46.4	6.9	15.8	21.2	—
16788	Nachos with beef, beans, cheese, tomatoes and onions	1	serving(s)	551	253.5	1576	59.1	137.5	20.4	90.8	32.6	41.9	9.4	—
6116	Pepperoni pizza	2	slice(s)	142	66.1	362	20.2	39.7	2.9	13.9	4.5	6.3	2.3	—
29601	Pizza with meat and vegetables, thin crust	2	slice(s)	158	81.4	386	16.5	36.8	2.7	19.1	7.7	8.1	2.2	—

PAGE KEY: A-6 = Breads/Baked Goods A-12 = Cereal/Rice/Pasta A-16 = Fruit A-20 = Vegetables/Legumes A-30 = Nuts/Seeds A-32 = Vegetarian A-34 = Dairy A-42 = Eggs A-42 = Seafood A-44 = Meats A-48 = Poultry
A-50 = Processed Meats A-50 = Beverages A-54 = Fats/Oils A-56 = Sweets A-58 = Spices/Condiments/Sauces A-62 = Mixed Foods/Soups/Sandwiches A-66 = Fast Food A-86 = Convenience A-88 = Baby Foods

Chol (mg)	Calc (mg)	Iron (mg)	Magn (mg)	Pota (mg)	Sodi (mg)	Zinc (mg)	Vit A (µg)	Thia (mg)	Vit E (mg α)	Ribo (mg)	Niac (mg)	Vit B$_6$ (mg)	Fola (µg)	Vit C (mg)	Vit B$_{12}$ (µg)	Sele (µg)
0	5	0.16	3.8	19.0	258.4	0.05	0	0.00	0.04	0.03	0.18	0.01	3.7	0.1	0	0.3
2	5	0.00	0.5	6.7	14.6	0.01	8.0	0.00	0.02	0.01	0.00	0.00	0.5	0.1	0	0.1
1	38	1.81	58.3	280.9	304.8	1.15	13.3	0.06	1.72	0.08	1.84	0.09	15.9	3.4	0.1	1.1
0	5	0.02	0.6	8.6	437.3	0.01	0	0.00	0.00	0.02	0.23	0.00	2.4	0.2	0.1	0.7
0	1	0.05	0.6	6.4	31.7	0.01	4.1	0.00	0.00	0.00	0.01	0.01	0.1	0.2	0	0
0	9	0.14	4.8	95.0	192.0	0.11	4.8	0.01	0.37	0.01	0.02	0.05	1.3	0.6	0	0.3
0	4	0.43	7.3	38.7	1018.9	0.08	0	0.01	0.00	0.03	0.72	0.04	3.3	0	0	0.1
0	5	0.20	1.2	8.2	97.5	0.01	0	0.00	—	0.01	0.11	0.03	0.2	0	0	—
0	5	0.30	11.0	40.5	689.9	0.01	0	0.01	0.00	0.01	0.22	0.01	1.4	0	0	0.2
0	23	1.24	28.9	536.8	268.6	0.36	—	0.08	0.52	0.08	1.64	0.20	23.2	32.0	0	1.0
4	74	0.20	8.8	97.5	221.3	0.25	—	0.04	—	0.11	0.25	0.02	3.1	0.5	0.2	—
0	6	0.30	0.7	45.4	55.6	0.01	0.3	0.00	0.00	0.01	0.03	0.00	0.5	0.7	0	0
0	0	0.00	—	—	0	—	0	—	—	—	—	—	—	0	—	—
0	1	0.03	0.7	10.9	0.7	0.01	0	0.00	0.00	0.00	0.00	0.00	0	0	0	0
0	1	0.09	0	2.3	0.1	0.00	0	0.00	0.00	0.00	0.00	0.00	0	0	0	5.0
0	0	0.07	—	2.3	0.7	—	—	0.07	—	0.07	0.07	—	—	0.3	0	—
41	68	1.86	58.1	539.7	510.6	1.50	31.5	0.07	4.11	0.22	9.57	0.43	26.6	5.1	0.3	13.6
120	26	1.70	32.4	419.7	500.9	2.67	—	0.09	0.01	0.24	6.87	0.40	15.0	7.9	0.3	19.5
38	333	3.01	52.5	447.6	510.6	1.98	—	0.28	0.01	0.30	1.92	0.19	122.9	8.2	0.3	15.9
45	85	3.65	37.9	475.0	756.0	3.52	17.8	0.38	0.80	0.29	5.33	0.39	69.1	23.4	2.1	28.3
74	31	1.68	20.5	248.3	547.8	0.89	25.6	0.32	1.28	0.24	2.55	0.18	38.4	4.0	0.3	17.5
71	29	2.90	—	613.0	292.0	—	—	0.15	0.51	0.17	4.70	—	—	17.0	0	15.0
74	69	3.25	35.8	391.7	816.6	3.63	69.1	0.21	1.25	0.30	3.80	0.21	48.6	1.3	1.8	27.9
41	68	1.86	58.1	539.7	510.6	1.50	31.5	0.07	4.11	0.22	9.57	0.43	26.6	5.1	0.3	13.6
17	249	2.78	28.0	294.0	739.2	1.42	47.6	0.29	1.05	0.33	2.84	0.14	61.6	15.3	0.4	18.6
23	190	1.04	13.5	75.6	469.3	0.86	58.3	0.11	0.43	0.15	0.89	0.02	21.6	2.4	0.1	9.2
103	26	2.20	—	149.0	600.0	—	—	0.05	—	0.17	4.30	—	—	0	—	29.0
42	28	1.19	22.7	299.7	620.5	1.32	81.0	0.07	1.11	0.14	5.28	0.36	16.2	22.5	0.2	12.0
109	47	1.97	40.0	489.3	492.1	2.13	—	0.11	0.00	0.20	9.81	0.57	16.1	14.0	0.3	22.6
156	423	8.79	110.3	1182.9	880.4	6.18	—	0.72	0.00	1.05	20.16	1.20	113.9	2.7	1.3	97.6
68	66	3.32	37.8	390.6	652.7	1.94	259.6	0.39	1.05	0.39	7.25	0.23	80.6	10.3	0.2	27.0
156	51	3.26	68.3	588.0	3208.6	3.75	31.7	0.15	0.58	0.36	16.68	0.88	24.4	2.0	0.5	36.1
42	20	0.82	23.9	202.5	246.5	1.13	—	0.05	0.62	0.09	4.06	0.25	15.8	2.5	0.2	10.7
27	42	2.83	50.6	636.8	864.8	2.36	—	0.15	0.01	0.22	3.18	0.19	58.1	10.3	0.6	7.3
7	41	0.53	9.0	162.9	20.7	0.18	47.7	0.06	—	0.05	0.24	0.11	24.3	29.4	0	0.6
90	63	0.64	19.8	194.4	198.0	2.45	34.2	0.05	—	0.04	1.74	0.10	31.8	1.7	3.6	24.4
27	266	1.07	38.5	251.4	276.3	1.26	—	0.07	0.02	0.16	1.27	0.17	44.6	59.3	0.2	6.0
0	28	1.74	41.8	298.4	149.9	0.76	0.5	0.07	—	0.08	0.53	0.06	47.4	0.8	0	0.5
9	218	1.83	38.4	163.9	472.9	1.24	—	0.41	0.00	0.35	2.85	0.09	96.2	1.6	0.4	38.4
73	146	2.66	61.3	222.9	885.7	3.43	0	0.10	0.10	0.16	3.00	0.26	95.7	0	1.2	36.7
103	38	2.77	33.7	196.0	833.6	1.34	41.6	0.33	1.60	0.18	4.17	0.27	97.0	3.4	0.3	22.0
66	26	1.46	23.4	248.2	849.7	1.40	29.2	0.10	1.62	0.18	6.28	0.28	23.4	12.0	0.2	19.9
0	13	0.65	11.4	178.0	26.9	0.21	59.0	0.03	—	0.05	0.56	0.08	38.3	24.0	0	0.4
0	60	1.91	35.7	212.8	297.7	1.34	0	0.10	0.92	0.06	0.49	0.49	72.6	9.7	0	3.0
65	50	1.96	63.2	427.7	907.2	1.50	38.9	0.15	4.32	0.14	13.22	0.58	42.1	7.5	0.3	23.0
90	38	2.95	40.1	493.2	495.6	6.60	—	0.08	1.29	0.28	8.03	0.21	28.3	1.4	2.9	30.4
68	233	2.22	40.1	420.1	433.6	2.70	—	0.21	0.21	0.29	3.06	0.22	50.4	15.0	0.8	21.4
34	310	2.06	40.0	258.0	784.0	2.06	180.0	0.27	0.72	0.43	2.18	0.08	64.0	0	0.5	30.6
15	35	2.10	20.1	283.6	1352.9	1.28	27.6	0.19	0.70	0.16	2.77	0.14	42.7	21.6	0.4	13.3
85	59	1.87	21.8	300.8	411.7	3.40	—	0.08	0.00	0.27	3.72	0.13	18.7	0.9	1.6	17.9
172	32	1.57	25.7	333.7	1052.5	1.82	49.8	0.49	5.39	0.36	2.88	0.31	21.1	8.0	0.8	30.0
154	948	7.32	242.4	1201.2	1862.4	10.68	259.0	0.29	7.71	0.81	6.39	1.09	148.8	16.0	2.6	44.1
28	129	1.87	17.0	305.3	533.9	1.03	105.1	0.26	—	0.46	6.09	0.11	73.8	3.3	0.4	26.1
36	258	3.14	31.6	352.3	971.7	2.02	49.0	0.37	1.13	0.37	3.77	0.19	64.8	15.6	0.6	22.8

TABLE A-1 **Table of Food Composition** *(continued)* (Computer code is for Cengage Diet Analysis program) (For purposes of calculations, use "0" for t, <1, <.1, <.01, etc.)

DA+ Code	Food Description	Quantity	Measure	Wt (g)	H₂0 (g)	Ener (kcal)	Prot (g)	Carb (g)	Fiber (g)	Fat (g)	Sat	Mono	Poly	Trans
												Fat Breakdown (g)		

Mixed Foods, Soups, Sandwiches—continued

DA+ Code	Food Description	Quantity	Measure	Wt (g)	H₂0 (g)	Ener (kcal)	Prot (g)	Carb (g)	Fiber (g)	Fat (g)	Sat	Mono	Poly	Trans
655	Potato salad	½	cup(s)	125	95.0	179	3.4	14.0	1.6	10.3	1.8	3.1	4.7	—
25109	Salisbury steaks with mushroom sauce	1	serving(s)	135	101.8	251	17.1	9.3	0.5	15.5	6.0	6.7	0.8	0
16637	Shrimp creole with rice	1	cup(s)	243	176.6	309	27.0	27.7	1.2	9.2	1.7	3.6	2.9	—
497	Spaghetti and meatballs with tomato sauce, prepared	1	cup(s)	248	174.0	330	19.0	39.0	2.7	12.0	3.9	4.4	2.2	—
28585	Spicy Thai noodles (pad Thai)	8	ounce(s)	227	73.3	221	8.9	35.7	3.0	6.4	0.8	3.3	1.8	—
33073	Stir fried pork and vegetables with rice	1	cup(s)	235	173.6	348	15.4	33.5	1.9	16.3	5.6	6.9	2.6	0
28588	Stuffed shells	2½	item(s)	249	157.5	243	15.0	28.0	2.5	8.1	3.1	3.0	1.3	
16821	Sushi with egg in seaweed	6	piece(s)	156	116.5	190	8.9	20.5	0.3	7.9	2.2	3.2	1.5	
16819	Sushi with vegetables and fish	6	piece(s)	156	101.6	218	8.4	43.7	1.7	0.6	0.2	0.1	0.2	
16820	Sushi with vegetables in seaweed	6	piece(s)	156	110.3	183	3.4	40.6	0.8	0.4	0.1	0.1	0.1	
25266	Sweet and sour pork	¾	cup(s)	249	205.9	265	29.2	17.1	1.0	8.1	2.6	3.5	1.5	0
16824	Tabouli, tabbouleh, or tabuli	1	cup(s)	160	123.7	198	2.6	15.9	3.7	14.9	2.0	10.9	1.6	—
25276	Three bean salad	½	cup(s)	99	82.2	95	1.9	9.7	2.6	5.9	0.8	1.4	3.5	0
160	Tuna salad	½	cup(s)	103	64.7	192	16.4	9.6	0	9.5	1.6	3.0	4.2	
25241	Turkey and noodles	1	cup(s)	319	228.5	270	24.0	21.2	1.0	9.2	2.4	3.5	2.3	
16794	Vegetable egg roll	2	item(s)	128	89.8	201	5.1	19.5	1.7	11.6	2.5	5.7	2.6	
16818	Vegetable sushi, no fish	6	piece(s)	156	99.0	226	4.8	49.9	2.0	0.4	0.1	0.1	0.1	
	Sandwiches													
1744	Bacon, lettuce, and tomato with mayonnaise	1	item(s)	164	97.2	341	11.6	34.2	2.3	17.6	3.8	5.5	6.7	—
30287	Bologna and cheese with margarine	1	item(s)	111	45.6	345	13.4	29.3	1.2	19.3	8.1	7.0	2.4	—
30286	Bologna with margarine	1	item(s)	83	33.6	251	8.1	27.3	1.2	12.1	3.7	5.0	2.1	—
16546	Cheese	1	item(s)	83	31.0	261	9.1	27.6	1.2	12.7	5.4	4.2	2.1	—
8789	Cheeseburger, large, plain	1	item(s)	185	78.9	564	32.0	38.5	2.6	31.5	12.5	10.2	1.0	1.8
8624	Cheeseburger, large, with bacon, vegetables, and condiments	1	item(s)	195	91.4	550	30.8	36.8	2.5	30.9	11.9	10.6	1.3	1.5
1745	Club with bacon, chicken, tomato, lettuce, and mayonnaise	1	item(s)	246	137.5	546	31.0	48.9	3.0	24.5	5.3	7.5	9.4	—
1908	Cold cut submarine with cheese and vegetables	1	item(s)	228	131.8	456	21.8	51.0	2.0	18.6	6.8	8.2	2.3	—
30247	Corned beef	1	item(s)	130	74.9	265	18.2	25.3	1.6	9.6	3.6	3.5	1.0	—
25283	Egg salad	1	item(s)	126	72.1	278	10.7	28.0	1.4	13.5	2.9	4.2	5.0	—
16686	Fried egg	1	item(s)	96	49.7	226	10.0	26.2	1.2	8.6	2.3	3.2	1.9	—
16547	Grilled cheese	1	item(s)	83	27.5	291	9.2	27.9	1.2	15.8	6.0	5.7	3.0	—
16659	Gyro with onion and tomato	1	item(s)	105	68.3	163	12.0	20.0	1.1	3.5	1.3	1.3	0.5	—
1906	Ham and cheese	1	item(s)	146	74.2	352	20.7	33.3	2.0	15.5	6.4	6.7	1.4	—
31890	Ham with mayonnaise	1	item(s)	112	56.3	271	13.0	27.9	1.9	11.6	2.8	4.0	4.0	—
756	Hamburger, double patty, large, with condiments and vegetables	1	item(s)	226	121.5	540	34.3	40.3	—	26.6	10.5	10.3	2.8	—
8793	Hamburger, large, plain	1	item(s)	137	57.7	426	22.6	31.7	1.5	22.9	8.4	9.9	2.1	—
8795	Hamburger, large, with vegetables and condiments	1	item(s)	218	121.4	512	25.8	40.0	3.1	27.4	10.4	11.4	2.2	—
25134	Hot chicken salad	1	item(s)	98	48.4	242	15.2	23.8	1.3	9.2	2.9	2.5	3.0	—
25133	Hot turkey salad	1	item(s)	98	50.1	224	15.6	23.8	1.3	6.9	2.3	1.6	2.5	—
1411	Hotdog with bun, plain	1	item(s)	98	52.9	242	10.4	18.0	1.6	14.5	5.1	6.9	1.7	—
30249	Pastrami	1	item(s)	134	71.2	328	13.4	27.8	1.6	17.7	6.1	8.3	1.2	—
16701	Peanut butter	1	item(s)	93	23.6	345	12.2	37.6	3.3	17.4	3.4	7.7	5.2	—
30306	Peanut butter and jelly	1	item(s)	93	24.2	330	10.3	41.9	2.9	14.7	2.9	6.5	4.4	—
1909	Roast beef submarine with mayonnaise and vegetables	1	item(s)	216	127.4	410	28.6	44.3	—	13.0	7.1	1.8	2.6	—
1910	Roast beef, plain	1	item(s)	139	67.6	346	21.5	33.4	1.2	13.8	3.6	6.8	1.7	—
1907	Steak with mayonnaise and vegetables	1	item(s)	204	104.2	459	30.3	52.0	2.3	14.1	3.8	5.3	3.3	—
25288	Tuna salad	1	item(s)	179	102.2	415	24.5	28.4	1.6	22.4	3.5	6.2	11.4	—
30283	Turkey submarine with cheese, lettuce, tomato, and mayonnaise	1	item(s)	277	168.0	529	30.4	49.4	3.0	22.8	6.8	6.0	8.6	—
31891	Turkey with mayonnaise	1	item(s)	143	74.5	329	28.7	26.4	1.3	11.2	2.6	2.6	4.8	—
	Soups													
25296	Bean	1	cup(s)	301	253.1	191	13.8	29.0	6.5	2.3	0.7	0.8	0.5	0
711	Bean with pork, condensed, prepared with water	1	cup(s)	253	215.9	159	7.3	21.0	7.3	5.5	1.4	2.0	1.7	—

PAGE KEY: A-6 = Breads/Baked Goods A-12 = Cereal/Rice/Pasta A-16 = Fruit A-20 = Vegetables/Legumes A-30 = Nuts/Seeds A-32 = Vegetarian A-34 = Dairy A-42 = Eggs A-42 = Seafood A-44 = Meats A-48 = Poultry
A-50 = Processed Meats A-50 = Beverages A-54 = Fats/Oils A-56 = Sweets A-58 = Spices/Condiments/Sauces A-62 = Mixed Foods/Soups/Sandwiches A-66 = Fast Food A-86 = Convenience A-88 = Baby Foods

Chol (mg)	Calc (mg)	Iron (mg)	Magn (mg)	Pota (mg)	Sodi (mg)	Zinc (mg)	Vit A (µg)	Thia (mg)	Vit E (mg α)	Ribo (mg)	Niac (mg)	Vit B_6 (mg)	Fola (µg)	Vit C (mg)	Vit B_{12} (µg)	Sele (µg)
85	24	0.81	18.8	317.5	661.3	0.38	40.0	0.09	—	0.07	1.11	0.17	8.8	12.5	0	5.1
60	74	1.94	23.8	314.5	360.5	3.45	—	0.10	0.00	0.27	3.95	0.13	20.8	0.7	1.6	17.4
180	102	4.68	63.2	413.1	330.5	1.72	94.8	0.29	2.06	0.11	4.75	0.21	75.3	12.9	1.2	49.3
89	124	3.70	—	665.0	1009.0	—	81.5	0.25	—	0.30	4.00	—	—	22.0	—	22.0
37	31	1.56	49.4	181.3	591.6	1.05	—	0.18	0.35	0.13	1.82	0.17	44.6	22.6	0.1	3.2
46	38	2.71	33.0	396.9	569.5	2.08	—	0.51	0.38	0.20	5.07	0.30	103.0	18.8	0.4	22.8
30	188	2.26	49.4	403.0	471.5	1.41	—	0.26	0.00	0.26	3.83	0.24	80.0	17.9	0.2	28.9
214	45	1.84	18.7	135.7	463.3	0.98	106.1	0.13	0.67	0.28	1.35	0.13	62.4	1.9	0.7	20.3
11	23	2.15	25.0	202.8	340.1	0.78	45.2	0.26	0.24	0.07	2.76	0.14	76.4	3.6	0.3	13.9
0	20	1.54	18.7	96.7	152.9	0.68	25.0	0.19	0.12	0.03	1.86	0.13	73.3	2.3	0	9.8
74	40	1.76	35.6	619.9	621.8	2.53	—	0.81	0.20	0.37	6.69	0.66	14.7	11.9	0.7	49.6
0	30	1.21	35.2	249.6	796.8	0.48	54.4	0.07	2.43	0.04	1.11	0.11	30.4	26.1	0	0.5
0	26	0.96	15.5	144.8	224.2	0.30	—	0.02	0.88	0.04	0.26	0.04	32.2	10.0	0	2.7
13	17	1.02	19.5	182.5	412.1	0.57	24.6	0.03	—	0.07	6.86	0.08	8.2	2.3	1.2	42.2
77	69	2.56	33.2	400.8	577.1	2.51	—	0.23	0.28	0.30	6.41	0.29	61.4	1.4	1.1	33.4
60	29	1.65	17.9	193.3	549.1	0.48	25.6	0.15	1.28	0.20	1.59	0.09	46.1	5.5	0.2	11.3
0	23	2.38	21.8	157.6	369.7	0.82	48.4	0.28	0.15	0.05	2.44	0.12	85.8	3.7	0	8.1
21	79	2.36	27.9	351.0	944.6	1.08	44.3	0.32	1.16	0.24	4.36	0.21	73.8	9.7	0.2	27.1
40	258	2.38	24.4	215.3	941.3	1.88	102.1	0.31	0.55	0.35	2.97	0.14	59.9	0.2	0.8	20.4
17	100	2.24	16.6	138.6	579.3	1.02	44.8	0.29	0.49	0.21	2.92	0.12	58.1	0.2	0.5	16.0
22	233	2.04	19.9	127.0	733.7	1.24	97.1	0.25	0.47	0.30	2.25	0.05	58.1	0	0.3	13.1
104	309	4.47	44.4	401.5	986.1	5.75	0	0.35	—	0.77	8.26	0.49	109.2	0	2.8	38.9
98	267	4.03	44.9	464.1	1314.3	5.20	0	0.33	—	0.67	8.25	0.47	95.6	1.4	2.4	6.6
71	157	4.57	46.7	464.9	1087.3	1.82	41.8	0.54	1.52	0.40	12.82	0.61	118.1	6.4	0.4	42.3
36	189	2.50	68.4	394.4	1650.7	2.57	70.7	1.00	—	0.79	5.49	0.13	86.6	12.3	1.1	30.8
46	81	3.04	19.5	127.4	1206.4	2.26	2.6	0.23	0.20	0.24	3.42	0.11	59.8	0.3	0.9	31.2
219	85	2.25	18.8	159.0	423.1	0.87	—	0.27	0.12	0.43	2.06	0.15	73.6	0.9	0.6	29.9
206	104	2.79	17.3	117.1	438.7	0.92	89.3	0.26	0.66	0.40	2.26	0.10	79.7	0	0.6	24.2
22	235	2.05	19.9	128.7	763.6	1.26	129.5	0.19	0.72	0.28	2.05	0.05	38.2	0	0.2	13.2
28	47	1.77	22.1	218.4	235.2	2.33	9.5	0.23	0.26	0.20	3.12	0.13	44.1	3.2	0.9	18.3
58	130	3.24	16.1	290.5	770.9	1.37	96.4	0.30	0.29	0.48	2.68	0.20	75.9	2.8	0.5	23.1
34	91	2.47	23.5	210.6	1097.6	1.13	5.6	0.57	0.50	0.26	3.80	0.25	60.5	2.2	0.2	20.2
122	102	5.85	49.7	569.5	791.0	5.67	0	0.36	—	0.38	7.57	0.54	76.8	1.1	4.1	25.5
71	74	3.57	27.4	267.2	474.0	4.11	0	0.28	—	0.28	6.24	0.23	60.3	0	2.1	27.1
87	96	4.92	43.6	479.6	824.0	4.88	0	0.41	—	0.37	7.28	0.32	82.8	2.6	2.4	33.6
39	115	1.88	20.0	172.4	505.1	1.19	—	0.23	0.28	0.22	4.84	0.19	46.4	0.5	0.2	19.9
37	114	1.99	21.3	189.0	494.5	1.07	—	0.22	0.28	0.20	4.26	0.22	46.4	0.5	0.2	23.4
44	24	2.31	12.7	143.1	670.3	1.98	0	0.23	—	0.27	3.64	0.04	48.0	0.1	0.5	26.0
51	80	3.02	22.8	182.2	1364.1	2.70	2.7	0.28	0.26	0.26	4.97	0.14	60.3	0.3	1.0	14.3
0	110	2.92	66.0	226.0	580.3	1.32	0	0.31	2.39	0.23	6.72	0.18	92.1	0	0	13.2
0	94	2.50	56.7	198.1	492.9	1.12	—	0.26	2.01	0.20	5.66	0.15	78.1	0.1	0	11.2
73	41	2.80	67.0	330.5	844.6	4.38	30.2	0.41	—	0.41	5.96	0.32	71.3	5.6	1.8	25.7
51	54	4.22	30.6	315.5	792.3	3.39	11.1	0.37	—	0.30	5.86	0.26	57.0	2.1	1.2	29.2
73	92	5.16	49.0	524.3	797.6	4.52	0	0.40	—	0.36	7.30	0.36	89.8	5.5	1.6	42.0
59	78	2.97	35.9	316.3	724.7	1.02	—	0.27	0.34	0.26	12.07	0.46	60.8	1.8	2.4	76.9
64	307	4.59	49.9	534.6	1759.0	2.74	74.8	0.49	1.19	0.65	3.91	0.31	121.9	10.5	0.6	42.1
67	100	3.46	34.3	304.6	564.9	3.00	5.7	0.28	0.74	0.32	6.84	0.46	64.4	0	0.3	40
5	79	3.05	61.8	588.8	689.0	1.41	—	0.27	0.02	0.15	3.63	0.23	140.1	3.6	0.2	7.9
3	78	1.89	43.0	371.9	883.0	0.96	43.0	0.08	1.08	0.03	0.52	0.03	30.4	1.5	0	7.8

Appendix A

DA+ Code	Food Description	Quantity	Measure	Wt (g)	H₂O (g)	Ener (kcal)	Prot (g)	Carb (g)	Fiber (g)	Fat (g)	Sat	Mono	Poly	*Trans*
											Fat Breakdown (g)			

Mixed Foods, Soups, Sandwiches—continued

DA+ Code	Food Description	Quantity	Measure	Wt (g)	H₂O (g)	Ener (kcal)	Prot (g)	Carb (g)	Fiber (g)	Fat (g)	Sat	Mono	Poly	*Trans*
713	Beef noodle, condensed, prepared with water	1	cup(s)	244	224.9	83	4.7	8.7	0.7	3.0	1.1	1.2	0.5	—
825	Cheese, condensed, prepared with milk	1	cup(s)	251	206.9	231	9.5	16.2	1.0	14.6	9.1	4.1	0.5	—
826	Chicken broth, condensed, prepared with water	1	cup(s)	244	234.1	39	4.9	0.9	0	1.4	0.4	0.6	0.3	—
25297	Chicken noodle soup	1	cup(s)	286	258.4	117	10.8	10.9	0.9	2.9	0.8	1.1	0.7	—
827	Chicken noodle, condensed, prepared with water	1	cup(s)	241	226.1	60	3.1	7.1	0.5	2.3	0.6	1.0	0.6	0
724	Chicken noodle, dehydrated, prepared with water	1	cup(s)	252	237.3	58	2.1	9.2	0.3	1.4	0.3	0.5	0.4	—
823	Cream of asparagus, condensed, prepared with milk	1	cup(s)	248	213.3	161	6.3	16.4	0.7	8.2	3.3	2.1	2.2	—
824	Cream of celery, condensed, prepared with milk	1	cup(s)	248	214.4	164	5.7	14.5	0.7	9.7	3.9	2.5	2.7	—
708	Cream of chicken, condensed, prepared with milk	1	cup(s)	248	210.4	191	7.5	15.0	0.2	11.5	4.6	4.5	1.6	—
715	Cream of chicken, condensed, prepared with water	1	cup(s)	244	221.1	117	3.4	9.3	0.2	7.4	2.1	3.3	1.5	—
709	Cream of mushroom, condensed, prepared with milk	1	cup(s)	248	215.0	166	6.2	14.0	0	9.6	3.3	2.0	1.8	0.1
716	Cream of mushroom, condensed, prepared with water	1	cup(s)	244	224.6	102	1.9	8.0	0	7.0	1.6	1.3	1.7	—
25298	Cream of vegetable	1	cup(s)	285	250.7	165	7.2	15.2	1.9	8.6	1.6	4.6	1.9	—
16689	Egg drop	1	cup(s)	244	228.9	73	7.5	1.1	0	3.8	1.1	1.5	0.6	—
25138	Golden squash	1	cup(s)	258	223.9	145	7.6	20.4	0.4	4.1	0.8	2.2	0.9	—
16663	Hot and sour	1	cup(s)	244	209.7	161	15.0	5.4	0.5	7.9	2.7	3.4	1.1	—
28054	Lentil chowder	1	cup(s)	244	202.8	153	11.4	27.7	12.6	0.5	0.1	0.1	0.2	0
28560	Macaroni and bean	1	cup(s)	246	138.8	146	5.8	22.9	5.1	3.7	0.5	2.2	0.6	0
714	Manhattan clam chowder, condensed, prepared with water	1	cup(s)	244	225.1	73	2.1	11.6	1.5	2.1	0.4	0.4	1.2	—
28561	Minestrone	1	cup(s)	241	185.4	103	4.5	16.8	4.8	2.3	0.3	1.4	0.4	0
717	Minestrone, condensed, prepared with water	1	cup(s)	241	220.1	82	4.3	11.2	1.0	2.5	0.6	0.7	1.1	—
28038	Mushroom and wild rice	1	cup(s)	244	199.7	86	4.7	13.2	1.7	0.3	0	0	0.2	0
828	New England clam chowder, condensed, prepared with milk	1	cup(s)	248	212.2	151	8.0	18.4	0.7	5.0	2.1	0.7	0.6	0
28036	New England style clam chowder	1	cup(s)	244	227.5	61	3.8	8.8	1.8	0.2	0.1	0	0	0
28566	Old country pasta	1	cup(s)	252	183.3	146	6.5	18.3	3.6	4.5	2.0	2.4	0.9	0
725	Onion, dehydrated, prepared with water	1	cup(s)	246	235.7	30	0.8	6.8	0.7	0	0	0	0	—
16667	Shrimp gumbo	1	cup(s)	244	207.2	166	9.5	18.2	2.4	6.7	1.3	2.9	2.0	—
28037	Southwestern corn chowder	1	cup(s)	244	217.8	98	4.9	17.0	2.4	0.5	0.1	0.1	0.2	0
30282	Soybean (miso)	1	cup(s)	240	218.6	84	6.0	8.0	1.9	3.4	0.6	1.1	1.4	—
25140	Split pea	1	cup(s)	165	119.7	72	4.5	16.0	1.6	0.3	0.1	0	0.2	0
718	Split pea with ham, condensed, prepared with water	1	cup(s)	253	206.9	190	10.3	28.0	2.3	4.4	1.8	1.8	0.6	—
726	Tomato vegetable, dehydrated, prepared with water	1	cup(s)	253	238.4	56	2.0	10.2	0.8	0.9	0.4	0.3	0.1	0
710	Tomato, condensed, prepared with milk	1	cup(s)	248	213.4	136	6.2	22.0	1.5	3.2	1.8	0.9	0.3	—
719	Tomato, condensed, prepared with water	1	cup(s)	244	223.0	73	1.9	16.0	1.5	0.7	0.2	0.2	0.2	—
28595	Turkey noodle	1	cup(s)	244	216.9	114	8.1	15.1	1.9	2.4	0.3	1.1	0.7	—
28051	Turkey vegetable	1	cup(s)	244	220.8	96	12.2	6.6	2.0	1.1	0.3	0.2	0.3	—
25141	Vegetable	1	cup(s)	252	228.1	82	5.2	16.5	4.5	0.3	0	0	0.1	0
720	Vegetable beef, condensed, prepared with water	1	cup(s)	244	224.0	76	5.4	9.9	2.0	1.9	0.8	0.8	0.1	—
28598	Vegetable gumbo	1	cup(s)	252	184.5	170	4.4	28.9	3.6	4.7	0.7	3.2	0.5	0
721	Vegetarian vegetable, condensed, prepared with water	1	cup(s)	241	222.7	67	2.1	11.8	0.7	1.9	0.3	0.8	0.7	—

Fast Food

Arby's

DA+ Code	Food Description	Quantity	Measure	Wt (g)	H₂O (g)	Ener (kcal)	Prot (g)	Carb (g)	Fiber (g)	Fat (g)	Sat	Mono	Poly	*Trans*
36094	Au jus sauce	1	serving(s)	85	—	43	1.0	7.0	0	1.3	0.4	—	—	0.4
751	Beef 'n cheddar sandwich	1	item(s)	195	—	445	22.0	44.0	2.0	21.0	6.0	—	—	1.0
9279	Cheddar curly fries	1	serving(s)	198	—	631	8.0	73.0	7.0	37.4	6.8	—	—	5.7

PAGE KEY: A-6 = Breads/Baked Goods A-12 = Cereal/Rice/Pasta A-16 = Fruit A-20 = Vegetables/Legumes A-30 = Nuts/Seeds A-32 = Vegetarian A-34 = Dairy A-42 = Eggs A-42 = Seafood A-44 = Meats A-48 = Poultry
A-50 = Processed Meats A-50 = Beverages A-54 = Fats/Oils A-56 = Sweets A-58 = Spices/Condiments/Sauces A-62 = Mixed Foods/Soups/Sandwiches A-66 = Fast Food A-86 = Convenience A-88 = Baby Foods

Chol (mg)	Calc (mg)	Iron (mg)	Magn (mg)	Pota (mg)	Sodi (mg)	Zinc (mg)	Vit A (μg)	Thia (mg)	Vit E (mg α)	Ribo (mg)	Niac (mg)	Vit B$_6$ (mg)	Fola (μg)	Vit C (mg)	Vit B$_{12}$ (μg)	Sele (μg)
5	20	1.07	7.3	97.6	929.6	1.51	12.2	0.06	1.22	0.05	1.03	0.03	19.5	0.5	0.2	7.3
48	289	0.80	20.1	341.4	1019.1	0.67	358.9	0.06	—	0.33	0.50	0.07	10.0	1.3	0.4	7.0
0	10	0.51	2.4	209.8	775.9	0.24	0	0.01	0.04	0.07	3.34	0.02	4.9	0	0.2	0
24	25	1.38	16.4	340.1	774.5	0.77	—	0.15	0.02	0.16	5.57	0.13	37.2	1.8	0.3	10.2
12	14	1.59	9.6	53.0	638.7	0.38	26.5	0.13	0.07	0.10	1.30	0.04	19.3	0	0	11.6
10	5	0.50	7.6	32.8	577.1	0.20	2.5	0.20	0.12	0.07	1.08	0.02	17.6	0	0.1	9.6
22	174	0.86	19.8	359.6	1041.6	0.91	62.0	0.10	—	0.27	0.88	0.06	29.8	4.0	0.5	8.0
32	186	0.69	22.3	310.0	1009.4	0.19	114.1	0.07	—	0.24	0.43	0.06	7.4	1.5	0.5	4.7
27	181	0.67	17.4	272.8	1046.6	0.67	178.6	0.07	—	0.25	0.92	0.06	7.4	1.2	0.5	8.0
10	34	0.61	2.4	87.8	985.8	0.63	163.5	0.02	—	0.06	0.82	0.01	2.4	0.2	0.1	7.0
10	164	1.36	19.8	267.8	823.4	0.79	81.8	0.10	1.01	0.29	0.62	0.05	7.4	0.2	0.6	6.0
0	17	1.31	4.9	73.2	775.9	0.24	9.8	0.05	0.97	0.05	0.50	0.00	2.4	0	0	2.9
1	80	1.20	17.5	340.7	787.9	0.56	—	0.12	1.05	0.18	3.32	0.12	39.5	10.7	0.3	4.3
102	22	0.75	4.9	219.6	729.6	0.48	41.5	0.02	0.29	0.19	3.02	0.05	14.6	0	0.5	7.6
4	262	0.78	42.4	542.2	515.6	0.88	—	0.16	0.52	0.30	1.14	0.16	32.7	12.5	0.7	6.0
34	29	1.24	19.5	373.3	1561.6	1.43	—	0.26	0.12	0.24	4.96	0.20	14.6	0.5	0.4	19.3
0	48	4.38	59.3	626.2	26.7	1.57	—	0.24	0.06	0.12	1.87	0.32	176.3	16.1	0	3.5
0	59	1.90	32.4	275.9	531.0	0.51	—	0.16	0.37	0.12	1.44	0.10	58.2	9.1	0	8.8
2	27	1.56	9.8	180.6	551.4	0.87	48.8	0.02	1.22	0.03	0.77	0.09	9.8	3.9	3.9	9.0
0	62	1.70	29.9	287.5	442.7	0.42	—	0.09	0.23	0.09	0.70	0.07	47.3	13.3	0	3.5
2	34	0.91	7.2	313.3	911.0	0.74	118.1	0.05	—	0.04	0.94	0.09	36.2	1.2	0	8.0
0	30	1.38	27.3	376.9	283.8	1.00	—	0.06	0.07	0.23	3.21	0.14	18.1	3.7	0.1	4.7
17	169	3.00	29.8	456.3	887.8	0.99	91.8	0.20	0.54	0.43	1.96	0.17	22.3	5.2	11.9	10.9
3	89	1.30	29.2	503.7	256.8	0.52	—	0.06	0.02	0.10	1.30	0.15	24.2	11.8	3.0	3.8
5	57	2.43	51.9	500.4	355.7	0.78	—	0.21	0.01	0.14	2.64	0.20	80.0	22.0	0.1	10.3
0	22	0.12	9.8	76.3	851.2	0.12	0	0.03	0.02	0.03	0.15	0.06	0	0.2	0	0.5
51	105	2.85	48.8	461.2	441.6	0.90	80.5	0.18	1.90	0.12	2.52	0.19	85.4	17.6	0.3	14.9
1	83	1.03	26.3	434.4	217.4	0.57	—	0.08	0.09	0.13	1.81	0.21	32.1	39.6	0.2	1.7
0	65	1.87	36.0	362.4	988.8	0.86	232.8	0.06	0.96	0.16	2.61	0.15	57.6	4.6	0.2	1.0
0	28	1.26	29.8	328.1	602.4	0.52	—	0.10	0.00	0.07	1.50	0.16	49.5	8.1	0	0.7
8	23	2.27	48.1	399.7	1006.9	1.31	22.8	0.14	—	0.07	1.47	0.06	2.5	1.5	0.3	8.0
0	20	0.60	10.1	169.5	334.0	0.20	10.1	0.06	0.43	0.09	1.26	0.06	12.7	3.0	0.1	2.0
10	166	1.36	29.8	466.2	711.8	0.84	94.2	0.09	0.44	0.31	1.35	0.15	5.0	15.6	0.6	9.2
0	20	1.31	17.1	273.3	663.7	0.29	24.4	0.04	0.41	0.07	1.23	0.10	0	15.4	0	6.1
26	28	1.40	24.1	223.8	395.9	0.75	—	0.21	0.01	0.12	2.85	0.15	45.0	7.1	0.1	13.7
21	38	1.48	25.0	423.4	348.7	0.99	—	0.09	0.01	0.09	3.64	0.27	23.9	10.6	0.2	10.3
0	40	2.08	39.1	681.5	670.3	0.67	—	0.16	0.00	0.09	2.70	0.26	36.3	22.4	0	2.2
5	20	1.09	7.3	168.4	773.5	1.51	190.3	0.03	0.58	0.04	1.00	0.07	9.8	2.4	0.3	2.7
0	56	1.85	38.9	360.2	518.4	0.64	—	0.18	0.64	0.08	1.77	0.17	57.6	21.9	0	4.1
0	24	1.06	7.2	207.3	814.6	0.45	171.1	0.05	1.39	0.04	0.90	0.05	9.6	1.4	0	4.3
0	0	—	—	—	1510.0	—	—	—	—	—	—	—	—	—	—	—
51	80	3.96	—	—	1274.0	—	—	—	—	—	—	—	—	1.8	—	—
0	80	3.24	—	—	1476.0	—	—	—	—	—	—	—	—	9.6	—	—

TABLE A-1 **Table of Food Composition** *(continued)* (Computer code is for Cengage Diet Analysis program) (For purposes of calculations, use "0" for t, <1, <.1, <.01, etc.)

DA+ Code	Food Description	Quantity	Measure	Wt (g)	H₂O (g)	Ener (kcal)	Prot (g)	Carb (g)	Fiber (g)	Fat (g)	Fat Breakdown (g)			
											Sat	Mono	Poly	*Trans*
Fast Food—continued														
34770	Chicken breast fillet sandwich, grilled	1	item(s)	233	—	414	32.0	36.0	3.0	17.0	3.0	—	—	0
36131	Chocolate shake, regular	1	serving(s)	397	—	507	13.0	83.0	0	13.0	8.0	—	—	0
36045	Curly fries, large size	1	serving(s)	198	—	631	8.0	73.0	7.0	37.0	7.0	—	—	6.0
36044	Curly fries, medium size	1	serving(s)	128	—	406	5.0	47.0	5.0	24.0	4.0	—	—	4.0
752	Ham 'n cheese sandwich	1	item(s)	167	—	304	23.0	35.0	1.0	7.0	2.0	—	—	0
36048	Homestyle fries, large size	1	serving(s)	213	—	566	6.0	82.0	6.0	37.0	7.0	—	—	5.0
36047	Homestyle fries, medium size	1	serving(s)	142	—	377	4.0	55.0	4.0	25.0	4.0	—	—	4.0
33465	Homestyle fries, small size	1	serving(s)	113	—	302	3.0	44.0	3.0	20.0	4.0	—	—	3.0
9249	Junior roast beef sandwich	1	item(s)	125	—	272	16.0	34.0	2.0	10.0	4.0	—	—	0
9251	Large roast beef sandwich	1	item(s)	281	—	547	42.0	41.0	3.0	28.0	12.0	—	—	2.0
39640	Market Fresh chicken salad with pecans sandwich	1	item(s)	322	—	769	30.0	79.0	9.0	39.0	10.0	—	—	0
39641	Market Fresh Martha's Vineyard salad, without dressing	1	serving(s)	330	—	277	26.0	24.0	5.0	8.0	4.0	—	—	0
34769	Market Fresh roast turkey and Swiss sandwich	1	serving(s)	359	—	725	45.0	75.0	5.0	30.0	8.0	—	—	1.0
9267	Market Fresh roast turkey ranch and bacon sandwich	1	serving(s)	382	—	834	49.0	75.0	5.0	38.0	11.0	—	—	1.0
39642	Market Fresh Santa Fe salad, without dressing	1	serving(s)	372	—	499	30.0	42.0	7.0	23.0	8.0	—	—	2.0
39650	Market Fresh Southwest chicken wrap	1	serving(s)	251	—	567	36.0	42.0	4.0	29.0	9.0	—	—	1.0
37021	Market Fresh Ultimate BLT sandwich	1	item(s)	294	—	779	23.0	75.0	6.0	45.0	11.0	—	—	1.0
750	Roast beef sandwich, regular	1	item(s)	154	—	320	21.0	34.0	2.0	14.0	5.0	—	—	1.0
36132	Strawberry shake, regular	1	serving(s)	397	—	498	13.0	81.0	0	13.0	8.0	—	—	0
2009	Super roast beef sandwich	1	item(s)	198	—	398	21.0	40.0	2.0	19.0	6.0	—	—	1.0
36130	Vanilla shake, regular	1	serving(s)	369	—	437	13.0	66.0	0	13.0	8.0	—	—	0
Auntie Anne's														
35371	Cheese dipping sauce	1	serving(s)	35	—	100	3.0	4.0	0	8.0	4.0	—	—	0
35353	Cinnamon sugar soft pretzel	1	item(s)	120	—	350	9.0	74.0	2.0	2.0	0	—	—	0
35354	Cinnamon sugar soft pretzel with butter	1	item(s)	120	—	450	8.0	83.0	3.0	9.0	5.0	—	—	0
35372	Marinara dipping sauce	1	serving(s)	35	—	10	0	4.0	0	0	0	0	0	0
35357	Original soft pretzel	1	serving(s)	120	—	340	10.0	72.0	3.0	1.0	0	—	—	0
35358	Original soft pretzel with butter	1	item(s)	120	—	370	10.0	72.0	3.0	4.0	2.0	—	—	0
35359	Parmesan herb soft pretzel	1	item(s)	120	—	390	11.0	74.0	4.0	5.0	2.5	—	—	—
35360	Parmesan herb soft pretzel with butter	1	item(s)	120	—	440	10.0	72.0	9.0	13.0	7.0	—	—	—
35361	Sesame soft pretzel	1	item(s)	120	—	350	11.0	63.0	3.0	6.0	1.0	—	—	0
35362	Sesame soft pretzel with butter	1	item(s)	120	—	410	12.0	64.0	7.0	12.0	4.0	—	—	0
35364	Sour cream and onion soft pretzel	1	item(s)	120	—	310	9.0	66.0	2.0	1.0	0	—	—	0
35366	Sour cream and onion soft pretzel with butter	1	item(s)	120	—	340	9.0	66.0	2.0	5.0	3.0	—	—	0
35373	Sweet mustard dipping sauce	1	serving(s)	35	—	60	0.5	8.0	0	1.5	1.0	—	—	0
35367	Whole wheat soft pretzel	1	item(s)	120	—	350	11.0	72.0	7.0	1.5	0	—	—	0
35368	Whole wheat soft pretzel with butter	1	item(s)	120	—	370	11.0	72.0	7.0	4.5	1.5	—	—	0
Boston Market														
34978	Butternut squash	¾	cup(s)	143	—	140	2.0	25.0	2.0	4.5	3.0	—	—	0
35006	Caesar side salad	1	serving(s)	71	—	40	3.0	3.0	1.0	20.0	2.0	—	—	1.5
35013	Chicken Carver sandwich with cheese and sauce	1	item(s)	321	—	700	44.0	68.0	3.0	29.0	7.0	—	—	0
34979	Chicken gravy	4	ounce(s)	113	—	15	1.0	4.0	0	0.5	0	—	—	0
35053	Chicken noodle soup	¾	cup(s)	283	—	180	13.0	16.0	1.0	7.0	2.0	—	—	0
34973	Chicken pot pie	1	item(s)	425	—	800	29.0	59.0	4.0	49.0	18.0	—	—	7.0
35054	Chicken tortilla soup with toppings	¾	cup(s)	227	—	340	12.0	24.0	1.0	22.0	7.0	—	—	0
35007	Cole slaw	¾	cup(s)	125	—	170	2.0	21.0	2.0	9.0	2.0	—	—	0
35057	Cornbread	1	item(s)	45	—	130	1.0	21.0	0	3.5	1.0	—	—	1.0
34980	Creamed spinach	¾	cup(s)	191	—	280	9.0	12.0	4.0	23.0	15.0	—	—	0
34998	Fresh vegetable stuffing	1	cup(s)	136	—	190	3.0	25.0	2.0	8.0	1.0	—	—	0
34991	Garlic dill new potatoes	¾	cup(s)	156	—	140	3.0	24.0	3.0	3.0	1.0	—	—	0
34983	Green bean casserole	¾	cup(s)	170	—	60	2.0	9.0	2.0	2.0	1.0	—	—	0
34982	Green beans	¾	cup(s)	91	—	60	2.0	7.0	3.0	3.5	1.5	—	—	0

PAGE KEY: A-6 = Breads/Baked Goods A-12 = Cereal/Rice/Pasta A-16 = Fruit A-20 = Vegetables/Legumes A-30 = Nuts/Seeds A-32 = Vegetarian A-34 = Dairy A-42 = Eggs A-42 = Seafood A-44 = Meats A-48 = Poultry
A-50 = Processed Meats A-50 = Beverages A-54 = Fats/Oils A-56 = Sweets A-58 = Spices/Condiments/Sauces A-62 = Mixed Foods/Soups/Sandwiches A-66 = Fast Food A-86 = Convenience A-88 = Baby Foods

Chol (mg)	Calc (mg)	Iron (mg)	Magn (mg)	Pota (mg)	Sodi (mg)	Zinc (mg)	Vit A (µg)	Thia (mg)	Vit E (mg α)	Ribo (mg)	Niac (mg)	Vit B_6 (mg)	Fola (µg)	Vit C (mg)	Vit B_{12} (µg)	Sele (µg)
9	90	3.06	—	—	913.0	—	—	—	—	—	—	—	—	10.8	—	—
34	510	0.54	—	—	357.0	—	—	—	—	—	—	—	—	5.4	—	—
0	80	3.24	—	—	1476.0	—	—	—	—	—	—	—	—	9.6	—	—
0	50	1.98	—	—	949.0	—	—	—	—	—	—	—	—	6.0	—	—
35	160	2.70	—	—	1420.0	—	—	—	—	—	—	—	—	1.2	—	—
0	50	1.62	—	—	1029.0	—	—	—	—	—	—	—	—	12.6	—	—
0	30	1.08	—	—	686.0	—	—	—	—	—	—	—	—	8.4	—	—
0	30	0.90	—	—	549.0	—	—	—	—	—	—	—	—	6.6	—	—
29	60	3.06	—	—	740.0	—	0	—	—	—	—	—	—	0	—	—
102	70	6.30	—	—	1869.0	—	0	—	—	—	—	—	—	0.6	—	—
74	180	4.32	—	—	1240.0	—	—	—	—	—	—	—	—	30.0	—	—
72	200	1.62	—	—	454.0	—	—	—	—	—	—	—	—	33.6	—	—
91	360	5.22	—	—	1788.0	—	—	—	—	—	—	—	—	10.2	—	—
109	330	5.40	—	—	2258.0	—	—	—	—	—	—	—	—	11.4	—	—
59	420	3.60	—	—	1231.0	—	—	—	—	—	—	—	—	36.6	—	—
88	240	4.50	—	—	1451.0	—	—	—	—	—	—	—	—	7.8	—	—
51	170	4.68	—	—	1571.0	—	—	—	—	—	—	—	—	16.8	—	—
44	60	3.60	—	—	953.0	—	0	—	—	—	—	—	—	0	—	—
34	510	0.72	—	—	363.0	—	—	—	—	—	—	—	—	6.6	—	—
44	70	3.78	—	—	1060.0	—	—	—	—	—	—	—	—	6.0	—	—
34	510	0.36	—	—	350.0	—	—	—	—	—	—	—	—	5.4	—	—
10	100	0.00	—	—	510.0	—	—	—	—	—	—	—	—	0	—	—
0	20	1.98	—	—	410.0	—	0	—	—	—	—	—	—	0	—	—
25	30	2.34	—	—	430.0	—	—	—	—	—	—	—	—	0	—	—
0	0	0.00	—	—	180.0	—	0	—	—	—	—	—	—	0	—	—
0	30	2.34	—	—	900.0	—	0	—	—	—	—	—	—	0	—	—
10	30	2.16	—	—	930.0	—	—	—	—	—	—	—	—	0	—	—
10	80	1.80	—	—	780.0	—	—	—	—	—	—	—	—	1.2	—	—
30	60	1.80	—	—	660.0	—	—	—	—	—	—	—	—	1.2	—	—
0	20	2.88	—	—	840.0	—	0	—	—	—	—	—	—	0	—	—
15	20	2.70	—	—	860.0	—	—	—	—	—	—	—	—	0	—	—
0	30	1.98	—	—	920.0	—	—	—	—	—	—	—	—	0	—	—
10	40	2.16	—	—	930.0	—	—	—	—	—	—	—	—	0	—	—
40	0	0.00	—	—	120.0	—	0	—	—	—	—	—	—	0	—	—
0	30	1.98	—	—	1100.0	—	0	—	—	—	—	—	—	0	—	—
10	30	2.34	—	—	1120.0	—	—	—	—	—	—	—	—	0	—	—
10	59	0.80	—	—	35.0	—	—	—	—	—	—	—	—	22.2	—	—
0	60	0.43	—	—	75.0	—	—	—	—	—	—	—	—	5.4	—	—
90	211	2.85	—	—	1560.0	—	—	—	—	—	—	—	—	15.8	—	—
0	0	0.00	—	—	570.0	—	0	—	—	—	—	—	—	0	—	—
55	0	1.07	—	—	220.0	—	—	—	—	—	—	—	—	1.8	—	—
115	40	4.50	—	—	800.0	—	—	—	—	—	—	—	—	1.2	—	—
45	123	1.32	—	—	1310.0	—	—	—	—	—	—	—	—	18.4	—	—
10	41	0.48	—	—	270.0	—	—	—	—	—	—	—	—	24.5	—	—
5	0	0.71	—	—	220.0	—	0	—	—	—	—	—	—	0	—	—
70	264	2.84	—	—	580.0	—	—	—	—	—	—	—	—	9.5	—	—
0	41	1.48	—	—	580.0	—	—	—	—	—	—	—	—	2.5	—	—
0	0	0.85	—	—	120.0	—	0	—	—	—	—	—	—	14.3	—	—
5	20	0.77	—	—	620.0	—	—	—	—	—	—	—	—	2.4	—	—
0	43	0.38	—	—	180.0	—	—	—	—	—	—	—	—	5.1	—	—

TABLE A-1 **Table of Food Composition** *(continued)*   (Computer code is for Cengage Diet Analysis program)   (For purposes of calculations, use "0" for t, <1, <.1, <.01, etc.)

DA+ Code	Food Description	Quantity	Measure	Wt (g)	H₂0 (g)	Ener (kcal)	Prot (g)	Carb (g)	Fiber (g)	Fat (g)	Sat	Mono	Poly	Trans
											Fat Breakdown (g)			

Fast Food—continued

DA+ Code	Food Description	Quantity	Measure	Wt (g)	H₂0 (g)	Ener (kcal)	Prot (g)	Carb (g)	Fiber (g)	Fat (g)	Sat	Mono	Poly	Trans
34984	Homestyle mashed potatoes	¾	cup(s)	221	—	210	4.0	29.0	3.0	9.0	6.0	—	—	0
34985	Homestyle mashed potatoes and gravy	1	cup(s)	334	—	225	5.0	33.0	3.0	9.5	6.0	—	—	0
34988	Hot cinnamon apples	¾	cup(s)	145	—	210	0	47.0	3.0	3.0	0	—	—	0
34989	Macaroni and cheese	¾	cup(s)	221	—	330	14.0	39.0	1.0	12.0	7.0	—	—	0.5
51193	Market chopped salad with dressing	1	item(s)	563	—	580	11.0	31.0	9.0	48.0	9.0	—	—	1.0
34970	Meatloaf	1	serving(s)	218	—	480	29.0	23.0	2.0	33.0	13.0	—	—	0
39383	Nestle Toll House chocolate chip cookie	1	item(s)	78	—	370	4.0	49.0	2.0	19.0	9.0	—	—	0
34965	Quarter chicken, dark meat, no skin	1	item(s)	134	—	260	30.0	2.0	0	13.0	4.0	—	—	0
34966	Quarter chicken, dark meat, with skin	1	item(s)	149	—	280	31.0	3.0	0	15.0	4.5	—	—	0
34963	Quarter chicken, white meat, no skin or wing	1	item(s)	173	—	250	41.0	4.0	0	8.0	2.5	—	—	0
34964	Quarter chicken, white meat, with skin and wing	1	item(s)	110	—	330	50.0	3.0	0	12.0	4.0	—	—	0
34968	Roasted turkey breast	5	ounce(s)	142	—	180	38.0	0	0	3.0	1.0	—	—	0
35011	Seasonal fresh fruit salad	1	serving(s)	142	—	60	1.0	15.0	1.0	0	0	0	0	0
51192	Spinach with garlic butter sauce	1	serving(s)	170	—	130	5.0	9.0	5.0	9.0	6.0	—	—	0
34969	Spiral sliced holiday ham	8	ounce(s)	227	—	450	40.0	13.0	0	26.0	10.0	—	—	0
35003	Steamed vegetables	1	cup(s)	136	—	50	2.0	8.0	3.0	2.0	0	—	—	0
35005	Sweet corn	¾	cup(s)	176	—	170	6.0	37.0	2.0	4.0	1.0	—	—	0
35004	Sweet potato casserole	¾	cup(s)	198	—	460	4.0	77.0	3.0	17.0	6.0	—	—	0
Burger King														
29731	Biscuit with sausage, egg, and cheese	1	item(s)	191	—	610	20.0	33.0	1.0	45.0	15.0	—	—	1.0
14249	Cheeseburger	1	item(s)	133	—	330	17.0	31.0	1.0	16.0	7.0	—	—	0.5
14251	Chicken sandwich	1	item(s)	219	—	660	24.0	52.0	4.0	40.0	8.0	—	—	2.5
3808	Chicken Tenders, 8 pieces	1	serving(s)	123	—	340	19.0	21.0	0.5	20.0	5.0	—	—	3.0
14259	Chocolate shake, small	1	item(s)	315	—	470	8.0	75.0	1.0	14.0	9.0	—	—	0
29732	Croissanwich with sausage and cheese	1	item(s)	106	37.2	370	14.0	23.0	0.5	25.0	9.0	12.7	3.3	2.0
14261	Croissanwich with sausage, egg, and cheese	1	item(s)	159	71.4	470	19.0	26.0	0.5	32.0	11.0	15.8	6.1	2.5
3809	Double cheeseburger	1	item(s)	189	—	500	30.0	31.0	1.0	29.0	14.0	—	—	1.5
14244	Double Whopper sandwich	1	item(s)	373	—	900	47.0	51.0	3.0	57.0	19.0	—	—	2.0
14245	Double Whopper with cheese sandwich	1	item(s)	398	—	990	52.0	52.0	3.0	64.0	24.0	—	—	2.5
14250	Fish Filet sandwich	1	item(s)	250	—	630	24.0	67.0	4.0	30.0	6.0	—	—	2.5
14255	French fries, medium, salted	1	serving(s)	116	—	360	4.0	41.0	4.0	20.0	4.5	—	—	4.5
14262	French toast sticks, 5 pieces	1	serving(s)	112	37.6	390	6.0	46.0	2.0	20.0	4.5	10.6	2.9	4.5
14248	Hamburger	1	item(s)	121	—	290	15.0	30.0	1.0	12.0	4.5	—	—	0
14263	Hash brown rounds, small	1	serving(s)	75	27.1	230	2.0	23.0	2.0	15.0	4.0	—	—	5.0
14256	Onion rings, medium	1	serving(s)	91	—	320	4.0	40.0	3.0	16.0	4.0	—	—	3.5
39000	Tendercrisp chicken sandwich	1	item(s)	286	—	780	25.0	73.0	4.0	43.0	8.0	—	—	4.0
37514	TenderGrill chicken sandwich	1	item(s)	258	—	450	37.0	53.0	4.0	10.0	2.0	—	—	0
14258	Vanilla shake, small	1	item(s)	296	—	400	8.0	57.0	0	15.0	9.0	—	—	0
1736	Whopper sandwich	1	item(s)	290	—	670	28.0	51.0	3.0	39.0	11.0	—	—	1.5
14243	Whopper with cheese sandwich	1	item(s)	315	—	760	33.0	52.0	3.0	47.0	16.0	—	—	1.5
Carl's Jr														
33962	Carl's bacon Swiss crispy chicken sandwich	1	item(s)	268	—	750	31.0	91.0	—	28.0	28.0	—	—	—
10801	Carl's Catch fish sandwich	1	item(s)	215	—	560	19.0	58.0	2.0	27.0	7.0	—	1.9	—
10862	Carl's Famous Star hamburger	1	item(s)	254	—	590	24.0	50.0	3.0	32.0	9.0	—	—	—
10785	Charbroiled chicken club sandwich	1	item(s)	270	—	550	42.0	43.0	4.0	23.0	7.0	—	2.9	—
10866	Charbroiled chicken salad	1	item(s)	437	—	330	34.0	17.0	5.0	7.0	4.0	—	1.0	—
10855	Charbroiled Santa Fe chicken sandwich	1	item(s)	266	—	610	38.0	43.0	4.0	32.0	8.0	—	—	—
10790	Chicken stars, 6 pieces	1	serving(s)	85	—	260	13.0	14.0	1.0	16.0	4.0	—	1.6	—
34864	Chocolate shake, small	1	serving(s)	595	—	540	15.0	98.0	0	11.0	7.0	—	—	—
10797	Crisscut fries	1	serving(s)	139	—	410	5.0	43.0	4.0	24.0	5.0	—	—	—
10799	Double Western Bacon cheeseburger	1	item(s)	308	—	920	51.0	65.0	2.0	50	21.0	—	6.6	—

Chol (mg)	Calc (mg)	Iron (mg)	Magn (mg)	Pota (mg)	Sodi (mg)	Zinc (mg)	Vit A (µg)	Thia (mg)	Vit E (mg α)	Ribo (mg)	Niac (mg)	Vit B$_6$ (mg)	Fola (µg)	Vit C (mg)	Vit B$_{12}$ (µg)	Sele (µg)
25	51	0.46	—	—	660.0	—	—	—	—	—	—	—	—	19.2	—	—
25	100	0.59	—	—	1230.0	—	—	—	—	—	—	—	—	24.9	—	—
0	16	0.28	—	—	15.0	—	—	—	—	—	—	—	—	0	—	—
30	345	1.65	—	—	1290.0	—	—	—	—	—	—	—	—	0	—	—
10	—	—	—	—	2010.0	—	—	—	—	—	—	—	—	—	—	—
125	140	3.77	—	—	970.0	—	—	—	—	—	—	—	—	1.8	—	—
20	0	1.32	—	—	340.0	—	—	—	—	—	—	—	—	0	—	—
155	0	1.52	—	—	260.0	—	0	—	—	—	—	—	—	0	—	—
155	0	2.14	—	—	660.0	—	0	—	—	—	—	—	—	0	—	—
125	0	0.89	—	—	480.0	—	0	—	—	—	—	—	—	0	—	—
165	0	0.78	—	—	960.0	—	0	—	—	—	—	—	—	0	—	—
70	20	1.80	—	—	620.0	—	0	—	—	—	—	—	—	0	—	—
0	16	0.29	—	—	20.0	—	—	—	—	—	—	—	—	29.5	—	—
20	—	—	—	—	200.0	—	—	—	—	—	—	—	—	—	—	—
140	0	1.73	—	—	2230.0	—	0	—	—	—	—	—	—	0	—	—
0	53	0.46	—	—	45.0	—	—	—	—	—	—	—	—	24.0	—	—
0	0	0.43	—	—	95.0	—	—	—	—	—	—	—	—	5.8	—	—
20	44	1.18	—	—	210.0	—	—	—	—	—	—	—	—	9.8	—	—
210	250	2.70	—	—	1620.0	—	89.9	—	—	—	—	—	—	0	—	—
55	150	2.70	—	—	780.0	—	—	0.24	—	0.31	4.17	—	—	1.2	—	—
70	64	2.89	—	—	1440.0	—	—	0.50	—	0.32	10.29	—	—	0	—	—
55	20	0.72	—	—	960.0	—	—	0.14	—	0.11	10.93	—	—	0	—	—
55	333	0.79	—	—	350.0	—	—	0.11	—	0.61	0.26	—	—	2.7	0	—
50	99	1.78	20.1	217.3	810.0	1.51	—	0.34	1.03	0.33	4.33	—	—	0	0.6	22.2
180	146	2.63	28.6	313.2	1060.0	2.08	—	0.38	1.66	0.51	4.72	0.28	—	0	1.1	38.0
105	250	4.50	—	—	1030.0	—	—	0.26	—	0.44	6.37	—	—	1.2	—	—
175	150	8.07	—	—	1090.0	—	—	0.39	—	0.59	11.05	—	—	9.0	—	—
195	299	8.08	—	—	1520.0	—	—	0.39	—	0.66	11.03	—	—	9.0	—	—
60	101	3.62	—	—	1380.0	—	—	—	—	—	—	—	—	3.6	—	—
0	20	0.71	—	—	590.0	—	0	0.15	—	0.48	2.30	—	—	8.9	—	—
0	60	1.80	21.3	124.3	440.0	0.57	—	0.31	0.98	0.19	2.88	0.05	—	0	0	13.7
40	80	2.70	—	—	560.0	—	—	0.25	—	0.28	4.25	—	—	1.2	—	—
0	0	0.36	—	—	450.0	—	0	0.11	0.83	0.06	1.35	0.17	—	1.2	—	—
0	100	0.00	—	—	460.0	—	0	0.14	—	0.09	2.32	—	—	0	—	—
75	79	4.43	—	—	1730.0	—	—	—	—	—	—	—	—	8.9	—	—
75	57	6.82	—	—	1210.0	—	—	—	—	—	—	—	—	5.7	—	—
60	348	0.00	—	—	240.0	—	—	0.11	—	0.63	0.21	—	—	2.4	0	—
51	100	5.38	—	—	1020.0	—	—	0.38	—	0.43	7.30	—	—	9.0	—	—
115	249	5.38	—	—	1450.0	—	—	0.38	—	0.51	7.28	—	—	9.0	—	—
80	200	5.40	—	—	1900.0	—	—	—	—	—	—	—	—	2.4	—	—
80	150	2.70	—	—	990.0	—	60.0	—	—	—	—	—	—	2.4	—	—
70	100	4.50	—	—	910.0	—	—	—	—	—	—	—	—	6.0	—	—
95	200	3.60	—	—	1330.0	—	—	—	—	—	—	—	—	9.0	—	—
75	200	1.80	—	—	880.0	—	—	—	—	—	—	—	—	30.0	—	—
100	200	3.60	—	—	1440.0	—	—	—	—	—	—	—	—	9.0	—	—
35	19	1.02	—	—	470.0	—	0	—	—	—	—	—	—	0	—	—
45	600	1.08	—	—	360.0	—	0	—	—	—	—	—	—	0	—	—
0	20	1.80	—	—	950.0	—	0	—	—	—	—	—	—	12.0	—	—
155	300	7.20	—	—	1730.0	—	—	—	—	—	—	—	—	1.2	—	—

TABLE A-1 **Table of Food Composition (continued)** (Computer code is for Cengage Diet Analysis program) (For purposes of calculations, use "0" for t, <1, <.1, <.01, etc.)

DA+ Code	Food Description	Quantity	Measure	Wt (g)	H₂O (g)	Ener (kcal)	Prot (g)	Carb (g)	Fiber (g)	Fat (g)	Fat Breakdown (g) Sat	Mono	Poly	Trans
Fast Food—continued														
14238	French fries, small	1	serving(s)	92	—	290	5.0	37.0	3.0	14.0	3.0	—	—	—
10798	French toast dips without syrup, 5 pieces	1	serving(s)	155	—	370	8.0	49.0	0	17.0	5.0	—	1.4	—
10802	Onion rings	1	serving(s)	128	—	440	7.0	53.0	3.0	22.0	5.0	—	0.8	—
34858	Spicy chicken sandwich	1	item(s)	198	—	480	14.0	48.0	2.0	26.0	5.0	—	—	—
34867	Strawberry shake, small	1	serving(s)	595	—	520	14.0	93.0	0	11.0	7.0	—	—	—
10865	Super Star hamburger	1	item(s)	348	—	790	41.0	52.0	3.0	47.0	14.0	—	—	—
38925	The Six Dollar burger	1	item(s)	429	—	1010	40.0	60.0	3.0	66.0	26.0	—	—	—
10818	Vanilla shake, small	1	item(s)	398	—	314	10.0	51.5	0	7.4	4.7	—	—	—
10770	Western Bacon cheeseburger	1	item(s)	225	—	660	32.0	64.0	2.0	30.0	12.0	—	4.8	—
Chick Fil-A														
38746	Biscuit with bacon, egg, and cheese	1	item(s)	163	—	470	18.0	39.0	1.0	26.0	9.0	—	—	3.0
38747	Biscuit with egg	1	item(s)	135	—	350	11.0	38.0	1.0	16.0	4.5	—	—	3.0
38748	Biscuit with egg and cheese	1	item(s)	149	—	400	14.0	38.0	1.0	21.0	7.0	—	—	3.0
38753	Biscuit with gravy	1	item(s)	192	—	330	5.0	43.0	1.0	15.0	4.0	—	—	4.0
38752	Biscuit with sausage, egg, and cheese	1	item(s)	212	—	620	22.0	39.0	2.0	42.0	14.0	—	—	3.0
38771	Carrot and raisin salad	1	item(s)	113	—	170	1.0	28.0	2.0	6.0	1.0	—	—	0
38761	Chargrilled chicken Cool Wrap	1	item(s)	245	—	390	29.0	54.0	3.0	7.0	3.0	—	—	0
38766	Chargrilled chicken garden salad	1	item(s)	275	—	180	22.0	9.0	3.0	6.0	3.0	—	—	0
38758	Chargrilled chicken sandwich	1	item(s)	193	—	270	28.0	33.0	3.0	3.5	1.0	—	—	0
38742	Chicken biscuit	1	item(s)	145	—	420	18.0	44.0	2.0	19.0	4.5	—	—	3.0
38743	Chicken biscuit with cheese	1	item(s)	159	—	470	21.0	45.0	2.0	23.0	8.0	—	—	3.0
38762	Chicken Caesar Cool Wrap	1	item(s)	227	—	460	36.0	52.0	3.0	10.0	6.0	—	—	0
38757	Chicken deluxe sandwich	1	item(s)	208	—	420	28.0	39.0	2.0	16.0	3.5	—	—	0
38764	Chicken salad sandwich on wheat bun	1	item(s)	153	—	350	20.0	32.0	5.0	15.0	3.0	—	—	0
38756	Chicken sandwich	1	item(s)	170	—	410	28.0	38.0	1.0	16.0	3.5	—	—	0
38768	Chick-n-Strip salad	1	item(s)	327	—	400	34.0	21.0	4.0	20.0	6.0	—	—	0
38763	Chick-n-Strips	4	item(s)	127	—	300	28.0	14.0	1.0	15.0	2.5	—	—	0
38770	Cole slaw	1	item(s)	128	—	260	2.0	17.0	2.0	21.0	3.5	—	—	0
38776	Diet lemonade, small	1	cup(s)	255	—	25	0	5.0	0	0	0	0	0	0
38755	Hashbrowns	1	serving(s)	84	—	260	2.0	25.0	3.0	17.0	3.5	—	—	1.0
38765	Hearty breast of chicken soup	1	cup(s)	241	—	140	8.0	18.0	1.0	3.5	1.0	—	—	0
38741	Hot buttered biscuit	1	item(s)	79	—	270	4.0	38.0	1.0	12.0	3.0	—	—	3.0
38778	IceDream, small cone	1	item(s)	135	—	160	4.0	28.0	0	4.0	2.0	—	—	0
38774	IceDream, small cup	1	serving(s)	227	—	240	6.0	41.0	0	6.0	3.5	—	—	0
38775	Lemonade, small	1	cup(s)	255	—	170	0	41.0	0	0.5	0	—	—	0
38777	Nuggets	8	item(s)	113	—	260	26.0	12.0	0.5	12.0	2.5	—	—	0
38769	Side salad	1	item(s)	108	—	60	3.0	4.0	2.0	3.0	1.5	—	—	0
38767	Southwest chargrilled salad	1	item(s)	303	—	240	25.0	17.0	5.0	8.0	3.5	—	—	0
40481	Spicy chicken cool wrap	1	serving(s)	230	—	380	30.0	52.0	3.0	6.0	3.0	—	—	0
38772	Waffle potato fries, small, salted	1	serving(s)	85	—	270	3.0	34.0	4.0	13.0	3.0	—	—	1.5
Cinnabon														
39572	Caramellata Chill w/whipped cream	16	fluid ounce(s)	480	—	406	10.0	61.0	0	14.0	8.0	—	—	0
39571	Cinnabon Bites	1	serving(s)	149	—	510	8.0	77.0	2.0	19.0	5.0	—	—	5.0
39570	Cinnabon Stix	5	item(s)	85	—	379	6.0	41.0	1.0	21.0	6.0	—	—	4.0
39567	Classic roll	1	item(s)	221	—	813	15.0	117.0	4.0	32.0	8.0	—	—	5.0
39568	Minibon	1	item(s)	92	—	339	6.0	49.0	2.0	13.0	3.0	—	—	2.0
39573	Mochalatta Chill w/whipped cream	16	fluid ounce(s)	480	—	362	9.0	55.0	0	13.0	8.0	—	—	0
39569	Pecanbon	1	item(s)	272	—	1100	16.0	141.0	8.0	56.0	10.0	—	—	5.0
Dairy Queen														
1466	Banana split	1	item(s)	369	—	510	8.0	96.0	3.0	12.0	8.0	—	—	0
38552	Brownie Earthquake®	1	serving(s)	304	—	740	10.0	112.0	0	27.0	16.0	—	—	0.5
38561	Chocolate chip cookie dough Blizzard,® small	1	item(s)	319	—	720	12.0	105.0	0	28.0	14.0	—	—	2.5
1464	Chocolate malt, small	1	item(s)	418	—	640	15.0	111.0	1.0	16.0	11.0	—	—	0.5
38541	Chocolate shake, small	1	item(s)	397	—	560	13.0	93.0	1.0	15.0	10.0	—	—	0.5
17257	Chocolate soft serve	½	cup(s)	94	—	150	4.0	22.0	0	5.0	3.5	—	—	0
1463	Chocolate sundae, small	1	item(s)	163	—	280	5.0	49.0	0	7.0	4.5	—	—	0

PAGE KEY: A-6 = Breads/Baked Goods A-12 = Cereal/Rice/Pasta A-16 = Fruit A-20 = Vegetables/Legumes A-30 = Nuts/Seeds A-32 = Vegetarian A-34 = Dairy A-42 = Eggs A-42 = Seafood A-44 = Meats A-48 = Poultry
A-50 = Processed Meats A-50 = Beverages A-54 = Fats/Oils A-56 = Sweets A-58 = Spices/Condiments/Sauces A-62 = Mixed Foods/Soups/Sandwiches A-66 = Fast Food A-86 = Convenience A-88 = Baby Foods

Chol (mg)	Calc (mg)	Iron (mg)	Magn (mg)	Pota (mg)	Sodi (mg)	Zinc (mg)	Vit A (µg)	Thia (mg)	Vit E (mg α)	Ribo (mg)	Niac (mg)	Vit B$_6$ (mg)	Fola (µg)	Vit C (mg)	Vit B$_{12}$ (µg)	Sele (µg)
0	0	1.08	—	—	170.0	—	0	—	—	—	—	—	—	21.0	—	—
3	0	0.00	—	—	470.0	—	0	0.25	—	0.23	2.00	—	—	0	—	—
0	20	0.72	—	—	700.0	—	0	—	—	—	—	—	—	3.6	—	—
40	100	3.60	—	—	1220.0	—	—	—	—	—	—	—	—	6.0	—	—
45	600	0.00	—	—	340.0	—	0	—	—	—	—	—	—	0	—	—
130	100	7.20	—	—	980.0	—	—	—	—	—	—	—	—	9.0	—	—
145	279	4.29	—	—	1960.0	—	—	—	—	—	—	—	—	16.7	—	—
30	401	0.00	—	—	234.0	—	0	—	—	—	—	—	—	0	—	—
85	200	5.40	—	—	1410.0	—	60.0	—	—	—	—	—	—	1.2	—	—
270	150	2.70	—	—	1190.0	—	—	—	—	—	—	—	—	0	—	—
240	80	2.70	—	—	740.0	—	—	—	—	—	—	—	—	0	—	—
255	150	2.70	—	—	970.0	—	—	—	—	—	—	—	—	0	—	—
5	60	1.80	—	—	930.0	—	0	—	—	—	—	—	—	0	—	—
300	200	3.60	—	—	1360.0	—	—	—	—	—	—	—	—	0	—	—
10	40	0.36	—	—	110.0	—	—	—	—	—	—	—	—	4.8	—	—
65	200	3.60	—	—	1020.0	—	—	—	—	—	—	—	—	6.0	—	—
65	150	0.72	—	—	620.0	—	—	—	—	—	—	—	—	30	—	—
65	80	2.70	—	—	940.0	—	—	—	—	—	—	—	—	6.0	—	—
35	60	2.70	—	—	1270.0	—	0	—	—	—	—	—	—	0	—	—
50	150	2.70	—	—	1500.0	—	—	—	—	—	—	—	—	0	—	—
80	500	3.60	—	—	1350.0	—	—	—	—	—	—	—	—	1.2	—	—
60	100	2.70	—	—	1300.0	—	—	—	—	—	—	—	—	2.4	—	—
65	150	1.80	—	—	880.0	—	—	—	—	—	—	—	—	0	—	—
60	100	2.70	—	—	1300.0	—	—	—	—	—	—	—	—	0	—	—
80	150	1.44	—	—	1070.0	—	—	—	—	—	—	—	—	6.0	—	—
65	40	1.44	—	—	940.0	—	—	—	—	—	—	—	—	0	—	—
25	60	0.36	—	—	220.0	—	—	—	—	—	—	—	—	36.0	—	—
0	0	0.36	—	—	5.0	—	0	—	—	—	—	—	—	15.0	—	—
5	20	0.72	—	—	380.0	—	—	—	—	—	—	—	—	0	—	—
25	40	1.08	—	—	900.0	—	—	—	—	—	—	—	—	0	—	—
0	60	1.80	—	—	660.0	—	0	—	—	—	—	—	—	0	—	—
15	100	0.36	—	—	80.0	—	—	—	—	—	—	—	—	0	—	—
25	200	0.36	—	—	105.0	—	—	—	—	—	—	—	—	0	—	—
0	0	0.36	—	—	10.0	—	0	—	—	—	—	—	—	15.0	—	—
70	40	1.08	—	—	1090.0	—	0	—	—	—	—	—	—	0	—	—
10	100	0.00	—	—	75.0	—	—	—	—	—	—	—	—	15.0	—	—
60	200	1.08	—	—	770.0	—	—	—	—	—	—	—	—	24.0	—	—
60	200	3.60	—	—	1090.0	—	—	—	—	—	—	—	—	3.6	—	—
0	20	1.08	—	—	115.0	—	0	—	—	—	—	—	—	1.2	—	—
46	—	—	—	—	187.0	—	—	—	—	—	—	—	—	—	—	—
35	—	—	—	—	530.0	—	—	—	—	—	—	—	—	—	—	—
16	—	—	—	—	413.0	—	—	—	—	—	—	—	—	—	—	—
67	—	—	—	—	801.0	—	—	—	—	—	—	—	—	—	—	—
27	—	—	—	—	337.0	—	—	—	—	—	—	—	—	—	—	—
46	—	—	—	—	252.0	—	—	—	—	—	—	—	—	—	—	—
63	—	—	—	—	600.0	—	—	—	—	—	—	—	—	—	—	—
30	250	1.80	—	—	180.0	—	—	—	—	—	—	—	—	15.0	—	—
50	250	1.80	—	—	350.0	—	—	—	—	—	—	—	—	0	—	—
50	350	2.70	—	—	370.0	—	—	—	—	—	—	—	—	1.2	—	—
55	450	1.80	—	—	340.0	—	—	—	—	—	—	—	—	2.4	—	—
50	450	1.44	—	—	280.0	—	—	—	—	—	—	—	—	2.4	—	—
15	100	0.72	—	—	75.0	—	—	—	—	—	—	—	—	0	—	—
20	200	1.08	—	—	140.0	—	—	—	—	—	—	—	—	0	—	—

TABLE A-1 Table of Food Composition (continued) (Computer code is for Cengage Diet Analysis program) (For purposes of calculations, use "0" for t, <1, <.1, <.01, etc.)

DA+ Code	Food Description	Quantity	Measure	Wt (g)	H₂0 (g)	Ener (kcal)	Prot (g)	Carb (g)	Fiber (g)	Fat (g)	Fat Breakdown (g) Sat	Mono	Poly	Trans
Fast Food—continued														
1462	Dipped cone, small	1	item(s)	156	—	340	6.0	42.0	1.0	17.0	9.0	4.0	3.0	1.0
38555	Oreo cookies blizzard, small	1	item(s)	283	—	570	11.0	83.0	0.5	21.0	10.0	—	—	2.5
38547	Royal Treats Peanut Buster® Parfait	1	item(s)	305	—	730	16.0	99.0	2.0	31.0	17.0	—	—	0
17256	Vanilla soft serve	½	cup(s)	94	—	140	3.0	22.0	0	4.5	3.0	—	—	0
	Domino's													
31606	Barbeque buffalo wings	1	item(s)	25	—	50	6.0	2.0	0	2.5	0.5	—	—	—
31604	Breadsticks	1	item(s)	30	—	115	2.0	12.0	0	6.3	1.1	—	—	—
37551	Buffalo Chicken Kickers	1	item(s)	24	—	47	4.0	3.0	0	2.0	0.5	—	—	—
37548	CinnaStix	1	item(s)	30	—	123	2.0	15.0	1.0	6.1	1.1	—	—	—
37549	Dot, cinnamon	1	item(s)	28	7.6	99	1.9	14.9	0.7	3.7	0.7	—	—	—
31605	Double cheesy bread	1	item(s)	35	—	123	4.0	13.0	0	6.5	1.9	—	—	—
31607	Hot buffalo wings	1	item(s)	25	—	45	5.0	1.0	0	2.5	0.5	—	—	—
	Domino's Classic hand tossed pizza													
31573	America's favorite feast, 12"	1	slice(s)	102	—	257	10.0	29.0	2.0	11.5	4.5	—	—	—
31574	America's favorite feast, 14"	1	slice(s)	141	—	353	14.0	39.0	2.0	16.0	6.0	—	—	—
37543	Bacon cheeseburger feast, 12"	1	slice(s)	99	—	273	12.0	28.0	2.0	13.0	5.5	—	—	—
37545	Bacon cheeseburger feast, 14"	1	slice(s)	137	—	379	17.0	38.0	2.0	18.0	8.0	—	—	—
37546	Barbeque feast, 12"	1	slice(s)	96	—	252	11.0	31.0	1.0	10.0	4.5	—	—	—
37547	Barbeque feast, 14"	1	slice(s)	131	—	344	14.0	43.0	2.0	13.5	6.0	—	—	—
31569	Cheese, 12"	1	slice(s)	55	—	160	6.0	28.0	1.0	3.0	1.0	—	—	0
31570	Cheese, 14"	1	slice(s)	75	—	220	8.0	38.0	2.0	4.0	1.0	—	—	0
37538	Deluxe feast, 12"	1	slice(s)	201	101.8	465	19.5	57.4	3.5	18.2	7.7	—	—	—
37540	Deluxe feast, 14"	1	slice(s)	273	138.4	627	26.4	78.3	4.7	24.1	10.2	—	—	—
31685	Deluxe, 12"	1	slice(s)	100	—	234	9.0	29.0	2.0	9.5	3.5	—	—	—
31694	Deluxe, 14"	1	slice(s)	136	—	316	13.0	39.0	2.0	12.5	5.0	—	—	—
31686	Extravaganzza, 12"	1	slice(s)	122	—	289	13.0	30.0	2.0	14.0	5.5	—	—	—
31695	Extravaganzza, 14"	1	slice(s)	165	—	388	17.0	40.0	3.0	18.5	7.5	—	—	—
31575	Hawaiian feast, 12"	1	slice(s)	102	—	223	10.0	30.0	2.0	8.0	3.5	—	—	—
31576	Hawaiian feast, 14"	1	slice(s)	141	—	309	14.0	41.0	2.0	11.0	4.5	—	—	—
31687	Meatzza, 12"	1	slice(s)	108	—	281	13.0	29.0	2.0	13.5	5.5	—	—	—
31696	Meatzza, 14"	1	slice(s)	146	—	378	17.0	39.0	2.0	18.0	7.5	—	—	—
31571	Pepperoni feast, extra pepperoni and cheese, 12"	1	slice(s)	98	—	265	11.0	28.0	2.0	12.5	5.0	—	—	—
31572	Pepperoni feast, extra pepperoni and cheese, 14"	1	slice(s)	135	—	363	16.0	39.0	2.0	17.0	7.0	—	—	—
31577	Vegi feast, 12"	1	slice(s)	102	—	218	9.0	29.0	2.0	8.0	3.5	—	—	—
31578	Vegi feast, 14"	1	slice(s)	139	—	300	13.0	40.0	3.0	11.0	4.5	—	—	—
	Domino's thin crust pizza													
31583	America's favorite, 12"	1	slice(s)	72	—	208	8.0	15.0	1.0	13.5	5.0	—	—	—
31584	America's favorite, 14"	1	slice(s)	100	—	285	11.0	20.0	2.0	18.5	7.0	—	—	—
31579	Cheese, 12"	1	slice(s)	49	—	137	5.0	14.0	1.0	7.0	2.5	—	—	—
31580	Cheese, 14"	1	slice(s)	68	27.0	214	8.8	19.0	1.4	11.4	4.6	2.9	2.5	—
31688	Deluxe, 12"	1	slice(s)	70	—	185	7.0	15.0	1.0	11.5	4.0	—	—	—
31697	Deluxe, 14"	1	slice(s)	94	—	248	10.0	20.0	2.0	15.0	5.5	—	—	—
31689	Extravaganzza, 12"	1	slice(s)	92	—	240	11.0	16.0	1.0	15.5	6.0	—	—	—
31698	Extravaganzza, 14"	1	slice(s)	123	—	320	14.0	21.0	2.0	20.5	8.0	—	—	—
31585	Hawaiian, 12"	1	slice(s)	71	—	174	8.0	16.0	1.0	9.5	3.5	—	—	—
31586	Hawaiian, 14"	1	slice(s)	100	—	240	11.0	21.0	2.0	13.0	5.0	—	—	—
31690	Meatzza, 12"	1	slice(s)	78	—	232	11.0	15.0	1.0	15.0	6.0	—	—	—
31699	Meatzza, 14"	1	slice(s)	104	—	310	14.0	20.0	2.0	20	8.0	—	—	—
31581	Pepperoni, extra pepperoni and cheese, 12"	1	slice(s)	68	—	216	9.0	14.0	1.0	14.0	5.5	—	—	—
31582	Pepperoni, extra pepperoni and cheese, 14"	1	slice(s)	93	—	295	13.0	20.0	1.0	19.0	7.5	—	—	—
31587	Vegi, 12"	1	slice(s)	71	—	168	7.0	15.0	1.0	9.5	3.5	—	—	—
31588	Vegi, 14"	1	slice(s)	97	—	231	10.0	21.0	2.0	13.5	5.0	—	—	—
	Domino's Ultimate deep dish pizza													
31596	America's favorite, 12"	1	slice(s)	115	—	309	12.0	29.0	2.0	17.0	6.0	—	—	—
31702	America's favorite, 14"	1	slice(s)	162	—	433	17.0	42.0	3.0	23.5	8.0	—	—	—
31590	Cheese, 12"	1	slice(s)	90	—	238	9.0	28.0	2.0	11.0	3.5	—	—	—
31591	Cheese, 14"	1	slice(s)	128	53.9	351	14.5	41.0	2.9	13.2	5.2	3.8	2.5	—
31589	Cheese, 6"	1	item(s)	215	—	598	22.9	68.4	3.9	27.6	9.9	—	—	—
31691	Deluxe, 12"	1	slice(s)	122	—	287	11.0	29.0	2.0	15.0	5.0	—	—	—

Chol (mg)	Calc (mg)	Iron (mg)	Magn (mg)	Pota (mg)	Sodi (mg)	Zinc (mg)	Vit A (µg)	Thia (mg)	Vit E (mg α)	Ribo (mg)	Niac (mg)	Vit B$_6$ (mg)	Fola (µg)	Vit C (mg)	Vit B$_{12}$ (µg)	Sele (µg)
20	200	1.08	—	—	130.0	—	—	—	—	—	—	—	—	1.2	—	—
40	350	2.70	—	—	430.0	—	—	—	—	—	—	—	—	1.2	—	—
35	300	1.80	—	—	400.0	—	—	—	—	—	—	—	—	1.2	—	—
15	150	0.72	—	—	70.0	—	—	—	—	—	—	—	—	0	—	—
26	10	0.36	—	—	175.5	—	—	—	—	—	—	—	—	0	—	—
0	0	0.72	—	—	122.1	—	—	—	—	—	—	—	—	0	—	—
9	0	0.00	—	—	162.5	—	—	—	—	—	—	—	—	0	—	—
0	0	0.72	—	—	111.4	—	—	—	—	—	—	—	—	0	—	—
0	6	0.59	—	—	85.7	—	—	—	—	—	—	—	—	0	—	—
6	40	0.72	—	—	162.3	—	—	—	—	—	—	—	—	0	—	—
26	10	0.36	—	—	254.5	—	—	—	—	—	—	—	—	1.2	—	—
22	100	1.80	—	—	625.5	—	—	—	—	—	—	—	—	0.6	—	—
31	140	2.52	—	—	865.5	—	—	—	—	—	—	—	—	0.6	—	—
27	140	1.80	—	—	634.0	—	—	—	—	—	—	—	—	0	—	—
38	190	2.52	—	—	900.0	—	—	—	—	—	—	—	—	0	—	—
20	140	1.62	—	—	600.0	—	—	—	—	—	—	—	—	0.6	—	—
27	190	2.16	—	—	831.5	—	—	—	—	—	—	—	—	0.6	—	—
0	0	1.80	—	—	110.0	—	0	—	—	—	—	—	—	0	—	—
0	0	2.70	—	—	150.0	—	0	—	—	—	—	—	—	0	—	—
40	199	3.56	—	—	1063.1	—	—	—	—	—	—	—	—	1.4	—	—
53	276	4.84	—	—	1432.2	—	—	—	—	—	—	—	—	1.8	—	—
17	100	1.80	—	—	541.5	—	—	—	—	—	—	—	—	0.6	—	—
23	130	2.34	—	—	728.5	—	—	—	—	—	—	—	—	1.2	—	—
28	140	1.98	—	—	764.0	—	—	—	—	—	—	—	—	0.6	—	—
37	190	2.70	—	—	1014.0	—	—	—	—	—	—	—	—	1.2	—	—
16	130	1.62	—	—	546.5	—	—	—	—	—	—	—	—	1.2	—	—
23	180	2.34	—	—	765.0	—	—	—	—	—	—	—	—	1.2	—	—
28	130	1.80	—	—	739.5	—	—	—	—	—	—	—	—	0	—	—
37	190	2.52	—	—	983.5	—	—	—	—	—	—	—	—	0	—	—
24	130	1.62	—	—	670.0	—	70.9	—	—	—	—	—	—	0	—	—
33	180	2.34	—	—	920.0	—	104.7	—	—	—	—	—	—	0	—	—
13	130	1.62	—	—	489.0	—	—	—	—	—	—	—	—	0.6	—	—
18	180	2.34	—	—	678.0	—	—	—	—	—	—	—	—	0.6	—	—
23	100	0.90	—	—	533.0	—	—	—	—	—	—	—	—	2.4	—	—
32	140	1.26	—	—	736.5	—	—	—	—	—	—	—	—	3.0	—	—
10	90	0.54	—	—	292.5	—	60.0	—	—	—	—	—	—	1.8	—	—
14	151	0.48	17.7	125.1	338.0	0.02	64.6	0.05	1.01	0.07	0.69	—	—	2.4	0.5	24.1
19	100	0.90	—	—	449.0	—	—	—	—	—	—	—	—	2.4	—	—
24	130	1.08	—	—	601.0	—	—	—	—	—	—	—	—	3.6	—	—
29	140	1.08	—	—	671.5	—	—	—	—	—	—	—	—	2.4	—	—
38	190	1.44	—	—	886.5	—	—	—	—	—	—	—	—	3.6	—	—
17	130	0.72	—	—	454.0	—	—	—	—	—	—	—	—	3.0	—	—
24	180	0.90	—	—	637.5	—	—	—	—	—	—	—	—	3.6	—	—
29	140	0.90	—	—	647.0	—	—	—	—	—	—	—	—	1.8	—	—
38	190	1.26	—	—	865.5	—	—	—	—	—	—	—	—	2.4	—	—
26	130	0.72	—	—	577.0	—	80.0	—	—	—	—	—	—	1.8	—	—
35	80	1.08	—	—	792.5	—	105.8	—	—	—	—	—	—	2.4	—	—
14	130	0.72	—	—	396.5	—	—	—	—	—	—	—	—	2.4	—	—
19	180	1.08	—	—	550.5	—	—	—	—	—	—	—	—	3.0	—	—
25	120	2.34	—	—	796.5	—	—	—	—	—	—	—	—	0.6	—	—
34	170	3.24	—	—	1110.0	—	—	—	—	—	—	—	—	0.6	—	—
11	110	1.98	—	—	555.5	—	70.0	—	—	—	—	—	—	0	—	—
18	189	3.78	32.0	209.9	718.1	1.75	99.8	0.29	1.13	0.31	5.44	—	—	0	0.6	45.6
36	295	4.67	—	—	1341.4	—	174.0	—	—	—	—	—	—	0.5	—	—
20	120	2.16	—	—	712.0	—	—	—	—	—	—	—	—	1.2	—	—

TABLE A-1 Table of Food Composition *(continued)* (Computer code is for Cengage Diet Analysis program) (For purposes of calculations, use "0" for t, <1, <.1, <.01, etc.)

DA+ Code	Food Description	Quantity	Measure	Wt (g)	H_2O (g)	Ener (kcal)	Prot (g)	Carb (g)	Fiber (g)	Fat (g)	Fat Breakdown (g) Sat	Mono	Poly	*Trans*
Fast Food—continued														
31700	Deluxe, 14"	1	slice(s)	156	—	396	15.0	42.0	3.0	20.0	7.0	—	—	—
31692	Extravaganzza, 12"	1	slice(s)	136	—	341	14.0	30.0	2.0	19.0	7.0	—	—	—
31701	Extravaganzza, 14"	1	slice(s)	186	—	468	20.0	43.0	3.0	25.5	9.5	—	—	—
31599	Hawaiian, 12"	1	slice(s)	114	—	275	12.0	30.0	2.0	13.0	5.0	—	—	—
31600	Hawaiian, 14"	1	slice(s)	162	—	389	17.0	43.0	3.0	18.0	6.5	—	—	—
31693	Meatzza, 12"	1	slice(s)	121	—	333	14.0	29.0	2.0	19.0	7.0	—	—	—
31703	Meatzza, 14"	1	slice(s)	167	—	458	19.0	42.0	3.0	25.0	9.5	—	—	—
31593	Pepperoni, extra pepperoni and cheese, 12"	1	slice(s)	110	—	317	13.0	29.0	2.0	17.5	6.5	—	—	—
31594	Pepperoni, extra pepperoni and cheese, 14"	1	slice(s)	155	—	443	18.0	42.0	3.0	24.0	9.0	—	—	—
31602	Vegi, 12"	1	slice(s)	114	—	270	11.0	30.0	2.0	13.5	5.0	—	—	—
31603	Vegi, 14"	1	slice(s)	159	—	380	15.0	43.0	3.0	18.0	6.5	—	—	—
31598	With ham and pineapple tidbits, 6"	1	item(s)	430	—	619	25.2	69.9	4.0	28.3	10.2	—	—	—
31595	With Italian sausage, 6"	1	item(s)	430	—	642	24.8	69.6	4.2	31.1	11.3	—	—	—
31592	With pepperoni, 6"	1	item(s)	430	—	647	25.1	68.5	3.9	32.0	11.7	—	—	—
31601	With vegetables, 6"	1	item(s)	430	—	619	23.4	70.8	4.6	28.7	10.1	—	—	—
	In-n-Out Burger													
34391	Cheeseburger with mustard and ketchup	1	serving(s)	268	—	400	22.0	41.0	3.0	18.0	9.0	—	—	0.5
34374	Cheeseburger	1	serving(s)	268	—	480	22.0	39.0	3.0	27.0	10.0	—	—	0.5
34390	Cheeseburger, lettuce leaves instead of buns	1	serving(s)	300	—	330	18.0	11.0	3.0	25.0	9.0	—	—	0
34377	Chocolate shake	1	serving(s)	425	—	690	9.0	83.0	0	36.0	24.0	—	—	1.0
34375	Double-Double cheeseburger	1	serving(s)	330	—	670	37.0	39.0	3.0	41.0	18.0	—	—	1.0
34393	Double-Double cheeseburger with mustard and ketchup	1	serving(s)	330	—	590	37.0	41.0	3.0	32.0	17.0	—	—	1.0
34392	Double-Double cheeseburger, lettuce leaves instead of buns	1	serving(s)	362	—	520	33.0	11.0	3.0	39.0	17.0	—	—	1.0
34376	French fries	1	serving(s)	125	—	400	7.0	54.0	2.0	18.0	5.0	—	—	0
34373	Hamburger	1	item(s)	243	—	390	16.0	39.0	3.0	19.0	5.0	—	—	0
34389	Hamburger with mustard and ketchup	1	serving(s)	243	—	310	16.0	41.0	3.0	10.0	4.0	—	—	0
34388	Hamburger, lettuce leaves instead of buns	1	serving(s)	275	—	240	13.0	11.0	3.0	17.0	4.0	—	—	0
34379	Strawberry shake	1	serving(s)	425	—	690	9.0	91.0	0	33.0	22.0	—	—	0.5
34378	Vanilla shake	1	serving(s)	425	—	680	9.0	78.0	0	37.0	25.0	—	—	1.0
	Jack in the Box													
30392	Bacon ultimate cheeseburger	1	item(s)	338	—	1090	46.0	53.0	2.0	77.0	30.0	—	—	3.0
1740	Breakfast Jack	1	item(s)	125	—	290	17.0	29.0	1.0	12.0	4.5	—	—	0
14074	Cheeseburger	1	item(s)	131	—	350	18.0	31.0	1.0	17.0	8.0	—	—	1.0
14106	Chicken breast strips, 4 pieces	1	serving(s)	201	—	500	35.0	36.0	3.0	25.0	6.0	—	—	6.0
37241	Chicken club salad, plain, without salad dressing	1	serving(s)	431	—	300	27.0	13.0	4.0	15.0	6.0	—	—	0
14064	Chicken sandwich	1	item(s)	145	—	400	15.0	38.0	2.0	21.0	4.5	—	—	2.5
14111	Chocolate ice cream shake, small	1	serving(s)	414	—	880	14.0	107.0	1.0	45.0	31.0	—	—	2.0
14073	Hamburger	1	item(s)	118	—	310	16.0	30.0	1.0	14.0	6.0	—	—	1.0
14090	Hash browns	1	serving(s)	57	—	150	1.0	13.0	2.0	10.0	2.5	—	—	3.0
14072	Jack's Spicy Chicken sandwich	1	item(s)	270	—	620	25.0	61.0	4.0	31.0	6.0	—	—	3.0
1468	Jumbo Jack hamburger	1	item(s)	261	—	600	21.0	51.0	3.0	35.0	12.0	—	—	1.5
1469	Jumbo Jack hamburger with cheese	1	item(s)	286	—	690	25.0	54.0	3.0	42.0	16.0	—	—	1.5
14099	Natural cut french fries, large	1	serving(s)	196	—	530	8.0	69.0	5.0	25.0	6.0	—	—	7.0
14098	Natural cut french fries, medium	1	serving(s)	133	—	360	5.0	47.0	4.0	17.0	4.0	—	—	5.0
1470	Onion rings	1	serving(s)	119	—	500	6.0	51.0	3.0	30.0	6.0	—	—	10
33141	Sausage, egg, and cheese biscuit	1	item(s)	234	—	740	27.0	35.0	2.0	55.0	17.0	—	—	6.0
14095	Seasoned curly fries, medium	1	serving(s)	125	—	400	6.0	45.0	5.0	23.0	5.0	—	—	7.0
14077	Sourdough Jack	1	item(s)	245	—	710	27.0	36.0	3.0	51.0	18.0	—	—	3.0
37249	Southwest chicken salad, plain, without salad dressing	1	serving(s)	488	—	300	24.0	29.0	7.0	11.0	5.0	—	—	0
14112	Strawberry ice cream shake, small	1	serving(s)	417	—	880	13.0	105.0	0	44.0	31.0	—	—	2.0

Chol (mg)	Calc (mg)	Iron (mg)	Magn (mg)	Pota (mg)	Sodi (mg)	Zinc (mg)	Vit A (µg)	Thia (mg)	Vit E (mg α)	Ribo (mg)	Niac (mg)	Vit B_6 (mg)	Fola (µg)	Vit C (mg)	Vit B_{12} (µg)	Sele (µg)
26	170	3.06	—	—	974.5	—	—	—	—	—	—	—	—	1.2	—	—
31	160	2.52	—	—	934.5	—	—	—	—	—	—	—	—	1.2	—	—
40	220	3.42	—	—	1260.0	—	—	—	—	—	—	—	—	1.2	—	—
19	150	1.98	—	—	717.0	—	—	—	—	—	—	—	—	1.2	—	—
26	210	2.88	—	—	1011.0	—	—	—	—	—	—	—	—	1.8	—	—
31	160	2.34	—	—	910.5	—	—	—	—	—	—	—	—	0	—	—
40	220	3.24	—	—	1230.0	—	—	—	—	—	—	—	—	0.6	—	—
27	150	2.16	—	—	840.5	—	86.5	—	—	—	—	—	—	0	—	—
37	220	3.06	—	—	1166.0	—	115.4	—	—	—	—	—	—	0.6	—	—
15	150	2.16	—	—	659.5	—	—	—	—	—	—	—	—	0.6	—	—
21	220	3.06	—	—	924.0	—	—	—	—	—	—	—	—	1.2	—	—
43	298	4.84	—	—	1497.8	—	—	—	—	—	—	—	—	1.5	—	—
45	302	4.89	—	—	1478.1	—	—	—	—	—	—	—	—	0.6	—	—
47	299	4.81	—	—	1523.7	—	167.9	—	—	—	—	—	—	0.6	—	—
36	307	5.10	—	—	1472.5	—	—	—	—	—	—	—	—	4.7	—	—
60	200	3.60	—	—	1080.0	—	—	—	—	—	—	—	—	12.0	—	—
60	200	3.60	—	—	1000.0	—	—	—	—	—	—	—	—	9.0	—	—
60	200	2.70	—	—	720.0	—	—	—	—	—	—	—	—	12.0	—	—
95	300	0.72	—	—	350.0	—	—	—	—	—	—	—	—	0	—	—
120	350	5.40	—	—	1440.0	—	—	—	—	—	—	—	—	9.0	—	—
115	350	5.40	—	—	1520.0	—	—	—	—	—	—	—	—	12.0	—	—
120	350	4.50	—	—	1160.0	—	—	—	—	—	—	—	—	12.0	—	—
0	20	1.80	—	—	245.0	—	0	—	—	—	—	—	—	0	—	—
40	40	3.60	—	—	650.0	—	—	—	—	—	—	—	—	9.0	—	—
35	40	3.60	—	—	730.0	—	—	—	—	—	—	—	—	12.0	—	—
40	40	2.70	—	—	370.0	—	—	—	—	—	—	—	—	12.0	—	—
85	300	0.00	—	—	280.0	—	—	—	—	—	—	—	—	0	—	—
90	300	0.00	—	—	390.0	—	—	—	—	—	—	—	—	0	—	—
140	308	7.38	—	540.0	2040.0	—	—	—	—	—	—	—	—	0.6	—	—
220	145	3.48	—	210.0	760.0	—	—	—	—	—	—	—	—	3.5	—	—
50	151	3.61	—	270.0	790.0	—	40.2	—	—	—	—	—	—	—	—	—
80	18	1.60	—	530.0	1260.0	—	—	—	—	—	—	—	—	1.1	—	—
65	280	3.35	—	560.0	880.0	—	—	—	—	—	—	—	—	50.4	—	—
35	100	2.70	—	240.0	730.0	—	—	—	—	—	—	—	—	4.8	—	—
135	460	0.47	—	840.0	330.0	—	—	—	—	—	—	—	—	0	—	—
40	100	3.60	—	250.0	600.0	—	0	—	—	—	—	—	—	0	—	—
0	10	0.18	—	190.0	230.0	—	0	—	—	—	—	—	—	0	—	—
50	150	1.80	—	450.0	1100.0	—	—	—	—	—	—	—	—	9.0	—	—
45	164	4.92	—	380.0	940.0	—	—	—	—	—	—	—	—	9.8	—	—
70	234	4.20	—	410.0	1310.0	—	—	—	—	—	—	—	—	8.4	—	—
0	20	1.42	—	1240.0	870.0	—	0	—	—	—	—	—	—	8.9	—	—
0	19	1.01	—	840.0	590.0	—	0	—	—	—	—	—	—	5.6	—	—
0	40	2.70	—	140.0	420.0	—	40.0	—	—	—	—	—	—	18.0	—	—
280	88	2.36	—	310.0	1430.0	—	—	—	—	—	—	—	—	0	—	—
0	40	1.80	—	580.0	890.0	—	—	—	—	—	—	—	—	0	—	—
75	200	4.50	—	430.0	1230.0	—	—	—	—	—	—	—	—	9.0	—	—
55	274	4.10	—	670.0	860.0	—	—	—	—	—	—	—	—	43.8	—	—
135	466	0.00	—	750.0	290.0	—	—	—	—	—	—	—	—	0	—	—

TABLE A-1 Table of Food Composition (continued) (Computer code is for Cengage Diet Analysis program) (For purposes of calculations, use "0" for t, <1, <.1, <.01, etc.)

DA+ Code	Food Description	Quantity	Measure	Wt (g)	H₂0 (g)	Ener (kcal)	Prot (g)	Carb (g)	Fiber (g)	Fat (g)	Sat	Mono	Poly	Trans
											Fat Breakdown (g)			

DA+ Code	Food Description	Quantity	Measure	Wt (g)	H₂0 (g)	Ener (kcal)	Prot (g)	Carb (g)	Fiber (g)	Fat (g)	Sat	Mono	Poly	Trans
Fast Food—continued														
14078	Ultimate cheeseburger	1	item(s)	323	—	1010	40.0	53.0	2.0	71.0	28.0	—	—	3.0
14110	Vanilla ice cream shake, small	1	serving(s)	379	—	790	13.0	83.0	0	44.0	31.0	—	—	2.0
	Jamba Juice													
31645	Aloha Pineapple smoothie	24	fluid ounce(s)	730	—	500	8.0	117.0	4.0	1.5	1.0	—	—	—
31646	Banana Berry smoothie	24	fluid ounce(s)	719	—	480	5.0	112.0	4.0	1.0	0	—	—	—
31656	Berry Lime Sublime smoothie	24	fluid ounce(s)	728	—	460	3.0	106.0	5.0	2.0	1.0	—	—	—
31647	Carribean Passion smoothie	24	fluid ounce(s)	730	—	440	4.0	102.0	4.0	2.0	1.0	—	—	—
38422	Carrot juice	16	fluid ounce(s)	472	—	100	3.0	23.0	0	0.5	0	—	—	—
31648	Chocolate Moo'd smoothie	24	fluid ounce(s)	634	—	720	17.0	148.0	3.0	8.0	5.0	—	—	—
31649	Citrus Squeeze smoothie	24	fluid ounce(s)	727	—	470	5.0	110.0	4.0	2.0	1.0	—	—	—
31651	Coldbuster smoothie	24	fluid ounce(s)	724	—	430	5.0	100.0	5.0	2.5	1.0	—	—	—
31652	Cranberry Craze smoothie	24	fluid ounce(s)	793	—	460	6.0	104.0	4.0	0.5	0	—	—	—
31654	Jamba Powerboost smoothie	24	fluid ounce(s)	738	—	440	6.0	105.0	6.0	1.0	0	—	—	—
38423	Lemonade	16	fluid ounce(s)	483	—	300	1.0	75.0	0	0	0	0	0	0
31657	Mango-a-go-go smoothie	24	fluid ounce(s)	690	—	440	3.0	104.0	4.0	1.5	0.5	—	—	—
38424	Orange juice, freshly squeezed	16	fluid ounce(s)	496	—	220	3.0	52.0	0.5	1.0	0	—	—	—
38426	Orange/carrot juice	16	fluid ounce(s)	484	—	160	3.0	37.0	0	1.0	0	—	—	—
31660	Orange-a-peel smoothie	24	fluid ounce(s)	726	—	440	8.0	102.0	5.0	1.5	0	—	—	—
31662	Peach Pleasure smoothie	24	fluid ounce(s)	720	—	460	4.0	108.0	4.0	2.0	1.0	—	—	—
31665	Protein Berry Pizzaz smoothie	24	fluid ounce(s)	710	—	440	20.0	92.0	5.0	1.5	0	—	—	—
31668	Razzmatazz smoothie	24	fluid ounce(s)	730	—	480	3.0	112.0	4.0	2.0	1.0	—	—	—
31669	Strawberries Wild smoothie	24	fluid ounce(s)	725	—	450	6.0	105.0	4.0	0.5	0	—	—	—
38421	Strawberry Tsunami smoothie	24	fluid ounce(s)	740	—	530	4.0	128.0	4.0	2.0	1.0	—	—	—
38427	Vibrant C juice	16	fluid ounce(s)	448	—	210	2.0	50.0	1.0	0	0	0	0	0
38428	Wheatgrass juice, freshly squeezed	1	ounce(s)	28	—	5	0.5	1.0	0	0	0	0	0	0
	Kentucky Fried Chicken (KFC)													
31850	BBQ baked beans	1	serving(s)	136	—	220	8.0	45.0	7.0	1.0	0	—	—	0
31853	Biscuit	1	item(s)	57	—	220	4.0	24.0	1.0	11.0	2.5	—	—	3.5
51223	Boneless Fiery Buffalo Wings	6	item(s)	211	—	530	30.0	44.0	3.0	26.0	5.0	—	—	2.5
39386	Boneless Honey BBQ Wings	6	item(s)	213	—	570	30.0	54.0	5.0	26.0	5.0	—	—	2.5
51224	Boneless Sweet & Spicy Wings	6	item(s)	203	—	550	30.0	50.0	3.0	26.0	5.0	—	—	2.5
31851	Cole slaw	1	serving(s)	130	—	180	1.0	22.0	3.0	10.0	1.5	—	—	0
31842	Colonel's Crispy Strips	3	item(s)	151	—	370	28.0	17.0	1.0	20.0	4.0	—	—	2.5
31849	Corn on the cob	1	item(s)	162	—	150	5.0	26.0	7.0	3.0	1.0	—	—	0
51221	Double Crunch sandwich	1	item(s)	213	—	520	27.0	39.0	3.0	29.0	5.0	—	—	1.5
3761	Extra Crispy chicken, breast	1	item(s)	162	—	370	33.0	10.0	2.0	22.0	5.0	—	—	1.5
3762	Extra Crispy chicken, drumstick	1	item(s)	60	—	150	12.0	4.0	0	10.0	2.5	—	—	1.0
3763	Extra Crispy chicken, thigh	1	item(s)	114	—	290	17.0	16.0	1.0	18.0	4.0	—	—	1.5
3764	Extra Crispy chicken, whole wing	1	item(s)	52	—	150	11.0	11.0	1.0	7.0	1.5	—	—	0
51218	Famous Bowls mashed potatoes with gravy	1	serving(s)	531	—	720	26.0	79.0	6.0	34.0	9.0	—	—	3.5
51219	Famous Bowls rice with gravy	1	serving(s)	384	—	610	25.0	67.0	5.0	27.0	8.0	—	—	2.5
31841	Honey BBQ chicken sandwich	1	item(s)	147	—	290	23.0	40.0	2.0	4.0	1.0	—	—	0
31833	Honey BBQ wing pieces	6	item(s)	157	—	460	27.0	26.0	3.0	27.0	6.0	—	—	2.0
10859	Hot wings pieces	6	piece(s)	134	—	450	26.0	19.0	2.0	30.0	7.0	—	—	2.0
42382	KFC Snacker sandwich	1	serving(s)	119	—	320	14.0	29.0	2.0	17.0	3.0	—	—	1.0
31848	Macaroni and cheese	1	serving(s)	136	—	180	8.0	18.0	0	8.0	3.5	—	—	1.0
31847	Mashed potatoes with gravy	1	serving(s)	151	—	140	2.0	20.0	1.0	5.0	1.0	—	—	0.5
10825	Original Recipe chicken, breast	1	item(s)	161	—	340	38.0	9.0	2.0	17.0	4.0	—	—	1.0
10826	Original Recipe chicken, drumstick	1	item(s)	59	—	140	13.0	3.0	0	8.0	2.0	—	—	0.5
10827	Original Recipe chicken, thigh	1	item(s)	126	—	350	19.0	7.0	1.0	27.0	7.0	—	—	1.0
10828	Original Recipe chicken, whole wing	1	item(s)	47	—	140	10.0	4.0	0	9.0	2.0	—	—	0.5
51222	Oven roasted Twister chicken wrap	1	item(s)	269	—	520	30.0	46.0	4.0	23.0	3.5	—	—	0
31844	Popcorn chicken, small or individual	1	item(s)	114	—	370	19.0	21.0	2.0	24.0	4.5	—	—	2.5
31852	Potato salad	1	serving(s)	128	—	180	2.0	22.0	2.0	9.0	1.5	—	—	0
10845	Potato wedges, small	1	serving(s)	102	—	250	4.0	32.0	3.0	12.0	2.0	—	—	1.5
31839	Tender Roast chicken sandwich with sauce	1	item(s)	236	—	430	37.0	29.0	2.0	18.0	3.5	—	—	0

PAGE KEY: A-6 = Breads/Baked Goods A-12 = Cereal/Rice/Pasta A-16 = Fruit A-20 = Vegetables/Legumes A-30 = Nuts/Seeds A-32 = Vegetarian A-34 = Dairy A-42 = Eggs A-42 = Seafood A-44 = Meats A-48 = Poultry
A-50 = Processed Meats A-50 = Beverages A-54 = Fats/Oils A-56 = Sweets A-58 = Spices/Condiments/Sauces A-62 = Mixed Foods/Soups/Sandwiches A-66 = Fast Food A-86 = Convenience A-88 = Baby Foods

Chol (mg)	Calc (mg)	Iron (mg)	Magn (mg)	Pota (mg)	Sodi (mg)	Zinc (mg)	Vit A (µg)	Thia (mg)	Vit E (mg α)	Ribo (mg)	Niac (mg)	Vit B$_6$ (mg)	Fola (µg)	Vit C (mg)	Vit B$_{12}$ (µg)	Sele (µg)
125	308	7.39	—	480.0	1580.0	—	—	—	—	—	—	—	—	0.6	—	—
135	532	0.00	—	750.0	280.0	—	—	—	—	—	—	—	—	0	—	—
5	200	1.80	60.0	1000.0	30.0	0.30	—	0.37	—	0.34	2.00	0.60	60.0	102.0	0	1.4
0	200	1.44	40.0	1010.0	115.0	0.60	—	0.09	—	0.25	0.80	0.70	24.0	15.0	0.2	1.4
5	200	1.80	16.0	510.0	35.0	0.30	—	0.06	—	0.25	6.00	0.70	140.0	54.0	0	1.4
5	100	1.80	24.0	810.0	60.0	0.30	—	0.09	—	0.25	5.00	0.50	100.0	78.0	0	1.4
0	150	2.70	80.0	1030.0	250.0	0.90	—	0.52	—	0.25	5.00	0.70	80.0	18.0	0	5.6
30	500	1.08	60.0	810.0	380.0	1.50	—	0.22	—	0.76	0.40	0.16	16.0	6.0	1.5	4.2
5	100	1.80	80.0	1170.0	35.0	0.30	—	0.37	—	0.34	1.90	0.60	100.0	180.0	0	1.4
5	100	1.08	60.0	1260.0	35.0	16.50	—	0.37	—	0.34	3.00	0.40	121.5	1302.0	0	1.4
0	250	1.44	16.0	500.0	50.0	0.30	—	0.03	—	0.25	5.00	0.60	120.0	54.0	0	1.4
0	1200	1.80	480.0	1070.0	45.0	16.50	—	5.55	—	6.12	68.00	7.40	640.0	288.0	10.8	77.0
0	20	0.00	8.0	200.0	10.0	0.00	—	0.03	—	0.17	14.00	1.80	320.0	36.0	0	0
5	100	1.08	24.0	780.0	50.0	0.30	—	0.15	—	0.25	5.00	0.70	120.0	72.0	0	1.4
0	60	1.08	60.0	990.0	0	0.30	—	0.45	—	0.13	2.00	0.20	160.0	246.0	0	0
0	100	1.80	60.0	1010.0	125.0	0.60	—	0.45	—	0.25	3.00	0.50	120.0	132.0	0	2.8
0	250	1.80	80.0	1380.0	160.0	0.90	—	0.45	—	0.42	2.00	0.50	140.0	240.0	0.6	1.4
5	100	0.72	32.0	740.0	60.0	0.30	—	0.06	—	0.25	4.00	0.60	80.0	18.0	0	1.4
0	1100	2.62	60.0	650.0	240.0	0.58	—	0.08	—	0.17	1.20	0.70	58.3	60.0	0	5.6
5	150	1.80	32.0	810.0	70.0	0.30	—	0.09	—	0.34	6.00	1.00	160.0	60.0	0	1.4
5	250	1.80	40.0	1050.0	180.0	0.90	—	0.12	—	0.34	0.80	0.40	40.0	60.0	0.6	1.4
5	100	1.08	24.0	480.0	10.0	0.30	—	0.06	—	0.34	14.00	1.80	320.0	90.0	0	1.4
0	20	1.08	40.0	720.0	0	0.30	—	0.30	—	0.10	1.60	0.40	80.0	678.0	0	0
0	0	1.80	8.0	80.0	0	0.00	0	0.03	—	0.03	0.40	0.04	16.0	3.6	0	2.8
0	100	2.70	—	—	730.0	—	—	—	—	—	—	—	—	1.2	—	—
0	40	1.80	—	—	640.0	—	—	—	—	—	—	—	—	0	—	—
65	40	1.80	—	—	2670.0	—	—	—	—	—	—	—	—	1.2	—	—
65	40	1.80	—	—	2210.0	—	—	—	—	—	—	—	—	1.2	—	—
65	60	1.80	—	—	2000.0	—	—	—	—	—	—	—	—	1.2	—	—
5	40	0.72	—	—	270.0	—	—	—	—	—	—	—	—	12.0	—	—
65	40	1.44	—	—	1220.0	—	0	—	—	—	—	—	—	1.2	—	—
0	60	1.08	—	—	10.0	—	—	—	—	—	—	—	—	6.0	—	—
55	100	2.70	—	—	1220.0	—	—	—	—	—	—	—	—	6.0	—	—
85	20	2.70	—	—	1020.0	—	—	—	—	—	—	—	—	1.2	—	—
55	0	1.44	—	—	300.0	—	0	—	—	—	—	—	—	0	—	—
95	20	2.70	—	—	700.0	—	—	—	—	—	—	—	—	—	—	—
45	20	1.08	—	—	340.0	—	—	—	—	—	—	—	—	0	—	—
35	200	5.40	—	—	2330.0	—	—	—	—	—	—	—	—	6.0	—	—
35	200	4.50	—	—	2130.0	—	—	—	—	—	—	—	—	6.0	—	—
60	80	2.70	—	—	710.0	—	—	—	—	—	—	—	—	2.4	—	—
140	40	1.80	—	—	970.0	—	—	—	—	—	—	—	—	21.0	—	—
115	40	1.44	—	—	990.0	—	—	—	—	—	—	—	—	1.2	—	—
25	60	2.70	—	—	690.0	—	—	—	—	—	—	—	—	2.4	—	—
15	150	0.72	—	—	800.0	—	—	—	—	—	—	—	—	1.2	—	—
0	40	1.44	—	—	560.0	—	—	—	—	—	—	—	—	1.2	—	—
135	20	2.70	—	—	960.0	—	—	—	—	—	—	—	—	6.0	—	—
70	20	1.08	—	—	340.0	—	—	—	—	—	—	—	—	0	—	—
110	20	2.70	—	—	870.0	—	—	—	—	—	—	—	—	1.2	—	—
50	20	1.44	—	—	350.0	—	0	—	—	—	—	—	—	1.2	—	—
60	40	6.30	—	—	1380.0	—	—	—	—	—	—	—	—	15.0	—	—
25	40	1.80	—	—	1110.0	—	0	—	—	—	—	—	—	0	—	—
5	0	0.36	—	—	470.0	—	—	—	—	—	—	—	—	6.0	—	—
0	20	1.08	—	—	700.0	—	0	—	—	—	—	—	—	0	—	—
80	80	2.70	—	—	1180.0	—	—	—	—	—	—	—	—	9.0	—	—

TABLE A-1 Table of Food Composition (*continued*) (Computer code is for Cengage Diet Analysis program) (For purposes of calculations, use "0" for t, <1, <.1, <.01, etc.)

DA+ Code	Food Description	Quantity	Measure	Wt (g)	H₂0 (g)	Ener (kcal)	Prot (g)	Carb (g)	Fiber (g)	Fat (g)	Fat Breakdown (g) Sat	Mono	Poly	*Trans*
Fast Food—continued														
	Long John Silver													
39392	Baked cod	1	serving(s)	101	—	120	22.0	1.0	0	4.5	1.0	—	—	0
3777	Batter dipped fish sandwich	1	item(s)	177	—	470	18.0	48.0	3.0	23.0	5.0	—	—	4.5
37568	Battered fish	1	item(s)	92	—	260	12.0	17.0	0.5	16.0	4.0	—	—	4.5
37569	Breaded clams	1	serving(s)	85	—	240	8.0	22.0	1.0	13.0	2.0	—	—	2.5
37566	Chicken plank	1	item(s)	52	—	140	8.0	9.0	0.5	8.0	2.0	—	—	2.5
39404	Clam chowder	1	item(s)	227	—	220	9.0	23.0	0	10.0	4.0	—	—	1.0
39398	Cocktail sauce	1	ounce(s)	28	—	25	0	6.0	0	0	0	0	0	0
3770	Coleslaw	1	serving(s)	113	—	200	1.0	15.0	3.0	15.0	2.5	1.8	4.1	0
39400	French fries, large	1	item(s)	142	—	390	4.0	56.0	5.0	17.0	4.0	—	—	5.0
3774	Fries, regular	1	serving(s)	85	—	230	3.0	34.0	3.0	10.0	2.5	—	—	3.0
3779	Hushpuppy	1	piece(s)	23	—	60	1.0	9.0	1.0	2.5	0.5	—	—	1.0
3781	Shrimp, batter-dipped, 1 piece	1	piece(s)	14	—	45	2.0	3.0	0	3.0	1.0	—	—	1.0
39399	Tartar sauce	1	ounce(s)	28	—	100	0	4.0	0	9.0	1.5	—	—	—
39395	Ultimate fish sandwich	1	item(s)	199	—	530	21.0	49.0	3.0	28.0	8.0	—	—	5.0
	McDonald's													
50828	Asian salad with grilled chicken	1	item(s)	362	—	290	31.0	23.0	6.0	10.0	1.0	—	—	0
2247	Barbecue sauce	1	item(s)	28	—	45	0	11.0	0	0	0	0	0	0
737	Big Mac hamburger	1	item(s)	219	—	560	25.0	47.0	3.0	30.0	10.0	—	—	1.5
29777	Caesar salad dressing	1	package(s)	44	—	150	1.0	5.0	0	13.0	2.5	—	—	—
38391	Caesar salad with grilled chicken, no dressing	1	serving(s)	278	230.6	181	26.4	10.5	3.1	6.0	2.9	1.7	0.8	0.2
38393	Caesar salad without chicken, no dressing	1	serving(s)	190	170.4	84	6.0	8.1	3.0	3.9	2.2	0.9	0.3	0.1
738	Cheeseburger	1	item(s)	119	—	310	15.0	35.0	1.0	12.0	6.0	—	—	1.0
29775	Chicken McGrill sandwich	1	item(s)	213	—	400	27.0	38.0	3.0	16.0	3.0	—	—	0
1873	Chicken McNuggets, 6 piece	1	serving(s)	96	—	250	15.0	15.0	0	15.0	3.0	—	—	1.5
3792	Chicken McNuggets, 4 piece	1	serving(s)	64	—	170	10.0	10.0	0	10.0	2.0	—	—	1.0
29774	Crispy chicken sandwich	1	item(s)	232	121.8	500	27.0	63.0	3.0	16.0	3.0	5.7	7.4	1.5
743	Egg McMuffin	1	item(s)	139	76.8	300	17.0	30.0	2.0	12.0	4.5	3.8	2.5	0
742	Filet-O-Fish sandwich	1	item(s)	141	—	400	14.0	42.0	1.0	18.0	4.0	—	—	1.0
2257	French fries, large	1	serving(s)	170	—	570	6.0	70.0	7.0	30.0	6.0	—	—	8.0
1872	French fries, small	1	serving(s)	74	—	250	2.0	30.0	3.0	13.0	2.5	—	—	3.5
33822	Fruit 'n Yogurt Parfait	1	item(s)	149	111.2	160	4.0	31.0	1.0	2.0	1.0	0.2	0.1	0
739	Hamburger	1	item(s)	105	—	260	13.0	33.0	1.0	9.0	3.5	—	—	0.5
2003	Hash browns	1	item(s)	53	—	140	1.0	15.0	2.0	8.0	1.5	—	—	2.0
2249	Honey sauce	1	item(s)	14	—	50	0	12.0	0	0	0	0	0	0
38397	Newman's Own creamy caesar salad dressing	1	item(s)	59	32.5	190	2.0	4.0	0	18.0	3.5	4.6	9.6	0
38398	Newman's Own low fat balsamic vinaigrette salad dressing	1	item(s)	44	29.1	40	0	4.0	0	3.0	0	1.0	1.2	0
38399	Newman's Own ranch salad dressing	1	item(s)	59	30.1	170	1.0	9.0	0	15.0	2.5	9.0	3.7	0
1874	Plain Hotcakes with syrup and margarine	3	item(s)	221	—	600	9.0	102.0	2.0	17.0	4.0	—	—	4.0
740	Quarter Pounder hamburger	1	item(s)	171	—	420	24.0	40.0	3.0	18.0	7.0	—	—	1.0
741	Quarter Pounder hamburger with cheese	1	item(s)	199	—	510	29.0	43.0	3.0	25.0	12.0	—	—	1.5
2005	Sausage McMuffin with egg	1	item(s)	165	82.4	450	20.0	31.0	2.0	27.0	10.0	10.9	4.6	0.5
50831	Side salad	1	item(s)	87	—	20	1.0	4.0	1.0	0	0	0	0	0
	Pizza Hut													
39009	Hot chicken wings	2	item(s)	57	—	110	11.0	1.0	0	6.0	2.0	—	—	0.3
14025	Meat Lovers hand tossed pizza	1	slice(s)	118	—	300	15.0	29.0	2.0	13.0	6.0	—	—	0.5
14026	Meat Lovers pan pizza	1	slice(s)	123	—	340	15.0	29.0	2.0	19.0	7.0	—	—	0.5
31009	Meat Lovers stuffed crust pizza	1	slice(s)	169	—	450	21.0	43.0	3.0	21.0	10.0	—	—	1.0
14024	Meat Lovers thin 'n crispy pizza	1	slice(s)	98	—	270	13.0	21.0	2.0	14.0	6.0	—	—	0.5
14031	Pepperoni Lovers hand tossed pizza	1	slice(s)	113	—	300	15.0	30.0	2.0	13.0	7.0	—	—	0.5
14032	Pepperoni Lovers pan pizza	1	slice(s)	118	—	340	15.0	29.0	2.0	19.0	7.0	—	—	0.5
31011	Pepperoni Lovers stuffed crust pizza	1	slice(s)	163	—	420	21.0	43.0	3.0	19.0	10.0	—	—	1.0
14030	Pepperoni Lovers thin 'n crispy pizza	1	slice(s)	92	—	260	13.0	21.0	2.0	14.0	7.0	—	—	0.5
10834	Personal Pan pepperoni pizza	1	slice(s)	61	—	170	7.0	18.0	0.5	8.0	3.0	—	—	1.0
10842	Personal Pan supreme pizza	1	slice(s)	77	—	190	8.0	19.0	1.0	9.0	3.5	—	—	1.0

Chol (mg)	Calc (mg)	Iron (mg)	Magn (mg)	Pota (mg)	Sodi (mg)	Zinc (mg)	Vit A (µg)	Thia (mg)	Vit E (mg α)	Ribo (mg)	Niac (mg)	Vit B$_6$ (mg)	Fola (µg)	Vit C (mg)	Vit B$_{12}$ (µg)	Sele (µg)
90	20	0.72	—	—	240.0	—	—	—	—	—	—	—	—	0	—	—
45	60	2.70	—	—	1210.0	—	—	—	—	—	—	—	—	2.4	—	—
35	20	0.72	—	—	790.0	—	—	—	—	—	—	—	—	4.8	—	—
10	20	1.08	—	—	1110.0	—	0	—	—	—	—	—	—	0	—	—
20	0	0.72	—	—	480.0	—	0	—	—	—	—	—	—	2.4	—	—
25	150	0.72	—	—	810.0	—	—	—	—	—	—	—	—	0	—	—
0	0	0.00	—	—	250.0	—	—	—	—	—	—	—	—	0	—	—
20	40	0.36	—	222.7	340.0	0.70	—	0.07	—	0.08	2.34	—	—	18.0	—	—
0	0	0.00	—	—	580.0	—	0	—	—	—	—	—	—	24.0	—	—
0	0	0.00	—	370.0	350.0	0.30	0	0.09	—	0.01	1.60	—	—	15.0	—	—
0	20	0.36	—	—	200.0	—	0	—	—	—	—	—	—	0	—	—
15	0	0.00	—	—	160.0	—	0	—	—	—	—	—	—	1.2	—	—
15	0	0.00	—	—	250.0	—	0	—	—	—	—	—	—	0	—	—
60	150	2.70	—	—	1400.0	—	—	—	—	—	—	—	—	4.8	—	—
65	150	3.60	—	—	890.0	—	—	—	—	—	—	—	—	54.0	—	—
0	0	0.00	—	55.0	260.0	—	—	—	—	—	—	—	—	0	—	—
80	250	4.50	—	400.0	1010.0	—	—	—	—	—	—	—	—	1.2	—	—
10	40	0.18	—	30.0	400.0	—	—	—	—	—	—	—	—	0.6	—	—
67	178	1.77	—	708.9	767.3	—	—	0.15	—	0.19	10.62	—	127.9	29.2	0.2	—
10	163	1.15	17.1	410.4	157.7	—	—	0.08	—	0.07	0.40	—	102.6	26.8	0	0.4
40	200	2.70	—	240.0	740.0	—	60.0	—	—	—	—	—	—	1.2	—	—
70	150	2.70	—	510.0	1010.0	—	—	—	—	—	—	—	—	6.0	—	—
35	20	0.72	—	240.0	670.0	—	—	—	—	—	—	—	—	1.2	—	—
25	0	0.36	—	160.0	450.0	—	—	—	—	—	—	—	—	1.2	—	—
60	80	3.60	62.6	526.6	1380.0	1.53	41.8	0.46	2.27	0.39	12.85	—	104.4	6.0	0.4	—
230	300	2.70	26.4	218.2	860.0	1.59	—	0.36	0.82	0.51	4.31	0.20	109.8	1.2	0.9	—
40	150	1.80	—	250.0	640.0	—	36.2	—	—	—	—	—	—	0	—	—
0	20	1.80	—	—	330.0	—	0	—	—	—	—	—	—	9.0	—	—
0	20	0.72	—	—	140.0	—	0	—	—	—	—	—	—	3.6	—	—
5	150	0.67	20.9	248.8	85.0	0.53	0	0.06	—	0.17	0.35	—	19.4	9.0	0.3	—
30	150	2.70	—	210.0	530.0	—	5.0	—	—	—	—	—	—	1.2	—	—
0	0	0.36	—	210.0	290.0	—	0	—	—	—	—	—	—	1.2	—	—
0	0	0.00	—	0	0	—	0	—	—	—	—	—	—	0	—	—
20	61	0.00	3.0	16.0	500.0	0.20	—	0.01	15.43	0.02	0.01	0.64	2.4	0	0.1	0.1
0	4	0.00	1.3	8.8	730.0	0.01	—	0.00	0.00	0.00	0.00	0.00	0	2.4	0	0
0	40	0.00	1.8	70.4	530.0	0.03	0	0.01	—	0.08	0.01	0.02	0.6	0	0	0.2
20	150	2.70	—	—	280.0	620.0	—	—	—	—	—	—	—	0	—	—
70	150	4.50	—	390.0	730.0	—	10.0	—	—	—	—	—	—	1.2	—	—
95	300	4.50	—	440.0	1150.0	—	100.0	—	—	—	—	—	—	1.2	—	—
255	300	3.60	29.7	282.2	950.0	2.01	—	0.43	0.82	0.56	4.83	0.24	—	0	1.2	—
0	20	0.72	—	—	10.0	—	—	—	—	—	—	—	—	15.0	—	—
70	0	0.36	—	—	450.0	—	—	—	—	—	—	—	—	0	—	—
35	150	1.80	—	—	760.0	—	—	—	—	—	—	—	—	6.0	—	—
35	150	2.70	—	—	750.0	—	—	—	—	—	—	—	—	6.0	—	—
55	250	2.70	—	—	1250.0	—	—	—	—	—	—	—	—	9.0	—	—
35	150	1.44	—	—	740.0	—	—	—	—	—	—	—	—	6.0	—	—
40	200	1.80	—	—	710.0	—	57.7	—	—	—	—	—	—	2.4	—	—
40	200	2.70	—	—	700.0	—	57.7	—	—	—	—	—	—	2.4	—	—
55	300	2.70	—	—	1120.0	—	—	—	—	—	—	—	—	3.6	—	—
40	200	1.44	—	—	690.0	—	58.0	—	—	—	—	—	—	2.4	—	—
15	80	1.44	—	—	340.0	—	38.5	—	—	—	—	—	—	1.4	—	—
20	80	1.86	—	—	420.0	—	—	—	—	—	—	—	—	3.6	—	—

TABLE A-1 **Table of Food Composition** *(continued)* (Computer code is for Cengage Diet Analysis program) (For purposes of calculations, use "0" for t, <1, <.1, <.01, etc.)

DA+ Code	Food Description	Quantity	Measure	Wt (g)	H₂0 (g)	Ener (kcal)	Prot (g)	Carb (g)	Fiber (g)	Fat (g)	Fat Breakdown (g)			
											Sat	Mono	Poly	*Trans*
Fast Food—continued														
39013	Personal Pan Veggie Lovers pizza	1	slice(s)	69	—	150	6.0	19.0	1.0	6.0	2.0	—	—	0.5
14028	Veggie Lovers hand tossed pizza	1	slice(s)	118	—	220	10.0	31.0	2.0	6.0	3.0	—	—	0.3
14029	Veggie Lovers pan pizza	1	slice(s)	119	—	260	10.0	30.0	2.0	12.0	4.0	—	—	0.3
31010	Veggie Lovers stuffed crust pizza	1	slice(s)	172	—	360	16.0	45.0	3.0	14.0	7.0	—	—	0.5
14027	Veggie Lovers thin 'n crispy pizza	1	slice(s)	101	—	180	8.0	23.0	2.0	7.0	3.0	—	—	0.5
39012	Wing blue cheese dipping sauce	1	item(s)	43	—	230	2.0	2.0	0	24.0	5.0	—	—	1.0
39011	Wing ranch dipping sauce	1	item(s)	43	—	210	0.5	4.0	0	22.0	3.5	—	—	0.5
	Starbucks													
38052	Cappuccino, tall	12	fluid ounce(s)	360	—	120	7.0	10.0	0	6.0	4.0	—	—	—
38053	Cappuccino, tall nonfat	12	fluid ounce(s)	360	—	80	7.0	11.0	0	0	0	0	0	0
38054	Cappuccino, tall soymilk	12	fluid ounce(s)	360	—	100	5.0	13.0	0.5	2.5	0	—	—	—
38059	Cinnamon spice mocha, tall nonfat w/o whipped cream	12	fluid ounce(s)	360	—	170	11.0	32.0	0	0.5	—	—	—	—
38057	Cinnamon spice mocha, tall w/whipped cream	12	fluid ounce(s)	360	—	320	10.0	31.0	0	17.0	11.0	—	—	—
38051	Espresso, single shot	1	fluid ounce(s)	30	—	5	0	1.0	0	0	0	0	0	0
38088	Flavored syrup, 1 pump	1	serving(s)	10	—	20	0	5.0	0	0	0	0	0	0
32562	Frappuccino bottled coffee drink, mocha	9½	fluid ounce(s)	298	—	190	6.0	39.0	3.0	3.0	2.0	—	—	—
32561	Frappuccino coffee drink, all bottled flavors	9½	fluid ounce(s)	281	—	190	7.0	35.0	0	3.5	2.5	—	—	—
38073	Frappuccino, mocha	12	fluid ounce(s)	360	—	220	5.0	44.0	0	3.0	1.5	—	—	—
38067	Frappuccino, tall caramel w/o whipped cream	12	fluid ounce(s)	360	—	210	4.0	43.0	0	2.5	1.5	—	—	—
38070	Frappuccino, tall coffee	12	fluid ounce(s)	360	—	190	4.0	38.0	0	2.5	1.5	—	—	—
39894	Frappuccino, tall coffee, light blend	12	fluid ounce(s)	360	—	110	5.0	22.0	2.0	1.0	0	—	—	—
38071	Frappuccino, tall espresso	12	fluid ounce(s)	360	—	160	4.0	33.0	0	2.0	1.5	—	—	—
39897	Frappuccino, tall mocha, light blend	12	fluid ounce(s)	360	—	140	5.0	28.0	3.0	1.5	0	—	—	—
39887	Frappuccino, tall Strawberries and Crème, w/o whipped cream	12	fluid ounce(s)	360	—	330	10.0	65.0	0	3.5	1.0	—	—	—
38063	Frappuccino, tall Tazo chai crème w/o whipped cream	12	fluid ounce(s)	360	—	280	10.0	52.0	0	3.5	1.0	—	—	—
38066	Frappuccino, tall Tazoberry	12	fluid ounce(s)	360	—	140	0.5	36.0	0.5	0	0	0	0	0
38065	Frappuccino, tall Tazoberry Crème	12	fluid ounce(s)	360	—	240	4.0	54.0	0.5	1.0	0	—	—	—
38080	Frappuccino, tall vanilla w/o whipped cream	12	fluid ounce(s)	360	—	270	10.0	51.0	0	3.5	1.0	—	—	—
39898	Frappuccino, tall white chocolate mocha, light blend	12	fluid ounce(s)	360	—	160	6.0	32.0	2.0	2.0	1.0	—	—	—
38074	Frappuccino, tall white chocolate w/o whipped cream	12	fluid ounce(s)	360	—	240	5.0	48.0	0	3.5	2.5	—	—	—
39883	Java Chip Frappuccino, tall w/o whipped cream	12	fluid ounce(s)	360	—	270	5.0	51.0	1.0	7.0	4.5	—	—	—
33111	Latte, tall w/nonfat milk	12	fluid ounce(s)	360	335.3	120	12.0	18.0	0	0	0	0	0	0
33112	Latte, tall w/whole milk	12	fluid ounce(s)	360	—	200	11.0	16.0	0	11.0	7.0	—	—	—
33109	Macchiato, tall caramel w/nonfat milk	12	fluid ounce(s)	360	—	170	11.0	30.0	0	1.0	0	—	—	—
33110	Macchiato, tall caramel w/whole milk	12	fluid ounce(s)	360	—	240	10.0	28.0	0	10.0	6.0	—	—	—
33107	Mocha coffee drink, tall nonfat, w/o whipped cream	12	fluid ounce(s)	360	—	170	11.0	33.0	1.0	1.5	0	—	—	—
38089	Mocha syrup	1	serving(s)	17	—	25	1.0	6.0	0	0.5	0	—	—	—
33108	Mocha, tall mocha w/whole milk	12	fluid ounce(s)	360	—	310	10.0	32.0	1.0	17.0	10.0	—	—	—
38042	Steamed apple cider, tall	12	fluid ounce(s)	360	—	180	0	45.0	0	0	0	0	0	0
38087	Tazo chai black tea, soymilk, tall	12	fluid ounce(s)	360	—	190	4.0	39.0	0.5	2.0	0	—	—	—
38084	Tazo chai black tea, tall	12	fluid ounce(s)	360	—	210	6.0	36.0	0	5.0	3.5	—	—	—
38083	Tazo chai black tea, tall nonfat	12	fluid ounce(s)	360	—	170	6.0	37.0	0	0	0	0	0	0
38076	Tazo iced tea, tall	12	fluid ounce(s)	360	—	60	0	16.0	0	0	0	0	0	0
38077	Tazo tea, grande lemonade	16	fluid ounce(s)	480	—	120	0	31.0	0	0	0	0	0	0

Chol (mg)	Calc (mg)	Iron (mg)	Magn (mg)	Pota (mg)	Sodi (mg)	Zinc (mg)	Vit A (µg)	Thia (mg)	Vit E (mg α)	Ribo (mg)	Niac (mg)	Vit B_6 (mg)	Fola (µg)	Vit C (mg)	Vit B_{12} (µg)	Sele (µg)
10	80	1.80	—	—	280.0	—	—	—	—	—	—	—	—	3.6	—	—
15	150	1.80	—	—	490.0	—	—	—	—	—	—	—	—	9.0	—	—
15	150	2.70	—	—	470.0	—	—	—	—	—	—	—	—	9.0	—	—
35	250	2.70	—	—	980.0	—	—	—	—	—	—	—	—	9.0	—	—
15	150	1.44	—	—	480.0	—	—	—	—	—	—	—	—	9.0	—	—
25	20	0.00	—	—	550.0	—	0	—	—	—	—	—	—	0	—	—
10	0	0.00	—	—	340.0	—	0	—	—	—	—	—	—	0	—	—
25	250	0.00	—	—	95.0	—	—	—	—	—	—	—	—	1.2	0	—
3	200	0.00	—	—	100.0	—	—	—	—	—	—	—	—	0	0	—
0	250	0.72	—	—	75.0	—	—	—	—	—	—	—	—	0	0	—
5	300	0.72	—	—	150.0	—	—	—	—	—	—	—	—	0	0	—
70	350	1.08	—	—	140.0	—	—	—	—	—	—	—	—	2.4	0	—
0	0	0.00	—	—	0	—	0	—	—	—	—	—	—	0	0	—
0	0	0.00	—	—	0	—	0	—	—	—	—	—	—	0	0	—
12	219	1.08	—	530.0	110.0	—	—	—	—	—	—	—	—	0	—	—
15	250	0.36	—	510.0	105.0	—	—	—	—	—	—	—	—	0	—	—
10	150	0.72	—	—	180.0	—	—	—	—	—	—	—	—	0	0	—
10	150	0.00	—	—	180.0	—	—	—	—	—	—	—	—	0	0	—
10	150	0.00	—	—	180.0	—	—	—	—	—	—	—	—	0	0	—
0	150	0.00	—	—	220.0	—	—	—	—	—	—	—	—	0	—	—
10	100	0.00	—	—	160.0	—	—	—	—	—	—	—	—	0	0	—
0	150	0.72	—	—	220.0	—	—	—	—	—	—	—	—	0	—	—
3	350	0.00	—	—	270.0	—	—	—	—	—	—	—	—	21.0	—	—
3	350	0.00	—	—	270.0	—	—	—	—	—	—	—	—	3.6	0	—
0	0	0.00	—	—	30.0	—	0	—	—	—	—	—	—	0	0	—
0	150	0.00	—	—	125.0	—	0	—	—	—	—	—	—	1.2	0	—
3	350	0.00	—	—	370.0	—	—	—	—	—	—	—	—	3.6	0	—
3	150	0.00	—	—	250.0	—	—	—	—	—	—	—	—	0	—	—
10	150	0.00	—	—	210.0	—	—	—	—	—	—	—	—	0	0	—
10	150	1.44	—	—	220.0	—	—	—	—	—	—	—	—	0	—	—
5	350	0.00	39.8	—	170.0	1.35	—	0.12	—	0.47	0.36	0.13	17.5	0	1.3	—
45	400	0.00	46.6	—	160.0	1.28	—	0.12	—	0.54	0.34	0.14	16.8	2.4	1.2	—
5	300	0.00	—	—	160.0	—	—	—	—	—	—	—	—	1.2	—	—
30	300	0.00	—	—	135.0	—	—	—	—	—	—	—	—	2.4	—	—
5	300	2.70	—	—	135.0	—	—	—	—	—	—	—	—	0	—	—
0	0	0.72	—	—	0	—	0	—	—	—	—	—	—	0	0	—
55	300	2.70	—	—	115.0	—	—	—	—	—	—	—	—	0	—	—
0	0	1.08	—	—	15.0	—	0	—	—	—	—	—	—	0	0	—
0	200	0.72	—	—	70.0	—	0	—	—	—	—	—	—	0	0	—
20	200	0.36	—	—	85.0	—	—	—	—	—	—	—	—	1.2	0	—
5	200	0.36	—	—	95.0	—	—	—	—	—	—	—	—	0	0	—
0	0	0.00	—	—	0	—	0	—	—	—	—	—	—	0	0	—
0	0	0.00	—	—	15.0	—	0	—	—	—	—	—	—	4.8	0	—

TABLE A-1 Table of Food Composition (continued) (Computer code is for Cengage Diet Analysis program) (For purposes of calculations, use "0" for t, <1, <.1, <.01, etc.)

DA+ Code	Food Description	Quantity	Measure	Wt (g)	H₂O (g)	Ener (kcal)	Prot (g)	Carb (g)	Fiber (g)	Fat (g)	Fat Breakdown (g) Sat	Mono	Poly	Trans
Fast Food—continued														
38045	Vanilla crème steamed nonfat milk, tall w/whipped cream	12	fluid ounce(s)	360	—	260	11.0	33.0	0	8.0	5.0	—	—	—
38046	Vanilla crème steamed soymilk, tall w/whipped cream	12	fluid ounce(s)	360	—	300	8.0	37.0	1.0	12.0	6.0	—	—	—
38044	Vanilla crème steamed whole milk, tall w/whipped cream	12	fluid ounce(s)	360	—	330	10.0	31.0	0	18.0	11.0	—	—	—
38090	Whipped cream	1	serving(s)	27	—	100	0	2.0	0	9.0	6.0	—	—	—
38062	White chocolate mocha, tall nonfat w/o whipped cream	12	fluid ounce(s)	360	—	260	12.0	45.0	0	4.0	3.0	—	—	—
38061	White chocolate mocha, tall w/whipped cream	12	fluid ounce(s)	360	—	410	11.0	44.0	0	20.0	13.0	—	—	—
38048	White hot chocolate, tall nonfat w/o whipped cream	12	fluid ounce(s)	360	—	300	15.0	51.0	0	4.5	3.5	—	—	—
38050	White hot chocolate, tall soymilk w/whipped cream	12	fluid ounce(s)	360	—	420	11.0	56.0	1.0	16.0	9.0	—	—	—
38047	White hot chocolate, tall w/whipped cream	12	fluid ounce(s)	360	—	460	13.0	50.0	0	22.0	15.0	—	—	—
	Subway													
15842	Cheese steak sandwich, 6", wheat bread	1	item(s)	250	—	360	24.0	47.0	5.0	10.0	4.5	—	—	0
40478	Chicken and bacon ranch sandwich, 6", white or wheat bread	1	serving(s)	297	—	540	36.0	47.0	5.0	25.0	10.0	—	—	0.5
38622	Chicken and bacon ranch wrap with cheese	1	item(s)	257	—	440	41.0	18.0	9.0	27.0	10.0	—	—	0.5
32045	Chocolate chip cookie	1	item(s)	45	—	210	2.0	30.0	1.0	10.0	6.0	—	—	0
32048	Chocolate chip M&M cookie	1	item(s)	45	—	210	2.0	32.0	0.5	10.0	5.0	—	—	0
32049	Chocolate chunk cookie	1	item(s)	45	—	220	2.0	30.0	0.5	10.0	5.0	—	—	0
4024	Classic Italian B.M.T. sandwich, 6", white bread	1	item(s)	236	—	440	22.0	45.0	2.0	21.0	8.5	—	—	0
15838	Classic tuna sandwich, 6", wheat bread	1	item(s)	250	—	530	22.0	45.0	4.0	31.0	7.0	—	—	0.5
15837	Classic tuna sandwich, 6", white bread	1	item(s)	243	—	520	21.0	43.0	2.0	31.0	7.5	—	—	0.5
16397	Club salad, no dressing and croutons	1	item(s)	412	—	160	18.0	15.0	4.0	4.0	1.5	—	—	0
3422	Club sandwich, 6", white bread	1	item(s)	250	—	310	23.0	45.0	2.0	6.0	2.5	—	—	0
4030	Cold cut combo sandwich, 6", white bread	1	item(s)	242	—	400	20.0	45.0	2.0	17.0	7.5	—	—	0.5
34030	Ham and egg breakfast sandwich	1	item(s)	142	—	310	16.0	35.0	3.0	13.0	3.5	—	—	0
3885	Ham sandwich, 6", white bread	1	item(s)	238	—	310	17.0	52.0	2.0	5.0	2.0	—	—	0
3888	Meatball marinara sandwich, 6", wheat bread	1	item(s)	377	—	560	24.0	63.0	7.0	24.0	11.0	—	—	1.0
4651	Meatball sandwich, 6", white bread	1	item(s)	370	—	550	23.0	61.0	5.0	24.0	11.5	—	—	1.0
15839	Melt sandwich, 6", white bread	1	item(s)	260	—	410	25.0	47.0	4.0	15.0	5.0	—	—	—
32046	Oatmeal raisin cookie	1	item(s)	45	—	200	3.0	30.0	1.0	8.0	4.0	—	—	0
16379	Oven-roasted chicken breast sandwich, 6", wheat bread	1	item(s)	238	—	330	24.0	48.0	5.0	5.0	1.5	—	—	0
32047	Peanut butter cookie	1	item(s)	45	—	220	4.0	26.0	1.0	12.0	5.0	—	—	0
4655	Roast beef sandwich, 6", wheat bread	1	item(s)	224	—	290	19.0	45.0	4.0	5.0	2.0	—	—	0
3957	Roast beef sandwich, 6", white bread	1	item(s)	217	—	280	18.0	43.0	2.0	5.0	2.5	—	—	0
16378	Roasted chicken breast, 6", white bread	1	item(s)	231	—	320	23.0	46.0	3.0	5.0	2.0	—	—	0
34028	Southwest steak and cheese sandwich, 6", Italian bread	1	item(s)	271	—	450	24.0	48.0	6.0	20.0	6.0	—	—	0
4032	Spicy Italian sandwich, 6", white bread	1	item(s)	220	—	470	20.0	43.0	2.0	25.0	9.5	—	—	0
4031	Steak and cheese sandwich, 6", white bread	1	item(s)	243	—	350	23.0	45.0	3.0	10.0	5.0	—	—	0
32050	Sugar cookie	1	item(s)	45	—	220	2.0	28.0	0.5	12.0	6.0	—	—	0
40477	Sweet onion chicken teriyaki sandwich, 6", white or wheat bread	1	serving(s)	281	—	370	26.0	59.0	4.0	5.0	1.5	—	—	0
38623	Turkey breast and bacon melt wrap with chipotle sauce	1	item(s)	228	—	380	31.0	20.0	9.0	24.0	7.0	—	—	0

PAGE KEY: A-6 = Breads/Baked Goods A-12 = Cereal/Rice/Pasta A-16 = Fruit A-20 = Vegetables/Legumes A-30 = Nuts/Seeds A-32 = Vegetarian A-34 = Dairy A-42 = Eggs A-42 = Seafood A-44 = Meats A-48 = Poultry
A-50 = Processed Meats A-50 = Beverages A-54 = Fats/Oils A-56 = Sweets A-58 = Spices/Condiments/Sauces A-62 = Mixed Foods/Soups/Sandwiches A-66 = Fast Food A-86 = Convenience A-88 = Baby Foods

Chol (mg)	Calc (mg)	Iron (mg)	Magn (mg)	Pota (mg)	Sodi (mg)	Zinc (mg)	Vit A (µg)	Thia (mg)	Vit E (mg α)	Ribo (mg)	Niac (mg)	Vit B$_6$ (mg)	Fola (µg)	Vit C (mg)	Vit B$_{12}$ (µg)	Sele (µg)
35	350	0.00	—	—	170.0	—	—	—	—	—	—	—	—	0	0	—
30	400	1.44	—	—	130.0	—	—	—	—	—	—	—	—	0	0	—
65	350	0.00	—	—	140.0	—	—	—	—	—	—	—	—	0	0	—
40	0	0.00	—	—	10.0	—	—	—	—	—	—	—	—	0	0	—
5	400	0.00	—	—	210.0	—	—	—	—	—	—	—	—	0	0	—
70	400	0.00	—	—	210.0	—	—	—	—	—	—	—	—	2.4	0	—
10	450	0.00	—	—	250.0	—	—	—	—	—	—	—	—	0	0	—
35	500	1.44	—	—	210.0	—	—	—	—	—	—	—	—	0	0	—
75	500	0.00	—	—	250.0	—	—	—	—	—	—	—	—	3.6	0	—
35	150	8.10	—	—	1090.0	—	—	—	—	—	—	—	—	18.0	—	—
90	250	4.50	—	—	1400.0	—	—	—	—	—	—	—	—	21.0	—	—
90	300	2.70	—	—	1680.0	—	—	—	—	—	—	—	—	9.0	—	—
15	0	1.08	—	—	150.0	—	—	—	—	—	—	—	—	0	—	—
10	20	1.00	—	—	100.0	—	—	—	—	—	—	—	—	0	—	—
10	0	1.00	—	—	100.0	—	—	—	—	—	—	—	—	0	—	—
55	150	2.70	—	—	1770.0	—	—	—	—	—	—	—	—	16.8	—	—
45	100	5.40	—	—	1030.0	—	—	—	—	—	—	—	—	21.0	—	—
45	100	3.60	—	—	1010.0	—	—	—	—	—	—	—	—	16.8	—	—
35	60	3.60	—	—	880.0	—	—	—	—	—	—	—	—	30.0	—	—
35	60	3.60	—	—	1290.0	—	—	—	—	—	—	—	—	13.8	—	—
60	150	3.60	—	—	1530.0	—	—	—	—	—	—	—	—	16.8	—	—
190	80	4.50	—	—	720.0	—	66.7	—	—	—	—	—	—	3.6	—	—
25	60	2.70	—	—	1375.0	—	—	—	—	—	—	—	—	13.8	—	—
45	200	7.20	—	—	1610.0	—	—	—	—	—	—	—	—	36.0	—	—
45	200	5.40	—	—	1590.0	—	—	—	—	—	—	—	—	31.8	—	—
45	150	5.40	—	—	1720.0	—	—	—	—	—	—	—	—	24.0	—	—
15	20	1.08	—	—	170.0	—	—	—	—	—	—	—	—	0	—	—
45	60	4.50	—	—	1020.0	—	—	—	—	—	—	—	—	18.0	—	—
15	20	0.72	—	—	200.0	—	—	—	—	—	—	—	—	0	—	—
20	60	6.30	—	—	920.0	—	—	—	—	—	—	—	—	18.0	—	—
20	60	4.50	—	—	900.0	—	—	—	—	—	—	—	—	13.8	—	—
45	60	2.70	—	—	1000.0	—	—	—	—	—	—	—	—	13.8	—	—
45	150	8.10	—	—	1310.0	—	—	—	—	—	—	—	—	21.0	—	—
55	60	2.70	—	—	1650.0	—	—	—	—	—	—	—	—	16.8	—	—
35	150	6.30	—	—	1070.0	—	—	—	—	—	—	—	—	13.8	—	—
15	0	0.72	—	—	140.0	—	—	—	—	—	—	—	—	0	—	—
50	80	4.50	—	—	1220.0	—	—	—	—	—	—	—	—	24.0	—	—
50	200	2.70	—	—	1780.0	—	—	—	—	—	—	—	—	6.0	—	—

TABLE A-1 Table of Food Composition (continued) (Computer code is for Cengage Diet Analysis program) (For purposes of calculations, use "0" for t, <1, <.1, <.01, etc.)

DA+ Code	Food Description	Quantity	Measure	Wt (g)	H₂O (g)	Ener (kcal)	Prot (g)	Carb (g)	Fiber (g)	Fat (g)	Sat	Mono	Poly	Trans
												Fat Breakdown (g)		

Fast Food—continued

DA+ Code	Food Description	Quantity	Measure	Wt (g)	H₂O (g)	Ener (kcal)	Prot (g)	Carb (g)	Fiber (g)	Fat (g)	Sat	Mono	Poly	Trans
15834	Turkey breast and ham sandwich, 6", white bread	1	item(s)	227	—	280	19.0	45.0	2.0	5.0	2.0	—	—	0
16376	Turkey breast sandwich, 6", white bread	1	item(s)	217	—	270	17.0	44.0	2.0	4.5	2.0	—	—	0
15841	Veggie Delite sandwich, 6", wheat bread	1	item(s)	167	—	230	9.0	44.0	4.0	3.0	1.0	—	—	0
16375	Veggie Delite, 6", white bread	1	item(s)	160	—	220	8.0	42.0	2.0	3.0	1.5	—	—	0
32051	White chip macadamia nut cookie	1	item(s)	45	—	220	2.0	29.0	0.5	11.0	5.0	—	—	0
	Taco Bell													
29906	7-Layer burrito	1	item(s)	283	—	490	17.0	65.0	9.0	18.0	7.0	—	—	1.0
744	Bean burrito	1	item(s)	198	—	340	13.0	54.0	8.0	9.0	3.5	—	—	0.5
749	Beef burrito supreme	1	item(s)	248	—	410	17.0	51.0	7.0	17.0	8.0	—	—	1.0
33417	Beef Chalupa Supreme	1	item(s)	153	—	380	14.0	30.0	3.0	23.0	7.0	—	—	0.5
34474	Beef Gordita Baja	1	item(s)	153	—	340	13.0	29.0	4.0	19.0	5.0	—	—	0
29910	Beef Gordita Supreme	1	item(s)	153	—	310	14.0	29.0	3.0	16.0	6.0	—	—	0.5
2014	Beef soft taco	1	item(s)	99	—	200	10.0	21.0	3.0	9.0	4.0	—	—	0
10860	Beef soft taco supreme	1	item(s)	135	—	250	11.0	23.0	3.0	13.0	6.0	—	—	0.5
34472	Chicken burrito supreme	1	item(s)	248	—	390	20.0	49.0	6.0	13.0	6.0	—	—	0.5
33418	Chicken Chalupa Supreme	1	item(s)	153	—	360	17.0	29.0	2.0	20.0	5.0	—	—	0
34475	Chicken Gordita Baja	1	item(s)	153	—	320	17.0	28.0	3.0	16.0	3.5	—	—	0
29909	Chicken quesadilla	1	item(s)	184	—	520	28.0	40.0	3.0	28.0	12.0	—	—	0.5
29907	Chili cheese burrito	1	item(s)	156	—	390	16.0	40.0	3.0	18.0	9.0	—	—	1.5
10794	Cinnamon twists	1	serving(s)	35	—	170	1.0	26.0	1.0	7.0	0	—	—	0
29911	Grilled chicken Gordita Supreme	1	item(s)	153	—	290	17.0	28.0	2.0	12.0	5.0	—	—	0
14463	Grilled chicken soft taco	1	item(s)	99	—	190	14.0	19.0	1.0	6.0	2.5	—	—	—
29912	Grilled Steak Gordita Supreme	1	item(s)	153	—	290	15.0	28.0	2.0	13.0	5.0	—	—	0
29904	Grilled steak soft taco	1	item(s)	128	—	270	12.0	20.0	2.0	16.0	4.5	—	—	0
29905	Grilled steak soft taco supreme	1	item(s)	135	—	235	13.0	21.0	1.0	11.0	6.0	—	—	—
2021	Mexican pizza	1	serving(s)	216	—	530	20.0	42.0	7.0	30.0	8.0	—	—	1.0
29894	Mexican rice	1	serving(s)	131	—	170	6.0	23.0	1.0	11.0	3.0	—	—	0
10772	Meximelt	1	serving(s)	128	—	280	15.0	22.0	3.0	14.0	7.0	—	—	0.5
2011	Nachos	1	serving(s)	99	—	330	4.0	32.0	2.0	21.0	3.5	—	—	2.0
2012	Nachos Bellgrande	1	serving(s)	308	—	770	19.0	77.0	12.0	44.0	9.0	—	—	3.0
2023	Pintos 'n cheese	1	serving(s)	128	—	150	9.0	19.0	7.0	6.0	3.0	—	—	0.5
34473	Steak burrito supreme	1	item(s)	248	—	380	18.0	49.0	6.0	14.0	7.0	—	—	0.5
33419	Steak Chalupa Supreme	1	item(s)	153	—	360	15.0	28.0	2.0	21.0	6.0	—	—	0
747	Taco	1	item(s)	78	—	170	8.0	13.0	3.0	10.0	3.5	—	—	0
2015	Taco salad with salsa, with shell	1	serving(s)	548	—	840	30.0	80.0	15.0	45.0	11.0	—	—	1.5
14459	Taco supreme	1	item(s)	113	—	210	9.0	15.0	3.0	13.0	6.0	—	—	0
748	Tostada	1	item(s)	170	—	230	11.0	27.0	7.0	10.0	3.5	—	—	0.5

Convenience Meals

DA+ Code	Food Description	Quantity	Measure	Wt (g)	H₂O (g)	Ener (kcal)	Prot (g)	Carb (g)	Fiber (g)	Fat (g)	Sat	Mono	Poly	Trans
	Banquet													
29961	Barbeque chicken meal	1	item(s)	281	—	330	16.0	37.0	2.0	13.0	3.0	—	—	—
14788	Boneless white fried chicken meal	1	item(s)	286	—	310	10.0	21.0	4.0	20.0	5.0	—	—	—
29960	Fish sticks meal	1	item(s)	207	—	470	13.0	58.0	1.0	20.0	3.5	—	—	—
29957	Lasagna with meat sauce meal	1	item(s)	312	—	320	15.0	46.0	7.0	9.0	4.0	—	—	—
14777	Macaroni and cheese meal	1	item(s)	340	—	420	15.0	57.0	5.0	14.0	8.0	—	—	—
1741	Meatloaf meal	1	item(s)	269	—	240	14.0	20.0	4.0	11.0	4.0	—	—	—
39418	Pepperoni pizza meal	1	item(s)	191	—	480	11.0	56.0	5.0	23.0	8.0	—	—	—
33759	Roasted white turkey meal	1	item(s)	255	—	230	14.0	30.0	5.0	6.0	2.0	—	—	—
1743	Salisbury steak meal	1	item(s)	269	196.9	380	12.0	28.0	3.0	24.0	12.0	—	—	—
	Budget Gourmet													
1914	Cheese manicotti with meat sauce entrée	1	item(s)	284	194.0	420	18.0	38.0	4.0	22.0	11.0	6.0	1.3	—
1915	Chicken with fettucini entrée	1	item(s)	284	—	380	20.0	33.0	3.0	19.0	10.0	—	—	—
3986	Light beef stroganoff entrée	1	item(s)	248	177.0	290	20.0	32.0	3.0	7.0	4.0	—	—	—
3996	Light sirloin of beef in herb sauce entrée	1	item(s)	269	214.0	260	19.0	30.0	5.0	7.0	4.0	2.3	0.3	—
3987	Light vegetable lasagna entrée	1	item(s)	298	227.0	290	15.0	36.0	4.8	9.0	1.8	0.9	0.6	—

PAGE KEY: A-6 = Breads/Baked Goods A-12 = Cereal/Rice/Pasta A-16 = Fruit A-20 = Vegetables/Legumes A-30 = Nuts/Seeds A-32 = Vegetarian A-34 = Dairy A-42 = Eggs A-42 = Seafood A-44 = Meats A-48 = Poultry A-50 = Processed Meats A-50 = Beverages A-54 = Fats/Oils A-56 = Sweets A-58 = Spices/Condiments/Sauces A-62 = Mixed Foods/Soups/Sandwiches A-66 = Fast Food A-86 = Convenience A-88 = Baby Foods

Chol (mg)	Calc (mg)	Iron (mg)	Magn (mg)	Pota (mg)	Sodi (mg)	Zinc (mg)	Vit A (µg)	Thia (mg)	Vit E (mg α)	Ribo (mg)	Niac (mg)	Vit B$_6$ (mg)	Fola (µg)	Vit C (mg)	Vit B$_{12}$ (µg)	Sele (µg)
25	60	2.70	—	—	1210.0	—	—	—	—	—	—	—	—	13.8	—	—
20	60	2.70	—	—	1000.0	—	—	—	—	—	—	—	—	13.8	—	—
0	60	4.50	—	—	520.0	—	—	—	—	—	—	—	—	18.0	—	—
0	60	2.70	—	—	500.0	—	—	—	—	—	—	—	—	13.8	—	—
15	20	0.72	—	—	160.0	—	—	—	—	—	—	—	—	0	—	—
25	250	5.40	—	—	1350.0	—	—	—	—	—	—	—	—	15.0	—	—
5	200	4.50	—	—	1190.0	—	5.9	—	—	—	—	—	—	4.8	—	—
40	200	4.50	—	—	1340.0	—	9.9	—	—	—	—	—	—	6.0	—	—
40	150	2.70	—	—	620.0	—	—	—	—	—	—	—	—	3.6	—	—
35	100	2.70	—	—	780.0	—	—	—	—	—	—	—	—	2.4	—	—
40	150	2.70	—	—	620.0	—	—	—	—	—	—	—	—	3.6	—	—
25	100	1.80	—	—	630.0	—	—	—	—	—	—	—	—	1.2	—	—
40	150	2.70	—	—	650.0	—	—	—	—	—	—	—	—	3.6	—	—
45	200	4.50	—	—	1360.0	—	—	—	—	—	—	—	—	9.0	—	—
45	100	2.70	—	—	650.0	—	—	—	—	—	—	—	—	4.8	—	—
40	100	1.80	—	—	800.0	—	—	—	—	—	—	—	—	3.6	—	—
75	450	3.60	—	—	1420.0	—	—	—	—	—	—	—	—	1.2	—	—
40	300	1.80	—	—	1080.0	—	—	—	—	—	—	—	—	0	—	—
0	0	0.37	—	—	200.0	—	0	—	—	—	—	—	—	0	—	—
45	150	1.80	—	—	650.0	—	—	—	—	—	—	—	—	4.8	—	—
30	100	1.08	—	—	550.0	—	14.6	—	—	—	—	—	—	1.2	—	—
40	100	2.70	—	—	530.0	—	—	—	—	—	—	—	—	3.6	—	—
35	100	2.70	—	—	660.0	—	—	—	—	—	—	—	—	3.6	—	—
35	120	1.44	—	—	565.0	—	29.2	—	—	—	—	—	—	3.6	—	—
40	350	3.60	—	—	1000.0	—	—	—	—	—	—	—	—	4.8	—	—
15	100	1.44	—	—	790.0	—	—	—	—	—	—	—	—	3.6	—	—
40	250	2.70	—	—	880.0	—	—	—	—	—	—	—	—	2.4	—	—
3	80	0.71	—	—	530.0	—	0	—	—	—	—	—	—	0	—	—
35	200	3.60	—	—	1280.0	—	—	—	—	—	—	—	—	4.8	—	—
15	150	1.44	—	—	670.0	—	—	—	—	—	—	—	—	3.6	—	—
35	200	4.50	—	—	1250.0	—	9.9	—	—	—	—	—	—	9.0	—	—
40	100	2.70	—	—	530.0	—	—	—	—	—	—	—	—	3.6	—	—
25	80	1.08	—	—	350.0	—	—	—	—	—	—	—	—	1.2	—	—
65	450	7.20	—	—	1780.0	—	—	—	—	—	—	—	—	12.0	—	—
40	100	1.08	—	—	370.0	—	—	—	—	—	—	—	—	3.6	—	—
15	200	1.80	—	—	730.0	—	—	—	—	—	—	—	—	4.8	—	—
																—
50	40	1.08	—	—	1210.0	—	0	—	—	—	—	—	—	4.8	—	—
45	80	1.44	—	—	1200.0	—	—	—	—	—	—	—	—	18.0	—	—
55	20	1.44	—	—	710.0	—	—	—	—	—	—	—	—	0	—	—
20	100	2.70	—	—	1170.0	—	—	—	—	—	—	—	—	0	—	—
20	150	1.44	—	—	1330.0	—	0	—	—	—	—	—	—	0	—	—
30	0	1.80	—	—	1040.0	—	0	—	—	—	—	—	—	0	—	—
35	150	1.80	—	—	870.0	—	0	—	—	—	—	—	—	0	—	—
25	60	1.80	—	—	1070.0	—	—	—	—	—	—	—	—	3.6	—	—
60	40	1.44	—	—	1140.0	—	0	—	—	—	—	—	—	0	—	—
85	300	2.70	45.4	484.0	810.0	2.29	—	0.45	—	0.51	4.00	0.22	30.7	0	0.7	—
85	100	2.70	—	—	810.0	—	—	0.15	—	0.42	6.00	—	—	0	—	—
35	40	1.80	38.9	280.0	580.0	4.71	—	0.17	—	0.36	4.28	0.27	18.9	2.4	2.5	—
30	40	1.80	57.7	540.0	850.0	4.81	—	0.15	—	0.29	5.53	0.37	38.4	6.0	1.6	—
15	283	3.03	78.5	420.0	780.0	1.39	—	0.22	—	0.45	3.13	0.32	74.8	59.1	0.2	—

TABLE A-1 Table of Food Composition (continued) (Computer code is for Cengage Diet Analysis program) (For purposes of calculations, use "0" for t, <1, <.1, <.01, etc.)

DA+ Code	Food Description	Quantity	Measure	Wt (g)	H₂0 (g)	Ener (kcal)	Prot (g)	Carb (g)	Fiber (g)	Fat (g)	Fat Breakdown (g) Sat	Mono	Poly	Trans
Convenience Meals—continued														
	Healthy Choice													
9425	Cheese French bread pizza	1	item(s)	170	—	340	22.0	51.0	5.0	5.0	1.5	—	—	—
9306	Chicken enchilada suprema meal	1	item(s)	320	251.5	360	13.0	59.0	8.0	7.0	3.0	2.0	2.0	—
3821	Familiar Favorites lasagna bake with meat sauce entrée	1	item(s)	255	—	270	13.0	38.0	4.0	7.0	2.5	—	—	—
13744	Familiar Favorites sesame chicken with vegetables and rice entrée	1	item(s)	255	—	260	17.0	34.0	4.0	6.0	2.0	2.0	2.0	—
9316	Lemon pepper fish meal	1	item(s)	303	—	280	11.0	49.0	5.0	5.0	2.0	1.0	2.0	—
9322	Traditional salisbury steak meal	1	item(s)	354	250.3	360	23.0	45.0	5.0	9.0	3.5	4.0	1.0	—
9359	Traditional turkey breasts meal	1	item(s)	298	—	330	21.0	50.0	4.0	5.0	2.0	1.5	1.5	—
	Stouffers													
2313	Cheese French bread pizza	1	serving(s)	294	—	380	15.0	43.0	3.0	16.0	6.0	—	—	—
11138	Cheese manicotti with tomato sauce entrée	1	item(s)	255	—	360	18.0	41.0	2.0	14.0	6.0	—	—	—
2366	Chicken pot pie entrée	1	item(s)	284	—	740	23.0	56.0	4.0	47.0	18.0	12.4	10.5	—
11116	Homestyle baked chicken breast with mashed potatoes and gravy entrée	1	item(s)	252	—	270	21.0	21.0	2.0	11.0	3.5	—	—	—
11146	Homestyle beef pot roast and potatoes entrée	1	item(s)	252	—	260	16.0	24.0	3.0	11.0	4.0	—	—	—
11152	Homestyle roast turkey breast with stuffing and mashed potatoes entrée	1	item(s)	273	—	290	16.0	30.0	2.0	12.0	3.5	—	—	—
11043	Lean Cuisine Comfort Classics baked chicken and whipped potatoes and stuffing entrée	1	item(s)	245	—	240	15.0	34.0	3.0	4.5	1.0	2.0	1.0	0
11046	Lean Cuisine Comfort Classics honey mustard chicken with rice pilaf entrée	1	item(s)	227	—	250	17.0	37.0	1.0	4.0	1.0	1.0	1.0	0
9479	Lean Cuisine Deluxe French bread pizza	1	item(s)	174	—	310	16.0	44.0	3.0	9.0	3.5	0.5	0.5	0
360	Lean Cuisine One Dish Favorites chicken chow mein with rice	1	item(s)	255	—	190	13.0	29.0	2.0	2.5	0.5	1.0	0.5	0
11054	Lean Cuisine One Dish Favorites chicken enchilada Suiza with Mexican-style rice	1	serving(s)	255	—	270	10.0	47.0	3.0	4.5	2.0	1.5	1.0	0
9467	Lean Cuisine One Dish Favorites fettucini alfredo entrée	1	item(s)	262	—	270	13.0	39.0	2.0	7.0	3.5	2.0	1.0	0
11055	Lean Cuisine One Dish Favorites lasagna with meat sauce entrée	1	item(s)	298	—	320	19.0	44.0	4.0	7.0	3.0	2.0	0.5	0
	Weight Watchers													
11164	Smart Ones chicken enchiladas suiza entrée	1	item(s)	255	—	340	12.0	38.0	3.0	10.0	4.5	—	—	—
39763	Smart Ones chicken oriental entrée	1	item(s)	255	—	230	15.0	34.0	3.0	4.5	1.0	—	—	—
11187	Smart Ones pepperoni pizza	1	item(s)	198	—	400	22.0	58.0	4.0	9.0	3.0	—	—	—
39765	Smart Ones spaghetti bolognese entrée	1	item(s)	326	—	280	17.0	43.0	5.0	5.0	2.0	—	—	—
31512	Smart Ones spicy szechuan style vegetables and chicken	1	item(s)	255	—	220	11.0	34.0	4.0	5.0	1.0	—	—	—
Baby Foods														
787	Apple juice	4	fluid ounce(s)	127	111.6	60	0	14.8	0.1	0.1	0	0	0	—
778	Applesauce, strained	4	tablespoon(s)	64	56.7	26	0.1	6.9	1.1	0.1	0	0	0	—
779	Bananas with tapioca, strained	4	tablespoon(s)	60	50.4	34	0.2	9.2	1.0	0	0	0	0	—
604	Carrots, strained	4	tablespoon(s)	56	51.7	15	0.4	3.4	1.0	0.1	0	0	0	—
770	Chicken noodle dinner, strained	4	tablespoon(s)	64	54.8	42	1.7	5.8	1.3	1.3	0.4	0.5	0.3	—
801	Green beans, strained	4	tablespoon(s)	60	55.1	16	0.7	3.8	1.3	0.1	0	0	0	—
910	Human milk, mature	2	fluid ounce(s)	62	53.9	43	0.6	4.2	0	2.7	1.2	1.0	0.3	—
760	Mixed cereal, prepared with whole milk	4	ounce(s)	113	84.6	128	5.4	18.0	1.5	4.0	2.2	1.2	0.4	—
772	Mixed vegetable dinner, strained	2	ounce(s)	57	50.3	23	0.7	5.4	0.8	0	—	—	0	—
762	Rice cereal, prepared with whole milk	4	ounce(s)	113	84.6	130	4.4	18.9	0.1	4.1	2.6	1.0	0.2	—
758	Teething biscuits	1	item(s)	11	0.7	44	1.0	8.6	0.2	0.6	0.2	0.2	0.1	—

Chol (mg)	Calc (mg)	Iron (mg)	Magn (mg)	Pota (mg)	Sodi (mg)	Zinc (mg)	Vit A (µg)	Thia (mg)	Vit E (mg α)	Ribo (mg)	Niac (mg)	Vit B$_6$ (mg)	Fola (µg)	Vit C (mg)	Vit B$_{12}$ (µg)	Sele (µg)
10	350	3.60	—	—	600.0	—	—	—	—	—	—	—	—	0	—	—
30	40	1.44	—	—	580.0	—	—	—	—	—	—	—	—	3.6	—	—
20	100	1.80	—	—	600.0	—	—	—	—	—	—	—	—	0	—	—
35	18	0.72	—	—	580.0	—	—	—	—	—	—	—	—	12.0	—	—
35	20	0.36	—	—	580.0	—	—	—	—	—	—	—	—	30.0	—	—
45	80	2.70	—	—	580.0	—	—	—	—	—	—	—	—	21.0	—	—
35	40	1.80	—	—	600.0	—	—	—	—	—	—	—	—	0	—	—
30	200	1.80	—	230.0	660.0	—	—	—	—	—	—	—	—	2.4	—	—
70	250	1.44	—	550.0	920.0	—	—	—	—	—	—	—	—	6.0	—	—
65	150	2.70	—	—	1170.0	—	—	—	—	—	—	—	—	2.4	—	—
55	20	0.72	—	490.0	770.0	—	0	—	—	—	—	—	—	0	—	—
35	20	1.80	—	800.0	960.0	—	—	—	—	—	—	—	—	6.0	—	—
45	40	1.08	—	490.0	970.0	—	—	—	—	—	—	—	—	3.6	—	—
25	40	1.16	—	500.0	650.0	—	—	—	—	—	—	—	—	3.6	—	—
30	64	0.38	—	370.0	650.0	—	—	—	—	—	—	—	—	0	—	—
20	150	2.70	—	300.0	700.0	—	—	—	—	—	—	—	—	15.0	—	—
25	40	0.72	—	380.0	650.0	—	—	—	—	—	—	—	—	2.4	—	—
20	150	0.72	—	350.0	510.0	—	—	—	—	—	—	—	—	2.4	—	—
15	200	0.72	—	290.0	690.0	—	0	—	—	—	—	—	—	0	—	—
30	250	1.47	—	610.0	690.0	—	—	—	—	—	—	—	—	2.4	—	—
40	200	0.72	—	—	800.0	—	—	—	—	—	—	—	—	2.4	—	—
35	40	0.72	—	—	790.0	—	—	—	—	—	—	—	—	6.0	—	—
15	200	1.08	—	401.0	700.0	—	69.1	—	—	—	—	—	—	4.8	—	—
15	150	3.60	—	—	670.0	—	—	—	—	—	—	—	—	9.0	—	—
10	40	1.44	—	—	890.0	—	—	—	—	—	—	—	—	0	—	—
0	5	0.72	3.8	115.4	3.8	0.03	1.3	0.01	0.76	0.02	0.10	0.03	0	73.4	0	0.1
0	3	0.12	1.9	45.4	1.3	0.01	0.6	0.01	0.36	0.02	0.04	0.02	1.3	24.5	0	0.2
0	3	0.12	6.0	52.8	5.4	0.04	1.2	0.01	0.36	0.02	0.08	0.04	3.6	10.0	0	0.4
0	12	0.20	5.0	109.8	20.7	0.08	320.9	0.01	0.29	0.02	0.25	0.04	8.4	3.2	0	0.1
10	17	0.40	9.0	89.0	14.7	0.32	70.4	0.03	0.12	0.04	0.44	0.04	7.0	0	0	2.4
0	23	0.40	12.0	87.6	3.0	0.12	10.8	0.02	0.04	0.04	0.20	0.02	14.4	0.2	0	0
9	20	0.02	1.8	31.4	10.5	0.10	37.6	0.01	0.04	0.02	0.10	0.01	3.1	3.1	0	1.1
12	249	11.82	30.6	225.7	53.3	0.80	28.4	0.49	—	0.65	6.54	0.07	12.5	1.4	0.3	
—	12	0.18	6.2	68.6	4.5	0.08	77.1	0.01	—	0.02	0.28	0.04	4.5	1.6	0	0.4
12	271	13.82	51.0	215.5	52.2	0.72	24.9	0.52	—	0.56	5.90	0.12	9.1	1.4	0.3	4.0
0	11	0.39	3.9	35.5	28.4	0.10	3.1	0.02	0.02	0.05	0.47	0.01	5.4	1.0	0	2.6

Appendix B

WHO: Nutrition Recommendations
Canada: Guidelines and Meal Planning

This appendix presents nutrition recommendations from the World Health Organization (WHO) and details for Canadians on the *Eating Well with Canada's Food Guide* and the *Beyond the Basics* meal-planning system.

Nutrition Recommendations from WHO

The World Health Organization (WHO) has assessed the relationships between diet and the development of chronic diseases. Its recommendations include:

▌ Energy: sufficient to support growth, physical activity, and a healthy body weight (BMI between 18.5 and 24.9) and to avoid weight gain greater than 11 pounds (5 kilograms) during adult life

▌ Total fat: 15 to 30 percent of total energy

▌ Saturated fatty acids: <10 percent of total energy

▌ Polyunsaturated fatty acids: 6 to 10 percent of total energy

▌ Omega-6 polyunsaturated fatty acids: 5 to 8 percent of total energy

▌ Omega-3 polyunsaturated fatty acids: 1 to 2 percent of total energy

▌ *Trans*-fatty acids: <1 percent of total energy

▌ Total carbohydrate: 55 to 75 percent of total energy

▌ Sugars: <10 percent of total energy

▌ Protein: 10 to 15 percent of total energy

▌ Cholesterol: <300 mg per day

▌ Salt (sodium): <5 g salt per day (<2 g sodium per day), appropriately iodized

▌ Fruits and vegetables: ≥400 g per day (about 1 pound)

▌ Total dietary fiber: >25 g per day from foods

▌ Physical activity: one hour of moderate-intensity activity, such as walking, on most days of the week

Eating Well with Canada's Food Guide

Figure B-1 presents the 2007 *Eating Well with Canada's Food Guide*. Additional publications, which are available from Health Canada through its website, provide many more details.

Search for "Canada's food guide" at Health Canada: www.hc-sc.gc.ca

FIGURE B-1 Eating Well with Canada's Food Guide.

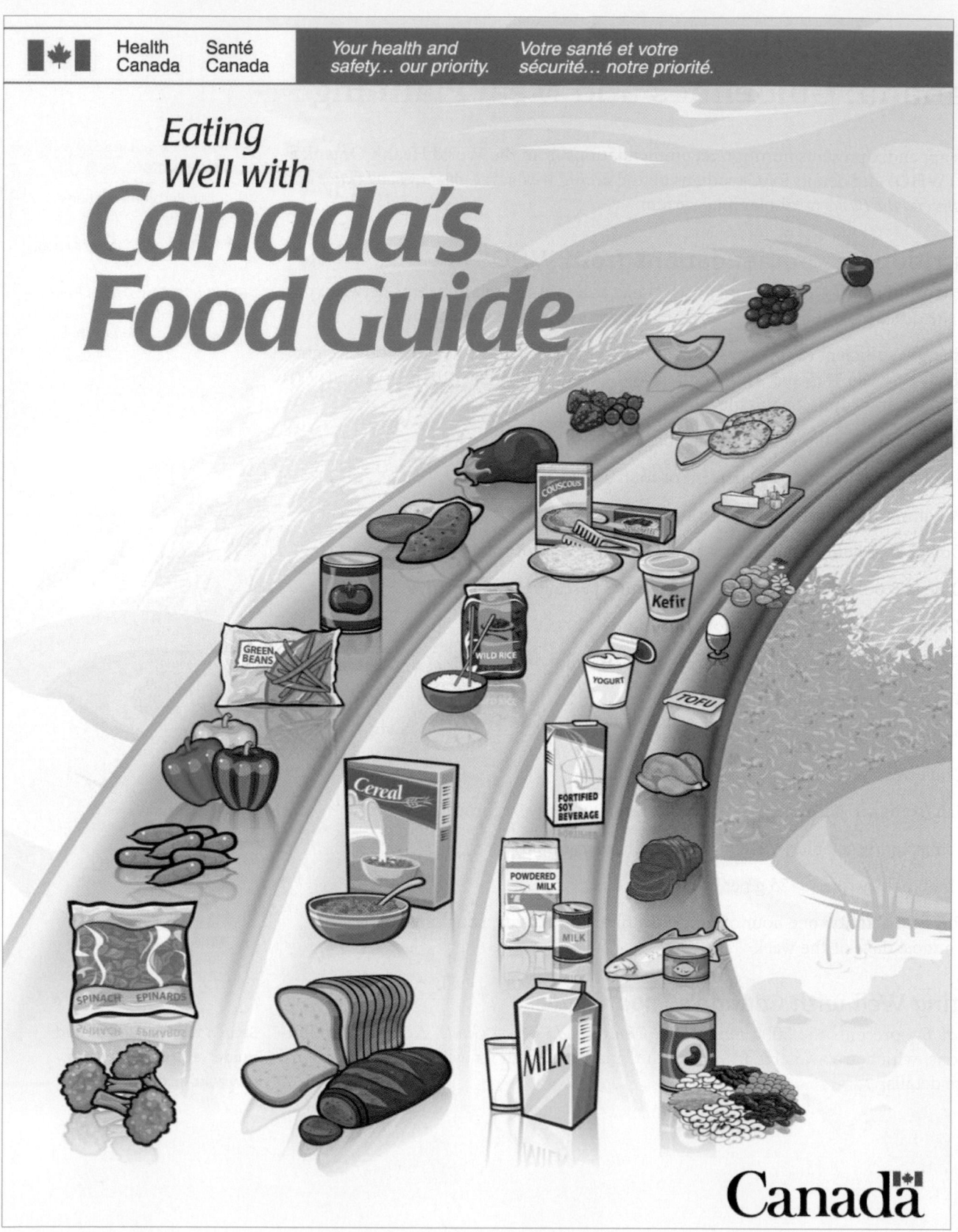

FIGURE B-1 **Eating Well with Canada's Food Guide. (*continued*)**

Recommended Number of *Food Guide Servings* per Day

	Children			Teens		Adults			
Age in Years	2-3	4-8	9-13	14-18		19-50		51+	
Sex	Girls and Boys			Females	Males	Females	Males	Females	Males
Vegetables and Fruit	4	5	6	7	8	7-8	8-10	7	7
Grain Products	3	4	6	6	7	6-7	8	6	7
Milk and Alternatives	2	2	3-4	3-4	3-4	2	2	3	3
Meat and Alternatives	1	1	1-2	2	3	2	3	2	3

The chart above shows how many Food Guide Servings you need from each of the four food groups every day.

Having the amount and type of food recommended and following the tips in *Canada's Food Guide* will help:

• Meet your needs for vitamins, minerals and other nutrients.

• Reduce your risk of obesity, type 2 diabetes, heart disease, certain types of cancer and osteoporosis.

• Contribute to your overall health and vitality.

FIGURE B-1 Eating Well with Canada's Food Guide. (*continued*)

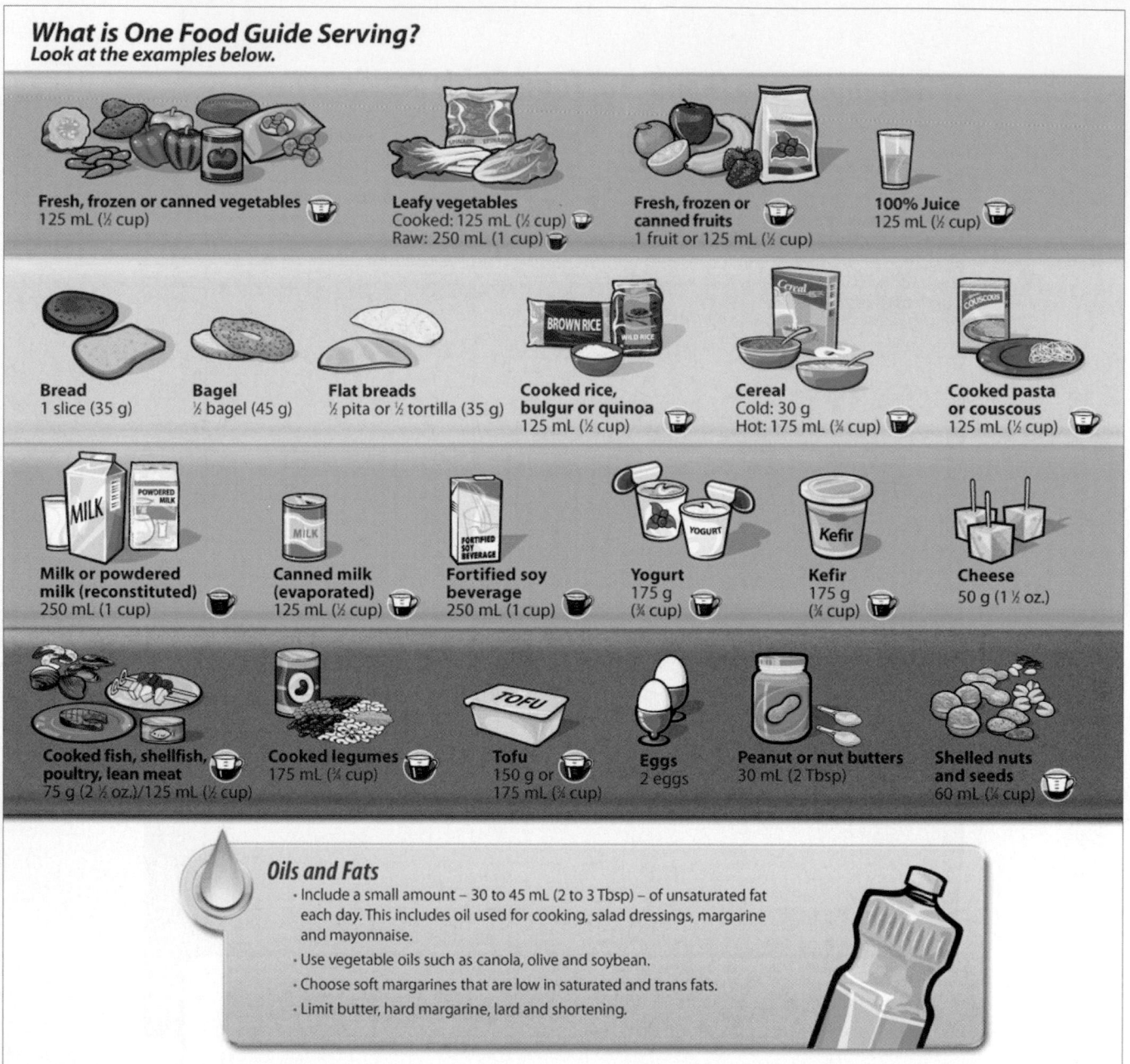

What is One Food Guide Serving?
Look at the examples below.

Fresh, frozen or canned vegetables
125 mL (½ cup)

Leafy vegetables
Cooked: 125 mL (½ cup)
Raw: 250 mL (1 cup)

Fresh, frozen or canned fruits
1 fruit or 125 mL (½ cup)

100% Juice
125 mL (½ cup)

Bread
1 slice (35 g)

Bagel
½ bagel (45 g)

Flat breads
½ pita or ½ tortilla (35 g)

Cooked rice, bulgur or quinoa
125 mL (½ cup)

Cereal
Cold: 30 g
Hot: 175 mL (¾ cup)

Cooked pasta or couscous
125 mL (½ cup)

Milk or powdered milk (reconstituted)
250 mL (1 cup)

Canned milk (evaporated)
125 mL (½ cup)

Fortified soy beverage
250 mL (1 cup)

Yogurt
175 g
(¾ cup)

Kefir
175 g
(¾ cup)

Cheese
50 g (1 ½ oz.)

Cooked fish, shellfish, poultry, lean meat
75 g (2 ½ oz.)/125 mL (½ cup)

Cooked legumes
175 mL (¾ cup)

Tofu
150 g or
175 mL (¾ cup)

Eggs
2 eggs

Peanut or nut butters
30 mL (2 Tbsp)

Shelled nuts and seeds
60 mL (¼ cup)

Oils and Fats
- Include a small amount – 30 to 45 mL (2 to 3 Tbsp) – of unsaturated fat each day. This includes oil used for cooking, salad dressings, margarine and mayonnaise.
- Use vegetable oils such as canola, olive and soybean.
- Choose soft margarines that are low in saturated and trans fats.
- Limit butter, hard margarine, lard and shortening.

FIGURE B-1 **Eating Well with Canada's Food Guide. (*continued*)**

Make each Food Guide Serving count...
wherever you are – at home, at school, at work or when eating out!

▸ **Eat at least one dark green and one orange vegetable each day.**
- Go for dark green vegetables such as broccoli, romaine lettuce and spinach.
- Go for orange vegetables such as carrots, sweet potatoes and winter squash.

▸ **Choose vegetables and fruit prepared with little or no added fat, sugar or salt.**
- Enjoy vegetables steamed, baked or stir-fried instead of deep-fried.

▸ **Have vegetables and fruit more often than juice.**

▸ **Make at least half of your grain products whole grain each day.**
- Eat a variety of whole grains such as barley, brown rice, oats, quinoa and wild rice.
- Enjoy whole grain breads, oatmeal or whole wheat pasta.

▸ **Choose grain products that are lower in fat, sugar or salt.**
- Compare the Nutrition Facts table on labels to make wise choices.
- Enjoy the true taste of grain products. When adding sauces or spreads, use small amounts.

▸ **Drink skim, 1%, or 2% milk each day.**
- Have 500 mL (2 cups) of milk every day for adequate vitamin D.
- Drink fortified soy beverages if you do not drink milk.

▸ **Select lower fat milk alternatives.**
- Compare the Nutrition Facts table on yogurts or cheeses to make wise choices.

▸ **Have meat alternatives such as beans, lentils and tofu often.**

▸ **Eat at least two Food Guide Servings of fish each week.***
- Choose fish such as char, herring, mackerel, salmon, sardines and trout.

▸ **Select lean meat and alternatives prepared with little or no added fat or salt.**
- Trim the visible fat from meats. Remove the skin on poultry.
- Use cooking methods such as roasting, baking or poaching that require little or no added fat.
- If you eat luncheon meats, sausages or prepackaged meats, choose those lower in salt (sodium) and fat.

Enjoy a variety of foods from the four food groups.

Satisfy your thirst with water!

Drink water regularly. It's a calorie-free way to quench your thirst. Drink more water in hot weather or when you are very active.

* Health Canada provides advice for limiting exposure to mercury from certain types of fish. Refer to www.hc-sc.gc.ca for the latest information.

Advice for different ages and stages...

Children

Following *Canada's Food Guide* helps children grow and thrive.

Young children have small appetites and need calories for growth and development.

- Serve small nutritious meals and snacks each day.

- Do not restrict nutritious foods because of their fat content. Offer a variety of foods from the four food groups.

- Most of all... be a good role model.

Women of childbearing age

All women who could become pregnant and those who are pregnant or breastfeeding need a multivitamin containing **folic acid** every day. Pregnant women need to ensure that their multivitamin also contains **iron**. A health care professional can help you find the multivitamin that's right for you.

Pregnant and breastfeeding women need more calories. Include an extra 2 to 3 Food Guide Servings each day.

Here are two examples:

- Have fruit and yogurt for a snack, or

- Have an extra slice of toast at breakfast and an extra glass of milk at supper.

Men and women over 50

The need for **vitamin D** increases after the age of 50.

In addition to following *Canada's Food Guide*, everyone over the age of 50 should take a daily vitamin D supplement of 10 µg (400 IU).

How do I count Food Guide Servings in a meal?

Here is an example:

Vegetable and beef stir-fry with rice, a glass of milk and an apple for dessert		
250 mL (1 cup) mixed broccoli, carrot and sweet red pepper	=	2 **Vegetables and Fruit** Food Guide Servings
75 g (2 ½ oz.) lean beef	=	1 **Meat and Alternatives** Food Guide Serving
250 mL (1 cup) brown rice	=	2 **Grain Products** Food Guide Servings
5 mL (1 tsp) canola oil	=	part of your **Oils and Fats** intake for the day
250 mL (1 cup) 1% milk	=	1 **Milk and Alternatives** Food Guide Serving
1 apple	=	1 **Vegetables and Fruit** Food Guide Serving

FIGURE B-1 Eating Well with Canada's Food Guide. (*continued*)

Eat well and be active today and every day!

The benefits of eating well and being active include:

- Better overall health.
- Lower risk of disease.
- A healthy body weight.
- Feeling and looking better.
- More energy.
- Stronger muscles and bones.

Be active

To be active every day is a step towards better health and a healthy body weight.

Canada's Physical Activity Guide recommends building 30 to 60 minutes of moderate physical activity into daily life for adults and at least 90 minutes a day for children and youth. You don't have to do it all at once. Add it up in periods of at least 10 minutes at a time for adults and five minutes at a time for children and youth.

Start slowly and build up.

Eat well

Another important step towards better health and a healthy body weight is to follow *Canada's Food Guide* by:

- Eating the recommended amount and type of food each day.
- Limiting foods and beverages high in calories, fat, sugar or salt (sodium) such as cakes and pastries, chocolate and candies, cookies and granola bars, doughnuts and muffins, ice cream and frozen desserts, french fries, potato chips, nachos and other salty snacks, alcohol, fruit flavoured drinks, soft drinks, sports and energy drinks, and sweetened hot or cold drinks.

Read the label

- Compare the Nutrition Facts table on food labels to choose products that contain less fat, saturated fat, trans fat, sugar and sodium.
- Keep in mind that the calories and nutrients listed are for the amount of food found at the top of the Nutrition Facts table.

Nutrition Facts

Per 0 mL (0 g)

Amount	% Daily Value
Calories 0	
Fat 0 g	0 %
Saturates 0 g	0 %
+ Trans 0 g	
Cholesterol 0 mg	
Sodium 0 mg	0 %
Carbohydrate 0 g	0 %
Fibre 0 g	0 %
Sugars 0 g	
Protein 0 g	

Vitamin A	0 %	Vitamin C	0 %
Calcium	0 %	Iron	0 %

Limit trans fat

When a Nutrition Facts table is not available, ask for nutrition information to choose foods lower in trans and saturated fats.

Take a step today...

✓ Have breakfast every day. It may help control your hunger later in the day.

✓ Walk wherever you can – get off the bus early, use the stairs.

✓ Benefit from eating vegetables and fruit at all meals and as snacks.

✓ Spend less time being inactive such as watching TV or playing computer games.

✓ Request nutrition information about menu items when eating out to help you make healthier choices.

✓ Enjoy eating with family and friends!

✓ Take time to eat and savour every bite!

For more information, interactive tools, or additional copies visit Canada's Food Guide on-line at: www.hc-sc.gc.ca

or contact:

Publications
Health Canada
Ottawa, Ontario K1A 0K9
E-Mail: publications@hc-sc.gc.ca
Tel.: 1-866-225-0709
Fax: (613) 941-5366
TTY: 1-800-267-1245

Également disponible en français sous le titre : Bien manger avec le Guide alimentaire canadien

This publication can be made available on request on diskette, large print, audio-cassette and braille.

Beyond the Basics: Meal Planning for Healthy Eating, Diabetes Prevention and Management

Beyond the Basics: Meal Planning for Healthy Eating, Diabetes Prevention and Management is Canada's system of meal planning.[1] Similar to the U.S. exchange system, *Beyond the Basics* sorts foods into groups and defines portion sizes to help people manage their blood glucose and maintain a healthy weight. Because foods that contain carbohydrate raise blood glucose, the food groups are organized into two sections—those that contain carbohydrate (presented in Table B-1) and those that contain little or no carbohydrate (shown in Table B-2). One portion from any of the food groups listed in Table B-1 provides about 15 grams of available carbohydrate (total carbohydrate minus fiber) and counts as one carbohydrate choice. Within each group, foods are identified as those to "choose more often" (generally higher in vitamins, minerals, and fiber) and those to "choose less often" (generally higher in sugar, saturated fat, or *trans* fat).

[1]The tables for the Canadian meal planning system are adapted from *Beyond the Basics: Meal Planning for Healthy Eating, Diabetes Prevention and Management,* copyright 2005, with permission of the Canadian Diabetes Association. Additional information is available from www.diabetes.ca.

Key:
● Choose more often
▲ Choose less often

TABLE B-1 Food Groups that Contain Carbohydrate

1 serving = 15 g carbohydrate or 1 carbohydrate choice

Food	Measure
Grains and starches: 15 g carbohydrate, 3 g protein, 0 g fat, 286 kJ (68 kcal)	
▲ Bagel, large (11 cm, 4.5")	¼
▲ Bagel, small (8 cm, 3")	½
▲ Bannock, fried	1.5" × 2.5"
● Bannock, whole grain baked	1.5" × 2.5"
● Barley, pearled, cooked	125 mL (½ c)
▲ Bread, white	1 slice (30 g)
● Bread, whole grain	1 slice (30 g)
● Bulghur, cooked	125 mL (½ c)
▲ Bun, hamburger or hotdog	½
▲ Cereal, flaked unsweetened	125 mL (½ c)
● Cream of Wheat, cooked	175 mL (¾ c)
● Red River, cooked	125 mL (½ c)
● Chapati (15 cm, 6")	1 (44 g)
● Corn	125 mL (½ c)
● Couscous, cooked	125 mL (½ c)
▲ Crackers, soda type	7
▲ Croutons	175 mL (¾ c)
● English muffin, whole grain	½ (28 g)
▲ French fries	10 (50 g)
● Millet, cooked	⅓ c (75 mL)

(continued)

TABLE B-1 **Food Groups that Contain Carbohydrate** (*continued*)

Food	Measure
Grains and starches: 15 g carbohydrate, 3 g protein, 0 g fat, 286 kJ (68 kcal)	
▲ Naan bread (15 cm, 6″)	¼
▲ Pancake (10 cm, 4″)	1
● Pasta, cooked	125 mL (½ c)
▲ Pita bread, white (15 cm, 6″)	½
● Pita bread, whole wheat (15 cm, 6″)	½
▲ Pizza crust (30 cm, 12″)	¹⁄₁₂ (90 g)
● Plantain, cooked, mashed	⅓ c (75 mL)
● Potatoes, boiled or baked	½ medium (84 g)
● Rice, white or brown, cooked	⅓ c (75 mL)
● Roti (15 cm, 6″)	1 (44 g)
● Soup, thick, chunky type	250 mL (1 c)
● Sweet potato, mashed	⅓ c (75 mL)
▲ Taco shells (13 cm, 5″)	2 (17 g)
● Tortilla, wheat flour (25 cm, 10″)	½
▲ Waffle (medium)	1 (39 g)
Fruits: 15 g carbohydrate, 1–2 g protein, 0 g fat, 269 kJ (64 kcal)	
● Apple	1 medium (138 g)
● Applesauce, unsweetened	125 mL (½ c)
● Banana	1 small or ½ large
● Blackberries	500 mL (2 c)
● Cherries	15 (102 g)
● Fruit, canned in juice	125 mL (½ c)
▲ Fruit, dried	60 mL (¼ c)
● Grapefruit	1 small
● Grapes	15 or ½ c (80 g)
● Kiwi	2 medium (150 g)
▲ Juice	125 mL (½ c)
● Mango	½ medium
● Melon	250 mL (1 c)
● Orange	1 medium
● Other berries	250 mL (1 c)
● Pear	1 medium
● Pineapple	175 mL (¾ c)
● Plum	2 medium
● Raspberries	500 mL (2 c)
● Strawberries	500 mL (2 c)

(continued)

Key:
● Choose more often
▲ Choose less often

TABLE B-1   Food Groups that Contain Carbohydrate (*continued*)

Food	Measure
Milk and alternatives: 15 g carbohydrate, 7 g protein, variable fat, 386–651 kJ (92–155 kcal)	
● Evaporated milk, canned	125 mL (½ c)
● Milk, fluid	250 mL (1 c)
● Milk powder, skim	60 mL (4 tbs)
● Soy beverage, flavoured	125 mL (½ c)
● Soy beverage, plain	250 mL (1 c)
● Soy yogourt, flavoured	75 mL (⅓ c)
● Yogourt, nonfat, plain	175 mL (¾ c)
● Yogourt, skim, artificially sweetened	250 mL (1 c)
Other choices (sweet foods and snacks): 15 g carbohydrate, variable protein and fat	
▲ Brownies, unfrosted	5 cm × 5 cm (2″ × 2″)
▲ Cake, unfrosted	5 cm × 5 cm (2″ × 2″)
▲ Cookies, arrowroot or gingersnap	3–4
▲ Jam, jelly, marmalade	15 mL (1 tbs)
● Milk pudding, skim, no sugar added	125 mL (½ c)
▲ Muffin, plain	1 small (45 g)
▲ Oatmeal granola bar	1 (28 g)
● Popcorn, low fat, air popped	750 mL (3 c)
▲ Pretzels, low fat, large	7
▲ Pretzels, low fat, sticks	30
▲ Sugar, white	15 mL (1 tbs, 3 tsp, or 3 packets)
▲ Syrup, honey, molasses	1 tbs (15 mL)

TABLE B-2   Food Groups That Contain Little or No Carbohydrate

Food	Measure
Vegetables: To encourage consumption, most vegetables are considered "free"	
▲ Artichokes, Jerusalem[a]	
● Asparagus	
● Beans, yellow or green	
● Bean sprouts	

(continued)

[a]These vegetables contain enough carbohydrate to be counted as one carbohydrate choice (15 g of available carbohydrate) when the portion size eaten is 1 cup (250 mL) or more.

TABLE B-2 Food Groups that Contain Little or No Carbohydrate (continued)

Food	Measure
Vegetables: To encourage consumption, most vegetables are considered "free"	
● Beets	
● Broccoli	
● Cabbage	
● Carrots	
● Cauliflower	
● Celery	
● Cucumber	
● Eggplant	
● Kale	
● Leeks	
● Mushrooms	
● Okra	
● Onions	
▲ Parsnips[a]	
▲ Peas[a]	
● Peppers	
● Rutabagas	
● Salad vegetables	
▲ Squash, Hubbard, pumpkin, spaghetti	
▲ Squash, acorn[a], butternut[a]	
● Tomatoes, fresh	
● Tomatoes, canned, regular	
▲ Tomatoes, canned, stewed[a]	
● Turnips	
Meat and alternatives: 0 g carbohydrate, 7 g protein, 3–5 g fat, 307 kJ (73 kcal)	
● Cheese, skim (<7% milk fat)	2.5 cm × 2.5 cm × 2.5 cm (1″ × 1″ × 1″)
● Cheese, light (<20% milk fat)	2.5 cm × 2.5 cm × 2.5 cm (1″ × 1″ × 1″)
▲ Cheese, regular (≥21% milk fat)	2.5 cm × 2.5 cm × 2.5 cm (1″ × 1″ × 1″)
● Cottage cheese (1–2% milk fat)	60 mL (¼ c)
● Egg	1 medium-large
● Fish, canned in oil or water	60 mL (¼ c)
● Fish, fresh or frozen, cooked	30 g (1 oz)
● Hummus	90 g (⅓ c)
● Legumes, cooked	125 mL (½ c)
● Meat, game, cooked	30 g (1 slice)
● Meat, ground, lean or extra lean, cooked	30 g (2 tbs)

(continued)

Key:
● Choose more often
▲ Choose less often

Key:
- ● Choose more often
- ▲ Choose less often

TABLE B-2 **Food Groups that Contain Little or No Carbohydrate** (*continued*)

Food	Measure
Meat and alternatives: 0 g carbohydrate, 7 g protein, 3–5 g fat, 307 kJ (73 kcal)	
● Meat, lean, cooked	30 g (1 slice)
● Meat, organ or tripe, cooked	30 g (1 slice)
● Meat, prepared, low fat	30 g (1–3 slices)
▲ Meat, prepared, regular fat	30 g (1–3 slices)
▲ Meat, regular, cooked	30 g (1–3 slices)
● Peameal/back bacon, cooked	30 g (1–2 slices)
● Poultry, ground, lean, cooked	30 g (2 tbs)
● Poultry, skinless, cooked	30 g (1 slice)
▲ Poultry/wings, skin on, cooked	45 g (2)
● Shellfish, cooked	30 g (3 medium)
● Tofu (soybean)	½ block (100 g)
● Vegetarian meat alternatives	30 g (1 oz)
Fats: 0 g carbohydrate, 0 g protein, 5 g fat, 189 kJ (45 kcal)	
▲ Avocado	34 g (⅙)
▲ Bacon	30 g (1 slice)
▲ Butter	5 mL (1 tsp)
▲ Cheese, spreadable	15 mL (1 tbs)
▲ Margarine, non-hydrogenated, regular	5 mL (1 tsp)
▲ Mayonnaise, light	15 mL (1 tbs)
▲ Nuts	15 mL (1 tbs)
▲ Oil, canola or olive	5 mL (1 tsp)
▲ Salad dressing, regular	5 mL (1 tsp)
▲ Seeds	15 g (1 tbs)
▲ Tahini	8 mL (½ tbs)
Extras: <5 g carbohydrate, 84 kJ (20 kcal)	

Broth
Coffee
Herbs and spices
Ketchup
Mustard
Relish
Sugar-free gelatin
Sugar-free soft drinks
Tea

Appendix C

Exchange Lists for Diabetes

Chapter 21 introduces the exchange system, and this appendix provides details from the 2008 Choose Your Foods: Exchange Lists for Diabetes. Appendix B presents Canada's meal-planning system.

Exchange lists can help people with diabetes to manage their blood glucose levels by controlling the amount and kinds of carbohydrates they consume. These lists can also help in planning diets for weight management by controlling kcalorie and fat intake.

The Exchange System

The exchange system sorts foods into groups by their proportions of carbohydrate, fat, and protein (Table C-1 on p. C-2). These groups may be organized into several exchange lists of foods (Tables C-2 through C-12 on pp. C-3–C-17). For example, the carbohydrate group includes these exchange lists:

▌ Starch

▌ Fruits

▌ Milk (fat-free, reduced-fat, and whole)

▌ Sweets, Desserts, and Other Carbohydrates

▌ Nonstarchy Vegetables

Then any food on a list can be "exchanged" for any other on that same list. Another group for alcohol has been included as a reminder that these beverages often deliver substantial carbohydrate and kcalories and therefore warrant their own list.

Serving Sizes

The serving sizes have been carefully adjusted and defined so that a serving of any food on a given list provides roughly the same amount of carbohydrate, fat, and protein, and, therefore, total energy. Any food on a list can thus be exchanged, or traded, for any other food on the same list without significantly affecting the diet's energy–nutrient balance or total kcalories. For example, a person may select 17 small grapes or ½ large grapefruit as one fruit exchange, and either choice would provide roughly 15 grams of carbohydrate and 60 kcalories. A whole grapefruit, however, would count as 2 fruit exchanges.

To apply the system successfully, users must become familiar with the specified serving sizes. A convenient way to remember the serving sizes and energy values is to keep in mind a typical item from each list (review Table C-1).

The Foods on the Lists

Foods do not always appear on the exchange list where you might first expect to find them. They are grouped according to their energy–nutrient contents rather than by their source (such as milks), their outward appearance, or their vitamin and mineral contents. For example, cheeses are grouped with meats (not milk) because, like meats, cheeses contribute energy from protein and fat but provide negligible carbohydrate.

For similar reasons, starchy vegetables such as corn, green peas, and potatoes are found on the Starch list with breads and cereals, not with the vegetables. Likewise, bacon is grouped with the fats and oils, not with the meats.

Diet planners learn to view mixtures of foods, such as casseroles and soups, as combinations of foods from different exchange lists. They also learn to interpret food labels with the exchange system in mind.

Controlling Energy, Fat, and Sodium

The exchange lists help people control their energy intakes by paying close attention to serving sizes. People wanting to lose weight can limit foods from the Sweets, Desserts, and Other Carbohydrates and Fats lists, and they might choose to avoid the Alcohol list altogether. The Free Foods list provide low-kcalorie choices.

By assigning items like bacon to the Fats list, the exchange lists alert consumers to foods that are unexpectedly high in fat. Even the Starch list specifies which grain products contain added fat (such as biscuits, cornbread, and waffles) by marking them with a symbol to indicate added fat (the symbols are explained in the table keys). In addition, the exchange lists encourage users to think of fat-free milk as milk and of whole milk as milk with added fat, and to think of lean meats as meats and of medium-fat and high-fat meats as meats with added fat. To that end, foods on the milk and meat lists are separated into categories based on their fat contents (review Table C-1). The Milk list is subdivided for fat-free, reduced fat, and whole; the meat list is subdivided for lean, medium fat, and high fat. The meat list also includes plant-based proteins, which tend to be rich in fiber. Notice that many of these foods (p. C-7) bear the symbol for "high fiber."

People wanting to control the sodium in their diets can begin by eliminating any foods bearing the "high sodium" symbol. In most cases, the symbol identifies foods that, in one serving, provide 480 milligrams or more of sodium. Foods on the Combination Foods or

TABLE C-1 The Food Lists

Lists	Typical Item/Portion Size	Carbohydrate (g)	Protein (g)	Fat (g)	Energy[a] (kcal)
Carbohydrates					
Starch[b]	1 slice bread	15	0–3	0–1	80
Fruits	1 small apple	15	—	—	60
Milk					
Fat-free, low-fat, 1%	1 c fat-free milk	12	8	0–3	100
Reduced-fat, 2%	1 c reduced-fat milk	12	8	5	120
Whole	1 c whole milk	12	8	8	160
Sweets, desserts, and other carbohydrates[c]	2 small cookies	15	varies	varies	varies
Nonstarchy vegetables	½ c cooked carrots	5	2	—	25
Meat and Meat Substitutes					
Lean	1 oz chicken (no skin)	—	7	0–3	45
Medium-fat	1 oz ground beef	—	7	4–7	75
High-fat	1 oz pork sausage	—	7	8+	100
Plant-based proteins	½ c tofu	varies	7	varies	varies
Fats	1 tsp butter	—	—	5	45
Alcohol	12 oz beer	varies	—	—	100

[a]The energy value for each exchange list represents an approximate average for the group and does not reflect the precise number of grams of carbohydrate, protein, and fat. For example, a slice of bread contains 15 grams of carbohydrate (60 kcalories), 3 grams protein (12 kcalories), and a little fat—rounded to 80 kcalories for ease in calculating. A ½ cup of vegetables (not including starchy vegetables) contains 5 grams carbohydrate (20 kcalories) and 2 grams protein (8 more), which has been rounded down to 25 kcalories.

[b]The Starch list includes cereals, grains, breads, crackers, snacks, starchy vegetables (such as corn, peas, and potatoes), and legumes (dried beans, peas, and lentils).

[c]The Sweets, Desserts, and Other Carbohydrates list includes foods that contain added sugars and fats such as sodas, candy, cakes, cookies, doughnuts, ice cream, pudding, syrup, and frozen yogurt.

Fast Foods lists that bear the symbol provide more than 600 milligrams of sodium. Other foods may also contribute substantially to sodium (consult Chapter 9 for details).

Planning a Healthy Diet

To obtain a daily variety of foods that provide healthful amounts of carbohydrate, protein, and fat, as well as vitamins, minerals, and fiber, the meal plan for adults and teenagers should include at least:

- Two to three servings of nonstarchy vegetables
- Two servings of fruits
- Six servings of grains (at least three of whole grains), beans, and starchy vegetables
- Two servings of low-fat or fat-free milk
- About 6 ounces of protein foods
- *Small* amounts of fat and sugar

The actual amounts are determined by age, gender, activity levels, and other factors that influence energy needs.

TABLE C-2 **Starch**

The Starch list includes bread, cereals and grains, starchy vegetables, crackers and snacks, and legumes (dried beans, peas, and lentils). 1 starch choice = 15 grams carbohydrate, 0–3 grams protein, 0–1 grams fat, and 80 kcalories.

NOTE: In general, one starch exchange is ½ cup cooked cereal, grain, or starchy vegetable; ⅓ cup cooked rice or pasta; 1 ounce of bread products; ¾ ounce to 1 ounce of most snack foods.

Bread

Food	Serving Size
Bagel, large (about 4 oz)	¼ (1 oz)
▽ Biscuit, 2½ inches across	1
Bread	
☺ reduced-kcalorie	2 slices (1½ oz)
white, whole-grain, pumpernickel, rye, unfrosted raisin	1 slice (1 oz)
Chapatti, small, 6 inches across	1
▽ Cornbread, 1¾ inch cube	1 (1½ oz)
English muffin	½
Hot dog bun or hamburger bun	½ (1 oz)
Naan, 8 inches by 2 inches	¼
Pancake, 4 inches across, ¼ inch thick	1
Pita, 6 inches across	½
Roll, plain, small	1 (1 oz)
▽ Stuffing, bread	⅓ cup
▽ Taco shell, 5 inches across	2

Bread (*continued*)

Food	Serving Size
Tortilla, corn, 6 inches across	1
Tortilla, flour, 6 inches across	1
Tortilla, flour, 10 inches across	⅓
▽ Waffle, 4-inch square or 4 inches across	1
Cereals and Grains	
Barley, cooked	⅓ cup
Bran, dry	
☺ oat	¼ cup
☺ wheat	½ cup
☺ Bulgur (cooked)	½ cup
Cereals	
☺ bran	½ cup
cooked (oats, oatmeal)	½ cup
puffed	1½ cups
shredded wheat, plain	½ cup

(continued)

Key:

☺ = More than 3 grams of dietary fiber per serving.

▽ = Extra fat, or prepared with added fat. (Count as 1 starch + 1 fat.)

🧂 = 480 milligrams or more of sodium per serving.

TABLE C-2 Starch (*continued*)

Cereals and Grains (*continued*)

Food	Serving Size
sugar-coated	½ cup
unsweetened, ready-to-eat	¾ cup
Couscous	⅓ cup
Granola	
low-fat	¼ cup
▽ regular	¼ cup
Grits, cooked	½ cup
Kasha	½ cup
Millet, cooked	⅓ cup
Muesli	¼ cup
Pasta, cooked	⅓ cup
Polenta, cooked	⅓ cup
Quinoa, cooked	⅓ cup
Rice, white or brown, cooked	⅓ cup
Tabbouleh (tabouli), prepared	½ cup
Wheat germ, dry	3 Tbsp
Wild rice, cooked	½ cup

Starchy Vegetables

Food	Serving Size
Cassava	⅓ cup
Corn	½ cup
on cob, large	½ cob (5 oz)
☺ Hominy, canned	¾ cup
☺ Mixed vegetables with corn, peas, or pasta	1 cup
☺ Parsnips	½ cup
☺ Peas, green	½ cup
Plantain, ripe	⅓ cup
Potato	
baked with skin	¼ large (3 oz)
boiled, all kinds	½ cup or ½ medium (3 oz)
▽ mashed, with milk and fat	½ cup
French fried (oven-baked)[a]	1 cup (2 oz)
☺ Pumpkin, canned, no sugar added	1 cup
Spaghetti/pasta sauce	½ cup

Starchy Vegetables (*continued*)

Food	Serving Size
☺ Squash, winter (acorn, butternut)	1 cup
☺ Succotash	½ cup
Yam, sweet potato, plain	½ cup

Crackers and Snacks[b]

Food	Serving Size
Animal crackers	8
Crackers	
▽ round-butter type	6
saltine-type	6
▽ sandwich-style, cheese or peanut butter filling	3
▽ whole-wheat regular	2–5 (¾ oz)
☺ whole-wheat lower fat or crispbreads	2–5 (¾ oz)
Graham cracker, 2½-inch square	3
Matzoh	¾ oz
Melba toast, about 2-inch by 4-inch piece	4
Oyster crackers	20
Popcorn	3 cups
▽ ☺ with butter	3 cups
☺ no fat added	3 cups
☺ lower fat	3 cups
Pretzels	¾ oz
Rice cakes, 4 inches across	2
Snack chips	
fat-free or baked (tortilla, potato), baked pita chips	15–20 (¾ oz)
▽ regular (tortilla, potato)	9–13 (¾ oz)

Beans, Peas, and Lentils[c]

The choices on this list count as 1 starch + 1 lean meat.

Food	Serving Size
☺ Baked beans	⅓ cup
☺ Beans, cooked (black, garbanzo, kidney, lima, navy, pinto, white)	½ cup
☺ Lentils, cooked (brown, green, yellow)	½ cup
☺ Peas, cooked (black-eyed, split)	½ cup
🥫 ☺ Refried beans, canned	½ cup

Key:

☺ = More than 3 grams of dietary fiber per serving.

▽ = Extra fat, or prepared with added fat. (Count as 1 starch + 1 fat.)

🥫 = 480 milligrams or more of sodium per serving.

[a]Restaurant-style French fries are on the Fast Foods list.
[b]For other snacks, see the Sweets, Desserts, and Other Carbohydrates list. For a quick estimate of serving size, an open handful is equal to about 1 cup or 1 to 2 ounces of snack food.
[c]Beans, peas, and lentils are also found on the Meat and Meat Substitutes list.

TABLE C-3 Fruits

Fruit[a]

The Fruits list includes fresh, frozen, canned, and dried fruits and fruit juices. 1 fruit choice = 15 grams carbohydrate, 0 grams protein, 0 grams fat, and 60 kcalories.

NOTE: In general, one fruit exchange is ½ cup canned or fresh fruit or unsweetened fruit juice; 1 small fresh fruit (4 ounces); 2 tablespoons dried fruit.

Food	Serving Size
Apple, unpeeled, small	1 (4 oz)
Apples, dried	4 rings
Applesauce, unsweetened	½ cup
Apricots	
canned	½ cup
dried	8 halves
☺ fresh	4 whole (5½ oz)
Banana, extra small	1 (4 oz)
☺ Blackberries	¾ cup
Blueberries	¾ cup
Cantaloupe, small	⅓ melon or 1 cup cubed (11 oz)
Cherries	
sweet, canned	½ cup
sweet fresh	12 (3 oz)
Dates	3
Dried fruits (blueberries, cherries, cranberries, mixed fruit, raisins)	2 Tbsp
Figs	
dried	1½
☺ fresh	1½ large or 2 medium (3½ oz)
Fruit cocktail	½ cup
Grapefruit	
large	½ (11 oz)
sections, canned	¾ cup
Grapes, small	17 (3 oz)
Honeydew melon	1 slice or 1 cup cubed (10 oz)
☺ Kiwi	1 (3½ oz)
Mandarin oranges, canned	¾ cup
Mango, small	½ (5½ oz) or ½ cup
Nectarine, small	1 (5 oz)
☺ Orange, small	1 (6½ oz)
Papaya	½ or 1 cup cubed (8 oz)
Peaches	
canned	½ cup
fresh, medium	1 (6 oz)
Pears	
canned	½ cup
fresh, large	½ (4 oz)
Pineapple	
canned	½ cup
fresh	¾ cup
Plums	
canned	½ cup
dried (prunes)	3
small	2 (5 oz)
☺ Raspberries	1 cup
☺ Strawberries	1¼ cup whole berries
☺ Tangerines, small	2 (8 oz)
Watermelon	1 slice or 1¼ cups cubes (13½ oz)

Fruit Juice

Food	Serving Size
Apple juice/cider	½ cup
Fruit juice blends, 100% juice	⅓ cup
Grape juice	⅓ cup
Grapefruit juice	½ cup
Orange juice	½ cup
Pineapple juice	½ cup
Prune juice	⅓ cup

Key:

☺ = More than 3 grams of dietary fiber per serving.

⚠ = Extra fat or prepared with added fat. (Count as 1 starch + 1 fat.)

🧂 = 480 milligrams or more of sodium per serving.

[a]The weight listed includes skin, core, seeds, and rind.

TABLE C-4 **Milk**

The Milk list groups milks and yogurts based on the amount of fat they have (fat-free/low fat, reduced fat, and whole). Cheeses are found on the Meat and Meat Substitutes list and cream and other dairy fats are found on the Fats list.

NOTE: In general, one milk choice is 1 cup (8 fluid ounces or ½ pint) milk or yogurt.

Milk and Yogurts

Food	Serving Size
Fat-free or low-fat (1%)	
1 fat-free/low-fat milk choice = 12 g carbohydrate, 8 g protein, 0-3 g fat, and 100 kcal.	
Milk, buttermilk, acidophilus milk, Lactaid	1 cup
Evaporated milk	½ cup
Yogurt, plain or flavored with an artificial sweetener	⅔ cup (6 oz)
Reduced-fat (2%)	
1 reduced-fat milk choice = 12 g carbohydrate, 8 g protein, 5 g fat, and 120 kcal.	
Milk, acidophilus milk, kefir, Lactaid	1 cup
Yogurt, plain	⅔ cup (6 oz)
Whole	
1 whole milk choice = 12 g carbohydrate, 8 g protein, 8 g fat, and 160 kcal.	
Milk, buttermilk, goat's milk	1 cup
Evaporated milk	½ cup
Yogurt, plain	8 oz

Dairy-Like Foods

Food	Serving Size	Count as
Chocolate milk		
fat-free	1 cup	1 fat-free milk + 1 carbohydrate
whole	1 cup	1 whole milk + 1 carbohydrate
Eggnog, whole milk	½ cup	1 carbohydrate + 2 fats
Rice drink		
flavored, low fat	1 cup	2 carbohydrates
plain, fat-free	1 cup	1 carbohydrate
Smoothies, flavored, regular	10 oz	1 fat-free milk + 2½ carbohydrates
Soy milk		
light	1 cup	1 carbohydrate + ½ fat
regular, plain	1 cup	1 carbohydrate + 1 fat
Yogurt		
and juice blends	1 cup	1 fat-free milk + 1 carbohydrate
low carbohydrate (less than 6 grams carbohydrate per choice)	⅔ cup (6 oz)	½ fat-free milk
with fruit, low-fat	⅔ cup (6 oz)	1 fat-free milk + 1 carbohydrate

TABLE C-5 **Sweets, Desserts, and Other Carbohydrates**

1 other carbohydrate choice = 15 grams carbohydrate, variable grams protein, variable grams fat, and variable kcalories.

NOTE: In general, one choice from this list can substitute for foods on the Starch, Fruits, or Milk lists.

Beverages, Soda, and Energy/Sports Drinks

Food	Serving Size	Count as
Cranberry juice cocktail	½ cup	1 carbohydrate
Energy drink	1 can (8.3 oz)	2 carbohydrates
Fruit drink or lemonade	1 cup (8 oz)	2 carbohydrates
Hot chocolate		
regular	1 envelope added to 8 oz water	1 carbohydrate + 1 fat
sugar-free or light	1 envelope added to 8 oz water	1 carbohydrate
Soft drink (soda), regular	1 can (12 oz)	2½ carbohydrates
Sports drink	1 cup (8 oz)	1 carbohydrate

Brownies, Cake, Cookies, Gelatin, Pie, and Pudding

Food	Serving Size	Count as
Brownie, small, unfrosted	1¼-inch square, ⅞ inch high (about 1 oz)	1 carbohydrate + 1 fat
Cake		
angel food, unfrosted	1/12 of cake (about 2 oz)	2 carbohydrates
frosted	2-inch square (about 2 oz)	2 carbohydrates + 1 fat
unfrosted	2-inch square (about 2 oz)	1 carbohydrate + 1 fat
Cookies		
chocolate chip	2 cookies (2¼ inches across)	1 carbohydrate + 2 fats
gingersnap	3 cookies	1 carbohydrate
sandwich, with crème filling	2 small (about ⅔ oz)	1 carbohydrate + 1 fat
sugar-free	3 small or 1 large (¾–1 oz)	1 carbohydrate + 1–2 fats
vanilla wafer	5 cookies	1 carbohydrate + 1 fat
Cupcake, frosted	1 small (about 1¾ oz)	2 carbohydrates + 1–1½ fats
Fruit cobbler	½ cup (3½ oz)	3 carbohydrates + 1 fat
Gelatin, regular	½ cup	1 carbohydrate
Pie		
commercially prepared fruit, 2 crusts	⅙ of 8-inch pie	3 carbohydrates + 2 fats
pumpkin or custard	⅛ of 8 inch pie	1½ carbohydrates + 1½ fats
Pudding		
regular (made with reduced-fat milk)	½ cup	2 carbohydrates
sugar-free or sugar- and fat-free (made with fat-free milk)	½ cup	1 carbohydrate

Candy, Spreads, Sweets, Sweeteners, Syrups, and Toppings

Food	Serving Size	Count as
Candy bar, chocolate/peanut	2 "fun size" bars (1 oz)	1½ carbohydrates + 1½ fats
Candy, hard	3 pieces	1 carbohydrate
Chocolate "kisses"	5 pieces	1 carbohydrate + 1 fat
Coffee creamer		
dry, flavored	4 tsp	½ carbohydrate + ½ fat
liquid, flavored	2 Tbsp	1 carbohydrate
Fruit snacks, chewy (pureed fruit concentrate)	1 roll (¾ oz)	1 carbohydrate
Fruit spreads, 100% fruit	1½ Tbsp	1 carbohydrate
Honey	1 Tbsp	1 carbohydrate
Jam or jelly, regular	1 Tbsp	1 carbohydrate

(continued)

TABLE C-5 **Sweets, Desserts, and Other Carbohydrates** (*continued*)

Candy, Spreads, Sweets, Sweeteners, Syrups, and Toppings—continued

Food	Serving Size	Count as
Sugar	1 Tbsp	1 carbohydrate
Syrup		
chocolate	2 Tbsp	2 carbohydrates
light (pancake type)	2 Tbsp	1 carbohydrate
regular (pancake type)	1 Tbsp	1 carbohydrate

Condiments and Sauces[a]

Food	Serving Size	Count as
Barbeque sauce	3 Tbsp	1 carbohydrate
Cranberry sauce, jellied	¼ cup	1½ carbohydrates
🥫 Gravy, canned or bottled	½ cup	½ carbohydrate + ½ fat
Salad dressing, fat-free, low-fat, cream-based	3 Tbsp	1 carbohydrate
Sweet and sour sauce	3 Tbsp	1 carbohydrate

Doughnuts, Muffins, Pastries, and Sweet Breads

Food	Serving Size	Count as
Banana nut bread	1-inch slice (1 oz)	2 carbohydrates + 1 fat
Doughnut		
cake, plain	1 medium (1½ oz)	1½ carbohydrates + 2 fats
yeast type, glazed	3¾ inches across (2 oz)	2 carbohydrates + 2 fats
Muffin (4 oz)	¼ muffin (1 oz)	1 carbohydrate + ½ fat
Sweet roll or Danish	1 (2½ oz)	2½ carbohydrates + 2 fats

Frozen Bars, Frozen Desserts, Frozen Yogurt, and Ice Cream

Food	Serving Size	Count as
Frozen pops	1	½ carbohydrate
Fruit juice bars, frozen, 100% juice	1 bar (3 oz)	1 carbohydrate
Ice cream		
fat-free	½ cup	1½ carbohydrates
light	½ cup	1 carbohydrate + 1 fat
no sugar added	½ cup	1 carbohydrate + 1 fat
regular	½ cup	1 carbohydrate + 2 fats
Sherbet, sorbet	½ cup	2 carbohydrates
Yogurt, frozen		
fat-free	⅓ cup	1 carbohydrate
regular	½ cup	1 carbohydrate + 0–1 fat

Granola Bars, Meal Replacement Bars/Shakes, and Trail Mix

Food	Serving Size	Count as
Granola or snack bar, regular or low-fat	1 bar (1 oz)	1½ carbohydrates
Meal replacement bar	1 bar (1⅓ oz)	1½ carbohydrates + 0-1 fat
Meal replacement bar	1 bar (2 oz)	2 carbohydrates + 1 fat
Meal replacement shake, reduced kcalorie	1 can (10–11 oz)	1½ carbohydrates + 0-1 fat
Trail mix		
candy/nut-based	1 oz	1 carbohydrate + 2 fats
dried fruit-based	1 oz	1 carbohydrate + 1 fat

Key:

🥫 = 480 milligrams or more of sodium per serving.

[a]You can also check the Fats list and Free Foods list for other condiments.

TABLE C-6 Nonstarchy Vegetables

The Nonstarchy Vegetables list includes vegetables that have few grams of carbohydrates or kcalories; starchy vegetables are found on the Starch list. 1 nonstarchy vegetable choice = 5 grams carbohydrate, 2 grams protein, 0 grams fat, and 25 kcalories.

NOTE: In general, one nonstarchy vegetable choice is ½ cup cooked vegetables or vegetable juice or 1 cup raw vegetables. Count 3 cups of raw vegetables or 1½ cups of cooked vegetables as one carbohydrate choice.

Nonstarchy Vegetables[a]

Amaranth or Chinese spinach	Kohlrabi
Artichoke	Leeks
Artichoke hearts	Mixed vegetables (without corn, peas, or pasta)
Asparagus	Mung bean sprouts
Baby corn	Mushrooms, all kinds, fresh
Bamboo shoots	Okra
Beans (green, wax, Italian)	Onions
Bean sprouts	Oriental radish or daikon
Beets	Pea pods
🖥 Borscht	☺ Peppers (all varieties)
Broccoli	Radishes
☺ Brussels sprouts	Rutabaga
Cabbage (green, bok choy, Chinese)	🖥 Sauerkraut
☺ Carrots	Soybean sprouts
Cauliflower	Spinach
Celery	Squash (summer, crookneck, zucchini)
☺ Chayote	Sugar pea snaps
Coleslaw, packaged, no dressing	☺ Swiss chard
Cucumber	Tomato
Eggplant	Tomatoes, canned
Gourds (bitter, bottle, luffa, bitter melon)	🖥 Tomato sauce
Green onions or scallions	🖥 Tomato/vegetable juice
Greens (collard, kale, mustard, turnip)	Turnips
Hearts of palm	Water chestnuts
Jicama	Yard-long beans

Key:

☺ = More than 3 grams of dietary fiber per serving.

🖥 = 480 milligrams or more of sodium per serving.

[a]Salad greens (like chicory, endive, escarole, lettuce, romaine, spinach, arugula, radicchio, watercress) are on the Free Foods list.

TABLE C-7 Meat and Meat Substitutes

The Meat and Meat Substitutes list groups foods based on the amount of fat they have (lean meat, medium-fat meat, high-fat meat, and plant-based proteins).

Lean Meats and Meat Substitutes

1 lean meat choice = 0 grams carbohydrate, 7 grams protein, 0–3 grams fat, and 100 kcalories.

Food	Amount
Beef: Select or Choice grades trimmed of fat: ground round, roast (chuck, rib, rump), round, sirloin, steak (cubed, flank, porterhouse, T-bone), tenderloin	1 oz
🖥 Beef jerky	1 oz
Cheeses with 3 grams of fat or less per oz	1 oz
Cottage cheese	¼ cup
Egg substitutes, plain	¼ cup
Egg whites	2
Fish, fresh or frozen, plain: catfish, cod, flounder, haddock, halibut, orange roughy, salmon, tilapia, trout, tuna	1 oz
🖥 Fish, smoked: herring or salmon (lox)	1 oz
Game: buffalo, ostrich, rabbit, venison	1 oz
🖥 Hot dog with 3 grams of fat or less per oz (8 dogs per 14 oz package) *Note: May be high in carbohydrate.*	1
Lamb: chop, leg, or roast	1 oz
Organ meats: heart, kidney, liver *Note: May be high in cholesterol.*	1 oz
Oysters, fresh or frozen	6 medium
Pork, lean	
🖥 Canadian bacon	1 oz
rib or loin chop/roast, ham, tenderloin	1 oz
Poultry, without skin: Cornish hen, chicken, domestic duck or goose (well-drained of fat), turkey	1 oz
Processed sandwich meats with 3 grams of fat or less per oz: chipped beef, deli thin-sliced meats, turkey ham, turkey kielbasa, turkey pastrami	1 oz
Salmon, canned	1 oz
Sardines, canned	2 medium
🖥 Sausage with 3 grams of fat or less per oz	1 oz
Shellfish: clams, crab, imitation shellfish, lobster, scallops, shrimp	1 oz
Tuna, canned in water or oil, drained	1 oz
Veal, lean chop, roast	1 oz

Medium-Fat Meat and Meat Substitutes

1 medium-fat meat choice = 0 grams carbohydrate, 7 grams protein, 4–7 grams fat, and 130 kcalories.

Food	Amount
Beef: corned beef, ground beef, meatloaf, Prime grades trimmed of fat (prime rib), short ribs, tongue	1 oz
Cheeses with 4–7 grams of fat per oz: feta, mozzarella, pasteurized processed cheese spread, reduced-fat cheeses,' string	1 oz
Egg *Note: High in cholesterol, so limit to 3 per week.*	1
Fish, any fried product	1 oz
Lamb: ground, rib roast	1 oz
Pork: cutlet, shoulder roast	1 oz
Poultry: chicken with skin; dove, pheasant, wild duck, or goose; fried chicken; ground turkey	1 oz
Ricotta cheese	2 oz or ¼ cup
🖥 Sausage with 4-7 grams of fat per oz	1 oz
Veal, cutlet (no breading)	1 oz

High-Fat Meat and Meat Substitutes

1 high-fat meat choice = 0 grams carbohydrate, 7 grams protein, 8+ grams fat, and 150 kcalories. These foods are high in saturated fat, cholesterol, and kcalories and may raise blood cholesterol levels if eaten on a regular basis. Try to eat 3 or fewer servings from this group per week.

Food	Amount
Bacon	
🖥 pork	2 slices (16 slices per lb or 1 oz each, before cooking)
🖥 turkey	3 slices (½ oz each before cooking)
Cheese, regular: American, bleu, brie, cheddar, hard goat, Monterey jack, queso, and Swiss	1 oz
▽ 🖥 Hot dog: beef, pork, or combination (10 per lb-sized package)	1
🖥 Hot dog: turkey or chicken (10 per lb-sized package)	1
Pork: ground, sausage, spareribs	1 oz
Processed sandwich meats with 8 grams of fat or more per oz: bologna, pastrami, hard salami	1 oz
🖥 Sausage with 8 grams fat or more per oz: bratwurst, chorizo, Italian, knockwurst, Polish, smoked, summer	1 oz

(continued)

Key:

☺ = More than 3 grams of dietary fiber per serving.

▽ = Extra fat, or prepared with added fat. (Count as 1 starch + 1 fat.)

🖥 = 480 milligrams or more of sodium per serving.

TABLE C-7 Meat and Meat Substitutes (*continued*)

Plant-Based Proteins

1 plant-based protein choice = variable grams carbohydrate, 7 grams protein, variable grams fat, and variable kcalories. Because carbohydrate content varies among plant-based proteins, you should read the food label.

Food	Serving Size	Count as
"Bacon" strips, soy-based	3 strips	1 medium-fat meat
☺ Baked beans	⅓ cup	1 starch + 1 lean meat
☺ Beans, cooked: black, garbanzo, kidney, lima, navy, pinto, white[a]	½ cup	1 starch + 1 lean meat
☺ "Beef" or "sausage" crumbles, soy-based	2 oz	½ carbohydrate + 1 lean meat
"Chicken" nuggets, soy-based	2 nuggets (1½ oz)	½ carbohydrate + 1 medium-fat meat
☺ Edamame	½ cup	½ carbohydrate + 1 lean meat
Falafel (spiced chickpea and wheat patties)	3 patties (about 2 inches across)	1 carbohydrate + 1 high-fat meat
Hot dog, soy-based	1 (1½ oz)	½ carbohydrate + 1 lean meat
☺ Hummus	⅓ cup	1 carbohydrate + 1 high-fat meat
☺ Lentils, brown, green, or yellow	½ cup	1 carbohydrate + 1 lean meat
☺ Meatless burger, soy-based	3 oz	½ carbohydrate + 2 lean meats
☺ Meatless burger, vegetable- and starch-based	1 patty (about 2½ oz)	1 carbohydrate + 2 lean meats
Nut spreads: almond butter, cashew butter, peanut butter, soy nut butter	1 Tbsp	1 high-fat meat
☺ Peas, cooked: black-eyed and split peas	½ cup	1 starch + 1 lean meat
🖥 ☺ Refried beans, canned	½ cup	1 starch + 1 lean meat
"Sausage" patties, soy-based	1 (1½ oz)	1 medium-fat meat
Soy nuts, unsalted	¾ oz	½ carbohydrate + 1 medium-fat meat
Tempeh	¼ cup	1 medium-fat meat
Tofu	4 oz (½ cup)	1 medium-fat meat
Tofu, light	4 oz (½ cup)	1 lean meat

Key:

☺ = More than 3 grams of dietary fiber per serving.

▽ = Extra fat, or prepared with added fat. (Add an additional fat choice to this food.)

🖥 = 480 milligrams or more of sodium per serving (based on the sodium content of a typical 3-oz serving of meat, unless 1 or 2 oz is the normal serving size).

[a]Beans, peas, and lentils are also found on the Starch list; nut butters in smaller amounts are found in the Fats list.

Appendix C

TABLE C-8 Fats

Fats and oils have mixtures of unsaturated (polyunsaturated and monounsaturated) and saturated fats. Foods on the Fats list are grouped together based on the major type of fat they contain. 1 fat choice = 0 grams carbohydrate, 0 grams protein, 5 grams fat, and 45 kcalories.

NOTE: In general, one fat exchange is 1 teaspoon of regular margarine, vegetable oil, or butter; 1 tablespoon of regular salad dressing.

When used in large amounts, bacon and peanut butter are counted as high-fat meat choices (see Meat and Meat Substitutes list). Fat-free salad dressings are found on the Sweets, Desserts, and Other Carbohydrates list. Fat-free products such as margarines, salad dressings, mayonnaise, sour cream, and cream cheese are found on the Free Foods list.

Monounsaturated Fats

Food	Serving Size
Avocado, medium	2 Tbsp (1 oz)
Nut butters (*trans* fat-free): almond butter, cashew butter, peanut butter (smooth or crunchy)	1½ tsp
Nuts	
almonds	6 nuts
Brazil	2 nuts
cashews	6 nuts
filberts (hazelnuts)	5 nuts
macadamia	3 nuts
mixed (50% peanuts)	6 nuts
peanuts	10 nuts
pecans	4 halves
pistachios	16 nuts
Oil: canola, olive, peanut	1 tsp
Olives	
black (ripe)	8 large
green, stuffed	10 large

Polyunsaturated Fats

Food	Serving Size
Margarine: lower-fat spread (30%–50% vegetable oil, *trans* fat-free)	1 Tbsp
Margarine: stick, tub (*trans* fat-free) or squeeze (*trans* fat-free)	1 tsp
Mayonnaise	
reduced-fat	1 Tbsp
regular	1 tsp
Mayonnaise-style salad dressing	
reduced-fat	1 Tbsp
regular	2 tsp
Nuts	
Pignolia (pine nuts)	1 Tbsp
walnuts, English	4 halves
Oil: corn, cottonseed, flaxseed, grape seed, safflower, soybean, sunflower	1 tsp
Oil: made from soybean and canola oil—Enova	1 tsp
Plant stanol esters	
light	1 Tbsp
regular	2 tsp

Polyunsaturated Fats—continued

Food	Serving Size
Salad dressing	
🔒 reduced-fat *Note: May be high in carbohydrate.*	2 Tbsp
🔒 regular	1 Tbsp
Seeds	
flaxseed, whole	1 Tbsp
pumpkin, sunflower	1 Tbsp
sesame seeds	1 Tbsp
Tahini or sesame paste	2 tsp

Saturated Fats

Food	Serving Size
Bacon, cooked, regular or turkey	1 slice
Butter	
reduced-fat	1 Tbsp
stick	1 tsp
whipped	2 tsp
Butter blends made with oil	
reduced-fat or light	1 Tbsp
regular	1½ tsp
Chitterlings, boiled	2 Tbsp (½ oz)
Coconut, sweetened, shredded	2 Tbsp
Coconut milk	
light	⅓ cup
regular	1½ Tbsp
Cream	
half and half	2 Tbsp
heavy	1 Tbsp
light	1½ Tbsp
whipped	2 Tbsp
whipped, pressurized	¼ cup
Cream cheese	
reduced-fat	1½ Tbsp (¾ oz)
regular	1 Tbsp (½ oz)
Lard	1 tsp
Oil: coconut, palm, palm kernel	1 tsp
Salt pork	¼ oz
Shortening, solid	1 tsp
Sour cream	
reduced-fat or light	3 Tbsp
regular	2 Tbsp

Key:

🔒 = 480 milligrams or more of sodium per serving.

TABLE C-9 Free Foods

A "free" food is any food or drink choice that has less than 20 kcalories and 5 grams or less of carbohydrate per serving.

▪ Most foods on this list should be limited to 3 servings (as listed here) per day. Spread out the servings throughout the day. If you eat all 3 servings at once, it could raise your blood glucose level.

▪ Food and drink choices listed here without a serving size can be eaten whenever you like.

Low Carbohydrate Foods

Food	Serving Size
Cabbage, raw	½ cup
Candy, hard (regular or sugar-free)	1 piece
Carrots, cauliflower, or green beans, cooked	¼ cup
Cranberries, sweetened with sugar substitute	½ cup
Cucumber, sliced	½ cup
Gelatin	
dessert, sugar-free	
unflavored	
Gum	
Jam or jelly, light or no sugar added	2 tsp
Rhubarb, sweetened with sugar substitute	½ cup
Salad greens	
Sugar substitutes (artificial sweeteners)	
Syrup, sugar-free	2 Tbsp

Modified Fat Foods with Carbohydrate

Food	Serving Size
Cream cheese, fat-free	1 Tbsp (½ oz)
Creamers	
nondairy, liquid	1 Tbsp
nondairy, powdered	2 tsp
Margarine spread	
fat-free	1 Tbsp
reduced-fat	1 tsp
Mayonnaise	
fat-free	1 Tbsp
reduced-fat	1 tsp
Mayonnaise-style salad dressing	
fat-free	1 Tbsp
reduced-fat	1 tsp
Salad dressing	
fat-free or low-fat	1 Tbsp
fat-free, Italian	2 Tbsp

Modified Fat Foods with Carbohydrate—continued

Food	Serving Size
Sour cream, fat-free or reduced-fat	1 Tbsp
Whipped topping	
light or fat-free	2 Tbsp
regular	1 Tbsp

Condiments

Food	Serving Size
Barbecue sauce	2 tsp
Catsup (ketchup)	1 Tbsp
Honey mustard	1 Tbsp
Horseradish	
Lemon juice	
Miso	1½ tsp
Mustard	
Parmesan cheese, freshly grated	1 Tbsp
Pickle relish	1 Tbsp
Pickles	
🧂 dill	1½ medium
sweet, bread and butter	2 slices
sweet, gherkin	¾ oz
Salsa	¼ cup
🧂 Soy sauce, light or regular	1 Tbsp
Sweet and sour sauce	2 tsp
Sweet chili sauce	2 tsp
Taco sauce	1 Tbsp
Vinegar	
Yogurt, any type	2 Tbsp

Drinks/Mixes

Any food on the list—without a serving size listed—can be consumed in any moderate amount.

🧂 Bouillon, broth, consommé

Bouillon or broth, low-sodium

Carbonated or mineral water

Club soda

(continued)

Key:

🧂 = 480 milligrams or more of sodium per serving.

TABLE C-9 **Free Foods** (*continued*)

Drinks/Mixes—continued	Seasonings
Cocoa powder, unsweetened (1 Tbsp)	Any food on this list can be consumed in any moderate amount.
Coffee, unsweetened or with sugar substitute	Flavoring extracts (for example, vanilla, almond, peppermint)
Diet soft drinks, sugar-free	Garlic
Drink mixes, sugar-free	Herbs, fresh or dried
Tea, unsweetened or with sugar substitute	Nonstick cooking spray
Tonic water, diet	Pimento
Water	Spices
Water, flavored, carbohydrate free	Hot pepper sauce
	Wine, used in cooking
	Worcestershire sauce

TABLE C-10 **Combination Foods**

Many foods are eaten in various combinations, such as casseroles. Because "combination" foods do not fit into any one choice list, this list of choices provides some typical combination foods.

Entrees

Food	Serving Size	Count as
🗋 Casserole type (tuna noodle, lasagna, spaghetti with meatballs, chili with beans, macaroni and cheese)	1 cup (8 oz)	2 carbohydrates + 2 medium-fat meats
🗋 Stews (beef/other meats and vegetables)	1 cup (8 oz)	1 carbohydrate + 1 medium-fat meat + 0–3 fats
Tuna salad or chicken salad	½ cup (3½ oz)	½ carbohydrate + 2 lean meats + 1 fat

Frozen Meals/Entrees

Food	Serving Size	Count as
🗋 😊 Burrito (beef and bean)	1 (5 oz)	3 carbohydrates + 1 lean meat + 2 fats
🗋 Dinner-type meal	generally 14–17 oz	3 carbohydrates + 3 medium-fat meats + 3 fats
🗋 Entrée or meal with fewer than 340 kcalories	about 8–11 oz	2–3 carbohydrates + 1–2 lean meats
Pizza		
🗋 cheese/vegetarian, thin crust	¼ of a 12 inch (4½-5 oz)	2 carbohydrates + 2 medium-fat meats
🗋 meat topping, thin crust	¼ of a 12 inch (5 oz)	2 carbohydrates + 2 medium-fat meats + 1½ fats
🗋 Pocket sandwich	1 (4½ oz)	3 carbohydrates + 1 lean meat + 1–2 fats
🗋 Pot pie	1 (7 oz)	2½ carbohydrates + 1 medium-fat meat + 3 fats

Salads (Deli-Style)

Food	Serving Size	Count as
Coleslaw	½ cup	1 carbohydrate + 1½ fats
Macaroni/pasta salad	½ cup	2 carbohydrates + 3 fats
🗋 Potato salad	½ cup	1½-2 carbohydrates + 1-2 fats

(*continued*)

Key:
😊 = More than 3 grams of dietary fiber per serving.
▽ = Extra fat, or prepared with added fat.
🗋 = 600 milligrams or more of sodium per serving (for combination food main dishes/meals).

TABLE C-10 Combination Foods (*continued*)

Soups

Food	Serving Size	Count as
🧂 Bean, lentil, or split pea	1 cup	1 carbohydrate + 1 lean meat
🧂 Chowder (made with milk)	1 cup (8 oz)	1 carbohydrate + 1 lean meat + 1½ fats
🧂 Cream (made with water)	1 cup (8 oz)	1 carbohydrate + 1 fat
🧂 Instant	6 oz prepared	1 carbohydrate
🧂 with beans or lentils	8 oz prepared	2½ carbohydrates + 1 lean meat
🧂 Miso soup	1 cup	½ carbohydrate + 1 fat
🧂 Oriental noodle	1 cup	2 carbohydrates + 2 fats
Rice (congee)	1 cup	1 carbohydrate
🧂 Tomato (made with water)	1 cup (8 oz)	1 carbohydrate
🧂 Vegetable beef, chicken noodle, or other broth-type	1 cup (8 oz)	1 carbohydrate

Key:
😊 = More than 3 grams of dietary fiber per serving.
▽ = Extra fat, or prepared with added fat.
🧂 = 600 milligrams or more of sodium per serving (for combination food main dishes/meals).

TABLE C-11 Fast Foods

The choices in the Fast Foods list are not specific fast-food meals or items, but are estimates based on popular foods. Ask the restaurant or check its website for nutrition information about your favorite fast foods.

Breakfast Sandwiches

Food	Serving Size	Count as
🧂 Egg, cheese, meat, English muffin	1 sandwich	2 carbohydrates + 2 medium-fat meats
🧂 Sausage biscuit sandwich	1 sandwich	2 carbohydrates + 2 high-fat meats + 3½ fats

Main Dishes/Entrees

Food	Serving Size	Count as
🧂 😊 Burrito (beef and beans)	1 (about 8 oz)	3 carbohydrates + 3 medium-fat meats + 3 fats
🧂 Chicken breast, breaded and fried	1 (about 5 oz)	1 carbohydrate + 4 medium-fat meats
Chicken drumstick, breaded and fried	1 (about 2 oz)	2 medium-fat meats
🧂 Chicken nuggets	6 (about 3½ oz)	1 carbohydrate + 2 medium-fat meats + 1 fat
🧂 Chicken thigh, breaded and fried	1 (about 4 oz)	½ carbohydrate + 3 medium-fat meats + 1½ fats
🧂 Chicken wings, hot	6 (5 oz)	5 medium-fat meats + 1½ fats

Oriental

Food	Serving Size	Count as
🧂 Beef/chicken/shrimp with vegetables in sauce	1 cup (about 5 oz)	1 carbohydrate + 1 lean meat + 1 fat
🧂 Egg roll, meat	1 (about 3 oz)	1 carbohydrate + 1 lean meat + 1 fat
Fried rice, meatless	½ cup	1½ carbohydrates + 1½ fats

(continued)

Key:
😊 = More than 3 grams of dietary fiber per serving.
▽ = Extra fat, or prepared with added fat.
🧂 = 600 milligrams or more of sodium per serving (for fast-food main dishes/meals)

TABLE C-11 Fast Foods (continued)

🧂 Meat and sweet sauce (orange chicken)	1 cup	3 carbohydrates + 3 medium-fat meats + 2 fats
🧂 ☺ Noodles and vegetables in sauce (chow mein, lo mein)	1 cup	2 carbohydrates + 1 fat

Pizza

Food	Serving Size	Count as
Pizza		
🧂 cheese, pepperoni, regular crust	⅛ of a 14 inch (about 4 oz)	2½ carbohydrates + 1 medium-fat meat + 1½ fats
🧂 cheese/vegetarian, thin crust	¼ of a 12 inch (about 6 oz)	2½ carbohydrates + 2 medium-fat meats + 1½ fats

Sandwiches

Food	Serving Size	Count as
🧂 Chicken sandwich, grilled	1	3 carbohydrates + 4 lean meats
🧂 Chicken sandwich, crispy	1	3½ carbohydrates + 3 medium-fat meats + 1 fat
Fish sandwich with tartar sauce	1	2½ carbohydrates + 2 medium-fat meats + 2 fats
Hamburger		
🧂 large with cheese	1	2½ carbohydrates + 4 medium-fat meats + 1 fat
regular	1	2 carbohydrates + 1 medium-fat meat + 1 fat
🧂 Hot dog with bun	1	1 carbohydrate + 1 high-fat meat + 1 fat
Submarine sandwich		
🧂 less than 6 grams fat	6-inch sub	3 carbohydrates + 2 lean meats
🧂 regular	6-inch sub	3½ carbohydrates + 2 medium-fat meats + 1 fat
Taco, hard or soft shell (meat and cheese)	1 small	1 carbohydrate + 1 medium-fat meat + 1½ fats

Salads

Food	Serving Size	Count as
🧂 ☺ Salad, main dish (grilled chicken type, no dressing or croutons)		1 carbohydrate + 4 lean meats
Salad, side, no dressing or cheese	Small (about 5 oz)	1 vegetable

Sides/Appetizers

Food	Serving Size	Count as
▽ French fries, restaurant style	small	3 carbohydrates + 3 fats
	medium	4 carbohydrates + 4 fats
	large	5 carbohydrates + 6 fats
🧂 Nachos with cheese	small (about 4½ oz)	2½ carbohydrates + 4 fats
🧂 Onion rings	1 serving (about 3 oz)	2½ carbohydrates + 3 fats

Desserts

Food	Serving Size	Count as
Milkshake, any flavor	12 oz	6 carbohydrates + 2 fats
Soft-serve ice cream cone	1 small	2½ carbohydrates + 1 fat

Key:

☺ = More than 3 grams of dietary fiber per serving.

▽ = Extra fat, or prepared with added fat.

🧂 = 600 milligrams or more of sodium per serving (for fast-food main dishes/meals).

TABLE C-12 Alcohol

1 alcohol equivalent = variable grams carbohydrate, 0 grams protein, 0 grams fat, and 100 kcalories.

NOTE: In general, one alcohol choice (½ ounce absolute alcohol) has about 100 kcalories. For those who choose to drink alcohol, guidelines suggest limiting alcohol intake to 1 drink or less per day for women, and 2 drinks or less per day for men. To reduce your risk of low blood glucose (hypoglycemia), especially if you take insulin or a diabetes pill that increases insulin, always drink alcohol with food. While alcohol, by itself, does not directly affect blood glucose, be aware of the carbohydrate (for example, in mixed drinks, beer, and wine) that may raise your blood glucose.

Alcoholic Beverage	Serving Size	Count as
Beer		
light (4.2%)	12 fl oz	1 alcohol equivalent + ½ carbohydrate
regular (4.9%)	12 fl oz	1 alcohol equivalent + 1 carbohydrate
Distilled spirits: vodka, rum, gin, whiskey, 80 or 86 proof	1½ fl oz	1 alcohol equivalent
Liqueur, coffee (53 proof)	1 fl oz	1 alcohol equivalent + 1 carbohydrate
Sake	1 fl oz	½ alcohol equivalent
Wine		
dessert (sherry)	3½ fl oz	1 alcohol equivalent + 1 carbohydrate
dry, red or white (10%)	5 fl oz	1 alcohol equivalent

The Exchange Lists are the basis of a meal planning system designed by a committee of the American Diabetes Association and the American Dietetic Association. While designed primary for people with diabetes and others who must follow special diets, the Exchange Lists are based on principles of good nutrition that apply to everyone. (c) 2008 by the American Diabetes Association and the American Dietetic Association.

Appendix D

Physical Activity and Energy Requirements

Chapter 6 described how to calculate your estimated energy requirements by using an equation that accounts for your gender, age, weight, height, and physical activity level. This appendix first helps you determine the correct physical activity factor to use in the equation, either by calculating your physical activity level or by guesstimating it. Then the appendix presents tables that provide a shortcut to estimating total energy expenditure.*

Calculating Your Physical Activity Level

To calculate your physical activity level, record all of your activities for a typical 24-hour day, noting the type of activity, the level of intensity, and the duration. Then, using a copy of Table D-1, find your activity in the first column (or an activity that is reasonably similar) and multiply the number of minutes spent on that activity by the factor in the third column. Put your answer in the last column and total the accumulated values for the day. Now add the subtotal of the last column to 1.1 (to account for basal energy and the thermic effect of food) as shown. This score indicates your level of physical activity. Using Table D-2, find the PA (physical activity) factor for your gender that correlates with your physical activity level score and use it in the energy equation presented on p. 144 of Chapter 6.

Guesstimating Your Physical Activity Level

As an alternative to recording your activities for a day, you can use the first two columns of Table D-3 to decide if your daily activity is sedentary, low active, active, or very active. Find the PA factor for your gender that correlates with your typical physical activity level and use it in the energy equation presented on p. 144.

Using a Shortcut to Estimate Total Energy Expenditure

The DRI committee has developed estimates of total energy expenditure based on the equations presented in Chapter 6. These estimates are presented in Table D-4 for women and Table D-5 for men. You can use these tables to estimate your energy requirements, the number of kcalories needed to maintain your current body weight. On the table appropriate for your gender, find your height in meters (or inches) in the left-hand column. Then follow the row across to find your weight in kilograms (or pounds). (If you can't find your exact height and weight, choose a value between the two closest ones.) Look down the column to find the number of kcalories that corresponds to your activity level.

Importantly, the values given in the tables are for 30-year-olds. Women 19 to 29 should add 7 kcalories per day for each year below age 30; older women should subtract 7 kcalories per day for each year above age 30. Similarly, men 19 to 29 should add 10 kcalories per day for each year below age 30; older men should subtract 10 kcalories per day for each year above age 30.

*This appendix, including the tables, is adapted from Committee on Dietary Reference Intakes, Food and Nutrition Board, Institute of Medicine, *Dietary Reference Intakes for Energy, Carbohydrate, Fiber, Fat, Fatty Acids, Cholesterol, Protein, and Amino Acids* (Washington, D.C.: National Academies Press, 2002).

Appendix D

TABLE D-1 Physical Activities and Their Scores

If your activity was equivalent to this . . .	Then list the number of minutes here and . . .	Multiply by this factor . . .	Add this column to get your physical activity level score:
Activities of Daily Living			
Gardening (no lifting)		0.0032	
Household tasks (moderate effort)		0.0024	
Lifting items continuously		0.0029	
Loading/unloading car		0.0019	
Lying quietly		0.0000	
Mopping		0.0024	
Mowing lawn (power mower)		0.0033	
Raking lawn		0.0029	
Riding in a vehicle		0.0000	
Sitting (idle)		0.0000	
Sitting (doing light activity)		0.0005	
Taking out trash		0.0019	
Vacuuming		0.0024	
Walking the dog		0.0019	
Walking from house to car or bus		0.0014	
Watering plants		0.0014	
Additional Activities			
Billiards		0.0013	
Calisthenics (no weight)		0.0029	
Canoeing (leisurely)		0.0014	
Chopping wood		0.0037	
Climbing hills (carrying 11-lb load)		0.0061	
Climbing hills (no load)		0.0056	
Cycling (leisurely)		0.0024	
Cycling (moderately)		0.0045	
Dancing (aerobic or ballet)		0.0048	
Dancing (ballroom, leisurely)		0.0018	
Dancing (fast ballroom or square)		0.0043	
Golf (with cart)		0.0014	
Golf (without cart)		0.0032	
Horseback riding (walking)		0.0012	
Horseback riding (trotting)		0.0053	
Jogging (6 mph)		0.0088	
Music (playing accordion)		0.0008	
Music (playing cello)		0.0012	
Music (playing flute)		0.0010	
Music (playing piano)		0.0012	
Music (playing violin)		0.0014	
Rope skipping		0.0105	

(continued)

Appendix D

TABLE D-1 Physical Activities and Their Scores (*continued*)

If your activity was equivalent to this . . .	Then list the number of minutes here and . . .	Multiply by this factor . . .	Add this column to get your physical activity level score:
Skating (ice)		0.0043	
Skating (roller)		0.0052	
Skiing (water or downhill)		0.0055	
Squash		0.0106	
Surfing		0.0048	
Swimming (slow)		0.0033	
Swimming (fast)		0.0057	
Tennis (doubles)		0.0038	
Tennis (singles)		0.0057	
Volleyball (noncompetitive)		0.0018	
Walking (2 mph)		0.0014	
Walking (3 mph)		0.0022	
Walking (4 mph)		0.0033	
Walking (5 mph)		0.0067	
Subtotal			
Factor for basal energy and the thermic effect of food			1.1
Your physical activity level score			

TABLE D-2 Physical Activity Level Scores and Their PA Factors

Physical Activity Level Score	Description	Men: PA Factor	Women: PA Factor
1.0 to 1.39	Sedentary	1.0	1.0
1.4 to 1.59	Low active	1.11	1.12
1.6 to 1.89	Active	1.25	1.27
1.9 and above	Very active	1.48	1.45

TABLE D-3 Physical Activity Equivalents and Their PA Factors

Description	Physical Activity Equivalents	Men: PA Factor	Women: PA Factor
Sedentary	Only those physical activities required for normal independent living	1.0	1.0
	Activities equivalent to walking at a pace of 2–4 mph for the following distances:		
Low active	1.5 to 3.0 miles/day	1.11	1.12
Active	3 to 10 miles/day	1.25	1.27
Very active	10 or more miles/day	1.48	1.45

Appendix D

TABLE D-4 Total Energy Expenditure (TEE in kCalories per Day) for Women 30 Years of Age[a] at Various Levels of Activity and Various Heights and Weights

Height m (in)	Physical Activity Level	Weight[b] kg (lb)					
1.45 (57)		38.9 (86)	45.2 (100)	52.6 (116)	63.1 (139)	73.6 (162)	84.1 (185)
		kCalories					
	Sedentary	1564	1623	1698	1813	1927	2042
	Low active	1734	1800	1912	2043	2174	2304
	Active	1946	2021	2112	2257	2403	2548
	Very active	2201	2287	2387	2553	2719	2886
1.50 (59)		41.6 (92)	48.4 (107)	56.3 (124)	67.5 (149)	78.8 (174)	90.0 (198)
		kCalories					
	Sedentary	1625	1689	1771	1894	2017	2139
	Low active	1803	1874	1996	2136	2276	2415
	Active	2025	2105	2205	2360	2516	2672
	Very active	2291	2382	2493	2671	2849	3027
1.55 (61)		44.4 (98)	51.7 (114)	60.1 (132)	72.1 (159)	84.1 (185)	96.1 (212)
		kCalories					
	Sedentary	1688	1756	1846	1977	2108	2239
	Low active	1873	1949	2081	2230	2380	2529
	Active	2104	2190	2299	2466	2632	2798
	Very active	2382	2480	2601	2791	2981	3171
1.60 (63)		47.4 (104)	55.0 (121)	64.0 (141)	76.8 (169)	89.6 (197)	102.4 (226)
		kCalories					
	Sedentary	1752	1824	1922	2061	2201	2340
	Low active	1944	2025	2168	2327	2486	2645
	Active	2185	2276	2396	2573	2750	2927
	Very active	2474	2578	2712	2914	3116	3318
1.65 (65)		50.4 (111)	58.5 (129)	68.1 (150)	81.7 (180)	95.3 (210)	108.9 (240)
		kCalories					
	Sedentary	1816	1893	1999	2148	2296	2444
	Low active	2016	2102	2556	2425	2594	2763
	Active	2267	2364	2494	2682	2871	3059
	Very active	2567	2678	2824	3039	3254	3469
1.70 (67)		53.5 (118)	62.1 (137)	72.3 (159)	86.7 (191)	101.2 (223)	115.6 (255)
		kCalories					
	Sedentary	1881	1963	2078	2235	2393	2550
	Low active	2090	2180	2345	2525	2705	2884
	Active	2350	2453	2594	2794	2994	3194
	Very active	2662	2780	2938	3166	3395	3623

(continued)

TABLE D-4 Total Energy Expenditure (TEE in kCalories per Day) for Women 30 Years of Age[a] at Various Levels of Activity and Various Heights and Weights (*continued*)

Height m (in)	Physical Activity Level	Weight[b] kg (lb)					
1.75 (69)		56.7 (125)	65.8 (145)	76.6 (169)	91.9 (202)	107.2 (236)	122.5 (270)
				kCalories			
	Sedentary	1948	2034	2158	2325	2492	2659
	Low active	2164	2260	2437	2627	2817	3007
	Active	2434	2543	2695	2907	3119	3331
	Very active	2758	2883	3054	3296	3538	3780
1.80 (71)		59.9 (132)	69.7 (154)	81.0 (178)	97.2 (214)	113.4 (250)	129.6 (285)
				kCalories			
	Sedentary	2015	2106	2239	2416	2593	2769
	Low active	2239	2341	2529	2731	2932	3133
	Active	2519	2634	2799	3023	3247	3472
	Very active	2855	2987	3172	3428	3684	3940
1.85 (73)		63.3 (139)	73.6 (162)	85.6 (189)	102.7 (226)	119.8 (264)	136.9 (302)
				kCalories			
	Sedentary	2083	2179	2322	2509	2695	2882
	Low active	2315	2422	2624	2836	3049	3262
	Active	2605	2727	2904	3141	3378	3615
	Very active	2954	3093	3292	3562	3833	4103
1.90 (75)		66.8 (147)	77.6 (171)	90.3 (199)	108.3 (239)	126.4 (278)	144.4 (318)
				kCalories			
	Sedentary	2151	2253	2406	2603	2800	2996
	Low active	2392	2505	2720	2944	3168	3393
	Active	2693	2821	3011	3261	3511	3760
	Very active	3053	3200	3414	3699	3984	4270
1.95 (77)		70.3 (155)	81.8 (180)	95.1 (209)	114.1 (251)	133.1 (293)	152.1 (335)
				kCalories			
	Sedentary	2221	2328	2492	2699	2906	3113
	Low active	2470	2589	2817	3053	3290	3526
	Active	2781	2917	3119	3383	3646	3909
	Very active	3154	3309	3538	3838	4139	4439

[a] For each year below 30, add 7 kcalories/day to TEE. For each year above 30, subtract 7 kcalories/day from TEE.
[b] These columns represent a BMI of 18.5, 22.5, 25, 30, 35, and 40, respectively.

TABLE D-5 Total Energy Expenditure (TEE in kCalories per Day) for Men 30 Years of Age[a] at Various Levels of Activity and Various Heights and Weights

Height m (in)	Physical Activity Level	Weight[b] kg (lb)					
1.45 (57)		38.9 (86)	47.3 (100)	52.6 (116)	63.1 (139)	73.6 (163)	84.1 (185)
		kCalories					
	Sedentary	1777	1911	2048	2198	2347	2496
	Low active	1931	2080	2225	2393	2560	2727
	Active	2127	2295	2447	2636	2826	3015
	Very active	2450	2648	2845	3075	3305	3535
1.50 (59)		41.6 (92)	50.6 (107)	56.3 (124)	67.5 (149)	78.8 (174)	90.0 (198)
		kCalories					
	Sedentary	1848	1991	2126	2286	2445	2605
	Low active	2009	2168	2312	2491	2670	2849
	Active	2215	2394	2545	2748	2951	3154
	Very active	2554	2766	2965	3211	3457	3703
1.55 (61)		44.4 (98)	54.1 (114)	60.1 (132)	72.1 (159)	84.1 (185)	96.1 (212)
		kCalories					
	Sedentary	1919	2072	2205	2376	2546	2717
	Low active	2089	2259	2401	2592	2783	2974
	Active	2305	2496	2646	2862	3079	3296
	Very active	2660	2887	3087	3349	3612	3875
1.60 (63)		47.4 (104)	57.6 (121)	64.0 (141)	76.8 (169)	89.6 (197)	102.4 (226)
		kCalories					
	Sedentary	1993	2156	2286	2468	2650	2831
	Low active	2171	2351	2492	2695	2899	3102
	Active	2397	2601	2749	2980	3210	3441
	Very active	2769	3010	3211	3491	3771	4051
1.65 (65)		50.4 (111)	61.3 (129)	68.1 (150)	81.7 (180)	95.3 (210)	108.9 (240)
		kCalories					
	Sedentary	2068	2241	2369	2562	2756	2949
	Low active	2254	2446	2585	2801	3017	3234
	Active	2490	2707	2854	3099	3345	3590
	Very active	2880	3136	3339	3637	3934	4232
1.70 (67)		53.5 (118)	65.0 (137)	72.3 (159)	86.7 (191)	101.2 (223)	115.6 (255)
		kCalories					
	Sedentary	2144	2328	2454	2659	2864	3069
	Low active	2338	2542	2679	2909	3139	3369
	Active	2586	2816	2961	3222	3483	3743
	Very active	2992	3265	3469	3785	4101	4417

(continued)

TABLE D-5 Total Energy Expenditure (TEE in kCalories per Day) for Men 30 Years of Age[a] at Various Levels of Activity and Various Heights and Weights (*continued*)

Height m (in)	Physical Activity Level	Weight[b] kg (lb)					
1.75 (69)		56.7 (125)	68.9 (145)	76.6 (169)	91.9 (202)	107.2 (236)	122.5 (270)
		kCalories					
	Sedentary	2222	2416	2540	2757	2975	3192
	Low active	2425	2641	2776	3020	3263	3507
	Active	2683	2927	3071	3347	3623	3900
	Very active	3108	3396	3602	3937	4272	4607
1.80 (71)		59.9 (132)	72.9 (154)	81.0 (178)	97.2 (214)	113.4 (250)	129.6 (285)
		kCalories					
	Sedentary	2301	2507	2628	2858	3088	3318
	Low active	2513	2741	2875	3132	3390	3648
	Active	2782	3040	3183	3475	3767	4060
	Very active	3225	3530	3738	4092	4447	4801
1.85 (73)		63.3 (139)	77.0 (162)	85.6 (189)	102.7 (226)	119.8 (264)	136.9 (302)
		kCalories					
	Sedentary	2382	2599	2718	2961	3204	3447
	Low active	2602	2844	2976	3248	3520	3792
	Active	2883	3155	3297	3606	3915	4223
	Very active	3344	3667	3877	4251	4625	4999
1.90 (75)		66.8 (147)	81.2 (171)	90.3 (199)	108.3 (239)	126.4 (278)	144.4 (318)
		kCalories					
	Sedentary	2464	2693	2810	3066	3322	3579
	Low active	2693	2948	3078	3365	3652	3939
	Active	2986	3273	3414	3739	4065	4390
	Very active	3466	3806	4018	4413	4807	5202
1.95 (77)		70.3 (155)	85.6 (180)	95.1 (209)	114.1 (251)	133.1 (293)	152.1 (335)
		kCalories					
	Sedentary	2547	2789	2903	3173	3443	3713
	Low active	2786	3055	3183	3485	3788	4090
	Active	3090	3393	3533	3875	4218	4561
	Very active	3590	3948	4162	4578	4993	5409

[a]For each year below 30, add 10 kcalories/day to TEE. For each year above 30, subtract 10 kcalories/day from TEE.
[b]These columns represent a BMI of 18.5, 22.5, 25, 30, 35, and 40, respectively.

Nutrition Assessment: Supplemental Information

Chapter 14 describes data from nutrition assessments that help health professionals evaluate patients' nutrition status and nutrient needs. This appendix provides additional information that may be useful for complete assessments.

Weight Gain during Pregnancy

Chapter 11 describes desirable weight-gain patterns during pregnancy. Figure E-1 shows prenatal weight-gain grids, which are used to plot the rate of weight gain during pregnancy.

Growth Charts

Health professionals evaluate physical development by monitoring the growth rates of children and comparing these rates with those on standard growth charts. Standard charts compare length or height to age, weight to age, weight to length, head circumference to age, and body mass index (BMI) to age. Although individual growth patterns vary, a child's growth curve will generally stay at about the same percentile throughout childhood. In children whose growth has been retarded, nutrition rehabilitation will ideally cause height and weight to increase to higher percentiles. In overweight children, the goal is for weight to remain stable as height increases, until weight becomes appropriate for height.

To evaluate growth in infants, an assessor uses a chart such as those in Figures E-2 through E-5. For example, the assessor follows these steps to determine the weight percentile:

▮ Select the appropriate chart based on age and gender.

▮ Locate the child's age along the horizontal axis on the bottom of the chart.

CONTENTS

Growth Charts

Measures of Body Fat and Lean Tissue

Nutritional Anemias

Reminder: The *body mass index (BMI)* is an index of a person's weight in relation to height, determined by dividing the weight in kilograms by the square of the height in meters:

$$BMI = \frac{Weight\ (kg)}{Height\ (m)^2}$$

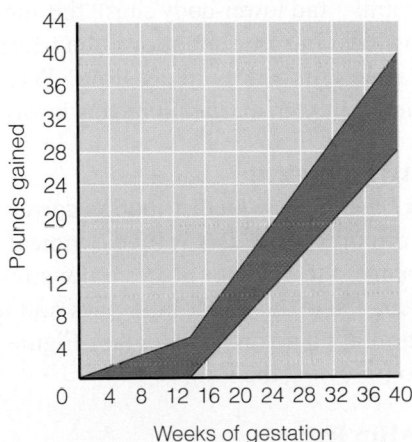

Normal-weight women should gain about 3½ pounds in the first trimester and just under 1 pound/week thereafter, achieving a total gain of 25 to 35 pounds by term.

Underweight women should gain about 5 pounds in the first trimester and just over 1 pound/week thereafter, achieving a total gain of 28 to 40 pounds by term.

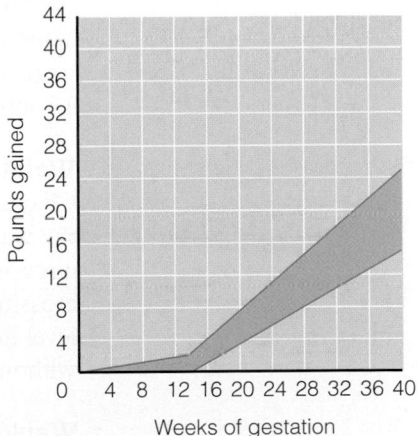

Overweight women should gain about 2 pounds in the first trimester and ⅔ pound/week thereafter, achieving a total gain of 15 to 25 pounds.

FIGURE E-1 **Recommended Prenatal Weight Gain Based on Prepregnancy Weight**

- Locate the child's weight in pounds or kilograms along the vertical axis.

- Mark the chart where the age and weight lines intersect, and read off the percentile.

Additional growth charts are available at **www.cdc.gov/ growthcharts**

For other measures, the assessor follows a similar procedure, using the appropriate chart. (When length is measured, use the chart for birth to 36 months; when height is measured, use the chart for 2 to 20 years.) Once all of the measures are plotted on growth charts, a skilled clinician can begin to interpret the data. Ideally, the height, weight, and head circumference should be in roughly the same percentile.

Head circumference is generally measured in children under 2 years of age. Since the brain grows rapidly before birth and during early infancy, extreme and chronic malnutrition during these times can impair brain development, curtailing the number of brain cells and the size of head circumference. Nonnutritional factors, such as certain disorders and genetic variation, can also influence head circumference.

Measures of Body Fat and Lean Tissue

Significant weight changes in both children and adults can reflect overnutrition or undernutrition with respect to energy and protein. To estimate the degree to which fat stores or lean tissues are affected by malnutrition, several anthropometric measurements are useful.

Skinfold Measures

Skinfold measures provide a good estimate of total body fat and a fair assessment of the fat's location. Most body fat lies directly beneath the skin, and the thickness of this subcutaneous fat correlates with total body fat. In some parts of the body, such as the back and the back of the arm over the triceps muscle, this fat is loosely attached. As illustrated in Figure E-6, an assessor can measure the thickness of the fat with calipers that apply a fixed amount of pressure. If a person gains body fat, the skinfold increases proportionately; if the person loses fat, it decreases. Measurements taken from central-body sites better reflect changes in fatness than those taken from upper sites (arm and back). Because subcutaneous fat may be thicker in one area than in another, skinfold measurements are often taken at three or four different places on the body (including upper-, central-, and lower-body sites); the sum of these measures is then compared to standard values. In some situations, the triceps skinfold measurement alone may be used because it is easily accessible. Triceps skinfold measures greater than 15 millimeters in men or 25 millimeters in women suggest excessive body fat.

Common sites for skinfold measures:
- Triceps
- Biceps
- Subscapular (below shoulder blade)
- Suprailiac (above hip bone)
- Abdomen
- Upper thigh

Waist Circumference

Chapter 6 explains how fat distribution correlates with health risks and points out that the waist circumference is a valuable indicator of abdominal fat. To measure waist circumference, the assessor places a nonstretchable tape around the person's body, crossing just above the upper hip bones and making sure that the tape remains on a level horizontal plane on all sides (see Figure E-7). The tape is tightened slightly but without compressing the skin.

Waist-to-Hip Ratio

The waist-to-hip ratio assesses abdominal obesity, but it offers no advantage over the waist circumference alone. To calculate the waist-to-hip ratio, divide the waistline measurement by the hip measurement. In general, women with a waist-to-hip ratio of 0.8 or greater and men with a waist-to-hip ratio of 0.9 or greater have an increased risk of developing diabetes and cardiovascular diseases.

The calculation of waist-to-hip ratio in a woman with a 28-inch waist and 38-inch hips is $28 \div 38 = 0.74$.

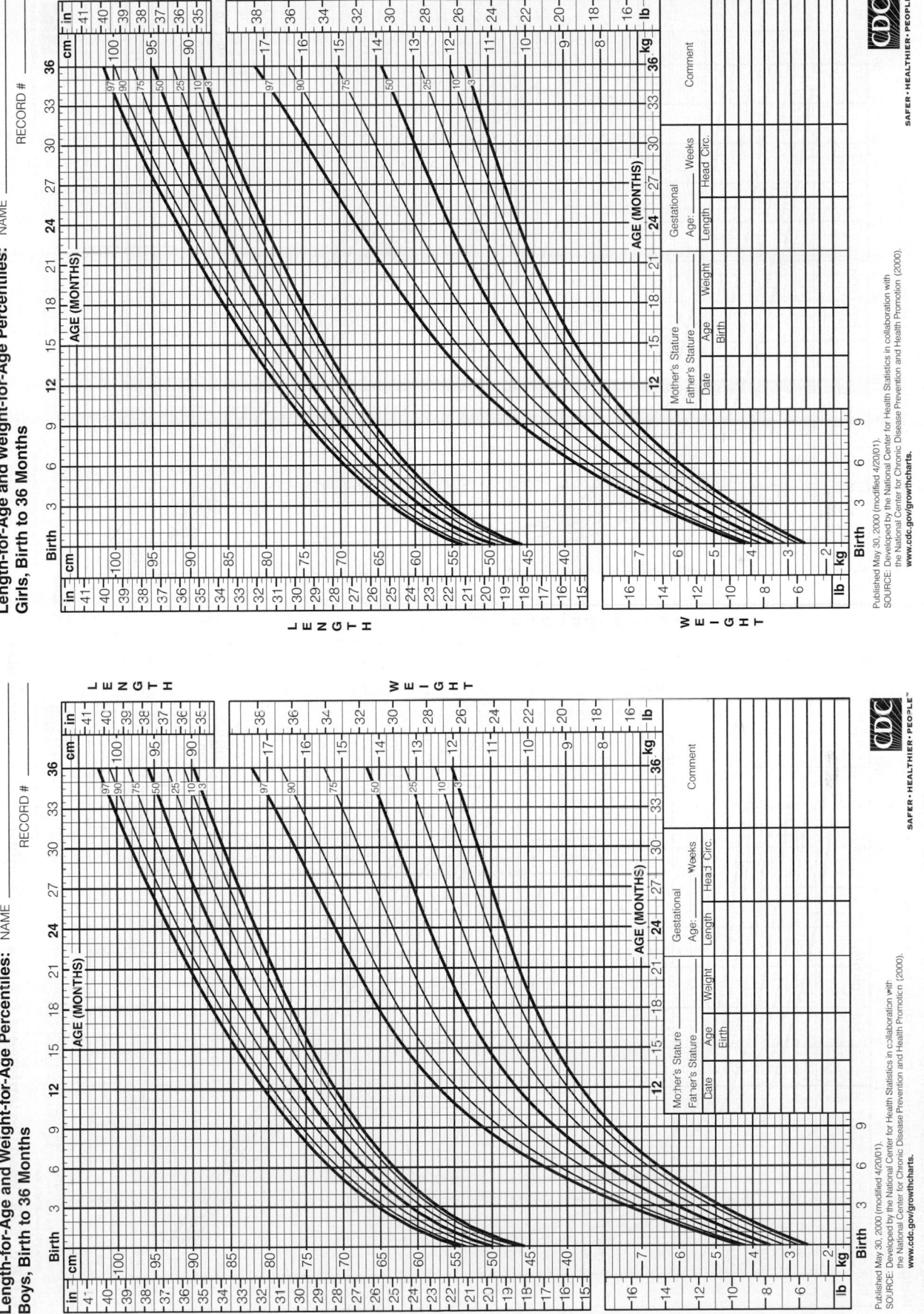

Length-for-Age and Weight-for-Age Percentiles: NAME
Girls, Birth to 36 Months

Published May 30, 2000 (modified 4/20/01).
SOURCE: Developed by the National Center for Health Statistics in collaboration with
the National Center for Chronic Disease Prevention and Health Promotion (2000).
www.cdc.gov/growthcharts.

Length-for-Age and Weight-for-Age Percentiles: NAME
Boys, Birth to 36 Months

Published May 30, 2000 (modified 4/20/01).
SOURCE: Developed by the National Center for Health Statistics in collaboration with
the National Center for Chronic Disease Prevention and Health Promotion (2000).
www.cdc.gov/growthcharts.

FIGURE E-2 Length-for-Age and Weight-for-Age Percentiles

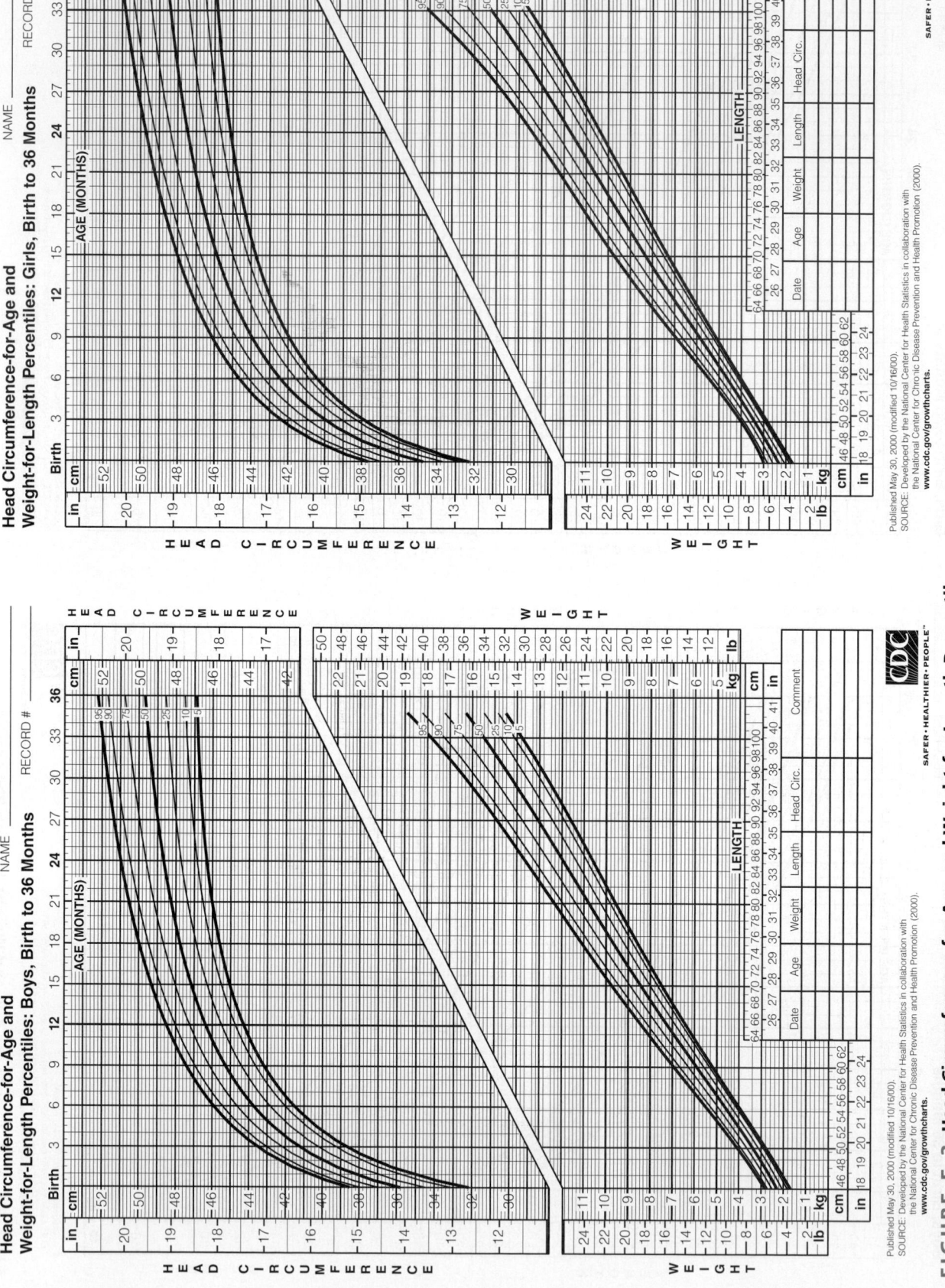

Head Circumference-for-Age and
Weight-for-Length Percentiles: Girls, Birth to 36 Months

Head Circumference-for-Age and
Weight-for-Length Percentiles: Boys, Birth to 36 Months

Published May 30, 2000 (modified 10/16/00).
SOURCE: Developed by the National Center for Health Statistics in collaboration with
the National Center for Chronic Disease Prevention and Health Promotion (2000).
www.cdc.gov/growthcharts.

FIGURE E-3 Head Circumference-for-Age and Weight-for-Length Percentiles

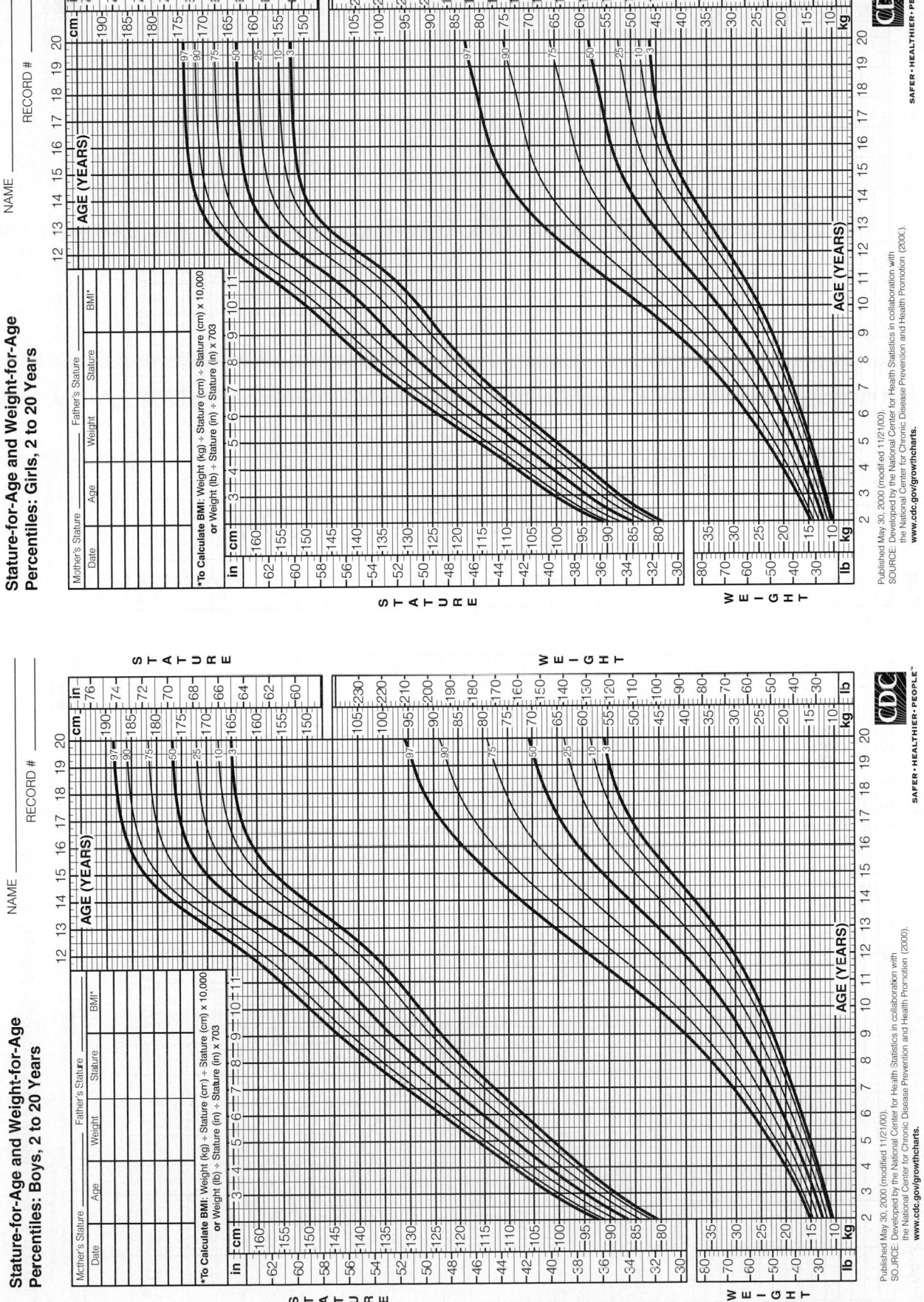

Stature-for-Age and Weight-for-Age Percentiles: Girls, 2 to 20 Years

Stature-for-Age and Weight-for-Age Percentiles: Boys, 2 to 20 Years

FIGURE E-4 Stature-for-Age and Weight-for-Age Percentiles

Published May 30, 2000 (modified 11/21/00).
SOURCE: Developed by the National Center for Health Statistics in collaboration with the National Center for Chronic Disease Prevention and Health Promotion (2000).
www.cdc.gov/growthcharts.

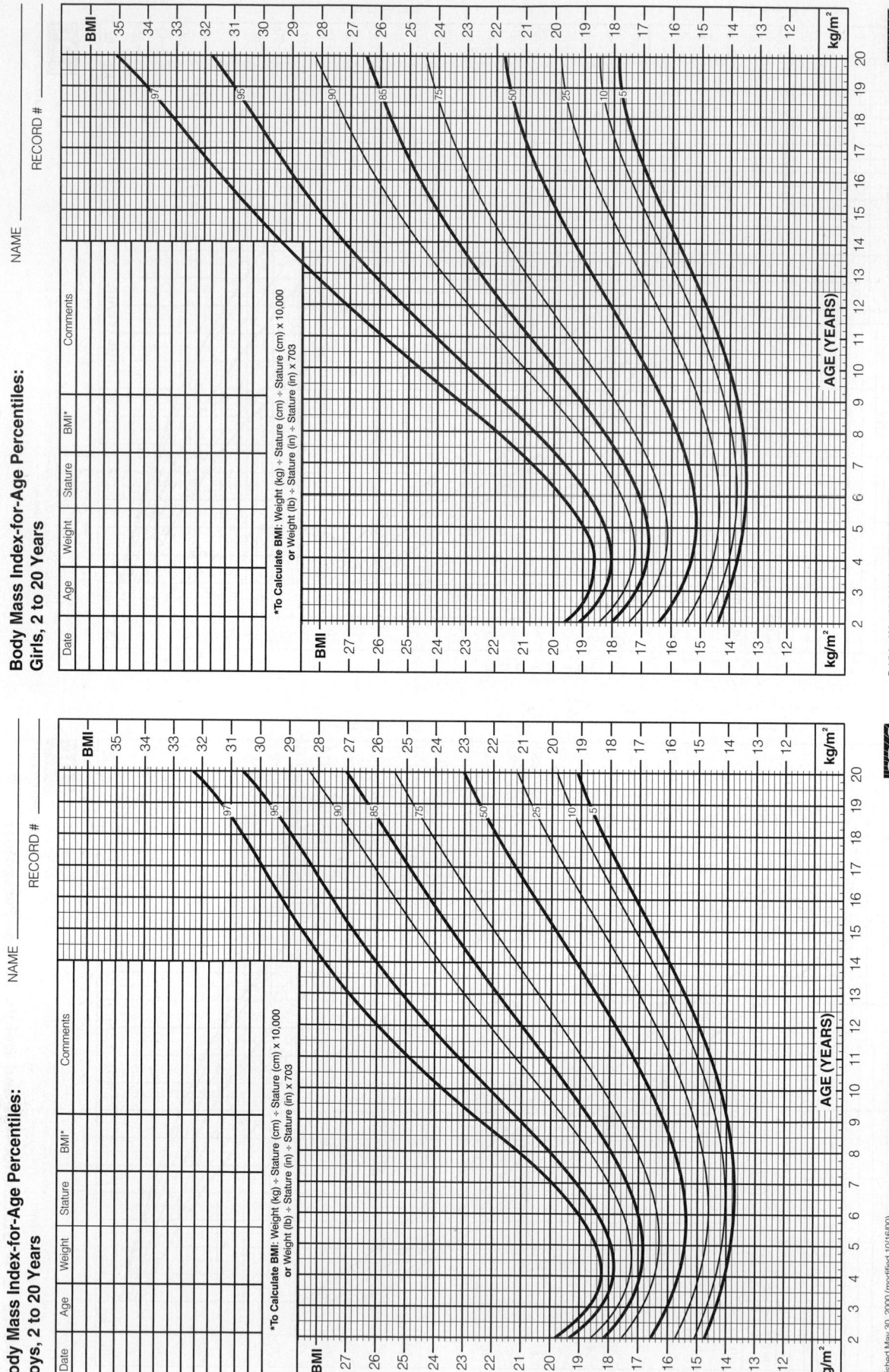

FIGURE E-5 Body Mass Index-for-Age Percentiles

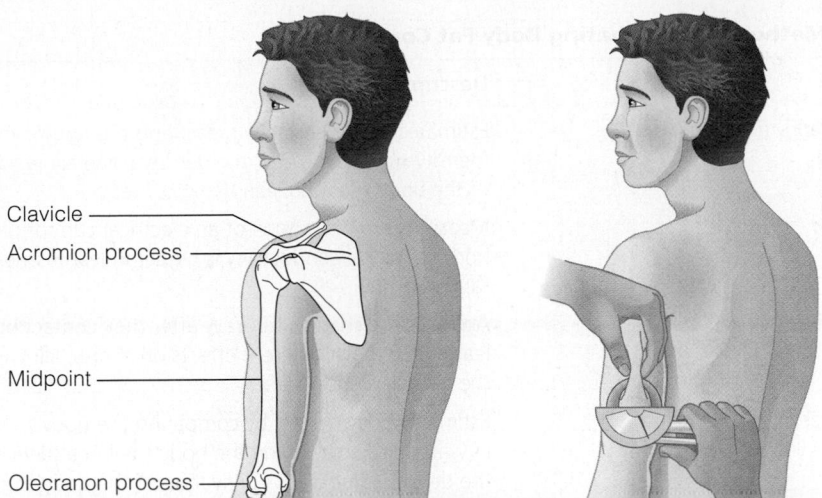

A. Find the midpoint of the arm:
   1. Ask the subject to bend his or her arm at the elbow and lay the hand across the stomach. (If he or she is right-handed, measure the left arm, and vice versa.)
   2. Feel the shoulder to locate the acromion process. It helps to slide your fingers along the clavicle to find the acromion process. The olecranon process is the tip of the elbow.
   3. Place a measuring tape from the acromion process to the tip of the elbow.

Divide this measurement by 2 and mark the midpoint of the arm with a pen.
B. Measure the skinfold:
   1. Ask the subject to let his or her arm hang loosely to the side.
   2. Grasp a fold of skin and subcutaneous fat between the thumb and forefinger slightly above the midpoint mark. Gently pull the skin away from the underlying muscle. (This step takes a lot of practice. If you want to be sure you don't have muscle as well as fat, ask the subject to

contract and relax the muscle. You should be able to feel if you are pinching muscle.)
   3. Place the calipers over the skinfold at the midpoint mark, and read the measurement to the nearest 1.0 millimeter in two to three seconds. (If using plastic calipers, align pressure lines, and read the measurement to the nearest 1.0 millimeter in two to three seconds.)
   4. Repeat steps 2 and 3 twice more. Add the three readings, and then divide by 3 to find the average.

FIGURE E-6 How to Measure the Triceps Skinfold

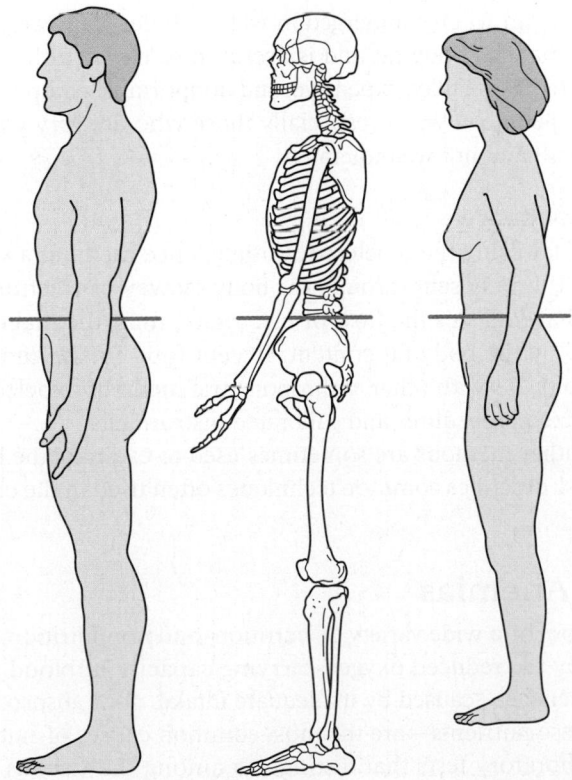

FIGURE E-7 How to Measure Waist Circumference

Place the measuring tape around the waist just above the bony crest of the hip. The tape runs parallel to the floor and is snug (but does not compress the skin). The measurement is taken at the end of normal expiration.

Source: National Institutes of Health Obesity Education Initiative, *Clinical Guidelines on the Identification, Evaluation, and Treatment of Overweight and Obesity in Adults* (Washington, D.C.: U.S. Department of Health and Human Services, 1998), p. 59.

TABLE E-1 Selected Methods for Estimating Body Fat Content

Method	Description
Air-displacement plethysmography (Bod Pod®)	Estimates body density by measuring the body's volume (density = mass/volume); the density value allows derivation of the body's fat and lean tissue contents
Bioelectrical impedance assay	Measures the magnitude of an electrical current passed through the body; electrical conductivity is higher in lean tissues than in fat tissue
Dual energy X-ray absorptiometry	Analyzes the change in X-rays after they contact body tissues; fat and lean tissues have different effects on X-rays, allowing quantification of the various tissues
Hydrodensitometry (underwater weighing)	Estimates body density by comparing the body's weight on land and in water or by measuring the body's volume (density = mass/volume); the density value allows derivation of the body's fat and lean tissue contents
Isotope dilution—deuterated water	Measures total body water content by analyzing the dilution of heavy water (water with a heavy form of hydrogen) in body tissues; allows an estimate of lean tissue and body fat content
Skinfolds	Estimates subcutaneous fat in several regions of the body by using calipers to measure skinfold thicknesses
Ultrasound	Estimates subcutaneous fat in several regions of the body by using ultrasound to measure skinfold thicknesses

Hydrodensitometry

To estimate body density using hydrodensitometry, the person is weighed twice—first on land and then again when submerged in water. Underwater weighing usually generates a good estimate of body fat and is useful in research, although the technique has drawbacks: it requires bulky, expensive, and nonportable equipment. Furthermore, submerging some people in water (especially those who are very young, very old, ill, or fearful) is difficult and not well tolerated.

Bioelectrical Impedance

To measure body fat using the bioelectrical impedance method, a very-low-intensity electrical current is briefly sent through the body by way of electrodes placed on the wrist and ankle. Fat impedes the flow of electricity; thus, the magnitude of the current is influenced by the body fat content. Recent food intake and hydration status can influence results. As with other anthropometric methods, bioelectrical impedance requires standardized procedures and calibrated instruments.

A number of other methods are sometimes used to estimate the body's content of body fat. Table E-1 describes common techniques often used in the clinical or research setting.

Nutritional Anemias

Anemia, a symptom of a wide variety of nutrition- and nonnutrition-related disorders, is characterized by the reduced oxygen-carrying capacity of blood. Iron, folate, and vitamin B_{12} deficiencies—caused by inadequate intake, poor absorption, or abnormal metabolism of these nutrients—are the most common causes of nutritional anemias. Table E-2 lists laboratory tests that distinguish among the various nutrition-related

TABLE E-2 Laboratory Tests Useful in Evaluating Nutrition-Related Anemias

Test or Test Result	What It Reflects
For Anemia (general)	
Hemoglobin (Hg)	Total amount of hemoglobin in the red blood cells (RBC)
Hematocrit (Hct)	Percentage of RBC in the total blood volume
Red blood cell (RBC) count	Number of RBC
Mean corpuscular volume (MCV)	RBC size; helps to determine if anemia is microcytic (iron deficiency) or macrocytic (folate or vitamin B_{12} deficiency)
Mean corpuscular hemoglobin concentration (MCHC)	Hemoglobin concentration within the average RBC; helps to determine if anemia is hypochromic (iron deficiency) or normochromic (folate or vitamin B_{12} deficiency)
Bone marrow aspiration	The manufacture of blood cells in different developmental states
For Iron-Deficiency Anemia	
↓ Serum ferritin	Early deficiency state with depleted iron stores
↓ Transferrin saturation	Progressing deficiency state with diminished transport iron
↑ Erythrocyte protoporphyrin	Later deficiency state with limited hemoglobin production
For Folate-Deficiency Anemia	
↓ Serum folate	Progressing deficiency state
↓ RBC folate	Later deficiency state
For Vitamin B_{12}-Deficiency Anemia	
↓ Serum vitamin B_{12}	Progressing deficiency state
↑ Serum methylmalonic acid	Vitamin B_{12} deficiency
Schilling test	Adequacy of vitamin B_{12} absorption

anemias. Some nonnutrition-related causes of anemia include massive blood loss, infections, hereditary blood disorders such as sickle-cell anemia, and chronic liver or kidney disease.

Assessment of Iron Status

Chapter 9 describes the progression of iron deficiency in detail, as well as the roles of some of the proteins involved in iron metabolism. This section describes the various tests that assess iron status, and Table E-3 provides acceptable values. Although other tests are more specific for detecting the early stages of iron deficiency, hemoglobin and hematocrit are most often used to detect iron-deficiency anemia because they are inexpensive and easily measured.

Reminder: Iron deficiency progresses as follows:
1. Iron stores diminish
2. Transport iron decreases
3. Hemoglobin production falls

Serum Ferritin

In the initial stage of iron deficiency, iron stores diminish. Iron is stored in the protein ferritin, which is located in the liver, spleen, and bone marrow. Serum ferritin values provide a noninvasive estimate of iron stores, because the ferritin levels in blood reflect the amounts stored in the tissues. Serum ferritin is not a reliable indicator of iron deficiency, however, because its concentrations are increased by infection, inflammation, alcohol consumption, and liver disease.

TABLE E-3 Criteria for Assessing Iron Status

Laboratory Test	Acceptable Values	Effect of Iron Deficiency
Serum ferritin	Male: 20-250 ng/mL Female: 10-120 ng/mL	Lower than normal
Serum iron	Male: 60-175 μg/dL Female: 50-170 μg/dL	Lower than normal
Total iron-binding capacity	250-450 μg/dL	Higher than normal
Transferrin saturation	Male: 20-50% Female: 15-50%	Lower than normal
Erythrocyte protoporphyrin	<70 μg/dL red blood cells	Higher than normal
Hemoglobin (Hb)	Male: 13.5-17.5 g/dL Female: 12.0-16.0 g/dL	Lower than normal
Hematocrit (Hct)	Male: 39-49% Female: 35-45%	Lower than normal
Mean corpuscular volume (MCV)	80-100 fL	Lower than normal

Note: ng = nanogram, μg = microgram, dL = deciliter, nmol = nanomole, fL = femtoliter
Sources: L. Goldman and D. Ausiello, coeditors, *Cecil Medicine* (Philadelphia: Saunders, 2008), pp. 2983–2991;
R. J. Wood and A. G. Ronnenberg, Iron, in M. E. Shils and coeditors, *Modern Nutrition in Health and Disease*
(Baltimore: Lippincott Williams & Wilkins, 2006), pp. 248–270.

Serum Iron and Total Iron-Binding Capacity (TIBC)

Early stages of iron deficiency are characterized by reduced levels of serum iron, which represent the amount of iron bound to transferrin, the iron transport protein. Total iron-binding capacity (TIBC) is a measure of the total amount of iron that the transferrin in blood can carry; thus, it is an indirect measure of the transferrin content of blood. During iron deficiency, the liver produces more transferrin in an effort to increase iron transport capacity, and therefore iron depletion is characterized by an increase in TIBC. TIBC reflects liver function as well as changes in iron metabolism.

Transferrin Saturation

The percentage of transferrin that is saturated with iron is an indirect measure derived from the serum iron and total iron-binding capacity measures, as follows:

$$\% \text{ Transferrin} = \frac{\text{serum iron}}{\text{total iron-binding capacity}} \times 100$$

During iron deficiency, transferrin saturation decreases. The transferrin saturation value is a useful indicator of iron status because it includes information about both the iron and transferrin content of blood.

Erythrocyte Protoporphyrin

The iron-containing molecule in hemoglobin is heme, which is formed from iron and protoporphyrin. Protoporphyrin accumulates in the blood when iron supplies are inadequate for the formation of heme. However, levels of protoporphyrin may increase when hemoglobin synthesis is impaired for other reasons, such as lead poisoning or inflammation.

Hemoglobin

When iron stores become depleted, hemoglobin production is impaired, and symptoms of anemia may eventually develop. Hemoglobin's usefulness in evaluating iron status is limited, however, because hemoglobin concentrations drop fairly late in the development of iron deficiency, and other nutrient deficiencies and medical conditions can also alter hemoglobin concentrations.

Hematocrit

The hematocrit is the percentage of the total blood volume occupied by red blood cells. To measure the hematocrit, a clinician spins the blood samples in a centrifuge to separate the red blood cells from the plasma. Low values indicate a reduced number or size of red blood cells. Although this test is not specific for iron status, it can help to detect the presence of iron-deficiency anemia.

Mean Corpuscular Volume (MCV)

The hematocrit value divided by the red blood cell count provides a measure of the average size of a red blood cell, referred to as the mean corpuscular volume (MCV). Such a measure helps to classify the type of anemia that is present. In iron deficiency, the red blood cells are smaller than average (microcytic cells).

Assessment of Folate and Vitamin B_{12} Status

Folate deficiency and vitamin B_{12} deficiency present a similar clinical picture—an anemia characterized by abnormally large, misshapen, and immature red blood cells (megaloblastic cells). Distinguishing between folate and vitamin B_{12} deficiency is essential, however, because their treatments differ. Giving folate to a person with vitamin B_{12} deficiency improves many of the test results indicative of vitamin B_{12} deficiency, but this would be a dangerous treatment because vitamin B_{12} deficiency causes nerve damage that folate cannot correct. Thus, inappropriate folate administration masks vitamin B_{12}-deficiency anemia, and nerve damage worsens. For this reason, it is critical to determine whether the anemia results from a folate deficiency or from a vitamin B_{12} deficiency. Several of the following assessment measures help to make this distinction.

Mean Corpuscular Volume (MCV)

As previously mentioned, MCV is a measure of red blood cell size. In folate and vitamin B_{12} deficiencies, the red blood cells are larger than average, or macrocytic. Macrocytic cells are not necessarily indicative of nutrient deficiency, however, as they may also result from a high alcohol intake, liver disease, and various medications.

Serum Folate and Vitamin B_{12} Levels

Analyses of serum folate and vitamin B_{12} levels are usually among the first tests conducted to determine the cause of macrocytic red blood cells. The presence of low serum levels of either nutrient is consistent with a deficiency of that nutrient, whereas adequate levels can help to rule out deficiency. Folate levels are not a specific measure of folate status, however; they may increase after folate consumption and decrease due to alcohol consumption, pregnancy, or use of anticonvulsants. The folate levels in red blood cells (called erythrocyte folate) correlate well with folate stores and can help to diagnose folate deficiency, but the more reliable testing methods are not widely available. Table E-4 shows the acceptable ranges for these tests.

TABLE E-4 Criteria for Assessing Folate and Vitamin B$_{12}$ Status

Laboratory Test	Acceptable Range	Effect of Folate or Vitamin B$_{12}$ Deficiency
Serum folate	3-16 ng/mL	Reduced in folate deficiency
Erythrocyte folate	140-628 ng/mL packed cells	Reduced in folate deficiency
Serum vitamin B$_{12}$	200-835 pg/mL	Reduced in vitamin B$_{12}$ deficiency
Serum methylmalonic acid	70-270 nmol/L	Increased in vitamin B$_{12}$ deficiency
Serum homocysteine	5-14 μmol/L	Increased in folate or vitamin B$_{12}$ deficiency

Note: ng = nanogram, pg = picogram, nmol = nanomole, μmol = micromole
Source: L. Goldman and D. Ausiello, coeditors, *Cecil Medicine* (Philadelphia: Saunders, 2008), p. 1238 and pp. 2983–2991.

Methylmalonic Acid and Homocysteine Levels

To determine whether a nutrient deficiency is present, clinicians can measure the levels of substances that accumulate when the functions of that nutrient are impaired. For example, blood levels of the amino acid homocysteine are usually increased by both folate and vitamin B$_{12}$ deficiency, because both nutrients are needed for its metabolism. Methylmalonic acid, a breakdown product of several amino acids, requires vitamin B$_{12}$ for its metabolism; hence, serum levels increase as a result of vitamin B$_{12}$ deficiency. Because methylmalonic acid levels are not influenced by folate status, this measure is useful in distinguishing between folate and vitamin B$_{12}$ deficiency.

Schilling Test

As Chapter 8 explains, vitamin B$_{12}$ deficiency most often results from malabsorption, not poor intake. The Schilling test can help to diagnose malabsorption of vitamin B$_{12}$: after the patient takes an oral dose of radioactive vitamin B$_{12}$, a urine test determines whether the vitamin B$_{12}$ was absorbed. The Schilling test is rarely performed at present, due to the difficulty in obtaining the chemical reagents needed for the test.

Antibodies to Intrinsic Factor

The presence of serum antibodies for intrinsic factor can help to confirm a diagnosis of pernicious anemia, an autoimmune disease characterized by destruction of the cells that produce intrinsic factor (a protein required for vitamin B$_{12}$ absorption; see Chapter 8). Serum antibodies to the parietal cells that produce and release intrinsic factor may also indicate pernicious anemia, but these antibodies may be present in various other conditions as well.

Cautions about Nutrition Assessment

The tests outlined in this appendix yield information that becomes meaningful only when they are conducted and interpreted by a skilled clinician. Potential sources of error may be introduced at any step, from the collection of samples to the analysis and reporting of data. Equipment must be regularly calibrated to ensure accuracy of measurements. In addition, the assessor must keep in mind that few tests may be specific to the nutrient of interest alone, and lab results may reflect physiological processes other than the ones being tested. Furthermore, because many tests are not sensitive enough to detect the early stages of deficiency, follow-up testing is often necessary to identify a nutrition problem.

Appendix F

Aids to Calculations

Many mathematical problems have been worked out in the "How to" sections of the text. These pages provide additional help and examples.

Conversion Factors

A *conversion factor* is a numerical factor—expressed as a ratio—that can convert a quantity expressed in one unit to another unit; for example, the factor may be used to convert pounds to kilograms or feet to inches. To create a conversion factor, an equality (such as *1 kilogram = 2.2 pounds*) is expressed as a fraction:

$$\frac{1 \text{ kg}}{2.2 \text{ lb}} \quad \text{and} \quad \frac{2.2 \text{ lb}}{1 \text{ kg}}.$$

Because a conversion factor has a value of *1* (the value in the numerator is equal to the value in the denominator), it can be used as a multiplier to change the *unit* of measure without changing the *value* of the measurement. To convert the units of a measurement, the fraction must have the desired unit in the numerator, as the unit in the denominator will cancel out the original unit.

Example 1 Convert the weight of 130 pounds to kilograms.

▪ Multiply 130 pounds by a conversion factor (fraction) that includes both pounds and kilograms and is arranged so that the desired unit (kilograms) is in the numerator:

$$130 \text{ lb} \times \frac{1 \text{ kg}}{2.2 \text{ lb}} = \frac{130 \text{ kg}}{2.2} = 59 \text{ kg}.$$

As this example shows, the unit for *pounds* cancels out, and the unit for *kilograms* remains in the solution.

Example 2 The food label on a bottle of apple juice shows the contents in both fluid ounces and liters. How many liters are contained in a bottle that holds 64 fluid ounces?

▪ Multiply 64 ounces by a conversion factor that includes both fluid ounces and liters, with the desired unit (liters) in the numerator:

$$64 \text{ fl oz} \times \frac{1 \text{ L}}{33.8 \text{ fl oz}} = \frac{64 \text{ L}}{33.8} = 1.89 \text{ L}.$$

Percentages

A percentage expresses a fraction that has 100 in the denominator; for example, the term *50 percent* is equivalent to the fraction 50/100. Similar to other fractions, percentages are used to express a *proportion* of the *whole;* therefore, the units in the numerator and denominator must be similar. Any fraction can be expressed in hundredths and converted to a percentage by dividing the numerator by the denominator and multiplying by 100:

$$1/4 = 0.25.$$
$$0.25 \times 100 = 25\%.$$

Example 3 Suppose your energy intake for the day is 2000 kcalories (kcal), and your recommended energy intake is 2400 kcalories. What percent of the recommended energy intake did you consume?

▮ Divide your intake by the recommended intake:

$$2000 \text{ kcal (your intake)} \div 2400 \text{ kcal (recommended intake)} = 0.83.$$

▮ Multiply by 100 to express the decimal as a percentage:

$$0.83 \times 100 = 83\%.$$

Example 4 A percentage can also be more than 100. Suppose your intake of vitamin C is 120 milligrams, and your RDA (male) is 90 milligrams. What percent of the RDA for vitamin C did you consume?

$$120 \text{ mg (your intake)} \div 90 \text{ mg (RDA)} = 1.33.$$
$$1.33 \times 100 = 133\%.$$

Weights and Measures

Length

1 meter (m) = 39 in
1 centimeter (cm) = 0.4 in
1 inch (in) = 2.5 cm
1 foot (ft) = 30 cm

Temperature

Boiling point	100°C	212°F	Boiling point
Body temperature	37°C	98.6°F	Body temperature
Freezing	0°C	32°F	Freezing
Celsius		Fahrenheit	

▮ To find degrees Fahrenheit (°F) when you know degrees Celsius (°C), multiply by 9/5 and then add 32.

▮ To find degrees Celsius (°C) when you know degrees Fahrenheit (°F), subtract 32 and then multiply by 5/9.

Volume

1 liter (L) = 1000 mL, 0.26 gal, 1.06 qt, 2.1 pt, or 33.8 fl. oz
1 milliliter (mL) = 1/1000 L or 0.03 fl. oz
1 gallon (gal) = 128 oz, 8 c, or 3.8 L
1 quart (qt) = 32 oz, 4 c, or 0.95 L
1 pint (pt) = 16 oz, 2 c, or 0.47 L
1 cup (c) = 8 oz, 16 tbs, about 250 mL, or 0.25 L
1 ounce (oz) = 30 mL
1 tablespoon (tbs) = 3 tsp or 15 mL
1 teaspoon (tsp) = 5 mL

Weight

1 kilogram (kg) = 1000 g or 2.2 lb
1 gram (g) = 1/1000 kg, 1000 mg, or 0.035 oz
1 milligram (mg) = 1/1000 g or 1000 μg

1 microgram (µg) = 1/1000 mg
1 pound (lb) = 16 oz, 454 g, or 0.45 kg
1 ounce (oz) = about 28 g

Energy
1 kilojoule (kJ) = 0.24 kcal
1 megajoule (MJ) = 240 kcal
1 kcalorie (kcal) = 4.2 kJ
1 g carbohydrate = 4 kcal = 17 kJ
1 g fat = 9 kcal = 37 kJ
1 g protein = 4 kcal = 17 kJ
1 g alcohol = 7 kcal = 29 kJ

Appendix G

Enteral Formulas

The large number of enteral formulas available allows patients to meet a wide variety of medical needs. The first step in narrowing the choice of formulas is to determine the patient's ability to digest and absorb nutrients. Table G-1 on pp. G-2 through G-3 lists examples of standard formulas for patients who can adequately digest and absorb nutrients, and Table G-2 on p. G-4 provides examples of elemental formulas for patients with a limited ability to digest or absorb nutrients. Each formula is listed only once, although a formula may have more than one use. A high-protein formula, for example, may also be a fiber-containing formula. Tables G-3 and G-4 on p. G-4 list modules that can be used to prepare modular formulas or enhance enteral formulas.

The information shown in this appendix reflects the literature provided by manufacturers and does not suggest endorsement by the authors. Manufacturers frequently add new formulas, discontinue old ones, and change formula composition. Consult the manufacturers' literature and websites for updates and additional examples of enteral formulas.* The following products are listed in this appendix:

- *Abbott Nutrition:* Glucerna 1.0 Cal, Jevity 1 Cal, Jevity 1.5 Cal, Nepro with Carb Steady, Optimental, Osmolite 1 Cal, Oxepa, Pivot 1.5 Cal, Polycose Powder, Promote, Promote with Fiber, Pulmocare, Suplena with Carb Steady

- *Nestlé Nutrition:* Compleat, Compleat Pediatric, Crucial, Diabetisource AC, Fibersource HN, Impact, Impact 1.5, Impact Glutamine, Isosource HN, MCT Oil, Microlipid, Novasource Renal, Nutren 1.0, Nutren 1.0 Fiber, Nutren 1.5, Nutren 2.0, Nutren Glytrol, Nutren Junior, Nutren Pulmonary, Nutren Replete, Nutren Replete Fiber, NutriHep, Peptamen, Peptamen Junior, Resource Beneprotein Instant Protein Powder, Vivonex Pediatric, Vivonex T.E.N.

---

*Sources for the information in this appendix: Abbott Nutrition, www.abbottnutrition.com; Nestlé Nutrition, www.nestle-nutrition.com.

TABLE G-1 Standard Formulas

Product	Volume to Meet 100% RDI[a] (mL)	Energy (kcal/mL)	Protein or Amino Acids (g/L)	Carbohydrate (g/L)	Fat (g/L)	Notes
Lactose-Free, Standard Formulas						
Compleat	1313	1.07	48	128	40	Blenderized formula, 6 g fiber/L
Nutren 1.0	1500	1.00	40	127	38	25% fat from MCT
Osmolite 1 Cal	1321	1.06	44	144	35	20% fat from MCT
Lactose-Free, Fiber-Enhanced Formulas						
Jevity 1 Cal	1321	1.06	44	155	35	14 g fiber/L
Nutren 1.0 Fiber	1500	1.00	40	127	38	14 g fiber/L
Promote with Fiber	1000	1.00	63	138	28	14 g fiber/L
Lactose-Free, High-kCalorie Formulas						
Jevity 1.5 Cal	1500	1.50	64	216	50	22 g fiber/L
Nutren 1.5	1000	1.50	60	169	68	50% fat from MCT
Nutren 2.0	750	2.00	80	196	104	75% fat from MCT
Lactose-Free, High-Protein Formulas						
Fibersource HN	1165	1.20	53	160	39	20% fat from MCT, 10 g fiber/L
Isosource HN	1165	1.20	53	160	39	20% fat from MCT, low residue
Promote	1000	1.00	63	130	26	20% fat from MCT, low residue
Specialized Formulas: Pediatric (1 to 10 years)						
Compleat Pediatric	Varies[b]	1.00	38	130	39	Blenderized formula, 7 g fiber/L
Nutren Junior	Varies[b]	1.00	30	110	50	21% fat from MCT
Specialized Formulas: Glucose Intolerance						
Diabetisource AC	1250	1.20	60	100	59	36% kcal from carbohydrate, 15 g fiber/L
Glucerna 1.0 Cal	1420	1.00	42	96	54	34% kcal from carbohydrate, 14 g fiber/L
Nutren Glytrol	1400	1.00	45	100	48	40% kcal from carbohydrate, 15 g fiber/L
Specialized Formulas: Immune System Support						
Impact	1500	1.00	56	130	28	Enriched with arginine, nucleic acids, and omega-3 fatty acids
Impact 1.5	1250	1.50	84	140	69	Same as above
Impact Glutamine	1000	1.30	78	150	43	Same as above, and enriched with glutamine

(continued)

TABLE G-1 **Standard Formulas** (*continued*)

Product	Volume to Meet 100% RDI[a] (mL)	Energy (kcal/mL)	Protein or Amino Acids (g/L)	Carbohydrate (g/L)	Fat (g/L)	Notes
Specialized Formulas: Chronic Kidney Disease (CKD)						
Nepro with Carb Steady	948	1.80	81	167	96	Low potassium, low phosphorus; to be used after dialysis has been instituted
Novasource Renal	1000	2.00	74	200	100	Low in electrolytes; to be used after dialysis has been instituted
Suplena with Carb Steady	948	1.80	45	205	96	Low in protein and electrolytes; for patients with CKD (stage 3 or 4)
Specialized Formulas: Respiratory Insufficiency						
Nutren Pulmonary	1000	1.50	68	100	95	55% kcal from fat, 40% fat from MCT
Oxepa	946	1.50	63	105	94	Enriched with omega-3 fatty acids and antioxidants; for mechanically ventilated patients
Pulmocare	947	1.50	63	106	93	55% kcal from fat, 20% fat from MCT, enriched with antioxidant nutrients
Specialized Formulas: Wound Healing						
Nutren Replete	1000	1.00	62	113	34	Enhanced with vitamins and minerals; for patients recovering from surgery, burns, and pressure ulcers
Nutren Replete Fiber	1000	1.00	62	113	34	Same as above; 14 g fiber/L

Note: MCT = Medium-chain triglycerides

[a]RDI = Reference Daily Intakes, which are labeling standards for vitamins, minerals, and protein. Consuming 100 percent of the RDI will meet the nutrient needs of most people using the product.
[b]Depends on age of child.

TABLE G-2 Elemental Formulas

Product	Volume to Meet 100% RDI[a] (mL)	Energy (kcal/mL)	Protein or Amino Acids (g/L)	Carbohydrate (g/L)	Fat (g/L)	Notes
Specialized Elemental Formula: Hepatic Insufficiency						
NutriHep	1000	1.50	40	290	21	Free amino acids, high in branched-chain amino acids and low in aromatic amino acids
Specialized Elemental Formulas: Immune System Support						
Crucial	1000	1.50	94	134	68	Enriched with arginine, omega-3 fatty acids, antioxidant nutrients, and zinc
Pivot 1.5 Cal	1500	1.50	94	172	51	Enriched with arginine, glutamine, omega-3 fatty acids, and antioxidant nutrients
Specialized Elemental Formulas: Malabsorption						
Optimental	1422	1.00	51	139	28	Contains MCT and arginine, enriched with antioxidants and omega-3 fatty acids
Peptamen	1500	1.00	40	127	39	70% fat from MCT
Vivonex T.E.N.	2000	1.00	38	210	3	Powder form, 100% free amino acids, enriched with glutamine
Specialized Elemental Formulas: Pediatric (1 to 10 years)						
Peptamen Junior	Varies[b]	1.00	30	138	38	60% fat from MCT
Vivonex Pediatric	Varies[b]	0.80	24	130	24	Powder form, 100% free amino acids

Note: MCT = Medium-chain triglycerides

[a]RDI = Reference Daily Intakes, which are labeling standards for vitamins, minerals, and protein. Consuming 100 percent of the RDI will meet the nutrient needs of most people using the product.
[b]Depends on age of child

TABLE G-3 Protein and Carbohydrate Modules

Product	Major Ingredient	Energy (kcal/g)	Nutrient Content (g/100 g)
Resource Beneprotein Instant Protein Powder	Whey protein	3.6	86 g protein
Polycose Powder	Hydrolyzed cornstarch	3.8	94 g carbohydrate

TABLE G-4 Fat Modules

Product	Major Ingredient	Energy (kcal/mL)	Fat Content (g/100 mL)
MCT Oil	Coconut oil	7.7	93
Microlipid	Safflower oil	4.5	50

Appendix H

Answers to Self-Check Questions

Chapter 1
1. d, 2. d, 3. c, 4. d, 5. c, 6. b,
7. c, 8. b, 9. c, 10. c

Chapter 2
1. b, 2. b, 3. c, 4. a, 5. a, 6. a,
7. b, 8. a, 9. d, 10. c

Chapter 3
1. b, 2. c, 3. a, 4. d, 5. a, 6. a,
7. d, 8. d, 9. b, 10. a

Chapter 4
1. c, 2. b, 3. d, 4. d, 5. a, 6. d,
7. c, 8. b, 9. c, 10. d

Chapter 5
1. b, 2. a, 3. d, 4. a, 5. c, 6. a,
7. d, 8. c, 9. d, 10. d

Chapter 6
1. d, 2. b, 3. c, 4. d, 5. d, 6. a,
7. b, 8. a, 9. b, 10. c

Chapter 7
1. b, 2. a, 3. a, 4. c, 5. b, 6. c,
7. d, 8. b, 9. c, 10. b

Chapter 8
1. d, 2. d, 3. c, 4. a, 5. d, 6. d,
7. b, 8. c, 9. c, 10. a

Chapter 9
1. d, 2. b, 3. a, 4. d, 5. c, 6. d,
7. a, 8. d, 9. d, 10. b

Chapter 10
1. b, 2. a, 3. b, 4. c, 5. d, 6. c,
7. d, 8. b, 9. a, 10. d

Chapter 11
1. a, 2. d, 3. c, 4. c, 5. b, 6. d,
7. b, 8. d, 9. b, 10. c

Chapter 12
1. c, 2. b, 3. a, 4. b, 5. d, 6. c,
7. d, 8. c, 9. b, 10. c

Chapter 13
1. d, 2. c, 3. d, 4. d, 5. d, 6. d,
7. b, 8. b, 9. a, 10. b

Chapter 14
1. b, 2. c, 3. d, 4. a, 5. b, 6. d,
7. a, 8. c, 9. c, 10. b

Chapter 15
1. d, 2. a, 3. a, 4. b, 5. d, 6. c,
7. b, 8. b, 9. d, 10. c

Chapter 16
1. c, 2. a, 3. b, 4. c, 5. d, 6. b,
7. a, 8. c, 9. a, 10. d

Chapter 17
1. a, 2. c, 3. b, 4. d, 5. a, 6. c,
7. d, 8. b, 9. b, 10. c

Chapter 18
1. b, 2. c, 3. d, 4. c, 5. d, 6. a,
7. b, 8. b, 9. a, 10. c

Chapter 19
1. d, 2. c, 3. b, 4. b, 5. a, 6. c,
7. a, 8. c, 9. a, 10. d

Chapter 20
1. d, 2. a, 3. b, 4. a, 5. b, 6. c,
7. d, 8. c, 9. b, 10. c

Chapter 21
1. b, 2. d, 3. d, 4. c, 5. a, 6. c,
7. d, 8. c, 9. a, 10. b

Chapter 22
1. a, 2. d, 3. c, 4. a, 5. d, 6. c,
7. d, 8. d, 9. b, 10. c

Chapter 23
1. a, 2. c, 3. b, 4. d, 5. c, 6. a,
7. b, 8. b, 9. c, 10. d

Chapter 24
1. d, 2. a, 3. b, 4. c, 5. d, 6. b,
7. c, 8. c, 9. d, 10. a

Chapter 25
1. d, 2. c, 3. c, 4. b, 5. d, 6. b,
7. b, 8. a, 9. a, 10. d

Glossary

2-in-1 solution: a parenteral solution that contains dextrose and amino acids but excludes lipids.

24-hour recall: a record of foods consumed in the previous 24 hours; sometimes modified to include foods consumed in a typical day.

abscesses (AB-sess-es): accumulations of pus.

Acceptable Daily Intake (ADI): the amount of an artificial sweetener that individuals can safely consume each day over the course of a lifetime without adverse effect.

Acceptable Macronutrient Distribution Ranges (AMDR): ranges of intakes for the energy-yielding nutrients that provide adequate energy and nutrients and reduce the risk of chronic disease.

acetaldehyde (ah-set-AL-deh-hide): an intermediate in alcohol metabolism that, in excess, can damage the body's tissues and interfere with cellular functions.

acetate (AH-seh-tate): the product of alcohol metabolism that is used as a source of energy by many tissues in the body.

acetone breath: a distinctive fruity odor on the breath of a person with ketosis.

acetyl CoA (ASS-eh-teel, or ah-SEET-il, coh-AY): a 2-carbon compound (acetate, or acetic acid) with a molecule of CoA attached to it.

achalasia (ack-ah-LAY-zhah): an esophageal disorder characterized by weakened peristalsis and impaired relaxation of the lower esophageal sphincter.

achlorhydria (AY-clor-HIGH-dree-ah): absence of gastric acid secretions.

acid-base balance: the balance maintained between acid and base concentrations in the blood and body fluids.

acidosis: acid accumulation in the blood and body fluids; depresses the central nervous system and can lead to disorientation and, eventually, coma.

acids: compounds that release hydrogen ions in a solution.

acquired immune deficiency syndrome (AIDS): the late stage of illness caused by infection with the human immunodeficiency virus (HIV); characterized by severe damage to immune function.

acute kidney injury: a condition associated with the abrupt decline of kidney function over a period of hours or days; potentially a cause of acute renal failure.

acute respiratory distress syndrome (ARDS): respiratory failure triggered by severe lung injury; a medical emergency that causes dyspnea and pulmonary edema and usually requires mechanical ventilation.

acute-phase response: changes in body chemistry resulting from infection, inflammation, or injury; characterized by alterations in plasma proteins.

added sugars: sugars and syrups added to a food for any purpose; also called *carbohydrate sweeteners*.

adequacy: the characteristic of a diet that provides all the essential nutrients, fiber, and energy necessary to maintain health and body weight.

Adequate Intakes (AI): a set of values that are used as guides for nutrient intakes when scientific evidence is insufficient to determine an RDA.

adipokines (AD-ih-poh-kynz): protein hormones made and released by adipose tissue (fat) cells.

adiponectin (AH-dih-poe-NECK-tin): a hormone produced by adipose cells that inhibits inflammation and improves insulin sensitivity.

adipose tissue: the body's fat, which consists of masses of fat-storing cells called adipose cells.

adolescence: the period of growth from the beginning of puberty until full maturity. Timing of adolescence varies from person to person.

advance medical directive: written or oral instruction regarding one's preferences for medical treatment to be used in the event of becoming incapacitated.

advanced glycation end products (AGEs): reactive compounds formed after glucose combines with protein; AGEs can damage tissues and lead to diabetic complications.

adverse reactions: unusual responses to food (including intolerances and allergies).

aerobic physical activity: activity in which the body's large muscles move in a rhythmic manner for a sustained period of time. Aerobic activity, also called *endurance activity*, improves cardiorespiratory fitness. Brisk walking, running, swimming, and bicycling are examples.

AIDS-defining illnesses: diseases and complications associated with the later stages of an HIV infection, including wasting, recurrent bacterial pneumonia, opportunistic infections, and certain cancers.

alcohol abuse: the continued use of alcohol despite the development of social, legal, or health problems. Individuals who abuse alcohol are called *alcoholics*.

alcoholic liver disease: liver disease that is related to excessive alcohol consumption. Disorders that may develop include fatty liver, hepatitis, and cirrhosis, which may lead to liver failure.

aldosterone (al-DOS-ter-own): a hormone secreted by the adrenal glands that stimulates the reabsorption of sodium by the kidneys; also regulates chloride and potassium concentrations.

alkalosis: excessive base in the blood and body fluids.

alpha-lactalbumin (lackt-AL-byoo-min): the chief protein in human breast milk.

alveoli (al-VEE-oh-lie): air sacs in the lungs (singular: *alveolus*).

Alzheimer's disease: a progressive, degenerative disease that attacks the brain and impairs thinking, behavior, and memory.

amenorrhea (ay-MEN-oh-REE-ah): the absence of or cessation of menstruation.

amino (a-MEEN-oh) **acids:** building blocks of protein. Each contains an amino group, an acid group, a hydrogen atom, and a distinctive side group, all attached to a central carbon atom.

amniotic (am-nee-OTT-ic) **sac:** the "bag of waters" in the uterus in which the fetus floats.

amylase (AM-uh-lace): an enzyme that splits amylose (a form of starch).

anabolic steroids: drugs related to the male sex hormone, testosterone, that stimulate the development of lean body mass.

anabolism (an-ABB-o-lism): energy-requiring reactions in which small molecules are put together to build larger ones.

anaerobic (AN-air-ROH-bic): not requiring oxygen.

anaphylactic (AN-ah-feh-LAC-tic) **shock:** a life-threatening whole-body allergic reaction to an offending substance.

androstenedione: a hormone made in the adrenal glands that serves as a precursor to the male hormone testosterone; falsely promoted as burning fat, building muscle, and slowing aging. Side effects include acne, aggressiveness, and liver enlargement.

anemia of chronic disease: anemia that develops in persons with chronic illness; may resemble iron-deficiency anemia even though iron stores are often adequate. Also called *anemia of chronic inflammation*.

anencephaly (AN-en-SEF-a-lee): an uncommon and always fatal type of neural tube defect; characterized by the absence of a brain.

aneurysm (AN-you-rih-zum): an abnormal enlargement or bulging of a blood vessel (usually an artery) caused by weakness in the blood vessel wall.

angina (an-JYE-nah or AN-ji-nah) **pectoris:** a condition caused by ischemia in the heart muscle that results in discomfort or dull pain in the chest region. The pain often radiates to the left shoulder and arm or to the back, neck, and lower jaw.

anorexia nervosa: an eating disorder characterized by a refusal to maintain a minimally normal body weight, self-starvation to the extreme, and a disturbed perception of body weight and shape.

anorexia: lack of appetite.

anthropometric (AN-throw-poe-MEH-trik): related to physical measurements of the human body, such as height, weight, body circumferences, and percentage of body fat.

antibodies: large proteins of the blood and body fluids, produced in response to invasion of the body by unfamiliar molecules (mostly proteins) called *antigens*. Antibodies inactivate the invaders and so protect the body.

antidiuretic hormone (ADH): a hormone released by the pituitary gland in response to high salt concentrations in the blood that stimulates the reabsorption of water by the kidneys.

antioxidant (anti-OX-ih-dant): a compound that protects other compounds from oxygen by itself reacting with oxygen. Antioxidants may be added to foods to delay or prevent rancidity.

anuria (ah-NOO-ree-ah): the absence of urine; clinically identified as urine output less than 50 mL/day.

anus (AY-nus): the terminal outlet of the GI tract.

aplastic anemia: anemia characterized by the inability of bone marrow to produce adequate numbers of blood cells. Causes include drug toxicity, viruses, and genetic defects.

appendix: a narrow blind sac extending from the beginning of the colon that stores lymph cells.

appetite: the psychological desire to eat; a learned motivation that is experienced as a pleasant sensation that accompanies the sight, smell, or thought of appealing foods.

arginine: a nonessential amino acid falsely promoted as enhancing the secretion of human growth hormone, the breakdown of fat, and the development of muscle.

artery: a vessel that carries blood away from the heart.

arthritis: inflammation of a joint, usually accompanied by pain, swelling, and structural changes.

artificial fats: zero-energy fat replacers that are chemically synthesized to mimic the sensory and cooking qualities of naturally occurring fats but are totally or partially resistant to digestion.

artificial sweeteners: noncarbohydrate, non-kcaloric synthetic sweetening agents; sometimes called *nonnutritive sweeteners*.

ascites (ah-SIGH-teez): an abnormal accumulation of fluid in the abdominal cavity.

ascorbic acid: one of the two active forms of vitamin C.

aspiration: drawing in by suction or breathing; a common complication of enteral feedings in which foreign material enters the lungs, often from GI secretions or the reflux of stomach contents.

atherosclerosis (ATH-er-oh-scler-OH-sis): a type of artery disease characterized by accumulations of lipid-containing material on the inner walls of arteries.

ATP or adenosine (ah-DEN-oh-seen) **triphosphate** (tri-FOS-fate): a common high-energy compound that contains three phosphate groups (joined by high-energy bonds).

atrophic gastritis (a-TRO-fik gas-TRI-tis): a condition characterized by chronic inflammation of the stomach accompanied by a diminished size and functioning of the mucosa and glands.

atrophy (AT-tro-fee): a decrease in size (for example, of a muscle) because of disuse.

autoimmune: an immune response directed against the body's own tissues.

bacterial cholangitis (KOH-lan-JYE-tis): bacterial infection involving the bile ducts.

bacterial overgrowth: excessive bacterial colonization of the stomach and small intestine; may be due to low gastric acidity, altered GI motility, mucosal damage, or contamination.

bacterial translocation: movement of bacteria across the intestinal mucosa, allowing access to body tissues.

balance: the dietary characteristic of providing foods of a number of types in proportion to each other, such that foods rich in some nutrients do not crowd out foods that are rich in other nutrients.

bariatric (BAH-ree-AH-trik) **surgery**: surgery that treats severe obesity.

Barrett's esophagus: a condition in which esophageal cells damaged by chronic exposure to stomach acid are replaced by cells that resemble those in the stomach or small intestine, sometimes becoming cancerous.

basal metabolic rate (BMR): the rate of energy use for metabolism under specified conditions: after a 12-hour fast and restful sleep, without any physical activity or emotional excitement, and in a comfortable setting. It is usually expressed as kcalories per kilogram of body weight per hour.

basal metabolism: the energy needed to maintain life when a person is at complete digestive, physical, and emotional rest.

bases: compounds that accept hydrogen ions in a solution.

behavior modification: the changing of behavior by the manipulation of *antecedents* (cues or environmental factors that trigger behavior), the behavior itself, and *consequences* (the penalties or rewards attached to behavior).

beneficence (be-NEF-eh-sense): the act of performing beneficial services rather than harmful ones.

beriberi: the thiamin-deficiency disease; characterized by loss of sensation in the hands and feet, muscular weakness, advancing paralysis, and abnormal heart action.

beta-carotene: a vitamin A precursor made by plants and stored in human fat tissue; an orange pigment.

BHA, BHT: preservatives commonly used to slow the development of "off" flavors, odors, and color changes caused by oxidation.

bicarbonate: an alkaline secretion of the pancreas; part of the pancreatic juice. (Bicarbonate also occurs widely in all cell fluids.)

bifidus (BIFF-id-us, by-FEED-us) **factors**: factors in colostrum and breast milk that favor the growth of the "friendly" bacterium *Lactobacillus* (lack-toe-ba-SILL-us) *bifidus* in the infant's intestinal tract. These bacteria prevent other, less desirable intestinal inhabitants from flourishing.

bile: an emulsifier that prepares fats and oils for digestion; made by the liver, stored in the gallbladder, and released into the small intestine when needed.

biliary system: the gallbladder and ducts that deliver bile from the liver and gallbladder to the small intestine.

binders: chemical compounds in foods that combine with nutrients (especially minerals) to form complexes the body cannot absorb.

binge eating disorder: an eating disorder whose criteria are similar to those of bulimia nervosa, excluding purging or other compensatory behaviors.

bioactive food components: compounds in foods, either nutrients or phytochemicals, that alter physiological processes in the body.

bioavailability: the rate and extent to which a nutrient is absorbed and used.

biopsy: removal of a tissue sample to determine the cause of an illness.

blastocyst (BLASS-toe-sist): the developmental stage of the zygote when it is about five days old and ready for implantation.

blenderized formulas: enteral formulas that are prepared by using a food blender to mix and puree whole foods.

body mass index (BMI): an index of a person's weight in relation to height; determined by dividing the weight (in kilograms) by the square of the height (in meters).

bolus (BOH-lus): the portion of food swallowed at one time.

bolus (BOH-lus) **feeding**: delivery of about 250 to 500 milliliters of formula in less than 20 minutes.

boron: a nonessential mineral that is falsely promoted as a "natural" steroid replacement.

bronchi (BRON-key), **bronchioles** (BRON-key-oles): the main airways of the lungs. The singular form of bronchi is *bronchus*.

buffalo hump: the accumulation of fatty tissue at the base of the neck.

buffers: compounds that can reversibly combine with hydrogen ions to help keep a solution's acidity or alkalinity constant.

built environment: the buildings, roads, utilities, homes, fixtures, parks, and all other manufactured entities that form the physical characteristics of a community.

bulimia (byoo-LEEM-ee-uh) **nervosa**: recurring episodes of binge eating combined with a morbid fear of becoming fat, usually followed by self-induced vomiting or purging.

C-reactive protein: an acute-phase protein released from the liver during acute inflammation or stress.

caffeine: a natural stimulant found in many common foods and beverages, including coffee, tea, and chocolate. High doses cause headaches, trembling, rapid heart rate, and other undesirable side effects.

calories: units by which energy is measured. Food energy is measured in *kilocalories* (1000 calories equal 1 kilocalorie).

cancer cachexia (ka-KEK-see-ah): a wasting syndrome associated with cancer that is characterized by anorexia, muscle wasting, weight loss, and fatigue.

cancers: diseases characterized by the uncontrolled growth of a group of cells, which can destroy adjacent tissues and spread to other areas of the body via lymph or blood.

candidiasis: a fungal infection on the mucous membranes of the oral cavity and elsewhere; usually caused by *Candida albicans*.

capillaries: small vessels that branch from an artery and connect arteries to veins. Oxygen, nutrients, and waste materials are exchanged across capillary walls.

carbohydrate-to-insulin ratio: the amount of carbohydrate that can be handled per unit of insulin. On average, every 15 grams of carbohydrate requires about 1 unit of rapid- or short-acting insulin.

carbohydrates: energy nutrients composed of monosaccharides.

carcinogenesis (CAR-sin-oh-JEN-eh-sis): the process of cancer development.

carcinogens (CAR-sin-oh-jenz or car-SIN-oh-jenz): substances that can cause cancer (the adjective is *carcinogenic*).

cardiac cachexia: a condition of severe malnutrition that develops in heart failure patients; characterized by weight loss and tissue wasting.

cardiac output: the volume of blood discharged by the heart each minute.

cardiopulmonary resuscitation (CPR): life-sustaining treatment that supplies oxygen and restores a person's ability to breathe and pump blood.

cardiorespiratory endurance: the ability to perform large-muscle dynamic exercise of moderate-to-high intensity for prolonged periods.

cardiovascular disease (CVD): a general term for all diseases of the heart and blood vessels.

cariogenic (KAH-ree-oh-JEN-ik): conducive to development of dental caries.

carnitine: a nonessential nonprotein amino acid made in the body from lysine that helps transport fatty acids across the mitochondrial membrane. Carnitine supposedly "burns" fat and spares glycogen during endurance events, but in reality it does neither.

casein (CAY-seen): the chief protein in cow's milk.

catabolism (ca-TAB-o-lism): energy-releasing reactions in which large molecules are broken down into smaller ones.

cataracts: thickenings of the eye lenses that impair vision and can lead to blindness.

cathartics (ca-THART-ics): strong laxatives.

catheter: a thin tube placed within a narrow lumen (such as a blood vessel) or body cavity; can be used to infuse or withdraw fluids or keep a passage open.

celiac (SEE-lee-ack) **disease:** an immune disorder characterized by an abnormal immune response to the dietary protein gluten; also called *gluten-sensitive enteropathy* or *celiac sprue*.

cellulite (SELL-you-light or SELL-you-leet): supposedly, a lumpy form of fat; actually, a fraud. Fatty areas of the body may appear lumpy, but the fat itself is the same as fat anywhere else in the body.

central obesity: excess fat around the trunk of the body; also called *abdominal fat* or *upper-body fat*.

central veins: the large-diameter veins located close to the heart.

cerebral cortex: the outer surface of the cerebrum, which is the largest part of the brain.

Certified Diabetes Educator (CDE): a health care professional who specializes in diabetes management education.

chemotherapy: the use of drugs to arrest or destroy cancer cells; these drugs are called *antineoplastic agents*.

cholecystectomy (KOH-leh-sis-TEK-toe-mee): surgical removal of the gallbladder.

cholecystitis (KOH-leh-sih-STY-tis): inflammation of the gallbladder, usually caused by obstruction of the cystic duct by gallstones.

cholelithiasis (KOH-leh-lih-THIGH-ah-sis): formation of gallstones.

choline: a nonessential nutrient that can be made in the body from an amino acid.

chromium picolinate (CROW-mee-um pick-oh-LYN-ate): a trace mineral supplement; falsely promoted as building muscle, enhancing energy, and burning fat.

chronic bronchitis (bron-KYE-tis): a lung disorder characterized by persistent inflammation and excessive secretions of mucus in the main airways of the lungs.

chronic diseases: diseases characterized by slow progression, long duration, and degeneration of body organs due in part to such personal lifestyle elements as poor food choices, smoking, alcohol use, and lack of physical activity.

chronic hypertension: in pregnant women, hypertension that is present and documented before pregnancy; in women whose prepregnancy blood pressure is unknown, the presence of sustained hypertension before 20 weeks of gestation.

chronic kidney disease: kidney disease characterized by gradual, irreversible deterioration of the kidneys; also known as *chronic renal failure*.

chronic obstructive pulmonary disease (COPD): a group of lung diseases characterized by persistent obstructed airflow through the lungs and airways; includes chronic bronchitis and emphysema.

chronological age: a person's age in years from his or her date of birth.

chylomicrons (kye-lo-MY-crons): the lipoproteins that transport lipids from the intestinal cells into the body.

chyme (KIME): the semiliquid mass of partly digested food expelled by the stomach into the duodenum (the top portion of the small intestine).

cirrhosis (sih-ROE-sis): an advanced stage of liver disease in which extensive scarring replaces healthy liver tissue, causing impaired liver function and liver failure.

claudication (CLAW-dih-KAY-shun): pain in the legs while walking; usually due to an inadequate supply of blood to muscles.

clear liquid diet: a diet that consists of foods that are liquid at room temperature and leave almost no residue (undigested material) in the intestines after digestion and absorption.

clinically severe obesity: a BMI of 40 or greater, or a BMI of 35 or greater with one or more serious conditions such as hypertension; also called *morbid obesity*.

closed feeding system: a delivery system in which the formula comes prepackaged in a container that can be attached directly to the feeding tube for administration.

CoA (coh-AY): coenzyme A; the coenzyme derived from the B vitamin pantothenic acid and central to energy metabolism.

coenzyme (co-EN-zime): a small molecule that works with an enzyme to promote the enzyme's activity; most contain B vitamins as part of their structure.

coenzyme Q_{10}: a lipid found in cells (mitochondria) shown to improve exercise performance in heart disease patients, but not effective in improving the performance of healthy athletes.

cognitive skills: as taught in behavior therapy, changes to conscious thoughts with the goal of improving adherence to lifestyle modifications; examples are problem-solving skills or the correction of false negative thoughts, termed *cognitive restructuring*.

cognitive therapy: psychological therapy aimed at changing undesirable behaviors by changing underlying thought processes contributing to these behaviors.

colectomy: removal of a portion or all of the colon.

collagen: the characteristic protein of connective tissue.

collaterals: blood vessels that enlarge to allow an alternative pathway for diverted blood.

colostomy (co-LAH-stoe-me): a surgical passage through the abdominal wall into the colon.

colostrum (co-LAHS-trum): a milk-like secretion from the breasts that is rich in protective factors and present during the first day or so after delivery, before milk appears.

complement: a group of plasma proteins that assist the activities of antibodies.

complementary proteins: two or more proteins whose amino acid assortments complement each other in such a way that the essential amino acids missing from one are supplied by the other.

conditionally essential amino acid: an amino acid that is normally nonessential but must be supplied by the diet in special circumstances when the need for it becomes greater than the body's ability to produce it.

conjugated linoleic acid: a naturally occurring *trans* fatty acid with 18 carbons and 2 double bonds; sometimes taken as a supplement to improve body composition.

continuous ambulatory peritoneal dialysis (CAPD): the most common method of peritoneal dialysis; involves frequent exchanges of dialysate, which remains in the peritoneal cavity throughout the day.

continuous feedings: slow delivery of formula at a constant rate over an 8- to 24-hour period.

continuous glucose monitoring: continuous monitoring of tissue glucose levels using a small sensor placed under the skin.

continuous parenteral nutrition: continuous administration of parenteral solutions over a 24-hour period.

continuous renal replacement therapy (CRRT): a slow, continuous method of removing solutes and/or fluids from blood by gently pumping blood across a filtration membrane over a prolonged time period.

cornea (KOR-nee-uh): the hard, transparent membrane covering the outside of the eye.

coronary heart disease (CHD): a chronic, progressive disease characterized by obstructed blood flow in the coronary arteries; also called *coronary artery disease*.

creatine (KREE-ah-tin): a nitrogen-containing compound that combines with phosphate to form the high-energy compound creatine phosphate (or phosphocreatine) in muscles.

cretinism (CREE-tin-ism): an iodine-deficiency disease characterized by mental and physical retardation.

critical pathways: coordinated programs of treatment that merge the care plans of different health practitioners; also called *clinical pathways*.

critical period: a finite period during development in which certain events occur that will have irreversible effects on later developmental stages; usually a period of rapid cell division.

Crohn's disease: an inflammatory bowel disease that usually occurs in the lower portion of the small intestine and the colon. Inflammation may pervade the entire intestinal wall.

cross-contamination: the contamination of food by bacteria that occurs when the food comes into contact with surfaces previously touched by raw meat, poultry, or seafood.

cryptosporidiosis (KRIP-toe-spor-ih-dee-OH-sis): a foodborne illness caused by the parasite *Cryptosporidium parvum*.

cultural competence: having an awareness and acceptance of one's own and others' cultures and possessing the skills that lead to effective interactions.

cyanosis (sigh-ah-NOH-sis): a bluish cast in the skin due to the color of deoxygenated hemoglobin.

cyclic parenteral nutrition: administration of parenteral solutions over an 8- to 16-hour period each day.

cystic fibrosis: an inherited disease that affects the transport of chloride across epithelial cell membranes and is characterized by the production of abnormally viscous exocrine secretions; often leads to respiratory illness and pancreatic insufficiency.

cystinuria (SIS-tin-NOO-ree-ah): an inherited disorder characterized by the elevated urinary excretion of several amino acids, including cystine.

cytokines (SIGH-toe-kines): signaling proteins produced by the body's cells; those produced by white blood cells regulate various aspects of immune function.

Daily Values: reference values developed by the FDA specifically for use on food labels.

dawn phenomenon: morning hyperglycemia that is caused by the early-morning release of growth hormone, which counteracts insulin's glucose-lowering effects.

deamination: removal of the amino group (NH_2) from a compound such as an amino acid.

debridement: the surgical removal of dead, damaged, or contaminated tissue resulting from burns or wounds; helps to prevent infection and hasten healing.

decision-making capacity: the ability to understand pertinent information and make appropriate decisions; known as *decision-making competency* within the legal system.

deep vein thrombosis: formation of a stationary blood clot (thrombus) in a deep vein, usually in the leg, which causes inflammation, pain, and swelling and is potentially fatal.

defibrillation: life-sustaining treatment in which an electronic device is used to shock the heart and reestablish a pattern of normal contractions. Defibrillation is used when the heart has arrhythmias or has experienced arrest.

deficient: in regard to nutrient intake, describes the amount below which almost all healthy people can be expected, over time, to experience deficiency symptoms.

dehydration: the loss of water from the body that occurs when water output exceeds water input. The symptoms progress rapidly from thirst, to weakness, to exhaustion and delirium, and end in death if not corrected.

denaturation (dee-nay-cher-AY-shun): the change in a protein's shape brought about by heat, acid, or other agents.

dental calculus: mineralized dental plaque, often associated with inflammation and bleeding.

dental caries (KAH-reez): infectious disease of the teeth that causes the gradual decay and disintegration of tooth structures.

dental plaque (PLACK): a film of bacteria and bacterial by-products that accumulates on the tooth surface.

dermatitis herpetiformis (DERM-ah-TYE-tis HER-peh-tih-FOR-mis): a gluten-sensitive disorder characterized by a severe skin rash.

DHA (docosahexaenoic [DOE-cosa-HEXA-ee-NO-ick] **acid):** an omega-3 fatty acid made from linolenic acid.

DHEA (dehydroepiandrosterone): a hormone made in the adrenal glands that serves as a precursor to the male hormone testosterone; falsely promoted as burning fat, building muscle, and slowing aging. Side effects include acne, aggressiveness, and liver enlargement.

diabetes (DYE-ah-BEE-teez) **mellitus:** a group of metabolic disorders characterized by hyperglycemia and disordered insulin metabolism.

diabetic coma: a coma that occurs in uncontrolled diabetes; may be due to diabetic ketoacidosis, the hyperosmolar hyperglycemic syndrome, or severe hypoglycemia.

diabetic nephropathy (neh-FRAH-pah-thee): damage to the kidneys that results from long-term diabetes.

diabetic neuropathy (nur-RAH-pah-thee): complications of diabetes that cause damage to nerves.

diabetic retinopathy (REH-tih-NAH-pah-thee): retinal damage that results from long-term diabetes.

dialysate (dye-AL-ih-sate): the solution used in dialysis to draw wastes and fluids from the blood.

dialysis (dye-AH-lih-sis): life-sustaining treatment in which a patient's blood is filtered using selective diffusion through a semipermeable membrane; substitutes for kidney function.

dialyzer (DYE-ah-LYE-zer): a machine used in hemodialysis to filter the blood; also called an *artificial kidney.*

diet manual: a resource that specifies the foods or preparation methods to include or exclude in modified diets and provides sample menus.

diet progression: a change in diet as a patient's tolerances permit.

dietary antioxidants: compounds typically found in plant foods that significantly decrease the adverse effects of oxidation on living tissues.

dietary fibers: a general term denoting in plant foods the polysaccharides cellulose, hemicellulose, pectins, gums and mucilages, as well as the nonpolysaccharide lignins, which are not digested by human digestive enzymes, although some are digested by GI tract bacteria.

dietary folate equivalents (DFE): the amount of folate available to the body from naturally occurring sources, fortified foods, and supplements, accounting for differences in bioavailability from each source.

Dietary Reference Intakes (DRI): a set of values for the dietary nutrient intakes of healthy people in the United States and Canada (used for planning and assessing diets).

dietary supplements: products that are added to the diet and contain any of the following ingredients: a vitamin; a mineral; an herb or other botanical; an amino acid; a metabolite; a constituent; or an extract.

dietetic technicians: persons who have completed a minimum of an associate's degree from an accredited college or university and an approved dietetic technician program.

dietetics: the application of nutrition principles to achieve and maintain optimal human health.

differentiation: the development of specific functions different from those of the original.

diffusion: movement of solutes from an area of high concentration to one of low concentration.

digestion: the process by which complex food particles are broken down to smaller, absorbable particles.

digestive system: all the organs and glands associated with the ingestion and digestion of food.

dipeptide: two amino acids bonded together.

disaccharides (dye-SACK-uh-rides): pairs of sugar units bonded together.

disclosure: the act of revealing pertinent information (e.g., about a treatment).

discretionary kcalorie allowance: the kcalories remaining in a person's energy allowance after consuming enough nutrient-dense foods to meet all nutrient needs for a day.

distributive justice: the equitable distribution of resources.

diuresis (DYE-uh-REE-sis): increased urine production.

diuretics (dye-yoo-RET-ics): medications that promote the excretion of water through the kidneys.

diverticulitis (dye-ver-tic-you-LYE-tis): an inflammation or infection involving diverticula.

diverticulosis (dye-ver-tic-you-LOH-sis): an intestinal condition characterized by the presence of small herniations (called diverticula) in the intestinal wall.

do-not-resuscitate (DNR) order: a request by a patient or surrogate to withhold cardiopulmonary resuscitation.

drink: an alcoholic beverage that contains about half an ounce of pure alcohol; equivalent to 12 ounces of beer, 5 ounces of wine, 10 ounces of wine cooler, or 1½ ounces of 80 proof distilled spirits.

dumping syndrome: a cluster of symptoms that result from the rapid emptying of an osmotic load from the stomach into the small intestine.

duodenum (doo-oh-DEEN-um, doo-ODD-num): the top portion of the small intestine.

durable power of attorney: a legal document (sometimes called a *health care proxy*) that gives legal authority to another *(a health care agent)* to make medical decisions in the event of incapacitation.

dysentery (DIS-en-terry): an infection of the gastrointestinal tract caused by an amoeba or bacterium that gives rise to severe diarrhea.

dyspepsia: a feeling of pain, bloating, or discomfort in the upper abdominal area, often called *indigestion;* a symptom of illness rather than a disease itself.

dysphagia (dis-FAY-ja): difficulty swallowing.

dyspnea (DISP-nee-ah): shortness of breath.

eating disorder: a disturbance in eating behavior that jeopardizes a person's physical and psychological health.

eclampsia (eh-KLAMP-see-ah): a severe complication during pregnancy in which seizures occur.

edema (eh-DEEM-uh): the swelling of body tissue caused by leakage of fluid from the blood vessels and accumulation of the fluid in the interstitial spaces.

eicosanoids (eye-KO-sa-noids): 20-carbon molecules derived from dietary fatty acids that help to regulate blood pressure, blood clotting, and other body functions.

electrolyte solutions: solutions that can conduct electricity.

electrolyte: a salt that dissolves in water and dissociates into charged particles called ions.

electron transport chain: the final pathway in energy metabolism that transports electrons from hydrogen to oxygen and captures the energy released in the bonds of a high-energy compound, ATP.

elemental formulas: enteral formulas that contain carbohydrates and proteins that are partially or fully hydrolyzed; also called *hydrolyzed, chemically defined,* or *monomeric formulas.*

embolism (EM-boh-lizm): the obstruction of a blood vessel by an embolus, causing sudden tissue death.

embolus (EM-boh-lus): an abnormal particle, such as a blood clot or air bubble, that travels in the blood.

embryo (EM-bree-oh): the developing infant from two to eight weeks after conception.

emergency shelters: facilities that are used to provide temporary housing.

emetic (em-ETT-ic): an agent that causes vomiting.

emphysema (EM-fih-ZEE-mah): a progressive lung condition characterized by the breakdown of the lungs' elastic structure and destruction of the walls of the bronchioles and alveoli, reducing the surface area involved in respiration.

empty-kcalorie foods: a popular term used to denote foods that contribute energy but lack protein, vitamins, and minerals.

emulsifier: a substance that mixes with both fat and water and that disperses the fat in the water, forming an emulsion.

end-stage renal disease (ESRD): an advanced stage of chronic kidney disease in which dialysis or a kidney transplant is necessary to sustain life.

endoscope: a flexible fiber optic tube used for viewing internal cavities of the body.

energy density: a measure of the energy a food provides relative to the amount of food (kcalories per gram).

energy metabolism: all the reactions by which the body obtains and expends the energy from food.

energy-yielding nutrients: the nutrients that break down to yield energy the body can use (carbohydrate, protein, and fat).

engorgement: overfilling of the breasts with milk.

enhanced water: water that is fortified with ingredients such as vitamins, minerals, protein, oxygen, or herbs.

enrichment: the addition to a food of nutrients to meet a specified standard; in the case of refined bread or cereal, the addition of thiamin, riboflavin, niacin, and folate in amounts approximately equivalent to, or higher than, those originally present, as well as iron in amounts to alleviate the prevalence of iron-deficiency anemia.

enteral (EN-ter-al) **nutrition:** the provision of nutrients using the GI tract, including the use of tube feedings and oral diets.

enteric coated: refers to medications or enzyme preparations that are coated to withstand gastric acidity and dissolve only at the higher pH of the small intestine.

enterostomy (EN-ter-AH-stoe-mee): an opening into the GI tract through the abdominal wall.

environmental tobacco smoke (ETS): the combination of exhaled smoke (mainstream smoke) and smoke from lighted cigarettes, pipes, or cigars (sidestream smoke) that enters the air and may be inhaled by other people.

enzymes: protein catalysts (compounds that facilitate chemical reactions without being changed in the process).

EPA (eicosapentaenoic [EYE-cosa-PENTA-ee-NO-ick] acid): an omega-3 fatty acid made from linolenic acid.

epiglottis (epp-ih-GLOTT-iss): cartilage in the throat that guards the entrance to the trachea and prevents fluid or food from entering it when a person swallows.

epinephrine: one of the stress hormones secreted whenever emergency action is needed; prescribed therapeutically to relax the bronchioles during allergy or asthma attacks.

epithelial (ep-i-THEE-lee-ul) **cells:** cells on the surface of the skin and mucous membranes.

epithelial tissue: tissue composing the layers of the body that serve as selective barriers between the body's interior and the environment.

ergogenic (ER-go-JEN-ick) **aids:** substances or techniques used in an attempt to enhance physical performance.

erythrocyte (er-REETH-ro-cite) hemolysis (he-MOLL-uh-sis): rupture of the red blood cells, caused by vitamin E deficiency.

erythrocyte protoporphyrin (PRO-toh-PORE-fe-rin): a precursor to hemoglobin.

erythropoiesis (eh-RIH-throh-poy-EE-sis): production of red blood cells within the bone marrow.

erythropoietin (eh-RIH-throh-POY-eh-tin): a hormone produced by kidney cells that stimulates red blood cell production.

esophageal (ee-SOF-a-GEE-al) **sphincter:** a sphincter muscle at the upper or lower end of the esophagus.

esophageal dysphagia: an inability to move food through the esophagus; usually caused by an obstruction or a motility disorder.

esophagus (ee-SOFF-ah-gus): the food pipe; the conduit from the mouth to the stomach.

essential amino acids: amino acids that the body cannot synthesize in amounts sufficient to meet physiological need (histidine, isoleucine, leucine, lysine, methionine, phenylalanine, threonine, tryptophan, and valine).

essential fatty acids: fatty acids that the body requires but cannot make in amounts sufficient to meet its physiological needs (linoleic and linolenic acids).

essential nutrients: nutrients a person must obtain from food because the body cannot make them for itself in sufficient quantities to meet physiological needs.

Estimated Average Requirements (EAR): the average daily nutrient intake levels estimated to meet the requirements of half of the healthy individuals in a given age and gender group; used in nutrition research and policymaking and as the basis on which RDA values are set.

Estimated Energy Requirement (EER): the dietary energy intake level that is predicted to maintain energy balance in a healthy adult of a defined age, gender, weight, and physical activity level consistent with good health.

ethical: in accordance with accepted principles of right and wrong.

ethnic diets: foodways and cuisines typical of national origins, races, cultural heritages, or geographic locations.

exclusive breastfeeding: an infant's consumption of human milk with no supplementation of any type (no water, no juice, no nonhuman milk, and no foods) except for vitamins, minerals, and medications.

exercise: planned, structured, and repetitive bodily movement that promotes or maintains physical fitness.

exocrine: pertains to external secretions, such as those of the mucous membranes or the skin. Opposite of *endocrine,* which pertains to hormonal secretions into the blood.

extracellular fluid: fluid residing outside the cells; includes the fluid between the cells *(interstitial fluid)*, plasma, and the water of structures such as the skin and bones.

fasting hyperglycemia: hyperglycemia that develops after an overnight fast of at least eight hours.

fat replacers: ingredients that replace some or all of the functions of fat in foods and may or may not provide energy.

fats: lipids that are solid at room temperature (70°F or 21°C).

fatty acids: organic compounds composed of a chain of carbon atoms with hydrogen atoms attached and an acid group at one end.

fatty liver: an accumulation of fat in liver tissue; also called *hepatic steatosis* (STEE-ah-TOE-sis).

fatty streaks: initial lesions of atherosclerosis that form on the artery wall, characterized by accumulations of foam cells, lipid material, and connective tissue.

female athlete triad: a potentially fatal triad of medical problems: disordered eating, amenorrhea, and osteoporosis.

fertility: the capacity of a woman to produce a normal ovum periodically and of a man to produce normal sperm; the ability to reproduce.

fetal alcohol spectrum disorders (FASD): a spectrum of physical, behavioral, and cognitive disabilities caused by prenatal alcohol exposure.

fetal alcohol syndrome (FAS): the cluster of symptoms seen in an infant or child whose mother consumed excessive alcohol during her pregnancy. FAS includes, but is not limited to, brain damage, growth retardation, mental retardation, and facial abnormalities.

fetus (FEET-us): the developing infant from eight weeks after conception until its birth.

fibrinogen (fye-BRIN-oh-jen): a liver protein that promotes blood clot formation.

filtrate: the substances that pass through the glomerulus and travel through the nephron's tubules, eventually forming urine.

fistulas (FIST-you-luz): abnormal passages between organs or tissues (or between an internal organ and the body's surface) that permit the passage of fluids or secretions.

fitness: the characteristics that enable the body to perform physical activity; more broadly, the ability to meet routine physical demands with enough reserve energy to rise to a physical challenge; the body's ability to withstand stress of all kinds.

flatulence: the condition of having excessive intestinal gas, which causes abdominal discomfort.

flexibility: the capacity of the joints to move through a full range of motion; the ability to bend and recover without injury.

flora: the bacteria that normally reside in a person's body.

fluid and electrolyte balance: maintenance of the necessary amounts and types of fluid and minerals in each compartment of the body fluids.

fluorapatite (floor-APP-uh-tite): the stabilized form of bone and tooth crystal, in which fluoride has replaced the hydroxy portion of hydroxyapatite.

fluorosis (floor-OH-sis): mottling of the tooth enamel from ingestion of too much fluoride during tooth development.

foam cells: swollen cells in the artery wall that accumulate lipids.

follicle (FOLL-i-cul): a group of cells in the skin from which a hair grows.

food allergy: an adverse reaction to food that involves an immune response; also called *food-hypersensitivity reactions.*

food aversions: strong desires to avoid particular foods.

food bank: a facility that collects and distributes food donations to authorized organizations feeding the hungry.

food cravings: deep longings for particular foods.

food frequency questionnaire: a survey of foods routinely consumed.

food group plan: a diet-planning tool that sorts foods into groups based on nutrient content and then specifies that people should eat certain amounts of food from each group.

food insecurity: limited or uncertain access to foods of sufficient quality or quantity to sustain a healthy and active life.

food intolerances: adverse reactions to foods or food additives that do not involve the immune system.

food pantries: programs that provide groceries to be prepared and eaten at home.

food poverty: hunger resulting from inadequate access to available food for various reasons, including inadequate resources, political obstacles, social disruptions, poor weather conditions, and lack of transportation.

food record: a detailed log of food eaten during a specified time period, usually several days; also called a *food diary.*

food recovery: collecting wholesome food for distribution to low-income people who are hungry.

food security: access to enough food to sustain a healthy and active life.

foodborne illness: illness transmitted to human beings through food and water, caused by either an infectious agent (foodborne infection) or a poisonous substance (food intoxication); commonly known as *food poisoning.*

foodways: the eating habits and culinary practices of a people, region, or historical period.

fortification: the addition to a food of nutrients that were either not originally present or present in insignificant amounts.

free radicals: highly reactive chemical forms that can cause destructive changes in nearby compounds, sometimes setting up a chain reaction.

French units: units of measure used to indicate a feeding tube's outer diameter; 1 French unit equals $\frac{1}{3}$ millimeter.

fructose: a monosaccharide; sometimes known as *fruit sugar.*

full liquid diet: a liquid diet that includes milk and other opaque liquids.

functional foods: whole or modified foods that contain bioactive food components believed to provide health benefits, such as reduced disease risks, beyond the benefits that their nutrients contribute.

futile: describes medical care that will not improve the medical circumstances of a patient.

galactose: a monosaccharide; part of the disaccharide lactose.

galactosemia (ga-LAK-toe-SEE-me-ah): an inherited disorder that affects galactose metabolism. Accumulated galactose causes damage to the liver, kidneys, and brain in untreated patients.

gallbladder: the organ that stores and concentrates bile; it contracts and squirts bile through the bile duct into the duodenum when signaled that fat is present there.

gallstones: stones that form in the gallbladder from crystalline deposits of cholesterol or bilirubin.

gamma-oryzanol: a plant sterol that supposedly provides the same physical responses as anabolic steroids without the adverse side effects; also known as ferulic acid, lerulate, or FRAC.

gangrene: death of tissue due to a deficient blood supply and/or infection.

gastrectomy (gah-STREK-ta-mee): the surgical removal of part of the stomach (partial gastrectomy) or the entire stomach (total gastrectomy).

gastric banding: a surgical means of producing weight loss by restricting stomach size with a constricting band.

gastric bypass: surgery that restricts stomach size and reroutes food from the stomach to the lower part of the small intestine; creates a chronic, lifelong state of malabsorption by preventing normal digestion and absorption of nutrients.

gastric decompression: the removal of the stomach contents (including swallowed saliva, stomach secretions, and gas) of patients with motility disorders or obstructions that prevent stomach emptying.

gastric glands: exocrine glands in the stomach wall that secrete gastric juice into the stomach.

gastric juice: the digestive secretion of the gastric glands containing a mixture of water, hydrochloric acid, and enzymes (principally pepsin and lipase).

gastric outlet obstruction: an obstruction that prevents the normal emptying of stomach contents into the duodenum.

gastric residual volume: the volume of formula remaining in the stomach from a previous feeding.

gastritis: inflammation of stomach tissue.

gastroesophageal reflux disease (GERD): condition characterized by the backward flow (reflux) of the stomach's acidic contents into the esophagus.

gastrointestinal (GI) tract: the digestive tract. The principal organs are the stomach and intestines.

gastrointestinal motility: spontaneous motion in the digestive tract accomplished by involuntary muscular contractions.

gastroparesis (GAS-troe-pah-REE-sis): delayed stomach emptying caused by nerve damage in stomach tissue.

gastrostomy (gah-STRAH-stoe-mee): an opening into the stomach through which a feeding tube can be passed.

gatekeeper: with respect to nutrition, a key person who controls other people's access to foods and thereby exerts a profound impact on their nutrition.

gene therapy: treatment for inherited disorders, in which DNA sequences are introduced into the chromosomes of affected cells, prompting the cells to express the protein needed to correct the disease.

genetic counseling: support for families at risk of genetic disorders; involves diagnosis of disease, identification of inheritance patterns within the family, and review of reproductive options.

gestation: the period of about 40 weeks (three trimesters) from conception to birth; the term of a pregnancy.

gestational diabetes: glucose intolerance with first onset or first recognition during pregnancy.

gestational hypertension: high blood pressure that develops in the second half of pregnancy and usually resolves after childbirth.

ghrelin (GRELL-in): a hormone produced primarily by the stomach cells that signals the hypothalamus of the brain to stimulate appetite and food intake.

gingiva (jin-JYE-va, JIN-jeh-va): the gums.

gingivitis (jin-jeh-VYE-tus): inflammation of the gums, characterized by redness, swelling, and bleeding.

ginseng: a plant whose extract supposedly boosts energy. Side effects of chronic use include headache, nervousness, and sleeplessness.

glands: cells or groups of cells that secrete materials for special uses in the body.

glomerular filtration rate (GFR): the rate at which filtrate is formed within the kidneys, normally about 125 mL/min.

glomerulus (gloh-MEHR-yoo-lus): a tuft of capillaries within the nephron that filters water and solutes from the blood as urine production begins (plural: *glomeruli*).

glucagon (GLOO-ka-gon): a hormone secreted by the pancreas in response to low blood glucose concentration that elicits release of glucose from storage.

gluconeogenesis (gloo-co-nee-oh-GEN-ih-sis): the making of glucose from a noncarbohydrate source.

glucose polymers: compounds that supply glucose, not as single molecules, but linked in chains somewhat like starch. The objective is to attract less water from the body into the digestive tract.

glucose: a monosaccharide common to all disaccharides and polysaccharides; also called *blood sugar* or *dextrose*.

glycated hemoglobin (HbA$_{1c}$): hemoglobin that has non-enzymatically attached to glucose; the level of HbA$_{1c}$ in blood helps to diagnose diabetes and evaluate long-term glycemic control. Also called *glycosylated hemoglobin*.

glycemic (gly-SEE-mic): pertaining to blood glucose.

glycemic index: a method of classifying foods according to their potential for raising blood glucose.

glycemic response: the extent to which a food raises the blood glucose concentration and elicits an insulin response.

glycerol (GLISS-er-ol): an organic compound, three carbons long, that can form the backbone of triglycerides and phospholipids.

glycogen (GLY-co-gen): a polysaccharide composed of glucose, made and stored by liver and muscle tissues of human beings and animals as a storage form of glucose.

glycolysis (gligh-COLL-ih-sis): the metabolic breakdown of glucose to pyruvate.

glycosuria (GLY-co-SOOR-ee-ah): the presence of glucose in the urine.

goiter (GOY-ter): an enlargement of the thyroid gland due to an iodine deficiency, malfunction of the gland, or overconsumption of a thyroid antagonist.

goitrogen (GOY-troh-jen): a substance (occurring naturally in such foods as cabbage, kale, brussels sprouts, cauliflower, broccoli, and kohlrabi) that enlarges the thyroid gland and causes *toxic goiter*.

gout (GOWT): a metabolic disorder characterized by elevated uric acid levels in the blood and urine and the deposition of uric acid in and around the joints, causing acute joint inflammation.

Hazard Analysis Critical Control Points (HACCP): a systematic plan to identify and correct potential microbial hazards in the manufacturing, distribution, and commercial use of food products; commonly referred to as "HASS-ip."

health care agent: a person given legal authority to make medical decisions for another in the event of incapacitation.

health claims: statements that characterize the relationship between a nutrient or other substance in food and a disease or health-related condition.

health: a range of states with physical, mental, emotional, spiritual, and social components. At a minimum, health means freedom from physical disease, mental disturbances, emotional distress, spiritual discontent, social maladjustment, and other negative states. At a maximum, health means *wellness*.

Healthy People: a national public health initiative under the jurisdiction of the U.S. Department of Health and Human Services (DHHS) that identifies the most significant preventable threats to health and focuses efforts toward eliminating them.

heart failure: a condition with various causes that is characterized by the heart's inability to pump adequate blood to the body's cells, resulting in fluid accumulation in the tissues; also called *congestive heart failure*.

heat stroke: an acute and life-threatening reaction to heat buildup in the body.

Helicobacter pylori (H. pylori): a species of bacterium that colonizes gastric mucosa; a primary cause of gastritis and peptic ulcer disease.

helper T cells: lymphocytes that have a specific protein called CD4 on their surfaces and therefore are also known as *CD4+ T cells;* these are the cells most affected in HIV infection.

hematocrit (hee-MAT-oh-crit): measurement of the volume of the red blood cells packed by centrifuge in a given volume of blood.

hematopoietic stem cell transplantation: transplantation of the stem cells that produce red blood cells and white blood cells; the stem cells are obtained from the bone marrow (*bone marrow transplantation*) or circulating blood.

hematuria (HE-mah-TOO-ree-ah): blood in the urine.

hemochromatosis (heem-oh-crome-a-TOH-sis): iron overload characterized by deposits of iron-containing pigment in many tissues, with tissue damage; it is usually caused by a hereditary defect in iron absorption.

hemodialysis (HE-moe-dye-AL-ih-sis): a treatment that removes fluids and wastes from the blood by passing the blood through a dialyzer.

hemofiltration: removal of fluid and solutes by pumping blood across a membrane; no osmotic gradients are created during the process.

hemoglobin: the oxygen-carrying protein of the red blood cells.

hemolytic (HE-moh-LIT-ick) **anemia:** the condition of having too few red blood cells as a result of erythrocyte hemolysis.

hemophilia (HE-moh-FEEL-ee-ah): an inherited bleeding disorder characterized by deficiency or malfunction of a plasma protein needed for clotting blood.

hemorrhagic (hem-oh-RAJ-ik) **disease:** the vitamin K–deficiency disease in which blood fails to clot.

hemorrhagic strokes: strokes caused by bleeding within the brain, which destroys or compresses brain tissue.

hepatic coma: loss of consciousness resulting from severe liver disease.

hepatic encephalopathy (en-sef-ah-LOP-ah-thee): a condition that develops in advanced liver disease that is characterized by altered neurological functioning, including changes in personality and behavior, reduced mental abilities, and disturbances in motor function.

hepatic portal vein: the vein that collects blood from the GI tract and conducts it to capillaries in the liver.

hepatic vein: the vein that collects blood from the liver capillaries and returns it to the heart.

hepatitis (hep-ah-TYE-tis): inflammation of the liver.

hepatomegaly (HEP-ah-toe-MEG-ah-lee): enlargement of the liver.

herpes simplex virus: a common virus that can cause blisterlike lesions on the lips and in the mouth.

hiatal hernia: a condition in which the upper portion of the stomach protrudes above the diaphragm; most cases are asymptomatic.

high food security: a form of food security characterized by no indications of food-access problems or limitations.

high-carbohydrate energy drinks: fruit-flavored commercial beverages used to restore muscle glycogen after exercise or as a pregame beverage.

high-density lipoproteins (HDL): the type of lipoproteins that transport cholesterol back to the liver from peripheral cells; composed primarily of protein.

high-quality proteins: dietary proteins containing all the essential amino acids in relatively the same amounts that human beings require.

high-risk pregnancy: a pregnancy characterized by risk factors that make it likely the birth will be surrounded by problems such as premature delivery, difficult birth, retarded growth, birth defects, and early infant death.

histamine-2 receptor blockers: a class of drugs that suppresses acid secretion by inhibiting receptors on acid-producing cells; commonly called *H2 blockers*. Examples include cimetidine (Tagamet), ranitidine (Zantac), and famotidine (Pepcid).

HIV-lipodystrophy (LIP-oh-DIS-tro-fee) syndrome: a collection of abnormalities in fat and glucose metabolism that may result from drug treatments for HIV infection; changes include body fat redistribution, abnormal blood lipid levels, and insulin resistance.

HMB (beta-hydroxy-beta-methylbutyrate): a metabolite of the branched-chain amine acid leucine. Some research shows that HMB may increase muscle mass and strength in untrained individuals.

homeostasis (HOME-ee-oh-STAY-sis): the maintenance of constant internal conditions (such as chemistry, temperature, and blood pressure) by the body's control systems.

hormones: chemical messengers secreted by a variety of glands in the body in response to altered conditions. Each travels to one or more target tissues or organs and elicits specific responses to restore normal conditions.

hourly sweat rate: the amount of weight lost plus fluid consumed during exercise per hour.

human immunodeficiency virus (HIV): the virus that causes acquired immune deficiency syndrome (AIDS). HIV destroys immune cells and progressively impedes the body's ability to fight infections and certain cancers.

hunger: the physiological need to eat, experienced as a drive to obtain food; an unpleasant sensation that demands relief.

hydrochloric acid (HCl): an acid composed of hydrogen and chloride atoms; normally produced by the gastric glands.

hydrogenation (high-dro-gen-AY-shun): a chemical process by which hydrogen atoms are added to monounsaturated or polyunsaturated fats to reduce the number of double bonds, making the fats more saturated (solid) and more resistant to oxidation; produces *trans*-fatty acids.

hyperactivity: inattentive and impulsive behavior that is more frequent and severe than is typical of others of a similar age; professionally called *attention deficit/hyperactivity disorder (ADHD)*.

hypercalcemia (HIGH-per-kal-SEE-me-ah): elevated serum calcium levels.

hypercalciuria (HIGH-per-kal-see-YOO-ree-ah): elevated urinary calcium levels.

hypercapnia (high-per-CAP-nee-ah): excessive carbon dioxide in the blood.

hyperglycemia: elevated blood glucose concentrations. Normal fasting plasma glucose levels are less than 100 mg/dL. Fasting plasma glucose levels between 100 and 125 mg/dL suggest prediabetes; values of 126 mg/dL and above suggest diabetes.

hyperinsulinemia: abnormally high levels of insulin in the blood.

hyperkalemia (HIGH-per-ka-LEE-me-ah): elevated serum potassium levels.

hypermetabolism: a higher-than-normal metabolic rate.

hyperosmolar hyperglycemic syndrome: extreme hyperglycemia associated with dehydration, hyperosmolar blood, and altered mental status; sometimes called the *hyperosmolar hyperglycemic nonketotic state*.

hyperoxaluria (HIGH-per-ox-ah-LOO-ree-ah): elevated urinary oxalate levels.

hyperphosphatemia (HIGH-per-fos-fa-TEE-me-ah): elevated serum phosphate levels.

hypertension: high blood pressure.

hyperthermia: an above-normal body temperature.

hypertonic formula: a formula with an osmolality greater than that of blood serum.

hypertriglyceridemia (HYE-per-try-gliss-er-rye-DEE-me-ah): high blood triglyceride levels.

hypertrophy (high-PURR-tro-fee): an increase in size (for example, of a muscle) in response to use.

hypoalbuminemia: low plasma albumin concentrations.

hypochlorhydria (HIGH-poe-clor-HIGH-dree-ah): abnormally low gastric acid secretions.

hypoglycemia: abnormally low concentrations of blood glucose. In diabetes, hypoglycemia is treated when plasma glucose levels fall below 70 mg/dL.

hypokalemia (HIGH-po-ka-LEE-me-ah): low serum potassium levels.

hyponatremia (HIGH-poh-na-TREE-mee-ah): a decreased concentration of sodium in the blood (*hypo* means "below"; *natrium* means "sodium"; *emia* means "blood").

hypothalamus (high-poh-THALL-uh-mus): a part of the brain that helps regulate many body balances, including fluid balance.

hypothermia: a below-normal body temperature.

hypoxemia (high-pock-SEE-me-ah): a low level of oxygen in the blood.

hypoxia (high-POCK-see-ah): a low amount of oxygen in body tissues.

ileocecal (ill-ee-oh-SEEK-ul) **valve:** the sphincter separating the small and large intestines.

ileostomy (ill-ee-AH-stoe-me): a surgical passage through the abdominal wall into the ileum.

ileum (ILL-ee-um): the last segment of the small intestine.

implantation: the stage of development in which the blastocyst embeds itself in the wall of the uterus and begins to develop; occurs during the first two weeks after conception.

inborn error of metabolism: an inherited trait (one that is present at birth) that causes the absence, deficiency, or malfunction of a protein that has a critical metabolic role.

inflammation: an immunological response to cellular injury characterized by an increase in white blood cells.

inflammatory response: a group of nonspecific immune responses to infection or injury.

informed consent: a patient's or caregiver's agreement to undergo a treatment that has been adequately disclosed. Persons must be mentally competent in order to make the decision.

inorganic: not containing carbon or pertaining to living things.

insoluble fibers: the tough, fibrous structures of fruits, vegetables, and grains; indigestible food components that do not dissolve in water.

insulin: a hormone secreted by the pancreas in response to high blood glucose that promotes cellular glucose uptake for use or storage.

insulin resistance: the condition in which a normal amount of insulin produces a subnormal effect in muscle, adipose, and liver cells, resulting in an elevated fasting glucose; a metabolic consequence of obesity that precedes type 2 diabetes.

intermittent claudication (claw-dih-KAY-shun): severe pain and weakness in the legs (especially the calves) caused by inadequate blood supply to the muscles; it usually occurs with walking and subsides during rest.

intermittent feedings: delivery of about 250 to 400 milliliters of formula over 20 to 40 minutes.

intestinal adaptation: the process of intestinal recovery following resection that leads to improved absorptive capacity.

intestinal flora: the bacterial inhabitants of the GI tract.

intestinal juice: the secretion of the intestinal glands; contains enzymes for the digestion of carbohydrate and protein and a minor enzyme for fat digestion.

intractable: not easily managed or controlled.

intractable vomiting: vomiting that is not easily managed or controlled.

intradialytic parenteral nutrition: the infusion of nutrients during hemodialysis, often providing amino acids, dextrose, lipids, and some trace minerals.

intrinsic factor: a factor inside the system; a compound that binds to vitamin B_{12} in the stomach and aids its absorption in the small intestine.

iron deficiency: the condition of having depleted iron stores.

iron-deficiency anemia: a blood iron deficiency characterized by small, pale red blood cells; also called *microcytic hypochromic anemia*.

iron overload: toxicity from excess iron.

irritable bowel syndrome: an intestinal disorder of unknown cause that disturbs the functioning of the lower bowel; symptoms include abdominal pain, flatulence, diarrhea, and constipation.

ischemia (iss-KEE-mee-a): inadequate blood supply within tissues due to obstructed blood flow in the arteries.

ischemic strokes: strokes caused by the obstruction of blood flow to brain tissue.

isotonic formula: a formula with an osmolality similar to that of blood serum (about 300 milliosmoles per kilogram).

jaundice (JAWN-dis): yellow discoloration of the skin and eyes due to an accumulation of bilirubin, a breakdown product of hemoglobin that normally exits the body via bile secretions.

jejunostomy (JEH-ju-NAH-stoe-mee): an opening into the jejunum through which a feeding tube can be passed.

jejunum (je-JOON-um): the first two-fifths of the small intestine beyond the duodenum.

Joint Commission: a nonprofit organization that sets standards for health care performance and safety and confers accreditation to health care organizations and programs that meet those standards.

Kaposi's (kah-POH-seez) **sarcoma:** a common cancer in HIV-infected persons that is characterized by lesions in the skin, lungs, and GI tract.

kcalorie (energy) control: management of food energy intake.

kcalorie counts: the determination of food energy (and often protein) consumed by patients for one or more days.

keratin (KERR-uh-tin): a water-insoluble protein; the normal protein of hair and nails.

ketoacidosis (KEY-toe-ass-ih-DOE-sis): an acidosis (lowering of blood pH) that results from the excessive production of ketone bodies.

ketones (KEY-tones): acidic, fat-related compounds formed from the incomplete breakdown of fat when carbohydrate is not available; technically known as *ketone bodies*.

ketonuria (KEY-toe-NOOR-ee-ah): the presence of ketone bodies in the urine.

kidney stones: crystalline masses that form in the urinary tract; also called *renal calculi* or *nephrolithiasis*.

kilocalorie (kcalorie or kcal): the amount of heat necessary to raise the temperature of 1 kilogram (kg) of water 1°C.

kwashiorkor (kwash-ee-OR-core or kwash-ee-or-CORE): a severe form of protein-energy malnutrition that occurs more frequently after 18 months of age. Kwashiorkor is characterized by failure to grow and develop, changes in the pigmentation of the hair and skin, edema, and fatty liver. Kwashiorkor is associated with inadequate protein intake and infections.

lactadherin (lack-tad-HAIR-in): a protein in breast milk that attacks diarrhea-causing viruses.

lactate: a compound produced during the breakdown of glucose in anaerobic metabolism.

carbohydrate loading: a regimen of moderate exercise, followed by eating a high-carbohydrate diet, which enables muscles to temporarily store glycogen beyond their normal capacity; also called *glycogen loading* or *glycogen supercompensation*.

lactation: production and secretion of breast milk for the purpose of nourishing an infant.

lacto-ovo vegetarians: people who include plant foods, milk or milk products, and eggs but exclude other animal-derived foods from their diets.

lacto-vegetarians: people who include plant foods and milk or milk products but exclude other animal-derived foods from their diets.

lactoferrin (lack-toe-FERR-in): protein in breast milk that binds iron and keeps it from supporting the growth of the infant's intestinal bacteria.

lactose: a disaccharide composed of glucose and galactose; commonly known as *milk sugar*.

lactose intolerance: a condition characterized by the inability to digest the lactose in milk products; symptoms include flatulence, abdominal discomfort, and diarrhea.

laparoscope: a narrow surgical telescope that is inserted into the abdominal cavity through a small incision. A video camera is usually attached so that the procedure can be viewed on a television monitor.

laparoscopic: pertaining to procedures that use a laparoscope for internal examination or surgery.

laparoscopic weight-loss surgery: a procedure in which surgeons gain access to the abdomen via several small incisions. A tiny video camera is inserted through one of the incisions and surgical instruments through the others. The surgeons watch their work on a large-screen monitor.

lapses: periods of returning to old habits.

large intestine or colon (COAL-un): the lower portion of intestine that completes the digestive process. Its segments are the ascending colon, the transverse colon, the descending colon, and the sigmoid colon.

lecithins: one type of phospholipid.

legumes (lay-GYOOMS, LEG-yooms): plants of the bean and pea family with seeds that are rich in protein compared with other plant-derived foods.

leptin: a hormone produced by fat cells under the direction of the (*ob*) gene. It decreases appetite and increases energy expenditure.

letdown reflex: the reflex that forces milk to the front of the breast when the infant begins to nurse.

life expectancy: the average number of years lived by people in a given society.

life span: the maximum number of years of life attainable by a member of a species.

limiting amino acid: an essential amino acid that is present in dietary protein in the shortest supply relative to the amount needed for protein synthesis in the body.

linoleic acid: a polyunsaturated fatty acid (with 2 points of unsaturation) that is essential for human beings.

linolenic acid: a polyunsaturated fatty acid (with 3 points of unsaturation) that is essential for human beings.

lipids: a family of compounds insoluble in water that includes triglycerides, phospholipids, and sterols.

lipomas (lih-POE-muz): benign tumors composed of fatty tissue.

lipoprotein lipase (LPL): an enzyme mounted on the surface of fat cells (and other cells) that hydrolyzes triglycerides in the blood into fatty acids and glycerol for absorption into the cells.

lipoproteins: clusters of lipids associated with proteins that serve as transport vehicles for lipids in the lymph and blood.

listeriosis: a serious foodborne infection that can cause severe brain infection or death in a fetus or newborn; caused by the bacterium *Listeria monocytogenes*, which is found in soil and water.

liver: the organ that manufactures bile.

living will: a written statement that specifies the medical procedures desired or not desired in the event that a person is unable to communicate or is incapacitated; also called an *advance medical directive*.

longevity: long duration of life.

low birthweight (LBW): a birthweight less than 5½ lb (2500 g); indicates probable poor health in the newborn and poor nutrition status of the mother during pregnancy.

low food security: a form of food insecurity characterized by reduced quality of diet with little or no indication of reduced food intake.

low-density lipoproteins (LDL): the type of lipoproteins derived from VLDL as cells remove triglycerides from them. LDL carry cholesterol and triglycerides from the liver to the cells of the body and are composed primarily of cholesterol.

low-microbial diet: a diet that contains foods that are unlikely to be contaminated with bacteria and other microbes.

low-risk pregnancy: a pregnancy characterized by factors that make it likely the birth will be normal and the infant healthy.

lymph (LIMF): the body fluid found in lymphatic vessels; consists of all the constituents of blood except red blood cells.

lymphatic system: a loosely organized system of vessels and ducts that conveys the products of digestion toward the heart.

macrosomia (MAK-roh-SOH-mee-ah): the condition of having an abnormally large body; in infants, refers to birth weights of 4000 grams (8 pounds 13 ounces) and above.

macrovascular complications: disorders that affect the large blood vessels, including the coronary arteries and arteries of the limbs.

macula: a small, oval, yellowish region in the center of the retina that provides the sharp, straight-ahead vision so critical to reading and driving.

macular (MACK-you-lar) **degeneration:** deterioration of the macular area of the eye that can lead to loss of central vision and eventual blindness.

major minerals: essential mineral nutrients required in the adult diet in amounts greater than 100 milligrams per day.

maleficence (mah-LEF-eh-sense): the act of doing evil or harm.

malignant (ma-LIG-nent): describes a cancerous cell or tumor, which can injure healthy tissue and spread cancer to other regions of the body.

malnutrition: any condition caused by deficient or excess energy or nutrient intake or by an imbalance of nutrients.

maltose: a disaccharide composed of two glucose units; sometimes known as *malt sugar*.

marasmus (ma-RAZZ-mus): the most common form of severe PEM before one year of age. Marasmus is characterized by generalized muscle wasting associated with extreme deprivation, or impaired absorption, of energy, protein, vitamins, and minerals.

marginal food security: a form of food security characterized by one or two indications of food-access problems but with little or no change in food intake.

mast cells: cells within connective tissue that produce and release histamine.

mastitis: infection of a breast.

mechanical ventilation: life-sustaining treatment in which a mechanical ventilator is used to substitute for a patient's failing lungs.

mechanically altered diets: diets that contain foods that have been mechanically altered in some way; foods may be liquid, pureed, ground, chopped, minced, or tenderized.

medical nutrition therapy: nutrition care provided by a registered dietitian; includes assessing nutrition status, diagnosing nutrition problems, developing nutrition care plans, and providing dietary counseling and nutrition education.

medium-chain triglycerides (MCT): triglycerides that contain fatty acids that are 8 to 10 carbons in length, do not require digestion, and can be absorbed in the absence of lipase or bile.

megaloblastic anemia: anemia characterized by large, immature red blood cells, as occurs in folate and vitamin B_{12} deficiency; also called *macrocytic anemia*.

metabolic stress: a disruption in the body's chemical environment due to the effects of disease or injury and characterized by changes in metabolic rate, heart rate, blood pressure, hormonal status, and nutrient metabolism.

metabolic syndrome: a cluster of interrelated clinical disorders, including obesity, insulin resistance, high blood pressure, and abnormal blood lipids, which together increase risk of diabetes and cardiovascular disease; also known as *insulin resistance syndrome* or *syndrome X*.

metabolism: the sum total of all the chemical reactions that go on in living cells.

metabolites: products of metabolism; compounds produced by a biochemical pathway.

metastasize (meh-TAS-tah-size): to spread from one part of the body to another; refers to cancer cells.

microalbuminuria: the presence of albumin (a blood protein) in the urine, a sign of diabetic nephropathy.

microcytic anemia: anemia characterized by small, hypochromic (pale) red blood cells, as occurs in iron deficiency.

microvascular complications: disorders that affect the small blood vessels and capillaries, including those in the retina and kidneys.

microvilli (MY-cro-VILL-ee or MY-cro-VILL-eye): tiny, hairlike projections on each cell of every villus that can trap nutrient particles and transport them into the cells (singular form: *microvillus*).

moderate-intensity physical activity: physical activity that requires some increase in breathing and/or heart rate and expends 3.5 to 7 kcalories per minute. Walking at a speed of 3 to 4.5 miles per hour (about 15 to 20 minutes to walk one mile) is an example.

moderation: providing enough, but not too much, of a substance.

modified diet: a diet that contains foods altered in texture, consistency, or nutrient content or that includes or omits specific foods; also called a *therapeutic diet*.

modular formulas: enteral formulas prepared in the hospital from *modules* that contain single macronutrients; used for people with unique nutrient needs.

molybdenum (mo-LIB-duh-num): a trace element.

monosaccharides (mon-oh-SACK-uh-rides): single sugar units.

monounsaturated fatty acid (MUFA): a fatty acid that has one point of unsaturation.

mouth: the oral cavity containing the tongue and teeth.

mucous membrane: membrane composed of mucus-secreting cells that lines the surfaces of body tissues.

mucus (MYOO-cuss): a mucopolysaccharide (a relative of carbohydrate) secreted by cells of the stomach wall that protects the cells from exposure to digestive juices (and other destructive agents). (The noun is *mucus*; the adjective is *mucous*.)

muscle endurance: the ability of a muscle to contract repeatedly within a given time without becoming exhausted.

muscle power: the efficiency of a muscle contraction, measured by force and time.

muscle strength: the ability of muscles to work against resistance.

mutation: an inheritable alteration in the DNA sequence of a gene.

myocardial (MY-oh-CAR-dee-al) **infarction** (in-FARK-shun), or **MI:** death of heart muscle caused by a sudden reduction in coronary blood flow; also called a *heart attack* or *cardiac arrest*.

myoglobin: the oxygen-carrying protein of the muscle cells

nasoduodenal (ND) feeding tube: a feeding tube that is placed into the duodenum via the nose.

nasoenteric feeding tube: a feeding tube that is placed into the GI tract via the nose.

nasogastric (NG) feeding tube: a feeding tube that is placed into the stomach via the nose.

nasojejunal (NJ) feeding tube: a feeding tube that is placed into the jejunum via the nose.

naturally occurring sugars: sugars that are not added to a food but are present as its original constituents, such as the sugars of fruit or milk.

nephron (NEF-ron): the functional unit of the kidneys, consisting of a glomerulus and tubules.

nephrotic (neh-FROT-ik) **syndrome:** a syndrome associated with kidney disorders that damage the glomerulus and cause urinary protein losses exceeding 3½ g/day.

nephrotoxic: toxic to the kidneys.

neural tube: the embryonic tissue that later forms the brain and spinal cord.

neural tube defects (NTD): malformations of the brain and/or spinal cord that occur during embryonic development.

neurofibrillary tangles: snarls of the threadlike strands that extend from the nerve cells, commonly found in the brains of people with Alzheimer's dementia.

neurons: nerve cells; the structural and functional units of the nervous system. Neurons initiate and conduct nerve transmissions.

neutropenia: a low white blood cell (neutrophil) count, which increases susceptibility to infection.

niacin equivalents (NE): the amount of niacin present in food, including the niacin that can theoretically be made from tryptophan, its precursor, present in the food.

night blindness: the slow recovery of vision after exposure to flashes of bright light at night; an early symptom of vitamin A deficiency.

nitric oxide: a compound produced by blood vessel cells that helps to regulate blood vessel function, including dilation and constriction.

nitrogen balance: the amount of nitrogen consumed (N in) as compared with the amount of nitrogen excreted (N out) in a given period of time.

nonessential amino acids: amino acids that the body can synthesize.

nonpathogenic: not capable of causing disease.

nursing bottle tooth decay: extensive tooth decay due to prolonged tooth contact with formula, milk, fruit juice, or other carbohydrate-rich liquid offered to an infant in a bottle.

nutrient claims: statements that characterize the quantity of a nutrient in a food.

nutrient density: a measure of the nutrients a food provides relative to the energy it provides (nutrient amounts per kcal).

nutrients: substances obtained from food and used in the body to provide energy and structural materials and to serve as regulating agents to promote growth, maintenance, and repair.

nutrition: the science of foods and the nutrients and other substances they contain, and of their ingestion, digestion, absorption, transport, metabolism, interaction, storage, and excretion.

nutrition care plans: strategies for meeting an individual's nutritional needs.

nutrition care process: a problem-solving method that dietetics professionals use to evaluate and treat nutrition-related problems.

nutrition prescriptions: specific instructions concerning nutrition care; also referred to as *diet orders* or *diet prescriptions*.

nutrition screening: a brief assessment of health-related variables to identify patients who are malnourished or at risk for nutrition

nutrition support: the delivery of nutrients using a feeding tube or intravenous infusions.

nutrition support teams: health care professionals responsible for the provision of nutrients by tube feeding or intravenous infusion.

nutritive sweeteners: sweeteners that yield energy, including both the sugars and the sugar alcohols.

obesity: overfatness with adverse health effects, as determined by reliable measures and interpreted with good medical judgment. Obesity is officially defined as a body mass index (BMI) of 30 or higher.

oils: lipids that are liquid at room temperature (70°F or 21°C).

olestra: a noncaloric synthetic fat made from sucrose and fatty acids; also known as *sucrose polyester*.

oliguria (OL-lih-GOO-ree-ah): an abnormally low amount of urine, often less than 400 mL/day.

omega-3 fatty acids: polyunsaturated fatty acids in which the endmost double bond is three carbons back from the end of the carbon chain (e.g., linolenic acid).

omega-6 fatty acid: a polyunsaturated fatty acid with its endmost double bond six carbons

back from the end of its carbon chain (e.g., linoleic acid).

oncotic pressure: the pressure exerted by fluid on one side of a membrane as a result of osmosis.

open feeding system: a delivery system that requires the formula to be transferred from its original packaging to a feeding container before being administered through the feeding tube.

opportunistic infections: infections from microorganisms that normally do not cause disease in healthy people but are damaging to persons with compromised immune function.

opsin (OP-sin): the protein portion of the visual pigment molecule.

oral glucose tolerance test: a test that evaluates a person's ability to tolerate an oral glucose load.

organic: carbon containing.

orogastric feeding tube: a feeding tube that is inserted into the stomach through the mouth.

oropharyngeal dysphagia (OR-oh-fah-ren-JEE-al diss-FAY-jee-ah): an inability to transfer food from the mouth and pharynx to the esophagus; usually caused by a neurological or muscular disorder.

osmolality (OZ-moe-LAL-ih-tee): the concentration of osmotically active solutes in a solution, expressed as milliosmoles (mOsm) per kilogram of solvent.

osmolarity: the concentration of osmotically active solutes in a solution, expressed as milliosmoles per liter of solution (mOsm/L). *Osmolality* (mOsm/kg) is an alternative term used to describe a solution's osmotic properties.

osmosis: movement of water across a membrane toward the side where solutes are more concentrated.

osteoarthritis: a painful, chronic disease of the joints that occurs when the cushioning cartilage in a joint breaks down; joint structure is usually altered, with loss of function; also called *degenerative arthritis*.

osteomalacia (os-tee-oh-mal-AY-shuh): a bone disease characterized by softening of the bones. Symptoms include bending of the spine and bowing of the legs. The disease occurs most often in adults with renal failure or malabsorption disorders.

osteoporosis (os-tee-oh-pore-OH-sis): literally, porous bones; reduced density of the bones, also known as *adult bone loss*.

overload: an extra physical demand placed on the body; an increase in the frequency, duration, or intensity of an activity.

overnutrition: overconsumption of food energy or nutrients sufficient to cause disease or increased susceptibility to disease; a form of malnutrition.

overweight: overfatness of a moderate degree; defined as a body mass index (BMI) of 25.0 through 29.9.

ovo-vegetarians: people who include plant foods and eggs but exclude other animal-derived foods from their diets.

ovum (OH-vum): the female reproductive cell, capable of developing into a new organism upon fertilization; commonly referred to as an egg.

oxidation (OKS-ee-day-shun): the process of a substance combining with oxygen.

pancreas: a gland that secretes digestive enzymes and juices into the duodenum and hormones that help to maintain glucose homeostasis into the blood.

pancreatic (pank-ree-AT-ic) **juice:** the exocrine secretion of the pancreas, containing enzymes for the digestion of carbohydrate, fat, and protein; released into the small intestine through the pancreatic duct.

paracentesis (pah-rah-sen-TEE-sis): a surgical puncture of a body cavity with an aspirator to draw out excess fluid.

parenteral (par-EN-ter-al) **nutrition:** the intravenous provision of nutrients that bypasses the GI tract.

pasteurization: heat processing of food that inactivates some, but not all, microorganisms in the food; not a sterilization process.

pathogenic: capable of causing disease.

pathogens (PATH-oh-jens): microorganisms capable of producing disease.

patient autonomy: a principle of self-determination, such that patients (or surrogate decision makers) are free to choose the medical interventions that are acceptable to them, even if they choose to refuse interventions that may extend their lives.

pellagra (pell-AY-gra): the niacin-deficiency disease. Symptoms include the "4 Ds": diarrhea, dermatitis, dementia, and, ultimately, death.

pepsin: a protein-digesting enzyme (gastric protease) in the stomach. It circulates as a precursor, pepsinogen, and is converted to pepsin by the action of stomach acid.

peptic ulcer: an open sore in the gastrointestinal mucosa; may develop in the esophagus, stomach, or duodenum.

percutaneous endoscopic gastrostomy (PEG): a nonsurgical technique for creating a gastrostomy under local anesthesia.

percutaneous endoscopic jejunostomy (PEJ): a nonsurgical technique for creating a jejunostomy. The tube can either be guided into the jejunum via a gastrostomy or passed directly into the jejunum (direct PEJ).

periodontal disease: disease that involves the connective tissue that supports the teeth.

periodontitis: inflammation or degeneration of the tissues that support the teeth.

periodontium: the tissues that support the teeth, including the gums, cementum (bonelike material covering the dentin layer of the tooth), periodontal ligament, and underlying bone.

peripheral blood smear: a circulating blood sample spread on a glass slide and stained for analysis under a microscope.

peripheral parenteral nutrition (PPN): the infusion of nutrient solutions into peripheral veins, usually a vein in the arm or back of the hand.

peripheral vascular disease: disorders characterized by impaired blood circulation in the limbs.

peripheral veins: the small-diameter veins that carry blood from the arms and legs.

peristalsis (peri-STALL-sis): successive waves of involuntary muscular contractions passing along the walls of the GI tract that push the contents along.

peritoneal (PEH-rih-toe-NEE-al) **dialysis:** a treatment that removes fluids and wastes from the blood by using the body's peritoneal membrane as a filter.

peritoneovenous (PEH-rih-toe-NEE-oh-VEE-nus) **shunt:** a surgical passage created between the peritoneum (membrane that surrounds the abdominal cavity) and the jugular vein to divert fluid and relieve ascites.

peritonitis: inflammation of the peritoneal membrane, which lines the abdominal cavity.

pernicious (per-NISH-us) **anemia:** a blood disorder that reflects a vitamin B_{12} deficiency caused by lack of intrinsic factor and characterized by large, immature red blood cells and damage to the nervous system.

persistent vegetative state: a vegetative mental state resulting from brain injury that persists for at least one month. Individuals lose awareness and the ability to think but retain noncognitive brain functions, such as motor reflexes and normal sleep patterns.

pH: the concentration of hydrogen ions. The lower the pH, the stronger the acid.

phagocytes (FAG-oh-sites): white blood cells (neutrophils and macrophages) that have the ability to engulf and destroy antigens.

pharynx (FAIR-inks): the passageway leading from the nose and mouth to the larynx and esophagus, respectively.

phenylketonuria (FEN-il-KEY-toe-NU-ree-ah) or **PKU:** an inherited disorder that affects the conversion of the essential amino acid phenylalanine to the amino acid tyrosine.

phlebitis (fleh-BYE-tiss): inflammation of a vein.

phospholipids: one of the three main classes of lipids; compounds that are similar to triglycerides but have choline (or another compound) and a phosphorus-containing acid in place of one of the fatty acids.

physical activity: bodily movement produced by muscle contractions that substantially increase energy expenditure.

physiological age: a person's age as estimated from physical health and probable life expectancy.

phytates: nonnutrient components of grains, legumes, and seeds that can bind minerals such as iron, zinc, calcium, and magnesium in insoluble complexes in the intestine.

phytochemicals (FIGH-toe-CHEM-ih-cals): compounds in plants that confer color, taste, and other characteristics.

pica (PIE-ka): a craving for nonfood substances; also known as geophagia (jee-oh-FAY-jee-uh) when referring to clay-eating behavior.

pigment: a molecule capable of absorbing certain wavelengths of light so that it reflects only those that we perceive as a certain color.

pituitary (pit-TOO-ih-tary) **gland:** in the brain, the "king gland" that regulates the operation of many other glands.

placenta (pla-SEN-tuh): an organ that develops inside the uterus early in pregnancy, in which maternal and fetal blood circulate in close proximity and exchange materials. The fetus receives nutrients and oxygen across the placenta; the mother's blood picks up carbon dioxide and other waste materials to be excreted via her lungs and kidneys.

plaque (PLACK): an accumulation of fatty deposits, smooth muscle cells, and fibrous connective tissue that develops in the artery walls in atherosclerosis. Plaque associated with atherosclerosis is known as atheromatous (ATH-er-OH-ma-tus) plaque.

plasminogen activator inhibitor-1: a protein that promotes blood clotting by inhibiting blood clot degradation within blood vessels.

polydipsia (POL-ee-DIP-see-ah): excessive thirst.

polypeptide: 10 or more amino acids bonded together.

polyphagia (POL-ee-FAY-jee-ah): excessive appetite or food intake.

polysaccharides: long chains of monosaccharide units arranged as starch, glycogen, or fiber.

polyunsaturated fatty acids (PUFA): fatty acids with two or more points of unsaturation.

polyuria (POL-ee-YOOR-ree-ah): excessive urine production.

portal hypertension: elevated blood pressure in the portal vein due to obstructed blood flow through the liver.

prebiotics: indigestible substances in foods that stimulate the growth of nonpathogenic bacteria within the large intestine.

precursors: compounds that can be converted into other compounds; with regard to vitamins, compounds that can be converted into active vitamins; also known as provitamins.

prediabetes: the condition in which blood glucose levels are higher than normal but not high enough to be diagnosed as diabetes.

preeclampsia (PRE-ee-KLAMP-see-ah): a condition characterized by hypertension and protein in the urine.

preformed vitamin A: vitamin A in its active form.

pregame meal: a meal eaten three to four hours before athletic competition.

prehypertension: blood pressure values that predict hypertension.

pressure gradient: the change in pressure over a given distance. In dialysis, a pressure gradient is created between the blood and the dialysate.

pressure sores: regions of damaged skin and tissue due to prolonged pressure on the affected area by an external object, such as a bed, wheelchair, or cast; also called decubitus (deh-KYU-bih-tus) ulcers.

probiotics: live bacteria provided in foods and dietary supplements for the purpose of preventing or treating disease.

protein digestibility: a measure of the amount of amino acids absorbed from a given protein intake.

protein isolates: proteins that have been isolated from foods.

protein turnover: the continuous breakdown and synthesis of body proteins involving the recycling of amino acids.

protein-energy malnutrition (PEM): a deficiency of protein and food energy; the world's most widespread malnutrition problem, including both marasmus and kwashiorkor.

proteins: compounds made from strands of amino acids composed of carbon, hydrogen, oxygen, and nitrogen atoms.

proteinuria (PRO-teen-NOO-ree-ah): loss of protein, especially albumin, in the urine; also known as albuminuria.

proton-pump inhibitors: a class of drugs that inhibits the enzyme that pumps hydrogen ions (protons) into the stomach. Examples include omeprazole (Prilosec) and lansoprazole (Prevacid).

puberty: the period in life in which a person becomes physically capable of reproduction.

purines (PYOO-reens): products of nucleotide metabolism that degrade to uric acid.

pyloric (pie-LORE-ic) **sphincter:** the circular muscle that separates the stomach from the small intestine and regulates the flow of partially digested food into the small intestine; also called pylorus or pyloric valve.

pyruvate (PIE-roo-vate): a 3-carbon compound that plays a key role in energy metabolism.

quality of life: a person's perceived physical and mental well-being.

radiation enteritis: inflammation of intestinal tissue caused by radiation therapy.

radiation therapy: the use of X-rays, gamma rays, or atomic particles to destroy cancer cells.

rancid: the term used to describe fats when they have deteriorated, usually by oxidation.

rebound hyperglycemia: hyperglycemia that results from the release of counterregulatory hormones following nighttime hypoglycemia; also called the Somogyi effect.

Recommended Dietary Allowances (RDA): a set of values reflecting the average daily amounts of nutrients considered adequate to meet the known nutrient needs of practically all healthy people in a particular life stage and gender group; a goal for dietary intake by individuals.

rectum: the muscular terminal part of the intestine, extending from the sigmoid colon to the anus.

refeeding syndrome: a condition that sometimes develops when a severely malnourished person is aggressively fed; characterized by electrolyte and fluid imbalances and hyperglycemia.

refined grain: a product from which the bran, germ, and husk have been removed, leaving only the endosperm.

reflux esophagitis: inflammation in the esophagus related to the reflux of acidic stomach contents.

registered dietitians (RDs): dietitians who have graduated from a university or college after completing a program of dietetics that has been accredited by the American Dietetic Association (or Dietitians of Canada). The dietitians must serve in an approved internship or coordinated program to practice the necessary skills, pass the association registration examination, and maintain competency through continuing education.

regular diet: a diet that includes all foods and meets the nutrient needs of healthy people; also called a *standard diet* or *house diet.*

renal (REE-nil): pertaining to the kidneys.

renal colic: the intense pain that occurs when a kidney stone passes through the ureter.

renal osteodystrophy: a bone disorder that develops in patients with chronic kidney disease as a consequence of the increased secretion of parathyroid hormone, reduced serum calcium, acidosis, and impaired vitamin D activation by the kidneys.

renal threshold: the blood concentration of a substance that exceeds the kidneys' capacity for reabsorption, causing the substance to be passed into the urine.

requirement: the lowest continuing intake of a nutrient that will maintain a specified criterion of adequacy.

resection: the surgical removal of part of an organ or body structure.

residue: material left in the intestine after digestion; includes mostly dietary fiber and undigested starches and proteins.

resistance training: the use of free weights, weight machines, or body weight to provide resistance for developing muscle strength, power, and endurance; also called *weight training.*

resistant starches: starches that escape digestion and absorption in the small intestine of healthy people.

resistin (re-ZIST-in): a hormone produced by adipose cells that induces insulin resistance.

respiratory stress: abnormal gas exchange between the air and blood, resulting in lower-than-normal oxygen levels and higher-than-normal carbon dioxide levels.

resting metabolic rate (RMR): a measure of the energy use of a person at rest in a comfortable setting; similar to the BMR but with less stringent criteria for recent food intake and physical activity (and consequently, slightly higher).

reticulocytes: immature red blood cells released into blood by bone marrow.

retina (RET-in-uh): the layer of light-sensitive nerve cells lining the back of the inside of the eye; consists of rods and cones.

retinol activity equivalents (RAE): a measure of vitamin A activity; the amount of retinol that the body will derive from a food containing preformed retinol or its precursor beta-carotene.

retinol-binding protein (RBP): the specific protein responsible for transporting retinol.

rheumatoid arthritis: a disease of the immune system involving painful inflammation of the joints and related structures.

rhodopsin (ro-DOP-sin): a light-sensitive pigment of the retina; contains the retinal form of vitamin A and the protein opsin.

ribose: a 5-carbon sugar falsely promoted as improving the regeneration of ATP and thereby the speed of recovery after high-intensity exercise.

rickets: the vitamin D–deficiency disease in children.

rooting reflex: a reflex that causes an infant to turn toward whichever cheek is touched in search of a nipple.

royal jelly: the substance produced by worker bees and fed to the queen bee; falsely promoted as enhancing stamina and reducing fatigue.

saliva: the secretion of the salivary glands; contains salivary amylase.

salivary glands: exocrine glands that secrete saliva into the mouth.

salts: compounds composed of charged particles (ions).

sarcopenia (SAR-koh-PEE-nee-ah): loss of skeletal muscle mass, strength, and quality.

saturated fatty acid: a fatty acid carrying the maximum possible number of hydrogen atoms (having no points of unsaturation).

scurvy: the vitamin C–deficiency disease.

sedentary: physically inactive (literally "sitting down a lot").

segmentation: a periodic squeezing or partitioning of the intestine by its circular muscles that both mixes and slowly pushes the contents along.

self-efficacy: a person's belief in his or her ability to succeed in an undertaking.

self-monitoring of blood glucose: home monitoring of blood glucose levels using a glucose meter.

semipermeable membrane: a membrane that allows some particles to pass through but not others.

senile dementia: the loss of brain function beyond the normal loss of physical adeptness and memory that occurs with aging.

senile plaques: clumps of the protein fragment beta-amyloid on the nerve cells, commonly found in the brains of people with Alzheimer's dementia.

sepsis: a whole-body inflammatory response caused by infection; characterized by symptoms similar to those of SIRS.

set-point theory: the theory that the body tends to maintain a certain weight by means of its own internal controls.

shock-wave lithotripsy: a nonsurgical procedure that uses high-amplitude sound waves to fragment gallstones or kidney stones.

shock: a severe reduction in blood flow that deprives the body's tissues of oxygen and nutrients; characterized by reduced blood pressure, raised heart and respiratory rates, and muscle weakness.

short bowel syndrome: the malabsorption syndrome that follows resection of the small intestine, resulting in inadequate absorptive capacity in the remaining intestine.

sinusoids: the small capillary-like passages that carry blood through liver tissue.

Sjögren's (SHOW-grenz) **syndrome:** an autoimmune disease characterized by the destruction of secretory glands, resulting in dry mouth and dry eyes.

skinfold measure: a clinical estimate of total body fatness in which the thickness of a fold of skin on the back of the arm (over the triceps muscle), below the shoulder blade (subscapular), or in other places is measured with a caliper.

sludge: literally, a semisolid mass. Biliary sludge is made up of mucus, cholesterol crystals, and bilirubin granules.

small intestine: a 10-foot length of small-diameter intestine that is the major site of digestion of food and absorption of nutrients. Its segments are the duodenum, jejunum, and ileum.

soaps: chemical compounds formed from fatty acids and positively charged minerals.

sodium bicarbonate: baking soda; an alkaline salt believed to neutralize blood lactic acid and thereby to reduce pain and enhance possible workload. Soda loading may cause intestinal bloating and diarrhea.

soluble fibers: indigestible food components that readily dissolve in water and often impart gummy or gel-like characteristics to foods.

soup kitchens: programs that provide prepared meals to be eaten on-site.

Special Supplemental Nutrition Program for Women, Infants, and Children (WIC): a high-quality, cost-effective health care and nutrition services program administered by the U.S. Department of Agriculture for low-income women, infants, and children who are nutritionally at risk. WIC provides supplemental foods, nutrition education, and referrals to health care and other social services.

specialized formulas: enteral formulas designed to meet the nutrient needs of patients with specific illnesses; also called *disease-specific formulas.*

sphincter (SFINK-ter): a circular muscle surrounding, and able to close, a body opening.

spina (SPY-nah) **bifida** (BIFF-ih-dah): one of the most common types of neural tube defects; characterized by the incomplete closure of the spinal cord and its bony encasement.

standard formulas: enteral formulas that contain mostly intact proteins and polysaccharides; also called *polymeric formulas.*

starch: a plant polysaccharide composed of glucose and digestible by human beings.

steatohepatitis (STEE-ah-to-HEP-ah-TYE-tis): liver inflammation that is associated with fatty liver.

steatorrhea (stee-AH-tor-REE-ah): excessive fat in the stools due to fat malabsorption; characterized by stools that are loose, frothy, and foul smelling due to a high fat content.

sterile: free of microorganisms, such as bacteria.

steroids (STARE-oids): medications used to reduce tissue inflammation, to suppress the immune response, or to replace certain steroid hormones in people who cannot synthesize them.

sterols: one of the main classes of lipids; includes cholesterol, vitamin D, and the sex hormones (such as testosterone).

stevia sweeteners: zero-kcalorie sweeteners derived from the sweetest part of the native South American stevia plant.

stoma (STOE-ma): a surgically created opening in a body tissue or organ.

stomach: a muscular, elastic, saclike portion of the digestive tract that grinds and churns swallowed food, mixing it with acid and enzymes to form chyme.

stress fractures: bone damage or breaks caused by stress on bone surfaces during exercise.

stress response: the chemical and physical changes that occur within the body during stress.

stricture: abnormal narrowing of a passageway; often due to inflammation, scarring, or a congenital abnormality.

stroke volume: the amount of oxygenated blood ejected from the heart toward body tissues at each beat.

stroke: a sudden injury to brain tissue resulting from impaired blood flow through an artery that supplies blood to the brain; also called a *cerebrovascular accident.*

structure-function claims: statements that describe how a product may affect a structure or function of the body; these statements do not require FDA authorization.

struvite (STROO-vite): crystals of magnesium ammonium phosphate.

subcutaneous (sub-cue-TAY-nee-us): beneath the skin.

subcutaneous fat: fat stored directly under the skin.

sucrose: a disaccharide composed of glucose and fructose; commonly known as *table sugar, beet sugar,* or *cane sugar.*

sugar alcohols: sugarlike compounds that are sweet to taste and yield 2 to 3 kcal per gram.

surrogate: a substitute; a person who takes the place of another.

sushi: vinegar-flavored rice and raw or cooked seafood, typically wrapped in seaweed and stuffed with colorful vegetables.

syringes: devices used for injecting medications. A syringe consists of a hypodermic needle attached to a hollow tube with a plunger inside.

systemic (sih-STEM-ic): relating to the entire body.

systemic inflammatory response syndrome (SIRS): a whole-body inflammatory response caused by severe illness or trauma; characterized by raised heart and respiratory rates, abnormal white blood cell counts, and elevated body temperature.

tannins: compounds in tea (especially black tea) and coffee that bind iron.

TCA cycle or tricarboxylic (try-car-box-ILL-ick) **acid cycle:** a series of metabolic reactions that break down molecules of acetyl CoA to carbon dioxide and hydrogen atoms; also called the *Kreb's cycle.*

teratogenic (ter-AT-oh-jen-ik): causing abnormal fetal development and birth defects.

thermic effect of food: an estimation of the energy required to process food (digest, absorb, transport, metabolize, and store ingested nutrients).

thrombosis (throm-BOH-sis): the formation or presence of a blood clot in blood vessels. A *coronary thrombosis* occurs in a coronary artery, and a *cerebral thrombosis* occurs in an artery that supplies blood to the brain.

thrombus: a blood clot formed within a blood vessel that remains attached to its place of origin.

tissue rejection: destruction of donor tissue by the recipient's immune system, which recognizes the donor cells as foreign.

tocopherol (tuh-KOFF-er-ol): a general term for several chemically related compounds, one of which has vitamin E activity.

Tolerable Upper Intake Levels (UL): a set of values reflecting the highest average daily nutrient intake levels that are likely to pose no risk of toxicity to almost all healthy individuals in a particular life stage and gender group.

tolerance level: the maximum amount of residue permitted in a food when a pesticide is used according to the label directions.

total nutrient admixture (TNA): a parenteral solution that contains dextrose, amino acids, and lipids; also called a *3-in-1 solution* or an *all-in-one solution.*

total parenteral nutrition (TPN): the infusion of nutrient solutions into a central vein.

trace minerals: essential mineral nutrients required in the adult diet in amounts less than 100 milligrams per day.

training: regular practice of an activity, which leads to physical adaptations of the body with improvement in flexibility, strength, or endurance.

trans-fatty acids: fatty acids in which the hydrogen atoms next to the double bond are on opposite sides of the carbon chain.

transferrin (trans-FERR-in): the body's iron-carrying protein.

transient ischemic attacks (TIAs): brief ischemic strokes that cause short-term neurological symptoms.

transnasal feeding tube: a feeding tube that is inserted through the nose.

traveler's diarrhea: nausea, vomiting, and diarrhea caused by consuming food or water contaminated by any of several organisms, most commonly *E. coli, Shigella, Campylobacter jejuni,* and *Salmonella.*

triglycerides (try-GLISS-er-rides): one of the main classes of lipids; the chief form of fat in foods and the major storage form of fat in the body; composed of glycerol with three fatty acids attached.

trimester: a period representing one-third of the term of gestation. A trimester is about 13 to 14 weeks.

tripeptide: three amino acids bonded together.

tube feedings: liquid formulas delivered through a tube placed in the stomach or intestine.

tubules: tubelike structures of the nephron that process filtrate during urine production. The tubules are surrounded by capillaries that reabsorb the substances retained by tubule cells.

tumor: an abnormal tissue mass that has no physiological function; also called a *neoplasm* (NEE-oh-plazm).

type 1 diabetes: the type of diabetes that accounts for 5 to 10 percent of diabetes cases and usually results from autoimmune destruction of pancreatic beta cells.

type 2 diabetes: the type of diabetes that accounts for 90 to 95 percent of diabetes cases and usually results from insulin resistance coupled with insufficient insulin secretion.

ulcerative colitis (ko-LYE-tis): an inflammatory bowel disease that involves the colon. Inflammation affects the mucosa and submucosa of the intestinal wall.

ultrafiltration: removal of fluids and solutes from blood by using pressure to transfer the blood across a semipermeable membrane.

umbilical (um-BIL-ih-cul) **cord:** the ropelike structure through which the fetus's veins and arteries reach the placenta; the route of nourishment and oxygen into the fetus and the route of waste disposal from the fetus.

undernutrition: underconsumption of food energy or nutrients severe enough to cause disease or increased susceptibility to disease; a form of malnutrition.

unsaturated fatty acid: a fatty acid with one or more points of unsaturation where hydrogen atoms are missing (includes monounsaturated and polyunsaturated fatty acids).

urea (yoo-REE-uh): the principal nitrogen-excretion product of protein metabolism.

urea kinetic modeling: a method of determining the adequacy of dialysis treatment by calculating the urea clearance from blood.

uremia (you-REE-me-ah): the accumulation of nitrogenous and various other waste products in the blood; often associated with symptoms that reflect impairments in multiple body systems (literally, "urine in the blood"). The term may also be used to indicate the toxic state that results when wastes are retained in the blood.

uremic syndrome: the cluster of symptoms associated with inadequate kidney function; the symptoms reflect fluid, electrolyte, and hormonal imbalances; altered heart function; neuromuscular disturbances; and other metabolic derangements.

USDA Food Guide: the USDA's food group plan for ensuring dietary adequacy that assigns foods to five major food groups.

uterus (YOO-ter-us): the womb, the muscular organ within which the infant develops before birth.

varices (VAH-rih-seez): abnormally dilated blood vessels (singular: *varix*).

variety (dietary): eating a wide selection of foods within and among the major food groups (the opposite of monotony).

vegans: people who exclude all animal-derived foods from their diets; also called *strict vegetarians* or *total vegetarians.*

vegetarians: people who include plant-based foods in their diets and eliminate some or all animal-derived foods.

vein: a vessel that carries blood back to the heart.

very low food security: a form of food insecurity characterized by multiple indications of disrupted eating patterns and reduced food intake.

very-low-density lipoproteins (VLDL): the type of lipoproteins made primarily by liver cells to transport lipids to various tissues in the body; composed primarily of triglycerides.

vigorous-intensity physical activity: physical activity that requires a large increase in breathing and/or heart rate and expends more than 7 kcalories per minute. Walking at a very brisk pace (>4.5 miles per hour) or running at a pace of at least 5 miles per hour are examples.

villi (VILL-ee or VILL-eye): fingerlike projections from the folds of the small intestine (singular form: *villus*).

visceral fat: fat stored within the abdominal cavity in association with the internal abdominal organs; also called *intra-abdominal fat.*

viscous: having a gel-like consistency.

vitamin A: a fat-soluble vitamin. Its three chemical forms are *retinol* (the alcohol form), *retinal* (the aldehyde form), and *retinoic acid* (the acid form).

vitamins: essential, nonkcaloric, organic nutrients needed in tiny amounts in the diet.

VO$_2$ max: the maximum rate of oxygen consumption by an individual (measured at sea level).

voluntary activities: the component of a person's daily energy expenditure that involves conscious and deliberate muscular work: walking, lifting, climbing, and other physical activities.

waist circumference: a measurement used to assess a person's abdominal fat.

wasting: the gradual atrophy (loss) of body tissues; associated with protein-energy malnutrition or chronic illness.

water balance: the balance between water intake and water excretion that keeps the body's water content constant.

water intoxication: the rare condition in which body water contents are too high. The symptoms may include confusion, convulsion, coma, and even death in extreme cases.

wean: to gradually replace breast milk with infant formula or other foods appropriate to an infant's diet.

weight cycling: repeated rounds of weight loss and subsequent regain, with reduced ability to lose weight with each attempt; also called *yo-yo dieting.*

wellness: maximum well-being; the top range of health states; the goal of the person who strives toward realizing his or her full potential physically, mentally, emotionally, spiritually, and socially.

wheat gluten (GLU-ten): a family of water-insoluble proteins in wheat; includes the gliadin (GLY-ah-din) fractions that are toxic to persons with celiac disease.

xerostomia (ZEE-roh-STOE-me-ah): dry mouth caused by reduced salivary flow.

zygote (ZY-goat): the product of the union of ovum and sperm; a fertilized ovum.

Index

C

Caffeine
  in adolescents, 344
  content of foods, Appendix A
  diuretic effect of, 227
  during lactation, 307
  before physical activity, 275, 283
  during pregnancy, 304
Calcium, 232–36, 238t
  adolescents' needs, 343
  in breast milk, 320
  in chronic kidney disease treatment, 606
  deficiency, 233, 343
  food sources of, 104t, 234–36
  intake recommendations, 234
  and kidney stone prevention, 612
  older adults' needs, 366
  and osteoporosis, 233
  pregnancy needs, 294–95
  in vegetarian diets, 103, 104t
Calcium balance, 233
Calcium oxalate stones, 610, 611–12
Calculation aids, Appendix F
Calories, 7
CAM (complementary and alternative medicine),
    425–28, 644, 652
Campylobacter bacterium, 123, 124t
Campylobacteriosis, 124t
Canada, nutrition guidelines, Appendix B
Cancer, 638–47, **639**
  antioxidants and risk of, 192, 210, 245, 253
  consequences of, 641–42
  development of, 638–39
  dietary fiber and risk of, 47
  environmental risk factors for, 639t
  fat intakes and risk of, 66
  nutritional risk factors, 639–41
  nutrition assessment checklist for, 654
  nutrition therapy for, 644–47
  oral, 483
  physical activity and risk of, 257
  side effects of surgery, 642
  supplements for, 644
  treatments for, 642–44, 651
  types of, 638
Cancer cachexia, **640,** 641, 644
Candidiasis, **649,** 649–50
Canned goods, safety of, 127t
CAPD (continuous ambulatory peritoneal dialysis),
    616, 617
Capillaries, **118**
Carbohydrate-controlled diets, 551–55
Carbohydrate counting, 552–54
Carbohydrate loading, **265**
Carbohydrate-modified diets
  for celiac disease, 511–13
  for cystic fibrosis, 509
  for lactose intolerance, 507
Carbohydrate requirements
  in acute stress, 624
  of children, 327
  during cirrhosis treatment, 530
  of diabetics, 551
  of older adults, 364–65
  for physical endurance, 263–66
  of physically active people, 276–78
  during pregnancy, 291–92
Carbohydrates, 33–51, **34.** *See also* Fiber; Glycemic
    index (GI); Starch
  balancing fats and proteins with, 171
  in digestive process, 108f, 115
  energy in, 7, 45
  excess, 138
  food label requirements, 25, 51
  food sources of, 49–51

glycemic index (GI), 46, 47, 55–56
  in heart-healthy diet, 577
  intake recommendations, 14t
  in parenteral solutions, 445
  and protein sparing, 99
  role in disease prevention, 46–49, 56–57
  supplements, 276
  types of, 34–37
Carbohydrate-to-insulin ratio, 556, **557**
Carcinogenesis, 638, **639**
Carcinogens, **639,** 640
Carcinomas, 638
Cardiac cachexia, **587**
Cardiac output, 262, **263,** 582
Cardiac sphincter, 107
Cardiopulmonary resuscitation (CPR), **657,** 658
Cardiorespiratory endurance, **260,** 262–63
Cardiorespiratory system, 262f
Cardiovascular disease (CVD), **350,** 569–88. *See
    also* Atherosclerosis; Coronary heart disease;
    Hypertension
  alcohol and risk of, 541
  carbohydrate intake recommendations, 46
  deaths resulting from, 569f
  dental disease as contributor to, 483
  development of, 65, 350–53, 569–70
  in diabetics, 547
  diet-drug interactions, 579
  fats and risk of, 66–69, 81, 82, 95
  heart failure, 586–87
  low-GI diets and, 57
  metabolic syndrome and risk of, 565, 566
  as multigene disorder, 408
  nutrition assessment checklist for, 588
  overweight/obesity as risk factor for, 149, 350,
    351
  physical activity and risk of, 257, 258,
    351–52
  protein excess and risk of, 95
  selenium deficiency and, 245
  stroke, 541, 569, 581
  vitamin E and risk of, 198
Cariogenic foods, **481,** 481–82
Carnitine, **282**
Carotenemia, 194
Carotenoids, **217t,** 219
Catabolism, **133,** 134f
Cataracts, 360, **361**
Catecholamines, 620
Cathartics, **153,** 157, 231
Catheters, 444, **445,** 447–48
CDC (Centers for Disease Control and Prevention),
    123, 247
CDEs (Certified Diabetes Educators), **549**
Celiac disease, **511,** 511–13
Cell differentiation, **191,** 191–92
Cell division, 208
Cellulite, 166, **167**
Cellulose, 36–37
Centers for Disease Control and Prevention (CDC),
    123, 247
Central obesity, **146,** 146–47
Central veins, **444**
Cerebral cortex, **361**
Certified Diabetes Educators (CDEs), **549**
Chain length, of fatty acids, 62
CHD. *See* Coronary heart disease
Cheese
  carbohydrates in, 51
  food safety precautions, 127t
  intake recommendations, 17
  iron in, 243
  protein in, 98
  saturated fat in, 71
Chemically defined formulas, 431
Chemotherapy, 642–43

Children, 326–41
  adverse drug effects in, 412
  calcium needs of, 232, 236
  cardiovascular disease development in,
    350–53
  diabetes prevention in, 351
  dietary guidelines for, 352t
  fat cell development in, 162–63
  food allergies in, 332–34
  food assistance programs for, 296–97
  food preferences of, 5, 334
  hunger/malnutrition in, 93–95, 330–31
  hyperactivity in, 334
  hypertension in, 351–52
  iodine deficiency in, 245
  iron deficiency in, 239, 240, 241
  iron poisoning in, 241
  lead poisoning in, 331–32
  mealtimes, 338–40
  nutrient/energy needs, 326–30
  nutrition assessment, 345, 391–92
  nutrition at school, 340–41
  obesity in, 335–38, 350–53
  selenium deficiency in, 245
  supplements for, 328
  vegan diets for, 103
  vitamin A deficiency/toxicity in, 192–93
  vitamin D deficiency in, 196, 198
Chiropractic treatment, **426,** 427
Chloride, 231, 238t
Choking hazards, 325, 339
Cholecalciferol, 195
Cholecystectomy, **535**
Cholecystitis, **533**
Cholelithiasis, 532, **533**
Cholesterol, dietary
  effect on blood cholesterol, 66–68
  food label requirements, 25
  food sources of, 67, 71, 575
  intake recommendations, 70, 575
Cholesterol, function and synthesis of, 65
Cholesterol gallstones, 533, 534
Choline, **64,** 209
Chromium, 247, 248t
Chromium picolinate, **282**
Chromosomes, 406, **407**
Chronic bronchitis, 626, **627**
Chronic disease, **12,** 518–19
Chronic hypertension, **301**
Chronic kidney disease, 602–7, **603**
Chronic obstructive pulmonary disease (COPD),
    626–29, **627**
Chronological age, **357**
Chylomicrons, **117,** 118, 120f
Chyme, **107**
Cigarette smoking
  as cause of chronic obstructive pulmonary
    disease, 627
  during lactation, 307–8
  during pregnancy, 302
  and risk of cardiovascular disease, 353, 571, 576
Cigar smoking, 576
Cirrhosis, 524–30, **525**
  causes of, 524
  consequences of, 525–27
  nutrition therapy for, 528–30
  treatment of, 527–28
Cis-fatty acids, **63**
CLA (conjugated linoleic acid), **63, 282,** 283
Claudication, **544,** 548
Cleaning guidelines, for food preparation,
    126, 127t
Clear liquid diet, **461,** 461–62, 463
Clinically severe obesity, 168, **169**
Closed feeding systems, **437**
Clostridium bacteria, 123, 124–25t

Daily Values for Food Labels

The Daily Values are standard values developed by the Food and Drug Administration (FDA) for use on food labels. Daily Values for protein, vitamins, and minerals reflect average allowances based on the RDA. Daily Values for nutrients and food components, such as fat and fiber, that do not have an established RDA but do have important relationships with health are based on recommended calculation factors as noted.

Nutrient	Amount
Protein[a]	50 g
Thiamin	1.5 mg
Riboflavin	1.7 mg
Niacin	20 mg NE
Biotin	300 µg
Pantothenic acid	10 mg
Vitamin B_6	2 mg
Folate	400 µg
Vitamin B_{12}	6 µg
Vitamin C	60 mg
Vitamin A	5000 IU[b]
Vitamin D	400 IU[b]
Vitamin E	30 IU[b]
Vitamin K	80 µg
Calcium	1000 mg
Iron	18 mg
Zinc	15 mg
Iodine	150 µg
Copper	2 mg
Chromium	120 µg
Selenium	70 µg
Molybdenum	75 µg
Manganese	2 mg
Chloride	3400 mg
Magnesium	400 mg
Phosphorus	1000 mg

Food Component	Amount	Calculation Factors
Fat	65 g	30% of kcalories
Saturated fat	20 g	10% of kcalories
Cholesterol	300 mg	Same regardless of kcalories
Carbohydrate (total)	300 g	60% of kcalories
Fiber	25 g	11.5 g per 1000 kcalories
Protein	50 g	10% of kcalories
Sodium	2400 mg	Same regardless of kcalories
Potassium	3500 mg	Same regardless of kcalories

Note: Daily Values were established for adults and children over 4 years old. The values for energy-yielding nutrients are based on 2000 kcalories a day.

[a]The Daily Values for protein vary for different groups of people: pregnant women, 60 g; nursing mothers, 65 g; infants under 1 year, 14 g; children 1 to 4 years, 16 g.

[b]Equivalent values for nutrients expressed as IU are: vitamin A, 1500 RAE (assumes a mixture of 40% retinol and 60% beta-carotene); vitamin D, 10 µg; vitamin E, 20 mg.

Glossary of Nutrient Measures

kcal: kcalories; a unit by which energy is measured (Chapter 1 provides more details).

g: grams; a unit of weight equivalent to about 0.03 ounces.

mg: milligrams; one-thousandth of a gram.

µg: micrograms; one-millionth of a gram.

IU: international units; an old measure of vitamin activity determined by biological methods (as opposed to new measures that are determined by direct chemical analyses). Many fortified foods and supplements use IU on their labels.
- For vitamin A, 1 IU = 0.3 µg retinol, 3.6 µg β-carotene, or 7.2 µg other vitamin A carotenoids.
- For vitamin D, 1 IU = 0.025 µg cholecalciferol.
- For vitamin E, 1 IU = 0.67 natural α-tocopherol (other conversion factors are used for different forms of vitamin E).

mg NE: milligrams niacin equivalents; a measure of niacin activity (Chapter 8 provides more details).
- 1 NE = 1 mg niacin.
  - = 60 mg tryptophan (an amino acid).

µg DFE: micrograms dietary folate equivalents; a measure of folate activity (Chapter 8 provides more details).
- 1 µg DFE = 1 µg food folate.
  - = 0.6 µg fortified food or supplement folate.
  - = 0.5 µg supplement folate taken on an empty stomach.

µg RAE: micrograms retinol activity equivalents; a measure of vitamin A activity (Chapter 8 provides more details).
- 1 µg RE = 1 µg retinol.
  - = 12 µg β-carotene.
  - = 24 µg other vitamin A carotenoids.

mmol: millimoles; one-thousanth of a mole, the molecular weight of a substance. To convert mmol to mg, multiply by the atomic weight of the substance.
- For sodium, mmol × 23 = mg Na.
- For chloride, mmol × 35.5 = mg Cl.
- For sodium chloride, mmol × 58.5 = mg NaCl.

Body Mass Index (BMI)

Height	18	19	20	21	22	23	24	25	26	27	28	29	30	31	32	33	34	35	36	37	38	39	40
									Body Weight (pounds)														
4'10"	86	91	96	100	105	110	115	119	124	129	134	138	143	148	153	158	162	167	172	177	181	186	191
4'11"	89	94	99	104	109	114	119	124	128	133	138	143	148	153	158	163	168	173	178	183	188	193	198
5'0"	92	97	102	107	112	118	123	128	133	138	143	148	153	158	163	168	174	179	184	189	194	199	204
5'1"	95	100	106	111	116	122	127	132	137	143	148	153	158	164	169	174	180	185	190	195	201	206	211
5'2"	98	104	109	115	120	126	131	136	142	147	153	158	164	169	175	180	186	191	196	202	207	213	218
5'3"	102	107	113	118	124	130	135	141	146	152	158	163	169	175	180	186	191	197	203	208	214	220	225
5'4"	105	110	116	122	128	134	140	145	151	157	163	169	174	180	186	192	197	204	209	215	221	227	232
5'5"	108	114	120	126	132	138	144	150	156	162	168	174	180	186	192	198	204	210	216	222	228	234	240
5'6"	112	118	124	130	136	142	148	155	161	167	173	179	186	192	198	204	210	216	223	229	235	241	247
5'7"	115	121	127	134	140	146	153	159	166	172	178	185	191	198	204	211	217	223	230	236	242	249	255
5'8"	118	125	131	138	144	151	158	164	171	177	184	190	197	203	210	216	223	230	236	243	249	256	262
5'9"	122	128	135	142	149	155	162	169	176	182	189	196	203	209	216	223	230	236	243	250	257	263	270
5'10"	126	132	139	146	153	160	167	174	181	188	195	202	209	216	222	229	236	243	250	257	264	271	278
5'11"	129	136	143	150	157	165	172	179	186	193	200	208	215	222	229	236	243	250	257	265	272	279	286
6'0"	132	140	147	154	162	169	177	184	191	199	206	213	221	228	235	242	250	258	265	272	279	287	294
6'1"	136	144	151	159	166	174	182	189	197	204	212	219	227	235	242	250	257	265	272	280	288	295	302
6'2"	141	148	155	163	171	179	186	194	202	210	218	225	233	241	249	256	264	272	280	287	295	303	311
6'3"	144	152	160	168	176	184	192	200	208	216	224	232	240	248	256	264	272	279	287	295	303	311	319
6'4"	148	156	164	172	180	189	197	205	213	221	230	238	246	254	263	271	279	287	295	304	312	320	328
6'5"	151	160	168	176	185	193	202	210	218	227	235	244	252	261	269	277	286	294	303	311	319	328	336
6'6"	155	164	172	181	190	198	207	216	224	233	241	250	259	267	276	284	293	302	310	319	328	336	345

Underweight	Healthy Weight	Overweight	Obese
(<18.5)	(18.5–24.9)	(25–29.9)	(≥30)

Find your height along the left-hand column and look across the row until you find the number that is closest to your weight. The number at the top of that column identifies your BMI. Chapter 6 describes how BMI correlates with disease risks and defines obesity. The area shaded in green represents healthy weight ranges.